Primer on the Metabolic Bone Diseases and Disorders of Mineral Metabolism

Ninth Edition

An Official Publication of the American Society for Bone and Mineral Research

Primer on the Metabolic Bone Diseases and Disorders of Mineral Metabolism

Ninth Edition

Editor-in-Chief

John P. Bilezikian, MD, PhD (hon)

Senior Associate Editors
Roger Bouillon, MD, PhD, FRCP
Thomas Clemens, PhD
Juliet Compston, OBE, MD, FRCP, FRCPE, FRCPath, FMedSci

Associate Editors
Douglas C. Bauer, MD
Peter R. Ebeling, AO, MBBS, MD, FRACP
Klaus Engelke, PhD
David Goltzman, MD
Theresa Guise, MD
Suzanne M. Jan de Beur, MD
Harald Jüppner, MD
Karen Lyons, PhD
Laurie McCauley, DDS, MS, PhD
Michael R. McClung, MD, FACP, FACE
Paul D. Miller, MD, FACP
Socrates E. Papapoulos, MD, PhD
G. David Roodman, MD, PhD
Clifford J. Rosen, MD
Ego Seeman, BSc, MBBS, FRACP, MD, AM
Rajesh V. Thakker, MD, ScD, FRCP, FRCPath, FMedSci, FRS
Michael P. Whyte, MD
Mone Zaidi, MD, PhD, FRCP

This edition first published 2019 © 2019 by American Society for Bone and Mineral Research

Edition History
First Edition published by American Society for Bone and Mineral Research © 1990 American Society for Bone and Mineral Research
Second Edition published by Raven Press, Ltd. © 1993 American Society for Bone and Mineral Research
Third Edition published by Lippincott-Raven Publishers © 1996 American Society for Bone and Mineral Research
Fourth Edition published by Lippincott Williams & Wilkins © 1999 American Society for Bone and Mineral Research
Fifth Edition published by American Society for Bone and Mineral Research © 2003 American Society for Bone and Mineral Research
Sixth Edition published by American Society for Bone and Mineral Research © 2006 American Society for Bone and Mineral Research
Seventh Edition published by American Society for Bone and Mineral Research © 2008 American Society for Bone and Mineral Research
Eighth Edition published by American Society for Bone and Mineral Research © 2013 American Society for Bone and Mineral Research

All rights reserved. No part of this publication may be reproduced, stored in a retrieval system, or transmitted, in any form or by any means, electronic, mechanical, photocopying, recording or otherwise, except as permitted by law. Advice on how to obtain permission to reuse material from this title is available at http://www.wiley.com/go/permissions.

The right of John P. Bilezikian to be identified as the author of the editorial material in this work has been asserted in accordance with law.

Registered Office
John Wiley & Sons, Inc., 111 River Street, Hoboken, NJ 07030, USA

Editorial Office
111 River Street, Hoboken, NJ 07030, USA

For details of our global editorial offices, customer services, and more information about Wiley products visit us at www.wiley.com.

Wiley also publishes its books in a variety of electronic formats and by print-on-demand. Some content that appears in standard print versions of this book may not be available in other formats.

Limit of Liability/Disclaimer of Warranty
While the publisher and authors have used their best efforts in preparing this work, they make no representations or warranties with respect to the accuracy or completeness of the contents of this work and specifically disclaim all warranties, including without limitation any implied warranties of merchantability or fitness for a particular purpose. No warranty may be created or extended by sales representatives, written sales materials or promotional statements for this work. The fact that an organization, website, or product is referred to in this work as a citation and/or potential source of further information does not mean that the publisher and authors endorse the information or services the organization, website, or product may provide or recommendations it may make. This work is sold with the understanding that the publisher is not engaged in rendering professional services. The advice and strategies contained herein may not be suitable for your situation. You should consult with a specialist where appropriate. Further, readers should be aware that websites listed in this work may have changed or disappeared between when this work was written and when it is read. Neither the publisher nor authors shall be liable for any loss of profit or any other commercial damages, including but not limited to special, incidental, consequential, or other damages.

Library of Congress Cataloging-in-Publication Data
Names: Bilezikian, John P., editor.
Title: Primer on the metabolic bone diseases and disorders of mineral metabolism.
Description: Ninth edition / editor-in-Chief, John P. Bilezikian, MD, PhD (hon) ; senior associate editors,
  Roger Bouillon, MD, PhD, FRCP, Tom Clemens, PhD, Juliet Compston, MD, FRCP, FRCPath, FMedSci ;
  associate editors, Doug Bauer, MD [and nine others]. | Hoboken, NJ : Wiley-Blackwell, 2019. |
  Includes bibliographical references and index. |
Identifiers: LCCN 2018038448 (print) | LCCN 2018038631 (ebook) | ISBN 9781119266570 (Adobe PDF) |
  ISBN 9781119266587 (ePub) | ISBN 9781119266563 (paperback)
Subjects: LCSH: Bones–Metabolism–Disorders. | Mineral metabolism–Disorders. | Bones–Diseases. |
  Minerals–Metabolism. | BISAC: SCIENCE / Life Sciences / Cytology.
Classification: LCC RC931.M45 (ebook) | LCC RC931.M45 P75 2019 (print) | DDC 616.7/16–dc23
LC record available at https://lccn.loc.gov/2018038448

Cover images: (Top row left to right) © Wiley; © Paul Gunning/Science Source; By kind permission of Alan Boyde; Courtesy of Kevin Mackenzie, Microscopy Facility University of Aberdeen; © GenScript USA Inc.; © Designua/Shutterstock; Reproduced with permission from Juliet Compston, M.D.; (Bottom row left to right) Courtesy of Prof. Ego Seeman; © designua/123RF; © SCOTT CAMAZINE/science source; © ZEPHYR/science source; (Floor) © DonNichols/Getty Images; (Frames) © iStock.com/JackF
Cover design by Wiley

Set in 9.5/11pt TrumpMediaeval by SPi Global, Pondicherry, India
Printed and bound in Singapore by Markono Print Media Pte Ltd

10 9 8 7 6 5 4 3 2 1

Contents

Contributors xi
Preface to the Ninth Edition of the *Primer*:
John P. Bilezikian xxiv
About ASBMR xxv
President's Preface: *Michael J. Econs* xxvi
About the Companion Website xxvii

Section I: Molecular and Cellular Determinants of Bone Structure and Function 1
Section Editor: Karen Lyons

1. Early Skeletal Morphogenesis in Embryonic Development 3
   Yingzi Yang

2. Endochondral Ossification 12
   Courtney M. Karner and Matthew J. Hilton

3. Local and Circulating Osteoprogenitor Cells and Lineages 20
   Naomi Dirckx and Christa Maes

4. Osteoblasts: Function, Development, and Regulation 31
   Elizabeth W. Bradley, Jennifer J. Westendorf, Andre J. van Wijnen, and Amel Dudakovic

5. Osteocytes 38
   Lynda F. Bonewald

6. Osteoclast Biology and Bone Resorption 46
   Hiroshi Takayanagi

7. Signal Transduction Cascades Controlling Osteoblast Differentiation 54
   David J.J. de Gorter, Gonzalo Sánchez-Duffhues, and Peter ten Dijke

8. The TGF-β Superfamily in Bone Formation and Maintenance 60
   Ce Shi and Yuji Mishina

9. Recent Developments in Understanding the Role of Wnt Signaling in Skeletal Development and Disease 68
   Zhendong A. Zhong, Nicole J. Ethen, and Bart O. Williams

10. Mechanotransduction in Bone Formation and Maintenance 75
    Whitney A. Bullock, Lilian I. Plotkin, Alexander G. Robling, and Fredrick M. Pavalko

11. The Composition of Bone 84
    Adele L. Boskey and Pamela G. Robey

12. Assessment of Bone Mass, Structure, and Quality in Rodents 93
    Jeffry S. Nyman and Deepak Vashishth

13. Skeletal Healing: Cellular and Molecular Determinants 101
    Alayna E. Loiselle and Michael J. Zuscik

14. Biomechanics of Fracture Healing 108
    Elise F. Morgan, Amira I. Hussein, and Thomas A. Einhorn

Section II: Skeletal Physiology 115
Section Editor: Ego Seeman

15. Human Fetal and Neonatal Bone Development 117
    Tao Yang, Monica Grover, Kyu Sang Joeng, and Brendan Lee

16. Skeletal Growth: a Major Determinant of Bone's Structural Diversity in Women and Men 123
    Ego Seeman

17. Racial Differences in the Acquisition and Age-related Loss of Bone Strength 131
    Shane A. Norris, Marcella D. Walker, Kate Ward, Lisa K. Micklesfield, and John M. Pettifor

18 Calcium, Vitamin D, and Other Nutrients During Growth **135**
Tania Winzenberg and Graeme Jones

19 Mechanical Loading and the Developing Skeleton **141**
Mark R. Forwood

20 Pregnancy and Lactation **147**
Christopher S. Kovacs and Henry M. Kronenberg

21 Menopause and Age-related Bone Loss **155**
Carlos M. Isales and Ego Seeman

Section III: Mineral Homeostasis 163
Section Editors: David Goltzman and Harald Jüppner

22 Regulation of Calcium Homeostasis **165**
Line Vautour and David Goltzman

23 Magnesium Homeostasis **173**
Aliya Aziz Khan, Asiya Sbayi, and Karl Peter Schlingmann

24 Fetal Calcium Metabolism **179**
Christopher S. Kovacs

25 FGF23 and the Regulation of Phosphorus Metabolism **187**
Kenneth E. White and Michael J. Econs

26 Gonadal Steroids **194**
Stavros C. Manolagas and Maria Schuller Almeida

27 Parathyroid Hormone **205**
Thomas J. Gardella, Robert A. Nissenson, and Harald Jüppner

28 Parathyroid Hormone-Related Protein **212**
John J. Wysolmerski and T. John Martin

29 Calcium-Sensing Receptor **221**
Geoffrey N. Hendy

30 Vitamin D: Production, Metabolism, Action, and Clinical Requirements **230**
Daniel D. Bikle, John S. Adams, and Sylvia Christakos

Section IV: Investigation of Metabolic Bone Diseases 241
Section Editors: Douglas C. Bauer and Klaus Engelke

31 Techniques of Bone Mass Measurement in Children with Risk Factors for Osteoporosis **243**
Nicola J. Crabtree and Leanne M. Ward

32 Standard Techniques of Bone Mass Measurement in Adults **252**
E. Michael Lewiecki, Paul D. Miller, and Nelson B. Watts

33 Advanced Techniques of Bone Mass Measurements and Strength in Adults **260**
Kyle K. Nishiyama, Enrico Dall'Ara, and Klaus Engelke

34 Magnetic Resonance Imaging of Bone **272**
Sharmila Majumdar

35 Trabecular Bone Score **277**
Barbara C. Silva and William D. Leslie

36 Reference Point Indentation **287**
Adolfo Diez-Perez and Joshua N. Farr

37 Biochemical Markers of Bone Turnover in Osteoporosis **293**
Pawel Szulc, Douglas C. Bauer, and Richard Eastell

38 Scintigraphy and PET in Metabolic Bone Disease **302**
Lorenzo Nardo, Paola A. Erba, and Benjamin L. Franc

39 Bone Histomorphometry in Clinical Practice **310**
Robert R. Recker and Carolina Aguiar Moreira

40 Diagnosis and Classification of Vertebral Fracture **319**
James F. Griffith and Harry K. Genant

41 FRAX: Assessment of Fracture Risk **331**
John A. Kanis, Eugene V. McCloskey, Nicholas C. Harvey, and William D. Leslie

Section V: Genetics of Bone 341
Section Editor: Rajesh V. Thakker

42 Introduction to Genetics 343
   Paul J. Newey, Michael P. Whyte, and Rajesh V. Thakker

43 Animal Models: Genetic Manipulation 351
   Karen Lyons

44 Animal Models: Allelic Determinants for Bone Mineral Density 359
   J. H. Duncan Bassett, Graham R. Williams, and Robert D. Blank

45 Transcriptional Profiling for Genetic Assessment 367
   Aimy Sebastian and Gabriela G. Loots

46 Approaches to Genetic Testing 373
   Christina Jacobsen, Yiping Shen, and Ingrid A. Holm

47 Human Genome-Wide Association Studies 378
   Douglas P. Kiel, Emma L. Duncan, and Fernando Rivadeneira

48 Translational Genetics of Osteoporosis: From Population Association to Individualized Assessment 385
   Bich Tran, Jacqueline R. Center, and Tuan V. Nguyen

Section VI: Osteoporosis 393
Section Editors: Paul D. Miller, Socrates E. Papapoulos, and Michael R. McClung

49 Osteoporosis: An Overview 395
   Michael R. McClung, Paul D. Miller, and Socrates E. Papapoulos

50 The Epidemiology of Osteoporotic Fractures 398
   Nicholas C. Harvey, Elizabeth M. Curtis, Elaine M. Dennison, and Cyrus Cooper

51 Fracture Liaison Service 405
   Piet Geusens, John A. Eisman, Andrea Singer, and Joop van den Bergh

52 Sex Steroids and the Pathogenesis of Osteoporosis 412
   Matthew T. Drake and Sundeep Khosla

53 Juvenile Osteoporosis 419
   Francis H. Glorieux and Craig Munns

54 Transplantation Osteoporosis 424
   Peter R. Ebeling

55 Premenopausal Osteoporosis 436
   Adi Cohen and Elizabeth Shane

56 Osteoporosis in Men 443
   Eric S. Orwoll and Robert A. Adler

57 Bone Stress Injuries 450
   Stuart J. Warden and David B. Burr

58 Inflammation-Induced Bone Loss in the Rheumatic Diseases 459
   Ellen M. Gravallese and Steven R. Goldring

59 Glucocorticoid-Induced Osteoporosis 467
   Kenneth Saag and Robert A. Adler

60 Human Immunodeficiency Virus and Bone 474
   Michael T. Yin and Todd T. Brown

61 Effects on the Skeleton from Medications Used to Treat Nonskeletal Disorders 482
   Nelson B. Watts

62 Diabetes and Fracture Risk 487
   Serge Ferrari, Nicola Napoli, and Ann Schwartz

63 Obesity and Skeletal Health 492
   Juliet Compston

64 Sarcopenia and Osteoporosis 498
   Gustavo Duque and Neil Binkley

65 Management of Osteoporosis in Patients with Chronic Kidney Disease 505
   Paul D. Miller

66 Other Secondary Causes of Osteoporosis 510
   Neveen A. T. Hamdy and Natasha M. Appelman-Dijkstra

67 Exercise for Osteoporotic Fracture Prevention and Management 517
   Robin M. Daly and Lora Giangregorio

- **68** Prevention of Falls 526
  Heike A. Bischoff-Ferrari

- **69** Nutritional Support for Osteoporosis 534
  Connie M. Weaver, Bess Dawson-Hughes, Rene Rizzoli, and Robert P. Heaney

- **70** Estrogens, Selective Estrogen Receptor Modulators, and Tissue-Selective Estrogen Complex 541
  Tobias J. de Villiers

- **71** Bisphosphonates for Postmenopausal Osteoporosis 545
  Andrea Giusti and Socrates E. Papapoulos

- **72** Denosumab 553
  Aline Granja Costa, E. Michael Lewiecki, and John P. Bilezikian

- **73** Parathyroid Hormone and Abaloparatide Treatment for Osteoporosis 559
  Felicia Cosman and Susan L. Greenspan

- **74** Combination Anabolic and Antiresorptive Therapy for Osteoporosis 567
  Joy N. Tsai and Benjamin Z. Leder

- **75** Strontium Ranelate and Calcitonin 573
  Leonardo Bandeira and E. Michael Lewiecki

- **76** Adverse Effects of Drugs for Osteoporosis 579
  Bo Abrahamsen and Daniel Prieto-Alhambra

- **77** Orthopedic Principles of Fracture Management 588
  Manoj Ramachandran and David G. Little

- **78** Adherence to Osteoporosis Therapies 593
  Stuart L. Silverman and Deborah T. Gold

- **79** Cost-Effectiveness of Osteoporosis Treatment 597
  Anna N.A. Tosteson

- **80** Future Therapies 603
  Michael R. McClung

Section VII: Metabolic Bone Diseases 611
Section Editors: Suzanne M. Jan de Beur and Peter R. Ebeling

- **81** Approach to Parathyroid Disorders 613
  John P. Bilezikian

- **82** Primary Hyperparathyroidism 619
  Shonni J. Silverberg, Francisco Bandeira, Jianmin Liu, Claudio Marcocci, and Marcella D. Walker

- **83** Familial States of Primary Hyperparathyroidism 629
  Andrew Arnold, Sunita K. Agarwal, and Rajesh V. Thakker

- **84** Non-Parathyroid Hypercalcemia 639
  Mara J. Horwitz

- **85** Hypocalcemia: Definition, Etiology, Pathogenesis, Diagnosis, and Management 646
  Anne L. Schafer and Dolores M. Shoback

- **86** Hypoparathyroidism 654
  Tamara Vokes, Mishaela R. Rubin, Karen K. Winer, Michael Mannstadt, Natalie E. Cusano, Harald Jüppner, and John P. Bilezikian

- **87** Pseudohypoparathyroidism 661
  Agnès Linglart, Michael A. Levine, and Harald Jüppner

- **88** Disorders of Phosphate Homeostasis 674
  Mary D. Ruppe and Suzanne M. Jan de Beur

- **89** Rickets and Osteomalacia 684
  Michaël R. Laurent, Nathalie Bravenboer, Natasja M. Van Schoor, Roger Bouillon, John M. Pettifor, and Paul Lips

- **90** Pathophysiology and Treatment of Chronic Kidney Disease–Mineral and Bone Disorder 695
  Mark R. Hanudel, Sharon M. Moe, and Isidro B. Salusky

- **91** Disorders of Mineral Metabolism in Childhood 705
  Thomas O. Carpenter and Nina S. Ma

- **92** Paget Disease of Bone 713
  Julia F. Charles, Ethel S. Siris, and G. David Roodman

- **93** Epidemiology, Diagnosis, Evaluation, and Treatment of Nephrolithiasis 721
  Murray J. Favus and David A. Bushinsky

- **94** Immobilization and Burns: Other Conditions Associated with Osteoporosis 730
  William A. Bauman, Christopher Cardozo, and Gordon L. Klein

Section VIII: Cancer and Bone 737
Section Editors: Theresa Guise and G. David Roodman

95 Mechanisms of Osteolytic and Osteoblastic Skeletal Lesions 739
    G. David Roodman and Theresa Guise

96 Clinical and Preclinical Imaging in Osseous Metastatic Disease 743
    Siyang Leng and Suzanne Lentzsch

97 Metastatic Tumors and Bone 752
    Julie A. Sterling and Rachelle W. Johnson

98 Myeloma Bone Disease and Other Hematological Malignancies 760
    Claire M. Edwards and Rebecca Silbermann

99 Osteogenic Osteosarcoma 768
    Yangjin Bae, Huan-Chang Zeng, Linchao Lu, Lisa L. Wang, and Brendan Lee

100 Skeletal Complications of Breast and Prostate Cancer Therapies 775
    Catherine Van Poznak and Pamela Taxel

101 Bone Cancer and Pain 781
    Denis Clohisy and Lauren M. MacCormick

102 Radiotherapy-Induced Osteoporosis 788
    Laura E. Wright

103 Skeletal Complications of Childhood Cancer 793
    Manasa Mantravadi and Linda A. DiMeglio

104 Medical Prevention and Treatment of Bone Metastases 799
    Catherine Handforth, Stella D'Oronzo, and Janet Brown

105 Radiotherapy of Skeletal Metastases 809
    Srinivas Raman, K. Liang Zeng, Oliver Sartor, Edward Chow, and Øyvind S. Bruland

106 Concepts and Surgical Treatment of Metastatic Bone Disease 816
    Kristy Weber and Scott L. Kominsky

Section IX: Sclerosing and Dysplastic Bone Diseases 823
Section Editor: Michael P. Whyte

107 Sclerosing Bone Disorders 825
    Michael P. Whyte

108 Fibrous Dysplasia 839
    Michael T. Collins, Alison M. Boyce, and Mara Riminucci

109 The Osteochondrodysplasias 848
    Fabiana Csukasi and Deborah Krakow

110 Ischemic and Infiltrative Disorders of Bone 853
    Michael P. Whyte

111 Tumoral Calcinosis – Dermatomyositis 861
    Nicholas J. Shaw

112 Genetic Disorders of Heterotopic Ossification: Fibrodysplasia Ossificans Progressiva and Progressive Osseous Heteroplasia 865
    Frederick S. Kaplan, Robert J. Pignolo, Mona Al Mukaddam, and Eileen M. Shore

113 Osteogenesis Imperfecta 871
    Joan C. Marini

114 Fibrillinopathies: Skeletal Manifestations of Marfan Syndrome and Marfan-Related Conditions 878
    Gary S. Gottesman and Michael P. Whyte

115 Hypophosphatasia and Other Enzyme Deficiencies Affecting the Skeleton 886
    Michael P. Whyte

Section X: Oral and Maxillofacial Biology and Pathology 891
Section Editor: Laurie McCauley

116 Craniofacial Morphogenesis 893
    Erin Ealba Bumann and Vesa Kaartinen

117 Development and Structure of Teeth and Periodontal Teeth 901
    Petros Papagerakis and Thimios Mitsiadis

118 Genetic Craniofacial Disorders Affecting the Dentition 911
    Yong-Hee Patricia Chun, Paul H. Krebsbach, and James P. Simmer

119 Pathology of the Hard Tissues of the Jaws 918
    Paul C. Edwards

120 Osteonecrosis of the Jaw 927
    Sotirios Tetradis, Laurie McCauley, and Tara Aghaloo

121 Alveolar Bone Homeostasis in Health and Disease **933**
Chad M. Novince and Keith L. Kirkwood

122 Oral Manifestations of Metabolic Bone Diseases **941**
Erica L. Scheller, Charles Hildebolt, and Roberto Civitelli

123 Dental Implants and Osseous Healing in the Oral Cavity **949**
Takashi Matsuura and Junro Yamashita

Section XI: Integrative Physiology of the Skeleton 957
Section Editors: Mone Zaidi and Clifford J. Rosen

124 Integrative Physiology of the Skeleton **959**
Clifford J. Rosen and Mone Zaidi

125 The Hematopoietic Niche and Bone **966**
Stavroula Kousteni, Benjamin J. Frisch, Marta Galan-Diez, and Laura M. Calvi

126 Adipocytes and Bone **974**
Clarissa S. Craft, Natalie K. Wee, and Erica L. Scheller

127 The Vasculature and Bone **983**
Marie Hélène Lafage-Proust and Bernard Roche

128 Immunobiology and Bone **992**
Roberto Pacifici and M. Neale Weitzmann

129 Cellular Bioenergetics of Bone **1004**
Wen-Chih Lee and Fanxin Long

130 Endocrine Bioenergetics of Bone **1012**
Patricia F. Ducy and Gerard Karsenty

131 Central Neuronal Control of Bone Remodeling **1020**
Hiroki Ochi, Paul Baldock, and Shu Takeda

132 Peripheral Neuronal Control of Bone Remodeling **1028**
Katherine J. Motyl and Mary F. Barbe

133 The Pituitary–Bone Axis in Health and Disease **1037**
Mone Zaidi, Tony Yuen, Wahid Abu-Amer, Peng Liu, Terry F. Davies, Maria I. New, Harry C. Blair, Alberta Zallone, Clifford J. Rosen, and Li Sun

134 Neuropsychiatric Disorders and the Skeleton **1047**
Madhusmita Misra and Anne Klibanski

135 Interactions Between Muscle and Bone **1055**
Marco Brotto

Index **1063**

Contributors

Bo Abrahamsen, PhD
Odense Patient Data Explorative Network, Department of Clinical Research
University of Southern Denmark, Odense;
and Department of Medicine, Holbæk Hospital, Holbæk, Denmark

Wahid Abu-Amer, MD
The Mount Sinai Bone Program, Department of Medicine
Icahn School of Medicine at Mount Sinai, New York, New York, USA

John S. Adams, MD
Orthopaedic Hospital Research Center; UCLA Clinical and Translational Science Institute; and the Departments of Orthopaedic Surgery, Medicine, and Molecular Cell and Developmental Biology, University of California Los Angeles, California, USA

Robert A. Adler, MD
Endocrinology and Metabolism Section, Hunter Holmes McGuire Veterans Affairs Medical Center; and Endocrine Division, Virginia Commonwealth University School of Medicine, Richmond, Virginia, USA

Sunita K. Agarwal, PhD
Metabolic Diseases Branch, The National Institute of Diabetes and Digestive and Kidney Diseases, National Institutes of Health, Bethesda, Maryland, USA

Tara Aghaloo, DDS, MD, PhD
Division of Diagnostic and Surgical Sciences
UCLA School of Dentistry, Los Angeles, California, USA

Mona Al Mukaddam, MD, MS
Center for Research in FOP and Related Disorders
Departments of Medicine and Orthopaedic Surgery
The University of Pennsylvania School of Medicine
Philadelphia, Pennsylvania, USA

Natasha M. Appelman-Dijkstra, MD, PhD
Department of Internal Medicine
Division of Endocrinology and Centre for Bone Quality
Leiden University Medical Center, The Netherlands

Andrew Arnold, MD
Division of Endocrinology and Metabolism;
Departments of Molecular Medicine and Medicine and Genetics; and Center for Molecular Oncology, University of Connecticut School of Medicine, Farmington, Connecticut, USA

Yangjin Bae, PhD
Department of Molecular and Human Genetics
Baylor College of Medicine, Texas Children's Hospital, Houston, Texas, USA

Paul Baldock, PhD
Bone and Mineral Research Program
Garvan Institute of Medical Research,
St Vincent's Hospital, Sydney,
New South Wales, Australia

Francisco Bandeira, MD, PhD, FACE
Division of Endocrinology, Diabetes and Metabolic Bone Diseases, Agamenon Magalhães Hospital, Brazilian Ministry of Health, University of Pernambuco Medical School, Recife, Brazil

Leonardo Bandeira, MD
Department of Medicine, College of Physicians and Surgeons
Columbia University Medical Center,
New York, New York, USA

Mary F. Barbe, PhD
Department of Anatomy and Cell Biology,
School of Medicine, Lewis Katz School of Medicine, Temple University, Philadelphia, Pennsylvania, USA

J.H. Duncan Bassett, BM BCh, PhD, FRCP
Molecular Endocrinology Laboratory
Department of Medicine, Imperial College London, London, UK

Douglas C. Bauer, MD
Departments of Medicine and Epidemiology & Biostatistics
University of California, San Francisco,
San Francisco, California, USA

Contributors

William A. Bauman, MD
Department of Veterans Affairs Rehabilitation Research and Development Service
National Center for the Medical Consequences of Spinal Cord Injury
James J. Peters Veterans Affairs Medical Center, Bronx, New York; and Departments of Medicine and Rehabilitation Medicine, Icahn School of Medicine at Mount Sinai, New York, New York, USA

Daniel D. Bikle, MD, PhD
Departments of Medicine and Dermatology
University of California San Francisco, Biomedical Sciences Graduate Program, San Francisco, California, USA

John P. Bilezikian, MD, PhD (hon)
Metabolic Bone Diseases Unit, Division of Endocrinology, Department of Medicine, College of Physicians and Surgeons, Columbia University, New York, New York, USA

Neil Binkley, MD
Department of Medicine, University of Wisconsin School of Medicine and Public Health; and Institute on Aging, University of Wisconsin, Madison, Wisconsin, USA

Heike A. Bischoff-Ferrari, MD, DrPH
Department of Geriatrics and Aging Research
University Hospital and University of Zurich, Switzerland

Harry C. Blair, MD
The Pittsburgh VA Medical Center and Departments of Pathology and of Cell Biology
University of Pittsburgh School of Medicine, Pittsburgh, Pennsylvania, USA

Robert D. Blank, MD, PhD
Department of Endocrinology, Metabolism, and Clinical Nutrition, Medical College of Wisconsin; and Clement J. Zablocki VAMC, Milwaukee, Wisconsin, USA

Lynda F. Bonewald, PhD
Indiana Center for Musculoskeletal Health;
Department of Anatomy and Cell Biology, and Department of Orthopaedic Surgery, Indiana University School of Medicine,
Indianapolis, Indiana, USA

Adele L. Boskey, PhD*
Research Institute, Hospital for Special Surgery and Department of Biochemistry and Graduate Field of Physiology, Biophysics and Systems Biology, Cornell University Medical and Graduate Medical Schools, New York, New York, USA

Roger Bouillon, MD, PhD
Laboratory of Clinical and Experimental Endocrinology, Department of Chronic Diseases, Metabolism and Ageing, KU Leuven, Leuven, Belgium

Alison M. Boyce, MD
Section on Skeletal Diseases and Mineral Homeostasis, National Institute of Dental and Craniofacial Research, National Institutes of Health, Department of Health and Human Services, Bethesda, Maryland, USA

Elizabeth W. Bradley, PhD
Department of Orthopedic Surgery, Mayo Clinic, Rochester, Minnesota, USA

Nathalie Bravenboer, PhD
Department of Clinical Chemistry, VU University Medical Center, MOVE Research Institute, Amsterdam, The Netherlands

Marco Brotto, BSN, MS, PhD
Bone-Muscle Collaborative Sciences, College of Nursing and Health Innovation, University of Texas at Arlington, Arlington, Texas, USA

Janet Brown, BMedSci, MBBS, MSc, MD, FRCP
Academic Unit of Clinical Oncology, University of Sheffield, Weston Park Hospital, Sheffield, UK

Todd T. Brown, MD, PhD
Division of Endocrinology, Diabetes, and Metabolism, Johns Hopkins University, Baltimore, Maryland, USA

Øyvind S. Bruland, MD, PhD
Department of Oncology, The Norwegian Radium Hospital, University of Oslo, Oslo, Norway

Whitney A. Bullock, BS
Department of Anatomy and Cell Biology,
Indiana University School of Medicine, Indianapolis, Indiana, USA

Erin Ealba Bumann, DDS, PhD, MS
Department of Oral and Craniofacial Sciences, University of Missouri-Kansas City School of Dentistry, Kansas City, Missouri, USA

David B. Burr, PhD
Department of Anatomy and Cell Biology, School of Medicine, Indiana University; and Department of Biomedical Engineering, Purdue School of Engineering and Technology, IUPUI, Indianapolis, Indiana, USA

David A. Bushinsky, MD
Department of Medicine, University of Rochester Medical Center, Rochester, New York, USA

Laura M. Calvi, MD
University of Rochester Multidisciplinary Neuroendocrinology Clinic, Rochester, New York, USA

Christopher Cardozo, MD
Department of Veterans Affairs Rehabilitation Research and Development Service
National Center for the Medical Consequences of Spinal Cord Injury

(*Deceased)

James J. Peters Veterans Affairs Medical Center, Bronx, New York; Departments of Medicine and Rehabilitation Medicine, Icahn School of Medicine at Mount Sinai, New York; Department of Pharmacological Sciences, Icahn School of Medicine at Mount Sinai, New York, New York, USA

Thomas O. Carpenter, MD
Yale University School of Medicine, New Haven, Connecticut, USA

Jacqueline R. Center, MBBS, MS, PhD
Division of Bone Biology, Garvan Institute of Medical Research; and St Vincent's Hospital Clinical School, UNSW Sydney, Sydney, New South Wales, Australia

Julia F. Charles, MD, PhD
Brigham and Women's Hospital, Boston, Massachusetts, USA

Edward Chow, MBBS, MSc, PhD, FRCPC
Department of Radiation Oncology, Odette Cancer Centre, Sunnybrook Health Sciences Centre, University of Toronto, Toronto, Ontario, Canada

Sylvia Christakos, PhD
Department of Microbiology, Biochemistry and Molecular Genetics, Rutgers, New Jersey Medical School, Newark, New Jersey, USA

Yong-Hee Patricia Chun, DDS, MS, PhD
Department of Periodontics; and Department of Cell Systems and Anatomy, University of Texas Health Science Center at San Antonio, San Antonio, Texas, USA

Roberto Civitelli, MD
Division of Bone and Mineral Diseases, Department of Internal Medicine, Washington University in St Louis, Missouri, USA

Thomas Clemens, PhD
Department of Orthopedic Surgery, Johns Hopkins School of Medicine, Baltimore, Maryland, USA

Denis Clohisy, MD
Orthopaedic Surgery, University of Minnesota, Minneapolis, Minnesota, USA

Adi Cohen, MD
Division of Endocrinology, Department of Medicine, College of Physicians and Surgeons, Columbia University, New York, New York, USA

Michael T. Collins, MD
Section of Skeletal Diseases and Mineral Homeostasis, National Institute of Dental and Craniofacial Research, National Institutes of Health, Department of Health and Human Services, Bethesda, Maryland, USA

Juliet Compston, OBE, MD, FRCP, FRCPE, FRCPath, FMedSci
Department of Medicine, Cambridge Biomedical Campus, University of Cambridge, Cambridge, UK

Cyrus Cooper, OBE, DL, FMedSci
MRC Lifecourse Epidemiology Unit, University of Southampton, Southampton General Hospital, Southampton, UK

Felicia Cosman, MD
Columbia College of Physicians and Surgeons, Columbia University; and Clinical Research Center, Helen Hayes Hospital, West Haverstraw, New York, New York, USA

Nicola J. Crabtree, PhD
Department of Endocrinology, Birmingham Women's and Children's Hospital, Birmingham, UK

Clarissa S. Craft, PhD
Division of Bone and Mineral Diseases, Department of Internal Medicine, Washington University, St Louis, Missouri, USA

Fabiana Csukasi, PhD
Orthopaedic Surgery, University of California at Los Angeles, California, USA

Elizabeth M. Curtis, MA, MB, BChir, MRCP
The MRC Lifecourse Epidemiology Unit, University of Southampton, Southampton General Hospital, Southampton, UK

Natalie E. Cusano, MD
Department of Medicine, Lenox Hill Hospital, New York, New York, USA

Stella D'Oronzo, MD
Academic Unit of Clinical Oncology, University of Sheffield, Weston Park Hospital, Sheffield, UK

Enrico Dall'Ara, PhD
Department of Oncology and Metabolism and INSIGNEO Institute for In Silico Medicine, University of Sheffield, Sheffield, UK

Robin M. Daly, PhD, FSMA
Institute for Physical Activity and Nutrition, School of Exercise and Nutrition Sciences, Deakin University, Geelong, Victoria, Australia

Terry F. Davies, MBBS, MD, FRCP, FACE
The Mount Sinai Bone Program, Department of Medicine, Icahn School of Medicine at Mount Sinai, New York, USA

Bess Dawson-Hughes, MD
USDA Nutrition Research Center at Tufts University, Boston, Massachusetts, USA

David J.J. de Gorter, PhD
Institute of Musculoskeletal Medicine, University Hospital Münster, Münster, Germany

Tobias J. de Villiers, MD, FRCOG
Department of Gynecology, Stellenbosch University, Mediclinic Panorama, Cape Town, South Africa

Elaine M. Dennison, MA, MB, BChir, MSc, PhD
The MRC Lifecourse Epidemiology Unit, University of Southampton, Southampton General Hospital, Southampton, UK

Adolfo Diez-Perez, MD, PhD
Hospital del Mar Institute of Medical Investigation and Department of Internal Medicine, Autonomous University of Barcelona, Barcelona, Spain

Linda A. DiMeglio, MD
Department of Pediatrics, Indiana University School of Medicine Indianapolis, Indiana, USA

Naomi Dirckx, PhD
Laboratory of Skeletal Cell Biology and Physiology (SCEBP), Skeletal Biology and Engineering Research Center (SBE), Department of Development and Regeneration, KU Leuven, Leuven, Belgium

Matthew T. Drake, MD, PhD
Department of Internal Medicine, Division of Endocrinology, Diabetes, and Metabolism, College of Medicine, Mayo Clinic, Rochester, Minnesota, USA

Patricia F. Ducy, PhD
Department of Pathology & Cell Biology, College of Physicians and Surgeons, Columbia University, New York, New York, USA

Amel Dudakovic, PhD
Department of Orthopedic Surgery, Mayo Clinic, Rochester, Minnesota, USA

Emma L. Duncan, MBBS, FRCP, FRACP, PhD
Royal Brisbane and Women's Hospital; Queensland University of Technology; and University of Queensland, Brisbane, Queensland, Australia

Gustavo Duque, MD, PhD, FRACP, FGSA
Australian Institute for Musculoskeletal Science (AIMSS), The University of Melbourne and Western Health; and Department of Medicine-Western Health, Melbourne Medical School, The University of Melbourne, St Albans, Victoria, Australia

Richard Eastell, MD, FRCP, FRCPath, FMedSci
Department of Oncology and Metabolism, University of Sheffield, Sheffield, UK

Peter R. Ebeling, AO, MBBS, MD, FRACP
Department of Medicine, School of Clinical Sciences, Monash University, Clayton, Victoria, Australia

Michael J. Econs, MD
Division of Endocrinology and Metabolism, Indiana University School of Medicine, Indianapolis, Indiana, USA
Department of Medicine, Indiana University School of Medicine, Indianapolis, IN, USA

Claire M. Edwards, PhD
Nuffield Department of Surgical Sciences; and Nuffield Department of Orthopaedics, Rheumatology and Musculoskeletal Sciences, Botnar Research Centre, University of Oxford, Oxford, UK

Paul C. Edwards, DDS, MSc, FRCD(C)
Department of Oral Pathology, Medicine and Radiology, Indiana University School of Dentistry, Indianapolis, Indiana, USA

Thomas A. Einhorn, MD
Department of Orthopaedic Surgery
NYU Langone Medical Center, New York University, New York, New York, USA

John A. Eisman, AO, MBBS, PhD, FRACP
Garvan Institute of Medical Research; Endocrinology, St Vincent's Hospital; School of Medicine Sydney, University of Notre Dame Australia; and UNSW Australia, Sydney, Australia

Klaus Engelke, PhD
Institute of Medical Physics, Friedrich-Alexander-Universität Erlangen-Nürnberg, Erlangen, Germany

Paola A. Erba, MD, PhD
Department of Translational Research and New Technology in Medicine, University of Pisa, Pisa, Italy

Nicole J. Ethen, PhD
Center for Cancer and Cell Biology and Program for Skeletal Disease and Tumor Microenvironment, Van Andel Research Institute, Grand Rapids, Michigan, USA

Joshua N. Farr, PhD
Division of Endocrinology and Kogod Center on Aging, Mayo Clinic College of Medicine, Rochester, Minnesota, USA

Murray J. Favus, MD
Department of Medicine, The University of Chicago, Chicago, Illinois, USA

Serge Ferrari, MD
Faculty of Medicine–Medicine Specialties, Geneva University Hospitals, Geneva, Switzerland

Mark R. Forwood, PhD
School of Medical Science and Menzies Health Institute Queensland, Griffith University, Gold Coast, Queensland, Australia

Contributors xv

Benjamin L. Franc, MD
Department of Radiology and Biomedical Imaging, Nuclear Medicine Section, University of California, San Francisco, California, USA

Benjamin J. Frisch, PhD
University of Rochester, Rochester, New York, USA

Marta Galan-Diez, PhD
Columbia University, New York, New York, USA

Thomas J. Gardella, PhD
Endocrine Unit, Department of Medicine, Harvard Medical School, Massachusetts General Hospital, Boston, Massachusetts, USA

Harry K. Genant, MD
Departments of Radiology, Orthopedic Surgery, Medicine, and Epidemiology, University of California at San Francisco, San Francisco, California, USA

Piet Geusens, MD, PhD
Department of Rheumatology, Maastricht University Medical Center, The Netherlands; and University Hasselt, Belgium

Lora Giangregorio, PhD
Schlegel-University of Waterloo Research Institute for Aging; and Department of Kinesiology, University of Waterloo, Waterloo, Ontario, Canada

Andrea Giusti, MD
Rheumatology Unit, Department of Locomotor System, La Colletta Hospital, Arenzano, Italy

Francis H. Glorieux, OC, MD, PhD
Shriners Hospital for Children-Canada and McGill University, Montréal, Québec, Canada

Deborah T. Gold, PhD
Departments of Psychiatry and Behavioral Sciences, Sociology, and Psychology and Neuroscience, Center for the Study of Aging and Human Development, Duke University Medical Center, Durham, North Carolina, USA

Steven R. Goldring, MD
Hospital for Special Surgery, Weill Medical College of Cornell University, New York, New York, USA

David Goltzman, MD
Department of Medicine, McGill University and McGill University Health Centre, Montréal, Québec, Canada

Gary S. Gottesman, MD, FAAP, FACMG
Center for Metabolic Bone Disease and Molecular Research, Shriners Hospitals for Children–St Louis, St Louis, Missouri, USA

Aline Granja Costa
Department of Medicine, Division of Endocrinology, Metabolic Bone Diseases Unit, College of Physicians and Surgeons, Columbia University, New York, New York, USA

Ellen M. Gravallese, MD
Division of Rheumatology; Departments of Medicine and Translational Research, Musculoskeletal Center of Excellence, University of Massachusetts Medical School, Worcester, Massachusetts, USA

Susan L. Greenspan, MD
Divisions of Geriatrics, Endocrinology and Metabolism, University of Pittsburgh; and Osteoporosis Prevention and Treatment Center, University of Pittsburgh Medical Center, Pittsburgh, Pennsylvania, USA

James F. Griffith, MB.BCh, BAO, MD, MRCP, FRCR, FHKCR, FHKAM
Department of Imaging and Interventional Radiology, The Chinese University of Hong Kong, Hong Kong

Monica Grover, MBBS
Stanford School of Medicine/Stanford Children's Hospital, Stanford, California, USA

Theresa Guise, MD
Department of Oncology, Medicine, and Pharmacology, Division of Endocrinology, Indiana University School of Medicine, Indianapolis, Indiana, USA

Neveen A.T. Hamdy, MD, MRCP
Department of Internal Medicine, Division of Endocrinology and Centre for Bone Quality, Leiden University Medical Center, The Netherlands

Catherine Handforth, MBChB
Academic Unit of Clinical Oncology, University of Sheffield, Weston Park Hospital, Sheffield, UK

Mark R. Hanudel, MD, MS
Department of Pediatrics, Division of Nephrology, David Geffen School of Medicine at UCLA, Los Angeles, California, USA

Nicholas C. Harvey, MA, MB, BChir, PhD, FRCP
MRC Lifecourse Epidemiology Unit, University of Southampton, Southampton General Hospital, Southampton, UK

Robert P. Heaney, MD*
Department of Medicine, Creighton University School of Medicine, Omaha, Nebraska, USA

Geoffrey N. Hendy, PhD*
Departments of Medicine, Physiology, and Human Genetics, McGill University, Montréal, Québec, Canada

(*Deceased)

Charles Hildebolt, PhD
Mallinckrodt Institute of Radiology, Washington University in St Louis, Missouri, USA

Matthew J. Hilton, PhD
Department of Orthopaedic Surgery, Duke Orthopaedic Cellular, Developmental, and Genome Laboratories, Duke University School of Medicine; and Department of Cell Biology, Duke University, Durham, North Carolina, USA

Ingrid A. Holm, MD, MPH
Divisions of Genetics and Genomics, Boston Children's Hospital; and Department of Pediatrics, Harvard Medical School, Boston, Massachusetts, USA

Mara J. Horwitz, MD
Division of Endocrinology and Metabolism, The University of Pittsburgh School of Medicine, Pittsburgh, Pennsylvania, USA

Amira I. Hussein, PhD
Department of Orthopaedic Surgery, Boston University Medical Center, Boston, Massachusetts, USA

Carlos M. Isales, MD, FACP
Department of Medicine, Augusta University, Augusta, Georgia, USA

Christina Jacobsen, MD, PhD
Divisions of Endocrinology, Genetics, and Genomics, Boston Children's Hospital; and Department of Pediatrics, Harvard Medical School, Boston, Massachusetts, USA

Suzanne M. Jan de Beur, MD
Division of Endocrinology and Metabolism, Department of Medicine, The Johns Hopkins School of Medicine, Baltimore, Maryland, USA

Kyu Sang Joeng, PhD
McKay Orthopaedic Research Laboratory, Department of Orthopaedic Surgery, Perelman School of Medicine, University of Pennsylvania, Philadelphia, Pennsylvania, USA

Rachelle W. Johnson, PhD
Center for Bone Biology, Department of Medicine, Division of Clinical Pharmacology, Vanderbilt University Medical Center; and Department of Cancer Biology, Vanderbilt University, Nashville, Tennessee, USA

Graeme Jones, MD, PhD
Menzies Institute for Medical Research, University of Tasmania, Hobart, Tasmania, Australia

Harald Jüppner, MD
Endocrine Unit and Pediatric Nephrology Unit, Departments of Medicine and Pediatrics, Harvard Medical School, Massachusetts General Hospital, Boston, Massachusetts, USA

Vesa Kaartinen, PhD
Department of Biologic and Materials Sciences, University of Michigan School of Dentistry, Ann Arbor, Michigan, USA

John A. Kanis, MD
Centre for Metabolic Bone Diseases, University of Sheffield Medical School, Sheffield, UK; and Mary McKillop Health Institute, Australian Catholic University, Melbourne, Victoria, Australia

Frederick S. Kaplan, MD
Center for Research in FOP and Related Disorders, Department of Orthopaedic Surgery, The University of Pennsylvania School of Medicine, Philadelphia, Pennsylvania, USA

Courtney M. Karner, PhD
Department of Orthopaedic Surgery, Duke Orthopaedic Cellular, Developmental, and Genome Laboratories, Duke University School of Medicine; and Department of Cell Biology, Duke University, Durham, North Carolina, USA

Gerard Karsenty, MD, PhD
Department of Genetics & Development, College of Physicians and Surgeons, Columbia University, New York, New York, USA

Aliya Aziz Khan, MD, FRCPC, FACP, FACE
Department of Medicine, Divisions of Endocrinology and Metabolism and Geriatric Medicine, McMaster University, Hamilton, Ontario, Canada

Sundeep Khosla, MD
Department of Internal Medicine, Division of Endocrinology, Diabetes, and Metabolism, College of Medicine, Mayo Clinic, Rochester, Minnesota, USA

Douglas P. Kiel, MD, MPH
Musculoskeletal Research Center, Institute for Aging Research, Hebrew SeniorLife; Department of Medicine, Beth Israel Deaconess Medical Center and Harvard Medical School; and Broad Institute of MIT and Harvard, Boston, Massachusetts, USA

Keith L. Kirkwood, DDS, PhD
Department of Oral Biology, University at Buffalo, The State University of New York, Buffalo, New York, USA

Gordon L. Klein, MD, MPH
Department of Orthopaedic Surgery, University of Texas Medical Branch, Galveston, Texas, USA

Anne Klibanski, MD
Neuroendocrine Unit, Massachusetts General Hospital and Harvard Medical School, Boston, Massachusetts, USA

Scott L. Kominsky, PhD
Departments of Orthopaedic Surgery and Oncology, Johns Hopkins University, Baltimore, Maryland, USA

Stavroula Kousteni, PhD
Columbia University, New York, New York, USA

Christopher S. Kovacs, MD
Faculty of Medicine—Endocrinology, Health Sciences Centre, Memorial University of Newfoundland, St. John's, Newfoundland, Canada

Deborah Krakow, MD
Orthopaedic Surgery and Human Genetics, University of California at Los Angeles, California, USA

Paul H. Krebsbach, DDS, PhD
UCLA School of Dentistry, Los Angeles, California, USA

Henry M. Kronenberg, MD
Endocrine Unit, Massachusetts General Hospital and Harvard Medical School, Boston, Massachusetts, USA

Marie Hélène Lafage-Proust, PU-PH
SAINBIOSE Inserm, Université de Lyon, Saint-Étienne, France

Michaël R. Laurent, MD, PhD
Centre for Metabolic Bone Diseases, University Hospitals Leuven, Leuven, Belgium; and Gerontology and Geriatrics, Department of Clinical and Experimental Medicine, KU Leuven, Leuven, Belgium

Benjamin Z. Leder, MD
Massachusetts General Hospital; and Harvard Medical School, Boston, Massachusetts, USA

Brendan Lee, MD, MPH
Department of Molecular and Human Genetics, Baylor College of Medicine, Texas Children's Hospital, Houston, Texas, USA

Wen-Chih Lee, PhD
Department of Orthopedic Surgery, Washington University School of Medicine, St Louis, Missouri, USA

Siyang Leng, MD
Division of Hematology and Oncology, Columbia University Medical Center, New York, New York, USA

Suzanne Lentzsch, MD, PhD
Division of Hematology and Oncology, Columbia University Medical Center, New York, New York, USA

William D. Leslie, MD, MSc
Department of Medicine, University of Manitoba, Winnipeg, Manitoba, Canada

Michael A. Levine, MD, MACE, FAAP, FACP
Division of Endocrinology and Diabetes, The Children's Hospital of Philadelphia; and Department of Pediatrics, University of Pennsylvania Perelman School of Medicine, Philadelphia, Pennsylvania, USA

E. Michael Lewiecki, MD
New Mexico Clinical Research & Osteoporosis Center, Albuquerque, New Mexico, USA

Agnès Linglart, MD, PhD
APHP, Hôpital Bicêtre Paris Sud, Service d'endocrinologie et diabétologie pour enfants, Centre de référence des Maladies Rares du métabolisme du calcium et du phosphate, Plateforme d'Expertise Paris Sud des maladies rares, filière OSCAR, Le Kremlin Bicêtre, France

Paul Lips, MD
Department of Internal Medicine, Endocrine Section, VU University Medical Center, Amsterdam, The Netherlands

David G. Little, PhD, FRACS (Orth)
Paediatrics and Child Health, University of Sydney, Sydney; and Orthopaedic Research and Biotechnology, The Children's Hospital at Westmead, Westmead, New South Wales, Australia

Jianmin Liu, MD, PhD
Department of Endocrine and Metabolic Diseases, Rui-jin Hospital, Shanghai Jiao-tong University School of Medicine, Shanghai, China

Peng Liu, MD, PhD
The Mount Sinai Bone Program, Department of Medicine, Icahn School of Medicine at Mount Sinai, New York, New York, USA

Alayna E. Loiselle, PhD
The Center for Musculoskeletal Research and The Department of Orthopaedics, The University of Rochester Medical Center, Rochester, New York, USA

Fanxin Long, PhD
Departments of Orthopedic Surgery, Medicine, and Developmental Biology, Washington University School of Medicine, St Louis, Missouri, USA

Gabriela G. Loots, PhD
Physical and Life Sciences, Lawrence Livermore National Laboratory, Livermore, California, USA

Linchao Lu
Department of Pediatrics, Section of Hematology/Oncology, Baylor College of Medicine, Texas Children's Hospital, Houston, Texas, USA

Karen Lyons, PhD
Department of Orthopaedic Surgery/Orthopaedic Hospital, University of California, Los Angeles, California, USA

Nina S. Ma, MD
Boston Children's Hospital/Harvard Medical School, Boston, Massachusetts, USA

Lauren M. MacCormick, MD
Orthopaedic Surgery, University of Minnesota, Minneapolis, Minnesota, USA

Christa Maes, PhD
Laboratory of Skeletal Cell Biology and Physiology (SCEBP), Skeletal Biology and Engineering Research Center (SBE), Department of Development and Regeneration, KU Leuven, Leuven, Belgium

Sharmila Majumdar, PhD
Department of Radiology and Biomedical Imaging; Department of Orthopedic Surgery; and Department of Bioengineering and Therapeutics Sciences, UCSF, San Francisco, California, USA

Michael Mannstadt, MD
Endocrine Unit, Massachusetts General Hospital and Harvard Medical School, Boston, Massachusetts, USA

Stavros C. Manolagas, MD, PhD
University of Arkansas for Medical Sciences and the Central Arkansas Veterans Healthcare System, Little Rock, Arkansas, USA

Manasa Mantravadi, MD, MS
Department of Pediatrics, Indiana University School of Medicine, Indianapolis, Indiana, USA

Claudio Marcocci, MD,
Department of Clinical and Experimental Medicine, University of Pisa, Pisa, Italy

Joan C. Marini, MD, PhD
The Eunice Kennedy Shriver National Institute of Child Health and Human Development, Section on Heritable Disorders of Bone and Extracellular Matrix, National Institutes of Health, Bethesda, Maryland, USA

T. John Martin, MD, DSc
St Vincent's Institute of Medical Research, Department of Medicine, University of Melbourne, Melbourne, Australia

Takashi Matsuura, DDS, PhD
Department of Oral Rehabilitation, Fukuoka Dental College, Fukuoka, Japan

Laurie McCauley, DDS, MS, PhD
Periodontics and Oral Medicine, University of Michigan School of Dentistry, Ann Arbor, Michigan, USA

Eugene V. McCloskey, MD, FRCPI
Centre for Metabolic Bone Diseases, University of Sheffield Medical School, Sheffield, UK

Michael R. McClung, MD, FACP, FACE
Institute of Health and Ageing, Australian Catholic University, Melbourne, Victoria, Australia; and Oregon Osteoporosis Center, Portland, Oregon, USA

Lisa K. Micklesfield, PhD
MRC Developmental Pathways for Health Research Unit, Department of Pediatrics, Faculty of Health Sciences, University of the Witwatersrand, Johannesburg, South Africa

Paul D. Miller, MD, FACP
Colorado Center for Bone Research at Panorama Orthopedics and Spine Center, Golden, Colorado, USA

Yuji Mishina, PhD
Department of Biologic and Materials Sciences, School of Dentistry, University of Michigan, Ann Arbor, Michigan, USA

Madhusmita Misra, MD, MPH
Department of Pediatrics, Harvard Medical School; and Pediatric Endocrinology, Massachusetts General Hospital for Children and Harvard Medical School, Boston, Massachusetts, USA

Thimios Mitsiadis, DDS, PhD
Institute of Oral Biology, Medical Faculty of the University of Zurich, Zurich, Switzerland

Sharon M. Moe, MD
Department of Medicine, Division of Nephrology, Indiana University School of Medicine, Indianapolis, Indiana, USA

Carolina Aguiar Moreira, MD, MACP, FACE
Department of Medicine & Section of Endocrinology (SEMPR), Federal University of Parana, Laboratory PRO, Fundacao Pro Renal, Curitiba, Paraná, Brazil

Elise F. Morgan, PhD
Department of Mechanical Engineering, Boston University; and Department of Orthopaedic Surgery, Boston University Medical Center, Boston, Massachusetts, USA

Katherine J. Motyl, PhD
Maine Medical Center Research Institute, Maine Medical Center, Scarborough, Maine, USA

Craig Munns, MBBS, PhD, FRACP
Sydney Medical School, University of Sydney, and the Institute of Endocrinology and Diabetes, The Children's Hospital at Westmead, Westmead, NSW, Australia

Nicola Napoli, MD, PhD
Campus Bio-Medico, University of Rome, Rome, Italy

Lorenzo Nardo, MD, PhD
Department of Radiology, University of California Davis, Sacramento, California, USA

Maria I. New, MD
The Mount Sinai Bone Program, Department of Medicine, Icahn School of Medicine at Mount Sinai, New York, New York, USA

Paul J. Newey, MBCHB, DPHIL
Division of Molecular and Clinical Medicine, University of Dundee, Dundee, UK

Tuan V. Nguyen, DSc, PhD
Division of Bone Biology, Garvan Institute of Medical Research; St Vincent's Hospital Clinical School, UNSW Australia; Centre for Health Technologies, School of Biomedical Engineering, University of Technology; and School of Medicine Sydney, University of Notre Dame Australia, Sydney, New South Wales, Australia

Kyle K. Nishiyama, PhD
Department of Medicine—Endocrinology, Columbia University Medical Center, New York, New York, USA

Robert A. Nissenson, PhD
Endocrine Research Unit, VA Medical Center, Departments of Medicine and Physiology, University of California, San Francisco, California, USA

Shane A. Norris, PhD
MRC Developmental Pathways for Health Research Unit, Department of Pediatrics, Faculty of Health Sciences, University of the Witwatersrand, Johannesburg, South Africa

Chad M. Novince, DDS, MSD, PhD
Department of Oral Health Sciences and the Center for Oral Health Research, Medical University of South Carolina, Charleston, South Carolina, USA

Jeffry S. Nyman, PhD
Department of Orthopaedic Surgery and Rehabilitation, Vanderbilt University Medical Center, Nashville, Tennessee, USA

Hiroki Ochi
Department of Physiology and Cell Biology, Tokyo Medical and Dental University, Tokyo, Japan

Eric S. Orwoll, MD
Division of Endocrinology, Diabetes, and Clinical Nutrition, Oregon Health and Science University, School of Medicine, Portland, Oregon, USA

Roberto Pacifici, MD
Division of Endocrinology, Metabolism and Lipids, Department of Medicine, and Immunology and Molecular Pathogenesis Program, Emory University, Atlanta, Georgia, USA

Petros Papagerakis, DDS, MSc, PhD
College of Dentistry; College of Medicine (Anatomy and Cell Biology); College of Pharmacy and Nutrition, Toxicology and Biomedical Engineering Graduate Programs, University of Saskatchewan, Saskatoon, SK, Canada; and School of Dentistry, Department of Orthodontics and Pediatric Dentistry, Center for Computational Medicine, University of Michigan, Ann Arbor, Michigan, USA

Socrates E. Papapoulos, MD, PhD
Center for Bone Quality, Department of Endocrinology and Metabolic Diseases, Leiden University Medical Center, Leiden, The Netherlands

Fredrick M. Pavalko, PhD
Department of Cellular and Integrative Physiology, Indiana University School of Medicine, Indianapolis, Indiana, USA

John M. Pettifor, MB BCh, PhDMed
MRC Developmental Pathways for Health Research Unit, Department of Pediatrics, Faculty of Health Sciences, University of the Witwatersrand, Johannesburg, South Africa

Robert J. Pignolo, MD, PhD
Division of Geriatric Medicine and Gerontology; and Division of Endocrinology, Diabetes, Metabolism, Nutrition, Department of Internal Medicine, Mayo Clinic College of Medicine, Rochester, Minnesota, USA

Lilian I. Plotkin, PhD
Indiana Center for Musculoskeletal Health (ICMH), Indianapolis, Indiana, USA

Daniel Prieto-Alhambra, MD, MSc(Oxf), PhD
Pharmaco- and Device Epidemiology, Oxford NIHR Biomedical Research Centre, Nuffield Department of Orthopaedics, Rheumatology, and Musculoskeletal Sciences (NDORMS), University of Oxford, Oxford, UK; and GREMPAL Research Group – Idiap Jordi Gol and CIBERFes, Universitat Autònoma de Barcelona and Instituto Carlos III (FEDER Research Funds), Barcelona, Spain

Manoj Ramachandran, BSc(Hons) MBBS(Hons) MRCS(Eng) FRCS(Tr&Orth)
Royal London Hospital, Barts Health NHS Trust, London, UK

Srinivas Raman, MD
Department of Radiation Oncology, Odette Cancer Centre, Sunnybrook Health Sciences Centre, University of Toronto, Toronto, Ontario, Canada

Robert R. Recker, MD, MACP, FACE
Division of Endocrinology, Osteoporosis Research Center, Creighton University School of Medicine, Omaha, Nebraska, USA

Mara Riminucci, PhD
Dipartimento di Medicina Molecolare, Sapienza Università, Rome, Italy

Fernando Rivadeneira, MD, PhD
Departments of Internal Medicine, Erasmus University Medical Center, Rotterdam, The Netherlands

Rene Rizzoli, MD
University Hospitals of Geneva, and Head of the Service of Bone Diseases, Geneva, Switzerland

Pamela G. Robey, PhD
Skeletal Biology Section, National Institute of Dental and Craniofacial Research, National Institutes of Health, Department of Health and Human Services, Bethesda, Maryland, USA

Alexander G. Robling, PhD
Department of Anatomy and Cell Biology, Indiana University School of Medicine, Indianapolis, Indiana, USA

Bernard Roche, PhD
SAINBIOSE Inserm, Université de Lyon, Saint-Étienne, France

G. David Roodman, MD, PhD
Division of Hematology and Oncology, Indiana University School of Medicine, Indianapolis, Indiana, USA

Clifford J. Rosen, MD
Maine Medical Center, Scarborough, Maine, USA

Mishaela R. Rubin, MD
Metabolic Bone Diseases Unit, Division of Endocrinology, Department of Medicine, College of Physicians and Surgeons, Columbia University, New York, New York, USA

Mary D. Ruppe, MD*
Division of Endocrinology, Department of Medicine, Houston Methodist Hospital, Houston, Texas, USA

Kenneth Saag, MD, MSc
Center for Outcomes and Effectiveness Research and Education (COERE); and Division of Clinical Immunology and Rheumatology, University of Alabama at Birmingham (UAB) School of Medicine, Birmingham, Alabama, USA

Isidro B. Salusky, MD
Department of Pediatrics, Division of Nephrology, David Geffen School of Medicine at UCLA, Los Angeles, California, USA

Gonzalo Sánchez-Duffhues, PhD
Department of Molecular Cell Biology and Oncode Institute, Leiden University Medical Center, Leiden, The Netherlands

Oliver Sartor, MD
Departments of Medicine and Urology, Tulane Medical School, New Orleans, Louisiana, USA

Asiya Sbayi, MPH Candidate
The George Washington University, Milken Institute School of Public Health, Washington DC, USA

Anne L. Schafer, MD
University of California San Francisco and San Francisco Department of Veterans Affairs Medical Center, San Francisco, California, USA

Erica L. Scheller, DDS, PhD
Division of Bone and Mineral Diseases, Department of Internal Medicine, Washington University, St Louis, Missouri, USA

Karl Peter Schlingmann, MD
Department of General Pediatrics, University Children`s Hospital, Münster, Germany

Maria Schuller Almeida, PhD
Center for Osteoporosis and Metabolic Bone Diseases, Department of Internal Medicine, Department of Orthopedic Surgery, University of Arkansas for Medical Sciences, Little Rock, Arkansas, USA

Ann Schwartz, PhD, MPH
Department of Epidemiology and Biostatistics, University of California, San Francisco, California, USA

Aimy Sebastian, PhD
Physical and Life Sciences, Lawrence Livermore National Laboratory, Livermore, California, USA

Ego Seeman, BSc, MBBS, FRACP, MD, AM
Departments of Endocrinology and Medicine, Austin Health, University of Melbourne; and the Mary MacKillop Institute for Health Research, Australian Catholic University, Melbourne, Victoria, Australia

Elizabeth Shane, MD
Division of Endocrinology, Department of Medicine, College of Physicians and Surgeons, Columbia University, New York, New York, USA

(*Deceased)

Nicholas J. Shaw, MB ChB, FRCPCH
Department of Endocrinology and Diabetes, Birmingham Children's Hospital, Birmingham, UK

Yiping Shen, PhD
Divisions of Genetics and Genomics, Boston Children's Hospital; and Department of Pediatrics, Harvard Medical School, Boston, Massachusetts, USA

Ce Shi, PhD, DDS
Department of Oral Pathology, School and Hospital of Stomatology, Jilin University, Changchun, Jilin, China; and Department of Biologic and Materials Sciences, School of Dentistry, University of Michigan, Ann Arbor, Michigan, USA

Dolores M. Shoback, MD
University of California San Francisco and San Francisco Department of Veterans Affairs Medical Center, San Francisco, California, USA

Eileen M. Shore, PhD
Center for Research in FOP and Related Disorders, Department of Orthopaedic Surgery, The University of Pennsylvania School of Medicine, Philadelphia, Pennsylvania, USA

Rebecca Silbermann, MD, MMS
Department of Medicine, Division of Hematology-Oncology, Indiana University School of Medicine, Indianapolis, Indiana, USA

Barbara C. Silva, MD, PhD
Division of Endocrinology, Felicio Rocho Hospital; and Santa Casa of Belo Horizonte Department of Medicine, UNI-BH, Belo Horizonte, Brazil

Shonni J. Silverberg, MD
Department of Medicine, Columbia University College of Physicians and Surgeons, New York, New York, USA

Stuart L. Silverman, MD, FACP, FACR
Department of Medicine, Division of Rheumatology, Cedars-Sinai Medical Center, UCLA David Geffen School of Medicine; and the OMC Clinical Research Center, Los Angeles, California, USA

James P. Simmer, DDS, PhD
Department of Biological and Materials Sciences, University of Michigan Dental Research Laboratory, Ann Arbor, Michigan, USA

Andrea Singer, MD
Departments of Obstetrics and Gynecology and Medicine, MedStar Georgetown University Hospital and Georgetown University Medical Center, Washington, DC, USA

Ethel S. Siris, MD
Department of Medicine, Columbia University, New York, New York, USA

Julie A. Sterling, PhD
Department of Veterans Affairs, Tennessee Valley Healthcare System; Center for Bone Biology, Department of Medicine, Division of Clinical Pharmacology, Vanderbilt University Medical Center; and Departments of Cancer Biology and Biomedical Engineering, Vanderbilt University, Nashville, Tennessee, USA

Li Sun, MD, PhD
The Mount Sinai Bone Program, Department of Medicine, Icahn School of Medicine at Mount Sinai, New York, New York, USA

Pawel Szulc, MD, PhD
INSERM UMR 1033, University of Lyon, Hospices Civils de Lyon, Lyon, France

Hiroshi Takayanagi, MD, PhD
Department of Immunology, Graduate School of Medicine and Faculty of Medicine, The University of Tokyo, Tokyo, Japan

Shu Takeda, MD, PhD
Division of Endocrinology, Toranomon Hospital, Tokyo, Japan

Pamela Taxel, MD
Department of Medicine, University of Connecticut Health Center, Farmington, Connecticut, USA

Peter ten Dijke, PhD
Department of Molecular Cell Biology and Oncode Institute, Leiden University Medical Center, Leiden, The Netherlands

Sotirios Tetradis, DDS, PhD
Division of Diagnostic and Surgical Sciences, UCLA School of Dentistry, Los Angeles, California, USA

Rajesh V. Thakker, MD, ScD, FRCP, FRCPath, FMedSci, FRS
University of Oxford, Nuffield Department of Clinical Medicine, OCDEM Churchill Hospital, Oxford, UK

Anna N.A. Tosteson, ScD
Multidisciplinary Clinical Research Center in Musculoskeletal Diseases, The Dartmouth Institute for Health Policy and Clinical Practice; and Department of Medicine, Geisel School of Medicine, Dartmouth College, Lebanon, New Hampshire, USA

Bich Tran, PhD
Centre for Big Data Research in Health, UNSW Sydney, Sydney, New South Wales, Australia

Joy N. Tsai, MD
Massachusetts General Hospital; and Harvard Medical School, Boston, Massachusetts, USA

Joop van den Bergh, MD, PhD
Department of Internal Medicine, VieCuri Medical Center for North Limburg, Venlo; and Department of Internal Medicine, NUTRIM School of Nutrition and Translational Research in Metabolism, Maastricht UMC, Maastricht, The Netherlands; Biomedical Research Center, University of Hasselt, Hasselt, Belgium

Catherine Van Poznak, MD
Department of Internal Medicine–Hematology/Oncology, University of Michigan, Ann Arbor, Michigan, USA

Natasja M. Van Schoor, PhD
Amsterdam Public Health Research Institute, Department of Epidemiology and Biostatistics, VU University Medical Center, Amsterdam, The Netherlands

Andre J. van Wijnen, PhD
Department of Orthopedic Surgery, Mayo Clinic, Rochester, Minnesota, USA

Deepak Vashishth, PhD
Department of Biomedical Engineering, Rensselaer Polytechnic Institute, Troy, New York, USA

Line Vautour, MD, FRCP(C)
Division of Endocrinology and Metabolism, McGill University Health Centre, Montréal, Québec, Canada

Tamara Vokes, MD
Department of Medicine, University of Chicago, Chicago, Illinois, USA

Marcella D. Walker, MD
Division of Endocrinology, Department of Medicine, Columbia University, College of Physicians and Surgeons, New York, New York, USA

Lisa L. Wang, MD
Department of Pediatrics, Section of Hematology/Oncology, Baylor College of Medicine, Texas Children's Hospital, Houston, Texas, USA

Kate Ward, PhD
MRC Lifecourse Epidemiology, University of Southampton, Southampton, UK

Leanne M. Ward, MD, FRCPC
Department of Pediatrics, Faculty of Medicine, University of Ottawa, Ottawa, Ontario, Canada

Stuart J. Warden, PT, PhD, FACSM
Department of Physical Therapy, School of Health and Human Sciences, Indiana University, Indianapolis, Indiana, USA

Nelson B. Watts, MD
Mercy Health, Osteoporosis and Bone Health Services, Cincinnati, Ohio, USA

Connie M. Weaver, PhD
Nutrition Science, Purdue University, West Lafayette, Indiana, USA

Kristy Weber, MD
Department of Orthopaedic Surgery, University of Pennsylvania, Philadelphia, Pennsylvania, USA

Natalie K. Wee, PhD
Division of Bone and Mineral Diseases, Department of Internal Medicine, Washington University, St Louis, Missouri, USA

M. Neale Weitzmann, PhD
Division of Endocrinology, Metabolism and Lipids, Department of Medicine, Emory University, Atlanta; Atlanta Department of Veterans Affairs Medical Center, Decatur, Georgia, USA

Jennifer J. Westendorf, PhD
Department of Orthopedic Surgery, Mayo Clinic, Rochester, Minnesota, USA

Kenneth E. White, PhD
Division of Endocrinology and Metabolism, Indiana University School of Medicine, Indianapolis, Indiana, USA

Michael P. Whyte, MD
Center for Metabolic Bone Disease and Molecular Research, Shriners Hospital for Children; and Division of Bone and Mineral Diseases, Department of Internal Medicine, Washington University School of Medicine at Barnes-Jewish Hospital, St Louis, Missouri, USA

Bart O. Williams, PhD
Center for Cancer and Cell Biology and Program for Skeletal Disease and Tumor Microenvironment, Van Andel Research Institute, Grand Rapids, Michigan, USA

Graham R. Williams, PhD
Molecular Endocrinology Laboratory, Department of Medicine, Imperial College London, London, UK

Karen K. Winer, MD
NICHD, National Institutes of Health, Bethesda, Maryland, USA

Tania Winzenberg, PhD
Menzies Institute for Medical Research, University of Tasmania; and Faculty of Health, University of Tasmania, Hobart, Tasmania, Australia

Laura E. Wright, PhD
Department of Medicine, Division of Endocrinology, Indiana University, Indianapolis, Indiana, USA

John J. Wysolmerski, MD
Section of Endocrinology and Metabolism, Department of Medicine, Yale School of Medicine, New Haven, Connecticut, USA

Junro Yamashita, DDS, MS, PhD
Department of Oral Rehabilitation, Fukuoka Dental College, Fukuoka, Japan

Tao Yang, PhD
Van Andel Research Institute, Grand Rapids, Michigan, USA

Yingzi Yang, PhD
Harvard School of Dental Medicine, Harvard Stem Cell Institute, Boston, Massachusetts, USA

Michael T. Yin, MD, MS
Division of Infectious Diseases, Columbia University Medical Center, New York, New York, USA

Tony Yuen, PhD
The Mount Sinai Bone Program, Department of Medicine, Icahn School of Medicine at Mount Sinai, New York, New York, USA

Mone Zaidi, MD, PhD, FRCP
The Mount Sinai Bone Program, Department of Medicine, Icahn School of Medicine at Mount Sinai, New York, New York, USA

Alberta Zallone, PhD
Department of Histology, University of Bari, Bari, Italy

Huan-Chang Zeng, PhD
Department of Molecular and Human Genetics, Baylor College of Medicine, Texas Children's Hospital, Houston, Texas, USA

K. Liang Zeng, MD
Department of Radiation Oncology, Odette Cancer Centre, Sunnybrook Health Sciences Centre, University of Toronto, Toronto, Ontario, Canada

Zhendong A. Zhong, PhD
Center for Cancer and Cell Biology and Program for Skeletal Disease and Tumor Microenvironment, Van Andel Research Institute, Grand Rapids, Michigan, USA

Michael J. Zuscik, PhD
The Center for Musculoskeletal Research and The Department of Orthopaedics, The University of Rochester Medical Center, Rochester, New York, USA

Preface to the Ninth Edition of the *Primer*

Over 30 years ago, pioneers of the ASBMR, in its early years of existence, had a remarkable vision, namely to create a *Primer on the Metabolic Bone Diseases and Disorders of Mineral Metabolism*, to provide 'a comprehensive, yet concise description of the clinical manifestations, pathophysiology, diagnostic approaches, and therapeutics of diseases that come under the rubric of bone and mineral disorders.' At that time, they pointed out that there was no repository of such information in specialty textbooks. The pioneers also had the remarkable vision to appoint Murray Favus as the inaugural Editor-in-Chief of the *Primer*. The first edition was published in 1990. With enviable longevity, Murray served as Editor-in-Chief of the *Primer* for the next 18 years, when Cliff Rosen assumed the role in 2008. Under Cliff's leadership, the *Primer* continued to reign as the undisputed source of key information in our field, not only for those being introduced to our specialty but also to those of us who periodically need to be refreshed. I am pleased and honored to have been selected to serve as the Editor-in-Chief of this ninth edition of the *Primer*.

The goals of the ninth edition are to continue to provide the most accurate, up-to-date evidence-based information on basic and clinical bone science to beginners and experts, in segments that are concise and eminently readable. Attesting to the vibrancy of our field, we have seen much change since the last edition was published in 2013. The revised chapters and the new ones reflect these changes. We have broadened the authorship to include our younger generation as well as greater international representation. Fully half of the 290 authors are new to the ninth edition. Many of them are our younger stars. One-third of the authorship is from outside the USA. These two points, namely younger and international representation, represent substantial increments over the eighth edition. The highlights of our 30-year history, as represented regularly in all nine editions of the *Primer*, are illustrated on the cover. They display great advances, framed and ready to be shown in a museum!

I am grateful to Juliet Compston and Roger Bouillon for returning as Senior Associate Editors of the ninth edition and to Tom Clemens for joining us at the Senior Associate Editor level. I am grateful also to the returning Section Editors who served in this capacity in the previous edition (Doug Bauer, Suzanne Jan de Beur, Theresa Guise, Karen Lyons, Laurie McCauley, Paul Miller, Socrates Papapoulos, Ego Seeman, Raj Thakker, and Mone Zaidi) and to the Section Editors who are new to this edition (Peter Ebeling, Klaus Engelke, David Goltzman, Harald Jüppner, Mike McClung, David Roodman, Cliff Rosen, and Mike Whyte). Kudos to Ann Elderkin, Executive Director of ASBMR, who took on this added administrative responsibility, when it became necessary, and to Katie Duffy, Publications Director of ASBMR, who worked tirelessly with me, literally from the moment she joined ASBMR. I am also grateful to the Publications Committee under Bob Jilka and its current Chair, Michael Mannstadt.

Many outstanding texts in our field have been published since 1990, when there were virtually none. The *Primer*, however, still stands tall as a unique resource for the broadest, most comprehensive, and easily readable text of them all. I hope all of you gain the knowledge, wisdom, and insights that are contained in these pages. As a result, your work, whether it is basic or clinical research, or patient care, or any combination will be enhanced every time you take the *Primer* down from your real or electronic bookshelf.

John P. Bilezikian, MD, PhD (hon)
College of Physicians and Surgeons
Columbia University, New York, New York, USA
July 2018

Meet Patricia, an ASBMR member.

The ASBMR helped Patricia with funding that allowed her to build her lab and get national recognition for her research on bone metastasis.

Did you know that the ASBMR membership represents a global network of scientists and clinician scientists? With nearly half of members located outside of North America, the ASBMR brings together the global bone and mineral research community to collaborate, advance their research and ultimately impact human health.

We're here to help you, too. Find your scientific home with the American Society for Bone and Mineral Research.

LEARN MORE ABOUT PATRICIA'S STORY AT

Patricia Juárez Camacho, Ph.D.
Assistant Professor, Center for Scientific Research and Higher Education at Ensenada (CICESE), Baja California, México
ASBMR member for 10 years

President's Preface

The ninth edition of the *Primer on the Metabolic Bone Diseases and Disorders of Mineral Metabolism* was developed during the 40th anniversary year of American Society for Bone and Mineral Research, a huge milestone for this seminal text that has introduced students and fellows to the field of bone, mineral, and musculoskeletal research since the first edition in 1990. We are grateful for the leadership of Editor-in-Chief John Bilezikian as well as Senior Associate Editors Tom Clemens, Juliet Compston, and Roger Bouillon and their outstanding team of 18 luminary Section Editors.

In this new edition, 11 sections capture the very cutting edge of research covering mineral homeostasis, osteoporosis, and other metabolic bone diseases, skeletal measurement technologies, genetics, and much, much more. The 135 chapters – 15 of them new for this edition – feature over 275 figures and almost 300 contributing authors from wide-ranging international research centers. Although the breadth of the *Primer* coverage is wide, John Bilezikian, the Associate Editors, and the Section Editors endeavored to condense essential materials into chapters with more compact reference lists, for easier reading and teaching.

The *Primer* represents the highest standards of collated scientific content and has evolved to include digital and print formats as well as a companion site at www.wiley.com/go/asbmrprimer, where researchers, instructors, clinicians, and students can download valuable teaching slides of tables and figures from the chapters. We hope that you will enjoy and value the extraordinary effort to capture the most current state of the field in the pages that follow.

Michael J. Econs, MD
Indiana University School of Medicine
Indianapolis, Indiana, USA
President, ASBMR
July 2018

About the Companion Website

This book is accompanied by a companion website:

www.wiley.com/go/asbmrprimer

The website includes:

- Videos from the ninth edition
- Editors' biographies
- Slide sets of all figures and tables from the book for downloading
- Useful website links

Section I
Molecular and Cellular Determinants of Bone Structure and Function

Section Editor: Karen Lyons

Chapter 1. Early Skeletal Morphogenesis in Embryonic Development   3
Yingzi Yang

Chapter 2. Endochondral Ossification   12
Courtney M. Karner and Matthew J. Hilton

Chapter 3. Local and Circulating Osteoprogenitor Cells and Lineages   20
Naomi Dirckx and Christa Maes

Chapter 4. Osteoblasts: Function, Development, and Regulation   31
Elizabeth W. Bradley, Jennifer J. Westendorf, Andre J. van Wijnen, and Amel Dudakovic

Chapter 5. Osteocytes   38
Lynda F. Bonewald

Chapter 6. Osteoclast Biology and Bone Resorption   46
Hiroshi Takayanagi

Chapter 7. Signal Transduction Cascades Controlling Osteoblast Differentiation   54
David J.J. de Gorter, Gonzalo Sánchez-Duffhues, and Peter ten Dijke

Chapter 8. The TGF-β Superfamily in Bone Formation and Maintenance   60
Ce Shi and Yuji Mishina

Chapter 9. Recent Developments in Understanding the Role of Wnt Signaling in Skeletal Development and Disease   68
Zhendong A. Zhong, Nicole J. Ethen, and Bart O. Williams

Chapter 10. Mechanotransduction in Bone Formation and Maintenance   75
Whitney A. Bullock, Lilian I. Plotkin, Alexander G. Robling, and Fredrick M. Pavalko

Chapter 11. The Composition of Bone   84
Adele L. Boskey and Pamela G. Robey

Chapter 12. Assessment of Bone Mass, Structure, and Quality in Rodents   93
Jeffry S. Nyman and Deepak Vashishth

Chapter 13. Skeletal Healing: Cellular and Molecular Determinants   101
Alayna E. Loiselle and Michael J. Zuscik

Chapter 14. Biomechanics of Fracture Healing   108
Elise F. Morgan, Amira I. Hussein, and Thomas A. Einhorn

Primer on the Metabolic Bone Diseases and Disorders of Mineral Metabolism, Ninth Edition. Edited by John P. Bilezikian.
© 2019 American Society for Bone and Mineral Research. Published 2019 by John Wiley & Sons, Inc.
Companion website: www.wiley.com/go/asbmrprimer

1

Early Skeletal Morphogenesis in Embryonic Development

Yingzi Yang

Harvard School of Dental Medicine, Harvard Stem Cell Institute, Boston, MA, USA

INTRODUCTION

Formation of the skeletal system is one of the hallmarks that distinguish vertebrates from invertebrates. In higher vertebrates (ie, birds and mammals), the skeletal system contains mainly cartilage and bone that are mesoderm-derived tissues and formed by chondrocytes and osteoblasts, respectively, during embryogenesis. A common mesenchymal progenitor cell also referred to as the osteochondral progenitor gives rise to both chondrocytes and osteoblasts. Skeletal development starts from mesenchymal condensation, during which mesenchymal progenitor cells aggregate at future skeletal locations. Because mesenchymal cells in different parts of the embryo are derived from different cell lineages, the locations of initial skeletal formation determine which of the three mesenchymal cell lineages contribute to the future skeleton. Neural crest cells from the branchial arches contribute to the craniofacial bone, the sclerotome compartment of the somites gives rise to most axial skeletons, and lateral plate mesoderm forms the limb mesenchyme, from which limb skeletons are derived (Fig. 1.1). Ossification is one of the most critical processes in skeletal development and this process is controlled by two major mechanisms: intramembranous and endochondral ossification. Osteochondral progenitors differentiate into osteoblasts to form the membranous bone during intramembranous ossification, whereas during endochondral ossification, osteochondral progenitors differentiate into chondrocytes instead to form a cartilage template of the future bone. The location of each skeletal element also determines its ossification mechanism and unique anatomic properties such as the shape and size. Importantly, the positional identity of each skeletal element is acquired early in embryonic development, even before mesenchymal condensation, through a process called pattern formation.

Cell–cell communication that coordinates cell proliferation, differentiation, and polarity plays a critical role in pattern formation. Patterning of the early skeletal system is controlled by several major signaling pathways that also regulate other pattern formation processes. These signaling pathways are mediated by morphogens including Wnts, Hedgehogs (Hhs), bone morphogenetic proteins (BMPs), fibroblast growth factors (FGFs), and Notch/Delta. Recently, the Turing model [1] of pattern formation that determines skeletal formation spatially and temporally has drawn increasing attention. In his seminar paper [1], Turing proposed an ingenious hypothesis that the patterns we observe during embryonic development arise in response to a spatial prepattern in morphogens. Cells would then respond to this prepattern by differentiating in a threshold-dependent way. Thus, Turing hypothesized that the patterns we see in nature, such as skeletal structures, are controlled by a self-organizing network of interacting morphogens. The Turing model has been successfully tested in limb skeletal patterning with combined computational modeling and experimental approaches [2–5].

EARLY SKELETAL PATTERNING

In the craniofacial region, neural crest cells are major sources of cells establishing the craniofacial skeleton [6]. It is the temporal and spatial-dependent reciprocal signaling between and among the neural crest cells and the epithelial cells (surface ectoderm, neural ectoderm,

---

Primer on the Metabolic Bone Diseases and Disorders of Mineral Metabolism, Ninth Edition. Edited by John P. Bilezikian.
© 2019 American Society for Bone and Mineral Research. Published 2019 by John Wiley & Sons, Inc.
Companion website: www.wiley.com/go/asbmrprimer

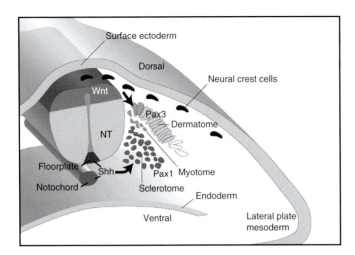

Fig. 1.1. Cell lineage contribution of chondrocytes and osteoblasts. Neural crest cells are born at the junction of dorsal neural tube and surface ectoderm. In the craniofacial region, neural crest cells from the branchial arches differentiate into chondrocytes and osteoblasts. In the trunk, axial skeletal cells are derived from the ventral somite compartment, sclerotome. Shh secreted from the notochord and floor plate of the neural tube induces the formation of sclerotome which expresses Pax1. Wnts produced in the dorsal neural tube inhibits sclerotome formation and induces dermomyotome that expresses Pax3. Cells from the lateral plate mesoderm will form the limb mesenchyme, from which limb skeletons are derived. Source: [16,17]. Reproduced with permission of Elsevier.

or endodermal cells) that ultimately establish the pattern of craniofacial skeleton formed by neural crest cells [7].

The most striking feature of axial skeleton patterning is the periodic organization of the vertebral columns along the anterior–posterior (A–P) axis. This pattern is established when somites, which are segmented mesodermal structures on either side of the neural tube and the underlying notochord, bud off at a defined pace from the anterior tip of the embryo's presomitic mesoderm (PSM) [8]. Somites give rise to axial skeleton, striated muscle, and dorsal dermis [9]. The repetitive and left–right symmetrical patterning of axial skeleton is controlled by a molecular oscillator or the segmentation clock that act in the PSM (Fig. 1.2A). The segmentation clock is operated by a traveling wave of gene expression (or cyclic gene expression) along the embryonic A–P axis, which is generated by an interacting molecular network of the Notch, Wnt/β-catenin, and FGF signaling pathways (Fig. 1.2B). Understanding molecular control of vertebrate segmentation has provided a conceptual framework to explain human diseases of the spine, such as congenital scoliosis [10].

The Notch signaling pathway mediates short-range communication between contacting cells [11]. The majority of cyclic genes are downstream targets of the Notch signaling pathway and code for Hairy/Enhancer of split (Hes) family members, Lunatic fringe (Lfng), and the Notch ligand Delta. The Wnt/β-catenin and FGF signaling pathways mediate long-range signaling across several cell diameters. Upon activation, β-catenin is stabilized and translocates to the nucleus where it binds Lef/Tcf factors and activates expression of downstream genes. Axin2, Dkk1, Dac1, and Nkd1 are Wnt-activated negative regulators that are rhythmically expressed in the PSM. The FGF signaling pathway is also activated periodically in the posterior PSM, indicated by the dynamic phosphorylation of ERK in the mouse PSM. FGF-negative feedback inhibitors, such as Sprouty homolog 2 and 4 (Spry2 and Spry4) and Dual specificity phosphatase 4 and 6 (Dusp4 and 6), are cyclically expressed. There are extensive cross-talks among these major oscillating signaling pathways. However, current studies suggest that none of the three signaling pathways individually acts as a global pacemaker. If there is no unidentified master pacemaker, it likely that each of the three pathways has the capacity to generate its own oscillations, while interactions among them allow efficient coupling and entrain them to each other.

The retinoic acid (RA) signaling controls somitogenesis by regulating the competence of PSM cells to undergo segmentation via antagonizing FGF signaling (Fig. 1.2A) [12]. RA signaling has an additional role in maintaining left–right bilateral symmetry of somites by buffering asymmetric signals that establish the left–right axis of the body, particularly Fgf8 [13].

The functional significance of the segmentation clock in human skeletal development is highlighted by congenital axial skeletal diseases. Abnormal vertebral segmentation (AVS) in humans is a relatively common malformation. For instance, mutations in NOTCH signaling components cause at least two human disorders, spondylocostal dysostosis (SCD, #277300, #608681, and #609813) and Alagille syndrome (AGS, OMIM #118450, and #610205), both of which exhibit vertebral column defects. However, the identified mutations explain only a minor fraction of congenital scoliosis cases. More work needs to be performed to elucidate the pathological mechanism underlying congenital and idiopathic scoliosis in human.

The formed somite is also patterned along the dorsal–ventral axis by cell signaling from the surface ectoderm, neural tube, and the notochord (Fig. 1.1). Ventralizing signals such as Sonic hedgehog (Shh) from the notochord and ventral neural tube is required to induce sclerotome formation on the ventral side [14,15], whereas Wnt signaling from the surface ectoderm and dorsal neural tube is required for the formation of dermomyotome on the dorsal side of the somite (Fig. 1.1) [16,17]. The sclerotome gives rise to the axial skeleton and the ribs. In the mouse mutant that lacks Shh function, the vertebral column and posterior ribs fail to form. The paired domain transcription factor Pax1 is expressed in the sclerotome and Shh is required to regulate its expression [18,19]. However, axial skeletal phenotypes in Pax1 mutant mice [20] were far less severe than those in the Shh mutants.

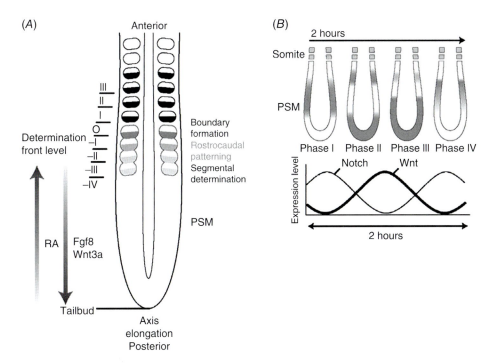

Fig. 1.2. Periodic and left–right symmetrical somite formation is controlled by signaling gradients and oscillations. (A) Somites form from the presomitic mesoderm (PSM) on either side of the neural tube in an anterior to posterior (A–P) wave. Each segment of the somite is also patterned along the A–P axis. Retinoic acid signaling controls the synchronization of somite formation on the left and right side of the neural tube. The most recent visible somite is marked by "0," whereas the region in the anterior PSM that is already determined to form somites is marked by a determination front that is determined by Fgf8 and Wnt3a gradients. This FGF signaling gradient is antagonized by an opposing gradient of retinoic acid. (B) Periodic somite formation (one pair of somite/2 hours) is controlled by a segmentation clock, the molecular nature of which is oscillated expression of signaling components in the Notch and Wnt pathway. Notch signaling oscillates out of phase with Wnt signaling.

Limb skeletons are patterned along the proximal–distal (P–D, shoulder to digit tip), anterior–posterior (A–P, thumb to little finger) and dorsal–ventral (D–V, back of the hand to palm) axes (Fig. 1.3) [21,22]. Along the P–D axis, the limb skeletons form three major segments: humerus or femur at the proximal end, radius and ulna or tibia and fibula in the middle and carpal/tarsal, metacarpal/metatarsal, and digits in the distal end. Along the A–P axis, the radius and ulna have distinct morphological features, as do each of the five digits. Patterning along the D–V limb axis also results in characteristic skeletal shapes and structures. For instance, the sesamoid processes are located ventrally whereas the knee patella forms on the dorsal side of the knee. The three-dimensional limb patterning events are regulated by three signaling centers in the early limb primordium, known as the limb bud, before mesenchymal condensation.

The apical ectoderm ridge (AER), a thickened epithelial structure formed at the distal tip of the limb bud, is the signaling center that directs P–D limb outgrowth (Fig. 1.3). Canonical Wnt signaling activated by Wnt3 induces AER formation, whereas BMP signaling leads to AER regression. FGF family members Fgf4, Fgf8, Fgf9, and Fgf17 are expressed specifically in the AER and Fgf8 alone is sufficient to mediate the function of AER. Fgf10, expressed in the presumptive limb mesoderm, is required for limb initiation and it also controls limb outgrowth by maintaining Fgf8 expression in the AER. It is interesting that exposure to the combined activities of distal signals (Wnt3a and Fgf8) and the proximal signal (RA) in the early limb bud or in culture maintains the potential to form both proximal and distal structures. As the limb bud grows, the proximal cells fall out of range of distal signals (Wnt3a and Fgf8) that act, in part, to keep the cells undifferentiated. Cells closer to the flank therefore differentiate and form proximal structures under the influence of proximal signals such as RA. The potential of distal mesenchymal cells becomes restricted over time to distal fates as they grow beyond the range of proximally produced RA [23,24]. Patterning of the limb bud progenitor cells into distinct segments along the P–D axis may also result in region-specific unique cellular properties such as cell sorting and aggregation behaviors that may direct their contribution toward specific skeletal elements such as the humerus or digits [25].

The second signaling center is the zone of polarizing activity (ZPA) which is a group of mesenchymal cells located at the posterior distal limb margin and immediately adjacent to the AER (Fig. 3.3B). When ZPA tissue is grafted to the anterior limb bud under the AER, it leads to digit duplication in mirror image of the endogenous ones [26]. Shh is expressed in the ZPA and is

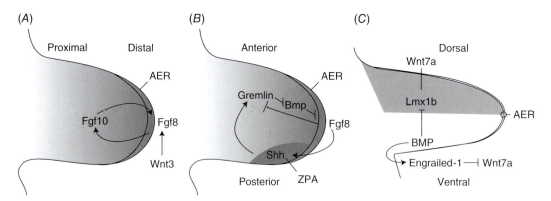

Fig. 1.3. Limb patterning and growth along the proximal–distal (P–D), anterior–posterior (A–P), and dorsal–ventral (D–V) axes are controlled by signaling interactions and feedback loops. (A) A signaling feedback loop between Fgf10 in the limb mesoderm and Fgf8 in the AER is required to direct P–D limb outgrowth. Wnt3 is required for AER formation. (B) Shh in the ZPA controls A–P limb patterning. A–P and P–D limb patterning and growth are also coordinated through a feedback loop between Shh and FGFs expressed in the AER. FGF signaling from the AER is required for Shh expression. Shh also maintains AER integrity by regulating Gremlin expression. Gremlin is a secreted antagonist of BMP signaling which promotes AER degeneration. The inhibitory feedback loop between Gremlin in the limb mesenchyme and FGFs in the AER is critical in terminating limb bud outgrowth. (C) D–V patterning of the limb is determined by Wnt7a and BMP signaling through regulating the expression of Lmx1b in the limb mesenchyme.

both necessary and sufficient to mediate ZPA activity in patterning digit identity along the A–P axis [27]. However, the A–P axis of the limb is established before Shh signaling. This pre-Shh A–P limb patterning is controlled by combined activities of Gli3, Alx4, and basic helix-loop-helix (bHLH) transcription factors dHand and Twist1. The Gli3 repressor form (Gli3R) and Alx4 establish the anterior limb territory by restricting dHand expression to the posterior limb, which in turn activates Shh expression [28,29]. The activity of dHand in the posterior limb is also antagonized by Twist1 via a dHand-Twist1 heterodimer. Recently, the zinc finger factors Sall4 and Gli3 have been found to cooperate for proper development of the A–P skeletal elements and also function upstream of Shh-dependent posterior skeletal element development [30].

Mutations in the human *TWIST1* gene cause Saethre–Chotzen syndrome (SCS, OMIM #101400), one of the most commonly inherited craniosynostosis conditions. The hallmarks of this syndrome are premature fusion of the calvarial bones and limb abnormalities. Mutations in the *GLI3* gene also cause limb malformations including Greig cephalopolysyndactyly syndrome (GCPS, OMIM #175700) and Pallister–Hall syndrome (PHS, OMIM #146510).

The third signaling center is the non-AER limb ectoderm that covers the limb bud. It sets up the D–V polarity of not only the ectoderm but also the underlying mesoderm (Fig. 1.3C) (review by [21,31]). Wnt and BMP signaling are required to control D–V limb polarity. *Wnt7a* is expressed specifically in the dorsal limb ectoderm and it activates the expression of *Lmx1b*, which encodes a dorsal-specific LIM homeobox transcription factor that determines the dorsal identity. *Wnt7a* expression in the ventral ectoderm is suppressed by En-1, which encodes a transcription factor that is expressed specifically in the ventral ectoderm. The BMP signaling pathway is also ventralizing in the early limb (Fig. 1.3C). It appears that the effects of BMP signaling are mediated by Msx1 and Msx2, two transcription factors that are also transcriptionally regulated by BMP signaling. The function of BMP signaling in the early limb ectoderm is upstream of En-1 in controlling D–V limb polarity [32]. However, when BMPRIA is specifically inactivated only in the mouse limb bud mesoderm, the distal limb is dorsalized without altering the expression of Wnt7a and En-1 in the limb ectoderm [33]. Thus, BMPs also have En-1-independent ventralization activity by directly signaling to the limb mesenchyme to inhibit *Lmx1b* expression.

Limb development is a coordinated three-dimensional event. Indeed, the three signaling centers interact with each other through interactions of the mediating signaling molecules. First, there is a positive feedback loop between Shh and FGFs expressed in the AER, which connects A–P limb patterning with P–D limb outgrowth (Fig. 1.3B) [21,22]. This positive feedback loop is antagonized by an FGF/Grem1 inhibitory loop that attenuates strong FGF signaling and terminates limb outgrowth signals in order to maintain a proper limb size [34]. Second, the dorsalizing signal Wnt7a is required for maintaining the expression of Shh that patterns the A–P axis [35,36]. Third, Wnt/β-catenin signaling has been found to be both distalizing and dorsalizing [37–39].

Identification of these interacting signaling networks in early limb patterning has provided a fertile ground to test the self-organizing Turing models [1] that simulate the pattern of digit formation in the limb. By combining experiments and modeling, a self-organizing Turing network implemented by BMP, Sox9, and Wnt has been found to drive digit specification. When modulated by morphogen gradients, the network is able to recapitulate the expression patterns of Sox9 in the wild type and in perturbation experiments [2]. Interestingly, the Turing model is also found to explain the dose effects of distal *Hox* genes in modulating the digit period or wavelength [3]. Progressive reduction in *Hoxa13* and *Hoxd11-Hoxd13* genes from the Gli3-null background results in progressively more severe polydactyly, displaying thinner and densely packed digits.

Recently, the generality and contribution of this Turing network implemented by BMP, Sox9, and Wnt to the morphological diversity of fins and limbs has been further explored [5]. Is has been suggested that the skeletal patterning of the catshark *Scyliorhinus canicula* pectoral fin is likely driven by a deeply conserved BMP–Sox9–Wnt Turing network. Therefore, the union of theory and experimentation is a powerful approach to not only identify and validate the minimal components of a network regulating digit pattern, but also to ask a new set of questions that will undoubtedly be answered as a result of the continued merging of disciplines.

EMBRYONIC CARTILAGE AND BONE FORMATION

The early patterning events determine where and when the mesenchymal cells condense, though the mechanism remains to be elucidated. Subsequently, osteochondral progenitors in the condensation form either chondrocytes or osteoblasts. Sox9 and Runx2, master transcription factors that are required for the determination of chondrocyte and osteoblast cell fates respectively [40,41], are both expressed in osteochondral progenitor cells, but Sox9 expression precedes that of Runx2 in the mesenchymal condensation in the limb [42]. Early Sox9-expressing cells give rise to both chondrocytes and osteoblasts regardless of ossification mechanisms [43]. Loss of Sox9 function in the limb leads to loss of mesenchymal condensation and Runx2 expression [42]. Coexpression of Sox9 and Runx2 is terminated upon chondrocyte and osteoblast differentiation when Sox9 and Runx2 expression is quickly segregated into chondrocytes and osteoblasts respectively. The mechanism controlling lineage-specific Sox9 and Runx2 expression is fundamental to the regulation of chondrocyte and osteoblast differentiation and the determination of ossification mechanisms. It is clear that cell–cell signaling, particularly those mediated by Wnts and Indian hedgehog (Ihh), are required for cell fate determination of chondrocytes and osteoblasts by controlling the expression of Sox9 and Runx2.

Active Wnt/β-catenin signaling is detected in the developing calvarium and perichondrium where osteoblasts differentiate through either intramembranous or endochondral ossification. Indeed, enhanced Wnt/β-catenin signaling enhanced bone formation and Runx2 expression, but inhibited chondrocyte differentiation and Sox9 expression [44–46]. Conversely, removal of β-catenin in osteochondral progenitor cells resulted in ectopic chondrocyte differentiation at the expense of osteoblasts during both intramembranous and endochondral ossification [46–48]. Therefore, during intramembranous ossification, Wnt/β-catenin signaling levels in the condensation are higher, which promotes osteoblast differentiation while inhibiting chondrocyte differentiation. During endochondral ossification, however, Wnt/β-catenin signaling in the condensation is initially lower, such that only chondrocytes can differentiate. Later, when Wnt/β-catenin signaling is upregulated in the periphery of the cartilage, osteoblasts will differentiate. It is likely that by manipulating Wnt signaling, mesenchymal progenitor cells, and perhaps even mesenchymal stem cells, can be directed to form only chondrocytes, which are needed to repair cartilage damage in osteoarthritis, or only form osteoblasts, which will lead to new therapeutic strategies to treat osteoporosis. These studies have provided new insights into tissue engineering that aims to fabricate cartilage or bone in vitro using mesenchymal progenitor cells or stem cells.

Ihh signaling is required for osteoblast differentiation by activating Runx2 expression only during endochondral bone formation [49]. Ihh is expressed in newly differentiated chondrocytes and Ihh signaling does not seem to affect chondrocyte differentiation from mesenchymal progenitors. However, when Hh signaling is inactivated in the perichondrium cells, they ectopically form chondrocytes that express Sox9 at the expense of Runx2. This is similar to what has been observed in the Osterix (Osx) mutant embryos, except that in the $Osx^{-/-}$ embryos, ectopic chondrocytes express both Sox9 and Runx2 [50], suggesting that Runx2 is not sufficient to inhibit Sox9 expression and chondrocyte differentiation. It is still not clear what controls Ihh-independent Runx2 expression during intramembranous ossification. One likely scenario is that the function of Ihh is compensated by Shh in the developing calvarium or Hh signaling is activated in a ligand-independent manner in the developing calvarium. Indeed, it has been recently found that in the rare human genetic disease progressive osseous heteroplasia (POH), which is caused by null mutations in GNAS that encodes Gαs, Hedgehog signaling is upregulated. Such activation of Hh signaling is independent of Hh ligands and is both necessary and sufficient to induce ectopic osteoblast cell differentiation in soft tissues [51]. Importantly, GNAS gain-of-function mutations upregulate Wnt/β-catenin signaling in osteoblast progenitor cells, resulting in their defective differentiation and fibrous dysplasia [52]. Therefore, studies of human genetic diseases identify Gαs as a key regulator of proper osteoblast differentiation through its maintenance of a balance between the Wnt/β-catenin and Hedgehog pathways.

Both Wnt/β-catenin and Ihh signaling pathways are required for endochondral bone formation. To understand which one acts first, a genetic epistatic test was carried out [53]. These studies found that β-catenin is required downstream of not just Ihh, but also Osx in promoting osteoblast maturation. By contrast, Ihh signaling is not required after Osx expression for osteoblast differentiation [54]. The sequential actions of Hh and Wnt signaling in osteoblast differentiation and maturation suggest that Hh and Wnt signaling need to be manipulated at distinct stages during fracture repair and tissue engineering.

BMPs are the transforming growth factor (TGF) superfamily members that were identified as secreted proteins able to promote ectopic cartilage and bone formation [55]. Unlike Ihh and Wnt signaling, BMP signaling promotes the differentiation of both osteoblast and chondrocyte differentiation from mesenchymal progenitors. The mechanisms underlying these unique activities of BMPs have been under intense investigation for the past two decades. During this time, our understanding of BMP action in chondrogenesis and osteogenesis has benefited greatly from molecular studies of BMP signal transduction [56]. Reducing BMP signaling by removing BMP receptors leads to impaired chondrocyte and osteoblast differentiation and maturation [57].

FGF ligands and FGF receptors (FGFR) are both expressed in the developing skeletal system. The significant role of FGF signaling in skeletal development was first identified by the discovery that achondroplasia (ACH, OMIM #100800), the most common form of skeletal dwarfism in humans, was caused by a missense mutation in FGFR3. Later, hypochondroplasia (HCH, OMIM #146000), a milder form of dwarfism and thanatophoric dysplasia (TD, OMIM #187600, and 187601), a more severe form of dwarfism, were also found to result from mutations in FGFR3. FGFR3 signaling acts to regulate the proliferation and hypertrophy of the differentiated chondrocytes. However, the function of FGF signaling in mesenchymal condensation and chondrocyte differentiation from progenitors remains to be elucidated as complete genetic inactivation of FGF signaling in mesenchymal condensation has not been achieved. Nevertheless, it is clear that FGF signaling acts in mesenchymal condensation to control osteoblast differentiation during intramembranous bone formation. Mutations in FGFR 1, 2 and 3 cause craniosynostosis (premature fusion of the cranial sutures). The craniosynostosis syndromes involving FGFR 1, 2, 3 mutations include Apert syndrome (AS, OMIM #101200), Beare-Stevenson cutis gyrata (OMIM #123790), Crouzon syndrome (CS, OMIM #123500), Pfeiffer syndrome (PS, OMIM #101600), Jackson-Weiss syndrome (JWS, OMIM #123150), Muenke syndrome (MS, OMIM #602849), crouzonodermoskeletal syndrome (OMIM #134934) and osteoglophonic dysplasia (OGD, OMIM #166250), a disease characterized by craniosynostosis, a prominent supraorbital ridge, and a depressed nasal bridge, as well as rhizomelic dwarfism and nonossifying bone lesions. All these mutations are autosomal dominant and many of them are activating mutations of FGF receptors. FGF signaling can promote or inhibit osteoblast proliferation and differentiation depending on the cell context. It does so either directly or through interaction with the Wnt and BMP signaling pathways.

Apart from having the right types of cells and proper size, cartilage and bone also have distinct morphologies which are required for their function. For example, the limb and long bones preferentially elongate along the P–D axis. It is well understood that Wnts can act as morphogens by forming gradients that specify distinct cell types in distinct spatial orders by inducing the expression of different target genes at threshold concentrations. In this regard, morphogen gradients provide quantitative information to generate a distinct pattern by coordinating cell proliferation and differentiation. Because the limbs are elongated organs instead of a three-dimensionally symmetrical ball, directional information has to be provided during limb and long bone elongation.

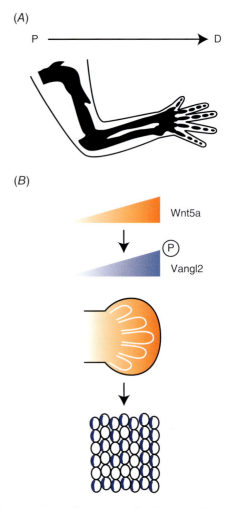

Fig. 1.4. Wnt5a gradient controls directional morphogenesis by regulating Vangl2 phosphorylation and asymmetrical localization. (A) Schematics of skeletons in a human limb that preferentially elongates along the proximal–distal (P–D) axis. (B) A model of a Wnt5a gradient controlling P–D limb elongation by providing a global directional cue. Wnt5a is expressed in a gradient (orange) in the developing limb bud and this Wnt5a gradient is translated into an activity gradient of Vangl2 by inducing different levels of Vangl2 phosphorylation (blue). In the distal limb bud of an E12.5 mouse embryo showing the forming digit cartilage, the Vangl2 activity gradient then induces asymmetrical Vangl2 localization (blue) and downstream polarized events.

Although the molecular mechanism underlying such directional morphogenesis was poorly understood in the past, there is evidence that alignment of the columnary chondrocytes of the growth plate might be regulated by planar cell polarity (PCP) during directional elongation of the formed cartilage [58]. PCP is an evolutionarily conserved pathway that is required in many directional morphogenetic processes including left–right asymmetry, neural tube closure, body axis elongation and brain wiring [59]. Recently, a major breakthrough has been made by demonstrating that newly differentiated chondrocytes in the developing long bones in the limb are polarized along the P–D axis. For the first time it was found with a definitive molecular marker, Vangl2 protein, a core regulatory component in the PCP pathway. Vangl2 protein is asymmetrically localized on the proximal side of the Sox9 positive chondrocytes, not in Sox9 negative interdigital mesenchymal cells [60]. Importantly, Vangl2 protein asymmetrical localization requires a Wnt5a signaling gradient. In the Wnt5a$^{-/-}$ mutant limb, the cartilage forms a ball-like structure and Vangl2 is symmetrically distributed on the cell membrane (Fig. 1.4). PCP mutations in the *WNT5a* and *ROR2* genes have been found in skeletal malformations such as the Robinow syndrome and brachydactyly type B1, which both exhibit short-limb dwarfisms [61–65]. In addition, mutations in PCP signaling components such as VANGL1 has been found in adolescent idiopathic scoliosis (AIS).

CONCLUSION

Skeletal formation is a process that has been perfected by nature in embryos during vertebrate evolution. Understanding the underlying molecular mechanisms of cartilage and bone formation in embryonic development will advance our knowledge of vertebrate embryonic morphogenesis in general. This knowledge will allow us to develop the strategy to promote skeletal tissue repair by endogenous cells or rejuvenate old skeletal tissues without having to use cells cultured in vitro. In addition, to use autologous cells and tissues or iPS (induced pluripotent stem) cells to repair bone and cartilage damaged during injury and disease, we require a more complete knowledge of skeletal development so that cartilage or bone can be fabricated using the body's own cells. Understanding skeletal development is indispensable for understanding pathological mechanisms of skeletal diseases, finding therapeutic targets, promoting consistent cartilage or bone repair in vivo, and eventually growing functional cartilage or bone in vitro.

ACKNOWLEDGMENT

I apologize to those authors whose work could not be cited directly because of space restrictions.

REFERENCES

1. Turing AM. The chemical basis of morphogenesis. Phil Trans R Soc Lond B. 1952;237(641):37–72.
2. Raspopovic J, Marcon L, Russo L, et al. Modeling digits. Digit patterning is controlled by a Bmp-Sox9-Wnt Turing network modulated by morphogen gradients. Science. 2014;345(6196):566–70.
3. Sheth R, Marcon L, Bastida MF, et al. Hox genes regulate digit patterning by controlling the wavelength of a Turing-type mechanism. Science. 2012;338(6113):1476–80.
4. Marcon L, Diego X, Sharpe J, et al. High-throughput mathematical analysis identifies Turing networks for patterning with equally diffusing signals. Elife. 2016;5.
5. Onimaru K, Marcon L, Musy M, et al. The fin-to-limb transition as the re-organization of a Turing pattern. Nature Comm. 2016;7:11582.
6. Santagati F, Rijli FM. Cranial neural crest and the building of the vertebrate head. Nat Rev Neurosci. 2003;4(10):806–18.
7. Helms JA, Cordero D, Tapadia MD. New insights into craniofacial morphogenesis. Development. 2005;132(5):851–61.
8. Hubaud A, Pourquie O. Signalling dynamics in vertebrate segmentation. Nat Rev Mol Cell Biol. 2014;15(11):709–21.
9. Bothe I, Ahmed MU, Winterbottom FL, et al. Extrinsic versus intrinsic cues in avian paraxial mesoderm patterning and differentiation. Dev Dyn. 2007;236(9):2397–409.
10. Pourquie O. Vertebrate segmentation: from cyclic gene networks to scoliosis. Cell. 2011;145(5):650–63.
11. Ilagan MX, Kopan R. SnapShot: notch signaling pathway. Cell. 2007;128(6):1246.
12. Moreno TA, Kintner C. Regulation of segmental patterning by retinoic acid signaling during Xenopus somitogenesis. Dev Cell. 2004;6(2):205–18.
13. Duester G. Retinoic acid regulation of the somitogenesis clock. Birth Defects Res C Embryo Today. 2007;81(2):84–92.
14. Fan CM, Tessier-Lavigne M. Patterning of mammalian somites by surface ectoderm and notochord: evidence for sclerotome induction by a hedgehog homolog. Cell. 1994;79(7):1175–86.
15. Johnson RL, Laufer E, Riddle RD, et al. Ectopic expression of Sonic hedgehog alters dorsal-ventral patterning of somites. Cell. 1994;79(7):1165–73.
16. Fan CM, Lee CS, Tessier-Lavigne M. A role for WNT proteins in induction of dermomyotome. Dev Biol. 1997;191(1):160–5.
17. Capdevila J, Johnson RL. Endogenous and ectopic expression of noggin suggests a conserved mechanism for regulation of BMP function during limb and somite patterning. Dev Biol. 1998;197(2):205–17.
18. Koseki H, Wallin J, Wilting J, et al. A role for Pax-1 as a mediator of notochordal signals during the dorsoventral specification of vertebrae. Development. 1993;119(3):649–60.
19. Chiang C, Litingtung Y, Lee E, et al. Cyclopia and defective axial patterning in mice lacking Sonic hedgehog gene function. Nature. 1996;383(6599):407–13.

20. Wallin J, Wilting J, Koseki H, et al. The role of Pax-1 in axial skeleton development. Development. 1994;120(5):1109–21.
21. Niswander L. Pattern formation: old models out on a limb. Nat Rev Genet. 2003;4(2):133–43.
22. Zeller R, Lopez-Rios J, Zuniga A. Vertebrate limb bud development: moving towards integrative analysis of organogenesis. Nat Rev Genet. 2009;10(12):845–58.
23. Cooper MK, Porter JA, Young KE, et al. Teratogen-mediated inhibition of target tissue response to Shh signaling. Science. 1998;280(5369):1603–7.
24. Rosello-Diez A, Ros MA, Torres M. Diffusible signals, not autonomous mechanisms, determine the main proximo-distal limb subdivision. Science. 2011;332(6033):1086–8.
25. Barna M, Niswander L. Visualization of cartilage formation: insight into cellular properties of skeletal progenitors and chondrodysplasia syndromes. Dev Cell. 2007;12(6):931–41.
26. Saunders JWJ, Gasseling MT. Ectoderm-mesenchymal interaction in the origin of wing symmetry. In: Fleischmajer R, Billingham RE (eds) *Epithelia-Mesenchymal Interactions*. Baltimore: Williams and Wilkins, 1968, pp. 78–97.
27. Riddle RD, Johnson RL, Laufer E, et al. Sonic hedgehog mediates the polarizing activity of the ZPA. Cell. 1993;75(7):1401–16.
28. Charite J, McFadden DG, Olson EN. The bHLH transcription factor dHAND controls Sonic hedgehog expression and establishment of the zone of polarizing activity during limb development. Development. 2000;127(11):2461–70.
29. Fernandez-Teran M, Piedra ME, Kathiriya IS, et al. Role of dHAND in the anterior-posterior polarization of the limb bud: implications for the Sonic hedgehog pathway. Development. 2000;127(10):2133–42.
30. Akiyama R, Kawakami H, Wong J, et al. Sall4-Gli3 system in early limb progenitors is essential for the development of limb skeletal elements. Proc Natl Acad Sci U S A. 2015;112(16):5075–80.
31. Tickle C. Patterning systems – from one end of the limb to the other. Dev Cell. 2003;4(4):449–58.
32. Lallemand Y, Nicola MA, Ramos C, et al. Analysis of Msx1; Msx2 double mutants reveals multiple roles for Msx genes in limb development. Development. 2005;132(13):3003–14.
33. Ovchinnikov DA, Selever J, Wang Y, et al. BMP receptor type IA in limb bud mesenchyme regulates distal outgrowth and patterning. Dev Biol. 2006;295(1):103–15.
34. Verheyden JM, Sun X. An Fgf/Gremlin inhibitory feedback loop triggers termination of limb bud outgrowth. Nature. 2008;454(7204):638–41.
35. Parr BA, McMahon AP. Dorsalizing signal Wnt-7a required for normal polarity of D-V and A-P axes of mouse limb. Nature. 1995;374(6520):350–3.
36. Yang Y, Niswander L. Interaction between the signaling molecules WNT7a and SHH during vertebrate limb development: dorsal signals regulate anteroposterior patterning. Cell. 1995;80(6):939–47.
37. Ten Berge D, Brugmann SA, Helms JA, et al. Wnt and FGF signals interact to coordinate growth with cell fate specification during limb development. Development. 2008;135(19):3247–57.
38. Hill TP, Taketo MM, Birchmeier W, et al. Multiple roles of mesenchymal beta-catenin during murine limb patterning. Development. 2006;133(7):1219–29.
39. Cooper KL, Hu JK, ten Berge D, et al. Initiation of proximal-distal patterning in the vertebrate limb by signals and growth. Science. 2011;332(6033):1083–6.
40. Zelzer E, Olsen BR. The genetic basis for skeletal diseases. Nature. 2003;423(6937):343–8.
41. Nakashima K, de Crombrugghe B. Transcriptional mechanisms in osteoblast differentiation and bone formation. Trends Genet. 2003;19(8):458–66.
42. Akiyama H, Chaboissier MC, Martin JF, et al. The transcription factor Sox9 has essential roles in successive steps of the chondrocyte differentiation pathway and is required for expression of Sox5 and Sox6. Genes Dev. 2002;16(21):2813–28.
43. Akiyama H, Kim JE, Nakashima K, et al. Osteo-chondro-progenitor cells are derived from Sox9 expressing precursors. Proc Natl Acad Sci U S A. 2005;102(41):14665–70.
44. Hartmann C, Tabin CJ. Dual roles of Wnt signaling during chondrogenesis in the chicken limb. Development. 2000;127(14):3141–59.
45. Guo X, Day TF, Jiang X, et al. Wnt/beta-catenin signaling is sufficient and necessary for synovial joint formation. Genes Dev. 2004;18(19):2404–17.
46. Day TF, Guo X, Garrett-Beal L, et al. Wnt/beta-catenin signaling in mesenchymal progenitors controls osteoblast and chondrocyte differentiation during vertebrate skeletogenesis. Dev Cell. 2005;8(5):739–50.
47. Hill TP, Spater D, Taketo MM, et al. Canonical Wnt/beta-catenin signaling prevents osteoblasts from differentiating into chondrocytes. Dev Cell. 2005;8(5):727–38.
48. Hu H, Hilton MJ, Tu X, et al. Sequential roles of Hedgehog and Wnt signaling in osteoblast development. Development. 2005;132(1):49–60.
49. St-Jacques B, Hammerschmidt M, McMahon AP. Indian hedgehog signaling regulates proliferation and differentiation of chondrocytes and is essential for bone formation. Genes Dev. 1999;13(16):2072–86.
50. Nakashima K, Zhou X, Kunkel G, et al. The novel zinc finger-containing transcription factor osterix is required for osteoblast differentiation and bone formation. Cell. 2002;108(1):17–29.
51. Regard JB, Malhotra D, Gvozdenovic-Jeremic J, et al. Activation of Hedgehog signaling by loss of GNAS causes heterotopic ossification. Nat Med. 2013;19(11):1505–12.
52. Regard JB, Cherman N, Palmer D, et al. Wnt/beta-catenin signaling is differentially regulated by Galpha proteins and contributes to fibrous dysplasia. Proc Natl Acad Sci U S A. 2011;108(50):20101–6.
53. Mak KK, Chen MH, Day TF, et al. Wnt/beta-catenin signaling interacts differentially with Ihh signaling in controlling endochondral bone and synovial joint formation. Development. 2006;133(18):3695–707.

54. Rodda SJ, McMahon AP. Distinct roles for Hedgehog and canonical Wnt signaling in specification, differentiation and maintenance of osteoblast progenitors. Development. 2006;133(16):3231–44.
55. Wozney JM. Bone morphogenetic proteins. Prog Growth Factor Res. 1989;1(4):267–80.
56. Derynck R, Zhang YE. Smad-dependent and Smad-independent pathways in TGF-beta family signalling. Nature. 2003;425(6958):577–84.
57. Yoon BS, Ovchinnikov DA, Yoshii I, et al. Bmpr1a and Bmpr1b have overlapping functions and are essential for chondrogenesis in vivo. Proc Natl Acad Sci U S A. 2005;102(14):5062–7.
58. Li Y, Dudley AT. Noncanonical frizzled signaling regulates cell polarity of growth plate chondrocytes. Development. 2009;136(7):1083–92.
59. Yang Y, Mlodzik M. Wnt-Frizzled/planar cell polarity signaling: cellular orientation by facing the wind (Wnt). Annu Rev Cell Dev Biol. 2015;31:623–46.
60. Gao B, Song H, Bishop K, et al. Wnt signaling gradients establish planar cell polarity by inducing Vangl2 phosphorylation through Ror2. Dev Cell. 2011;20(2):163–76.
61. Minami Y, Oishi I, Endo M, et al. Ror-family receptor tyrosine kinases in noncanonical Wnt signaling: their implications in developmental morphogenesis and human diseases. Dev Dyn.239(1):1–15.
62. Person AD, Beiraghi S, Sieben CM, et al. WNT5A mutations in patients with autosomal dominant Robinow syndrome. Dev Dyn. 2010;239(1):327–37.
63. van Bokhoven H, Celli J, Kayserili H, et al. Mutation of the gene encoding the ROR2 tyrosine kinase causes autosomal recessive Robinow syndrome. Nat Genet. 2000;25(4):423–6.
64. Schwabe GC, Tinschert S, Buschow C, et al. Distinct mutations in the receptor tyrosine kinase gene ROR2 cause brachydactyly type B. Am J Hum Genet. 2000;67(4):822–31.
65. DeChiara TM, Kimble RB, Poueymirou WT, et al. Ror2, encoding a receptor-like tyrosine kinase, is required for cartilage and growth plate development. Nat Genet. 2000;24(3):271–4.

2
Endochondral Ossification

Courtney M. Karner[1,2] and Matthew J. Hilton[1,2]

[1]*Department of Orthopaedic Surgery, Duke Orthopaedic Cellular, Developmental, and Genome Laboratories, Duke University School of Medicine, Durham, NC, USA*
[2]*Department of Cell Biology, Duke University, Durham, NC, USA*

INTRODUCTION

A defining feature of vertebrates is the presence of a mineralized skeleton. Aside from giving animals their characteristic shape, the skeleton provides diverse functions including protecting internal organs, supporting body mass and movement, production of blood cells, calcium storage, and endocrine signaling. The skeleton is comprised primarily of two tissues, cartilage and bone, which are formed embryonically by chondrocytes and osteoblasts respectively. During skeletal development, these specialized cells are derived from a common mesenchymal progenitor of either neural crest origin in the craniofacial region or mesodermal origin for bones formed elsewhere in the body. Bones develop via two distinct mechanisms: intramembranous or endochondral ossification. Intramembranous ossification is responsible for forming specific parts of the skull and the clavicle, whereby mesenchymal progenitors differentiate directly into osteoblasts responsible for secreting bone matrix. In contrast, endochondral ossification, the process responsible for generating most of the skeleton, requires a cartilage intermediate before forming bone. Here we will discuss the major cellular events of endochondral ossification: chondrogenesis, chondrocyte hypertrophy, and osteoblast differentiation, as well as important molecular mediators governing each of these processes.

CHONDROGENESIS AND CHONDROCYTE HYPERTROPHY DURING ENDOCHONDRAL OSSIFICATION

Bones within the limbs serve as the model for endochondral ossification. Development of the limb skeleton initiates during embryogenesis via the migration of multipotent mesenchymal progenitors from the lateral plate mesoderm into the developing limb field. These progenitors rapidly proliferate, expanding the limb bud, followed by the formation of condensations ultimately giving rise to cartilage anlagen (Fig. 2.1*A,B*). During the condensation phase, mesenchymal progenitors express multiple cell adhesion related molecules such as: *N-cadherin* (*Ncad*), *N-cam* (*Ncam1*), and *tenascin C* (*Tnc*) aiding in mesenchymal cell compaction. Cells within condensations undergo differentiation to generate mature chondrocytes (cartilage cells), a process known as chondrogenesis (Fig. 2.1*A,B*). Newly formed chondrocytes take on a characteristic round shape, continue to proliferate, and begin producing an extracellular matrix rich in type II, type IX, and type XI collagens (COL2A1, COL9A1, COL11A1) and the proteoglycan, aggrecan (ACAN). As cartilage rudiments continue to grow, chondrocytes nearest the epiphyseal ends maintain their round appearance and reduce their proliferative index, whereas chondrocytes near the center of rudiments

---

Primer on the Metabolic Bone Diseases and Disorders of Mineral Metabolism, Ninth Edition. Edited by John P. Bilezikian.
© 2019 American Society for Bone and Mineral Research. Published 2019 by John Wiley & Sons, Inc.
Companion website: www.wiley.com/go/asbmrprimer

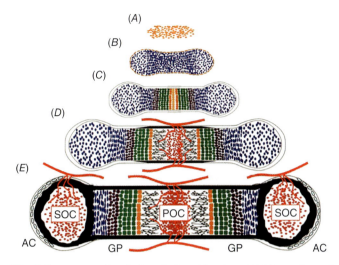

Fig. 2.1. Stages of endochondral ossification. (A) Mesenchymal condensation (orange cells = mesenchymal progenitors). (B) Chondrogenesis (blue cells = chondrocytes; orange cells = perichondrial progenitors. (C) Chondrocyte hypertrophy (blue cells = epiphyseal round chondrocytes and flat columnar chondrocytes; purple cells = prehypertrophic chondrocytes; green cells = hypertrophic chondrocytes; orange cells = late stage hypertrophic chondrocytes. (D) Formation of the primary ossification center (POC) (all cells colored as described above; blood vessels and marrow cells in red; osteoblasts and bone matrix in black). (E) Formation of secondary ossification center (SOC) separates the articular chondrocytes (AC) from growth plate (GP) chondrocytes (light blue cells = articular chondrocytes; all other cells/tissues as previously described).

enhance their rate of proliferation, adopt a flattened appearance, and align into columns, driving longitudinal growth of the cartilage elements (Fig. 2.1C). A combination of chondrogenesis and chondrocyte proliferation establishes the early cartilaginous skeleton, which serves as the template for endochondral bone development.

Calcification and ossification of the endochondral skeleton begins with chondrocyte hypertrophy. During this process, columnar chondrocytes located at the center of growing cartilage rudiments, also known as prehypertrophic and hypertrophic chondrocytes, undergo further differentiation after exiting the cell cycle. Hypertrophic differentiation consists of genetic programs responsible for dramatically increasing chondrocyte cell size, switching the production of type II collagen to type X collagen (COL10A1), and inducing factors responsible for calcification and vascularization of the cartilage matrix such as ALP and VEGF respectively (Fig. 2.1C). Hypertrophic chondrocytes express transcriptional regulators and a myriad of growth factors that not only coordinate the hypertrophic chondrocyte differentiation process, but also induce osteoblast differentiation of surrounding perichondrial cells and promote vascularization of the calcified cartilage by surrounding blood vessels, establishing a marrow cavity and primary ossification center (POC) (Fig. 1.1D). Late stage hypertrophic chondrocytes secrete the catabolic enzyme matrix metalloprotease 13 (MMP13), which helps to degrade the cartilage matrix. Previously coined as terminal hypertrophic chondrocytes, these cells were thought to undergo exclusively a form of programmed cell death; however, lineage tracing studies recently showed that many hypertrophic chondrocytes undergo transdifferentiation into the osteoblast lineage. The combination of calcified cartilage degradation and hypertrophic chondrocyte transdifferentiation provides both a scaffold and a cell source for the generation of bone within the POC. Concominant with the ossification process directly associated with cartilage, osteoblasts derived from perichondrial cells and perivascular mesenchymal progenitors also utilize the degrading cartilage as a scaffold for further bone formation. The continuous processes of chondrocyte proliferation, hypertrophy, calcification, vascularization, cartilage matrix degradation, transdifferentiation, and bone formation drive embryonic and postnatal endochondral bone growth.

During early postnatal endochondral ossification, round chondrocytes maintained near the epiphyseal ends of bones undergo a maturation process similar to chondrocytes during embryonic skeletogenesis. Epiphyseal chondrocytes hypertrophy, generate a calcified matrix, degrade the matrix, undergo apoptosis and/or transdifferentiation, and eventually are replaced by invading vasculature and osteoblasts to create the secondary ossification center (SOC) (Fig. 2.1D,E). This SOC serves an important support role within weight-bearing articulating joints and separates the only two areas of remaining cartilage within the adult endochondral skeleton: the articular cartilage (AC) and growth plate (GP) cartilage (Fig. 2.1E). As cartilage growth and turnover decreases in the postnatal or adult skeleton, the contribution of cartilage to bone formation dramatically decreases, ultimately terminating the process of endochondral ossification.

MOLECULAR MEDIATORS OF CARTILAGE DEVELOPMENT

Specific transcriptional regulators are critical in establishing the cartilage phase of the endochondral skeleton. Several Sry-box (SOX) factors are required for chondrogenesis and early cartilage development. The master regulator of cartilage development, SOX9, is expressed in mesenchymal progenitors, osteochondral progenitors, and immature chondrocytes. SOX9 controls cell morphology at the mesenchyme to chondrocyte transition [1], while also directly regulating the expression of *Col2a1*, *Col9a1*, *Col11a1*, *Acan*, and other cartilage-related genes [2]. Much of the transcriptional regulation imposed by SOX9 occurs via interactions with other SOX factors, specifically SOX5 and SOX6, which together form the SOX trio. Mouse genetic studies in which *Sox9* is either removed from the germline or specifically within the limb mesenchyme highlight the requirement for SOX9 in forming organized condensations capable of undergoing chondrogenesis [2, 3]. Interestingly, Sox5-/- Sox6-/- double mutant mice develop normal mesenchymal cell condensations, but subsequently show impaired chondrogenesis, columnar chondrocyte disruption, and failure to maintain the chondrocyte phenotype, even though *Sox9* expression is maintained [4]. Similarly, oseteochondral progenitor cell deletion of *Sox9* results in normal condensations, but mutants subsequently develop a severe chondrodysplasia phenotype. These mouse genetic studies underscore the critical and sequential roles for SOX9, SOX5, and SOX6 during cartilage development.

Numerous developmental signaling pathways are critical during chondrogenesis and early phases of endochondral ossification. Factors such as the BMPs play key roles in the compaction of mesenchymal cells and shaping of condensations [1]. Both the BMP and related TGFβ pathways induce *Sox9* expression to promote chondrogenesis and cartilage development. Conditional mutant mouse models with *Bmpr1a* floxed alleles deleted in osteochondral progenitors of a *Bmpr1b*-/- background display a severe generalized chondrodysplasia in which formation of mesenchymal condensations and chondrogenic rudiments fail because of a lack of *Sox9*, *Sox5*, and *Sox6* expression [5]. SMAD proteins are intracellular mediators of BMP signaling. Genetic removal of *Smad1* and *Smad5* floxed alleles in osteochondral progenitors results in reduced condensation size, more compacted cells with less cartilage matrix, decreased chondrocyte proliferation, and an increased incidence of immature chondrocyte cell death, a phenotype slightly less severe than that of BMP receptor mutant mice. These data suggest that whereas BMPs mostly exert their prochondrogenic functions via SMAD activation, alternative signaling mechanisms important for BMP-mediated regulation of endochondral ossification likely exist (Fig. 2.2) [6]. The TGFβ pathway signals through the TGFβ-related SMADS, SMAD2, and SMAD3. TGFβ1, TGFβ2, and TGFβ3 are each sufficient to induce chondrogenesis and cartilage matrix synthesis (Fig. 2.2); however, individual mutant mice have not defined a requisite role

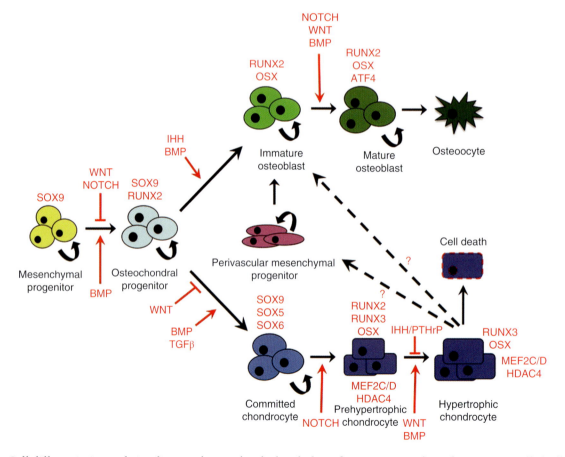

Fig. 2.2. Cell differentiation and signaling regulators of endochondral ossification. Mesenchymal progenitor cells (yellow) differentiate into osteochondral progenitors (light blue) before committing to either the osteoblast (shades of green; osteoblastogenesis) or chondrocyte lineages (shades of blue; chondrogenesis). Osteoblast differentiation proceeds from immature osteoblasts to mature osteoblasts before becoming an osteocyte. Chondrocyte differentiation proceeds from committed chondrocytes to prehypertrophic and hypertrophic chondrocytes before undergoing cell death or transdifferentiation into the osteoblast lineage (pink cells = perivascular mesenchyme progenitors). The NOTCH, BMP, TGFβ, IHH, and WNT pathways play important roles in regulating chondrogenesis, chondrocyte hypertrophy, and osteoblast differentiation during endochondral ossificaiton. Question marks indicate unknown molecular mechanisms.

for any one TGFβ ligand in regulating chondrogenesis. Analyses of Smad2-/- and Tgfβr2 or Smad3 conditional mutant mice in which floxed alleles were deleted from mesenchymal or osteochondral progenitors have also yet to uncover a requisite role for TGFβ signaling during overt chondrogenesis [7–9]; however, important components of the pathway that may be critical in promoting chondrogenesis have yet to be scrutinized. As an opposition to BMP/TGFβ signaling, the WNT and NOTCH signaling pathways antagonize the formation of mesenchymal condensations and inhibit chondrogenic differentiation. Activation of WNT signaling in mesenchymal progenitors using a stabilized β-catenin floxed allele (β-catenin$^{ex3/ex3}$) inhibits Sox9, resulting in impaired condensation formation and suppression of chondrogenesis, ultimately leading to a failure of endochondral ossification (Fig. 2.2) [10]. Conversely, genetic deletion of β-catenin floxed alleles in mesenchymal or osteochondral progenitors resulting in WNT pathway loss of function leads to enhanced Sox9 expression and accelerated chondrogenesis at the expense of osteoblastogenesis, indicating that WNT/β-catenin signaling regulates a critical fate switch in osteochondral progenitors by inhibiting the chondrogenic fate (Fig. 2.2) [10, 11]. Similar to WNT effects on chondrogenesis, NOTCH signaling suppresses condensation formation and chondrogenic differentiation. Activation of NOTCH signaling via conditional overexpression of the NOTCH intracellular domain (NICD) within mesenchymal progenitors suppresses expression of Sox9, Sox5, and Sox6, disrupts formation of mesenchymal condensations, and completely blocks endochondral ossification, an effect that is overturned via the simultaneous genetic removal of the NOTCH nuclear effector, RBPjk. Analyses of mutant mice in which only Rbpjk floxed alleles were deleted in mesenchymal progenitors showed an increase in chondrogenesis and chondrogenic gene expression, indicating the requirement for RBPjk-dependent NOTCH signaling as a regulator of the pace of chondrogenesis and endochondral ossification (Fig. 2.2) [12]. Similar, although less severe, effects on chondrogenesis were also observed in gain- and loss-of-function studies for the NOTCH target genes, Hes1 and Hes5, indicating that NOTCH regulation of chondrogenesis occurs at least partially via a HES-mediated mechanism [13]. Each of these pathways exhibits distinct functions in space and time to regulate mesenchymal condensations and chondrogenesis; however, it is likely all intersect on a common transcriptional program governed by the SOX trio.

Chondrocyte hypertrophy is coordinated by balancing the expression and activities of specific transcriptional regulators including: runt-related transcription factor 2 (RUNX2), runt-related transcription factor 3 (RUNX3), osterix (OSX), myocyte enhancer factor 2c (MEF2C), myocyte enhancer factor 2d (MEF2D), histone deacetylase 4 (HDAC4), and SOX9. RUNX2 is expressed in prehypertrophic and hypertropic chondrocytes and regulates Col10a1, Alpl, Vegf, and Mmp13 expression, coordinating hypertrophic differentiation with calcification, vascularization, and catabolism of the hypertrophic cartilage matrix (Fig. 2.2). Assessments of various Runx2-/- isoform mutants, as well as conditional mutant mice in which Runx2 floxed alleles were removed from osteochondral progenitors, show a significant delay or absence of hypertrophic differentiation, mineralization, and vascularization of cartilage elements [14, 15]. Interestingly, combined Runx2-/-; Runx3-/- double mutants exhibit a more robust blockade in hypertrophic differentiation leading to a complete failure of endochondral ossification, showing an important and potentially redundant role for RUNX3 in cartilage development [16]. OSX is expressed in prehypertrophic and hypertrophic chondrocytes and is another critical regulator of chondrocyte hypertrophy, calcification, and catabolism of calcified cartilage functioning downstream of RUNX2 and RUNX3 (Fig. 2.2). Germline deletion of Osx or conditional removal of Osx floxed alleles in mesenchymal or osteochondral progenitors results in severely delayed chondrocyte hypertrophy, failed matrix calcification, and an inability to catabolize the cartilage matrix [17]. The latter effects are probably manifest because of the direct transcriptional regulation of Mmp13 by OSX. HDAC4 is yet another critical transcriptional regulator expressed in hypertrophic chondrocytes, but not osteoblasts (Fig. 2.2). Germline deletion of Hdac4 results in accelerated chondrocyte hypertrophy, cartilage calicifiation, and advanced endochondral ossification, whereas transgenic overexpression of Hdac4 suppresses chondrocyte hypertrophy and endochondral ossification, likely because of the chromatin and transcriptional regulation of Runx2 imposed by HDAC4 [18]. MEF2C and MEF2D are related transcription factors also expressed in prehypertrophic and hypertrophic chondrocytes (Fig. 2.2). Conditional deletion of Mef2c alone or in combination with Mef2d from mesenchymal or osteochondral progenitors results in failed chondrocyte hypertrophy, cartilage vascularization, and endochondral ossification reminiscent of Runx2 mutants [19]. Genetic interactions between HDAC4 and MEF2C were shown by deleting either a single allele of Mef2c in a Hdac4-/- background or deleting a single Hdac4 allele in a Mef2c+/- background resulting in normalization of their respective mutant phenotypes. Molecular studies further determined that both factors converge on the regulation of Runx2 in controlling chondrocyte hypertrophy, calcification, vascularization, and overall endochondral ossification [19]. Finally, SOX9 not only regulates chondrogenic gene expression important for inducing and maintaining immature chondrocytes, but also coordinates the onset of hypertrophy (Fig. 2.2). Forced expression of Sox9 in hypertrophic chondrocytes delays chondrocyte maturation and inhibits both calcification and vascularization of the hypertrophic cartilage [20], whereas cartilage-specific loss of Sox9 showed roles for SOX9 in (i) maintaining early hypertrophic chondrocytes via appropriate transcriptional regulation of Col10a1 and (ii) acting as a counterbalance to RUNX2 and OSX activities, preventing osteoblast differentiation of chondrocytes and excessive endochondral ossification [21].

Multiple signaling molecules regulate chondrocyte hypertrophy in both direct and indirect manners. Indian hedgehog (IHH) and PTHrP form a negative feedback loop critical for coordinating chondrocyte hypertrophy and endochondral ossification. IHH is secreted from prehypertrophic chondrocytes and directly regulates PTHrP in epiphyseal chondrocytes, which in turn signals to its receptor, PTHrP-R, on prehypertrophic chondrocytes to antagonize the pace of chondrocyte hypertrophy (Fig. 2.2). Germline deletions of *Ihh* and *Ptrhp* both exhibit reduced chondrocyte proliferation and precocious chondrocyte hypertrophy; however, *Pthrp* mutants generate accelerated bone formation whereas *Ihh* mutants fail to form bone because of a critical function of IHH in osteoblast differentiation from perichondral progenitors (see later) [22–24]. Cartilage specific deletions of *Smoothened* (*Smo*), a critical cell surface protein mediating hedgehog signaling, aided in uncoupling IHH effects on chondrocyte hypertrophy and endochondral ossification from perichodrial bone formation because mutants developed accelerated chondrocyte hypertrophy without defects in perichondrial bone formation [25]. A critical intracellular function of IHH signaling is to antagonize GLI3 repressor function while activating other GLI family members (GLI1 and GLI2). Analyses of $Ihh^{-/-}$; $Gli3^{-/-}$ mutant mice identified the requirement for GLI3 in mediating IHH actions on PTHrP, chondrocyte hypertrophy, and endochondral ossification, but not on vascularization of the cartilage or perichondrial osteoblast differentiation [26, 27]. Finally, PTHrP induces dephosphorylation of HDAC4 in turn decreasing HDAC4 and 14-3-3 protein interactions in the cytoplasm to promote nuclear translocation of HDAC4 and repression of MEF2C transcriptional activation of *Runx2*, thereby establishing an IHH/PTHrP-HDAC4-MEF2C-RUNX2 molecular regulation of chondrocyte hypertrophy and endochondral ossification [28]. Antagonistic to the IHH/PTHrP pathway, both BMP and WNT signaling promote chondrocyte hypertrophy (Fig. 2.2). Conditional removal of *Smad1* and *Smad5* floxed alleles from osteochondral progenitors [6] or *Bmpr1a* floxed alleles from osteochondral progenitors in a $Bmpr1b^{+/-}$ background [29] defined an extracellular to intracellular signaling cascade leading to BMP-mediated chondrocyte proliferation, survival, hypertrophy, and proper endochondral ossification; a cascade culminating in SMAD-mediated regulation of IHH signaling and transcriptional control of *Runx2*. Similarly, conditional deletion of β-catenin floxed alleles in mesenchymal and osteochondral progenitors showed the role WNT/β-catenin plays in promoting chondrocyte hypertrophy, calcification, vascularization, and endochondral ossification via β-catenin induced degradation and antagonism of SOX9 combined with transcriptional activation of *Osx* (Fig. 2.2) [10, 11, 30, 31]. NOTCH signaling also promotes chondrocyte hypertrophy and endochondral ossification (Fig. 2.2), via alternative mechanisms, however. First, both in vivo and in vitro lines of evidence support NOTCH suppression of chondrocyte proliferation and cell cycle exit during early chondrocyte hypertrophy [32].

Second, conditional deletion of *Notch1* and *Notch2* floxed alleles from mesenchymal progenitors or RBPjk floxed alleles from mesenchymal progenitors, osteochondral progenitors, or chondrocytes leads to delayed onset of chondrocyte hypertrophy and cartilage matrix catabolism, whereas overexpression of NICD in chondrocytes promotes these processes [32–34]. Molecular dissection of the pathway indicates that NOTCH indirectly promotes chondrocyte hypertrophy in an RBPjk-dependent manner via HES/HEY-mediated downregulation of *Sox9*, *Col2a1*, *Acan*, and other chondrogenic genes [13], while simultaneously inducing cartilage catabolism and turnover of growth plate cartilage via RBPjk-mediated induction of numerous catabolic genes including *Mmp13* [32, 35]. Each of the aforementioned transcriptional regulators and signaling pathways impart profound effects on chondrocyte hypertrophy and maturation during endochondral ossification. However, the identification of signals positively or negatively regulating hypertrophic chondrocyte transdifferentiation into osteoblast lineage cells (mesenchymal progenitors or committed osteoblasts) remains to be elucidated (Fig. 2.2).

OSTEOBLAST DIFFERENTIATION AND BONE FORMATION

Osteoblasts are responsible for producing and secreting a combination of extracellular proteins that comprise the bone matrix. These include copious amounts of type 1 collagen (COL1A1) and noncollagenous matrix proteins including ALPL, integrin-linked bone sialoprotein (IBSP), and osteocalcin (BGLAP), which serve as markers of distinct stages of osteoblast differentiation in addition to regulating diverse aspects of bone matrix mineralization. The process of osteoblast differentiation begins with condensation of multipotent mesenchymal progenitors, specification of osteochondral progenitors, formation of committed preosteoblasts, differentiation into mature functional osteoblasts, and finally encasement in bone matrix to form osteocytes (Fig. 2.2). Unlike osteoblasts that secrete the bone matrix, osteocytes function as mechanosensory cells that transduce mechanical loads into biochemical signals to regulate osteoblast differentiation and bone formation. Recent studies highlight the diverse cellular sources that can give rise to osteoblasts, including perichondrial cells, perivascular mesenchymal progenitors (pericytes) brought in during vascular invasion, circulating progenitors, hypertrophic chondrocytes [36, 37], and other mesenchymal cells within the bone marrow (see also Chapters 3, 4, and 5). Extensive studies over the past few decades have identified a number of transcription factors and developmental signals that regulate osteoblast differentiation. Differentiating osteoblasts are characterized by the expression of master transcriptional regulators including *Sox9* (expressed in mesenchymal progenitors), *Runx2* (expressed in osteochondral progenitors and mature osteoblasts),

Osterix (*Osx*, preosteoblasts and mature osteoblasts), and *Atf4* (mature osteoblasts) (Fig. 2.2). RUNX2 is indispensable for osteoblast differentiation and promotes the differentiation of osteochondral progenitors into preosteoblasts as well as the proper function of mature osteoblasts. Homozygous deletion of *Runx2* in mice results in a complete loss of osteoblasts. Like RUNX2, OSX is required for osteoblast differentiation and bone formation but functions downstream of *Runx2*. OSX is required for the transition from the preosteoblast to a functional mature osteoblast. Homozygous deletion of *Osx* in mice results in a thickened perichondrium at the diaphysis because of a failure of osteoblast differentiation. OSX is also crucial for postnatal osteoblast differentiation and function. ATF4 is required for terminal differentiation and regulates bone-forming activities in mature osteoblasts. Homozygous deletion of *Atf4* results in delayed osteoblast differentiation and decreased bone formation. ATF4 is dispensable for *Runx2* and *Osx* expression but coregulates *Ibsp* and *Bglap* expression with RUNX2. ATF4 promotes amino acid uptake to facilitate protein synthesis and bone matrix production by osteoblasts. This appears to be the primary function of ATF in osteoblasts as a high protein diet can correct the bone phenotypes of $Atf4^{-/-}$ mice.

Developmental signals such as NOTCH, IHH, WNT, and BMP are required at different stages and play unique roles in osteoblast differentiation. NOTCH signaling plays an important role in maintaining an osteochondral progenitor pool to provide osteoblasts throughout life. NOTCH signaling maintains the osteochondral progenitor pool by inhibiting RUNX2 transcriptional activity and preventing osteoblast differentiation (Fig. 2.2). Loss of this blockade through genetic inhibition of NOTCH signaling in mesenchymal progenitors results in exuberant differentiation and bone formation early in life with a severe reduction of bone mass later because of exhaustion of the progenitor pool. In committed osteoblasts, forced NOTCH activation inhibits terminal differentiation and stimulates osteoblast activity and bone formation by expanding the number of active osteoblasts resulting in sclerotic bone formation [41]. However, genetic ablation of NOTCH signaling in committed osteoblasts results in no discernable phenotype, underscoring the uncertainty of a physiological role for NOTCH signaling in mature osteoblasts. Conversely, Hedgehog signaling is required to initiate osteoblast differentiation. IHH, expressed in prehypertrophic and hypertrophic chondrocytes, signals to adjacent perichondrial cells to initiate osteoblast differentiation by regulating *Runx2* and *Osx* expression (Fig. 2.2). The WNT pathway similarly promotes osteoblast differentiation but functions downstream of IHH (Fig. 2.2) (also see Chapter 9). The WNT transcriptional effector β-catenin (encoded by *Catnnb1*) is required for osteoblast differentiation. Genetic deletion of *Catnnb1* in mesenchymal progenitors abolishes osteoblast formation and results in ectopic cartilage formation. Both β-catenin dependent and independent WNT signaling are required for progression from the *Runx2* positive progenitor to the *Osx* positive preosteoblast and from the preosteoblast to the mature osteoblast stages [11, 31, 44, 45]. Recent studies implicate cellular metabolism as a target of WNT regulation during differentiation. WNT stimulates glucose uptake, which favors osteoblast differentiation by increasing RUNX2 expression and activity [46, 47]. In mature osteoblasts, WNT stimulates glutamine catabolism which increases ATF4 translation to stimulate osteoblast activity and terminal differentiation [48]. Like WNT signaling, BMPs play multiple roles in regulating osteoblast differentiation [49]. BMP signaling directly regulates both *Runx2* and *Osx* expression and is required to form preosteoblasts (Fig. 2.2) (see also Chapter 8). Later, BMP promotes differentiation by suppressing proliferation in preosteoblasts and stimulating osteoblast activity. BMP signaling ultimately regulates osteoblast activity and bone formation by increasing ATF4 protein expression downstream of the unfolded protein response [50]. Combined, these signals cooperate in an elaborate and elegant web to coordinate endochondral ossification.

CONCLUSION

Here we have provided a general overview of endochondral ossification with a focus on the major cellular events (chondrogenesis, chondrocyte hypertrophy, and osteoblast differentiation), transcription factors (SOX trio, RUNX2, OSX, HDAC4, MEF2C/D, and ATF4), and signaling pathways (BMP/TGFB, WNT, IHH/PTHrP, and NOTCH) governing each stage of the process. In particular, we highlighted critical murine studies using sophisticated genetic approaches to determine the function(s) for many of the transcriptional regulators and signaling molecules important in coordinating proper development of the endochondral skeleton.

ACKNOWLEDGMENTS

This work was supported in part by the following United States National Institute of Health R01 grants (AR057022 and AR063071 to MJH) and funds from the Department of Orthopaedic Surgery at Duke University School of Medicine. Because of space constraints, we would like to acknowledge and apologize to the many authors whose important works were unable to be cited.

REFERENCES

1. Barna M, Niswander L. Visualization of cartilage formation: insight into cellular properties of skeletal progenitors and chondrodysplasia syndromes. Dev Cell. 2007;12(6):931–41.
2. Bi W, Deng JM, Zhang Z, et al. Sox9 is required for cartilage formation. Nat Genet. 1999;22(1):85–9.

3. Akiyama H, Chaboissier MC, Martin JF, et al. The transcription factor Sox9 has essential roles in successive steps of the chondrocyte differentiation pathway and is required for expression of Sox5 and Sox6. Genes Dev. 2002;16(21):2813–28.
4. Smits P, Li P, Mandel J, Zhang Z, et al. The transcription factors L-Sox5 and Sox6 are essential for cartilage formation. Dev Cell. 2001;1(2):277–90.
5. Yoon BS, Ovchinnikov DA, Yoshii I, et al. Bmpr1a and Bmpr1b have overlapping functions and are essential for chondrogenesis in vivo. Proc Natl Acad Sci U S A. 2005;102(14):5062–7.
6. Retting KN, Song B, Yoon BS, et al. BMP canonical Smad signaling through Smad1 and Smad5 is required for endochondral bone formation. Development. 2009;136(7):1093–104.
7. Baffi MO, Slattery E, Sohn P, et al. Conditional deletion of the TGF-beta type II receptor in Col2a expressing cells results in defects in the axial skeleton without alterations in chondrocyte differentiation or embryonic development of long bones. Dev Biol. 2004;276(1):124–42.
8. Nomura M, Li E. Smad2 role in mesoderm formation, left-right patterning and craniofacial development. Nature. 1998;393(6687):786–90.
9. Spagnoli A, O'Rear L, Chandler RL, et al. TGF-beta signaling is essential for joint morphogenesis. J Cell Biol. 2007;177(6):1105–17.
10. Hill TP, Spater D, Taketo MM, et al. Canonical Wnt/beta-catenin signaling prevents osteoblasts from differentiating into chondrocytes. Dev Cell. 2005;8(5):727–38.
11. Day TF, Guo X, Garrett-Beal L, et al. Wnt/beta-catenin signaling in mesenchymal progenitors controls osteoblast and chondrocyte differentiation during vertebrate skeletogenesis. Dev Cell. 2005;8(5):739–50.
12. Dong Y, Jesse AM, Kohn A, et al. RBPjkappa-dependent Notch signaling regulates mesenchymal progenitor cell proliferation and differentiation during skeletal development. Development. 2010;137(9):1461–71.
13. Rutkowski TP, Kohn A, Sharma D, et al. HES factors regulate specific aspects of chondrogenesis and chondrocyte hypertrophy during cartilage development. J Cell Sci. 2016;129(11):2145–55.
14. Chen H, Ghori-Javed FY, Rashid H, et al. Runx2 regulates endochondral ossification through control of chondrocyte proliferation and differentiation. J Bone Miner Res. 2014;29(12):2653–65.
15. Otto F, Thornell AP, Crompton T, et al. Cbfa1, a candidate gene for cleidocranial dysplasia syndrome, is essential for osteoblast differentiation and bone development. Cell. 1997;89(5):765–71.
16. Yoshida CA, Yamamoto H, Fujita T, et al. Runx2 and Runx3 are essential for chondrocyte maturation, and Runx2 regulates limb growth through induction of Indian hedgehog. Genes Dev. 2004;18(8):952–63.
17. Nakashima K, Zhou X, Kunkel G, et al. The novel zinc finger-containing transcription factor osterix is required for osteoblast differentiation and bone formation. Cell. 2002;108(1):17–29.
18. Vega RB, Matsuda K, Oh J, et al. Histone deacetylase 4 controls chondrocyte hypertrophy during skeletogenesis. Cell. 2004;119(4):555–66.
19. Arnold MA, Kim Y, Czubryt MP, et al. MEF2C transcription factor controls chondrocyte hypertrophy and bone development. Dev Cell. 2007;12(3):377–89.
20. Hattori T, Muller C, Gebhard S, et al. SOX9 is a major negative regulator of cartilage vascularization, bone marrow formation and endochondral ossification. Development. 2010;137(6):901–11.
21. Dy P, Wang W, Bhattaram P, Wang Q, et al. Sox9 directs hypertrophic maturation and blocks osteoblast differentiation of growth plate chondrocytes. Dev Cell. 2012;22(3):597–609.
22. Karaplis AC, Luz A, Glowacki J, et al. Lethal skeletal dysplasia from targeted disruption of the parathyroid hormone-related peptide gene. Genes Dev. 1994;8(3):277–89.
23. St-Jacques B, Hammerschmidt M, McMahon AP. Indian hedgehog signaling regulates proliferation and differentiation of chondrocytes and is essential for bone formation. Genes Dev. 1999;13(16):2072–86.
24. Vortkamp A, Lee K, Lanske B, et al. Regulation of rate of cartilage differentiation by Indian hedgehog and PTH-related protein. Science. 1996;273(5275):613–22.
25. Long F, Zhang XM, Karp S, et al. Genetic manipulation of hedgehog signaling in the endochondral skeleton reveals a direct role in the regulation of chondrocyte proliferation. Development. 2001;128(24):5099–108.
26. Hilton MJ, Tu X, Cook J, et al. Ihh controls cartilage development by antagonizing Gli3, but requires additional effectors to regulate osteoblast and vascular development. Development. 2005;132(19):4339–51.
27. Koziel L, Wuelling M, Schneider S, et al. Gli3 acts as a repressor downstream of Ihh in regulating two distinct steps of chondrocyte differentiation. Development. 2005;132(23):5249–60.
28. Kozhemyakina E, Cohen T, Yao TP, et al. Parathyroid hormone-related peptide represses chondrocyte hypertrophy through a protein phosphatase 2A/histone deacetylase 4/MEF2 pathway. Mol Cell Biol. 2009;29(21):5751–62.
29. Yoon BS, Pogue R, Ovchinnikov DA, et al. BMPs regulate multiple aspects of growth-plate chondrogenesis through opposing actions on FGF pathways. Development. 2006;133(23):4667–78.
30. Akiyama H, Lyons JP, Mori-Akiyama Y, et al. Interactions between Sox9 and beta-catenin control chondrocyte differentiation. Genes Dev. 2004;18(9):1072–87.
31. Hu H, Hilton MJ, Tu X, et al. Sequential roles of Hedgehog and Wnt signaling in osteoblast development. Development. 2005;132(1):49–60.
32. Kohn A, Dong Y, Mirando AJ, et al. Cartilage-specific RBPjkappa-dependent and -independent Notch signals regulate cartilage and bone development. Development. 2012;139(6):1198–212.
33. Hilton MJ, Tu X, Wu X, et al. Notch signaling maintains bone marrow mesenchymal progenitors by suppressing osteoblast differentiation. Nat Med. 2008;14(3):306–14.

34. Mead TJ, Yutzey KE. Notch pathway regulation of chondrocyte differentiation and proliferation during appendicular and axial skeleton development. Proc Natl Acad Sci U S A. 2009;106(34):14420–5.
35. Liu Z, Chen J, Mirando AJ, et al. A dual role for NOTCH signaling in joint cartilage maintenance and osteoarthritis. Sci Signal. 2015;8(386):ra71.
36. Yang L, Tsang KY, Tang HC, et al. Hypertrophic chondrocytes can become osteoblasts and osteocytes in endochondral bone formation. Proc Natl Acad Sci U S A. 2014;111(33):12097–102.
37. Zhou X, von der Mark K, Henry S, et al. Chondrocytes transdifferentiate into osteoblasts in endochondral bone during development, postnatal growth and fracture healing in mice. PLoS Genet. 2014;10(12):e1004820.
38. Yang X, Matsuda K, Bialek P, et al. ATF4 is a substrate of RSK2 and an essential regulator of osteoblast biology; implication for Coffin-Lowry Syndrome. Cell. 2004;117(3):387–98.
39. Ducy P, Zhang R, Geoffroy V, et al. Osf2/Cbfa1: a transcriptional activator of osteoblast differentiation. Cell. 1997;89(5):747–54.
40. Elefteriou F, Benson MD, Sowa H, et al. ATF4 mediation of NF1 functions in osteoblast reveals a nutritional basis for congenital skeletal dysplasiae. Cell Metab. 2006;4(6):441–51.
41. Engin F, Yao Z, Yang T, et al. Dimorphic effects of Notch signaling in bone homeostasis. Nat Med. 2008;14(3):299–305.
42. Tu X, Chen J, Lim J, et al. Physiological notch signaling maintains bone homeostasis via RBPjk and Hey upstream of NFATc1. PLoS Genet. 2012;8(3):e1002577.
43. Tu X, Joeng KS, Long F. Indian hedgehog requires additional effectors besides Runx2 to induce osteoblast differentiation. Dev Biol. 2012;362(1):76–82.
44. Guo X, Day TF, Jiang X, et al. Wnt/beta-catenin signaling is sufficient and necessary for synovial joint formation. Genes Dev. 2004;18(19):2404–17.
45. Tu X, Joeng KS, Nakayama KI, et al. Noncanonical Wnt signaling through G protein-linked PKCdelta activation promotes bone formation. Dev Cell. 2007;12(1):113–27.
46. Wei J, Shimazu J, Makinistoglu MP, et al. Glucose uptake and Runx2 synergize to orchestrate osteoblast differentiation and bone formation. Cell. 2015;161(7):1576–91.
47. Esen E, Chen J, Karner CM, et al. WNT-LRP5 signaling induces Warburg effect through mTORC2 activation during osteoblast differentiation. Cell Metab. 2013;17(5):745–55.
48. Karner CM, Esen E, Okunade AL, et al. Increased glutamine catabolism mediates bone anabolism in response to WNT signaling. J Clin Invest. 2015;125(2):551–62.
49. Bandyopadhyay A, Tsuji K, Cox K, et al. Genetic analysis of the roles of BMP2, BMP4, and BMP7 in limb patterning and skeletogenesis. PLoS Genet. 2006;2(12):e216.
50. Saito A, Ochiai K, Kondo S, et al. Endoplasmic reticulum stress response mediated by the PERK-eIF2(alpha)-ATF4 pathway is involved in osteoblast differentiation induced by BMP2. J Biol Chem. 2011;286(6):4809–18.

3

Local and Circulating Osteoprogenitor Cells and Lineages

Naomi Dirckx and Christa Maes

Laboratory of Skeletal Cell Biology and Physiology, Skeletal Biology and Engineering Research Center, Department of Development and Regeneration, KU Leuven, Leuven, Belgium

INTRODUCTION

Mature, bone-forming osteoblasts represent principle mediators of skeletal development, growth, and repair by being responsible for bone matrix deposition and mineralization. During adult bone homeostasis, continual bone remodeling is mediated by the tightly balanced activities of bone-resorbing osteoclasts and bone-rebuilding osteoblasts, thereby ensuring proper bone maintenance as well as regulation of calcium and phosphate homeostasis. Cells of the osteoblast lineage also contribute to the regulation of hematopoiesis, by constituting essential components of the HSC niches within the bone and BM environment. In addition, osteoblasts play important roles in the control of whole-body energy metabolism.

To fulfill their key functions, osteoblasts need to differentiate from mesenchymal progenitors typically residing within the stromal BM environment. When committed to the osteoblast lineage, osteoprogenitors further differentiate into matrix-producing osteoblasts characterized by abundant expression of the prime bone matrix constituent collagen type I (Col1), and next into mineralizing osteoblasts typically expressing osteocalcin (OCN) (Fig. 3.1). Ultimately, osteoblast lineage cells can undergo apoptosis, become flattened quiescent bone lining cells (BLCs), or become matrix-embedded osteocytes.

Accordingly, indispensable aspects of bone formation — whether during development, growth, remodeling, or repair — include the recruitment and engagement of progenitors with osteogenic potential, their migration towards and attachment on the bone surface at sites in need of bone formation, and their proper differentiation and activation into functional osteoblasts [1]. A better understanding of the endogenous osteogenic progenitor cells present in the bone and BM environment and in the circulation will therefore be of vital importance for the development of osteoanabolic therapies for widespread low bone mass disorders such as osteoporosis, and to intervene therapeutically in situations of compromised fracture healing.

In recent years, an impressive body of work has increased our knowledge on the localization and characteristics of osteogenic progenitors, although the univocal identification of specific stem/progenitor cell subsets in bone by unique markers is still awaited. Localizing such cells in vivo by immunohistochemical staining is accordingly difficult. However, the increasing availability of transgenic mice carrying (constitutively active or tamoxifen-inducible) Cre recombinase constructs under the control of a variety of gene promoters, and the existence of a broad variety of reporter mice, made it possible to mark and trace cell populations characterized by expression of specific markers over time and space in a controlled way and visualize them in vivo. The use of these mice has started to shed light on the potential candidate cell (sub-) populations constituting sources of skeletal stem cells (SSCs), multipotent mesenchymal stem or progenitor cells (MSPCs), and osteoprogenitors functioning in bone development, homeostasis, and fracture repair, as will be discussed in this chapter. Although much of this knowledge is being derived from murine models, which will constitute the main focus of this overview, the existence and characteristics of human counterparts will be indicated.

Primer on the Metabolic Bone Diseases and Disorders of Mineral Metabolism, Ninth Edition. Edited by John P. Bilezikian.
© 2019 American Society for Bone and Mineral Research. Published 2019 by John Wiley & Sons, Inc.
Companion website: www.wiley.com/go/asbmrprimer

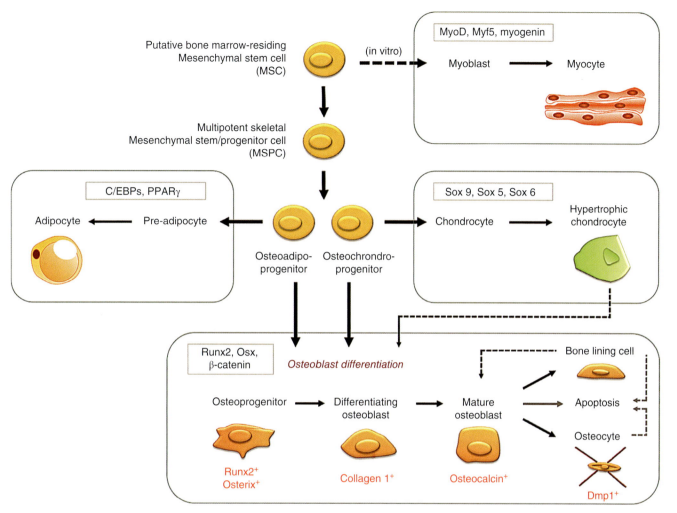

Fig. 3.1. Differentiation of osteoprogenitor cells and lineages. The bone marrow harbors putative mesenchymal stem cells (MSCs) that can give rise to osteoblasts, adipocytes, chondrocytes and, at least in vitro, myocytes. The three skeletal lineages can descend from multipotent skeletal mesenchymal stem or progenitor cells (MSPCs), also often called bone marrow stromal cells (BMSCs). Lineage differentiation of the precursors is controlled by the indicated respective sets of transcription factors; Runx2, Osx, and β-catenin mediate osteoblast differentiation and functioning. Osteoprogenitors in vivo, at least in mice, can be derived from an osteo-chondro-progenitor intermediate (particularly in fetal development) or an osteoadipogenic bipotent progenitor cell (found in postnatal bone). Progressive osteoblast differentiation is characterized by changes in gene expression typifying specific stages or subsets of osteoblast lineage cells as indicated. Mature osteoblasts ultimately undergo apoptosis, flatten to form a layer of bone lining cells (BLCs) on the bone surface, or become embedded within the bone as osteocytes. Recent studies indicate that in addition to the classical, lineage-committed osteoprogenitors, hypertrophic chondrocytes may also constitute a source of osteoblasts during development and growth. In addition, quiescent BLCs can convert into mature osteoblasts in postnatal bone upon induction of active bone formation.

IN SEARCH OF THE SKELETAL STEM CELL

While the quest for the alleged, yet so far unspecified, SSC is vigorously ongoing and a topic of extensive research driven especially by the high hopes for clinical applicability in bone regenerative purposes [2], a prevailing challenge in the field remains to adopt a unifying terminology and classification based on molecular marker determinants for the myriads of multipotent mesenchymal progenitor cell populations currently studied [3–5]. Here, as shown in Fig. 3.1, we indicate as multipotent skeletal MSPCs those cells that can be derived from BM and exhibit the potential to differentiate along the adipogenic, chondrogenic, and osteogenic lineages in vivo and in vitro; such cells are themselves thought to be descending from a putative BM-residing mesenchymal stem cell (MSC) that can also differentiate along the myogenic lineage, at least in vitro (see Fig. 3.1). These populations are most likely contained within the broader and heterogeneous cell population generically designated as BM stromal cells (BMSCs).

The history of MSPCs within bone originates from 1968, with the discovery that transplantation of boneless fragments of BM could induce heterotopic ossification at the graft site [6]. In the 1970s, Friedenstein and colleagues discovered the existence of a cell type distinct from hematopoietic cells within the BM environment that had the ability to adhere to plastic culture recipients and form colonies when seeded at low density. These cells had a fibroblastic morphology with clonogenic potential (defined as the colony-forming unit-fibroblasts [CFU-F]) and cells derived from a single colony were able to differentiate along the osteogenic, chondrogenic, and adipogenic lineages in vitro and in vivo (reviewed by [7]). Such multipotent mesenchymal cells were later more generically adopted within the name "mesenchymal stem cells (MSCs)" as cells that are derived from embryonic mesoderm, but with an occurrence that is not restricted to the BM environment yet in fact encompasses various postnatal connective tissues. Importantly, however, in order to be designated a "stem cell," a cell needs to meet two criteria: self-renewal and multipotency. Notwithstanding the ability of extensive proliferation, the term MSC in the abovementioned sense has become a topic of controversy and debate, particularly because of the common lack of evidence of a universal self-renewing capacity after serial passaging or transplantation and the lack of classification based on tissue specificity [3, 7]. It is therefore more desirable to signify MSC as "multipotent mesenchymal stromal cells," while denoting those cells derived from BM specifically as BMSCs or SSCs, depending on their characteristics [4].

In terms of marker expression, for human MSCs, the criteria listed in [5] include the presence of expression of CD105, CD73, and CD90, and lack of CD45, CD34, CD14 or CD11b, CD79alpha or CD19, and HLA-DR surface molecules [5]. For BMSCs or SSCs, consensus markers are as yet to be defined. In the 1990s, studies by Simmons and colleagues led to the discovery that a monoclonal antibody termed STRO-1, reacting with an unidentified cell surface antigen, could be used to isolate human BMSCs enriched with a clonogenic population able to form CFU-Fs, transfer a functional hematopoietic microenvironment in vitro, and differentiate in vitro into multiple mesenchymal lineages, including smooth muscle cells, adipocytes, chondrocytes, and osteoblasts [8]. Later on, these authors optimized the isolation protocol by adding antibodies against vascular cell adhesion molecule-1 (VCAM-1/CD106) for positive selection [9]. Sacchetti and colleagues defined melanoma-associated cell adhesion molecule (MCAM/CD146) as a marker for stromal progenitors in human BM with the capacity to self-renew, regenerate bone and stroma, and establish a hematopoietic microenvironment in vivo [10]. For murine BM, stem cell biologists have used expression of stem cell antigen-1 (Sca-1) to designate the MSPC population, defining them as Sca-1$^+$Lin$^-$CD45$^-$CD31$^-$ cells, with Lineage (Lin) and CD45 as hematopoietic markers and CD31 as the classical endothelial cell (EC) marker for negative selection [11].

Clearly, although extensive research has been performed, the distinct and incontrovertible identification of the stem/progenitor cells of bone, as well as their functional characterization, remains to be accomplished. The ultimate goal is to reach a linear branching tree that defines the skeletal stem cell hierarchy into a similar extent as the organization of the well-defined hematopoietic system with HSCs progressively giving rise to the different specified blood cell lineages. Very recently, advances towards this challenging goal were reported, with novel approaches and markers identifying the murine SSC [12, 13]. These studies retrieved a CD45$^-$Ter119$^-$Tie2$^-$AlphaV$^+$Thy$^-$6C3$^-$CD105$^-$CD200$^+$ stem cell population [12] and so-called osteochondroreticular (OCR) cells expressing Gremlin (a BMP antagonist) [13] from the metaphyseal bone regions close to the growth plate, suggesting their intensive involvement in growth and bone formation.

OSTEOPROGENITOR CELLS IN BONE DEVELOPMENT, GROWTH, AND HOMEOSTASIS

Osteoblasts are derived from mesenchymal cells that originate in condensing mesenchyme during embryogenesis or reside in the BM environment in the adult [14]. Being the driving engine for skeletal growth in the long bones, the cartilaginous growth plates consist of vigorously proliferating and matrix-producing chondrocytes that ultimately become hypertrophic, before being replaced by bone tissue. Osteoblasts settle on cartilage remnants and deposit bone matrix in close interplay with neovascularization of the developing and growing endochondral bone [15].

The earliest known markers for bona fide osteoprogenitors, classically used to denote immature yet committed cells of the osteoblast lineage, are Runt-related transcription factor 2 (Runx2) and the zinc finger-containing transcription factor Osterix (Osx). Although both transcription factors are also expressed in hypertrophic and prehypertrophic chondrocytes, respectively, they are absolutely indispensable for mesenchymal osteogenic precursor cells to differentiate along the osteoblast lineage and function as bone-forming cells (Fig. 3.1). This was first shown by the finding that a full knockout of Runx2 in mice resulted in a cartilaginous skeleton with complete absence of osteoblast differentiation and bone formation (for references, see [14]). Haploinsufficiency of Runx2 in mice led to hypoplastic clavicles and delayed closure of the fontanelles, which are typical characteristics of cleidocranial dysplasia, caused by RUNX2 mutations in humans. Deletion of Osx in mice also led to a phenotype of total absence of bone formation. The finding that Osx null mice displayed expression of Runx2 while Runx2 null mice lacked Osx expression led to the positioning of Osx downstream of Runx2 (Fig. 3.1) (see [14]).

Beside their importance during embryonic bone development, both transcription factors also play crucial roles in postnatal bone homeostasis. For Runx2, a function beyond development was initially suggested by its postnatal expression, including in fully differentiated osteoblasts, and because it is a major regulator of Ocn expression [16]. Expression of a dominant-negative form of Runx2 in osteoblasts led to virtual absence of Ocn expression and impaired postnatal bone formation [17]. Conditional disruption of Osx in maturing osteoblasts (by using Col1a1-Cre/CreERt2 mice) led to osteopenia caused by impaired osteoblast differentiation in adult mice [18, 19], and postnatal deletion of Osx in adult bones by using tamoxifen-inducible CAG-CreERt;Osx(flox/−) transgenic mice abolished the maturation, morphology, and function of osteocytes [20]. Notably, these mice also showed accumulation of unresorbed calcified cartilage remnants below the growth plate and an almost complete absence of BM cells leading to a decrease in osteoclast number [21]. These findings classified Osx as a vital and multifunctional player in postnatal bone growth and homeostasis.

The promoter regions of both the Runx2 and Osx genes have been used to create Runx2-Cre [22], Osx-Cre [23], and Osx-CreErt [24] transgenic mouse models for cell-specific recombination of target genes or reporters, which are being extensively used in the field. These model systems made it possible to investigate these osteoprogenitor subsets in detail during embryonic bone formation and adult bone remodeling and repair. During embryogenesis, tracing experiments with tamoxifen-inducible Osx-CreERt mice combined with a LacZ reporter revealed the capacity for Osx-expressing osteoprogenitors to invade, as pericytes or in close vicinity of blood vessels, the hypertrophic cartilage region during the formation of the primary ossification center, and give rise to differentiated, trabecular bone-forming osteoblasts inside the developing bone [24]. In contrast, the more mature osteoblasts that already expressed Col1 while residing in the perichondrium stayed in these outer layers and formed the cortical bone [24].

The use of these models also helped reveal the key role of β-catenin, the prime downstream transcriptional mediator of the canonical Wnt signaling pathway [25], in the osteoblast lineage specification of Osx-expressing cells, and evidenced their plasticity and/or multipotent nature in development and adult bone homeostasis. During embryogenesis, Osx-Cre-mediated removal of β-catenin led the osteoprogenitors to convert to a chondrocyte fate [23]. This finding is in line with the notion that during development and in early postnatal life, common osteochondro-progenitor cells exist that can give rise to both cell types (see Fig. 3.1), as has been documented using lineage tracing strategies employing Col2, aggrecan (Acan), and Sox9 gene promoters driving tamoxifen-inducible CreERt expression [24, 26]. In addition, differentiated growth plate chondrocytes have been strongly suggested to have the ability to transdifferentiate into functional osteoblasts, as based on Acan-CreERt- and ColX-CreERt-driven fate-mapping studies [27, 28] (Fig. 3.1).

Later in life, osteolineage cells appear to share a common progenitor with marrow adipocytes (Fig. 3.1).

Postnatal deletion of functional β-catenin in Osx-expressing cells induces a cell fate shift towards adipocytes instead of their classic differentiation into osteoblasts [29]. These findings were ground-breaking initial findings in the currently enfolding endeavors towards understanding the origin of BM adipocytes, and marked Osx+ cells, previously considered committed osteoprogenitors, as bipotent precursors that can under certain circumstances switch to an adipogenic fate. Additional studies highlighted this latter finding, by describing a similar phenotype of increased BM adiposity upon Osx-Cre-driven targeting of the genes encoding VEGF [30] or the heterotrimeric G protein subunit $G_s\alpha$ [31]. Conversely, inactivation of Foxo1, -3, and -4 (members of the FOXO family of transcription factors that attenuate Wnt/β-catenin signaling by diverting β-catenin from TCF- to FOXO-mediated transcription) resulted in a reduced adipogenic differentiation of Osx+ cells, associated with decreased adiposity in the aged BM, an increased osteoblast number, and a high bone mass that was maintained in old age [32].

Adult bone undergoes constant remodeling with a central role for osteoblasts and osteoclasts communicating through several cytokines and growth factors, and osteocytes playing important remodeling–regulating roles. However, the actual source for active matrix-depositing osteoblasts during adult bone homeostasis has as yet not been absolutely defined. BMSCs or MSPCs in the BM are potential candidates for populating the osteogenic pool in the bone resorption pit, as discussed in the next section. The recruitment of such osteogenic progenitors towards the bone surfaces may be coupled to the resorption phase of the remodeling cycle through the osteoclast-mediated release of growth factors that were stored in the bone matrix and that are chemo-attractive for the osteogenic cells [1, 33]. Yet, a bone surface-associated cell that increasingly comes into the picture as a potential source of active osteoblasts in adult bone is the bone lining cell (BLC). BLCs are descendants from mature osteoblasts characterized by a flat morphology and distinct location on the bone surface (see Fig. 3.1) that could represent a pool of quiescent, osteolineage-committed precursors playing major roles during tissue turnover or repair. The idea that BLCs become reactivated upon intermittent administration of PTH had been postulated already two decades ago [34]. Interestingly, recent lineage-tracing approaches confirmed the conversion of inactive BLCs into mature, cuboidal osteoblasts upon PTH treatment in mice [35]. Direct reactivation of BLCs was also shown to contribute to the acute increase in osteoblast numbers after sclerostin antibody treatment [36] and to the transient increase in bone formation observed after acute exposure to high dose whole-body γ-radiation [37]. Conversely, BLC activation was inhibited by glucocorticoids, known inducers of osteoporosis [38]. This latter study employed a mouse model in which active osteoblasts, but not BLCs, could be genetically ablated by ganciclovir administration [38]. In the subsequent recovery period, rapid bone formation was observed, through lineage tracing, to be executed by former BLCs that had switched to a cuboidal cell shape and regained bone-forming activity. Isolation of these cells by FACS revealed that BLCs express cell surface markers characteristic of MSPCs that are largely absent in osteoblasts, including Sca-1 and Leptin Receptor (LepR) [38].

PERIVASCULAR PROGENITORS WITH OSTEOGENIC POTENTIAL IN THE BM MICROENVIRONMENT

As outlined above, BMSCs or MSPCs in the BM environment are believed to be prime sources of new osteoblasts during adult bone homeostasis and remodeling. Such skeletal stem/progenitor cells have been identified as a heterogenic cell population that can adhere to plastic, form CFU-Fs, and has the potential to differentiate into osteoblasts, chondrocytes, and adipocytes in vitro. However, their identification and nature in vivo remain largely unclear. Recent work has increasingly suggested that osteoprogenitor lineages (or fractions thereof) constitute in fact pericytes in bone, with mesenchymal progenitor cells residing preferentially in close association with the walls of skeletal blood vessels (Fig. 3.2). Moreover, several publications have identified the skeletal vasculature as a niche for progenitors with both osteogenic and adipogenic potential as well as the capacity to support HSCs and/or organize a functional hematopoietic environment (Fig. 3.2).

In humans, CD146 has been identified as a marker for such self-renewing, multipotent SSCs, physically residing as perivascular cells (or adventitial reticular cells) on the outer endothelial lining of BM sinusoids, and acting as organizers of the local BM microenvironment [4, 39, 40]. In mice, several candidate markers have been

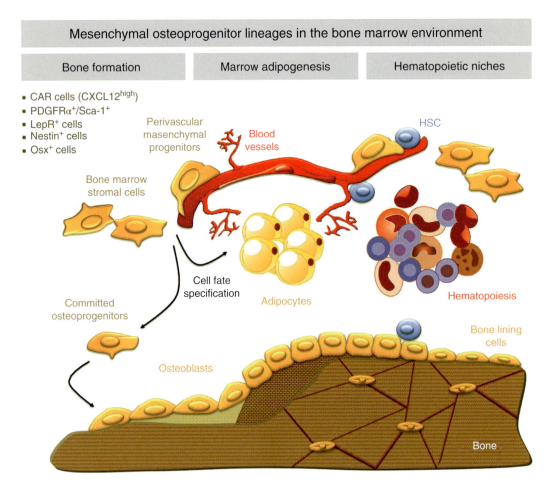

Fig. 3.2. Mesenchymal osteoprogenitor lineages in the BM environment. Recent studies using genetically modified mouse models have indicated cell populations characterized by the expression of a number of specific marker genes (see top left corner) to correspond to (subsets of) skeletal mesenchymal stem or progenitor cells (MSPCs), usually residing in perivascular locations, and/or bone marrow stromal cells (BMSCs). Besides their osteogenic potential, ie the capacity to differentiate into functional, bone-forming osteoblasts (left and bottom parts of the scheme), several of these cell populations are multi- or bipotent and can also give rise to marrow adipocytes (middle portion of the scheme). In addition, many of these cell populations are able to support hematopoiesis, such as by constituting functional niches for hematopoietic stem cells (HSCs) that are thought to reside in osteoblastic, vascular, and perivascular niches in the bone and BM environment (right part of the scheme).

associated with (likely partly overlapping populations of) such osteogenic mesenchymal progenitors based on extensive research using sophisticated genetic models, as briefly listed below. For more extensive and specialized reviews, see [41–43].

Interestingly, several of these populations are not only pivotal to the growth and lifelong turnover of bone but have also been documented to be activated after injury and contribute to bone repair in fracture models. These findings support the notion that these perivascular progenitor populations in fact constitute tissue-resident reserve pools of osteogenic cells, capable of reconstituting functional bone-forming cells mediating endogenous tissue repair, and thereby determining the native regeneration capacity of bone (also see next section).

Chemokine C-X-C motif ligand 12-abundant reticular cells

Chemokine C-X-C motif ligand 12 (CXCL12), also known as stromal cell-derived factor (SDF)-1, is a ligand for the CXCR4 receptor expressed on HSCs and plays important roles in HSC maintenance and hematopoiesis. Stromal cells characterized by strong CXCL12 expression were found to be perivascular reticular cells residing adjacent to BM sinusoids and in the endosteum [41, 43]. These so-called CXCL12-abundant reticular cells (CAR) cells represent key components of the niche for HSCs, as was shown by their selective ablation in vivo (through the use of a CXCL12-DTR-GFP transgene responsive to diphtheria toxin [DT] administration) [44] and by investigating conditional CXC12 knockout models [45]. CAR cells also constitute a pool of osteoadipogenic progenitors; BMSCs from CAR cell-depleted mice showed significantly reduced osteogenic and adipogenic differentiation potential in vitro, and CXCL-12-GFP+ cells in vivo express both the osteogenic transcription factors Runx2 and Osx, and the adipogenic transcription factor PPARγ [44]. These findings support the notion that perivascular, multipotent adipogenic–osteogenic progenitor cells secreting large amounts of hematopoietic cytokines such as CXCL12, stem cell factor (SCF), and angiopoietin-1 (Ang-1) are critical to maintain the BM niche and to regulate quiescence and differentiation of HSCs (Fig. 3.2).

PαS (PDGFRα+ Sca-1+) cells

PDGFRα+ Sca-1+ CD45- TER119- (PαS) nonhematopoietic mesenchymal progenitor cells in adult murine BM reside in the arterial perivascular space in vivo and are highly enriched for CFU-Fs [46]. In vitro, clonogenic colonies had the capacity to differentiate along the chondrogenic, adipogenic, and osteogenic lineages and remarkably also the endothelial lineage. In vivo, cotransplantation of uncultured GFP+ PαS cells with HSCs into lethally irradiated donor mice showed that PαS cells could engraft and repopulate the hematopoietic niche. Moreover, GFP+ cells in the BM gave rise to Ocn+ osteoblasts, perilipin+ adipocytes, CXCL12-, and Ang-1-expressing reticular perivascular cells, and, importantly, to PαS cells themselves, suggesting their self-renewal in vivo [46].

Leptin receptor+ cells

Another marker that identifies perivascular mesenchymal progenitors in adult murine bone is leptin receptor (LepR). LepR+ cells were reported to locate around BM sinusoids and arterioles, to be highly enriched for mesenchymal stromal cells, and to account for 94% of the CFU-Fs in the BM [47]. LepR+ cells were able to form bone, cartilage, and adipocytes in vitro as well as upon transplantation in vivo, and they express the hematopoietic niche factors SCF and CXCL12 [47]. Moreover, LepR+ cell-derived SCF is essential for HSC maintenance [48]. Although LepR+ cells do not or rarely contribute to osteogenic and chondrogenic lineages during development and growth, their descendants do contribute to a large fraction of osteoblasts, BLCs, osteocytes, as well as adipocytes in adulthood. Moreover, although LepR+ perivascular stromal cells in normal adult BM were mostly quiescent, they proliferated after injury and constitute a major source of osteoblasts and adipocytes in bone regenerated after irradiation or fracture [47, 49].

Nestin+ cells

Nestin (Nes), an intermediate filament protein previously known particularly as a neural stem cell marker, has in recent years also come into the picture as a marker for mesenchymal stromal cells in murine BM. A Nes-GFP transgene was found to be expressed within the BM stroma and by perivascular cells around skeletal arterioles. Isolated Nes-GFP+ stromal cells included all BM CFU-F activity and formed nonadherent multipotent "mesenspheres" that self-renewed and spontaneously differentiated into osteoblasts, chondrocytes, and adipocytes in vitro. Serial heterotopic transplantations in phosphocalcic ceramic ossicles also showed the self-renewal capacity, the ability to differentiate into osteogenic lineages and the association with hematopoietic activity of Nes-GFP+ cells in vivo [50]. In fact, HSCs commonly localize adjacent to Nes-GFP+ cells, which express high levels of SCF, CXCL12, and Ang-1. Moreover, Nes-GFP+ cells were proven to constitute an essential HSC niche component, as in vivo Nes+ cell depletion (in DT-treated Nes-CreERt/ iDTR mice) rapidly reduced the HSC content in the BM as well as the homing of hematopoietic progenitors into bones of lethally irradiated mice [50].

During fetal endochondral ossification and in early postnatal bones, Nes+ cells constitute a heterogeneous population that is associated with the vasculature and comprises a range of cells in the osteoblast, stromal, and endothelial lineages [51]. Later, in adult bones, Nes+ cells appear to be quiescent early cells with osteogenic potential, shown by in vivo fate mapping to contribute to osteoblasts, osteocytes, and chondrocytes. Moreover, the proliferation and osteoblastic differentiation of Nes+ cells markedly increased upon PTH treatment [50].

Osx+ cells

Pulse-chase studies have followed the fate of Osx-expressing cells labeled at various stages in life. During bone development, Osx+ cells originating in the fetal perichondrium contribute to stromal cells and mature osteoblasts inside the bone [24]. Later on, these cells are likely replaced, with new waves of Osx+ cells originating in early postnatal life, giving rise to BMSCs for prolonged periods of time, and contributing to tissue regeneration after injury [49]. In contrast, Osx+ cells marked in adult mice did not contribute much to the BM stroma [52]. Overall, the data suggest that labeling of Osx+ cells is mostly transient and that the Osx+ population does not represent a durable, self-renewing pool of osteoprogenitors in adult bone [49, 52]. Interestingly, at least a fraction of Osx+ cells is pericytic during embryonic bone development, postnatal growth, and fracture repair [24, 49, 53]. In addition, ample studies have indicated that the Osx+ population contains bipotent osteoadipo-progenitors [29–31] and contributes to the regulation of hematopoiesis, especially B-lymphopoiesis [45, 54].

OSTEOPROGENITOR RECRUITMENT IN FRACTURE HEALING

Generally, bone repair is a rapid and efficient process, which largely recapitulates the processes of embryonic bone development. This "re-generation" process at the site of the defect results in newly formed bone that is basically indistinguishable from the original, uninjured bone. Microfractures and stabilized fractures principally heal through intramembranous ossification, with MSPCs or osteogenic progenitors differentiating directly into osteoblasts. Semistabilized fractures largely heal through endochondral ossification, a process involving the formation of a soft fibrocartilaginous callus intermediate, followed by osteoblastogenesis and the deposition of bone matrix bridging the fracture site. In both processes the initial hematoma that releases a myriad of growth factures and cytokines is indispensable for proper fracture healing because it initiates proliferation and recruitment of the necessary stem and progenitor cells to mediate the subsequent repair [1]. The major sources for mesenchymal cell recruitment to the fracture site are the periosteum and BMSCs, although some reports also stress the importance of MSPCs from the circulation and contributions by muscle-derived progenitor cells.

The periosteum

The periosteum, a composite tissue lining the outer surface of the bone that provides a niche for multipotent osteochondro-progenitor cells, exhibits a remarkable regenerative capacity for de novo bone formation after injury [55]. Ample studies revealed that disruption of the periosteum impairs fracture healing, that isolated periosteum-derived cells display unique properties in terms of expansion capacity, mesenchymal stem/progenitor marker expression, and bone-forming potential, and that the periosteum bears enormous therapeutic promise as a cell source for bone tissue engineering purposes [55]. Lineage-tracing studies in mice support the idea that a subset of periosteal cells provides the principal source of chondrocytes and osteoblasts during endochondral fracture repair. Such periosteal stem/progenitor cells that respond to the bone fracture have for instance been marked by virtue of their expression of Prx1, a very early fetal limb bud mesenchymal cell marker that in adulthood becomes abundantly expressed in the periosteal cambium layer [56], or α-smooth muscle actin (α-SMA) [57]. In fact, a very elegant lineage tracing strategy using a combined transplantation and injury model in mice, showed that the periosteum, endosteum, and BM contain pools of stem/progenitor cells with distinct properties: whereas periosteum and BM/endosteum both gave rise to osteoblasts within the callus during fracture repair, the periosteum was the major source of chondrocytes in endochondral bone healing processes [58].

BMSCs

Many studies have also highlighted the capacity of BM aspirates or BMSCs to enhance fracture healing in experimental bone regeneration studies. Although there is little evidence and many limitations, BMSCs are being used in clinical trials for craniofacial bone regeneration, distal tibial nonunion, and osteonecrosis of the femoral head, with promising results [59]. Some of the limitations include the poor definition of the cells and the lack of mechanistic insights, aspects that are being addressed by basic research using mouse models. Fate mapping studies in transgenic mice (described previously) already allocated a role in fracture healing for Osx+ and LepR+ cells and the Grem1+ OCR population, contributing to both the chondrogenic and osteogenic lineages during endochondral bone repair, yet it must be noted that these cell populations are not confined to the BM and locate to the periosteum as well [13, 47, 49]. Another stromal marker, myxovirus resistance-1 (Mx1), delineates a pool of progenitor cells located at the endosteal surface and within the BM, which respond to fracture by proliferation and migration to the site of injury, supplying new osteoblasts during fracture healing. These Mx1+ cells appeared to supply the majority of osteogenic cells in the regeneration process and are largely osteolineage-restricted [52].

CIRCULATING OSTEOGENIC PRECURSOR CELLS

The existence in the blood stream of MSPCs with osteogenic potential, or circulating osteoprogenitors or other osteoblast lineage populations — here collectively termed circulating osteogenic precursor cells (COPs) — and their actual function and contribution to new bone formation is still a topic of controversy and debate.

On the one hand, COPs have been isolated based on plastic adherence and marker analysis from human and experimental animal peripheral blood, although their frequency was extremely low, especially in humans [60, 61]. Nevertheless, the isolated adherent fibroblast-like cells displayed osteogenic differentiation potential in vitro and in vivo upon ectopic (subcutaneous) transplantation [60]. Circulating MSPC-like cells have also been isolated from umbilical cord blood [62]. A number of experimental studies have further provided evidence in favor of the existence of COPs, and of their ability to access bone formation sites from the circulation. Experimental approaches typically used in such studies include parabiosis experiments, in which a conjoint pair of mice is created that shares a common circulatory system (see later), and transplantation experiments using BM or MSPC populations. For instance, Otsuru and colleagues showed the existence of COPs by transplanting BM from GFP-expressing transgenic mice into recipients undergoing BMP2-induced bone formation in muscle-implanted collagen pellets, because $GFP^+;Ocn^+$ osteoblastic cells were found to contribute to the newly formed ectopic bone [63]. The COPs were characterized to be $CD45^-$ $CD44^+$ $CXCR4^+$, thus expressing receptors for osteopontin and CXCL12/SDF-1 and capable of differentiating into osteoblasts in vitro and in vivo [64].

On the other hand, plastic-adherent MSPCs generally exhibit poor homing to noninjured skeletal tissues in systemic infusion studies, suggesting that MSPCs may generally not be physiologically circulating [65]. During pubertal growth in adolescent boys and upon fracture in adult patients, however, significant increases in the number of COPs have been reported [66, 67]. Still, it could be argued that (pathology-associated) COPs could be the result of disruption of the integrity of the bone tissue, thereby releasing BMSCs or related cell populations into the circulation, or that cells with osteogenic potential retrieved from the peripheral blood may in fact represent hematopoietic- or EC-derived populations, given that several such cell types have been shown to be able to revert to an osteoblastic fate under certain conditions [65, 68].

Overall, the prevailing notion appears to be that COPs may participate particularly in bone formation in circumstances of highly active osteoanabolic responses, fracture healing, and/or ectopic/heterotopic ossification. Several experimental animal models support this idea (for review, see [65]). For instance, Kumagai and colleagues created a parabiosis model by joining a wild-type mouse and a syngeneic mouse constitutively expressing GFP [69]. After the induction of a fibular fracture in the wild-type partner, GFP^+ cells also expressing alkaline phosphatase (ALP) were found in the fracture callus, albeit still at relatively low frequency, suggesting that osteogenic progenitor cells had been recruited through the circulation, homed to the fracture site, and contributed to skeletal repair [69].

Despite the likely relatively small contribution of COPs to normal physiology and even to bone formation in circumstances of pathology and repair, the therapeutic prospects towards the enhancement of failing fracture healing are extensive. For instance, the administration of exogenous MSPCs that are primed to home to the site of the defect through molecular or genetic modifications, or stimulation of the mobilization and recruitment to the repair tissue of endogenous osteoprogenitors by means of pharmacological interventions, could be efficient and patient-friendly ways to enhance compromised bone healing or treat nonunions. With regard to the former strategy, interesting work indicated that overexpression of integrin α4 or administration of a peptidomimetic ligand of α4β1 coupled to a bone-seeking agent (alendronate) increased the homing of BMSCs to bone, leading to encouraging results in osteopenic mouse models [70, 71]. Concerning the latter approach, endogenous progenitor recruitment to the site of the defect may be enhanced by exploiting or amplifying mechanisms that mediate normal tissue regeneration. The CXCL12/SDF-1–CXCR4 axis has been implicated in homing mechanisms of a variety of stem cell populations, and may be the final common pathway for mobilization of BM-derived progenitor cells by injury, inflammation, or relative hypoxia [64, 65]. In a mouse model of segmental bone defects, a combination of the CXCR4 antagonist AMD3100 and IGF-1 provided significant augmentation of bone repair, likely by combined effects on MSPC mobilization and proliferation [72]. In another study, administration of AMD3100 significantly increased the number of endothelial and osteogenic progenitors in the circulation and improved repair of femoral fractures [73]. The potential of endogenous stem/progenitor mobilization to enhance bone regeneration may be magnified by such strategies that stimulate both osteogenesis and angiogenesis at the fracture site, because both processes are vital for bone healing [15]. The discovery of $CD34^+$ progenitor cells in the circulation could as such be highly promising for cell-based therapies of fracture healing and tissue engineering, because they show multilineage differentiation potential into ECs and osteoblasts [74].

ACKNOWLEDGMENTS

Research in the Laboratory for Skeletal Cell Biology and Physiology (SCEBP) is supported by grants to CM, from the European Research Council (ERC Starting Grant 282131 under the European Union's Seventh Framework Programme, FP7), the Fund for Scientific Research of

Flanders (FWO) and the University of Leuven (KU Leuven). ND is supported by a doctoral fellowship of the Flemish government agency for Innovation by Science and Technology (IWT).

REFERENCES

1. Dirckx N, Van Hul M, Maes C. Osteoblast recruitment to sites of bone formation in skeletal development, homeostasis, and regeneration. Birth Defects Res C Embryo Today. 2013;99:170–91.
2. Caplan AI, Bruder SP. Mesenchymal stem cells: building blocks for molecular medicine in the 21st century. Trends Mol Med. 2001;7:259–64.
3. Bianco P. "Mesenchymal" stem cells. Annu Rev Cell Dev Biol. 2014;30:677–704.
4. Bianco P, Robey PG. Skeletal stem cells. Development. 2015;142:1023–7.
5. Dominici M, Le Blanc K, Mueller I, et al. Minimal criteria for defining multipotent mesenchymal stromal cells. The International Society for Cellular Therapy position statement. Cytotherapy. 2006;8:315–17.
6. Tavassoli M, Crosby WH. Transplantation of marrow to extramedullary sites. Science. 1968;161:54–6.
7. Bianco P, Robey PG, Simmons PJ. Mesenchymal stem cells: revisiting history, concepts, and assays. Cell Stem Cell. 2008;2:313–19.
8. Gronthos S, Graves SE, Ohta S, et al. The STRO-1+ fraction of adult human bone marrow contains the osteogenic precursors. Blood. 1994;84:4164–73.
9. Gronthos S, Zannettino A, Hay S, et al. Molecular and cellular characterization of highly purified stromal stem cells derived from human bone marrow. J Cell Sci. 2003;116:1827–35.
10. Sacchetti B, Funari A, Michienzi S, et al. Self-renewing osteoprogenitors in bone marrow sinusoids can organize a hematopoietic microenvironment. Cell. 2007;131:324–36.
11. Short BJ, Brouard N, Simmons PJ. Prospective isolation of mesenchymal stem cells from mouse compact bone. Methods Mol Biol. 2009;482:259–68.
12. Chan CK, Seo EY, Chen JY, et al. Identification and specification of the mouse skeletal stem cell. Cell. 2015;160:285–98.
13. Worthley DL, Churchill M, Compton JT, et al. Gremlin 1 identifies a skeletal stem cell with bone, cartilage, and reticular stromal potential. Cell. 2015;160:269–84.
14. Long F. Building strong bones: molecular regulation of the osteoblast lineage. Nature Rev. 2011;13:27–38.
15. Maes C. Role and regulation of vascularization processes in endochondral bones. Calcif Tissue Int.2013;92:307–23.
16. Ducy P, Zhang R, Geoffroy V, et al. Osf2/Cbfa1: a transcriptional activator of osteoblast differentiation. Cell. 1997;89:747–54.
17. Ducy P, Starbuck M, Priemel M, et al. A Cbfa1-dependent genetic pathway controls bone formation beyond embryonic development. Genes Dev. 1999;13:1025–36.
18. Baek WY, Lee MA, Jung JW, et al. Positive regulation of adult bone formation by osteoblast-specific transcription factor osterix. J Bone Miner Res. 2009;24:1055–65.
19. Baek WY, de Crombrugghe B, Kim JE. Postnatally induced inactivation of Osterix in osteoblasts results in the reduction of bone formation and maintenance. Bone. 2010;46:920–8.
20. Zhou X, Zhang Z, Feng JQ, et al. Multiple functions of Osterix are required for bone growth and homeostasis in postnatal mice. Proc Natl Acad Sci U S A. 2010;107: 12919–24.
21. Zhou X, Zhang Z, Feng JQ, et al. Multiple functions of Osterix are required for bone growth and homeostasis in postnatal mice. Proc Natl Acad Sci U S A. 2010;107: 12919–24.
22. Rauch A, Seitz S, Baschant U, et al. Glucocorticoids suppress bone formation by attenuating osteoblast differentiation via the monomeric glucocorticoid receptor. Cell Metab. 2010;11:517–31.
23. Rodda SJ, McMahon AP. Distinct roles for Hedgehog and canonical Wnt signaling in specification, differentiation and maintenance of osteoblast progenitors. Development. 2006;133:3231–44.
24. Maes C, Kobayashi T, Selig MK, et al. Osteoblast precursors, but not mature osteoblasts, move into developing and fractured bones along with invading blood vessels. Dev Cell. 2010;19:329–44.
25. Baron R, Kneissel M. WNT signaling in bone homeostasis and disease: from human mutations to treatments. Nat Med. 2013;19:179–92.
26. Ono N, Ono W, Nagasawa T, et al. A subset of chondrogenic cells provides early mesenchymal progenitors in growing bones. Nat Cell Biol. 2014;16:1157–67.
27. Yang L, Tsang KY, Tang HC, et al. Hypertrophic chondrocytes can become osteoblasts and osteocytes in endochondral bone formation. Proc Natl Acad Sci U S A. 2014;111:12097–102.
28. Zhou X, von der Mark K, Henry S, et al. Chondrocytes transdifferentiate into osteoblasts in endochondral bone during development, postnatal growth and fracture healing in mice. PLoS Genet. 2014;10:e1004820.
29. Song L, Liu M, Ono N, et al. Loss of wnt/beta-catenin signaling causes cell fate shift of preosteoblasts from osteoblasts to adipocytes. J Bone Miner Res. 2012;27:2344–58.
30. Liu Y, Berendsen AD, Jia S, et al. Intracellular VEGF regulates the balance between osteoblast and adipocyte differentiation. J Clin Invest. 2012;122:3101–13.
31. Sinha P, Aarnisalo P, Chubb R, et al. Loss of Gsalpha early in the osteoblast lineage favors adipogenic differentiation of mesenchymal progenitors and committed osteoblast precursors. J Bone Miner Res. 2014;29:2414–26.
32. Iyer S, Ambrogini E, Bartell SM, et al. FOXOs attenuate bone formation by suppressing Wnt signaling. J Clin Invest. 2013;123:3409–19.
33. Tang Y, Wu X, Lei W, et al. TGF-beta1-induced migration of bone mesenchymal stem cells couples bone resorption with formation. Nature Med. 2009;15:757–65.
34. Dobnig H, Turner R. Evidence that intermittent treatment with parathyroid hormone increases bone

formation in adult rats by activation of bone lining cells. Endocrinology. 1995;136:3632–38.
35. Kim SW, Pajevic PD, Selig M, et al. Intermittent parathyroid hormone administration converts quiescent lining cells to active osteoblasts. J Bone Miner Res. 2012;27:2075–84.
36. Kim SW, Lu Y, Williams EA, et al. Sclerostin antibody administration converts bone lining cells into active osteoblasts. J Bone Miner Res. 2016;
37. Turner RT, Iwaniec UT, Wong CP, et al. Acute exposure to high dose gamma-radiation results in transient activation of bone lining cells. Bone. 2013;57:164–73.
38. Matic I, Matthews BG, Wang X, et al. Quiescent bone lining cells are a major source of osteoblasts during adulthood. Stem Cells. 2016;34:2930–42.
39. Sacchetti B, Funari A, Michienzi S, et al. Self-renewing osteoprogenitors in bone marrow sinusoids can organize a hematopoietic microenvironment. Cell. 2007;131: 324–36.
40. Crisan M, Yap S, Casteilla L, et al. A perivascular origin for mesenchymal stem cells in multiple human organs. Cell Stem Cell. 2008;3:301–13.
41. Boulais PE, Frenette PS. Making sense of hematopoietic stem cell niches. Blood. 2015;125:2621–9.
42. Morrison SJ, Scadden DT. The bone marrow niche for haematopoietic stem cells. Nature. 2014;505:327–34.
43. Ono N, Kronenberg HM. Bone repair and stem cells. Curr Opin Genet Dev. 2016;40:103–107.
44. Omatsu Y, Sugiyama T, Kohara H, et al. The essential functions of adipo-osteogenic progenitors as the hematopoietic stem and progenitor cell niche. Immunity. 2010;33:387–99.
45. Greenbaum A, Hsu YM, Day RB, et al. CXCL12 in early mesenchymal progenitors is required for haematopoietic stem-cell maintenance. Nature. 2013;495:227–30.
46. Morikawa S, Mabuchi Y, Kubota Y, et al. Prospective identification, isolation, and systemic transplantation of multipotent mesenchymal stem cells in murine bone marrow. J Exp Med. 2009;206:2483–96.
47. Zhou BO, Yue R, Murphy MM, et al. Leptin-receptor-expressing mesenchymal stromal cells represent the main source of bone formed by adult bone marrow. Cell Stem Cell. 2014;15:154–68.
48. Ding L, Saunders TL, Enikolopov G, et al. Endothelial and perivascular cells maintain haematopoietic stem cells. Nature. 2012;481:457–62.
49. Mizoguchi T, Pinho S, Ahmed J, et al. Osterix marks distinct waves of primitive and definitive stromal progenitors during bone marrow development. Developmental Cell. 2014;29:340–9.
50. Mendez-Ferrer S, Michurina TV, Ferraro F, et al. Mesenchymal and haematopoietic stem cells form a unique bone marrow niche. Nature. 2010;466:829–34.
51. Ono N, Ono W, Mizoguchi T, et al. Vasculature-associated cells expressing nestin in developing bones encompass early cells in the osteoblast and endothelial lineage. Developmental Cell. 2014;29:330–9.
52. Park D, Spencer JA, Koh BI, et al. Endogenous bone marrow MSCs are dynamic, fate-restricted participants in bone maintenance and regeneration. Cell Stem Cell. 2012;10:259–72.
53. Kusumbe AP, Ramasamy SK, Adams RH. Coupling of angiogenesis and osteogenesis by a specific vessel subtype in bone. Nature. 2014;507:323–8.
54. Wu JY, Purton LE, Rodda SJ, et al. Osteoblastic regulation of B lymphopoiesis is mediated by Gs(alpha)-dependent signaling pathways. Proc Natl Acad Sci U S A. 2008;105:16976–81.
55. Colnot C, Zhang X, Knothe Tate ML. Current insights on the regenerative potential of the periosteum: molecular, cellular, and endogenous engineering approaches. J Orthop Res. 2012;30:1869–78.
56. Kawanami A, Matsushita T, Chan YY, et al. Mice expressing GFP and CreER in osteochondro progenitor cells in the periosteum. Biochem Biophys Res Commun. 2009;386:477–82.
57. Matthews BG, Grcevic D, Wang L, et al. Analysis of alphaSMA-labeled progenitor cell commitment identifies notch signaling as an important pathway in fracture healing. J Bone Miner Res. 2014;29:1283–94.
58. Colnot C. Skeletal cell fate decisions within periosteum and bone marrow during bone regeneration. J Bone Miner Res. 2009;24:274–82.
59. Grayson WL, Bunnell BA, Martin E, et al. Stromal cells and stem cells in clinical bone regeneration. Nat Rev Endocrinol. 2015;11:140–50.
60. Kuznetsov SA, Mankani MH, Gronthos S, et al. Circulating skeletal stem cells. J Cell Biol. 2001;153: 1133–40.
61. Kuznetsov SA, Mankani MH, Leet AI, et al. Circulating connective tissue precursors: extreme rarity in humans and chondrogenic potential in guinea pigs. Stem Cells. 2007;25:1830–9.
62. Rosada C, Justesen J, Melsvik D, et al. The human umbilical cord blood: a potential source for osteoblast progenitor cells. Calcif Tissue Int. 2003;72:135–42.
63. Otsuru S, Tamai K, Yamazaki T, et al. Bone marrow-derived osteoblast progenitor cells in circulating blood contribute to ectopic bone formation in mice. Biochem Biophys Res Commun. 2007;354:453–8.
64. Otsuru S, Tamai K, Yamazaki T, et al. Circulating bone marrow-derived osteoblast progenitor cells are recruited to the bone-forming site by the CXCR4/stromal cell-derived factor-1 pathway. Stem Cells. 2008;26:223–34.
65. Pignolo RJ, Kassem M. Circulating osteogenic cells: implications for injury, repair, and regeneration. J Bone Miner Res. 2011;26:1685–93.
66. Alm JJ, Koivu HM, Heino TJ, et al. Circulating plastic adherent mesenchymal stem cells in aged hip fracture patients. J Orthop Res. 2010;28:1634–42.
67. Eghbali-Fatourechi GZ, Lamsam J, Fraser D, et al. Circulating osteoblast-lineage cells in humans. N Engl J Med. 2005;352:1959–66.
68. Medici D, Shore EM, Lounev VY, et al. Conversion of vascular endothelial cells into multipotent stem-like cells. Nat Med. 2010;16:1400–6.
69. Kumagai K, Vasanji A, Drazba JA, et al. Circulating cells with osteogenic potential are physiologically mobilized

into the fracture healing site in the parabiotic mice model. J Orthop Res. 2008;26:165–75.
70. Guan M, Yao W, Liu R, et al. Directing mesenchymal stem cells to bone to augment bone formation and increase bone mass. Nat Med. 2012;18:456–62.
71. Kumar S, Ponnazhagan S. Bone homing of mesenchymal stem cells by ectopic alpha 4 integrin expression. Faseb J. 2007;21:3917–27.
72. Kumar S, Ponnazhagan S. Mobilization of bone marrow mesenchymal stem cells in vivo augments bone healing in a mouse model of segmental bone defect. Bone. 2012;50:1012–18.
73. Toupadakis CA, Granick JL, Sagy M, et al. Mobilization of endogenous stem cell populations enhances fracture healing in a murine femoral fracture model. Cytotherapy. 2013;15:1136–47.
74. Matsumoto T, Kawamoto A, Kuroda R, et al. Therapeutic potential of vasculogenesis and osteogenesis promoted by peripheral blood CD34-positive cells for functional bone healing. Am J Pathol. 2006;69:1440–57.

4

Osteoblasts: Function, Development, and Regulation

Elizabeth W. Bradley, Jennifer J. Westendorf, Andre J. van Wijnen, and Amel Dudakovic

Department of Orthopedic Surgery, Mayo Clinic, Rochester, MN, USA

CELL BIOLOGY OF OSTEOBLASTS

Osteoblasts are derived from a variety of progenitor populations, including bone marrow, neural crest, and periosteal cells. Osteoblasts produce extracellular matrix proteins and paracrine factors that together support formation of bone tissue. Osteoblasts survive approximately 2 weeks unless they become bone-lining cells or osteocytes [1]. Bone-lining osteoblasts reside on nonmineralized bone surfaces (osteoid), have a single nucleus, and are basophilic. Osteoblasts have abundant endoplastic reticulum to support the production of collagens (predominantly type I collagen) that will eventually become mineralized. Osteoblasts that become embedded in the mineralized matrix are called osteocytes. They reside in lacunae and sense mechanical forces on the skeleton.

OSTEOBLAST FUNCTION

Collagen production and extracellular matrix mineralization

The major function of osteoblasts is to produce the organic constituents of the bone extracellular matrix that facilitate its mineralization by inorganic compounds (Fig. 4.1). The resulting organic and inorganic matrix forms a composite material that resists both stress and strain and that adapts to the loads under which it is placed (Wolff's law).

Type I collagen comprises over 90% of the organic bone matrix. In normal physiological conditions, osteoblasts produce type I collagen fibers in a highly organized fashion such that they are aligned in parallel to each other and in staggered arrays to form lamellar bone. A 90-degree juxtaposition of subsequent collagen fiber layers provides matrix tensile strength with anisotropic biomechanical properties. The pores and spaces between the fibers are sites of mineral nucleation. During periods of rapid growth or in disease settings, the collagen fibers are arranged randomly to form woven bone with isotropic mechanical properties. Collagen fibril assembly, as well as the timing and degree of mineralization, are controlled by other osteoblast-produced proteins, including other collagens, bone alkaline phosphatase (ALPL), gamma carboxylated glutamic acids (osteocalcin/bone gla protein [BGLAP], matrix gla protein [MGP]), small integrin binding ligand with N-linked glycoproteins (SIBLING proteins) (eg, osteopontin/secreted sialoprotein 1 [SSP1], integrin binding bone sialoprotein [IBSP], matrix extracellular phosphoglycoprotein [MEPE], dentin sialophosphoprotein [DMP1], osteonectin/secreted protein-acidic&cysteine-rich [SPARC]), small lipid proteoglycans (eg, biglycan/BGN), and numerous cytokines and hormones.

The inorganic phase of the bone matrix produced by osteoblasts is a calcium-phosphate-hydroxide salt called hydroxyapatite $[Ca_{10}(PO_4)_6(OH)_2]$. Matrix mineralization occurs via active transport of matrix vesicles and activation of phosphatases [2]. The mineral component produced by osteoblasts confers compressive strength and varies depending on environmental factors such as diet. Bone is the major physiological source of calcium and phosphate, but is also a depot for toxic chemicals (eg, lead). Toxins can be incorporated into the hydroxyapatite structure and are released again during bone remodeling.

Primer on the Metabolic Bone Diseases and Disorders of Mineral Metabolism, Ninth Edition. Edited by John P. Bilezikian.
© 2019 American Society for Bone and Mineral Research. Published 2019 by John Wiley & Sons, Inc.
Companion website: www.wiley.com/go/asbmrprimer

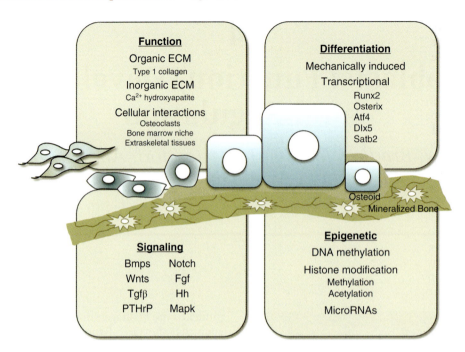

Fig. 4.1. Osteoblasts are mesenchymal cells present on bone surfaces. Their major function is to produce collagenous and noncollagenous proteins of the skeletal elements, but they also coordinate physiological processes both within skeletal and extraskeletal tissues. Osteoblast differentiation and function is regulated by numerous cytokines and growth factors, mechanical stimulation, key transcription factors as well as epigenetic control.

Interactions of osteoblasts with other cell types

Bone is a dynamic tissue that is constantly remodeled by the coordinated actions of osteoblasts and osteoclasts within the bone-remodeling unit [3]. Osteoblasts influence osteoclast differentiation via expression of Receptor Activator of Nuclear Factor Kappa B Ligand (RANKL/TNFSF11) and monocyte colony stimulating factor (M-CSF-1), molecules that bind cognate receptors (RANK and CSFR) on osteoclasts and facilitate their differentiation. Osteoblasts also produce a decoy receptor of RANK, called osteoprotegerin (OPG/TNFRSF11B) that blocks this interaction. Other direct interactions between osteoblasts and osteoclasts that are active in the bone remodeling unit involve cell surface proteins, such as ephrins B2 and B4 (EPHB2 and EPHB4). Osteoclasts in turn produce several cytokines including WNTs, BMP6, and TGFβ to recruit osteoblast progenitors to the bone surface [4]. Thus, bone resorption is coupled to bone formation to support normal bone homeostasis. Uncoupling of these processes leads to altered bone mass.

Osteoblasts lining the bone surface generate an extracellular environment that both directly and indirectly supports nonskeletal hematopoietic cells [5]. Osteoblasts also produce hormones that enter the circulation and affect distant tissues. For example, extraskeletal effects of nonglycosylated osteocalcin and the phosphaturic hormone FGF23 have been well documented [6].

DEVELOPMENT

Osteogenic phenotype commitment of MSCs and osteoblast maturation

Mesenchymal stromal/stem cells (MSCs) are considered the developmental precursors of osteoblasts. Yet, as an isolated entity they represent a heterogeneous cell type with a number of common properties and cell surface markers [7, 8]. MSCs are multipotent fibroblast-like cells that are capable of self-renewal and differentiation into a variety of cells, including osteoblasts, chondrocytes, adipocytes, and myocytes. MSCs isolated from bone marrow aspirates account for only 0.001–0.01% of all cells [9]. Bone marrow derived MSCs are related to perivascular MSCs (pericytes) that can be isolated from many tissues, including adipose tissue, synovial membrane, dental pulp, endometrium, umbilical cord, and other tissues [8]. The commitment of MSCs into the osteoblast lineage is controlled by several molecular mechanisms including signaling pathways (eg, BMP, WNT, FGF), transcription factors (eg, RUNX2, SP7/Osterix), and epigenetic mechanisms (eg, histone modifications and microRNAs). Mechanical forces also promote MSC differentiation into osteoblasts.

Osteoblast growth and differentiation has been extensively studied with cultured osteoblasts derived from calvarial bone formed by membranous ossification. Differentiation of mesenchymal stem cells into the osteogenic lineage is followed by a progressive series of maturation events in which preosteoblasts mature to osteoblasts and ultimately to osteocytes. The successive stages of osteoblastogenesis in vitro are characterized by the temporal expression of a series of well-characterized biomarkers that reflect the developmental steps required for bone formation in vivo. During proliferative stages, osteoblasts carry out a cell cycle program to duplicate the chromosomes (eg, by inducing histone proteins like HIST2H4A) and to enlarge cytoarchitecture in preparation of cell division. When osteoblasts reach a postproliferative state after about 3–4 days, they begin to produce vast quantities of collagens (eg, COL1A1) that are cross-linked by hydroxylation. Formation of the collagen extracellular matrix, which proceeds from days 3 to 10 in culture, permits deposition of a number of noncollagenous proteins that are either modified by various glycan moieties (eg, SSP1, IBSP, SPARC) or carboxylated in a vitamin K dependent manner (eg, BGLAP, MGP). Deposition of these proteins occurs concomitantly with production of alkaline phosphatase (ALP), one of the most commonly assayed classical bone markers that begins to peak between 10 and 14 days in culture. The negative charges of noncollagenous proteins are thought to support the deposition of divalent calcium cations and hydroxyapatite containing mineral crystals. These are visualized in culture by Alizarin Red or Von Kossa staining methods and is most prominent between 14 and 28 days in culture. In these final stages of osteoblast maturation, osteoblasts may convert into osteocytes that express the unique matrix markers MEPE and DMP1, as well as the WNT-inhibitor sclerostin (SOST).

Signaling pathways controlling osteoblast differentiation and function

Mouse genetic studies combined with identification of human mutations established the critical role for BMP signaling in skeletogenesis and osteoblast differentiation [10]. BMPs promote osteoblast commitment and proliferation, as well as induce bone formation by enhancing the expression and activities of Sp7/Osterix, Runx2, and Dlx5 [11]. BMP signaling enhances expression of RANKL and suppresses expression of OPG [12]. Due to their osteogenic potential, BMP2 and BMP7 are used clinically to enhance bone grafting [10].

Tgfβ, another key member of the TGFβ/BMP/GDF superfamily, plays a role in coupling bone resorption and bone formation and directly impacts osteoblast differentiation and function. Tgfβ enhances proliferation and migration of osteoprogenitor cells and survival of mature osteoblasts, but diminishes in vitro mineralization [11]. Gain-of-function mutations result in Camurati–Engelmann disease that alters bone density and is recapitulated in mice [11].

Mutations in Wnt-related genes, such as coreceptors LRP5/6 and the soluble antagonist Sclerostin/SOST, alter osteoblast activity and are corroborated by mouse genetic and in vitro studies [13]. Canonical Wnt signaling enhances osteoblast lineage allocation, proliferation, matrix production, survival, and osteoclast coupling [11, 13].

PTH and the related peptide PTHrP (PTHLH) regulate skeletogenesis and osteoblast function. PTH has both catabolic and anabolic effects dependent on dose and periodicity [14]. Consistent systemic levels of PTH act to stimulate RANKL expression, enhancing osteoclastogenesis and bone resorption [14]. In contrast, intermittent PTH treatment has anabolic effects by enhancing osteoblast commitment, proliferation, and reactivation of bone lining cells [14]. Similarly, PTHrP increases osteoblast progenitor proliferation and enhances osteoblast differentiation and survival [15].

Mutations affecting Notch signaling also affect skeletal homeostasis and osteoblast function [16]. Notch controls osteoblast development in a stage-dependent fashion. By repressing Runx2 expression in mesenchymal progenitors, Notch signaling impairs osteoblast differentiation [17]. In contrast, Notch promotes the expansion of preosteoblasts, but inhibits terminal differentiation [17]. Notch signaling also regulates OPG expression and facilitates maintenance of the bone marrow niche [17].

Gain-of-function mutations in FGF receptors (FGFR2 and FGFR3) cause craniosynostosis and/or skeletal dysplasias [18]. Fgf signaling is complex, but overall acts to promote osteoblast proliferation, differentiation, and survival. FGF2/FGFR1 interactions are highly mitogenic and control cell cycle related proteins, whereas subsequent signaling in postproliferative cells may involve FGFR2 and FGFR3. Osteogenic effects of FGF2 may be achieved by modulating Runx2 levels, phosphorylation via MAPK, and transcriptional stimulation of osteoblast specific genes [18]. Beyond its effects on Runx2, members of the MAPK family also have important roles in governing osteoblast function by regulating the expression and activity of other downstream factors including Osterix, Dlx5, and Rsk2 [19].

Disruption of Hedgehog (eg, Ihh, Shh) signaling also causes skeletal dysplasias. Hh is required for optional osteoblast function during intramembranous ossification and differentiation from perichondrial cells during endochondral ossification; however, Hh signaling is linked to age-related bone loss because it increases osteoclastogenesis by enhancing expression of RANKL [20].

A number of additional signaling pathways are crucial for osteoblast differentiation and function. PI3K/Akt-dependent signaling is activated by many different ligands, including Bmp2 and insulin/Igf1. This pathway functions to promote expression of Runx2, osteoblast differentiation, and survival. A growing body of evidence has identified factors that adversely affect osteoblast differentiation and function, including adipokines, senescence-associated proteins, reactive oxygen species, reduced autophagy, and glucocorticoids.

Mechanical signals promoting osteogenic differentiation

All cells within the body, including osteoblasts, respond to mechanical signals. Mechanical cues that act upon skeletal cells include hydrostatic pressure, shear stress induced by fluid flow, substrate strain, stiffness and topography, and electromagnetic fields. Osteoblasts within the skeleton are subjected to constant high frequency, low intensity mechanical stimulation during normal activities. Furthermore, the importance of mechanical stimulation on bone formation is illustrated by decreased bone mass associated with unloading conditions, such as immobilization during prolonged bed rest, microgravity experienced during space flight, or experimental models for bone loading and unloading [21]. At the cellular level, increased mechanical stimulation enhances osteogenic lineage-commitment of MSCs, enhances preosteoblast proliferation and increases matrix production by mature osteoblasts [21](see also Chapter 10).

Mechanical cues that facilitate osteoblast differentiation and function are recognized and propagated by mechanoreceptors on cell surfaces. Mechanoreceptors include integrins, cadherins, focal adhesions, and connexins. Primary cilia also detect mechanical changes in extracellular environment. Activation of mechanoreceptors induces many intracellular signaling pathways, including MAPKs, Wnt signaling, G-protein coupled cascades, and release of intracellular calcium to increase osteoblast activity (see also Chapters 5 and 10). Estrogen facilitates the response of osteoblasts to mechanotransduction [22].

REGULATION

Transcriptional control of osteoblast differentiation

Several transcription factors (TF) are essential for the differentiation of MSCs into osteoblasts. Together they control expression of lineage-specific genes though sequence-directed binding to DNA. Although numerous transcription factors have been implicated in osteogenesis, this section will focus on TFs that have been studied in the most detail over the years.

Many of the pro-osteogenic signaling pathways rely on the activity of Runx2 (Cbfa1/Aml3/Pebp2αA), which is considered the master regulator of osteoblast differentiation because Runx2 null mutations prevent bone formation in vivo. Specifically, Runx2 deletion prevents the differentiation of osteoblasts from MSCs and suppresses both intramembranous and endochondral bone development and mineralization. Elevated expression of Runx2 increases osteoclastogenesis [23], presumably by raising expression of RANKL in osteoblasts. Runx2 haploinsufficiency causes cleidocranial dysplasia, which is characterized by the absence of clavicles and incomplete development of cranial bones [24]. The interaction of Runx2 with coactivators and corepressors can either enhance or suppress the expression osteoblastic genes to ensure proper osteoblast differentiation [25].

The second crucial osteoblast-specific TF is Sp7 (Osterix), which is downstream of Runx2. As with Runx2, deletion of the mouse Sp7 gene prevents osteoblast differentiation and bone formation [26]. Functionally, Sp7 interacts with NFATc1 and a variety of transcriptional cofactors to stimulate the expression of genes required for osteoblast differentiation [27].

A third key TF that contributes to osteoblast differentiation is ATF4, which forms a complex with Runx2 to stimulate the expression of osteogenic genes [28]. Lack of ATF4 delays bone formation and causes low bone mass. Disruption of ATF4 transcriptional activity is linked to Coffin–Lowry syndrome, which is characterized by mental retardation and skeletal abnormalities [29]. Beyond Runx2, Sp7/Osterix, and ATF4, there are many other TFs that contribute to osteoblast differentiation, including Dlx3, Dlx5, Msx2, Tcf7/Lef1, Satb2, as well as steroid hormone receptors such as vitamin D receptor (Vdr), glucocorticoid receptor (GR/NR3C1), and estrogen receptor-α (Erα/ESR1). The combined biological activities of these TFs permit coordinate commitment of MSCs into the osteogenic lineage, but also control the differentiation process and the termination of the bone-forming program.

Epigenetic transcriptional mechanisms controlling osteoblast differentiation

Transcriptional regulation of gene expression in osteoblasts is controlled by epigenetic events that are important for normal osteoblast differentiation and function. Two major epigenetic mechanisms that control gene expression and function of osteoblasts are transcription-related chromatin modifications (DNA methylation and histone posttranslational modifications). These chemical changes to DNA and histones modulate the accessibility of TFs and RNA polymerase II to DNA. Additional post-transcriptional mechanisms (microRNAs and long-noncoding RNAs) control mRNA stability and translation and are considered epigenetic.

DNA methylation at the fifth carbon of cytosine promotes gene silencing. This modification is actively regulated in osteoblasts and changes during differentiation and in response to external stimuli. Differentiation of MSCs into osteoblasts is accompanied by progressive DNA methylation of stem cell genes [30]. Hypomethylation in promoters of osteogenic genes occurs as precursor cells differentiate into osteoblasts. For example, reduction in CpG methylation of the osteocalcin promoter enhances its expression [31]. Active DNA demethylation promotes osteogenic differentiation in vitro and in vivo [32]. Thus, the silencing of stem cell genes by DNA hypermethylation and the activation of osteoblast-specific genes by DNA hypomethylation is a fundamental epigenetic mechanism that mediates osteogenic differentiation.

Core histone proteins (H2A, H2B, H3, and H4) undergo posttranslational modifications at several chemically labile amino acids (eg, lysine, arginine, and serine). Some histone modifications are associated with gene activation (eg, trimethylation at H3 lysine 4 [H3K4me3] or acetylation at lysine 27 [H3K27ac]), whereas others are more generally linked to gene silencing (eg, trimethylation at H3 lysine 27 [H3K27me3]). These and a plethora of other histone modifications collectively control gene expression. Changes in this "histone code" can modulate the osteogenic commitment of progenitor cells by modulating the accessibility of transcription factors such as Runx2 [33].

Acetylation of histones by acetyltransferases (eg, p300 and WDR5) is generally associated with gene activation during osteogenesis and these enzymes play crucial roles in osteogenic gene activation [34]. Removal of acetyl groups by histone deacetylases (HDACs) also contributes to osteoblastogenesis [35]. While enhancing global acetylation of histones promotes osteoblasts differentiation in vitro [36], alterations in these epigenetic marks negatively affect osteoblasts in vivo by reducing the number of osteoprogenitor cells [37].

Methylation of histones is also important in the differentiation and function of osteoblasts. Methylation of H3K27 promotes adipogenic differentiation, whereas demethylation promotes osteogenic differentiation of progenitor cells [38, 39]. Pro-osteogenic effects of inactivating the H3K27 methyltransferase Ezh2 are evident in vivo [40]. Furthermore, genetic mutation of the H3K36 methyltransferase WHSC1, which is linked to Wolf–Hirshhorn syndrome, causes craniofacial defects and growth abnormalities [41]. Consistent with this result, deletion of the NO66 gene, a histone H3K4/H3K36 demethylase, enhances bone formation by enhancing the expression of several osteogenic markers, including SP7 [42]. Thus, histone modifications play a critical role in osteoblast differentiation.

Epigenetic posttranscriptional mechanisms controlling osteoblast differentiation

Posttranslational control of gene expression during bone cell growth and differentiation is mediated by microRNAs (miRNAs). These small noncoding RNAs (<24 nucleotides long) are produced by the RNA processing enzyme Dicer and bind to specific sequences in the 3′ untranslated regions (3′ UTRs) of their targets to control mRNA translation and/or degradation.

There is strong genetic and mechanistic evidence that miRNA function is important for normal skeletal development and osteoblast function in vivo. Genetic loss of Dicer during postnatal bone maturation results in a prominent increase in cortical bone volume, which is thought to be due to increased collagen expression [43]. There is clear evidence that miRNAs in general control osteoblast activity in cell culture, albeit few studies have specifically examined miRNA effects in vivo. Two representative studies showed that some miRNAs block either alternative cell lineages or inhibitors that suppress osteogenesis. For example, the myogenic miR-133 suppressed osteogenic commitment and the depletion of these molecules enhances osteogenic differentiation [44]. Conversely, upregulation of miR-218 suppressed inhibitors of WNT and BMP signaling to stimulate osteogenesis [45].

The different functions of miRNAs in skeletal development and disease have been well-documented [46]. Beyond osteoblasts, these RNAs have key roles in the growth, differentiation, and function of mesenchymal stem cells, osteoclasts, chondrocytes, adipocytes, and myoblasts. One important theme that emerged from studies on miRNAs in osteoblasts is that they act through a "double-negative principle" (eg, inhibiting repressors) to stimulate principal osteogenic signaling pathways (eg, Wnt signaling).

Similar to miRNAs, long noncoding RNAs (lncRNAs) are RNA transcripts longer than 200 nucleotides that do not encode proteins. While lncRNAs were initially considered random nonspecific transcripts, they are now known to play an essential role in controlling nuclear structure, gene expression, and cell differentiation [47]. While there are only a few studies that have assessed the role of lncRNAs in osteogenic differentiation, at least two lncRNAs (ANCR and HoxA-AS3) alter the function of Ezh2 and heterochromatization in osteoblast progenitors through direct interaction [48, 49]. Disruption of the complex that these lncRNAs form with Ezh2 enhances osteogenic differentiation, similar to the finding that Ezh2 inhibition itself promotes bone formation in vitro and in vivo [39, 40]. LncRNA-H19 has been shown to be pro-osteogenic in MSCs by suppressing mRNA and protein expression of TGF-β1 [50]. The mechanism of action of this lncRNA is complex. It involves the expression of miR-675, which is expressed within an exon of lncRNA-H19, and also alters the expression of HDAC4 and HDAC5. As such, the proposed lncRNAs/miR-675/HDAC4/HDAC5 axis represents an intricate and pleiotropic pathway that attenuates TGF-β1 signaling and induces osteogenic commitment of MSCs.

CONCLUSION

Osteoblasts are mesenchymally-derived cells that produce many factors essential for mineralization of the extracellular matrix. Osteoblasts coordinate bone remodeling and homeostasis with osteoclasts through paracrine signaling events. Within osteoblast nuclei, numerous transcription factors drive expression of tissue-specific genes in a temporal and highly regulated fashion. Epigenetic events coordinated by histone or DNA-modifying proteins and noncoding RNAs amplify and promote retention of tissue-specific functions. Understanding these events is essential for understanding bone degeneration and regeneration, and permits design of strategies that may prevent or mitigate bone-related disorders.

REFERENCES

1. Jilka RL, Weinstein RS, Bellido T, et al. Osteoblast programmed cell death (apoptosis): modulation by growth factors and cytokines. J Bone Miner Res. 1998;13(5): 793–802.
2. Millan JL. The role of phosphatases in the initiation of skeletal mineralization. Calcif Tissue Int. 2013;93(4): 299–306.
3. Sims NA, Martin TJ. Coupling the activities of bone formation and resorption: a multitude of signals within the basic multicellular unit. Bonekey Rep. 2014;3:481.
4. Pederson L, Ruan M, Westendorf JJ, et al. Regulation of bone formation by osteoclasts involves Wnt/Bmp signaling and the chemokine spingosine-1 phosphate. Proc Natl Acad Sci U S A. 2008;105:20674–9.
5. Morrison SJ, Scadden DT. The bone marrow niche for haematopoietic stem cells. Nature. 2014;505(7483): 327–34.
6. Karsenty G, Olson EN. Bone and muscle endocrine functions: unexpected paradigms of inter-organ communication. Cell. 2016;164(6):1248–56.
7. Russell KC, Phinney DG, Lacey MR, et al. In vitro high-capacity assay to quantify the clonal heterogeneity in trilineage potential of mesenchymal stem cells reveals a complex hierarchy of lineage commitment. Stem Cells. 2010;28(4):788–98.
8. Lv FJ, Tuan RS, Cheung KM, et al. Concise review: the surface markers and identity of human mesenchymal stem cells. Stem Cells. 2014;32(6):1408–19.
9. Pittenger MF, Mackay AM, Beck SC, Jaiswal RK, Douglas R, Mosca JD, et al. Multilineage potential of adult human mesenchymal stem cells. Science. 1999;284(5411): 143–7.
10. Salazar VS, Gamer LW, Rosen V. BMP signalling in skeletal development, disease and repair. Nat Reviews Endocrinol. 2016;12(4):203–21.
11. Wu M, Chen G, Li YP. TGF-beta and BMP signaling in osteoblast, skeletal development, and bone formation, homeostasis and disease. Bone Res. 2016;4:16009.
12. Chau JF, Leong WF, Li B. Signaling pathways governing osteoblast proliferation, differentiation and function. Histol Histopathol. 2009;24(12):1593–606.
13. Monroe DG, McGee-Lawrence ME, Oursler MJ, et al. Update on Wnt signaling in bone cell biology and bone disease. Gene. 2012;492(1):1–18.
14. Silva BC, Bilezikian JP. Parathyroid hormone: anabolic and catabolic actions on the skeleton. Curr Opin Pharmacol. 2015;22:41–50.
15. Augustine M, Horwitz MJ. Parathyroid hormone and parathyroid hormone-related protein analogs as therapies for osteoporosis. Curr Osteoporos Rep. 2013;11(4):400–6.
16. Zanotti S, Canalis E. Notch signaling in skeletal health and disease. Eur J Endocrinol 2013;168(6):R95–103.
17. Engin F, Lee B. NOTCHing the bone: insights into multi-functionality. Bone. 2010;46(2):274–80.
18. Ornitz DM, Marie PJ. Fibroblast growth factor signaling in skeletal development and disease. Genes Dev. 2015;29(14):1463–86.
19. Greenblatt MB, Shim JH, Glimcher LH. Mitogen-activated protein kinase pathways in osteoblasts. Ann Rev Cell Dev Biol 2013;29:63–79.
20. Yang J, Andre P, Ye L, et al. The Hedgehog signalling pathway in bone formation. Int J Oral Sci 2015;7(2): 73–9.
21. Robling AG, Turner CH. Mechanical signaling for bone modeling and remodeling. Crit Rev Eukaryot Gene Expr 2009;19(4):319–38.
22. Klein-Nulend J, van Oers RF, Bakker AD, et al. Bone cell mechanosensitivity, estrogen deficiency, and osteoporosis. J Biomech 2015;48(5):855–65.
23. Komori T. Regulation of bone development and maintenance by Runx2. Front Biosci. 2008;13:898–903.
24. Jaruga A, Hordyjewska E, Kandzierski G, et al. Cleidocranial dysplasia and RUNX2-clinical phenotype-genotype correlation. Clin Genet. 2016;6(10):12812.
25. Schroeder TM, Jensen ED, Westendorf JJ. Runx2: a master organizer of gene transcription in developing and maturing osteoblasts. Birth Defects Res C Embryo Today. 2005;75(3):213–25.
26. Nakashima K, Zhou X, Kunkel G, et al. The novel zinc finger-containing transcription factor osterix is required for osteoblast differentiation and bone formation. Cell. 2002;108(1):17–29.
27. Koga T, Matsui Y, Asagiri M, et al. NFAT and Osterix cooperatively regulate bone formation. Nat Med. 2005;11(8):880–5.
28. Xiao G, Jiang D, Ge C, et al. Cooperative interactions between activating transcription factor 4 and Runx2/Cbfa1 stimulate osteoblast-specific osteocalcin gene expression. J Biol Chem. 2005;280(35):30689–96.
29. Yang X, Matsuda K, Bialek P, et al. ATF4 is a substrate of RSK2 and an essential regulator of osteoblast biology; implication for Coffin-Lowry Syndrome. Cell. 2004;117(3): 387–98.
30. Dansranjavin T, Krehl S, Mueller T, et al. The role of promoter CpG methylation in the epigenetic control of stem cell related genes during differentiation. Cell Cycle. 2009;8(6):916–24.
31. Villagra A, Gutierrez J, Paredes R, et al. Reduced CpG methylation is associated with transcriptional activation of the bone-specific rat osteocalcin gene in osteoblasts. J Cell Biochem. 2002;85(1):112–22.
32. Thaler R, Maurizi A, Roschger P, et al. Anabolic and anti-resorptive modulation of bone homeostasis by the epigenetic modulator sulforaphane, a naturally occurring isothiocyanate. J Biol Chem. 2016;291(13); 6754–71.
33. Meyer MB, Benkusky NA, Sen B, et al. Epigenetic plasticity drives adipogenic and osteogenic differentiation of marrow-derived mesenchymal stem cells. J Biol Chem. 2016;291(34);17829–47.
34. Jun JH, Yoon WJ, Seo SB, et al. BMP2-activated Erk/MAP kinase stabilizes Runx2 by increasing p300 levels and histone acetyltransferase activity. J Biol Chem. 2010;285(47):36410–9.
35. Bradley EW, Carpio LR, van Wijnen AJ, et al. Histone deacetylases in bone development and skeletal disorders. Physiol Rev. 2015;95(4):1359–81.

36. Dudakovic A, Evans JM, Li Y, et al. Histone deacetylase inhibition promotes osteoblast maturation by altering the histone H4 epigenome and reduces Akt phosphorylation. J Biol Chem. 2013;288(40):28783–91.
37. McGee-Lawrence ME, Bradley EW, Dudakovic A, et al. Histone deacetylase 3 is required for maintenance of bone mass during aging. Bone. 2013;52(1):296–307.
38. Hemming S, Cakouros D, Isenmann S, et al. EZH2 and KDM6A act as an epigenetic switch to regulate mesenchymal stem cell lineage specification. Stem Cells. 2014;32(3):802–15.
39. Dudakovic A, Camilleri ET, Xu F, et al. Epigenetic control of skeletal development by the histone methyltransferase Ezh2. J Biol Chem. 2015;290(46): 27604–17.
40. Dudakovic A, Camilleri ET, Riester SM, et al. Enhancer of Zeste Homolog 2 inhibition stimulates bone formation and mitigates bone loss due to ovariectomy in skeletally mature mice. J Biol Chem. 2016;10:740571.
41. Lee YF, Nimura K, Lo WN, et al. Histone H3 lysine 36 methyltransferase Whsc1 promotes the association of Runx2 and p300 in the activation of bone-related genes. PloS one. 2014;9(9).
42. Chen Q, Sinha K, Deng JM, et al. Mesenchymal deletion of histone demethylase NO66 in mice promotes bone formation. J Bone Miner Res 2015;30(9):1608–17.
43. Gaur T, Hussain S, Mudhasani R, et al. Dicer inactivation in osteoprogenitor cells compromises fetal survival and bone formation, while excision in differentiated osteoblasts increases bone mass in the adult mouse. Dev Biol. 2010;340(1):10–21.
44. Li Z, Hassan MQ, Volinia S, et al. A microRNA signature for a BMP2-induced osteoblast lineage commitment program. Proc Natl Acad Sci U S A. 2008;105(37): 13906–11.
45. Taipaleenmaki H, Farina NH, van Wijnen AJ, et al. Antagonizing miR-218-5p attenuates Wnt signaling and reduces metastatic bone disease of triple negative breast cancer cells. Oncotarget. 2016;7(48):79032–46.
46. van Wijnen AJ, van de Peppel J, van Leeuwen JP, et al. MicroRNA functions in osteogenesis and dysfunctions in osteoporosis. Curr Osteoporos Rep. 2013;11(2): 72–82.
47. Hassan MQ, Tye CE, Stein GS, et al. Non-coding RNAs: epigenetic regulators of bone development and homeostasis. Bone. 2015;81:746–56.
48. Zhu L, Xu PC. Downregulated LncRNA-ANCR promotes osteoblast differentiation by targeting EZH2 and regulating Runx2 expression. Biochem Biophys Res Commun. 2013;432(4):612–7.
49. Zhu XX, Yan YW, Chen D, et al. Long non-coding RNA HoxA-AS3 interacts with EZH2 to regulate lineage commitment of mesenchymal stem cells. Oncotarget. 2016;23(10):11538.
50. Huang Y, Zheng Y, Jia L, et al. Long noncoding RNA H19 promotes osteoblast differentiation via TGF-beta1/Smad3/HDAC signaling pathway by deriving miR-675. Stem Cells 2015;33(12):3481–92.

5
Osteocytes

Lynda F. Bonewald

Indiana Center for Musculoskeletal Health; Department of Anatomy and Cell Biology, and Department of Orthopaedic Surgery, Indiana University School of Medicine, Indianapolis, IN, USA

INTRODUCTION

In the adult skeleton, osteocytes make up over 90–95% of all bone cells compared with 4–6% osteoblasts and around 1–2% osteoclasts. These cells are regularly dispersed throughout the mineralized matrix, connected to each other and cells on the bone surface through dendritic processes generally radiating towards the bone surface and the blood supply. The dendritic processes travel through the bone in tiny canals called canaliculi (250–300nm) whereas the cell body is encased in a lacuna (15–20μm) (see Figs 5.1, 5.2, and 5.3). Osteocytes are thought to function as a network of sensory cells mediating the effects of mechanical loading through this extensive lacunocanalicular network. Not only do these cells communicate with each other and with cells on the bone surface, but their dendritic processes can extend past the bone surface into the bone marrow and vascular spaces. Osteocytes have long been thought to respond to mechanical strain to send signals of resorption or formation, and evidence is accumulating to show that this is a major function of these cells. The number of functions attributed to these cells has been expanding [1] and includes regulation of phosphate homeostasis; therefore, the osteocyte network functions as an endocrine gland [2]. Defective osteocyte function may play a role in a number of bone diseases, especially glucocorticoid-induced bone fragility and osteoporosis in the adult, and in nonbone disease such as chronic kidney disease and skeletal and cardiac muscle function [3].

OSTEOCYTE ONTOGENY

Osteoprogenitor cells reside in the bone marrow before differentiating into plump, polygonal osteoblasts on the bone surface (for reviews see [4,5]). By an unknown mechanism some of these cells are destined to become osteocytes, whereas some become lining cells, and some undergo apoptosis [6]. Osteoblasts, osteoid-osteocytes, and osteocytes may play distinct roles in the initiation and regulation of bone matrix mineralization. Bordier first proposed that osteoid-osteocytes are major regulators of this process [7]. Osteoid-osteocytes actively make matrix while simultaneously calcifying it (see Figs 5.1, 5.2, and 5.3).

Whereas numerous markers for osteoblasts have been identified such as Cbfa1/Runx2, Osterix/Sp7, alkaline phosphatase, collagen type 1, etc., few markers have been available for osteocytes until recently (for review see [5]). In 1996, the markers described for osteocytes were limited to low or no alkaline phosphatase, high casein kinase II, high osteocalcin protein expression and high CD44 as compared with osteoblasts. Now markers for osteocytes include E11/gp38, phosphate-regulating neutral endopeptidase on the chromosome X (Phex), Dentin Matrix Protein 1 (DMP1), matrix extracellular phosphoglycoprotein (MEPE), fibroblast growth factor (FGF23), ORP150, and sclerostin along with a list of "osteoclast" specific markers such as Cathepsin K and TRAP (Table 5.1).

Some of these markers are also expressed in osteoblasts at specific stages of differentiation. The identification of these markers revealed new osteocyte functions

Primer on the Metabolic Bone Diseases and Disorders of Mineral Metabolism, Ninth Edition. Edited by John P. Bilezikian.
© 2019 American Society for Bone and Mineral Research. Published 2019 by John Wiley & Sons, Inc.
Companion website: www.wiley.com/go/asbmrprimer

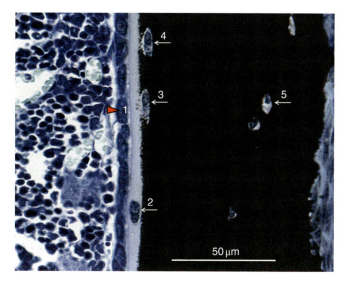

Fig. 5.1. Histological section of tetrachrome-stained murine cortical bone showing osteoblast to osteocyte differentiation. 1, matrix producing osteoblast; 2, osteoid-osteocyte; 3, embedding osteocyte; 4, newly embedded osteocyte; 5, mature osteocyte. From this histological section one would assume only the lacunae are the porosities within bone. However, as can be observed in Figs 5.2 and 5.3, the osteocyte canaliculi provide extensive porosities within the mineralized bone matrix. The osteocyte lacunocanalicular surface is extensive and is a source of calcium and other factors.

Fig. 5.3. The osteocyte lacunocanalicular network is intimately associated with the blood vessel network in the bone matrix. The white marker points to an osteocyte lacuna intimately associated with the blood vessel.

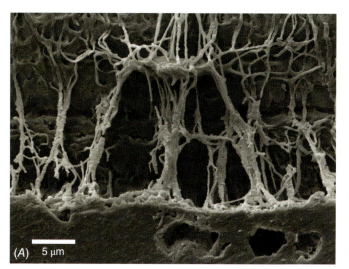

Fig. 5.2. The embedding osteocyte retains its connectivity with cells on the bone surface. (A) Acid-etched plastic-embedded murine cortical bone. With this technique, resin fills the lacunocanalicular system, osteoid, and marrow, but cannot penetrate mineral. Mild acid is used to remove the mineral leaving behind a resin cast relief. Note the canaliculi connecting the lacunae with the bone surface at the bottom of the image. (B) A transmission electron microscopy image showing a fully embedded osteocyte and an osteoid-osteocyte becoming surrounded by mineral (white). The osteoid is black and the osteoblasts are at the bottom of the image.

Table 5.1. Osteocyte markers.

Marker	Expression	Function
E11/gp38	Early, embedding cell	Dendrite formation
CD44	More highly expressed in osteocytes compared with osteoblasts	Hyaluronic acid receptor associated with E11 and linked to cytoskeleton
Fimbrin	All osteocytes	Dendrite branching?
Phex	Early and late osteocytes	Phosphate metabolism
OF45/MEPE	Mature osteocytes	Inhibitor of bone formation/regulator of phosphate metabolism
DMP1	Early and mature osteocytes	Phosphate metabolism and mineralization
Sclerostin	Mature embedded osteocyte	Inhibitor of bone formation
FGF23	Early and mature osteocytes	Induces hypophosphatemia
"Osteoclast" specific genes	Mature osteocytes	Remove calcium under calcium demanding conditions
ORP150	Mature osteocytes	Protection from hypoxia

(Table 5.1). Osteocytes also have distinct cytoskeletal markers compared with osteoblasts, including the actin-bundling proteins, villin, alpha-actinin and fimbrin, with strong expression of fimbrin at branching points in osteocyte dendrites [8]. It is likely that these actin reorganizing proteins play a role in osteocyte cell body movement within lacunae and in the retraction and extension of dendritic processes [9]. CapG and destrin have been shown to be more highly expressed in the embedding osteocyte than in osteoblasts [10].

Promoters for specific markers have been used to drive GFP to follow osteoblast to osteocyte differentiation in vivo. Collagen type 1-GFP is strongly expressed in both osteoblasts and osteocytes, osteocalcin-GFP is expressed in a few osteoblastic cells lining the endosteal bone surface and in scattered osteocytes, and the osteocyte-selective tracer, the 8 kilobase (kb) DMP1 promoter driving GFP showed selective expression in early osteocytes [11]. These promoters have also been used to drive Cre-recombinases to perform targeted deletion of genes at different stages of osteoblast to osteocyte differentiation. This genetic approach has significantly advanced our knowledge regarding the functions of osteocytes.

OSTEOCYTES AS ORCHESTRATORS OF BONE (RE)MODELING

Evidence is mounting that osteocytes can conduct and control both bone resorption and bone formation. Some of the earliest data supporting the theory that osteocytes can send signals to initiate bone resorption were observations that isolated avian osteocytes can support osteoclast formation and activation in the absence of osteotropic factors [12] as can the osteocyte-like cell line, MLO-Y4 [13]. It was suggested that expression of RANKL along exposed osteocyte dendritic processes provides a potential means for osteocytes within bone to stimulate osteoclast precursors at the bone surface. Deletion of RANKL in osteocytes using Dmp1-Cre resulted in increased bone formation; neutralizing antibodies against RANKL are now successfully being used to treat osteoporosis [3].

One of the major means by which osteocytes may support osteoclast activation and formation is through their death and expression of RANKL while dying. Osteocyte apoptosis is an orderly process that can occur at sites of microdamage; it is proposed that dying osteocytes send signals to osteoclasts for their removal and the repair of microdamage. The expression of antiapoptotic and proapoptotic molecules in osteocytes surrounding microcracks was mapped and it was found that expression of proapoptotic molecules is elevated in osteocytes immediately at the microcrack locus, whereas antiapoptotic molecules are expressed 1–2 mm from the microcrack [14]. Therefore, those osteocytes that do not undergo apoptosis are prevented from doing so by protective mechanisms, whereas those destined for removal by osteoclasts undergo apoptosis. Targeted ablation of osteocytes was performed using the 10 kb Dmp1 promoter to drive expression of the diptheria toxin receptor in mice [15]. Injection of a single dose of diphtheria toxin eliminated approximately 70% of osteocytes in cortical bone, leading to dramatic osteoclast activation. Therefore, viable osteocytes are necessary to prevent osteoclast activation and maintain bone mass (Fig. 5.4).

Osteocytes can also send signals to osteoblasts to inhibit new bone formation. A marker for the late, embedded osteocyte is sclerostin, which is coded by the gene *Sost*. This protein is a negative regulator of bone formation, targeting the osteoblast through inhibition of the Wnt/β-catenin pathway. Inhibition of sclerostin activity leads to increased bone mass; therefore, therapeutics targeting sclerostin such as neutralizing antibody or small inhibitory molecules are the focus of much

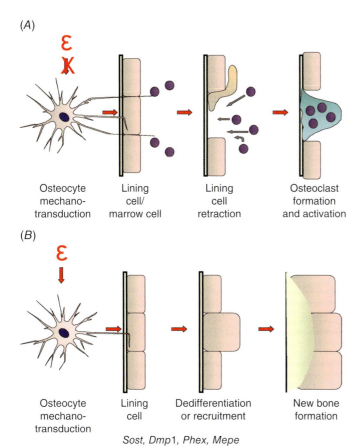

Fig. 5.4. Osteocytes as orchestrators of bone (re)modeling. Osteocytes play a role in bone formation and mineralization as promoters of mineralization through Dmp1 and Phex and inhibitors of mineralization and bone formation such as Sost/sclerostin and MEPE/OF45 which are highly expressed in osteocytes (A). Osteocytes are prodigious secretors of prostaglandin in response to loading which also plays a role in bone formation. These supporters and inhibitors of bone formation and mineralization are most likely exquisitely balanced to maintain equilibrium in order to maintain bone mass. Osteocytes also appear to play a major role in the regulation of osteoclasts, by both inhibiting and activating osteoclastic resorption. With loading, the osteocytes send signals inhibiting osteoclast activation (B) [15]. In contrast, compromised, hypoxic, apoptotic or dying osteocytes, especially with unloading, appear to send signals such as RANKL to osteoclasts/preosteoclasts on the bone surface to initiate resorption. Therefore, osteocytes within the bone regulate bone formation and mineralization and inhibit osteoclastic resorption, while having the capacity also to send signals of osteoclast activation under specific conditions.

effort (for review see [16]). Clinical trials to date have yielded very favorable results with dramatic bone formation with few adverse effects [17].

The osteocyte-like cell line MLO-Y4 not only supports osteoclast formation, but also osteoblast differentiation, and surprisingly, mesenchymal stem cell differentiation. Since these observations were made using a cell line, they have also been validated in vivo (for review see [2]).

These data support the hypothesis that osteocytes possess the unique capacity to regulate all phases of bone remodeling.

OSTEOCYTE CELL DEATH AND APOPTOSIS

It had been proposed that the major purpose of osteocytes is to die, thereby releasing signals of resorption for bone repair. However, it is now clear that excessive osteocyte cell death can lead to a compromised skeleton. Osteocyte cell death can occur in association with pathological conditions, such as osteoporosis and osteoarthritis, leading to increased skeletal fragility [18]. Such fragility is considered to be caused by the loss of the ability of osteocytes to sense microdamage and/or to initiate repair. Several conditions result in osteocyte cell death, such as oxygen deprivation during immobilization, withdrawal of estrogen as with ovariectomy or menopause, and glucocorticoid treatment [18]. TNFα and Interleukin-1 (IL-1) have been reported to increase with estrogen deficiency and to induce osteocyte apoptosis (for review see [5]).

Just as many agents have been shown to induce osteocyte cell death, many agents have been shown to reduce or inhibit osteoblast and osteocyte apoptosis. These include estrogen, selective estrogen receptor modulators, bisphosphonates, calcitonin, CD40 Ligand, Calbindin-D28k, Monocyte Chemotactic Proteins MCP1 and 3, and mechanical loading in the form of fluid flow shear stress which releases prostaglandin (for review see [5]). More recently it has been shown that muscle-secreted factors will protect osteocytes [20]. These agents do not interfere with osteocyte cell death that occurs during the repair of microdamage. The death process and consequently the resorption signals sent by dying osteocytes in an aging or glucocorticoid treated skeleton may be distinct from those in a normal, healthy skeleton in response to microdamage. It will be important to identify and characterize these differences.

Rather than undergoing apoptosis under stressful conditions, osteocytes can enter a state of self-preservation called autophagy where the cell "eats itself" (ie, "auto-phagy") to maintain viability until a favorable environment returns. Glucocorticoids can induce this state in osteocytes [21] and high or low glucocorticoid dose determines whether the osteocyte will undergo apoptosis or autophagy, respectively [22].

OSTEOCYTE MODIFICATION OF THEIR MICROENVIRONMENT

Almost 100 years ago it was proposed that osteocytes may resorb their lacunar wall [23]. The term "osteolytic osteolysis" was initially used to describe the enlarged lacunae in patients with hyperparathyroidism and later in immobilized rats. One must keep in mind that enlarged lacunae caused by osteolytic osteolysis are distinct from the enlarged lacunae in bone from patients with renal osteodystrophy, because the latter is most likely caused by defective mineralization during embedding of the osteoid-osteocyte (for review see [5,24]). Osteolytic osteolysis has a negative connotation, because it was confused with osteoclastic bone resorption. When resorption "pits" were not observed using primary avian osteocytes seeded onto dentin slices, it was concluded that osteocytes cannot remove mineralized matrix. However, removal of mineral by osteocytes would not be detectable using this approach because these cells are within a lacuna and do not form the characteristic sealed osteoclast resorption lacuna that rapidly decalcifies bone.

In addition to enlargement of the lacunae, osteocytes alter the perilacunar matrix. The term "osteocyte halos" was used to describe perilacunar demineralization in rickets [25] and periosteocytic lesions in X-linked hypophosphatemic rickets [26]. Glucocorticoids, in addition to effects on osteocyte apoptosis/autophagy, appear to compromise the metabolism and function of the osteocyte, causing them to enlarge their lacunae and remove mineral from the perilacunar space, thereby generating "halos" of hypomineralized bone [27]. Glucocorticoid may therefore alter or compromise the metabolism and function of the osteocyte, not just induce cell death.

Over three decades ago, it was suggested that the osteocyte not only has matrix destroying capability but also can form new matrix [28]. Osteocyte lacunae were shown to uptake tetracycline, indicating the ability to calcify or form bone. Moreover, mice undergoing calcium-demanding conditions such as lactation remove their perilacunar matrix and upon weaning replace this matrix [29]. Therefore, the osteocyte is capable of both adding and removing mineral from its surroundings, suggestive of perilacunar remodeling.

The capacity of osteocytes to deposit or remove mineral from lacunae and canaliculi has important implications with regards to (i) mineral homeostasis, (ii) magnitude of fluid shear stress applied to the cell, and (iii) mechanical properties of bone. The surface area of the osteocyte lacuno-canalicular system is several orders of magnitude greater than the bone surface area; therefore, removal of only a few angstroms of mineral would have significant effects on systemic ion levels. Enlargement of the lacunae and canaliculi would reduce bone fluid flow shear stress, thereby reducing mechanical loading on the osteocyte. As holes in a material act as stress concentrators, enlargement of lacunae would enhance this effect in bone. Therefore, changes in lacunar size and matrix properties could have dramatic effects on bone properties and quality in addition to osteocyte function.

MECHANOSENSATION AND TRANSDUCTION

The skeleton is able to adapt continually to mechanical loading by the process of adaptive remodeling where new bone is added to withstand increased amounts of loading and bone is removed in response to unloading or disuse. The parameters for inducing bone formation or resorption in vivo are fairly well characterized. Frequency, intensity, and timing of loading are all important parameters. Bone mass is influenced by peak applied strain [30] and bone formation rate is related to loading rate [31]. When rest periods are inserted, the loaded bone shows increased bone formation rates compared with bone subjected to a single bout of mechanical loading, and improved bone structure and strength is greatest if loading is applied in shorter versus longer increments [32].

The major challenge in the field of mechanotransduction has been to translate these wellcharacterized in vivo parameters to in vitro cell culture models. Theoretical models and experimental studies suggest that flow of bone fluid within the osteocyte lacunocanalicular network is driven by extravascular pressure as well as applied cyclic mechanical loading of osteocytes [33]. Mechanical forces applied to bone cause fluid flow through the canaliculi surrounding the osteocyte to induce shear stress and deformation of the cell membrane. It has also been proposed that mechanical information is relayed by primary cilia, a flagellar-like structure found on every cell or most likely through a combination of means to sense mechanical strain [34]. Theoretical modeling predicts osteocyte wall and membrane shear stresses resulting from peak physiological loads in-vivo to be in the range of 8–30 dynes/cm^3 [33]. In vivo experiments have been performed showing an estimated 5 Pa which is within the 8–30 dynes/cm^3 peak shear stress along the osteocyte membrane [35] validating the use of this magnitude of strain in vitro.

It has been shown that the Wnt/β-catenin signaling pathway plays a role in mechanosensation (for review see [36]), in addition to its role in skeletal growth and development (for review see Chapter 9 [37]), but also has recently been shown to be important for mechanosensation in osteocytes. This pathway is activated in osteocytes within an hour of anabolic loading. Mice with reduced β-catenin do not respond to anabolic load and are more susceptible to stress and unloading. The Wnt inhibitor sclerostin most likely prevents or reduces the effects of anabolic loading on the skeleton through effects in osteocytes. As stated above, therapeutics targeting this pathway such as antisclerostin antibodies are proving to be useful not only in the treatment of osteoporosis but also in other conditions of bone loss and fracture healing. It is most likely that the Wnt/β-catenin pathway will continue to provide direction for future anabolic therapies.

ROLE OF GAP JUNCTIONS AND HEMICHANNELS IN OSTEOCYTE COMMUNICATION

Osteocytes communicate intracellularly through gap junctions, transmembrane channels which connect the cytoplasm of two adjacent cells, through which molecules with molecular weights less than 1 kD$_a$ can pass. Gap junction channels are formed by proteins known as connexins, and Cx43 is the major connexin in bone cells. Much of mechanotransduction in bone is thought to be mediated through gap junctions. Primary osteocytes and MLO-Y4 osteocyte-like cells [38] express large amounts of Cx43, suggesting that Cx43 has functions in addition to being a component of gap junctions. It was shown that connexins can form and function as hemichannels, unapposed halves of gap junction channels. Hemichannels directly serve as the conduit for the exit of intracellular PGE$_2$ into the osteocyte microenvironment in response to fluid flow shear stress [39] and function as essential transducers of the antiapoptotic effects of bisphosphonates [40]. Hemichannels are now one of several types of openings or channels to the extracellular bone fluid; others include calcium, ion, voltage, and stretch activated channels, in addition to others (see review [41]). Therefore, gap junctions at the connecting tips of dendrites appear to mediate a form of intracellular communication whereas hemichannels along the dendrite and the cell body appear to mediate a form of extracellular communication between osteocytes. Deletion of Cx43 in osteoblasts and osteocytes in vivo has a negative effect on the skeleton and Cx43 is necessary to mediate the positive effects of anabolic load and to decrease the negative effects of skeletal unloading [42]. In vivo studies also suggest that hemichannels play a major role in osteocyte survival, endocortical bone resorption and remodeling, and periosteal apposition [43].

THE POTENTIAL ROLE OF OSTEOCYTES IN BONE DISEASE

Osteocytes most likely play a role not only in bone disease but also in disease of other organs. Bone diseases include osteoporosis, bone loss with aging, hypophosphatemic rickets, and certain genetic conditions. Examples of nonbone disease include chronic kidney disease, sarcopenia and cachexia, and cardiac disease. A number of therapeutics have been developed for treatment of these diseases that target proteins specifically or selectively expressed by osteocytes, such as antisclerostin, anti-RANKL, and anti-FGF23 (for review see [3]).

Osteoid-osteocytes play a role in phosphate homeostasis. Once the osteoblast begins to embed in osteoid, molecules such as Phex and Dmp1 are elevated (see Table 5.1). Autosomal recessive hypophosphatemic rickets is caused by mutations in *Dmp1* [44]. *Dmp1* null mice have a similar phenotype to *hyp* mice carrying a *Phex* mutation, that of osteomalacia and rickets caused by elevated FGF23 levels in osteocytes [44,45]. The osteocyte lacunocanalicular system should therefore be viewed as an endocrine organ regulating phosphate metabolism through the Phex-Dmp1-Mepe-FGF23 axis. A molecule expressed in the mature osteocyte is MEPE which also plays a role in regulating FGF23, potentially through Phex [46]. The unraveling of the interactions of these molecules should lead to insights into diseases of hyper- and hypophosphatemia (see Chapter 00).

The connectivity and structure of the osteocyte lacunocanalicular system most likely plays a role in bone disease. Osteocyte dendricity may change with static and dynamic bone formation and has been shown to be disrupted in bone disease [47]. In osteoporotic bone there is disorientation of the canaliculi as well as a marked decrease in connectivity that correlates with severity of disease. In contrast, in osteoarthritic bone, although a decrease in connectivity is observed, orientation is intact. In osteomalacic bone, the osteocytes appear viable with high connectivity, but the processes are distorted and the network chaotic [47]. A reduction in osteocyte number and connectivity is also observed with aging [48]. Variability in complexity and number of dendrites and canaliculi could have a dramatic effect on osteocyte function and viability and on the mechanical properties of bone.

As described previously, physiological osteocyte cell death is necessary for bone repair. Pathological osteocyte cell death is responsible for disease. Osteocyte cell death may be responsible for some forms of osteonecrosis, especially glucocorticoid-induced osteonecrosis. Osteonecrotic bone is "dead" bone containing empty osteocyte lacunae that does not remodel, but can persist for years. Proposed mechanisms responsible for osteonecrosis include an early "mechanical theory" where osteoporosis and the accumulation of unhealed trabecular microcracks result in fatigue fractures, and the "vascular theory" where ischemia is caused by microscopic fat emboli. A newer theory postulates that agents inducing osteocyte cell death result in dead bone that does not remodel [18]. The health, viability, and capacity of osteocytes to regulate their own deaths most likely play a highly significant role in the maintenance and integrity of bone. Bone loss in osteoporosis may be due in part to pathological rather than physiological osteocyte cell death [6]. It will be important to develop therapeutics that maintain normal osteocyte viability while allowing the physiological osteocyte cell death responsible for normal bone repair.

CONCLUSION

In conclusion, although there has been a great deal of progress in identifying the functions of osteocytes and how they mediate these functions, it is most likely that osteocytes utilize undiscovered specific molecules to regulate bone (re)modeling. With the dramatic increases or maintenance of bone mass being observed with neutralizing antibody to sclerostin, an osteocyte selective marker [16,17], greater effort to identify additional markers and to unravel the mysteries surrounding osteocyte function is warranted. It is likely that new functions will be discovered for these cells, making them a prime target of investigation, not only to understand basic bone physiology, but also to understand and treat bone disease.

ACKNOWLEDGMENT

The author's work in osteocyte biology was supported by the National Institutes of Health NIAMS PO1AR-46798 and now NIA PO1AG-039355.

REFERENCES

1. Bonewald LF. The amazing osteocyte. J Bone Miner Res. 2011;26:229–38.
2. Dallas SL, Prideaux M, Bonewald LF. The osteocyte: an endocrine cell... and more. Endocr Rev. 2013;34:658–90.
3. Bonewald L. The role of the osteocyte in bone and nonbone disease. Endocrinol Metab Clin North Am 2017;46:1–18.
4. Franz-Odendaal TA, Hall BK, Witten PE. Buried alive: how osteoblasts become osteocytes. Dev Dyn. 2006;235:176–90.
5. Bonewald L. Osteocytes. In: Marcus R, Feldman D, Nelson D, Roden C (eds) *Osteoporosis* (3rd ed.). Cambridge, MA: Academic Press, 2007, pp. 169–90.
6. Manolagas SC. Birth and death of bone cells: basic regulatory mechanisms and implications for the pathogenesis and treatment of osteoporosis. Endocr Rev. 2000;21:115–37.
7. Bordier PJ, Miravet L, Ryckerwaert A, et al. Morphological and morphometrical characteristics of the mineralization front. A vitamin D regulated sequence of bone remodeling. In: Meunier PJ (ed.) *Bone Histomorphometry*. Paris: Armour Montagu, 1976, pp. 335–54.
8. Tanaka-Kamioka K, Kamioka H, Ris H, et al. Osteocyte shape is dependent on actin filaments and osteocyte processes are unique actin-rich projections. J Bone Miner Res. 1998;13:1555–68.
9. Dallas SL, Bonewald LF. Dynamics of the transition from osteoblast to osteocyte. Ann N Y Acad Sci. 2010;1192:437–43.
10. Guo D, Keightley A, Guthrie J, et al. Identification of osteocyte-selective proteins. Proteomics. 2010;10:3688–98.
11. Kalajzic I, Braut A, Guo D, et al. Dentin matrix protein 1 expression during osteoblastic differentiation, generation of an osteocyte GFP-transgene. Bone. 2004;35:74–82.
12. Tanaka K, Yamaguchi, Y., Hakeda, Y. Isolated chick osteocytes stimulate formation and bone-resorbing activity of osteoclast-like cells. J Bone Miner Metab. 1995;13:61–70.
13. Zhao S, Zhang YK, Harris S, et al. MLO-Y4 osteocyte-like cells support osteoclast formation and activation. J Bone Miner Res. 2002;17:2068–79.
14. Verborgt O, Tatton NA, Majeska RJ, et al. Spatial distribution of Bax and Bcl-2 in osteocytes after bone fatigue: complementary roles in bone remodeling regulation? J Bone Miner Res. 2002;17:907–14.
15. Tatsumi S, Ishii K, Amizuka N, et al. Targeted ablation of osteocytes induces osteoporosis with defective mechanotransduction. Cell Metab. 2007;5:464–75.
16. Paszty C, Turner CH, Robinson MK. Sclerostin: a gem from the genome leads to bone-building antibodies. J Bone Miner Res. 2010;25:1897–904.
17. Cosman F, Crittenden DB, Adachi JD, et al. Romosozumab treatment in postmenopausal women with osteoporosis. N Engl J Med. 2016;375:1532–43.
18. Weinstein RS, Nicholas RW, Manolagas SC. Apoptosis of osteocytes in glucocorticoid-induced osteonecrosis of the hip. J Clin Endocrinol Metab. 2000;85:2907–12.
19. Fonseca H, Moreira-Goncalves D, Esteves JL, et al. Voluntary exercise has long-term in vivo protective effects on osteocyte viability and bone strength following ovariectomy. Calcif Tissue Int. 2011;88:443–54.
20. Brotto M, Bonewald, LF. Bone and muscle: interactions beyond mechanical. Bone. 2015;80;109–114.
21. Xia X, Kar R, Gluhak-Heinrich J, et al. Glucocorticoid-induced autophagy in osteocytes. J Bone Miner Res. 2010;25:2479–88.
22. Jia J, Yao W, Guan M, et al. Glucocorticoid dose determines osteocyte cell fate. FASEB J. 2011;25:3366–76.
23. Recklinghausen FV. *Untersuchungen über rachitis and osteomalacia*. Jena: Gustav Fischer, 1910.
24. Qing H, Bonewald LF. Osteocyte remodeling of the perilacunar and pericanalicular matrix. Int J Oral Sci. 2009;1:59–65.
25. Heuck F. Comparative investigations of the function of osteocytes in bone resorption. Calcif Tissue Res. 1970(Suppl):148–9.
26. Marie PJ, Glorieux FH. Relation between hypomineralized periosteocytic lesions and bone mineralization in vitamin D-resistant rickets. Calcif Tissue Int. 1983;35:443–8.
27. Lane NE, Yao W, Balooch M, et al. Glucocorticoid-treated mice have localized changes in trabecular bone material properties and osteocyte lacunar size that are not observed in placebo-treated or estrogen-deficient mice. J Bone Miner Res. 2006;21:466–76.
28. Baud CA, Dupont DH. The fine structure of the osteocyte in the adult compact bone. In: Breese SSJ (ed.) *Electron Microscopy*, vol. 2. New York: Academic Press, 1962.
29. Qing H, Ardeshirpour L, Pajevic PD, et al. Demonstration of osteocytic perilacunar/canalicular remodeling in mice during lactation. J Bone Miner Res. 2012;27:1018–29.
30. Rubin C. Skeletal strain and the functional significance of bone architecture. Calcif Tissue Int. 1984;36:S11–S18.
31. Turner CH, Forwood MR, Otter MW. Mechanotransduction in bone: do bone cells act as sensors of fluid flow? FASEB J. 1994;8:875–8.
32. Robling AG, Hinant FM, Burr DB, et al. Shorter, more frequent mechanical loading sessions enhance bone mass. Med Sci Sports Exerc. 2002;34:196–202.
33. Weinbaum S, Cowin SC, Zeng Y. A model for the excitation of osteocytes by mechanical loading-induced bone fluid shear stresses. J Biomech. 1994;27:339–60.
34. Bonewald LF. Mechanosensation and transduction in osteocytes. Bonekey Osteovision. 2006;3:7–15.

35. Price C, Zhou X, Li W, et al. Real-time measurement of solute transport within the lacunar-canalicular system of mechanically loaded bone: direct evidence for load-induced fluid flow. J Bone Miner Res. 2011;26:277–85.
36. Duan P, Bonewald LF. The role of the wnt/beta-catenin signaling pathway in formation and maintenance of bone and teeth. Int J Biochem Cell Biol. 2016;77:23–9.
37. Baron R, Kneissel M. WNT signaling in bone homeostasis and disease: from human mutations to treatments. Nat Med. 2013;19:179–92.
38. Kato Y, Windle JJ, Koop BA, et al. Establishment of an osteocyte-like cell line, MLO-Y4. J Bone Miner Res. 1997;12:2014–23.
39. Cherian PP, Siller-Jackson AJ, Gu S, et al. Mechanical strain opens connexin 43 hemichannels in osteocytes: a novel mechanism for the release of prostaglandin. Mol Biol Cell. 2005;16:3100–6.
40. Plotkin LI, Manolagas SC, Bellido T. Transduction of cell survival signals by connexin-43 hemichannels. J Biol Chem. 2002;277:8648–57.
41. Klein-Nulend J, Bonewald LF. The osteocyte. In: Bilezikian JP, Raisz LG (ed.) *Principles of Bone Biology*. Cambridge, MA: Academic Press, 2008.
42. Plotkin LI, Speacht TL, Donahue HJ. Cx43 and mechanotransduction in bone. Curr Osteoporos Rep. 2015;13:67–72.
43. Xu H, Gu S, Riquelme MA, et al. Connexin 43 channels are essential for normal bone structure and osteocyte viability. J Bone Miner Res. 2015;30:550–62.
44. Feng JQ, Ward LM, Liu S, et al. Loss of DMP1 causes rickets and osteomalacia and identifies a role for osteocytes in mineral metabolism. Nat Genet. 2006;38:1310–15.
45. Liu S, Lu Y, Xie Y, et al. Elevated levels of FGF23 in dentin matrix protein 1 (DMP1) null mice potentially explain phenotypic similarities to Hyp Mice. J Bone Min Res. 2006;21:S51.
46. Liu S, Rowe PS, Vierthaler L, et al. Phosphorylated acidic serine-aspartate-rich MEPE-associated motif peptide from matrix extracellular phosphoglycoprotein inhibits phosphate regulating gene with homologies to endopeptidases on the X-chromosome enzyme activity. J Endocrinol. 2007;192:261–7.
47. Knothe Tate ML, Adamson JR, Tami AE, et al. The osteocyte. Int J Biochem Cell Biol. 2004;36:1–8.
48. Xie Y, Dusevich V, Hulbert M, et al. Age-related changes in osteocyte connectivity and bone structure in a murine model of aging. J Bone Min Res. 2014; 29 (Suppl 1).

6

Osteoclast Biology and Bone Resorption

Hiroshi Takayanagi

Department of Immunology, Graduate School of Medicine and Faculty of Medicine,
The University of Tokyo, Tokyo, Japan

CELL BIOLOGY OF THE OSTEOCLAST

Pathological bone loss, regardless of etiology, invariably represents an increase in the rate at which the skeleton is degraded by osteoclasts relative to its formation by osteoblasts. Thus, prevention of conditions such as osteoporosis requires an understanding of the molecular mechanisms of bone resorption.

The osteoclast, the exclusive bone resorptive cell (Fig. 6.1), is a polykaryon derived from the monocyte/macrophage lineage cell of hematopoietic origin. The multinucleated osteoclasts can be generated in vitro from mononuclear monocyte/macrophage precursors resident in a number of tissues [1]. There is, however, general agreement that the principal physiological osteoclast precursor is the bone marrow monocyte/macrophage lineage precursor cell. Two cytokines are essential for osteoclastogenesis, the first being receptor activator of nuclear factor-κB ligand (RANKL) [1,2]. and the second being M-CSF, also designated CSF-1 [3]. These two proteins exist as both membrane-bound and soluble forms. While soluble M-CSF plays a major role, RANKL functions more strongly in a membrane-bound form than the soluble one (eg, secreted by activated T cells) [2]. M-CSF and RANKL are mainly produced by osteoblast-lineage cells including osteoblasts and osteocytes, and thus differentiation of osteoclasts from their mononuclear precursors requires the presence of these nonhematopoietic, bone-residing cells [1,4,5]. Of these, osteocytes were shown to be the major source of RANKL in the adult bone remodeling stage [4,5]. RANKL, a member of the TNF superfamily cytokines, is the key osteoclastogenic cytokine, because osteoclast formation and activation requires its presence and osteoclastogenesis is completely abrogated in the RANKL deficiency in mice and humans [2,6]. Although several studies suggested the possibility of RANKL-independent osteoclastogenesis, there is no evidence for any molecules that substitute for RANKL in vivo [7]. M-CSF contributes to the proliferation, survival, and differentiation of osteoclast precursors, as well as the survival and cytoskeletal rearrangement required for efficient bone resorption [3]. The discovery of RANKL was preceded by identification of its physiological inhibitor osteoprotegerin (OPG), to which it binds with higher affinity than its receptor RANK [2,6]. In contrast, M-CSF is a moiety long known to regulate the broader biology of myeloid cells, including osteoclasts [3].

Our understanding of how osteoclasts resorb bone derives from two major types of studies: biochemical and genetic [8–10]. The unique osteoclastogenic properties of RANKL permit generation of pure populations of osteoclasts in culture, and hence the performance of meaningful biochemical and molecular experiments that provide insights into the molecular mechanisms by which osteoclasts develop and resorb bone. Further evidence has come from our capacity to generate mice lacking specific genes, plus the positional cloning of genetic abnormalities in people with abnormal osteoclast function [9,10]. Key to the resorptive event is the capacity of the osteoclast to form a microenvironment between itself and the underlying bone matrix (Fig. 6.2A). This compartment, which is isolated from the general extracellular space, is acidified by an electrogenic proton pump (H^+-ATPase) and a Cl^- channel to a pH of ~4.5 [8]. The acidified milieu mobilizes the mineralized component of bone (hydroxyapatite, a calcium phosphate mineral), exposing its organic matrix, consisting largely of type I collagen that is subsequently degraded by lysosomal enzymes such as cathepsin K. The critical role that the proton pump, Cl^- channel, and cathepsin K play in osteoclast action is underscored by the fact

---

Primer on the Metabolic Bone Diseases and Disorders of Mineral Metabolism, Ninth Edition. Edited by John P. Bilezikian.
© 2019 American Society for Bone and Mineral Research. Published 2019 by John Wiley & Sons, Inc.
Companion website: www.wiley.com/go/asbmrprimer

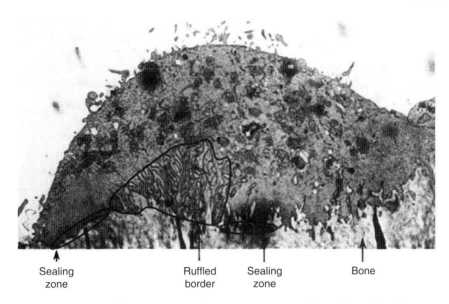

Fig. 6.1. The osteoclast as a resorptive cell. Transmission electron microscopy of a multinucleated primary rat osteoclast on bone. Note the extensive ruffled border, close apposition of the cell to bone, and the partially degraded matrix between the sealing zones. Courtesy of H. Zhao.

that diminished function of each results in a human disease of excess bone mass, namely osteopetrosis or pyknodysostosis [8,9,11]. Degraded protein fragments are endocytosed and transported in undefined vesicles to the basolateral surface of the cell, where they are discharged into the surrounding intracellular fluid [12,13].

The above model of bone degradation clearly depends on physical intimacy between the osteoclast and bone matrix, a role provided by integrins. Integrins are αβ heterodimers with long extracellular and single transmembrane domains [3]. In most instances, the integrin cytoplasmic region is relatively short, consisting of 40–70 amino acids. Integrins are the principal cell-matrix attachment molecules which mediate osteoclastic bone recognition. αvβ3 is the principal integrin mediating bone resorption and this heterodimer recognizes the amino acid motif Arg-Gly-Asp (RGD), which is present in a variety of bone-residing proteins such as osteopontin and bone sialoprotein [3]. Thus, osteoclasts attach to and spread on these substrates in an RGD-dependent manner and, most importantly, competitive ligands that prevent integrin engagement arrest bone resorption in vivo. Proof of the pivotal role that αvβ3 has in the resorptive process came with the generation of the β3 integrin knockout mouse, which develops a progressive increase in bone mass because of osteoclast dysfunction [3,9].

Bone resorption also requires a polarization event in which the osteoclast delivers effector molecules like HCl and cathepsin K into the resorptive microenvironment. Osteoclasts are characterized by a unique cytoskeleton, which mediates the resorptive process. Specifically, when the cell contacts bone, it generates two polarized structures, which enable it to degrade skeletal tissue. In the first instance, a subset of acidified vesicles containing specific cargo, including cathepsin K and other MMPs, are transported, probably through microtubules and actin, to the bone-apposed plasma membrane, to which they fuse in a manner not currently understood, but which may involve PLEKHM1 [11,14]. Insertion of these vesicles into the plasmalemma results in formation of a villous structure, unique to the osteoclast, called the ruffled membrane. This resorptive organelle contains the abundant H^+ transporting machinery to create the acidified microenvironment, whereas the accompanying exocytosis serves as the means by which cathepsin K is secreted (Fig. 6.2B).

In addition to inducing ruffled membrane formation, contact with bone also prompts the osteoclast to polarize its fibrillar actin into a circular structure known as the "actin ring" [11,15]. A separate "sealing zone" surrounds and isolates the acidified resorptive microenvironment in the active cell, but its composition is almost completely unknown. The actin ring, like the ruffled membrane, is a hallmark of the degradative capacity of the osteoclast, because structural abnormalities in either occur in conditions of arrested resorption [15]. In most cells, such as fibroblasts, matrix attachment prompts formation of stable structures known as focal adhesions, which mediate contact and formation of actin stress fibers. In keeping with the substitution of the actin ring for stress fibers in osteoclasts, these cells form podosomes instead of focal adhesions. Podosomes, which in resorbing osteoclasts are present in the actin ring, consist of an actin core surrounded by αvβ3 and associated cytoskeletal proteins [3,9]. Consistently, the integrin β3 subunit knockout ($β3^{-/-}$) osteoclast forms abnormal ruffled membranes in vivo and the mutant osteoclasts fail to spread when plated on the mineralized matrix [3,9].

48 Osteoclast Biology and Bone Resorption

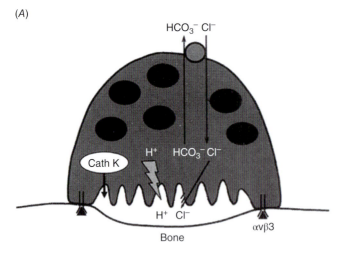

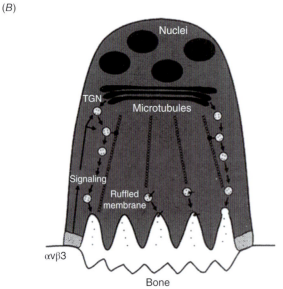

Fig. 6.2. Mechanism of osteoclastic bone resorption. (A) The osteoclast adheres to bone through the integrin αvβ3, creating a sealing zone, into which is secreted hydrochloric acid and acidic proteases such as cathepsin K, MMP9, and MMP13. The acid is generated by the combined actions of a vacuolar H^+-ATPase; it couples a chloride channel and a basolateral chloride–bicarbonate exchanger. Carbonic anhydrase converts CO_2 into H^+ and HCO_3^- (data not shown). (B) Integrin engagement results in signals that target acidifying vesicles containing specific cargo (black dots) to the bone-apposed face of the cell. Fusion of the vesicles with the plasma membrane generates a polarized cell capable of secreting the acid and proteases required for bone resorption.

SIGNALS THAT REGULATE OSTEOCLAST FUNCTION

Osteoclast bone-resorbing function is dependent on the multiple factors regulating cytoskeletal reorganization, acidification, matrix degradation, and the survival of osteoclasts. Therefore, osteoclast function is impaired in humans or mice deficient in the molecules involved in these processes such as αvβ3 integrin, vacuolar ATPase, Cl channel, and cathepsin K. Bone resorbing activity is measured by the resorption pit assay on the bone or dentine slices. This activity is known to be stimulated by M-CSF, IL-1, and RANKL as well as integrin αvβ3 [1,3]. The initial αvβ3 integrin signaling event involves the protooncogene c-Src, which is also stimulated by M-CSF binding to c-Fms. c-Src, acting as a kinase and an adaptor protein, regulates formation of lamellipodia and disassembly of podosomes, indicating that c-Src controls formation of resorptive organelles of the cell, such as the ruffled membrane, and also arrests migration on the bone surface [3,9]. There is continuing debate surrounding the molecules which link c-Src to the cytoskeleton, one proposal being that the focal adhesion kinase family member Pyk2, acts in concert with c-Cbl, a protooncogene and ubiquitin ligase [16]. A second strong candidate is Syk, a nonreceptor tyrosine kinase that is recruited to the active conformation of αvβ3 in osteoclasts in a c-Src-dependent manner [17] where it targets Vav3 [3], a member of the large family of guanine nucleotide exchange factors (GEFs) that convert Rho GTPases from their inactive GDP to their active GTP conformation.

M-CSF, IL-1, and RANKL promote osteoclast survival and bone resorbing activity, partly sharing downstream signaling pathways with αvβ3 integrin. The signal transduction cascades regulating osteoclast function are summarized in Fig. 6.3. However, it has not been well understood how these cytokines and signaling molecules distinctly contribute to the various processes of bone resorption including cytoskeletal reorganization, decalcification, matrix degradation, and the survival of osteoclasts. Exploration of the detailed signaling pathways governing osteoclast function will be an important issue to be pursued in the future.

SIGNALS THAT REGULATE OSTEOCLAST DIFFERENTIATION

Various molecules were identified to be essential for osteoclast differentiation based on findings obtained from osteopetrotic mice [9,10]. Since osteoclast differentiation was found to be driven mainly by RANKL in the presence of M-CSF, the signal transduction pathways regulating osteoclastogenesis has been extensively studied in the context of RANKL signaling. Binding of RANKL to its receptor RANK induces activation of TNF receptor-associated factor (TRAF) 6 and transforming growth factor β-activated kinase 1 (TAK1), which stimulate NF-κB and MAPKs, including JNK and p38 [10,18]. RANKL stimulation results in the upregulation of the activator protein 1 (AP-1) complex through JNK and NF-κB as well as cyclic AMP-responsive element-binding protein (CREB) activated by Ca^{2+}/calmodulin-dependent kinase (CaMK) IV (Fig. 6.4).

Signals that Regulate Osteoclast Differentiation 49

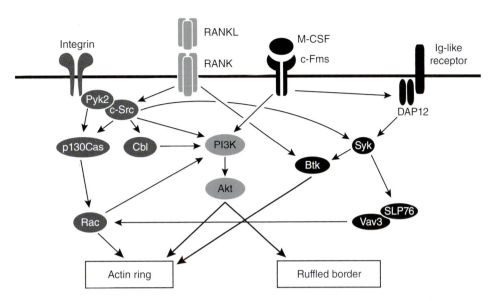

Fig. 6.3. Signaling pathways implicated in cytoskeletal organization in osteoclasts. In concert with RANKL, M-CSF, and ITAM signaling, activation of the $\alpha v \beta 3$ integrin organizes the osteoclast cytoskeleton through c-Src, which transmits intracellular signals to the cytoskeleton mainly by activating c-Cbl, Rac and Syk. PI3K/Akt signaling also plays a central role in organizing the cytoskeleton of the osteoclast.

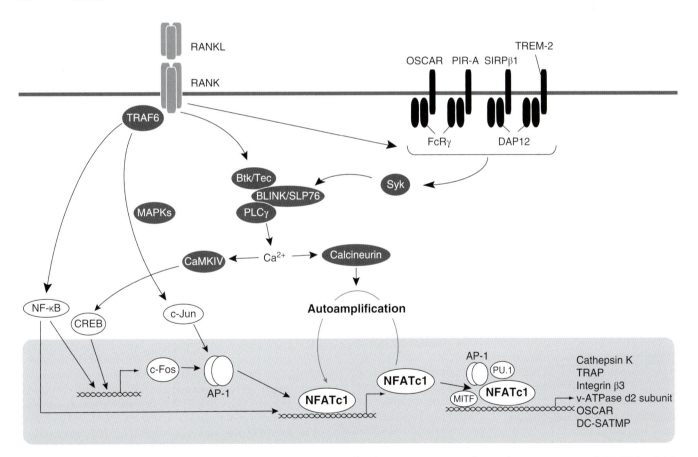

Fig. 6.4. Signaling cascades during osteoclastogenesis. RANKL binding to RANK results in the recruitment of TRAF6, which activates MAPKs and NF-κB. RANKL also stimulates the induction of c-Fos, a component of AP-1. NF-κB and AP-1 are important for the robust induction of NFATc1. Costimulatory receptors (OSCAR, PIR-A, SIRPβ1 and TREM-2) associated with the ITAM-harboring adaptors, FcRγ and DAP12, stimulate the activation of Syk and subsequent induction of calcium signaling, which is critical for the activation and autoamplification of NFATc1. NFATc1 controls expression of a variety of osteoclast-specific genes in cooperation with other osteoclastogenic transcription factors.

RANKL stimulation cooperates with its costimulatory receptors to activate calcium signaling. Costimulatory receptors are immunoglobulin-like receptors such as osteoclast-associated receptor (OSCAR) and triggering receptor expressed on myeloid cells (TREM)-2, which associate with the adaptor molecules DNAX-activating protein 12 (DAP12) and FcRγ [19,20]. These adaptor molecules contain immunoreceptor tyrosine-based activation motifs (ITAMs), which upon phosphorylation recruits the spleen tyrosine kinase (Syk). Activation of Tec-family kinases (Tec and Btk) by RANK leads to the formation of the osteoclastogenic complex composed of Tec kinases, B-cell linker protein (BLNK)/SH2-containing leukocyte protein of 76 kDa (SLP76), and phospholipase Cγ (PLCγ), which induces the release of calcium from the endoplasmic reticulum, resulting in calcium oscillation in the osteoclast precursor cells [21].

Shortly after RANKL stimulation, NF-κB induces the initial induction of nuclear factor of activated T cells, cytoplasmic 1 (NFATc1), the master regulator of osteoclast differentiation [22]. NFATc1 is continuously activated by calcium oscillation during osteoclastogenesis and binds to its own promoter with AP-1 complex, resulting in the robust induction of NFATc1 (autoamplification of NFATc1) [23]. Thus, mature osteoclasts express a very high level of NFATc1, which directly controls the promoter of a number of osteoclast-specific genes, such as cathepsin K, tartrate-resistant acid phosphatase (TRAP), and the calcitonin receptor, in cooperation with other transcription factors such as AP-1, CREB, PU.1, and microphthalmia-associated transcription factor (MITF). Autoamplification of NFATc1 depends on TRAF6, NF-κB, AP-1, and calcium signaling and the deficiency of any one of these signals results in impaired NFATc1 induction and osteoclastogenesis. The essential role of NFATc1 in osteoclastogenesis has been proven by genetic experiments [23–25].

Studies using osteoclast lineage-specific conditional knockout mice generated by the Cre-loxP system have provided convincing evidence for essential molecules for osteoclasts. However, it is important to carefully check the deletion efficiency and the timing of Cre-mediated recombination, because deletion efficiency is dependent on a combination of flox mice and the Cre strain. A failure of deletion or a very low level of deletion in osteoclasts has been reported in certain Cre-expressing strains such as LysM-Cre [24].

The previous text might suggest that numerous mutations in many genes linked to the osteoclast are likely to have been discovered in humans. In fact, few such genetic changes have been defined, with fewer than 50% of those reported being in patients with osteopetrosis caused by defects in the chloride channel that modulates osteoclast acid secretion (Fig. 6.2). Rare reports link deficiencies in RANKL, RANK, the proton pump, or carbonic anhydrase II to osteopetrosis, whereas decreased cathepsin K function leads to pyknodysostosis [6]. In contrast, RANK activation manifests as osteolytic bone disease, whereas OPG deficiency leads to a severe form of high turnover osteoporosis.

FACTORS REGULATING OSTEOCLAST FORMATION AND/OR FUNCTION

Proteins

In addition to the two key osteoclastogenic cytokines M-CSF and RANKL, a number of other proteins play important roles in osteoclast biology, either in physiological and/or pathophysiological circumstances.

As discussed earlier, OPG, a high-affinity ligand for RANKL that acts as a soluble inhibitor of RANKL, is secreted by cells of mesenchymal origin, both basally and in response to other regulatory signals, including cytokines and bone-targeting steroids [2,26]. Proinflammatory cytokines suppress OPG expression while simultaneously enhancing that of RANKL, with the net effect being a marked increase in osteoclast formation and function. Genetic deletion of OPG in both mice and humans leads to profound osteoporosis, whereas overexpression of the molecule under the control of a hepatic promoter results in severe osteopetrosis [6,26]. Together, these observations indicate that skeletal and perhaps circulating OPG modulates the bone resorptive activity of RANKL and helps to explain the increased bone loss in clinical situations accompanied by increased levels of TNF, IL-1, PTH, or PTH-related protein (PTHrP). Serum PTH levels are increased in hyperparathyroidism of whatever etiology, whereas PTHrP is secreted by metastatic lung and breast carcinoma [27,28]. TNF antibodies or a soluble TNF receptor-IgG fusion protein potently suppress the bone loss in disorders of inflammatory osteolysis such as rheumatoid arthritis [29]. In addition to the ability to induce RANKL, the inflammatory cytokine synergizes with RANKL in a unique manner. RANKL and TNF each activate a number of key downstream effector pathways, leading to nuclear localization of a range of osteoclastogenic transcription factors [10].

T-cell cytokines including interferon-γ (IFN-γ), IL-4, and IL-10 are potent suppressors of osteoclast formation [18,30]. These findings seem to be in conflict with other in vivo observations in which activated T-cell immune responses are associated with enhanced bone resorption. This is partly because T cells are divided into several subsets with distinct cytokine production. The detailed studies revealed that Th1 cells (producing IFN-γ) and Th2 cells (producing IL-4) inhibit osteoclast formation, but Th17 cells stimulate osteoclastogenesis through IL-17-mediated induction of RANKL as well as induction of other inflammatory cytokines [31]. IFN-γ treatment of children with osteopetrosis ameliorates the disease [32], but IFN-γ was used to rescue the immunodeficiency associated with osteopetrosis not to increase the osteoclast number. It is necessary to interpret the in vivo data by taking into consideration various conditions in addition to the in vitro effect of each cytokine to osteoclastogenesis.

Many additional studies have implicated a range of other soluble factors and cytokines in the regulation of the osteoclast. These include a range of interleukins, granulocyte macrophage colony-stimulating factor (GM-CSF), IFN-β, stromal cell-derived factor 1 (SDF-1), macrophage inflammatory protein-1α (MIP-1α), monocyte chemoattractant protein 1 (MCP-l), transforming growth factor β (TGF-β), various Toll-like receptor ligands, Wnt ligands, and semaphorins [18,33,34].

Small molecules

1,25-dihydroxyvitamin D has all the characteristics of a steroid hormone, including a high-affinity nuclear receptor that binds as a heterodimer with the retinoid X receptor to regulate transcription of a set of specific target genes. This active form of vitamin D, generated by successive hydroxylation in the liver and kidney, is a well-established stimulator of bone resorption when present at supraphysiological levels. Studies over many years have indicated that this steroid hormone increases mesenchymal cell transcription of the RANKL gene, while diminishing that of OPG [35]. Separately, 1,25-dihydroxyvitamin D suppresses synthesis of the pro-osteoclastogenic hormone PTH and enhances calcium uptake from the gut [35]. Taken together, the two latter effects would seem to be antiresorptive, but many studies in humans indicate the net osteolytic action resulting from high levels of this steroid hormone, suggesting that its ability to stimulate osteoclast function overrides any bone anabolic actions.

Loss of estrogen (E_2), most often observed in the context of menopause, is a major reason for the development of significant bone loss in aging. Interestingly, it is now clear that estrogen is the main sex steroid regulating bone mass in both men and women [36]. The mechanisms by which estrogen mediates its osteolytic effects are still incompletely understood, but significant advances have been made over the last decade. The original hypothesis, now considered to be only part of the explanation, is that decreased serum E_2 led to increased production, by circulating macrophages, of osteoclastogenic cytokines such as IL-6, TNF, and IL-1. These molecules act on stromal cells and osteoclast precursors to enhance bone resorption by regulating expression of pro- (RANKL, M-CSF) and anti- (OPG) osteoclastogenic cytokines (in the case of mesenchymal cells) and by synergizing with RANKL itself (in the case of myeloid osteoclast precursors). However, recent studies suggest other targets of E_2 such as T cells and osteoclast lineage cells [36–38]. Further studies are needed to evaluate the relative contribution of E_2 effects on these multiple cell types.

Both endogenous glucocorticoids and their synthetic analogs, which have been and continue to be a major mainstay of immunosuppressive therapy, are members of a third steroid hormone family having a major impact on bone biology [39]. One consequence of their chronic mode of administration is severe osteoporosis arising from decreased bone formation and resorption with the latter absolutely decreased (low turnover osteoporosis). The majority of evidence focuses on the osteoblast as the prime target with the steroid increasing apoptosis of these bone-forming cells. However, numerous human studies document a rapid initial decrease in bone resorption, suggesting that the ostcoclast and/or its precursors may also be targets. The molecular basis for this latter finding is unclear. However, because osteoblasts are a requisite part of the resorptive cycle, one consequence of their long-term diminution could be decreased osteoclast formation and/or function secondary to lower levels of RANKL and/or M-CSF production. Alternatively, glucocorticoids have been shown to decrease osteoclast apoptosis [40].

A wide range of clinical information shows that excess prostaglandins stimulate bone loss, but once again, the cellular basis has not been established. Prostaglandins target stromal and osteoblastic cells, stimulating expression of RANKL and suppressing that of OPG [41]. This increase in the RANKL/OPG ratio, observed in a variety of human studies, is sufficient of itself to explain the clinical findings of increased osteoclastic activity. However, highlighting again the dilemma of interpreting in vitro studies, there have been a number of studies in which prostaglandins regulate osteoclastogenesis per se in murine cell culture. Phosphoinositides play distinct and important roles in organization of the osteoclast cytoskeleton [42]. Binding of M-CSF or RANKL to their cognate receptors, c-Fms and RANK, or activation of αυβ3, recruits phosphoinositol-3-kinase (PI3K) to the plasma membrane, where it converts membrane-bound phosphatidylinositol 4,5-bisphosphate into phosphatidylinositol 3,4,5-trisphosphate (Fig. 6.3). The latter compound is recognized by specific motifs in a wide range of cytoskeletally active proteins, and thus PI3K plays a central role in organizing the cytoskeleton of the osteoclast, including its ruffled membrane [42]. Akt is a downstream target of PI3K and plays an important role in osteoclast function, particularly by mediating RANKL and/or M-CSF-stimulated proliferation and/or survival [42].

Cell–cell interactions in bone marrow

Recent evidence has indicated that a number of additional cell types are important for osteoclast biology in a variety of situations. First, as discussed previously, T cells play a key role in estrogen deficiency bone loss but also are important in a range of inflammatory diseases, most notably rheumatoid arthritis [18] and periodontal disease [43]; here the Th17 subset likely secretes TNF and IL-17, a newly described osteoclastogenic cytokine [31]. Given that both osteoclast precursors and the various lymphocyte subsets, such as T, B, and NK cells, arise from the same stem cell, it is not surprising that some of the same receptors and ligands that mediate the immune process also govern the maturation of osteoclast precursors and the capacity of the mature cell to degrade bone. This interface has given rise to the new discipline of osteoimmunology, which promises to provide important and exciting findings in the future [10,18,33,44].

Second, whereas it is well established that mesenchymal cells are major mediators of cytokine and prostaglandin action on osteoclasts, it has become clear recently that cells of the same lineage, residing on cortical and trabecular bone, comprise the site of a HSC niche [44–46]. Specifically, HSCs reside close to osteoblasts as a result of multiple interactions involving receptors and ligands on both cells types [47]. Furthermore, the mesenchymally derived cells secrete both membrane-bound and soluble factors that contribute to survival and proliferation of multipotent osteoclast precursors, as well as molecules that influence osteoclast formation and function. Both committed osteoblasts and the numerous stromal cells in bone marrow produce a range of proteins both basally and in response to hormones and growth factors, resulting in modulation of the capacity of HSCs to become functional osteoclasts.

Third, cancer cells facilitate their infiltration into the marrow cavity by stimulating osteoclast formation and function. An initial stimulus is PTHrP generation by lung and breast cancer cells [27,28,48], thus enhancing mesenchymal production of RANKL and M-CSF, whereas decreasing that of OPG and possibly chemotactic factors. The resulting increase in matrix dissolution releases bone-residing cytokines and growth factors that, feeding back on the cancer cells, increase their growth and/or survival. This loop has been termed "the vicious cycle" [27]. Multiple myeloma seems to use a different but related strategy, namely secretion of MIP-1α and MCP-1, both of which are chemotactic and proliferative for osteoclast precursor [49,50]. The latter compound has been reported to be secreted by osteoclasts in response to RANKL and enhances osteoclast formation [8]. It seems likely that future studies will uncover additional molecules mediating bone loss in metastatic disease.

CONCLUSION

Osteoclasts act in and elicit changes in the complex bone microenvironment. These cells have a crucial role in maintaining bone health, and they participate in essential reciprocal interactions with osteoblasts, osteocytes, and immune cells to maintain homeostasis. Although many of the key factors that regulate osteoclast formation and activity have been identified, there are many remaining unknowns. We still do not understand how the primary modulators of osteoclastogenesis function or dysfunction in healthy bone and in disease states. The mechanisms underlying different responses of osteoclasts in different bones or in different bone compartments are not well understood. The role of aging in the accumulation of epigenetic changes that affect the activity of osteoclasts has not been elucidated. Answers to these and other questions will expand therapeutic options for the treatment of bone loss in osteoporosis and metabolic bone disease.

ACKNOWLEDGMENT

The author is indebted to Dr F. Patrick Ross, who wrote the previous version of this chapter, upon which this update is based.

REFERENCES

1. Suda T, Takahashi N, Udagawa N, et al. Modulation of osteoclast differentiation and function by the new members of the tumor necrosis factor receptor and ligand families. Endocr Rev. 1999;20:345–57.
2. Theill LE, Boyle WJ, Penninger JM. RANK-L and RANK: T cells, bone loss, and mammalian evolution. Annu Rev Immunol. 2002;20:795–823.
3. Ross FP, Teitelbaum SL. αvβ3 and macrophage colony stimulating factor: partners in osteoclast biology. lmmunol Rev. 2005;208:88–105.
4. Nakashima T, Hayashi M, Fukunaga T, et al. Evidence for osteocyte regulation of bone homeostasis through RANKL expression. Nat Med. 2011;17:1231–4.
5. Xiong J, Onal M, Jilka RL, et al. Matrix-embedded cells control osteoclast formation. Nat Med. 2011;17:1235–41.
6. Nakashima T, Hayashi M, Takayanagi H. New insights into osteoclastogenic signaling mechanisms. Trends Endocrinol Metab. 2012;23:582–90.
7. Tsukasaki M, Hamada K, Okamoto K, et al. LOX fails to substitute for RANKL in osteoclastogenesis. J Bone Miner Res. 2017;32:434–9.
8. Boyle WJ, Simonet WS, Lacey DL. Osteoclast differentiation and activation. Nature. 2003;423:337–42.
9. Teitelbaum SL, Ross FP. Genetic regulation of osteoclast development and function. Nat Rev Genet. 2003;4:638–49.
10. Takayanagi H. Osteoimmunology: shared mechanisms and crosstalk between the immune and bone systems. Nat Rev Immunol. 2007;7:292–304.
11. Sobacchi C, Schulz A, Coxon FP, et al. Osteopetrosis: genetics, treatment and new insights into osteoclast function. Nat Rev Endocrinol. 2013;9:522–36.
12. Salo J, Lehenkari P, Mulari M, et al. Removal of osteoclast bone resorption products by transcytosis. Science. 1997;276:270–3.
13. Stenbeck G. Horton MA. Endocytic trafficking in actively resorbing osteoclasts. J Cell Sci. 2004;117:827–36.
14. Teitelbaum SL, Abu-Amer Y, Ross FP. Molecular mechanisms of bone resorption. J Cell Biochem. 1995;59:1–10.
15. Vaananen HK, Zhao H, Mulari M, et al. The cell biology of osteoclast function. J Cell Sci. 2000;113:377–81.
16. Horne WC, Sanjay A, Bruzzaniti A, et al. The role(s) of Src kinase and Cbl proteins in the regulation of osteoclast differentiation and function. Immunol Rev. 2005;208:106–25.
17. Zou W, Kitaura H, Reeve J, et al. Syk, c-Src, the αvβ3 integrin, and ITAM immunoreceptors, in concert, regulate osteoclastic bone resorption. J Cell Biol. 2007;176: 877–88.

18. Takayanagi H. Osteoimmunology and the effects of the immune system on bone. Nat Rev Rheumatol. 2009;5:667–76.
19. Koga T, Inui M, Inoue K, et al. Costimulatory signals mediated by the ITAM motif cooperate with RANKL for bone homeostasis. Nature. 2004;428:758–63.
20. Mócsai A, Humphrey MB, Van Ziffle JA, et al. The immunomodulatory adapter proteins DAP12 and Fc receptor γ-chain (FcRγ) regulate development of functional osteoclasts through the Syk tyrosine kinase. Proc Natl Acad Sci U S A. 2004;101:6158–63.
21. Shinohara M, Koga T, Okamoto K, et al. Tyrosine kinases Btk and Tec regulate osteoclast differentiation by linking RANK and ITAM signals. Cell 2008;132:794–806.
22. Takayanagi H, Kim S, Koga T, et al. Induction and activation of the transcription factor NFATc1 (NFAT2) integrate RANKL signaling in terminal differentiation of osteoclasts. Dev Cell. 2002;3:889–901.
23. Asagiri M, Sato K, Usami T, et al. Autoamplification of NFATc1 expression determines its essential role in bone homeostasis. J Exp Med. 2005;202:1261–9.
24. Aliprantis AO, Ueki Y, Sulyanto R, et al. NFATc1 in mice represses osteoprotegerin during osteoclastogenesis and dissociates systemic osteopenia from inflammation in cherubism. J Clin Invest. 2008;118:3775–89.
25. Winslow MM, Pan M, Starbuck M, et al. Calcineurin/NFAT signaling in osteoblasts regulates bone mass. Dev Cell. 2006;10:771–82.
26. Simonet WS, Lacey DL, Dunstan CR, et al. Osteoprotegerin: a novel secreted protein involved in the regulation of bone density. Cell. 1997;89:309–19.
27. Clines GA, Guise TA. Hypercalcaemia of malignancy and basic research on mechanisms responsible for osteolytic and osteoblastic metastasis to bone. Endocr Relat Cancer. 2005;12:549–83.
28. Martin TJ. Manipulating the environment of cancer cells in bone: a novel therapeutic approach. J Clin Invest. 2002;110:1399–401.
29. Zwerina J, Redlich K, Schett G, et al. Pathogenesis of rheumatoid arthritis: targeting cytokines. Ann NY Acad Sci. 2005;1051:716–29.
30. Takayanagi H, Ogasawara K, Hida S, et al. T-cell-mediated regulation of osteoclastogenesis by signalling cross-talk between RANKL and IFN-γ. Nature 2000;408:600–5.
31. Sato K, Suematsu A, Okamoto K, et al. Th17 functions as an osteoclastogenic helper T cell subset that links T cell activation and bone destruction. J Exp Med. 2006;203:2673–82.
32. Key LL, Rodriguiz RM, Willi SM, et al. Long-term treatment of osteopetrosis with recombinant human interferon gamma. N Engl J Med. 1995;332:1594–99.
33. Lorenzo Jl, Horowitz M, Choi Y. Osteoimmunology: interactions of the bone and immune system. Endocr Rev. 2008;29:403–40.
34. Hayashi M, Nakashima T, Taniguchi M, et al. Osteoprotection by semaphorin 3A. Nature. 2012;485:69–74.
35. Takahashi N, Udagawa N, Suda T. Vitamin D endocrine system and osteoclasts. Bonekey Rep. 2014;3:495.
36. Syed F, Khosla S. Mechanisms of sex steroid effects on bone. Biochem Biophys Res. Commun. 2005;328:688–96.
37. Pacifici R. Role of T cells in ovariectomy induced bone loss-revisited. J Bone Miner Res. 2012;27:231–9.
38. Nakamura T, Imai Y, Matsumoto T, et al. Estrogen prevents bone loss via estrogen receptor α and induction of Fas ligand in osteoclasts. Cell. 2007;130:811–23.
39. Canalis E, Bilezikian JP, Angeli A, Giustina A. Perspectives on glucocorticoid-induced osteoporosis. Bone. 2004;34:593–8.
40. Weinstein RS, Chen J-R, Powers CC, et al. Promotion of osteoclast survival and antagonism of bisphosphonate-induced osteoclast apoptosis by glucocorticoids. J Clin Invest. 2002;109:1041–8.
41. Kobayashi T, Narumiya S. Function of prostanoid receptors: studies on knockout mice. Prostaglandins Other Lipids Mediat. 2002;68–69:557–73.
42. Golden LH, Insogna KL. The expanding role of PI3-kinase in bone. Bone. 2004;34:3–12.
43. Hienz SA, Paliwal S, Ivanovski S. Mechanisms of bone resorption in periodontitis. J Immunol Res. 2015:615486.
44. Takayanagi H. New developments in osteoimmunology. Nat Rev Rheumatol. 2012;8:684–9.
45. Suda T, Arai F, Hirao A. Hematopoietic stem cells and their niche. Trends Immunol. 2005;26:428–33.
46. Morrison SJ, Scadden DT. The bone marrow niche for haematopoietic stem cells. Nature. 2014;505:327–34.
47. Taichman RS. Blood and bone: two tissues whose fates are intertwined to create the hematopoietic stem-cell niche. Blood. 2005;105:2631–9.
48. Bendre M, Gaddy D, Nicholas RW, et al. Breast cancer metastasis to bone: it is not all about PTHrP. Clin Orthop Relat Res. 2003;415 Suppl: S39–45.
49. Hata H. Bone lesions and macrophage inflammatory protein-1 alpha (MIP-1α) in human multiple myeloma. Leuk Lymphoma. 2005;46:967–72.
50. Kim MS, Day CJ, Morrison NA. MCP-1 is induced by receptor activator of nuclear factor-κB ligand, promotes human osteoclast fusion, and rescues granulocyte macrophage colony-stimulating factor suppression of osteoclast formation. J Biol Chem. 2005;280:16163–9.

7

Signal Transduction Cascades Controlling Osteoblast Differentiation

David J.J. de Gorter[1], Gonzalo Sánchez-Duffhues[2], and Peter ten Dijke[2]

[1]*Institute of Musculoskeletal Medicine, University Hospital Münster, Münster, Germany*
[2]*Department of Molecular Cell Biology and Oncode Institute, Leiden University Medical Centre, Leiden, The Netherlands*

INTRODUCTION

Mesenchymal stem cells (MSCs) are pluripotent cells located in the bone marrow, muscles, and fat that potentially can differentiate into all mesenchymal tissues. Differentiation towards these cell lineages is controlled by a multitude of cytokines, which regulate the expression of cell-lineage-specific sets of transcription factors. Osteoblasts and chondrocytes are thought to differentiate from a common mesenchymal precursor, the osteochondrogenic precursor. The osteoblastic differentiation process can be divided into several stages, including proliferation, extracellular matrix deposition, matrix maturation, and mineralization (Fig. 7.1).

RUNX2 AND OSTERIX TRANSCRIPTION FACTORS

An essential event in osteoblast differentiation, and a point of convergence of many signal transduction pathways involved, is the activation of the transcription factor Runx2 (also known as Cbfa1 or AML3). Runx2 is the master switch for osteoblast differentiation and interacts with many transcriptional activators and repressors thereby controlling the expression of a variety of osteoblast-specific genes including type I Collagen (Col1), alkaline phosphatase (ALP), osteopontin (OPN), osteonectin (ON), and osteocalcin (OC). Notably, *Runx2*-deficient mice completely lack osteoblasts and produce a cartilaginous skeleton that is completely devoid of mineralized matrix [1]. In humans, insertions, deletions, and mutations leading to translational stop codons in the DNA-binding domain or in the C-terminal transactivating region of *Runx2* underlie the rare skeletal disorder cleidocranial dysplasia (CCD). CCD is characterized by defective development of the cranial bones and the complete or partial absence of the collar bones, emphasizing the importance of Runx2 in bone formation [2].

Importantly, Runx2 also regulates the expression of the transcription factor Osterix, encoded by the *Osx (Sp7)* gene [3]. Similar to mice deficient in *Runx2*, *Osx*-/- mice lack osteoblasts. Osterix can interact with nuclear factor for activated T cells-2 (NFAT2), which cooperates with Osterix in controlling transcription of target genes such as Col1, thereby promoting osteoblast function [4]. Since nuclear localization of NFAT transcription factors is regulated by the Ca^{2+}-calcineurin pathway, signaling cascades which modulate intracellular Ca^{2+} levels can potentially control Osterix-mediated osteoblast differentiation via NFAT activation.

Other transcription factors that are involved in osteoblast differentiation are homeobox proteins such as Msx2, Dlx-3, Dlx-5, Dlx-6, and members of the AP-1 family such as Fos, Fra, and ATF4 (Fig. 7.1). However, deficiency of these genes does not result in complete loss of osteoblasts like in *Runx2*-/- and *Osx*-/- mice, pointing at a facilitatory role in osteoblastogenesis.

BMP SIGNALING

BMPs belong to the TGF-β superfamily and were originally identified as the active components in bone extracts capable of inducing bone formation at ectopic sites. BMPs are

---

Primer on the Metabolic Bone Diseases and Disorders of Mineral Metabolism, Ninth Edition. Edited by John P. Bilezikian.
© 2019 American Society for Bone and Mineral Research. Published 2019 by John Wiley & Sons, Inc.
Companion website: www.wiley.com/go/asbmrprimer

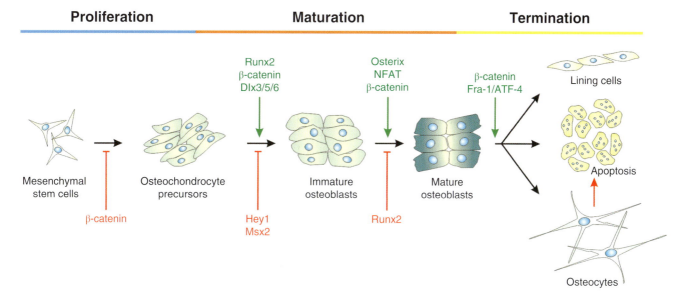

Fig. 7.1. Schematic model of MSC differentiation toward the osteoblastic lineage and the impact of transcriptional regulators in this process.

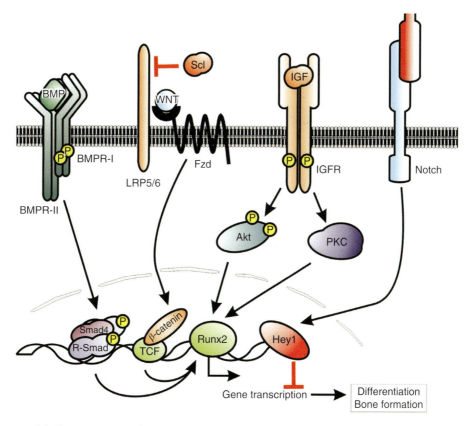

Fig. 7.2. Schematic model illustrating signaling pathways controlling Runx2-mediated osteoblast differentiation. Source: [5,15].

expressed in skeletal tissue and are required for skeletal development and maintenance of adult bone homeostasis, and play an important role in fracture healing [5].

BMPs bind as dimers to type-I and type-II serine/threonine receptor kinases, forming an oligomeric complex (Fig. 7.2). Upon oligomerization, the constitutively active type-II receptors phosphorylate and consequently activate the type-I receptors which subsequently phosphorylate the intracellular signaling mediators BMP receptor-regulated Smads, Smad1, -5, and -8, at their extreme C-termini.

The receptor-regulated Smads then associate with the Co-Smad, Smad4, and translocate into the nucleus, where they, together with other transcription factors, control the expression of target genes (Fig. 7.2) [5]. For example, Runx2 interacts with Smad1 and -5, and cooperates in controlling BMP-induced osteoblast-specific gene expression and osteogenic differentiation [6]. Interestingly, a CCD-mutation resulting in expression of a truncated Runx2 mutant displays impaired Smad1 interaction and inhibited BMP-induced ALP activity [7]. Furthermore, BMP-2 was found to induce Runx2-mediated expression of Osterix [3]. Besides Smad1, -5, and -8 phosphorylation, BMPs also induce the activation of noncanonical cascades involving the p38, JNK, and ERK MAPK pathways, which also modulate the expression of osteogenic genes [5].

Conditional knockout mice deficient in specific BMP members in bone display skeletal defects and several naturally occurring mutations in BMPs or their receptors underlie inherited skeletal disorders [8]. For example, fibrodysplasia ossificans progressiva, in which bone is progressively formed at ectopic sites, has been linked to a heterozygous activating mutation in the BMP type-I receptor ALK2 [9]. Interestingly, this mutation enables the receptor to induce Smad1, -5, and -8 signaling and enhanced heterotopic ossification in response to activins, a different subset of the TGF-β superfamily that normally inhibit bone formation [9].

TGF-β SIGNALING

TGF-β is one of the most abundant cytokines in the bone matrix and plays a major role in the development and maintenance of the skeleton, affecting both cartilage and bone metabolism [10]. Interestingly, TGF-β can have both positive and negative effects on bone formation depending on the context and concentration [10,11].

TGF-β signals via a similar mechanism as the related BMPs. However, upon binding to its specific type-I and type-II receptors, TGF-β induces activation of Smad2 and -3 [10]. Mice deficient in Smad3 display low bone mass disorders, partially caused by increased osteoblast apoptosis [12]. As is the case for Smad1 and -5, Runx2 also interacts with Smad3 and cooperates in regulating TGF-β-induced transcription [11]. This interaction requires a functional Runx2 C-terminal domain, because the truncated Runx2 mutant derived of a CCD patient is unable to interact with Smad3 [7,11]. The effects of TGF-β/Smad3 signaling on the function of Runx2 depend on the cell type and promoter context [11].

Interestingly, only TGF-β2 deficient mice, but not those lacking TGF-β1 or TGF-β3, exhibit severe skeletal abnormalities [13]. In humans, the Loeys–Dietz syndrome is caused by mutations in either TGF-β1, TGF-β2, the TGF-β type-I or type-II receptors, or Smad3, and leads to, besides cardiovascular malformations, skeletal overgrowth [14].

WNT SIGNALING

WNTs are secreted glycoproteins that transduce their signals via 7-transmembrane spanning receptors of the frizzled family and co-receptors low-density lipoprotein receptor-related protein (LRP)-5 and -6 to β-catenin (Fig. 7.2) [15]. In the absence of WNT ligand, β-catenin forms a complex with Adenomatous Polyposis Coli (APC), Axin, Glycogen Synthase Kinase 3 (GSK3), and casein kinase I (CK1). This complex facilitates the phosphorylation and proteosomal degradation of β-catenin. In the presence of WNT ligand, this complex dissociates leading to the accumulation of cytoplasmic β-catenin and its translocation into the nucleus, where it initiates the transcription of target genes via complex formation with TCF/Lef1 transcription factors [15](Fig. 7.2). Conditional deletion of β-catenin led osteochondroprogenitor cells to differentiate into chondrocytes instead of osteoblasts during both intramembranous and endochondral ossification, whereas ectopic WNT signaling enhanced osteoblast differentiation [16–18] (Fig. 7.1). Moreover, WNT signaling in osteocytes plays an essential role in controlling normal bone homeostasis, whereas mice deficient in β-catenin expression in osteocytes display progressive bone loss [19]. Furthermore, loss- or gain-of-function mutations in the LRP-5 gene have been associated with low or high bone mass diseases caused by decreased or increased osteoblast activity, respectively [20]. In addition, mutations in *SOST*, the gene encoding the osteocyte-derived WNT antagonist sclerostin, underlie the rare high bone mass disorders sclerosteosis and Van Buchem disease [21]. All these findings show the importance of WNT signaling in controlling bone formation.

HEDGEHOG SIGNALING

Hedgehogs (Hhs), of which there are three in mammals (ie, sonic, Indian, and desert hedgehog), are critically important for bone development and homeostasis. Cellular responses to the Hh signal are controlled by two transmembrane proteins, the 12-transmembrane-spanning protein Patched-1 (Ptch) and the 7-transmembrane-spanning receptor Smoothened (Smo). The latter has homology to G protein-coupled receptors and transduces the Hh signal. In the absence of Hh, Ptch maintains Smo in an inactive state. With the binding of Hh, Ptch inhibition of Smo is released and intracellular signaling is initiated [22]. The transcriptional response to Hh signaling is mediated by three closely related zinc finger transcription factors termed Gli proteins, Gli1, Gli2, and Gli3, each with different roles and distinct set of target genes. Gli2 functions mainly as a transcriptional activator. In the absence of Hh, Gli3 is processed into a repressor of transcription. In the presence of Hh, however, full length Gli3 translocates into the nucleus which has transcriptional activation properties. Gli1 acts only as transcriptional activator and is induced by Hh signals [22]. Ihh regulates osteoblast differentiation via a Gli2-mediated increase in expression and function of Runx2 [23]. In addition, Hh proteins also activate a so-called noncanonical pathway via Rho-GTPases. It was found that inhibition of the canonical Hh-Gli signaling and a subsequent upregulation of the noncanonical Hh-RhoA signaling pathway in osteoprogenitor cells results in growth retardation and osteopenia by impairing osteoblast differentiation [24].

A number of human diseases affecting bone development have been linked to mutations in Hh family members. For example, heterotopic ossification in progressive osseous heteroplasia (POH) is caused by enhanced Hh signaling, caused by mutations in GNAS that lose their ability to counteract Hh activity [25]. In the endochondral skeleton, Indian hedgehog (Ihh) was found to be indispensable for osteoblast development, because mice deficient in Ihh completely lack osteoblasts in bones formed by endochondral ossification [26].

NOTCH SIGNALING

Notch proteins are transmembrane receptors that control cell-fate decisions and regulate osteoblastic differentiation. Binding of the transmembrane Notch ligands Delta, Serrate, and Lag2 to Notch receptors induces cleavage of the Notch extracellular domain near the transmembrane region [27]. The resulting membrane-associated Notch is then cleaved by Presenilin generating the Notch intracellular domain (NICD), which then translocates into the nucleus. Here the NCID forms a complex with members of the CSL family of DNA binding proteins, which recruits coactivators to drive transcription of target genes [27]. Notably, various mutations affecting the Notch pathway lead to skeletal developmental disorders [28].

Opposite effects on osteoblast differentiation have been described for Notch target genes, including Hes1 and Hey1, which interact with Runx2 [29,30]. Consequently, the role of Notch signalling in animal models is still under debate and seems to be dependent on the differentiation stage of the osteoprogenitor cells [31,32](Figs 7.1 and 7.2). Deletion of the Notch transcriptional effector RBPjk in skeletal progenitors highlighted the relevance of Notch signalling in bone marrow MSCs in fracture healing. Inhibition of Notch signalling in mature osteoblasts did not affect fracture healing [33]. However, inducible overactivation of Notch signalling in mature osteoblasts was shown to have an anabolic effect on bone, thereby promoting fracture healing and preventing osteoporosis in mouse models [34].

OTHER SIGNALING PATHWAYS

Furthermore, other signaling cascades can modulate osteoblast activity. One of these is induced by PTH and its related peptide PTHrP. PTH(rP) signals via the 7-transmembrane G protein coupled receptor PTHR1 and upon ligand binding, several intracellular signaling pathways can be activated, including the cAMP/protein kinase A (PKA) and PKC pathways. Interestingly, whereas intermittent PTH administration induces bone formation, continuous treatment of PTH leads to bone loss. In humans, loss of function in PTH1R has been linked to Blomstrand lethal osteochondrodysplasia [35], characterized by advanced maturation and premature ossification of the skeleton. Different mechanisms have been suggested to explain the anabolic and catabolic effects of PTH; PTH may have diverse effects on the proliferation, commitment, differentiation, or apoptosis of the osteoblasts.

Finally, also various growth factors, including IGF-1 and FGFs, can affect osteoblast function by activating their specific receptor tyrosine kinases (RTKs). Activation of most RTKs results in activation of the phosphatidylinositol 3-kinase (PI3K)-Akt and Ras-ERK MAP kinase pathways (Fig. 7.2). Interestingly, Akt1/Akt2 double-knockout mice show a phenotype resembling that of IGF-1 receptor-deficient mice, which includes impaired bone development [36]. Many human craniosynostosis disorders have been linked to activating mutations in FGF receptors [37]. FGFs affect both chondrogenesis and osteogenesis, and induce proliferation of immature osteoblasts via the Ras/ERK MAPK pathway and via PKC stimulate Runx2 activity [37]. Disruption of FGFR2 signaling in skeletal tissues results in skeletal dwarfism and decreased bone density [38].

CONCLUSION

Because many of the signaling pathways mentioned here are activated subsequently or simultaneously, the ultimate effects they have on the osteoblast differentiation process is highly dependent on which signaling molecules are activated or inhibited, the magnitude and the duration of the responses, and the differentiation stage of the responding cells. Besides that regulation of Runx2 and Osterix activity are points of convergence of many signal transduction cascades, there is also a high degree of cross-talk between various pathways, adding additional degrees of complexity and providing further fine-tuning of the differentiation process. For example, TGF-β can inhibit BMP-induced osteoblast differentiation; however, under specific conditions TGF-β can also promote BMP-induced osteoblast differentiation [39]. Apart from its C-terminal phosphorylation by BMP type-I receptors, Smad proteins can be phosphorylated by MAP kinases and GSK-3 activated by RTKs and WNTs, resulting in cytoplasmic retention and proteosomal degradation and inhibition of signaling [5]. β-catenin-TCF/Lef1 can interact with Smad1 and -3 proteins to cooperate in inducing gene transcription [40], and Hh signaling is required for accurate β-catenin-mediated Wnt signaling in osteoblasts [18]. Thus, the combined action of the signal transduction pathways induced by bone promoting cytokines determines commitment of MSCs towards the osteoblast lineage and the efficiency of bone formation.

ACKNOWLEDGMENTS

We apologize to all authors whose primary work could not be cited owing to space constraints. DJJdG is supported by the "Innovative Medizinische Forschung" of the Medical Faculty of Münster University. Bone research at the lab of PTD is supported by the LeDucq Foundation and Cancer Genomics Centre Netherlands. GSD is supported by Netherlands CardioVascular Research Initiative: the Dutch Heart Foundation, Dutch Federation of University Medical Centers, the Netherlands Organization for Health Research and Development, and the Royal Netherlands Academy of Sciences (CVON-RECONNECT).

REFERENCES

1. Otto F, Thornell AP, Crompton T, et al. Cbfa1, a candidate gene for cleidocranial dysplasia syndrome, is essential for osteoblast differentiation and bone development. Cell. 1997;89(5):765–71.
2. Mundlos S. Cleidocranial dysplasia: clinical and molecular genetics. J Med Genet. 1999;36(3):177–82.
3. Nakashima K, Zhou X, Kunkel G, et al. The novel zinc finger-containing transcription factor osterix is required for osteoblast differentiation and bone formation. Cell. 2002;108(1):17–29.
4. Koga T, Matsui Y, Asagiri M, et al. NFAT and Osterix cooperatively regulate bone formation. Nat Med. 2005;11(8):880–5.
5. Sanchez-Duffhues G, Hiepen C, Knaus P, et al. Bone morphogenetic protein signaling in bone homeostasis. Bone. 2015;80:43–59.
6. Lee KS, Kim HJ, Li QL, et al. Runx2 is a common target of transforming growth factor beta1 and bone morphogenetic protein 2, and cooperation between Runx2 and Smad5 induces osteoblast-specific gene expression in the pluripotent mesenchymal precursor cell line C2C12. Mol Cell Biol. 2000;20(23):8783–92.
7. Zhang YW, Yasui N, Ito K, et al. A RUNX2/PEBP2alpha A/CBFA1 mutation displaying impaired transactivation and Smad interaction in cleidocranial dysplasia. Proc Natl Acad Sci USA. 2000;97(19):10549–54.
8. Salazar VS, Gamer LW, Rosen V. BMP signalling in skeletal development, disease and repair. Nat Rev Endocrinol. 2016;12(4):203–21.
9. de Gorter DJJ, Sanchez-Duffhues G, ten Dijke P. Promiscuous signaling of ligands via mutant ALK2 in fibrodysplasia ossificans progressiva. Receptor Clin Invest. 2016;3:e1356.
10. Janssens K, ten Dijke P, Janssens S, et al. Transforming growth factor-beta1 to the bone. Endocr Rev. 2005;26(6):743–74.
11. Alliston T, Choy L, Ducy P, et al. TGF-beta-induced repression of CBFA1 by Smad3 decreases cbfa1 and osteocalcin expression and inhibits osteoblast differentiation. EMBO J. 2001;20(9):2254–72.
12. Borton AJ, Frederick JP, Datto MB, et al. The loss of Smad3 results in a lower rate of bone formation and osteopenia through dysregulation of osteoblast differentiation and apoptosis. J Bone Miner Res. 2001;16(10):1754–64.
13. Wu M, Chen G, Li YP. TGF-beta and BMP signaling in osteoblast, skeletal development, and bone formation, homeostasis and disease. Bone Res. 2016;4:16009.
14. Loeys BL, Chen J, Neptune ER, et al. A syndrome of altered cardiovascular, craniofacial, neurocognitive and skeletal development caused by mutations in TGFBR1 or TGFBR2. Nat Genet. 2005;37(3):275–81.
15. Clevers H, Nusse R. Wnt/beta-catenin signaling and disease. Cell. 2012;149(6):1192–205.
16. Day TF, Guo X, Garrett-Beal L, et al. Wnt/beta-catenin signaling in mesenchymal progenitors controls osteoblast and chondrocyte differentiation during vertebrate skeletogenesis. Dev Cell. 2005;8(5):739–50.
17. Hill TP, Spater D, Taketo MM, et al. Canonical Wnt/beta-catenin signaling prevents osteoblasts from differentiating into chondrocytes. Dev Cell. 2005;8(5):727–38.
18. Hu H, Hilton MJ, Tu X, et al. Sequential roles of Hedgehog and Wnt signaling in osteoblast development. Development. 2005;132(1):49–60.
19. Kramer I, Halleux C, Keller H, et al. Osteocyte Wnt/beta-catenin signaling is required for normal bone homeostasis. Mol Cell Biol. 2010;30(12):3071–85.
20. Boyden LM, Mao J, Belsky J, et al. High bone density due to a mutation in LDL-receptor-related protein 5. N Engl J Med. 2002;346(20):1513–21.
21. ten Dijke P, Krause C, de Gorter DJ, et al. Osteocyte-derived sclerostin inhibits bone formation: its role in bone morphogenetic protein and Wnt signaling. J Bone Joint Surg Am. 2008;90(Suppl 1):31–5.
22. Hooper JE, Scott MP. Communicating with Hedgehogs. Nat Rev Mol Cell Biol. 2005;6(4):306–17.
23. Shimoyama A, Wada M, Ikeda F, et al. Ihh/Gli2 signaling promotes osteoblast differentiation by regulating Runx2 expression and function. Mol Biol Cell. 2007;18(7):2411–8.
24. Yuan X, Cao J, He X, Serra R, et al. Ciliary IFT80 balances canonical versus non-canonical hedgehog signalling for osteoblast differentiation. Nat Commun. 2016;7:11024.
25. Regard JB, Malhotra D, Gvozdenovic-Jeremic J, et al. Activation of Hedgehog signaling by loss of GNAS causes heterotopic ossification. Nat Med. 2013;19(11):1505–12.
26. Long F, Chung UI, Ohba S, et al. Ihh signaling is directly required for the osteoblast lineage in the endochondral skeleton. Development. 2004;131(6):1309–18.
27. Ehebauer M, Hayward P, Martinez-Arias A. Notch signaling pathway. SciSTKE. 2006;2006(364):cm7.
28. Chen S, Lee BH, Bae Y. Notch signaling in skeletal stem cells. Calcif Tissue Int. 2014;94(1):68–77.
29. Hilton MJ, Tu X, Wu X, et al. Notch signaling maintains bone marrow mesenchymal progenitors by suppressing osteoblast differentiation. Nat Med. 2008;14(3):306–14.
30. McLarren KW, Lo R, Grbavec D, et al. The mammalian basic helix loop helix protein HES-1 binds to and modu-

lates the transactivating function of the runt-related factor Cbfa1. J Biol Chem. 2000;275(1):530–8.
31. Zanotti S, Smerdel-Ramoya A, Stadmeyer L, et al. Notch inhibits osteoblast differentiation and causes osteopenia. Endocrinology. 2008;149(8):3890–9.
32. Engin F, Yao Z, Yang T, et al. Dimorphic effects of Notch signaling in bone homeostasis. Nat Med. 2008;14(3):299–305.
33. Wang C, Inzana JA, Mirando AJ, et al. NOTCH signaling in skeletal progenitors is critical for fracture repair. J Clin Invest. 2016;126(4):1471–81.
34. Liu P, Ping Y, Ma M, et al. Anabolic actions of Notch on mature bone. Proc Natl Acad Sci U S A. 2016;113(15):E2152–61.
35. Zhang P, Jobert AS, Couvineau A, et al. A homozygous inactivating mutation in the parathyroid hormone/parathyroid hormone-related peptide receptor causing Blomstrand chondrodysplasia. J Clin Endocrinol Metab. 1998;83(9):3365–8.
36. Peng XD, Xu PZ, Chen ML, et al. Dwarfism, impaired skin development, skeletal muscle atrophy, delayed bone development, and impeded adipogenesis in mice lacking Akt1 and Akt2. Genes Dev. 2003;17(11):1352–65.
37. Ornitz DM. FGF signaling in the developing endochondral skeleton. Cytokine Growth Factor Rev. 2005;16(2):205–13.
38. Yu K, Xu J, Liu Z, et al. Conditional inactivation of FGF receptor 2 reveals an essential role for FGF signaling in the regulation of osteoblast function and bone growth. Development. 2003;130(13):3063–74.
39. de Gorter DJJ, van Dinther M, Korchynskyi O, et al. Biphasic effects of transforming growth factor beta on bone morphogenetic protein-induced osteoblast differentiation. J Bone Miner Res. 2011;26(6):1178–87.
40. Labbe E, Letamendia A, Attisano L. Association of Smads with lymphoid enhancer binding factor 1/T cell-specific factor mediates cooperative signaling by the transforming growth factor-beta and wnt pathways. Proc Natl Acad Sci USA. 2000;97(15):8358–63.

8
The TGF-β Superfamily in Bone Formation and Maintenance

Ce Shi[1,2] and Yuji Mishina[2]

[1]*Department of Oral Pathology, School and Hospital of Stomatology, Jilin University, Changchun, Jilin, China*
[2]*Department of Biologic and Materials Sciences, School of Dentistry, University of Michigan, Ann Arbor, MI, USA*

INTRODUCTION

BMPs were originally identified for their ability to induce ectopic bone formation [1]. Subsequent studies have demonstrated that the TGF-β superfamily, which includes BMPs, possess pleiotropic functions. Phenotypic analyses of transgenic and gene knockout models reveal diverse functions of TGF-β superfamily members and their downstream signaling effectors. These studies also emphasize the highly context-dependent contribution of TGF-β superfamily members in the skeletal system (Fig. 8.1). Detailed reviews of the mechanisms of signaling and the roles of TGF-β and BMP pathways in skeletal tissues are available in recent comprehensive reviews [2]. This chapter summarizes some of the skeletal phenotypes found in genetically modified animals and discusses how they relate to human conditions.

BASICS OF TGF-β AND BMP SIGNALING

The TGF-β superfamily is the largest class of cytokines in vertebrates. A detailed review of the mechanics of signal transduction is available [2]. In brief, TGF-β superfamily ligands signal as dimers by binding to specific serine/threonine kinase receptors. Ligand binding triggers formation of heterotetrameric complexes composed of type I and type II receptors. There are two major arms: a TGF-β/activin branch and a BMP branch. The majority of ligands in the superfamily fall into one of these branches. TGF-β/activin ligands bind to specific sets of type I/type II receptors. Ligand binding triggers phosphorylation of the receptors, leading to creation of a docking site that enables phosphorylation of the intracellular proteins Smad2 and Smad3. Phosphorylated Smad2 and Smad3 then enter the nucleus, where they act as transcriptional regulators in conjunction with Smad4. Ligands in the BMP branch act through a distinct but structurally related set of receptors to activate the intracellular mediators Smad1, -5, and -9, which act in conjunction with Smad4. These proteins regulate expression of a distinct set of genes. In addition to these canonical pathways, both TGF-β/activin and BMP ligands trigger a variety of noncanonical pathways. The relative contributions of Smad-mediated and noncanonical pathways to the effects of TGF-β/activin and BMP signaling in skeletal tissues is poorly understood.

TGF-β AND BONE DEVELOPMENT

TG/KO phenotypes of ligands

There are three subtypes of TGF-β in mammals (TGF-β1, -2, and -3). These ligands bind to receptor complexes (discussed later in this chapter) that activate the intracellular transducers and transcriptional regulators Smad2 and -3. A second group of ligands, the activins, which bind to receptor complexes distinct from those that bind TGF-β ligands, also activate Smad2 and -3. Very little is known about the role of activins in skeletal development [2]. However, previous studies have reported that TGF-β1, -2, and -3 contribute to skeletogenesis.

Primer on the Metabolic Bone Diseases and Disorders of Mineral Metabolism, Ninth Edition. Edited by John P. Bilezikian.
© 2019 American Society for Bone and Mineral Research. Published 2019 by John Wiley & Sons, Inc.
Companion website: www.wiley.com/go/asbmrprimer

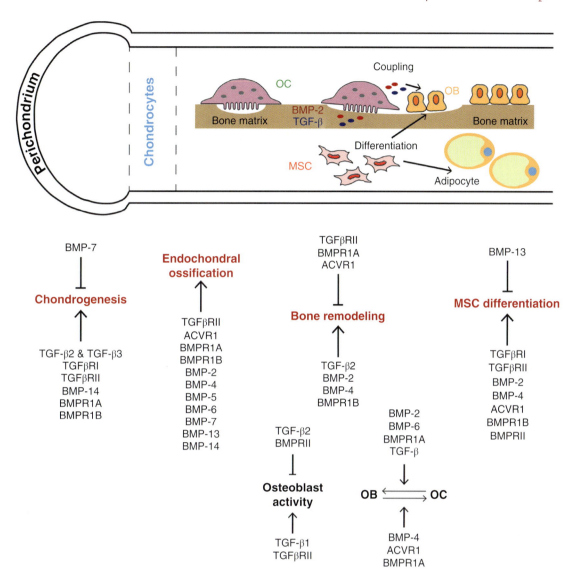

Fig. 8.1. TGF-β/BMP ligands and receptors play multiple roles within different processes during bone development including mesenchymal stem cell (MSC) differentiation, chondrogenesis, endochondral ossification, and bone remodeling. In early chondrogenesis, MSCs condense to form cartilage primordia. In the bones that undergo endochondral ossification, TGF-β/BMP members regulate this process both directly and indirectly in part through FGF, Indian hedgehog (IHH), and PTHrP signaling. Bone remodeling is the consequence of a balance between osteoblastic bone formation activity and osteoclastic bone resorption activity. During bone remodeling, cross-talk between osteoblast (OB) and osteoclast (OC) needs to be well orchestrated. MSCs are the major sources of osteoblasts in the bone marrow. Additionally, MSCs can differentiate into adipocytes. Arrows (↑) indicate positive regulation, while lines with a flat end (⊥) indicate negative regulation.

TGF-β1

Tgfb1$^{-/-}$ mice exhibit loss of mature osteoblasts in association with reduced alkaline phosphatase (ALP) activity, despite normal osteoclast function and number [3]. Fourier transform-infrared imaging (FTIR) analyses reveal reduction of collagen maturity, but no alteration in mineral content or crystallinity in the growth plate or metaphyses [4]. In contrast, all measurements are reduced in the secondary ossification center and cortical bone compartments in *Tgfb1*$^{-/-}$ mice [4].

TGF-β2

Tgfb2$^{-/-}$ mice exhibit reduced bone size in long bones and many craniofacial bones ([2] and citations therein). The mutant mice die soon after birth with a wide range of developmental defects including cleft palate. Overexpression of *Tgfb2* in osteoblasts driven by the osteocalcin (*Ocn*) promoter results in defective mineralization in long bones [5]. The transgenic mice show a higher bone formation rate, suggesting that an imbalance between osteoblastic and osteoclastic activity leads to bone loss that mimics high-turnover osteoporosis [5].

TGF-β3

Tgfb3⁻/⁻ mice show no overt abnormalities in the skeleton. However, *Tgfb2⁻/⁻;Tgfb3⁻/⁻* and *Tgfb2⁻/⁻;Tgfb3⁺/⁻* mutant embryos show a more severe reduction in rib cages than is seen in *Tgfb2⁻/⁻* mice ([2] and citations therein).

TG/KO phenotypes of receptors and signaling molecules

TGF-β1, -2, and -3 bind to the TGF-β receptor type II (TGFβRII) to form a heterotetramer along with ALK-5 (TGFβR1). TGF-β ligand–receptor complexes phosphorylate Smad2 and Smad3, leading to their nuclear transport allowing for regulation of target gene expression.

TGFβR1/ALK-5

Mesenchyme-specific knockout of *Tgfbr1* using *Dermo1*-Cre results in short and wide long bones, associated with reduced bone collars and trabecular bone compartments [6]. Histological assessment reveals that TGFβR1/ALK-5 signaling is critical for proliferation and differentiation of the perichondrium [6].

TGFβRII

Early limb-mesenchyme-specific knockout of *Tgfbr2* using *Prx*-Cre results in short limbs, fusions of phalangeal joints, and thinner skull vaults due to less proliferation and differentiation of osteoblasts [7]. An increase of chondrogenesis in the mutant embryos suggests that TGFβRII limits chondrogenesis in the interzone, and that activity is required for the formation of phalangeal joints [7].

Disease connections

It is reported that in humans some polymorphisms in the gene encoding TGF-β1 (*TGFB1*) are associated with an osteoporotic pathology. One single base deletion in intron 8 of *TGFB1* that likely affects splicing is associated with an osteoporosis phenotype, and an association of the TT genotype (T(816-20)-C mutation) in *TGFB1* with higher bone mass is also reported [8]. It has been reported that mutations in the pro-region of TGF-β1 are causes of Camurati-Engelmann disease, an autosomal dominant bone dysplasia [9]. Mutations in genes encoding multiple components of the TGFβ pathway, including *TGFB1*, *TGFB2*, *TGFBR1*, and *TGFBR2* have been shown to lead to Marfan syndrome and related connective tissue disorders, such as Loeys-Dietz syndrome. These conditions are caused by elevated TGFβ signaling, and skeletal manifestations include long bone overgrowth and joint laxity [10].

TGF AND BONE MAINTENANCE

TGF and coupling

Bone remodeling is a process comprised of bone formation mediated by osteoblasts and bone resorption mediated by osteoclasts. The cells responsible for remodeling are located within the BMU. To maintain bone mass during adulthood, the bone formed in each BMU must be replaced precisely with the amount removed by resorption. This stimulation of osteoblast activity in response to resorption is termed "coupling." There are four main classes of osteoclast-derived coupling factors: matrix-derived signals released during bone resorption, factors synthesized and secreted by mature osteoclasts, factors expressed on the osteoclast cell membrane, and topographic changes effected by the osteoclasts on the bone surface.

The mature domain of TGF-β ligands (25 kDa) is secreted into the extracellular matrix as a latent form that is noncovalently associated with a 75 to 80 kDa portion of the precursor (pro-region); these domains together bind to latent TGF-β binding proteins (LTBPs). Therefore, large amounts of latent TGF-β are stored in bone matrix, and active TGF-β is released during bone resorption. The active form of TGF-β1 released from bone matrix prompts migration of bone marrow stromal cells to the site of resorption. These facts suggest that TGF-β1 is one of the crucial factors to couple bone resorption and formation to maintain bone mass. An analysis of skeletal phenotypes supports a role for TGF-βs in osteoclast activity [11].

BMPS AND BONE DEVELOPMENT

TG/KO phenotypes of ligands

The BMPs comprise a group of more than 15 ligands, and is the largest TGF-β subfamily. Based on sequence similarities and their diversity of functions, this subfamily can be further subdivided into BMP-2/-4, BMP-5/-6/-7/-8, BMP-9/-10, and BMP-12/-13/-14 (growth and differentiation factor, GDF-7/-6/-5, respectively) subgroups. The roles of BMP pathway components in bone formation have been summarized in a recent review [12]. The key findings are summarized here.

BMP-2

An osteoblast-specific disruption of *Bmp2* using 3.6 kilobase (kb) *Col1a1*-Cre leads to thinner bones in mice with increased brittleness. Loss of *Bmp2* in limb mesenchyme leads to a greatly increased rate of fracture owing to defects in the periosteum and ability to mount a reparative response ([12] and citations therein). Chondrocyte-specific disruption of *Bmp2* or both *Bmp2* and *Bmp4* leads to severe defects in chondrocyte proliferation and maturation during endochondral bone formation, while chondrocyte-specific disruption of only *Bmp4* causes minor changes in chondrocyte maturation [13]. This indicates that BMP-2 has a crucial and nonredundant role in chondrocyte proliferation and maturation during endochondral bone development, as well as in fracture repair.

BMP-4

Limb-mesenchyme-specific knockouts of *Bmp2* and *Bmp4* demonstrate that these ligands function redundantly to control osteogenesis in the limb [14]. Osteoblastic differentiation and osteogenesis are defective in *Bmp2* and *Bmpr4* conditional (*Prx1*-Cre) knockout mice, but these defects can be rescued by one functional allele of either *Bmp2* or *Bmp4* ([12] and citations therein). Consistent with an anabolic role for BMPs in promotion of bone remodeling, overexpression of *Bmp4* in osteoblasts results in an increase of osteoclastogenesis and a reduction in bone mass [15].

BMP-5/-6/-7

Mutation in *Bmp5* is associated with a wide range of skeletal defects, including reductions in long bone width, the size of several vertebral processes, and an overall lower body mass [16].

BMP-6 is highly expressed in hypertrophic chondrocytes. However, $Bmp6^{-/-}$ mice show only a slight delay in the ossification of the sternum and a reduction in the size of long bones, which is slightly exacerbated in *Bmp5*/*Bmp6* double mutants [17]. Compound knockout mice lacking one allele of *Bmp2* and both alleles of *Bmp6* ($Bmp2^{+/-}$;$Bmp6^{-/-}$) exhibit moderate growth retardation, showing a reduction in trabecular bone volume with suppressed bone formation, while the single deficient mice ($Bmp2^{+/-}$ or $Bmp6^{-/-}$) do not [18]. Thus, there is a considerable functional overlap at the level of BMP ligands, and the combination of BMP-2 and BMP-6 plays a pivotal role in bone formation.

Bmp7 homozygous null mice exhibit an early postnatal lethality mutation that is associated with various developmental defects: holes in the basisphenoid bone and xiphoid cartilage, retarded ossification of bones, fused ribs and vertebrae, underdeveloped neural arches of the lumbar and sacral vertebrae, and polydactyly of the hindlimbs [19]. On the other hand, conditional deletion of *Bmp7* from the limb has only minor consequences ([12] and citations therein). These divergent findings suggest that BMP-7 may have redundant roles with other BMPs in bone formation and may have a more prominent role in axial than in appendicular elements.

BMP-13/GDF-6 and BMP-14/GDF-5/CDMP-1

Mutations in the BMP-13, -14, and -15 subgroups are associated with fusions of joints and defects in cranial suture formation that are not seen in mice lacking other BMP family members. Mutation in *Bmp13* causes defects at multiple sites, including joint fusions in the wrist and ankle, fusions of carpal and tarsal bones, cartilage defects in the middle ear, and the absence of the coronal suture [20]. *Bmp13* knockout mice have accelerated coronal suture fusion, indicating an inhibitory role of BMP-13 in osteogenic differentiation [21]. *Bmp13* knockout mice also exhibit shorter dermal flat bones in the skull and shorter digits [22]. It is also shown that a hindlimb enhancer in the *Bmp13* locus has a striking correlation with known changes in hindlimb digit length and musculature that have evolved during the transition to bipedal locomotion in the human lineage [22]. The joints altered in $Bmp13^{-/-}$ mice are distinct from those altered in *Bmp14* mutants, and *Bmp13*/*Gdf6*;*Bmp14*/*Gdf5* double mutants show additional defects.

BMP-14/GDF-5 has a fundamental role in limb development, where it controls the size of the initial cartilaginous condensations as well as the coordination of bone and joint formation. Mutations of *Bmp14*/*Gdf5* in mice cause brachypodism, reduction of digit number, fusion of some bones in the wrist and ankle, ankylosis of the knee joint and malformation with early onset osteoarthritis of the elbow joint [23]. Transgenic mice expressing *Bmp14*/*Gdf5* under the control of a *Col11a2* promoter show extensive cartilage overgrowth and complete absence of joints [24]. How members of this subgroup promote joint formation on the one hand, but promote chondrogenesis on the other, is not understood.

TG/KO phenotypes of receptors

BMP ligands transduce their signals through complexes composed of type I and type II serine/threonine kinase receptors. BMPs bind to three distinct type I receptors, called activin receptor-like kinase 2 (ALK-2, also known as activin receptor type I, ACVR1), ALK-3 (also known as BMP receptor type IA, BMPR1A), and ALK-6 (also known as BMP receptor type IB, BMPR1B). BMP receptor type II (BMPRII), activin receptor type II (ACVRII), and activin receptor type IIB (ACVRIIB) serve as type II receptors for BMPs.

ACVR1/ALK-2

Deletion of *Acvr1* using a 3.2-kb *Col1*-CreER increases bone mass in association with suppression of *Sost* and *Dkk1* [25]. Mice deficient for *Acvr1* in chondrocytes, achieved using a *Col2*-Cre driver, exhibit a shortened cranial base and hypoplastic cervical vertebrae [26]. Activation experiments showed that BMPR1A, BMPR1B, and ACVR1 are all able to promote chondrogenesis. However, deletion of each individual receptor only results in mild skeletal defects or defects restricted to isolated skeletal elements [26,27]. *Bmpr1a*/*Bmpr1b* double mutant mice, *Acvr1*/*Bmpr1a* double mutant mice, and *Acvr1*/*Bmpr1b* double mutant mice exhibit generalized chondrodysplasia that is much more severe than any of the corresponding mutant strains [26,27]. Unlike compound mutant mice for *Bmpr1a* and *Bmpr1b*, compound mutant mice for *Avcr1* and *Bmpr1b* can develop cartilage primordia and subsequent bones through endochondral ossification [26], suggesting that BMP signaling through ACVR1 plays a relatively minor role compared with other type I receptors during chondrogenesis.

BMPR1A/ALK-3

Deletion of *Bmpr1a* using a 3.2-kb *Col1*-CreER increases bone mass due to decreased osteoblast activity and more dramatically decreased osteoclast activity [28]. The constitutively active form of *Bmpr1a* (*caBmpr1a*) is associated with a partial rescue of the bone phenotype of *Bmpr1a*-deficient mice [28]. Conditional disruption of *Bmpr1a* using *Dmp1*-Cre demonstrates an increased bone mass concomitant with accelerated cell proliferation and *Sost* reduction [29,30]. When *Bmpr1a* is conditionally disrupted in osteoclasts using a *Cathepsin K* (*CtsK*) promoter, bone mass increases [31].

Disruption of *Bmpr1a* in chondrocytes demonstrates impairment of articular cartilage and growth plate cartilage, resulting in decreased bone size and bone mass [27]. The loss of both BMPR1A and BMPR1B blocks chondrocyte condensation, proliferation, differentiation, survival, and function due to impaired *Sox* expression [27]. These mouse models demonstrate that BMP signaling is essential for almost every step during endochondral bone development.

The constitutively active form of *Bmpr1a* (*caBmpr1a*) in neural crest cells results in craniosynostosis in mice [28,32], as well as bone and cartilage defects of the nasomaxillary complex, due to an increased level of cell death in skeletal primordia [32,33].

BMPR1B/ALK-6

Unlike in ACVR1 and BMPR1A mice, mice homozygous null for *Bmpr1b* are viable. In *Bmpr1b*-deficient mice, proliferation of prechondrogenic cells and differentiation of chondrocytes are markedly reduced, leading to reduced lengths in the phalangeal region [34]. In addition, bone mass in the mutant mice is decreased, in association with compromised osteoblastic differentiation of bone marrow mesenchymal progenitors [35].

In *Bmpr1b* and *Bmp7* double mutant mice, severe skeletal defects are observed in the forelimbs and hindlimbs [34]. Since BMP-7 binds efficiently to both BMPR1B and ACVR1, it is conceivable that BMPR1B and ACVR1 play important synergistic or overlapping roles in cartilage and bone formation in vivo. Conditional disruption of *Bmpr1a* driven by *Col2*-Cre and *Bmpr1b* homozygous null double mutant mice exhibit a dramatic decrease in the size of skeletal primordia (i.e. chondrodysplasia) due to a reduction of proliferation and an increase in apoptosis around E12.5 to E16.5 [27]. This suggests a possible functional compensation mechanism between BMPR1A and BMPR1B in chondrocytes during early cartilage development in growth plates [27].

BMPRII

Deletion of *Bmpr2* using *Prx1*-Cre results in normal bone development both at embryonic stages and at birth, probably due to compensation by the other type II receptors, ACVRII and ACVRIIB, suggesting that BMPRII is not required for endochondral ossification in the limb. However, the mutant mice have increased bone mass at 2 months after birth [36]. In this mouse model, BMP signaling is unchanged, whereas activin signaling is impaired, leading to increased osteoblast activity. Activins (which bind to type I activin receptors and lead to phosphorylation of Smad2 and -3) and BMPs (which bind to ACVR1, BMPR1A, and BMPR1B and lead to phosphorylation of Smad1 and -5) can all transduce their signals through receptor complexes that contain ACVRII and ACVRIIB. This study therefore suggests that type II receptor segregation and/or competition could be a generalized mechanism by which BMP and activin signaling interact.

Disease connection

Given the important roles of BMP signaling in chondrogenesis and osteogenesis, mutations in BMP ligands or receptors have been identified as the basis of a wide range of skeletal disorders in humans.

Fibrodysplasia ossificans progressiva (FOP)

FOP is an extremely rare and debilitating genetic disorder characterized by congenital malformations of the halluces (big toes) and by progressive heterotopic endochondral ossification in predictable anatomical patterns. The classic phenotype is caused by a mutation (617G>A; R206H) in *ACVR1* that accounts for at least 98% of classic presentations [37].

Further studies have identified new mutations including c.982G>A (p.G328R), c.1124G>C (p.R375P), c.590_592delCTT, P197_F198delinsL, and c.619C>G (Q207E) [37]. Recent studies have shown that these mutations in *ACVR1* lead to altered responsiveness to activin. Activin ligands typically bind to AVCR1 but do not induce signal transduction. The FOP mutations lead to a structural alteration that enables activin to activate BMP signaling through ACVR1 [38]. Thus, in addition to competition of activins and BMPs for type II receptors, the occupation of ACVR1 by either activin or BMP ligands appears to be a second mechanism by which BMP and activin pathways interact.

Osteoarthritis (OA)

OA is a disease involving degeneration of the articular cartilage in synovial joints, such as the knee, hip, and hand. Associations between polymorphisms in *BMP5* and *BMP2* and OA have been found, suggesting that variability in gene expression is a susceptibility factor for the disease. The strongest evidence for a role in OA susceptibility comes from a SNP in the 5′ untranslated region (5′-UTR) of *BMP14/GDF5*, rs143383, associated with OA [39]. rs143383 is a C/T transition and the OA-associated T allele of rs143383 has been shown to produce less mRNA transcript compared with the ancestral C allele, indicating that reduced *BMP14/GDF5* expression is likely to be the mechanism through which this OA susceptibility locus is working.

Brachydactyly (BD)

BD is a shortening of the hands/feet due to small or missing metacarpals/metatarsals and/or phalanges. Depending on the affected phalanges, five different types of brachydactylies are categorized (BDA to BDE) including seven subgroups (BDA1 to BDA7).

Mutations in *BMP14/GDF5* have been linked to isolated traits of different types of brachydactyly including BDA1, BDA2, and BDC [40]. Dominant-negative mutations in *BMPR1B* and a specific missense mutation in *BMP14/GDF5* are known to cause isolated brachydactyly type BDA2 due to a loss of interaction between BMPR1B and BMP-14/GDF-5 [41]. A mutation in *BMPR1B* (R486Q) is associated with either BDA2 or a BDC/symphalangism (SYM1)-like phenotype [42]. Duplication of a regulatory element that affects the expression of *BMP2* is associated with BDA2 [43].

Symphalangism (SYM)

SYM is an uncommon condition characterized by fusion of the joints of the fingers or toes. Activating mutations in *BMP14/GDF5* result in increased chondrogenic activity as described for proximal SYM and the multiple synostoses syndrome 2 [44]. A mutation in *BMP14* (R438L) leads to a loss of receptor-binding specificity, causing the SYM phenotype [41].

BMPS AND BONE MAINTENANCE

BMPs and coupling

BMP-2 is one of the latent growth factors in the bone matrix that stimulates osteoclast activity ([2] and citations therein). *Bmp6* is highly expressed in osteoclasts, compared with macrophages, and is identified as being an osteoclast-derived coupling factor [45].

Conditional disruption of *Bmpr1a* in osteoclasts using a *CtsK* promoter increases bone mass and bone formation rate [31], suggesting that BMP signaling in osteoclasts negatively regulates osteoblast function through activation of downstream target genes within osteoclasts.

BMPs and bone quality

Conditional disruption of *Bmp2* in osteoblasts using *Osx*-Cre results in an alteration of mechanical properties at whole bone and material levels, and a decrease of cortical mineral/matrix ratio (MMR) in the long bones [46]. In *Bmp5*-deficient adult animals, the femurs are significantly weaker due to their smaller cross-sectional geometry. The cross-sectional geometry is diminished in 4-week-old animals and is accompanied by an apparent increase in material strength, resulting in comparable structural integrity in the weanling animals. However, the femurs are notably strong for the size of the animals, indicated by the level of in vivo mechanical loading [47].

BMP-12 is also thought to play a role in the structural integrity of bone [48]. *Bmp12* deficiency is associated with elevated cortical bone material properties that compensate for decreased geometric properties, thereby preserving bone structural integrity.

Conditional deletion of *Bmpr1a* using a 3.2-kb *Col1*-CreER results in a significant increase in an MMR in the trabecular compartments, but not the cortical compartments of femurs [49]. Collagen cross-link ratios are increased in the trabecular compartment of male mutant mice [49]. In addition, weight-bearing exercise increases trabecular bone volume, cortical collagen fibril diameters, and ductility and toughness in mutant bones [50]. These results indicate that BMP signaling mediated by BMPR1A contributes to the biomechanics quality of bone, together with bone mass.

CONCLUSION

TGF-β superfamily members play complex, critical, and diverse roles in bone formation and maintenance. Studies using transgenic or knockout mouse models have revealed that ligands within specific subgroups often exhibit overlapping functions; other studies outlined in this chapter have shown that TGF-βs/activins and BMPs have distinct functions in chondrogenesis and bone formation/remodeling. These studies are revealing that competition between TGF-βs/activins and BMP pathways at the level of receptor utilization plays a major role in regulating bone formation in vivo. Additional studies will help in understanding the mechanisms underlying how alterations in these pathways lead to related diseases in humans, which will then provide insights into new strategies for the treatment of these diseases.

ACKNOWLEDGMENTS

We apologize to those authors whose work could not be cited directly due to space limitations. Work in the Mishina laboratory is supported by R01DE020843.

REFERENCES

1. Urist MR. Bone: formation by autoinduction. Science. 1965;150:893–9.
2. Wu M, Chen G, Li YP. TGF-beta and BMP signaling in osteoblast, skeletal development, and bone formation, homeostasis and disease. Bone Res. 2016;4:16009.
3. Geiser AG, Zeng QQ, Sato M, et al. Decreased bone mass and bone elasticity in mice lacking the transforming growth factor-beta1 gene. Bone. 1998;23:87–93.
4. Atti E, Gomez S, Wahl SM, et al. Effects of transforming growth factor-beta deficiency on bone development: a Fourier transform-infrared imaging analysis. Bone. 2002;31:675–84.
5. Erlebacher A, Derynck R. Increased expression of TGF-beta 2 in osteoblasts results in an osteoporosis-like phenotype. J Cell Biol. 1996;132:195–210.
6. Matsunobu T, Torigoe K, Ishikawa M, et al. Critical roles of the TGF-beta type I receptor ALK5 in perichondrial formation and function, cartilage integrity, and osteoblast differentiation during growth plate development. Dev Biol. 2009;332:325–38.
7. Seo HS, Serra R. Deletion of Tgfbr2 in Prx1-cre expressing mesenchyme results in defects in development of the long bones and joints. Dev Biol. 2007;310:304–16.
8. Langdahl BL, Carstens M, Stenkjaer L, et al. Polymorphisms in the transforming growth factor beta 1 gene and osteoporosis. Bone. 2003;32:297–310.
9. Kinoshita A, Saito T, Tomita H, et al. Domain-specific mutations in TGFB1 result in Camurati-Engelmann disease. Nat Genet. 2000;26:19–20.
10. Verstraeten A, Alaerts M, Van Laer L, et al. Marfan syndrome and related disorders: 25 years of gene discovery. Hum Mutat. 2016;37:524–31.
11. Tang Y, Wu X, Lei W, et al. TGF-beta1-induced migration of bone mesenchymal stem cells couples bone resorption with formation. Nat Med. 2009;15:757–65.
12. Salazar VS, Gamer LW, Rosen V. BMP signalling in skeletal development, disease and repair. Nat Rev Endocrinol. 2016;12:203–21.
13. Shu B, Zhang M, Xie R, et al. BMP2, but not BMP4, is crucial for chondrocyte proliferation and maturation during endochondral bone development. J Cell Sci. 2011;124:3428–40.
14. Bandyopadhyay A, Tsuji K, Cox K, et al. Genetic analysis of the roles of BMP2, BMP4, and BMP7 in limb patterning and skeletogenesis. PLoS Genet. 2006;2:e216.
15. Okamoto M, Murai J, Yoshikawa H, et al. Bone morphogenetic proteins in bone stimulate osteoclasts and osteoblasts during bone development. J Bone Miner Res. 2006;21:1022–33.
16. Kingsley DM, Bland AE, Grubber JM, et al. The mouse short ear skeletal morphogenesis locus is associated with defects in a bone morphogenetic member of the TGF beta superfamily. Cell. 1992;71:399–410.
17. Solloway MJ, Dudley AT, Bikoff EK, et al. Mice lacking Bmp6 function. Dev Genet. 1998;22:321–39.
18. Kugimiya F, Kawaguchi H, Kamekura S, et al. Involvement of endogenous bone morphogenetic protein (BMP) 2 and BMP6 in bone formation. J Biol Chem. 2005;280:35704–12.
19. Jena N, Martin-Seisdedos C, McCue P, et al. BMP7 null mutation in mice: developmental defects in skeleton, kidney, and eye. Exp Cell Res. 1997;230:28–37.
20. Settle SH Jr, Rountree RB, Sinha A, et al. Multiple joint and skeletal patterning defects caused by single and double mutations in the mouse Gdf6 and Gdf5 genes. Dev Biol. 2003;254:116–30.
21. Clendenning DE, Mortlock DP. The BMP ligand Gdf6 prevents differentiation of coronal suture mesenchyme in early cranial development. PLoS One. 2012;7:e36789.
22. Indjeian VB, Kingman GA, Jones FC, et al. Evolving new skeletal traits by cis-regulatory changes in bone morphogenetic proteins. Cell. 2016;164:45–56.
23. Masuya H, Nishida K, Furuichi T, et al. A novel dominant-negative mutation in Gdf5 generated by ENU mutagenesis impairs joint formation and causes osteoarthritis in mice. Hum Mol Genet. 2007;16:2366–75.
24. Tsumaki N, Nakase T, Miyaji T, et al. Bone morphogenetic protein signals are required for cartilage formation and differently regulate joint development during skeletogenesis. J Bone Miner Res. 2002;17:898–906.
25. Kamiya N, Kaartinen VM, Mishina Y. Loss-of-function of ACVR1 in osteoblasts increases bone mass and activates canonical Wnt signaling through suppression of Wnt inhibitors SOST and DKK1. Biochem Biophys Res Commun. 2011;414:326–30.
26. Rigueur D, Brugger S, Anbarchian T, et al. The type I BMP receptor ACVR1/ALK2 is required for chondrogenesis during development. J Bone Miner Res. 2015;30:733–41.
27. Yoon BS, Ovchinnikov DA, Yoshii I, et al. Bmpr1a and Bmpr1b have overlapping functions and are essential for chondrogenesis in vivo. Proc Natl Acad Sci USA 2005;102:5062–7.
28. Kamiya N, Ye L, Kobayashi T, et al. BMP signaling negatively regulates bone mass through sclerostin by inhibiting the canonical Wnt pathway. Development. 2008;135:3801–11.
29. Lim J, Shi Y, Karner CM, et al. Dual function of Bmpr1a signaling in restricting preosteoblast proliferation and stimulating osteoblast activity in mouse. Development. 2016;143:339–47.
30. Kamiya N, Shuxian L, Yamaguchi R, et al. Targeted disruption of BMP signaling through type IA receptor (BMPR1A) in osteocyte suppresses SOST and RANKL, leading to dramatic increase in bone mass, bone mineral density and mechanical strength. Bone. 2016;91:53–63.
31. Okamoto M, Murai J, Imai Y, et al. Conditional deletion of Bmpr1a in differentiated osteoclasts increases osteoblastic bone formation, increasing volume of remodeling bone in mice. J Bone Miner Res. 2011;26:2511–22.
32. Komatsu Y, Yu PB, Kamiya N, et al. Augmentation of Smad-dependent BMP signaling in neural crest cells causes craniosynostosis in mice. J Bone Miner Res. 2013;28:1422–33.

33. Hayano S, Komatsu Y, Pan H, et al. Augmented BMP signaling in the neural crest inhibits nasal cartilage morphogenesis by inducing p53-mediated apoptosis. Development. 2015;142:1357–67.
34. Yi SE, Daluiski A, Pederson R, et al. The type I BMP receptor BMPRIB is required for chondrogenesis in the mouse limb. Development. 2000;127:621–30.
35. Shi C, Iura A, Terajima M, et al. Deletion of BMP receptor type IB decreased bone mass in association with compromised osteoblastic differentiation of bone marrow mesenchymal progenitors. Sci Rep. 2016;6:24256.
36. Lowery JW, Intini G, Gamer L, et al. Loss of BMPR2 leads to high bone mass due to increased osteoblast activity. J Cell Sci. 2015;128:1308–15.
37. Kaplan FS, Xu M, Seemann P, et al. Classic and atypical fibrodysplasia ossificans progressiva (FOP) phenotypes are caused by mutations in the bone morphogenetic protein (BMP) type I receptor ACVR1. Hum Mutat. 2009;30:379–90.
38. Hatsell SJ, Idone V, Wolken DM, et al. ACVR1R206H receptor mutation causes fibrodysplasia ossificans progressiva by imparting responsiveness to activin A. Sci Transl Med. 2015;7:303ra137.
39. Miyamoto Y, Mabuchi A, Shi D, et al. A functional polymorphism in the 5′ UTR of GDF5 is associated with susceptibility to osteoarthritis. Nat Genet. 2007;39:529–33.
40. Ploger F, Seemann P, Schmidt-von Kegler M, et al. Brachydactyly type A2 associated with a defect in proGDF5 processing. Hum Mol Genet. 2008;17:1222–33.
41. Seemann P, Schwappacher R, Kjaer KW, et al. Activating and deactivating mutations in the receptor interaction site of GDF5 cause symphalangism or brachydactyly type A2. J Clin Invest. 2005;115:2373–81.
42. Lehmann K, Seemann P, Boergermann J, et al. A novel R486Q mutation in BMPR1B resulting in either a brachydactyly type C/symphalangism-like phenotype or brachydactyly type A2. Eur J Hum Genet. 2006;14:1248–54.
43. Dathe K, Kjaer KW, Brehm A, et al. Duplications involving a conserved regulatory element downstream of BMP2 are associated with brachydactyly type A2. Am J Hum Genet. 2009;84:483–92.
44. Dawson K, Seeman P, Sebald E, et al. GDF5 is a second locus for multiple-synostosis syndrome. Am J Hum Genet. 2006;78:708–12.
45. Pederson L, Ruan M, Westendorf JJ, et al. Regulation of bone formation by osteoclasts involves Wnt/BMP signaling and the chemokine sphingosine-1-phosphate. Proc Natl Acad Sci U S A. 2008;105:20764–9.
46. McBride SH, McKenzie JA, Bedrick BS, et al. Long bone structure and strength depend on BMP2 from osteoblasts and osteocytes, but not vascular endothelial cells. PLoS One. 2014;9:e96862.
47. Mikic B, Van der Meulen MC, Kingsley DM, et al. Mechanical and geometric changes in the growing femora of BMP-5 deficient mice. Bone. 1996;18:601–7.
48. Maloul A, Rossmeier K, Mikic B, et al. Geometric and material contributions to whole bone structural behavior in GDF-7-deficient mice. Connect Tissue Res. 2006;47:157–62.
49. Zhang Y, McNerny EG, Terajima M, et al. Loss of BMP signaling through BMPR1A in osteoblasts leads to greater collagen cross-link maturation and material-level mechanical properties in mouse femoral trabecular compartments. Bone. 2016;88:74–84.
50. Iura A, McNerny EG, Zhang Y, et al. Mechanical loading synergistically increases trabecular bone volume and improves mechanical properties in the mouse when BMP signaling is specifically ablated in osteoblasts. PLoS One. 2015;10:e0141345.

9

Recent Developments in Understanding the Role of Wnt Signaling in Skeletal Development and Disease

Zhendong A. Zhong, Nicole J. Ethen, and Bart O. Williams

Center for Cancer and Cell Biology and Program for Skeletal Disease and Tumor Microenvironment, Van Andel Research Institute, Grand Rapids, MI, USA

INTRODUCTION

Wnt/β-catenin signaling plays a number of key roles in regulating embryogenesis, organogenesis, cell fate determination, and differentiation. Our knowledge of Wnt signaling is still expanding, with recent work identifying several mechanisms by which the pathway is fine-tuned at several levels, including the activity of Wnt as a ligand and the regulation of Wnt receptor protein levels. We are becoming more and more aware of the complexity of Wnt regulation of bone homeostasis, and we are trying to manipulate this pathway to cure human skeletal diseases.

WNT/B-CATENIN SIGNALING CORE PATHWAY

Wnt signaling is initiated by a conserved Wnt family of secreted glycolipoproteins, either through a β-catenin-dependent route (known as canonical Wnt signaling) or a β-catenin-independent route (noncanonical Wnt signaling) [1]. This review focuses on the canonical pathway (Fig. 9.1). Under normal conditions, the amount of β-catenin in the cytoplasm is maintained in a steady state. Excess β-catenin is removed by a "destruction complex" containing Axin, the adenomatous polyposis coli (Apc) protein, and a serine/threonine protein kinase, glycogen synthase kinase 3 (GSK3). GSK3 phosphorylates β-catenin and targets it for ubiquitin-dependent degradation. When a Wnt molecule engages a receptor complex that contains a member of the Frizzled family of seven-transmembrane receptors and either LRP5 or LRP6 (low-density lipoprotein receptor-related protein), this ligand–receptor interaction induces phosphorylation of the cytoplasmic tail of Lrp5/6, creating a binding site for Axin. The recruitment of Axin to Lrp5/6 at the plasma membrane inactivates the destruction complex because it fails to recruit β-TrCP (the beta-transducin repeat-containing E3 ubiquitin protein ligase) for ubiquitination, thus blocking the degradation of β-catenin. As β-catenin levels increase in the cytoplasm, it is translocated to the nucleus, where it complexes with members of the lymphoid enhancer-binding factor/T-cell factor (LEF/TCF) family of DNA-binding proteins to activate the transcription of its target genes.

REGULATION OF WNT PRODUCTION AND SECRETION

There are 19 genes in humans and mice that encode for members of the Wnt family of secreted glycoproteins, all of which exhibit different expression patterns during skeletal development [2]. Wnts may be functionally divided into two groups according to their downstream pathways. One group consists of the canonical Wnts, which are dependent on LRP5/6 and can stabilize β-catenin and induce LEF/TCF downstream targets. A recent in vitro survey showed that in at least one context 14 out of 19 human Wnts were

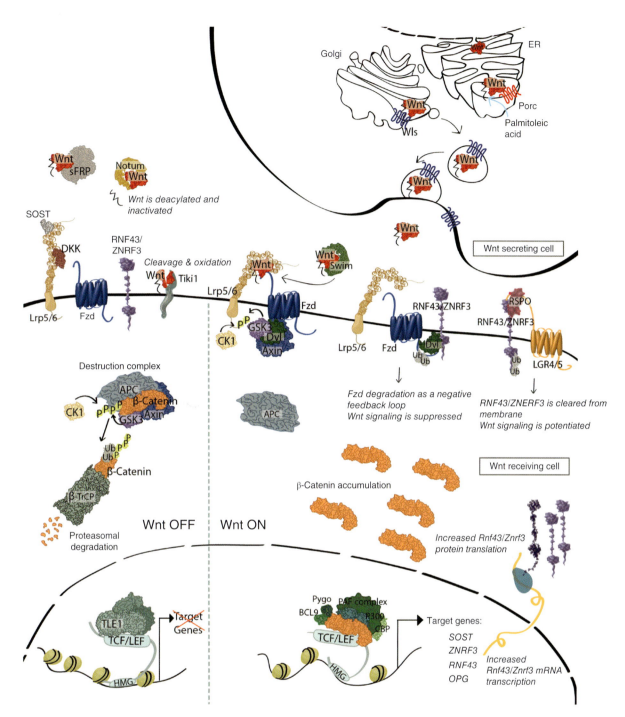

Fig. 9.1. Overview of the Wnt/β-catenin signaling pathway. During Wnt ligand biogenesis in "Wnt-secreting cells," the endoplasmic reticulum (ER) localized Porcupine adds palmitoleic acid to Wnts. This modication is required for recognition by the protein Wntless (Wls) which then facilitates transport of Wnt to the plasma membrane for secretion. In addition, this lipid modification is required for Wnt to interact with Frizzled. In "Wnt-receiving cells," when Wnt signaling is "off" in the absence of Wnt ligand engaging receptors, a multiprotein complex ("destruction complex"), which includes GSK3, Axin, and APC, is formed to facilitate CK1-primed and GSK3-dependent phosphorylation of β-catenin, targeting β-catenin for proteolytic degradation via the E3-ubiquitin ligase β-TrCP. Without β-catenin binding, the transcription factor TCF suppresses target genes by binding its target DNAs with its high mobility group (HMG) domain and complexing with the chromatin repressor TLE1 (or Groucho in *Drosophila*). Wnt signaling can also be suppressed by posttranslational regulation of the ligand (via inhibitors that include sFRP, Notum, and Tiki1) or via the availability or levels of receptors (by SOST, DKK, RNF43/ZNRF3, and others). Wnt signaling is turned "on" when Wnt proteins bind to the receptor complex that includes either Lrp5 or Lrp6 and a member of the Fzd family. Activation of the Wnt receptor complex leads to the activation of Dvl and the phosphorylation of the cytoplasmic domain of Lrp5/6, resulting in the recruitment of Axin to the plasma membrane. This inhibits the degradation of β-catenin, and consequently β-catenin can accumulate in the cytoplasm and enter the nucleus where it binds to members of the LEF/TCF family and activates target gene transcription via the recruitment of factors such as BCL9, Pygo, CBP, P300, and Parafibromin (a component of the PAF complex). Among Wnt target genes, *RNF43* and *ZNRF3* were recently shown to play a critical role in the negative feedback loop by clearing Fzd receptors on the cell membrane through ubiquitination. A potent Wnt agonist, RSPO can form a complex consisting of RNF43/ZNRF3/LGR/RSPO to remove RNF43/ZNRF3 from the cell membrane and to potentiate Wnt signaling by increasing Fzd availability on the cell membrane. Source: [1,25]. Reproduced with permission of Elsevier.

able to induce LRP6 phosphorylation (a proximal marker for Wnt-induced canonical signaling) and β-catenin-dependent signaling [3]. The other group contains the so-called noncanonical Wnts, which do not directly signal through β-catenin and are not the focus of this chapter. However, some Wnts are able to activate either pathway under different circumstances, ie, in different cell types or with various receptors. For example, Wnt5a activates noncanonical Wnt signaling and inhibits canonical Wnt signaling with Ror2 receptor expression [4]. On the other hand, Wnt5a can activate canonical Wnt signaling in the presence of Frizzled4 (Fzd4) and Lrp5 (not Lrp6), which is similar to how Norrin stabilizes β-catenin [5].

Wnts are highly hydrophobic, containing a conserved serine residue that is palmitoylated in the endoplasmic reticulum of a Wnt-producing cell by the acyl transferase, Porcupine [6]. The crystal structure of the *Xenopus* Wnt8 in complex with a cysteine-rich domain (CRD) of mouse Fzd8 confirmed that palmitoylation of this serine residue is required for its secretion and function [7]. In addition, Wnt secretion requires an endoplasmic reticulum (ER) resident protein, Wntless, to escort the palmitoylated Wnt protein to the cell surface and release it from Wnt-producing cells [6]. So far, Porcupine and Wntless have proven to be indispensable for the normal secretion of all mammalian Wnts.

WNT GENE MUTATIONS AND SKELETAL DISEASE

Alterations in a number of Wnt genes have been associated with skeletal diseases. For example, a homozygous loss of function Wnt3 mutation was identified as the cause of tetra-amelia syndrome (limblessness) in humans [8]. Also, Wnt10b loss of function mutations are associated with split-hand/foot malformation, and Wnt16 loss is associated with low cortical bone mass [9]. In mice, global deletions of Wnt genes have varied skeletal phenotypes, with Wnt1 and Wnt3a knockouts showing the most severe and earliest phenotypes [10]. Wnt10b null mice show age-dependent and progressive bone loss, presumably due to a reduction of the mesenchymal progenitors that give rise to osteoblasts that support bone growth [10]. On the other hand, transgenic mice that ectopically express Wnt10b in brown adipose tissue and bone marrow show increased bone mass and strength [10]. Wnt16 overexpression in osteoblasts protected cortical bone by reducing osteoclast activity, thus helping prevent cortical bone fractures [11]. However, osteoblast-specific expression of human Wnt16 in mice increased both cortical and trabecular bone mass and structure [12]. This indicates that either the Wnt16 expression pattern or compensation from other Wnts played a role in the disparity between Wnt16 knockout and ectopic expression mouse models.

Wntless/porcupine in skeletal tissues

To overcome potential redundancy among the 19 Wnts so that the function of Wnts originating from cells of the osteochondral lineage could be assessed, we and others took advantage of the requirement of the Wntless gene (*Wls*) for secretion of any Wnts (thus making any *Wls*-deficient cell functionally null for the ability to secrete any Wnt ligands) and have characterized several mouse models with conditional deletion of *Wls* in the osteoblast. When *Wls* is deleted in mature osteoblasts in vivo, the mice display significant defects in postnatal skeletal maintenance [13]. When *Wls* is deleted in both osteoblasts and chondrocytes, cartilage development is affected and a more severe skeletal phenotype is observed [13].

Wnt signaling is activated in many types of human cancer and therefore has emerged as a target for inhibition during cancer treatment [14]. Several such approaches are aimed at blocking the ability of Wnt ligands to activate their cognate receptors. For example, a series of Porcupine inhibitors are being evaluated in clinical trials to treat several types of human cancer [15]. Since these Porcupine inhibitors act similarly to the *Wls* knockouts to systemically block secretion of all Wnts, we may expect the same detrimental effects, such as loss of bone mass and increased risk of fracture, when patients are treated with drugs that inhibit Wnt signaling. In fact, this has already been reported in the context of inhibitors that block the pathway by interfering with Frizzled activation [16]. Thus, it will be important to proactively implement strategies to mitigate deleterious effects on the skeleton when treating patients with therapies that systemically inhibit Wnt signaling.

Secreted frizzled-related proteins

Secreted frizzled-related proteins (Sfrps) are a family of four proteins that contain CRDs homologous to the Wnt-binding sites of the Frizzled (Fzd) receptors. The interaction between an Sfrp and a Wnt can prevent Wnt ligands from binding to Fzd receptors and will consequently decrease Wnt signaling [17]. In some contexts, however, an Sfrp could facilitate Wnt signaling, perhaps by facilitating transport through the extracellular space [18]. Recently, homozygous loss of function mutations in Sfrp4 were found to be the cause of Pyle disease, a rare bone disorder characterized by genu valgum (knock knees), cortical bone thinning, and other bone defects (OMIM #265900). Sfrp4$^{-/-}$ mice have higher trabecular bone mass but significantly reduced cortical bone thickness in the calvarium and long bones, which recapitulates the skeletal phenotypes of human Pyle disease. The different effects on trabecular and cortical bone were suggested to be caused by the fact that *Sfrp4* knockout in cortical bones increases both canonical and noncanonical Wnt signaling; the activation of noncanonical Wnt signaling might cause the cortical bone thinness through increased BMP signaling and subsequent increased sclerostin (SOST) expression [19]. Independent groups have confirmed the cortical and trabecular bone phenotypes in Sfrp4−/− mice, and they report that Sfrp4 deletion could prevent age-related trabecular bone loss and other abnormalities [20,21]. In addition, osteoblast-specific overexpression of *Sfrp4* is associated with low trabecular bone mass [22].

Notum and Tiki

Notum was recently identified as an extracellular Wnt deacylase and a highly conserved, secreted, feedback antagonist of Wnt [23]. It essentially counteracts Porcupine by removing the palmitoleic acid group and thus inactivating Wnts. A group of Notum inhibitors was developed by Lexicon Pharmaceuticals and they were shown to specifically stimulate cortical (but not trabecular) bone formation and leave soft tissues phenotypically unaltered [24]. Members of the Tiki family also act directly on the Wnt ligand to inactivate it, via proteolysis-induced oxidation-oligomerization [23]. More detailed examination of potential role(s) for these proteins in skeletal development and disease are undoubtedly underway.

Lrp5 and Lrp6

The roles of Lrp5 and Lrp6 have been the subject of numerous recent reviews (for example, see [25]) and thus will only be briefly summarized here. Alterations in the *LRP5* gene, which encodes a Wnt coreceptor, were identified as being associated with low or high bone mass in humans in the early 2000s [25]. These pioneering discoveries encouraged scientists to invest significant effort into studying the mechanisms underlying Wnt signaling in bone. One of the resulting models posits that LRP5 signals within the enterochromaffin cells of the duodenum to regulate serotonin expression; secretion of serotonin then regulates osteoblast differentiation systemically [26]. However, several groups have independently shown that Lrp5 exerts its role within osteoblasts to regulate bone formation in mice [13]. The explanation for these different observations is not yet clear. A loss of function mutation in LRP6 (a close homolog of LRP5) was shown to be associated with osteoporosis and metabolic syndrome in human patients [27], while loss of function mutations in Lrp6 (or those that otherwise reduced activity) are associated with multiple skeletal dysmorphologies and low bone mass [25].

Dickkopfs and sclerostin

Wnt signaling activity is inhibited by several proteins that block signaling by binding to the Lrp5/6 component and blocking the ability of Wnt ligands to interact with Lrp5/6. Three members of the Dickkopf family (Dkk1, -2, and -4) act via this mechanism [28]. Heterozygosity for a germline-inactivating allele of Dkk1 in mice leads to high bone mass [10], while complete lack of Dkk1 results in embryonic lethality caused by failure to form head structures anterior to the midbrain as well as developing polydactylylism [10]. Paradoxically, Dkk2-deficient mice display decreased bone mass which may be due to a requirement to reduce Wnt activity at later stages of osteoblast differentiation [10]. In vitro loss of Dkk4 leads to increased osteoblastogenesis [10]. Antibodies that block Dkk1 function are being evaluated in clinical trials for efficacy against multiple myeloma, a disease that has been associated with elevated Dkk1 levels [29]. In addition, recent studies have suggested that dual inhibition of Dkk1 and Sost via functionally blocking antibodies (described in more detail later in this chapter) may be a relevant strategy to treating osteoporosis [30].

Sclerosteosis is an autosomal recessive disease associated with progressive skeletal overgrowth early in life [31]. Patients with a related disorder, Van Buchem disease, have similar, but slightly milder, symptoms. In both cases, dysregulation of the *Sost* gene is the underlying genetic mechanism for the disorder. Patients with sclerosteosis have inactivating mutations in Sost, while Van Buchem patients have a homozygous deletion of a 52 kilobase (kb) region that harbors an enhancer for *Sost* gene expression [32]. The realization that Sost was a secreted protein that worked via binding to Lrp5/6 and inhibiting Wnt signaling, coupled with the relatively specific expression of Sost from osteocytes, were key factors to support the development of therapies to inhibit Sost function to treat osteoporosis [31]. The most advanced of these approaches is romosozumab, an anti-Sost antibody developed by Amgen and UCB. The results of a phase III study showing the efficacy of romosozumab in treating postmenopausal osteoporosis were recently published [33]. At least two other similar anti-Sost-based approaches are being pursued. Blosozumab (developed by Eli Lilly [34]) and BPS804 (originally developed by Novartis but now being pursued by Mereo BioPharma [34]) are also being evaluated for efficacy in the treatment of skeletal diseases.

Lrp4

LRP4 is a member of the low-density lipoprotein (LDL) receptor family whose extracellular domain closely resembles those found in LRP5 and LRP6 and has been identified as a Wnt/β-catenin signaling antagonist [35]. Homozygous missense mutations in LRP4 in humans are the underlying genetic cause of some types of sclerosteosis and of Cenani-Lenz syndrome, an autosomal recessive congenital disorder affecting distal limb development [35]. Osteoblast-specific *Lrp4* deletion in mice or anti-Lrp4 antibody treatment in rats inhibits Lrp5/6, causing high bone mass phenotypes that resemble those of human patients who have mutant SOST, a well-known Wnt antagonist. Osteoblast-specific *Lrp4* conditional knockout mice and sclerosteosis patients with *LRP4* mutations (but not osteoclast-specific Lrp4-deficient mice) have dramatically higher serum SOST, suggesting that osteoblastic Lrp4 can retain SOST and prevent its secretion from bone tissue [36–38].

Frizzleds

The Frizzled receptors in a seven-transmembrane spanning family were first identified as Wnt receptors in 1996 [39], but their specific roles in skeletal development have not been studied in as much detail as Lrp5/6. Two mouse strains missing individual Fzd genes (*Fzd8* and *Fzd9*) show osteopenic phenotypes due to increased osteoclastogenesis or decreased bone formation, respectively [40,41]. This might explain the low bone density in patients with Williams-Beuren syndrome, who have an FZD9 deletion [42]. However, in the global knockout animals, the fact that Fzd8 (together with Fzd4) controls kidney development while Fzd9 regulates B-cell and hippocampal development might have affected the bone phenotypes [43–45]. Bone-specific Fzd deletions will provide more convincing evidence for further characterizing the function of each Frizzled in bone.

REGULATION OF FRIZZLED RECEPTOR STABILITY AT THE CELL SURFACE

The stability and cell surface levels of the 10 mammalian Frizzleds are regulated primarily by the activity of two single-pass transmembrane ubiquitin E3 ligases, Ring finger 43 (Rnf43) and Zinc and ring finger 3 (Znrf3), that apparently are dedicated to this purpose. Rnf43 and Znfr3 directly ubiquitinate Frizzled proteins on cytoplasmic lysine residues to target them for internalization and degradation. This process also reduces membrane levels of Lrp5 and Lrp6. Rnf43 and Znrf3 levels at the plasma membrane are themselves under tight regulatory control via the actions of the four mammalian members of the R-spondin family of secreted proteins [46]. R-spondins bind both to Rnf43/Znrf3 and either Lgr4, -5, or -6 (leucine-rich repeat containing G-protein-coupled receptor 4, 5, or 6) stimulating autoubiquitination of Rnf43/Znrf3, and resulting in the internalization and degradation of the complex. Thus, the ultimate result of exposure to R-spondin is the reduction of Rnf43/Znrf3 levels and increases in Frizzled and Lrp5/6 levels at the cell surface, thereby sensitizing cells to Wnt-mediated activation. Dishevelled (Dvl), an essential component that is recruited to Frizzled upon Wnt activation to form the Wnt receptor complex signalosome, also plays a role as an adaptor for Rnf43 and Znrf3 to form an RNF43/ZNRF3/Fzd/Dvl complex that facilitates Fzd degradation [47].

Several genetically engineered mouse models lacking members of the R-spondin family display significant skeletal abnormalities [48]. For example, severe malformations can be found in the limbs of embryos homozygous for a hypomorphic allele of R-Spo2 [48]. More severe hindlimb truncations were observed in embryos lacking both R-Spo3 and R-Spo2, suggesting a redundant function of these genes in bone development [48]. While approximately half of mice homozygous for a null allele of *Lgr4* die perinatally, assessment of the animals that survived until adulthood revealed a significant low bone mass phenotype, with suppressed bone formation and elevated bone resorption [48]. Lgr4 has been also found to directly interact with RANKL in negatively regulating osteoclast differentiation and apoptosis [48]. In addition, Lgr6-deficient mice have regeneration defects in both nails and bones [48].

SUMMARY

The development of therapies to treat osteoporosis based on activating the Wnt signaling pathway hold great promise. One limitation of anti-sclerostin therapy in clinical trials is that bone formation only lasts for 2–3 months and then the anabolic effect wanes [30]. One potential explanation is that Wnt signaling induces several transcriptional targets that act as negative feedback inhibitors to blunt prolonged activation of the pathway. One such negative feedback inhibitor, Dkk1, was shown to contribute to this blunting of response in that simultaneous administration of antibodies that neutralized Sclerostin and Dkk1 (or exposure to a bispecific Dkk1/Sost-blocking antibody) resulted in an increased anabolic response [30]. Importantly, it was noted that additional feedback inhibitors were almost certainly still acting to reduce the long-term response in this context. Thus, gaining further insight into how Wnt ligands and receptors are regulated may increase the efficacy of these clinically relevant agents and uncover additional mechanisms by which the pathway can be safely manipulated to treat osteoporosis.

ACKNOWLEDGMENTS

We apologize to any of our colleagues whose work we were unable to reference due to the strict limitations on the number of references. In many cases this necessitated citing review articles to which we refer readers for additional references. Work in the Williams Laboratory is supported by the Van Andel Research Institute and by NIH/NIAMS grant AR053237.

REFERENCES

1. Clevers H, Nusse R. Wnt/beta-catenin signaling and disease. Cell. 2012;149(6):1192–205.
2. Tan SH, Senerath-Yapa K, Chuyng MT, et al. Wnts produced by Osterix-expressing osteolineage cells regulate their proliferation and differentiation. Proc Natl Acad Sci USA. 2014;111(49):E5262–71.

3. Najdi R, Proffitt K, Sprowl S, et al. A uniform human Wnt expression library reveals a shared secretory pathway and unique signaling activities. Differentiation. 2012;84(2):203–13.
4. Minami Y, Oishi I, Endo M, et al. Ror-family receptor tyrosine kinases in noncanonical Wnt signaling: their implications in developmental morphogenesis and human diseases. Dev Dyn. 2010;239(1):1–15.
5. Mikels AJ, Nusse R. Purified Wnt5a protein activates or inhibits beta-catenin-TCF signaling depending on receptor context. PLoS Biol. 2006;4(4):e115.
6. Ke J, Xu HE, Williams BO. Lipid modification in Wnt structure and function. Curr Opin Lipidol. 2013;24(2): 129–33.
7. Janda CY, Waghray D, Levin AM, et al. Structural basis of Wnt recognition by Frizzled. Science. 2012;337(6090): 59–64.
8. Al-Qattan MM. WNT pathways and upper limb anomalies. J Hand Surg Eur Vol. 2011;36(1):9–22.
9. Gori F, Lerner U, Ohlsson C, et al. A new WNT on the bone: WNT16, cortical bone thickness, porosity and fractures. Bonekey Rep. 2015;4:669.
10. Maupin KA, Droscha CJ, Williams BO. A comprehensive overview of skeletal phenotypes associated with alterations in Wnt/beta-catenin signaling in humans and mice. Bone Res. 2013;1(1):27–71.
11. Moverare-Skrtic S, Henning P, Liu X, et al. Osteoblast-derived WNT16 represses osteoclastogenesis and prevents cortical bone fragility fractures. Nat Med. 2014;20(11): 1279–88.
12. Alam I, Alkhouli M, Gerard-O'Riley RL, et al. Osteoblast-specific overexpression of human WNT16 increases both cortical and trabecular bone mass and structure in mice. Endocrinology. 2016;157(2):722–36.
13. Zhong Z, Ethen NJ, Williams BO. WNT signaling in bone development and homeostasis. Wiley Interdiscip Rev Dev Biol. 2014;3(6):489–500.
14. Rey JP, Ellies DL. Wnt modulators in the biotech pipeline. Dev Dyn. 2010;239(1):102–14.
15. Lum L, Clevers H. Cell biology. The unusual case of Porcupine. Science. 2012;337(6097):922–3.
16. Messersmith WA, Cohen S, Shahda H, et al. Phase 1b study of WNT inhibitor vantictumab (VAN, human monoclonal antibody) with nab-paclitaxel (Nab-P) and gemcitabine (G) in patients (pts) with previously untreated stage IV pancreatic cancer (PC). Ann Oncol. 2016;27(6):207–42.
17. Cruciat CM, Niehrs C. Secreted and transmembrane wnt inhibitors and activators. Cold Spring Harb Perspect Biol. 2013;5(3):a015081.
18. Mii Y, Taira M. Secreted Frizzled-related proteins enhance the diffusion of Wnt ligands and expand their signalling range. Development. 2009;136(24):4083–8.
19. Simsek Kiper PO, Saito H, Gori F, et al. Cortical-bone fragility—insights from sFRP4 deficiency in Pyle's disease. N Engl J Med. 2016;374(26):2553–62.
20. Haraguchi R, Kitazawa R, Mori K, et al. sFRP4-dependent Wnt signal modulation is critical for bone remodeling during postnatal development and age-related bone loss. Sci Rep. 2016;6:25198.
21. Mastaitis J, Eckersdorff M, Min S, et al. Loss of SFRP4 alters body size, food intake, and energy expenditure in diet-induced obese male mice. Endocrinology. 2015;156(12):4502–10.
22. Nakanishi R, Akiyama H, Kimur H, et al. Osteoblast-targeted expression of Sfrp4 in mice results in low bone mass. J Bone Miner Res. 2008;23(2):271–7.
23. Zhang X, He X. Methods for studying Wnt protein modifications/inactivations by extracellular enzymes, Tiki and Notum. Methods Mol Biol. 2016;1481:29–38.
24. Tarver JE Jr, Pabba PK, Barbosa J, et al. Stimulation of cortical bone formation with thienopyrimidine based inhibitors of Notum pectinacetylesterase. Bioorg Med Chem Lett. 2016;26(6):1525–8.
25. Joiner DM, Ke J, Zhong Z, et al. LRP5 and LRP6 in development and disease. Trends Endocrinol Metab. 2013;24(1):31–9.
26. Williams BO, Insogna KL. Where Wnts went: the exploding field of Lrp5 and Lrp6 signaling in bone. J Bone Miner Res. 2009;24(2):171–8.
27. Mani A, Radhakrishman J, Wang H, et al. LRP6 mutation in a family with early coronary disease and metabolic risk factors. Science. 2007;315(5816):1278–82.
28. Niehrs C. Function and biological roles of the Dickkopf family of Wnt modulators. Oncogene. 2006;25(57): 7469–81.
29. Zhou F, Meng S, Song H, et al. Dickkopf-1 is a key regulator of myeloma bone disease: opportunities and challenges for therapeutic intervention. Blood Rev. 2013;27(6):261–7.
30. Florio M, Gunasekaran K, Stolina K, et al. A bispecific antibody targeting sclerostin and DKK-1 promotes bone mass accrual and fracture repair. Nat Commun. 2016;7:11505.
31. Williams BO. Insights into the mechanisms of sclerostin action in regulating bone mass accrual. J Bone Miner Res. 2014;29(1):24–8.
32. Nassar K, Rachidi W, Janani S, et al. Van Buchem's disease. Joint Bone Spine. 2016;83(6):737–8.
33. Cosman F, Gilchrist N, McClung M, et al. A phase 2 study of MK-5442, a calcium-sensing receptor antagonist, in postmenopausal women with osteoporosis after long-term use of oral bisphosphonates. Osteoporos Int. 2016;27(1):377–86.
34. Lewiecki EM. Role of sclerostin in bone and cartilage and its potential as a therapeutic target in bone diseases. Ther Adv Musculoskelet Dis. 2014;6(2):48–57.
35. Shen C, Xiong WC, Mei L. LRP4 in neuromuscular junction and bone development and diseases. Bone. 2015;80:101–8.
36. Fijalkowski I, Geets E, Steenackers E, et al. A novel domain-specific mutation in a sclerosteosis patient suggests a role of LRP4 as an anchor for sclerostin in human bone. J Bone Miner Res. 2016;31(4):874–81.
37. Xiong L, Jung JU, Wu H, et al. Lrp4 in osteoblasts suppresses bone formation and promotes osteoclastogenesis and bone resorption. Proc Natl Acad Sci U S A. 2015;112(11):3487–92.
38. Chang MK, Kramer I, Huber T, et al. Disruption of Lrp4 function by genetic deletion or pharmacological

blockade increases bone mass and serum sclerostin levels. Proc Natl Acad Sci U S A. 2014;111(48):E5187–95.
39. Wang Y, Chang H, Rattner A, et al. Frizzled receptors in development and disease. Curr Top Dev Biol. 2016;117:113–39.
40. Albers J, Keller J, Baranowsky A, et al. Canonical Wnt signaling inhibits osteoclastogenesis independent of osteoprotegerin. J Cell Biol. 2013;200(4):537–49.
41. Albers J, Schulze J, Beil FT, et al., Control of bone formation by the serpentine receptor Frizzled-9. J Cell Biol. 2011;192(6):1057–72.
42. Francke U. Williams-Beuren syndrome: genes and mechanisms. Hum Mol Genet. 1999;8(10):1947–54.
43. Ye X, Wang Y, Rattner A, et al. Genetic mosaic analysis reveals a major role for frizzled 4 and frizzled 8 in controlling ureteric growth in the developing kidney. Development. 2011;138(6):1161–72.
44. Zhao C, Avilés C, Abel RA, et al. Hippocampal and visuospatial learning defects in mice with a deletion of frizzled 9, a gene in the Williams syndrome deletion interval. Development. 2005;132(12):2917–27.
45. Ranheim EA, Kwan HC, Reya T, et al. Frizzled 9 knock-out mice have abnormal B-cell development. Blood. 2005;105(6):2487–94.
46. de Lau W, Peng WC, Gros P, et al. The R-spondin/Lgr5/Rnf43 module: regulator of Wnt signal strength. Genes Dev. 2014;28(4):305–16.
47. Jiang X, Charlat O, Zamponi R, et al. Dishevelled promotes Wnt receptor degradation through recruitment of ZNRF3/RNF43 E3 ubiquitin ligases. Mol Cell. 2015;58(3):522–33.
48. Knight MN, Hankenson KD. R-spondins: novel matricellular regulators of the skeleton. Matrix Biol. 2014;37:157–61.

10
Mechanotransduction in Bone Formation and Maintenance

Whitney A. Bullock[1], Lilian I. Plotkin[2], Alexander G. Robling[1], and Fredrick M. Pavalko[3]

[1]*Department of Anatomy and Cell Biology, Indiana University School of Medicine, Indianapolis, IN, USA*
[2]*Indiana Center for Musculoskeletal Health (ICMH), Indianapolis, IN, USA*
[3]*Department of Cellular and Integrative Physiology, Indiana University School of Medicine, Indianapolis, IN, USA*

INTRODUCTION

Among the most important regulators of bone shape, strength, architecture, and overall quality is the type, duration, and magnitude of mechanical loads that are placed on the skeleton. Osteocytes, which comprise 90% to 95% of all the cells in bone, and are dispersed throughout the mineralized tissue, are the primary sensors and regulators of bone's response to its mechanical loading environment [1]. Detection of mechanical stimuli, and transformation of mechanical signals into biochemical responses, is facilitated by the vast network of cytoplasmic processes that extend from the osteocyte cell body throughout the lacuna–canalicular network in mineralized bone. Cell surface molecules and structures (integrins/cadherins, ion channels, G-protein coupled receptors, primary cilia) and intracellular signaling pathways (Wnt/β-catenin, mitogen-activated protein kinases, tyrosine kinases, cGMP/cAMP pathways) detect and transduce mechanical signals from osteocyte "sensor cells" into altered bone remodeling by "effector cells" (osteoblasts, osteoclasts, bone lining). Several fundamental processes that control skeletal adaptation to mechanical loading have been identified and integrated with signaling pathways, leading to a better understanding of bone tissue engineering and physiology [2], creating new opportunities for clinical/pharmaceutical and exercise strategies to promote skeletal health.

OSTEOGENIC MECHANICAL STIMULI

This chapter considers the types of cells (mechanosensors) and forces likely to impact bone formation/remodeling and then examine the cellular and molecular mechanisms regulating the biological responses of cells to those forces.

Mechanosensors in bone

Cells within bone detect, coordinate, and mediate mechanical loading on the skeleton. Key questions include: (i) how the cells embedded within and on the surfaces of bone sense and respond to externally applied mechanical loads; (ii) whether the mechanical stimulus is direct tissue strain from loading/bending of bone, or an indirect consequence of the load (ie, load-induced fluid flow over the surfaces of bone cells); and (iii) how cells translate a mechanical signal detected at the cell surface into an appropriate sequence of biochemical changes inside the cell that results in a coordinated anabolic or catabolic response.

The cells most likely to function as the "mechanosensors" in bone are the osteocytes. Intuitive considerations and abundant experimental evidence supports the assignment of osteocytes as the bone cells most directly responsible for detecting mechanical signals and

Primer on the Metabolic Bone Diseases and Disorders of Mineral Metabolism, Ninth Edition. Edited by John P. Bilezikian.
© 2019 American Society for Bone and Mineral Research. Published 2019 by John Wiley & Sons, Inc.
Companion website: www.wiley.com/go/asbmrprimer

orchestrating the skeletal response to those signals. Osteocytes within individual lacunae are perfectly positioned as mechanosensors because they are distributed uniformly throughout the cortical and trabecular bone and are connected by a system of canals. This allows osteocytes to function in a network capable of sensing load-induced signals virtually anywhere in the skeleton. Long cytoplasmic processes (~50/cell) interconnect the cells comprising this vast network to each other via gap junctions, which facilitate cell–cell communication. However, these same properties make osteocytes unlikely candidates for effector cells that directly add or remove bone from bone surfaces. Although osteocytes can mediate localized removal of mineralized matrix and calcium release from their own lacunae, this localized activity has little effect on the size, shape, and structural properties of the skeleton. Osteoblasts, bone lining cells, and osteoclasts function as the "effector" cells that add or remove bone under the direction of signals from the osteocyte network.

To understand how physical stimuli affect bone cells, we must consider the vastly different microenvironments that different bone cells inhabit. Stromal mesenchymal stem cells (MSCs) for instance reside in the bone marrow space of long bones, have minimal extracellular space surrounding them, and no rigid substrate to contact. In contrast, osteoblasts, osteocytes, and bone lining cells attached to the hard surfaces of bone, experience higher surface strains as well as movement of fluid over their surfaces. The physical environment of the osteocyte is truly unique. Osteocytes encase themselves in mineralized matrix. Importantly, they create a small fluid-filled, form-fitting cavity within their lacuna and extend cell processes that course through the matrix in narrow canaliculi. The canaliculi average about 260 nm in diameter and the space between the projection membrane and the canalicular wall is very narrow (~80 nm), creating the potential for high-velocity fluid movement driven by load-induced pressures or matrix strains. Finite element modeling indicates that typical macroscopic-level strains on bone can result in greatly amplified strains at the microscopic level of the perilacunar space [3]. For years the strongest argument favoring the osteocyte network as the primary mechanosensory system in bone was based on these characteristics. Fortunately, experimental support for this hypothesis now exists. Osteocytes are more sensitive and respond differently to fluid flow-induced shear stress compared with osteoblasts in vitro [4,5], and experiments in mice have demonstrated a key functional role for osteocytes in detecting disruptions to normal mechanical loading. A transgenic mouse was created that enabled efficient ablation of the majority of osteocytes without affecting osteoblasts and osteoclasts [6]. These mice were protected from disuse-induced bone loss, demonstrating a key role for the osteocyte network in sensing the loss of normal skeletal loading. Although inflammatory effects, including enhanced osteoclast formation, following induction of such massive osteocyte death, complicate this analysis [7], there is an overwhelming consensus that osteocytes are the primary mechanosensory cells in bone.

Fluid flow and tissue strain in bone

There is a strong argument in favor of mechanically generated oscillatory fluid movement within the lacunocanalicular network being the most physiologically relevant cell-level mechanical stimulus. Although tissue strains produced by bending the bone matrix can induce mechanical signals detected by osteocytes, the vast array of cytoplasmic processes of the osteocyte positions these cells to sense fluid flow. Indeed, the ability of the extracellular domains of integrins expressed on the long cytoplasmic osteocyte processes appear to be capable of "amplifying" relatively small fluid forces into disproportionally large biochemical responses inside osteocytes [8]. Osteocytes are physically connected to the extracellular glycocalyx within the canalicular walls via integrin heterodimers. Subtle changes in the conformation of integrin extracellular domains are transduced to the cell interior via short cytoplasmic tails. Growing evidence suggests that integrin-associated signaling and cytoskeletal proteins transmit mechanical signals directly to the nucleus to alter gene expression.

Experimental limitations must be considered when evaluating mechanical sensory mechanisms and the relative contributions of fluid flow and substrate strain in osteocytes. First, most published studies are based on in vitro cell culture experiments using osteoblasts or osteocyte-like cell lines because primary osteocytes are difficult to isolate in sufficient numbers for experimental study. Determining the contribution of tissue-level strains in bone as biologically relevant osteocyte stimuli is also challenging [9]. Measurements of tissue strain collected on the periosteal surface of long bones indicate that the peak strain generated during strenuous activity is approximately 3000 µε. When bone cells grown on two-dimensional culture substrates are exposed to 3000 µε strain no detectable response can be measured. In fact, greater than 10,000 µε must be applied before most in vitro models will yield a measurable response. Application of 10,000 µε is beyond the yield point of bone in vivo and fracture occurs before sufficient tissue strain stimulates bone cells. However, application of loads generating ~2000 µε on a bone's surface may generate local strains up to 30,000 µε on the osteocyte lacunar wall. This difference in macro- versus micro-level strain is possibly a consequence of stress-concentrating effects of the voids produced by osteocyte lacunae [8].

Much less controversial is the impact of shear stress generated as extracellular fluid moves across osteocyte surfaces from areas of higher to lower pressure when a long bone is bent during loading (Fig. 10.1). Fluid shear stress potently stimulates osteocytes and other bone cells in vitro. Enhanced transport of growth factors and nutrients, removal of waste products of metabolism

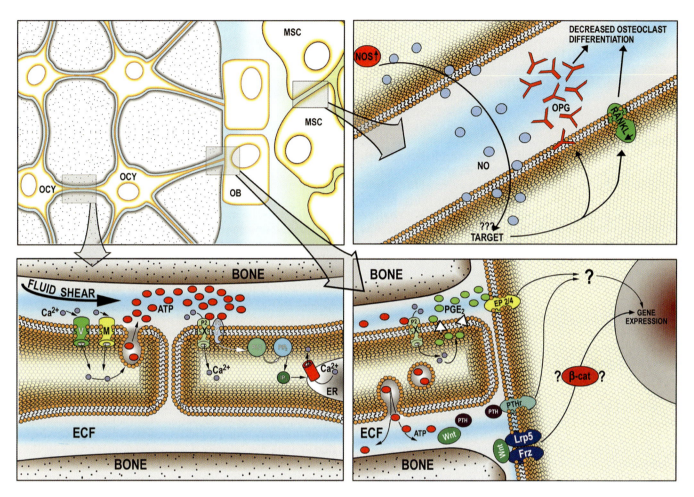

Fig. 10.1. Fluid shear stress-induced signaling in bone. ATP = adenosine triphosphate; ECF = extracellular fluid; ER = endoplasmic reticulum; MSC, mesenchymal stem cell; NO = nitric oxide; NOS = nitric oxide synthase; OB = osteoblast; OCY = osteocyte; PGE_2 = prostaglandin E_2; OPG = osteoprotegerin; RANKL = RANK ligand.

(chemotransport), and generation of electrical streaming potentials may also affect osteocyte responses to fluid flow. Chemotransport also affects the response of bone cells to fluid flow independent of mechanical signal transduction [10].

An intriguing model for osteocyte mechanotransduction has been developed by the Schaffler and Weinbaum laboratories [8] based upon microscopic-level studies of osteocytes in bone pieces. Their model considers the observation that the cytoplasmic processes appear to be suspended from the canalicular wall via integrin adhesion complexes attached to tethering structures made from the glycocalyx. They suggest that fluid flow through the space between the membrane and the glycocalyx wall "deflects" these tethering structures, resulting in deformation referred to as "hoop strain." Deformation of these tethering structure results in a strain amplification that is radial, like a hoop. As a result of this amplification, estimates suggest that local cell membrane strains may be 10- to 100-fold higher than grossly measured tissue strains.

MOLECULAR BASIS FOR MECHANOTRANSDUCTION IN BONE

Having considered both the physical environment and the mechanical forces to which osteocytes are exposed, how do osteocytes respond appropriately to relevant mechanical stimuli? This complex question must consider the contributions of membrane proteins, lipids, intracellular signals, and cytoskeleton, all of which convert mechanical forces into biochemical signals that ultimately alter gene expression. No aspect of mechanotransduction in bone may be more fundamental than—or as controversial as—identification of the predominant cell surface mechanoreceptor. It is probable that no single mechanoreceptor protein or structure is singularly responsible for sensing and transducing mechanical stimuli into altered bone formation. Fortunately, excellent progress is being made and great potential exists to therapeutically manipulate mechanotransduction pathways to treat human bone mass diseases.

Bone cell mechanoreceptors can be classified into broad categories: (i) integrins; (ii) G-protein coupled receptors/receptor tyrosine kinases; and (iii) ion channels/connexin hemichannels. A role for the primary cilium is also gaining acceptance. Inhibition of a specific protein's ability to blunt normal mechanotransduction is the standard of proof that a protein is involved in vitro. However, most in vitro studies fail to recapitulate the complex three-dimensional environment of osteoblasts and osteocytes in vivo, and it is more challenging to determine which proteins are essential in vivo. While it is likely that any predominant mechanoreceptor would be activated within milliseconds of stimulation, temporal and quantitative assessment of mechanical signaling activity in real time in living cells within bone is extremely difficult.

Integrins/focal adhesions

Integrins as mechanosensors at specialized sites of cell-matrix adhesion, known as "focal adhesions," have been intensively studied [11]. Integrin-containing focal adhesions mediate both cell-matrix adhesion and transmembrane signaling. Responses to strain and fluid flow mediated by integrins in bone cells, initially studied in osteoblast-like cells, and more recently in osteocytes, include changes in intracellular tension via reorganization of the actin cytoskeleton, increased MAPK and tyrosine kinase activity, release of paracrine signaling molecules (prostaglandins and nitric oxide), and altered gene expression. These findings led to the "mechanosome hypothesis" which postulated that protein complexes (mechanosomes) capable of associating with cell adhesion molecules (integrins and cadherins) are "launched" from the membrane to the nucleus in response to mechanical stimuli, leading to alteration of gene expression [12]. DNA-binding proteins localized in adhesion molecule-associated complexes are translocated to the nucleus in response to mechanical loading, thereby transferring mechanical information from adhesion complexes at the cell surface to target genes [13]. Experimental support is based largely on the effects of deleting/inhibiting specific mechanosome proteins on signaling and load-induced bone formation.

Evidence that integrins are activated by fluid flow is compelling and comes both from direct demonstration of changes in integrin conformation and from studies using integrin-specific inhibitory antibodies and function-inhibiting Arg-Gly-Asp (RGD) peptides. Thus, focal adhesions exhibit many properties implicating them as primary mechanosensory structures [14]. Importantly, integrins have two interrelated roles: a structural (load-bearing) function in which they link (indirectly) to the actin cytoskeleton, and a signal transduction/scaffolding function in which they bind and organize proteins that have significant downstream effects on the cell. Although cytoskeletal reorganization received considerable initial attention as a mechanism for transducing mechanical signals, many biochemical responses to mechanical stimulation in bone cells do not require an intact cytoskeleton [15]. Conformational changes induced by mechanical stimuli expose binding sites in the cytoplasmic tails of integrin subunits. These are binding sites for kinases (focal adhesion kinase [FAK], proline rich kinase-2 [Pyk2], Src kinase) and adaptor proteins (eg, paxillin, α-actinin, Nmp4, zyxin, p130^{cas}, among many others) that can stimulate signaling cascades and translocate mechanosome complexes from the cell membrane into the nucleus in a multi-stop process. This process involves detection of a mechanical signal, activation of membrane-associated proteins, formation of mechanosome complexes at the membrane, and then their translocation to the nucleus where they alter the transcription of genes that regulate bone remodeling (Fig. 10.2).

G-protein and receptor tyrosine kinase signaling in mechanotransduction

Osteocyte responses to mechanical load include activation of G-protein coupled receptors (GPCRs). GPCRs are a large class of surface receptors stimulated by a wide range of ligands. The most common method of monitoring GPCR activity is though measuring hydrolysis of the G proteins that they activate. The Frangos lab showed that fluid shear stimulates G-protein signaling and that pharmacological inhibition of G proteins inhibits activation of downstream signals [16]. The temporal and spatial advantages of fluorescence resonance energy transfer (FRET) microscopy allowed them to show that a conformational change occurs in the parathyroid hormone 1 receptor (PTH1R) and B$_2$ bradykinin receptor within milliseconds of applying fluid shear stress [17]. This response did not require the presence of either receptor's ligand and could be modulated by changing membrane fluidity, showing that these receptors exhibit the characteristics required of direct cell surface sensors of mechanical stimuli.

Molecular interplay between receptor tyrosine kinase (RTK) signaling and integrins was shown by the Bikle laboratory who reported that signaling through the insulin-like growth factor 1 receptor (IGF1R) is required for optimal transduction of fluid shear-induced signaling and is also regulated by direct interaction with integrins [18]. This led to their proposing a model in which the bone-forming effects of IGF-1 are not only mediated by interactions with IGF1R but can simultaneously be modulated by mechanical stimulation of integrins and pathways downstream of integrins. The Pilz lab showed that fluid shear-mediated activation of nitric oxide, cGMP, and protein kinase G (PKG) signaling in osteoblasts was dependent on Src activation downstream of integrins [19]. This process involved PKG II dependent Src dephosphorylation via a fluid shear-induced PKG-II/Src/Shp mechanosome. Whether GPCRs and RTKs are able to sense mechanical stimuli directly, or do so primarily via cross-communication with integrin mechanosensors, remains to be determined.

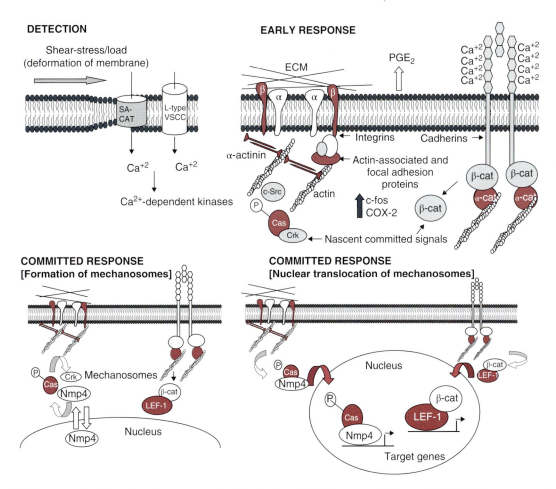

Fig. 10.2. Multi-step activation of mechanosome signaling in bone. ECM = extracellular matrix; SA-CAT = stress-activated cation channel; PGE$_2$, prostaglandin E$_2$; VSCC = voltage sensitive calcium channel.

Gap junctions/ion channels in mechanotransduction

Gap junction channels formed by connexins (Cx) mediate communication among osteocytes and between osteocytes and cells on the bone surface [20]. All bone cells express connexins, Cx43 being the most abundant. In vitro and in vivo studies showed that Cx43 protein levels and function are increased by mechanical stimulation, and, conversely, lack of mechanical signals decreases Cx43 expression.

Cx43 also forms hemichannels that enable communication between cells and the extracellular environment [21] and are opened by mechanical stimulation [20], although their function is still under debate. It has been proposed that hemichannels are required for adenosine triphosphate (ATP) and prostaglandin release induced by mechanical stimulation. However, another study showed that activation of purinergic receptors increases prostaglandin E$_2$ (PGE$_2$) release independently of hemichannels, and that blockade of P2X7 receptors, which are activated by ATP, prevents PGE$_2$ release from osteoblastic and osteocytic cell lines. Further, it was recently shown that osteoblastic cells derived from Cx43$^{-/-}$ mice still respond to mechanical stimulation by releasing PGE$_2$ [20]. This later study proposes that hemichannels formed by pannexin1, a transmembrane protein with topology similar to connexins [22], rather than Cx43 are responsible for channel activity induced by ATP in osteoblastic cells.

In vitro studies, together with early evidence showing the existence of gap junctions between osteocytes and osteoblasts on the bone surface [23], led to the hypothesis that Cx43 is required for the response to mechanical stimulation in vivo. This was proven in mice lacking Cx43 in pre-osteoblasts, osteoblasts, and osteocytes, which exhibit an attenuated anabolic response to mechanical stimulation on the tibial endocortical surface [20]. However, all studies in which periosteal bone formation was investigated showed, surprisingly, an enhanced response to mechanical loading in the absence of Cx43 in osteochondral progenitors, osteoblasts, and osteocytes. The molecular basis for this effect remains unknown.

Mechanotransduction can be modulated by extracellular nucleotides [24]. ATP and uridine triphosphate (UTP) are released by osteoblasts and osteocytes subjected to mechanical stimulation in vitro, and can activate P2X ligand-gated ion channels and metabotropic P2Y GPCRs [25]. Receptor

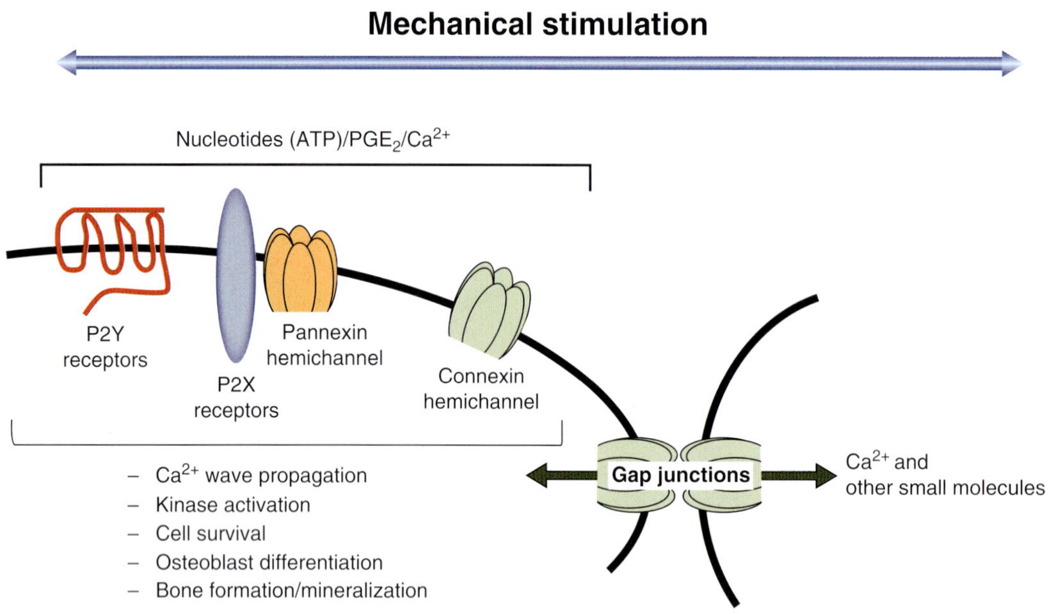

Fig. 10.3. Activation of gap junction signaling by mechanical stimulation in bone.

activation leads to various responses depending on the nucleotide and type of receptor, including Ca^{2+} wave propagation, intracellular kinase activation, and modulation of osteoblast function and survival. In particular, P2X7-deficient mice exhibit a deficient anabolic response to mechanical stimulation in vivo, suggesting an important role for these channels in mechanotransduction [26]. Further, a complex formed by P2X7 receptors and pannexin1 channels mediates ATP release induced by mechanical stimulation in vitro [27]. The relevance for this complex in transmission of mechanical signals in vivo remains to be determined (Fig. 10.3).

Ion channels sensitive to mechanical stimulation are also found in osteoblastic cells [28,29]. Among them, gadolinium-sensitive stretch-activated cation channels are involved in the response to stretching in vitro. In addition, transient receptor potential channels and voltage-gated calcium channels (VGCCs) mediate the propagation of calcium waves following mechanical stimulation in osteoblastic cells. Interestingly, osteoblast to osteocyte differentiation is accompanied by changes in the type of VGCC from L type in osteoblasts to T type in osteocytes, a change associated with the different sensitivity of the two cell types to mechanical stimulation [30]. The role of VGCC channels in mechanotransduction has been demonstrated in vivo, as inhibition of L-type channels in rats blocks the anabolic response to loading [28].

BIOCHEMICAL RESPONSES TO MECHANICAL STIMULI

Detection of a molecular response to mechanical stimulation does not prove that it is important for mechanotransduction in bone. Determining which are important to regulating bone responses to mechanical stimulation and which are merely tangential is a critical question if we are to effectively target mechanotransduction pathways for therapeutic benefit. The following signaling pathways have been verified in vivo to affect the skeleton.

Wnt/β-catenin signaling

The Wnt pathway has been identified as one of the most important mediators of bone cell mechanotransduction (Fig. 10.4). In brief, Wnt ligands bind Lrp5/6 receptors, which colocalize with Frizzled. This in turn phosphorylates Axin, resulting in dissociation of the Axin/GS3κβ complex and accumulation of β-catenin. Translocation of β-catenin to the nucleus occurs, where it coordinates lymphoid enhancer-binding factor/T-cell factor (LEF/TCF) binding to DNA (see Chapter 9). Altered canonical Wnt signaling leads to disruptions in bone mass. Several gain of function mutations in the Wnt coreceptor Lrp5 increase bone mass (eg, A214V and G171V) [31–33], while loss of function mutations induce osteopenic effects on the skeleton. Moreover, both gain- and loss-of-function mutations in Lrp5 have profound effects on the efficacy of mechanical stimulation in bone [32,34]. The downstream Wnt effector molecule β-catenin is also important for bone mechanotransduction. Deletion of β-catenin in mice dramatically decreases bone mass and strongly increases the risk of fracture [35]. Constitutive activation of β-catenin increases bone mass, although activation of both alleles increases porosity and decreases bone strength [36]. Female mice haploinsufficient for β-catenin lose bone in response to hindlimb suspension [37], while mice with constitutive activation are protected from disuse-induced bone loss.

The Wnt pathway is regulated by inhibitors such as Sost/sclerostin and Dkk. Sclerostin binds Lrp5/6 and prohibits a

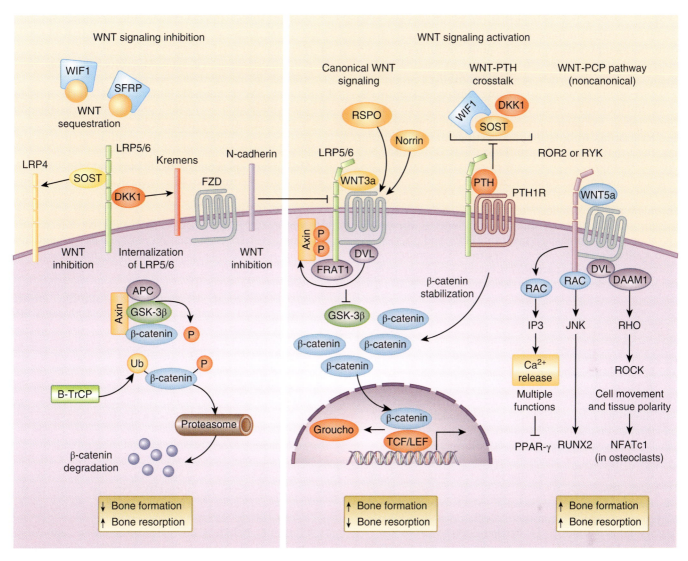

Fig. 10.4. Wnt/β-catenin signaling. Source: [46]. Reproduced with permission of Springer Nature.

subset of Wnt proteins from binding and activating the pathway. Upon mechanical stimulation of bone (exercise/loading), sclerostin is decreased, and Sost expression levels are inversely correlated to areas of high strain in bone [38]. Importantly, when load-induced sclerostin downregulation is prevented by overexpressing the Sost gene in transgenic mice, load-induced bone formation was reduced by 70% to 85% [39]. Lrp4, a membrane-bound receptor related to Lrp5/6, is crucial for the recruitment of Sost to Lrp5/6. Three different point mutations in Lrp4 have been identified in individuals with sclerosteosis-like high bone mass syndrome. These mutations prevent the binding of Sost to Lrp4, and, consequently, Sost is unable to inhibit Lrp5/6, resulting in decreased Wnt inhibition in bone.

Prostaglandin and NO signaling

Osteoblasts and osteocytes secrete PGE_2 and upregulate expression of cyclooxygenase 2 (COX2) in response to substrate strain and fluid flow [40]. COX2 is necessary for PGE_2 production, and inhibition of COX2 using inhibitors such as NS398 blocked PGE_2 release in vitro. PGE_2, a lipid generated from arachidonic acid via COX enzymes that functions like a hormone, is secreted in response to a number of stimuli including PTH, estrogen, and mechanical loading. Vigorous exercise in humans results in rapid release of PGE_2 from loaded lower limb bones [41]. In mice, mechanical loading upregulated COX2 mRNA and protein within minutes [42]. A functional role for PGE_2 signaling on the skeleton in response to mechanical loading was shown in vivo by pharmacological inhibition of both COX1 and COX2 using indomethacin, or via selective inhibition of COX2 alone using NS398 [43]. In both cases, inhibition of PGE_2 production prior to loading reduced the osteogenic response. PGE_2 secretion studies from bone cells in vitro suggest PGE_2 is released through Cx43 hemichannels or from purinergic P2X7 protein complexes [44]. Downstream of PGE_2 release, autocrine and/or paracrine effects may be mediated via Ep receptors.

Nitric oxide, a free radical that diffuses like a gas very rapidly through the plasma membrane, is released from osteoblasts and osteocytes in response to fluid shear and strain. Depletion of NO in vivo prior to loading in rats using nitric oxide synthase (NOS) inhibitors attenuates the osteogenic response to loading. Unlike WT animals, mice null for inducible NOS (iNOS$^{-/-}$), failed to reform bone that is lost during hindlimb unloading upon returning to normal ambulation, suggesting that the cellular response to mechanical stimulation in bone requires NO signaling [45].

In conclusion, the mechanical loads that are placed on the skeleton have a critical role in bone remodeling throughout life. Orchestrated by osteocytes as mechanosensors, mechanotransduction in bone is driven by the action of osteoblasts, bone lining cells, and osteoclasts. The molecular pathways that control mechanotransduction in bone represent promising targets for pharmacological manipulation to enhance skeletal health.

REFERENCES

1. Bonewald LF. The amazing osteocyte. J Bone Min Res. 2011;26:229–38.
2. Robling AG, Castillo AB, Turner CH. Biomechanical and molecular regulation of bone remodeling. Annu Rev Biomed Eng. 2006;293:455–98.
3. Bonivtch AR, Bonewald LF, Nicolella DP. Tissue strain amplification at the osteocyte lacuna: a microstructural finite element analysis. J Biomech. 2011;40:2199–206.
4. Ponik SM, Triplett JW, Pavalko FM. Osteoblasts and osteocytes respond differently to oscillatory and undirectinal fluid flow profiles. J Cell Biochem. 2007;100:794–807.
5. Klein-Nulend J, van der PA, Semeins CM, et al. Sensitivity of osteocytes to biomechanical stress in vitro. FASEB J. 1995;9:441–5.
6. Tatsumi S, Amizuka IK, Li M, et al. Targeted ablation of osteocytes induces osteoporosis with defective mechanotransduction. Cell Metab. 2007;5:464–75.
7. Komori T. Mouse models for the evaluation of osteocyte functions. J Bone Metab. 2014;21:55–60.
8. Wang Y, McNamara LM, Schaffler MB, et al. Strain amplification and integrin based signalilng in osteocytes. J Musculoskelet Neuronal Interact. 2008;8:332–4.
9. Stern AR, Nicolella DP. Measurment and estimation of osteocyte mechanical strain. Bone. 2013;54:191–5.
10. Donahue TL, Haut TR, Yellowley CE, et al. Mechanosensitivity of bone cells to oscillating fluid flow induced shear stress may be modulated by chemotransport. J Biomech. 2003;36:363–71.
11. Dubash AD, Menold MM, Samson T, et al. Focal adhesions: new angles on an old structure. Int Rev Cell Mol Biol. 2009;277:1–65.
12. Pavalko FM, Norvell SM, Burr DB, et al. A model for mechanotransduction in bone cells: the load-bearing mechanosomes. J Cell Biochem. 2003;88:104–12.
13. Bidwell, JP, Pavalko FM. The load-bearing mechanosome revisited. Clin Rev Bone Miner Metab. 2010;8:213–23.
14. Yavropoulou MP, Yovos JG. The molecular basis of bone mechanotransduction. J. Musculoskelet Neuronal Interact. 2016;16:221–36.
15. Norvell SM, Ponik SM, Bowen DK, et al. Fluid shear stress induction of COX-2 protein and prostaglandin release in cultured MC3T3-E1 osteoblasts does not require intact microfilaments or microtubules. J Appl Physiol. 2004;96:957–66.
16. McAllister TN, Frangos JA. Steady and transient fluid shear stress stimulate NO release in osteoblasts through distinct biochemical pathways. J Bone Miner Res. 1999;14:930–6.
17. Zhang YL, Frangos JA, Chachisvilis M. Mechanical stimulus alters conformation of type 1 parathyroid hormone receptor in bone cells. Am J Physiol Cell Physiol. 2009;296:1391–9.
18. Tahimic CG, Long RK, Kubota T, et al. Regulation of ligand and shear stress-induced insulin-like growth factor 1 (IGF1) signaling by the integrin pathway. J Biol Chem. 2016;291(15):8140–9.
19. Rangaswami H, Schwappacher R, Marathe N, et al. Cyclic GMP and protein kinase G control a Src-containing mechanosome in osteoblasts. Sci Signal. 2010;3(153):ra91.
20. Plotkin LI, Speacht TL, Donahue,H.J. Cx43 and mechanotransduction in bone. Curr Osteoporos Rep. 2015;13:67–72.
21. Goodenough DA, Paul DL. Beyond the gap: functions of unpaired connexon channels. Nat Rev Mol Cell Biol. 2003;4:285–94.
22. Penuela S, Gehi R, Laird DW. The biochemistry and function of pannexin channels. Biochim Biophys Acta. 2013;1828:15–22.
23. Plotkin LI, Bellido T. Beyond gap junctions: Connexin43 and bone cell signaling. Bone 2013;52:157–66.
24. Turner CH, Robling AG. Exercise as an anabolic stimulus for bone. Curr Pharm Des. 2004;10:2629–41.
25. Orriss IR. The role of purinergic signaling in the musculoskeletal system. Auton Neurosci. 2015;191:124–34.
26. Li J, Liu D, Ke HZ, et al. The P2X7 nucleotide receptor mediates skeletal mechanotransduction. J Biol Chem. 2005;280:42952–9.
27. Seref-Ferlengez Z, Maung S, Schaffler MB, et al. P2X7R-Panx1 complex impairs bone mechanosignaling under high glucose levels associated with type-1 diabetes. PLoS One. 2016;11:e0155107.
28. Thompson WR, Rubin CT, Rubin J. Mechanical regulation of signaling pathways in bone. Gene 2012;503:179–93.
29. Lieben L, Carmeliet G. The Involvement of TRP channels in bone homeostasis. Front Endocrinol (Lausanne). 2012;3:99.
30. Lu XL, Huo B, Chiang V, et al. Osteocytic network is more responsive in calcium signaling than osteoblastic network under fluid flow. J. Bone Miner Res. 2012;27:563–74.
31. Robinson JA, Chatterjee-Kishore M, Yaworsky PJ, et al. Wnt/beta-catenin signaling is a normal physiological

response to mechanical loading in bone. J Biol Chem. 2006;281:31720–8.
32. Saxon LK, Jackson BF, Sugiyama T, et al. Analysis of multiple bone responses to graded strains above functional levels, and to disuse, in mice in vivo show that the human Lrp5 G171V high bone mass mutation increases the osteogenic response to loading but that lack of Lrp5 activity reduces it. Bone. 2011;49:184–93.
33. Niziolek PJ, Warman ML, Robling AG. Mechanotransduction in bone tissue: the A214V and G171V mutations in Lrp5 enhance load-induced osteogenesis in a surface-selective manner. Bone. 2012; 51:459–65.
34. Sawakami K, Robling AG, Ai M, et al. The Wnt co-receptor LRP5 is essential for skeletal mechanotransduction but not for the anabolic bone response to parathyroid hormone treatment. J Biol Chem. 2006;281:23698–711.
35. Kramer I, Halleux C, Keller H, et al. Osteocyte Wnt/beta-catenin signaling is required for normal bone homeostasis. Mol Cell Biol. 2010;30(12):3071–85.
36. Chen S, Feng J, Bao Q, et al. Adverse effects of osteocytic constitutive activation of ß-catenin on bone strength and bone growth. J Bone Miner Res. 2015;30(7): 1184–94.
37. Maurel DB, Duan P, Farr J, et al. Beta-catenin haplo insufficient male mice do not lose bone in response to hindlimb unloading. PLoS One. 2016;11(7):e0158381.
38. Robling AG, Niziolek PJ, Baldridge LA, et al. Mechanical stimulation of bone in vivo reduces osteocyte expression of Sost/sclerostin. J Biol Chem. 2008;283(9):5866–75.
39. Tu X, Rhee Y, Condon KW, et al. Sost downregulation and local Wnt signaling are required for the osteogenic response to mechanical loading. Bone. 2012;50(1): 209–17.
40. Burger EH, Klein-Nulend J. Microgravity and bone cell mechanosensitivity. Bone. 1998;22(suppl 5):S127S–30.
41. Thorsen K, Kristoffersson AO, Lerner UH, et al. In situ microdialysis in bone tissue. Stimulation of prostaglandin E2 release by weight-bearing mechanical loading. J Clin Invest. 1996;98(11):2446–9.
42. Tanaka SM, Sun HB, Roeder RK, et al. Osteoblast responses one hour after load-induced fluid flow in a three-dimensional porous matrix. Calcif Tissue Int. 2005;76(4):261–71.
43. Li J, Burr DB, Turner CH. Suppression of prostaglandin synthesis with NS-398 has different effects on endocortical and periosteal bone formation induced by mechanical loading. Calcif Tissue Int. 2002;70(4):320–9.
44. Burra S, Jiang JX. Connexin 43 hemichannel opening associated with prostaglandin E(2) release is adaptively regulated by mechanical stimulation. Commun Integr Biol. 2009;2:239–40.
45. Saura M, Tarin C, Zaragoza C. Recent insights into the implication of nitric oxide in osteoblast differentiation and proliferation during bone development. Sci World J. 2010;10:624–32.
46. Baron R, Kneissel M. WNT signaling in bone homeostasis and disease: from human mutations to treatments. Nat Med. 2013;19:179–92.

11
The Composition of Bone

Adele L. Boskey[1] and Pamela G. Robey[2]

[1]*Research Institute, Hospital for Special Surgery; and Department of Biochemistry and Graduate Field of Physiology, Biophysics, and Systems Biology, Cornell University Medical and Graduate Medical Schools, New York, NY, USA*
[2]*Skeletal Biology Section, National Institute of Dental and Craniofacial Research, National Institutes of Health, Department of Health and Human Services, Bethesda, MD, USA*

INTRODUCTION

Bone comprises the largest proportion of the body's connective tissue mass. Unlike most other connective tissue matrices, bone matrix is physiologically mineralized and is unique in that it is constantly regenerated throughout life as a consequence of bone turnover. Bone as an organ is made up of the cartilaginous joints, the calcified cartilage in the growth plate (in developing individuals), the marrow space, and the cortical and cancellous mineralized structures. Bone as a tissue consists of mineralized and nonmineralized (osteoid) components of the cortical and cancellous regions of long and flat bones. There are three cell types on (and in) bone tissue: (i) the bone-forming osteoblasts, which when engulfed in mineral become (ii) osteocytes; and (iii) the bone-destroying osteoclasts. Each of these cells communicates with one another by either direct cell contact or through signaling molecules, and respond to each other. The detailed properties of these cells have been discussed in numerous publications; see for example the review by Teti [1] or the text by Bilezikian and colleagues [2]. This chapter focuses on the extracellular matrix, which is synthesized primarily by osteoblasts and osteocytes but also contains proteins adsorbed from the circulation. The preponderance of bone is mineral and extracellular matrix (ECM). Information on the gene and protein structure and potential function of bone ECM constituents has exploded during the last few decades. This information has been described in great detail in several recent texts [2,3], to which the reader is referred for specific references, which are too numerous to be listed adequately here. This chapter summarizes the composition of bone and the salient features of the classes of bone matrix proteins. The tables list specific details for the individual ECM components.

BONE AS A COMPOSITE

Bone is a composite material whose ECM consists of mineral, collagen, water, noncollagenous proteins, and lipids in decreasing proportion (depending on age, species, and site). These components contribute to both the mechanical and metabolic functions of bone. Understanding of some of the biological functions of these components has come from mouse models and analyses of healthy and diseased human tissues, and from cell culture studies.

Mineral

The mineral phase of bone is a nanocrystalline, highly substituted analog of the naturally occurring mineral, hydroxylapatite $[Ca_{10}(PO_4)_6(OH)_2]$. The major substitutions are with carbonate, magnesium, and acid phosphate, along with other trace elements, the content of which depends on diet and environment. Although the precise chemical nature of the initial mineral formed has been debated [4–6],recent data suggest the first mineral deposits are disordered and heterogeneous [5]. It is well accepted, however, that the majority of the "biomineral" present in bones is apatitic [7] and that this mineral is aligned with the collagen fibrils. The physical and chemical properties of this

---

Primer on the Metabolic Bone Diseases and Disorders of Mineral Metabolism, Ninth Edition. Edited by John P. Bilezikian.
© 2019 American Society for Bone and Mineral Research. Published 2019 by John Wiley & Sons, Inc.
Companion website: www.wiley.com/go/asbmrprimer

Table 11.1. Characteristics of collagen-related genes and proteins found in bone matrix.

Protein/Gene	Function	Disease/Animal Model/Phenotype
Type I: 17q21.23, 7q21.3-22 [α1(I)$_2$α2(I)] [α1(I)$_3$]	Serves as scaffolding, binds and orients other proteins that nucleate hydroxyapatite deposition	Human mutations: osteogenesis imperfecta (OMIM #166210, 166200, 610854, 259420, 166220) Mouse models: oim mouse; mov 13 mouse; brittle mouse; Amish mouse; bones hypermineralized and mechanically weak, mineral crystals small, some mineral outside collagen
Type X: 6q21-22.3 [α1(X)$_3$]	Present in hypertrophic cartilage of the growth plate, but does not appear to regulate matrix mineralization	Human mutations: Schmid metaphyseal chondrodysplasia (OMIM #120110) Knockout mouse: no apparent skeletal phenotype
Type III: 2q24.3-31 [α1(III)]$_3$	Present in bone in trace amounts, may regulate collagen fibril diameter, its paucity in bone may explain the large diameter size of bone collagen fibrils	Human mutations in type III: different forms of vascular Ehlers-Danlos syndrome and abnormal collagen type I folding (OMIM #130050) Mouse model: disrupted trabecular bone formation
Type V: 9q34.2-34.3;2q24.3-31, 9q34.2-34.3 [α1(V)$_2$α2(V)] [α1(V)α2(V)α3(V)]	Present in bone in trace amounts, may regulate collagen fibril diameter, their paucity in bone may explain the large diameter size of bone collagen fibrils	Mutations in type V α1 or α2 (OMIM #120215, 120190) Mouse model: disrupted fibril arrangement

mineral have been determined by a variety of techniques including chemical analyses, X-ray diffraction, vibrational spectroscopy, energy dispersive electron analysis, nuclear magnetic resonance, small angle scattering, and transmission and atomic force microscopy [8].

The functions of the mineral are to strengthen the collagen composite, providing more mechanical resistance to the tissue, and also to serve as a source of calcium, phosphate, and magnesium ions for mineral homeostasis. For physicochemical reasons, usually it is the smallest mineral crystals that are lost during remodeling. Thus, in osteoporosis, it is not surprising that the larger, more perfect crystals persist within the matrix [9] contributing to the brittle nature of osteoporotic bone. When remodeling is impaired, as in osteopetrosis, the mineral crystals remain small relative to age-matched controls [9].

Collagen

The basic building block of the bone matrix fiber network is type I collagen, which is a triple helical molecule containing two identical α1(1) chains and a structurally similar, but genetically different, α1(2) chain [2]. Collagen α chains are characterized by a Gly-X-Y repeating triplet (where X is usually proline, and Y is often hydroxyproline) and by several posttranslational modifications including: (i) hydroxylation of certain lysyl or prolyl residues; (ii) glycosylation of the hydroxylysine with glucose or galactose residues or both; (iii) addition of mannose at the propeptide termini; and (iv) formation of intra- and intermolecular covalent cross-links that differ from those found in soft connective tissues. Measurement of these bone-derived collagen cross-links in urine has proven to be a good measure of bone resorption [10]. Bone matrix proper consists predominantly of type I collagen; however, trace amounts of type III and V and fibril-associated collagens (Table 11.1) may be present during certain stages of bone formation and may regulate collagen fibril diameter.

Noncollagenous proteins

Noncollagenous proteins (NCPs) compose 10% to 15% of the total bone protein content. NCPs are multifunctional, having roles in organizing the ECM, coordinating cell-matrix and mineral-matrix interactions, and regulating the mineralization process. The multifunctionality can be attributed to their protein structure, as a large number of these proteins are intrinsically disordered (mostly random coils) and thus can bind to many partners [11]. Knowledge of their specific functions has come from studies of the isolated proteins in solution, from analyses of mice in which the proteins are ablated (knocked out [KO]), or overexpressed, characterization of human diseases in which these proteins have mutations, and studies using appropriate cell cultures. Tables 11.2–11.7 summarize the gene and protein structures, and the functions of these protein families.

Serum-derived proteins

Approximately one-fourth of the total NCP content is exogenously derived (Table 11.2). This fraction is largely composed of serum-derived proteins, such as albumin and α$_2$-HS-glycoprotein, which are acidic in character and bind to bone matrix because of their affinity for hydroxyapatite. Although these proteins are not endogenously synthesized, they may exert effects on matrix

Table 11.2. Gene and protein characteristics of serum proteins found in bone matrix.

Protein/Gene	Function	Disease/Animal Model/Phenotype
Albumin: 2q11-13 69 kDa, nonglycosylated, one sulfhydryl, 17 disulfide bonds, high affinity hydrophobic binding pocket	Inhibits hydroxyapatite crystal growth	
α2HS glycoprotein: 3q27-29 Precursor protein of fetuin, cleaved to form A and B chains that are disulfide linked, Ala-Ala and Pro-Pro repeat sequences, N-linked oligosaccharides, cystatin-like domains	Promotes endocytosis, has opsonic properties, chemoattractant for monocytic cells, bovine analog (fetuin) is a growth factor; inhibits calcification	Knockout mouse: adult ectopic calcification

mineralization and bone cell proliferation. For example, α_2-HS-glycoprotein, the human analog of fetuin, when ablated in mice causes ectopic calcification [12] suggesting that the protein is a mineralization inhibitor. The remainder of the exogenous fraction is composed of growth factors and a large variety of other molecules present in trace amounts, which influence local bone cell activity [1,2].

On a mole-to-mole basis, bone-forming cells synthesize and secrete as many molecules of NCPs as of collagen. These molecules can be classified into four general (and sometimes overlapping) groups: (i) proteoglycans; (ii) glycosylated proteins; (iii) glycosylated proteins with potential cell attachment activities; and (iv) γ-carboxylated (gla) proteins. The physiological roles for individual bone protein constituents are not well defined; however, they may participate not only in regulating the deposition of mineral but also in the control of osteoblastic and osteoclastic metabolism.

Proteoglycans

Proteoglycans are macromolecules that contain acidic polysaccharide side chains (glycosaminoglycans) attached to a central core protein. Bone matrix contains several members of this family [2] (Table 11.3).

During initial stages of bone formation, the large chondroitin sulfate proteoglycan, versican, and the glycosaminoglycan, hyaluronan (which is not attached to a protein core), are highly expressed and may delineate areas that will become bone. With continued osteogenesis, versican is replaced by two small chondroitin sulfate proteoglycans, decorin and biglycan, composed of tandem repeats of a leucine-rich repeat (LRR) sequence. Decorin has been implicated in the regulation of collagen fibrillogenesis and is distributed predominantly in the ECM space of connective tissues and in bone, whereas biglycan tends to be found in pericellular locales. A heparan sulfate proteoglycan, perlecan, is involved in limb patterning, and is found surrounding chondrocytes in the growth plate, whereas the glypican family of cell surface-associated heparan sulfate proteoglycans also affect skeletal growth. In addition, there are other small leucine-rich proteoglycans (SLRPs) in bone [13], including osteoglycin (mimecan), keratocan, osteoadherin, lumican, asporin, and fibromodulin. Although their exact physiological functions are not known, these proteoglycans are assumed to be important for the integrity of most connective tissue matrices. Deletion of the *biglycan* gene, for example, leads to a significant decrease in the development of trabecular bone, indicating that it is a positive regulator of bone formation. Deletion of the *epiphican* gene, or the *epiphican* and *biglycan* genes together, causes shortening of the femur during growth and early onset osteoarthritis [14]. Other functions might arise from the ability of these proteoglycans to bind and modulate the activity of the growth factors in the extracellular space, thereby influencing cell proliferation and differentiation [1].

Glycosylated proteins

Glycosylated proteins with diverse functions abound in bone. One of the hallmarks of bone formation is the synthesis of high levels of alkaline phosphatase (Table 11.4).

Alkaline phosphatase, a glycoprotein enzyme, is primarily bound to the cell surface through a phosphoinositol linkage, but is cleaved from the cell surface and found within mineralized matrix. The function of alkaline phosphatase in bone cell biology has been the matter of much speculation and remains undefined. Mice lacking tissue nonspecific alkaline phosphatase have impaired mineralization, suggesting the importance of this enzyme for mineral deposition [15].

The most abundant NCP produced by bone cells is osteonectin [16,17], a phosphorylated glycoprotein accounting for ~2% of the total protein of developing bone in most animal species. Osteonectin is transiently produced in nonbone tissues that are rapidly proliferating, remodeling, or undergoing profound changes in tissue architecture, and is also found constitutively expressed in certain types of epithelial cells, cells associated with the skeleton, and in platelets. Osteonectin, along with thrombospondin-2 (TSP-2) and periostin are members of the class of "matricellular proteins," each of which has a role in bone cell proliferation and differentiation; with some role in regulating mineralization. Tetranectin (which is important for wound healing), tenascin

Table 11.3. Gene and protein characteristics: glycosaminoglycan-containing molecules in bone.

Protein/Gene	Function	Disease/Animal Model/Phenotype
Aggrecan: 15q26.1 ~2.5×10^6 intact protein, ~180–370,000 kDa core, ~100 CS chains of 25 kDa, and some KS chains of similar size, G1, G2, and G3 globular domains with hyaluronan-binding sites, EGF and CRP-like sequences	Matrix organization, retention of water and ions, resilience to mechanical forces	Human mutation: spondyloepiphyseal dysplasia (OMIM #155760, 608361) and premature growth cessation (OMIM #165800) Mouse models: brachymorphic mouse; accelerated growth plate calcification, cartilage matrix deficiency, shortened stature Nanomelic chick (mutation): abnormal bone shape
Versican (PG-100): 5q12-14 ~1×10^6 intact protein, ~360 kDa core, ~12 CS chains of 45 kDa, G1 and G3 globular domains with hyaluronan binding sites, EGF and CRP-like sequences	Regulates chondrogenesis; may "capture" space that is destined to become bone	Human mutation: Wagner syndrome (an ocular disorder) (OMIM #143200)
Decorin (class 1 LRR): 12q21.33 ~130 kDa intact protein, ~38–45 kDa core with 10 leucine-rich repeat sequences, 1 CS chain of 40 kDa	Binds to collagen and may regulate fibril diameter, binds to TGF-β and may modulate activity, inhibits cell attachment to fibronectin	Human mutation: congenital stromal corneal dystrophy (OMIM #610048) Mouse knockout: no apparent skeletal phenotype although collagen fibrils are abnormal; Decorin (DCN)/Biglycan (BGN) double knockout – progeroid form of Ehlers-Danlos syndrome
Biglycan (class 1 LRR): Xq27 ~270 kDa intact protein, ~38–45 kDa core protein with 12 leucine-rich repeat sequences, 2 CS chains of 40 kDa	Binds to collagen, TGF-β, and other growth factors; pericellular environment, a genetic determinant of peak bone mass	Human mutation: thoracic aortic aneurysms and dissections (OMIM #615291) Knockout mouse: osteopenia; thin bones, decreased mineral content, increased crystal size; short stature
Asporin (class 1 LRR): 9q22.31 67 kDa, most likely few GAG chains	Regulates collagen structure	Human polymorphisms: associated with osteoarthritis (OMIM #608135) and intervertebral disk degeneration (OMIM #603932)
Fibromodulin (class 2 LRR): 1q32.1 59 kDa intact protein, 42 kDa core protein, one N-linked KS chain	Binds to collagen, may regulate fibril formation, binds to TGF-β	Mouse model: fewer tendon fibril bundles Fmod/Bgn double knockout mice: joint laxity and formation of supernumery sesmoid bones
Osteoadherin (class 2 LRR)/ osteomodulin 85 kDa intact protein, 47 kDa core protein, rich in KS, RGD sequence	Expression restricted to mineralized tissues, may mediate cell attachment and play a role in endochondral bone formation	
Lumican (class 2 LRR): 12q21.33 70–80 kDa intact protein, 37 kDa core proetin	Binds to collagen, may regulate fibril formation and growth	Lum/Fmod double knockout mouse: ectopic calcification and a variant of Ehlers-Danlos syndrome (OMIM #130000)
Perlecan: 1p36.12 Five domain heparan sulfate proteoglycan, core protein 400 kDa	Interacts with matrix components to regulate cell signaling; cephalic development	Human mutations: short stature, dystrophy of epiphyseal cartilage; dyssegmental dysplasia, Silverman-Handmaker type Transgenic mice with mutated perlecan: Schwartz-Jampel syndrome (OMIM #142461); mice have impaired mineralization and misshapen skeletons and joint abnormalities Knockout mice: phenotype resembling thanatophoric dysplasia (TD) type I (OMIM #187600)
Glypican: Xq26 Lipid-linked heparan sulfate proteoglycan, 14 conserved cysteine residues	Regulates BMP-SMAD signaling; regulates cell development	Human mutation: Simpson-Golabi-Behmel syndrome (OMIM #300037) Knockout mouse: delayed endochondral ossification, impaired osteoclast development
Osteoglycin/mimecan (class 3 LRR): 9q22 299 aa precursor, 105 aa mature protein, no GAG in bone, keratan sulfate in other tissues	Binds to TGF-β, regulates collagen fibrillogenesis	
Hyaluronan: multi-gene complex Multiple proteins associated outside of the cell, structure unknown	May work with versican molecule to capture space destined to become bone	

CRP = C-reactive protein; EGF = epidermal growth factor; TGF = transforming growth factor.

Table 11.4. Gene and protein characteristics of glycoproteins in bone matrix.

Protein/Gene	Function	Disease/Animal Models/Phenotype
Alkaline phosphatase (bone-liver-kidney isozyme): 1p34-36.1 Two identical subunits of ~80 kDa, disulfide bonded, tissue-specific posttranslational modifications	Potential Ca^{2+} carrier, hydrolyzes inhibitors of mineral deposition such as pyrophosphates, increases local phosphate concentration	Human mutations: hypophosphatasia (OMIM #171760) (decreased activity) TNAP knockout mouse: growth impaired; decreased mineralization
Osteonectin: 5q31.3-q32 ~35–45 kDa, intramolecular disulfide bonds, α helical amino terminus with multiple low-affinity Ca^{2+} binding sites, two EF hand high-affinity Ca^{2+} sites, ovomucoid homology, glycosylated, phosphorylated, tissue-specific modifications	Regulates collagen organization; may mediate deposition of hydroxyapatite, binds to growth factors, may influence cell cycle, positive regulator of bone formation	Knockout mouse: severe osteopenia, decreased trabecular connectivity; decreased mineral content; increased crystal size
Tetranectin: 3p22-p21.3 21 kDa protein composed of four identical subunits of 5.8 kDa, sequence homologies with a sialoprotein receptor and G3 domain of aggrecan	Binds to plasminogen, may regulate matrix mineralization	Knockout mouse: no long bone phenotype, spinal deformity (kyphosis, increased curvature of the thoracic spine), increased mineralization in implant model
Tenascin-C: 9q33.1 Hexameric structure, six identical chaines of 320 kDA, Cys rich, EGF-like repeats, FN type III repeats	Interferes with cell–FN interactions	Knockout mouse: no apparent skeletal phenotype
Tenascin-X: 6p21.33 Hexameric with 5 N-linked glycosylation sites and multiple EGF and 40 FN type III repeats	Regulates cell-matrix interactions	Human mutation and knockout mouse: Ehlers-Danlos II phenotype with hyperextensible skin, hypermobile joints, and tissue fragility (OMIM #600985)
Secreted phosphoprotein 2: 2q37.1 24-kDa secreted phosphoprotein, shares sequence homology with members of the cystatin family of thiol protease inhibitors	Associates with regulators of mineralization in serum, may regulate thiol proteases in bone, may have a role in inhibiting calcification	

EGF = epidermal growth factor; FN = fibronectin;

(which regulates organization of ECM) [18], and secreted phosphoprotein 24 (which regulates bone morphogenetic protein expression) along with periostin [19] are other glycoproteins found in the bone matrix.

Small integrin-binding ligand, N-glycosylated proteins, and other glycoproteins with cell attachment activity

All connective tissue cells interact with their extracellular environment in response to stimuli that direct or coordinate (or both) specific cell functions, such as migration, proliferation, and differentiation (Tables 11.5 and 11.6). These particular interactions involve cell attachment through transient or stable focal adhesions to extracellular macromolecules, which are mediated by cell surface receptors that subsequently transduce intracellular signals. Bone cells synthesize at least 12 proteins that may mediate cell attachment: members of the small integrin-binding ligand, N-glycosylated protein (SIBLING) family (osteopontin [OPN], bone sialoprotein, dentin matrix protein-1, dentin sialophosphoprotein, and matrix extracellular phosphoprotein [MEPE]), type I collagen, fibronectin, thrombospondin(s) (predominantly TSP-2 with lower levels of TSP-1, -3, and -4 and cartilage oligomeric matrix protein [COMP]), vitronectin, fibrillin, BAG-75, and osteoadherin (which is also a proteoglycan). Many of these proteins are phosphorylated and/or sulfated, and all contain RGD (Arg-Gly-Asn), the cell attachment consensus sequence that binds to the integrin class of cell surface molecules. However, in some cases, cell attachment seems to be independent of RGD, indicating the presence of other sequences or mechanisms of cell attachment [2]. Thrombospondin(s), fibronectin, vitronectin, fibrillin, and osteopontin are expressed in many tissues. Certain types of epithelial cells synthesize bone sialoprotein, and it is highly enriched in bone and is expressed by hypertrophic chondrocytes, osteoblasts, osteocytes, and osteoclasts. In bone, the expression of bone sialoprotein correlates with the appearance of mineral [20]. The bone sialoprotein knockout at an early age has impaired new bone formation, while adult knockout mice have shorter stature,

Table 11.5. Gene and protein characteristics of SIBLINGs (small integrin-binding ligands, N-glycosylated proteins).

Protein/Gene	Function	Disease/Animal Models/Phenotype
Osteopontin: 4q21 ~44–75 kDa, polyaspartyl stretches, no disulfide bonds, glycosylated, phosphorylated, RGD located 2/3 from the N terminal	Binds to cells, may regulate mineralization, may regulate proliferation, inhibits nitric oxide synthase, may regulate resistance to viral infection	Knockout mouse: decreased crystal size; increased mineral content; not subject to osteoclastic remodeling
Bone sialoprotein: 4q21 ~46–75 kDa, polyglutamyl stretches, no disulfide bonds, 50% carbohydrate, tyrosine-sulfated, RGD near the C terminus	Binds to cells, may initiate mineralization, regulates turnover	Knockout mouse: delayed endochondral ossification, adults have shorter stature, lower bone turnover and higher trabecular bone mass
DMP-1: 4q21 513 amino acids predicted; serine-rich, acidic, RGD 2/3 from N terminus	Regulator of biomineralization; regulates osteocyte function	Human mutations: dentinogenesis imperfecta and hypophosphatemia (OMIM #600980) Knockout mouse: undermineralized with craniofacial and growth plate abnormalities and defective osteocyte function
Dentin sialophosphoprotein: 4q21.3 Gene produces three proteins, dentin sialoprotein, dentin phosphophoryn, and dentin glycoprotein. All have RGD sites; dentin phosphophoryn is highly phosphorylated	Regulation of biomineralization	Human mutations: dentinal dysplasias and dentinogenesis imperfecta; no bone disease (OMIM #125485) Knockout mouse: thinner bones at 9 months, no significant other bone phenotype, and severe dentin abnormalities
MEPE: 4q21.1 525 amino acids, two N-glycosylation motifs, a glycosaminoglycan-attachment site, an RGD cell-attachment motif, and phosphorylation motifs	Regulation of biomineralization; regulation of PHEX (phosphaturic hormone) activity	Humans: association with oncogenic osteomalacia Knockout mouse: increased bone mass and resistance to ovariectomy-induced bone loss

Table 11.6. Gene and protein characteristics of other RGD-containing glycoproteins.

Protein/Gene	Function	Disease/Animal Models
Thrombospondins (1-4, COMP): 15Q-1, 6q27, 1q21-24, 5q13, 19p13.1 ~450 kDa molecules, three identical disulfide linked subunits of ~150–180 kDa, homologies to fibrinogen, properdin, EGF, collagen, von Willebrand, *Plasmodium falciparum* and calmodulin, RGD at the C terminal globular domain	Cell attachment (but usually not spreading), binds to heparin, platelets, types I and V collagens, thrombin, fibrinogen, laminin, plasminogen and plasminogen activator inhibitor, histidine-rich glycoprotein	Human COMP mutation: pseudoachondroplasia (OMIM #600310) TSP-2 knockout mouse: large collagen fibrils, thickened bones; spinal deformities
Fibronectin: 2q34 ~400 kDa with two nonidentical subunits of ~200 kDa, composed of type I, II, and III repeats, RGD in the 11th type III repeat 2/3 from N terminus	Binds to cells, fibrin, heparin, gelatin, collagen	Knockout mouse: lethal prior to skeletal development
Vitronectin: 17q11 ~70 kDa, RGD close to N terminus, homology to somatomedin B, rich in cysteines, sulfated, phosphorylated	Cell attachment protein, binds to collagen, plasminogen, and plasminogen activator inhibitor, and to heparin	
Fibrillin 1 and 2: 15q21.1, 5q23-q31 350 kDa, EGF-like domains, RGD, cysteine motifs	May regulate elastic fiber formation	Human fibrillin 1 mutations: Marfan syndrome (OMIM #134797) Human fibrillin 2 mutations: congenital contractural arachnodactyly (OMIM #121050).

EGF = epidermal growth factor.

lower bone turnover, and higher bone mass, which may be associated with the elevated expression of other SIBLING proteins (MEPE and OPN) [21]. In solution, bone sialoprotein can function as an hydroxyapatite nucleator, is found in association with bone acidic glycoprotein-75 in mineralization foci [22], and is upregulated during mineralization in culture [23].

Both osteopontin and bone sialoprotein are known to anchor osteoclasts to bone, and in addition to supporting cell attachment, bind Ca^{2+} with extremely high affinity through polyacidic amino acid sequences. Each SIBLING protein regulates hydroxyapatite formation in solution [11], and their knockouts have phenotypes that can be correlated with these in vitro functions. It is not immediately clear why there are such a plethora of RGD-containing proteins in bone; however, the pattern of expression varies from one RGD protein to another, as does the pattern of the different integrins that bind to these proteins. This variability indicates that cell-matrix interactions change as a function of maturational stage, suggesting that they also may play a role in osteoblastic maturation. Their posttranslational modifications also vary, suggesting that these modifications may determine their in situ functions [11,24].

Gla-containing proteins

At least five bone-matrix NCPs, matrix gla protein (MGP), osteocalcin (bone gla protein [BGP]), periostin (also a bone matrix glycoprotein), and gla-rich protein (GRP) [25], all of which are made endogenously, and protein S (made primarily in the liver but also made by osteogenic cells) are posttranslationally modified by the action of vitamin K-dependent γ-carboxylases (Table 11.7). The dicarboxylic glutamyl (gla) residues enhance calcium binding. MGP and GRP, also called unique cartilage matrix protein (UCMA) are found in many connective tissues, whereas osteocalcin is more bone-specific, and periostin is made in all connective tissues that respond to load [26]. The physiological roles of these proteins are still under investigation; MGP, GRP, osteocalcin, and periostin may function in the control of mineral deposition and remodeling. MGP-deficient mice develop calcification in extraskeletal sites such as the aorta [27], implying that it is an inhibitor of mineralization. GRP knockout mice, in contrast, had no phenotype [25]. Expression of MGP in blood vessels of the MGP-deficient mice prevents calcification, whereas expression in osteoblasts prevents mineralization [28]. Osteocalcin seems to be involved in regulating bone turnover. Osteocalcin-deficient mice are reported to have increased BMD compared with normal [29] but, with age, the mineral properties did not show the changes that occurred in age-matched controls, suggesting a role for osteocalcin in osteoclast recruitment [30]. In human bone, osteocalcin is concentrated in osteocytes, and its release may be a signal in the bone turnover cascade. Osteocalcin measurements in serum have proved valuable as a marker of bone turnover in metabolic disease states [9]. In contrast, uncarboxylated osteocalcin was reported to be a hormone involved in the regulation of energy and glucose metabolism in mice [31], but was not found to have this function in human studies [32]. Periostin senses load and regulates periodontal and vascular calcification [33]. Periostin-deficient mice also have increased vascular calcifications.

Table 11.7. Gene and protein characteristics of γ-carboxy glutamic acid-containing proteins in bone matrix.

Protein/Gene	Function	Disease/Animal Model/Phenotype
Matrix Gla protein: 12p13.1 ~15 kDa, five gla residues, one disulfide bridge, phosphoserine residues	May function in cartilage metabolism, a negative regulator of mineralization	Human mutations: Keutel syndrome (OMIM #245150), excessive cartilage calcification Knockout mouse: excessive cartilage calcification
Osteocalcin: 1q25-31 ~5 kDa, one disulfide bridge, gla residues located in α helical region	May regulate activity of osteoclasts and their precursors, may mark the turning point between bone formation and resorption, suggested to be a hormone	Knockout mouse: osteopetrotic - thickened bones, decreased crystal size, increased mineral content
Gla-rich protein–UCMA: NC_000010.11	Modifies calcium availability; accumulates at sites of pathological calcification and inhibits vascular and valvular calcification, function might be associated with prevention of calcium-induced signaling pathways and direct mineral binding to inhibit crystal formation/maturation	Knockout mouse: no phenotype Zebrafish KO: skeletal malformations
Protein S: 3p11-q11.2 ~72 kDa	Primarily a liver product, but may be made by osteogenic cells	Human mutations: protein deficiency with osteopenia (OMIM #076080)

Table 11.8. Effects of bone matrix molecules on mineralization in vitro.

Promote or Support Apatite Formation	Inhibit Mineralization	Dual Function (Nucleate and Inhibit)	No Published Effect
Bone sialoprotein	Aggrecan	Biglycan	Decorin
Type I collagen	α2-HS glycoprotein	Osteonectin	Lumican
Proteolipid (matrix vesicle nucleational core)	Matrix gla protein (MGP)	Fibronectin	Mimecan
BAG-75	Osteocalcin	Bone sialoprotein Osteopontin	Tetranectin
Alkaline phosphatase		MEPE	Thrombospondin
Phospho-1			
Osteoadherin			

MEPE = matrix extracellular phosphoprotein.

OTHER COMPONENTS

The previous sections summarize the major components of bone ECM, but there are other minor components that affect the properties of the tissue. For example, there are numerous enzymes that are important for processing the ECM components. Some of these are cell-associated; some are found in the ECM. Readers are referred to other reviews [2,34–37] for more details. Growth factors sequestered in bone regulate cell-matrix interactions and cell function [1,2]. Water accounts for ~10% of the weight of bone, depending on species and bone age. Water is important for cell and matrix nutrition, for control of ion flux, and for maintenance of the collagen structure, because type I collagen contains the bulk of the tissue water.

Lipids make up <2% of the dry weight of bone; however, they have some significant effects on bone properties [38]. The lipid droplets on the surface of collagen fibrils in cortical bone that appear to be associated with the mineral have been identified as triglycerides [39]. Animals lacking lipids and lipid enzymes may have skeletal phenotypes; for example, the dwarfed neutral sphingomyelinase-deficient mouse [40], the fro/fro mouse that mimics severe osteogenesis imperfecta and has a chemically induced mutation in sphingomyelinase [41], and the caveolin-1-deficient mouse that has increased bone mass and increased stiffness [42]. Conditional phospholipase A2 knockout mice were reported to have increased age-related bone loss [43], implying a role for this phospholipid-cleaving enzyme in bone turnover.

Each of the components in the organic matrix of bone influences the mechanism of mineral deposition. Some promote mineralization on the collagen matrix, some inhibit the formation and/or growth of mineral crystals, and some are multifunctional, promoting in some cases and inhibiting in others. The known effects on hydroxyapatite formation in solution for each of the extracellular matrix components discussed in this chapter are summarized in Table 11.8.

REFERENCES

1. Teti A. Bone development: overview of bone cells and signaling. Curr Osteoporos Rep. 2011;9(4):264–73.
2. Bilezikian JP, Raisz LG, Martin TJ. *Principles of Bone Biology* (3rd ed.). San Diego: Academic Press, 2008.
3. Johnson ML. How rare bone diseases have informed our knowledge of complex diseases. Bonekey Rep. 2016;5:839.
4. Grynpas MD, Omelon S. Transient precursor strategy or very small biological apatite crystals? Bone. 2007;41(2):162–4.
5. Dey A, Bomans PH, Müller FA, et al. The role of prenucleation clusters in surface-induced calcium phosphate crystallization. Nat Mater. 2010;9(12):1010–14.
6. Akiva A, Kerschnitzki M, Pinkas I, et al. Mineral formation in the larval zebrafish tail bone occurs via an acidic disordered calcium phosphate phase. J Am Chem Soc. 2016;138(43):14481–7.
7. Campi G, Ricci A, Guagliardi A, et al. Early stage mineralization in tissue engineering mapped by high resolution X-ray microdiffraction. Acta Biomater. 2012;8(9):3411–8.
8. Boskey AL. Organic and inorganic matrices. In: Wnek GE, Bowlin GL (eds) *Encyclopedia of Biomaterials and Biomedical Engineering* (2nd ed.). New York: Informa, 2008, pp. 2039–53.
9. Boskey AL. Osteoporosis and osteopetrosis. In: Baeuerlein E, Bchrens P, Epple M (eds) *Biomineralization in Medicine*, vol. 7. New York: Wiley, 2007, pp 59–75.
10. Garnero P. Biomarkers for osteoporosis management: utility in diagnosis, fracture risk prediction and therapy monitoring. Mol Diagn Ther. 2008;12(3):157–70.
11. Boskey AL, Villarreal-Ramirez E. Intrinsically disordered proteins and biomineralization. Matrix Biol. 2016;52–54:43–59.
12. Schafer C, Heiss A, Schwarz A, et al. The serum protein alpha 2-Heremans-Schmid glycoprotein/fetuin-A is a systemically acting inhibitor of ectopic calcification. J Clin Invest 2003;112(3):357–66.

13. Nikitovic D, Aggelidakis J, Young MF, et al. The biology of small leucine-rich proteoglycans in bone pathophysiology. J Biol Chem. 2012;287(41):33926–33.
14. Nuka S, Zhou W, Henry SP, et al. Phenotypic characterization of epiphycan-deficient and epiphycan/biglycan double-deficient mice. Osteoarthritis Cartilage. 2010;18(1):88–96.
15. Anderson HC, Sipe JB, Hessle L, et al. Impaired calcification around matrix vesicles of growth plate and bone in alkaline phosphatase-deficient mice. Am J Pathol. 2004;164(3):841–7.
16. Rosset EM, Bradshaw AD. SPARC/osteonectin in mineralized tissue. Matrix Biol. 2016;52–54:78–87.
17. Delany AM, Hankenson KD. Thrombospondin-2 and SPARC/osteonectin are critical regulators of bone remodeling. J Cell Commun Signal. 2009;3(3–4):227–38.
18. Kimura H, Akiyama H, Nakamura T, et al. Tenascin-W inhibits proliferation and differentiation of preosteoblasts during endochondral bone formation. Biochem Biophys Res Commun. 2007;356(4):935–41.
19. Kii I, Nishiyama T, Li M, et al. Incorporation of tenascin-C into the extracellular matrix by periostin underlies an extracellular meshwork architecture. J Biol Chem. 2010;285(3):2028–39.
20. Paz J, Wade K, Kiyoshima T, et al. Tissue- and bone cell-specific expression of bone sialoprotein is directed by a 9.0 kb promoter in transgenic mice. Matrix Biol. 2005;24(5):341–52.
21. Bouleftour W, Boudiffa M, Wade-Gueye NM, et al. Skeletal development of mice lacking bone sialoprotein (BSP)—impairment of long bone growth and progressive establishment of high trabecular bone mass. PLoS One. 2014;9(5):e95144.
22. Huffman NT, Keightley JA, Chaoying C, et al. Association of specific proteolytic processing of bone sialoprotein and bone acidic glycoprotein-75 with mineralization within biomineralization foci. J Biol Chem 2007;282(36):26002–13.
23. Egusa H, Kayashima H, Miura J, et al. Comparative analysis of mouse-induced pluripotent stem cells and mesenchymal stem cells during osteogenic differentiation in vitro. Stem Cells Dev. 2014;23(18):2156–69.
24. Prasad M, Butler WT, Qin C. Dentin sialophosphoprotein in biomineralization. Connect Tissue Res. 2010;51(5):404–17.
25. Cancela ML, Conceição N, Laizé V. Gla-rich protein, a new player in tissue calcification? Adv Nutr. 2012;3(2):174–81.
26. Coutu DL, Wu JH, Monette A, et al. Periostin, a member of a novel family of vitamin K-dependent proteins, is expressed by mesenchymal stromal cells. J Biol Chem. 2008;283(26):17991–8001.
27. Luo G, Ducy P, McKee MD, et al. Spontaneous calcification of arteries and cartilage in mice lacking matrix GLA protein. Nature. 1997;386(6620):78–81.
28. Murshed M, Schinke T, McKee MD, et al. Extracellular matrix mineralization is regulated locally; different roles of two gla-containing proteins. J Cell Biol. 2004;165(5):625–30.
29. Ducy P, Desbois C, Boyce B, et al. Increased bone formation in osteocalcin-deficient mice. Nature. 1996;382(6590):448–52.
30. Boskey AL, Gadaleta S, Gundberg C, et al. Fourier transform infrared microspectroscopic analysis of bones of osteocalcin-deficient mice provides insight into the function of osteocalcin. Bone. 1998;23(3):187–96.
31. Lee NK, Sowa H, Hinoi E, et al. Endocrine regulation of energy metabolism by the skeleton. Cell. 2007;130(3):456–69.
32. Kumar R, Vella A. Carbohydrate metabolism and the skeleton: picking a bone with the beta-cell. J Clin Endocrinol Metab. 2011;96(5):1269–71.
33. Bonnet N, Standley KN, Bianchi EN, et al. The matricellular protein periostin is required for sost inhibition and the anabolic response to mechanical loading and physical activity. J Biol Chem. 2009;284(51):35939–50.
34. Trackman PC. Diverse biological functions of extracellular collagen processing enzymes. J Cell Biochem. 2005;96(5):927–37.
35. Ge G, Greenspan DS. Developmental roles of the BMP1/TLD metalloproteinases. Birth Defects Res C Embryo Today. 2006;78(1):47–68.
36. Yadav MC, Simao AM, Narisawa S, et al. Loss of skeletal mineralization by the simultaneous ablation of PHOSPHO1 and alkaline phosphatase function: a unified model of the mechanisms of initiation of skeletal calcification. J Bone Miner Res. 2011;26(2):286–97.
37. Mebarek S, Abousalham A, Magne D, et al. Phospholipases of mineralization competent cells and matrix vesicles: roles in physiological and pathological mineralizations. Int J Mol Sci. 2013;14(3):5036–129.
38. Goldberg M, Boskey AL. Lipids and biomineralizations. Prog Histochem Cytochem. 1996;31(2):1–187.
39. Mroue KH, Xu J, Zhu P, et al. Selective detection and complete identification of triglycerides in cortical bone by high-resolution (1)H MAS NMR spectroscopy. Phys Chem Chem Phys. 2016;18(28):18687–91.
40. Stoffel W, Jenke B, Block B, et al. Neutral sphingomyelinase 2 (smpd3) in the control of postnatal growth and development. Proc Natl Acad Sci USA. 2005;102(12):4554–9.
41. Aubin I, Adams CP, Opsahl S, et al. A deletion in the gene encoding sphingomyelin phosphodiesterase 3 (Smpd3) results in osteogenesis and dentinogenesis imperfecta in the mouse. Nat Genet. 2005;37(8):803–5.
42. Rubin J, Schwartz Z, Boyan BD, et al. Caveolin-1 knockout mice have increased bone size and stiffness. J Bone Miner Res. 2007;22(9):1408–18.
43. Ramanadham S, Yarasheski KE, Silva MJ, et al. Age-related changes in bone morphology are accelerated in group VIA phospholipase A2 (iPLA2β)-null mice. Am J Path. 2008;172(4):868–81.

12

Assessment of Bone Mass, Structure, and Quality in Rodents

Jeffry S. Nyman[1] and Deepak Vashishth[2]

[1]*Department of Orthopaedic Surgery and Rehabilitation, Vanderbilt University Medical Center, Nashville, TN, USA*
[2]*Department of Biomedical Engineering, Rensselaer Polytechnic Institute Troy, NW, USA*

INTRODUCTION

Rodents are commonly used in preclinical studies that assess the ability of signaling pathways or new therapies to regulate bone mass, cortical structure, trabecular architecture, bone quality, and, ultimately, to maintain or increase fracture resistance. This assessment is possible because of the widely available tools capable of measuring multiple bone characteristics in vivo and ex vivo. Moreover, the organization of bone at the macroscale, microscale, and nanoscale is similar in rodents and humans: bone shapes of the appendicular and axial skeleton are homologous despite the difference in locomotion; trabeculae are located at the ends of long bones and the centrum of vertebrae; and at the ultrastructural level of organization, bone matrix is composed of hydrated type I collagen organized into fibrils with noncollagenous proteins and nanocrystals of calcium phosphate (hydroxyapatite with carbonate substitutions) existing within and outside the fibrils [1,2]. While there are some differences in the relative fractions of the matrix constituents and in the collagen cross-linking pattern between human and rodent bone [3], the primary difference is the lack of an extensive Haversian system in rodent bone, although a few osteons are present [4]. Although rodents do not necessarily phenocopy certain musculoskeletal diseases, a number of age-related changes in bone are similar between humans and rodents [3]. Moreover, ovariectomy (OVX) in rodents mimics the effects of human menopause on bone mass and trabecular architecture, thus providing a model of post-menopausal osteopenia/osteoporosis [5,6].

Unlike humans, wild-type mice and rats do not experience low-energy, fragility fractures as they age or upon gonadectomy. Thus, the definitive way to determine whether a treatment improves fracture resistance in a rodent study is the mechanical testing of bones, namely the lumbar vertebra and femur, ex vivo. Prior to this destructive technique, rodent bones are typically subjected to X-ray analysis to characterize effects on mass, cortical structure, trabecular architecture, and mineral density. Following the gold standard assessment of bone mass in patients, DXA scanners can longitudinally measure areal bone mineral density (aBMD) for specific bones or the entire skeleton of rodents. There are now in vivo micro-computed tomography (μCT) scanners that can assess cross-sectional geometry of cortical bone and three-dimensional architecture of trabecular bone and do so with low radiation exposure to the animal.

This chapter reviews the standard imaging techniques that assess bone mass, structure, and architecture in rodent studies as well as the common mechanical testing techniques used to determine multiple biomechanical properties of rodent bones. Together these tools can provide a comprehensive picture of bone density, structure, and quality to provide insight into treatment and targeted signaling pathways that affect fracture resistance. Emphasis is given to practical considerations when evaluating rodent bones in preclinical studies as there are published guidelines on the technical aspects of μCT imaging [7] and mechanical testing [8,9] of rodent bone.

Primer on the Metabolic Bone Diseases and Disorders of Mineral Metabolism, Ninth Edition. Edited by John P. Bilezikian.
© 2019 American Society for Bone and Mineral Research. Published 2019 by John Wiley & Sons, Inc.
Companion website: www.wiley.com/go/asbmrprimer

DUAL-ENERGY X-RAY ABSORPTIOMETRY

Although the widely used PIXImus (GE Lunar, Madison, WI, USA) is no longer manufactured, new small animal DXA scanners with high spatial resolution (~50 μm instead of ~180 μm pixel size) are available to researchers, typically as part of a digital X-ray imaging cabinet. The basic principles remain the same: X-ray beams are passed through the rodent at two different photon energies (eg, 28 and 48 keV), and the difference in X-ray attenuation between hard and soft tissue at the two energy levels provides the means to estimate bone mass and density, lean mass, and fat mass [10]. As with clinical DXA imaging of patients [11], consistent positioning of the anesthetized rodent ensures that the orientation of the bones does not vary across animals and scan sessions. Otherwise, variable bone orientation would alter the resulting data. Moreover, with routine calibration of the DXA imaging system, measurements of bone mineral content in grams and aBMD (g/cm^2 as the content within selected area) are relatively precise, with an intraoperator error of ~2% among repeated scans [12].

The advantages of DXA in preclinical studies are: (i) the low cost; (ii) the low radiation exposure and short scan time allowing for weekly measurements from numerous animals; (iii) additional assessment of body composition (fat); and (iv) measurement of multiple regions of interest, such as the lumbar vertebra and distal femur, from a single image. The latter requires precise selection of the region of interest (ie, the selected area). The disadvantages of DXA primarily stem from being a two-dimensional projection method in that: (i) treatment-related changes in trabecular bone cannot be distinguished from changes in cortical bone; (ii) imaging artifacts due to obesity, osteophytes, or other degenerative changes in the spine, and implants (eg, osmotic minipump), can alter the aBMD measurement; and (iii) there is insufficient sensitivity to directly assess changes in cortical thickness or porosity. As with any X-ray-based technique, DXA is also insensitive to nonmineral attributes of bone, namely type I collagen.

MICRO-COMPUTED TOMOGRAPHY

With the availability of high-resolution μCT scanners for in vivo imaging of small animals and ex vivo imaging of small bone samples (eg, mouse L6 vertebra), μCT assessment of cortical bone structure, trabecular architecture, and volumetric bone mineral density is now the gold standard for establishing a bone phenotype in rodents. Initially, pQCT scanners were used to analyze ex vivo bone structure and volumetric BMD (vBMD) of cortical and trabecular compartments in rat studies at nominal resolutions between 70 and 120 μm. Now scanners can image the entire skeleton of a rodent or isolated limbs in vivo at nominal resolutions of ~35 and ~10 μm, respectively. There are also ex vivo scanners capable of achieving submicron resolution without a synchrotron radiation beamline; and thus, nano-CT imaging of lacunar pores and small vascular channels in rodent bones is now practical [13] if not widely available. The primary advantage of μCT imaging over DXA is the ability for three-dimensional assessment of structural and architectural properties, but the disadvantage is the higher cost related to the production of the instrument, maintenance of the scanner, and labor. Still, μCT, whether ex vivo or in vivo, has become the most commonly reported technique to assess skeletal phenotypes in rodents.

Practical considerations

There are a number of trade-offs in μCT imaging. With longitudinal in vivo imaging, which is a sensitive way to establish drug-induced or OVX-induced changes from baseline scans [14,15], radiation exposure needs to be minimized so as to avoid undue biological effects [16]. This should be accomplished by both reducing the radiation exposure of each scan and limiting the total number of in vivo scans for each animal. Thus, forgoing a global assessment of skeletal changes, isolated scans of the hindlimb can provide high-resolution images of the proximal tibia and distal femur as well as the midshaft of either bone. Such images are sufficient to resolve trabecular architecture and cortical structure but not lacunar or small vascular pores in rodent bones. Coregistering serial scans can be a challenge, especially in rapidly growing animals, but when achievable [17,18] can provide volumetric dynamic measurements of both bone formation and resorption in vivo in rodents [19].

The ability to resolve small structural features depends on spatial resolution, which is determined by nominal resolution (ie, voxel size) and contrast/noise ratio (CNR) (ie, graininess of the image). Achieving a small voxel size with high CNR requires a long scan time because spatial resolution depends on the number of X-ray projections per rotation, integration time, beam intensity, and the averaging of multiple acquisitions per rotation. High-resolution scans also generate extremely large data sets, requiring large disk space to store the image stacks and large amounts of computational power for reconstruction and analysis. Thus, there is a practical limit to what can be resolved in typical μCT scans of rodent bone. Fortunately, images with 12 and 20 μm isotropic voxels and moderate CNR are sufficient to provide accurate structural measurements of trabeculae and cortices from adult mouse and rat bones, respectively. Achieving such nominal resolutions with relatively short scan times is possible when the bone is positioned relatively closer to the X-ray source than the detector, taking advantage of the magnification effect of the cone-shaped beam. This also reduces the diameter of the field of view and hence the region of bone that can be imaged. However, a narrow region (metaphysis including growth plate or the midshaft of the diaphysis) is usually sufficient to accurately measure structure, architecture, and density. Although not useful for density measurements, phase contrast-enhanced μCT imaging is emerging as an improved way to detect small pores in rodent cortical bone by disproportionately highlighting the edges or surfaces of objects of highly different densities [20].

Other considerations for proper ex vivo μCT imaging include immersing the bone in an appropriate solution or buffer for any subsequent analyses, selecting the field of view, employing validated beam hardening corrections during image reconstruction, and calibrating the scanner to a hydroxyapatite phantom. While scanning dry bone actually provides better contrast, bones need to be fully hydrated when mechanically tested (ie, dry bone is brittle [21]) and drying prior to histology distorts marrow morphology. To avoid the potentially confounding effect of dehydration/rehydration (ie, collagen contracts when dry [22]), bones are best imaged immersed in PBS when μCT precedes biomechanical testing of the same samples. Moreover, the conversion of linear attenuation to hydroxyapatite density (mgHA/cm^3) may assume the bone is in a liquid. If properties related to bone geometry are sought for biomechanical analysis, then the rodent bone (L6 vertebra, femur, etc.) should be axially aligned with the long axis of the specimen tube or orthogonal to the X-rays. This is to ensure the image cross-section (XY plane) matches the orientation of the bone when loaded either in compression or in bending. The angle of the bone in the image stack can be digitally realigned post-scanning, but the process shortens the usable Z-length of the scan. Since the X-ray source is polychromatic (X-rays at multiple photon energies), beam hardening artifacts can confound the measurement of vBMD. The use of filters built into the scanner between the X-ray source and sample can narrow the range of energy levels such that a beam harden correction—an empirical conversion typically offered by manufacturers—can minimize the reported effect of bone size on vBMD measurements [23]. This of course requires weekly calibration tests against the hydroxyapatite phantom and occasional recalibration as the spectrum of the X-ray source drifts.

Segmentation and analysis

There are published guidelines on evaluating cortical and trabecular bone by μCT imaging (selecting volumes of interest, noise filtering, segmenting, as well as property reporting and meaning of each property) [7], and evaluations of bones prior to mechanical testing are discussed here. For long bones undergoing flexural mechanical testing, the central portion of the diaphysis, where the load or moment is applied, is evaluated. There are no published studies that systematically investigated how much of the diaphysis should be evaluated, but as a general rule of thumb, the axial length of the midshaft evaluation can be between 1 and 3 mm for mouse bones and 3 and 5 mm for rat bones. The key considerations are to evaluate the diaphysis region subjected to the maximum moment and to ensure the long axis of bone is aligned with the z axis of the scan. The rotation of the bone within the XY plane (ie, cross-section) is not as critical because the invariant known as the minimum principal of the moment of inertia (I_{min}) coincides with the moment of inertia for the bone orientation in which the direction of loading is directed along the anterior–posterior axis (I_{a-p})[24] (Fig. 12.1). The moment of inertia (also known as the second moment of area) characterizes the distribution of bone tissue to resist bending [25]. The polar moment of inertia characterizes the structural resistance to torsion.

Since key structural properties such as I_{min} and the distance between the centroid and the outer most bone surface in the anterior–posterior direction (c_{min}) are determined from binary segmented images (bone is the object analyzed and medullary canal is the background), contouring to select the ROI is actually not necessary as long as isolated trabeculae are not present or negligible in the medullary space. However, contouring the cortex of the midshaft is useful for assessing apparent vBMD, medullary volume, total cross-sectional area (Tt.Ar), and cortical porosity. There are various edge detection algorithms that can be used to fit contours to the periosteal and endosteal surfaces (eg, snakes [26], dual thresholding [27], attenuation profile [28], polar segmentation [29]). Although the techniques are automated or semiautomated, the fit of the contour to the surface, especially the endosteum, which can undergo resorption creating cortical pores and "trabecularization" (transitional zone)[30], should be checked.

COMPRESSION TESTS OF VERTEBRAL BODIES

Compression testing of lumbar vertebra (typically L5 or L6) assesses whether a treatment or signaling pathway affects whole bone strength at a clinically relevant site. In particular, it is sensitive to OVX and treatment effects on trabecular bone [31–34]. Before testing, the end plates are typically removed to provide nearly parallel surfaces. Using a material testing system with a linear actuator, the vertebral body (VB) is loaded at a specified rate in displacement control (eg, 3 mm/min) between two compression platens. These platens can have a slightly rough surface to prevent slippage. One platen can sit on a moment relief since the cut surfaces may not be exactly parallel, thereby minimizing inadvertent shear deformation. During the quasi-static test, the load cell sensor having the appropriate capacity (maximum of 100 and 1000 N for mouse and rat bones, respectively) measures the force, and the linear variable displacement transducer (LVDT), which is attached to the actuator, measures the displacement. From the resulting force versus displacement curve, stiffness is the slope of the linear portion of the curve and whole bone strength is the maximum or peak force. Identifying the failure point can be rather difficult in compression tests as the bone can compact after

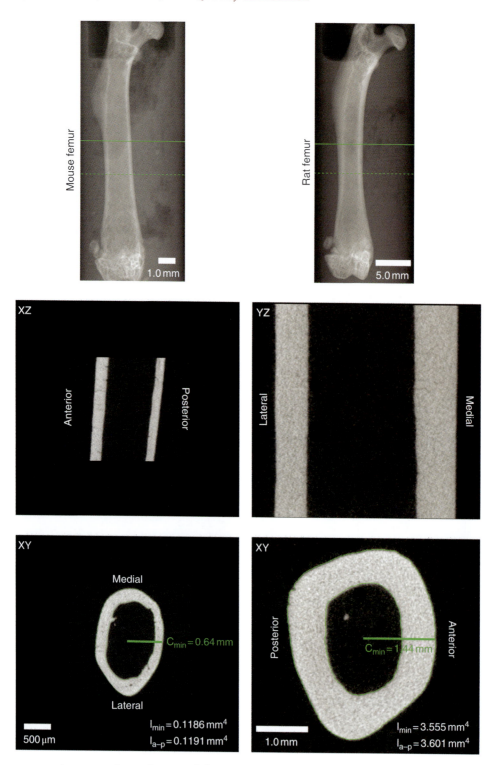

Fig. 12.1. Micro-computed tomography evaluation of the rodent femur midshaft. By aligning the long axis of the femur with the scanner axis, the proper cross-sectional properties can be determined without having to realign the image stack. Typically, a short segment of the midshaft corresponding to the point of loading is evaluated.

failure, thereby causing the force to stay above zero. Therefore, work to fracture (area under the curve) is typically not determined, although work to maximum force is a useful indicator of how well the VB dissipates energy.

BENDING TESTS OF LONG BONES

Rodent long bones are amenable to flexural testing, namely three-point (3 pt) and four-point (4 pt) bending. As discussed by Jepsen et al. [9], 3 pt is not as technically challenging as 4 pt, and, in general, mechanical properties from 3 pt bending tests are more widely reported. Thus, descriptions of this test and the mechanical property calculations follow.

The hydrated bone, most commonly the femur, is placed on the lower contact points (rounded) with a specified span (eg, 8 mm for mouse femur and 16 mm for rat femur). The orientation of the bone (eg, medial forward and anterior down) must be consistent across animals. A preload of 0.5 N (mouse) and 1 N (rat) can be applied to prevent rotation of the bone during the initial loading phase. In the center of the span, the upper contact point (rounded) engages the midpoint of the diaphysis), the region evaluated by μCT. For the radius, this is the point of curvature [35]. As with the compression test of the lumbar VB, the long bone is loaded at a specified rate in displacement control, and a force versus displacement curve is generated (Fig. 12.2). Usually, but not always, the bone snaps in two, thus giving an unequivocal failure point. Structural properties are stiffness (or rigidity), maximum force (or maximum moment) endured by the bone, and work to fracture (or span-adjusted work to fracture) [36,37]. To determine post-yield displacement, an indicator of brittleness, the yield point (proportional limit when elastic deformation transitions to permanent or plastic deformation) can be defined by the 0.2% offset or by 10% to 15% loss in secant stiffness (Fig. 12.2). These measurements and the mechanical properties of bone are affected by state of hydration and loading rates. Thus, these parameters should be kept constant among different test groups for comparison [9]. Span can be adjusted to match the bone size, but then rigidity, maximum moment, and span-adjusted work to fracture should be reported [38].

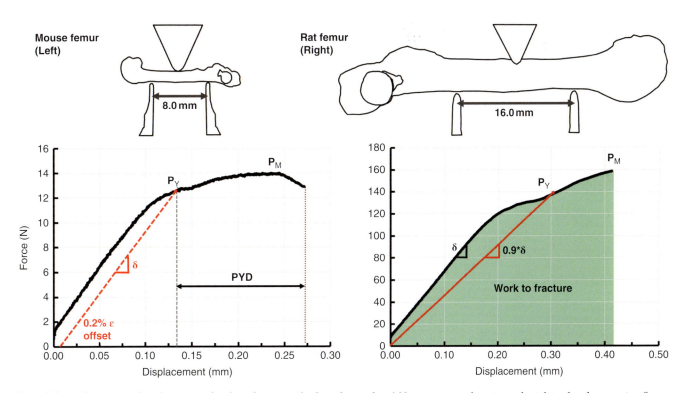

Fig. 12.2. Three-point bending test of rodent femurs. The long bone should be consistently oriented within the three-point fixture (eg, medial side forward and anterior side down), and the upper loading point should contact the midpoint that was evaluated by μCT. The flexural test generates a force versus displacement curve from which the stiffness (δ), yield force (P_Y), maximum force (P_M), post-yield displacement (PYD), and work to fracture (area under the force versus displacement curve) can be determined. There are two different ways to determine the yield point. Either one can be used for mouse or rat long bones. For clarity, PYD is arbitrarily depicted in the mouse curve and work to fracture in the rat curve.

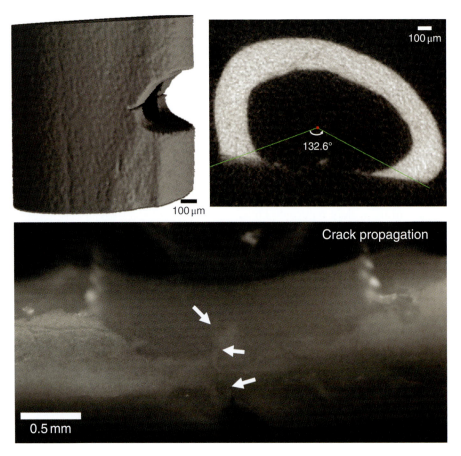

Fig. 12.3. Fracture toughness testing of rodent bone. The starter crack is created by rubbing a razor blade coated with a dimond solution within the notch (3D μCT rendering). The notch angle is determined from the cross-section of the notch (2Θ). When loaded in three-point bending, a crack propogates from the micro-notch (note the bone is kept hydrated during the test).

Using the geometric characteristics from prior μCT evaluations of the midshaft, namely I_{min}/c_{min} or I_{a-p}/c_{a-p}, material properties of the cortical bone can be *estimated* as follows [unit of measurement] at the *apparent* level:

$$\text{Modulus} = \text{stiffness} \times \text{span}^3/I_{min}/48 \, [\text{GPa}]$$

$$\text{Strength} = \text{maximum force} \times \text{span} \times c_{min}/I_{min}/4 \, [\text{MPa}]$$

$$\text{Toughness} = 3 \times \text{work to fracture}/\text{span}/\text{Ct.Ar} \, [\text{MJ/m}^3]$$

The calculation of toughness [39] is an alternative to the area under the stress versus strain curve. To derive the flexural formulas, the material is assumed to be homogenous and to behave in an elastic, not plastic, manner. Again, these calculations can only estimate material properties because rodent bones are not uniform in geometry and not long enough to minimize shear deformation.

Unlike work to fracture or toughness, fracture toughness (K_c) provides a measure of the resistance that the bone material provides against fracture. Fracture toughness relates to changes only at the material level such as changes in the composition, proportion, or interrelationships between mineral, protein (cross-linking, amount, distribution), and water content of bone, and the porosity and organization of matrix (orientation, number, size of fiber bundles, lamellae) that are often altered with development, aging, exercise, and pharmaceutical treatments.

Determining the fracture toughness of rodent bone involves introducing a notch in the mid-diaphyseal region (Fig. 12.3) such that the maximum moment occurs at the notch during 3 pt bending tests [40]. The notch should be sharp, so typically after introducing the notch with a thin diamond-embedded wafer blade, it is sharpened with a razor and diamond solution. Notched bones are loaded at a low displacement rate (0.06 mm/min) until fracture in displacement control (available on most mechanical testing machines), and the resulting load–displacement curve is used to calculate fracture toughness (see following equations) at initiation (ie, initiation toughness) and maximum load (ie, propagation toughness). Human and large animal bones (eg, cows, pigs, dogs) can readily be machined into standardized geometries to reduce variations in measurements in material properties [41,42]. Because fracture toughness testing involves the introduction of a defect (such as a microcrack) and measurements related to this defect, fracture toughness properties show less variation than those from traditional strength tests and often require a smaller sample size. To account for inherent tissue heterogeneity and to reduce errors associated with variations in bone geometry (eg, femur), μCT-based measurements are recommended to estimate the geometric parameters required to calculate fracture toughness [39,43] by the following equations:

$$\left[k = F_b * \frac{P_c * S * R_o}{\pi\left(R_o^4 - R_i^4\right)} * \sqrt{\pi * R_m * \Theta}\right]$$

$$F_b = \left(1 + \frac{t}{2R_m}\right)\left[A_b + B_b\left(\frac{\Theta}{\pi}\right) + C_b\left(\frac{\Theta}{\pi}\right)^2 + D_b\left(\frac{\Theta}{\pi}\right)^3 + E_b\left(\frac{\Theta}{\pi}\right)^4\right]$$

$$A_b = 0.65133 - 0.5774\xi - 0.3427\xi^2 - 0.0681\xi^3$$

$$B_b = 1.879 + 4.795\xi + 2.343\xi^2 - 0.6197\xi^3$$

$$C_b = -9.779 - 38.14\xi - 6.611\xi^2 + 3.972\xi^3$$

$$D_b = 34.56 + 129.9\xi + 50.55\xi^2 + 3.374\xi^3$$

$$E_b = -30.82 - 147.69\xi - 78.38\xi^2 - 15.54\xi^3$$

$$\xi = \log\left(\frac{t}{R_m}\right)$$

where:

F_b = geometric factor for an edge-cracked cylindrical pipe
P_c = maximum load (propagation toughness via maximum load method) or load at yield (initiation toughness; a secant line with a 5% lower slope than the elastic modulus was plotted on the load deformation curve, and its intersection with the curve was used to determine the load at initiation)
S = span length
R_o = periosteal radius of cortical shell
R_i = endosteal radius of cortical shell
R_m = mean radius of cortical shell
Θ = half-crack angle at crack initiation
t = cortical thickness

Unlike the crack initiation point, K_c calculated at maximum load provides a more comprehensive measurement of bone's resistance to fracture [44] and has a smaller standard deviation (assuming consistent notching) compared with initiation or other methods of determining bone fragility including strength tests [39].

CONCLUSION

DXA, μCT, and mechanical testing (compression or three-point bending) are commonly used to determine how genetic manipulations or drug treatments affect bone in rodent studies. In vivo imaging with DXA and μCT can longitudinally assess changes in bone mineral density, trabecular architecture, and cortical structure. However, ultimately, to know whether a manipulation or treatment affects the ability of bone to resist fracture, ex vivo destructive, mechanical testing is necessary. Using proper techniques and knowing the limitations of each technology can ensure proper interpretation of the observed differences in bone properties among experimental groups.

ACKNOWLEDGMENTS

We thank Sasidhar Uppuganti for his assistance in generating the figures.

REFERENCES

1. Sroga GE, Vashishth D. Effects of bone matrix proteins on fracture and fragility in osteoporosis. Curr Osteoporos Rep. 2012;10(2):141–50.
2. Reznikov N, Shahar R, Weiner S. Bone hierarchical structure in three dimensions. Acta Biomater. 2014;10(9):3815–26.
3. Nyman JS. Age-related changes to bone structure and quality in rodent models. In: Ram J, Conn PM (eds) Conn's Handbook of Models for Human Aging (2nd ed.). London: Academic Press, 2017, pp 919–36.
4. Zhou H, Chernecky R, Davies JE. Deposition of cement at reversal lines in rat femoral bone. J Bone Mineral Res. 1994;9(3):367–74.
5. Jee WS, Yao W. Overview: animal models of osteopenia and osteoporosis. J Musculoskelet Neuronal Interact. 2001;1(3):193–207.
6. Komori T. Animal models for osteoporosis. Eur J Pharmacol. 2015;759:287–94.
7. Bouxsein ML, Boyd SK, Christiansen BA, et al. Guidelines for assessment of bone microstructure in rodents using micro-computed tomography. J Bone Miner Res. 2010;25(7):1468–86.
8. Turner CH, Burr DB. Basic biomechanical measurements of bone: a tutorial. Bone. 1993;14(4):595–608.
9. Jepsen KJ, Silva MJ, Vashishth D, et al. Establishing biomechanical mechanisms in mouse models: practical guidelines for systematically evaluating phenotypic changes in the diaphyses of long bones. J Bone Miner Res. 2015;30(6):951–66.
10. Blake GM, Fogelman I. Technical principles of dual energy x-ray absorptiometry. Semin Nucl Med. 1997;27(3):210–28.
11. Lilley J, Walters BG, Heath DA, et al. In vivo and in vitro precision for bone density measured by dual-energy X-ray absorption. Osteoporos Int. 1991;1(3):141–6.
12. Nagy TR, Clair AL. Precision and accuracy of dual-energy X-ray absorptiometry for determining in vivo body composition of mice. Obes Res. 2000;8(5):392–8.
13. Langer M, Peyrin F. 3D X-ray ultra-microscopy of bone tissue. Osteoporos Int. 2016;27(2):441–55.
14. Boyd SK, Davison P, Müller R, et al. Monitoring individual morphological changes over time in ovariectomized rats by in vivo micro-computed tomography. Bone. 2006;39(4):854–62.

15. Campbell GM, Tiwari S, Grundmann F, et al. Three-dimensional image registration improves the long-term precision of in vivo micro-computed tomographic measurements in anabolic and catabolic mouse models. Calcif Tissue Int. 2014;94(3):282–92.
16. Longo AB, Sacco SM, Salmon PL, et al. Longitudinal use of micro-computed tomography does not alter microarchitecture of the proximal tibia in sham or ovariectomized Sprague-Dawley rats. Calcif Tissue Int. 2016;98(6):631–41.
17. Altman AR, Tseng W-J, de Bakker CMJ, et al. Quantification of skeletal growth, modeling, and remodeling by in vivo micro computed tomography. Bone. 2015;81:370–9.
18. Lu Y, Boudiffa M, Dall'Ara E, et al. Development of a protocol to quantify local bone adaptation over space and time: quantification of reproducibility. J Biomech. 2016;49(10):2095–9.
19. de Bakker CMJ, Altman AR, Tseng W-J, et al. μCT-based, in vivo dynamic bone histomorphometry allows 3D evaluation of the early responses of bone resorption and formation to PTH and alendronate combination therapy. Bone. 2015;73:198–207.
20. Pratt IV, Belev G, Zhu N, et al. In vivo imaging of rat cortical bone porosity by synchrotron phase contrast micro computed tomography. Phys Med Biol. 2015;60(1):211–32.
21. Nyman JS, Roy A, Shen X, et al. The influence of water removal on the strength and toughness of cortical bone. J Biomech. 2006;39(5):931–8.
22. Vesper EO, Hammond MA, Allen MR, et al. Even with rehydration, preservation in ethanol influences the mechanical properties of bone and how bone responds to experimental manipulation. Bone. 2017;97:49–53.
23. Fajardo RJ, Cory E, Patel ND, et al. Specimen size and porosity can introduce error into microCT-based tissue mineral density measurements. Bone. 2009;44(1):176–84.
24. Uppuganti S, Granke M, Makowski AJ, et al. Age-related changes in the fracture resistance of male Fischer F344 rat bone. Bone. 2016;83:220–32.
25. Jepsen KJ, Pennington DE, Lee YL, et al. Bone brittleness varies with genetic background in A/J and C57BL/6 J inbred mice. 2001;16(10):1854–62.
26. Maksimovic R, Stankovic S, Milovanovic D. Computed tomography image analyzer: 3D reconstruction and segmentation applying active contour models—'snakes'. Int J Med Inform. 2000;58–59:29–37.
27. Buie HR, Campbell GM, Klinck RJ, et al. Automatic segmentation of cortical and trabecular compartments based on a dual threshold technique for in vivo micro-CT bone analysis. Bone. 2007;41(4):505–15.
28. Zebaze R, Ghasem-Zadeh A, Mbala A, et al. A new method of segmentation of compact-appearing, transitional and trabecular compartments and quantification of cortical porosity from high resolution peripheral quantitative computed tomographic images. Bone. 2013;54(1):8–20.
29. Manhard MK, Uppuganti S, Granke M, et al. MRI-derived bound and pore water concentrations as predictors of fracture resistance. Bone. 2016;87:1–10.
30. Zebaze R, Seeman E. Cortical bone: a challenging geography. J Bone Min Res. 2015;30:30–8.
31. Kousteni S, Chen JR, Bellido T, et al. Reversal of bone loss in mice by nongenotropic signaling of sex steroids. Science. 2002;298(5594):843–6.
32. Wang F, Wang P-X, Wu X-L, et al. Deficiency of adiponectin protects against ovariectomy-induced osteoporosis in mice. PLoS One. 2013;8(7):e68497.
33. Li X, Niu Q-T, Warmington KS, et al. Progressive increases in bone mass and bone strength in an ovariectomized rat model of osteoporosis after 26 weeks of treatment with a sclerostin antibody. Endocrinology. 2014;155(12):4785–97.
34. Amugongo SK, Yao W, Jia J, et al. Effects of sequential osteoporosis treatments on trabecular bone in adult rats with low bone mass. Osteoporos Int. 2014;25(6):1735–50.
35. Schriefer JL, Robling AG, Warden SJ, et al. A comparison of mechanical properties derived from multiple skeletal sites in mice. J Biomech. 2005;38(3):467–75.
36. Willinghamm MD, Brodt MD, Lee KL, et al. Age-related changes in bone structure and strength in female and male BALB/c mice. Calcif Tissue Int. 2010;86(6):470–83.
37. Creecy A, Uppuganti S, Merkel AR, et al. Changes in the fracture resistance of bone with the progression of type 2 diabetes in the ZDSD rat. Calcif Tissue Int. 2016;99(3):289–301.
38. Brodt MD, Ellis CB, Silva MJ. Growing C57Bl/6 mice increase whole bone mechanical properties by increasing geometric and material properties. J Bone Miner Res. 1999;14(12):2159–66.
39. Ritchie RO, Koester KJ, Ionova S, et al. Measurement of the toughness of bone: a tutorial with special reference to small animal studies. Bone. 2008;43(5):798–812.
40. Poundarik AA, Diab T, Sroga GE, et al. Dilatational band formation in bone. Proc Natl Acad Sci U S A. 2012;109(47):19178–83.
41. Vashishth D, Behiri JC, Bonfield W. Crack growth resistance in cortical bone: concept of microcrack toughening. J Biomech. 1997;30(8):763–9.
42. Granke M, Makowski AJ, Uppuganti S, et al. Prevalent role of porosity and osteonal area over mineralization heterogeneity in the fracture toughness of human cortical bone. J Biomech. 2016;49(13):2748–55.
43. Vashishth D. Small animal bone biomechanics. Bone. 2008;43(5):794–7.
44. Vashishth D. Rising crack-growth-resistance behavior in cortical bone: implications for toughness measurements. J Biomech. 2004;37(6):943–6.

13

Skeletal Healing: Cellular and Molecular Determinants

Alayna E. Loiselle and Michael J. Zuscik

The Center for Musculoskeletal Research and The Department of Orthopaedics, The University of Rochester Medical Center, Rochester, NY, USA

INTRODUCTION

The process of skeletal repair is essential for resolution of: (i) orthopaedic trauma that has caused bony disjunction; or (ii) surgical interventions that are intended to create bony injury with the aim of inducing a repair response. Understanding the cellular and molecular basis of this healing process has been the focus of intense research both in humans and animal models over the past 25 years, with this work largely driven by the need to develop therapeutic strategies to enable or enhance healing of fibrous nonunions, critically-sized defects, or other situations of impaired healing. Combined, failed, or delayed healing impacts up to 10% of all fracture patients seen clinically [1] and can result from multiple factors including comminution, inadequate fixation, infection, tumor, hypoxia/poor blood supply, metabolic dysfunction, and other chronic comorbid diseases [2]. Overall, research efforts have led to a general understanding of the molecular and genetic control over the inflammatory, cellular and tissue processes that are required for healing, which are generally conserved across species and are similar in structurally distinct skeletal elements. This chapter provides a concise and up to date overview of our understanding of the skeletal healing process at the cellular and molecular level, a discussion of a few key situations that complicate healing, and a summary of therapeutic modalities that are either in development or employed clinically to enhance repair or facilitate healing in nonunion situations.

It should be noted that since 2000, study of the biology and pathophysiology of bone healing has grown into a robust field, with >5200 primary citations and >550 clinical or scientific reviews in the published literature. Since our need to accommodate space limitations has precluded inclusion of numerous seminal contributions, we sincerely apologize to authors whose work could not be directly quoted in this overview.

PROCESS OF SKELETAL HEALING

Cellular contribution to healing tissues

The fracture healing process requires the coordinated activity of several different cell types including inflammatory cells, chondro- and osteoprogenitors, chondrocytes, osteoblasts, and osteoclasts. Healing events that occur in various vertebrate species are similar, except that relative to humans the pace of repair is generally accelerated in smaller animals/rodents. Thus, the schematic presented in Fig. 13.1, which depicts the unique morphogenesis of reparative tissue during the phases of bone fracture healing, provides a benchmark for the description of the healing process in general. This process begins immediately after fracture and usually involves both intramembranous and endochondral ossification [3]. The trauma that induces the fracture initially results in hematoma formation at the injury site. Hematoma-associated cytokines including TNF-α and IL-1, -6, -11, and -18 [3] lead to recruitment and infiltration of inflammatory cells to the fracture site, which themselves potentiate the inflammatory environment and induce the secondary recruitment of key mesenchymal stem cell (MSC) populations. These MSCs may derive from several niches, including bone marrow, muscle, periosteum, and possibly the general circulation [4–6]. While there is

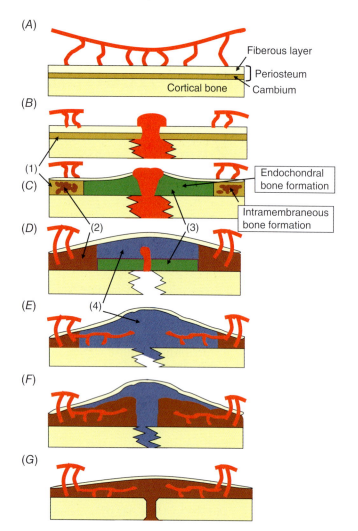

Fig. 13.1. Tissue morphogenesis during bone repair. (A) Periosteum is a well-microvascularized tissue (vessels in red) consisting of an outer fibrous layer and an inner cambium layer. The cambium layer contains abundant stem/progenitor cells that can differentiate into bone and cartilage. (B) Following fracture or osteotomy, the blood supply is disrupted at the defect and a blood clot (hematoma) forms near the disjunction. (C) Progenitor cells residing in the periosteum are recruited to differentiate into osteoblasts to facilitate intramembranous bone formation where intact blood supply is preserved and chondrocytes to facilitate endochondral bone formation adjacent to the fracture where the tissue is hypoxic. In this panel, osteogenic tissue is labeled (1) with newly mineralized tissue labeled (2). Tissues supporting chondrogenesis are labeled (3). (D) Intramembranous bone formation proceeds with robust matrix mineralization (2) where the blood supply is present distal to the fracture site. Endochondral bone formation proceeds simultaneously with chondrogenic tissue supporting a growing population of chondrocytes that comprise the hypertrophic cartilage (4). (E) Cartilage tissue continues to mature, ultimately encompassing the callus nearest the fracture site. Revascularization of the callus also ensues. (F) Chondrocytes in the hypertrophic cartilage undergo terminal differentiation and the matrix is progressively mineralized, expanding the portion of the callus that is comprised of woven bone (brown). (G) The remodeling process proceeds with osteoclasts and osteoblasts facilitating the conversion of woven bone into lamellar bone to ultimately support re-establishment of the appropriate anatomical shape.

much debate about which of these populations are the most critical for initiating repair, data suggest that a key participant is the periosteal progenitor cell that responds to the inflammation by entering the osteogenic or chondrogenic lineage [3]. Endochondral bone formation takes place closest to the fracture site where the oxygen tension is low and vascularity is disrupted. Intramembranous bone formation, on the other hand, occurs distal to the disjunction where intact vasculature remains present. The mechanical stability of the fractured bone markedly affects the fate of the progenitor cells, with stabilized fractures healing with virtually no evidence of cartilage whereas nonstabilized fractures produce abundant cartilage at the fracture site [7].

Given that the periosteum represents a primary source of MSCs that contribute to bone repair, understanding its structure/function is critical for dissecting the tissue and cellular dynamics of the healing process. Overall, periosteum is a vascularized connective tissue that covers the outer surface of cortical bone. It can be separated into two distinct layers: an outer layer that contains fibroblasts and Sharpey's fibers (which facilitate connection to the underlying cortical bone) and an inner layer referred to as the cambium, which contains multipotent MSCs and osteoprogenitor cells that contribute to normal bone growth, healing, and regeneration [8]. It is known that the cambium layer in children is much thicker and better vascularized than in adults, facilitating faster healing.

Once periosteal MSCs have committed to the chondrogenic or osteogenic lineage, chondrocyte and osteoblast differentiation takes place. Directly overlying the site of the fracture, the ends of the original bone have decreased perfusion due to disrupted vascularity. In this central hypoxic region, MSCs differentiate into chondrocytes and endochondral bone formation is initiated. This is consistent with the concept that hypoxia is a critical inducer of chondrogenesis [9]. The tissue that forms as the cell population expands is referred to as the callus, and differentiation of MSCs into chondrocytes occurs directionally within the callus with the process starting in the most central avascular region. While these centrally positioned MSCs persist in the callus area directly overlying the fracture site, chondrocytes that differentiate radially recapitulate the maturation process that occurs in the growth plate, including phases of proliferation, hypertrophy, and terminal differentiation [3]. The calcified cartilage, which acts a template for primary bone formation, is populated by the most terminally differentiated hypertrophic chondrocytes that contribute to the mineralization of the tissue. While many terminally differentiated chondrocytes residing in the calcified cartilage matrix undergo apoptosis, recent work has demonstrated that cartilage-derived cells remain in mature bony tissue during healing and can contribute to new bone formation, taking on a $Col1^+$ phenotype [10]. Concomitant with this transition is a phase of osteoclast-driven remodeling to remove the remaining cartilaginous callus in response to M-CSF, RANKL, and osteoprotegerin (OPG) [11].

Maximal resorption occurs when the ratio of OPG to RANKL expression is at its lowest. Distal to the fracture site and flanking the chondrocytes undergoing endochondral ossification is the location where intramembranous ossification occurs. This process, which proceeds in the zone of injury where blood supply has been better preserved, is characterized by the differentiation of periosteal cells into osteoblasts, which directly lay down new mineral without a cartilage intermediate. As mentioned, the better the fracture is fixed (minimizing instability), the greater the ratio of intramembranous to endochondral ossification in the overall healing process.

Fractures are considered healed when bone stability has been restored by the formation of new bone that bridges the area of fracture. However, this initial woven bone matrix is replaced by organized lamellar bone through a second remodeling process that is the critical final step in achieving an anatomically correct skeletal element. Again, this process is governed by osteoclasts, which become dominant in this final stage due to the induction of IL-1 and TNF and the subsequent expansion of the functional osteoclast population via RANKL in the remodeling callus [11]. Similarly, the initial cortical bone is remodeled and replaced at the fracture site, where necrosis occurred secondary to loss of vascularization due to the injury. The completion of this final remodeling phase results in an anatomically correct element with biomechanical stability that matches the pre-fracture state.

Gene expression profile during fracture healing

Given that the bone repair process is dependent on a combination of endochondral and intramembranous ossification followed by osteoclast remodeling, the genetic profile of the healing tissue is stage-dependent and reflective of the differentiation of these cells. Since the endochondral healing process recapitulates the events that occur during skeletal development, it is no surprise that the genetic profile partially reflects the profile seen in the growth plate chondrocyte hypertrophic program. Overlying this is the genetic profile of osteoblast differentiation that occurs during intramembranous bone formation and in the process of cartilaginous callus conversion into woven bone. Regarding the endochondral process, mesenchymal cell condensation coincides with the expression of early markers of cartilage formation that include Sox-9 and type II collagen [12]. As chondrocyte differentiation ensues, there is a significant increase in cell volume that is associated with the expression of hypertrophy-associated genes that include *type X collagen*, *MMP9* and *MMP13*, *osteocalcin*, and *Indian hedgehog (Ihh)* [3]. Partially in response to the upregulation of hypoxia-inducible factor-1α (HIF-1α), terminally mature chondrocytes contribute to revascularization via expression of VEGF [13] and also may initiate the remodeling process via the induction of osteoclast formation/activity via expression of RANKL [11]. Regarding the intramembranous process, markers of osteoblast differentiation are detected including type I collagen, osteopontin, and osteocalcin. Osteoblasts also contribute to callus revascularization by producing VEGF [13]. The differentiation process in these cells is driven by the expression of Runx2 [14], a transcription factor required for mineralization. By establishing the temporal and regional gene expression pattern during the healing process, a benchmark has emerged that facilitates monitoring healing rates that may be delayed or accelerated depending on comorbidities such as aging or diabetes, or in therapeutic interventions such as BMP-2 treatment.

Molecular control of fracture healing

The molecular signaling pathways involved in the initiation of fracture repair are only superficially understood. Although animals and humans have only very limited capacity to regenerate damaged tissues, it has long been suspected that postnatal bone repair involving endochondral ossification recapitulates some of the essential pathways/factors in limb development [3]. Regarding the developmental process, the most notable regulators are the bone morphogenetic proteins that belong to the TGF-β superfamily, Ihh, the mammalian homologs of Wingless in *Drosophila* (Wnt) proteins, FGFs, and IGFs. Relative to fracture repair, BMP-2 expression has been observed in early periosteal callus just a few days following cortical bone fracture, while elimination of BMP-2 in the limb disrupts the initiation of postnatal fracture healing [15]. Consistent with this, antagonism of the BMP receptor BMPR1A primarily affects the periosteal response. BMPR1A antagonism early after fracture results in delayed formation of the cartilaginous callus, a transition to fibrous tissue, and impaired bony bridging [16]. Moreover, recent work suggests that BMP-2 induction depends in large part on hypoxia, with suppression of hypoxia resulting in a nonunion-like phenotype [17]. Taken together, there is clearly an essential role of BMP-2 in bone repair, driving its use in a number of bone-healing situations (discussed later in this chapter). Evidence has also emerged indicating that Wnt/β-catenin signaling is required to drive osteoblast differentiation in the fracture callus, implicating this pathway as an important participant in the callus mineralization process [18]. Recently, *Sostdc1*, a paralog of *Sclerostin*, has been identified as a marker of osteochondral progenitors, with loss of *Sostdc1* resulting in enhanced bone formation and remodeling [19], suggesting that *Sostdc1* acts to maintain periosteal stem cell quiescence and identifying inhibition of *Sostdc1* as a translational approach to improve fracture repair. Hedgehogs are likely important during the cartilage differentiation phase of healing, and contribute to the repair process [20]. Similarly FGFs [21] and IGFs [22] also play a role during skeletal healing. Overall, studies are ongoing in the fracture healing field with the aim of fully characterizing the role of these pathways and factors in the adult fracture healing process.

In addition to cell differentiation processes that recapitulate limb development, genes that are involved in injury and inflammatory responses during bone repair play key roles in endochondral bone repair. For example, during the inflammatory phase, a constellation of cytokines drive MSC commitment to the chondrogenic and osteogenic lineages as described earlier. Later, during the endochondral healing phase, a turnover of mineralized cartilage occurs that sets the stage for primary woven bone formation. As mentioned, this initial remodeling process coincides with the upregulation of M-CSF, RANKL, OPG, and TNF [3], implicating these factors as critical for the transition from cartilage to bone. During the second remodeling phase when woven bone is converted into lamellar bone, TNF and IL-1 and -6 expression is upregulated, implicating these factors in the recruitment of osteoclasts that are critical for this final remodeling step [11]. Supporting this idea, IL-6 knockout mice display delayed callus mineralization and maturation during repair of femoral osteotomy [23]. Importantly, suppression of inflammation at both remodeling phases is necessary to facilitate progression of healing and provide protection of reparative tissues from chronic inflammatory insult [3] through the production of immunosuppressive paracrine factors [24]. In summary, there is a clear involvement of pro- and anti-inflammatory mediators in the bone repair process.

Several studies have also shown that cyclooxygenase activity is involved in normal bone metabolism and suggest that nonsteroidal anti-inflammatory drugs (NSAIDs) have a negative impact on bone repair. The most compelling data implicating COX function during skeletal healing comes from genetic models that demonstrate a critical role for COX-2. While COX-2$^{-/-}$ mice develop normally, bone repair is impaired in adult knockout mice following fracture [25]. Defective healing in this model occurs at the early inflammatory phase and persists into the reparative phase, including delayed chondrogenesis and persistent mesenchyme at the fracture site. Recent work has delineated the function of COX-2 in specific cell populations during fracture healing in a mouse model. Loss of COX-2 in mesenchymal progenitors (Prx1$^+$ cells) suppresses bone formation, while loss of COX-2 specifically in chondrocytes inhibits the conversion of cartilage to bone [26].

Critical to successful bone repair is the revascularization of injured tissues to provide oxygen, facilitate nutrient/metabolic waste management, and deliver a population of precursor cells of hematopoietic origin that may contribute to healing. As mentioned, support for angiogenesis during the repair process is thought to be modulated by VEGFs and their cognate receptors VEGFR1 and VEGFR2. It has been demonstrated that exogenous administration of VEGF during mouse femur fracture and allograft healing enhances vascular ingrowth into the callus and accelerates repair, providing a basis for VEGF to be in the conversation as a potential therapeutic agent [27].

CONDITIONS THAT IMPAIR FRACTURE HEALING AND THERAPEUTIC MODALITIES

The normal progression of the fracture healing process can be significantly compromised by several physiological, pathological, and environmental factors including aging, diabetes, and cigarette smoking. Clinical data provide evidence for this and basic research has begun to reveal the details of the underlying biological basis in some cases. A brief discussion follows of three of the most important conditions that are documented to impair the process of skeletal healing.

Aging

While it has been known for more than 30 years that the rate of fracture healing is reduced with aging [28], minimal progress has been made towards understanding the mechanisms involved. It has also been suggested that development of nonunion in the aging population is a significant clinical problem [29]. Several mechanisms have been proposed to explain reduced/delayed fracture healing in the elderly, with recent work suggesting impairment of macrophage function, increased systemic inflammation ("inflammaging"), and reduction in size and differentiation potential of progenitor populations [30]. Additionally, the normal upregulation of BMP-2, Ihh, and various Wnts during chondrocyte maturation, and osteoblast differentiation, is reduced in aged mice, further impairing the progression of healing in senescence [31]. Enhanced adipogenic potential at the expense of chondrogenesis and osteogenesis or altered competency to support osteoclastogenesis at various stages of healing, may contribute [31]. There has also been an association between aging and a decrement in endothelial cells and the factors/pathways that modulate them, suggesting that impaired blood vessel formation in aging could also affect healing [31]. Ongoing efforts aim to address the relative mechanistic contribution of these and other processes to impaired skeletal healing in the elderly.

Diabetes

Documented clinical findings establish that fracture healing is impaired in patients with type 1 and type 2 diabetes [32]. Consistent with this, animal models of streptozotocin-induced type 1 diabetes show impaired healing evidenced by reduced mesenchymal cell proliferation in the early callus, reduced matrix deposition (collagen), and reduced biomechanical properties in the healed fracture [33]. Additionally, diabetes-related overexpression of TNF leads to accelerated loss of cartilage in the callus due to increased chondrocyte apoptosis [34]. While it is not known if the impaired healing is the result of hypoinsulinemia or hyperglycemia/formation of advanced glycation end-products, insulin treatment to normalize blood glucose in a diabetic murine femur fracture model has been shown to reverse the deficit in healing [35]. In a diabetic rat fracture model, local intramedullary delivery of insulin to the fracture site, which does not provide systemic management of glucose, reversed the healing deficit at both early (mesenchymal cell proliferation and chondrogenesis) and late (mineralization and biomechanical strength) time points [36]. This suggests a direct anabolic effect of insulin on cells at the fracture site. The mechanisms of impaired healing in type 2 diabetes are less clear. Recent work in a murine tibia fracture model demonstrated impaired healing associated with decreased woven bone formation and increased callus adiposity [37]. This was not recapitulated in a femur fracture model in obese mice [38], suggesting that obesity alone is not deleterious unless accompanied by fulminant type 2 diabetes. Interestingly, metformin treatment to normalize blood glucose levels further impaired fracture healing in diabetic rats [39], so simple rescue of altered systemic glucose management is itself not therapeutic. This is likely to be an active area of research given the growing number of people who are obese and have type 2 diabetes worldwide.

Cigarette smoking

Clinically, smoking has been shown to have a negative impact on skeletal healing following long bone fracture [29]. Little is known about the mechanism underlying the deficits in skeletal repair, with mesenchymal cell condensation and the process of chondrogenesis hypothesized to be important targets of cigarette smoke. The most widely studied molecule in the context of smoking and bone healing is nicotine, which has been shown to inhibit distraction osteogenesis [40] and fracture healing in rabbits [41]. Conversely, cigarette smoke, but not nicotine, was found to affect mechanical strength in a rat fracture model [42]. Recently, another class of molecules in cigarette smoke, polycyclic aromatic hydrocarbons, has been implicated in impaired tibial fracture healing in the mouse via activation of the aryl hydrocarbon receptor [43]. In general, however, a full characterization of the healing process in smokers or animal models of smoke exposure is necessary and work to identify which components of cigarette smoke are responsible for its effect(s) is important if the underlying molecular mechanisms are to be fully understood.

MOLECULAR THERAPIES TO ENHANCE BONE HEALING

Currently, the only US Food and Drug Administration (FDA) approved molecular therapy for bone healing is BMP-2. As mentioned previously, the underlying rationale for its therapeutic potential is based on the finding that elimination of BMP-2 in the limb disrupts the initiation of postnatal fracture healing [15], establishing its essential role in the process. Thus, it is no surprise that several animal studies have identified a positive effect of BMP-2, or activation of its signaling pathway, on various bone healing situations. As its use has gained attention, clinical data have emerged that support the use of BMP-2 in clinical situations; for example, recombinant BMP-2 delivered in a collagen sponge with cancellous autograft aids in the healing of tibial diaphyseal fractures [44]. Spine fusion patients were also found to have better neck disability and arm pain scores 24 months postoperatively when an INFUSE bone graft (collagen sponge impregnated with BMP-2) was used [45]. Despite the positive outcomes of these and other studies, it should be noted that the clinical and cost effectiveness of BMP-2 in skeletal repair situations remains an open debate [46].

There has been a focus on developing molecular therapies to enhance healing or induce repair in nonunion situations using agents known for their bone anabolic capability: parathyroid hormone (PTH) and activators of the Wnt/β-catenin pathway. Because PTH is an FDA-approved treatment for enhancing bone mass in osteoporosis patients, its repurposing as a candidate therapy for fracture healing in patients with nonunion has been proposed. In addition to several case reports, recent studies in humans [47] and animals [48] show compelling evidence of positive actions of PTH on bone healing. Regarding the mechanism of PTH action in this context, recent findings demonstrate that PTH treatment can suppress fibrous tissue formation and promote osteogenesis via enhancement of angiogenesis [49]. Regarding the modulation of Wnt signaling, either genetic or molecular (via Wnt3a) enhancement of β-catenin signaling in osteo- and chondroprogenitors is known to accelerate fracture healing in the mouse [50]. Given that clinical trials are currently ongoing to test antibody blockade of the Wnt decoy receptors Sclerostin and Dickkopf-1 as bone anabolic strategies in osteoporosis [51], testing of these agents in fracture repair in humans is gaining momentum [52]. Overall, advances identifying PTH and modulation of Wnt signaling as viable strategies in bone healing, set the stage for these established bone anabolic therapies as treatment options for accelerating fracture repair or alleviating/reversing the development of fracture nonunion.

REFERENCES

1. Tzioupis C, Giannoudis PV. Prevalence of long-bone non-unions. Injury. 2007;38(suppl 2):S3–9.
2. Garrison KR, Shemilt I, Donell S, et al. Bone morphogenetic protein (BMP) for fracture healing in adults. Meta-analysis review. Cochrane Database Syst Rev. 2010(6):CD006950.
3. Einhorn TA, Gerstenfeld LC. Fracture healing: mechanisms and interventions. Nat Rev Rheumatol. 2015;11(1):45–54.

4. Abou-Khalil R, Yang F, Lieu S, et al. Role of muscle stem cells during skeletal regeneration. Stem Cells. 2015;33(5):1501–11.
5. Wang C, Inzana JA, Mirando AJ, et al. NOTCH signaling in skeletal progenitors is critical for fracture repair. J Clin Invest. 2016;126(4):1471–81.
6. Wang T, Zhang X, Bikle DD. Osteogenic differentiation of periosteal cells during fracture healing. J Cell Physiol. 2017;232(5):913–21.
7. Morgan EF, Salisbury Palomares KT, Gleason RE, et al. Correlations between local strains and tissue phenotypes in an experimental model of skeletal healing. J Biomech. 2010;43(12):2418–24.
8. Orwoll ES. Toward an expanded understanding of the role of the periosteum in skeletal health. J Bone Miner Res. 2003;18(6):949–54.
9. Provot S, Zinyk D, Gunes Y, et al. Hif-1alpha regulates differentiation of limb bud mesenchyme and joint development. J Cell Biol. 2007;177(3):451–64.
10. Zhou X, von der Mark K, Henry S, et al. Chondrocytes transdifferentiate into osteoblasts in endochondral bone during development, postnatal growth and fracture healing in mice. PLoS Genet. 2014;10(12):e1004820.
11. El-Jawhari JJ, Jones E, Giannoudis PV. The roles of immune cells in bone healing; what we know, do not know and future perspectives. Injury. 2016;47(11):2399–406.
12. Gerstenfeld LC, Cullinane DM, Barnes GL, et al. Fracture healing as a post-natal developmental process: molecular, spatial, and temporal aspects of its regulation. J Cell Biochem. 2003;88(5):873–84.
13. Hu K, Olsen BR. Osteoblast-derived VEGF regulates osteoblast differentiation and bone formation during bone repair. J Clin Invest. 2016;126(2):509–26.
14. McGee-Lawrence ME, Carpio LR, Bradley EW, et al. Runx2 is required for early stages of endochondral bone formation but delays final stages of bone repair in Axin2-deficient mice. Bone. 2014;66:277–86.
15. Tsuji K, Bandyopadhyay A, Harfe BD, et al. BMP2 activity, although dispensable for bone formation, is required for the initiation of fracture healing. Nat Genet. 2006;38(12):1424–9.
16. Morgan EF, Pittman J, DeGiacomo A, et al. BMPR1A antagonist differentially affects cartilage and bone formation during fracture healing. J Orthop Res. 2016;34(12):2096–105.
17. Muinos-Lopez E, Ripalda-Cemborain P, Lopez-Martinez T, et al. Hypoxia and reactive oxygen species homeostasis in mesenchymal progenitor cells define a molecular mechanism for fracture nonunion. Stem Cells. 2016;34(9):2342–53.
18. Xu H, Duan J, Ning D, et al. Role of Wnt signaling in fracture healing. BMB Rep. 2014;47(12):666–72.
19. Collette NM, Yee CS, Hum NR, et al. Sostdc1 deficiency accelerates fracture healing by promoting the expansion of periosteal mesenchymal stem cells. Bone. 2016;88:20–30.
20. Kazmers NH, McKenzie JA, Shen TS, et al. Hedgehog signaling mediates woven bone formation and vascularization during stress fracture healing. Bone. 2015;81:524–32.
21. Du X, Xie Y, Xian CJ, Chen L. Role of FGFs/FGFRs in skeletal development and bone regeneration. J Cell Physiol. 2012;227(12):3731–43.
22. Sheng MH, Lau KH, Baylink DJ. Role of osteocyte-derived insulin-like growth factor I in developmental growth, modeling, remodeling, and regeneration of the bone. J Bone Metab. 2014;21(1):41–54.
23. Yang X, Ricciardi BF, Hernandez-Soria A, et al. Callus mineralization and maturation are delayed during fracture healing in interleukin-6 knockout mice. Bone. 2007;41(6):928–36.
24. Montespan F, Deschaseaux F, Sensebe L, et al. Osteodifferentiated mesenchymal stem cells from bone marrow and adipose tissue express HLA-G and display immunomodulatory properties in HLA-mismatched settings: implications in bone repair therapy. J Immunol Res. 2014;2014:230346.
25. Zhang X, Schwarz EM, Young DA, et al. Cyclooxygenase-2 regulates mesenchymal cell differentiation into the osteoblast lineage and is critically involved in bone repair. J Clin Invest. 2002;109(11):1405–15.
26. Huang C, Xue M, Chen H, et al. The spatiotemporal role of COX-2 in osteogenic and chondrogenic differentiation of periosteum-derived mesenchymal progenitors in fracture repair. PLoS One. 2014;9(7):e100079.
27. Hankenson KD, Gagne K, Shaughnessy M. Extracellular signaling molecules to promote fracture healing and bone regeneration. Adv Drug Deliv Rev. 2015;94:3–12.
28. Skak SV, Jensen TT. Femoral shaft fracture in 265 children. Log-normal correlation with age of speed of healing. Acta Orthop Scand. 1988;59(6):704–7.
29. Foulke BA, Kendal AR, Murray DW, et al. Fracture healing in the elderly: a review. Maturitas. 2016;92:49–55.
30. Gibon E, Lu L, Goodman SB. Aging, inflammation, stem cells, and bone healing. Stem Cell Res Ther. 2016;7:44.
31. Gruber R, Koch H, Doll BA, et al. Fracture healing in the elderly patient. Exp Gerontol. 2006;41(11):1080–93.
32. Sellmeyer DE, Civitelli R, Hofbauer LC, et al. Skeletal metabolism, fracture risk, and fracture outcomes in type 1 and type 2 diabetes. Diabetes. 2016;65(7):1757–66.
33. Beam HA, Parsons JR, Lin SS. The effects of blood glucose control upon fracture healing in the BB Wistar rat with diabetes mellitus. J Orthop Res. 2002;20(6):1210–16.
34. Kayal RA, Siqueira M, Alblowi J, et al. TNF-alpha mediates diabetes-enhanced chondrocyte apoptosis during fracture healing and stimulates chondrocyte apoptosis through FOXO1. J Bone Miner Res. 2010;25(7):1604–15.
35. Kayal RA, Alblowi J, McKenzie E, et al. Diabetes causes the accelerated loss of cartilage during fracture repair which is reversed by insulin treatment. Bone. 2009;44(2):357–63.
36. Gandhi A, Beam HA, O'Connor JP, et al. The effects of local insulin delivery on diabetic fracture healing. Bone. 2005;37(4):482–90.
37. Brown ML, Yukata K, Farnsworth CW, et al. Delayed fracture healing and increased callus adiposity in a C57BL/6J murine model of obesity-associated type 2 diabetes mellitus. PLoS One. 2014;9(6):e99656.
38. Histing T, Andonyan A, Klein M, et al. Obesity does not affect the healing of femur fractures in mice. Injury. 2016;47(7):1435–44.

39. La Fontaine J, Chen C, Hunt N, et al. Type 2 Diabetes and metformin influence on fracture healing in an experimental rat model. J Foot Ankle Surg. 2016;55(5):955–60.
40. Ma L, Zheng LW, Cheung LK. Inhibitory effect of nicotine on bone regeneration in mandibular distraction osteogenesis. Front Biosci. 2007;12:3256–62.
41. Raikin SM, Landsman JC, Alexander VA, et al. Effect of nicotine on the rate and strength of long bone fracture healing. Clin Orthop. 1998;353:231–7.
42. Skott M, Andreassen TT, Ulrich-Vinther M, et al. Tobacco extract but not nicotine impairs the mechanical strength of fracture healing in rats. J Orthop Res. 2006;24(7):1472–9.
43. Kung MH, Yukata K, O'Keefe RJ, et al. Aryl hydrocarbon receptor-mediated impairment of chondrogenesis and fracture healing by cigarette smoke and benzo(a)pyrene. J Cell Physiol. 2012;227(3):1062–70.
44. Swiontkowski MF, Aro HT, Donell S, et al. Recombinant human bone morphogenetic protein-2 in open tibial fractures. A subgroup analysis of data combined from two prospective randomized studies. J Bone Joint Surg Am. 2006;88(6):1258–65.
45. Baskin DS, Ryan P, Sonntag V, et al. A prospective, randomized, controlled cervical fusion study using recombinant human bone morphogenetic protein-2 with the CORNERSTONE-SR allograft ring and the ATLANTIS anterior cervical plate. Spine. 2003;28(12):1219–24.
46. Poon B, Kha T, Tran S, et al. Bone morphogenetic protein-2 and bone therapy: successes and pitfalls. J Pharm Pharmacol. 2016;68(2):139–47.
47. Zhang D, Potty A, Vyas P, et al. The role of recombinant PTH in human fracture healing: a systematic review. J Orthop Trauma. 2014;28(1):57–62.
48. Takahata M, Awad HA, O'Keefe RJ, et al. Endogenous tissue engineering: PTH therapy for skeletal repair. Cell Tissue Res. 2012;347(3):545–52.
49. Dhillon RS, Xie C, Tyler W, et al. PTH-enhanced structural allograft healing is associated with decreased angiopoietin-2-mediated arteriogenesis, mast cell accumulation, and fibrosis. J Bone Miner Res. 2013;28(3):586–97.
50. Minear S, Leucht P, Jiang J, et al. Wnt proteins promote bone regeneration. Sci Transl Med. 2010;2(29):29ra30.
51. Ke HZ, Richards WG, Li X, Ominsky MS. Sclerostin and Dickkopf-1 as therapeutic targets in bone diseases. Endocr Rev. 2012;33(5):747–83.
52. Florio M, Gunasekaran K, Stolina M, et al. A bispecific antibody targeting sclerostin and DKK-1 promotes bone mass accrual and fracture repair. Nat Commun. 2016;7:11505.

14
Biomechanics of Fracture Healing

Elise F. Morgan[1,2], Amira I. Hussein[2], and Thomas A. Einhorn[3]

[1]*Department of Mechanical Engineering, Boston University, Boston, MA, USA*
[2]*Department of Orthopaedic Surgery, Boston University Medical Center, Boston, MA, USA*
[3]*Department of Orthopaedic Surgery, NYU Langone Medical Center, New York University, New York, NY, USA*

INTRODUCTION

Fracture healing involves a dynamic interplay of biological processes that when properly executed restore form and function to the injured bone. This chapter presents a biomechanical description of fracture healing, with an emphasis on methods of assessing the extent of healing—as defined principally by the extent of regain of mechanical function—and on the role of the local mechanical environment. Fracture healing is often classified as either primary or secondary fracture healing, where the former is characterized by direct cortical reconstitution and the latter involves substantial periosteal callus formation. The techniques for assessing healing that are presented in this chapter apply equally well to primary and secondary healing; however, the overviews of the biomechanical stages of fracture healing and the mechanobiology of fracture healing are largely specific to secondary healing. We also note that this chapter does not include a discussion of the biomechanics of fracture fixation, as this topic has been extensively reviewed elsewhere [1–3].

BIOMECHANICAL ASSESSMENT OF FRACTURE HEALING

In the laboratory setting, the mechanical properties of a healing bone are commonly assessed by mechanical tests that load the bone in torsion or in three-point bending. Tension and compression tests are less common. The choice of the type of test is dictated by technical as well as physiological considerations. For example, bending and torsion are logical choices when studying fracture healing in long bones, because these bones experience bending and torsional moments in vivo, and because these tests can be relatively robust [4]. However, whereas torsion tests subject every cross-section of the callus to the same torque, three-point bending creates a nonuniform bending moment throughout the callus. As a result, failure of the callus during a three-point bend test does not necessarily occur at the weakest cross-section of the callus.

Regardless of the type of mechanical test, the outcome measures that can be obtained are the strength, stiffness, rigidity, and toughness of the healing bone (Fig. 14.1). For torsion tests, an additional parameter, twist to failure, can be used as a measure of the ductility of the callus. Although strength, a measure of the force or moment that causes failure, can only be measured once for a given callus, it is possible to obtain more than one measure of stiffness and rigidity. Multistage testing protocols have been reported that apply nondestructive loads to the callus in planes or in loading modes that are different from those used for the stage of the test in which the callus is loaded to failure. With these protocols, it is possible to quantify the bending stiffness in multiple planes [5] or the torsional as well as compressive stiffness [6].

The mechanical properties illustrated in Fig. 14.1 are structural, rather than material, properties. *Material properties* describe the intrinsic mechanical behavior of a particular type of material (tissue), such as woven bone, fibrocartilage, or granulation tissue. The *structural properties* of a fracture callus depend on the material properties of the individual callus tissues as well as the spatial arrangement of the tissues and the overall geometry of the callus. While it is possible to use measurements of callus geometry together with those of structural properties to gain some insight into callus tissue material properties,

---

Primer on the Metabolic Bone Diseases and Disorders of Mineral Metabolism, Ninth Edition. Edited by John P. Bilezikian.
© 2019 American Society for Bone and Mineral Research. Published 2019 by John Wiley & Sons, Inc.
Companion website: www.wiley.com/go/asbmrprimer

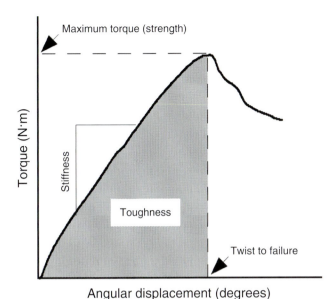

Fig. 14.1. Representative torque–twist curve for a mouse tibia 21 days post fracture. The curve is annotated to show definitions of basic biomechanical parameters. Torsional rigidity is computed by multiplying the torsional stiffness by the gage length. Analogous definitions hold for bending tests.

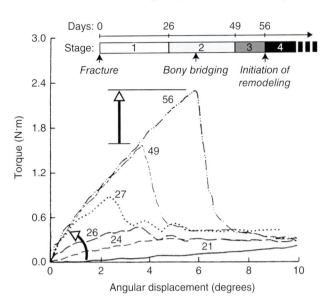

Fig. 14.2. The four biomechanical stages of secondary fracture healing. The graph contains torque–twist curves for healing rabbit tibias at various time points (in days) post fracture. The duration (in days) of each biomechanical stage of healing is depicted by the shaded bar across the top. The increase in callus stiffness that occurs between stages 1 and 2 is indicated by the curved open-headed arrow, while the increase in callus strength from stage 2 to stage 3 is indicated by the straight open-headed arrow. Source: [9].

true measurement of these material properties requires direct testing of individual callus tissues [7,8].

BIOMECHANICAL STAGES OF FRACTURE HEALING

White and colleagues used the results of torsion tests performed on healing rabbit tibiae at multiple time points (Fig. 14.2) to define four biomechanical stages of secondary fracture healing [9]. Stage 1 is characterized by extremely low callus stiffness and strength, and failure during the torsion test occurs at the original fracture line. Stage 2 corresponds to a notable increase in callus stiffness and, to a lesser extent, strength. However, it is not until stage 3 that failure during the torsion test occurs at least partly outside of the original fracture line. This stage is also characterized by an increase in callus strength compared with stage 2. Finally, in stage 4, failure during the torsion test occurs in the intact bone rather than through the original fracture line. Although fracture healing is commonly described in terms of four biological phases (inflammation, soft callus formation, bony callus formation, and remodeling), these phases do not map onto the four biomechanical stages in a one-to-one manner (Fig. 14.2). Stage 1 corresponds to the inflammatory phase, yet stage 2 encompasses the soft callus phase as well as the first part of the bony callus phase. It is the occurrence of bony bridging of the fracture line that is responsible for the increase in stiffness observed in stage 2. The transition from stage 3 to stage 4 roughly corresponds to the start of the remodeling phase.

If the bony callus is sufficiently large, the rigidity and strength of the callus during stage 3 can exceed that of the intact bone. Even though the callus tissues at this stage are not as rigid or as strong as those of well-mineralized lamellar bone, the larger cross-sectional area and moments of inertia of the callus as compared with the intact bone can overcompensate for the inferior material properties. However, though robust, the callus at this point in the healing progression is also mechanically inefficient. Through remodeling, the callus is able to retain sufficient mechanical integrity with less mass.

Results of several studies further illustrate the biomechanical consequences of individual biological phases of healing. For example, intermittent PTH(1–34) treatment has been shown to increase callus strength ([10] and citations therein) primarily as a result of enhanced chondrogenesis [11]. However, while PTH treatment leads to an increase in callus size, a slight decrease in the fraction of the callus that is comprised of mineralized tissue was observed [11], suggesting that the mechanical enhancement results purely from modulation of callus geometry (Fig. 14.3A). In contrast, evidence suggests that strontium ranelate can increase the mineralized fraction of the callus in ovariectomized animals [12]. The biomechanical importance of the extent of bony bridging (stage 2), and in particular, the extent of outer cortical bridging was demonstrated in a study of the effect of lovastatin treatment on fracture healing [13]. With respect to the later stages of healing, bisphosphonate treatment has been shown to enhance callus strength through inhibition of callus remodeling, resulting

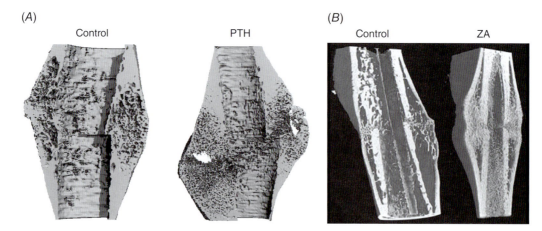

Fig. 14.3. (A) Longitudinal cut-away views of 3D micro-computed tomography reconstructions of representative saline- (control) and PTH-treated murine fracture calluses at 14 days post fracture. (B) Longitudinal cross-sections of the fracture callus and cortex at 6 weeks post fracture in rats treated with saline (control) and zolendronic acid (ZA) beginning 2 weeks after fracture. (A) Source: [16]. Reproduced with permission of Springer. (B) Source: [10] (images not to scale). Reproduced with permission of John Wiley & Sons.

in a larger callus and larger proportion of mineralized tissue (Fig. 14.3B) [14,15]. Hegde and colleagures in a recent review further discuss the biomechanical consequences of bisphosphonates and PTH as well as other common osteoporosis medications on fracture [16].

NONINVASIVE ASSESSMENT OF FRACTURE HEALING

While mechanical tests provide the gold standard measures of healing in laboratory studies of fracture healing, clinical assessment of healing requires noninvasive methods. Multiple noninvasive approaches to quantifying callus stiffness, whether in axial loading or bending, have been reported, and the clinical feasibility of several has been demonstrated. Typically, these measurements rely on measuring the displacement across the fracture gap or the pin-to-pin displacement under a known force or bending moment [17]. If an external fixator is present, it is necessary to consider only the fraction of the applied load that is borne by the callus as opposed to the fixator. From these approaches, quantitative criteria for healing have been suggested. For example, it has been proposed: (i) that a fracture can be considered healed when the bending stiffness (the ratio of the applied bending moment to the angular displacement) exceeds a certain threshold (15 N·m/degree in the case of human tibia fractures) [18]; (ii) that "healing time" can be defined as the time required to achieve bony bridging of the callus (though assessment of bridging by radiographs is subjective) [19]; and (iii) that in distraction osteogenesis, external fixation can be removed when the fraction of the axial force borne by the fixator is less than 10% [20].

Other noninvasive methods of assessing healing provide surrogate, rather than direct, measures of callus mechanical properties and include acoustic emission [21], resonant frequency [22], ultrasound [23–25], and CT imaging. Direct comparisons of CT and standard radiographic analyses have indicated that the former can yield comparable or better predictions of callus compressive strength [26], bending strength [13], and torsional strength and stiffness [27] and more definitive diagnoses of healing progression [28] and of nonunions [29]. No consensus currently exists, however, as to which CT-derived measures, or combinations of measures, best predict callus strength and stiffness for a range of types of fractures and/or bony defects.

Importantly, the vast majority of noninvasive approaches to monitoring healing focus on callus stiffness and not callus strength. While noninvasive measures of stiffness may provide valuable information about the healing process, a method to evaluate strength would be more clinically meaningful as it would theoretically provide information regarding the ability to bear weight and carry loads. In this respect, acoustic methods may pose a considerable advantage, as analysis of ultrasonic wave propagation across the fracture gap can be used to detect bony bridging of the gap. Another viable approach is CT-based finite element analysis, in which CT images are used to construct a finite element model of the callus. This approach was demonstrated for estimating callus stiffness [30]. However, this approach requires two key types of input for accurate estimates: (i) the elastic and failure properties of the callus tissues; and (ii) the types of loads and/or displacements that the callus is subjected to in vivo. As mentioned earlier, direct measurements of the material properties of callus tissues have been reported [8]. Other studies have also made substantial progress in using techniques such as inverse dynamics analysis to estimate the loads that bone defects experience during gait (eg, [31]).

In parallel, a number of investigators have pushed for better standardization of the presently available clinical methods for assessing fracture healing [32]. The common clinical approaches include plain film radiography and assessments of pain, weight bearing, and/or palpation of fracture sites. While there are no universally recognized

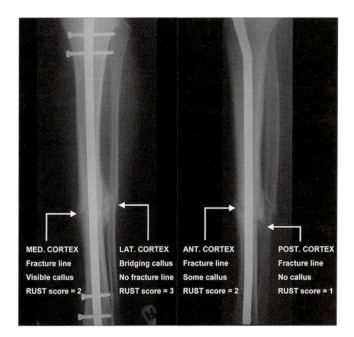

Fig. 14.4. An application of RUST scores to each of the four cortices visible on the anterior–posterior and lateral radiographs of a tibial shaft fracture. The RUST score for this callus was 8 out of a maximum of 12. The score for each cortex was: medial cortex (RUST = 2), lateral cortex (RUST = 3), anterior cortex (RUST = 2), and posterior cortex (RUST = 1). Source: [33] (images not to scale). Reproduced with permission of Lippincott Williams & Wilkins.

guidelines to evaluate radiographic union, recently standardized radiographic scoring systems such as that applied to tibia fractures (Radiographic Union Score for Tibial, RUST) [33] were introduced to provide a quantitative approach to the assessment of fracture healing. Whereas earlier scoring systems were based on different combinations of the presence of callus, callus bridging, bone formation, visibility of fracture lines, and/or remodeling, RUST assigns scores to each of the four cortices visible on the anterior–posterior and lateral radiographs, reasoning that cortical continuity is correlated to callus strength. Thus, in RUST, a score is assigned based on: "no callus," "callus," and "remodeled callus, fracture not visible" (Fig. 14.4). In a modified version of RUST (mRUST) the "callus" category is further split into "callus present" and "callus bridging." RUST scores offer a reliable and reproducible tool to assess tibial fracture healing, which led to examining the use of RUST at other anatomical sites such as the distal femur [34]. However, a comparison of these scores with biomechanical properties of fracture calluses has yet to be published.

MECHANOBIOLOGY OF FRACTURE HEALING

Fracture healing is one of the most frequently employed scenarios for studies of the effects the local mechanical environment on skeletal tissue differentiation. Mechanical loading of a fracture callus occurs most commonly as a consequence of weight bearing; however, dynamization, or applied micromotion, of the fracture gap has also been investigated. Results of these studies have shown that the effects of loading depend heavily on the mode [35], rate [36], and magnitude of loading [37], as well as gap size [37] and the time during healing at which the dynamization is enacted (eg, [38]). Application of cyclic compressive displacements can enhance healing through increased callus formation and more rapid ossification and bridging [39]. However, the benefits of applied cyclic compressive displacements appear to be limited to displacements that induce an interfragmentary strain (defined as the ratio of applied displacement to the gap size) of 7% or less [35,37]. Moreover, dynamization of the fracture gap appears to be detrimental in the very early stage of healing [38], yet beneficial during later stages [40,41].

As evidenced by the success of distraction osteogenesis in both experimental and clinical settings, application of successive tensile displacements across an osteotomy gap can also promote bone formation. In contrast to the effects of cyclic compressive loading, however, bone formation in distraction osteogenesis occurs primarily via intramembranous ossification. These characteristics of distraction also appear to hold when the tensile displacements are applied for only 2 days at a time, followed by shortening of the osteotomy gap to its original length [42], but not when the tensile displacements are applied in a true oscillatory manner (eg, 1 to 10 Hz frequency) [35]. The effects of shear or transverse movement at the fracture site are controversial [43]. Studies investigating the use of a bending motion to an osteotomy gap reported formation of cartilage rather than bone within the gap [44].

In parallel with some of the earlier experimental investigations summarized earlier, and Perren and Cordey [45] proposed the interfragmentary strain theory, which states that only tissue that is capable of withstanding the present value of interfragmentary strain can form in the fracture gap. This theory is consistent with observations that granulation tissue forms initially in the gap, followed by cartilage and then bone. The successive formation of each type of tissue further reduces the interfragmentary strain that occurs as a result of the applied load and allows a stiffer tissue to form next.

The interfragmentary strain theory presents an oversimplified description of the mechanical environment within the fracture gap in that it uses one scalar (interfragmentary strain) to describe a multiaxial strain field that varies as a function of position within the gap. More recent models of the mechanobiology of skeletal tissue differentiation have sought to account for this complexity by considering the distributions of local mechanical stimuli present throughout the fracture gap (Fig. 14.5A–C) [46–48] and the interplay between osteogenesis and angiogenesis (Fig. 14.5D) [49]. Carter and colleagues have proposed that different combinations of hydrostatic pressure and tensile strain promote the formation of different skeletal tissues [46], while Claes and Heigle have postulated that these two

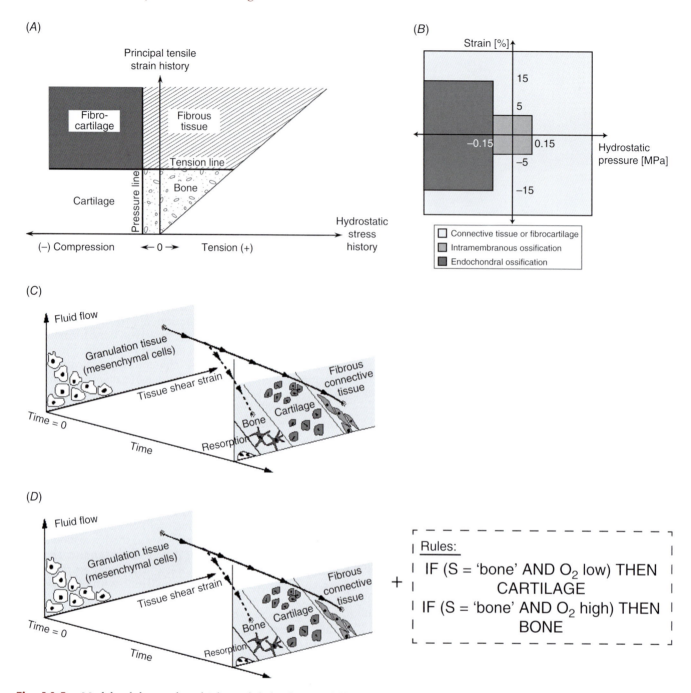

Fig. 14.5. Models of the mechanobiology of skeletal tissue differentiation by: (A) Carter and colleagues; (B) Claes and Heigle; (C) Lacroix and Prendergast; and (D) Checa and Prendergast. (A) Source: [46]. Reproduced with permission of Wolter Kluwer. (B) Source: [47]. Reproduced with permission of Elsevier. (C) Source: [48]. Reproduced with permission of Elsevier. (D) Source: [49]. Reproduced with permission of Springer.

stimuli regulate intramembranous versus endochondral ossification [47]. Shefelbine and colleagues adapted this model to also include bone resorption and tissue failure [30]. Lacroix and Prendergast have instead proposed that the two key stimuli are shear strain and fluid flow [48]. Direct comparison of these models' predictions to histological analyses of bone healing, and also experimental measurement of local mechanical stimuli such as shear strain within a bone defect, suggests that the most accurate predictions are those based on shear strain and fluid flow [44,50]. However, each of these theories is unable to predict certain histological features of the fracture healing process [48,50] indicating that the definitive role of the local mechanical environment in modulating healing has yet to be elucidated fully.

SUMMARY

An essential outcome in fracture healing is restoration of sufficient mechanical integrity to allow weight bearing and activities of daily living. Thus, biomechanical analyses of fracture healing are critical for thorough assessment of the repair process. At present, the biomechanical progression of secondary fracture healing is well characterized, and standardized in vitro methods of quantifying the extent of healing have been established. Noninvasive methods of measuring the regain of bone stiffness have also been reported; however, development of noninvasive methods of measuring the regain of strength has lagged behind. Studies to date on the effects of mechanical factors indicate that it is possible to augment healing via mechanical loading, and the growing body of literature in this area suggests that further enhancements in healing may be possible. Thus, an understanding of the biomechanics of fracture healing can be applied not only to the assessment of healing but also to development of new repair strategies.

REFERENCES

1. Moss DP, Tejwani NC. Biomechanics of external fixation: a review of the literature. Bull NYU Hosp Joint Dis. 2007;65(4):294–9.
2. Bong MR, Kummer FJ, Koval KJ, et al. Intramedullary nailing of the lower extremity: biomechanics and biology. J Am Acad Orthop Surg. 2007;15(2):97–106.
3. Bottlang M, Schemitsch CE, Nauth A, et al. Biomechanical concepts for fracture fixation. J Orthop Trauma. 2015;29(suppl 12):S28–33.
4. Steiner M, Volkheimer D, Meyers N, et al. Comparison between different methods for biomechanical assessment of ex vivo fracture callus stiffness in small animal bone healing studies. PLoS One. 2015;10(3):e0119603.
5. Foux A, Black RC, Uhthoff HK. Quantitative measures for fracture healing: an in-vitro biomechanical study. J Biomech Engin. 1990;112(4):401–6.
6. Tsiridis E, Morgan EF, Bancroft JM, et al. Effects of OP-1 and PTH in a new experimental model for the study of metaphyseal bone healing. J Orthop Res. 2007;25(9):1193–203.
7. Leong PL, Morgan EF. Measurement of fracture callus material properties via nanoindentation. Acta Biomater. 2008;4(5):1569–75.
8. Manjubala I, Liu Y, Epari DR, et al. Spatial and temporal variations of mechanical properties and mineral content of the external callus during bone healing. Bone. 2009;45(2):185–92.
9. White AA 3rd, Panjabi MM, Southwick WO. The four biomechanical stages of fracture repair. J Bone Joint Surg. 1977;59(2):188–92.
10. Ellegaard M, Kringelbach T, Syberg S, et al. The effect of PTH (1-34) on fracture healing during different loading conditions. J Bone Miner Res. 2013;28(10):2145–55.
11. Kakar S, Einhorn TA, Vora S, et al. Enhanced chondrogenesis and Wnt-signaling in parathyroid hormone treated fractures. J Bone Miner Res. 2007;22(12):1903–12.
12. Li YF, Luo E, Feng G, et al. Systemic treatment with strontium ranelate promotes tibial fracture healing in ovariectomized rats. Osteoporos Int. 2010;21(11):1889–97.
13. Nyman JS, Munoz S, Jadhav S, et al. Quantitative measures of femoral fracture repair in rats derived by microcomputed tomography. J Biomech. 2009;42(7):891–7.
14. Little DG, McDonald M, Bransford R, et al. Manipulation of the anabolic and catabolic responses with OP-1 and zoledronic acid in a rat critical defect model. J Bone Miner Res. 2005;20(11):2044–52.
15. Hao Y, Wang X, Wang L, et al. Zoledronic acid suppresses callus remodeling but enhances callus strength in an osteoporotic rat model of fracture healing. Bone. 2015;81:702–11.
16. Hegde V, Jo J, Andreopoulou P, et al. Effect of osteoporosis medications on fracture healing. Osteoporos Int. 2016;27(3):861–71.
17. Hente R, Cordey J, Perren SM. In vivo measurement of bending stiffness in fracture healing. Biomed Engin Online. 2003;2:8.
18. Richardson JB, Cunningham JL, Goodship AE, et al. Measuring stiffness can define healing of tibial fractures. J Bone Joint Surg Br. 1994;76(3):389–94.
19. Claes LE, Cunningham JL. Monitoring the mechanical properties of healing bone. Clin Orthop Relat Res. 2009;467(8):1964–71.
20. Aarnes GT, Steen H, Ludvigsen P, et al. In vivo assessment of regenerate axial stiffness in distraction osteogenesis. J Orthop Res. 2005;23(2):494–8.
21. Watanabe Y, Takai S, Arai Y, et al. Prediction of mechanical properties of healing fractures using acoustic emission. J Orthop Res. 2001;19(4):548–53.
22. Tower SS, Beals RK, Duwelius PJ. Resonant frequency analysis of the tibia as a measure of fracture healing. J Orthop Trauma. 1993;7(6):552–7.
23. Gerlanc M, Haddad D, Hyatt GW, et al. Ultrasonic study of normal and fractured bone. Clin Orthop Relat Res. 1975;111:175–80.
24. Glinkowski W, Gorecki A. Clinical experiences with ultrasonometric measurement of fracture healing. Technol Health Care. 2006;14(4–5):321–33.
25. Dodd SP, Miles AW, Gheduzzi S, et al. Modelling the effects of different fracture geometries and healing stages on ultrasound signal loss across a long bone fracture. Comput Methods Biomech Biomed Engin. 2007;10(5):371–5.
26. Jamsa T, Koivukangas A, Kippo K, et al. Comparison of radiographic and pQCT analyses of healing rat tibial fractures. Calcif Tissue Int. 2000;66(4):288–91.
27. Nazarian A, Pezzella L, Tseng A, et al. Application of structural rigidity analysis to assess fidelity of healed fractures in rat femurs with critical defects. Calcif Tissue Int. 2010;86(5):397–403.
28. Grigoryan M, Lynch JA, Fierlinger AL, et al. Quantitative and qualitative assessment of closed fracture healing using computed tomography and conventional radiography. Acad Radiol. 2003;10(11):1267–73.

29. Kuhlman JE, Fishman EK, Magid D, et al. Fracture nonunion: CT assessment with multiplanar reconstruction. Radiology. 1988;167(2):483–8.
30. Shefelbine SJ, Simon U, Claes L, et al. Prediction of fracture callus mechanical properties using micro-CT images and voxel-based finite element analysis. Bone. 2005;36(3):480–8.
31. Histing T, Kristen A, Roth C, et al. In vivo gait analysis in a mouse femur fracture model. J Biomech. 2010;43(16):3240–3.
32. Morshed S. Current options for determining fracture union. Adv Med. 2014;2014:Article ID 708574.
33. Whelan DB, Bhandari M, Stephen D, et al. Development of the radiographic union score for tibial fractures for the assessment of tibial fracture healing after intramedullary fixation. J Trauma Acute Care Surg. 2010;68(3):629–32.
34. Litrenta J, Tornetta P III, Mehta S, et al. Determination of radiographic healing: an assessment of consistency using RUST and modified RUST in metadiaphyseal fractures. J Orthop Trauma. 2015;29(11):516–20.
35. Augat P, Merk J, Wolf S, Claes L. Mechanical stimulation by external application of cyclic tensile strains does not effectively enhance bone healing. J OrthopTtrauma. 2001;15(1):54–60.
36. Wolf S, Augat P, Eckert-Hubner K, et al. Effects of high-frequency, low-magnitude mechanical stimulus on bone healing. Clin Orthop. 2001;385:192–8.
37. Claes L, Augat P, Suger G, et al. Influence of size and stability of the osteotomy gap on the success of fracture healing. J Orthop Res. 1997;15(4):577–84.
38. Claes L, Blakytny R, Gockelmann M, et al. Early dynamization by reduced fixation stiffness does not improve fracture healing in a rat femoral osteotomy model. J Orthop Res. 2009;27(1):22–7.
39. Goodship AE, Kenwright J. The influence of induced micromovement upon the healing of experimental tibial fractures. J Bone Joint Surg Br. 1985;67(4):650–5.
40. Boerckel JD, Uhrig BA, Willett NJ, et al. Mechanical regulation of vascular growth and tissue regeneration in vivo. Proc Natl Acad Sci U S A. 2011;108(37):E674–80.
41. Gardner MJ, van der Meulen MC, Demetrakopoulos D, et al. In vivo cyclic axial compression affects bone healing in the mouse tibia. J Orthop Res. 2006;24(8):1679–86.
42. Claes L, Augat P, Schorlemmer S, et al. Temporary distraction and compression of a diaphyseal osteotomy accelerates bone healing. J Orthop Res. 2008;26(6):772–7.
43. Schell H, Epari DR, Kassi JP, et al. The course of bone healing is influenced by the initial shear fixation stability. J Orthop Res. 2005;23(5):1022–8.
44. Miller GJ, Gerstenfeld LC, Morgan EF. Mechanical microenvironments and protein expression associated with formation of different skeletal tissues during bone healing. Biomech Model Mechanobiol. 2015;14(6):123—53.
45. Perren SM, Cordey J. The concept of interfragmentary strain. In: Uhthoff HK (ed.) *Current Concepts of Internal Fixation of Fractures*. Berlin: Springer, 1980, pp. 63–77.
46. Carter DR, Beaupre GS, Giori NJ, et al. Mechanobiology of skeletal regeneration. Clin Orthop. 1998(suppl 355):S41–55.
47. Claes LE, Heigele CA. Magnitudes of local stress and strain along bony surfaces predict the course and type of fracture healing. J Biomech. 1999;32(3):255–66.
48. Lacroix D, Prendergast PJ. A mechano-regulation model for tissue differentiation during fracture healing: analysis of gap size and loading. J Biomech. 2002;35(9):1163–71.
49. Checa S, Prendergast PJ. A mechanobiological model for tissue differentiation that includes angiogenesis: a lattice-based modeling approach. Ann Biomed Engin. 2009; 37(1):129–45.
50. Isaksson H, Wilson W, van Donkelaar CC, et al. Comparison of biophysical stimuli for mechano-regulation of tissue differentiation during fracture healing. J Biomech. 2006;39(8):1507–16.

Section II
Skeletal Physiology

Section Editor: Ego Seeman

Chapter 15. Human Fetal and Neonatal Bone Development   117
Tao Yang, Monica Grover, Kyu Sang Joeng, and Brendan Lee

Chapter 16. Skeletal Growth: a Major Determinant of Bone's Structural Diversity in Women and Men   123
Ego Seeman

Chapter 17. Racial Differences in the Acquisition and Age-related Loss of Bone Strength   131
Shane A. Norris, Marcella D. Walker, Kate Ward, Lisa K. Micklesfield, and John M. Pettifor

Chapter 18. Calcium, Vitamin D, and Other Nutrients During Growth   135
Tania Winzenberg and Graeme Jones

Chapter 19. Mechanical Loading and the Developing Skeleton   141
Mark R. Forwood

Chapter 20. Pregnancy and Lactation   147
Christopher S. Kovacs and Henry M. Kronenberg

Chapter 21. Menopause and Age-related Bone Loss   155
Carlos M. Isales and Ego Seeman

Primer on the Metabolic Bone Diseases and Disorders of Mineral Metabolism, Ninth Edition. Edited by John P. Bilezikian.
© 2019 American Society for Bone and Mineral Research. Published 2019 by John Wiley & Sons, Inc.
Companion website: www.wiley.com/go/asbmrprimer

15
Human Fetal and Neonatal Bone Development

Tao Yang[1], Monica Grover[2], Kyu Sang Joeng[3], and Brendan Lee[4]

[1]*Van Andel Research Institute, Grand Rapids, MI, USA*
[2]*Stanford School of Medicine/Stanford Children's Hospital, Stanford, CA, USA*
[3]*McKay Orthopaedic Research Laboratory, Department of Orthopaedic Surgery, Perelman School of Medicine, University of Pennsylvania, Philadelphia, PA, USA*
[4]*Department of Molecular and Human Genetics, Baylor College of Medicine, Texas Children's Hospital, Houston, TX, USA*

INTRODUCTION

Our understanding of human bone development, especially that occurring in utero, has been greatly accelerated through the analysis of animal models. However, direct studies of human bone development are still invaluable because pathological and genetic findings from human bone disorders have been extremely important for generating novel hypotheses, validating model organism studies, and uncovering new mechanisms for bone development. Moreover, animal models cannot recapitulate all human conditions. In this chapter, we will focus on human data related to the physiology of fetal and neonatal bone development and the intrinsic and extrinsic factors that lead to fetal and neonatal bone disorders.

PHYSIOLOGY OF FETAL AND NEONATAL BONE DEVELOPMENT

At the beginning of human fetal development (8 weeks post fertilization), the patterning of the skeleton has been largely determined. Compared to the earliest fetus, newborns are approximately 12 times longer in body length (30 mm versus 360 mm in crown–rump length). Hence, the major theme of bone development during the fetal stage is very rapid growth. For example, the rate of femur elongation during gestation between 16 and 41 weeks is 0.35 mm per day [1]. Ossification is an important component of bone development and growth, and it involves the coordination of osteoblast differentiation, matrix production, mineralization, and vasculogenesis. Studies showed that the majority of bones commence ossification during the first several weeks of the fetal stage, and that there is a sequential appearance of ossification centers in each individual bone. For example, the ossification of clavicle, humerus, and mandible occurs during the embryonic stage (6 weeks or 7 weeks). In contrast, the ossification of the talus or cuboid starts late at 28 weeks or after birth [2].

Secondary ossification centers (SOCs) develop in the cartilage epiphysis of the long bones and are initiated from the formation of cartilage canals, which are the invaginated perichondrium invading the epiphysis center. This process brings mesenchymal cells and vasculature into the epiphysis for bone formation [3]. In the distal part of the femur, the vascular invasion, along with the initiation of cartilage canal formation in the epiphysis, begins during weeks 8 to 10 of gestation. By 14 weeks (several months prior to SOC development), a complex vascular system within the canals is fully developed [4]. Most SOCs appear between the late embryonic stage and a few years after birth, except in the clavicles, where SOCs do not develop until 18 to 20 years of age [5].

In order to maintain bone shape while accommodating rapid growth, the progression of ossification must be tightly coupled with bone resorption, and this is mediated by osteoclasts acting both inside and outside the bone. This bone modeling–remodeling process begins during the fetal period and becomes prominent by the fourth and fifth months of gestation [2].

Primer on the Metabolic Bone Diseases and Disorders of Mineral Metabolism, Ninth Edition. Edited by John P. Bilezikian.
© 2019 American Society for Bone and Mineral Research. Published 2019 by John Wiley & Sons, Inc.
Companion website: www.wiley.com/go/asbmrprimer

To adapt to the rapid growth and ossification of the fetal skeleton, a fetus requires a large quantity of building blocks including proteins and minerals. These substances are transported against a concentration gradient across the placenta from the mother. More than 150 g of calcium and 70 g of phosphorus per kg of fetal body weight are transferred via active transport during the third trimester [6]. The actual steps of mineral transport are not completely understood. It has been proposed that calcium is transported across the placenta via a three-step model. TRPV6, a voltage-dependent calcium channel, is present on the maternal side of the placenta. Some studies have suggested that this channel transfers calcium to calbindin D9K, which is an intracellular binding protein in trophoblastic cells. Finally, calcium is transferred via PMCA3, a plasma membrane calcium–ATPase protein on the basolateral membrane to the fetal bloodstream [7–9]. Transport of phosphorus across the placenta is less well understood but NaPi-IIb, a sodium-dependent inorganic phosphorus transporter, is believed to play an important role in transplacental phosphorus transport [10].

The primary hormone responsible for the active transport of minerals across the placenta to the fetus is PTHrP [11]. The fetus, placenta, umbilical cord and breast tissue produce this hormone. Mice lacking PTHrP exhibit lethal skeletal dysplasia characterized by premature mineralization of all bones that are formed through an endochondral process. In the placenta, PTHrP is known to act through a receptor distinct from PTH1R [12]. On the other hand, PTH and vitamin D, the two critical hormones for maintaining calcium and phosphorus homeostasis in adults, are present at low levels in fetal serum which may be a response to high serum calcium levels [13]. PTH is important in the mineralization of fetal bones but not in active transport of calcium across the placenta. Similarly, fetal vitamin D does not have a major role in mineral transport although maternal vitamin D deficiency has been associated with congenital rickets [14]. The Institute of Medicine (IOM) recommends optimal maternal vitamin D levels of >20 ng/mL (>50 nmol/L) to be maintained during pregnancy [15]. Recent data also suggest that maternal vitamin D status during pregnancy influences bone mass in offspring at young adulthood [16]. In contrast, high maternal $1,25(OH)_2$ vitamin D levels can cross the placenta and lead to reduced fetal bone mass and neonatal lethality as suggested in an animal model [17].

Calcitonin, on the other hand, may not play a major role in fetal bone development as seen in calcitonin or calcitonin gene related peptide ablated mice [18]. Other hormones that play a role in skeletal health during adulthood, such as growth hormone and cortisol, have been shown to influence birth weight and weight gain during infancy. Furthermore, levels of growth hormone and cortisol were found to be determinants of prospective bone loss rate. This is compatible with the hypothesis that environmental influences during intrauterine life may alter sensitivity of the skeleton to growth hormone and cortisol [19].

After birth, the skeleton maintains a fast rate of growth and requires substantial mineral input to support bone development. Different from the fetus, whose calcium level is higher than that in its mother's serum, the newborn exhibits a reduced calcium level, rapidly reaching a base level because the placental source is removed; concomitantly, PTH level rapidly rises. The neonate becomes dependent on intestinal calcium absorption and the primary hormones responsible for maintaining serum calcium levels are PTH and vitamin D. Skeletal calcium is stored and renal calcium is reabsorbed to maintain these serum levels. Premature birth (resulting in lack of transfer of calcium during third trimester) along with associated comorbidities predisposes neonates to metabolic bone disease. Also, small for gestational age status, maternal vitamin D deficiency, and maternal diabetes make the calcium nadir more dramatic. During infancy, a phase of rapid bone mineralization, vitamin D deficiency can lead to rickets and hypocalcemia. Hence, the American Academy of Pediatrics recommends supplementing all infants with 600 IU of vitamin D daily.

EXTRINSIC FACTORS THAT AFFECT FETAL/NEONATAL BONE DEVELOPMENT

Nutritional influences

Maternal nutrition during pregnancy influences fetal nutrition. Childhood bone mass has been shown to be associated with maternal protein, calcium, phosphorus and vitamin B_{12} levels in the first trimester [20]. In addition, neonatal bone mass was found to be strongly and positively associated with birth weight, birth length and placental weight, after adjusting for sex and gestational age, indicating the importance of maternal nutrition during pregnancy [21]. There are also studies suggesting that genetic influences on BMD and adult bone size may be modified by undernutrition in utero [22].

Mechanical influences

Fetal movement in utero is a form of mechanical stimulation against resistance, which leads to mineral accretion. The importance of muscle–bone interaction (probably regulated by a network of osteocytes) in utero is evident in newborns with muscular disease or hypotonia as they have lower BMD [23]. Physiological osteoporosis of infancy, a condition of decreased cortical density, presents within 6 months following birth. Although it is mainly attributed to expansion of the bone marrow cavity size [24], lack of resistance in movement after delivery may also be a contributing factor [23]. Whether this is of clinical significance is controversial.

Environmental influences

Approximately 1 in 1000 live births is affected with axial skeletal defects. Many toxins and drugs have been implicated in its etiology including retinoic acid, valproic acid, arsenic, and carbon monoxide. These can lead to vertebral body defects such as block vertebra and nonsegmented hemivertebra. Uncontrolled maternal diabetes mellitus can cause fetal skeletal defects, specifically caudal dysgenesis, along with neonatal hypocalcemia by mechanisms not completely understood [25]. Rats with uncontrolled diabetes have been shown to have decreased calbindin mRNA in the placenta and this could explain decreased calcium transport across the placenta [26]. In addition, maternal smoking has been associated with decreased numbers of ossification centers, and maternal alcohol consumption affects calciotropic hormones and can thus cause fetal bone defects [25].

Other influences

Seasonal variation has been shown to influence newborn BMC, possibly due to the effect on maternal vitamin D levels. Prematurity and small for gestational age status are also associated with increased risk of rickets and osteoporosis due to multiple factors including hypoxia, immobility and decreased mineral supply/intake. In addition, gender and race appear to play a role. In some studies, BMC for male newborns was higher than for females and higher for African-American newborns than for white newborns [27].

Epigenetic contributions

Epigenetic regulation has been found to affect placental transfer of nutrients important for skeletal development and growth. WNT2 is highly expressed in human placenta, regulates placenta development, and supports nutrient transfer from mother to fetus. A high level of DNA methylation on the WNT2 promoter in placental tissue (but not in the fetus) is an epigenetic variation associated with fetal body size [28].

Moreover, environmental or nutrient factors also profoundly influence the risk of some skeletal diseases in later life by shaping the epigenetic landscape of the genome during fetal or neonatal stages. Epidemiologic studies have indicated a strong association between infant birth weight and adult bone mass as well as between infant birth weight/height and adult proximal femoral geometry/risk of hip fracture [29–31]. It was reported that poor nutrition in utero or in the neonatal stage can induce chronic epigenetic effects that alter the expression of bone development–related genes, such as those coding for the insulin-like growth factor 2 (IGF$_2$) and leptin etc. [32].

INHERITED FETAL/NEONATAL BONE DISORDERS

Multiple signaling and metabolic pathways are involved in fetal bone development, and the identification of human mutations has served as a major guide for uncovering these signaling pathways and mechanisms. Although genetically related dysregulation of these pathways can eventually lead to human skeletal diseases, many of them are difficult to diagnose in newborns. This is because the milder spectrum of diseases may not cause pronounced deformity of the skeleton, and the clinical consequences of abnormal bone mass, such as fracture, may not be evident given the relatively mild mechanical loading in the fetal or neonatal stages. Here, we have selected several examples of severe bone diseases that underscore key developmental processes affecting fetal and neonatal bone development to review.

Defects in bone matrix production

OI is a group of inborn bone diseases in humans characterized by brittle bone. The most severe forms of OI can lead to bone fractures and lethality in fetuses and neonates. Etiologies of these severe OIs are related to abnormal production, posttranslational modification, or metabolism of fibrillar collagens, especially type I collagen, which are the major content of bone matrix produced from osteoblast lineage. For example, dominantly inherited point mutations in COL1A1 and COL1A2, encoding the proα1(I) and proα2(I) chains of type I collagen, lead to posttranslational overmodification of collagen chains and severe forms of OI (types II and III) [33]. More recently described recessive mutations in genes important for modification or trafficking of type I collagen also cause OI. The expanding list of OI-related genes (and the corresponding gene products) includes CRTAP (cartilage associated protein) [34], LEPRE1 (prolyl 3 hydroxylase 1) [35], PPIB (cyclophilin B) [36], FKBP10 (FK506 binding protein 10) [37], SERPINH1 (heat shock protein 47) [38] and SERPINF1 (pigment epithelial derived factor) [39], BMP1(bone morphogenetic protein-1/Tolloid) [40], WNT1 (MMTV integration site 1) [41], etc.

Defects in mineral homeostasis

Recessive inactivating mutations of the calcium sensing receptor gene (CASR) are the cause of neonatal severe primary hyperparathyroidism (NSHPT) [42,43]. This disease is characterized by extreme hypercalcemia and severe neonatal hyperparathyroidism, including demineralization of the skeleton, respiratory distress and parathyroid hyperplasia. Without prompt parathyroidectomy of the affected infants, NSHPT is usually lethal. In contrast, familial hypocalciuric hypercalcemia (FHH), caused by haploinsufficiency of CASR, affords a much milder hypocalcemia and does not exhibit the complexity of hyperparathyroidism.

Defects in mineral deposition

Perinatal and infantile hypophosphatasia is a pernicious inborn metabolic disease manifesting in utero with profound hypomineralization that results in caput membraneceum, deformed or shortened limbs, and rapid death due to respiratory failure. Infantile hypophosphatasia is caused by recessive mutations in the gene encoding tissue-nonspecific isoenzyme of alkaline phosphatase (*TNSALP*), a glycoprotein localized to the plasma membranes of osteoblasts and chondrocytes that hydrolyzes monophosphate esters at an alkaline pH optimum [44]. The deficiency of TSNALP activity leads to extracellular accumulation of inorganic pyrophosphate (PP_i) which potently inhibits growth of hydroxyapatite crystal and causes severe hypomineralization in the infant's bone [45]. Haploinsufficiency of *TNSALP* also causes hypophosphatasia, but in a milder manner, usually diagnosed later in life.

Defects in osteoclastic function

Infantile malignant osteopetrosis (IMO) is a group of severe autosomal recessive osteopetrosis. The affected bones become very brittle, although bone mass is markedly higher than normal. IMO arises in the fetal stage, thus fractures of the clavicle can be found during delivery and frequent bone fractures occur during infancy. The affected infants suffer from hypocalcemia. Moreover, due to defective osteoclastic function, the bone marrow space, which accommodates hematopoiesis, is gradually diminished. Hence, if not properly treated in the first year, most affected infants develop anemia and thrombocytopenia because of encroachment of bone on marrow [46]. Genetically, IMO is caused by mutations in the genes important for osteoclast activity. The bone resorption of osteoclasts primarily relies on the acidification of bone resorption lacunae. Hence, defects in the machinery for acid secretion, such as that caused by mutations in either *CLCN7* or *OSTM1* (*CLCN7* encodes the chloride channel 7 which complexes with and is stabilized by the *OSTM1* gene product, the osteopetrosis-associated transmembrane protein 1) or in *TRCIRG1* (encoding T cell immune regulator 1, a subunit of a vacuolar proton pump) have been identified in the osteoclast-rich IMO patients [47–50].

Defects in cranial suture closure and osteogenesis

The skull of neonates is composed of separate cranial bones connected by fibrous cranial sutures (fontanels). These sutures provide flexibility for the skull to facilitate its passage through the birth canal without damaging the infant's brain. Moreover, cranial sutures contain osteogenic mesenchymal cells, serving as important sites for cranial bone growth to adapt to the rapid brain growth of infancy [51,52]. The fusion of cranial bones normally starts after infancy and completes by adulthood. Disorders characterized by delayed or premature closure of cranial sutures are not rare in newborns. Cleidocranial dysplasia (CCD) patients have persistently open and unossified skull sutures. This is caused by a haploinsufficiency of *Runx2*, a master gene that regulates multiple steps of osteoblast differentiation [53–55]. In contrast, premature suture closure leads to craniosynostosis, which can severely restrain growth of the skull, thus leading to increased intracranial pressure that can severely impair neural development [51]. The etiologies of craniosynostosis include dominant activating mutations in the FGF receptors (*FGFR1, 2* and *3*) [56–59] or haploinsufficiency of *TWIST1* [60,61]. Mutations in *MSX2* [62], *EFNB1* [63], *Gli3* [64], *RAB23* [65], *POR* [66], and *RECQL4* [67] have also been identified in some rare types of craniosynostosis.

SUMMARY

Overall, fetal and neonatal bone development is a dynamic and complicated process orchestrated by multiple intrinsic or extrinsic factors. Mutations in the genes important for skeletal cell differentiation and function, or for extracellular matrix production, modifications, and mineralization, can cause dramatic abnormality of the fetal/neonatal skeleton. Moreover, fetal/neonatal bone development is greatly influenced by maternal nutrition and health, hormones, toxins, and in utero environment. Accumulating data suggest that aside from affecting fetal bone development, the exposure to these external factors in early life can epigenetically influence postnatal bone homeostasis. This underscores that better understanding of the physiological processes of early bone development may also help optimize bone health throughout life.

REFERENCES

1. Salle BL, Rauch F, Travers R, et al. Human fetal bone development: histomorphometric evaluation of the proximal femoral metaphysis. Bone. 2002;30(6):823–8.
2. Gardner E. *Osteogenesis in the Human Embryo and Fetus. Development and Growth*. Elsevier BV, 1971, p. 77–118.
3. Blumer MJ, Schwarzer C, Perez MT, et al. Identification and location of bone-forming cells within cartilage canals on their course into the secondary ossification centre. J Anat. 2006;208(6):695–707.
4. Burkus JK, Ganey TM, Ogden JA. Development of the cartilage canals and the secondary center of ossification in the distal chondroepiphysis of the prenatal human femur. Yale J Biol Med. 1993;66(3):193–202.
5. Zoetis T, Tassinari MS, Bagi C, et al. Species comparison of postnatal bone growth and development. Birth Defects Res B Dev Reprod Toxicol. 2003;68(2):86–110.
6. Neer R, Berman M, Fisher L, et al. Multicompartmental Analysis of Calcium Kinetics in Normal Adult Males. J Clin Invest. 1967;46(8):1364–79.
7. Bianco SDC, Peng J-B, Takanaga H, et al. Marked Disturbance of Calcium Homeostasis in Mice With Targeted Disruption of the Trpv6 Calcium Channel Gene. J Bone Miner Res. 2006;22(2):274–85.

8. Belkacemi L, Bédard I, Simoneau L, et al. Calcium channels, transporters and exchangers in placenta: a review. Cell Calcium. 2005;37(1):1–8.
9. Suzuki Y, Kovacs CS, Takanaga H, et al. Calcium Channel TRPV6 Is Involved in Murine Maternal-Fetal Calcium Transport. J Bone Miner Res. 2008;23(8):1249–56.
10. Shibasaki Y, Etoh N, Hayasaka M, et al. Targeted deletion of the tybe IIb Na+-dependent Pi-co-transporter, NaPi-IIb, results in early embryonic lethality. Biochem Biophys Res Commun. 2009;381(4):482–6.
11. Kovacs CS. Bone development and mineral homeostasis in the fetus and neonate: roles of the calciotropic and phosphotropic hormones. Physiol Rev. 2014;94(4):1143–218.
12. Kovacs CS, Lanske B, Hunzelman JL, et al. Parathyroid hormone-related peptide (PTHrP) regulates fetal-placental calcium transport through a receptor distinct from the PTH/PTHrP receptor. Proc Natl Acad Sci U S A. 1996;93(26):15233–8.
13. Salle BL, Glorieux FH, Delvin EE. Perinatal Vitamin D Metabolism. Biol Neonate. 1988;54(4):181–7.
14. Mahon P, Harvey N, Crozier S, et al. Low maternal vitamin D status and fetal bone development: Cohort study. J Bone Miner Res. 2009;25(1):14–9.
15. Ross AC, Manson JE, Abrams SA, et al. The 2011 report on dietary reference intakes for calcium and vitamin D from the Institute of Medicine: what clinicians need to know. J Clin Endocrinol Metab. 2011;96(1):53–8.
16. Zhu K, Whitehouse AJ, Hart PH, et al. Maternal vitamin D status during pregnancy and bone mass in offspring at 20 years of age: a prospective cohort study. J Bone Miner Res. 2014;29(5):1088–95.
17. Lieben L, Stockmans I, Moermans K, et al. Maternal hypervitaminosis D reduces fetal bone mass and mineral acquisition and leads to neonatal lethality. Bone. 2013;57(1):123–31.
18. McDonald KR, Fudge NJ, Woodrow JP, et al. Ablation of calcitonin/calcitonin gene-related peptide-alpha impairs fetal magnesium but not calcium homeostasis. Am J Physiol Endocrinol Metab. 2004;287(2):E218–26.
19. Dennison EM, Syddall HE, Rodriguez S, et al. Polymorphism in the growth hormone gene, weight in infancy, and adult bone mass. J Clin Endocrinol Metab. 2004;89(10):4898–903.
20. Heppe DH, Medina-Gomez C, Hofman A, et al. Maternal first-trimester diet and childhood bone mass: the Generation R Study. Am J Clin Nutr. 2013;98(1):224–32.
21. Godfrey K, Walker-Bone K, Robinson S, et al. Neonatal Bone Mass: Influence of Parental Birthweight, Maternal Smoking, Body Composition, and Activity During Pregnancy. J Bone Miner Res. 2001;16(9):1694–703.
22. Dennison EM, Arden NK, Keen RW, et al. Birthweight, vitamin D receptor genotype and the programming of osteoporosis. Paediatr Perinat Epidemiol. 2001;15(3):211–9.
23. Land C, Schoenau E. Fetal and postnatal bone development: reviewing the role of mechanical stimuli and nutrition. Best Pract Res Clin Endocrinol Metab. 2008;22(1):107–18.
24. Rauch F, Schoenau E. Changes in Bone Density During Childhood and Adolescence: An Approach Based on Bone's Biological Organization. J Bone Miner Res. 2001;16(4):597–604.
25. Alexander PG, Tuan RS. Role of environmental factors in axial skeletal dysmorphogenesis. Birth Defects Res C Embryo Today. 2010;90(2):118–32.
26. Husain SM, Birdsey TJ, Glazier JD, et al. Effect of Diabetes Mellitus on Maternofetal Flux of Calcium and Magnesium and Calbindin9K mRNA Expression in Rat Placenta. Pediatr Res. 1994;35(3):376–80.
27. Namgung R, Tsang RC. Factors affecting newborn bone mineral content: in utero effects on newborn bone mineralization. Proc Nutr Soc. 2000;59(01):55–63.
28. Ferreira JC, Choufani S, Grafodatskaya D, et al. WNT2 promoter methylation in human placenta is associated with low birthweight percentile in the neonate. Epigenetics. 2011;6(4):440–9.
29. Cooper C, Fall C, Egger P, et al. Growth in infancy and bone mass in later life. Ann Rheum Dis. 1997;56(1):17–21.
30. Javaid MK, Lekamwasam S, Clark J, et al. Infant growth influences proximal femoral geometry in adulthood. J Bone Miner Res. 2006;21(4):508–12.
31. Cooper C, Eriksson JG, Forsen T, et al. Maternal height, childhood growth and risk of hip fracture in later life: a longitudinal study. Osteoporosis Int. 2001;12(8):623–9.
32. Heijmans BT, Tobi EW, Stein AD, et al. Persistent epigenetic differences associated with prenatal exposure to famine in humans. Proc Natl Acad Sci U S A. 2008;105(44):17046–9.
33. Marini JC, Forlino A, Cabral WA, et al. Consortium for osteogenesis imperfecta mutations in the helical domain of type I collagen: regions rich in lethal mutations align with collagen binding sites for integrins and proteoglycans. Hum Mutat. 2007;28(3):209–21.
34. Morello R, Bertin TK, Chen Y, et al. CRTAP Is Required for Prolyl 3- Hydroxylation and Mutations Cause Recessive Osteogenesis Imperfecta. Cell. 2006;127(2):291–304.
35. Cabral WA, Chang W, Barnes AM, et al. Prolyl 3-hydroxylase 1 deficiency causes a recessive metabolic bone disorder resembling lethal/severe osteogenesis imperfecta. Nat Genet. 2007;39(3):359–65.
36. van Dijk FS, Nesbitt IM, Zwikstra EH, et al. PPIB Mutations Cause Severe Osteogenesis Imperfecta. Am J Hum Genet. 2009;85(4):521–7.
37. Alanay Y, Avaygan H, Camacho N, et al. Mutations in the Gene Encoding the RER Protein FKBP65 Cause Autosomal-Recessive Osteogenesis Imperfecta. Am J Hum Genet. 2010;86(4):551–9.
38. Christiansen HE, Schwarze U, Pyott SM, et al. Homozygosity for a Missense Mutation in SERPINH1, which Encodes the Collagen Chaperone Protein HSP47, Results in Severe Recessive Osteogenesis Imperfecta. Am J Hum Genet. 2010;86(3):389–98.
39. Becker J, Semler O, Gilissen C, et al. Exome Sequencing Identifies Truncating Mutations in Human SERPINF1

in Autosomal-Recessive Osteogenesis Imperfecta. Am J Hum Genet. 2011;88(3):362–71.
40. Asharani PV, Keupp K, Semler O, et al. Attenuated BMP1 function compromises osteogenesis, leading to bone fragility in humans and zebrafish. Am J Hum Genet. 2012;90(4):661–74.
41. Laine CM, Joeng KS, Campeau PM, et al. WNT1 mutations in early-onset osteoporosis and osteogenesis imperfecta. N Engl J Med. 2013;368(19): 1809–16.
42. Bai M, Pearce SH, Kifor O, et al. In vivo and in vitro characterization of neonatal hyperparathyroidism resulting from a de novo, heterozygous mutation in the Ca2+-sensing receptor gene: normal maternal calcium homeostasis as a cause of secondary hyperparathyroidism in familial benign hypocalciuric hypercalcemia. J Clin Invest. 1997;99(1):88–96.
43. Pollak MR, Brown EM, Chou Y-HW, et al. Mutations in the human Ca2+-sensing receptor gene cause familial hypocalciuric hypercalcemia and neonatal severe hyperparathyroidism. Cell. 1993;75(7):1297–303.
44. Weiss MJ, Cole DE, Ray K, et al. A missense mutation in the human liver/bone/kidney alkaline phosphatase gene causing a lethal form of hypophosphatasia. Proc Natl Acad Sci U S A. 1988;85(20):7666–9.
45. Whyte MP. Physiological role of alkaline phosphatase explored in hypophosphatasia. Ann N Y Acad Sci. 2010;1192(1):190–200.
46. Stark Z, Savarirayan R. Osteopetrosis. Orphanet J Rare Dis. 2009;4(1):5.
47. Lange PF, Wartosch L, Jentsch TJ, et al. ClC-7 requires Ostm1 as a β-subunit to support bone resorption and lysosomal function. Nature. 2006;440(7081):220–3.
48. Vezzoni P, Frattini A, Orchard PJ, et al. Nat Genet. 2000;25(3):343–6.
49. Kornak U, Kasper D, Bösl MR, et al. Loss of the ClC-7 Chloride Channel Leads to Osteopetrosis in Mice and Man. Cell. 2001;104(2):205–15.
50. Ramírez A, Faupel J, Goebel I, et al. Identification of a novel mutation in the coding region of the grey-lethal gene OSTM1 in human malignant infantile osteopetrosis. Hum Mutat. 2004;23(5):471–6.
51. Morriss-Kay GM, Wilkie AOM. Growth of the normal skull vault and its alteration in craniosynostosis: insights from human genetics and experimental studies. J Anat. 2005;207(5):637–53.
52. Opperman LA. Cranial sutures as intramembranous bone growth sites. Dev Dyn. 2000;219(4):472–85.
53. Mundlos S, Otto F, Mundlos C, et al. Mutations Involving the Transcription Factor CBFA1 Cause Cleidocranial Dysplasia. Cell. 1997;89(5):773–9.
54. Lee B, Thirunavukkarasu K, Zhou L, et al. Missense mutations abolishing DNA binding of the osteoblast-specific transcription factor OSF2/CBFA1 in cleidocranial dysplasia. Nat Genet. 1997;16(3):307–10.
55. Otto F, Thornell AP, Crompton T, et al. Cbfa1, a Candidate Gene for Cleidocranial Dysplasia Syndrome, Is Essential for Osteoblast Differentiation and Bone Development. Cell. 1997;89(5):765–71.
56. Muenke M, Schell U, Hehr A, et al. A common mutation in the fibroblast growth factor receptor 1 gene in Pfeiffer syndrome. Nat Genet. 1994;8(3):269–74.
57. Meyers GA, Orlow SJ, Munro IR, et al. Fibroblast growth factor receptor 3 (FGFR3) transmembrane mutation in Crouzon syndrome with acanthosis nigricans. Nat Genet. 1995;11(4):462–4.
58. Rutland P, Pulleyn LJ, Reardon W, et al. Identical mutations in the FGFR2 gene cause both Pfeiffer and Crouzon syndrome phenotypes. Nat Genet. 1995;9(2):173–6.
59. Bellus GA, Gaudenz K, Zackai EH, et al. Identical mutations in three different fibroblast growth factor receptor genes in autosomal dominant craniosynostosis syndromes. Nat Genet. 1996;14(2):174–6.
60. Howard TD, Paznekas WA, Green ED, et al. Mutations in TWIST, a basic helix–loop–helix transcription factor, in Saethre-Chotzen syndrome. Nat Genet. 1997;15(1):36–41.
61. Ghouzzi VE, Merrer ML, Perrin-Schmitt F, et al. Mutations of the TWIST gene in the Saethre-Chotzene syndrome. Nat Genet. 1997;15(1):42–6.
62. Jabs EW, Müller U, Li X, et al. A mutation in the homeodomain of the human MSX2 gene in a family affected with autosomal dominant craniosynostosis. Cell. 1993;75(3):443–50.
63. Wieland I, Jakubiczka S, Muschke P, et al. Mutations of the Ephrin-B1 Gene Cause Craniofrontonasal Syndrome. Am J Hum Genet. 2004;74(6):1209–15.
64. Vortkamp A, Gessler M, Grzeschik K-H. GLI3 zinc-finger gene interrupted by translocations in Greig syndrome families. Nature. 1991;352(6335):539–40.
65. Jenkins D, Seelow D, Jehee FS, et al. RAB23 Mutations in Carpenter Syndrome Imply an Unexpected Role for Hedgehog Signaling in Cranial-Suture Development and Obesity. Am J Hum Genet. 2007;80(6):1162–70.
66. Flück CE, Tajima T, Pandey AV, et al. Mutant P450 oxidoreductase causes disordered steroidogenesis with and without Antley-Bixler syndrome. Nat Genet. 2004;36(3): 228–30.
67. Mendoza-Londono R, Lammer E, Watson R, et al. Characterization of a New Syndrome That Associates Craniosynostosis, Delayed Fontanel Closure, Parietal Foramina, Imperforate Anus, and Skin Eruption: CDAGS. Am J Hum Genet. 2005;77(1):161–8.

16

Skeletal Growth: a Major Determinant of Bone's Structural Diversity in Women and Men

Ego Seeman[1,2]

[1]*Departments of Endocrinology and Medicine, Austin Health, University of Melbourne, Melbourne, VIC, Australia*
[2]*Mary MacKillop Institute for Health Research, Australian Catholic University, Melbourne, VIC, Australia*

VARIANCES IN BONE TRAIT ARE ESTABLISHED EARLY IN LIFE

The variance of bone traits like bone mass is large; 1 SD is about 10% to 15% of the mean [1]. Thus, individuals at the 95th and 5th percentiles for bone mass and size differ by ~50% in that trait. The variance in the rate of bone loss in adulthood is about an order of magnitude less (1 SD = 1% of the mean). So the difference in the percentile location of a trait at the completion of growth is likely to be an important determinant of bone strength and fracture risk in advanced age, and perhaps more important than differences in rates of bone loss for many years [2].

Intrauterine growth

Differences in bone structure appear during intrauterine life but there is no evidence that ranking in femur length is established in utero; on the contrary, the location of an individual's femur length in a given quartile varies throughout gestation such that under 10% of individuals have femur length at birth in the same quartile as found in earlier stages of gestation [3] (Fig. 16.1).

Postnatal growth

Several studies report associations between bone traits at birth and adulthood but others suggest that trait variances are established within the first 2 years of life [4,5]. For example, morphology at 6 months or beyond, *not* at birth, predicts morphology at puberty and in adulthood. Tracking appears to become established during the first year of life. Crown-heel length (CHL) at 6 months or later, but not at birth, predicts height, bone size, mass, and strength almost two decades later [6,7] (Fig. 16.2). CHL and height track from 6 months of age onwards through adolescence to adulthood. This also applies to total and regional bone mass and size, tibial and radial cross-sectional area, and indices of bending and compressive strength first measured at 11.5 years of age and then during 7 years to maturity at 18 years of age.

Tracking occurs through puberty without change in percentile location. For example, Loro et al. reported that the percentile ranking of traits at Tanner stage 2 was unchanged during 3 years and 60% to 90% of the variance at maturity was accounted for by the variance before puberty [8]. An individual with a larger vertebral or femoral shaft cross section, or higher vertebral volumetric bone mineral density (vBMD) or femoral cortical area,

---

Primer on the Metabolic Bone Diseases and Disorders of Mineral Metabolism, Ninth Edition. Edited by John P. Bilezikian.
© 2019 American Society for Bone and Mineral Research. Published 2019 by John Wiley & Sons, Inc.
Companion website: www.wiley.com/go/asbmrprimer

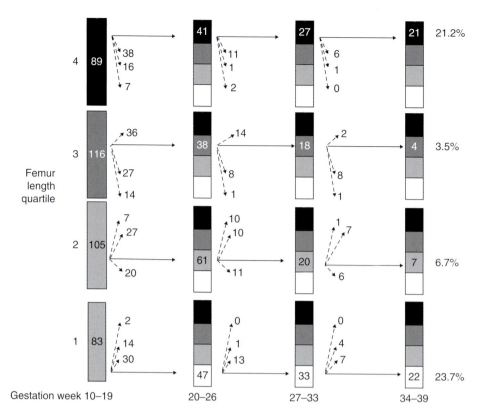

Fig. 16.1. The proportion of 412 fetal femurs whose length remained within the same quartile during gestation was 13% ($n = 54$). Percentages on the far right refer to the percentage tracking within a given quartile. The numbers to the right of the bars give examples of the disposition of individuals' femur length from their baseline quartile location through gestation. The numbers are fetuses that kept their quartile (solid array) or deviated (dashed line) from quartile 1 (white), quartile 2 (light gray), quartile 3 (dark gray), and quartile 4 (black). Source: adapted from [3] with permission from John Wiley & Sons.

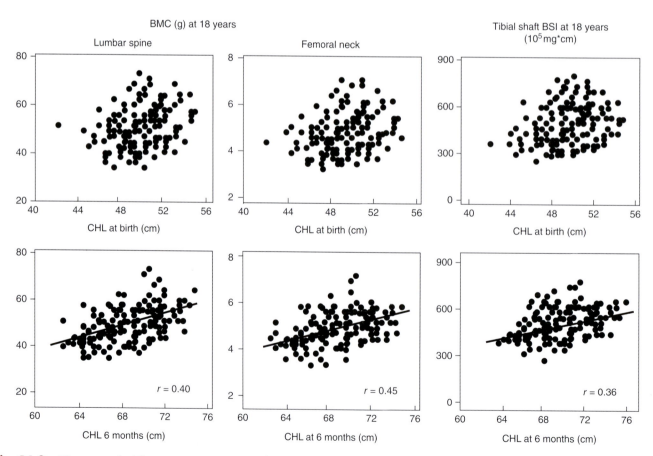

Fig. 16.2. Upper panels. There was no association between crown–heel length (CHL) at birth and lumbar spine or femoral neck bone mineral content (BMC) or tibia shaft bending strength index (BSI) at 18 years of age. These associations appeared at 6 months (lower panels). Source: adapted from [5] with permission from John Wiley & Sons.

than their peers of the same age before puberty retained this relative position to maturity.

Healthy premenopausal daughters of women with fractures have structural abnormalities at the corresponding site suggesting that the location of an individual's skeletal trait relative to others, and the familial resemblance of bone traits at maturity, are established during growth and perhaps early in postnatal life [9].

BONE SIZE, SHAPE, AND MICROARCHITECTURE

Bone modeling assembles bone size and its shape according to a genetic program; fetal limb buds grown in vitro develop the shape of the proximal femur [10]. Nevertheless, environmental factors influence bone morphology. Differences in bone size in the playing arm and non-playing arm of adolescent tennis players attest to the ability of periosteal apposition to model bone structure in response to loading during growth [11]; comparable effects in adulthood have never been reported.

Prepubertal growth

In prepubertal girls, tibial cross-sectional shape is already elliptical at 10 years of age [12]. During 2 years, periosteal apposition increased ellipticity by adding twice the amount of bone anteriorly and posteriorly than medially and laterally (Fig. 16.3). Consequently, estimates of bending strength increase more in the anteroposterior than mediolateral direction. Greater periosteal apposition on the anterior and posterior surfaces than medial and lateral surfaces creates the elliptical shape of the tibia and demonstrates how strength is optimized and mass minimized by modeling and remodeling being point-specific, and so modifying the spatial distribution of the material rather than using more material. If cortical thickness increased by the same amount of periosteal apposition at each point around the perimeter of a cross section, the amount of bone producing the same increase in bending resistance would be four-fold more than observed.

This is further illustrated by the heterogeneity in femoral neck (FN) shape [13] (Fig. 16.4). At the junction with the shaft, the size and ellipticity of FN cross section is greatest. Cross-sectional area lessens and becomes more circular moving proximally. This diversity in total

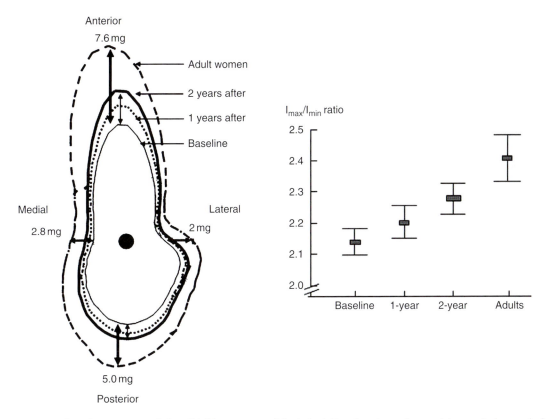

Fig. 16.3. Bone mass distribution around the tibial bone center (black dot). Focal periosteal apposition varied at each degree around the bone perimeter. More bone was deposited at the anteroposterior (AP) regions than at the mediolateral (ML) regions, increasing the ellipticity and bending strength more along the AP axis (I_{max}) than the ML axis (I_{min}). Source: adapted from [12] with permission from Oxford University Press.

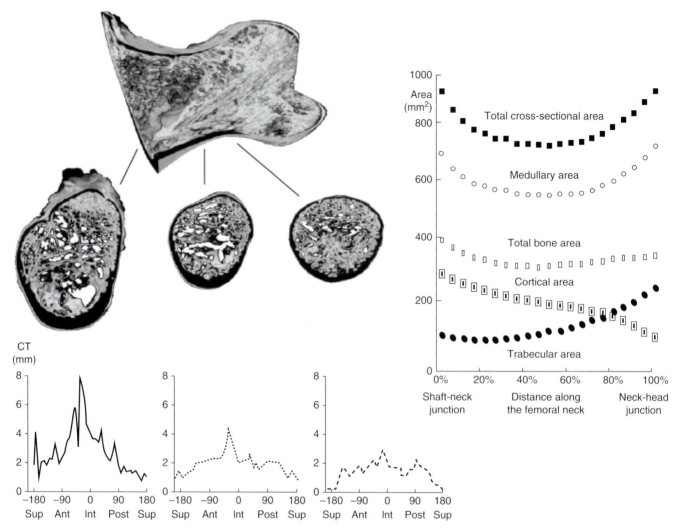

Fig. 16.4. Bone area of adjacent slices along the femoral neck (FN) is constant but is distributed differently around and along each slice. Cortical bone dominates adjacent to the femoral shaft and varies in thickness around a cross section, being thickest inferiorly. Moving proximally, cortical bone decreases and trabecular bone increases. The FN is more circular, the cortical thicknesses vary less, and most of the bone is trabecular at the junction between the FN and head. CT, cortical thickness. Source: adapted from [13] with permission from Elsevier.

cross-sectional area and shape, from cross section to cross section, is achieved using the *same* amount of material. The constant amount of material is distributed differently in space at each cross section using void volume.

The same amount of bone is assembled as a larger cross section with mainly cortical bone adjacent to the femoral shaft, and most of this is distributed inferiorly. More proximally, the proportion of bone that is cortical decreases while the proportion that is trabecular increases reciprocally. Cortices become thinner and, at the femoral head junction, most bone is trabecular and the cortex is thin and evenly distributed around the perimeter.

Differing periosteal apposition at each point around the bone perimeter is accompanied by concurrent resorption on the endosteal surfaces of the bone. In tubular bones lightness is achieved by endocortical resorption, which excavates the marrow cavity, shifting the thickening cortex outward; distance from from the neutral axis increases the bone's resistance to bending [14].

Wang et al. reported that the amount of bone deposited during 2 years on the periosteal surface in prepubertal children with larger tibial cross sections was no different to the amount deposited on the periosteal surface of smaller cross sections [12]. Although counterintuitive, these data suggest that larger bones deposit *less* bone *relative* to their starting cross-sectional size than smaller bones because resistance to bending requires less material to be deposited upon the periosteal surface of a larger than smaller bone.

Larger bones also excavate a larger medullary canal by higher rates of endocortical resorption so larger bones are relatively lighter. Larger bones also have higher cortical porosity [15]. Both features result in bones with a lower apparent vBMD. Individuals with smaller tubular bone cross sections assemble them with *more* mass relative to their size, forming a bone with a higher apparent vBMD. In bones with a smaller cross section, the liability to fracture due to slenderness is offset by more periosteal apposition relative to their starting cross-sectional size and excavation of a smaller medullary canal and lower cortical porosity so vBMD is higher. Asians have smaller bones with a lower cortical porosity and higher vBMD than whites [16].

So, a high peak vBMD is *not* the result of increased bone formation (mass has a high energy cost), it is the result of *reduced* bone resorption. (The resorption is not followed by formation and is modeling, not remodeling.) Similarly, a lower vBMD is the result of more bone resorption not less bone formation.

PUBERTY AND THE APPEARANCE OF SEX DIFFERENCES IN BONE MORPHOLOGY

Stature is a heterogeneously assembled trait. Growth in height is the result of appendicular and axial growth [17]. Appendicular and axial growth are rapid at birth and slow precipitously after birth. Around 1 year of age, appendicular growth velocity accelerates and remains about twice that of axial growth velocity until puberty, so most of the growth in height before puberty is due to growth of the legs (Fig. 16.5).

At puberty, appendicular growth decelerates while axial growth accelerates. Sex differences in bone length, width, mass, and strength emerge largely during puberty.

During the first 2 years of puberty (11 to 13 years in girls and 13 to 15 years in boys) the contribution of axial and appendicular growth to the standing height is similar (7.7 versus 7.4 cm in girls and 8.5 versus 8.0 cm in boys), while late in puberty the increase in standing height is derived more from axial than appendicular growth (4.5 versus 1.5 cm in both sexes).

Males may have a 1- to 2-year longer prepubertal growth than females because puberty occurs later in males resulting in greater sex differences in leg than trunk length [18].

During puberty, periosteal apposition increases bone width while endocortical resorption enlarges the medullary cavity [19] (Fig. 16.6). Net cortical thickness increases because periosteal apposition is greater than endocortical resorption. In girls, periosteal apposition decelerates earlier and the medullary cavity contracts. The net effect of cessation of periosteal apposition and medullary contraction in girls is the construction of a bone with a smaller total cross-sectional size and medullary size but similar cortical thickness to boys.

At the metaphyses of long bones, vBMD of the trabecular compartment remains constant from 5 years of age to young adulthood in both sexes. At this region sex differences in BMC, vBMD, and cross-sectional size emerge after puberty; males are reported to have thicker trabeculae and higher bone volume fraction (BV/TV) [8].

As the size of the vertebral body increases during growth, and the amount of bone within it also increases, there is no increase in vBMD before puberty [9]. At puberty trabecular vBMD increases in both sexes and in African-Americans and whites, and is due to an increase in trabecular thickness not number. The increase is race-specific but no different by sex [20] (Fig. 16.7). The vertebral body cross section is ~15% larger in boys than girls before puberty and ~25% greater at maturity but there is no sex difference in trabecular number or thickness. That is, the sex differences in morphology

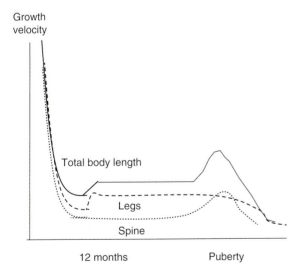

Fig. 16.5. The heterogeneity of growth in stature. After birth, growth velocity of the total body slows due to slowing of both axial and appendicular growth. At 1 year of age, growth velocity of leg length accelerates and remains higher than growth velocity of the spine until puberty. At puberty, growth velocity declines in the legs but accelerates at the spine, then late in puberty longitudinal growth slows and ceases. Source: [19]. Reproduced with permission.

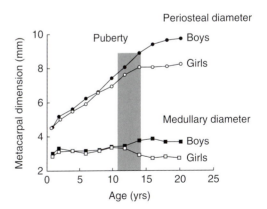

Fig. 16.6. Periosteal diameter of the metacarpal bone does not differ before puberty in boys and girls. During puberty, periosteal diameter increases in boys and ceases to expand in girls, whereas medullary diameter remains fairly constant in boys but contracts in girls. Source: adapted from [19].

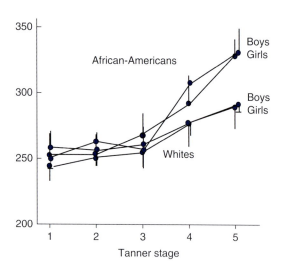

Fig. 16.7. Before puberty, trabecular vBMD is no different by sex or race. vBMD then increases at Tanner stage 3, similarly by sex within a race but more greatly in African-Americans than whites. Source: adapted from [20] with permission from Elsevier.

are in size, not apparent density; vertebral total cross-sectional area but neither vertebral height nor trabecular density differ before puberty.

GROWTH OF METAPHYSES AND FRACTURES IN CHILDHOOD

The incidence of fracture, especially at the distal metaphysis of radius, peaks at 10 to 12 years of age in girls and 12 to 14 years in boys, coinciding with the pubertal growth spurt [21]. The rate of linear growth peaks earlier than that of bone mass. In particular, growth of the distal radius is more rapid at the distal metaphysis, and at this site the longitudinal growth outpaces bone formation upon the surfaces of trabeculae emerging from the growth plate, delaying their coalescence [22]. Because of the rapidity of longitudinal growth at the distal growth plate, this delay in "corticalization" of trabeculae results in a transitory phase of porosity [23]. vBMD of the metaphysis decreases during puberty due to this transitory phase of porosity, which predisposes to fracture. By contrast, vBMD of the diaphysis of long bone continues to increase during puberty.

In late puberty, there is slowing of longitudinal bone growth. Trabeculae coalesce as bone formation proceeds on trabecular surfaces. Cortical porosity decreases and matrix mineralization increases, resulting in an increase in cortical vBMD. Exposure to sex steroids in peri- and postpubertal girls may enhance the consolidation of metaphyseal cortex at the endocortical surface, decreasing the residual cortical porosity sooner than in males [24,25].

The wider metaphysis must be modeled to fit the relatively slender diaphysis during longitudinal growth. Unlike the diaphysis where bone diameter increases by periosteal apposition, the metaphyseal cortex is resorbed from the periosteal surface while bone formation occurs upon the endocortical surface by trabecular coalescence. This modeling by periosteal resorption and deposition of bone on the endocortical surface by corticalization of trabeculae differs from the growth taking place at the diaphyseal cortex, perhaps making this region more susceptible to fracture due to stress concentration and buckling during compressive stress in a fall onto the outstretched hand.

Trabecular density at the metaphyseal region in females is independent of age in childhood and adolescence. We observed no increase in trabecular number, thickness, or separation at the distal metaphyses of radius or tibia in girls aged 5 to 18 years. By contrast, trabecular vBMD in males increases during puberty, due to increased trabecular thickness. Thus, in young adulthood, males have thicker trabeculae but similar trabecular number relative to females [26]. This sex difference may have implications in later life when bone resorption occurs. In females, thin trabeculae are more readily perforated.

EFFECT OF ILLNESSES ON BONE MORPHOLOGY IS MATURATIONAL STAGE-SPECIFIC

The effects of illness during growth depend on the maturational stage at the time of disease exposure, not just the "severity" of the illness. Longitudinal growth is more rapid in the appendicular than axial skeleton before and during early puberty, so illness may produce greater deficits in appendicular morphology. For example, disease affecting radial growth before puberty and especially during early puberty compromises the gain in bending strength. Illness during late puberty may produce greater deficits in the axial than the appendicular morphology while illness post puberty is unlikely to produce deficits in bone size [27]. This regional specificity in growth and the effects of illness are obscured by the study of standing height alone or BMD alone.

Diseases leading to sex hormone deficiency in females during puberty produce loss of sexual dimorphism in leg length as estrogen deficiency allows continued growth in females so the epiphyses do not fuse and periosteal apposition continues. Periosteal apposition continues, increasing bone width, whereas endocortical apposition fails to occur, so cortical thickness is reduced but only modestly and the bone is wider. By contrast, the pubertal growth spurt in trunk length may be affected, creating a shorter but wider vertebral body and a shorter sitting height.

Delayed puberty in males reduces periosteal apposition, producing a narrower bone with a thinner cortex, while appendicular growth in length continues, producing a longer, more slender bone with a thinner cortex.

This predisposes to greater fragility in males than females: greater bone diameter in females with delayed puberty produces a biomechanical advantage and less cortical thinning because the lack of endocortical apposition is offset by continued periosteal apposition [28–30].

CONCLUSION

Skeletal fragility in advanced age has its antecedence in growth because the variance in bone traits achieved during growth is an order of magnitude greater than rates of loss during aging. Factors modifying skeletal morphology such as exercise and nutrition are likely to be best during growth. Recent advances in imaging allow quantification of the material composition and microstructure of bone and now open doors to quantifying deficits in microstructural determinants of bone strength that predispose persons to risk of fractures during growth and when they are adults.

REFERENCES

1. Seeman E. Growth in bone mass and size—Are racial and gender differences in bone mineral density more apparent than real? J Clin Endocrinol Metab. 1998;83(5):1414–9.
2. Hui SL, Slemenda CW, Johnston CC. The contribution of bone loss to postmenopausal osteoporosis. Osteoporos Int. 1990;1(1):30–4.
3. Bjornerem A, Johnsen SL, Nguyen TV, et al. The Shifting Trajectory of Growth in Femur Length During Gestation. J Bone Miner Res. 2010;25(5):1029–33.
4. Pietilainen KH, Kaprio J, Rasanen M, et al. Tracking of body size from birth to late adolescence: contribution of birth length, birth weight, duration of gestation, parents' body size, and twinship. Am J Epidemiol. 2001;154(1):21–9.
5. Wang Q, Alen M, Nicholson P, et al. Growth patterns at distal radius and tibial shaft in pubertal girls: a 2-year longitudinal study. J Bone Miner Res. 2005;20(6):954–61.
6. Ruff C. Growth tracking of femoral and humeral strength from infancy through late adolescence. Acta Paediatr. 2005;94(8):1030–7.
7. Cheng S, Volgyi E, Tylavsky FA, et al. 2009. Trait-specific tracking and determinants of body composition: A 7-year follow-up study of pubertal growth in girls. BMC Med. 7:5.
8. Loro ML, Sayre J, Roe TF, et al. Early identification of children predisposed to low peak bone mass and osteoporosis later in life. J Clin Endocrinol Metab. 2000;85(10):3908–18.
9. Seeman E, Hopper JL, Bach LA, et al. Reduced bone mass in daughters of women with osteoporosis. N Engl J Med. 1989;320(9):554–8.
10. Murray PDF, Huxley JS. Self-differentiation in the grafted limb bud of the chick. J Anat. 1925;59:379–84.
11. Lanyon LE. Control of bone architecture by functional load bearing. J Bone Miner Res. 1992;7(suppl 2):S369–75.
12. Wang Q, Cheng S, Alen M et al.;Finnish Calex Study Group. Bone's structural diversity in adult females is established before puberty. J Clin Endocrinol Metab. 2009;94(5):1555–61.
13. Zebaze RM, Jones A, Welsh F, et al. Femoral neck shape and the spatial distribution of its mineral mass varies with its size: Clinical and biomechanical implications. Bone. 2005;37(2):243–52.
14. Ruff CB, Hayes WC. Subperiosteal expansion and cortical remodeling of the human femur and tibia with aging. Science. 1982;217(4563):945–8.
15. Bjornerem A, Bui Q, Ghasem-Zadeh A, et al. Fracture risk and height: An association partly accounted for by cortical porosity of relatively thinner cortices. J Bone Miner Res. 2013;28(9):2017–26.
16. Boutroy S, Walker MD, Liu XS, et al. Lower cortical porosity and higher tissue mineral density in Chinese American versus white women. J Bone Miner Res. 2014;29(3):551–61.
17. Karlberg J. The infancy-childhood growth spurt. Acta Paediatr Scand Suppl. 1990;367:111–8.
18. Tanner JM, Whitehouse RH. Clinical longitudinal standards for height, weight, height velocity, weight velocity, and stages of puberty. Arch Dis Child. 1976;51:170–9.
19. Garn SM. The course of bone gain and the phases of bone loss. Orth Clin North Am. 1972;3(3):503–20.
20. Gilsanz V, Boechat MI, Roe TF, et al. Gender differences in vertebral body sizes in children and adolescents. Radiology. 1994;190(3):673–7.
21. Cooper C, Dennison E, Leufkins H, et al. Epidemiology of childhood fractures in Britain: a study using the general practice research database. J Bone Miner Res. 2004;19(12):1976–81.
22. Cadet ER, Gafni RI, McCarthy EF, et al. Mechanisms responsible for longitudinal growth of the cortex: coalescence of trabecular bone into cortical bone. J Bone Joint Surg. 2003;85-A(9):1739–48.
23. Wang Q, Wang X-F, Iuliano-Burns S, et al. Rapid Growth Produces Transient Cortical Weakness: A Risk Factor for Metaphyseal Fractures During Puberty. J Bone Miner Res. 2010;25(7):1521–6.
24. Wang Q, Nicholson PH, Suuriniemi M, et al. Relationship of sex hormones to bone geometric properties and mineral density in early pubertal girls. J Clin Endocrinol Metab. 2004;89(4):1698–703.
25. Rauch F, Schoenau E. Peripheral quantitative computed tomography of the distal radius in young subjects—new reference data and interpretation of results. J Musculoskelet Neuronal Interact. 2005;5(2):119–26.
26. Khosla S, Riggs BL, Atkinson EJ, et al. Effects of sex and age on bone microstructure at the ultradistal radius: a population-based noninvasive in vivo assessment. J Bone Miner Res 2006;21(1):124–31.
27. Seeman E, Karlsson MK, Duan Y. On exposure to anorexia nervosa, the temporal variation in axial and appendicular skeletal development predisposes

to site-specific deficits in bone size and density: A cross-sectional study. J Bone Miner Res. 2000;15(11):2259–65.

28. Kirmani S, Christen D, van Lenthe GH, et al. Bone structure at the distal radius during adolescent growth. J Bone Miner Res. 2008;24:1033–42.

29. Havill LM, Mahaney MC, L Binkley T, et al. Effects of genes, sex, age, and activity on BMC, bone size, and areal and volumetric BMD. J Bone Miner Res. 2007;22(5):737–46.

30. Schoenau E, Neu CM, Rauch F, et al. Gender-specific pubertal changes in volumetric cortical bone mineral density at the proximal radius. Bone. 2002;31(1): 110–3.

17

Racial Differences in the Acquisition and Age-related Loss of Bone Strength

Shane A. Norris[1], Marcella D. Walker[2], Kate Ward[3], Lisa K. Micklesfield[1], and John M. Pettifor[1]

[1]MRC Developmental Pathways for Health Research Unit, Department of Pediatrics, Faculty of Health Sciences, University of the Witwatersrand, Johannesburg, South Africa
[2]Division of Endocrinology, Department of Medicine, Columbia University, College of Physicians and Surgeons, New York, NY, USA
[3]MRC Lifecourse Epidemiology, University of Southampton, Southampton, UK

INTRODUCTION

Childhood and adolescence are critically important periods in the establishment of peak bone mass, macro- and microstructure and the impact of age- and menopause-related bone loss on bone fragility. Peak bone mass, along with macro- and microstructure, is achieved by the end of the second or third decade of life [1]. Racial differences in the tempo of growth of the axial and appendicular skeleton and differences in the age of onset of puberty and duration of pubertal growth contribute to the racial differences in axial and appendicular morphology and the risk of fractures during the pre- and peripubertal years, as well as determining racial differences in fracture risk in advanced age [2–5]. This chapter addresses several of these issues and underscores the value of studying racial dimorphism as a means of understanding bone fragility.

RACIAL DIFFERENCES IN FRACTURE RATES IN CHILDHOOD AND ADOLESCENCE

Racial differences in fracture rates vary among adolescents. In South Africa, fracture rates in blacks during childhood and adolescence are less than half those in whites of comparable age [4]. Fracture rates are also lower in African-American, Hispanic and Asian individuals than in US white children and adolescents aged 6 to 17 [6]. Understanding factors that contribute to racial heterogeneity in fracture rates is fundamental to identifying risk factors for fracture.

RACIAL DIFFERENCES IN FRACTURE RATES IN ADULTHOOD

The highest incidence of hip fractures is reported in Scandinavia, with an age-standardized hip fracture rate of 532 per 100,000 per annum in Norwegian women. The lowest fracture rates (per 100,000 persons) are reported in Africa, at 4.1 in women in Cameroon [7,8], 4.5 for black South African women [5], and 173 in Chinese women [9]. Even within a racial group, heterogeneity in hip fracture rates is common. For example, among Han Chinese, hip fracture rates vary between Beijing, Hong Kong, and Taiwan [8]. Within countries, the age-standardized incidence of hip fracture in men is approximately half that documented in women.

In the United States (US), hip fractures are lowest in African-Americans and Asians and intermediate or similar in Hispanic compared to non-Hispanic white populations [10,11]. Vertebral fracture rates are similar or higher in Asians than whites [11,12]. African-Americans have lower rates of vertebral fracture than whites [12–14]. Vertebral fracture risk is lower in Mexicans than in whites [15].

Primer on the Metabolic Bone Diseases and Disorders of Mineral Metabolism, Ninth Edition. Edited by John P. Bilezikian.
© 2019 American Society for Bone and Mineral Research. Published 2019 by John Wiley & Sons, Inc.
Companion website: www.wiley.com/go/asbmrprimer

RACIAL DIFFERENCES IN BONE MASS AND GEOMETRY IN CHILDREN

Racial differences in BMD are found in children and adolescents. Data derived from black children and adolescents from South Africa and Gambia do not show consistently greater BMD at all skeletal sites as documented in the US [16]. In South Africa, black children have higher BMD at the hip but not at other sites, whereas in Gambia, black children had lower radial bone mineral content (BMC) than UK white children [17–20].

These data illustrate the difficulty in generalizing about any racial group, because site-specific and geographic differences in BMD and BMC exist. Similarly, when comparing 9-year-old black South African children to white and African-American children of a similar age, after adjusting for age, sex and body size, whole body BMC was higher in South African than US children [17].

Studies using central QCT indicate that, regardless of sex, race has differing effects on axial and appendicular skeletal morphology [21]. In part, this may be due to differences in the duration of prepubertal growth, which is dominated by appendicular growth and is earlier in Asians than whites but later in African-Americans than whites. A later and shorter pubertal growth period influences axial growth and so accounts for similar trunk length in Asians and whites but shorter trunk length in African-Americans than whites. Although precise details by ethnic groups still require further research, the timing of puberty is likely to be an important determinant of racial differences in morphology [22,23].

Vertebrae are shorter in African-Americans but not different in diameter. They build higher trabecular density due to thicker and more connected trabecular plates than whites. The increase from prepubertal to postpubertal values is greater in African-American than in white individuals (34% versus 11%, respectively) [21,24], features that may protect African-Americans from fractures as age advances.

In the appendicular skeleton, the later onset of puberty may partly account for the greater bone length and width in African-American children [21]. On pQCT [25–28], 10-year-old African-American and Hispanic children have higher volumetric BMD (vBMD) than whites because of larger cortical area even after adjusting for age, sex, tibial length and tibia muscle cross-sectional area [26]. Leonard et al. [28] support the view that racial differences are likely to be maturation-dependent. After adjusting for covariates, polar section modulus was 13.4% higher in African-Americans than whites in Tanner stage 1 children, but 2.5% higher in Tanner stage 5 adolescents.

Racial differences in pQCT-derived vBMD and bone geometry are likely to be site-specific. South African data have shown racial differences in bone size and strength in 13-year-old black compared to white children at the 38% tibia, a predominantly cortical site, but not at predominantly trabecular metaphyseal regions [25]. These findings are consistent with differences in BMD between South African black and white adults and children found using DXA at the hip, a site containing both cortical and trabecular bone, but not at the lumbar spine (LS), a predominantly trabecular site.

Using HRpQCT and micro-finite element analysis (µFEA), Misra and colleagues reported that African-American girls had greater cortical perimeter and area, trabecular thickness and vBMD, and stiffness at the distal radius than their white and Asian peers [29]. These deficits in strength and microstructure at the radius are associated with greater risk of fractures [30].

RACIAL DIFFERENCES IN BONE MASS AND GEOMETRY IN ADULTHOOD

Data from the Study of Osteoporotic Fractures (SOF) indicate that age-adjusted mean BMD at the proximal femur in Tobago Afro-Caribbean and African-American women was 21% to 31% and 13% to 23% higher, respectively, compared to US whites [31]. These differences may be due, in part, to differences in bone size and body weight. Higher body weight is associated with higher BMD [32].

Racial differences in BMD are attenuated when bone size and/or weight is taken into account. For example, the differences in BMD between Asian and non-Asian women in SOF were attenuated after adjustment for weight [31]. In the study of women's health across the nation (SWAN), unadjusted LS and femoral neck (FN) BMD was highest in African-American, lowest in Japanese- and Chinese-American and intermediate in white pre- and early perimenopausal women [33]. However, adjusted for weight, differences between Asian and white women in BMD are attenuated whereas differences between African-Americans and other racial groups persist.

Data derived using HRpQCT and µFEA provide insights into racial differences in fracture risk. Putman et al. reported that African-American women from the SWAN study had larger bones than whites, with greater area and total vBMD at both skeletal sites (perhaps because of lower intracortical porosity or higher matrix mineral density). African-Americans had higher trabecular vBMD at the radius and higher cortical vBMD at the tibia with higher cortical area, thickness, and volumes at both sites, and lower cortical porosity, but only at the tibia. These microstructural differences led to greater estimated bone stiffness and failure load at both skeletal sites in African-Americans, consistent with their lower risk of fracture than white women [34].

Premenopausal Chinese-American women have smaller bone size, thicker and denser cortical bone due to lower cortical porosity, higher matrix mineral density, greater trabecular thickness, and more trabecular plates than rods (ie, than white women [35,36]), features conferring greater stiffness in Chinese-American women than their white peers, despite smaller bone size.

Chinese women are endowed with a more robust skeleton than whites. Postmenopausal Chinese-Americans have smaller bone with relatively thicker and denser cortical bone (due to lower porosity and higher matrix mineral density), and more trabecular plates [37]. Similarly, more robust microstructural features as measured by HRpQCT are reported in Australian-Chinese versus white women [38] and Chinese women in Hong Kong. Advantages in Asian women are not restricted to the peripheral skeleton. Similarly, Chinese women have greater FN cortical vBMD and cortical thickness [39].

US Asian young adult men also have smaller bones, thicker and denser cortices (due to lower porosity and higher matrix mineral density), and more platelike trabeculae. These features led to similar estimates of bone strength compared to US white men [40]. US Hispanic men have greater cortical thickness and lower cortical porosity and perhaps higher matrix mineral density than their white peers leading to similar mechanical competence despite smaller bone size [41].

RACIAL DIFFERENCES IN BONE LOSS AND REMODELING IMBALANCE

Bone loss occurs because remodeling becomes unbalanced; less bone is deposited than removed by each remodeling event leading to microstructural deterioration. The negative balance is exacerbated when the remodeling rate increases in midlife in women and later in men. Details of racial differences in remodeling balance and remodeling rate are not well described but, in general, it appears that remodeling is slower in Asians than whites [42]. This area needs more research, but it is hypothesized that the amount of bone lost in African-Americans is similar to whites and thus any differences in morphology late in life are largely due to racial differences in peak bone structure. An alternative hypothesis is that if remodeling is slower, then the negative BMU balance should be less in African-Americans if net bone loss is not different by race.

SUMMARY

Racial differences in skeletal strength in advanced age are the net result of the amount of bone mass and microstructure fashioned by modeling and remodeling during growth and by unbalanced remodeling during aging. Differences in fracture risk can be better understood by studying racial and sex differences in bone size, shape, microstructure and material composition. Less variance in these bone qualities in racial groups within the US than outside the US, particularly in Africa, may signal secular trends in pre- and pubertal growth or the impact of environmental factors. Comparator studies examining racial and sex differences in skeletal growth and later deterioration will enhance our understanding of the global epidemiology of fractures.

REFERENCES

1. Baxter-Jones AD, Faulkner RA, Forwood MR, et al. Bone mineral accrual from 8 to 30 years of age: An estimation of peak bone mass. J Bone Miner Res. 2011;26:1729–39.
2. Clark EM, Ness AR, Bishop NJ, et al. Association between bone mass and fractures in children: a prospective cohort study. J Bone Miner Res. 2006;21:1489–95.
3. Chevalley T, Bonjour JP, van Rietbergen B, et al. Fractures during Childhood and Adolescence in Healthy Boys: Relation with Bone Mass, Microstructure, and Strength. J Clin Endocrinol Metab. 2011;96(10):3134–42.
4. Thandrayen K, Norris SA, Pettifor JM. Fracture rates in urban South African children of different ethnic origins: The Birth to Twenty Cohort. Osteoporos Int. 2009;20:47–52.
5. Solomon L. Osteoporosis and fracture of the femoral neck in the South African Bantu. J Bone Joint Surg Br. 1968;50:2–13.
6. Wren TA, Shepherd JA, Kalkwarf HJ, et al. Racial disparity in fracture risk between white and nonwhite children in the United States. J Pediatr. 2012;161(6):1035–40.
7. Dhanwal DK, Cooper C, Dennison EM. Geographic variation in osteoporotic hip fracture incidence: the growing importance of asian influences in coming decades. J Osteoporos. 2010;2010:757102.
8. Cheng SY, Levy AR, Lefaivre KA, et al. Geographic trends in incidence of hip fractures: a comprehensive literature review. Osteoporos Int. 2011;22:2575–86.
9. Kanis JA, Odén A, McCloskey EV, et al; IOF Working Group on Epidemiology and Quality of Life. A systematic review of hip fracture incidence and probability of fracture worldwide. Osteoporos Int. 2012;23(9):2239–56.
10. Maggi S, Kelsey JL, Litvak J, et al. Incidence of hip fractures in the elderly: a cross-national analysis. Osteoporos Int 1991;1:232–41.
11. Barrett-Connor E, Siris ES, Wehren LE, et al. Osteoporosis and fracture risk in women of different ethnic groups. J Bone Miner Res. 2005;20(2):185–94.
12. Bow CH, Cheung E, Cheung CL, et al. Ethnic difference of clinical vertebral fracture risk. Osteoporos Int. 2012;23:879–85.
13. Taylor AJ, Gary LC, Arora T, et al. Clinical and demographic factors associated with fractures among older Americans. Osteoporos Int. 2011;22:1263–74.
14. Jacobsen SJ, Cooper C, Gottlieb MS, et al. Hospitalization with vertebral fracture among the aged: a national population-based study, 1986-1989. Epidemiology. 1992;3:515–18.
15. Bauer RL, Deyo RA. Low risk of vertebral fracture in Mexican American women. Arch Intern Med. 1987;147:1437–9.

16. Nelson DA, Simpson PM, Johnson CC, et al. The accumulation of whole body skeletal mass in third- and fourth-grade children: effects of age, gender, ethnicity, and body composition. Bone. 1997;20(1):73–8.
17. Micklesfield LK, Norris SA, Nelson DA, et al. Comparisons of body size, composition, and whole body bone mass between North American and South African children. J Bone Miner Res. 2007;22:1869–77.
18. Micklesfield LK, Norris SA, van der Merwe L, et al. Comparison of site-specific bone mass indices in South African children of different ethnic groups. Calcif Tissue Int. 2009;85:317–25.
19. Vidulich L, Norris SA, Cameron N, et al. Differences in bone size and bone mass between black and white 10-year-old South African children. Osteoporos Int. 2006;17:433–40.
20. Prentice A, Laskey MA, Shaw J, et al. Bone mineral content of Gambian and British children aged 0–36 months. Bone Miner. 1990;10:211–24.
21. Gilsanz V, Skaggs DL, Kovanlikaya A, et al. Differential effect of race on the axial and appendicular skeletons of children. J Clin Endocrinol Metab. 1998;83:1420–7.
22. Bass S, Delmas PD, Pearce G, et al. The differing tempo of growth in bone size, mass, and density in girls is region- specific. J Clin Invest. 1999;104(6):795–804.
23. Parent AS, Teilmann G, Juul A, et al. The timing of normal puberty and the age limits of sexual precocity: variations around the world, secular trends, and changes after migration. Endocr Rev. 2003;24(5):668–93.
24. Kleerekoper M, Nelson DA, Flynn MJ, et al. Comparison of radiographic absorptiometry with dual-energy x-ray absorptiometry and quantitative computed tomography in normal older white and black women. J Bone Miner Res. 1994;9:1745–9.
25. Micklesfield LK, Norris SA, Pettifor JM. Determinants of bone size and strength in 13-year-old South African children: The influence of ethnicity, sex and pubertal maturation. Bone. 2011;48:777–85.
26. Wetzsteon RJ, Hughes JM, Kaufman BC, et al. Ethnic differences in bone geometry and strength are apparent in childhood. Bone. 2009;44:970–5.
27. Pollock NK, Laing EM, Taylor RG, et al. Comparisons of trabecular and cortical bone in late adolescent black and white females. J Bone Miner Metab. 2011;29:44–53.
28. Leonard MB, Elmi A, Mostoufi-Moab S, et al. Effects of Sex, Race, and Puberty on Cortical Bone and the Functional Muscle Bone Unit in Children, Adolescents, and Young Adults. J Clin Endocrinol Metab. 2010;95: 1681–9.
29. Misra M, Ackerman KE, Bredella MA, et al. Racial Differences in Bone Microarchitecture and Estimated Strength at the Distal Radius and Distal Tibia in Older Adolescent Girls: a Cross-Sectional Study. J Racial Ethn Health Disparities. 2017;4(4):587–98.
30. Määttä M, Macdonald HM, Mulpuri K, et al. Deficits in distal radius bone strength, density and microstructure are associated with forearm fractures in girls: an HR-pQCT study. Osteoporos Int. 2015;26(3):1163–74.
31. Nam HS, Kweon SS, Choi JS, et al. Racial/ethnic differences in bone mineral density among older women. J Bone Miner Metab. 2013;31(2):190–8.
32. Reid IR. Relationships among body mass, its components, and bone. Bone. 2002;31(5):547–55.
33. Finkelstein JS, Lee ML, Sowers M, et al. Ethnic variation in bone density in premenopausal and early perimenopausal women: effects of anthropometric and lifestyle factors J Clin Endocrinol Metab. 2002;87(7): 3057–67.
34. Putman MS, Yu EW, Lee H, et al. Differences in skeletal microarchitecture and strength in African-American and white women J Bone Miner Res. 2013;28(10): 2177–85.
35. Liu XS, Walker MD, McMahon DJ, et al. Better skeletal microstructure confers greater mechanical advantages in Chinese-American women versus white women. J Bone Miner Res. 2011;26:1783–92.
36. Walker MD, McMahon DJ, Udesky J, et al. Application of high-resolution skeletal imaging to measurements of volumetric BMD and skeletal microarchitecture in Chinese-American and white women: explanation of a paradox. J Bone Miner Res. 2009;24:1953–9.
37. Walker MD, Liu XS, Zhou B, et al. Pre- and postmenopausal differences in bone microstructure and mechanical competence in Chinese-American and white women. J Bone Miner Res. 2013;28(6):1308–18.
38. Wang XF, Wang Q, Ghasem-Zadeh A, et al. Differences in macro- and microarchitecture of the appendicular skeleton in young Chinese and white women. J Bone Miner Res. 2009;24:1946–52.
39. Walker MD, Saeed I, McMahon DJ, et al. Volumetric bone mineral density at the spine and hip in Chinese American and White women. Osteoporos Int. 2012;23(10): 2499–506.
40. Kepley AL, Nishiyama KK, Zhou B, et al. Differences in bone quality and strength between Asian and Caucasian young men. Osteoporos Int. 2017;28(2):549–58.
41. Walker MD, Kepley A, Nishiyama K, et al. Cortical microstructure compensates for smaller bone size in young Caribbean Hispanic versus non-Hispanic white men. Osteoporos Int. 2017;28(7):2147–54.
42. Duan Y, Seeman E. Bone fragility in Asian and Caucasian men. Ann Acad Med Singapore. 2002;31(1):54–66.

18

Calcium, Vitamin D, and Other Nutrients During Growth

Tania Winzenberg[1,2] and Graeme Jones[1]

[1]*Menzies Institute for Medical Research, University of Tasmania, Hobart, TAS, Australia*
[2]*Faculty of Health, University of Tasmania, Hobart, TAS, Australia*

INTRODUCTION

BMD in later life is a function of peak bone mass and the rate of subsequent bone loss [1]. Childhood is potentially an important time to intervene because modeling suggests that a 10% increase in peak bone mass will delay the onset of osteoporosis by 13 years [2]. In addition, low BMD in childhood is a risk factor for childhood fractures [3] suggesting that optimizing age-appropriate bone mass could also have a more immediate benefit on childhood fracture rates. This chapter reviews key nutritional influences on childhood bone development.

CALCIUM

It is widely accepted that an adequate calcium intake in childhood is important for bone development, though the results of observational and intervention studies are mixed [4]. Evidence with regards to effects of dairy and/or calcium intake on fracture is also mixed. In case-control studies low dairy intake has been found to be associated with increased fracture risk in 11- to 13-year-old boys [5] but no other group [6], and in a meta-analysis of case-control studies no association between calcium intake and fracture was observed [6]. However, low calcium/dairy intake has been found to be associated with recurrent fracture in both sexes [7,8].

High levels of calcium intake for children are recommended in many developed countries. Current WHO recommendations based on North American and western European data are from 300 to 400 mg/day for infants, 400 to 700 mg/day for children, and 1300 mg/day for adolescents [9]. Modeling of data from calcium balance studies in 348 children [10] suggests that there is a calcium threshold below which skeletal calcium accumulation was related to intake, but above which skeletal accumulation remained constant. This varied with age to up to 1730 mg in 9- to 17-year-olds. A similar threshold of about 1300 mg was described in girls aged 12 to 15 years [11]. However the relationship between short-term calcium balance studies and achieving bone outcome improvements from longer term calcium supplementation is open to question. In a meta-analysis of randomized controlled trials (RCTs) [12,13], bone outcomes were no different above or below calcium intakes of 1400 mg/day, casting doubt on the clinical relevance of the balance studies' results.

This meta-analysis also [12,13] found that calcium supplementation had no effect on BMD at the femoral neck (FN) or lumbar spine (LS). Supplementation had a small effect on total body (TB) BMC but this did not persist once supplementation ceased. There was a small persistent effect on upper limb BMD, equivalent to a 1.7 percentage point greater increase in BMD in the supplemented compared to the control group which might reduce the absolute risk of fracture at the peak childhood fracture incidence by at most 0.2% per annum (p.a.). Thus, the small increase in bone density in the upper limb from increasing intake from an average 700 mg/day to 1200 mg/day is unlikely to result in a clinically significant decrease in fracture risk. Furthermore the evidence did not suggest that increasing the duration of supplementation led to increasing effects. In the meta-analysis, the effect size did not vary with baseline calcium intakes, down to a level of <600 mg/day. A subsequent RCT targeting children (mean age 12 years) with an habitual calcium intake

<650 mg/day resulted in greater increases in TB BMC (2.3%) and total hip (TH) and LS BMD (2.5 and 2.2% respectively) in children supplemented with an average of 555 mg calcium/day after 18 months but as in the meta-analysis, the effects did not persist once supplements ceased [14]. Short-term benefits of supplementing Gambian toddlers (aged 12 to 18 months) for 18 months for distal forearm BMD (5% to 6% greater gain) also dissipated within 12 months of supplementation ceasing [15]. Because the meta-analysis only included placebo-controlled trials, some RCTs of dairy products were not included but qualitatively the results of these studies were similar with at best small to moderate short-term effects at varying sites [16,17] which did not persist after supplementation ceased in those studies reporting such data [16]. One study reported a larger effect but the intervention group had substantially greater vitamin D intake which may have led to an overestimate of the effect of calcium [18].

VITAMIN D

Vitamin D deficiency is common in children, especially in late adolescence [19]. Observational data on associations between vitamin levels and fracture are sparse and with inconsistent results, at least partly because some studies have long lag times between fracture events and assessment of vitamin D levels [20]. The effectiveness of vitamin D supplementation for improving bone density in children was examined in a meta-analysis of six RCTs [21,22]. In all children, vitamin D supplementation had no statistically significant effects on TB BMC, hip BMD, or forearm BMD and effect sizes were small (standardized mean difference [SMD] 0.10 or less at all three sites). There was a trend to a small effect on LS BMD (SMD +0.15, [95%CI –0.01 to +0.31], $p=0.07$). However, in studies in which the mean baseline serum vitamin D level of the children was low (<35 nmol/L), there were significant effects on TB BMC and LS BMD. These were approximately equivalent to a 2.6% and 1.7% percentage point greater increase from baseline in the supplemented group. RCTs published after this review are broadly consistent with the meta-analysis results, with no effect in studies of replete children but larger (but not statistically significant) effect sizes in pilot studies of children with baseline vitamin D <25 nmol/L [19]. No studies were powered for fracture and it is not known if effects accumulate with ongoing supplementation. Nonetheless, the data suggest that vitamin D supplementation of deficient children could result in clinically useful improvements, particularly if future trials demonstrate that effects accumulate with ongoing supplementation.

FRUIT AND VEGETABLES

Fruit and vegetable intake is postulated to have effects on bone through mechanisms including induction of mild metabolic alkalosis, vitamin K, vitamin C, antioxidants, and phytoestrogens, though phytoestrogens alone have little effect on bone turnover in children [23]. Observational data support a positive relationship between fruit and vegetable intake and bone outcomes in children. In a cross-sectional study, in 8-year-old children [24] urinary potassium was positively associated with both fruit and vegetable intake and BMD. In addition, girls at Tanner stage 2 [25] who consumed ≥3 servings of fruit and vegetables daily had higher bone area, lower urinary calcium excretion, and lower parathyroid hormone levels than those consuming <3 servings daily, though there were no differences in BMD or bone turnover markers. In other cross-sectional studies, 12-year-old girls consuming high amounts of fruit had higher heel BMD than moderate fruit consumers [26] and in adolescent boys and girls [27] fruit intake was positively associated with spine size-adjusted BMC (SA-BMC) and, in boys, with FN SA-BMC. In Chinese adolescents, fruit but not vegetable intake was associated with BMD and BMC at the total body, LS, and hip [28]. Longitudinal data also suggest benefits. Over 7 years, fruit and vegetable intake independently predicted TB BMC in boys but not girls [29]. In children aged 10 to 15 years [30], over 1 year girls increasing fruit intake had a 4.7% greater increase in stiffness index (SI) (measured by QUS) than those who did not, girls increasing vegetable intake had a 3.6% greater increase, and boys increasing vegetable intake had a 2.4% greater increase. In children followed from age 3.8 to 7.8 years, a dietary pattern characterized by a high intake of dark-green and deep-yellow vegetables was associated with high bone mass [31]. Fruit and vegetable intake in children can be increased by 0.3 to 0.99 servings per day by dietary intervention [32]. Further research is needed to confirm if bone health is changed by clinically significant amounts by such increases.

DIET IN PREGNANCY

Nutritional influences on childhood bone development may begin in utero, and because of in utero programming, such influences may affect both early skeletal development and the acquisition of bone mass throughout childhood. However, RCT data are limited.

A systematic review of RCTs of calcium supplementation during pregnancy to improve neonatal bone density identified few studies [33]. One trial demonstrated that either 600 mg or 300 mg of calcium daily resulted in higher neonatal bone density of the ulna, radius, fibula, and tibia (measured by X-ray). However, pooled analysis of two studies did not demonstrate an effect on TB BMD and single studies found no effect on midshaft radius or lumbar spine BMD. Lastly, a RCT in pregnant adolescents compared 1200 mg/day of calcium from either calcium-supplemented orange juice/calcium carbonate supplements or dairy foods with no-intervention controls. The dairy but not the calcium supplementation group had higher total body calcium than controls, which may have been due to the higher vitamin D content of the dairy foods [34]. Given the limited data, it remains unclear whether improving maternal calcium intake in pregnancy is beneficial for in utero bone development.

Observational evidence regarding the effect of vitamin D status on bone outcomes is conflicting [35] possibly because of the diversity in study methodologies and populations which also precludes pooling of the observational literature. One small intervention study ($n=76$, of whom only 19 received vitamin D supplementation) reported no statistically significant effect of vitamin D supplementation to mothers on neonatal forearm BMC measured by single-photon absorptiometry (SPA) but the study was underpowered and at high risk of bias. Published after this review, in observational data from the Danish National Birth Cohort, predicted 25(OH)D levels were not associated with forearm fracture risk in childhood, but taking >10μg/day of vitamin D in midpregnancy was (unexpectedly) associated with increased risk of forearm fracture in girls [36]. This study is limited by the lack of data on confounders in offspring and the fact that, for most women, serum 25(OH)D was predicted rather than measured. Nonetheless this finding underlines the need for strong RCT evidence, such as that provided by the recently published Maternal Vitamin D Osteoporosis Study (MAVIDOS) [37]. This is a large ($n=1134$) multicenter, double-blind randomized placebo-controlled trial of 1000IU of vitamin D_3 orally in pregnant women with serum 25(OH)D of 25 to 100nmol/L. Its primary outcome was neonatal whole-body BMC by DXA within 2 weeks of birth, and this did not differ significantly between intervention and placebo groups (61.6g [95%CI 60.3 to 62.8] versus 60.5 [95% CI 59.3 to 61.7] respectively). An interaction between season and birth and intervention group suggests that there may be a benefit on whole-body BMC for babies born in winter, but this is not definitive. Thus the best available evidence to date suggests that vitamin D supplementation in pregnancy is not an effective population health intervention for improving bone acquisition in utero. However, the results of the planned follow-up of these children at 4 years of age are of considerable interest, given the potential both for beneficial epigenetic effects but also detrimental impacts on childhood fracture [36].

Data on effects of other supplements and nutritional factors in pregnancy on childhood bone outcomes are sparse. In an RCT, zinc supplementation in pregnancy in a disadvantaged area in a developing country resulted in increased fetal femur diaphysis length [38]. Other evidence comes from observational data. Maternal folate intake at 32 weeks was positively associated with spinal SA-BMC after adjusting for children's weight and height [39] in 9-year-olds, and maternal red blood cell folate at 28 weeks gestation with spine BMD in 6-year-olds [40].

Dietary intake of magnesium, phosphorus, potassium, and protein of mothers during the third trimester of pregnancy has been shown to be positively associated, and maternal fat intake negatively associated, with bone density in their children at age 8 [41]. In the same children at age 16, FN and LS BMD associations with magnesium density and fat density in the maternal diet in pregnancy persisted. LS BMD was also positively associated with maternal milk intake and calcium and phosphorus density. With all significant nutrients in the same model, fat density remained negatively associated for the FN and LS, whereas magnesium density remained positively associated for the FN [42]. In another cohort, first-trimester maternal protein intake was positively and carbohydrate intake negatively associated with TB BMC at age 6 years [43] with no associations with maternal fat, magnesium, or folate. In a third cohort, maternal magnesium intake at 32 weeks gestation was positively associated with TB BMC and BMD at age 9 years until adjusted for child's height, and maternal intake of potassium was positively associated with spinal BMC and BMD, until adjusted for child's weight [39]. Principal component analysis of maternal diet in the same study identified a pattern of a high intake of fruit and vegetables and wholemeal bread, pasta, and rice and low intake of processed foods which was quantified by a "prudent diet score." A high score was associated with higher TB and lumbar BMC and BMD [44]. Similarly, in a cohort of rural Indian mother–child pairs, milk product, pulses, and fruit intake were all positively associated with spine BMD in children at age 6 years [40].

Though limited, these data support the need for further research into nutritional interventions in pregnancy.

BREASTFEEDING

Generally, infants fed human milk have lower bone accretion compared to formula-fed infants, possibly because of the low vitamin D content and decreasing phosphorus content of human milk with continued lactation [45]. However, data on the long-term effects of breastfeeding on bone health in children born at term suggest that this initial lower bone accretion is temporary, and catch-up growth occurs later in childhood. These include data from an RCT of infant feeding comparing two different formulas and breastfeeding, in which initial differences in BMC accretion did not persist past 12 months of age [46], as well as longitudinal observational data. In 8-year-old children [47], breastfed children had higher FN, LS, and TB BMD compared with bottle-fed children, and the effect was most marked in children breastfed for >3 months. In 7- to 9-year-old children, being breastfed was not associated with broadband ultrasound attenuation (BUA) or speed of sound (SOS), but in breastfed children duration of breastfeeding was positively associated with metacarpal diameter [48]. The effects of prolonged breastfeeding may differ from that of shorter exposures and differ between sexes though this is not certain. In one observational study breastfeeding for more than 7 months resulted in lower TB BMD, LS bone area, and LS BMC at age 32 years in males but not females [49] but breastfeeding duration was not associated with bone density outcomes at age 4 in a different cohort [50].

Other observational studies with bone measures at younger ages [51,52] did not demonstrate associations between breastfeeding and bone density. However, in a retrospective study, premenopausal women who had been breastfed for >3 months had greater cortical thickness at the radius and a trend towards greater cortical area and cortical BMC at the radius, but not at other sites [53]. Importantly, breastfeeding was protective for childhood fractures in a longitudinal study of prepubertal children [54] and in a case-control study of children aged 4 to 15 years [8], though this was not observed in a longitudinal study of fracture risk from birth to 18 years [55].

SALT

Urinary sodium excretion has been shown to be associated with urinary calcium excretion in girls [56–58] though not with an acute sodium chloride load [58]. Despite this, in the few studies assessing bone outcomes in children urinary sodium excretion has not in turn been shown to be associated with bone density [24,57], though dietary sodium intake was associated with size-adjusted bone area but not BMC in a cross-sectional study of 10-year-old girls [59]. Urinary sodium has also been shown to be associated with a high bone turnover state in adolescent boys [60]. Whether high dietary sodium intake adversely affects other bone outcomes in children is uncertain. Initially, more longitudinal studies are needed to determine if sodium intake does in fact have a clinically important effect on bone in children.

SOFT DRINKS AND MILK AVOIDANCE

Carbonated beverage consumption has been linked with decreased BMD in girls but not boys [61,62] and with increased fracture risk in both sexes. Low milk intake and a higher consumption of carbonated beverages were independent fracture risk factors in children with recurrent fractures [8]. Other studies have reported increased fracture risk with higher cola intake but not non-cola carbonated beverage intake [63,64]. It is unclear if this effect is due to milk replacement—two studies have demonstrated that associations between fracture risk [5], pQCT measures [65], and cola drinks persist after adjustment for milk intake, suggesting independent effects. Milk avoidance also appears to have deleterious effects on children's bone. Prepubertal children who avoid milk have lower TB BMC and areal BMD [66] as well as an increased risk of childhood fracture [6]. There are limited and conflicting data examining whether milk consumption is beneficial in the long-term for reducing fractures. Low milk consumption in childhood has been associated with higher risk of combined hip, wrist, and spine fracture in adult women [67] but in another long-term cohort, for each additional glass of milk consumed daily between age 13 and 18 years, there was a 9% increase in hip fracture risk in males and no association in women after age 50 years [68]. The effect in males appeared in part to be caused by the effect of dairy intake on height.

CONCLUSION

In conclusion, there is increasing evidence linking a number of nutritional factors with children's bone development. Calcium supplementation has been investigated to the greatest extent, but its effects are of limited public health significance. This makes the exploration of other nutritional approaches of key importance.

ACKNOWLEDGMENTS

Graeme Jones receives an NHMRC Practitioner Fellowship, which supported this work.

REFERENCES

1. Hansen MA, Overgaard K, Riis BJ, et al. Role of peak bone mass and bone loss in postmenopausal osteoporosis: 12 year study. BMJ. 1991;303(6808):961–4.
2. Hernandez CJ, Beaupre GS, Carter DR. A theoretical analysis of the relative influences of peak BMD, age-related bone loss and menopause on the development of osteoporosis. Osteoporos Int. 2003;14(10):843–7.
3. Clark EM, Tobias JH, Ness AR. Association between bone density and fractures in children: a systematic review and meta-analysis. Pediatrics. 2006;117(2):e291–7.
4. Lanou AJ, Berkow SE, Barnard ND. Calcium, dairy products, and bone health in children and young adults: a reevaluation of the evidence. Pediatrics. 2005;115(3):736–43.
5. Ma D, Jones G. Soft drink and milk consumption, physical activity, bone mass, and upper limb fractures in children: a population-based case-control study. Calcif Tissue Int. 2004;75(4):286–91.
6. Handel MN, Heitmann BL, Abrahamsen B. Nutrient and food intakes in early life and risk of childhood fractures: a systematic review and meta-analysis. Am J Clin Nutr. 2015;102(5):1182–95.
7. Goulding A, Grant AM, Williams SM. Bone and body composition of children and adolescents with repeated forearm fractures. J Bone Miner Res. 2005;20(12):2090–6.
8. Manias K, McCabe D, Bishop N. Fractures and recurrent fractures in children; varying effects of environmental factors as well as bone size and mass. Bone. 2006;39(3):652–7.
9. World Health Organization and Food and Agriculture Organization of the United Nations. *Vitamin and Mineral Requirements in Human Nutrition* (2nd ed.). World Health Organization and Food and Agriculture Organization of the United Nations, 2004.
10. Matkovic V, Heaney RP. Calcium balance during human growth: evidence for threshold behavior. Am J Clin Nutr. 1992;55(5):992–6.

11. Jackman LA, Millane SS, Martin BR, et al. Calcium retention in relation to calcium intake and postmenarcheal age in adolescent females. Am J Clin Nutr. 1997;66(2):327–33.
12. Winzenberg T, Shaw K, Fryer J, et al. Effects of calcium supplementation on bone density in healthy children: meta-analysis of randomised controlled trials. BMJ. 2006;333(7572):775.
13. Winzenberg TM, Shaw K, Fryer J, et al. Calcium supplementation for improving bone mineral density in children. Cochrane Database Syst Rev. 2006; (2):CD005119.
14. Lambert HL, Eastell R, Karnik K, et al. Calcium supplementation and bone mineral accretion in adolescent girls: an 18-mo randomized controlled trial with 2-y follow-up. Am J Clin Nutr. 2008;87(2):455–62.
15. Umaretiya PJ, Thacher TD, Fischer PR, et al. Bone mineral density in Nigerian children after discontinuation of calcium supplementation. Bone. 2013;55(1):64–8.
16. Winzenberg T, Jones G. Cost-effectiveness of nutritional interventions for bone health in children and young adults – what is known and where are the gaps? In: Watson RR, Gerald JK, Preedy VR (eds.) *Nutrients, Dietary Supplements, and Nutriceuticals: Cost Analysis versus Clinical Benefits*. Springer/Humana Press, 2011.
17. Zhang ZQ, Ma XM, Huang ZW, et al. Effects of milk salt supplementation on bone mineral gain in pubertal Chinese adolescents: a 2-year randomized, double-blind, controlled, dose-response trial. Bone. 2014;65:69–76.
18. Chan GM, Hoffman K, McMurry M. Effects of dairy products on bone and body composition in pubertal girls. J Pediatr. 1995;126(4):551–6.
19. Winzenberg T, Jones G. Vitamin D and bone health in childhood and adolescence. Calcif Tissue Int. 2013;92(2):140–50.
20. Moon RJ, Harvey NC, Davies JH, et al. Vitamin D and skeletal health in infancy and childhood. Osteoporos Int. 2014;25(12):2673–84.
21. Winzenberg T, Powell S, Shaw KA, et al. Effects of vitamin D supplementation on bone density in healthy children: systematic review and meta-analysis. BMJ. 2011;342:c7254.
22. Winzenberg TM, Powell S, Shaw KA, et al. Vitamin D supplementation for improving bone mineral density in children. Cochrane Database Syst Rev. 2010;10:CD006944.
23. Jones G, Dwyer T, Hynes K, et al. A randomized controlled trial of phytoestrogen supplementation, growth and bone turnover in adolescent males. Eur J Clin Nutr. 2003;57(2):324–7.
24. Jones G, Riley MD, Whiting S. Association between urinary potassium, urinary sodium, current diet, and bone density in prepubertal children. Am J Clin Nutr. 2001;73(4):839–44.
25. Tylavsky FA, Holliday K, Danish R, et al. Fruit and vegetable intakes are an independent predictor of bone size in early pubertal children. Am J Clin Nutr. 2004;79(2):311–7.
26. McGartland CP, Robson PJ, Murray LJ, et al. Fruit and vegetable consumption and bone mineral density: the Northern Ireland Young Hearts Project. Am J Clin Nutr. 2004;80(4):1019–23.
27. Prynne CJ, Mishra GD, O'Connell MA, et al. Fruit and vegetable intakes and bone mineral status: a cross sectional study in 5 age and sex cohorts. Am J Clin Nutr. 2006;83(6):1420–8.
28. Li JJ, Huang ZW, Wang RQ, et al. Fruit and vegetable intake and bone mass in Chinese adolescents, young and postmenopausal women. Public Health Nutr. 2013;16(1):78–86.
29. Vatanparast H, Baxter-Jones A, Faulkner RA, et al. Positive effects of vegetable and fruit consumption and calcium intake on bone mineral accrual in boys during growth from childhood to adolescence: the University of Saskatchewan Pediatric Bone Mineral Accrual Study. Am J Clin Nutr. 2005;82(3):700–6.
30. Hirota T, Kusu T, Hirota K. Improvement of nutrition stimulates bone mineral gain in Japanese school children and adolescents. Osteoporos Int. 2005;16(9):1057–64.
31. Wosje KS, Khoury PR, Claytor RP, et al. Dietary patterns associated with fat and bone mass in young children. Am J Clin Nutr. 2010;92(2):294–303.
32. Knai C, Pomerleau J, Lock K, et al. Getting children to eat more fruit and vegetables: a systematic review. Prev Med. 2006;42(2):85–95.
33. Buppasiri P, Lumbiganon P, Thinkhamrop J, et al. Calcium supplementation (other than for preventing or treating hypertension) for improving pregnancy and infant outcomes. Cochrane Database Syst Rev. 2015(2):CD007079.
34. Chan GM, McElligott K, McNaught T, et al. Effects of dietary calcium intervention on adolescent mothers and newborns: A randomized controlled trial. Obstet Gynecol. 2006;108(3 Pt 1):565–71.
35. Harvey NC, Holroyd C, Ntani G, et al. Vitamin D supplementation in pregnancy: a systematic review. Health Technol Assess. 2014;18(45):1–190.
36. Petersen SB, Strom M, Maslova E, et al. Predicted vitamin D status during pregnancy in relation to offspring forearm fractures in childhood: a study from the Danish National Birth Cohort. Br J Nutr. 2015;114(11):1900–8.
37. Cooper C, Harvey NC, Bishop NJ, et al. Maternal gestational vitamin D supplementation and offspring bone health (MAVIDOS): a multicentre, double-blind, randomised placebo-controlled trial. Lancet Diabetes Endocrinol. 2016;4(5):393–402.
38. Merialdi M, Caulfield LE, Zavaleta N, et al. Randomized controlled trial of prenatal zinc supplementation and fetal bone growth. Am J Clin Nutr. 2004;79(5):826–30.
39. Tobias JH, Steer CD, Emmett PM, et al. Bone mass in childhood is related to maternal diet in pregnancy. Osteoporos Int. 2005;16(12):1731–41.
40. Ganpule A, Yajnik CS, Fall CH, et al. Bone mass in Indian children—relationships to maternal nutritional status and diet during pregnancy: the Pune Maternal Nutrition Study. J Clin Endocrinol Metab. 2006;91(8):2994–3001.
41. Jones G, Riley MD, Dwyer T. Maternal diet during pregnancy is associated with bone mineral density in children: a longitudinal study. Eur J Clin Nutr. 2000;54(10):749–56.

42. Yin J, Dwyer T, Riley M, et al. The association between maternal diet during pregnancy and bone mass of the children at age 16. Eur J Clin Nutr. 2010;64(2):131–7.
43. Heppe DH, Medina-Gomez C, Hofman A, et al. Maternal first-trimester diet and childhood bone mass: the Generation R Study. Am J Clin Nutr. 2013;98(1):224–32.
44. Cole ZA, Gale CR, Javaid MK, et al. Maternal dietary patterns during pregnancy and childhood bone mass: a longitudinal study. J Bone Miner Res. 2009;24(4):663–8.
45. Specker B. Nutrition influences bone development from infancy through toddler years. J Nutr. 2004;134(3):691S–5S.
46. Specker BL, Beck A, Kalkwarf H, et al. Randomized trial of varying mineral intake on total body bone mineral accretion during the first year of life. Pediatrics. 1997;99(6):E12.
47. Jones G, Riley M, Dwyer T. Breastfeeding in early life and bone mass in prepubertal children: a longitudinal study. Osteoporos Int. 2000;11(2):146–52.
48. Micklesfield L, Levitt N, Dhansay M,. Maternal and early life influences on calcaneal ultrasound parameters and metacarpal morphometry in 7- to 9-year-old children. J Bone Miner Metab. 2006;24(3):235–42.
49. Pirila S, Taskinen M, Viljakainen H, et al. Infant milk feeding influences adult bone health: a prospective study from birth to 32 years. PLoS One. 2011;6(4):e19068.
50. Harvey NC, Robinson SM, Crozier SR, et al. Breastfeeding and adherence to infant feeding guidelines do not influence bone mass at age 4 years. Br J Nutr. 2009;102(6):915–20.
51. Kurl S, Heinonen K, Jurvelin JS, et al. Lumbar bone mineral content and density measured using a Lunar DPX densitometer in healthy full-term infants during the first year of life. Clin Physiol Funct Imaging. 2002;22(3):222–5.
52. Young RJ, Antonson DL, Ferguson PW, et al. Neonatal and infant feeding: effect on bone density at 4 years. J Pediatr Gastroenterol Nutr. 2005;41(1):88–93.
53. Laskey MA, de Bono S, Smith EC, et al. Influence of birth weight and early diet on peripheral bone in premenopausal Cambridge women: a pQCT study. J Musculoskelet Neuronal Interact. 2007;7(1):83.
54. Ma DQ, Jones G. Clinical risk factors but not bone density are associated with prevalent fractures in prepubertal children. J Paediatr Child Health. 2002;38(5):497–500.
55. Jones IE, Williams SM, Goulding A. Associations of birth weight and length, childhood size, and smoking with bone fractures during growth: evidence from a birth cohort study. Am J Epidemiol. 2004;159(4):343–50.
56. O'Brien KO, Abrams SA, Stuff JE, et al. Variables related to urinary calcium excretion in young girls. J Pediatr Gastroenterol Nutr. 1996;23(1):8–12.
57. Matkovic V, Ilich JZ, Andon MB, et al. Urinary calcium, sodium, and bone mass of young females. Am J Clin Nutr. 1995;62(2):417–25.
58. Duff TL, Whiting SJ. Calciuric effects of short-term dietary loading of protein, sodium chloride and potassium citrate in prepubescent girls. J Am Coll Nutr. 1998;17(2):148–54.
59. Hoppe C, Molgaard C, Michaelsen KF. Bone size and bone mass in 10-year-old Danish children: effect of current diet. Osteoporos Int. 2000;11(12):1024–30.
60. Jones G, Dwyer T, Hynes KL, et al. A prospective study of urinary electrolytes and bone turnover in adolescent males. Clin Nutr. 2007;26(5):619–23.
61. Whiting S, Heaky A, Psiuk S, et al. Relationship between carbonated and other low nutrient dense beverages and bone mineral content of adolescents. Nutr Res. 2001;21:1107–15.
62. McGartland C, Robson PJ, Murray L, et al. Carbonated soft drink consumption and bone mineral density in adolescence: the Northern Ireland Young Hearts project. J Bone Miner Res. 2003;18(9):1563–9.
63. Wyshak G, Frisch RE. Carbonated beverages, dietary calcium, the dietary calcium/phosphorus ratio, and bone fractures in girls and boys. J Adolesc Health. 1994;15(3):210–5.
64. Petridou E, Karpathios T, Dessypris N, et al. The role of dairy products and non alcoholic beverages in bone fractures among schoolage children. Scand J Soc Med. 1997;25(2):119–25.
65. Libuda L, Alexy U, Remer T, et al. Association between long-term consumption of soft drinks and variables of bone modeling and remodeling in a sample of healthy German children and adolescents. Am J Clin Nutr. 2008;88(6):1670–7.
66. Black RE, Williams SM, Jones IE, et al. Children who avoid drinking cow milk have low dietary calcium intakes and poor bone health. Am J Clin Nutr. 2002;76(3):675–80.
67. Kalkwarf HJ, Khoury JC, Lanphear BP. Milk intake during childhood and adolescence, adult bone density, and osteoporotic fractures in US women. Am J Clin Nutr. 2003;77(1):257–65.
68. Feskanich D, Bischoff-Ferrari HA, Frazier AL, et al. Milk consumption during teenage years and risk of hip fractures in older adults. JAMA Pediatr. 2014;168(1):54–60.

19
Mechanical Loading and the Developing Skeleton

Mark R. Forwood

School of Medical Science and Menzies Health Institute Queensland, Griffith University, Gold Coast, QLD, Australia

INTRODUCTION

The basic morphology of the skeleton is determined genetically, but its final mass and architecture result from epigenetic mechanisms sensitive to mechanical factors. When subjected to loading, fracture resistance depends on bone mass, material properties, geometry, and tissue quality [1]. To determine if an intervention in children could reduce skeletal fragility in adults would require understanding of its effects on each mechanical determinant. Yet it is only recently that studies have moved beyond areal bone mineral density (aBMD) to include volumetric density and geometry [2]. None are yet able to determine if childhood adaptations translate into anti-fracture efficacy in adults, but the latter variables do provide surrogates of bone strength. There is convincing evidence that growing bone has greater capacity to respond to increased mechanical loading than the adult skeleton [2–9]. The question is what to prescribe, what is the window of opportunity, and does adherence to the prescription optimize bone strength so that the risk of fracture is reduced in adulthood and old age?

WHAT IS THE WINDOW OF OPPORTUNITY?

Osteoporotic fractures occur because reduced bone mass decreases the safety factor for skeletal loading. This results from age-related bone loss and/or failure to achieve optimal peak bone mass [10,11]. During childhood and adolescence the skeleton undergoes rapid change due to growth, modeling, and remodeling, the processes that maximize bone accrual. It is generally believed that bone mass increases substantially during adolescence, reaching a plateau, being peak bone mass (PBM), in the late teen or young adult years [12]. Until recently the timing of this event was still disputed. Some studies suggested that it was reached by 20 years of age [13,14], but others concluded that PBM was attained in the third decade of life [15]. Longitudinal bone mineral accrual data measured from childhood (age 8) to young adulthood (up to age 30) demonstrate that PBM occurs by the end of the second, or very early in the third, decade of life [16] (Fig. 19.1). On a regional basis the lower limb achieves a plateau earliest, within 1 year of peak height velocity (PHV), and the lumbar spine (LS) the latest, some 4 years after PHV. Importantly, depending on the skeletal site, 33% to 46% of the adult BMC was accrued over a 4-year period of adolescent growth surrounding PHV. In females this total accrual represents double the amount of bone mineral that will subsequently be lost during the postmenopausal years from age 50 to 80 years [17].

Although the variance in PBM is genetically determined [10] it is influenced by mechanical factors, such as exercise. The evidence, while not universal [18,19], is mounting that the peripubertal period is more advantageous than post puberty to elicit an adaptive response in growing bones [5,8,9,20–27]. This observation is exemplified in unilateral racquet sports in which the differences between bones of dominant and non-dominant arms are greatest in players who started training before puberty, compared to those who started playing during adulthood [8,9].

The consequence of physical activity on the skeleton of adults is conservation, not acquisition [3]. So factors that optimize PBM must be implemented during growth,

---

Primer on the Metabolic Bone Diseases and Disorders of Mineral Metabolism, Ninth Edition. Edited by John P. Bilezikian.
© 2019 American Society for Bone and Mineral Research. Published 2019 by John Wiley & Sons, Inc.
Companion website: www.wiley.com/go/asbmrprimer

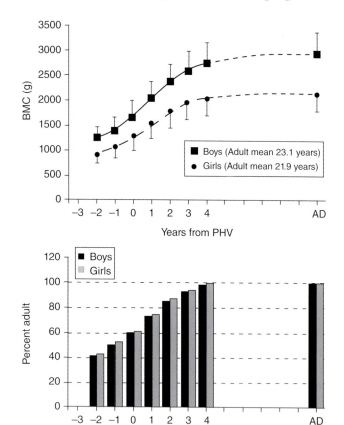

Fig. 19.1. Bone mineral accrual (total body BMC) from preadolescence (2 years before peak height velocity—PHV) to adulthood (AD). When controlled for maturity, total body BMC reached a plateau by 7 years post PHV, representing about 18.8 and 20.5 years of age in girls and boys, respectively [16]. Data from [16].

when the stimulus of mechanical loading can augment modeling processes that are already active and the surfaces have a greater proportion of active bone cells. There is also evidence that the increase in estrogen levels in males and females during adolescence augments the amount of functional estrogen receptor alpha (ERα) available to facilitate strain-related responses in bone [28,29], explaining enhanced sensitivity to physical activity during early puberty [19,21,24–27].

CHARACTERISTICS OF AN EFFECTIVE LOADING PRESCRIPTION

If fracture prevention is a consequence of an activity program then we need to know the characteristics of mechanical stimuli that effect skeletal adaptation. We know that static, or isometric, loading provides minimal adaptive stimulus to bone [30–32] and can even inhibit normal appositional growth [33]. Activation of new bone formation also requires that a threshold magnitude of loading is exceeded [34], but that there is an interaction between strain rate and amplitude of loading that modulates this threshold [32,34–36]. The moderating effect of strain rate occurs because bone tissue is viscoelastic and interstitial fluid mechanics underlie transduction of applied dynamic loads into bone cell responses [37]. For external loads, such as those experienced during exercise, these responses are optimal in a range of loading frequencies up to about 2.0 Hz [36]. In terms of physical activity, exercises that create relatively high strain rates will be more adaptive than loads applied gently, or in which the load is held constant for a period of time. That is, jumping exercises will create a greater osteogenic effect than simply walking or doing isometric strength exercises.

A key characteristic of loading is that very few loading cycles are required to elicit adaptive responses [36–38]. The loading effect saturates relatively quickly so that increasing the duration of loading beyond about 40 cycles per day has little additional effect [31]. Cardiovascular health notwithstanding, long exercise sessions will therefore have diminishing returns in terms of growing strong bones. Children do not need to participate in long periods of exercise, nor disrupt their normal schedule of activities, or inactivity, to provide an adequate adaptive stimulus to their skeleton. Moreover, a given physical activity will be more osteogenic if it is divided into shorter bouts with rest periods in between [39,40]. This is because the sensitivity of the bone cells to the loading stimulus returns after a period of rest. For example, new bone formation was 80% greater in tibias subjected to 4 bouts of 90 cycles per day for 2 weeks, compared to one bout of 360 cycles [39]. When a similar program was extended to 16 weeks, groups that received 4 bouts of 90 cycles per day had significantly greater bending strength in the loaded ulnas compared to those receiving a single bout of 360 cycles [40]. In ulnas loaded 4 times per day, the increased strength was attributed to greater geometric adaptations that resist axial bending (Fig. 19.2) [40].

These mechanical loading principles have been adopted increasingly in physical activity interventions [18,24–27,41–44]. But stronger bones can also be achieved in children who undertake greater levels of normal physical activity than sedentary children [22] or compete in high-impact sports like gymnastics [45,46]. Although numerous controlled trials have applied some of these principles to maximize modeling in growing bone [2,21,41,43], the Healthy Bones II [25,26] and "Action Schools! BC" programs [18,24,42] in Vancouver were specifically designed around principles to optimize the osteogenic index [47] of a practical and sustainable activity intervention. In "Action Schools! BC," the bone-loading component of the program included an extra 15 minutes of simple activities for 5 days per week, and "bounce at the bell" [42] in which just 3 minutes of variegated jumping activities were implemented 3 times per day (at each school bell) for 4 days per week. During initial trials, the program induced an increase in bone mass (BMC) at the LS and femoral neck (FN) of about 2% in boys and girls.

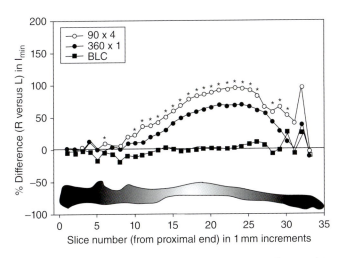

Fig. 19.2. When daily loading was divided into 4 bouts of 90 cycles per day, instead of a single bout of 360 cycles, ulnas showed significantly greater increases in the geometric property responsible for resistance to bending (CSMI). The graph illustrates the difference between loaded and unloaded limbs (%) for the CSMI of loaded animals, age-matched controls (AMC), and baseline controls (BLC). Source: [40]. Reproduced with permission from Elsevier.

Controlled trials using programs like "Action Schools! BC" have increased bone strength, estimated using pQCT, in the distal tibia [18], and produced increases of 2% to 4% in spine and total body BMC [24] of prepubertal boys, and BMC and section modulus of the FN (an index of bending strength) in peripubertal girls [24]. By optimizing the osteogenic index this program is able to achieve modest but significant increases in parameters of bone strength that are similar to other studies that involve more intensive bone-loading programs [2,21,41,43,48].

PEAK BONE MASS OR PEAK BONE STRENGTH?

Skeletal adaptation to mechanical loading must increase bone strength without unduly increasing the metabolic cost of locomotion. This design gives rise to a trade-off between the goals of strength and lightness. Efficient adaptation, therefore, cannot simply increase bone mass but must effect efficient increases in bone geometry (Fig. 19.3). These adaptations are often independent of the material properties. The contribution of altered bone geometry to fracture risk is unappreciated by clinical assessment using DXA because it cannot distinguish geometry and density, nor cortical and cancellous bone [1,49]. The resolution of DXA is also too low to detect small changes in bone dimensions that can provide substantial increases in bone strength. An excellent example of this characteristic is the modest 5% increase in the aBMD of rat ulnas after 16 weeks of axial loading 3 times per day [40]. This is starkly contrasted by an incredible 64% increase in

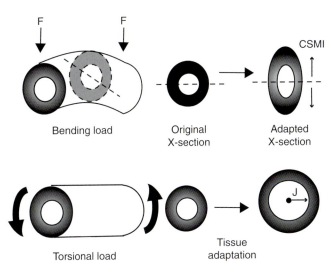

Fig. 19.3. The aim of skeletal adaptation is not to increase bone mass per se, but to improve resistance to increased loads efficiently. Increased strength is therefore achieved without jeopardizing lightness. Modeling on appropriate bone surfaces improves the biomechanical determinants for enhanced bending or torsional strength (CSMI), or polar moment of inertia (J), respectively. These changes improve bending and torsional strength, disproportionately to the change in BMC or BMD, and hence can be misinterpreted by DXA. Source: [1]. Reproduced with permission from Springer.

ultimate breaking strength. Such a marked discrepancy occurs because the new bone formation occurs at the periosteal surface where a relatively small increase in bone apposition provides a disproportionate mechanical advantage at the locations of greatest strain. That is, the small amount of bone is strategically placed away from the axis of bending where it has an exponential effect to resist bending loads (Fig. 19.2 and 19.3). Nonetheless, the vast majority of physical activity studies have relied upon measures of BMD or BMC, derived from DXA, to assess the adaptive response. The use of hip structural analysis, derived from DXA, has provided some mechanical indices to assess maturation of bone strength [22]. But the increasing use of micro–computed tomography (μCT) in animal studies, and pQCT in children [2,9,18,50,51], specifically allows adaptations in density and geometry to be distinguished and proves that increased bone strength, not mass, is the goal of adaptation [9,51]. The important question to address is whether these changes are maintained into adulthood and old age.

PERSISTENCE OF CHILDHOOD BONE ADAPTATION

During childhood and adolescence it is clear that the mechanical component of physical activity effects adaptations in the growing skeleton. The increases in bone mass (BMD) are relatively modest, between 1 and 5%, but may be manifested by geometric adaptations that

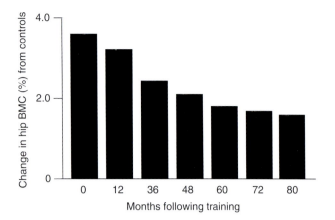

Fig. 19.4. Results of a school-based jumping intervention for 7 months in children aged ~8 years [41]. At the end of the 7-month program (0 months), hip BMC was significantly greater than controls (3.6%). When followed for up to 8 years of detraining, the exercise group retained greater hip BMC (1.4%) after controlling for baseline age and change in height, weight, and sports participation [54]. Source: adapted from [41,54].

augment bone strength by a significantly greater degree. It is practically impossible to prove that the adaptations of childhood prevent osteoporotic fracture in old age. Confounding variables associated with retrospective studies, such as self-selection for physique, reduce the certainty that a given activity created long-lasting skeletal protection. But a small number of studies have now traversed childhood and followed adolescents into adulthood. Early studies of this phenomenon suggested that the gains in bone mass from childhood training were lost in adulthood [52,53]. Some of these studies were cross-sectional, or started relatively late in adolescence, when it is harder to control for confounding variables such as maturity. Longitudinal studies from childhood, through adolescence, report a sustained effect on BMC accrual up to 14 years after training, or cessation of a physical activity intervention [41,54,55] (Fig. 19.4). In addition physically active children achieve greater bone mass (BMC) during adolescence than their sedentary peers [20,22], and their higher BMC is maintained into adulthood [56].

Retention of childhood bone mass into adulthood may be modest but preservation of skeletal architecture and geometry, underlying bone strength, could be more significant. The distinction between preservation of bone mass or architecture is elegantly illustrated by lifelong preservation of adapted bone structure after short-term exercise in rapidly growing rodents [57]. Forearm training started at 5 weeks of age, with a short 7-week exercise program, following which animals were limited to cage activity for up to 92 weeks (2 years of age, equivalent to senescence for rodents). Increases in bone mass induced by exercise (aBMD and BMC) were not retained into adulthood but there was long-term preservation of bone structural changes, which was manifested in superior strength and fatigue life in the trained animals [57]. Similar retention of exercise-induced structural adaptation from childhood has been observed following cessation of gymnastics [55] and racquet sports in male and female tennis players who had started training during childhood (about 10 years of age) [58,59]. Up to 3 years after retirement a greater bone mass was retained at the age of 30, modeled into an architecture that provided greater indices of bone strength, such as cortical area and CSMI. These data support the argument that physical activity during childhood can effect structural adaptations in the skeleton that persist well into adulthood. The greater strength afforded by these structural changes could reduce the risk of fracture in adults, more than that predicted by bone mass alone.

CONCLUSION

Compared to adults, the skeleton of children and adolescents is capable of greater structural adaptations in response to the mechanical stimulus of physical activity. The evidence is persuasive that the peripubertal period is more advantageous to effect adaptive responses. To grow healthy bones, physical activity should not consist of static or isometric exercises, but should incorporate repetitive cyclical loads that include a range of strain magnitudes and directions, such as running and jumping. Because only a few cycles of loading are required to elicit an adaptive response, distributed bouts of loading that incorporate rest periods are more osteogenic than single sessions of long duration. These parameters of loading have been translated into feasible public health interventions that have achieved improved bone mass and strength in children and adolescents. Those architectural adaptations can persist into adulthood and be translated into a lower risk of fracture.

REFERENCES

1. Forwood MR. Mechanical effects on the skeleton: are there clinical implications? Osteoporos Int. 2001;12:77–83.
2. Heinonen A, Sievänen H, Kannus P, et al. High-impact exercise and bones of growing girls: a 9-month controlled trial. Osteoporos Int. 2000;11:1010–17.
3. Forwood MR, Burr DB. Physical activity and bone mass: exercises in futility? Bone Miner. 1993;21:89–112.
4. Jarvinen TL, Pajamaki I, Sievanen H, et al. Femoral neck response to exercise and subsequent deconditioning in young and adult rats. J Bone Miner Res. 2003;18:1292–9.
5. Rantalainen T, Weeks BK, Nogueira RC, et al. Effects of bone-specific physical activity, gender and maturity on tibial cross-sectional bone material distribution: a cross-sectional pQCT comparison of children and young adults aged 5-29 years. Bone 2015;72:101–8.
6. Rubin CT, Bain SD, McLeod KJ. Suppression of the osteogenic response in the aging skeleton. Calcif Tissue Int. 1992;50:306–13.

7. Turner CH, Takano Y, Owan I. Aging changes mechanical loading thresholds for bone formation in rats. J Bone Miner Res. 1995;10:1544–9.
8. Kannus P, Haapasalo H, Sankelo M, et al. Effect of starting age of physical activity on bone mass in the dominant arm of tennis and squash players. Ann Intern Med. 1995;123:27–31.
9. Kontulainen S, Sievanen H, Kannus P, et al. Effect of long-term impact-loading on mass, size, and estimated strength of humerus and radius of female racquet-sports players: a peripheral quantitative computed tomography study between young and old starters and controls. J Bone Miner Res. 2002;17:2281–9.
10. Ferrari S, Rizzoli R, Slosman D, et al. Familial resemblance for bone mineral mass is expressed before puberty. J Clin Endocrinol Metab. 1998;83:358–61.
11. Hui SL, Slemenda CW, Johnston CC. The contribution of bone loss to post menopausal osteoporosis. Osteoporos Int. 1990;1:30–4.
12. Faulkner RA, Bailey DA. Osteoporosis: a pediatric concern? In: Daly RM, Petit MA (eds) *Optimizing Bone Mass and Strength: The Role of Physical Activity and Nutrition During Growth*. Vol. 51, Medicine and Sport Science. Karger, 2007, pp. 1–12.
13. Bachrach LK, Hastie T, Wang M-C, et al. Bone mineral acquisition in healthy Asian, Hispanic, Black and Caucasian youth: A longitudinal study. J Clin Endocrinol Metab. 1999;84:4702–12.
14. Faulkner RA, Bailey DA, Drinkwater DT, et al. Bone densitometry in Canadian children 8-17 years of age. Calcif Tissue Int. 1996;59:344–51.
15. Recker EE, Davies KM, Hinders SM, et al. Bone gain in young adult women. JAMA 1992;268:2403–8.
16. Baxter-Jones ADG, Faulkner RA, Forwood MR, et al. Bone mineral accrual from 8 to 30 years of age: An estimation of peak bone mass. J Bone Miner Res. 2011;26:1729–39.
17. Arlot M, Sornay-Rendu E, Garnero P, et al. Apparent pre- and postmenopausal bone loss evaluated by DXA at different skeletal sites in women: The OFELY cohort. J Bone Miner Res. 1997;12:683–90.
18. Macdonald HM, Kontulainen SA, Khan KM, et al. Is a school-based physical activity intervention effective for increasing tibial bone strength in boys and girls? J Bone Miner Res. 2007;22:434–46.
19. Sundberg M, Gardsell P, Johnell O, et al. Peripubertal moderate exercise increases bone mass in boys but not in girls: a population-based intervention study. Osteoporos Int. 2001;12:230–8.
20. Bailey DA, McKay HA, Mirwald RL, et al. A six-year longitudinal study of the relationship of physical activity to bone mineral accrual in growing children: the university of Saskatchewan bone mineral accrual study. J Bone Miner Res. 1999;14:1672–9.
21. Bradney M, Pearce G, Naughton G, et al. Moderate exercise during growth in prepubertal boys: changes in bone mass, size, volumetric density, and bone strength: a controlled prospective study. J Bone Miner Res. 1998;13:1814–21.
22. Forwood MR, Baxter-Jones AD, Beck TJ, et al. Physical activity and strength of the proximal femur during the adolescent growth spurt: a longitudinal analysis. Bone. 2006;38:576–83.
23. Kannus P, Haapasalo H, Sankelo M, et al. Effect of starting age of physical activity on bone mass in the dominant arm of tennis and squash players. Ann Intern Med. 1995;123:27–31.
24. MacDonald HM, Kontulainen S, Petit M, et al. Does a novel school-based physical activity model benefit femoral neck bone strength in pre- and early pubertal children? Osteoporos Int. 2008;9:1445–56.
25. Mackelvie KJ, McKay HA, Khan KM, et al. A school-based exercise intervention augments bone mineral accrual in early pubertal girls. J Pediatr. 2001;139:501–8.
26. MacKelvie KJ, Petit MA, Khan KM, et al. Bone mass and structure are enhanced following a 2-year randomized controlled trial of exercise in prepubertal boys. Bone. 2004;34:755–64.
27. Petit MA, McKay HA, MacKelvie KJ, et al. A randomized school-based jumping intervention confers site and maturity-specific benefits on bone structural properties in girls: a hip structural analysis study. J Bone Miner Res. 2002;17:363–72.
28. Damien E, Price JS, Lanyon LE. Mechanical strain stimulates osteoblast proliferation through the estrogen receptor in males as well as females. J Bone Miner Res. 2000;15:2169–77.
29. Zaman G, Jessop HL, Muzylak M, et al. Osteocytes use estrogen receptor alpha to respond to strain but their ERalpha content is regulated by estrogen. J Bone Miner Res. 2006;21:1297–306.
30. Hert J, Liskova M, Landa J. Reaction of bone to mechanical stimuli. 1. Continuous and intermittent loading of tibia in rabbit. Folia Morphol (Praha). 1971;19:290–300.
31. Rubin CT, Lanyon LE. Regulation of bone formation by applied dynamic loads. J Bone Joint Surg Am. 1984;66:397–402.
32. Turner CH, Owan I, Takano Y. Mechanotransduction in bone: role of strain rate. Am J Physiol 1995;269:E438–42.
33. Robling AG, Duijvelaar KM, Geevers JV, et al. Modulation of appositional and longitudinal bone growth in the rat ulna by applied static and dynamic force. Bone. 2001;29:105–13.
34. Turner C, Forwood M, Rho J, et al. Mechanical Loading Thresholds for Lamellar and Woven Bone Formation. J Bone Miner Res. 1994;9:87–97.
35. O'Connor JA, Lanyon LE, MacFie H. The influence of strain rate on adaptive bone remodelling. J Biomech. 1982;15:767–81.
36. Turner CH, Forwood MR, Otter MW. Mechanotransduction in bone: do bone cells act as sensors of fluid flow? Faseb J. 1994;8:875–8.
37. Rubin CT, Lanyon LE. Kappa Delta Award paper. Osteoregulatory nature of mechanical stimuli: function as a determinant for adaptive remodeling in bone. J Orthop Res. 1987;5:300–10.

38. Rubin CT, Lanyon LE. Regulation of bone mass by mechanical strain magnitude. Calcif Tissue Int. 1985;37:411–17.
39. Robling AG, Burr DB, Turner CH. Partitioning a daily mechanical stimulus into discrete loading bouts improves the osteogenic response to loading. J Bone Miner Res. 2000;15:1596–602.
40. Robling AG, Hinant FM, Burr DB, et al. Improved bone structure and strength after long-term mechanical loading is greatest if loading is separated into short bouts. J Bone Miner Res. 2002;17:1545–54.
41. Fuchs RK, Bauer JJ, Snow CM. Jumping improves hip and lumbar spine bone mass in prepubescent children: a randomized controlled trial. J Bone Miner Res. 2001;16:148–56.
42. McKay HA, MacLean L, Petit M, et al. "Bounce at the Bell": a novel program of short bouts of exercise improves proximal femur bone mass in early pubertal children. Br J Sports Med. 2005;39:521–6.
43. Linden C, Ahlborg HG, Besjakov J, et al. A school curriculum-based exercise program increases bone mineral accrual and bone size in prepubertal girls: two-year data from the pediatric osteoporosis prevention (POP) study. J Bone Miner Res. 2006;21:829–35.
44. Weeks BK, Young CM, Beck BR. Eight months of regular in-school jumping improves indices of bone strength in adolescent boys and girls: the POWER PE study. J Bone Miner Res. 2008;23:1002–11.
45. Faulkner RA, Forwood MR, Beck TJ, et al. Strength indices of the proximal femur and shaft in prepubertal female gymnasts. Med Sci Sports Exerc. 2003;35:513–8.
46. Erlandson MC, Kontulainen SA, Chilibeck PD, et al. Bone mineral accrual in 4- to 10-year-old precompetitive, recreational gymnasts: a 4-year longitudinal study. J Bone Miner Res. 2011;26:1313–20.
47. Turner CH, Robling AG. Designing exercise regimens to increase bone strength. Exerc Sport Sci Rev. 2003;31:45–50.
48. MacKelvie KJ, McKay HA, Petit MA, et al. Bone mineral response to a 7-month randomized controlled, school-based jumping intervention in 121 prepubertal boys: associations with ethnicity and body mass index. J Bone Miner Res. 2002;17:834–44.
49. Prentice A, Parsons TJ, Cole TJ. Uncritical use of bone mineral density in absorptiometry may lead to size-related artifacts in the identification of bone mineral determinants. Am J Clin Nutr. 1994;60:837–2.
50. Daly RM, Saxon L, Turner CH, et al. The relationship between muscle size and bone geometry during growth and in response to exercise. Bone. 2004;34:281–7.
51. Macdonald HM, Cooper DM, McKay HA. Anterior-posterior bending strength at the tibial shaft increases with physical activity in boys: evidence for non-uniform geometric adaptation. Osteoporos Int. 2009;20:61–70.
52. Karlsson MK, Linden C, Karlsson C, et al. Exercise during growth and bone mineral density and fractures in old age. Lancet. 2000;355:469–70.
53. Nordström A, Olsson T, Nordström P. Bone gained from physical activity and lost through detraining: a longitudinal study in young males. Osteoporos Int. 2004;16:835–41.
54. Gunter K, Baxter-Jones AD, Mirwald RL, et al. Impact exercise increases BMC during growth: an 8-year longitudinal study. J Bone Miner Res. 2008;23:986–93.
55. Erlandson MC, Kontulainen SA, Chilibeck PD, et al. 2012 Higher premenarcheal bone mass in elite gymnasts is maintained into young adulthood after long-term retirement from sport: a 14-year follow-up. J Bone Miner Res. 2012;27:104–10.
56. Jackowski SA, Kontulainen SA, Cooper DM, et al. Adolescent physical activity and bone strength at the proximal femur in adulthood. Med Sci Sports Exerc. 2014;46:736–44.
57. Warden SJ, Fuchs RK, Castillo AB, et al. Exercise when young provides lifelong benefits to bone structure and strength. J Bone Miner Res. 2007;22:251–9.
58. Haapasalo H, Kontulainen S, Sievanen H, et al. Exercise-induced bone gain is due to enlargement in bone size without a change in volumetric bone density: a peripheral quantitative computed tomography study of the upper arms of male tennis players. Bone. 2000;27:351–7.
59. Kontulainen S, Kannus P, Haapasalo H, et al. Good maintenance of exercise-induced bone gain with decreased training of female tennis and squash players: a prospective 5-year follow-up study of young and old starters and controls. J Bone Miner Res. 2001;16:195–201.

20
Pregnancy and Lactation

Christopher S. Kovacs[1] and Henry M. Kronenberg[2]

[1]Faculty of Medicine—Endocrinology, Health Sciences Centre, Memorial University of Newfoundland, St. John's, NL, Canada
[2]Endocrine Unit, Massachusetts General Hospital and Harvard Medical School, Boston, MA, USA

INTRODUCTION

Pregnancy and lactation place demands on women to provide sufficient calcium and other minerals for the fetus and neonate, respectively. Despite similar magnitudes of demand for minerals, different hormone-mediated adaptations are invoked in each of these reproductive periods (Fig. 20.1). Detailed references on this subject are available in a more extensive review [1].

PREGNANCY

The developing human skeleton accretes about 30g of calcium and 20g of phosphate by term, and about 80% of both during the third trimester [2]. The calcium demand appears to be largely met by a doubling of maternal intestinal calcium absorption, mediated by $1,25(OH)_2D_3$ and other factors.

Mineral ions and calciotropic hormones

Normal pregnancy results in characteristic alterations in serum chemistries and calciotropic hormones [1]. The total serum calcium falls because of a decline in serum albumin, but the ionized calcium (the physiologically important fraction) remains constant. Serum phosphate and magnesium levels are normal.

In studies of women from North America and Europe, the "intact" PTH level falls to the low end of the normal range during the first trimester, but increases to mid-normal by term. In contrast, PTH did not suppress in some studies of women from Asia and Africa, which may reflect lower intakes of calcium and vitamin D, and higher oxalate intakes.

Total and free $1,25(OH)_2D_3$ increase to reach 2 to 3 times baseline values by the third trimester. The rise in $1,25(OH)_2D_3$ is not driven by PTH, because PTH falls while $1,25(OH)_2D_3$ increases. Also, $1,25(OH)_2D_3$ increases threefold or more in pregnant rodents despite lacking PTH or parathyroids. The maternal kidneys and not the placenta account for most of the rise in $1,25(OH)_2D_3$ during pregnancy. This has been confirmed by animal studies [1] and findings of an anephric woman, who displayed very low endogenous $1,25(OH)_2D_3$ levels before and during pregnancy [3]. The renal 1α-hydroxylase may be upregulated by such factors as PTHrP, estradiol, prolactin, and placental lactogen.

PTHrP levels progressively increase from the first trimester and may contribute to the rise in $1,25(OH)_2D_3$ and suppression of PTH [1]. Serum calcitonin levels are also increased, and there are significant changes in estradiol, prolactin, placental lactogen, IGF_1, and other hormones. Each of these may have effects on calcium and bone metabolism during pregnancy.

Intestinal absorption of calcium

Studies using stable calcium isotopes have confirmed that intestinal calcium absorption is doubled from as early as 12 weeks of gestation in women; this appears to be a major maternal adaptation to meet the fetal need for calcium. It may largely represent a $1,25(OH)_2D_3$ mediated increase in active and passive absorption, although intestinal calcium absorption appears to increase before

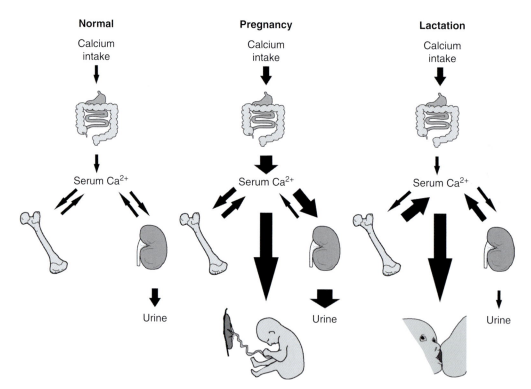

Fig. 20.1. Schematic illustration contrasting calcium homeostasis in human pregnancy and lactation, as compared to normal. The thickness of arrows indicates a relative increase or decrease with respect to the normal and nonpregnant state. Although not illustrated, the serum (total) calcium is decreased during pregnancy, whereas the ionized calcium remains normal during both pregnancy and lactation. Source: adapted from [47]. Reproduced with permission of The Endocrine Society.

free $1,25(OH)_2D_3$ levels change. Also, studies in rodents indicate that intestinal calcium absorption increases during pregnancy despite severe vitamin D deficiency or absence of the vitamin D receptor [4–6]. Prolactin, placental lactogen, and other factors also stimulate intestinal calcium absorption in rodents, independently of $1,25(OH)_2D_3$ [7, 8]. The increase in calcium absorption early in pregnancy contributes to a positive calcium balance in most women [9], which may be utilized when the peak fetal demand occurs during the third trimester.

Renal handling of calcium

The 24-hour urine calcium excretion increases as early as the 12th week of gestation and often exceeds the normal range [1]. Because fasting urine calcium values are normal or low, the increased 24-hour urine calcium reflects the increased intestinal absorption of calcium (absorptive hypercalciuria).

Skeletal calcium metabolism

Animal models indicate that histomorphometric parameters of bone turnover are increased during pregnancy and that bone mineral content may increase or decrease, influenced in part by calcium intake [1, 10, 11].

Comparable histomorphometric data are not available for women, but one small study indicated that bone resorption parameters were increased at 8 to 10 weeks in 15 women who electively terminated pregnancy [12].

Most human studies have examined changes in biochemical markers of bone formation and resorption during pregnancy. These studies are fraught with confounding variables that may artifactually raise or lower the values [1]. Given these limitations, many studies have reported that urinary and serum markers of bone resorption are increased from early to mid-pregnancy. Conversely, serum markers of bone formation are often reduced in early or mid-pregnancy, rising to normal or above before term. Total alkaline phosphatase rises early in pregnancy, largely because of contributions from the placenta; it is not a useful marker of bone formation in pregnancy.

The available data are consistent with a possible modest increase in bone resorption that begins as early as the 10th week. There is comparatively little maternal–fetal calcium transfer occurring then, as compared to the peak rate during the third trimester. One might anticipate that markers of bone resorption should increase further in the third trimester, but this has not been consistently observed.

Changes in skeletal calcium content have been assessed by sequential areal bone density (aBMD) measurements. These studies are potentially confounded by changes in body composition, weight, and skeletal volumes during normal pregnancy. Moreover, to avoid fetal radiation exposure, DXA measurements have usually been obtained 1 to 8 months before planned pregnancies, and then again 1 to 6 weeks after delivery [1]. These studies have reported zero change to as much as a 5% decrease in lumbar spine (LS) aBMD, and smaller changes at appendicular sites. In many of the studies a small net loss of aBMD observed at the postpartum measurement was confounded by bone loss from lactation (see "Lactation/Skeletal Calcium Metabolism").

It seems certain that any acute changes in bone metabolism during pregnancy do not cause long-term changes in skeletal calcium content or strength. More than five dozen epidemiological studies have found that parity is associated with a neutral or protective effect on bone density or fracture risk; very few studies suggest a contrary view [1].

Osteoporosis in pregnancy

Women occasionally present with fragility fractures and low BMD during or shortly after pregnancy [13]. In a few cases, low bone density or a genetic cause of fragility was confirmed prior to pregnancy. Some women may experience excessive resorption of calcium from the skeleton, especially if calcium intake is insufficient for maternal and fetal requirements. The apparently increased rate of bone resorption during normal pregnancy may contribute to fracture risk, because a high rate of bone turnover is an independent risk factor for fractures. During lactation, additional changes in mineral metabolism occur that may further increase fracture risk (see "Lactation/Skeletal Calcium Metabolism" and "Osteoporosis of Lactation").

Focal, transient osteoporosis of the hip is a rare, self-limited form of pregnancy-associated osteoporosis [13]. It is probably not a manifestation of altered calciotropic hormone levels or mineral balance during pregnancy, but rather might be a consequence of local factors. These patients present with unilateral or bilateral hip pain, limp, and/or hip fracture in the third trimester. There is objective evidence of reduced aBMD of the symptomatic femoral head and neck. MRI reveals increased water content of the femoral head and the marrow cavity, and a joint effusion may be present. The symptoms and radiological findings usually resolve within 2 to 6 months postpartum.

Primary hyperparathyroidism

Although probably a rare condition, primary hyperparathyroidism during pregnancy has been historically associated with an alarming rate of adverse outcomes in the fetus and neonate, including a 30% rate of spontaneous abortion or stillbirth [1, 14]. The adverse postnatal outcomes are thought to result from suppression of the fetal and neonatal parathyroid glands; this suppression may occasionally be prolonged for months after birth or permanent. Surgical correction of primary hyperparathyroidism during the second trimester has been almost universally recommended. Several case series have found elective surgery to be well tolerated and to dramatically reduce the rate of adverse events when compared to the earlier cases reported in the literature. Many of the women in those early cases had a severe form of primary hyperparathyroidism that is not often seen today (symptomatic, with nephrocalcinosis and renal insufficiency). Although mild, asymptomatic primary hyperparathyroidism during pregnancy has been followed conservatively with successful outcomes, complications continue to occur, so that, in the absence of definitive data, surgery during the second trimester remains the most common recommendation [15].

Familial hypocalciuric hypercalcemia

Although familial hypocalciuric hypercalcemia (FHH) has not been reported to adversely affect the mother during pregnancy, the maternal hypercalcemia has caused fetal and neonatal parathyroid suppression with subsequent tetany, even in babies that carry the FHH mutation [16]. The absorptive hypercalciuria also shifts the relationship between urinary calcium and creatinine, thereby making Ca/Cr nomograms unreliable to diagnose FHH during pregnancy [17, 18].

Hypoparathyroidism and pseudohypoparathyroidism

Early in pregnancy, some hypoparathyroid women have fewer hypocalcemic symptoms and require less supplemental calcium [1]. This is consistent with a limited role for PTH in the pregnant woman, and suggests that an increase in $1,25(OH)_2D_3$ and/or increased intestinal calcium absorption will occur in the absence of PTH. However, it is clear from other case reports that some pregnant hypoparathyroid women require increased $1,25(OH)_2D_3$ replacement in order to avoid worsening hypocalcemia [1]. It is important to maintain a normal ionized or albumin-corrected calcium because maternal hypocalcemia due to hypoparathyroidism can result in intrauterine fetal hyperparathyroidism and fetal death. Late in pregnancy, hypercalcemia can occur unless the calcitriol dosage is substantially reduced or discontinued. This effect may be mediated by the increasing levels of PTHrP in the maternal circulation.

In limited case reports of pseudohypoparathyroidism, pregnancy has been noted to normalize the serum calcium level, reduce the PTH level by half, and increase the $1,25(OH)_2D_3$ level two- to threefold [19]. Why this occurs remains unclear, just as it is unexplained in normal pregnant women.

Vitamin D deficiency and insufficiency

Maternal 25OHD levels do not change significantly as a result of pregnancy [1]. Consequently, pregnant women do not require higher intakes of vitamin D to maintain a set 25OHD level. There are as yet no large randomized trials that have examined the effects of vitamin D deficiency or insufficiency on human pregnancy. However, available data from small clinical trials of vitamin D supplementation, observational studies, and case reports suggest that, consistent with animal studies, vitamin D deficiency is not associated with any worsening of maternal calcium homeostasis, and that the fetus will have a normal serum calcium and fully mineralized skeleton at term (this topic is reviewed in detail in [14]). Randomized trials have found that maternal vitamin D supplementation during pregnancy increases maternal and cord blood 25OHD levels without altering cord blood calcium or anthropometric parameters.

Low calcium intake

Because absorptive hypercalciuria typically occurs during pregnancy, this may be viewed as evidence that calcium intake normally exceeds maternal requirements. A randomized intervention suggested that calcium supplementation may benefit bone density only of those women with very low dietary intakes of calcium [20]. If calcium intake is insufficient to meet the combined requirements of mother and fetus, then maternal skeletal resorption must occur.

Low calcium intake has been associated with an increased risk of preeclampsia. Calcium supplementation reduces the risk of preeclampsia when the dietary calcium intake is very low, whereas there is no effect with adequate dietary calcium intake.

LACTATION

The average daily loss of calcium in breast milk is 210 mg, although losses of 500 to 1000 mg calcium have been documented in women nursing twins or triplets. A temporary demineralization of the skeleton appears to be the main mechanism by which lactating humans meet the calcium demand. It is not mediated by PTH or $1,25(OH)_2D_3$, but may be stimulated by PTHrP in the setting of a fall in estradiol levels.

Mineral ions and calciotropic hormones

The mean ionized calcium level of exclusively breastfeeding women increases slightly but remains within the normal range, whereas serum phosphate levels increase and may exceed the normal range. The rise in serum phosphate is caused by increased flux of phosphate into the blood from skeletal resorption in the setting of decreased renal phosphate excretion.

Intact PTH is low or undetectable in most exclusively breastfeeding women, and then it rises to normal or above after weaning. But as with pregnancy, some studies of women from Asia and Africa have found that PTH does not suppress and may even increase. Maternal free and bound $1,25(OH)_2D_3$ levels fall to normal within days of parturition and remain there. Calcitonin falls to normal values after the first 6 weeks. Calcitonin protects the rodent skeleton from excessive resorption during lactation [10], but whether it does so in women is unknown.

PTHrP levels are significantly higher in lactating women than in nonpregnant controls. The source of PTHrP is the breast, which excretes it into milk at concentrations up to 10,000 times those found in the blood of patients with hypercalcemia of malignancy or normal human controls. Animal models have confirmed that mammary tissue is the main source of PTHrP during lactation [21], and that PTHrP secretion is regulated by the calcium-sensing receptor found in the breast [22]. Circulating PTHrP causes resorption of calcium from the maternal skeleton (through osteoclast activity and osteocytic osteolysis), renal tubular reabsorption of calcium, and (indirectly) suppression of PTH. Deletion of the PTHrP gene from mammary tissue at the onset of lactation resulted in more modest losses of bone mineral content during lactation in mice [21]. In humans, PTHrP levels correlate with the amount of BMD lost, negatively with PTH levels, and positively with the ionized calcium levels of lactating women [23–25]. Furthermore, observations in aparathyroid women provide additional evidence of the impact of PTHrP (see 'Lactation/Hypoparathyroidism and Pseudohypoparathyroidism').

Intestinal absorption of calcium

Intestinal calcium absorption decreases to the nonpregnant rate from the increased rate of pregnancy.

Renal handling of calcium

In humans, the glomerular filtration rate (GFR) falls during lactation, and renal calcium excretion is typically reduced to levels as low as 50 mg per 24 hours. Tubular reabsorption of calcium must be increased, perhaps mediated by PTHrP.

Skeletal calcium metabolism

Histomorphometric data from animals consistently show increased bone turnover during lactation. Losses of 25% to 35% or more of bone mineral and ash weight result occur in rodents after 2 to 3 weeks of lactation [1]. These losses result from both osteoclast-mediated bone resorption and osteocytic osteolysis [1], with osteolysis prevented by deleting the PTH/PTHrP receptor from osteocytes [26].

Comparative histomorphometric data are lacking for humans; in place of that, bone formation and resorption markers have been assessed in cross-sectional and prospective studies. A reduced GFR and contracted intravascular volume may confound such measurements. Urinary and serum markers of bone resorption are increased two- to threefold during lactation and are higher than the levels attained in the third trimester. Serum markers of bone formation are generally high and are increased over the levels observed during the third trimester. Total alkaline phosphatase falls immediately postpartum due to loss of the placental fraction, but may still remain above normal due to the bone-specific fraction. These findings are consistent with a marked increase in bone turnover.

Serial measurements of aBMD by DXA during lactation have shown a fall of 3% to 10.0% in bone mineral content after 2 to 6 months of lactation at trabecular sites (LS, hip, femur, and distal radius), with smaller losses at cortical sites, including whole body [1]. These aBMD changes are in accord with studies in rats, mice, and primates, in which the skeletal resorption has been shown to occur largely at trabecular and to a lesser degree at endocortical surfaces. The loss occurs at a peak rate of 1% to 3% *per month*, far exceeding the rate of 1% to 3% per year that can occur in women after menopause who are considered to be losing bone rapidly. Loss of mineral from the maternal skeleton appears to be a normal, hormonally programmed consequence of lactation. Several studies have demonstrated that calcium supplementation does not reduce the amount of bone lost during lactation [27–30]. The lactational decrease in BMD correlates with the amount of calcium lost in the breast milk [31].

The mechanisms controlling the rapid loss of skeletal calcium content are not well understood. The reduced estradiol levels of lactation are important but are unlikely to be the sole explanation. Six months of acute estrogen deficiency in reproductive age women, induced by gonadotropin-releasing hormone (GnRH) agonist therapy, leads to only 1% to 4% losses in trabecular (but not cortical) aBMD, increased urinary calcium excretion, and suppression of $1,25(OH)_2D_3$ and PTH [1]. This contrasts with lactating women who are not as estrogen deficient but who lose more aBMD (at both trabecular and cortical sites), have normal (as opposed to low) $1,25(OH)_2D_3$ levels, and have reduced (as opposed to increased) urinary calcium excretion. The difference between isolated GnRH-induced estrogen deficiency and lactation is caused in part by the effects of PTHrP, which stimulates osteoclasts and osteocytes to resorb mineral. PTHrP and estrogen deficiency are both induced by suckling and high prolactin levels, and their combined effects are greater than either alone during lactation (Fig. 20.2).

The bone density losses of lactation appear to be substantially reversed by 6 to 12 months after weaning [1, 28, 32], although the speed and completeness of recovery may differ by skeletal site and technique used [33, 34]. This corresponds to a gain in bone density of 0.5% to 2% per month in a woman who has weaned her infant. The mechanism for this restoration of bone density is uncertain. In the long term, the consequences of lactation-induced

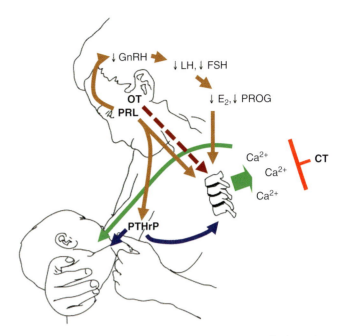

Fig. 20.2. The breast–brain–bone circuit controls lactation. Suckling and prolactin (PRL) both inhibit the hypothalamic gonadotropin-releasing hormone (GnRH) pulse center, which in turn suppresses the gonadotropins (luteinizing hormone [LH] and follicle-stimulating hormone [FSH]), leading to low levels of the ovarian sex steroids (estradiol [E_2] and progesterone [PROG]). Prolactin may also have direct effects on its receptor in bone cells. PTHrP production and release from the breast are stimulated by suckling, prolactin, low estradiol, and the calcium receptor. PTHrP enters the bloodstream and combines with systemically low estradiol levels to markedly upregulate bone resorption and osteocytic osteolysis. Increased bone resorption releases calcium and phosphate into the bloodstream, which then reach the breast ducts and are actively pumped into the breast milk. PTHrP also passes into milk at high concentrations, but whether swallowed PTHrP plays a role in regulating calcium physiology of the neonate is uncertain. In addition to stimulating milk ejection, oxytocin (OT) may directly affect osteoblast and osteoclast function (dashed line). Calcitonin (CT) may inhibit skeletal responsiveness to PTHrP and low estradiol. Source: adapted from [48]. Reproduced with permission of Springer Science and Business Media B.V.

depletion of bone mineral appear clinically unimportant. Over five dozen epidemiologic studies of pre- and postmenopausal women have found no adverse effect, or even a protective effect, of lactation on peak bone mass, bone density, or hip fracture risk [1].

Osteoporosis of lactation

Rarely, women suffer fragility fractures during lactation, and osteoporotic readings are confirmed by DXA. Like osteoporosis in pregnancy, this may represent a coincidental, unrelated disease. Some women were confirmed to have low bone density prior to pregnancy or even a genetic cause of skeletal fragility. Alternatively, some cases may represent an exacerbation of the normal degree of skeletal demineralization that occurs during lactation, and a continuum of changes that may have begun during pregnancy. For example, excessive PTHrP release from the breasts, combined with estrogen deficiency, could conceivably cause excessive bone resorption, osteoporosis, and fractures. Sustained high PTHrP levels have been found in a few cases of lactational osteoporosis [35]. Also, in most cases where vertebral compression fractures occurred during lactation, bone density spontaneously increased after weaning, consistent with temporary lactation-induced bone loss [36].

Hypoparathyroidism and pseudohypoparathyroidism

Levels of calcitriol and calcium supplementation required for treatment of hypoparathyroid women fall early in the postpartum period, especially in women who breastfeed, and hypercalcemia may occur if the calcitriol dosage is not substantially reduced or stopped [37]. This is consistent with PTHrP reaching the maternal circulation in amounts sufficient to allow stimulation of $1,25(OH)_2D_3$ synthesis and normalization of calcium homeostasis [38].

The effect of lactation on pseudohypoparathyroidism has not been documented. Because these patients have renal but not skeletal resistance to PTH, it is possible that lactation may lead to increased skeletal resorption compared to normal when the effects of PTH and PTHrP are combined.

Vitamin D deficiency and insufficiency

During lactation, maternal 25OHD levels are unchanged [39, 40] because very little vitamin D or 25OHD passes into breast milk. The available data from small clinical trials, observational studies, and case reports indicate that lactation proceeds normally regardless of vitamin D status, and breast milk calcium content is unaffected by vitamin D deficiency or by supplementation with doses as high as 6400 IU per day [1]. This is likely because maternal calcium homeostasis is dominated by skeletal resorption induced by estrogen deficiency and PTHrP. It is the neonate who will suffer the consequences of being born of a vitamin D deficient mother, and especially if exclusively breastfed without supplements or sunlight exposure [14].

Low calcium intake

The calcium content of milk appears to be largely derived from skeletal resorption; consequently, low calcium intake does not alter breast milk calcium content nor does it accentuate maternal bone loss during lactation [41–44].

ADOLESCENT PREGNANCY AND LACTATION

Adolescent pregnancy and lactation do not reduce peak bone mass, as previously feared [45]. In a National Health and Nutrition Examination Survey (NHANES III) analysis of 819 women aged 20 to 25, women who had been pregnant as adolescents had the same BMD as nulliparous women and women who had been pregnant as adults [46]. Women who breastfed as adolescents had higher BMD than women who had not breastfed and nulliparous women [46].

IMPLICATIONS

During pregnancy and lactation, novel regulatory systems specific to these settings complement the usual regulators of calcium homeostasis. The fetal calcium demand is met in large part by a doubling of intestinal calcium absorption, an adaptation that may not be fully explained by an observed increase in $1,25(OH)_2D_3$ concentrations. In comparison, skeletal calcium resorption is a dominant mechanism by which calcium is supplied to the breast milk, aided by renal calcium conservation. These changes during lactation appear to be driven by PTHrP in association with estrogen deficiency, and are independent of calcium intake.

The rapidity of calcium regain by the skeleton of the lactating woman occurs through a mechanism that is not understood. Although it is apparent that some women can experience fragility fractures as a consequence of pregnancy or lactation, for most women these adaptations in calcium and bone metabolism occur silently and without apparent long-term adverse consequences.

REFERENCES

1. Kovacs CS. Maternal Mineral and Bone Metabolism During Pregnancy, Lactation, and Post-Weaning Recovery. Physiol Rev. 2016;96(2):449–547.
2. Trotter M, Hixon BB. Sequential changes in weight, density, and percentage ash weight of human skeletons from an early fetal period through old age. Anat Rec. 1974;179(1):1–18.
3. Turner M, Barre PE, Benjamin A, et al. Does the maternal kidney contribute to the increased circulating 1,25-dihydroxyvitamin D concentrations during pregnancy? Miner Electrolyte Metab. 1988;14:246–52.
4. Halloran BP, DeLuca HF. Calcium transport in small intestine during pregnancy and lactation. Am J Physiol. 1980;239:E64–E8.
5. Brommage R, Baxter DC, Gierke LW. Vitamin D-independent intestinal calcium and phosphorus absorption during reproduction. Am J Physiol. 1990;259:G631–G8.

6. Fudge NJ, Kovacs CS. Pregnancy up-regulates intestinal calcium absorption and skeletal mineralization independently of the vitamin D receptor. Endocrinology. 2010;151(3):886–95.
7. Pahuja DN, DeLuca HF. Stimulation of intestinal calcium transport and bone calcium mobilization by prolactin in vitamin D-deficient rats. Science. 1981;214:1038–9.
8. Mainoya JR. Effects of bovine growth hormone, human placental lactogen and ovine prolactin on intestinal fluid and ion transport in the rat. Endocrinology. 1975;96:1165–70.
9. Heaney RP, Skillman TG. Calcium metabolism in normal human pregnancy. J Clin Endocrinol Metab. 1971;33(4):661–70.
10. Woodrow JP, Sharpe CJ, Fudge NJ, et al. Calcitonin plays a critical role in regulating skeletal mineral metabolism during lactation. Endocrinology. 2006;147(9):4010–21.
11. Kirby BJ, Ardeshirpour L, Woodrow JP, et al. Skeletal recovery after weaning does not require PTHrP. J Bone Miner Res. 2011;26(6):1242–51.
12. Purdie DW, Aaron JE, Selby PL. Bone histology and mineral homeostasis in human pregnancy. Br J Obstet Gynaecol. 1988;95(9):849–54.
13. Kovacs CS, Ralston SH. Presentation and management of osteoporosis presenting in association with pregnancy or lactation. Osteoporos Int. 2015;26(9):2223–41.
14. Kovacs CS. Bone Development and Mineral Homeostasis in the Fetus and Neonate: Roles of the Calciotropic and Phosphotropic Hormones. Physiol Rev. 2014;94(4):1143–218.
15. Schnatz PF, Curry SL. Primary hyperparathyroidism in pregnancy: evidence-based management. Obstet Gynecol Surv. 2002;57(6):365–76.
16. Thomas AK, McVie R, Levine SN. Disorders of maternal calcium metabolism implicated by abnormal calcium metabolism in the neonate. Am J Perinatol. 1999;16(10):515–20.
17. Walker A, Fraile JJ, Hubbard JG. "Parathyroidectomy in pregnancy"—a single centre experience with review of evidence and proposal for treatment algorithim. Gland Surg. 2014;3(3):158–64.
18. Morton A. Altered calcium homeostasis during pregnancy may affect biochemical differentiation of hypercalcaemia. Intern Med J. 2004;34(11):655–6; author reply 6–7.
19. Breslau NA, Zerwekh JE. Relationship of estrogen and pregnancy to calcium homeostasis in pseudohypoparathyroidism. J Clin Endocrinol Metab. 1986;62:45–51.
20. Koo WW, Walters JC, Esterlitz J, et al. Maternal calcium supplementation and fetal bone mineralization. Obstet Gynecol. 1999;94(4):577–82.
21. VanHouten JN, Dann P, Stewart AF, et al. Mammary-specific deletion of parathyroid hormone-related protein preserves bone mass during lactation. J Clin Invest. 2003;112(9):1429–36.
22. Mamillapalli R, VanHouten J, Dann P, et al. Mammary-specific ablation of the calcium-sensing receptor during lactation alters maternal calcium metabolism, milk calcium transport, and neonatal calcium accrual. Endocrinology. 2013;154(9):3031–42.
23. Kovacs CS, Chik CL. Hyperprolactinemia caused by lactation and pituitary adenomas is associated with altered serum calcium, phosphate, parathyroid hormone (PTH), and PTH-related peptide levels. J Clin Endocrinol Metab. 1995;80(10):3036–42.
24. Dobnig H, Kainer F, Stepan V, et al. Elevated parathyroid hormone-related peptide levels after human gestation: relationship to changes in bone and mineral metabolism. J Clin Endocrinol Metab. 1995;80(12):3699–707.
25. Sowers MF, Hollis BW, Shapiro B, et al. Elevated parathyroid hormone-related peptide associated with lactation and bone density loss. JAMA. 1996;276(7):549–54.
26. Qing H, Ardeshirpour L, Pajevic PD, et al. Demonstration of osteocytic perilacunar/canalicular remodeling in mice during lactation. J Bone Miner Res. 2012;27(5):1018–29.
27. Kolthoff N, Eiken P, Kristensen B, et al. Bone mineral changes during pregnancy and lactation: a longitudinal cohort study. Clin Sci (Lond). 1998;94(4):405–12.
28. Polatti F, Capuzzo E, Viazzo F, et al. Bone mineral changes during and after lactation. Obstet Gynecol. 1999;94(1):52–6.
29. Kalkwarf HJ, Specker BL, Bianchi DC, et al. The effect of calcium supplementation on bone density during lactation and after weaning. N Engl J Med. 1997;337(8):523–8.
30. Cross NA, Hillman LS, Allen SH, et al. Changes in bone mineral density and markers of bone remodeling during lactation and postweaning in women consuming high amounts of calcium. J Bone Miner Res. 1995;10(9):1312–20.
31. Laskey MA, Prentice A, Hanratty LA, et al. Bone changes after 3 mo of lactation: influence of calcium intake, breast-milk output, and vitamin D-receptor genotype. Am J Clin Nutr. 1998;67(4):685–92.
32. Sowers M. Pregnancy and lactation as risk factors for subsequent bone loss and osteoporosis. J Bone Miner Res. 1996;11(8):1052–60.
33. Bjornerem A, Ghasem-Zadeh A, Wang X, et al. Irreversible Deterioration of Cortical and Trabecular Microstructure Associated with Breastfeeding. J Bone Miner Res. 2017;32(4):681–7.
34. Brembeck P, Lorentzon M, Ohlsson C, et al. Changes in cortical volumetric bone mineral density and thickness, and trabecular thickness in lactating women postpartum. J Clin Endocrinol Metab. 2015;100(2):535–43.
35. Reid IR, Wattie DJ, Evans MC, et al. Post-pregnancy osteoporosis associated with hypercalcaemia. Clin Endocrinol (Oxf). 1992;37(3):298–303.
36. Phillips AJ, Ostlere SJ, Smith R. Pregnancy-associated osteoporosis: does the skeleton recover? Osteoporos Int. 2000;11(5):449–54.
37. Caplan RH, Beguin EA. Hypercalcemia in a calcitriol-treated hypoparathyroid woman during lactation. Obstet Gynecol. 1990;76(3 Pt 2):485–9.
38. Mather KJ, Chik CL, Corenblum B. Maintenance of serum calcium by parathyroid hormone-related peptide during lactation in a hypoparathyroid patient. J Clin Endocrinol Metab. 1999;84(2):424–7.

39. Kent GN, Price RI, Gutteridge DH, et al. Human lactation: forearm trabecular bone loss, increased bone turnover, and renal conservation of calcium and inorganic phosphate with recovery of bone mass following weaning. J Bone Miner Res. 1990;5:361–9.
40. Sowers M, Zhang D, Hollis BW, et al. Role of calciotrophic hormones in calcium mobilization of lactation. Am J Clin Nutr. 1998;67(2):284–91.
41. Prentice A. Calcium in pregnancy and lactation. Annu Rev Nutr. 2000;20:249–72.
42. Prentice A, Jarjou LM, Cole TJ, et al. Calcium requirements of lactating Gambian mothers: effects of a calcium supplement on breast-milk calcium concentration, maternal bone mineral content, and urinary calcium excretion. Am J Clin Nutr. 1995;62(1):58–67.
43. Prentice A, Jarjou LM, Stirling DM, et al. Biochemical markers of calcium and bone metabolism during 18 months of lactation in Gambian women accustomed to a low calcium intake and in those consuming a calcium supplement. J Clin Endocrinol Metab. 1998;83(4):1059–66.
44. Prentice A, Yan L, Jarjou LM, et al. Vitamin D status does not influence the breast-milk calcium concentration of lactating mothers accustomed to a low calcium intake. Acta Paediatr. 1997;86(9):1006–8.
45. Bezerra FF, Mendonca LM, Lobato EC, et al. Bone mass is recovered from lactation to postweaning in adolescent mothers with low calcium intakes. Am J Clin Nutr. 2004;80(5):1322–6.
46. Chantry CJ, Auinger P, Byrd RS. Lactation among adolescent mothers and subsequent bone mineral density. Arch Pediatr Adolesc Med. 2004;158(7):650–6.
47. Kovacs CS, Kronenberg HM. Maternal-fetal calcium and bone metabolism during pregnancy, puerperium, and lactation. Endocr Rev. 1997;18(6):832–72.
48. Kovacs CS. Calcium and bone metabolism during pregnancy and lactation. J Mammary Gland Biol Neoplasia. 2005;10(2):105–18.

21

Menopause and Age-related Bone Loss

Carlos M. Isales[1] and Ego Seeman[2,3]

[1]Department of Medicine, Augusta University, Augusta, GA, USA
[2]Departments of Endocrinology and Medicine, Austin Health, University of Melbourne, Melbourne, VIC, Australia
[3]Mary MacKillop Institute for Health Research, Australian Catholic University, Melbourne, VIC, Australia

INTRODUCTION—BONE MODELING AND REMODELING

This chapter focuses on two of the major factors that compromise bone strength in women: the occurrence of menopause and the effects of advancing age. Another important factor is peak bone strength achieved during growth. This is dealt with in Chapter 19.

Two cellular processes determine the macro- and microstructural configuration of bone. Bone modeling refers to the deposition of bone upon a quiescent bone surface that has not undergone prior bone resorption. The deposition of bone upon an existing surface alters the size and shape of bone. An example of formative bone modeling is periosteal apposition during growth and the slow continued periosteal apposition that occurs in young adulthood and only modestly during advancing age [1]. Although modeling is commonly regarded as being formative, modeling of bone may also be resorptive. Under this circumstance, resorption takes place on a quiescent surface and is not followed by bone formation. An example of this process is the excavation of the marrow cavity by resorptive excavation upon the endocortical surface of bone during growth [2].

Bone remodeling is different. It is carried out by the cellular machinery of the BMU. The BMU refers to groups of cells, the executive cells being osteoclasts and osteoblasts that respectively resorb and deposit volumes of bone at the same location. Bone remodeling does not alter the size or shape of bone. When the BMU is activated on a bone surface, a highly choreographed cellular response is initiated characterized by recruitment of osteoclasts within a bone remodeling compartment. The resorptive phase of remodeling lasts around 3 weeks and is followed by a reversal or quiescent phase believed to be the time during which coupling is initiated with subsequent differentiation of osteoblast precursors into mature bone-forming osteoblasts which deposit bone during the formation phase of remodeling. This formation phase lasts about 3 months [3–5]. Primary mineralization is rapid, occurring shortly after matrix deposition, but secondary mineralization takes many months if not years to reach completion [6–8].

Bone remodeling is balanced during young adulthood with equal volumes of bone resorbed and formed so that no permanent bone loss or microstructural deterioration occurs. There is a temporary or transient deficit in bone matrix and mineral volumes caused by the delay in the initiation of bone formation (the reversal phase) and slowness of the formation phase. This deficit is focally transient and is formed by excavated cavities, by cavities containing osteoid without mineral, and cavities containing osteoid that has undergone primary but not secondary mineralization or incomplete secondary mineralization. This reversible deficit is referred to as the remodeling "transient" [5]. It is globally ever present because remodeling sites are being completed at some locations while others are generated at other locations, but it is focally transient because the excavated matrix is replaced when bone formation and its mineralization is eventually completed. The size of this ever present deficit is a function of the remodeling rate; the deficit is larger in the trabecular compartment because this compartment is more vigorously remodeled—there are more sites in various stages of the remodeling cycle per unit volume of bone matrix than there are remodeling the cortical compartment.

SURFACE AREA/BONE MATRIX VOLUME CONFIGURATION

The effect of bone remodeling partly depends upon the surface area/bone matrix volume configuration of the skeleton. Trabecular bone and bone of the transitional compartment, situated between trabecular and cortical bone, has a large surface area. The trabecular matrix is fashioned as thin platelike structures of around 150 μm diameter. This configuration is an advantage because remodeling can easily excavate matrix, remove microdamage, and replace it with an equal volume of bone. However, when remodeling becomes unbalanced, especially early in menopause when large numbers of deep resorption cavities are excavated, these trabeculae may perforate and become disconnected, resulting in loss of bone strength in excess of the bone strength lost if there is only thinning of trabeculae (by reduced bone formation as found in males) [1, 9–12]. Thus, this surface area/matrix volume configuration of trabecular bone becomes a liability when remodeling becomes unbalanced. Trabecular bone has the larger surface area but accounts for only about 20% of bone volume. Because of its greater surface area, trabecular bone is more rapidly lost than cortical bone when remodeling becomes unbalanced.

Cortical bone accounts for 80% of total bone volume and is enveloped by the periosteal envelope externally, the intracortical surfaces of the canals traversing it and the endocortical surface lining the marrow cavity on its "inside" [1, 5, 13]. Therefore, cortical bone matrix volume resides "inside" the periosteum and "outside" – the intracortical and endocortical components of the endosteal (inner) envelope.

Remodeling is uncommon upon the periosteal envelope. It occurs mainly at points upon the three (endocortical, intracortical, trabecular) components of the endosteal envelope. Cortical bone is therefore remodeled by BMUs active upon the intracortical surfaces lining the many canals traversing its matrix and by BMUs active upon the endocortical surface. As the cortical bone matrix volume is large, it is less accessible to being remodeled than trabecular bone and so it is lost more slowly than trabecular bone. With time, this fourfold larger bone matrix volume accounts for more of the total bone loss than the more rapid loss of the smaller total trabecular bone matrix volume [14].

Unbalanced remodeling at points upon the intracortical canals enlarges them focally. Those traversing cortex nearer the medullary canal enlarge and coalesce forming irregularly shaped pores as seen in cross section. The enlarging canals produce loss of cortical bone matrix volume which increases in the intracortical surface area providing more surface for remodeling to be initiated upon. Thus, intracortical remodeling becomes self-perpetuating: a decreasing total cortical bone matrix volume is now being eroded by more unbalanced remodeling events—more bone matrix volume is lost from an ever decreasing cortical bone matrix volume.

The cavitation within the inner part of the cortex adjacent to the medullary canal thins the cortex from within, fragmenting it. The cortical fragments look like trabeculae and can be mistaken for them, producing a seeming increase in the number of "trabeculae" in what appears to be the "medullary" canal. This common error results in several misconceptions. The intracortical intracortical porosity is "seen" as medullary canal and so results in underestimation of cortical porosity because the porosity has been relegated to the medullary canal. The cortical fragments are seen as trabecular bone. This overestimates trabecular density and so blunts the age-related decrease in trabecular bone density. Moreover, the loss of cortical bone with age and across menopause is exaggerated because the cortical fragments "seen" as trabecular bone are no longer included in the calculation of cortical matrix volume in postmenopausal women. Both the underestimation of cortical porosity and overestimation of trabecular density result in an underestimate in the loss of bone strength during advancing age.

AGE-RELATED BONE LOSS

Age-related bone loss is the result of a reduction in the volume of bone formed by each BMU. The mechanisms responsible for this reduction in bone formation are not understood but may involve a reduction in osteoblast numbers, their work capacity, or life span. Osteoblasts are replenished from a stem cell pool, the bone marrow derived mesenchymal stem cells (MSCs). Aging is accompanied by a decrease in MSC number and differentiation capacity in both mice [15, 16] and humans [17–19], resulting in decreased bone formation. This decrease in MSC osteoblastic differentiation is accompanied by an increase in MSC differentiation to adipocytes, resulting in a fatty infiltrated bone marrow [20–22]. Bone marrow fat also increases in postmenopausal women and estrogen replacement decreases marrow adiposity [23]. The mechanism of age-related decline in MSC function may relate in part to an increase in cell damage from ROS [24–27]. Further discussion of this subject is outside the brief of this chapter. Whatever the mechanism, this reduction in bone formation leads to a reduction in mean wall thickness as age advances [28].

Osteoclasts are derived from hematopoietic precursor stem cells. These stem cells also exhibit decreased proliferative and differentiation capacity [29]. Bone resorption by the cells of the BMU also decrease so the volume of bone resorbed decreases but does so less than the decrease in the volume of bone formed, producing the negative BMU balance that is the necessary and sufficient cause of bone loss and microstructural deterioration [30].

MENOPAUSAL BONE LOSS

Estrogen appears to be the main steroid hormone modulating bone mass in both men and women, and circulating estrogen levels correlate with fracture risk [31].

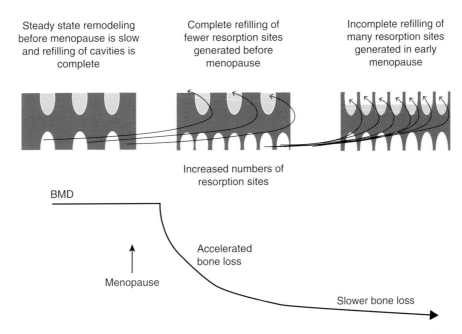

Fig. 21.1. Loss of bone after menopause. Before menopause, remodeling is slow with small numbers of remodeling sites in their formation phase and others in their resorption phase. At menopause the number of remodeling sites in their resorptive phase increases while the fewer numbers in their formation phase before menopause now enter their refilling phase so BMD rapidly declines. Later in menopause a new steady state is achieved at the higher remodelling rate so that now more similar numbers of remodelling sites are being excavated and incompletely refilled (see text).

At the cellular level two estrogen receptor subtypes ERα and ERβ are widely expressed in multiple bone cell types including osteocytes, osteoblasts, and osteoclasts [32]. Decreased estrogen levels induce osteocyte apoptosis and increased release of inflammatory cytokines such as TNF-α and interleukins 1 and 7, which promote osteoclastogenesis [32].

Estrogen itself may have antioxidant properties and loss of estrogen may induce MSC damage through increased levels of ROS thus impairing osteoblast differentiation [24, 26, 27]. Thus the net effect of decreased estrogen levels at the cellular level is increased osteocyte and osteoblast apoptosis, decreased osteoclast apoptosis, and increased RANKL-mediated differentiation [33].

Menopause-related bone loss is a result of four distinct mechanisms [34–36]: (i) a decline in the net amount of bone deposited by each BMU; (ii) a transitory increase in the volume of bone resorbed by each BMU; (iii) an increase in the rate of bone remodeling—greater numbers of BMUs remodel the skeleton and each resorbs more bone and deposits less because sex hormone deficiency increases the life span of osteoclasts and reduces the life span of osteoblasts [33]; and (iv) a reduction in periosteal apposition associated with continued intracortical, endocortical, and trabecular bone loss.

As estrogen levels decrease in women between the ages of 40 and 50 there is an increase in the number of BMUs. This increase in BMUs in their resorptive phase results in enlargement of the remodeling space deficit in matrix and mineral. The appearance of many BMUs in their resorptive phase excavating bone is matched by the fewer BMUs initiated *before* menopause only now entering their formation phase. This perturbation of surface level remodeling is responsible for the accelerated decline in BMD associated with the menopause (Fig. 21.1).

As the postmenopausal state continues, those large numbers of BMUs initiated in early menopause move into their reversal then refilling phase but now refilling is incomplete because of the increased resorption depth and reduced osteoblastic refilling. The worsening negative BMU balance produces continued bone loss but more slowly than in early menopause because remodeling at the surface level has returned to steady state at a higher rate of remodeling. The loss of bone is now driven only by the rapidity of remodeling and the negative BMU balance, not the perturbation of remodeling at the surface level occurring in early menopause.

Because BMUs are present on bone surfaces and trabecular bone has a larger surface area, initial bone loss occurs more rapidly—a greater proportion of the smaller trabecular than cortical bone volume is remodeled and lost after menopause. However, despite the slower remodeling of cortical bone, the slower remodeling of the larger cortical bone volume results in more cortical than trabecular bone loss as women transition from pre- to perimenopausal, peri- to postmenopausal, and then advance in the postmenopausal years. Cortical bone loss accounts for ~70% of all peri- and postmenopausal bone loss [14, 37].

Increased bone remodeling results in an increase in bone fragility in several ways. The resorption pits act as stress risers while loss of trabeculae and loss of connectivity

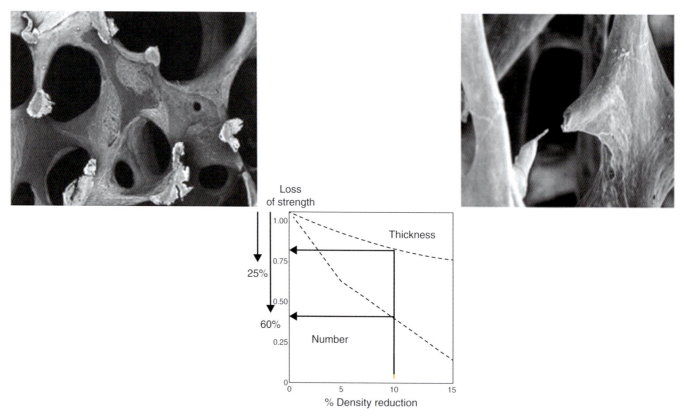

Fig. 21.2. Upper panels: trabecular platelike structure (left) and loss of trabecular connectivity (right). Perforation of trabeculae results in loss of connectivity which reduces strength more greatly than trabecular thinning (lower panel). Source: adapted from [12]. Reproduced with permission of John Wiley and Sons.

reduce bone strength [38] (Fig. 21.2). Cortical porosity reduces bone strength as a seventh power function of the porosity; the loss of strength by increasing the porosity of an already porous structure like trabecular bone reduces strength to the third power [39]. By contrast, in men, trabecular remodeling results in thinning rather than perforation, a process that compromises trabecular strength less [12].

As trabeculae are lost with continued remodeling, the cortical compartment becomes the main source of bone loss leading to increased cortical porosity and "trabecularization" of cortical bone as discussed previously [40]. Increased intracortical bone surface area facilitates continued high remodeling predisposing to appendicular fractures. High remodeling also reduces matrix mineral density as older more mineralized bone is replaced by newer less mineralized bone. This results in a more heterogeneously mineralized bone and decreased strength (Fig. 21.3).

Periosteal apposition is believed to increase as an adaptive response to compensate for the loss of strength produced by endocortical bone loss, so there will be no *net* loss of bone, no cortical thinning, and no loss of bone strength [41]. In a prospective study of over 600 women, Szulc et al. reported that endocortical bone loss occurred in premenopausal women with concurrent periosteal apposition [42]. Because periosteal apposition was less than endocortical resorption, the cortices thinned but there was no *net* bone loss because the thinner cortex was now distributed around a larger perimeter, conserving total bone mass. Moreover, resistance to bending increased despite bone loss and cortical thinning because this same amount of bone was now distributed further from the neutral axis (Fig. 21.4).

During the perimenopausal period endocortical resorption increased yet periosteal apposition decreased. The cortices thinned but bending strength remained unchanged despite bone loss and cortical thinning because periosteal apposition was still sufficient to shift the thinning cortex outwards. Bone fragility emerged only after menopause when accelerated endocortical bone resorption and deceleration in periosteal apposition produced further cortical thinning. As periosteal apposition was now minimal, there was little outward displacement of the thinning cortex so cortical area now declined as did resistance to bending. During aging, both increasing endocortical bone resorption and reduced periosteal apposition cause *net* bone loss, alterations in the distribution of the remaining bone, and the emergence of bone fragility [43].

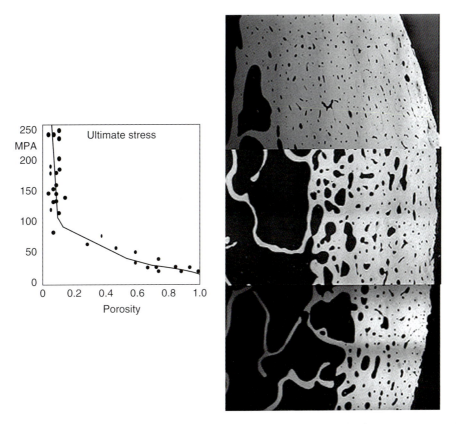

Fig. 21.3. Cortical porosity in specimens from 27-year-old, 70-year-old, and 80-year-old cadavers progressing from top, illustrating thinning of the cortex from within by intracortical remodeling. This is associated with a decline in ultimate stress. MPA = megapascal. Source: adapted from [14]. Reproduced with permission of Elsevier.

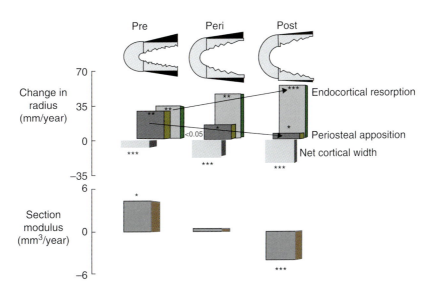

Fig. 21.4. The amount of bone resorbed by endocortical resorption increases with age. The amount deposited by periosteal apposition decreases. The net effect is a decline in cortical thickness. In premenopausal women, the thinner cortex is displaced radially, increasing section modulus (Z). In perimenopausal women Z does not decrease despite cortical thinning because periosteal apposition still produces radial displacement. In postmenopausal women, Z decreases because endocortical resorption continues, periosteal apposition declines, and little radial displacement occurs. Source: adapted from [42]. Reproduced with permission of John Wiley and Sons.

SUMMARY

Bone loss resulting in bone fragility was thought to occur through two distinct mechanisms and lead to type I (postmenopausal) or type II (age-related or involutional) osteoporosis [44]. Increased understanding of the underlying cellular mechanisms involved in bone loss reveals a common mechanism—the appearance of a negative BMU balance, the necessary and sufficient cause of bone loss, and microstructural deterioration producing bone fragility. After menopause, the negative BMU balance worsens, perhaps transiently but with rapid remodeling; there is accelerated loss of trabecular and cortical bone. The greater rapidity of remodeling compromises both compartments of bone: trabecular bone is lost more rapidly, but in absolute terms the slower loss of the larger volume of cortical bone results in greater loss of cortical than trabecular bone and increased risk for fractures of the axial and appendicular skeleton.

REFERENCES

1. Seeman E, Delmas PD. Bone quality—the material and structural basis of bone strength and fragility. N Engl J Med. 2006;354(21):2250–61.
2. Seeman E. Structural basis of growth-related gain and age-related loss of bone strength. Rheumatology (Oxford). 2008;47(Suppl 4):iv2–8.
3. Parfitt AM. The physiologic and clinical significance of bone histomorphometric data. In: Recker R (ed.) *Bone Histomorphometry Techniques and Interpretation.* Boca Raton: CRC Press, 1983, pp 142–223.
4. Hattner R, Epker BN, Frost HM. Suggested sequential mode of control of changes in cell behaviour in adult bone remodelling. Nature. 1965;206(983):489–90.
5. Parfitt AM. Osteonal and hemi-osteonal remodeling: the spatial and temporal framework for signal traffic in adult human bone. J Cell Biochem. 1994;55(3):273–86.
6. Akkus O, Adar F, Schaffler MB. Age-related changes in physicochemical properties of mineral crystals are related to impaired mechanical function of cortical bone. Bone. 2004;34(3):443–53.
7. Akkus O, Polyakova-Akkus A, Adar F, et al. Aging of microstructural compartments in human compact bone. J Bone Miner Res. 2003;18(6):1012–19.
8. Boskey AL. Variations in bone mineral properties with age and disease. J Musculoskelet Neuronal Interact. 2002;2(6):532–4.
9. Aaron JE, Makins NB, Sagreiya K. The microanatomy of trabecular bone loss in normal aging men and women. Clin Orthop Relat Res. 1987;(215):260–71.
10. Aaron JE, Shore PA, Shore RC, et al. Trabecular architecture in women and men of similar bone mass with and without vertebral fracture: II. Three-dimensional histology. Bone. 2000;27(2):277–82.
11. Hordon LD, Raisi M, Aaron JE, et al. Trabecular architecture in women and men of similar bone mass with and without vertebral fracture: I. Two-dimensional histology. Bone. 2000;27(2):271–6.
12. van der Linden JC, Homminga J, Verhaar JA, et al. Mechanical consequences of bone loss in cancellous bone. J Bone Miner Res. 2001;16(3):457–65.
13. Bala Y, Zebaze R, Seeman E. Role of cortical bone in bone fragility. Curr Opin Rheumatol. 2015;27(4):406–13.
14. Zebaze RM, Ghasem-Zadeh A, Bohte A, et al. Intracortical remodelling and porosity in the distal radius and post-mortem femurs of women: a cross-sectional study. Lancet. 2010;375(9727):1729–36.
15. Yukata K, Xie C, Li TF, et al. Aging periosteal progenitor cells have reduced regenerative responsiveness to bone injury and to the anabolic actions of PTH 1-34 treatment. Bone. 2014;62:79–89.
16. Zhang W, Ou G, Hamrick M, et al. Age-related changes in the osteogenic differentiation potential of mouse bone marrow stromal cells. J Bone Miner Res. 2008;23(7):1118–28.
17. Stolzing A, Jones E, McGonagle D, et al. Age-related changes in human bone marrow-derived mesenchymal stem cells: consequences for cell therapies. Mech Ageing Dev. 2008;129(3):163–73.
18. Stolzing A, Scutt A. Age-related impairment of mesenchymal progenitor cell function. Aging Cell. 2006;5(3):213–24.
19. Zhou S, Greenberger JS, Epperly MW, et al. Age-related intrinsic changes in human bone-marrow-derived mesenchymal stem cells and their differentiation to osteoblasts. Aging Cell. 2008;7(3):335–43.
20. Devlin MJ, Rosen CJ. The bone-fat interface: basic and clinical implications of marrow adiposity. Lancet Diabetes Endocrinol. 2015;3(2):141–7.
21. Di Iorgi N, Rosol M, Mittelman SD, et al. Reciprocal relation between marrow adiposity and the amount of bone in the axial and appendicular skeleton of young adults. J Clin Endocrinol Metab. 2008;93(6):2281–6.
22. Rosen CJ, Bouxsein ML. Mechanisms of disease: is osteoporosis the obesity of bone? Nat Clin Pract Rheumatol. 2006;2(1):35–43.
23. Limonard EJ, Veldhuis-Vlug AG, van Dussen L, et al. Short-Term Effect of Estrogen on Human Bone Marrow Fat. J Bone Miner Res. 2015;30(11):2058–66.
24. Almeida M, Han L, Martin-Millan M, et al. Oxidative stress antagonizes Wnt signaling in osteoblast precursors by diverting beta-catenin from T cell factor- to forkhead box O-mediated transcription. J Biol Chem. 2007;282(37):27298–305.
25. El Refaey M, Watkins CP, Kennedy EJ, et al. Oxidation of the aromatic amino acids tryptophan and tyrosine disrupts their anabolic effects on bone marrow mesenchymal stem cells. Mol Cell Endocrinol. 2015;410:87–96.
26. Kousteni S. FoxOs: Unifying links between oxidative stress and skeletal homeostasis. Curr Osteoporos Rep. 2011;9(2):60–6.
27. Manolagas SC. From estrogen-centric to aging and oxidative stress: a revised perspective of the pathogenesis of osteoporosis. Endocr Rev. 2010;31(3):266–300.

28. Lips P, Courpron P, Meunier PJ. Mean wall thickness of trabecular bone packets in the human iliac crest: changes with age. Calcif Tissue Res. 1978;26(1):13–7.
29. Tuljapurkar SR, McGuire TR, Brusnahan SK, et al. Changes in human bone marrow fat content associated with changes in hematopoietic stem cell numbers and cytokine levels with aging. J Anat. 2011;219(5):574–81.
30. Parfitt AM, Mathews CH, Villanueva AR, et al. Relationships between surface, volume, and thickness of iliac trabecular bone in aging and in osteoporosis. Implications for the microanatomic and cellular mechanisms of bone loss. J Clin Invest. 1983;72(4):1396–409.
31. Cauley JA. Estrogen and bone health in men and women. Steroids. 2015;99(Pt A):11–15.
32. Khalid AB, Krum SA. Estrogen receptors alpha and beta in bone. Bone. 2016;87:130–5.
33. Khosla S, Oursler MJ, Monroe DG. Estrogen and the skeleton. Trends Endocrinol Metab. 2012;23(11):576–81.
34. Seeman E. Age- and menopause-related bone loss compromise cortical and trabecular microstructure. J Gerontol A Biol Sci Med Sci. 2013;68(10):1218–25.
35. Seeman E. Growth and Age-Related Abnormalities in Cortical Structure and Fracture Risk. Endocrinol Metab (Seoul). 2015;30(4):419–28.
36. Seeman E, Martin TJ. Co-administration of antiresorptive and anabolic agents: a missed opportunity. J Bone Miner Res. 2015;30(5):753–64.
37. Bjornerem A, Ghasem-Zadeh A, Wang X, et al. Irreversible Deterioration of Cortical and Trabecular Microstructure Associated with Breastfeeding. J Bone Miner Res. 2017;32(4):681–7.
38. Hernandez CJ, Gupta A, Keaveny TM. A biomechanical analysis of the effects of resorption cavities on cancellous bone strength. J Bone Miner Res. 2006;21(8):1248–55.
39. Schaffler MB, Burr DB. Stiffness of compact bone: effects of porosity and density. J Biomech. 1988;21(1):13–16.
40. Brockstedt H, Kassem M, Eriksen EF, et al. Age- and sex-related changes in iliac cortical bone mass and remodeling. Bone. 1993;14(4):681–91.
41. Balena R, Shih MS, Parfitt AM. Bone resorption and formation on the periosteal envelope of the ilium: a histomorphometric study in healthy women. J Bone Miner Res. 1992;7(12):1475–82.
42. Szulc P, Seeman E, Duboeuf F, et al. Bone fragility: failure of periosteal apposition to compensate for increased endocortical resorption in postmenopausal women. J Bone Miner Res. 2006;21(12):1856–63.
43. Seeman E. Periosteal bone formation—a neglected determinant of bone strength. N Engl J Med. 2003;349(4):320–3.
44. Eastell R, Riggs BL. Treatment of osteoporosis. Obstet Gynecol Clin North Am. 1987;14(1):77–88.

Section III
Mineral Homeostasis
Section Editors: David Goltzman and Harald Jüppner

Chapter 22. Regulation of Calcium Homeostasis 165
Line Vautour and David Goltzman

Chapter 23. Magnesium Homeostasis 173
Aliya Aziz Khan, Asiya Sbayi, and Karl Peter Schlingmann

Chapter 24. Fetal Calcium Metabolism 179
Christopher S. Kovacs

Chapter 25. FGF23 and the Regulation of Phosphorus Metabolism 187
Kenneth E. White and Michael J. Econs

Chapter 26. Gonadal Steroids 194
Stavros C. Manolagas and Maria Schuller Almeida

Chapter 27. Parathyroid Hormone 205
Thomas J. Gardella, Robert A. Nissenson, and Harald Jüppner

Chapter 28. Parathyroid Hormone-Related Protein 212
John J. Wysolmerski and T. John Martin

Chapter 29. Calcium-Sensing Receptor 221
Geoffrey N. Hendy

Chapter 30. Vitamin D: Production, Metabolism, Action, and Clinical Requirements 230
Daniel D. Bikle, John S. Adams, and Sylvia Christakos

Primer on the Metabolic Bone Diseases and Disorders of Mineral Metabolism, Ninth Edition. Edited by John P. Bilezikian.
© 2019 American Society for Bone and Mineral Research. Published 2019 by John Wiley & Sons, Inc.
Companion website: www.wiley.com/go/asbmrprimer

22

Regulation of Calcium Homeostasis

Line Vautour[1] and David Goltzman[2]

[1]Division of Endocrinology and Metabolism, McGill University Health Centre, Montréal, QC, Canada
[2]Department of Medicine, McGill University and McGill University Health Centre, Montréal, QC, Canada

CALCIUM DISTRIBUTION

Total body distribution

In adults, the body contains approximately 1000 g of calcium (Ca), of which 99% is located in the mineral phase of bone as the hydroxyapatite crystal $[Ca_{10}(PO4)_6(OH)_2]$. The crystal plays a key role in the mechanical weight-bearing properties of bone and serves as a ready source of Ca to support a number of Ca-dependent biological systems and to maintain blood ionized Ca within the normal range. The remaining 1% of total body Ca is located in the blood, extracellular fluid (ECF), and soft tissues. In the serum, total Ca is 10^{-3} M and is the most frequent measurement of serum Ca level.

Cell levels

The concentration of Ca in the cytoplasm is about 10^{-6} M. In the ECF the Ca concentration is 10^{-3} M which creates a 1000-fold gradient across the plasma membrane that favors Ca entry into the cell. There is also an electrical charge across the plasma membrane of about 50 mV with the cell interior negative. Thus, the chemical and electrical gradients across the plasma membrane favor Ca entry which the cell must defend against to preserve cell viability. Ca-induced cell death is largely prevented by several mechanisms including: extrusion of Ca from the cell by adenosine triphosphate-dependent energy driven Ca pumps and Ca channels; Na–Ca exchangers; and binding of intracellular Ca by proteins located in the cytoplasm, endoplasmic reticulum (ER), and mitochondria. Ca binding to ER and mitochondrial sites buffer intracellular Ca and can be mobilized to maintain cytoplasmic Ca levels and to create pulsatile peaks of Ca to mediate membrane receptor signaling that regulate a variety of biological systems.

Blood levels

Of the total Ca in the blood, the free or ionized fraction (45%) is the biologically functional portion of total Ca and can be measured clinically; 45% of the total is bound to albumin in a pH-dependent manner and the remaining 10% exists as a complex with anions including phosphate (PO_4) and citrate [1]. Although only the ionized Ca is available to move into cells and activate cellular processes, most clinical laboratories report total serum Ca concentrations. Concentrations of total Ca in normal serum generally range between 8.5 and 10.5 mg/dL (2.12 to 2.62 mM) and levels above this are considered to be hypercalcemic. The reference range for ionized Ca is 4.65 to 5.25 mg/dL (1.16 to 1.31 mM). When protein concentrations fluctuate, especially albumin concentrations, total Ca levels may vary, whereas the ionized Ca may remain relatively stable. Dehydration or hemoconcentration during venipuncture may elevate serum albumin and falsely elevate total serum Ca. Such elevations in total Ca, when albumin levels are increased, can be "corrected" by subtracting 0.8 mg/dL from the total Ca for every 1.0 g/dL by which the serum albumin concentration is >4 g/dL. Conversely, when albumin levels are low, total Ca can be corrected by adding 0.8 mg/dL for every 1.0 g/dL by which the albumin is <4 g/dL. Even in the presence of a normal serum albumin, changes in blood pH can alter the equilibrium constant of the albumin-Ca^{2+} complex, with acidosis reducing the binding and alkalosis enhancing it. Consequently, a major shift in serum protein or pH

Primer on the Metabolic Bone Diseases and Disorders of Mineral Metabolism, Ninth Edition. Edited by John P. Bilezikian.
© 2019 American Society for Bone and Mineral Research. Published 2019 by John Wiley & Sons, Inc.
Companion website: www.wiley.com/go/asbmrprimer

CALCIUM BALANCE

The ECF concentration of Ca is tightly maintained within a rather narrow range because of the importance of the Ca ion to numerous cellular functions including cell division, cell adhesion and plasma membrane integrity, protein secretion, muscle contraction, neuronal excitability, glycogen metabolism, and coagulation.

The skeleton, the gut, and the kidney each play a major role in assuring Ca homeostasis. Overall, in a typical individual, if 1000 mg of Ca are ingested in the diet per day, approximately 200 mg will be absorbed. Approximately 10 g of Ca will be filtered daily through the kidney and most will be reabsorbed, with about 200 mg being excreted in the urine. The normal 24 hour excretion of Ca may however vary between 100 and 300 mg per day (2.5 to 7.5 mmol/day). The skeleton, a storage site of about 1 kg of Ca, is the major Ca reservoir in the body. Ordinarily, as a result of normal bone turnover, approximately 500 mg of Ca is released from bone per day and the equivalent amount is accreted per day (Fig. 22.1).

HORMONAL REGULATION OF CALCIUM HOMEOSTASIS

Overview of integrated regulation of calcium homeostasis

Tight regulation of the ECF calcium concentration is maintained through the action of Ca-sensitive cells which modulate the production of hormones [2–5]. These hormones act on specific cells in bone, gut, and kidney which can respond by altering fluxes of Ca to maintain ECF Ca. Thus a reduction in ECF Ca stimulates release of PTH from the parathyroid glands in the neck (Fig. 22.2). This hormone can then act to enhance bone resorption and liberate both Ca and phosphate from the skeleton. PTH has also been reported to increase release of the phosphaturic hormone, FGF23, from mature osteoblasts and osteocytes [6]. PTH can augment Ca reabsorption in the kidney and at the same time reduce phosphate reabsorption producing phosphaturia. Hypocalcemia and PTH itself can both stimulate the conversion of the inert metabolite of vitamin D, 25-hydroxyvitamin D (25OHD), to the active moiety 1,25-dihydroxyvitamin D [1,25(OH)$_2$D] [7] which in turn will augment intestinal Ca absorption, and to a lesser extent renal phosphate reabsorption. The net effect

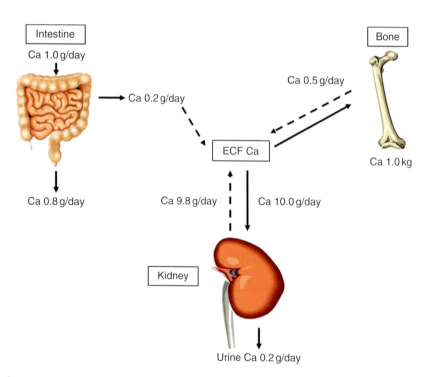

Fig. 22.1. Calcium balance. On average, in a typical adult, approximately 1 g of elemental calcium (Ca^{2+}) is ingested per day. Of this, about 200 mg/day will be absorbed and 800 mg/day excreted. Approximately 1 kg of Ca is stored in bone and about 500 mg/day is released by resorption and deposited during bone formation when bone turnover is in balance. Of the 10 g of Ca filtered through the kidney per day only about 200 mg or less appears in the urine, the remainder being reabsorbed.

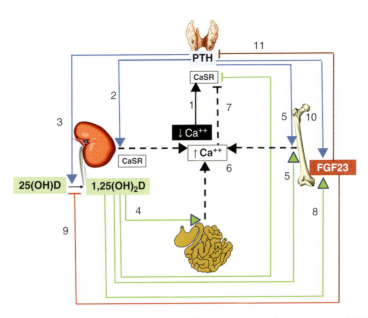

Fig. 22.2. Hormonal regulation of ECF Ca homeostasis. Decreased (↓) ECF Ca results in increased PTH release from the parathyroid glands via the CaSR (1). The increased PTH can enhance Ca reabsorption from the kidney (2), and with decreased ECF Ca, CaSR is not activated to cause calciuria. PTH can also increase conversion of 25OHD to 1,25(OH)$_2$D in the kidney (3). The 1,25(OH)$_2$D produced can increase intestinal absorption of Ca (4), and PTH and 1,25(OH)$_2$D can resorb bone and increase Ca release from bone (5). The net result is normalization (→) of ECF Ca and inhibition (⊣) of further PTH release (7). 1,25(OH)$_2$D can also stimulate FGF23 release from bone (8) which in turn can inhibit further renal 1,25(OH)$_2$D production from 25OHD (9). PTH may also stimulate FGF23 (10) which may then limit further PTH release (11).

of the mobilization of Ca from bone, the increased absorption of Ca from the gut, and the increased reabsorption of filtered Ca along the nephron is to restore the ECF Ca to normal and to inhibit further production of PTH and 1,25(OH)$_2$D. Furthermore, released FGF23 can also reduce 1,25(OH)$_2$D [8] and has been reported to decrease PTH production therefore further ensuring that Ca homeostasis is restored.

When the ECF Ca is raised above the normal range, the opposite sequence of events occurs, ie, diminished PTH secretion caused by stimulation of the parathyroid Ca-sensing receptor (CaSR) and diminished renal 1,25(OH)$_2$D production. In addition, a direct calciuric effect of hypercalcemia on the kidney can occur mediated by the renal CaSR in the cortical thick ascending limb (CTAL) of Henle's loop. Therefore, the effect of suppressing the release of PTH and production of 1,25(OH)$_2$D and of stimulating renal CaSR results in diminished skeletal Ca release, decreased intestinal Ca absorption, and reduced renal tubular Ca reabsorption respectively, and restores the elevated ECF Ca to normal.

Regulation of hormone production

PTH production

A major regulator of parathyroid gland secretion of PTH is ECF Ca. The parathyroid glands detect ECF Ca via a CaSR, which is a Gq protein-coupled receptor [9] Thus a decrease in ECF Ca is sensed by the CaSR in parathyroid chief cells resulting in an acute increase in PTH secretion. The relationship between ECF Ca and PTH secretion is governed by a steep inverse sigmoidal curve which is characterized by a maximal secretory rate at low ECF Ca, a midpoint or "set point" which is the level of ECF Ca which half-maximally suppresses PTH, and a minimal secretory rate at high ECF Ca [10, 11]. Sustained hypocalcemia can eventually lead to parathyroid cell proliferation [12] and an increased total secretory capacity of the parathyroid gland. ECF Ca, acting via the CaSR, therefore functions as a hormone in modulating PTH release and parathyroid cell function. 1,25-dihydroxyvitamin D$_3$ (1,25(OH)$_2$D$_3$) reduces PTH synthesis and parathyroid cell proliferation [13]. Molecular events in PTH secretion and CaSR function are found in Chapters 26 and 28.

Vitamin D production and metabolism

Vitamin D produced in the skin or ingested in the diet is 25-hydroxylated in liver and the 25OHD metabolite thus formed is then converted in the kidney and, to a lesser extent, in other tissues [14] to the active form, 1,25(OH)$_2$D, by a mitochondrial enzyme, 25OHD-1α hydroxylase [CYP27B1 or 1α(OH)ase] The renal production of 1,25(OH)$_2$D is stimulated by hypocalcemia, hypophosphatemia, and elevated PTH levels. The renal 1α(OH)ase is potently inhibited by 1,25(OH)$_2$D as part of a negative feedback loop. Furthermore, 1,25(OH)$_2$D can also stimulate a renal 24-hydroxylase enzyme [CYP24A1 or 24(OH)ase], converting 25OHD into 24,25(OH)$_2$D, and thereby reducing the substrate of the renal 1α(OH)ase, and converting 1,25(OH)$_2$D into 1,24,25(OH)$_3$D, which is then metabolized to calcitroic acid (1-hydroxy-23-carboxy-vitamine D3) or 23,25OHD-26,23-lactone, two inactive forms of vitamin D. FGF23 is a potent inhibitor of the renal 1α(OH)ase and can also stimulate the renal 24(OH)ase thereby participating in the reduction of circulating 1,25(OH)$_2$D concentrations. In this way, FGF23 acts as a counter regulatory hormone for 1,25(OH)$_2$D effects on mineral homeostasis. The molecular details of the vitamin D metabolic pathway are described in Chapter 29.

FGF23 production

FGF23 may be stimulated by both local and systemic factors. In humans, high dietary phosphorus increases and low dietary phosphorus decreases serum FGF23 [15–18]; but the changes are modest, and there is a lag period between phosphate loading and elevations of FGF23 [17]. Serum Ca-mediated increases in serum FGF23 were shown, in mice, to require a threshold level of serum phosphorus and, likewise, phosphate-elicited increases in FGF23 were markedly blunted if serum Ca was below a threshold level [19]. PTH can increase FGF23 but ambient concentrations of 1,25(OH)2D appear to supersede the effects of PTH on regulating this hormone [20]. Thus 1,25(OH)$_2$D appears to be the most important physiological stimulus for FGF23 production and FGF23 and 1,25(OH)$_2$D participate in a bone–kidney endocrine loop, in which 1,25(OH)$_2$D stimulates FGF23 production and FGF23 suppresses 1,25(OH)$_2$D levels.

PTH, 1,25(OH)$_2$D calcium, and FGF23 actions in target tissues to regulate calcium homeostasis

Intestinal Ca transport

Net intestinal Ca absorption can be determined by the external balance technique in which a diet of known composition with a known amount of Ca is ingested, and urine Ca excretion and fecal Ca loss are measured. Negative balance occurs when net absorption declines to about 200 mg Ca per day (5.0 mmol). The portion of dietary Ca absorbed varies with age and amount of Ca ingested, and may range from 20 to 60%. Rates of net Ca absorption are high in growing children, during growth spurts in adolescence, and during pregnancy and lactation. The efficiency of Ca absorption increases during prolonged dietary Ca restriction to absorb the greatest portion of that ingested. Net absorption declines with age in men and women, and so increased Ca intake is required to compensate for the lower absorption rate. Fecal Ca losses vary between 100 and 200 mg per day (2.5 to 5.0 mmol). Fecal Ca is composed of unabsorbed dietary Ca and Ca contained in intestinal, pancreatic, and biliary secretions. Secreted Ca is not regulated by hormones or serum Ca.

About 90% of absorbed Ca occurs in the large surface area of the duodenum and jejunum. Increased Ca requirements stimulate expression of the epithelial Ca active transport system in duodenum, ileum, and colon sufficient to increase fractional Ca absorption from 20% to 45% in older men and women and to 55% to 70% in children and young adults. 1,25(OH)$_2$D$_3$ increases the efficiency of the gut to absorb dietary Ca. Intestinal epithelial Ca transport includes both an energy-dependent, cell-mediated saturable active process that is largely regulated by 1,25(OH)$_2$D, and a passive, diffusional paracellular path of absorption that is driven by transepithelial electrochemical gradients. Active Ca absorption accounts for absorption of 10% to 15% of a dietary load [21]. Active transcellular intestinal absorption involves three sequential cellular steps: first, a rate-limiting step involving transfer of luminal Ca into the intestinal cell, via the epithelial apical Ca channel of the transient receptor potential vanilloid (TRPV) family, TRPV6; second, transport of Ca across the cell via a channel-associated protein, calbindin-D9K; and finally extrusion of Ca across the basolateral membrane into ECF by an energy-requiring process via the basolateral Ca ATPase system, PMCA1b [21, 22]. Reductions in dietary Ca intake can increase PTH secretion and 1,25(OH)$_2$D production. Increased 1,25(OH)$_2$D can then increase expression of these proteins resulting in enhanced fractional Ca absorption and compensation for the dietary reduction [23] (see Chapter 29). This cell-mediated pathway involving the TRPV6 Ca channel is saturable with a K_t (1/2 maximal transport) of 1.0 mM.

By regulating claudin 2 and claudin 12, which form paracellular calcium channels, 1,25(OH)$_2$D may also reroute Ca through paracellular epithelial cell junctions. However, passive diffusion increases linearly with luminal Ca concentration and during high dietary Ca intake, 1,25(OH)$_2$D is suppressed and passive paracellular transport accounts for almost all absorption. Causes of increased and decreased intestinal Ca absorption are listed in Table 22.1.

Renal Ca handling

The kidney plays a central role in ensuring Ca balance, and PTH has a major role in fine-tuning this renal function [24–26] along with ECF Ca per se. Multiple influences on Ca handling are listed in Table 22.2 (Chapters 26, 28, and 29 contain descriptions of the molecular actions of PTH on the kidney). PTH has little effect on modulating Ca fluxes in the proximal tubule where 65% of the filtered Ca is reabsorbed, coupled to the bulk transport of solutes such as sodium and water [25]. In this nephron region, PTH can stimulate the 1α(OH)ase, leading to increased synthesis of 1,25(OH)$_2$D [27]. A reduction in ECF Ca can itself stimulate 1,25(OH)$_2$D production but whether this occurs via the CaSR is presently unknown. Finally PTH can also inhibit Na and HCO$_3$- reabsorption in the proximal tubule by inhibiting the apical type 3 Na$^+$/H$^+$ exchanger [28], and the basolateral Na$^+$/K$^+$-ATPase [29], and can

Table 22.1. Conditions that increase or decrease intestinal Ca absorption.

Increased Ca Absorption	Decreased Ca Absorption
Increased renal 1,25(OH)$_2$D production Growth Pregnancy (may also include increased placental production) Lactation Primary hyperparathyroidism Idiopathic hypercalciuria Phosphate-wasting disorders, such as those caused by NPT2a or NPT2c mutations	**Decreased renal 1,25(OH)$_2$D production** Vitamin D deficiency Chronic renal insufficiency Hypoparathyroidism Aging 25-hydroxylase deficiency (CYP2R1 mutation) Vitamin D-dependent rickets type I (hereditary pseudo-vitamin D deficient rickets) (CYP27B1 mutation)
Decreased renal 1,25(OH)$_2$D metabolism Idiopathic infantile hypercalcemia (CYP24A1 mutations)	**Resistance to 1,25(OH)$_2$D** Vitamin D-dependent rickets type II (hereditary vitamin D resistant rickets) (VDR mutation)
Increased extra-renal 1,25(OH)$_2$D production Sarcoid and other granulomatous diseases B-cell lymphoma	**Normal 1,25(OH)$_2$D production** Glucocorticoid excess Hyperthyroidism

Table 22.2. Hormones and conditions that regulate urine Ca excretion via increasing or decreasing glomerular filtration and/or tubular reabsorption.

Decreased Ca Excretion	Increased Ca Excretion
Decreased glomerular filtration Hypocalcemia Hypomagnesemia Renal insufficiency	**Increased glomerular filtration** Hypercalcemia
Increased tubular reabsorption Hypocalcemia ECF volume contraction Thiazide diuretics Phosphate administration Metabolic alkalosis PTH — as in hyperparathyroidism and pseudohypoparathyroidism Parathyroid hormone-related peptide (PTHrP) Gitelman syndrome Familial hypocalciuric hypercalcemia (inactivating CaSR mutation)	**Decreased tubular reabsorption** Hypercalcemia ECF volume expansion Loop diuretics Phosphate deprivation Metabolic acidosis Cyclosporin A Dent disease Bartter syndrome Autosomal dominant hypocalcemia (activating CaSR mutation)

inhibit apical Na+/PO$_4$- cotransport by inhibiting the type II Na+-dependent phosphate cotransporters NaPi-IIa and NaPi-IIc.

About 20% of filtered Ca is reabsorbed in the cortical thick ascending limb of the loop of Henle (CTAL) and 15% is reabsorbed in the distal convoluted tubule (DCT). At both sites PTH binds to the PTH receptor (PTHR) [30, 31], and enhances Ca reabsorption. In the CTAL, at least, this appears to occur by increasing the activity of the Na/K/2Cl cotransporter that drives NaCl reabsorption and stimulates paracellular Ca and Mg reabsorption [32].

The CaSR is also resident in the CTAL [33], where increased ECF Ca activates phospholipase A2, thereby reducing the activity of the Na/K/2Cl cotransporter and of an apical K channel, and diminishing paracellular Ca reabsorption. Consequently, a raised ECF Ca antagonizes the effect of PTH in this nephron segment and ECF Ca can in fact participate in this way in the regulation of its own homeostasis. Inhibition of NaCl reabsorption and loss of NaCl in the urine may contribute to the volume depletion observed in severe hypercalcemia. ECF Ca may therefore act in a manner analogous to "loop" diuretics such as furosemide. 1,25(OH)$_2$D$_3$ has a direct effect on renal Ca handling through stimulation of CaSR. It remains controversial whether 1,25(OH)$_2$D$_3$ plays a direct role in enhancing tubular Ca reabsorption in humans although it appears to do so in mouse models.

In the DCT, PTH can also influence [21] luminal Ca transfer into the renal tubule cell via TRPV5, translocation of Ca across the cell from apical to basolateral surface involving proteins such as calbindin-D28K, and finally active extrusion of Ca from the cell into the blood via a Na+/Ca exchanger, designated NCX1. PTH markedly stimulates Ca reabsorption in the DCT primarily by augmenting NCX1 activity via a cyclic AMP-mediated mechanism.

Bone remodeling and mineralization, and Ca homeostasis

In bone, the PTHR is localized on cells of the osteoblast phenotype which are of mesenchymal origin [34] but not on osteoclasts which are of hematogenous origin. A major physiological role of PTH appears to be to maintain normal Ca homeostasis by enhancing release of the cytokine, receptor activator of NFkB ligand (RANKL) [35]. RANKL binds to its receptor, RANK, on osteoclast precursors and osteoclasts, enhancing the formation of mature osteoclasts from precursors and increasing the resorptive activity of existing osteoclasts especially in cortical bone. PTH may also reduce the osteoblastic protein, osteoprotegerin, which binds to RANKL, forming an inactive complex, and preventing it from binding to RANK, thus reducing osteoclastic activity [36, 37]. It has also been suggested that PTH can acutely release mineral at the bone surface in an osteoclast-independent manner by modifying its solubility [38]. PTH may also have anabolic effects via its action on osteoblastic cells, mainly on trabecular bone.

Vitamin D is essential for normal mineralization of bone that may be caused by an indirect effect of enhancing intestinal calcium and phosphate absorption and maintaining these ions within a range that facilitates hydroxyapatite deposition in bone matrix. A major direct function of $1,25(OH)_2D$ on bone appears to be to enhance mobilization of Ca stores when dietary Ca is insufficient to maintain a normal ECF Ca [39]. As with PTH [40], $1,25(OH)_2D$ enhances osteoclastic bone resorption by binding to receptors on cells of the osteoblastic lineage and increasing the RANKL/OPG ratio to enhance the proliferation, differentiation, and activation of the osteoclastic system from its monocytic precursors [41]. High levels of $1,25(OH)_2D$ may also inhibit mineralization. Endogenous and exogenous $1,25(OH)_2D_3$ have also been reported to have an anabolic role in vivo [42, 43].

Bone formation and resorption are discussed in detail in Section I, and Chapters 25 to 29. The molecular basis for physiological and pathological states of bone turnover is detailed in Chapters 67, 70, 73–76 and 79 and in Section VI.

Temporal sequence of regulation of calcium homeostasis

The elevation in circulating PTH level in response to hypocalcemia enhances distal renal tubular Ca reabsorption within minutes. There may also be PTH-independent "buffering" of ECF Ca by bone through incompletely understood mechanisms, perhaps involving the CaSR, that rapidly returns ECF Ca to its baseline value after induction of hypocalcemia [44]. Therefore, a short period of hypocalcemia may be corrected exclusively through increased renal conservation of Ca and mobilization of Ca from bone. PTH-induced osteoclastic bone resorption likely occurs after several hours to days. PTH-induced stimulation of renal synthesis of $1,25(OH)_2D$ from 25OHD requires several hours [45] and more prolonged hypocalcemia with more prolonged exposure to elevated PTH, may involve a $1,25(OH)_2D$-mediated augmentation of intestinal Ca absorption, as well as $1,25(OH)_2D$-mediated release of Ca from bone.

Generally, an elevation in circulating PTH in response to hypocalcemia is adequate to restore normocalcemia within minutes to a few hours. There are various clinical conditions, however (eg, markedly low Ca intake or vitamin D deficiency), in which greater and more prolonged increases in PTH levels are required to restore and maintain normocalcemia. This can be achieved through a temporal graded series of responses of the parathyroid glands to low ECF Ca and/or associated deficiency of $1,25(OH)_2D$ [46]. Thus, after the initial release of stored PTH in response to hypocalcemia, which occurs within seconds and lasts for 60 to 90 min, there is decreased intracellular degradation of PTH within 20 to 30 min [47], enhanced PTH gene expression over hours and, eventually, increased parathyroid cell proliferation over weeks to months or more [48]. In this situation, large increases in circulating PTH and in parathyroid cellular mass can occur, e.g. in patients with chronic kidney disease and severe hyperparathyroidism.

REFERENCES

1. Walser M. Ion association: VI. Interactions between calcium, magnesium, inorganic phosphate, citrate, and protein in normal human plasma. J Clin Invest. 1961;40:723–30.
2. Parfitt AM, Kleerekoper M. Clinical disorders of calcium, phosphorus and magnesium metabolism. In: Maxwell MH, Kleeman CR (eds) *Clinical Disorders of Fluid and Electrolyte Metabolism* (3rd ed.). New York: McGraw-Hill, 1980, p. 947.
3. Stewart AF, Broadus AE. Mineral metabolism. In: Felig P, Baxter ID, Broadus AE, et al. (eds) *Endocrinology and Metabolism* (2nd ed.). New York: McGraw-Hill, 1987, p. 1317.
4. Bringhurst FR, Demay MB, Kronenberg HM. Hormones and disorders of mineral metabolism. In: Wilson JD, Foster DW, Kronenberg HM, et al. (eds) *Williams Textbook of Endocrinology* (9th ed.). Philadelphia: Saunders, 1998, p. 1155.
5. Brown EM. Physiology of calcium homeostasis. In: Bilezikian JP, Marcus R, Levine MA (eds) *The Parathyroids: Basic and Clinical Concepts* (2nd ed.). San Diego: Academic Press, 2001, p. 167.
6. Lavi-Moshayoff V, Wasserman G, Meir T, et al. PTH increases FGF23 gene expression and mediates the high-FGF23 levels of experimental kidney failure: a bone parathyroid feedback loop. Am J Physiol Renal Physiol. 2010;299(4):F882–9.
7. Fraser DR, Kodicek E. Regulation of 25-hydroxycholecalciferol-1-hydroxylase activity in kidney by parathyroid hormone. Nat New Biol. 1973;241(110):163–6.
8. Bai XY, Miao D, Goltzman D, et al. The autosomal dominant hypophosphatemic rickets R176Q mutation in fibroblast growth factor 23 resists proteolytic cleavage and enhances in vivo biological potency. J Biol Chem. 2003;278(11):9843–9.
9. Brown EM, Gamba G, Riccardi D, et al. Cloning and characterization of an extracellular Ca(2+)-sensing receptor from bovine parathyroid. Nature. 1993; 366(6455):575–80.
10. Potts JT Jr, Juppner H. Parathyroid hormone and parathyroid hormone-related peptide in calcium homeostasis, bone metabolism, and bone development: the proteins, their genes, and receptors. In: Avioli LV, Krane SM.(eds) *Metabolic Bone Disease* (3rd ed.). New York: Academic Press, 1997, p. 51.

11. Grant FD, Conlin PR, Brown EM. Rate and concentration dependence of parathyroid hormone dynamics during stepwise changes in serum ionized calcium in normal humans. J Clin Endocrinol Metab. 1990;71(2):370–8.
12. Kremer R, Bolivar I, Goltzman D, et al. Influence of calcium and 1,25-dihydroxycholecalciferol on proliferation and proto-oncogene expression in primary cultures of bovine parathyroid cells. Endocrinology. 1989;125(2):935–41.
13. Goltzman D, Miao D, Panda DK, et al. Effects of calcium and of the vitamin D system on skeletal and calcium homeostasis: lessons from genetic models. J Steroid Biochem Mol Biol. 2004;89–90(1–5):485–9.
14. Nguyen-Yamamoto L, Karaplis AC, St-Arnaud R, et al. Fibroblast growth factor 23 regulation by systemic and local osteoblast-synthesized 1,25-dihydroxyvitamin D. J Am Soc Nephrol. 2017; 28(2): 586–97
15. Antoniucci DM, Yamashita T, Portale AA. Dietary phosphorus regulates serum fibroblast growth factor-23 concentrations in healthy men. J Clin Endocrinol Metab. 2006;91(8):3144–49.
16. Saito H, Maeda A, Ohtomo S, et al. Circulating FGF-23 is regulated by 1alpha,25-dihydroxyvitamin D3 and phosphorus in vivo. J Biol Chem. 2005;280(4):2543–9.
17. Perwad F, Azam N, Zhang MY, et al. Dietary and serum phosphorus regulate fibroblast growth factor 23 expression and 1,25-dihydroxyvitamin D metabolism in mice. Endocrinology. 2005;146(12):5358–64.
18. Scanni R, von Rotz M, Jehle S, et al. The human response to acute enteral and parenteral phosphate loads. J Am Soc Nephrol. 2014;25(12):2730–9
19. Quinn SJ, Thomsen AR, Pang L, et al. Interactions between calcium and phosphorus in the regulation of the production of fibroblast growth factor 23 in vivo. Am J Physiol Endocrinol Metab. 2013;304(3):E310–20.
20. Favus MF. Intestinal absorption of calcium, magnesium and phosphorus. In: Coe FL, Favus MJ (eds) *Disorders of Bone and Mineral Metabolism*. New York: Raven, 1992, p. 57.
21. Hoenderop JG, Nilius B, Bindels RJM. Calcium absorption across epithelia. Physiol Rev. 2005;85(1):373–422.
22. Van de Graaf SF, Boullart I, Hoenderop JG, et al. Regulation of the epithelial Ca^{2+} channels TRPV5 and TRPV6 by $1\alpha,25$-dihydroxy Vitamin D3 and dietary Ca^{2+}. J Steroid Biochem Molec Biol. 2004;89–90(1–5):303–8.
23. Lieben L, Benn BS, Ajibade D, et al. Trpv6 mediates intestinal calcium absorption during calcium restriction and contributes to bone homeostasis. Bone. 2010;47(2): 301–8.
24. Friedman PA, Gesek FA. Cellular calcium transport in renal epithelia: measurement, mechanisms, and regulation. Physiol Rev. 1995;75(3):429–71.
25. Nordin BE, Peacock M. Role of kidney in regulation of plasma-calcium. Lancet. 1969;2(7633):1280–3.
26. Rouse D, Suki WN. Renal control of extracellular calcium. Kidney Int. 1990;38(4):700–8.
27. Brenza HL, Kimmel-Jehan C, Jehan F, et al. Parathyroid hormone activation of the 25-hydroxyvitamin D3-1α-hydroxylase gene promoter. Proc Natl Acad Sci U S A. 1998;95(4):1387–91.
28. Azarani A, Goltzman D, Orlowski J. Parathyroid hormone and parathyroid hormone-related peptide inhibit the apical Na+/H+ exchanger NHE-3 isoform in renal cells (OK) via a dual signaling cascade involving protein kinase A and C. J Biol Chem. 1995;270(34):20004–10.
29. Derrickson BH, Mandel LJ. Parathyroid hormone inhibits Na(+)-K(+)-ATPase through Gq/G11 and the calcium-independent phospholipase A2. Am J Physiol. 1997;272(6 Pt 2):F781–8.
30. Jüppner H, Abou-Samra AB, Freeman M, et al. A G protein-linked receptor for parathyroid hormone and parathyroid hormone-related peptide. Science. 1991; 254(5034):1024–6.
31. Abou-Samra AB, Jüppner H, Force T, et al. Expression cloning of a common receptor for parathyroid hormone and parathyroid hormone-related peptide from rat osteoblast-like cells: a single receptor stimulates intracellular accumulation of both cAMP and inositol triphosphates and increases intracellular free calcium. Proc Natl Acad Sci U S A. 1992;89(7):2732–6.
32. de Rouffignac C, Quamme GA. Renal magnesium handling and its hormonal control. Physiol Rev. 1994;74(2):305–22.
33. Hebert SC. Extracellular calcium-sensing receptor: implications for calcium and magnesium handling in the kidney. Kidney Int. 1996;50(6):2129–39.
34. Rouleau MF, Mitchell J, Goltzman D. In vivo distribution of parathyroid hormone receptors in bone: Evidence that a predominant osseous target cell is not the mature osteoblast. Endocrinology. 1988;123(1):187–91.
35. Kim S, Yamazaki M, Shevde NK, et al. Transcriptional control of receptor activator of nuclear factor-kappaB ligand by the protein kinase A activator forskolin and the transmembrane glycoprotein 130-activating cytokine, oncostatin M, is exerted through multiple distal enhancers. Mol Endocrinol. 2007;21(1): 197–214.
36. Huang JC, Sakata T, Pfleger LL, et al. PTH differentially regulates expression of RANKL and OPG. J Bone Miner Res. 2004;19(2):235–44.
37. Boyce BF, Xing L. Functions of RANKL/RANK/OPG in bone modeling and remodeling. Arch Biochem Biophys. 2008;473(2):139–46.
38. Talmage DW, Talmage RV. Calcium homeostasis: how bone solubility relates to all aspects of bone physiology. J Musculoskelet Neuronal Interact. 2007;7(2):108–12.
39. Li YC, Pirro AE, Amling M, et al. Targeted ablation of the vitamin D receptor: an animal model of vitamin D-dependent rickets type II with alopecia. Proc Natl Acad Sci U S A. 1997;94(18):9831–5.
40. Lee SK, Lorenzo JA. Parathyroid hormone stimulates TRANCE and inhibits osteoprotegerin messenger ribonucleic acid expression in murine bone marrow cultures: correlation with osteoclast-like cell formation. Endocrinology. 1999;140(8):3552–61.
41. Takahashi N, Udagawa N, Takami M, et al. Cells of bone: osteoclast generation. In: Bilezikian JP, Raisz LG, Rodan GA (eds) *Principles of Bone Biology* (2nd ed.). San Diego: Academic Press, 2002, p. 109.

42. Panda DK, Miao D, Bolivar I, et al. Inactivation of the 25-dihydroxyvitamin D 1α-hydroxylase and vitamin D receptor demonstrates independent effects of calcium and vitamin D on skeletal and mineral homeostasis. J Biol Chem. 2004;279(16):16754–66.
43. Xue Y, Karaplis AC, Hendy GN, et al. Exogenous 1,25-dihydroxyvitamin D3 exerts a skeletal anabolic effect and improves mineral ion homeostasis in mice which are homozygous for both the 1alpha-hydroxylase and parathyroid hormone null alleles. Endocrinology. 2006;147(10):4801–10.
44. Lewin E, Wang W, Olgaard K. Rapid recovery of plasma ionized calcium after acute induction of hypocalcaemia in parathyroidectomized and nephrectomized rats. Nephrol Dial Transplant. 1999;14(3):604–9.
45. Sommerville BA, Maunder E, Ross R, et al. Effect of dietary calcium and phosphorus depletion on vitamin D metabolism and calcium binding protein in the growing pig. Horm Metab Res. 1985;17(2):78–81.
46. Brown EM. Extracellular Ca2+ sensing, regulation of parathyroid cell function, and role of Ca2+ and other ions as extracellular (first) messengers. Physiol Rev. 1991;71(2):371–411.
47. Morrissey JJ, Hamilton JW, MacGregor RR, et al. The secretion of parathormone fragments 34–84 and 37–84 by dispersed porcine parathyroid cells. Endocrinology. 1980;107(1):164–71.
48. Naveh-Many T, Rahamimov R, Livni N, et al. Parathyroid cell proliferation in normal and chronic renal failure rats. The effects of calcium, phosphate, and vitamin D. J Clin Invest. 1995;96(4):1786–93.

23

Magnesium Homeostasis

Aliya Aziz Khan[1], Asiya Sbayi[2], and Karl Peter Schlingmann[3]

[1]*Department of Medicine, Divisions of Endocrinology and Metabolism and Geriatric Medicine, McMaster University, Hamilton, ON, Canada*
[2]*The George Washington University, Milken Institute School of Public Health, Washington DC, USA*
[3]*Department of General Pediatrics, University Children`s Hospital, Münster, Germany*

INTRODUCTION

Magnesium (Mg^{2+}) is essential for a vast number of cellular processes, including energy metabolism, protein and nucleic acid synthesis, and in the maintenance of the electrical potential of nervous tissues and cell membranes [1]. It is a cofactor for a number of enzymes and plays a key role in energy metabolism. Approximately 99% of the total body Mg^{2+} is intracellular, with the remainder present in the extracellular fluid [2]. About 90% of total body Mg^{2+} is found in bone, muscle, and soft tissues with 0.3% present in the serum, about a third being protein-bound [3, 4]. Roughly 10% is complexed as salts (bicarbonate, citrate, phosphate, sulfate) with approximately 60% of the magnesium being biologically active and available in the free, ionized form [1, 5]. Serum ionized Mg^{2+} is maintained in a tight normal reference range through the actions of the kidneys, bowel, and bone [6–9].

HYPOMAGNESEMIA

The symptoms of hypomagnesemia depend on its severity and rate of onset. Symptoms include fatigue and leg cramps in mild hypomagnesemia, and seizures, coma, and death in severely deficient states [2]. Hypomagnesemia during pregnancy can contribute to malformations in the developing fetus [10]. Hypomagnesemia may be caused by decreased intake, decreased intestinal absorption, increased losses, or redistribution of Mg^{2+} [11].

Mg^{2+} is widely present in all food groups [11]. Common causes of intestinal hypomagnesemia are decreased absorption caused by malabsorption, short bowel syndrome, severe vomiting, diarrhea, or steatorrhea [4]. An inherited disorder impairing intestinal magnesium absorption is familial hypomagnesemia with secondary hypocalcemia (fHSH) caused by mutations in *TRPM6* (Transient Receptor Potential Cation Channel Subfamily M Member 6). *TRPM6* is involved in the formation of apical magnesium-permeable ion channels in intestine and kidney. Recessive mutations therefore not only result in defective active, transcellular magnesium uptake in the intestine, but also impaired renal magnesium conservation [12, 13].

Long-term proton pump inhibitor (PPI) use may result in an acquired form of hypomagnesemia most likely caused by decreased intestinal absorption or losses [14, 15]. This may be caused by inhibition of TRPM6 mediated active transportation of Mg^{2+} caused by alterations in intestinal pH. However, the exact mechanism leading to hypomagnesemia still needs to be elucidated [16].

Renal magnesium losses can be differentiated from intestinal losses by evaluating the fractional excretion of magnesium (FEMg):

$$FEMg = [(\text{urine magnesium} \times \text{plasma creatinine})/(0.7 \times \text{plasma magnesium} \times \text{urine creatinine})] \times 100\%$$

If FEMg is more than 4% in a patient with hypomagnesemia it is consistent with renal magnesium wasting. An extrarenal cause of magnesium loss is likely to be present if FEMg is less than 2% [17]. Falsely reduced FEMg can occur in the presence of low glomerular filtration rates as well as severe hypomagnesemia. Therefore, it may be necessary to supplement magnesium before detecting renal magnesium wasting by measuring FEMg from spot urine samples [18]. In absolute terms a urinary magnesium excretion of more than 1 mmol/day in the presence of hypomagnesemia would be consistent with renal magnesium wasting [18].

A number of drugs may result in hypomagnesemia by promoting renal magnesium excretion including diuretics, both thiazide and furosemide, antibiotics and antimycotics (foscarnet, amphotericin B, and aminoglycosides), anticancer medications (ie, platinum derivatives such as cisplatin, carboplatin), immunosuppressants (rapamycin and calcineurin inhibitors such as tacrolimus and cyclosporine A), and also EGF-receptor inhibitors (cetuximab). The latter might specifically interfere with transcellular magnesium transport in the distal tubule.

HYPERMAGNESEMIA

In the presence of hypermagnesemia, the fractional excretion of Mg^{2+} is enhanced in order to maintain normal serum Mg^{2+}. In the presence of a decline in renal function with a GFR less than 30 mL/min, the excretion of magnesium becomes impaired and serum magnesium levels begin to rise [19]. Impaired renal clearance of Mg^{2+} also occurs in familial hypocalciuric hypercalcemia (see later) as well as in the presence of lithium therapy [20]. In contrast to an increased oral intake of Mg^{2+}, that is by ingestion of antacids, cathartics, or laxatives, that rarely results in hypermagnesemia, parenteral administrations of Mg^{2+} are able to produce clinically relevant hypermagnesemic states [21]. The resulting hypermagnesemia may in turn cause hypocalcemia caused by the inhibition of PTH release [22]. Clinically, hypermagnesemia is associated with gastrointestinal symptoms including nausea and vomiting. In addition, electrocardiographic changes can develop with prolonged QRS, PR, and QT intervals and can result in complete heart block and shock. Neurological symptoms include confusion and coma.

MAGNESIUM ABSORPTION

Intestinal magnesium absorption

Mg^{2+} is mainly absorbed in the small bowel (jejunum and ileum) with some absorption occurring in the colon [23, 24]. Under physiological conditions, approximately 30% to 40% of orally ingested Mg^{2+} is absorbed [25]. However, the amount absorbed from the bowel can increase to 80% in the presence of deficiency. Mg^{2+} absorption occurs via two different pathways: a saturable, active transcellular transport and a nonsaturable passive paracellular pathway [26]. At low intraluminal concentrations, Mg^{2+} is primarily absorbed via the active transcellular route whereas with rising intraluminal concentrations the passive paracellular pathway gains importance. The saturation kinetics of the transcellular pathway indicate a limited active transport capacity. The two transport systems together yield a curvilinear kinetic of transepithelial Mg^{2+} transport. The defect in intestinal Mg^{2+} absorption observed in children with *TRPM6* defects (see later) points to a critical role of this apically located Mg^{2+}-permeable ion channel for active transcellular Mg^{2+} absorption.

Renal magnesium conservation

Following absorption from the bowel, Mg^{2+} enters the blood stream and is filtered within the renal glomeruli. Approximately 95% to 99% of filtered Mg^{2+} is reabsorbed along the kidney tubule [27] so that under physiological, normomagnesemic conditions, 3% to 5% of filtered Mg^{2+} is finally excreted in the urine. The mechanisms of transepithelial Mg^{2+} transport along the kidney tubule significantly vary depending on the different tubular segments involved.

A minor portion (15% to 20%) of filtered Mg^{2+} is already reabsorbed in the proximal convoluted tubule (PCT). Interestingly, this nephron segment is capable of reabsorbing up to 70% of filtered Mg^{2+} in the neonate. The increased paracellular permeability of the PCT for Mg^{2+} responsible for this phenomenon disappears with maturation of this tubular segment [28].

The vast majority of filtered Mg^{2+} (~70%) is reabsorbed in the thick ascending limb of Henle's loop (TAL). Mg^{2+} reabsorption in the TAL is passive and paracellular in nature and occurs together with calcium through specialized tight junctions. These tight junctions are composed of a specific set of proteins of the claudin family that, on one hand, seal the paracellular space for water and electrolytes, but, on the other hand, allow for the selective passage of ions. In the TAL, tight junctions are formed, amongst others, by the claudin proteins claudin-16 and claudin-19 that play a key role in regulating paracellular Ca^{2+} and Mg^{2+} transport [29]. Mutations affecting these two proteins result in impaired paracellular reabsorption of both Ca^{2+} and Mg^{2+} causing familial hypomagnesemia with hypercalciuria and nephrocalcinosis (FHHNC) [30]. Affected patients exhibit hypomagnesemia and develop nephrocalcinosis in childhood caused by the presence of hypercalciuria [31]. Unfortunately, patients almost uniformly develop chronic renal failure mainly in the second decade of life [32].

Paracellular Ca^{2+} and Mg^{2+} transport in the TAL is regulated by the action of the basolaterally located calcium sensing receptor (CaSR). The CaSR senses extracellular Ca^{2+} as well as Mg^{2+} concentrations in the distal nephron as well as in other tissues and thereby plays an essential role in Ca^{2+} and Mg^{2+} homeostasis [33]. Concerning renal tubular magnesium handling, PTH not only increases Mg^{2+} reabsorption in the cortical TAL by enhancing paracellular permeability, but also increases transcellular Mg^{2+} reabsorption in the DCT [34, 35].

In the parathyroid, the CaSR is responsible for adjusting the rate of PTH synthesis and release to the extracellular levels of Ca^{2+} and Mg^{2+}. The CaSR bears multiple low-affinity cation binding sites in its extracellular domain allowing for a cooperative interaction with multiple cations in a millimolar concentration range [36]. Therefore, Mg^{2+} and Ca^{2+} are both able to activate the CaSR and affect PTH synthesis and secretion [34]. In addition, intracellular Mg^{2+} is involved in the activation of adenylate cyclase and in intracellular signaling of cyclic AMP [37]. Activation of the CaSR by Mg^{2+} results in stimulation of phospholipase C and A2 and inhibition of cellular cAMP with inhibition of PTH release [38]. In the kidney, CaSR activation decreases paracellular sodium, calcium, and magnesium transport, resulting in a renal loss of these cations. Hereditary disorders may result from either activating or inactivating CaSR mutations. Heterozygous, activating mutations lead to autosomal-dominant hypocalcemia (ADH). Patients may present with hypocalcemic cerebral seizures or muscle spasms. Inappropriately low PTH levels (caused by activation of the CaSR in the parathyroid gland) lead to the diagnosis of primary hypoparathyroidism. The hypocalcemia is associated with hypomagnesemia in a significant number of affected patients [39]. In addition, patients may develop a clinically relevant degree of renal salt and water losses caused by inhibition of active transcellular NaCl reabsorption in the TAL that is reflected by a laboratory profile similar to Bartter syndrome [19].

Inactivating mutations on one or two alleles lead to familial hypocalciuric hypercalcemia and neonatal severe hyperparathyroidism (NSHPT), respectively. Serum PTH levels are inappropriately high and renal Ca^{2+} and Mg^{2+} excretions are markedly reduced. In addition to symptomatic hypercalcemia, patients with NSHPT also exhibit mild hypermagnesemia [20].

Although only 5–10% of filtered Mg^{2+} is reabsorbed in the distal convoluted tubule (DCT), active transcellular Mg^{2+} transport in this segment is of critical importance for determining the final urinary Mg^{2+} excretion because there is no significant Mg^{2+} transport in the collecting duct [30, 40].

Though transepithelial magnesium transport in the DCT is far from being completely understood, molecular genetic studies in patients with different forms of hereditary hypomagnesemia have provided critical insight into the mechanisms and regulation of underlying transport processes.

Molecular genetic studies in patients with hypomagnesemia with secondary hypocalcemia lead to the identification of TRPM6 as a critical component of active transcellular Mg^{2+} transport in intestine and kidney [12, 13]. TRPM6 is thought to be involved in the formation of apically-located Mg^{2+} permeable ion channels. Through these ion channels, Mg^{2+} enters the epithelial cell driven by the membrane potential. Recessive loss of function mutations in TRPM6 result in the development of severe hypomagnesemia and cerebral seizures during infancy. In addition to hypomagnesemia, patients also display suppressed PTH levels and consecutive hypocalcemia. The suppression of PTH is thought to result from a block of PTH synthesis and secretion in the presence of profound hypomagnesemia [41]. This paradoxical inhibition of the parathyroid involves intracellular signaling pathways of the CaSR with an increase in the activity of inhibitory G alpha subunits [42].

In addition, PTH-induced release of Ca^{2+} from bone is substantially impaired in hypomagnesemia [21, 22, 43]. Intracellular Mg^{2+} is a cofactor of adenylate cyclase and decreases in intracellular ionized Mg^{2+} result in a resistance to PTH [44–46]. The hypocalcemia is resistant to treatment with Ca^{2+} or vitamin D, but rapidly responds to Mg^{2+} supplementation.

Unfortunately, the molecular nature of basolateral Mg^{2+} export is still unknown. However, a number of molecular studies in patients with rare forms of hereditary hypomagnesemia could show that the transcellular transport of Mg^{2+} in the DCT is highly dependent on membrane potential and cellular energy content. Examples are a dominant negative mutation in the gamma subunit of basolateral Na-K-ATPase that physiologically influences the affinities for sodium and ATP [47] or mutations in *KCNA1* encoding the Kv1.1 potassium channel thought to be involved in the generation of the apical membrane potential in DCT cells [48].

A more recent study identified mutations in a transmembrane protein, *CNNM2* (cyclin M2), that potentially represents a basolaterally expressed sensor for interstitial Mg^{2+} concentrations [49]. Finally, Mg^{2+} transport in the DCT was found to be hormonally regulated by EGF (epidermal growth factor) that acts via basolaterally located receptors [50]. A mutation disrupting the basolateral sorting of this EGF receptor leads to impaired membrane trafficking of TRPM6, which finally results in reduced Mg^{2+} reabsorption [49].

Besides these hereditary disorders leading to renal Mg^{2+} wasting, hypomagnesemia may be a result of a number of diverse clinical conditions and also a clinically relevant side effect of a multitude of medical treatments: polyuria per se may result in decreased renal tubular reabsorption of Mg^{2+}. Diuretics, antibiotics, calcineurin inhibitors, and epidermal growth factor receptor antagonists may also decrease renal tubular reabsorption of Mg^{2+} [4]. These drugs have been shown to downregulate renal Mg^{2+} transport proteins including TRPM6 ion channels in the DCT and thereby induce urinary magnesium losses [21, 22, 43].

Deficiencies in intracellular Mg^{2+} may develop in the presence of a normal serum Mg^{2+} [51, 52]. Intracellular Mg^{2+} may be a key regulator of serum PTH [3].

CONCLUSION

Ca^{2+} and Mg^{2+} homeostases are closely linked. Our understanding of Mg^{2+} homeostasis has significantly advanced by increased understanding of the pathophysiology of inherited disorders resulting in hypomagnesemia (Table 23.1). Hypomagnesemia in turn can result in hypocalcemia and requires careful assessment and correction in order to normalize serum Ca^{2+}. Our understanding of the hormonal regulation of Mg^{2+} is still incomplete and is an area of active research. Mg^{2+} homeostasis is maintained by the bowel, kidney, and bone and is essential for cardiovascular, neuromuscular, and skeletal health.

Table 23.1. Inherited disorders of magnesium homeostasis.

Disorder	Inheritance	Genetic Abnormality	Clinical Features
Hypomagnesemia with secondary hypocalcemia (HSH)	Autosomal recessive	TRPM6 (transient receptor potential cation channel)	Severe hypomagnesemia with hypocalcemia; hypoparathyroidism, and neurological symptoms of seizures, tetany, or muscle spasm
Familial hypomagnesemia with hypercalciuria and nephrocalcinosis (FHHNC)	Autosomal recessive	CLDN16, CLDN19	Renal Ca^{2+} and Mg^{2+} wasting, polyuria/polydipsia, cramps, tremors, and convulsions; renal impairment
Autosomal dominant hypoparathyroidism (ADH)	Autosomal recessive	CASR (Ca^{2+}-sensing receptor, CaSR)	Hypocalcemia, hypomagnesemia
Gitelman syndrome	Autosomal dominant	SLC12A3 (sodium chloride cotransporter (NCC))	Hypokalemic alkalosis, secondary hyperaldosteronism, hypomagnesemia, and hypocalciuria
EAST (epilepsy, ataxia, sensorineural deafness, and renal tubulopathy) syndrome	Autosomal recessive	KCNJ10 (Kir4.1 K^+-channel)	Salt wasting, hypomagnesemia, epilepsy, ataxia, sensorineural deafness, mental retardation
Isolated dominant hypomagnesemia	Autosomal dominant, de novo	FXYD2 (Na/K-ATPase gamma subunit)	Hypomagnesemia, seizures, hypocalciuria
		HNF1B (hepatocyte nuclear factor 1 beta)	HNF1b nephropathy, MODY 5 diabetes, hypomagnesemia
		CNNM2 (cyclin M2)	Hypomagnesemia, seizures, intellectual disability
		KCNA1 (Kv1.1, K^+-channel)	Muscle cramps, tetany, muscle weakness, tremor, hypomagnesemia
Isolated recessive hypomagnesemia	Autosomal recessive	EGF (epidermal growth factor)	Hypomagnesemia, seizures

REFERENCES

1. Elin RJ. Magnesium metabolism in health and disease. Dis Mon. 1988;34:161–218.
2. Navarro-Gonzalez JF, Mora-Fernandez C, Garcia-Perez J. Clinical implications of disordered magnesium homeostasis in chronic renal failure and dialysis. Semin Dial. 2009;22:37–44.
3. Pironi L, Malucelli E, Guidetti M, et al. The complex relationship between magnesium and serum parathyroid hormone: a study in patients with chronic intestinal failure. Magnes Res. 2009;22:37–43.
4. Hoorn EJ, Zietse R. Disorders of calcium and magnesium balance: a physiology-based approach. Pediatr Nephrol. 2013;28:1195–206.
5. Speich M, Bousquet B, Nicolas G. Reference values for ionized, complexed, and protein-bound plasma magnesium in men and women. Clin Chem. 1981;27:246–8.
6. Ferre S, Hoenderop JG, Bindels RJ. Sensing mechanisms involved in $Ca2+$ and $Mg2+$ homeostasis. Kidney Int. 2012;82:1157–66.
7. Al-Azem H, Khan AA. Hypoparathyroidism. Best Pract Res Clin Endocrinol Metab. 2012;26:517–22.
8. Reilly RF, Ellison DH. Mammalian distal tubule: physiology, pathophysiology, and molecular anatomy. Physiol Rev. 2000;80:277–313.
9. Glaudemans B, Knoers NV, Hoenderop JG, et al. New molecular players facilitating Mg(2+) reabsorption in the distal convoluted tubule. Kidney Int. 2010;77:17–22.
10. Schlegel RN, Cuffe JS, Moritz KM, et al. Maternal hypomagnesemia causes placental abnormalities and fetal and postnatal mortality. Placenta. 2015;36:750–8.
11. Steen O, Khan A. Role of magnesium in parathyroid physiology. In: Brandi ML, Brown EM (eds) *Hypoparathyroidism*. Milan: Springer, 2015, pp. 61–7.
12. Schlingmann KP, Weber S, Peters M, et al. Hypomagnesemia with secondary hypocalcemia is caused by mutations in TRPM6, a new member of the TRPM gene family. Nat Genet. 2002;31:166–70.
13. Walder RY, Landau D, Meyer P, et al. Mutation of TRPM6 causes familial hypomagnesemia with secondary hypocalcemia. Nat Genet. 2002;31:171–4.

14. Hess MW, Hoenderop JG, Bindels RJ, et al. Systematic review: hypomagnesaemia induced by proton pump inhibition. Aliment Pharmacol Ther. 2012;36: 405–13.
15. Hoorn EJ, van der Hoek J, de Man RA, et al. A case series of proton pump inhibitor-induced hypomagnesemia. Am J Kidney Dis. 2010;56:112–6.
16. Cundy T, Dissanayake A. Severe hypomagnesaemia in long-term users of proton-pump inhibitors. Clin Endocrinol (Oxf). 2008;69:338–41.
17. Elisaf M, Panteli K, Theodorou J, et al. Fractional excretion of magnesium in normal subjects and in patients with hypomagnesemia. Magnes Res 1997;10:315–20.
18. Sutton RA, Domrongkitchaiporn S. Abnormal renal magnesium handling. Miner Electrolyte Metab. 1993; 19:232–40.
19. Watanabe S, Fukumoto S, Chang H, et al. Association between activating mutations of calcium-sensing receptor and Bartter's syndrome. Lancet. 2002;360:692–4.
20. Marx SJ, Attie MF, Levine MA, et al. The hypocalciuric or benign variant of familial hypercalcemia: clinical and biochemical features in fifteen kindreds. Medicine (Baltimore). 1981;60:397–412.
21. Hoorn EJ, Walsh SB, McCormick JA, et al. The calcineurin inhibitor tacrolimus activates the renal sodium chloride cotransporter to cause hypertension. Nat Med. 2011;17:1304–9.
22. Nijenhuis T, Vallon V, van der Kemp AW, et al. Enhanced passive Ca2+ reabsorption and reduced Mg2+ channel abundance explains thiazide-induced hypocalciuria and hypomagnesemia. J Clin Invest. 2005;115:1651–8.
23. Graham LA, Caesar JJ, Burgen AS. Gastrointestinal absorption and excretion of Mg 28 in man. Metabolism. 1960;9:646–59.
24. Brannan PG, Vergne-Marini P, Pak CY, et al. Magnesium absorption in the human small intestine. Results in normal subjects, patients with chronic renal disease, and patients with absorptive hypercalciuria. J Clin Invest. 1976;57:1412–8.
25. Quamme GA. Recent developments in intestinal magnesium absorption. Curr Opin Gastroenterol. 2008;24:230–5.
26. Fine KD, Santa Ana CA, Porter JL, et al. Intestinal absorption of magnesium from food and supplements. J Clin Invest. 1991;88:396–402.
27. Dimke H, Hoenderop JG, Bindels RJ. Hereditary tubular transport disorders: implications for renal handling of Ca2+ and Mg2+. Clin Sci (Lond). 2009;118:1–18.
28. Lelievre-Pegorier M, Merlet-Benichou C, Roinel N, de RC. Developmental pattern of water and electrolyte transport in rat superficial nephrons. Am J Physiol. 1983;245:F15–F21.
29. Hou J, Goodenough DA. Claudin-16 and claudin-19 function in the thick ascending limb. Curr Opin Nephrol Hypertens. 2010;19:483–8.
30. Ferre S, Hoenderop JJ, Bindels RJ. Role of the distal convoluted tubule in renal Mg2+ handling: molecular lessons from inherited hypomagnesemia. Magnes Res. 2011;24:S101–S108.
31. Weber S, Schneider L, Peters M, et al. Novel paracellin-1 mutations in 25 families with familial hypomagnesemia with hypercalciuria and nephrocalcinosis. J Am Soc Nephrol. 2001;12:1872–81.
32. Konrad M, Hou J, Weber S, et al. CLDN16 genotype predicts renal decline in familial hypomagnesemia with hypercalciuria and nephrocalcinosis. J Am Soc Nephrol. 2008;19:171–81.
33. Alfadda TI, Saleh AM, Houillier P, et al. Calcium-sensing receptor 20 years later. Am J Physiol Cell Physiol. 2014;307:C221–C231.
34. Vetter T, Lohse MJ. Magnesium and the parathyroid. Curr Opin Nephrol Hypertens. 2002;11:403–10.
35. Wittner M, Mandon B, Roinel N, de RC, et al. Hormonal stimulation of Ca2+ and Mg2+ transport in the cortical thick ascending limb of Henle's loop of the mouse: evidence for a change in the paracellular pathway permeability. Pflugers Arch. 1993;423:387–96.
36. Hebert SC. Extracellular calcium-sensing receptor: implications for calcium and magnesium handling in the kidney. Kidney Int. 1996;50:2129–39.
37. Grubbs RD, Maguire ME. Magnesium as a regulatory cation: criteria and evaluation. Magnesium. 1987;6: 113–27.
38. Chang W, Pratt S, Chen TH, et al. Coupling of calcium receptors to inositol phosphate and cyclic AMP generation in mammalian cells and Xenopus laevis oocytes and immunodetection of receptor protein by region-specific antipeptide antisera. J Bone Miner Res. 1998;13:570–80.
39. Pearce SH, Williamson C, Kifor O, et al. A familial syndrome of hypocalcemia with hypercalciuria due to mutations in the calcium-sensing receptor. N Engl J Med. 1996;335:1115–22.
40. Bindels RJ. 2009 Homer W. Smith Award: Minerals in motion: from new ion transporters to new concepts. J Am Soc Nephrol. 2010;21:1263–9.
41. Anast CS, Mohs JM, Kaplan SL, et al. Evidence for parathyroid failure in magnesium deficiency. Science. 1972;177:606–8.
42. Quitterer U, Hoffmann M, Freichel M, et al. Paradoxical block of parathormone secretion is mediated by increased activity of G alpha subunits. J Biol Chem. 2001;276:6763–9.
43. Groenestege WM, Thebault S, van der Wijst J, et al. Impaired basolateral sorting of pro-EGF causes isolated recessive renal hypomagnesemia. J Clin Invest. 2007;117:2260–7.
44. Mune T, Yasuda K, Ishii M, et al. Tetany due to hypomagnesemia induced by cisplatin and doxorubicin treatment for synovial sarcoma. Intern Med. 1993;32:434–7.
45. Mori S, Harada S, Okazaki R, et al. Hypomagnesemia with increased metabolism of parathyroid hormone and reduced responsiveness to calcitropic hormones. Intern Med. 1992;31:820–4.
46. Mihara M, Kamikubo K, Hiramatsu K, et al. Renal refractoriness to phosphaturic action of parathyroid hormone in a patient with hypomagnesemia. Intern Med. 1995;34:666–9.
47. Arystarkhova E, Sweadner KJ. Splice variants of the gamma subunit (FXYD2) and their significance in regu-

lation of the Na, K-ATPase in kidney. J Bioenerg Biomembr. 2005;37:381–6.
48. Glaudemans B, van der Wijst J, Scola RH, et al. A missense mutation in the Kv1.1 voltage-gated potassium channel-encoding gene KCNA1 is linked to human autosomal dominant hypomagnesemia. J Clin Invest. 2009;119:936–42.
49. Stuiver M, Lainez S, Will C, et al. CNNM2, encoding a basolateral protein required for renal Mg2+ handling, is mutated in dominant hypomagnesemia. Am J Hum Genet. 2011;88:333–43.
50. Thebault S, Alexander RT, Tiel Groenestege WM, et al. EGF increases TRPM6 activity and surface expression. J Am Soc Nephrol. 2009;20:78–85.
51. Dyckner T, Wester PO. The relation between extra- and intracellular electrolytes in patients with hypokalemia and/or diuretic treatment. Acta Med Scand. 1978;204:269–82.
52. Rob PM, Bley N, Dick K, et al. Magnesium deficiency after renal transplantation and cyclosporine treatment despite normal serum-magnesium detected by a modified magnesium-loading-test. Transplant Proc. 1995;27:3442–3.

24

Fetal Calcium Metabolism

Christopher S. Kovacs

Faculty of Medicine — Endocrinology, Health Sciences Centre, Memorial University of Newfoundland, St John's, NL, Canada

INTRODUCTION

Our understanding of how fetal mineral homeostasis is regulated has derived in part from studies of cord blood chemistries and hormone levels, and from both normal and abnormal fetuses that died at birth. However, much of human regulation of fetal mineral homeostasis must be inferred from studies in animals. Some observations in animals may not apply to humans. This chapter briefly reviews existing human and animal data; a recent lengthier review should be consulted for detailed references [1].

Fetal mineral metabolism has adapted to maintain a high extracellular level of calcium (and other minerals) that is physiologically appropriate for fetal tissues, and to provide sufficient calcium (and other minerals) to fully mineralize the skeleton before birth. Mineralization occurs rapidly in late gestation, such that a human accretes 80% of its required 30g of calcium and 20g of phosphorus in the third trimester, whereas a rat accretes 95% of the required 12.5mg of calcium in the last 5 days of its 3-week gestation.

MINERAL IONS AND CALCIOTROPIC HORMONES

Human and other mammalian fetuses consistently maintain serum calcium (both total and ionized) at significantly higher values than in the mother during late gestation. Similarly, serum phosphate is significantly elevated whereas serum magnesium is minimally elevated above maternal concentrations.

These increased serum mineral concentrations have physiological importance. Maintaining the fetal calcium above the maternal level is required to achieve normal mineralization of the fetal skeleton, as discussed below. Survival to the end of gestation is unaffected by significant hypocalcemia in animal models. However, survival after birth may be aided by a high blood calcium level in utero. The serum calcium declines 20% to 30% after birth in humans [2–4] and 40% in rodents [5, 6] before increasing to adult values over the succeeding 24 to 48 hours. A lower fetal blood calcium may predispose to an even lower trough level being reached after birth, thereby increasing the risk of tetany and death.

The high level of calcium in the fetal circulation is robustly maintained despite maternal hypocalcemia from a variety of causes. For example, serum calcium is normal in fetal mice lacking the vitamin D receptor (*Vdr* null fetuses) and also in pups born of severely vitamin D–deficient rodents [1]. Similarly, the cord blood calcium is normal in human babies with severe vitamin D deficiency (25-hydroxyvitamin D [25OHD] levels of 10nmol/L versus 138nmol/L in offspring of vitamin D–treated mothers) [7], whereas children with naturally occurring deletions in the VDR do not present with hypocalcemia or rickets until their second year of life [1].

Calciotropic hormone levels are maintained at levels that differ from the adult. These differences appear to reflect the relatively different roles that these hormones play in the fetus, and are not an artifact of altered metabolism or clearance of hormones. Intact PTH levels are much lower than maternal PTH levels near term. However, PTH is important for fetal development because fetal mice lacking parathyroids or PTH are hypocalcemic and have undermineralized skeletons [8–11]. Circulating 1,25-dihydroxyvitamin D (1,25-D) levels are also low, owing to suppression of the fetal 1α-hydroxylase by high serum calcium and phosphate, increased 24-hydroxylation of 1,25-D and 25OHD, and low PTH. 1,25-D appears to be relatively unimportant for fetal mineral homeostasis, because several vitamin D

Primer on the Metabolic Bone Diseases and Disorders of Mineral Metabolism, Ninth Edition. Edited by John P. Bilezikian.
© 2019 American Society for Bone and Mineral Research. Published 2019 by John Wiley & Sons, Inc.
Companion website: www.wiley.com/go/asbmrprimer

deficiency models, 1α-hydroxylase null pigs, and *Vdr* null mice, all have normal serum mineral concentrations and fully mineralized skeletons (reviewed in detail in [1]). Fetal calcitonin levels are higher than maternal levels, but calcitonin is not required for fetal calcium homeostasis [12].

Cord blood levels of PTHrP are up to 15-fold higher than simultaneous PTH levels at term. PTHrP is produced in many tissues and plays multiple roles during embryonic and fetal development [1]. Absence of PTHrP (in *Pthrp* null fetuses) leads to abnormal endochondral bone development, modest hypocalcemia [13], and reduced placental calcium transfer [13, 14]. These *Pthrp* null fetuses have secondary hyperparathyroidism [9] but the blood calcium is reduced to the maternal level, confirming that PTH does not make up for lack of PTHrP in maintaining a normal fetal calcium concentration. Conversely, PTHrP does not make up for absence of PTH, given that fetuses lacking either parathyroids or PTH are hypocalcemic and have no compensatory increase in PTHrP [8–10].

The role (if any) of the sex steroids in fetal skeletal development and mineral accretion is uncertain. Mice lacking estrogen receptor alpha and beta or the aromatase appear normal at birth and develop altered skeletal metabolism postnatally, but the fetal skeleton has not been examined [15–19].

FETAL PARATHYROIDS

Intact parathyroid glands are required for maintenance of normal fetal calcium, magnesium, and phosphate levels. Lack of parathyroids causes the fetal blood calcium to fall below the maternal level in mice [8, 9], whereas lack of either PTH or PTHrP causes the fetal calcium to fall to the maternal level [10]. Fetal parathyroids and PTH are also required for normal accretion of mineral by the skeleton, and may be required for regulation of placental mineral transfer. Studies in fetal lambs have indicated that the fetal parathyroids may contribute to mineral homeostasis by producing both PTH and PTHrP, whereas detailed study of rats indicated that the fetal parathyroids produce only PTH. Whether human fetal parathyroids produce PTH alone, or PTH and PTHrP together, is unclear.

CALCIUM SENSING RECEPTOR (CASR)

The parathyroid CaSR regulates the serum calcium level in adults by inhibiting PTH, but it does not appear to set the high serum calcium level of fetuses. Instead, the CaSR likely suppresses fetal PTH in response to the high fetal blood calcium [20]. On the other hand, inactivating mutations of the CaSR (*Casr* null fetuses) disrupt fetal homeostasis by inducing hyperparathyroidism with increased serum calcium, 1,25-D, and bone turnover, and this results in a lower skeletal calcium content [20].

The CaSR is also expressed within human and murine placentas [21], and may play some role in regulating placental mineral transfer. *Casr* null fetuses have a reduced rate of placental calcium transfer, but whether this is a direct consequence of the loss of placental CaSR is unknown [20].

FETAL KIDNEYS AND AMNIOTIC FLUID

Fetal kidneys partly regulate calcium homeostasis by adjusting the relative reabsorption and excretion of calcium, magnesium, and phosphate in response to the filtered load and other regulatory factors, such as PTHrP and PTH. The fetal kidneys also synthesize 1,25-D, but because *Vdr* null fetal mice and severely vitamin D deficient rodent fetuses show no alteration in serum minerals or skeletal mineral content [1], it appears likely that production of 1,25-D by the fetal kidneys is relatively unimportant.

Renal calcium handling during fetal life may have minimal importance because calcium excreted by the kidneys is not permanently lost. Fetal urine is the major source of fluid and solute in amniotic fluid, and with swallowing the excreted calcium is made available again to the fetus.

PLACENTAL MINERAL ION TRANSPORT

The bulk of placental calcium, phosphorus, and magnesium transfer occurs late in gestation at a rapid rate. Active transport is necessary for the fetal requirements to be met; only placental calcium transfer has been studied in detail. Analogous to calcium transfer across the intestinal mucosa, it has been theorized that calcium enters calcium-transporting cells through channels in maternal-facing basement membranes, is carried across these cells by calcium binding proteins, and is then actively extruded at the fetal-facing basement membranes by Ca^{2+}-ATPase.

Data from animal models indicates that a normal rate of maternal-to-fetal calcium transfer can usually be maintained despite the presence of maternal hypocalcemia or maternal hormone deficiencies, such as aparathyroidism, vitamin D deficiency, and absence of the VDR [1]. Whether the same is true for human pregnancies is less certain. A "normal" rate of maternal–fetal calcium transfer does not necessarily imply that the fetus is unaffected by maternal hypocalcemia. Instead, it is an indication of the resilience of the fetal–placental unit to be able to extract the required amount of calcium from a hypocalcemic maternal circulation.

Fetal regulation of placental calcium transfer has been studied in a number of different animal models. Thyroparathyroidectomy in fetal lambs results in a reduced rate of calcium transfer across isolated, perfused placentas, suggesting that the parathyroids regulate placental calcium transfer [22]. In contrast, mice lacking parathyroids as a consequence of ablation of the *Hoxa3* gene have a normal rate of placental calcium transfer [8]. The discrepancy between these findings in lambs and mice may be because of whether or not the parathyroids are an important source of PTHrP in the circulation, as discussed previously. Studies in fetal lambs and in *Pthrp* null fetal mice are in agreement that PTHrP, and in particular mid-molecular forms of PTHrP, stimulate placental calcium transfer [14, 23, 24]. There is also evidence that PTH is expressed within murine placenta from where it may stimulate placental transfer of calcium and other cations [10]. Conversely calcitonin and 1,25-D are not required for placental calcium transport [12, 25].

FETAL SKELETON

A complete cartilaginous skeleton with digits and intact joints is present by the eighth week of gestation in humans. Primary ossification centers form in the vertebrae and long bones between the eighth and 12th weeks, but it is not until the third trimester that the bulk of mineralization occurs. At the 34th week of gestation, secondary ossification centers form in the femurs, but otherwise most epiphyses are cartilaginous at birth, with secondary ossification centers appearing in other bones in the neonate and child [26].

The skeleton must undergo substantial growth and be sufficiently mineralized by the end of gestation to support the organism, but as in the adult, the fetal skeleton participates in the regulation of mineral homeostasis. Calcium accreted by the fetal skeleton can be subsequently resorbed to help maintain the concentration of calcium in the blood, and this resorption can become more pronounced with severe maternal hypocalcemia [1] or in *Casr* null fetal mice [20]. Functioning fetal parathyroid glands are needed for normal skeletal mineral accretion, and both hypoparathyroidism and hyperparathyroidism reduce the skeletal mineral at term.

Further comparative study of fetal mice lacking parathyroids, PTH, or PTHrP has shown the interlocking roles of PTH and PTHrP in regulating the development and mineralization of the fetal skeleton [27]. PTHrP is produced locally in proliferating chondrocytes and perichondrium from where it directs the development of the cartilaginous scaffold that is later broken down and transformed into bone [28]. PTHrP is also expressed in preosteoblasts and osteoblasts from where it acts in an autocrine and paracrine manner to stimulate osteoblast function [29]. PTH acts from the systemic circulation to stimulate osteoblasts and contribute to maintaining the fetal blood calcium and magnesium at levels that facilitate mineralization. In the absence of PTHrP, a severe chondrodysplasia results from rapid differentiation and early apoptosis of chondrocytes [13]; accelerated and abnormal calcification also occurs, resulting in an apparently normal mineral content [9, 13]. Secondary hyperparathyroidism enables the *Pthrp* null growth plates to maintain normal expression of collagen α1(I) and collagenase-3, upregulated expression of osteocalcin and osteopontin, and increased mineralization [13, 30]. However, when the PTH/PTHrP receptor (PTH1R) is ablated (*Pthr1* null fetuses), the resultant phenotype combines the *Pthrp* null chondrodysplasia with decreased osteoblast function, as evidenced by reduced expression of collagenase-3, osteocalcin, and osteopontin, and reduced mineralization [30, 31].

In the absence of parathyroids or PTH, the chondrocytic aspects of endochondral bone formation proceed normally but the bone compartment is significantly undermineralized at term [9–11]. Whether defects in osteoblast function are present when PTH is absent is unclear, because the expression of osteoblast-specific genes is normal in aparathyroid *Hoxa3* null fetuses but inconsistently reduced in *Pth* null fetuses [9–11]. Blood calcium and magnesium are significantly reduced in fetuses lacking parathyroids or PTH, and so lack of PTH may impair mineralization simply by reducing the amount of mineral presented to the skeletal surface and to osteoblasts.

FETAL RESPONSE TO MATERNAL HYPERPARATHYROIDISM

In humans, maternal primary hyperparathyroidism has been associated with adverse fetal outcomes, including spontaneous abortion and stillbirth, which are thought to result from suppression of the fetal parathyroid glands [32]. Since PTH cannot cross the placenta [33, 34], fetal parathyroid suppression may result from increased calcium flux across the placenta to the fetus, facilitated by maternal hypercalcemia. The suppressed parathyroid function may persist for months after birth and has even been permanent [35, 36]. Similar suppression of fetal parathyroids occurs when the mother has hypercalcemia from familial hypocalciuric hypercalcemia [37, 38]. Chronic elevation of the maternal serum calcium in mice results in suppression of the fetal PTH level [20], but fetal survival is not notably affected by this.

FETAL RESPONSE TO MATERNAL HYPOPARATHYROIDISM

Maternal hypoparathyroidism during human pregnancy can cause fetal hyperparathyroidism. This is characterized by fetal parathyroid gland hyperplasia, generalized skeletal demineralization, subperiosteal bone resorption, bowing of the long bones, osteitis fibrosa cystica, rib and limb fractures, low birth weight, spontaneous abortion, stillbirth, and neonatal death [1]. Similar skeletal findings have been reported in the fetuses and neonates of women with pseudohypoparathyroidism, renal tubular acidosis, and chronic renal failure [1]. These changes in human skeletons differ from what has been found in animal models of maternal hypoparathyroidism, in which the fetal skeleton and the blood calcium are generally normal.

FETAL RESPONSE TO MATERNAL VITAMIN D DEFICIENCY

25OHD readily crosses the placenta, such that its cord blood concentrations are usually within 80% to 100% of the maternal value at term [1].

As mentioned earlier, fetal calcium metabolism and skeletal mineral content are normal in animal models of severe vitamin D deficiency, and in *Vdr* null mice [1]. The available but limited human data indicate that calcium homeostasis and skeletal mineral content in human fetuses may also be unaffected by severe vitamin D deficiency and absence of VDR or calcitriol.

These data include the finding of normal ash weight, skeletal mineral content (by atomic absorption spectroscopy), and absence of radiological signs of rickets, in severely vitamin D deficient infants who died of obstetrical accidents [39]. Observational data of infants with severe vitamin D deficiency or genetic absence of 1α-hydroxylase or VDR show that hypocalcemia and rickets do not normally develop or become recognized until at least several months after birth, with the peak incidence occurring in the second year [1]. Studies of bone mineral density in newborns have not shown any association with vitamin D sufficiency [1]. Recent randomized trials have examined the use of up to 4000IU of vitamin D per day during pregnancy and found no effect of supplementation on cord blood calcium or skeletal parameters in the newborns [40–42].

The available animal and human data predict that human fetuses will have a normal skeleton and serum calcium despite severe vitamin D deficiency. After birth, vitamin D deficient neonates and infants are at risk for hypocalcemia followed by the development of rickets [1]. Maternal vitamin D deficiency may have little or no effect on fetal calcium homeostasis because secondary hyperparathyroidism in the mother minimizes the hypocalcemia, and because placental transport of calcium does not require calcitriol or VDR.

However, recent associational studies have questioned whether the fetal skeleton is truly normal with maternal vitamin D insufficiency. These studies examined associations between single measurements of maternal serum 25OHD during pregnancy and various skeletal outcomes in the fetus, neonate, or child. In none of these was an association found with birth weight, cord blood calcium, skeletal lengths, and bone mineral density [43–46]. In one study a 25OHD level below 28nmol/L was reportedly associated with a slightly shorter knee-heel length, but the difference was not statistically significant after correcting for gestational age [43]. A second study found that maternal 25OHD levels *below* 50nmol/L were associated with greater distal metaphyseal cross-sectional area in the femur and concluded that this was evidence of early rickets [44]. But a third study found that maternal 25OHD levels *above* 42.6nmol/L were associated with greater metaphyseal cross-sectional area in the tibia, and concluded that this meant stronger bone [45]. The latter pair of studies exemplify how subjective the interpretations can be, with greater metaphyseal cross-sectional area considered an adverse effect in one study and a benefit in another.

A well-publicized study by Javaid and colleagues found no associations between maternal serum 25OHD and serum calcium or anthropometric parameters of the neonates at birth or 9 months of age [46]. However, a maternal serum 25OHD level below 27.5nmol/L during pregnancy was associated with a modestly lower bone mineral content in offspring at *9 years of age* compared with offspring of mothers whose 25OHD levels had been >50nmol/L. These findings support the theory that vitamin D exposure during fetal development programs childhood peak bone mass [47]. However, a study that was 20 times larger found no significant associations between maternal 25OHD during pregnancy and offspring bone mass at age 9 years [48].

These associational studies are confounded by factors which predispose to low maternal 25OHD levels, including obesity, lower socioeconomic status, poorer nutrition, lack of exercise, and prenatal care or vitamin supplementation, etc. Therefore, is lower 25OHD simply a marker of a less healthy pregnant woman? In the study by Javaid and colleagues, much time elapsed between birth (when no effect was observed) and 9 years of age. Does low 25OHD in utero really program lower bone mineral content at 9 years of age? Or does lower maternal 25OHD in late pregnancy serve as a surrogate signal for lower socioeconomic status, poorer nutrition, and other factors in the mother that will remain unchanged and continue to be shared with the child?

Ultimately, associational studies are hypothesis-generating and do not prove causality, and so clinical trials of vitamin D supplementation with bone mineral content at birth as the endpoint are needed to determine if vitamin D supplementation during pregnancy confers any skeletal benefit on the fetus, neonate, infant, or child.

The recent MAVIDOS study did report, in a subgroup analysis, a possible benefit of vitamin D supplementation during pregnancy on the bone mineral content of winter-born babies only [49]. However, DXA was performed up to 14 days after birth, by which time the neonatal skeleton should have accreted 1400mg of calcium [1]. Rather than confirming a fetal role for calcitriol, the result may simply reflect the postnatal role of calcitriol to stimulate intestinal calcium absorption and thereby mineral accretion.

INTEGRATED FETAL CALCIUM HOMEOSTASIS

The evidence discussed in the preceding sections suggests the following summary models.

Calcium sources

The main flux of calcium and other minerals is across the placenta and into fetal bone, but calcium is also made available to the fetal circulation through several routes (Fig. 24.1). The kidneys reabsorb calcium; calcium excreted by the kidneys into the urine and amniotic fluid may be swallowed and reabsorbed; calcium is also resorbed from the developing skeleton. Some calcium returns to the maternal circulation (backflux). The maternal skeleton is a potential source of mineral, and it may be compromised in mineral deficiency states in order to provide to the fetus.

Blood calcium regulation

Fetal blood calcium is set at a level higher than the maternal value through the actions of PTHrP and PTH acting in concert. The CaSR suppresses PTH in response to the high calcium level, but the low level of PTH is required for maintaining the blood calcium and facilitating mineral accretion by the skeleton. 1,25-D concentrations remain low, suppressed by low PTH, high blood calcium and phosphate, and increased 24-hydroxylation. The parathyroids may play a central role by producing PTH and PTHrP, or may produce PTH alone. PTHrP and PTH are produced by the placenta, whereas PTHrP is also produced by many other fetal tissues.

PTH and PTHrP, both present in the fetal circulation, independently and additively regulate the fetal blood calcium. Neither hormone can make up for absence of the other: if one is missing the blood calcium is reduced, and if both are missing the blood calcium is reduced even further. PTH upregulates in the absence of PTHrP but PTHrP does not upregulate when parathyroids or PTH are absent.

PLACENTAL CALCIUM TRANSFER

Placental calcium transfer is regulated by both PTHrP and PTH, and the placenta (and possibly the parathyroids) is likely an important source of both hormones. The CaSR may also be involved in some aspect of calcium sensing within the calcium-transporting cells of the placenta.

SKELETAL MINERALIZATION

PTH and PTHrP have separate roles with respect to skeletal development and mineralization (Fig. 24.2). PTH normally acts systemically to direct the mineralization of bone matrix by maintaining blood calcium at the adult level, and by direct actions on osteoblasts within the bone matrix. In contrast, PTHrP acts locally to direct endochondral bone development and osteoblast function, and acts from outside of bone to affect skeletal development and mineralization by contributing to the regulation of the blood calcium and placental calcium transfer. PTH may have the more dominant effect to maintain skeletal mineral accretion because the absence of PTH leads to an undermineralized skeleton whereas absence of PTHrP does not.

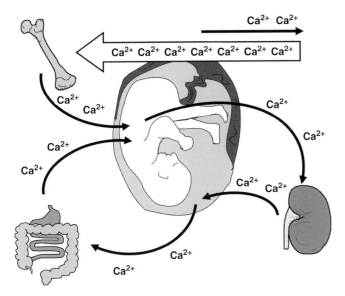

Fig. 24.1. Circulation of mineral within the fetal–placental unit. Calcium is represented here but these statements apply to phosphorus (phosphate) and magnesium as well. At the top right, the main flux of mineral is across the placenta and through the fetal circulation into bone; however, some mineral returns to the maternal circulation (backflux). On the bottom right, the fetal kidneys filter the blood and excrete mineral into urine, which in turn makes up much of the volume of amniotic fluid. On the bottom left, amniotic fluid is swallowed and its mineral content can be absorbed by the fetal intestines, thereby restoring it to the circulation. The renal–amniotic–intestinal loop is likely a minor component for fetal mineral homeostasis. On the top left, although the net flux of mineral is into bone, some mineral is resorbed from the developing skeleton to re-enter the fetal circulation. If placental delivery of mineral is deficient, fetal secondary hyperparathyroidism ensues, which causes more substantial resorption of mineral from the fetal skeleton, reduced skeletal mineral content, and possible fractures occurring in utero or during the birthing process. Source: [50]. Reproduced with permission of Elsevier.

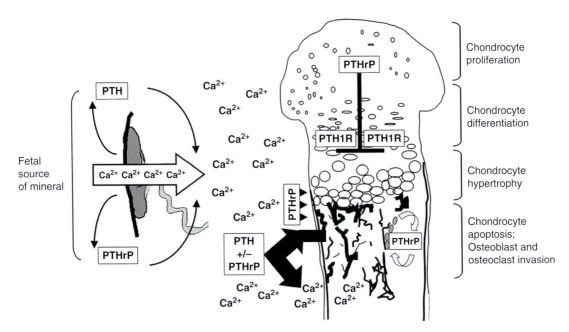

Fig. 24.2. Relative roles of PTH, PTHrP, and 1,25-D during fetal life. The placenta is the main source of mineral. PTH and PTHrP are expressed within the placenta but may also act on it from systemic sources. PTHrP stimulates calcium and possibly magnesium transfer; PTH also regulates calcium transfer and the expression of cation transporters. The regulators of placental phosphorus transfer are unknown. Within the endochondral skeleton, PTHrP is produced by proliferating chondrocytes and perichondrial cells (arrowheads) from which sites it acts on prehypertrophic chondrocytes (where the PTH1R is expressed) in order to delay their differentiation into hypertrophic chondrocytes. Hypertrophic chondrocytes undergo apoptosis, vascular invasion occurs, chondroclasts and osteoclasts resorb the cartilaginous matrix, and osteoblasts lay down primary spongiosa. PTHrP is also produced within preosteoblasts and osteoblasts from where it acts in a paracrine and autocrine fashion to stimulate bone formation (semicircular arrows). During fetal life PTHrP and PTH both regulate the fetal blood calcium, magnesium, and phosphorus, the concentrations of which are maintained above ambient maternal levels in order to facilitate mineralization. The regulation of blood calcium, placental calcium transfer, endochondral bone formation, and skeletal mineralization do not require 1,25-D or VDR during fetal life. Source: [50]. Reproduced with permission of Elsevier.

REFERENCES

1. Kovacs CS. Bone development and mineral homeostasis in the fetus and neonate: roles of the calciotropic and phosphotropic hormones. Physiol Rev. 2014;94(4):1143–218.
2. Loughead JL, Mimouni F, Tsang RC. Serum ionized calcium concentrations in normal neonates. Am J Dis Child. 1988;142:516–8.
3. David L, Anast CS. Calcium metabolism in newborn infants. The interrelationship of parathyroid function and calcium, magnesium, and phosphorus metabolism in normal, sick, and hypocalcemic newborns. J Clin Invest. 1974;54:287–96.
4. Schauberger CW, Pitkin RM. Maternal-perinatal calcium relationships. Obstet Gynecol. 1979;53:74–6.
5. Garel JM, Barlet JP. Calcium metabolism in newborn animals: the interrelationship of calcium, magnesium, and inorganic phosphorus in newborn rats, foals, lambs, and calves. Pediatr Res. 1976;10:749–54.
6. Krukowski M, Smith JJ. pH and the level of calcium in the blood of fetal and neonatal albino rats. Biol Neonate. 1976;29:148–61.
7. Brooke OG, Brown IR, Bone CD, et al. Vitamin D supplements in pregnant Asian women: effects on calcium status and fetal growth. Br Med J. 1980;280:751–4.
8. Kovacs CS, Manley NR, Moseley JM, et al. Fetal parathyroids are not required to maintain placental calcium transport. J Clin Invest. 2001;107(8):1007–15.
9. Kovacs CS, Chafe LL, Fudge NJ, et al. PTH regulates fetal blood calcium and skeletal mineralization independently of PTHrP. Endocrinology. 2001;142(11):4983–93.
10. Simmonds CS, Karsenty G, Karaplis AC, et al. Parathyroid hormone regulates fetal-placental mineral homeostasis. J Bone Miner Res. 2010;25(3):594–605.
11. Miao D, He B, Karaplis AC, et al. Parathyroid hormone is essential for normal fetal bone formation. J Clin Invest. 2002;109(9):1173–82.
12. McDonald KR, Fudge NJ, Woodrow JP, et al. Ablation of calcitonin/calcitonin gene related peptide-α impairs fetal magnesium but not calcium homeostasis. Am J Physiol Endocrinol Metab. 2004;287(2):E218–E26.
13. Karaplis AC, Luz A, Glowacki J, et al. Lethal skeletal dysplasia from targeted disruption of the parathyroid hormone-related peptide gene. Genes Dev. 1994;8:277–89.
14. Kovacs CS, Lanske B, Hunzelman JL, et al. Parathyroid hormone-related peptide (PTHrP) regulates fetal-placental

calcium transport through a receptor distinct from the PTH/PTHrP receptor. Proc Natl Acad Sci U S A. 1996;93:15233–8.
15. Mueller SO, Korach KS. Estrogen receptors and endocrine diseases: lessons from estrogen receptor knockout mice. Curr Opin Pharmacol. 2001;1(6):613–9.
16. Vidal O, Lindberg MK, Hollberg K, et al. Estrogen receptor specificity in the regulation of skeletal growth and maturation in male mice. Proc Natl Acad Sci U S A. 2000;97(10):5474–9.
17. Windahl SH, Andersson G, Gustafsson JA. Elucidation of estrogen receptor function in bone with the use of mouse models. Trends Endocrinol Metab. 2002;13(5):195–200.
18. Lubahn DB, Moyer JS, Golding TS, et al. Alteration of reproductive function but not prenatal sexual development after insertional disruption of the mouse estrogen receptor gene. Proc Natl Acad Sci U S A. 1993;90(23): 11162–6.
19. Couse JF, Curtis SW, Washburn TF, et al. Analysis of transcription and estrogen insensitivity in the female mouse after targeted disruption of the estrogen receptor gene. Mol Endocrinol. 1995;9(11):1441–54.
20. Kovacs CS, Ho-Pao CL, Hunzelman JL, et al. Regulation of murine fetal-placental calcium metabolism by the calcium-sensing receptor. J Clin Invest. 1998;101:2812–20.
21. Kovacs CS, Chafe LL, Woodland ML, et al. Calcitropic gene expression suggests a role for intraplacental yolk sac in maternal-fetal calcium exchange. Am J Physiol Endocrinol Metab. 2002;282(3):E721–E32.
22. Care AD, Caple IW, Abbas SK, et al. The effect of fetal thyroparathyroidectomy on the transport of calcium across the ovine placenta to the fetus. Placenta. 1986;7:417–24.
23. Care AD, Abbas SK, Pickard DW, et al. Stimulation of ovine placental transport of calcium and magnesium by mid-molecule fragments of human parathyroid hormone-related protein. Exp Physiol. 1990;75:605–8.
24. Rodda CP, Kubota M, Heath JA, et al. Evidence for a novel parathyroid hormone-related protein in fetal lamb parathyroid glands and sheep placenta: comparisons with a similar protein implicated in humoral hypercalcaemia of malignancy. J Endocrinol. 1988;117:261–71.
25. Kovacs CS, Woodland ML, Fudge NJ, et al. The vitamin D receptor is not required for fetal mineral homeostasis or for the regulation of placental calcium transfer. Am J Physiol Endocrinol Metab. 2005;289(1):E133–E44.
26. Moore KL, Persaud TVN, Torchia MG. *The Developing Human: Clinically Oriented Embryology* (9th ed.). Philadelphia, PA: Saunders/Elsevier; 2013.
27. Simmonds CS, Kovacs CS. Role of parathyroid hormone (PTH) and PTH-related protein (PTHrP) in regulating mineral homeostasis during fetal development. Crit Rev Eukaryot Gene Expr. 2010;20(3):235–73.
28. Karsenty G. Chondrogenesis just ain't what it used to be. J Clin Invest. 2001;107(4):405–7.
29. Miao D, He B, Jiang Y, et al. Osteoblast-derived PTHrP is a potent endogenous bone anabolic agent that modifies the therapeutic efficacy of administered PTH 1-34. J Clin Invest. 2005;115(9):2402–11.
30. Lanske B, Divieti P, Kovacs CS, et al. The parathyroid hormone/parathyroid hormone-related peptide receptor mediates actions of both ligands in murine bone. Endocrinology. 1998;139:5192–204.
31. Lanske B, Karaplis AC, Lee K, et al. PTH/PTHrP receptor in early development and Indian hedgehog-regulated bone growth. Science. 1996;273:663–6.
32. Schnatz PF, Curry SL. Primary hyperparathyroidism in pregnancy: evidence-based management. Obstet Gynecol Surv. 2002;57(6):365–76.
33. Northrop G, Misenhimer HR, Becker FO. Failure of parathyroid hormone to cross the nonhuman primate placenta. Am J Obstet Gynecol. 1977;129:449–53.
34. Garel JM, Dumont C. Distribution and inactivation of labeled parathyroid hormone in rat fetus. Horm Metab Res. 1972;4:217–21.
35. Bruce J, Strong JA. Maternal hyperparathyroidism and parathyroid deficiency in the child, with account of effect of parathyroidectomy on renal function, and of attempt to transplant part of tumor. Q J Med. 1955;24:307–19.
36. Better OS, Levi J, Grief E, et al. Prolonged neonatal parathyroid suppression. A sequel to asymptomatic maternal hyperparathyroidism. Arch Surg. 1973;106:722–4.
37. Powell BR, Buist NR. Late presenting, prolonged hypocalcemia in an infant of a woman with hypocalciuric hypercalcemia. Clin Pediatr (Phila). 1990;29:241–3.
38. Thomas BR, Bennett JD. Symptomatic hypocalcemia and hypoparathyroidism in two infants of mothers with hyperparathyroidism and familial benign hypercalcemia. J Perinatol. 1995;15:23–6.
39. Maxwell JP, Miles LM. Osteomalacia in China. J Obstet Gynaecol Br Empire. 1925;32(3):433–73.
40. Hollis BW, Johnson D, Hulsey TC, et al. Vitamin D supplementation during pregnancy: double-blind, randomized clinical trial of safety and effectiveness. J Bone Miner Res. 2011;26(10):2341–57.
41. Wagner CL, McNeil R, Hamilton SA, et al. A randomized trial of vitamin D supplementation in 2 community health center networks in South Carolina. Am J Obstet Gynecol. 2013;208(2):137.e1–13.
42. Grant CC, Stewart AW, Scragg R, et al. Vitamin D during pregnancy and infancy and infant serum 25-hydroxyvitamin D concentration. Pediatrics. 2014; 133(1):e143–53.
43. Morley R, Carlin JB, Pasco JA, et al. Maternal 25-hydroxyvitamin D and parathyroid hormone concentrations and offspring birth size. J Clin Endocrinol Metab. 2006;91(3):906–12.
44. Mahon P, Harvey N, Crozier S, et al. Low maternal vitamin D status and fetal bone development: cohort study. J Bone Miner Res. 2010;25(1):14–9.
45. Viljakainen HT, Saarnio E, Hytinantti T, et al. Maternal vitamin D status determines bone variables in the newborn. J Clin Endocrinol Metab. 2010;95(4):1749–57.
46. Javaid MK, Crozier SR, Harvey NC, et al. Maternal vitamin D status during pregnancy and childhood bone mass at age 9 years: a longitudinal study. Lancet. 2006;367(9504):36–43.

47. Cooper C, Westlake S, Harvey N, et al. Review: developmental origins of osteoporotic fracture. Osteoporos Int. 2006;17(3):337–47.
48. Lawlor DA, Wills AK, Fraser A, et al. Association of maternal vitamin D status during pregnancy with bone-mineral content in offspring: a prospective cohort study. Lancet. 2013;381(9884):2176–83.
49. Cooper C, Harvey NC, Bishop NJ, et al. Maternal gestational vitamin D supplementation and offspring bone health (MAVIDOS): a multicentre, double-blind, randomised placebo-controlled trial. Lancet Diabetes Endocrinol. 2016;4(5):393–402.
50. Kovacs CS. Fetal control of calcium and phosphate homeostasis. In: Thakker RV, Whyte MP, Eisman JA, et al. (eds) *Genetics of Bone Biology and Skeletal Disease* (2nd ed.). San Diego: Academic Press/Elsevier; 2017, pp. 324–48.

25
FGF23 and the Regulation of Phosphorus Metabolism

Kenneth E. White[1] and Michael J. Econs[1,2]

[1]*Division of Endocrinology and Metabolism, Indiana University School of Medicine, Indianapolis, IN, USA*
[2]*Department of Medicine, Indiana University School of Medicine, Indianapolis, IN, USA*

INTRODUCTION

Phosphorus is required to maintain skeletal integrity, as well as for key intracellular processes including nucleic acid synthesis, ATP production, and as substrates for kinase and phosphatase activity. The systemic regulation of phosphorus is maintained through endocrine feedback loops involving the intestines, kidney, and bone. Disorders of phosphate (Pi) homeostasis, in concert with powerful in vitro and in vivo studies have shown that FGF23 is central to the control of renal Pi and vitamin D homeostasis. Although the molecular mechanisms are unique to each disorder, elevated FGF23 is associated with syndromes manifested by hypophosphatemia with paradoxically low or normal 1,25(OH)$_2$ vitamin D (1,25(OH)$_2$D), and include: autosomal dominant hypophosphatemic rickets (ADHR), X-linked hypophosphatemic rickets (XLH), tumor-induced osteomalacia (TIO), and autosomal recessive hypophosphatemic rickets (ARHR1–3). Heritable disorders of hyperphosphatemia and elevated 1,25(OH)$_2$D, such as tumoral calcinosis (TC), are associated with reduced FGF23 activity. These collective findings have provided unique insight into the activity of FGF23 on renal Pi and vitamin D metabolism.

PHOSPHATE METABOLISM

Phosphate is distributed in the soft tissues of the body, both in inorganic form and as a component of organic molecules, including nucleic acids, membrane phospholipids, and other phosphoproteins. These nonmineralized phosphates comprise the minor portion of total body content, with the remaining 80% stored as hydroxyapatite in the bone matrix. This is the primary reserve, which is mobilized to regulate blood phosphate levels. The small intestine absorbs phosphate through passive mechanisms and via the type II sodium–phosphate cotransporter NPT2b (SLC34A2), expressed in the microvilli [1]. NPT2b is up-regulated by 1,25-D [2]. The absorbed phosphate is then either excreted or resorbed by the kidney and/or deposited in bone. The kidney is the major organ regulating more acute blood phosphate concentrations. Seventy percent of filtered phosphate is reabsorbed within the proximal tubule where the NPT2a (encoded by *SLC34A1*) and NPT2c (encoded by *SLC34A3*) transporters are localized in the apical membrane [3]. Transport of phosphate from the tubule lumen to across the renal proximal tubular cell is largely unidirectional. As shown through mouse genetics, Npt2a is responsible for 70% of phosphate transport and Npt2c compromising the balance [4]. In humans, NPT2c may play a larger role in phosphate handling because loss of function mutations in *SLC34A3* cause hereditary hypophosphatemic rickets with hypercalciuria [5, 6].

PTH is a central regulator of calcium balance, with hypocalcemia stimulating its secretion. PTH increases the expression of the proximal tubule 25(OH) vitamin D 1-αhydroxylase (encoded by *CYP27B1*) [7], the enzyme that produces the active form of vitamin D (1,25(OH)$_2$D) and increases calcium reabsorption in the renal distal convoluted tubule (DCT). In addition to its effects on calcium, PTH has been very well characterized as a key hormonal regulator of serum phosphate. After PTH

Primer on the Metabolic Bone Diseases and Disorders of Mineral Metabolism, Ninth Edition. Edited by John P. Bilezikian.
© 2019 American Society for Bone and Mineral Research. Published 2019 by John Wiley & Sons, Inc.
Companion website: www.wiley.com/go/asbmrprimer

delivery in vitro or in vivo, NPT2a protein expression in the proximal tubule is reduced [8]. This effect results from relatively rapid internalization of NPT2a and Npt2c and their lysosomal degradation [9]. These effects are mirrored in patients with defects in PTH, because patients with hyperparathyroidism develop renal phosphate wasting, and those with hypoparathyroidism have increased renal phosphate reabsorption. To regulate NPT2a expression, PTH signals through the type 1 PTH receptor (PTHR1) via PKA and PKC, as well as through the MAPK pathway [10]. $1,25(OH)_2D$ also contributes to control of serum phosphate by increasing intestinal calcium and phosphate absorption, and at high concentrations, increasing bone phosphate mobilization through increasing osteoclast activity. The endocrine effects of PTH and vitamin D on the kidney and the intestine, respectively, maintain phosphate balance while preserving calcium homeostasis. Evidence supports that PTH and vitamin D production could be influenced by FGF23, adding further levels of regulation to this system (see later).

THE *FGF23* GENE AND PROTEIN

The *FGF23* gene resides on human chromosome 12p13 (mouse chromosome 6), is comprised of three coding exons, and contains an open reading frame of 251 residues [11]. The tissue with the highest FGF23 expression is bone. FGF23 mRNA is observed in osteoblasts, osteocytes, flattened bone-lining cells, and osteoprogenitor cells [12]. Quantitative PCR showed that FGF23 mRNA was most highly expressed in long bone, followed by thymus, brain, and heart. Immunoblot analyses revealed that wild type FGF23 is secreted as a full length 32 kDa species, as well as cleavage products of 12 and 20 kDa [13, 14]. Cleavage of FGF23 occurs within a subtilisin-like proprotein convertase (SPC) proteolytic site ($_{176}RXXR_{179}/S_{180}AE$), that separates the conserved FGF-like N-terminal domain from the variable C-terminal tail.

FGF23 ACTIVITY

FGF23 has overlapping function with PTH to reduce renal Pi reabsorption, but has opposite effects on $1,25(OH)_2D$. FGF23 delivery leads to renal Pi wasting through the downregulation of both Npt2a and Npt2c [15]. Normally, hypophosphatemia is a strong positive stimulator for increasing serum $1,25(OH)_2D$. However, patients with TIO, ADHR, XLH, and ARHR manifest hypophosphatemia with paradoxically low or inappropriately normal $1,25(OH)_2D$. In mice, the expression of the vitamin D 1α(OH)ase enzyme and the catabolic 24(OH) ase (encoded by *CYP24A1*) are reduced, and elevated, respectively, when the animals are exposed to FGF23 [13]. Thus, the effects of FGF23 on the renal vitamin D metabolic enzymes is responsible for the reductions in $1,25(OH)_2D$ in the setting of persistent hypophosphatemia in ADHR, XLH, TIO, and ARHR patients.

REGULATION OF FGF23 IN VIVO

In humans, dietary Pi supplementation increased FGF23 whereas Pi restriction and the addition of Pi binders suppressed serum FGF23 [16], supporting that FGF23 plays a role in maintenance of Pi homeostasis. In animal studies, the FGF23 response to serum Pi has been more dramatic than in the human studies. Mice given high and low Pi diets produce the expected correlations between FGF23 and dietary Pi intake [17].

Vitamin D has important regulatory effects on FGF23. In mice, injections of 20–200 ng $1,25(OH)_2D$ led to dose-dependent increases in serum FGF23 [18]. These changes in FGF23 occurred before changes in serum Pi, indicating that FGF23 is directly regulated by vitamin D. Physiologically, this would be consistent with results examining the role of FGF23 in vitamin D metabolism. FGF23 has been shown to downregulate the 1α(OH)ase mRNA [15, 18], thus because $1,25(OH)_2D$ rises in the blood as a product of 1α(OH)ase activity, vitamin D would then stimulate FGF23, which would complete the feedback loop and downregulate 1α(OH)ase expression.

FGF23 RECEPTORS

FGF23 is a member of a unique class of FGFs including FGF19 and FGF21 that are endocrine, as opposed to paracrine/autocrine factors. FGF23 requires the coreceptor αKlotho (αKL) for bioactivity. αKL-null mice have severe calcifications as well as markedly elevated serum Pi [19], which parallels *Fgf23*-null mice [20, 21], and that of TC patients. However, both the αKL-null and *Fgf23*-null mice have more extreme phenotypes than that observed in patients. Importantly, these defects in the αKL-null and *Fgf23*-null mice can be ameliorated with a low Pi diet to reduce serum Pi [22]. In parallel with *Fgf23*-null mice, αKL-null mice have increased Npt2a in the proximal tubule, indicating that the hyperphosphatemia is secondary to increased renal reabsorption of Pi.

αKL is produced as several isoforms. Membrane bound KL (mKL) is a 130 kDa single-pass transmembrane protein characterized by a large extracellular domain and a very short (10 residue) intracellular domain that does not possess signaling capabilities [23]. The mKL protein can also be cleaved extracellularly near the transmembrane domain, giving rise to a circulating form of αKL (110 kDa) [24].

The most likely mechanism for FGF23 signaling through αKL is the recruitment of canonical FGF receptors (FGFRs) to form heteromeric complexes. One group has identified a specific complex between FGFR1c and αKL [25] and another found that FGFR3c and FGFR4 were also involved. Signaling appears to be through MAPK cascades [26]. FGFR1 may be more important for phosphate homeostasis, whereas FGFR3c and FGFR4 may be more relevant to vitamin D status [27]. Within the kidney, αKL localizes to the distal tubule. However, FGF23 mediates its effects on NPT2a, NPT2c, and vitamin D within the proximal tubule [13, 15]. Acute delivery of FGF23 results in p-ERK1/2 signaling in the renal DCT [28], therefore the mechanisms underlying a local DCT–PT axis in the kidney after FGF23 delivery are unclear.

SERUM ASSAYS

FGF23 is measured in the circulation via several assays. One extensively used assay is a "C-terminal" FGF23 ELISA with both the capture and detection antibodies binding C-terminal to the FGF23 $_{176}$RXXR$_{179}$/S cleavage site [29]. This assay thus recognizes full-length FGF23 as well as C-terminal proteolytic fragments ("cFGF23"). In a study with a large number of controls and TIO patients, this ELISA was used to test the levels of FGF23 in TIO and XLH [29], and showed that serum FGF23 is detectable in normal individuals. The mean FGF23 was greater than 10-fold elevated in TIO patients, and rapidly fell after tumor resection. Importantly, many XLH patients (13 out of 21) had elevated FGF23 compared with controls [29], and in those with "normal" FGF23, these levels may be "inappropriately normal" in the setting of hypophosphatemia. Mouse and human "intact" FGF23 ("iFGF23") ELISA assays have been developed that span the $_{176}$RXXR$_{179}$/S$_{180}$ SPC site and thus recognize N- and C-terminal portions of FGF23 [30]. The assays detect a mean iFGF23 circulating concentration of 29 pg/mL in normal individuals [30]. The results of the cFGF23 and iFGF23 assays generally agree with regard to the relative ranges of FGF23 concentrations in XLH and in TIO patients, and that FGF23 is elevated in most XLH patients.

FGF23-ASSOCIATED SYNDROMES

Disorders associated with increased FGF23 bioactivity

ADHR (OMIM #193100)

Importantly, ADHR is distinguished from other hereditary hypophosphatemias by having either early or delayed onset with variable expressivity [31]. The ADHR mutations replace arginine (R) residues at positions 176 or 179 with glutamine (Q) or tryptophan (W) within the FGF23 subtilisin-like proprotein convertase (SPC) cleavage site, $_{176}$RXXR$_{179}$/S$_{180}$ [11, 13, 32] (Table 25.1). Following insertion of the ADHR mutations into wild type FGF23, FGF23 secreted from mammalian cells was primarily full length (32 kDa), active polypeptide, as opposed to the 32 kDa and cleavage products typically observed for wild type FGF23 expression [14].

TIO

TIO is an acquired disorder of isolated renal Pi wasting that is associated with tumors. TIO patients present with similar biochemistries because patients with ADHR [33] and osteomalacia can be found upon bone biopsy. Clinical symptoms include muscle weakness, fatigue, and bone pain [33]. Insufficiency fractures are common and proximal muscle weakness can become severe [33]. FGF23 is elevated in patients with TIO [29, 30] and tumors that cause TIO have a dramatic overexpression of FGF23 mRNA [32]. Surgical resection of the tumor results in rapid decreases in serum FGF23 [29].

XLH (OMIM #307800)

X-linked hypophosphatemic rickets is caused by inactivating mutations in *PHEX* (phosphate-regulating gene with homologies to endopeptidases on the X chromosome) [34]. PHEX is a member of the M13 family of membrane-bound metalloproteases, and shows the highest expression in bone cells such as osteoblasts, osteocytes, and odontoblasts in teeth [35].

Reports have established that FGF23 is elevated in many XLH patients [29, 30]. Although it was initially thought that PHEX might cleave FGF23, this is not the case [3]. Instead, FGF23 mRNA expression is markedly increased in *Hyp* mice (mouse model of XLH) bone [17, 36]. The elevated FGF23 mRNA levels indicate that the increase in serum FGF23 in XLH is caused by overproduction by skeletal cells, as opposed to a decreased rate of FGF23 degradation by cell surface proteases after secretion into the circulation. At present, the PHEX substrate is unknown.

ARHR types 1–3

ARHR1 (OMIM #241520)

Dentin Matrix Protein-1 (*DMP1*), a member of the Small Integrin-Binding LIgand, N-linked Glycoprotein (SIBLING) family, is highly expressed in osteocytes. Both *Dmp1*-null mice and patients with ARHR1 manifest rickets and osteomalacia with isolated renal Pi wasting associated with elevated FGF23. Mutational analyses revealed that an ARHR1 family carried a mutation that ablated the *DMP1* start codon, and a second family exhibited a deletion in the DMP1 C-terminus [37]. Mutations have also been identified in *DMP1* splicing sites, which likely result in nonfunctional protein [38]. Mechanistic studies using the *Dmp1*-null mouse show that loss of DMP1 causes defective osteocyte maturation, associated with elevated FGF23 expression and pathological changes in bone mineralization [37]. Importantly, *Dmp1*-null mice are biochemical phenocopies of the *Hyp* mouse, and patients with ARHR1 and XLH (as well as the *Dmp1*-null and *Hyp* mice) share a unique bone histology characterized by distinctive periosteocytic lesions [37]. Thus, these findings suggest that PHEX may also have a role in osteocyte maturation in a parallel pathway to DMP1 that leads to overexpression of FGF23.

ARHR2 (OMIM #613312)

ARHR2 is caused by mutations in the ectonucleotide pyrophosphatase/phosphodiesterase-1 (*ENPP1*) gene that controls physiological mineralization and pathological chondrocalcinosis by generating inorganic pyrophosphate. Studies support that ENPP1 may regulate osteoblastic differentiation in an extracellular phosphate-independent manner [39]. Therefore, similar to DMP1, loss of function *ENPP1* mutations could potentially result in an early osteocyte differentiation defect and overexpression of FGF23.

ARHR3 (OMIM #259775)

Loss of function mutations in the kinase family with sequence similarity 20, member C (FAM20C), give rise to ARHR3, also known as Raine syndrome. When the mutations arise in the conserved C-terminal domain of FAM20C Raine syndrome is generally lethal. However, some mutations have been described in surviving Raine's patients [40–42]. Patients may present with craniofacial malformation and osteosclerosis of the skull and long bones. In one case a R408W FAM20C substitution caused a hypophosphatemic rickets phenotype because of high serum iFGF23 and renal phosphate wasting [43]. Another study revealed both sclerosing and hypophosphatemic rickets phenotypes in a Raine's case [44]. FAM20C is a casein kinase and phosphorylates-secreted proteins containing an S-X-E motif. FAM20C has higher activity in the presence of the related gene FAM20A [45]. *Fam20c* knockout mice display a significant rise in FGF23 and renal phosphate wasting [46]. In vitro analyses found the SIBLING family members DMP1 and OPN are substrates for FAM20c phosphorylation. The increased iFGF23 in some Raine syndrome patients indicated that direct interactions between FAM20C and FGF23 may occur. The Fam20C S-X-E phosphorylation motif is found at the FGF23 SPC cleavage site R_{176}-H_{177}-T_{178}-R_{179}/S_{180}-A_{181}-E_{182}. In vitro studies showed that the FAM20C-mediated phosphorylation of FGF23 S_{180} blocked O-glycosylation by GALNT3 at residue T_{178}, which promoted furin proteolysis of FGF23 [47]. Exposure of FGF23 to a FAM20C carrying a Raine mutation showed reduced FGF23 phosphorylation and some stabilization of FGF23, explaining the increased iFGF23 and low serum phosphate in some ARHR3 patients [47]. Studies are needed to test additional FAM20C mutants to determine why the development of severe osteosclerosis in most ARHR3 patients occurs compared with the rachitic bone phenotype observed Fam20c-null mice.

Other heritable disorders involving elevated FGF23

In addition to the disorders described above, FGF23 is also upregulated in several bone dysplasias that manifest documented isolated renal Pi wasting. These disorders include: McCune Albright syndrome (OMIM #174800) caused by somatic activating mutations in G_s; opsismodysplasia (OMIM#258480); osteoglophonic dysplasia (OMIM#166250) which is caused by activating mutations in FGFR1; and epidermal nevus syndrome (MIM #163200).

Therapy for FGF23-mediated hypophosphatemic disorders

Although there are no approved therapies for FGF23-mediated hypophosphatemic disorders, clinicians frequently use high dose calcitriol and phosphate, which improves lower extremity deformities and reduces bone pain [48]. This therapy also decreases dental abscesses, but does not reduce enthesopathy [49]. FGF23 neutralizing antibody, burosumab, has been used in clinical trials in XLH. Results show persistent increases in tubular maximum resorption of phosphate/glomerular filtration rate (TmP/GMR), serum phosphorus, and 1,25 dihydroxyvitamin D concentrations [50]. Additional data on improvement in rickets and growth in children will be forthcoming.

Disorders associated with reduced FGF23 bioactivity

Familial TC (OMIM #211900) is an autosomal recessive disorder characterized by dental abnormalities, as well as soft tissue periarticular and vascular calcification [40]. Biochemical abnormalities include hyperphosphatemia, increased % tubular reabsorption of phosphate and inappropriately normal or elevated 1,25$(OH)_2$D. Calcium and PTH are usually within the normal ranges, although PTH may be suppressed. Hyperostosis–hyperphosphatemia syndrome (HHS) is a rare metabolic disorder characterized by a biochemical profile that is identical to TC, with localized hyperostosis [41].

TC/HHS caused by GALNT3 mutations

The first gene identified for heritable TC was UDP-N-acetyl-alpha-D-galactosamine: polypeptide N-acetylgalactosaminyl transferase-3 (*GALNT3*) [42]. GALNT3 is expressed in the Golgi and initiates *O*-linked glycosylation of nascent proteins. These TC patients were originally reported to manifest serum FGF23 levels approximately 30-fold above the normal mean when assessed with the C-terminal FGF23 ELISA [42]. Importantly, it was subsequently shown that the TC patients did indeed have elevated C-terminal FGF23. However, the same individuals had low FGF23 when measured with the intact FGF23 ELISA (Table 25.1) [43]. These findings were then confirmed by showing that loss of GALNT3 resulted in the production of nonfunctional FGF23 protein caused by intracellular degradation [44]. FGF23 is *O*-glycosylated on specific residues within the $_{176}$RH$\underline{T}$R$_{179}$/S$_{180}$ site (at threonine 178), thus the lack of glycosylation at this residue is thought to destabilize intact active FGF23 [41].

HHS was also found to be caused by inactivating mutations in *GALNT3* [41], and these patients also manifest inappropriate cFGF23/iFGF23 ELISA ratios (Table 25.1).

Table 25.1. Summary of heritable and acquired disorders involving FGF23.

Disorder	Mutated Gene	Mutation Consequence	Relationship to FGF23	Effect on Serum Pi	Effect on Serum 1,25-D	iFGF23 ELISA conc.	cFGF23 ELISA conc.
ADHR	FGF23	Gain of function	Stabilize full-length, active FGF23	↓	↔	↔ or ↑	↔ or ↑
XLH	PHEX	Loss of function	Increased FGF23 production in osteocytes	↓	↔	↔ or ↑	↔ or ↑
ARHR1	DMP1	Loss of function	Increased FGF23 production in osteocytes	↓	↔	↔ or ↑	↔ or ↑
ARHR2	ENPP1	Loss of function	Increased FGF23 production	↓	↔	↔ or ↑	↔
ARHR3	FAM20C	Loss of function	Increased iFGF23	↓	↔	↑	↑
TIO	–	–	FGF23 overproduced by tumor	↓	↔ or ↓	↔ or ↑	↔ or ↑
TC/HHS	FGF23 or GALNT3	Loss of function	Destabilize full-length, active FGF23	↑	↔ or ↑	↓	↑
TC	αKLOTHO	Loss of function	Decreased FGF23-dependent signaling	↑	↔ or ↑	↑	↑

ADHR = autosomal dominant hypophosphatemic rickets; ARHR1–3 = autosomal recessive hypophosphatemic rickets 1–3; HHS = hyperostosis–hyperphosphatemia syndrome; TC = tumoral calcinosis; TIO = tumor-induced osteomalacia; XLH = X-linked hypophosphatemic rickets.

Indeed, some of the HHS mutations are the same as those that result in TC, indicating that genetic background may influence disease phenotype and/or that TC and HHS may represent a spectrum of severity within the same disease.

TC caused by FGF23 mutations

TC can also be caused by recessive, inactivating mutations in the *FGF23* gene [45, 46, 51]. These mutations have all been missense mutations (S71G, M96T, S129F) within the FGF23 N-terminal FGF-like domain. The TC alterations destabilize FGF23, as supported by the findings that the TC patients with *FGF23* mutations have the same FGF23 ELISA pattern as GALNT3-TC patients, ie, markedly elevated C-terminal concentrations, in concert with low intact values [45, 46], and the fact that these mutants are cleaved before cellular secretion [45, 46, 51]. Thus, the common denominator in GALNT3-TC and FGF23-TC is the lack of production of intact FGF23. This lack of intact FGF23 then results in elevation of serum Pi through increased renal reabsorption, which in turn results in elevated secretion of nonfunctional FGF23 fragments through a positive feedback cycle.

TC caused by Klotho mutations

αKlotho (αKL) is a coreceptor for FGF23, and was therefore tested as a candidate gene for TC in a 13-year-old female with hypothesized end-organ defects in renal FGF23 bioactivity. This patient manifested hyperphosphatemia, hypercalcemia, elevated PTH, elevated iFGF23, and cFGF23 [47] (approximately 100-550-fold elevation of the normal means), as well as ectopic calcifications in the heel and brain. She had normal pubertal development, and her disease paralleled α*KL*-null mice with regard to ectopic calcifications, and dramatic elevation of circulating FGF23 [25]. This patient had a novel recessive mutation in a highly conserved residue (Histidine193Arginine, or H193R) in the extracellular domain of αKL (KL1 domain). Mutant KL expression was markedly reduced compared with that of wild type αKL, which resulted in a striking reduction in the ability of αKL to mediate FGF23-dependent signaling [47]. Thus, an inactivating H193R αKL mutation results in a TC phenotype and shows that αKL is required for FGF23 bioactivity.

Chronic kidney disease

FGF23 is elevated in patients with chronic kidney disease (CKD) and recent studies indicate that this FGF23 is biologically active [52]. One report has shown that higher FGF23 levels are a predictor of increased progression of renal disease in patients with nondiabetic CKD [53]. Other reports show an association between high FGF23 concentrations and left ventricular hypertrophy in patients with CKD [54]. Furthermore, epidemiological studies show higher mortality in renal and nonrenal patients with higher FGF23 concentrations [55]. Although Klotho is not expressed in the heart, the extremely high iFGF23 levels observed in patients with CKD may allow binding to FGF receptors. Studies indicate that FGF23 can bind FGFR4 without Klotho to activate calcineurin/NFAT signaling and *FGFR4* blockade protects rats from FGF23-induced cardiac hypertrophy [56]. Moreover, FGFR4 gain of function mutations result in cardiac hypertrophy in mice [56]. If these data are confirmed by additional studies, it will be critical to determine the

FGF23 concentration needed to cause clinically relevant cardiac problems. For example, patients with FGF23-mediated hypophosphatemia, such as XLH, do not appear to manifest clinically apparent cardiac disease. It is possible therefore that cardiac disease only occurs with extreme elevation in FGF23, as found in end stage renal disease, but additional data are needed to confirm this.

REFERENCES

1. Hilfiker H, Hattenhauer O, Traebert M, et al. Characterization of a murine type II sodium-phosphate cotransporter expressed in mammalian small intestine. Proc Natl Acad Sci U S A. 1998;95(24):14564–9.
2. Xu H, Bai L, Collins JF, et al. Age-dependent regulation of rat intestinal type IIb sodium-phosphate cotransporter by 1,25-(OH)(2) vitamin D(3). Am J Physiol Cell Physiol. 2002;282(3):C487–93.
3. Tenenhouse HS. Regulation of phosphorus homeostasis by the type IIa Na/phosphate cotransporter. Annu Rev Nutr. 2005;25:197–214.
4. Biber J, Hernando N, Forster I, et al. Regulation of phosphate transport in proximal tubules. Pflugers Arch. 2009;458(1):39–52.
5. Bergwitz C, Roslin NM, Tieder M, et al. SLC34A3 mutations in patients with hereditary hypophosphatemic rickets with hypercalciuria predict a key role for the sodium-phosphate cotransporter NaPi-IIc in maintaining phosphate homeostasis. Am J Hum Genet. 2006;78(2):179–92.
6. Lorenz-Depiereux B, Benet-Pages A, et al. Hereditary hypophosphatemic rickets with hypercalciuria is caused by mutations in the sodium-phosphate cotransporter gene SLC34A3. Am J Hum Genet. 2006;78(2):193–201.
7. Puschett JB, Fernandez PC, Boyle IT, et al. The acute renal tubular effects of 1,25-dihydroxycholecalciferol. Proc Soc Exp Biol Med Soc Exp Biol Med. 1972;141(1):379–84.
8. Bacic D, Lehir M, Biber J, et al. The renal Na+/phosphate cotransporter NaPi-IIa is internalized via the receptor-mediated endocytic route in response to parathyroid hormone. Kidney Int. 2006;69(3):495–503.
9. Pfister MF, Lederer E, Forgo J, et al. Parathyroid hormone-dependent degradation of type II Na+/Pi cotransporters. J Biol Chem. 1997;272(32):20125–30.
10. Traebert M, Volkl H, Biber J, et al. Luminal and contraluminal action of 1-34 and 3-34 PTH peptides on renal type IIa Na-P(i) cotransporter. Am J Physiol Renal Physiol. 2000;278(5):F792–8.
11. ADHRConsortium. Autosomal dominant hypophosphataemic rickets is associated with mutations in FGF23. Nat Genet. 2000;26(3):345–8.
12. Riminucci M, Collins MT, Fedarko NS, et al. FGF-23 in fibrous dysplasia of bone and its relationship to renal phosphate wasting. J Clin Invest. 2003;112(5):683–92.
13. Shimada T, Mizutani S, Muto T, et al. Cloning and characterization of FGF23 as a causative factor of tumor-induced osteomalacia. Proc Natl Acad Sci U S A. 2001;98(11):6500–5.
14. White KE, Carn G, Lorenz-Depiereux B, et al. Autosomal-dominant hypophosphatemic rickets (ADHR) mutations stabilize FGF-23. Kidney Int. 2001;60(6):2079–86.
15. Larsson T, Marsell R, Schipani E, et al. Transgenic mice expressing fibroblast growth factor 23 under the control of the alpha1(I) collagen promoter exhibit growth retardation, osteomalacia, and disturbed phosphate homeostasis. Endocrinology. 2004;145(7):3087–94.
16. Burnett SM, Gunawardene SC, Bringhurst FR, et al. Regulation of C-terminal and intact FGF-23 by dietary phosphate in men and women. J Bone Miner Res. 2006;21(8):1187–96.
17. Perwad F, Azam N, Zhang MY, et al. Dietary and serum phosphorus regulate fibroblast growth factor 23 expression and 1,25-dihydroxyvitamin D metabolism in mice. Endocrinology. 2005;146(12):5358–64.
18. Shimada T, Hasegawa H, Yamazaki Y, et al. FGF-23 is a potent regulator of vitamin D metabolism and phosphate homeostasis. J Bone Miner Res. 2004;19(3):429–35.
19. Tsujikawa H, Kurotaki Y, Fujimori T, et al. Klotho, a gene related to a syndrome resembling human premature aging, functions in a negative regulatory circuit of vitamin D endocrine system. Mol Endocrinol. 2003;17(12):2393–403.
20. Shimada T, Kakitani M, Yamazaki Y, et al. Targeted ablation of Fgf23 demonstrates an essential physiological role of FGF23 in phosphate and vitamin D metabolism. J Clin Invest. 2004;113(4):561–8.
21. Sitara D, Razzaque MS, Hesse M, et al. Homozygous ablation of fibroblast growth factor-23 results in hyperphosphatemia and impaired skeletogenesis, and reverses hypophosphatemia in Phex-deficient mice. Matrix Biol. 2004;23(7):421–32.
22. Segawa H, Yamanaka S, Ohno Y, et al. Correlation between hyperphosphatemia and type II Na-Pi cotransporter activity in klotho mice. Am J Physiol Renal Physiol. 2007;292(2):F769–79.
23. Matsumura Y, Aizawa H, Shiraki-Iida T, et al. Identification of the human klotho gene and its two transcripts encoding membrane and secreted klotho protein. Biochem Biophys Res Commun. 1998;242(3):626–30.
24. Imura A, Iwano A, Tohyama O, et al. Secreted Klotho protein in sera and CSF: implication for post-translational cleavage in release of Klotho protein from cell membrane. FEBS Lett. 2004;565(1–3):143–7.
25. Urakawa I, Yamazaki Y, Shimada T, et al. Klotho converts canonical FGF receptor into a specific receptor for FGF23. Nature. 2006;444(7120):770–4.
26. Kurosu H, Ogawa Y, Miyoshi M, et al. Regulation of fibroblast growth factor-23 signaling by klotho. J Biol Chem. 2006;281(10):6120–3.
27. Li H, Martin A, David V, et al. Compound deletion of Fgfr3 and Fgfr4 partially rescues the Hyp mouse phenotype. Am J Physiol Endocrinol Metab. 2011;300(3):E508–17.
28. Farrow EG, Davis SI, Summers LJ, et al. Initial FGF23-mediated signaling occurs in the distal convoluted tubule. J Am Soc Nephrol. 2009;20(5):955–60.
29. Jonsson KB, Zahradnik R, Larsson T, et al. Fibroblast growth factor 23 in oncogenic osteomalacia and X-linked hypophosphatemia. N Engl J Med. 2003;348(17):1656–63.

30. Yamazaki Y, Okazaki R, Shibata M, et al. Increased circulatory level of biologically active full-length FGF-23 in patients with hypophosphatemic rickets/osteomalacia. J Clin Endocrinol Metab. 2002;87(11):4957–60.
31. Econs MJ, McEnery PT. Autosomal dominant hypophosphatemic rickets/osteomalacia: clinical characterization of a novel renal phosphate-wasting disorder. J Clin Endocrinol Metab. 1997;82(2):674–81.
32. White KE, Jonsson KB, Carn G, et al. The autosomal dominant hypophosphatemic rickets (ADHR) gene is a secreted polypeptide overexpressed by tumors that cause phosphate wasting. J Clin Endocrinol Metab. 2001;86(2):497–500.
33. Ryan EA, Reiss E. Oncogenous osteomalacia. Review of the world literature of 42 cases and report of two new cases. Am J Med. 1984;77(3):501–12.
34. HYP-Consortium. A gene (PEX) with homologies to endopeptidases is mutated in patients with X-linked hypophosphatemic rickets. The HYP Consortium. Nat Genet. 1995;11(2):130–6.
35. Beck L, Soumounou Y, Martel J, et al. Pex/PEX tissue distribution and evidence for a deletion in the 3′ region of the Pex gene in X-linked hypophosphatemic mice. J Clin Invest. 1997;99(6):1200–9.
36. Liu S, Guo R, Simpson LG, et al. Regulation of fibroblastic growth factor 23 expression but not degradation by PHEX. J Biol Chem. 2003;278(39):37419–26.
37. Feng JQ, Ward LM, Liu S, et al. Loss of DMP1 causes rickets and osteomalacia and identifies a role for osteocytes in mineral metabolism. Nat Genet. 2006; 38(11):1310–5.
38. Lorenz-Depiereux B, Bastepe M, Benet-Pages A, et al. DMP1 mutations in autosomal recessive hypophosphatemia implicate a bone matrix protein in the regulation of phosphate homeostasis. Nat Genet. 2006;38(11): 1248–50.
39. Nam HK, Liu J, Li Y, et al. Ectonucleotide pyrophosphatase/phosphodiesterase-1 (Enpp1) regulates osteoblast differentiation. J Biol Chem. 2011;286(55):39059–71.
40. Prince MJ, Schaeffer PC, Goldsmith RS, et al. Hyperphosphatemic tumoral calcinosis: association with elevation of serum 1,25-dihydroxycholecalciferol concentrations. Ann Intern Med. 1982;96(5):586–91.
41. Frishberg Y, Topaz O, Bergman R, et al. Identification of a recurrent mutation in GALNT3 demonstrates that hyperostosis-hyperphosphatemia syndrome and familial tumoral calcinosis are allelic disorders. J Mol Med. 2005;83(1):33–8.
42. Topaz O, Shurman DL, Bergman R, et al. Mutations in GALNT3, encoding a protein involved in O-linked glycosylation, cause familial tumoral calcinosis. Nat Genet. 2004;36(6):579–81.
43. Garringer HJ, Fisher C, Larsson TE, et al. The role of mutant UDP-N-acetyl-alpha-D-galactosamine-polypeptide N-acetylgalactosaminyltransferase 3 in regulating serum intact fibroblast growth factor 23 and matrix extracellular phosphoglycoprotein in heritable tumoral calcinosis. J Clin Endocrinol Metab. 2006;91(10): 4037–42.
44. Frishberg Y, Ito N, Rinat C, et al. Hyperostosis-hyperphosphatemia syndrome: a congenital disorder of O-glycosylation associated with augmented processing of fibroblast growth factor 23. J Bone Miner Res. 2007;22(2):235–42.
45. Benet-Pages A, Orlik P, Strom TM, et al. An FGF23 missense mutation causes familial tumoral calcinosis with hyperphosphatemia. Hum Mol Genet. 2005;14(3): 385–90.
46. Larsson T, Yu X, Davis SI, et al. A novel recessive mutation in fibroblast growth factor-23 causes familial tumoral calcinosis. J Clin Endocrinol Metab. 2005;90(4):2424–7.
47. Ichikawa S, Imel EA, Kreiter ML, et al. A homozygous missense mutation in human KLOTHO causes severe tumoral calcinosis. J Clin Invest. 2007;117(9): 2684–91.
48. Carpenter TO, Imel EA, Holm IA, et al. A clinician's guide to X-linked hypophosphatemia. J Bone Miner Res. 2011;26(7):1381–8.
49. Connor J, Olear EA, Insogna KL, et al. Conventional therapy in adults with X-linked hypophosphatemia: effects on enthesopathy and dental disease. J Clin Endocrinol Metab. 2015;100(10):3625–32.
50. Carpenter TO, Imel EA, Ruppe MD, et al. Randomized trial of the anti-FGF23 antibody KRN23 in X-linked hypophosphatemia. J Clin Invest. 2014;124(4):1587–97.
51. Araya K, Fukumoto S, Backenroth R, et al. A novel mutation in fibroblast growth factor 23 gene as a cause of tumoral calcinosis. J Clin Endocrinol Metab. 2005;90(10):5523–7.
52. Shimada T, Urakawa I, Isakova T, et al. Circulating fibroblast growth factor 23 in patients with end-stage renal disease treated by peritoneal dialysis is intact and biologically active. J Clin Endocrinol Metab. 2010;95(2):578–85.
53. Fliser D, Kollerits B, Neyer U, et al. Fibroblast growth factor 23 (FGF23) predicts progression of chronic kidney disease: the Mild to Moderate Kidney Disease (MMKD) Study. J Am Soc Nephrol. 2007;18(9):2600–8.
54. Mirza MA, Larsson A, Melhus H, et al. Serum intact FGF23 associate with left ventricular mass, hypertrophy and geometry in an elderly population. Atherosclerosis. 2009;207(2):546–51.
55. Gutierrez OM, Mannstadt M, Isakova T, et al. Fibroblast growth factor 23 and mortality among patients undergoing hemodialysis. N Engl J Med. 2008;359(6):584–92.
56. Grabner A, Amaral AP, Schramm K, et al. Activation of cardiac fibroblast growth factor receptor 4 causes left ventricular hypertrophy. Cell Metabol. 2015;22(6): 1020–32.

26
Gonadal Steroids

Stavros C. Manolagas[1] and Maria Schuller Almeida[2]

[1]University of Arkansas for Medical Sciences and the Central Arkansas Veterans Healthcare System, Little Rock, AR, USA
[2]Center for Osteoporosis and Metabolic Bone Diseases, Department of Internal Medicine, Department of Orthopedic Surgery, University of Arkansas for Medical Sciences, Little Rock, AR, USA

INTRODUCTION

Estrogens and androgens influence the growth of the skeleton during puberty and help to maintain skeletal assets during adulthood. A decline of estrogen levels at menopause and both estrogens and androgens in elderly males contributes to the development of osteoporosis. In this chapter, we will briefly review current understanding of the actions of estrogens and androgens at the molecular, cellular, and whole tissue level in bone; evidence that distinct cellular and molecular targets are responsible for their actions in cancellous versus cortical bone; their role in bone growth during puberty and the maintenance of the skeleton during adulthood; the contribution of estrogen and androgen deficiency to the pathogenesis of osteoporosis in females and males; and the role of natural or synthetic sex steroids in the management of osteoporosis. For a more detailed and comprehensive treatise of the subject the reader should consult the recent review article by Almeida and colleagues [1].

HORMONE BIOSYNTHESIS

Both estrogens and androgens are derived from C19 metabolites of cholesterol and are synthesized in the gonads and the adrenals [2]. The most abundant estrogen in women is 17β-estradiol (E_2) and is synthesized in the ovaries. Men also produce E2; 15% is made and secreted by the testes and 85% by peripheral aromatization of testosterone. Testosterone is the main circulating androgen and is made by the Leydig cells of the testicles (95%) and the adrenals (5%). Testosterone acts in its target cells unmodified or after conversion to the more potent dihydrotestosterone (DHT). The bioavailability of both estrogens and androgens is limited because of their binding to sex hormone-binding globulin (SHBG), albumin, or other proteins; with only 1% to 5% being biologically active [3] (Fig. 26.1).

After menopause, the circulating levels of estrogens are generally lower than in men of the same age. The circulating levels of testosterone decrease only marginally (~1%/year) during aging in men, whereas the total E_2 level remains constant. However, SHBG increases moderately in elderly men, resulting in a decrease of bioavailable testosterone and E_2 [4].

MOLECULAR MECHANISM OF ACTION

The effects of estrogens and androgens on bone are mediated by the estrogen receptors (ER) α and β (known as NR3A1 and NR3A2) and the androgen receptor (AR) (known as NR3C4), respectively [5]. Upon binding of the cognate ligand, the receptors form homo-dimers that bind to palindromic nucleotide sequences, called hormone response elements (EREs or AREs) (Fig. 26.2A). Several coregulators form multiprotein complexes with the receptors and can either activate or repress their transcriptional activity [6]. In addition to direct DNA binding, these receptors can bind to chromatin and regulate transcription indirectly via tethering with other transcription factors, like NF-kB and AP1 (Fig. 26.2B).

In addition to nuclear initiated actions, estrogens or androgens bind to subsets of their cognate receptors localized outside the nucleus, either on the cell membrane or in the cytosol [7]. Binding of these ligands to receptors localized on the membrane initiates signal transduction cascades that ultimately activate cytoplasmic kinases,

---

Primer on the Metabolic Bone Diseases and Disorders of Mineral Metabolism, Ninth Edition. Edited by John P. Bilezikian.
© 2019 American Society for Bone and Mineral Research. Published 2019 by John Wiley & Sons, Inc.
Companion website: www.wiley.com/go/asbmrprimer

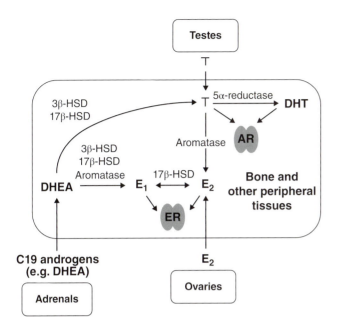

Fig. 26.1. Biosynthesis of estrogens and androgens in the gonads and peripheral tissues. T = testosterone; E_1 = estrone; E_2 = 17β-estradiol; DHT = dihydrotestosterone; DHEA = dehydroepiandrosterone; AR = androgen receptor; ER = estrogen receptor; HSD = hydroxysteroid dehydrogenase.

such as ERKs, PI3K, and JNKs. Activated kinases, in turn, phosphorylate substrate proteins and transcription factors which then modulate gene transcription (Fig. 26.2C,D) [8]. Notably, a cell membrane-impermeable E2 conjugate that can selectively activate nonnuclear initiated actions of the ERα, but has no effects on the nuclear-initiated actions of this receptor, replicates the beneficial effects of estrogens on cortical bone and the cardiovascular system in mice without affecting reproductive organs, such as the uterus and the breast [9, 10].

CELLULAR TARGETS

ERα is present in myeloid precursors and mature osteoclasts, as well as mesenchymal progenitors and their progeny, which includes chondrocytes, osteoblasts, and osteocytes. The AR is also expressed in mesenchymal progenitors and their progeny and in very low levels in osteoclasts [11]. Both ERα and AR are expressed in other cell types in the bone microenvironment, such as B and T lymphocytes, as well as muscle cells.

In the last few years, generation of genetic mouse models with cell-specific deletion of these receptors has allowed the evaluation of their biological role in the different cell types, thereby providing more clear understanding of the role of their ligands on bone in vivo [12]. It has been elucidated that the osteoclast ERα mediates the protective effect of estrogens on the cancellous, but not the cortical, bone compartment in females [11–13]. In contradistinction to estrogens, the antiresorptive effects of androgens on cancellous bone are exerted indirectly via osteoblasts and osteocytes [11, 14]. Furthermore, in cortical bone, estrogens protect against resorption in both females and males, at least in part, via ERα-mediated actions (upon aromatization of androgens to estrogens in males) on mesenchymal progenitors [13] (Fig. 26.3).

Estrogen deficiency increases osteocyte apoptosis in humans and rodents [15–17]; this phenomenon has been spatially associated with regional bone resorption in some studies, but not others [18, 19]. Nonetheless, mice with ERα deletion in osteocytes do not exhibit increased osteoclast number or bone resorption, indicating that if estrogens suppress osteoclast number by controlling osteocyte lifespan or function, these effects must be indirect [20].

ERα is required for the adaptive response of bone to mechanical loading [21]. Notably, this particular effect is independent of ligand binding and is mediated via the ERα expressed in osteoblast progenitors (Fig. 26.3). Indeed, the ERα in these progenitors promotes cortical bone accrual at the periosteum in both females and males, independently of ligand binding [22]. Moreover, mice with conditional deletion of the ERα in osteoblast progenitors fail to exhibit the expected increase in bone formation and periosteal bone accrual in response to mechanical loading, whereas mice with ERα deletion in mature osteoblasts/osteocytes have normal loading response. Global deletion of ERβ in mice potentiates the response of cortical bone to mechanical loading, suggesting that ERβ and ERα may exert opposite effects [23, 24].

Loss of estrogens or androgens increases B lymphocyte number and this promotes osteoclast formation, by contributing osteoclastogenic cytokines. In support of this contention, B-cell-derived RANKL contributes to ovariectomy-induced cancellous bone loss in mice [25], whereas RANKL produced by osteocytes is required for the increase in B cells and the bone loss caused by estrogen deficiency [26]. T-cell-derived RANKL, on the other hand, has no effect on the ovariectomy-induced loss of bone. AR deletion from osteoblastic cells or from B lymphocytes also increases B-cell number, indicating that androgens may suppress B-cell numbers directly and indirectly. In contrast, estrogens suppress B-cell number indirectly because ERα deletion from B lymphocytes does not alter the number of these cells or bone mass [27].

Finally, it remains unclear whether the ERα or the AR present in muscle cells indirectly affects skeletal homeostasis [28, 29].

PUTATIVE GENE TARGETS

Studies with cell lines and primary cell cultures have suggested over the years a long list of putative target genes of estrogens and androgens in osteoclasts and osteoblasts (including IL-1β, IL-6, IL-7, TNFα, M-CSF, RANKL, OPG, and prostaglandins) produced by bone marrow stromal cells, T and B lymphocytes, macrophages, and dendritic

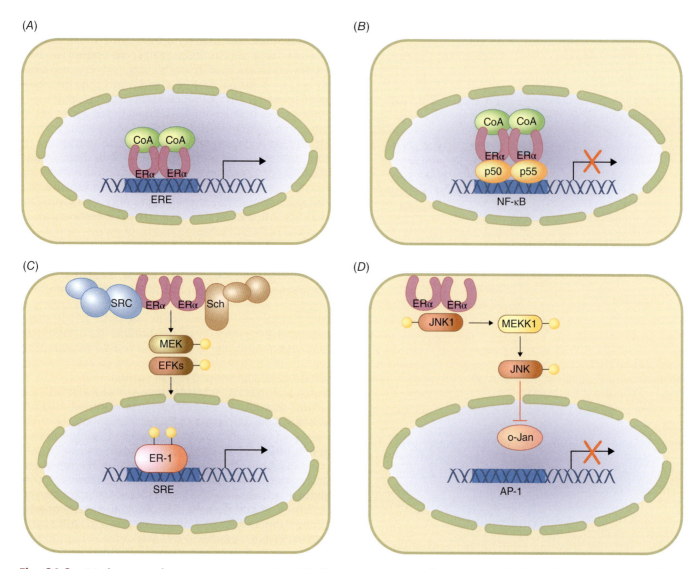

Fig. 26.2. Mechanisms of estrogen receptor action. (A) Classic genomic signaling in which the ligand-activated receptor dimer attaches to estrogen response elements (ERE) on DNA, and activates or represses transcription. (B) ERE-independent genomic signaling pathway in which the ligand-activated receptor binds to other transcription factors (eg, p50 and p65 subunits of NF-κB) and prevents them from binding to their response elements. (C, D) Nongenotropic mode of action in which the ligand-activated receptor (in the plasma membrane) activates cytoplasmic kinases which in turn cause the phosphorylation of substrate proteins and transcription factors (eg, Elk-1 and c-jun) that positively (C) or negatively (D) regulate transcription.

cells [30, 31]. However, functional evidence that any of these genes are biologically relevant targets of estrogens or androgens on bone remains elusive.

A study of bone marrow cells from pre- and postmenopausal women suggested that estrogen deficiency may increase RANKL and decrease osteoprotegerin (OPG) production by osteoblast progenitors, and T and B cells [32]. Nonetheless, estrogens do not seem to suppress the transcription of the RANKL gene directly and RANKL production by osteoblast progenitors or osteoblasts is not a major contributor to osteoclastogenesis during remodeling in mice [33]. In addition, the levels of circulating OPG in pre- and postmenopausal women are not different [34].

Recent work from the authors' laboratory, using microarray analysis of cells isolated from mice with conditional ERα deletion, has suggested that a critical target gene for the direct antiresorptive effect of estrogens on cancellous bone is the calcium-binding protein S100A8. S100A8 is produced in macrophages and osteoclasts and stimulates osteoclast formation via the Toll-like receptor 4 and activation of NF-kB. CXCL12 – a chemoattractant with a seminal function in the bone marrow niche [35] – as well as the matrix metalloproteinase 13 (MMP13), which promotes osteoclast fusion independent of its enzymatic activity [36] are, on the other hand, likely gene targets of the indirect effects of estrogens on bone resorption and in particular the antiresorptive effects of

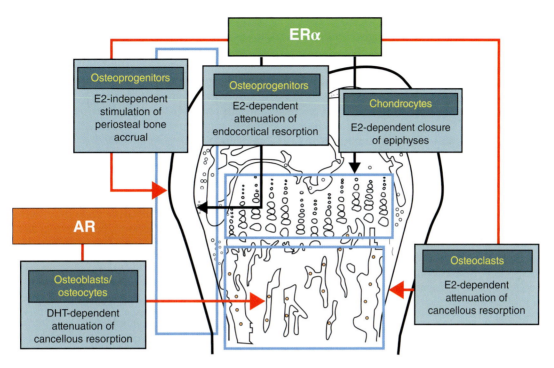

Fig. 26.3. Site-specific effects of ERα and AR on different bone compartments. The cortical and the cancellous bone compartments as well as the growth plate of a long bone are depicted in the respective blue boxes. The cells responsible for the particular action along with their dependency in each compartment are deduced from the respective cell-specific ERα deletion murine models [1]. The ERα in osteoblast progenitors is required for optimal cortical bone accrual at the periosteum and mediates a protective effect of estrogens against endocortical bone resorption. Estrogens act directly on osteoclasts to protect against the loss of cancellous bone mass in females. In contrast, androgens acting via the AR in osteoblasts preserve cancellous bone mass in males. Estrogens act directly on chondrocytes to promote epiphyseal closure in both sexes.

estrogens (and androgens) on cortical bone. BMP3b (also known as GDF10) – an inhibitor of BMP2 signaling – may be a seminal target of the effect of ERα signaling on the periosteal response to mechanical loading.

EFFECTS ON SKELETAL DEVELOPMENT AND GROWTH

Mammalian skeletons grow in length at the epiphysial growth plate by the process of endochondral bone formation, whereby calcified cartilage made by chondrocytes is resorbed by osteoclasts and gradually replaced by mineralized bone made by osteoblasts. Simultaneously, bones expand radially, their cortices thicken, and the medullary cavities become larger as a result of bone formation at the periosteum and increased resorption at the endosteal surface [37]. The marrow cavity enlargement is greater in males than in females.

At the initiation of puberty, low levels of estrogens and perhaps androgens are responsible for a spurt in linear bone growth in both sexes. This effect results from the stimulation of endochondral bone formation [37]. At the end of puberty, high levels of estrogens are essential for the closure of the epiphyses and the cessation of linear growth in either sex. In parallel to the acceleration of linear growth, pubertal boys and girls experience an accelerated enlargement of the outer perimeter and further widening of the medullary cavity of long bones. These changes lead to bigger bones in boys than in girls primarily caused by a larger increase in periosteal bone formation, placing the cortex further away from the neutral axis. Puberty starts earlier in girls, but lasts longer in boys. This accounts for part of the differences in the size of the skeleton between the two sexes; and it is the main reason why men have stronger bones.

Linear growth

Linear skeletal growth is dependent on the growth hormone (GH)/insulin-like growth factor (IGF) axis. The effects of androgens and estrogens on the pubertal growth spurt are mediated via effects on the GH secretion patterns, whereas estrogens also have direct effects on hepatic IGF-1 release [37]. Estrogens are essential for the pubertal bone changes in both girls and boys. Thus, aromatase-deficient females and males lacking estrogens do not exhibit a growth spurt and do not undergo closure of their epiphyses [38]. Estrogen replacement in patients with aromatase deficiency enhances bone growth. In addition, administration of E_2 to pre- or early pubertal boys increases longitudinal growth [39]. Furthermore, overexpression of the aromatase gene in both sexes accelerates growth and leads to premature closure of the epiphyses [40]. The importance of estrogens for the male skeleton is further highlighted by the absence of a pubertal growth spurt in a man with a loss of function mutation of ERα [41].

It is unclear whether androgens have a significant effect on linear growth. In support of the contention that androgens may not play a significant role in linear bone growth, serum testosterone was high normal in the man with the ERα mutation or the aromatase-deficient men and women [38, 41]. On the other hand, men with androgen insensitivity syndrome caused by AR mutations have intermediate height between normal males and females [42]. Moreover, administration of DHT to boys with delayed growth, or to growing rats and rabbits, stimulates longitudinal bone growth, supporting a role of androgens in this process [37]. To date, there are no consistent data in the murine model of the effects of androgens, estrogens, or their receptors, on pubertal bone growth spurt.

Periosteal expansion

The greater periosteal expansion during puberty in boys as compared with girls has been ascribed to the higher levels of androgens in the male. Nevertheless, administration of E_2 to an aromatase-deficient young man increased bone size, and this was presumed to be the result of increased periosteal apposition [43]. Consistent with this finding, studies in rodents have revealed that both estrogens and androgens, acting via their respective receptors, are involved in expansion of the periosteum in growing males. Thus, male mice with a mutation or deletion of the AR exhibit reduced periosteal expansion [44]. DHT may be the androgen responsible for periosteal expansion in the male. Thus, male mice lacking 5α-reductase type I exhibit reduced cortical thickness [45]. However, deletion of ERα, but not ERβ, also causes decreased periosteal bone formation, and decreased femoral width in male mice [46]. Pharmacological inhibition of E_2 synthesis by an aromatase inhibitor further reduces periosteal expansion in orchidectomized male mice, indicating that the estrogens needed for radial bone growth are derived from peripheral tissues [47]. In line with the need for both AR and ERα for optimal periosteal expansion in males, periosteal circumference is lower in mice lacking both receptors, as compared with mice lacking only one of the receptors [48].

Unlike the situation in males, estrogens restrain radial bone growth in females, as evidenced by the observation that ovariectomy of growing rats or mice increases periosteal expansion during early, but not late, puberty [47]. Moreover, female mice lacking ERβ exhibit increased periosteal circumference [49]. Whether the opposite effect of estrogens on periosteal apposition in males versus females also occurs in humans is unknown.

Cessation of longitudinal bone growth at the end of puberty is caused by a decline in the replication of chondrocytes in the proliferative zone of the growth plate. This results in reduced cartilage synthesis at the distal end of the growth plate, as well as reduced replacement of the cartilage by bone at the proximal end. Consequently, the growth plate closes. The closure of the growth plate at the end of puberty is clearly mediated by E_2 in both men and women as indicated by failure of growth plate closure and continued longitudinal growth in the man lacking ERα [41], and in aromatase deficient men and women [38, 50]. Moreover, chondrocyte-specific deletion of ERα results in failure of growth plate closure indicating a direct suppressive effect of E_2 on chondrocyte proliferation [51]. In agreement with this finding, ovariectomy increased the number of proliferating chondrocytes in the proliferative zone of growing rats [52].

To conclude this section, both estrogens and androgens stimulate pubertal growth spurt, at least in part, by stimulating the secretion of growth hormone and IGF-1. Estrogens stimulate IGF-1 release by direct actions on the liver. In females, the unliganded ERα is required for optimal periosteal expansion. Estrogens, on the other hand, decrease periosteal expansion and help to maintain the endosteal perimeter, via direct actions on bone cells. In males, both AR and ERα are required for optimal cortical bone expansion, but the cells responsible for the effects of AR are unknown.

EFFECTS ON SKELETAL MAINTENANCE

After bone development and growth is completed and bones have achieved adult sizes and shapes, skeletal mass is determined by the balance of the amount of bone resorbed and formed during remodeling by the teams of osteoclasts and osteoblasts assembled in the basic multicellular units (Fig. 26.4).

During adulthood, periosteal expansion continues in both sexes, but is greater in men, whereas endosteal resorption is greater in women [1]. Because osteoblasts and osteoclasts are short-lived cells, balanced remodeling depends on the timely supply and lifespan of these two cells. Too many osteoclasts relative to the need for remodeling or too few osteoblasts relative to the need for cavity repair lead to loss of bone mass and are the seminal pathogenetic changes responsible for most acquired metabolic bone diseases [30, 53].

The rate of remodeling during adulthood is slower than it is during growth and varies, depending on the mechanical strains experienced in different bones (and even within different areas of a particular bone), and thereby the chance of microdamage. Remodeling is apparently orchestrated and targeted to sites that are in need of repair by osteocytes [54]. Importantly, apoptotic, old, or dysfunctional osteocytes signal their live neighbors to initiate remodeling and this mechanism may lead to bone loss [55].

Estrogens slow bone remodeling by restraining bone resorption in cancellous bone and the endocortical surface. As it was discussed earlier, these effects result from the ability of estrogens to decrease the birth rate of

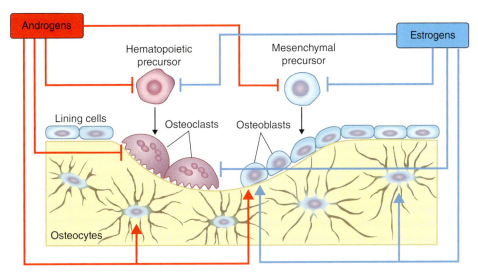

Fig. 26.4. Schematic representation of the remodeling process and the effects of estrogens and androgens. Osteoclasts and osteoblasts are derived from hematopoietic and mesenchymal precursors, respectively. During bone remodeling, bone matrix excavated by osteoclasts is replaced with new matrix produced by osteoblasts. Both estrogens and androgens influence the generation and lifespan of osteoclasts and osteoblasts, as well as the lifespan of osteocytes. Negative and positive effects of sex steroids on the generation and survival of the cells are depicted, by bookends and arrowheads.

osteoclasts and shorten the lifespan of osteoclast progenitors and mature osteoclasts by stimulating apoptosis [56] (Fig. 26.4). Estrogens or androgens, on the other hand, decrease the apoptosis of osteoblasts and osteocytes [53, 57]. Conversely, estrogen or androgen deficiency causes loss of bone associated with an increase in the bone remodeling rate, increased osteoclast and osteoblast numbers, and increased resorption and formation (albeit unbalanced).

SEX STEROID DEFICIENCY AND THE DEVELOPMENT OF OSTEOPOROSIS

It is widely appreciated that osteoporosis is a multifactorial disease of both sexes in which a decline in ovarian or testicular function are only one of several other progressive and cumulative pathologies. Low bone mass is only one of many risk factors responsible for osteoporotic fractures – the clinical manifestation of the disease. Old age is by far the most critical predictor of osteoporotic fractures [58].

Both women and men lose bone as a result of age, but men are less likely to develop osteoporosis than women for two reasons. First, men gain more bone during puberty, and second they lose less bone during aging because unlike women, men do not experience an abrupt loss of estrogens.

In either sex, bone loss begins within 10 years after the achievement of peak bone mass, which occurs during the third decade of life. This early loss is independent of changes in sex steroid levels [59]. At menopause, the loss of cancellous bone in the spine accelerates. The accelerated cancellous bone loss caused by menopause results predominantly from trabecular perforation and loss of connectivity. Histological analysis of cancellous bone biopsies suggests that loss of estrogens gives rise to more aggressive, so called "killer osteoclasts" [60]. Evidence from mice with osteoclast-specific ERα deletion suggests that in cancellous bone this is the result of the prolongation of osteoclast lifespan, caused by the loss of the direct proapoptotic effects of estrogens on osteoclasts [61, 62]. Bone loss in elderly men is associated with trabecular thinning rather than perforation [63].

The accelerated rate of bone loss after menopause is followed within 5 to 10 years by a slower phase that also occurs in men and occurs primarily in cortical bone. Over the age of 65 in women, bone loss is primarily cortical, not trabecular, and most fractures over the age of 65 occur predominantly at cortical sites [64]. After the age of 80, 90% of the decline of bone mass is cortical. High resolution peripheral quantitative CT measurements (HR-pQCT) have revealed that the majority of the loss of cortical bone in women between the ages of 50 and 80 is caused by increased cortical porosity [65–68]. Notably, cortical porosity cannot be detected by DXA BMD, the routine measurement used to diagnose patients with osteoporosis and determine their fracture risk.

Like humans, mice exhibit a progressive loss of bone mass and strength with advancing age, but unlike women, they do not undergo menopause [13, 69]. Of note, the effects of ovariectomy in adult female mice are transient and the ovariectomy-induced increases in osteoclastogenesis and osteoblastogenesis return to baseline in fewer than 2 months [70]. Similar to humans, mice exhibit an age-dependent decrease in cancellous bone mass and cortical thickness as well as increased cortical porosity [13, 71]. The decrease in cortical thickness and the increase in cortical porosity in mice is caused by increased bone resorption as well as decreased osteoblast number and bone formation. The age-dependent decrease in bone formation is well documented in both elderly women and men by the decreased wall width – the histological hallmark of decreased osteoblast work output [72, 73]. Therefore, the evidence from humans and mice alike strongly suggests that the loss of bone mass in old age is caused by a combination of increased resorption and decreased bone formation. As will be detailed in the next section, the increase in resorption in the murine model is caused by age-related mechanisms per se, not estrogen deficiency [13, 71].

Decline of estrogens at menopause increases periosteal bone apposition, in line with the evidence in mice that periosteal expansion is attenuated as a result of estrogen signaling via the ERα of osteoblast progenitors [74, 75]. The mechanical benefit of the outward expansion of cortical bone in postmenopausal women, however, is outweighed by the simultaneous increase in endosteal bone resorption. It is presently unknown whether the decreasing estrogen levels of menopause attenuate the ERα-dependent sensitivity of bone to mechanical loading.

Men do not experience "andropause" and in the majority of elderly men total testosterone levels are maintained above those found in symptomatic hypogonadism [76]. Instead, elderly men experience a mild to moderate increase in SHBG, which results in a small decline of bioavailable levels of testosterone or estradiol [76, 77]. The preponderance of epidemiologic evidence indicates that bioavailable or free E2, but not total or free testosterone, are positively associated with BMD. Strong support of the evidence that low free estradiol rather than low free testosterone contributes to the low BMD in elderly men is provided from genetic evidence in one man and one woman with loss of function of ERα mutations [78, 79] as well as men with aromatase deficiency [80]. Patients with AR mutations and androgen insensitivity syndrome, on the other hand, have only a very mild bone phenotype [81] and estrogen administration to these patients restores BMD. Consistent with the genetic evidence in humans, recent findings from the authors' work with AR and ERα-deficient mice suggest that estrogens (derived from androgen aromatization) are the sex steroid responsible for the protection of cortical bone mass in both sexes, whereas nonaromatizable androgens are responsible for the protection of cancellous bone in males [11, 13]. This is consistent with results of human studies showing that estrogens account for 70% of the protective effect of sex steroids on bone resorption in men [82] and congruent with the fact that almost 80% of the skeleton is cortical.

CONTRIBUTION OF SEX STEROID DEFICIENCY TO THE AGE-DEPENDENT SKELETAL INVOLUTION

Because of the menopause in women and the decrease of both androgens and estrogens in elderly men, it has been difficult, if not impossible, to dissect the contribution of sex steroid deficiency to the effects of aging in the involution of the human skeleton. Unlike humans, rodents do not experience menopause and androgen levels do not seem to decrease with age in male mice [56]. Surgical removal of the gonads after the age of 4 to 5 months, nonetheless, replicate the effects of estrogen or androgen deficiency in humans on both cancellous and cortical bone mass [69].

Mechanistic studies and genetic evidence from the mouse strongly suggests that several aging-related changes intrinsic to bone are practically identical to those responsible for the effects of aging in all other tissues and organs [83]. These changes are the primary culprits in the development of osteoporosis later in life. Age-related changes in other organs, for example declining ovarian function, are contributory [1, 84, 85]. Previously, the age-related mechanisms that have been implicated in the pathogenesis of involutional osteoporosis include mitochondrial dysfunction, oxidative stress, a decrease in the histone deacetylase Sirt1, activation of the FoxO transcription factors, senescence of mesenchymal stem cells and osteocytes, and declining autophagy. Importantly, all such mechanisms adversely affect the skeleton independently of sex. Moreover, all features of skeletal aging in mice, including the loss of cancellous and cortical bone mass as well as the development of increased cortical porosity, are independent of sex steroid deficiency [13] (Fig. 26.5).

Furthermore, the mechanisms responsible for the effects of sex steroid deficiency in both female and male mice are evidently distinct from the mechanisms of aging. Indeed, an increase in H_2O_2 generation in cells of the osteoclast lineage seems responsible for the loss of cortical bone caused by acute sex steroid deficiency, but not by aging [13]. On the other hand, an increase in H_2O_2 generation in the mitochondria of cells of osteoblast progenitors (and their descendants) with old age seems to be partially responsible for the age-dependent loss of cortical bone. An age-related change in cells of the osteoblast lineage, for example senescence causing increased production of osteoclastogenic cytokines, may also contribute to the increased cortical porosity. In any case, because osteonal remodeling is not as well organized in rats and mice as in humans, future work is needed before one can extrapolate directly this preclinical evidence to human cortical bone.

ROLE IN THE TREATMENT OF OSTEOPOROSIS

Hormone replacement therapy with estrogens alone or in combination with progesterone was the main treatment of osteoporosis for over 50 years. During the last 10 years, however, this is no longer the case because of: (i) the recognition of the serious side effects associated with natural estrogen-based therapies in the uterus, breast, and the cardiovascular system as well as their decreased efficacy in older women with time since menopause [86, 87]; and (ii) the availability of alternative and more potent antiresorptive agents, such as the bisphosphonates

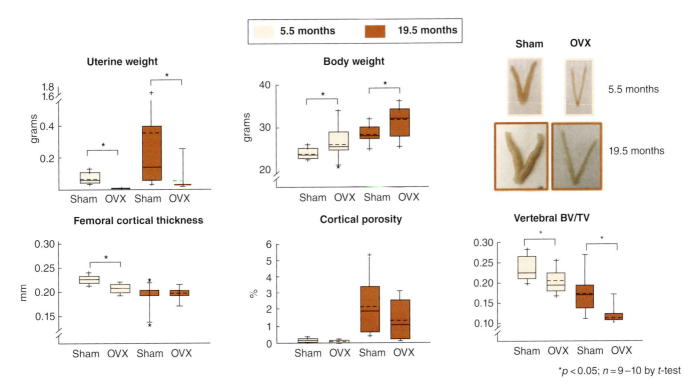

Fig. 26.5. The effects of aging on cortical bone are independent of estrogen status. Mice were either sham operated (n = 10) or ovariectomized (5.5 months, n = 10; 19.5 months, n = 9) for 6 weeks. Boxes depict values from the 25th to 75th quartiles, the middle line depicts the mean, and the vertical whiskers show the 10th and 90th percentiles; values outside this range are plotted as dots. Femoral cortical thickness was determined by microCT in the midshaft region and cortical porosity in the distal metaphysis. Cancellous bone (bone volume/tissue volume, BV/TV) was measured in the fifth lumbar vertebra. *$p < 0.05$ by two-way ANOVA.

and the anti-RANKL antibody, denosumab. However, initiation of hormone replacement therapy soon after menopause for the management of menopausal symptoms and limited duration of this therapy might have a safer benefit-to-harm ratio [88].

Selective estrogen receptor modulators (SERMs), such as raloxifene and lasofoxifene, remain a useful alternative in women with high breast cancer risk and/or contraindications to the other drugs. However, their efficacy is lower than natural estrogens and even lower as compared with the available alternative antiresorptive agents. Currently, there is no clear understanding of how SERMs act as agonists in some tissues and antagonists in others. This situation hinders rational design of an ideal SERM that can be used long term in the prevention and treatment of osteoporosis without any of the side effects associated with estrogens, including the adverse effects on clotting mechanisms and venous thromboembolic events [89, 90].

Testosterone has not been found to be efficacious in male osteoporosis and there are concerns for potential cardiovascular risk [91, 92]. Selective AR modulators (SARMs) have been shown in preclinical studies to retain the anabolic efficacy of androgens on bone and muscle while acting as partial antagonists in the prostate [93]. However, the development of SARMs as therapeutics has been hindered by evidence that, as a class, they may have adverse effects on the heart. Therefore, the future of such compounds is at this stage uncertain.

It is the authors' expectation that elucidation of the estrogen target genes and deeper understanding of the age-related mechanisms causing skeletal involution may lead to new, more effective, and safer therapies; even ones that could simultaneously treat more than one disease of old age.

REFERENCES

1. Almeida M, Laurent MR, Dubois V, et al. Estrogens and androgens in skeletal physiology and pathophysiology. Physiol Rev. 2017;97(1):135–87.
2. Vanderschueren D, Laurent MR, Claessens F, et al. Sex steroid actions in male bone. Endocr Rev. 2014;35(6):906–60.
3. Laurent MR, Vanderschueren D. Reproductive endocrinology: functional effects of sex hormone-binding globulin variants. Nat Rev Endocrinol. 2014;10(9):516–7.
4. Khosla S, Melton LJ, III, Atkinson EJ, et al. Relationship of serum sex steroid levels and bone turnover markers with bone mineral density in men and women: a key role for bioavailable estrogen. J Clin Endocrinol Metab. 1998;83(7):2266–74.

5. Beato M, Klug J. Steroid hormone receptors: an update. Hum Reprod Update. 2000;6(3):225–36.
6. Bevan C, Parker M. The role of coactivators in steroid hormone action. Exp Cell Res. 1999;253(2):349–56.
7. Hammes SR, Levin ER. Extranuclear steroid receptors: nature and actions. Endocr Rev. 2007;28(7):726–41.
8. Kousteni S, Han L, Chen J-R, et al. Kinase-mediated regulation of common transcription factors accounts for the bone-protective effects of sex steroids. J Clin Invest. 2003;111:1651–64.
9. Bartell SM, Han L, Kim HN, et al. Non-nuclear-initiated actions of the estrogen receptor protect cortical bone mass. Mol Endocrinol. 2013;27(4):649–56.
10. Chambliss KL, Wu Q, Oltmann S, et al. Non-nuclear estrogen receptor alpha signaling promotes cardiovascular protection but not uterine or breast cancer growth in mice. J Clin Invest. 2010;120(7):2319–30.
11. Ucer S, Iyer S, Bartell SM, et al. The effects of androgens on murine cortical bone do not require AR or ERalpha signaling in osteoblasts and osteoclasts. J Bone Miner Res. 2015;30(7):1138–49.
12. Manolagas SC, O'Brien CA, Almeida M. The role of estrogen and androgen receptors in bone health and disease. Nat Rev Endocrinol. 2013;9(12):699–712.
13. Ucer S, Iyer S, Kim H-N, et al. The effects of aging and sex steroid deficiency on the murine skeleton are independent and mechanistically distinct. J Bone Min Res. 2016;S30.
14. Sinnesael M, Jardi F, Deboel L, et al. The androgen receptor has no direct antiresorptive actions in mouse osteoclasts. Mol Cell Endocrinol. 2015;411:198–206.
15. Kousteni S, Bellido T, Plotkin LI, et al. Nongenotropic, sex-nonspecific signaling through the estrogen or androgen receptors: dissociation from transcriptional activity. Cell. 2001;104:719–30.
16. Tomkinson A, Gevers EF, Wit JM, et al. The role of estrogen in the control of rat osteocyte apoptosis. J Bone Miner Res. 1998;13(8):1243–50.
17. Tomkinson A, Reeve J, Shaw RW, et al. The death of osteocytes via apoptosis accompanies estrogen withdrawal in human bone. J Clin Endocrinol Metab. 1997;82(9):3128–35.
18. Emerton KB, Hu B, Woo AA, et al. Osteocyte apoptosis and control of bone resorption following ovariectomy in mice. Bone. 2010;46(3):577–83.
19. Moriishi T, Maruyama Z, Fukuyama R, et al. Overexpression of Bcl2 in osteoblasts inhibits osteoblast differentiation and induces osteocyte apoptosis. PLoS One. 2011;6(11):e27487.
20. Li Y, Li A, Yang X, et al. Ovariectomy-induced bone loss occurs independently of B cells. J Cell Biochem. 2007;100(6):1370–5.
21. Lee K, Jessop H, Suswillo R, et al. Endocrinology: bone adaptation requires oestrogen receptor-alpha. Nature. 2003;424(6947):389.
22. Almeida M, Iyer S, Martin-Millan M, et al. Estrogen receptor-alpha signaling in osteoblast progenitors stimulates cortical bone accrual. J Clin Invest. 2013;123(1):394–404.
23. Saxon LK, Galea G, Meakin L, et al. Estrogen receptors alpha and beta have different gender-dependent effects on the adaptive responses to load bearing in cancellous and cortical bone. Endocrinology. 2012;153(5):2254–66.
24. Saxon LK, Robling AG, Castillo AB, et al. The skeletal responsiveness to mechanical loading is enhanced in mice with a null mutation in estrogen receptor-beta. Am J Physiol Endocrinol Metab. 2007;293(2):E484–E491.
25. Onal M, Xiong J, Chen X, et al. Receptor activator of nuclear factor kappaB ligand (RANKL) protein expression by B lymphocytes contributes to ovariectomy-induced bone loss. J Biol Chem. 2012;287(35):29851–60.
26. Fujiwara Y, Piemontese M, Liu Y, et al. RANKL produced by osteocytes is required for the increase in B cells and bone loss caused by estrogen deficiency in mice. J Biol Chem. 2016.
27. Henning P, Ohlsson C, Engdahl C, et al. The effect of estrogen on bone requires ERalpha in nonhematopoietic cells but is enhanced by ERalpha in hematopoietic cells. Am J Physiol Endocrinol Metab. 2014;307(7):E589–E595.
28. Carson JA, Manolagas SC. Effects of sex steroids on bones and muscles: similarities, parallels, and putative interactions in health and disease. Bone. 2015;80:67–78.
29. Ophoff J, Van PK, Callewaert F, et al. Androgen signaling in myocytes contributes to the maintenance of muscle mass and fiber type regulation but not to muscle strength or fatigue. Endocrinology. 2009;150(8):3558–66.
30. Manolagas SC, Kousteni S, Jilka RL. Sex steroids and bone. Recent Prog Horm Res. 2002;57:385–409.
31. Weitzmann MN, Pacifici R. Estrogen deficiency and bone loss: an inflammatory tale. J Clin Invest. 2006;116(5):1186–94.
32. Eghbali-Fatourechi G, Khosla S, Sanyal A, et al. Role of RANK ligand in mediating increased bone resorption in early postmenopausal women. J Clin Invest. 2003;111(8):1221–30.
33. Xiong J, Onal M, Jilka RL, et al. Matrix-embedded cells control osteoclast formation. Nature Medicine. 2011;17(10):1235–41.
34. Clowes JA, Riggs BL, Khosla S. The role of the immune system in the pathophysiology of osteoporosis. Immunol Rev. 2005;208:207–27.
35. Iyer S, Kim HN, Ucer S, et al. From conditional ER alpha deletion mouse models to novel gene targets of the antiresorptive effects of estrogens. J Bone Miner Res. 2016;31(Suppl 1):S128.
36. Fu J, Li S, Feng R, et al. Multiple myeloma-derived MMP-13 mediates osteoclast fusogenesis and osteolytic disease. J Clin Invest. 2016;125(5):1759–72.
37. van der Eerden BC, Karperien M, Wit JM. Systemic and local regulation of the growth plate. Endocr Rev. 2003;24(6):782–801.
38. Jones ME, Boon WC, McInnes K, et al. Recognizing rare disorders: aromatase deficiency. Nat Clin Pract Endocrinol Metab. 2007;3(5):414–21.
39. Caruso-Nicoletti M, Cassorla F, Skerda M, et al. Short term, low dose estradiol accelerates ulnar growth in boys. J Clin Endocrinol Metab. 1985;61(5):896–8.
40. Stratakis CA, Vottero A, Brodie A, et al. The aromatase excess syndrome is associated with feminization of both sexes and autosomal dominant transmission of aberrant

P450 aromatase gene transcription. J Clin Endocrinol Metab. 1998;83(4):1348–57.
41. Smith EP, Boyd J, Frank GR, et al. Estrogen resistance caused by a mutation in the estrogen- receptor gene in a man. N Engl J Med. 1994;331(16):1056–61.
42. Quigley CA, De Bellis A, Marschke KB, et al. Androgen receptor defects: historical, clinical, and molecular perspectives. Endocr Rev. 1995;16(3):271–321.
43. Bouillon R, Bex M, Vanderschueren D, et al. Estrogens are essential for male pubertal periosteal bone expansion. J Clin Endocrinol Metab. 2004;89(12):6025–9.
44. Callewaert F, Boonen S, Vanderschueren D. Sex steroids and the male skeleton: a tale of two hormones. Trends Endocrinol Metab. 2010;21(2):89–95.
45. Windahl SH, Andersson N, Borjesson AE, et al. Reduced bone mass and muscle strength in male 5alpha-reductase type 1 inactivated mice. PLoS ONE. 2011;6(6):e21402.
46. Sims NA, Dupont S, Krust A, et al. Deletion of estrogen receptors reveals a regulatory role for estrogen receptors-beta in bone remodeling in females but not in males. Bone. 2002;30(1):18–25.
47. Callewaert F, Venken K, Kopchick JJ, et al. Sexual dimorphism in cortical bone size and strength but not density is determined by independent and time-specific actions of sex steroids and IGF-1: evidence from pubertal mouse models. J Bone Miner Res. 2010;25(3):617–26.
48. Callewaert F, Venken K, Ophoff J, et al. Differential regulation of bone and body composition in male mice with combined inactivation of androgen and estrogen receptor-alpha. FASEB J. 2009;23(1):232–40.
49. Windahl SH, Vidal O, Andersson G, et al. Increased cortical bone mineral content but unchanged trabecular bone mineral density in female ERbeta(-/-) mice. J Clin Invest. 1999;104(7):895–901.
50. Santen RJ, Brodie H, Simpson ER, et al. History of aromatase: saga of an important biological mediator and therapeutic target. Endocr Rev. 2009;30(4):343–75.
51. Borjesson AE, Lagerquist MK, Liu C, et al. The role of estrogen receptor alpha in growth plate cartilage for longitudinal bone growth. J Bone Miner Res. 2010;25(12):2690–700.
52. Tajima Y, Yokose S, Kawasaki M, et al. Ovariectomy causes cell proliferation and matrix synthesis in the growth plate cartilage of the adult rat. Histochem J. 1998;30(7):467–72.
53. Manolagas SC. Birth and death of bone cells: basic regulatory mechanisms and implications for the pathogenesis and treatment of osteoporosis. Endocr Rev. 2000;21(2):115–37.
54. Bonewald LF. The amazing osteocyte. J Bone Miner Res. 2011;26(2):229–38.
55. Xiong J, O'Brien CA. Osteocyte RANKL: new insights into the control of bone remodeling. J Bone Miner Res. 2012;27(3):499–505.
56. Manolagas SC, Parfitt AM. For whom the bell tolls: distress signals from long-lived osteocytes and the pathogenesis of metabolic bone diseases. Bone. 2013;54(2):272–8.
57. Jilka RL, Hangoc G, Girasole G, et al. Increased osteoclast development after estrogen loss: mediation by interleukin-6. Science. 1992;257:88–91.
58. Hui SL, Slemenda CW, Johnston CC, Jr. Age and bone mass as predictors of fracture in a prospective study. J Clin Invest. 1988;81(6):1804–9.
59. Khosla S, Melton LJ, III, Riggs BL. The unitary model for estrogen deficiency and the pathogenesis of osteoporosis: is a revision needed? J Bone Miner Res. 2011;26(3):441–51.
60. Parfitt AM, Mundy GR, Roodman GD, et al. A new model for the regulation of bone resorption, with particular reference to the effects of bisphosphonates. J Bone Miner Res. 1996;11:150–9.
61. Nakamura T, Imai Y, Matsumoto T, et al. Estrogen prevents bone loss via estrogen receptor alpha and induction of Fas ligand in osteoclasts. Cell. 2007;130(5):811–23.
62. Martin-Millan M, Almeida M, Ambrogini E, et al. The estrogen receptor alpha in osteoclasts mediates the protective effects of estrogens on cancellous but not cortical bone. Mol Endocrinol. 2010;24(2):323–34.
63. Manolagas SC, Almeida M. Gone with the Wnts: beta-catenin, T-cell factor, forkhead box O, and oxidative stress in age-dependent diseases of bone, lipid, and glucose metabolism. Mol Endocrinol. 2007;21(11):2605–14.
64. Riggs BL, Wahner HW, Dunn WL, et al. Differential changes in bone mineral density of the appendicular and axial skeleton with aging: relationship to spinal osteoporosis. J Clin Invest. 1981;67(2):328–35.
65. Zebaze RM, Ghasem-Zadeh A, Bohte A, et al. Intracortical remodelling and porosity in the distal radius and post-mortem femurs of women: a cross-sectional study. Lancet. 2010;375(9727):1729–36.
66. Nicks KM, Amin S, Atkinson EJ, et al. Relationship of age to bone microstructure independent of areal bone mineral density. J Bone Miner Res. 2012;27(3):637–44.
67. Bala Y, Zebaze R, Seeman E. Role of cortical bone in bone fragility. Curr Opin Rheumatol. 2015;27(4):406–13.
68. Bala Y, Zebaze R, Ghasem-Zadeh A, et al. Cortical porosity identifies women with osteopenia at increased risk for forearm fractures. J Bone Miner Res. 2014;29(6):1356–62.
69. Almeida M, Han L, Martin-Millan M, et al. Skeletal involution by age-associated oxidative stress and its acceleration by loss of sex steroids. J Biol Chem. 2007;282(37):27285–97.
70. Jilka RL, Takahashi K, Munshi M, et al. Loss of estrogen upregulates osteoblastogenesis in the murine bone marrow: evidence for autonomy from factors released during bone resorption. J Clin Invest. 1998;101:1942–50.
71. Jilka RL, O'Brien CA, Roberson PK, et al. Dysapoptosis of osteoblasts and osteocytes increases cancellous bone formation but exaggerates cortical porosity with age. J Bone Miner Res. 2014;29(1):103–17.
72. Parfitt AM, Villanueva AR, Foldes J, et al. Relations between histologic indices of bone formation: implications for the pathogenesis of spinal osteoporosis. J Bone Miner Res. 1995;10:466–73.
73. Han ZH, Palnitkar S, Rao DS, et al. Effects of ethnicity and age or menopause on the remodeling and turnover of iliac bone: Implications for mechanisms of bone loss. J Bone Miner Res. 1997;12(4):498–508.

74. Ucer, S, Iyer, S, Han, Li, et al. H$_2$O$_2$ generated in the mitochondria of osteoclasts is required for the loss of cortical bone mass caused by estrogen or androgen deficiency, but not aging. J Bone Miner Res. 2015; 30(Suppl 1):S26.
75. Ahlborg HG, Johnell O, Turner CH, et al. Bone loss and bone size after menopause. N Engl J Med. 2003;349(4):327–34.
76. Bhasin S, Pencina M, Jasuja GK, et al. Reference ranges for testosterone in men generated using liquid chromatography tandem mass spectrometry in a community-based sample of healthy nonobese young men in the Framingham Heart Study and applied to three geographically distinct cohorts. J Clin Endocrinol Metab. 2011;96(8):2430–9.
77. Wu FC, Tajar A, Beynon JM, et al. Identification of late-onset hypogonadism in middle-aged and elderly men. N Engl J Med. 2010;363(2):123–35.
78. Huhtaniemi IT, Pye SR, Limer KL, et al. Increased estrogen rather than decreased androgen action is associated with longer androgen receptor CAG repeats. J Clin Endocrinol Metab. 2009;94(1):277–84.
79. Travison TG, Shackelton R, Araujo AB, et al. Frailty, serum androgens, and the CAG repeat polymorphism: results from the Massachusetts Male Aging Study. J Clin Endocrinol Metab. 2010;95(6):2746–54.
80. Smith EP, Specker B, Bachrach BE, et al. Impact on bone of an estrogen receptor-alpha gene loss of function mutation. J Clin Endocrinol Metab. 2008;93(8):3088–96.
81. Vanderschueren D, Venken K, Ophoff J, et al. Clinical review: sex steroids and the periosteum – reconsidering the roles of androgens and estrogens in periosteal expansion. J Clin Endocrinol Metab. 2006;91(2):378–82.
82. Falahati-Nini A, Riggs BL, Atkinson EJ, et al. Relative contributions of testosterone and estrogen in regulating bone resorption and formation in normal elderly men. J Clin Invest. 2000;106(12):1553–60.
83. Lopez-Otin C, Blasco MA, Partridge L, et al. The hallmarks of aging. Cell. 2013;153(6):1194–217.
84. Manolagas SC. From estrogen-centric to aging and oxidative stress: a revised perspective of the pathogenesis of osteoporosis. Endocr Rev. 2010;31(3):266–300.
85. Almeida M, O'Brien CA. Basic biology of skeletal aging: role of stress response pathways. J Gerontol A Biol Sci Med Sci. 2013;68(10):1197–208.
86. Rossouq JE, Anderson GL, Prentice RL, et al. Risks and benefits of estrogen plus progestin in healthy postmenopausal women: principal results from the Women's Health Initiative randomized controlled trial. JAMA. 2002;288(3):321–33.
87. Estrogen and progestogen use in postmenopausal women: 2010 position statement of The North American Menopause Society. Menopause. 2010;17(2):242–55.
88. Sassarini J, Lumsden MA. Oestrogen replacement in postmenopausal women. Age Ageing. 2015;44(4):551–8.
89. Cummings SR, Ensrud K, Delmas PD, et al. Lasofoxifene in postmenopausal women with osteoporosis. N Engl J Med. 2010;362(8):686–96.
90. Cummings SR, McClung M, Reginster JY, et al. Arzoxifene for prevention of fractures and invasive breast cancer in postmenopausal women. J Bone Miner Res. 2011;26(2):397–404.
91. Basaria S, Coviello AD, Travison TG, et al. Adverse events associated with testosterone administration. N Engl J Med. 2010;363(2):109–22.
92. Basaria S, Davda MN, Travison TG, et al. Risk factors associated with cardiovascular events during testosterone administration in older men with mobility limitation. J Gerontol A Biol Sci Med Sci. 2013;68(2):153–60.
93. Rosen J, Negro-Vilar A. Novel, non-steroidal, selective androgen receptor modulators (SARMs) with anabolic activity in bone and muscle and improved safety profile. J Musculoskelet Neuronal Interact. 2002;2(3):222–4.

27

Parathyroid Hormone

Thomas J. Gardella[1], Robert A. Nissenson[2], and Harald Jüppner[1,3]

[1]Endocrine Unit, Department of Medicine, Harvard Medical School, Massachusetts General Hospital, Boston, MA, USA
[2]Endocrine Research Unit, VA Medical Center, Departments of Medicine and Physiology, University of California, San Francisco, CA, USA
[3]Pediatric Nephrology Unit, Department of Pediatrics, Harvard Medical School, Massachusetts General Hospital, Boston, MA, USA

INTRODUCTION

The parathyroid glands first appear during evolution with the movement of animals from an aquatic environment to a terrestrial environment deficient in calcium. Maintenance of adequate levels of blood ionized calcium (1.1 to 1.3 mM) is required for normal neuromuscular function, bone mineralization, and many other physiological processes. Chief cells in the parathyroid gland secrete PTH in response to very small decrements in ionized calcium in order to maintain the normocalcemic state. As discussed later, PTH accomplishes this task by binding to its receptor, the PTH/PTHrP receptor (PTH1R), in target cells of bone and kidney, promoting bone resorption, and releasing calcium from the skeletal reservoir. PTH furthermore reduces urinary calcium losses and increases phosphate excretion, and it enhances intestinal calcium absorption albeit indirectly through the renal production of the active vitamin D metabolite $1,25(OH)_2$ vitamin D.

Blood ionized calcium and $1,25(OH)_2$ vitamin D contribute to the negative feedback inhibition of PTH secretion, whereas serum phosphate increases PTH secretion. Fibroblast growth factor 23 (FGF23) is the third hormone contributing to the regulation of calcium and phosphate homeostasis; it promotes renal phosphate excretion and reduces the circulating levels of $1,25(OH)_2$ vitamin D thus diminishing intestinal calcium absorption. The interplay between serum calcium, PTH, FGF23, $1,25(OH)_2$ vitamin D, and phosphate permit blood ionized calcium levels to be maintained within very narrow limits over a wide range of dietary calcium intake. This chapter summarizes our current understanding of the biology of PTH secretion and action (for an historical perspective on the field, see [1]).

PTH AND PTH-RELATED PEPTIDE (PTHrP)

Mammalian PTH is synthesized as a pre-pro-peptide comprising 115 amino acids, but only the single chain 84 amino acid full-length polypeptide, PTH(1-84), is secreted by the parathyroid glands; very limited expression has also been detected in the rodent hypothalamus and thymus. Normal development of the parathyroid glands depends on the transcription factor GCMB (Gcm2 in rodents) as well as several other upstream proteins, including SOX3, the transcription factor cascade Hoxa3-Pax1/9-Eya, GATA3, the transcription factor Tbx1, and the Shh-Bmp4 signaling network [2, 3]. Mutations in GATA3 [4, 5], GCMB [6–9], and one of the signaling proteins at the calcium-sensing receptor, namely Gα11 [10–14], can be a cause of isolated hypoparathyroidism, with hypocalcemia ranging from mild to severe [5–7, 15–20].

PTH shares homology with PTH-related protein (PTHrP), a polypeptide of 141 amino acids that is genetically and functionally distinct from PTH (Fig. 27.1). PTHrP was originally identified as the humoral mediator of hypercalcemia of malignancy [21–23], but was soon discovered to play a key role in development, particularly of the skeleton where it regulates maturation of the growth plate during endochondral bone formation. PTHrP

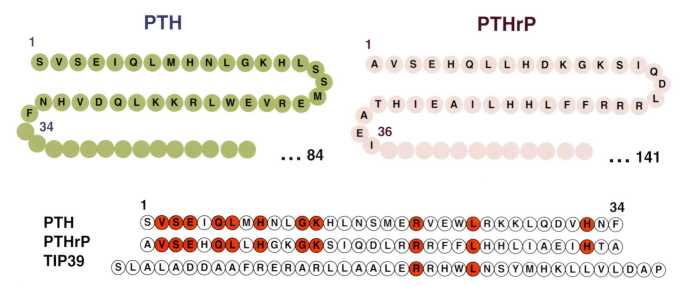

Fig. 27.1. Amino acid sequences of the amino-terminal portions of PTH, PTHrP, and intact TIP39; through alternative splicing two additional PTHrP variants can be generated, namely PTHrP(1–139) and PTHrP(1–173)..

also regulates branching morphogenesis in developing mammary glands, and likely contributes to calcium homeostasis during pregnancy and lactation [24, 25].

Both PTH and PTHrP mediate their biological actions through the same receptor, the PTH1R, which is a member of the class B subgroup of the larger protein superfamily of G protein coupled receptors (GPCRs) [24–26]. The shared amino acid sequence homology between PTH and PTHrP is strongest in the N-terminal portions of the ligands, where eight of the first 13 residues are identical; homology then diminishes and is all but absent after residue 32. This pattern of shared N-terminal amino acid homology correlates with the location of key determinants of receptor interaction, as for both ligands the N-terminal (1-34) portions contain all known sites of contact required for high affinity binding and signal transduction responses at the PTH1R. Synthetic or recombinant PTH(1-34) or PTHrP(1-36) peptides thus exhibit binding and signaling responses at the PTH1R that are generally indistinguishable from those induced by the corresponding longer-length intact peptides, PTH(1-84) or PTHrP(1-41), and indeed PTH(1-34) is used widely not only in the laboratory, but also as a PTH1R-targeted therapeutic for osteoporosis [27].

A functionally distinct peptide that shares weak but detectable homology with the N-terminal regions of PTH and PTHrP is TIP39 (tuberoinfundibular peptide of 39 amino acids) [28]. TIP39 is expressed in the brain, where it contributes to nociception, and testes, where it contributes to spermatogenesis. TIP39 does not interact efficiently with PTH1R, but rather with a structurally related GPCR, called the PTH2 receptor (PTH2R). The genes encoding PTH, PTHrP, and TIP39 are distinct in chromosomal location, but are similar in terms of exonic structure, suggesting they were derived from a common ancestral precursor gene (Fig. 27.2).

PTH SYNTHESIS AND SECRETION

In humans, the gene encoding PTH resides on the short arm of chromosome 11 [26]. In the parathyroid gland, PTH gene expression is limited to the chief cells of the parathyroid gland. The PTH protein is synthesized as a pre-pro precursor molecule of 115 amino acids, with the first 25 amino acids comprising a pre-sequence and the subsequent 6 amino acids comprising a pro-sequence, each of which is removed during protein processing to yield the intact PTH(1-84) polypeptide [29]. The pre-sequence functions as a signal sequence that directs the nascent polypeptide chain to the machinery of the endoplasmic reticulum (ER) that transports the chain across the ER membrane into the lumen. The pre-sequence is removed during this process.

Once produced and packaged into secretory vesicles within the parathyroid chief cells, the PTH(1-84) molecule is subject to two alternative fates, likely depending in part on the prevailing blood calcium concentrations. The mature hormone may be secreted into the circulation through a classical exocytotic mechanism, or it may be cleaved by calcium-sensitive proteases present within the secretory vesicles. This cleavage can result in the accumulation of carboxyl-terminal PTH fragments that lack the amino-terminal residues, which are critical for PTH1R interaction, and are thus inactive with respect to responses mediated by that receptor [26]. Cleavage of circulating PTH(1-84) into carboxyl-fragments can also occur in peripheral tissues, such as the liver and kidney [30]. Historically, cleavage of PTH(1-84) has been viewed as a mechanism for inactivation of the hormone, but data do exist to suggest that the carboxyl-terminal PTH fragments can exert some biological effects, for example on bone cells in culture, although the broader signifi-

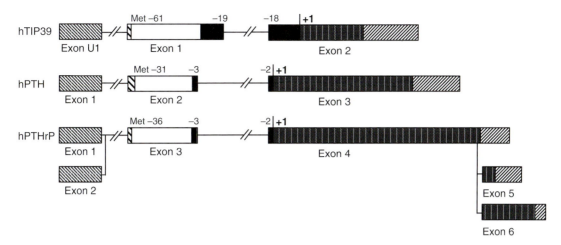

Fig. 27.2. Diagrammatic structures of the genes encoding human PTH, PTHrP, and TIP39. Boxed areas represent exons (the 5′ end of exon U1 in the TIP39 gene is not known). White boxes denote pre-sequences, black boxes are pro-sequences, gray stippled boxes are mature protein sequences, and striped boxes are noncoding regions. The small striped boxes preceding the white boxes denote untranslated exonic sequences. The positions of the initiator methionines are also indicated. +1 represents the relative position of the beginning of the secreted protein Source: [59]. Reproduced with permission of Oxford University Press.

cance of these effects, and the receptor mechanisms involved remain undefined [31].

REGULATION OF PTH SECRETION

The major physiological function of the parathyroid glands is to act as a "calciostat," sensing the prevailing blood ionized calcium level and adjusting the secretion of PTH accordingly. The relationship between ionized calcium and PTH secretion is a steep sigmoidal one, allowing significant changes in PTH secretion in response to very small changes in blood ionized calcium. The midpoint of this curve ("set-point") is a reflection of the sensitivity of the parathyroid gland secretion response to suppression by extracellular calcium.

Alteration in blood ionized calcium affects the secretion of PTH(1-84) by multiple mechanisms. Short-term increases in extracellular ionized calcium produce increased levels of intracellular free calcium in the parathyroid cell, resulting in the activation of calcium-sensitive proteases in secretory vesicles. As a result, there is increased cleavage of PTH(1-84) into carboxyl-terminal fragments. Increased extracellular calcium also reduces the release of stored PTH from secretory vesicles, although the molecular details of this regulation are not well defined. Long-term decreases in blood ionized calcium (eg, chronic dietary calcium deficiency, resistance towards PTH) result in increased PTH mRNA expression and in the number of PTH-secreting parathyroid cells.

Extracellular calcium is "measured" at the surface of the parathyroid cell through a calcium-sensing receptor (CaSR) that is abundantly expressed at the plasma membrane of these cells [32]. Increased levels of extracellular calcium suppress PTH secretion, whereas diminished levels increase PTH secretion. Unlike intracellular calcium-binding proteins, which have an affinity for free calcium in the nanomolar range (consistent with intracellular levels of free calcium), CaSR binds free calcium with an affinity in the millimolar range. The CaSR is a class C GPCR, and contains calcium-binding elements in its large extracellular domain. Binding of calcium (or calcimimetic agents such as cinacalcet) to the CaSR triggers activation of heterotrimeric G proteins comprising the alpha-subunits Gαq and Gα11, and (to a lesser extent) Gαi, which results in stimulation of phospholipase C and inhibition of adenylyl cyclase, respectively [33, 34]. This results in an increase in intracellular free calcium and a decrease in cyclic AMP levels in parathyroid cells. By mechanisms that are not fully defined, activation of these signaling pathways suppresses the synthesis and secretion of PTH. When blood ionized calcium falls, there is less intracellular signaling by the CaSR on the plasma membrane of parathyroid cells and PTH secretion consequently increases. The essential role of the CaSR can best be observed in humans bearing loss-of-function mutations in the CaSR gene. In the heterozygous state, such mutations cause familial hypocalciuric hypercalcemia (FHH), characterized by inappropriately elevated levels of PTH in the face of hypercalcemia [35–37]. These individuals are quantitatively resistant to the suppressive effect of calcium on PTH secretion because of reduced function of the CaSR; this disorder usually requires no surgical intervention. In the homozygous state, patients display a severe increase in PTH secretion with life-threatening hypercalcemia (neonatal severe hyperparathyroidism), which can be controlled with bisphosphonates and possibly with calcimetics, but usually requires removal of all parathyroid glands during infancy. Mice with homozygous and heterozygous disruption of the CaSR gene display similar phenotypes [38]. Of interest, in mice, combined deletion of Gαq and Gα11, the two major G proteins associated with the CaSR, also results in neonatal severe hyperparathyroidism, confirming the role of these closely related signaling proteins in CaSR signaling [39]. Heterozygous point mutations in the CaSR that induce constitutive signaling can be a cause of autosomal dominant hypocalcemia (ADH1) [40, 41]; similar findings were made in different families with Gα11 mutations (AHD2) [10–14].

MECHANISM OF ACTION OF PTH

Both PTH and PTHrP are thought to bind and activate the PTH1R via a similar mechanism (Fig. 27.3). As for the other class B GPCRs, each of which binds a peptide ligand that is similar in size to PTH(1-34), including calcitonin, glucagon, glucagon like peptide-1, corticotropin-releasing factor (CRF), secretin, and vasoactive intestinal peptide, the PTH1R binds its ligand, PTH or PTHrP, via a two-component mechanism [26, 42]. This mechanism thus involves initial docking interactions between the 15-34 helical domain of the ligand and the large (~160 amino acids) amino-terminal extracellular domain of the receptor, which enable subsequent signaling interactions to occur between the N-terminal (1-14) portion of the ligand and the heptahelical transmembrane domain region of the receptor, and these latter interactions induce the conformational changes that lead to receptor activation and coupling to a hetero-trimeric G protein, which, for the PTH1R typically contains the Gαs subunit that mediates signaling via increases in cellular cyclic AMP (Fig. 27.3).

Although both PTH and PTHrP follow this general mode of binding and activation, there are subtle differences in the interactions that could conceivably contribute to the capacity of this single receptor to mediate the distinct biological actions of two structurally distinct ligands [43, 44].

The PTH1R can be traced backward in evolutionary time to fish; in fact, the zebrafish genome contains, in addition to the PTH1R and the PTH2R orthologs, a third receptor, the PTH3R, that is more closely related to the fish PTH1R than to the fish PTH2R. The evolutionary conservation of structure and function suggests important biological roles for these receptors, even in fish, which lack discrete parathyroid glands but produce two molecules that are closely related to mammalian PTH.

In target cells, PTH ligand activation of the PTH1R results in coupling of the activated receptor to several intracellular signaling pathways [26]. The most important pathway is receptor-mediated activation of Gαs signaling, which results in stimulation of adenylyl cyclase with subsequent increases in intracellular levels of cyclic AMP, which in turn activates protein kinase A. The important role that this signaling pathway plays in PTH biology is highlighted by the renal resistance to this hormone that is observed in patients with pseudohypoparathyroidism type Ia (PHPIA), in which function or expression of Gαs in proximal renal tubular cells is diminished [45, 46] or in patients with acrodysostosis who have mutations in PRKAR1A, the regulatory subunit of PKA, which disrupts cyclic AMP binding and hence kinase activation [47, 48]. The effects of PTH on the expression of key genes that regulate bone resorption and bone formation (eg, RANKL, SOST) is mediated at least in part through the cyclic AMP pathway [49, 50].

The PTH1R also couples to Gαq/α11, resulting in the activation of phospholipase C with consequent activation

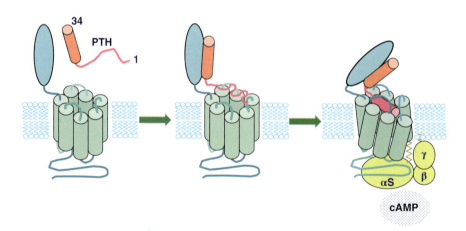

Fig. 27.3. Two-site model of ligand interaction used by the PTH1R. PTH(1-34) binds to the PTH1R via initial docking interactions between the 15-34 helical domain of the ligand and the large (~160 amino acids) amino-terminal extracellular domain of the receptor (oval), and this docking enables subsequent signaling interactions to occur between the N-terminal (1–14) portion of the ligand and the extracellularly exposed surface of the heptahelical transmembrane domain region (TMD) of the receptor. The TMD interactions induce conformational changes in the receptor that underlie receptor activation and coupling to hetero-trimeric G proteins. The PTH1R typically couples to G protein trimers containing the Gs-alpha subunit which mediates signaling via cyclic AMP.

of protein kinase C and increased intracellular free calcium. This signaling pathway appears to play an important, but less well understood, role in the action of PTH in kidney and bone [51–53]. Binding of PTH to the PTH1R also recruits the adaptor protein β-arrestin to the plasma membrane [26, 54]. As for most GPCRs, β-arrestin plays a key role in terminating PTH1R signaling by mediating receptor desensitization and internalization, but it may also promote signaling, as through the extracellular kinase (ERK)-1/2 pathway, and thus may contribute to target cell responses to PTH [55].

Recent findings in cell culture assays have uncovered possible new variations in PTH1R signaling mechanisms. First, the ligand-activated PTH receptor has classically been thought to mediate signaling exclusively from the plasma membrane and to terminate signaling upon internalization to endosomal vesicles, which typically occurs within minutes of initial ligand binding. Fluorescent colocalization microscopy analyses and kinetic FRET-based methods that assess intermolecular proximities, recently revealed that the PTH1R remains associated with the ligand, Gαs and adenylyl cyclase during endocytosis, and moreover that these internalized complexes can be functional [56–58]. These findings suggest that the temporal and/or spatial domain at which PTH receptor signaling occurs in a given target cell may be an important determinant of that target cell's overall response.

REFERENCES

1. Potts JT, Gardella TJ. Progress, paradox, and potential: parathyroid hormone research over five decades. Ann N Y Acad Sci. 2007;1117:196–208.
2. Zajac JD, Danks JA. The development of the parathyroid gland: from fish to human. Curr Opin Nephrol Hyperten. 2008;17(4):353–6.
3. Gordon J, Patel SR, Mishina Y, et al. Evidence for an early role for BMP4 signaling in thymus and parathyroid morphogenesis. Dev Biol. 2010;339(1):141–54.
4. Van Esch H, Groenen P, Nesbit MA, et al. GATA3 haploinsufficiency causes human HDR syndrome. Nature. 2000;406(6794):419–22.
5. Grigorieva IV, Mirczuk S, Gaynor KU, et al. Gata3-deficient mice develop parathyroid abnormalities due to dysregulation of the parathyroid-specific transcription factor Gcm2. J Clin Invest. 2010;120(6):2144–55.
6. Ding C, Buckingham B, Levine MA. Familial isolated hypoparathyroidism caused by a mutation in the gene for the transcription factor GCMB. J Clin Invest. 2001;108(8):1215–20.
7. Mannstadt M, Bertrand G, Muresan M, et al. Dominant-negative GCMB mutations cause an autosomal dominant form of hypoparathyroidism. J Clin Endocrinol Metab. 2008;93(9):3568–76.
8. Thomee C, Schubert SW, Parma J, et al. GCMB mutation in familial isolated hypoparathyroidism with residual secretion of parathyroid hormone. J Clin Endocrinol Metab. 2005;90(5):2487–92.
9. Tomar N, Bora H, Singh R, et al. Presence and significance of a R110W mutation in the DNA-binding domain of GCM2 gene in patients with isolated hypoparathyroidism and their family members. Eur J Endocrinol. 2010;162(2):407–21.
10. Nesbit MA, Hannan FM, Howles SA, et al. Mutations affecting G-protein subunit alpha11 in hypercalcemia and hypocalcemia. N Engl J Med. 2013;368(26):2476–86.
11. Mannstadt M, Harris M, Bravenboer B, et al. Germline mutations affecting Galpha11 in hypoparathyroidism. N Engl J Med. 2013;368(26):2532–4.
12. Li D, Opas EE, Tuluc F, et al. Autosomal dominant hypoparathyroidism caused by germline mutation in GNA11: phenotypic and molecular characterization. J Clin Endocrinol Metab. 2014;99(9):E1774–83.
13. Piret SE, Gorvin CM, Pagnamenta AT, et al. Identification of a G-protein subunit-alpha11 gain-of-function mutation, Val340Met, in a family with autosomal dominant hypocalcemia type 2 (ADH2). J Bone Min Res. 2016;31(6):1207–14.
14. Tenhola S, Voutilainen R, Reyes M, et al. Impaired growth and intracranial calcifications in autosomal dominant hypocalcemia caused by a GNA11 mutation. Eur J Endocrinol. 2016;175(3):211–8.
15. Sunthornthepvarakul T, Churesigaew S, Ngowngarmratana S. A novel mutation of the signal peptide of the preproparathyroid hormone gene associated with autosomal recessive familial isolated hypoparathyroidism. J Clin Endocrinol Metab. 1999;84(10):3792–6.
16. Arnold A, Horst SA, Gardella TJ, et al. Mutation of the signal peptide-encoding region of the preproparathyroid hormone gene in familial isolated hypoparathyroidism. J Clin Invest. 1990;86:1084–7.
17. Zahirieh A, Nesbit MA, Ali A, et al. Functional analysis of a novel GATA3 mutation in a family with the hypoparathyroidism, deafness, and renal dysplasia syndrome. J Clin Endocrinol Metab. 2005;90(4):2445–50.
18. Adachi M, Tachibana K, Asakura Y, et al. A novel mutation in the GATA3 gene in a family with HDR syndrome (hypoparathyroidism, sensorineural deafness and renal anomaly syndrome). J Pediatr Endocrinol Metab. 2006;19(1):87–92.
19. Datta R, Waheed A, Shah GN, et al. Signal sequence mutation in autosomal dominant form of hypoparathyroidism induces apoptosis that is corrected by a chemical chaperone. Proc Natl Acad Sci U S A. 2007;104(50):19989–94.
20. Lee S, Mannstadt M, Guo J, et al. A homozygous [Cys25] PTH(1-84) mutation that impairs PTH/PTHrP receptor activation defines a novel form of hypoparathyroidism. J Bone Min Res. 2015;30(10):1803–13.
21. Suva LJ, Winslow GA, Wettenhall RE, et al. A parathyroid hormone-related protein implicated in malignant hypercalcemia: cloning and expression. Science. 1987;237:893–6.
22. Strewler GJ, Stern PH, Jacobs JW, et al. Parathyroid hormone-like protein from human renal carcinoma cells. Structural and functional homology with parathyroid hormone. J Clin Invest. 1987;80:1803–7.
23. Mangin M, Webb AC, Dreyer BE, et al. Identification of a cDNA encoding a parathyroid hormone-like peptide

from a human tumor associated with humoral hypercalcemia of malignancy. Proc Natl Acad Sci U S A. 1988; 85:597–601.
24. Wysolmerski JJ. Parathyroid hormone-related protein. In: DeGroot LJ, Jameson JL (eds) *Endocrinology* (7th ed.). Philadelphia: W.B. Saunders, 2016, pp. 991–1003.
25. Kovacs CS. Fetal mineral homeostasis. In: Glorieux F (ed.) *Pediatric Bone: Biology and Diseases*. San Diego: Academic Press, 2012. p. 247–75.
26. Gardella TJ, Jüppner H, Brown EM, et al. Parathyroid hormone and parathyroid hormone receptor type 1 in the regulation of calcium and phosphate homeostasis and bone metabolism. In: DeGroot LJ, Jameson JL (eds) *Endocrinology* (7th ed.). Philadelphia: W.B. Saunders, 2016, pp. 969–90.
27. Baron R, Hesse E. Update on bone anabolics in osteoporosis treatment: rationale, current status, and perspectives. J Clin Endocrinol Metab. 2012;97(2):311–25.
28. Usdin TB, Hoare SR, Wang T, et al. TIP39: a new neuropeptide and PTH2-receptor agonist from hypothalamus. Nat Neurosci. 1999;2(11):941–3.
29. Kemper B, Habener JF, Mulligan RC, et al. Preproparathyroid hormone: a direct translation product of parathyroid messenger RNA. Proc Natl Acad Sci U S A. 1974;71(9):3731–5.
30. D'Amour P. Metabolism and measurement of parathyroid hormone. In: Bilezikian J (ed.) *The Parathyroids: Basic and Clinical Concepts*. San Diego: Academic Press; 2015, pp. 245–52.
31. Murray TM, Rao LG, Divieti P, et al. Parathyroid hormone secretion and action: evidence for discrete receptors for the carboxyl-terminal region and related biological actions of carboxyl-terminal ligands. Endocr Rev. 2005;26(1):78–113.
32. Brown EM, Gamba G, Riccardi D, et al. Cloning and characterization of an extracellular Ca^{2+}-sensing receptor from bovine parathyroid. Nature. 1993;366:575–80.
33. Brown EM, MacLeod RJ. Extracellular calcium sensing and extracellular calcium signaling. Physiol Rev. 2001;81(1):239–97.
34. Chang W, Chen TH, Pratt S, et al. Amino acids in the second and third intracellular loops of the parathyroid Ca^{2+}-sensing receptor mediate efficient coupling to phospholipase C. J Bio Chem. 2000;275(26):19955–63.
35. Pearce SH, Williamson C, Kifor O, et al. A familial syndrome of hypocalcemia with hypercalciuria due to mutations in the calcium-sensing receptor. N Engl J Med. 1996;335(15):1115–22.
36. Pollak MR, Brown EM, WuChou YH, et al. Mutations in the human Ca^{2+}-sensing receptor gene cause familial hypocalciuric hypercalcemia and neonatal severe hyperparathyroidism. Cell. 1993;75:1297–303.
37. Pollak MR, Seidman CE, Brown EM. Three inherited disorders of calcium sensing. Medicine. 1996;75(3):115–23.
38. Ho C, Conner DA, Pollak M, et al. A mouse model for familial hypocalciuric hypercalcemia and neonatal severe hyperparathyroidism. Nature Genet. 1995;11:389–94.
39. Wettschureck N, Lee E, Libutti SK, et al. Parathyroid-specific double knockout of Gq and G11 alpha-subunits leads to a phenotype resembling germline knockout of the extracellular Ca^{2+}-sensing receptor. Mol Endocrinol. 2007;21(1):274–80.
40. Diaz R, Brown E. Familial hypocalciuric hypercalcemia and other disorders due to calcium-sensing receptor mutations. In: DeGroot L, Jameson J (eds) *Endocrinology. 2.* (5th ed.). Philadelphia: W.B. Saunders, 2005, pp. 1595–609.
41. Hu J, Spiegel AM. Structure and function of the human calcium-sensing receptor: insights from natural and engineered mutations and allosteric modulators. J Cell Mol Med. 2007;11(5):908–22.
42. Cheloha RW, Gellman SH, Vilardaga JP, et al. PTH receptor-1 signalling-mechanistic insights and therapeutic prospects. Nat Rev Endocrinol. 2015;11(12):712–24.
43. Pioszak AA, Parker NR, Gardella TJ, et al. Structural basis for parathyroid hormone-related protein binding to the parathyroid hormone receptor and design of conformation-selective peptides. J Bio Chem. 2009;284(41):28382–91.
44. Dean T, Vilardaga JP, Potts JT, Jr., et al. Altered selectivity of parathyroid hormone (PTH) and PTH-related protein (PTHrP) for distinct conformations of the PTH/PTHrP receptor. Mol Endocrinol. 2008;22(1):156–66.
45. Yu S, Yu D, Lee E, et al. Variable and tissue-specific hormone resistance in heterotrimeric Gs protein alpha-subunit ($G_s\alpha$) knockout mice is due to tissue-specific imprinting of the $G_s\alpha$ gene. Proc Natl Acad Sci U S A. 1998;95(15):8715–20.
46. Weinstein LS, Liu J, Sakamoto A, et al. Minireview: GNAS: normal and abnormal functions. Endocrinology. 2004;145(12):5459–64.
47. Linglart A, Menguy C, Couvineau A, et al. Recurrent PRKAR1A mutation in acrodysostosis with hormone resistance. N Engl J Med. 2011;364(23):2218–26.
48. Linglart A, Fryssira H, Hiort O, et al. PRKAR1A and PDE4D mutations cause acrodysostosis but two distinct syndromes with or without GPCR-signaling hormone resistance. J Clin Endocrinol Metab. 2012;97(12): E2328–38.
49. Fu Q, Manolagas SC, O'Brien CA. Parathyroid hormone controls receptor activator of NF-kappaB ligand gene expression via a distant transcriptional enhancer. Mol Cell Biol. 2006;26(17):6453–68.
50. Keller H, Kneissel M. SOST is a target gene for PTH in bone. Bone. 2005;37(2):148–58.
51. Pfister MF, Forgo J, Ziegler U, et al. cAMP-dependent and -independent downregulation of type II Na-Pi cotransporters by PTH. Am J Physiol. 1999;276:F720–5.
52. Guo J, Liu M, Yang D, et al. Phospholipase C signaling via the parathyroid hormone (PTH)/PTH-related peptide receptor is essential for normal bone responses to PTH. Endocrinology. 2010;151(8):3502–13.
53. Guo J, Song L, Liu M, et al. Activation of a non-cAMP/PKA signaling pathway downstream of the PTH/PTHrP receptor is essential for a sustained hypophosphatemic response to PTH infusion in male mice. Endocrinology. 2013;154(5):1680–9.
54. Vilardaga JP, Frank M, Krasel C, et al. Differential conformational requirements for activation of G proteins

and the regulatory proteins arrestin and G protein-coupled receptor kinase in the G protein-coupled receptor for parathyroid hormone (PTH)/PTH-related protein. J Bio Chem. 2001;276(36):33435–43.
55. Bianchi EN, Ferrari SL. Beta-arrestin2 regulates parathyroid hormone effects on a p38 MAPK and NFkappaB gene expression network in osteoblasts. Bone. 2009;45(4):716–25.
56. Ferrandon S, Feinstein TN, Castro M, et al. Sustained cyclic AMP production by parathyroid hormone receptor endocytosis. Nat Chem Biol. 2009;5(10):734–42.
57. Feinstein TN, Wehbi VL, Ardura JA, et al. Retromer terminates the generation of cAMP by internalized PTH receptors. Nat Chem Biol. 2011;7(5):278–84.
58. Gidon A, Al-Bataineh MM, Jean-Alphonse FG, et al. Endosomal GPCR signaling turned off by negative feedback actions of PKA and v-ATPase. Nat Chem Biol. 2014;10(9):707–9.
59. John M, Arai M, Rubin D, et al. Identification and characterization of the murine and human gene encoding the tuberoinfundibular peptide of 39 residues (TIP39). Endocrinology. 2002;143:1047–57.

28

Parathyroid Hormone-Related Protein

John J. Wysolmerski[1] and T. John Martin[2]

[1]*Section of Endocrinology and Metabolism, Department of Medicine, Yale School of Medicine, New Haven, CT, USA*
[2]*St Vincent's Institute of Medical Research, Department of Medicine, University of Melbourne, Melbourne, Australia*

INTRODUCTION

Fuller Albright first postulated that tumors associated with hypercalcemia might elaborate a PTH-like humor [1]. Work in the 1980s and 1990s provided a detailed biochemical description of humoral hypercalcemia of malignancy (HHM), which allowed the identification and characterization of PTHrP [2, 3]. We now know that PTHrP and PTH are related molecules that stimulate the same Type 1 PTH/PTHrP receptor (PTH1R) [4]. PTHrP normally serves local autocrine, paracrine, or intracrine functions but, in patients with HHM, PTHrP circulates and mimics the systemic actions of PTH. Chapter 84 will discuss malignancy-associated hypercalcemia in detail. This chapter will review the physiology of PTHrP and its role in pathophysiology apart from hypercalcemia.

THE PTHrP GENE

The human PTHrP gene (gene symbol, *PTHLH*) encompasses 9 exons and 3 promoters on the short arm of chromosome 12 (Fig. 28.1) [2, 5, 6]. Alternative splicing at the 3′ end of the gene generates different mRNAs coding for translation products of 139, 141, or 173 amino acids. Additional splicing of the 5′ end of the gene allows each different 3′ coding sequence to have a series of different 5′ untranslated sequences. The physiological significance of these various PTHrP transcripts remains unclear, although some cells preferentially transcribe specific 3′ coding variants [2, 3].

The *PTHLH* and *PTH* genes share structural elements and sequence homology identifying them as members of a gene family that also includes tuberoinfundibular peptide of 39 residues (TIP-39) and additional PTH/PTHrP-like genes in lower vertebrates [2, 3, 5–7]. The human *PTH* and *PTHLH* genes likely arose through duplication of a common ancestor that gave rise to this gene family (Fig. 28.1). Sequence homology at the amino-terminal portion of both genes generates peptides that share eight of the first 13 amino acids and a high degree of secondary structure over the next 21 amino acids, although the two genes and peptides diverge beyond this point [2]. Given that the first 34 amino acids of both peptides are sufficient for their ability to activate the common PTH1R, this primary and secondary sequence homology is the basis for their overlapping biological effects.

Cells in almost every organ express PTHrP mRNA, especially during development [2, 3, 8, 9]. Many different hormones and growth factors regulate the transcription and/or stability of PTHrP mRNA. As with PTH, the calcium-sensing receptor (CaSR) has been found to regulate PTHrP gene expression in many cells [10,11]. Another common theme is the observation that mRNA levels are induced by mechanical deformation [12]. Recent evidence shows that noncoding RNAs are also important regulators of *PTHLH* gene expression and may help to modify chromatin and organize its three-dimensional structure to coordinate cell-specific gene expression [13]. For example, in chondrocytes, a long, noncoding RNA encoded by a cis regulatory element on chromosome 12q can interact with the *PTHLH* gene locus to regulate its transcription within long chromatin loop structures formed between chromosome 12p and 12q [14].

Primer on the Metabolic Bone Diseases and Disorders of Mineral Metabolism, Ninth Edition. Edited by John P. Bilezikian.
© 2019 American Society for Bone and Mineral Research. Published 2019 by John Wiley & Sons, Inc.
Companion website: www.wiley.com/go/asbmrprimer

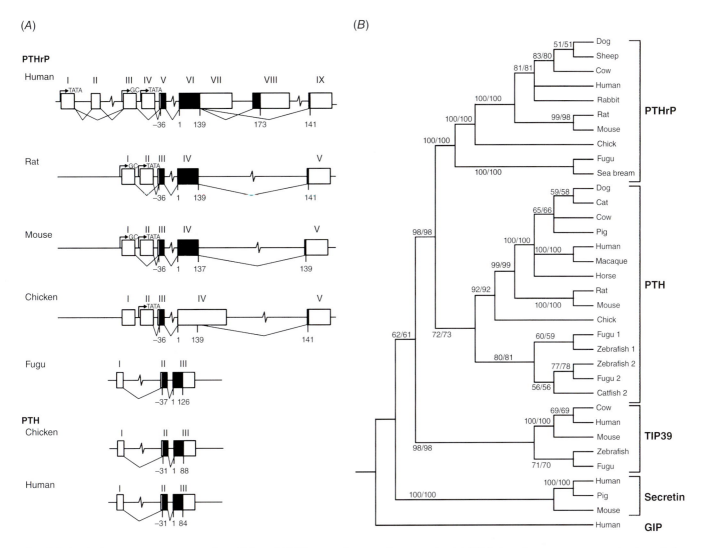

Fig. 28.1. (A) Genomic structure of PTHrP and PTH genes from various species highlighting the common organization of the genes. White boxes represent noncoding exons and black boxes represent the coding sections. Source: [2]. Reproduced with permission. (B) The phylogentic relationship between PTH, PTHrP, and TIP-39 peptides. PTH and PTHrP are most closely related and split from a common ancestor related to TIP-39, Secretin and GIP. Source: [75]. Reproduced with permission of Oxford University Press.

PTHrP IS A POLYHORMONE

Similar to the pro-opiomelanocortin (POMC) gene, the primary translation product of PTHrP undergoes posttranslational processing to generate a series of biologically active peptides [2, 15]. The details of cell-specific PTHrP processing and the biological significance of the different PTHrP peptides are not entirely clear, but several PTHrP fragments have been defined. PTHrP(1–36) is secreted from many cell types, and longer forms of PTHrP containing the amino-terminus are secreted from keratinocytes and mammary epithelial cells, and circulate in patients with cancer and during lactation [16–18]. PTHrP peptides containing the amino-terminus interact with the classical PTH/PTHrP receptor (see below). Mid-region peptides starting at amino acid 38 and extending variably to amino acids 94 through 101 have also been described [15, 19]. Mid-region PTHrP has been described to stimulate placental calcium transport, modulate renal bicarbonate handling, and it contains nuclear localization signals (see later) [2, 20, 21]. Finally, C-terminal PTHrP fragments consisting of amino acids 107-138 and 109-138 have been described to inhibit osteoclast function and stimulate osteoblast proliferation [2, 22].

PTHrP RECEPTORS

The amino-terminal portions of both PTH and PTHrP bind to and activate the same seven transmembrane-containing, G-protein coupled receptor (GPCR), termed the PTH1R [4, 23, 24]. The PTH1R is a member of class B of the large family of GPCRs and, like the *PTHLH* gene, the *PTH1R* gene is one of several related PTH receptor genes. While mid-region and C-terminal portions of PTHrP have biological activity, no receptors have been identified for these peptides.

Most studies in vitro suggest that PTHrP and PTH are equipotent at triggering identical signaling events and biological effects downstream of the PTH1R. This is also true when amino-terminal fragments of PTH and PTHrP are infused into animals [2, 4, 23, 24]. However, human subjects subjected to continuous infusion of the two peptides for 72 hours were found to become hypercalcemic with lower doses of PTH(1–34) than PTHrP(1–36) [25]. In these same studies, PTHrP was also less potent than PTH at stimulating renal $1,25(OH)_2D$ production. This may be explained by physical differences in the binding of the two peptides to different conformational states of the receptor, so that PTH remains engaged with the receptor longer than PTHrP. As a result, the duration of cAMP production is longer for stimulation with PTH(1–34) than it is for PTHrP(1–36) [23, 26]. This idea is also supported by crystal structures of the extracellular domain of the PTH1R that suggest that PTHrP(1–36) may not fit into the binding pocket of the receptor as tightly as PTH(1–34) (Fig. 28.2) [23, 27]. Thus, the human PTH1R may respond somewhat differently to PTH and PTHrP, and this might help to explain differences in the biochemical profiles of HHM and hyperparathyroidism (see Chapter 82).

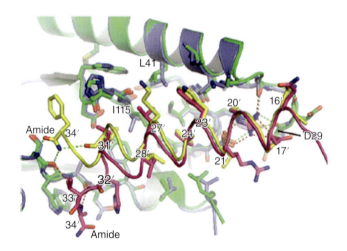

Fig. 28.2. Three-dimensional model of PTHrP (magenta) or PTH (yellow) binding to the extracellular domain (ECD) of the Type 1 PTH/PTHrP receptor (PTH1R). Numbers refer to the respective amino acids of each peptide. Selected side-chains are shown as sticks and the hydrogen bonds between PTHrP and the ECD are shown as red dashed lines whereas the hydrogen bonds between PTH and the ECD are shown as green dashed lines. Note that the helical structure of both peptides within the binding pocket is identical from amino acids 16 through 28. However, after that point they diverge and the longer helix in PTH fits into the binding pocket more tightly. Source: [27]. Reproduced with permission.

NUCLEAR PTHrP

Nuclear localization sequences (NLS) between amino acids 84–93 allow PTHrP to shuttle into and out of the nucleus in a regulated fashion. This process requires binding to microtubules and a specific shuttle protein known as importin β1, which allows PTHrP to transit the nuclear pore [22, 28]. Nuclear export is facilitated by a related shuttle protein known as CRM1. The nuclear trafficking of PTHrP is not completely understood but phosphorylation at Thr^{85} by the cell-cycle-regulated, cyclin-dependent kinase, $p34^{cdc2}$ regulates nuclear import in a cell-cycle-dependent fashion [22, 28]. The function(s) of nuclear PTHrP is unclear, but it can bind RNA and localizes to the nucleolus suggesting that PTHrP might be involved in regulating RNA trafficking, ribosomal dynamics, and/or protein translation [22, 28]. In cell lines, nuclear PTHrP regulates proliferation and/or apoptosis and, in vivo, replacement of the endogenous mouse PTHrP gene with mutant versions that cannot enter the nucleus caused widespread cellular senescence, growth retardation, and early death [22, 28–30]. Thus, nuclear PTHrP may be of fundamental importance to a variety of cell types.

PHYSIOLOGICAL FUNCTIONS OF PTHrP

Like other growth factors or cytokines, PTHrP has a myriad of functions in many different cell types. The reader is referred to more comprehensive reviews for a complete discussion of the various functions of PTHrP [2, 3, 8, 9, 24, 31]. What follows is a brief outline of selected areas where PTHrP has been well documented to have physiological effects in intact organisms.

The skeleton

Results from animal models and human mutations in the *PTH1R* have clearly documented that amino-terminal PTHrP coordinates the rate of chondrocyte differentiation and maintains the proper architecture of the growth plate, both of which support the orderly growth of long bones during development [2, 31]. Disruption of the *PTHLH* or *PTH1R* genes accelerates chondrocyte differentiation and causes a lethal form of short-limbed dwarfism, whereas transgenic overexpression of PTHrP or a constitutively active PTH1R within growth plate chondrocytes produces opposite effects [2, 31–33]. These and other observations defined a canonical pathway (Fig. 28.3) whereby PTHrP is secreted by immature chondrocytes at the top of the growth plate in response to Indian Hedgehog (IHH) produced by differentiating hypertrophic chondrocytes. PTHrP, in turn, acts on its receptor located on proliferating and prehypertrophic cells to slow their rate of differentiation into hypertrophic cells. In this manner,

Physiological Functions of PTHrP

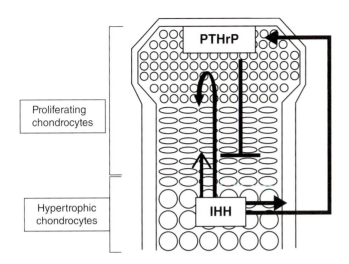

Fig. 28.3. PTHrP and Indian hedgehog (IHH) act as part of a negative feedback loop regulating chondrocyte proliferation and differentiation. The chondrocyte differentiation program proceeds from undifferentiated chondrocytes at the end of the bone, to proliferative chondrocytes within the columns, and then to prehypertrophic and terminally differentiated hypertrophic chondrocytes nearest the primary spongiosum. PTHrP is made by undifferentiated and proliferating chondrocytes at the ends of long bones. It acts through the PTH1R on proliferating and prehypertrophic chondrocytes to delay their differentiation, maintain their proliferation, and delay the production of IHH, which is made by hypertrophic cells. IHH, by contrast, increases the rate of chondrocyte proliferation and differentiation and stimulates the production of PTHrP at the ends of the bone. IHH also acts on perichondrial cells in order to generate osteoblasts in the bone collar. Source: [76]. Reproduced with permission of John Wiley & Sons.

IHH and PTHrP act in a local negative feedback loop to regulate the rate of chondrocyte differentiation.

In recent years, it has become clear that the IHH-PTHrP pathway is part of a complicated web of signaling events that affects chondrocyte proliferation and differentiation in the growth plate [34]. IHH increases PTHrP expression at the ends of the growth plate by antagonizing the activity of the transcription factor, Gli3 [35, 36], whereas stimulation of mechanistic target of rapamycin complex 1 (mTORC1) has been shown to upregulate PTHrP expression in chondrocytes by activating Gli2 [37]. PTHrP acts on the PTH1R to stimulate G_s, cAMP production, and PKA activity which, in turn, mediates a series of downstream events including the phosphorylation of SOX9, the inhibition of p57 expression, the induction of Gli3, Bcl-2, and cyclin D1 expression, and, eventually, the phosphorylation and degradation of Runx2 and Runx3, transcription factors necessary for chondrocyte differentiation [31, 34]. PTHrP also modulates chondrocyte differentiation by promoting the movement of histone deacetylase 4 (HDAC4) into the nucleus, which, in turn, regulates a network of transcription factors such as Zfp521, MEF2 and Runx2 [2, 38–40]. PTHrP expression persists within round chondrocytes at the top of the growth plate in adult life (mice do not close their epiphyses) and regulated disruption of the *Pthlh* gene in these cells during postnatal life causes their abnormal differentiation, leading to loss of the growth plate [41]. This study raised the intriguing possibility that reductions in PTHrP expression or signaling might contribute to growth plate closure during adolescence in humans.

PTHrP also has important osteoanabolic functions. Heterozygous PTHrP-null mice develop osteopenia with increasing age [42]. In addition, selective deletion of the PTHrP gene from osteoblasts results in a decreased bone mass, reduced bone formation, and mineral apposition, and a reduction in the formation and survival of osteoblasts [43]. Studies in PTHrP(1–84) knock-in mice have suggested that at least some of the anabolic effects of PTHrP on osteoblast differentiation and function may be mediated through nuclear actions of PTHrP to inhibit activity of the cell cycle inhibitor, $p27^{kip1}$, and to promote the nuclear activity of Bmi-1 [29].

Finally, PTHrP is prominently expressed at the insertions sites of ligaments and tendons into bone, which are called entheses [44]. PTHrP is induced by mechanical loading of entheses, which upregulates RANKL production to induce osteoclast formation [44, 45]. These periosteal osteoclasts serve to model the bone surface during growth and development, and also erode the cortical surface to create the root system by which the tendons and ligaments are anchored into the bone. This process of periosteal bone resorption is also necessary for the migration of ligament insertion sites along the surface of the bone because longitudinal growth occurs away from the joint.

Mammary gland

The mammary gland forms as a bud-like invagination of epidermal cells that grow into a developing fatty stroma as a branching tube to become the mammary duct system [46]. In mice and in humans, epithelial cells in the nascent mammary bud produce PTHrP, which interacts with the PTH1R expressed on surrounding mesenchymal cells [47,48]. This interaction is necessary for proper differentiation of the dense mammary mesenchyme that surrounds the embryonic mammary bud so that these mesenchymal cells can maintain the mammary fate of the epithelial cells, initiate outgrowth of the duct system, and stimulate the formation of the specialized epidermis that comprises the nipple.

PTHrP is also produced by breast epithelial cells during lactation and is secreted into the maternal circulation, where it participates in the regulation of systemic calcium metabolism [17, 18, 49]. The maternal skeleton is an important source of calcium for milk production and elevated rates of bone resorption and rapid bone loss are well documented in both nursing women and rodents [50]. Elevated levels of PTHrP correlate with bone loss during lactation in humans, and circulating levels of PTHrP correlate directly with rates of bone resorption and inversely with bone mass in mice [17, 49]. In addition, mammary-specific disruption of the *Pthlh* gene during lactation reduces circulating PTHrP levels, lowers bone turnover, and

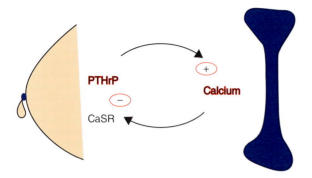

Fig. 28.4. The breast and the skeleton communicate during lactation in order to provide a steady supply of calcium for milk production. The lactating breast secretes PTHrP into the systemic circulation during lactation. PTHrP interacts with the PTH1R in bone cells in order to increase the rate of bone resorption and liberate skeletal calcium stores. Mammary epithelial cells in the lactating breast express the CaSR and suppress PTHrP production in response to increased delivery of calcium, defining a classical endocrine negative feedback loop between breast and bone.

preserves bone mass [18]. These data show that the lactating breast secretes PTHrP into the circulation to increase bone resorption. The lactating breast also expresses the CaSR, which signals to suppress PTHrP secretion in response to an increase in calcium delivery to the breast [11, 51]. These interactions define a classical endocrine negative feedback loop, whereby mammary cells secrete PTHrP to mobilize calcium from the bone. Calcium, in turn, feeds back to inhibit further PTHrP secretion from the breast. In this way, the breast communicates with the skeleton to regulate the mobilization of calcium stores to ensure a steady supply for milk production (Fig. 28.4).

Large amounts of PTHrP are secreted into milk, although its exact function in milk remains unclear [18, 51, 52]. Genetic removal of PTHrP from the mammary gland does not alter milk calcium levels and PTHrP-null mammary epithelial cells are able to transport calcium in vitro at the same rate as wild-type mammary epithelial cells [18]. However, experiments in intact mice have shown a dose-responsive, inverse relationship between milk PTHrP levels and neonatal ash calcium content [51]. The mechanisms by which milk PTHrP might modulate neonatal calcium and/or bone metabolism remain unexplored. Nevertheless, given that milk PTHrP levels vary inversely with maternal calcium availability and milk calcium content, these observations suggest that PTHrP may serve as a metabolic messenger to entrain neonatal bone and mineral metabolism to the mother's ability to supply calcium in milk.

Placenta

During pregnancy, calcium must be actively transported across the placenta from mother to fetus. Furthermore, circulating calcium concentrations in the fetus are higher than in the mother, so that calcium must be transported against a gradient [50]. In PTHrP$^{-/-}$ mice, this gradient is abolished and PTHrP-deficient fetuses are relatively hypocalcemic, suggesting that fetal PTHrP is important in mediating placental calcium transport from the mother [21, 50]. The circulating PTHrP in fetal mice is partly derived from the placenta and placental PTHrP production is regulated by the CaSR [2, 50, 53]. Experiments in sheep and mice have shown that mid-region PTHrP, not the amino-terminal portion, stimulates placental calcium transport suggesting that this action of PTHrP is not mediated by the PTH1R [20,21].

Smooth muscle and the cardiovascular system

Mechanical deformation increases the expression of PTHrP in many smooth muscle cell beds [2,3]. In turn, PTHrP acts through the PTH1R to relax the muscle that has been stretched. In the stomach, bladder, or uterus, this autocrine/paracrine feedback loop may facilitate gradual filling. In the vasculature, PTHrP is induced by vasoconstrictive agents as well as stretch and acts as a vasodilator to resistance vessels.

Secreted amino-terminal PTHrP inhibits the proliferation of vascular smooth muscle cells (VSMCs) by activating the PTH1R. However, mid-region and C-terminal portions of PTHrP act in the nucleus to stimulate the proliferation of VSMCs by regulating the levels of p27^{kip1} [54]. It is not clear how these differing actions of PTHrP are balanced in vivo but several studies suggest that PTHrP plays an important role in the response of VSMCs to injury and may contribute to the development of a neointima after angioplasty [54, 55].

Teeth

Tooth eruption relies on the formation of osteoclasts over the crown of the tooth in order to resorb the overlying bone. At the same time, bone formation at the base of the tooth propels it upward out of the dental crypt. Just before the onset of eruption, PTHrP is produced by stellate reticulum cells and signals to dental follicle cells to drive the differentiation of osteoclasts above the crypt. In the absence of PTHrP, osteoclasts do not appear, eruption fails to occur, and teeth become impacted within the surrounding bone [56].

Pancreatic islets

PTHrP is expressed by all four neuroendocrine cell types within the pancreatic islets. In beta-cells, it is stored within secretory granules and is coreleased with insulin [3,57,58]. PTHrP stimulates the PTH1R to induce the proliferation of mouse and human beta-cells [57, 58]. Overexpression of PTHrP in beta-cells in vivo causes increased islet mass, hyperinsulinemia, and hypoglycemia [57, 58]. Furthermore, daily subcutaneous injections of PTHrP(1–36) increase the proliferation of beta-cells in islets in adult mice and improve glucose tolerance [59]. These observations suggest that PTHrP administration might prove useful to maintain islet cell mass and treat diabetes.

PTHrP IN DISEASE

The most established roles for PTHrP in disease states are to cause hypercalcemia in patients with cancer and to increase bone resorption around osteolytic bone metastases [2, 10]. These topics are discussed in Chapter 84. A short discussion follows of other conditions in which PTHrP may contribute to pathophysiology or have therapeutic applications.

Skeletal dysplasias

Loss-of-function mutations in the *PTH1R* gene cause Blomstrand's chondrodysplasia, which is associated with bone abnormalities that mimic the PTHrP knockout mouse and cause fetal demise [2, 31, 60]. Gain-of-function mutations in the *PTH1R* gene cause Jansen metaphyseal chondrodysplasia. This is a form of short-limbed dwarfism that results from inhibition of chondrocyte differentiation caused by overactive PTH1R signaling [2, 31, 60]. Finally, partial loss-of-function mutations in the *PTHLH* gene have been shown to cause brachydactyly type E, a syndrome that includes short stature, shortened metacarpals and metatarsals, and learning disabilities [14,61].

Cancer

Many studies have suggested that PTHrP modulates the proliferation, differentiation, and/or survival of cancer cells from a variety of different tumor types in vitro [2, 3, 62]. However, fewer studies address the effects of PTHrP on tumor cell growth in animal models in vivo or in human cancers. Anti-PTHrP antibodies have been shown to prevent tumor growth in murine models of renal cell carcinoma [63]. Recent data have also suggested that PTHrP is an important survival factor for p53-deficient osteosarcomas [64]. The best documented role for PTHrP may be in breast cancer, although studies have shown conflicting results [3]. Some case series suggest that PTHrP predicts a more aggressive clinical course and/or the occurrence of bone metastases in patients with breast cancer [3]. However, in a large well-controlled study, Henderson and colleagues found that PTHrP expression was an independent predictor of a more benign clinical course [65]. Conflicting results have also been reported in two studies employing transgenic models of breast cancer. Mammary-specific disruption of the *Pthlh* gene increased the incidence of tumors in MMTV-Neu mice, but slowed tumor growth and reduced metastases in MMTV-PyMT mice [66, 67]. Breast cancer is heterogeneous and the opposing results of these studies suggest that the molecular context of transformation may be critical for determining PTHrP's actions. Nevertheless, large genome-wide association studies (GWAS) have shown the *PTHLH* gene to be a breast cancer susceptibility locus [68], underscoring the clinical importance of PTHrP's actions in breast cancer.

Emerging data suggest that PTHrP may contribute to metabolic derangements in cancer-associated cachexia. Sato and colleagues originally showed that passive immunization against PTHrP reversed weight loss and greatly prolonged survival in murine xenograft tumor models [69]. Subsequently, Kir and colleagues showed that PTHrP caused adipocyte "browning," hypermetabolism, and cachexia in a murine model of lung cancer [70]. Recently, a clinical study of cancer patients from Korea showed correlations between weight loss and elevated levels of circulating PTHrP, raising the possibility that PTHrP might also contribute to cachexia in human cancer [71].

Osteoporosis

Although PTHrP has not been shown to be involved in the pathophysiology of osteoporosis, translational studies showed that, like PTH(1–34), intermittent injections of PTHrP(1–36) were able to increase bone mass in humans [72]. Recently, a synthetic analog of PTHrP(1–34) called abaloparatide has been shown to increase bone density and prevent both vertebral and nonvertebral fractures in postmenopausal women with osteoporosis, albeit requiring much higher doses than PTH(1–34) [73]. Studies in vitro show that abaloparatide, like native PTHrP(1–36), may favor a conformational state of the PTH1R associated with transient receptor activation and less cAMP production, perhaps explaining its lower incidence of hypercalcemia as compared with teriparatide [73, 74].

REFERENCES

1. Mallory TB. Case records of the Massachusetts General Hospital. Case #27461. N Eng J Med. 1941;225:789–91.
2. Martin TJ. Parathyroid hormone-related protein, its regulation of cartilage and bone development, and role in treating bone diseases. Physiol Rev. 2016;96(3):831–71.
3. Wysolmerski JJ. Parathyroid hormone-related protein: an update. J Clin Endocrinol Metab. 2012;97(9):2947–56.
4. Juppner H, Abou-Samra AB, Freeman M, et al. A G protein-linked receptor for parathyroid hormone and parathyroid hormone-related peptide. Science. 1991;254(5034):1024–6.
5. Mangin M, Webb AC, Dreyer BE, et al. Identification of a cDNA encoding a parathyroid hormone-like peptide from a human tumor associated with humoral hypercalcemia of malignancy. Proc Natl Acad Sci U S A. 1988;85(2):597–601.
6. Suva LJ, Winslow GA, Wettenhall RE, et al. A parathyroid hormone-related protein implicated in malignant hypercalcemia: cloning and expression. Science. 1987;237(4817):893–6.

7. Guerreiro PM, Renfro JL, Power DM, et al. The parathyroid hormone family of peptides: structure, tissue distribution, regulation, and potential functional roles in calcium and phosphate balance in fish. Am J Physiol Regul Integr Comp Physiol. 2007;292(2):R679–96.
8. Philbrick WM, Wysolmerski JJ, Galbraith S, et al. Defining the roles of parathyroid hormone-related protein in normal physiology. Physiol Rev. 1996;76(1): 127–73.
9. Strewler GJ. The physiology of parathyroid hormone-related protein. N Engl J Med. 2000;342(3):177–85.
10. Chattopadhyay N. Effects of calcium-sensing receptor on the secretion of parathyroid hormone-related peptide and its impact on humoral hypercalcemia of malignancy. Am J Physiol Endocrinol Metab. 2006;290(5): E761–70.
11. VanHouten J, Dann P, McGeoch G, et al. The calcium-sensing receptor regulates mammary gland parathyroid hormone-related protein production and calcium transport. J Clin Invest. 2004;113(4):598–608.
12. Philbrick WM, Wysolmerski JJ, Galbraith S, et al. Defining the roles of parathyroid hormone-related protein in normal physiology. Physiol Rev. 1996;76:127–73.
13. Bohmdorfer G, Wierzbicki AT. Control of chromatin structure by long noncoding RNA. Trends Cell Biol. 2015;25(10):623–32.
14. Maass PG, Rump A, Schulz H, et al. A misplaced lncRNA causes brachydactyly in humans. J Clin Invest. 2012;122(11):3990–4002.
15. Orloff JJ, Reddy D, de Papp AE, et al. Parathyroid hormone-related protein as a prohormone: posttranslational processing and receptor interactions. Endocr Rev. 1994;15(1):40–60.
16. Burtis WJ, Brady TG, Orloff JJ, et al. Immunochemical characterization of circulating parathyroid hormone-related protein in patients with humoral hypercalcemia of cancer. N Engl J Med. 1990;322(16):1106–12.
17. Sowers MF, Hollis BW, Shapiro B, et al. Elevated parathyroid hormone-related peptide associated with lactation and bone density loss. JAMA. 1996;276(7):549–54.
18. VanHouten JN, Dann P, Stewart AF, et al. Mammary-specific deletion of parathyroid hormone-related protein preserves bone mass during lactation. J Clin Invest. 2003;112(9):1429–36.
19. Soifer NE, Dee KE, Insogna KL, et al. Parathyroid hormone-related protein. Evidence for secretion of a novel mid-region fragment by three different cell types. J Biol Chem. 1992;267(25):18236–43.
20. Care AD, Abbas SK, Pickard DW, et al. Stimulation of ovine placental transport of calcium and magnesium by mid-molecule fragments of human parathyroid hormone-related protein. Exp Physiol. 1990;75(4):605–8.
21. Kovacs CS, Lanske B, Hunzelman JL, et al. Parathyroid hormone-related peptide (PTHrP) regulates fetal-placental calcium transport through a receptor distinct from the PTH/PTHrP receptor. Proc Natl Acad Sci U S A. 1996;93(26):15233–8.
22. Jans DA, Thomas RJ, Gillespie MT. Parathyroid hormone-related protein (PTHrP): a nucleocytoplasmic shuttling protein with distinct paracrine and intracrine roles. Vitam Horm. 2003;66:345–84.
23. Cheloha RW, Gellman SH, Vilardaga JP, et al. PTH receptor-1 signalling-mechanistic insights and therapeutic prospects. Nat Rev Endocrinol. 2015;11(12):712–24.
24. Gensure RC, Gardella TJ, Juppner H. Parathyroid hormone and parathyroid hormone-related peptide, and their receptors. Biochem Biophys Res Commun. 2005;328(3):666–78.
25. Horwitz MJ, Tedesco MB, Sereika SM, et al. Continuous PTH and PTHrP infusion causes suppression of bone formation and discordant effects on 1,25(OH)2 vitamin D. J Bone Miner Res. 2005;20(10):1792–803.
26. Dean T, Vilardaga JP, Potts JT, Jr., et al. Altered selectivity of parathyroid hormone (PTH) and PTH-related protein (PTHrP) for distinct conformations of the PTH/PTHrP receptor. Mol Endocrinol. 2008;22(1):156–66.
27. Pioszak AA, Parker NR, Gardella TJ, et al. Structural basis for parathyroid hormone-related protein binding to the parathyroid hormone receptor and design of conformation-selective peptides. J Biol Chem. 2009;284(41):28382–91.
28. Fiaschi-Taesch NM, Stewart AF. Minireview: parathyroid hormone-related protein as an intracrine factor – trafficking mechanisms and functional consequences. Endocrinology. 2003;144(2):407–11.
29. Miao D, Su H, He B, et al. Severe growth retardation and early lethality in mice lacking the nuclear localization sequence and C-terminus of PTH-related protein. Proc Natl Acad Sci U S A. 2008;105(51):20309–14.
30. Toribio RE, Brown HA, Novince CM, et al. The midregion, nuclear localization sequence, and C terminus of PTHrP regulate skeletal development, hematopoiesis, and survival in mice. FASEB J. 2010;24(6):1947–57.
31. Kronenberg HM. PTHrP and skeletal development. Ann N Y Acad Sci. 2006;1068:1–13.
32. Lanske B, Karaplis AC, Lee K, et al. PTH/PTHrP receptor in early development and Indian hedgehog-regulated bone growth. Science. 1996;273(5275):663–6.
33. Weir EC, Philbrick WM, Amling M, et al. Targeted overexpression of parathyroid hormone-related peptide in chondrocytes causes chondrodysplasia and delayed endochondral bone formation. Proc Natl Acad Sci U S A. 1996;93(19):10240–5.
34. Marino R. Growth plate biology: new insights. Curr Opin Endocrinol Diabetes Obes. 2011;18(1):9–13.
35. Hilton MJ, Tu X, Cook J, et al. Ihh controls cartilage development by antagonizing Gli3, but requires additional effectors to regulate osteoblast and vascular development. Development. 2005;132(19):4339–51.
36. Koziel L, Wuelling M, Schneider S, et al. Gli3 acts as a repressor downstream of Ihh in regulating two distinct steps of chondrocyte differentiation. Development. 2005;132(23):5249–60.
37. Yan B, Zhang Z, Jin D, et al. mTORC1 regulates PTHrP to coordinate chondrocyte growth, proliferation and differentiation. Nat Commun. 2016;7:11151.
38. Correa D, Hesse E, Seriwatanachai D, et al. Zfp521 is a target gene and key effector of parathyroid hormone-related peptide signaling in growth plate chondrocytes. Dev Cell. 2010;19(4):533–46.

39. Kozhemyakina E, Cohen T, Yao TP, et al. Parathyroid hormone-related peptide represses chondrocyte hypertrophy through a protein phosphatase 2A/histone deacetylase 4/MEF2 pathway. Mol Cell Biol. 2009;29(21): 5751–62.
40. Seriwatanachai D, Densmore MJ, Sato T, et al. Deletion of Zfp521 rescues the growth plate phenotype in a mouse model of Jansen metaphyseal chondrodysplasia. FASEB J. 2011;25(9):3057–67.
41. Hirai T, Chagin AS, Kobayashi T, et al. Parathyroid hormone/parathyroid hormone-related protein receptor signaling is required for maintenance of the growth plate in postnatal life. Proc Natl Acad Sci U S A. 2011;108(1):191–6.
42. Amizuka N, Karaplis AC, Henderson JE, et al. Haploinsufficiency of parathyroid hormone-related peptide (PTHrP) results in abnormal postnatal bone development. Dev Biol. 1996;175(1):166–76.
43. Miao D, He B, Jiang Y, et al. Osteoblast-derived PTHrP is a potent endogenous bone anabolic agent that modifies the therapeutic efficacy of administered PTH 1-34. J Clin Invest. 2005;115(9):2402–11.
44. Chen X, Macica C, Nasiri A, et al. Mechanical regulation of PTHrP expression in entheses. Bone. 2007;41(5): 752–9.
45. Wang M, VanHouten JN, Nasiri AR, et al. PTHrP regulates the modeling of cortical bone surfaces at fibrous insertion sites during growth. J Bone Miner Res. 2013;28(3):598–607.
46. Hens JR, Wysolmerski JJ. Key stages of mammary gland development: molecular mechanisms involved in the formation of the embryonic mammary gland. Breast Cancer Res. 2005;7(5):220–4.
47. Foley J, Dann P, Hong J, et al. Parathyroid hormone-related protein maintains mammary epithelial fate and triggers nipple skin differentiation during embryonic breast development. Development. 2001;128(4):513–25.
48. Wysolmerski JJ, Cormier S, Philbrick WM, et al. Absence of functional type 1 parathyroid hormone (PTH)/PTH-related protein receptors in humans is associated with abnormal breast development and tooth impaction. J Clin Endocrinol Metab. 2001;86(4):1788–94.
49. VanHouten JN, Wysolmerski JJ. Low estrogen and high parathyroid hormone-related peptide levels contribute to accelerated bone resorption and bone loss in lactating mice. Endocrinology. 2003;144(12):5521–9.
50. Kovacs CS. Maternal mineral and bone metabolism during pregnancy, lactation, and post-weaning recovery. Physiol Rev. 2016;96(2):449–547.
51. Mamillapalli R, VanHouten J, Dann P, et al. Mammary-specific ablation of the calcium-sensing receptor during lactation alters maternal calcium metabolism, milk calcium transport, and neonatal calcium accrual. Endocrinology. 2013;154(9):3031–42.
52. Budayr AA, Halloran BP, King JC, et al. High levels of a parathyroid hormone-like protein in milk. Proc Natl Acad Sci U S A. 1989;86(18):7183–5.
53. Kovacs CS, Ho-Pao CL, Hunzelman JL, et al. Regulation of murine fetal-placental calcium metabolism by the calcium-sensing receptor. J Clin Invest. 1998;101(12): 2812–20.
54. Fiaschi-Taesch N, Sicari BM, Ubriani K, et al. Cellular mechanism through which parathyroid hormone-related protein induces proliferation in arterial smooth muscle cells: definition of an arterial smooth muscle PTHrP/p27kip1 pathway. Circ Res. 2006;99(9):933–42.
55. Ishikawa M, Akishita M, Kozaki K, et al. Expression of parathyroid hormone-related protein in human and experimental atherosclerotic lesions: functional role in arterial intimal thickening. Atherosclerosis. 2000;152(1):97–105.
56. Philbrick WM, Dreyer BE, Nakchbandi IA, et al. Parathyroid hormone-related protein is required for tooth eruption. Proc Natl Acad Sci U S A. 1998;95(20): 11846–51.
57. Vasavada RC, Cavaliere C, D'Ercole AJ, et al. Overexpression of parathyroid hormone-related protein in the pancreatic islets of transgenic mice causes islet hyperplasia, hyperinsulinemia, and hypoglycemia. J Biol Chem. 1996;271(2):1200–8.
58. Vasavada RC, Wang L, Fujinaka Y, et al. Protein kinase C-zeta activation markedly enhances beta-cell proliferation: an essential role in growth factor mediated beta-cell mitogenesis. Diabetes. 2007;56(11):2732–43.
59. Williams K, Abanquah D, Joshi-Gokhale S, et al. Systemic and acute administration of parathyroid hormone-related peptide(1-36) stimulates endogenous beta cell proliferation while preserving function in adult mice. Diabetologia. 2011;54(11):2867–77.
60. Juppner H, Schipani E. Receptors for parathyroid hormone and parathyroid hormone-related peptide: from molecular cloning to definition of diseases. Curr Opin Nephrol Hypertens. 1996;5(4):300–6.
61. Thomas-Teinturier C, Pereda A, Garin I, et al. Report of two novel mutations in PTHLH associated with brachydactyly type E and literature review. Am J Med Genet A. 2016;170(3):734–42.
62. Luparello C. Parathyroid hormone-related protein (PTHrP): a key regulator of life/death decisions by tumor cells with potential clinical applications. Cancers (Basel). 2011;3(1):396–407.
63. Talon I, Lindner V, Sourbier C, et al. Antitumor effect of parathyroid hormone-related protein neutralizing antibody in human renal cell carcinoma in vitro and in vivo. Carcinogenesis. 2006;27(1):73–83.
64. Walia MK, Ho PM, Taylor S, et al. Activation of PTHrP-cAMP-CREB1 signaling following p53 loss is essential for osteosarcoma initiation and maintenance. Elife. 2016;5.
65. Henderson MA, Danks JA, Slavin JL, et al. Parathyroid hormone-related protein localization in breast cancers predict improved prognosis. Cancer Res. 2006;66(4): 2250–6.
66. Fleming NI, Trivett MK, George J, et al. Parathyroid hormone-related protein protects against mammary tumor emergence and is associated with monocyte infiltration in ductal carcinoma in situ. Cancer Res. 2009;69(18): 7473–9.

67. Li J, Karaplis AC, Huang DC, et al. PTHrP drives breast tumor initiation, progression, and metastasis in mice and is a potential therapy target. J Clin Invest. 2011;121(12):4655–69.
68. Ghoussaini M, Fletcher O, Michailidou K, et al. Genome-wide association analysis identifies three new breast cancer susceptibility loci. Nat Genet. 2012;44(3):312–8.
69. Sato K, Yamakawa Y, Shizume K, et al. Passive immunization with anti-parathyroid hormone-related protein monoclonal antibody markedly prolongs survival time of hypercalcemic nude mice bearing transplanted human PTHrP-producing tumors. J Bone Miner Res. 1993;8(7):849–60.
70. Kir S, White JP, Kleiner S, et al. Tumour-derived PTH-related protein triggers adipose tissue browning and cancer cachexia. Nature. 2014;513(7516):100–4.
71. Hong N, Yoon HJ, Lee YH, et al. Serum PTHrP predicts weight loss in cancer patients independent of hypercalcemia, inflammation, and tumor Burden. J Clin Endocrinol Metab. 2016;101(3):1207–14.
72. Horwitz MJ, Augustine M, Khan L, et al. A comparison of parathyroid hormone-related protein (1-36) and parathyroid hormone (1-34) on markers of bone turnover and bone density in postmenopausal women: the PrOP study. J Bone Miner Res. 2013;28(11):2266–76.
73. Miller PD, Hattersley G, Riis BJ, et al. Effect of abaloparatide vs placebo on new vertebral fractures in postmenopausal women with osteoporosis: a randomized clinical trial. JAMA. 2016;316(7):722–33.
74. Hattersley G, Dean T, Corbin BA, et al. Binding selectivity of abaloparatide for PTH-type-1-receptor conformations and effects on downstream signaling. Endocrinology. 2016;157(1):141–9.
75. Papasani MR, Gensure RC, Yan YL, et al. Identification and characterization of the zebrafish and fugu genes encoding tuberoinfundibular peptide 39. Endocrinology. 2004;145(11):5294–304.
76. Kronenberg HM. The role of the perichondrium in fetal bone development. Ann N Y Acad Sci. 2007;1116:59–64.

29

Calcium-Sensing Receptor

Geoffrey N. Hendy

Departments of Medicine, Physiology, and Human Genetics, McGill University, Montréal, QC, Canada

INTRODUCTION

The extracellular ionized calcium (Ca^{2+}_o) concentration is maintained within a normal range of 1.1 to 1.3 mM [1]. Ca^{2+}_o has numerous roles, for example as a cofactor for clotting factors and other proteins, and modulating neuronal excitability [1]. Salts of calcium and phosphorus provide the mineral phase of the skeleton. In contrast, the resting cytosolic calcium concentration (Ca^{2+}_i), is ~100 nM, 10,000-fold lower than that of Ca^{2+}_o [2]. Ca^{2+}_i is a key intracellular second messenger, regulating cellular motility, differentiation, proliferation, and apoptosis as well as muscle contraction and hormonal secretion [2]. All Ca^{2+}_i derives from Ca^{2+}_o. Therefore, maintaining Ca^{2+}_o at a constant level ensures that calcium is available for its intracellular roles.

In mammals, a homeostatic system comprising the PTH-secreting parathyroid glands, thyroidal calcitonin (CT)-secreting C-cells, kidney, bone, and intestines [1] maintains the near constancy of Ca^{2+}_o. Key components of this system are the cells that sense small perturbations in Ca^{2+}_o from its normal value and respond so as to return Ca^{2+}_o to normal. The parathyroid glands play key roles in this process by secreting PTH in response to hypocalcemia, which then increases renal tubular reabsorption of Ca^{2+}, contributes to net release of Ca^{2+} from bone and enhances intestinal Ca^{2+} absorption by increasing renal synthesis of the active vitamin D metabolite, 1,25-dihydroxyvitamin D [1,25(OH)$_2$D] [1].

This chapter describes the properties and functions of the calcium-sensing receptor (CaSR), a G-protein coupled receptor (GPCR) that plays a central role in Ca^{2+}_o homeostasis by virtue of its ability to sense Ca^{2+}_o [1, 3]. The CaSR provides the principal mechanism in parathyroid cells, C-cells, and several nephron segments in the kidney, as well as in bone and intestine, for measuring the level of Ca^{2+}_o. It serves as the body's calciostat to modulate the functions of those cell types listed above that participate in Ca^{2+}_o homeostasis.

STRUCTURE AND FUNCTION OF THE CaSR

Protein sequence alignment and phylogenetic analysis provide evidence of a functional association of the CaSR, cloned initially from parathyroid gland [3], with the vertebrate skeleton that has an ancient origin [4]. Thus, the CaSR originated before the migration of vertebrates from the oceans onto dry land and maintains stability of Ca^{2+}_o in bony and cartilaginous fishes [4].

The CaSR is a member of GPCR family C, other members of which are metabolic glutamate receptors (mGluRs), γ-aminobutyric acid B (GABA$_B$) receptors, taste and pheromone receptors and the amino acid and cation-sensing family C receptor group 6 member A (GPRC6A) [5, 6].

CASR gene and properties of the CaSR

CASR gene

The single copy *CASR* gene maps to chromosome 3q in humans. Two promoters (P1 and P2) drive transcription from exon 1A and exon 1B, respectively, and the alternative transcripts splice to the common exon 2 that has the translation initiation ATG codon (Fig. 29.1). Exons 2–7 encode the CaSR protein of 1078 amino acids. Both promoters of the *CASR* have vitamin D response elements (VDREs) and cis-acting elements responsive to inflammatory cytokines interleukin-1β and interleukin-6 as well as the parathyroid-specific transcription factor, glial cells missing-2 (GCM2) [7, 8] (Fig. 29.1). The CaSR's expression is commonly downregulated in various forms of hyperparathyroidism although the precise basis for this is unknown.

Primer on the Metabolic Bone Diseases and Disorders of Mineral Metabolism, Ninth Edition. Edited by John P. Bilezikian.
© 2019 American Society for Bone and Mineral Research. Published 2019 by John Wiley & Sons, Inc.
Companion website: www.wiley.com/go/asbmrprimer

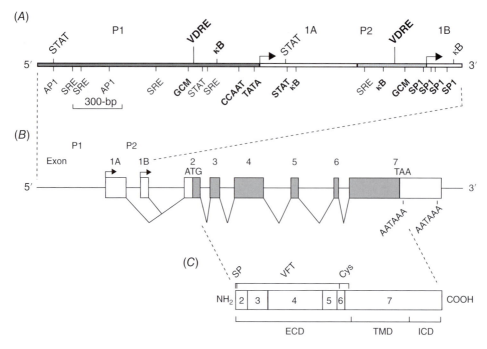

Fig. 29.1. Calcium-sensing receptor: gene, mRNA and protein. (A) *CASR* gene has two promoters, P1 and P2, gray bars, upstream of exons 1A and 1B, white bars, respectively. Arrows show transcription start sites. CCAAT and TATA boxes, and SP-1 sites driving transcription of exon 1A and 1B, respectively (bold). Cis-acting elements are shown. VDRE = vitamin D response element; κB = kappa-B element responsive to nuclear factor kappa-light-chain-enhancer of activated B cells; STAT = signal transducer and activator of transcription; GCM = glial cells missing; AP1 = activator protein 1; SRE = serum response element. Bold: those shown to be functionally active. Roman: those predicted but either not functionally active or not yet evaluated. Not all predicted cis-acting elements are shown. (B) Exon/intron organization of the *CASR* gene. Exons are drawn to scale introns are not. White bars: mRNA untranslated (exons; 1A, 1B, part of 2, part of 7). Gray bars: mRNA protein coding (exons; part of 2, 3-6, part of exon 7). ATG = initiation codon; TAA = stop codon; AATAAA = polyadenylation signals. Alternative splicing of exons 1A and 1B to exon 2 is shown. (C) CaSR protein: 1078 amino acid (aa) protein encoded by exons 2–7. SP = signal peptide; VFT = Venus flytrap domain; Cys = cysteine rich domain; ECD = extracellular domain; TMD = transmembrane domain; ICD = intracellular domain.

CaSR protein

The human CaSR has a 19-amino-acid signal peptide targeting it to the endoplasmic reticulum [9, 10] and the mature protein has an ~600 amino acid extracellular domain (ECD), a 250 amino acid transmembrane domain (TMD) and a 216 amino acid cytoplasmic tail [9, 11] (Fig. 29.1). In the endoplasmic reticulum the receptor dimerizes via both covalent and noncovalent interactions, and undergoes immature and mature glycosylation before trafficking to the cell surface in its mature dimeric active state [9, 11, 12]. The CaSR interacts with several diverse proteins that may aid its trafficking to the cell surface, maintain its expression there, facilitate its signaling and its endocytosis and recycling to the cell surface [13, 14].

The ECD of each CaSR monomer has a bilobed Venus flytrap-like (VFT) structure. Immediately after the VFT there is a cysteine-rich domain and then a linker sequence before the TMD [6, 11]. There is marked cooperativity of binding of Ca^{2+} with the receptor and several Ca^{2+} binding sites have been proposed, most within the VFT domain [6, 9]. Binding sites for aromatic amino acids have been proposed with L-amino acids acting in an allosteric fashion to promote activation by mineral ions [6, 15].

CaSR activation

Until recently, knowledge of CaSR structure and activation was based upon ECD crystal structures of the related mGluRs in active and inactive conformations [6, 9, 16]. Recently, the crystal structure of the CaSR ECD has been elucidated to provide insights into the binding to the VFT of Ca^{2+} and Mg^{2+} and amino acids, and subsequent conformational changes [17, 18].

The crystal structure of the entire ECD (the VFT and the cysteine-rich domain) of the CaSR has been determined in the resting and active conformations [17]. Initial binding of an amino acid, in this case L-tryptophan (L-Trp), was essential for the fully active structure to be achieved upon subsequent cation (Ca^{2+}) binding (Fig. 29.2). Multiple novel binding sites were revealed for Ca^{2+} and PO_4^{3-} ions with both being crucial for structural integrity of the receptor. Ca^{2+} ions stabilize the active state and PO_4^{3-} ions reinforce the inactive conformation [17]. In a second study, the crystal structure of the ECD VFT (without the cys-rich domain) of human CaSR was determined in the closed state bound with Mg^{2+} [18]. A novel high-affinity co-agonist, a tryptophan derivative, bound in the hinge region between LB1 and LB2 of the VFT, was identified in the crystal [18].

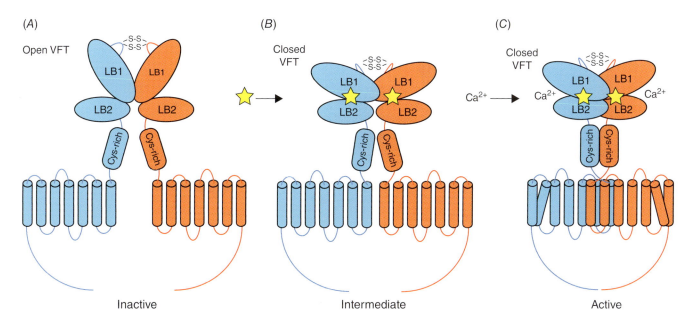

Fig. 29.2. Model of ligand activation of the CaSR dimer at the cell surface. (A) Intermolecular disulfide bonds link lobes 1 (LB1) of the Venus fly trap (VFT) domain of each protomer (monomer) and in the absence of any ligand maintain the VFT formed by LB1 and LB2 in an open (inactive) conformation. (B) An aromatic amino acid (eg, L-tryptophan or a derivative) binds within the cleft between LB1 and LB2 causing VFT closure and a rotation about the dimer interface. (C) Ca^{2+} binds within the VFT causing an extended homodimer interface to form involving not only the LB1 but also the LB2 and the cysteine-rich domain (Cys-rich) of each protomer. This is predicted to bring about reconfiguration of some of the transmembrane α-helices so that intracellular loops and part of the C-tail can productively contact G proteins triggering cell signaling.

The Ca^{2+} ion is directly involved in receptor activation by stabilizing the unique homodimer interface between the membrane proximal LB2 of the VFT and the cysteine-rich domains in the active state [17] (Fig. 29.2). Ca^{2+} enhances L-Trp binding to the CaSR ECD reinforcing the active conformation of the receptor. Exactly how these ligand-activated conformational changes alter the structure of the heptahelical TMD initiating coupling to G proteins remains to be elucidated.

Binding of Ca^{2+}_o and activation of CaSR-mediated signaling

CaSR ligands

Besides Ca^{2+} and Mg^{2+} the CaSR has many other orthosteric ligands that bind within the ECD including polycations, Sr^{2+}, Ba^{2+}, Gd^{3+}, and La^{3+}, and charged polyamines, such as spermine and spermidine, β-amyloid peptides, and aminoglycoside antibiotics such as neomycin [6]. Ionic strength negatively modulates, whereas pH and acid-base status, and the L-α-amino acids, phenylalanine, tryptophan, and histidine positively modulate sensitivity of the CaSR to Ca^{2+} [19]. The polycationic agonists have been termed type I agonists and can activate the receptor in the absence of Ca^{2+}. In contrast, type II agonists require some Ca^{2+} (or equivalent type I agonist) to be present to activate the CaSR [20]. Activation by amino acids and cations may serve to coordinate nutrient and mineral metabolism [21]. With respect to binding by anions, such as phosphate, that stabilize the inactive conformation, it can be noted that metabolic balances of Ca^{2+} and PO_4^{3-} are linked through hormones such as PTH, $1,25(OH)_2D$ and the phosphatonin, fibroblast growth factor-23 (FGF23), which control the homeostasis of both ions [22, 23]. Previously, genetic variants of the CASR have been linked to serum phosphate concentrations in GWAS and other association studies [24, 25]. In addition, the CaSR may regulate phosphate homeostasis independently of PTH, $1,25(OH)_2D$, and FGF23 [26, 27]. Further studies will be required to evaluate the CaSR as a "phosphate sensor."

Intracellular signaling by the CaSR

At the plasma membrane the CaSR can couple to the G proteins $G_{q/11}$, $G_{i/o}$, $G_{12/13}$, and in rare cases G_s [19]. The respective signaling pathways activated are PLC, increasing diacyl glycerol and inositol trisphosphate production, inhibition of adenylate cyclase or activating phosphodiesterase, activation of Rho kinases, and in rare cases activation of adenylate cyclase [19] (Fig. 29.3). CaSR-mediated activation of PKC diminishes stimulation of PLC by the receptor, conferring a negative feedback upon it, primarily by phosphorylating a key PKC site (T888) in the CaSR's C-tail [19].

CaSR expression

Ca^{2+}_o fluctuations just below or above the normal range do not alter CaSR expression in the parathyroid gland and kidney given the requirement to maintain the

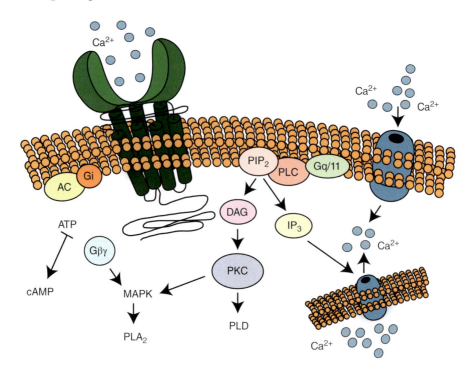

Fig. 29.3. Signaling pathways activated by the CaSR. The most well-documented interactions of the receptor are with $G_{q/11}$ and G_i (shown), but coupling to $G_{12/13}$, G_o and G_s occurs in some cell types (not shown). The stimulated CaSR couples to Gq/11 causing phospholipase C (PLC)-mediated cleavage of phosphatidylinositol-4,5-bisphosphate (PIP_2) to form 1,2-diacylglycerol (DAG), and inositol 1,4,5-trisphosphate (IP_3) production with intracellular Ca^{2+} mobilization from endoplasmic reticulum stores followed by influx of extracellular Ca^{2+} via plasma membrane Ca^{2+} channels. The activated CaSR also couples to G_i causing inhibition of cAMP formation. By increasing the intracellular Ca^{2+} concentration and phospholipid metabolites such as DAG, the stimulated CaSR activates the serine/threonine kinase protein kinase C (PKC). Both conventional and atypical isoforms of PKC can be involved. Via PLC and PKC activation, the activated CaSR also stimulates phospholipase A_2 (PLA_2) with arachidonic acid production and phospholipase D (PLD) with phosphatidic acid formation. The mitogen-activated protein kinase (MAPK) family includes the extracellular-regulated protein kinases $ERK_{1/2}$ (p44/p42), p38, and c-Jun N-terminal kinase (JNK). Both $G_{i/o}$ and $G_{q/11}$ coupling has been implicated in $ERK_{1/2}$ phosphorylation by the stimulated CaSR that can also activate p38 and JNK.

CaSR — the calciostat — at a constant level at these Ca^{2+}_o concentrations. In normal rats no evidence was found for Ca^{2+}_o modulation of parathyroid gland CaSR expression [28, 29]. However, in some situations there is evidence that increasing Ca^{2+}_o itself can upregulate the CaSR [7, 30]. The mechanism(s) involved are unknown. An additional element of regulation of CaSR by Ca^{2+}_o relates to the ability of the ligand to stimulate trafficking of the CaSR from intracellular stores to the plasma membrane by so-called agonist driven insertional secretion [13]. Expression of the CaSR at the cell surface increases whereas overall CasR expression may not necessarily be changed.

Active vitamin D, 1,25(OH)$_2$D, upregulates the CaSR [29]. This occurs by the ligand-activated VDR with its partner retinoid X receptor (RXR) transactivating the CASR gene via VDREs located in both promoters P1 and P2 of the gene [31] (Fig. 29.1). As activation of the CaSR upregulates expression of the VDR [32] this would potentiate vitamin D action via increased VDR occupancy further enhancing CaSR expression.

CaSR allosteric modulators: PAMs and NAMs

Calcimimetics are low molecular weight positive allosteric modulators (PAMs). One calcimimetic, cinacalcet, is used to suppress severe secondary hyperparathyroidism in patients with end-stage renal disease on hemodialysis [33]. Clinically, calcimimetics are often combined with VDR activators to achieve therapeutic endpoints such as acceptable reduction in serum PTH levels and attenuation of vascular complications [34]. For more on calcimimetic and calcilytic drugs for treating bone and mineral-related disorders the interested reader is referred to references [35] and [36].

PAMs bind in the TMD of the receptor and stabilize the receptor in a more active state. Phenylalkylamine calcimimetics like cinacalcet bind in the TMD in a cavity that overlaps a putative Ca^{2+} binding site and the extracellular loops [37, 38]. Calcilytics like the phenylalkylamine NPS-2143 that are negative allosteric modulators (NAMs) also bind within the TMD and induce a more inactivate conformation. The pockets for cinacalcet and NPS-2143 overlap to an extent with some of the receptor amino acids contacted being common and others not [37].

The CaSR is subject to biased signaling, the phenomenon by which distinct ligands stabilize the receptor conformation such that it activates a preferred downstream signaling pathway [38]. There is the potential for CaSR allosteric modulators that cause biased signaling and promote selective activity in different tissues to be useful for specific disease states [37, 38].

ROLES OF THE CaSR IN TISSUES MAINTAINING Ca^{2+}_o HOMEOSTASIS

Parathyroid

Activation of the CaSR on parathyroid chief cells suppresses PTH secretion [20], PTH gene expression by modulating PTH mRNA stability [39], and parathyroid cell proliferation [22], and stimulates PTH degradation [40]. Dicer is an endoribonuclease that processes microRNAs affecting expression of specific proteins. Experiments in parathyroid-specific Dicer1 knockout mice indicated that the normal response to reductions in Ca^{2+}_o of increased PTH synthesis and release was Dicer dependent [41]. Thus, inhibition of CaSR downstream signaling is not just caused by reversal or relaxation of CaSR activation, but rather CaSR inhibition by hypocalcemia is an active process that requires involvement of microRNAs and the proteins they control by a mechanism yet to be elucidated.

The actual mechanism whereby the activated CaSR inhibits PTH secretion is unclear. Activating $G_{q/11}$ is essential, because mice with knockout of both of these G proteins in the parathyroid gland have severe hyperparathyroidism similar to that present in mice homozygous for global knockout of the CaSR [42]. In humans, the parathyroid expresses predominantly GNA11 rather than GNAQ [43] and patients with either heterozygous GNA11 inactivating or activating mutations present with familial hypocalciuric hypercalcemia type 2 (FHH2) or autosomal dominant hypocalcemia type 2 (ADH2) similar to FHH1 and ADH1 caused by heterozygous loss-of-function or gain-of-function mutations in the CASR, respectively [43, 44]. FHH3 is caused by inactivating mutations in the adaptor-related protein complex 2 sigma subunit 1 (AP2S1) gene that encodes the σ-subunit of the ubiquitously expressed AP2 complex [45]. The AP2 complex is a component of clathrin-coated vesicles and facilitates the endocytosis of plasma membrane proteins including GPCRs [46].

Downstream signaling pathways involved in Ca^{2+}_o-regulated PTH release may include products of the 12- and 15-lipoxygenase pathways of arachidonate metabolism [47] and/or ERK1/2 [48]. A further downstream mechanism may involve rearrangement of the cytoskeleton to block access of the PTH-containing secretory vesicles to the plasma membrane [49].

The CaSR-mediated change in PTH gene expression is the result of a change in PTH mRNA stability rather than in PTH gene transcription. High Ca^{2+}_o recruits a degradation complex comprising both endo- and exoribonucleases to the 3′ untranslated region of the PTH mRNA to destabilize it [39]. Ca^{2+}_o also exerts a posttranslational effect on PTH expression by promoting cleavage of intact PTH stored within secretory vesicles to yield fragments comprising the mid- and COOH-terminal region, some of which are only truncated at the far NH_2-terminus [40].

Individuals homozygous for inactivating CASR mutations [1] or homozygous CASR knockout mice [50] exhibit marked parathyroid cellular hyperplasia. The CaSR-mediated inhibition of parathyroid cellular proliferation results in part from induction of the cyclin-dependent kinase inhibitor, $p21^{WAF1}$, and downregulation of the growth factor, TGF-α, and its receptor, the EGFR [51].

C-Cells

Studies in CaSR knockout mice documented the mediatory role of the CaSR in high Ca^{2+}_o-stimulated CT secretion from thyroidal C-cells [52]. Activation by the CaSR of a nonselective cation channel that depolarizes the cells activates voltage-sensitive Ca^{2+} channels producing increases in the cytosolic Ca^{2+} concentration stimulating granule exocytosis [53].

Kidney

In rat kidney the CaSR is found along nearly the entire nephron [54, 55], with highest levels of CaSR at the basolateral surface of the cells of the cortical thick ascending limb (cTAL) [56]. In the proximal tubule, the CaSR suppresses PTH-induced phosphaturia [57] and enhances VDR expression [58]. The latter may participate in the direct, high Ca^{2+}_o-elicited lowering of circulating $1,25(OH)_2D_3$ levels.

In the cTAL the CaSR mediates the inhibitory action of high peritubular Ca^{2+}_o on Ca^{2+} and Mg^{2+} reabsorption [59]. The CaSR inhibits the activity of the Na-K-2Cl (NKCC2) cotransporter that contributes to the generation of the lumen-positive, transepithelial potential gradient that is suggested to drive paracellular reabsorption of Ca^{2+} and Mg^{2+} in the cTAL [55]. NKCC2 activity is inhibited by severe CaSR activating mutants in renal cells [60] that present clinically as type 5 of the Bartter syndromes in which dysregulated epithelial transport of metal ions across the cTAL has been ascribed to mutations of one or other of critical ion transporters and channels [61]. The hypercalcemia-induced hypercalciuria has distinct CaSR-mediated components: inhibition of PTH release, which then reduces renal Ca^{2+} reabsorption, and direct suppression of reabsorption of Ca^{2+} in the cTAL [56, 62].

In the distal convoluted tubule, the basolateral CaSR stimulates the activity of the apical uptake channel, TRPV5, providing the influx mechanism for transcellular Ca^{2+} reabsorption [57]. In addition, Ca^{2+}_o increases the expression of TRPV5, calbindin D28K, the basolateral calcium pump, PMCA1b, and the sodium–calcium exchanger, NCX1 [63]. The CaSR-induced stimulation of acid secretion by the intercalated cells of the cortical collecting duct protects against the nephrolithiasis that occurs in hypercalciuric mice with knockout of TRPV5 [64]. The apical CaSR in the inner medullary collecting duct (IMCD) has been suggested to defend against calcium stone formation by inhibiting vasopressin-stimulated reabsorption of water in the IMCD when Ca^{2+}_o in the final urine is high, thereby diluting urinary Ca^{2+} [65].

Intestine

The CaSR is found on the basal surface of the small intestinal epithelial cells, within the crypts of the large and small intestines, and in the enteric nervous system [66]. The GI tract *per se* has the ability to sense Ca^{2+}_o. Hypercalcemia reduces the absorption of dietary Ca^{2+} [67] and dietary and/or blood Ca^{2+} have direct actions on the expression of the intestinal apical uptake channel, TRPV6, calbindin D9K, and PMCA1b in mice lacking the *Cyp27b1* gene that cannot make $1,25(OH)_2D$ [68].

The CaSR serves as a GI nutrient sensor, monitoring levels of both minerals and amino acids in the luminal contents so that appropriate adjustments in the digestive process can be made [19, 69]. The CaSR in the enteric nervous system, which regulates secretomotor functions of the GI tract, could contribute to hypo- and hypercalcemia enhancing and decreasing, respectively, GI motility. Activation of the CaSR in the colon markedly reduces fluid secretion, suggesting the use of calcimimetics to treat diarrheal states [70].

Bone and cartilage

The CaSR mediates the stimulatory effects of high Ca^{2+}_o on important parameters of preosteoblast and osteoblast function in vitro, such as proliferation and chemotaxis [71], and differentiation and mineralization [72]. In vivo conditional knockout of the CaSR in cells of the osteoblast lineage produced mice that had small poorly mineralized skeletons and died after several weeks, supporting a key role for the CaSR in osteoblasts [73].

While preosteoclast-like cells and cultured osteoclasts express the CaSR [74], the expression is less apparent in multinucleated osteoclasts [75]. In vivo the CaSR seems to play a permissive/stimulatory role in the generation of osteoclasts [74, 76]. The CaSR is also necessary to achieve the full calcemic and the osteoanabolic action of PTH in vivo in mice [76–78]. However, very high levels of Ca^{2+}_o suppress osteoclast activity and stimulate their apoptosis [74].

Cartilage cells including the hypertrophic chondrocytes of the growth plate, which are key for endochondral bone formation, express the CaSR [75]. Conditional knockout of the CaSR in chondrocytes of mice results in an embryonic lethal phenotype with death before day 14 of embryonic life (E14), confirming an essential role of the CaSR in chondrogenesis [73].

The CaSR in breast and placenta

CaSR expression increases markedly in the breast during lactation and returns to baseline after the termination of breastfeeding [79, 80]. In breast epithelial cells, particularly the milk-producing alveoli, the CaSR is expressed on the basolateral side opposite to the apical surface facing the milk. During lactation the CaSR suppresses the secretion of PTHrP and stimulates the transport of Ca^{2+} into breast milk. Decrease in the maternal serum Ca^{2+} level stimulates PTHrP secretion into the milk and into the maternal circulation. The increase in systemic PTHrP concentrations stimulates bone resorption releasing additional Ca^{2+} into the circulation for transport into the milk. Ligand activation of the breast epithelial cell basolateral CaSR stimulates transport of Ca^{2+} into milk by the apical calcium-ATPase pump, PMCA2 [79, 80]. Thus, during lactation the inverse relationship between Ca^{2+}_o and PTHrP release is like that between serum Ca^{2+} and PTH release from the parathyroid gland.

Therefore, during lactation the CaSR in normal breast cells coordinates a feedback loop that matches the transport of calcium into milk and maternal calcium metabolism to the supply of calcium. During malignant transformation, a switch in CaSR G protein usage (from G_i to G_s) occurs such that an increase in Ca^{2+}_o levels stimulates PTHrP production and release. Thus, the normal feedback loop is converted into a feed-forward "vicious" cycle in breast cancer cells that may promote osteolytic skeletal metastases [80].

During intrauterine life the placenta provides adequate quantities of Ca^{2+} to the developing fetal skeleton (during the third trimester in humans). Ca^{2+} is pumped transcellularly using the TRPV6, calbindin-D9K, and PMCA machinery utilized by other Ca^{2+}-transporting epithelia. The CaSR in human placenta is expressed in trophoblasts, cytotrophoblasts, and syncytiotrophoblasts [81]. In the mouse Ca^{2+} is pumped transcellularly in the placental yolk sac. Fetal $Casr^{-/-}$ mice are hypercalcemic, and have elevated PTH levels because of defective sensing to Ca^{2+} levels by the fetal parathyroid glands, leading to increased bone resorption. Placental transport in the $Casr^{-/-}$ fetuses was less than that of wild type and heterozygous fetuses [82]. The CaSR promotes placental transport of Ca^{2+} in a PTHrP-dependent manner as knockout of PTHrP (in $Pthlh^{-/-}$ mice) decreases Ca^{2+} transport to a similar level to that of $Casr^{-/-}$ fetuses.

In addition to the roles that it plays in tissues participating in Ca^{2+}_o homeostasis, the CaSR is expressed in and modulates the functions of numerous other cells uninvolved in mineral ion homeostasis [83].

REFERENCES

1. Brown EM. Clinical lessons from the calcium-sensing receptor. Nat Clin Pract Endocrinol Metab. 2007;3: 122–33.
2. Berridge MJ, Bootman MD, Roderick HL. Calcium signaling: dynamics, homeostasis and remodelling. Nat Rev Mol Cell Biol. 2003;4:517–29.
3. Brown EM, Gamba G, Riccardi D, et al. Cloning and characterization of an extracellular Ca(2+)-sensing receptor from bovine parathyroid. Nature. 1993;366:575–80.
4. Herberger AL, Loretz CA. Vertebrate extracellular calcium-sensing receptor evolution: selection in relation to life history and habitat. Comp Biochem Physiol Part D Genomics Proteomics. 2013;8(1):86–94.
5. Bräuner-Osborne H, Wellendorph P, Jensen AA. Structure, pharmacology and therapeutic prospects of family C G-protein coupled receptors. Curr Drug Targets. 2007;8:169–84.
6. Zhang C, Miller CL, Gorkhali R, et al. Molecular basis of the extracellular ligands mediated signaling by the calcium sensing receptor. Front Physiol. 2016;7:441.
7. Hendy GN, Canaff L. Calcium-sensing receptor gene: regulation of expression. Front Physiol. 2016;7:394.
8. Hendy GN, Canaff L. Calcium-sensing receptor, proinflammatory cytokines and calcium homeostasis. Semin Cell Dev Biol. 2016;49:37–43.
9. Hu J, Spiegel AM. Structure and function of the human calcium-sensing receptor: insights from natural and engineered mutations and allosteric modulators. J Cell Mol Med. 2007;11(5):908–22.
10. Pidasheva S, Canaff L, Simonds WF, et al. Impaired cotranslational processing of the calcium-sensing receptor due to signal peptide missense mutations in familial hypocalciuric hypercalcemia. Hum Mol Genet. 2005;14(12):1679–90.
11. Hendy GN, Canaff L, Cole DE. The CASR gene: alternative splicing and transcriptional control, and calcium-sensing receptor (CaSR) protein: structure and ligand binding sites. Best Pract Res Clin Endocrinol Metab. 2013;27(3):285–301.
12. Pidasheva S, Grant M, Canaff L, et al. Calcium-sensing receptor dimerizes in the endoplasmic reticulum: biochemical and biophysical characterization of CASR mutants retained intracellularly. Hum Mol Genet. 2006;15(14):2200–9.
13. Breitwieser GE. The calcium sensing receptor life cycle: trafficking, cell surface expression, and degradation. Best Pract Res Clin Endocrinol Metab. 2013;27(3):303–13.
14. Ray K. Calcium-sensing receptor: trafficking, endocytosis, recycling, and importance of interacting proteins. Prog Mol Biol Transl Sci. 2015;132:127–50.
15. Conigrave AD, Quinn SJ, Brown EM. L-amino acid sensing by the extracellular Ca2+-sensing receptor. Proc Natl Acad Sci U S A. 2000;97(9):4814–9.
16. Zajickova K, Vrbikova J, Canaff L, et al. Identification and functional characterization of a novel mutation in the calcium-sensing receptor gene in familial hypocalciuric hypercalcemia: modulation of clinical severity by vitamin D status. J Clin Endocrinol Metab. 2007;92(7): 2616–23.
17. Geng Y, Mosyak L, Kurinov I, et al. Structural mechanism of ligand activation in human calcium-sensing receptor. Elife. 2016; 5. pii: e13662.
18. Zhang C, Zhang T, Zou J, et al. Structural basis for regulation of human calcium-sensing receptor by magnesium ions and an unexpected tryptophan derivative co-agonist. Sci Adv. 2016;2(5):e1600241.
19. Conigrave AD, Ward DT. Calcium-sensing receptor (CaSR): pharmacological properties and signaling pathways. Best Pract Res Clin Endocrinol Metab. 2013;27(3):315–31.
20. Nemeth EF, Steffey ME, Hammerland LG, et al. Calcimimetics with potent and selective activity on the parathyroid calcium receptor. Proc Natl Acad Sci U S A. 1998;95(7):4040–5.
21. Conigrave AD, Mun HC, Brennan SC. Physiological significance of L-amino acid sensing by extracellular Ca(2+)-sensing receptors. Biochem Soc Trans. 2007; 35(Pt 5):1195–8.
22. Brown EM. Role of the calcium-sensing receptor in extracellular calcium homeostasis. Best Pract Res Clin Endocrinol Metab. 2013;27(3):333–43.
23. Tyler Miller R. Control of renal calcium, phosphate, electrolyte, and water excretion by the calcium-sensing receptor. Best Pract Res Clin Endocrinol Metab. 2013;27(3):345–58.
24. Kestenbaum B, Glazer NL, Köttgen A, et al. Common genetic variants associate with serum phosphorus concentration. J Am Soc Nephrol. 2010; 21(7):1223–32.
25. Babinsky VN, Hannan FM, Youhanna SC, et al. Association studies of calcium-sensing receptor (CaSR) polymorphisms with serum concentrations of glucose and phosphate, and vascular calcification in renal transplant recipients. PLoS One. 2015;10(3):e0119459.
26. Quinn SJ, Thomsen AR, Pang JL, et al. Interactions between calcium and phosphorus in the regulation of the production of fibroblast growth factor 23 in vivo. Am J Physiol Endocrinol Metab. 2013;304(3):E310–20.
27. Ba J, Brown D, Friedman PA. Calcium-sensing receptor regulation of PTH-inhibitable proximal tubule phosphate transport. Am J Physiol Renal Physiol. 2003;285(6): F1233–43.
28. Rogers KV, Dunn CK, Conklin RL, et al. Calcium receptor messenger ribonucleic acid levels in the parathyroid glands and kidney of vitamin D-deficient rats are not regulated by plasma calcium or 1,25-dihydroxyvitamin D3. Endocrinology. 1995;136(2):499–504.
29. Brown AJ, Zhong M, Finch J, et al. Rat calcium-sensing receptor is regulated by vitamin D but not by calcium. Am J Physiol. 1996;270(3 Pt 2):F454–60.
30. Chakrabarty S, Wang H, Canaff L, et al. Calcium sensing receptor in human colon carcinoma: interaction with Ca(2+) and 1,25-dihydroxyvitamin D(3). Cancer Res. 2005;65(2):493–8.
31. Canaff L, Hendy GN. Human calcium-sensing receptor gene. Vitamin D response elements in promoters P1

and P2 confer transcriptional responsiveness to 1,25-dihydroxyvitamin D. J Biol Chem. 2002;277:30337–50.
32. Rodriguez ME, Almaden Y, Cañadillas S, et al. The calcimimetic R-568 increases vitamin D receptor expression in rat parathyroid glands. Am J Physiol Renal Physiol. 2007;292(5):F1390–5.
33. Block GA, Martin KJ, de Francisco AL, et al. Cinacalcet for secondary hyperparathyroidism in patients receiving hemodialysis. N Engl J Med. 2004;350(15):1516–25.
34. Drüeke TB. Calcimimetics and outcomes in CKD. Kidney Int Suppl. (2011). 2013;3(5):431–435.
35. Nemeth EF, Shoback D. Calcimimetic and calcilytic drugs for treating bone and mineral-related disorders. Best Pract Res Clin Endocrinol Metab. 2013;27(3):373–84.
36. Nemeth EF, Goodman WG. Calcimimetic and calcilytic drugs: feats, flops, and futures. Calcif Tissue Int. 2016;98(4):341–58.
37. Leach K, Gregory KJ, Kufareva I, et al. Towards a structural understanding of allosteric drugs at the human calcium-sensing receptor. Cell Res. 2016;26(5):574–92.
38. Leach K, Conigrave AD, Sexton PM, et al. Towards tissue-specific pharmacology: insights from the calcium-sensing receptor as a paradigm for GPCR (patho)physiological bias. Trends Pharmacol Sci. 2015;36(4):215–25.
39. Naveh-Many T. Minireview: the play of proteins on the parathyroid hormone messenger ribonucleic acid regulates its expression. Endocrinology. 2010;151(4):1398–402.
40. Kawata T, Imanishi Y, Kobayashi K, et al. Direct in vitro evidence of extracellular Ca^{2+}-induced amino-terminal truncation of human parathyroid hormone (1-84) by human parathyroid cells. J Clin Endocrinol Metab. 2005;90(10):5774–8.
41. Shilo V, Ben-Dov IZ, Nechama M, et al. Parathyroid-specific deletion of dicer-dependent microRNAs abrogates the response of the parathyroid to acute and chronic hypocalcemia and uremia. FASEB J. 2015;29(9):3964–76.
42. Wettschureck N, Lee E, Libutti SK, et al. Parathyroid-specific double knockout of Gq and G11 alpha-subunits leads to a phenotype resembling germline knockout of the extracellular Ca^{2+}-sensing receptor. Mol Endocrinol. 2007;21:274–80.
43. Nesbit MA, Hannan FM, Howles SA, et al. Mutations affecting G-protein subunit α11 in hypercalcemia and hypocalcemia. N Engl J Med. 2013;368(26):2476–86.
44. Hendy GN, Guarnieri V, Canaff L. Calcium-sensing receptor and associated diseases. Prog Mol Biol Transl Sci. 2009;89:31–95.
45. Nesbit MA, Hannan FM, Howles SA, et al. Mutations in AP2S1 cause familial hypocalciuric hypercalcemia type 3. Nat Genet. 2013;45(1):93–7.
46. Hendy GN, Cole DE. Ruling in a suspect: the role of AP2S1 mutations in familial hypocalciuric hypercalcemia type 3. J Clin Endocrinol Metab. 2013;98(12):4666–9.
47. Bourdeau A, Moutahir M, Souberbielle J, et al. Effects of lipoxygenase products of arachidonate metabolism on parathyroid hormone secretion. Endocrinology. 1994;135:1109–12.
48. Corbetta S, Lania A, Filopanti M, et al. Mitogen-activated protein kinase cascade in human normal and tumoral parathyroid cells. J Clin Endocrinol Metab. 2002;87(5):2201–5.
49. Quinn SJ, Kifor O, Kifor I, et al. Role of the cytoskeleton in extracellular calcium-regulated PTH release. Biochem Biophys Res Commun. 2007;354:8–13.
50. Ho C, Conner DA, Pollak MR, et al. A mouse model of human familial hypocalciuric hypercalcemia and neonatal severe hyperparathyroidism. Nat Genet. 1995;11:389–94.
51. Cozzolino M, Lu Y, Finch J, et al. p21WAF1 and TGF-alpha mediate parathyroid growth arrest by vitamin D and high calcium. Kidney Int. 2001;60:2109–17.
52. Fudge NJ, Kovacs CS. Physiological studies in heterozygous calcium sensing receptor (CaSR) gene-ablated mice confirm that the CaSR regulates calcitonin release in vivo. BMC Physiol. 2004; 4:5.
53. McGehee DS, Aldersberg M, Liu KP, et al. Mechanism of extracellular Ca^{2+} receptor-stimulated hormone release from sheep thyroid parafollicular cells. J Physiol (Lond). 1997;502:31–44.
54. Graca JA, Schepelmann M, Brennan SC, et al. Comparative expression of the extracellular calcium-sensing receptor in the mouse, rat, and human kidney. Am J Physiol Renal Physiol. 2016;310(6):F518–33.
55. Riccardi D, Valenti G. Localization and function of the renal calcium-sensing receptor. Nat Rev Nephrol. 2016;12(7):414–25.
56. Loupy A, Ramakrishnan SK, Wootla B, et al. PTH-independent regulation of blood calcium concentration by the calcium-sensing receptor. J Clin Invest. 2012;122(9):3355–67.
57. Topala CN, Schoeber JP, Searchfield LE, et al. Activation of the Ca^{2+}-sensing receptor stimulates the activity of the epithelial Ca^{2+} channel TRPV5. Cell Calcium. 2009;45:331–9.
58. Maiti A, Beckman MJ. Extracellular calcium is a direct effecter of VDR levels in proximal tubule epithelial cells that counter-balances effects of PTH on renal Vitamin D metabolism. J Steroid Biochem Mol Biol. 2007;103:504–8.
59. Ba J, Friedman PA. Calcium-sensing receptor regulation of renal mineral ion transport. Cell Calcium. 2004;35:229–37.
60. Carmosino M, Gerbino A, Hendy GN, et al. NKCC2 activity is inhibited by the Bartter's syndrome type 5 gain-of-function CaR-A843E mutant in renal cells. Biol Cell. 2015;107(4):98–110.
61. Hebert SC. Bartter syndrome. Curr Opin Nephrol Hypertens. 2003;12(5):527–32.
62. Kantham L, Quinn SJ, Egbuna OI, et al. The calcium-sensing receptor (CaSR) defends against hypercalcemia independently of its regulation of parathyroid hormone secretion. Am J Physiol Endocrinol Metab. 2009;297:E915–23.
63. Thebault S, Hoenderop JG, Bindels RJ. Epithelial Ca^{2+} and Mg^{2+} channels in kidney disease. Adv Chronic Kidney Dis. 2006;13:110–7.

64. Renkema KY, Velic A, Dijkman HB, et al. The calcium-sensing receptor promotes urinary acidification to prevent nephrolithiasis. J Am Soc Nephrol. 2009;20:1705–13.
65. Sands JM, Naruse M, Baum M, et al. Apical extracellular calcium/polyvalent cation-sensing receptor regulates vasopressin-elicited water permeability in rat kidney inner medullary collecting duct. J Clin Invest. 1997;99:1399–405.
66. Chattopadhyay N, Cheng I, Rogers K, et al. Identification and localization of extracellular Ca(2+)-sensing receptor in rat intestine. Am J Physiol. 1998;274:G122–30.
67. Krishnamra N, Angkanaporn K, Deenoi T. Comparison of calcium absorptive and secretory capacities of segments of intact or functionally resected intestine during normo-, hypo-, and hyper-calcemia. Can J Physiol Pharmacol. 1994;72:764–70.
68. Li YC, Bolt MJ, Cao LP, et al. Effects of vitamin D receptor inactivation on the expression of calbindins and calcium metabolism. Am J Physiol Endocrinol Metab. 2001;281(3):E558–64.
69. Geibel JP, Hebert SC. The functions and roles of the extracellular Ca2+-sensing receptor along the gastrointestinal tract. Annu Rev Physiol. 2009;71:205–17.
70. Geibel J, Sritharan K, Geibel R, et al. Calcium-sensing receptor abrogates secretagogue- induced increases in intestinal net fluid secretion by enhancing cyclic nucleotide destruction. Proc Natl Acad Sci U S A. 2006;103:9390–7.
71. Chattopadhyay N, Yano S, Tfelt-Hansen J, et al. Mitogenic action of calcium-sensing receptor on rat calvarial osteoblasts. Endocrinology. 2004;145:3451–62.
72. Dvorak MM, Siddiqua A, Ward DT, et al. Physiological changes in extracellular calcium concentration directly control osteoblast function in the absence of calciotropic hormones. Proc Natl Acad Sci U S A. 2004;101:5140–5.
73. Chang W, Tu C, Chen TH, et al. The extracellular calcium-sensing receptor (CaSR) is a critical modulator of skeletal development. Sci Signal. 2008;1:ra1.
74. Mentaverri R, Yano S, Chattopadhyay N, et al. The calcium sensing receptor is directly involved in both osteoclast differentiation and apoptosis. Faseb J. 2006;20:2562–4.
75. Chang W, Tu C, Chen T-H, et al. Expression and signal transduction of calcium-sensing receptors in cartilage and bone. Endocrinology 1999;140:5883–93.
76. Shu L, Ji J, Zhu Q, et al. The calcium-sensing receptor mediates bone turnover induced by dietary calcium and parathyroid hormone in neonates. J Bone Miner Res. 2011;26:1057–71.
77. Goltzman D, Hendy GN. The calcium-sensing receptor in bone — mechanistic and therapeutic insights. Nat Rev Endocrinol. 2015;11(5):298–307.
78. Santa Maria C, Cheng Z, Li A, et al. Interplay between CaSR and PTH1R signaling in skeletal development and osteoanabolism. Semin Cell Dev Biol. 2016;49:11–23.
79. VanHouten J, Dann P, McGeoch G, et al. The calcium-sensing receptor regulates mammary gland parathyroid hormone-related protein production and calcium transport. J Clin Invest. 2004;113:598–608.
80. VanHouten JN, Wysolmerski JJ. The calcium-sensing receptor in the breast. Best Pract Res Clin Endocrinol Metab. 2013;27(3):403–14.
81. Bradbury RA, Sunn KL, Crossley M, et al. Expression of the parathyroid Ca(2+)-sensing receptor in cytotrophoblasts from human term placenta. J Endocrinol. 1998;156:425–30.
82. Kovacs CS, Ho-Pao CL, Hunzelman JL, et al. Regulation of murine fetal-placental calcium metabolism by the calcium-sensing receptor. J Clin Invest. 1998;101:2812–20.
83. Magno AL, Ward BK, Ratajczak T. The calcium-sensing receptor: a molecular perspective. Endocr Rev. 2011;32(1):3–30.

30

Vitamin D: Production, Metabolism, Action, and Clinical Requirements

Daniel D. Bikle[1], John S. Adams[2,3,4], and Sylvia Christakos[5]

[1]Departments of Medicine and Dermatology, University of California San Francisco, Biomedical Sciences Graduate Program, San Francisco, CA, USA
[2]Departments of Orthopaedic Surgery, Medicine, and Molecular Cell and Developmental Biology, University of California, Los Angeles, CA, USA
[3]Orthopaedic Hospital Research Center, University of California, Los Angeles, CA, USA
[4]UCLA Clinical and Translational Science Institute, University of California, Los Angeles, CA, USA
[5]Department of Microbiology, Biochemistry and Molecular Genetics, Rutgers, New Jersey Medical School, Newark, NJ, USA

VITAMIN D_3 PRODUCTION

Vitamin D_3 is produced from 7-dehydrocholesterol (7-DHC) (Fig. 30.1). Although irradiation of 7-DHC was known to produce previtamin D_3 (pre-D_3; which subsequently undergoes a temperature-dependent rearrangement of the triene structure to form vitamin D_3, lumisterol, and tachysterol), the physiological regulation of this pathway was not well understood until the studies of Holick and colleagues [1–3]. They showed that the formation of pre-D_3 under the influence of solar or UVB irradiation (maximal effective wavelength between 290 and 310) is relatively rapid and reaches a maximum within hours. Both the degree of epidermal pigmentation and the intensity of exposure correlate with the time required to achieve this maximal concentration of pre-D_3 but do not alter the maximal level achieved. Although pre-D_3 levels reach a maximum level, the biologically inactive metabolites lumisterol and tachysterol accumulate with continued UV exposure. Thus, prolonged exposure to sunlight does not produce toxic amounts of D_3 because of the photoconversion of pre-D_3 to lumisterol and tachysterol. By absorbing UV irradiation, melanin in the epidermis can act as a sunscreen to reduce the effectiveness of sunlight in cutaneous vitamin D_3 synthesis. Sunlight exposure increases melanin production and so provides another mechanism by which excess D_3 production can be prevented. As noted, the intensity of UV irradiation is also important for vitamin D_3 production and is dependent on latitude. In Edmonton, Canada (52°N) very little D_3 is produced in exposed skin from mid-October to mid-April, while in San Juan (18°N) the skin is able to produce D_3 all year long [4]. Clothing and commercially available sunscreens effectively prevent D_3 production.

VITAMIN D METABOLISM

Vitamin D_3 and its plant-derived counterpart vitamin D_2 (collectively referred to as vitamin D) are by themselves biologically inert for most of their actions. After cutaneous synthesis and transport into the general circulation, vitamin D disappears from the serum within a week [5]. Vitamin D is bound to the serum vitamin D binding protein (DBP) and ferried in the circulation to sites of storage (mainly fat and muscle) [6] and to the tissues, primarily the liver, where the conversion to the prohormone 25-hydroxyvitamin D (25OHD; Fig. 30.2), occurs. There are a number of cytochrome P450 enzymes capable of converting vitamin D to 25OHD, of which CYP2R1 appears to be the most important, accounting for up to 50% of the total 25OHD synthesized [7]. These enzymes exhibit a high capacity for substrate vitamin D, and release product 25OHD back into the circulation and not into the bile.

Primer on the Metabolic Bone Diseases and Disorders of Mineral Metabolism, Ninth Edition. Edited by John P. Bilezikian.
© 2019 American Society for Bone and Mineral Research. Published 2019 by John Wiley & Sons, Inc.
Companion website: www.wiley.com/go/asbmrprimer

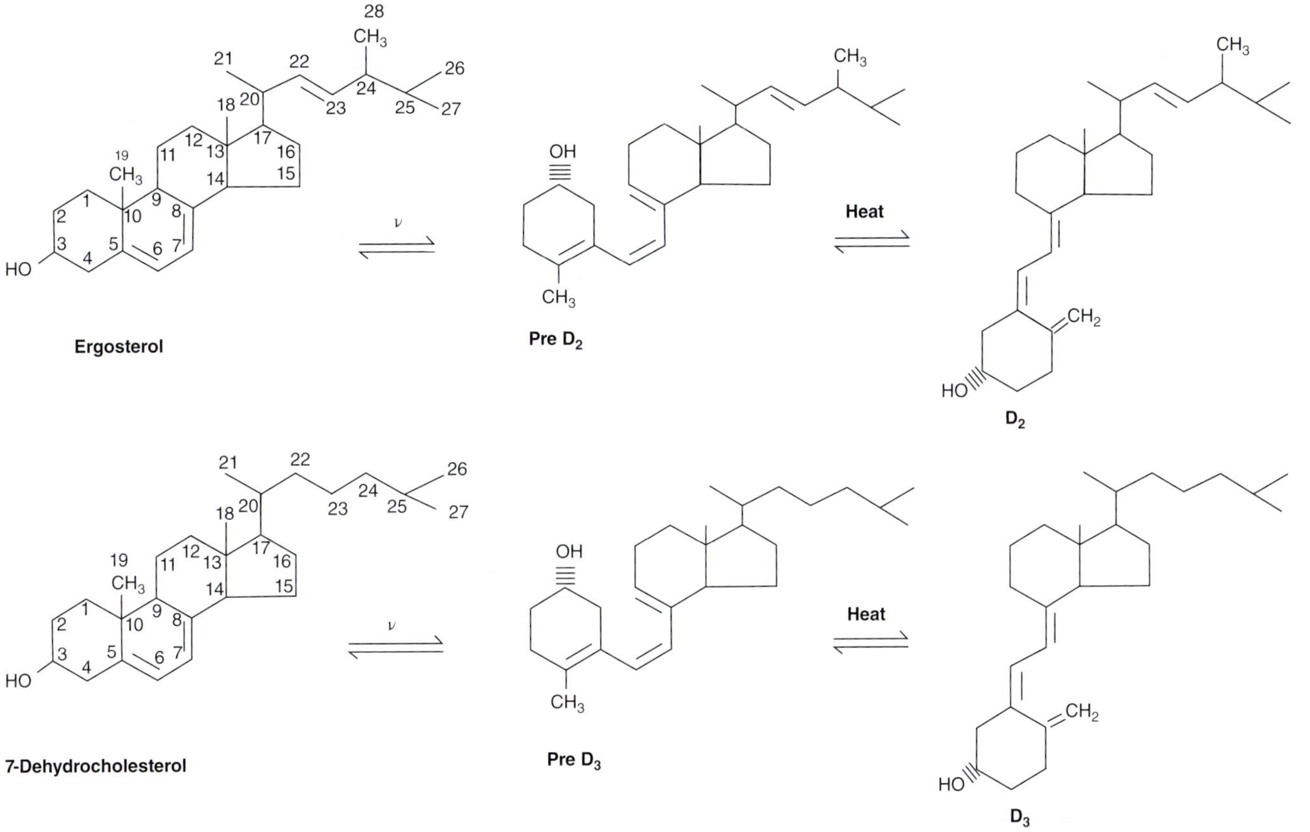

Fig. 30.1. The photolysis of ergosterol and 7-dehydrocholesterol to vitamin D_2 (ergocalciferol) and vitamin D_3 (cholecalciferol), respectively. An intermediate is formed after photolysis, which undergoes a thermal-activated isomerization to the final form of vitamin D. The rotation of the A-ring puts the 3β-hydroxyl group into a different orientation with respect to the plane of the A-ring during production of vitamin D.

As such, the serum 25OHD level is the most reliable indicator of whether too little or too much vitamin D is entering the host [8]. 25OHD is biologically inert unless present in intoxicating concentrations in the blood owing to the ingestion of large amounts of vitamin D. Otherwise, it must be converted via CYP27B1 (25OHD-1α-hydroxylase) to 1,25-dihydroxyvitamin D (1,25-(OH)$_2$D), the specific, naturally-occurring ligand for the vitamin D receptor (VDR) (Fig. 30.2). CYP27B1 is a heme-containing, inner mitochondrial membrane-embedded, cytochrome P450 mixed function oxidase requiring molecular oxygen and a source of electrons for biological activity. The proximal renal tubular epithelial cell is the richest source of CYP27B1 and responsible for generating the relatively large amounts of 1,25-(OH)$_2$D that are required to achieve the endocrine functions of the hormone in mineral ion homeostasis. This enzyme is also encountered in a number of extrarenal sites, including the placenta, immune cells, and a variety of normal and malignant epithelia [9], where it functions to provide 1,25-(OH)$_2$D for intracrine or paracrine access to the VDR in these and neighboring cells. As discussed below, the VDR has an extraordinarily broad distribution among human tissues. There are four major recognized means of regulating CYP27B1: (i) controlling the availability of substrate 25OHD; (ii) controlling the amount of CYP27B1; (iii) controlling the amount and activity of the catabolic CYP24A1 (the 24-hydroxylase, encoded by a structural gene distinct from the CYP27B1); and (iv) altering the activity of the hydroxylases by cofactor availability.

For the kidney, substrate 25OHD for the CYP27B1 is provided by the endocytic internalization of filtered, megalin/cubulin-bound DBP carrying 25OHD into the proximal tubular cell from the urinary side of that cell. Endocrine regulation of the CYP27B1 in the proximal nephron is principally controlled at the level of transcription with circulating PTH and FGF-23 being the major stimulator and inhibitor of CYP27B1 gene expression, respectively (see later). Downstream of their respective, plasma membrane-anchored receptors, PTH and FGF-23 modulate the intracellular cyclic AMP and kinase signaling pathways, respectively. In human tissues including the kidney, the CYP24A1, like the CYP27B1, is a mitochondrial P450, and serves not only to limit the amount of 1,25(OH)$_2$D leaving the kidney for distant target tissues by accelerating its catabolism to

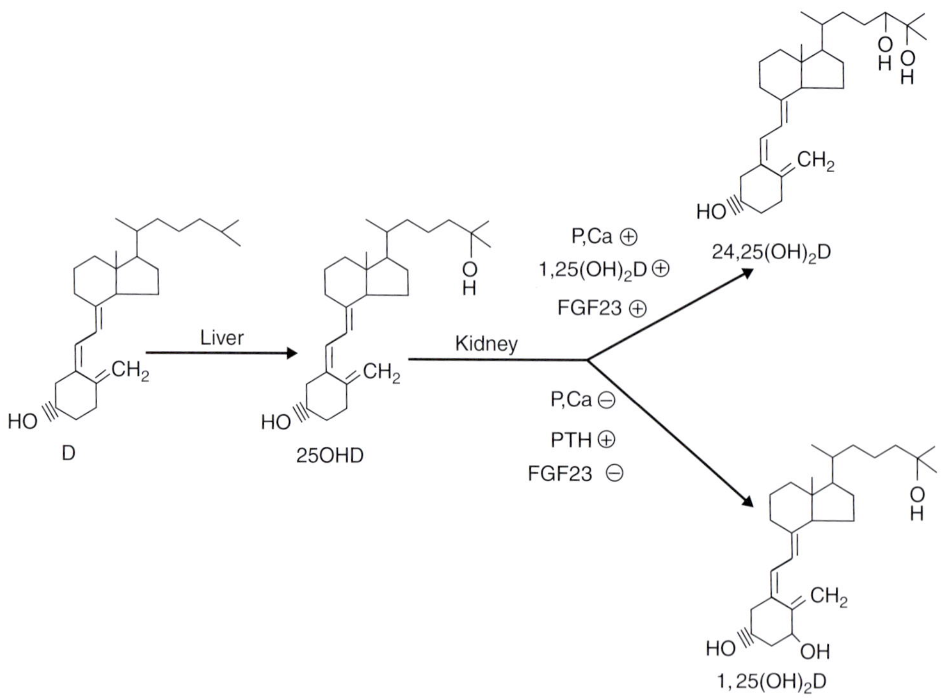

Fig. 30.2. The metabolism of vitamin D. The liver converts vitamin D to 25OHD. The kidney converts 25OHD to $1,25(OH)_2D_3$ and $24,25(OH)_2D_3$. Control of metabolism is exerted primarily at the level of the kidney, where low serum phosphorus, low serum calcium, low FGF23, and high parathyroid hormone (PTH) levels favor production of $1,25(OH)_2D_3$, whereas high serum phosphorus, calcium, FGF23, and $1,25(OH)_2D_3$ and low PTH favor $24,25(OH)_2D_3$ production.

$1,24,25\text{-}(OH)_3D$ but also by shunting available substrate 25OHD away from CYP27B1. In both cases, the 24-hydroxylated products are degraded by the same enzyme to side chain-cleaved, water-soluble catabolites. The CYP24A1 gene is under stringent transcriptional control by $1,25\text{-}(OH)_2D$ itself, providing a robust means of proximate, negative feedback regulation of the amount of $1,25\text{-}(OH)_2D$ made in and released from the synthetic tissue of origin [10]. By comparison, the activity of some of the extrarenal, intracrine/paracrine-acting CYP27B1, such as that in keratinocytes and disease-activated macrophages, appears to be primarily governed by the availability of extracellular substrate 25OHD to the enzyme, stimulatory cytokines, such as tumor necrosis factor-alpha (TNF-α), interferon-gamma (IFN-γ), and toll-like receptor (TLR) activation. In human macrophage proliferative, granuloma-forming diseases $1,25\text{-}(OH)_2D$ may be produced in excess, spilling into the circulation and causing hypercalcemia. It is postulated that this is in part because of the expression of a catalytically nonfunctional splice variant product of the catabolic CYP24A1 gene [11–13]. Also contrary to the renal CYP27B1, the extrarenal CYP27B1 in the macrophage and keratinocyte is (i) immune to control by either PTH or FGF-23, (ii) susceptible to induction by TLR ligands shed by microbial agents, and (iii) upregulated by nontraditional electron donors like nitric oxide [9].

TRANSPORT OF VITAMIN D IN THE BLOOD

With the exception of the skin, where it can be both produced and act locally as just described, in order for the hormone $1,25\text{-}(OH)_2D$ to reach any of its target tissues, it must be able to escape its synthetic site in the skin or its absorption site in the gut and be transported to tissues expressing one of the vitamin D-25-hydroxylase genes. In order to be activated to its hormonal form, 25OHD must travel and gain access to the interior of cells expressing CYP27B1. Once synthesized, $1,25(OH)_2D$ must be able to gain access to the VDR in an intra-, para- or endocrine mode in order for the genomic actions of the sterol hormone to be realized. As noted previously, the serum DBP is the specific chaperone for vitamin D and its metabolites in the serum [14]. It has a high capacity (less than 5% saturated with vitamin D metabolites in humans) and is bound with high affinity (nM range) by vitamin Ds, particularly 25-hydroxylated metabolites [11]. DBP is produced mainly in the liver and is freely filterable across the glomerulus into the urine. DBP has a serum half-life of 2.5 to 3.0 days, indicating that it must be largely reclaimed from the urine once filtered. Reclamation from the urine is achieved by DBP being bound by the endocytic, LDL-like coreceptor molecules megalin and cubulin embedded in the luminal membrane of the proximal renal tubular epithelial cell. Lack of DBP in animals or humans may be embryonically lethal.

INTERNALIZATION OF VITAMIN D METABOLITES

Once bound to DBP and shuttled to sites of metabolism, action and/or catabolism, vitamin D metabolites must gain access to the interior of their target cell and arrive safely at their intracellular destination. With the exception of kidney epithelial cells, parathyroid gland, and placenta where megalin/cubulin-mediated uptake of 25OHD is dominant, it is the free fraction of 25OHD and 1,25-(OH)$_2$D, representing <1% of the total 25OHD and 1,25-(OH)$_2$D in the general circulation [12, 13], that appears to have the best opportunity for crossing the plasma membrane of target cells. Still unclear is whether free vitamin D metabolites enter cells in vivo by simple and/or facilitated diffusion.

MOLECULAR MECHANISM OF ACTION

The mechanism of action of the active form of vitamin D, 1,25-(OH)$_2$D, is similar to that of other steroid hormones. The intracellular mediator of 1,25-(OH)$_2$D function is the vitamin D receptor (VDR). 1,25-(OH)$_2$D is bound stereospecifically with high affinity (0.1 pM range) but low capacity by the VDR, a protein that has extensive homology with other members of the superfamily of nuclear hormone receptors including receptors for steroid, thyroid, and retinoid hormones. VDR functions as a heterodimer with the retinoid X receptor (RXR) for activation for vitamin D target genes. Once formed, the 1,25-(OH)$_2$D-VDR-RXR heterodimeric complex interacts with specific DNA sequences (vitamin D response elements; VDREs) in and around target genes resulting in either activation or repression of transcription [15–18]. Although the exact sequence of these direct repeats is quite variable from gene to gene, VDRE consensus sequence consists of two direct repeats of the hexanucleotide sequence GGGTGA separated by three nucleotide pairs. The molecular mechanisms and recruited regulatory proteins involved in VDR-mediated transcription after binding of the 1,25-(OH)$_2$D-VDR-RXR heterodimeric complex to DNA are now beginning to be defined [17, 19–23]. These regulatory factors include: (i) TFIIB; (ii) several TATA binding protein associated factors (TAFs); (iii) the p160 coactivators, the DRIP (vitamin D receptor interacting protein; also known as the mediator complex coactivator complex) that functions in recruitment of RNA polymerase II; (iv) transcription factors YY1 and CCAAT enhancer binding proteins β and α; (v) steroid receptor activator-1, 2, and 3 (SRC-1, SRC-2, and SRC-3), all of which have histone acetylase (HAT) activity; and (vi) SWI/SNF complexes that also remodel chromatin. In addition, genome-wide studies of VDR binding sites have indicated that although many of the VDR regulatory regions are located in proximal promoters of target genes, most are situated many kilobases upstream and downstream and in intronic and exonic sites. VDR binding to these sites has been reported to be largely, but not exclusively [24], dependent on activation by 1,25-(OH)$_2$D [25]. These genome-wide studies have provided a new perspective on mechanisms involved in the cell-specific regulation of gene expression by the VDR in both a ligand-dependent and nondependent mode [26]. Recently, recombinant mouse and human bacterial artificial chromosomes (BACs) harboring the VDR have been introduced into mice [27]. This approach represents an important advance that will enable future studies related to the identification of tissue specific determinants and the consequence of selective mutations in mouse and human VDR-directed gene targets in vivo [27].

REGULATION OF CALCIUM AND PHOSPHATE METABOLISM

The classical, endocrine actions of 1,25-(OH)$_2$D involve its regulation of calcium and phosphate flux across three target tissues: bone, gut, and kidney. In this regard for the mineral homeostatic, endocrine actions of vitamin D, 1,25-(OH)$_2$D acts in concert with two peptide hormones, PTH and FGF23 (Fig. 30.3). In each case feed-forward and feedback regulatory loops are operative. PTH is the major stimulator of 1,25-(OH)$_2$D production in the kidney. In turn, the 1,25(OH)$_2$D-activated VDR, and perhaps the unliganded VDR [24], suppresses PTH production directly via a transcriptional mechanism and indirectly by increasing serum calcium levels. Calcium acts via the calcium sensing receptor (CaSR) in the parathyroid gland to suppress PTH release. 1,25-(OH)$_2$D increases the levels of CaSR in the parathyroid gland just as calcium increases the VDR in the parathyroid gland further enhancing the negative influence of calcium and 1,25(OH)$_2$D on PTH secretion. The parathyroid gland also expresses CYP27B1, enabling circulating 25(OH)D to regulate PTH secretion by providing substrate for local 1,25-(OH)$_2$D synthesis. On the other hand, FGF23, a hormone made largely by osteocytes in bone [28], inhibits 1,25-(OH)$_2$D production in the kidney by increasing expression of CYP24A1, whereas 1,25-(OH)$_2$D stimulates FGF23 production in bone. 1,25-(OH)$_2$D also upregulates klotho-α, and the loss of klotho results in induction of CYP27B1. Klotho- and FGF23-deficient mice exhibit homologous phenotypes (including hyperphosphatemia) and increased synthesis of 1,25-(OH$_2$)D [29].

CLASSICAL TARGET TISSUES

Bone

Whether 1,25-(OH)$_2$D acts directly on bone or whether the antirachitic effects of 1,25-(OH)$_2$D are indirect, owing to 1,25-(OH)$_2$D promotion of intestinal calcium

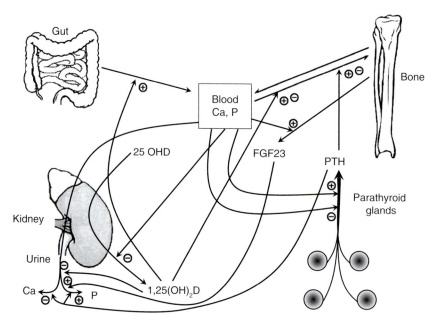

Fig. 30.3. Classical actions: bone mineral homeostasis mineral feedback loops. 1,25(OH)$_2$D interacts with other hormones, in particular FGF23 and PTH, to regulate calcium and phosphate homeostasis. As noted in the legend to Fig. 30.2, FGF23 inhibits whereas PTH stimulates 1,25(OH)$_2$D production by the kidney. In turn, 1,25(OH)$_2$D inhibits PTH production but stimulates that of FGF23.

and phosphorus absorption with secondary increased incorporation of calcium and phosphorus into bone, has been a matter of debate. VDR null mice develop secondary hyperparathyroidism, hypocalcemia, and rickets after weaning [30, 31]. However, when VDR null mice are fed a rescue diet containing high levels of calcium, phosphorus, and lactose, serum ionized calcium and PTH levels are normalized and rickets/osteomalacia is prevented; these results suggest that the major effect of 1,25-(OH)$_2$D$_3$ in the hypocalcemic state is the provision of calcium and phosphate to bone from the stimulated intestinal absorption of calcium and phosphorous [32]. In addition, transgenic expression of VDR in the intestine of VDR null mice results in normalization of serum calcium, bone density, and bone volume [33]. However, studies using CYP27B1 null mice and CYP27B1/VDR double knockout mice showed that when hypocalcemia and secondary hyperparathyroidism are prevented by a rescue diet in vivo, not all changes in osteoblast number, mineral apposition rate, and bone volume are rescued, suggesting direct skeletal effects of the 1,25-(OH)$_2$D-VDR system [34]. In vitro 1,25-(OH)$_2$D can stimulate osteoblast differentiation and osteoclastogenesis [35]. Runx2, a transcriptional regulator of osteoblast differentiation to mature bone-forming cells, is regulated by 1,25-(OH)$_2$D [36, 37]. Transgenic mice overexpressing VDR in osteoblastic cells have increased bone formation, further indicating direct effects of 1,25(OH)$_2$D on bone [38]. On the other hand, the effect of 1,25-(OH)$_2$D on osteoclastogenesis is indirect. Stimulation of osteoclast formation by 1,25-(OH)$_2$D involves upregulation of RANKL by 1,25-(OH)$_2$D in osteoblastic cells and requires cell-to-cell contact between osteoblasts and osteoclast precursors [39].

The osteoclastogenesis inhibitory factor, osteoprotegerin (OPG), is a decoy receptor for RANKL that antagonizes RANKL action thus blocking osteoclastogenesis. OPG is downregulated by 1,25-(OH)$_2$D [39]. 1,25-(OH)$_2$D has also been reported to stimulate the production of the calcium-binding proteins osteocalcin and osteopontin (OPN), a potent mineralization inhibitor [40, 41], in osteoblasts. It has been suggested that induction of OPN is one mechanism whereby 1,25-(OH)$_2$D$_3$ inhibits bone matrix mineralization to preserve normal serum calcium levels during a negative calcium balance [42, 43].

Intestine

When the demand for calcium increases as a consequence of (i) a diet deficient in calcium, (ii) demand for skeletal growth, (iii) pregnancy, and (iv) lactation, the synthesis of 1,25-(OH)$_2$D is increased. VDR null mice display a major defect in intestinal calcium absorption, suggesting that a principal action of 1,25-(OH)$_2$D$_3$ is to maintain calcium homeostasis via maximizing the efficiency of intestinal calcium absorption [44–47]. Intestinal calcium absorption is accounted for by two different modes of calcium transport. The first is a saturable process which is mainly transcellular. The second is a diffusional mode which is (i) nonsaturating, (ii) requires a lumenal free calcium concentration >2 to 6mM, and (iii) is paracellular (ie, between neighboring intestinal epithelial cells) in nature. 1,25-(OH)$_2$D has been reported to affect both the transcellular and the paracellular path [45–47]. The transcellular process is comprised of three 1,25-(OH)$_2$D-regulated steps: the entry of calcium across the brush border membrane, intracellular diffusion, and the energy requiring extrusion of calcium across the basolateral membrane [45–47]. The calcium-binding protein, calbindin, whose expression is induced by 1,25-(OH)$_2$D in the intestine, acts to facilitate the diffusion of calcium through the cell interior, directing calcium to the basolateral membrane. Interestingly, studies in calbindin-D$_{9k}$ null mice showed no change in 1,25-(OH)$_2$D$_3$-mediated intestinal calcium absorption and in serum calcium levels compared with wild-type mice [44, 45, 48]; this observation provides evidence that calbindin alone is not responsible for 1,25-(OH)$_2$D-mediated intestinal calcium absorption.

The rate of calcium entry into the enterocyte is increased by 1,25-(OH)$_2$D. It has been suggested that the calcium-selective channel TRPV6, which is colocalized with calbindin in the intestine and is induced by 1,25-(OH)$_2$D$_3$, plays a key role in vitamin D-dependent calcium entry into the enterocyte [49, 50]. Moreover, transgenic mice overexpressing TRPV6 in the intestine develop hypercalcemia, hypercalciuria, and soft tissue calcification, further suggesting a direct role for TRPV6 in the process of intestinal calcium transport [51]. Studies in calbindin-D$_{9k}$/TRPV6 double knockout mice have shown that intestinal calcium absorption in response to low dietary calcium or to 1,25-(OH)$_2$D is least efficient in the absence of both proteins; therefore it is postulated that calbindin-D$_{9k}$ and TRPV6 act together to promote calcium absorption, especially in the calcium-deprived state [52]. 1,25(OH)$_2$D also affects calcium extrusion from the enterocyte via stimulation of the plasma membrane calcium pump (PMCA) [46]. Although the duodenum has been a focus of research related to 1,25-(OH)$_2$D$_3$-mediated active calcium absorption, 1,25-(OH)$_2$D$_3$ regulation of active calcium transport in ileum, cecum, and colon has also been reported [43, 45]. VDR, TRPV6, and calbindin-D$_{9k}$ are present in all segments of the small and large intestine [43, 45]. Transgenic expression of VDR restricted to the ileum, cecum, and colon of VDR null mice prevented VDR dependent rickets [43]. In addition, disturbed calcium metabolism was noted in mice that lack VDR in the large intestine [53]. These findings suggest that although calcium is absorbed more rapidly in the duodenum, the distal segments of the intestine contribute significantly to 1,25-(OH)$_2$D$_3$-mediated regulation of calcium homeostasis, especially in states of calcium deprivation.

In addition to intestinal calcium absorption, 1,25-(OH)$_2$D also acts to enhance intestinal phosphorus absorption. Although the mechanisms involved have been a matter of debate, it has been suggested that 1,25(OH)$_2$D$_3$ stimulates the active transport of phosphorus and FGF23 inhibits it by suppressing 1,25(OH)$_2$D$_3$ synthesis [54, 55].

Kidney

A third target tissue involved in 1,25-(OH)$_2$D-mediated mineral homeostasis is the kidney. 1,25-(OH)$_2$D has been reported to enhance the actions of PTH on calcium transport in the distal tubule by increasing PTH receptor expression action in distal tubule cells [56]. 1,25-(OH)$_2$D also induces the synthesis of the calbindins in the distal tubules [43]. Similar to studies in the intestine, an apical (lumenal) calcium channel, TRPV5, which is colocalized with the calbindins and induced by 1,25-(OH)$_2$D, has been identified in the distal convoluted tubule and distal connecting tubules [49]. Calbindin-D$_{28k}$ was reported to associate directly with TRPV5 and to control TRPV5-mediated calcium influx [57]. Thus, 1,25-(OH)$_2$D affects calcium transport in the distal tubule by enhancing the action of PTH and by inducing TRPV5 and the calbindins. Another important effect of 1,25-(OH)$_2$D in the kidney is the inhibition of the CYP27B1 and the induction of the CYP24A1 [58]. Beside effects on calcium transport in the distal nephron and modulation of CYP27B1 and CYP24A1 in tubular epithelium, depending on the parathyroid and FGF23 status of the host, 1,25-(OH)$_2$D has been reported to increase or decrease renal phosphate reabsorption via control of the sodium dependent phosphate cotransporter in the proximal tubule [59]; the FGF23-directed, PTH-amplified phosphaturic response requires klotho-α as a co-receptor for FGF23 signaling.

NONCLASSICAL TARGET TISSUES

Parathyroids

The parathyroid glands are an important target of 1,25-(OH)$_2$D. As discussed previously, 1,25(OH)$_2$D inhibits the synthesis and secretion of PTH and prevents the proliferation of PTH-producing cells in the gland in order to maintain normal parathyroid status [60, 61]. It has also been shown that 1,25-(OH)$_2$D upregulates VDR-directed calcium sensing receptor (CaSR) transcription [62], suggesting that 1,25-(OH)$_2$D sensitizes the parathyroid gland to calcium inhibition. Parathyroid cells also express the CYP27B1; as such, local production of 1,25-(OH)$_2$D as well as circulating 1,25(OH)$_2$D may contribute to regulation of PTH production and secretion.

Immune cells

Nonclassical regulation of immune responses by 1,25-(OH)$_2$D was first reported over 30 years ago with disease-activated lymphocytes found to express the VDR and macrophages to make 1,25-(OH)$_2$D [63, 64]. Recent studies have shown that 1,25-(OH)$_2$D regulates both innate and adaptive immunity, but in opposite directions, namely promoting the former while repressing the latter. Innate immunity encompasses the ability of the host immune system to recognize and respond to an offending antigen. In silico screening of the subhuman and human primate genome, but not lower mammals (eg, mice) revealed the insertion of an Alu repeat element bearing a canonical vitamin D response element in the promoter of the human gene for cathelicidin, whose product LL37 is an antimicrobial peptide capable of killing bacteria [65]. Subsequent investigations confirmed the ability of 1,25-(OH)$_2$D [66] and its precursor 25OHD [67] to induce expression of cathelicidin in cells of the monocyte/macrophage and epidermal lineage [67, 68], highlighting the potential for intracrine induction of antimicrobial responses in cells that also express the 25OHD-activating enzyme, CYP27B1. Reinforcing these events is the ability of locally generated 1,25-$(OH)_2D$ to escape the confines of the cell in which it is made to act on neighboring VDR-expressing monocytes to promote their maturation to mature macrophages [69], thus acting as a feed-forward signal to further enhance the innate immune response.

The adaptive immune response is generally defined by T- and B-lymphocytes and their ability to produce cytokines and immunoglobulins, respectively, to specifically combat the source of the antigen presented to them by innate immune cells. Contrary to the role of locally produced 1,25-$(OH)_2D$ to promote the innate immune response, the hormone exerts a generalized dampening effect on lymphocyte function. With respect to B cells, 1,25-$(OH)_2D$ suppresses proliferation and immunoglobulin production, and retards the differentiation of B-lymphocyte precursors to mature plasma cells. With regard to T cells, 1,25-$(OH)_2D$ inhibits their proliferation in response to proinflammatory cytokines [70]. In sum, the collective, concerted action of 1,25-$(OH)_2D$ is to promote the host's response to an invading pathogen while simultaneously acting to limit what might be an overzealous immune response to that pathogen. Perhaps the best example of 25OHD-dependent local production of 1,25-$(OH)_2D$ exerting immune tolerance is that imparted by the mother to the fetus. By 12 weeks of gestation, maternal total serum 1,25-$(OH)D_2$ levels are already twice that of a nonpregnant control and continue to increase such that by term 1,25-$(OH)_2D$ in mother and baby can be greater than 300 pg/mL [71, 72]. In contrast to the pathological granuloma-forming disease states noted previously, supraphysiological concentrations of 1,25-$(OH)_2D$ attained during pregnancy occur without the development of dysregulated calcium homeostasis in either the baby or mother; the mechanism underlying the "resistance" to the calcium mobilizing actions of 1,25-$(OH)_2D$ in pregnancy are not understood. Moreover, a recent study by Mirzakhani and colleagues [73] indicated that a low maternal 25OHD level in the early stages of pregnancy was more likely to result in preeclampsia and premature delivery.

Pancreas

The pancreas was one of the first nonclassical target tissues in which the VDR was identified [74]. Although 1,25-$(OH)_2D$ has been reported to play a role in insulin secretion, the exact mechanisms remain unclear. Autoradiographic data and immunocytochemical studies have localized the VDR and calbindin-D_{28k}, respectively, in pancreatic beta cells [75, 76]. Studies using calbindin-D_{28k} null mice indicate that calbindin-D_{28k} can modulate depolarization-stimulated insulin release by regulating intracellular calcium [77]. In addition to modulating insulin release, calbindin-D_{28k} can protect against cytokine-mediated destruction of beta cells by buffering calcium [78]. Recent studies have also noted that proliferation of VDR-expressing pancreatic stellate cells under the influence of paracrine growth factors produced by pancreatic cancer cells can be antagonized by 1,25-$(OH)_2D$, suggesting that treatment with nonhypercalcemia-causing analogues of 1,25-$(OH)_2D$ could be used in therapy of ductal adenocarcinoma of the pancreas [79].

Epidermis and hair follicles

The keratinocyte, like the kidney tubular epithelial cells, expresses both the VDR and CYP27B1 enabling it to produce as well as respond to 1,25$(OH)_2D$ [80]. The expression of VDR and CYP27B1 are highest in the basal layer (stratum basale) of the epidermis where the epidermal stem cells are located. 1,25$(OH)_2D$ induced differentiation of these skin stem cells is marked by the rise in involucrin and transglutaminase expression in the stratum spinosum [81], filaggrin and loricrin in the stratum granulosum [82], and synthesis of long chain fatty acids in the stratum granulosum. Loricrin and involucrin are major protein components of the permeability barrier formed within the stratum corneum, whereas the lipids produced and packaged into lamellar bodies in the stratum granulosum are secreted into the protein network of the permeability barrier, effectively waterproofing the epidermis and providing for its resistance to foreign pathogens. The mechanisms by which 1,25$(OH)_2D$ alters keratinocyte proliferation and differentiation are multiple including induction of the proteins and lipid synthesizing enzymes referred to above, the induction of the calcium-sensing receptor that enables calcium signaling by these cells, and stimulation of the E-cadherin/catenin complex enabling both cell adhesion and migration as well as forming a key signaling complex in the plasma membrane [83].

Alopecia is a well-known feature of VDR mutations in humans as well as mice [84], but is not characteristic of vitamin D deficiency or CYP27B1 mutations. Two transacting proteins, hairless (Hr) and ß-catenin, appear to interact with VDR to regulate hair follicle (HF) cycling. Their inactivating mutations produce phenocopies of the VDR-null animal with regard to HF cycling. Loss of stem cells from the HF bulge is found in both VDR and β-catenin null mice [85, 86]. HF cycling represents the best example by which VDR regulates a physiological process that is independent of its ligand, 1,25$(OH)_2D$, and so points to a novel mechanism of action for this transcriptional regulator.

NUTRITIONAL CONSIDERATIONS

Serum 25OHD levels provide a useful surrogate for assessing vitamin D status, as the conversion of vitamin D to 25OHD is less well controlled (ie, primarily substrate dependent) than the subsequent conversion of 25OHD to

1,25(OH)$_2$D. Unlike 25OHD levels, 1,25-(OH)$_2$D levels remain well within the reference range until extremes of vitamin D deficiency occur, because secondary hyperparathyroidism efficiently increases 1,25-(OH)$_2$D production from precursor 25OHD; as such, 1,25-(OH)$_2$D levels do not provide a useful index for assessing the initial stages of vitamin D deficiency. Levels of 25OHD below 10 ng/mL or 25 nM are associated with a high prevalence of rickets or osteomalacia. However, there is a growing consensus that these lower limits of normal are too low. Recently an expert panel for the Institute of Medicine (IOM) recommended that a level of 20 ng/mL (50 nM) was sufficient for 97.5% of the otherwise healthy population and levels up to 50 ng/mL (125 nM) were safe [87]. A level of 600 IU vitamin D was thought to be sufficient to meet these goals for otherwise healthy individuals between the ages of 1 and 70 years, although up to 4000 IU vitamin D daily was considered safe [87]. Contrary to the IOM, the Endocrine Society considered the lower end of these recommendations to be too low and the upper end too restrictive [88]. At least with respect to the lower recommended levels of vitamin D supplementation, the IOM guidelines are unlikely to correct vitamin D deficiency in individuals with obesity, dark complexions, limited capacity for sunlight exposure, or malabsorption. Because of lack of compelling data from randomized controlled trials, the IOM did not make recommendations for nonskeletal effects of vitamin D. Moreover, these guidelines fail to consider the likelihood that it is the free concentration of these metabolites and not the total concentration that is biologically active in most nonrenal tissues. Furthermore, a large body of cell, animal, as well as epidemiological association data in humans focused on all-cause mortality, support a large range of beneficial actions for maintaining 25OHD greater than 20 ng/mL [89, 90].

VITAMIN D TREATMENT STRATEGIES

Adequate sunlight exposure is the most cost effective means of obtaining vitamin D. Whole body exposure to a 0.5 minimal erythema dose of sunlight (ie, half the dose required to produce a slight reddening of the skin) has been calculated to provide the equivalent of 10,000 IU vitamin D$_3$ [91]. UVB radiation to the arms and legs achieved in 5 to 10 minutes on a bright summer day in light-skinned individuals has been calculated to be the equivalent of 3000 IU vitamin D$_3$ [8]. Although sunlight exposure remains a viable option for those unable or unwilling to consider oral supplementation of vitamin D, concerns regarding the association between sunlight and skin cancer and/or solar aging of the skin have limited this approach. While studies have shown that on average for every incremental increase of 100 IU of vitamin D$_3$ supplemented daily, the 25OHD level will rise by 0.5 to 1 ng/mL [91, 92], much higher doses are likely to be required by obese individuals with an increased volume of distribution of vitamin D to fat or those with intestinal malabsorption (including after bariatric surgery). A number of studies suggest that 700 to 800 IU is the lower limit of vitamin D supplementation required to prevent fractures and falls, although as noted the IOM has concluded that 600 IU suffices. With the exception of wild-caught salmon, other fish products such as cod liver oil, and UVB-treated mushrooms (enriched in vitamin D$_2$), unfortified food contains little vitamin D. Milk and other fortified beverages typically contain 100 IU per 8 oz serving. Vitamin D$_2$ is less potent than vitamin D$_3$ in increasing the total serum 25OHD, but is at least equipotent in increasing the free 25OHD [93]; this is explained in part because 25OHD$_2$ is bound less avidly than 25OHD$_3$ by the serum DBP with the free fraction being rapidly cleared from the circulation. Currently there is no clear clinical indication for measuring serum 25OHD$_2$ and 25OHD$_3$ separately [93]. Toxicity (hypercalciuria and/or hypercalciuria) caused by vitamin D supplementation has not been observed at doses less than 10,000 IU per day [94].

Finally, it is important to emphasize the clinical reality that 25OHD-deficiency is not the same as 1,25-(OH)$_2$D deficiency as is observed in patients with chronic renal failure (CRF). 25OHD-deficient subjects with normal renal function develop compensatory increases in PTH that significantly increase their serum 1,25-(OH)$_2$D. On the other hand, CRF subjects frequently harbor low serum levels of both 25OHD and 1,25-(OH)$_2$D; the former as a consequence of deficient cutaneous production and insufficient oral supplementation with vitamin D and the latter owing to compromised renal 1,25-(OH)$_2$D production. As such, CRF patients benefit from the coincident replacement of both 25OHD and 1,25-(OH)$_2$D (or an analog of 1,25-(OH)$_2$D).

REFERENCES

1. Holick MF, McLaughlin JA, Clark MB, et al. Factors that influence the cutaneous photosynthesis of previtamin D3. Science. 1981;211:590–93.
2. Holick MF, McLaughlin JA, Clark MB, et al. Photosynthesis of previtamin D3 in human and the physiologic consequences. Science. 1980;210:203–5.
3. Holick MF, Richtand NM, McNeill SC, et al. Isolation and identification of previtamin D3 from the skin of exposed to ultraviolet irradiation. Biochemistry. 1979;18:1003–8.
4. Webb AR, Kline L, Holick MF Influence of season and latitude on the cutaneous synthesis of vitamin D3: exposure to winter sunlight in Boston and Edmonton will not promote vitamin D3 synthesis in human skin. J Clin Endocrinol Metab. 1988;67:373–8.
5. Adams JS, Clemens TL, Parrish JA, et al. Vitamin-D synthesis and metabolism after ultraviolet irradiation of normal and vitamin-D-deficient subjects. N Engl J Med. 1982;306:722–5.
6. Heaney RP, Horst RL, Cullen DM, et al. Vitamin D3 distribution and status in the body. J Am Coll Nutr. 2009;28:252–6.

7. Thacher TD, Levine MA. CYP2R1 mutations causing vitamin D-deficiency rickets. J Steroid Biochem Mol Biol. 2017;173:333–6.
8. Holick MF. Vitamin D deficiency. N Engl J Med. 2007;357:266–81.
9. Hewison M, Burke F, Evans KN, et al. Extra-renal 25-hydroxyvitamin D3-1alpha-hydroxylase in human health and disease. J Steroid Biochem Mol Biol. 2007;103:316–21.
10. Zierold C, Darwish HM, DeLuca HF. Two vitamin D response elements function in the rat 1,25-dihydroxyvitamin D 24-hydroxylase promoter. J Biol Chem. 1995;270:1675–8.
11. Liang C, Cooke N. Vitamin D-binding protein In: Feldman D, Pike JW, Glorieux F (eds) *Vitamin D* (2nd ed.). San Diego: Academic Press, 2005, pp. 117–34.
12. Nielson CM, Jones KS, Chun RF, et al. Osteoporotic Fractures in Men (MrOS) Research Group. Free 25-hydroxyvitamin D: impact of vitamin D binding protein assays on racial-genotypic associations. J Clin Endocrinol Metab. 2016;101(5):2226–34.
13. Nielson CM, Jones KS, Bouillon R. Osteoporotic Fractures in Men (MrOS) Research Group. Role of assay type in determining free 25-hydroxyvitamin D levels in diverse populations. N Engl J Med. 2016;374(17):1695–6.
14. Cooke NE, Haddad JG. Vitamin D binding protein (Gc-globulin). Endocr Rev. 1989;10:294–307.
15. Christakos S, Dhawan P, Liu Y, et al. New insights into the mechanisms of vitamin D action. J Cell Biochem. 2003;88:695–705.
16. DeLuca HF. Overview of general physiologic features and functions of vitamin D. Am J Clin Nutr. 2004;80:1689S–96S.
17. Rachez C, Freedman LP. Mechanisms of gene regulation by vitamin D(3) receptor: a network of coactivator interactions. Gene. 2000;246:9–21.
18. Sutton AL, MacDonald PN. Vitamin D: more than a "bone-a-fide" hormone. Mol Endocrinol. 2003;17:777–91.
19. Christakos S, Dhawan P, Peng X, et al. New insights into the function and regulation of vitamin D target proteins. J Steroid Biochem Mol Biol. 2007;103:405–10.
20. Dhawan P, Peng X, Sutton AL, et al. Functional cooperation between CCAAT/enhancer-binding proteins and the vitamin D receptor in regulation of 25-hydroxyvitamin D3 24-hydroxylase. Mol Cell Biol. 2005;25:472–87.
21. Guo B, Aslam F, van Wijnen AJ, et al. YY1 regulates vitamin D receptor/retinoid X receptor mediated transactivation of the vitamin D responsive osteocalcin gene. Proc Natl Acad Sci U S A. 1997;94:121–6.
22. Raval-Pandya M, Dhawan P, Barletta F, et al. YY1 represses vitamin D receptor-mediated 25-hydroxyvitamin D(3)24-hydroxylase transcription: relief of repression by CREB-binding protein. Mol Endocrinol. 2001;15:1035–46.
23. Seth-Vollenweider T, Joshi S, Dhawan P, et al. Novel mechanism of negative regulation of 1,25-dihydroxyvitamin D3-induced 25-hydroxyvitamin D 24-hydroxylase (Cyp24a1) transcription. Epigenetic modification involving cross-talk between protein-arginine methyltransferase 5 and the SWI/SNF complex. J Biol Chem. 2014;289:33958–70.
24. Lee SM, Pike JW. The vitamin D receptor functions as a transcription regulator in the absence of 1,25-dihydroxyvitamin D(3). J Steroid Biochem Mol Biol. 2016;Nov;164:265–70.
25. Pike JW, Meyer MB, Benkusky NA, et al. Genomic determinants of vitamin D-regulated gene expression. Vitamins Hormones. 2016;100:21–44.
26. Lips P, Welsh J, Demay M, et al. Highlights from the 18th workshop on vitamin D, Delft, The Netherlands, April 21–24, 2015. J Steroid Biochem Mol Biol. 2016;164:1–3.
27. Lee SM, Bishop KA, Goellner JJ, et al. Mouse and human BAC transgenes recapitulate tissue- specific expression of the vitamin D receptor in mice and rescue the VDR-null phenotype. Endocrinology. 2014;155:2064–76.
28. Feng JQ, Ward LM, Liu S, et al. Loss of DMP1 causes rickets and osteomalacia and identifies a role for osteocytes in mineral metabolism. Nat Genet. 2006;38:1310–5.
29. Imura A, Tsuji Y, Murata M, et al. alpha-Klotho as a regulator of calcium homeostasis. Science. 2007;316:1615–18.
30. Li YC, Pirro AE, Amling M, et al. Targeted ablation of the vitamin D receptor: an animal model of vitamin D-dependent rickets type II with alopecia. Proc Natl Acad Sci U S A. 1997;94:9831–5.
31. Yoshizawa T, Handa Y, Uematsu Y, et al. Mice lacking the vitamin D receptor exhibit impaired bone formation, uterine hypoplasia and growth retardation after weaning. Nat Genet. 1997;16:391–6.
32. Amling M, Priemel M, Holzmann T, et al. Rescue of the skeletal phenotype of vitamin D receptor-ablated mice in the setting of normal mineral ion homeostasis: formal histomorphometric and biomechanical analyses. Endocrinology. 1999;140:4982–7.
33. Xue Y, Fleet JC. Intestinal vitamin D receptor is required for normal calcium and bone metabolism in mice. Gastroenterology. 2009;136:1317–1327, e1311–12.
34. Panda DK, Miao D, Bolivar I, et al. Inactivation of the 25-hydroxyvitamin D 1alpha-hydroxylase and vitamin D receptor demonstrates independent and interdependent effects of calcium and vitamin D on skeletal and mineral homeostasis. J Biol Chem. 2004;279:16754–66.
35. Raisz LG, Trummel CL, Holick MF, et al. 1,25-dihydroxycholecalciferol: a potent stimulator of bone resorption in tissue culture. Science. 1972;175:768–9.
36. Drissi H, Pouliot A, Koolloos C, et al. 1,25-(OH)2-vitamin D3 suppresses the bone-related Runx2/Cbfa1 gene promoter. Exp Cell Res. 2002;274:323–33.
37. Meyer MB, Benkusky NA, Lee CH, et al. Genomic determinants of gene regulation by 1,25-dihydroxyvitamin D3 during osteoblast-lineage cell differentiation. J Biol Chem. 2014;289:19539–54.
38. Gardiner EM, Baldock PA, Thomas GP, et al. Increased formation and decreased resorption of bone in mice with elevated vitamin D receptor in mature cells of the osteoblastic lineage. Faseb J. 2000;14:1908–16.
39. Yasuda H, Shima N, Nakagawa N, et al. Osteoclast differentiation factor is a ligand for osteoprotegerin/osteoclastogenesis-inhibitory factor and is identical

to TRANCE/RANKL. Proc Natl Acad Sci U S A. 1998;95:3597–2.
40. Price PA, Baukol SA. 1,25-Dihydroxyvitamin D3 increases synthesis of the vitamin K-dependent bone protein by osteosarcoma cells. J Biol Chem. 1980;255:11660–3.
41. Prince CW, Butler WT. 1,25-dihydroxyvitamin D3 regulates the biosynthesis of osteopontin, a bone-derived cell attachment protein, in clonal osteoblast-like osteosarcoma cells. Coll Relat Res. 1987;7:305–13.
42. Lieben L, Masuyama R, Torrekens S, et al. Normocalcemia is maintained in mice under conditions of calcium malabsorption by vitamin D-induced inhibition of bone mineralization. J Clin Invest. 2012;122;1803–15.
43. Christakos S, Dhawan P, Verstuyf A, et al. Vitamin D: metabolism, molecular mechanism of action and pleiotropic effects. Physiol Rev. 2016;96:365–408.
44. Akhter S, Kutuzova GD, Christakos S, et al. Calbindin D9k is not required for 1,25-dihydroxyvitamin D3-mediated Ca2+ absorption in small intestine. Arch Biochem Biophys. 2007;460:227–32.
45. Christakos S. Recent advances in our understanding of 1,25-dihydroxyvitamin D_3 regulation of intestinal calcium absorption. Arch Biochem Biophys. 2012;523:73–6.
46. Wasserman RH, Fullmer CS. Vitamin D and intestinal calcium transport: facts, speculations and hypotheses. J Nutr. 1995;125:1971S–9S.
47. Fleet JC, Schoch RD. Molecular mechanisms for regulation of intestinal calcium absorption by vitamin D and other factors Crit Rev Clin Lab Sci. 2010;47:181–95.
48. Kutuzova GD, Akhter S, Christakos S, et al. Calbindin D(9k) knockout mice are indistinguishable from wild-type mice in phenotype and serum calcium level. Proc Natl Acad Sci U S A. 2006;103:12377–81.
49. Hoenderop JG, Nilius B, Bindels RJ. Epithelial calcium channels: from identification to function and regulation. Pflugers Arch. 2003;446:304–8.
50. Peng JB, Chen XZ, Berger UV, et al. Molecular cloning and characterization of a channel-like transporter mediating intestinal calcium absorption. J Biol Chem. 1999;274:22739–46.
51. Cui M, Li Q, Johnson R, Fleet JC. Villin promoter-mediated transgenic expression of transient receptor potential cation channel, subfamily V, member 6 (TRPV6) increases intestinal calcium absorption in wild-type and vitamin D receptor knockout mice. J Bone Miner Res. 2012;27:2097–107.
52. Benn BS, Ajibade D, Porta A, et al. Active intestinal calcium transport in the absence of transient receptor potential vanilloid type 6 and calbindin-D9k. Endocrinology. 2008;149:3196–205.
53. Reyes-Fernandez PC, Fleet JC. Compensatory changes in calcium metabolism accompany the loss of vitamin D receptor (VDR) from the distal intestine and kidney of mice. J Bone Miner Res. 2015;31:143–51.
54. Williams KB, DeLuca HF. Characterization of intestinal phosphate absorption using a novel in vivo method. Am J Physiol Endocrinol Metab. 2007;292:E1917–21.
55. Sabbagh Y, O'Brien SP, Song W, et al. Intestinal npt2b plays a major role in phosphate absorption and homeostasis. J Am Soc Nephrol. 2009;20:2348–58.
56. Sneddon WB, Barry EL, Coutermarsh BA, et al. Regulation of renal parathyroid hormone receptor expression by 1, 25-dihydroxyvitamin D3 and retinoic acid. Cell Physiol Biochem. 1998;8:261–77.
57. Lambers TT, Mahieu F, Oancea E, et al. Calbindin-D28k dynamically controls TRPV5 mediated Ca2+ transport. Embo J. 2006;25: 2978–88.
58. Omdahl JL, Bobrovnikova EA, Choe S, et al. Overview of regulatory cytochrome P450 enzymes of the vitamin D pathway. Steroids. 2001;66:381–9.
59. Kaneko I, Segawa H, Furutani J, et al. Hypophosphatemia in vitamin D receptor null mice: effect of rescue diet on the developmental changes in renal Na+-dependent phosphate cotransporters. Pflugers Arch J Physiol. 2011;461:77–90.
60. Demay MB, Kiernan MS, DeLuca HF, et al. Sequences in the human parathyroid hormone gene that bind the 1,25-dihydroxyvitamin D3 receptor and mediate transcriptional repression in response to 1,25-dihydroxyvitamin D3. Proc Natl Acad Sci U S A. 1992;89:8097–101.
61. Martin KJ, Gonzalez EA. Vitamin D analogs: actions and role in the treatment of secondary hyperparathyroidism. Semin Nephrol. 2004;24:456–9.
62. Canaff L, Hendy GN. Human calcium-sensing receptor gene. Vitamin D response elements in promoters P1 and P2 confer transcriptional responsiveness to 1,25-dihydroxyvitamin D. J Biol Chem. 2002;277:30337–50.
63. Provvedini DM, Tsoukas CD, Deftos LJ, et al. 1,25-dihydroxyvitamin D3 receptors in human leukocytes. Science. 1983;221:1181–3.
64. Adams JS, Sharma OP, Gacad MA, et al. Metabolism of 25-hydroxyvitamin D3 by cultured pulmonary alveolar macrophages in sarcoidosis. J Clin Invest. 1983;72: 1856–60.
65. Wang TT, Nestel FP, Bourdeau V, et al. Cutting edge: 1,25-dihydroxyvitamin D3 is a direct inducer of antimicrobial peptide gene expression. J Immunol. 2004;173: 2909–12.
66. Gombart AF, Borregaard N, Koeffler HP. Human cathelicidin antimicrobial peptide (CAMP) gene is a direct target of the vitamin D receptor and is strongly up-regulated in myeloid cells by 1,25-dihydroxyvitamin D3. Faseb J. 2005;19:1067–77.
67. Weber G, Heilborn JD, Chamorro Jimenez CI, et al. Vitamin D induces the antimicrobial protein hCAP18 in human skin. J Invest Dermatol. 2005;124:1080–2.
68. Schauber J, Dorschner RA, Coda AB, et al. Injury enhances TLR2 function and antimicrobial peptide expression through a vitamin D-dependent mechanism. J Clin Invest. 2007;117:803–11.
69. Kreutz M, Andreesen R, Krause SW, et al. 1,25-dihydroxyvitamin D3 production and vitamin D3 receptor expression are developmentally regulated during differentiation of human monocytes into macrophages. Blood. 1993;82:1300–7.
70. Bruce D, Ooi JH, Yu S, et al. Vitamin D and host resistance to infection? Putting the cart in front of the horse. Exp Biol Med. 2010;235:921–7.
71. Hollis BW, Johnson D, Hulsey TC, et al. Vitamin D supplementation during pregnancy: double-blind,

71. randomized clinical trial of safety and effectiveness. J Bone Miner Res. 2011;26(10):2341–57. Erratum in: J Bone Miner Res. 2011;26(12):3001.
72. Walker VP, Zhang X, Rastegar I, et al. Cord blood vitamin D status impacts innate immune responses. J Clin Endocrinol Metab. 2011;96(6):1835–43.
73. Mirzakhani H, Litonjua AA, McElrath TF, et al. Early pregnancy vitamin D status and risk of preeclampsia. J Clin Invest. 2016;126(12):4702–15.
74. Christakos S, Norman AW. Studies on the mode of action of calciferol. XVIII. Evidence for a specific high affinity binding protein for 1,25 dihydroxyvitamin D3 in chick kidney and pancreas. Biochem Biophys Res Commun. 1979;89:56–63.
75. Clark SA, Stumpf WE, Sar M, et al. Target cells for 1,25 dihydroxyvitamin D3 in the pancreas. Cell Tissue Res. 1980;209:515–20.
76. Morrissey RL, Bucci TJ, Richard B, et al. Calcium-binding protein: its cellular localization in jejunum, kidney and pancreas. Proc Soc Exp Biol Med. 1975;149:56–60.
77. Sooy K, Schermerhorn T, Noda M, et al. Calbindin-D(28k) controls [Ca(2+)](i) and insulin release. Evidence obtained from calbindin-d(28k) knockout mice and beta cell lines. J Biol Chem. 1999;274:34343–9.
78. Rabinovich A, Suarez-Pinzon WL, Sooy K, et al. Expression of calbindin-D28k in a pancreatic islet beta cell line protects against cytokine-induced apoptosis and necrosis. Endocrinology. 2001;142:3649–55.
79. Sherman MH, Yu RT, Engle DD, et al. Vitamin D receptor-mediated stromal reprogramming suppresses pancreatitis and enhances pancreatic cancer therapy. Cell. 2014;159:80–93.
80. Bikle DD, Nemanic MK, Whitney JO, et al. Neonatal human foreskin keratinocytes produce 1,25-dihydroxyvitamin D3. Biochemistry. 1986;25:1545–8.
81. Su MJ, Bikle DD, Mancianti ML, et al. 1,25-dihydroxyvitamin D3 potentiates the keratinocyte response to calcium. J Biol Chem. 1994;269:14723–9.
82. Hawker NP, Pennypacker SD, Chang SM, et al. Regulation of human epidermal keratinocyte differentiation by the vitamin D receptor and its coactivators DRIP205, SRC2, and SRC3. J Invest Dermatol. 2007;127:874.
83. Bikle DD. Vitamin D and the skin: physiology and pathophysiology. Rev Endocr Metab Disord. 2012;13(1):3–19.
84. Malloy PJ, Pike JW, Feldman D. The vitamin D receptor and the syndrome of hereditary 1,25-dihydroxyvitamin D-resistant rickets. Endocr Rev. 1999;20:156–88.
85. Bikle DD, Elalieh H, Chang S, et al. Development and progression of alopecia in the vitamin D receptor null mouse. J Cell Physiol. 2006;207:340–53.
86. Cianferotti L, Cox M, Skorija K, et al. Vitamin D receptor is essential for normal keratinocyte stem cell function. Proc Natl Acad Sci U S A. 2007;104:9428–33.
87. Ross AC, Manson JE, Abrams SA, et al. The 2011 report on dietary reference intakes for calcium and vitamin D from the Institute of Medicine: what clinicians need to know. J Clin Endocrinol Metab. 2011;96:53–8.
88. Holick MF, Binkley NC, Bischoff-Ferrari HA, et al. Evaluation, treatment, and prevention of vitamin D deficiency: an Endocrine Society clinical practice guideline. J Clin Endocrinol Metab. 2011;96:1911–30.
89. Adams JS, Ramin J, Rafison B, et al. Redefining human vitamin D sufficiency: back to the basics. Bone Res. 2013;1(1):2–10.
90. Bikle DD. Extraskeletal actions of vitamin D. Ann N Y Acad Sci. 2016;1376(1):29–52.
91. Vieth R. Vitamin D supplementation, 25-hydroxyvitamin D concentrations, and safety. Am J Clin Nutr. 1999;69:842–56.
92. Heaney RP, Davies KM, Chen TC, et al. Human serum 25-hydroxycholecalciferol response to extended oral dosing with cholecalciferol. Am J Clin Nutr. 2003;77:204–10
93. Chun RF, Hernandez I, Pereira R, et al. Differential responses to vitamin D2 and Vitamin D3 are associated with variations in free 25-hydroxyvitamin D. Endocrinology. 2016;157:3420–30.
94. Hathcock JN, Shao A, Vieth R, et al. Risk assessment for vitamin D. Am J Clin Nutr. 2007;85:6–18.

Section IV
Investigation of Metabolic Bone Diseases

Section Editors: Douglas C. Bauer and Klaus Engelke

Chapter 31. Techniques of Bone Mass Measurement in Children with Risk Factors for Osteoporosis 243
Nicola J. Crabtree and Leanne M. Ward

Chapter 32. Standard Techniques of Bone Mass Measurement in Adults 252
E. Michael Lewiecki, Paul D. Miller, and Nelson B. Watts

Chapter 33. Advanced Techniques of Bone Mass Measurements and Strength in Adults 260
Kyle K. Nishiyama, Enrico Dall'Ara, and Klaus Engelke

Chapter 34. Magnetic Resonance Imaging of Bone 272
Sharmila Majumdar

Chapter 35. Trabecular Bone Score 277
Barbara C. Silva and William D. Leslie

Chapter 36. Reference Point Indentation 287
Adolfo Diez-Perez and Joshua N. Farr

Chapter 37. Biochemical Markers of Bone Turnover in Osteoporosis 293
Pawel Szulc, Douglas C. Bauer, and Richard Eastell

Chapter 38. Scintigraphy and PET in Metabolic Bone Disease 302
Lorenzo Nardo, Paola A. Erba, and Benjamin L. Franc

Chapter 39. Bone Histomorphometry in Clinical Practice 310
Robert R. Recker and Carolina Aguiar Moreira

Chapter 40. Diagnosis and Classification of Vertebral Fracture 319
James F. Griffith and Harry K. Genant

Chapter 41. FRAX: Assessment of Fracture Risk 331
John A. Kanis, Eugene V. McCloskey, Nicholas C. Harvey, and William D. Leslie

Primer on the Metabolic Bone Diseases and Disorders of Mineral Metabolism, Ninth Edition. Edited by John P. Bilezikian.
© 2019 American Society for Bone and Mineral Research. Published 2019 by John Wiley & Sons, Inc.
Companion website: www.wiley.com/go/asbmrprimer

31

Techniques of Bone Mass Measurement in Children with Risk Factors for Osteoporosis

Nicola J. Crabtree[1] and Leanne M. Ward[2]

[1]Department of Endocrinology, Birmingham Women's and Children's Hospital, Birmingham, UK
[2]Department of Pediatrics, Faculty of Medicine, University of Ottawa, Ottawa, ON, Canada

INTRODUCTION

Understanding normal patterns of bone growth and mineral accrual is important for optimizing bone health in children and reducing the risk of osteoporotic fractures in later life. At the same time, there are a number of potentially harmful assaults on bone health in childhood including the impact of serious chronic diseases and/or genetic disorders such as OI.

The detrimental effects of chronic diseases are multifactorial, with the most significant threats being exogenous glucocorticoid therapy and reduced mobility; poor nutrition, delayed puberty, and inflammatory cytokines are other potential modifiers of skeletal strength. Ultimately it is important to establish to what extent bone strength has been compromised by a chronic or genetic condition and to estimate the child's current and future risk of fragility fracture. In addition, it is also important to identify otherwise healthy children with recurrent fractures who have an underlying bone fragility condition. It is in these two arenas that the use of pediatric BMD quantification techniques is beneficial as part of the overall bone health assessment. At the same time it is important to note that the diagnosis of osteoporosis cannot be made on the basis of BMD alone; rather, the diagnosis rests largely on the presence of a clinically significant fracture history [1, 2], with BMD Z-scores and their subsequent clinical course providing further information about the child's overall skeletal status and bone health trajectory.

In this chapter, the various techniques for assessing bone mass and density are discussed, along with issues in their interpretation and their clinical utility in the course of routine care.

DUAL-ENERGY X-RAY ABSORPTIOMETRY

DXA is the most widely used quantitative bone imaging technique in pediatric practice, but many aspects of its use, and the clinical interpretation of the data obtained, remain contentious [3–11]. It provides estimates of bone size in two dimensions, and bone mass within that envelope, with the value of bone mass adjusted for size being "areal" BMD (g/cm^2, aBMD) [12]. The reference values for aBMD provided by the machine manufacturer, which increase in a similar way to height and weight during childhood and adolescence, clearly indicate that DXA is not measuring true volumetric bone density, but a composite measure of bone size and mass. This is not necessarily a disadvantage because bone size, especially in the long bones that in children are most prone to fracture, is an important predictor of bone strength [13].

Measurement sites for DXA in children are typically the lumbar spine (LS) (L1 to L4 or L2 to L4) and total body less head (TBLH) (Fig. 31.1*A,B*) where precision is similar to that achieved in adults [14]. The forearm and proximal femur have been used in some studies, as has the lateral distal femur when spine deformity or hip or knee contractures preclude accurate ascertainment of spine or

---

Primer on the Metabolic Bone Diseases and Disorders of Mineral Metabolism, Ninth Edition. Edited by John P. Bilezikian.
© 2019 American Society for Bone and Mineral Research. Published 2019 by John Wiley & Sons, Inc.
Companion website: www.wiley.com/go/asbmrprimer

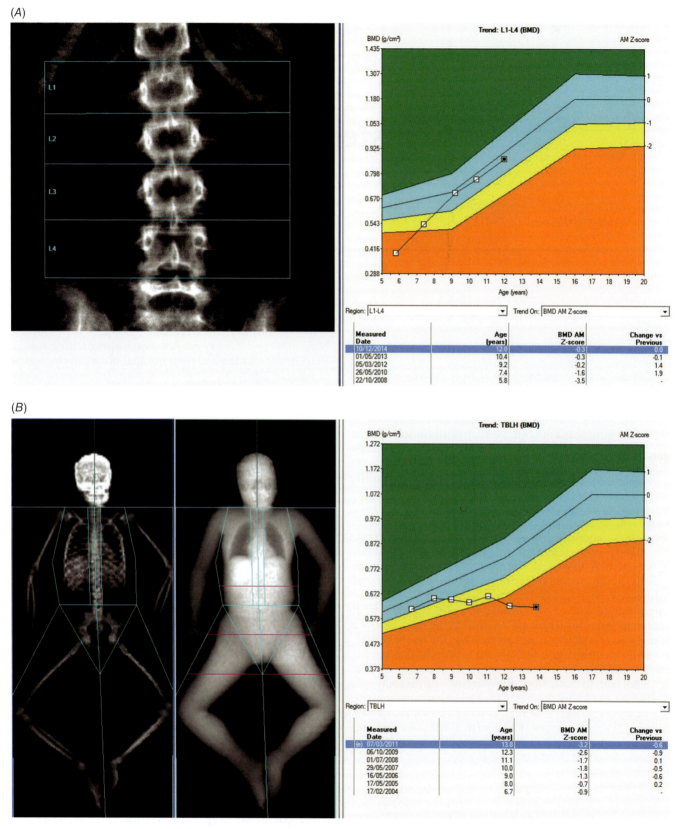

Fig. 31.1. (A) 12-year-old child with OI illustrating significant restitution of BMD Z-scores following 3 years' treatment with an intravenous bisphosphonate. (B) 13-year-old child with Duchenne muscular dystrophy and progressive deterioration in BMD Z-scores on glucocorticoid therapy, with loss of ambulation at age 11. TBLH BMC also showed a decline (from 588 g at age 11 to 568 g at age 12) consistent with true bone loss (as opposed to failure to accrue at a normal rate).

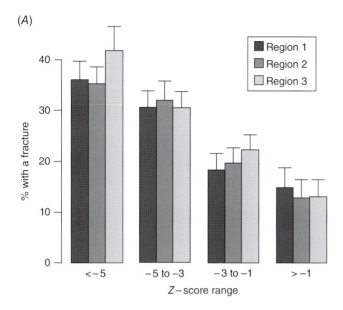

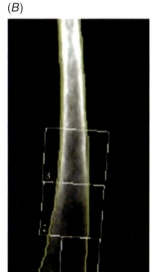

Region[8]	BMD (g/cm²)	BMD (9)	Area (cm¹)
1	0.201	1.00	6.60
2	0.473	4.30	9.09
3	0.577	5.06	7.47

Reference data from: Zemel et al. (Journal of clinical densitometry 2009 12(2): 207-18)

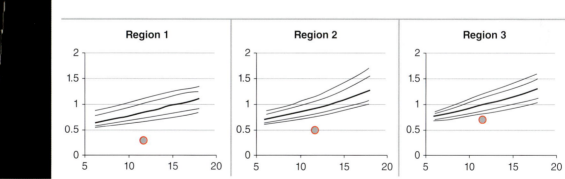

Fig. 31.2. (A) Measurement of BMD in three regions at the lateral distal femur, as first described by Henderson et al. (B) Lateral distal femur aBMD Z-score result in an 11-year-old boy with Duchenne muscular dystrophy and osteoporosis (multiple vertebral fractures).

total body aBMD. Normative reference data are available for spine, femoral neck, total body, and lateral distal femur [5, 7, 15–18]. Lateral distal femur aBMD is particularly useful in children with neuromuscular disorders who are more comfortable lying on their sides; low values at this site have been shown in children with motor disabilities to be associated with an increased risk of long bone fractures [18] (Fig. 31.2A,B). It should be noted that inconsistencies have been reported among various reference databases that have led to clinically relevant differences in aBMD Z-scores, with the disparity among different Z-scores varying by as much as 2.0; this observation makes the use of an aBMD Z-score cut-off challenging as part of the diagnosis of osteoporosis in children. On the other hand, the relationship between spine aBMD and vertebral fractures is highly consistent, underscoring that the use of an aBMD Z-score as a continuous variable predictor of vertebral fracture in the context of research studies remains valid [2, 19].

The advantages of using DXA in children to quantify aBMD are the short scan time, low radiation dose, and general widespread availability. DXA measures of aBMD in healthy children are predictive of fracture risk both at the measurement site (in the forearm) and elsewhere; total body aBMD less head, adjusted for weight, height, and bone area, is the measure that has been found in a prospective cohort study at age 9.9 years to be most strongly associated with fracture risk over the following 2 years [20]. However, there are no data for the predictive value of aBMD by DXA at other ages in apparently healthy children. On the other hand, numerous studies in children with glucocorticoid-treated illnesses have shown that spine aBMD Z-scores predict vertebral fractures at different time points during the course of their illnesses; simply put, the lower the spine aBMD Z-score, the more likely a child is to sustain a vertebral fracture [21–25].

There are many medical conditions in which DXA may be useful (Table 31.1) [26], other requests come when a child has recurrent fractures in the absence of an obvious underlying predisposition.

Unfortunately, children with chronic diseases and OI often have substantially reduced stature compared with their age-matched peers and the use of DXA for aBMD quantification is potentially flawed in this instance [27–29]. This is because the measurement of aBMD relies on the 2D areal projection of a 3D object and is intrinsically related to the size of the object (bone) which is inherently related to the size of the child. Thus DXA can systematically overestimate the bone density of a tall child and underestimate bone density in a short child [4]. The failure to account for delayed growth and maturation is a common cause of misinterpretation of pediatric DXA results [30, 31].

Increasingly, the size limitation of DXA is recognized and understood and a number of approaches have been developed to account for this problem through mathematical and statistical size adjustment techniques [32–39]. The most frequently used techniques are discussed in the following paragraphs.

Bone mineral apparent density

There are two commonly used approaches to estimate bone mineral apparent density (BMAD): either the Kröger technique [40], where the vertebrae are considered to be a stack of cylindrical volumes, or the Carter technique [32], where the spine is considered to be a stack of cuboid volumes. An approximation of bone volume is calculated using the projected bone area from the DXA scan and an estimate of bone depth. Using the Kröger technique the projected bone width is used to estimate bone depth, whereas with the Carter technique the square root of the projected bone area is used to estimate bone depth. From an estimation of bone volume an approximate volumetric bone density (BMAD) (g/cm^3) can then be calculated [32, 40]. BMAD is relatively independent of bone size and, along with aBMD for age, has also been shown to be related to fracture risk in children [41].

Bone mineral content for height or height for age

Reporting bone mineral content (BMC) for height is the simplest of all of the size adjustment methods because it requires no assumptions about bone size. Although this size adjustment has not been related to fracture risk, it has been shown to be useful in comparing populations with diminished stature [28]. It also correlates with estimated bone strength as measured by pQCT in children (this technique is described in "Quantitative computed tomography (QCT)") [36]. Recently a more sophisticated height-based size adjustment has been proposed which uses height for age to predict LS or whole body BMC and aBMD. Deviations from predicted values are referred to as height for age Z-scores (HAZ) [7]. Zemel and colleagues demonstrated that adjustments using HAZ were less biased compared to other size adjustment techniques in relation to height and age and postulated that they could be used to evaluate the effect of short or tall stature [7]. Currently, this technique has not been related to fracture risk.

Table 31.1. Example groups considered to be at increased risk of fracture in whom DXA scanning is indicated.

Primary bone diseases	Osteogenesis imperfecta (OI)
	Idiopathic juvenile osteoporosis
Neuromuscular diseases/ immobilization	Cerebral palsy
	Muscular dystrophies
	Spinal cord injury
Inflammatory conditions	Crohn's disease
	Cystic fibrosis
	Juvenile rheumatic disorders, eg,
	• idiopathic arthritis
	• vasculitis
	• dermatomyositis
	• systemic lupus erythematosus
Nutritional and/or endocrine disorders	Anorexia nervosa
	Celiac disease
	Cushing syndrome
	Hyperparathyroidism
Others	Leukemia
	Thalassemia major
	Sickle cell anemia
	Solid organ transplantation
	Galactosemia
	Glycogen storage disease
	Ehlers–Danlos syndrome

Allometric approach

The allometric approach or "Mølgaard" model, so named after the Danish physician who first introduced this method in children, provides a three-stage assessment to explain reduced bone mineralization in a sick child. The model assesses height for age, bone area for height, and BMC for bone area. These three steps then correspond to three different causes of reduced bone mass: short bones, narrow bones, and light bones [37]. The important diagnostic value of this approach is that it provides both geometric and densitometric information to give a better understanding of the bones' underlying fragility. For example, smaller and thinner bones have been shown to have reduced bone strength by Binkley and colleagues in children with cerebral palsy [42].

Mechanostat functional model

The mechanostat or functional model uses an alternative approach to size adjustment based on the principles proposed by Harold Frost in that bone strength is highly correlated to muscle strength [43]. The model was first applied as a diagnostic tool by Schöenau and colleagues using pQCT measures of bone strength and muscle mass to distinguish different diseases and risk factors for reduced bone strength [44]. The application of the two-stage algorithm was extended to DXA using the assumptions that BMC could act as a surrogate for bone strength and lean body mass as a surrogate for muscle load [45–47]. The two stages of assessment are: (i) whether the child has sufficient muscle for their height, and (ii) whether they have sufficient bone for that muscle. This leads to four outcomes: (i) appropriate muscle mass for their height and appropriate bone mass for their muscle mass; (ii) primary bone defect, where the child has sufficient muscle for their height but insufficient bone mass for their muscle; (iii) primary muscle defect where the child has reduced muscle for their height but sufficient bone for their muscle mass; and (iv) a mixed muscle and bone defect where muscle and bone are both reduced.

Currently, there is no consensus as to the best method of size adjustment, or which of the size adjustment techniques will best predict future fracture risk. In a retrospective study in children with chronic diseases, with and without fracture, BMAD using the Kröger method was found to have the best discriminative power for vertebral fractures [27]. In contrast, for long bone fractures the mechanostat approach was the most successful of all the size adjustment techniques [27]. However, in general, any of the size adjustment techniques will improve the estimation of aBMD by DXA in children with short stature [27, 28]. In cases of delayed puberty, bone age (by hand X-ray according to the method of Greulich and Pyle) can be substituted for chronological age [48].

REPORTING aBMD BY DXA

Interpretation of the DXA scan results depends on the clinical context. A diagnosis of osteoporosis should not be made on the basis of an aBMD measurement in isolation [11]. Terminology such as "low bone density for chronological age" may be used if the Z-score is below −2.0 [8]. If a child presents with recurrent fractures but no clinically apparent underlying disease despite an extensive bone health assessment (as outlined in the Osteoporosis Diagnosis and Treatment Algorithm presented by Ward and colleagues [19]), and aBMD is within the expected range, reassurance can be offered. If there is an underlying problem, additional follow-up will be required. Such a clinical setting would be a child with apparently mild OI with aBMD in the normal range who may have occult fractures of thoracic vertebrae; such findings have also been reported in nephrotic syndrome [49]. Normal bone mass by DXA does not, therefore, preclude the presence of vertebral fractures that merit intervention. Low aBMD in the presence of vertebral or recurrent long bone fractures resulting in loss of independent mobility and chronic bone pain should prompt evaluation of the need for active intervention, as should falling bone mass in conditions that predispose to fracture [1].

In diseases and disorders in childhood where fracture risk is increased, initiation of bone mass measurement by DXA is typically part of the routine bone health assessment, and the monitoring interval should reflect the severity of the disease. That said, monitoring for signs of overt bone fragility (such as vertebral fractures) is even more important than aBMD by DXA, because low-trauma fractures provide the starting point for considering whether a child is a candidate for osteoporosis therapy [19]. There is little evidence to support monitoring at intervals of less than 6 months in any setting [14]. It does not seem appropriate at this time to suggest that measurements should be any more frequent in children receiving such therapy [26].

DIAGNOSTIC POTENTIAL OF DXA

The International Society for Clinical Densitometry (ISCD) has published an updated position statement on the use of aBMD by DXA in pediatrics [11]. The guidelines recommend using DXA of the LS and TBLH as part of the assessment process in children with primary or secondary bone disease and that follow-up should be no more frequent than 6 monthly. They also recommend that results should be size-adjusted in children of reduced stature and reported as "low bone mineral content (BMC) or bone mineral density for chronological age" if the BMC or aBMD Z-score is less than or equal to −2.0 [8]. However, the guidelines clearly state that a diagnosis of osteoporosis must not be made purely on densitometric criteria. The newly agreed definition of osteoporosis in children states that:

1. The finding of one or more vertebral compression fractures is indicative of osteoporosis in the absence of local disease or high-energy trauma, *regardless of the aBMD Z-score.*
2. In the absence of vertebral fractures, the diagnosis of osteoporosis is indicated by both the presence of a clinically significant fracture and aBMD ≤ −2.0, where clinically significant fractures are either:
   a. two or more long bone fractures by age 10 years;
   b. three or more long bone fractures at any age up to 19 years [1]

However, an important caveat to this aspect of the definition (stated on page 4 of the ISCD 2014 guidelines) is that "an aBMD/BMC Z score of > −2.0 does not preclude the possibility of skeletal fragility and increased fracture risk."

Practically speaking, this means that in any child with risk factors for skeletal fragility (such as a chronic illness that is known to be associated with an increased frequency of fractures relative to the healthy population, or in children with OI), a "normal" aBMD does not rule out the diagnosis of osteoporosis. These concepts are reiterated in the "Osteoporosis Diagnosis and Treatment" algorithm provided by Ward and colleagues [19].

VERTEBRAL FRACTURE ASSESSMENT

The dichotomy between normal aBMD and vertebral fracture and the often occult presentation of theses fractures has led to the increasing desire to include an assessment for vertebral fracture alongside standard bone densitometry studies [19]. Until recently vertebral fracture assessment (VFA) by DXA was considered inferior to plain X-rays for the diagnosis of vertebral fracture [50]. However, with the development of new higher-resolution scanners, VFA by DXA is proving to be an attractive imaging tool for children (Fig. 31.3A,B). Compared to lateral radiographs, it affords the child a significant (approximately threefold) reduction in radiation dose and is available at the time of routine DXA scanning [51–54].

OTHER TECHNIQUES

Quantitative computed tomography

Aside from the size-related issues of DXA, one of the other main limitations of the technique is that it measures a composite of trabecular and cortical bone. One technique which has the potential to overcome both the size-related issues and separately measure the two bone types is QCT. Axial QCT has been available since the late 1970s but has gained renewed interest in pediatric bone density assessment. The development of spiral and multislice CT

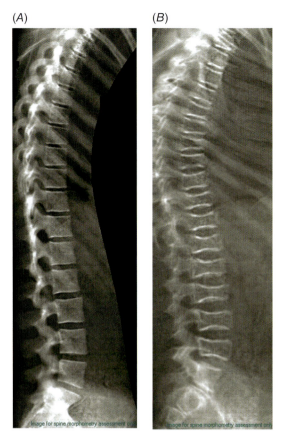

Fig. 31.3. (A) Example of vertebral fracture assessment (VFA) acquired by DXA of a 14-year-old child with anorexia nervosa—no evidence of vertebral fracture. (B) Example of VFA acquired by DXA of an 11-year-old child with idiopathic juvenile osteoporosis—multiple vertebral fractures (images acquired from a GE Lunar iDXA scanner).

scanners allows the rapid, precise acquisition of volumetric scans of the spine, hip, and peripheral skeleton [55, 56]. As such, QCT can be applied to central or peripheral skeletal sites. However, radiation exposure is much lower from dedicated pQCT scanners than axial QCT scanners and is comparable to or even less than DXA [57].

Other advantages to pQCT are that, in addition to quantification of true volumetric BMD, muscle and bone geometry can also be ascertained including muscle and cortical cross-sectional area as well as periosteal and endosteal circumferences. This has enhanced our understanding of the ways in which muscle–bone geometry in growing children changes in response to chronic diseases and recovery (such as in pediatric leukemia, arthritis, and Crohn's disease [58–60]. Additionally, pQCT-measured parameters such as volumetric BMD, cortical area, and strength-strain index have also been associated to fractures in children [13]. The most recent development in QCT has been the introduction of dedicated high-resolution scanning (HRpQCT) [61]. HRpQCT has sufficient spatial resolution (130 μm) to enable visualization and quantification of individual trabeculae; direct measures of trabecular number and estimates of thickness and separation are given [62]. In a study, HRpQCT highlighted bone strength and structural deficits in children with distal forearm fractures resulting from mild trauma [63].

There are however limitations which need to be considered; even with modern day scanners, axial QCT (though not peripheral) still affords a relatively higher radiation than DXA [57]. QCT, pQCT, and HRpQCT are only available at a few specialized centres around the world and require highly trained personnel to both acquire and interpret the outcomes. As such, the method remains primarily a research tool apart from in those centres with expertise where reference data are available [56].

Magnetic resonance imaging

The most recently developed technique is MRI. It can be applied to the peripheral or axial skeleton. The whole bone or specific regions can be measured [64, 65]. Dedicated MRI techniques can be used to measure trabecular bone architecture (ie, apparent trabecular bone volume to total volume [appBV/TV], trabecular number [appTb.N], trabecular thickness [appTb.Th] and trabecular separation [appTb.Sp]) and cortical bone architecture [ie, cortical volume, total volume, section modulus (Z), and polar moment of inertia (J)] [66]. The main advantage of MRI is that it provides a volumetric measure without the use of ionizing radiation; additionally, imaging in multiple anatomical planes is possible without having to reposition the subject and simultaneous scanning of several limbs is also feasible. However, the limitations of MRI are the long scanning times in a relatively confined environment making it difficult for some children to tolerate (particularly young children but also older children who find the confined environment challenging), and as with QCT techniques there are limited reference and fracture data for widespread use. To date, MRI has been used only in a few adult research protocols. Currently, there are no published data relating MRI-acquired bone parameters to low-trauma fracture risk in either adults or children. As such, its applicability in clinical practice is limited.

Developing techniques

Because of the limited number of studies, other techniques such as trabecular bone score (TBS), hip axis length (HAL), hip strength analysis (HSA), and QUS are currently not recommended as diagnostic tools for the assessment of fracture risk in children [8].

CONCLUSION

DXA is the most widely used method of measuring aBMD in children and it provides precise results with very low doses of ionizing radiation. The basic rationale for DXA-based aBMD examinations is their support in identifying patients at increased risk of fracture. DXA has some important limitations, in particular issues with interpreting results in children in whom the dependency on bone size is an issue, and to date there is no consensus on whether size correction should be applied and which method is optimal. This issue is obviated to some extent by the current recommendation that the diagnosis of osteoporosis in children rests largely on the presence of a clinically significant fracture history; in fact, in children with known risk factors for bone fragility, a BMD Z-score cut-off to diagnosis osteoporosis is no longer required given observations that such children can sustain low-trauma long bone or vertebral fractures with aBMD Z-scores > –2 [1].

Although alternative techniques such as QCT, pQCT, HRqCT, and MRI are available to pediatrics, their main use is still limited to research studies rather than clinical diagnostic purposes. The clinical utility of these more sophisticated techniques merits further study through the conduct of studies which assess CT- and MRI-based skeletal parameters in relationship to fractures.

ACKNOWLEDGMENTS

Dr. Leanne Ward is supported by a University of Ottawa Research Chair Award in Pediatric Bone Health and a Children's Hospital of Eastern Ontario Capacity Building Award. Dr. Nicola Crabtree was supported in part by a National Institutes of Health Research Clinical Development Fellowship (HCS/P10/009). The views expressed are those of the authors and not necessarily those of the NHS, the NIHR, or the Department of Health.

REFERENCES

1. Bishop N, Arundel P, Clark E, et al. Fracture prediction and the definition of osteoporosis in children and adolescents: the ISCD 2013 Pediatric Official Positions. J Clin Densitom. 2014;17(2):275–80.
2. Ma J, Siminoski K, Alos N, et al. The choice of normative pediatric reference database changes spine bone mineral density Z-scores but not the relationship between bone mineral density and prevalent vertebral fractures. J Clin Endocrinol Metab. 2015;100(3):1018–27.
3. Van Rijn RR, van Kuijk C. Bone density in children. Semin Musculoskel Radiol. 2002;6(3):233–9.
4. Fewtrell M, Ahmed SF, Allgrove J, et al. Bone densitometry in children assessed by dual X-ray absorptiometry: uses and pitfalls. Arch Dis Child. 2003;88(9):795–8.
5. Kalkwarf HJ, Zemel BS, Gilsanz V, et al. The bone mineral density in childhood study: bone mineral content and density according to age, sex, and race. J Clin Endocrinol Metab. 2007;92(6):2087–99.
6. Kocks J, Ward K, Mughal Z, et al. Z-score comparability of bone mineral density reference databases for children. J Clin Endocrinol Metab. 2010;95(10):4652–9.

7. Zemel BS, Leonard MB, Kelly A, et al. Height adjustment in assessing dual energy x-ray absorptiometry measurements of bone mass and density in children. J Clin Endocrinol Metab. 2010;95(3):1265–73.
8. Crabtree NJ, Arabi A, Bachrach LK, et al. Dual-energy X-ray absorptiometry interpretation and reporting in children and adolescents: the revised 2013 ISCD Pediatric Official Positions. J Clin Densitom. 2014;17(2):225–42.
9. Short DF, Zemel BS, Gilsanz V, et al. Fitting of bone mineral density with consideration of anthropometric parameters. Osteoporos Int. 2011;22(4):1047–57.
10. Katzman DK, Bachrach LK, Carter DR, et al. Clinical and anthropometric correlates of bone mineral acquisition in healthy adolescent girls. J Clin Endocrinol Metab. 1991;73(6):1332–9.
11. Gordon CM, Leonard MB, Zemel BS, International Society for Clinical Densitometry. 2013 Pediatric Position Development Conference: executive summary and reflections. J Clin Densitom. 2014;17(2):219–24.
12. Crabtree N, Ward K. Bone Densitometry: Current Status and Future Perspective. Endocr Dev. 2015;28:72–83.
13. Kalkwarf HJ, Laor T, Bean JA. Fracture risk in children with a forearm injury is associated with volumetric bone density and cortical area (by peripheral QCT) and areal bone density (by DXA). Osteoporos Int. 2011;22(2):607–16.
14. Shepherd JA, Wang L, Fan B, et al. Optimal monitoring time interval between DXA measures in children. J Bone Miner Res. 2011;26(11):2745–52.
15. Crabtree NJ, Shaw NJ, Bishop NJ, et al. Amalgamated Reference Data for Size-Adjusted Bone Densitometry Measurements in 3598 Children and Young Adults-the ALPHABET Study. J Bone Miner Res. 2017;32(1):172–80.
16. Ward KA, Ashby R, Roberts S, et al. United Kingdom reference data for the hologic QDR discovery dual energy X-ray absorptiometry scanner in healthy children aged 6–17 years. Arch Dis Child. 2007;92(1):53–9.
17. Zemel BS, Kalkwarf HJ, Gilsanz V, et al. Revised reference curves for bone mineral content and areal bone mineral density according to age and sex for black and non-black children: results of the bone mineral density in childhood study. J Clin Endocrinol Metab. 2011;96(10):3160–9.
18. Henderson RC, Berglund LM, May R, et al. The relationship between fractures and DXA measures of BMD in the distal femur of children and adolescents with cerebral palsy or muscular dystrophy. J Bone Miner Res. 2010;25(3):520–6.
19. Ward LM, Konji VN, Ma J. The management of osteoporosis in children. Osteoporos Int. 2016;27(7):2147–79.
20. Clark EM, Ness AR, Bishop NJ, et al. Association between bone mass and fractures in children: a prospective cohort study. J Bone Miner Res. 2006;21(9):1489–95.
21. Halton J, Gaboury I, Grant R, et al. Advanced vertebral fracture among newly diagnosed children with acute lymphoblastic leukemia: results of the Canadian Steroid-Associated Osteoporosis in the Pediatric Population (STOPP) research program. J Bone Miner Res. 2009;24(7):1326–34.
22. Alos N, Grant RM, Ramsay T, et al. High incidence of vertebral fractures in children with acute lymphoblastic leukemia 12 months after the initiation of therapy. J Clin Oncol. 2012;30(22):2760–7.
23. Cummings EA, Ma J, Fernandez CV, et al. Incident Vertebral Fractures in Children With Leukemia During the Four Years Following Diagnosis. J Clin Endocrinol Metab. 2015;100(9):3408–17.
24. Halton J, Gaboury I, Grant R, et al. Advanced vertebral fracture among newly diagnosed children with acute lymphoblastic leukemia: results of the Canadian Steroid-Associated Osteoporosis in the Pediatric Population (STOPP) research program. J Bone Miner Res. 2009;24(7):1326–34.
25. LeBlanc CM, Ma J, Taljaard M, et al. Incident Vertebral Fractures and Risk Factors in the First Three Years Following Glucocorticoid Initiation Among Pediatric Patients With Rheumatic Disorders. J Bone Miner Res. 2015;30(9):1667–75.
26. Bianchi ML, Leonard MB, Bechtold S, et al. Bone health in children and adolescents with chronic diseases that may affect the skeleton: the 2013 ISCD Pediatric Official Positions. J Clin Densitom. 2014;17(2):281–94.
27. Crabtree NJ, Hogler W, Cooper MS, et al. Diagnostic evaluation of bone densitometric size adjustment techniques in children with and without low trauma fractures. Osteoporos Int. 2013;24(7):2015–24.
28. Fewtrell M, Gordon I, Biassoni L, et al. Dual X-ray absorptiometry (DXA) of the lumbar spine in a clinical paediatric setting: does the method of size-adjustment matter? Bone. 2005;37(3):413–9.
29. Leonard MB, Propert KJ, Zemel BS, et al. Discrepancies in pediatric bone mineral density reference data: potential for misdiagnosis of osteopenia. J Pediatr. 1999;135:182–8.
30. Gafni RI, Baron J. Overdiagnosis of osteoporosis in children due to misinterpretation of dual-energy X-ray absorptiometry (DEXA). J Pediatr. 2004;144(2):253–7.
31. Leonard MB, Zemel BS. Current concepts in pediatric bone disease. Pediatr Clin North Am. 2002;49(1):143–73.
32. Carter DR, Bouxsein ML, Marcus R. New approaches for interpreting projected bone densitometry data. J Bone Miner Res. 1992;7(2):137–45.
33. Ellis KJ, Shypailo RJ, Hardin DS, et al. Z-Score prediction model for assessment of bone mineral content in pediatric diseases. J Bone Miner Res. 2001;16(9):1658–64.
34. Horlick M, Wang J, Peirson RN, et al. Prediction models for evaluation of total-body bone mass with dual-energy X-ray absorptiometry among children and adolescents. Pediatrics. 2004;114(3):337–45.
35. Kroger H, Kotaniemi A, Kroger L, et al. Development of bone mass and bone density of the spine and femoral neck—a prospective study of 65 children and adolescents. Bone Miner. 1993;23:171–82.
36. Leonard MB, Shults J, Elliot DM, et al. Interpretation of whole body dual energy X-ray absorptiometry measures in children: comparison with peripheral quantitative computed tomography. Bone. 2004;34:1044–52.

37. Molgaard C, Thomsen BL, Prentice A, et al. Whole body bone mineral content in healthy children and adolescents. Arch Dis Child. 1997;76:9–15.
38. Schoenau E, Schwahn B, Rauch F. The muscle-bone relationship: methods and management - perspectives in glycogen storage disease. Eur J Pediatr. 2002;161:50–2.
39. Warner JT, Cowan FJ, Dunstan FDJ, et al. Measured and predicted bone mineral content in healthy boys and girls aged 6-18 years: adjustment for body size and puberty. Acta Paediatr. 1998;87:244–9.
40. Kroger H, Kotaniemi A, Vainio P, et al. Bone densitometry of the spine and femur in children by dual-energy X-ray absorptiometry. Bone Miner. 1992;17:75–85.
41. Goulding A, Jones IE, Taylor RW, et al. More broken bones: a 4-year double cohort study of young girls with and without distal forearm fractures. J Bone Miner Res. 2000;15(10):2011–8.
42. Binkley TL, Johnson J, Vogel L, et al. Bone measurements by peripheral quantitative computed tomography (pQCT) in children with cerebral palsy. J Pediatr. 2005;147:791–6.
43. Frost HM. The Mechanostat: a proposed pathogenic mechanism of osteoporoses and the bone mass effects of mechanical and nonmechanical agents. Bone Miner. 1987;2:73–85.
44. Schoenau E, Neu CM, Beck B, et al. Bone mineral content per muscle cross-sectional area as an index of the functional muscle-bone unit. J Bone Miner Res. 2002;17(6):1095–101.
45. Crabtree NJ, Kibirige MS, Fordham JN, et al. The relationship between lean body mass and bone mineral content in paediatric health and disease. Bone. 2004;35:965–72.
46. Hogler W, Briody JN, Woodhead HJ, et al. Importance of lean mass in the interpretation of total body densitometry in children and adolescents. J Pediatr. 2003;143:81–8.
47. Pludowski P, Lebiedowski M, Lorenc RS. Evaluation of practical use of bone assessments based on DXA-derived hand scans in diagnosis of skeletal status in healthy and diseased children. J Clin Densitom. 2005;8(1):48–56.
48. Greulich WW, Pyle SI. *Radiographic Atlas of Skeletal Development of the Hand and Wrist* (2nd ed.). Stanford University Press, 1959.
49. Sbrocchi AM, Rauch F, Matzinger M, et al. Vertebral fractures despite normal spine bone mineral density in a boy with nephrotic syndrome. Pediatr Nephrol. 2011;26(1):139–42.
50. Mayranpaa MK, Helenius I, Valta H, et al. Bone densitometry in the diagnosis of vertebral fractures in children: accuracy of vertebral fracture assessment. Bone. 2007;41(3):353–9.
51. Adiotomre E, Summers L, Allison A, et al. Diagnostic accuracy of DXA compared to conventional spine radiographs for the detection of vertebral fractures in children. Eur Radiol. 2017;27(5):2188–99.
52. Crabtree N, Chapman S, Hogler W, et al. Is vertebral fracture assessment by DXA more useful in a high fracture risk paediatric population than in a low-risk screening population? Bone Abstracts. 2013;2.
53. Diacinti D, Pisani D, D'Avanzo M, et al. Reliability of vertebral fractures assessment (VFA) in children with osteogenesis imperfecta. Calcif Tissue Int. 2015;96(4):307–12.
54. Kyriakou A, Shepherd S, Mason A, et al. A critical appraisal of vertebral fracture assessment in paediatrics. Bone. 2015;81:255–9.
55. Adams JE. Quantitative computed tomography. Eur J Radiol. 2009;71(3):415–24.
56. Adams JE, Engelke K, Zemel BS, et al., International Society of Clinical Densitometry. Quantitative computer tomography in children and adolescents: the 2013 ISCD Pediatric Official Positions. J Clin Densitom. 2014;17(2):258–74.
57. Damilakis J, Adams JE, Guglielmi G, et al. Radiation exposure in X-ray-based imaging techniques used in osteoporosis. Eur Radiol. 2010;20(11):2707–14.
58. Mostoufi-Moab S, Brodsky J, Isaacoff EJ, et al. Longitudinal assessment of bone density and structure in childhood survivors of acute lymphoblastic leukemia without cranial radiation. J Clin Endocrinol Metab. 2012;97(10):3584–92.
59. Burnham JM, Shults J, Dubner SE, et al. Bone density, structure, and strength in juvenile idiopathic arthritis: importance of disease severity and muscle deficits. Arthritis Rheum. 2008;58(8):2518–27.
60. Dubner SE, Shults J, Baldassano RN, et al. Longitudinal assessment of bone density and structure in an incident cohort of children with Crohn's disease. Gastroenterology. 2009;136(1):123–30.
61. Tjong W, Kazakia GJ, Burghardt AJ, et al. The effect of voxel size on high-resolution peripheral computed tomography measurements of trabecular and cortical bone microstructure. Med Phys. 2012;39(4):1893–903.
62. Bouxsein ML. Bone structure and fracture risk: do they go arm in arm? J Bone Miner Res. 2011;26(7):1389–91.
63. Farr JN, Amin S, Melton LJ, 3rd, et al. Bone strength and structural deficits in children and adolescents with a distal forearm fracture resulting from mild trauma. J Bone Miner Res. 2014;29(3):590–9.
64. Daly RM, Saxon L, Turner CH, et al. The relationship between muscle size and bone geometry during growth and in response to exercise. Bone. 2004;34:281–7.
65. Hogler W, Blimkie CJ, Cowell CT, et al. A comparison of bone geometry and cortical density at the mid-femur between prepuberty and young adulthood using magnetic resonance imaging. Bone. 2003;33(5):771–8.
66. Hong J, Hipp JA, Mulkern RV, et al. Magnetic resonance imaging measurements of bone density and cross-sectional geometry. Calcif Tissue Int. 2000;66(1):74–8.

32

Standard Techniques of Bone Mass Measurement in Adults

E. Michael Lewiecki[1], Paul D. Miller[2], and Nelson B. Watts[3]

[1]New Mexico Clinical Research & Osteoporosis Center, Albuquerque, NM, USA
[2]Colorado Center for Bone Research at Panorama Orthopedics and Spine Center, Golden, CO, USA
[3]Mercy Health, Osteoporosis and Bone Health Services, Cincinnati, OH, USA

INTRODUCTION

DXA was first approved for the measurement of BMD in clinical practice by the United States Food and Drug Administration (FDA) in 1988. Six years later, in 1994, the WHO released criteria for diagnosing osteoporosis according to variation of a patient's BMD from the mean BMD of a young-adult reference population [1] expressed as a value that is now called a T-score (Table 32.1). These two developments, followed by the approval of alendronate for the treatment of osteoporosis in 1995, launched the modern era of osteoporosis management. For the first time, osteoporosis could be diagnosed before a fracture occurred and medication to reduce fracture risk became widely available. DXA has emerged as the dominant technology for measuring BMD because of: (i) a robust correlation between DXA-measured BMD and bone strength in biomechanical studies [2]; (ii) epidemiologic studies showing a strong relationship between fracture risk and DXA BMD [3]; and (iii) randomized clinical trials with drug therapy selecting subjects based on DXA BMD as one of their randomization criteria [4] with excellent accuracy and precision [5] and low radiation [6]. DXA is now used to diagnose osteoporosis, assess fracture risk, and monitor changes in BMD over time. In recent years, DXA technology has been adapted to measurement of more than BMD (eg, detecting vertebral fractures, body composition testing, trabecular bone score); other technologies for assessing bone strength, including QUS, have also been developed. This is a review of the major clinical applications of DXA and QUS.

DXA TECHNOLOGY

A "central" DXA system (one that measures BMD at the LS and hips) consists of a table to support the patient, a radiation source (usually beneath the patient), a radiation detector (usually above the patient), and computer software that creates images of bone and soft tissue and analyzes the data to provide a quantitative result. "Peripheral" DXA (pDXA) uses the same technology with a smaller and more portable instrument to measure BMD at a peripheral skeletal site, such as the calcaneus or radius. The X-ray source emits two distinct energy levels. It is the difference in attenuation of these two beams passing through body tissues of variable composition that allows the instrument to generate a quantitative measure of BMD and soft tissue. The original DXA scanners used a pencil beam of X-rays with a single detector that scanned in a rectilinear fashion across the anatomical site with scan times of 5 to 10 minutes. Subsequently, fan-beam machines were introduced with detectors in an array, which resulted in shorter scan times of less than 1 minute per skeletal site and improved image quality. The radiation dose from DXA examinations is extremely low, in the range of 1 to 10 microsieverts (µSv) for a spine and hip

---

Primer on the Metabolic Bone Diseases and Disorders of Mineral Metabolism, Ninth Edition. Edited by John P. Bilezikian.
© 2019 American Society for Bone and Mineral Research. Published 2019 by John Wiley & Sons, Inc.
Companion website: www.wiley.com/go/asbmrprimer

Table 32.1. World Health Organization (WHO) classification of bone mineral density [1].

WHO Classification	T-score
Normal	−1.0 or greater
Low bone mass (osteopenia)	Between −1.0 and −2.5
Osteoporosis	−2.5 or less

The T-score compares the patient's BMD to that of a young-adult reference population and is used to classify BMD in postmenopausal women and men age 50 years and older. It is calculated with the following equation, where SD = standard deviation, g = grams, cm = centimeter. The T-score is properly expressed as a number with one decimal place and no units [15].

$$T\text{-score} = \frac{(\text{patient's BMD in g/cm}^2) - (\text{mean young-adult BMD in g/cm}^2)}{(1\ SD\ \text{of young-adult BMD in g/cm}^2)}$$

A Z-score, which is not used with the WHO classification, compares the patient's BMD to an age-, ethnicity-, and sex-matched reference population. Z-scores, rather than T-scores, are preferred for reporting the results for premenopausal women, men under age 50, and children, using the following equation.

$$Z\text{-score} = \frac{(\text{patient's BMD in g/cm}^2) - \text{mean age/race/sex-adjusted BMD in g/cm}^2}{(1\ SD\ \text{of age/race/sex-adjusted BMD in g/cm}^2)}$$

examination; this is less than or comparable to the daily dose from natural background radiation (~7 µSv/day) [7].

QUALITY STANDARDS

DXA systems manufactured in recent years are generally very reliable. However, unexpected changes in calibration may not be detected unless instrument stability is assessed regularly; errors of scan acquisition, analysis, interpretation, and reporting are common and potentially harmful to patients [8–13]. In order to guide DXA facilities toward the achievement of high-quality DXA and to assist referring clinicians and patients in determining whether high-quality DXA is being performed, the International Society for Clinical Densitometry (ISCD) has developed certification courses in bone densitometry, established standards for the assessment of skeletal health, and recently released DXA Best Practices [14] (Table 32.2). DXA Best Practices are largely based on the findings of ISCD Position Development Conferences, where an international panel of experts has reviewed the best available evidence for clinically relevant topics and made recommendations that become ISCD Official Positions [15]. DXA technologists and interpreters can be trained in high-quality DXA through courses in bone densitometry; familiarity with a basic skill set can be demonstrated by passing a certification test in bone densitometry and maintaining a valid certification [14]. Facility accreditation, offered by organizations such as the ISCD, Ontario Association of Radiologists, Canadian Association of Radiologists, and the Brazilian College of Radiology and Brazilian Association of Bone Health Assessment and Metabolism, provides the highest level of assurance that essential elements for quality bone density testing have been implemented at a DXA facility.

MEASUREMENT OF BONE MINERAL DENSITY

Indications for BMD testing vary according to societal priorities, availability of testing facilities, and financial concerns depending on local requirements. The ISCD Official Positions recommend that BMD testing be considered according to criteria listed in Table 32.3. DXA and pDXA systems generate a 2D projection of bone, resulting in an areal BMD expressed as grams per square centimeter (g/cm^2). This value is then converted to a T-score (comparison with young-adult mean BMD) for diagnostic classification or a Z-score for comparison with an age-, ethnicity-, and sex-matched population (Table 32.1) through the use of appropriate reference data. The WHO and the ISCD recommend that the international standard for T-score calculation for the hip in women and men is the female, white, 20- to 29-year-old Third National Health and Nutrition Examination Survey (NHANES III) database, with the ISCD stating that application of the recommendation may vary according to local requirements [15]. Each manufacturer still uses their own specific reference population database for T-score calculation at the spine and wrist; this may lead to discrepancies among results if different manufacturer DXA machines are used for these specific skeletal sites. It should be noted that software with many DXA systems in clinical use may calculate T-scores using a female reference database for women and a male database for men, and that in some countries an ethnicity-matched

Table 32.2. DXA best practices.

Scan acquisition and analysis
1.1. At least one practicing DXA technologist, and preferably all, has a valid certification in bone densitometry.
1.2. Each DXA technologist has access to the manufacturer's manual of technical standards and applies these standards for BMD measurement.
1.3. Each DXA facility has detailed Standard Operating Procedures for DXA performance that are updated when appropriate and available for review by all key personnel.
1.4. The DXA facility must comply with all applicable radiation safety requirements.
1.5. Spine phantom BMD measurement is performed at least once weekly to document stability of DXA performance over time. BMD values must be maintained within a tolerance of ± 1.5%, with a defined ongoing monitoring plan that defines a correction approach when the tolerance has been exceeded.
1.6. Each DXA technologist has performed *in-vivo* precision assessment according to standard methods and the facility LSC has been calculated.
1.7. The LSC for each DXA technologist should not exceed 5.3% for the lumbar spine, 5.0% for the total proximal femur, 6.9% for the femoral neck.

Interpretation and reporting
2.1. At least one practicing DXA interpreter, and preferably all, has a valid certification in bone densitometry.
2.2. The DXA manufacturer and model are noted on the report.
2.3. The DXA report includes a statement regarding scan factors that may adversely affect acquisition/analysis quality and artifacts/confounders, if present.
2.4. The DXA report identifies the skeletal site, region of interest, and body side for each technically valid BMD measurement.
2.5. There is a single diagnosis reported for each patient, not a different diagnosis for each skeletal site measured.
2.6. A fracture risk assessment tool is used appropriately.
2.7. When reporting differences in BMD with serial measurements, only those changes that meet or exceed the LSC are reported as a change.

LSC = least significant change.
Source: [14]. Reproduced with permission of Elsevier.

Table 32.3. International Society for Clinical Densitometry indications for BMD testing.

- Women age 65 years and older
- Postmenopausal women younger than 65 years of age when a risk factor for bone mass is present, such as low body weight, prior fracture, high-risk medication use, disease or condition associated with bone loss
- Women during the menopausal transition with clinical risk factors for fracture, such as low body weight, prior fracture, or high-risk medication use
- Men age 70 years and older
- Men younger than 70 years of age, when a risk factor for low bone mass is present, such as low body weight, prior fracture, high-risk medication use, disease or condition associated with bone loss
- Adults with a fragility fracture
- Adults with a disease or condition associated with low bone mass or bone loss
- Adults taking medications associated with low bone mass or bone loss
- Anyone being considered for pharmacologic therapy
- Anyone being treated to monitor treatment effect
- Anyone not receiving therapy in whom evidence of bone loss would lead to treatment

Source: [15]. Reproduced with permission of Elsevier.

database for *T*-scores may be used for both men and women. Careful review of the DXA printout will reveal which of the many options for databases has been selected. Central DXA systems measure BMD at the LS (ideally L1 to L4, with operator exclusion of up to two vertebral bodies from analysis when confounding structural abnormalities are present) and hip; the forearm should be measured under some circumstances, such as when a valid measurement cannot be obtained at the LS and/or hip, the patient's body weight exceeds the limit of the DXA table, or when hyperparathyroidism is present [15]. pDXA systems measure BMD at peripheral skeletal sites, such as the calcaneus and radius.

As with all biological measurements, there is some inherent variability in DXA measurements (ie, two measurements on the same day are likely to be different, even though BMD has not changed). Quantitative comparison of serial BMD measurement requires performance of precision assessment and calculation of the least significant change (LSC), the smallest change in BMD that is statistically significant, usually with a 95% CI. There are well-established guidelines for when and how to perform precision assessment, and how to use serial BMD in managing patients with osteoporosis [15, 16].

DIAGNOSIS OF OSTEOPOROSIS

A *T*-score –2.5 or below at the LS, femoral neck (FN), or total proximal femur (measured by DXA), or 33% (⅓) radius (measured by DXA or pDXA), using the lowest *T*-score of these skeletal regions of interest, is consistent with a diagnosis of osteoporosis by WHO criteria. There is only one diagnostic classification for a patient, not a different diagnosis for each skeletal site. Other skeletal sites and other technologies for BMD measurement cannot be used for diagnostic classification, with the exception of QCT *T*-scores at the FN and total proximal femur calculated from 2D projections [15]. Because other skeletal disorders, such as osteomalacia, may also be associated with *T*-scores in this range, it is incumbent on the clinician to evaluate the patient to exclude non-osteoporosis causes of low BMD.

Other means of diagnosing osteoporosis have been considered. The US National Bone Health Alliance (NBHA) recommends (applicable in the United States) three different pathways for diagnosing osteoporosis: (i) the WHO criteria; (ii) the occurrence of a hip fracture, osteopenia-associated vertebral, proximal humerus, pelvis, or some wrist fractures; or (iii) Fracture Risk Assessment Tool (FRAX) 10-year probability of major osteoporotic fracture is ≥20% or the 10-year probability of hip fracture is ≥3%, consistent with the US National Osteoporosis Foundation (NOF) thresholds for initiating pharmacological therapy to reduce fracture risk [17]. This scheme for diagnosing osteoporosis may not be appropriate in other countries.

ASSESSMENT OF FRACTURE RISK

Fracture risk approximately doubles for every SD decrease in BMD, with a predictive ability that is similar to or better than that of blood pressure in predicting stroke and cholesterol in predicting coronary events [18]. Combining BMD with clinical risk factors for fracture (eg, previous fracture, chronic glucocorticoid therapy, advancing age) predicts fracture risk better than BMD or clinical risk factors alone [19]. For this reason, fracture risk assessment tools such as FRAX [20], the Garvan Fracture Risk Calculator [21], and the Canadian Association of Radiologists and Osteoporosis Canada Risk Assessment Tool (CAROC) [22] have been developed for use in clinical practice. These tools can provide helpful information for making clinical decisions and may be incorporated into treatment guidelines [23].

VERTEBRAL FRACTURE ASSESSMENT

Vertebral fracture assessment (VFA) is imaging of the spine by DXA for the detection of vertebral fractures (Fig. 32.1). Fractures of the spine may cause chronic back pain, reduced pulmonary function, loss of height, kyphosis, abdominal discomfort, disability, loss of independence, and increased risk of death [24]. Vertebral fractures are the most common type of fragility fracture, with only about one-third being clinically recognized [25]. The finding of a previously unrecognized vertebral fracture may change diagnostic classification (eg, from osteopenia by WHO criteria to osteoporosis based on fracture) and assessment of fracture risk (to a higher level), and alter treatment decisions (leading to a decision to treat) [24]. The NOF treatment guidelines recommend initiating pharmacological therapy to reduce fracture risk when there is a vertebral fracture, regardless of BMD [26]. Indications for VFA are provided in Table 32.4.

VFA compares favorably with standard spine radiography in reliably and accurately diagnosing vertebral fractures. In a study of women age ≥65 years, the sensitivity of VFA for diagnosing moderate and severe vertebral fractures was 87% to 93%, with a specificity of 93% to 95% [27]. VFA did not perform well for diagnosing mild vertebral fractures in the presence of scoliosis or moderate to severe osteoarthritis, possibly because of differences in image resolution; however, mild vertebral fractures are

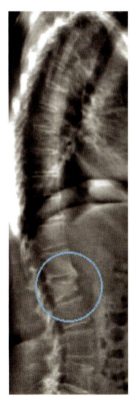

Fig. 32.1. Vertebral fracture identified by VFA.

Table 32.4. Indications for lateral spine imaging by standard radiography or VFA by DXA from International Society for Clinical Densitometry Official Positions.

When T-score is < –1.0 and one or more of the following is present:
- Women age ≥70 years or men ≥ age 80 years
- Historical height loss >4 cm 1.5 inches)
- Self-reported but undocumented prior vertebral fracture
- Glucocorticoid therapy equivalent to ≥5 mg of prednisone or equivalent per day for ≥3 months

Source: [15]. Reproduced with permission of Elsevier.

less clinically significant than moderate or severe vertebral fractures for predicting the risk of future fractures [28]. Although VFA may poorly visualize vertebral levels T4 to T7, this is not a major clinical concern because vertebral fractures are infrequent in the upper thoracic spine [29]. Compared with standard spine radiography, VFA provides greater patient convenience (ie, it can be performed at the time of BMD testing by DXA), smaller radiation doses, less parallax effects that may distort X-ray images, and lower cost.

HIP GEOMETRY

The DXA-generated 2D image of the hip can be analyzed with proprietary software such as Hip Structural Analysis (HSA, Hologic, Bedford, MA, USA) and Advanced Hip Assessment (AHA, GE Lunar, Madison, WI, USA) to measure macro-architecture of the hip and derive parameters aimed at providing a better estimation of bone strength and fracture risk than BMD alone. Direct geometrical measurements include hip axis length (HAL), neck–shaft angle (NSA), and outer diameter (OD); calculated parameters include cross-sectional area (CSA), CSMI, buckling ratio (BR), and section modulus (SM). All of these have been associated with bone strength [30]. At the 2015 ISCD Position Development Conference, the clinical utility of these seven values for estimating fracture risk was evaluated [30]. It was concluded that HAL (defined in 1993 as the distance measured along the FN axis from the base of the greater trochanter to the inner pelvic rim [31], Fig. 32.2) was associated with hip fracture risk in postmenopausal women, but not in men, independent of BMD; CSA, OD, SM, BR, CSMI, and NSA should not be used to assess hip fracture risk in clinical practice. It was also recommended that none of the hip geometry parameters be used in clinical practice to initiate treatment and none be used to monitor the effects of treatment. A separate analysis of the Manitoba Bone Density Database found a relative increase in hip fracture probability of 4.7% for every millimeter that HAL is above the sex-specific average and a relative decrease in hip fracture probability of 3.8% for every millimeter that

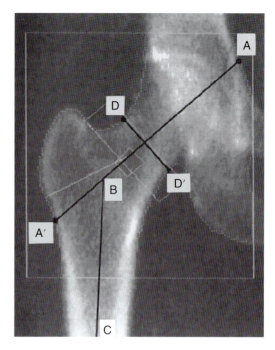

Fig. 32.2. Hip axis length (HAL), AA′, the distance from the base of the greater trochanter to the inner pelvic rim along the FN axis. Angle ABC, neck/shaft angle, the angle formed between the femoral neck and the shaft of the femur; DD′, neck width, the shortest distance within the femoral neck region of interest, perpendicular to the femoral neck axis. Source: [31]. Reproduced with permission of John Wiley & Sons.

HAL is below the sex-specific average. Despite encouraging evidence for clinical applications of HAL, it is currently challenging to apply HAL measurements to making treatment decisions because of the absence of reference data for average HAL in appropriate populations and lack of data showing treatment effect in patients selected for treatment based on HAL [32].

BODY COMPOSITION

The three major tissue components of the body (fat mass, lean mass, and bone mineral mass) can be measured with high precision and low scanning time using a single whole body DXA scan [33]. This positions DXA as a potentially attractive technology for clinical applications that include assessment and management of nutritional disorders, management of diseases that alter body tissue components, interventions for the treatment of sarcopenia, and evaluation of the effects of diet and training in recreational and elite athletes. After consideration of the best available medical evidence, which is limited with regard to clinical utility, the ISCD developed indications for DXA total body composition testing (Table 32.5).

There are numerous quality issues that must be addressed in performing body composition studies with DXA [34]. At this time, there are no available total body phantoms that serve as absolute reference standards

Table 32.5. Indications for DXA body composition testing with regional analysis from the International Society for Clinical Densitometry.

1. DXA total body composition with regional analysis can be used in the following conditions:
   a. In patients living with HIV to assess fat distribution in those using antiretroviral agents associated with a risk of lipoatrophy (currently stavudine [d4T] and zidovudine [ZDV, AZT]). Quality of evidence: good.
   b. In obese patients undergoing bariatric surgery (or medical, diet, or weight loss regimens with anticipated large weight loss) to assess fat and lean mass changes when weight loss exceeds approximately 10%. The impact on clinical outcomes is uncertain. Quality of evidence: poor.
   c. In patients with muscle weakness or poor physical functioning to assess fat and lean mass. The impact on clinical outcomes is uncertain. Quality of evidence: fair.
2. Pregnancy is a contraindication to DXA body composition. Limitations in the use of clinical DXA for total body composition or BMD are weight over the table limit, recent administration of contrast material, and/or artifact. Radiopharmaceutical agents may interfere with accuracy of results using systems from some DXA manufacturers. Quality of evidence: fair.

Source: [34]. Reproduced with permission of Elsevier.

for soft tissue composition or bone mineral mass, and no phantom has been identified to remove systematic differences in body composition measurements with different DXA manufacturers. Each DXA technologist should perform in vivo precision assessment, analogous to what is done for BMD measurements, for all body composition measures of interest, using patients who are representative of the clinic's patient population. The ISCD Official Positions state that the minimum acceptable precision for an individual technologist is 3%, 2%, and 2% for total fat mass, total lean mass, and percent fat mass, respectively [34].

QUANTITATIVE ULTRASOUND

QUS devices operate with inaudible high-frequency sound waves in the ultrasonic range, typically between 0.1 and 1.0 megahertz (MHz), produced and detected by means of high-efficiency piezoelectric transducers. There are substantial technical differences among QUS systems, with variable frequencies, different transducer sizes, and sometimes measuring different regions of interest, even at the same skeletal site. The calcaneus is the skeletal site most often tested, although other bones, including the radius, tibia, and finger phalanges, may be used. QUS devices typically measure the speed of sound (SOS) and broadband ultrasound attenuation (BUA); proprietary values, such as "quantitative ultrasound index" (QUI) with the Hologic Sahara or "stiffness index" with the GE Healthcare Achilles Express, may then be calculated and reported. Values obtained from calculations using ultrasound parameters may be used to generate an estimated BMD and a T-score. A QUS T-score is not the same as a DXA T-score, because different properties of bone are being measured and different reference databases are used; thus, QUS-derived T-scores cannot be used with the WHO classification.

The ISCD Official Positions state that the calcaneus is the only validated skeletal site with QUS for the management of patients with osteoporosis [34]. Validated calcaneus QUS devices predict fragility fractures in postmenopausal women and men age ≥65 years. Where central DXA is not available, pharmacological therapy to reduce fracture risk may be initiated when fracture probability is high, as assessed by QUS at the calcaneus, using device-specific thresholds and clinical risk factors for fracture. QUS measurements cannot be used to diagnose osteoporosis with the WHO criteria and cannot be used to monitor the effects of osteoporosis treatment.

SUMMARY

DXA is the most versatile technology for managing patients with osteoporosis. It is widely available at modest cost, producing data that can be used for assessing fracture risk, diagnosing osteoporosis, and monitoring the skeletal effects of treatment. Non-BMD applications of DXA include assessments of hip geometry, diagnosis of vertebral fractures, analysis of body composition, and trabecular bone score. Effective use of DXA in clinical practice depends on DXA facilities, technologists, and interpreters adhering to established quality standards. QUS devices have a more limited role in managing osteoporosis, but may be clinically useful in locations where DXA is not available.

DISCLOSURES

E. Michael Lewiecki has received institutional grant/research support from Amgen, Merck, and Eli Lilly; he serves on scientific advisory boards for Amgen, Merck, Eli Lilly, Radius Health, Shire, and Alexion; he is on speakers' bureaus for Shire and Alexion. Paul D. Miller has received research grants from Alexion, Amgen, Boehringer Ingelheim, Immunodiagnostics, Eli Lilly, Merck, Merck Serrano, National Bone Health Alliance,

Novartis, Radius Pharma, Roche Diagnostics, Regeneron, Daiichi Sankyo, Inc. Ultragenyx; he serves on Scientific Advisory Boards for Amgen, AgNovos, Eli Lilly, Merck, Radius Pharma, Roche, and Ultragenyx. Nelson B. Watts serves on advisory boards or as a consultant for AbbVie, Amgen, Radius, and Sanofi and is on speakers' bureaus for Amgen and Radius.

REFERENCES

1. Assessment of fracture risk and its application to screening for postmenopausal osteoporosis. Report of a WHO Study Group. World Health Organ Tech Rep Ser. 1994;843:1–129.
2. Lotz JC, Cheal EJ, Hayes WC. Fracture prediction for the proximal femur using finite element models: Part I—Linear analysis. J Biomechan Eng. 1991;113:353–60.
3. Nielson CM, Marshall LM, Adams AL, et al. BMI and fracture risk in older men: the osteoporotic fractures in men study (MrOS). J Bone Miner Res. 2011;26(3): 496–-502.
4. Cranney A, Tugwell P, Wells G, et al. Meta-analyses of therapies for postmenopausal osteoporosis. I. Systematic reviews of randomized trials in osteoporosis: introduction and methodology. Endocr Rev. 2002;23(4): 497–507.
5. Mazess R, Chesnut CH, III, McClung M, et al. Enhanced precision with dual-energy X-ray absorptiometry. CalcifTissue Int. 1992;51(1):14–17.
6. Njeh CF, Fuerst T, Hans D, et al. Radiation exposure in bone mineral density assessment. Appl Radiat Isot. 1999;50(1):215–36.
7. Damilakis J, Adams JE, Guglielmi G, et al. Radiation exposure in X-ray-based imaging techniques used in osteoporosis. Eur Radiol. 2010;20(11):2707–14.
8. Lewiecki EM, Lane NE. Common mistakes in the clinical use of bone mineral density testing. Nat Clin Pract Rheumatol. 2008;4(12):667–74.
9. Lewiecki EM, Binkley N, Petak SM. DXA Quality Matters. J Clin Densitom. 2006;9(4):388–92.
10. Watts NB. Fundamentals and pitfalls of bone densitometry using dual-energy X-ray absorptiometry (DXA). Osteoporos Int. 2004;15(11):847–54.
11. Messina C, Bandirali M, Sconfienza LM, et al. Prevalence and type of errors in dual-energy x-ray absorptiometry. Eur Radiol. 2015;25(5):1504–11.
12. Kim TY, Schafer AL. Variability in DXA Reporting and Other Challenges in Osteoporosis Evaluation. JAMA Intern Med. 2016;176(3):393–95.
13. Fenton JJ, Robbins JA, Amarnath AL, et al. Osteoporosis Overtreatment in a Regional Health Care System. JAMA Intern Med. 2016;176(3):391–93.
14. Lewiecki EM, Binkley N, Morgan SL, et al., International Society for Clinical Densitometry. Best Practices for Dual-Energy X-ray Absorptiometry Measurement and Reporting: International Society for Clinical Densitometry Guidance. J Clin Densitom. 2016;19(2): 127–40.
15. Shepherd JA, Schousboe JT, Broy SB, et al. Executive Summary of the 2015 ISCD Position Development Conference on Advanced Measures From DXA and QCT: Fracture Prediction Beyond BMD. J Clin Densitom. 2015;18(3):274–86.
16. Lenchik L, Kiebzak GM, Blunt BA, International Society for Clinical Densitometry, Position Development Panel, Scientific Advisory Committee. What is the role of serial bone mineral density measurements in patient management? J Clin Densitom. 2002;5(Suppl):S29–S38.
17. Siris ES, Adler R, Bilezikian J, et al. The clinical diagnosis of osteoporosis: a position statement from the National Bone Health Alliance Working Group. Osteoporos Int. 2014;25(5):1439–43.
18. Marshall D, Johnell O, Wedel H. Meta-analysis of how well measures of bone mineral density predict occurrence of osteoporotic fractures. BMJ. 1996;312(7041):1254–9.
19. Kanis JA, on behalf of the World Health Organization Scientific Group. *Assessment of Osteoporosis at the Primary Health-Care Level.* Technical Report.Sheffield, UK: University of Sheffield, 2007.
20. Kanis JA, Hans D, Cooper C, et al. Interpretation and use of FRAX in clinical practice. Osteoporos Int. 2011;22(9):2395–411.
21. Garvan Institute. Bone Fracture Risk Calculator. 2011. https://www.garvan.org.au/bone-fracture-risk (accessed May 2018).
22. Leslie WD, Berger C, Langsetmo L, et al. Construction and validation of a simplified fracture risk assessment tool for Canadian women and men: results from the CaMos and Manitoba cohorts. Osteoporos Int. 2011;22(6):1873–83.
23. Leslie WD, Lix LM. Comparison between various fracture risk assessment tools. Osteoporos Int. 2014;25(1):1–21.
24. Lewiecki EM, Laster AJ. Clinical applications of vertebral fracture assessment by dual-energy X-ray absorptiometry. J Clin Endocrinol Metab. 2006;91(11):4215–22.
25. Cooper C, O'Neill T, Silman A. The epidemiology of vertebral fractures. European Vertebral Osteoporosis Study Group. Bone. 1993;14(suppl 1):S89–97.
26. Cosman F, de Beur SJ, LeBoff MS, et al. Clinician's Guide to Prevention and Treatment of Osteoporosis. Osteoporos Int. 2014;25(10):2359–81.
27. Schousboe JT, DeBold CR. Reliability and accuracy of vertebral fracture assessment with densitometry compared to radiography in clinical practice. Osteoporos Int. 2006;17(2):281–9.
28. Black DM, Arden NK, Palermo L, et al. Prevalent vertebral deformities predict hip fractures and new vertebral fractures but not wrist fractures. J Bone Miner Res. 1999;14:821–8.
29. Van der Klift M, De Laet CE, McCloskey EV, et al. The incidence of vertebral fractures in men and women: the Rotterdam Study. J Bone Miner Res. 2002;17(6):1051–6.
30. Broy SB, Cauley JA, Lewiecki EM, et al. Fracture Risk Prediction by Non-BMD DXA Measures: the 2015 ISCD

Official Positions Part 1: Hip Geometry. J Clin Densitom. 2015;18(3):287–308.
31. Faulkner KG, Cummings SR, Black D, et al. Simple measurement of femoral geometry predicts hip fracture: The study of osteoporotic fractures. J Bone Miner Res. 1993;8:1211–17.
32. Lewiecki EM. Clinical vignettes: using non-BMD measurements in clinical practice. Clin Rev Bone Miner Metab. 2016;14(1):50–4.
33. Albanese CV, Diessel E, Genant HK. Clinical applications of body composition measurements using DXA. J Clin Densitom. 2003;6(2):75–85.
34. Shepherd JA, Baim S, Bilezikian JP, et al. Executive summary of the 2013 International Society for Clinical Densitometry Position Development Conference on Body Composition. J Clin Densitom. 2013;16(4): 489–95.

33

Advanced Techniques of Bone Mass Measurements and Strength in Adults

Kyle K. Nishiyama[1], Enrico Dall'Ara[2], and Klaus Engelke[3]

[1]*Department of Medicine—Endocrinology, Columbia University Medical Center, New York, NY, USA*
[2]*Department of Oncology and Metabolism and INSIGNEO Institute for In Silico Medicine, University of Sheffield, Sheffield, UK*
[3]*Institute of Medical Physics, Friedrich-Alexander-Universität Erlangen-Nürnberg, Erlangen, Germany*

INTRODUCTION

BMD is one of the most important predictors of fracture risk. However, BMD is a surrogate parameter and its assessment does not differentiate pathophysiological causes for low BMD. Areal bone mineral density (aBMD) as measured by DXA has limitations to characterize bone status under treatment. 3D QCT imaging provides a true physical bone mineral density (vBMD) and is the basis for advanced measurements of bone morphology, bone strength, trabecular structure, and cortical porosity, which provide extended insight into bone pathophysiology and may overcome limitations in the use of aBMD. HRpQCT using dedicated peripheral CT scanners provides improved spatial resolution compared to QCT obtained from whole body clinical CT scanners but is limited to the distal arms and legs. Finite element analysis (FEA) is an add-on technique to estimate bone strength from QCT or HRpQCT scans. Technical characteristics and clinical applications of these techniques are the topics of this chapter.

METHODS

QCT of spine and hip

CT is an X-ray-based technique that provides a spatial distribution of the X-ray absorption coefficient, which after normalization to the absorption of water and air is defined as CT value and is measured in Hounsfield Units (HU).

For QCT, CT values are calibrated to BMD. Traditionally simultaneous calibration, obtained from a reference phantom containing known concentrations of hydroxyapatite (HA), that is positioned below the patient [1] has been used (Fig. 33.1). The imaging process consists of a survey radiograph, also called a scout scan or topogram, to locate the scan range, followed by acquisition of the tomographic data called projections. From the acquired projections, a stack of axial CT images is reconstructed (Fig. 33.1). One important parameter of the reconstruction process is the so-called kernel that determines the ratio of noise to spatial resolution (Fig. 33.2). For final image analysis, QCT data are usually transferred to an external workstation where the specific bone of interest such as the vertebral body or proximal femur is segmented. QCT is a true 3D method; trabecular and cortical bone compartments can be assessed separately. BMD is measured as physical BMD in g/cm^3 (vBMD) compared to a projected or areal density in g/cm^2 as measured by DXA [2].

Recommended volumes of interest (VOIs) for BMD analysis in the spine are L1 + L2. For QCT of the hip the combination of femoral neck (FN), trochanter and intertrochanter should be assessed (Fig. 33.2) [1,3]. CTXA (computed tomography X-ray absorptiometry) is a technique to simulate DXA-type projectional images from QCT data for measurement of aBMD in the hip (Fig. 33.1). Table 33.1 lists typical acquisition and reconstruction parameters for spine and hip QCT that have been widely used in the literature. Radiation exposure levels are also listed.

Precision characteristics are shown in Table 33.2. With advanced, largely automated 3D image analysis

---

Primer on the Metabolic Bone Diseases and Disorders of Mineral Metabolism, Ninth Edition. Edited by John P. Bilezikian.
© 2019 American Society for Bone and Mineral Research. Published 2019 by John Wiley & Sons, Inc.
Companion website: www.wiley.com/go/asbmrprimer

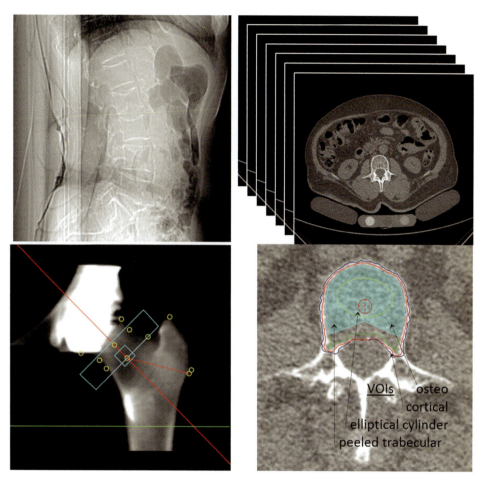

Fig. 33.1. QCT: lateral scout view of the lumbar spine (top left); stack of reconstructed CT images of the lumbar spine with calibration phantom positioned below patient (top right); CTXA of the hip (bottom left); axial multiplanar reformation of segmented vertebra with analysis VOIs (bottom right).

algorithms, QCT precision errors are comparable to DXA. Accuracy data have not been published recently. One major contributor to the accuracy error of trabecular BMD is marrow fat that artificially lowers BMD [4]. This is the consequence of the single-energy technique, which assumes that the trabecular compartment is a two material—water and HA—mixture, and neglects fat [2]. This so-called fat error decreases with lower kV settings [4]. As already shown 30 years ago, dual-energy QCT largely reduces the fat error but at the expense of higher radiation exposure or higher precision errors. Interestingly, the advanced dual-energy CT techniques nowadays implemented in all high-end CT scanners have not been applied to osteodensitometry so far.

Accuracy errors of cortical BMD or thickness are primarily caused by the limited spatial resolution of clinical CT, resulting in partial volume artifacts and associated blurring of the cortex in the reconstructed CT images. As a result, the segmented cortex in the CT image is thicker and its BMD is lower than the true cortex. These accuracy errors increase with decreasing cortical thickness, which is rather low in the vertebral body (<~0.5 mm) and in the FN (<~1 mm). Obviously they also depend on the spatial resolution of the image (Fig. 33.2). Several cortical segmentation techniques have been proposed [5–7] and it is important to understand their respective impact on cortical measurements. In order to reduce this impact of segmentation, the use of cortical BMC has been suggested, which to a certain degree counterbalances the inaccurate decrease in cortical BMD and increase in cortical thickness.

HRpQCT

HRpQCT is a special QCT technique to assess BMD and trabecular bone structure at peripheral sites such as the distal radius and tibia. So far, only one company produces two versions of HRpQCT scanners. The first-generation device has an isotropic voxel size of 82 μm and the second-generation of 61 μm (Table 33.1, Fig. 33.3). The second-generation device scans a greater axial length (10.2 mm versus 9.0 mm) in a shorter time (2.0 min versus 2.8 min). Further details on technical differences between

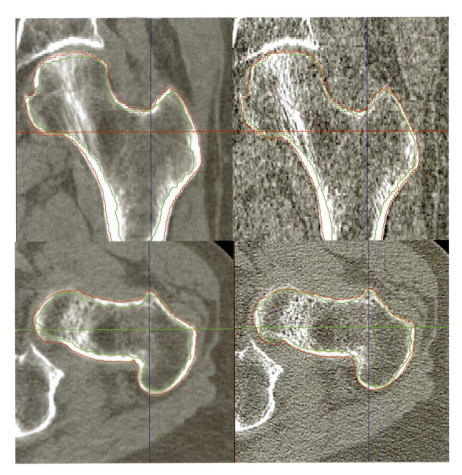

Fig. 33.2. QCT of the hip. Cortical segmentation of two reconstructions shown in coronal (top) and axial (bottom) multiplanar reformations from same scan using a smooth kernel (left), which results in lower noise and a sharp kernel (right) that provides higher spatial resolution. The same segmentation algorithm was used but average cortical thickness is 20% lower and average cortical BMD 15% higher in the sharp kernel reconstruction. Adapted from Engelke 2008 and Engelke 2015.

Table 33.1. Typical reconstruction and acquisition parameters for spine and hip QCT used in most of the recent clinical trials in osteoporosis and for HRpQCT as specified by the manufacturer; adapted from Engelke 2016 and ICRP 2007 [69].

Parameter	QCT Spine	QCT Hip	HRpQCT1	HRpQCT2
X-ray tube voltage (kV)	120[a]		60	68
Exposure: acquisition time multiplied by X-ray tube current	100 mAs[b]	170 mAs[b]	2.8 min × 900 µA	2.0 min × 1460 µA
Pitch	1[c]		41 µm	30 µm
Anatomical coverage	L1 + L2	1 cm above head to 2–3 cm below lesser trochanter	Distal radius and distal tibia	Distal radius, distal tibia, and up to knee and elbow in some individuals
Scan time	<30 s for CT scanners with ≥16 detector rows		2.8 min	2.0 min
Reconstructed field of view (FoV) [cm]	200/400		126 mm	140 mm
Reconstructed slice thickness	1 mm[d]		82 µm	61 µm
In plane pixel size	0.2/0.4 cm² for a FoV of 200/400 mm (matrix: 512 × 512 voxels)		82 µm	61 µm
Reconstruction kernel	Standard body kernel		Modified Feldkamp	Modified Feldkamp
Radiation exposure (mSv)[e]	m: 1.0; f: 1.6	m: 1.8; f: 2.0	<0.003	<0.005

[a] In the spine 80 kV may be an alternative (see text).
[b] Reference mAs values modulated by automatic exposure control.
[c] Exact value depends on scanner and detector characteristics.
[d] Slight variations exist among scanners, for example most GE scanners use a slice thickness of 1.25 mm instead of 1 mm.
[e] m = males; f = females; Monte Carlo calculations performed with Impact Dose version 2.2 (CT Imaging GmbH, Erlangen, Germany). Tissue weighing factors according to ICRP 103 [70]; acquisition parameters as shown in the table; automated exposure control activated; spine: 10 cm scan length; hip 15 cm scan length.

Methods

Modality	Compartment/Parameter	Precision Error [%]
QCT	Total hip	0.8–1.3/0.8–2.0/1.2–2.0[a] [39,71–73][c]
	FN	1.3–3.3/1.1–4.6/1.1–2.9[a] [39,71–73][c]
	Spine	0.5–0.7/1.0–1.7/1.8–2.0[a] [74,75][c]
HRpQCT	Distal radius	0.4–0.5/0.4–13.0/0.5–3.9[a] [13,76]
	Distal tibia	0.3–0.7/0.3–6.2/0.5–3.5[a] [13,76]
QCT FEA	Femur	1.9/1.6–6.4[b] [77,78]
HRpQCT FEA	Distal radius	3.3–4.4/2.8–5.0[b] [79]
	Distal tibia	2.1–3.7/2.5–2.9[b] [79]

Table 33.2. Precision errors for QCT, HRpQCT and FEA.

[a] Values for integral/cortical/trabecular compartments.
[b] Values for stiffness/strength.
[c] Some of the original publications reported just interoperator reanalysis precision. In this case the numbers shown in the table were increased by 0.5% to consider repositioning [71].

the two scanners have been published recently [8]. Most information on HRpQCT scanner technology [9, 10] is available for the first-generation device. Precision values are shown in Table 33.2.

Briefly, cortical and trabecular regions can be segmented using semi-automatic or automatic methods [11–13]. vBMD in mg/cm^3 can be determined for the whole bone and for separate regions. In addition to vBMD, a morphometric analysis can be used to assess the microstructure of the trabecular network (Fig. 33.3). For the first-generation scanners, average trabecular thickness (Tb.Th) and separation (Tb.Sp) are derived from bone volume fraction (BV/TV) and average number of trabeculae (Tb.N). BV/TV is determined from BMD of the trabecular compartment by assuming a density of 1200 mg/cm^3 for fully mineralized bone. Tb.N is directly measured using ridge extraction methods. The second-generation scanners now use direct quantification of the trabecular bone indices [8]. The accuracy of the measurements compared to micro–CT in cadaver specimens has also been published [11, 14].

Different cortical thickness (Ct.Th) measurements are provided by the scanner manufacturer software. One value is derived as volume of cortical bone volume divided by outer bone surface. The other one is directly measured using distance transform methods. Cortical porosity (Ct.Po) is calculated as percentage of void voxels

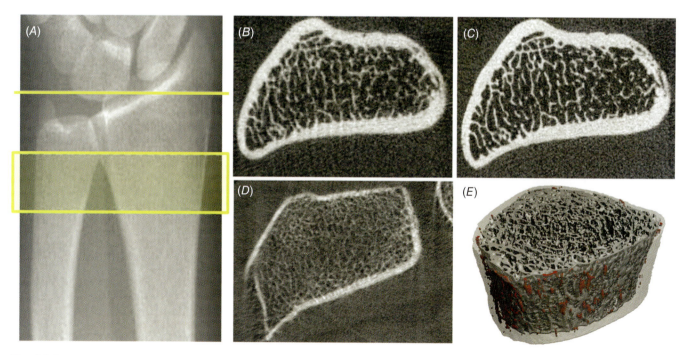

Fig. 33.3. HRpQCT of the radius. (A) Scout view with reference line and region to be scanned. One slice of the radius for the same individual scanned with the first- (B) and second- (C) generation scanners. (D) Image with motion and ring artifacts. (E) 3D representation of the distal tibia with the cortex shown in transparent gray and pores highlighted in red.

in the cortex [11] after thresholding the image (Fig. 33.3). Another method [15], available as external software, uses density information to estimate pores by assuming fully mineralized bone to be 1200 mg/cm^3. Both techniques showed strong correlations with synchrotron CT [16], but absolute values differ significantly.

FEA

Three-dimensional QCT or HRpQCT images can be used to estimate bone mechanical properties noninvasively by subject-specific FEA. After segmentation of the periosteal bone surface, the bone volume is discretized into small elements (meshing). Physically meaningful material properties are assigned to each element (constitutive law), and a particular loading scenario is simulated (boundary conditions). A computer program solves the equilibrium equations and provides local mechanical properties in each point of the meshed volume (Fig. 33.4).

Homogenized FEA (hFEA) is used for QCT scans of the spine and hip from whole body clinical CT scanners where spatial resolution is too low for identification of individual trabeculae. In hFEA each element includes several voxels. Material properties (eg, Young's modulus) are assigned according to the average BMD value of this element using phenomenological laws obtained by independent mechanical testing [17]. The material is usually considered as linear or nonlinear isotropic because trabecular orientation cannot be measured because of limited spatial resolution, although attempts to add anisotropy have been reported recently [18,19]. Typically, 1 to 3 mm in side length, linear hexahedral or quadratic tetrahedral elements are used. The modeled part usually covers a large portion of the total bone, such as the vertebral body or the proximal femur.

In contrast to hFEA, voxel FEA (vFEA) [20] is used for HRpQCT, which allows segmentation of the trabecular microstructure. In line with the HRpQCT scanning protocol, only a small portion (~10 mm) of the distal forearm

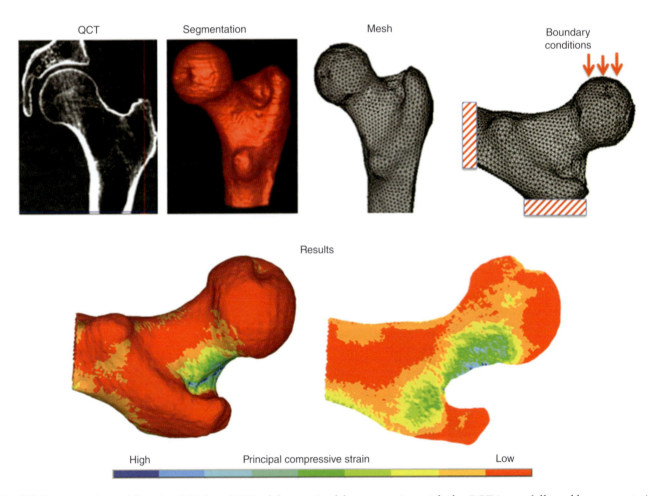

Fig. 33.4. Example workflow for QCT-based FEA of the proximal femur, starting with the QCT image followed by segmentation to extract the periosteal surface, by the meshing with tetrahedrons, the assignment of material properties (not shown) and the assignment of the boundary conditions to simulate a fall onto the greater trochanter. The color plots of the results represent the distribution of local principal compressive strains in the 3D model (bottom left) and in a frontal section (bottom right). Images kindly provided by M. Qasim, INSIGNEO Institute for In Silico Medicine, University of Sheffield.

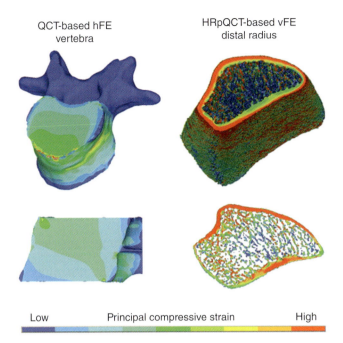

Fig. 33.5. QCT-based hFEA of a vertebra (left) and HRpQCT-based vFEA of a distal radius section (right). In both cases axial compression was simulated. The distribution of the principal compressive strains is shown in color plots of the 3D models (top), of a sagittal section of the vertebral body (bottom left) and of a cross section of the distal radius (bottom right). Images kindly provided by A.M. Campos Marin and M.C. Costa, INSIGNEO Institute for In Silico Medicine, University of Sheffield.

or tibia is modeled. For vFEA the image is binarized to separate background from bone voxels, which are directly converted into linear hexahedral elements. These are small enough to assume local material homogeneity and isotropy.

Under simulated loading conditions, hFEA and vFEA outcomes are mechanical properties such as deformations and stress maps with one data point per element, which can be visualized as 3D color plots (Fig. 33.5). The models can further be used to estimate bone mechanical properties according to different failure criteria, by evaluating those elements that are loaded above their yield or failure properties. In the case of hFEA these are estimated from local BMD [21], and in the case of vFEA from a threshold in the strain values of the bulk material [22]. Assuming a linear relationship between stress and strain in each element, linear analyses can estimate integral measures such as stiffness directly or failure load (bone strength) indirectly. Nonlinear analyses include more complex failure criteria and can provide a direct estimation of bone strength and fracture location. Accuracy of predicted bone strength has been evaluated ex vivo in a number of well-controlled experiments with cadaver bones, but almost no performance measures have been published for in vivo FEA (Table 33.2).

CLINICAL APPLICATIONS

With the advanced 3D imaging techniques described previously, bone can be characterized in much finer detail than with DXA: separation of trabecular, cortical, and subcortical compartments, assessment of cortical thickness and porosity and of trabecular bone architecture, and the determination of local mechanical properties and integral bone strength. Clinical expectations are improved fracture prediction and associated with it increased sensitivity and specificity of identification of patients at high fracture risk. In addition, decisions made before or during treatment with anti-osteoporotic drugs should be put on firmer ground: which treatment option, in particular anabolic versus antiresorptive, should be selected? How can monitoring of treatment efficacy be improved? When to stop or to switch treatment? Finally, what is the role of the advanced imaging modalities to improve the identification of causes for secondary osteoporosis or of specific phenotypes that may require specific treatments? So far the advanced 3D imaging techniques have largely been used in clinical trials and research and only rarely in patient clinical care.

Recent reviews have summarized the state of the art with respect to fracture prediction and monitoring [1, 3, 20, 23]. QCT and FEA of the spine and hip are mature techniques that have been endorsed by the International Society for Clinical Densitometry (ISCD) for fracture prediction and for monitoring age- and treatment-related BMD and strength changes in adults [1, 3]. As a result of its novelty, official endorsements do not yet exist for HRpQCT but a number of reviews have addressed topics of clinical applications including fracture prediction and longitudinal monitoring [9, 10, 24]. Earlier recommendations of the ISCD [1] also included the single-slice pQCT for adults, nowadays mostly used in children (Chapter 31).

The advanced techniques cannot be used for the diagnosis of osteoporosis, which by convention is based on the WHO schema and therefore on DXA aBMD T-scores. T-scores can be calculated for any technique but usually T-scores of different modalities are not comparable because the fracture risk gradient, which describes the dependence of fracture risk on T-score, differs among techniques. One exception is CTXA [25], which after proper normalization with DXA scans provides DXA-equivalent hip T-scores from QCT hip scans. Although the diagnosis of osteoporosis remains the realm of DXA, QCT-based techniques can be used for fracture prediction if normative data and fracture risk gradients have been determined [1].

A large variety of normative data have been published for the old single-slice QCT of the spine that was used before the development of spiral CT [1]. Volumetric CT spine and hip data sets exist for an age-stratified (20 to 97 years) random sample of Rochester, MN, residents [26], a community highly characteristic of the US white population. QCT [26–30] and FEA analyses of the spine and hip have been published. For HRpQCT BMD, structure and strength reference data have been published for Canadian [31,32], US [33], and European [34,35] populations.

QCT

In contrast to DXA, QCT provides regional 3D BMD distributions. Differential BMD and BMC effects in cortical, subcortical, and trabecular BMD increase understanding of pathophysiology of different anti-osteoporotic treatments. Together with measurements of cortical thickness and of bone volume to potentially identify periosteal apposition, a much more detailed understanding of treatment effects has been achieved. Second, QCT is less affected by degenerative changes in the spine, in particular when trabecular BMD is assessed. Changes of trabecular BMD, either age-related or under treatment, are much higher than changes of integral BMD and changes of aBMD. In combination with least significant change (LSC) values comparable to DXA, this makes QCT the ideal tool to monitor BMD changes in the vertebrae, which is important in men and woman beyond age 50, a population with a rapid increase in spinal fractures compared to younger age. It also explains the improved risk prediction for incident vertebral fractures in men of QCT versus DXA (ROC-AUC: 0.83 versus 0.76, $p<0.05$) [36]). QCT data for fracture prediction in women [37] were consistent with those found in men.

In the hip, the benefit of QCT lies in the separate assessment of the cortex, which adds independent information to BMD [5, 38, 39] for fracture prediction. However, as shown in Table 33.1, radiation exposure of QCT is higher than for DXA, which is about 5 μSv for a hip or spine scan. Exposure of a QCT scan is comparable to the annual background radiation. Therefore, a QCT scan should only be performed if clinical benefits compared to DXA can be expected. According to ISCD recommendations, DXA should be used when both DXA and QCT are available in order to limit radiation exposure [40].

Third, QCT is the basis of FEA and other advanced techniques such as statistical parameter mapping at the hip and spine. Thus, standard QCT analysis is always available in these cases. Also, dedicated high-resolution CT techniques have been developed for the spine to obtain estimates of trabecular structure parameters [41]. Finally, clinical CT scans may be the basis for improved and more widespread identification of patients at high fracture risk. This new approach is called opportunistic screening.

Opportunistic screening

Opportunistic screening describes a variety of different techniques applied to routine CT scans of the abdomen, pelvis, or chest to screen for patients at high (and potentially also for low) BMD and fracture risk [3]. It addresses one of the major bottlenecks in osteoporosis: the identification of patients at high risk for fracture. Two principal strategies are discussed. In the first scenario, opportunistic screening serves as a screening tool for DXA. Its outcome is a rough categorization into low, medium, and high fracture risk. Patients with high fracture risk are followed up with established diagnostic procedures including DXA. The second strategy bypasses DXA. A treatment decision is based on fracture risk determined directly from vBMD, strength, or DXA-equivalent T-scores of the hip derived from CTXA, of course in conjunction with clinical risk factors. Both scenarios are convenient for the patient and cost effective because the dual use of clinical CT scans eliminates DXA scans and associated logistics. An additional benefit of the dual use of CT scans is the possibility to use a lateral scout view or a lateral projection of the spinal column to assess fractures. However, there are a number of challenges.

One problem of opportunistic screening is the absence of a calibration phantom in routine clinical scans. There are a number of options to address this difficulty, for example asynchronous external calibration, where a calibration phantom is measured separately from the patient. Calibration equations obtained from the phantom analysis are applied to the subsequent patient scans [42], a technique also used in DXA or HRpQCT. Another option is the use of an internal calibration technique based on the CT values of internal soft tissues such as muscle and fat and their mutual relations to calculate BMD. A third approach currently discussed in the literature is the direct use of CT values without any BMD calibration [43, 44]. Pros and cons of these solutions have been summarized recently [3]. Interested readers should closely follow further progress as this is an area of widespread research.

Another problem of opportunistic screening is the frequent use of contrast in clinical CT scans that increases CT values. In the hip, an increase in CTXA aBMD of $0.032\,mg/cm^2$ was reported [45]. This may not affect the ability to distinguish subjects with high and low BMD but, as indicated by recent reviews [46], further studies are needed to better quantify the impact of contrast agents in terms of concentration, bolus, differential effects in the spine and hip, and options to correct the effects.

HRpQCT

Currently HRpQCT is predominantly used as a research tool. The availability of HRpQCT is still limited; only about 65 scanners are in use worldwide (end of 2017). The second-generation device addresses many of the challenges of the first-generation one. However, so far most clinical data are only available for the first-generation device, which has been used in studies examining differences between populations with and without fractures, and for longitudinal monitoring [9,10,24]. In the past, the clinical relevance of BMD assessments at the forearm has been questioned [1,47], because age-related changes are higher at the spine. Risk gradients for hip and vertebral fractures are higher for BMD measurements at the hip and spine, respectively, and anti-osteoporotic treatment has been less effective to increase radius BMD and to reduce forearm fracture compared to spine and hip.

In two recent studies, quantification of trabecular structure in vivo, now possible with HRpQCT, contributed moderately to fracture risk independently of DXA aBMD of the forearm [48, 49] but results were not adjusted for integral, trabecular, or cortical vBMD as measured by HRpQCT. Thus the added value of the assessment of trabecular architecture for fracture risk prediction still has to be proven. The second domain of HRpQCT, the measurement of cortical parameters, is another focus of current research. For both women [50] and men [51], cortical architecture is associated with severity of vertebral fractures. The measurement of cortical porosity also seems to be relevant in patients with diabetes type 2, who despite normal aBMD and increased trabecular BMD have higher fracture risk than controls. As shown by HRpQCT, these patients have reduced cortical vBMD [52] and, in the case of fragility fractures, increased cortical porosity [53].

Differential effects on cortical and trabecular structure induced by a variety of anti-osteoporotic drugs have been summarized recently [24]: "Responses to therapies were treatment-specific and divergent effects in cortical and trabecular bone with antiresorptive or anabolic agents were observed". Often, effects were stronger in the tibia than in the radius. Results for cortical thickness and porosity depended on segmentation and revealed challenges to monitor changes, particularly in instances of endocortical or periosteal resorption or apposition.

More consistent results will require standardization. 3D matching of baseline and follow-up scans should also improve sensitivity to quantify longitudinal changes [54]. Another important area of standardization is the exact location of the analysis VOI, which depends on the location of the reference line positioned in the scout scan and the distance between reference line and analysis VOI. There is ongoing discussion whether this distance should be fixed or depend on limb length.

FEA

Ex vivo studies have shown that FEA predicts bone strength more accurately than BMD measured with QCT or DXA for both femur and vertebrae [23]. In these validation studies, loading conditions between models and experiments were matched and FE models were only challenged to predict failure load (bone strength) in a chosen loading scenario. The prediction of fracture risk in vivo is much more complicated as a priori loading conditions are unknown. In fact, several prospective and cross-sectional in vivo studies showed that FEA at the hip was not superior to aBMD measured by DXA or vBMD measured by QCT to classify fractured from nonfractured patients or to predict fractures [55]. However, there is evidence that in the spine FEA is superior to DXA for vertebral fracture prediction [36].

In most cases, fracture events are caused by a combination of a weak bone and overloading (ie, a fall). Therefore, FE models should focus not only on accurate prediction of bone strength, but also on estimation of applied loads. One possibility is to integrate varying loading conditions in the FEA process, which adds a stochastic element to the mechanistic models. QCT-based hFEA of the femur has been used to identify the critical loading scenarios by simulating several physiological [56] and pathological (eg, fall) loading conditions [57]. For the most critical loading scenario, FEA better classified patients at risk for femoral fractures compared to DXA aBMD of the hip (5% increase in area under the curve [AUC]).

A 5% AUC increase may not justify the added radiation exposure of the required CT scan when compared to DXA. However, the results encourage exploration of further measures to increase the predictive power of FEA: more accurate estimations of subject-specific overloading forces, more realistic boundary conditions in case of impact from falls, and development of better models to calculate the bone response to impact loading and large deformations.

Treatment-related increases in bone strength in osteoporotic patients are usually higher than BMD increases [55]. FEA includes information about the BMD distribution and the bone geometry when calculating strength. In contrast, DXA only measures integral BMD changes. QCT provides information on integral, compartmental (trabecular and cortical), and even local BMD change. For peripheral sites, HRpQCT can add information on microarchitecture changes over time, but FEA can predict how changes of that heterogeneous distribution affect bone strength. For example, the deposition of newly formed bone in the cortex increases bending or torsional strength more than bone deposition in the center, that is, in the trabecular compartment. Under treatment or age-related normalized (ie, percentage) trabecular BMD or BMC changes are usually higher than cortical changes; however, it has recently been shown that absolute BMC changes, that is, an increase in BMC in grams, can be higher in the cortex than in the trabecular compartment [58, 59]. Although the impact on strength has not been investigated this effect may at least partly explain the discrepancy between treatment-related strength and BMD increases.

Limitations of QCT and HRpQCT imaging techniques such as limited spatial resolution and radiation exposure higher than DXA for spine and hip QCT, or frequent motion artifacts for HRpQCT also affect corresponding FEA. One additional limitation of current hFEA is the assumption of local isotropy due to the impossibility of resolving the trabecular architecture from CT scans of the spine and hip. Therefore the material mapping is based on the heterogeneous BMD distribution. This may explain the similar trends found for FEA-predicted strength and BMD changes in treated patients, although strength changes are usually larger.

Numerous QCT-based hFEA methods have been developed over the past decades to estimate vertebral and femoral strength. Changes in segmentation, meshing, constitutive models, and boundary conditions affect the outcome of the FE models although the clinical impact is less obvious. Recently, two different FE techniques gave almost identical results in a pharmaceutical trial [60, 61]. Nevertheless, standardization of FEA techniques should be pursued. Another limitation is the need for expert modelers for running complex FEA and for data post-processing. HRpQCT-based vFE models can be run through the user-friendly software package developed by the manufacturer. However, this usually means that the potential of FEA is not fully exploited.

Special topics

Statistical parameter mapping

Statistical parameter mapping (SPM) denotes a number of related techniques to create maps of morphological and biomechanical variations (see [3] for a summary). These approaches are opposite to FEA, which integrates all available information into one measure of strength. SPM is based on spatial registration of so-called feature maps extracted from the images, such as the local distribution of cortical thickness. A combination of linear and nonlinear registrations is used to compare bone features between individuals, between groups of individuals, or relative to a feature map of a reference population. This anatomical normalization enables a statistical comparison of the feature for each voxel of the considered VOI. SPM has been used to study the association of the BMD and cortical thickness distribution with hip fractures in elderly women. SPM has also been used to monitor the effect of anti-osteoporotic treatment. One problem with SPM is the interpretation of the resulting maps. The relevance of information from single voxels remains unclear, which in essence results in a clustering of voxels containing similar information. Still, the interpretation of the results remains challenging.

Bone lesions

Bone metastases and in particular focal bone lesions are a major problem for cancer patients. The evaluation of bone stability in the presence of lesions is paramount for appropriate treatment decisions. Current semiquantitative criteria to define impaired stability are based on simple scoring systems [62]. Recently, more advanced methods have been developed by combining CT imaging, image processing, and/or structural engineering analyses to improve strength prediction of bones with lytic lesions. CT-based structural rigidity analysis (CTRA) [63] improved sensitivity and specificity of femoral fracture prediction in patients with metastatic lesions when compared to the standard Mirels scoring system [64].

Another approach combined low- and high-resolution CT images to study early weakening of vertebral bodies in patients with multiple myeloma [65]. A special CT protocol was applied to increase the spatial resolution in order to add information of the trabecular architecture to density. Although hFE is the method of choice to evaluate bone strength, so far only ex vivo validation studies using bones with real [66] or simulated [67, 68] lesions have been performed. There is a large potential to apply the advanced imaging methods reviewed in this chapter to the field of bone lesions to address pending clinical demands.

REFERENCES

1. Engelke K, Adams JE, Armbrecht G, et al. Clinical Use of Quantitative Computed Tomography and Peripheral Quantitative Computed Tomography in the Management of Osteoporosis in Adults: The 2007 ISCD Official Positions. J Clin Densitom. 2008;11:123–62.
2. Kalender W, Engelke K, Fuerst T, et al. ICRU Report 81: Quantitative Aspects Of Bone Densitometry. Journal of the ICRU. 2009;9.
3. Engelke K, Lang T, Khosla S, et al. Clinical Use of Quantitative Computed Tomography-Based Advanced Techniques in the Management of Osteoporosis in Adults: the 2015 ISCD Official Positions-Part III. J Clin Densitom. 2015;18:393–407.
4. Glüer CC, Genant HK. Impact of marrow fat on accuracy of quantitative CT. J Comput Assist Tomogr. 1989;13:1023–35.
5. Treece GM, Gee AH. Independent measurement of femoral cortical thickness and cortical bone density using clinical CT. Med Image Anal. 2015;20:249–64.
6. Prevrhal S, Engelke K, Kalender WA. Accuracy limits for the determination of cortical width and density: the influence of object size and CT imaging parameters. Phys Med Bio. 1999;44:751–64.
7. Museyko O, Gerner B, Engelke K. Cortical bone thickness estimation in CT images: a model-based approach without profile fitting. Third MICCAI Workshop on Computational Methods and Clinical Applications for Spine Imaging. Lecture Notes in Computer Science, vol. 9402. Munich: Springer, 2015, pp 66–75.
8. Manske SL, Zhu Y, Sandino C, et al. Human trabecular bone microarchitecture can be assessed independently of density with second generation HR-pQCT. Bone. 2015;79:213–21.
9. Burghardt AJ, Link TM, Majumdar S. High-resolution computed tomography for clinical imaging of bone microarchitecture. Clin Orthop Relat Res. 2011;469:2179–93.
10. Nishiyama KK, Shane E. Clinical imaging of bone microarchitecture with HR-pQCT. Curr Osteoporos Rep. 2013;11:147–55.
11. Nishiyama KK, Macdonald HM, Buie HR, et al. Postmenopausal women with osteopenia have higher cortical porosity and thinner cortices at the distal radius and tibia than women with normal aBMD: an in vivo HR-pQCT study. J Bone Miner Res. 2010;25:882–90.

12. Buie HR, Campbell GM, Klinck RJ, et al. Automatic segmentation of cortical and trabecular compartments based on a dual threshold technique for in vivo micro-CT bone analysis. Bone. 2007;41:505–15.
13. Burghardt AJ, Buie HR, Laib A, et al. Reproducibility of direct quantitative measures of cortical bone microarchitecture of the distal radius and tibia by HR-pQCT. Bone. 2010;47:519–28.
14. Zhou B, Wang J, Yu YE, et al. High-resolution peripheral quantitative computed tomography (HR-pQCT) can assess microstructural and biomechanical properties of both human distal radius and tibia: Ex vivo computational and experimental validations. Bone. 2016;86: 58–67.
15. Zebaze R, Ghasem-Zadeh A, Mbala A, et al. A new method of segmentation of compact-appearing, transitional and trabecular compartments and quantification of cortical porosity from high resolution peripheral quantitative computed tomographic images. Bone. 2013;54:8–20.
16. Jorgenson BL, Buie HR, McErlain DD, et al. A comparison of methods for in vivo assessment of cortical porosity in the human appendicular skeleton. Bone. 2015;73:167–75.
17. Helgason B, Perilli E, Schileo E, et al. Mathematical relationships between bone density and mechanical properties: a literature review. Clin Biomech (Bristol, Avon). 2008;23:135–46.
18. Larsson D, Luisier B, Kersh ME, et al. Assessment of transverse isotropy in clinical-level CT images of trabecular bone using the gradient structure tensor. Ann Biomed Eng. 2014;42:950–9.
19. Marangalou JH, Ito K, van Rietbergen B. A novel approach to estimate trabecular bone anisotropy from stress tensors. Biomech Model Mechanobiol. 2016;14: 39–48.
20. van Rietbergen B, Ito K. A survey of micro-finite element analysis for clinical assessment of bone strength: the first decade. J Biomech. 2015;48:832–41.
21. Dall'Ara E, Luisier B, Schmidt R, et al. A nonlinear QCT-based finite element model validation study for the human femur tested in two configurations in vitro. Bone. 2013;52:27–38.
22. Pistoia W, van Rietbergen B, Lochmuller EM, et al. Estimation of distal radius failure load with micro-finite element analysis models based on three-dimensional peripheral quantitative computed tomography images. Bone. 2002;30:842–8.
23. Zysset PK, Dall'ara E, Varga P, et al. Finite element analysis for prediction of bone strength. Bonekey Rep. 2013;2:386.
24. Lespessailles E, Hambli R, Ferrari S. Osteoporosis drug effects on cortical and trabecular bone microstructure: a review of HR-pQCT analyses. Bonekey Rep. 2016;5:836.
25. Cann CE, Adams JE, Brown JK, et al. CTXA hip—an extension of classical DXA measurements using quantitative CT. PLoS One. 2014;9:e91904.
26. Riggs BL, Melton LJ, 3rd, Robb RA, et al. Population-based study of age and sex differences in bone volumetric density, size, geometry, and structure at different skeletal sites. J Bone Miner Res. 2004;19:1945–54.
27. Nicks KM, Amin S, Melton LJ, 3rd, et al. Three-dimensional structural analysis of the proximal femur in an age-stratified sample of women. Bone. 2013;55:179–88.
28. Kaneko M, Ohnishi I, Matsumoto T, et al. Prediction of proximal femur strength by a quantitative computed tomography-based finite element method–Creation of predicted strength data of the proximal femur according to age range in a normal population. Mod Rheumatol. 2016;26:151–5.
29. Keaveny TM, Kopperdahl DL, Melton LJ, 3rd, et al. Age-dependence of femoral strength in white women and men. J Bone Miner Res. 2010;25:994–1001.
30. Lang TF, Sigurdsson S, Karlsdottir G, et al. Age-related loss of proximal femoral strength in elderly men and women: the Age Gene/Environment Susceptibility Study—Reykjavik. Bone. 2012;50:743–8.
31. Burt LA, Macdonald HM, Hanley DA, et al. Bone microarchitecture and strength of the radius and tibia in a reference population of young adults: an HR-pQCT study. Arch Osteoporos. 2014;9:183.
32. Macdonald HM, Nishiyama KK, Kang J, et al. Age-related patterns of trabecular and cortical bone loss differ between sexes and skeletal sites: a population-based HR-pQCT study. J Bone Miner Res. 2011;26:50–62.
33. Khosla S, Riggs BL, Atkinson EJ, et al. Effects of sex and age on bone microstructure at the ultradistal radius: a population-based noninvasive in vivo assessment. J Bone Miner Res. 2006;21:124–31.
34. Dalzell N, Kaptoge S, Morris N, et al. Bone micro-architecture and determinants of strength in the radius and tibia: age-related changes in a population-based study of normal adults measured with high-resolution pQCT. Osteoporos Int. 2009;20:1683–94.
35. Vilayphiou N, Boutroy S, Sornay-Rendu E, et al. Age-related changes in bone strength from HR-pQCT derived microarchitectural parameters with an emphasis on the role of cortical porosity. Bone. 2016;83:233–40.
36. Wang X, Sanyal A, Cawthon PM, et al. Prediction of new clinical vertebral fractures in elderly men using finite element analysis of CT scans. J Bone Miner Res. 2012;27:808–16.
37. Kopperdahl DL, Aspelund T, Hoffmann PF, et al. Assessment of incident spine and hip fractures in women and men using finite element analysis of CT scans. J Bone Miner Res. 2014;29:570–80.
38. Poole KE, Treece GM, Ridgway GR, et al. Targeted regeneration of bone in the osteoporotic human femur. PLoS One. 2011;6:e16190.
39. Museyko O, Bousson V, Adams J, et al. QCT of the proximal femur-which parameters should be measured to discriminate hip fracture? Osteoporos Int. 2016;27:1137–47.
40. Engelke K, Lang T, Khosla S, et al. Clinical Use of Quantitative Computed Tomography (QCT) of the Hip in the Management of Osteoporosis in Adults: the 2015 ISCD Official Positions-Part I. J Clin Densitom. 2015;18:338–58.

41. Graeff C, Campbell GM, Pena J, et al. Administration of romosozumab improves vertebral trabecular and cortical bone as assessed with quantitative computed tomography and finite element analysis. Bone. 2015;81:364–9.
42. Brown JK, Timm W, Bodeen G, et al. Asynchronously Calibrated Quantitative Bone Densitometry. J Clin Densitom. 2017;20:216–25.
43. Emohare O, Dittmer A, Morgan RA, et al. Osteoporosis in acute fractures of the cervical spine: the role of opportunistic CT screening. J Neurosurg Spine. 2015;23:1–7.
44. Lee SJ, Binkley N, Lubner MG, et al. Opportunistic screening for osteoporosis using the sagittal reconstruction from routine abdominal CT for combined assessment of vertebral fractures and density. Osteoporos Int. 2016;27:1131–6.
45. Ziemlewicz TJ, Maciejewski A, Binkley N, et al. Direct Comparison of Unenhanced and Contrast-Enhanced CT for Opportunistic Proximal Femur Bone Mineral Density Measurement: Implications for Osteoporosis Screening. AJR Am J Roentgenol. 2016;206:694–8.
46. Brett A, Brown K. Quantitative computed tomography and opportunistic bone density screening by dual use of computed tomography scans. J Orthopaedic Translation. 2015;3:178–84.
47. Clowes JA, Eastell R, Peel NF. The discriminative ability of peripheral and axial bone measurements to identify proximal femoral, vertebral, distal forearm and proximal humeral fractures: a case control study. Osteoporos Int. 2005;16:1794–802.
48. Sornay-Rendu E, Boutroy S, Munoz F, et al. Alterations of cortical and trabecular architecture are associated with fractures in postmenopausal women, partially independent of decreased BMD measured by DXA: the OFELY study. J Bone Miner Res. 2007;22:425–33.
49. Vilayphiou N, Boutroy S, Sornay-Rendu E, et al. Finite element analysis performed on radius and tibia HR-pQCT images and fragility fractures at all sites in postmenopausal women. Bone. 2010;46:1030–7.
50. Sornay-Rendu E, Cabrera-Bravo JL, Boutroy S, et al. Severity of vertebral fractures is associated with alterations of cortical architecture in postmenopausal women. J Bone Miner Res. 2009;24:737–43.
51. Szulc P, Boutroy S, Vilayphiou N, et al. Cross-sectional analysis of the association between fragility fractures and bone microarchitecture in older men: the STRAMBO study. J Bone Miner Res. 2011;26:1358–67.
52. Burghardt AJ, Issever AS, Schwartz AV, et al. High-resolution peripheral quantitative computed tomographic imaging of cortical and trabecular bone microarchitecture in patients with type 2 diabetes mellitus. J Clin Endocrinol Metab. 2010;95:5045–55.
53. Patsch JM, Burghardt AJ, Yap SP, et al. Increased cortical porosity in type 2 diabetic postmenopausal women with fragility fractures. J Bone Miner Res. 201328:313–24.
54. Nishiyama KK, Pauchard Y, Nikkel LE, et al. Longitudinal HR-pQCT and image registration detects endocortical bone loss in kidney transplantation patients. J Bone Miner Res. 2015;30:554–61.
55. Zysset P, Qin L, Lang T, et al. Clinical Use of Quantitative Computed Tomography-Based Finite Element Analysis of the Hip and Spine in the Management of Osteoporosis in Adults: the 2015 ISCD Official Positions-Part II. J Clin Densitom. 2015;18:359–92.
56. Falcinelli C, Schileo E, Balistreri L, et al. Multiple loading conditions analysis can improve the association between finite element bone strength estimates and proximal femur fractures: A preliminary study in elderly women. Bone. 2014;67:71–80.
57. Qasim M, Farinella G, Zhang J, et al. Patient-specific finite element estimated femur strength as a predictor of the risk of hip fracture: the effect of methodological determinants. Osteoporos Int. 2016;27:2815–22.
58. Engelke K, Fuerst T, Dardzinski B, et al. Odanacatib treatment affects trabecular and cortical bone in the femur of postmenopausal women: results of a two-year placebo-controlled trial. J Bone Miner Res. 2015;30:30–8.
59. Genant H, Bolognese M, Mautalen C, et al. Romosozumab Administration Is Associated with Significant Improvements in Lumbar Spine and Hip Volumetric Bone Mineral Density and Content Compared with Teriparatide. Ann Rheum Dis. 2014;73:172.
60. Keaveny TM, McClung MR, Genant HK, et al. Femoral and vertebral strength improvements in postmenopausal women with osteoporosis treated with denosumab. J Bone Miner Res. 2014;29:158–65.
61. Zysset P, Pahr D, Engelke K, et al. Comparison of proximal femur and vertebral body strength improvements in the FREEDOM trial using an alternative finite element methodology. Bone. 2015;81:122–30.
62. Mirels H. Metastatic disease in long bones: A proposed scoring system for diagnosing impending pathologic fractures. Clin Orthop Relat Res. 1989; (249):256–64.
63. Villa-Camacho JC, Iyoha-Bello O, Behrouzi S, et al. Computed tomography-based rigidity analysis: a review of the approach in preclinical and clinical studies. Bonekey Rep. 2014;3:587.
64. Damron TA, Nazarian A, Entezari V, et al. CT-based Structural Rigidity Analysis Is More Accurate Than Mirels Scoring for Fracture Prediction in Metastatic Femoral Lesions. Clin Orthop Relat Res. 2016;474:643–51.
65. Borggrefe J, Giravent S, Thomsen F, et al. Association of QCT Bone Mineral Density and Bone Structure With Vertebral Fractures in Patients With Multiple Myeloma. J Bone Miner Res. 2015;30:1329–37.
66. Yosibash Z, Plitman Mayo R, Dahan G, et al. Predicting the stiffness and strength of human femurs with real metastatic tumors. Bone. 2014;69:180–90.
67. Derikx LC, van Aken JB, Janssen D, et al. The assessment of the risk of fracture in femora with metastatic lesions: comparing case-specific finite element analyses with predictions by clinical experts. J Bone Joint Surg Br. 2012;94:1135–42.
68. Tanck E, van Aken JB, van der Linden YM, et al. Pathological fracture prediction in patients with

metastatic lesions can be improved with quantitative computed tomography based computer models. Bone. 2009;45:777–83.
69. Engelke K. Quantitative Computed Tomography (QCT) – Current Status and New Developments. J Clin Densitom. 2017;20(3):309–21.
70. ICRP. The 2007 Recommendations of the International Commission on Radiological Protection. ICRP publication 103. Ann ICRP. 2007;37:1–332.
71. Lang TF, Keyak JH, Heitz MW, et al. Volumetric quantitative computed tomography of the proximal femur: precision and relation to bone strength. Bone. 1997;21:101–8.
72. Li W, Sode M, Saeed I, et al. Automated registration of hip and spine for longitudinal QCT studies: integration with 3D densitometric and structural analysis. Bone. 2006;38:273–9.
73. Yang L, Burton AC, Bradburn M, et al. Distribution of bone density in the proximal femur and its association with hip fracture risk in older men: the osteoporotic fractures in men (MrOS) study. J Bone Miner Res. 2012;27:2314–24.
74. Engelke K, Mastmeyer A, Bousson V, et al. Reanalysis precision of 3D quantitative computed tomography (QCT) of the spine. Bone. 2009;44:566–72.
75. Lang TF, Li J, Harris ST, et al. Assessment of vertebral bone mineral density using volumetric quantitative CT. J Comput Assist Tomogr. 1999;23:130–7.
76. MacNeil JA, Boyd SK. Improved reproducibility of high-resolution peripheral quantitative computed tomography for measurement of bone quality. Med Eng Phys. 2008;30:792–9.
77. Carpenter RD, Saeed I, Bonaretti S, et al. Inter-scanner differences in in vivo QCT measurements of the density and strength of the proximal femur remain after correction with anthropomorphic standardization phantoms. Med Eng Phys. 2014;36:1225–32.
78. Cody DD, Hou FJ, Divine GW, et al. Short term in vivo precision of proximal femoral finite element modeling. Ann Biomed Eng. 2000;28:408–14.
79. Kawalilak CE, Kontulainen SA, Amini MA, et al. In vivo precision of three HR-pQCT-derived finite element models of the distal radius and tibia in postmenopausal women. BMC Musculoskelet Disord. 2016;17:389.

34
Magnetic Resonance Imaging of Bone

Sharmila Majumdar

*Department of Radiology and Biomedical Imaging; Department of Orthopedic Surgery;
and Department of Bioengineering and Therapeutics Sciences, UCSF, San Francisco, CA, USA*

INTRODUCTION

Three-dimensional (3D) imaging techniques that reveal bone structure are emerging as important contenders for defining bone quality, at least partially. Techniques such as micro-CT (µCT) have recently been developed and provide high-resolution images of the trabecular and cortical bone micro- and macroarchitecture. This method is routinely used in specimen evaluation and has recently been extended to in vivo animal and human extremity imaging. Another recent development in the assessment of trabecular bone and cortical bone structure is the use of MRI, a nonionizing technique that makes it possible to obtain noninvasive "bone biopsies" at multiple anatomical sites.

MRI OF TRABECULAR BONE

Trabecular bone consists of a network of rodlike elements interconnected by platelike elements, immersed in bone marrow composed partly of water and partly of fat. Magnetic susceptibility of trabecular bone is substantially different from that of bone marrow. This gives rise to susceptibility gradients at the bone–marrow interface. Magnetic inhomogeneity arising from these susceptibility gradients depends on the static magnetic field strength, number of bone–bone marrow interfaces, and size of individual trabeculae [1–3]. These effects cause dephasing of spins and signal decay at a rate known as $T2^*$. In a voxel partly occupied by bone and partly by marrow, the static inhomogeneity induced intravoxel dephasing of spins leads to signal cancellation within the voxel. $T2^*$ methods have been used to quantify trabecular bone and these measures have been related to bone strength, osteoporotic status, and therapeutic response [4].

Beside the tissue composition, the small dimensions of the trabecular elements (~100 µm) require very high imaging resolutions. The suitability of an MRI method (acquisition and analysis) for depicting bone microstructure depends on its ability to yield images with high enough signal in a reasonable acquisition time and its ability to derive trabecular structural measurements from the images accurately and reproducibly. The three competing factors to be considered in high-resolution MRI (hr-MRI) are signal-to-noise ratio, spatial resolution, and imaging time. Spatial resolution and signal-to-noise ratio (SNR) are both directly related to imaging time but are inversely related to each other. Recent technique developments in trabecular bone MRI technique reflect all these considerations and have been aiming for increasing SNR and accelerating total acquisition times.

Magnetic resonance (MR) pulse sequences can be broadly classified into spin-echo and gradient-echo sequences. Ideally 3D spin-echo (SE) sequences are better suited for imaging of trabecular bone microarchitecture than gradient-echo (GE)-based sequences because they are less sensitive to the thickening of trabeculae owing to susceptibility differences. However, GE sequences can be employed with short repetition time (TR) because of their higher SNR efficiency and can thus acquire a 3D volume in shorter scan time and avoid patient motion artifacts [5, 6]. 3D SE type pulse sequences with variable flip angle such as rapid SE excitation (RASEE) [7, 8], large-angle SE imaging [9], and subsequently fast 3D large-angle SE imaging (FLASE) [10] and a new fully balanced steady state 3D SE pulse sequence have also been developed [11]. The choice of pulse sequence for trabecular bone imaging is still a topic of active research. Availability of the sequences at multiple centers, their

---

Primer on the Metabolic Bone Diseases and Disorders of Mineral Metabolism, Ninth Edition. Edited by John P. Bilezikian.
© 2019 American Society for Bone and Mineral Research. Published 2019 by John Wiley & Sons, Inc.
Companion website: www.wiley.com/go/asbmrprimer

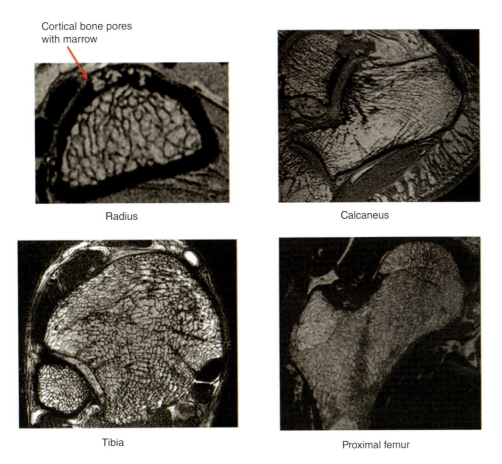

Fig. 34.1. Representative MR images showing trabecular bone in the radius, calcaneus, tibia, and femur. Cortical bone in the radius shows evidence of marrow-filled cortical porosity. Images were acquired at 3 T using a General Electric Signa scanner.

robustness, and total imaging time versus the anatomical coverage are typical considerations.

SNR is linearly proportional to the static magnetic field strength, perhaps making 3 T preferable over 1.5 T magnets. Phan and colleagues imaged the trabecular microarchitecture in 40 cadaveric calcaneus specimens at 1.5 T and 3 T and compared with μCT as gold standard [12]; they found that correlations between trabecular structural parameters derived from 3 T MR images and μCT were significantly higher ($p < 0.05$) than correlations between structural parameters obtained from 1.5 T MRI and μCT. Preliminary experiments conducted on a 7 T GE Signa scanner yielded a twofold increase in SNR for hr-MRI of trabecular bone [13, 14]. Fig. 34.1 shows an example of an hr-MR image obtained at 3 T in the radius, calcaneus, tibia, and femur, where the bone marrow is bright and trabecular bone is depicted as dark striations. Image such as these can be analyzed to derive structural measures of microarchitecture.

The most common structural measures analogous to quantitative histomorphometry derived from MR images include apparent (app.) bone volume/total volume (app. BV/TV), trabecular number (app.Tb.N), trabecular spacing (app.Tb.Sp), and trabecular thickness (app. Tb.Th) [15, 16] and require the images to be subdivided into a bone and marrow component or binarized. Because the MR images are not acquired at true microscopic resolutions, Majumdar and colleagues [17] described these measures derived from MR images as "apparent" measurements, which although obtained in the limited-resolution regime are highly correlated to the "true" structure. Binarization of a MR image is not a trivial task, mainly because of partial volume effects; multiple techniques have been developed which operate directly on the grayscale image. Recognizing the fuzzy nature of the images caused by partial volume effects, Saha and Wehrli [18] applied a fuzzy distance transform (FDT) technique for computing trabecular thickness and observed an improved robustness in the computation against loss of resolution. Digital topological analysis techniques have also been applied to quantify the number of surface and curved edges, junctions, and interiors in the trabecular network [19].

MRI OF CORTICAL BONE

MRI has been used to image cortical bone as well. Specifically for the proximal femur, the ability of MRI to align the image plane perpendicular to the femoral neck is a great advantage and enables more accurate acquisition of the cortical architecture [20]. MRI allows the visualization of soft tissues such as bone marrow and thus a quantification of the amount of cortical porosity that contains bone marrow.

Using advanced MRI methods with ultra short echo times (UTE), the bone water content in the microscopic pores of the haversian and the lacuno-canalicular systems of cortical bone can be quantified. A smaller water fraction is also bound to collagen and the matrix substrate and imbedded in the crystal structure of the mineral [21]. These micropores have usually a very small size in the order of a few micrometers and are thus difficult to visualize but the quantification of bone water using MRI could potentially provide a surrogate measure of bone porosity without resolving these individual small pores.

RELATIONSHIP OF MR-DERIVED STRUCTURE MEASURES TO BONE STRENGTH, FRACTURE, OSTEOPOROTIC STATUS, AND RESPONSE TO THERAPY

Several studies relating the measures of trabecular structure obtained using MRI to measures of bone strength in vitro have been conducted [17, 22–24]. Relationships between whole bone strength and bone structure measures have been demonstrated in radii (Hudalmeier) and in the proximal femur [25, 26].

hr-MR images of the distal radius were obtained at 1.5 T in premenopausal normal, postmenopausal normal, and postmenopausal osteoporotic women [27]. Significant differences were evident in spinal BMD, radial trabecular BMD, trabecular bone volume fraction, Tb.Sp, and Tb.N between the postmenopausal nonfracture and the postmenopausal osteoporotic subjects. Tb.Sp and Tb.N showed moderate correlation with radial trabecular BMD but correlated poorly with radial cortical BMC.

Distance transformation techniques were applied to the 3D image of the distal radius of postmenopausal patients and structural indices such as app.Tb.N, app.Tb.Th, and app.Tb.Sp were determined without model assumptions [28]. A new metric index, the apparent intra-individual distribution of separations (app.Tb.Sp.SD), was introduced. It was found that app.Tb.Sp.SD discriminates fracture subjects from nonfracture patients as well as DXA measurements of the radius and the spine, but not as well as DXA of the hip. MR-derived measures of trabecular bone architecture in the distal radius [29] and calcaneus [30] were obtained in 20 subjects with hip fractures and 19 age-matched postmenopausal controls, in addition to BMD measures at the hip (DXA) and the distal radius (pQCT). Measures of app. Tb.Sp and app. Tb.N in the distal radius showed significant ($p < 0.05$) differences between the two groups, as did hip BMD measures. However, radial trabecular BMD measures showed only a marginal difference ($p = 0.05$). In the calcaneus, significant differences between both patient groups were obtained using morphological parameters.

Sagittal MR images of the calcaneus were obtained in 50 men (26 patients with osteoporosis and 24 age-matched healthy control subjects) [31]; structural parameters, especially connectivity parameters, showed significant differences between control subjects and patients ($p < 0.05$).

In addition, in vivo images have also been combined with micro-finite element analysis in a limited set of subjects. Newitt and colleagues [32] studied subjects in two groups: postmenopausal women with normal BMD ($n = 22$, mean age 58 ± 7 years) and postmenopausal women with spine or femur BMD –1 SD to –2.5 SD below young normal ($n = 37$, mean age 62 ± 11 years). Anisotropy of trabecular bone microarchitecture, as measured by the ratios of the mean intercept length (MIL) values (MIL1/MIL3, etc.), and the anisotropy in elastic modulus (E1/E3, etc.), were greater in the osteopenic group.

Ninety-one postmenopausal osteoporotic women were followed for 2 years ($n = 46$ for nasal spray calcitonin, $n = 45$ for placebo) [33]. MRI measurements of trabecular structure were obtained at distal radius and calcaneus in addition to DXA-BMD at spine/hip/wrist/calcaneus (obtained yearly). MRI assessment of trabecular microarchitecture at individual regions of the distal radius revealed preservation (no significant loss) in the treated group compared with significant deterioration in the placebo control group.

Trabecular bone structure of the tibia was studied in 10 men with severe, untreated hypogonadism and age- and race-matched eugonadal men. Two composite topological indices were determined: the ratio of surface voxels (representing plates) to curve voxels (representing rods), which is higher when architecture is more intact; and the erosion index, a ratio of parameters expected to increase upon architectural deterioration to those expected to decrease, which is higher when deterioration is greater. The surface/curve ratio was 36% lower ($p = 0.004$) and the erosion index was 36% higher ($p = 0.003$) in the hypogonadal men than in the eugonadal men [34]. In contrast, BMD of the spine and hip was not significantly different between the two groups. After 24 months of testosterone treatment, BMD of the spine increased 7.4% ($p < 0.001$) and that of the total hip increased 3.8% ($p = 0.008$). Architectural parameters assessed by MRI also changed: the surface/curve ratio increased 11% ($p = 0.004$) and the topological erosion index decreased 7.5% ($p = 0.004$) [35].

Until recently, in vivo MRI of trabecular microarchitecture was limited to peripheral sites (such as distal tibia and femur, radius, calcaneus) because of SNR limitations. However, the main sites of osteoporotic fractures are nonperipheral regions such as the vertebral bodies (spine) and the proximal femur (hip). hr-MRI has only recently been applied to the proximal femur [36] by using

SNR efficient sequences, high magnetic field strength (3 T), and phased array coils.

APPLICATIONS USING MR-DERIVED CORTICAL BONE MEASURES

Although significant work has been done using MRI to measure trabecular bone structure, there is relatively little work on the macroarchitectural geometry of the cortex which may play an equally important role for bone strength. Gomberg and colleagues [20] investigated cortical shell geometry of the femur, further expanding the potential role of MRI in characterizing bone. In one study, images of the distal radius and the distal tibia of 49 postmenopausal osteopenic women (age 56±3.7) were acquired with both HRpQCT and MRI [37]. It was found that the amount of cortical porosity did not vary greatly between subjects but the type of cortical pore containing marrow versus not containing marrow varied highly between subjects.

Bone water quantification measurements have been previously conducted in sheep and human cadaveric specimens and the method's sensitivity to distinguish subjects of different age and disease state has been evaluated [38]. The data were compared with areal and volumetric BMD from DXA and pQCT respectively. The bone water content was calibrated with the aid of an external reference (10% H_2O in D_2O doped with 27 mmol/L $MnCl_2$) which was attached anteriorly to the subject's tibial midshaft. Excellent agreement ($R^2=0.99$) was found in the specimen between the water displaced by using D_2O exchange and water measured with respect to the reference sample. Measurements in vivo revealed that the bone water content was increased 65% in the postmenopausal group compared to the premenopausal group [39]. Patients with renal osteodystrophy had 135% higher bone water content than the premenopausal group whereas conventional BMD measurements showed an opposite behavior, with much smaller group differences.

SUMMARY

Imaging trabecular and cortical microarchitecture, characterizing the features of trabecular and cortical bone, has been an area of fertile and ongoing research. Beyond relating microarchitecture to the biomechanical properties of bone in specimens, advances have been made to extend these measures in vivo in human subjects. In this context the relationship between age, fracture status, and even post-therapeutic response has been studied. New advances in peripheral CT, MR (nonionizing, peripheral sites, calcaneus and femur) are ongoing and evolving at a rapid pace and with the establishment of robust analysis methodologies and normative databases have the potential for further clinical utilization in the coming years.

REFERENCES

1. Majumdar S, Thomasson D, Shimakawa A, et al. Quantitation of the susceptibility difference between trabecular bone and bone marrow: experimental studies. Magn Reson Med. 1991;22(1):111–27.
2. Weisskoff RM, Zuo CS, Boxerman JL, et al. Microscopic susceptibility variation and transverse relaxation: theory and experiment. Magn Reson Med. 1994;31(6):601–10.
3. Ford JC, Wehrli FW, Chung HW. Magnetic field distribution in models of trabecular bone. Magn Reson Med. 1993;30(3):373–9.
4. Link TM, Majumdar S, Augat P, et al. Proximal femur: assessment for osteoporosis with T2* decay characteristics at MR imaging. Radiology. 1998;209(2):531–6.
5. Majumdar S, Link TM, Augat P, et al. Trabecular bone architecture in the distal radius using magnetic resonance imaging in subjects with fractures of the proximal femur. Magnetic Resonance Science Center and Osteoporosis and Arthritis Research Group. Osteoporos Int. 1999;10(3):231–9.
6. Newitt DC, Van Rietbergen B, Majumdar S. Processing and Analysis of In Vivo High-Resolution MR Images of Trabecular Bone for Longitudinal Studies: Reproducibility of Structural Measures and Micro-Finite Element Analysis Derived Mechanical Properties. Osteoporos Int. 2002;13:278–87.
7. Jara H, Wehrli FW, Chung H, et al. High-resolution variable flip angle 3D MR imaging of trabecular microstructure in vivo. Magn Reson Med. 1993;29(4):528–39.
8. Bogdan AR, Joseph PM. RASEE: a rapid spin-echo pulse sequence. Magn Reson Imaging. 1990;8(1):13–9.
9. DiIorio G, Brown JJ, Borrello JA, et al. Large angle spin-echo imaging. Magn Reson Imaging. 1995;13(1):39–44.
10. Ma J, Wehrli FW, Song HK. Fast 3D large-angle spin-echo imaging 3D FLASE. Magn Reson Med. 1996;35(6):903–10.
11. Krug R, Han ET, Banerjee S, et al. Fully balanced steady-state 3D-spin-echo (bSSSE) imaging at 3 Tesla. Magn Reson Med. 2006;56(5):1033–40.
12. Phan CM, Matsuura M, Bauer JS, et al. Trabecular bone structure of the calcaneus: comparison of MR imaging at 3.0 and 1.5 T with micro-CT as the standard of reference. Radiology. 2006;239(2):488–96.
13. Zuo J, Bolbos R, Hammond K, et al. Reproducibility of the quantitative assessment of cartilage morphology and trabecular bone structure with magnetic resonance imaging at 7 T. Magn Reson Imaging. 2008;26(4):560–6.
14. Krug R, Carballido-Gamio J, Banerjee S, et al. In vivo bone and cartilage MRI using fully-balanced steady-state free-precession at 7 tesla. Magn Reson Med. 2007;58(6):1294–8.
15. Parfitt AM, Mathews CH, Villanueva AR, et al. Relationships between surface, volume, and thickness of iliac trabecular bone in aging and in osteoporosis. Implications for the microanatomic and cellular mechanisms of bone loss. J Clin Invest. 1983;72(4):1396–409.

16. Parfitt AM. Assessment of trabecular bone status. Henry Ford Hosp Med J. 1983;31(4):196–8.
17. Majumdar S, Newitt D, Mathur A, et al. Magnetic resonance imaging of trabecular bone structure in the distal radius: relationship with X-ray tomographic microscopy and biomechanics. Osteoporos Int. 1996;6(5):376–85.
18. Saha PK, Wehrli FW. Measurement of trabecular bone thickness in the limited resolution regime of in vivo MRI by fuzzy distance transform. IEEE Trans Med Imaging. 2004;23(1):53–62.
19. Gomberg BR, Saha PK, Song HK, et al. Topological analysis of trabecular bone MR images. IEEE Trans Med Imaging. 2000;19(3):166–74.
20. Gomberg BR, Saha PK, Wehrli FW. Method for cortical bone structural analysis from magnetic resonance images. Acad Radiol. 2005;12(10):1320–32.
21. Timmins PA, Wall JC. Bone water. Calcif Tissue Res. 1977;23(1):1–5.
22. Hwang SN, Wehrli FW, Williams JL. Probability-based structural parameters from three-dimensional nuclear magnetic resonance images as predictors of trabecular bone strength. Med Phys. 1997;24(8):1255–61.
23. Pothuaud L, Laib A, Levitz P, et al. Three-dimensional-line skeleton graph analysis of high-resolution magnetic resonance images: a validation study from 34-microm-resolution microcomputed tomography. J Bone Miner Res. 2002;17(10):1883–95.
24. Majumdar S, Kothari M, Augat P, et al. High-resolution magnetic resonance imaging: Three-dimensional trabecular bone architecture and biomechanical properties. Bone. 1998;22:445–54.
25. Ammann P, Rizzoli R. Bone strength and its determinants. Osteoporos Int. 2003;14(suppl 3):S13–8.
26. Link TM, Bauer J, Kollstedt A, et al. Trabecular bone structure of the distal radius, the calcaneus, and the spine: which site predicts fracture status of the spine best? Invest Radiol. 2004;;39(8):487–97.
27. Majumdar S, Genant H, Grampp S, et al. Correlation of trabecular bone structure with age, bone mineral density and osteoporotic status: in vivo studies in the distal radius using high resolution magnetic resonance imaging. J Bone Miner Res. 1997;12:111–8.
28. Laib A, Newitt DC, Lu Y, et al. New model-independent measures of trabecular bone structure applied to in vivo high-resolution MR images. Osteoporos Int. 2002;13(2):130–6.
29. Majumdar S, Link T, Augat P, et al. Trabecular bone architecture in the distal radius using MR imaging in subjects with fractures of the proximal femur. Osteoporos Int. 1999;10:231–9.
30. Link TM, Majumdar S, Augat P, et al. In vivo high resolution MRI of the calcaneus: differences in trabecular structure in osteoporosis patients. J Bone Miner Res. 1998;13(7):1175–82.
31. Boutry N, Cortet B, Dubois P, et al. Trabecular bone structure of the calcaneus: preliminary in vivo MR imaging assessment in men with osteoporosis. Radiology. 2003;227(3):708–17.
32. Newitt DC, Majumdar S, van Rietbergen B, et al. In vivo assessment of architecture and micro-finite element analysis derived indices of mechanical properties of trabecular bone in the radius. Osteoporos Int. 2002;13(1):6–17.
33. Chesnut CH, 3rd, Majumdar S, Newitt DC, et al. Effects of salmon calcitonin on trabecular microarchitecture as determined by magnetic resonance imaging: results from the QUEST study. J Bone Miner Res. 2005;20(9):1548–61.
34. Benito M, Gomberg B, Wehrli FW, et al. Deterioration of trabecular architecture in hypogonadal men. J Clin Endocrinol Metab. 2003;88(4):1497–502.
35. Benito M, Vasilic B, Wehrli FW, et al. Effect of testosterone replacement on trabecular architecture in hypogonadal men. J Bone Miner Res. 2005;20(10):1785–91.
36. Krug R, Banerjee S, Han ET, et al. Feasibility of in vivo structural analysis of high-resolution magnetic resonance images of the proximal femur. Osteoporos Int. 2005;16(11):1307–14.
37. Goldenstein J, Kazakia G, Majumdar S. In vivo evaluation of the presence of bone marrow in cortical porosity in postmenopausal osteopenic women. Ann Biomed Eng. 2010;38(2):235–46.
38. Techawiboonwong A, Song HK, Wehrli FW. In vivo MRI of submillisecond T(2) species with two-dimensional and three-dimensional radial sequences and applications to the measurement of cortical bone water. NMR Biomed. 2008;21(1):59–70.
39. Techawiboonwong A, Song HK, Leonard MB, et al. Cortical bone water: in vivo quantification with ultrashort echo-time MR imaging. Radiology. 2008;248(3):824–33.

35
Trabecular Bone Score

Barbara C. Silva[1] and William D. Leslie[2]

[1]Division of Endocrinology, Felicio Rocho Hospital; and Santa Casa of Belo Horizonte,
Department of Medicine, UNI-BH, Belo Horizonte, Brazil
[2]Department of Medicine, University of Manitoba, Winnipeg, MB, Canada

INTRODUCTION

Trabecular bone score (TBS), a gray-level texture measure derived from lumbar spine (LS) DXA images, improves fracture risk prediction beyond that provided by the combination of BMD by DXA and clinical risk factors [1]. In 2012, the US Food and Drug Administration (FDA) cleared the software used to calculate TBS (TBS iNsight; Medimaps Group, Geneva, Switzerland), which is now commercially available, with the labeling: "TBS is derived from the texture of the DXA image and has been shown to be related to bone microarchitecture and fracture risk. This data provides information independent of BMD value… The TBS score can assist the health care professional in assessment of fracture risk…." More recently, Task Forces of the International Society for Clinical Densitometry (ISCD) and the European Society for Clinical and Economic Aspects of Osteoporosis and Osteoarthritis (ESCEO) developed guidelines on how to use TBS in clinical practice [2, 3]. This chapter briefly reviews technical aspects of TBS, major studies that have examined its ability to predict fracture risk, and the use of TBS in clinical practice.

TBS: TECHNICAL ASPECTS

TBS was developed as an index of skeletal architecture, computed from an experimental variogram of 2D projection images. In its first description, TBS was derived from simulated 2D projections of micro-CT (μCT) images [4] and was subsequently adapted for DXA [5]. TBS (unitless) is defined as the initial slope of the variogram, calculated as the sum of the squared gray-level differences between pixels at a specific distance. In general, a more homogeneous 2D DXA image, with many gray-level variations of small amplitude, generates a high TBS that is correlated with finely textured (well-structured) bone. A DXA image with reduced pixel value variations of higher amplitude produces a low TBS, which is related to coarsely textured (deteriorated) bone structure [5].

Ex vivo studies reported significant unadjusted correlations between TBS and trabecular microarchitecture assessed by μCT [4–7]. The associations between TBS and 3D bone parameters examined in vivo have shown mixed results. TBS derived from LS DXA scans was moderately correlated (R^2~0.4 to 0.6) with integral and trabecular volumetric BMD measures by QCT of the LS and femur [8]. In multiple linear regression models, TBS was independently associated with trabecular microstructure indices assessed by μCT of iliac crest biopsies (partial R^2 0.3 to 0.7) [9]. In contrast, LS TBS explained little of the variance in trabecular microarchitecture assessed by HRpQCT of the distal radius and tibia [8, 10–12], and, in a multiple regression model, none of the HRpQCT indices were associated with TBS [8]. Regardless of the structural properties assessed by TBS, its clinical utility derives from its demonstrated ability to predict fractures.

CLINICAL ASSESSMENT OF TBS

TBS and fracture risk: cross-sectional studies

Most cross-sectional studies have demonstrated that lower TBS is associated with vertebral, hip, and overall osteoporotic fractures in older women and men [13–26].

In general, each SD decrease in TBS resulted in an odds ratio (OR) of 1.3 to 3.8 for fragility fractures ([1, 2, 27] provide detailed reviews). In a number of these reports, TBS was associated with fracture risk even after adjusting for DXA BMD.

TBS and fracture risk: longitudinal studies

Most longitudinal studies, summarized in Table 35.1, have shown that TBS predicts fracture risk in women and men >40 years old [28–40]. A large clinical DXA registry in the Canadian province of Manitoba contributed several retrospective cohort studies examining the association between TBS and incident osteoporotic fracture. The first report evaluated 29,407 women aged >50 years, of whom 1668 had incident major osteoporotic fractures (MOF), including 439 clinical vertebral and 293 hip fractures, during average follow-up of 4.7 years [28]. In models adjusted for age, LS BMD, and clinical risk factors, each SD decline in TBS conferred a 17% greater risk of MOF (hazard ratio [HR] 1.17; 95% CI 1.09 to 1.25), a 14% increase risk of vertebral fracture (HR 1.14; 95% CI 1.03 to 1.26), and a 47% greater risk of hip fracture (HR 1.47; 95% CI 1.30 to 1.67).

Using the French OFELY cohort, Boutroy and colleagues studied 560 postmenopausal Caucasian women for a mean follow up period of 8.0 years [30]. Ninety-four women sustained an osteoporotic fracture. Both LS BMD and TBS were lower in women with than women without incident fractures. In models controlling for age, body weight and prevalent fracture, fracture prediction was similar for LS BMD (OR = 1.30; 95% CI 1.06 to 1.58) and TBS (OR = 1.34; 95% CI 1.04 to 1.73), but lower than for total hip BMD (OR 1.99; 95% CI 1.52 to 2.62).

In another European trial, 556 postmenopausal elderly women (mean age 76.1 years) from the prospective Swiss Evaluation of the Methods of Measurement of Osteoporotic Fracture Risk (SEMOF) study were evaluated [31]. During mean follow-up of 2.7 years, 52 women sustained a clinical fragility fracture. Women with fractures were older, more likely to have a prevalent fracture, and had lower TBS values and BMD T-scores at all sites than women without incident fractures. TBS predicted fractures in models adjusted for age and BMI (HR/SD 2.01; 95% CI 1.54 to 2.63), and in those further adjusted for the lowest BMD (HR/SD 1.87; 95% CI 1.38 to 2.54). The combination of LS BMD and TBS predicted incident fractures (area under the curve [AUC] 0.71) better than LS BMD alone (AUC 0.62; $p=0.03$).

TBS was also evaluated in 1007 postmenopausal women >50 years old from three European centers of the Osteoporosis and Ultrasound Study (OPUS) [32]. Over 6 years, 82 participants reported low-energy fractures, and 46 women had incident vertebral fractures on radiographs. Compared with nonfractured subjects, women with incident fractures were older, and had lower TBS values and BMD at all sites. In unadjusted models, each SD decline in TBS conferred a 62% greater risk of incident clinical fragility fractures (OR 1.62; 95% CI 1.30 to 2.01), and a 54% greater risk of vertebral fractures (OR 1.54; 95% CI 1.17 to 2.03).

TBS and incident fractures were also assessed in 665 Japanese women >50 years old (mean age 64.1 years) followed for a median period of 10 years [33]. Incident morphometric vertebral fractures ($n=140$) were identified by vertebral fracture assessment (VFA) in 92 women. Compared with women without fractures, those with fractures had a lower LS BMD and TBS at baseline. The TBS difference between groups was substantially attenuated, but still significant, after adjusting for confounding variables. In models adjusted for age and LS BMD, each SD decline in TBS conferred a 54% greater risk of vertebral fracture (OR 1.54; 95% CI 1.17 to 2.02). The combination of TBS and LS BMD predicted fracture no better than BMD alone.

The relationship between TBS and incident fractures was explored in men. Leslie and colleagues studied 3620 men >50 years (mean age 67.7) from the province of Manitoba [34]. After average 4.5 years follow-up, incident MOFs were observed in 183 men, including 91 with clinical vertebral fractures and 46 with hip fractures. Men with incident fractures had lower TBS and BMD T-scores at the LS and femoral neck (FN), and higher Fracture Risk Assessment Tool (FRAX) scores than men without fractures. TBS was a predictor of MOF (HR/SD 1.22; 95% CI 1.05 to 1.41) and hip fracture (HR/SD 1.60; 95% CI 1.21 to 2.11) in models adjusted for clinical FRAX score and osteoporosis treatment. After further adjustment for FN (HR/SD 1.36; 95% CI 1.01 to 1.83) or LS BMD (HR/SD 1.44; 95% CI 1.07 to 1.94), TBS remained a predictor of hip fracture, but not MOFs. No association was seen between TBS and clinical vertebral fractures.

Schousboe and colleagues assessed TBS and incident fractures in 5863 community-dwelling men ≥65 years of age enrolled in the MrOS cohort [35]. During a mean follow-up of 10 years, 448 men sustained a MOF and 181 experienced a hip fracture. TBS was a predictor of MOFs (HR/SD 1.31; 95% CI 1.20 to 1.43) and hip fractures (HR/SD 1.24; 95% CI 1.08 to 1.49) in models adjusted for FRAX probabilities with BMD. The ability of TBS to predict MOF and hip fractures remained significant after further adjustment for prevalent morphometric vertebral fracture.

TBS and incident fractures were also examined in 1872 community-dwelling Japanese men ≥65 years of age (mean age 73) from the Fujiwara-kyo Osteoporosis Risk in Men (FORMEN) study [36]. Over median follow-up of 4.5 years 22 participants sustained 23 MOFs. TBS and BMD at all sites were lower, and FRAX scores were higher, in men with incident MOFs than in those without fractures. In unadjusted models, TBS was a predictor of MOFs, with an OR of 1.89 (95% CI 1.28, 2.81) for each SD decline in TBS. The combination of FRAX score and TBS predicted MOFs no better than FRAX alone.

Finally, a meta-analysis of 14 prospective population-based cohorts that included 17,809 men and women from diverse countries confirmed that TBS can predict MOFs

Table 35.1. Summary of longitudinal studies.

Reference	Study Subjects	Mean Age ± SD (years)	Mean Follow-up (years)	Outcome Measure (number of subjects)	Adjustments	HR or OR per SD decrease in TBS (95% CI)
Su et al., 2016 [38]	Community-dwelling men ($n=1923$) and women ($n=1950$) ≥65 years, from Mr. OS and Ms. OS Hong Kong study	Men: 72.3 ± 4.9 Women: 72.5 ± 5.3	9.9 8.8	MOF ($n=126$) MOF ($n=215$)	FRAX score	HR = 1.38 (1.15–1.65) HR = 1.32 (1.13–1.54)
McCloskey et al., 2016 [39]	17,809 men and women (59% women), meta-analysis of 14 international population-based cohorts	72 (range: 40–90)	6.1	MOF ($n=1109$) Hip Fx ($n=298$)	Age, time since baseline and FRAX probability of MOF (with BMD) Age, time since baseline and FRAX probability of hip Fx (with BMD)	HR = 1.32 (1.24–1.41) HR = 1.28 (1.13–1.45)
Hans et al., 2011 [28]	29,407 women ≥50 years (98% white)	65.4 ± 9.5	4.7	MOF ($n=1668$) assessed in health service records by fracture codes Clinical vertebral Fx ($n=439$) Hip Fx ($n=239$)	Age, LS BMD, and a combination of clinical risk factors[a]	HR = 1.17 (1.09–1.25) HR = 1.14 (1.03–1.26) HR = 1.47 (1.30–1.67)
Boutroy et al., 2013 [30]	560 postmenopausal white women	66.2 ± 7.9	8.0	Clinical and radiographic vertebral Fx and fragility Fx at any site ($n=94$) (confirmed by radiographs), except head, toes, and fingers	Age, weight, and prevalent fracture at baseline	OR = 1.34 (1.04–1.73)
Iki et al., 2014 [33]	665 Japanese women >50 years old	64.1 ± 8.1	8.3	Vertebral Fx ($n=92$) (by VFA)	Age, LS BMD, and prevalent vertebral deformity	OR = 1.52 (1.16–2.00)
Briot et al., 2013 [32]	1007 postmenopausal white women >50 years old	65.9 ± 6.9	6.0	Clinical osteoporotic Fx ($n=82$) (peripheral and clinical vertebral fractures), by self-report, confirmed by radiographs Vertebral Fx ($n=46$) (by radiographs)	None	OR = 1.62 (1.30–2.01) OR = 1.54 (1.17–2.03)
Leslie et al., 2014 [29][c]	33,352 women aged 40 to 100 years (98% white)	63.2 ± 10.8	4.7	MOF ($n=1872$) assessed in health service records by fracture codes 1754 deaths	Clinical risk factors[b] and LS BMD	HR = 1.17 (1.11–1.23) HR = 1.26 (1.19–1.32)

(Continued)

Table 35.1. (Continued)

Reference	Study Subjects	Mean Age ± SD (years)	Mean Follow-up (years)	Outcome Measure (number of subjects)	Adjustments	HR or OR per SD decrease in TBS (95% CI)
Popp et al., 2016 [31]	556 postmenopausal elderly women	76.1 ± 3.0	2.7	Clinical fragility fractures (n=52) (20 forearm, 6 hip, 10 clinical vertebral, 9 humerus, 2 pelvis, 3 ankle, 1 clavicle, and 1 elbow)	Age, BMI, and lowest BMD	HR = 1.87 (1.38–2.54)
McCloskey et al., 2015 [40]c	33,352 women aged 40 to 100 years (98% white)	63.2 ± 10.8	4.7	At least one MOF (n=1639) excluding hip Fx (assessed in health service records by fracture codes) Hip Fx (n=306) 1754 (5.3%) deaths	Age, time since baseline, FN BMD, and clinical risk factors (BMI, previous fracture, smoking, glucocorticoids, rheumatoid arthritis, secondary osteoporosis, and alcohol use)	HR = 1.18 (1.12–1.24) HR = 1.23 (1.09–1.38) HR = 1.20 (1.14–1.26)
Leslie et al., 2014 [34]	3620 men >50 years (98% white)	67.6 ± 9.8	4.5	MOF (n=183) assessed in health service records by fracture codes Clinical vertebral Fx (n=91) Hip Fx (n=46)	Clinical FRAX score, osteoporosis treatment, and LS BMD	HR = 1.08 (0.92–1.26) HR = 1.02 (0.81–1.27) HR = 1.44 (1.07–1.94)
Iki et al., 2015 [36]	1872 community-dwelling Japanese men ≥65 years	73 ± 5.1	4.5 (median)	MOF (n=22) identified by interviews or mail and telephone surveys	FRAX score	OR = 1.76 (1.16–2.67)
Schousboe et al., 2016 [35]	5863 community-dwelling men ≥65 years	73.7 ± 5.9	10	MOF (n=448) identified by mail and confirmed by radiographs Hip Fx (n=181)	FRAX with BMD 10-year fracture risks and prevalent radiographic vertebral Fx	HR = 1.27 (1.17–1.39) HR = 1.20 (1.05–1.39)

a Clinical risk factors: ADG (ambulatory diagnostic groups) comorbidity score, rheumatoid arthritis, chronic obstructive pulmonary disease, diabetes, substance abuse, body mass index, prior osteoporotic fracture, systemic corticosteroid use in the last year, and osteoporosis treatment in the last year.
b Clinical risk factors: BMI, previous fracture, chronic obstructive pulmonary disease (smoking proxy), glucocorticoid use >90 days, rheumatoid arthritis, secondary osteoporosis, and high alcohol use.
c These studies included the same study population.
BMD = bone mineral density; BMI = body mass index; FN = femoral neck; Fx = fracture; FRAX = Fracture Risk Assessment Tool; HR = hazard ratio; LS = lumbar spine; MOF: major osteoporotic fracture (hip, clinical spine, forearm, and humerus); OR = odds ratio; TBS = trabecular bone score; VFA = vertebral fracture assessment.

and hip fractures in both women and men after adjusting for FRAX probabilities [39]. This study is detailed in the next section.

In summary, there is consistent clinical evidence that TBS is a predictor of fracture risk in postmenopausal women and older men, even after adjusting for BMD, FRAX probabilities, and/or clinical risk factors. The association of TBS with BMD slightly improved fracture discrimination compared with BMD alone in some of these studies [28, 31, 32, 34]. At the time of this review, no fracture data have been published for premenopausal women or younger men.

Combination of TBS with FRAX to improve fracture risk prediction

In the Manitoba cohort (33,352 women aged 40 to 100 years followed for 4.7 years), TBS was established to be a predictor of MOFs after adjusting for FRAX variables including FN BMD and after accounting for the TBS-associated death hazard [29]. Compared with high TBS (90th percentile), low TBS (10th percentile) increased the risk of MOF by 1.5- to 1.6-fold. A subsequent analysis developed a method to adjust FRAX-based probabilities based upon TBS [40]. During the observation period, 1754 subjects died, 1639 sustained a non-hip MOF, and 306 sustained a hip fracture. After controlling for age, time since baseline, FN BMD, and FRAX clinical risk factors, TBS was a predictor of mortality, non-hip MOF, and hip fracture with respective HRs per SD decrease in TBS of 1.20 (95% CI 1.14 to 1.26), 1.18 (95% CI 1.12 to 1.24), and 1.23 (95% CI 1.09 to 1.38). The 10-year probabilities of fracture were calculated with and without TBS, confirming an incremental improvement in fracture prediction when TBS was used in combination with FRAX. There was a significant interaction between TBS and age, with TBS having a larger impact on 10-year fracture probability from FRAX in younger women, and less in older women.

Finally, the TBS adjustment derived from the Manitoba study described [40] was applied to the meta-analysis of 14 prospective population-based cohorts from North America, Asia, Australia, and Europe to calculate TBS-adjusted FRAX probabilities [39]. The meta-analysis included 17,809 men and women (mean age 72 years old; 59% women), followed for a mean time of 6.1 years. FRAX probabilities of MOFs and hip fractures were estimated at baseline using individual-level data in each cohort and country-specific FRAX models. During follow-up, 298 participants sustained at least one hip fracture and 1109 had one or more MOFs. In models adjusted for age, time since baseline, and FRAX probabilities, TBS was able to predict MOF (HR/SD 1.32; 95% CI 1.24 to 1.41) and hip fracture (HR/SD 1.28; 95% CI 1.13 to 1.45). Effects were similar between men and women ($p > 0.10$ for sex interaction). Risk stratification for MOF and hip fracture was slightly greater for TBS-adjusted FRAX probability than for FRAX probability without TBS. Similar results were seen in a study of 1923 community-dwelling men, and 1950 women ≥65 years, from the Mr. OS and Ms. OS Hong Kong study [38]. The HRs/SD decrease in FRAX score for MOF were 1.58 (95% CI 1.40 to 1.79) for men, and 1.35 (95% CI 1.23 to 1.48) for women. The respective HRs/SD decrease in TBS-adjusted FRAX score were 1.65 (95% CI 1.45 to 1.86), and 1.39 (95% CI 1.27 to 1.53). Based on these results, TBS can now be entered in the internet-based FRAX calculator, allowing for the calculation of TBS-adjusted 10-year probability of MOF and hip fracture (see later in this chapter).

TBS and fracture risk assessment in special conditions

A growing number of studies have assessed TBS in various conditions known to increase fracture risk. In general, compared with control subjects, TBS was lower in patients with diabetes [41–43], on long-term glucocorticoid exposure [44–47], with primary hyperparathyroidism [48, 49], kidney transplant recipients [50], with thalassemia major [51], with Ehlers–Danlos syndrome [52], and with acromegaly [53]. TBS was also assessed in subjects with rheumatoid arthritis [54, 55], with anorexia nervosa [56], and in patients on hemodialysis [57].

Diabetes

TBS and incident fractures were examined in 29,407 women ≥50 years from the Province of Manitoba, including 2356 with diabetes (mostly type 2 diabetes) [41]. Compared with controls, diabetic women had higher baseline BMDs at all sites, but lower TBS, even after adjusting for multiple confounding variables. Over 4.7 years of follow-up, the incidence of MOF was greater in patients with diabetes (7.4%, $n = 175$) than in nondiabetic women (5.5%, $n = 1493$; $p < 0.001$). TBS predicted MOF independent of BMD in women with diabetes (HR = 1.27; 95% CI 1.10 to 1.46) and similar to those without diabetes (HR = 1.31; 95% CI 1.24 to 1.38).

Additional studies have confirmed that, despite greater BMD measurements, patients with type 2 diabetes have lower TBS values than controls [42, 43]. Another study [58], which enrolled 99 postmenopausal women with type 2 diabetes and 107 nondiabetic controls, found a greater prevalence of morphometric vertebral fractures in diabetics (34.3%, $n = 34$) than in controls (18.7%, $n = 20$; $p = 0.01$). Vertebral fractures were associated with lower values of TBS (AUC 0.69, $p < 0.0001$) and FN BMD (AUC 0.63, $p < 0.004$).

Finally, TBS was examined in 119 patients with type 1 diabetes (59 males, 60 premenopausal females; mean age 43.4 years) and 68 matched healthy controls [59]. TBS was similar between diabetics and nondiabetic controls, but lower in patients with diabetes with prevalent clinical fractures ($n = 24$) than in controls. Using a multivariate model, TBS ($p = 0.049$) and HbA_1c ($p = 0.036$) were independently associated with prevalent fractures in patients with type 1 diabetes.

Long-term glucocorticoid exposure

Two independent studies have shown that TBS differentiated subjects according to chronic glucocorticoid (GC) exposure. Paggiosi and colleagues studied 484 women (ages 55 to 79 years), allocated into one of three groups: 64 women taking prednisolone ≥5 mg/day for >3 months; 141 women who had sustained a recent MOF; and 279 healthy population-based women [44]. Compared with healthy women, those with fractures had lower age-adjusted LS BMD and TBS Z-scores. In contrast, women on GC had similar age-adjusted LS BMD, but lower adjusted TBS Z-scores ($p<0.001$) than healthy controls. TBS (AUC = 0.721), but not LS BMD (AUC = 0.572), was able to discriminate between GC-treated and GC-naïve women.

Leib and colleagues also assessed TBS, BMD, and osteoporotic fractures in 416 individuals (mean age 63.4 years; 72 men) taking GCs (prednisone ≥5 mg/day, for ≥3 months) compared with 1104 sex-, age-, and BMI-matched controls [47]. Prevalent osteoporotic fractures were present in 16.3% of cases and in 13.1% of GC-naïve subjects ($p=0.16$). TBS and BMD Z-scores at the hip sites, but not LS BMD, were lower in the GC group than in controls. In GC-naïve subjects, both TBS and LS BMD were able to differentiate between patients with and without fracture. In contrast, in the GC group, TBS (but not LS BMD) was able to distinguish between individuals with and without fracture. Using a multivariate model, each SD decrease in TBS conferred a 51% greater risk of prevalent fracture (OR 1.51; 95% CI 1.23 to 1.86).

EFFECT OF DIFFERENT THERAPIES ON TBS

The effect of different anti-osteoporotic therapies on TBS has been evaluated mostly in postmenopausal women and in women with breast cancer treatment-induced bone loss. Despite good short-term in vivo precision (range 1.1% to 2.1% [28,32,60–62]), on-treatment increases exceeding the least significant change (LSC) are less frequently observed with TBS than for LS BMD. Indeed, the majority of studies have shown that changes in TBS due to anti-osteoporotic agents tend to be much smaller than those observed in LS BMD.

Overall, treatment with bisphosphonates leads to a slight increase or maintenance of TBS in postmenopausal women treated for up to 3 to 4 years [61–63]. In a study of women >50 years old from the Manitoba database, 534 women initiating antiresorptive agents with high adherence (86% bisphosphonates, 10% raloxifene, 4% calcitonin) were compared with 1150 untreated women followed for a mean period of 3.7 years [63]. In untreated subjects, there was a similar decrease in TBS and LS BMD (−0.31%/year versus −0.36%/year). Among treated women, the mean increase in TBS was much lower than for LS BMD (+0.2%/year versus +1.86%/year). Similarly, in 54 postmenopausal women, 3-year treatment with zoledronic acid resulted in an increase of 1.41% in TBS and 9.58% in LS BMD [61]. Treatment of 60 postmenopausal women with denosumab for 1 year gave nonsignificant increases in TBS relative to baseline [64].

In contrast with antiresorptive therapies, treatment with teriparatide, an anabolic agent, might lead to greater improvements in TBS [62, 65]. In an open-label trial [62], postmenopausal women were treated with teriparatide ($n=65$) or ibandronate ($n=122$) for 2 years. Teriparatide led to greater increases in TBS than ibandronate (+4.3 versus +0.3%; $p<0.0001$). Another nonrandomized retrospective study assessed changes in TBS in 390 subjects (72 men) ≥40 years old treated for an average of 1.7 years with one of the following: calcium + vitamin D, alendronate, risedronate, testosterone, denosumab, or teriparatide [65]. Consistent with previous studies, LS BMD increases were greater than with TBS for all pharmacological therapies (+4.1% to +8.8% in BMD versus +1.4% to +3.6% in TBS). The greatest improvement in TBS was observed in the group of 30 women treated with teriparatide (+3.6%).

TBS was also assessed in women with breast cancer on estrogen-deprivation therapies with or without concurrent use of bisphosphonates [66–69]. Overall, treatment with estrogen-deprivation therapies for ~2 years resulted in unadjusted decreases in TBS (−2.1 to −2.35%) and in LS BMD (−1.7 to −6.4%). The concomitant use of zoledronic acid or risedronate preserved TBS and increased LS BMD [67,69].

TBS IN CLINICAL PRACTICE

Considerations regarding TBS measurement

TBS is clinically measured at the LS with specialized software (TBS iNsight, Med-Imaps, France) that uses the same region of interest as for conventional BMD measurement. A result is given for each evaluable vertebra (from L1 to L4) and for the overall region of interest, and can be performed at the time of the DXA acquisition or retrospectively. LS DXA images obtained from the current generation of fan-beam densitometers (Hologic Delphi, QDR 4500, and Discovery; GE/Lunar Prodigy and iDXA) are suitable for TBS analyses. Of note, TBS values may not be comparable across different DXA machines or scan modes [60, 70, 71].

There is evidence that TBS is unaffected by the presence of ossifications that typically overestimate LS DXA BMD. Data from two independent studies found that neither osteoarthritic changes in elderly women nor lumbar syndesmophytes in men with spondyloarthritis influence TBS results [72, 73]. In contrast, TBS is affected by excessive abdominal soft tissue, which degrades image texture and reduces TBS values [74–76]. This tissue effect on TBS explains why the original TBS algorithm, which had been optimized for women, gave paradoxically lower TBS in men than in women because adiposity in men tends to be more abdominal. The algorithm has since been updated to address these technical issues, and became applicable to both women and men. Under TBS version 2.x, the BMI dependency has been eliminated for GE/Lunar scanners, but has not yet been fully addressed with Hologic devices [77]. As a result of this BMI/abdominal adiposity dependence, the manufacturers of the TBS software recommend against its use in individuals with BMI outside of the range 15 to 37 kg/m^2.

Use of TBS to guide clinical decisions

Recently published guidance documents from the ISCD and the ESCEO support the use of TBS to assess fracture risk in postmenopausal women and in men >50 years [2, 3]. In addition, TBS was proposed for use in postmenopausal women with type 2 diabetes to predict MOFs.

There is no consensus regarding what represents normal versus abnormal TBS. The TBS manufacturer has proposed that TBS values ≥1.350 be considered normal while TBS between 1.200 and 1.350 is consistent with "partially degraded" bone and TBS ≤1.200 indicates "degraded" bone [78]. A low TBS is associated with greater risk of fracture, but a TBS threshold to initiate treatment has not been identified and the ISCD has recommended against the use of TBS as a single measurement to guide treatment decisions [2]. Instead, the ISCD has suggested that using FRAX probabilities adjusted for TBS could assist in treatment decisions [2]. In 2016, the US FDA cleared for clinical use the TBS-adjusted FRAX tool to estimate 10-year probabilities of MOF and hip fracture.

As previously mentioned, the online calculator on the FRAX website has the option to "adjust with TBS" when BMD is entered in the calculator and TBS is available, providing a TBS-adjusted 10-year fracture probability of MOF and hip fracture. This adjustment contains a "TBS × age" interaction term which reflects the declining strength of the TBS adjustment on FRAX with increasing age. In general, the use of TBS to adjust the FRAX score has a greater clinical impact in those patients who are close to the intervention threshold when risk is determined from FRAX in the absence of TBS (Fig. 35.1).

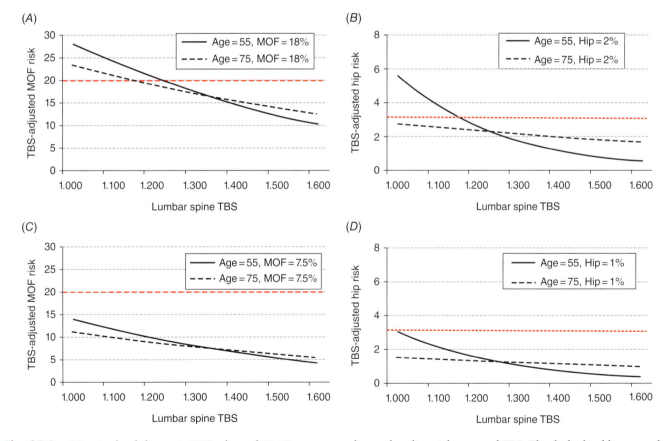

Fig. 35.1. Magnitude of change in TBS-adjusted FRAX score according to baseline risk, age, and TBS. The dashed red lines are the National Osteoporosis Foundation (NOF) treatment thresholds. The black lines represent the TBS-adjusted FRAX risk for a given woman of 55 years of age (solid lines) or a woman of 75 years of age (dashed lines) according to different TBS values. Notice that TBS has a greater effect in younger (55 years) than in older (75 years) women, in those with greater unadjusted fracture probabilities (panels *A* and *B*), compared with lower unadjusted fracture probabilities (panels *C* and *D*). TBS = trabecular bone score; MOF = major osteoporotic fracture.

Given the much smaller changes in TBS than in BMD in patients treated with bisphosphonates, and because of the lack of studies demonstrating that on-treatment changes in TBS are associated with fracture risk, the ISCD recommended against use of TBS for monitoring patients on bisphosphonate therapy [2]. The role of TBS for monitoring patients on teriparatide, denosumab, or newer anti-osteoporotic drugs remains uncertain.

REFERENCES

1. Silva BC, Leslie WD, Resch H, et al. Trabecular bone score: a noninvasive analytical method based upon the DXA image. J Bone Miner Res. 2014;29(3):518–30.
2. Silva BC, Broy SB, Boutroy S, et al. Fracture Risk Prediction by Non-BMD DXA Measures: the 2015 ISCD Official Positions Part 2: Trabecular Bone Score. J Clin Densitom. 2015;18(3):309–30.
3. Harvey NC, Gluer CC, Binkley N, et al. Trabecular bone score (TBS) as a new complementary approach for osteoporosis evaluation in clinical practice. Bone. 2015;78:216–24.
4. Pothuaud L, Carceller P, Hans D. Correlations between grey-level variations in 2D projection images (TBS) and 3D microarchitecture: applications in the study of human trabecular bone microarchitecture. Bone. 2008;42(4):775–87.
5. Hans D, Barthe N, Boutroy S, et al. Correlations between trabecular bone score, measured using anteroposterior dual-energy X-ray absorptiometry acquisition, and 3-dimensional parameters of bone microarchitecture: an experimental study on human cadaver vertebrae. J Clin Densitom. 2011;14(3):302–12.
6. Winzenrieth R, Michelet F, Hans D. Three-Dimensional (3D) Microarchitecture Correlations with 2D Projection Image Gray-Level Variations Assessed by Trabecular Bone Score Using High-Resolution Computed Tomographic Acquisitions: Effects of Resolution and Noise. J Clin Densitom. 2013;16(3):287–96.
7. Roux JP, Wegrzyn J, Boutroy S, et al. The predictive value of trabecular bone score (TBS) on whole lumbar vertebrae mechanics: an ex vivo study. Osteoporos Int. 2013;24(9):2455–60.
8. Silva BC, Walker MD, Abraham A, et al. Trabecular bone score is associated with volumetric bone density and microarchitecture as assessed by central QCT and HRpQCT in Chinese American and white women. J Clin Densitom. 2013;16(4):554–61.
9. Muschitz C, Kocijan R, Haschka J, et al. TBS reflects trabecular microarchitecture in premenopausal women and men with idiopathic osteoporosis and low-traumatic fractures. Bone. 2015;79:259–66.
10. Silva BC, Boutroy S, Zhang C, et al. Trabecular bone score (TBS)—a novel method to evaluate bone microarchitectural texture in patients with primary hyperparathyroidism. J Clin Endocrinol Metab. 2013;98(5): 1963–70.
11. Popp AW, Buffat H, Eberli U, et al. Microstructural parameters of bone evaluated using HR-pQCT correlate with the DXA-derived cortical index and the trabecular bone score in a cohort of randomly selected premenopausal women. PLoS One. 2014;9(2):e88946.
12. Amstrup AK, Jakobsen NF, Moser E, et al. Association between bone indices assessed by DXA, HR-pQCT and QCT scans in post-menopausal women. J Bone Miner Metab. 2016;34(6):638–45.
13. Pothuaud L, Barthe N, Krieg MA, et al. Evaluation of the potential use of trabecular bone score to complement bone mineral density in the diagnosis of osteoporosis: a preliminary spine BMD-matched, case-control study. J Clin Densitom. 2009;12(2):170–6.
14. Winzenrieth R, Dufour R, Pothuaud L, et al. A retrospective case-control study assessing the role of trabecular bone score in postmenopausal Caucasian women with osteopenia: analyzing the odds of vertebral fracture. Calcif Tissue Int. 2010;86(2):104–9.
15. Rabier B, Heraud A, Grand-Lenoir C, et al. A multicentre, retrospective case-control study assessing the role of trabecular bone score (TBS) in menopausal Caucasian women with low areal bone mineral density (BMDa): Analysing the odds of vertebral fracture. Bone. 2010;46(1):176–81.
16. Del Rio LM, Winzenrieth R, Cormier C, et al. Is bone microarchitecture status of the lumbar spine assessed by TBS related to femoral neck fracture? A Spanish case-control study. Osteoporos Int. 2013;24(3):991–8.
17. Krueger D, Fidler E, Libber J, et al. Spine trabecular bone score subsequent to bone mineral density improves fracture discrimination in women. J Clin Densitom. 2014;17(1):60–5.
18. Lamy O, Krieg MA, Stoll D, et al. The OsteoLaus Cohort Study: Bone mineral density, micro-architecture score and vertebral fracture assessment extracted from a single DXA device in combination with clinical risk factors improve significantly the identification of women at high risk of fracture. Osteologie. 2012;21:77–82.
19. Vasic J, Petranova T, Povoroznyuk V, et al. Evaluating spine micro-architectural texture (via TBS) discriminates major osteoporotic fractures from controls both as well as and independent of site matched BMD: the Eastern European TBS study. J Bone Miner Metab. 2014;32(5):556–62.
20. Leib E, Winzenrieth R, Lamy O, et al. Comparing bone microarchitecture by trabecular bone score (TBS) in Caucasian American women with and without osteoporotic fractures. Calcif Tissue Int. 2014;95(3):201–8.
21. Ayoub ML, Maalouf G, Bachour F, et al. DXA-based variables and osteoporotic fractures in Lebanese postmenopausal women. Orthop Traumatol Surg Res. 2014;100(8):855–8.
22. Nassar K, Paternotte S, Kolta S, et al. Added value of trabecular bone score over bone mineral density for identification of vertebral fractures in patients with areal bone mineral density in the non-osteoporotic range. Osteoporos Int. 2014;25(1):243–9.
23. Touvier J, Winzenrieth R, Johansson H, et al. Fracture discrimination by combined bone mineral density

(BMD) and microarchitectural texture analysis. Calcif Tissue Int. 2015;96(4):274–83.
24. Leib E, Winzenrieth R, Aubry-Rozier B, et al. Vertebral microarchitecture and fragility fracture in men: a TBS study. Bone. 2014;62:51–5.
25. Ayoub ML, Maalouf G, Cortet B, et al. Trabecular Bone Score and Osteoporotic Fractures in Obese Postmenopausal Women. J Clin Densitom. 2016;19(4): 544–5.
26. Jain RK, Narang DK, Hans D, et al. Ethnic Differences in Trabecular Bone Score. J Clin Densitom. 2017;20(2): 172–9.
27. Leslie WD, Binkley N. Spine bone texture and the trabecular bone score (TBS). In: Patel VB, Preedy VR (eds) *Biomarkers in Disease: Methods, Discoveries and Applications.* Biomarkers in Bone Disease. Springer; 2016.
28. Hans D, Goertzen AL, Krieg MA, et al. Bone microarchitecture assessed by TBS predicts osteoporotic fractures independent of bone density: the Manitoba study. J Bone Miner Res. 2011;26(11):2762–9.
29. Leslie WD, Johansson H, Kanis JA, et al. Lumbar spine texture enhances 10-year fracture probability assessment. Osteoporos Int. 2014;25(9):2271–7.
30. Boutroy S, Hans D, Sornay-Rendu E, et al. Trabecular bone score improves fracture risk prediction in non-osteoporotic women: the OFELY study. Osteoporos Int. 2013;24(1):77–85.
31. Popp AW, Meer S, Krieg MA, et al. Bone mineral density (BMD) and vertebral trabecular bone score (TBS) for the identification of elderly women at high risk for fracture: the SEMOF cohort study. Eur Spine J. 2016;25(11):3432–8.
32. Briot K, Paternotte S, Kolta S, et al. Added value of trabecular bone score to bone mineral density for prediction of osteoporotic fractures in postmenopausal women: The OPUS study. Bone. 2013;57(1):232–6.
33. Iki M, Tamaki J, Kadowaki E, et al. Trabecular bone score (TBS) predicts vertebral fractures in Japanese women over 10 years independently of bone density and prevalent vertebral deformity: the Japanese Population-Based Osteoporosis (JPOS) cohort study. J Bone Miner Res. 2014;29(2):399–407.
34. Leslie WD, Aubry-Rozier B, Lix LM, et al. Spine bone texture assessed by trabecular bone score (TBS) predicts osteoporotic fractures in men: the Manitoba Bone Density Program. Bone. 2014;67:10–14.
35. Schousboe JT, Vo T, Taylor BC, et al; Osteoporotic Fractures in Men MrOS Study Research Group. Prediction of Incident Major Osteoporotic and Hip Fractures by Trabecular Bone Score (TBS) and Prevalent Radiographic Vertebral Fracture in Older Men. J Bone Miner Res. 2016;31(3):690–7.
36. Iki M, Fujita Y, Tamaki J, et al. Trabecular bone score may improve FRAX® prediction accuracy for major osteoporotic fractures in elderly Japanese men: the Fujiwara-kyo Osteoporosis Risk in Men (FORMEN) Cohort Study. Osteoporos Int. 2015;26(6):1841–8.
37. Su Y, Leung J, Hans D, et al. Added clinical use of trabecular bone score to BMD for major osteoporotic fracture prediction in older Chinese people: the Mr. OS and Ms. OS cohort study in Hong Kong. Osteoporos Int. 2017;28(1):151–60.
38. Su Y, Leung J, Hans D, et al. The added value of trabecular bone score to FRAX® to predict major osteoporotic fractures for clinical use in Chinese older people: the Mr. OS and Ms. OS cohort study in Hong Kong. Osteoporos Int. 2017;28(1):111–17.
39. McCloskey EV, Oden A, Harvey NC, et al. A Meta-Analysis of Trabecular Bone Score in Fracture Risk Prediction and Its Relationship to FRAX. J Bone Miner Res. 2016;31(5):940–8.
40. McCloskey EV, Oden A, Harvey NC, et al. Adjusting fracture probability by trabecular bone score. Calcif Tissue Int. 2015;96(6):500–9.
41. Leslie WD, Aubry-Rozier B, Lamy O, et al. TBS (Trabecular Bone Score) and Diabetes-Related Fracture Risk. J Clin Endocrinol Metab. 2013;98(2):602–9.
42. Dhaliwal R, Cibula D, Ghosh C, et al. Bone quality assessment in type 2 diabetes mellitus. Osteoporos Int. 2014;25(7):1969–73.
43. Kim JH, Choi HJ, Ku EJ, et al. Trabecular bone score as an indicator for skeletal deterioration in diabetes. J Clin Endocrinol Metab. 2015;100(2):475–82.
44. Paggiosi MA, Peel NF, Eastell R. The impact of glucocorticoid therapy on trabecular bone score in older women. Osteoporos Int. 2015;26(6):1773–80.
45. Eller-Vainicher C, Morelli V, Ulivieri FM, et al. Bone quality, as measured by trabecular bone score in patients with adrenal incidentalomas with and without subclinical hypercortisolism. J Bone Miner Res. 2012;27(10): 2223–30.
46. Koumakis E, Avouac J, Winzenrieth R, et al. Trabecular bone score in female patients with systemic sclerosis: comparison with rheumatoid arthritis and influence of glucocorticoid exposure. J Rheumatol. 2015;42(2):228–35.
47. Leib ES, Winzenrieth R. Bone status in glucocorticoid-treated men and women. Osteoporos Int. 2016;27(1): 39–48.
48. Eller-Vainicher C, Filopanti M, Palmieri S, et al. Bone quality, as measured by trabecular bone score, in patients with primary hyperparathyroidism. Eur J Endocrinol. 2013;169(2):155–62.
49. Romagnoli E, Cipriani C, Nofroni I, et al. "Trabecular Bone Score" (TBS): An indirect measure of bone microarchitecture in postmenopausal patients with primary hyperparathyroidism. Bone. 2013;53(1):154–9.
50. Naylor KL, Lix LM, Hans D, et al. Trabecular bone score in kidney transplant recipients. Osteoporos Int. 2016;27(3):1115–21.
51. Baldini M, Ulivieri FM, Forti S, et al. Spine bone texture assessed by trabecular bone score (TBS) to evaluate bone health in thalassemia major. Calcif Tissue Int. 2014; 95(6):540–6.
52. Eller-Vainicher C, Bassotti A, Imeraj A, et al. Bone involvement in adult patients affected with Ehlers-Danlos syndrome. Osteoporos Int. 2016;27(8):2525–31.
53. Hong AR, Kim JH, Kim SW, et al. Trabecular bone score as a skeletal fragility index in acromegaly patients. Osteoporos Int. 2016;27(3):1123–9.

54. Breban S, Briot K, Kolta S, et al. Identification of rheumatoid arthritis patients with vertebral fractures using bone mineral density and trabecular bone score. J Clin Densitom. 2012;15(3):260–6.
55. Kim D, Cho SK, Kim JY, et al. Association between trabecular bone score and risk factors for fractures in Korean female patients with rheumatoid arthritis. Mod Rheumatol. 2016;26(4):540–5.
56. Donaldson AA, Feldman HA, O'Donnell JM, et al. Spinal Bone Texture Assessed by Trabecular Bone Score in Adolescent Girls With Anorexia Nervosa. J Clin Endocrinol Metab. 2015;100(9):3436–42.
57. Brunerova L, Ronova P, Veresova J, et al. Osteoporosis and Impaired Trabecular Bone Score in Hemodialysis Patients. Kidney Blood Press Res. 2016;41(3):345–54.
58. Zhukouskaya VV, Eller-Vainicher C, Gaudio A, et al. The utility of lumbar spine trabecular bone score and femoral neck bone mineral density for identifying asymptomatic vertebral fractures in well-compensated type 2 diabetic patients. Osteoporos Int. 2016;27(1):49–56.
59. Neumann T, Lodes S, Kastner B, et al. Trabecular bone score in type 1 diabetes—a cross-sectional study. Osteoporos Int. 2016;27(1):127–33.
60. Krueger D, Libber J, Binkley N. Spine Trabecular Bone Score Precision, a Comparison Between GE Lunar Standard and High-Resolution Densitometers. J Clin Densitom. 2015;18(2):226–32.
61. Popp AW, Guler S, Lamy O, et al. Effects of zoledronate versus placebo on spine bone mineral density and microarchitecture assessed by the trabecular bone score in postmenopausal women with osteoporosis: a three-year study. J Bone Miner Res. 2013;28(3):449–54.
62. Senn C, Gunther B, Popp AW, et al. Comparative effects of teriparatide and ibandronate on spine bone mineral density (BMD) and microarchitecture (TBS) in postmenopausal women with osteoporosis: a 2-year open-label study. Osteoporos Int. 2014;25(7):1945–51.
63. Krieg MA, Aubry-Rozier B, Hans D, et al. Effects of antiresorptive agents on trabecular bone score (TBS) in older women. Osteoporos Int. 2013;24(3):1073–8.
64. Petranova T, Sheytanov I, Monov S, et al. Denosumab improves bone mineral density and microarchitecture and reduces bone pain in women with osteoporosis with and without glucocorticoid treatment. Biotechnol Biotechnol Equip. 2014;28(6):1127–37.
65. Di Gregorio S, Del Rio L, Rodriguez-Tolra J, et al. Comparison between different bone treatments on areal bone mineral density (aBMD) and bone microarchitectural texture as assessed by the trabecular bone score (TBS). Bone. 2015;75:138–43.
66. Kalder M, Hans D, Kyvernitakis I, et al. Effects of Exemestane and Tamoxifen treatment on bone texture analysis assessed by TBS in comparison with bone mineral density assessed by DXA in women with breast cancer. J Clin Densitom. 2014;17(1):66–71.
67. Kalder M, Kyvernitakis I, Albert US, et al. Effects of zoledronic acid versus placebo on bone mineral density and bone texture analysis assessed by the trabecular bone score in premenopausal women with breast cancer treatment-induced bone loss: results of the ProBONE II substudy. Osteoporos Int. 2015;26(1):353–60.
68. Pedrazzoni M, Casola A, Verzicco I, et al. Longitudinal changes of trabecular bone score after estrogen deprivation: effect of menopause and aromatase inhibition. J Endocrinol Invest. 2014;37(9):871–4.
69. Prasad C, Greenspan SL, Vujevich KT, et al. Risedronate may preserve bone microarchitecture in breast cancer survivors on aromatase inhibitors: A randomized, controlled clinical trial. Bone. 2016;90:123–6.
70. Bandirali M, Di Leo G, Messina C, et al. Reproducibility of trabecular bone score with different scan modes using dual-energy X-ray absorptiometry: a phantom study. Skeletal Radiol. 2015;44(4):573–6.
71. Chen W, Slattery A, Center J, et al. The Effect of Changing Scan Mode on Trabecular Bone Score Using Lunar Prodigy. J Clin Densitom. 2016;19(4):502–6.
72. Kolta S, Briot K, Fechtenbaum J, et al. TBS result is not affected by lumbar spine osteoarthritis. Osteoporos Int. 2014;25(6):1759–64.
73. Wildberger L, Boyadzhieva V, Hans D, et al. Impact of lumbar syndesmophyte on bone health as assessed by bone density (BMD) and bone texture (TBS) in men with axial spondyloarthritis. Joint Bone Spine. 2017;84(4):463–6.
74. Kim JH, Choi HJ, Ku EJ, et al. Regional body fat depots differently affect bone microarchitecture in postmenopausal Korean women. Osteoporos Int. 2016;27(3):1161–8.
75. Langsetmo L, Vo TN, Ensrud KE, et al.; Osteoporotic Fractures in Men Research Group. The Association Between Trabecular Bone Score and Lumbar Spine Volumetric BMD Is Attenuated Among Older Men With High Body Mass Index. J Bone Miner Res. 2016;31(10):1820–6.
76. Looker AC, Sarafrazi Isfahani N, Fan B, et al. Trabecular bone scores and lumbar spine bone mineral density of US adults: comparison of relationships with demographic and body size variables. Osteoporos Int. 2016;27(8):2467–75.
77. Mazzetti G, Berger C, Leslie W, et al. Densitometer-specific differences in the correlation between body mass index and lumbar spine trabecular bone score. J Clin Densitom. 2017;20(2):233–8.
78. Cormier C, Lamy O, Poriau S. *TBS in Routine Clinical Practice: Proposals of Use*. For the Medimaps Group. 2012: http://www.roentgen-baden.at/home/wp-content/uploads/TBS%20in%20der%20klinischen%20Routine.pdf (accessed May 2018).

36
Reference Point Indentation

Adolfo Diez-Perez[1] and Joshua N. Farr[2]

[1]Hospital del Mar Institute of Medical Investigation and Department of Internal Medicine,
Autonomous University of Barcelona, Barcelona, Spain
[2]Division of Endocrinology and Kogod Center on Aging, Mayo Clinic College of Medicine,
Rochester, MN, USA

INTRODUCTION

Fracture is the clinical event in osteoporosis that induces morbidity and mortality. It results from the application of a load (trauma) to a bone that exceeds the bone's mechanical resistance (strength). Therefore, deterioration in bone strength decreases the ability to resist even moderate- or low-level impact to the bone, which results in fracture, the hallmark of osteoporosis.

Bone strength depends on three basic components: (i) amount of mineral; (ii) architecture (both at the macro- and microscopic levels); and (iii) tissue characteristics. In clinical practice, the first component can be measured by bone densitometry, most commonly with DXA. The second component is easily assessed at a macroscopic level (bone geometry) by common imaging techniques. However, for the study of microscopic architecture, more sophisticated imaging methods are needed, as reviewed in other sections of this primer. With regards to CT, the better the image resolution, the more accurate the assessment of bone microarchitecture [1]. Finite element analysis of these spatial images can simulate the ability of a bone to resist an impact force, providing a mathematical approach to the assessment of bone strength [2]. A recent method, more accessible to daily clinical practice, is the trabecular bone score (TBS) analysis of structure based on DXA images [3].

The third component, tissue characteristics, is complex and only partially understood. Micro- and nanoscale hierarchical elements, such as collagen maturity, tissue hydration, noncollagen proteins (osteopontin, osteocalcin), sacrificial bonds, microdamage, osteon characteristics, and regional cortical microporosity, play a role along with other factors. The final combination of these elements determines bone tissue mechanical performance [4–6]. Assessing this third component of bone at the tissue level has traditionally required invasive techniques to acquire bone samples and sophisticated analytical approaches to evaluate the samples, making these studies cumbersome and limited to small numbers of samples and only a few specialized research laboratories. Recently, however, a minimally invasive technique called reference point indentation (RPI) [7] has been developed as a novel, feasible measurement of the mechanical properties of bone at the tissue level.

TWO MODALITIES OF RPI: CYCLIC AND IMPACT

The two RPI techniques, cyclic reference point indentation (cRPI) and impact reference point indentation (iRPI), differ in their loading rates and patterns and use of test probes that have slightly different diameters. Accordingly, correlations between parameters obtained by iRPI and cRPI are generally weak or moderate [8,9] and tend to be lower as bone quality decreases [9]. Both techniques analyze the resistance of the outer cortex of a weightbearing bone (ie, the tibia) to the penetration of a conic tip: the deeper the penetration, the less resistant is the bone. However, the bone tissue elements that determine this resistance to penetration are not fully understood and, indeed, likely differ between cRPI (a repeated-loading, fatigue-like challenge) and iRPI (a high-speed, monotonic-like test).

The first series of RPI devices was based on the cyclic microindentation technique, which was originally used

---

Primer on the Metabolic Bone Diseases and Disorders of Mineral Metabolism, Ninth Edition. Edited by John P. Bilezikian.
© 2019 American Society for Bone and Mineral Research. Published 2019 by John Wiley & Sons, Inc.
Companion website: www.wiley.com/go/asbmrprimer

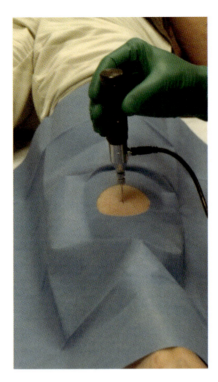

Fig. 36.1. Image of the OsteoProbe impact Reference Point Indenter (Active Life Scientific Inc., Santa Barbara, CA, USA), a handheld microindentation instrument designed for in vivo measurements of the bone material strength index (BMSi) in humans at the midshaft of the anterior tibia. Reproduced from [21] with permission.

in clinical studies both ex vivo on cadaver specimens and in vivo in humans [10] and also more recently (and commonly) in preclinical animal studies. Both the cRPI and iRPI techniques have been described in detail [11]. For cRPI, biomechanical parameters can be calculated, including pre- and post-yield properties and dissipated energy, although the most often reported data from this technique are indentation distances (ie, total indentation distance [TID] and indentation distance increase [IDI]) into the bone, which are inversely related to the material properties of the underlying tissue [12]. Since the conception of this technology, several cRPI instruments have been developed with evolving changes to both hardware and software. The latest version (termed the BioDent®) utilizes repeated, cyclic microindentations applied to the bone over a period of seconds at relatively high frequency, yet low force (2 to 10N). More recently, an iRPI device has been developed for clinical studies (termed the OsteoProbe; Fig. 36.1), which differs from the cRPI devices not only with regards to the intrinsic machinery but also in the measurement of bone tissue material properties. Thus, it is not surprising, as noted earlier, that weak to moderate correlations have been found between the two instruments [8]. In contrast to the BioDent, the OsteoProbe is a handheld device with an impact mechanism, a displacement transducer, and a stainless steel probe that has a 90-degree conical tip. With the probe residing on the bone surface, the iRPI measurement is actuated by compressing the outer housing unit of the OsteoProbe, compressing an internal spring until a force of 10N is achieved, at which time a trigger mechanism initiates an impact (with a maximum force of 40N) into the bone. The distance the probe moves (ie, the IDI) relative to its penetration into a methyl methacrylate reference phantom is defined as the bone material strength index (BMSi). Thus, the farther the probe indents the bone (ie, higher the IDI), the lower or "worse" the bone material properties (ie, BMSi).

PRECLINICAL STUDIES IN ANIMALS

Several preclinical studies have been performed using cRPI in various animal models, including mice [13], rats [14–16], pigs [17], and dogs [18]. To date, while the work in dogs and pigs has focused on examining treatment effects of bone-altering medications and aging on bone material properties, much of the rodent cRPI data has focused on measurement reproducibility and longitudinal changes in untreated animals in order to establish the feasibility of the technique and to help guide the design of future studies. In addition, studies using cRPI in rats modeling diabetes [15] or chronic kidney disease [16] have provided evidence that bone material properties may be altered by these conditions, although more work is needed to establish the underlying mechanisms responsible for these observations.

Overall, the available data suggest that bone material properties derived from cRPI are improved in response to treatment with raloxifene [18] and decline with age [17]. Further, the variability of cRPI measurements tends to be higher in animals [13, 14], which is perhaps related to their smaller bone size. In addition, mounting evidence suggests that some animals (ie, dogs and rats) have lower cRPI parameters as compared with humans [14, 18], although the reason for this observation is currently unclear. Additionally, studies have shown that cRPI parameters (eg, TID and IDI) are consistently lower at follow-up relative to baseline, even in young, healthy animals that have not received any treatments [13]. While the underlying causes for these findings remain incompletely understood, it is possible that the technique itself causes detrimental alterations in the underlying bone material properties of these animals that persist over time. Clearly, additional preclinical studies are needed to establish optimal protocols for preclinical testing and to determine whether cRPI is a reliable technology for assessing in vivo bone material properties in various animal models.

CLINICAL STUDIES IN HUMANS

Accumulating experience in utilizing microindentation technology in various clinical studies around the world has established that RPI is a safe, minimally invasive method for in vivo assessment of BMSi in humans. However, it should be noted that there are still several important unresolved questions regarding the RPI technology that have been discussed at length elsewhere [11, 19]. Nevertheless, when performed correctly, RPI technology holds promise as a novel, innovative approach for measuring clinically relevant indices of bone material properties in vivo in humans. The first two studies in humans used cRPI to show that indices of bone material properties are worse in relatively small cohorts of hip fracture [10] and atypical femoral fracture [20] patients, independently of BMD. Since the time of these initial studies, the research community has transitioned towards the utilization of iRPI in clinical studies.

The first clinical study used iRPI to cross-sectionally examine BMSi in 60 postmenopausal women: 30 diagnosed with type 2 diabetes mellitus (T2DM), and 30 age-matched nondiabetic controls [21]. Data from this study revealed that compared to controls, postmenopausal women with T2DM had significantly lower BMSi: both unadjusted (−11.7%) and following adjustment for BMI (−10.5%), suggesting that impaired bone material properties may contribute to the increased fracture risk observed in these patients [22]. In addition, data from this study showed that the average glycated hemoglobin level over the previous 10 years was negatively correlated with BMSi in women with T2DM, thus highlighting the potential detrimental effects of prolonged hyperglycemia on bone quality [21]. A subsequent cross-sectional study in 16 postmenopausal women with T2DM also found that BMSi was lower (−9.2%) in patients with T2DM as compared to 19 nondiabetic controls. Further, this study reported that increased skin autofluorescence (a crude index of AGEs) was negatively associated with BMSi in women with T2DM, suggesting that accumulation of AGEs may contribute to diabetic skeletal fragility. However, future adequately powered longitudinal studies are needed to confirm these findings.

Additional studies using iRPI have explored measures of BMSi in several different populations of human subjects. For example, Sosa and colleagues [23] compared 42 Norwegian women with 32 Spanish women to establish whether BMSi at least in part explains the reported difference in fracture risk between these populations (ie, Spanish women have ~50% lower risk of hip fracture). Interestingly, on average, Norwegian women had significantly lower BMSi, but paradoxically higher hip BMD as compared to Spanish women, suggesting that worse bone material properties and perhaps factors related to greater stature and/or altered hip geometry could be contributing to the differences in fracture risk between these populations. In another cross-sectional study, Malgo and colleagues [24] measured BMSi in 90 patients with low bone mass (ie, DXA T-score < −1.0) either with ($n=63$) or without ($n=27$) fragility fractures and found that despite comparable BMD values, BMSi was significantly lower in fracture patients as compared to nonfracture controls regardless of the presence of osteopenia or osteoporosis. In addition to fragility fractures, it is also noteworthy that impaired bone material properties may contribute to the incidence of stress fractures because BMSi has been shown to be significantly reduced in women with stress fractures as compared to controls [25]. Furthermore, recent cross-sectional studies in patients infected with the HIV virus [26] or patients who received a kidney transplant [27] have provided evidence that bone material properties may be worse in these populations, potentially contributing to their increased fracture risk.

It should be noted that much of the published RPI data have been obtained cross-sectionally in relatively small cohorts of human subjects. Indeed, to date, there remain very few longitudinal iRPI data in humans. However, in the only longitudinal study published thus far, Mellibovsky and colleagues [28] measured BMSi (at baseline, 7 weeks, and 20 weeks) in a mixed cohort of patients within 4 weeks of initiating glucocorticoid therapy who then received either calcium + vitamin D alone or with risedronate, denosumab, or teriparatide. The data from this study showed that BMSi was reduced over time in the calcium + vitamin D group and unchanged in the risedronate group, whereas it was significantly improved (even after only 7 weeks of therapy) in both the denosumab and teriparatide groups. While intriguing, these findings should be considered preliminary given the small sample size and mixed cohort of patients, which resulted in groups that differed with respect to several important covariates (eg, sex, age, BMD, fracture history, glucocorticoid dosage). It should be noted that this experiment was planned as a proof-of-concept study on the sensitivity of the iRPI technique to detect short-term changes in bone material properties. For this reason, a rapidly deteriorating condition (high dose of glucocorticoids) was chosen and several therapeutic interventions were tested. Perhaps surprising was the finding that significant changes in BMSi were observed by the first follow-up microindentation assessment (ie, in only 7 weeks). Interestingly, this may suggest a rapid, remodeling-independent effect of the interventions on the underlying skeletal components involved in the tissue resistance to indentation. However, these findings should be considered preliminary given the proof-of-concept study design. Nevertheless, the longitudinal iRPI data obtained as part of this study will be important toward the design and execution of future longitudinal trials using microindentation technology. Such studies are direly needed to expand on our current knowledge of iRPI, which is largely based on small, cross-sectional studies.

WHAT DOES RPI MEASURE?

The limited experimental data available associate cortical porosity, water content, and anisotropy with resistance to penetration in the cRPI test [6] and both cRPI and iRPI measurements have been correlated with tissue mineral density and the accumulation of AGEs [7]. These findings indicate that the combined effect of these (and likely other) components of bone tissue at a micro- and nanoscopic levels influences RPI measurements, although in different proportions and combinations for each technique.

One possibility is that RPI measurements on the tibia are correlated with tissue-level mechanical properties throughout the skeleton. In cadaver experiments, cRPI was a significant predictor of femoral bone strength [29]. Furthermore, in surgical specimens from patients with hip fracture, cRPI measurements at the femoral neck have shown significantly worse material bone properties than controls [30], similar to previous tibia microindentation reports in living patients with fracture [10]. Finally, clinical studies have consistently shown that both cRPI and iRPI measurements are poorly correlated with areal BMD measured by DXA [10, 20, 21, 23, 24, 30], and therefore capture independent information on bone material properties. This suggests added value of RPI for the prediction of bone fragility [29, 30], in addition to the correlation between RPI variables and tissue density [9].

Although the relationship between the variables measured by RPI and traditional biomechanical tests has been explored, further research is needed to understand exactly which mechanical properties are measured by microindentation. Indeed, RPI testing conditions are highly divergent from the typical, most often quasi-static, tests used in biomechanical laboratories. Correlations between cRPI and toughness have been demonstrated in some experiments [31, 32] whereas more limited iRPI data have shown no correlation [9], and the translation of RPI parameters to traditional mechanical testing variables has been problematic under various experimental conditions [33].

The bottom line is that the merit of searching for evidence that RPI is comparable to the usual quasi-static tests is debatable because they test notably different biomechanical properties. It has been suggested that, while traditional mechanical tests seem to be representative of structural components such as porosity, RPI appears to better capture tissue composition elements, such as AGEs [9]. Moreover, in vivo testing of the material properties of bone may significantly differ from ex vivo experiments as shown, for example, by the different range of values obtained by cRPI measurement in both conditions [11]. Several limitations of traditional biomechanical testing techniques also exist [34]. For example, a number of alendronate experiments have demonstrated detrimental effects on mechanical characteristics of bone in traditional laboratory experiments under various conditions [35–37], whereas clinical use of the drug consistently reduces fractures [38]. This example highlights the complexity of this type of testing as well as the debatable value of attempting direct translation of the results into clinically helpful information in living patients [39].

POTENTIAL ADVANTAGES OF RPI TECHNOLOGY

Although comparison to traditional mechanical tests has yielded some insights, the "gold standard" for judging the value of RPI testing is its capacity to predict fractures, which is the clinical hallmark of skeletal fragility. The possibility has been raised that iRPI might offer more physiologically relevant information on fracture propensity, compared to biomechanical testing [9]. Therefore, RPI, and particularly iRPI, must be considered in light of in vivo studies in clinical populations of different types in order to establish the value added to bone health and therapeutic interventions assessment [40].

One aspect to be emphasized is the potential ability of iRPI to detect drug-induced changes in tissue mechanical properties before completion of even one bone remodeling cycle. The longitudinal study in glucocorticoid-treated patients showed these effects as soon as 7 weeks after initial exposure to treatment with glucocorticoids and anti-osteoporosis medications [41]. Furthermore, some treatments have cell-independent effects on bone toughness by increasing matrix-bound water, and modifications in collagen nanomorphology also have been described [41]. In dogs, material-level mechanical properties, measured by cRPI, showed improvement after just 6 months of treatment [18].

These results raise a thought-provoking possibility. To date, our understanding of bone physiology and disease has been almost exclusively limited to the bone-remodeling paradigm. Accordingly, the effects of treatments have been assessed from this perspective. Instruments for measuring bone health have provided information on the amount of mineral (density) and, more recently, structure, both of which are highly dependent on the remodeling balance. However, patients with osteoporosis do not follow a homogeneous path to decreased bone strength. For example, patients with diabetes mellitus [21] or HIV infection [26], or those with fractures at near-normal BMD levels [24], show distinct clinical features that complicate the assessment of their fracture risk using the available methods that are clinically accessible. Whether RPI, and particularly iRPI, is a clinically feasible tool for the assessment of these nondensity, nonarchitectural components of bone strength remains to be determined. The preliminary results suggest the potential to add meaningful information to what can be measured in current clinical practice, particularly in patient groups in which fracture propensity has not yet been reasonably explained.

FUTURE NEEDS AND DIRECTIONS

Crucial needs for further investigation of RPI include the following: (i) methodological aspects such as the selection of outliers, inter- and intra-observed variability, and standardization of the measurement method require further development and implementation; (ii) reference values must be established, with the potential caveat that different populations may have different normal ranges [23]; and (iii) the independent ability of RPI in predicting fractures must be evaluated in prospective studies that are sufficiently powered.

CONCLUSION

In conclusion, if RPI techniques are to have a role in our clinical approach to the evaluation of osteoporosis, further research is needed to show that a comprehensive, individualized evaluation of bone health including RPI adds useful information to that obtained by currently available measurement techniques.

REFERENCES

1. Jorgenson BL, Buie HR, McErlain DD, et al. A comparison of methods for in vivo assessment of cortical porosity in the human appendicular skeleton. Bone. 2015;73:167–75.
2. Kopperdahl DL, Aspelund T, Hoffmann PF, et al. Assessment of incident spine and hip fractures in women and men using finite element analysis of CT scans. J Bone Miner Res. 2014;29(3):570–80.
3. Hans D, Goertzen AL, Krieg MA, et al. Bone microarchitecture assessed by TBS predicts osteoporotic fractures independent of bone density: the Manitoba study. J Bone Miner Res. 2011;26(11):2762–9.
4. Fantner GE, Hassenkam T, Kindt JH, et al. Sacrificial bonds and hidden length dissipate energy as mineralized fibrils separate during bone fracture. Nat Mater. 2005;4(8):612–6.
5. Bala Y, Seeman E. Bone's Material Constituents and their Contribution to Bone Strength in Health, Disease, and Treatment. Calcif Tissue Int. 2015;97(3):308–26.
6. Nyman JS, Granke M, Singleton RC, et al. Tissue-Level Mechanical Properties of Bone Contributing to Fracture Risk. Curr Osteoporos Rep. 2016;14(4):138–50.
7. Hansma P, Yu H, Schultz D, et al. The tissue diagnostic instrument. Rev Sci Instrum. 2009;80(5):054303.
8. Karim L, Van Vliet M, Bouxsein ML. Comparison of cyclic and impact-based reference point indentation measurements in human cadaveric tibia. Bone. 2018;106:90–5.
9. Abraham AC, Agarwalla A, Yadavalli A, et al. Microstructural and compositional contributions towards the mechanical behavior of aging human bone measured by cyclic and impact reference point indentation. Bone. 2016;87:37–43.
10. Diez-Perez A, Guerri R, Nogues X, et al. Microindentation for in vivo measurement of bone tissue mechanical properties in humans. J Bone Miner Res. 2010;25(8):1877–85.
11. Allen MR, McNerny EM, Organ JM, et al. True Gold or Pyrite: A Review of Reference Point Indentation for Assessing Bone Mechanical Properties In Vivo. J Bone Miner Res. 2015;30(9):1539–50.
12. Hansma P, Turner P, Drake B, et al. The bone diagnostic instrument II: indentation distance increase. Rev Sci Instrum. 2008;79(6):064303.
13. Srisuwananukorn A, Allen MR, Brown DM, et al. In vivo reference point indentation measurement variability in skeletally mature inbred mice. Bonekey Rep. 2015;4:712.
14. Allen MR, Newman CL, Smith E, et al. Variability of in vivo reference point indentation in skeletally mature inbred rats. J Biomech. 2014;47(10):2504–7.
15. Reinwald S, Peterson RG, Allen MR, et al. Skeletal changes associated with the onset of type 2 diabetes in the ZDF and ZDSD rodent models. Am J Physiol Endocrinol Metab. 2009;296(4):E765–74.
16. Newman CL, Moe SM, Chen NX, et al. Cortical bone mechanical properties are altered in an animal model of progressive chronic kidney disease. PLoS One. 2014;9(6):e99262.
17. Rasoulian R, Raeisi Najafi A, Chittenden M, et al. Reference point indentation study of age-related changes in porcine femoral cortical bone. J Biomech. 2013;46(10):1689–96.
18. Aref M, Gallant MA, Organ JM, et al. In vivo reference point indentation reveals positive effects of raloxifene on mechanical properties following 6 months of treatment in skeletally mature beagle dogs. Bone. 2013;56(2):449–53.
19. Jepsen KJ, Schlecht SH. Biomechanical mechanisms: resolving the apparent conundrum of why individuals with type II diabetes show increased fracture incidence despite having normal BMD. J Bone Miner Res. 2014;29(4):784–6.
20. Guerri-Fernandez RC, Nogues X, Quesada Gomez JM, et al. Microindentation for in vivo measurement of bone tissue material properties in atypical femoral fracture patients and controls. J Bone Miner Res. 2013;28(1):162–8.
21. Farr JN, Drake MT, Amin S, et al. In vivo assessment of bone quality in postmenopausal women with type 2 diabetes. J Bone Miner Res. 2014;29(4):787–95.
22. Melton LJ, Leibson CL, Achenbach SJ, et al. Fracture risk in type 2 diabetes: update of a population-based study. J Bone Miner Res. 2008;23:1334–42.
23. Duarte Sosa D, Vilaplana L, Guerri R, et al. Are the High Hip Fracture Rates Among Norwegian Women Explained by Impaired Bone Material Properties? J Bone Miner Res. 2015;30(10):1784–9.
24. Malgo F, Hamdy NA, Papapoulos SE, et al. Bone material strength as measured by microindentation in vivo is decreased in patients with fragility fractures independently of bone mineral density. J Clin Endocrinol Metab. 2015;100(5):2039–45.
25. Duarte Sosa D, Fink Eriksen E. Women with previous stress fractures show reduced bone material strength. Acta Orthop. 2016;87(6):626–31.
26. Guerri-Fernandez R, Molina D, Villar-Garcia J, et al. Brief Report: HIV Infection Is Associated With Worse Bone Material Properties, Independently of Bone Mineral Density. J Acquir Immune Defic Syndr. 2016;72(3):314–8.
27. Perez-Saez MJ, Herrera S, Prieto-Alhambra D, et al. Bone density, microarchitecture and tissue quality long-term after kidney transplant. Transplantation. 2017;101(6):1290–4.
28. Mellibovsky L, Prieto-Alhambra D, Mellibovsky F, et al. Bone Tissue Properties Measurement by Reference Point

28. ...Indentation in Glucocorticoid-Induced Osteoporosis. J Bone Miner Res. 2015;30(9):1651–6.
29. Abraham AC, Agarwalla A, Yadavalli A, et al. Multiscale Predictors of Femoral Neck In Situ Strength in Aging Women: Contributions of BMD, Cortical Porosity, Reference Point Indentation, and Nonenzymatic Glycation. J Bone Miner Res. 2015;30(12):2207–14.
30. Jenkins T, Coutts LV, D'Angelo S, et al. Site-Dependent Reference Point Microindentation Complements Clinical Measures for Improved Fracture Risk Assessment at the Human Femoral Neck. J Bone Miner Res. 2016; 31(1):196–203.
31. Granke M, Coulmier A, Uppuganti S, et al. Insights into reference point indentation involving human cortical bone: sensitivity to tissue anisotropy and mechanical behavior. J Mech Behav Biomed Mater. 2014;37:174–85.
32. Gallant MA, Brown DM, Organ JM, et al. Reference-point indentation correlates with bone toughness assessed using whole-bone traditional mechanical testing. Bone. 2013;53(1):301–5.
33. Krege JB, Aref MW, McNerny E, et al. Reference point indentation is insufficient for detecting alterations in traditional mechanical properties of bone under common experimental conditions. Bone. 2016;87:97–101.
34. Wallace RJ, Pankaj P, Simpson AH. Major source of error when calculating bone mechanical properties. J Bone Miner Res. 2014;29(12):2697.
35. Allen MR, Reinwald S, Burr DB. Alendronate reduces bone toughness of ribs without significantly increasing microdamage accumulation in dogs following 3 years of daily treatment. Calcif Tissue Int. 2008;82(5): 354–60.
36. Bala Y, Depalle B, Farlay D, et al. Bone micromechanical properties are compromised during long-term alendronate therapy independently of mineralization. J Bone Miner Res. 2012;27(4):825–34.
37. Bajaj D, Geissler JR, Allen MR, et al. The resistance of cortical bone tissue to failure under cyclic loading is reduced with alendronate. Bone. 2014;64:57–64.
38. Mackey DC, Black DM, Bauer DC, et al. Effects of antiresorptive treatment on nonvertebral fracture outcomes. J Bone Miner Res. 2011;26(10):2411–8.
39. Allen MR, Burr DB. Bisphosphonate effects on bone turnover, microdamage, and mechanical properties: what we think we know and what we know that we don't know. Bone. 2011;49(1):56–65.
40. Farr JN, Amin S, Khosla S. Regarding "True Gold or Pyrite: A Review of Reference Point Indentation for Assessing Bone Mechanical Properties In Vivo". J Bone Miner Res. 2015;30(12):2325–6.
41. Gallant MA, Brown DM, Hammond M, et al. Bone cell-independent benefits of raloxifene on the skeleton: a novel mechanism for improving bone material properties. Bone. 2014;61:191–200.

37

Biochemical Markers of Bone Turnover in Osteoporosis

Pawel Szulc[1], Douglas C. Bauer[2], and Richard Eastell[3]

[1]*INSERM UMR 1033, University of Lyon, Hospices Civils de Lyon, Lyon, France*
[2]*Departments of Medicine and Epidemiology & Biostatistics, University of California, San Francisco, San Francisco, CA, USA*
[3]*Department of Oncology and Metabolism, University of Sheffield, Sheffield, UK*

INTRODUCTION

Bone turnover is characterized by bone resorption and bone formation [1]. Both activities are coupled at a BMU (bone remodeling unit, BRU). During bone resorption, mineral dissolution and bone matrix catabolism by osteoclasts result in the formation of resorptive cavity and the release of bone matrix components. Then, during bone formation, osteoblasts synthesize bone matrix which fills in the resorption cavity and undergoes mineralization.

Bone turnover markers (BTMs) comprise markers of bone formation and of bone resorption (Table 37.1). Serum or plasma PINP and CTX-I are the referent markers of bone formation and resorption [2]. PINP is derived from posttranslational cleavage of type I procollagen. Serum PINP originates primarily from bone, does not show circadian variation, and increases during bone formation–stimulating therapy. CTX-I is a product of breakdown of type I collagen. It is specific for bone and decreases during antiresorptive treatment. CTX-I occurs in its native (α) and β-isomerized forms which undergo racemization (D- and L-forms). Bone alkaline phosphatase (ALP) and TRACP5b are enzymes reflecting the metabolic activity of osteoblasts and osteoclasts. Other BTMs are bone matrix components released during bone formation or resorption.

ANALYTICAL AND PREANALYTICAL VARIABILITY

The analytical variability (assessed by the intra-assay and interassay coefficients of variation [CV]) depends on the BTM, the assay, and the technician's expertise [3]. Monoclonal antibodies permit specific measurements of the BTMs. Automated analyzers permit a rapid, convenient, automated, and precise measurement of BTMs [4].

The preanalytical variability comprises many factors (Table 37.2). Circadian rhythm influences BTM variability, mainly serum CTX-I that has its peak after midnight and nadir in the afternoon [5, 6]. Food intake influences bone resorption. The postprandial drop in blood CTX-I may be stimulated by glucose inducing the intestinal synthesis of glucagon-like peptide 2 [7]. Therefore, blood for the measurement of CTX-I should be collected in the fasting state in the morning. For urinary BTMs, a spot sample corrected for urine creatinine (preferably second morning void) may be collected conveniently. However, the 24-hour collection best reflects the overall bone turnover because the excretion of BTMs per mg of urinary creatinine is artifactually overestimated in the case of low creatinine excretion due to sarcopenia. The amount of BTMs per glomerular filtrate volume assumes that their glomerular filtration is the same as that of creatinine and there is no tubular reuptake of the BTM.

Primer on the Metabolic Bone Diseases and Disorders of Mineral Metabolism, Ninth Edition. Edited by John P. Bilezikian.
© 2019 American Society for Bone and Mineral Research. Published 2019 by John Wiley & Sons, Inc.
Companion website: www.wiley.com/go/asbmrprimer

Table 37.1. Biochemical bone turnover markers.

Bone formation
  Osteocalcin (OC)
  Bone alkaline phosphatase (bone ALP)
  N-terminal propeptides of type I procollagen (PINP)
  C-terminal propeptides of type 1 procollagen (PICP)

Bone resorption
  C-terminal cross-linking telopeptides of type I collagen (CTX-I)
  N-terminal cross-linking telopeptides of type I collagen (NTX-I)
  C-terminal cross-linking telopeptide of type I collagen generated
  by matrix metalloproteinases (CTX-MMP, ICTP)
  Helical peptide 620-633 of the α1 chain
  Deoxypyridinoline (DPD)
  Isoform 5b of tartrate resistant acid phosphatase (TRACP5b)

Table 37.2. Determinants of the preanalytical variability of bone turnover.

Modifiable determinants
Circadian variation
Menstrual variation
Seasonal variation
Fasting and food intake (particularly serum CTX-I)
Exercise and physical activity
Lifestyle factors (smoking, alcohol abuse)

Determinants that cannot be easily modified
Age
Sex
Menopausal status
Vitamin D deficit and secondary hyperparathyroidism
Short- and long-term day-to-day variation
Diseases characterized by an acceleration of bone turnover
  Primary hyperparathyroidism
  Thyrotoxicosis
  Acromegaly
  Paget disease
  Bone metastases
  Hypogonadism (depending on the severity)

Diseases characterized by a dissociation of bone turnover
  Cushing's disease
  Multiple myeloma

Diseases characterized by a low bone turnover
  Hypothyroidism
  Hypoparathyroidism
  Hypopituitarism
  Growth hormone deficit

Renal impairment (depending on stage)
Recent fracture
Depression
HIV infection

Chronic diseases associated with limited mobility
  Stroke
  Hemiplegia
  Dementia
  Alzheimer disease
  Sarcopenia

Medications
  Oral corticosteroids
  Inhaled corticosteroids (only osteocalcin)
  Aromatase inhibitors (anti-aromatases)
  Oral contraceptives
  Gonadoliberin agonists
  Antiepileptic drugs
  Thiazolidinediones
  Nucleoside reverse transcriptase inhibitors (mainly tenofovir)
  Protease inhibitors
  Heparin
  Vitamin K antagonists

Bone metabolism is influenced by the vitamin D and calcium status. BTMs are increased in the institutionalized and home-bound elderly who have lower 25OHD and higher PTH levels versus those who are ambulatory. The vitamin D–deficient elderly also have evident seasonal variation of BTM levels.

BTM levels are strongly influenced by a major fracture [8, 9]. Osteocalcin decreases during the first hours after fracture because of stress-related cortisol secretion. Subsequently, bone formation and resorption increase, reflecting the healing of the fracture. BTM levels may be increased by 50% to 100% for 4 months after fracture, and then decrease for at least 1 year.

CONDITIONS THAT IMPACT BTM MEASUREMENTS

Primary hyperparathyroidism (PHPT)

PHPT is characterized by the accelerated bone turnover [10]. In early normocalcemic PHPT BTMs are in the normal range [11]. In the asymptomatic hypercalcemic PHPT BTMs are on average at the upper limit of the normal range [12]. In both groups, BTMs remain stable. In the symptomatic PHPT, all BTMs are increased and correlated positively with serum PTH. After surgery a decrease in PTH and calcium levels is followed by a reduction in bone resorption and, later on, a slow decrease in bone formation [10]. In PHPT patients, antiresorptive drugs (bisphosphonates) suppress BTM levels but not PTH [10].

Paget disease

BTMs (mainly bone ALP, PINP, NTX-I) are elevated in active untreated Paget disease [13]. However, normal BTM levels do not exclude Paget disease. BTM levels are correlated with scintigraphic activity of the disease, being higher in the polyostotic versus monostotic disease. BTMs decrease after treatment with bisphosphonates, most promptly after zoledronic acid. NTX-I decreases first. Later on, PINP and NTX-I correlate with the residual activity of the disease. Bone ALP, PINP, and urinary NTX-I are the most valuable BTMs in the clinical assessment of Paget disease. Bone ALP is a sensitive marker for Paget disease relapse. However, total ALP is inexpensive and useful in the assessment of Paget disease in patients without liver pathology.

Diabetes mellitus

Data on BTMs in diabetes mellitus (DM) are inconsistent. The inconsistencies may be related to the type of DM (type 1 [T1DM] versus type 2 [T2DM]), duration, degree of metabolic control, accompanying factors, renal impairment, and treatments. Recently diagnosed T1DM may be associated with low BTM levels which normalize during treatment, but data are inconsistent [14]. In T2DM serum osteocalcin levels are slightly decreased, whereas data for other BTMs are discordant [15].

Renal osteodystrophy

Renal osteodystrophy is an alteration of bone morphology in the late stage of chronic kidney disease – mineral and bone disorder (CKD-MBD). It presents several clinical forms with high or low bone turnover. BTMs not influenced by renal function are useful in CKD [16]. Bone ALP is degraded by the liver. Intact PINP is taken up mainly by liver endothelial cells. Both may be useful as bone formation markers in CKD. By contrast, PINP monomers accumulate, leading to disproportionate increase in total PINP. TRAP5b may be useful for assessment of bone resorption. It reflects osteoclast number and activity. Its blood levels are not affected by renal function. BTM levels correlate moderately with histomorphometric parameters of bone turnover and fracture risk [17, 18]. Other BTMs accumulate proportionally to kidney function loss and are not useful in CKD.

Bone metastases

BTM levels are increased in patients with solid tumor bone metastases and correlate with their spread [19, 20]. Among available BTMs, ICTP and native α-α-CTX-I appear to be the most sensitive markers of bone involvement [19, 21]. In cancer patients without known bone metastases, increased BTM levels indicate a higher probability of the presence of bone metastases in bone scintigraphy [22]. In cancer patients with known bone metastases, higher BTM levels are associated with higher risk of skeletal-related events (SREs) and higher mortality [23].

In studies of the treatments of bone metastases (pharmaceutical agents, radiotherapy), BTMs may be used as early markers of development and progression of bone metastases or SREs. In clinical practice, BTMs may be used for the early detection of bone metastases and for identification of candidates for chemotherapy combined with antiresorptive treatment. In patients with bone metastases antiresorptive treatment promptly decreases BTM levels. However, higher BTM levels (pre- or on-treatment) predict SREs, bone disease progression, and death.

Multiple myeloma

In multiple myeloma (MM) proliferation of a plasma cell clone leads to rapid bone resorption which results in osteolysis and high levels of bone resorption markers (ICTP, CTX-I, NTX-I, TRAP5b) [24]. Parallel bone formation suppression leads to low levels of osteocalcin and bone ALP. Bone resorption markers are increased in newly diagnosed MM, especially in its diffuse or lytic forms [25]. Some MM treatments (bortezomib) decrease bone resorption and increase bone formation. Antiresorptive therapy also reduces bone resorption. By contrast, BTMs do not decrease in treatment-refractory MM and increase further during relapse. Serum ICTP and CTX-I may rise 3 to 6 months prior to MM relapse [24, 25]. Interpretation of BTM levels should be cautious in MM patients with poor renal function.

Medications

The effect of glucocorticoids (GCs) on bone depends on their type, route of administration, absorption, treatment duration, and the underlying disease. Endogenous and exogenous GCs inhibit bone formation [26]. The decrease in osteocalcin level is rapid and followed by a delayed and milder decrease in PICP and PINP. Data on bone resorption are less consistent which may be related to the underlying disease. Patients with endogenous cortisol excess have higher bone resorption [27]. GCs are usually used in inflammatory diseases associated with high bone resorption induced by inflammatory cytokines. In these patients GCs directly stimulate bone resorption and reduce it indirectly by inhibiting inflammation. Low-dose prednisone (5 mg/day) lowers bone formation but not resorption. Inhaled GCs decrease osteocalcin levels in a dose- and drug-dependent manner without effect on other BTMs [28]. Withdrawal of chronic GC therapy is followed by an increase in osteocalcin levels without change in the levels of other BTMs.

Aromatase inhibitors reduce the residual secretion of estrogens. GnRH agonists inhibit the secretion of sex steroids. Both groups of drugs lead to accelerated bone turnover. The expected increase in bone turnover is prevented by concomitant antiresorptive treatment.

Premenopausal women taking oral contraceptives have lower BTMs; however, changes in BTM levels slightly depend on the composition of combined oral contraceptives [29].

BTMs AND REFERENCE VALUES

Reference intervals for BTMs reported from several studies vary because of differences in age and sex of subjects and inclusion criteria. However, data from healthy, premenopausal women are quite consistent and offer the possibility to develop a *T*-score approach similar to BMD [30, 31].

BONE TURNOVER RATE AND BONE LOSS

In young adults, the quantity of bone replaced by formation is roughly equal to that removed by resorption. However, after menopause and in diseases associated with rapid bone loss, bone resorption increases but bone formation does not match resorption and there is a net bone loss. Among postmenopausal women and older men, higher baseline BTM levels are associated with faster bone loss. Thus, bone turnover rate seems to determine the subsequent bone loss. However, for a given BTM level, there is a large scatter of values of bone loss [32]. Thus, BTMs cannot be used for the prediction of accelerated bone loss at the individual level in clinical practice.

BONE TURNOVER RATE AND FRACTURE RISK

In some, not all, prospective studies of untreated individuals, increased BTM levels predict fractures independently of age, BMD, and prior fracture [33–35]. This pattern has been found in postmenopausal and elderly women, but not in men or frail elderly. BTM levels predict major osteoporotic fractures (vertebra, hip, multiple fractures), but not minor peripheral fractures. BTMs predict fractures during short-term follow-up (<5 years) but not in longer studies. In various studies fracture risk was predicted mainly by urine bone resorption markers and bone ALP. In a recent meta-analysis PINP and CTX-I predicted fracture [36]. However, this analysis is based on poorly controlled models not adjusted for BMD.

High bone turnover is associated with lower BMD, faster bone loss, cortical thinning, higher cortical porosity, and poor trabecular connectivity [37]. The remaining bone sustains higher stress, leading to rapid fatigue of bone tissue and further loss of its strength. Resorption cavities trigger stress risers, leading to local weakening of a trabecula [38]. Shorter periods between remodeling cycles leave less time for the posttranslational modifications of bone matrix proteins (cross-linking, β-isomerization of type I collagen).

BTMs may help identify women who benefit the most from anti-osteoporotic treatment. However, data on BTMs and fracture risk should be interpreted cautiously. BTM results may be inaccurate with fluctuation in renal function and improper collection. In patients with sarcopenia urinary bone resorption markers may be overestimated and reflect the risk of falling rather than bone status.

Before widespread clinical use or incorporation in risk prediction models, the use of BTMs for fracture prediction requires further standardization concerning time of collection of samples, choice of BTM, definition of clinically valid thresholds, and duration of follow-up for which BTMs may be valid.

BONE TURNOVER RATE AND MONITORING

BTMs reflect the metabolic effect of drugs on bone turnover, and may help to establish optimal dose and predict treatment-related increase in BMD and reduction in fracture risk. Thus, BTMs have the potential to be helpful in clinical management of osteoporosis.

Metabolic effect

Changes in BTM levels depend on the mechanism of action of the drug. Antiresorptive drugs (bisphosphonates, denosumab) rapidly decrease bone resorption. As bone formation continues in BMUs activated before treatment, bone formation markers are stable for a couple of weeks and decrease when osteoblasts fill in the lower number of BMUs formed after the beginning of the treatment. BMD increases rapidly during the early period, when bone resorption is reduced and bone formation is stable. The decrease in BTMs depends on the degree and the rapidity of inhibition of bone resorption determined by the mechanism of action of the drug, its dose, and route of administration, eg, intravenous bisphosphonates or subcutaneous denosumab decrease BTM levels faster than orally administered agents.

Cathepsin K (CatK), cysteine protease expressed by osteoclasts, degrades collagen under acidic conditions [39, 40]. CatK inhibitors reduce the catabolism of type I collagen. During bone resorption, collagen is degraded first by MMPs, then by CatK. MMPs release CTX-MMP [22]. CatK degrades CTT-MMP to release CTX-I and NTX-I. Thus, CatK inhibitors decrease the levels of CTX-I and NTX-I, whereas CTX-MMP is not catabolized and its serum level increases. CatK inhibitors reduce the activity of osteoclasts, not their number. Therefore, TRACP5b level, which reflects osteoclast number, decreases during treatment with bisphosphonates or denosumab, but not during CatK inhibitor treatment [41, 42]. As the osteoclasts stimulate osteoblast recruitment and differentiation [43], bone formation remains relatively stable during treatment with CatK inhibitors.

Bone formation–stimulating drugs, eg, recombinant hPTH(1–34) (teriparatide) and PTH(1–84), induce a rapid increase in bone formation (mainly PINP) followed by an increase in bone resorption [44, 45]. Abaloparatide (an analog of PTH-related peptide) increases PINP similarly to teriparatide [46]. After 1 month of treatment, PINP levels decreased and were lower than those observed during teriparatide treatment. CTX-I levels were similar to those found in the placebo group.

This phase, when bone formation increases and resorption is still low, is called the "anabolic window." During this phase, BMD increases rapidly, mainly in trabecular bone.

Humanized sclerostin monoclonal antibody induces a rapid, dose-dependent increase in bone formation and a milder, transient decrease in serum CTX-I [47]. The first months of treatment are characterized by the largest dissociation between increased bone formation and decreased bone resorption followed by a rapid increase in BMD which slows later on [48].

Strontium ranelate slightly increases bone ALP and slightly lowers serum CTX-I at the beginning of therapy and then both plateau throughout treatment [49].

BTMs and start of treatment for osteoporosis

BTMs help to establish the optimal dose of anti-osteoporotic drugs because the treatment-related changes in BTMs are more rapid compared with BMD. Higher doses of antiresorptive agents are associated with lower steady state BTM level and greater increases in BMD. Transdermal 17β-estradiol, SERMs, and oral bisphosphonates induce a dose-dependent decrease in bone resorption (maximal after 3 months) and bone formation (maximal after 6 months), followed by an increase in BMD. The first dose of treatment with denosumab or intravenous bisphosphonates decreases bone resorption dose-dependently [50]. Bone formation–stimulating PTH(1–84) induces a dose-dependent increase in bone formation [51].

BTMs may also be useful for the assessment of the therapeutic equivalence of various doses of the same drug. A similar decrease in BTM levels in two groups of patients treated with different regimens of the same drug suggests that both regimens have similar efficacy [52, 53].

BTM levels and therapeutic efficacy of anti-osteoporotic treatment

BTMs (mainly PINP) allow detection of the response to antiresorptive therapy and between-person response to treatment [54]. Relative reduction of the nonspine fracture risk induced by alendronate was greater in women with higher baseline PINP levels [55]. Relative fracture risk reduction induced by risedronate or teriparatide did not depend on pretreatment BTM level [56, 57]. However, because untreated women with higher BTM levels had higher fracture incidence, more fractures were avoided in women with high bone turnover.

The early treatment-induced decrease in BTM levels is associated with long-term increase in BMD and antifracture efficacy of antiresorptive therapy [58, 59]. For a given decrease in BTM levels and for a given on-treatment BTM level, the vertebral fracture incidence was similar in the active treatment and placebo groups [59, 60]. Early teriparatide-induced increase in BTMs is correlated positively with subsequent increase in BMD, mainly trabecular volumetric BMD, but not with fracture risk [61].

BTM levels after discontinuation of antifracture treatments

Bisphosphonates are accumulated in bone and not metabolized. Thus, the lower the overall dose of bisphosphonates, the sooner the BTMs return to baseline. After the withdrawal of alendronate administered for several years, BTMs increased and BMD decreased, but slowly [62]. Hormone replacement therapy, denosumab, and CatK inhibitors do not accumulate in bone. Their withdrawal is followed by a rapid increase in the BTMs, which may attain values exceeding pretreatment levels [63–65]. This increase is followed by a decrease in BMD. Withdrawal of PTH(1–84) after 1 year of treatment was followed by a return of BTM levels to baseline values and a decrease in trabecular volumetric BMD [66].

After denosumab withdrawal, BTMs return to or even surpass pretreatment levels within months, and retreatment with the same agent 12 months after discontinuation rapidly reduces BTM levels to the values observed in patients treated continuously [64]. After teriparatide withdrawal followed by 12 months without treatment, retreatment with the same agent increased BTM levels [67]. However, the increase was much lower compared with the initial treatment in the treatment-naïve subjects.

BTMs and treatment monitoring at the individual level

The goal of monitoring of antiresorptive therapy is to assess the degree of decrease in bone turnover. Changes in BTM level that exceed the least significant change (determined by interassay CV of BTM assay and within-subject CV) exceed random BTM variability and likely represent a true biological effect of treatment. Ideally, the on-treatment BTM level should be below the mean for premenopausal women. If BTMs are not available before starting therapy, a BTM value below the young healthy mean level (e.g. 35 μg/L for P1NP) is associated with adequate response [68].

Poor adherence during anti-osteoporotic treatment leads to a higher fracture risk [69]. The better the adherence, the greater is the average decrease in bone turnover, but it is unclear if the association is strong enough to be clinically useful [70]. Of note, persistence with antiresorptive treatment was significantly better in women who received positive information, corresponding to a substantial decrease in the NTX-I level [71, 72].

Available data are not sufficient to evaluate whether BTM measurement may be useful for the identification of patients at high risk of atypical subtrochanteric fracture or osteonecrosis of the jaw [73, 74]. Similarly, it is not possible to establish a threshold value of BTMs warranting discontinuation or reinitiation of treatment.

BTMs IN MEN

In boys, the growth spurt starts later and lasts longer than in girls. Young men enter the phase of consolidation (formation of peak BMD after growth arrest) later than women. At the age of 20 to 25 years, men have BTM levels higher than women because they have more active bone turnover in longer and wider bones. Then BTMs decrease and attain their lowest levels at about 50 years of age [75–77]. After age 60, bone formation is stable or increases slightly, whereas bone resorption increases. In older men, urinary DPD and serum CTX-MMP increase whereas serum CTX-I is stable, which reflects relative activities of various enzymes involved in the degradation of collagen [76, 78].

Men with high bone turnover have lower BMD and poor cortical microarchitecture [76, 79]. It indicates that in men increased bone resorption determines age-related bone loss. Elderly men with high BTM levels have faster bone loss; however, the association is weak [80–82]. Large prospective cohort studies showed that BTMs do not predict fractures in elderly men [80, 81]. In a nested case-control study, increased CTX-MMP level was associated with a higher incidence of clinical fracture [83].

BTMs and anti-osteoporotic treatment in men

Testosterone replacement therapy (TRT) decreases bone resorption, if normal testosterone level has been achieved [84, 85]. Bone resorption decreases promptly, but decrease in urinary excretion per mg creatinine may be masked by the increase in muscle mass. Bone formation increases at the beginning of TRT (direct stimulation), levels off, and finally decreases because of the slowdown of bone turnover.

In osteoporotic men, hypogonadal men, HIV-infected men, and men after stroke or cardiac transplantation, bisphosphonates and denosumab decrease BTM levels to a similar degree as in women [86–91]. However, in men receiving androgen–deprivation therapy for prostate cancer, denosumab decreased BTM levels less than in women [92].

In men, teriparatide increased bone formation (PINP) after 1 month and bone resorption after 3 months of treatment [93]. After teriparatide withdrawal, BTMs decreased [94]. In men treated with alendronate for 6 months, teriparatide increased bone formation (but less than in men treated with teriparatide alone) and (more weakly) serum NTX-I concentration which returned to baseline [95]. Sclerostin monoclonal antibody and CatK inhibitors induce in men changes in BTMs similar to those observed in women [96, 97].

CONCLUSION

BTMs improve our understanding of the relationship between bone turnover, BMD, bone fragility, and the effect of anti-osteoporotic treatment (biological mechanism, time course, antifracture efficacy). Data on BTMs show that the rate of bone turnover (spontaneous or modified by therapy) is independently associated with bone fragility. From a clinical point of view, BTM measurement may help to identify postmenopausal women at high risk of fracture and improve persistence with antiresorptive treatment. Thus, the use of BTMs may improve the cost-effectiveness of anti-osteoporotic treatment.

REFERENCES

1. Marti J, Seeman E. Bone remodelling: its local regulation and the emergence of bone fragility. Best Pract Res Clin Endocrinol Metab. 2008;22:701–22.
2. Vasikaran S, Cooper C, Eastell R, et al. International Osteoporosis Foundation and International Federation of Clinical Chemistry and Laboratory Medicine Position on bone marker standards in osteoporosis. Clin Chem Lab Med. 2011;49(8):1271–4.
3. Schafer AL, Vittinghoff E, Ramachandran R, et al. Laboratory reproducibility of biochemical markers of bone turnover in clinical practice. Osteoporos Int. 2010;21:439–45.
4. Garnero P, Vergnaud P, Hoyle N. Evaluation of a fully automated serum assay for total N-terminal propeptide of type I collagen in postmenopausal osteoporosis. Clin Chem. 2008;54:188–96.
5. Szulc P, Delmas PD. Biochemical markers of bone turnover: potential use in the investigation and management of postmenopausal osteoporosis. Osteoporos Int. 2008;19:1683–704.
6. Qvist P, Christgau C, Pedersen BJ, et al. Circadian variation in the serum concentration of C-terminal telopeptide of type I collagen (serum CTx): effects of gender, age, menopausal status, posture, daylight, serum cortisol, and fasting. Bone. 2002;31:57–61.
7. Yavropoulou MP, Tomos K, Tsekmekidou X, et al. Response of biochemical markers of bone turnover to oral glucose load in diseases that affect bone metabolism. Eur J Endocrinol. 2011;164:1035–41.
8. Ivaska KK, Gerdhem P, Akesson K, et al. Effect of fracture on bone turnover markers: a longitudinal study comparing marker levels before and after injury in 113 elderly women. J Bone Miner Res. 2007;22:1155–64.
9. Stoffel K, Engler H, Kuster M, et al. Changes in biochemical markers after lower limb fractures. Clin Chem. 2007;53:131–4.
10. Costa AG, Bilezikian JP. Bone turnover markers in primary hyperparathyroidism. J Clin Densitom. 2013;16:22–7.
11. Lowe H, McMahon DJ, Rubin MR, et al. Normocalcemic primary hyperparathyroidism: further characterization of a new clinical phenotype. J Clin Endocrinol Metab. 2007;92:3001–5.
12. Silverberg SJ, Gartenberg F, Jacobs TP, et al. Longitudinal measurements of bone density and biochemical indices in untreated primary hyperparathyroidism. J Clin Endocrinol Metab. 1995;80:723–8.

13. Al Nofal AA, Altayar O, BenKhadra K, et al. Bone turnover markers in Paget's disease of the bone: A Systematic review and meta-analysis. Osteoporos Int. 2015;26:1875–91.
14. Starup-Linde J, Vestergaard P. Biochemical bone turnover markers in diabetes mellitus - A systematic review. Bone. 2016;82:69–78.
15. Gilbert MP, Pratley RE. The impact of diabetes and diabetes medications on bone health. Endocr Rev. 2015;36:194–213.
16. Mazzaferro S, Tartaglione L, Rotondi S, et al. News on biomarkers in CKD-MBD. Semin Nephrol. 2014;34:598–611.
17. Sprague SM, Bellorin-Font E, Jorgetti V,. Diagnostic Accuracy of Bone Turnover Markers and Bone Histology in Patients With CKD Treated by Dialysis. Am J Kidney Dis. 2016;67:559–66.
18. Malluche HH, Davenport DL, Cantor T, et al. Bone mineral density and serum biochemical predictors of bone loss in patients with CKD on dialysis. Clin J Am Soc Nephrol. 2014;9:1254–62.
19. Voorzanger-Rousselot N, Juillet F, Mareau E, et al. Association of 12 serum biochemical markers of angiogenesis, tumour invasion and bone turnover with bone metastases from breast cancer: a crossectional and longitudinal evaluation. Br J Cancer. 2006;95:506–14.
20. Leeming DJ, Koizumi M, Qvist P, et al. Serum N-Terminal Propeptide of Collagen Type I is Associated with the Number of Bone Metastases in Breast and Prostate Cancer and Correlates to Other Bone Related Markers. Biomark Cancer. 2011;3:15–23.
21. Leeming DJ, Delling G, Koizumi M, et al. Alpha CTX as a biomarker of skeletal invasion of breast cancer: immunolocalization and the load dependency of urinary excretion. Cancer Epidemiol Biomarkers Prev. 2006;15:1392–5.
22. Lumachi F, Basso SM, Camozzi V, et al. Bone turnover markers in women with early stage breast cancer who developed bone metastases. A prospective study with multivariate logistic regression analysis of accuracy. Clin Chim Acta. 2016;460:227–30.
23. López-Carrizosa MC, Samper-Ots PM, Pérez AR. Serum C-telopeptide levels predict the incidence of skeletal-related events in cancer patients with secondary bone metastases. Clin Transl Oncol. 2010;12:568–73.
24. Pecoraro V, Roli L, Germagnoli L, et al. The prognostic role of bone turnover markers in multiple myeloma patients: The impact of their assay. A systematic review and meta-analysis. Crit Rev Oncol Hematol. 2015;96:54–66.
25. Ting KR, Brady JJ, Hameed A, et al. Clinical utility of C-terminal telopeptide of type 1 collagen in multiple myeloma. Br J Haematol. 2016;173:82–8.
26. Dovio A, Perazzolo L, Osella G, et al. Immediate fall of bone formation and transient increase of bone resorption in the course of high-dose, short-term glucocorticoid therapy in young patients with multiple sclerosis. J Clin Endocrinol Metab. 2004;89:4923–8.
27. Chiodini I, Carnevale V, Torlontano M, et al. Alterations of bone turnover and bone mass at different skeletal sites due to pure glucocorticoid excess: study in eumenorrheic patients with Cushing's syndrome. J Clin Endocrinol Metab. 1998;83:1863–7.
28. Richy F, Bousquet J, Ehrlich GE, et al. Inhaled corticosteroids effects on bone in asthmatic and COPD patients: a quantitative systematic review. Osteoporos Int. 2003;14:179–90.
29. Herrmann M, Seibel MJ. The effects of hormonal contraceptives on bone turnover markers and bone health. Clin Endocrinol (Oxf). 2010;72:571–83.
30. Glover SJ, Garnero P, Naylor K, et al. Establishing a reference range for bone turnover markers in young, healthy women. Bone. 2008;42:623–30.
31. Michelsen J, Wallaschofski H, Friedrich N, et al. Reference intervals for serum concentrations of three bone turnover markers for men and women. Bone. 2013;57:399–404.
32. Rogers A, Hannon RA, Eastell R. Biochemical markers as predictors of rates of bone loss after menopause. J Bone Miner Res. 2000;15:1398–404.
33. Vasikaran S, Eastell R, Bruyère O, et al. Markers of bone turnover for the prediction of fracture risk and monitoring of osteoporosis treatment: a need for international reference standards. Osteoporos Int. 2011;22:391–420.
34. Daele PLA van, Seibel MJ, Burger H, et al. Case-control analysis of bone resorption markers, disability, and hip fracture risk: The Rotterdam study. Br Med J. 1996;312:482–3.
35. Garnero P, Hausher E, Chapuy MC, et al. Markers of bone resorption predict hip fracture in elderly women: The Epidos prospective study. J Bone Miner Res 1996;11:1531–8.
36. Johansson H, Odén A, Kanis JA, et al. A meta-analysis of reference markers of bone turnover for prediction of fracture. Calcif Tissue Int. 2014;94:560–7.
37. Bouxsein ML, Delmas PD. Considerations for development of surrogate endpoints for antifracture efficacy of new treatments in osteoporosis: a perspective. J Bone Miner Res. 2008;23:1155–67.
38. Dempster DW. The contribution of trabecular architecture to cancellous bone quality. J Bone Miner Res. 2000;15:20–23.
39. Bone HG, McClung MR, Roux C, et al. Odanacatib, a cathepsin-K inhibitor for osteoporosis: a two-year study in postmenopausal women with low bone density. J Bone Miner Res 2010;25:937–47.
40. Eastell R, Nagase S, Ohyama M, et al. Safety and efficacy of the cathepsin K inhibitor ONO-5334 in postmenopausal osteoporosis: The OCEAN study. J Bone Miner Res. 2011;26:1303–12.
41. Garnero P, Ferreras M, Karsdal MA, et al. The type I collagen fragments ICTP and CTX reveal distinct enzymatic pathways of bone collagen degradation. J Bone Miner Res. 2003;18:859–67.
42. Eastell R, Christiansen C, Grauer A, et al. Effects of denosumab on bone turnover markers in postmenopausal osteoporosis. J Bone Miner Res. 2011;26:530–7.
43. Pederson L, Ruan M, Westendorf JJ, et al. Regulation of bone formation by osteoclasts involves Wnt/BMP

signaling and the chemokine sphingosine-1-phosphate. Proc Natl Acad Sci U S A. 2008;105:20764–9.
44. Glover SJ, Eastell R, McCloskey EV, et al. Rapid and robust response of biochemical markers of bone formation to teriparatide therapy. Bone. 2009;45:1053–8.
45. Greenspan SL, Bone HG, Ettinger MP, et al. Effect of recombinant human parathyroid hormone (1-84) on vertebral fracture and bone mineral density in postmenopausal women with osteoporosis: a randomized trial. Ann Intern Med. 2007;146:326–39.
46. Miller PD, Hattersley G, Riis BJ,. Effect of Abaloparatide vs Placebo on New Vertebral Fractures in Postmenopausal Women With Osteoporosis: A Randomized Clinical Trial. JAMA. 2016;316:722–33.
47. Padhi D, Jang G, Stouch B, et al. Single-dose, placebo-controlled, randomized study of AMG 785, a sclerostin monoclonal antibody. J Bone Miner Res. 2011;26:19–26.
48. Cosman F, Crittenden DB, Adachi JD, et al. Romosozumab Treatment in Postmenopausal Women with Osteoporosis. N Engl J Med. 2016;375:1532–43.
49. Meunier PJ, Roux C, Seeman E, et al. The effects of strontium ranelate on the risk of vertebral fracture in women with postmenopausal osteoporosis. N Engl J Med. 2004;350:459–68.
50. McClung MR, Lewiecki EM, Cohen SB, et al. Denosumab in postmenopausal women with low bone mineral density. N Engl J Med. 2006;354:821–31.
51. Hodsman AB, Hanley DA, Ettinger MP, et al. Efficacy and safety of human parathyroid hormone-(1-84) in increasing bone mineral density in postmenopausal osteoporosis. J Clin Endocrinol Metab. 2003;88:5212–20.
52. Delmas PD, Benhamou CL, Man Z, et al. Monthly dosing of 75 mg risedronate on 2 consecutive days a month: efficacy and safety results. Osteoporos Int. 2008;19:1039–45.
53. Rizzoli R, Greenspan SL, Bone G 3rd, et al. Two-year results of once-weekly administration of alendronate 70 mg for the treatment of postmenopausal osteoporosis. J Bone Miner Res. 2002;17:1988–96.
54. Bell KJ, Hayen A, Glasziou P, et al. Potential Usefulness of BMD and Bone Turnover Monitoring of Zoledronic Acid Therapy Among Women With Osteoporosis: Secondary Analysis of Randomized Controlled Trial Data. J Bone Miner Res. 2016;31:1767–73.
55. Bauer DC, Garnero P, Hochberg MC, et al. Pretreatment levels of bone turnover and the antifracture efficacy of alendronate: the fracture intervention trial. J Bone Miner Res. 2006;21:292–9.
56. Seibel MJ, Naganathan V, Barton I, et al. Relationship between pretreatment bone resorption and vertebral fracture incidence in postmenopausal osteoporotic women treated with risedronate. J Bone Miner Res. 2004;19:323–9.
57. Delmas PD, Licata AA, Reginster JY, et al. Fracture risk reduction during treatment with teriparatide is independent of pretreatment bone turnover. Bone. 2006;39:237–43.
58. Bauer DC, Black DM, Garnero P, et al. Change in bone turnover and hip, non-spine, and vertebral fracture in alendronate-treated women: the fracture intervention trial. J Bone Miner Res. 2004;19:1250–8.
59. Eastell R, Hannon RA, Garnero P, et al. Relationship of early changes in bone resorption to the reduction in fracture risk with risedronate: review of statistical analysis. J Bone Miner Res. 2007;22:1656–60.
60. Reginster JY, Sarkar S, Zegels B, et al. Reduction in PINP, a marker of bone metabolism, with raloxifene treatment and its relationship with vertebral fracture risk. Bone. 2004;34:344–51.
61. Chen P, Satterwhite JH, Licata AA, et al. Early changes in biochemical markers of bone formation predict BMD response to teriparatide in postmenopausal women with osteoporosis. J Bone Miner Res. 2005;20:962–70.
62. Black DM, Schwartz AV, Ensrud KE, et al. Effects of continuing or stopping alendronate after 5 years of treatment: the Fracture Intervention Trial Long-term Extension (FLEX): a randomized trial. JAMA. 2006;296:2927–38.
63. Sornay-Rendu E, Garnero P, Munoz F, et al. Effect of withdrawal of hormone replacement therapy on bone mass and bone turnover: the OFELY study. Bone. 2003;33:159–66.
64. Miller PD, Bolognese MA, Lewiecki EM, et al. Effect of denosumab on bone density and turnover in postmenopausal women with low bone mass after long-term continued, discontinued, and restarting of therapy: a randomized blinded phase 2 clinical trial. Bone. 2008;43:222–9.
65. Eastell R, Nagase S, Small M, et al. Effect of ONO-5334 on bone mineral density and biochemical markers of bone turnover in postmenopausal osteoporosis: 2-year results from the OCEAN study. J Bone Miner Res. 2014;29:458–66.
66. Black DM, Bilezikian JP, Ensrud KE, et al. One year of alendronate after one year of parathyroid hormone (1-84) for osteoporosis. N Engl J Med. 2005;353:555–65.
67. Finkelstein JS, Wyland JJ, Leder BZ, et al. Effects of teriparatide retreatment in osteoporotic men and women. J Clin Endocrinol Metab. 2009;94:2495–501.
68. Naylor KE, Jacques RM, Paggiosi M, et al. Response of bone turnover markers to three oral bisphosphonate therapies in postmenopausal osteoporosis: the TRIO study. Osteoporos Int. 2016;27:21–31.
69. Siris ES, Harris ST, Rosen CJ, et al. Adherence to bisphosphonate therapy and fracture rates in osteoporotic women: relationship to vertebral and nonvertebral fractures from 2 US claims databases. Mayo Clin Proc. 2006;81:1013–22.
70. Eastell R, Vrijens B, Cahall DL, et al. Bone turnover markers and bone mineral density response with risedronate therapy: Relationship with fracture risk and patient adherence. J Bone Miner Res. 2011;26:1662–9.
71. Clowes JA, Peel NF, Eastell R. The impact of monitoring on adherence and persistence with antiresorptive treatment for postmenopausal osteoporosis: a randomized controlled trial. J Clin Endocrinol Metab. 2004;89:1117–23.
72. Delmas PD, Vrijens B, Eastell R, et al. Effect of monitoring bone turnover markers on persistence with risedronate treatment of postmenopausal osteoporosis. J Clin Endocrinol Metab. 2007;92:1296–304.

73. Baim S, Miller PD. Assessing the clinical utility of serum CTX in postmenopausal osteoporosis and its use in predicting risk of osteonecrosis of the jaw. J Bone Miner Res. 2009;24:561–74.
74. Visekruna M, Wilson D, McKiernan FE. Severely suppressed bone turnover and atypical skeletal fragility. J Clin Endocrinol Metab. 2008;93:2948–52.
75. Fatayerji D, Eastell R. Age-related changes in bone turnover in men. J Bone Miner Res. 1999;14:1203–10.
76. Szulc P, Garnero P, Munoz F, et al. Cross-sectional evaluation of bone metabolism in men. J Bone Miner Res. 2001;16:1642–50.
77. Khosla S, Melton LJ 3rd, Atkinson EJ, et al. Relationship of serum sex steroid levels and bone turnover markers with bone mineral density in men and women: a key role for bioavailable estrogen. J Clin Endocrinol Metab. 1998;83:2266–74.
78. Chandani AK, Scariano JK, Glew RH, et al. Bone mineral density and serum levels of aminoterminal propeptides and cross-linked N-telopeptides of type I collagen in elderly men. Bone. 2000;26:513–18.
79. Chaitou A, Boutroy S, Vilayphiou N, et al. Association between bone turnover rate and bone microarchitecture in men: the STRAMBO study. J Bone Miner Res. 2010;25:2313–23.
80. Bauer DC, Garnero P, Harrison SL, et al. Biochemical markers of bone turnover, hip bone loss, and fracture in older men: the MrOS study. J Bone Miner Res. 2009;24:2032–8.
81. Szulc P, Montella A, Delmas PD. High bone turnover is associated with accelerated bone loss but not with increased fracture risk in men aged 50 and over: the prospective MINOS study. Ann Rheum Dis. 2008;67:1249–55.
82. Dennison E, Eastell R, Fall CH, et al. Determinants of bone loss in elderly men and women: a prospective population-based study. Osteoporos Int. 1999;10:384–91.
83. Meier C, Nguyen TV, Center JR, et alA. Bone resorption and osteoporotic fractures in elderly men: the dubbo osteoporosis epidemiology study. J Bone Miner Res. 2005;20:579–87.
84. Amory JK, Watts NB, Easley KA, et al. Exogenous testosterone or testosterone with finasteride increases bone mineral density in older men with low serum testosterone. J Clin Endocrinol Metab. 2004;89:503–10.
85. Wang C, Swerdloff RS, Iranmanesh A, et al. Effects of transdermal testosterone gel on bone turnover markers and bone mineral density in hypogonadal men. Clin Endocrinol (Oxf). 2001;54:739–50.
86. Orwoll ES, Binkley NC, Lewiecki EM, et al. Efficacy and safety of monthly ibandronate in men with low bone density. Bone. 2010;46:970–6.
87. Boonen S, Orwoll ES, Wenderoth D, et al. Once-weekly risedronate in men with osteoporosis: results of a 2-year, placebo-controlled, double-blind, multicenter study. J Bone Miner Res. 2009;24:719–25.
88. Bolland MJ, Grey AB, Horne AM, et al. Annual zoledronate increases bone density in highly active antiretroviral therapy-treated human immunodeficiency virus-infected men: a randomized controlled trial. J Clin Endocrinol Metab. 2007;92:1283–8.
89. Orwoll E, Ettinger M, Weiss S, et al. Alendronate for the treatment of osteoporosis in men. N Engl J Med. 2000;343:604–10.
90. Boonen S, Reginster JY, Kaufman JM, et al. Fracture risk and zoledronic acid therapy in men with osteoporosis. N Engl J Med. 2012;367:1714–23.
91. Orwoll E, Teglbjærg CS, Langdahl BL, et al. A randomized, placebo-controlled study of the effects of denosumab for the treatment of men with low bone mineral density. J Clin Endocrinol Metab. 2012;97:3161–9.
92. Smith MR, Egerdie B, Hernández Toriz N, et al. Denosumab in men receiving androgen-deprivation therapy for prostate cancer. N Engl J Med. 2009;361:745–55.
93. Orwoll ES, Scheele WH, Paul S, et al. The effect of teriparatide [human parathyroid hormone (1-34)] therapy on bone density in men with osteoporosis. J Bone Miner Res. 2003;18:9–17.
94. Leder BZ, Neer RM, Wyland JJ, et al. Effects of teriparatide treatment and discontinuation in postmenopausal women and eugonadal men with osteoporosis. J Clin Endocrinol Metab. 2009;94:2915–21.
95. Finkelstein JS, Leder BZ, Burnett SM, et al. Effects of teriparatide, alendronate, or both on bone turnover in osteoporotic men. J Clin Endocrinol Metab. 2006;91:2882–7.
96. Padhi D, Allison M, Kivitz AJ, et al. Multiple doses of sclerostin antibody romosozumab in healthy men and postmenopausal women with low bone mass: a randomized, double-blind, placebo-controlled study. J Clin Pharmacol. 2014;54:168–78.
97. Anderson MS, Gendrano IN, Liu C, et al. Odanacatib, a selective cathepsin K inhibitor, demonstrates comparable pharmacodynamics and pharmacokinetics in older men and postmenopausal women. J Clin Endocrinol Metab. 2014;99:552–60.

38
Scintigraphy and PET in Metabolic Bone Disease

Lorenzo Nardo[1], Paola A. Erba[2], and Benjamin L. Franc[3]

[1] Department of Radiology, University of California Davis, Sacramento, CA, USA
[2] Department of Translational Research and New Technology in Medicine, University of Pisa, Pisa, Italy
[3] Department of Radiology and Biomedical Imaging, Nuclear Medicine Section, University of California, San Francisco, CA, USA

INTRODUCTION

Nuclear medicine offers a broad set of radiotracers and imaging systems to reliably and noninvasively assess for metabolic musculoskeletal diseases.

Radiotracers

Radiotracers relevant to the study of osseous metabolic disease can be separated into bone-seeking agents and those agents that do not target specific bone components. The use of bone-seeking radiotracers is based on the concept that pathophysiologic processes affecting bone usually result in increased local bone turnover, which, in turn, results in increased radiotracer uptake. Bone-seeking radiotracers include ^{99m}technetium [^{99m}Tc] methylene diphosphonate (MDP) or hydroxymethylene diphosphonate (HDP), [^{18}F] sodium fluoride (^{18}F-NaF) [1], and more recently developed radiotracers such as ^{64}Cu-labeled arginyl-glycyl-aspartic peptides [2]. ^{99m}Tc-MDP/HDP and ^{18}F-NaF are the only radiotracers clinically approved for use in humans. Following intravenous injection these radiotracers bind to the mineral phase of the bone, a process known as chemisorption. Unbound tracer remaining in the blood pool is quickly cleared by the kidneys, resulting in a high target-to-background ratio in the skeleton within 30 to 45 minutes in the case of ^{18}F-NaF, and 120 to 240 minutes in the case of ^{99m}Tc-MDP/HDP. ^{18}F-NaF and ^{99m}Tc-MDP/HDP are essentially surrogates of both bone perfusion and bone turnover: increased bone perfusion or turnover results in increased uptake [3]. In addition, other factors that influence the amount of bone uptake include local acid–base status, vitamins, hormones, innervation, and drugs [4].

Rather than targeting mineralization, ^{64}Cu-labeled arginyl-glycyl-aspartic peptides [2] specifically target αvβ3, an integrin expressed by osteoclasts [2]. ^{64}Cu-labeled arginyl-glycyl-aspartic peptides binding αvβ3 provide a means to quantify numbers of osteoclasts and may prove useful in monitoring the response of osteoclast-mediated disease to therapy, although they are not yet routinely used clinically.

Nonspecific radiotracers—those agents that do not target pathophysiologic processes specific to bone—include ^{18}F-FDG. ^{18}F-FDG is a marker of glucose consumption and accumulates in both physiologic and pathologic situations. Specific patient preparation is required prior to ^{18}F-FDG injection, including one or more days of limited physical activity and at least 6 hours of fasting. Ideally, the patient's blood glucose level, as determined by standard fingerstick method, should be in the range of 80 to 120 mg/dL. ^{18}F-FDG (555 to 740 MBq) is administered intravenously according to the patient's weight. It distributes within the body based on relative tissue metabolism. ^{18}F-FDG remaining in the blood pool is quickly cleared by the kidneys and, to some extent, by the liver, resulting in a high target-to-background ratio in highly metabolic areas within 45 to 50 minutes.

Primer on the Metabolic Bone Diseases and Disorders of Mineral Metabolism, Ninth Edition. Edited by John P. Bilezikian.
© 2019 American Society for Bone and Mineral Research. Published 2019 by John Wiley & Sons, Inc.
Companion website: www.wiley.com/go/asbmrprimer

Imaging systems

The different scanners used in nuclear medicine can be divided into two major categories: single-photon imaging systems (gamma cameras), and imaging systems that coincidently detect two photons emitted from a positron annihilation (PET systems).

In the case of gamma cameras, the patient is administered a single gamma photon-emitting radiotracer intravenously and is scanned subsequently to obtain planar or tomographic images, the latter in the case of systems capable of single-photon emission computed tomography (SPECT). Typically 20 to 30 minutes (5 to 7 bed positions, 3 to 5 minutes each) are required for planar image acquisition, whereas 10 to 15 minutes are required to obtain tomographic images of a limited area (about 30 to 40 cm). In the case of systems allowing serial acquisition of SPECT and X-ray CT, referred to as SPECT-CT systems, CT images can be utilized for anatomic correlation as well as to correct SPECT images for various physical factors such as soft tissue attenuation of lower energy gamma photons (140 KeV in the case of [^{99m}Tc]), photon scatter, and collimator blurring.

PET systems are generally more sensitive than gamma camera-based techniques, in part because of their lower soft tissue attenuation of the higher energy (511 KeV) gamma photons emitted upon annihilation of the positron. Because of their ring design and component requirements, PET scanners are more expensive than gamma cameras. Depending on the radiotracer imaged, the PET technique can expose patients to higher radiation doses than gamma imaging. For example, the whole body radiation exposure for an imaging study to evaluate bone remodeling is 0.024 mSv/MBq for PET using ^{18}F-NaF, versus 0.0057 mSv/MBq for gamma imaging (planar or SPECT) using ^{99m}Tc-MDP [5]. PET is obtained in 6 to 8 bed positions of 3 to 4 minutes each, depending on the patient's height, for a total acquisition time of approximately 35 to 45 minutes. PET imaging systems are routinely integrated with either CT or MRI systems with the dual purpose of attenuation correction of the PET emission images and anatomic correlation. The standard CT acquisition parameters typically utilize a voltage close to 130 kV and a current of 110 mA with 3.75 axial slice placement. Sequence selection for whole body MRI includes a T2-weighted sequence with fat suppression, such as short tau inversion recovery (STIR) in the coronal plane. Whole body diffusion-weighted imaging (DWI), a nonenhanced functional MRI technique, can be added to evaluate tissue microstructure. When specific attention to the pelvis or lower extremities is required, sequences should include T1-weighted axial images over those regions. Spine imaging should include T1-weighted and STIR sequences in the sagittal and coronal planes. All of the above sequences can be acquired in about 35 to 45 minutes with contemporaneous acquisition of PET images. PET-MRI requires a lower radiation dose than PET-CT. However, the smaller field of view provided by PET-MRI may not be suitable for some applications, and current algorithms for attenuation correction using data from MRI vary in their accuracy near the bone–soft tissue interface.

Most recently a new PET scanner has been created: the EXPLORER PET-CT scanner. The EXPLORER scanner allows image acquisition of the whole body in a single bed position. When compared to standard PET-CT, EXPLORER decreases the radiation exposure and increases sensitivity and spatial resolution. In addition, EXPLORER allows for entire body dynamic acquisition. However, this sophisticated technology is available only in a few institutions at the present time.

APPLICATIONS OF NUCLEAR MEDICINE TECHNIQUES IN METABOLIC BONE DISEASE

Osteoporosis

Osteoporosis is a skeletal disorder characterized by decreased bone density and quality, resulting in bone fragility and fractures. Given the availability of effective treatments, early diagnosis of osteoporosis is crucial to slow its evolution and reduce the risk of complications. Although several nonspecific findings on nuclear bone scan imaging might suggest the diagnosis of osteoporosis, such as reduced bone to soft tissue uptake, decreased resolution of vertebral body endplates, and relatively increased skull uptake of the radiopharmaceutical, the value of nuclear bone scan imaging is predominantly in the assessment for osteoporotic fractures and other potential complications (Fig. 38.1). Bone scan can be used to select patients who would benefit from kyphoplasty treatment when MRI is contraindicated; specifically bone scan has a higher specificity than either plain radiographs or CT [6]. When directly compared to radiographs and CT, bone scan performs better in distinguishing painful osteoporotic vertebral fractures, which can benefit from kyphoplasty [6]. When compared to MRI, bone scan performs better only for older fractures [7]; the use of MRI is recommended for more acute fractures because it may take up to 12 days for the bone scan to become abnormal [8].

One of the main advantages of bone scanning is its ability to provide evaluation of the entire skeleton in a single and rapid imaging test. The patient needs to lie on the table for only 20 minutes to have the whole body scanned.

Bone scan is also an appropriate technique to assess the age of a fracture. Abnormal uptake at the site of closed and nonmanipulated fractures can be detected in 80% of cases within 24 hours and in 95% of cases within 72 hours [9]. The uptake at the site of fracture returns to a level similar to background after 5 to 7 months, on average, but ranging up to 3 years depending on fracture site and patient age. For example, long bone fractures tend to normalize in radiopharmaceutical uptake earlier than vertebral fractures; in older patients elevated ^{99m}Tc-MDP uptake tends to appear later and to persist longer [9].

Using ^{18}F-fluorodeoxyglucose (^{18}F-FDG), radiolabeled glucose uptake is typically high in malignancy, and PET-CT can help distinguish a pathological vertebral fracture

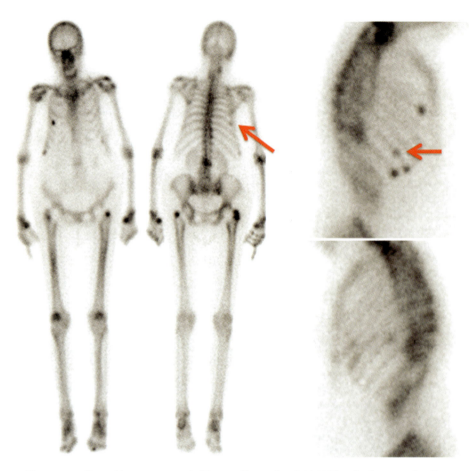

Fig. 38.1. A 59-year old woman affected by osteoporosis (T-score [femur] = –2.9). Diffusely decreased radiotracer uptake throughout the skeleton with relatively increased uptake in the skull associated with multiple bilateral rib fractures (arrows), better assessed on oblique views. In addition, photopenic areas in the bilateral proximal femurs are consistent with bilateral hip arthroplasty related to prior femoral fractures.

from osteoporotic fracture. Acute vertebral fractures that originated from osteoporosis or preclinical osteoporosis tend to have no pathologically increased FDG uptake, whereas high FDG uptake is characteristic for malignant and inflammatory processes [10].

The role of nuclear medicine is not limited to diagnosis of fracture caused by osteoporosis but extends to the assessment of side effects from drugs used in the treatment of osteoporosis. For example, bone scan can be used to assess for fractures related to antiresorptive therapy [11]. Bisphosphonate or denosumab therapy may result in low-energy subtrochanteric and proximal diaphyseal femoral fractures with a characteristic appearance on bone scan. These fractures often are subtle and missed on initial radiographs [12]; however, because of the risk of delayed diagnosis of displaced complete fracture, early assessment for such fractures using a highly sensitive technique, such as scintigraphy or MRI, is recommended.

Paget disease of the bone

Bone scan is more sensitive than radiography in detecting Paget disease [13, 14]. Only when 30% to 50% of the bone has been resorbed can the lesion be seen on radiography. An additional advantage of bone scan over radiography is that it allows for assessment of the entire skeleton. A limitation of bone scan is the difficulty in differentiating between osteolytic and osteoblastic disease [14]. It has been reported that the amount of radioactivity correlates with both bone deformation on radiographs and increased pain [14]. To assist in the diagnosis of Paget disease, relatively specific signs have been described on both radiographs and bone scans [15]. "Picture frame sign" represents cortical thickening associated with radiolucency of the center of an enlarged, flattened vertebral body. "Ivory sign" is typical of the sclerotic phase and characterized by an enlarged and homogeneously dense vertebral body. "Mickey Mouse" sign describes increased uptake in the vertebral body, posterior elements, and spinous process (Fig. 38.2). This same sign has also been termed the "T-sign" or "champagne glass" sign [16].

The most commonly encountered differential diagnoses for Paget disease are malignant metastasis and fibrous dysplasia. MRI and ^{18}F-FDG PET-CT may help to differentiate metastatic disease from Paget disease. In the lytic phase, a Paget lesion has preserved fatty marrow signal on MRI because the destruction of the bone is caused by resorption and not infiltration as in metastatic disease [17]. However, in the sclerotic phase this MRI feature is not helpful because the signal would be low in all sequences [18]. Because ^{18}F-FDG uptake is often normal or only slightly

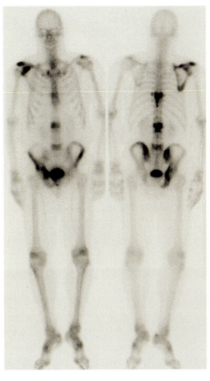
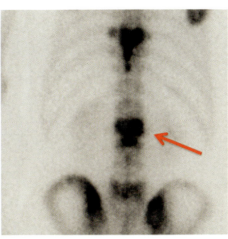

Fig. 38.2. Polyostotic Paget disease. Multifocal radiotracer uptake is seen in the right scapula, several thoracic and lumbar vertebrae, and right acetabulum. Note the specific pattern of uptake in the lumbar vertebral body and posterior elements (arrow), which forms an inverted triangular pattern, previously reported as the "Mickey Mouse" sign (left-hand image courtesy of R. Flavell, MD).

increased in pagetic lesions, ^{18}F-FDG PET-CT has some additional value in this differential diagnosis in that it is mildy avid when alkaline phosphatase is elevated, whereas metastases are usually hypermetabolic. However, on occasion, marked FDG uptake can occur and incidental Paget's disease rarely could cause false-positive scans when assessing for metastatic disease [19].

Paget disease and fibrous dysplasia are frequently difficult to differentiate on the basis of the bone scan because uptake may be intense in both diseases. However, clinical data and other radiologic techniques can help; young age, and ground glass appearance are more typical of fibrous dysplasia. When the head is involved, fibrous dysplasia tends to spare the skull, whereas Paget disease often involves skull bones.

Paget disease presents a pitfall also in the use of new PET-CT radiotracers such as ^{68}Ga-PSMA [20] and ^{11}C-fluorocholine [21]. These radiotracers have been used to assess for recurrent or metastatic prostate cancer. However, there are several examples in the literature that demonstrate their nonspecific uptake in Paget disease. ^{18}F-NaF PET-CT has been used to quantify bone turnover in Paget disease with particular attention to the assessment of response to bisphosphonate treatment [22], however this technique is not used routinely in clinical practice.

Hyperparathyroidism

In the diagnosis of hyperparathyroidism nuclear medicine alone often cannot indicate the precise type of hyperparathyroidism, for which other radiological, pathologic, and clinical data need to be obtained. ^{99m}Tc-sestamibi scans have played a key role in the diagnosis of parathyroid adenoma. A scintigraphic scan is performed after injection of 740 to 1110 MBq of ^{99m}Tc-sestamibi or ^{99m}Tc-tetrofosmin at 10 to 30 minutes (immediate timepoint) and 90 to 180 minutes (late timepoint) in the anterior view, including neck and upper chest [23]. Additional oblique views and SPECT images may be obtained to increase the sensitivity of the scan and to accurately evaluate the anatomic localization. Recently, new radiotracers have been investigated for localizing parathyroid adenoma with promising results. For example ^{18}F-fluorocholine and ^{11}C-methionine can detect parathyroid lesions in the case of a negative sestamibi scan [24, 25]. These new radiotracers are reliable second-line imaging techniques to enable minimally invasive parathyroidectomy [26]. Notably, the production of ^{18}F-fluorocholine does not require an onsite cyclotron as in the case of ^{11}C-methionine, facilitating its broader use.

The typical bone scan pattern of hyperparathyroidism is represented by generalized increased uptake in the bone, especially long bones, skull, and mandible, and decreased or absent renal activity (Fig. 38.3). This clinical scenario has been described as a "super scan" or "beautiful scan" because of the high contrast between bones and background. Other radiologic signs associated with hyperparathyroidism are the "rosary beads sign," related to increased uptake at the costochondral junctions, and the "tie sign," caused by prominent uptake in the sternum. Hyperparathyroidism has been also described in association with brown tumors, soft tissue calcifications

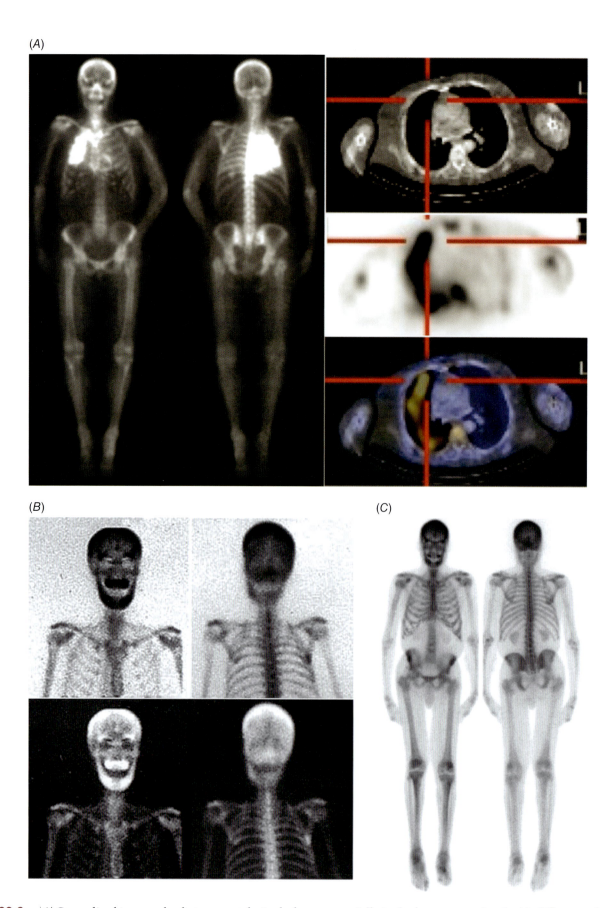

Fig. 38.3. (A) Generalized increased radiotracer uptake in the bones, especially in the femurs, associated with diffuse uptake in the right lung in a patient with severe hyperparathyroidism. There is faint uptake in the kidneys, consistent with a "super scan" (courtesy of R. Flavell, MD). (B) Characteristic pattern of prominent skull and mandibular uptake in a patient affected by longstanding primary hyperparathyroidism. (C) Diffuse increased uptake throughout the skeleton with minimal excreted activity. Prominent skull, costochondral, and lower extremity uptake representing features of hyperparathyroidism.

(Fig. 38.3A), fractures, and subchondral bone resorption. Brown tumors are also known as osteitis fibrosa cystica or osteoclastomas and result from normal bone marrow replacement with hemorrhage and granulation tissue. These tumors can be visualized with whole body ^{99m}Tc-sestamibi scintigraphy, however some reports have described ^{18}F-FDG PET-CT as a more sensitive technique [27]. Soft tissue calcifications, also known as metastatic calcifications, have been described in the soft tissue of both upper and lower extremities [28], lungs [29], thyroid, and stomach [30]. Resolution of these findings has been reported upon parathyroidectomy [30]. The increased risk of fractures in hyperparathyroidism is related to the bone loss induced by calcium reabsorption. In particular, subchondral bone resorption is often better appreciated with pinhole acquisitions at the radial aspect of the phalanges, clavicles, and pubic bones.

Hypertrophic osteoarthropathy

Hypertrophic osteoarthropathy can be idiopathic or secondary to malignancy. This condition is characterized by periosteal reaction, more prominent in the long bones. The bone scan appearance is that of linear uptake in parallel lines along the cortex of the diaphysis of the radii, tibias, and femurs [31] (Fig. 38.4).

Renal osteodystrophy

Renal osteodystrophy refers to the set of findings seen in chronic renal dysfunction, which include those related to osteomalacia (adult), rickets (children), and hyperparathyroidism. The typical appearance on a bone scan is that of a "beautiful scan" without physiological distribution of radiotracer in the bladder. However, this appearance is not specific for renal osteodystrophy.

A diagnosis of adynamic bone disease is indicated by poor uptake of tracer.

Osteomalacia and rickets

Osteomalacia and rickets are characterized by defective mineralization of bone, resulting in fragile bones.

Findings on bone scan in the setting of osteomalacia can be normal, especially in the early stages. In the later stages, there is increased radiotracer uptake in the metaphysis and in the ossification centers of the long bones resulting in the "chicken bone appearance". The typical scintigraphic appearance of osteomalacia shares metabolic features with hyperparathyroidism including increased bone-to-soft tissue uptake ratio, increased uptake in long bones and skull, beading of the costochondral junctions, and the

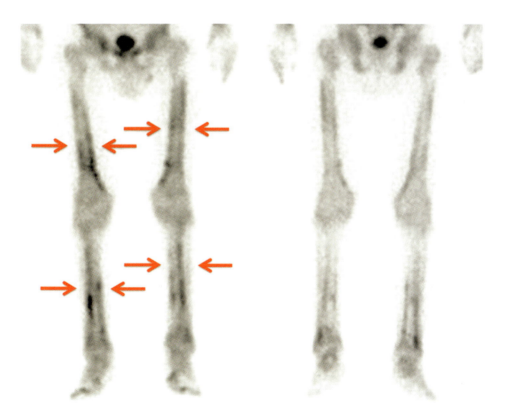

Fig. 38.4. Lower extremity images, obtained 2 hours after administration of ^{99}Tc-HDP, demonstrating linear uptake along the cortex and periosteum of both lower extremities with "tram tracks" (arrows) characteristic of hypertrophic osteoarthropathy.

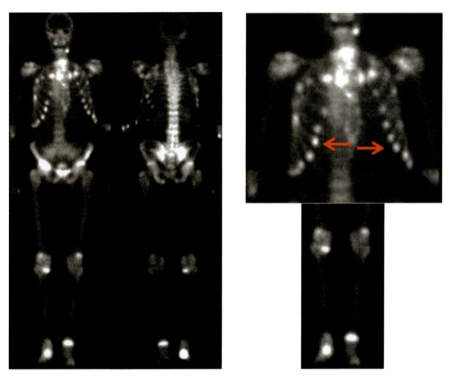

Fig. 38.5. 99Tc-MDP bone scan showing multiple foci of uptake in the costochondral (arrows) and costovertebral, sacroiliac, and large extremity joints, more prominent in knees and ankles. These findings are suggestive of osteomalacia.

"tie sternum" appearance (Fig. 38.5) [32]. "Pathologic" fractures have also been described in association with osteomalacia and may be mistaken for metastasis.

Bone scan is not diagnostic for osteomalacia but the presence of metabolic features associated with fractures and/or pseudofractures may suggest this condition in the setting of clinical symptoms and signs (weakness, myalgia, bone pain) and blood biochemistry abnormalities. In addition bone scanning is more sensitive than radiography in the detection of pseudofractures [33].

Oncogenic osteomalacia is a rare form of osteomalacia but deserves a special note for its clinical significance. This condition is also known as tumor-induced osteomalacia and usually associated with mesenchymal tumors [34]. If there is suspicion of tumor-induced osteomalacia, whole body scanning with ^{18}F-FDG PET-CT is recommended to locate the primary cancer [35, 36]. ^{18}F-FDG [35] and the somatostatin receptor imaging agents gallium-68 (^{68}Ga)-DOTATATE [37], ^{68}Ga-DOTANOC [38], and ^{68}Ga-DOTATOC have been reported to detect the primary tumor in these cases. Somatostatin receptor imaging appears to be more sensitive than ^{18}F-FDG in some reports [39], however further investigation is required.

CONCLUSION

Nuclear medicine provides several imaging modalities critical for the evaluation of several metabolic bone disorders. These imaging techniques will continue to be important in evaluating the evolution of metabolic bone disorders, assessing their causes, and monitoring for side effects related to their treatment.

REFERENCES

1. Beheshti M, Mottaghy FM, Payche F, et al. (18)F-NaF PET/CT: EANM procedure guidelines for bone imaging. Eur J Nucl Med Mol Imaging. 2015;42(11):1767–77.
2. Sprague JE, Kitaura H, Zou W, et al. Noninvasive imaging of osteoclasts in parathyroid hormone-induced osteolysis using a 64Cu-labeled RGD peptide. J Nucl Med. 2007;48(2):311–8.
3. Wong KK, Piert M. Dynamic bone imaging with 99mTc-labeled diphosphonates and 18F-NaF: mechanisms and applications. J Nucl Med. 2013;54(4):590–9.
4. Van den Wyngaert T, Strobel K, et al. The EANM practice guidelines for bone scintigraphy. Eur J Nucl Med Mol Imaging. 2016;43(9):1723–38.
5. Segall G, Delbeke D, Stabin MG, et al. SNM practice guideline for sodium 18F-fluoride PET/CT bone scans 1.0. J Nucl Med. 2010;51(11):1813–20.
6. Tang ZB, Lei Z, Yang HL, et al. Value of bone scan imaging in determining painful vertebrae of osteoporotic vertebral compression fractures patients with contraindications to MRI. Orthop Surg. 2012;4(3):172–6.
7. Masala S, Schillaci O, Massari F, et al. MRI and bone scan imaging in the preoperative evaluation of painful vertebral fractures treated with vertebroplasty and kyphoplasty. In vivo. 2005;19(6):1055–60.

8. Spitz J, Lauer I, Tittel K, et al. Scintimetric evaluation of remodeling after bone fractures in man. J Nucl Med. 1993;34(9):1403–9.
9. Matin P. The appearance of bone scans following fractures, including immediate and long-term studies. J Nucl Med. 1979;20(12):1227–31.
10. Schmitz A, Risse JH, Textor J, et al. FDG-PET findings of vertebral compression fractures in osteoporosis: preliminary results. Osteoporos Int. 2002;13(9):755–61.
11. Bush LA, Chew FS. Subtrochanteric Femoral Insufficiency Fracture in Woman on Bisphosphonate Therapy for Glucocorticoid-Induced Osteoporosis. Radiol Case Rep. 2009;4(1):261.
12. Patel RN, Ashraf A, Sundaram M. Atypical Fractures Following Bisphosphonate Therapy. Semin Musculoskelet Radiol. 2016;20(4):376–81.
13. Fogelman I, Carr D. A comparison of bone scanning and radiology in the assessment of patients with symptomatic Paget's disease. Eur J Nucl Med. 1980;5(5):417–21.
14. Vellenga CJ, Pauwels EK, Bijvoet OL, et al. Untreated Paget disease of bone studied by scintigraphy. Radiology. 1984;153(3):799–805.
15. Estrada WN, Kim CK. Paget's disease in a patient with breast cancer. J Nucl Med. 1993;34(7):1214–6.
16. van Heerden BB, Prins MJ. [The value of pinhole collimator imaging in the scintigraphic analysis of vertebral diseases]. S Afr Med J. 1989;75(6):280–3.
17. Sundaram M, Khanna G, El-Khoury GY. T1-weighted MR imaging for distinguishing large osteolysis of Paget's disease from sarcomatous degeneration. Skeletal Radiol. 2001;30(7):378–83.
18. Sundaram M. Imaging of Paget's disease and fibrous dysplasia of bone. J Bone Miner Res. 2006;21 (suppl 2):P28–30.
19. Cook GJ, Maisey MN, Fogelman I. Fluorine-18-FDG PET in Paget's disease of bone. J Nucl Med. 1997;38(9):1495–7.
20. Bourgeois S, Gykiere P, Goethals L, et al. Aspecific Uptake of 68GA-PSMA in Paget Disease of the Bone. Clin Nucl Med. 2016;41(11):877–8.
21. Giovacchini G, Samanes Gajate AM, Messa C, et al. Increased C-11 choline uptake in pagetic bone in a patient with coexisting skeletal metastases from prostate cancer. Clin Nucl Med. 2008;33(11):797–8.
22. Cook GJ, Blake GM, Marsden PK, et al. Quantification of skeletal kinetic indices in Paget's disease using dynamic 18F-fluoride positron emission tomography. J Bone Miner Res. 2002;17(5):854–9.
23. Hindie E, Ugur O, Fuster D, et al. 2009 EANM parathyroid guidelines. Eur J Nucl Med Mol Imaging. 2009;36(7):1201–16.
24. Chun IK, Cheon GJ, Paeng JC, et al. Detection and Characterization of Parathyroid Adenoma/Hyperplasia for Preoperative Localization: Comparison Between (11)C-Methionine PET/CT and (99m)Tc-Sestamibi Scintigraphy. Nucl Med Mol Imaging. 2013;47(3):166–72.
25. Weber T, Gottstein M, Schwenzer S, et al. Is C-11 Methionine PET/CT Able to Localise Sestamibi-Negative Parathyroid Adenomas? World J Surg. 2017;41(4):980–5.
26. Kluijfhout WP, Pasternak JD, Drake FT, et al. Use of PET tracers for parathyroid localization: a systematic review and meta-analysis. Langenbecks Arch Surg. 2016;401(7):925–35.
27. Gahier Penhoat M, Drui D, Ansquer C, et al. Contribution of 18-FDG PET/CT to brown tumor detection in a patient with primary hyperparathyroidism. Joint Bone Spine. 2017;84(2):209–12.
28. Niemann KE, Kropil F, Hoffmann MF, et al. A 23-year-old patient with secondary tumoral calcinosis: Regression after subtotal parathyroidectomy: A case report. Int J Surg Case Rep. 2016;23:56–60.
29. Ando T, Mochizuki Y, Iwata T, et al. Aggressive pulmonary calcification developed after living donor kidney transplantation in a patient with primary hyperparathyroidism. Transplant Proc. 2013;45(7):2825–30.
30. Hwang GJ, Lee JD, Park CY, et al. Reversible extraskeletal uptake of bone scanning in primary hyperparathyroidism. J Nucl Med. 1996;37(3):469–71.
31. Terry DW, Jr., Isitman AT, Holmes RA. Radionuclide bone images in hypertrophic pulmonary osteoarthropathy. Am J Roentgenol Radium Ther Nucl Med. 1975;124(4):571–6.
32. Fogelman I, McKillop JH, Bessent RG, et al. The role of bone scanning in osteomalacia. J Nucl Med. 1978;19(3):245–8.
33. Fogelman I, McKillop JH, Greig WR, et al. Pseudofracture of the ribs detected by bone scanning. J Nucl Med. 1977;18(12):1236–7.
34. Chakraborty PP, Bhattacharjee R, Mukhopadhyay S, et al. 'Rachitic rosary sign' and 'tie sign' of the sternum in tumour-induced osteomalacia. BMJ Case Rep. 2016;2016.
35. Kaneuchi Y, Hakozaki M, Yamada H, et al. Missed causative tumors in diagnosing tumor-induced osteomalacia with (18)F-FDG PET/CT: a potential pitfall of standard-field imaging. Hell J Nucl Med. 2016;19(1):46–8.
36. Jain AS, Shelley S, Muthukrishnan I, et al. Diagnostic importance of contrast enhanced (18)F-fluorodeoxyglucose positron emission computed tomography in patients with tumor induced osteomalacia: Our experience. Indian J Nucl Med. 2016;31(1):14–19.
37. Breer S, Brunkhorst T, Beil FT, et al. 68Ga DOTA-TATE PET/CT allows tumor localization in patients with tumor-induced osteomalacia but negative 111In-octreotide SPECT/CT. Bone. 2014;64:222–7.
38. Bhavani N, Reena Asirvatham A, Kallur K, et al. Utility of Gallium-68 DOTANOC PET/CT in the localization of Tumour-induced osteomalacia. Clin Endocrinol (Oxf). 2016;84(1):134–40.
39. Agrawal K, Bhadada S, Mittal BR, et al. Comparison of 18F-FDG and 68Ga DOTATATE PET/CT in localization of tumor causing oncogenic osteomalacia. Clin Nucl Med. 2015;40(1):e6–10.

39

Bone Histomorphometry in Clinical Practice

Robert R. Recker[1] and Carolina Aguiar Moreira[2]

[1]Division of Endocrinology, Osteoporosis Research Center, Creighton University School of Medicine, Omaha, NE, USA
[2]Department of Medicine & Section of Endocrinology (SEMPR), Federal University of Parana, Laboratory PRO, Fundacao Pro Renal, Curitiba, Paraná, Brazil

INTRODUCTION

Histological examination of undecalcified transilial bone biopsy specimens is a valuable and well-established clinical and research tool for studying the etiology, pathogenesis, and treatment of metabolic bone diseases. In this chapter, we will review the underlying organization and function of bone cells; identify a set of basic structural and kinetic histomorphometric variables; outline an approach to interpretation of findings, with examples from a range of metabolic bone diseases; describe techniques for obtaining, processing, and analyzing transilial biopsy specimens; identify clinical situations in which bone histomorphometry can be useful; and relate histomorphometric measures to data from other methods for assessing bone properties and bone physiology.

ORGANIZATION AND FUNCTION OF BONE CELLS

Intermediary organization of the skeleton

In what he termed the intermediary organization (IO) of the skeleton, Frost [1] described four discrete functions of bone cells: growth, modeling, remodeling, and fracture repair. Although each involves osteoclasts and osteoblasts, the coordinated outcomes differ greatly. Growth elongates the skeleton; modeling shapes it during growth; remodeling removes and replaces bone tissue; and fracture repair heals sites of structural failure.

The remodeling IO, which predominates during adult life, is the focus of this chapter. Coordinated groups of bone cells (ie, osteoclasts, osteoblasts, osteocytes, and lining cells—see Chapters 3 and 4) comprise the basic BMUs that carry out bone remodeling. Basic structural units (BSUs) are the packets of new bone that BMUs form [2]. Nearly all adult-onset metabolic bone disease involves derangement of the remodeling IO.

Bone remodeling process

Remodeling occurs in cancellous and haversian (cortical) bone. The first step is activation of osteoclast precursors to form osteoclasts that begin to excavate a cavity. After removal of about 0.05 mm³ of bone tissue, the site remains quiescent for a short time. Then, activation of osteoblast precursors occurs at the site, and the excavation is refilled. The average length of time required to complete the remodeling cycle is approximately 6 months [3], about 4 weeks for resorption and the rest for formation.

The healthy bone remodeling system accesses the required building materials within a favorable physiologic milieu to replace fully a packet of aged, microdamaged bone tissue with new, mechanically competent bone. However, overuse can overwhelm the capacity of the system to repair microdamage (stress fractures in military recruits are an example). The healthy bone remodeling system modifies bone architecture to meet changing mechanical needs. The system also promptly reduces the mass of underused bone (the bone loss of extended bedrest, paralysis, or space travel are examples). All bone loss occurs through bone remodeling. The bone

---

Primer on the Metabolic Bone Diseases and Disorders of Mineral Metabolism, Ninth Edition. Edited by John P. Bilezikian.
© 2019 American Society for Bone and Mineral Research. Published 2019 by John Wiley & Sons, Inc.
Companion website: www.wiley.com/go/asbmrprimer

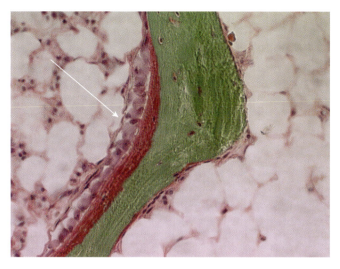

Fig. 39.1. A normal bone-forming surface. Unmineralized osteoid is covered with plump osteoblasts, as identified by the arrow.

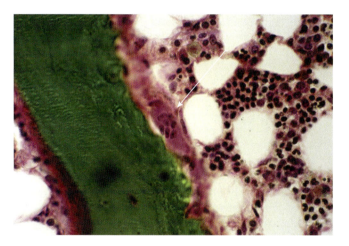

Fig. 39.2. A normal bone-resorbing surface. The arrow locates a multinucleated osteoclast in a Howship's lacuna.

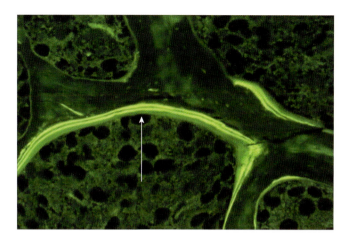

Fig. 39.3. The arrow identifies a mineralizing surface containing fluorescent double labels.

remodeling system responds to nutritional and humoral as well as mechanical influences. Figs 39.1, 39.2, and 39.3 present representative photomicrographs from human transiliac biopsy specimens. An extensive atlas has also been published [4].

BASIC HISTOMORPHOMETRIC VARIABLES

Bone biopsy specimens are ordinarily obtained at the transiliac site and shipped to specialized laboratories for processing and microscopic analysis. Of the dozens of measurements and calculations that have been devised, we provide here descriptions of several frequently used variables. Nomenclature was approved by a committee of the American Society of Bone and Mineral Research [5].

Structural features from the transilial biopsy

Core width (C.Wi)

This represents the distance (in mm) between periosteal surfaces, at the point of biopsy.

Cortical bone

Cortical width (Ct.Wi) is the combined thickness (in mm) of both cortices. Cortical porosity (Ct.Po) is the area of intracortical voids (haversian canals) as percent of total cortical area.

Cancellous bone

Cancellous bone volume (BV/TV) is the percent of entire marrow area occupied by cancellous (trabecular) bone. Wall thickness (W.Th) is the mean distance in µm between resting cancellous surfaces (ie, surfaces without osteoid or Howship's lacunae) and corresponding cement lines.

Trabecular thickness (Tb.Th) is the mean distance across individual trabeculae in µm, and trabecular separation (Tb.Sp) is the mean distance in µm between trabeculae. Trabecular number (Tb.N) per millimeter is calculated as (BV/TV)/Tb.Th. These variables can be used to evaluate trabecular connectivity [6]. Other measures of trabecular connectivity include the ratio of nodes to free ends [7], star volume [8, 9], and trabecular bone pattern factor (TBPf) [10]. Eroded surface (ES/BS) is the percent of cancellous surface occupied by Howship's lacunae, with and without osteoclasts. Osteoblast surface (Ob.S/BS) and osteoclast surface (Oc.S/BS) identify the percent of cancellous surface occupied by osteoblasts and osteoclasts, respectively. Osteoid surface (OS/BS) is the percent of cancellous surface with unmineralized osteoid, with and without osteoblasts. Osteoid thickness (O.Th) is the mean thickness in µm of the osteoid on cancellous surfaces.

Kinetic features

A fluorochrome labeling agent, taken orally on a strict schedule before biopsy, deposits a fluorescent double label at sites of active mineralization and allows rates of change to be determined [11].

Mineralizing surface (MS/BS) is the percent of trabecular surface that is mineralizing and thus labeled. The most accurate version of MS/BS includes surfaces with a double label plus one-half of those with a single label [12]. Clear definition of MS/BS is crucial, because it is used to calculate bone formation rates, bone formation periods, and mineralization lag time.

Mineral appositional rate (MAR) is the rate in μm/day at which new bone mineral is being added to cancellous surfaces. MAR represents distance between labels at doubly labeled surfaces divided by the marker interval (span in days between the midpoints of each labeling period). This and all measurements of thickness must be corrected for obliquity (ie, the randomness of the angle between the plane of the section and the plane of the cancellous surface) by use of a scaling factor [6].

Activation frequency (Ac.f) is the probability that a new remodeling cycle will begin at any point on the cancellous bone surface. Bone formation rates (BFR/BV and BFR/BS) are estimates of cancellous bone volume (in mm^3/mm^3/year) and cancellous bone surface (in $mm^3 mm^2$/mm/year), respectively, that are being replaced annually; BFR/BS = Ac.f × W.Th [13].

Formation period (FP) is the time in years required to complete a new cancellous BSU.

Mineralization lag time (Mlt) is the interval in days between osteoid formation and mineralization. The most accurate version of Mlt is calculated as O.Th/MAR × MS/OS (mineralizing surface/osteoid surface).

Microcrack density (Cr.d.) is the number of microcracks per area of mineralized bone area ($\#/mm^2$), and microcrack length (Cr.L) is the average length (in mm) of visualized microcracks [14].

Apoptosis of any bone cells—osteoblasts, osteocytes, osteoclasts—can be quantified (as percent of total in each case). Apoptosis is identified by using special stains [15].

INTERPRETATION OF FINDINGS

Reference data

In 1988, Recker et al. [3] published the results of a study to establish reference values for histomorphometric variables in postmenopausal white women. The 34 healthy subjects were evenly distributed into three age groups: 45 to 54, 55 to 64, and 65 to 74 years. A comparative study of 12 blacks and 13 whites, aged 19 to 46 years, has also been published [16]. In addition, Glorieux et al. [17] reported histomorphometric data from 58 white subjects in children, adolescents, and young adults. Others have published reference databases [17–22]. Bone histomorphometry findings may vary widely among healthy individuals which makes it difficult to establish normal values. Features such as age, gender, and race have an important influence.

Replacement of normal marrow elements

A variety of hematopoietic cells and fat cells normally occupy the marrow space at the transilial site. If these normal marrow elements are displaced by fibrous tissue, clumps of tumor cells, or sheets of abnormal hematopoietic cells, it will be obvious to the histomorphometrist. The biopsy preparations described are unsuitable for hematologic diagnosis because of the time required to generate a report (typically, at least 4 weeks).

Adipocytes in bone marrow

There is growing interest in the relationship between bone marrow fat (BMF) and bone [23,24] metabolism, because osteoblasts and adipocytes differentiate from the same mesenchymal stem cells (MSC). BMF can be measured by MRI and also by bone histomorphometry in which the adipocyte volume, perimeter, and density can be quantified. Enhanced adipogenesis in the bone marrow of osteoporotic patients is inversely correlated with trabecular bone volume. Further, higher BMF is associated with reduced bone formation [24,25].

Cortical bone deficit

Both the angle of the biopsy and site-to-site variation in cortical thickness at the biopsy site influence Ct.Wi. Nevertheless, low bone density at the lumbar spine and/or proximal femur is often reflected in low values for Ct.Wi [26]. Evidence of trabeculation of the cortex (ie, formation of a transitional zone with characteristic coarse trabeculae) indicates that cortical bone once present in the area adjacent to the marrow space has been lost [27].

Cancellous bone deficit

Low BV/TV indicates a cancellous bone deficit. Generalized trabecular thinning (decreased Tb.Th) and/or complete loss of trabecular elements (poor trabecular connectivity) may contribute to this deficit. Low Tb.N with high Tb.Sp characterizes bone that is more fragile than its overall mass would suggest [28].

Altered bone remodeling

Ac.f is an indicator of overall level of remodeling activity in cancellous bone. Values for Ac.f correlate with excretion of bone resorption markers ($r = 0.71$, Recker, unpublished).

In biopsy specimens from ostensibly healthy women we have yet to see a case in which the subject was compliant with the fluorochrome labeling protocol but label could not be found in cancellous areas. However, a recent paper from our laboratory, cited earlier, reports three cases of no label (ie, zero Ac.f) among women with untreated osteoporosis [18]. Further, extensive attention has been directed at osteoporosis drugs that may suppress remodeling to the extent that microdamage repair is not adequate. This has led to questions concerning the histomorphometric definition of abnormally low remodeling rates [29]. These authors concluded that absence of fluorochrome label in a human transiliac biopsy is evidence of abnormal reduction in remodeling.

Abnormal osteoid morphology

The difference in the arrangement of osteoid (collagen) fibers between lamellar and woven bone is readily apparent. Lamellar bone contains collagen fibers arranged in layers, and woven bone contains collagen fibers arranged in a random fashion. Woven bone is generally associated with an intense stimulus to rapid bone formation, as in Paget disease or renal osteodystrophy. It can also occur in osteitis fibrosa. In OI, collagen abnormalities result in production of variable amounts of woven bone, but may be subtle enough to escape detection. The presence of woven bone can be suspected in stained sections by the presence of increased numbers of osteocytes, and by the hint of randomly arranged collagen. However, the best way to identify woven bone is by using polarizing lenses and light microscopy of unstained sections.

Accumulation of unmineralized osteoid

Parfitt has described the complex relationships between dynamic indices of bone formation and static indices of osteoid accumulation [13]. Increases in OS/BS, O.Th, and Mlt indicate failure of osteoid to mineralize normally. If mineralization is arrested completely, no double label will be seen, and Mlt is unmeasurable [30].

FINDINGS IN METABOLIC BONE DISEASE

In Table 39.1, we identify key histomorphometric findings that characterize representative types of metabolic bone disease.

Postmenopausal osteoporosis

Osteoporosis in postmenopausal women is characterized by a cortical bone deficit with trabeculation of endocortical bone and a cancellous bone deficit with poor trabecular connectivity. Decreases in Tb.Th are modest, and dynamic measures vary widely [31, 32]. Median Ac.f. remains high in specimens from women with postmenopausal osteoporosis, but varies widely, from low to normal to high bone turnover [18].

Glucocorticoid-induced osteoporosis

Early in treatment, Ac.f is increased; later, Ac.f, MAR, and MS/BS are all decreased. In femoral specimens from patients with glucocorticoid-induced osteonecrosis, abundant apoptotic osteocytes and lining cells have been reported [33].

Primary hyperparathyroidism

Primary hyperparathyroidism leads to a cortical bone deficit, with increased Ct.Po and trabeculation of endocortical bone [34]. Ct.Po correlates positively with fasting serum PTH [35]. BV/TV is generally preserved, and normal cancellous bone architecture is maintained [36, 37]. Osteoid with a woven appearance and peritrabecular fibrosis are also found [38].

Hypogonadism

Hypogonadism in both women and men increases Ac.f and leads to deficits of both cortical bone and trabecular bone. At low levels of BV/TV and/or Tb.Th, loss of trabecular connectivity occurs [39].

Hypovitaminosis D osteopathy

Vitamin D depletion of any etiology leads to hypovitaminosis D osteopathy (HVO). Parfitt [30] describes three stages. In HVOi ("pre-osteomalacia") Ac.f and OS/BS are increased, but O.Th is not. Accumulation of unmineralized osteoid characterizes both HVOii and HVOiii (osteomalacia), with Mlt and O.Th clearly increased, ie, Mlt >100 days and O.Th >12.5 μm after correction for obliquity. Some double labels can be seen in HVOii, but not in HVOiii. A cortical bone deficit also characterizes advanced HVO, secondary hyperparathyroidism in response to reduced serum ionized calcium is usual, and fibrous tissue is frequently seen in the marrow. Hepatic enzyme–inducing anticonvulsant drugs have been most clearly associated with these problems [40, 41].

Hypophosphatemic osteopathy

Phosphate depletion of any etiology also leads to osteomalacia, with histomorphometric findings similar to those of advanced HVO [30]. These cases involve defects in phosphorus metabolism manifest as defects in renal tubular reabsorption of phosphorus and increase in FGF23. However, most cases are not the result of a primary renal

Table 39.1. Patterns of key histomorphometric findings that characterize several types of metabolic bone disease.

	Marrow Spaces	Cortical Bone	Cancellous Bone	Bone Remodeling	Osteoid Morphology	Osteoid Mineralization
Postmenopausal osteoporosis	—	Cortical bone deficit with endocortical trabeculation	Cancellous bone deficit with poor trabecular connectivity	Ac.f generally increased, but varies widely	—	—
Glucocorticoid-induced osteoporosis	—	Cortical bone deficit	Cancellous bone deficit	Early, increased Ac.f; later, decreased Ac.f	—	—
Primary hyperparathyroidism	Peritrabecular fibrosis may be seen	Cortical bone deficit, increased Ct.Po, endocortical trabeculation	Typically unremarkable	Increased Ac.f	Woven bone may be seen	—
Hypogonadism (males and females)	—	Cortical bone deficit	Cancellous bone deficit, sometimes with poor trabecular connectivity	Increased Ac.f	—	—
Hypovitaminosis D osteopathy	Fibrous tissue may be seen	—	—	Early, increased Ac.f	—	Early, increased OS/BS; later, increased Mlt and O.Th; double label may be absent
Hypophosphatemic osteopathy	Fibrous tissue may be seen	—	—	—	—	Increased Mlt and O.Th; double label may be absent
Renal osteodystrophy (high turnover type)	Fibrous tissue may be seen	Endocortical trabeculation	Osteoblast, osteocyte, and trabecular abnormalities	Markedly increased remodeling activity	Woven bone may be seen	Increased OS/BS
Renal osteodystrophy (low turnover type)	—	—	—	Markedly decreased remodeling activity	—	Increased OS/BS (osteomalacic type); decreased OS/BS (adynamic type)
Renal osteodystrophy (mixed type)	Fibrous tissue may be seen	—	Variable BV/TV	Patchy remodeling activity	Irregular, woven bone and osteoid may be seen	Increased OS/BS and O.Th

tubular abnormality, but instead are caused by an abnormality in plasma phosphorus homeostasis [42]. Secondary hyperparathyroidism occurs variably. Transilial biopsy can be quite useful to assess the efficacy of treatment.

Gastrointestinal bone disease

Evidence of HVO has been reported in a variety of absorptive and digestive disorders [43]. However, these conditions also may promote deficiency of calcium and other nutrients. Bone histomorphometry may also reflect the results of treatment (ie, corticosteroids or surgery). Parfitt describes a histomorphometric profile of low bone turnover, often with evidence of HVO, and secondary hyperparathyroidism in these patients [30]. Studies have demonstrated a significant deterioration of bone microarchitecture in premenopausal women with newly diagnosed celiac disease using HRpQCT and improvement after 12 months of a gluten-free diet [43,44].

Renal osteodystrophy

At least three patterns of histomorphometric findings have been described among patients with end-stage renal disease (ESRD): high bone turnover with osteitis fibrosa (hyperparathyroid bone disease); low bone turnover (including osteomalacic and adynamic subtypes); and mixed osteodystrophy with high bone turnover, altered bone formation, and accumulation of unmineralized osteoid [45–48] (see Chapter 90).

Transilial bone biopsy remains a useful "gold standard" on which to base decisions about treatment of bone disease in ESRD [45]. A dramatic example is the evaluation of bone pain and fractures in a chronic dialysis patient with hypercalcemia. If the biopsy shows high bone turnover and osteitis fibrosa, partial parathyroidectomy may be indicated. However, if the biopsy shows low turnover, then parathyroidectomy is contraindicated. The same biopsy can also help determine the extent of vitamin D deprivation and indicate the adequacy of vitamin D treatment.

OBTAINING THE SPECIMEN

In this section, we outline the procedures for obtaining bone biopsy specimens, processing them, and carrying out histomorphometric analysis. For greater detail, we recommend another publication [49].

Fluorochrome labeling

In clinical settings, tetracycline antibiotics are the only suitable fluorochrome labeling agents [11]. Demeclocycline (150 mg, four times daily) or tetracycline hydrochloride (250 mg, four times daily) are commonly used. The double-labeling process involves two dosing periods, and close adherence to the dosing schedule is crucial. A schedule of 3 days on, 14 days off, 3 days on, and 5 to 14 days off before biopsy (abbreviated as 3–14-3–5) produces good results, with a marker interval of 17 days [12]. Tetracyclines must be taken on an empty stomach. Thus, oral intake must be avoided for at least 1 hour before and after each dose.

Biopsy procedure

Specimens for histomorphometric examination require use of a trephine with inner diameter of ≥7.5 mm. The teeth should be sharpened (and reconditioned, if necessary) after every two or three procedures. Transilial bone biopsy is performed in an outpatient minor surgery facility using sterile procedures and monitoring vital signs. The patient should be off aspirin for at least 3 days and have nothing orally for 4 hours. A second biopsy should always be on the side opposite the first; there is thus a practical limit of two transilial biopsy specimens per patient. The gowned patient lies in the supine position on the surgical table, and midazolam (2.5 to 5 mg) is given through a forearm intravenous catheter.

The biopsy site is ~2 cm posterior to the anterior-superior spine, which is ~2 cm inferior to the iliac crest. The skin, subcutaneous tissues, and periosteum on both sides of the ilium are infiltrated with local anesthetic. The periosteum is accessed by a 2-cm skin incision and blunt dissection. The trephine is inserted and advanced with steady, gentle pressure and a deliberate pace. The specimen—an intact, unfractured core with both cortices and the intervening cancellous bone—is transferred into a 20-mL screw-cap vial containing 70% ethanol, unless an unfixed specimen is required. The bony defect is then packed with Surgicel. After local pressure to facilitate hemostasis, the wound is closed with three to five stitches and covered by a pressure dressing. Follow-up care is specified clearly (ie, dressing in place and absolutely dry for 48 hours; then a daily shower is allowed; no bathing or strenuous physical activity until suture removal, 1 week after the procedure). The procedure produces localized aching for about 2 days and a small scar at the site.

Patients typically describe feeling something "like a cramp" as the trephine advances. The biopsy procedure described here rarely evokes more than mild discomfort. Although bleeding during the procedure is typically minimal, there is risk of bleeding in some situations (eg, liver disease, hemodialysis, or medications that compromise hemostasis). Local bruising sometimes occurs, but hematoma is uncommon. In an early survey, physicians who were obtaining transilial biopsy specimens reported adverse events in 0.7% of 9131 biopsy specimens, that is, 22 with hematomas, 17 with pain for >7 days, 11 with transient neuropathy, 6 with wound infection, 2 with fracture, and 1 with osteomyelitis. No cases of death or permanent disability were reported [50].

SPECIMEN PROCESSING AND ANALYSIS

Specimen handling and processing

For routine histomorphometry, the bone biopsy specimen should remain in 70% ethanol for at least 48 hours for proper fixation. This solution is suitable for shipping and long-term storage at room temperature. The specimen vials should be filled to capacity with 70% ethanol for shipping, handling, and storage. Steps in laboratory processing include dehydrating, de-fatting, embedding, sectioning, mounting, de-plasticizing, staining, and microscopic examination.

After proper trimming, the tissue block is sectioned parallel to the long axis of the biopsy core. Two or more sets of sections are obtained at intervals of 400 μm, beginning 35% to 40% into the embedded specimen. Unstained sections 8 to 10 μm thick are used to examine osteoid morphology and to measure fluorochrome-labeled surfaces. Sections 5 to 7 μm thick stained with toluidine blue are used to measure wall thickness. Sections 5 μm thick with Goldner's stain [51] are used for other histomorphometric measurements. In some circumstances von Kossa staining is used to document the presence and extent of mineralization of bone tissue.

Microscopy

Our histomorphometry laboratory uses an interactive image analysis system (BIOQUANT OSTEO 2016 v 6.1.60 mp, BIOQUANT Image Analysis Corporation, Nashville, TN, USA). A digital camera mounted on the microscope presents the microscopic images on-screen, and measurements are made using a mouse. Fluorescent light at a wavelength of 350 nm is used to examine fluorochrome labels (see Fig. 39.3).

INDICATIONS FOR BONE BIOPSY AND HISTOMORPHOMETRY

The purpose of bone histomorphometry in the clinical setting is to gather information (ie, to establish a diagnosis, clarify a prognosis, or evaluate adherence or response to treatment) on which to base informed clinical decisions. However, the number of clinical indications for this procedure is limited. Although clinicians can manage most metabolic bone diseases, including osteoporosis, without the aid of a bone biopsy, there are some situations in which bone biopsy after fluorochrome labeling is appropriate, as outlined in Table 39.2.

Bone histomorphometry has been, and remains, crucial for assessing the mechanism of action, safety, and efficacy of new bone-active agents. Testing of every new bone-active treatment should include bone biopsy in at least a subset of subjects. Trabecular bone histomorphometry provides a method for examining both bone properties and bone physiology. Although cortical bone histomorphometry offers less information than trabecular bone, more recently cortical bone has attracted attention of the researchers. It is the most abundant type of bone of the skeleton and its porosity has been linked to bone strength [53].

Table 39.2. Examples of clinical situations in which bone histomorphometry can provide useful information*.

1. When there is excessive skeletal fragility in unusual circumstances
2. When a mineralizing defect is suspected
3. To evaluate adherence to treatment in a malabsorption syndrome
4. To characterize the bone lesion in renal osteodystrophy
5. To diagnose and assess response to treatment in vitamin D–resistant osteomalacia and similar disorders
6. When a rare metabolic bone disease is suspected

*Adapted from [52] with permission.

ACKNOWLEDGMENTS

The authors thank Susan Bare for assistance in describing technical methods and preparing digital photomicrographs.

DISCLOSURES

Dr. Recker has received research funding from Merck, Radius Pharmaceuticals, Grunenthal, and Lilly.

REFERENCES

1. Frost HM. *Intermediary Organization of the Skeleton*. Boca Raton: CRC Press, 1986.
2. Frost HM. *Bone Remodeling and its Relationship to Metabolic Bone Diseases*. Springfield Illinois: Charles C. Thomas, 1973.
3. Recker RR, Kimmel DB, Parfitt AM, et al. Static and tetracycline-based bone histomorphometric data from 34 normal postmenopausal females. J Bone Miner Res. 1988;3:133–44.
4. Malluche HH, Faugere MC. In: Malluche HH, Faugere MC (eds) *Atlas of Mineralized Bone Histology*. New York: Karger, 1986.
5. Dempster DW, Compston JE, Drezner MK, et al. Standardized nomenclature, symbols, and units for bone histomorphometry: A 2012 update of the report of the ASBMR Histomorphometry Nomenclature Committee. J Bone Miner Res. 2013;28(1):2–17.
6. Parfitt AM. The physiologic and clinical significance of bone histomorphometric data. In: Recker RR (ed.) *Bone Histomorphometry: Techniques and Interpretation*. Boca Raton: CRC Press, 1983, pp 143–224.

7. Garrahan NJ, Mellish RWE, Compston JE. A new method for the two-dimensional analysis of bone structure in human iliac crest biopsies. J Microsc. 1986;142:341–9.
8. Vesterby A, Gundersen HJG, Melsen F. Star volume of marrow space and trabeculae of the first lumbar vertebra: sampling efficiency and biological variation. Bone. 1989;10:7–13.
9. Vesterby A, Gundersen HJG, Melsen F, et al. Marrow space star volume in the iliac crest decreases in osteoporotic patients after continuous treatment with fluoride, calcium, and vitamin D2 for five years. Bone. 1991;12:33–7.
10. Hahn M, Vogel M, Pompesius-Kempa M, et al. Trabecular bone pattern factor: a new parameter for simple quantification of bone microarchitecture. Bone. 1992;13:327–30.
11. Frost HM. Measurement of human bone formation by means of tetracycline labelling. Can J Biochem Physiol. 1969;41:331–42.
12. Schwartz MP, Recker RR. The label escape error: determination of the active bone-forming surface in histologic sections of bone measured by tetracycline double labels. Metab Bone Dis Relat Res. 1982;4:237–41.
13. Parfitt AM. Physiologic and pathogenetic significance of bone histomorphometric data. In: Coe FL, Favus M (eds) *Disorders of Bone and Mineral Metabolism* (2nd ed.). Philadelphia: Lippincott Williams & Wilkins, 2002, pp 469–85.
14. Chapurlat RD, Arlot M, Burt-Pichat B, et al. Microcrack frequency and bone remodeling in postmenopausal osteoporotic women on long-term bisphosphonates: a bone biopsy study. J Bone Miner Res. 2007;22:1502–9.
15. Jilka RL, Weinstein RS, Parfitt AM, et al. Quantifying osteoblast and osteocyte apoptosis: challenges and rewards. J Bone Miner Res. 2007;22:1492–505.
16. Weinstein RS, Bell NH. Diminished rates of bone formation in normal black adults. N Engl J Med. 1988;319:1698–701.
17. Glorieux FH, Travers R, Taylor A, et al. Normative data for iliac bone histomorphometry in growing children. Bone. 2000;26:103–9.
18. Recker R, Lappe J, Davies KM, et al. Bone remodeling increases substantially in the years after menopause and remains increased in older osteoporosis patients. J Bone Miner Res. 2004;19:1628–33.
19. Parfitt AM, Travers R, Rauch F, et al. Structural and cellular changes during bone growth in healthy children. Bone. 2000;27:487–94.
20. Cosman F, Morgan D, Nieves J, et al. Resistance to bone resorbing effects of PTH in black women. J Bone Miner Res. 1997;12:958–66.
21. Han Z-H, Palnitkar S, Rao DS, et al. Effects of ethnicity and age or menopause on the remodeling and turnover of iliac bone: implications for mechanisms of bone loss. J Bone Miner Res. 1997;12:498–508.
22. Dahl E, Nordal KP, Halse J, et al. Histomorphometric analysis of normal bone from the iliac crest of Norwegian subjects. Bone Miner. 1988;3:369–77.
23. Blake GM, Griffith JF, Yeung DKW, et al. Effect of increasing vertebral marrow fat content on BMD measurement, T-score status and fracture risk prediction by DXA. Bone. 2009;44:495–501.
24. Verma S, Rajaratnam JH, Denton J, et al. Adipocytic proportion of bone marrow is inversely related to bone formation in osteoporosis. J Clin Pathol. 2002;55(9):693–8.
25. Justesen J, Stenderup K, Ebbesen EN, et al. Adipocyte tissue volume in bone marrow is increased with aging and in patients with osteoporosis. Biogerontology. 2001;2(3):165–71.
26. Cosman R, Schnitzer MB, McCann PD, et al. Relationships betwen quantitative histological measurements and noninvasive assessments of bone mass. Bone. 1992;13:237–42.
27. Keshawarz NM, Recker RR. Expansion of the medullary cavity at the expense of cortex in postmenopausal osteoporosis. Metab Bone Dis Relat Res. 1984;5:223–8.
28. Parfitt AM. Age-related structural changes in trabecular and cortical bone: cellular mechanisms and biomechanical consequences. Calcif Tissue Int. 1984;36(suppl 1):S123–8.
29. Recker RR, Kimmel DB, Dempster D, et al. Issues in modern bone histomorphometry. Bone. 2011;49:955–64.
30. Parfitt AM. Osteomalacia and related disorders. In: Avioli LV, Krane SM (eds) *Metabolic Bone Disease and Clinically Related Disorders*. Boston: Academic Press, 1998, pp 327–86.
31. Kimmel DB, Recker RR, Gallagher JC, et al. A comparison of iliac bone histomorphometric data in postmenopausal osteoporotic and normal subjects. Bone Miner. 1990;11(2):217–35.
32. Recker RR, Barger-Lux MJ. Bone remodeling findings in osteoporosis. In: Marcus R, Feldman D, Kelsey J (eds) *Osteoporosis* (2nd ed.). San Diego: Academic Press, 2001.
33. Weinstein RS, Nicholas RW, Manolagas SC. Apoptosis of osteocytes in glucocorticoid-induced osteonecrosis of the hip. J Clin Endocrinol Metab. 2000;85:2907–12.
34. Ericksen E. Primary hyperparathyroidism: Lessons from bone histomorphometry. J Bone Miner Res. 2002; 17:S2:N95–N97.
35. van Doorn L, Lips P, Netelenbos JC, et al. Bone histomorphometry and serum concentrations of intact parathyroid hormone (PTH(1-84)) in patients with primary hyperparathyroidism. Bone Miner. 1993;23:233–42.
36. Parisien M, Mellish RWE, Silverberg SJSE, et al. Maintenance of cancellous bone connectivity in primary hyperparathyroidism: trabecular strut analysis. J Bone Miner Res. 1992;7:913–9.
37. Uchiyama T, Tanizawa T, Ito A, et al. Microstructure of the trabecula and cortex of iliac bone in primary hyperparathyroidism patients determined using histomorphometry and node-strut analysis. J Bone Miner Res. 1999;17:283–8.
38. Monier-Faugere M-C, Langub MC, Malluche HH. Bone biopsies: a modern approach. In: Avioli LV, Krane SM (eds) *Metabolic Bone Disease and Clinically Related Disorders* (3rd ed.). San Diego: Academic Press, 1998, pp 237–73.
39. Audran M, Chappard D, Legrand E, et al. Bone microarchitecture and bone fragility in men: DXA and histomorphometry in humans and in the orchidectomized rat model. Calcif Tissue Int. 2001;69:214–7.

40. Pack AM, Morrell MJ. Epilepsy and bone health in adults. Epilepsy Behav. 2004;5:S24–S29.
41. Fitzpatrick LA. Pathophysiology of bone loss in patients receiving anticonvulsant therapy. Epilepsy Behav 2004;5:S3–S15.
42. Antoniucci DM, Yamashita T, Portale AA. Dietary phosphorus regulates serum fibroblast growth factor-23 concentrations in healthy men. J Clin Endocrinol Metab. 2006;91:3144–9.
43. Arnala I, Kemppainen T, Kroger H, et al. Bone histomorphometry in celiac disease. Ann Chir Gynaecol. 2001; 90:100–4.
44. Ott SM, Tucci JR, Heaney RP, et al. Hypocalciuria and abnormalities in mineral and skeletal homeostasis in patients with celiac sprue without intestinal symptoms. Endocrinol Metab. 1997;4:201–6.
45. Pecovnik BB, Bren A. Bone histomorphometry is still the golden standard for diagnosing renal osteodystrophy. Clin Nephrol. 2000;54:463–9.
46. Parker CR, Blackwell PJ, Freemont AJ, et al. Biochemical measurements in the prediction of histologic subtype of renal transplant bone disease in women. Am J Kidney Dis. 2002;40(385):396.
47. Elder G. Pathophysiology and recent advances in the management of renal osteodystrophy. J Bone Miner Res. 2002;17:2094–105.
48. Malluche HH, Langub MC, Monier-Faugere MC. Pathogenisis and histology of renal osteodystrophy. J Bone Miner Res. 1997;7:S184–S187.
49. Recker RR, Barger-Lux MJ. Transilial bone biopsy. In: Bilezikian JP, Raisz L, Rodan GA (eds) *Principles of Bone Biology* (2nd ed.). San Diego: Academic Press, 2001, pp 1625–34.
50. Rao DS, Matkovic V, Duncan H. Transiliac bone biopsy: complications and diagnostic value. Henry Ford Hosp Med J. 1980;28:112–8.
51. Goldner J. A modification of the Masson trichrome technique for routine laboratory purposes. Am J Pathol. 1938;14:237–43.
52. Kimmel DB, Jee WSW. Measurements of area, perimeter, and distance: Details of data collection in bone histomorphometry. In: Recker RR (ed.) *Bone Histomorphometry: Techniques and Interpretation.* Boca Raton: CRC Press, 1983, pp 80–108.
53. Zebaze R, Seeman E. Cortical bone: a challenging geography. J Bone Miner Res. 2015;30(1):24–9.

40

Diagnosis and Classification of Vertebral Fracture

James F. Griffith[1] and Harry K. Genant[2]

[1] Department of Imaging and Interventional Radiology, The Chinese University of Hong Kong, Hong Kong
[2] Departments of Radiology, Orthopedic Surgery, Medicine, and Epidemiology, University of California at San Francisco, San Francisco, CA, USA

SIGNIFICANCE OF VERTEBRAL FRACTURE

Vertebral fracture is the most common and usually the first osteoporotic fracture to occur, being present in 15% of women aged 50 to 59 years, 15% of men aged 69 to 81 years, and 50% of women aged more than 85 years [1, 2]. Accurate recognition of vertebral fracture is essential to clinical evaluation, determination of population prevalence, fracture risk, and evaluating treatment [3]. Nearly half of all vertebral fractures occur in patients with low BMD (T-score at or below −1.0; osteopenia) rather than osteoporosis (T-score at or below −2.5) because low BMD is much more common than osteoporosis [4]. The presence of a nontraumatic vertebral fracture provides indisputable evidence of reduced bone strength, ie, osteoporosis.

Vertebral fracture risk is 20% in the year following incident (ie, new) vertebral fracture. However, this relative risk is fourfold greater in those with a severe, rather than a mild, fracture and threefold greater in those with multiple, rather than a single, vertebral fracture [5]. Vertebral fracture is also associated with reduced quality of life, reduced self-esteem, and increased mortality, particularly from pulmonary disease and cancer [6]. Early recognition of vertebral fracture and appropriate treatment of osteoporosis significantly reduces the occurrence of new vertebral and nonvertebral fractures [7].

Despite the clear clinical importance of vertebral fractures, they remain underdiagnosed in clinical practice [8]. This is because the typical clinical symptoms of back pain and restricted movement are usually attributed to degenerative change [9]. Also, many vertebral fractures evident on imaging are not reported. Radiologists and clinicians who review imaging studies are urged to clearly report the presence of vertebral fractures [3]. Ambiguous terminology (eg, "vertebral collapse," "compressed vertebral body," "loss of vertebral height," "wedging of vertebral body," "wedge deformity," "biconcavity," or "codfish deformity") should be avoided in isolation, or used in conjunction with the term "vertebral fracture." Proper identification of vertebral fracture is critically important also for research in that over- or under-reporting by an inexperienced reader can significantly skew research findings [10].

DETECTION OF VERTEBRAL FRACTURE

Clinical detection

Only about one in four vertebral fractures is recognized as a distinct clinical event because symptoms are often mild and nonspecific [11]. Probably the most effective clinical discriminators of incident vertebral fracture are a measured height loss of >2 cm or a recalled height loss of >4 cm [12].

Spinal radiography

Osteoporotic vertebral fractures usually occur in the presence of reduced bone density. On radiography, one can accurately distinguish between unequivocally normal bone and unequivocally osteoporotic bone. However, bone in most subjects is neither unequivocally normal nor unequivocally osteoporotic. Radiographic detection

---

Primer on the Metabolic Bone Diseases and Disorders of Mineral Metabolism, Ninth Edition. Edited by John P. Bilezikian.
© 2019 American Society for Bone and Mineral Research. Published 2019 by John Wiley & Sons, Inc.
Companion website: www.wiley.com/go/asbmrprimer

320 Diagnosis and Classification of Vertebral Fracture

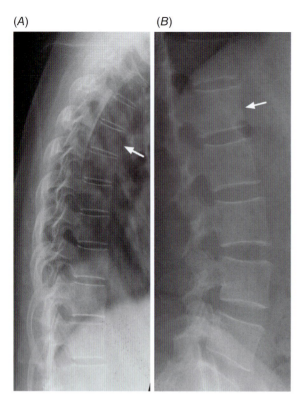

Fig. 40.1. (A) Lateral radiograph of normal thoracic spine in a subject with low BMD. There is minor physiological anterior wedging of the T6 vertebral body (arrow). (B) Lateral radiograph of normal lumbar spine. There is slight physiological wedging of the L1 vertebral body (arrow).

of reduced bone density and osteoporosis is subjective and dependent on radiographic technique, equipment, and patient body habitus. Additional radiographic signs of osteoporosis such as cortical thinning and porosity, as well as trabecular rarefaction, are helpful to recognize osteoporosis although appreciation is heavily dependent on the observer's experience.

A standard radiographic protocol of the thoracic and lumbar spine should visualize the C7 to S1 vertebrae on finely collimated properly positioned antero-posterior (AP) and lateral views with the X-ray beam centered at T7 and L3 for the thoracic and lumbar spines respectively and a focus–film distance of 100 cm (Fig. 40.1A,B). The upper thoracic vertebrae are not clearly seen on lateral thoracic spine radiographs but, fortunately, isolated osteoporotic fractures are uncommon in this region. On the lateral projection, the spine must be parallel to the film so that the vertebral endplates at the level of the central X-ray beam are seen as a single, dense, well-defined cortical line. As a result of the divergent X-ray beam the endplates distant from the centering point appear slightly concave and must not be mistaken for vertebral fractures. The AP projection is helpful mainly in scoliotic patients though mild vertebral fractures may be overlooked on this projection. The typical effective dose of ionizing radiation from a single lateral and AP projection of the thoracic spine is 0.3 mSv and 0.4 mSv respectively, whereas for the lumbar spine it is 0.3 mSv and 0.7 mSv respectively. For comparison, a 16-hour return transatlantic flight would amount to 0.07 mSv background radiation [13].

Problems in the diagnosis of vertebral fracture on radiographs

Although there is good agreement in the radiological diagnosis of moderate and severe vertebral fractures, there is considerable contention in the diagnosis of mild vertebral fractures. The difficulty in defining a mild vertebral fracture radiographically is a reflection of six potential pitfalls as follows:

1. Mild physiological wedging may be misinterpreted as a mild fracture but is a normal and entirely necessary feature of the thoracic and lumbar vertebral bodies as the spine changes from a thoracic kyphosis to a lumbar lordosis. Normal spinal curvature dictates that vertebral bodies are slightly anteriorly wedge shaped in the thoracic and upper lumbar spine, and slightly posteriorly wedge shaped in the lower lumbar region (Fig. 40.2A,B) [14]. The degree of wedging is dependent on sagittal spinal curvature.

2. Short vertebral body height (SVH) is a common feature of increasing age and spondylosis in the absence of osteoporosis (Fig. 40.3). In adult women, the combined height of the anterior aspects of the vertebral bodies from T4 to L5 decreases by about 1.5 mm per year while the combined middle and posterior heights decline by about 1.2 mm per year [15]. Although SVH refers to reduction in vertebral height of up to about 20% of expected height, differentiating SVH from a mild vertebral fracture is probably the most contentious and difficult area in vertebral fracture diagnosis. Most evidence suggests that isolated SVH is not associated with low BMD or vertebral fracture [16].

3. Nonfracture vertebral deformity including SVH is most commonly the result of degenerative- or stress-related remodeling of the vertebral body. It can also be a feature of an uncommon spinal developmental or acquired disorder known as Scheuermann disease. This condition, observed during late adolescence and probably stress-induced, is characterized by endplate irregularity/indentation of the thoracic or lumbar vertebrae (Fig. 40.4A,B). The disorder may involve only one or two vertebrae or longer segments of the spine. It is typically associated with reduced vertebral height, sometimes with an increased anteroposterior vertebral body diameter, and usually with small disks and premature degeneration. The deformity is largely irreversible and its persistence into later life makes distinction from some forms of osteoporotic vertebral fracture difficult, especially if there are no earlier spinal radiographs for comparison.

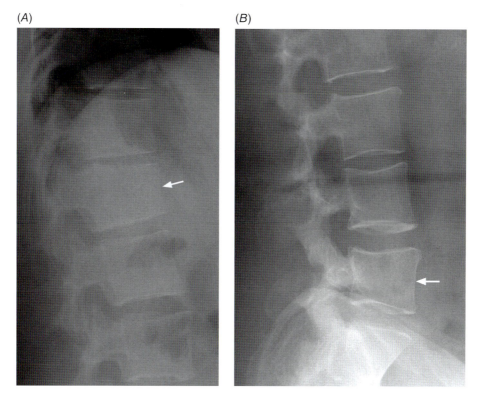

Fig. 40.2. (A) Anterior physiological wedging of the L1 vertebral body (arrow). (B) Posterior physiological wedging of the L5 vertebral body (arrow).

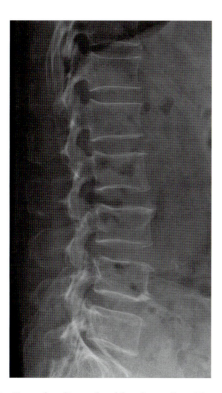

Fig. 40.3. Lateral radiograph of lumbar spine. There is "short vertebral height" with all of the lumbar vertebrae reduced in height by 10% to 15% compared to expected vertebral body height. Note associated spondylosis with marginal osteophytosis.

4. Degenerative-type scoliosis is common in the elderly and may lead to obliquity of vertebral bodies and side-to side discrepancy in vertebral body height. On the lateral projection, this obliquity produces a spurious biconcave outline to the vertebral endplates which may be misinterpreted as a vertebral fracture. On the AP projection, the vertebral bodies are shortened on the concave side and of normal height or even elongated on the convex side of the curve. This scoliotic wedging, providing it is predominantly one-sided and commensurate with the severity of scoliosis, should not be misinterpreted as a vertebral fracture.

5. Schmorl's nodes are discrete indentations of the endplates caused by remote intervertebral disk herniation. Most Schmorl's nodes are small and have well-defined sclerotic borders, unlike osteoporotic vertebral fractures. Occasionally, medium-sized or large Schmorl's nodes may be misinterpreted as an endplate fracture (Fig. 40.4A,C).

6. Cupid's bow deformity is a developmental abnormality arising from a focal deficiency of the cartilage endplate and, perhaps, enlarged nucleus pulposus, so-called megalonucleus, most frequently affecting the inferior endplates of the fourth and fifth lumbar vertebral bodies [17] (Fig. 40.4A,D). It leads to characteristic concave endplate depressions seen radiographically resembling a "Cupid's bow" on the AP projection and indentation of the posterior two-thirds of the inferior endplate on a lateral projection simulating an endplate fracture.

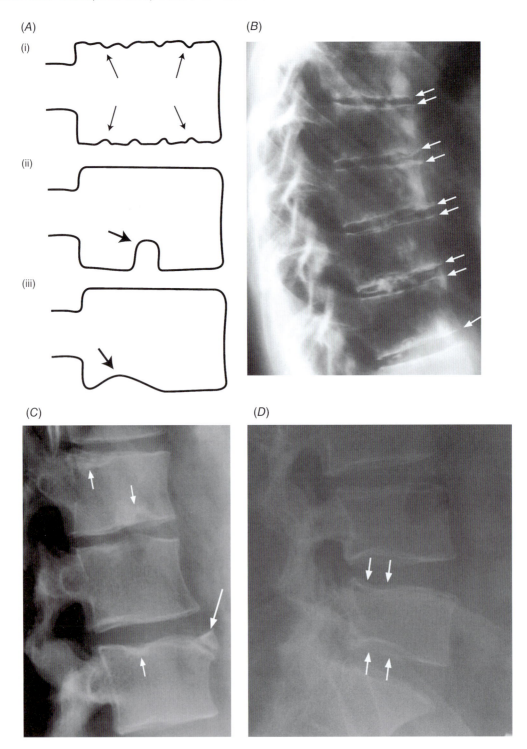

Fig. 40.4. (A) Schematic diagram showing endplate impressions caused by (i) Scheuermann disease, (ii) Schmorl's node, and (iii) Cupid's bow deformity. (B) Lateral radiograph midthoracic region showing diffuse endplate irregularity (arrows), narrowing of the intervertebral disks, and elongated AP diameter of the vertebral bodies, all consistent with Scheurmann disease. (C) Lateral radiograph of lumbar spine showing medium-sized Schmorl's nodes of the superior and inferior endplates (small arrows). There is also a limbus vertebra present (long arrow). (D) Lateral radiograph lower lumbar region showing smooth indentation of inferior and superior endplates of L5 (arrows) caused by Cupid's bow deformity.

Defining a vertebral fracture

In an effort to produce definable, reproducible, and objective methods of detecting vertebral fracture, several methods have been developed and refined to diagnose and grade severity of vertebral fracture on radiographs. These methods, which have also been applied to DXA vertebral fracture assessment (VFA) and CT images, can be broadly considered as either qualitative, semiquantitative (SQ), or quantitative in approach.

Semiquantitative (SQ) assessment

SQ analysis involves evaluation of spinal radiographs by an experienced reader without prior measurement of vertebral heights. The most widely used SQ approach is that of Genant et al. [18, 19] (Fig. 40.5). Vertebral fractures are graded from 1 (mild) to 3 (severe), and incident fractures are defined as an increase of one grade or more on follow-up radiographs (Fig. 40.6). Grade 1 (mild) vertebral fracture corresponds to a ~20% to 25% reduction in anterior, middle, and/or posterior height compared to the expected height of the vertebral body based on experience (Fig. 40.5). Grade 2 (moderate) vertebral fracture is a ~25% to 40% reduction in vertebral height (Fig. 40.5), and grade 3 (severe) vertebral fracture is a ~>40% reduction in vertebral height (Fig. 40.5). The approximation symbol (~) is applied because the height reductions are visually estimated rather than measured directly. Additionally, other morphologic changes of endplate buckling or bowing and cortical fractures are factored into the diagnosis, particularly in distinguishing mild vertebral fracture from nonfracture vertebral deformity. A spinal deformity index (SDI) can be assigned to each patient by summating the SQ scores for the T4 to L4 vertebrae [20].

Grading severity of vertebral fracture recognizes the incremental nature of vertebral fracture and enables fracture progression from mild to moderate or moderate to severe to be meaningfully described on follow-up radiographs (Fig. 40.6). The more severe the vertebral fracture, the greater the deterioration in bone architectural parameters [21].

The SQ method has excellent inter-reader reliability among trained observers. Agreement between each of three readers and a consensus reading yields a kappa score of 0.84 to 0.96 [22]. SQ analysis is much quicker to perform than other methods of vertebral fracture assessment, is easy to implement in clinical practice, and is suited to both epidemiologic research studies and clinical practice. For follow-up studies, serial radiographs should be viewed together in temporal order to fully appreciate changing vertebral morphology. Although visual assessment methods of vertebral fracture detection are potentially more subjective than morphometric analysis, they do allow the experienced reader to address issues such as nonosteoporotic deformity. SQ analysis is also better suited to deal with errors introduced by radiographic technique such as varying projection and magnification effects, which clearly would influence serial vertebral body measurements. The SQ method is a practical and reproducible method of VFA when performed by trained and experienced readers [23].

Quantitative morphometry (QM)

Vertebral QM is used in a research rather than a clinical setting [24]. The two main advantages of vertebral morphometry over other methods are that (i) it can in theory be undertaken by inexperienced or nonmedical research staff, and (ii) it provides an objective measure of loss of vertebral height on serial images. Although the methodology is straightforward to define and readily describable, its application in practice requires training and is often rather subjective. Further, serial assessments are subject to moderate errors related to radiographic projectional differences.

The margins of each vertebral body from T4 to L4 are identified by six points on the upper and lower endplates – one for each corner and one for each of the endplate midpoints (Fig. 40.7). Marginal osteophytes and Schmorl's nodes are not included.

Placement of the six points can be manual or automated. The automated method comprises computerized vertebral boundary recognition of a digitalized radiograph or other image. Automated point placement is checked and adjusted, if necessary, by a trained reader. The anterior (A_H), middle (M_H), and posterior (P_H) vertebral heights are measured.

Vertebral height ratios are used to define vertebral shape with A_H/P_H reflecting anterior wedging, M_H/P_H reflecting endplate concavity, and $P_H/P_{H'}$ of the adjacent normal vertebrae reflecting posterior compression [25]. Prevalent vertebral fracture is defined as a reduction in one or more of the three vertebral height ratios (A_H/P_H, M_H/P_H, or $P_H/P_{H'}$) of greater than 20% or 3 SD from the mean of a reference population. Incident vertebral fracture is defined as a reduction in one of the three height ratios (A_H/P_H, M_H/P_H, or $P_H/P_{H'}$) greater than 15% to 20% with or without a 3 to 4 mm minimal decrease compared to baseline [26]. Although the reproducibility of QM is good in normal subjects and when repeated on the same images (interobserver coefficient of variation of ~2%), it is poorer in the very elderly and in those with osteoporotic fractures (interobserver coefficient of variation of 6.3% for M_H) [27].

Positioning QM reference points is partly subjective, especially for the vertebral midpoints where obliquity of the radiographic beam often produces a double endplate contour. Even with good radiographic technique, a mild degree of scoliosis will lead to the endplate being visualized slightly en face. In such situations, observer experience will influence reference point placement for baseline and sequential imaging examinations. Small differences in reference point placement on follow-up radiographs can result in misdiagnosis of incident vertebral fracture

Grade 0: normal, unfractured vertebra.

Grade 0.5: uncertain or questionable fracture with borderline 20% reduction in anterior, middle, or posterior heights relative to the same or adjacent vertebrae.

Grade 1: mid fracture with approximately 20–25% reduction in anterior, middle, or posterior heights relative to the same or adjacent vertebrae.

Grade 2: moderate fracture with approximately 25–40% reduction in anterior, middle, or posterior heights relative to the same or adjacent vertebrae.

  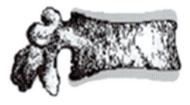

Grade 3: severe fracture with approximately >40% reduction in anterior, middle, or posterior heights relative to the same or adjacent vertebrae.

Fig. 40.5. Schematic diagram of Genant SQ analysis of vertebral fracture severity. Adapted from [19] with permission from John Wiley & Sons.

by QM. QM also does not distinguish between vertebral fracture and nonfracture vertebral deformity (such as SVH or physiological wedging). Although there is good concordance between QM and SQ methods in the detection of moderate or severe vertebral fractures, the concordance between both methods for detection of mild vertebral fractures is poor. In most instances, this is the result of a false positive diagnosis of mild fractures by QM [27]. Therefore, all fractures identified by vertebral morphometry should be confirmed by an expert reader [27].

Defining normative databases for QM is fraught with difficulty, as is defining the threshold used for vertebral

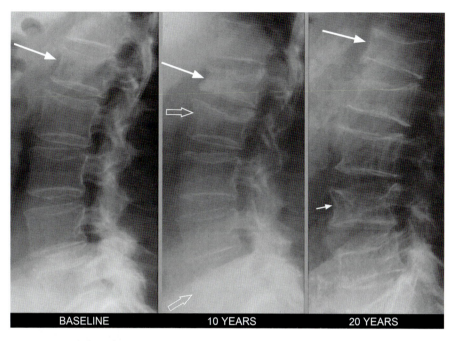

Fig. 40.6. Lateral radiographs of lumbar spine showing progression of vertebral fractures in a given subject akin to "vertebral cascade." At baseline, there is a mild vertebral fracture of L1 (long arrow). Ten years later, this L1 vertebral fracture is moderate in severity (long arrow) and there are new fractures of L2 and L3 (open arrows). Twenty years later, there is a new fracture of L4 (short arrow).

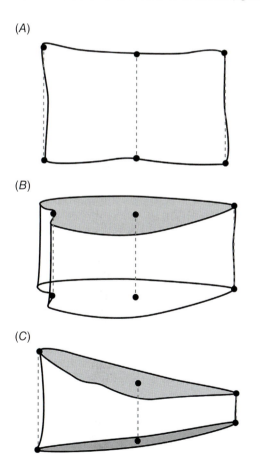

Fig. 40.7. Schematic diagram showing reference point placement for (A) normal vertebral body, (B) obliquely depicted vertebral body, and (C) anterior wedge fracture.

fracture diagnosis, because the quality of reference data and, more particularly, the thresholds chosen will greatly influence fracture detection rate [3]. Vertebral morphometry is probably most suited to evaluating (i) large longitudinal studies and (ii) individual vertebrae in DXA VFA where large normative databases are available. More sophisticated automated models for vertebral morphometry have been developed using statistical model–based and other approaches [28].

Algorithm-based qualitative (ABQ) assessment

The ABQ method, as the name implies, emphasizes a qualitative assessment of vertebral fracture and relies more on detection of vertebral endplate abnormalities related to fracture than loss of vertebral body height. The ABQ method categorizes vertebrae as either (i) normal, (ii) osteoporotic fracture, or (iii) nonosteoporotic deformity or SVH. The diagnosis of an osteoporotic vertebral fracture requires evidence of vertebral endplate fracture ± loss of expected vertebral height but with no minimum threshold for apparent reduction in vertebral height [29]. If a fracture line of the cortical margin is also visible radiographically, this provides clear-cut evidence that there is a fracture present and it is likely to be of recent origin. When one or more vertebral heights (anterior, middle, or posterior) is shorter than expected but without specific endplate abnormalities of fracture (such as altered bone texture due to trabecular microfracture), this is designated as a nonosteoporotic deformity.

DXA

Imaging vertebral fractures using modern DXA technology is known as vertebral fracture assessment (VFA) (Fig. 40.8). VFA has several advantages over radiography including greater convenience as it can be performed at the same time as DXA and on the same equipment, with lower radiation dose (less than 5% of spine radiography) and low cost [13, 30]. Combining prevalent vertebral fracture status with BMD enhances fracture risk prediction of both vertebral and nonvertebral fractures [31].

Following image acquisition, manual or automated vertebral morphometry known as MXA is available [2]. The vertebral body is demarcated by four or six reference points, and vertebral body height, height ratios, and average height calculated automatically (Fig. 40.9) and compared with normative reference data [3]. Superimposition of these baseline reference points on follow-up VFA spine images allows ready comparison of baseline and follow-up VFA [3]. VFA in children may be problematic because consistent definition of the vertebral outline is not always possible [32].

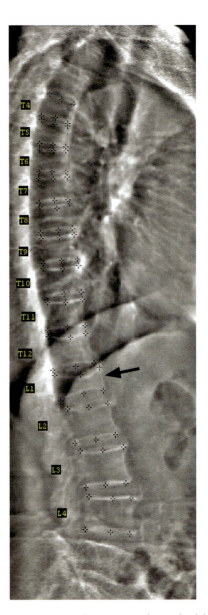

Fig. 40.9. DXA VFA with margins of vertebral bodies from T4 to L4 outlined by six reference points. There is a mild fracture of the L1 vertebral body (arrow).

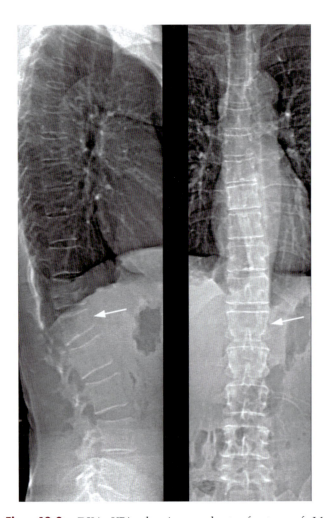

Fig. 40.8. DXA VFA showing moderate fracture of L1 vertebral body (arrow).

Although morphometric analysis is routinely undertaken on VFA, visual inspection using the Genant SQ method is the International Society of Clinical Densitometry (ISCD) recommendation for diagnosing and grading vertebral fracture severity on VFA (Table 40.1). SQ analysis of vertebral fracture on VFA compares well with radiography [19, 33] for detecting moderate or severe vertebral fractures [34]. Because the X-ray beam is orthogonal, rather than divergent, image distortion is less than on radiographs. However, VFA is subject to the same difficulty as radiography in assessing vertebral fracture when scoliosis is present. VFA will have an increasingly important role in vertebral fracture diagnosis. The added

Table 40.1. ISCD Recommendations for screening for vertebral fractures using VFA http://www.iscd.org/visitors/positions/OfficialPositionsText.cfm.

- Consider VFA when the results may influence clinical management
- Postmenopausal women with low bone mass (osteopenia) by BMD criteria, *plus* any one of the following:
  - Age greater than or equal to 70 years
  - Historical height loss greater than 4 cm (1.6 in)
  - Prospective height loss greater than 2 cm (0.8 in)
  - Self-reported vertebral fracture (not previously documented)
  - Two or more of the following:
    - age 60 to 69 years
    - self-reported prior nonvertebral fracture
    - historical height loss of 2 to 4 cm
    - chronic systemic diseases associated with increased risk of vertebral fractures (for example, moderate to severe chronic obstructive pulmonary or airway disease [COPD/COAD], seropositive rheumatoid arthritis, Crohn's disease)
- Men with low bone mass (osteopenia) by BMD criteria, *plus* any one of the following:
  - Age 80 years or older
  - Historical height loss greater than 6 cm (2.4 in)
  - Prospective height loss greater than 3 cm (1.2 in)
  - Self-reported vertebral fracture (not previously documented)
  - Two or more of the following:
    - age 70 to 79 years
    - self-reported prior nonvertebral fracture
    - historical height loss of 3 to 6 cm
    - on pharmacologic androgen deprivation therapy or following orchiectomy
    - chronic systemic diseases associated with increased risk of vertebral fractures (for example, moderate to severe COPD or COAD, seropositive rheumatoid arthritis, Crohn's disease)
- Women or men on chronic glucocorticoid therapy (equivalent to 5 mg or more of prednisone daily for 3 months or longer).
- Postmenopausal women or men with osteoporosis by BMD criteria, if documentation of one or more vertebral fractures will alter clinical management

information provided by VFA can be incorporated into the Fracture Risk Assessment Tool (FRAX) model along with DXA BMD data and clinical risk factors to improve individual 10-year prediction of developing a major osteoporotic fracture [35].

OTHER IMAGING METHODS

CT

The ease of sagittal reconstructions with multidetector CT (MDCT) allows the spine to be evaluated on all thoracic or abdominal CT studies undertaken for nonspinal clinical indications (Fig. 40.10), enabling the fortuitous identification of vertebral fractures [11]. The CT scout views should also be scrutinized routinely because these usually include a greater length of the spine than that covered by axial sections [36, 37]. The major limitation to the more widespread primary, rather than fortuitous, use of CT for vertebral fracture diagnosis is the lack of availability and the radiation dose involved, although the latter will improve with techniques such as iterative CT which dramatically reduce radiation dose [38].

MRI

Because vertebral fractures are typically incremental in development, with occasional mild stepwise progression in severity, high sensitivity and specificity with radiographic assessment may not be achievable. A vertebral fracture is diagnosed on radiography when there is at least 20% loss of vertebral height or visible cortical/endplate fracture. This approach clearly overlooks a significant number of mild vertebral fractures. MRI can help solve this problem by demonstrating marrow edema in even the mildest of acute vertebral fractures. Marrow edema, in the absence of marrow infiltration, is a sensitive sign of acute or subacute vertebral fracture even when no fracture is visible radiographically.

Determining the age of a vertebral fracture is often difficult radiographically in the absence of previous radiographs. Lack of a cortical fracture line, the presence of reparative sclerosis, and marginal osteophytosis point to a chronic fracture. However, these features are insensitive and chronic fractures may refracture. The presence and degree of edema on sagittal T2-weighted fat-suppressed MR images is a reliable guide as to presence and acuteness of a vertebral fracture [39]. These acute

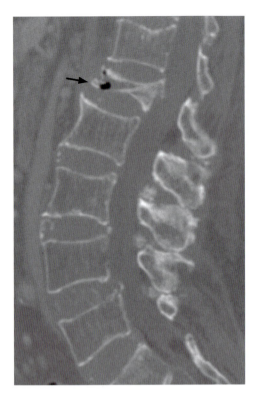

Fig. 40.10. Sagittal reconstruction of abdominal CT data set (bone window) showing moderate generalized osteopenia and severe fracture of L1 (arrow) vertebral body with intravertebral gas.

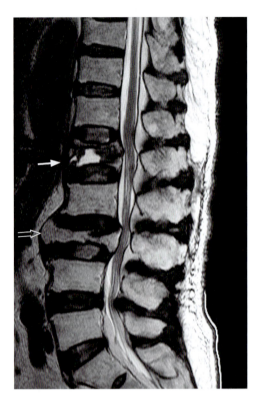

Fig. 40.11. T2-weighted sagittal MR image showing typical osteoporotic-type fracture of T12 vertebral body with preservation of some marrow fat and fluid-filled cavity within vertebral body (closed arrow). There is a chronic fracture of the L3 vertebral body with fat occupying the marrow cavity (open arrow).

fractures are more likely to respond to percutaneous vertebroplasty or kyphoplasty [39].

DIFFERENTIATING OSTEOPOROTIC FRACTURE FROM NEOPLASTIC FRACTURE

The spine is the most common site for both skeletal metastases and osteoporotic fracture. Almost one-third of vertebral fractures in patients with a known primary malignancy are caused by osteoporosis and not neoplasia [40]. In the acute stage, the marrow cavity of an osteoporotic fractured vertebral body may be filled with blood and fluid. This is gradually reabsorbed and replaced by granulation and fibroblastic tissue. This reparative tissue is reabsorbed over time with restoration of normal fatty marrow. It is thus not difficult to differentiate a chronic osteoporotic vertebral fracture in which the marrow is filled with fat from a pathological fracture caused by neoplasia (Fig. 40.11). Difficulty arises when differentiating acute/subacute osteoporotic fracture from neoplastic fracture.

Table 40.2 outlines the most helpful imaging signs to discriminate between acute/subacute osteoporotic fracture and neoplastic fracture [39]. A combination of these signs should be applied with suitable weighting being given to their relative discriminatory power. Residual marrow fat within the fractured vertebral body is a particularly helpful sign of osteoporotic fracture. Fluid in a cavity within the vertebral body is another helpful sign, being present in more (about 40%) osteoporotic than neoplastic fractures (about 6%) [41] (Fig. 40.11). Fluid accumulates in a fracture cavity adjacent to the endplates and is equivalent to the vacuum cleft seen on radiographs or CT. Gas within a vacuum cleft arises from compression/decompression forces leading to the release of nitrogen [42]. Fluid in the vertebral body on MRI and gas on radiographs both indicate a medium to large cavity within the vertebral body and provide good evidence that a vertebral body is not filled with neoplastic or other soft tissue. Intravenous contrast medium enhancement is not a useful discriminatory criterion because acute/subacute vertebral fractures enhance, as do most neoplastic-related fractures [39]. In case of doubt, standard MRI can be supplemented by diffusion-weighted [41] or chemical shift MRI [44]. The authors do not use these techniques and consider CT to be a more useful adjunct investigation. In practice, one can usually distinguish accurately between osteoporotic and neoplastic vertebral fractures on imaging grounds alone without the need for percutaneous biopsy.

Table 40.2. Useful imaging signs that help distinguish osteoporotic from neoplastic vertebral fracture.

1. Preservation of some marrow fat signal within marrow***
2. No involvement of pedicles or posterior elements***
3. Fluid or gas within vertebral body**
4. T1-hypointense fracture line within vertebral body (often near fractured endplate)**
5. Lack of discrete soft tissue mass***
6. Only minimal or mild paravertebral soft tissue swelling**
7. Absence of epidural mass**
8. Fracture not occurring above T4 level***
9. Posterior located triangular fracture fragment*
10. Nonconvex posterior cortical margin*
11. Evidence of metastases elsewhere in spine*
12. Near complete fatty marrow of adjacent vertebrae*
13. Radiographic evidence of osteopenia*
14. Preservation of trabeculae within fractured vertebral body on CT*** (Fig. 40.12)

The relative usefulness of these signs is indicated with *** = most useful, ** = less useful and * = least useful signs. A known history of malignancy is moderately helpful. If doubt still exists after standard MRI, one can proceed to diffusion-weighted imaging or chemical shift imaging [39].

and a high level of observer experience are key to the reliable diagnosis of vertebral fracture. VFA is being utilized increasingly for vertebral fracture identification. Density and structural parameters obtained by volumetric QCT can potentially predict vertebral compressive strength and these parameters, together with nonlinear finite element analysis (FEA), provide a potential for more efficiently identifying patients at risk of vertebral fracture. MRI can detect even minor acute or subacute vertebral fracture or refracture, gauge fracture age, and distinguish between osteoporotic and neoplastic fracture with greater sensitivity and specificity than any other imaging technique.

REFERENCES

1. Melton LJ, 3rd, Lane AW, Cooper C, et al. Prevalence and incidence of vertebral deformities. Osteoporos Int. 1993;3:113–19.
2. Karlsson MK, Kherad M, Hasserius R, et al. Characteristics of Prevalent VertebralFractures Predict New Fractures in Elderly Men. J Bone Joint Surg Am. 2016;98(5):379–85.
3. Lenchik L, Rogers LF, Delmas PD, et al. Diagnosis of osteoporotic vertebral fractures: importance of recognition and description by radiologists. AJR Am J Roentgenol. 2004;183:949–58.
4. Siris ES, Miller PD, Barrett-Connor E, et al. Identification and fracture outcomes of undiagnosed low bone mineral density in postmenopausal women: results from the National Osteoporosis Risk Assessment. JAMA. 2001;286:2815–22.
5. Lindsay R, Silverman SL, Cooper C, et al. Risk of new vertebral fracture in the year following a fracture. JAMA. 2001;285:320–3.
6. Lips P, van Schoor NM. Quality of life in patients with osteoporosis. Osteoporos Int. 2005;16:447–55.
7. Ensrud KE, Schousboe JT. Clinical practice. Vertebral fractures. N Engl J Med. 2011;364:1634–42.
8. Delmas PD, van de Langerijt L, Watts NB, et al.; IMPACT Study Group. Underdiagnosis of vertebral fractures is a worldwide problem: the IMPACT study. J Bone Miner Res. 2005;20:557–63.
9. Cooper C, Atkinson EJ, O'Fallon WM, et al. Incidence of clinically diagnosed vertebral fractures: a population-based study in Rochester, Minnesota, 1985-1989. J Bone Miner Res. 1992;7:221–7.
10. Li EK, Tam LS, Griffith JF, et al. High prevalence of asymptomatic vertebral fractures in Chinese women with systemic lupus erythematosus. J Rheumatol. 2009;36:1646–52.
11. Adams JE, Lenchik L, Roux C, et al. Radiological Assessment of Vertebral Fracture. International Osteoporosis Foundation Vertebral Fracture Initiative Resource Document Part II, page 1–49. 2010. http://www.iofbonehealth.org/health-professionals/educational-tools-and-slide-kits/vertebral-fracture-teaching-program.html

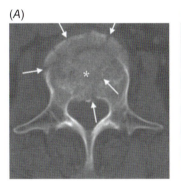 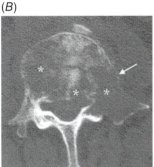

Fig. 40.12. Axial CT images of (A) compressive osteoporotic fracture showing preservation of vertebral trabeculae (*) with several cortical- and trabecular-based fracture lines (arrows). (B) Neoplastic fracture showing several lytic areas caused by metastatic infiltration (*) with marked expansion of the left pedicle (arrow).

CONCLUSION

Vertebral fractures occurring with minimal or no trauma provide indisputable evidence of reduced bone strength, irrespective of BMD. Because vertebral fractures are usually the first insufficiency fracture to occur, recognition and appropriate treatment at this relatively early stage can reduce future fracture risk, patient pain, and deformity. Radiographic SQ assessment is the usual standard for detecting vertebral fracture. Good radiographic technique

12. Siminoski K, Warshawski RS, Jen H, et al. The accuracy of historical height loss for the detection of vertebral fractures in postmenopausal women. Osteoporos Int. 2006;17:290–6.
13. Damilakis J, Adams JE, Guglielmi G, et al. Radiation exposure in X-ray-based imaging techniques used in osteoporosis. Eur Radiol. 2010;20:2707–14.
14. Masharawi Y, Salame K, Mirovsky Y, et al. Vertebral body shape variation in the thoracic and lumbar spine: characterization of its asymmetry and wedging. Clin Anat. 2008;21:46–54.
15. Diacinti D, Acca M, D'Erasmo E, et al. Aging changes in vertebral morphometry. Calcif Tissue Int. 1995;57:426–9.
16. Ferrar L, Roux C, Reid DM, et al. Prevalence of non-fracture short vertebral height is similar in premenopausal and postmenopausal women: the osteoporosis and ultrasound study. Osteoporos Int. 2012;23(3):1035–40.
17. Chan KK, Sartoris DJ, Haghighi P, et al. Cupid's bow contour of the vertebral body: evaluation of pathogenesis with bone densitometry and imaging-histopathologic correlation. Radiology. 1997;202:253–6.
18. Genant HK, Jergas M, Palermo L, et al. Comparison of semiquantitative visual and quantitative morphometric assessment of prevalent and incident vertebral fractures in osteoporosis The Study of Osteoporotic Fractures Research Group. J Bone Miner Res. 1996;11:984–96.
19. Genant HK, Wu CY, van Kuijk C, et al. Vertebral fracture assessment using a semiquantitative technique. J Bone Miner Res. 1993;8:1137–48.
20. Genant HK, Siris E, Crans GG, et al. Reduction in vertebral fracture risk in teriparatide-treated postmenopausal women as assessed by spinal deformity index. Bone 2005;37:170–4.
21. Genant HK, Delmas PD, Chen P, et al. Severity of vertebral fracture reflects deterioration of bone microarchitecture. Osteoporos Int. 2007;18:69–76.
22. Wu CY, Li J, Jergas M, et al. Comparison of semiquantitative and quantitative techniques for the assessment of prevalent and incident vertebral fractures. Osteoporos Int. 1995;5:354–70.
23. Buehring B, Krueger D, Checovich M, et al. Vertebral fracture assessment: impact of instrument and reader. Osteoporos Int. 2010;21:487–94.
24. Guglielmi G, Diacinti D, van Kuijk C, et al. Vertebral morphometry: current methods and recent advances. Eur Radiol. 2008;18:1484–96.
25. Grados F, Fechtenbaum J, Flipon E, et al. Radiographic methods for evaluating osteoporotic vertebral fractures. Joint Bone Spine. 2009;76:241–7.
26. Eastell R, Cedel SL, Wahner HW, et al. Classification of vertebral fractures. J Bone Miner Res. 1991;6:207–15.
27. Grados F, Roux C, de Vernejoul MC, et al. Comparison of four morphometric definitions and a semiquantitative consensus reading for assessing prevalent vertebral fractures. Osteoporos Int. 2001;12:716–22.
28. Roberts MG, Pacheco EM, Mohankumar R, et al. Detection of vertebral fractures in DXA VFA images using statistical models of appearance and a semi-automatic segmentation. Osteoporos Int. 2010;21:2037–46.
29. Jiang G, Eastell R, Barrington NA, et al. Comparison of methods for the visual identification of prevalent vertebral fracture in osteoporosis. Osteoporos Int. 2004;15:887–96.
30. Gallacher SJ, Gallagher AP, McQuillian C, et al. The prevalence of vertebral fracture amongst patients presenting with non-vertebral fractures. Osteoporos Int. 2007;18:185–92.
31. Siris ES, Genant HK, Laster AJ, et al. Enhanced prediction of fracture risk combining vertebral fracture status and BMD. Osteoporos Int. 2007;18:761–70.
32. Mayranpaa MK, Helenius I, Valta H, et al. Bone densitometry in the diagnosis of vertebral fractures in children: Accuracy of vertebral fracture assessment. Bone. 2007;41:353–9.
33. Schousboe JT, Debold CR. Reliability and accuracy of vertebral fracture assessment with densitometry compared to radiography in clinical practice. Osteoporos Int. 2006;17:281–9.
34. Fuerst T, Wu C, Genant HK, et al. Evaluation of vertebral fracture assessment by dual X-ray absorptiometry in a multicenter setting. Osteoporos Int. 2009;20:1199–205.
35. Kanis JA, Hans D, Cooper C, et al.; Task Force of the FRAX Initiative. Interpretation and use of FRAX in clinical practice. Osteoporos Int. 2011;22(9):2395–411.
36. Takada M, Wu CY, Lang TF, et al. Vertebral fracture assessment using the lateral scoutview of computed tomography in comparison with radiographs. Osteoporos Int. 1998;8:197–203.
37. Samelson EJ, Christiansen BA, Demissie S, et al. Reliability of vertebral fracture assessment using multidetector CT lateral scout views: the Framingham Osteoporosis Study. Osteoporos Int. 2011;22:1123–31.
38. Krug R, Burghardt AJ, Majumdar S, et al. High-resolution imaging techniques for the assessment of osteoporosis. Radiol Clin North Am. 2010;48:601–21.
39. Fornasier VL, Czitrom AA. Collapsed vertebrae: a review of 659 autopsies. Clin Orthop Relat Res. 1978;(131):261–5.
40. Griffith JF, Guglielmi G. Vertebral fracture. Radiol Clin North Am. 2010;48:519–529.
41. Baur A, Stäbler A, Arbogast S, et al. Acute osteoporotic and neoplastic vertebral compression fractures: fluid sign at MR imaging. Radiology. 2002;225:730–5.
42. Malghem J, Maldague B, Labaisse MA, et al. Intravertebral vacuum cleft: changes in content after supine positioning. Radiology. 1993;187:483–7.
43. Karchevsky M, Babb JS, Schweitzer ME. Can diffusion-weighted imaging be used to differentiate benign from pathologic fractures? A meta-analysis. Skeletal Radiol. 2008;37:791–5.
44. Ragab Y, Emad Y, Gheita T, et al. Differentiation of osteoporotic and neoplastic vertebral fractures by chemical shift {in-phase and out-of phase} MR imaging. Eur J Radiol. 2009;72:125–33.

41

FRAX: Assessment of Fracture Risk

John A. Kanis[1,2], Eugene V. McCloskey[1], Nicholas C. Harvey[3], and William D. Leslie[4]

[1]Centre for Metabolic Bone Diseases, University of Sheffield Medical School, Sheffield, UK
[2]Mary McKillop Health Institute, Australian Catholic University, Melbourne, VIC, Australia
[3]MRC Lifecourse Epidemiology Unit, University of Southampton, Southampton General Hospital, Southampton, UK
[4]Department of Medicine, University of Manitoba, Winnipeg, MB, Canada

INTRODUCTION

A major objective of fracture risk assessment is to enable the targeting of interventions to those at need and avoid unnecessary treatment in those at low risk of fracture. Historically, fracture risk assessment was largely based on the measurement of BMD, because osteoporosis is defined operationally in terms of bone mass [1, 2]. Whereas BMD forms a central component in the assessment of risk, the accuracy of risk prediction is improved by taking into account other readily measured indices of fracture risk, particularly those that add information to that provided by BMD. Several risk prediction models have been developed, but the most widely used is the Fracture Risk Assessment Tool (FRAX).

INPUT AND OUTPUT

FRAX is a computer-based algorithm (http://www.sheffield.ac.uk/FRAX) first released in 2008 [3, 4]. The algorithm, intended for primary care, calculates fracture probability from easily obtained clinical risk factors (CRFs) in men aged 40 years or more and postmenopausal women. The output of FRAX is the 10-year probability of a major osteoporotic fracture (hip, clinical spine, proximal humerus, or forearm fracture) and the 10-year probability of hip fracture (Fig. 41.1).

Fracture probability is derived from the risk of fracture as well as the risk of death. Fracture risk is calculated from age, body mass index, and dichotomized risk factors comprising prior fragility fracture, parental history of hip fracture, current tobacco smoking, long-term oral glucocorticoid use, rheumatoid arthritis, excessive alcohol consumption, and other causes of secondary osteoporosis. Femoral neck (FN) BMD can be optionally input to enhance fracture risk prediction. Apart from rheumatoid arthritis and long-term use of glucocorticoids, the other secondary causes of osteoporosis considered (Table 41.1) are conservatively assumed to contribute to increased fracture risk because of low BMD.

The relationships between risk factors and fracture risk have been constructed using information derived from the primary data of population-based cohorts from around the world, including centers from North America, Europe, Asia, and Australia, based on a series of meta-analyses to identify clinical risk factors for fracture that provide independent information on fracture risk [3]. The use of primary data for the model construct permits the determination of the predictive importance in a multivariable context of each of the risk factors, as well as interactions between risk factors, and thereby optimizes the accuracy with which fracture risk can be computed [5, 6]. The fracture risk assessment with the combined use of these clinical risk factors with and without the use of BMD has been validated in independent cohorts with a similar geographic distribution with a follow-up in excess of 1 million patient years [7–9].

Fracture probability is computed taking into account both the risk of fracture and the risk of death. The inclusion of the death hazard is important because those with a high immediate likelihood of death are less likely to

---

Primer on the Metabolic Bone Diseases and Disorders of Mineral Metabolism, Ninth Edition. Edited by John P. Bilezikian.
© 2019 American Society for Bone and Mineral Research. Published 2019 by John Wiley & Sons, Inc.
Companion website: www.wiley.com/go/asbmrprimer

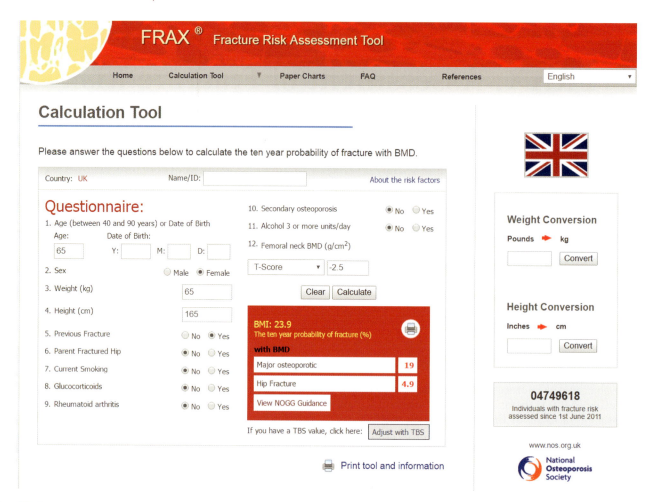

Fig. 41.1. Screen page for input of data and format of results in the UK version of the FRAX tool (UK model, version 3.10. http://www.shef.ac.uk/FRAX). Reproduced with permission of the Centre for Metabolic Bone Diseases, University of Sheffield Medical School, UK.

fracture than individuals with longer life expectancy (Fig. 41.2). This is highlighted by the downturn in 10-year fracture probability after age 80 to 85 years whereas fracture incidence rates continue to rise. In addition, some of the risk factors affect the risk of death as well as the risk of fracture. Examples include increasing age, low BMI, low BMD, glucocorticoids, and smoking. Other risk tools calculate the risk of fracture without considering the possibility of death [9, 10]. Fracture probability varies markedly in different regions of the world [11, 12]. Thus, the FRAX models are calibrated to those countries where the epidemiology of fracture and death is known. Models are currently available for 63 countries in 31 languages covering 80% of the world population [13]. Ethnicity-specific models are available in the US and Singapore.

FRAX has been widely used for the assessment of patients since the launch of the website in 2008, and currently makes about 11,000 calculations per working day. When uptake is expressed per million of the population, highest usage is seen for Slovenia, Switzerland, the US, Belgium, New Zealand, and the UK [13]. Following regulatory review by the US Food and Drug Administration (FDA), FRAX was incorporated into DXA scanners to provide FRAX probabilities at the time of DXA scanning. For those without internet access, handheld calculators and smartphone applications have been developed by the International Osteoporosis Foundation (http://itunes.apple.com/us/app/frax/id370146412?mt=8 and https://play.google.com/store/apps/details?id=com.inkrypt.clients.iof.frax&hl=en_GB)

PERFORMANCE CHARACTERISTICS

For the purpose of risk assessment, a characteristic of major importance is the ability of a technique to predict the outcome of interest. This is traditionally expressed as the increase in relative risk per SD unit change in risk score—termed the gradient of risk. The gradients of risk for incident fractures are shown in Table 41.2 for the use of the clinical risk factors alone, FN BMD, and the combination.

Table 41.1. Secondary causes of osteoporosis associated with an increase in fracture risk. Adapted from [3] with permission of the Centre for Metabolic Bone Diseases, University of Sheffield Medical School, UK.

Secondary cause	Example
Glucocorticoids	Any dose, by mouth, for 3 months or more
	High doses of inhaled glucocorticoids
	Cushing's disease
Rheumatoid arthritis	
Chronic liver disease	Alcoholism
Untreated hypogonadism	Bilateral oophorectomy or orchidectomy
	Anorexia nervosa
	Chemotherapy for breast cancer
	Tamoxifen in premenopausal women
	Aromatase inhibitors
	GnRH inhibitors for prostate cancer
	Hypopituitarism
Prolonged immobility	Spinal cord injury
	Parkinson's disease
	Stroke
	Muscular dystrophy
	Ankylosing spondylitis
Organ transplantation	
Type 1 and 2 diabetes	
Thyroid disorders	Untreated hyperthyroidism
	Overtreated hypothyroidism
Gastrointestinal disease	Crohn's disease
	Ulcerative colitis
Chronic obstructive pulmonary disease	
Osteogenesis imperfecta in adults	

The use of clinical risk factors alone provides a gradient of risk (GR) that lies between 1.4 and 2.1, depending upon age and the type of fracture predicted. These gradients are comparable to the use of BMD alone to predict fractures [15, 16], indicating that clinical risk factors alone are of value in fracture risk prediction and might be used, therefore, in the many countries where DXA facilities are sparse [17]. Nevertheless, there are substantial gains in the use of the clinical risk factors in conjunction with BMD, particularly in the case of hip fracture prediction. At the age of 50 years, for example, the GR with BMD alone is 3.7/SD, but with the addition of clinical risk factors it is 4.2/SD. Although the improvement in GR with the addition of BMD appears relatively modest, particularly in the case of other osteoporotic fractures, it should be recognized that gradients of risk are not necessarily multiplicative. For example, at the age of 70 years, BMD alone gives a GR of 2.8/SD for hip fracture. For the clinical risk factors the GR is 1.8/SD. If these two tests were totally independent, the combined GR would be $\sqrt{(2.8^2 + 1.8^2)} = 3.3$. The observed GR (2.9) falls short of the theoretical upper limit, because there is a significant correlation between the clinical risk factor score and BMD. Overall, the predictive value compares very favorably with other risk engines such as the Gail score for breast cancer [18].

LIMITATIONS

FRAX should not be considered as a gold standard in patient assessment, but rather as a reference platform. The same argument applies to BMD testing. Thus, the result should not be uncritically used in the management of patients without an appreciation of its limitations as well as its strengths. In some instances, limitations (eg, to experts in bone disease) are perceived as strengths to others (eg, primary care physicians).

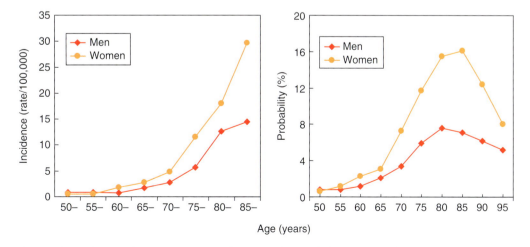

Fig. 41.2. Incidence of hip fracture by age in women from Sweden and the corresponding 10-year probability of hip fracture. Source: [14]. Reproduced with permission of Springer Science+Business Media B.V.

Table 41.2. Gradients of risk (HR/per SD change in risk score) (with 95% CI) with the use of BMD at the femoral neck, clinical risk factors, or the combination [7]. Outcomes comprised (a) hip fracture and (b) other fractures at sites associated with osteoporosis. Reproduced with permission of Springer Science + Business Media B.V.

	Gradient of risk		
Age (Years)	Clinical Risk Factors Alone	BMD Only	Clinical Risk Factors + BMD
(a) Hip fracture			
50	2.05 (1.58–2.65)	3.68 (2.61–5.19)	4.23 (3.12–5.73)
60	1.95 (1.63–2.33)	3.07 (2.42–3.89)	3.51 (2.85–4.33)
70	1.84 (1.65–2.05)	2.78 (2.39–3.23)	2.91 (2.56–3.31)
80	1.75 (1.62–1.90)	2.28 (2.09–2.50)	2.42 (2.18–2.69)
90	1.66 (1.47–1.87)	1.70 (1.50–1.93)	2.02 (1.71–2.38)
(b) Other osteoporotic fractures			
50	1.41 (1.28–1.56)	1.19 (1.05–1.34)	1.44 (1.30–1.59)
60	1.48 (1.39–1.58)	1.28 (1.18–1.39)	1.52 (1.42–1.62)
70	1.55 (1.48–1.62)	1.39 (1.30–1.48)	1.61 (1.54–1.68)
80	1.63 (1.54–1.72)	1.54 (1.44–1.65)	1.71 (1.62–1.80)
90	1.72 (1.58–1.88)	1.56 (1.40–1.75)	1.81 (1.67–1.97)

Table 41.3. Percentage adjustment of 10-year probabilities of a hip fracture or a major osteoporotic fracture by age according to dose of glucocorticoids.

Dose	Prednisolone equivalent (mg/day)	Age (years)						All ages
		40	50	60	70	80	90	
Hip fracture								
Low	<2.5	−40	−40	−40	−40	−30	−30	−35
Medium[a]	2.5–7.5							
High	≥7.5	+25	+25	+25	+20	+10	+10	+20
Major osteoporotic fracture								
Low	<2.5	−20	−20	−15	−20	−20	−20	−20
Medium[a]	2.5–7.5							
High	≥7.5	+20	+20	+15	+15	+10	+10	+15

[a] No adjustment.
Source: [22]. Reproduced with permission of Springer Science + Business Media B.V.

Assessment with FRAX takes no account of current or prior treatment though the effect of treatment on the estimated fracture probability is modest, related in part to treatment-induced increases in BMD [19]. FRAX also takes no account of dose-responses for several risk factors. For example, two or more prior vertebral fractures carry a much higher risk than a single prior fracture [20]. A prior clinical vertebral fracture carries an approximately twofold higher risk than other prior fractures. Dose-responses are also evident for glucocorticoid use, smoking, and alcohol consumption [3, 21]. Because it is not possible to model all such scenarios with the FRAX algorithm, these limitations should temper clinical judgement.

In the case of glucocorticoids, simple arithmetic procedures have been formulated that can be applied to conventional FRAX estimates of probabilities of hip fracture and a major osteoporotic fracture to adjust the probability assessment with knowledge of the dose of glucocorticoids (Table 41.3) [22]. For example, an individual at the age of 60 years taking a high dose of glucocorticoids and with a probability of a major fracture of 18% would have the FRAX estimate uplifted by 15% giving a revised probability of 21% (18×1.15). In contrast, if the patient were exposed to a low dose, the revised estimate would be 15%.

A further limitation is that the FRAX algorithm uses T-scores for FN BMD from DXA and QCT at the hip and does not accommodate other sites or technologies. The lumbar spine (LS) is frequently measured by DXA for the assessment of patients and, indeed, is incorporated into many clinical guidelines [23–26]. It is also the site favored for monitoring treatment. There is, therefore, much interest in the incorporation into FRAX of measurements at the

LS. Whereas this is not possible at present, some guidance is available in cases where there is a large discordance between the *T*-score at the FN and the LS [27, 28]. It has been proposed that the FRAX estimate for a major fracture is increased/decreased by one-tenth for each rounded *T*-score difference between the LS and FN. An example is a case in which the *T*-score for FN BMD is –2.2 SD with a FRAX-calculated fracture probability of 19% and a *T*-score of –3.5 SD at the LS. In this scenario, the *T*-score discordance is 1.3 SD (3.5 to 2.2). If the figure is rounded off (to 1.0 SD), the estimated probability with the inclusion of lumbar BMD is upward revised by 10% (19 × 1.10) to 21%.

Additionally, arithmetic procedures have been proposed that can be applied to conventional FRAX estimates of probabilities of hip fracture and a major fracture to adjust the probability assessment with knowledge of trabecular bone score (TBS) [29, 30], hip axis length [31], falls history [32], and immigration status [33].

INTERVENTION AND ASSESSMENT THRESHOLDS

The use of FRAX in clinical practice demands a consideration of the fracture probability at which to intervene, both for treatment (an intervention threshold) and for BMD testing (assessment thresholds). A general approach is shown in Fig. 41.3 [34]. The management process begins with the assessment of fracture probability and the categorization of fracture risk on the basis of age, sex, BMI, and the clinical risk factors. On this information alone, some patients at high risk may be offered treatment without recourse to BMD testing. Many guidelines recommend treatment in the absence of information on BMD in women with a previous fragility fracture (a prior vertebral or hip fracture

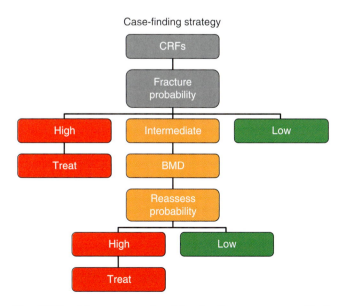

Fig. 41.3. Management algorithm for the assessment of individuals at risk of fracture. Source: [34]. Reproduced with permission of Springer Science + Business Media B.V.

in North America) [24, 26, 35]. Many physicians would also perform a BMD test, but frequently this is for reasons other than to decide on intervention, for example as a baseline to monitor treatment. There will be other instances where the probability will be so low that a decision not to treat can be made without BMD. An example might be the well woman at menopause with no clinical risk factors. Thus, not all individuals require a BMD test. The size of the intermediate category in Fig. 41.3 will vary in different countries. In the US this will be a large category, whereas in a large number of countries with limited or no access to densitometry [17] the size of the intermediate group will necessarily be small. In other countries (eg, the UK), where provision for BMD testing is suboptimal, the intermediate category will lie between the two extremes.

Since its release, FRAX has been incorporated into more than 80 guidelines wordwide [36]. The setting of universal intervention thresholds is problematic from an international perspective because the risk of fracture, the cost of fracture, the cost of treatment, reimbursement, and willingness to pay vary in different countries. Thus, probability-based guidelines differ in detail. Guidelines variously use an age-dependent fracture probability or a fixed probability threshold applied to all relevant ages [36]. Examples of each are provided in the guidelines of the UK and the US, illustrated in the following paragraphs in the case of postmenopausal women.

GUIDELINES IN THE UK

The UK guidance for the identification of individuals at high fracture risk, developed by the National Osteoporosis Guideline Group (NOGG) [34, 37], recommends that postmenopausal women with a prior fragility fracture may be considered for intervention without the necessity for a BMD test. In women without a fragility fracture but a FRAX risk factor, the intervention threshold set by NOGG is at the age-specific fracture probability equivalent to women with a prior fragility fracture. The same intervention threshold is applied to men, because the effectiveness and cost-effectiveness of intervention in men is broadly similar to that in women for equivalent risk [38, 39].

The NOGG management strategy considers two additional thresholds (Fig. 41.4):

- a threshold probability below which neither treatment nor a BMD test should be considered (lower assessment threshold),
- a threshold probability above which treatment may be recommended irrespective of BMD (upper assessment threshold).

In other words, some patients at high risk may be offered treatment without recourse to BMD testing. Conversely, some patients at low risk would not warrant treatment or a BMD test. An attraction of the approach is that efficient use is made of BMD testing. For example, the NOGG strategy requires only 3.5 scans at the age of 50 years to identify

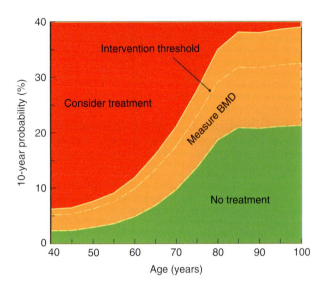

Fig. 41.4. Assessment guidelines of the National Osteoporosis Guideline Group based on the 10-year probability of a major fracture (%). The dotted line denotes the intervention threshold, which increases with age. Where assessment is made in the absence of BMD, a BMD test is recommended for individuals where the probability assessment lies in the orange-shaded region Source: [34]. Reproduced with permission of Springer Science+Business Media B.V.

one case of hip fracture, whereas the former guidelines of the Royal College of Physicians required 14. The lower number of BMD tests means that the acquisition costs for identifying a fracture case and the total costs (acquisition and treatment) per fracture averted are also lower [40].

The justification for this parsimonious approach derives from the correlation between fracture probability (without BMD) and BMD. It has been shown in several studies that the reclassification of high risk to low risk (and vice versa, from low risk to high risk) with the incorporation of BMD in the FRAX calculation is more or less confined to those close to an intervention threshold. Indeed, more than 80% of women that are reclassified have probabilities without BMD within 5% of the treatment threshold. This figure rises to about 95% for probabilities without BMD within 10% of the treatment threshold. Thus, a BMD test is only required in individuals whose fracture probability lies within about 10% of the intervention threshold, avoiding unnecessary testing in about 50% of eligible women without a prior fragility fracture [41]. Initial analysis of the SCOOP study indicates the effectiveness of this approach [42].

GUIDELINES IN THE US

In the US, the guideline of the National Osteoporosis Foundation recommends treatment for postmenopausal women who have had a prior spine or hip fracture and for women with a BMD at or below a T-score of −2.5 SD [23–25, 35]. Thus, as for the UK, it recognizes high-risk groups in whom treatment can be recommended without FRAX; however, the high-risk groups differ (a prior fragility fracture in the UK). Conversely, the guideline does not recommend treatment in postmenopausal women with a T-score of >−1.0 SD. In the UK, the exclusion group is women without a clinical risk factor. Thus, FRAX in the US becomes relevant only in women with a T-score between −1 and −2.5 SD. Treatment is recommended in such patients in whom the 10-year probability of a major fracture equals or exceeds 20% or where the 10-year probability of a hip fracture equals or exceeds 3%.

Thus, there are similarities and differences with the UK and US guidelines. The similarities are that both select high-risk patients in whom treatment is recommended and low-risk groups in whom treatment is not recommended (though the categories differ somewhat). The major difference lies in the intervention thresholds (fixed or age-dependent) and the use of BMD in the intermediate group (all women in the US after age 65 years and many younger menopausal women, assessment thresholds in the UK). Neither approach is right or wrong, but both are imperfect and will remain so until fracture risk prediction can be improved still further.

GUIDELINES WITHOUT BMD TESTING

Many regions of the world have little or no access to BMD testing [17]. In these circumstances, FRAX without BMD can be used [43, 44]. The clinical risk factors used in FRAX are not totally independent of BMD. Indeed, there is a weak but significant correlation between the clinical risk factor score for hip fracture (assessed without BMD) and FN BMD. This indicates that the selection of individuals with the use of FRAX, without knowledge of BMD, will preferentially select those with low BMD, and that the higher the fracture probability, the lower will be the BMD. When a fixed intervention threshold is chosen, the high-risk group has a BMD that is approximately 1 SD lower than in the low-risk group [31, 41].

OTHER APPLICATIONS OF FRAX

FRAX (without BMD) has been recommended as a screening tool to detect osteoporosis [45, 46]. Its use for this purpose is inappropriate, because other tools are more sensitive [47, 48].

Several studies have examined the relationship between FRAX-based probability of fracture and efficacy of bone-active agents used in the treatment of osteoporosis. In some, but not all instances, efficacy (relative risk reduction) has been shown to be greater in patients with higher baseline fracture probabilities [49]. This has implications for targeting treatments to high-risk patients in that the dividend in terms of fractures saved is amplified. This also has implications for health economic assessment

Table 41.4. Comparative features of QFracture and FRAX.

	QFracture	FRAX
Externally validated	Yes (UK only)	Yes, internationally
Calibrated	Yes (hip only)	Yes
Applicability	UK	63 countries
Falls as an input variable	Yes	No
BMD as an input variable	No	Yes
Prior fracture as an input variable	Yes	Yes
Family history as an input variable	Yes	Yes
Outcome	Hip, forearm, spine, shoulder	Hip, forearm, spine, humerus
Outcome metric	Incidence	Probability

Source: [47]. Reproduced with permission of Springer Science+Business Media B.V.

and conventional meta-analyses of interventions used in osteoporosis [50].

OTHER RISK ASSESSMENT TOOLS

As well as the FRAX tool, other fracture risk calculators are available online; these include the Garvan fracture risk calculator and QFracture [9, 10]. Both QFracture and FRAX have been approved by the National Institute for Health and Care Excellence (NICE) for use in the UK [51] and FRAX is approved by the FDA (https://www.itnonline.com/content/fda-clears-hologic-bone-densitometer-risk-calculator). Their comparative features are summarized in Table 41.4. The QFracture tool is based on a UK prospective open cohort study of routinely collected data from 357 General Practices on over 2 million men and women aged 30 to 85 years (www.qfracture.org). Like the FRAX tool, it takes into account history of smoking, alcohol, glucocorticoid use, parental history (of hip fracture or osteoporosis), and multiple secondary causes of osteoporosis. Unlike FRAX, it also includes a history of falls (yes/no over an unspecified time frame). It has been internally validated (ie, from a stratum of the same population) for hip fracture and also externally validated, but only from GP records in the UK [52]. It is poorly calibrated for major osteoporotic fracture [47].

CONCLUSION

FRAX represents a significant advance in the assessment of fracture risk both in women and in men, and allows the tailoring of pharmacologic interventions to high-risk subjects. Although FRAX does not define intervention thresholds, which depend on country-specific considerations, it provides a platform to assess fracture probability which is needed to make rational treatment decisions by clinicians and public health agencies. The tool is, however, far from perfect, but better than BMD alone. The widespread use of and interest in FRAX, and its adoption into management guidelines, have fueled interest as to how models can be improved and extended to other countries and, in particular, how the limitations of FRAX should temper clinical judgement.

COMPETING INTERESTS

The authors have no competing interests with regard to the content of this chapter. Professor Kanis is the principal architect of FRAX but has no financial interest. The financial proceeds of the sales of FRAX go to the International Osteoporosis Foundation.

REFERENCES

1. World Health Organization. Assessment of fracture risk and its application to screening for postmenopausal osteoporosis. Technical Report Series 843. Geneva: WHO, 1994.
2. Kanis JA, McCloskey EV, Johansson H, et al. A reference standard for the description of osteoporosis. Bone. 2008;42:467–75.
3. Kanis JA on behalf of the World Health Organization Scientific Group. Assessment of osteoporosis at the primary health-care level. Technical Report. WHO Collaborating Centre, University of Sheffield, UK, 2008. Available at https://www.sheffield.ac.uk/FRAX/pdfs/WHO_Technical_Report.pdf, accessed 17 May 2018.
4. Kanis JA, Johnell O, Oden A, et al. FRAX™ and the assessment of fracture probability in men and women from the UK. Osteoporos Int. 2008;19:385–97.
5. De Laet C, Oden A, Johansson H, et al. The impact of the use of multiple risk indicators for fracture on case-finding strategies: a mathematical approach. Osteoporos Int. 2005;16:313–18.
6. Kanis JA, Johnell O, Oden A, et al. Ten-year risk of osteoporotic fracture and the effect of risk factors on screening strategies. Bone. 2002;30:251–8.

7. Kanis JA, Oden A, Johnell O, et al. The use of clinical risk factors enhances the performance of BMD in the prediction of hip and osteoporotic fractures in men and women. Osteoporos Int. 2007;18:1033–46.
8. Leslie WD, Lix LM, Johansson H, et al.; Manitoba Bone Density Program. Independent clinical validation of a Canadian FRAX tool: Fracture prediction and model calibration. J Bone Miner Res. 2010;25:2350–8.
9. Hippisley-Cox J, Coupland C. Predicting risk of osteoporotic fracture in men and women in England and Wales: prospective derivation and validation of QFractures Scores. Br Med J. 2009;339:b4229.
10. Nguyen ND, Frost SA, Center JR, et al. Development of prognostic nomograms for individualizing 5-year and 10-year fracture risks. Osteoporos Int. 2008;19:1431–44.
11. Kanis JA, Johnell O, De Laet C, et al. International variations in hip fracture probabilities: implications for risk assessment. J Bone Miner Res. 2002;17:1237–44.
12. Kanis JA, Odén A, McCloskey EV, et al. A systematic review of hip fracture incidence and probability of fracture worldwide. Osteoporos Int. 2012;23:2239–56.
13. Kanis JA, Johansson H, Oden A, et al.; the Epidemiology and Quality of Life Working Group of IOF. Worldwide uptake of FRAX. Arch Osteoporos. 2014;9:166.
14. Kanis JA, Johnell O, Oden A, et al. Long-term risk of osteoporotic fracture in Malmo. Osteoporos Int. 2000;11:669–74.
15. Johnell O, Kanis JA, Oden A, et al. Predictive value of bone mineral density for hip and other fractures. J Bone Miner Res. 2005;20:1185–1194. Erratum J Bone Miner Res. 22:774.
16. Marshall D, Johnell O, Wedel H. Meta-analysis of how well measures of bone mineral density predict occurrence of osteoporotic fractures. Br Med J. 1996;312:1254–9.
17. Kanis JA, Johnell O. Requirements for DXA for the management of osteoporosis in Europe. Osteoporosis Int. 2004;16:229–38.
18. Chlebowski RT, Anderson GL, Lane DS, et al. Predicting risk of breast cancer in postmenopausal women by hormone receptor status. J Natl Cancer Inst. 2007;99:1695–705.
19. Leslie WD, Lix LM, Johansson H, et al. Does osteoporosis therapy invalidate FRAX for fracture prediction? J Bone Miner Res. 2012;27:1243–51.
20. Delmas PD, Genant HK, Crans GG, et al. Severity of prevalent vertebral fractures and the risk of subsequent vertebral and nonvertebral fractures: results from the MORE trial. Bone. 2003;33:522–32.
21. Van Staa TP, Leufkens HG, Abenhaim L, et al. Use of oral corticosteroids and risk of fractures. J Bone Miner Res. 2000;15:993–1000.
22. Kanis JA, Johansson H, Oden A, et al. Guidance for the adjustment of FRAX according to the dose of glucocorticoids. Osteoporos Int. 2011;22:809–16.
23. Baim S, Binkley N, Bilezikian JP, et al. Official Positions of the International Society for Clinical Densitometry and executive summary of the 2007 ISCD Position Development Conference. J Clin Densitom. 2008;11:75–91.
24. Dawson-Hughes B. A revised clinician's guide to the prevention and treatment of osteoporosis. J Clin Endocrinol Metab. 2008;93:2463–5.
25. Cosman F, de Beur SJ, LeBoff MS, et al. Clinician's guide to prevention and treatment of osteoporosis. Osteoporos Int. 2014;25:2359–81.
26. Papaioannou A, Morin S, Cheung AM, et al. 2010 clinical practice guidelines for the diagnosis and management of osteoporosis in Canada: summary. CMAJ. 2010;182:1864–73.
27. Leslie WD, Lix LM, Johansson H, et al. Spine-hip discordance and fracture risk assessment: a physician-friendly FRAX enhancement. Osteoporos Int. 2011;22:839–47.
28. Johansson H, Kanis JA, Odén A, et al. Impact of femoral neck and lumbar spine BMD discordances on FRAX probabilities in women: a meta-analysis of international cohorts. Calcif Tiss Int. 2014;95:428–35.
29. Leslie WD, Johansson H, Kanis JA, et al. Lumbar spine texture enhances ten-year fracture probability assessment. Osteoporos Int. 2014;25:2271–7.
30. McCloskey EV, Odén A, Harvey NC, et al. A meta-analysis of trabecular bone score in fracture risk prediction and its dependence on FRAX. J Bone Miner Res. 2016;31:940–8.
31. Leslie WD, Lix LM, Morin SN, et al. Adjusting hip fracture probability in men and women using hip axis length: The Manitoba Bone Density Database. J Clin Densitom. 2015;19:326–31.
32. Masud T, Binkley N, Boonen S, et al. on behalf of the FRAX Position Conference members. Can falls and frailty be used in FRAX? J Clin Densitom. 2011;14:194–204.
33. Johansson H, Odén A, Lorentzon M, et al. Is the Swedish FRAX model appropriate for immigrants to Sweden? Osteoporos Int. 2015;26:2617–22.
34. Kanis JA, McCloskey EV, Johansson H, et al.;the National Osteoporosis Guideline Group. Case finding for the management of osteoporosis with FRAX® - Assessment and intervention thresholds for the UK. Osteoporos Int. 2008;19:1395–408. Erratum Osteoporos Int. 2009;20:499–502.
35. Grossman JM, Gordon R, Ranganath VK, et al. American College of Rheumatology 2010 recommendations for the prevention and treatment of glucocorticoid-induced osteoporosis. Arthritis Care Res (Hoboken). 2010;62:1515–26.
36. Kanis JA, Harvey NC, Cooper C, et al.; the Advisory Board of the National Osteoporosis Guideline Group. A systematic review of intervention thresholds based on FRAX. Arch Osteoporos. 2016;11(1):25.
37. Compston J, Cooper A, Cooper C, et al., on behalf of the National Osteoporosis Guideline Group (NOGG). Guidelines for the diagnosis and management of osteoporosis in postmenopausal women and men from the age of 50 years in the UK. Maturitas. 2009;62:105–8.
38. Kanis JA, Stevenson M, McCloskey EV, et al. Glucocorticoid-induced osteoporosis: a systematic review and cost-utility analysis. Health Technol Assess. 2007;11:1–256.

39. Tosteson AN, Melton LJ 3rd, Dawson-Hughes B, et al.; National Osteoporosis Foundation Guide Committee. Cost-effective osteoporosis treatment thresholds: The United States perspective. Osteoporos Int. 2008;19: 437–47.
40. Johansson H, Kanis JA, Oden A, et al. A comparison of case finding strategies in the UK for the management of hip fractures. Osteoporos Int. 2011;23:907–15.
41. Johansson H, Oden A, Johnell O, et al. Optimisation of BMD measurements to identify high risk groups for treatment – A test analysis. J Bone Miner Res. 2004;19:906–13.
42. Shepstone L, Lenaghan E, Cooper C, et al. Screening in the community to reduce fractures in older women (SCOOP): A randomised controlled trial. Lancet. 2018; 391(10122):741–7.
43. Leslie WD, Morin S, Lix LM, et al. Fracture risk assessment without bone density measurement in routine clinical practice. Osteoporos Int. 2012;23:75–85.
44. Kanis JA, McCloskey E, Johansson H, et al. FRAX® with and without BMD. Calcif Tissue Int. 2012;90:1–13.
45. Scottish Intercollegiate Guidelines Network (SIGN). Management of osteoporosis and the prevention of fragility fractures. Edinburgh: SIGN publication no. 142. 2015. Available at http://www.sign.ac.uk, accessed 11 May 2015.
46. Nelson HD, Haney EM, Chou R, et al. Screening for osteoporosis: systematic review to update the 2002 US Preventive Services Task Force Recommendation. Evidence Syntheses No. 77. AHRQ Publication No. 10-05145-EF-1. 2010.
47. Kanis JA, Compston J, Cooper C, et al. SIGN. Guidelines for Scotland: BMD Versus FRAX Versus QFracture. Calcif Tissue Int. 2016;98:417–25.
48. Kanis JA, McCloskey EV, Harvey NC, et al. Intervention thresholds and the diagnosis of osteoporosis. J Bone Miner Res. 2015;30:1747–53.
49. Kanis JA, Oden A, Johansson H, et al. FRAX® and its applications to clinical practice. Bone 2009;44:734–43.
50. Ström O, Borgström F, Kleman M, et al. FRAX and its applications in health economics - cost-effectiveness and intervention thresholds using bazedoxifene in a Swedish setting as an example. Bone. 2010;47:430–7.
51. National Institute for Health and Care Excellence. NICE Clinical Guideline 146. Osteoporosis: assessing the risk of fragility fracture. London, UK, 2014. https://www.nice.org.uk/guidance/cg146, accessed 18 May 2015.
52. Collins GS, Mallett S, Altman DG. Predicting risk of osteoporotic and hip fracture in the United Kingdom: Prospective independent and external validation of QFractureScores. BMJ. 2011;342:d3651.

Section V
Genetics of Bone

Section Editor: Rajesh V. Thakker

Chapter 42. Introduction to Genetics   343
Paul J. Newey, Michael P. Whyte, and Rajesh V. Thakker

Chapter 43. Animal Models: Genetic Manipulation   351
Karen Lyons

Chapter 44. Animal Models: Allelic Determinants for Bone Mineral Density   359
J. H. Duncan Bassett, Graham R. Williams, and Robert D. Blank

Chapter 45. Transcriptional Profiling for Genetic Assessment   367
Aimy Sebastian and Gabriela G. Loots

Chapter 46. Approaches to Genetic Testing   373
Christina Jacobsen, Yiping Shen, and Ingrid A. Holm

Chapter 47. Human Genome-Wide Association Studies   378
Douglas P. Kiel, Emma L. Duncan, and Fernando Rivadeneira

Chapter 48. Translational Genetics of Osteoporosis: From Population Association to Individualized Assessment   385
Bich Tran, Jacqueline R. Center, and Tuan V. Nguyen

Primer on the Metabolic Bone Diseases and Disorders of Mineral Metabolism, Ninth Edition. Edited by John P. Bilezikian.
© 2019 American Society for Bone and Mineral Research. Published 2019 by John Wiley & Sons, Inc.
Companion website: www.wiley.com/go/asbmrprimer

42

Introduction to Genetics

Paul J. Newey[1], Michael P. Whyte[2], and Rajesh V. Thakker[3]

[1]*Division of Molecular and Clinical Medicine, University of Dundee, Dundee, UK*
[2]*Center for Metabolic Bone Disease and Molecular Research, Shriners Hospital for Children; and Division of Bone and Mineral Diseases, Department of Internal Medicine, Washington University School of Medicine at Barnes-Jewish Hospital, St. Louis, MO, USA*
[3]*University of Oxford, Nuffield Department of Clinical Medicine, OCDEM Churchill Hospital, Oxford, UK*

INTRODUCTION

Many mineral metabolism and skeletal diseases have a genetic basis, which may be a germline single gene abnormality (ie, a monogenic or Mendelian disorder), a somatic single gene defect (ie, a postzygotic mosaic disorder), or several contributing genetic variants (ie, an oligogenic or polygenic disorder) [1]. Genetic mutations causing Mendelian diseases usually have a large effect (ie, penetrance), whereas oligogenic or polygenic disorders summate smaller effects, sometimes with contributions from the environment (ie, a multifactorial disorder) [1–3].

INHERITANCE

Inheritance of monogenic mineral metabolism and skeletal diseases occurs as one of six traits: (i) autosomal dominant (eg, familial benign hypercalcemia due to mutations involving Ca^{2+} sensing, and the most common forms of osteogenesis imperfecta due to defects in the genes encoding type I collagen); (ii) autosomal recessive (eg, vitamin D-dependent rickets types I and II from mutations of the renal 1 alpha hydroxylase and vitamin D receptor genes, respectively); (iii) X-linked recessive (eg, Dent disease involving chloride channel 5); (iv) X-linked dominant (eg, X-linked hypophosphatemia from mutations of a phosphate endopeptidase on the X chromosome [PHEX] gene); (v) Y-linked (eg, azoospermia and oligospermia); and (vi) non-Mendelian mitochondrial defects (eg, hypoparathyroidism in Kearns-Sayre syndrome and mitochondrial encephalopathy, lactic acidosis, and stroke [MELAS] syndrome). Sporadic postzygotic mosaicism accounts for fibrous dysplasia including McCune-Albright syndrome. Inheritance of polygenic disorders may be suspected by familial occurrence, but the phenotype and transmission pattern may be complex partly because of environmental factors (eg, in osteoporosis, osteoarthritis, or hypercalciuria) [3, 4].

Herein, we briefly review mineral metabolism and skeletal diseases with a genetic element underpinning their identification and characterization based on clinical evaluation including a detailed medical history and physical examination, and their mode of inheritance based on appropriate interpretation using genetic testing.

CLINICAL APPROACH

The diagnosis and treatment of hereditary mineral metabolism and skeletal diseases requires clinical skill [4–8]. Initial and follow-up patient evaluations may involve one or many diagnostic tools and choices from a wide array of treatment approaches. Mutation analysis, like biochemical testing, skeletal imaging, and histopathological assessments requires physician interpretation that can recognize and deal with inherent uncertainties and limitations [4–14]. Moreover, as

---

Primer on the Metabolic Bone Diseases and Disorders of Mineral Metabolism, Ninth Edition. Edited by John P. Bilezikian.
© 2019 American Society for Bone and Mineral Research. Published 2019 by John Wiley & Sons, Inc.
Companion website: www.wiley.com/go/asbmrprimer

molecular testing is increasingly available from commercial laboratories, judicious and cost-effective use of such technology is required and comes from experience with patients.[1]

Medical history and physical examination

Diagnosis and treatment of dysplastic and metabolic bone diseases, including those that are heritable, begins by acquiring information that can come from the patient's stated medical history and the findings from thorough physical examination [4–8,15]. The importance of the medical history cannot be overemphasized. It determines whether any of the many adverse exogenous factors that complicate metabolic bone disease and some skeletal dysplasias will be uncovered. A questionnaire may be a beginning, but is hardly a substitute. Only by talking with his/her patient will the physician sense how knowledgeable this individual might be and judge the value of the information. Subsequently, the medical history should be reported as a narrative to capture the clinical problem(s). Paramount is orderly accumulation, documentation, and consideration of the information acquired directly from patients. This effort helps to disclose potentially important medical records, guide the physical examination and laboratory studies, including any genetic testing, and choose safe and effective therapy.

Most genetically-based metabolic and dysplastic bone diseases are chronic conditions. The "history of present illness" may be lengthy, but provides infrastructure for diagnosis and therapy. Critical clues concerning etiology and pathogenesis should emerge perhaps with a glimpse at prognosis. Here the physician can learn if previous medical records, radiographs, laboratory tests, etc., can help in diagnosis and prognostication. Have the signs and symptoms been lifelong, or have they begun recently prompting very different diagnostic considerations and interventions? Has the patient been compliant with medical care; if not, will therapy be safe? Here, the physician can also provide the basis for sound clinical research. Physical examination can show a considerable variety of findings for diagnosis including skeletal deformities especially common and unique in children, but that are also complications, including skeletal deformities, requiring attention for successful treatment. The diagnosis of a genetic bone disease may emanate from recognition of a single physical finding; for example, blue or gray sclerae (osteogenesis imperfecta), large café-au-lait spots (McCune-Albright syndrome), premature loss of deciduous teeth (hypophosphatasia), hallux valgus (fibrodysplasia ossificans progressiva), alopecia (some patients with vitamin D-dependent rickets, type II), brachydactyly (pseudohypoparathyroidism, type IA), or numerous surgical scars (multiple endocrine neoplasia [MEN] syndromes) [4–8]. For some genetic bone diseases, a constellation of physical findings suggests the diagnosis; for example, rickets featuring craniotabes at birth and soon after a rachitic rosary (enlargement of the costochondral junctions) appearing during the first year of life. Childhood-onset rickets causes bowed legs, short stature, and flared wrists and ankles from metaphyseal widening. There can be Harrison's groove (rib cage ridging from diaphragmatic pull producing a horizontal depression along the lower border of the chest at the costal insertions of the diaphragm). Although weight bearing typically bows rachitic lower limbs, knock-knee deformity may instead occur, especially if the rachitic disturbance occurs during the adolescent growth spurt. In adults, skeletal deformation originating in childhood can cause much of the morbidity from metabolic bone disease. Bowing of the lower limbs predisposes to osteoarthritis, especially in the knees. Without a complete physical examination, these important problems may go unnoticed.

Family medical history for determining mode of disease inheritance

Many mineral metabolic and skeletal disorders have a monogenic etiology. This may be suspected because of an early age of onset, occurrence of other abnormalities consistent with a syndromic disease, or a family history suggestive of the disorder. The "family history" concerning such diseases is vital for revealing the mode of inheritance [1, 2, 4–8]. For example, consanguinity that is apparent, or inapparent involving geographic isolation and a "founder" mutation, can be an important clue for autosomal recessive conditions, whereas the autosomal dominant disorders may be disclosed by prior or prospective study of relatives (eg, familial benign [hypocalciuric] hypercalcemia or osteogenesis imperfecta). Inborn errors of vitamin D bioactivation or resistance are rare, and the family history may be key to considering one because of national or ethnic background. Furthermore, significant information can come from screening studies to identify "carriers" and to treat or counsel affected relatives who may provide important clues to the patient's future complications and prognosis. Medical records from living or deceased affected family members may establish the diagnosis, guide prognostication, and indicate a safe and effective treatment. To report this history as "negative" without first establishing the basis for the information might be misleading. If the patient is adopted, he/she is less likely to give useful details. Knowing the family size is essential before dismissing possible transmission of a heritable disorder. The only child of only children, or

---

[1]"The more resources we have, and the more complex they are, the greater are the demands upon our clinical skill. These resources are calls upon judgment and not substitutes for it. Do not, therefore, scorn clinical examination; learn it sufficiently to get from it all it holds, and gain in it the confidence it merits." Sir F.M.R. Walshe (1881–1973), *Canadian Medical Association Journal* 1952;67:395.

the patient from a disrupted family, is not as likely to disclose a heritable disorder as someone from a large, cohesive kindred.

In autosomal dominant disease, the affected person (proband, propositus (♂), proposita (♀), or sometimes index case) usually has one affected parent (unless a "new mutation" sporadic case), and the disease occurs in both sexes and is transmitted by either the father or mother. In autosomal recessive diseases, which can affect both sexes, the proband is born to parents who are usually asymptomatic "carriers" and sometimes related (ie, consanguineous). In X-linked recessive diseases, usually only males are affected, parents are unaffected yet the mother is an asymptomatic carrier, and there is no male to male transmission. In X-linked dominant diseases, both males and females can be affected, although the females are often more mildly and variably affected than males, and 50% of offspring (girls and boys) from an affected woman will have the disease, and 100% of the daughters but 0% of the sons of an affected man will have the disease. In Y-linked diseases, only males are affected and unless representing a sporadic case they have an affected father (patrilineal inheritance) and all sons of an affected man will have the disease. Mitochondrial inherited disorders (non-Mendelian) can affect both sexes, being transmitted only by an affected mother in her mitochondrial, not genomic, DNA. The small volume of sperm precludes them from contributing mitochondria to the zygote. Thus, all mitochondrial DNA in people is inherited through the maternal line in the egg (ie, matrilineal inheritance). These patterns of inheritance may be complicated by: (i) nonpenetrance or variable expression in autosomal dominant disorders (eg, in MEN1); (ii) imprinting whereby expression of an autosomal dominant disorder is conditioned by whether it is maternally or paternally transmitted (eg, pseudohypoparathyroidism type 1A versus pseudopseudohypoparathyroidism); (iii) anticipation, whereby some dominant disorders become more severe (or have earlier onset) in successive generations; (iv) pseudodominant inheritance of autosomal recessive disorders reflecting repeated consanguineous marriages in successive generations often within small/medium sized populations; and (v) mosaicism in which an individual has two or more populations of cells with different genotypes because of postzygotic mutations during their development from a single fertilized egg (eg, McCune-Albright syndrome). In the special circumstance of germline mosaicism within eggs or sperm arising from somatic mutation during gametogenesis, there may be diagnostic and recurrence risk confusion because of seemingly unaffected parents having multiple affected offspring suggesting autosomal recessive inheritance, but actually reflecting an autosomal dominant disorder (eg, osteogenesis imperfecta type II) [16]. Hence, these inheritance patterns, which can help to diagnose a genetic disorder and identify individuals at risk, can come from a detailed family history [2, 4–8].

GENETIC TESTS, THEIR CLINICAL UTILITY AND INTERPRETATION

Value of genetic testing

Establishing a disorder's genetic etiology creates potential benefits for the patient and relatives (Table 42.1). These include: appropriate clinical investigations and treatment; prognostication; screening for associated features not initially apparent; genetic counseling; testing of first-degree asymptomatic relatives; identification of relatives not at risk of having or transmitting the disease; and facilitating preconception genetic counseling and/or prenatal genetic testing. Establishing the genetic basis of any disease should improve diagnosis and lead to new insights concerning pathogenesis useful for choosing or developing therapeutic agents [17]. An excellent example is romosozumab, a monoclonal antibody to the inhibitor of bone formation sclerostin, which represents a treatment for osteoporosis and was discovered by identifying mutations of the *SOST* gene in patients with rare sclerosing bone dysplasias, sclerosteosis, and van Buchem disease [17–19].

Pretest considerations—which test?

Several factors must be considered before requesting genetic testing. These include the phenotype of the patient, the likely mode of inheritance, the potential genetic etiology (eg, aneuploidy, copy number variation [CNV], or single

Table 42.1. Value of genetic testing in clinical practice of patients with metabolic bone disease.

Benefits for the patient
Allow appropriate investigation and treatment of disease
Ability to undertake screening/surveillance for associated features that may not be clinically apparent
Prognostic information regarding disease course

Benefits for first degree relatives and/or progeny
Identify first degree family members who may be at risk of disease or those at risk of passing on to their progeny
Identify family members who do not harbor the genetic abnormality thereby alleviating the anxiety and burden of disease from them and/or their progeny
Where appropriate, to enable preconception genetic counseling
Where appropriate, to enable prenatal diagnosis

Academic/research benefits
Improved molecular characterization and understanding of respective disorder
Development of novel therapeutic targets/pathways

Potential future clinical benefits
Advent of personalized medicine—matching therapies to genetic defects

gene defect), and availability of additional pedigree members if necessary to aid diagnosis. Indeed, DNA sequencing of 'trios' (ie, both parents and the affected offspring) may identify autosomal recessive compound heterozygous or de novo mutation(s) in the patient not appreciated without parental samples [20]. Selecting the appropriate genetic test(s) will increase the likelihood of achieving a genetic diagnosis. Although most genetic skeletal disorders are monogenic in origin and will typically require high-resolution DNA sequencing for detection, some feature marked chromosomal abnormalities (eg, aneuploidy in Turner syndrome, or CNVs such as deletion on chromosome 15 in Prader–Willi syndrome), whose detection requires alternative approaches [21]. Furthermore, for some monogenic syndromes, causal whole or partial gene deletions may go undetected by direct sequencing methods. Thus, it is crucial to know that negative genetic testing does not exclude genetic disease, but rather may reflect: (i) an alternative genetic etiology to the one being tested; (ii) limitations of the genetic methodology employed (ie, inadequate resolution or coverage); or (iii) incorrect assumptions regarding the clinical phenotype or mode of inheritance. Sequential or simultaneous genetic tests (Table 42.2) may therefore be required, as briefly reviewed below.

Detection of chromosomal abnormalities, CNVs, and mutations causing disease

Karyotyping

Karyotyping is frequently the first cytogenetic test for major chromosomal abnormalities including aneuploidy (abnormal number of chromosomes) or large insertions, deletions, duplications, inversions, or reciprocal translocations [21, 22]. Karyotype analysis is usually performed by high-resolution G-banding (Giemsa staining) of at least 20 metaphase nuclei prepared from peripheral blood leucocytes. Occasionally, diseased tissues are studied directly to identify mosaicism (eg, fibroblasts). However, the resolution of G-band karyotype analysis is limited to ~5 to 10 megabases (Mb) of DNA, and cannot identify smaller abnormalities (eg, CNVs) [21].

Fluorescence in situ hybridization

FISH utilizes DNA probes that hybridize, as visualized by fluorescence microscopy, to specific target regions on metaphase chromosomes [21, 22]. This may identify: (i) a chromosomal deletion by absence of probe binding; (ii) duplications by additional probe binding; or (iii) translocations or inversions by probes binding to aberrant chromosomal regions. The molecular resolution of FISH is typically ~50 kilobases (kb) to 2 Mb. Newer FISH-based methods allow simultaneous evaluation of several regions of interest, including whole chromosome "painting probes" (termed multiplex FISH [M-FISH] and spectral karyotyping [SKY]), where each chromosome is labeled a different color [21, 22].

Multiplex ligation-dependent probe amplification (MLPA)

Multiplex ligation-dependent probe amplification (MLPA) is a PCR-based method that detects complete or partial gene deletions by using a pool of custom-designed probes to amplify specific genomic regions of interest [21].

Microarray comparative genomic hybridization

Microarray comparative genomic hybridization (aCGH) detects small chromosomal abnormalities (ie, CNVs) [21, 23]. In aCGH, the patient's fragmented DNA sample (labeled green) mixes with fragmented normal DNA (labeled red) prior to applying the samples to the array platform containing immobilized control DNA fragments for competitive hybridization [21–23]. Automated measuring of red-green fluorescence identifies deletions (appearing as an excess of red) and duplications (appearing as an excess of green) in the patient sample. aCGH resolution depends on the number and distance between the DNA clones (probes) selected for the array. However, all people harbor many small CNVs (eg, five to 10) without discernable adverse impact on health, whilst several sometimes pathogenic CNVs do not cause disease in all individuals (ie, reduced penetrance).

Single nucleotide polymorphism arrays

SNP arrays detect CNVs as well as genome-wide genotyping. For example, deletions (or uniparental disomy) spanning several adjacent SNPs included on the array may reveal loss of heterozygosity (LOH), whilst copy number gains (eg, duplication) may be indicated by increased numbers of different genotypes [21].

Single gene testing by Sanger sequencing (first generation sequencing)

Single gene testing by Sanger sequencing (first generation sequencing) uses PCR to amplify DNA fragments up to 1000 bp long, after which a single primer is utilized to initiate a reaction containing DNA polymerase to add nucleotides according to the complementary DNA strand, whilst randomly incorporating terminator nucleotides which harbor nucleotide-specific fluorescent labels. The resulting mix of DNA fragments of varying length is resolved by gel electrophoresis, and their DNA sequence is established from the chromatogram. The high fidelity of DNA polymerase results in a high degree of accuracy such that Sanger sequencing achieves a base accuracy of >99.99%, and remains the gold standard for DNA sequencing [24, 25].

Next generation sequencing (second generation sequencing)

Next generation sequencing (NGS) (second generation sequencing) represents a paradigm shift for both the investigation and diagnosis of genetic disease, and

Table 42.2. Examples of genetic tests, their molecular resolution, and utility.

Genetic Test	Resolution	Abnormalities Detected	Additional Notes
Detection of chromosomal abnormalities including copy number variations (CNVs)			
Karyotype: G-banding (typsine-Giemsa staining)	5–10 Mb	Aneuploidy Large chromosomal deletions, duplications, translocations, inversions, insertions	Limited resolution Requirement to study many cells to detect mosaicism
Fluorescence in situ hybridization (FISH)	50 kb to 2 Mb (dependent on size of probes employed)	Structural chromosomal abnormalities (eg, microdeletions, translocations)	Labor-intensive Low resolution limits its use Unsuitable where unknown genetic etiology
Multiplex ligation-dependent probe amplification (MLPA)	Probe-dependent 50–70 nucleotides Single exon deletion or duplication possible	CNVs including (partial) gene deletions or duplications)	Low cost, technically simple method Simultaneous evaluation of multiple genomic regions Not suitable for genome-wide approaches Not suitable for analysis of single cells
Array-comparative genomic hybridization (aCGH)	10 kb (high resolution) 1 Mb (low resolution) (Dependent on probes set)	Genome-wide CNVs	Inability to detect balanced translocations Useful for detection of low level mosaicism
Single nucleotide polymorphism (SNP) array	~50–400 kb (Dependent on probe set)	Genome-wide detection of SNP genotypes CNVs	Inability to detect balanced translocation Useful for detection of low level mosaicism Detection of copy number neutral regions or absence of heterozygosity (ie, due to uniparental disomy)
Detection of monogenic disorders (and CNVs)			
First generation sequencing (Sanger)			
Single gene test	Single nucleotide (exonic regions and intron/exon boundaries of candidate gene)	Single nucleotide variants (SNVs) Small insertions or deletions ("indels")	Relative high cost/base May miss large deletions/duplications Unsuitable where unknown genetic etiology
Next generation sequencing			
Disease-targeted gene panels	Single nucleotide (exonic regions and intron/exon boundaries of candidate genes)	Single nucleotide variants (SNVs) Small insertions or deletions ('indels')	May lack complete coverage of exomic regions (may require Sanger sequencing to fill in "gaps") Increased likelihood of identifying variants of uncertain significance (VUS) as number of genes increases Unsuitable where unknown genetic etiology
Whole exome sequencing (WES)	Single nucleotide (all exonic regions and intron/exon boundaries)	SNVs *Small insertions or deletions ("indels") CNVs	Not all exons may be covered/captured Difficulties with GC-rich regions and presence of homologous regions/pseudogenes Bioinformatic expertise required for data analysis High likelihood of incidental findings and VUSs Detection of CNVs requires additional data analysis (ie loss of heterozygosity mapping across exonic regions) Suitable for disease associated gene-discovery
Whole genome sequencing (WGS)	Single nucleotide	SNVs Small insertions or deletions ("indels") CNVs (Translocations/rearrangements)	Relative high cost Large data sets generated and complex data analysis requiring bioinformatic expertise Very high likelihood of incidental findings and VUSs CNV analysis possible but may present specific challenges Suitable for disease-associated gene discovery

*Small indels may not be captured.

increasingly contributes to routine clinical care. The principles of NGS do not differ markedly from those of Sanger sequencing. However, in NGS incorporation of the fluorescently labeled nucleotide does not terminate the reaction. Thus, the sequence is established through successive cycles of nucleotide addition captured by an automated camera, thereby enabling simultaneous sequencing of millions of DNA fragments ("massively parallel sequencing") [24, 25]. These principles form the basis for the three most widely employed uses of NGS: whole genome sequencing (WGS), whole exome sequencing (WES), and disease-targeted gene panels.

WGS determines the DNA sequence of the entire genome including coding and noncoding regions, and can identify single nucleotide variants (SNVs), small insertions or deletions ("indels"), and CNVs. Mutations can be identified in both coding and regulatory regions (eg, enhancers, promoters) [24, 25]. Limitations of WGS include difficulties sequencing certain regions of the genome (ie, GC-rich regions), whilst highly homologous regions of DNA (eg, due to pseudogenes) may pose difficulties aligning (and interpreting) sequence reads to the reference genome.

WES analyzes the 1% to 2% of the genome that encodes the ~20,000 protein-coding genes (ie, the "exome"), which are expected to harbor most of the disease-associated mutations [24, 25]. Indeed, WES has been the mainstay of highly successful disease gene discovery over the past decade, including for mineral metabolic and skeletal diseases [1,2,24]. However, WES can also detect large CNVs (eg, by LOH mapping). WES uses probe sets to capture only the exomic regions, and thus may not be complete because of missing exons, or capture may be uneven due to SNPs or indels. Here too, GC-rich and highly homologous regions may present difficulties, whilst the lack of coverage of noncoding regions will miss mutations in regulatory regions.

Disease-targeted sequencing is the most widely utilized NGS method in the clinical arena because it can be configured to simultaneously analyze collections of genes (eg, <10 to >150 genes) associated with a specific disorder [24–27]. Such NGS disease-targeted panels have been established for osteogenesis imperfecta and other low bone mass phenotypes, hypophosphatemic rickets, and skeletal dysplasias including those featuring high bone mass. Potential advantages include reduced cost, simplified data analysis, and rapid simultaneous evaluation of multiple genes compared with sequential analysis of genes using Sanger sequencing. However, its limitations include knowing the potential causative genes—it is unsuitable for "gene-discovery" studies.

Genetic tests to detect mosaicism

Improved genome-wide genetic testing (including aCGH, SNP arrays, and NGS) now also provides far greater sensitivity for detecting low levels of mosaicism (eg, 5% for SNP array) [16]. However, choosing the best test depends on the clinical picture as well as the suspected type of mutation (eg, aneuploidy, CNV, SNV, "indel"), and its likely extent and tissue distribution [16, 28]. Often circulating lymphocyte DNA will suffice, but testing other affected tissue(s) may be required to detect somatic mosaic mutations, as illustrated by identification of *GNAS1* and *AKT1* mutations causing McCune-Albright and Proteus syndromes, respectively [16,29].

Genetic tests for prenatal diagnosis

Prenatal genetic testing may be undertaken at preimplantation or prenatal stages, and its implementation will depend on the clinical scenario.

Preimplantation genetic diagnosis

Preimplantation genetic diagnosis uses a single cell from the embryo taken several days after in vitro fertilization (IVF) to detect major chromosomal abnormalities (by aCGH or FISH) or single gene defects (conventional PCR and sequencing). Thus, embryos are screened for genetic disorders prior to establishing pregnancy.

Prenatal genetic testing

Prenatal genetic testing is used once pregnancy is established to identify fetuses at risk of genetic disease. Typically, this involves invasive methods such as chorionic villous sampling (CVS) or amniocentesis to obtain cells for karyotyping, FISH, aCGH, and DNA sequencing. Detection of cell-free circulating fetal DNA in the maternal circulation offers the potential for noninvasive prenatal genetic diagnosis (NIPD) and/or testing (NIPT) [30].

Approximately 10% to 20% of cell-free DNA in the maternal circulation arises from the placenta, and the ability to detect this fetal DNA (cffDNA) provides the basis for NIPD/NIPT, which involves a maternal blood sample at an appropriate stage of gestation (eg, after 10 weeks). NIPD/NIPT primarily screens for aneuploidy (eg, for Down, Edwards, Patau, and Turner syndromes) and fetal sex determination (eg, important for X-linked disorders). However, it can also detect monogenic disorders, although this is limited to paternally inherited mutations or those arising de novo, because it cannot distinguish abnormalities present in the mother (due to the presence of maternal cell-free DNA in the sample) [30].

Data interpretation and incidental findings

Establishing a genetic diagnosis may benefit the patient and family, but it can also present clinical and ethical challenges [31–34]. For example, multiple gene testing or whole genome approaches employing NGS may reveal variants of uncertain significance (VUSs), whose relevance to the clinical phenotype is ambiguous, as well as incidental genetic abnormalities. Hence, the possibility of identifying ambiguous or incidental results should be part of informed consent prior to genetic testing. Indeed, recent studies have demonstrated that many variants reported as pathogenic may instead be benign, or far less penetrant than hitherto recognized. Any misclassification may therefore have a substantial adverse impact for patients and family members. Thus, caution is required in data interpretation. Clinicians and patients alike must appreciate uncertainties from genetic testing and its limitations. Finally, accurate reporting of genetic and phenotype data will improve disease-specific mutation databases and facilitate clinical research [32,33].

ACKNOWLEDGMENTS

PJN is supported by a Chief Scientist Office (CSO) and NHS Research Scotland (NRS) Fellowship (UK). MPW is supported by Shriners Hospitals for Children and The Clark and Mildred Cox Inherited Metabolic Bone Disease Research Fund at the Barnes-Jewish Hospital Foundation, St. Louis, MO, USA. RVT is supported by the Medical Research Council (UK), Wellcome Trust (UK), and National Institute of Health Research (NIHR), UK.

REFERENCES

1. Newey PJ, Gorvin CM, Whyte MP, et al. Introduction to genetics of skeletal and mineral diseases. In: Thakker RV, Whyte MP, Eisman J, et al. (eds). *Genetics of Bone Biology and Skeletal Disease* (2nd ed.). San Diego: Elsevier-Academic Press, 2018, pp. 3–23.
2. Bonafe L, Cormier-Daire V, Hall C, et al. Nosology and classification of genetic skeletal disorders: 2015 revision. Am J Med Genet A. 2015;167A(12):2869–92.
3. Estrada K, Styrkarsdottir U, Evangelou E, et al. Genome-wide meta-analysis identifies 56 bone mineral density loci and reveals 14 loci associated with risk of fracture. Nat Genet. 2012;44(5):491–501.
4. Rosen RJ. *Primer on the Metabolic Bone Diseases and Disorders of Mineral Metabolism* (8th ed.). Ames, IA: American Society for Bone and Mineral Research/Wiley-Blackwell, 2013.
5. Coe FL, Favus MJ. *Disorders of Bone and Mineral Metabolism* (2nd ed.). Philadelphia: Lippincott Williams & Wilkins, 2002.
6. Avioli LV, Krane SM. *Metabolic Bone Disease and Clinically Related Disorders* (3rd ed.). San Diego: Academic Press, 1997.
7. Scriver CR, Beaudet AL, Sly WS, et al. *The Metabolic and Molecular Bases of Inherited Disease* (8th ed.). New York: McGraw-Hill Medical, 2001.
8. McKusick V. Online Mendelian Inheritance in Man (OMIM). http://www.ncbi.nlm.nih.gov/omim (accessed May 2018).
9. Edeiken J, Dalinka MK, Karasick D. *Edeiken's Roentgen Diagnosis of Diseases of Bone* (4th ed.). Baltimore: Williams & Wilkins, 1990.
10. Lachman RS. *Taybi and Lachman's Radiology of Syndromes, Metabolic Disorders and Skeletal Dysplasias* (5th ed.). St. Louis: Mosby, 2007.
11. Revell PA. *Pathology of Bone*. Berlin: Springer-Verlag, 1986.
12. Greenfield GB. *Radiology of Bone Diseases* (5th ed.) Philadelphia: Lippincott Williams & Wilkins, 1990.
13. Resnick DL, Kransdorf MJ. *Bone and Joint Imaging* (3rd ed.). Philadelphia: Saunders, 2004.
14. Resnick D. *Diagnosis of Bone and Joint Disorders* (4th ed.). Philadelphia: Saunders, 2002.
15. LeBlond RF, Brown DD, Suneja M. *Degowin's Diagnostic Examination* (10th ed.). New York: McGraw-Hill Education, 1999.
16. Biesecker LG, Spinner NB. A genomic view of mosaicism and human disease. Nat Rev Genet. 2013; 14(5):307–20.
17. Rivadeneira F, Mäkitie O. Osteoporosis and bone mass disorders: from gene pathways to treatments. Trends Endocrinol Metab. 2016;27(5):262–81.
18. Brunkow ME, Gardner JC, Van Ness J, et al. Bone dysplasia sclerosteosis results from loss of the SOST gene product, a novel cystine knot-containing protein. Am J Hum Genet. 2001;68(3):577–89.
19. Balemans W, Patel N, Ebeling M, et al. Identification of a 52 kb deletion downstream of the SOST gene in patients with van Buchem disease. J Med Genet. 2002; 39(2):91–7.
20. Goldstein DB, Allen A, Keebler J, et al. Sequencing studies in human genetics: design and interpretation. Nat Rev Genet. 2013;14(7):460–70.
21. Gijsbers AC, Ruivenkamp CA. Molecular karyotyping: from microscope to SNP arrays. Horm Res Paediatr. 2011;76(3):208–13.
22. Dave BJ, Sanger WG. Role of cytogenetics and molecular cytogenetics in the diagnosis of genetic imbalances. Semin Pediatr Neurol. 2007;14(1):2–6.
23. Kharbanda M, Tolmie J, Joss S. How to use… microarray comparative genomic hybridisation to investigate developmental disorders. Arch Dis Child Educ Pract Ed. 2015;100(1):24–9.
24. Lazarus S, Zankl A, Duncan EL. Next-generation sequencing: a frameshift in skeletal dysplasia gene discovery. Osteoporos Int. 2014;25(2):407–22.
25. Falardeau F, Camurri MV, Campeau PM. Genomic approaches to diagnose rare bone disorders. Bone. 2017; 102:5–14.
26. Rehm HL. Disease-targeted sequencing: a cornerstone in the clinic. Nat Rev Genet. 2013;14(4):295–300.
27. Árvai K, Horváth P, Balla B, et al. Next-generation sequencing of common osteogenesis imperfecta-related genes in clinical practice. Sci Rep. 2016;6:28417.
28. Cohen AS, Wilson SL, Trinh J, et al. Detecting somatic mosaicism: considerations and clinical implications. Clin Genet. 2015;87(6):554–62.
29. Lindhurst MJ, Sapp JC, Teer JK, et al. A mosaic activating mutation in AKT1 associated with the Proteus syndrome. N Engl J Med. 2011;365(7):611–19.
30. van den Veyver IB, Eng CM. Genome-wide sequencing for prenatal detection of fetal single-gene disorders. Cold Spring Harb Perspect Med. 2015;5(10):pii:a023077

31. Manrai AK, Funke BH, Rehm HL, et al. Genetic misdiagnoses and the potential for health disparities. N Engl J Med. 2016;375(7):655–65.
32. MacArthur DG, Manolio TA, Dimmock DP, et al. Guidelines for investigating causality of sequence variants in human disease. Nature. 2014;508(7497):469–76.
33. Richards S, Aziz N, Bale S, et al; ACMG Laboratory Quality Assurance Committee. Standards and guidelines for the interpretation of sequence variants: a joint consensus recommendation of the American College of Medical Genetics and Genomics and the Association for Molecular Pathology. Genet Med. 2015;17(5):405–24.
34. Lek M, Karczewski KJ, Minikel EV, et al; Exome Aggregation Consortium. Analysis of protein-coding genetic variation in 60,706 humans. Nature. 2016;536 (7616):285–91.

# 43

Animal Models: Genetic Manipulation

Karen Lyons

Department of Orthopaedic Surgery/Orthopaedic Hospital, University of California, Los Angeles, CA, USA

INTRODUCTION

Studies involving genetically manipulated mice have contributed enormously to the identification of genes controlling skeletal development and elucidation of their mechanisms of action. Genetically modified mice have facilitated the production of animal models of human diseases, cell lineage studies, and dissection of distinct functions at specific stages of differentiation within a single cell lineage. Public repositories of genetically modified embryonic stem (ES) cells and mice have greatly facilitated genetic studies. The adaptation of CRISPR/Cas9 technology to introduce defined mutations has made the production of genetically modified animals increasingly more feasible.

RESOURCES AND REPOSITORIES

Several large-scale government-sponsored research programs have provided access to targeting vectors, ES cells, or mice harboring a large number of mutated alleles. Such programs include the Knockout Mouse Project (KOMP, USA), EUCOMM (Europe), North American Conditional Mouse Mutagenesis Project (NorCOMM, Canada), and Texas A&M Institute for Genomic Medicine (TIGM, USA). Alleles generated by these consortia are coordinated by the International Knockout Mouse Consortium (IKMC), which publishes a searchable list of all available vectors, ES cell clones, and mice. The IKMC is a component of the International Mouse Phenotyping Consortium (IMPC) (http://www.mousephenotype.org, accessed May 2018) and the searchable database is accessible through their website. The goal of the IMPC is to produce loss of function alleles for 20,000 known and predicted genes in the mouse genome. Some of the alleles have been created by gene targeting. Others have been generated through gene-trapping technologies. Yet others have been generated through large-scale chemical mutagenesis screens. Modified ES cells obtained from all of these projects can greatly simplify and accelerate the process of generating a mutant mouse strain. In this chapter, we focus on tissue-specific gene targeting, the most widely used approach for analyzing gene function in skeletal tissues.

OVEREXPRESSION OF TARGET GENES

The first widely used approach to study gene function in vivo was to produce transgenic mice that overexpress target genes. Several promoters have been well characterized and widely used to drive gene expression in skeletal tissues.

Chondrocytes

The most widely used cartilage-specific promoter is derived from the mouse *pro aI(II) collagen (Col2a1)* gene. This promoter drives high levels of expression beginning after the condensation stage in appendicular elements, but prior to condensation in the sclerotomal compartment [1]. The *Col11a2* promoter has also been used, although some of these promoters also drive expression in perichondrium and osteoblasts [2].

Osteoblasts/osteocytes

The most commonly used promoter to drive overexpression in osteoblasts is a 2.3-kilobase (kb) proximal promoter fragment from the rat or mouse *Col1a1* gene. Strong activity is seen in fetal and adult mature osteoblasts and osteocytes [3]. The 3.6-kb proximal *Col1a1* promoter drives strong expression at an earlier stage of differentiation (preosteoblasts), but is also expressed in nonosseous tissues, including tendon, skin, muscle, and brain [4, 5]. Expression is driven by 3.5- to 3.9-kb human osteocalcin promoter fragments in a large proportion of mature osteoblasts and osteocytes [6], and have been very widely used. The identification of an osteocyte-specific promoter to drive transgene expression has been a challenge. A 10-kb *dentin matrix protein 1* (*DMP1*) promoter has been described, but this promoter can potentially drive gene expression in osteoblasts and myoblasts in addition to osteocytes; an 8-kb DMP1 promoter appears to have higher specificity for osteocytes (reviewed in [7]).

Tendon and ligament

Tendon patterning and differentiation has been difficult to study genetically owing to a lack of tendon-specific promoters. *Scleraxis* (*Scx*) encodes a transcription factor expressed in developing tendons and ligaments. A *Scx* promoter capable of driving tendon-specific gene expression has not yet been characterized. Scx-Cre strains (discussed later in this chapter) provide a potential strategy for inducing expression of Cre-inducible transgenes; however, expression has also been reported in cartilage and nonskeletal tissue [8].

Osteoclasts

A variety of promoters drive high levels of gene expression in osteoclasts and their progenitors. These include *CD11b*, expressed in monocytes, macrophages, and along the osteoclast differentiation pathway from mononucleated progenitor cells and into mature osteoclasts [9], and *TRACP*, expressed in mature osteoclasts and their precursors [10].

Advantages and disadvantages of overexpression approaches

The major advantages of the transgenic approach are that it is straightforward, inexpensive, and transgenic mice often show obvious phenotypes. Furthermore, transgenic strains in which marker genes such as *LacZ*, *GFP*, and/or *ALP* are expressed under the control of tissue-specific promoters allow visualization of specific cell types in vivo, and permit their isolation with a resolution not possible using other methods [11]. The transgenic overexpression approach can also be used to introduce a dominant negative gene product in order to block a gene activity (eg, [12]). Finally, transgenic approaches have been used to ablate a specific cell population by driving overexpression of diptheria toxin receptor; this approach provided one of the first clear demonstrations of the essential role of osteocytes in mechanotransduction [13].

A major caveat of the transgenic approach is that overexpression models often yield nonphysiological levels of gene expression, confounding interpretations of the role of the gene under normal conditions. On the other hand, the use of transgenic techniques to overexpress dominant negative variants or natural antagonists leads to loss of function and thus targets pathways in their normal physiological contexts.

A major limitation of the transgenic approach is that there are relatively few well-characterized promoters for skeletal tissues. Moreover, the site of transgene integration can have major consequences on tissue specificity and levels of expression. This can be exploited to examine dose-dependent effects, but care must be taken to assess levels and sites of expression.

Finally, overexpression of genes may confer embryonic lethality, precluding the establishment of stable transgenic lines. This can be overcome by the generation of inducible transgenic strains. The most widely used method to generate inducible transgenic strains is to modify the transgene so that its expression is repressed by a strong transcriptional stop sequence flanked by LoxP sites located upstream of the transgene coding sequence. By crossing the above transgenic strain to one expressing Cre recombinase (discussed later), transgene expression can be activated in a tissue-specific or inducible manner in skeletal tissues [14].

GENE TARGETING

The most widely used technique for genetic manipulation is gene targeting in mouse ES cells. Basic methodologies are described in review articles [15]. Several large-scale government-sponsored research programs provide access to targeting vectors, ES cells, or mice harboring a large number of mutated alleles. These reagents can greatly simplify and accelerate the process of generating a mutant mouse strain. The efforts of these consortia are coordinated by the IKMC, which maintains a searchable list of all available vectors, ES cells, and mice. The IKMC is a component of the IMPC.

The Jackson Laboratories Mouse Genome Informatics website (www.informatics.jax.org, accessed May 2018) provides information on all published genetic modifications, results of ongoing phenotypic analysis, and links to publically available ES and mouse resources. Because all genetic modifications are included on this website, one can request strains that have not been deposited into a public repository by contacting the investigator directly.

Advantages and disadvantages of gene targeting

Phenotypes caused by global loss of function provide direct insight into the physiological roles of the ablated gene and can be used to generate animal models of human disease. Moreover, novel actions of targeted genes can be revealed. However, a complication of the global knockout approach is that deletion of genes essential for early development may lead to early lethality. A second issue is that it may be difficult to dissect direct effects on skeletal tissues from indirect ones related to roles in metabolic, cardiovascular, neuroendocrine, or other systems. An important consideration is that global knockout stains usually retain the selection cassette used to screen the ES colonies. On occasion, this leads to effects on neighboring genes. For these reasons, global knockout models may not always recapitulate the phenotypes seen in humans bearing loss of function alleles for a particular gene.

Tissue-specific and inducible knockout and overexpression

The ability to ablate or overexpress genes in a tissue-specific manner has revolutionized our understanding of the functions of genes and pathways in specific skeletal cell types and lineages, and at specific stages of development. Several methods can be used to achieve tissue-specific gene knockout or activation; these rely on site-specific recombinases derived from bacteriophage (Cre) or yeast (Flp) [16]. Cre and Flp recombine DNA at specific target sites (Fig. 43.1). Two mouse lines are required. For the Cre-loxP system, these are the "floxed" strain, in which the region of the gene targeted for

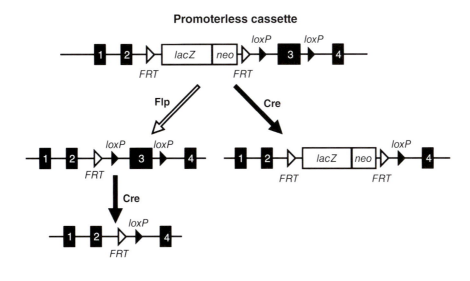

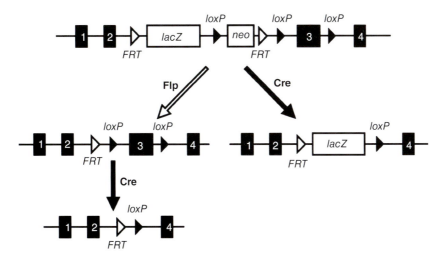

Fig. 43.1. Schematic of knockout-first alleles generated by the IKMC. The targeting strategies used by the IKMC rely on the identification of a "critical" exon. The generation of the knockout-first allele is flexible and can produce reporter knockouts, conditional knockouts, and null alleles following exposure to site-specific recombinases Cre and Flp. Source: [51]. Reproduced with permission of SpringerNature.

deletion is flanked by loxP sites, and a second transgenic line in which Cre recombinase is expressed under the control of an inducible and/or tissue-specific promoter. In mice carrying both the floxed gene and the Cre transgene, Cre deletes the sequence flanked by loxP sites. The loxP sites are usually placed in introns and generally do not interfere with the normal function of the gene. Hence, the floxed target gene usually functions normally except in tissues where Cre is expressed. Figure 43.1 shows a typical IKMC vector design and various targeting strategies for the generation of conditional alleles from ES cells obtained through this database.

Conditional alleles have also been used to achieve site-specific inducible overexpression. In this case, a transgene is generated under the control of a strong ubiquitious promoter such as CAG [17]. The expression of the transgene is prevented by a strong transcriptional stop signal, flanked by loxP sites. The gene is activated when Cre catalyzes excision of the stop signal [18]. Engin and colleagues provides an example of this approach [19].

Drivers for gene recombination

As discussed, the site-specific recombinase most commonly used to catalyze gene recombination is Cre. Most studies have employed constitutively active forms of Cre. However, ligand-regulated forms that enable temporal control of gene activity are also available. The most popular strategy employs versions of Cre fused to a ligand-binding domain from a mutant estrogen receptor (ER). The ER domain recognizes the synthetic estrogen antagonist 4-OH tamoxifen (T), but is insensitive to endogenous β-estradiol. In the absence of T, the Cre-ER(T) fusion protein is retained in the cytoplasm. Binding of T to the ER domain permits the fusion protein to enter the nucleus and catalyze recombination. A complete list of published Cre lines can be found on the Mouse Genome Informatics site (www.informatics.jax.org/home/recombinase, accessed May 2018). A few of the constitutive and inducible Cre lines most widely used to study skeletal biology, and some promising newcomers, are discussed in the following list and outlined in Table 43.1. Examination of the table reveals the variety of methodologies used to generate tissue-specific Cre drivers, including transgenic mice harboring traditional promoter-driven constructs, and the generation of transgenic strains in which Cre has been inserted into a gene of interest on a bacterial artificial chromosome (BAC). In other cases, the Cre cassette has been inserted into an endogenous locus.

Uncondensed mesenchyme and mesenchymal condensations

Prrxl-Cre drives expression in early uncondensed limb and head mesenchyme [20]. Dermol-Cre expresses Cre in mesenchymal condensations [21]. A Sox9-Cre knock-in strain drives Cre-mediated excision in precursors of osteoblasts and chondrocytes in these condensations [22].

Cartilage

The most widely used Cre strain for cartilage is *Col2a1-Cre* [1]. Promoter activity seems to be restricted to chondrocytes in most studies, but there is a window during chondrogenesis when *Col2a1-Cre* is expressed in perichondrium. *Col10a1-Cre* lines permit gene deletion in hypertrophic chondrocytes [23, 24].

Tools that permit inducible recombination in cartilage are available. Several groups generated transgenic lines using CreER(T) under the control of the *Col2a1* promoter. These permit efficient ablation in prenatal cartilage and up to 2 weeks after birth; a *Col2a1-CreERT2* line is efficient in adults (eg, [25, 26]. Aggrecan-CreERT2 mice, in which CreERT2 has been knocked into the aggrecan locus, exhibit robust Cre expression in prenatal and adult growth plates and articular cartilage, as well as fibrocartilage [27].

There are no Cre lines that drive Cre-mediated recombination specifically in articular cartilage. However, *Gdf5-Cre* mice drive expression in the joint interzone, synovial tissues, and articular cartilage [28]. A *Prg4GFPCreERT2* line, in which GFP and Cre-ERT2 have been introduced into the *Prg4* locus, drives expression in superficial zone cartilage in young mice, but descendants of these cells span the depth of articular cartilage in adults [29].

Osteoblasts/osteocytes

A 3.6-kb *Col1a1* promoter drives high levels of Cre expression in osteoblasts but also targets tendon and fibrous cells types in the suture, skin, and several organs [4]. *Col1a1-Cre* lines (2.3 kb) show more restricted expression to mature osteoblasts, but ectopic expression in brain is noted in some lines [3, 5]. Osteocalcin-Cre drives excision in mature osteoblasts but is not activated until just before birth [6]. Several Cre lines based on the *DMP1* promoter have been used to target osteocytes. As discussed above, lines containing 10 kb or longer *DMP1* proximal promoter regions have been generated (reviewed in [7]). These exhibit expression in mature osteoblasts in addition to osteocytes. An 8-kb DMP1-Cre strain is reported to have more osteocyte-selective expression [30]. Several inducible Cre transgenic strains have been developed for bone. The transcription factor *Osterix (Osx1)* is expressed in osteoblast precursors. A Cre-GFP fusion protein whose expression is regulated by doxycycline has been inserted into the *Osx1* locus on a BAC transgene [31]. This construct targets osteoblast precursors. Inducible 2.3-kb *Col1a1-CreERT* and 8-kb *Dmp1-cre/ERT2* strains have been described [32].

Osteoclasts

Several Cre strains permit ablation in myeloid cells. These include LysMcre mice, in which Cre has been introduced into the *M Lysozyme* locus [33], and a strain in which the

Table 43.1. Commonly used Cre drivers and lineage tracers for Cre activity for skeletal analysis.

	Description	Primary Sites of Expression
Precondensed mesenchyme		
Dermo1-Cre (Twist2-Cre) Twist2$^{tm1.1(cre)Dor}$	Cre knockin that abolishes function of *Twist2* allele	Craniofacial and splanchnic mesenchyme; skeletal condensations
Prrx1-Cre Tg(Prrx1-Cre)1Cjt	Transgene insertion	Limb bud mesenchyme; skull vault
Prrx1-CreER-GFP Tg(Prrx1-cre/ERT2,-GFP)1Smkm	Transgene insertion Inducible, GFP activity tracer	Limb bud mesenchyme; skull vault
Cartilage		
Sox9-Cre Sox9$^{tm3(cre)Crm}$	IRES-Cre cassette inserted into 3' UTR of Sox9	Skeletal condensations Other sites: intestine, testes, etc.
Sox9-Cre-ERT2	IRES-Cre-ERT2 cassette inserted into 3' UTR of *Sox9*	Skeletal condensations Other sites: intestine, testes, etc.
Col2a1-Cre Tg(Col2a1-cre)1Bhr	Transgene insertion	Differentiating chondrocytes in appendicular elements; sclerotome Other sites: heart valves, notochord
Col2a1-CreERT Tg(Col2a1-cre/ERT) KA3Smac	Transgene insertion Inducible; not responsive in adults	Tamoxifen-inducible in differentiating chondrocytes
Col2a1-CreERT2 Tg(Col2a1-cre/ERT2)1Dic	Transgene insertion Inducible in embryos and adults	Tamoxifen-inducible in differentiating chondrocytes
Aggrecan-CreERT2 Acan$^{tm1(cre/ERT2)Crm}$	Cre-ERT2 knockin into 3' UTR of *Acan* locus Inducible in embryos and adults	Growth plate and articular cartilage; ligament; fibrocartilage of meniscus; trachea; intervertebral disk
GDF5-Cre Tg(Gdf5-cre-ALPP)1Kng	Transgene insertion of a BAC encoding *Gdf5* modified by insertion of a Cre-IRES-ALPP cassette into the translational start site	Joint interzone in embryos; derivatives contribute to synovium and articular cartilage
Prg4-Cre-ERT2 Prg4$^{tm1(GFP/cre/ERT2)Abl}$	GFP/Cre-ERT2 cassette inserted into translation initiation codon of the Prg4 gene; abolishes function of the modified Prg4 locus Inducible in embryos and adults	Superficial zone cartilage; descendant cells contribute extensively to full thickness articular cartilage
FSP1-Cre Tg(S100a4-cre)1Egn	Transgene insertion	Broadly in a subset of fibroblasts; fibrous periosteum/perichondrium
Bone		
Osx-Cre g(Sp7-tTA,tetO-EGFP/cre)1Amc	A tetracycline-regulated transactivator (tTA) upstream of a tetracycline-responsive element (tetO)-controlled EGFP/Cre fusion protein placed in exon 1 of a BAC construct encoding Sp7 (Osterix) Inducible in embryos and adults	Osteoprogenitors; the transgene itself impacts bone mass and causes a craniofacial defect
Col1a1-Cre Tg(Col1a1-cre)1Kry	Transgene insertion. 2.3-kb promoter	Osteoblasts throughout differentiation
Col1a1-Cre-ERT2 Tg(Col1a1-cre/ERT2)1Crm	Transgene insertion. 2.3-kb promoter Inducible	Osteoblasts throughout differentiation
OCN-Cre Tg(BGLAP-cre)1Clem	Transgene insertion	Mature osteoblasts
DMP1-Cre Tg(Dmp1-cre)1Jqfe	Transgene insertion. 14-kb promoter	Odonotblasts, osteoblasts, osteocytes
DMP1-Cre-ERT2 Tg(Dmp1-cre/ERT2) D77Pdp	Transgene insertion. 10-kb promoter Inducible in embryos and adults	Mature osteoblasts and osteocytes
Osteoclasts		
LysM-Cre Lyz2$^{tm1(cre)Ifo}$	A Cre inserted into the ATG start site of the endogenous *Lyz2* gene	Myeloid lineage cells
CD11b-Cre Tg(ITGAM-cre)2781Gkl	Transgene insertion	Myeloid lineaage cells

(Continued)

	Description	Primary Sites of Expression
Reporters for detecting Cre activity		
R26R Gt(ROSA)26Sor[tm1Sor]	Targeted insertion of a floxed neo cassette	LacZ expressed in all cells where Cre is active and their descendants
R26R-Confetti Gt(ROSA)26Sor[tm1(CAG-Brainbow2.1)Cle]	Targeted insertion of a strong CAGG promoter, a loxP site, a PGK-Neo-pA cassette (serving as a transcriptional roadblock), and the Brainbow 2.1 construct inserted between exons 1 and 2 of the Gt(ROSA)26Sor locus	Individual cells and their descendants can be tracked upon Cre activation
mTmG Gt(ROSA)26Sor[tm4(ACTB-tdTomato,-EGFP)Luo]	Targeted insertion of a CAGG promoter driving expression of loxP-flanked DsRed protein (tdTomato) followed by a membrane-tagged EGFP protein	Red fluorescence in all tissues; EGFP in Cre-expressing cells and their descendants

EGFP, enhanced green fluorescent protein; GRP, green fluorescent protein; UTR, untranslated region.

CD11b promoter drives Cre expression in macrophages and osteoclasts [34], and transgenic lines expressing Cre under the control of the *TRAPC* and *Ctsk* promoters [35].

Considerations when using inducible knockouts

The most significant advantage of the Cre-loxP system is its flexibility, permitting exploration of gene function in distinct tissues at multiple time points. However, there are caveats. Depending on the Cre line used, excision of the target gene may not be complete; this is highly dependent on the floxed allele. Floxed alleles also vary with respect to the kinetics of Cre-mediated recombination. This must be borne in mind when attempting to compare phenotypes caused by excision of different genes using the same Cre transgenic line.

Cre transgenic strains based on identical promoters but generated in different laboratories can exhibit different specificities and efficiencies. Furthermore, the promoters most widely used for analysis of skeletal phenotypes will exhibit expression in multiple skeletal cell types, and for many, in nonskeletal cell types as well. For this reason, every study should include controls to verify the extent and location of Cre-mediated recombination of the floxed line of interest. Lineage controls and activity reporters are discussed in the next section.

The presence of a drug selection cassette in a floxed allele can have a major effect on expression levels of the targeted allele or neighboring genes even in the absence of the Cre transgene; for example, *scleraxis* knockouts, in which the presence of the drug selection cassette led to embryonic lethality by day 9.5 of gestation [36]. In striking contrast, mice homozygous for a *scleraxis* null allele in which the drug selection cassette was removed are viable as adults [37].

With respect to inducible models, both doxycycline and tamoxifen can have effects on cartilage, bone, and osteoclasts independently of target gene deletion. Even the low doses of tamoxifen used to catalyze Cre-ER(T)-mediated excision may have effects on bone [38]. Thus, a control group of Cre-negative mice treated with the inducer may need to be included to examine the impact of this variable on the mutant phenotype under study.

LINEAGE TRACING AND ACTIVITY REPORTERS

Genetically modified mice have permitted the determination of cell lineage relationships and the relative contributions of cells from various sources to a given organ with unprecedented resolution. These studies rely on strains that carry a floxed reporter gene, such as *LacZ* or *GFP*. For example, the R26R strain carries a floxed LacZ cassette introduced into the ROSA26 locus. When bred to a Cre-expressing strain, all cells in which Cre is expressed, and all of their descendants, express LacZ. R26R mice have been widely used to test the specificity and efficiency of Cre-expressing transgenic lines, although use of this strain in bone is limited by the fact that osteoblasts express endogenous LacZ. More recently, multiple lines have been generated that permit live imaging of expression of fluorescent proteins in specific organelles (cell membrane, nucleus, etc.) in a tissue-specific and inducible manner [39].

Major insights have been made using R26R and other lineage tracers to study osteoprogenitors. These are discussed in several comprehensive reviews [11, 40]. An example is the demonstration that immature osteoblasts move into developing bone along with invading blood vessels [41]. A large number of reporters have been used recently as lineage tracers that mark distinct skeletal populations.

Transgenic reporter lines can also be used to monitor signaling pathway activity. For example, Wnt pathway activity has been monitored in vivo using TOPGAL mice to track ß-catenin activity during endochondral bone formation [42]. Tools are also available to monitor canonical BMP pathway activity in vivo [43, 44].

CRISPR/CAS9 GENOMIC ENGINEERING

In the last 5 years, the advancement of CRISPR/Cas9 technology has revolutionized genomic engineering. CRISPR/Cas9 is a genome editing technique in which the generation of 20-nucleotide (nt) guide sequences determines the site of genomic editing. Error-prone DNA repair mechanisms are exploited to introduce insertions or deletions. The technology can also be used to generate specific point mutations or conditional alleles. When inserting *loxP* sites, in addition to the efficiency of the technique, CRISPR/Cas9 guide sequences can be injected into the zygote (rather than in ES cells) to directly generate mutant offspring. Thus gene-modified mice can be produced in as few as 4 weeks, compared with homologous recombination using ES cells, which typically requires 9 to 12 months [45]. Detailed methodology for generating mutant animals using CRISPR/Cas9 is evolving but can be found in Yang and colleagues [45]. Additionally, there are many core facilities and commercial groups that cost effectively generate mouse models using CRISPR/Cas9 technology.

GENERAL CONSIDERATIONS

With all of the successes in genetic manipulation, the real bottleneck is phenotyping. Many knockout strains exhibit no obvious phenotypes due to functional redundancy; the creation of double or even triple knockouts may be necessary. Phenotype is dependent on genetic and environmental factors. Inbred strains vary considerably in their peak BMD. ES cells from the 129 strain were the first to be derived and have been the most frequently used. However, 129 mice exhibit abnormal immunological characteristics [46]. Germline competent ES cell lines from 129, C56Bl/6, and C3H are available commercially from multiple sources. These strains have different BMD profiles [47] and different responses to mechanical loading [48] that must be taken into consideration when interpreting skeletal phenotypes. Moreover, housing conditions, food intake, intestinal microbiota, and metabolic stress can have significant impacts on metabolic parameters and BMD [49, 50]. Caution must be taken when extrapolating findings in mouse models to functions in humans. Biomechanical loading and hormonal effects on bones are clearly different in mice and humans. Moreover, linear growth in humans ceases after epiphyseal closure, whereas in mice the growth plate does not fuse. Nonetheless, similarities greatly outweigh the differences, and genetic models are likely to play an increasingly prominent role in every aspect of research in skeletal biology.

ACKNOWLEDGMENTS

The author acknowledges funding from NIH (AR052686 and AR044528).

REFERENCES

1. Ovchinnikov DA, Deng JM, Ogunrinu G, et al. Col2a1-directed expression of Cre recombinase in differentiating chondrocytes in transgenic mice. Genesis. 2000;26: 145–6.
2. Horiki M, Imamura T, Okamoto M, et al. Smad6/Smurf1 overexpression in cartilage delays chondrocyte hypertrophy and causes dwarfism with osteopenia. J Cell Biol. 2004;165:433–45.
3. Dacquin R, Starbuck M, Schinke T, et al. Mouse alpha1(I)-collagen promoter is the best known promoter to drive efficient Cre recombinase expression in osteoblast. Dev Dyn. 2002;224:245–51.
4. Liu F, Woitge HW, Braut A, et al. Expression and activity of osteoblast-targeted Cre recombinase transgenes in murine skeletal tissues. Int J Dev Biol. 2004;48:645–53.
5. Scheller EL, Leinninger GM, Hankenson KD, et al. Ectopic expression of Col2.3 and Col3.6 promoters in the brain and association with leptin signaling. Cells Tissues Organs. 2011;194:268–73.
6. Zhang M, Xuan S, Bouxsein ML, et al. Osteoblast-specific knockout of the insulin-like growth factor (IGF) receptor gene reveals an essential role of IGF signaling in bone matrix mineralization. J Biol Chem. 2002;277:44005–12.
7. Kalajzic I, Matthews BG, Torreggiani E, et al. In vitro and in vivo approaches to study osteocyte biology. Bone. 2013;54:296–306.
8. Sugimoto Y, Takimoto A, Hiraki Y, et al. Generation and characterization of ScxCre transgenic mice. Genesis. 2013;51:275–83.
9. Ferron M, Vacher J. Targeted expression of Cre recombinase in macrophages and osteoclasts in transgenic mice. Genesis. 2005;41:138–45.
10. Kim JH, Kim K, Kim I, et al. Role of CrkII signaling in RANKL-induced osteoclast differentiation and function. J Immunol. 2016;196:1123–31.
11. Roeder E, Matthews BG, Kalajzic I. Visual reporters for study of the osteoblast lineage. Bone. 2016;92:189–95.
12. Hiramatsu K, Iwai T, Yoshikawa H, et al. Expression of dominant negative TGF-beta receptors inhibits cartilage formation in conditional transgenic mice. J Bone Miner Metab. 2011;29:493–500.
13. Tatsumi S, Ishii K, Amizuka N, et al. Targeted ablation of osteocytes induces osteoporosis with defective mechanotransduction. Cell Metab. 2007;5:464–75.
14. Liu Z, Chen J, Mirando AJ, et al. A dual role for NOTCH signaling in joint cartilage maintenance and osteoarthritis. Sci Signal. 2015;8:ra71.
15. Bouabe H, Okkenhaug K. Gene targeting in mice: a review. Methods Mol Biol. 2013;1064:315–36.

16. Birling MC, Gofflot F, Warot X. Site-specific recombinases for manipulation of the mouse genome. Methods Mol Biol. 2009;561:245–63.
17. Niwa H, Yamamura K, Miyazaki J. Efficient selection for high-expression transfectants with a novel eukaryotic vector. Gene. 1991;108:193–9.
18. Saunders TL. Inducible transgenic mouse models. Methods Mol Biol. 2011;693:103–15.
19. Engin F, Yao Z, Yang T, et al. Dimorphic effects of Notch signaling in bone homeostasis. Nature Med. 2008;14:299–305.
20. Logan M, Martin JF, Nagy A, et al. Expression of Cre recombinase in the developing mouse limb bud driven by a Prx1 enhancer. Genesis. 2002;33:77–80.
21. Yu K, Xu J, Liu Z, et al. Conditional inactivation of FGF receptor 2 reveals an essential role for FGF signaling in the regulation of osteoblast function and bone growth. Development. 2003;130:3063–74.
22. Akiyama H, Kim JE, Nakashima K, et al. Osteo-chondroprogenitor cells are derived from Sox9 expressing precursors. Proc Natl Acad Sci U S A. 2005;102:14665–70.
23. Golovchenko S, Hattori T, Hartmann C, et al. Deletion of beta catenin in hypertrophic growth plate chondrocytes impairs trabecular bone formation. Bone. 2013;55:102–12.
24. Yang L, Tsang KY, Tang HC, et al. Hypertrophic chondrocytes can become osteoblasts and osteocytes in endochondral bone formation. Proc Natl Acad Sci U S A. 2014;111:12097–102.
25. Grover J, Roughley PJ. Generation of a transgenic mouse in which Cre recombinase is expressed under control of the type II collagen promoter and doxycycline administration. Matrix Biol. 2006;25:158–65.
26. Chen M, Lichtler AC, Sheu TJ, et al. Generation of a transgenic mouse model with chondrocyte-specific and tamoxifen-inducible expression of Cre recombinase. Genesis. 2007;45:44–50.
27. Henry SP, Jang CW, Deng JM, et al. Generation of aggrecan-CreERT2 knockin mice for inducible Cre activity in adult cartilage. Genesis. 2009;47:805–14.
28. Koyama E, Shibukawa Y, Nagayama M, et al. A distinct cohort of progenitor cells participates in synovial joint and articular cartilage formation during mouse limb skeletogenesis. Dev Biol. 2008;316:62–73.
29. Kozhemyakina E, Zhang M, Ionescu A, et al. Identification of a Prg4-expressing articular cartilage progenitor cell population in mice. Arthritis Rheumatol. 2015;67:1261–73.
30. Bivi N, Condon KW, Allen MR, et al. Cell autonomous requirement of connexin 43 for osteocyte survival: consequences for endocortical resorption and periosteal bone formation. J Bone MIner Res. 2012;27:374–89.
31. Rodda SJ, McMahon AP. Distinct roles for Hedgehog and canonical Wnt signaling in specification, differentiation and maintenance of osteoblast progenitors. Development. 2006;133:3231–44.
32. Kim SW, Pajevic PD, Selig M, et al. Intermittent parathyroid hormone administration converts quiescent lining cells to active osteoblasts. J Bone Mineral Res. 2012;27:2075–84.
33. Clausen BE, Burkhardt C, Reith W, et al. Conditional gene targeting in macrophages and granulocytes using LysMcre mice. Transgenic Res. 1999;8:265–77.
34. Nakamura T, Imai Y, Matsumoto T, et al. Estrogen prevents bone loss via estrogen receptor alpha and induction of Fas ligand in osteoclasts. Cell. 2007;130:811–23.
35. Chiu WS, McManus JF, Notini AJ, et al. Transgenic mice that express Cre recombinase in osteoclasts. Genesis. 2004;39:178–85.
36. Brown D, Wagner D, Li X, et al. Dual role of the basic helix-loop-helix transcription factor scleraxis in mesoderm formation and chondrogenesis during mouse embryogenesis. Development. 1999;126:4317–29.
37. Murchison ND, Price BA, Conner DA, et al. Regulation of tendon differentiation by scleraxis distinguishes force-transmitting tendons from muscle-anchoring tendons. Development. 2007;134:2697–708.
38. Starnes LM, Downey CM, Boyd SK, et al. Increased bone mass in male and female mice following tamoxifen administration. Genesis. 2007;45:229–35.
39. Abe T, Fujimori T. Reporter mouse lines for fluorescence imaging. Dev Growth Differ. 2013;55:390–405.
40. Scott RW, Underhill TM. Methods and strategies for lineage tracing of mesenchymal progenitor cells. Methods Mol Biol. 2016;1416:171–203.
41. Maes C, Kobayashi T, Selig MK, et al. Osteoblast precursors, but not mature osteoblasts, move into developing and fractured bones along with invading blood vessels. Dev Cell. 2010;19:329–44.
42. Day TF, Guo X, Garrett-Beal L, et al. Wnt/beta-catenin signaling in mesenchymal progenitors controls osteoblast and chondrocyte differentiation during vertebrate skeletogenesis. Dev Cell. 2005;8:739–50.
43. Monteiro RM, de Sousa Lopes SM, Korchynskyi O, et al. Spatio-temporal activation of Smad1 and Smad5 in vivo: monitoring transcriptional activity of Smad proteins. J Cell Sci. 2004;117:4653–63.
44. Javier AL, Doan LT, Luong M, et al. Bmp indicator mice reveal dynamic regulation of transcriptional response. PLoS One. 2012;7:e42566.
45. Yang H, Wang H, Jaenisch R. Generating genetically modified mice using CRISPR/Cas-mediated genome engineering. Nat Protoc. 2014;9:1956–68.
46. McVicar DW, Winkler-Pickett R, Taylor LS, et al. Aberrant DAP12 signaling in the 129 strain of mice: implications for the analysis of gene-targeted mice. J Immunol. 2002;169:1721–8.
47. Rosen CJ, Beamer WG, Donahue LR. Defining the genetics of osteoporosis: using the mouse to understand man. Osteoporos Int. 2001;12:803–10.
48. Li J, Liu D, Ke HZ, et al. The P2X7 nucleotide receptor mediates skeletal mechanotransduction. J Biol Chem. 2005;280:42952–9.
49. Nagy TR, Krzywanski D, Li J, et al. Effect of group vs. single housing on phenotypic variance in C57BL/6J mice. Obes Res. 2002;10:412–5.
50. Li JY, Chassaing B, Tyagi AM, et al. Sex steroid deficiency-associated bone loss is microbiota dependent and prevented by probiotics. J Clin Invest. 2016;126:2049–63.
51. Osterwalder M, Galli A, Rosen B, et al. Dual RMCE for efficient re-engineering of mouse mutant alleles. Nat Methods. 2010;7:893–5.

44
Animal Models: Allelic Determinants for Bone Mineral Density

J. H. Duncan Bassett[1], Graham R. Williams[1], and Robert D. Blank[2]

[1]Molecular Endocrinology Laboratory, Department of Medicine, Imperial College London, London, UK
[2]Department of Endocrinology, Metabolism, and Clinical Nutrition, Medical College of Wisconsin; and Clement J. Zablocki VAMC, Milwaukee, WI, USA

INTRODUCTION

There is a large body of work studying the impact of allelic variation on skeletal phenotypes in mice. It is beyond the scope of this chapter to provide encyclopedic coverage of this literature. Rather, examples will be used to illustrate specific aspects of the investigations. The review will encompass three broad areas. First, the phenotypes studied in mice will be described. Second, general themes emerging from the data will be summarized. Finally, expected directions for future work will be explored.

It is important for readers who are not familiar with laboratory mice to recognize that there are many specialized genetic resources available for mouse work. The material in this chapter assumes cursory familiarity with specially bred mouse strains. One of us (RDB) has previously reviewed these and genetically engineered mice for the nonspecialist [1]. It is also important to note at the outset that the emphasis here is on variation existing within established mouse strains and stocks, not the rapidly growing collection of knockout and knockin mice with interesting skeletal phenotypes. These are obviously of great importance, but beyond the scope of this chapter.

PHENOTYPES

Most phenotypes of interest to bone biologists are "complex traits." By complex traits, we mean properties that are determined by the combined influence of multiple genes and environmental conditions. Often, but not invariably, complex traits are also quantitative, meaning that there is a continuum of possible trait values rather than a categorical classification of the trait. Genetic loci contributing to quantitative traits are called quantitative trait loci or QTLs. Two properties of traits make them attractive objects of genetic study. The first is that they are highly heritable, or that much of the variation in the trait can be attributed to genetics. In human studies, this is determined by twin studies and recurrence rates. In mouse studies, this is determined in backcrosses, intercrosses, or analysis of specially bred mice such as recombinant inbred lines. The second criterion is that the assay used for phenotyping is precise, that is, highly reproducible. Less precise assays, although not precluding study, make it more difficult because larger sample sizes are needed to distinguish the genetic signal from the noise attending a less precise phenotyping assay. Another desirable, but not essential, property is that phenotyping should be quick and technically easy to perform, as genetic mapping studies require large samples. An example of such a rapid-throughput phenotyping assay is that used by the Origins of Bone and Cartilage (OBCD) Consortium, which has undertaken skeletal phenotype screening of mutants generated at the Sanger Institute as part of the International Knockout Mouse Consortium (IKMC) [2, 3] (Fig. 44.1). In addition to the genetic desiderata for phenotypes, it is also important to consider their potential utility in guiding interpretation of the biology. Thus, phenotypes that are difficult to measure reliably may still be worth pursuing if they provide insight that cannot be gained otherwise.

Primer on the Metabolic Bone Diseases and Disorders of Mineral Metabolism, Ninth Edition. Edited by John P. Bilezikian.
© 2019 American Society for Bone and Mineral Research. Published 2019 by John Wiley & Sons, Inc.
Companion website: www.wiley.com/go/asbmrprimer

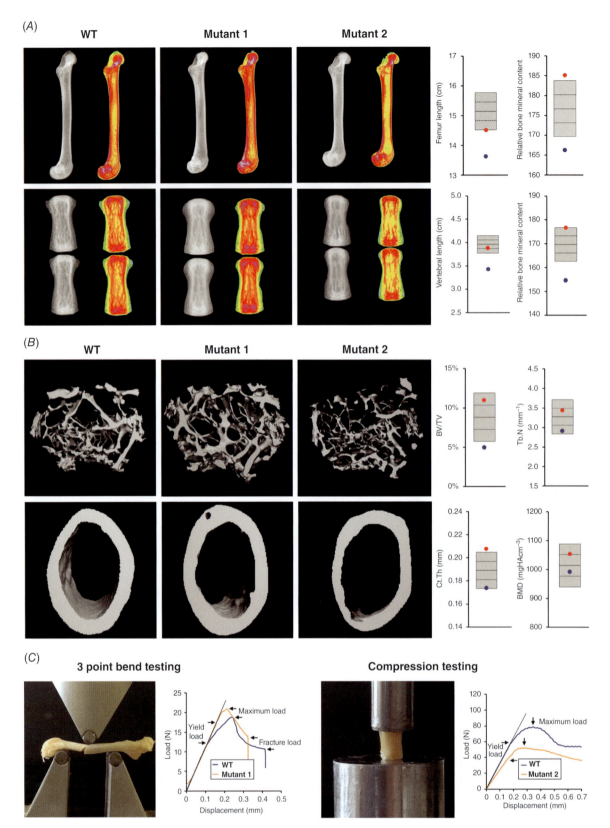

Fig. 44.1. Rapid throughput skeletal phenotyping as performed by the Origin of Bone Cartilage Disease (OBCD) Consortium. (*A*) X-ray microradiography images of femur and fifth to sixth tail vertebrae from wild-type (WT) and mutant mice with high (mutant 1) or low bone mass (mutant 2). Low bone mineral content in pseudo-colored microradiography images is represented by green/yellow colors and high bone mineral content by red/pink colors. Mean femur and vertebral lengths and bone mineral content are shown compared with reference data derived from >250 female wild-type mice of identical C57BL/6 genetic background at 16 weeks of age. Graphs show mean values from mutant 1 (red points) and mutant 2 (blue points) in comparison with reference mean ± 2 standard deviations (gray box). (*B*) Micro-CT images of trabecular and cortical bone from wild-type and mutant mice. Mean trabecular bone volume/total volume (BV/TV) and cortical thickness parameters are shown. (*C*) Femur three-point bend and vertebral compression analysis with load–displacement curves illustrating biomechanical parameters. (*A*) Reproduced with permission of Freudenthal 2016, http://joe.endocrinology-journals.org/content/231/1/R31/F4.expansion.html, licensed under CC BY 3.0. (*B,C*) Reproduced with permission of Logan 2016, https://www.ncbi.nlm.nih.gov/pmc/articles/PMC5064764/figure/fig4/, Licensed under CC BY 3.0.

Bone mineral density

The contemporary era of mouse bone genetics began with the demonstration that inbred mouse strains differ in apparent volumetric BMD, using pQCT scanning [4]. Subsequent studies (eg, [5, 6]) have used both pQCT and DXA technologies to map quantitative trait loci for volumetric and areal BMD in mice, respectively.

Trabecular structure

Improvements in micro-CT technology have allowed trabecular structure, a phenotype previously assayable only by histomorphometry, to be studied with sufficiently high throughput for genetic analyses (eg, [7]). While micro-CT makes it feasible to analyze hundreds of specimens for trabecular structure, in practice these phenotypes have been studied primarily to characterize congenic mice bred on the basis of another phenotype, as these investigations require far smaller sample sizes.

Dimensions

Both lengths and cross-sectional dimensions of long bones have been widely studied in mice (eg, [8–11]), using a variety of methods. Measurement of bone geometry greatly enhances interpretation of biomechanical tests, and therefore is included in virtually all studies that include biomechanical phenotypes. It is worth noting that the robustness of the genetic signals obtained for bone geometry typically exceeds those for any other class of traits, manifested by higher linkage statistics for bone dimensions than for bone density or mechanical performance.

Mechanical performance

The ability to perform mechanical tests represents one of the great advantages of the mouse model system. Various aspects of mechanical performance have been widely studied (eg, [9, 11–14]. There are several distinct aspects of mechanical performance, and while each has been mapped successfully, the robustness of each for genetic analysis varies. In general, both strength (yield or maximum load or stress) performs best among these, reflecting its better reproducibility relative to displacement (or strain) or energy (or toughness) [15]. It is important to recall that the domains of mechanical performance are distinct, and that each is important to function in vivo.

Gene expression

Relatively inexpensive microarrays allow genome-wide measurement of message abundance. The abundance of a specific message may itself be considered a phenotype, and mapped genetically [16]. When considered in conjunction with pleiotropic traditional phenotypes, expression QTLs (eQTLs) offer the promise of improved identification of the gene underlying the QTL (eg, [17]). eQTLs typically explain a larger fraction of the genetic variance in gene expression than phenotypic QTLs do, in some cases approaching 50%. Chief among the possible reasons for the impressive performance of eQTLs is that the phenotype being assessed—gene expression—is far less removed physiologically from the causative variant than is the case for a "clinical" or physiological phenotype, thus the influences of adaptation and feedback are restricted.

Dynamic phenotypes

All the phenotypes considered above are "snapshots" obtained at a single time. It is also possible to map genes for changes in a trait, either as a consequence of time or in response to an intervention. In mice, this has been done for post-maturity change in BMD [18] and modeling in response to mechanical loading [19].

Principal components and other composite phenotypes

The phenotypes considered above are not independent, as each has some biological overlap with at least one other trait, and it is useful to attempt to extract the unique information from each. One approach to this challenge is to apply principal component (PC) analysis to the data [20]. PC analysis transforms the original phenotypes to an equal number of orthogonal PCs, each defined as a specific linear combination of the original phenotypes. While PCs have been used to study bone phenotypes in mice (eg, [2, 9, 21], they suffer from two important limitations. First, as they are linear combinations of directly measured phenotypes chosen algorithmically, they defy intuitive biological interpretation. Second, because the PCs are dependent on the specific phenotypes contributing to them, PCs studied by different investigative teams cannot readily be compared.

The genetics of other composite phenotypes have been studied as well. A particularly interesting example is mandibular shape [22]. The method used in this study employs software that converts the positions of multiple anatomical landmarks to normalized distances from the corresponding mean landmark positions. This approach allows a useful mathematical framework for comparing morphological differences independently of size.

THEMES OF EXISTING DATA

Heritability

It has long been recognized that inbred mouse strains have distinctive, reproducible phenotypes. This is true for volumetric BMD [4], areal BMD [6], architectural features [23], various aspects of biomechanical performance [23, 24], and responsiveness to mechanical loading and unloading [25, 26]. In mice, all of these traits are highly heritable, a prerequisite for successfully mapping the responsible genes.

Covariation

Bone properties are interrelated. Larger bones are stronger bones, more heavily mineralized bones are stiffer bones, and areal BMD is dependent on both bone size and mineralization. It is natural to ask which traits used in mapping are more "important" or "informative." There is no simple answer to the question, and several research groups have studied the interrelationship among bone phenotypes in detail (eg, [14, 27, 28]). The simplest way to gauge the interdependence of phenotypes is to construct a correlation matrix, in which the correlation of each trait with every other trait is tabulated. These uniformly reveal that some traits are largely redundant with each other, for example stiffness and maximum load, and therefore provide partially redundant information. This redundancy is a motivation for performing PC analysis, as discussed earlier. More interesting from a biological perspective is the observed negative correlation between mineralization and long bone cross-sectional size (eg, [14, 28]). Recent work has demonstrated that chromosome substitution strains can disrupt the physiological feedback between bone architecture and material properties, with consequent impairment of mechanical properties [29]. The relationship provides powerful evidence supporting the mechanostat model of the skeleton [30, 31].

Pleiotropy

Pleiotropy is the property of a single genetic locus affecting multiple traits. Unsurprisingly, many bone genes and loci identified in mouse experiments display pleiotropy. Pleiotropy is best appreciated in experiments that have studied multiple traits in the same population. Pleiotropy is commonly observed among mechanical and geometric phenotypes at a single site (eg, [8, 32]), as well as between measures of the same phenotype at different anatomical sites (eg, [10, 33]).

One possible interpretation of the data is that the observed pleiotropy reflects covariation among the traits, as discussed in the previous section. The extreme version of this view implies that the traits being studied are only approximations of an underlying, fundamental set of traits that cannot be measured directly with current assessment methods, but can be approximated by existing phenotyping assays.

An alternative interpretation instead focuses on the large mechanistic gap that exists between most phenotypes and the proteins encoded by the QTLs. According to this formulation, a primary task for bone biologists is to fill in how differences in the level of expression or activity of specific proteins cascades through integrative physiology to impact the skeleton. This is a daunting task, entailing detailed studies of protein function in multiple tissues in isolation, and additionally of dissecting the cross-talk among tissues to understand the whole organism physiology. An elegant in vivo approach to address this difficult problem is to employ cre-lox gene targeting techniques to conditionally overexpress or delete genes of interest in individual bone cell lineages. This can be achieved by crossing mice harboring floxed alleles for the gene of interest with mice expressing cre-recombinase specifically in chondrocytes, osteoblasts, osteoclasts, or osteocytes.

The fraction of variance explained by allelic variation serves as a rough surrogate for the extent to which the physiological impact of the polymorphism can be buffered by compensatory mechanisms. As a trivial example, consider a polymorphism that affects the activity of albumin transcription. Such a polymorphism would be expected to have a large impact on the abundance of albumin mRNA, a lesser effect on serum albumin concentration, even less impact on serum calcium, and negligible impact on serum phosphate. At each step in this hierarchy, additional regulatory feedback loops contribute to the ultimate phenotype.

Clustering

QTLs identified in intercrosses of inbred mouse strains, while accounting for much phenotypic variability, are only imprecisely located. The next step in identifying the responsible gene(s) is often construction of nested congenic strains, in which a donor chromosome segment is crossed into a recipient strain. In this way, the phenotypic consequence of substituting only a short genomic region can be assessed independently of the genetic contributions of other QTLs. Recombination, or crossing over, within the donor segment allows more precise chromosomal localization of the QTL. The result of several such experiments (eg, [33–35]) has been that the donor segment contained not one, but multiple, closely linked QTLs for bone phenotypes. Physical linkage of genes contributing to a common set of functions is an evolutionary mechanism for keeping compatible alleles together (for review, see [36]). This mechanism is believed to underlie the emergence of sex chromosomes, and may be operating with regard to the skeleton as well.

Sex limitation

Multiple research groups have reported QTLs that affect only males, only females, or have significantly greater effects in one sex than the other (eg, [11,35,37]). The most important implication of these findings is that the genes underlying such QTLs are involved in pathways that include sex hormone signaling, or that display cross-talk with sex hormone signaling pathways. The additional mechanistic insight arising from sex-specificity is a powerful tool in identifying the causative gene and an equally powerful tool for investigating the more general question of how sex hormones act on the skeleton.

Intersite discordance

One of the most important insights gained from mice is that the genetic bases of cortical and trabecular bone properties are distinct (eg, [7]), even though some QTLs affect both the cortical and trabecular compartments. In contrast, long bone length QTLs tend to affect multiple sites similarly (eg, [10]). This finding demonstrates that cortical bone and trabecular bone are subject to different physiological feedback. A common interpretation is that cortical bone is more responsive to regulation related to mechanical loading, while trabecular bone is more responsive to metabolic signals.

Gene networks

A powerful, complementary approach to examining individual mRNA abundance is to identify sets of coexpressed genes, or "modules." This allows identification of key genes in various cell types and the determination of which genes within a module are likely to be key drivers of the module's expression. This approach has been successfully used in osteoblasts from the 96 inbred strains comprising the hybrid mouse diversity panel to define an osteoblast-specific module that is highly correlated with traditional osteoblast markers [38]. These authors identified Maged1 (encoding melanoma antigen, family D, member 1) and Pard6g (encoding partitioning defective 6, homolog γ) as key regulators of osteoblast differentiation and maturation. A similar analysis in osteoclasts identified Asxl2 (encoding additional sex combs-like 2) as a key regulator of osteoclast maturation [39]. Several other studies using this strategy have been published recently [40, 41], including the recent identification of a novel gene network that regulates macrophage multinucleation and osteoclastogenesis [42] (Fig. 44.2).

Experimental design and genetic architecture

The genetic architecture of skeletal phenotypes (and all other phenotypes thus far examined in mice) is dependent on the design of the breeding system used [43]. Methods relying on short breeding protocols and phenotypic analyses conducted in the context of a defined genetic background typically find loci with major large effects and strong epistasis. In contrast, those based on breaking up linkage relationships show more relevant loci with much smaller contributions to phenotype and do not suggest that epistasis plays an important role in mediating skeletal phenotypes.

Concordance with human data

Not only are the sequences of genes conserved across species, but so are linkage relationships. Therefore, it has been possible to develop a detailed comparative genetic map of the human and the mouse. If one knows the location of a gene in one organism, then its position is also known in the other. Armed with this knowledge, it is possible to ask whether the genes that contribute to bone properties in one species also have an impact in the other. There is substantial overlap between mouse and human genes for BMD [44]. This analysis did not include any of the BMD-related phenotypes that have also been studied in mice, as most of these have not been amenable to measurement in humans.

The mouse data and the human data are complementary in important ways. The human data allow identification of specific DNA sequence variants that are associated with BMD, either because they have a functional significance themselves or because they are in linkage disequilibrium with functional variants. However, the analysis is limited to common sequence variants, so that the contribution of rare variants to BMD is not addressed. Consequently, human studies account for only a very small fraction of the phenotypic variability. In mice, the fraction of the variance explained is about 5- to 10-fold greater than in humans, but the localization of the responsible loci is much less precise than in humans. Moreover, the genetic linkage studies in mice are performed in experimental crosses in which only a small number of alleles are considered. The greater fraction of the variance explained in the mouse studies is likely due in part to the capturing of all the contributions of genetic differences to the phenotype, unlike the situation in the human studies. In addition, since the mouse linkage peaks include a greater fraction of the genome, they may also contain multiple genes, as discussed earlier. Nevertheless, the high concordance between the human and mouse QTL for BMD is a critical validation of the mouse as a model organism for studying bone genetics.

The above notwithstanding, it is important to recognize that mice are not a perfect model, as differences of body size and other features result in important differences between humans and mice. One obvious consequence of the body size difference is the absence of Haversian systems in mouse cortical bone. Thus, there are aspects of human bone architecture that cannot be studied in mice.

FUTURE DIRECTIONS

In addition to the recognized limitations of mice in reproducing human bone biology, it is also important to recognize that there are important genetic limitations that attend the use of mouse models. These include the extent to which studies have been performed in highly inbred animals, the limited genetic variation being studied, and

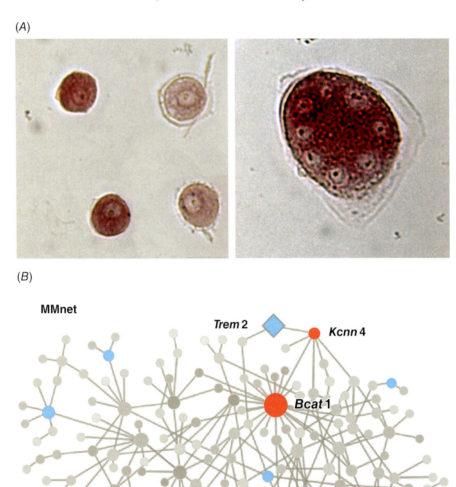

Fig. 44.2. Identification of a gene network controlling macrophage multinucleation and osteoclastogenesis. (A) Differentiation of human peripheral blood monocytes (left) to multinucleated osteoclasts (right) stained red with tartrate resistant acid phosphatase. (B) Macrophage multinucleation gene network (MMnet) showing master and key genes that control MMnet and regulate osteoclastogenesis. Source: [42]. Reproduced with permission of Elsevier.

the poor genetic resolving power achievable in most experimental crosses, which only extend two or three generations.

In order to overcome these limitations, the collaborative cross and the related diversity outcross are ongoing efforts to produce mouse resources for genetic mapping that overcome the genetic limitations of currently available mouse resources [45, 46]. The objective is to generate a large series of eight-progenitor recombinant inbred (RI) strains. The progenitors are chosen to include more than 80% of the known interstrain diversity among present inbred mouse strains. F1 animals produced from such strains are isogenic, outbred, and possess haplotype blocks whose length approaches those of natural outbred populations. Combinational mating among these inbred strains will allow enough distinct genotypes to apply genome-wide association methods as are used in human studies. The short haplotype blocks will allow localization of functionally important genetic variation to short genomic segments. Thus, the advantages of isogeneity will be maintained, including the need to perform genotyping only once, and the ability to estimate phenotype from the pooling of multiple animals sharing a common genotype. The first collaborative cross RI strains have been bred, and hundreds more are approaching completion. These strains will provide a powerful resource for future mouse studies of bone.

Another approach to the limited resolving power of short mouse breeding experiments is to study advanced intercross mice [47]. With every generation of breeding, additional recombination, or crossing-over, occurs. This shortens the lengths of chromosome segments that have been inherited from a specific ancestor. For this reason, advanced intercross lines have improved genetic resolving power relative to F2 mice, with genetic resolving power increasing as a function of the number of generations of breeding. The challenge of using advanced intercross mice, however, is that statistical analysis of the data requires accounting for family structure, and is therefore more computationally difficult than for F2 or backcross experiments. Some work using advanced intercross populations to study skeletal phenotypes has already been published (eg, [33]). Such studies will no doubt increase in frequency in the years ahead.

A third approach can be described as genotype-driven rather than phenotype-driven. Rather than using existing variation as a starting point, this approach generates new mutations and screens for the presence of a skeletal phenotype. New mutants can be generated blindly via chemical mutagenesis or gene trapping or in a targeted fashion via genome editing. Success of this strategy depends on the robustness and scalability of the phenotyping protocol [2,48]. Two of us (GRW, JHB) have recently reviewed the application of this approach to skeletal phenotypes [3].

Of course, the usefulness of mouse models in advancing the study of bone genetics in the future will depend most on the ability to integrate functional, structural, and mechanical elements in innovative experiments. The genetic tools available in the mouse will allow resourceful investigators to continue to learn new biology that can be applied to improving the human condition.

REFERENCES

1. Blank RD. *UpToDate*. Waltham: Wolters Kluwer, 2010.
2. Bassett JH, Gogakos A, White JK, et al. Rapid-throughput skeletal phenotyping of 100 knockout mice identifies 9 new genes that determine bone strength. PLoS Genet. 2012;8(8):e1002858.
3. Freudenthal B, Logan J, Sanger Institute Mouse Pipelines, et al. Rapid phenotyping of knockout mice to identify genetic determinants of bone strength. J Endocrinol. 2016;231(1):R31–46.
4. Beamer WG, Donahue LR, Rosen CJ, et al. Genetic variability in adult bone density among inbred strains of mice. Bone. 1996;18(5):397–403.
5. Beamer WG, Shultz KL, Churchill GA, et al. Quantitative trait loci for bone density in C57BL/6J and CAST/EiJ inbred mice. Mamm Genome. 1999;10(11):1043–9.
6. Klein RF, Mitchell SR, Phillips TJ, et al. Quantitative trait loci affecting peak bone mineral density in mice. J Bone Miner Res. 1998;13(11):1648–56.
7. Bouxsein ML, Uchiyama T, Rosen CJ, et al. Mapping quantitative trait loci for vertebral trabecular bone volume fraction and microarchitecture in mice. J Bone Miner Res. 2004;19(4):587–99.
8. Volkman SK, Galecki AT, Burke DT, et al. Quantitative trait loci that modulate femoral mechanical properties in a genetically heterogeneous mouse population. J Bone Miner Res. 2004;19(9):1497–505.
9. Koller DL, Schriefer J, Sun Q, et al. Genetic effects for femoral biomechanics, structure, and density in C57BL/6J and C3H/HeJ inbred mouse strains. J Bone Miner Res. 2003;18(10):1758–65.
10. Kenney-Hunt JP, Wang B, Norgard EA, et al. Pleiotropic patterns of quantitative trait loci for 70 murine skeletal traits. Genetics. 2008;178(4):2275–88.
11. Saless N, Litscher SJ, Lopez Franco GE, et al. Quantitative trait loci for biomechanical performance and femoral geometry in an intercross of recombinant congenic mice: restriction of the Bmd7 candidate interval. FASEB J. 2009;23(7):2142–54.
12. Li X, Masinde G, Gu W, et al. Chromosomal regions harboring genes for the work to femur failure in mice. Funct Integr Genomics. 2002;1(6):367–74.
13. Li X, Masinde G, Gu W, et al. Genetic dissection of femur breaking strength in a large population (MRL/MpJ x SJL/J) of F2 mice: single QTL effects, epistasis, and pleiotropy. Genomics. 2002;79(5):734–40.
14. Saless N, Lopez Franco GE, Litscher S, et al. Linkage mapping of femoral material properties in a reciprocal intercross of HcB-8 and HcB-23 recombinant mouse strains. Bone. 2010;46(5):1251–9.
15. Leppanen OV, Sievanen H, Jarvinen TL. Biomechanical testing in experimental bone interventions—may the power be with you. J Biomech. 2008;41(8):1623–31.
16. Cookson W, Liang L, Abecasis G, et al. Mapping complex disease traits with global gene expression. Nat Rev Genet. 2009;10(3):184–94.
17. Farber CR, van Nas A, Ghazalpour A, et al. An integrative genetics approach to identify candidate genes regulating BMD: combining linkage, gene expression, and association. J Bone Miner Res. 2009;24(1):105–16.
18. Szumska D, Benes H, Kang P, et al. A novel locus on the X chromosome regulates post-maturity bone density changes in mice. Bone. 2007;40(3):758–66.
19. Kesavan C, Mohan S, Srivastava AK, et al. Identification of genetic loci that regulate bone adaptive response to mechanical loading in C57BL/6J and C3H/HeJ mice intercross. Bone. 2006;39(3):634–43.
20. Pearson K. On lines and planes of closest fit to systems of points in space. Philos Mag. 1901;2:559–72.
21. Saless N, Litscher SJ, Vanderby R, et al. Linkage mapping of principal components for femoral biomechanical performance in a reciprocal HCB-8 x HCB-23 intercross. Bone. 2011;48(3):647–53.
22. Klingenberg CP, Leamy LJ, Cheverud JM. Integration and modularity of quantitative trait locus effects on geometric shape in the mouse mandible. Genetics. 2004;166(4):1909–21.
23. Turner CH, Hsieh YF, Muller R, et al. Genetic regulation of cortical and trabecular bone strength and

microstructure in inbred strains of mice. J Bone Miner Res. 2000;15(6):1126–31.
24. Jepsen KJ, Pennington DE, Lee YL, et al. Bone brittleness varies with genetic background in A/J and C57BL/6J inbred mice. J Bone Miner Res. 2001;16(10):1854–62.
25. Akhter MP, Cullen DM, Pedersen EA, et al. Bone response to in vivo mechanical loading in two breeds of mice. Calcif Tissue Int. 1998;63(5):442–9.
26. Judex S, Donahue LR, Rubin C. Genetic predisposition to low bone mass is paralleled by an enhanced sensitivity to signals anabolic to the skeleton. FASEB J. 2002;16(10):1280–2.
27. Jepsen KJ, Akkus OJ, Majeska RJ, et al. Hierarchical relationship between bone traits and mechanical properties in inbred mice. Mamm Genome. 2003;14(2):97–104.
28. Jepsen KJ, Hu B, Tommasini SM, et al. Genetic randomization reveals functional relationships among morphologic and tissue-quality traits that contribute to bone strength and fragility. Mamm Genome. 2007;18(6–7):492–507.
29. Smith LM, Bigelow EM, Nolan BT, et al. Genetic perturbations that impair functional trait interactions lead to reduced bone strength and increased fragility in mice. Bone. 2014;67:130–8.
30. Frost HM. The Utah paradigm of skeletal physiology: an overview of its insights for bone, cartilage and collagenous tissue organs. J Bone Miner Metab. 2000;18(6):305–16.
31. Frost HM. From Wolff's law to the Utah paradigm: insights about bone physiology and its clinical applications. Anat Rec. 2001;262(4):398–419.
32. Volkman SK, Galecki AT, Burke DT, et al. Quantitative trait loci for femoral size and shape in a genetically heterogeneous mouse population. J Bone Miner Res. 2003;18(8):1497–505.
33. Norgard EA, Jarvis JP, Roseman CC, et al. Replication of long-bone length QTL in the F9-F10 LG,SM advanced intercross. Mamm Genome. 2009;20(4):224–35.
34. Beamer WG, Shultz KL, Ackert-Bicknell CL, et al. Genetic dissection of mouse distal chromosome 1 reveals three linked BMD QTLs with sex-dependent regulation of bone phenotypes. J Bone Miner Res. 2007;22(8):1187–96.
35. Edderkaoui B, Baylink DJ, Beamer WG, et al. Genetic regulation of femoral bone mineral density: complexity of sex effect in chromosome 1 revealed by congenic sublines of mice. Bone. 2007;41(3):340–5.
36. Charlesworth B. The evolution of chromosomal sex determination. Novartis Found Symp. 2002;244:207–19; discussion 20–4, 53–7.
37. Orwoll ES, Belknap JK, Klein RF. Gender specificity in the genetic determinants of peak bone mass. J Bone Miner Res. 2001;16(11):1962–71.
38. Calabrese G, Bennett BJ, Orozco L, et al. Systems genetic analysis of osteoblast-lineage cells. PLoS Genet. 2012;8(12):e1003150.
39. Farber CR, Bennett BJ, Orozco L, et al. Mouse genome-wide association and systems genetics identify Asxl2 as a regulator of bone mineral density and osteoclastogenesis. PLoS Genet. 2011;7(4):e1002038.
40. Wang L, Lu W, Zhang L, et al. Trps1 differentially modulates the bone mineral density between male and female mice and its polymorphism associates with BMD differently between women and men. PLoS One. 2014;9(1):e84485.
41. Mesner LD, Ray B, Hsu YH, et al. Bicc1 is a genetic determinant of osteoblastogenesis and bone mineral density. J Clin Invest. 2014;124(6):2736–49.
42. Kang H, Kerloc'h A, Rotival M, et al. Kcnn4 is a regulator of macrophage multinucleation in bone homeostasis and inflammatory disease. Cell Rep. 2014;8(4):1210–24.
43. Buchner DA, Nadeau JH. Contrasting genetic architectures in different mouse reference populations used for studying complex traits. Genome Res. 2015;25(6):775–91.
44. Ackert-Bicknell CL, Karasik D, Li Q, et al. Mouse BMD quantitative trait loci show improved concordance with human genome-wide association loci when recalculated on a new, common mouse genetic map. J Bone Miner Res. 2010;25(8):1808–20.
45. Churchill GA, Airey DC, Allayee H, et al. The Collaborative Cross, a community resource for the genetic analysis of complex traits. Nat Genet. 2004;36(11):1133–7.
46. Chesler EJ. Out of the bottleneck: the Diversity Outcross and Collaborative Cross mouse populations in behavioral genetics research. Mamm Genome. 2014;25(1–2):3–11.
47. Darvasi A, Soller M. Advanced intercross lines, an experimental population for fine genetic mapping. Genetics. 1995;141(3):1199–207.
48. Ruffoni D, Kohler T, Voide R, et al. High-throughput quantification of the mechanical competence of murine femora—a highly automated approach for large-scale genetic studies. Bone. 2013;55(1):216–21.

45
Transcriptional Profiling for Genetic Assessment

Aimy Sebastian and Gabriela G. Loots
Physical and Life Sciences, Lawrence Livermore National Laboratory, Livermore, CA, USA

INTRODUCTION

Genome-wide expression profiling allows us to study the molecular basis of biological systems and identify disease biomarkers. Since the development of gene expression microarrays in the late 1990s, it has been the technology of choice for large-scale gene expression studies. Microarrays enabled us to simultaneously measure the expression of thousands of genes in one experimental sample. However, microarrays have several limitations including the dependence on predesigned probes, high background noise due to nonspecific probe binding, limited ability to measure transcripts with low abundance, and inability to accurately measure genes with high expression due to signal saturation. Recent advances in RNA sequencing (RNA-seq) technology have enabled us to overcome many of these limitations. RNA-seq is the direct sequencing of transcripts expressed in a cell/tissue using high-throughput sequencing approaches. This technology eliminates the need for predesigned probes. RNA-seq also allows the detection of high and low expressed genes, novel transcripts, novel isoforms, allele-specific expression, single nucleotide variants, deletion and insertions.

Detailed information about gene expression patterns and regulatory networks involved in skeletal development and remodeling is crucial to completely understand bone structure, function, and homeostasis, and for the development of appropriate therapeutic strategies for various skeletal diseases. While genome-wide expression profiling using microarrays and RNA-seq have not yet been explored to their full potential in the field of bone biology, work over the last decade has greatly enhanced knowledge of skeletal development and remodeling. Several studies have used genome-wide expression profiling as a tool for furthering the understanding of bone development and metabolism, biomechanical effects on bone, age- and anatomical location-dependent differences in skeletal gene expression, and molecular changes associated with bone diseases.

GENE EXPRESSION PROFILING OF SKELETAL CELLS AND BONE METABOLISM

Bone is made up of mainly three types of cells: osteoblasts, osteocytes, and osteoclasts. Osteoblasts are responsible for the synthesis and mineralization of bone matrix. Osteoblasts that get trapped in the bone matrix during bone formation become osteocytes. Osteocytes form long channels called canaliculi that allow them to communicate among themselves and with other cells present on the bone surfaces. Osteoclasts are large multinucleated cells that reside on bone surfaces and function primarily to resorb bone through acidification and proteolytic degradation of the mineralized bone matrix.

Over the years, much work has been done in identifying genes driving osteoblast and osteoclast differentiation and function. Several studies have utilized microarrays to profile gene expression changes during differentiation of various cell lines [1–4] and primary cultures derived from humans and rodents [5–8] towards osteoblast and osteoclast linages, in vitro. Notably, Kulterer and colleagues examined the gene expression profiles of mesenchymal stem cells (MSCs) differentiating into mature osteoblasts and identified gene expression patterns that characterize different stages of osteogenic differentiation: proliferation, matrix maturation, and mineralization [6]. Takayanagi

Primer on the Metabolic Bone Diseases and Disorders of Mineral Metabolism, Ninth Edition. Edited by John P. Bilezikian.
© 2019 American Society for Bone and Mineral Research. Published 2019 by John Wiley & Sons, Inc.
Companion website: www.wiley.com/go/asbmrprimer

and colleagues [8] and Ishida and colleagues [3] profiled gene expression changes during osteoclast differentiation and identified several key regulators of osteoclastogenesis including NFATC1. These studies have significantly enhanced our understanding of osteoblastogenesis, osteoclastogenesis, and bone metabolism. However, the gene expression profiles obtained from these in vitro studies may not accurately describe the changes occurring in vivo. Therefore, genes identified in these studies should be validated in appropriate in vivo models.

Osteocytes are the most abundant cell type in bones and constitute ~95% of all bone cells. Osteocytes are thought to be mechanosensors and may regulate the activity of both osteoblasts and osteoclasts in response to bone tissue strain and biomechanical load. Paic and colleagues [9] profiled gene expression in mouse calvarial osteoblasts and osteocytes, and identified genes differentially expressed between these two cell types. To purify osteoblasts and osteocytes from calvaria they utilized dual GFP reporter mice in which osteocytes expressed GFP (topaz) driven by the DMP1 promoter and osteoblasts expressed GFP (cyan) driven by 2.3 kilobase (kb) of the Col1a1 promoter. $Col2.3^{Cyan+}$ (osteoblasts) and $Dmp1^{Topaz+}$ (preosteocytes and osteocytes) cells were isolated from neonatal calvaria by FACS sorting cells based on GFP expression. Microarray-based expression profiling of $Col2.3^{Cyan+}$ and $Dmp1^{Topaz+}$ cells identified 385 genes differentially expressed between these two cell populations. Genes differentially expressed between osteoblasts and osteocytes include several known regulators of bone metabolism including members of the transforming growth factor β (*TGF-β*) family, bone morphogenetic proteins (*Bmps*), insulin-like growth factors (*Igfs*), and fibroblast growth factors (*Fgfs*). Interestingly, several genes associated with muscle development and function including *Myh11*, *Acta1*, *Tnnt2*, *Tnni1*, *Myoz2*, *Tnnt3*, *Tnnc2*, *Tnnt1*, *Actn2*, and *Tpm2* were identified as enriched in osteocytes. The data generated by Paic and colleagues highlight several novel regulators of both osteoblast and osteocyte function [9].

CHANGES IN GENE EXPRESSION PROFILES DURING BONE DEVELOPMENT

Several studies have used microarrays or RNA-seq to study global gene expression dynamics during skeletal development [10–13]. Long bones and bones at the base of the skull are formed through a process called endochondral ossification, which involves the replacement of a cartilaginous template by bone [14]. Endochondral ossification is also responsible for the postnatal longitudinal growth of long bones in the epiphyseal (growth) plate. During endochondral ossification, chondrocytes proliferate, undergo hypertrophy, mineralize their matrix, and die, attracting blood vessels and osteoblasts that remodel the cartilaginous template into bone [14].

To determine the molecular drivers of specific stages of endochondral bone formation James and colleagues [12] performed microarray-based analyses of gene expression changes during mouse endochondral ossification. Mouse embryonic tibias were segmented and designated into three zones: (i) zone I, which contains proliferating and resting cells; (ii) zone II, which mostly contains prehypertrophic and hypertrophic chondrocytes; and (iii) zone III, which contains both the most mature hypertrophic chondrocytes and the mineralized portion of the tibias. Their study identified 6185 probes differentially regulated between zone I and II, 8134 probes differentially regulated between zone II and III, and 7220 probes differentially regulated between zone I and III. Of these probes, 834, 1482, and 1027 probe sets were unique to zone I versus II, II versus III, and I versus III comparisons, respectively. Overall, early-stage chondrocyte markers including Sox9 and Col2a1 exhibited higher expression in zone I and lower expression in zone III, whereas molecules involved in processes such as matrix turnover and osteoblast and osteoclast differentiation (*Mmp13*, *Tnfsf11*, *Ibsp*, *Dmp1*, *Spp1*, *Runx2*, etc.) showed high expression in zone III. This study suggests that endochondral ossification is transcriptionally regulated by both zone-specific genes and genes with broad expression patterns [12].

Using microarrays, Taher and colleagues [11] identified genes differentially expressed at five stages of limb development (E9.5 to 13.5), during fore- and hindlimb patterning. Compared with whole embryo samples, 3520 genes were found to be upregulated in the limb during different stages of development. These included 855 transcripts upregulated specifically in the forelimb and 511 in the hindlimb, at least in one developmental time point. These fore- and hindlimb-specific genes may dictate the formation of unique skeletal structures in the fore- and hindlimb.

James and colleagues [12] and Taher and colleagues [11] used highly heterogeneous tissue samples in their studies. Such studies may not capture signal that comes exclusively from a subset of cells as the surrounding cells may exhibit opposite expression profiles, and may fail to uncover some cell type/stage-specific regulators of complex processes such as bone development. In a recent study, Li and colleagues [13] profiled the transcriptome of 217 individual cells from mouse postnatal growth plate region using single cell RNA-seq. Single cell RNA-seq enables the gene expression profiling of individual cells. In their study, Li and colleagues identified more than 9000 genes with confident expression in at least 10 single cell samples. Among the 217 cells analyzed, 13 cells did not show significant expression of cartilage matrix proteins Col9a, Comp, or Col10a. These cells were considered as outliers and the remaining 204 cells were considered as cells that originated from the growth plate.

Next, Li and colleagues ordered these 204 cells computationally, based on their gene expression patterns, to generate a pseudo-temporal order as these cells may represent different stages of growth plate development, and determined the beginning and end of the timeline based on the expression levels of the known hypertrophic marker *Col10a*. Subsequently, more than 600 genes that were dynamically regulated in these temporally ordered cells were identified and these genes were grouped into six clusters based on the similarity in their temporal expression profiles. A gene ontology (GO) analysis showed that hypertrophic differentiation-related GO terms were enriched in clusters of constantly upregulated and transiently upregulated genes. A cluster of genes showing delayed downregulation was enriched for steroid hormone signaling and nonhypertrophic chondrocyte-associated genes. Genes showing transient downregulation were enriched for transcription and translational activity-associated GO terms. Adopting approaches like the one presented by Li and colleagues [13] will allow the systematic discovery of cell type/stage-specific genes and potential signal pathways that regulate complex biological processes such as bone development and metabolism.

AGE- AND LOCATION-DEPENDENT CHANGES IN SKELETAL GENE EXPRESSION

Genes expressed in the skeleton may show different expression patterns depending on the age (young versus old), anatomical location and origin (endochondral versus intramembranous ossification) of the cells. Genome-wide expression profiling of juvenile and adult mice calvaria performed by Aalami and colleagues [15] provides insights into age-dependent differences in skeletal gene expression. These researchers isolated calvaria from 6- and 60-day-old male mice and profiled gene expression changes using microarrays. They identified 1324 differentially expressed genes; 976 genes with increased expression in the juvenile calvaria and 357 with increased expression in the adult calvaria. Several known regulators of osteogenesis including osteogenic transcription factor *Runx2*, *Bmp2*, *Col1*, *Dmp1*, and *Ptn* showed an elevated expression in juvenile calvaria while osteocalcin showed a higher expression in adult calvaria. This study suggests that the juvenile calvaria have an increased osteogenic potential compared with adult calvaria.

In another study, Rawlinson and colleagues [16] investigated how the gene expression profiles of bones from different skeletal sites differ in relation to their origin and location. Using microarrays, Rawlinson and colleagues profiled the transcriptome of rat limb and skull bones, and osteoblasts derived from skull and limb [16]. The majority of the genes analyzed did not show differential expression between skull and limb bone suggesting that the gene expression in these skeletal tissues is largely similar irrespective of their origin and location. However, 1236 genes were found to be significantly differentially expressed between skull and limb. Several genes known to be associated with osteogenesis including *Opg*, *Pthr1*, *Lrp5*, *Sost*, *Ibsp*, *Bmp3*, and *Cthrc1* were upregulated in the skull compared with limb, while genes such as *Wnt16*, *β-catenin*, *Bmp5*, and *Comp* showed high expression in the limb. Rawlinson and colleagues also identified several transcription factors including *Sp7*, *Vdr*, *Tcf7*, *Dlx5*, *Twist1*, and *Fos* as upregulated in the skull bone, whereas transcription factors *Sox6*, *Gata1*, *Gata3*, *Cited4*, and several members of the Hox family were found to be upregulated in the limb. Most of the differences observed in skull and limb gene expression profiles were lost in skull and limb bone-derived osteoblasts cultured in vitro. However, 249 genes were found to be significantly differentially expressed between skull and limb bone-derived osteoblasts. This finding suggests that transcriptome profiles of the bone differ based on the anatomical location and these differences are established not only by the differences in the local mechanical microenvironment but also by the developmental origin of the cells.

The studies discussed suggest that genes associated with bone development and metabolism are differentially expressed in functionally distinct skeletal sites, and as a function of age. These spatial and temporal differences in gene expression must be considered when generalizing results from different genome-wide expression profiling studies.

PROFILING MECHANICAL LOADING-INDUCED GENE EXPRESSION CHANGES

Mechanical stimuli activate bone formation and increase bone mineral density [17]. Microarray- and RNA-seq-based gene expression profiling studies have identified several genes and signaling pathways responsible for mechanical loading-induced bone formation. Mantila Roosa and colleagues [18] evaluated loading-induced gene expression in rat ulna over a time course of 4 hours to 32 days. The right forelimb was loaded axially for 3 minutes per day while the left forearm served as the contralateral control. This study identified 1051 genes that were differentially expressed in at least one time point in response to loading. Mantila Roosa and colleagues categorized these genes into three clusters: genes upregulated early in the time course, genes upregulated during matrix formation, and genes downregulated during matrix formation. Several chemokines, calcium signaling genes, matrix proteins, and AP-1 transcription factors *Fosl1* and *Junb* were identified as early response genes. A number of extracellular matrix genes, growth factors, ion channels, and solute carriers were identified as up- or downregulated during matrix formation. Interestingly, several muscle-related genes (*Acta1*, *Myocd*, *Myl1*, *Myplf*, *Tnni2*, *Tnnt3*, *Tpm2*, etc.) were identified as downregulated in loaded ulna. Other important bone metabolism genes differentially regulated during matrix formation include *Bmp2*, *Tgfb1*, *Pthr1*, *Vdr*, *Wif1*, *Wisp1* (upregulated genes), *Bmpr1b*, *Grem1*, *Tgfbr3*, *Chrdl1*, and Wnt signaling pathway inhibitors *Sost* and *Sfrp4* (downregulated genes).

In a recent study, Kelly and colleagues [19] employed RNA-seq to investigate load-induced transcriptional changes in cortical and cancellous bone. Left tibias of 10-week-old mice were subjected to a single session of mechanical loading and the gene expression changes in loaded cortical and cancellous bone after 3 and 24 hours compared with nonloaded contralateral controls were identified. Three hours after loading, 43 genes were differentially expressed in cortical bone and 18 genes in cancellous bone. Eleven genes including *Wnt1*, *Wnt7b*, *Timp1*, *Ptgs2*, and *Opg* were upregulated in both cortical and cancellous bone. The cortical bone also showed an upregulation of *Wnt10b* while *Lrp5* was modestly decreased. Wnt pathway inhibitors *Sost* and *Dkk1* were downregulated in loaded cancellous bone. Twenty-four hours after loading, 58 genes were differentially expressed in cortical bone and 32 genes in cancellous bone. Twelve genes, including *Ptn*, *Vcan*, and *Cthrc1* were identified as differentially regulated in both cortical and cancellous bone. *Wnt1* and *Wnt10b* remained upregulated in cortical bone 24 hours after loading. Several regulators of muscle development and function including *Myh4*, *Tnnc2*, *Tnnt3*, *Actn3*, *Myl1*, *Myh2*, *Mylpf*, and *Tnni2* were downregulated in cancellous bone 24 hours after loading. The data generated by Kelly and colleagues provide novel insights into differential response of cortical and cancellous bone to mechanical loading.

These studies have identified several regulators of load-induced bone formation. It is important to note that Paic and colleagues [9] identified several of the muscle-related genes downregulated in response to loading as genes enriched in osteocytes compared with osteoblasts. Further investigation is required to understand the functions of these muscle-related genes in bone.

GENE EXPRESSION PROFILES IN OSTEOPOROSIS

Osteoporosis (OP) is a disease characterized by decreased bone mass, increased bone fragility, and increased risk of fractures. Understanding the molecular mechanisms that contribute to OP will open new avenues for therapeutic intervention. Hopwood and colleagues [20] used microarrays to identify candidate OP genes by comparing gene expression in bone from individuals with fracture of the neck of the proximal femur (OP) with that from age-matched individuals with osteoarthritis (OA), and control (CTL) individuals with no known bone pathology. This study identified 150 genes differentially expressed in OP, of which 75 genes had known or suspected roles in bone metabolism. Several genes involved in promoting myelomonocytic/osteoclast precursor differentiation and osteoclast function, including *TREM2*, *ANXA2*, *SCARB2*, *CCL3*, *CD14*, *ST14*, *CCR1*, *ADAM9*, *PTK9*, and *CCL2*, showed an elevated expression in OP bone compared with OA and CTL. The expression profile for these genes is consistent with increased osteoclast numbers and activity observed in OP bone.

In another study, Reppe and colleagues [21] investigated the relationship between gene expressions in transiliacal bone biopsies and BMD (in hip and spine) in 84 postmenopausal women. Using microarrays, they identified all the genes expressed in the bone samples. Among almost 23,000 expressed transcripts, four genes (*ACSL3*, *NIPSNAP3B*, *ABCA8*, *DLEU2*) showed an inverse correlation to BMD while four genes including *C1ORF61*, *DKK1*, *SOST* showed a positive correlation.

Xiao and colleagues [22] used ovariectomized (OVX) rats, a model for osteoporosis, to study OP-associated gene expression changes. Using microarrays, they examined the gene expression in bone marrow mesenchymal stromal cells (BMSCs) isolated from the bone marrow of four different experimental groups: juvenile (7-week-old), adult (7-month-old), OP (7-month-old OVX), and aged (>2-year-old) rats. Comparisons between OP and adult rats identified 195 up- and 109 downregulated transcripts. OP caused an increase in the expression of lipid metabolism genes (*Alox5*, *Baat*, *Sult4a1*, *Lpl*, etc.) and genes involved in cell growth and maintenance (*A2m*, *Alpl*, *Crabp2*, *Cdkn2b*, etc.), and downregulation of genes such as *Npy*, *Cd24*, *Ramp3*, *Marcksl1*, *Wnt4*, and *Adrb3*. Adult versus aged comparison identified 62 up- and 86 downregulated transcripts and juvenile versus adult rats identified 120 up- and 80 downregulated genes. OP versus aged comparison identified 14 genes (*Mmp8*, *Braf*, *Inhbp*, *Pgr*, *Slc26a1*, *Sp1*, etc.) upregulated in OP and six genes (*Prlpb*, *Iilrn*, *Plpcb*, *Loc171569*, *Ramp3*, *Mip*) downregulated in OP.

MicroRNAs (miRNAs) are short (~20 to 25 nucleotides long), noncoding RNAs that play a major role in posttranscriptional gene regulation; they play a key role in regulating bone development and metabolism [23, 24]. In a recent study, An and colleagues [25] investigated the changes in miRNA profiles in OVX compared with sham-operated mice and identified nine miRNAs differentially expressed in OVX including miR-127, miR-133a, miR-133b, miR-136, miR-206, and miR-378. They further studied the functional role two of these miRNAs, miR-127 and miR -136, and found that these miRNAs may contribute to bone loss by suppressing osteoblast differentiation and osteocyte function and survival, while promoting osteoclast differentiation.

Teriparatide or PTH(1–34) is the only US Food and Drug Administration (FDA) approved anabolic agent for the treatment of OP; however, while intermittent administration increases bone formation, continuous infusion of PTH results in bone loss. To understand the molecular basis for these opposing biological effects, Onyia and colleagues [26] examined gene expression in the distal femurs of rats receiving either intermittent (once-daily subcutaneous injection) or continuous (subcutaneous infusion) PTH(1–34) treatment for 1 week. Both modes of PTH treatment resulted in differential regulation of 22 genes that were similarly regulated in both magnitude and direction. Intermittent treatment regulated 19 unique genes whereas 173 genes were specific to continuous treatment. The genes uniquely changed by intermittent PTH treatment included *Icam2*, *Igfbp6*, *Pspn*,

Sparcl1, Cpe (upregulated) and *Spc18, Nup54, Nrbp, Pcoln3* (downregulated). Genes uniquely regulated by continuous PTH treatment included *Omd, Thbs4, Fn1, Ibsp, Alpl* (upregulated) and *Esm1, Fmo1, Gpt, Ass* (downregulated). Further studies on genes regulated by intermittent PTH may reveal new insights into PTH-mediated anabolic response.

Here we discussed some key findings from genomic profiling in human and animal models with altered bone metabolism; however, it is important to note that numerous other studies have also used microarrays or RNA-seq to profile gene expression changes in OP [27–29]. These studies have identified several novel candidates that can be further explored as potential therapeutic targets for OP.

GENE EXPRESSION PROFILING DURING FRACTURE HEALING

Fractures are among the most common traumatic injuries in humans and OP-related fractures are a major health care problem due to huge hospitalization and rehabilitation expenses. Using microarrays, Niikura and colleagues [30] compared gene expression profiles of atrophic nonunion and standard closed healing fractures in rats at 3, 7, 10, 14, 21, and 28 days post fracture. They identified 559 genes upregulated and 462 genes downregulated in nonunion fractures compared with standard healing fractures. This study showed that BMPs and their antagonists are involved in both normal and abnormal fracture healing. Expression levels of several BMP family members including *Bmp2, Bmp3, Bmp4, Bmp6,* and *Bmp7* and BMP antagonists such as *Nog, Drm,* and *Bambi* were significantly lower in nonunions compared with standard healing fractures, at several time points. Their study suggests that a transcriptional balance between BMP/BMP antagonists may be crucial for promoting 'healthy' fracture healing and targeting BMP signaling may be an effective strategy in treating nonunion fractures [30].

Waki and colleagues [31] investigated the differential regulation of miRNAs in healing femoral shaft fractures and nonhealing fractures, in rats. They performed a microarray analysis of miRNA samples from each group on post-fracture day 14 and identified 317 miRNAs with high expression in healing femoral shaft fractures compared with nonhealing fractures. Of these 317 miRNAs, miR-140-3p, miR-181a-5p, and miR-451a have previously been reported to be involved in the regulation of inflammatory responses and miR-140-5p, miR-181a-5p, miR-181d-5p, and miR-451a have been reported to be involved in the regulation of skeletal development. Together, these gene expression studies further our knowledge of the genomic regulation of fracture healing. However, functional studies are still required to determine the precise role of each of these genes during fracture healing.

CONCLUSION

This chapter has highlighted a few key genome-wide expression studies covering several topics in bone biology that have greatly benefited from gene expression profiling. These studies have identified several novel regulators of skeletal development and metabolism and, potential therapeutic targets for treating bone diseases. But these putative candidates still require significant validation through traditional experimental approaches, which will continue to be the bottleneck of scientific discoveries. Regardless, these studies have helped in expanding our knowledge of skeletal biology.

ACKNOWLEDGMENTS

This work was performed under the auspices of the US Department of Energy by Lawrence Livermore National Laboratory under Contract DE-AC52-07NA27344.

REFERENCES

1. Balint E, Lapointe D, Drissi H, et al. Phenotype discovery by gene expression profiling: mapping of biological processes linked to BMP-2-mediated osteoblast differentiation. J Cell Biochem. 2003;89(2):401–26.
2. Korchynskyi O, Dechering KJ, Sijbers AM, et al. Gene array analysis of bone morphogenetic protein type I receptor-induced osteoblast differentiation. J Bone Miner Res. 2003;18(7):1177–85.
3. Ishida N, Hayashi K, Hoshijima M, et al. Large scale gene expression analysis of osteoclastogenesis in vitro and elucidation of NFAT2 as a key regulator. J Biol Chem. 2002;277(43):41147–56.
4. Sambandam Y, Blanchard JJ, Daughtridge G, et al. Microarray profile of gene expression during osteoclast differentiation in modelled microgravity. J Cell Biochem. 2010;111(5):1179–87.
5. Roman-Roman S, Garcia T, Jackson A, et al. Identification of genes regulated during osteoblastic differentiation by genome-wide expression analysis of mouse calvaria primary osteoblasts in vitro. Bone. 2003;32(5):474–82.
6. Kulterer B, Friedl G, Jandrositz A, et al. Gene expression profiling of human mesenchymal stem cells derived from bone marrow during expansion and osteoblast differentiation. BMC Genomics. 2007;8:70.
7. Cappellen D, Luong-Nguyen NH, Bongiovanni S, et al. Transcriptional program of mouse osteoclast differentiation governed by the macrophage colony-stimulating factor and the ligand for the receptor activator of NFkappa B. J Biol Chem. 2002;277(24):21971–82.
8. Takayanagi H, Kim S, Koga T, et al. Induction and activation of the transcription factor NFATc1 (NFAT2) integrate RANKL signaling in terminal differentiation of osteoclasts. Dev Cell. 2002;3(6):889–901.

9. Paic F, Igwe JC, Nori R, et al. Identification of differentially expressed genes between osteoblasts and osteocytes. Bone. 2009;45(4):682–92.
10. Wang Y, Middleton F, Horton JA, et al. Microarray analysis of proliferative and hypertrophic growth plate zones identifies differentiation markers and signal pathways. Bone. 2004;35(6):1273–93.
11. Taher L, Collette NM, Murugesh D, et al. Global gene expression analysis of murine limb development. PLoS One. 2011;6(12):e28358.
12. James CG, Stanton LA, Agoston H, et al. Genome-wide analyses of gene expression during mouse endochondral ossification. PLoS One. 2010;5(1):e8693.
13. Li J, Luo H, Wang R, et al. Systematic reconstruction of molecular cascades regulating GP development using single-cell RNA-seq. Cell Rep. 2016;15(7):1467–80.
14. Mackie EJ, Ahmed YA, Tatarczuch L, et al. Endochondral ossification: how cartilage is converted into bone in the developing skeleton. Int J Biochem Cell Biol. 2008;40(1):46–62.
15. Aalami OO, Nacamuli RP, Salim A, et al. Differential transcriptional expression profiles of juvenile and adult calvarial bone. Plast Reconstr Surg. 2005;115(7):1986–94.
16. Rawlinson SC, McKay IJ, Ghuman M, et al. Adult rat bones maintain distinct regionalized expression of markers associated with their development. PLoS One. 2009;4(12):e8358.
17. Galli C, Passeri G, Macaluso GM. Osteocytes and WNT: the mechanical control of bone formation. J Dent Res. 2010;89(4):331–43.
18. Mantila Roosa SM, Liu Y, Turner CH. Gene expression patterns in bone following mechanical loading. J Bone Miner Res. 2011;26(1):100–12.
19. Kelly NH, Schimenti JC, Ross FP, et al. Transcriptional profiling of cortical versus cancellous bone from mechanically-loaded murine tibiae reveals differential gene expression. Bone. 2016;86:22–29.
20. Hopwood B, Tsykin A, Findlay DM, et al. Gene expression profile of the bone microenvironment in human fragility fracture bone. Bone. 2009;44(1):87–101.
21. Reppe S, Refvem H, Gautvik VT, et al. Eight genes are highly associated with BMD variation in postmenopausal Caucasian women. Bone. 2010;46(3):604–12.
22. Xiao Y, Fu H, Prasadam I, et al. Gene expression profiling of bone marrow stromal cells from juvenile, adult, aged and osteoporotic rats: with an emphasis on osteoporosis. Bone. 2007;40(3):700–15.
23. Murata K, Ito H, Yoshitomi H, et al. Inhibition of miR-92a enhances fracture healing via promoting angiogenesis in a model of stabilized fracture in young mice. J Bone Miner Res. 2014;29(2):316–26.
24. Inose H, Ochi H, Kimura A, et al. A microRNA regulatory mechanism of osteoblast differentiation. Proc Natl Acad Sci U S A. 2009;106(49):20794–9.
25. An JH, Ohn JH, Song JA, et al. Changes of microRNA profile and microRNA-mRNA regulatory network in bones of ovariectomized mice. J Bone Miner Res. 2014;29(3):644–56.
26. Onyia JE, Helvering LM, Gelbert L, et al. Molecular profile of catabolic versus anabolic treatment regimens of parathyroid hormone (PTH) in rat bone: an analysis by DNA microarray. J Cell Biochem. 2005;95(2):403–18.
27. Ayturk UM, Jacobsen CM, Christodoulou DC, et al. An RNA-seq protocol to identify mRNA expression changes in mouse diaphyseal bone: applications in mice with bone property altering Lrp5 mutations. J Bone Miner Res. 2013;28(10):2081–93.
28. Taylor S, Ominsky MS, Hu R, et al. Time-dependent cellular and transcriptional changes in the osteoblast lineage associated with sclerostin antibody treatment in ovariectomized rats. Bone. 2016;84:148–59.
29. Zhu M, Zhang J, Dong Z, et al. The p27 pathway modulates the regulation of skeletal growth and osteoblastic bone formation by parathyroid hormone-related peptide. J Bone Miner Res. 2015;30(11):1969–79.
30. Niikura T, Hak DJ, Reddi AH. Global gene profiling reveals a downregulation of BMP gene expression in experimental atrophic nonunions compared to standard healing fractures. J Orthop Res. 2006;24(7):1463–71.
31. Waki T, Lee SY, Niikura T, et al. Profiling microRNA expression during fracture healing. BMC Musculoskelet Disord. 2016;17(1):83.

46
Approaches to Genetic Testing

Christina Jacobsen[1], Yiping Shen[2], and Ingrid A. Holm[2]

[1]*Divisions of Endocrinology, Genetics, and Genomics, Boston Children's Hospital; and Department of Pediatrics, Harvard Medical School, Boston, MA, USA*
[2]*Divisions of Genetics and Genomics, Boston Children's Hospital; and Department of Pediatrics, Harvard Medical School, Boston, MA, USA*

INTRODUCTION

Genetic testing is the analysis of human DNA, RNA, chromosomes, proteins, and certain metabolites in order to detect heritable disease-related genotypes, mutations, phenotypes, or karyotypes for clinical purposes [1, 2]. In practice, genetic testing mainly involves looking at an individual's DNA (genes and genome) for variants that may be the underlying cause of a clinical condition, and this chapter will focus on genetic testing of DNA. As knowledge about genetic disorders of bone has expanded greatly over the past few years, the number of disorders for which genetic testing is commercially available has increased. In this chapter we will discuss the types of genetic testing available, what testing is available for disorders of bone, and suggest an approach to genetic testing in individuals with a disorder of bone.

OVERVIEW OF THE TYPES OF GENETIC TESTING AVAILABLE

There are many ways in which an individual's genome can vary from "normal." Based on the impact on genome structure, variants can be classified as either small or large scale. Laboratory molecular techniques have been developed to specifically test the different types of variants. Here we briefly discuss the types of genetic testing available in molecular diagnostic laboratories to detect different types of variants associated with a variety of genetic disorders.

Types of variants

Small-scale variants

Base-pair substitutions

The replacement of one nucleotide base by another is the most abundant variant type in the human genome. The vast majority of single nucleotide variants are located in intergenic regions (stretches of DNA that contain few or no genes) and intronic regions (stretches of DNA within a gene that is removed by RNA splicing). Those located in the coding regions (exons) can be further classified based on the effect on the amino acid sequence of the protein.

Synonymous variants cause no change in the final protein product as they do not change the amino acid. In most cases, they are thought not to have clinical consequences, although the consequences of synonymous variants can be hard to predict.

Nonsynonymous variants are those that result in changes in the amino acid coded for. When the nucleotide change results in the replacement of one amino acid for another it is called a *missense variant*; when the nucleotide change results in the gain of a stop codon it is called a *nonsense mutation*. Sometimes a nucleotide change results in the loss of a stop codon (ie, a stop codon changes to a codon that inserts an amino acid). A nonsynonymous variant that causes a significant impact on protein structure may have clinical consequences.

Splicing variants are variants at the splicing junctions (the first two or last two nucleotides at the beginning or end of the exon, respectively) and results in an alteration in exon splicing. This can lead to exon skipping and the

---

Primer on the Metabolic Bone Diseases and Disorders of Mineral Metabolism, Ninth Edition. Edited by John P. Bilezikian.
© 2019 American Society for Bone and Mineral Research. Published 2019 by John Wiley & Sons, Inc.
Companion website: www.wiley.com/go/asbmrprimer

loss of an exon in the mRNA. If the number of nucleotides in the skipped (deleted) exon is not a multiple of three, it results in a frameshift and a downstream change in the amino acid sequence usually leading to a new stop codon downstream, resulting in a truncated protein. This type of variant often has significant clinical consequences.

Indels

These are the insertion or deletion of one or several nucleotides. When the number of nucleotides is a multiple of three it usually results in an in-frame deletion or insertion of one amino acid. When the number of nucleotides is not a multiple of three, it results in an out-of-frame deletion or duplication, a frameshift, and a new stop codon downstream. Out-of-frame indels usually have larger impact on protein structure and function, often with clinical consequences.

Repeat expansions

There are areas within the genome that contain repeating sequences in both coding and noncoding regions, and an increase in the number of repeats can occur. This type of variant is known to be associated with a limited number of disease-causing genes, but constitutes an important category of mutation and disease mechanism.

Epigenetic variants

Some disorders are caused by changes in the epigenetic modification pattern such as methylation, which can affect gene expression and cause disease.

Methods to detect small scale changes in the DNA

Sanger sequencing has been the most effective method for detecting most small-scale variants. Genotyping methods are useful in detecting targeted mutations but are less often used. For repeat expansions, PCR-based assays, and occasionally Southern-blot-based assays, are necessary to assess the number of repeats. For epigenetic mutations, to detect the methylation status of disease-related genes, methylation-specific PCR or multiplex ligation-dependent probe amplification (MLPA) methods are often used. MLPA is a form of multiplex PCR used not only to detect methylation status but also to detect mutations and gene deletions/duplications. Multiple genomic targets are amplified using a common primer pair located at the outside ends of the multiple oligonucleotide probe pairs specific to the genomic targets. Since PCR amplification of the genomic targets only occurs when the two probes hybridize to their target, probes that are unbound will not be amplified. Thus, the strength of amplification signal reflects the amount of genomic target (inferred as copy number) available for hybridization and ligation. MLPA is used to detect methylation status by using probes at the methylation loci and treating the genomic DNA with methylation-sensitive restriction enzymes that disrupt the methylated DNA preventing the ligation, hence reducing the amplification signal.

Large-scale variants

This type of variant affects at least one exon of a gene or a larger genomic segment and cannot be reliably detected by conventional PCR and Sanger sequencing-based assays.

Copy number variants

Copy number variants (CNVs) are imbalanced structure variants. Genomic DNA copy number gain or loss has been known for many years, but CNVs are now known to be much more abundant than previously appreciated, with some resulting in significant clinical relevance. CNVs are the second most frequent mutation type associated with genetic disorders. Microarray-based genomic profiling technologies have enabled the effective detection of CNVs in a genome-wide manner with much improved sensitivity, resolution, and reproducibility. This has resulted in their being recommended as the first tier genetic test in many clinical scenarios.

Balanced structure variants

Translocations and inversions represent another type of large-scale variant. Until very recently, balanced genomic variants could only be detected by conventional cytogenetic approaches but with limited resolution. Microarray techniques are not able to detect balanced variants but can be used to detect cryptic imbalances undetected in apparently balanced rearrangements.

Loss of heterozygosity (LOH)

LOH results from deletion of one allele, uniparental origin of both alleles (often referred to as uniparental disomy), or consanguinity. Microarray platforms are able to detect large segments of LOH in a genome-wide manner. Next generation sequencing (NGS) data can also easily provide detailed genotyping information for the whole genome, providing a much more complete picture of LOH in the whole genome.

Evolving approaches of DNA-based genetic testing

The nature of the variant dictates what method should be used for genetic testing. Conventional approaches deal with one mutation or one gene and one patient a time. Clinicians make a preliminary diagnosis and order the gene test that is most likely to explain a patient's clinical condition. Depending on the known mutation spectrum of the disease gene, a molecular diagnostic laboratory will either use Sanger sequencing-based whole gene or gene panel tests often complemented by CNV testing using MLPA or quantitative PCR to detect potential exonic deletion or duplication.

Whole genome chromosomal microarray (CMA) analysis for CNV and LOH has enhanced clinical utility compared with conventional cytogenetic techniques, which justifies its use as the first tier test for patients with complex or unknown genetic conditions.

NGS-based whole exome or whole genome testing generates sequence data by massive parallel sequencing of clonally amplified or single DNA molecules. Recent advances in NGS technologies are making testing more cost-effective with a reasonable turnaround time. NGS is poised to bring a paradigm shift in genetic testing. The premise for using whole exome instead of whole genome sequencing for Mendelian disorders is that protein coding sequences constitute about 1% of the human genome but harbor about 85% of the known mutations that cause human diseases. However, given the fact that large-scale variants can be simultaneously detected by a whole genome-based test, but not easily by whole exome sequencing, it is predicted that when the cost for whole genome sequencing becomes acceptable for routine testing, it will become the method of choice for both research and clinical diagnostics. Many challenges lie ahead, particularly in the area of data interpretation, which often involves database searching, segregation analysis, bioinformatic prediction, and functional demonstration. Even though NGS will eventually replace many current genetic testing methods, conventional sequencing, genotyping, and CNV detecting technologies are still useful for validating and confirming variants detected by NGS.

GENETIC TESTS AVAILABLE FOR SKELETAL DISORDERS

As knowledge about genetic disorders of bone has expanded greatly over the past few years, the number of disorders for which genetic testing is commercially available has also increased. Unfortunately, there has been a lag between the discovery of causative genes on a research basis and the availability of commercial diagnostic testing [3]. However, genetic testing is available for a large number of skeletal diseases commonly seen in the clinic.

Metabolic bone disease

Genetic testing for metabolic disorders of bone can be useful for diagnostic purposes. There are a number of genetic causes of metabolic bone disease for which there is genetic testing available.

Familial hypophosphatemic rickets

Familial hypophosphatemic rickets (FHR) is the most common form of heritable rickets. Clinically, FHR presents in childhood with the typical signs of rickets, including bowing of the legs and growth delay [4]. X-linked hypophosphatemic rickets (XLH) is by far the most common form of FHR, and is due to mutations in the phosphate-regulating gene with homology to endopeptidases located on the X chromosome (PHEX) [4]. XLH is inherited in an X-linked dominant manner and thus there are more affected females than males (2:1 ratio) and no male to male transmission. Other forms are inherited in an autosomal dominant or autosomal recessive manner. Genetic testing for all forms is available and can be used to differentiate between the forms when it is not clear from an inheritance pattern [5].

Vitamin D-related disorders

There are two genetic defects in the vitamin D pathway that cause rickets. Vitamin D-dependent rickets type I (VDDR-I), also known as 1-α-hydroxylase deficiency, and vitamin D-dependent rickets type II (VDDR-II), also called vitamin D-resistant rickets [6]. VDDR-I is due to mutations in the 1-α-hydroxylase gene *CYP27B* and VDDR-II is due to mutations in the vitamin D receptor, *VDR*. Both genetic tests are available clinically.

Hypophosphatasia

Hypophosphatasia is an inherited bone disorder characterized by rickets in childhood and osteomalacia in adulthood due to a defect in mineralization of bone and teeth [7]. Hypophosphatasia is due to mutations in *ALPL*, the gene encoding the alkaline phosphatase, tissue-nonspecific isozyme (TNSALP). Therapy is now available to treat hypophosphatasia, so confirming a diagnosis through genetic testing can ensure patients receive appropriate treatment [8, 9].

Skeletal dysplasias

Genetic testing for skeletal dysplasias can be used to confirm a clinical diagnosis, or to make a diagnosis in cases where a clinical diagnosis is not immediately clear but several different disorders are under consideration. Here we highlight some of the more common skeletal dysplasias for which there is genetic testing available.

Osteogenesis imperfecta

OI is characterized by low bone mass and increased fracture risk. Most patients with OI have mutations in the type I collagen genes, *COL1A1* and *COL1A2*, or in genes encoding proteins that participate in the assembly, modification, and secretion of type I collagen [10].

Genetic testing in OI generally serves two purposes. In children with multiple fractures but who do not have significant short stature or deformities, genetic testing can be used to make the diagnosis of OI. In children with the more severe forms of OI that are diagnosed on clinical exam, genetic testing can confirm the diagnosis of OI and differentiate between dominantly and recessively inherited OI [10]. Confirming the diagnosis of OI by genetic testing is helpful as there is pharmacological treatment available [11, 12].

Achondroplasia and other FGFR3-related disorders

Achondroplasia is a relatively common skeletal dysplasia [13]. Although the diagnosis of achondroplasia is typically made based on the physical examination and radiographs, genetic testing for the disorder is useful to confirm the diagnosis. Nearly all cases of achondroplasia are caused by one of two mutations in the same nucleotide of the fibroblast growth factor receptor 3 (*FGFR3*) gene. Achondroplasia is inherited in an autosomal dominant manner and approximately 80% of cases are new mutations [14, 15]. Mutations in *FGFR3* are responsible for several other conditions. Of these, hypochondroplasia is the most common and presents as a milder form of achondroplasia [16].

Multiple epiphyseal dysplasia

Multiple epiphyseal dysplasia (MED) is a skeletal dysplasia characterized by abnormal epiphyses of the long bones [17]. MED can be both dominantly and recessively inherited [18]. While a clinical diagnosis of MED is typically made based on symptoms and abnormal epiphyses seen on radiographs, genetic testing can be very useful in confirming the diagnosis and differentiating between the dominant and recessive forms. Dominant MED is a disorder with genetic heterogeneity, that is, mutations in several different genes can cause the disorder, including *COMP*, *COL9A1*, *COL9A2*, *COL9A3*, and *MATN3* [18]. Given this scenario, gene panels including all known causative genes are very useful.

WHEN TO ORDER GENETIC TESTING

The initial evaluation of suspected metabolic bone disorders should include X-rays and laboratory tests. In the case of a suspected skeletal dysplasia, the evaluation starts with a comprehensive physical examination and a skeletal radiographic series to characterize the skeletal findings.

Once a preliminary diagnosis of a genetic condition is made, the decision to order genetic testing is influenced by several factors, including the diagnostic and treatment necessity, the need for reproductive counseling, patient and family desire for testing, and cost and payment issues. Prior to ordering any genetic test, the provider needs to determine if the desired test is clinically available. In the United States, all genetic testing must be performed in a Clinical Laboratory Improvement Amendments (CLIA) certified clinical laboratory for both legal and reimbursement reasons [19].

The most common clinical situation where genetic testing is ordered is when the genetic testing is required to make a diagnosis or to confirm the diagnosis. Genetic testing is especially useful if a course of treatment is available and dependent on the diagnosis. In these cases, genetic testing will provide a clear benefit for the care of the affected patient.

Often, if the diagnosis is unclear, a staged approach to testing, while perhaps prolonging the time to final diagnosis, can potentially conserve resources. This involves requesting one genetic test at a time and moving on to other genetic testing only if the initial tests are negative. Frequently, one blood sample can almost always provide enough DNA for a laboratory to carry out multiple sequencing analyses, thus preventing repetitive blood draws. However, in cases where diagnosis will affect treatment and delaying therapy is not optimal, ordering multiple tests at once may benefit the patient.

Reproductive counseling is another reason that genetic testing may be useful in a clinical setting. It is not uncommon for skeletal disorders to be diagnosed clinically based on family history, physical exam, and radiographic evidence. While the diagnosis may not be in question, knowledge of the specific mutation causing the phenotype may be useful for future reproductive decisions. In the pediatric setting, parents of an affected child may desire prenatal testing for future pregnancies. In this situation, the patient's exact mutation needs to be known to determine if the parents are carriers (or affected themselves if the mutation is dominant). In the case of recessive disorders, carrier testing is appropriate for adult family members but should not be performed on minor children until they reach maturity and can consent to their own testing [20].

Patients or family members may request or refuse genetic testing for various reasons. In cases where genetic testing is needed for initial diagnosis or to confirm a clinical diagnosis, a full discussion of the risks and benefits should be undertaken with the patient (or parents if the patient is a minor). This discussion should include the reason(s) why genetic testing is needed, how the results, whether a mutation is found or not, will affect the patient's immediate treatment, and how the results may affect the patient's care in the future. Discussion of these issues can help alleviate many misunderstandings and anxieties around genetic testing [21]. In the United States, the 2008 Genetic Information Nondiscrimination Act (GINA) bars discrimination by health insurance providers and employers due to genetic testing results, although GINA does not protect from genetic discrimination in other areas such as long-term care or life insurance [22]. Finally, the discussion should include the limits of genetic testing, particularly when a negative test will not exclude a clinical diagnosis, which is often the case. Available genetic tests may only detect mutations in a small number of clinically diagnosed patients, depending on the disorder. Patients and families may consider a negative test result "the final answer" and need to understand that a negative genetic test does not always exclude a given clinical diagnosis.

The cost of genetic testing can affect the ability to order genetic tests in different settings. Genetic tests can be among the most expensive laboratory-based tests available, ranging from hundreds to tens of thousands of dollars. Third party payers are becoming increasingly reluctant to pay for the costs of genetic testing, particularly in the absence of documented clinical necessity. In the pediatric population, genetic testing for diagnostic purposes is typically covered by private insurance companies as long as the patient's pertinent history and physical findings are well documented. However, families with high deductible plans should be warned that the cost of one genetic test may be greater than the entire deductible amount, requiring a single large payment. Carrier testing may not be covered, except when done for prenatal testing purposes. As genetic tests are ordered more frequently, institutions such as hospitals may limit the number of tests that can be ordered or require that tests be ordered only from laboratories that bill the patients' insurance company directly. This can effectively put testing for some disorders out of the reach of patients and providers. These increasing costs place a responsibility on providers to ensure that genetic tests are only ordered under appropriate circumstances where there is a clear benefit for the patient.

The increasing availability of genetic testing for all types of skeletal disorders can be of great benefit to patients and providers. Genetic testing can be useful for diagnosis and subsequent treatment as well as for reproductive counseling for patients and families. However, as with all medical testing, genetic testing has risks and benefits and may not be appropriate for every clinical situation.

REFERENCES

1. National Human Genome Research Institute. *Promoting Safe and Effective Genetic Testing in the United States*. 1997. http://www.genome.gov/10001733 (accessed May 2018).
2. Holtzman NA, Watson MS. Promoting safe and effective genetic testing in the United States. Final report of the Task Force on Genetic Testing. J Child Fam Nurs. 1999;2(5):388–90.
3. Das S, Bale SJ, Ledbetter DH. Molecular genetic testing for ultra rare diseases: models for translation from the research laboratory to the CLIA-certified diagnostic laboratory. Genet Med. 2008;10(5):332–6.
4. Imel EA, Carpenter TO. A practical clinical approach to paediatric phosphate disorders. Endocr Dev. 2015;28: 134–61.
5. Carpenter TO, Imel EA, Holm IA, et al. A clinician's guide to X-linked hypophosphatemia. J Bone Miner Res. 2011;26(7):1381–8.
6. Allgrove J, Shaw NJ. A practical approach to vitamin D deficiency and rickets. Endocr Dev. 2015;28:119–33.
7. Whyte MP. Hypophosphatasia—aetiology, nosology, pathogenesis, diagnosis and treatment. Nat Rev Endocrinol. 2016;12(4):233–46.
8. Whyte MP, Madson KL, Phillips D, et al. Asfotase alfa therapy for children with hypophosphatasia. JCI Insight. 2016;1(9):e85971.
9. Whyte MP, Rockman-Greenberg C, Ozono K, et al. Asfotase alfa treatment improves survival for perinatal and infantile hypophosphatasia. J Clin Endocrinol Metab. 2016;101(1):334–42.
10. Trejo P, Rauch F. Osteogenesis imperfecta in children and adolescents—new developments in diagnosis and treatment. Osteoporos Int. 2016;27(12):3427–37.
11. Rijks EB, Bongers BC, Vlemmix MJ, et al. Efficacy and safety of bisphosphonate therapy in children with osteogenesis imperfecta: a systematic review. Horm Res Paediatr. 2015;84(1):26–42.
12. Dwan K, Phillipi CA, Steiner RD, et al. Bisphosphonate therapy for osteogenesis imperfecta. Cochrane Database Syst Rev. 2016;10:CD005088.
13. Stevenson DA, Carey JC, Byrne JL, et al. Analysis of skeletal dysplasias in the Utah population. Am J Med Genet A. 2012;158A(5):1046–54.
14. Carter EM, Davis JG, Raggio CL. Advances in understanding etiology of achondroplasia and review of management. Curr Opin Pediatr. 2007;19(1):32–7.
15. Shirley ED, Ain MC. Achondroplasia: manifestations and treatment. J Am Acad Orthop Surg. 2009;17(4): 231–41.
16. Xue Y, Sun A, Mekikian PB, et al. FGFR3 mutation frequency in 324 cases from the International Skeletal Dysplasia Registry. Mol Genet Genomic Med. 2014;2(6): 497–503.
17. Anthony S, Munk R, Skakun W, et al. Multiple epiphyseal dysplasia. J Am Acad Orthop Surg. 2015;23(3):164–72.
18. Briggs MD, Chapman KL. Pseudoachondroplasia and multiple epiphyseal dysplasia: mutation review, molecular interactions, and genotype to phenotype correlations. Hum Mutat. 2002;19(5):465–78.
19. Rivers PA, Dobalian A, Germinario FA. A review and analysis of the clinical laboratory improvement amendment of 1988: compliance plans and enforcement policy. Health Care Manage Rev. 2005;30(2):93–102.
20. Borry P, Fryns JP, Schotsmans P, et al. Carrier testing in minors: a systematic review of guidelines and position papers. Eur J Hum Genet. 2006;14(2):133–8.
21. Henneman L, Timmermans DR, Van Der Wal G. Public attitudes toward genetic testing: perceived benefits and objections. Genet Test. 2006;10(2):139–45.
22. Payne PW Jr, Goldstein MM, Jarawan H, et al. Health insurance and the Genetic Information Nondiscrimination Act of 2008: implications for public health policy and practice. Public Health Rep. 2009;124(2):328–31.

47
Human Genome-Wide Association Studies

Douglas P. Kiel[1], Emma L. Duncan[2], and Fernando Rivadeneira[3]

[1]Musculoskeletal Research Center, Institute for Aging Research, Hebrew SeniorLife; Department of Medicine, Beth Israel Deaconess Medical Center and Harvard Medical School; Associate Member, Broad Institute of MIT and Harvard, Boston, MA, USA
[2]Royal Brisbane and Women's Hospital, Queensland University of Technology; and University of Queensland, Brisbane, QLD, Australia
[3]Department of Internal Medicine, Erasmus University Medical Center, Rotterdam, The Netherlands

INTRODUCTION

Gene mapping in both common and rare diseases has changed enormously in recent times. Technological changes (particularly high-throughput microarray genotyping and, more recently, massively parallel sequencing) and newer statistical approaches have resulted in genome-wide association studies (GWAS) with unprecedented success in identifying loci underlying common traits and diseases. This has meant an explosion of new discoveries in skeletal genetics, as for many other human diseases. Findings from all published GWAS are recorded on a web site maintained by the National Genome Research Institute (https://www.ebi.ac.uk/gwas/, accessed May 2018).

HERITABILITY OF BONE MINERAL DENSITY

Heritability is the proportion of the total variance of a trait due to genetic factors. Twin and family studies have demonstrated that BMD, one of the most commonly studied bone phenotypes, is highly heritable (0.6 to 0.8), as are bone geometry (0.3 to 0.7), and bone ultrasound measures (0.4 to 0.5) [1–3]. Risk of fracture is more heterogeneous, with lower but still significant heritability (0.3 to 0.5) for fractures in general and for hip fracture; there is a trend for greater heritability for fractures occurring at younger ages [4]. Segregation studies suggested that BMD is a polygenic trait, with many genes each contributing a small amount to the overall variance of the trait and with skeletal site specificity [5, 6].

Before the GWAS era the field of genetics of osteoporosis and fracture had been confined to a very large number of genome-wide linkage and candidate gene association studies. With few exceptions, the majority of these studies were inadequately-powered (small sample size) studies generating controversial and very frequently irreproducible results, as was highlighted in a retrospective review of 150 candidate gene regions for osteoporosis using current standards to scrutinize genetic associations [7].

GENOME-WIDE ASSOCIATION STUDIES

General principles

GWAS use high-throughput microarrays to genotype simultaneously in one or more individuals hundreds of thousands and even millions of the most common forms of genetic variation, SNPs. This powerful approach interrogates the entire genome for associations between the variants and the phenotypes, without prior assumptions as to the underlying cause of a particular disease. From a statistical point of view, the very large number of associations tested across the genome requires an adjustment for multiple testing, resulting in a very stringent definition of genome-wide significance (conventionally, $p < 5 \times 10^{-8}$ to account for the roughly 1 million independent tests covering common variants). Replication across independent cohorts is then sought – for example, through GWAS meta-analyses – a compulsory step to validate significant findings.

Primer on the Metabolic Bone Diseases and Disorders of Mineral Metabolism, Ninth Edition. Edited by John P. Bilezikian.
© 2019 American Society for Bone and Mineral Research. Published 2019 by John Wiley & Sons, Inc.
Companion website: www.wiley.com/go/asbmrprimer

The conduct of GWAS meta-analyses requires careful attention to quality control. Important considerations include the extent of missing data on variants across studies and the numbers of individuals with missing data on genotyping. Large amounts of missing data may reflect technical problems with the quality of the DNA or the platform used. Because of the millions of genotype–phenotype association tests that are performed in typical GWAS, quantile-quantile plots comparing the number of expected p-values of a given magnitude to those that are observed are used to indicate whether the study has generated more significant results than expected by chance. If the number of observed p-values systematically exceeds the number expected under the null hypothesis of no association, this may have resulted from unappreciated population stratification or cryptic relatedness of the individuals in the study. On the other hand, when the number of observed p-values exceeds those expected at the most significant end of the range, this may indicate robust, large-effect, susceptibility loci. Nevertheless, in most cases, independent replication of genetic association study findings is necessary to reduce the probability of making a type I error.

Imputation

In addition to genotyped variants, GWAS typically "impute" the genotypes of millions more SNPs, resulting in a very large final data set of uniform content for analysis between studies. Imputation is based on the presence of haplotypes in the genome—stretches of DNA (blocks) where surrounding SNPs lying on a chromosomal strand are inherited together more often than should occur by chance. This phenomenon is known as linkage disequilibrium. Once the genotype of one SNP within the haplotype is known, the genotypes of other SNPs on the haplotype can be inferred.

Increasingly large, free, publically accessible databases are available to perform imputation [8, 9]. The HapMap project developed a human haplotype map using genotyped data from samples drawn from populations with diverse geographic ancestry [10], providing ethnically-appropriate resources for SNP imputation and tag-SNP project design. More recently, the 1000 Genomes Project (1000GP) (http://www.1000genomes.org/, accessed May 2018) has characterized over 95% of variants in genomic regions accessible to current high-throughput sequencing technologies that have a minor allele frequency (MAF) of 1% or higher in each of five major population groups (populations in or with ancestry from Europe, East Asia, South Asia, West Africa, and the Americas). The 1000GP is also cataloguing lower frequency alleles (in the range of 0.1%) because these low-frequency alleles are often found in coding regions [11]. Recently, the Haplotype Reference Consortium (HRC) comprising 64,976 haplotypes at 39,235,157 SNPs (including 1000GP data), provides a reference panel enabling researchers to perform accurate and reliable imputation at an even lower frequency (0.02%). As a measure of the depth of these resources, currently 96% of genetic variation due to common variants (and 73% of that due to rare variants) can be captured after genotyping performed using a standard microarray platform (genotyping fewer than a million SNPs) with subsequent imputation [12].

Imputation is also critical for meta-analysis of GWAS, to create a large uniform set of SNPs from data generated by multiple individual cohorts and different genotyping platforms (discussed further later in this chapter).

Power and sample size considerations

Individually, most studies are not sufficiently powered to interrogate so many SNPs and be able to detect association with complex traits at genome-wide significance, particularly given the small contribution of each individual variant to the overall phenotype. There is a very close correlation ($r^2 > 0.9$) between number of loci identified and GWAS size, almost irrespective of the phenotype in question and its heritability [13]. This is true for both dichotomous and quantitative traits. To maximize power when the sample size may be limited, some have shown that using the extremes of a continuous phenotype may have advantages over using the entire range of values [14]. Thus, the GWAS era has spawned a dramatic change in the collaborative efforts across the scientific community; and in osteoporosis, as in many other diseases, cohorts across the world have developed consortia that cultivate the necessary trust and collegiality required to undertake large meta-analyses of the GWAS studies from individual cohorts.

Meta-analysis: combining data from different clinical groups

Careful phenotyping harmonization

One of the challenges of these collaborative GWAS efforts includes proper harmonization of phenotypes to minimize measurement heterogeneity. There have been efforts to collect the best phenotypic measures as part of the PhenX project (https://www.phenx.org/, accessed May 2018). When measurement differences arise across cohorts such as DXA manufacturer differences, these may sometimes be tolerable when the individual study GWAS analysis is then meta-analyzed with other cohorts. If the phenotypes are widely different across cohorts, use of Z-scores to "standardize" measures represents another approach to harmonization. Further efforts to minimize heterogeneity require standardization of data analysis to account for potential confounding, methods of minimizing the effects of possible population stratification, and appropriate quality metrics for the handling of genotyping data that come from a variety of genotyping platforms.

Combining data from different platforms

A major challenge in combining data from many cohorts is that each cohort may have been genotyped on a different microarray platform, resulting in only a small number of genotyped SNPs common to all studies. However, as detailed above this pool of common SNPs can be vastly increased by using imputed data, resulting in a much large common data set and providing for a much more comprehensive assessment of genomic variation.

Combining data from different ethnic groups

As the science of GWAS has matured, there has been a growing appreciation of the value of examining genetic associations in a variety of ethnic groups because limiting studies to participants of European ancestry is insufficient for fully uncovering the variants underlying disease populations of other ancestries [15]. Furthermore, inclusion of other ethnicities in GWAS offers the possibility of improving the resolution of fine-mapping of causal variants, as the underlying difference in linkage disequilibrium across ethnic groups provides an opportunity to amplify the signal of association for a causal variant. Methods have been developed to leverage these multiethnic studies in meta-analyses [16]. Taking advantage of the growing number of individuals of non-European background included in reference databases, imputation can be used to impute variants across multiple ethnic groups.

SKELETAL PHENOTYPES

The field of skeletal genetics has expanded beyond the early focus on the BMD phenotype exclusively. While most studies do leverage the availability of BMD for meta-analyses, there has been some evolution from the use of areal BMD using DXA to volumetric measures of bone density using QCT [17, 18], quantitative heel ultrasound [19], and fracture [20].

GWAS SUCCESS IN THE MUSCULOSKELETAL FIELD

The first GWAS for DXA-derived BMD and hip geometry traits was performed in the Framingham Study using a genotyping platform with only 100,000 SNPs in a sample of only 1141 men and women. This yielded no genome-wide significant findings using the strict p-value criterion of 5×10^{-8} [21]. Since that time, as the sample sizes of the GWAS meta-analyses has grown, the number of genome-wide significant loci has dramatically increased for BMD (Fig. 47.1).

In the most recent GWAS meta-analysis from the Genetic Factors of Osteoporosis (GEFOS) Consortium, using DXA-derived BMD of the hip and spine from 17 genome-wide association studies and 32,961 individuals of European and East Asian ancestry, the top-associated SNPs with BMD at either the femoral neck or lumbar spine were then tested for replication in 50,933 independent subjects [22]. These independent samples for replication came from a previously funded consortium called Genetic Markers for Osteoporosis (GENOMOS), which had produced several of the largest candidate gene meta-analyses prior to the advent of GWAS efforts. The availability of fracture phenotypes in the replication cohorts also permitted the GEFOS meta-analysis to include the testing of the top BMD associations for their association with low-trauma fracture in 31,016 cases and 102,444 controls. This meta-analysis identified 56 loci (32 novel) associated with BMD at a genome-wide significant level ($p < 5 \times 10^{-8}$). The overall results of this meta-analysis are displayed in Fig. 47.2, referred to as a "Manhattan" plot, as each tower of points indicate $-\log_{10}$ p-values. Each point represents a p-value for a SNP–phenotype association meta-analysis result.

When the loci associated with BMD were then tested for association with fracture, it was found that 14 BMD loci were also associated with fracture ($p < 5 \times 10^{-4}$), of which six reached $p < 5 \times 10^{-8}$ including 18p11.21 (C18orf19), 7q21.3 (SLC25A13), 11q13.2 (LRP5), 4q22.1 (MEPE), 2p16.2 (SPTBN1), and 10q21.1 (DKK1).

These and similar genetic discovery studies have focused primarily on "common variants" with allele frequency of 5% or greater. Newly available genome sequencing and better imputation of genotyped cohorts spawned an effort to characterize less common and noncoding variants affecting BMD. In a study by Zheng and colleagues, novel noncoding genetic variants with large effects on BMD (n_{total} 553,236) and fracture (n_{total} 5508,253) were identified [23]. In fact, a low-frequency, noncoding variant near a novel locus, EN1, was confirmed with an effect size fourfold larger than the mean of previously reported common variants for BMD of the lumbar spine (rs11692564(T), MAF = 1.6%, replication effect size = +0.20 standard deviations, p-value for the meta-analysis = 2×10^{-14}). This variant's positive effect on BMD was accompanied by a decreased risk of fracture comparing 598,742 fracture cases with 409,511 controls (odds ratio = 0.85; $p = 2 \times 10^{-11}$). Such a novel finding with BMD and fracture associations suggests a previously unrecognized target for future osteoporosis drug discovery.

The largest GWAS meta-analysis for skeletal phenotypes to date was recently reported by investigators using data from the UK Biobank Study [24]. This GWAS focused on estimated BMD from heel ultrasounds and comprised 140,623 individuals of European descent (75,275 females and 65,349 males) and identified 307 conditionally independent SNPs attaining genome-wide significance at 203 loci (of which 153 were novel) and altogether explaining 11.8% of the variance. These included the majority of SNPs previously associated with DXA-derived BMD, as well as 308 novel loci, including many containing genes that have not been previously implicated in bone physiology. This study highlights the value of expanding the sample size of GWAS and the great potential for identifying

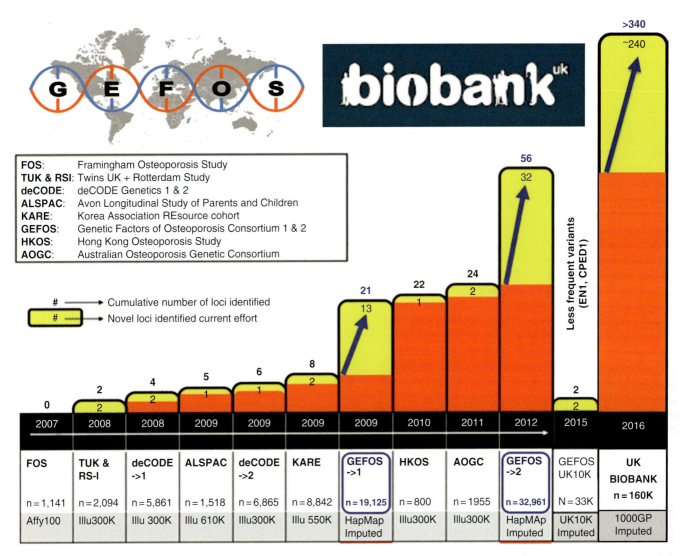

Fig. 47.1. An illustration of how increasing sample sizes of GWAS over the years have resulted in larger number of genome-wide significant associations with BMD.

new biology as a next step towards driving these discoveries in the direction of clinical applications (eg, drug targets). This is well illustrated by the pathway analyses derived from the GWAS signals that have identified gene clusters characterized by factors from major pathways critical to bone biology like RANK-RANKL-OPG, mesenchymal stem cell differentiation, endochondral ossification, and Wnt signaling (among others). In addition to the identification of factors in well-established biological pathways, the hypothesis-free GWAS approach has revealed completely new biology. An example of this is the case of *FAM210A*, one of the strongest loci associated with fracture risk across the different GEFOS efforts, as well as for at least 30 of the 56 BMD loci, containing genes underlying the GWAS signals for which nothing is known regarding their potential role in bone biology [22].

One of the greatest advancements in this direction is well characterized by the discovery of *WNT16* as a critical molecule in bone biology using the GWAS approach. From its initial identification as a BMD locus [22], several subsequent GWAS performed in premenopausal women [25], wrist BMD [23], total body BMD in children and adults [26], cortical thickness from pQCT of the tibia [27], and quantitative ultrasound of the heel [19], all confirmed the importance of *WNT16* across multiple skeletal traits. The regulatory mechanisms of *WNT16* have been recently elucidated by a functional study in murine models [28] showing that Wnt16 KO mice have reduced cortical bone thickness and increased cortical bone porosity (but not trabecular bone mass), which leads to spontaneous nonvertebral fractures in these mice. Recent work generating Wnt16 KO mice demonstrated that the phenotypic presentation of reduced cortical bone thickness and increased cortical bone porosity resulting in skeletal fragility is the consequence of osteoclastogenesis inhibition via noncanonical Wnt pathway activation, while

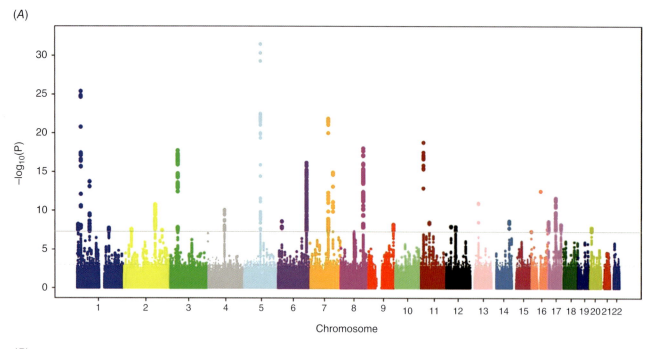

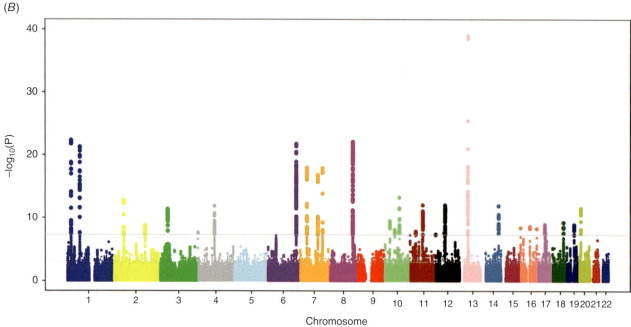

Fig. 47.2. Manhattan plots from the most recent GWAS meta-analysis of (A) lumbar spine and (B) femoral neck BMD from the GEFOS Consortium [22]. The plot displays the genome-wide significant associations between all of the SNPs that were imputed using the "HapMap2" data. Each point on the plot represents a *p*-value for association with lumbar spine and femoral neck BMD from the meta-analysis using a fixed-effects model. The significant loci for each skeletal site are not completely overlapping.

also exerting an indirect effect by inducing OPG expression in osteoblasts [28]. Interestingly, overexpression of WNT16 has recently been shown to increase trabecular bone mass independently of estrogen action [29].

These are only a few "proof of concept" examples of the potential derived from employing genetic information for the identification of drug targets. There are more discoveries awaiting scientific inquiry considering that these examples represent only a fraction of the identified loci implicating new biology. In fact, nearly all current osteoporosis agents, either in clinical use or in advanced clinical trials, target known BMD-associated genes that lie in biological pathways key to bone metabolism [30]. A recent study demonstrated how known genetic associations are successful predictors of drug mechanisms and success in clinical development. This is particularly the case in the musculoskeletal (BMD), metabolic, and hematological fields, which supports the use of genetics for drug–target indications [31]. Some existing osteoporosis drugs target molecules discovered by genetic investigations of monogenic skeletal disorders (eg, sclerostosis and van Buchem disease) [32]. Since several of the loci identified through GWAS overlap with human monogenic conditions presenting with severe skeletal fragility [33], GWAS findings have a huge potential to identify novel drug targets for osteoporosis.

Another important clinical implication emerges for the diagnosis of individuals with unresolved familial forms of osteoporosis. Different variants in the same gene (ranging from rare mutations to common polymorphisms) can be responsible for both Mendelian monogenic and common complex conditions [33]. Therefore, genes responsible for monogenic forms of osteoporosis can also harbor common variants influencing BMD variation (and other osteoporosis-related traits) in the general population. Conversely, common polymorphisms discovered by GWAS of osteoporosis-related traits in the general population can point to genes potentially responsible for monogenic familial forms of osteoporosis.

While GWAS in the field of osteoporosis have identified large numbers of novel loci, the translation of these discoveries into palpable clinical applications remains hampered by the difficulty in pinpointing the actual genes contributing to the GWAS signals. With few exceptions, most of the genes claimed as "underlying the GWAS signals" have been labeled as such by allocating a gene to a genetic signal based on physical distance and current knowledge of biology. This practice may actually constrain new discovery by only looking "under the lamp post." The follow-up of these GWAS discoveries requires robust functional evidence linking the human genome sequence to the function of its regulatory elements. The Encyclopedia of DNA Elements (ENCODE) Project [34] has started a revolution by going beyond the "linear" configuration of regulatory elements, challenging the previous context of boundaries and gaps between genes. The knowledge derived from the ENCODE Project provides new approximations to the understanding of diverse processes of gene regulation, including chromatin interactions, epigenetic states, promoter activities, enhancer binding, and nuclear lamina occupancy to name a few.

NOVEL APPLICATIONS OF GWAS META-ANALYSES SUMMARY DATA

With the increasing availability of summary level data from large GWAS meta-analyses that are often made publicly available, there are several novel applications that can be leveraged to stimulate new scientific understanding and hypotheses. First, summary level statistics can be used to find potential pleiotropy, where individual variants may be implicated as being associated with more than one phenotype. Second, several methods have been developed that are able to use summary data from GWAS meta-analyses to estimate genetic correlation between traits. The method of linkage disequilibrium (LD) score regression considers the effects of all SNPs, including those that do not reach genome-wide significance [35]. Finally, Mendelian randomization can be used to infer causality using observational data. This is accomplished by using a set of genetic risk alleles identified from a GWAS meta-analysis as being significantly associated with a given risk exposure (eg, BMD) that randomly assort at the time of gamete formation. By using these random alleles in a given individual to test for association with another trait (eg, fracture), one is able to estimate association free of confounding [36].

The immense wealth of information generated by GWAS has huge potential to change our thinking about biological science and human health—from understanding disease etiology to the identification of novel therapeutic targets to risk prediction, prognostication, and personalized management for affected individuals. The challenge is to realize the potential of GWAS and ensure that these results are embraced by the medical and scientific communities.

REFERENCES

1. Karasik D, Dupuis J, Cupples LA, et al. Bivariate linkage study of proximal hip geometry and body size indices: the Framingham study. Calcif Tissue Int. 2007;81:162–73.
2. Karasik D, Myers RH, Hannan MT, et al. Mapping of quantitative ultrasound of the calcaneus bone to chromosome 1 by genome-wide linkage analysis. Osteoporos Int. 2002;13:796–802.
3. Karasik D, Myers RH, Cupples LA, et al. Genome screen for quantitative trait loci contributing to normal variation in bone mineral density: the Framingham Study. J Bone Miner Res. 2002;17:1718–27.
4. Michaelsson K, Melhus H, Ferm H, et al. Genetic liability to fractures in the elderly. Arch Intern Med. 2005;165:1825–30.
5. Hofman A, Brusselle GG, Darwish Murad S, et al. The Rotterdam Study: 2016 objectives and design update. Eur J Epidemiol. 2015;30:661–708.
6. Khosla S, Shane E. A crisis in the treatment of osteoporosis. J Bone Miner Res. 2016;31:1485–7.
7. Richards JB, Kavvoura FK, Rivadeneira F, et al. Collaborative meta-analysis: associations of 150 candidate genes with osteoporosis and osteoporotic fracture. Ann Intern Med. 2009;151:528–37.
8. de Bakker PI, Ferreira MA, Jia X, et al. Practical aspects of imputation-driven meta-analysis of genome-wide association studies. Hum Mol Genet. 2008;17:R122–8.
9. Marchini J, Howie B, Myers S, et al. A new multipoint method for genome-wide association studies by imputation of genotypes. Nat Genet. 2007;39:906–13.

10. Manolio TA, Brooks LD, Collins FS. A HapMap harvest of insights into the genetics of common disease. J Clin Invest. 2008;118:1590–605.
11. 1000 Genomes Consortium. A map of human genome variation from population-scale sequencing. Nature. 2010;467:1061–73.
12. Yang J, Bakshi A, Zhu Z, et al. Genetic variance estimation with imputed variants finds negligible missing heritability for human height and body mass index. Nat Genet. 2015;47:1114–20.
13. Visscher PM, Brown MA, McCarthy MI, et al. Five years of GWAS discovery. Am J Hum Genet. 2012;90:7–24.
14. Duncan EL, Danoy P, Kemp JP, et al. Genome-wide association study using extreme truncate selection identifies novel genes affecting bone mineral density and fracture risk. PLoS Genet. 2011;7:e1001372.
15. Rosenberg NA, Huang L, Jewett EM, et al. Genome-wide association studies in diverse populations. Nat Rev Genet. 2010;11:356–66.
16. Morris AP. Transethnic meta-analysis of genomewide association studies. Genet Epidemiol. 2011;35:809–22.
17. Paternoster L, Lorentzon M, Lehtimaki T, et al. Genetic determinants of trabecular and cortical volumetric bone mineral densities and bone microstructure. PLoS Genet. 2013;9:e1003247.
18. Nielson CM, Liu CT, Smith AV, et al. Novel genetic variants associated with increased vertebral volumetric BMD, reduced vertebral fracture risk, and increased expression of SLC1A3 and EPHB2. J Bone Miner Res. 2016;31:2085–97.
19. Moayyeri A, Hsu YH, Karasik D, et al. Genetic determinants of heel bone properties: genome-wide association meta-analysis and replication in the GEFOS/GENOMOS consortium. Hum Mol Genet. 2014;23:3054–68.
20. Trajanoska K, Morris JA, Oei L, et al. Assessment of the genetic and clinical determinants of fracture risk: a Mendelian randomization approach. BMJ. 2018 (in press).
21. Kiel DP, Demissie S, Dupuis J, et al. Genome-wide association with bone mass and geometry in the Framingham Heart Study. BMC Med Genet. 2007;8(suppl 1):S14.
22. Estrada K, Styrkarsdottir U, Evangelou E, et al. Genome-wide meta-analysis identifies 56 bone mineral density loci and reveals 14 loci associated with risk of fracture. Nat Genet. 2012;44:491–501.
23. Zheng HF, Forgetta V, Hsu YH, et al. Whole-genome sequencing identifies EN1 as a determinant of bone density and fracture. Nature. 2015;526:112–17.
24. Kemp JP, Morris JA, Medina-Gómez M, et al. *Genome-wide association study of bone mineral density in the UK Biobank Study identifies over 376 loci associated with osteoporosis.* Presented at American Society for Human Genetics, Vancouver, CA, October 2016.
25. Koller DL, Ichikawa S, Lai D, et al. Genome-wide association study of bone mineral density in premenopausal European-American women and replication in African-American women. J Clin Endocrinol Metab. 2010;95:1802–9.
26. Medina-Gomez C, Heppe DH, Yin JL, et al. Bone mass and strength in school-age children exhibit sexual dimorphism related to differences in lean mass: the Generation R Study. J Bone Miner Res. 2016;31:1099–106.
27. Zheng HF, Tobias JH, Duncan E, et al. WNT16 influences bone mineral density, cortical bone thickness, bone strength, and osteoporotic fracture risk. PLoS Genet. 2012;8:e1002745.
28. Moverare-Skrtic S, Henning P, Liu X, et al. Osteoblast-derived WNT16 represses osteoclastogenesis and prevents cortical bone fragility fractures. Nat Med. 2014;20:1279–88.
29. Moverare-Skrtic S, Wu J, Henning P, et al. The bone-sparing effects of estrogen and WNT16 are independent of each other. Proc Natl Acad Sci U S A. 2015;112:14972–7.
30. Richards JB, Zheng HF, Spector TD. Genetics of osteoporosis from genome-wide association studies: advances and challenges. Nat Rev Genet. 2012;13:576–88.
31. Nelson MR, Tipney H, Painter JL, et al. The support of human genetic evidence for approved drug indications. Nat Genet. 2015;47:856–60.
32. Balemans W, Van Den Ende J, Freire Paes-Alves A, et al. Localization of the gene for sclerosteosis to the van Buchem disease-gene region on chromosome 17q12-q21. Am J Hum Genet. 1999;64:1661–9.
33. Rivadeneira F, Makitie O. Osteoporosis and bone mass disorders: from gene pathways to treatments. Trends Endocrinol Metab. 2016;27:262–81.
34. Maurano MT, Humbert R, Rynes E, et al. Systematic localization of common disease-associated variation in regulatory DNA. Science. 2012;337:1190–5.
35. Bulik-Sullivan BK, Loh PR, Finucane HK, et al. LD Score regression distinguishes confounding from polygenicity in genome-wide association studies. Nat Genet. 2015;47:291–5.
36. Smith GD, Ebrahim S. Mendelian randomization: prospects, potentials, and limitations. Int J Epidemiol. 2004;33:30–42.

48

Translational Genetics of Osteoporosis: From Population Association to Individualized Assessment

Bich Tran[1], Jacqueline R. Center[2], and Tuan V. Nguyen[2,3]

[1]Centre for Big Data Research in Health, UNSW Sydney, Sydney, NSW, Australia
[2]Division of Bone Biology, Garvan Institute of Medical Research; and St Vincent's Hospital Clinical School, UNSW Sydney, Sydney, NSW, Australia
[3]Centre for Health Technologies, School of Biomedical Engineering, University of Technology; and School of Medicine Sydney, University of Notre Dame Australia, Sydney, NSW, Australia

BACKGROUND

Osteoporosis is a condition characterized by low BMD and micro-architectural deterioration leading to increased fracture risk. Osteoporosis affects up to 40% postmenopausal women and 15% elderly men of white background [1]. From the age of 50 years, approximately 44% of women and 25% of men will sustain a fracture during their remaining lifetime [2]. While fracture can occur in any bones, hip fracture is the most serious event because it is associated with increased risks of subsequent fractures [3], premature mortality [4], and incurs social and economic cost [5]. Up to 24% women and 38% men will die within the first 3 months after experiencing a hip fracture [6]. Given the ongoing aging of the population worldwide, it is expected that the burden of osteoporosis and osteoporosis-related fractures will become more pronounced in the near future.

GENETICS OF OSTEOPOROSIS

Osteoporosis is a complex phenotype because its risk is determined by environmental and genetic factors, and possibly their interactions. A key measure of genetic contribution to a phenotype is the index of heritability, which is defined as the extent to which genetic individual differences contribute to individual differences in the observed phenotype. Because the between-individual difference is often quantified in terms of variance, the index of heritability can be operationally interpreted as the proportion of phenotypic variance attributable to genetic factors.

Twins and family studies have shown that the heritability of BMD ranged between 60% and 80% depending on sites measured [7, 8]. For fracture, the index of heritability ranged from 25% to 48% [9–11], of which the index was higher for fractures occurring before 70 years of age [12]. Other markers of bone phenotypes are also heritable. For example, between 59% and 73% of the variance of QU measurements are attributable to genetic factors [13]; and 58% to 69% for bone turnover markers [14].

The heritability of bone phenotypes has motivated several studies to search for putative genes. Initially, the identification of genes associated with bone phenotypes has started with studies of monogenic syndromes such as osteogenesis imperfecta caused by defects in *COL1A1* and *COL1A2* [15] or osteoporosis pseudoglioma syndrome (OPPG) which is related to *LRP5* [16]. However, this study approach has not always been successful, mainly due to the complexity in developing osteoporosis and/or methodological problems. Subsequent efforts on mapping genes mostly focused on two common approaches: candidate genes and genome-wide studies.

Candidate gene studies

This approach compares allele frequency of genetic variant(s) in osteoporotic and nonosteoporotic persons. Studies using this approach usually select variants in or near gene(s) whose role in the pathophysiology of osteoporosis is known. Findings from some candidate gene studies are summarized in Table 48.1.

These studies are simple and the interpretation is straightforward; however it suffers from some shortcomings. The selection of nonosteoporotic persons can be a challenge, because osteoporosis could develop in later life. Thus, using an arbitrary cut-off in BMD at study entry to define osteoporosis may not fully account for the time effect. Furthermore, any significant association found in these analyses may not necessarily imply a causative relationship, but that could be due to linkage disequilibrium between studied variant(s) and the disease-causing variant(s), or population stratification.

Genome-wide studies

While the candidate gene approach focuses on biologically plausible genes, the genome-wide approach is a hypothesis-free design that scans hundreds of thousands of variants across the entire genome using tagging SNPs. Tagging SNPs are defined on the basis of linkage disequilibrium where SNPs are more likely to be inherited together when they are physically close to each other on a chromosomal region (ie, haplotype). Knowledge of haplotype blocks developed by the HapMap Project facilitates the scan of the entire genome by only genotyping a number of tagging SNPs, then imputing to a much greater number of other variants. Thus, genome-wide studies can overcome weaknesses of the candidate gene design by providing a holistic representation of genes contributing to the susceptibility of disease.

The genome-wide approach in osteoporosis research has been undertaken using two strategies: linkage analysis and association analysis. *Genome-wide linkage studies* identify variants that cosegregate in families with affected members. The principle of this study design is based on the assumption that loci close to each other are more likely to be passed on to offspring. Some linkage analyses have successfully mapped QTL associated with bone phenotypes (Table 48.2). *Genome-wide association studies* (GWAS) use a similar analytic approach to candidate gene association studies and test for differences in allele frequency between osteoporotic and nonosteoporotic persons. Although GWAS can overcome the biased effect in candidate gene studies because of the hypothesis-free design, there is a possibility of generating false positive findings as a result of multiple testings [17].

Following the first GWAS in osteoporosis published in 2007 [18], two other GWAS were published in 2008 [19,20] that identified several genetic variants associated with BMD. Some of these were located in or near osteoporosis-related genes (*ESR1*, *OPG*, *TNFSF11*, *LRP5*); novel loci on chromosomes 1p36 and 6p21 were also

Table 48.1. Genes associated with BMD identified from candidate genes association studies.

Gene	Gene Name	Location
ARHGEF3	Rho guanine nucleotide exchange factor 3	3p14-p21
COL11A1	Collagen type I alpha 1	17q21.33
CYP19A1	Cytochrome P450, family 19, subfamily A, polypeptide 1	15q21.1
DBP	D site of albumin promoter (albumin D-box) binding protein	19q13.3
ESR1	Estrogen receptor 1	6q25.1
ESR2	Estrogen receptor 2	14q
FNLB	Filamin B	3p14.3
FOXC2	Forkhead box C2	16q24.4
ITGA1	Integrin, alpha 1	5q11.2
LRP4	LDL receptor-related protein 4	11p11.2
LRP5	LDL receptor-related protein 5	11q13.4
MHC	Major histocompatibility complex	6p21
MTHFR	5, 10-methylenetetrahydrofolate reductase	1p36.3
PTH	Parathyroid hormone	11p15.3-p15.1
RHOA	Ras homologue gene family, member A	3p21.3
SFRP1	Secreted frizzled-related protein 1	8p12-p11.1
SOST	Sclerosteosis	17q11.2
SPP1	Secreted phosphoprotein 1 (osteopontin)	4q21-q25
TNFSF11	Tumor necrosis factor ligand superfamily, member 11 (RANKL)	13q14
TNFRSF11A	Tumor necrosis factor ligand superfamily, member 11a, NFκB activator (RANKL)	18q22.1
TNFRSF11B	Tumor necrosis factor ligand superfamily, member 11b (OPG)	8q24
VDR	Vitamin D receptor	12q13.11
WNT10B	Wingless-type MMTv integration site family, member 10B	12q13
ZBTB40	Zinc finger and BTB domain-containing protein 40	1p36

Table 48.2. Quantitative trait loci (QTL) associated with BMD identified from linkage studies.

Study	Phenotype	Locus/Marker	LOD Score
Deng et al., 2002 [51]	Spinal BMD	4q32, 7p22 (IL6, TWIST), 12q24 (IGF1, TBX3, TBX5)	2.2–2.3
	Wrist BMD	4q32	2.53
Karasik et al., 2002 [52]	FN BMD	6p21, 21q22	2.9–2.4
	LS BMD	12q24	2.1
	Trochanteric BMD	21q22, 21qter	2.3–3.1
	Ward BMD	8q24	2.1
Niu et al., 1999 [53]	Forearm BMD	2p21 (CALM2), 2p23 (STK, POMC)	2.15
Devoto et al., 1998 [54]	Hip BMD	1p36	3.51
	Spinal BMD	2p23-24	2.07
Willaert et al., 2008 [55]	Spinal BMD	1p36	3.1
Kaufman et al., 2008 [56]	LS BMD	11q12 (LRP5), 17q21 (COL1A1, SOST), 22q11	2.6–3.6
	FN BMD	13q12 (RANK)	2.7

FN = femoral neck; LS = lumbar spine.

suggested as being at a genome-wide significant threshold. These studies also suggested some loci associated with fracture risk (1p36, 2p16, OPG, MHC, LRP4, LRP5, TNFRSF11A).

Among more than 90 genetic variants suggested from GWAS, results from two meta-analyses of GWAS showed that variants in 1p36, ESR1, LRP4, LRP5, TNFSF11, SOST, and TNFRSF11A were associated with BMD [21, 22], and variants in LRP5, SOST, and TNFRSF11A were associated with fracture risk [22] (Table 48.3).

Whole genome sequencing

Despite novel findings of genetic variants associated with BMD from GWAS, only a small proportion of the heritability was explained by the discovered variants, giving rise to the idea of "missing heritability." As expected with other multifactorial diseases, GWAS can only identify common variants, each with small effect size, and that the missing heritability is partly caused by the rare variants that are poorly captured in almost GWAS by design. Recent technological advances in whole genome sequencing (WGS) and whole exome sequencing (WES) has enabled a more thorough scan of exon regions encoding proteins involved in bone physiology.

A study that combined data from the UK10K and 1000 Genomes projects has identified a novel candidate gene, EN1, encoding homeobox protein engrailed-1, which is related to BMD and fracture risk [23]. Genetic variant in the noncoding region of this gene (rs11692564) has a minor allele frequency of 1.6%, but the allele was associated with 0.2 SD higher lumbar spine BMD, and a 15% lower odds of fracture [23]. Another successful application of WGS is the discovery of LGR4, a member of the G-protein coupled receptor (GPCR) superfamily, which has various roles in brain and bone development. A rare mutation in the noncoding region of this gene was found to be associated with BMD and fracture risk [24]. Another study using WGS technology found two rare mutations in the coding regions of COL1A2 (p.Gly496Ala and p.Gly703Ser) associated with between 0.5 and 0.9 SD lower BMD and increased fracture risk [25].

Pathways

Genetic variants identified by GWAS and WGS mostly relate to the three known pathways that are involved in bone formation and bone remodeling: RANK-RANKL-OPG (TNFRSF11B, TNFRSF11A, TNFSF11), Wnt-β-catenin (LRP5, LRP4, SOST), and endochondral ossification.

1. The RANK-RANKL-OPG pathway is encoded by TNFRSF11A, TNFSF11, and TNFRS11B, respectively. RANK and osteoprotegerin (OPG) are members of the tumor necrosis receptor family, and RANKL is a member of the TNF family. Binding of RANKL to RANK stimulates the formation and differentiation of osteoclasts regulating bone resorption. An inhibitor of this pathway, OPG, blocks the binding of RANKL to RANK, thereby preventing the bone resorption. Genetic variants within TNFRSF11A, TNFRSF11, and TNFRS11B have been found to be associated with BMD and fracture risk in both candidate gene studies and GWAS [20, 26].

2. The Wnt-β-catenin signaling pathway contributes to the process of bone formation by regulating the differentiation and proliferation of osteoblasts, and bone mineralization, which is crucial for bone maintenance and fracture healing. The pathway is activated when there is a binding between membrane-spanning frizzled receptor proteins and LRP5/LRP6, which transfers signal into the nucleus thereby controlling gene expression. LRP5, which encodes one of the important elements of this pathway, has been identified as a candidate gene for BMD and fracture risk [27]. An inhibitor of this pathway, SOST, preventing the binding of Wnt to LRP5, is also a candidate gene for osteoporosis research [26].

Table 48.3. Genetic variants associated with fracture risk identified from genome-wide association studies and meta-analyses.*

SNP	Position	Gene	Allele	Allele Frequency	Odds Ratio and 95% CI	P-value
rs7524102	1p36		A	0.83	1.12 (1.05–2.30)	8.4×10^{-4}
rs6696981	1p36		G	0.87	1.15 (1.07–1.25)	2.4×10^{-4}
rs3130340	6p21	MHC	T	0.80	1.09 (1.02–1.16)	0.008
rs9479055	6q25	ESR1	C	0.36	1.05 (1.00–1.11)	0.06
rs4870044	6q25	ESR1	T	0.28	1.02 (0.97–1.09)	0.14
rs1038304	6q25	ESR1	G	0.47	1.04 (0.99–1.10)	0.11
rs6929137	6q25	ESR1	A	0.30	1.05 (0.99–1.10)	0.12
rs1999805	6q25	ESR1	C	0.44	1.03 (0.97–1.08)	0.35
rs6993813	8q24	OPG	C	0.51	1.06 (1.00–1.11)	0.04
rs6469804	8q24	OPG	A	0.52	1.05 (1.00–1.11)	0.052
rs9594738	13q14	RANKL	T	0.57	1.04 (0.98–1.11)	0.23
rs9594759	13q14	RANKL	T	0.63	1.02 (0.97–1.07)	0.52
rs11898505	2p16		G	0.69	1.11 (1.05–1.17)	1.8×10^{-4}
rs3018362	18p21	RANK	A	0.37	1.08 (1.02–1.14)	0.005
rs2306033	11p11	LRP4	G	0.87	1.11 (1.03–1.19)	0.007
rs7935346	11p11	LRP4	G	0.78	1.08 (1.01–1.14)	0.02
rs4233949	2p16.2	SPTBN1	G	0.63	1.06 (1.04–1.08)	2.6×10^{-8}
rs6532023	4q22.1	MEPE/SPP1	G	0.67	1.06 (1.04–1.09)	1.7×10^{-8}
rs4727338	7q21.3	SLC25A13	G	0.32	1.08 (1.05–1.10)	5.9×10^{-11}
rs1373004	1q21.1	MBL2/DKK1	T	0.13	1.10 (1.06–1.13)	9.0×10^{-8}
rs3736228	11q13.2	LRP5	T	0.15	1.09 (1.06–1.13)	1.4×10^{-8}
rs4796995	18p11.21	FAMB210A	G	0.39	1.08 (1.06–1.10)	8.8×10^{-13}
rs6426749	1p36.12	ZBTB40	G	0.83	1.07 (1.04–1.10)	3.6×10^{-6}
rs7521902	1p36.12	WNT4	A	0.27	1.09 (1.06–1.13)	1.4×10^{-7}
rs430727	3p22.1	CTNNB1	T	0.47	1.06 (1.03–1.08)	2.9×10^{-7}
rs6959212	7p14.1	STARD3NL	T	0.33	1.05 (1.02–1.07)	7.2×10^{-5}
rs3801387	7q31.31	WNT16	A	0.74	1.06 (1.04–1.08)	2.7×10^{-7}
rs7851693	9q34.11	FUBP3	G	0.37	1.05 (1.02–1.07)	3.5×10^{-5}
rs163879	11p14.1	DCDC5	T	0.36	1.05 (1.03–1.07)	3.3×10^{-5}
rs1286083	14q32.12	RPS6KA5	T	0.81	1.05 (1.03–1.08)	7.2×10^{-5}
rs4792909	17q21.31	SOST	G	0.62	1.07 (1.04–1.10)	6.9×10^{-6}
rs227584	17q21.31	C17orf53	A	0.67	1.05 (1.03–1.07)	4.1×10^{-5}

*This list was compiled from a previous GWAS [20] and a recent meta-analysis of GWAS [57].

3. *Endochondral ossification* is a major process of bone formation, where osteoblasts deposit collagen and noncollagenous proteins on a cartilaginous template, which are then mineralized [28]. GWAS have identified several candidate genes in this pathway, including genes involved in the establishment of the cartilage growth plate (*PTHLH*, *SOX6*) allowing for the development of endochondral bone, cartilage matrix (*SOX9*), and ossification through the deposit of mineral (*RUNX2*) and osteoblast differentiation (*SP7*) [28].

The identification of genetic variants for either BMD or fracture risk has many applications. The genes that are associated with BMD and bone fracture together with their encoded proteins can be potential targets in the development of new treatments for bone disease. In the clinical setting, genetic variants can lead to new diagnostics for the early detection of high-risk individuals. Because genes likely interact with intermediate risk factors to elevate an individual's risk of fracture, genetic variants can be used to form the basis for the promotion of effective prevention. Another potential use of genetic variants is to predict the response to therapies.

CLINICAL APPLICATION 1: INDIVIDUALIZED RISK ASSESSMENT

BMD has been used as a primary tool for fracture risk assessment and decisions concerning therapy. However, BMD alone cannot reliably predict an individual who is (or is not) going to sustain a fracture. It has been estimated that less than 40% of fracture cases occur in osteoporotic patients [29]. On the other hand, among those who sustained a fracture, almost 60% had BMD above the osteoporotic cut-off point (T-score less than –2.5). In other words, more than half of individuals with low BMD were "resistant to fracture." The situation in elderly men is similar: 70% of men with low BMD did not sustain a fracture, and among fracture cases, 77% occurred in those with nonosteoporotic BMD levels. This finding suggests that non-BMD factors may improve fracture risk assessment.

There are a number of predictive models of fracture based on an individual risk profile [30–32]. However, these models have low sensitivity and high specificity and their predictive performance is often modest with the area under the receiver-operating characteristic curve (AUC) ranging between 0.70 and 0.80 [31, 32]. None of the existing models use genetic markers. Although the Fracture Risk Assessment Tool (FRAX) model uses family history of fracture as a risk factor in the risk assessment, no genetic variants have been incorporated into the model. Thus, there is room for genetic research of osteoporosis to improve the prognostic accuracy of these models using genetic variants.

Using genetic factors for predicting fracture risk has some advantages. First, because of the time-invariant characteristics, it is easier to estimate its effect size and to incorporate its information into a model. Second, the independent association between genetic variants and fracture risk could potentially improve the predictive value. Third, although there is uncertain about the genetic therapy for those at high risk, genetic testing predicting fracture risk could stratify those with high risk from low risk and help allocate effective and efficient interventions.

However, given the modest effect size of genetic variants, the utility of any single variant in prediction is low, and a profiling of multiple variants may be helpful. In a semi-simulated study [33], a profiling of up to 50 genetic variants was shown to improve the fracture prediction by 11% of AUC. In a recent study [34], a genetic profiling estimated from 62 common BMD-associated variants was significantly associated with fracture risk, and by incorporating the profile into the existing Garvan Fracture Risk Calculator, the reclassification of fracture versus nonfracture was significantly improved. In the Osteoporotic Fractures in Men (MrOS) study cohort, a genetic profiling of 63 SNPs was also associated with lower BMD and greater risk of total fracture [35]. Two studies in postmenopausal women of Korean background found that a genetic profiling of 39 SNPs could increase the precision of nonvertebral fracture prediction and help to define the risk threshold [36], while 35 risk alleles were significantly associated with the risk of vertebral fracture [37] in patients on bisphosphonate. These latest findings suggest that genetic profiling is useful in the identification of high-risk individuals, and that genetic profiling could help realize the personalized fracture risk assessment paradigm.

CLINICAL APPLICATION 2: PHARMACOGENETICS

Response to osteoporosis treatments is highly variable. The standard deviation of BMD change induced by antiresorptive drugs is up to twice that of the mean rate of change. As a result, while the majority of patients benefits from the treatment, up to 10% of patients apparently lose bone [38]. Along with patient characteristics such as age, gender, ethnicity, and concomitant disease, there is evidence that genetic factors are associated with variation in drug response.

There are more than 20 studies reporting the association between genetic variants and response to antiresorptive drugs. These studies mostly focus on candidate genes with plausible functions such as *ER*, *VDR*, *COL1A1*, and *LRP5*, and the common outcome is BMD change. Some key findings of these pharmacogenetic studies are highlighted as follows.

1. *Response to hormone-replacement therapy (HRT).* HRT has been used in the treatment of osteoporosis to help increase BMD and thus reduce fracture risk. However, 8% of women on this therapy did not respond [39]. Some studies, but not all, that investigated the differential effects of HRT on BMD change suggested that genetic variants in *ER* and *VDR* are associated with variation in response to treatment [40].

2. *Response to selective estrogen receptor modulators.* Among patients on raloxifene, the *B* allele of the *VDR* gene was associated with a greater increase in BMD than the *b* allele [41]. Among those on combined alendronate and raloxifene, there was no significant association between genetic variants in *VDR* and BMD change. These results could be due to the interaction between *VDR* and various antiresorptive drug therapies regulating BMD change.

3. *Response to bisphosphonates.* Oral bisphosphonates are considered first-line treatment for osteoporosis. Several randomized controlled trials (RCTs) have consistently shown that bisphosphonates can help increase BMD, prevent bone loss, and reduce fracture risk [42, 43], including hip fracture risk [44]. However, response to this therapy varies among patients. Findings from Palomba and colleagues [45] suggested that among postmenopausal women on alendronate, the *b* allele of the *VDR Bsm-I* had greater increase in BMD than the *B* allele. Genetic variants in *COL1A1* were found to be associated with response to bisphosphonate therapy. One RCT of 108 perimenopausal women with osteopenia on cyclical etidronate showed that the SS genotype of *Sp1* was associated with increased femoral neck BMD, while the s allele was associated with decreased BMD [46].

4. *Osteonecrosis of the jaw (ONJ).* Bisphosphonates not only have been prescribed for increased BMD and reduced fracture risk, they are also used in the management of patients with advanced cancers that have metastasized to the bone. In this latter case, very high doses of bisphosphonates are given intravenously, and this could result in the development of ONJ. A systematic review of published articles between 2004 and 2006 reported there have been 368 cases of ONJ associated with high doses of bisphosphonates; of those ~95% occurred in patients with myeloma or breast cancer [47]. Findings from a GWAS suggested that four genetic variants (rs1934951,

rs1934980, rs1341162, rs17110453) in *CYP2C8* were associated with the risk of ONJ and the relative risk ranged from 10 to 13 [48]. Another GWAS showed that genetic variants in *RBMS3* were significantly associated with a sixfold increase in ONJ risk [49].

Finding genes associated with drug response is a challenging task. Most studies to date have been largely based on candidate genes, which could be prone to poor replication and lack of statistical power. Given the hypothesis that the variability in drug response is determined by multiple genes, each with modest effect size, and given that the estimated number of common variants is ~10 million [34], the probability of a randomly selected common variant being associated with drug response is very low, probably in the order of 1/100,000 or 0.000001. Even if there is a priori biological justification, this probability may be around 0.001; because the likelihood of this is generally low, the probability of a true association between a genetic variant and drug response is also low. Using the Bayesian approach and a prior probability of association set at 0.001 and 0.000001 (corresponding to that expected for a candidate gene and for a random SNP in a GWAS, respectively), most previous associations between genetic variants and BMD change have a probability of a false positive rate of more than 0.20, which means that most of these associations are probably not "true." However, a genetic profiling may be a better alternative to define the association between genetic variation and response to antiosteoporotic therapies.

CONCLUSION

There is strong evidence for the contribution of genetic factors to the variation of BMD and fracture risk. There is also evidence that the variation in response to osteoporosis therapies is attributable to genetic factors. However, all the variants identified from GWAS explain less than 10% of the total heritability of BMD [50]. It is likely that both bone phenotypes and drug responses are modified by multiple genes (and possibly interactions between genes), but it is not clear which genes are involved and what the mode of inheritance is. With current methodology, it is unlikely that we will completely understand the causes of fracture, and why some people sustain fracture and others do not. Nevertheless, knowledge of the underlying mechanisms generated from genetic research of osteoporosis will definitely inform the development of the next therapeutic generation. One approach that incorporates genetic information and clinical risk factors to improve the predictive performance could become a tool assisting in making medical decisions.

This chapter has reviewed evidence for the contribution of genetic factors to BMD, fracture risk, and response to therapies. It is now possible to find modifiable risk factors accounting for a substantial number of cases, and thus to reduce the risk. Newly identified genetic variants in combination with clinical risk factors can help improve the accuracy of individualized prognosis of fracture, and segregate individuals at high risk from those with lower risk to reduce the burden of osteoporosis in the general community. Recent advances in genetic technologies such as whole genome sequencing, together with implementing high performance data mining platforms and rapidly reduced costs for genotyping, will help in the discovery of more rare variants which have not been captured previously. This will not only increase the chance of detecting true associations, but will also decrease the chance of false positive findings. These approaches will potentially have impacts on the future direction of drug discovery and development, and the possibility of personalized regimens for fracture risk prevention.

REFERENCES

1. Melton LJ 3rd. The prevalence of osteoporosis. J Bone Miner Res. 1997;12(11):1769–71.
2. Nguyen ND, Ahlborg HG, Center JR, et al. Residual lifetime risk of fractures in women and men. J Bone Miner Res. 2007;22(6):781–8.
3. Center JR, Bliuc D, Nguyen TV, et al. Risk of subsequent fracture after low-trauma fracture in men and women. JAMA. 2007;297(4):387–94.
4. Center JR, Nguyen TV, Schneider D, et al. Mortality after all major types of osteoporotic fracture in men and women: an observational study. Lancet. 1999;353(9156): 878–82.
5. Randell A, Sambrook PN, Nguyen TV, et al. Direct clinical and welfare costs of osteoporotic fractures in elderly men and women. Osteoporos Int. 1995;5(6):427–32.
6. Hindmarsh DM, Hayen A, Finch CF, et al. Relative survival after hospitalisation for hip fracture in older people in New South Wales, Australia. Osteoporos Int. 2009;20(2):221–9.
7. Pocock NA, Eisman JA, Hopper JL, et al. Genetic determinants of bone mass in adults. A twin study. J Clin Invest. 1987;80(3):706–10.
8. Young D, Hopper JL, Nowson CA, et al. Determinants of bone mass in 10- to 26-year-old females: a twin study. J Bone Miner Res. 1995;10(4):558–67.
9. Andrew T, Antioniades L, Scurrah KJ, et al. Risk of wrist fracture in women is heritable and is influenced by genes that are largely independent of those influencing BMD. J Bone Miner Res. 2005;20(1):67–74.
10. Deng HW, Chen WM, Recker S, et al. Genetic determination of Colles' fracture and differential bone mass in women with and without Colles' fracture. J Bone Miner Res. 2000;15(7):1243–52.
11. Kannus P, Palvanen M, Kaprio J, et al. Genetic factors and osteoporotic fractures in elderly people: prospective 25 year follow up of a nationwide cohort of elderly Finnish twins. BMJ. 1999;319(7221):1334–7.
12. Michaelsson K, Melhus H, Ferm H, et al. Genetic liability to fractures in the elderly. Arch Intern Med. 2005; 165(16):1825–30.

13. Howard GM, Nguyen TV, Harris M, et al. Genetic and environmental contributions to the association between quantitative ultrasound and bone mineral density measurements: a twin study. J Bone Miner Res. 1998;13(8):1318–27.
14. Wagner H, Melhus H, Pedersen NL, et al. Genetic influence on bone phenotypes and body composition: a Swedish twin study. J Bone Miner Metab. 2013;31(6):681–9.
15. Pope FM, Nicholls AC, McPheat J, et al. Collagen genes and proteins in osteogenesis imperfecta. J Med Genet. 1985;22(6):466–78.
16. Gong Y, Slee RB, Fukai N, et al. LDL receptor-related protein 5 (LRP5) affects bone accrual and eye development. Cell. 2001;107(4):513–23.
17. Pearson TA, Manolio TA. How to interpret a genome-wide association study. JAMA. 2008;299(11):1335–44.
18. Kiel DP, Demissie S, Dupuis J, et al. Genome-wide association with bone mass and geometry in the Framingham Heart Study. BMC Med Genet. 2007;8(suppl 1):S14.
19. Richards JB, Rivadeneira F, Inouye M, et al. Bone mineral density, osteoporosis, and osteoporotic fractures: a genome-wide association study. Lancet. 2008;371:1505–12.
20. Styrkarsdottir U, Halldorsson BV, Gretarsdottir S, et al. Multiple genetic loci for bone mineral density and fractures. N Engl J Med. 2008;358(22):2355–65.
21. Rivadeneira F, Styrkarsdottir U, Estrada K, et al. Twenty bone-mineral-density loci identified by large-scale meta-analysis of genome-wide association studies. Nat Genet. 2009;41(11):1199–206.
22. Richards JB, Kavvoura FK, Rivadeneira F, et al. Collaborative meta-analysis: associations of 150 candidate genes with osteoporosis and osteoporotic fracture. Ann Intern Med. 2009;151(8):528–37.
23. Zheng HF, Forgetta V, Hsu YH, et al. Whole-genome sequencing identifies EN1 as a determinant of bone density and fracture. Nature. 2015;526(7571):112–17.
24. Styrkarsdottir U, Thorleifsson G, Sulem P, et al. Nonsense mutation in the LGR4 gene is associated with several human diseases and other traits. Nature. 2013;497(7450):517–20.
25. Styrkarsdottir U, Thorleifsson G, Eiriksdottir B, et al. Two rare mutations in the COL1A2 gene associate with low bone mineral density and fractures in Iceland. J Bone Miner Res. 2016;31(1):173–9.
26. Styrkarsdottir U, Halldorsson BV, Gretarsdottir S, et al. New sequence variants associated with bone mineral density. Nat Genet. 2009;41(1):15–17.
27. Richards JB, Rivadeneira F, Inouye M, et al. Bone mineral density, osteoporosis, and osteoporotic fractures: a genome-wide association study. Lancet. 2008;371(9623):1505–12.
28. Clark GR, Duncan EL. The genetics of osteoporosis. Br Med Bull. 2015;113(1):73–81.
29. Nguyen ND, Eisman JA, Center JR, et al. Risk factors for fracture in nonosteoporotic men and women. J Clin Endocrinol Metab. 2007;92(3):955–62.
30. Kanis JA. Assessment of fracture risk. In: Rosen CJ (ed.) *Primer on the Metabolic Bone Diseases and Disorders of Mineral Metabolism* (7th edn). Washington: American Society for Bone and Mineral Research, 2008, pp. 170–3.
31. Nguyen ND, Frost SA, Center JR, et al. Development of a nomogram for individualizing hip fracture risk in men and women. Osteoporos Int. 2007;18(8):1109–17.
32. Nguyen ND, Frost SA, Center JR, et al. Development of prognostic nomograms for individualizing 5-year and 10-year fracture risks. Osteoporos Int. 2008;19(10):1431–44.
33. Tran BN, Nguyen ND, Nguyen VX, et al. Genetic profiling and individualized prognosis of fracture. J Bone Miner Res. 2011;26(2):414–19.
34. Ho-Le TP, Center JR, Eisman JA, et al. Prediction of bone mineral density and fragility fracture by genetic profiling. J Bone Miner Res. 2016;32:285–93.
35. Eriksson J, Evans DS, Nielson CM, et al. Limited clinical utility of a genetic risk score for the prediction of fracture risk in elderly subjects. J Bone Miner Res. 2015;30(1):184–94.
36. Lee SH, Lee SW, Ahn SH, et al. Multiple gene polymorphisms can improve prediction of nonvertebral fracture in postmenopausal women. J Bone Miner Res. 2013;28(10):2156–64.
37. Lee SH, Cho EH, Ahn SH, et al. Prediction of future osteoporotic fracture occurrence by genetic profiling: a 6-year follow-up observational study. J Clin Endocrinol Metab. 2016;101(3):1215–24.
38. Francis RM. Non-response to osteoporosis treatment. J Br Menopause Soc. 2004;10(2):76–80.
39. Rosen CJ, Kessenich CR. The pathophysiology and treatment of postmenopausal osteoporosis. An evidence-based approach to estrogen replacement therapy. Endocrinol Metab Clin North Am. 1997;26(2):295–311.
40. van Meurs JB, Schuit SC, Weel AE, et al. Association of 5′ estrogen receptor alpha gene polymorphisms with bone mineral density, vertebral bone area and fracture risk. Hum Mol Genet. 2003;12(14):1745–54.
41. Palomba S, Numis FG, Mossetti G, et al. Raloxifene administration in post-menopausal women with osteoporosis: effect of different BsmI vitamin D receptor genotypes. Hum Reprod. 2003;18(1):192–8.
42. Cranney A, Tugwell P, Adachi J, et al. Meta-analyses of therapies for postmenopausal osteoporosis. III. Meta-analysis of risedronate for the treatment of postmenopausal osteoporosis. Endocr Rev. 2002;23(4):517–23.
43. Cranney A, Wells G, Willan A, et al. Meta-analyses of therapies for postmenopausal osteoporosis. II. Meta-analysis of alendronate for the treatment of postmenopausal women. Endocr Rev. 2002;23(4):508–16.
44. Nguyen ND, Eisman JA, Nguyen TV. Anti-hip fracture efficacy of bisphosphonates: a Bayesian analysis of clinical trials. J Bone Miner Res. 2006;21(1):340–9.
45. Palomba S, Orio F Jr, Russo T, et al. BsmI vitamin D receptor genotypes influence the efficacy of antiresorptive treatments in postmenopausal osteoporotic women. A 1-year multicenter, randomized and controlled trial. Osteoporos Int. 2005;16(8):943–52.
46. Qureshi AM, Herd RJ, Blake GM, et al. COLIA1 Sp1 polymorphism predicts response of femoral neck bone density to cyclical etidronate therapy. Calcif Tissue Int. 2002;70(3):158–63.

47. Woo SB, Hellstein JW, Kalmar JR. Narrative [corrected] review: bisphosphonates and osteonecrosis of the jaws. Ann Intern Med. 2006;144(10):753–61.
48. Sarasquete ME, Garcia-Sanz R, Marin L, et al. Bisphosphonate-related osteonecrosis of the jaw is associated with polymorphisms of the cytochrome P450 CYP2C8 in multiple myeloma: a genome-wide single nucleotide polymorphism analysis. Blood. 2008;112:2709–12.
49. Nicoletti P, Cartsos VM, Palaska PK, et al. Genomewide pharmacogenetics of bisphosphonate-induced osteonecrosis of the jaw: the role of RBMS3. Oncologist. 2012;17(2):279–87.
50. Estrada K, Styrkarsdottir U, Evangelou E, et al. Genome-wide meta-analysis identifies 56 bone mineral density loci and reveals 14 loci associated with risk of fracture. Nat Genet. 2012;44(5):491–501.
51. Deng HW, Xu FH, Huang QY, et al. A whole-genome linkage scan suggests several genomic regions potentially containing quantitative trait loci for osteoporosis. J Clin Endocrinol Metab. 2002;87(11):5151–9.
52. Karasik D, Myers RH, Cupples LA, et al. Genome screen for quantitative trait loci contributing to normal variation in bone mineral density: the Framingham Study. J Bone Miner Res. 2002;17(9):1718–27.
53. Niu T, Chen C, Cordell H, et al. A genome-wide scan for loci linked to forearm bone mineral density. Hum Genet. 1999;104(3):226–33.
54. Devoto M, Shimoya K, Caminis J, et al. First-stage autosomal genome screen in extended pedigrees suggests genes predisposing to low bone mineral density on chromosomes 1p, 2p and 4q. Eur J Hum Genet. 1998;6(2):151–7.
55. Willaert A, Van Pottelbergh I, Zmierczak H, et al. A genome-wide linkage scan for low spinal bone mineral density in a single extended family confirms linkage to 1p36.3. Eur J Hum Genet. 2008;16(8):970–6.
56. Kaufman JM, Ostertag A, Saint-Pierre A, et al. Genome-wide linkage screen of bone mineral density (BMD) in European pedigrees ascertained through a male relative with low BMD values: evidence for quantitative trait loci on 17q21-23, 11q12-13, 13q12-14, and 22q11. J Clin Endocrinol Metab. 2008;93(10):3755–62.
57. Estrada K, Styrkarsdottir U, Evangelou E, et al. Genome-wide meta-analysis identifies 56 bone mineral density loci and reveals 14 loci associated with risk of fracture. Nat Genet. 2012;44(5):491–501.

Section VI
Osteoporosis

Section Editors: Paul D. Miller, Socrates E. Papapoulos, and Michael R. McClung

Chapter 49. Osteoporosis: An Overview   395
Michael R. McClung, Paul D. Miller, and Socrates E. Papapoulos

Chapter 50. The Epidemiology of Osteoporotic Fractures   398
Nicholas C. Harvey, Elizabeth M. Curtis, Elaine M. Dennison, and Cyrus Cooper

Chapter 51. Fracture Liaison Service   405
Piet Geusens, John A. Eisman, Andrea Singer, and Joop van den Bergh

Chapter 52. Sex Steroids and the Pathogenesis of Osteoporosis   412
Matthew T. Drake and Sundeep Khosla

Chapter 53. Juvenile Osteoporosis   419
Francis H. Glorieux and Craig Munns

Chapter 54. Transplantation Osteoporosis   424
Peter R. Ebeling

Chapter 55. Premenopausal Osteoporosis   436
Adi Cohen and Elizabeth Shane

Chapter 56. Osteoporosis in Men   443
Eric S. Orwoll and Robert A. Adler

Chapter 57. Bone Stress Injuries   450
Stuart J. Warden and David B. Burr

Chapter 58. Inflammation-Induced Bone Loss in the Rheumatic Diseases   459
Ellen M. Gravallese and Steven R. Goldring

Chapter 59. Glucocorticoid-Induced Osteoporosis   467
Kenneth Saag and Robert A. Adler

Chapter 60. Human Immunodeficiency Virus and Bone   474
Michael T. Yin and Todd T. Brown

Chapter 61. Effects on the Skeleton from Medications Used to Treat Nonskeletal Disorders   482
Nelson B. Watts

Chapter 62. Diabetes and Fracture Risk   487
Serge Ferrari, Nicola Napoli, and Ann Schwartz

Chapter 63. Obesity and Skeletal Health   492
Juliet Compston

Chapter 64. Sarcopenia and Osteoporosis   498
Gustavo Duque and Neil Binkley

Chapter 65. Management of Osteoporosis in Patients with Chronic Kidney Disease   505
Paul D. Miller

Chapter 66. Other Secondary Causes of Osteoporosis   510
Naveen A.T. Hamdy and Natasha M. Appelman-Dijkstra

Chapter 67. Exercise for Osteoporotic Fracture Prevention and Management   517
Robin M. Daly and Lora Giangregorio

Chapter 68. Prevention of Falls   526
Heike A. Bischoff-Ferrari

Chapter 69. Nutritional Support for Osteoporosis   534
Connie M. Weaver, Bess Dawson-Hughes, Rene Rizzoli, and Robert P. Heaney

(Continued)

Primer on the Metabolic Bone Diseases and Disorders of Mineral Metabolism, Ninth Edition. Edited by John P. Bilezikian.
© 2019 American Society for Bone and Mineral Research. Published 2019 by John Wiley & Sons, Inc.
Companion website: www.wiley.com/go/asbmrprimer

Chapter 70. Estrogens, Selective Estrogen Receptor Modulators, and Tissue Selective Estrogen Complex   541
Tobias J. de Villiers

Chapter 71. Bisphosphonates for Postmenopausal Osteoporosis   545
Andrea Giusti and Socrates E. Papapoulos

Chapter 72. Denosumab   553
Aline Granja Costa, E. Michael Lewiecki, and John P. Bilezikian

Chapter 73. Parathyroid Hormone and Abaloparatide Treatment for Osteoporosis   559
Felicia Cosman and Susan L. Greenspan

Chapter 74. Combination Anabolic and Antiresorptive Therapy for Osteoporosis   567
Joy N. Tsai and Benjamin Z. Leder

Chapter 75. Strontium Ranelate and Calcitonin   573
Leonardo Bandeira and E. Michael Lewiecki

Chapter 76. Adverse Effects of Drugs for Osteoporosis   579
Bo Abrahamsen and Daniel Prieto-Alhambra

Chapter 77. Orthopedic Principles of Fracture Management   588
Manoj Ramachandran and David G. Little

Chapter 78. Adherence to Osteoporosis Therapies   593
Stuart L. Silverman and Deborah T. Gold

Chapter 79. Cost-Effectiveness of Osteoporosis Treatment   597
Anna N.A. Tosteson

Chapter 80. Future Therapies   603
Michael R. McClung

49

Osteoporosis: An Overview

Michael R. McClung[1], Paul D. Miller[2], and Socrates E. Papapoulos[3]

[1] Institute of Health and Ageing, Australian Catholic University, Melbourne, VIC,
Australia and Oregon Osteoporosis Center, Portland, OR, USA
[2] Colorado Center for Bone Research at Panorama Orthopedics and Spine Center, Golden, CO, USA
[3] Center for Bone Quality, Department of Endocrinology and Metabolic Diseases,
Leiden University Medical Center, Leiden, The Netherlands

This section of the *Primer* includes updated and current knowledge about the clinical aspects and management of osteoporosis in postmenopausal women, men, young women, and various forms of secondary osteoporosis. Comparing and contrasting this information with that contained in the first edition of the *Primer* in 1991 tells the story of how much we have learned about this disorder and its treatment over the past 25 years. Dual energy X-ray absorptiometry (DXA) had been introduced in 1987, providing an important tool for clinical research, including clinical trials of bisphosphonates and other drugs. With DXA, the epidemiology of postmenopausal and age-related bone loss has been extensively studied in many populations. The relationship between areal BMD and fracture risk was documented [1]. High resolution imaging techniques have provided glimpses into the microarchitectural changes of both trabecular and cortical skeletal compartments, providing new insights into the structural deficits that occur with osteoporosis. Clinical risk factors for fracture are now well known and have been evaluated in multiple meta-analyses. Of particular importance, previous fractures, especially recent vertebral fractures, have been identified as robust predictors of subsequent fracture [2].

A clinical definition of osteoporosis as a "systemic disease characterized by low bone mass and microarchitectural deterioration of bone tissue, with a consequent increase in bone fragility and susceptibility to fracture" was provided in 1993 [3]. This was followed by an operational diagnosis of osteoporosis in postmenopausal women defined as an areal BMD T-score in the femoral neck of −2.5 or lower [4]. In 2004, a report of the Surgeon General of the USA identified osteoporosis as one of the nation's most important health problems [5].

In 1991, estrogen and injectable salmon calcitonin were available as treatments for postmenopausal osteoporosis. In 1990, etidronate therapy for 2 years had been shown to reduce vertebral fracture risk [6], but this drug was never licensed as an osteoporosis treatment in the USA. Meanwhile, the promise of sodium fluoride as a bone growth stimulating agent was doomed by the documentation of an increased risk of nonvertebral fracture [7]. Since then, beginning with alendronate in 1995, several drugs of different classes with widely different mechanisms of action have been licensed, based on large clinical trials documenting significant and often substantial reductions in the risk of vertebral fracture in women with postmenopausal osteoporosis [8]. Some agents have also reduced the risk of hip fracture by 40% to 50% whereas our ability to reduce the incidence of nonvertebral fracture is more limited with relative reductions in risk of 20% to 35%. The full effects on vertebral fracture risk reduction are evident within 12 months of beginning treatment, whereas hip fracture risk is reduced after 1 to 3 years by treatment with potent antiremodeling agents. With bisphosphonates and denosumab, protection from fractures persists as long as therapy is administered. Drugs are available for oral (daily, weekly, monthly dosing) or parenteral (subcutaneous, intravenous) administration, some with dosing intervals of 6 months or longer, which many patients find to be very convenient and which improve short-term persistence with therapy. Teriparatide, by activating bone remodeling, stimulates the formation of new bone, particularly in the trabecular compartment where the structure and strength are substantially improved. In general, our drugs are quite well tolerated with few serious safety concerns.

Primer on the Metabolic Bone Diseases and Disorders of Mineral Metabolism, Ninth Edition. Edited by John P. Bilezikian.
© 2019 American Society for Bone and Mineral Research. Published 2019 by John Wiley & Sons, Inc.
Companion website: www.wiley.com/go/asbmrprimer

With availability of these treatments for postmenopausal osteoporosis, clinical guidelines appeared in the late 1990s, with recommendations for treatment initially based on a history of fracture or BMD, When it became apparent that combining independent risk factors, especially age, prior fracture, and BMD, improved risk prediction, several risk predicting algorithms were developed, and some of these have been incorporated into clinical guidelines with the concept being that the benefit:risk profile of therapy is enhanced when treatment is directed toward a patient who is at high risk for fracture [8]. There is strong consensus, based on solid clinical trial evidence, that postmenopausal women, and probably older men, who have experienced fragility fractures of the spine and hip are definite candidates for pharmacological therapy, irrespective of other risk factors. Patients with nonhip, nonspine fractures are also at higher risk for fracture and deserve, at least, to be evaluated for other risk factors and as potential candidates for therapy.

Most serious and chronic illnesses have adverse skeletal effects, as do many of the drugs used to treat these conditions. The mechanisms of bone loss and structural decay of many of these diseases have been elucidated whereas the pathogenesis of increased fracture risk with some diseases, such as type 2 diabetes, and some drugs, including proton pump inhibitors, remain unclear.

There is no question that remarkable progress has been made in our field since 1991. We can make the diagnosis of osteoporosis, identify patients at high risk for fracture, prevent bone loss if we need or choose to and prevent, although not eliminate, the risk of serious osteoporosis-related fractures. Yet, despite these advances and the acquisition of so much knowledge and so many tools, the field of osteoporosis has recently been described as being "in crisis" by leading experts [9]. Despite the breadth and depth of this knowledge among osteoporosis specialists, our strategies and recommendations are not being applied in daily practice by our primary care colleagues and are either not known or not believed by patients. Great concern exists about rare side effects, especially with bisphosphonates, squelching any interest in osteoporosis treatment by patients. Trust among the general public of medical science and advice by physicians has markedly diminished. It is widely held among patients that osteoporosis drugs are only minimally effective, if at all, and are more likely to cause harm than benefit. In contrast, there is a strong misperception that nutrition, exercise, and "natural" supplements are both effective and sufficient for treating osteoporosis. The number of prescriptions for bisphosphonates, the most commonly used treatment for osteoporosis, decreased by 50% from 2008 to 2012 despite a flood of baby boomers reaching age 65 [10]. This decrease was not made up for by increased use of other drugs, meaning that the total number of people treated for osteoporosis declined substantially. The proportion of patients receiving bisphosphonates after a hip fracture, the most obvious and least controversial group of patients to treat, decreased from 40% to 21% between 2001 and 2011, despite robust evidence that therapy reduces subsequent fractures and mortality [11,12]. The decline in age-adjusted hip fracture incidence observed in the USA during the 2000s appears to have plateaued.

There is hope on the horizon. Finite element analysis (FEA) of routine CT scans of the hip and spine provides accurate in vivo measurement of skeletal strength. The utility of FEA to predict fracture risk has been documented, but whether changes in FEA estimates of strength in response to therapy correlate with fracture protection is not yet known. The ambitious Foundation for the National Institutes of Health (FNIH) project to compile individual patient data from all of the large Phase 3 osteoporosis fracture trials, including CT scans acquired in those studies, may provide the opportunity to explore the relationship between changes in FEA and fracture risk. An important outcome of this project could be the validation of an effective surrogate for fracture risk in patients on osteoporosis therapy, perhaps streamlining and simplifying clinical trials required for the evaluation and registration of new drugs.

The elucidation of new molecular pathways for the control of bone metabolism have led to the development of denosumab which became available in 2010, and to abaloparatide and romosozumab which have just completed Phase 3 fracture trials [13,14]. These studies are particularly important because they both incorporated sequential therapy regimens of an anabolic agent followed by an antiremodeling drug for the primary study endpoints. The studies showed for the first time that such a sequence results in more effective fracture risk reduction than beginning treatment with an antiremodeling drug.

The anticipated availability of abaloparatide and romosozumab in 2017 will be the first new treatments for osteoporosis since denosumab was approved in 2010. The clinical development of odanacatib, a very promising novel agent, was derailed by an unexpected side effect found in adjudicated analyses of a very large Phase 3 fracture trial. There are no other drugs in late stage clinical development. As a result, we will soon reach a plateau of treatment options on which we will reside for several years.

The chapters in this Section collectively summarize the current state of osteoporosis management, and it is a very good story. However, even with the new drugs and improved assessment tools, the current "crisis" — the disparity between the number of patients who should be treated and who are currently being evaluated and treated — will not be solved. Rather than more tools, we need new initiatives and strategies to re-engage patients, our physician colleagues, health systems, and payors to partner with us to identify and to treat the appropriate patients — those at high risk for fracture. Where the story of osteoporosis goes from here will depend on how successful we are in doing that.

REFERENCES

1. Black DM, Cummings SR, Genant HK, et al. Axial and appendicular bone density predict fractures in older women. J Bone Miner Res. 1992;7:633–8.
2. Kanis JA, Johnell O, De Laet C, et al. A meta-analysis of previous fracture and subsequent fracture risk. Bone. 2004;35:375–82.
3. Anonymous. Consensus development conference: diagnosis, prophylaxis and treatment of osteoporosis. Am J Med 1993;94:646–50.
4. World Health Organization. *Assessment of fracture risk and its application to screening for postmenopausal osteoporosis: technical report series 843.* Geneva: WHO, 1994.
5. US Department of Health and Human Services. *Bone Health and Osteoporosis: A Report of the Surgeon General.* US Department of Health and Human Services, Office of the Surgeon General: Rockville, MD, 2004.
6. Watts NB, Harris ST, Genant HK, et al. Intermittent cyclical etidronate treatment of postmenopausal osteoporosis. N Engl J Med. 1990;323:73–9.
7. Riggs BL, Hodgson SF, O'Falton WM, et al. Effect of fluoride treatment on fracture rate in postmenopausal women with osteoporosis. N Engl J Med. 1990;332:802–9.
8. Black DM, Rosen CJ. Postmenopausal osteoporosis. N Engl J Med. 2016;374:2096–7.
9. Khosla S, Shane E. A crisis in the treatment of osteoporosis. J Bone Miner Res. 2016;31:1485–7.
10. Wysowski DK, Greene P. Trends in osteoporosis treatment with oral and intravenous bisphosphonates in the United States, 2002–2012. Bone. 2013;57:423–8.
11. Solomon DH, Johnston SS, Boytsov NN, et al. Osteoporosis medication use after hip fracture in U.S. patients between 2002 and 2011. J Bone Miner Res. 2014;29:1929–37.
12. Lyles KW, Colón-Emeric CS, Magaziner JS, et al.; HORIZON Recurrent Fracture Trial. Zoledronic acid and clinical fractures and mortality after hip fracture. N Engl J Med. 2007;357:1799–809.
13. Cosman F, Miller PD, Williams GC, et al. Eighteen months of treatment with subcutaneous abaloparatide followed by 6 months of treatment with alendronate in postmenopausal women with osteoporosis: results of the ACTIVExtend Trial. Mayo Clin Proc. 2017;92:200–10.
14. Cosman F, Crittenden DB, Adachi JD, et al. Romosozumab treatment in postmenopausal women with osteoporosis. N Engl J Med. 2016;375:1532–43.

50
The Epidemiology of Osteoporotic Fractures

Nicholas C. Harvey, Elizabeth M. Curtis, Elaine M. Dennison, and Cyrus Cooper

The MRC Lifecourse Epidemiology Unit, University of Southampton, Southampton General Hospital, Southampton, UK

INTRODUCTION

Osteoporosis is a skeletal disease characterized by low bone mass and microarchitectural deterioration of bone tissue with a consequent increase in bone fragility and susceptibility to fracture [1]. The term osteoporosis was first introduced in France and Germany during the last century. It means "porous bone," and initially implied a histological diagnosis, but was later refined to mean bone that was normally mineralized, but reduced in quantity. Clinically, osteoporosis has been difficult to define: a focus on bone mineral density may not encompass all the risk factors for fracture, whereas a fracture-based definition will not enable identification of at risk populations. In 1994 the World Health Organization convened to resolve this issue [2], operationally defining osteoporosis in terms of bone mineral density (BMD), and previous fracture. The development of the FRAX algorithm [3], a web-based tool which uses clinical risk factors plus or minus BMD, has enabled fracture risk assessment strategies based on an individual's absolute risk of major osteoporotic or hip fracture over the next 10 years [4]. This approach has the advantage of incorporating risk factors that are partly independent of BMD, such as age and previous fracture, and thus allows decisions regarding commencement of therapy to be made more readily. Osteoporosis-related fractures have a huge impact economically, in addition to their effect on health: the cost to the US economy is around $17.9 billion per annum, with the burden to the UK being almost £4 billion (Table 50.1 summarizes fracture impact across the European Union) [5].

FRACTURE EPIDEMIOLOGY

The 2004 report from the US Surgeon General highlighted the enormous burden of osteoporosis-related fractures [6]. An estimated 10 million Americans over 50 years old have osteoporosis and there are around 1.5 million fragility fractures each year. Another 34 million Americans are at risk of the disease. A study of British fracture occurrence indicates that population risk is similar in the UK [7], with one in two women aged 50 years expected to have an osteoporosis-related fracture in their remaining lifetime; the figure for men is one in five.

Fracture incidence in the community is bimodal, showing peaks in youth and in the very elderly [8, 9]. In young people, fractures of long bones predominate, usually after substantial trauma, and are more frequent in males than females. In this group the question of bone strength rarely arises, although there are now data suggesting that this may not be entirely irrelevant as a risk factor [10]. Over the age of 50 years, fracture incidence in women begins to climb steeply, so that rates become twice those in men. This peak was historically thought to be mainly caused by hip and distal forearm fracture, but as Fig. 50.1 shows, vertebral fracture, when ascertained from radiographs rather than clinical presentation, can now be shown to make a significant contribution [11].

Hip fracture

Incidence and prevalence

In most populations hip fracture incidence increases exponentially with age (Fig. 50.1). Above 50 years of age, there is

---

Primer on the Metabolic Bone Diseases and Disorders of Mineral Metabolism, Ninth Edition. Edited by John P. Bilezikian.
© 2019 American Society for Bone and Mineral Research. Published 2019 by John Wiley & Sons, Inc.
Companion website: www.wiley.com/go/asbmrprimer

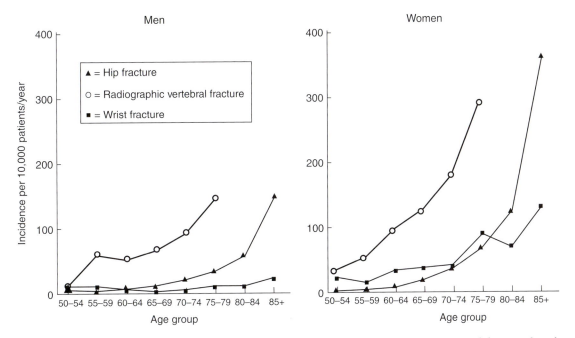

Table 50.1. Impact of osteoporosis-related fractures across Europe.

	Hip	Spine	Wrist
Lifetime risk (%)			
Women	23	29	21
Men	11	14	5
Cases/year	620,000	810,000	574,000
Hospitalization (%)	100	2–10	5
Relative survival	0.83	0.82	1.00

Costs: all sites combined ~ Euro 39 billion.
Source: [5]. Reproduced with permission of Springer.

Fig. 50.1. Radiographic vertebral, hip and wrist fracture incidence by age and gender [7,11]. Source: [7]. Reproduced with permission of Elsevier.

a female to male incidence ratio of around two to one [8]. Overall, around 98% of hip fractures occur among people aged 35 years or over, and 80% occur in women (because there are more elderly women than men). Worldwide there were an estimated 1.66 million hip fractures in 1990 [12]; about 1.19 million in women and 463,000 in men. The majority occur after a fall from standing height or less and 90% occur in people over 50 years old [13]. Hip fractures are seasonal, with an increase in winter in temperate countries, but their occurrence mainly indoors would imply that this increase is not simply caused by slipping on icy pavements: other possible causes may include slowed neuromuscular reflexes and lower light in winter weather. The direction of falling is important, and a fall directly onto the hip (sideways) is more likely to cause a fracture than falling forwards [13]. Factors such as ethnicity, geographic location, and socioeconomic status (Fig. 50.2) have all been shown to influence hip fracture incidence [8].

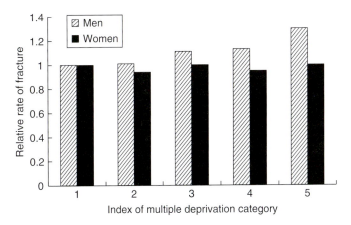

Fig. 50.2. Incidence of hip fracture by fifths of socioeconomic deprivation in the UK (Multiple Deprivation Category 5 = most deprivation). Source: [8]. Reproduced with permission of Elsevier.

Mortality and morbidity

Hip fracture mortality is higher in men than women, increases with age [14], and is greater for those with coexisting illnesses and poor prefracture functional status. There are around 31,000 excess deaths within 6 months of the approximately 300,000 hip fractures that occur annually in the USA. About 8% of men and 3% of women aged over 50 years die whilst hospitalized for their fracture. In the UK, the 12-month survival post-hip fracture for men is 63.3% versus 90.0% expected, and for women is 74.9% versus 91.1% expected [7]. The risk of death is greatest immediately after the fracture and decreases gradually over time. The cause of death is not usually directly attributable to the fracture itself, but to other chronic diseases, which lead both to the fracture and to the reduced life expectancy. Recent data suggest that elevated mortality persists for up 10 years after hip fracture, and reduced survival is associated with all types of fracture except for minor fractures where mortality was increased only for those of 75 years old or older [15].

As with mortality, hip fractures contribute most to osteoporosis-associated disability. Patients are prone to developing acute complications such as pressure sores, bronchopneumonia, and urinary tract infections. Perhaps the most important long-term outcome is impairment of the ability to walk. Fifty percent of those ambulatory before the fracture are unable to walk independently afterwards. Age is an important determinant of outcome, with 14% of 50- to 55-year-old hip fracture victims being discharged to nursing homes, versus 55% of those over 90 years old [16].

Vertebral fracture

Incidence and prevalence

Data from the European Vertebral Osteoporosis Study (EVOS) have shown that the age-standardized population prevalence across Europe was 12.2% for men and 12.0% for women aged 50 to 79 years [17]. Historically it was believed that vertebral fractures were more common in men than women, but the EVOS data suggest that this is not the case at younger ages: the prevalence of deformities in 50 to 60 year olds is similar, if not higher in men, possibly because of a greater incidence of trauma. The majority of vertebral fractures in elderly women occur through normal activities such as lifting, rather than through falling.

Many vertebral fractures are asymptomatic, and there is disagreement about the radiographic definition of deformities in those patients who do present. Thus, in studies using radiographic screening of populations, the incidence of all vertebral deformities has been estimated to be three times that of hip fracture, with only one third of these coming to medical attention. Data from EVOS have allowed accurate assessment of radiographically determined vertebral fractures in a large population. At age 75 to 79 years, the incidence of vertebral fractures so-defined was 13.6 per 1000 person-years for men and 29.3 per 1000 person-years for women [11]. This compares with 0.2 per 1000 person-years for men and 9.8 per 1000 person-years in 75 to 84 year olds where the fractures were defined by clinical presentation in an earlier study from Rochester, Minnesota [18]. The overall age-standardized incidence in EVOS was 10.7 per 1000 person-years in women and 5.7 per 1000 person-years in men.

Mortality and morbidity

Vertebral fractures are associated with increased mortality well beyond a year postfracture [15, 19], with comorbid conditions contributing significantly to the decreased relative survival. The impairment of survival following a vertebral fracture also markedly worsens as time from diagnosis of the fracture increases. This is in contrast to the pattern of survival for hip fractures. In the UK General Practice Research Database (GPRD) study, the observed survival in women 12 months after vertebral fracture was 86.5% versus 93.6% expected. At 5 years, survival was 56.5% observed and 69.9% expected [7]. The major clinical consequences of vertebral fracture are back pain, kyphosis, and height loss. Quality of life (QUALEFFO) scores decrease as the number of vertebral fractures increase [20].

Distal forearm fracture

Incidence and prevalence

Wrist fractures show a different pattern of occurrence to hip and vertebral fractures, with a gradual increase in rates with age [8]. Rates are higher in women than men at older ages, with an incidence of 39.7 per 10,000 person-years and 8.9 per 10,000 person-years respectively in the UK for individuals aged 50 years or older [8].

Mortality and morbidity

Wrist fractures do not appear to increase mortality [7]. Although wrist fractures may impact on some activities such as writing or meal preparation, overall few patients are completely disabled, despite over half reporting only fair to poor function at 6 months [16].

Clustering of fractures in individuals

Epidemiological studies suggest that patients with different types of fragility fractures are at increased risk of developing other types of fracture. A meta-analysis of 15,259 men and 44,902 women across 11 population cohorts showed that a history of prior fracture was associated with an 86% increase in risk of any new fracture, with a similar risk ratio for osteoporotic fracture and hip fracture [21]. Other studies have shown even higher risk ratios for specific fractures, for example data from EVOS showed that prevalent vertebral deformity predicts incident hip fracture with a rate ratio of 2.8 to 4.5, and this increases with the number of vertebral deformities [22]. Although not characterized in detail, there is evidence that risk of subsequent fracture is particularly high in the period immediately after the index fracture [23].

Fracture Epidemiology 401

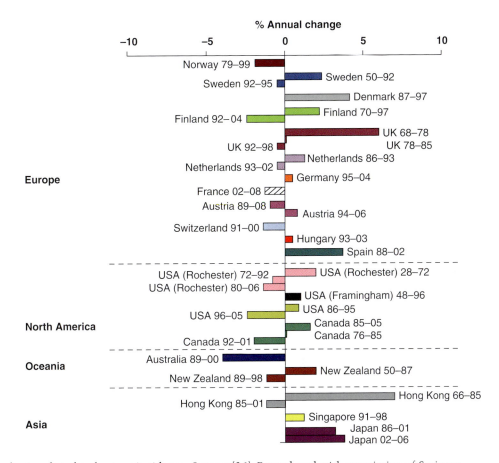

Fig. 50.3. Secular trends in hip fracture incidence. Source: [26]. Reproduced with permission of Springer.

Time trends and future projections

There are currently estimated to be 901 million people aged 60 years or over globally, comprising 12% of the world population. Although Europe has the greatest percentage of its population aged 60 years or older (24%), rapid aging in other parts of the world means that by 2050 all major areas of the world except Africa will have nearly a quarter or more of their populations aged 60 years or older [24]. This growth in world population and increasingly aging demographic will substantially impact the number of hip fractures globally in coming decades, with a conservative estimate of the annual number of hip fractures increasing from 1.66 million in 1990 to 6.26 million in 2050, with the latter figure potentially over 20 million when known secular trends are considered [12, 25]. Alterations in age- and sex-adjusted incidence rates have been documented most robustly for hip fracture (Fig. 50.3) and in many developed countries, such rates appeared to have plateaued or decreased in the last 10 to 20 years, following a rise in preceding years; however, in the developing world, age-and sex-specific rates are still rising in many areas [26].

Recent work in the UK has shown no overall change in fracture rates (1990 to 2012), but substantial variation by individual fracture site [27]. Whilst the burden of osteoporosis can be assessed in terms of consequent fracture, there is merit in identifying the number of individuals at high fracture risk. Using this approach, it has been estimated that in 2010 there were 21 million men and 137 million women aged 50 years or older at high fracture risk, and that this number is expected to double by 2040, with the increase predominantly borne by Asia, Africa, and Latin America [28], as shown in Fig. 50.4.

Geography

There is variation in the incidence of hip fracture within populations of a given ethnicity and gender, documented within [8] and across countries [29]. Within Europe the range of variation is approximately 11-fold [30], with the differences not explained by variation in activity levels, smoking, obesity, alcohol consumption, or migration status. The EVOS study showed a threefold difference in the prevalence of vertebral deformities between countries, with the highest rates in Scandinavia. The differences were not as great as those observed for hip fracture in Europe, and some of the differences could be explained by levels of physical activity and body mass index [17]. The greater incidence of fracture with increasingly northerly latitude may implicate vitamin D status [31, 32], although such associations are often confounded by dietary and behavioural factors, and ethnic differences in bone

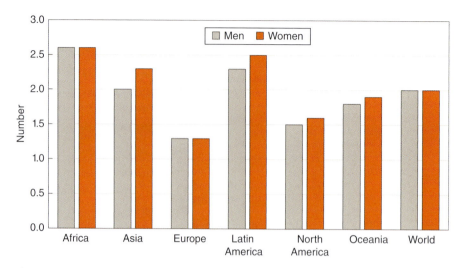

Fig. 50.4. Number of men and women at high fracture risk in 2040 relative to 2010, by world region. Source: [28]. Reproduced with permission of Springer.

mineral density, bone geometry, and microarchitecture may also underlie global variation in fracture incidence [8].

Low bone mass in children

There has been considerably less investigation of the role of bone fragility in childhood fractures, probably because of the perception that the primary determinant of fracture in this age group is trauma. There are also considerable difficulties in reaching a definition of "osteoporosis" in a growing skeleton, when there is no straightforward relationship between BMD and fracture risk. Thus, the consensus view is that the term "low bone mass for age (with or without fractures)" is used rather than "osteoporosis." Most evidence comes from two large European studies which describe the epidemiology of fractures in childhood [9, 33]. In Malmo, Sweden, the overall incidence of fracture was 212 per 10,000 girls and 257 per 10,000 boys, with 27% of girls and 42% of boys sustaining a fracture between birth and 16 years of age. Fractures of the distal radius occurred most commonly, followed by fractures of the phalanges of the hand [33]. A follow-up study in Malmo a decade later found the incidence of fracture had decreased by almost 10% since the original study [34].

A similar pattern was found in the UK Clinical Practice Research Datalink (CPRD) [9]. The overall incidence of fracture was 137 per 10,000 person-years, with fractures being more common in boys than girls with an incidence of 169 per 10,000 and 103 per 10,000 person-years respectively. Based on these data, 30% of UK boys and 19% of girls are expected to sustain a fracture before their 18th birthday. Again, the most common fracture site in both sexes was the radius/ulna with a rate of 29.7 per 10,000 person-years. There were marked differences in fracture rates by ethnicity, with rates in white children being over twice those in black children, with children of South Asian origin at intermediate risk (Fig. 50.5). Fracture incidence also varied by geographic location within the UK, a finding likely related to differences in socioeconomic status and ethnic mix, amongst other factors [9]. The greatest fracture incidence corresponded with the age of entry to puberty (14.5 years in males and 11.5 years in females), a time when there is discordance between velocity of height gain and accrual of volumetric bone density [35, 36]. Childhood obesity has been associated with increased fracture risk [10]; whilst increased physical activity has been associated with increased bone mass in childhood, such activity also increases exposure to trauma [37], and children with fractures tend to have lower BMD than do their nonfracture peers [10].

Early life influences on adult fragility fracture

The importance of bone mineral accrual in childhood and achievement of adequate peak bone mass (PBM) in early adulthood has been emphasized in recent work, showing that PBM is a major determinant osteoporosis risk in later life [38]. Over the last 20 years, evidence has accrued that the early environment may have long-term influences on future bone health. This phenomenon of "developmental plasticity," whereby a single genotype may lead to different phenotypes, dependent upon the prevailing environmental milieu, is well established in the natural world, and potentially mediated by epigenetic mechanisms [39]. There is a growing body of epidemiological evidence that a poor intrauterine environment leads to lower bone mass in adult life, both at peak bone mass and in older age, and also to increased risk of adult hip fracture [39–42]. A key early determinant of skeletal development may be maternal 25(OH)-vitamin D status, and recently a multicentre randomized, placebo-controlled, double-blind trial of vitamin D supplementation during pregnancy has shown potential benefits for offspring bone mass at birth amongst those babies delivered in the

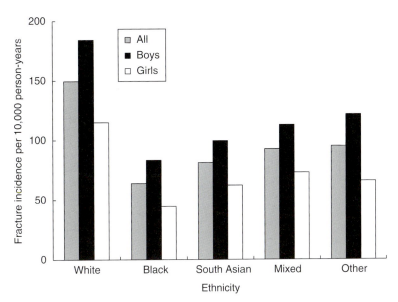

Fig. 50.5. Fracture incidence per 10,000 person-years amongst UK children, 1988 to 2012, by ethnicity. Source: [9]. Reproduced with permission of Elsevier.

winter months [43]. Mechanistic insight has come from studies of perinatal epigenetic markers, with sites within the retinoid X-receptor-A gene (critical to the action of 1,25(OH)2-vitamin D and other nuclear hormones) associated with offspring bone mass and estimated maternal free 25(OH)-vitamin D status [44]. This novel area of research may ultimately lead to innovative strategies aimed at improving bone health in children, with a subsequent reduction in the burden of osteoporotic fracture in future generations.

CONCLUSION

Osteoporosis is a disease that places a huge burden on individuals, healthcare systems, and societies as a whole. Many risk factors for inadequate peak bone mass, excessive involutional loss, and fracture have been characterized, and coupled with new pharmacological therapies, we are now in a position to develop novel preventative and therapeutic strategies aimed at preventing osteoporotic fractures, both for the entire population and for those at highest risk.

ACKNOWLEDGMENTS

We would like to thank the Medical Research Council (UK), Arthritis Research UK, the National Osteoporosis Society (UK), the International Osteoporosis Foundation, and the European Union Network in Male Osteoporosis for funding this work.

REFERENCES

1. Consensus development conference: diagnosis, prophylaxis, and treatment of osteoporosis. Am J Med. 1993; 94(6):646–50.
2. Kanis JA. Assessment of fracture risk and its application to screening for postmenopausal osteoporosis: synopsis of a WHO report. WHO Study Group. Osteoporos Int. 1994;4(6):368–81.
3. Kanis JA. Assessment of osteoporosis at the primary health care level. WHO Scientific Group Technical Report. Geneva: World Health Organization, 2007.
4. Kanis JA, Harvey NC, Cooper C, et al. A systematic review of intervention thresholds based on FRAX: a report prepared for the National Osteoporosis Guideline Group and the International Osteoporosis Foundation. Arch Osteoporos. 2016;11(1):25.
5. Svedbom A, Hernlund E, Ivergard M, et al. Osteoporosis in the European Union: a compendium of country-specific reports. Arch Osteoporos. 2013;8(1–2):137.
6. *Bone Health and Osteoporosis: A Report of the Surgeon General*. Rockville, MD: Office of the Surgeeon General, 2004.
7. van Staa TP, Dennison EM, Leufkens HG, et al. Epidemiology of fractures in England and Wales. Bone. 2001;29(6):517–22.
8. Curtis EM, van der Velde R, Moon RJ, et al. Epidemiology of fractures in the United Kingdom 1988–2012: variation with age, sex, geography, ethnicity and socioeconomic status. Bone. 2016;87:19–26.
9. Moon RJ, Harvey NC, Curtis EM, et al. Ethnic and geographic variations in the epidemiology of childhood fractures in the United Kingdom. Bone. 2016;85:9–14.

10. Goulding A, Jones IE, Taylor RW, et al. More broken bones: a 4-year double cohort study of young girls with and without distal forearm fractures. J Bone Miner Res. 2000;15(10):2011–18.
11. Felsenberg D, Silman AJ, Lunt M, et al. of vertebral fracture in europe: results from the European Prospective Osteoporosis Study (EPOS). J Bone Miner Res. 2002;17(4):716–24.
12. Cooper C, Campion G, Melton LJ. Hip fractures in the elderly: a world-wide projection. Osteoporos Int. 1992;2(6):285–9.
13. Blain H, Masud T, Dargent-Molina P, et al. A comprehensive fracture prevention strategy in older adults: The European Union Geriatric Medicine Society (EUGMS) Statement. J Nutr Health Aging. 2016;20(6):647–52.
14. Klop C, Welsing PM, Cooper C, et al. Mortality in British hip fracture patients, 2000–2010: a population-based retrospective cohort study. Bone. 2014;66:171–7.
15. Bliuc D, Nguyen ND, Milch VE, et al. Mortality risk associated with low-trauma osteoporotic fracture and subsequent fracture in men and women. JAMA. 2009;301(5):513–21.
16. Chrischilles EA, Butler CD, Davis CS, et al. A model of lifetime osteoporosis impact. Arch Intern Med. 1991;151(10):2026–32.
17. O'Neill TW, Felsenberg D, Varlow J, et al. The prevalence of vertebral deformity in European men and women: the European Vertebral Osteoporosis Study. J Bone Miner Res. 1996;11(7):1010–18.
18. Cooper C, Atkinson EJ, O'Fallon WM, et al. Incidence of clinically diagnosed vertebral fractures: a population-based study in Rochester, Minnesota, 1985–1989. J Bone Miner Res. 1992;7(2):221–7.
19. Cooper C, Atkinson EJ, Jacobsen SJ, et al. Population-based study of survival after osteoporotic fractures. Am J Epidemiol. 1993;137(9):1001–5.
20. Oleksik A, Lips P, Dawson A, et al. Health-related quality of life in postmenopausal women with low BMD with or without prevalent vertebral fractures. J Bone Miner Res. 2000;15(7):1384–92.
21. Kanis JA, Johnell O, De Laet C, et al. A meta-analysis of previous fracture and subsequent fracture risk. Bone. 2004;35(2):375–82.
22. Ismail AA, Cockerill W, Cooper C, et al. Prevalent vertebral deformity predicts incident hip though not distal forearm fracture: results from the European Prospective Osteoporosis Study. Osteoporos.Int. 2001;12(2):85–90.
23. Johnell O, Kanis JA, Oden A, et al. Fracture risk following an osteoporotic fracture. Osteoporos Int. 2004;15(3):175–9.
24. *World Population Prospects: The 2015 Revision.* New York: United Nations, 2015.
25. Gullberg B, Johnell O, Kanis JA. World-wide projections for hip fracture. Osteoporos Int. 1997;7(5):407–13.
26. Cooper C, Cole ZA, Holroyd CR, et al. Secular trends in the incidence of hip and other osteoporotic fractures. Osteoporos Int. 2011;22(5):1277–88.
27. van der Velde RY, Wyers CE, Curtis EM, et al. Secular trends in fracture incidence in the UK between 1990 and 2012. Osteoporos Int. 2016;27(11):3197–206.
28. Oden A, McCloskey EV, Kanis JA, et al. Burden of high fracture probability worldwide: secular increases 2010–2040. Osteoporos Int. 2015;26(9):2243–8.
29. Kanis JA, Oden A, McCloskey EV, et al. A systematic review of hip fracture incidence and probability of fracture worldwide. Osteoporos Int. 2012;23(9):2239–56.
30. Elffors I, Allander E, Kanis JA, et al. The variable incidence of hip fracture in southern Europe: the MEDOS Study. Osteoporos Int. 1994;4(5):253–63.
31. Oden A, Kanis JA, McCloskey EV, et al. The effect of latitude on the risk and seasonal variation in hip fracture in Sweden. J Bone Miner Res. 2014;29(10):2217–23.
32. Wahl DA, Cooper C, Ebeling PR, et al. A global representation of vitamin D status in healthy populations. Arch Osteoporos. 2012;7:155–72.
33. Landin LA. Fracture patterns in children. Analysis of 8,682 fractures with special reference to incidence, etiology and secular changes in a Swedish urban population 1950–1979. Acta Orthop Scand Suppl. 1983;202:1–109.
34. Tiderius CJ, Landin L, Duppe H. Decreasing incidence of fractures in children: an epidemiological analysis of 1,673 fractures in Malmo, Sweden, 1993–1994. Acta Orthop Scand. 1999;70(6):622–6.
35. Walsh JS, Paggiosi MA, Eastell R. Cortical consolidation of the radius and tibia in young men and women. J Clin Endocrinol Metab. 2012;97(9):3342–8.
36. Holroyd CR, Osmond C, Barker D, et al. Placental size is associated differentially with postnatal bone size and density. J Bone Miner Res. 2016;31(10):1855–64.
37. Clark EM, Ness AR, Tobias JH. Vigorous physical activity at age 9 increases the risk of childhood fractures, despite increasing bone mass. Rheumatology. 2007;46(Supplement 1):i3–i4.
38. Hernandez CJ, Beaupre GS, Carter DR. A theoretical analysis of the relative influences of peak BMD, age-related bone loss and menopause on the development of osteoporosis. Osteoporos Int. 2003;14(10):843–7.
39. Harvey N, Dennison E, Cooper C. Osteoporosis: a life-course approach. J Bone Miner Res. 2014;29(9):1917–25.
40. Cooper C, Eriksson JG, Forsen T, et al. Maternal height, childhood growth and risk of hip fracture in later life: a longitudinal study. Osteoporos Int. 2001;12(8):623–9.
41. Javaid MK, Eriksson JG, Kajantie E, et al. Growth in childhood predicts hip fracture risk in later life. Osteoporos Int. 2011;22(1):69–73.
42. Baird J, Kurshid MA, Kim M, et al. Does birthweight predict bone mass in adulthood? A systematic review and meta-analysis. Osteoporos Int. 2011;22(5):1323–34.
43. Cooper C, Harvey NC, Bishop NJ, et al. Maternal gestational vitamin D supplementation and offspring bone health (MAVIDOS): a multicentre, double-blind, randomised placebo-controlled trial. Lancet Diabetes Endocrinol. 2016;4(5):393–402.
44. Harvey NC, Sheppard A, Godfrey KM, et al. Childhood bone mineral content is associated with methylation status of the RXRA promoter at birth. J Bone Miner Res. 2014;29(3):600–7.

51
Fracture Liaison Service

Piet Geusens[1], John A. Eisman[2], Andrea Singer[3], and Joop van den Bergh[4]

[1]Department of Rheumatology, Maastricht University Medical Center,
the Netherlands and University Hasselt, Belgium
[2]Garvan Institute of Medical Research; Endocrinology, St Vincent's Hospital; School of Medicine
Sydney, University of Notre Dame Australia, UNSW Australia, Sydney, Australia
[3]Departments of Obstetrics and Gynecology, and Medicine, MedStar Georgetown University
Hospital and Georgetown University Medical Center, Washington, DC, USA
[4]Department of Internal Medicine, VieCuri Medical Center for North Limburg, Venlo, The Netherlands;
Department of Internal Medicine, NUTRIM School of Nutrition and Translational Research
in Metabolism, Maastricht UMC, Maastricht, The Netherlands; Biomedical Research Center,
University of Hasselt, Hasselt, Belgium

INTRODUCTION

After the age of 50 years, a fracture is the clinical expression of a decrease in bone quality, and — in frail elderly — often accompanied by decreased neuromuscular performance, both of which have a multifactorial pathophysiology. Repeat fractures contribute substantially to the overall fracture burden [1]. A fracture results in morbidity, an increased risk of subsequent fracture, and additionally after major fractures, and, in elderly also after minor fractures, an increased mortality risk. In spite of this evidence, there is still a large investigation gap in patients with a recent fracture, and an even bigger treatment gap related to preventing subsequent fractures in high-risk patients [2, 3].

A recent fracture is a window of opportunity to "capture the fracture" for secondary fracture prevention: the patient presents with the clinical problem of a fracture, and the "opportunity" to prevent fractures in the future. Subsequent fracture prevention immediately following the acute fracture healing should therefore be implemented, as in patients with other diseases with acute events, such as acute myocardial infarction, thromboembolism, cerebrovascular accident, arthritis, etc., in whom evaluation and treatment is recommended after an acute event in order to prevent further events in high risk patients [4].

In this chapter we review the evidence that a multidisciplinary structured program at a Fracture Liaison Service (FLS) is the most appropriate approach for the prevention of subsequent fractures in patients older than 50 years at high risk for additional fractures (Fig. 51.1).

We review the time-dependent risk of subsequent fractures, the diagnostic and therapeutic decisions to be considered based on a 5-step decision plan, including case finding, risk evaluation, differential diagnosis, therapy and follow up, and the implementation of a multidisciplinary organization and structured approach for secondary fracture prevention at the FLS (Fig. 51.2).

EPIDEMIOLOGY

The rationale for "capturing the fracture" and ensuring that these individuals are investigated and treated is based on the clear evidence for each fragility fracture signaling not only risk for further fractures, but also risk of premature mortality in both men and women [5–9]. These increased risks are important in view of the additional evidence from randomized controlled trials (RCT) that osteoporosis medications can approximately halve the risk of further fractures and, from both epidemiological and some RCT data, that osteoporosis treatment may reduce the risk of premature mortality [8, 10–12].

Primer on the Metabolic Bone Diseases and Disorders of Mineral Metabolism, Ninth Edition. Edited by John P. Bilezikian.
© 2019 American Society for Bone and Mineral Research. Published 2019 by John Wiley & Sons, Inc.
Companion website: www.wiley.com/go/asbmrprimer

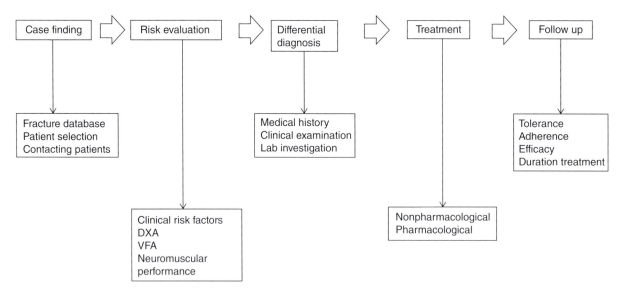

Fig. 51.1. Components of a FLS in a 5-step decision plan. Source: [23]. Reproduced with permission of BMJ.

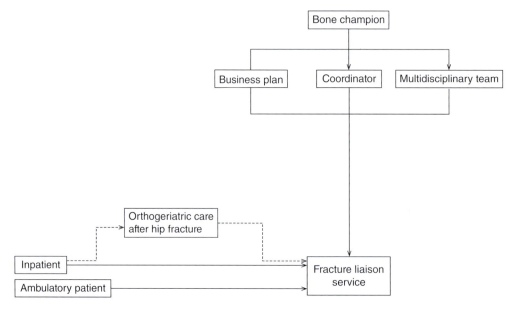

Fig. 51.2. Administrative and clinical organization of a FLS [2, 3, 20, 22–25, 44, 46].

The evidence that fragility fractures signal increased risk comes from multiple epidemiological cohort studies. These show that the risk of a first fracture is approximately twice as high in women than in men of the same age. However, after a single fragility fracture event, the risk of another fracture approximately doubles in women and increases approximately fourfold in men [6, 13]. Moreover, the increased risk is predominantly in the first 5 years after a fracture or 10 years following a hip fracture [6, 13].

The evidence of premature mortality also comes from epidemiological studies. As well as increased mortality post-hip fracture, after all types of fragility fractures, especially major or axial fractures, and even after minor, distal fractures in older individuals, mortality is increased approximately twofold in women and two- to threefold in men, especially in the first 5 years [5]. Comorbidities and surgery explain only up to 50% of this mortality [14].

Osteoporosis therapy may reduce mortality, by 28% in one RCT of intravenous zoledronate after hip fracture, and is supported by a meta-analysis of other RCTs suggesting a 10% reduction in mortality with osteoporosis therapies. Cohort studies have also reported marked reductions in mortality in individuals treated for osteoporosis, albeit with potential confounding by immortal time and healthy complier biases [15].

PHENOTYPE OF PATIENTS WITH A RECENT FRACTURE ATTENDING THE FLS

Components of the phenotype of patients attending the FLS have been reported in many publications [2, 3, 16–20]. The surveys varied widely in terms of patient selection and reported patient characteristics, such as the proportion of men, mean age, and fracture location [20]. The studies varied also in the types of additional assessments, and, when reported, there was a high variability in the presence of osteoporosis, prevalent vertebral fracture, fall risk, and in known and newly diagnosed contributors to secondary osteoporosis and other metabolic bone disorders [21]. In a survey of 834 FLS patients, 20% had only bone-related risks, 40% had both bone- and fall-related risks, 20% had only fall-related risk, and in 20% no risk factors could be identified [18]. These results show the heterogeneous phenotype of patients invited and attending the FLS. The presence of a combination of bone- and fall-related risk factors in most patients indicates that these risk factors should be evaluated in order to determine as accurately as possible the patient phenotype [2, 3, 19–21].

CASE FINDING

The principal goal of the FLS is to facilitate patient care by inclusive case finding; that is, to automatically include all patients with a low-trauma/fragility fracture to provide them with the appropriate intervention to prevent avoidable fracture-related complications, readmissions, and secondary fractures [2, 3, 22–25]. This would include identification of the following:

- males and females, age 50 years and older,
- fracture locations: hip, femur, wrist, elbow, shoulder, clavicle, pelvis, vertebrae, tibia/fibula, and ankle (all fractures except fingers, toes, face and skull),
- low trauma fractures: defined as fractures sustained with fall(s) from standing height or less,
- patients who sustain a moderate to high trauma fracture are also at increased risk for subsequent fracture, and could warrant inclusion [26].

While capturing 100% of patients at risk for secondary fractures is the ultimate goal [2, 3, 23–25], the journey toward best practice may be incremental, and goals may be subdivided into more manageable components [3]. As an example, a new FLS program may initially focus on inpatient hip and other major fracture patients. After assessing clinical volume within the system, availability of staffing for the program, and refining the workflow and data reporting processes, the scope of the program should be progressively expanded to include all inpatient fractures, followed by outpatient fractures, and finally, proactive screening or identification of previously unrecognized vertebral fractures [3, 22]. Other FLS units have started with ambulant patients after a nonvertebral fracture [18], or focused on elderly frail patients with a hip fracture as part of a comprehensive orthogeriatric care program [27]. It is essential to work with the information technology department of the FLS facility to ensure that all fracture patients are appropriately identified. Defined patient populations can be captured on the basis of diagnostic (ICD-10) or procedure (CPT) coding and through the electronic health record or billing system/claims data [22]. Automatically generated lists that do not depend on individual referrals are best. Additional methods that may be used for case finding include admitting office lists, emergency department admissions, operating room schedules, referrals from other clinicians, and radiology reports.

RISK EVALUATION

Simple assessment by age plus fracture history has good predictive value for all fractures, but risk profiles differ for first and subsequent fractures [28]. A number of factors influence subsequent fracture risk in patients presenting at the FLS. Additional assessments that can capture this risk include a detailed evaluation of medical history, medication use, clinical risk factors for bone fragility and fall risk, and laboratory investigations, together with assessment of BMD and vertebral fracture assessment (VFA) using DXA or lateral spine radiographs, when the fracture is a nonvertebral-nonhip fracture [29].

The reported prevalence of previously known diseases/medications associated with metabolic bone disorders in most studies with systematic evaluation of medical history, medication use, and clinical risk factors was 25% to 30% in both men and women [19].

DXA assessment is informative in all patients with a recent fracture for diagnosis and for follow up. However, in elderly frail patients with a recent low-trauma hip fracture, the inability to perform a DXA does not preclude starting treatment. In the NORA study, fracture rates were highest in women with BMD-defined osteoporosis albeit assessed at the heel, finger, or forearm [30]. However, they accounted for only 18% of the osteoporotic fractures (wrist or forearm, rib, spine, or hip) and 26% of the hip fractures. Based on systematic BMD assessment, 30% to 40% of patients with a fracture at the FLS had osteoporosis [29].

Since most patients at the FLS have a BMD in the osteopenic range, the combination of BMD measurement with vertebral imaging (VFA or spine radiographs) helps to identify patients with increased fracture risk, provides a more accurate fracture history, and can be used during follow up to identify incident vertebral fractures. The prevalence of one or more vertebral fractures in patients with a nonvertebral fracture is reported to be 20% to 25%, even in those with osteopenia or even normal BMD [31].

Epidemiological studies have identified a number of risk factors for falling. Roughly 15% of falls have a sole recognizable cause such as Parkinson's disease or syncope; a similar percentage can be attributed to an external event that would result in most people falling, and multiple interacting factors are responsible for the remaining falls [32]. A targeted history and physical examination, covering potential home hazards, cognitive and visual impairment, functional limitations, medications, orthostatic hypotension, and gait and balance abnormalities, can be used to identify risk factors for falls [33].

Thirteen externally validated algorithms designed to predict osteoporotic fracture risk are currently available to clinicians and researchers. Most of these tools are feasible in clinical practice. FRAX, QFracture, and Garvan are the most extensively studied tools. FRAX was evaluated in a larger number of countries. Adding BMD to FRAX increases the area under the receiver–operator curves (AUC) for hip fractures in both men and women. Studies with QFracture present the highest AUCs [34]. Garvan may be more useful in patients with a history of multiple falls and multiple previous fractures, and predicts all types of fragility fractures [35]. These algorithms have not been studied in the FLS setting in patients with a recent fracture; FRAX is the exception, and it has been reported that it underestimates the risk of recurrent fragility fractures after an incident fracture, particularly in individuals younger than 65 years [36]. These algorithms can be used to aid risk evaluation, but do not replace clinical decision making [37].

DIFFERENTIAL DIAGNOSIS

Previously unknown and subclinical metabolic bone disorders have been documented in 27% of FLS patients [21]. In most guidelines it is therefore recommended to perform standard laboratory tests for detection of contributors to underlying metabolic bone diseases, including calcium, phosphate, albumin, creatinine, alkaline phosphatase, 25OH vitamin D, TSH, and testosterone in men, with additional testing of PTH, ESR, or other tests on indication [21]. With these standard tests, detection of newly diagnosed secondary osteoporosis or other metabolic bone disorders (excluding vitamin D deficiency) ranged from 10% in patients aged 50 to 60 years with minor fractures, up to 35% of men and women age 80 or older with major or hip fractures, regardless of BMD level [21,29]. When known and new contributors to secondary osteoporosis or other metabolic bone disorders are combined, 45% to 70% of the patients at the FLS have one or more secondary factors for bone fragility [21].

THERAPY AND FOLLOW-UP

Patient education is recommended, but on its own is insufficient to increase treatment [22, 23]. All patients at the FLS should be counseled on lifestyle (moderate alcohol intake and smoking cessation), nutrition (including adequate calcium and vitamin D intake), physical activity, fall prevention, and medications [2, 3, 23–25]. Addressable contributors to secondary osteoporosis or other metabolic bone disorders should be corrected [21].

Most guidelines and reimbursement criteria include the results of BMD and/or a prevalent hip or vertebral fracture for treatment decisions [2, 3, 23–25, 38]. In some guidelines, the FRAX calculator is considered sufficient to make treatment decisions when the calculated fracture risk is high [37]. As a fracture is a risk factor for any subsequent fracture, the choice of drug treatment is based on the spectrum of demonstrated fracture prevention (which is broadest for alendronate, risedronate, zoledronate, and denosumab). The route of administration (oral, i.v., s.c.) allows adaptation to the needs and tolerance of the patient. Antiresorptive drugs will most often be the first choice. The osteoanabolic agent teriparatide can be considered in severe osteoporosis.

Major challenges with all therapy for osteoporosis are initiation of treatment and longer term adherence [2, 3, 23–25]. RCTs indicate that a designated coordinator significantly improves the initiation and adherence to treatment [39, 40]. Risk communication and shared decision making in the care of patients with osteoporosis may have a positive influence on adherence [41, 42].

IMPLEMENTATION

The key steps and components for successful FLS development and implementation are detailed in Table 51.1. Specific components and strategies are highlighted in this section.

The initial driver in establishing an FLS program is the bone health champion, who is generally a physician, but an administrative leader or advanced practice provider can also serve in this essential role [43]. The bone health champion solicits support for the benefits of the FLS model of care, delivers the business case supporting development, handles wider negotiations around the politics of service development, and manages the multidisciplinary stakeholder group. He/she also provides leadership in the development of relevant clinical protocols.

The core of the FLS program is also built on the FLS coordinator who is crucial for the success of this care model [22–25]. The FLS coordinator has primary responsibility for providing clinical assessment of osteoporosis, nonpharmacological management, development and delivery of patient and family education, and case management including appointment coordination and utilization [22–25, 44]. Although not a requirement, the FLS coordinator is typically an advanced practice provider (nurse practitioner or physician assistant), and therefore also has the ability to initiate treatment. All FLS coordinators are responsible for following up with the patient to ensure treatment adherence [22–25, 44].

In addition to the bone health champion/physician lead and FLS coordinator described above, centralized workflow with key personnel to support such functions is necessary for implementation and sustainability of the

Table 51.1. Key steps and components for successful FLS program development and implementation [2, 3, 20, 22–25, 44, 46].

Champions and key personnel	• Bone health champion • Administrative champion • FLS coordinator
Establish a vision and mission statement	Example: to enhance the care of patients with a low trauma fracture and close the care gap for this group at high risk of secondary fractures
Establish core objectives to achieve the mission	• Institute inclusive case finding • Employ evidence-based assessment • Treat in accordance with appropriate guidelines • Improve adherence with therapy
Conduct a baseline audit	• Helps define the extent of the care gap and need for service implementation • Provides an idea of the number of fractures within the system and informs decisions regarding the scope of the problem and realistic goals
Establish measurable goals	
Create a sound business plan with realistic expectations	Business considerations: • Income at risk with current care model • Pay for performance/quality metrics • FLS expenses • New income sources • Covered lives
Identify key stakeholders/ multidisciplinary team	Includes but is not limited to: • Orthopedics • Neurosurgery • Radiology/Interventional Radiology • Emergency Department • Hospitalists • Endocrinology • Rheumatology • Primary Care • Geriatrics • Obstetrics/Gynecology • Physical Medicine and Rehabilitation/Physical Therapy • Pharmacy • Information Technology
IT infrastructure	• Regional "at risk" database • Inreach/outreach supported by the electronic health record • Robust performance reporting

program. Ideally, the nurse or administrative navigator can manage the "at risk" database to identify all patients at risk for a secondary fracture, conduct outreach by phone or mail, and assist with scheduling tests and appointments [22, 44]. Added job responsibilities may include assisting with patient education and insurance verification for prescriptions or testing [22].

Since osteoporosis bridges many specialties and the FLS program touches many departments or services, it is imperative that a team of key stakeholders be assembled to help establish policy, sponsor the mission, and develop the model of care [2, 3, 19, 22–25, 44, 45]. This is performed by presenting relevant and targeted information that conveys how the program will benefit members of the multidisciplinary team and their patients [46].

FLS programs in closed healthcare settings and in single payer healthcare systems have been shown to reduce costs [47–49]. It is more challenging to convince the leadership of an open healthcare system that the return on investment in a FLS program is justified. The importance of developing a solid business case and gaining the support of a hospital or system administrative champion or governing body cannot be underestimated [44].

The pathway to a successful project plan includes a number of fundamental factors — the need, the product or service itself, the business model, and the team. The need indicates the size of the problem. In this case, a baseline audit that shows the number of fracture patients within the system and the pre-FLS care gap is highly informative and important. The product or service description has been detailed elsewhere in this chapter, but should focus on how it addresses the need, what competitive advantage it offers, and should identify the strategic fit within the institution or system. Although providing a detailed business model is beyond the scope of this chapter, consideration must be given to demonstrating how the service will earn a return on investment and that the business risks are fully understood.

Resources, including templates for business plans and a return on investment (ROI) calculator, are available and may provide a starting point for plan development [50]. Finally, the FLS team, with their experience, expertise, and understanding of the market, gives credibility. It should be noted that US healthcare reform is transforming the healthcare system from fee for service to paying for quality, outcomes, and care coordination. Payments to a hospital for providing high quality care and avoiding readmission penalties are examples of potential additional justification for an FLS program [43].

FLS programs should start by using best practices, but need to be flexible and creative [2, 3, 23–25]. Leadership should anticipate reevaluation and evolution as the delivery system matures.

REFERENCES

1. Langsetmo L, Goltzman D, Kovacs CS, et al.; CaMos Research Group. Repeat low-trauma fractures occur frequently among men and women who have osteopenic BMD. J Bone Miner Res. 2009;24(9):1515–22.
2. Eisman JA, Bogoch ER, Dell R, et al.; ASBMR Task Force on Secondary Fracture Prevention. Making the first fracture the last fracture: ASBMR task force report on secondary fracture prevention. J Bone Miner Res. 2012;27(10):2039–46.
3. Akesson K, Marsh D, Mitchell PJ, et al.; IOF Fracture Working Group. Capture the fracture: a best practice framework and global campaign to break the fragility fracture cycle. Osteoporos Int. 2013;24(8):2135–52.
4. Majumdar SR. Implementation research in osteoporosis: an update. Curr Opin Rheumatol. 2014;26(4):453–7.
5. Center JR, Nguyen TV, Schneider D, et al. Mortality after all major types of osteoporotic fracture in men and women: an observational study. Lancet 1999;353(9156):878–82.
6. Center JR, Bliuc D, Nguyen TV, et al. Risk of subsequent fracture after low-trauma fracture in men and women. JAMA 2007;297(4):387–94.
7. Nguyen ND, Ahlborg HG, Center JR, et al. Residual lifetime risk of fractures in women and men. J Bone Miner Res 2007;22(6):781–8.
8. Center JR, Bliuc D, Nguyen ND, et al. Osteoporosis medication and reduced mortality risk in elderly women and men. J Clin Endocrinol Metab 2011;96:1006–14.
9. Bliuc D, Nguyen ND, Nguyen TV, et al. Compound risk of high mortality following osteoporotic fracture and refracture in elderly women and men. J Bone Miner Res 2013;28(11):2317–24.
10. Bolland MJ, Grey AB, Gamble GD, et al. Effect of osteoporosis treatment on mortality: a meta-analysis. J Clin Endocrinol Metab 2010;95:1174–81.
11. Lyles KW, Colon-Emeric CS, Magaziner JS, et al. Zoledronic acid and clinical fractures and mortality after hip fracture. N Engl J Med 2007;357(18):1799–809.
12. Eriksen EF, Lyles KW, Colon-Emeric CS, et al. Antifracture efficacy and reduction of mortality in relation to timing of the first dose of zoledronic acid after hip fracture. J Bone Miner Res 2009;24(7):1308–13.
13. Bliuc D, Alarkawi D, Nguyen TV, et al. Risk of subsequent fractures and mortality in elderly women and men with fragility fractures with and without osteoporotic bone density: the Dubbo Osteoporosis Epidemiology Study. J Bone Miner Res 2015;30(4):637–46.
14. Smith T, Pelpola K, Ball M, et al. Pre-operative indicators for mortality following hip fracture surgery: a systematic review and meta-analysis. Age Ageing. 2014;43(4):464–71.
15. Jones M, Fowler R. Immortal time bias in observational studies of time-to-event outcomes. J Crit Care 2016; 36:195–9.
16. Mitchell PJ, Chem C. Secondary prevention and estimation of fracture risk. Best Pract Res Clin Rheumatol. 2013;27(6):789–803.
17. van Helden S, van Geel AC, Geusens PP, et al. Bone and fall-related fracture risks in women and men with a recent clinical fracture. J Bone Joint Surg Am. 2008;90(2):241–8.
18. Huntjens KM, van Geel TA, van Helden S, et al. The role of the combination of bone and fall related risk factors on short-term subsequent fracture risk and mortality. BMC Musculoskelet Disord. 2013;14:121.
19. Blain H, Masud T, Dargent-Molina P, et al.; EUGMS Falls and Fracture Interest Group.; European Society for Clinical and Economic Aspects of Osteoporosis and Osteoarthritis (ESCEO), Osteoporosis Research and Information Group (GRIO), and International osteoporosis Foundation (IOF). A Comprehensive Fracture Prevention Strategy in Older Adults: The European Union Geriatric Medicine Society (EUGMS) Statement. J Nutr Health Aging. 2016;20(6):647–52.
20. Huntjens KM, Kosar S, van Geel TA, et al. Risk of subsequent fracture and mortality within 5 years after a non-vertebral fracture. Osteoporos Int. 2010;21(12):2075–82.
21. Bours SP, van den Bergh JP, van Geel TA, et al. Secondary osteoporosis and metabolic bone disease in patients 50 years and older with osteoporosis or with a recent clinical fracture: a clinical perspective. Curr Opin Rheumatol. 2014;26(4):430–9.
22. Miller AN, Lake AF, Emory CL. Establishing a fracture liaison service: an orthopaedic approach. J Bone Joint Surg Am. 2015;97(8):675–81.
23. Lems WF, Dreinhöfer KE, Bischoff-Ferrari H, et al. EULAR-EFORT recommendations for management of patients older than 50 years with a fragility fracture and prevention of subsequent fractures. Ann Rheum Dis. 2017;76(5):802–10.
24. American Academy of Orthopaedic Surgeons. Management of Hip Fractures in the Elderly. Secondary Management of Hip Fractures in the Elderly, 2014. http://www.aaos.org/research/guidelines/HipFxGuideline.pdf
25. British Orthopaedic Association. *The Care of Patients with Fragility Fracture*, 2007. http://www.fractures.com/pdf/BOA-BGS-Blue-Book.pdf
26. Warriner AH, Patkar NM, Yun H, et al. Minor, major, low-trauma, and high-trauma fractures: what are the

subsequent fracture risks and how do they vary? Curr Osteoporos Rep. 2011;9(3):122–8.
27. Prestmo A, Hagen G, Sletvold O, et al. Comprehensive geriatric care for patients with hip fractures: a prospective, randomised, controlled trial. Lancet. 2015;385 (9978):1623–33.
28. Watts NB; GLOW Investigators. Insights from the Global Longitudinal Study of Osteoporosis in Women (GLOW). Nature Rev Endocrinol. 2014;10:412–22.
29. van den Bergh JP, van Geel TA, Geusens PP. Osteoporosis, frailty and fracture: implications for case finding and therapy. Nat Rev Endocrinol. 2012;8:163–72.
30. Siris ES, Brenneman SK, Barrett-Connor E, et al. The effect of age and bone mineral density on the absolute, excess, and relative risk of fracture in postmenopausal women aged 50–99: results from the National Osteoporosis Risk Assessment (NORA). Osteoporos Int. 2006;17(4):565–74.
31. Gallacher SJ, Gallagher AP, McQuillian C, et al. The prevalence of vertebral fracture amongst patients presenting with non-vertebral fractures. Osteoporos Int. 2007;18:185–92.
32. La Grow SJ, Robertson MC, Campbell AJ, et al. Reducing hazard related falls in people 75 years and older with significant visual impairment: how did a successful program work? Inj Prev. 2006;12:296–301.
33. Kwan E, Straus SE. Assessment and management of falls in older people. CMAJ 2014;186:E610–21.
34. Marques A, Ferreira RJ, Santos E, et al. The accuracy of osteoporotic fracture risk prediction tools: a systematic review and meta-analysis. Ann Rheum Dis. 2015;74: 1958–67.
35. van Geel TA, Eisman JA, Geusens PP, et al. The utility of absolute risk prediction using FRAX® and Garvan Fracture Risk Calculator in daily practice. Maturitas 2014;77:174–9.
36. Roux S, Cabana F, Carrier N, et al. The World Health Organization Fracture Risk Assessment Tool (FRAX) underestimates incident and recurrent fractures in consecutive patients with fragility fractures. J Clin Endocrinol Metab. 2014;99:2400–8.
37. Kanis JA, Harvey NC, Cooper C, et al. A systematic review of intervention thresholds based on FRAX. Arch Osteoporos. 2016;11:25.
38. Appelman-Dijkstra NM, Papapoulos SE. Prevention of incident fractures in patients with prevalent fragility fractures: current and future approaches. Best Pract Res Clin Rheumatol 2013;27(6):805–20.
39. Majumdar SR, Beaupre LA, Harley CH, et al. Use of a case manager to improve osteoporosis treatment after hip fracture: results of a randomized controlled trial. Arch Intern Med 2007;167(19):2110–5.
40. Morrish DW, Beaupre LA, Bell NR, et al. Facilitated bone mineral density testing versus hospital-based case management to improve osteoporosis treatment for hip fracture patients: additional results from a randomized trial. Arthritis Rheum 2009;61(2): 209–15.
41. Fried TR. Shared decision making — finding the sweet spot. N Engl J Med. 2016;374(2):104–6.
42. Lewiecki EM. Risk communication and shared decision making in the care of patients with osteoporosis. J Clin Densitom 2010;13(4):335–45.
43. http://capturethefracture.org/sites/default/files/2014-IOF-CTF-FLS_toolkit.pdf
44. Kaiser Permanente — adapted with permission from "Healthy Bones FLS Care Management Policies and Procedures" 11-20-2011. Personal communication Richard Dell, MD and Denise Greene MSN, FNP-BC.
45. Curtis JR, Silverman SL. Commentary: the five Ws of a Fracture Liaison Service: why, who, what, where and how? In osteoporosis we reap what we sow. Curr Osteoporos Rep 2013;11:365–8.
46. Marsh D, Åkesson K, Beaton DE, et al. and the IOF CSA Fracture Working Group. Coordinator-based systems for secondary prevention in fragility fracture patients. Osteoporos Int 2011;22: 2051–2065.
47. Newman ED, Ayoub WT, Starkey RH, et al. Osteoporosis disease management in a rural health care population: hip fracture reduction and reduced costs in postmenopausal women after 5 years. Osteoporos Int. 2003;14(2):146–51.
48. Dell RM, Greene D, Anderson D, et al. Osteoporosis disease management: what every orthopaedic surgeon should know. J Bone Joint Surg Am. 2009;91(Suppl 6):79–86.
49. Dell R, Greene D, Schelkun SR, et al. Osteoporosis disease management: the role of the orthopaedic surgeon. J Bone Joint Surg Am. 2008;90(Suppl 4):188–94.
50. NBHA Fracture Prevention Central website at: http://www.nbha.org/fpc

52

Sex Steroids and the Pathogenesis of Osteoporosis

Matthew T. Drake and Sundeep Khosla

Department of Internal Medicine, Division of Endocrinology, Diabetes, and Metabolism,
College of Medicine, Mayo Clinic, Rochester, MN, USA

INTRODUCTION

Significant bone loss occurs with aging in both men and women, leading to alterations in skeletal microarchitecture and increased fracture incidence [1]. Much work has now significantly enhanced our understanding of the role that sex steroids (primarily estrogen and testosterone) play in the development and progression of osteoporosis in both men and women.

CHANGES IN BONE MASS AND STRUCTURE WITH AGING

The composite DXA data shown in Fig. 52.1 shows that at the menopausal transition, women undergo rapid trabecular bone loss [2]. While somewhat variable in length, this period of accelerated bone loss extends for approximately 5 to 10 years, with loss of ~20% to 30% of trabecular bone, but only 5% to 10% of cortical bone. Following this initial phase of rapid bone loss, a second phase of slow and continuous bone loss becomes predominant. In this phase, which extends throughout the remaining female life span, cortical and trabecular bone loss occur at approximately equal rates. In comparison, men from middle life onwards show slow progressive trabecular and cortical bone loss nearly equivalent to the latter phase observed in postmenopausal women. However, because men do not have a menopausal equivalent, the early accelerated bone loss observed in women does not occur; thus, overall loss of both trabecular and cortical bone is less pronounced in men.

Studies using QCT have challenged the prevailing notion that there is relative preservation of skeletal integrity in both men and women from the end of puberty until middle age. QCT distinguishes trabecular and cortical components much more precisely than DXA imaging, which only provides areal bone mineral density (aBMD) assessment. Cross-sectional QCT studies show that in the spine (composed primarily of trabecular bone), large decreases (-55% in women and -45% in men) in volumetric BMD (vBMD) occur beginning in the third decade (Fig. 52.2) [3]. In contrast, cortical vBMD (assessed at the distal radius) shows little change in either sex until midlife. Thereafter, roughly linear declines in cortical bone occur in both sexes, although cumulative decreases are greater in women than men (28% versus 18%, $p < 0.01$), reflecting the period of rapid bone loss observed in early menopause. Importantly, these cross-sectional findings have been confirmed by longitudinal studies [4] in which vBMD was followed at both the distal radius and tibia, with substantial trabecular bone loss found to start shortly after completion of puberty in both men and women, an age range during which sex steroid levels are considered to be normal. The relative contribution of bone loss during these years to future skeletal fragility remains undetermined.

In addition to changes in bone mass that occur with aging, changes in bone cross-sectional area also occur in both sexes. Despite a net decrease in cortical area and thickness resulting from endocortical resorption, concomitant outward cortical displacement caused by ongoing periosteal apposition has been shown in both men and women. This net outward displacement increases bone strength for bending stresses, and partially offsets the decrease in bone strength caused by cortical thinning [5].

Primer on the Metabolic Bone Diseases and Disorders of Mineral Metabolism, Ninth Edition. Edited by John P. Bilezikian.
© 2019 American Society for Bone and Mineral Research. Published 2019 by John Wiley & Sons, Inc.
Companion website: www.wiley.com/go/asbmrprimer

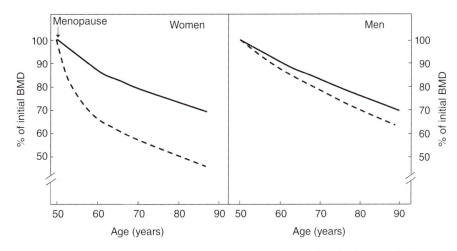

Fig. 52.1. Patterns of age-related bone loss in women and men. Dashed lines, trabecular bone; solid lines, cortical bone. The figure is based on multiple cross-sectional and longitudinal studies using DXA. Source: [2]. Reprinted with permission of Elsevier.

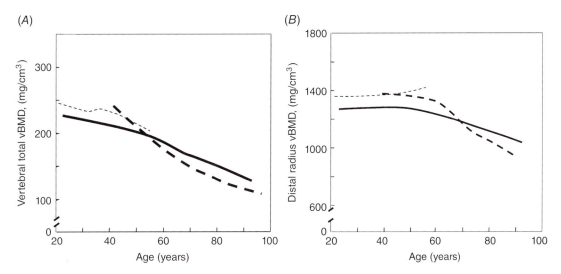

Fig. 52.2. (A) Values for vBMD (mg/cm³) of the total vertebral body in a population sample of women and men from Rochester, MN, USA between the ages of 20 and 97 years. Thin-dashed line, premenopausal women; thick-dashed line, postmenopausal women; solid line, men. (B) Values for cortical vBMD at the distal radius in the same cohort. Line coding as in (A). All changes with age were significant ($p < 0.05$). Source: [3]. Reprinted with permission of John Wiley & Sons.

When taken together, these changes in both bone quantity and structure with aging in both women and men lead to increases in annual osteoporotic fracture incidence (Fig. 52.3). Thus, distal forearm fractures in women rise markedly around the time of menopause and plateau ~15 years after menopause. Similarly, vertebral fracture incidence also begins to rise with menopausal onset. Whereas wrist fracture incidence plateaus, however, vertebral fracture incidence continues to increase for the duration of the female lifespan. Hip fracture rates in women initially parallel those of vertebral fractures, but rise markedly later in life, as shown.

In contrast, distal forearm fractures remain uncommon in men throughout life. Cross-sectional studies of the distal forearm in both sexes using pQCT show that whereas women undergo both trabecular loss and increased trabecular spacing with aging, men begin young adult life with relatively thicker trabeculae and primarily undergo trabecular thinning rather than loss with aging [6]. Thus, in addition to on average having larger bones than women, elderly men also have comparatively more trabeculae at the distal forearm, likely contributing to the rarity of wrist fractures. Although delayed by about one decade relative to women, vertebral and hip fracture incidence in men is roughly equivalent to that in women, with the decade delay again likely reflecting the lack of a male menopause and the associated rapid skeletal loss observed during this period in women.

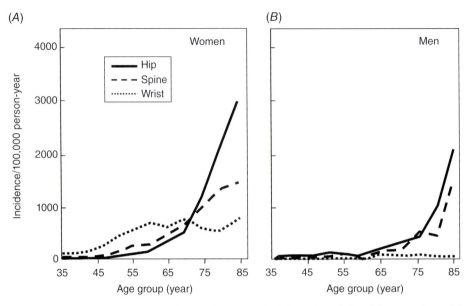

Fig. 52.3. Age specific incidence rates for proximal femur (hip), vertebral (spine), and distal forearm (wrist) factures in women (A) and men (B) from Rochester, MN, USA. Source: [55]. Reproduced with permission of Elsevier.

SEX STEROIDS AND BONE LOSS IN WOMEN

The relationship between diminishing estrogen levels in women caused by ovarian failure and the development of postmenopausal osteoporosis has been recognized for seven decades [7]. Relative to premenopausal levels, serum estradiol (E2) levels decrease by 85% to 90%, and serum estrone (E1; a fourfold weaker estrogen) declines by 65% to 75% over the menopausal transition [8]. Temporally associated with this decline are changes in both bone formation and resorption rates. Whereas coupled bone remodeling maintains bone formation and resorption rates approximately equivalent before menopause, menopausal onset heralds an increase in the BMU activation frequency rate, extension of the resorption period [9], and shortening of the formation period [10]. As assessed by biochemical markers, bone resorption at the menopause increases by 90%, whereas bone formation increases by only 45% [11]. This net bone remodeling imbalance leads to the accelerated phase of bone loss detailed above, and a net efflux of skeletal-derived calcium to the extracellular fluid. As a result, compensatory mechanisms to limit the development of hypercalcemia are employed, including increased renal calcium clearance [12], decreased intestinal calcium absorption [13], and partial suppression of PTH secretion [14]. Together, however, these compensatory mechanisms contribute to a net negative total body calcium balance from skeletal losses. Importantly, these compensatory effects appear directly related to estrogen deficiency, because estrogen repletion, at least in early menopause, leads to preservation of both renal calcium reclamation and intestinal calcium absorption [15].

The effects of estrogen on modulating skeletal metabolism at the cellular and molecular level remain the subject of active study. Estrogen plays a central role in osteoblast biology, where it promotes bone marrow stromal cell differentiation toward the osteoblast lineage, increases preosteoblast to osteoblast differentiation, and limits both osteoblast and osteocyte apoptosis [16, 17]. In addition, estrogen increases osteoblastic production of growth factors (IGF-1 and TGF-α) and procollagen synthesis [18, 19]. Estrogen also suppresses serum [20], bone marrow plasma [21], and bone mRNA [22] levels of the potent Wnt signaling inhibitor, sclerostin. Moreover, mice with LRP5 mutations that confer resistance to sclerostin are at least partially protected from ovariectomy-induced bone loss [23]. Collectively, this data suggests that estrogen may mediate some of its skeletal protective effects via inhibition of sclerostin production to thereby maintain bone formation [24]. In addition, recent studies also show that sclerostin can increase RANKL and inhibit production of the soluble RANKL decoy receptor osteoprotegerin (OPG) by osteocytes [25], and that Wnt signaling can directly inhibit osteoclast formation [26]. Thus, inhibition of sclerostin production by estrogen may be important not only for estrogen effects on maintaining bone formation but also on its antiresorptive effects.

Estrogen also has other important effects on suppressing bone resorption. Both in vitro and in vivo, estrogen suppresses production of RANKL, the central molecule in osteoclast development, from bone marrow stromal/osteoblast precursor cells, T cells, and B cells [27]. Furthermore, estrogen increases OPG production by osteoblast lineage cells [28]. Additional estrogen-suppressible cytokines produced by osteoblasts and bone marrow mononuclear cells, including interleukin-1

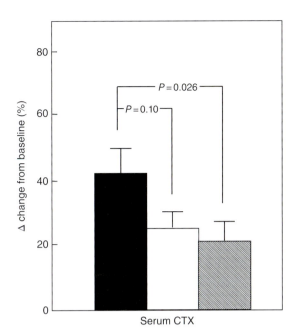

Fig. 52.4. The proportional change (%) in means for the bone resorption marker serum CTX between the baseline value and those on the final two intervention days. Note that both anakinra and etanercept both reduced the rise in serum CTX by approximately one-half. p values as assessed by t-test for the difference in serum CTX after anakinra or etanercept treatment and the control group are shown in brackets. Solid bar = control; open bar = anakinra; hatched bar = etanercept. Source: [33]. Reproduced with permission of John Wiley & Sons.

(IL-1), IL-6, TNF-α, M-CSF, and prostaglandins also appear to play central roles in mediating bone resorption [29–32]. In support of this role for estrogen, pharmacological blockade of either IL-1 or TNF-α activity partially blunted a rise in bone resorption markers in early postmenopausal women induced to undergo acute estrogen withdrawal (Fig. 52.4) [33].

In addition to influencing osteoclast development, estrogen both directly and indirectly promotes apoptosis of both osteoclast lineage cells and mature osteoclasts. In osteoclast lineage cells, estrogen induces direct apoptosis through a decrease in c-jun activity, thereby limiting activator protein-1 (AP-1)-dependent transcription [34]. The importance of direct estrogen effects on osteoclast apoptosis has been shown in mice in which estrogen receptor alpha had been selectively deleted in osteoclasts, in which the pro-apoptotic effects of estrogen deficiency on osteoclasts were lost [35, 36].

Although a role for follicle-stimulating hormone (FSH) in regulating postmenopausal bone loss based on human epidemiological [37] and rodent [38] studies has been postulated, more recent direct human interventional studies have provided strong evidence that FSH does not regulate bone resorption in either women [39, 40] or men [41].

Like estrogen, testosterone has a primary effect on bone to limit resorption, although at least part of this effect likely derives from testosterone aromatization to estrogen [42]. In vitro, testosterone can both stimulate osteoblast proliferation [43] albeit weakly, and limit osteoblast apoptosis [10]. Whereas testosterone likely plays a role in increasing bone formation (and perhaps periosteal apposition) in women, there is at present little data suggesting a role for testosterone in maintaining postmenopausal skeletal integrity.

SEX STEROIDS AND BONE LOSS IN MEN

Relative to postmenopausal women, elderly men on average lose one-half as much bone and sustain one-third as many fragility fractures [14]. Although men do not undergo a hormonal menopausal equivalent, substantial changes in biologically available sex steroid levels occur over the male life span primarily as a result of the more than twofold age-related increase in sex hormone–binding globulin (SHBG) levels [44]. Whereas SHBG-bound circulating sex steroids have limited capacity to reach target tissues, free (1% to 3% of total) and albumin-associated (35% to 55% of total) sex steroids are biologically available. As a result of this rise in SHBG levels, bioavailable estrogen and testosterone levels decline an average of 47% and 64%, respectively, over the male life span, as shown in a cross-sectional study of 346 men between the ages of 23 and 90 in Rochester, MN, USA [44].

Although testosterone is the predominant sex steroid in men, evidence from both cross-sectional [45, 46] and longitudinal [47] studies shows that male BMD at various skeletal sites better correlates with circulating bioavailable E2 than testosterone levels. Although suggestive, the associations described above do not provide direct evidence for a causal role of estrogen in maintenance of the aging male skeleton. To address the relative roles of estrogen and testosterone in skeletal maintenance, Falahati-Nini and colleagues [42] used pharmacological suppression of endogenous estrogen and testosterone production (through treatment with both a GnRH agonist and an aromatase inhibitor) in elderly men. Physiological estrogen (E) and testosterone (T) levels were maintained by placement of topical estrogen and testosterone patches, and baseline markers of bone resorption and formation were obtained. Subjects were then randomized to one of four groups: group A (−T,−E) discontinued both patches; group B (−T,+E) discontinued only the testosterone patch; group C (+T,−E) discontinued only the estrogen patch; and group D (+T,+E) continued both patches. As such, this direct human interventional study allowed for changes in bone metabolism resulting from either estrogen or testosterone to be determined, because suppression of endogenous sex steroid production was maintained for the entire intervention period.

As shown in Fig. 52.5A, the significant increases in bone resorption observed in group A (-T,-E) were completely prevented by treatment with both testosterone

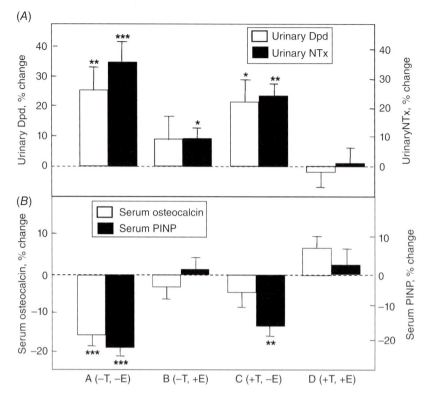

Fig. 52.5. Percent changes in (A) bone resorption markers (urinary deoxypyridinoline [Dpd] and N-telopeptide of type I collagen [NTX]) and (B) bone formation markers (serum osteocalcin and N-terminal extension peptide of type I collagen [P1NP]) in a group of elderly men (mean age, 68 years) made acutely hypogonadal and treated with an aromatase inhibitor (group A), estrogen alone (group B), testosterone alone (group C), or both estrogen and testosterone (group D). Significance for change from baseline: *, $p<0.05$; **, $p<0.01$; ***, $p<0.001$. Source: [42].

and estrogen (group D). Whereas estrogen alone was almost completely able to prevent the rise in bone resorption (group B). Testosterone alone, however, was much less potent (group C). By comparison (Fig. 52.5B), the marked decreases in bone formation observed with dual sex steroid deficiency in group A were completely prevented by continuation of estrogen and testosterone (group D). Interestingly, serum osteocalcin levels were only slightly diminished with either estrogen or testosterone alone, whereas levels of serum aminoterminal propeptide of type I collagen (PlNP) were sustained with estrogen (group B) but not testosterone. In summary, these results are consistent with a dominant role for estrogen in maintenance of skeletal integrity in aging men. These findings were recently confirmed by Finkelstein and colleagues [48] who suppressed endogenous gonadal steroid production in healthy men aged 20-50 years with a GnRH agonist and treated study subjects with increasing doses of testosterone, either in the presence or absence of aromatase blockade, for 16 weeks. Consistent with the prior study [42], estrogen was again determined to be the dominant regulator of both bone resorption and bone structural changes, as assessed by both central QCT and high resolution peripheral QCT.

OSTEOPOROSIS AND NONSEX STEROID HORMONE CHANGES WITH AGING

In addition to the effects of sex steroids, it is important to recognize that nonsex hormonal changes also occur with aging in both men and women. These include reductions in the production of growth factors important for osteoblast differentiation and function. Thus with aging, both the frequency and amplitude of growth hormone secretion is diminished [49], leading to decreased hepatic production of IGF-1 and IGF-2, an effect that may contribute to decreased bone formation with aging [50]. Additionally, aging is associated with increased levels of the IGF inhibitory binding protein, IGFBP-2, which also correlates inversely with bone mass in the elderly [51]. Finally, it is likely that intrinsic changes occur in osteoblast and perhaps osteoclast lineage cells with aging [52]. For example, the role of cellular senescence in contributing to effects of age on multiple tissues has recently been recognized [53]. Importantly, Farr and colleagues [54] have found that in mice, aging is associated with the development of senescence in multiple cell types in the bone microenvironment, particularly osteocytes. These changes, which are likely independent of changes in sex steroids or other hormonal factors, are the focus of ongoing animal and human studies.

REFERENCES

1. Riggs BL, Khosla S, Melton LJ. Sex steroids and the construction and conservation of the adult skeleton. Endocr Rev. 2002;23:279–302.
2. Khosla S, Riggs BL. Pathophysiology of age-related bone loss and osteoporosis. Endocrinol Metab Clin North Am. 2005;34:1015–30.
3. Riggs BL, Melton LJ, Robb RA, et al. Population-based study of age and sex differences in bone volumetric density, size, geometry, and structure at different skeletal sites. J Bone Miner Res. 2004;19:1945–54.
4. Riggs BL, Melton LJ, Robb RA, et al. A population-based assessment of rates of bone loss at multiple skeletal sites: evidence for substantial trabecular bone loss in young adult women and men. J Bone Miner Res. 2008;23:205–14.
5. Seeman E. From density to structure: growing up and growing old on the surfaces of bone. J Bone Miner Res. 1997;12:509–21.
6. Khosla S, Riggs BL, Atkinson EJ, et al. Effects of sex and age on bone microstructure at the ultradistal radius: a population-based noninvasive in vivo assessment. J Bone Miner Res. 2006;21:124–31.
7. Albright F, Smith PH, Richardson AM. Postmenopausal osteoporosis. JAMA. 1941;116:2465–74.
8. Khosla S, Atkinson EJ, Melton LJ, et al. Effects of age and estrogen status on serum parathyroid hormone levels and biochemical markers of bone turnover in women: a population-based study. J Clin Endocrinol Metab. 1997;82:1522–7.
9. Hughes DE, Dai A, Tiffee JC, et al. Estrogen promotes apoptosis of murine osteoclasts mediated by TGF-beta. Nat Med. 1996;2:1132–6.
10. Manolagas SC. Birth and death of bone cells: basic regulatory mechanisms and implications for the pathogenesis and treatment of osteoporosis. Endocr Rev. 2000;21:115–37.
11. Garnero P, Sornay-Rendu E, Chapuy M, et al. Increased bone turnover in late postmenopausal women is a major determinant of osteoporosis. J Bone Miner Res. 1996;11:337–49.
12. Young MM, Nordin BEC. Effects of natural and artificial menopause on plasma and urinary calcium and phosphorus. Lancet. 1967;2:118–20.
13. Gennari C, Agnusdei D, Nardi P, et al. Estrogen preserves a normal intestinal responsiveness to 1,25-dihydroxyvitamin D3 in oophorectomized women. J Clin Endocrinol Metab. 1990;71:1288–93.
14. Riggs BL, Khosla S, Melton LJ. A unitary model for involutional osteoporosis: estrogen deficiency causes both type I and type II osteoporosis in postmenopausal women and contributes to bone loss in aging men. J Bone Miner Res. 1998;13:763–73.
15. McKane WR, Khosla S, Burritt MF, et al. Mechanism of renal calcium conservation with estrogen replacement therapy in women in early postmenopause — a clinical research center study. J Clin Endocrinol Metab. 1995;80:3458–64.
16. Chow J, Tobias JH. Colston KW, et al. Estrogen maintains trabecular bone volume in rats not only by suppression of bone resorption but also by stimulation of bone formation. J Clin Invest. 1992;89:74–8.
17. Qu Q, Perala-Heape M, Kapanen A, et al. Estrogen enhances differentiation of osteoblasts in mouse bone marrow culture. Bone. 1998;22:201–9.
18. Ernst M, Heath JK, Rodan GA. Estradiol effects on proliferation, messenger ribonucleic acid for collagen and insulin-like growth factor-I, and parathyroid hormone-stimulated adenylate cyclase activity in osteoblastic cells from calvariae and long bones. Endocrinology. 1989;125:825–33.
19. Oursler MJ, Cortese C, Keeting PE, et al. Modulation of transforming growth factor-beta production in normal human osteoblast-like cells by 17beta-estradiol and parathyroid hormone. Endocrinology. 1991;129:3313–20.
20. Mirza FS, Padhi ID, Raisz LG, et al. Serum sclerostin levels negatively correlate with parathyroid hormone levels and free estrogen index in postmenopausal women. J Clin Endocrinol Metab. 2010;95:1991–7.
21. Modder UI, Roforth MM, Hoey K, et al. Effects of estrogen on osteoprogenitor cells and cytokines/bone-regulatory factors in postmenopausal women. Bone. 2011;49:202–7.
22. Farr JN, Roforth MM, Fujita K, et al. Effects of age and estrogen on skeletal gene expression in humans as assessed by RNA sequencing. PLoS One. 2015;10:e0138347.
23. Niziolek PJ, Bullock W, Warman ML, et al. Missense mutations in LRP5 associated with high bone mass protect the mouse skeleton from disuse- and ovariectomy-induced osteopenia. PLos One. 2015;10:e0140775.
24. Khastgir G, Studd J, Holland N, et al. Anabolic effect of estrogen replacement on bone in postmenopausal women with osteoporosis: histomorphometric evidence in a longitudinal study. J Clin Endocrinol Metab. 2001;86:289–95.
25. Tu X, Delgado-Calle J, Condon KW, et al. Osteocytes mediate the anabolic actions of canonical Wnt/β-catenin signaling in bone. Proc Natl Acad Sci USA. 2015;112:E478–E86.
26. Weivoda MM, Ruan M, Hachfeld CM, et al. Wnt signaling inhibits osteoclast differentiation by activating canonical and noncanonical cAMP/PKA pathways. J Bone Miner Res. 2016;31:65–75.

27. Eghbali-Fatourechi C, Khosla S, Sanyal A, et al. Role of RANK ligand in mediating increased bone resolution in early postmenopausal women. J Clin Invest. 2003;111:1221–30.
28. Hofbauer LC, Khosla S, Dunstan CR, et al. Estrogen stimulates gene expression and protein production of osteoprotegerin in human osteoblastic cells. Endocrinology. 1999;140:4367–70.
29. Jilka RL, Hangoc G, Girasole G, et al. Increased osteoclast development after estrogen loss: mediation by interleukin-6. Science. 1992;257:88–91.
30. Ammann P, Rizzoli R, Bonjour J, et al. Transgenic mice expressing soluble tumor necrosis factor-receptor are protected against bone loss caused by estrogen deficiency. J Clin Invest. 1997;99:1699–703.
31. Tanaka S, Takahashi N, Udagawa N, et al. Macrophage colony-stimulating factor is indispensable for both proliferation and differentiation of osteoclast progenitors. J Clin Invest. 1993;91:257–63.
32. Kawaguchi H, Pilbeam CC, Vargas SJ, et al. Ovariectomy enhances and estrogen replacement inhibits the activity of bone marrow factors that stimulate prostaglandin production in cultured mouse calvariae. J Clin Invest. 1995;96:539–48.
33. Charatcharoenwitthaya N, Khosla S, Atkinson EJ, et al. Effect of blockade of TNF-α and interleukin-1 action on bone resorption in early postmenopausal women. J Bone Miner Res. 2007;22:724–9.
34. Srivastava S, Toraldo G, Weitzmann MN, et al. Estrogen decreases osteoclast formation by downregulating receptor activator of NF-kB ligand (RANKL)-induced JNK activation. J Biol Chem. 2001;276:8836–40.
35. Nakamura T, Imai Y, Matsumoto T, et al. Estrogen prevents bone loss via estrogen receptor alpha and induction of fas ligand in osteoclasts. Cell. 2007;130:811–23.
36. Martin-Millan M, Almeida M, Ambrogini E, et al. The estrogen receptor-alpha in osteoclasts mediates the protective effects of estrogens on cancellous but not cortical bone. Mol Endocrinol. 2010;24:323–34.
37. Ebeling PR, Atley LM, Guthrie JR, et al. Bone turnover markers and bone density across the menopausal transition. J Clin Endocrinol Metab. 1996;81:3366–71.
38. Sun L, Peng Y, Sharrow AC, et al. FSH directly regulates bone mass. Cell. 2006;125:247–60.
39. Drake MT, McCready LK, Hoey KA, et al. Effects of suppression of follicle-stimulating hormone secretion on bone resorption markers in postmenopausal women. J Clin Endocrinol Metab. 2010;95:5063–8.
40. Omodei U, Mazziotti G, Donarini G, et al. Effects of recombinant follicle-stimulating hormone on bone turnover markers in infertile women undergoing in vitro fertilization procedure. J Clin Endocrinol Metab. 2013;98:330–6.
41. Uihlein AV, Finkelstein JS, Lee H, et al. FSH suppression does not affect bone turnover in eugonadal men. J Clin Endocrinol Metab. 2014;99:2510–15
42. Falahati-Nini A, Riggs BL, Atkinson EJ, et al. Relative contributions of testosterone and estrogen in regulating bone resorption and formation in normal elderly men. J Clin Invest. 2000;106:1553–60.
43. Kasperk CH, Wergedal JE, Farley JR, et al. Androgens directly stimulate proliferation of bone cells in vitro. Endocrinology. 1989;124:1576–8.
44. Khosla S, Melton LJ, Atkinson EJ, et al. Relationship of serum sex steroid levels and bone turnover markers with bone mineral density in men and women: a key role for bioavailable estrogen. J Clin Endocrinol Metab. 1998;83:2266–74.
45. Slemenda CW, Longcope C, Zhou L, et al. Sex steroids and bone mass in older men: Positive associations with serum estrogens and negative associations with androgens. J Clin Invest. 1997;100:1755–9.
46. Szulc P, Munoz F, Claustrat B, et al. Bioavailable estradiol may be an important determinant of osteoporosis in men: the MINOS study. J Clin Endocrinol Metab. 2001;86:192–9.
47. Khosla S. Melton LJ, Atkinson EJ, et al. Relationship of serum sex steroid levels to longitudinal changes in bone density in young versus elderly men. J Clin Endocrinol Metab. 2001;86:3555–61.
48. Finkelstein JS, Lee H, Leder BZ, et al. Gonadal steroid-dependent effects on bone turnover and bone mineral density in men. J Clin Invest. 2016;126:1114–25.
49. Ho KY, Evans WS, Blizzard RM, et al. Effects of sex and age on the 24-hour profile of growth hormone secretion in man: importance of endogenous estradiol concentrations. J Clin Endocrinol Metab. 1987;64:51–8.
50. Boonen S, Mohan S, Dequeker J, et al. Downregulation of the serum stimulatory components of the insulin-like growth factor (IGF) system (IGF-I, IGF-II, IGF binding protein [BP]-3, and IGFBP-5) in age-related (type II) femoral neck osteoporosis. J Bone Miner Res. 1999;14:2150–8.
51. Amin S, Riggs BL, Atkinson EJ, et al. A potentially deleterious role of IGFBP-2 on bone density in aging men and women. J Bone Miner Res. 2004;19:1075–83.
52. Moerman EJ, Teng K, Lipschitz DA, et al. Aging activates adipogenic and suppresses osteogenic programs in mesenchymal marrow stroma/stem cells: the role of PPAR-gamma2 transcription factor and TGF-beta/BMP signaling pathways. Aging Cell. 2004;3:379–89.
53. Tchkonia T, Zhu Y, Van Deuersen J, et al. Cellular senescence and the senescent secretory phenotype: therapeutic opportunities. J Clin Invest. 2013;123:966–72.
54. Farr JN, Fraser DG, Wang H, et al. Identification of senescent cells in the bone microenvironment. J Bone Miner Res. 2016;31(11):1920–9.
55. Cooper C, Melton LJ. Epidemiology of osteoporosis. Trends Endocrinol Metab. 1992;3:224–9.

53

Juvenile Osteoporosis

Francis H. Glorieux[1] and Craig Munns[2]

[1]*Shriners Hospital for Children-Canada and McGill University, Montreal, QC, Canada*
[2]*Sydney Medical School, University of Sydney, and the Institute of Endocrinology and Diabetes, The Children's Hospital at Westmead, Westmead, NSW, Australia*

INTRODUCTION

Taken literally, the phrase "juvenile osteoporosis" means "osteoporosis in children and adolescents," and thus does not refer to any particular form of osteoporosis in this age group. However, in the scientific literature and in clinical practice, the term "juvenile osteoporosis" is usually used to refer to idiopathic juvenile osteoporosis (IJO). This chapter therefore discusses IJO as a primary disease rather than the entirety of largely secondary osteoporotic conditions that may occur in young people.

Osteoporosis in childhood and adolescence may result from mutations in genes principally affecting the amount and quality of the fibrous component of bone, presenting clinically as osteogenesis imperfecta (OI), or secondary to a spectrum of diverse conditions, such as prolonged immobilization, and chronic inflammatory disease. Bone loss may be worsened by treatment with anticonvulsants and/or steroids, but may also improve as the underlying condition improves. Life-threatening diseases such as leukemia also may present with osteoporotic fractures, particularly of the vertebrae. It is clearly important to exclude such causes of osteoporosis. If no underlying cause can be detected, IJO is said to be present.

IJO was first described as a separate entity by Dent and Friedman [1] five decades ago. According to the classical description, IJO is a self-limiting disease that develops in a prepubertal, previously healthy child, leads to metaphyseal and vertebral compression fractures, and is characterized radiologically by radiolucent areas in the metaphyses of long bones, dubbed "neo-osseous osteoporosis" [2]. This implies a disorder of trabecular bone architecture and mass, possibly related to changes in the hormonal milieu around the time of the growth spurt.

It is clear, however, that there are many children and adolescents who have low bone mass (defined as a body size-adjusted BMC or BMD measured by DXA at the spine or total body that is >2 SD below the mean, ie, a Z-score below −2) and who sustain recurrent fractures after minimal trauma, but whose clinical findings do not correspond to the classical description of Dent and Friedman.

In the Pediatric Position Statements of the International Society for Clinical Densitometry published in 2007, these patients would simply fulfill the diagnosis of "osteoporosis." Thus, it may be useful to distinguish "classical IJO" (for patients whose presentation is similar to the description of Dent and Friedman) from "osteoporosis in the wider sense" (for patients who do not match the description of Dent and Friedman, but nevertheless have unexplained fractures with low bone mass).

Most reviews on the topic state that IJO is an extremely rare disease, because fewer than 200 patients have been mentioned in the literature with this diagnosis. This is probably because few patients present with the classical picture. However, most clinicians who see children and adolescents with fractures could probably list a few of their patients who have osteoporosis without recognizable etiology. In our clinical settings, IJO is ~10 times less common than OI.

PATHOPHYSIOLOGY

The etiology of IJO is unknown. One study found a normal rise in serum osteocalcin in six IJO patients after calcitriol was administered orally, which was postulated to indicate "normal osteoblast function" [3]. However, the fact that osteoblasts in this test released normal

---

Primer on the Metabolic Bone Diseases and Disorders of Mineral Metabolism, Ninth Edition. Edited by John P. Bilezikian.
© 2019 American Society for Bone and Mineral Research. Published 2019 by John Wiley & Sons, Inc.
Companion website: www.wiley.com/go/asbmrprimer

amounts of osteocalcin into the circulation does not necessarily mean that they also deposited matrix on the bone surface in a normal fashion. Early histomorphometric reports on IJO were limited to static methods to quantify bone metabolism, described single cases [4–8], or did not have adequate control data. No conclusive picture emerged from these reports.

More recent studies using dynamic histomorphometry showed that IJO was characterized by a markedly reduced activation frequency and therefore low remodeling activity [9]. In addition, the amount of bone formed at each remodeling site was abnormally low. No evidence was found for increased bone resorption. Interestingly, the bone formation defect was limited to bone surfaces that were exposed to the bone marrow environment; no abnormalities were detected in intracortical and periosteal surfaces [10]. In one study, bone matrix composition in IJO and controls was studied by infrared imaging. Increased heterogeneity for mineral-to-matrix and collagen maturity ratios were found, reflecting a slower mineralization process [11].

These results suggested that, in IJO, impaired osteoblast performance decreases the ability of cancellous bone to adapt to the increasing mechanical needs during growth. This results in load failure at sites where cancellous bone is essential for stability. The initial trigger of the decrease in osteoblast performance remains nevertheless elusive. Three reports have indicated that heterozygous mutations in the low-density lipoprotein receptor-related protein 5 (LRP5) can result in low bone mass with fractures in some children [12–14]. However, no mutations in LRP5 and LRP6 could be identified in 10 IJO children evaluated by a single center, pointing to likely heterogeneity in IJO [15]. Next generation sequencing studies may help revealing causative genes.

CLINICAL FEATURES

Classical IJO typically develops in a prepubertal (mostly between 8 and 12 years of age), previously healthy child of either sex [16]. However, in one series, 21 children who were presented as having IJO were recorded as having a mean age at onset of 7 years with a range of 1 to 13 years [17].

Symptoms generally begin with an insidious onset of pain in the lower back, hips, and feet, and difficulty walking. Knee and ankle pain, and fractures of the lower extremities may be present, as well as diffuse muscle weakness. Vertebral compression fractures are frequent, resulting in a short back (Fig. 53.1). Long bone fractures, mostly at metaphyseal sites, may occur. Physical examination may be entirely normal or show thoracolumbar kyphosis or kyphoscoliosis, pigeon chest deformity, loss of height, deformities of the long bones, and a limp.

Fig. 53.1. Lateral lumbar spine radiograph of a 10-year-old girl with idiopathic juvenile osteoporosis. Compression fractures of all vertebral bodies and severe osteoporosis are evidence. At the time of this radiograph, lumbar spine areal BMD Z-score was –4.9 (at a height Z-score of –2.5).

RADIOLOGICAL FEATURES

Children with fully expressed classical IJO present with generalized osteopenia and collapsed or biconcave vertebrae. Disc spaces may be widened asymmetrically because of wedging of the vertebral bodies. Long bones usually have normal diameter and cortical width, unlike the thin, gracile bones of children with OI. The typical radiographical finding in IJO is neoosseous osteoporosis, a radiolucent band at sites of newly formed metaphyseal bone. This localized metaphyseal weakness can give rise to fractures, often at the distal tibias and adjacent to the knee and hip joints. Nevertheless, "neoosseous osteoporosis" is not a prerequisite for diagnosing IJO.

BIOCHEMICAL FINDINGS

Biochemical studies of bone and mineral metabolism have not detected any consistent abnormality in children with IJO [18, 19]; bone resorption was increased in a young woman before pregnancy and increased further, with associated bone loss (25% decrease in spine BMD) and vertebral crush fractures [20].

BONE BIOPSY

Iliac bone biopsies show low trabecular bone volume but largely preserved core width (ie, a normal outer size of the biopsy specimen) and cortical width [9, 10]. Tetracycline double labeling shows a low extent of mineralizing surface (a sign of decreased remodeling activity) and low mineral apposition rate (a sign of weakness of the individual osteoblast team at a remodeling site). There is no indication of a mineralization defect. Osteoclasts are normal in appearance and number.

DIFFERENTIAL DIAGNOSIS

The diagnosis of IJO is made by the exclusion of known etiologies for low bone mass and fractures. The list of conditions that may be associated with bone fragility in children and adolescents is shown in Table 53.1. The exclusion of most of these disorders is usually not difficult. The most frequent diagnostic problem facing a clinician is probably to separate IJO from OI type I.

Table 53.2 presents the typical distinguishing features between IJO and OI type I. Apart from bone fragility and low bone mass, most patients with OI type I have associated extraskeletal connective tissue signs, such as blue or grey scleral hue, dentinogenesis imperfecta, joint hyperlaxity, and Wormian bones (on skull X-rays). However, the extraskeletal involvement can be absent or too subtle to be clinically recognizable in some OI patients. In this situation, genetic analysis of the genes that encode the two collagen type I α-chains *(COL1A1* and *COL1A2)* can be helpful. Mutations affecting a glycine residue in either gene or those leading to a quantitative defect in *COL1A2* expression are diagnostic of OI. DNA-based collagen type I analysis now detects mutations in more than 99% of those individuals where mutations exist. LRP5 sequencing may be informative in 10% to 15% of cases [12, 13].

An iliac bone biopsy, preferably after tetracycline double labeling, may also contribute to clarifying the diagnosis. Microscopically, a "lack of activity" is usually noted in IJO, whereas there is "hypercellularity" in OI. In histomorphometric terms, this translates into low activation frequency and bone surface-based remodeling parameters in IJO and an increase in these values in OI. Also, hyperosteocytosis is a common feature in OI, whereas the amount of osteocytes appears to be normal in IJO.

Fractures and low bone mass may also occur in healthy prepubertal children. Indeed, during late prepuberty and early puberty, fracture rates are almost as high as in postmenopausal women [21–23]. Similar to IJO, such fractures frequently involve metaphyseal bone sites, especially the distal radius. This may reflect problems in the adaptation of the skeleton, in particular the metaphyseal cortex, to the increasing mechanical needs during growth [24]. Growing children and adolescents who have had a few forearm fractures and who have borderline low areal BMD at the spine are frequently encountered in pediatric bone clinics. We propose that classical IJO should only be diagnosed when vertebral compression fractures are present (with or without extremity fractures).

Table 53.1. Forms of osteoporosis in children, according to current literature.

I Primary
Osteogenesis imperfecta
Idiopathic juvenile osteoporosis

II. Secondary
Endocrine disorders
    Cushing syndrome
    Thyrotoxicosis
    Anorexia nervosa

Inflammatory disorders
    Juvenile arthritis
    Dermatomyositis
    Systemic lupus erythematosis
    Inflammatory bowel disease
    Cystic fibrosis
    Chronic hepatitis

Malabsorption syndromes
    Biliary atresia

Inborn errors of metabolism
    Homocystinuria
    Glycogen storage disease type 1

Immobilization
    Cerebral palsy
    Duchenne dystrophy

Hematology/oncology
    Acute lymphoblastic leukemia
    Thalassemia
    Severe congenital neutropenia

TREATMENT

There is no treatment with proven benefit to the patient. The effect of any kind of medical intervention is difficult to judge in IJO, because the disease is rare, has a variable course, and is said to resolve without treatment. Long-term outcome studies are lacking, however.

Given the current enthusiasm for pediatric bisphosphonate therapy, many IJO patients probably are receiving treatment with such drugs. The most commonly used compounds are pamidronate [25] and zoledronic acid [26]. The latter is now favored as more convenient to administer. We would normally restrict such an intervention to those children with multiple

Table 53.2. Differential diagnosis between idiopathic juvenile osteoporosis (IJO) and osteogenesis imperfecta (OI) type I.

	IJO	OI type I
Family history	Negative	Often positive
Onset	Late prepubertal	Birth or soon after
Duration	1-5 year	Lifelong
Clinical finding	Metaphyseal fractures	Long bone diaphyseal fractures
	No signs of connective tissue involvement	Blue sclerae, joint hyperlaxity, sometimes abnormal dentition
	Abnormal gait	
Growth rate	Normal	Normal or low
Radiological finding	Vertebral compression fractures	Vertebral compression fractures
	Long bones: predominantly metaphyseal involvement "neoosseous osteoporosis"	"Narrow bones" [low diameter]
	No Wormian bones	Wormian bones (skull)
Bone biopsy	Decreased bone turnover	Increased bone turnover
	Normal amount of osteocytes	Hyperosteocytosis
Genetic testing	Negative	Mutations affecting collagen type I in most patients

vertebral crush fractures, who may also experience debilitating chronic bone pain.

A number of case reports have described increasing BMD and clinical improvement after treatment with bisphosphonates was started [27–29]. Medical therapies should complement orthopedic and rehabilitative measures such as physiotherapy in all such cases. Review at 6-month intervals is also warranted in children not receiving bisphosphonates. Changes in the shape of the spine should be monitored carefully, and early referral to a specialist pediatric spine surgeon should be made in any progressive cases.

PROGNOSIS

The disease process appears to be active only in growing children, and spontaneous recovery is the rule after 3 to 5 years of evolution [17]. However, in some of the most severe cases reported to date, deformities and severe functional impairment persisted, which left them wheelchair bound with cardiorespiratory abnormalities. Preventing such deformities with attendant loss of function should be the focus of attention during the active phase of the disease.

REFERENCES

1. Dent CE, Friedman M. Idiopathic juvenile osteoporosis. Q J Med. 1965;34:177–210.
2. Dent CE. Osteoporosis in childhood. Postgrad Med J. 1977;53:450–7.
3. Bertelloni S, Baroncelli GI, Di Nero G, et al. Idiopathic juvenile osteoporosis: evidence of normal osteoblast function by 1,25-dihydroxyvitamin D3 stimulation test. Calcif Tissue Int. 1992;51:20–3.
4. Cloutier MD, Hayles AB, Riggs BL, et al. Juvenile osteoporosis: report of a case including a description of some metabolic and microradiographic studies. Pediatrics. 1967;40:649–55.
5. Gooding CA, Ball JH. Idiopathic juvenile osteoporosis. Radiology. 1969;93:1349–50.
6. Jowsey J, Johnson KA. Juvenile osteoporosis: bone findings in seven patients. J Pediatr. 1972;81:511–17.
7. Smith R. Idiopathic osteoporosis in the young. J Bone Joint Surg Br. 1980;62-B:417–27.
8. Evans RA, Dunstan CR, Hills E. Bone metabolism in idiopathic juvenile osteoporosis: a case report. Calcif Tissue Int. 1983;35:5–8.
9. Rauch F, Travers R, Norman ME, et al. Deficient bone formation in idiopathic juvenile osteoporosis: a histomorphometric study of cancellous iliac bone. J Bone Miner Res. 2000;15:957–63.
10. Rauch F, Travers R, Norman ME, et al. The bone formation defect in idiopathic juvenile osteoporosis is surface-specific. Bone. 2002;31:85–9.
11. Garcia I, Chiodo V, Ma Y, et al. Evidence of altered matrix composition in iliac crest biopsies from patients with idiopathic juvenile osteoporosis. Connect Tissue Res. 2016;57:28–37.
12. Toomes C, Bottomley HM, Jackson RM, et al. Mutations in LRP5 or FZD4 underlie the common familial exudative vitreoretinopathy locus on chromosome 11q. Am J Hum Genet. 2004;74:721–30.
13. Hartikka H, Makitie O, Mannikko M, et al. Heterozygous mutations in the LDL receptor-related protein 5 (LRP5) gene are associated with primary osteoporosis in children. J Bone Miner Res. 2005;20:783–9.
14. Fahiminiya S, Majewski J, Roughley P, et al. Whole-exome sequencing reveals a heterozygous LRP5 mutation in a 6-year-old boy with vertebral compression fractures and low trabecular bone density. Bone. 2013;57:41–6.

15. Franceschi R, Vincenzi M, Camilot M, et al. Idiopathic juvenile osteoporosis: clinical experience from a single centre and screening of LRP5 and LRP6 genes. Calcif Tissue Int. 2015;96:575–9.
16. Teotia M, Teotia SP, Singh RK. Idiopathic juvenile osteoporosis. Am J Dis Child. 1979;133:894–900.
17. Smith R. Idiopathic juvenile osteoporosis: experience of twenty-one patients. Br J Rheumatol. 1995;34:68–77.
18. Saggese G, Bertelloni S, Baroncelli GI, et al. Serum levels of carboxyterminal propeptide of type I procollagen in healthy children from 1st year of life to adulthood and in metabolic bone diseases. Eur J Pediatr. 1992;151:764–8.
19. Saggese G, Bertelloni S, Baroncelli GI, et al. Mineral metabolism and calcitriol therapy in idiopathic juvenile osteoporosis. Am J Dis Child. 1991;145:457–62.
20. Black AJ, Reid R, Reid DM, et al. Effect of pregnancy on bone mineral density and biochemical markers of bone turnover in a patient with juvenile idiopathic osteoporosis. J Bone Miner Res. 2003;18:167–71.
21. Landin LA. Epidemiology of children's fractures. J Pediatr Orthop B. 1997;6:79–83.
22. Cooper C, Dennison EM, Leufkens HG, et al. Epidemiology of childhood fractures in Britain: a study using the general practice research database. J Bone Miner Res. 2004;19:1976–81.
23. Khosla S, Melton W III, Dekutoski MB, et al. Incidence of childhood distal forearm fractures over 30 years: a population-based study. JAMA. 2003;290:1479–85.
24. Rauch F, Neu C, Manz F, et al. The development of metaphyseal cortex-implications for distal radius fractures during growth. J Bone Miner Res. 2001;16:1547–55.
25. Baroncelli GI, Vierucci F, Bertelloni S, et al. Pamidronate treatment stimulates the onset of recovery phase reducing fracture rate and skeletal deformities in patients with idiopathic juvenile osteoporosis: comparison with untreated patients. J Bone Miner Metab. 2013;31:533–43.
26. Ooi HL, Briody J, Biggin A, et al. Intravenous zoledronic acid given every 6 months in childhood osteoporosis. Horm Res Paediatr. 2013;80:179–84.
27. Hoekman K, Papapoulos SE, Peters AC, et al. Characteristics and bisphosphonate treatment of a patient with juvenile osteoporosis. J Clin Endocrinol Metab. 1985;61:952–6.
28. Brumsen C, Hamdy NA, Papapoulos SE. Long-term effects of bisphosphonates on the growing skeleton. Studies of young patients with severe osteoporosis. Medicine (Baltimore). 1997;76:266–83.
29. Kauffman RP, Overton TH, Shiflett M, et al. Osteoporosis in children and adolescent girls: case report of idiopathic juvenile osteoporosis and review of the literature. Obstet Gynecol Surv. 2001;56:492–504.

54

Transplantation Osteoporosis

Peter R. Ebeling

Department of Medicine, School of Clinical Sciences, Monash University, Clayton, VIC, Australia

INTRODUCTION

Improved survival rates necessitate a greater awareness of long-term complications of transplantation such as fractures and osteoporosis [1, 2]. Both the presence of preexisting bone disease and the type of transplant will determine the requirement for, and duration of, posttransplant therapy (Table 54.1).

PREEXISTING BONE DISEASE

Chronic kidney disease

Chronic kidney disease–mineral and bone disorder (CKD–MBD) predominates and includes secondary hyperparathyroidism (SHPT), low turnover bone disease (osteomalacia, adynamic bone disease, or aluminum bone disease), osteoporosis, mixed bone disease, and β_2-microglobulin amyloidosis. In addition, hypogonadism, both in men and women, metabolic acidosis, and medications (loop diuretics, heparin, warfarin, glucocorticoids, or immunosuppressive agents) also adversely affect bone health. Patients with CKD who have low BMD and bone turnover markers in the upper half of the normal premenopausal range are at the highest risk of fracture [3].

Adynamic bone disease needs exclusion before treatment with bisphosphonates. It is characterized by a scarcity of bone cells, reduced osteoid thickness, and a low bone formation rate on bone histomorphometry [4]. High serum phosphate and FGF-23 levels, can override the stimulatory effect of PTH in early CKD and cinacalcet, calcium, and calcitriol use may also reduce bone turnover. Bone histomorphometry is the best method to evaluate bone remodeling in CKD, whereas the combination of a low bone-specific alkaline phosphatase level with a slightly increased or normal PTH level is less specific.

In hemodialysis patients, low BMD and fractures at all skeletal sites are common. Vertebral fracture prevalence is as high as 21% and the relative risk of hip fracture is increased two- to 14-fold. Fracture risk is increased with older age, female sex, white race [5], hemodialysis duration [6], diabetic nephropathy, peripheral vascular disease [7], low spine BMD, and low bone turnover.

Congestive heart failure

Osteoporotic BMD affects up to 40% of patients with congestive heart failure (CHF), with a 2.5-fold increase in fracture risk in one study [8]. In another study of patients awaiting heart transplantation, LS osteopenia was found in 43%, and osteoporosis in 7% [9]. Mild renal insufficiency, vitamin D deficiency, SHPT and increased bone resorption markers, and loop diuretics may contribute.

End-stage liver disease

Osteoporosis and fractures commonly accompany chronic liver disease and low BMD can be found in the majority of patients undergoing liver transplantation. Osteoporosis at the spine or hip has been reported in 11% to 52% of patients awaiting liver transplantation [1, 10]. Low body mass index (BMI) before liver transplant, cholestatic liver disease and older age are important risk factors [11, 12] for osteoporosis.

Chronic respiratory failure

Osteoporosis may be most common in patients awaiting lung transplantation, affecting up to 61% [13]. Hypoxia, hypercapnia, smoking, low BMI, and glucocorticoids (GC) all contribute [13]. Fragility fractures are extremely

---

Primer on the Metabolic Bone Diseases and Disorders of Mineral Metabolism, Ninth Edition. Edited by John P. Bilezikian.
© 2019 American Society for Bone and Mineral Research. Published 2019 by John Wiley & Sons, Inc.
Companion website: www.wiley.com/go/asbmrprimer

Table 54.1. Risk factors for posttransplant bone loss and fractures.

Contributing Factors	Mechanisms
Aging	**Low pretransplant BMD**
Low body mass index	
Hypogonadism	
Calcium and vitamin D deficiency	
Tobacco	
Alcohol abuse	
Cholestasis (liver disease)	
Organ failure (heart, lung, liver, kidney)	
Pancreatic insufficiency (cystic fibrosis)	
Physical inactivity	
High dose prednisone	**Decreased bone formation**
	Direct effect
	Decreased gonadal function
	Reduced intestinal and renal calcium transport
Calcineurin inhibitors	**Increased bone resorption**
Cyclosporine or FK506	Decreased renal function and 1,25(OH)$_2$D
	Increased PTH secretion
	Possible direct effect
Calcineurin inhibitor	**Decreased bone formation**
Sirolimus	Possible direct effect

common in cystic fibrosis because additional risk factors (pancreatic insufficiency, vitamin D deficiency, calcium malabsorption, hypogonadism, genetic factors, and inactivity) exist.

Candidates for stem cell transplantation

Bone loss in stem cell transplantation (SCT) recipients is related both to the underlying diseases and to chemotherapeutic drugs. These include GC-induced decreases in bone formation and serum 1,25-(OH)$_2$D$_3$, as well as hypogonadism secondary to the effects of high-dose chemotherapy, total body irradiation (TBI), and GCs. Women are particularly sensitive to the adverse effects of TBI and chemotherapy on gonadal function. Ovarian insufficiency occurs in the majority [14, 15]. In men, testosterone levels decline acutely after bone marrow transplantation (BMT) then return to normal in most men [16–18]. Long-term impairment of spermatogenesis with elevated FSH occurs in 47% of men [14, 15]. In patients studied after chemotherapy but before BMT, osteopenia was present in 24% and osteoporosis in 4% [18]. Chemotherapy-induced increases in cytokines, including macrophage chemoattractant protein-1 (MCP-1), are associated with increased bone resorption [19].

Candidates for intestinal transplantation

Osteoporosis occurs in 36% of candidates for intestinal transplantation, with age and duration of parenteral nutrition being significant risk factors. On average, spine and hip BMD Z scores are −1.5 [20].

SKELETAL EFFECTS OF IMMUNOSUPPRESSIVE DRUGS

Glucocorticoids

GC exposure varies with the organ transplanted and the number of rejection episodes. High doses commonly prescribed immediately after transplantation are rapidly weaned. Doses increase during rejection episodes. The highest GC-associated rates of bone loss are in the first 3 to 12 months posttransplant. Trabecular sites are predominantly affected. The use of calcineurin inhibitors and more recent immunosuppressive regimens has both limited GC use and slowed rates of posttransplant bone loss.

However, even small doses of GC are associated with marked increases in fracture risk in epidemiological studies [21] and reduce bone formation, whereas receptor activator for NFkB-ligand (RANK-L) is upregulated. The immediate posttransplant period is characterized by high bone remodeling and increased bone resorption. Hyperparathyroidism results from GC-induced reductions of intestinal and renal calcium absorption. Early GC withdrawal after kidney transplantation has been associated with a 31% fracture risk reduction [22], including fractures associated with hospitalization, and with improved BMD parameters 1-year posttransplantation. However, a recent retrospective cohort study that compared two cohorts of patients undergoing kidney transplantation 5 years apart, found a lower incidence of fractures in more recent trials, despite less GC withdrawal [23]. Steroid-sparing or withdrawal has the potential to improve bone loss posttransplantation. However, this needs to be balanced against the potential risk of higher rates of rejection.

Calcineurin inhibitors

Cyclosporine (CsA) has independent adverse effects to increase bone turnover [24]. Although CsA treatment could result in high bone turnover after transplantation, it is reassuring that kidney transplant patients receiving CsA without GCs [25, 26] do not lose bone and fractures are also reduced [22]. Tacrolimus (FK506), another calcineurin inhibitor, also causes trabecular bone loss in the rat [19]. Both cardiac [27] and liver [28] transplant recipients sustained rapid bone loss with tacrolimus. However, tacrolimus may cause less bone loss in humans than CsA [29, 30], and may also protect the skeleton by reducing GC use.

Other immunosuppressive agents

Limited information is available regarding the effects of other immunosuppressive drugs on BMD and bone metabolism. However, azathioprine, sirolimus (rapamycin), mycophenelate mofetil and daclizumab may also protect the skeleton by reducing GC use. In vitro studies suggest rapamycim inhibits osteoblast proliferation and differentiation [31], but more clinical data are required.

MANAGEMENT OF TRANSPLANTATION OSTEOPOROSIS

Diagnostic strategies

Before organ transplantation

All candidates going onto the waiting list for organ transplantation should have bone densitometry, by DXA of the hip and spine. Spinal X-rays should be performed to diagnose prevalent fractures. Any secondary causes of osteoporosis should be identified and treated. Common secondary causes include hyperparathyroidism, hypogonadism, smoking, use of loop diuretics, low dietary calcium intake, and vitamin D deficiency (<20 ng/mL).

Vitamin D deficiency [32] should be corrected and all patients should receive adequate calcium and vitamin D (1000 to 1300 mg of calcium and at least 800 IU of vitamin D per day). Replacement doses of vitamin D may need to be higher (2000 IU of vitamin D per day), but should be selected to achieve a 25(OH)D concentration ≥30 ng/mL. Patients with CKD should be evaluated for CKD–MBD and for adynamic bone disease and secondary hyperparathyroidism, in particular.

After organ transplantation

Risk factors for posttransplant bone loss and fractures are shown in Table 54.1. Bone loss is most rapid immediately after transplantation. Fractures often occur in the first year after transplantation and may affect patients with either low or normal pretransplant BMD. Therefore, the majority of patients may benefit from treatment instituted immediately after transplantation, with the exception of patients with CKD–MBD and adynamic bone disease. Patients who present after being transplanted months or years before should also be assessed for treatment.

Vitamin D deficiency is common posttransplantation and in long-term graft recipients. Vitamin D status is partly determined by demographic and lifestyle factors and deficiency is associated with poorer general health, lower serum albumin levels, and even decreased survival in these groups [32].

Most therapeutic trials have focused on the use of active vitamin D metabolites and antiresorptive drugs, particularly oral and intravenous bisphosphonates. Hormone therapy with estrogen ± progestin helps protect the skeleton in women receiving liver, lung, and bone marrow transplantation. Because amenorrhea is a common sequela of BMT in premenopausal women, they should receive HRT. However, it does not prevent bone loss after BMT. Hypogonadism is common in male cardiac and bone marrow transplant recipients, caused by chronic illness and hypothalamic-pituitary-adrenal suppression by GCs and CsA. Testosterone levels fall immediately after transplantation and normalize 6 to 12 months later. However, testosterone treatment alone does not prevent bone loss after cardiac transplantation or BMT in men.

Recent studies examined prevention of bone loss after transplantation (Table 54.2).

Kidney transplantation

Cross-sectional studies of patients evaluated several years after kidney transplantation have reported osteoporosis in 17% to 49% at the spine, 11% to 56% at the FN and 22% to 52% at the radius [1]. There is a correlation between cumulative GC dose and BMD. Rates of bone loss are greatest in the first 6 to 18 months after transplantation [33], and range from 4% to 9% at the spine and 5% to 8% at the hip. Bone loss has not been consistently related to sex, patient age, cumulative GC dose, rejection episodes, activity level, or PTH levels. Studies examining BMD after the first year or two do not consistently show ongoing bone loss. However, BMD remains low up to 20 years after transplantation. SHPT and low $1,25(OH)_2D$ levels also often persist [3, 34]. Fractures affect appendicular sites (hips, long bones, ankles, feet) more commonly than axial sites (spine and ribs) [34]. Women and patients transplanted for diabetic nephropathy are at particularly increased risk of fractures. The majority of fractures occur within the first 3 years. However, fractures continue to increase over time [35].

Prevention and treatment

Calcium and vitamin D supplementation alone does not prevent bone loss in renal transplant patients [36]. Bisphosphonates reduce bone loss after kidney transplantation [37]. The effects of bisphosphonates on bone loss and fractures during the first year after kidney transplantation was examined in a meta-analysis of 11 studies and 780 patients [38]. There was an increase of approximately 3% in both FN and LS BMD, and an overall reduction in fractures, but no significant reduction in vertebral fractures. Two more recent meta-analyses both confirmed the results of earlier studies, and showed improved BMD at the FN and LS. However, there was no difference in fracture incidence [39, 40]. Adynamic bone disease remains a concern with the use of bisphosphonates. Their use has been associated with an increase in biopsy proven adynamic bone disease, although the effect of this finding on fracture incidence remains

Table 54.2. Randomized controlled trials using vitamin D analogues or bisphosphonates for prevention of bone loss after heart, lung, liver, and bone marrow transplantation.

Transplant Type	First Author (Year)	n	Duration	Treatment Regimen	Control Regimen	Findings/Summary
Heart and lung	Sambrook (2000) [52]	65	24 months	**Calcitriol** 0.5–0.75 µg for 12 months or 24 months Calcium 600 mg/day	Placebo Calcium 600 mg/day	**BMD:** FN (but not LS) bone loss was attenuated in the calcitriol groups at 12 months. LS bone loss was similar among all three groups. **Fracture:** Not powered.
Lung (CF)	Aris (2000) [48]	37	24 months	**Pamidronate** 30 mg i.v. q 3 months Calcium 1000 mg/day Vitamin D 800 IU/day	Calcium 1000 mg/day Vitamin D 800 IU/day	**BMD:** LS and TH BMD increased significantly more in the pamidronate group versus controls. **Fracture:** No difference.
Heart	Shane (2004) [9]	149[a]	12 months	**Alendronate** 10 mg/day or **Calcitriol** 0.5 µg/day Calcium 945 mg/day Vitamin D 1000 IU/day	Non-randomized reference group	**BMD:** Similar small losses at LS and TH in both groups. Significantly less bone loss at LS and TH than reference group. **Fracture:** No difference.
Heart	Gil-Fraguas (2005) [57]	87		**Alendronate** 10 mg/day Calcitonin 200 IU/day		**BMD:** Less bone loss from FN in the alendronate group. **Fracture:** Fewer vertebral fractures than in calcitonin group (6 versus 15).
Heart	Fahrleitner-Pammer (2009) [58]	35		**Ibandronate** 2 mg i.v. q 3 months Calcium 1000 mg/day Vitamin D 400 IU/day		**BMD:** Bone loss from LS and FN prevented in ibandronate group. **Fracture:** Fewer vertebral fractures than in control group (2 versus 17).
Liver and multivisceral	Hommann (2002) [64]	36	12 months	**Ibandronate** 2 mg i.v. q 3 months Calcium 1000 mg/day Vitamin D 1000 IU/day	Calcium 1000 mg/day Vitamin D 1000 IU/day	**BMD:** LS, FN, and forearm BMD decreased initially in both groups. Reversal of bone loss with ibandronate observed after 12 months.
Liver	Ninkovic (2002) [6]	99	12 months	**Pamidronate** 60 mg i.v. given once before transplantation	No treatment	**BMD:** Significant, comparable bone loss at FN in pamidronate and control groups. **Fracture:** No difference.
Liver	Crawford (2006) [65]	62	12 months	**Zoledronic acid** 4 mg i.v. administered within 7 days of transplantation and at months 1, 3, 6, and 9 after transplant Calcium 600 mg/day Vitamin D 1000 IU/day	Placebo Calcium 600 mg/day Vitamin D 1000 IU/day	**BMD:** At 3 months, difference in bone loss from baseline was decreased in ZA group versus placebo. At 12 months, the differences in % bone loss was less. **Fracture:** Not powered.
Liver	Bodinghauer (2007) [66]	69		**Zoledronic acid** 4 mg i.v. 1–6, 9, and 12 months Calcium 600 mg/day Vitamin D 1000 IU/day		**BMD:** Less bone loss from LS (but not FN) in ZA group. **Fracture:** Fewer vertebral fractures than in control group (4 versus 11).
Liver	Monegal (2009) [67]	79		**Pamidronate** 90 mg i.v. 0 and 3 months Calcium 1000 mg/day Vitamin D 16,000 IU/day		**BMD:** Increase in LS in pamidronate group. Decrease FN in both groups. **Fracture:** More fractures in pamidronate group (15 versus 3).
Liver	Kaemmerer (2010) [68]	74		**Ibandronate** 2 mg i.v. q 3 months Calcium 1000 mg/day Vitamin D 800–1000 IU/day		**BMD:** Increase LS and less bone loss from FN in ibandronate group. **Fractures:** fewer fractures in ibandronate group (2 versus 8).

(Continued)

Table 54.2. (Continued)

Transplant Type	First Author [Year]	n	Duration	Treatment Regimen	Control Regimen	Findings/Summary
BMT	Tauchmanova (2003) [81]	34	12 months	**Risedronate** 5 mg/day Calcium 1 g/day Vitamin D 800 IU/day	Calcium 1 g/day Vitamin D 800 IU/day	**BMD:** LS BMD significantly increased in risedronate group at 6 and 12 months and decreased in the control group at 6 months. FN BMD decreased significantly in control group at 6 months only.
BMT	Tauchmanova (2005) [82]	32	12 months	**Zoledronic acid** 4 mg i.v. administered at 1, 2, and 3 months Calcium 500 mg/day Vitamin D 400 IU/day	Calcium 500 mg/day Vitamin D 400 IU/day	**BMD:** LS and FN BMD significantly increased in ZA group and did not change in the control group at 12 months.
BMT	Kananen (2005) [73]	99	12 months	**Pamidronate** 60 mg i.v. administered before transplant and 1, 2, 3, 6, and 9 months after transplant Calcium 1000 mg/day Vitamin D 800 IU/day Estrogen — women Testosterone — men	Calcium 1000 mg/day Vitamin D 800 IU/day Estrogen — women Testosterone — men	**BMD:** At 12 months, difference in bone loss from baseline at LS and TH was decreased in pamidronate group versus no infusion. No difference in bone loss from baseline at the FN. **Fracture:** Not powered.
BMT	Grigg (2006) [83]	116	24 months	**Pamidronate** 90 mg i.v. administered before transplant and every month after transplant for 12 months Calcium 1000 mg/day Calcitriol 0.25 µg/day for 24 months	Calcium 1000 mg/day Calcitriol 0.25 µg/day	**BMD:** At 12 months, difference in bone loss from baseline was decreased at LS, FN, and TH in pamidronate group versus no infusion. At 24 months, the difference in bone loss from baseline was only significant at the TH (3.9%). **Fracture:** Not powered.
BMT	Tauchmanova (2006) [86]	55	12 months	**Estrogen** 2 mg/day **Risedronate** 35 mg/week **Zoledronate** 4 mg i.v. at 0, 1, and 2 months Calcium 1000 mg/day Vitamin D 800 IU/day	Calcium 1000 mg/day Vitamin D 800 IU/day	**BMD:** At 12 months, bone loss occurred with calcium and vitamin D. LS BMD increased with risedronate and zoledronate, but FN BMD increased only with zoledronate. **Fractures:** Not powered.
BMT	Hari (2013) [85]	61	12 months	**Zoledronate** 4 mg i.v. at 0, 3, and 6 months Calcium 1000 mg/day Vitamin D 400–500 IU/day	Calcium 1000 mg/day Vitamin D 400–500 IU/day	**BMD:** Improvements in LS and FN BMD at 12 months. **Fractures:** Not powered.
BMT	Lu (2016) [87]	78	12 months	**Ibandronate** 3 mg i.v. at 0, 3, 6, and 9 months Calcium 500 mg/day Vitamin D 800 IU/day	Calcium 500 mg/day Vitamin D 800 IU/day	**BMD:** At 12 months, LS BMD was stable and higher than control group, whereas FN BMD decreased in both groups.

[a]Number randomized to alendronate or calcitriol, 27 prospectively recruited nonrandomized patients served as a reference group.
BMT = bone marrow transplantation; CF = cystic fibrosis; ZA = zoledronic acid.

uncertain [41]. In the meta-analyses discussed above, bisphosphonate therapy was superior to active vitamin D in preserving BMD. However, the use of either therapy was beneficial compared with no treatment. An unexpected finding was a reduction in the risk of graft rejection associated with bisphosphonate therapy. The efficacy and safety of bisphosphonates in those with an estimated glomerular filtration rate (eGFR) of more than 30 mL/min remains unclear. Similarly, no consensus exists about duration of treatment. However, given that bone loss is greatest in the first 12 months, benefits are likely to be greatest in this period.

Denosumab is a fully human monoclonal antibody that inhibits RANKL, and decreases the differentiation and activity of osteoclasts, reduces bone resorption, and increases BMD. In a large study of osteoporotic women [42], denosumab improved BMD and decreased fracture risk, and was safe in those with reduced eGFR including CKD stages 3 to 4. However, its use in an end-stage CKD cohort was associated with severe hypocalcemia [43]. Its safety postkidney transplant was assessed in an open-label prospective study of 90 patients, where denosumab was administered at baseline and 6 months [44]. There was an increase in BMD and a decrease in bone turnover markers, with no differences in serum calcium or eGFR. Denosumab is a potential alternative for reducing posttransplantation bone loss. A trial using fractures as a primary end-point is now required to compare treatment with oral or parenteral bisphosphonates with denosumab.

Kidney–pancreas transplantation

Severe osteoporosis complicates kidney–pancreas transplants in recipients with type 1 diabetes, occurring in 23% and 58% at the LS and FN, respectively. Vertebral or nonvertebral fractures were documented in 45% [1]. Other retrospective studies have documented a fracture prevalence of 26% to 49% up to 8.3 years after transplantation [45].

A prospective study addressed osteoporosis and secondary hyperparathyroidism in simultaneous kidney-pancreas recipients before and 4 years after transplant. Prior to transplantation, 68% had hyperparathyroidism. After 6 months, bone loss of 6.0% and 6.9% occurred at both LS and FN sites, respectively, and fractures were related to low pretransplant FN BMD [46].

Lung transplantation

The prevalence of osteoporosis is as high as 73% in lung transplantation recipients. During the first year after lung transplantation, rates of bone loss at the LS and FN range from 2% to 5% [1]. Fracture rates are also high during the first year, ranging from 18% to 37%. Bone turnover is also increased [47]. Repeated doses of intravenous pamidronate prevented LS and FN bone loss in lung transplantation recipients [48, 49].

Cardiac transplantation (CT)

The most rapid rate of bone loss occurs in the first year posttransplant. Spinal BMD declines by 6% to 10% during the first 6 months, whereas FN BMD falls by 6% to 11% in the first year, and stabilizes thereafter in most cases. BMD declines at the largely cortical proximal radius site over the second and third years, perhaps reflecting posttransplant SHPT. Vitamin D deficiency and testosterone deficiency (in men) are associated with more severe bone loss. Testosterone levels fall immediately after CT and normalize after 6 to 12 months. Some studies have found correlations between GC dose and bone loss. Vertebral fracture incidence ranges from 33% to 36% during the first 1 to 3 years after CT [50, 51].

Prevention and treatment
Vitamin D and calcitriol

Calcium and vitamin D alone do not prevent bone loss after CT [1]. Early studies showed calcitriol was effective at reducing bone loss, particularly at the FN, after CT [52]. Another study compared rates of bone loss in patients randomized to receive calcitriol (0.5 μg/day) or two cycles of etidronate during the first 6 months after CT or lung transplantation [53]. Significant and similar bone loss (3% to 8%) occurred at the spine and FN in both treatment groups, but was less than in historical controls [52, 53]. Other studies observed that CT recipients randomized to either alphacalcidol or cyclic etidronate sustained considerable bone loss at the spine and FN during the first year after transplantation [1], whereas another study of calcitriol [54] found no protective benefit. Thus, data regarding calcitriol and prevention of post-CT bone loss are inconsistent. Monitoring of serum and urine calcium levels is also required.

Intranasal calcitonin

One small study showed spinal BMD was higher 1 to 3 years, but not 7 years after CT in those treated with intranasal salmon calcitonin [55].

Testosterone

Because low posttransplant testosterone concentrations are often transient, only hypogonadal men should receive testosterone therapy.

Bisphosphonates

An open-label study of a single intravenous dose of pamidronate (60 mg) followed by four cycles of etidronate (400 mg every 3 months) and daily low-dose calcitriol (0.25 μg), prevented spinal and FN bone loss and reduced fracture rates in CT recipients compared with historical controls [56]. Compared with calcitonin (200 IU/day), alendronate (10 mg/day) treatment reduced hip bone loss and resulted in fewer vertebral fractures [57]. In a small study of 35 men post-CT, intravenous ibandronate (2 mg every 3 months) prevented spine and hip bone loss and resulted in fewer morphometric vertebral fractures [58].

In the largest study, where 149 patients were randomized immediately after CT to receive either alendronate (10 mg/day) or calcitriol (0.25 µg twice daily) for one year, bone loss at the spine and hip was prevented by both regimens compared with a prospectively recruited, nonrandomized reference group who received only calcium and vitamin D [59]. After one year of treatment withdrawal, BMD did not change in either the former alendronate or calcitriol group, but bone resorption increased in the calcitriol group [60]. This suggests that antiresorptive therapy may be discontinued one year posttransplant in CT recipients. However, these patients still require observation to ensure that BMD remains stable.

Exercise

Resistance exercise significantly improved LS BMD after lung [61] and heart [62] transplantation when used alone and in combination with alendronate.

Liver transplantation

Bone loss and fracture rates after liver transplantation are highest in the first 6 to 12 months. Spine BMD declines by 2% to 24% during the first year in earlier studies. In more recent studies rates of bone loss have been lower, or absent. Fracture rates range from 24% to 65% with the ribs and vertebrae being most common. Women with primary biliary cirrhosis have the most severe preexisting bone disease and the greatest risk. Older age, pre-liver transplantation spinal and FN BMD, and pre-liver transplantation vertebral fractures predict post-liver transplantation fractures in one recent prospective study [1, 8].

Prevention and treatment

Both oral and intravenous bisphosphonates are effective in reducing post-liver transplantation bone loss. However, a single pamidronate infusion failed to prevent post-liver transplantation bone loss [63]. A randomized trial of intravenous ibandronate in liver [64] transplant recipients found a significant protective effect on BMD at one year. In a randomized, double-blind trial, 62 liver transplantation recipients received treatment with either infusions of 4 mg zoledronic acid (ZA), or saline within 7 days of transplantation and again at 1, 3, 6, and 9 months post-liver transplantation [65]. All patients also received calcium and vitamin D. ZA significantly prevented bone loss from the LS, FN, and total hip by 3.8% to 4.7%, with differences being greatest at 3 months post-liver transplantation. At 12 months post-liver transplantation, differences only remained significant at the total hip. Similar findings were identified in another study using 4 mg intravenous ZA at 1 to 6, 9, and 12 months post-liver transplantation. Bone loss from the spine (but not hip) was prevented and fewer vertebral fractures occurred in the ZA group [66]. Vitamin D deficiency should be corrected before giving bisphosphonates post-liver transplantation to prevent hypocalcemia.

One study using two intravenous doses of pamidronate (90 mg) at baseline and 3 months post-liver transplantation showed an increase in spinal BMD in the pamidronate group, but more fractures in the pamidronate group [67]. Intravenous ibandronate given every 3 months post-liver transplantation resulted in increases in spinal BMD and less bone loss from the FN, and fewer fractures in the ibandronate group [68]. Two studies examined effects of alendronate on bone after liver transplantation. An uncontrolled, prospective study of 136 liver transplantation patients showed alendronate prevented bone loss in patients with osteopenia and led to an increase in BMD at the spine and FN in patients with osteoporosis over 4 years [8]. Another study of 59 liver transplantation patients used historical controls to examine effects of alendronate combined with calcium and calcitriol 0.5 µg daily [69]. Increases in spinal, FN, and total hip BMD at 12 months were higher than in historical controls.

Small bowel transplantation

Small bowel transplantation (SBT) is being used increasingly for severe inflammatory bowel disease. It may also include concomitant liver, pancreas, and stomach transplantation. In a cross-sectional study of 81 patients who had SBT 2.2 years previously, BMD at the spine, total hip, and FN was reduced by about 0.8 SD compared with age- and sex-matched controls with similar small bowel diseases. Long-term SBT recipients are at risk of both osteoporosis (44%) and fractures (20%) [20]. In a small longitudinal study of nine patients, significant bone loss occurred at both the spine (2.6%) and the total hip and FN (by about 15%) 1.3 years after SBT [70]. A larger longitudinal study ($n=24$) documented acceleration ($p=0.025$) of bone loss after SBT with a decline of 13.4% (FN), 12.7% (total hip), and 2.1% (spine) over 2.5 years. Alendronate reduced ($p<0.05$), but did not prevent bone loss [20].

Stem cell transplantation

Stem cell transplantation (SCT) is the treatment of choice for patients with many hematological malignancies, the majority of whom will survive for many years. However, up to 29% and 52% of survivors have osteopenia at the spine or FN, respectively [1]. Osteoporosis is more common at proximal femur sites. The pathogenesis of post-SCT osteoporosis is complex, relating both to effects of treatment and effects on the bone marrow stromal cell compartment [71,72]. Cytokines induced by cytotoxic chemotherapy drive increases in bone resorption, whereas bone formation decreases [1, 73] resulting in early, rapid bone loss. In addition to osteoporosis, osteomalacia and avascular necrosis may also occur.

Dramatic bone loss from the proximal femur occurs within the first 12 months of allogeneic SCT [1, 74, 75]. Spinal bone loss is less. Most studies suggest that little additional bone loss occurs after this time. Studies of long-term survivors of SCT have shown that losses from the proximal femur are not regained [76]. After autologous SCT, bone loss from the proximal femur is less (about 4%), but persists at 2 years, whereas spine BMD returns to baseline [77].

Bone loss after SCT is related to both cumulative GC exposure and duration of CsA exposure [74]. There may also be a direct effect of graft-versus-host disease (GVHD) itself on bone cells. Abnormal cellular or cytokine-mediated bone marrow function may affect bone turnover and BMD after SCT [1]. Both myeloablative treatment and SCT stimulate early cytokine release, including MCP-1 [19]. SCT also has adverse effects on bone marrow osteoprogenitors. Osteocyte viability is decreased after SCT and bone marrow stromal cells are damaged by high-dose chemotherapy, TBI, GCs and CsA, reducing osteoblastic differentiation [78]. In this regard, fibroblastic colony-forming units (CFU-F) are reduced for up to 12 years after SCT [1].

Avascular necrosis develops in 10% to 20% of allo-SCT survivors, a median of 12 months after SCT [72, 79]. GC treatment of chronic GVHD inducing osteoblast apoptosis is the most important risk factor. Avascular necrosis appears to be related to decreased numbers of bone marrow CFU-F colonies in vitro [79], and may be facilitated by the deficit in bone marrow stromal stem cell regeneration post-SCT [78].

Prevention and treatment

Vitamin D treatment reduces episodes of overall chronic GVHD post-SCT by about 64% over 12 months [80]. Risedronate or intravenous ZA given 12 months after SCT prevent spinal and proximal femoral bone loss [81, 82]. ZA effects may be related to increased osteoblast numbers post-SCT because increases in ex vivo growth of CFU-F have been shown.

Two randomized trials utilized intravenous pamidronate to prevent bone loss after SCT. The first studied 99 allogeneic SCT recipients, randomized to received calcium and vitamin D daily, hormone therapy with estrogen in females or testosterone in men, or the same treatments plus intravenous 60 mg pamidronate infusions before and 1, 2, 3, 6, and 9 months post-SCT [73]. In the pamidronate group, spine BMD remained stable but decreased in the control group. Total hip BMD and FN BMD decreased by 5.1% and 4.2%, respectively, in the pamidronate group and by 7.8% and 6.2%, respectively, in the control group at 12 months. Thus, pamidronate reduced bone loss more than in those treated with calcium, vitamin D, and sex steroid replacement alone.

A larger randomized, multicentre open-label 12-month prospective study compared intravenous pamidronate (90 mg/month) beginning before conditioning versus no pamidronate [83]. All 116 patients also received calcitriol (0.25 μg/day) and calcium, which were continued for a further year. Pamidronate significantly reduced bone loss at the spine, FN, and total hip at 12 months. However, BMD of the FN and total hip was still 2.8% and 3.5% lower than baseline, respectively, after pamidronate. Only the BMD benefit at the total hip remained significant between the two groups at 24 months. Benefits of pamidronate therapy were restricted to patients receiving an average daily prednisolone dose of more than 10 mg.

A small uncontrolled, prospective study of a single 4 mg ZA infusion in allogeneic BMT patients with either osteoporosis or rapid bone loss postallogeneic BMT [84] showed reduced bone loss at the spine and FN. A larger study of 61 patients showed three 4 mg doses of ZA reduced bone loss at the spine and FN by 9% and 8.2%, respectively, compared with calcium and vitamin D [85]. Another multiarm study of 55 women undergoing allogenic STC showed only three 4 mg doses of ZA reduced bone loss at both spine and FN BMD by 12.9% and 9.6%, respectively, compared with calcium and vitamin D [86]. An open-label prospective randomized controlled study of 3 mg intravenous ibandronate every 3 months after allo-SCT showed it reduced spinal bone loss by 4.2% at 12 months, but did not reduce bone loss at either the FN or total hip sites compared with calcium and vitamin D alone [87]. Currently, trial data show a beneficial effect of only ZA at the spine and proximal femur after allogeneic SCT. To date, no studies have used denosumab.

CONCLUSION

Pretransplantation bone disease and immunosuppressive therapy result in rapid bone loss and increased fracture rates, soon after transplantation. There is increased bone resorption and decreased bone formation. In the late posttransplant period, with weaning of GC doses, bone formation begins to increase. However, the underlying high bone remodeling results in osteoporosis. Although rates of bone loss and fractures reported in recent studies are lower than those of 10 years ago, they remain too high. Transplant candidates should be assessed and pretransplantation bone disease should be treated. Preventive therapy initiated in the immediate posttransplantation period is indicated in patients with osteopenia or osteoporosis, because further bone loss will occur immediately after transplantation. Long-term organ transplant recipients should also have bone mass measurement and treatment of osteoporosis.

A meta-analysis showed treatment with a bisphosphonate or active vitamin D metabolite during the first year after solid organ transplantation is associated with a 50% reduction in the number of subjects with fractures and 76% fewer vertebral fractures [38]. Bisphosphonate treatment was associated with a 47% reduction in the number of subjects with fractures [38]. Overall, bisphosphonates are the most promising approach for the prevention and treatment of transplantation osteoporosis.

Active vitamin D metabolites may have additional benefits in reducing hyperparathyroidism, particularly after kidney transplantation. Potential new agents for transplantation osteoporosis include anabolic agents that stimulate bone formation, namely PTH (1-34) or teriparatide, and the potent antiresorptive drug, human antibodies to RANKL (denosumab). PTH (1-34) and other PTH1 receptor agonists may have a specific role after BMT in stimulating marrow stromal stem cell differentiation into the osteoblast lineage and reducing adipogenesis [88, 89].

Several issues remain regarding the administration of bisphosphonates for transplantation bone disease, including the optimal route of administration and duration of therapy. Treatment may only need to be given for one year after cardiac transplantation, but its optimal duration is less clear after other transplants. Another special consideration in using bisphosphonates in kidney transplant recipients is adynamic bone disease. Large multicentre trials comparing treatment with oral or parenteral bisphosphonates and calcitriol, and commencing at the time of transplantation that are powered to detect differences in fracture rates are recommended here. Trials of denosumab after transplantation should also be encouraged. Much has been learnt about transplantation osteoporosis. Armed with this information, it is now critical to act to prevent and treat this disabling disease.

ACKNOWLEDGMENT

I thank Dr Elizabeth Shane for her mentorship in this area.

REFERENCES

1. Cohen A, Sambrook P, Shane E. Management of bone loss after organ transplantation. J Bone Miner Res. 2004;19(12):1919–32.
2. Cohen A, Shane E. Osteoporosis after solid organ and bone marrow transplantation. Osteoporos Int. 2003;14(8):617–30.
3. Nickolas TL, Cremers S, Zhang A, et al. Discriminants of prevalent fractures in chronic kidney disease. J Am Soc Nephrol. 2011;22(8):1560–72.
4. Gal-Moscovici A, Sprague SM. Osteoporosis and chronic kidney disease. Semin Dialysis. 2007;20(5):423–30.
5. Stehman-Breen CO, Sherrard DJ, et al. Risk factors for hip fracture among patients with end-stage renal disease. Kidney Int. 2000;58(5):2200–5.
6. Alem AM, Sherrard DJ, Gillen DL, et al. Increased risk of hip fracture among patients with end-stage renal disease. Kidney Int. 2000;58(1):396–9.
7. Ball AM, Gillen DL, Sherrard D, et al. Risk of hip fracture among dialysis and renal transplant recipients. JAMA. 2002;288(23):3014–8.
8. Majumdar S, Ezekowitz JA, Lix LM, et al. Heart failure is a clinically and densitometrically independent and novel risk factor for major osteoporotic fractures: population-based cohort study of 45,509 subjects. J Bone Miner Res. 2011;26 (Suppl 1):S10.
9. Shane E, Mancini D, Aaronson K, et al. Bone mass, vitamin D deficiency and hyperparathyroidism in congestive heart failure. Am J Med. 1997;103:197–207.
10. Monegal A, Navasa M, Guanabens N, et al. Bone disease after liver transplantation: a long-term prospective study of bone mass changes, hormonal status and histomorphometric characteristics. Osteoporos Int. 2001;12(6):484–92.
11. Millonig G, Graziadei IW, Eichler D, et al. Alendronate in combination with calcium and vitamin D prevents bone loss after orthotopic liver transplantation: a prospective single-center study. Liver Transpl. 2005;11:960–6.
12. Ninkovic M, Love SA, Tom B, et al. High prevalence of osteoporosis in patients with chronic liver disease prior to liver transplantation. Calcif Tissue Int. 2001;69(6):321–6.
13. Tschopp O, Boehler A, Speich R, et al. Osteoporosis before lung transplantation: association with low body mass index, but not with underlying disease. Am J Transplant. 2002;2(2):167–72.
14. Keilholz U, Max R, Scheibenbogen C, et al. Endocrine function and bone metabolism 5 years after autologous bone marrow/blood-derived progenitor cell transplantation. Cancer. 1997;79(8):1617–22.
15. Tauchmanova L, Selleri C, Rosa GD, et al. High prevalence of endocrine dysfunction in long-term survivors after allogeneic bone marrow transplantation for hematologic diseases. Cancer. 2002;95(5):1076–84.
16. Valimaki M, Kinnunen K, Volin L, et al. A prospective study of bone loss and turnover after allogeneic bone marrow transplantation: effect of calcium supplementation with or without calcitonin. Bone Marrow Transplant. 1999;23:355–61.
17. Kananen K, Volin L, Laitinen K, et al. Prevention of bone loss after allogeneic stem cell transplantation by calcium, vitamin D, and sex hormone replacement with or without pamidronate. J Clin Endocrinol Metab. 2005;90:3877–85.
18. Schulte C, Beelen D, Schaefer U, et al. Bone loss in long-term survivors after transplantation of hematopoietic stem cells: a prospective study. Osteoporosis Int. 2000;11:344–53.
19. Quach JM, Askmyr M, Jovic T, et al. Myelosuppressive therapies significantly increase pro-inflammatory cytokines and directly cause bone loss. J Bone Miner Res. 2015;30(5):886–97.
20. Resnick JI, Gupta N, Wagner J, et al. Skeletal integrity and visceral transplantation. Am J Transplant. 2010;10(10):2331–40.
21. Van Staa TP, Leufkens HG, Abenhaim L, et al. Use of oral corticosteroids and risk of fractures. J Bone Miner Res. 2000;15:993–1000.
22. Nikkel LE, Mohan S, Zhang A, et al. Reduced fracture risk with early corticosteroid withdrawal after kidney transplant. Am J Transplant. 2012;12(3):649–59.

23. Perrin P, Kiener C, Javier RM, et al. Recent changes in chronic kidney disease-mineral and bone disorders (CKD-MBD) and associated fractures after kidney transplantation. Transplantation. 2016:101(8);1897–905.
24. Epstein S. Post-transplantation bone disease: the role of immunosuppressive agents on the skeleton. J Bone Miner Res 1996;11:1–7.
25. Ponticelli C, Aroldi A. Osteoporosis after organ transplantation. Lancet. 2001;357(9268):1623.
26. McIntyre HD, Menzies B, Rigby R, et al. Long-term bone loss after renal transplantation: comparison of immunosuppressive regimens. Clin Transplant. 1995;9(1):20–4.
27. Stempfle HU, Werner C, Echtler S, et al. Rapid trabecular bone loss after cardiac transplantation using FK506 (tacrolimus)-based immunosuppression. Transplant Proc. 1998;30(4):1132–3.
28. Park KM, Hay JE, Lee SG, et al. Bone loss after orthotopic liver transplantation: FK 506 versus cyclosporine. Transplant Proc. 1996;28(3):1738–40.
29. Goffin E, Devogelaer JP, Depresseux G, et al. Osteoporosis after organ transplantation. Lancet. 2001;357(9268):1623.
30. Monegal A, Navasa M, Guanabens N, et al. Bone mass and mineral metabolism in liver transplant patients treated with FK506 or cyclosporine A. Calcif Tiss Int. 2001;68:83–6.
31. Singha UK, Jiang Y, Yu S, et al. Rapamycin inhibits osteoblast proliferation and differentiation in MC3T3-E1 cells and primary mouse bone marrow stromal cells. J Cell Biochem. 2007;103(2):434–46.
32. Stein EM, Shane E. Vitamin D in organ transplantation. Osteoporos Int. 2011;22(7):2107–18.
33. Julian BA, Laskow DA, Dubovsky J, et al. Rapid loss of vertebral bone density after renal transplantation. N Engl J Med. 1991;325:544–50.
34. Ramsey-Goldman R, Dunn JE, Dunlop DD, et al. Increased risk of fracture in patients receiving solid organ transplants. J Bone Miner Res. 1999;14(3):456–63.
35. Sprague SM, Josephson MA. Bone disease after kidney transplantation. Sem Nephrol. 2004;24:82–90.
36. Wissing KM, Broeders N, Moreno-Reyes R, et al. A controlled study of vitamin D_3 to prevent bone loss in renal-transplant patients receiving low doses of steroids. Transplantation. 2005;79:108–15.
37. Palmer SC, Strippoli GFM, McGregor DO. Interventions for preventing bone disease in kidney transplant recipients. Cochrane Database Syst Rev. 2007;(3):CD005015.
38. Stein EM, Ortiz D, Jin Z, et al. Prevention of fractures after solid organ transplantation: a meta-analysis. J Clin Endocrinol Metab. 2011;96, 3457–65.
39. Toth-Manikowski SM, Francis JM, Gautam A, et al. Outcomes of bisphosphonate therapy in kidney transplant recipients: a systematic review and meta-analysis. Clin Transplant. 2016;30:1090–6.
40. Wang J, Yao M, Xu JH, et al. Bisphosphonates for prevention of osteopenia in kidney-transplant recipients: a systematic review of randomized controlled trials. Osteoporos Int. 2016;27:1683–90.
41. Coco M, Glicklich D, Faugere MC, et al. Prevention of bone loss in renal transplant recipients: a prospective, randomized trial of intravenous pamidronate. J Am Soc Nephrol. 2003;14:2669–76.
42. Jamal SA, Ljunggren O, Stehman-Breen C, et al. Effects of denosumab on fracture and bone mineral density by level of kidney function. J Bone Miner Res. 2011;26:1829–35.
43. Dave V, Chiang CY, Booth J, et al. Hypocalcemia post denosumab in patients with chronic kidney disease stage 4–5. Am J Nephrol. 2015;41:129–37.
44. Bonani M, Frey D, Brockmann J, et al. Effect of twice-yearly denosumab on prevention of bone mineral density loss in de novo kidney transplant recipients: a randomized controlled trial. Am J Transplant. 2016;16:1882–91.
45. Chiu MY, Sprague SM, Bruce DS, et al. Analysis of fracture prevalence in kidney-pancreas allograft recipients. J Am Soc Nephrol. 1998;9(4):677–83.
46. Smets, YFC, De Fijter JW, Ringers J, et al. Long-term follow-up study on bone mineral density and fractures after simultaneous pancreas-kidney transplantation. Kidney Int. 2004;66(5):2070–6.
47. Shane E, Papadopoulos A, Staron RB, et al. Bone loss and fracture after lung transplantation. Transplantation. 1999;68:220–7.
48. Aris RM, Lester GE, Renner JB, et al. Efficacy of pamidronate for osteoporosis in patients with cystic fibrosis following lung transplantation. Am J Respir Crit Care Med. 2000;162(3 Pt 1):941–6.
49. Trombetti A, Gerbase MW, Spiliopoulos A, et al. Bone mineral density in lung-transplant recipients before and after graft: prevention of lumbar spine posttransplantation-accelerated bone loss by pamidronate. J Heart Lung Transplant. 2000;19(8):736–43.
50. Shane E, Rivas M, Staron RB, et al. Fracture after cardiac transplantation: a prospective longitudinal study. J. Clin. Endocrinol. Metab. 1996;81:1740–6.
51. Leidig-Bruckner G, Hosch S, Dodidou P, et al. Frequency and predictors of osteoporotic fractures after cardiac or liver transplantation: a follow-up study. Lancet. 2001;357:342–7.
52. Sambrook P, Henderson NK, Keogh A, et al. Effect of calcitriol on bone loss after cardiac or lung transplantation. J Bone Miner Res. 2000;15(9):1818–24.
53. Henderson K, Eisman J, Keogh A, et al. Protective effect of short-term calcitriol or cyclical etidronate on bone loss after cardiac or lung transplantation. J Bone Miner Res. 2001;16(3):565–71.
54. Stempfle HU, Werner C, Echtler S, et al. Prevention of osteoporosis after cardiac transplantation: a prospective, longitudinal, randomized, double-blind trial with calcitriol. Transplantation. 1999;68(4):523–30.
55. Kapetanakis EI, Antonopoulos AS, Antoniou TA, et al. Effect of long-term calcitonin administration on steroid-induced osteoporosis after cardiac transplantation. J Heart Lung Transplant. 2005;24(5):526–32.
56. Bianda T, Linka A, Junga G, et al. Prevention of osteoporosis in heart transplant recipients: a comparison of calcitriol with calcitonin and pamidronate. Calcif Tissue Int. 2000;67:116–21.

57. Gil-Fraguas L, Jodar E, Martinez G, et al. Evolution of bone density after heart transplantation: influence of anti-resorptive therapy. J Bone Miner Res. 2005;20 (Suppl):S439–40.
58. Fahrleitner-Pammer A, Piswanger-Soelkner JC, Pieber TR, et al. Ibandronate prevents bone loss and reduces vertebral fracture risk in male cardiac transplant patients: a randomized double-blind, placebo-controlled trial. J Bone Miner Res. 2009;24:1335–44.
59. Shane E, Addesso V, Namerow PB, et al. Alendronate versus calcitriol for the prevention of bone loss after cardiac transplantation. New Engl J Med. 2004;350: 767–76.
60. Cohen A, Addesso V, McMahon DJ, et al. Discontinuing antiresorptive therapy one year after cardiac transplantation: effect on bone density and bone turnover Transplantation. 2006;81:686–91.
61. Mitchell MJ, Baz MA, Fulton MN, et al. Resistance training prevents vertebral osteoporosis in lung transplant recipients. Transplantation. 2003;76:557–62.
62. Braith RW, Magyari PM, Fulton MN, et al. Comparison of calcitonin versus calcitonin and resistance exercise as prophylaxis for osteoporosis in heart transplant recipients. Transplantation. 2006;81:1191–5.
63. Ninkovic M, Love S, Tom BD, et al. Lack of effect of intravenous pamidronate on fracture incidence and bone mineral density after orthotopic liver transplantation. J Hepatol. 2002;37(1):93–100.
64. Hommann M, Abendroth K, Lehmann G, et al. Effect of transplantation on bone: osteoporosis after liver and multivisceral transplantation. Transplant Proc. 2002;34(6):2296–8.
65. Crawford, BAL, Kam C, Pavlovic J, et al. Zoledronic acid prevents bone loss after liver transplantation: a randomized, double-blind, placebo-controlled trial. Annals Intern Med. 2006;144(4):239–48.
66. Bodingbauer M, Wekerle T, Pakrah B, et al. Prophylactic bisphosphonate treatment prevents bone fractures after liver transplantation. Am J Transplant. 2007;7(7): 1763–9.
67. Monegal A, Guañabens N, Suárez MJ, et al. Pamidronate in the prevention of bone loss after liver transplantation: a randomized controlled trial. Transpl Int. 2009;22(2): 198–206.
68. Kaemmerer D, Lehmann G, Wolf G, et al. Treatment of osteoporosis after liver transplantation with ibandronate. Transpl Int. 2010;23(7):753–9.
69. Karasu Z, Kilic M, Tokat Y. The prevention of bone fractures after liver transplantation: experience with alendronate treatment. Transplant Proc. 2006;38:1448–52.
70. Awan KS, Wagner JM, Martin D, et al. Bone loss following small bowel transplantation. J Bone Miner Res. 2007;22(S1):Abstract T497, S356.
71. Banfi A, Podesta M, Fazzuoli L, et al. High-dose chemotherapy shows a dose-dependent toxicity to bone marrow osteoprogenitors: a mechanism for post-bone marrow transplantation osteopenia. Cancer. 2001;92(9):2419–28.
72. Lee WY, Cho SW, Oh ES, et al. The effect of bone marrow transplantation on the osteoblastic differentiation of human bone marrow stromal cells. J Clin Endocrinol Metab. 2002;87(1):329–35.
73. Kananen K, Volin L, Laitinen K, et al. Prevention of bone loss after allogeneic stem cell transplantation by calcium, vitamin D, and sex hormone replacement with or without pamidronate. J Clin Endocrinol Metab. 2005;90:3877–85.
74. Ebeling P, Thomas D, Erbas B, et al. Mechanism of bone loss following allogeneic and autologous hematopoeitic stem cell transplantation. J Bone Miner Res. 1999;14: 342–50.
75. Ebeling PR Bone Disease After Bone Marrow Transplantation. In: Compston J, Shane E (eds) Bone Disease of Organ Transplantation. New York: Elsevier, 2005, pp. 339–52.
76. Lee WY, Kang MI, Baek KH, et al. The skeletal site-differential changes in bone mineral density following bone marrow transplantation: 3-year prospective study. J Korean Med Sci. 2002;17(6):749–54.
77. Gandhi MK, Lekamwasam S, Inman I, et al. Significant and persistent loss of bone mineral density in the femoral neck after haematopoietic stem cell transplantation: long-term follow-up of a prospective study. Br J Haematol. 2003;121:462–8.
78. Ebeling PR. Is defective osteoblast function responsible for bone loss from the proximal femur despite pamidronate therapy? J Clin Endocrinol Metab. 2005;90:4414–6.
79. Tauchmanova L, De Rosa G, Serio B, et al. Avascular necrosis in long-term survivors after allogeneic or autologous stem cell transplantation: a single center experience and a review. Cancer. 2003;97:2453–61.
80. Caballero-Velázquez T, Montero I, Sánchez-Guijo F, et al.; GETH (Grupo Español de Trasplante Hematopoyético). Immunomodulatory effect of vitamin D after allogeneic stem cell transplantation: results of a prospective multicenter clinical trial. Clin Cancer Res. 2016;22(23):5673–81.
81. Tauchmanova L, Selleri C, Esposito M, et al. Beneficial treatment with risedronate in long-term survivors after allogeneic stem cell transplantation for hematological malignancies. Osteoporos Int. 2003;14:1013–9.
82. Tauchmanova L, Ricci P, Serio B, et al. Short-term zoledronic acid treatment increases bone mineral density and marrow clonogenic fibroblast progenitors after allogeneic stem cell transplantation. J Clin Endocrinol Metab. 2005;90:627–34.
83. Grigg AP, Shuttleworth P, Reynolds J, et al. Pamidronate reduces bone loss after allogeneic stem cell transplantation. J Clin Endocrinol Metab. 2006;91:3835–43.
84. D'Souza AB, Grigg AP, Szer J, et al. Zoledronic acid prevents bone loss after allogeneic haemopoietic stem cell transplantation. Int Med J. 2006;36:600–3.
85. Hari P, DeFor TE, Vesole DH, et al. Intermittent zoledronic acid prevents bone loss in adults after allogeneic hematopoietic cell transplantation. Biol Blood Marrow Transplant. 2013;19(9):1361–7.
86. Tauchmanovà L, De Simone G, Musella T, et al. Effects of various antireabsortive treatments on bone mineral

density in hypogonadal young women after allogeneic stem cell transplantation. Bone Marrow Transplant. 2006;37(1):81–8.
87. Lu H, Champlin RE, Popat U, et al. Ibandronate for the prevention of bone loss after allogeneic stem cell transplantation for hematologic malignancies: a randomized-controlled trial. Bonekey Rep. 2016;5:843.
88. Rickard DJ, Wang FL, Rodriguez-Rojas AM, et al. Intermittent treatment with parathyroid hormone (PTH) as well as a non-peptide small molecule agonist of the PTH1 receptor inhibits adipocyte differentiation in human bone marrow stromal cells. Bone. 2006;39(6):1361–72.
89. Chan GK, Miao D, Deckelbaum R, et al. Parathyroid hormone-related peptide interacts with bone morphogenetic protein 2 to increase osteoblastogenesis and decrease adipogenesis in pluripotent C3H10T mesenchymal cells. Endocrinology. 2003;144:5511–20.

55

Premenopausal Osteoporosis

Adi Cohen and Elizabeth Shane

Division of Endocrinology, Department of Medicine, College of Physicians and Surgeons,
Columbia University, New York, NY, USA

INTRODUCTION

In this chapter we will discuss issues specific to the diagnosis, clinical evaluation, and management of premenopausal women who present with low-trauma fracture and/or low BMD.

PREMENOPAUSAL WOMEN WITH A HISTORY OF LOW-TRAUMA FRACTURE

The diagnosis of osteoporosis in premenopausal women is most secure when there is a history of low-trauma fracture(s). A fracture (excluding fracture of the face, skull, or digits) that occurs with trauma equivalent to a fall from a standing height or less may be a sign of decreased bone strength, regardless of BMD.

Several studies have shown that fractures before menopause predict postmenopausal fractures [1–3]. In the Study of Osteoporotic Fractures (SOF), women with a history of premenopausal fracture were 35% more likely to fracture during the early postmenopausal years than women without a history of premenopausal fracture [1]. These findings suggest that certain life-long traits, such as fall frequency, neuromuscular protective response to falls, bone mass, or various aspects of bone quality can affect life-long fracture risk [2].

PREMENOPAUSAL WOMEN WITH LOW BMD

In postmenopausal women, BMD assessment by DXA is a cornerstone of fracture risk prediction models used for therapeutic decision making because of the wealth of longitudinal observational and interventional studies correlating DXA findings with fracture incidence in this population (see Chapter 32). In contrast, in premenopausal women, incidence and prevalence of fracture is much lower [1, 2, 4], and such longitudinal data is not available. Thus, the relationship between BMD and fracture risk is unclear.

Several studies have shown that young women with low BMD are at higher risk for fractures than young women with normal BMD [5,6]. Premenopausal women with Colles fractures have been found to have significantly lower BMD at the nonfractured radius [7], lumbar spine, and femoral neck [8] than controls without fractures. Stress fractures in female military recruits and athletes are associated with lower BMD than controls [5, 6, 9]. In addition, high resolution imaging and transiliac bone biopsy studies have found that healthy, normally menstruating, premenopausal women with unexplained low BMD and no fractures have similar microarchitectural disruption to a comparable cohort of premenopausal women with low trauma fractures (Fig. 55.1) [10–12], suggesting that very low BMD may represent a presymptomatic phase of osteoporosis in this group, as it does in postmenopausal women.

Even though these cross-sectional studies suggest a relationship between DXA BMD and bone strength in premenopausal women, the lack of prospective data and the very low incidence of premenopausal fractures lead to distinct recommendations for BMD interpretation in the premenopausal years. Screening BMD is not recommended [13, 14], and BMD measurement should not be used as the sole guide for diagnosis and treatment of osteoporosis. In premenopausal women, the World Health Organization criteria for diagnosis of osteoporosis and osteopenia do not apply to, and generally should not be used to categorize, BMD measurements. The International Society for Clinical Densitometry (ISCD) recommends using Z scores (comparison to an age-matched reference population), to categorize BMD

Primer on the Metabolic Bone Diseases and Disorders of Mineral Metabolism, Ninth Edition. Edited by John P. Bilezikian.
© 2019 American Society for Bone and Mineral Research. Published 2019 by John Wiley & Sons, Inc.
Companion website: www.wiley.com/go/asbmrprimer

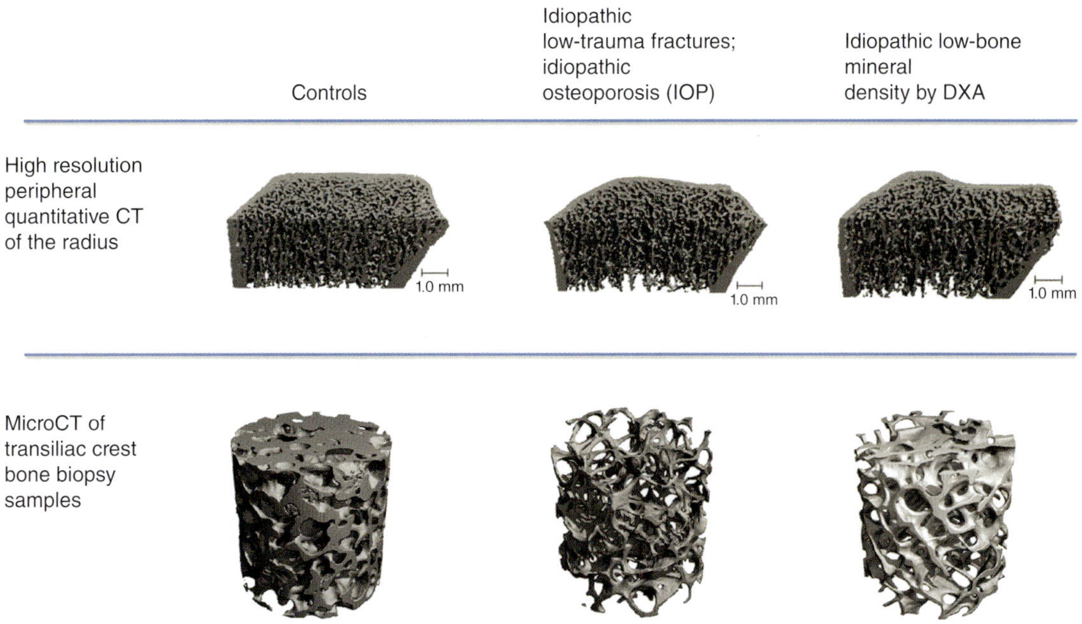

Fig. 55.1. Premenopausal women with idiopathic low bone mineral density by DXA have deficient bone microstructure in comparison to healthy premenopausal controls and similar bone microstructure in comparison to premenopausal women with idiopathic low trauma fractures, as shown by high resolution peripheral QCT of the radius and microCT of transiliac crest bone biopsy. Adapted from Cohen et al. [10] and Cohen et al. [12].

measurements in premenopausal women. Young women with BMD Z scores at or below −2.0 should be categorized as having BMD that is "below expected range for age" and those with Z scores above −2.0 should be categorized as having BMD that is "within the expected range for age" [13,14]. Because Z scores rather than T scores are used, the diagnostic categories of "osteoporosis" and "osteopenia" based upon T scores alone should not be applied to premenopausal women. An exception to these recommendations occurs in perimenopausal women, in whom T scores may be used [13, 14].

The International Osteoporosis Foundation (IOF) recommends use of Z score less than −2 to define low bone mass in children, adolescents, those under 20 years, and some over 20 years in the context of delayed puberty. In contrast to the ISCD, the IOF recommends use of T scores in those aged 20 to 50 years and suggests use of T score less than −2.5 to define osteoporosis, particularly in those with known secondary causes or in the context of low-trauma fractures that provide evidence of bone fragility [15].

SPECIAL ISSUES RELATED TO BMD INTERPRETATION IN PREMENOPAUSAL WOMEN

1. Although the majority of bone mass acquisition occurs during adolescence, BMD may continue to increase slightly between ages 20 and 30 [16] (Chapter 16). Thus, very young women with low BMD measurements may not have yet achieved peak bone mass.
2. There are expected changes in bone mass associated with both pregnancy and lactation (Chapter 20). At the lumbar spine, longitudinal studies document losses of 3% to 5% over a pregnancy and 3% to 10% over a 6-month period of lactation [17], with recovery of bone mass expected over 6 to 12 months thereafter [17–20]. Therefore, when interpreting a low BMD measurement in a premenopausal woman, the clinician must take the timing of recent pregnancy and lactation into account.
3. Pregnancy and lactation associated osteoporosis: pregnancy and lactation may represent a particularly vulnerable time for the skeleton for some women. Although it is rare, premenopausal osteoporosis may first present with low trauma fracture(s), most commonly vertebral fractures, occurring in the last trimester of pregnancy or during lactation [21]. Women with pregnancy- and lactation-associated osteoporosis still require an evaluation for potential secondary causes of osteoporosis (see below).

SECONDARY CAUSES OF OSTEOPOROSIS IN PREMENOPAUSAL WOMEN

Most premenopausal women with low-trauma fractures or low BMD have an underlying disorder or medication exposure that has interfered with bone mass accrual during adolescence and/or has caused excessive bone loss

Table 55.1. Secondary causes of osteoporosis in premenopausal women.

Premenopausal amenorrhea (eg, hypothalamic amenorrhea, pituitary diseases, medications)
Anorexia nervosa
Cushing syndrome
Hyperthyroidism
Primary hyperparathyroidism
Hypercalciuria
Vitamin D, calcium, and/or other nutrient deficiency
Gastrointestinal malabsorption (celiac disease, inflammatory bowel disease, cystic fibrosis, postoperative states)
Rheumatoid arthritis, systemic lupus erythematosus, other inflammatory conditions
Renal disease
Liver disease
Diabetes mellitus
Alcoholism

Connective tissue diseases:
• Osteogeneis imperfecta
• Marfan syndrome and Ehlers–Danlos syndrome
• Hypophosphatasia
Other rare conditions (eg, hemochromatosis, Gaucher disease, mastocytosis, thalassemia)

Medications:
• Glucocorticoids
• Immunosuppressants (eg, cyclosporine)
• Antiepileptic drugs (particularly cytochrome P450 inducers such as phenytoin, carbamazepine)
• Cancer chemotherapy
• GnRH agonists (when used to suppress ovulation)
• Heparin
Idiopathic osteoporosis

after reaching peak bone mass. In a population study from Olmstead County, Minnesota, USA, 90% of men and women aged 20 to 44 with osteoporotic fractures were found to have a secondary cause [22]. In contrast, several case series of young women with osteoporosis evaluated in tertiary centers report that only 50% have secondary causes [23, 24], likely reflecting referral bias of more obscure cases to specialists.

Potential secondary causes are listed in Table 55.1. Many of these are discussed elsewhere in this Primer. The main goal of the evaluation of a premenopausal woman with low-trauma fractures or low BMD is to identify any secondary cause, and to institute specific treatment for that cause if it is correctable. Often this can be accomplished by a detailed history and physical examination, although an exhaustive biochemical evaluation may be necessary.

IDIOPATHIC OSTEOPOROSIS

Premenopausal women with osteoporosis, in whom no definable cause can be found after a detailed evaluation, are said to have idiopathic osteoporosis (IOP). IOP is predominantly reported in white people, and family history of osteoporosis is common [22, 24, 25]. Mean age at diagnosis is 35 years. Multiple vertebral and/or nonvertebral fractures may occur over 5 to 15 years; alternatively, there may be a single, major osteoporotic fracture, such as a low trauma spine, hip, or long bone fracture [22, 24, 26]. Studies of bone microarchitecture in premenopausal women with IOP have shown that women with unexplained low BMD and those with low trauma fractures have comparably abnormal bone microstructure, with thinner cortices and thinner, more widely spaced and heterogeneously distributed trabeculae [10, 12]. Bone turnover, as assessed by tetracycline-labeled transiliac bone biopsies, was heterogeneous, with very low, normal, and high remodeling observed, suggesting diverse pathogeneses could account for the microarchitectural deterioration. Women with high bone turnover had a biochemical pattern resembling idiopathic hypercalciuria (mildly increased 24-hour urinary calcium and higher serum $1,25(OH)_2D$ concentrations compared with controls). In women with low bone turnover, microstructural deficits were more profound, serum IGF-1 concentrations were higher and osteoblasts appeared to synthesize less bone matrix per remodeling site, suggesting osteoblast resistance to IGF-1 [10,26].

EVALUATION OF THE PREMENOPAUSAL WOMAN WITH OSTEOPOROSIS

Premenopausal women with low BMD (Z score ≤ −2.0 or T score < −2.5), and those with a low-trauma fracture regardless of whether their BMD is frankly low, should undergo a thorough evaluation for secondary causes of bone loss. Identification of a contributing condition often helps to guide management of the affected individual.

A careful medical history is essential, including information about family history, fractures, kidney stones or family history of kidney stones, oligo- or amenorrhea or other history consistent with premenopausal estrogen deficiency, timing of recent pregnancies and lactation, dieting and exercise behavior, subtle gastrointestinal symptoms, and medications, including over-the-counter supplements. During the physical examination, look for signs of Cushing syndrome, thyrotoxicosis, or connective tissue disorders (eg, blue sclerae in some forms of osteogenesis imperfecta or joint hypermobility in Ehlers–Danlos syndrome).

The laboratory evaluation (Table 55.2) should be aimed at identifying secondary causes such as hyperthyroidism, hyperparathyroidism, Cushing syndrome, early menopause, renal or liver disease, celiac disease, malabsorption, and idiopathic hypercalciuria. In those with a clinical presentation or family history suggesting a genetic condition, such as osteogenesis imperfecta, Gaucher disease or Ehlers-Danlos syndrome, genetic testing can be considered. Follow-up bone density testing may help to distinguish those with stable low BMD from those with ongoing bone loss who may be at higher short-term risk of fracture. Transiliac crest bone biopsy after

Table 55.2. Laboratory evaluation.

Initial laboratory evaluation
- Complete blood count
- Electrolytes, renal function
- Serum calcium, phosphate
- Serum albumin, transaminases, total alkaline phosphatase
- Serum TSH
- Serum 25-hydroxyvitamin D
- 24-hour urine for calcium and creatinine

Additional laboratory evaluation
- Estradiol, LH, FSH, prolactin
- PTH
- 1,25-dihydroxyvitamin D
- 24-hour urine for free cortisol
- Iron/total iron binding capacity, ferritin
- Celiac screen
- Serum/urine protein electrophoresis
- Erythrocyte sedimentation rate or C-reactive protein
- Tryptase
- Bone turnover markers
- Transiliac crest bone biopsy
- Genetic testing

double tetracycline labeling may be useful in certain clinical scenarios when it is necessary to examine bone remodeling at the tissue level, rule out osteomalacia, differentiate between different types of renal osteodystrophy, or complete an examination for rare secondary causes.

MANAGEMENT ISSUES

General measures

For all patients, one should recommend a set of general measures that generally benefit bone health: adequate weightbearing exercise [27,28], nutrition (protein, calories, calcium, vitamin D; see Chapters 18), and lifestyle modifications (smoking cessation, avoidance of excess alcohol). However, in our clinical experience, instituting these measures does not result in major increases in BMD measurements.

In the authors' opinion, pharmacological therapy is rarely justified for premenopausal women with isolated low BMD and no history of fractures, in whom there is no identifiable secondary cause, particularly if the Z score is above −3.0. Low BMD in such young women may be caused by genetic low peak bone mass, or to past insults to the growing or adult skeleton (nutritional deficiency, alcohol excess, medications, estrogen deficiency) that are no longer operative. A small study with a mean of 3 years of follow-up suggests a low short-term risk of fracture: Peris and colleagues reported slight BMD improvement and no further fractures in women with unexplained osteoporosis managed with only calcium (total intake of 1500 mg/day), vitamin D (400 to 800 IU/daily), and exercise [29]. Bone density should be remeasured after 1 or 2 years to confirm that it is stable and identify patients with ongoing bone loss.

In women with low BMD or low trauma fractures and a known secondary cause, address the underlying cause if possible. Bone density benefits have been shown in the context of intervention for several such secondary causes in premenopausal women. Women with estrogen deficiency should receive estrogen [30] (unless contraindicated), women with anorexia nervosa should be treated with the goal of nutritional rehabilitation and weight gain [31], those with celiac disease should begin a gluten free diet [32, 33], and those with primary hyperparathyroidism may benefit from parathyroidectomy [34] (see Chapter 82). Although data are lacking in premenopausal women, those with idiopathic hypercalciuria may benefit from thiazide diuretics [35].

In some women, it is not possible to address or alleviate the secondary cause directly. Premenopausal women requiring long-term glucocorticoids and those being treated for breast cancer may require pharmacological therapy to prevent excessive bone loss or fractures. Treatment options include antiresorptive drugs, such as estrogen, bisphosphonates and denosumab, or anabolic agents such as teriparatide. SERMs, such as raloxifene, should not be used to treat bone loss in menstruating women because they block estrogen action on bone and may lead to further bone loss [36, 37].

Bisphosphonates

Bisphosphonates have been shown to improve BMD or prevent bone loss in premenopausal women with various conditions, including glucocorticoid therapy, breast cancer therapy, pregnancy and lactation associated fractures, anorexia nervosa, cystic fibrosis, and thalassemia [38–47]; in some cases, premenopausal women were studied specifically [38–43, 46, 47]. Readers are also referred to a recent review of treatment studies for premenopausal osteoporosis by Ferrari and colleagues [48].

Large randomized trials are scarce and the United States Food and Drug Administration has approved oral bisphosphonates only for premenopausal women on glucocorticoids. Because bisphosphonates accumulate in the maternal skeleton, cross the placenta, accumulate in the fetal skeleton [49], and cause toxic effects in pregnant rats [50], they should be used with caution in women who are intending pregnancy or could become pregnant. While several reports document normal pregnancies and fetal outcomes in women receiving bisphosphonates [43, 51–53], the potential for fetal abnormalities should be considered when prescribing bisphosphonates for a premenopausal woman.

Because there are so few data regarding the long-term efficacy and safety of bisphosphonates in young women, the decision to initiate treatment must be made on a case by case basis with consideration of individual fracture risk, and with a plan for the shortest possible duration of use. In general, bisphosphonates should be reserved for those with fragility fractures or ongoing bone loss.

Human PTH(1–34)

There are even fewer data on the effects of teriparatide or PTH(1–34) in premenopausal women, but this medication has been studied in women with medication-induced amenorrhea [54], IOP [55], anorexia nervosa [56], pregnancy and lactation associated osteoporosis [57], and those on glucocorticoids [58]. In young women treated with the GnRH analog nafarelin for endometriosis, spine BMD declined by 4.9%, whereas those treated with PTH(1–34) 40 μg daily together with nafarelin had an increase of 2.1% ($p<0.001$) [54]. It is not clear whether these results would apply to premenopausal women with normal gonadal status. A recent study comparing teriparatide and alendronate for glucocorticoid-induced osteoporosis included some premenopausal women. Overall, teriparatide was associated with significantly greater increases in lumbar spine and total hip BMD and resulted in significantly fewer incident vertebral fractures than alendronate [58]. The BMD responses were similar in premenopausal women as in men and postmenopausal women, but no fractures occurred in either premenopausal group.

In an observational study of teriparatide 20 μg daily for 18 to 24 months in 21 premenopausal women with IOP, BMD increased by $10.8\% \pm 8.3\%$ at the lumbar spine, $6.2\% \pm 5.6\%$ at the total hip and $7.6\% \pm 3.4\%$ at the femoral neck (all $p<0.001$) [55]. However, among this unique cohort, a small subset with very low baseline bone turnover had little or no increase in BMD on this medication [55]. Because the long-term effects of teriparatide in young women are not known, use of this medication should be reserved for those at highest risk for fracture or those who are experiencing recurrent fractures. In young women less than 25 years of age, documentation of fused epiphyses is recommended before consideration of teriparatide treatment, because continued bone growth is considered a contraindication to use of this medication.

Few data are available to guide treatment options for premenopausal women after teriparatide cessation. One study documented BMD gain in premenopausal women who resumed menses after cessation of both long acting GnRH analog and PTH(1–34) [59]. However, in a study of 13 premenopausal women with idiopathic osteoporosis and normal gonadal function followed for 2.0 ± 0.6 years after teriparatide cessation, BMD declined $4.2\% \pm 3.9\%$ at the spine although it remained stable at the hip [60]. This finding suggests that women with IOP will require antiresorptive treatment to prevent bone loss after teriparatide.

Denosumab

Denosumab is currently approved for the treatment of osteoporosis in postmenopausal women and men at high risk for fracture. Although denosumab may have some advantages in premenopausal women because of its shorter half-life relative to bisphosphonates and lack of skeletal accumulation, the efficacy and safety of this medication have not been defined in this population. Denosumab, as marketed for osteoporosis, has been assigned a designation of pregnancy category X; animal studies indicate that denosumab may cause fetal harm.

Glucocorticoid-induced osteoporosis in premenopausal women

Bisphosphonates are approved for prevention and treatment of glucocorticoid-induced osteoporosis in premenopausal women. However, relatively few premenopausal women participated in the relevant large registration trials for bisphosphonates in glucocorticoid-induced osteoporosis and none of the premenopausal women in those trials fractured [61–64]. Guidelines from the American College of Rheumatology suggest that bisphosphonates or teriparatide could be considered for premenopausal women of childbearing potential with a history of fragility fracture, if there is glucocorticoid exposure of at least 7.5 mg of prednisone or equivalent per day for 3 months or more [65].

CONCLUSION

Premenopausal woman with low-trauma fracture(s) or low BMD (Z score ≤ –2.0) should have a thorough evaluation for secondary causes of osteoporosis and bone loss. In most, a secondary cause can be found, the most common being glucocorticoid excess, anorexia nervosa, premenopausal estrogen deficiency, and gastrointestinal malabsorption. Where possible, identification and treatment of the underlying cause should be the focus of management. Although pharmacological therapy is rarely justified in premenopausal women, those with an ongoing cause of bone loss and those who have had or continue to have low-trauma fractures may require pharmacological intervention, such as bisphosphonates or teriparatide. Few high quality clinical trials exist to provide guidance and there are no data that such therapeutic intervention reduces the risk of future fractures.

REFERENCES

1. Hosmer WD, Genant HK, Browner WS Fractures before menopause: a red flag for physicians. Osteoporos Int. 2002;13:337–41.
2. Wu F, Mason B, Horne A, et al. Fractures between the ages of 20 and 50 years increase women's risk of subsequent fractures. Arch Intern Med. 2002;162:33–6.
3. Honkanen R, Tuppurainen M, Kroger H, et al. Associations of early premenopausal fractures with subsequent fractures vary by sites and mechanisms of fractures. Calcif Tissue Int. 1997;60:327–31.

4. Thompson PW, Taylor J, Dawson A. The annual incidence and seasonal variation of fractures of the distal radius in men and women over 25 years in Dorset, UK. Injury. 2004;35:462–6.
5. Lauder TD, Dixit S, Pezzin LE, et al. The relation between stress fractures and bone mineral density: evidence from active-duty Army women. Arch Phys Med Rehabil. 2000;81:73–9.
6. Lappe J, Davies K, Recker R, et al. Quantitative ultrasound: use in screening for susceptibility to stress fractures in female army recruits. J Bone Miner Res. 2005; 20:571–8.
7. Wigderowitz CA, Cunningham T, Rowley DI, et al. Peripheral bone mineral density in patients with distal radial fractures. J Bone Joint Surg Br. 2003;85:423–5.
8. Hung LK, Wu HT, Leung PC, et al. Low BMD is a risk factor for low-energy Colles' fractures in women before and after menopause. Clin Orthop Relat Res 2005:219–25.
9. Myburgh KH, Hutchins J, Fataar AB, et al. Low bone density is an etiologic factor for stress fractures in athletes. Ann Intern Med. 1990;113:754–9.
10. Cohen A, Dempster D, Recker R, et al. Abnormal bone microarchitecture and evidence of osteoblast dysfunction in premenopausal women with idiopathic osteoporosis. J Clin Endocrinol Metab. 2011;96:3095.
11. Cohen A, Lang TF, McMahon DJ, et al. Central QCT reveals lower volumetric bmd and stiffness in premenopausal women with idiopathic osteoporosis, regardless of fracture history. J Clin Endocrinol Metab. 2012;97:4244–52.
12. Cohen A, Liu XS, Stein EM, et al. Bone microarchitecture and stiffness in premenopausal women with idiopathic osteoporosis. J Clin Endocrinol Metab. 2009;94:4351–60.
13. Lewiecki EM, Gordon CM, Baim S, et al. International Society for Clinical Densitometry 2007 Adult and Pediatric Official Positions. Bone. 2008;43:1115–21.
14. Schousboe JT, Shepherd JA, Bilezikian JP, et al. Executive summary of the 2013 International Society for Clinical Densitometry Position Development Conference on bone densitometry. J Clin Densitom. 2013;16:455–66.
15. Ferrari S, Bianchi ML, Eisman JA, et al. Osteoporosis in young adults: pathophysiology, diagnosis, and management. Osteoporos Int. 2012;23:2735–48.
16. Recker RR, Davies KM, Hinders SM, et al. Bone gain in young adult women. JAMA. 1992;268:2403–8.
17. Karlsson MK, Ahlborg HG, Karlsson C Maternity and bone mineral density. Acta Orthop. 2005;76:2–13.
18. Kolthoff N, Eiken P, Kristensen B, et al. Bone mineral changes during pregnancy and lactation: a longitudinal cohort study. Clin Sci (Lond) 1998;94:405–12.
19. Kovacs CS. Maternal mineral and bone metabolism during pregnancy, lactation, and post-weaning recovery. Physiol Rev. 2016;96:449–547.
20. Sowers M, Corton G, Shapiro B, et al. Changes in bone density with lactation. JAMA. 1993;269:3130–5.
21. Kovacs CS, Ralston SH. Presentation and management of osteoporosis presenting in association with pregnancy or lactation. Osteoporos Int. 2015;26:2223–41.
22. Khosla S, Lufkin EG, Hodgson SF, et al. Epidemiology and clinical features of osteoporosis in young individuals. Bone. 1994;15:551–5.
23. Moreira Kulak CA, Schussheim DH, McMahon DJ, et al. Osteoporosis and low bone mass in premenopausal and perimenopausal women. Endocr Pract. 2000;6: 296–304.
24. Peris P, Guanabens N, Martinez de Osaba MJ, et al. Clinical characteristics and etiologic factors of premenopausal osteoporosis in a group of Spanish women. Semin Arthritis Rheum. 2002;32:64–70.
25. Kulak CAM, Schussheim DH, McMahon DJ, et al. Osteoporosis and low bone mass in premenopausal and perimenopausal women. Endocr Pract. 2000;6:296–304.
26. Cohen A, Recker RR, Lappe J, et al. Premenopausal women with idiopathic low-trauma fractures and/or low bone mineral density. Osteoporos Int. 2012;23:171–82.
27. Wallace BA, Cumming RG. Systematic review of randomized trials of the effect of exercise on bone mass in pre- and postmenopausal women. Calcif Tissue Int. 2000;67:10–18.
28. Mein AL, Briffa NK, Dhaliwal SS, et al. Lifestyle influences on 9-year changes in BMD in young women. J Bone Miner Res. 2004;19:1092–8.
29. Peris P, Monegal A, Martinez MA, et al. Bone mineral density evolution in young premenopausal women with idiopathic osteoporosis. Clin Rheumatol 2007;26:958–61.
30. Liu SL, Lebrun CM. Effect of oral contraceptives and hormone replacement therapy on bone mineral density in premenopausal and perimenopausal women: a systematic review. Br J Sports Med. 2006;40:11–24.
31. Miller KK, Lee EE, Lawson EA, et al. Determinants of skeletal loss and recovery in anorexia nervosa. J Clin Endocrinol Metab. 2006;91:2931–7.
32. Ciacci C, Maurelli L, Klain M, et al. Effects of dietary treatment on bone mineral density in adults with celiac disease: factors predicting response. Am J Gastroenterol. 1997;92:992–6.
33. McFarlane XA, Bhalla AK, Robertson DA. Effect of a gluten free diet on osteopenia in adults with newly diagnosed coeliac disease. Gut. 1996;39:180–4.
34. Lumachi F, Camozzi V, Ermani M, et al. Bone mineral density improvement after successful parathyroidectomy in pre- and postmenopausal women with primary hyperparathyroidism: a prospective study. Ann N Y Acad Sci. 2007;1117:357–61.
35. Adams JS, Song CF, Kantorovich V. Rapid recovery of bone mass in hypercalciuric, osteoporotic men treated with hydrochlorothiazide. Ann Intern Med. 1999;130: 658–60.
36. Powles TJ, Hickish T, Kanis JA, et al. Effect of tamoxifen on bone mineral density measured by dual-energy x-ray absorptiometry in healthy premenopausal and postmenopausal women. J Clin Oncol. 1996;14:78–84.
37. Vehmanen L, Elomaa I, Blomqvist C, et al. Tamoxifen treatment after adjuvant chemotherapy has opposite effects on bone mineral density in premenopausal patients depending on menstrual status. J Clin Oncol. 2006;24:675–80.

38. Fuleihan Gel H, Salamoun M, Mourad YA, et al. Pamidronate in the prevention of chemotherapy-induced bone loss in premenopausal women with breast cancer: a randomized controlled trial. J Clin Endocrinol Metab. 2005;90:3209–14.
39. Golden NH, Iglesias EA, Jacobson MS, et al. Alendronate for the treatment of osteopenia in anorexia nervosa: a randomized, double-blind, placebo-controlled trial. J Clin Endocrinol Metab. 2005;90:3179–85.
40. Miller KK, Grieco KA, Mulder J, et al. Effects of risedronate on bone density in anorexia nervosa. J Clin Endocrinol Metab. 2004;89:3903–6.
41. Nakayamada S, Okada Y, Saito K, et al. Etidronate prevents high dose glucocorticoid induced bone loss in premenopausal individuals with systemic autoimmune diseases. J Rheumatol. 2004;31:163–6.
42. Nzeusseu Toukap A, Depresseux G, Devogelaer JP, et al. Oral pamidronate prevents high-dose glucocorticoid-induced lumbar spine bone loss in premenopausal connective tissue disease (mainly lupus) patients. Lupus. 2005;14:517–20.
43. O'Sullivan SM, Grey AB, Singh R, et al. Bisphosphonates in pregnancy and lactation-associated osteoporosis. Osteoporos Int. 2006;17:1008–12.
44. Conwell LS, Chang AB. Bisphosphonates for osteoporosis in people with cystic fibrosis. Cochrane Database Syst Rev. 2012;4:CD002010.
45. Skordis N, Ioannou YS, Kyriakou A, et al. Effect of bisphosphonate treatment on bone mineral density in patients with thalassaemia major. Pediatr Endocrinol. 2008;Rev 6 Suppl 1:144–8.
46. Okada Y, Nawata M, Nakayamada S, et al. Alendronate protects premenopausal women from bone loss and fracture associated with high-dose glucocorticoid therapy. J Rheumatol. 2008;35:2249–54.
47. Yeap SS, Fauzi AR, Kong NC, et al. A comparison of calcium, calcitriol, and alendronate in corticosteroid-treated premenopausal patients with systemic lupus erythematosus. J Rheumatol. 2008;35:2344–7.
48. Ferrari S, Bianchi ML, Eisman JA, et al. Osteoporosis in young adults: pathophysiology, diagnosis, and management. Osteoporos Int. 2012;23:2735–48.
49. Patlas N, Golomb G, Yaffe P, et al. effects of bisphosphonates on fetal skeletal ossification and mineralization in rats. Teratology. 1999;60:68–73.
50. Minsker DH, Manson JM, Peter CP. Effects of the bisphosphonate, alendronate, on parturition in the rat. Toxicol Appl Pharmacol. 1993;121:217–23.
51. Biswas PN, Wilton LV, Shakir SA. Pharmacovigilance study of alendronate in England. Osteoporos Int. 2003;14:507–14.
52. Chan B, Zacharin M. Maternal and infant outcome after pamidronate treatment of polyostotic fibrous dysplasia and osteogenesis imperfecta before conception: a report of four cases. J Clin Endocrinol Metab. 2006;91: 2017–20.
53. Levy S, Fayez I, Taguchi N, et al. Pregnancy outcome following in utero exposure to bisphosphonates. Bone. 2009;44:428–30.
54. Finkelstein JS, Klibanski A, Arnold AL, et al. Prevention of estrogen deficiency-related bone loss with human parathyroid hormone-(1-34): a randomized controlled trial. JAMA. 1998;280:1067–73.
55. Cohen A, Stein EM, Dempster D, et al. In: Premenopausal Women with Idiopathic Osteoporosis, Baseline Bone Turnover Predicts Response to Teriparatide. American Society for Bone and Mineral Research, San Diego, CA, USA, 2011.
56. Fazeli PK, Wang IS, Miller KK, et al. Teriparatide increases bone formation and bone mineral density in adult women with anorexia nervosa. J Clin Endocrinol Metab. 2014;99:1322–9.
57. Choe EY, Song JE, Park KH, et al. Effect of teriparatide on pregnancy and lactation-associated osteoporosis with multiple vertebral fractures. J Bone Min Metab. 2012;30:596–601.
58. Langdahl BL, Marin F, Shane E, et al. Teriparatide versus alendronate for treating glucocorticoid-induced osteoporosis: an analysis by gender and menopausal status. Osteoporos Int. 2009;20:2095–104.
59. Finkelstein JS, Arnold AL Increases in bone mineral density after discontinuation of daily human parathyroid hormone and gonadotropin-releasing hormone analog administration in women with endometriosis. J Clin Endocrinol Metab. 1999;84:1214–19.
60. Cohen A, Kamanda-Kosseh M, Recker RR, et al. Bone density after teriparatide discontinuation in premenopausal idiopathic osteoporosis. J Clin Endocrinol Metab. 2015;100:4208–14.
61. Adachi JD, Bensen WG, Brown J, et al. Intermittent etidronate therapy to prevent corticosteroid-induced osteoporosis. N Engl J Med. 1997;337:382–7.
62. Saag KG, Emkey R, Schnitzer TJ, et al. Alendronate for the prevention and treatment of glucocorticoid-induced osteoporosis. Glucocorticoid-Induced Osteoporosis Intervention Study Group. N Engl J Med. 1998;339:292–9.
63. Wallach S, Cohen S, Reid DM, et al. Effects of risedronate treatment on bone density and vertebral fracture in patients on corticosteroid therapy. Calcif Tissue Int. 2000;67:277–85.
64. Langdahl BL, Marin F, Shane E, et al. Teriparatide versus alendronate for treating glucocorticoid-induced osteoporosis: an analysis by gender and menopausal status. Osteoporos Int. 2009;20;2095–104.
65. Buckley L, Guyatt G, Fink HA, et al. American College of Rheumatology guideline for the prevention and treatment of glucocorticoid-induced osteoporosis. Arthritis Care Res. 2017;69:1095–1110.

56

Osteoporosis in Men

Eric S. Orwoll[1] and Robert A. Adler[2]

[1]*Division of Endocrinology, Diabetes, and Clinical Nutrition, Oregon Health and Science University, School of Medicine, Portland, OR, USA*
[2]*Endocrinology and Metabolism Section, Hunter Holmes McGuire Veterans Affairs Medical Center, Richmond; Endocrine Division, Virginia Commonwealth University School of Medicine, Richmond, VA, USA*

INTRODUCTION

Osteoporosis in men is an important public health problem and effective diagnostic, preventive, and treatment strategies have been developed [1]. Moreover, the study of osteoporosis in men has revealed male–female differences that in turn have fostered a greater understanding of bone biology in general. Despite this tremendous increase in knowledge, osteoporosis in men is frequently not detected or treated.

SKELETAL DEVELOPMENT

Bone mass accumulation in males occurs gradually during childhood and accelerates dramatically during adolescence. Peak bone mass is closely tied to pubertal development, and male–female differences in the skeleton appear during adolescence [2]. The rapid increase in bone mass occurs somewhat later in boys than girls; the majority of the increase has occurred by an average age of 16 years in girls and age 18 years in boys. Moreover, whereas trabecular BMD accumulation is similar in boys and girls, boys generally develop thicker cortices and larger bones, even when adjusted for body size. These differences may provide important biomechanical advantages that in part underlie the lower fracture risk observed in men later in life. The reasons for these sexual differences in skeletal development are unclear but are related at least in part to differences in sex steroid action (androgens may stimulate periosteal bone formation and bone expansion), growth factor concentrations, and mechanical forces exerted on bone (eg, by greater muscle action or activity). Sex-specific effects of a variety of genetic loci have been reported in animals and humans. Despite these average sex differences, there is wide variation in bone mass and structure in men after adjustment for body size and there is considerable overlap with the range of similar measures in women.

EFFECTS OF AGING ON THE SKELETON IN MEN

As in women, aging is associated with large changes in bone mass and architecture in men [3]. Trabecular bone loss (eg, in the vertebrae and proximal femur) occurs during midlife and accelerates in later life. The magnitude of these changes is less than those in women. In men there is more trabecular thinning and less trabecular drop-out than in women. Endocortical bone loss with resulting cortical thinning takes place in long bones, and cortical porosity increases with age, but those processes may be accompanied by a concomitant increase in periosteal bone expansion that tends to preserve the breaking strength of bone [4]. In general, the pattern of age-related bone loss in men is similar in men and women, but in men the rate of loss is slower and there is no analog to the accelerated phase of bone loss associated with the menopause in women. In the elderly of both sexes, the rate of bone loss accelerates with increasing age.

Primer on the Metabolic Bone Diseases and Disorders of Mineral Metabolism, Ninth Edition. Edited by John P. Bilezikian.
© 2019 American Society for Bone and Mineral Research. Published 2019 by John Wiley & Sons, Inc.
Companion website: www.wiley.com/go/asbmrprimer

FRACTURE EPIDEMIOLOGY

Fractures are common in men, although data in men are derived primarily from the study of white populations. The incidence of fracture is bimodal, with a peak incidence in adolescence and mid-adulthood, a lower incidence between 40 and 60 years, and a dramatic increase after the age of 70 year (Fig. 56.1) [5]. The types of fractures sustained in younger and older men are different, with long bone fractures being common in younger men, whereas vertebral and hip fractures predominate in the elderly. These differences suggest that the etiologies of fractures at these two periods of life are distinct. In younger men, trauma appears to play a larger role, whereas in older men skeletal fragility and fall propensity are likely to be major factors.

The exponential increase in fracture incidence as men age is as dramatic as the similar increase in women, but it begins 5–10 years later in life. This delay, combined with the longer life expectancy in women, underlies the greater burden of osteoporotic fractures in women. The age-adjusted incidence of hip fracture in men is one-quarter to one-third that in women, and 20% to 25% of hip fractures occur in men [3]. The consequences of fracture in men are at least as great as in women, and in fact, elderly men seem to be more likely to die and to suffer disability than women after a hip fracture. Older men suffer lower rates of long bone fractures than do women [6]. Vertebral fractures are common in men [7], but are frequently not clinically recognized [8, 9]. In younger men, the prevalence of vertebral fracture is actually greater in men than in women, at least in part the result of higher rates of spinal trauma experienced by men. Although there are inadequate data, the epidemiology of fracture in men seems to be dramatically influenced by both race and geography [10, 11]. For instance, compared to white men, black men have a much lower likelihood of fractures, and Asian men have a lower likelihood of suffering hip fracture. Much more information is needed concerning these differences and their causation.

Over the last few decades the incidence of fractures has apparently been changing in both men and women. In Western societies the rate of hip fracture increased dramatically until about 10 years ago, but after that it began to decline, particularly in women but also in men [12, 13]. The reasons for the change are not clear but may be related to increased efforts in screening and treatment of osteoporosis, a greater prevalence of obesity, reduced smoking, etc. In contrast, recent data suggest that the rates of fracture are increasing quickly in Asian societies [14], potentially because of urbanization and other cultural changes. These divergent trends emphasize the importance of environmental influences on fracture causation. Importantly, anticipated increases in the size of older populations will result in a greater number of fractures despite gradual reductions in fracture incidence.

CAUSES OF OSTEOPOROSIS IN MEN

Fractures in men are related to a variety of risk factors. Certainly, skeletal fragility makes fracture more likely. This trait is most commonly measured as reduced BMD, but almost certainly has other components (biomechanically important alterations in bone geometry, material properties, etc.). Aging and a previous history of fracture are independently associated with a higher probability of future fracture, and men of lower weight have a higher fracture risk [3, 15]. Finally, falling becomes much more common with increasing age in men, and falls are strongly associated with increased fracture risk [16].

The causation of osteoporosis in men is commonly heterogeneous, and most osteoporotic men have several factors that contribute to the disease. One-half to two-thirds of men with osteoporosis have multiple risk factors, including other medical conditions, medications, or lifestyle issues that result in bone loss and fragility (Table 56.1) [3, 10]. Important comorbidities include alcohol abuse, glucocorticoid excess, and hypogonadism. An important fraction of osteoporotic men, however, have idiopathic disease. In a recent study, risk factors had a striking effect on prediction of hip fractures in men, when added to DXA [17]. Some of these risk factors are in addition to those in Fracture Risk Assessment tool (FRAX) and include hypoglycemic agents, Parkinson disease, poor mobility, and tricyclic antidepressants.

Idiopathic osteoporosis

Osteoporosis of unknown etiology can present in men of any age [18] but its presentation (usually with vertebral fracture) is most dramatic in younger men who are otherwise unlikely to be affected by osteoporosis. A low bone formation rate has been found to be more common in

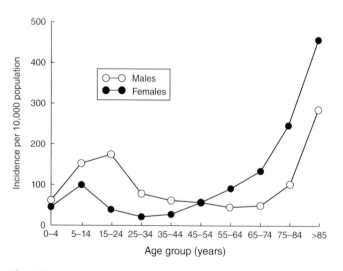

Fig. 56.1. Average annual fracture incidence rate per 10,000 population in Leicester, UK, by age group and by sex.

Table 56.1. Causes of osteoporosis and bone loss in men.

Primary
   Aging
   Idiopathic
Secondary
   Hypogonadism
   Glucocorticoid excess
   Alcoholism, tobacco use
   Renal insufficiency
   Gastrointestinal, hepatic disorders, malabsorption
   Hyperparathyroidism
   Anticonvulsant
   Thyrotoxicosis
   Chronic respiratory disorders
   Anemias, hemoglobinopathies
   Immobilization
   Osteogenesis imperfecta
   Homocystinuria
   Systemic mastocytosis
   Neoplastic diseases and chemotherapy
   Rheumatoid arthritis

these patients. Several possible etiologies have been considered. Most prominent among them are genetic factors, because BMD and the risk of fracture are highly heritable. The specific genes that may be responsible are uncertain.

Hypogonadism

Sex steroids are clearly important for skeletal health in men, both during growth and the attainment of peak bone mass as well as in the maintenance of bone strength in adults [18]. Hypogonadism is associated with low BMD and the development of hypogonadism results in increased bone remodeling and rapid bone loss (at least in the early phases of hypogonadism). Testosterone replacement increases BMD in hypogonadal men. One of the most important causes of severe hypogonadism is androgen deprivation therapy for prostate cancer (ADT); in this situation, bone loss is rapid, and the risk of fractures is almost 20% after 5 years of ADT [19]. Gonadal function and sex steroid levels decline with age in men, and it has been postulated that the decline may be an important risk factor in age-related bone loss and fracture risk, but the strength of this association remains somewhat unclear.

Although both are important, the relative roles of estrogens and androgens in skeletal physiology in men are uncertain [20–22]. Estrogen is essential for normal bone development in young men, as evidenced by immature delay in development and low bone mass in men with aromatase deficiency and their reversal with estrogen therapy. Moreover, estrogen is correlated with bone remodeling, BMD, and rate of BMD loss in older men, apparently more strongly than is testosterone [23]. However, testosterone is independently related to indices of bone resorption and formation and may stimulate periosteal bone [24–26]. Low levels of estradiol have been clearly linked to increased fracture risk in older men [27, 28]. Testosterone levels appear less strongly related to fracture but may have an effect, particularly at very low concentrations [29]. High sex hormone–binding globulin (SHBG) concentrations have also been linked to increased fracture propensity [28]. The relative roles of estrogen, androgen, and SHBG must be better defined, and how measures of their levels can be used in clinical situations must be clarified.

EVALUATION OF OSTEOPOROSIS IN MEN

Guidelines for the evaluation of osteoporosis in men are not well validated, but comprehensive recommendations have recently emerged [30].

BMD measurements

BMD measures are at least as effective in men as in women in predicting the risk of future fractures [31]. In light of the prevalence of osteoporosis and the high incidence of fractures in men, BMD measures are performed too infrequently. Two groups of men would benefit from BMD testing:

- men older than 50 years of age who have suffered a fracture, including those with vertebral deformity. Younger men who suffer low-trauma fractures should also be assessed,
- men who have known secondary causes of bone loss should have BMD determined. These include men treated with glucocorticoids or other medications associated with osteoporosis, men with hypogonadism of any cause including those treated with androgen deprivation therapy, or men who have alcoholism. Many other risk factors may also prompt BMD measures (Table 56.1).

Screening BMD measures in older men have been recommended (eg, >70 years of age) [32] and recent cost-effectiveness analysis suggested that screening may be appropriate at this age, especially as treatment costs have declined [33, 34]. Nonetheless, the US Preventive Services Task Force has found insufficient evidence to recommend screening DXAs for men, based simply on age [35].

The presence of reduced BMD in men is commonly quantified with T-scores using a grading system parallel to that used in women (BMD T-score −1.0 to −2.5 = low bone mass; BMD T-score < −2.5 = osteoporosis). Although whether BMD measurements in men should be interpreted using T-scores based on a male-specific reference range or using the same reference range used in women has been controversial, analyses of large population level data suggest that the BMD–fracture risk association is the same in men and women and the use of female specific references ranges for both sexes has become common. The use of FRAX or other fracture risk calculators provides an important means of incorporating risk factors other than BMD (eg, age, BMI, previous fracture history) into an estimation of fracture risk and the selection of men for therapy. However, there are no studies in which men only identified by FRAX have been shown to respond to treatment in a manner similar to those identified by DXA and/or previous fragility fracture.

Men who have been selected for androgen deprivation therapy deserve special note because the risk of bone loss and fractures is clearly increased, especially in the first 5 year after sex steroid deficiency is induced [19, 36]. Men starting high-dose glucocorticoid therapy present the same challenges and should be similarly managed. When antiandrogen (or glucocorticoid) therapy is begun, a BMD assessment is appropriate. If it is normal, routine preventive measures are reasonable. A repeat BMD measurement should be performed 1 to 2 years later. If BMD is reduced at the onset of therapy, more aggressive preventive measures should be considered (eg, bisphosphonate therapy). In men with osteoporosis even before antiandrogen (or glucocorticoid) therapy is begun, pharmacological approaches to prevent further bone loss or fractures are warranted.

Clinical evaluation

The clinical evaluation of men found to have low BMD should include a careful history and physical examination designed to identify any factors that may contribute to deficits in bone mass. Attention should be paid to lifestyle factors, nutrition (especially calcium, vitamin D, and protein nutrition), alcohol and tobacco abuse, activity level, and family history. A history of previous fracture should be identified, and fall risk should be assessed. This information should be used to formulate recommendations for prevention and treatment.

Laboratory testing

In a man undergoing an evaluation for osteoporosis, laboratory testing is intended to identify correctable causes of bone loss. Appropriate tests are shown in Table 56.2. While such testing has been advocated [30, 37] the value of such testing has not been proven [38].

OSTEOPOROSIS PREVENTION IN MEN

The essentials of fracture prevention in men are similar to those in women. In early life, excellent nutrition and exercise seem to have positive effects on bone mass.

Table 56.2. Evaluation of osteoporosis in men: laboratory tests.

Serum calcium, phosphate, creatinine, alkaline phosphatase, liver function
Complete blood count
Serum 25(OH) vitamin D
Serum total testosterone
24-hour urine calcium, creatinine, and sodium
Targeted diagnostic testing in men with signs, symptoms, or other indications of secondary disorders
When an etiology is not apparent after the above, additional testing may be appropriate, including: calculated free or bioavailable testosterone, serum protein electrophoresis with free κ and λ light chains and/or urine protein electrophoresis, tissue transglutaminase antibodies (for celiac disease), thyroid function tests, and PTH levels

These principles and the avoidance of lifestyle factors known to be associated with bone loss (Table 56.1) remain important throughout life. Calcium and vitamin D probably provide beneficial effects on bone mass and fractures in men as in women. Recent Institute of Medicine recommendations for men include 1000 mg/day of calcium for those 30 to 70 years of age and 1200 mg/day for those older than 70 years of age, with suggested vitamin D intakes of 600 IU/day until age 70 and 800 IU/day thereafter [39]. A recent approach to vitamin D nutrition likely applies to men [40]. In those at risk for falls (eg, with reduced strength, poor balance, previous falls), attempts to increase strength and balance may be beneficial.

TREATMENT OF OSTEOPOROSIS IN MEN

Ensuring adequate calcium and vitamin D intake and appropriate physical activity are essential foundations for preserving and enhancing bone mass in men who have osteoporosis. Secondary causes of osteoporosis should be identified and treated. In addition, there are pharmacological therapies that have been shown to enhance BMD, and in some cases, reduce fracture risk in men. Although the available data are not as extensive as in women, these therapies seem to be as effective in increasing BMD and in reducing fracture risk in men. The treatment indications for these drugs are similar in men and women. There are no long-term treatment studies in men, but it has been suggested that recommendations for women also apply to men [41]. Contrary to most studies in women, one retrospective analysis of a large male database did not find an association of bisphosphonate therapy duration with the most feared side effect, atypical femoral fracture [42].

Idiopathic and age-related osteoporosis

Alendronate, risedronate, ibandronate, zoledronate, teriparatide, and denosumab are effective in improving BMD [43–45] regardless of age or gonadal function. Although the trials are relatively small, therapy is also apparently effective in reducing vertebral fracture risk. Zoledronate reduced the incidence of vertebral fracture in men [46]. Zoledronate also reduced the risk of recurrent fracture in men and women after hip fracture [47], and although the independent effects in men could not be reliably ascertained the effects sizes on fracture risk were similar in men and women. Well-powered studies to evaluate the antifracture effectiveness of therapy in men are clearly warranted.

Glucocorticoid-induced osteoporosis

Bisphosphonate therapy (eg, alendronate, risedronate) is effective in improving BMD and, although the data are not extensive, also probably reduces fracture [48, 49]. In one study [50], teriparatide was more effective than alendronate in reducing fractures in men and women with glucocorticoid-induced osteoporosis.

Hypogonadal osteoporosis

Bisphosphonate, denosumab, and teriparatide therapy are effective in increasing BMD in hypogonadal men. Moreover, bisphosphonate and denosumab treatment can prevent the bone loss after androgen deprivation therapy for prostate cancer, and denosumab reduces vertebral fracture risk in these men [51]. Testosterone replacement therapy results in increases in serum levels of both estradiol and testosterone and improves BMD in men with established hypogonadism [52, 53], but whether fracture risk is reduced is unknown. In older men with less severe, age-related reductions in gonadal function, the usefulness of testosterone is less certain. Intramuscular or transdermal testosterone is associated with an increase in BMD and muscle strength in older men with low testosterone levels [52, 53], but its impact on fracture risk has not been examined. Moreover, the long-term risks of testosterone therapy in older men are unknown. Congruent with published guidelines [30], testosterone replacement therapy is appropriate for management of hypogonadal symptoms in older men, but the treatment of osteoporosis in an older man with low testosterone levels is most confidently undertaken with an osteoporosis drug for which there are more data demonstrating fracture risk reduction (eg, a bisphosphonate or teriparatide). A new agent, romosozumab, has been shown [54] to increase spine and hip bone density in men. Its approval and clinical use are pending.

REFERENCES

1. Adler RA. Update on osteoporosis in men. Best Pract Res Clin Endocrinol Metab. 2018; doi 10.1016/j.beem.2018.05.007.
2. Seeman E. Sexual dimorphism in skeletal size, density, and strength. J Clin Endocrinol Metab. 2001;86(10) 4576–84.
3. Orwoll ES, Vanderschueren D, Boonen S. Osteoporosis in men: epidemiology, pathophysiology, and clinical characterization In: Marcus R, Feldman D, Dempster DW, Luckey M, Cauley JA, eds. *Osteoporosis*. 4th ed. San Diego, CA: Academic Press; 2013: pp 757–802.
4. Seeman E. Pathogenesis of bone fragility in women and men. Lancet. 2002;359(9320):1841–50.
5. Donaldson LJ, Cook A, Thomson RG. Incidence of fractures in a geographically defined population. J Epidemiol Comm Health. 1990;44:241–5.
6. Ismail AA, Pye SR, Cockerill WC, et al. Incidence of limb fracture across Europe: results from the European prospective osteoporosis study (EPOS). Osteoporos Int. 2002;13(7):565–71.
7. Schousboe JT. Epidemiology of vertebral fractures. J Clin Densitom. 2016;19(1):8–22.
8. Briot K, Fechtenbaum J, Roux C. Clincal relevance of vertebral fractures in men. J Bone Miner Res. 2016;31(8):1497–9.
9. European Prospective Osteoporosis Study (EPOS) Group, Felsenberg D, Silman AJ, Lunt M, et al. Incidence of vertebral fracture in Europe: results from the European prospective osteoporosis study (EPOS). J Bone Miner Res. 2002;17(4):716–24.
10. Amin S, Felson DT. Osteoporosis in men. Rheum Dis Clin North Am. 2001;27(1):19–47.
11. Schwartz AV, Kelsey JL, Maggi S, et al. International variation in the incidence of hip fractures: cross-national project on osteoporosis for the World Health Organization Program for Research on Aging. Osteoporosis Int. 1999;9(3):242–53.
12. Cooper C, Cole ZA, Holroyd CR, et al.; IOF CSA Working Group on Fracture Epidemiology. Secular trends in the incidence of hip and other osteoporotic fractures. Osteoporos Int. 2011;22(5):1277–88.
13. Leslie WD, O'Donnell S, Jean S, et al. Osteoporosis Surveillance Expert Working Group. Trends in hip fracture rates in Canada. JAMA. 2009;302(8):883–9.
14. Xia WB, He SL, Xu L, et al. Rapidly increasing rates of hip fracture in Beijing, China. J Bone Miner Res. 2012;27(1):125–9.
15. Nguyen TV, Eisman JA, Kelly PJ, et al. Risk factors for osteoporotic fractures in elderly men. Am J Epidemiol. 1996;144(3):255–63.
16. Chan BKS, Marshall LM, Lambert LC, et al. The risk of non-vertebral and hip fracture and prevalent falls in older men: The MrOS Study. J Bone Miner Res. 2005;20(Suppl 1):S385.
17. Cauley JA, Cawthon PM, Peters KE, et al., Osteoporotic Fractures in Men (MrOS) Study Research Group. Risk factors for hip fracture in older men: The Osteoporotic in Men Study (MrOS). J Bone Miner Res. 2016;31(10): 1810–19.
18. Vanderschueren D, Boonen S, Bouillon R. Osteoporosis and osteoportic fractures in men: a clinical perspective. Baillieres Best Pract Res Clin Endocrinol Metab. 2000;14(2):299–315.

19. Shahinian VB, Kuo YF, Freeman JL, et al. Risk of fracture after androgen deprivation for prostate cancer. N Engl J Med. 2005;352(2):154–64.
20. Khosla S, Oursler MJ, Monroe DG. Estrogen and the skeleton. Trends in Endocrinol Metab. 2012;23(11): 576–81.
21. Vanderschueren D, Laurent MR, Claessens F, et al. Sex steroid actions in male bone. Endocr Rev. 2014;35(6): 906–60.
22. Orwoll ES. Men, bone and estrogen: unresolved issues. Osteoporos Int. 2003;14(2):93–8.
23. Weber TJ. Battle of the sex steroids in the male skeletono: and the winner is… J Clin Invest. 2016;126(3):829–32.
24. Finkelstein JS, Lee H, Leder BZ, et al. Gonadal steroid-dependent effects on bone turnover and bone mineral density in men. J Clin Invest. 2016;126(3):1114–25.
25. Wiren KM, Zhang XW, Olson DA, et al. Androgen prevents hypogonadal bone loss via inhibition of resorption mediated by mature osteoblasts/osteocytes. Bone. 2012;51(5):835–46.
26. Khosla S. New insights into androgen and estrogen receptor regulation of the male skeleton. J Bone Miner Res. 2015;30(7):1134–7.
27. Mellström D, Vandenput L, Mallmin H, et al. Older men with low serum estradiol and high serum SHBG have an increased risk of fractures. J Bone Miner Res. 2008;23(10):1552–1560.
28. LeBlanc ES, Nielson CM, Marshall LM, et al. Osteoporotic Fractures in Men Study Group. The effects of serum testosterone, estradiol, and sex hormone binding globulin levels on fracture risk in older men. J Clin Endocrinol Metab. 2009;94(9):3337–46.
29. Fink HA, Ewing SK, Ensrud KE, et al. Association of testosterone and estradiol deficiency with osteoporosis and rapid bone loss in older men. J Clin Endocrinol Metab. 2006;91(10):3908–15.
30. Watts NB, Adler RA, Bilezikian JP, et al. Endocrine Society. Osteoporosis in men: an Endocrine Society clinical practice guideline. J Clin Endocrinol Metab. 2012;97(6):1802–22.
31. Nguyen ND, Pongchaiyakul C, Center JR, et al. Identification of high-risk individuals for hip fracture: a 14-year prospective study. J Bone Miner Res. 2005;20(11):1921–8.
32. Schousboe JT, Shepherd JA, Bilezikian JP, et al. Executive summary of the 2013 International Society for Clinical Densitometry Position Development Conference on bone densitometry. J Clin Densitom. 2013;16(4): 455–66.
33. Schousboe JT, Taylor BC, Fink HA, et al. Cost-effectiveness of bone densitometry followed by treatment of osteoporosis in older men. JAMA. 2007;298(6): 629–37.
34. Dawson-Hughes B, Tosteson AN, Melton LJ 3rd, et al. National Osteoporosis Foundation Guide Committee. Implications of absolute fracture risk assessment for osteoporosis practice guidelines in the USA. Osteoporosis International. 2008;19(4):449–58.
35. Preventive Services Task Force US. Screening for osteoporosis: U.S. preventive services task force recommendation statement. Ann Intern Med. 2011;154(5):356–64.
36. Abrahamsen B, Nielsen M, Brixen K, et al. Fracture risk is increased in Danish Men with prostate cancer: a nation-wide register study. Abstracts of the 29th Annual Meeting of the American Society for Bone and Mineral Research, Honolulu, HI, 2007.
37. Ryan CS, Petkov VI, Adler RA. Osteoporosis in men: the value of laboratory testing. Osteoporos Int. 2011; 22(6):1845–53.
38. Fink HA, Litwack-Harrison S, Taylor BC, et al. Osteoporotic Fractures in Men (MrOS) Study Group. Clinical utility of routine laboratory testing to identify possible secondary causes in older men with osteoporosis: the Osteoporotic Fractures in Men (MrOS) Study. Osteoporos Int. 2016;27(1):331–8.
39. Institute of Medicine (U. S.) Committee to Review Dietary Reference Intakes for Vitamin D and Calcium. Ross AC, Taylor CL, Yaktine AL, et al. (eds). *Dietary Reference Intakes for Calcium and Vitamin D*. Washington, DC: National Academies Press, 2011.
40. Fuleihan Gel-H, Bouillon R, Clarke B, et al. Serum 25-Hydroxyvitamin D levels: variability, knowledge gaps, and the concept of a desirable range. J Bone Miner Res. 2015;30(7):1119–33.
41. Adler RA, El-Hajj Fuleihan G, Bauer DC, et al. Managing osteoporosis in patients on long-term bisphosphonate treatment: report of a task force of the American Society for Bone and Mineral Research. J Bone Miner Res. 2016;31(1):16–35.
42. Safford MM, Barasch A, Curtis JR, et al. Bisphosphonates and hip and nontraumatic subtrochanteric femoral fractures in the Veterans Health Administration. J Clin Rheumatol. 2014;20(7):357–62.
43. Orwoll E, Ettinger M, Weiss S, et al. Alendronate for the treatment of osteoporosis in men. N Engl J Med. 2000;343(9):604–10.
44. Orwoll ES, Scheele WH, Paul S, et al. The effect of teriparatide [Human Parathyroid Hormone (1-34)] therapy on bone density in men with osteoporosis. J Bone Miner Res. 2003;18(1):9–17.
45. Orwoll E, Teglbjærg CS, Langdahl BL, et al. A randomized, placebo-controlled study of the effects of denosumab for the treatment of men with low bone mineral density. J Clin Endocrinol Metab. 2012;97(9):3161–9.
46. Boonen S, Reginster JY, Kaufman JM, et al. Fracture risk and zoledronic acid therapy in men with osteoporosis. N Engl J Med. 2012;367(18):1714–23.
47. Lyles KW, Colón-Emeric CS, Magaziner JS, et al.; for the HORIZON Recurrent Fracture Trial. Zoledronic acid in reducing clinical fracture and mortality after hip fracture. N Engl J Med. 2007;357:nihpa40967.
48. Adachi JD, Bensen WG, Brown J, et al. Intermittent etidronate therapy to prevent corticosteroid-induced osteoporosis. N Engl J Med. 1997;337(6):382–7.
49. Reid DM, Hughes RA, Laan RF, et al. Efficacy and safety of daily residronate in the treatment of corticosteroid-

induced osteoporosis in men and women: a randomized trial. J Bone Miner Res. 2000;15:1006–13.
50. Saag KG, Zanchetta JR, Devogelaer JP, et al. Effects of teriparatide versus alendronate for treating glucocorticoid-induced osteoporosis: thirty-six-month results of a randomized, double-blind, controlled trial. Arthritis Rheum. 2009;60(11):3346–55.
51. Smith MR, Egerdie B, Hernández Toriz N, et al.; Denosumab HALT Prostate Cancer Study Group. Denosumab in men receiving androgen-deprivation therapy for prostate cancer. N Engl J Med. 2009;361(8):745–55.
52. Page ST, Amory JK, Bowman FD, et al. Exogenous testosterone (T) alone or with finasteride increases physical performance, grip strength, and lean body mass in older men with low serum T. J Clin Endocrinol Metab. 2005;90(3):1502–10.
53. Snyder PJ, Kopperdahl DL, Stephens-Shields AJ, et al. Effect of testosterone treatment on volumetric bone density and strength in older men with low testosterone: a controlled clinical trial. JAMA Intern Med. 2017;177(4):471–9.
54. Lewiecki EM, Blicharski T, Goemaere S, et al. A phase 3 randomized placebo-controlled trail to evaluate efficacy and safety of romosozumab in men with osteoporosis. J Clin Endocrinol Metab. 2018; doi 10.1210/jc.2017-02163.

57

Bone Stress Injuries

Stuart J. Warden[1] and David B. Burr[2]

[1]*Department of Physical Therapy, School of Health and Human Sciences, Indiana University, Indianapolis, IN, USA*
[2]*Department of Anatomy and Cell Biology, School of Medicine, Indiana University; and Department of Biomedical Engineering, Purdue School of Engineering and Technology, IUPUI, Indianapolis, IN, USA*

INTRODUCTION

The skeleton is well designed to withstand habitual and most incidental loads; however, as with all structures exposed to repeated loading, bone is subject to overuse and fatigue. Skeletal fatigue occurs when repetitive loads below the fracture threshold create microscopic structural damage that reduces bone's stiffness. In 1855, Prussian military physician Breithaupt [1] first described an overuse syndrome presenting as "painful swollen feet associated with marching." The pathology was later referred to as a "march fracture" and radiologically identified as a metatarsal stress fracture.

We now know that stress fractures represent part of a pathology continuum which includes stress reactions, stress fractures and, ultimately, complete bone fracture [2]. Stress reactions present on MRI as increased bone turnover associated with periosteal and/or marrow edema identified by MRI, whereas stress fractures have the addition of a radiographically discernable fracture line. Continued loading of a bone that is mechanically compromised by a stress fracture can ultimately lead to a relatively low-trauma complete bone fracture.

The pathology continuum from stress reaction to stress fracture and complete bone fracture is captured by the term bone stress injury (BSI). BSI can be defined as an inability of a bone to withstand repetitive loading resulting in progressive stiffness loss due to structural damage, and localized tenderness and bone pain [2].

PATHOPHYSIOLOGY

The mechanism responsible for BSIs remains theoretical; however, there is consensus that it involves an imbalance between load-induced microdamage formation and its removal (Fig. 57.1) [3]. Mechanical loading results in bone deformation (expressed in terms of strain), with the amount of strain depending on the magnitude of the applied load and ability of the bone to resist deformation. Microscopic damage (termed microdamage) can develop in response to repeated subfailure bone deformation, with the threshold for damage formation depending on the interaction between the number of bone strain cycles, strain magnitude, and rate at which the strain is introduced [4].

Microdamage is a normal phenomenon that helps dissipate energy to avoid complete fracture and serves as a stimulus for targeted remodeling (Fig. 57.2). The latter refers to site-specific remodeling targeted towards areas of damage (likely directed by osteocyte apoptosis [5]), and contrasts the hormonally driven nontargeted (stochastic) remodeling responsible for releasing calcium into the circulation [6]. Targeted remodeling involves activation of a remodeling unit, which maintains homeostasis between damage formation and its removal, preserving bone tissue properties, and enabling a bone to adapt over time to changing demands.

BSIs develop when there is an imbalance between microdamage formation and its removal. An imbalance can occur when: (i) loading produces microdamage at a rate exceeding the normal remodeling rate or (ii) the normal

---

Primer on the Metabolic Bone Diseases and Disorders of Mineral Metabolism, Ninth Edition. Edited by John P. Bilezikian.
© 2019 American Society for Bone and Mineral Research. Published 2019 by John Wiley & Sons, Inc.
Companion website: www.wiley.com/go/asbmrprimer

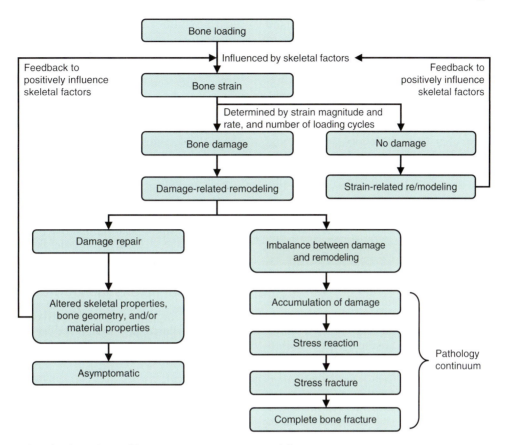

Fig. 57.1. Proposed pathophysiology of bone stress injuries. Source: [3].

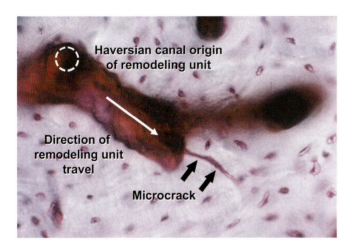

Fig. 57.2. Bone microdamage and targeted remodeling. A microcrack is visible in this basic fuchsin stained histological section of cortical bone. A stimulus (possibly osteocyte apoptosis) has triggered targeted remodeling by a remodeling unit which is advancing toward the damage from a nearby Haversian canal. Source: [50]. Reproduced with permission of Elsevier.

rate of remodeling is suppressed. These mechanisms are thought to contribute to BSIs in healthy, physically active individuals (ie, athletes and military personnel) and older people taking antiremodeling drugs for osteoporosis.

In the scenario of healthy, physically active individuals, remodeling normally removes damage approximately as fast as it occurs and a reserve exists whereby additional remodeling units can be activated in response to increased damage formation. Thus, changes in loading can generally be tolerated. However, remodeling is time dependent and, if insufficient time is given to adapt to a new mechanical stimulus, progressively more damage may form due to a positive feedback loop between remodeling and damage formation. Resorption precedes formation in remodeling so that an increase in the number of active remodeling units reduces local bone mass and energy absorbing capacity, potentiating further damage formation. Accumulating microdamage may combine or coalesce to initiate the BSI pathology continuum.

In the scenario of patients taking antiremodeling drugs, BSIs present as atypical femoral (stress) fractures (AFFs) in the subtrochanteric region and diaphysis of the femur. The pathophysiology is believed to involve damage accumulation due to impaired osteoclast-mediated removal [7].

EPIDEMIOLOGY

BSIs are rare in the population as a whole; however, incidence is increased in the most physically active. The incidence of BSIs is highest within the military, occurring at a rate of 3.2 to 5.7 cases per 1000 person years [8, 9]. New recruits account for the majority of cases, with an incidence over five times greater than in individuals who are further along in their military careers. Up to 5% to 10% of new recruits develop a BSI during the 2 to 3 months of initial basic training [8–10]. In contrast, the one-year prospective incidence of BSIs in athletes is 5% or less [10, 11] with 1.5 BSIs occurring per 100,000 athlete-exposures in high school athletes and between one- and two-thirds of runners having a previous history of a BSI [12–14]. The incidence of atypical femoral fractures in people taking bisphosphonates is very low (≤0.5 cases per 1000 person-years) [15].

RISK FACTORS

BSIs occur relatively infrequently, but are a cause for concern due to their associated morbidity. In athletes, BSIs cause loss of participation and are a source of frustration, particularly when occurring in the lead up to a major competition. In the military, BSIs cost millions of dollars each year due to lost training time, medical expenses, and trainee attrition. In older people, AFFs have high morbidity and usually require surgical intervention, which is associated with its own morbidity and costs.

To reduce the morbidity associated with BSIs, there is a need to identify risk factors for their development (Fig. 57.3). Risk factor identification is important not only for prevention of first-time injury, but also to reduce the risk of recurrence. A history of a BSI is the single largest risk factor for future injury, increasing BSI risk by fivefold [16].

Because BSIs occur when loading at a specific site repetitively exceeds the threshold for microdamage formation, risk factors can be grouped into two categories — factors that modify (i) the load applied to a bone and (ii) the ability of a bone to resist load without damage accumulation.

Factors that modify the load applied to a bone

BSIs occur in response to repetitive loading and, thus, heightened participation in physical activities raises risk. However, risk varies according to activity type. Activities

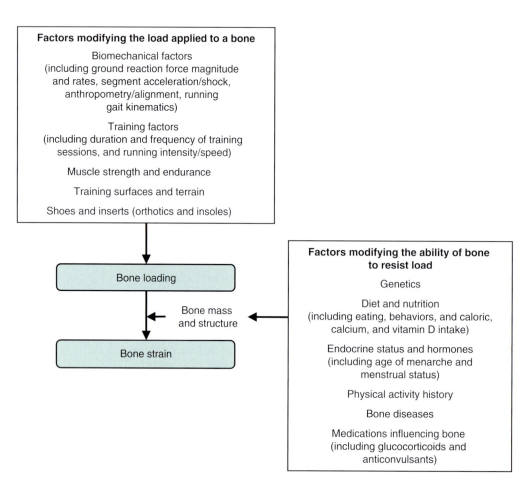

Fig. 57.3. Risk factors for bone stress injuries.

Table 57.1. Sites for stress fractures and the commonly associated activities.

Skeletal Site	Stress Fracture Site	Common Associated Sport or Activity
Scapula	Coracoid process	Trapshooting
	Body	Running with hand-held weights
Humerus	Diaphysis	Throwing, racquet sports, arm wrestling
Ulna	Diaphysis	Racquet sports (especially tennis), gymnastics, volleyball, weightlifting
	Olecranon	Throwing/pitching
Ribs	First	Throwing/pitching
	Second to tenth	Rowing, kayaking
Lumbar spine	Pars interarticularis	Gymanstics, ballet, cricket fast bowling, volleyball, springboard diving
Pelvis	Pubic ramus	Distance running, ballet
Femur	Neck	Distance running, jumping, ballet
	Diaphysis	Distance running
Patella		Running, hurdling
Tibia	Plateau	Running
	Diaphysis	Running, ballet
	Medial malleolus	Basketball, running
Fibula		Running, aerobics, race walking, ballet
Talus	Lateral process	Pole vaulting
Navicular		Sprinting, middle distance running, hurdling, long/triple jump, football
Metatarsals	General	Running, ballet, marching
	Base of the second	Ballet
	Fifth	Tennis, ballet
Sesamoid bones of the foot		Running, ballet, basketball, skating

involving high magnitude loads introduced over short periods of time (ie, sprinting) increase risk due to high strain magnitudes and rates, whereas activities involving high numbers of load repetitions (ie, long distance running, marching) heighten risk due to more cyclic fatigue. Because different activities load different parts of the skeleton, activity type influences site-specific BSI susceptibility (Table 57.1).

Site-specific skeletal loading is influenced not only by activity type, but also by how the specific activity is performed. Subject-specific biomechanics contribute to BSI risk and can be dichotomized into those related to abnormal forces and those related to abnormal motions. Increased ground reaction force magnitudes and rates, and high accelerations applied to the normally aligned skeleton, can increase BSI risk by elevating bone loading [2,17,18]. Alternatively, normal forces applied to a malaligned skeleton can alter strain distribution within a bone leading to increased and abnormal loading of a less accustomed site.

The influence of load magnitude, rate, and number of cycles is most important when an activity-related loading is initiated too rapidly or aggressively. Increased activity intensity or speed increases load magnitudes and their rates of introduction [19], whereas increased duration and/or frequency of training sessions increase the total number of bone loading cycles. In the absence of a change in the load bearing capacity of a bone, large changes in activity, and subsequent bone loading, may contribute to damage accumulation and BSI generation.

Evidence from military studies confirms that large changes in loading increase BSI risk. Military recruits with a shorter history of regular physical activity prior to standardized basic training (ie, those with larger changes in physical activity levels) are at greater risk of developing a BSI [20–22]. Most athletes do not introduce changes in their bone loading environment to the same extremes or at the same rate as many military recruits; however, progression remains an important means of improving activity performance. Incrementing training too rapidly or frequently relative to an athlete's usual activities is thought to be central to disrupting the balance between bone microdamage formation and damage repair.

Activity progression may independently contribute to BSI development, but the associated risk may be compounded by muscle factors. An intimate mechanical relationship exists between muscle and bone, and the general consensus is that muscle protects against BSIs, particularly those occurring in the lower extremity, by dissipating impact forces through eccentric contractions. When muscles are dysfunctional (weakened, fatigued, or altered in their activation patterns), their ability to attenuate loads becomes compromised, leading to increased skeletal loading. For instance, fatigue results in a decrease in shock attenuation [23], increase in loading rates and peak accelerations [24, 25], and increase in bone strain magnitudes and rates [26, 27]. In addition, fatigue can lead to altered kinematics, which may increase bone strain at less accustomed bone sites [28].

Factors influencing the ability of bone to resist load

The amount and rate of strain engendered when load is applied to a bone is dependent upon the ability of the bone to resist deformation. For a given applied load, less rigid bones experience greater strain and at a faster rate than more rigid bones and, thus, are more susceptible to microdamage and BSI formation.

Skeletal features influencing bone rigidity include the quality, amount, and distribution of bone material present. There is preliminary evidence bone material properties contribute to BSI risk [29] and good prospective evidence for contributions of bone mass and structure [22,30–32]. As such, it is important to consider modifiable factors that may contribute to these skeletal characteristics.

A previous history of physical activity is protective against the development of a BSI [20–22] likely due to bone mechanoadaptation and improved load bearing tolerance. Adaptation increases bone rigidity in the direction of loading, with the net result being a reduction in bone strain and increase in fatigue life [33, 34].

Females are at double the risk of BSI than males [10,16], with the heightened risk likely due to skeletal consequences of interrelationships between energy availability, menstrual function and bone mass — otherwise known as the female athlete triad. While one or more components of the triad may be present, low energy availability appears to be central [35]. Low energy availability results from low dietary energy intake and/or excessive energy expenditure. The menstrual and skeletal changes associated with low energy availability reduce the ability of bone to resist load and/or impair its ability to repair microdamage. The net result is heightened BSI risk, particularly in elite female long distance runners.

Bone properties and BSI risk may also be influenced by calcium and vitamin D. Prospective studies indicate roles for both in BSIs [36–38] and a large trial showed daily calcium and vitamin D supplementation to reduce BSI incidence by 20% in female Navy recruits with prior suboptimal baseline daily calcium intake [39].

Hormonal changes that cause bone loss, or medications that attempt to control bone loss, can also alter mechanical properties. The use of antiremodeling agents to control bone loss in osteoporosis can be associated with increased tissue brittleness that can increase the risk for some types of fracture. In the case of AFFs, this increased risk appears to escalate with longer duration of bisphosphonate use [15, 40].

Finally, the presence of other systemic bone diseases can impact the ability of bone to resist load and contribute to BSI risk, and should be considered in the differential diagnosis of presenting individuals. In particular, hypophosphatasia manifesting in middle age (ie, adult hypophosphatasia) can initially present as recurrent, slowly healing metatarsal BSIs followed by painful, debilitating, proximal femur fractures or pseudofractures [41]. The BSIs seen in cases of hypophosphatasia are best screened by the finding of a low total serum alkaline phosphatase and often have elevated serum phosphorus. Other cases, such as tumor-induced osteomalacia (TIO) and/or hypophosphatemic vitamin D resistant rickets, mostly associated with FGF 23-induced renal phosphate wasting, also can present with BSIs. Even though BSIs may be associated with each of these conditions, the pathogenesis of each is clearly distinct.

DIAGNOSIS

BSIs present with activity-related pain. At early stages, the pain may be described as a mild diffuse ache occurring after a specific amount of bone loading activity. The pain does not tend to subside with activity and only abates once bone loading is ceased. With progression of the pathology, the pain may become more severe and localized, occur at an earlier stage of activity, persist for longer periods following activity completion, and begin to be present during activities of daily living. At advanced stages, activity restriction occurs and any associated inflammatory response may contribute to resting and night pain.

On physical examination the most obvious feature of a BSI is localized bony tenderness. Certain bones (such as the tibia, fibula, and metatarsals) lend themselves well to palpation because of the absence of overlying muscle. Direct palpation is obviously not possible at deeper sites (such as the femur and pars interarticularis of the spine), with symptoms at these sites possibly being provoked by specific bone loading tests.

In individuals displaying signs and symptoms of a BSI, imaging can aid in making a more definitive diagnosis. Plain radiographs remain the first line of imaging for BSIs because of their low cost and wide availability; however, they are limited by their planar nature and low spatial resolution which contribute to extremely low sensitivity. Computed tomography also lacks sensitivity, but may be utilized in specific cases where demonstration of a fracture line may affect treatment. Bone scintigraphy has high sensitivity, but is limited by low specificity and the use of extremely high ionizing radiation doses. Of the imaging modalities currently available, MRI is the modality of choice because of its superior contrast resolution, non-use of ionizing radiation, and high sensitivity and specificity [42].

CLASSIFICATION

Numerous classification systems have been reported for grading BSIs to guide management decisions and determine prognosis [43]. While there is no universally accepted system, most consider pain, and the anatomical location and imaging appearance of the BSI.

Management 455

Table 57.2. Low- and high-risk bone stress injuries (BSIs).

Low-Risk BSIs	High-Risk BSIs and Typical Management Other Than Rest
Posteromedial tibia	Subtrochanteric or diaphyseal femur in elderly — intramedullary rod fixation of incomplete fractures (with cortical lucency)
Fibula/lateral malleolus	Femoral neck — nondisplaced: initial bed rest for 1 week, then gradual weight bearing; displaced: surgical fixation
Femoral shaft	Anterior cortex of the tibia — nonweight bearing on crutches for 6 to 8 weeks or intramedullary rod fixation
Pelvis	Medial malleolus — nonweight bearing cast immobilization for 6 weeks or surgical fixation
Calcaneus	Talus (lateral process) — nonweight bearing cast immobilization for 6 weeks or surgical excision of fragment
Diaphysis of 2nd to 4th metatarsals	Navicular — nonweight bearing cast immobilization for 6 to 8 weeks or surgery
	Proximal diaphysis of the 5th metatarsal — cast immobilization or percutaneous screw fixation
	Base of 2nd metatarsal — nonweight bearing for 2 weeks; partial weight bearing for 2 weeks
	Sesamoid bone of the foot — nonweight bearing for 4 weeks

Table 57.3. Bone stress injury grade according to MRI appearance.

Grade	Management
1	Periosteal surface: mild to moderate edema on T2-weighted images Marrow: normal on T1- and T2-weighted images
2	Periosteal surface: moderate to severe edema on T2-weighted images Marrow: edema on T2-weighted images
3	Periosteal surface: moderate to severe edema on T2-weighted images Marrow: edema on T1- and T2-weighted images
4	Periosteal surface: moderate to severe edema on T2-weighted images Marrow: edema on T1- and T2-weighted images Clearly visible fracture line

Source: [47]. Reproduced with permission of Sage Publications.

Pain can be used to guide management, but does not appear to inform prognosis. Imaging changes consistent with a BSI are common in asymptomatic individuals and have not been found to predict subsequent symptoms, even if intensive physical activity continues [44–46]. Thus, precautionary imaging and management of presymptomatic BSI changes is not indicated for routine (ie, low-risk) BSIs. The same is not true for high-risk BSIs where risk of progression to complete fracture is greater.

BSIs can be dichotomized into either low- or high-risk groups according to their location (Table 57.2). Low-risk BSIs predominantly occur on the compression side of a bone's bending axis, and typically recover with a low incidence of complications and without the need for aggressive intervention (such as surgery and/or prolonged modified weight bearing). High-risk BSIs often occur on the tension side of a bone's bending axis, and present treatment challenges demanding specific attention because they are prone to delayed- or non-union and/or are at greater high risk for progression to complete fracture. This is typical of AFFs which occur on the lateral (tension) cortex of the femoral diaphysis, have a high risk for progression, and often do not to heal quickly.

BSIs can also be categorized as either low- or high-grade according to their imaging appearance. A seminal MRI grading system for BSIs is shown in Table 57.3 [47] with more recent grading systems yielding only slight variations. Grades 1 and 2 BSIs on the grading system can be grouped as low-grade BSIs, whereas grades 3 and 4 can be categorized as high-grade BSIs [48].

Combining knowledge regarding the anatomical location and imaging grade of BSIs provides a basis to formulate management and prognosis. Generally speaking, BSIs at: (i) low-risk sites with a low grade have the most favorable outcome and shortest time to activity return; (ii) low-risk sites, but with a high grade will require a longer time to return to activity; and (iii) high-risk sites need to be carefully managed and can be expected to take longer for return to activity regardless of grade.

MANAGEMENT

Low-risk BSIs

Management of low-risk BSIs is relatively straightforward from the standpoint that they readily heal without complication. The overarching goal is to return the individual to their preinjury level of function in the shortest time possible without compromising tissue level healing. However, the high recurrence rate of BSIs indicates that management needs to involve identification and modification of potential risk factors for future BSIs.

Low-risk BSIs can be managed with a two-phase approach consisting of modified activity followed by its gradual resumption. Temporary discontinuation of the inciting activity is initially required. The early goal is for the individual to be pain-free during and after activities of daily living. If a pain-free normal gait cannot be achieved, partial or nonweight bearing using assistive gait devices or a period in a walking boot may be considered. However, progression to unassisted pain-free gait should be sought as soon as possible.

Once the individual is completely pain free for 5 consecutive days during usual daily activities, a graduated loading program consisting of progressive, appropriate loading can be commenced [2]. Appropriate loading can be defined as loading that does not provoke BSI symptoms either during or after completion of an activity. Provocation of symptoms indicates excessive loading for the stage of healing and the need for activity titration and slower activity progression.

During the early stage of BSI healing, techniques aimed at accelerating tissue-level healing may be considered, such as the introduction of low-intensity pulsed ultrasound or use of PTH or antisclerostin antibody therapy. Such approaches have promise in stimulating osteogenesis, but considering low-risk BSIs usually readily heal, clinical utility may be best limited to the management of high-risk, delayed- or nonunited BSIs.

Physical conditioning can be maintained during early management via activities such as cycling, swimming, deep water running, and antigravity treadmill training. Also, early consideration should be given to the identification of potential subject-specific causative risk factors. A detailed activity history is important, including review of usual and recent changes in activity type, frequency, duration, and intensity. Similarly, initial assessment and management of potential biomechanical factors influencing bone loading can be explored, including static posture and alignment, muscle strength and endurance, joint range, and dynamic mechanics. A full dietary history should be taken, with particular attention paid towards the possible presence of deficits in energy, calcium, and vitamin D intake. Bone health can be assessed, and a detailed menstrual history taken in females, remembering that diagnosis of a BSI may be the first time issues associated with the female athlete triad are identified.

As recovery and resumption of activity progresses, biomechanical techniques aimed at reducing the magnitude and rate of bone loading may be considered, particularly in those with a history of recurrent BSIs. For instance, a number of gait retraining techniques are currently being investigated as means to reduce loading during running, including: (i) the use of accelerometer-based biofeedback to encourage reductions in loading magnitude and rate; (ii) increasing stride rate to reduce stride length and, subsequently, vertical excursion and velocity of the center of mass, and ground reaction forces and tibial accelerations; and (iii) modifying initial contact to encourage a more forefoot, rather than heel or rearfoot strike pattern [49]. However, these gait retraining techniques should not be taken lightly as inducing a change in gait may alter injury risk at an alternative site.

High-risk BSIs

High-risk BSIs present challenges because they are: (i) more difficult to diagnose resulting in delays in diagnosis; (ii) prone to delayed- or nonunion; and/or (iii) at high risk for progression to complete fracture. Specific sites for high-risk BSIs and their typical management are detailed in Table 57.2. Management ranges from prolonged modified activity to nonweight bearing with or without a cast and/or surgical fixation. Factors determining management choice include BSI location, presence or absence of a cortical defect on imaging, and duration of symptoms and/or the pathology.

Return to activity following a high-risk BSI typically requires a greater degree of tissue-level healing compared with a low-risk BSI to minimize the risk of pathology progression. In most cases, gradual return to inciting physical activities can occur when imaging studies are consistent with cortical bridging or healing and the individual is asymptomatic on weight bearing and palpation. However, at some locations (ie, tarsal navicular) or in the absence of a cortical defect at initial presentation, repeat imaging is not informative and return to activity decisions are based on the absence of symptoms following completion of the initial mandatory management. As per low-risk BSIs, it is important to assess for and address potential risk factors for BSIs during management of a high-risk BSI.

CONCLUSION

BSIs result from disruption of the homeostasis between microdamage formation and its removal, and remain a source of concern because of the morbidity they cause and high rate of recurrence. Risk for a BSI relates to both the load being applied to a bone and the ability of the bone to resist load, with the former being most amenable to intervention. While most BSIs readily heal with a period of modified loading and progressive return to activity, the high recurrence rate of BSIs signals a need to address underlying reasons for their occurrence. In particular, there is a need to look beyond changes in loading as the sole cause of BSIs. Interventions aimed at reducing loads being applied to the skeleton may include techniques to reduce impact-related forces and increase the strength and/or endurance of local musculature. Similarly, malalignments and abnormal movement patterns should be explored and addressed. Also, the ability of the skeleton to resist load should not be ignored despite being more difficult to intervene. In particular, female long-distance runners exhibiting signs and/or symptoms of the female athlete triad need appropriate management and individuals taking bisphosphonates should be aware of a potential BSI with the onset of thigh pain.

REFERENCES

1. Breithaupt J. Zur Pathologie des menschlichen Fußes. Med Zeittung. 1855;24:169–75.
2. Warden SJ, Davis IS, Fredericson M. Management and prevention of bone stress injuries in long-distance runners. J Orthop Sports Phys Ther. 2014;44(10):749–65.

3. Warden SJ, Burr DB, Brukner PD. Stress fractures: pathophysiology, epidemiology, and risk factors. Curr Osteoporos Rep. 2006;4(3):103–9.
4. Burr DB, Forwood MR, Fyhrie DP, et al. Bone microdamage and skeletal fragility in osteoporotic and stress fractures. J Bone Miner Res. 1997;12:6–15.
5. Verborgt O, Gibson GJ, Schaffler MB. Loss of osteocyte integrity in association with microdamage and bone remodeling after fatigue in vivo. J Bone Miner Res. 2000;15(1):60–7.
6. Burr DB. Targeted and nontargeted remodeling. Bone. 2002;30(1):2–4.
7. Ettinger B, Burr DB, Ritchie RO. Proposed pathogenesis for atypical femoral fractures: lessons from materials research. Bone. 2013;55(2):495–500.
8. Lee D. Stress fractures, active component, U.S. Armed Forces, 2004–2010. MSMR. 2011;18(5):8–11.
9. Waterman BR, Gun B, Bader JO, et al. Epidemiology of lower extremity stress fractures in the United States Military. Mil Med. 2016;181(10):1308–13.
10. Wentz L, Liu PY, Haymes E, et al. Females have a greater incidence of stress fractures than males in both military and athletic populations: a systemic review. Mil Med. 2011;176(4):420–30.
11. Tenforde AS, Sayres LC, McCurdy ML, et al. Identifying sex-specific risk factors for stress fractures in adolescent runners. Med Sci Sports Exerc. 2013;45(10):1843–51.
12. Bennell KL, Malcolm SA, Thomas SA, et al. Risk factors for stress fractures in female track-and-field athletes: a retrospective analysis. Clin J Sports Med. 1995;5(4):229–35.
13. Changstrom BG, Brou L, Khodaee M, et al. Epidemiology of stress fracture injuries among US high school athletes, 2005–2006 through 2012–2013. Am J Sports Med. 2015;43(1):26–33.
14. Kelsey JL, Bachrach LK, Procter-Gray E, et al. Risk factors for stress fracture among young female cross-country runners. Med Sci Sports Exerc. 2007;39(9):1457–63.
15. Shane E, Burr D, Abrahamsen B, et al. Atypical subtrochanteric and diaphyseal femoral fractures: second report of a task force of the American Society for Bone and Mineral Research. J Bone Miner Res. 2014;29(1):1–23.
16. Wright AA, Taylor JB, Ford KR, et al. Risk factors associated with lower extremity stress fractures in runners: a systematic review with meta-analysis. Br J Sports Med. 2015;49(23):1517–23.
17. Meardon SA, Willson JD, Gries SR, et al. Bone stress in runners with tibial stress fracture. Clin Biomech. 2015;30(9):895–902.
18. Popp KL, McDermott W, Hughes JM, et al. Bone strength estimates relative to vertical ground reaction force discriminates women runners with stress fracture history. Bone. 2017;94:22–8.
19. Hamill J, Bates BT, Knutzen KM, et al. Variations in ground reaction force parameters at different running speeds. Hum Mov Sci. 1983;2:47–56.
20. Lappe JM, Stegman MR, Recker RR. The impact of lifestyle factors on stress fractures in female Army recruits. Osteoporos Int. 2001;12(1):35–42.
21. Milgrom C, Simkin A, Eldad A, et al. Using bone's adaptation ability to lower the incidence of stress fractures. Am J Sports Med. 2000;28:245–51.
22. Cosman F, Ruffing J, Zion M, et al. Determinants of stress fracture risk in United States Military Academy cadets. Bone. 2013;55(2):359–66.
23. Mercer JA, Bates BT, Dufek JS, et al. Characteristics of shock attenuation during fatigued running. J Sports Sci. 2003;21(11):911–9.
24. Clansey AC, Hanlon M, Wallace ES, et al. Effects of fatigue on running mechanics associated with tibial stress fracture risk. Med Sci Sports Exerc. 2012;44(10):1917–23.
25. Mizrahi J, Verbitsky O, Isakov E. Fatigue-related loading imbalance on the shank in running: a possible factor in stress fractures. Ann Biomed Eng. 2000;28:463–9.
26. Fyhrie DP, Milgrom C, Hoshaw SJ, et al. Effect of fatiguing exercise on longitudinal bone strain as related to stress fracture in humans. Ann Biomed Eng. 1998;26:660–5.
27. Milgrom C, Radeva-Petrova DR, Finestone A, et al. The effect of muscle fatigue on in vivo tibial strains. J Biomech. 2007;40(4):845–50.
28. Yoshikawa T, Mori S, Santiesteban AJ, et al. The effects of muscle fatigue on bone strain. J Exp Biol. 1994;188:217–33.
29. Duarte Sosa D, Fink Eriksen E. Women with previous stress fractures show reduced bone material strength. Acta Orthop. 2016:1–6.
30. Beck TJ, Ruff CB, Shaffer RA, et al. Stress fracture in military recruits: gender differences in muscle and bone susceptibility factors. Bone. 2000;27:437–44.
31. Bennell KL, Malcolm SA, Thomas SA, et al. Risk factors for stress fractures in track and field athletes: a twelve-month prospective study. Am J Sports Med. 1996;24:810–8.
32. Giladi M, Milgrom C, Simkin A, et al. Stress fractures and tibial bone width: a risk factor. J Bone Joint Surg. 1987;69B:326–9.
33. Warden SJ, Hurst JA, Sanders MS, et al. Bone adaptation to a mechanical loading program significantly increases skeletal fatigue resistance. J Bone Miner Res. 2005;20:809–16.
34. Warden SJ, Mantila Roosa SM, Kersh ME, et al. Physical activity when young provides lifelong benefits to cortical bone size and strength in men. Proc Natl Acad Sci U S A. 2014;111:5337–42.
35. Nattiv A, Loucks AB, Manore MM, et al.; American College of Sports Medicine. American College of Sports Medicine position stand. The female athlete triad. Med Sci Sports Exerc. 2007;39(10):1867–82.
36. Davey T, Lanham-New SA, Shaw AM, et al. Low serum 25-hydroxyvitamin D is associated with increased risk of stress fracture during Royal Marine recruit training. Osteoporos Int. 2016;27(1):171–9.
37. Nieves JW, Melsop K, Curtis M, et al. Nutritional factors that influence change in bone density and stress fracture risk among young female cross-country runners. PM R. 2010;2(8):740–50.

38. Ruohola JP, Laaksi I, Ylikomi T, et al. Association between serum 25(OH)D concentrations and bone stress fractures in Finnish young men. J Bone Miner Res. 2006;21(9):1483–8.
39. Lappe J, Cullen D, Haynatzki G, et al. Calcium and vitamin D supplementation decreases incidence of stress fractures in female navy recruits. J Bone Miner Res. 2008;23(5):741–9.
40. Dell RM, Adams AL, Greene DF, et al. Incidence of atypical nontraumatic diaphyseal fractures of the femur. J Bone Miner Res. 2012;27(12):2544–50.
41. Whyte MP. Hypophosphatasia – aetiology, nosology, pathogenesis, diagnosis and treatment. Nat Rev Endocrinol. 2016;12(4):233–46.
42. Wright AA, Hegedus EJ, Lenchik L, et al. Diagnostic accuracy of various imaging modalities for suspected lower extremity stress fractures: a systematic review with evidence-based recommendations for clinical practice. Am J Sports Med. 2016;44(1):255–63.
43. Miller T, Kaeding CC, Flanigan D. The classification systems of stress fractures: a systematic review. Phys Sportsmed. 2011;39(1):93–100.
44. Bergman AG, Fredericson M, Ho C, et al. Asymptomatic tibial stress reactions: MRI detection and clinical follow-up in distance runners. Am J Roentgenol. 2004;183(3):635–8.
45. Kiuru MJ, Niva M, Reponen A, et al. Bone stress injuries in asymptomatic elite recruits: a clinical and magnetic resonance imaging study. Am J Sports Med. 2005;33(2):272–6.
46. Niva MH, Mattila VM, Kiuru MJ, et al. Bone stress injuries are common in female military trainees: a preliminary study. Clin Orthop Relat Res. 2009;467(11):2962–9.
47. Fredericson M, Bergman AG, Hoffman KL, et al. Tibial stress reaction in runners. Correlation of clinical symptoms and scintigraphy with a new magnetic resonance imaging grading system. Am J Sports Med. 1995;23(4):472–81.
48. Chen YT, Tenforde AS, Fredericson M. Update on stress fractures in female athletes: epidemiology, treatment, and prevention. Curr Rev Musculoskelet Med. 2013;6(2):173–81.
49. Davis IS, Futrell E. Gait retraining: altering the fingerprint of gait. Phys Med Rehabil Clin N Am. 2016;27(1):339–55.
50. Warden SJ, Burr DB, Brukner PD. Repetitive stress pathology: bone. In: Magee DJ, Zachazewski JE, Quillen WS (eds) *Pathology and Intervention in Musculoskeletal Rehabilitation*. St Louis: Saunders Elsevier, 2009, pp. 685–705.

58

Inflammation-Induced Bone Loss in the Rheumatic Diseases

Ellen M. Gravallese[1] and Steven R. Goldring[2]

[1] Division of Rheumatology; Departments of Medicine and Translational Research, Musculoskeletal Center of Excellence, University of Massachusetts Medical School, Worcester, MA, USA
[2] Hospital for Special Surgery, Weill Medical College of Cornell University, New York, NY, USA

INTRODUCTION

The inflammatory joint diseases include a diverse group of disorders that share in common the presence of inflammatory and destructive changes that adversely affect the structure and function of articular and periarticular tissues. In many of these disorders, the inflammatory processes that target the joint tissues may affect extra-articular tissues and organs, and in addition, there may be generalized effects on systemic bone remodeling. Attention will focus on rheumatoid arthritis (RA), systemic lupus erythematosus (SLE), and seronegative spondyloarthritis, which includes ankylosing spondylitis (AS), reactive arthritis (formerly designated Reiter syndrome), arthritis of inflammatory bowel disease, juvenile-onset spondyloarthropathy, and psoriatic arthritis. The discussion of seronegative spondyloarthritis will be limited to AS, which is the prototypical spondyloarthritis.

In RA and SLE, the synovial lining is the initial site of the inflammatory process. Under physiological conditions, the synovium forms a thin membrane that lines the surface of the joint cavity and is responsible for generation of the synovial fluid that contributes to joint lubrication and nutrition for the chondrocytes that populate the articular cartilage. In patients with RA and SLE, the synovium becomes a site of an intense immune-mediated inflammatory process that results in synovial proliferation and production of inflammatory cytokines and soluble mediators that are responsible for the clinical signs of joint inflammation [1–3]. In patients with RA this inflammatory process ultimately leads to destruction of the joint tissues. Although the clinical signs of inflammation in SLE and RA are similar, the synovitis associated with SLE characteristically does not lead to direct destruction of articular cartilage and bone. Of interest Nzeusseu Toukap and colleagues [4] have reported that patients with SLE exhibit a synovial gene profile that is distinct from the patterns observed in RA and osteoarthritis, suggesting that differential inflammatory and immunological processes are involved in the pathogenesis and biological activity of the synovium in these conditions. The potential mechanisms involved in the differential effects on bone resorption in SLE and RA will be reviewed in this chapter.

Despite the absence of destructive changes, joint deformities (referred to as Jaccoud arthropathy) do develop in patients with SLE, but these have been attributed to alterations in the integrity of periarticular tendons and connective tissues rather than destruction of the articular cartilage and bone [5]. However, recent studies employing ultrasonographic techniques and computed tomography have shown that a small percentage of patients with SLE (so-called rhupus syndrome) do exhibit evidence of joint erosions [6].

Synovial inflammation is also present in the seronegative spondyloarthropathies. Unlike the pattern of joint inflammation in RA and SLE, the joint inflammation is most often oligoarticular and asymmetrical, involving distal as well as proximal joints, and, importantly, the axial skeleton also is affected. Anatomical and histopathological analyses have established that the entheses, which are the sites of tendon or ligament attachments to bone, are the initial sites of inflammation in the spondyloarthropathies [7]. Subsequently, the extension of the inflammatory process to the synovium and joint margins and the development of synovial pannus may be accompanied by the development of marginal joint erosions. In contrast to the findings in RA, the inflammatory process

---

Primer on the Metabolic Bone Diseases and Disorders of Mineral Metabolism, Ninth Edition. Edited by John P. Bilezikian.
© 2019 American Society for Bone and Mineral Research. Published 2019 by John Wiley & Sons, Inc.
Companion website: www.wiley.com/go/asbmrprimer

in the spondyloarthropathies may be associated with calcification and ossification at the enthesis and eventual bony ankylosis of the joint [8, 9]. A similar inflammatory process may affect the axial skeleton, leading to enhanced bone formation, the formation of so-called syndesmophytes, and fusion or ankylosis of adjacent vertebrae [8, 10, 11].

RHEUMATOID ARTHRITIS

RA is a systemic inflammatory disorder characterized by symmetrical polyarthritis. Four major forms of pathological skeletal remodeling can be observed in this disorder, including focal marginal articular erosions, subchondral bone loss, periarticular osteopenia, and systemic osteoporosis [1–3]. The focal marginal erosions are the radiological hallmark of RA. Histopathological examination of these sites of focal bone loss reveals the presence of inflamed synovial tissue that has attached to the bone surface, forming a mantle or covering referred to as "pannus." The interface between the pannus and adjacent bone is frequently lined by resorption lacunae containing mono- and multinucleated cells with phenotypic features of authentic osteoclasts, thus implicating osteoclasts as the principal cell type responsible for the focal synovial resorptive process [12, 13]. Similar sites of focal bone loss are present on the endosteal surface of the subchondral bone, and cells with phenotypic features of osteoclasts also are present on these bone surfaces [12, 13]. Erosion of the subchondral bone at these sites contributes to cartilage destruction by providing access to the deep zones of the articular cartilage, which is subject to degradation by the invading inflammatory tissue [14]. These regions of subchondral bone erosion frequently conform to sites of so-called bone marrow edema visualized by MRI. Histological analysis of the bone marrow in these regions reveals that the bone marrow has been replaced by a fibrovascular stroma populated by inflammatory cells [15]. Importantly, the presence of bone marrow lesions is strongly predictive of the subsequent development of local bone erosions at these sites [16, 17].

More definitive evidence implicating osteoclasts in the pathogenesis of focal articular bone erosions has come from the use of genetic approaches in which investigators have induced inflammatory arthritis with features of RA in mice lacking the ability to form osteoclasts [14, 18, 19]. In these models, the inability to form osteoclasts results in protection from focal articular bone resorption despite the presence of extensive synovial inflammation.

The propensity of the synovial lesion in RA to induce osteoclast-mediated bone resorption can be attributed to the production by cells within the inflamed tissue of a wide variety of products with the capacity to recruit osteoclast precursors and induce their differentiation and activation. These include a host of chemokines, as well as receptor activator of NF-κB ligand (RANKL), TNF, IL-1, IL-6, IL-11, IL-15, IL-17, M-CSF, prostaglandins, and parathyroid hormone-related peptide [3, 20, 21]. Among these products, particular attention has focused on RANKL, which is produced by both synovial fibroblasts and T cells within the synovial tissue [22–25]. The critical role of this cytokine in the pathogenesis of focal bone erosions is suggested by the observation that blocking the activity of RANKL in animal models of RA with osteoprotegerin (OPG) results in marked attenuation of articular bone erosions [23, 26, 27]. The pivotal role of RANKL in the resorptive process is further supported by results obtained in genetic models of inflammatory arthritis in which deletion of RANKL [14] or disruption of its signaling pathway [19] protects animals from articular bone erosions. More recently, blockade of RANKL with denosumab, a monoclonal antibody that blocks RANKL activity, was shown to significantly reduce articular bone erosions in a group of patients with RA, providing further evidence that osteoclasts and osteoclast-mediated bone resorption represent a rational therapeutic target for preventing articular bone destruction in RA [28–31]. Interestingly, bisphosphonates, which demonstrate beneficial effects in protecting from systemic bone loss in RA, have not been effective in reducing focal joint destruction [32], with the exception of a publication in which the investigators used a protocol involving the sequential administration of zoledronic acid [32, 33]. Although there may be limitations with respect to the use of these agents to prevent joint destruction, as discussed later in this chapter, there clearly is a role for bisphosphonates in treating and preventing systemic bone loss in RA [33]. In addition, several recent studies have also shown beneficial effects of targeting RANKL with denosumab in the prevention of systemic bone loss in patients with RA, including patients concomitantly treated with glucocorticoids [34, 35].

The RA synovium is also a source of inhibitors of osteoclastogenesis, including interferon-γ, interferon-α/β, IL-4, IL-10, IL-12, IL-18, and possibly also IL-23 [36, 37]. Many of these factors are produced by T cells, which are a major cellular constituent of the RA synovium. Synovial T cells also are a major source of pro-osteoclastogenic factors, including RANKL and TNF-α but the presence of progressive focal bone erosions and systemic bone loss indicate that the inhibitory effects of these cytokines is not sufficient to prevent bone loss.

The discovery of a new class of autoantibodies in patients with RA has provided a link between autoimmunity and osteoclast-mediated bone resorption. These autoantibodies are directed against multiple citrullinated proteins and have been defined as anticitrullinated protein antibodies (ACPAs). Certain ACPAs recognize citrullinated vimentin on the cell surface membrane of osteoclasts and their precursors, and binding of the antibodies results in enhanced osteoclastogenesis [38]. Of interest, the presence of ACPA is associated with bone loss even before the onset of clinical joint inflammation and, importantly, there is an additive effect of ACPA and rheumatoid factor on marginal joint erosion [39, 40].

An additional striking feature of the focal marginal and subchondral bone loss in patients with RA is the virtual absence of bone repair. Matzelle and colleagues [41] have demonstrated in an animal model of RA that once inflammation almost completely resolves at erosion sites, osteoblast precursors line eroded surfaces, mature, and lay down bone to repair erosions. The lack of repair of erosions in patients with RA may thus reflect the persistence of low levels of inflammation in joints, or may indicate a paucity of osteoblast precursor cells to produce bone [41].

Diarra and colleagues [42] have also provided insights into the mechanism involved in the uncoupling of bone resorption and formation in this form of inflammatory arthritis. They demonstrated that cells in the inflamed RA synovial tissue produced dickkopf-1 (DKK-1), the inhibitor of the wingless (Wnt) signaling pathway that plays a critical role in osteoblast-mediated bone formation. Studies by Walsh and colleagues [43] have confirmed these observations and identified additional Wnt family antagonists, including members of the DKK and secreted Frizzled-related protein families in RA synovium. In the Diarra studies, synovial fibroblasts, endothelial cells, and chondrocytes were the principal sources of DKK-1 [42], and TNF-α was shown to be a potent inducer of DKK-1, thus implicating this proinflammatory cytokine in the impaired bone formation at sites of bone erosion. A somewhat surprising observation in the Diarra study was that inhibition of DKK-1 with a blocking antibody produced beneficial effects not only on bone formation, but also suppressed osteoclast-mediated bone resorption [42]. The effects on suppression of bone resorption were attributed to downregulation of RANKL production by the inflamed synovium and upregulation of OPG [44].

In addition to production of DKK-1, synovial fibroblasts also produce sclerostin in response to TNF-α [45]. In follow-up studies, Chen and colleagues showed that treatment of human TNF transgenic mice (hTNFtg mice) with inflammatory arthritis using antibodies that blocked sclerostin activity protected the mice from focal, periarticular, and systemic bone loss [46]. More recently, Wehmeyer and colleagues [47], using a similar hTNFtg mouse model, reported that treatment of these mice with antisclerostin antibodies exacerbated joint inflammation and bone loss. They went on to demonstrate that sclerostin inhibits downstream signaling pathways by which TNF induces inflammation, and these pathways are enhanced in the presence of sclerostin inhibition. Further studies will be needed to resolve this apparent discrepancy regarding the role of sclerostin in regulating joint inflammation and bone remodeling.

Focal marginal bone erosions are the radiographic hallmark of RA, but the earliest skeletal feature of RA is the development of periarticular osteopenia. Of importance, there is evidence that the juxta-articular bone loss has high predictive value with respect to the subsequent development of marginal joint erosions in the hand [48–50]. There are few studies examining the histopathological changes associated with peri-articular osteopenia. Shimizu and colleagues [51] examined the peri-articular bone obtained from a series of patients with RA undergoing joint arthroplasty and observed evidence of both increased bone resorption and formation based on histomorphometric analysis. Examination of the bone marrow in the juxta-articular tissues frequently reveals the presence of focal accumulations of inflammatory cells, including lymphocytes and macrophages, and these cells are a likely source of cytokines and related proinflammatory mediators that could adversely affect bone remodeling [21]. Immobilization and reduced mechanical loading are additional factors that have been implicated in the pathogenesis of peri-articular bone loss.

The final skeletal feature of RA is the presence of generalized osteoporosis. Numerous studies have documented that patients with RA have lower BMD and an increased risk of fracture compared with disease controls [52–55]. The presence in patients with RA of multiple confounding factors that influence bone remodeling has made it difficult to define in any given patient the underlying pathogenic mechanism responsible for the reduced bone mass. These include the effects of sex, age, nutritional state, level of physical activity, disease duration and severity, and the use of medications such as glucocorticoids that can adversely affect bone remodeling. Lodder and colleagues [53] evaluated the relationship between bone mass and disease activity in a cohort of patients with RA with low to moderate disease activity and observed that disease activity was a significant contributory factor to systemic bone loss, supporting the earlier observations made by several other investigators. Solomon and colleagues [56] examined the relationship between focal bone erosions and generalized osteoporosis in a cohort of postmenopausal women with RA. Although they observed an association between low hip BMD and joint erosions, the association disappeared after multivariable adjustment, suggesting that the relationship between erosions and BMD is complex and influenced by multiple disease- and treatment-related factors.

Several different approaches have been utilized to gain insights into the mechanism responsible for systemic bone loss in RA, including histomorphometric analysis of bone biopsies, measurement of urinary and serum biomarkers of bone remodeling, and the assessment of serum cytokine levels. Earlier studies employing histomorphometric analysis suggested that the decrease in bone mass was attributable to depressed bone formation [57]. In contrast, Gough and colleagues [58], as well as several other investigators, observed the presence of increased bone resorption based on assessment of urinary markers. Of interest, Garnero and colleagues [59] observed that a high urinary CTx-1 level (a marker of bone resorption) predicted risk of radiographic progression of joint damage independent of rheumatoid factor or erythrocyte sedimentation rate. Several groups of investigators have used the indices of bone remodeling and/or serum cytokine levels to assess the effects of treatment interventions on focal articular and systemic bone loss in RA patients [35, 60–63]. Results indicate that suppression of signs of inflammation and improved functional status are reflected in improvement in the level and pattern of bone remodeling indices.

The disturbance in systemic bone remodeling in RA has been attributed to the adverse effects of proinflammatory cytokines that are released into the circulation from sites of synovial inflammation. These cytokines act in a manner similar to endocrine hormones to regulate systemic bone remodeling. Although the serum levels of multiple osteoclastogenic cytokines are elevated in RA patients, particular attention has been focused on the levels of RANKL and OPG. Geusens and others have reported results from patients with RA and shown that circulating OPG/RANKL in early RA predicted subsequent bone destruction over an 11-year follow-up [63–65]. In another study, Vis and colleagues [63] showed that anti-TNF therapy with infliximab was accompanied by decreased systemic bone loss and these effects correlated with a fall in serum RANKL levels. There also is evidence that cytokines and mediators released from inflamed joints can adversely affect systemic bone formation. This conclusion is supported by the observations of Diarra and colleagues [42] who detected elevated levels of DKK-1, an inhibitor of bone formation, in the sera of patients with RA. Similar findings have been reported by other authors [66, 67]. Of interest, in the Diarra study, the authors observed that levels of DKK-1 were not increased compared with controls in patients with AS, which is associated with focal increases in periarticular bone formation, as discussed in the following section.

These studies and the related investigations described in the preceding discussion highlight the importance of monitoring patients with RA for evidence of systemic bone loss and for the institution of early therapeutic interventions that have been shown to reduce the long-term risks of fracture and disability. Similar approaches should be considered in patients with SLE and related forms of inflammatory arthritis who also are at risk for the development of systemic osteoporosis and fracture.

ANKYLOSING SPONDYLITIS

As described, AS is characterized by inflammation in the entheses, as well as the synovial lining of peripheral and sacroiliac joints [68]. Examination of the synovial lesion reveals many of the same features as the RA synovium, including synovial lining hyperplasia, lymphocytic infiltration, and pannus formation. In contrast to the pattern of articular bone remodeling in RA, the inflammatory process in patients with AS may be accompanied by evidence of increased bone formation [9]. This is particularly the case at sites of entheseal inflammation, such as ligament and tendon insertion sites, especially in the spine. To investigate the mechanism responsible for the enhanced bone formation, Braun and coworkers [69] obtained biopsies from the sacroiliac joints of patients with AS. They noted the presence of dense infiltrates of lymphocytes, similar to infiltrates in RA synovial lesions, but unlike in the RA synovium they also detected foci of endochondral ossification. Using in situ hybridization, they noted the presence of increased expression of TGF-β_2 mRNA in these regions and speculated that the upregulation of this growth factor could be responsible for the enhanced bone formation. Bleil and colleagues [70] examined tissues from facet joints of patients with AS and noted the presence of foci of new bone formation at contacts between the inflammatory tissue and cartilage. These zones contained cells expressing RUNX-2 and type I collagen, consistent with an osteoblastic phenotype. Lories and colleagues [71] analyzed synovial tissues from a series of patients with AS or RA and noted the presence of elevated levels of BMP-2 and BMP-6 in tissues from both patient populations. They speculated that the differential pattern of new bone formation in AS could be related to the localization of the inflammatory process to the periosteal bone at the entheses. They have extended these observations to studies in DBA/1 mice that spontaneously develop an inflammatory arthritis, as well as enthesial bone formation that recapitulates the excessive bone formation characteristic of AS. They showed that systemic delivery of noggin, a BMP antagonist, attenuated new bone formation [72]. Direct evidence supporting a role for BMPs in the new bone formation associated with AS was provided by immunohistochemical analysis of enthesial biopsies from patients with AS, demonstrating the presence of phosphorylated Smad-1/5, consistent with local activation of the BMP signaling pathway [72].

Maksymowych and coworkers [73] examined the relationship between inflammation and new bone formation in the axial skeleton in patients with AS. MRI and radiographs of the spine revealed that new syndesmophytes developed more frequently at the vertebral body margins with inflammation than in sites without inflammation. These observations are supported by a more recent study by Ramiro and colleagues [74] who showed that erosions and sclerosis on radiographs preceded the development of syndesmophytes at the same site. Earlier studies indicated that syndesmophytes continued to develop despite resolution of the inflammation with anti-TNF therapy [75, 76]. However, a more recent study by Maas and colleagues [77] showed a reduction in spinal radiographic progression in patients with AS receiving prolonged treatment with TNF-α inhibitors.

Despite the tendency of patients with AS to produce excessive bone formation at sites of inflammation, many individuals exhibit evidence of spinal osteopenia [78, 79]. This has been attributed to the adverse effects of immobilization that results from spinal ankylosis, although decreased bone density also has been detected in patients even in the absence of bony ankylosis [79–81]. These authors and others have suggested that, as in the other forms of inflammatory arthritis, the bone loss is related to the adverse effects of inflammation on systemic bone remodeling. This conclusion is supported by the observations of Rossini and colleagues [82], who noted that the presence of higher serum levels of DKK-1 in patients with AS was associated with low BMD and a higher prevalence of vertebral fractures. These observations support the importance of the anatomical site of inflammation in rheumatic diseases for the outcome for bone and the distinct microenvironments that exist in and around bone [43, 83].

SYSTEMIC LUPUS ERYTHEMATOSUS

SLE, similar to RA, is a systemic inflammatory disease that, in addition to targeting joint structures, may be associated with widespread extra-articular organ damage. Although the pattern and distribution of joint inflammation in SLE and RA are similar, the joint inflammation in SLE most often does not result in extensive articular bone erosions or cartilage destruction [84]. Joint deformity and subluxation do occur, but these have been attributed primarily to ligamentous laxity related to persistent periarticular soft tissue inflammation. This pattern of arthritis (Jaccoud arthropathy) is characterized by the presence of "hook" erosions that occur on the radial aspect of metacarpal bones [84]. These local bone changes are distinct from the marginal erosions seen in RA. A similar pattern of joint deformity has been described in other inflammatory disorders, including rheumatic fever and sarcoid, so the condition is not unique to SLE.

As described earlier, analysis of the transcriptome in synovial tissue from patients with SLE reveals a gene profile that exhibits substantial differences from the patterns observed in RA [4]. The most prominent finding in the SLE synovial tissue is the upregulation of interferon-inducible genes. A type I interferon signature has been reported in the peripheral blood cells from patients with SLE compared with disease control subjects [85, 86]. Of interest, both interferon-α and -β have been shown to inhibit osteoclastogenesis in vitro, and the upregulation of these genes and their products in the SLE synovium could contribute to protection from the development of osteoclast-mediated bone erosions [87]. Studies by Mensah and colleagues [88] in the (NZB×NZW)F1 mouse model of SLE provide experimental support implicating synovial-derived interferon-α in the inhibition of osteoclast-mediated bone erosion in SLE. They showed that interferon-α shifted osteoclast precursors towards myeloid dendritic cell differentiation and away from osteoclast formation both in vivo and in vitro. Similar to patients with RA, SLE patients also are at risk for the development of systemic osteoporosis and associated fragility fractures [89–91]. In addition to the adverse effects of chronic inflammation on bone remodeling, additional factors including use of glucocorticoid therapy, renal impairment, and vitamin D insufficiency likely contribute to the low bone mass.

ACKNOWLEDGMENTS

SG has a research grant from Boehringer Ingelheim and EG has research grants from AbbVie, Inc. and Eli Lilly and Company.

REFERENCES

1. Baum R, Gravallese EM. Bone as a target organ in rheumatic disease: impact on osteoclasts and osteoblasts. Clin Rev Allergy Immunol. 2016;51(1):1–15.
2. Goldring SR. Inflammatory signaling induced bone loss. Bone. 2015;80:143–9.
3. Schett G. Effects of inflammatory and anti-inflammatory cytokines on the bone. Eur J Clin Invest. 2011;10:1–6.
4. Nzeusseu Toukap A, Galant C, Theate I, et al. Identification of distinct gene expression profiles in the synovium of patients with systemic lupus erythematosus. Arthritis Rheum. 2007;56(5):1579–88.
5. Santiago MB, Galvao V. Jaccoud arthropathy in systemic lupus erythematosus: analysis of clinical characteristics and review of the literature. Medicine (Baltimore). 2008;87(1):37–44.
6. Piga M, Saba L, Gabba A, et al. Ultrasonographic assessment of bone erosions in the different subtypes of systemic lupus erythematosus arthritis: comparison with computed tomography. Arthritis Res Ther. 2016;18(1):222.
7. McGonagle D, Tan AL, Moller Dohn U, et al. Microanatomic studies to define predictive factors for the topography of periarticular erosion formation in inflammatory arthritis. Arthritis Rheum. 2009;60(4):1042–1051.
8. Lories RJ, Luyten FP, de Vlam K. Progress in spondylarthritis. Mechanisms of new bone formation in spondyloarthritis. Arthritis Res Ther. 2009;11(2):221.
9. Lories RJ, Schett G. Pathophysiology of new bone formation and ankylosis in spondyloarthritis. Rheum Dis Clin North Am. 2012;38(3):555–67.
10. Braun J, Baraliakos X, Golder W, et al. Analysing chronic spinal changes in ankylosing spondylitis: a systematic comparison of conventional x rays with magnetic resonance imaging using established and new scoring systems. Ann Rheum Dis. 2004;63(9):1046–55.
11. Benjamin M, McGonagle D. The enthesis organ concept and its relevance to the spondyloarthropathies. Adv Exp Med Biol. 2009;649:57–70.
12. Bromley M, Woolley DE. Chondroclasts and osteoclasts at subchondral sites of erosion in the rheumatoid joint. Arthritis Rheum. 1984;27(9):968–75.
13. Gravallese EM, Harada Y, Wang JT, et al. Identification of cell types responsible for bone resorption in rheumatoid arthritis and juvenile rheumatoid arthritis. Am J Pathol. 1998;152(4):943–51.
14. Pettit AR, Ji H, von Stechow D, Muller R, et al. TRANCE/RANKL knockout mice are protected from bone erosion in a serum transfer model of arthritis. Am J Pathol. 2001;159(5):1689–99.
15. Jimenez-Boj E, Nobauer-Huhmann I, Hanslik-Schnabel B, et al. Bone erosions and bone marrow edema as defined by magnetic resonance imaging reflect true bone marrow inflammation in rheumatoid arthritis. Arthritis Rheum. 2007;56(4):1118–24.
16. Boyesen P, Haavardsholm EA, van der Heijde D, et al. Prediction of MRI erosive progression: a comparison of

17. Hetland ML, Ejbjerg B, Horslev-Petersen K, et al. MRI bone oedema is the strongest predictor of subsequent radiographic progression in early rheumatoid arthritis. Results from a 2-year randomised controlled trial (CIMESTRA). Ann Rheum Dis. 2009;68(3):384–90.
18. Redlich K, Hayer S, Ricci R, et al. Osteoclasts are essential for TNF-alpha-mediated joint destruction. J Clin Invest. 2002;110:1419–27.
19. Li P, Schwarz EM, O'Keefe RJ, et al. RANK signaling is not required for TNFalpha-mediated increase in CD11(hi) osteoclast precursors but is essential for mature osteoclast formation in TNFalpha-mediated inflammatory arthritis. J Bone Miner Res. 2004;19(2):207–13.
20. Walsh NC, Crotti TN, Goldring SR, et al. Rheumatic diseases: the effects of inflammation on bone. Immunol Rev. 2005;208:228–51.
21. Schett G, Gravallese E. Bone erosion in rheumatoid arthritis: mechanisms, diagnosis and treatment. Nat Rev Rheumatol. 2012;8(11):656–64.
22. Gravallese EM, Goldring SR. Cellular mechanisms and the role of cytokines in bone erosions in rheumatoid arthritis. Arthritis Rheum. 2000;43(10):2143–51.
23. Kong YY, Feige U, Sarosi I, et al. Activated T cells regulate bone loss and joint destruction in adjuvant arthritis through osteoprotegerin ligand. Nature. 1999;402(6759):304–9.
24. Pettit AR, Walsh NC, Manning C, et al. RANKL protein is expressed at the pannus-bone interface at sites of articular bone erosion in rheumatoid arthritis. Rheumatology (Oxford). 2006;45(9):1068–76.
25. Romas E, Bakharevski O, Hards DK, et al. Expression of osteoclast differentiation factor at sites of bone erosion in collagen-induced arthritis. Arthritis Rheum. 2000;43(4):821–6.
26. Redlich K, Hayer S, Maier A, et al. Tumor necrosis factor-a-mediated joint destruction is inhibited by targeting osteoclasts with osteoprotegerin. Arthritis Rheum. 2002;46:785–92.
27. Romas E, Gillespie MT, Martin TJ. Involvement of receptor activator of NFkappaB ligand and tumor necrosis factor-alpha in bone destruction in rheumatoid arthritis. Bone. 2002;30(2):340–6.
28. Cohen SB, Dore RK, Lane NE, et al. Denosumab treatment effects on structural damage, bone mineral density, and bone turnover in rheumatoid arthritis: a twelve-month, multicenter, randomized, double-blind, placebo-controlled, phase II clinical trial. Arthritis Rheum. 2008;58(5):1299–309.
29. Deodhar A, Dore RK, Mandel D, et al. Denosumab-mediated increase in hand bone mineral density associated with decreased progression of bone erosion in rheumatoid arthritis patients. Arthritis Care Res. 2010;62(4):569–74.
30. Takeuchi T, Tanaka Y, Ishiguro N, et al. Effect of denosumab on Japanese patients with rheumatoid arthritis: a dose-response study of AMG 162 (Denosumab) in patients with RheumatoId arthritis on methotrexate to Validate inhibitory effect on bone Erosion (DRIVE)-a 12-month, multicentre, randomised, double-blind, placebo-controlled, phase II clinical trial. Ann Rheum Dis. 2016;75(6):983–90.
31. Yue J, Griffith JF, Xiao F, et al. Repair of bone erosion in rheumatoid arthritis by denosumab: a high-resolution peripheral quantitative computed tomography study. Arthritis Care Res. 2017;69(8):1156–63.
32. Goldring SR, Gravallese EM. Bisphosphonates: environmental protection for the joint? Arthritis Rheum. 2004;50(7):2044–7.
33. Jarrett SJ, Conaghan PG, Sloan VS, et al. Preliminary evidence for a structural benefit of the new bisphosphonate zoledronic acid in early rheumatoid arthritis. Arthritis Rheum. 2006;54(5):1410–14.
34. Dore RK, Cohen SB, Lane NE, et al. Effects of denosumab on bone mineral density and bone turnover in patients with rheumatoid arthritis receiving concurrent glucocorticoids or bisphosphonates. Ann Rheum Dis. 2010;69(5):872–5.
35. Miller PD, Wagman RB, Peacock M, et al. Effect of denosumab on bone mineral density and biochemical markers of bone turnover: six-year results of a phase 2 clinical trial. J Clin Endocrinol Metab. 2011;96(2):394–402.
36. Takayanagi H, Ogasawara K, Hida S, et al. T-cell-mediated regulation of osteoclastogenesis by signalling cross-talk between RANKL and IFN-gamma. Nature. 2000;408(6812):600–5.
37. Takayanagi H, Sato K, Takaoka A, et al. Interplay between interferon and other cytokine systems in bone metabolism. Immunol Rev. 2005;208:181–93.
38. Harre U, Georgess D, Bang H, et al. Induction of osteoclastogenesis and bone loss by human autoantibodies against citrullinated vimentin. J Clin Invest. 2012;122(5):1791–802.
39. Hecht C, Englbrecht M, Rech J, et al. Additive effect of anti-citrullinated protein antibodies and rheumatoid factor on bone erosions in patients with RA. Ann Rheum Dis. 2015;74(12):2151–6.
40. Hecht C, Schett G, Finzel S. The impact of rheumatoid factor and ACPA on bone erosion in rheumatoid arthritis. Ann Rheum Dis. 2015;74(1):e4.
41. Matzelle MM, Gallant MA, Condon KW, et al. Resolution of inflammation induces osteoblast function and regulates the Wnt signaling pathway. Arthritis Rheum. 2012;64(5):1540–50.
42. Diarra D, Stolina M, Polzer K, et al. Dickkopf-1 is a master regulator of joint remodeling. Nat Med. 2007;13(2):156–63.
43. Walsh NC, Reinwald S, Manning CA, et al. Osteoblast function is compromised at sites of focal bone erosion in inflammatory arthritis. J Bone Miner Res. 2009;24(9):1572–85.
44. Goldring SR, Goldring MB. Eating bone or adding it: the Wnt pathway decides. Nat Med. 2007;13(2):133–4.
45. Heiland GR, Zwerina K, Baum W, et al. Neutralisation of Dkk-1 protects from systemic bone loss during inflammation and reduces sclerostin expression. Ann Rheum Dis. 2010;69(12):2152–9.

46. Chen XX, Baum W, Dwyer D, et al. Sclerostin inhibition reverses systemic, periarticular and local bone loss in arthritis. Ann Rheum Dis. 2013;72(10):1732–6.
47. Wehmeyer C, Frank S, Beckmann D, et al. Sclerostin inhibition promotes TNF-dependent inflammatory joint destruction. Sci Transl Med. 2016;8(330):330ra335.
48. Goldring SR. Periarticular bone changes in rheumatoid arthritis: pathophysiological implications and clinical utility. Ann Rheum Dis. 2009;68(3):297–9.
49. Hoff M, Haugeberg G, Odegard S, et al. Cortical hand bone loss after 1 year in early rheumatoid arthritis predicts radiographic hand joint damage at 5-year and 10-year follow-up. Ann Rheum Dis. 2009;68(3):324–9.
50. Stewart A, Mackenzie LM, Black AJ, et al. Predicting erosive disease in rheumatoid arthritis. A longitudinal study of changes in bone density using digital X-ray radiogrammetry: a pilot study. Rheumatology (Oxford). 2004;43(12):1561–4.
51. Shimizu S, Shiozawa S, Shiozawa K, et al. Quantitative histologic studies on the pathogenesis of periarticular osteoporosis in rheumatoid arthritis. Arthritis Rheum. 1985;28(1):25–31.
52. Haugeberg G, Orstavik RE, Kvien TK. Effects of rheumatoid arthritis on bone. Curr Opin Rheumatol. 2003;15(4):469–75.
53. Lodder MC, de Jong Z, Kostense PJ, et al Bone mineral density in patients with rheumatoid arthritis: relation between disease severity and low bone mineral density. Ann Rheum Dis. 2004;63(12):1576–80.
54. van Staa TP, Geusens P, Bijlsma JW, et al. Clinical assessment of the long-term risk of fracture in patients with rheumatoid arthritis. Arthritis Rheum. 2006;54(10):3104–12.
55. Vis M, Haavardsholm EA, Boyesen P, et al. High incidence of vertebral and non-vertebral fractures in the OSTRA cohort study: a 5-year follow-up study in postmenopausal women with rheumatoid arthritis. Osteoporos Int. 2011;22(9):2413–9.
56. Solomon DH, Finkelstein JS, Shadick N, et al. The relationship between focal erosions and generalized osteoporosis in postmenopausal women with rheumatoid arthritis. Arthritis Rheum. 2009;60(6):1624–1631.
57. Compston JE, Vedi S, Croucher PI, et al. Bone turnover in non-steroid treated rheumatoid arthritis. Ann Rheum Dis. 1994;53(3):163–6.
58. Gough A, Sambrook P, Devlin J, et al. Osteoclastic activation is the principal mechanism leading to secondary osteoporosis in rheumatoid arthritis. J Rheumatol. 1998;25(7):1282–9.
59. Garnero P, Landewe R, Boers M, et al. Association of baseline levels of markers of bone and cartilage degradation with long-term progression of joint damage in patients with early rheumatoid arthritis: the COBRA study. Arthritis Rheum. 2002;46(11):2847–56.
60. Barnabe C, Hanley DA. Effect of tumor necrosis factor alpha inhibition on bone density and turnover markers in patients with rheumatoid arthritis and spondyloarthropathy. Semin Arthritis Rheum. 2009;39(2):116–22.
61. Seriolo B, Paolino S, Sulli A, et al. Bone metabolism changes during anti-TNF-alpha therapy in patients with active rheumatoid arthritis. Ann N Y Acad Sci. 2006;1069:420–7.
62. Syversen SW, Haavardsholm EA, Boyesen P, et al. Biomarkers in early rheumatoid arthritis: longitudinal associations with inflammation and joint destruction measured by magnetic resonance imaging and conventional radiographs. Ann Rheum Dis. 2011;69(5):845–50.
63. Vis M, Havaardsholm EA, Haugeberg G, et al. Evaluation of bone mineral density, bone metabolism, osteoprotegerin and receptor activator of the NFkappaB ligand serum levels during treatment with infliximab in patients with rheumatoid arthritis. Ann Rheum Dis. 2006;65(11):1495–1499.
64. Geusens PP, Landewe RB, Garnero P, et al. The ratio of circulating osteoprotegerin to RANKL in early rheumatoid arthritis predicts later joint destruction. Arthritis Rheum. 2006;54(6):1772–7.
65. van Tuyl LH, Voskuyl AE, Boers M, et al. Baseline RANKL:OPG ratio and markers of bone and cartilage degradation predict annual radiological progression over 11 years in rheumatoid arthritis. Ann Rheum Dis. 2011;69(9):1623–8.
66. Garnero P, Tabassi NC, Voorzanger-Rousselot N. Circulating dickkopf-1 and radiological progression in patients with early rheumatoid arthritis treated with etanercept. J Rheumatol. 2008;35(12):2313–15.
67. Wang SY, Liu YY, Ye H, et al. Circulating Dickkopf-1 is correlated with bone erosion and inflammation in rheumatoid arthritis. J Rheumatol. 2011;38(5):821–7.
68. Kehl AS, Corr M, Weisman MH. Review. Enthesitis: new insights into pathogenesis, diagnostic modalities, and treatment. Arthritis Rheumatol. 2016;68(2):312–22.
69. Braun J, Bollow M, Neure L, et al. Use of immunohistologic and in situ hybridization techniques in the examination of sacroiliac joint biopsy specimens from patients with ankylosing spondylitis. Arthritis Rheum. 1995;4:499–505.
70. Bleil J, Maier R, Hempfing A, et al. Granulation tissue eroding the subchondral bone also promotes new bone formation in ankylosing spondylitis. Arthritis Rheumatol. 2016;68(10):2456–65.
71. Lories RJ, Derese I, Ceuppens JL, et al. Bone morphogenetic proteins 2 and 6, expressed in arthritic synovium, are regulated by proinflammatory cytokines and differentially modulate fibroblast-like synoviocyte apoptosis. Arthritis Rheum. 2003;48(10):2807–18.
72. Lories RJ, Derese I, Luyten FP. Modulation of bone morphogenetic protein signaling inhibits the onset and progression of ankylosing enthesitis. J Clin Invest. 2005;115(6):1571–9.
73. Maksymowych WP, Chiowchanwisawakit P, Clare T, et al. Inflammatory lesions of the spine on magnetic resonance imaging predict the development of new syndesmophytes in ankylosing spondylitis: evidence of a relationship between inflammation and new bone formation. Arthritis Rheum. 2009;60(1):93–102.

74. Ramiro S, van Tubergen A, van der Heijde D, et al. Brief report: erosions and sclerosis on radiographs precede the subsequent development of syndesmophytes at the same site: a twelve-year prospective followup of patients with ankylosing spondylitis. Arthritis Rheumatol. 2014;66(10):2773–9.
75. van der Heijde D, Landewe R, Baraliakos X, et al. Radiographic findings following two years of infliximab therapy in patients with ankylosing spondylitis. Arthritis Rheum. 2008;58(10):3063–70.
76. van der Heijde D, Landewe R, Einstein S, et al. Radiographic progression of ankylosing spondylitis after up to two years of treatment with etanercept. Arthritis Rheum. 2008;58(5):1324–31.
77. Maas F, Arends S, Brouwer E, et al. Reduction in spinal radiographic progression in ankylosing spondylitis patients receiving prolonged treatment with tumor necrosis factor inhibitors. Arthritis Care Res. 2017;69(7):1011–9.
78. Geusens P, De Winter L, Quaden D, et al. The prevalence of vertebral fractures in spondyloarthritis: relation to disease characteristics, bone mineral density, syndesmophytes and history of back pain and trauma. Arthritis Res Ther. 2015;17:294.
79. Geusens P, Vosse D, van der Linden S. Osteoporosis and vertebral fractures in ankylosing spondylitis. Curr Opin Rheumatol. 2007;19(4):335–9.
80. Will R, Bhalla A, Palmer R, et al. Osteoporosis in early ankylosing spondylitis; a primary pathological event? Lancet. 1989;23:1483–5.
81. Ralston SH, Urquhart GD, Brzeski M, et al. Prevalence of vertebral compression fractures due to osteoporosis in ankylosing spondylitis. BMJ. 1990;300(6724):563–5.
82. Rossini M, Viapiana O, Idolazzi L, et al. Higher level of Dickkopf-1 is associated with low bone mineral density and higher prevalence of vertebral fractures in patients with ankylosing spondylitis. Calcif Tissue Int. 2016;98(5):438–45.
83. Walsh DA. Angiogenesis in osteoarthritis and spondylosis: successful repair with undesirable outcomes. Curr Opin Rheumatol. 2004;16(5):609–15.
84. Ostendorf B, Scherer A, Specker C, et al. Jaccoud's arthropathy in systemic lupus erythematosus: differentiation of deforming and erosive patterns by magnetic resonance imaging. Arthritis Rheum. 2003;48(1):157–65.
85. Bennett L, Palucka AK, Arce E, et al. Interferon and granulopoiesis signatures in systemic lupus erythematosus blood. J Exp Med. 2003;197(6):711–23.
86. Crow MK. Type I interferon in the pathogenesis of lupus. J Immunol. 2014;192(12):5459–68.
87. Coelho LF, Magno de Freitas Almeida G, Mennechet FJ, et al. Interferon-alpha and -beta differentially regulate osteoclastogenesis: role of differential induction of chemokine CXCL11 expression. Proc Natl Acad Sci U S A. 2005;102(33):11917–22.
88. Mensah KA, Mathian A, Ma L, et al. Mediation of nonerosive arthritis in a mouse model of lupus by interferon-alpha-stimulated monocyte differentiation that is nonpermissive of osteoclastogenesis. Arthritis Rheum. 2011;62(4):1127–37.
89. Alele JD, Kamen DL. The importance of inflammation and vitamin D status in SLE-associated osteoporosis. Autoimmun Rev. 2010;9(3):137–9.
90. Wang X, Yan S, Liu C, et al. Fracture risk and bone mineral density levels in patients with systemic lupus erythematosus: a systematic review and meta-analysis. Osteoporos Int. 2016;27(4):1413–23.
91. Bultink IE, Lems WF. Lupus and fractures. Curr Opin Rheumatol. 2016;28(4):426–32.

59

Glucocorticoid-Induced Osteoporosis

Kenneth Saag[1] and Robert A. Adler[2]

[1]Center for Outcomes and Effectiveness Research and Education (COERE); and Division of Clinical Immunology and Rheumatology, University of Alabama at Birmingham (UAB) School of Medicine, Birmingham, AL, USA

[2]Endocrinology and Metabolism Section, Hunter Holmes McGuire Veterans Affairs Medical Center; and Endocrine Division, Virginia Commonwealth University School of Medicine, Richmond, VA, USA

EPIDEMIOLOGY

Drug-induced osteoporosis is the most common secondary cause of bone loss leading to fractures, and glucocorticoids constitute the class of drug most typically associated with this form of osteoporosis. While glucocorticoid use may be supplanted by new more directed therapies, such as biologics, oral glucocorticoids are still used chronically (defined as 3 months or more) by up to 1% of the population [1]. Population-based epidemiologic studies demonstrate that up to 40% of individuals using long-term glucocorticoids will experience fractures due to glucocorticoid-induced osteoporosis (GIOP) [2, 3].

Changes in bone density and fracture rates

Bone loss occurs rapidly when glucocorticoids are initiated, with up to 20% loss seen within the first year [2, 4]. In subsequent years, this rate of bone loss slows to 1% to 3% per year. Glucocorticoids affect both bone quantity measured by BMD and bone quality; thus fractures are the outcome of greatest concern. Fractures can occur at any skeletal site but are most prevalent in sites richest in trabecular bone such as the hip trochanter and lumbar spine. Due to the direct toxic effects of glucocorticoids on osteocytes and osteoblasts, an increased fracture risk may occur soon after the onset of glucocorticoid use and at a greater bone density than that seen in postmenopausal osteoporosis [5]. Indeed, fracture risk is increased in the first 3 months of glucocorticoid use [6]. Randomized controlled trials show a 15% incidence rate of morphometrically defined vertebral fractures in persons on an average prednisone dose of 10 mg per day who were not assigned to receive osteoporosis therapy [7, 8].

Dose effects and routes of glucocorticoid administration

The association of glucocorticoids with bone loss is mostly linear and a threshold beneath which the effects of glucocorticoids are deleterious to bone has not been identified [9]. Doses exceeding 20 mg per day prednisone equivalent may be particularly deleterious to bone. Cumulative glucocorticoid dose has its greatest effect on BMD, while peak dose may directly affect osteocyte and osteoblast function leading to fractures, or potentially osteonecrosis, with relatively well-preserved bone mass. While safer to bone than oral preparations, even inhaled glucocorticoids used for respiratory illnesses may adversely affect bone [10]. It is estimated that more than 7 years use of high-dose inhaled glucocorticoid will lead to a one SD, or 10%, loss of BMD [11, 12]. Controlled release budesonide is less well absorbed from the gut and exerts most of its effects topically in the gastrointestinal tract, but there are still some systemic effects on bone [13]. Topical glucocorticoids applied to the skin do not have measurable effects on bone except at high-dose use in children or when placed on mucous membranes or surfaces with very thin skin such as the scrotum. The effects of glucocorticoids abate once the therapy is discontinued and fracture risk reverts towards normal.

Primer on the Metabolic Bone Diseases and Disorders of Mineral Metabolism, Ninth Edition. Edited by John P. Bilezikian.
© 2019 American Society for Bone and Mineral Research. Published 2019 by John Wiley & Sons, Inc.
Companion website: www.wiley.com/go/asbmrprimer

Populations at high risk for glucocorticoid-induced osteoporosis

Beyond glucocorticoid dose, glucocorticoid effects on bone are mediated by many factors including age, sex, race, menopausal status, concomitant illnesses, and comedications used to treat conditions for which glucocorticoids are prescribed. While postmenopausal women are at greatest risk for fracture while on glucocorticoids, men and premenopausal women experience fractures as well. For example, among younger women with systemic lupus erythematosus there was over 12% risk of fracture in over 6000 person-years of follow-up [14]. Rheumatoid arthritis (RA) and chronic obstructive lung disease (COPD) are among the diseases for which glucocorticoids are used most commonly. The patterns of glucocorticoid use vary considerably, with RA use being at lower dose and longer term and COPD use being more episodic and at higher dose. RA directly increases bone loss through the underlying disease process causing both a generalized and localized bone loss around joints, independent of glucocorticoid use. Moreover, there is some debate about possible protective effects to bone in RA, given the reduction in proinflammatory cytokines deleterious to bone and improvements in weight-bearing physical activity, also beneficial to bone, seen in RA patients treated with glucocorticoids [15]. Glucocorticoids and RA are both independent significant risk factors for fracture [16]. In COPD, chronic smoking, low body mass index (common in emphysema), and diminished sunlight exposure confound the effects of glucocorticoids on bone. Somewhat less prevalent diseases with great glucocorticoid-associated morbidity are polymyalgia rheumatica and the related disorder temporal arteritis [17]. These conditions, particularly temporal arteritis, with a typical starting prednisone dose of 60 mg daily, affect older, predominately white adults. Fracture outcomes are common and in some cases more serious than the underlying morbidity or condition for which they are prescribed.

PATHOGENESIS

Direct effects on bone

What separates GIOP from most other types of osteoporosis is that bone formation is severely impaired with the onset of excess glucocorticoid status. Indeed, biochemical markers of bone formation such as osteocalcin decline rapidly following initiation of oral, inhaled, or even intra-articular glucocorticoids [18]. A mouse model of GIOP [19, 20] demonstrated that glucocorticoids reduce osteoblast and osteocyte differentiation and lifespan via increased apoptosis. Changes in osteoblast formation may be mediated by the PPARγ transcription factor [21], and glucocorticoid suppression of the Wnt signaling pathway [22] in osteoblasts may lead to decreased function. Key histological changes in bone reflecting these pathways are reduced bone area and osteoid, fewer osteoblasts, and prolongation of the reversal phase of the bone remodeling cycle [23]. Prior to the direct effect on bone, the 11β hydroxysteroid dehydrogenase (11β-HSD) shunt may alter the concentrations of active (cortisol) and inactive (cortisone) glucocorticoid. It has been postulated that the 11β-HSD enzymes are responsible for the clinical finding of variable sensitivity to a given dose of exogenous glucocorticoid.

It is well established that growth is adversely affected in children treated with exogenous glucocorticoids, and there is evidence that growth hormone secretion may be decreased [24]. New studies demonstrate a potential interaction with IGF-1 secretion, as well [25].

Effects on osteoclasts are more controversial. Clinically, there is evidence, as manifested by increased bone resorption markers, of increased osteoclastic activity in the first several months of exogenous glucocorticoid exposure [26]. Glucocorticoids increase RANKL and decrease osteoprotegerin [27]. However, much of the increased osteoclastic activity may be a reflection of the inflammatory disorder for which glucocorticoids have been prescribed [28]. As the underlying disorder, such as RA, responds to treatment, inflammation is reduced, as are markers of bone resorption. Indeed, there is also evidence that osteoclastic activity is diminished by glucocorticoids [28].

Indirect effects of glucocorticoids

Also controversial have been what the effects of glucocorticoids are on calcium balance and parathyroid activity. While there is evidence that glucocorticoids lead to decreased calcium absorption in the gut and increased urinary calcium excretion [29], the clinical impact of these effects is unclear. For example, there is no evidence that PTH is elevated by these effects. In normal men given 50 mg of prednisone daily for about 4 months, there was no change in serum calcium or PTH [30]. While serum 25OHD levels are often considered low in patients on glucocorticoids [31], there is less evidence that such levels are different from the general population.

Glucocorticoids have effects on muscle, leading to the classic loss of muscle mass noted in patients with endogenous or exogenous glucocorticoid excess. Proteins are the source of amino acids for gluconeogenesis stimulated by glucocorticoids. Decreased lower body strength is clearly a risk factor for falls and fracture. Hypogonadism is also noted in patients with Cushing syndrome and in those taking exogenous glucocorticoids, and may have deleterious effects on bone indirectly (eg, decreased testosterone and muscle) or directly (eg, a decrease of estrogen's antiresorptive effect). Histological studies show that direct bone effects of glucocorticoids are most important, but the indirect effects on calcium loss, muscle loss, and diminished sex steroids may play some role in fracture risk [23, 32]. Finally, both the inflammatory effects of the underlying disorder for which glucocorticoids are prescribed plus the increased fracture risk of the underlying disorder per se add to the markedly high probability that a given patient on glucocorticoids will fracture, compared to age-matched peers not on such medications.

EVALUATION

History and physical examination

GIOP illustrates the importance of careful medical history taking and physical examination for the evaluation of osteoporosis. Most health care systems are fragmented, so that each patient may be treated by multiple clinicians; oral or other systemic glucocorticoids may be prescribed by many different generalists or specialists without other clinicians being aware of the prescription. Thus, glucocorticoid medication reconciliation is a major part of the assessment of any patient and may reveal why a given patient has had a low trauma fracture. In addition, the underlying disorder for which glucocorticoids have been prescribed may be a risk factor for osteoporosis and fracture (see earlier in this chapter). The clinician must assess the details of glucocorticoid administration and estimate the length of treatment. As described earlier, increased fracture risk has been found in patients taking as little as 7.5 mg of prednisolone equivalent for 3 months [9]. A brief dietary history may help the clinician decide which patients need calcium supplementation. Physical examination should include measurement of height (with calculation of change from maximum attained) and may reveal generalized weakness or decreased lower body strength. The inability to rise from a chair without using arms and hands is an easy method to detect such weakness. Stigmata of glucocorticoid excess such as muscle wasting, central obesity, moon facies, bruising, violaceous abdominal striae, and so-called buffalo hump may be found. The absence of such findings does not mean the patient is free of fracture risk.

Laboratory studies are helpful for the diagnosis of endogenous Cushing syndrome, such as the overnight dexamethasone suppression test. For patients on exogenous glucocorticoids, testing beyond what is done for most osteoporosis patients may be determined by the underlying condition. Most osteoporosis patients require a measure of serum and urinary calcium, renal function, and vitamin D status (25OHD), whereas patients with sarcoidosis may also need measurement of 1,25-dihydroxyvitamin D. Patients with RA or COPD may need laboratory or functional assessments of disease activity as a method to estimate duration of glucocorticoid treatment. The 25OHD level will help the clinician determine whether vitamin D supplementation is needed and if so, how much. While serum bone turnover markers are affected by endogenous and exogenous glucocorticoids, it is not established that they should be measured in most patients. Early after starting glucocorticoids, reflections of bone resorption such as serum CTx may be elevated, and measures of bone formation such as serum P1NP may be decreased. The changes in these turnover markers vary with the timing of glucocorticoid administration as well as with other factors such menopausal state and the underlying glucocorticoid-requiring illness. Thus, for a given patient, the results of such testing are unlikely to affect the decision as to whether to start osteoporosis therapy or not.

DXA is indicated for patients at risk for GIOP. However, fracture may occur at greater bone density than in patients with postmenopausal osteoporosis [5]. The American College of Rheumatology (ACR) guideline on GIOP recommends using a fracture risk calculator, such as the Fracture Risk Assessment Tool (FRAX), in addition to DXA for determining fracture risk and choice of management [33]. This guideline categorized patients into low, medium, and high risk based on glucocorticoid dose, DXA results, and 10-year fracture risk prediction by FRAX. FRAX can be calculated using a free website (www.shef.ac.uk/FRAX/, accessed May 2018) by entering the patient's age, sex, weight, height, several risk factors including glucocorticoid exposure, and femoral neck BMD. In the latest iteration of the Guideline, an adjustment can be made for the degree of glucocorticoid exposure.

Other modalities of bone assessment hold some promise for better estimating fracture risk, although in some cases ionizing radiation exposure and/or cost may prevent widespread use. Quantitative computed tomography with finite element analysis of the hip [34] and high-resolution quantitative computed tomography of the vertebrae [35] appear to be very sensitive to changes induced by glucocorticoids. Magnetic resonance imaging of GIOP patients may hold some promise for fracture risk assessment [36]. Reference point indentation of cortical bone (tibia) [37] may reflect glucocorticoid effect. Recently, trabecular bone score, a bone texture measurement of spine DXA data, has been used in patients treated for GIOP [38]. Just how these newer techniques will augment current management based on history, physical examination, DXA, and the FRAX calculation remains to be determined.

TREATMENT AND PREVENTION

Calcium and vitamin D

Sufficient calcium and vitamin D are necessary but generally insufficient by themselves for most long-term glucocorticoid users to maintain adequate bone health. An intake of 1200 to 1500 mg/day of elemental calcium achieved through dietary and supplemental sources may partially attenuate bone loss. About 800 IU of vitamin D can be provided via calcium supplements that also contain cholecalciferol but it is also available through multivitamins or as stand-alone supplements. For those who are D insufficient when commencing glucocorticoids (defined as a 25OHD level of 30 ng/mL or less), higher dose ergocalciferol or cholecalciferol, given intermittently is recommended. Activated forms of vitamin D preserve or improve BMD significantly compared with placebo but are less efficacious than bisphosphonates in GIOP [39, 40]. Weight-bearing physical activity and other traditional osteoporosis risk factor modifications may have a modest effect on bone health.

Clinical trials in GIOP have included patients who are both newly initiating glucocorticoids (prevention) and those who are longer term users (treatment). Interpretation of these studies is further confounded by the diverse diseases for which glucocorticoids are administered as well as the varying ages and other factors affecting baseline bone health prior to glucocorticoid use.

Bisphosphonates

Clinical trials specific to glucocorticoid use have led to the international registration and widespread use of three bisphosphonate: alendronate, risedronate, and zoledronic acid. Data also show efficacy of etidronate [41, 42], pamidronate [43–46], and ibandronate [47], although these agents are either unapproved by regulatory agencies or much less commonly utilized. In a placebo-controlled clinical trial, alendronate at 5 or 10 mg/day led to small but significant increases in bone mass at the spine and hip of 1% to 4% of patients over a 2-year period among both people initiating and continuing glucocorticoids [8]. In the subgroup of women who were premenopausal, 5 mg of daily alendronate appeared largely equivalent to 10 mg/day in its effects on bone mass. All study subjects in both treatment groups also received supplemental calcium and vitamin D. The finding of relative preservation of bone mass from baseline at both the lumbar spine and total hip in the group receiving placebo and calcium and vitamin D, may be partially attributed to approximately one-third men and one-third of persons enrolled who had well-preserved bone mass at baseline. It suggests that in persons at low risk for bone loss, calcium and vitamin D constitute a reasonable initial strategy [48]. An extension to the original alendronate clinical trial demonstrated a reduced risk of morphometric vertebral fractures [49]. There were more nonserious upper gastrointestinal adverse events seen with alendronate compared with placebo.

Risedronate was studied in two separate investigations: one among a population newly initiating glucocorticoids [7] and one among more chronic users [50]. As expected, there was more bone loss or smaller gains in BMD overall in the prevention study, but risedronate preserved or increased bone density significantly more than the placebo. Greater gains in BMD at all three primary measurement sites were seen in the study of longer term glucocorticoid users. A post hoc combination of these two studies showed a significant reduction in fractures at the spine [51]. These primary studies of alendronate and risedronate used a daily dose, although a weekly dose of alendronate looked fairly equivalent to daily therapy [52]. Weekly bisphosphonate therapy has become the standard of care in GIOP due to greater ease of use.

The largest GIOP study was conducted using an annual 5 mg infusion of zoledronic acid in an active comparator study with risedronate [53]. A BMD entry criterion was not used, thus the population of patients studied was at somewhat lower fracture risk compared with prior GIOP studies. In both the prevention and treatment populations, spine BMD gains were significantly greater in the group receiving zoledronic acid compared with those receiving risedronate. There was no significant difference in fracture risk between the treatment groups.

A meta-analysis of bisphosphonate use in GIOP demonstrated fracture risk reductions that were similar in magnitude to those seen in postmenopausal osteoporosis. Since GIOP is a low turnover state, emerging questions about the rare adverse effects, such as atypical femoral fracture and osteonecrosis of the jaw that have been associated with bisphosphonates, are of added concern in GIOP. The number of postmenopausal women on prednisone needed to treat with a bisphosphonate to prevent one fracture is low, between 8 and 26 [54]. Further, discontinuing bisphosphonates in GIOP patients may result in a heightened rate of bone loss [55]. Thus, indications for taking a bisphosphonate "drug holiday," when such therapy is used to prevent GIOP is uncertain and not evidence-based.

Teriparatide

Since GIOP is largely a problem of bone formation, an agent that is directed to the osteoblast has considerable biological plausibility for GIOP prevention and treatment. The anabolic agent teriparatide was compared with daily alendronate in patients who were either longer term or new glucocorticoid users [56]. Teriparatide use led to greater gains in BMD at both the spine and hip compared with alendronate. Although this study (and all other GIOP studies) was not powered in the primary analyses for a fracture endpoint, patients randomized to teriparatide had significantly fewer vertebral fractures using a semiquantitative method. Teriparatide may be a preferred agent in glucocorticoid users at very high fracture risk, tempered by the inconvenience of a daily subcutaneous injection and added cost. Although the GIOP teriparatide study was a 3-year study that showed continued efficacy and safety [57], the normal period of use is only 2 years, creating a need for other therapies in many patients who often have long-term use of glucocorticoids and need more protracted treatment. Anabolic therapy should be followed by antiresorptive treatment to fully mineralize new bone and consolidate BMD gains.

Denosumab

Due to the increase in RANKL observed in GIOP, there is a biological rationale for use of denosumab, a monoclonal antibody to RANKL in this setting. In an active comparator trial of denosumab versus risedronate, denosumab led to greater gains in BMD at the spine and hip, but there were no significant differences in fracture rates [58].

Denosumab was generally well tolerated without an increase in infection rates, even among a small subgroup of patients who were using either biologics or immunosuppressive drugs.

Other therapeutic options

The selective estrogen receptor modulator, raloxifene, attenuated bone loss at the spine in a small randomized controlled trial [59]. It must be used cautiously in populations at higher risk for thromboembolic events, such as some patients with certain connective tissue diseases, such as systemic lupus erythematosis. Testosterone therapy also had some benefit in GIOP among men, but its use should be restricted to those who need testosterone for gonadal insufficiency [60].

Calcitonin has been used in GIOP with modest benefit [61] seen in some but not all studies. Calcitonin use is largely in disfavor as an osteoporosis therapeutic agent due to its weak antiresportive benefits, absence of nonspine fracture reduction overall, and new safety concerns with regard to a possible malignancy signal.

Treatment guidelines

Guidelines from the ACR [33] and other groups with a vested interest in glucocorticoid use have disseminated recommendations for bone health protection among beginning and chronic glucocorticoid users. Most guidelines advocate therapy when glucocorticoid use is anticipated to be long term and at more than a minimal dose (typically 5 mg/day prednisone equivalent). A key distinction of GIOP treatment guidelines from other forms of osteoporosis relate to populations of patients in whom traditional bone treatments would not be recommended, such as those with normal or minimally reduced bone mass on entry and persons, such as premenopausal women, in whom traditional osteoporosis therapies would not be indicated. Of particular interest is guidance for premenopausal women of child-bearing potential who are initiating or maintaining long-term glucocorticoids. This has special relevance given the long-term skeletal retention of bisphosphonates, in particular, and uncertainty about effects of bisphosphonates on a future developing fetus. In general, guidelines from both the ACR and International Osteoporosis Foundation/European Calcified Tissue Society (IOF-ECTS) [62] have left this decision up to the treating clinician, given the paucity of data on safety considerations in this setting, but there appears to be real potential for fracture risk with chronic glucocorticoids even among younger women.

REFERENCES

1. Mudano A, Allison J, Hill J, et al. Variations in glucocorticoid induced osteoporosis prevention in a managed care cohort. J Rheumatol. 2001;28(6):1298–305.
2. Kanis JA, Johnell O, Oden A, et al. Ten year probabilities of osteoporotic fractures according to BMD and diagnostic thresholds. Osteoporos Int. 2001;12(12):989–95.
3. van Staa T, Leufkens H, Abenhaim L, et al. Use of oral corticosteroids and risk of fractures. J Bone Miner Res. 2000;15(6):993–1000.
4. Bijlsma JW, Saag KG, Buttgereit F, et al. Developments in glucocorticoid therapy. Rheum Dis Clin North Am. 2005;31(1):1–17, vii.
5. van Staa TP, Laan RF, Barton IP, et al. Bone density threshold and other predictors of vertebral fracture in patients receiving oral glucocorticoid therapy. Arthritis Rheum. 2003;48(11):3224–9.
6. van Staa T, Leukens HGM, Abenhaim L, et al. Use of oral glucocorticoids and risk of fractures. J Bone Miner Res. 2000;15:952–6.
7. Cohen S, Levy RM, Keller M, et al. Risedronate therapy prevents corticosteroid-induced bone loss: A twelve-month, multicenter, randomized, double-blind, placebo-controlled, parallel-group study. Arthritis Rheum. 1999;42(11):2309–18.
8. Saag KG, Emkey R, Schnitzer T, et al. Alendronate for the treatment and prevention of glucocorticoid-induced osteoporosis. N Engl J Med. 1998;339:292–9.
9. van Staa TP, Leukens HGM, Abenhaim L, et al. Oral corticosteroids and fracture risk: relationship to daily and cumulative doses. Rheumatology (Oxford). 2000;39:1383–9.
10. Sutter SA, Stein EM. The skeletal effects of inhaled glucocorticoids. Curr Osteoporos Rep. 2016;14(3):106–13.
11. Israel E, Banerjee TR, Fitzmaurice GM, et al. Effects of inhaled glucocorticoids on bone density in premenopausal women. N Engl J Med. 2001;345:941–7.
12. Wong CA, Walsh LJ, Smith CJP, et al. Inhaled corticosteroid use and bone-mineral density in patients with asthma. Lancet. 2000;355:1399–403.
13. Schoon EJ, Bollani S, Mills PR, et al. Bone mineral density in relation to efficacy and side effects of budesonide and prednisolone in Crohn's disease. Clin Gastroenterol Hepatol. 2005;3(2):113–21.
14. Ramsey-Goldman R, Dunn JE, Huang C-F, et al. Frequency of fractures in women with systemic lupus erythematosus: comparison with United States population data. Arthritis Rheum. 1999;42:882–90.
15. Verhoeven MC, Boers M. Limited bone loss due to corticosteroids; a systematic review of prospective studies in rheumatoid arthritis and other diseases. J Rheumatol. 1997;24:1495–503.
16. Cooper C, Coupland C, Mitchell M. Rheumatoid arthritis, corticosteroid therapy and hip fracture. Ann Rheum Dis. 1995;54:49–52.
17. Proven A, Gabriel SE, Orces C, et al. Glucocorticoid therapy in giant cell arteritis: duration and adverse outcomes. Arthritis Rheum. 2003;49(5):703–8.
18. Kauh E, Mixson L, Malice MP, et al. Prednisone affects inflammation, glucose tolerance, and bone turnover within hours of treatment in healthy individuals. Eur J Endocrinol. 2012;166(3):459–67.

19. O'Brien CA, Jia D, Plotkin LI, et al. Glucocorticoids act directly on osteoblasts and osteocytes to induce their apoptosis and reduce bone formation and strength. Endocrinology. 2004;145(4):1835–41.
20. Weinstein RS, Jia D, Powers CC, et al. The skeletal effects of glucocorticoid excess override those of orchidectomy in mice. Endocrinology. 2004;145(4):1980–7.
21. Ito S, Suzuki N, Kato S, et al. Glucocorticoids induce the differentiation of a mesenchymal progenitor cell line, ROB-C26 into adipocytes and osteoblasts, but fail to induce terminal osteoblast differentiation. Bone. 2007;40(1):84–92.
22. Guanabens N, Gifre L, Peris P. The role of Wnt signaling and sclerostin in the pathogenesis of glucocorticoid-induced osteoporosis. Curr Osteoporos Rep. 2014;12(1):90–7.
23. Adler RA, Weinstein RS, Saag KG. Glucocorticoid-induced osteoporosis. In: Marcus R, Feldman D, Dempster DW, et al. (eds) Osteoporosis (4th edn). Boston: Academic Press, 2013, pp. 1191–223.
24. Mazziotti G, Giustina A. Glucocorticoids and the regulation of growth hormone secretion. Nat Rev Endocrinol. 2013;9(5):265–76.
25. Mazziotti G, Formenti AM, Adler RA, et al. Glucocorticoid-induced osteoporosis: pathophysiological role of GH/IGF-I and PTH/VITAMIN D axes, treatment options and guidelines. Endocrine. 2016;54(3):603–11.
26. Ton FN, Gunawardene SC, Lee H, et al. Effects of low-dose prednisone on bone metabolism. J Bone Miner Res. 2005;20(3):464–70.
27. Kondo T, Kitazawa R, Yamaguchi A, et al. Dexamethasone promotes osteoclastogenesis by inhibiting osteoprotegerin through multiple levels. J Cell Biochem. 2008;103(1):335–45.
28. Teitelbaum SL. Glucocorticoids and the osteoclast. Clin Exp Rheumatol. 2015;33(4 suppl 92):S37–9.
29. Gennari C. Differential effect of glucocorticoids on calcium absorption and bone mass. Br J Rheumatol. 1993;32(suppl 2):11–4.
30. Pearce G, Tabensky DA, Delmas PD, et al. Corticosteroid-induced bone loss in men. J Clin Endocrinol Metab. 1998;83(3):801–6.
31. Davidson ZE, Walker KZ, Truby H. Clinical review: do glucocorticosteroids alter vitamin D status? A systematic review with meta-analyses of observational studies. J Clin Endocrinol Metab. 2012;97(3):738–44.
32. Weinstein RS. Glucocorticoid-induced osteoporosis and osteonecrosis. Endocrinol Metab Clin North Am. 2012;41(3):595–611.
33. Buckley L, Guyatt G, Fink HA, et al. American College of Rheumatology guideline for the prevention and treatment of glucocorticoid-induced osteoporosis. Arth Rheum. 2017;69(8):1521–37.
34. Lian KC, Lang TF, Keyak JH, et al. Differences in hip quantitative computed tomography (QCT) measurements of bone mineral density and bone strength between glucocorticoid-treated and glucocorticoid-naive postmenopausal women. Osteoporos Int. 2005;16(6):642–50.
35. Graeff C, Marin F, Petto H, et al. High resolution quantitative computed tomography-based assessment of trabecular microstructure and strength estimates by finite-element analysis of the spine, but not DXA, reflects vertebral fracture status in men with glucocorticoid-induced osteoporosis. Bone. 2013;52(2):568–77.
36. Chang G, Rajapakse CS, Regatte RR, et al. 3 Tesla MRI detects deterioration in proximal femur microarchitecture and strength in long-term glucocorticoid users compared with controls. J Magn Reson Imaging. 2015;42(6):1489–96.
37. Mellibovsky L, Prieto-Alhambra D, Mellibovsky F, et al. Bone tissue properties measurement by reference point indentation in glucocorticoid-induced osteoporosis. J Bone Miner Res. 2015;30(9):1651–6.
38. Saag KG, Agnusdei D, Hans D, et al. Trabecular bone score in patients with chronic glucocorticoid therapy-induced osteoporosis treated with alendronate or teriparatide. Arthritis Rheumatol. 2016;68(9):2122–8.
39. de Nijs RN, Jacobs JW, Algra A, et al. Prevention and treatment of glucocorticoid-induced osteoporosis with active vitamin D3 analogues: a review with meta-analysis of randomized controlled trials including organ transplantation studies. Osteoporos Int. 2004;15(8):589–602.
40. Richy F, Ethgen O, Bruyere O, et al. Efficacy of alphacalcidol and calcitriol in primary and corticosteroid-induced osteoporosis: a meta-analysis of their effects on bone mineral density and fracture rate. Osteoporos Int. 2004;15(4):301–10.
41. Campbell IA, Douglas JG, Francis RM, et al. Five year study of etidronate and/or calcium as prevention and treatment for osteoporosis and fractures in patients with asthma receiving long term oral and/or inhaled glucocorticoids. Thorax. 2004;59(9):761–8.
42. Adachi JD, Bensen WG, Brown J, et al. Intermittent etidronate therapy to prevent corticosteroid-induced osteoporosis. N Engl J Med. 1997;337(6):382–7.
43. Ringe JD, Dorst A, Faber H, et al. Intermittent intravenous ibandronate injections reduce vertebral fracture risk in corticosteroid-induced osteoporosis: results from a long-term comparative study. Osteoporos Int. 2003;14(10):801–7.
44. Gallacher SJ, Fenner JAK, Anderson K, et al. Intravenous pamidronate in the treatment of osteoporosis associated with corticosteroid dependent lung disease: an open pilot study. Thorax. 1992;47:932–6.
45. Reid IR, King AR, Alexander CJ, et al. Prevention of steroid-induced osteoporosis with (3-amino-1-hydroxypropylidene)-1, 1-bisphosphonate (APD). Lancet. 1988;1:143–6.
46. Valkema R, Vismans F-JE, Papapoulos SE, et al. Maintained improvement in calcium balance and bone mineral content in patients with osteoporosis treated with the bisphosphonate APD. Bone Miner. 1989;5:183–92.
47. Boutsen Y, Jamart J, Esselnickx W, et al. Primary prevention of glucocorticoid-induced osteoporosis with intravenous pamidronate and calcium: a prospective controlled 1-year study comparing a single infusion, an

infusion given once every 3 months, and calcium alone. J Bone Miner Res. 2001;16(1):104–12.
48. Buckley LM, Leib ES, Cartularo KS, et al. Calcium and vitamin D3 supplementation prevents bone loss in the spine secondary to low-dose corticosteroids in patients with rheumatoid arthritis. Ann Intern Med. 1996;125:961–8.
49. Adachi R, Saag K, Emkey R, et al. Effects of alendronate for two years on BMD and fractures in patients receiving glucocorticoids. Arthritis Rheum. 2001;44:202–11.
50. Reid DM, Hughes R, Laan RF. Efficacy and safety of daily risedronate in the treatment of corticosteroid-induced osteoporosis in men and women: a randomized trial. J Bone Miner Res. 2000;15:1006–13.
51. Wallach S, Cohen S, Reid DM, et al. Effects of risedronate treatment on bone density and vertebral fracture in patients on corticosteroid therapy. Calcif Tissue Int. 2000;67:277–85.
52. Stoch SA, Saag KG, Greenwald M, et al. Once-weekly oral alendronate 70 mg in patients with glucocorticoid-induced bone loss: a 12-month randomized, placebo-controlled clinical trial. J Rheumatol. 2009;36(8): 1705–14.
53. Reid DM, Devogelaer JP, Saag K, et al. Zoledronic acid and risedronate in the prevention and treatment of glucocorticoid-induced osteoporosis (HORIZON): a multicentre, double-blind, double-dummy, randomised controlled trial. Lancet. 2009;373(9671):1253–63.
54. Sambrook PN. Corticosteroid osteoporosis: practical implications of recent trials. J Bone Miner Res. 2000;15(9):1645–49.
55. Emkey R, Delmas PD, Goemaere S, et al. Changes in bone mineral density following discontinuation or continuation of alendronate therapy in glucocorticoid-treated patients: a retrospective, observational study. Arthritis Rheum. 2003;48(4):1102–8.
56. Saag KG, Shane E, Boonen S, et al. Teriparatide or alendronate in glucocorticoid-induced osteoporosis. N Engl J Med. 2007;357(20):2028–39.
57. Saag KG, Zanchetta JR, Devogelaer JP, et al. Effects of teriparatide versus alendronate for treating glucocorticoid-induced osteoporosis: thirty-six-month results of a randomized, double-blind, controlled trial. Arthritis Rheum. 2009;60(11):3346–55
58. Saag KG, Wagman RB, Geusens P, et al. Effect of denosumab compared with risedronate in glucocorticoid-treated individuals: results from the 12-month primary analysis of a randomized, double-blind, active-controlled study. Arthritis Rheumatol. 2016;68(suppl 10):abstract 2L.
59. Mok CC, Ying KY, To CH, et al. Raloxifene for prevention of glucocorticoid-induced bone loss: a 12-month randomised double-blinded placebo-controlled trial. Ann Rheum Dis. 2011;70(5):778–84.
60. Tracz MJ, Sideras K, Bolona ER, et al. Testosterone use in men and its effects on bone health. A systematic review and meta-analysis of randomized placebo-controlled trials. J Clin Endocrinol Metab. 2006;91(6): 2011–6.
61. Adachi JD, Bensen WG, Bell MJ, et al. Salmon calcitonin nasal spray in the prevention of corticosteroid-induced osteoporosis. Br J Rheumatol. 1997;36(2):255–7.
62. Lekamwasam S, Adachi JD, Agnusdei D, et al. A framework for the development of guidelines for the management of glucocorticoid-induced osteoporosis. Osteoporos Int. 2012;23(9):2257–76.

60

Human Immunodeficiency Virus and Bone

Michael T. Yin[1] and Todd T. Brown[2]

[1]*Division of Infectious Diseases, Columbia University Medical Center, New York, NY, USA*
[2]*Division of Endocrinology, Diabetes, and Metabolism, Johns Hopkins University, Baltimore, MD, USA*

INTRODUCTION

With effective antiretroviral therapy (ART), persons living with human immunodeficiency virus (HIV) have life expectancies that are approaching that of the general population, with the main cause of death shifted from opportunistic infections to non-acquired immunodeficiency syndrome (non-AIDS) illnesses related to aging, such as malignancy, cardiovascular disease, and liver and renal failure [1–3]. In the United States, almost half of persons living with HIV are over the age of 50 [4]. Osteoporosis and fracture has been increasingly recognized as a HIV-associated morbidity. This chapter provides an overview of the epidemiology, pathogenesis, screening, and management of HIV-associated bone loss and fractures, with an emphasis on the role of ART.

EPIDEMIOLOGY OF OSTEOPOROSIS AND FRACTURE

A meta-analysis in 2006 by Brown and colleagues estimated that HIV-infected adults were three times more likely to have osteoporosis than uninfected, age-matched controls [5]. Since then several other cross-sectional studies have corroborated those findings in different study populations. The availability of large study cohorts has also provided evidence of a higher incidence of fractures in HIV-infected adults compared with age- and sex-matched controls or population-based controls (relative risk [RR] = 1.56 all fractures; RR = 1.36 fragility fractures) [6], with the risk increasing in older age [7]. Another important observation from these studies is that exposure to antiretrovirals is associated with lower BMD and greater fracture risk [8, 9].

PUTATIVE MECHANISMS

HIV associated bone loss is most likely a multifactorial problem which includes traditional risk factors that may occur more commonly in HIV-infected individuals, direct and indirect effects of HIV infection, and the effect of antiretrovirals (Fig. 60.1). HIV-infected persons are more likely to be smokers, use opiates, be exposed to glucocorticoids, have gonadal insufficiency, and lower body weight, all of which are known risk factors for osteoporosis and fracture. However, the effect of HIV on BMD or fracture remains significant after controlling for these traditional risk factors, suggesting that the effect of HIV on bone cannot solely be explained by traditional risk factors. Some in vitro studies have demonstrated that HIV proteins have direct effects on activating osteoclasts and inhibiting osteoblasts [10–12], and the direct effect of HIV on bone was confirmed in a transgenic rat model of HIV infection [13]. Increasingly, there is an understanding that HIV infection, even with effective ART, is associated with a chronically heightened state of immune activation that may result from persistent low-grade viral replication or from translocation of microbial peptides from the intestines that trigger an immune response [14]. Studies have linked cellular markers of immune activation with increases in surrogate markers of cardiovascular disease [15–18], and some studies have shown an association with T-cell activation and lower BMD [19–21]; however, the data are not consistent across all studies.

Role of antiretroviral therapy

The natural history of HIV infection has been transformed by the availability of combination ART (cART, previously referred to as highly active antiretroviral

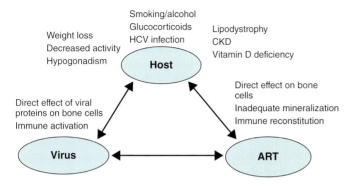

Fig. 60.1. Multifactorial etiology of bone loss in HIV-infected individuals. ART = antiretroviral therapy; CKD = chronic kidney disease; HCV = hepatitis C virus.

therapy or HAART) in the mid-1990s. Over the past 20 years, cART has evolved, such that the complexity and metabolic toxicities have been greatly diminished. Modern ART generally consist of two nucleoside/nucleotide reverse transcriptase inhibitors (NRTIs) along with either a non-NRTI (NNRTI), protease inhibitor (PI), or integrase stand transfer inhibitor (INSTI or integrase inhibitor). Novel NRTI-sparing regimens are also emerging. There are specific bone toxicities to certain ART medication, but this is also rapidly evolving with the availability of newer agents.

The timing of cART initiation has also evolved. Previously, it was suggested that cART be started only when the CD4 cell count declined below a certain low threshold (eg, 350 cells/mm^3). However, based on strong observational and clinical trial data, it is currently recommended not to defer cART initiation, but to start as soon as HIV is diagnosed [22, 23]. In general, cART initiation is associated with salutary effects on lean body mass, gonadal function, and systemic inflammation; however, the effect of cART initiation on BMD is different. Through many different studies utilizing DXA measure before and after initiation of different cART regimens, it is now recognized that first 1–2 years after cART initiation is the most significant and dynamic phase of bone loss among HIV-infected persons, and most likely confers the majority of the risk for fracture. Areal BMD by DXA usually decreases 2% to 4% within the first year after initiation of ART, with most of the bone loss occurring within the first 6 months at the lumbar spine and first 12 months at the hip [24–26]. BMD usually stabilizes and increases after that point; however mean BMD does not return to pretreatment BMD. For the few patients who fail their initial regimen, ART initiation with a second-line treatment is associated with similar bone loss [27]

During the first year after cART initiation, bone turnover markers also increase dramatically, with increases in bone resorption markers preceding the compensatory increase in bone formation markers, suggestive of a catabolic window [28]. Some antiretroviral regimens are associated with more bone loss than others. Among first-line agents for treatment of HIV, tenofovir disoproxil fumarate (TDF) is associated with a 1% to 2% greater declines in BMD than with NRTIs [8, 9, 25] or tenofovir alafenamide (TAF), a prodrug of TDF that has lower plasma concentrations of tenofovir [29]. Protease inhibitors are also associated with greater bone loss than NNRTIs [25] and integrase inhibitors [24], although some protease inhibitors seem to be less problematic than others. Despite these slight differences between ART regimens, bone loss has been observed in every study of ART initiation, regardless of the regimen. This has led some to hypothesize that it is partially immune-mediated, due to the immune reconstitution of T-cell numbers and function with ART [30, 31]. This notion is supported by the observation that lower CD4 cells counts and higher HIV RNA levels prior to cART initiation is associated with a greater degree of bone loss with cART initiation [32].

TDF is of special interest not only because it is one of the most commonly utilized antiretrovirals but also because of its utilization in the prevention of infection in uninfected individuals (pre-exposure prophylaxis or PrEP) and in the treatment of hepatitis B infection. In studies of men and women who are uninfected receiving PrEP with TDF or TDF plus emtricitabine, BMD decreases 1% to 2% within the first year [33, 34]. Fortunately, after cessation of PrEP, BMD returns back to baseline levels [35]; therefore, in young adults who use TDF-based PrEP only for a few years while at higher risk of HIV infection, should not suffer negative long-term effects on bone. Despite ample evidence of the negative effects of TDF on BMD, the mechanism is still uncertain. Potential mechanisms include: (i) a direct negative effect on osteoblasts, as suggested by in vitro studies [11, 36]; (ii) inadequate mineralization because TDF is known to cause proximal tubular toxicity leading to varying degrees of phosphaturia in 30% of patients, although phosphatemia and osteomalacia are extremely rare; and (iii) secondary hyperparathyroidism caused by functional vitamin D deficiency [37, 38]. Protease inhibitors have also been associated with bone loss with the putative mechanisms related to either: (i) direct effects of protease inhibitors on inhibiting osteoblastic activity or enhancing osteoclast formation [39–42]; or (ii) indirect effects by affecting vitamin D metabolism through the suppression of 25- and 1α-hydroxylase in hepatocytes and monocytes [43]. Unfortunately, there are no definitive bone biopsy studies to clarify mechanisms.

SCREENING

Based upon data demonstrating a higher risk of fracture independent of traditional osteoporosis risk factors, there is recognition that some risk stratification for prevention of bone loss or fracture is indicated in older HIV-infected adults [44–47]. Most guidelines favor screening with DXA in all HIV-infected postmenopausal women and men 50 years and over, regardless of whether or not other risk factors are present. There have been no studies to assess provider acceptance or adherence to these guidelines, or of the cost-benefit ratio for implementation. European HIV society guidelines also promote the use of the Fracture Risk Assessment Tool (FRAX) for risk stratification in HIV-infected persons. However, some studies have indicated that FRAX using clinical risk factors underestimates fracture risk in HIV-infected adults [48]. Calibration can improve with the addition of HIV as a cause of secondary osteoporosis in the calculation when FRAX is utilized without a BMD measurement.

MANAGEMENT

Nonpharmacological management

As in the general population, assessment of secondary causes of osteoporosis should be the first step in management of low BMD in the HIV-infected population. Of special emphasis in an HIV-infected patient is: (i) the assessment of renal phosphate wasting (generally by measuring the fractional excretion of phosphate, or the ratio of tubular maximum reabsorption of phosphate to glomerular filtration rate) in TDF-treated persons; (ii) the assessment of 25OHD because certain antiretrovirals like efavirenz may increase 25OHD catabolism [49]; and (iii) the assessment of hypogonadism in men. Given that sex hormone-binding globulin levels are higher in HIV-infected men compared with HIV-uninfected men [50], assessment of hypogonadism in HIV-infected men should use a reliable assay of free testosterone.

Optimization of calcium and vitamin D intake is recommended in the HIV-infected population, as is counseling regarding smoking cessation and alcohol abuse. As HIV-infected individuals have a high burden of risk factors that may lead to falls including peripheral neuropathy, polypharmacy, symptoms of imbalance, and muscle wasting, a fall risk assessment should be done. For those at risk of falling, referral to physical therapy for strength and balance training and minimization of environmental factors leading to falls are recommended.

Bone-specific pharmacological management

Guidelines for bone-specific treatment in HIV-infected patients should generally follow those of the underlying general population. For patients who have fractured or are at very high risk of fracture (eg, in the United States, a previous history of fragility fracture, osteoporosis by T-score, or in those with osteopenia a10-year FRAX score ≥20% for all major osteoporotic fracture or ≥3% for the hip), treatment with bisphosphates should be considered first-line therapy. In randomized controlled studies in HIV-infected adults, treatment with alendronate or zoledronic acid has been shown to be well tolerated, safe, and effective in increasing BMD [51, 52]. Teriparatide has also been used in patients who have failed bisphosphonate therapy and has been reported in case studies, but there are no clinical trials in the HIV-infected population. Denosumab has not been utilized in HIV-infected individuals because of concern for increased risk of infection, although in patients with high CD4 counts on ART, this is usually not a concern.

Switching cART to improve bone heath

Recent studies also suggest that there may be significant gains in BMD from coming off TDF or protease inhibitors, agents which have been most clearly linked with bone loss. Switching from TDF to raltegravir, an integrase inhibitor, resulted in a 2.5% or 3.0% increase in BMD at the total hip and lumbar spine, respectively [53]. Substitution of TDF for abacavir is another option [54], although the changes may not be as robust. A newly approved prodrug of tenofovir, called TAF, leads to less renal and bone toxicity. In three recently published studies in slightly different study populations, a switch from TDF to TAF resulted in a 1.1% to 2.5% increase in BMD [55–57]. While there are fewer studies that examine the effects on BMD after switching from a protease inhibitor to another agent, one study found an increase in hip BMD in patients switching from lopinavir/ritonavir to raltegravir [58].

The decision of whether to switch a patient off TDF or protease inhibitors is multifaceted and includes considerations related to cost, availability, and toxicities of alternatives, other comorbid conditions, and patient preferences. If a patient has known osteoporosis or history of fracture, or evidence of urinary phosphate wasting, the authors of this chapter would recommend a switch off TDF and protease inhibitors based on skeletal considerations. Other considerations may include additional risk factors for fracture, older age, osteopenia, hepatitis C (HCV), concomitant use medications adversely affecting bone metabolism, or family history of fracture. We recommend reassessment of BMD 1 year after the switch and consideration for bone-specific treatment, such as bisphosphonates, if the BMD remains stable or continues to decrease.

Reducing bone loss during ART initiation

Since the most dynamic period of bone loss in HIV-infected persons occurs around the time of cART initiation, several strategies have also been investigated to mitigate that bone loss. The first strategy involves avoidance of TDF or protease inhibitors in the cART regimen in patients with osteoporosis or previous fracture, or in patients with many risk factors for osteoporosis and fracture (eg, older age, HCV coinfection, glucocorticoid use). There are many excellent choices for ART initiation that have high efficacy, low pill burden, and low toxicity so that the avoidance of TDF and protease inhibitors is usually possible. However, there is no consensus yet on which regimens are considered the best for bone health, although they are likely to be integrase inhibitor-based. Another strategy involves supplementation of ART with vitamin D and calcium. One randomized clinical trial demonstrated that daily administration of vitamin D3 (4000 IU) and calcium carbonate (1000 mg) resulted in 50% less decrease in BMD and increase in bone turnover markers levels than placebo at 48 weeks after initiation of TDF/emtricitabine/efavirenz, a commonly utilized once-daily cART regimen [59]. Whether the same benefit will be evident with other dosages of vitamin D and calcium, and when combined with other initiating ART regimens, remains to be seen. Lastly, a recent study demonstrated the efficacy of a single dose of intravenous zoledronic acid 5 mg to prevent bone loss associated with initiation of ART [60]. The choice of which strategy to utilize depends on the availability of antiretroviral regimens and other medications, especially in resource-limited settings where there are typically only one first-line and one second-line antiretroviral regimen available.

SPECIAL POPULATIONS

Children and adolescents

Approximately one-fourth of the 37 million people living with HIV worldwide are children and adolescents under the age of 24 years. Children and adolescents who acquire HIV infection either perinatally or during adolescence through sexual transmission have the greatest cumulative lifetime exposure to the direct and indirect negative effects of HIV infection and ART on bone metabolism (Fig. 60.2). Fracture risk may be increased in adulthood, even in comparison with adults infected later in life, since they may not achieve optimal peak bone mass, a key determinant of osteoporosis and fracture risk later in life [61]. A study of 20- to 25-year-old HIV-infected men, whether infected perinatally or during adolescence, found lower BMD by DXA and markedly abnormal trabecular and cortical microarchitecture at the radius and tibia compared with uninfected controls. This suggested that young adults infected with HIV at an early age have lower peak bone mass and compromised bone strength [62]. Many cross-sectional and longitudinal studies have been conducted in perinatally infected children, several of which included healthy controls [63–68], but have varied in their approach to account for skeletal size and maturational delays, including age, sex, height, weight, BMI, race, pubertal stage, and bone age [63, 65, 69, 70]. Despite variable adjustment techniques for body size or growth retardation, most studies found that measures of bone mass are reduced in children and adolescents with HIV [71].

TDF is just becoming available in certain resource-limited countries and is being increasingly used in adolescents to improve tolerability and adherence. Although TDF has been approved for use in children as young as 2 years, there remains concern about its effect on developing bones in children. Data on its effects on bone acquisition are inconclusive. Similarly, the impact of TDF exposure in utero on fetal bone development and implications for neonatal growth is an area of ongoing interest; while some studies have found an association between maternal TDF exposure and decreased infant growth [72, 73] and others have not [74, 75].

Hepatitis C virus infection

HCV infection is more common than HIV infection, and HIV/HCV coinfection is relatively common among HIV-infected adults. Recent data indicate that HIV/HCV coinfection is an independent risk factor for incident fractures [76–80], with an overall increase in relative risk of fracture of 2.05 (1.75, 2.41) for HIV/HCV coinfection compared with HIV monoinfection [9]. Several well-controlled studies using DXA and QCT found significantly lower areal and volumetric BMD in HIV/HCV coinfection, sug-

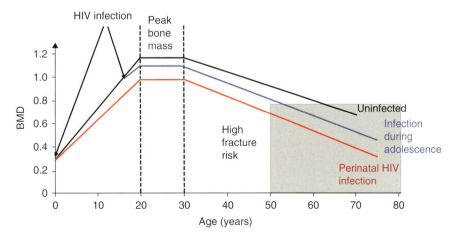

Fig. 60.2. Potential negative effects of HIV infection (perinatal or during adolescence) on bone. BMD = bone mineral density. Source: [93]. Reproduced with permission of Oxford University Press.

gesting that true deficits in bone strength contribute to the higher fracture rates [81–83]. HCV monoinfection is also associated with low BMD and osteoporosis [84–86], even in the absence of cirrhosis [87].

Whereas the clinical impact of HIV/HCV coinfection on fracture risk is quite clear, the complex mechanisms of how chronic viral coinfections affect bone metabolism require further clarification, especially the role of increased inflammation and immune activation. Studies have demonstrated higher levels of T-cell activation in HIV/HCV coinfected individuals than HIV monoinfected individuals [88, 89], possibly augmented by increased microbial translocation in HIV/HCV coinfected individuals [90]. This may result in enhanced bone resorption and greater bone loss and fracture among HIV/HCV coinfected individuals. Treatment of HCV monoinfection with interferon and ribavirin also decreased inflammatory and bone turnover markers and increased BMD [91], suggesting that HCV-associated bone loss is at least partially mediated by increased inflammation and bone turnover and modulated by eradication of HCV infection. Fracture risk was noted to decrease in postmenopausal women who had successfully treated HCV monoinfection [92]. With the availability of oral, noninterferon-requiring, direct-acting antiviral agents (DAAs) for HCV therapy, there is tremendous enthusiasm for widespread detection and treatment of HCV-infected individuals. Currently, there are no data yet on the effect of DAAs on bone metabolism, turnover, and BMD, but this could potentially be the most important single intervention for improvement of bone health in HCV-infected and HIV/HCV coinfected individuals.

SUMMARY

Recognition of the negative consequences of HIV infection, antiretroviral initiation, and HCV coinfection on bone health in the last 15 years is a testament to the incredible advances in treatment of HIV and longer survival of HIV-infected persons. The care of HIV-infected patients with osteoporosis and fracture presents unique challenges and opportunities for interdisciplinary collaboration and research.

REFERENCES

1. Lohse N, Hansen AB, Pedersen G, et al. Survival of persons with and without HIV infection in Denmark, 1995–2005. Ann. Intern. Med. 2007;146(2):87–95.
2. Samji H, Cescon A, Hogg RS, et al. Closing the gap: increases in life expectancy among treated HIV-positive individuals in the United States and Canada. PloS One. 2013;8(12):e81355.
3. Smith CJ, Ryom L, Weber R, et al. Trends in underlying causes of death in people with HIV from 1999 to 2011 (D:A:D): a multicohort collaboration. Lancet. 2014;384(9939):241–8.
4. High KP, Brennan-Ing M, Clifford DB, et al. HIV and aging: state of knowledge and areas of critical need for research. A report to the NIH Office of AIDS Research by the HIV and Aging Working Group. J Acquir Immune Defic Syndr. 2012;60(suppl 1):S1–18.
5. Brown TT, Qaqish RB. Antiretroviral therapy and the prevalence of osteopenia and osteoporosis: a meta-analytic review. AIDS. 2006;20(17):2165–74.
6. Shiau S, Broun EC, Arpadi SM, et al. Incident fractures in HIV-infected individuals: a systematic review and meta-analysis. AIDS. 2013;27(12):1949–57.
7. Triant VA, Brown TT, Lee H, et al. Fracture prevalence among human immunodeficiency virus (HIV)-infected versus non-HIV-infected patients in a large U.S. healthcare system. J Clin Endocrinol Metab. 2008;93(9):3499–504. Epub 2008/07/03.
8. Bedimo R, Maalouf NM, Zhang S, et al. Osteoporotic fracture risk associated with cumulative exposure to tenofovir and other antiretroviral agents. AIDS. 2012;26(7):825–31.
9. Hansen AB, Gerstoft J, Kronborg G, et al. Incidence of low and high-energy fractures in persons with and without HIV infection: a Danish population-based cohort study. AIDS. 2012;26(3):285–93.
10. Gibellini D, De Crignis E, Ponti C, et al. HIV-1 triggers apoptosis in primary osteoblasts and HOBIT cells through TNFalpha activation. J Med Virol. 2008;80(9):1507–14.
11. Cotter EJ, Malizia AP, Chew N, et al. HIV proteins regulate bone marker secretion and transcription factor activity in cultured human osteoblasts with consequent potential implications for osteoblast function and development. AIDS Res Hum Retroviruses. 2007;23(12):1521–30.
12. Gibellini D, De Crignis E, Ponti C, et al. HIV-1 Tat protein enhances RANKL/M-CSF-mediated osteoclast differentiation. Biochem Biophys Res Commun. 2010;401(3):429–34.
13. Vikulina T, Fan X, Yamaguchi M, et al. Alterations in the immuno-skeletal interface drive bone destruction in HIV-1 transgenic rats. Proc Natl Acad Scie U S A. 2010;107(31):13848–53.
14. Brenchley JM, Price DA, Schacker TW, et al. Microbial translocation is a cause of systemic immune activation in chronic HIV infection. Nat Med. 2006;12(12):1365–71.
15. Karim R, Mack WJ, Kono N, et al. T-cell activation, both pre- and post-HAART levels, correlates with carotid artery stiffness over 6.5 years among HIV-infected women in the WIHS. J Acquir Immune Defic Syn. 2014;67(3):349–56.
16. Kaplan RC, Sinclair E, Landay AL, et al. T cell activation predicts carotid artery stiffness among HIV-infected women. Atherosclerosis. 2011;217(1):207–13.
17. Fitch KV, Srinivasa S, Abbara S, et al. Noncalcified coronary atherosclerotic plaque and immune activation in HIV-infected women. J Infect Dis. 2013;208(11):1737–46.
18. Pereyra F, Lo J, Triant VA, et al. Increased coronary atherosclerosis and immune activation in HIV-1 elite controllers. AIDS. 2012;26(18):2409–12.

19. Erlandson KM, O'Riordan M, Labbato D, et al. Relationships between inflammation, immune activation, and bone health among HIV-infected adults on stable antiretroviral therapy. J Acquir Immune Defic Syn. 2014;65(3):290–8.
20. Gazzola L, Bellistri GM, Tincati C, et al. Association between peripheral T Lymphocyte activation and impaired bone mineral density in HIV-infected patients. J Transl Med. 2013;11:51.
21. Manavalan JS, Arpadi S, Tharmarajah S, et al. Abnormal bone acquisition with early-life HIV infection: role of immune activation and senescent osteogenic precursors. J Bone Miner Res. 2016;31(11):1988–96.
22. US Department of Health and Human Services. Guidelines for the use of antiretroviral agents in adults and adolescents living with HIV. 2016. https://aidsinfo.nih.gov/guidelines/html/1/adult-and-adolescent-treatment-guidelines/0/ (accessed May 2018).
23. European AIDS Clinical Society (EACS). Guidelines Version 8.1 October 2016. 2016. http://www.eacsociety.org/files/guidelines_8.1-english.pdf (accessed May 2018).
24. Brown TT, Moser C, Currier JS, et al. Changes in bone mineral density after initiation of antiretroviral treatment with tenofovir disoproxil fumarate/emtricitabine plus atazanavir/ritonavir, darunavir/ritonavir, or raltegravir. J Infect Dis. 2015;212(8):1241–9.
25. McComsey GA, Kitch D, Daar ES, et al. Bone mineral density and fractures in antiretroviral-naive persons randomized to receive abacavir-lamivudine or tenofovir disoproxil fumarate-emtricitabine along with efavirenz or atazanavir-ritonavir: Aids Clinical Trials Group A5224s, a substudy of ACTG A5202. J Infect Dis. 2011;203(12):1791–801.
26. Stellbrink HJ, Orkin C, Arribas JR, et al. Comparison of changes in bone density and turnover with abacavir-lamivudine versus tenofovir-emtricitabine in HIV-infected adults: 48-week results from the ASSERT study. Clin infect Dis. 2010;51(8):963–72.
27. Martin A, Moore C, Mallon PW, et al. Bone mineral density in HIV participants randomized to raltegravir and lopinavir/ritonavir compared with standard second line therapy. AIDS. 2013;27(15):2403–11.
28. van Vonderen MG, Mallon PW, Murray B, et al. *Changes in bone biomarkers in antiretroviral naïve HIV-infected men randomised to nevirapine/lopinavir/ritonavir (NVP/LPV/r) or zidovudine/lamivudine/lopinavir/ritonavir (AZT/3TC/LPV/r) help explain limited loss of bone mineral density over first 12 months after antiretroviral therapy (ART) initiation.* Conference on Retroviruses and Opportunistic Infections, Boston, MA2011.
29. Sax PE, Wohl D, Yin MT, et al. Tenofovir alafenamide versus tenofovir disoproxil fumarate, coformulated with elvitegravir, cobicistat, and emtricitabine, for initial treatment of HIV-1 infection: two randomised, double-blind, phase 3, non-inferiority trials. Lancet. 2015;385(9987):2606–15.
30. Ofotokun I, Titanji K, Vikulina T, et al. Role of T-cell reconstitution in HIV-1 antiretroviral therapy-induced bone loss. Nat Commun. 2015;6:8282.
31. Ofotokun I, Titanji K, Vunnava A, et al. Antiretroviral therapy induces a rapid increase in bone resorption that is positively associated with the magnitude of immune reconstitution in HIV infection. AIDS. 2016;30(3):405–14.
32. Grant PM, Kitch D, McComsey GA, et al. Low baseline CD4+ count is associated with greater bone mineral density loss after antiretroviral therapy initiation. Clin Infect Dis. 2013;57(10):1483–8.
33. Liu AY, Vittinghoff E, Sellmeyer DE, et al. Bone mineral density in HIV-negative men participating in a tenofovir pre-exposure prophylaxis randomized clinical trial in San Francisco. PloS One. 2011;6(8):e23688.
34. Mulligan K, Glidden DV, Anderson PL, et al. Effects of emtricitabine/tenofovir on bone mineral density in HIV-negative persons in a randomized, double-blind, placebo-controlled trial. Clin Infect Dis. 2015;61(4):572–80.
35. Mirembe BG, Kelly CW, Mgodi N, et al. Bone mineral density changes among young, healthy African women receiving oral tenofovir for HIV preexposure prophylaxis. J Acquir Immune Defic Syn. 2016;71(3):287–94.
36. Grigsby IF, Pham L, Mansky LM, et al. Tenofovir treatment of primary osteoblasts alters gene expression profiles: implications for bone mineral density loss. Biochem Biophys Res Commun. 2010;394(1):48–53.
37. Havens PL, Kiser JJ, Stephensen CB, et al. Association of higher plasma vitamin D binding protein and lower free calcitriol levels with tenofovir disoproxil fumarate use and plasma and intracellular tenofovir pharmacokinetics: cause of a functional vitamin D deficiency? Antimicrob Agents Chemother. 2013;57(11):5619–28.
38. Havens PL, Stephensen CB, Hazra R, et al. Vitamin D3 decreases parathyroid hormone in HIV-infected youth being treated with tenofovir: a randomized, placebo-controlled trial. Clin Infect Dis. 2012;54(7):1013–25.
39. Modarresi R, Xiang Z, Yin M, et al. WNT/beta-catenin signaling is involved in regulation of osteoclast differentiation by human immunodeficiency virus protease inhibitor ritonavir: relationship to human immunodeficiency virus-linked bone mineral loss. Am J Pathol. 2009;174(1):123–35.
40. Yin MT, Modarresi R, Shane E, et al. Effects of HIV infection and antiretroviral therapy with ritonavir on induction of osteoclast-like cells in postmenopausal women. Osteoporos Int. 2011;22(5):1459–68.
41. Santiago F, Oguma J, Brown AM, et al. Noncanonical Wnt signaling promotes osteoclast differentiation and is facilitated by the human immunodeficiency virus protease inhibitor ritonavir. Biochem Biophys Res Comm. 2012;417(1):223–30.
42. Malizia AP, Cotter E, Chew N, et al. HIV protease inhibitors selectively induce gene expression alterations associated with reduced calcium deposition in primary human osteoblasts. AIDS Res Hum Retroviruses. 2007;23(2):243–50.
43. Cozzolino M, Vidal M, Arcidiacono MV, et al. HIV-protease inhibitors impair vitamin D bioactivation to 1,25-dihydroxyvitamin D. AIDS. 2003;17(4):513–20.

44. Brown TT, Mallon PW. Editorial: Working towards an understanding of bone disease in HIV. Curr Opin HIV AIDS. 2016;11(3):251–2.
45. Hoy J, Young B. Do people with HIV infection have a higher risk of fracture compared with those without HIV infection? Curr Opin HIV AIDS. 2016;11(3):301–5.
46. Brown TT, Hoy J, Borderi M, et al. Recommendations for evaluation and management of bone disease in HIV. Clin Infect Dis. 2015;60(8):1242–51.
47. Hileman CO, Eckard AR, McComsey GA. Bone loss in HIV: a contemporary review. Curr Opin Endocrinol Diabetes Obesity. 2015;22(6):446–51.
48. Yin MT, Shiau S, Rimland D, et al. Fracture prediction with modified-FRAX in older HIV-infected and uninfected men. J Acquir Immune Defic Syn. 2016;72(5):513–20.
49. Brown TT, McComsey G. Association between initiation of antiretroviral therapy with efavirenz and decrease in 25-hydroxyvitamin D. Antiviral Ther. 2010;15(3):425–9.
50. Monroe AK, Dobs AS, Palella FJ, et al. Morning free and total testosterone in HIV-infected men: implications for the assessment of hypogonadism. AIDS Res Ther. 2014;11(1):6.
51. McComsey GA, Kendall MA, Tebas P, et al. Alendronate with calcium and vitamin D supplementation is safe and effective for the treatment of decreased bone mineral density in HIV. AIDS. 2007;21(18):2473–82.
52. Bolland MJ, Grey AB, Horne AM, et al. Annual zoledronate increases bone density in highly active antiretroviral therapy-treated human immunodeficiency virus-infected men: a randomized controlled trial. J Clin Endocrinol Metab. 2007;92(4):1283–8.
53. Bloch M, Tong WW, Hoy J, et al. Switch from tenofovir to raltegravir increases low bone mineral density and decreases markers of bone turnover over 48 weeks. HIV Med. 2014;15(6):373–80.
54. Negredo E, Domingo P, Perez-Alvarez N, et al. Improvement in bone mineral density after switching from tenofovir to abacavir in HIV-1-infected patients with low bone mineral density: two-centre randomized pilot study (OsteoTDF study). J Antimicrob Chemother. 2014;69(12):3368–71.
55. Gallant J, Brunetta J, Crofoot G, et al. Efficacy and safety of switching to a single-tablet regimen of elvitegravir/cobicistat/emtricitabine/tenofovir alafenamide (E/C/F/TAF) in HIV-1/hepatitis B coinfected adults. J Acquir Immune Defic Syn. 2016;73(3):294–8.
56. Mills A, Crofoot G Jr, McDonald C, et al. Tenofovir alafenamide versus tenofovir disoproxil fumarate in the first protease inhibitor-based single-tablet regimen for initial HIV-1 therapy: a randomized phase 2 study. J Acquir Immune Defic Syn. 2015;69(4):439–45.
57. Pozniak A, Arribas JR, Gathe J, et al. Switching to tenofovir alafenamide, coformulated with elvitegravir, cobicistat, and emtricitabine, in HIV-infected patients with renal impairment: 48-week results from a single-arm, multicenter, open-label phase 3 study. J Acquir Immune Defic Syn. 2016;71(5):530–7.
58. Curran A, Martinez E, Saumoy M, et al. Body composition changes after switching from protease inhibitors to raltegravir: SPIRAL-LIP substudy. AIDS. 2012;26(4):475–81.
59. Overton ET, Chan ES, Brown TT, et al. Vitamin D and calcium attenuate bone loss with antiretroviral therapy initiation: a randomized trial. Ann Intern Med. 2015;162(12):815–24.
60. Ofotokun I, Titanji K, Lahiri CD, et al. A single-dose zoledronic acid infusion prevents antiretroviral therapy-induced bone loss in treatment-naive HIV-infected patients: a phase IIb trial. Clin Infect Dis. 2016;63(5):663–71.
61. Heaney RP, Abrams S, Dawson-Hughes B, et al. Peak bone mass. Osteoporos Int. 2000;11(12):985–1009.
62. Yin MT, Lund E, Shah J, et al. Lower peak bone mass and abnormal trabecular and cortical microarchitecture in young men infected with HIV early in life. AIDS. 2014;28(3):345–53.
63. Pitukcheewanont P, Safani D, Church J, et al. Bone measures in HIV-1 infected children and adolescents: disparity between quantitative computed tomography and dual-energy X-ray absorptiometry measurements. Osteoporos Int. 2005;16(11):1393–6.
64. Stagi S, Bindi G, Galluzzi F, et al. Changed bone status in human immunodeficiency virus type 1 (HIV-1) perinatally infected children is related to low serum free IGF-I. Clin Endocrinol. 2004;61(6):692–9.
65. Jacobson DL, Lindsey JC, Gordon CM, et al. Total body and spinal bone mineral density across Tanner stage in perinatally HIV-infected and uninfected children and youth in PACTG 1045. AIDS. 2010;24(5):687–96.
66. Zuccotti G, Vigano A, Gabiano C, et al. Antiretroviral therapy and bone mineral measurements in HIV-infected youths. Bone. 2010;46(6):1633–8.
67. Mulligan K, Harris DR, Emmanuel P, et al. Low bone mass in behaviorally HIV-infected young men on antiretroviral therapy: Adolescent Trials Network Study 021B. Clin Infect Dis. 2012;55(3):461–8.
68. Negredo E, Domingo P, Ferrer E, et al. Peak bone mass in young HIV-infected patients compared with healthy controls. J Acquir Immune Defic Syndr. 2014;65(2):207–12.
69. Mora S, Zamproni I, Giacomet V, et al. Analysis of bone mineral content in horizontally HIV-infected children naive to antiretroviral treatment. Calcif Tissue Int. 2005;76(5):336–40.
70. Palchetti CZ, Szejnfeld VL, de Menezes Succi RC, et al. Impaired bone mineral accrual in prepubertal HIV-infected children: a cohort study. Braz J Infect Dis. 2015;19(6):623–30.
71. Arpadi SM, Shiau S, Marx-Arpadi C, et al. Bone health in HIV-infected children, adolescents and young adults: a systematic review. J AIDS Clin Res. 2014;5(11):374.
72. Ransom CE, Huo Y, Patel K, et al. Infant growth outcomes after maternal tenofovir disoproxil fumarate use during pregnancy. J Acquir Immune Defic Syn. 2013;64(4):374–81.

73. Siberry GK, Williams PL, Mendez H, et al. Safety of tenofovir use during pregnancy: early growth outcomes in HIV-exposed uninfected infants. AIDS. 2012;26(9):1151–9.
74. Gibb DM, Kizito H, Russell EC, et al. Pregnancy and infant outcomes among HIV-infected women taking long-term ART with and without tenofovir in the DART trial. PLoS Med. 2012;9(5):e1001217.
75. Vigano A, Mora S, Giacomet V, et al. In utero exposure to tenofovir disoproxil fumarate does not impair growth and bone health in HIV-uninfected children born to HIV-infected mothers. Antiviral Ther. 2011;16(8):1259–66.
76. Yin MT, Shi Q, Hoover DR, et al. Fracture incidence in HIV-infected women: results from the Women's Interagency HIV Study. AIDS. 2010;24(17):2679–86.
77. Young B, Dao CN, Buchacz K, et al. Increased rates of bone fracture among HIV-infected persons in the HIV Outpatient Study (HOPS) compared with the US general population, 2000-2006. Clin Infect Dis. 2011;52(8):1061–8.
78. Hansen AB, Gerstoft J, Kronborg G, et al. Incidence of low- and high-energy fractures in persons with and without HIV-infection: a Danish population-based cohort study. AIDS. 2012;26(3):285–93.
79. Collin F, Duval X, Le Moing V, et al. Ten-year incidence and risk factors of bone fractures in a cohort of treated HIV1-infected adults. AIDS. 2009;23(8):1021–4.
80. Lo Re V 3rd, Volk J, Newcomb CW, et al. Risk of hip fracture associated with hepatitis C virus infection and hepatitis C/HIV coinfection. Hepatology. 2012;56(5):1688–98.
81. Bedimo R, Cutrell J, Zhang S, et al. Mechanisms of bone disease in HIV and hepatitis C virus: impact of bone turnover, tenofovir exposure, sex steroids and severity of liver disease. AIDS. 2016;30(4):601–8.
82. Walker Harris V, Sutcliffe CG, Araujo AB, et al. Hip bone geometry in HIV/HCV-co-infected men and healthy controls. Osteoporos Int. 2012;23(6):1779–87.
83. Lo Re V 3rd, Lynn K, Stumm ER, et al. Structural bone deficits in HIV/HCV-coinfected, HCV-monoinfected, and HIV-monoinfected women. J Infect Dis. 2015;212(6):924–33.
84. Loria I, Albanese C, Giusto M, et al. Bone disorders in patients with chronic liver disease awaiting liver transplantation. Transplant Proc. 2010;42(4):1191–3.
85. Carey EJ, Balan V, Kremers WK, et al. Osteopenia and osteoporosis in patients with end-stage liver disease caused by hepatitis C and alcoholic liver disease: not just a cholestatic problem. Liver Transpl. 2003;9(11):1166–73.
86. Chen CC, Wang SS, Jeng FS, et al. Metabolic bone disease of liver cirrhosis: is it parallel to the clinical severity of cirrhosis? J Gastroenterol Hepatol. 1996;11(5):417–21.
87. Schiefke I, Fach A, Wiedmann M, et al. Reduced bone mineral density and altered bone turnover markers in patients with non-cirrhotic chronic hepatitis B or C infection. World J Gastroenterol. 2005;11(12):1843–7.
88. Kovacs A, Karim R, Mack WJ, et al. Activation of CD8 T cells predicts progression of HIV infection in women coinfected with hepatitis C virus. J Infect Dis. 2010;201(6):823–34.
89. Kovacs A, Al-Harthi L, Christensen S, et al. CD8(+) T cell activation in women coinfected with human immunodeficiency virus type 1 and hepatitis C virus. J Infect Dis. 2008;197(10):1402–7.
90. Balagopal A, Philp FH, Astemborski J, et al. Human immunodeficiency virus-related microbial translocation and progression of hepatitis C. Gastroenterology. 2008;135(1):226–33.
91. Redondo-Cerezo E, Casado-Caballero F, Martin-Rodriguez JL, et al. Bone mineral density and bone turnover in non-cirrhotic patients with chronic hepatitis C and sustained virological response to antiviral therapy with peginterferon-alfa and ribavirin. Osteoporos Int. 2014;25(6):1709–15.
92. Arase Y, Suzuki F, Suzuki Y, et al. Virus clearance reduces bone fracture in postmenopausal women with osteoporosis and chronic liver disease caused by hepatitis C virus. J Med Virol. 2010;82(3):390–5.
93. Orwoll ES, Klein RF. Osteoporosis in men. Endocr Rev. 1995;16(1):87–116.

61

Effects on the Skeleton from Medications Used to Treat Nonskeletal Disorders

Nelson B. Watts

Mercy Health, Osteoporosis and Bone Health Services, Cincinnati, OH, USA

INTRODUCTION

Iatrogenic bone loss caused by therapies for nonskeletal diseases is a growing and important cause of osteoporosis. The skeletal effects of glucocorticoids, progestins, excess thyroid hormone, chemotherapy, and calcineurin inhibitors is described in other sections of the this book. Medications covered in this chapter include aromatase inhibitors, androgen deprivation therapy, thiazolidenediones, canagliflozin, proton pump inhibitors, and heparin—medications that have adverse effects on bone and fracture risk (Table 61.1)—as well as thiazide diuretics and β-adrenergic blocking agents that have beneficial effects (Table 61.2).

DRUGS ASSOCIATED WITH BONE LOSS AND FRACTURE RISK

Antihormonal drugs

Androgen deprivation therapy

Reducing levels of testosterone is beneficial in many cases of prostate cancer. Androgen deprivation therapy (ADT) can be accomplished with surgery or medications including gonadotropin-releasing hormone (GnRH) analog therapy and antiandrogenic agents, including cyproterone acetate, flutamide, and bicalutamide. ADT is associated with increased bone loss [1, 2] and fractures [2, 3].

Men receiving long-term ADT should undergo baseline assessment of fracture risk using BMD measurement and clinical risk factors, with repeat BMD measurement at 1 to 2 years as clinically indicated. Bone protective therapy should be started in men with a history of hip or vertebral fracture and/or a T-score ≤ –2.5. Additionally, fracture-reducing therapy has been recommended in men with a low T-score (–1.0 to –2.5) and a 10-year risk of ≥3% for hip fracture or ≥20% for major osteoporotic fracture as assessed by the Fracture Risk Assessment Tool (FRAX) [4].

A number of interventions have been shown to have beneficial effects on BMD in men treated with ADT. These include raloxifene, toremifene, risedronate, pamidronate, zoledronic acid, alendronate, and denosumab [5]. A reduction in vertebral fracture risk has been shown for the use of toremifene and denosumab [6, 7].

Aromatase inhibitors

Aromatase inhibitors (AIs) block peripheral conversion of androgens to estrogen, reducing endogenous estrogen by 80% to 90%. They have largely replaced the selective estrogen receptor modulator, tamoxifen, as the preferred treatment to reduce the risk of recurrence in women with early-stage estrogen receptor-positive breast cancer. The most commonly used AIs are exemestane, anastrazole, and letrozole [8].

In contrast to tamoxifen, which is bone-protective in postmenopausal women (but has the opposite effect in premenopausal women), AIs increase rates of bone loss [9] and fracture risk [10]. Interpretation of studies of the effects of AIs on bone is complicated by the use of tamoxifen as a comparator in many studies and also, in some, the use of tamoxifen before AI therapy. In addition, there are no comparative data on the effects of different AIs on BMD and fracture rates. Nevertheless, the existing biomarker data indicate that all AIs increase bone turnover.

Prevention of bone loss associated with AI therapy has been demonstrated with intravenous zoledronic acid [11],

---

Primer on the Metabolic Bone Diseases and Disorders of Mineral Metabolism, Ninth Edition. Edited by John P. Bilezikian.
© 2019 American Society for Bone and Mineral Research. Published 2019 by John Wiley & Sons, Inc.
Companion website: www.wiley.com/go/asbmrprimer

Table 61.1. Medications associated with bone loss and/or increased fracture risk.

Glucocorticoids
Thyroxine in supraphysiological doses
Calcineurin inhibitors
Medications that reduce sex steroids
    Androgen deprivation therapy
    Aromatase inhibitors
Antidiabetic agents
    Canagliflozin
    Thiazolidenediones
Acid-suppressing medications
    H_2 receptor blockers
    Proton pump inhibitors
Antiepileptic drugs (often used off-label for other conditions)
Selective serotonin receptor reuptake inhibitors
Heparin

Table 61.2. Medications that may have bone benefits.

Thiazide and other proximally-acting diuretics
Beta-adrenergic blockers
Statins

oral risedronate [12], and denosumab [13, 14], although data on fracture reduction are lacking.

It seems reasonable to advise risk assessment, including BMD measurements, in all postmenopausal women treated with AIs and to perform repeat BMD measurements at 1- to 2-year intervals in those at moderate risk based on age, BMD, and other clinical risk factors. In the absence of data on the antifracture efficacy of bone protective therapy, the indications for intervention are not clearly defined [15].

Antidiabetic agents

Canagliflozin

Canagliflozin is one of several inhibitors of sodium glucose cotransporter 2 (SGLT2), lowering blood glucose levels by promoting urinary glucose excretion. Prospective studies showed small but statistically significant BMD loss in the total hip (but not at other sites) [16] and an increase in fractures, primarily at peripheral sites [17]. The incidence of fractures was small but the increase was seen as early as 6 weeks, too soon to be explained by changes in BMD, suggesting that nonskeletal factors, such as falling, may be responsible. More research is needed on canagliflozin and other SGLT2 inhibitors to determine the significance of these findings and whether or not this is a class effect.

Thiazolidinediones

Thiazolidinediones (TZDs) are ligands for PPARγ used to treat type 2 diabetes. Activation of PPARγ increases marrow adiposity, increases insulin sensitivity, and suppresses bone formation [18]. The mechanisms by which suppression of bone formation occurs have not been fully established but may include inhibition of the Wnt/ß-catenin signaling pathway, inhibition of osteoblast differentiation genes including *Runx2* and *osterix*, and suppression of insulin-like growth factor production [19].

The effects of TZDs in humans are particularly relevant in view of the increased risk of fracture associated with type 2 diabetes. In observational studies and clinical trials, TZDs have been shown to increase rates of bone loss [20] and fracture risk [21]. No studies have been done to see if any of the current bone active agents will protect against the negative skeletal effects of TZDs; therefore, it seems prudent to avoid TZD use in patients at high risk of fractures.

Acid-suppressive mediations

Increased risk of fracture has been reported in individuals treated with acid-suppressive medications including H_2 receptor blockers [22], but primarily with proton pump inhibitors (PPIs) [23, 24]. Overall, current data support an association between acid-suppressive medication and fracture, although the limitations of observational studies, particularly the effects of potential but unmeasured confounding factors, have to be recognized.

Inhibition of the osteoclastic proton pump would be expected to have beneficial skeletal effects, so the increased fracture risk is counterintuitive. Whether PPIs have a causal effect or an unrelated association is not clear. PPIs do not affect calcium absorption [25] or cause increased rates of bone loss [26]. Recent data suggest that PPI users are at increased risk of falling [27, 28] which could be due, at least in part, to the underlying disorders where PPIs are prescribed, rather than a direct effect of PPI use.

Given the uncertainty of whether PPIs directly increase fracture risk (and lack of a clear countermeasure), it is reasonable to try to avoid their use in patients at high risk of fracture, understanding that this may not always be possible. It is unknown whether current treatments for osteoporosis will reduce fracture risk caused by PPI use.

Antiepileptic drugs

An association between antiepileptic drugs (AEDs), increased rates of bone loss, and fracture risk has been reported [29, 30]. The underlying pathogenesis is unclear; vitamin D deficiency, trauma during seizures, increased risk of falling, and comedications including glucocorticoids may all contribute. In a few patients with severe vitamin D deficiency, osteomalacia or rickets may be present. Currently, there are insufficient data to distinguish between the skeletal effects of specific AED regimens.

Management guidelines to prevent and treat bone disease in AED users have been proposed, although at present these lack a robust evidence base. Routine prophylaxis of vitamin D deficiency should be considered in high-risk individuals (eg, elderly or institutionalized); higher than normal doses of vitamin D may be required in patients taking some AEDs, and in such cases, calcium supplements should also be considered. Routine bone densitometry in all AED users cannot be justified at present, although BMD should be measured in those who present with fracture or have other clinical risk factors. Treatment of established osteoporosis in this population has not been specifically evaluated.

Selective serotonin receptor uptake inhibitors

Selective serotonin receptor uptake inhibitors (SSRIs) are widely prescribed as antidepressants as well as used for other indications. Depression itself has been implicated as a risk factor for fracture, but SSRIs appear to increase fracture risk further [31]; SSRIs have also been show to increase fracture risk in postmenopausal women who are not depressed [32]. SSRI use has been associated with increased rates of bone loss in older women [33] and low bone mass in adolescents [34] and in men [35].

The negative skeletal effects of SSRIs appear due at least in part to actions in the Wnt signaling pathway [36]. As with PPIs, there do not appear to be any effective countermeasures against the increased fracture risk of SSRIs; therefore, it is reasonable to try to avoid their use in patients at high risk of fracture, understanding that this may not always be possible.

Heparin

Long-term heparin therapy, which is used as prophylaxis against thromboembolism in high-risk women during pregnancy, is associated with an increased risk of reduced BMD, increased rates of bone loss, and increased fracture risk [37–39], although the mechanisms responsible for bone loss have not been established. The use of low molecular weight heparin and of newer antithrombotic agents such as fondaparinux may be associated with fewer adverse skeletal effects, but osteoporosis and fractures have been reported with enoxaparin use [40]. Calcium and vitamin D supplements are often advocated but, in common with other antiresorptive regimens, have not been formally evaluated in this situation.

DRUGS THAT MAY PROTECT AGAINST OSTEOPOROSIS

Beta-adrenergic blockers

A protective effect of ß-adrenergic blocker therapy on fracture risk has been reported [41–43]. This is a side benefit worth noting; however, using ß-adrenergic blocker therapy to reduce fracture risk is not advisable.

Thiazide diuretics

Users of thiazide diuretics appear to have a reduced risk of hip fractures [44], and prospective studies have shown improvement in BMD [45, 46]. Increased renal calcium reabsorption is thought to play a role. Thiazide diuretics may be useful in the management of osteoporosis in patients with hypercalciuria.

Statins

Statins inhibit the enzyme 3-hydroxy-3-methyl-glutaryl-coenzyme A reductase in the mevalonate pathway, thus reducing cholesterol biosynthesis but also preventing the prenylation of guanosine triphosphate (GTP) binding proteins, thus inhibiting osteoclast activity. Beneficial skeletal effects of statins in animals have been shown in vitro and in vivo [47], but studies in humans have produced conflicting results. A meta-analysis of the effect of statins on fracture showed no benefit [48]. A meta-analysis of the effect of statins on BMD showed a small but statistically significant benefit [49], but a double-blind randomized placebo-controlled trial using clinically relevant doses of atorvastatin showed no effect on BMD or biochemical indices of bone metabolism [50].

REFERENCES

1. Greenspan SL, Coates P, Sereika SM, et al. Bone loss after initiation of androgen deprivation therapy in patients with prostate cancer. J Clin Endocrinol Metab. 2005;90(12): 6410–17.
2. Morgans AK, Fan KH, Koyama T, et al. Bone complications among prostate cancers survivors: long-term follow-up from the prostate cancer outcomes study. Prostate Cancer Prostatic Dis. 2014;17:338–42.
3. Wang A, Obertova Z, Brown C, et al. Risk of fracture in men with prostate cancer on androgen deprivation therapy: a population-based study in New Zealand. BMJ Cancer. 2015;15:837.
4. Saylor PJ, Smith MR. Adverse effects of androgen deprivation therapy: defining the problem and promoting health among men with prostate cancer. J Natl Compr Cancer Network. 2010;8:211–23.

5. Garg A, Leitzel K, Ali S, et al. Antiresorptive therapy in the management of cancer treatment-induced bone loss. Curr Osteoporos Rep. 2015;13:73–7.
6. Smith MR, Morton RA, Barnette KG, et al. Toremifene to reduce fracture risk in men receiving androgen deprivation therapy for prostate cancer. J Urol. 2010;184:1316–21.
7. Smith MR, Egerdie B, Hernandez Toriz N, et al. Denosumab in men receiving androgen-deprivation therapy for prostate cancer. N Engl J Med. 2009;361:1745–55.
8. McCloskey E. Effects of third-generation aromatase inhibitors on bone. Eur J Cancer. 2006;42:1044–51.
9. Chebowski RT, Haque R, Hedlin H, et al. Benefit/risk for adjuvant breast cancer therapy with tamoxifen or aromatase inhibitor use by age and race/ethnicity. Breast Cancer Res Treat. 2015;154:609–16.
10. Bouvard B, Soulie P, Hoppe E, et al. Fracture incidence after 3 years of aromatase inhibitor therapy. Ann Oncol. 2014;25:843–7.
11. Majithia N, Atherton PJ, Lafky JM, et al. Zoledronic acid for treatment of osteopenia and osteoporosis in women with primary breast cancer undergoing adjuvant aromatase inhibitor therapy: a 5-year follow-up. Support Care Cancer. 2016;24:1219–26.
12. Greenspan SL, Vujevich KT, Brufsky A, et al. Prevention of bone loss with risedronate in breast cancer survisors: a randomized, controlled clinical trial. Osteoporos Int. 2015;26:1857–64.
13. Ellis GK, Bone HG, Chebowski R, et al. Effect of denosumab on bone mineral density in women receiving adjuvant aromatase inhibitors for non-metastatic breast cancer: subgroup analysis of a Phase 3 study. Breast Cancer Res Treat. 2009;118:81–7.
14. Gnant M, Pfeiler G, Dubsky PC, et al. Adjuvant denosumab in breast cancer (ABSCG-18): a multicentre, randomised, double-blind, placebo-controlled trial. Lancet. 2015;386:433–43.
15. Coleman RE, Rathbone E, Brown JE. Management of cancer treatment-induced bone loss. Nature Rev Rheumatol. 2013;9:365–74.
16. Bilezikian JP, Watts NB, Usiskin K, et al. Evaluation of bone mineral density and bone biomarkers in patients with type 2 diabetes treated with canagliflozin. J Clin Endocrinol Metab. 2016;101:44–51.
17. Watts NB, Bilezikian JP, Usiskin K, et al. Effects of canagliflozin on fracture risk in patients with type 2 diabetes mellitus. J Clin Endocrinol Metab. 2016;101:157–66.
18. Ali AA, Weinstein RS, Stewart SA, et al. Rosiglitazone causes bone loss in mice by suppressing osteoblast differentiation and bone formation. Endocrinology. 2005;146:1226–35.
19. Lecka-Czernik B, Acker-Bicknell C, Adamo ML, et al. Activation of peroxisome proliferator-activated receptor gamma (PPAR gamma) by rosiglitzaone suppresses components of the insulin-like growth factor regulatory system in vitro and in vivo. Endocrinology. 2007;148:903–11.
20. Grey A, Bolland M, Gamble G, et al. The peroxisome-proliferator-activated receptor gamma agonist rosiglitazone decreased bone formation and bone mineral density in healthy postmenopausal women: a randomized controlled trial. J Clin Endocrinol Metab. 2007;92:1305–10.
21. Loke YK, Singh S, Furberg CD. Long-term use of thiazolidinediones and fractures in type 2 diabetes: a meta-analysis. CMAJ. 2009;180:32–9.
22. Grisso JA, Kelscy JL, O'Brien LA, et al. Risk factors for hip fracture in men. Am J Epidemiol. 1997;145:786–93.
23. Vestergaard P, Rejnmark L, Mosekilde L. Proton pump inhibitors, histamine H2 receptor antagonists and other antacid medications and the risk of fracture. Calcif Tissue Res. 2006;79:76–83.
24. van der Hoorn MM, Tett SE, de Vries OJ, et al. The effect of dose and type of proton pump inhibitor use on the risk of fractures and osteoporosis treatment in Australian women: a prospective cohort study. Bone. 2015;675:682.
25. O'Connell MB, Madden DM, Murray AM, et al. Effects of proton pump inhibitors on calcium carbonate absorption in women: a randomized crossover trial. Am J Med. 2005;120:778–81.
26. Solomon DH, Diem SJ, Ruppert K, et al. Bone mineral density changes among women initiating proton pump inhibitors or H2 receptor antagonists: a SWAN cohort study. J Bone Miner Res. 2015;30:232–9.
27. Lewis JR, Barre D, Zhu K, et al. Long-term proton pump inhibitor therapy and falls and fractures in elderly women: a prospective cohort study. J Bone Miner Res. 2014;39:2489–97.
28. Thaler HW, Sterke CS, van der Cammen TJ. Association of proton pump inhibitor use with recurrent falls and risk of fractures in older women: a study of medication use in older fallers. J Nutr. 2016;20:77–81.
29. Lee RH, Lyles KW, Colon-Emeric CS. A review of the effect of anticonvulsant medications on bone mineral density and fracture risk. Am J Geriatr Pharmacother. 2010;8:34–46.
30. Beerhorst K, van der Kruijs SJ, Verschuure P, et al. Bone disease during chronic antiepileptic drug therapy: general versus specific risk factors. J Neurol Sci. 2015;331:19–25.
31. Lateigne A, Sheu YH, Sturmer T, et al. Serotonin-norepinephrine reuptake inhibitor and selective serotonin reuptake inhibitor use and risk of fractures: a new-user cohort study among US adults aged 50 years and older. CNS Drugs. 2015;29:245–52.
32. Sheu YH, Lateigne A, Sturmer T, et al. SSRI use and risk of fractures among postmenopausal women without mental disorders. Injury Prev. 2015;21:397–403.
33. Diem SJ, Blackwell TL, Stone KL, et al. Use of antidepressants and rates of hip bone loss in older women: the Study of Osteoporotic Fractures. Arch Intern Med. 2007;167:1240–5.
34. Feuer AJ, Demmer RT, Thai A, et al. Use of selective serotonin reuptake inhibitors and bone mass in adolescents: an NHANES study. Bone. 2015;78:28–33.
35. Haney EM, Chan BK, Diem SJ, et al. Association of low bone mineral density with selective serotonin reuptake inhibitors in older men. Arch Intern Med. 2007;167:1256–1.
36. Warden SJ, Robling AG, Haney EM, et al. The emerging role of serotonin (5-hydroxytryptamine) in the skeleton

and its mediation of the skeletal effects of low-density lipoprotein receptor-related protein 5 (LRP5). Bone. 2010;46:4–12.
37. deSweit M, Ward P, Fidler A, et al. Prolonged heparin therapy in pregnancy causes bone demineralisation. Br J Obstet Gynaecol. 1983;90:1129–34.
38. Dalhman T. Osteoporotic fractures and the recurrence of thromboembolism during pregnancy and the puerperium in 184 women undergoing thromboprophylaxis with heparin. Am J Obstet Gynecol. 1993;168:1265–70.
39. Barbour LA, Kick S, Steiner J, et al. A prospective study of heparin-induced osteoporosis in pregnancy using bone densitometry. Am J Obstet Gynecol. 1994;170:862–9.
40. Ozdemir D, Tam AA, Dirikoc A, et al. Postpartum osteoporosis and vertebral fractures in two patients treated with enoxaparin during pregnancy. Osteoporos Int. 2015;26:415–18.
41. Bonnet N, Gadois C, McCloskey E, et al. Protective effect of beta blockers in postmenopausal women: influence on fractures, bone density, micro and macroarchitecture. Bone. 2007;40:1209–16.
42. Solomon DH, Mogun H, Garneau K, et al. The risk of fractures in older adults using antihypertensive medications. J Bone Miner Res. 2011;26:1561–7.
43. Yang S, Nguyen ND, Center JR, et al. Association between beta blocker use and fracture risk: the Dubbo Osteoporosis Epidemiology Study. Bone. 2011;48: 451–5.
44. LaCroix AZ, Weinpahl J, White LR, et al. Thiazide diuretic agents and the incidence of hip fracture. N Engl J Med. 1990;322:286–90.
45. LaCroix AZ, Ott SM, Ichikawa L, et al. Low dose hydrochlorothiazide and preservation of bone mineral density in older adults: a randomized double-blind placebo-controlled trial. Ann Intern Med. 2000;133:516–26.
46. Bolland MJ, Ames RW, Horne AM, et al. The effect of treatment with a thiazide diuretic for 4 years on bone mineral density in normal postmenopausal women. Osteoporos Int. 2007;18:479–86.
47. Mundy G, Garret R, Harris S, et al. Stimulation of bone formation in rodents by statins. Science. 1999;286: 1946–9.
48. Nguyen D, Wang CY, Eisman JA, et al. On the association between statins and fractures: a Bayesian consideration. Bone. 2007;40:813–20.
49. Uzzan B, Cohen RM, Nicolas P, et al. Effect of statins on bone mineral density: a meta-analysis of clinical studies. Bone. 2007;40:1581–7.
50. Bone HG, Kiel DP, Lindsay RL, et al. Effects of atorvastatin on bone in postmenopausal woman with dyslipidemia: a double-blind, placebo-controlled, dose-ranging trial. J Clin Endocrinol Metab. 2007;92:4671–7.

62

Diabetes and Fracture Risk

Serge Ferrari[1], Nicola Napoli[2], and Ann Schwartz[3]

[1]Faculty of Medicine–Medicine Specialties, Geneva University Hospitals, Geneva, Switzerland
[2]Campus Bio-Medico, University of Rome, Rome, Italy
[3]Department of Epidemiology and Biostatistics, University of California, San Francisco, CA, USA

INTRODUCTION

Diabetes is a group of metabolic diseases characterized by hyperglycemia resulting from defects in insulin secretion, insulin action, or both. Fragility fractures are increasingly recognized as another complication of diabetes. This chapter discusses the pathophysiology, epidemiology, clinical assessment, and pharmacology of diabetes-associated bone disease.

PATHOPHYSIOLOGY OF BONE FRAGILITY

The pathophysiology of bone fragility in diabetes is complex, with some alterations being common to type 1 and 2 diabetes, while other are distinct, pertaining to the different ages of onset and underlying mechanisms of diabetes development [1, 2].

Type 1 diabetes (T1D) is an autoimmune disease, also called juvenile diabetes, characterized by β-cell destruction and absolute insulin deficiency. Insulin stimulates osteoblast proliferation by enhancing Runx2 activity, promotes collagen synthesis, and increases glucose uptake [3]. Streptozotocin-induced diabetic mice and nonobese diabetic (NOD) mice, the most common animal models used for T1D, have shown that insulin deficiency is associated with low bone turnover, low trabecular and cortical BMD, and decreased bone strength [4]. Moreover T1D is associated with low IGF-1 levels [5, 6], a major stimulus of osteoblastic functions. Added to that, the proinflammatory state that is characteristic of this disease, the altered nutritional status, particularly a poor calcium and/or protein intake, and decreased levels of physical activity in children and adolescents who develop T1D may all contribute to the development of low peak bone mass [7]. Hence in children with diabetes, osteocalcin and P1NP levels are inversely proportional to the level of chronic hyperglycemia, as assessed by glycated hemoglobin (HbA$_{1C}$) levels [7]. Accordingly, deficits in both trabecular and cortical bone volume have been documented by HRpQCT in young adults with T1D, in particular those with severe disease characterized by microvascular complications [8]. In these patients, there is also evidence of bone marrow fat infiltration using micro-MRI [6]. Note, however, that a bone biopsy study in young adults with relatively well-controlled T1D failed to document decreased bone-forming indices compared with controls [9], but showed increased levels of non-enzymatic collagen cross-linking by pentosidine and other AGEs [10], which may alter bone material properties.

Type 2 diabetes (T2D) is generally adult-onset and the most common form of diabetes. T2D patients are usually obese and insulin-resistant with normal or high BMD (favored by increased load, raised estrogen levels, and hypersinulinemia). However, β-cell function declines and glucose control deteriorates with time, determining a state of chronic hyperglycemia causing organ damage and increasing risk of complications, including bone impairment. Similarly to type 1, T2D has been associated with low bone turnover, assessed by both by biochemical markers and on bone biopsies [11], as well as trabecular and cortical deficits, with notably increased cortical porosity, particularly among subjects with fragility fractures and microvascular complications [12]. Moreover, in vivo microindentation testing of the tibia has shown decreasd bone mineral strength (BMS) in 60 postmenopausal women including 30 patients diagnosed with T2D for >10 years [13]. As in type 1, the latter may reflect the reduction in enzymatic collagen cross-links and the accumulation of AGEs in bone. Another potential mechanism by which hyperglycemia and AGEs may negatively impact on bone strength is by altering osteoblast and osteocyte functions [2]. Hence both in vitro and in vivo studies have indicated increased sclerostin levels in this condition [14, 15]. In addition, a fine balance exists between adipogenesis and osteoblastogenesis, which are mainly regulated by Wnt signaling and the PPARγ pathways. Mouse models of diabetes display increased PPARγ2 in bone tissue, reduced bone formation, and increased marrow adiposity [16]. Overweight postmenopausal women with T2D present an inverse correlation between marrow adipose tissue and BMD [17].

Eventually, diabetic patients present a loss of incretin effects. The incretin effect depends primarily on two peptides, glucose-dependent insulinotropic polypeptide (GIP) and glucagon-like peptide 1 (GLP-1). GLP-1 receptors are present on bone marrow stromal cells and osteoblasts [18], and GLP-1 inhibits mesenchymal stem cell differentiation into adipocytes. GLP-1 receptor knockout mice have decreased cortical bone mass due to increased osteoclast number and activity and impaired mechanical and material properties. Conversely, in animal models, administration of GLP-1 increases bone formation both in control and in streptozocin-induced diabetes or fructose-induced insulin-resistant rats, suggesting an insulin-independent action. In another study, GLP-1 administration in a T2D animal model lowered sclerostin serum levels and increased osteocalin levels [19].

EPIDEMIOLOGY OF FRACTURES

Study of the epidemiology of fracture risk in diabetes has developed with several large case-control studies from Denmark and the Women's Health Initiative (WHI) indicating that adults with diabetes have a 0.5- to twofold increased risk of osteoporotic fractures [20, 21]. In T1D this risk is actually larger and present throughout life in both men and women [22], but with an exponential increase in hip fracture risk that appears after 40 to 50 years of age, ie, 15 years earlier than in the nondiabetic population. A recent meta-analysis indicated an up to fivefold increased risk of hip fractures in T1D [23]. In T2D the risk is lower, but increases up to twofold with duration of the disease and insulin use [24, 25], probably as a marker of disease duration and severity and also because of the associated increased risk of falls due to proprioceptive problems and hypoglycemic episodes, and poor glycemic control (ie, higher HbA_{1C} levels) [26].

EVALUATION OF FRACTURE RISK

In T1D, the lower areal bone mineral density (aBMD) and/or the possibility to adjust Fracture Risk Assessment Tool (FRAX) estimates of fracture probability for the presence of diabetes (as a secondary cause) means fracture risk is similar to that of the nondiabetic population [1]. In contrast, in T2D, both aBMD, which on average is 5% to 10% higher than in the normal population [21, 27], and FRAX, which has currently not been calibrated for T2D, underestimate by 30% to 50% the actual fracture risk [28, 29]. Hence the 10-year incidence of nonspine (particularly hip) fractures appears to be equivalent at −2 T-scores among T2D to the fracture incidence at −2.5 T-scores among nondiabetics [28]. In this context, evaluation of the spine trabecular bone score—which is slightly lower among diabetics compared with controls—appears to be a weaker (than aBMD) but independent risk factor for fractures [30].

SKELETAL EFFECTS OF GLUCOSE-LOWERING MEDICATIONS

Thiazolidinediones

Activation of PPARγ by thiazolidinediones (TZDs) improves insulin sensitivity but also has negative effects on bone. TZDs reduce bone formation through shifts in mesenchymal stem cells which favors adipocyte over osteoblast differentiation and increases osteoclastogenesis through the recruitment of osteoclasts from hematopoietic stem cells and increased RANKL production in mesenchymal cells [31]. TZDs may also increase osteocyte apoptosis through a distinct pathway dependent on G-protein coupled receptor 40 [32]. Awareness of the potential effects of glucose-lowering medications on skeletal health is relatively new, resulting from the discovery that TZD use increases fracture risk. Publication of the ADOPT trial results in 2006 provided the first clinical evidence of increased fracture risk [33], confirming concerns from animal [34] and clinical [35] studies that reported increased bone loss with TZD use. A meta-analysis of 22 randomized clinical trials found a doubling of fracture risk with TZD use in women but no increased risk in men [36]. However, observational studies in older age groups indicate that risk in men may be elevated [37].

A meta-analysis of 18 randomized trials with a median duration of 48 weeks found more rapid bone loss at the total hip and spine of ~1% with TZD use [38]. It is not known if excess bone loss continues at a similar level with longer term TZD use. Interestingly, the same meta-analysis could not identify a consistent effect of TZD use on bone turnover markers. Limited evidence suggests that more rapid bone loss does not continue after TZD withdrawal but the initial excess bone loss with TZD use is not regained [38]. In an observational study, fracture risk in women returned to baseline levels within 2 years after TZD withdrawal [39].

Medical care guidelines for T2D patients recommend avoidance of TZD therapy in those with higher fracture risk [40]. Outstanding questions remain regarding the effects of longer term TZD use on bone and fracture risk.

Incretin-based therapies

GLP-1, an incretin gut hormone, is secreted in response to a meal, stimulates insulin secretion by the pancreas, and is rapidly degraded by dipeptidyl peptidase 4 (DPP-4) [31]. GLP-1 receptor agonists and DPP-4 inhibitors, targeting this pathway, have been successfully developed as therapies for T2D. Gut hormones, primarily GLP-2 and GIP, also affect bone metabolism, lowering bone resorption in response to feeding. Thus, it is possible that incretin-based therapies might have positive effects on bone. An initial meta-analysis of randomized clinical trials ($n = 28$) of DPP-4 inhibitors reported a protective effect on fractures reported as adverse events [41]. However, a more recent, larger meta-analysis ($n = 62$) that included 722 fractures found no difference in fracture risk, comparing DPP-4 inhibitors with controls [42]. Evidence regarding the fracture effects of GLP-1 receptor agonists is more limited. The largest published meta-analysis (14 trials and 38 fracture events) found no increased risk for GLP-1 receptor agonists as a class [43]. When analyzed separately, exenatide was associated with increased risk while liraglutide was associated with lower risk. However, there is no clear biological basis for this difference, and results from a large observational study indicate that liraglutide, exenatide, and GLP-1 receptor agonists as a class were not associated with fracture risk [44]. Clarification of the fracture effects of the GLP-1 receptor agonists will require larger studies, particularly additional randomized clinical trials that monitor fracture events.

Sodium glucose cotransporter-2 inhibitors

Sodium glucose cotransporter 2 (SGLT-2) inhibitors prevent renal reabsorption of glucose, leading to increased urinary glucose excretion and reduced blood glucose [45]. Several therapies based on this mechanism of glucose-lowering are currently available, including cangliflozin, dapagliflozin, and empagliflozin. Empagliflozin also reduces cardiovascular events in high-risk diabetes patients [46]. This class of medications is unlikely to have a direct effect on bone as SGLT-2 is not present in bone, but indirect effects are possible through alterations in calcium and phosphate homeostasis and secondary hyperparathyroidism. SGLT-2 inhibitors also cause weight loss, associated with bone loss and fracture, and result in more frequent episodes of volume-depletion adverse events, possibly contributing to fall risk. In a randomized trial, canagliflozin use for 2 years was associated with greater bone loss at the total hip (−1.2%), compared with placebo [47].

A well-designed meta-analysis of nine randomized trials that included adjudication of fractures ($n = 246$) reported as adverse events found increased fracture risk with canagliflozin use (hazard ratio 1.32; 1.00, 1.74) [48]. The increased risk was driven by interim results from the CANVAS trial (151 fractures), conducted among patients at higher risk for cardiovascular disease. In contrast, for empagliflozin there was no difference in fracture risk compared with placebo based on the EmpaReg trial with median follow-up of 3.1 years [46].

Clinical guidelines currently recommend avoidance of SGLT-2 inhibitors in T2D patients with a higher risk of fracture [40]. Further studies are needed to clarify the effects of SGLT-2 inhibitors on fracture risk as a class and individually.

Insulin, metformin, and sulfonylureas

In observational studies, insulin therapy has been identified as a risk factor for fracture [31]. Insulin is unlikely to have direct negative effects on bone strength, but use of insulin is associated with more frequent hypoglycemia and falls, and patients using insulin are more likely to have other comorbidities that increase fracture risk. In a randomized comparison of metformin and sulfonylurea in the ADOPT trial, fracture rates based on adverse event reports did not differ across these two therapies [49].

EFFICACY AND SAFETY OF OSTEOPOROSIS THERAPIES IN DIABETIC PATIENTS

Currently, fracture prevention approaches for older adults with diabetes are the same as those for the general population, including consideration of pharmacological therapy for those at higher risk. Diabetes is characterized by lower bone turnover but increased fracture risk, leading to concerns that suppression of bone turnover with antiresorptive therapy may not be effective for fracture prevention in patients with diabetes [50]. Available evidence is limited but suggests similar efficacy of osteoporosis therapies in T2D patients [51–54]. Post hoc analyses of two randomized trials of raloxifene found a reduction in vertebral fractures in diabetic as well as nondiabetic women [53, 54]. Alendronate [51] and risedronate [52] improved bone density in women with diabetes, compared with placebo. Randomized trial results are not yet available for fracture efficacy of the bisphosphonates. Observational studies based in large cohorts suggest that bisphosphonate effects on fracture are similar for those with and without diabetes [55, 56]. For anabolic therapies, a small observational study found that teriparatide prevented bone loss and fracture in diabetic as well as nondiabetic patients [57].

Appreciation of the effects of the skeleton on energy metabolism led to concerns that reducing bone turnover through use of antiresorptive therapies might increase risk of diabetes [58]. However, a post hoc analysis of randomized trials of these therapies found no increased risk of progression to diabetes or increase in fasting glucose levels [59].

REFERENCES

1. Hough FS, Pierroz DD, Cooper C, et al. Mechanisms in endocrinology: mechanisms and evaluation of bone fragility in type 1 diabetes mellitus. Eur J Endocrinol. 2016;174:R127–38.
2. Napoli N, Chandran M, Pierroz DD, et al. Mechanisms of diabetes mellitus-induced bone fragility. Nat Rev Endocrinol. 2017;13(4):208–19.
3. Napoli N, Strollo R, Paladini A, et al. The alliance of mesenchymal stem cells, bone, and diabetes. Int J Endocrinol. 2014;2014:690783.
4. Botolin S, McCabe LR. Bone loss and increased bone adiposity in spontaneous and pharmacologically induced diabetic mice. Endocrinology. 2007;148:198–205.
5. Sorensen JS, Birkebaek NH, Bjerre M, et al. Residual beta-cell function and the insulin-like growth factor system in Danish children and adolescents with type 1 diabetes. J Clin Endocrinol Metab. 2015;100:1053–61.
6. Abdalrahaman N, McComb C, Foster JE, et al. Deficits in trabecular bone microarchitecture in young women with type 1 diabetes mellitus. J Bone Miner Res. 2015;30:1386–93.
7. Maggio AB, Ferrari S, Kraenzlin M, et al. Decreased bone turnover in children and adolescents with well controlled type 1 diabetes. J Pediatr Endocrinol Metab. 2010;23:697–707.
8. Shanbhogue VV, Hansen S, Frost M, et al. Bone geometry, volumetric density, microarchitecture, and estimated bone strength assessed by HR-pQCT in adult patients with type 1 diabetes mellitus. J Bone Miner Res. 2015;30:2188–99.
9. Armas LA, Akhter MP, Drincic A, et al. Trabecular bone histomorphometry in humans with type 1 diabetes mellitus. Bone. 2012;50:91–6.
10. Farlay D, Armas LA, Gineyts E, et al. Nonenzymatic glycation and degree of mineralization are higher in bone from fractured patients with type 1 diabetes mellitus. J Bone Miner Res. 2016;31:190–5.
11. Manavalan JS, Cremers S, Dempster DW, et al. Circulating osteogenic precursor cells in type 2 diabetes mellitus. J Clin Endocrinol Metab. 2012;97:3240–50.
12. Shanbhogue VV, Hansen S, Frost M, et al. Compromised cortical bone compartment in type 2 diabetes mellitus patients with microvascular disease. Eur J Endocrinol. 2016;174:115–24.
13. Farr JN, Drake MT, Amin S, et al. In vivo assessment of bone quality in postmenopausal women with type 2 diabetes. J Bone Miner Res. 2014;29:787–95.
14. Tanaka K, Yamaguchi T, Kanazawa I, et al. Effects of high glucose and advanced glycation end products on the expressions of sclerostin and RANKL as well as apoptosis in osteocyte-like MLO-Y4-A2 cells. Biochem Biophys Res Commun. 2015;461:193–9.
15. Ardawi MS, Akhbar DH, Alshaikh A, et al. Increased serum sclerostin and decreased serum IGF-1 are associated with vertebral fractures among postmenopausal women with type-2 diabetes. Bone. 2013;56:355–62.
16. Botolin S, Faugere MC, Malluche H, et al. Increased bone adiposity and peroxisomal proliferator-activated receptor-gamma2 expression in type I diabetic mice. Endocrinology. 2005;146:3622–31.
17. Baum T, Yap SP, Karampinos DC, et al. Does vertebral bone marrow fat content correlate with abdominal adipose tissue, lumbar spine bone mineral density, and blood biomarkers in women with type 2 diabetes mellitus? J Magn Reson Imaging. 2012;35:117–24.
18. Sanz C, Vazquez P, Blazquez C, et al. Signaling and biological effects of glucagon-like peptide 1 on the differentiation of mesenchymal stem cells from human bone marrow. Am J Physiol Endocrinol Metab. 2010;298: E634–43.
19. Kim JY, Lee SK, Jo KJ, et al. Exendin-4 increases bone mineral density in type 2 diabetic OLETF rats potentially through the down-regulation of SOST/sclerostin in osteocytes. Life Sci. 2013;92:533–40.
20. Vestergaard P, Rejnmark L, Mosekilde L. Relative fracture risk in patients with diabetes mellitus, and the impact of insulin and oral antidiabetic medication on relative fracture risk. Diabetologia. 2005;48:1292–9.
21. Bonds DE, Larson JC, Schwartz AV, et al. Risk of fracture in women with type 2 diabetes: the Women's Health Initiative Observational Study. J Clin Endocrinol Metab. 2006;91:3404–10.
22. Weber DR, Haynes K, Leonard MB, et al. Type 1 diabetes is associated with an increased risk of fracture across the life span: a population-based cohort study using The Health Improvement Network (THIN). Diabetes Care. 2015;38:1913–20.
23. Shah VN, Shah CS, Snell-Bergeon JK. Type 1 diabetes and risk of fracture: meta-analysis and review of the literature. Diabet Med.2015;32:1134–42.
24. Forsen L, Meyer HE, Midthjell K, et al. Diabetes mellitus and the incidence of hip fracture: results from the Nord-Trondelag Health Survey. Diabetologia. 1999;42:920–5.
25. Napoli N, Strotmeyer ES, Ensrud KE, et al. Fracture risk in diabetic elderly men: the MrOS study. Diabetologia. 2014;57:2057–65.
26. Li CI, Liu CS, Lin WY, et al. Glycated hemoglobin level and risk of hip fracture in older people with type 2 diabetes: a competing risk analysis of Taiwan Diabetes Cohort Study. J Bone Miner Res. 2015;30:1338–46.
27. de Liefde II, van der Klift M, de Laet CE, et al. Bone mineral density and fracture risk in type-2 diabetes mellitus: the Rotterdam Study. Osteoporos Int. 2005;16:1713–20.
28. Schwartz AV, Vittinghoff E, Bauer DC, et al. Association of BMD and FRAX score with risk of fracture in older adults with type 2 diabetes. JAMA. 2011;305:2184–92.
29. Giangregorio LM, Leslie WD, Lix LM, et al. FRAX underestimates fracture risk in patients with diabetes. J Bone Miner Res. 2012;27:301–8.
30. Leslie WD, Aubry-Rozier B, Lamy O, Hans D. TBS (trabecular bone score) and diabetes-related fracture risk. J Clin Endocrinol Metab. 2013;98:602–9.
31. Meier C, Schwartz AV, Egger A, Lecka-Czernik B. Effects of diabetes drugs on the skeleton. Bone. 2016;82:93–100.

32. Mieczkowska A, Basle MF, Chappard D, Mabilleau G. Thiazolidinediones induce osteocyte apoptosis by a GPR40-dependent mechanism. J Biol Chem. 2012;287(28):23517–26.
33. Kahn SE, Haffner SM, Heise MA, et al. Glycemic durability of rosiglitazone, metformin, or glyburide monotherapy. New Engl J Med. 2006;355:2427–43.
34. Rzonca SO, Suva LJ, Gaddy D, et al. Bone is a target for the antidiabetic compound rosiglitazone. Endocrinology. 2004;145:401–6.
35. Schwartz A, Sellmeyer D, Vittinghoff E, et al. Thazolidinedione (TZD) use and change in bone density in older diabetic adults. Diabetes. 2005;54:A41.
36. Zhu ZN, Jiang YF, Ding T. Risk of fracture with thiazolidinediones: an updated meta-analysis of randomized clinical trials. Bone. 2014;68:115–23.
37. Colhoun HM, Livingstone SJ, Looker HC, et al. Hospitalised hip fracture risk with rosiglitazone and pioglitazone use compared with other glucose-lowering drugs. Diabetologia. 2012;55:2929–37.
38. Billington EO, Grey A, Bolland MJ. The effect of thiazolidinediones on bone mineral density and bone turnover: systematic review and meta-analysis. Diabetologia. 2015;58:2238–46.
39. Schwartz AV, Chen H, Ambrosius WT, et al. Effects of TZD use and discontinuation on fracture rates in ACCORD Bone Study. J Clin Endocrinol Metab. 2015;100:4059–66.
40. American Diabetes Association. Standards of medical care in diabetes—2015. Diabetes Care. 2015;38(suppl 1).
41. Monami M, Dicembrini I, Antenore A, et al. Dipeptidyl peptidase-4 inhibitors and bone fractures: a meta-analysis of randomized clinical trials. Diabetes Care. 2011;34:2474–6.
42. Fu J, Zhu J, Hao Y, et al. Dipeptidyl peptidase-4 inhibitors and fracture risk: an updated meta-analysis of randomized clinical trials. Sci Rep. 2016;6:29104.
43. Su B, Sheng H, Zhang M, et al. Risk of bone fractures associated with glucagon-like peptide-1 receptor agonists' treatment: a meta-analysis of randomized controlled trials. Endocrine. 2015;48(1):107–15.
44. Driessen JH, Henry RM, van Onzenoort HA, et al. Bone fracture risk is not associated with the use of glucagon-like peptide-1 receptor agonists: a population-based cohort analysis. Calcif Tissue Int. 2015;97:104–12.
45. Egger A, Kraenzlin ME, Meier C. Effects of incretin-based therapies and SGLT2 inhibitors on skeletal health. Curr Osteoporos Rep 2016;14:345–50.
46. Zinman B, Wanner C, Lachin JM, et al. Empagliflozin, cardiovascular outcomes, and mortality in type 2 diabetes. New Engl J Med. 2015;373:2117–28.
47. Bilezikian JP, Watts NB, Usiskin K, et al. Evaluation of bone mineral density and bone biomarkers in patients with type 2 diabetes treated with canagliflozin. J Clin Endocrinol Metab. 2016;101:44–51.
48. Watts NB, Bilezikian JP, Usiskin K, et al. Effects of canagliflozin on fracture risk in patients with type 2 diabetes mellitus. J Clin Endocrinol Metab. 2016;101:157–66.
49. Kahn SE, Zinman B, Lachin JM, et al. Rosiglitazone associated fractures in type 2 diabetes: an analysis from ADOPT. Diabetes Care. 2008;31:845–51.
50. Schwartz AV. Efficacy of osteoporosis therapies in diabetic patients. Calcif Tissue Int. 2017;100(2):165–73.
51. Keegan TH, Schwartz AV, Bauer DC, et al. Effect of alendronate on bone mineral density and biochemical markers of bone turnover in type 2 diabetic women: the fracture intervention trial. Diabetes Care. 2004;27:1547–53.
52. Inoue D, Muraoka R, Okazaki R, et al. Efficacy and safety of risedronate in osteoporosis subjects with comorbid diabetes, hypertension, and/or dyslipidemia: a post hoc analysis of phase III trials conducted in Japan. Calcif Tissue Int. 2016;98:114–22.
53. Johnell O, Kanis JA, Black DM, et al. Associations between baseline risk factors and vetebral fracture risk in the Multiple Outcomes of Raloxifene Evaluation (MORE) study. J Bone Miner Res. 2004;19:764–72.
54. Ensrud KE, Stock JL, Barrett-Connor E, et al. Effects of raloxifene on fracture risk in postmenopausal women: the Raloxifene Use for the Heart Trial. J Bone Miner Res. 2008;23:112–20.
55. Vestergaard P, Rejnmark L, Mosekilde L. Are antiresorptive drugs effective against fractures in patients with diabetes? Calcif Tissue Int. 2011;88:209–14.
56. Abrahamsen B, Rubin KH, Eiken PA, et al. Characteristics of patients who suffer major osteoporotic fractures despite adhering to alendronate treatment: a National Prescription Registry Study. Osteoporos Int. 2013;24:321–8.
57. Schwartz AV, Pavo I, Alam J, et al. Teriparatide in patients with osteoporosis and type 2 diabetes. Bone. 2016;91:152–8.
58. Ferron M, Wei J, Yoshizawa T, et al. Insulin signaling in osteoblasts integrates bone remodeling and energy metabolism. Cell. 2010;142:296–308.
59. Schwartz AV, Schafer AL, Grey A, et al. Effects of antiresorptive therapies on glucose metabolism: results from the FIT, HORIZON-PFT, and FREEDOM trials. J Bone Miner Res. 2013;28:1348–54.

63
Obesity and Skeletal Health

Juliet Compston

Department of Medicine, Cambridge Biomedical Campus, University of Cambridge, Cambridge, UK

INTRODUCTION

In the past, obesity was believed to be associated with reduced fracture risk as a result of higher BMD and the protective effect of subcutaneous tissue on the impact of falls. However, recent studies have challenged this perception and have demonstrated that fractures in obese individuals make a substantial contribution to the overall fracture burden in the older population. In view of the rapidly rising prevalence of obesity in many parts of the world, this contribution is set to increase, emphasizing the need for a better understanding of the pathogenesis of fractures in obese individuals and the development of effective preventive strategies.

EPIDEMIOLOGY OF FRACTURES IN THE OBESE

Relationship between body mass index and fracture

The inverse relationship between BMI and fracture risk is well documented and is principally, although not exclusively, mediated through the positive association between BMI and BMD [1]. The steepest gradient of risk is observed at BMI values below $20\,kg/m^2$, and only small decreases in fracture risk are seen with BMI increases above $25\,kg/m^2$. However, the relationship between BMI and fracture risk is to some extent site-specific and the inverse relationship described above does not apply to all fractures.

Early evidence that obesity is not protective against fractures was provided in an audit of a Fracture Liaison Service in the United Kingdom, in which it was found that 28% of postmenopausal women presenting with an incident clinical fracture in a 2-year period were obese [2]. This was further explored in the Global Longitudinal study of Osteoporosis in Women (GLOW), a large, prospective, observational, multinational study of postmenopausal women. During 2 years of follow-up, fractures in obese women accounted for 23% and 22% of all prevalent and incident fractures, respectively [3]. Subsequently, other studies have confirmed the substantial contribution of fractures in obese individuals to the fracture burden in the older population. Thus, in women enrolled in the Study of Osteoporotic Fractures (SOF), 37.5% of obese women sustained a clinical nonvertebral fracture during a median of 11 years follow-up compared with 44% of nonobese women [4]. In the Million Women Study, 40% of hip fractures occurred in women who were obese or overweight, and in the Osteoporotic Fractures in Men Study (MrOS), 13% of hip fractures and 17% of nonvertebral fractures occurred in obese men [5, 6]. Failure to observe these associations in earlier studies may reflect the growing prevalence of obesity in recent years together with a decline in the proportion of underweight individuals in the population.

Site-specificity

Further analysis of these studies has clearly demonstrated that the relationship between BMI and fracture is site-specific, high BMI being associated with an increased risk of fractures at some sites whilst being protective at others. Thus increased risk of ankle, upper leg (excluding hip), lower leg, and humerus fractures has been reported in obese individuals in several studies, whilst protection by obesity against hip and wrist fracture has emerged as a consistent finding [3, 7–12]. Data on the relationship between BMI and spine fracture are somewhat conflicting, possibly reflecting differences in the definition of vertebral fracture and the inclusion, in some studies, only of clinical vertebral fractures [13, 14].

Primer on the Metabolic Bone Diseases and Disorders of Mineral Metabolism, Ninth Edition. Edited by John P. Bilezikian.
© 2019 American Society for Bone and Mineral Research. Published 2019 by John Wiley & Sons, Inc.
Companion website: www.wiley.com/go/asbmrprimer

The relationship between fracture site and BMI, body weight, and height was investigated in an analysis of 52,939 postmenopausal women followed up for 3 years in GLOW [15]. BMI was inversely related to hip, clinical spine, and wrist fractures, whilst there was a positive association between BMI and ankle fractures. Interestingly, the association between rib and pelvic fracture with BMI was nonlinear and U-shaped, the risk being higher in underweight and obese women than in those of intermediate weight. Linear height was inversely associated with fractures of the clavicle and shoulder/upper arm, but no relationship with BMI or body weight was seen. In a meta-analysis of data from 398,610 women aged 20 to 105 years followed for a mean of 5.7 years, the inverse association between BMI and hip, distal forearm, and other osteoporotic fractures was again observed. Additionally, low BMI was protective against lower leg fracture and high BMI was associated with increased risk of upper arm fracture [16]. Ankle fractures were not included in this meta-analysis.

Taken together, these studies indicate clear differences in the relationship between BMI and fracture at different sites, the most consistent evidence supporting a positive association between BMI and ankle, lower leg, and humerus fractures and a negative one for hip and wrist fractures. Differences between studies may partly be explained by differences in study design, the prevalence of obesity in populations studied, geographic variations in fracture incidence, and inaccuracies associated with self-reporting of fracture. Possible reasons for the differences in fracture risk at different sites include the protective effect of soft tissue padding at some sites, particularly the hip, and differences in the direction and impact of falls between obese and nonobese subjects. For example, reduced mobility associated with obesity may predispose to falling backwards or sideways rather than forwards, and the protective responses to falling may also be impaired in obese subjects. In the case of ankle fractures, greater introversion/extroversion stresses in obese individuals may be relevant.

MORBIDITY AND MORTALITY OF FRACTURES IN THE OBESE

Although increased risk of fracture nonunion has been consistently reported in obese individuals, and some studies have also reported increased length of stay in hospital, post-fracture mortality does not appear to be increased and may even be decreased when compared with nonobese individuals [17]. The latter finding is consistent with the so-called "obesity paradox," in which obesity has been associated with reduced mortality in a number of chronic disease states including cardiovascular disease and pulmonary disease. However, rehabilitation after fracture may be prolonged, and in the GLOW study self-reported quality of life and functional status were significantly reduced after fracture in obese postmenopausal women with fracture when compared with nonobese women with fracture [18].

PATHOGENESIS OF FRACTURES IN THE OBESE

Clinical risk factors

Risk factors for fracture in the obese are similar in some respects to those in nonobese people, but also exhibit some differences. In GLOW, the prevalence of previous fracture, parental hip fracture, and glucocorticoid therapy was not significantly different between obese and nonobese women who suffered an incident fracture, but two or more falls in the past 2 years and use of arms to assist rising from a sitting position were both significantly more common in the obese women. Furthermore, certain comorbidities, including self-reported asthma, emphysema, osteoarthritis, and diabetes were more common in obese women with fracture [3].

Falls and sarcopenia

Falls have not been well studied in the obese population but are likely to make a significant contribution to the increased risk of fractures. Reduced mobility and increased intramuscular adipose tissue may predispose to increased risk of falls and, as noted earlier, the direction of falling and lack of protective responses to falling may have adverse effects on the consequences of falling. Current definitions of sarcopenic obesity are unsatisfactory and do not adequately encompass function. However, there is some evidence for a positive association between intramuscular adipose tissue and risk of falls. In the Health, Ageing, and Body Composition Study, the risk of incident hip and all clinical fractures was significantly higher in women in the highest tertile of thigh muscle adipose tissue when compared with those in the lower two tertiles [19–21].

Bone mineral density

Although BMD is often normal or only mildly osteopenic in obese postmenopausal women who fracture, this may reflect appropriate adaptation to increased weight-bearing and does not necessarily indicate greater bone strength than that of a lighter individual with lower BMD. However, in obese older postmenopausal women in the SOF cohort who were closely matched for age and BMI, those who sustained an incident nonvertebral fracture had significantly lower BMD at the hip and spine when compared with their obese counterparts who did not fracture [4]. Thus lower BMD may play a role in the pathogenesis of fractures in the obese. The complex interactions between bone and adipose tissue are currently the focus of much research, and whether obesity leads to alterations in bone material and matrix composition has yet to be explored. Finally, other factors may contribute to increased bone fragility in obese individuals including vitamin D insufficiency, secondary hyperparathyroidism, reduced physical activity and, in men, hypogonadism [22–26].

Bone turnover and structure

Relatively little is known about the effects of obesity on bone turnover and structure. Reduced serum levels of bone turnover have been reported in obese older adults and there is also evidence of reduced rates of bone loss in obese women during the menopause [27, 28]. Increased cortical thickness in the radius and tibia have been reported, albeit with a reduction in tibial cortical volumetric BMD in one study [27, 29]. Finally, in a recent study conducted in older postmenopausal women, Sundh and colleagues reported a positive correlation between tibial subcutaneous fat and cortical porosity and an inverse association between BMI and the bone material strength index as measured by microindentation [30]. Further studies are required, but the available evidence suggests that despite the higher BMD and cortical thickness in obese subjects, bone quality and possibly strength may be compromised.

CLINICAL MANAGEMENT

Fracture risk assessment

The Fracture Risk Assessment Tool (FRAX) has been shown to perform reasonably well in the assessment of fracture risk in obese postmenopausal women, in spite of their higher BMI and BMD. In a study of older postmenopausal women recruited into the SOF, tests of discrimination and calibration showed a similar performance of FRAX in obese and nonobese women, both for hip fracture and for major osteoporotic fracture [31]. In another study, it was shown that the performance of FRAX was not affected by variations in body composition [32]. However, the fracture sites most commonly affected in the obese are not included in the output of FRAX, and underestimation of these fractures is therefore likely.

Lifestyle measures

Although not formally tested, it seems reasonable to advocate similar lifestyle measures in obese women at risk of fracture to those recommended in nonobese women. These include avoiding tobacco use, moderating alcohol intake, and ensuring appropriate levels of physical activity. Falls risk assessment and advice should be given high priority, given the increased risk of falls in the obese. An adequate calcium intake should be maintained, preferably by dietary means or, if not practicable, with calcium supplements. There is some evidence that obese subjects require higher doses of vitamin D than the nonobese to maintain adequate serum levels of 25-hydroxyvitamin D [33].

Weight loss

Weight loss in obese adults is beneficial for many aspects of health but has adverse effects on the skeleton. A number of studies have demonstrated that weight loss is associated with increased rates of bone loss and increased fracture risk. Different patterns of fracture are seen with unintentional and intentional weight loss, the former being associated with increased risk of hip, upper limb, and spine fracture whereas the latter has been associated with increased risk of lower leg fracture but decreased risk of hip and central body (hip, spine, and pelvis) fracture [34–38]. In the GLOW trial, unintentional weight loss in postmenopausal women was associated with increased risk of hip, clinical spine, and clavicle fractures within 1 year of weight loss, this association persisting for the 5-year study duration [39] (Figure 63.1). In addition, the cumulative risk of wrist, rib, and pelvis

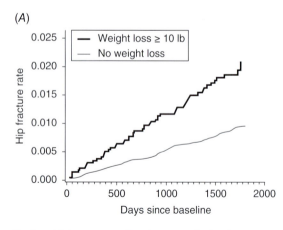
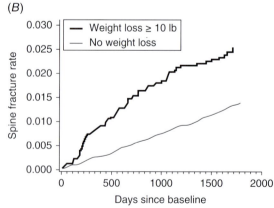

Fig. 63.1. Kaplan–Meier curves showing cumulative fracture rates of (A) the hip and (B) the spine over 5 years, by baseline weight loss in postmenopausal women. Source: [39]. Reproduced with permission of John Wiley & Sons.

fractures was also increased at 5 years. Bone loss in this situation may partially reflect physiological adaptation to reduced mechanical loading, but other factors including comorbidities in individuals with unintentional weight loss are also likely to contribute. There is some evidence that exercise programs reduce bone loss associated with weight loss, although the feasibility and safety of this approach in older frail populations has not been established [40].

Pharmacological intervention

Undertreatment of individuals at high risk of fragility fractures is well recognized and there is evidence that this trend is even more marked in the obese. In the GLOW trial, only 27% of obese women sustaining an incident fracture received anti-osteoporosis medication, compared with 41% of nonobese and 57% of underweight women with incident fracture [3]. The low treatment rate may reflect, at least in part, the perception that fractures in obese women are not true fragility fractures because of their higher BMD.

Pivotal clinical trials of antiosteoporosis medication have included relatively few obese subjects and those that have participated have generally had low BMD with or without previous fracture. The evidence base for treating obese individuals is therefore weak and decisions in clinical practice are mainly based on the antifracture efficacy of interventions in the nonobese. There is, however, some evidence that approved pharmacological interventions are less effective against nonvertebral fractures in overweight and obese women. In older, postmenopausal community-based women not selected on the basis of low BMD and/or fracture, clodronate therapy significantly reduced osteoporotic fractures in nonobese, but not obese, women [41]. Subgroup analyses of pivotal clinical trials have also been conducted to investigate efficacy of interventions in overweight and obese women. In the HORIZON Pivotal Fracture Trial, the reduction in vertebral fracture risk was significantly greater in obese than in nonobese women treated with zoledronic acid, but no interaction was seen for BMI and reduction in nonvertebral fracture risk [42]. In subgroup analyses of the FREEDOM study, significant reduction in nonvertebral fractures in women taking denosumab was seen only in women with a BMI <25 kg/m^2, although reductions in vertebral fracture were unaffected by BMI [43].

Since reduced effectiveness against nonvertebral fractures has been reported for several treatments in non-obese women with osteopenia as opposed to osteoporosis [44, 45], it is possible that the higher BMD in obese women with fracture contributes to a relative lack of responsiveness at nonvertebral sites. This in turn raises the possibility that higher doses of bisphosphonates or other drugs might be required in obese individuals to achieve full effectiveness.

SUMMARY

Fractures in obese postmenopausal women make a contribution to the overall fracture burden in this population. The pathophysiology of these fractures is incompletely understood but increased risk of, and altered biomechanics of, falling are likely to be important contributors and may explain the site-specificity of BMI effects on fracture risk. Establishing effective strategies to reduce fracture risk in obese postmenopausal women is an important priority for future research.

REFERENCES

1. De Laet C, Kanis JA, Oden A, et al. Body mass index as a predictor of fracture risk: a meta-analysis. Osteoporos Int. 2005;16:1330–8.
2. Premaor MO, Pilbrow L, Tonkin C, et al. Obesity and fractures in postmenopausal women. J Bone Miner Res. 2010;25:292–7.
3. Compston JE, Watts NB, Chapurlat R, et al. Obesity is not protective against fracture in postmenopausal women: GLOW. Am J Med. 2011;124:1043–50.
4. Premaor MO, Ensrud K, Lui L, et al. Study of Osteoporotic Fractures. Risk factors for nonvertebral fracture in obese older women. J Clin Endocrinol Metab. 2011;96:2414–21.
5. Armstrong ME, Spencer EA, Cairns BJ, et al. Body mass index and physical activity in relation to the incidence of hip fracture in postmenopausal women. J Bone Miner Res. 2011;26:1330–8.
6. Nielson CM, Marshall LM, Adams AL, et al. BMI and fracture risk in older men: the Osteoporotic Fractures in Men Study (MrOS). J Bone Miner Res. 2011;26:496–502.
7. Bergkvist D, Hekmat K, Svensson T, et al. Obesity in orthopedic patients. Surg Obes Relat Dis. 2009; 5:670–2.
8. Gnudi S, Emanuela S, Lisi L. Relationship of body mass index with main limb fragility fractures in postmenopausal women. J Bone Miner Metab. 2009;27:479–84.
9. King CM, Hamilton GA, Cobb M, et al. Association between ankle fractures and obesity. J Foot Ankle Surg. 2012; 5:543–7.
10. Prieto-Alhambra D, Premaor MO, Fina Avilés F, et al. The association between fracture and obesity is site-dependent: a population-based study in postmenopausal women. J Bone Miner Res. 2012; 27:294–300.
11. Ong T, Sahota O, Tan W, et al. A United Kingdom perspective on the relationship between body mass index (BMI) and bone health: a cross sectional analysis of data from the Nottingham Fracture Liaison Service. Bone. 2014;59:207–10.
12. Lacombe J, Cairns BJ, Green J, et al; Million Women Study collaborators. The effects of age, adiposity, and physical activity on the risk of seven site-specific fractures in postmenopausal women. J Bone Miner Res. 2016;31:1559–68.

13. Pirro M, Fabbriciani G, Leli C, et al. High weight or body mass index increase the risk of vertebral fractures in postmenopausal osteoporotic women. J Bone Miner Metab. 2010;28:88–93.
14. Laslett LL, Just Nee Foley SJ, Quinn SJ, et al. Excess body fat is associated with higher risk of vertebral deformities in older women but not in men: a crosssectional study. Osteoporos Int. 2012;23:67–74.
15. Compston JE, Flahive J, Hosmer DW, et al. Relationship of weight, height, and body mass index with fracture risk at different sites in postmenopausal women: the Global Longitudinal study of Osteoporosis in Women (GLOW). J Bone Miner Res. 2014;29:487–93.
16. Johansson H, Kanis JA, Oden A, et al. A meta-analysis of the association of fracture risk and body mass index in women. J Bone Miner Res. 2014;29:223–33.
17. Prieto-Alhambra D, Premaor MO, Avilés FF, et al. Relationship between mortality and BMI after fracture: a population-based study of men and women aged ≥40 years. J Bone Miner Res. 2014;29(8):1737–44.
18. Compston JE, Flahive J, Hooven FH, et al.; GLOW Investigators. Obesity, health-care utilization, and health-related quality of life after fracture in postmenopausal women: Global Longitudinal Study of Osteoporosis in Women (GLOW). Calcif Tissue Int. 2014;94:223–31.
19. Inacio M, Ryan AS, Bair WN, et al. Gluteal muscle composition differentiates fallers from nonfallers in community dwelling older adults. BMC Geriatr. 2014;14:37.
20. Schafer AL, Vittinghoff E, Lang TF, et al. Fat infiltration of muscle, diabetes, and clinical fracture risk in older adults. Health, Aging, and Body Composition (Health ABC) Study. J Clin Endocrinol Metab. 2010;95:E368–72.
21. Lang T, Cauley JA, Tylavsky F, et al.; Health ABC Study. Computed tomographic measurements of thigh muscle cross-sectional area and attenuation coefficient predict hip fracture: the health, aging, and body composition study. J Bone Miner Res. 2010;25:513–19.
22. Cheng S, Massaro JM, Fox CS, et al. Adiposity, cardiometabolic risk, and vitamin D status: the Framingham Heart Study. Diabetes. 2010;59:242–8.
23. Earthman CP, Beckman LM, Masodkar K, et al. The link between obesity and low circulating 25-hydroxyvitamin D concentrations: considerations and implications. Int J Obes (Lond). 2012;36:96–107.
24. Bolland MJ, Grey AB, Ames RW, et al. Fat mass is an important predictor of parathyroid hormone levels in postmenopausal women. Bone. 2006;38:317–21.
25. Grethen E, McClintock R, Gupta CE, et al. Vitamin D and hyperparathyroidism in obesity. J Clin Endocrinol Metab. 2011;96:1320–6.
26. Orwoll E, Lambert LC, Marshall LM, et al. Endogenous testosterone levels, physical performance, and fall risk in older men. Arch Int Med 2006;166:2124–31.
27. Evans AL, Paggiosi MA, Eastell R, et al. Bone density, microstructure and strength in obese and normal weight men and women in younger and older adulthood. J Bone Miner Res. 2015;30:920–8.
28. Sowers MR, Zheng H, Greendale GA, et al. Changes in bone resorption across the menopause transition: effects of reproductive hormones, body size, and ethnicity. J Clin Endocrinol Metab. 2013;98:2854–63.
29. Sukumar D, Schlussel Y, Riedt CS, et al. Obesity alters cortical and trabecular bone density and geometry in women. Osteoporos Int. 2011;22:635–45.
30. Sundh D, Rudäng R, Zoulakis M, et al. A high amount of local adipose tissue is associated with high cortical porosity and low bone material strength in older women. J Bone Miner Res. 2016;31:749–57.
31. Premaor M, Parker RA, Cummings SR, et al. Predictive value of FRAX for fracture in obese older women. J Bone Miner Res. 2013;28:188–95.
32. Leslie WD, Orwoll ES, Nielson CM, et al. Estimated lean mass and fat mass differentially affect femoral bone density and strength index but are not FRAX independent risk factors for fracture. J Bone Miner Res. 2014;29:2511–19.
33. Gallagher JC, Yalamanchili V, Smith LM. The effect of vitamin D supplementation on serum 25(OH)D in thin and obese women. J Steroid Biochem Mol Biol. 2013;136:195–200.
34. Langlois JA, Harris T, Looker AC, et al. Weight change between age 50 years and old age is associated with risk of hip fracture in white women aged 67 years and older. Arch Intern Med. 1996;156:989–94.
35. Ensrud KE, Cauley J, Lipschutz R, et al. Weight change and fractures in older women. Study of Osteoporotic Fractures Research Group. Arch Intern Med. 1997;157:857–63.
36. Langlois JA, Mussolino ME, Visser M, et al. Weight loss from maximum body weight among middle-aged and older white women and the risk of hip fracture: the NHANES I epidemiologic follow-up study. Osteoporos Int. 2001;12:763–8.
37. Ensrud KE, Ewing SK, Stone KL, et al. Intentional and unintentional weight loss increase bone loss and hip fracture risk in older women. J Am Geriatr Soc. 2003;51:1740–7.
38. Crandall C, Yildiz V, Wactawski-Wenda J, et al. Postmenopausal weight change and fracture incidence: posthoc findings from the Women's Health Initiative Observational Study and Clinical Trials. BMJ. 2015;350:h25.
39. Compston JE, Wyman A, FitzGerald G, et al. Increase in fracture risk following unintentional weight loss in postmenopausal women: the Global Longitudinal Study of Osteoporosis in Women. J Bone Miner Res. 2016;31:1466–72.
40. Shah K, Armamento-Villareal R, Parimi N, et al. Exercise training in obese older adults prevents increase in bone turnover and attenuates decrease in hip bone mineral density induced by weight loss despite decline in bone-active hormones. J Bone Miner Res. 2011;26:2851–9.
41. McCloskey EV, Johansson H, Oden A, et al. Ten-year fracture probability identifies women who will benefit from clodronate therapy—additional results from a double-blind, placebo-controlled randomised study. Osteoporos Int. 2009;20:811–17.

42. Eastell R, Black DM, Boonen S, et al. HORIZON Pivotal Fracture Trial. Effect of once-yearly zoledronic acid five milligrams on fracture risk and change in femoral neck bone mineral density. J Clin Endocrinol Metab. 2009;94:3215–25.
43. McClung M, Boonen S, Torring O, et al. Effect of denosumab treatment on the risk of fractures in subgroups of women with postmenopausal osteoporosis. J Bone Miner Res. 2012;27:211–18.
44. Cummings SR, Black DM, Thompson DE, et al. Effect of alendronate on risk of fracture in women with low bone density but without vertebral fractures: results from the Fracture Intervention Trial. JAMA. 1998;280:2077–82.
45. Silverman SL, Christiansen C, Genant HK, et al. Efficacy of bazedoxifene in reducing new vertebral fracture risk in postmenopausal women with osteoporosis: results from a 3-year, randomized, placebo-, and active-controlled clinical trial. J Bone Miner Res. 2008;23:1923–34.

64

Sarcopenia and Osteoporosis

Gustavo Duque[1] and Neil Binkley[2]

[1]Australian Institute for Musculoskeletal Science (AIMSS), The University of Melbourne and Western Health; and Department of Medicine-Western Health, Melbourne Medical School, The University of Melbourne, St. Albans, VIC, Australia
[2]Department of Medicine, University of Wisconsin School of Medicine and Public Health; and Institute on Aging, University of Wisconsin, Madison, WI, USA

COMMON MECHANISMS IN SARCOPENIA AND OSTEOPOROSIS

Although sarcopenia—defined as the age-related decline in muscle mass, strength, and function—and osteoporosis, which could similarly be defined as the age-related decline in bone mass and strength, are generally considered as separate processes, their phenotypes are increasingly identified as co-occurring with aging [1]. These syndromes may cause a dual impact and thereby lead to increased risk for falls, fractures [2], and frailty with concomitant increase in morbidity and mortality [1]. Indeed, some have suggested that osteoporosis and sarcopenia are the result of the same processes reflected in bone and muscle, respectively.

In this regard, knowledge about the pathophysiological mechanisms of these concurrent syndromes has rapidly grown in recent years [3] (Fig. 64.1). Osteoblasts and myocytes derive from a similar lineage in which mesenchymal stem cells (MSCs) are the common parent. Therefore, current theory contends that MSC pathogenicity may be at the crux of these disorders, especially in the context of physiological senescence and lipotoxicity [4].

An MSC is a type of pluripotent cell that can differentiate into various cellular lineages and thus is essential for organ growth and repair [5]. Bone and skeletal muscle loss with advancing age can be attributed to alterations in genetic, mechanical, systemic, and local factors that alter the stromal microenvironment in which MSCs reside and the communication between muscle and bone (Fig. 64.2). Consequently, with aging, MSC homeostasis is altered causing decreased capacity for self-renewal and differentiation, in which decreases in osteoblastogenic potential and myoblastogenic potential in favor of increased adipogenesis have been observed [6, 7]. This process of cell senescence can be exemplified by the fact that MSCs obtained from older adults tend to be larger in size compared with those of a younger population, with a higher tendency to differentiate into adipocytes [4].

Furthermore, many factors also have the capacity to disturb the microenvironments of the MSCs and their daughter cells. Some key examples of these factors are low-grade inflammation, changes in adipokine concentration [8], low levels of growth/differentiation factors [9, 10], and increased apoptosis of osteoblasts and osteocytes [11]. In addition, excess adiposity within the bone marrow microenvironment and muscle fibers enhances inflammatory and adipogenic signals in the MSC microenvironment that further alter MSC homeostasis. The net result of this can lead to bone and skeletal muscle loss.

CLINICAL IMPLICATIONS OF SARCOPENIA AND OSTEOPOROSIS

It is intuitive that many older individuals, especially frailer ones, would suffer from both osteoporosis and sarcopenia, increasing their risks and complications, which are exacerbated even further by comorbid conditions [12]. The combination of these two diseases exacerbates negative health outcomes and has been described as a "hazardous duet" adding the propensity of falls associated with sarcopenia to the vulnerability of bones in those with osteoporosis [13].

Primer on the Metabolic Bone Diseases and Disorders of Mineral Metabolism, Ninth Edition. Edited by John P. Bilezikian.
© 2019 American Society for Bone and Mineral Research. Published 2019 by John Wiley & Sons, Inc.
Companion website: www.wiley.com/go/asbmrprimer

Osteosarcopenia contributes to an even higher risk of falls, fractures, institutionalization, and poorer quality of life [2, 12, 14]. Consequently, fracture prevention approaches in older adults must include assessment of physical function and, ideally, muscle mass to evaluate whether sarcopenia is also present. At this point in time, a multitude of consensus definitions for sarcopenia exists, almost all of which include measurement of lean mass and some assessment of physical function. It is to be expected that clinical application of approaches considering osteosarcopenia will be facilitated by a single consensus definition of sarcopenia. In individuals with both conditions (ie, osteosarcopenia), it is essential that planned interventions should address the strength of not only bone but also muscle.

It was recently advocated that the diagnosis of osteoporosis be expanded to include individuals at increased risk for fracture [15, 16]. At face value, this is a reasonable proposition as many older adults with osteopenia or even normal BMD sustain "osteoporosis-related" fractures [17]. Thus, diagnosing osteoporosis and treating only those with a BMD T-score of -2.5 or less is not adequate to detect many people who will subsequently fracture. Conversely, some individuals with osteoporosis based upon BMD alone are at low fracture risk; clearly, moving beyond a BMD T-score-based approach to guide fracture risk therapeutic initiation is appropriate. However, continued focus on a diagnosis of "osteoporosis," or even on approaches for osteoporosis therapeutic intervention thresholds based upon estimated fracture risk, diverts attention from the other concomitant problems such as sarcopenia, and diverts clinical attention from the important clinical consequence, fragility fracture risk.

Indeed, it is osteoporosis, in concert with age-related loss of muscle mass and quality (ie, sarcopenia), and often combined with other age-related morbidities that negatively impact ambulation such as neuropathy, reduced balance, impaired vision, polypharmacy, osteoarthritis, and others conditions leading to increased falls risk, that ultimately causes the vast majority of "osteoporosis-related" fractures. That these age-related fragility fractures result from much more than simply compromised bone mass is exemplified by the fact that approximately one in six fragility fractures occur in those with normal proximal femur BMD [17]. Whether such individuals truly have bone loss and microarchitectural deterioration (ie, osteoporosis) is unclear. Of great clinical consequence, it is unknown if those who sustain "osteoporosis-related" fractures despite normal BMD will benefit from classic bone-directed pharmacological treatments. Moreover, the multitude of rigorous prospective studies of effective bone-directed pharmacological agents demonstrate only an approximate 50% to 80% reduction in vertebral fractures and an approximate 35% reduction in nonvertebral fractures [18–21]. Thus, the current best available therapies directed solely at improving bone do not prevent a large number of fragility fractures. These fractures continue to occur as other contributing factors lead to falls, generating a force on

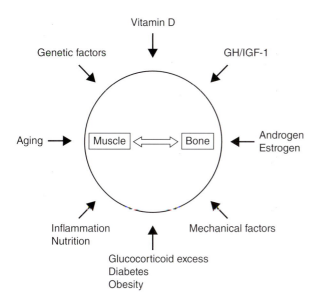

Fig. 64.1. Muscle and bone are affected by multiple factors. Source: [3]. Reproduced with permission of John Wiley & Sons.

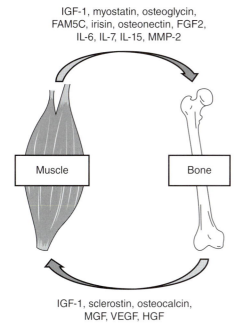

Fig. 64.2. Endocrine and autocrine communication between muscle and bone. FAM5C = family with sequence similarity 5, member, C; FGF2 = fibroblast growth factor 2; HGF = hepatocyte growth factor; IL = interleukin; MGF = mechano growth factor; MMP-2 = matrix metalloproteinase-2; VEGF = vascular endothelial growth factor. Source: [3]. Reproduced with permission of John Wiley & Sons.

Bone and muscle are interconnected tissues; not only because of the adjacent nature of their surfaces, but also chemically and metabolically. Since osteoporosis and sarcopenia are closely linked, the term osteosarcopenia has recently been proposed to describe the disease state of individuals suffering from both conditions.

the skeleton that exceeds the load-bearing capacity of even bone with normal mass.

It is obvious that factors in addition to bone loss contribute to the age-related increase in fracture risk as fragility fractures increase dramatically with advancing age, but bone mass does not have a comparable decline [22]. A major contributor to this increase in fracture risk is falls. Falls become common with advancing age and approximately 95% of hip fractures result from a fall [23]. Indeed, fall risk factors such as prior falls and slow gait speed predict hip fracture independent of BMD [24, 25]. Thus, it is unsurprising that poor physical performance is associated with increased hip fracture risk [26]. Such observations point to the importance of including sarcopenia as an integral consideration in efforts to reduce fragility fracture risk.

Sarcopenia has been defined as the "age-associated loss of skeletal muscle mass and function … a complex syndrome associated with muscle mass loss alone or in conjunction with increased fat mass" [27]. This decline in muscle mass/function becomes common with advancing age [28] and is associated with impaired walking, falls, and fractures.

Sarcopenia is rarely diagnosed clinically, in large part reflecting the absence of a universally accepted diagnostic definition. To fill this void, several recent consensus conferences have proposed definitions, all of which include measurement of muscle mass and muscle function [27, 29, 30]. The recent definition proposed by the Foundation of the NIH Sarcopenia Project importantly integrates consideration of sarcopenic obesity (in essence, too much fat for the amount of muscle present) [31] by suggesting a unique approach to define low muscle mass as appendicular (leg + arm) lean mass divided by BMI [30]. Essentially, this is muscle mass/height-corrected weight. Whether this approach will ultimately be included in a consensus definition of sarcopenia remains to be determined.

Considering sarcopenic obesity as a risk factor for fragility fracture may seem counterintuitive for some, as obesity has historically been thought of as being protective against fragility fractures by increasing mechanical load [32, 33]. However, that the relationship between fat and bone is much more complex than simple weight-bearing is increasingly being recognized [4, 34, 35]. Moreover, infiltration of fat into muscle, or intramuscular adipose tissue (IMAT), leads to muscle dysfunction [36]. It is therefore not surprising that some studies find obesity to increase fracture risk [37]. As such, consideration of obesity in fragility fracture risk estimation is appropriate.

Direct support for the importance of nonbone-related factors in fracture risk is apparent given the inclusion of demographic, lifestyle, and medical history factors included in fracture risk calculators such as the Fracture Risk Assessment Tool (FRAX) and Garvan [38, 39]. Additionally, evidence that these calculators are not currently comprehensive and require continued evolution was recently provided by the implementation of a trabecular bone score (TBS) adjustment to the FRAX

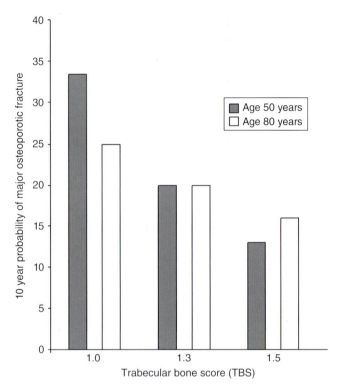

Fig. 64.3. Effect of age on the trabecular bone score (TBS) adjustment in FRAX. This figure depicts the effect of low or high TBS on FRAX-estimated 10-year risk of major osteoporosis-related fracture in two individuals (age 50 and 80 years) whose nonadjusted risk is 21%. It is apparent that bone microarchitectural status, as estimated by TBS, has a much less pronounced effect on fracture risk at an older age indicating that nonbone factors, likely falls, play a greater role in fractures in older adults.

calculator. The TBS is bone text assessment software applied to lumbar spine DXA scans that can serve as a surrogate for bone microarchitecture, independent of BMD [40]. The effect of the TBS adjustment on fracture risk calculation further confirms that nonbone factors are part of the cause for increased fracture risk with age. Specifically, in a study conducted by McCloskey and colleagues [41], lower TBS values exerted a less profound effect on fracture risk in older individuals (Fig. 64.3), indicating that bone fragility contributes less to fracture risk in more elderly people. A logical explanation for this observation is that falls and other age-related morbidities play a greater role while bone fragility plays a lesser role with advancing age.

Overall, and considering that fragility fractures are considered part of a larger syndrome that involves bone, muscle, and fat, the term "dysmobility syndrome" has been proposed as a syndrome that encompasses osteosarcopenia, obesity, and other fracture risk factors as an approach to improve identification and ultimately treatment of older adults to reduce their risk for falls and fractures [42]. This approach is analogous to the widely recognized metabolic syndrome in which various conditions (eg, hypertension, hyperlipidemia, etc.)

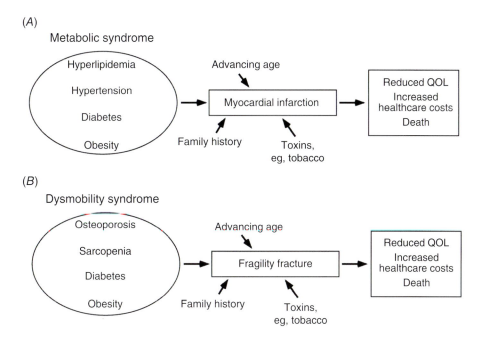

Fig. 64.4. Analogous situation of metabolic syndrome and dysmobility syndrome. (A) Metabolic syndrome is familiar to clinicians as being a group of conditions, predisposing individuals with the syndrome to the development of cardiovascular disease. Each component of the syndrome is a therapeutic target, with optimization of each contributing to reduced risk. (B) Similarly, dysmobility syndrome consists of conditions predisposing to fragility fracture. Each component of dysmobility syndrome (others should also be considered, eg, osteoarthritis, neurological disease) should ideally be optimized to reduce fracture risk. This approach directly acknowledges that other conditions contribute to what is currently recognized as "osteoporosis-related" fracture. This paradigm directly acknowledges that the "failure" to prevent all fractures is not just linked to bone mass, but to not considering other conditions that contribute to elevated fracture risk. A comprehensive approach that appropriately intervenes with each of these components is required to optimally reduce fracture risk. QOL = quality of life.

are recognized contributors to the adverse outcome of cardiovascular disease (Fig. 64.4).

It is a logical hypothesis, as yet untested, that patients would accept an approach to reduce their fracture risk similar to the approaches currently routinely taken to reduce risk following myocardial infarction. However, a critical factor will be education to inform the lay public that fragility fractures are often indicative of multisystem treatable problems, not simply "I fell." Moreover, educational efforts are sorely needed that emphasize the adverse consequences related to fracture, specifically mortality, and likely of equal, or greater, importance, loss of independence which occurs following fragility fracture. The latter is not only devastating to the quality of life for a community-dwelling individual, but also has huge societal impact due to the cost of long-term care.

In addition to recognizing this concept, approaches to the clinical identification of individuals with dysmobility syndrome will be required. Diagnosis of dysmobility syndrome could be made by a combination of slow gait speed (<0.8 m/s), BMD T-score −2.5 or less, low appendicular lean mass as measured by DXA (4.45 and 7.26 kg/m^2 for women and men, respectively), low grip strength (<20 and <30 kg for women and men, respectively), and high total body fat mass (obesity).

It is certainly appropriate that fragility fracture risk factors other than those originally proposed be considered in the dysmobility syndrome concept as they have a deleterious effect on bone and muscle. Two logical candidates (among others) to be included are diabetes mellitus and osteoarthritis. Diabetes is recognized to have adverse consequences upon bone and muscle [43, 44] and to increase fracture risk above that predicted by FRAX [45]. Similarly, some studies find osteoarthritis to be associated with increased risk for sarcopenia, osteoporosis, falls, and fractures [46, 47]. Although the dysmobility syndrome concept seems a logical approach to fracture risk reduction, perhaps increasing care of osteosarcopenia can be a step toward more comprehensive approaches to reducing fragility fracture risk.

OSTEOSARCOPENIA, DYSMOBILITY, AND ADVERSE HEALTH OUTCOMES

Osteosarcopenia has been associated with poor outcomes in older persons. For example, a cross-sectional study in 680 Australian subjects (mean age = 79 years, 65% female) showed a significantly higher prevalence of falls and fractures in the osteosarcopenic group [2]. In addition, osteosarcopenia is also associated with higher risk of frailty [1] and mortality [48].

The clinical consequences of the dysmobility syndrome have also been assessed. Clynes and colleagues, in

398 older adults in a Herfordshire cohort, found dysmobility to be associated prospectively with falls, but interestingly not with fractures [49]. Data from the Osteoporotic Fractures in Men Study (MrOS) followed 5826 older men for a mean of 6.2 years and found those with dysmobility to be more likely than those without to sustain major osteoporosis-related fractures and hip fractures, with hazard ratios of ~3 and ~2, respectively [50]. In this report, the diagnosis of dysmobility was an independent risk factor for fracture even after adjustment for FRAX score. Finally, Looker evaluated 2975 National Health and Nutrition Examination Survey (NHANES) participants aged 50+ and found dysmobility to be associated with increased mortality risk, with highest hazard ratios (~3) for those with dysmobility who were 50 to 69 years of age [51].

In summary, both osteosarcopenia and dysmobility are associated with adverse health outcomes that could be prevented if muscle and bone loss are recognized and effective therapeutic interventions are implemented.

COMBINED THERAPEUTIC INTERVENTIONS TO INCREASE MUSCLE AND BONE MASS

Therapies for osteosarcopenia should have a beneficial effect not only on bone and muscle mass, but also on physical function and strength. Intuitively, increasing muscle mass should increase muscle strength and restore bone loss from disuse. However, most of the results from clinical trials targeting both muscle and bone have been inconclusive because pharmacological strategies inducing isolated myocyte hypertrophy have not translated into increased muscle power or better bone mass [52].

Physical exercise has a beneficial effect on bone and muscle, with additional beneficial effects of decreasing BMI and fat infiltration in muscle and bone and potentially improving balance [53]. As would be expected some studies find that exercise decreases falls and fractures through a combined effect on bone, muscle, and mobility [54]. In addition, whole body vibration has also demonstrated a combined beneficial effect in terms of improving muscle and bone mass [55]. However, a beneficial effect of vibration therapy on muscle function requires further study to identify effective approaches regarding vibration intensity, duration, and frequency of use.

Nutritional supplementation is likely to be a key component of treatment approaches for osteosarcopenia/dysmobility. In this regard, protein supplements may also have a beneficial effect on bone and muscle. Protein supplements, optimally administered right after physical exercise, may improve muscle mass and function and improve bone mass [56–58]. However, protein supplements alone have not shown any antifracture effect. Additionally, vitamin D deficiency is associated with low muscle and bone mass. Vitamin D supplementation may have a direct effect on both muscle and bone [54, 57–59]. Serum concentrations above 60 nmol/L (24 ng/mL) are associated with a lower prevalence of falls and fractures [60], an effect that could potentially be explained by the dual effect of vitamin D on bone and muscle. Well-designed studies of nutritional supplementation, including requiring subjects to be low in whatever nutrient is being studied, are necessary to clarify the role of supplementation in osteosarcopenia treatment.

Recently, antibodies against myostatin or activin receptor type IIA and IIB (eg, bimagrumab) have been tested in phase II clinical trials [61]. A proof-of-concept, randomized, placebo-controlled, double-blind, multicentre, phase II study, in 365 patients aged 75 years or older who had fallen in the past year, tested the effect of the humanized monoclonal antimyostatin antibody LY2495655 on appendicular lean mass as measured by DXA with physical performances as secondary outcomes. The beneficial effects of this antibody were limited to a reduction in fat mass, a modest increase in lean mass, and a limited improvement in functional measures of muscle power [62].

Although changes in bone mass were not assessed in this study, experiments using animal models suggest that antimyostatin antibody could have a dual effect on bone and muscle mass. This evidence has allowed the suggestion that these medications could become the "holy grail" for muscle bone and fat [63].

In summary, combined therapies that target muscle, bone, and fat seem to be a very reasonable approach to the treatment of osteosarcopenia and dysmobility. It is essential that fracture prevention approaches and post-fracture management include assessment and treatment of muscle and bone loss, thereby reducing the risk of falls and fractures. There is solid evidence that nonpharmacological treatment can have dual effects on bone and muscle. However, so far, no pharmacological treatment has demonstrated clinical benefits on both tissues.

REFERENCES

1. Drey M, Sieber CC, Bertsch T, et al; FiAT intervention group.Osteosarcopenia is more than sarcopenia and osteopenia alone. Aging Clin Exp Res. 2016;28:895–9.
2. Huo YR, Suriyaarachchi P, Gomez F, et al. Comprehensive nutritional status in sarco-osteoporotic older fallers. J Nutr Health Aging. 2015;19:474–80.
3. Kawao N, Kaji H. Interactions between muscle tissues and bone metabolism. J Cell Biochem. 2015;116: 687–695
4. Demontiero O, Boersma D, Suriyaarachchi P, et al. Clinical outcomes of impaired muscle and bone interactions. Clinic Rev Bone Miner Metab. 2014;12:86–92
5. Ormsbee MJ, Prado CM, Ilich JZ, et al. Osteosarcopenic obesity: the role of bone, muscle, and fat on health. J Cachexia Sarcopenia Muscle. 2014;5(3):183–92
6. Zhou S, Greenberger JS, Epperly MW, et al. Age-related intrinsic changes in human bone-marrow-derived mesenchymal stem cells and their differentiation to osteoblasts. Aging Cell. 2008;7(3):335–43.

7. Stolzing A, Jones E, McGonagle D, et al. Age-related changes in human bone marrow-derived mesenchymal stem cells: consequences for cell therapies. Mech Ageing Dev. 2008;129(3):163–73.
8. Magni P, Dozio E, Galliera E, et al. Molecular aspects of adipokine-bone interactions. Curr Mol Med. 2010;10(6):522–32.
9. Jiang SS, Chen CH, Tseng KY, et al. Gene expression profiling suggests a pathological role of human bone marrow-derived mesenchymal stem cells in aging-related skeletal diseases. Aging. 2011;3(7):672–84.
10. Benisch P, Schilling T, Klein-Hitpass L, et al. The transcriptional profile of mesenchymal stem cell populations in primary osteoporosis is distinct and shows overexpression of osteogenic inhibitors. PLoS One. 2012;7(9):e45142.
11. Almeida M. Aging mechanisms in bone. Bonekey Rep. 2012;1:ii.
12. Huo YR, Suriyaarachchi P, Gomez F, et al. Phenotype of osteosarcopenia in older individuals with a history of falling. J Am Med Dir Assoc. 2015;16:290–95
13. Crepaldi G, Maggi S. Sarcopenia and osteoporosis: a hazardous duet. J Endocrinol Invest. 2015;28:66–68.
14. Levinger I, Phu S, Duque G. Sarcopenia and osteoporotic fractures. Clin Rev Bone Miner Metab. 2016;14:38–44.
15. Siris ES, Adler R, Bilezikian J, et al. The clinical diagnosis of osteoporosis: a position statement from the National Bone Health Alliance Working Group. Osteoporos Int. 2014;25(5):1439–43.
16. Siris ES, Boonen S, Mitchell PJ, et al. What's in a name? What constitutes the clinical diagnosis of osteoporosis? Osteoporos Int. 2012;23(8):2093–7.
17. Schuit SCE, van der Klift M, Weel AE, et al. Fracture incidence and association with bone mineral density in elderly men and women: the Rotterdam study. Bone. 2004;34:195–202.
18. Black DM, Cummings SR, Karpf DB, et al. Randomised trial of effect of alendronate on risk of fracture in women with existing vertebral fractures. Fracture Intervention Trial Research Group. Lancet. 1996;348(9041):1535–41.
19. McClung MR, Geusens P, Miller PD, et al. Effect of risedronate on the risk of hip fracture in elderly women. N Engl J Med. 2001;344:333–40.
20. Lyles KW, Colón-Emeric CS, Magaziner JS, et al. Zoledronic acid and clinical fractures and mortality after hip fracture. N Engl J Med. 2007;357(18):1799–809.
21. Neer RM, Arnaud CD, Zanchetta JR, et al. Effect of parathyroid hormone (1-34) on fractures and bone mineral density in postmenopausal women with osteoporosis. N Engl J Med. 2001;344(19):1434–41.
22. Kelly TL, Wilson KE, Heymsfield SB. Dual energy X-ray absorptiometry body composition reference values from NHANES. PLoS One. 2009;4(9):e7038.
23. Parkkari J, Kannus P, Palvanen M, et al. Majority of hip fractures occur as a result of a fall and impact on the greater trochanter of the femur: a prospective controlled hip fracture study with 206 consecutive patients. Calcif Tissue Int. 1999;65(3):183–7.
24. Dargent-Molina P, Favier F, Grandjean H, et al. Fall-related factors and risk of hip fracture: the EPIDOS prospective study. Lancet. 1996;348(9021):145–9.
25. Nguyen ND, Pongchaiyakul C, Center JR, et al. Identification of high-risk individuals for hip fracture: a 14-year prospective study. J Bone Miner Res. 2005;20:1921–8.
26. Cawthon PM, Fullman RL, Marshall L, et al. Physical performance and risk of hip fractures in older men. J Bone Miner Res. 2008;23(7):1037–44.
27. Fielding RA, Vellas B, Evans WJ, et al. Sarcopenia: an undiagnosed condition in older adults. Current consensus definition: prevalence, etiology, and consequences. International Working Group on Sarcopenia. J Am Med Dir Assoc. 2011;12(4):249–56.
28. Looker AC, Wang CY. Prevalence of reduced muscle strength in older U.S. adults: United States, 2011–2012. NCHS Data Brief. 2015;179:1–8.
29. Cruz-Jentoft AJ, Baeyens JP, Bauer JM, et al. Sarcopenia: European consensus on definition and diagnosis: Report of the European Working Group on Sarcopenia in Older People. Age Ageing. 2010;39(4):412–23.
30. Studenski SA, Peters KW, Alley DE, et al. The FNIH Sarcopenia Project: rationale, study description, conference recommendations, and final estimates. J Gerontol A Biol Sci Med Sci. 2014;69(5):547–58.
31. Ilich JZ, Kelly OJ, Inglis JE, et al. Interrelationship among muscle, fat, and bone: connecting the dots on cellular, hormonal, and whole body levels. Ageing Res Rev. 2014;15:51–60.
32. Cummings SR, Nevitt MC, Browner WS, et al. Risk factors for hip fracture in white women. Study of Osteoporotic Fractures Research Group. N Engl J Med. 1995;332(12):767–73.
33. Roy DK, O'Neill TW, Finn JD, et al. Determinants of incident vertebral fracture in men and women: results from the European Prospective Osteoporosis Study (EPOS). Osteoporos Int. 2003;14(1):19–26.
34. Rosen CJ, Bouxsein ML. Mechanisms of disease: is osteoporosis the obesity of bone. Nat Clin Pract Rheumatol. 2006;2:35–43.
35. Rosen CJ, Klibanski A. Bone, fat and body composition: evolving concepts in the pathogenesis of osteoporosis. Am J Med. 2009;122:409–14.
36. Addison O, Marcus RL, Lastayo PC, et al. Intermuscular fat: a review of the consequences and causes. Int J Endocrinol. 2014;2014:309570.
37. Compston JE, Watts NB, Chapurlat R, et al. Obesity is not protective against fracture in postmenopausal women: GLOW. Am J Med. 2011;124:1043–50.
38. Kanis JA, McCloskey EV, Johansson H, et al. European guidance for the diagnosis and management of osteoporosis in postmenopausal women. Osteoporos Int. 2008;19(4):399–428.
39. Nguyen ND, Frost SA, Center JR, et al. Development of prognostic nomograms for individualizing 5-year and 10-year fracture risks. Osteoporos Int. 2008;19(10):1431–44.
40. Silva BC, Leslie WD, Resch H, et al. Trabecular bone score: a noninvasive analytical method based upon the DXA image. J Bone Miner Res. 2014;29(3):518–30.

41. McCloskey EV, Odén A, Harvey NC, et al. Adjusting fracture probability by trabecular bone score. Calcif Tissue Int. 2015;96(6):500–9.
42. Binkley N, Krueger D, Buehring B. What's in a name revisited: should osteoporosis and sarcopenia be considered components of "dysmobility syndrome?". Osteoporos Int. 2013;24(12):2955–9.
43. Schwartz AV. Epidemiology of fractures in type 2 diabetes. Bone. 2016;82:2–8.
44. Wang J, You W, Jing Z, et al. Increased risk of vertebral fracture in patients with diabetes: a meta-analysis of cohort studies. Int Orthop. 2016;40(6):1299–307.
45. Giangregorio LM, Leslie WD, Lix LM, et al. FRAX underestimates fracture risk in patients with diabetes. J Bone Miner Res. 2012;27:301–8.
46. Prieto-Alhambra D, Nogues X, Javaid MK, et al. An increased rate of falling leads to a rise in fracture risk in postmenopausal women with self-reported osteoarthritis: a prospective multinational cohort study (GLOW). Ann Rheum Dis. 2013;72(6):911–17.
47. Zasadzka E, Borowicz AM, Roszak M, et al. Assessment of the risk of falling with the use of timed up and go test in the elderly with lower extremity osteoarthritis. Clin Interv Aging. 2015;10:1289–98.
48. Pasco JA, Mohebbi M, Holloway KL, et al. Musculoskeletal decline and mortality: prospective data from the Geelong Osteoporosis Study. J Cachexia Sarcopenia Muscle. 2017;8(3):482–9.
49. Clynes MA, Edwards MH, Buehring B, et al. Definitions of sarcopenia: associations with previous falls and fracture in a population sample. Calcif Tissue Int. 2015;97(5):445–52.
50. Buehring B, Hansen KE, Lewis BL, et al. Osteoporotic Fractures in Men (MrOS) Study Research Group. Dysmobility Syndrome Independently Increases Fracture Risk in the Osteoporotic Fractures in Men (MrOS) Prospective Cohort Study. J Bone Miner Res. 2018; doi: 10.1002/jbmr.3455 [epub ahead of print].
51. Looker AC. Dysmobility syndrome and mortality risk in US men and women age 50 years and older. Osteoporos Int. 2015;26(1):93–102.
52. Girgis CM, Mokbel N, Digirolamo DJ. Therapies for musculoskeletal disease: can we treat two birds with one stone? Curr Osteoporos Rep. 2014;12(2):142–53.
53. Phu S, Boersma D, Duque G. Exercise and sarcopenia. J Clin Densitom. 2015;18(4):488–92.
54. Daly RM, Duckham RL, Gianoudis J. Evidence for an interaction between exercise and nutrition for improving bone and muscle health. Curr Osteoporos Rep. 2014;12:219–26.
55. Luo X, Zhang J, Zhang C, et al. The effect of whole-body vibration therapy on bone metabolism, motor function, and anthropometric parameters in women with postmenopausal osteoporosis. Disabil Rehabil. 2017;39:2315–23
56. Genaro PS, Martini LA. Effect of protein intake on bone and muscle mass in the elderly. Nutr Rev. 2010;68:616–23.
57. Rizzoli R, Stevenson JC, Bauer JM, et al. The role of dietary protein and vitamin D in maintaining musculoskeletal health in postmenopausal women: a consensus statement from the European Society for Clinical and Economic Aspects of Osteoporosis and Osteoarthritis (ESCEO). Maturitas. 2014;79:122–32.
58. Bischoff-Ferrari H, Stähelin HB, Walter P. Vitamin D effects on bone and muscle. Int J Vitam Nutr Res. 2011;81(4):264–72.
59. Gunton JE, Girgis CM, Baldock PA, et al. Bone muscle interactions and vitamin D. Bone. 2015;80:89–94.
60. Rizzoli R, Boonen S, Brandi ML, et al. Vitamin D supplementation in elderly or postmenopausal women: a 2013 update of the 2008 recommendations from the European Society for Clinical and Economic Aspects of Osteoporosis and Osteoarthritis (ESCEO). Curr Med Res Opin. 2013;29:305–13.
61. Bradley L, Yaworsky PJ, Walsh FS. Myostatin as a therapeutic target for musculoskeletal disease. Cell Mol Life Sci. 2008;65:2119–24.
62. Becker C, Lord SR, Studenski SA, et al. STEADY Group. Myostatin antibody (LY2495655) in older weak fallers: a proof-of-concept, randomised, phase 2 trial. Lancet Diabetes Endocrinol. 2015;3(12):948–57.
63. Buehring B, Binkley N. Myostatin—the holy grail for muscle, bone, and fat? Curr Osteoporos Rep. 2013;11(4):407–14.

65

Management of Osteoporosis in Patients with Chronic Kidney Disease

Paul D. Miller

Colorado Center for Bone Research at Panorama Orthopedics and Spine Center,
Golden, CO, USA

INTRODUCTION

Chronic kidney disease is associated with increased risk for fragility fractures. At each stage of CKD, fracture risk is higher than in postmenopausal populations with similar bone mineral densities and increased age without CKD [1–10]. The National Kidney Foundation has classified five stages of CKD based on the glomerular filtration rate (GFR) and/or proteinuria [11, 12].

Stage I: Kidney damage with normal or increased GFR (>90 mL/min).
Stage II: Kidney damage with normal or decreased GFR (60 to 89 mL/min).
Stage III: Moderate decrease in GFR (30 to 59 mL/min).
Stage IV: Severe decrease in GFR (15 to 29 mL/min).
Stage V: Kidney failure (GFR <15 mL/min or on dialysis).

For stage I–II, kidney damage is defined as pathological abnormalities or markers of damage including abnormalities of urine, blood tests, or imaging studies. The proteinuria portion of the definitions for CKD includes persistent as opposed to intermittent proteinuria, a transient situation that can occur with heavy exercise or fever.

FRACTURE RISK IN CHRONIC KIDNEY DISEASE

Fracture risk approximately doubles at equivalent age and BMD as soon as the GFR is 60 mL/min or below (eg, stage IIIA CKD). Fracture risk increases progressively as renal function deteriorates. The mechanisms behind the greater skeletal fragility as GFR declines are multifactorial and include the following.

1. Secondary hyperparathyroidism, which often progresses as the GFR declines [13–17].
2. Hyperphosphatemia, which may directly or indirectly alter mineralization of bone [18, 19].
3. Elevated FGF-23, which, in addition to its dominant role in phosphorus homeostasis, may have important effects on bone mineralization [20, 21].
4. Decrease in serum levels of 1,25-dihydroxyvitamin D, which is predominantly produced in the kidney (even with normal precursor 25OHD) [11, 12, 15].
5. Chronic metabolic acidosis, which affects bone resorption as well as mineralization [22].
6. Sarcopenia, with a subsequent greater risk for falling [23].

Morbidity and mortality after fragility fractures in CKD are also much higher in subjects with CKD, even after adjusting for age, BMD, and BMI, for multifactorial reasons. CKD, in particular stage IV–V, is accompanied by multisystem disease, especially cardiovascular disease, which adds to the higher risk for death following a fracture in these patients. In the populations used to develop the Fracture Risk Assessment Tool (FRAX), there were too few patients with seriously impaired renal function to evaluate the relationship between GFR or estimated GFR (eGFR) and fracture risk.

Before the Kidney Disease Improving Global Outcome (KDIGO) working group coined the term CKD-MBD (chronic kidney disease–mineral and bone disorder) to embrace the systemic pathology that accompanies altered

---

Primer on the Metabolic Bone Diseases and Disorders of Mineral Metabolism, Ninth Edition. Edited by John P. Bilezikian.
© 2019 American Society for Bone and Mineral Research. Published 2019 by John Wiley & Sons, Inc.
Companion website: www.wiley.com/go/asbmrprimer

bone turnover in this population, the classification of the bone diseases accompanying CKD was defined by quantitative bone histomorphometry [24]. These histomorphometric classifications are still scientific and valid. Quantitative histomorphometry requires double tetracycline labeling to define dynamic bone turnover parameters, which have specific criteria for the various types of renal osteodystrophy (adynamic, hyperparathyroid, mixed renal bone disease, osteomalacia, osteoporosis) [25–27]. In contrast, CKD-MBD is difficult to define in clinical practice and does not have specific International Classification of Diseases, Ninth Revision (ICD-9) or ICD-10 diagnostic codes. While it is known that the biochemical abnormalities accompanying CKD-MBD alter bone turnover or mineralization, which can influence bone strength, the primary clinical challenge is to differentiate among the diseases accompanying CKD—all of which may have a low BMD and/or be associated with fragility (including hip) fractures [27–29]. The challenge for physicians managing fragility fractures in patients with CKD is discriminating fractures due to osteoporosis from fractures due to the traditional bone diseases accompanying CKD.

DIAGNOSIS OF OSTEOPOROSIS IN PATIENTS WITH RENAL IMPAIRMENT

Stage I–III CKD

There are a number reasons why the WHO criteria for diagnosing osteoporosis can be used across the spectrum of stage I–III CKD (GFR 30 to >90 mL/min). First is that fact that such patients were included among those identified by WHO diagnostic criteria in the registration trials for the treatment of PMO. In these trials, some form of renal function assessment was used to enroll subjects. Exclusion criteria ranged from: (i) a baseline serum creatinine concentration below 1.27 mg/dL for the alendronate trials; (ii) 1.1 times the upper limit of normal serum creatinine concentration for the risedronate registration trials; (iii) a serum creatinine concentration of <2.4 mg/dL for the ibandronate trials; or (iv) eGFR >30 mL/min for the zoledronic acid and denosumab trials [30]. Second, the measurable derangements in bone and mineral metabolism suggesting the presence of CKD-MBD, such as secondary hyperparathyroidism or hyperphosphatemia, are less pronounced at a GFR >30 mL/min, unless there are also nonrenal related causes of secondary hyperparathyroidism. Third, neither serum PTH nor serum phosphorus were systematically randomly measured in the whole populations constituting the osteoporosis registration trials. The biology of bone is clearly in a different milieu with CKD patients than in patients with osteoporosis without CKD since measurable changes in molecules that affect bone metabolism (PTH, FGF-23, serum phosphorus) may be seen in early, even stage II, CKD [11, 20]. However, to what extent these early rises in serum phosphorus regulatory peptides have on the clinical profiling between osteoporosis and CKD-MBD remains elusive.

Stage IV–V CKD

By the time patients progress to stage IV–V CKD, the derangements in bone metabolism become so dominant that the WHO criteria for the diagnosis of osteoporosis or use of FRAX without a clinical adjustment for the greater fracture risk than calculated by FRAX alone become invalid. The most recent International Society for Clinical Densitometry (ISCD) position development conference that dealt with how to incorporate nonvalidated (by the original WHO data) data into FRAX still did not have enough data to add CKD to the clinical risk log to calculate fracture risk [31]. Nevertheless, clinical recognition of the higher fracture risk that is observed in severe (stage IV–V) CKD is useful because it emphasizes the additional risk that CKD-MBD derangements in bone metabolism add to management decisions intended to reduce risk.

DXA underestimates the fracture risk in stage IV–V CKD [14]. However, more sensitive methodologies for measuring cortical bone of the radius or tibia, such as pQCT or its high resolution counterpart HRpQCT, perform better (receiver-operating curve 0.78) than does 2D DXA at discriminating between fractured and nonfractured patients with stage IV–V CKD [4, 11, 12]. This greater differentiation may be related to the improved capacity of high-resolution 3D modalities to define bone size, bone microarchitecture, and cortical porosity than 2D DXA measurements.

BONE TURNOVER MARKERS IN CHRONIC KIDNEY DISEASE

In clinical practice, several serum biochemical bone turnover markers (BTMs), both resorption and formation, can be measured [32, 33]. Studies suggest that certain serum BTM levels, including serum PTH, may aid in the discrimination among the heterogeneous forms of renal bone disease [34]. There are two markers that are not cleared by the kidney: an osteoclast cellular number marker, tartrate-resistant acid phosphatase (TRAP5b), and the osteoblast activity marker, bone-specific alkaline phosphatase (BSAP). Changes in serum levels of the osteoblast-derived marker P1NP correlated with an improvement in BMD and/or bone microarchitecture in the pooled clinical teriparatide trial data [35]. P1NP is currently assessed by two assays: one measures the intact (monomer and trimer) form of P1NP (Roche Diagnostics, Mannheim, Germany) while the other measures only the trimer form of P1NP (Immunodiagnostic Systems, Gaithersburg, MD, USA). The trimer form is not cleared by the kidney, while the intact molecule is cleared by the kidney. Currently there are insufficient data to know if these differences influence P1NP clearance enough to influence clinical utilization of the intact P1NP in determining osteoblast activity.

BONE BIOPSY FOR QUANTITATIVE PURPOSES

Transiliac bone biopsy done with prior double tetracycline labeling is the most sensitive and specific means of discrimination among the various renal bone diseases [27, 28], and, by exclusion, making the diagnosis of osteoporosis. The science underpinning quantitative histomorphometry is rooted in robust data sets defining normal and abnormal bone turnover. Whereas hyperparathyroid bone parameters have a spectrum of histomorphometry according to the severity and longevity of the hyperparathyroid disorder, osteomalacia has a clear set of criteria required for its definition. Adynamic bone disease is generally considered to be a turnover disorder best defined by the absence of any single or double tetracycline labels [15].

TREATMENT OF OSTEOPOROSIS IN PATIENTS WITH RENAL IMPAIRMENT

Osteoporosis management and treatment should for stage I–III CKD should be the same as that for people without CKD, as long as there are no biochemical markers suggestive of the presence of CKD-MBD. The registration trials for all of the US Food and Drug Administration (FDA) approved pharmacological therapies for osteoporosis contained individuals across the spectrum of stage I–III CKD [12, 16]. The major limitation in understanding the role of FDA-approved pharmacological drugs for osteoporosis in patients with severe (stage IV–V) CKD is the lack of evidence for fracture risk reduction in these patients, with the exception of a few post hoc analyses in small subsets of registered cohorts.

Antiresorptive agents

Bisphosphonates and denosumab are the most widely used antiresorptive agents for osteoporosis, and therefore these two agents will be focused here. Their mechanisms of action and skeletal effects are discussed elsewhere.

Bisphosphonates are cleared by the kidney both by filtration and active proximal tubular secretion. The amount of bisphosphonate retained in the skeleton is likely a function of the baseline remodeling space, the chronic rate of bone turnover, and the GFR [36]. Approximately 50% of the absorbed dose of oral bisphosphonates and of the administered dose of intravenous bisphosphonates is excreted by the kidney. Oral bisphosphonates have never been shown to have renal toxicity, while intravenous bisphosphonates, especially zoledronic acid, may acutely reduce the GFR via a tubular lesion that mimics acute tubular necrosis. Intravenous ibandronate has not been shown in either clinical trials or post-marketing reports to have a negative effect on the kidney, but there are no head-to-head studies comparing the renal effects of these two bisphosphonates in normal, healthy subjects and in patients with impaired GFR. Even zoledronic acid, when administered more slowly than the registered label advises (15 minutes), seems safe in clinical experience, even in those patients with impaired GFR. Data from the zoledronic acid cancer trials provide an insight that the potential renal damage with this drug is related to dose and rate of infusion [37]. From a pharmacokinetic profile, renal damage appears to be related to the C_{max} (the peak concentration of the drug) rather than to the total dose administered. Nevertheless, because of the renal clearance and lack of clinical trial data in subjects with GFR <30 mL/min, bisphosphonates carry either a warning or a contraindication label for use in patients with GFR <30 to 35 mL/min. In that regard, use of bisphosphonates is an off-label use in patients with stage IV–V CKD, but, if used, intravenous doses should be administered very slowly (60 minutes).

In separate post hoc analysis from the pooled risedronate registration studies and the alendronate fracture intervention trials, both of these oral bisphosphonates were used in approximately 600 patients per trial (300 treated and 300 placebo) to treat subjects with PMO and eGFR using Cockgroft-Gault equations of below 30 mlL/min and down to 15 mL/min. In those patients with impaired renal function, both bisphosphonates reduced the incidence of either morphometric vertebral fractures or all clinical fractures significantly compared with placebo over an average of 2.6 years, without any change in renal function [36].

Denosumab, the anti-RANKL antibody, is metabolized in the reticuloendothelial system, is not excreted through the kidney, and is not nephrotoxic [38]. As a result, denosumab is not contraindicated in patients with impaired renal function. In post hoc analyses from the original registration trial, denosumab significantly increased BMD and reduced incident vertebral fractures in subjects down to a GFR of 15 mL/min [39]. Although the pharmacokinetics of denosumab are not affected by renal function [40], the skeletal effects of the drug in patients on dialysis have not been adequately evaluated. Because the risk of hypocalcemia with denosumab appears to be increased in patients with severe renal impairment, even more careful attention to the vitamin D status of these patients is warranted.

Anabolic agents

Teriparatide as well as abaloparatide are the two anabolic agents currently approved for the treatment of postmenopausal osteoporosis. Teriparatide is also approved for male and glucocorticoid-induced osteoporosis [41]. The teriparatide registration trial, like other registration trials for osteoporosis, did not randomize subjects with known stage IV–V CKD but had a small subset of patents who had an eGFR down to 30 mL/min [41, 42]. Similar increases in BMD and P1NP were observed across tertiles of eGFR. Fracture numbers were too small to have power for statistical analysis across these three tertiles. There were no changes in renal function as assessed by changes in serum creatinine or serum calcium concentrations as a function of eGFR during the registration trial with the approved 20 µg/day or the higher 40 µg/day doses of teriparatide. While 24-hour urine calcium excretion increased on average ~50 mg/day more than with placebo, there was no greater risk of clinical nephrolithiasis, though preexisting kidney stones were an exclusionary criteria for trial randomization.

There is a paucity of data on the effect of teriparatide in subjects with stage IV–V CKD or in subjects with adynamic renal bone disease. A single case of biopsy-proven renal adynamic bone disease treated with teriparatide showed the appearance of new bone formation on paired bone biopsy [43]. The potential use of teriparatide in known adynamic bone disease is predicated based on the knowledge that teriparatide can increase bone turnover and improve bone microarchitecture, that a strong correlation exists between increases in BMD and fracture risk reduction with teriparatide, and that this is a disease without other treatment options [43]. Currently there are no published data on the effect of abaloparatide across the spectrum of CKD [44].

CONCLUSION

The management of patients with osteoporosis or fragility fractures who have stage I–III CKD due to age-related reductions in GFR and without suggestions of CKD-MBD should not differ from that of patients without reductions in eGFR. In patients with stage IV–V CKD who have fragility fractures, the first management step is making the correct diagnosis. Biochemical markers of bone turnover, in particular serum PTH and tissue-specific alkaline phosphatase, may provide differentiation among biopsy-proven adynamic and hyperparathyroid bone disease and/or osteomalacia. Diagnosis of osteoporosis in stage IV–V CKD is made only after excluding other forms of metabolic bone disease. The exclusion of renal adynamic bone disease is particularly important because use of antiresorptive agents may not be beneficial in persons with no bone turnover to begin with. The diagnosis of adynamic bone disease is best made by quantitative histomorphometry, a clinical science that is underutilized. There is a great need to gain knowledge and evidence for the effects of registered therapies for osteoporosis in very high-risk stage IV–V CKD subjects who have sustained a low trauma fracture.

REFERENCES

1. Ensrud KE, Lui LY, Taylor BC, et al; Osteoporotic Fractures Research Group. Renal function and risk of hip and vertebral fractures in older women. Arch Intern Med. 2007;167(2):133–9.
2. Dukas L, Schacht E, Stahelin HB. In elderly men and women treated for osteoporosis a low creatinine clearance <65 ml/min is a risk factor for falls and fractures. Osteoporosis Int. 2005;16(12):1683–90.
3. Fried LF, Biggs ML, Shlipak MG, et al. Association of kidney function with incident hip fracture in older adults. J Am Soc Nephrol. 2007;18(1):282–6.
4. Nickolas TL, Cremers S, Zhang A, et al. Discriminants of prevalent fractures in chronic kidney disease. J Am Soc Nephrol. 2011;22(8):1560–7.
5. Nickolas TL, Leonard MB, Shane E. Chronic kidney disease and bone fracture: a growing concern. Kidney Int. 2008;2(6):1–11.
6. Alem AM, Sherrard DJ, Gillen DL, et al. Increased risk of hip fracture among patients with end-stage renal disease. Kidney Int. 2000;58(1):396–9.
7. Ball AM, Gillen DL, Sherrard D, et al. Risk of hip fracture among dialysis and renal transplant recipients. JAMA. 2002;288(23):3014–18.
8. Jadoul M, Albert JM, Akiba T, et al. Incidence and risk factors for hip or other bone fractures among hemodialysis patients in the Dialysis Outcomes and Practice Patterns Study. Kidney Int. 2006;70(7):1358–66.
9. Stehman-Breen CO, Sherrard DJ, Alem AM, et al. Risk factors for hip fracture among patients with end-stage renal disease. Kidney Int. 2000;58(5):2200–5.
10. Ensrud KE. Fracture risk in CKD. Clin J Am Soc Nephrol. 2013;8(8):1282–3.
11. Kidney Disease: Improving Global Outcomes (KDIGO) CKD–MBD Working Group. KDIGO clinical practice guideline for the diagnosis, evaluation, prevention, and treatment of chronic kidney disease–mineral and bone disorder (CKD–MBD). Kidney Int. 2009;76(suppl 113):S1–130.
12. Zangeneh F, Clarke BL, Hurley DL, et al. Chronic kidney disease mineral and bone disorder: what the endocrinologist needs to know. Endocrine Pract. 2013;20(5):500–16.
13. Jamal S, Miller PD. Secondary and tertiary hyperparathyroidism. J Clin Densitom. 2013;16(1):64–8.
14. Bucur RC, Panjwani DD, Turner L, et al. Low bone mineral density and fractures in stages 3-5 CKD: an updated systematic review and meta-analysis. Osteoporos Int. 2015;26(2):449–58.
15. Andress DL, Sherrard DJ. The osteodystrophy of chronic renal failure. In: Schrier RW (ed.) *Diseases of the Kidney and Urinary Tract*. Philadelphia: Lippincott Williams and Wilkins, 2003, pp 2735–68.
16. Gal-Moscovici A, Sprague SM. Osteoporosis and chronic kidney disease. Semin Dial. 2007;20(5):423–30.
17. Lamb EJ, Vickery S, Ellis AR. Parathyroid hormone, kidney disease, evidence and guidelines. Ann Clin Biochem. 2007;44(1):1–4.
18. Fang Y, Ginsburg C, Sugatani T, et al. Early chronic kidney disease-mineral bone disorder stimulates vascular calcification. Kidney Int. 2013;85(1):1–9.
19. Hruska KA, Saab G, Mattew S, et al. Renal osteodystrophy, phosphate homeostasis, and vascular calcification. Semin Dial. 2007;20(4):309–15.
20. Jüppner H, Wolf M, Salusky IB. FGF-23: more than a regulator of renal phosphate handling? J Bone Miner Res. 2010;25:2091–7.

21. Bonewald L, Johnson ML. Osteocytes, mechanosensing and Wnt signaling. Bone. 2008;42(4):606–15.
22. Gambaro G, Croppi E, Coe F, et al; Consensus Conference Group. Metabolic diagnosis and medical prevention of calcium nephrolithiasis and its systemic manifestations: a consensus statement. J Nephrol. 2016;29(6): 715–34.
23. Carrero JJ, Johansen KL, Lindholm B, et al. Screening for muscle wasting and dysfunction in patients with chronic kidney disease. Kidney Int. 2016;90(1):53–66.
24. Parfitt AM, Drezner M, Glorieux F, et al. Bone histomorphometry: standardization of nomenclature, symbols, and units. Report of the ASBMR Histomorphometry Nomenclature Committee. J Bone Miner Res. 1987;2(6): 595–610.
25. Frost HM. Tetracycline-based histological analysis of bone remodeling. Calcif Tissue Res. 1969;3(3):211–37.
26. Hitt O, Jaworski ZF, Shimizu AG, et al. Tissue-level bone formation rates in chronic renal failure, measured by means of tetracycline bone labeling. Can J Physiol Pharmacol. 1970;48(12):824–8.
27. Parfitt AM. Renal bone disease: a new conceptual framework for the interpretation of bone histomorphometry. Curr Opin Nephrol Hyperten. 2003;12(4):387–403.
28. Miller PD. The role of bone biopsy in chronic kidney disease. Clin J Am Soc Nephrol. 2008;3(suppl 3): S140–50.
29. Trueba D, Sawaya BP, Mawad H, et al. Bone biopsy: indications, techniques, and complications. Semin Dial. 2003;16(4):341–5.
30. Miller PD. Osteoporosis in patients with chronic kidney disease: diagnosis, evaluation and management. In: Basow DS (ed.) *Up-to-Date*. Waltham, MA: UpToDate, 2013.
31. International Society for Clinical Densitometry. 2010 *Official Positions of the ISCD/IOF on the Interpretation and Use of FRAX in Clinical Practice*. http://www.iscd.org/official-positions/2010-official-positions-iscd-iof-frax/ (accessed May 2018).
32. Eastell R, Hannon RA. Biomarkers of bone health and osteoporosis risk. Proc Nutr Soc. 2008;67(2):157–62.
33. Civitelli R, Armamento-Villareal R, Napoli N. Bone turnover markers: understanding their value in clinical trials and clinical practice. Osteoporos Int. 2009;20(6):843–51.
34. Coen G, Ballanti P, Bonnucci E, et al. Renal osteodystrophy in predialysis and hemodialysis patients: comparison of histologic patterns and diagnostic predictively of intact PTH. Nephron. 2002;91(1):103–11.
35. Krege J, Jennifer Meyer Harris, Lane N, et al. PINP as a biological response marker during teriparatide treatment for osteoporosis. Osteoporos Int. 2014;25(9): 2159–71.
36. Miller PD. The kidney and bisphosphonates. Bone. 2011;49(1):77–81.
37. Saad F, Gleason DM, Murray R, et al.; Zoledronic Acid Prostate Cancer Study Group. A randomized, placebo-controlled trial of zoledronic acid in patients with hormone-refractory metastatic prostate carcinoma. J Natl Cancer Inst. 2002;94(19):1458–68.
38. Miller PD. A review of the efficacy and safety of denosumab in postmenopausal women with osteoporosis. Ther Adv Musculoskelet Dis. 2011;3(6):271–82.
39. Jamal SA, Ljunggren O, Stehman-Breen C, et al. The effects of denosumab on fracture and bone mineral density by level of kidney function. J Bone Miner Res. 2011;26(8):1829–35.
40. Block GA, Bone HG, Fang L, et al. A single-dose study of denosumab in patients with various degrees of renal impairment. J Bone Miner Res. 2012;27(7):1471–9.
41. Neer RM, Arnaud CD, Zanchetta JR, et al. Effect of parathyroid hormone (1-34) on fractures and bone mineral density in postmenopausal women with osteoporosis N Engl J Med. 2011;344(19):1434–41.
42. Miller PD, Schwartz EN, Chen P, et al. Teriparatide in postmenopausal women with osteoporosis and impaired renal function. Osteoporosis Int. 2007;18(1):59–68.
43. Palcu P, Dion N, Ste-Marie LG, et al. Teriparatide and bone turnover and formation in a hemodialysis patient with low-turnover bone disease: a case report. Am J Kidney Dis. 2015;65(6):933–6.
44. Miller PD, Hattersley G, Riis BJ, et al. Effect of abaloparatide vs placebo on new vertebral fractures in postmenopausal women with osteoporosis: a randomized clinical trial. JAMA. 2016;316(7):722–33.

66

Other Secondary Causes of Osteoporosis

Neveen A. T. Hamdy and Natasha M. Appelman-Dijkstra

Department of Internal Medicine, Division of Endocrinology and Centre for Bone Quality, Leiden University Medical Center, The Netherlands

INTRODUCTION

Bone loss is an inevitable consequence of aging, starting some years before the menopause, accelerating after its onset and continuing throughout life in both men and women. "Secondary causes of osteoporosis" represent the collection of heterogeneous underlying diseases and medications that may contribute to bone loss and increase bone fragility through a number of different mechanisms independently of age or estrogen deficiency. It is important to identify secondary causes of osteoporosis in clinical practice, as these are often associated with the more severe bone loss, and may also be associated with increased bone fragility by altering bone quality, independently of changes in bone mass. A number of these secondary causes may also be treatable and thus reversible [1, 2].

Over the past decade there has been increasing awareness of the impact of these secondary factors on the fracture risk profile of patients with osteoporosis [3]. As a result of the implemented policy of screening for secondary causes in the evaluation of osteoporosis in the fracture liaison services, these causes have been identified in about 50% to 80% of men [4], in more than 50% of premenopausal women [5], and in up to 30% of postmenopausal women evaluated for osteoporosis [6]. Interestingly, these secondary causes of osteoporosis were found to be prevalent in patients who had sustained a recent fracture independently of BMD measurements, whether these were in the osteoporosis, osteopenia, or normal range [7].

Secondary causes of osteoporosis are numerous, ranging from easily identifiable specific disease states such as systemic inflammatory disorders, hematological disorders, and endocrine disorders, to the use of medication, particularly glucocorticoids [2]. More "occult" conditions such as vitamin D deficiency, hypercalciuria, and hyperparathyroidism are frequently encountered causes of unexpected bone loss, which can only be diagnosed by a high degree of suspicion, and easily confirmed by appropriate investigations [6, 8].

A number of secondary causes of osteoporosis are individually discussed elsewhere in this book. This chapter focuses on systemic inflammatory diseases other than rheumatic diseases, systemic mastocytosis, and endocrine disorders as underlying contributory mechanism for bone loss and/or increased bone fragility in patients with osteoporosis.

OSTEOPOROSIS ASSOCIATED WITH SYSTEMIC INFLAMMATORY DISORDERS

The RANKL to osteoprotegerin (OPG) ratio is the primary determinant of osteoclastogenesis and thus of the maintenance of bone mass [9, 10]. In inflammatory disorders, T-cell activation leads to increased expression of T-cell-derived RANKL, which stimulates bone resorption by enhancing all steps of osteoclastogenesis [11, 12]. Bone loss resulting from underlying disease activity may be further exacerbated by the use of glucocorticoids to control the inflammatory process, which primarily affect bone remodeling by decreasing osteoblast number and function and by inhibiting OPG expression [13]. However, in systemic inflammatory disorders, the ultimate effect of glucocorticoids on bone remodeling is the result of the fine balance between their beneficial effect when successful in controlling the inflammatory process and their harmful effect, especially when high doses are required to control inflammation [14].

Primer on the Metabolic Bone Diseases and Disorders of Mineral Metabolism, Ninth Edition. Edited by John P. Bilezikian.
© 2019 American Society for Bone and Mineral Research. Published 2019 by John Wiley & Sons, Inc.
Companion website: www.wiley.com/go/asbmrprimer

Inflammatory arthritis

Rheumatoid arthritis, discussed elsewhere in this book, represents the prototype of a systemic inflammatory disorder, in which inflammation triggers the increased expression of RANKL from activated T cells, resulting in localized bone loss in the form of periarticular erosions, and in generalized bone loss in the form of osteoporosis [15, 16]. The use of glucocorticoids and/or biologic drugs, both of which interfere with the inflammatory process, have been shown to be associated with preservation of bone mass [14, 17].

Inflammatory bowel diseases

In Crohn's disease, the pathophysiology of osteoporosis is multifactorial, including the effect of inflammatory cytokines mediating disease activity (IL-6, IL-1, TNF-α), intestinal malabsorption due to disease activity or intestinal resection, the use of glucocorticoids, inability to achieve peak bone mass when the disease presents in childhood, malnutrition, immobilization, low BMI, smoking, and hypogonadism [18–20]. Ileum resection has been identified as one of the most significant risk factor for osteoporosis, followed by age, which is of relevance in predicting overall lifetime risk as Crohn's disease peaks in the second and third decade of life, so that osteoporosis potentially becomes clinically significant only as patients grow older [21]. Patients with Crohn's disease are indeed relatively young, and the exact relationship between the host of factors potentially deleterious to the skeleton remains unclear. Opinion is thus still divided on the prevalence of fractures and on bone loss in the long term in this inflammatory disorder [22–27]. Maintaining a vitamin D-replete status contributes to the maintenance of bone mass. As in the case of rheumatoid arthritis, the current use of anti-TNF-α agents as first-line therapy not only significantly improves control of disease activity, but also counteracts the deleterious effects of cytokine-driven disease activity on the skeleton [28], although additional use of antiresorptive agents may be necessary in some cases. Concern about the intestinal absorption of bisphosphonates has not been substantiated by pharmacodynamic studies which have shown that in patients with reasonably well-controlled Crohn's disease, the nitrogen-containing bisphosphonate alendronate is adequately absorbed from the gut and retained in the skeleton despite underlying chronic inflammatory gut changes and/or gut resection [29]. A randomized placebo-controlled trial conducted in 131 patients with Crohn's disease and osteopenia confirmed these findings by demonstrating a significant increase in bone mass at the lumbar spine and a decrease in biochemical markers of bone turnover in patients treated with risedronate 35 mg once weekly for 2 years compared with placebo-treated patients. However, fracture risk was overall low in the whole population studied, and no difference could be observed between active and placebo groups [30].

A meta-analysis of the efficacy and safety of treatment with bisphosphonates in inflammatory bowel disease further showed that these agents were effective and well tolerated in the treatment of low BMD in these patients, although a beneficial effect on fracture risk remains to be established [31].

Chronic obstructive pulmonary disease

In chronic obstructive pulmonary disease (COPD), proinflammatory cytokines, particularly TNF-α, are major contributory factors to the pathophysiology of the disease process [32]. Elevated inflammatory markers reflect not only severity of lung disease but also the likelihood of increased risk for comorbidities, particularly cardiovascular disease, diabetes, and osteoporosis [33, 34]. A high prevalence of osteoporosis was thus observed within the first year after diagnosis among 2699 COPD patients from the UK General Practice Research Database (GPRD) [35]. Data from over 9500 subjects from the Third National Health and Nutrition Examination Survey conducted in the United States between 1988 and 1994, also showed that airflow obstruction was associated with increased odds of osteoporosis compared with no airflow obstruction and that these odds increased with increased severity of airways obstruction [36]. Neither BMD alone nor Fracture Risk Assessment Tool (FRAX) score was found to be predictive for the presence of vertebral fractures in COPD patients of different grades of severity [36]. Loss of bone mass is associated with increased excretion of collagen breakdown products, suggesting a protein catabolic state which may not only lead to bone loss, but also to loss of skeletal muscle mass and function and progressive disability [37]. In addition to the effects of COPD on bone mass, there is evidence for COPD-related changes in bone microstructure and material properties [38, 39]. Continuous users of systemic glucocorticoids are more than twice as likely to have one or more vertebral fractures compared with nonusers [40]. In a large case-controlled study including more than 100,000 cases from the GPRD, an association between inhaled corticosteroids at daily doses equivalent to >1600 μg beclomethasone and increased fracture risk disappeared after adjustment for disease severity, suggesting that it is disease severity rather than inhaled corticosteroids that contributed to the increased fracture risk in chronic obstructive airways disorders [41].

Factors other than chronic inflammation and corticosteroid use that also contribute to bone loss and increased fracture risk in COPD include vitamin D deficiency or insufficiency, reduced skeletal muscle mass and strength, immobilization, low BMI and changes in body composition, hypogonadism, reduced levels of insulin-like growth factors, smoking, increased alcohol intake, and genetic factors [42].

The morbidity associated with vertebral fractures is particularly high in patients with COPD as vertebral fractures are associated with restrictive changes in pulmonary function, with significant decreases in forced expiratory volume, and with up to 9% reduction of predicted lung vital capacity for each additional thoracic vertebral compression fracture [43]. Osteoporosis and increased fracture risk remain, however, largely underdiagnosed and untreated in patients with COPD of various severity, despite their high fracture rate [44, 45].

OSTEOPOROSIS ASSOCIATED WITH MASTOCYTOSIS

In all forms of mastocytosis, the proximity of the mast cell to bone remodeling surfaces and the production by this cell of a large number of chemical mediators and cytokines capable of modulating bone turnover translates in skeletal involvement, ranging from severe osteolysis to significant osteosclerosis, with osteoporosis being the most frequently observed pathology [46–48]. Bone loss is also exacerbated by the use of glucocorticoids. A clinical hallmark of mastocytosis-associated osteoporosis is bone pain, which is often severe and resistant to conventional analgesia, particularly in cases of extensive bone marrow involvement and of rapidly progressive disease. This clinical feature should raise suspicion for this secondary cause for osteoporosis, particularly when associated with systemic manifestations of enhanced mast cell activity such as flushes and gastrointestinal symptoms [46–48]. Skeletal abnormalities other than osteoporosis include lytic changes, sclerotic changes or a mixed pattern of both, predominantly observed in the axial skeleton. Osteoporosis may also be the sole presentation of bone marrow mastocytosis in which case the osteoporotic process may be severe and progressive [49, 50]. Bone marrow mastocytosis has been identified as an important "occult" secondary cause of osteoporosis in men because it is diagnosed on bone biopsy in up to 9% of men with "idiopathic osteoporosis" [50]. A high prevalence of fractures has been reported in patients with indolent mastocytosis [48, 51]. However, bone turnover markers may be high, low, or normal in systemic mastocytosis, and none of the evaluated markers have been found to be consistently predictive for fracture, probably because of the complex effects of the various chemical mediators on bone remodeling. Serum tryptase may be very high, usually associated with diffuse sclerosis, or mildly elevated, or normal [48]. The measurement of the 24-hour urine excretion of N-methyl histamine has been suggested to represent a valuable noninvasive surrogate to bone marrow biopsies in establishing the diagnosis of mastocytosis and in evaluating the degree of mast cell load [50–52]. The diagnosis of systemic mastocytosis can be, however, only confidently established by the histological examination of bone marrow biopsies demonstrating pathognomonic bone marrow infiltration with large numbers of morphologically abnormal mast cells, individually or in aggregates of more than 15 cells [46, 50].

There is evidence for a beneficial effect of bisphosphonates on bone loss and risk of fracture [53–57]. Whereas mast cell cytoreductive therapy, used in the more severe and aggressive forms of systemic mastocytosis, may potentially prevent bone loss and decrease fracture risk, there are no available data on the effect of these agents on bone turnover markers, bone loss, or fracture risk [58].

OSTEOPOROSIS ASSOCIATED WITH ENDOCRINE DISORDERS

Hyperparathyroidism and hypogonadism, discussed elsewhere in this book, are the most common secondary causes of osteoporosis of endocrine origin, with correction of the endocrine disturbance leading to improvement in bone health. An improvement in BMD, bone microarchitecture, cortical thickness, and bone strength, associated with a decrease in fracture risk, is thus observed in hyperparathyroidism after curative parathyroidectomy [59].

Hyperthyroidism

Normal thyroid function is essential for normal skeletal development, linear growth, and the achievement of peak bone mass. In children, thyroid hormone deficiency is associated with impaired skeletal development and delayed bone age and the reverse is observed in hyperthyroidism. Excess thyroid hormone is often associated with increased bone turnover, potentially leading to significant bone loss and increased fracture risk, particularly in postmenopausal women. Subclinical hyperthyroidism may also be associated with a generalized increase in fracture risk, and thyroid hormone supplementation has been shown to be potentially associated with increased fracture risk even if the free T4 levels remain in the normal range [60, 61]. Whether the increased fracture risk observed in hyperthyroidism is related to the low thyroid-stimulating hormone (TSH) levels or to high physiological levels of free thyroxine remains, therefore, a matter of debate [61–63].

Disturbances in growth hormone secretion

Growth hormone (GH) and IGF-1 play an important role in skeletal growth and metabolism throughout the life span of an individual. IGF-1 mediates skeletal growth by promoting endochondral ossification and by enhancing osteoblast formation and differentiation [63, 64]. Upregulation of IGF-1-mediated pathways in osteoblasts results in widening of cortical bone and in an increase in the length of long bones, but has no significant effect on trabecular bone. IGF-1 is also a coregulator of the differentiation of osteoclasts through the induction of RANKL expression. GH also exerts direct effects on bone by stimulating proliferation, and, to some extent, differentiation of osteoblasts [63, 64]. Both GH and IGF-1 thus have anabolic effects on bone. Untreated GH deficiency is characterized by low bone turnover, decreased BMD, and increased fracture risk [65, 66]. Recombinant human GH (rhGH) replacement therapy has been shown to increase bone turnover, resulting in an initial decline in BMD, followed by a small increase of 1% to 2% in the first 2 years of therapy because of a positive balance in favor of bone formation, which has been shown to persist for up to 15 years of continuous supplementation [67–70]. In acromegaly, GH overproduction is associated with increased bone turnover and a high prevalence of often silent vertebral fractures, which correlate with the duration of the disease and the height of serum IGF-1 levels [71–73]. Radiological vertebral fractures can be documented in as many as a third of patients with acromegaly who demonstrate no significant decreases in BMD, suggesting that it is a decrease in bone quality rather than a decrease in bone mass that is responsible for the observed increase in bone fragility [71–73]. BMD measurement is thus of limited value in the assessment of fracture risk in acromegaly. Taking into account that fracture risk is increased in the presence of prevalent fractures, that these factures are mostly occult in these patients, and that although treatment of acromegaly does improve bone turnover, fracture risk persists after cure [73, 74], it seems reasonable to recommended that all patients with acromegaly should be evaluated for the presence of vertebral fractures by means of conventional plain radiography of the spine, as well as after surgical or medical cure, to allow timely use of antiresorptive agents should that become necessary. Attention should also be given to the correction of other hormone deficiencies such as hypogonadism, which may further contribute to fracture risk if left untreated [73–76].

Hypercorticolism: Cushing syndrome

Glucocorticoid-induced osteoporosis, discussed elsewhere in this book, is the most common form of secondary osteoporosis, largely caused by the exogenous use of glucocorticoids for the treatment of a variety of inflammatory and autoimmune disorders. The endogenous excessive production of glucocorticoids is a much less common cause of hypercorticolism, most commonly due to an adrenocorticotropic hormone (ACTH) producing pituitary adenoma, and less commonly to overproduction of cortisol by an adrenal adenoma or carcinoma. Glucocorticoid excess increases bone loss but also directly increases bone fragility independently of changes in BMD. Up to 50% of patients with endogenous hypercortisolism experience predominantly vertebral fractures. These fractures may be the presenting symptom of the hypercortisolism or they may be silent, only identified on conventional spinal radiography. Fracture risk is associated with disease severity and duration [77, 78]. In children with Cushing syndrome, the deleterious effects of hypercortisolism on the skeleton may be severe and associated with growth failure, decreased longitudinal growth, and failure to achieve peak bone mass, often in the absence of significant changes in BMD. It is imperative that fracture risk is evaluated in all children and adults with hypercorticolism, and conventional radiographs of the spine are strongly recommended to evaluate the presence of vertebral fractures. Recovery of bone loss is slow after disease cure, and bone fragility may persist, particularly in the presence of prevalent fractures. Similar to exogenous glucocorticoid-induced osteoporosis, patients with endogenous hypercorticolism benefit from treatment with bone-modulating agents, particularly if at high risk for fractures.

WHO NEEDS TO BE SCREENED FOR SECONDARY CAUSES FOR OSTEOPOROSIS?

The high prevalence of potentially reversible secondary causes of osteoporosis, which may be identified with a sensitivity of 92% by cost-effective laboratory investigations [8], dictates that the majority of patients with osteoporosis require a basic battery of laboratory tests before the start of treatment. Such investigations include a full blood count, serum biochemistry panel, 24-hour urine calcium excretion, and 25-hydroxyvitamin D measurements. Secondary causes of osteoporosis should be particularly sought in young patients, premenopausal women, men under the age of 65, in all patients with unexpected or severe osteoporosis, in those with accelerated bone loss, and in those experiencing bone loss under treatment with conventional osteoporosis therapy. Further laboratory tests should be requested to confirm or exclude hypogonadism, thyrotoxicosis, celiac disease, hypercortisolism, mastocytosis, and multiple myeloma. If suspicion remains high, or in the case of fragility fractures in the presence of a normal BMD, a bone marrow biopsy is indicated to diagnose a nonsecretory myeloma, mastocytosis, or another bone marrow abnormality as a potential secondary cause of osteoporosis. A double tetracycline-labeled transiliac bone biopsy may be very rarely further indicated if a diagnosis is still not reached or an occult mineralization defect is suspected.

CONCLUSION

Secondary causes of osteoporosis are very common, particularly in premenopausal women and in men with osteoporosis, while also being the cause of accelerated bone loss in postmenopausal and age-related osteoporosis. In addition to representing significant comorbidity in specific disease entities such as systemic inflammatory disorders, malignant disease, bone marrow disorders, and endocrinopathies, secondary osteoporosis is also commonly associated with often silent disturbances in calcium homeostasis such as vitamin D deficiency, hypercalciuria, and hyperparathyroidism, all of which are easily detectable by standard laboratory testing. The ubiquitous nature of "secondary osteoporosis" suggests that diverse medical disciplines need to better interact to meet some of the challenges presented by osteoporosis as a chronic comorbidity of specific disease entities. Screening for secondary causes of osteoporosis should represent an intrinsic part of the optimal management of any patient with osteoporosis.

REFERENCES

1. Painter SE, Kleerekoper M, Camacho PM. Secondary osteoporosis: a review of the recent evidence. Endocr Pract. 2006;12:436–45.
2. Mirza F, Canalis E. Secondary osteoporosis: pathophysiology and management. Eur J Endocrinol. 2015;173: R131–51.
3. Holm JP, Hyldstrup L, Jensen JB. Time trends in osteoporosis risk factor profiles: a comparative analysis of risk factors, comorbidities, and medications over twelve years. Endocrine. 2016;54:241–55.
4. Fink HA, Litwack-Harrison S, Taylor BC, et al.; Osteoporotic Fractures in Men (MrOS) Study Group. Clinical utility of routine laboratory testing to identify possible secondary causes in older men with osteoporosis: the Osteoporotic Fractures in Men (MrOS) Study. Osteoporos Int. 2016;27:331–8.
5. Peris P, Guanabens N, Martinez de Osaba MJ, et al. Clinical characteristics and etiologic factors of premenopausal osteoporosis in a group of Spanish women. Semin Arthritis Rheum. 2002;32:64–70.
6. Gabaroi DC, Peris P, Monegal A, et al. Search for secondary causes in postmenopausal women with osteoporosis. Menopause. 2010;17:135–9.
7. Malgo F, Appelman-Dijkstra NM, Termaat MF, et al. High prevalence of secondary factors for bone fragility in patients with a recent fracture independently of BMD. Arch Osteoporos. 2016;11:12.
8. Tannenbaum C, Clark J, Schwartzman K, et al. Yield of laboratory testing to identify secondary contributors to osteoporosis in otherwise healthy women. J Clin Endocrinol Metab. 2002;87:4431–7.
9. Boyle WJ, Simonet WS, Lacey DL. Osteoclast differentiation and activation. Nature. 2003;423:337–42.
10. Walsh MC, Kim N, Kadono Y, et al. Osteoimmunology: interplay between the immune system and bone metabolism. Annu Rev Immunol. 2006;24:33–6.
11. Teitelbaum SL. Osteoclasts: culprits in inflammatory osteolysis. Arthritis Res Ther. 2006;8:201.
12. Boyce BF, Schwartz EM, Xing L. Osteoclast precursors: cytokine stimulated immunomodulators of inflammatory bone disease. Curr Opin Rheumatol. 2006;18:427–32.
13. Hofbauer LC, Gori F, Riggs BL, et al. Stimulation of osteoprotegerin ligand and inhibition of osteoprotegerin production by glucocorticoids in human osteoblastic lineage cells: potential paracrine mechanisms of glucocorticoid-induced osteoporosis Endocrinology. 1999;140:4382–9.
14. Cooper C, Bardin T, Brandi ML, et al. Balancing benefits and risks of glucocorticoids in rheumatic diseases and other inflammatory joint disorders: new insights from emerging data. An expert consensus paper from the European Society for Clinical and Economic Aspects of Osteoporosis and Osteoarthritis (ESCEO). Aging Clin Exp Res. 2016;28:1–16.
15. Kong YY, Feige U, Sarosi I, et al. Activated T cells regulate bone loss and joint destruction in adjuvant arthritis through osteoprotegerin ligand. Nature. 1999;402:304–9.
16. Shimizu T, Takahata M, Kimura-Suda H, et al. Autoimmune arthritis deteriorates bone quantity and quality of peri articular bone in a mouse model of rheumatoid arthritis. Osteoporos Int. 2017;28(2):709–18.
17. Zerbini CAF, Clark P, Mendez-Sanchez L, et al.; IOF Chronic Inflammation and Bone Structure (CIBS) Working Group. Biologic therapies and bone loss in rheumatoid arthritis. Osteoporos Int. 2017;28(2):429–46.
18. Laakso S, Valta H, Verkasalo M, et al. Compromised peak bone mass in patients with inflammatory bowel disease- a prospective study. J Pediatr. 2014;164:1436–43.
19. Moschen AR, Kaser A, Enrich B, et al. The RANKL/OPG system is activated in inflammatory bowel disease and relates to the state of bone loss. Gut. 2005;54:479–87.
20. Compston J. Osteoporosis in inflammatory bowel disease. Gut. 2003;52:63–4.
21. van Hogezand RA, Banffer D, Zwinderman AH, et al. Ileum resection is the most predictive factor for osteoporosis in patients with Crohn's disease. Osteoporos Int. 2006;17:535–42.
22. Bernstein CN, Blanchard JF, Leslie W, et al. The incidence of fracture among patients with inflammatory bowel disease. A population-based cohort study. Ann Intern Med. 2000;133:795–9.
23. Loftus EV, Crowson CS, Sandborn WJ, et al. Long-term fracture risk in patients with Crohn's disease: a population-based study in Olmsted County, Minnesota. Gastroenterology. 2002;123:468–75.
24. Vestergaard P, Mosekilde L. Fracture risk in patients with celiac disease, Crohn's disease, and ulcerative colitis: a nationwide follow-up study of 16,416 patients in Denmark. Am J Epidemiol. 2002;156:1–10.
25. van Staa TP, Cooper C, Brosse LS, et al. Inflammatory bowel disease and the risk of fracture. Gastroenterology. 2003;125:1591–7.
26. Jahnsen J, Falch JA, Mowinckel, et al. Bone mineral density in patients with inflammatory bowel disease: a population-based prospective two-year follow-up study. Scand J Gastroenterol. 2004;39:145–53.
27. Targownik LE, Bernstein CN, Leslie WD. Risk factors and management of osteoporosis in inflammatory bowel disease. Curr Opin Gastroenterol. 2014;30:168–74.
28. Mauro M, Radovic V, Armstrong D. Improvement of lumbar bone mass after infliximab therapy in Crohn's disease patients. Can J Gastroenterol. 2007;21:637–42.
29. Cremers SC, van Hogezand R, Banffer D, et al. Absorption of the oral bisphosphonate alendronate in osteoporotic patients with Crohn's disease. Osteoporos Int. 2005;16: 1727–30.

30. van Bodegraven AA, Bravenboer N, Witte BI, et al.; Dutch Initiative on Crohn and Colitis (ICC). Treatment of bone loss in osteopenic patients with Crohn's disease: a double-blind, randomised trial of oral risedronate 35 mg once weekly or placebo, concomitant with calcium and vitamin D supplementation. Gut. 2014;63:1424–30.
31. Melek J, Sakuraba A. Efficacy and safety of medical therapy for low bone mineral density in patients with inflammatory bowel disease: a meta-analysis and systematic review. Clin Gastroenterol Hepatol. 2014;12: 32–44.
32. Franciosi LG, Page CP, Celli BR, et al. Markers of disease severity in chronic obstructive pulmonary disease. Pulm Pharmacol Ther. 2006;19:189–99.
33. Gan WQ, Man SF, Senthilselvan A, et al. Association between chronic obstructive pulmonary disease and systemic inflammation: a systematic review and a meta-analysis. Thorax. 2004;59:574–80.
34. Sevenoaks MJ, Stockley RA. Chronic obstructive pulmonary disease, inflammation and co-morbidity: a common inflammatory phenotype? Respir Res. 2006;7: 70–8.
35. Soriano JB, Visick GT, Muellerova H, et al. Patterns of comorbidities in newly diagnosed COPD and asthma in primary care. Chest. 2005;128:2099–107.
36. Sin DD, Man JP, Man SF. The risk of osteoporosis in Caucasian men and women with obstructive airways disease. Am J Med. 2003;114:10–14.
37. Ogura-Tomomatsu H, Asano K, Tomomatsu K, et al. Predictors of osteoporosis and vertebral fractures in patients presenting with moderate-to-severe chronic obstructive lung disease. COPD. 2012;9:332–7.
38. Bolton CE, Ionexcu AA, Shiels KM, et al. Associated loss of fat-free mass and bone mineral density in chronic obstructive pulmonary disease. Am J Respir Crit Care Med. 2004;170:1286–93.
39. Misof BM, Moreira CA, Klaushofer K, et al. Skeletal implications of chronic obstructive pulmonary disease. Curr Osteoporos Rep. 2016;14:49–53.
40. McEvoy CE, Ensrud KE, Bender E, et al. Association between corticosteroid use and vertebral fractures in older men with chronic obstructive pulmonary disease. Am J Resp Crit Care Med. 1998;157:704–9.
41. de Vries F, van Staa TP, Bracke MSGM, et al. Severity of obstructive airway disease and risk of osteoporotic fracture. Eur Respir J. 2005;25:879–84.
42. Ionescu AA, Schoon E. Osteoporosis in chronic obstructive pulmonary disease. Eur Respir J. 2003;22(suppl 46): S64–75.
43. Schlaich C, Minne HW, Bruckner T, et al. Reduced pulmonary function in patients with spinal osteoporotic fractures. Osteoporos Int. 1998;8:261–7.
44. Morden NE, Sullivan SD, Bartle B, et al. Skeletal health in men with chronic lung disease: rates of testing, treatment, and fractures. Osteoporos Int. 2011;22:1855–62.
45. Graat-Verboom L, van den Borne BE, Smeenk FW, et al. Osteoporosis in COPD outpatients based on bone mineral density and vertebral fractures. J Bone Miner Res. 2011;26:561–8.
46. Valent P, Akin C, Escribano L, et al. Standards and standardization in mastocytosis: consensus statements on diagnostics, treatment recommendations and response criteria. Eur J Clin Invest. 2007;37:435–53.
47. Barete S, Assous N, de Gennes C, et al. Systemic mastocytosis and bone involvement in a cohort of 75 patients Ann Rheum Dis. 2010;69:1838–41.
48. Rossini M, Zanotti R, Orsolini G, et al. Prevalence, pathogenesis, and treatment options for mastocytosis-related osteoporosis. Osteoporos Int. 2016;27:2411–21.
49. Lidor C, Frisch B, Gazit D, et al. Osteoporosis as the sole presentation of bone marrow mastocytosis. J Bone Miner Res. 1990;5:871–6.
50. Brumsen C, Papapoulos SE, Lentjcs EG, et al. A potential role for the mast cell in the pathogenesis of idiopathic osteoporosis in men. Bone. 2002;31:556–61.
51. van der Veer E, van der Goot W, de Monchy JG, et al. High prevalence of fractures and osteoporosis in patients with indolent systemic mastocytosis. Allergy. 2012;67:431–8.
52. Oranje AP, Mulder PG, Heide R, et al. Urinary N-methylhistamine as an indicator of bone marrow involvement in mastocytosis. Clin Exp Dermatol 2002;27:502–6.
53. Marshall A, Kavanagh RT, Crisp AJ. The effect of pamidronate on lumbar spine bone density and pain in osteoporosis secondary to systemic mastocytosis. Br J Rheumatol. 1997;36:393–6.
54. Brumsen C, Hamdy NA, Papapoulos SE. Osteoporosis and bone marrow mastocytosis: dissociation of skeletal responses and mast cell activity during long-term bisphosphonate therapy. J Bone Miner Res. 2002;17: 567–9.
55. Grey A, Bolland MJ, Horne A, et al. Five years of antiresorptive activity after a single dose of zoledronate—results from a randomized double-blind placebo-controlled trial. Bone. 2012;50:1389–93.
56. Reid IR, Black DM, Eastell R, et al. Reduction in the risk of clinical fractures after a single dose of zoledronic acid 5 milligrams. J Clin Endocrinol Metab. 2013;98:557–63.
57. Rossini M, Zanotti R, Viapiana O, et al. Zoledronic acid in osteoporosis secondary to mastocytosis. Am J Med. 2014;127:1127–60.
58. Laroche M, Bret J, Brouchet A, et al. Clinical and densitometric efficacy of the association of interferon alpha and pamidronate in the treatment of osteoporosis in patients with systemic mastocytosis. Clin Rheumatol. 2007;26:242–3.
59. Hansen S, Hauge EM, Rasmussen L, et al. Parathyroidectomy improves bone geometry and microarchitecture in female patients with primary hyperparathyroidism: a one-year prospective controlled study using high-resolution peripheral quantitative computed tomography. J Bone Miner Res. 2012;27:1150–8.
60. Bassett JH, Williams GR. Critical role of the hypothalamic–pituitary–thyroid axis in bone. Bone. 2008;43: 418–26.
61. Bauer DC, Ettinger B, Nevitt MC, et al.; Study of Osteoporotic Fractures Research Group. Risk for fracture in women with low serum levels of thyroid-stimulating hormone. Ann Intern Med. 2001;134:561–8.

62. Abe E, Marians RC, Yu W, et al. TSH is a negative regulator of skeletal remodeling. Cell. 2003;115:151–62.
63. Canalis E, Giustina A, Bilezikian JP. Mechanisms of anabolic therapies for osteoporosis. N Engl J Med. 2007;357: 905–16.
64. Ohlsson C, Bengtsson BA, Isaksson OG, et al. Growth hormone and bone. Endocr Rev. 1998;19:55–79.
65. Rosen T, Hansson T, Granhed H, et al. Reduced bone mineral content in adult patients with growth hormone deficiency. Acta Endocrinol (Copenh). 1993;129:201–6.
66. Bouillon R, Koledova E, Bezlepkina O, et al. Bone status and fracture prevalence in Russian adults with childhood-onset growth hormone deficiency. J Clin Endocrinol Metab. 2004;89:4993–8.
67. Janssen YJH, Hamdy NA, Frolich M, et al. Skeletal effects of two-years of treatment with low physiological doses of recombinant human growth hormone (GH) in patients with adult-onset GH deficiency. J Clin Endocrinol Metab. 1998;83:2143–8.
68. Gotherstrom G, Svensson J, Koranyi J, et al. A prospective study of 5 years of GH replacement therapy in GH-deficient adults: sustained effects on body composition, bone mass, and metabolic indices. J Clin Endocrinol Metab. 2001;86:4657–65.
69. Appelman-Dijkstra NM, Claessen KMJA, Hamdy NAT, et al. Effects of up to 15 years treatment with recombinant human GH (rhGH) on bone metabolism of adults with growth hormone deficiency (GHD): the Leiden Cohort Study. Clin Endocrinol. 2014;81:727–35.
70. Elbornsson M, Gotherstrom G, Bosaeus I, et al. Fifteen years of GH replacement increases bone mineral density in hypopituitary patients with adult-onset GH deficiency. Eur J Endocrinol. 2012;166:787–95.
71. Bonadonna S, Mazziotti G, Nuzzo M, et al. Increased prevalence of radiological spinal deformities in active acromegaly: a cross-sectional study in postmenopausal women. J Bone Miner Res. 2005;20:1837–44.
72. Wassenaar MJ, Biermasz NR, Hamdy NA, et al. High prevalence of vertebral fractures despite normal bone mineral density in patients with long-term controlled acromegaly. Eur J Endocrinol. 2011;164:475–83.
73. Claessen KM, Kroon HM, Pereira AM, et al. Progression of vertebral fractures despite long-term biochemical control of acromegaly: a prospective follow-up study. J Clin Endocrinol Metab. 2013;98:4808–15.
74. Mazziotti G, Bianchi A, Porcelli T, et al. Vertebral fractures in patients with acromegaly: a 3-year prospective study. J Clin Endocrinol Metab. 2013;98:3402–10.
75. Diamond T, Nery L, Posen S. Spinal and peripheral bone mineral densities in acromegaly: the effects of excess growth hormone and hypogonadism. Ann Intern Med. 1989;111:567–73.
76. Lesse GP, Fraser WD, Farquharson R, et al. Gonadal status is an important determinant of bone density in acromegaly. Clin Endocrinol. 1998;48:59–65.
77. Vestergaard P, Lindholm J, Jørgensen JO, et al. Increased risk of osteoporotic fractures in patients with Cushing's syndrome. Eur J Endocrinol. 2002;146:51–6.
78. Trementino L, Appolloni G, Ceccoli L, et al. Bone complications in patients with Cushing's syndrome: looking for clinical, biochemical, and genetic determinants. Osteoporos Int. 2014;25:913–21.

67

Exercise for Osteoporotic Fracture Prevention and Management

Robin M. Daly[1] and Lora Giangregorio[2]

[1]Institute for Physical Activity and Nutrition, School of Exercise and Nutrition Sciences, Deakin University, Geelong, VIC, Australia
[2]Schlegel-University of Waterloo Research Institute for Aging; and Department of Kinesiology, University of Waterloo, Waterloo, ON, Canada

INTRODUCTION

Exercise can improve bone health during childhood and adolescence, and in the adult years attenuate bone loss, increase or preserve muscle mass, strength, and power, and reduce the risk of falls, all of which may reduce fracture risk. However, not all exercise modalities are effective at improving all fracture risk factors (fall risk, fall impact, bone strength). A fundamental principle in exercise physiology is specificity; physiological adaptations are closely coupled to the mode, intensity, and volume of exercise, or, that exercise should be tailored to the desired outcome. For exercise to be most effective at reducing both fall and fracture risk, the type(s) of exercises and doses must be tailored to an individual's needs and therapeutic goals, and informed by high-quality evidence. This chapter will summarize current evidence with regard to the role of exercise for the prevention and management of osteoporosis, falls, and fractures.

THEORETICAL BASIS AND PRINCIPLES OF LOADING IMPORTANT TO BONE

Bone adapts to changes in its loading environment via modeling and remodeling, controlled via a negative feedback system in which bone cells detect and alter the mass, structure, and strength of bone in response to its loading requirements. Several theories have been proposed to describe this feedback system. For example, Frost's "mechanostat" theory proposes that bones have an adaptation set point, or "minimum effective strain" (MES), such that loading-induced strains above (or below) this set point will cause bone formation (or resorption) leading to an increase (or decrease) in bone strength [1]. However, Frost's theory does not specify the loading characteristics necessary to stimulate bone cells to elicit an adaptive skeletal response. Nevertheless, many animal studies have shown that the skeletal response of bone to mechanical loading is influenced by key loading characteristics:

1. Intermittent dynamic loads (eg, jumping, skipping) are more osteogenic than low impact or static activities [2].
2. High magnitude loads that are applied rapidly are effective at eliciting a positive skeletal response [3, 4].
3. Bone cells accommodate to customary loads and thus novel or diverse loading patterns are more stimulating than repetitive loads [5, 6].
4. Relatively few loading cycles (repetitions) are needed to elicit an adaptive response (if an adequate load is applied), and continual loading diminishes the capacity for bone to respond [6, 7].
5. Short loading bouts with rest periods are more effective than continuous loading [8].

These mechanical loading principles have formed the basis for human intervention trials that have quantified the effects of different types and dosages of exercise on bone health at various stages throughout the life span. While this chapter will focus on the role of exercise in

---

Primer on the Metabolic Bone Diseases and Disorders of Mineral Metabolism, Ninth Edition. Edited by John P. Bilezikian.
© 2019 American Society for Bone and Mineral Research. Published 2019 by John Wiley & Sons, Inc.
Companion website: www.wiley.com/go/asbmrprimer

middle-aged and older adults, it is important to acknowledge that the growing years represent a critical period during which exercise can enhance the mass, structure, and strength of bone. Several reviews and meta-analyses of physical activity interventions in children have reported that targeted weight-bearing exercise programs incorporating dynamic and diverse impact loading activities can enhance bone mineral accrual, structure, and strength, with the greatest benefits apparent during the prepubertal and early pubertal years [9–11]. However, questions still remain as to whether such skeletal adaptations persist into later life and translate into antifracture efficacy.

EXERCISE FOR OSTEOPOROSIS, FALLS, AND FRACTURE PREVENTION

Clinical practice guidelines recommend exercise as a strategy to reduce the risk of fractures [12, 13], but there have been no large-scale, long-term, and adequately powered randomized controlled trials (RCTs) to address this question as it would require a sample size of approximately 7000 high-risk persons to be followed for at least 5 years [14]. The highest level of evidence comes from meta-analyses of RCTs in which fracture was not the primary outcome. For instance, a 2013 systematic review and meta-analysis of clinical trials (with and without randomization) in adults aged 45 years and older found that exercise reduced overall fracture number (10 trials) by 51% (relative risk [RR], 0.49; 95% CI, 0.31 to 0.76) and vertebral fracture number (three trials) by 44% (RR, 0.56; 95% CI, 0.30 to 1.04) [15]. While there are limitations associated with this study, including the low number of trials included, the wide confidence intervals, and evidence of publication bias, a meta-analysis of 15 RCTs including 3136 participants also found that exercise reduced the risk of fall-related fractures by 40% in adults aged ≥50 years (RR, 0.60; 95% CI, 0.45 to 0.84), with little evidence of publication bias [16]. While there is high-level evidence from other meta-analyses that exercise, particularly high challenging balance training, can reduce the rate of falls (and *injurious* falls) by approximately 20% to 40% in community-dwelling older adults [17–19], balance training has no effect on *bone mineral density* [20]. For many older adults, the risk of falling increases when they undertake a concurrent motor or cognitive task such as walking while carrying objects, which is referred to as the "dual task paradigm" [21]. Emerging evidence indicates that dual task training, which involves exercise whilst performing a secondary cognitive or motor task (eg, dancing in time to music), can improve dual task performance and reduce falls [22, 23]. Therefore, the types of exercise that have been shown to be most effective for reducing fall risk are different to those that have been shown to alter bone strength. For fracture prevention, exercise prescription needs to specify the type and dose (frequency, intensity, and duration) that can target both fall and fracture risk factors, and a multimodal program may be needed to include exercises that address multiple risk factors.

Walking is an exercise that can improve aerobic fitness, body composition, and cardiometabolic health, but there is little evidence from RCTs that it counteracts bone loss [24]. Walking is a customary activity for most people and imparts relatively low-magnitude loads (strains) on bones that are unlikely to exceed the MES threshold to stimulate an adaptive skeletal response. Similarly, swimming and cycling have no effect on bone health, even though they incorporate forceful muscle contractions [25, 26]. While there is some evidence that brisk and hill walking, or walking in combination with other weight-bearing impact exercises or weighted vests, may slow bone loss [27, 28], others have reported that frequent walking is associated with an increased risk of falls [28] and fracture [29]. Exercise interventions that emphasized walking, without including exercises targeting balance, were also found to be less effective in reducing falls [17, 30]. Thus, walking as a single intervention to prevent osteoporosis, falls, or fractures is not recommended.

Weight-bearing impact exercise programs that include moderate to high impact (more than two to three times body weight) and novel or diverse multidirectional activities have been shown to maintain or improve hip and spine BMD in premenopausal women, and to a lesser extent in postmenopausal women and older men (for review refer to [31]). Although the gains in BMD are modest (~1% to 3%), relatively few multidirectional impact loads (~10 to 100 jumps per day, three to seven times per week) appear necessary to stimulate an osteogenic response [32–36]. Recent advances in imaging technology have allowed researchers to explore whether impact loading can improve cortical and trabecular density within the hip at sites of focal weakness, including the superior region of the femoral neck, with some encouraging findings [32] (Fig. 67.1). Nevertheless, questions still remain regarding the safety and efficacy of high, novel, or diverse impact loading for older or osteoporotic individuals given the mixed findings reported in the literature and the fact older people may experience pain from comorbidities such as osteoarthritis, which may influence long-term adherence. While further research is still needed, there are encouraging findings from a 12-month RCT in 88 women aged 50 to 66 years with mild knee osteoarthritis, which showed that a progressively implemented high impact training program did not affect the biochemical composition of cartilage, but did improve femoral neck bone mass [37].

Progressive resistance training (PRT) is the most effective mode of exercise to improve muscle mass, size, and strength, including in individuals who are frail or have a history of fracture [38, 39]. However, there are heterogeneous findings with regard to the effects of PRT alone on hip and spine BMD in postmenopausal women and older men [40]. Based on the available evidence, the most effective PRT programs are those that applied moderate to high loads (70% to 85% of maximal strength), incorporated the principle of progressive overload, were performed at least

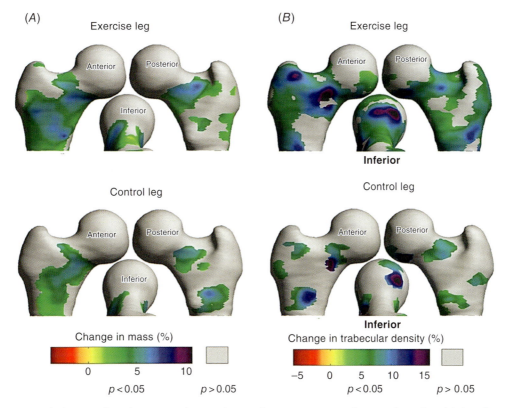

Fig. 67.1. Changes in (A) cortical surface mass density (cortical mass per unit of cortical area) and (B) endocortical trabecular density in the exercise leg and control leg following a 12-month unilateral exercise program in which healthy older men performed 50 multidirectional jumps. Data are expressed as a percentage change from preintervention values. 3D color maps are displayed across an average right proximal femur in anterior, posterior, and inferior anatomical views. Areas where there were no statistically significant changes are shown in gray. Source: [32]. Reproduced with permission of John Wiley & Sons.

twice per week, and which specifically targeted muscles attached to or near the hip and spine [12, 31, 41].

Resistance training is also often prescribed to improve functional outcomes (eg, balance, gait, mobility) and prevent falls, but, as reviewed by others, there are mixed findings from RCTs in older adults [30, 42]. Most PRT programs are designed to increase muscle strength through slow and controlled movements, but many common daily tasks (eg, a rapid step to regain balance) require rapid coordinated and dynamic contractions within 50 to 200 ms, which is less than the time needed to achieve peak muscle force (~400 to 600 ms). High-velocity PRT (power training), which involves rapid concentric muscle contractions, has been shown to be effective for improving functional performance, including movement speed and muscle force [43], even if a low external resistance (40% of maximal strength) is used [44]. Lower limb muscle power (ie, ability to produce force quickly), declines earlier and more rapidly with increasing age compared with muscle strength [45]. Furthermore, there is evidence that osteoporosis is associated with a preferential atrophy of type II fast twitch muscle fibers, which is proportional to the degree of bone loss [46]. Power training can target type II muscle fibers, and may have a positive effect on bone strength due to the high rate of loading [47], but this requires further investigation.

Based on the current evidence, exercise programs targeting multiple fall and fracture risk factors in adults without osteoporosis should be multimodal in design, that is, include a combination of moderate to high-intensity PRT or power training, weight-bearing impact, and functional balance and mobility training [18, 40]. A summary of the key recommendations is presented in Table 67.1. The inclusion of PRT in a multimodal exercise program may have the advantage of preventing diet-induced bone and muscle loss in individuals on a weight loss program [48]. However, for the greatest benefits, progression of exercise intensity or volume over time is necessary. Finally, it is important to ensure an adequate intake of calcium, vitamin D, and protein, as there is evidence that the benefits of exercise on bone and muscle health may be attenuated with inadequate intakes [49].

EXERCISE FOR OSTEOPOROSIS AND FRACTURE MANAGEMENT

Currently there is little to no direct evidence that exercise can prevent fractures in people with established osteoporosis, but there is indirect evidence that exercise can influence fracture risk via outcomes along the causal

Table 67.1. Exercise prescription recommendations unique to the prevention and management of osteoporosis, falls, and fractures.*

Type	Frequency	Intensity	Dose	Exercises/Precautions
Progressive resistance training	≥2 days per week	Start with slow and controlled movements Progress to 70–85% of 1-RM (5–7/8 on Borg 0–10-point RPE scale or hard/very hard) Consider progressing to high-velocity resistance and functional training for lower extremities to improve muscle power (light to moderate loads, 30–70% of 1-RM)	≥8 exercises 2–3 sets 8–12 repetitions 1–3-minute rest between sets	Exercises: squats, lunges, hip abduction/adduction, leg press, thoracic/lumbar extension, plantar/dorsiflexion, abdominal/postural exercises, lateral pulldown/bent over row, wall/counter/floor push up, triceps dips, and lateral shoulder raises Emphasize exercises performed in standing (weight bearing); clinical judgment is needed regarding the safety of lifting weights higher than shoulder height; use spine sparing strategies to avoid spine flexion or twisting
Weight-bearing impact exercise	4–7 times per week	Moderate to high impact activities (>2–4 BW), as tolerated Increase height of jumps, add weights/weighted vest, change direction of movements For sedentary or moderate/high-risk individuals, start with low impact exercises (see precautions)	50–100 jumps per session (3–5 sets, 10–20 repetitions) 1–2-minute rest between sets For high-risk individuals, aim to progress to 50 repetitions or as part of short bouts (≥10 minutes) of weight-bearing exercise	Multidirectional activities: jumping, bounding, skipping, hopping, bench stepping, and drop jumps It is advisable that moderate/high-risk individuals perform low impact only, or progress impact magnitude and direction with caution
Challenging balance/mobility	Accumulate at least 2–3 hours per week of activity that includes challenging balance activities	Must be challenging (close to limit of balance)	Incorporate into daily activities or combine with PRT or impact exercise (eg, balance for 10–30 seconds while waiting for kettle to boil)	Include static and dynamic movements: reduce base of support, shift weight to limits of stability (eg, leaning/reaching), perturb center of mass, stepping over obstacles, alter surface (foam mats), multisensory activities (eg, reduce vision), and dual tasking. Consider tai chi For individuals with impaired balance or high fracture risk, start with static and progress to dynamic balance exercises

* In accordance with most national physical activity guidelines, adults should accumulate ≥150 minutes per week of moderate to vigorous intensity physical activity. To realistically accomplish all of the above therapeutic goals, one could combine activities (eg, lunges as a leg-strengthening exercise that also challenges balance, and a step class that includes impact exercise and moderate/vigorous aerobic challenge and simultaneously challenges balance).
BW = body weight; PRT = progressive resistance training; 1-RM = one repetition maximum; RPE = rating of perceived exertion.

pathway to fractures in older adults, including BMD, falls, and spinal alignment. To date, few trials have examined the efficacy of exercise on BMD in individuals with osteoporosis, vertebral fractures, or secondary osteoporosis, largely because individuals with osteoporosis are often on medications influencing bone metabolism. However, several trials over 12 to 18 months in older women or men with osteopenia and/or osteoporosis or falls risk factors have reported that supervised, multimodal exercise programs incorporating moderate- to high-intensity PRT, impact, and balance exercise training can maintain or improve BMD and increase muscle mass, strength, and function [50–53]. Furthermore, a 16-year prospective study comparing 59 osteopenic women who self-selected to a multimodal exercise group with 46 osteopenic controls reported that bone loss was attenuated by 3.1% to 4.3% at the hip and spine in the exercise group, and this was associated with a 49% fracture risk reduction (RR, 0.51; 95% CI, 0.23 to 0.97; $p < 0.05$) [54]. Only two studies have examined the effects of exercise on BMD in individuals with vertebral fractures. One trial involving home-based resistance and aerobic training with stretching reported no beneficial effect of exercise on BMD [55], whereas another reported a positive effect of exercises for spinal extensors and core muscles on lumbar spine BMD after 52 weeks [56].

In light of the limited evidence, an international consensus process was conducted (termed Too Fit to Fracture) to develop exercise recommendations for individuals with osteoporosis or vertebral fractures [13]. Given the larger body of evidence examining the effects of exercise on bone health in postmenopausal women and older men, the consensus was that the effects of exercise on BMD in older adults is site- and exercise mode-specific, and should combine dynamic, weight-bearing, aerobic exercise with PRT and balance exercises. The recommendations also discouraged aerobic exercise to the exclusion of PRT and balance training. Individuals at high risk of fracture, such as those with vertebral fractures, should emphasize moderate over vigorous aerobic exercise, and form and alignment over intensity when it comes to PRT. The safety of moderate or high impact exercise in individuals with established osteoporosis is unknown. It is also unknown whether individuals with established osteoporosis can improve BMD with exercise, therefore the therapeutic goal should be to prevent bone loss.

The Too Fit to Fracture initiative also considered the efficacy of exercise for fall prevention in individuals with osteoporosis. As described earlier, the anti-fall efficacy of exercise for falls prevention in community-dwelling older adults is well established, and this is not expected to be different in individuals at moderate or high risk of falls or fractures. For example, exercise effectively reduced falls by 53% and 45% in individuals with Parkinson disease and cognitive impairment, respectively [17]. However, there is a need to consider safety precautions during physical activity, exercise, or everyday transitions for high-risk individuals. Fractures attributable to exercise or study assessments have been reported (eg, fractured costal cartilage during prone exercise, or fractured rib when rolling from supine to prone) [57]. For frail or medically complex older adults, such as those in residential care settings, or upon discharge from hospital, the fall prevention efficacy of exercise is less well established, and exercise interventions may need to be part of a multifactorial falls prevention program [17, 58]. For the elderly in long-term care facilities, the findings from a 2013 meta-analysis of 12 RCTs revealed that combined resistance and balance training programs were effective at preventing falls (RR, 0.71; 95% CI, 0.55 to 0.90), with the strongest effects observed with long-term (>6 months) programs with a frequency of at least two to three sessions per week [59].

Individuals at moderate to high risk of fracture also need to consider fall and fracture prevention during exercise and transitions [13]. Fall prevention strategies during exercise include: wearing shoes with good traction; avoiding fast movements or changes in direction, especially on hard or slippery surfaces; having a support object available during exercises that challenge balance; or using an assistive aid if balance is impaired. Spine sparing strategies are encouraged to reduce the risk of vertebral fractures. The types of movements that are the riskiest are those that involve rapid, repetitive, weighted, sustained, or end-range forward flexion or twisting of the spine. Examples of spine sparing strategies include: hip hinge (ie, flexing at the hips and knees while bringing the hips posterior to the base of support and maintaining the head over the base of support); step-to-turn; avoiding lifting from/lowering to floor; slow, controlled twist, not to end of range of motion; balance loads on either side of body; supporting trunk when flexing; and holding weight close to body, and not overhead. Individuals with pain due to osteoporotic vertebral fractures may benefit from 15 to 20 minutes of supine lying at intervals throughout the day, which promotes extension of the spine and stretching of the chest and front shoulder muscles. An alternative is lying in a prone position, which may encourage spinal extension and flexibility of the hip flexors. When sitting, individuals with pain or vertebral fractures should be encouraged to sit erect with proper lumbar support, and to avoid sitting for prolonged periods. Figure 67.2 provides some examples of beginner exercises for individuals at moderate to high risk of falls and fracture.

Hyperkyphosis can occur with vertebral fractures, weak back extensor muscles, habitual poor posture, or other chronic conditions (eg, Parkinson disease, ankylosing spondylitis) and may contribute to impaired lower extremity function or balance recovery and an increased risk of falls or fracture [60, 61]. For example, hyperkyphosis was found to be associated with an increased risk of nonspine fractures, even after adjustment for risk factors (eg, vertebral fractures) [62]. Although it has been postulated that a hyperkyphotic posture increases the risk of vertebral fractures, one study failed to find an association after controlling for prevalent vertebral fractures [63]. It may be that spinal alignment during various movements, and not just the presence of hyperkyphosis, contributes to vertebral fracture risk. Although some fractures of the spine occur during falls, many occur during daily activities (eg, bending to tie shoes). Nevertheless, there is some evidence that exercise interventions that target back extensor muscle endurance or strength may improve posture in individuals with hyperkyphosis [64, 65]. Whether the therapeutic goal for individuals with osteoporosis should be to improve back extensor endurance or back extensor strength is debatable, but a focus on back extensor muscles is agreed upon [13]. In summary, individuals at moderate or high risk of fracture should consider spine sparing strategies, fall prevention exercises, and exercises targeting back extensor muscles, in addition to PRT, balance exercises, and weight-bearing aerobic physical activity.

It has been hypothesized that whole body vibration (WBV) (eg, standing or exercising on a vibrating platform) stimulates mechanical loading and reflexive muscular contractions, and may increase muscle and bone strength, and prevent falls. However, meta-analyses of RCTs have reported inconsistent effects on BMD; heterogeneity may

Fig. 67.2. Example beginner exercises for thoracic and lumbar extensor muscles (*A*, arm and leg lengthener; *B*, bird dog), and beginner lower extremity functional strengthening exercises (*C*, body weight squat; *D*, step-up). Exercises should be tailored to ability and require instruction to ensure proper form and alignment.

be due to differences in WBV frequency, intensity, or cumulative dose, body position (eg, standing versus semiflexed knee), type of vibration, participant age, or methodological quality of the study [66, 67], or that positive findings have been spurious. While some studies suggest WBV may improve balance and fall risk [66], most of the WBV studies have focused on postmenopausal women, and thus there is little evidence regarding its benefits or harms in individuals with osteoporosis. One 24-month trial in elderly, osteopenic women reported no significant effect of 10 minutes of daily WBV on BMD [68]. Clinicians considering whether to recommend WBV should consider the cost/benefit (eg, financial cost, time commitment, and potential risks of WBV) to the client, and know that the evidence is equivocal and that not all WBV protocols have similar effects. Currently, there is not enough evidence to strongly recommend it for individuals with osteoporosis, and it should not supplant a multimodal exercise program.

CONCLUSION

Exercise is the only strategy that has the potential to improve all modifiable fracture risk parameters (fall risk, fall impact, bone strength), if it is tailored to each individual's needs and the right type and dose is prescribed. For the *prevention* of osteoporosis and falls in community-dwelling healthy adults, multimodal programs including targeted PRT (or power training), weight-bearing impact activities, and challenging balance and mobility training are most effective for improving hip and spine BMD, and muscle mass, strength, power, and function. Regular walking has modest or no effect on bone or muscle mass or function, and the evidence for whole body vibration is inconclusive. For people with a previous low trauma fragility fracture, or who are deconditioned, have comorbid conditions, kyphosis, poor posture, poor trunk muscle control/strength, and/or impaired mobility, a multimodal program is also recommended but with a focus on challenging balance and mobility training, trunk postural exercises, and spine sparing strategies in addition to PRT and weight-bearing (low impact) aerobic physical activity. For these individuals and those with osteoporosis, supervision and coaching on good alignment and correct technique is particularly important when initiating and/or progressing an exercise program.

REFERENCES

1. Frost HM. Bone's mechanostat: a 2003 update. Anat Rec A Discov Mol Cell Evol Biol. 2003;275(2):1081–101.
2. Lanyon LE, Rubin CT. Static vs dynamic loads as an influence on bone remodelling. J Biomech. 1984;17(12):897–905.
3. O'Connor JA, Lanyon LE, MacFie H. The influence of strain rate on adaptive bone remodelling. J Biomech. 1982;15(10):767–81.
4. Rubin CT, Lanyon LE. Regulation of bone mass by mechanical strain magnitude. Calcif Tissue Int. 1985;37(4):411–17.
5. Lanyon LE, Goodship AE, Pye CJ, et al. Mechanically adaptive bone remodelling. J Biomech. 1982;15(3):141–54.
6. Rubin CT, Lanyon LE. Regulation of bone formation by applied dynamic loads. J Bone Joint Surg Am. 1984;66(3):397–402.
7. Umemura Y, Ishiko T, Yamauchi T, et al. Five jumps per day increase bone mass and breaking force in rats. J Bone Miner Res. 1997;12(9):1480–5.
8. Robling AG, Hinant FM, Burr DB, et al. Improved bone structure and strength after long-term mechanical loading is greatest if loading is separated into short bouts. J Bone Miner Res. 2002;17(8):1545–54.
9. Daly RM. The effect of exercise on bone mass and structural geometry during growth. Med Sport Sci. 2007;51:33–49.
10. Nogueira RC, Weeks BK, Beck BR. Exercise to improve pediatric bone and fat: a systematic review and meta-analysis. Med Sci Sports Exerc. 2014;46(3):610–21.
11. Tan VP, Macdonald HM, Kim S, et al. Influence of physical activity on bone strength in children and adolescents: a systematic review and narrative synthesis. J Bone Miner Res. 2014;29(10):2161–81.
12. Beck BR, Daly RM, Singh MA, et al. Exercise and Sports Science Australia (ESSA) position statement on exercise prescription for the prevention and management of osteoporosis. J Sci Med Sport. 2017;20(5):438–45.
13. Giangregorio LM, Papaioannou A, Macintyre NJ, et al. Too Fit to Fracture: exercise recommendations for individuals with osteoporosis or osteoporotic vertebral fracture. Osteoporos Int. 2014;25(3):821–35.
14. Moayyeri A. The association between physical activity and osteoporotic fractures: a review of the evidence and implications for future research. Ann Epidemiol. 2008;18(11):827–35.
15. Kemmler W, Haberle L, von Stengel S. Effects of exercise on fracture reduction in older adults: a systematic review and meta-analysis. Osteoporos Int. 2013;24(7):1937–50.
16. Zhao R, Feng F, Wang X. Exercise interventions and prevention of fall-related fractures in older people: a meta-analysis of randomized controlled trials. Int J Epidemiol 2017;46(1):149–61.
17. Sherrington C, Michaleff ZA, Fairhall N,.et al. Exercise to prevent falls in older adults: an updated systematic review and meta-analysis. Br J Sports Med 2017;51(24):1750–8.
18. El-Khoury F, Cassou B, Charles MA, et al. The effect of fall prevention exercise programmes on fall induced injuries in community dwelling older adults: systematic review and meta-analysis of randomised controlled trials. BMJ. 2013;347:f6234.
19. Gillespie LD, Robertson MC, Gillespie WJ, et al. Interventions for preventing falls in older people living in the community. Cochrane Database Syst Rev. 2012(9):CD007146.
20. Duckham RL, Masud T, Taylor R, et al. Randomised controlled trial of the effectiveness of community group and home-based falls prevention exercise programmes on bone health in older people: the ProAct65+ bone study. Age Ageing. 2015;44(4):573–9.
21. Woollacott M, Shumway-Cook A. Attention and the control of posture and gait: a review of an emerging area of research. Gait Posture. 2002;16(1):1–14.
22. Booth V, Hood V, Kearney F. Interventions incorporating physical and cognitive elements to reduce falls risk in cognitively impaired older adults: a systematic review.

JBI Database System Rev Implement Rep. 2016;14(5): 110–35.
23. Trombetti A, Hars M, Herrmann FR, et al. Effect of music-based multitask training on gait, balance, and fall risk in elderly people: a randomized controlled trial. Arch Intern Med. 2011;171(6):525–33.
24. Ma D, Wu L, He Z. Effects of walking on the preservation of bone mineral density in perimenopausal and postmenopausal women: a systematic review and meta-analysis. Menopause. 2013;20(11):1216–26.
25. Stewart AD, Hannan J. Total and regional bone density in male runners, cyclists, and controls. Med Sci Sports Exerc. 2000;32(8):1373–7.
26. Taaffe DR, Snow-Harter C, Connolly DA, et al. Differential effects of swimming versus weight-bearing activity on bone mineral status of eumenorrheic athletes. J Bone Miner Res. 1995;10(4):586–93.
27. Borer KT, Fogleman K, Gross M, et al. Walking intensity for postmenopausal bone mineral preservation and accrual. Bone. 2007;41(4):713–21.
28. Ebrahim S, Thompson PW, Baskaran V, et al. Randomized placebo-controlled trial of brisk walking in the prevention of postmenopausal osteoporosis. Age Ageing. 1997;26(4):253–60.
29. Nikander R, Gagnon C, Dunstan DW, et al. Frequent walking, but not total physical activity, is associated with increased fracture incidence: a 5-year follow-up of an Australian population-based prospective study (AusDiab). J Bone Miner Res. 2011;26(7):1638–47.
30. Sherrington C, Tiedemann A, Fairhall N, et al. Exercise to prevent falls in older adults: an updated meta-analysis and best practice recommendations. N S W Public Health Bull. 2011;22(3–4):78–83.
31. Taaffe D, Daly RM, Suominen H, et al. Physical activity and exercise in the maintenance of the adult skeleton and the prevention of osteoporotic fractures. In: *Marcus R, Feldman D, Dempster D, et al. (eds) Osteoporosis* (4th ed.). Amsterdam: Elsevier Publisher, 2013, pp 683–719.
32. Allison SJ, Poole KE, Treece GM, et al. The Influence of high-impact exercise on cortical and trabecular bone mineral content and 3D distribution across the proximal femur in older men: a randomized controlled unilateral intervention. J Bone Miner Res. 2015;30(9): 1709–16.
33. Bailey CA, Brooke-Wavell K. Optimum frequency of exercise for bone health: randomised controlled trial of a high-impact unilateral intervention. Bone. 2010;46(4): 1043–9.
34. Bassey EJ, Ramsdale SJ. Increase in femoral bone density in young women following high-impact exercise. Osteoporos Int. 1994;4(2):72–5.
35. Heinonen A, Sievanen H, Kannus P, et al. Effects of unilateral strength training and detraining on bone mineral mass and estimated mechanical characteristics of the upper limb bones in young women. J Bone Miner Res. 1996;11(4):490–501.
36. Niu K, Ahola R, Guo H, et al. Effect of office-based brief high-impact exercise on bone mineral density in healthy premenopausal women: the Sendai Bone Health Concept Study. J Bone Miner Metab. 2010;28(5):568–77.
37. Multanen J, Nieminen MT, Hakkinen A, et al. Effects of high-impact training on bone and articular cartilage: 12-month randomized controlled quantitative MRI study. J Bone Miner Res. 2014;29(1):192–201.
38. Borde R, Hortobagyi T, Granacher U. Dose-response relationships of resistance training in healthy old adults: a systematic review and meta-analysis. Sports Med. 2015;45(12):1693–720.
39. Stewart VH, Saunders DH, Greig CA. Responsiveness of muscle size and strength to physical training in very elderly people: a systematic review. Scand J Med Sci Sports. 2014;24(1):e1–10.
40. Zhao R, Zhao M, Xu Z. The effects of differing resistance training modes on the preservation of bone mineral density in postmenopausal women: a meta-analysis. Osteoporos Int. 2015;26(5):1605–18.
41. Giangregorio LM, Macintyre NJ, Thabane L, et al. Exercise for improving outcomes after osteoporotic vertebral fracture. Cochrane Database Syst Rev. 2013;(1):CD008618.
42. Orr R, Raymond J, Fiatarone Singh M. Efficacy of progressive resistance training on balance performance in older adults : a systematic review of randomized controlled trials. Sports Med. 2008;38(4):317–43.
43. Steib S, Schoene D, Pfeifer K. Dose-response relationship of resistance training in older adults: a meta-analysis. Med Sci Sports Exerc. 2010;42(5):902–14.
44. Reid KF, Martin KI, Doros G, et al. Comparative effects of light or heavy resistance power training for improving lower extremity power and physical performance in mobility-limited older adults. J Gerontol A Biol Sci Med Sci. 2015;70(3):374–80.
45. Reid KF, Fielding RA. Skeletal muscle power: a critical determinant of physical functioning in older adults. Exerc Sport Sci Rev. 2012;40(1):4–12.
46. Terracciano C, Celi M, Lecce D, et al. Differential features of muscle fiber atrophy in osteoporosis and osteoarthritis. Osteoporos Int. 2013;24(3):1095–100.
47. Stengel SV, Kemmler W, Pintag R, et al. Power training is more effective than strength training for maintaining bone mineral density in postmenopausal women. J Appl Physiol. 2005;99(1):181–8.
48. Villareal DT, Chode S, Parimi N, et al. Weight loss, exercise, or both and physical function in obese older adults. N Engl J Med. 2011;364(13):1218–29.
49. Daly RM, Duckham RL, Gianoudis J. Evidence for an interaction between exercise and nutrition for improving bone and muscle health. Curr Osteoporos Rep. 2014;12(2):219–26.
50. Bolton KL, Egerton T, Wark J, et al. Effects of exercise on bone density and falls risk factors in post-menopausal women with osteopenia: a randomised controlled trial. J Sci Med Sport. 2012;15(2):102–9.
51. Gianoudis J, Bailey CA, Ebeling PR, et al. Effects of a targeted multimodal exercise program incorporating high-speed power training on falls and fracture risk factors in older adults: a community-based randomized controlled trial. J Bone Miner Res. 2014;29(1):182–91.
52. Kukuljan S, Nowson CA, Sanders KM, et al. Independent and combined effects of calcium-vitamin D3 and exercise on bone structure and strength in older men: an 18-

month factorial design randomized controlled trial. J Clin Endocrinol Metab. 2011;96(4):955–63.
53. Watson SL, Weeks BK, Weis LJ, Harding AT, Horan SA, Beck BR. High-Intensity Resistance and Impact Training Improves Bone Mineral Density and Physical Function in Postmenopausal Women With Osteopenia and Osteoporosis: The LIFTMOR Randomized Controlled Trial. J Bone Miner Res. 2018 Feb;33(2):211–220.
54. Kemmler W, Bebenek M, Kohl M, et al. Exercise and fractures in postmenopausal women. Final results of the controlled Erlangen Fitness and Osteoporosis Prevention Study (EFOPS). Osteoporos Int. 2015;26(10):2491–9.
55. Papaioannou A, Adachi JD, Winegard K, et al. Efficacy of home-based exercise for improving quality of life among elderly women with symptomatic osteoporosis-related vertebral fractures. Osteoporos Int. 2003;14(8):677–82.
56. Wang XF, Xu B, Ye XY, et al. [Effects of different treatments on patients with osteoporotic fracture after percutaneous kyphoplasty]. Zhongguo Gu Shang. 2015;28(6): 512–16.
57. Gold DT, Shipp KM, Pieper CF, et al. Group treatment improves trunk strength and psychological status in older women with vertebral fractures: results of a randomized, clinical trial. J Am Geriatr Soc. 2004;52(9):1471–8.
58. Papaioannou A, Santesso N, Morin SN, et al. Recommendations for preventing fracture in long-term care. CMAJ. 2015;187(15):1135–44, E450–61.
59. Silva RB, Eslick GD, Duque G. Exercise for falls and fracture prevention in long term care facilities: a systematic review and meta-analysis. J Am Med Dir Assoc. 2013;14(9):685–9, e2.
60. Granacher U, Gollhofer A, Hortobagyi T, et al. The importance of trunk muscle strength for balance, functional performance, and fall prevention in seniors: a systematic review. Sports Med. 2013;43(7):627–41.
61. Katzman WB, Vittinghoff E, Kado DM. Age-related hyperkyphosis, independent of spinal osteoporosis, is associated with impaired mobility in older community-dwelling women. Osteoporos Int. 2011;22(1):85–90.
62. Kado DM, Miller-Martinez D, Lui LY, et al. Hyperkyphosis, kyphosis progression, and risk of non-spine fractures in older community dwelling women: the study of osteoporotic fractures (SOF). J Bone Miner Res. 2014;29(10):2210–16.
63. Katzman WB, Vittinghoff E, Kado DM, et al. Thoracic kyphosis and rate of incident vertebral fractures: the Fracture Intervention Trial. Osteoporos Int. 2016;27(3):899–903.
64. Bansal S, Katzman WB, Giangregorio LM. Exercise for improving age-related hyperkyphotic posture: a systematic review. Arch Phys Med Rehabil. 2014;95(1): 129–40.
65. Sinaki M, Itoi E, Wahner HW, et al. Stronger back muscles reduce the incidence of vertebral fractures: a prospective 10 year follow-up of postmenopausal women. Bone. 2002;30(6):836–41.
66. Ma C, Liu A, Sun M, et al. Effect of whole-body vibration on reduction of bone loss and fall prevention in postmenopausal women: a meta-analysis and systematic review. J Orthop Surg Res. 2016;11:24.
67. Oliveira LC, Oliveira RG, Pires-Oliveira DA. Effects of whole body vibration on bone mineral density in postmenopausal women: a systematic review and meta-analysis. Osteoporos Int. 2016;27(10):2913–33.
68. Kiel DP, Hannan MT, Barton BA, et al. Low-magnitude mechanical stimulation to improve bone density in persons of advanced age: a randomized, placebo-controlled trial. J Bone Miner Res. 2015;30(7):1319–28.

68

Prevention of Falls

Heike A. Bischoff-Ferrari

Department of Geriatrics and Aging Research, University Hospital and University of Zurich, Switzerland

INTRODUCTION

Close to 75% of hip and non-hip fractures occur among seniors age 65 years and older [1]. Notably, the primary risk factor for a hip fracture is a fall, and over 90% of all fractures occur after a fall [2]. Thus, critical for the understanding and prevention of fractures at later age is their close relationship with muscle weakness [3] and falling [4, 5]. In fact, antiresorptive treatment alone may not reduce fractures among individuals 80 years and older in the presence of nonskeletal risk factors for fractures despite an improvement in bone metabolism [6]. This chapter will review the epidemiology of falls, and their importance in regard to fracture risk. Finally, fall prevention strategies and how these translate into fracture reduction are evaluated based on data from randomized controlled trials.

EPIDEMIOLOGY AND COST OF FALLS

Each year one out of three persons age 65 years and older, and one out of two aged 80 years and older experience at least one fall [7]. Nine percent of falls require an emergency room visit [8], and serious injuries occur with 10% to 15% of falls, resulting in fractures in 5% and hip fractures in 1% to 2% [4]. Additionally, falls are an independent determinant of functional decline and 40% of all nursing home admissions are due to a fall [9]. Moreover, the primary risk factor for a hip fracture is a fall, and over 90% of all fractures occur after a fall [3]. Recurrent fallers may have close to a fourfold increased odds of sustaining a fall-related fracture compared with individuals with a single fall [10]. As the number of seniors aged 65 and older is predicted to increase from 25% to 40% by 2030 [11], the number of fall-related fractures will increase substantially. Notably, even today 75% of fractures occur among seniors aged 65 years and older [1], and by 2050 the worldwide incidence in hip fractures is expected to increase by 240% among women and 310% among men [12]. Because of the increasing proportion of older individuals, annual costs from all fall-related injuries in the United States in persons 65 years or older were projected to increase from US$20.3 billion in 1994 to US$32.4 billion in 2020, including medical, rehabilitation, and hospital costs, and the costs of morbidity and mortality [13]. In fact, fall injuries are among the 20 most expensive medical conditions amounting to US$34 billion annually of direct medical costs for fall injuries [14]. Thus, therapeutic interventions that are effective in fall prevention are urgently needed.

FALL DEFINITION AND INCLUSION OF FALL RISK IN FRACTURE RISK PREDICTION

Buchner and colleagues created a useful fall definition for the common database of the FICSIT (Frailty and Injuries: Cooperative Studies of Intervention Techniques) trials [15]. Falls were defined as "unintentionally coming to rest on the ground, floor, or other lower level." Coming to rest against furniture or a wall was not counted as a fall [15]. Because falls tend to be forgotten if not associated with significant injury [16], accurately assessing fall frequency is challenging. Thus, high-quality fall assessment requires a prospective ascertainment of falls and their circumstances, ideally in short time periods (<3 months) [16], and supported by a diary [17].

Fall reports may be in the form of postcards, phone calls, hotline, or diary/calendar, although the usefulness and comprehensiveness of different ascertainment methods need further study [18]. One study among community-dwelling seniors suggested that retrospective 3-month

---

Primer on the Metabolic Bone Diseases and Disorders of Mineral Metabolism, Ninth Edition. Edited by John P. Bilezikian.
© 2019 American Society for Bone and Mineral Research. Published 2019 by John Wiley & Sons, Inc.
Companion website: www.wiley.com/go/asbmrprimer

recall of falls based on phone calls resulted in underreporting of falls by as much as 25% compared with daily calendars [17]. However, reliable use of diary calendars may not be possible in pre-frail seniors or frail seniors after hip fracture; these people may need closer in-person follow-up with monthly phone calls to assess fall events comprehensively [19].

Notably, fall assessment has not been standardized across randomized controlled trials (RCTs) or large epidemiologic data sets [20], which prevented falls from being included in the WHO Fracture Risk Assessment Tool (FRAX) tool (http://www.shef.ac.uk/FRAX/, accessed May 2018) that estimates the probability of a major osteoporotic fracture in the next 10 years [20]. Thus, FRAX may underestimate fracture risk among seniors with frequent falls [21]. Based on one Australian cohort study, the Garvan nomogram has been developed as an alternative fracture prediction tool that includes falling as a risk factor of fracture (www.fractureriskcalculator.com, accessed May 2018). In a comparative assessment, however, the predictive accuracy of the two tools showed similar performance in postmenopausal women and a possible advantage of the Garvan nomogram over FRAX among men [22]. One explanation for a similar predictive accuracy between FRAX and the Garvan nomogram may be the relatively long time interval of fall assessment in the Garvan nomogram (fall recall in the last 12 months), which may lead to the underreporting of falls not associated with significant injury [16]. The incorporation of falls as a risk factor for fracture in the FRAX algorithm is currently being evaluated within the ongoing European DO-HEALTH trial.

FALL MECHANICS AND RISK OF FRACTURE

Mechanistically, the circumstances [2] and the direction [23] of a fall determine the type of fracture, whereas bone density and factors that attenuate a fall, such as better strength or better padding, critically determine whether a fracture will take place when the faller lands on a certain bone [24]. Moreover, falling may affect bone density indirectly by resulting decreased mobility and self-restriction of activities [25]. It is well known that falls may lead to psychological trauma known as fear of falling [26]. After their first fall, about 30% of persons develop fear of falling resulting in self-restriction of activities, and decreased quality of life [25]. Key strategies to overcome fear of falling include optimization of vision and hearing, eliminating fall hazards at home, and stopping medications that may promote dizziness and weakness, plus improving self-confidence by strength and balance training plus nutritional supplements that support muscle health (eg, vitamin D, whey protein; see later in this chapter) [27]. Figure 68.1 illustrates the fall–fracture construct that describes the complexity of osteoporosis prevention introduced by nonskeletal risk factors for fractures among older individuals.

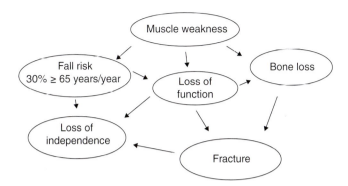

Fig. 68.1. Nonskeletal fall–fracture construct.

In support of the concept that falling is a key determinant of fracture risk, antiresorptive treatment alone may not reduce fractures among individuals 80 years and older in the presence of nonskeletal risk factors for fractures despite an improvement in bone metabolism [6]. Further, consistent with the understanding that factors unrelated to bone are at play in fracture epidemiology, the circumstances of different fractures are strikingly different. Hip fractures tend to occur in less active individuals falling indoors from a standing height with little forward momentum; they tend to fall sideways or straight down on their hip [24]. Other nonvertebral fractures, such as distal forearm or humerus fractures, tend to occur among more active older individuals who are more likely to be outdoors and have a greater forward momentum when they fall [28].

Supporting the notion that bone should not to be viewed in isolation, fracture risk due to falling is increased among individuals with osteoarthritis of the weight-bearing joints despite having increased bone density compared with controls [29]. One prospective study found that prevalent knee pain due to osteoarthritis increased the risk of falling by 26% and the risk of hip fracture twofold [30]. Another study suggested that fall force and soft tissue thickness are predictive of hip fracture risk independent of femoral strength estimated from BMD [31].

SARCOPENIA AND FALLS

Falling is a widely accepted serious consequence of sarcopenia. After being coined by Rosenberg as "paucity of flesh", the definition of sarcopenia was initially related to loss of muscle mass alone [32]. Clinical observations in ageing cohorts then emerged suggesting that muscle mass loss alone may not predict future strength decline [33, 34]. These important observations supported the rationale for use of composite definitions of sarcopenia, which required both low muscle mass and the presence of decreased strength/gait performance to define sarcopenia [33]. A recent 3-year prospective study among 445 community-dwelling seniors age 65 and older compared seven published operational definitions of sarcopenia [35–41] and two related definitions [42] for their validity in predicting the prospective rate of falls in sarcopenic versus nonsarcopenic individuals [43]. The rate of falls was best predicted by the Baumgartner definition [35] based on low appendicular lean mass (ALM) alone (relative risk [RR], 1.54; 95% CI, 1.09 to 2.18) and the Cruz-Jentoft composite definition [39] of low ALM plus low gait speed or decreased grip strength (RR, 1.82; 95% CI, 1.24 to 2.69). Notably, with the same cut-off for low ALM, the additional requirement of decreased function in the Cruz-Jentoft definition increased the prediction of the rate of falls among sarcopenic individuals from a relative risk of 1.54 (Baumgartner) to 1.82 (Cruz-Jentoft) and reduced the respective prevalence of sarcopenia from 11% (Baumgartner) to 7.1% (Cruz-Jentoft) [43]. The authors point out that based on these findings, the moderate gain in fall rate prediction by the composite endpoint of reduced ALM and reduced function (Cruz-Jentoft) is counterbalanced by the low percentage of individuals identified with sarcopenia by this definition [43]. These individuals likely represent a progressed disease stage and may therefore miss out on early treatment opportunities for sarcopenia and fall prevention.

Ongoing efforts are in place to reach a consensus on an internationally accepted operational definition for sarcopenia. These efforts are supported by the establishment in 2016 of a code in the International Classification of Diseases and Related Health Problems, 10th Revision (ICD-10) for sarcopenia (M62.84).

RISK FACTORS FOR FALLS

Falls are a hallmark of aging and becoming frail, and falls are often heralded by a decline of muscle mass and function (sarcopenia [38, 39]), onset of gait instability, visual impairment or its correction by multifocal glasses, drug treatment with antidepressants, anticonvulsants/barbiturates, or benzodiazepines, weakness, cognitive impairment, vitamin D deficiency, poor mental health, home hazards, or often a combination of several risk factors. Some studies indicate that falls due to snow and ice may play an important role in the seasonality of fractures because older persons are more likely to slip and fall during such periods [44]. On the other hand, hip fractures, which mostly occur indoors [45], may be less affected by snow and ice with a smaller seasonal swing compared with distal forearm, humerus, and ankle fractures [46].

The seemingly inseparable relationship of falls to worsening health status and the complexity of factors involved in falling has led to pessimism on the part of physicians when faced with falling, especially recurrent falling. However, there is a growing body of literature that should encourage the standardized assessment of falls and application of fall prevention strategies for fracture prevention.

FALL PREVENTION STRATEGIES

Fall prevention by risk factor reduction has been tested in a number of approaches. Multifactorial approaches—such as medical and occupational therapy assessment or adjustment in medications, behavioral instructions, and exercise programs—have been demonstrated in the PROFET (Prevention of Falls in the Elderly Trial) [47] and FICSIT trials [48]. Multifactorial approaches may be especially useful in high-risk populations for falls, such as older individuals in care institutions [49].

Many studies demonstrate that simple weight-bearing exercise programs improve gait speed, muscle strength, and balance in community-dwelling and frail seniors, which translates into a fall reduction of 25% to 50% [47, 48]. As falls are the primary risk factor for fractures, the rationale is that these interventions should also protect against fractures, although this needs confirmation in a large clinical study (also see Chapter 68).

Significant limitations of exercise programs are their cost and high implementation time. To overcome this barrier, exercise as a strategy of fall prevention may be applied at a smaller expense when the program is instructed but unsupervised. Such a home exercise program reduced falls in a randomized trial by Campbell and colleagues among community-dwelling elderly women age 80 and older [50]. Consistently, a simple unsupervised exercise home program, instructed during acute care after hip fracture repair, reduced falls significantly by 25% over 12 months follow-up among senior hip fracture patients with a mean age of 84 years [19]. Although not powered for a fracture endpoint, there was a suggestion that the unsupervised home exercise program contributed to a reduction in repeat fractures among acute hip fracture patients (relative fracture rate difference was –56% for the exercise home program versus control; 95% CI, –82% to +9%; $p = 0.08$) [19].

As a precaution of exercise programs in frail seniors with poor balance, increased mobility may lead to an increased opportunity to fall and fracture. Tai chi has been successful in reducing falls among healthy older individuals [51], and physically inactive community-dwelling older individuals [52], while frail older individuals [53] and fallers [54] may not benefit as much. Furthermore, tai chi may not improve bone density [55] and fracture prevention has not been explored as an endpoint with tai chi intervention programs.

As an extension of exercise, programs that support dual tasking may be of great value for fall prevention. Earlier studies suggested that fall risk is increased in seniors unable to walk while talking ("stops walking when talking" test [56, 57]). Thus, dual tasking assessments may best identify those at the greatest risk of falling, and programs that improve dual tasking may be useful in fall prevention at older age. This concept was tested in a trial that showed that a music-based multitask exercise program improved gait and balance, and reduced fall risk significantly by 39% in community-dwelling seniors [58].

Fall prevention strategies with evidence for fracture reduction

Two interventions among older individuals resulted in both fall and fracture reduction. One is *cataract surgery* with limited evidence from one trial; 306 women aged over 70, with cataracts, were randomized to expedited (~4-week wait) or routine (12-month wait) surgery. Over a 12-month follow-up, the rate of falling was reduced by 34% in the expedited group (rate ratio, 0.66; 95% CI, 0.45 to 0.96) accompanied by a significantly lower number of persons with a new fracture (p = 0.04) [59].

The other intervention is *vitamin D* supplementation. Mechanistically, muscle weakness is an important risk factor for falls and is a prominent feature of the clinical syndrome of vitamin D deficiency [60]. Thus, muscle weakness due to vitamin D deficiency may plausibly mediate fracture risk through an increased susceptibility to falls. The vitamin D receptor (VDR) is expressed in human muscle tissue, as suggested in several but not all studies [61]. Vitamin D bound to its nuclear receptor in muscle tissue may lead to de novo protein synthesis [62], followed by a relative increase in the diameter and number of fast type II muscle fibers [62]. Notably, fast type II muscle fibers decline with age relative to slow type I muscle fibers, resulting in an increased propensity to fall. Moreover, supplementation with 4000 IU vitamin D compared with placebo increased the number of VDRs in muscle tissue, and the number and diameter of type II muscle fibers, among postmenopausal women [62].

Finally, it is important to note that vitamin D may address several components of the fall–fracture construct, including strength [8], balance [63], lower extremity function [64], falling [65], bone density [66, 67], the risk of hip and nonvertebral fractures [68, 69], and the risk of nursing home admission [70] (see Fig. 68.1).

With evidence from several double-blind RCTs summarized in two trial-based and one pooled individual data meta-analyses, supplementation with 800 IU vitamin D per day should reduce the risk of falls (−19%) [65], hip fractures (−30%) [68, 69], and nonvertebral fractures (−14%) [68, 69] among seniors at risk of vitamin D deficiency. Notably, for both endpoints, lower doses were not effective for fall (<700 IU/day) [65] and fracture (<792 IU/day) [69] prevention. Extending to more comprehensive meta-analyses of any vitamin D dose, route of application (oral daily, bolus, intramuscular), and trial quality with respect to both study design (blinded and open designs) and fall ascertainment (any interval, prospective and retrospective), vitamin D supplementation still reduced fall risk significantly in most peer-reviewed meta-analyses (see Table 68.1), including those that focused on trials that tested active vitamin D metabolites. The most recent 2014 comprehensive meta-analysis by Bolland and colleagues included 25 clinical trials and documented only a small nonsignificant 5% benefit for fall prevention with vitamin D [71]. The main inconsistency in this recent meta-analysis [71] compared with the trials presented in Table 68.1, is the inclusion of several trials that cannot be considered reliable indicators of true treatment efficacy. Bolland and colleagues included several open-design trials and trials that did not define how falls were assessed and/or did not assess falls prospectively, or falls were imputed from fracture data. Also, the authors included trials where vitamin D was a component of a multifactorial intervention, or was applied to seniors with unstable health such as those in acute hospital care or those who had suffered a stroke or heart failure. Further, they included several trials with

Table 68.1. Meta-analyses of clinical trials testing the impact of vitamin D supplementation on fall prevention.

Year	Authors	% Fall Reduction With Vitamin D	Effect Size (95% CI)
2004	Bischoff-Ferrari et al. [77]	−22% with supplemental and active vitamin D (only double-blind RCTs)	OR = 0.78 (0.64–0.92)
2007	Jackson et al. [78]	−12% with vitamin D_3 supplementation	RR = 0.88 (0.78–1.00)
2008	O'Donnell et al. [79]	−34% with active vitamin D	OR = 0.66 (0.44–0.98)
2008	Richy et al. [80]	−21% with active vitamin D	RR = 0.79 (0.64–0.96)
2009	Bischoff-Ferrari et al. [65]	−19% with 700–1000 IU/day and +10% with 200–600 IU/day of vitamin $D_{3/2}$ supplementation (only double-blind RCTs)	RR = 0.81 (0.71–0.92) RR = 1.10 (0.89–1.35)
2010	Kalyani et al. [81]	−14% with 200–100 IU/day	RR = 0.86 (0.79–0.93)
2010	Cameron et al. [82]	−28% with vitamin D supplementation	RaR = 0.72 (0.55–0.95) (rate)
2011	Michael et al. [83]	−17% with vitamin D supplementation	RR = 0.83 (0.77–0.89)
2011	Murad et al. [84]	−14% with vitamin D supplementation	OR = 0.86 (0.77–0.96)
2014	Bolland et al. [71]	−5% with vitamin D supplementation	RR = 0.95 (0.90–1.00) (any quality trial)

OR = odds ratio; RaR = rate ratio; RR = relative risk.

oral or injected unphysiologically large bolus doses of 300,000 IU vitamin D or more, which may not be considered equivalent to daily or lower bolus dosing. In summary, the latest meta-analysis of Bolland and colleagues does not invalidate the overall finding that vitamin D at a daily dose of 800 IU/day reduces the risk of falling among seniors at risk for vitamin D deficiency, both in the community and institutions [65].

However, 800 IU vitamin D per day may not reduce falls among seniors who are vitamin D-replete. A 2-year placebo-controlled trial published in 2015 by Uusi-Rasi and colleagues from Finland tested 800 IU vitamin D per day in vitamin D-replete community-dwelling women aged 70 to 80 years with a fall in the previous year [72]. In the intent to treat analyses, vitamin D maintained femoral neck BMD and increased tibial trabecular density marginally, but did not reduce falls. The lack of benefit in the vitamin D group may be explained by the fact that none of the participants were vitamin D-deficient at baseline. The placebo group started with mean 25OHD levels of 27.5 ng/mL and stayed at 27.5 ng/mL at the 2-year follow-up exam. The group that received vitamin D started at 25.1 ng/mL and reached 37.0 ng/mL at the 2-year follow-up. In contrast, 160 Brazilian postmenopausal women with a fall in the past year and low 25OHD levels were treated with placebo or 1000 IU of vitamin D_3 for 9 months [64]. The vitamin D_3 group started at a 25OHD level of 15.0 ng/mL and reached 27.5 ng/mL on treatment. The placebo group started at 16.9 ng/mL and dropped to 13.8 ng/mL during the trial. Over the 9-month intervention period, the risk of falling in the placebo group was almost twice that in the vitamin D-treated group (RR, 1.95; 95% CI, 1.23 to 3.08) [64].

Supporting a concept that fall prevention benefits of vitamin D supplementation are primarily seen in vitamin D-deficient seniors, a recent meta-analysis of 17 RCTs suggested that a benefit of vitamin D on lower extremity strength may be seen primarily among those with vitamin D deficiency, defined as 25OHD <10 ng/mL [73].

Regarding higher bolus doses of vitamin D, recent clinical trials do not support a benefit among seniors at high risk of falling. In the most recent trial published in 2016, participants were randomized into three treatment groups: the first received a standard dose of 24,000 IU of vitamin D per month, the second received 60,000 IU of vitamin D per month, and the third received 24,000 IU of vitamin D plus 300 μg of calcifediol per month. The 200 enrolled participants had fallen at least once in the 12 months leading up to the study, were on average 78 years old, 58% were vitamin D-deficient (25OHD <20 ng/mL), and all lived independently at home [74]. Of the 200 participants, 60.5% (121 of 200) fell during the 12-month treatment period. The two monthly high-dose groups did not improve lower extremity function more than those having the standard monthly dose of 24,000 IU vitamin D and had higher percentages of participants who fell (66.9% and 66.1%, respectively) compared with the 24,000 IU group (47.9%). Participants in the 24,000 IU vitamin D group (equivalent to 800 IU/day) experienced the most improved lower extremity function and also had the fewest number of falls. A consistent pattern was seen by achieved 25OHD blood levels. The best functional improvement and fewest falls were observed at the lower 25OHD quartile range of 21.3 to 30.3 ng/mL, while no functional benefit plus most falls were observed at the highest achieved 25OHD quartile range (44.7 to 98.9 ng/mL). Notably, for the dosages examined in this study, it turned out that the two higher doses were most likely to achieve the detrimental highest quartile at 6 months and 12 months follow-up, independent of starting level. On the other hand, participants in the standard dose of 24,000 IU vitamin D group were most likely to achieve the optimal lower replete range of 21.3 to 30.3 ng/mL, and none of the standard group participants reached the undesirable highest quartile range of >45 ng/mL.

High oral doses of vitamin D and fall prevention have been evaluated in two other trials with prospective fall assessment [18, 75] and one trial that assessed falls retrospectively [76]. In one trial among 173 frail seniors after acute hip fracture, vitamin D 2000 IU/day versus 800 IU/day did not improve lower extremity function or reduce falls over a 12-month follow-up (+28%; 95% CI, –4% to +68%) [19]. Notably, the achieved mean 25OHD level at 12 months was 44.6 ng/mL in the 2000 IU/day group compared with 35.4 ng/mL in the 800 IU/day group [19]. In another trial with 2256 senior women at high risk for hip fracture, an annual bolus of 500,000 IU vitamin D versus placebo increased the risk of falling (RR, 1.15; 95% CI, 1.02 to 1.30) [75]. The bolus group achieved a 25OHD level of 48 ng/mL at 1 month and 36 ng/mL at 3 months follow-up, the timeframe where most falls occurred in the trial [75]. Notably, in both of these trials, high-dose vitamin D shifted participants to a mean 25OHD level overlapping the highest achieved quartile in the 2016 trial outlined earlier [18, 75]. In a third trial of 2686 community-dwelling seniors age 65 to 85 years, a bolus of 100,000 IU every 4 months for 5 years, reduced the risk of any new fracture significantly by 22%, but did not reduce the risk of falling (RR, 0.93; 95% CI, 0.76 to 1.14) [76]. However, falls were only assessed retrospectively for the last year of follow-up and the treatment group shifted to a mean 25OHD level of 30 ng/mL [76].

Notably, the physiology behind a possible detrimental effect of high bolus doses of vitamin D on muscle function and falls remains unclear and needs further investigation. Two ongoing trials (VITAL and DO-HEALTH) use 2000 IU vitamin D per day. These trials will provide important opportunities to verify and expand higher dose findings to other endpoints.

CONCLUSION

Fall risk reduction is a significant component of fracture prevention at older age and the public health impact of falls is significant. Falls can be reduced by a number of interventions with exercise and vitamin D offering efficacy, as established in several RCTs, extending to fracture reduction in some of the same trials. In order to study falls and the fall–fracture risk profile from different interventions and cohort studies better, fall definition and ascertainment need to be standardized in fracture ascertainment cohorts.

REFERENCES

1. Melton LJ 3rd, Crowson CS, O'Fallon WM. Fracture incidence in Olmsted County, Minnesota: comparison of urban with rural rates and changes in urban rates over time. Osteoporos Int. 1999;9:29–37.
2. Cummings SR, Nevitt MC. Non-skeletal determinants of fractures: the potential importance of the mechanics of falls. Study of Osteoporotic Fractures Research Group. Osteoporos Int. 1994;4(suppl 1):67–70.
3. Cummings SR, Nevitt MC, Browner WS, et al. Risk factors for hip fracture in white women. Study of Osteoporotic Fractures Research Group. N Engl J Med. 1995;332:767–73.
4. Centers for Disease Control and Prevention (CDC). Fatalities and injuries from falls among older adults—United States, 1993–2003 and 2001–2005. MMWR Morb Mortal Wkly Rep. 2006;55:1221–4.
5. Schwartz AV, Nevitt MC, Brown BW Jr, et al. Increased falling as a risk factor for fracture among older women: the study of osteoporotic fractures. Am J Epidemiol. 2005;161:180–5.
6. McClung MR, Geusens P, Miller PD, et al. Effect of risedronate on the risk of hip fracture in elderly women. Hip Intervention Program Study Group. N Engl J Med. 2001;344:333–40.
7. Tinetti ME. Prevention of falls and fall injuries in elderly persons: a research agenda. Prev Med. 1994;23:756–62.
8. Bischoff HA, Stahelin HB, Dick W, et al. Effects of vitamin D and calcium supplementation on falls: a randomized controlled trial. J Bone Miner Res. 2003;18:343–51.
9. Tinetti ME, Williams CS. Falls, injuries due to falls, and the risk of admission to a nursing home. N Engl J Med. 1997;337:1279–84.
10. Pluijm SM, Smit JH, Tromp EA, et al. A risk profile for identifying community-dwelling elderly with a high risk of recurrent falling: results of a 3-year prospective study. Osteoporos Int. 2006;17:417–25.
11. Eberstadt N, Groth H. *Europe's Coming Demographic Challenge: Unlocking the Value of Health*. Washington: American Enterprise Institute for Health Policy Research, 2007.
12. Gullberg B, Johnell O, Kanis JA. World-wide projections for hip fracture. Osteoporos Int. 1997;7:407–13.
13. Englander F, Hodson TJ, Terregrossa RA. Economic dimensions of slip and fall injuries. J Forensic Sci. 1996;41:733–46.
14. Stevens JA, Corso PS, Finkelstein EA, et al. The costs of fatal and nonfatal falls among older adults. Injury Prev. 2006;12:290–5.
15. Buchner DM, Hornbrook MC, Kutner NG, et al. Development of the common data base for the FICSIT trials. J Am Geriatr Soc. 1993;41:297–308.
16. Cummings SR, Nevitt MC, Kidd S. Forgetting falls. The limited accuracy of recall of falls in the elderly. J Am Geriatr Soc. 1988;36:613–16.
17. Hannan MT, Gagnon MM, Aneja J, et al. Optimizing the tracking of falls in studies of older participants: comparison of quarterly telephone recall with monthly falls calendars in the MOBILIZE Boston Study. Am J Epidemiol. 2010;171:1031–6.
18. Teister CJ, Chocano-Bedoya PO, Orav EJ, et al. Which fall ascertainment method captures most falls in pre-frail and frail seniors? Am J Epidemiol. 2018; doi: 10.1093/aje/kwy113 [epub ahead of print].
19. Bischoff-Ferrari HA, Dawson-Hughes B, Platz A, et al. Effect of high-dosage cholecalciferol and extended physiotherapy on complications after hip fracture: a randomized controlled trial. Arch Intern Med. 2010;170:813–20.
20. Kanis JA, Borgstrom F, De Laet C, et al. Assessment of fracture risk. Osteoporos Int. 2005;16:581–9.
21. Masud T, Binkley N, Boonen S, et al; FPDC Members. Official positions for FRAX(R) clinical regarding falls and frailty: can falls and frailty be used in FRAX(R)? From Joint Official Positions Development Conference of the International Society for Clinical Densitometry and International Osteoporosis Foundation on FRAX(R). J Clinical Densitom. 2011;14:194–204.
22. Sandhu SK, Nguyen ND, Center JR, et al. Prognosis of fracture: evaluation of predictive accuracy of the FRAX algorithm and Garvan nomogram. Osteoporos Int. 2010;21:863–71.
23. Nguyen ND, Frost SA, Center JR, et al. Development of a nomogram for individualizing hip fracture risk in men and women. Osteoporos Int. 2007;17:17.
24. Nevitt MC, Cummings SR. Type of fall and risk of hip and wrist fractures: the study of osteoporotic fractures. The Study of Osteoporotic Fractures Research Group. J Am Geriatr Soc. 1993;41:1226–34.
25. Vellas BJ, Wayne SJ, Romero LJ, et al. Fear of falling and restriction of mobility in elderly fallers. Age Ageing. 1997;26:189–93.
26. Arfken CL, Lach HW, Birge SJ, et al. The prevalence and correlates of fear of falling in elderly persons living in the community. Am J Public Health. 1994;84:565–70.
27. Bischoff-Ferrari HA. Three steps to unbreakable bones: the 2011 World Osteoporosis Day Report. Nyon, Switzerland: International Osteoporosis Foundation, 2011. https://wwwiofbonehealthorg/news/three-steps-unbreakable-bones-world-osteoporosis-day (accessed May 2018).
28. Graafmans WC, Ooms ME, Hofstee HM, et al. Falls in the elderly: a prospective study of risk factors and risk profiles. Am J Epidemiol. 1996;143:1129–36.

29. Arden NK, Nevitt MC, Lane NE, et al. Osteoarthritis and risk of falls, rates of bone loss, and osteoporotic fractures. Study of Osteoporotic Fractures Research Group. Arthritis Rheum. 1999;42:1378–85.
30. Arden NK, Crozier S, Smith H, et al. Knee pain, knee osteoarthritis, and the risk of fracture. Arthritis Rheum. 2006;55:610–15.
31. Dufour AB, Roberts B, Broe KE, et al. The factor-of-risk biomechanical approach predicts hip fracture in men and women: the Framingham Study. Osteoporos Int. 2011;23(2):513–20.
32. Rosenberg IH. Sarcopenia: origins and clinical relevance. J Nutr. 1997;127:S990–1.
33. Visser M, Deeg DJ, Lips P, et al. Skeletal muscle mass and muscle strength in relation to lower-extremity performance in older men and women. J Am Geriatr Soc. 2000;48:381–6.
34. Visser M, Schaap LA. Consequences of sarcopenia. Clin Geriatr Med. 2011;27:387–99.
35. Baumgartner RN, Koehler KM, Gallagher D, et al. Epidemiology of sarcopenia among the elderly in New Mexico. Am J Epidemiol. 1998;147:755–63.
36. Delmonico MJ, Harris TB, Lee JS, et al. Alternative definitions of sarcopenia, lower extremity performance, and functional impairment with aging in older men and women. J Am Geriatr Soc. 2007;55:769–74.
37. Delmonico MJ, Harris TB, Visser M, et al. Longitudinal study of muscle strength, quality, and adipose tissue infiltration. Am J Clin Nutr. 2009;90:1579–85.
38. Fielding RA, Vellas B, Evans WJ, et al. Sarcopenia: an undiagnosed condition in older adults. Current consensus definition: prevalence, etiology, and consequences. International Working Group on Sarcopenia. J Am Med Dir Assoc. 2011;12:249–56.
39. Cruz-Jentoft AJ, Baeyens JP, Bauer JM, et al. Sarcopenia: European consensus on definition and diagnosis: Report of the European Working Group on Sarcopenia in Older People. Age Ageing. 2010;39:412–23.
40. Muscaritoli M, Anker SD, Argiles J, et al. Consensus definition of sarcopenia, cachexia and pre-cachexia: joint document elaborated by Special Interest Groups (SIG) "cachexia-anorexia in chronic wasting diseases" and "nutrition in geriatrics." Clin Nutr. 2010;29:154–9.
41. Morley JE, Abbatecola AM, Argiles JM, et al. Sarcopenia with limited mobility: an international consensus. J Am Med Dir Assoc. 2011;12:403–9.
42. Studenski SA, Peters KW, Alley DE, et al. The FNIH sarcopenia project: rationale, study description, conference recommendations, and final estimates. J Gerontol A Biol Sci Med Sci. 2014;69:547–58.
43. Bischoff-Ferrari HA, Orav JE, Kanis JA, et al. Comparative performance of current definitions of sarcopenia against the prospective incidence of falls among community-dwelling seniors age 65 and older. Osteoporos Int. 2015;26:2793–802.
44. Ralis ZA. Epidemic of fractures during period of snow and ice. BMJ. 1981;282:603–5.
45. Carter SE, Campbell EM, Sanson-Fisher RW, et al. Accidents in older people living at home: a community-based study assessing prevalence, type, location and injuries. Aust N Z J Public Health. 2000;24:633–6.
46. Bischoff-Ferrari HA, Orav JE, Barrett JA, et al. Effect of seasonality and weather on fracture risk in individuals 65 years and older. Osteoporos Int 2007;24:24.
47. Close J, Ellis M, Hooper R, et al. Prevention of falls in the elderly trial (PROFET): a randomized controlled trial. Lancet. 1999;353:93–7.
48. Province MA, Hadley EC, Hornbrook MC, et al. The effects of exercise on falls in elderly patients. A preplanned meta-analysis of the FICSIT Trials. Frailty and Injuries: Cooperative Studies of Intervention Techniques. JAMA. 1995;273:1341–7.
49. Oliver D, Connelly JB, Victor CR, et al. Strategies to prevent falls and fractures in hospitals and care homes and effect of cognitive impairment: systematic review and meta-analyses. BMJ. 2007;334:82.
50. Campbell AJ, Robertson MC, Gardner MM, et al. Randomised controlled trial of a general practice programme of home based exercise to prevent falls in elderly women. BMJ. 1997;315:1065–9.
51. Gillespie LD, Robertson MC, Gillespie WJ, et al. Interventions for preventing falls in older people living in the community. Cochrane Database Syst Rev. 2012:CD007146.
52. Li F, Harmer P, Fisher KJ, et al. Tai chi and fall reductions in older adults: a randomized controlled trial. J Gerontol A Biol Sci Med Sci. 2005;60:187–94.
53. Wolf SL, Sattin RW, Kutner M, et al. Intense tai chi exercise training and fall occurrences in older, transitionally frail adults: a randomized, controlled trial. J Am Geriatr Soc. 2003;51:1693–701.
54. Voukelatos A, Cumming RG, Lord SR, et al. A randomized, controlled trial of tai chi for the prevention of falls: the Central Sydney tai chi trial. J Am Geriatr Soc. 2007;55:1185–91.
55. Lee MS, Pittler MH, Shin BC, et al. Tai chi for osteoporosis: a systematic review. Osteoporos Int. 2008;19:139–46.
56. Lundin-Olsson L, Nyberg L, Gustafson Y. "Stops walking when talking" as a predictor of falls in elderly people. Lancet. 1997;349:617.
57. de Hoon EW, Allum JH, Carpenter MG, et al. Quantitative assessment of the stops walking while talking test in the elderly. Arch Phys Med Rehabil. 2003;84:838–42.
58. Trombetti A, Hars M, Herrmann FR, et al. Effect of music-based multitask training on gait, balance, and fall risk in elderly people: a randomized controlled trial. Arch Intern Med. 2011;171:525–33.
59. Harwood RH, Foss AJ, Osborn F, et al. Falls and health status in elderly women following first eye cataract surgery: a randomised controlled trial. Br J Ophthalmol. 2005;89:53–9.
60. Schott GD, Wills MR. Muscle weakness in osteomalacia. Lancet. 1976;1:626–9.
61. Bischoff-Ferrari HA. Relevance of vitamin D in muscle health. Rev Endocr Metab Disord. 2012;13:71–7.

62. Ceglia L, Niramitmahapanya S, da Silva Morais M, et al. A randomized study on the effect of vitamin D3 supplementation on skeletal muscle morphology and vitamin D receptor concentration in older women. J Clin Endocrinol Metab. 2013;98:E1927–35.
63. Pfeifer M, Begerow B, Minne HW, et al. Effects of a short-term vitamin D and calcium supplementation on body sway and secondary hyperparathyroidism in elderly women. J Bone Miner Res. 2000;15:1113–18.
64. Cangussu LM, Nahas-Neto J, Orsatti CL, et al. Effect of isolated vitamin D supplementation on the rate of falls and postural balance in postmenopausal women fallers: a randomized, double-blind, placebo-controlled trial. Menopause. 2016;23(3):267–74.
65. Bischoff-Ferrari HA, Dawson-Hughes B, Staehelin HB, et al. Fall prevention with supplemental and active forms of vitamin D: a meta-analysis of randomised controlled trials. BMJ. 2009;339:b3692.
66. Dawson-Hughes B, Harris SS, Krall EA, et al. Effect of calcium and vitamin D supplementation on bone density in men and women 65 years of age or older. N Engl J Med. 1997;337:670–6.
67. Bischoff-Ferrari HA, Dietrich T, Orav EJ, et al. Positive association between 25-hydroxy vitamin D levels and bone mineral density: a population-based study of younger and older adults. Am J Med. 2004;116:634–9.
68. Bischoff-Ferrari HA, Willett WC, Wong JB, et al. Prevention of nonvertebral fractures with oral vitamin D and dose dependency: a meta-analysis of randomized controlled trials. Arch Intern Med. 2009;169:551–61.
69. Bischoff-Ferrari HA, Willett WC, Orav EJ, et al. A pooled analysis of vitamin D dose requirements for fracture prevention. N Engl J Med. 2012;367:40–9.
70. Visser M, Deeg DJ, Puts MT, et al. Low serum concentrations of 25-hydroxyvitamin D in older persons and the risk of nursing home admission. Am J Clin Nutr. 2006;84:616–22, quiz 71–2.
71. Bolland MJ, Grey A, Gamble GD, et al. Vitamin D supplementation and falls: a trial sequential meta-analysis. Lancet Diabetes Endocrinol. 2014;2:573–80.
72. Uusi-Rasi K, Patil R, Karinkanta S, et al. Exercise and vitamin D in fall prevention among older women: a randomized clinical trial. JAMA Intern Med. 2015;17(5):703–11.
73. Stockton KA, Mengersen K, Paratz JD, et al. Effect of vitamin D supplementation on muscle strength: a systematic review and meta-analysis. Osteoporos Int. 2011;22:859–71.
74. Bischoff-Ferrari HA, Dawson-Hughes B, Orav EJ, et al. Monthly high-dose vitamin D treatment for the prevention of functional decline: a randomized clinical trial. JAMA Intern Med. 2016;176:175–83.
75. Sanders KM, Stuart AL, Williamson EJ, et al. Annual high-dose oral vitamin D and falls and fractures in older women: a randomized controlled trial. JAMA. 2010;303:1815–22.
76. Trivedi DP, Doll R, Khaw KT. Effect of four monthly oral vitamin D3 (cholecalciferol) supplementation on fractures and mortality in men and women living in the community: randomised double blind controlled trial. BMJ. 2003;326:469.
77. Bischoff-Ferrari HA, Dawson-Hughes B, Willett WC, et al. Effect of vitamin D on falls: a meta-analysis. JAMA. 2004;291:1999–2006.
78. Jackson C, Gaugris S, Sen SS, et al. The effect of cholecalciferol (vitamin D3) on the risk of fall and fracture: a meta-analysis. QJM. 2007;100:185–92.
79. O'Donnell S, Moher D, Thomas K, et al. Systematic review of the benefits and harms of calcitriol and alfacalcidol for fractures and falls. J Bone Miner Metab. 2008;26:531–42.
80. Richy F, Dukas L, Schacht E. Differential effects of D-hormone analogs and native vitamin D on the risk of falls: a comparative meta-analysis. Calcif Tissue Int. 2008;82:102–7.
81. Kalyani RR, Stein B, Valiyil R, et al. Vitamin D treatment for the prevention of falls in older adults: systematic review and meta-analysis. J Am Geriatr Soc. 2010;58:1299–310.
82. Cameron ID, Murray GR, Gillespie LD, et al. Interventions for preventing falls in older people in nursing care facilities and hospitals. Cochrane Database Syst Rev. 2010:CD005465.
83. Michael YL, Whitlock EP, Lin JS, et al. Primary care-relevant interventions to prevent falling in older adults: a systematic evidence review for the U.S. Preventive Services Task Force. Ann Intern Med. 2011;153:815–25.
84. Murad MH, Elamin KB, Abu Elnour NO, et al. Clinical review: the effect of vitamin D on falls: a systematic review and meta-analysis. J Clin Endocrinol Metab. 2011;96:2997–3006.

69

Nutritional Support for Osteoporosis

Connie M. Weaver[1], Bess Dawson-Hughes[2], Rene Rizzoli[3], and Robert P. Heaney[4]

[1]Nutrition Science, Purdue University, West Lafayette, IN, USA
[2]USDA Nutrition Research Center at Tufts University, Boston, MA, USA
[3]University Hospitals of Geneva, and Head of the Service of Bone Diseases, Geneva, Switzerland
[4]Department of Medicine, Creighton University School of Medicine, Omaha, NB, USA

INTRODUCTION

Nutrients are essential to the viability of all cells, including those in bone. However, it is the whole diet rather than individual nutrients that determines many factors that influence bone, including nutrition adequacy of all essential nutrients, the presence or absence of inhibitors to absorption and utilization of individual nutrients, the energy available for growth and maintenance of bone and adiposity, and acid–base balance. Diet and some lifestyle choices around the world lead to shortages of some nutrients that are particularly important to bone (ie, calcium, protein, and vitamin D). Our ability to accurately link and quantify the role of individual nutrients or whole diet in building and maintaining bone is handicapped by methodological limitations in assessing dietary intakes and the time lag for seeing consequences of diet on bone. Nutrition is an important component for treating those with osteoporosis as addressed elsewhere in this book. Diet, including dietary supplements, may be the most important complement to drugs for combination therapy. However, the more important role of diet is preventive. The cumulative effect of diet over the life span influences development of peak bone mass and its subsequent maintenance. Osteoporosis has been called a pediatric disorder because adult peak bone mass is largely determined during childhood.

ROLE OF DIET IN BUILDING PEAK BONE MASS

Rapid skeletal growth occurs in infancy and adolescence. During growth, there is a high demand for nutrients. For bone mineral matrix formation, calcium, phosphorus, and magnesium are particularly important. Vitamin D status is important for active calcium absorption across the gut. Many nutrients are important for collagen synthesis, including protein, copper, zinc, and iron. A National Osteoporosis Foundation (NOF) position paper of a systematic review of predictors of peak bone mass gave grades for the strength of evidence for various nutrients, dietary patterns, and physical activities [1] (Table 69.1). The concern for low peak bone mass is risk of fracture, especially later in life. Fracture risk is also of concern in childhood, particularly during the period of relatively low BMD when bone consolidation lags behind growth [2].

Meeting nutrient needs is easier during infancy through breastfeeding or carefully developed infant formulas. With the exception of its vitamin D content, the nutrient profile of breast milk is relatively constant and nearly independent of the diet of mothers [3]. In contrast, the pubertal growth spurt occurs at a life stage where diet becomes increasingly influenced by peers. This is an extremely important period for development of peak

---

Primer on the Metabolic Bone Diseases and Disorders of Mineral Metabolism, Ninth Edition. Edited by John P. Bilezikian.
© 2019 American Society for Bone and Mineral Research. Published 2019 by John Wiley & Sons, Inc.
Companion website: www.wiley.com/go/asbmrprimer

Table 69.1. Evidence grades for lifestyle predictors of peak bone mass.

Lifestyle Factor	Grade*
Macronutrients	
Fat	D
Protein	C
Micronutrients	
Calcium	A
Vitamin D	B
Micronutrients other than calcium and vitamin D	D
Food patterns	
Dairy	B
Fiber	C
Fruits and vegetables	C
Detriment of cola and caffeinated beverages	C
Infant nutrition	
Duration of breastfeeding	D
Breastfeeding versus formula feeding	D
Enriched formula	D
Special nutrition issues	
Detriment of alcohol	D
Detriment of smoking	C
Physical activity and exercise	
Effect on bone mass and density	A
Effect on bone structural outcomes	B

Source: [1]. Reproduced with permission of SpringerNature.

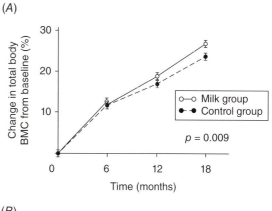

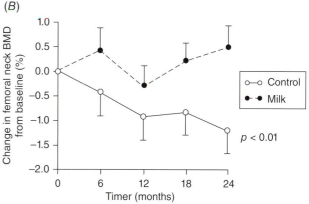

Fig. 69.1. Milk supplementation in RCTs was shown to (A) increase bone accrual in growing girls and (B) reduce bone loss in postmenopausal women. BMC = bone mineral content; BMD = bone mineral density. (A) Source: [5]. Reproduced with permission of BMJ. (B) Source: [15]. Reproduced with permission of SpringerNature.

bone mass. During the 4 years surrounding peak bone mass accretion, ~40% of peak bone mass is acquired. Peak bone mass velocity determined from a longitudinal study in white boys and girls was 409 g/day in boys and 325 g/day in girls [2]. During peak bone mass accrual, controlled feeding studies on a range of calcium intakes in black and white girls showed that calcium intake explained 12.3% of the variance in skeletal calcium retention compared with the 13.7% explained by race, whereas a measure of sexual maturity explained an additional 4% [3]. The large contribution of just one nutrient emphasizes the importance of nutrition at this life stage. It also shows the large genetic influence on bone as race is a crude marker of genotype [4].

In addition to providing raw materials for growth, diet can alter regulators of growth that affect bone accretion. In a randomized controlled trial (RCT) where early pubertal girls drank a pint of milk a day, serum IGF-1 increased, which was thought to be partly causal for the increase in BMD in the intervention group relative to the control group [5] (Fig. 69.1A). Of the factors measured, serum IGF-1 was the greatest predictor of calcium retention after calcium intake in adolescent white boys; calcium intake predicted 21.7% and serum IGF-1 predicted 11.5% of calcium retention [6]. Diet can alter timing of menarche, perhaps through modulating growth hormones. In a study of girls 7.9 years of age who were randomized to products fortified with dairy minerals and followed for 16 years until approximate peak mass had been achieved [7], those who had received the mineral complex, although only for 1 year, achieved menarche almost 5 months earlier on average than the control group. Earlier menarche with longer exposure to estrogen resulted in greater bone accretion at six skeletal sites and with higher estimated bone strength [8].

ROLE OF NUTRITION IN MAINTAINING BONE MASS

Bone mass is ultimately determined by genetics as modified by current and past mechanical loading and limited or permitted by nutrition. The genetic potential cannot be reached or maintained if intake and absorption of essential nutrients are insufficient.

Calcium is the principal cation of bone mineral. Bone constitutes a very large nutrient reserve for calcium, which, over the course of evolution, acquired a secondary, structural function that is responsible for its importance for osteoporosis. Bone strength varies as the approximate second power of bone structural density. Accordingly, any decrease in bone mass produces a corresponding decrease in bone strength. The aggregate total of bone resorptive activity is controlled systemically by PTH, which in turn responds to the demands of extracellular fluid calcium ion homeostasis and not to the structural need for bone mass. Whenever absorbed calcium intake is insufficient to meet the demands of growth and/or the drain of cutaneous and excretory losses, resorption will be stimulated and bone mass will be reduced as the body scavenges the calcium released in bone resorption. Reserves are designed to be used in times of need, so such use would normally be temporary. Sustained, unbalanced withdrawals deplete the reserves and thereby reduce bone strength. The principal skeletal role of calcium once peak bone mass has been achieved is to offset obligatory losses of calcium through sweat, desquamated skin, and excreta. In addition to depleting or limiting bone mass, low calcium intake in older age directly causes fragility through the PTH-stimulated increase in bone remodeling. Resorption pits on trabeculae cause applied loads to shift to adjacent bone, leading to increased strain locally. In this way excessive remodeling is itself a fragility factor, separately from its effect on bone mass. When adequate calcium is absorbed, PTH-stimulated remodeling decreases immediately [9], and with it, fragility.

Vitamin D is acquired from the diet and from skin synthesis upon exposure to ultraviolet B rays. The best clinical indicator of vitamin D status is the serum 25OHD level. Serum 25OHD levels decline with aging because skin synthesis of vitamin D declines with age. Calcium absorption efficiency declines with age, likely due to vitamin D insufficiency and to loss of intestinal vitamin D receptors.

DIETARY PATTERNS

Food

Recommended food patterns around the world attempt to meet national nutrient requirements and to promote health and reduce risk of disease. The food group most associated with bone health is the dairy group. This food group provides between 20% and 75% of recommended calcium, protein, phosphorus, magnesium, and potassium. Recommended intakes of dairy products around the world are two to three servings per day. Intakes on average are much less than the recommended levels for most populations. Dairy intakes in much of the world have been low persistently, possibly related to a high incidence of lactose maldigestion. Milk consumption has declined over the last half century in the United States, concurrent with increased consumption of soft drinks [10]. Adequacy of milk intake has been associated with adequacy of a number of nutrients in children including calcium, potassium, magnesium, zinc, iron, riboflavin, vitamin A, folate, and vitamin D [11]. Alternative sources to replace this whole package of nutrients are not typically consumed in sufficient amounts to replace milk [12]. Milk consumption within various cultures has been positively associated with bone health. High dairy-consuming regions have better bone measures than low dairy-consuming regions in Yugoslavia [13] and China [14]. A 2-year RCT of milk (1200mg calcium/day) in 173 postmenopausal Chinese women reduced loss of femoral neck BMD relative to a control group [15] (Fig. 69.1B). Milk avoiders have higher risk of fracture than milk drinking counterparts in children [16] and adults [17]. Retrospective studies show that milk drinking in childhood is inversely associated with risk of hip fracture later in life [18]. A European panel concluded that most individuals can tolerate 12 g lactose in a single dose [19]. A National Institute of Health Consensus Conference concluded that the majority of people who self-identify as lactose malabsorbers do not have clinical lactose intolerance [20]. Unnecessary avoidance of dairy due to perceived lactose intolerance can predispose individuals to decrease bone accrual [21].

Other food groups have also been associated with bone health by affecting acid–base balance—fruits and vegetables positively and cereal grains, meats, fish, and poultry negatively. Sulfur-rich amino acids from the grain and meat groups favor an acidic ash that increases calciuria, whereas fruits and vegetables favor an alkaline ash largely because they are metabolized to bicarbonate. Despite its positive contribution to acid ash, protein has a net positive effect on bone. The hypercalciuria seen with increased protein intake is offset by increased calcium absorption [22]. Protein intake negatively predicts age-related bone loss and protein supplements decrease fracture rates in the elderly [23]. A protein and calcium interaction has been identified through retrospective analysis of a RCT [24]. Subsequently, subjects in the Framingham Study with the highest tertile of animal protein intake who consumed <800mg calcium/day had 2.8 times the risk of hip fracture as those in the lowest tertile of protein ($p=0.02$), whereas higher protein intake was associated with decreased fracture incidence when calcium intakes were >800mg/day [25]. In a dose–response study of potassium citrate in older men and women, calcium balance improved with increasing potassium citrate intake due to decreased urinary calcium with no change in fractional calcium absorption [26]. Some vegetables and herbs have the ability to decrease bone resorption, but the effect is independent of the alkaline load or potassium content [27].

Dietary salt is the largest predictor of urinary calcium excretion [28]. Sodium and calcium share transport proteins in the kidney. In adolescence, less sodium, and consequently less calcium, is excreted by black compared with white girls, presumably because of racial differences in renal transport [29].

Bioactive dietary constituents

There is increasing interest in bioactive constituents that can replace or reduce the dose of drug therapies for ameliorating bone loss or promoting bone health in a preventive manner. Typically, these bioactive constituents must be added to the food supply or taken as dietary supplements as effective doses may not be achievable in amounts naturally present in foods.

One category of dietary bioactive constituents that has been studied more than other categories is the flavonoids. Soybean isoflavones have been studied the most but there is little support for their amelioration of postmenopausal bone loss [30]. Flavonoids from plant sources other than soy that have promise in in vitro and in animal models are those in dried plums and blueberries. Osteoporosis and other chronic diseases are being considered as inflammatory disorders. To the extent that flavonoids can modify reactive oxygen species and redox status in bone cells involved in the regulation of bone turnover and survival of osteoblasts, osteoclasts, and osteocytes, they may have a protective role in the diet. There is much to learn about the nature and dose of bioactive constituents, their mechanisms of action, and under what conditions they may be effective.

Certain carbohydrates and fibers, which are fermented in the lower gut, are of interest for their mineral-enhancing intestinal absorption capacity with subsequent benefits to bone [31]. Upon fermentation of these prebiotics by gut microbiota, short chain fatty acids are produced that may solubilize minerals and influence microbiota, increasing the proportion of bifidobacteria.

SKELETAL EFFECTS OF CALCIUM AND VITAMIN D SUPPLEMENTATION

Bone mineral density

Calcium supplementation in adults causes small increases in BMD of 0.54% to 1.19% [32]. In one trial, calcium from food (milk powder) and supplement sources had similar effects on BMD in older postmenopausal women [33]. Higher 25OHD levels have been associated with higher hip BMD. Supplementation with vitamin D alone modestly reduces rates of bone loss in older adults, as does the combination of calcium and vitamin D.

Muscle strength, balance, and falling

The role of vitamin D in preventing fractures includes its effects on muscle performance, balance, and risk of falling. Higher serum 25OHD levels are associated with better lower extremity function and 25OHD levels <50 nmol/L (or 20 ng/mL) have been associated with more rapid declines in physical performance. However, results of vitamin D intervention studies have been variable. In a meta-analysis of 17 trials, supplemental vitamin D had no significant effect on lower extremity muscle strength except in individuals with very low starting serum 25OHD levels (<25 nmol/L or 10 ng/mL) [34]. In a meta-analysis of 30 trials, vitamin D supplementation had a positive impact on muscle strength. The mechanism(s) by which vitamin D influences muscle performance/strength are not well established, but may be mediated through vitamin D receptors in muscle.

Vitamin D supplementation appears to have a favorable effect on balance in older adults. The amplitude of sway in the medial–lateral direction is a strong predictor of falling. In two independent trials, 800 IU of vitamin D_3 plus 1000 mg of calcium per day compared with calcium alone, reduced sway by up to 28% over periods of 2 and 12 months [35, 36]. Mechanisms by which vitamin D affects balance have not been defined.

The effect of vitamin D supplementation on fall risk appears to be dose-dependent. Higher dose trials (700 to 1000 IU/day) showed a risk reduction, whereas lower dose trials (200 to 600 IU/day) did not [37]. The magnitude of the risk reduction in the higher dose trials averaged 34% [37]. In more recent trials, treatment with 800 IU of vitamin D_3 per day for 2 years had no effect on risk of falling in older women with a mean starting 25OHD level of 67.5 nmol/L [38]. In contrast, treatment with 1000 IU of vitamin D_3 per day significantly reduced the number of falls in older women with a lower starting 25OHD level of 37.5 nmol/L [39]. Finally, treatment with 2000 IU when compared with 800 IU per day did not reduce fall risk in elderly acute hip fracture patients [40], suggesting that 800 IU/day is adequate for fall risk reduction. The serum 25OHD level needed to reduce falls was estimated to be 60 nmol/L (or 24 ng/mL) [40].

Fractures

A meta-analysis of 29 trials in 63,897 men and women aged 50 years and older found that supplementation with calcium alone overall had a marginal effect on fracture risk (relative risk [RR], 0.90; 95% CI, 0.80 to 1.00) [32]. Among seniors with usual calcium intakes <700 mg/day, however, the risk reduction was significant (RR, 0.80; 95% CI, 0.71 to 0.89) [32].

Chung and colleagues synthesized the results of 16 trials examining the effect of vitamin D with or without calcium supplementation on fracture risk [41]. Combined vitamin D and calcium supplementation significantly reduced fracture risk (RR, 0.88; 95% CI, 0.78 to 0.99), with most of the benefit occurring in institutionalized rather than community-dwelling settings. A more recent trial-level meta-analysis found that calcium with vitamin D reduced total fractures by 14% and hip fractures by 39% [42]. An individual subject-level meta-analysis examined the impact of vitamin D on fracture risk [43]. This analysis included 31,022 persons (mean age 76 years, 91% women) and 1111 incident hip fractures. Overall, there was a nonsignificant 10% reduction in hip fracture incidence (hazard ratio [HR], 0.90; 95% CI, 0.80 to 1.01). When examined by quartiles of actual intake (the product of dose administered and compliance), however, hip fracture risk was significantly reduced only in the highest quartile (median vitamin D intake 800IU/day). Corresponding serum 25OHD levels in the highest and lowest quartiles, in the subset measured, were >61 and <30 nmol/L, respectively. Consistent with the level that appears to be needed to reduce fall risk, an achieved 25OHD level of about 60 nmol/L appears to be sufficient to reduce risk of fractures.

Role in pharmacotherapy

In trials testing the antifracture efficacy of antiresorptive and anabolic therapies, calcium and vitamin D have been given to both the control and intervention groups. Thus the antifracture efficacy has been established only in calcium- and vitamin D-replete patients. One cannot conclude that these drugs would have the same efficacy in calcium- and vitamin D-deficient patients.

Safety

Bolland and colleagues reported that calcium supplement use without coadministered vitamin D may increase risk of myocardial infarction [44]; however, subsequent reports have challenged that observation [45]. A position statement from the NOF and American Society for Preventive Cardiology concluded that there was grade B level evidence of no relation between the level of calcium and vitamin D supplementation and cardiovascular disease risk [46].

A detailed report from the Women's Health Initiative revealed a 17% increase in renal stones in the calcium and vitamin D group compared with placebo-treated women [47]. Individuals with high calcium intake from food sources do not share this risk and may in fact have reduced risk of nephrolithiasis [48]. It is prudent to obtain calcium from food sources to the greatest extent possible. The IOM and others have identified no risk associated with vitamin D supplementation that achieves serum 25OHD levels up to 125 nmol/L (or 50 ng/mL) [49].

Calcium and vitamin D recommendations

Calcium intake recommendations vary globally. The IOM recommends, for ages 51 to 70 years, 1000 mg/day for men and 1200 mg/day for women; it recommends 1200 mg/day for everyone aged 70 years or more [49]. Calcium supplementation should be recommended only when the requirement cannot be met with food. Calcium from calcium carbonate is better absorbed when taken with a meal. Absorption from all supplements is more efficient in doses up to 500 mg than from higher doses. Individuals requiring more than 500 mg/day from supplements should take it in divided doses.

The vitamin D intake recommendations of the IOM are, for ages 51 to 70 years, 15 µg (600 IU) per day, and for age 71 years and older, 20 µg (800 IU) per day [49]. The IOM recommends a level of 50 nmol/L to meet the needs of 97.5% of the population [49]. The evidence cited above suggests that a serum 25OHD level of 60 nmol/L is needed to maximally reduce risk of falls and fractures. The Endocrine Society, the International Osteoporosis Foundation, and other organizations focused on osteoporosis patients recommend a level of 75 nmol/L. Vitamin D is available in two forms, plant-derived ergocalciferol (D_2) and animal-derived cholecalciferol (D_3). Vitamin D_3 increases serum 25OHD levels more efficiently than vitamin D_2. Moreover, vitamin D_2 is not accurately measured in all 25OHD assays. For these reasons, vitamin D_3, when available, is the preferred form for clinical use.

CONCLUSION

Bone health rests on a combination of mechanical loading and adequate intakes of a broad array of macro- and micronutrients. Three important essential nutrients for bone health are calcium, vitamin D, and protein. Most diets inadequate in one key nutrient will be inadequate in several. Optimal protection of bone requires a diet rich in all the essential nutrients. Mononutrient supplementation regimens will often be inadequate to ensure optimal nutritional protection of bone health. Some bioactive ingredients may improve bone health by reducing chronic inflammation.

REFERENCES

1. Weaver CM, Gordon CM, Janz KF, et al. The National Osteoporosis Foundation's position statement on peak bone mass development and lifestyle factors: a systematic review and implementation recommendations. Osteoporos Int. 2016;27(4):1281–386.
2. Bailey DA, McKay HA, Mirwald RL, et al. A six-year longitudinal study of the relationship of physical activity to bone mineral accrual in growing children: the University of Saskatchewan bone mineral accrual study. J Bone Miner Res. 1999;14:1672–9.
3. Braun M, Palacios C, Wigertz K, et al. Racial differences in skeletal calcium retention in adolescent girls on a range of controlled calcium intakes. Am J Clin Nutr. 2007;85:1657–63.
4. Walker MD, Novotny R, Bilezikian JP, et al. Race and diet interactions in the acquisition, maintenance, and loss of bone. J. Nutr. 2008;138:S1256–60.
5. Cadogan J, Eastell R, Jones N, et al. Milk intake and bone mineral acquisition in adolescent girls: Randomized, controlled intervention trial. BMJ. 1997;315:1255–60.
6. Hill K, Braun MM, Kern M, et al. Predictors of calcium retention in adolescent boys. J. Clin. Endocrin. Metab. 2008;93(12):4743–8.

7. Chevalley T, Rizzoli R, Hans D, et al. Interaction between calcium intake and menarcheal age on bone mass gain: An eight-year follow-up study from prepuberty to postmenarche. J Clin Endocrinol Metab. 2005;90:44–51.
8. Chevalley T, Bonjour JP, van Rietbergen B, et al. Fractures in healthy females followed from childhood to early adulthood are associated with later menarcheal age and with impaired bone microstructure at peak bone mass. J Clin Endocrinol Metab. 2012;97(11):4174–81.
9. Wastney ME, Martin BR, Peacock M, et al. Changes in calcium kinetics in adolescent girls induced by high calcium intake. J Clin Endocrinol Metab. 2000;85:4470–5.
10. US Department of Health and Human Services and US Department of Agriculture. *Dietary Guidelines for Americans, 2005* (6th ed.). Washington, DC: US Government Printing Office, 2005.
11. Ballow C, Kuester S, Gillespie C. Beverage choices affect adequacy of children's nutrient intakes. Arch Pediatr Adolesc Med. 2000;154:1148–52.
12. Gao X, Wilde PE, Lichtenstein AH, et al. Meeting adequate intake for dietary calcium without dairy foods in adolescents aged 9 to 18 years (National Health and Nutrition Examination Survey 2001–2002). J Am Diet Assoc. 2006;106:1759–65.
13. Matkovic V, Kostial K, Siminovic I, et al. Bone status and fracture rates in two regions of Yugoslavia. Am J Clin Nutr. 1979;32:540–9.
14. Hu J-F, Zbao X-H, Jia J-B, et al. Dietary calcium and bone density among middle-aged and elderly women in China. Am J Clin Nutr. 1993;58:219–27.
15. Chee WSS, Suriah AR, Chan SP, et al. The effect of milk supplementation on bone mineral density in postmenopausal Chinese women living in Malaysia. Osteoporos Inter. 2003;14:828–34.
16. Goulding A, Rockell JE, Black RE, et al. Children who avoid drinking cow's milk are at increased risk for prepubertal bone fractures. J Am Diet Assoc. 2004;104:250–3.
17. Honkanen R, Kroger H, Alhava E, et al. Lactose intolerance associated with fractures in weight-bearing bones in Finnish women aged 38–57 years. Bone. 1997;21:473–7.
18. Kalkwarf HJ, Khoury JC, Lanphear BP. Milk intake during childhood and adolescence, adult bone density, and osteoporotic fractures in US women. Am J Clin Nutr. 2003;77:257–65.
19. EFSA Panel on Dietetic Products, Nutrition and Allergies. Scientific opinion on lactose thresholds in lactose intolerance and galactosaemia Eur Food Safety Authority J 2010;8(9):1777.
20. Suchy FJ, Brannon PM, Carpenter TO, et al. NIH consensus development conference statement: lactose intolerance and health. NIH Consens State Sci Statements. 2010;27(2):1–27.
21. Matlik L, Savaiano D, McCabe G, et al. Perceived milk intolerance is related to bone mineral content in 10- to 13-year-old female adolescents. Pediatrics. 2007;120:3669.
22. Kerstetter JE, O'Brien KO, Caseria DM, et al. The impact of dietary protein on calcium absorption and kinetic measures of bone turnover in women. J Clin Endocrinol Metab. 2005;90:26–31.
23. Bonjour JP. Dietary protein: an essential nutrient for bone health. J Am Coll Nutr. 2005;24:S526–36.
24. Dawson-Hughes B, Harris SS. Calcium intake influences the association of protein intake with rates of bone loss in elderly men and women. Am J Clin Nutr. 2002;75(4):773–9.
25. Sahni S, Cupples A, McLean RR, et al. Protective effect of high protein and calcium intake on the risk of hip fracture in the Framingham offspring cohort. J Bone Miner Res. 2010;25:2770–6.
26. Moseley KF, Weaver CM, Appel L, et al. Potassium citrate supplementation results in sustained improvement in calcium balance in older men and women. J Bone Miner Res. 2013;28:497–504.
27. Muhlbauer RC, Lozano A, Reinli A. Onion and a mixture of vegetables, salads, and herbs affect bone resorption in the rat by a mechanism independent of their base exceeds. J Bone Miner Res. 2002;17:1230–6.
28. Nordin BE, Need AG, Morris HA, et al. The nature and significance of the relationship between urinary sodium and urinary calcium in women. J Nutr. 1993;123:1615–22.
29. Wigertz K, Palacios C, Jackman LA. et al. Racial differences in calcium retention in response to dietary salt in adolescent girls. Am J Clin Nutr. 2005;81:845–50.
30. North American Menopause Society (NAMS). The role of soy isoflavones in menopausal health: report of The North American Menopause Society/Wulf H. Utian Translational Science Symposium in Chicago, IL. Menopause. 2011;18(7):732–53.
31. Jakeman SA, Henry CN, Martin BR, et al. Soluble corn fiber increases bone retention in postmenopausal women in a dose-dependent manner: a randomized crossover trial. Am J Clin Nutr. 2016;104(3):837–43.
32. Tang BM, Eslick GD, Nowson C, et al. Use of calcium or calcium in combination with vitamin D supplementation to prevent fractures and bone loss in people aged 50 years and older: a meta-analysis. Lancet. 2007;370:657–66.
33. Prince R, Devine A, Dick I, et al. The effects of calcium supplementation (milk powder or tablets) and exercise on bone density in postmenopausal women. J Bone Miner Res. 1995;10:1068–75.
34. Beaudart C, Buckinx F, Rabenda V, et al. The effects of vitamin D on skeletal muscle strength, muscle mass, and muscle power: a systematic review and meta-analysis of randomized controlled trials. J Clin Endocrinol Metab. 2014;99(11):4336–45.
35. Bischoff HA, Stahelin HB, Dick W, et al. Effects of vitamin D and calcium supplementation on falls: a randomized controlled trial. J Bone Miner Res. 2003;18:343–351.
36. Pfeifer M, Begerow B, Minne HW, et al. Effects of a short-term vitamin D and calcium supplementation on body sway and secondary hyperparathyroidism in elderly women. J Bone Miner Res. 2000;15:1113–18.

37. Bischoff-Ferrari HA. Authors' reply. BMJ 2011;342:d2608.
38. Uusi-Rasi K, Patil R, Karinkanta S, et al. Exercise and vitamin D in fall prevention among older women: a randomized clinical trial. JAMA Intern Med. 2015;175:703–11.
39. Cangussu LM, Nahas-Neto J, Orsatti CL, et al. Effect of isolated vitamin D supplementation on the rate of falls and postural balance in postmenopausal women fallers: a randomized, double-blind, placebo-controlled trial. Menopause. 2016;23:267–74.
40. Bischoff-Ferrari HA, Dawson-Hughes B, Orav EJ, et al. Monthly high-dose vitamin D treatment for the prevention of functional decline: a randomized clinical trial. JAMA Intern Med. 2016;176(2):175–83.
41. Chung M, Lee J, Terasawa T, et al. Vitamin D with or without calcium supplementation for prevention of cancer and fractures: an updated meta-analysis for the U.S. Preventive Services Task Force. Ann Intern Med. 2011;155:827–38.
42. Weaver CM, Dawson-Hughes B, Lappe JM, et al. Erratum and additional analyses re: calcium plus vitamin D supplementation and the risk of fractures: an updated meta-analysis from the National Osteoporosis Foundation. Osteoporos Int. 2016;27:2643–6.
43. Bischoff-Ferrari HA, Willett WC, Orav EJ, et al. A pooled analysis of vitamin D dose requirements for fracture prevention. N Engl J Med. 2012;367:40–9.
44. Bolland MJ, Avenell A, Baron JA, et al. Effect of calcium supplements on risk of myocardial infarction and cardiovascular events: meta-analysis. BMJ. 2010;341:c3691.
45. Harvey NC, Biver E, Kaufman J-M, et al. The role of calcium supplementation in healthy musculoskeletal ageing. Osteopor Int. 2017;28(2):447–62.
46. Kopecky SL, Bauer DC, Gulati M, et al. Lack of evidence linking calcium with or without vitamin D supplementation to cardiovascular disease in generally healthy adults: a clinical guideline from the National Osteoporosis Foundation and American Society for Preventive Cardiology. Ann Intern Med. 2016;165(12):867–8.
47. Wallace RB, Wactawski-Wende J, O'Sullivan MJ, et al. Urinary tract stone occurrence in the Women's Health Initiative (WHI) randomized clinical trial of calcium and vitamin D supplements. Am J Clin Nutr. 2011;94:270–7.
48. Curhan GC, Willett WC, Speizer FE, et al. Comparison of dietary calcium with supplemental calcium and other nutrients as factors affecting the risk for kidney stones in women [see comments]. Ann Intern Med. 1997;126:497–504.
49. Institute of Medicine (IOM). *Dietary Reference Intakes for Calcium and Vitamin D*. In. Washington, DC: National Academies Press, 2011.

70

Estrogens, Selective Estrogen Receptor Modulators, and Tissue-Selective Estrogen Complex

Tobias J. de Villiers

Department of Gynecology, Stellenbosch University, Mediclinic Panorama, Cape Town, South Africa

ESTROGENS

Fuller Albright published his observations on the causal relationship between menopausal estrogen deficiency and osteoporosis in 1941 and he introduced the concept of menopausal hormone therapy (MHT) for the prevention of osteoporosis [1].

In 1996, the Postmenopausal Estrogen/Progestin Intervention (PEPI) trial was the first, large, randomized controlled trial (RCT) utilizing dual X-ray absorptiometry to show that BMD in early menopause is conserved at the hip and spine by intervention with estrogen therapy [2]. It was suggested by observational data that MHT prevented fractures, but no RCT with a fracture endpoint had been conducted.

In 2002, the results of the Women's Health Initiative (WHI) study, a large RCT, gave concrete proof of the antifracture efficacy of MHT. In the WHI estrogen and progestogen arm, reductions in the rates of fractures were reported as hazard ratios (HRs): hip fracture 0.66 (95% CI, 0.45 to 0.98), clinical vertebral fracture 0.66 (95% CI, 0.44 to 0.98), and nonvertebral fracture 0.77 (95% CI, 0.69 to 0.86) [3]. In the WHI estrogen-only arm, HRs were: hip fracture 0.61 (95% CI, 0.41 to 0.91), clinical vertebral fracture 0.62 (95% CI, 0.42 to 0.93), and total fractures 0.79 (95% CI, 0.63 to 0.79) [4]. These results are remarkable as only clinical fractures were recorded, without routine X-rays to detect morphometric vertebral fractures. Furthermore, the population studied was at low risk of fracture. This resulted in the magnitude of antifracture efficacy of MHT being understated. MHT has proven efficacy against all the major classes of osteoporotic fractures (vertebral and nonvertebral including hip). HRT is virtually in a class of its own in offering fracture protection in patients with osteopenia [5].

MHT also offers exceptional extraskeletal benefits such as the most effective treatment for menopause-associated vasomotor symptoms, the improvement of vaginal and urological health, possible benefits for cardiovascular health, and lowered all-cause mortality (in patients initiated on therapy between the ages of 50 and 60 years or within 10 years of menopause), as well as a reduction in risk of colorectal carcinoma. Reported risks were breast cancer, coronary arterial disease, and increased thrombotic events [3].

Based on these provisional results of the WHI study and a reported poor benefit/risk profile (excess risk of 19 events per 10,000 person-years), the uptake of MHT in the world of osteoporosis remained low [6]. This reduced MHT to the role of treatment for vasomotor symptoms at the lowest effective dose for the shortest possible time. In the United States, the additional approved indication of prevention of osteoporosis remained, but perceived restrictions on the duration of use made this approval pointless as any benefit on bone is rapidly lost after cessation of therapy.

In recent years major menopause societies have recommended a less restrictive view of MHT for the indication of fracture prevention. This is embodied in the latest Global Consensus Statement (GCS) by all the major international menopause societies [7]:

- MHT, including tibolone, can be initiated in postmenopausal women at risk of fracture or osteoporosis before the age of 60 years or within 10 years after menopause.

Primer on the Metabolic Bone Diseases and Disorders of Mineral Metabolism, Ninth Edition. Edited by John P. Bilezikian.
© 2019 American Society for Bone and Mineral Research. Published 2019 by John Wiley & Sons, Inc.
Companion website: www.wiley.com/go/asbmrprimer

- Initiation of MHT after the age of 60 years for the indication of fracture prevention is considered second-line therapy and requires individually calculated benefit/risk, compared with other approved drugs.
- If MHT is elected, the lowest effective dose should be used.
- Duration of treatment should be consistent with the treatment goals of the individual, and the benefit/risk profile needs to be individually reassessed annually. This is important in view of new data indicating longer duration of vasomotor symptoms (VMS) in some women.

These more enlightened recommendations are based on the final as well as follow-up results from the WHI, which have shown a much better benefit/risk profile and is reflected in the following statements of the GCS [8]:

- If MHT is initiated within the window of opportunity (age group 50 to 60 years or within 10 years of menopause) at least a neutral or possible positive effect on the cardiovascular system is evident as well as reduced all-cause mortality.
- Short duration estrogen therapy alone has been shown to reduce breast cancer detection rate whereas estrogen plus progestogen may increase the detection rate after 5 years of treatment. The risk of breast cancer attributable to MHT is rare. It equates to an incidence of <1.0 per 1000 women per year of use. This is similar to or lower than the increased risk associated with common factors such as sedentary lifestyle, obesity, and alcohol consumption [7].
- The risk of venous thromboembolism (VTE) and ischemic stroke increases with oral MHT, although the absolute risk of stroke with initiation of MHT before age 60 years is rare. Observational studies and a meta-analysis point to a probable lower risk of VTE and possibly stroke with transdermal therapy (0.05 mg twice weekly or lower) compared with oral therapy [7].

In making these recommendations, the GCS took into account specific limitations of the WHI such as the age at initiation of MHT, the dose and type of MHT used, as well as inconsistencies in the statistical analysis as reported [9, 10].

MHT is often avoided by nongynecological physicians because of being uncertain about the general principles governing the use of MHT. Clear guidelines as provided by the GCS are aimed at overcoming this obstacle. These principles are [7]:

- The type and route of administration of MHT should be consistent with treatment goals, patient preference, and safety issues and should be individualized.
- The dosage should be titrated to the lowest appropriate and most effective dose.
- Estrogen as a single systemic agent is appropriate in women after hysterectomy but concomitant progestogen is required in the presence of a uterus for endometrial protection with the exception that CE can be combined with BZA for uterine protection.

The chapter on MHT in the previous (seventh) edition of this publication concluded that "estrogen is indicated for relief of menopausal symptoms; for women who are candidates for estrogen either short term or long term, bone effects may be considered a side benefit." It is the opinion of this author that we have moved away from this very restrictive view to the wider indications for MHT in osteoporosis as advocated by the GCS. In our quest to reduce the burden of osteoporosis-related fractures, we need to use all the means at our disposal. MHT is a powerful tool in the younger menopausal women to reduce all types of osteoporosis-related fractures, even in women with osteopenia. This is important in view of more restricted use of approved bone-specific medications based on changing side effect profiles. This is illustrated by recommendations that oral bisphosphonates generally be restricted to 5 years of continuous use. This opens the opportunity for MHT in younger women followed by a bisphosphonate if MHT is stopped.

SELECTIVE ESTROGEN RECEPTOR MODULATORS

SERMs are a complex group of synthetic drugs that act as either estrogen receptor (ER) agonists or antagonists in a tissue-specific way. The ideal SERM will act as an ER agonist on the cardiovascular system, bone, vagina, and bladder, while acting as an ER antagonist on the breast and endometrium [11].

The only SERM approved in the United States for osteoporosis prevention and treatment is raloxifene, a second generation SERM at an oral daily dose of 60 mg. Raloxifene was shown to reduce the risk of vertebral fracture by 34% to 51% in a large RCT, in spite of a very modest increase in BMD [12]. It failed to show a protective effect on nonvertebral fractures (including hip fracture). Raloxifene reduces the risk of invasive ER-positive breast cancer by 76% [13]. Raloxifene is as effective as tamoxifene (a first generation SERM) in the prevention of breast cancer in nonosteoporotic women [14]. The Raloxifene Use for The Heart (RUTH) trial failed to show that raloxifene offers protection against coronary heart disease (in patients at high risk of coronary heart disease). Compared with placebo, participants on raloxifene therapy were at a higher risk of venous thrombotic events (HR, 1.44) and fatal stroke (HR, 1.49) [15]. Raloxifene does not cause endometrial stimulation or result in uterine bleeding [16]. Unlike estrogen, raloxifene does not treat the vasomotor symptoms of menopause and may cause hot flushes. Raloxifene is typically used in patients at risk of vertebral fracture and breast cancer. The use of raloxifene is limited because of lack of evidence for protection against hip and other nonvertebral fractures as well as side effects and safety concerns as described earlier in this chapter.

Several third generation SERMs were in clinical development but most failed to reach registration. Lasofoxifene 5 mg daily per mouth significantly reduced vertebral fracture, nonvertebral fracture, ER-positive breast cancer, major coronary heart disease, and the absolute risk of stroke, compared with placebo [17]. In spite of these very favorable results, regulatory approval by the US authorities was withheld with the request for more data as there were 25 more deaths in a lower dose lasofoxifene group (0.25 mg) than in the placebo group ($p=0.05$). Although regulatory approval for lasofoxifene was given in Europe, the sponsor stopped further development of the drug; there is interest at present to resume development of it.

BZA is a third generation SERM. It is approved outside the United States as monotherapy for osteoporosis treatment. BZA has been shown to preserve BMD and to reduce bone turnover. In postmenopausal women with osteoporosis, BZA has demonstrated significant protection against new vertebral fractures and against nonvertebral fractures in women at higher risk of fracture [18]. BZA was generally safe and well tolerated in phase III studies over 7 years with neutral effects on the breast and an excellent endometrial safety profile [19]. An increased risk of VTE was confirmed as an expected class effect of SERMs.

TISSUE-SELECTIVE ESTROGEN COMPLEX: CONJUGATED ESTROGEN/BAZEDOXIFENE

When using MHT in women with a uterus, stimulation of the endometrium by estrogen needs to be opposed by a progestin. The concept of tissue-selective estrogen complex (TSEC) is based on the desire to oppose the endometrial stimulation of estrogen with the SERM BZA. Pairing CE 0.45 mg with BZA 20 mg results in a blended tissue-selective activity profile that protects the endometrium, alleviates vasomotor symptoms, prevents bone loss, has a neutral effect on the breast, and does not compound the risk of VTE [20]. Compared with placebo, over 24 months, TSEC increased BMD by 3.61% and was significantly better than raloxifene at 18 and 24 months [21]. No trials with fracture outcomes have been done. TSEC thus provides a progestin-free alternative to traditional estrogen-progestin therapy MHT in women with a uterus. CE 0.45 mg/BZA 20 mg has been approved in the United States for menopausal symptom relief and osteoporosis prevention for postmenopausal women with a uterus. This option is attractive to both women with a uterus and nongynecologists, as the therapy does not result in uterine bleeding, as is often the case in traditional estrogen-progestin MHT.

REFERENCES

1. Albright F, Smith PH, Richardson AM. Postmenopausal osteoporosis. JAMA. 1941;161:2465–74.
2. The Writing Group for the PEPI. Effects of hormone therapy on bone mineral density: results from the postmenopausal estrogen/ progestin interventions (PEPI) trial. JAMA. 1996;276:1389–96.
3. Rossouw JE, Anderson GL, Prentice RL, et al.; Writing Group for the Women's Health Initiative Investigators. Risks and benefits of estrogen plus progestin in healthy postmenopausal women: principal results from the Women's Health Initiative randomized controlled trial. JAMA. 2002;288:321–33.
4. Anderson GL, Limacher M, Assaf AR, et al., for the Women's Health Initiative Steering Committee. Effects of conjugated equine estrogen in postmenopausal women with hysterectomy: the Women's Health Initiative randomized controlled trial. JAMA. 2004;291:1701–12.
5. De Villiers TJ. The role of menopausal hormone therapy in the management of osteoporosis. Climacteric. 2015;18:19–21.
6. Cauley JA, Robbins J, Chen Z, et al.; Women's Health Initiative Investigators. Effects of estrogen plus progestin on risk of fracture and bone mineral density: the Women's Health Initiative randomized trial. JAMA. 2003;290:1729–38.
7. De Villiers TJ, Hall JE, Pinkerton JV, et al. Revised Global Consensus Statement on Menopausal Hormone Therapy. Climacteric. 2016;19:313–15.
8. Manson JE, Chlebowski RT, Stefanick ML, et al. Menopausal hormone therapy and health outcomes during the intervention and extended poststopping phases of the Women's Health Initiative randomized trials. JAMA. 2013;310:1353–68.
9. De Villiers TJ. Estrogen and bone: have we completed a full circle? Climacteric. 2014;17:1–4.
10. De Villiers TJ, Stevenson J. The WHI: the effect of hormone replacement therapy on fracture prevention. Climacteric. 2012;15:263–6.
11. De Villiers TJ. Clinical issues regarding cardiovascular disease and selective estrogen receptor modulators in postmenopausal women. Climacteric. 2009;12:108–11.
12. Ettinger B, Black DM, Mitlack BH, et al. Reduction of vertebral fracture risk in postmenopausal women with osteoporosis treated with raloxifene: results from a 3-year randomized clinical trial. Multiple Outcomes of Raloxifene Evaluation (MORE) Investigators. JAMA. 1999;282:637–45.
13. Martino S, Cauley JA, Barrett-Connor E; CORE Investigators. Continuing outcomes relevant to Evista; breast cancer incidence in postmenopausal osteoporotic women in a randomized trial of raloxifene. J Natl Cancer Inst. 2004;96:1751–61.
14. Bevers TB. The STAR trial: evidence for raloxifene as a breast cancer risk reduction agent for postmenopausal patients. J Natl Compr Canc Netw. 2007;5:719–24.
15. Barret-Connor E, Mosca L, Collins P, et al. Effects of raloxifene on cardiovascular events and breast cancer in postmenopausal women. N Engl J Med. 2006;355:125–37.
16. Neven P, Lunde T, Benedetti-Panici P, et al. A multicentre randomised trial to compare uterine safety of raloxifene with a continuous combined hormone replacement therapy containing oestradiol and norethisterone acetate. Br J Obstet Gynaecol. 2003;110:157–67.
17. Cummings SR, Ensrud K, Delmas P, et al., for the PEARL Study Investigators. Lasofoxifene in postmenopausal

women with osteoporosis. N Engl J Med. 2010;362: 686–96.
18. De Villiers TJ, Chines AA, Palacios S, et al. Safety and tolerability of bazedoxifene in postmenopausal women with osteoporosis: results of a 5-year, randomized, placebo-controlled phase 3 trial. Osteoporos Int. 2011;22: 567–76.
19. Palacios S, Silverman SL, de Villiers TJ, et al., on behalf of the Bazedoxifene Study Group. A 7-year randomized, placebo-controlled trial assessing the long-term efficacy and safety of bazedoxifene in postmenopausal women with osteoporosis: effects on bone density and fracture. Menopause. 2015;22:806–13.
20. Komm BS, Mirkin S, Jenkins SN. Development of conjugated estrogens/bazedoxifene, the first tissue selective estrogen complex (TSEC) for management of menopausal hot flashes and postmenopausal bone loss. Steroids. 2014;90:71–81.
21. Lindsay R, Gallagher JC, Kagan R, et al. Efficacy of tissue-selective estrogen complex of bazedoxifene/conjugated estrogens for osteoporosis prevention in at-risk postmenopausal women. Fertil Steril. 2009;92:1045–52.

71

Bisphosphonates for Postmenopausal Osteoporosis

Andrea Giusti[1] and Socrates E. Papapoulos[2]

[1]*Rheumatology Unit, Department of Locomotor System, La Colletta Hospital, Arenzano, Italy*
[2]*Center for Bone Quality, Department of Endocrinology and Metabolic Diseases, Leiden University Medical Center, Leiden, The Netherlands*

INTRODUCTION

Bisphosphonates (BPs) are synthetic compounds that have high affinity for calcium crystals, concentrate selectively in the skeleton, and decrease bone resorption. The first BP was synthesized in the 19th century but their relevance to medicine was recognized in the 1960s, and they were first given to patients with osteoporosis in the early 1970s. Currently, alendronate, ibandronate, risedronate, and zoledronic acid are approved for the treatment of osteoporosis worldwide, while other BPs are also available in some countries.

PHARMACOLOGY

BPs are synthetic analogs of inorganic pyrophosphate in which the oxygen atom that connects the two phosphates is replaced by a carbon atom (Fig. 71.1A). This substitution renders BPs resistant to biological degradation and suitable for clinical use. BPs have two additional side chains (R1 and R2), attached to the carbon atom, that allowed the synthesis of a large number of analogs with different pharmacological properties (Fig. 71.1B). A hydroxyl substitution at R1 enhances the affinity of BPs for calcium crystals, while the presence of a nitrogen atom in R2 determines their potency and mechanism of action. The whole molecule is responsible for the action of BPs on bone resorption and probably for their affinity for bone mineral [1, 2].

The intestinal absorption of BPs is poor (less than 1%) and decreases further in the presence of food, calcium, or other minerals that bind them. Oral BPs should be given in the fasting state 30 to 60 minutes before meals with water. BPs are cleared rapidly from the circulation; about 50% of the administered dose concentrates in the skeleton, primarily at active remodeling sites, while the rest is excreted unmetabolized in urine. Skeletal uptake depends on the rate of bone turnover, renal function, and affinity for bone mineral [3]. The capacity of the skeleton to retain BPs is large and saturation of binding sites with the doses used in the treatment of osteoporosis is impossible even if these are given for a very long time. After exerting their action on bone resorption at the bone surface, BPs are embedded in bone, where they remain for a long time and are pharmacologically inactive. The elimination of BPs from the body is multiexponential. The calculated terminal half-life of elimination from the skeleton can be as long as 10 years, and pamidronate has been detected in the urine of patients for up to 8 years after discontinuation of treatment. The slow release of BPs from the skeleton is probably responsible for the slow rate of reversal of their effect on bone turnover following cessation of treatment, which differentiates them from all other anti-osteoporotic medications. The speed of reversal of the effect may be different among BPs depending on their pharmacological properties, particularly their affinity for bone mineral.

The decrease in bone resorption by BPs is followed by a slower decrease in the rate of bone formation, due to the coupling of the two processes, so that a new steady state at a lower rate of bone turnover is reached 3 to 6 months after the start of treatment. This level of bone turnover remains constant during the whole period of treatment, up to 10 years in clinical studies, demonstrating that the accumulation of BP in the skeleton is not associated with a

---

Primer on the Metabolic Bone Diseases and Disorders of Mineral Metabolism, Ninth Edition. Edited by John P. Bilezikian.
© 2019 American Society for Bone and Mineral Research. Published 2019 by John Wiley & Sons, Inc.
Companion website: www.wiley.com/go/asbmrprimer

546 *Bisphosphonates for Postmenopausal Osteoporosis*

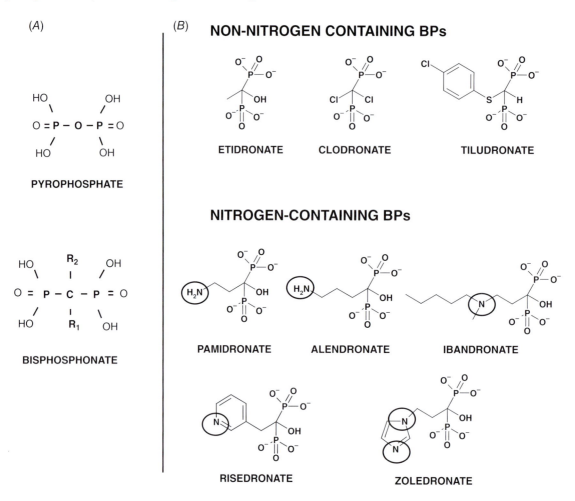

Fig. 71.1. (A) Structure of pyrophosphate and geminal bisphosphonate. (B) Structure of clinically used bisphosphonates (acid forms are depicted).

cumulative effect on bone turnover. In addition to decreasing the rate of bone turnover to premenopausal levels, BPs preserve or may improve trabecular and cortical architecture, correct the hypomineralization of osteoporotic bone, increase areal mineral density, and may reduce the rate of osteocyte apoptosis. The relevant clinical outcome of these actions is the decrease in the risk of fractures [Fig. 71.2].

At the cellular level BPs inhibit the activity of osteoclasts [1–3]. BPs bound to bone hydroxyapatite are released in the acidic environment of the resorption lacunae under the osteoclasts and are taken up by them. BPs without a nitrogen atom in their molecule incorporate into adenosine triphosphate (ATP) and generate metabolites that induce osteoclast apoptosis. Nitrogen-containing BPs (N-BPs) induce changes in the cytoskeleton of osteoclasts, leading to their inactivation and potentially apoptosis. This action is mainly the result of inhibition of farnesyl pyrophosphate synthase (FPPS), an enzyme of the mevalonate biosynthetic pathway. FPPS is responsible for the formation of isoprenoid metabolites required for the prenylation of small guanosine triphosphatases (GTPases) that are important for the cytoskeletal integrity and function of osteoclasts. There is a close relation between the degree of inhibition of FPPS and the antiresorptive potencies of N-BPs. In addition, the inhibition of FPPS by N-BPs leads to accumulation of isopentenyl pyrophosphate, a metabolite immediately upstream of FPPS, which reacts with adenosine monophosphate (AMP) leading to the production of a new metabolite that induces osteoclast apoptosis.

ANTIFRACTURE EFFICACY

All BPs given daily in adequate doses reduce significantly the risk of vertebral fractures by 35% to 65% [4–10] (Fig. 71.3). As also illustrated in Fig. 71.3, the incidence of vertebral fractures varied greatly among placebo-treated patients in different clinical trials. Thus, results obtained in different studies should not be used to compare efficacy of individual BPs; for this head-to-head studies with fracture endpoints are needed. The overall efficacy and consistency of daily oral BPs in reducing the risk of vertebral fractures

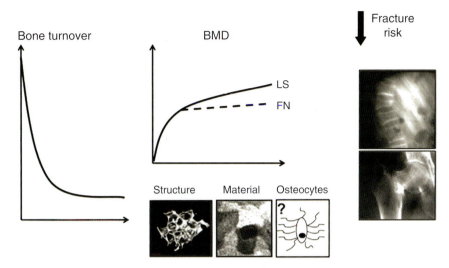

Fig. 71.2. The effects of bisphosphonates on bone metabolism and strength in osteoporosis. BMD = bone mineral density; FN = femoral neck; LS = lumbar spine.

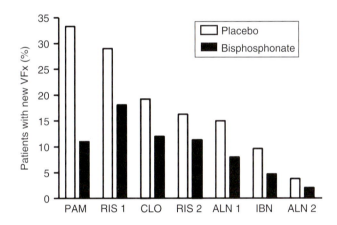

Fig. 71.3. Incidence of vertebral fractures (VFx) in patients with osteoporosis treated with daily oral placebo or bisphosphonate after 3 years. PAM = pamidronate [4]; RIS 1 = risedronate (VERT study [8]); CLO = clodronate [10]; RIS 2 = risedronate (VERT study [7]); ALN 1 = alendronate (FIT 1 study [5]); IBN = ibandronate (BONE study [9]); ALN 2 = alendronate (FIT 2 study [6]).

has been demonstrated by meta-analyses of randomized controlled trials (RCTs) for alendronate and risedronate [11–13]. In studies in which radiographs were taken annually (eg, the Vertebral Efficacy with Risedronate Therapy (VERT) study with risedronate), the effect of the BP in reducing the risk of vertebral fractures was already evident after 1 year, demonstrating a rapid protection of skeletal integrity [8]. This was also shown for moderate and severe vertebral fractures with ibandronate [14] and for clinical vertebral fractures with alendronate [15].

The efficacy of daily oral BPs in reducing the risk of nonvertebral fractures was evaluated in a number of RCTs. It should be noted that definitions and adjudication procedures of nonvertebral fractures were different among clinical trials. Two meta-analyses of the Cochrane Collaboration reported an overall reduction in the risk of nonvertebral fractures in women with osteoporosis of 23% (relative risk [RR], 0.77; 95% CI, 0.64 to 0.92) with alendronate and 20% (RR, 0.80; 95% CI, 0.72 to 0.90) with risedronate [12, 13]. The corresponding reductions in the risk of hip fractures were 53% (RR, 0.47; 95% CI, 0.26 to 0.85) with alendronate and 26% (RR, 0.74; 95% CI, 0.59 to 0.94) with risedronate. These estimates are in agreement with earlier published meta-analyses [16, 17]. With daily ibandronate, a reduction (69%) in the risk of nonvertebral fractures was reported only in a population at high risk (femoral neck BMD T-score < –3.0), by post hoc analysis [9]. As with vertebral fractures, the effect of BPs on nonvertebral fractures was observed early after the start of treatment.

Daily administration of oral BPs is inconvenient and may also be associated with gastrointestinal adverse effects. These reduce adherence to treatment and can diminish the therapeutic response [18]. To overcome these problems, once-weekly formulations, the sum of seven daily doses, have been developed for alendronate and risedronate and were shown to significantly improve patients' adherence to treatment while maintaining the pharmacodynamic response of daily treatment [19, 20]. Recently, new formulations of once-weekly risedronate (slow-release tablets administered after breakfast) and alendronate (effervescent tablets and liquid formulations), potentially improving gastric tolerability, have also become available. Daily and weekly BPs are pharmacologically equivalent and should be considered to be continuous administration, while the term intermittent or cyclical administration should be reserved for treatments with drug-free intervals longer than 2 weeks [3].

Intermittent administration

Results of early attempts to give BPs intermittently to patients with osteoporosis were equivocal but a meta-analysis of studies with cyclical etidronate showed a significant reduction in the risk of vertebral but not of nonvertebral fractures [21]. The efficacy of intermittent administration of N-BPs was explored in studies with ibandronate. These indicated that dose and dosing intervals are important determinants of the response to intermittent BP therapy, which in turn depends on the safety and tolerability of the administered dose [22]. An oral ibandronate preparation given once monthly and an intravenous preparation given once every 3 months, providing higher annual cumulative doses than the daily regimen, were shown to significantly reduce the risk of nonvertebral fractures by 38% compared with the oral daily dose [23, 24]. A once-monthly oral preparation of risedronate is also available.

The efficacy of intermittent administration of zoledronic acid, the most potent N-BP, in reducing the risk of osteoporotic fractures was examined in the Health Outcomes and Reduced Incidence with Zoledronic Acid Once Yearly (HORIZON) trial, in which postmenopausal women with osteoporosis were randomized to receive 15-minute infusions of zoledronic acid 5 mg or placebo once-yearly [25]. Compared with placebo, zoledronic acid reduced the incidence of vertebral fractures by 70%, of hip fractures by 41%, and of nonvertebral fractures by 25% after 3 years. The effect of zoledronic acid on vertebral fractures was significant already at 1 year. In a second controlled study, zoledronic acid infusions given within 90 days after surgical repair of a hip fracture decreased the rate of new clinical fractures by 35% and significantly improved patient survival (28% reduction in all-cause mortality) [26]. Epidemiologic studies reported also survival benefits in patients treated with oral BPs [27–29].

LONG-TERM EFFECTS ON BONE FRAGILITY

Skeletal fragility on long-term BP therapy has been examined in the extensions of four clinical trials for up to 10 years [30–35]. None of these extension studies was specifically designed to assess antifracture efficacy, but rather safety and efficacy on surrogate endpoints as well as the consistency of the effect of BPs over longer periods were evaluated. In all studies the incidence of nonvertebral fractures was constant over time. In the Fracture Intervention Trial (FIT) extension (FLEX), women who received on average alendronate for 5 years were randomized to placebo (ALN/PBO) or alendronate (5 or 10 mg/day) (ALN/ALN), and were followed for another 5 years [32]. Continuation of alendronate treatment led to further modest increases in spine BMD and stabilization of hip BMD, whereas a slow progressive decrease of total hip BMD was reported in patients who received placebo. At the end of the 10-year observation period, the incidence of nonvertebral and hip fractures in the ALN/PBO group was similar to that of the ALN/ALN group, while the incidence of clinical vertebral fractures was significantly decreased in the ALN/ALN group (2% versus 5%). In a post hoc analysis, women without vertebral fracture who entered the extension trial with a femoral neck BMD T-score of < -2.5 and continued alendronate use showed a significant 50% reduction in the risk of nonvertebral fractures during the 5-year extension [34]. Similar BMD and fracture data were reported in the first extension of the HORIZON trial in which patients treated with zoledronic acid for 3 years were randomized to three additional years of zoledronic acid or placebo [33]. Treatment for a further 3 years (total 9 years) had no additional benefit [35]. Taken together, the findings of the long-term extension studies of BPs are reassuring and indicate that prolonged exposure of bone tissue to BPs maintains the effect of treatment and does not compromise bone strength. Whether continuation of treatment offers additional antifracture benefits is not unequivocally established but, within the limitations of the studies, data strongly suggest that patients at increased risk of fractures can benefit from continuing treatment.

In line with these considerations, a task force of the American Society for Bone and Mineral Research (ASBMR) suggested reassessment of fracture risk in women after 5 years treatment with oral BP or 3 years treatment with intravenous BP [34]. In women at high risk (eg, older women, those with low hip BMD T-score, and those with previous major osteoporotic fractures or who fracture during therapy), continuation of BP treatment for up to 10 years (oral) or 6 years (intravenous) with periodic evaluations should be considered. In women not at high risk, discontinuation of treatment after 5 or 3 years can be considered for a period of 2 to 3 years (a so-called "drug holiday"). This approach is based on rather limited evidence and does not replace clinical judgment.

SPECIAL ISSUES RELATED TO TREATMENT OF OSTEOPOROSIS WITH BISPHOSPHONATES

Excessive suppression of bone remodeling

There have been concerns that the long-term decrease of bone remodeling by BPs may compromise bone strength, leading to increased bone fragility. Numerous studies in different animal models with N-BPs given at a wide range of doses and time intervals have consistently shown preservation or improvement of bone strength. Earlier reports of potential compromise of the biomechanical competence of bone due to increases in microdamage accumulation in bone biopsies of healthy dogs treated with high BP doses were not substantiated by later animal and human studies [36, 37].

In human controlled studies of osteoporosis the incidence of nonvertebral fractures was not increased with long-term therapy, and bone turnover markers increased after cessation of treatment indicating metabolically active bone. In addition, an analysis of the FIT data showed that higher decreases of bone turnover were associated with

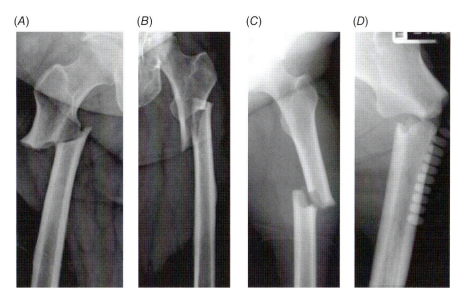

Fig. 71.4. Radiographs of patients with atypical subtrochanteric (ST) and femoral shaft (FS) fractures treated or not with bisphosphonates (BPs). (A,B) Atypical ST fractures in patients treated with BPs. (C) Atypical FS fracture that occurred in a patient who had discontinued BP 5 years before the fracture. (D) Atypical ST fracture in a man who has never been treated with a BP. Source: [47]. Reproduced with permission of Elsevier.

larger decreases in the incidence of nonvertebral and hip fractures [38], a finding supported by the previously mentioned analysis of the ibandronate studies [24]. Moreover, in studies of patients treated with BPs followed by teriparatide, early significant increases in bone markers have been reported, indicating that BP-treated bone can readily respond to stimuli [39]. This conclusion is further supported by a study of zoledronic acid treatment in patients previously treated with alendronate which showed that alendronate-treated bone reacted normally to an acute BP load, as provided by zoledronic acid, indicating that metabolic activity was preserved [40]. Similar results have been also reported in animal and human studies of transitioning from BPs to denosumab [41–43]. Finally, in bone biopsies of patients in the FLEX study there was no increase in bone matrix mineralization between 5 and 10 years of treatment with alendronate [44]; similarly the increase in bone matrix mineralization observed after 3 years of treatment with risedronate did not increase after a further 2 years of treatment [45]. Thus, in postmenopausal women with osteoporosis, long-term BP treatment does not lead to abnormally high mineralization of bone (sometimes called "hypermineralization").

Atypical fractures of the femur

During the past 10 years there has been growing concern about the potential relationship between low-energy subtrochanteric/diaphyseal femoral fractures, deemed to be atypical for patients with osteoporosis, and long-term use of BPs. Atypical fractures of the femur are often preceded by prodromal pain, can be bilateral, and healing may be delayed. A task force of the ASBMR has proposed criteria for the identification and diagnosis of atypical femoral fractures [46]. These fractures are rare (about 1% of all femoral fractures), and occur more frequently in patients treated with BPs than in untreated patients [46, 47] (Fig. 71.4). A causal association between BPs and atypical fractures has not been established, but it appears that the risk rises with increasing duration of exposure.

Osteonecrosis of the jaw

Osteonecrosis of the jaw (ONJ) is defined as exposed bone in the mandible, maxilla, or both that persists for at least 8 weeks in the absence of previous irradiation or metastases in the jaw. It has been reported mainly in patients with malignant diseases receiving high intravenous doses of BPs. The background incidence in the population and its pathogenesis are poorly defined, and a causal relation with BPs has not been established. In patients with osteoporosis treated with BPs ONJ is rare, its estimated incidence being between 1:10,000 and <1:100,000 patient-years, and appears to increase with the length of treatment [48]. In two RCTs of yearly intravenous zoledronic acid, two adjudicated cases of ONJ were reported among 9892 patients after 3 years, one in the placebo-treated group and one in the zoledronic acid -treated group [25, 26].

ADVERSE EFFECTS

BPs are relatively safe compounds and their benefits outweigh their potential risks. Specific adverse effects related to the use of BPs in osteoporosis include gastrointestinal toxicity associated with the oral, particularly daily, use of

N-BPs and symptoms related to an acute phase reaction, mainly after first exposure to intravenous N-BPs. Gastrointestinal toxicity appears to be higher with some generic preparations of oral BPs, resulting in significantly poorer adherence and effectiveness [49]. Case reports have suggested a relationship between oral N-BP treatment and esophageal cancer but this was not confirmed in analyses of large databases, while a reduction in the incidence of gastric and colon cancer was reported in alendronate users [29, 50]. The kidney is the principal route of BP elimination and use of BP is not indicated in patients with impaired renal function. Renal toxicity of intravenous BP is not a concern provided that the indications for treatment and instructions for administration are closely followed. A significant increase in the incidence of atrial fibrillation was observed in one [25] but not in another study [26] of patients receiving zoledronic acid compared with those receiving placebo. A biological explanation for this effect is not apparent and further analyses of clinical trials with alendronate, ibandronate, and risedronate did not confirm such association.

CONCLUSION

BPs are considered first-line therapy for postmenopausal women with osteoporosis due to their favorable benefit to harm balance and low cost. The selection of a BP for the treatment of an individual patient should be based on the evidence of its efficacy and risk profile as well as on the patient's preferences.

DISCLOSURES

A. Giusti has received consulting fees from bisphosphonate manufacturers: Abiogen, Merck & Co, and Chiesi. S Papapoulos has received research support and/or honoraria from bisphosphonate manufacturers: Merck & Co, Novartis, Procter & Gamble, and Roche/GSK.

REFERENCES

1. Russell RG, Watts NB, Ebetino FH, et al. Mechanisms of action of bisphosphonates: similarities and differences and their potential influence on clinical efficacy. Osteoporos Int. 2008;19(6):733–59.
2. Papapoulos SE. Bisphosphonates: how do they work? Best Pract Res Clin Endocrinol Metab. 2008;22(5):831–47.
3. Cremers SC, Pillai G, Papapoulos SE. Pharmacokinetics/pharmacodynamics of bisphosphonates: use for optimisation of intermittent therapy for osteoporosis. Clin Pharmacokinet. 2005;44(6):551–70.
4. Brumsen C, Papapoulos SE, Lips P, et al. Daily oral pamidronate in women and men with osteoporosis: a 3-year randomized placebo-controlled clinical trial with a 2-year open extension. J Bone Miner Res. 2002;17(6):1057–64.
5. Black DM, Cummings SR, Karpf DB, et al. Randomised trial of effect of alendronate on risk of fracture in women with existing vertebral fractures. Fracture Intervention Trial Research Group. Lancet. 1996;348(9041):1535–41.
6. Cummings SR, Black DM, Thompson DE, et al. Effect of alendronate on risk of fracture in women with low bone density but without vertebral fractures: results from the Fracture Intervention Trial. JAMA. 1998;280(24):2077–82.
7. Harris ST, Watts NB, Genant HK, et al. Effects of risedronate treatment on vertebral and nonvertebral fractures in women with postmenopausal osteoporosis: a randomized controlled trial. Vertebral Efficacy with Risedronate Therapy (VERT) Study Group. JAMA. 1999;282(14):1344–52.
8. Reginster J, Minne HW, Sorensen OH, et al. Randomized trial of the effects of risedronate on vertebral fractures in women with established postmenopausal osteoporosis. Vertebral Efficacy with Risedronate Therapy (VERT) Study Group. Osteoporos Int. 2000;11(1):83–91.
9. Chesnut CH 3rd, Skag A, Christiansen C, et al.; Oral Ibandronate Osteoporosis Vertebral Fracture Trial in North America and Europe (BONE). Effects of oral ibandronate administered daily or intermittently on fracture risk in postmenopausal osteoporosis. J Bone Miner Res. 2004;19(8):1241–9.
10. McCloskey E, Selby P, Davies M, et al. Clodronate reduces vertebral fracture risk in women with postmenopausal or secondary osteoporosis: results of a double-blind, placebo-controlled 3-year study. J Bone Miner Res. 2004;19(5):728–36.
11. Cranney A, Guyatt G, Griffith L, et al.; Osteoporosis Methodology Group and The Osteoporosis Research Advisory Group. Meta-analyses of therapies for postmenopausal osteoporosis. IX: Summary of meta-analyses of therapies for postmenopausal osteoporosis. Endocr Rev. 2002;23(4):570–8.
12. Wells G, Cranney A, Peterson J, et al. Risedronate for the primary and secondary prevention of osteoporotic fractures in postmenopausal women. Cochrane Database Syst Rev. 2008;(1):CD004523.
13. Wells GA, Cranney A, Peterson J, et al. Alendronate for the primary and secondary prevention of osteoporotic fractures in postmenopausal women. Cochrane Database Syst Rev. 2008;(1):CD001155.
14. Felsenberg D, Miller P, Armbrecht G, et al. Oral ibandronate significantly reduces the risk of vertebral fractures of greater severity after 1, 2, and 3 years in postmenopausal women with osteoporosis. Bone. 2005;37(5):651–4.
15. Black DM, Thompson DE, Bauer DC et al.; Fracture Intervention Trial. Fracture risk reduction with alendronate in women with osteoporosis: the Fracture Intervention Trial. FIT Research Group. J Clin Endocrinol Metab. 2000;85(11):4118–24.
16. Papapoulos SE, Quandt SA, Liberman UA, et al. Meta-analysis of the efficacy of alendronate for the prevention of hip fractures in postmenopausal women. Osteoporos Int. 2005;16(5):468–74.

17. Nguyen ND, Eisman JA, Nguyen TV. Anti-hip fracture efficacy of biophosphonates: a Bayesian analysis of clinical trials. J Bone Miner Res. 2006;21(2):340–9.
18. Seeman E, Compston J, Adachi J, et al. Non-compliance: the Achilles' heel of anti-fracture efficacy. Osteoporos Int. 2007;18(6):711–19.
19. Schnitzer T, Bone HG, Crepaldi G, et al. Therapeutic equivalence of alendronate 70 mg once-weekly and alendronate 10 mg daily in the treatment of osteoporosis. Alendronate Once-Weekly Study Group. Aging (Milano). 2000;12(1):1–12.
20. Brown JP, Kendler DL, McClung MR, et al. The efficacy and tolerability of risedronate once a week for the treatment of postmenopausal osteoporosis. Calcif Tissue Int. 2002;71(2):103–11.
21. Cranney A, Guyatt G, Krolicki N, et al.; Osteoporosis Research Advisory Group (ORAG). A meta-analysis of etidronate for the treatment of postmenopausal osteoporosis. Osteoporos Int. 2001;12(2):140–51.
22. Papapoulos SE, Schimmer RC. Changes in bone remodelling and antifracture efficacy of intermittent bisphosphonate therapy: implications from clinical studies with ibandronate. Ann Rheum Dis. 2007;66(7):853–8.
23. Reginster JY, Adami S, Lakatos P, et al. Efficacy and tolerability of once-monthly oral ibandronate in postmenopausal osteoporosis: 2 year results from the MOBILE study. Ann Rheum Dis. 2006;65(5):654–61.
24. Cranney A, Wells GA, Yetisir E, et al. Ibandronate for the prevention of nonvertebral fractures: a pooled analysis of individual patient data. Osteoporos Int. 2009;20(2):291–7.
25. Black DM, Delmas PD, Eastell R, et al.; HORIZON Pivotal Fracture Trial. Once-yearly zoledronic acid for treatment of postmenopausal osteoporosis. N Engl J Med. 2007;356(18):1809–22.
26. Lyles KW, Colón-Emeric CS, Magaziner JS, et al.; HORIZON Recurrent Fracture Trial. Zoledronic acid and clinical fractures and mortality after hip fracture. N Engl J Med. 2007;357(18):1799–809.
27. Kranenburg G, Bartstra JW, Weijmans M, et al. Bisphosphonates for cardiovascular risk reduction: a systematic review and meta-analysis. Atherosclerosis. 2016;252:106–15.
28. Lee P, Ng C, Slattery A, et al. Preadmission bisphosphonate and mortality in critically ill patients. J Clin Endocrinol Metab. 2016;101(5):1945–53.
29. Pazianas M, Abrahamsen B, Eiken PA, et al. Reduced colon cancer incidence and mortality in postmenopausal women treated with an oral bisphosphonate—Danish National Register Based Cohort Study. Osteoporos Int. 2012;23(11):2693–701.
30. Bone HG, Hosking D, Devogelaer JP, et al.; Alendronate Phase III Osteoporosis Treatment Study Group. Ten years' experience with alendronate for osteoporosis in postmenopausal women. N Engl J Med. 2004;350(12):1189–99.
31. Mellström DD, Sörensen OH, Goemaere S, et al. Seven years of treatment with risedronate in women with postmenopausal osteoporosis. Calcif Tissue Int. 2004;75(6):462–8.
32. Black DM, Schwartz AV, Ensrud KE, et al.; FLEX Research Group. Effects of continuing or stopping alendronate after 5 years of treatment: the Fracture Intervention Trial Long-term Extension (FLEX): a randomized trial. JAMA. 2006;296(24):2927–38.
33. Black DM, Reid IR, Boonen S, et al. The effect of 3 versus 6 years of zoledronic acid treatment of osteoporosis: a randomized extension to the HORIZON-Pivotal Fracture Trial (PFT). J Bone Miner Res. 2012;27(2):243–54.
34. Adler RA, El-Hajj Fuleihan G, Bauer DC, et al. Managing osteoporosis in patients on long-term bisphosphonate treatment: report of a task force of the American Society for Bone and Mineral Research. J Bone Miner Res. 2016;31(10):1910.
35. Black DM, Reid IR, Cauley JA, et al. The effect of 6 versus 9 years of zoledronic acid treatment in osteoporosis: a randomized second extension to the HORIZON-Pivotal Fracture Trial (PFT). J Bone Miner Res. 2015;30(5):934–44.
36. Allen MR, Iwata K, Phipps R, et al. Alterations in canine vertebral bone turnover, microdamage accumulation, and biomechanical properties following 1-year treatment with clinical treatment doses of risedronate or alendronate. Bone. 2006;39(4):872–9.
37. Chapurlat RD, Arlot M, Burt-Pichat B, et al. Microcrack frequency and bone remodeling in postmenopausal osteoporotic women on long-term bisphosphonates: a bone biopsy study. J Bone Miner Res. 2007;22(10):1502–9.
38. Bauer DC, Black DM, Garnero P, et al.; Fracture Intervention Trial Study Group. Change in bone turnover and hip, non-spine, and vertebral fracture in alendronate-treated women: the fracture intervention trial. J Bone Miner Res. 2004;19(8):1250–8.
39. Miller PD, Delmas PD, Lindsay R, et al.; Open-label Study to Determine How Prior Therapy with Alendronate or Risedronate in Postmenopausal Women with Osteoporosis Influences the Clinical Effectiveness of Teriparatide Investigators. Early responsiveness of women with osteoporosis to teriparatide after therapy with alendronate or risedronate. J Clin Endocrinol Metab. 2008;93(10):3785–93.
40. McClung M, Recker R, Miller P, et al. Intravenous zoledronic acid 5 mg in the treatment of postmenopausal women with low bone density previously treated with alendronate. Bone. 2007;41(1):122–8.
41. Kostenuik PJ, Smith SY, Samadfam R, et al. Effects of denosumab, alendronate, or denosumab following alendronate on bone turnover, calcium homeostasis, bone mass and bone strength in ovariectomized cynomolgus monkeys. J Bone Miner Res. 2015;30(4):657–69.
42. Roux C, Hofbauer LC, Ho PR, et al. Denosumab compared with risedronate in postmenopausal women suboptimally adherent to alendronate therapy: efficacy and safety results from a randomized open-label study. Bone. 2014;58:48–54.
43. Kendler DL, Roux C, Benhamou CL, et al. Effects of denosumab on bone mineral density and bone turnover in postmenopausal women transitioning from alendronate therapy. J Bone Miner Res. 2010;25(1):72–81.

44. Roschger P, Lombardi A, Misof BM, et al. Mineralization density distribution of postmenopausal osteoporotic bone is restored to normal after long-term alendronate treatment: qBEI and sSAXS data from the fracture intervention trial long-term extension (FLEX). J Bone Miner Res. 2010;25(1):48–55.
45. Borah B, Dufresne TE, Ritman EL, et al. Long-term risedronate treatment normalizes mineralization and continues to preserve trabecular architecture: sequential triple biopsy studies with micro-computed tomography. Bone. 2006;39(2):345–52.
46. Shane E, Burr D, Abrahamsen B, et al. Atypical subtrochanteric and diaphyseal femoral fractures: second report of a task force of the American Society for Bone and Mineral Research. J Bone Miner Res. 2014;29(1):1–23.
47. Giusti A, Hamdy NA, Dekkers OM, et al. Atypical fractures and bisphosphonate therapy: a cohort study of patients with femoral fracture with radiographic adjudication of fracture site and features. Bone. 2011;48(5):966–71.
48. Khan AA, Morrison A, Hanley DA, et al.; International Task Force on Osteonecrosis of the Jaw. Diagnosis and management of osteonecrosis of the jaw: a systematic review and international consensus. J Bone Miner Res. 2015;30(1):3–23.
49. Kanis JA, Reginster JY, Kaufman JM, et al. A reappraisal of generic bisphosphonates in osteoporosis. Osteoporos Int. 2012;23(1):213–21.
50. Abrahamsen B, Pazianas M, Eiken P, et al. Esophageal and gastric cancer incidence and mortality in alendronate users. J Bone Miner Res. 2012;27(3):679–86.

72

Denosumab

Aline Granja Costa[1], E. Michael Lewiecki[2], and John P. Bilezikian[1]

[1]*Department of Medicine, Division of Endocrinology, Metabolic Bone Diseases Unit, College of Physicians and Surgeons, Columbia University, New York, NY, USA*
[2]*New Mexico Clinical Research & Osteoporosis Center, Albuquerque, NM, USA*

INTRODUCTION

Denosumab (Prolia; Amgen, Inc., Thousand Oaks, CA, USA) is a fully human monoclonal antibody that inhibits the receptor activator of RANKL. It is a relatively recent addition to the antiresorptive drugs available for the treatment of osteoporosis. It was approved by the US Food and Drug Administration (FDA) and by the European Medicines Agency (EMA) in 2010 to reduce fracture incidence in postmenopausal women with osteoporosis at high risk for fragility fractures. Later, in 2012, denosumab was approved for men with osteoporosis at high risk for fracture to increase bone mass. Other indications are to increase bone mass in women at high risk for fracture receiving adjuvant aromatase inhibitor therapy for breast cancer; and to increase bone mass in men at high risk for fracture receiving androgen deprivation therapy for nonmetastatic prostate cancer [1]. It is important to point out that in Europe denosumab is not approved for use in women receiving aromatase inhibitors to increase bone mass.

MECHANISM OF ACTION

Denosumab is a highly specific inhibitor of RANKL, a powerful bone-resorbing cytokine and the principal regulator of osteoclastic bone resorption (Fig. 72.1). By preventing binding of RANKL to its cognate receptor, RANK, on mature osteoclasts and their progenitors, bone resorption is markedly reduced [2]. Additionally, reduced bone resorption with denosumab administration reduces bone formation and thus leads to an overall reduction in bone turnover [3–5]. The consequences of this mechanism are substantial increases in BMD in the lumbar spine and more modest increases in the femoral neck, total proximal femur, and one-third radius. Fracture risk is reduced at the spine, hip, and other nonvertebral sites.

PHARMACODYNAMICS, PHARMACOKINETICS, AND METABOLISM

The effects of denosumab on bone turnover markers (BTMs) are substantial. CTx, a bone resorption marker, falls over a short period of time after the subcutaneous injection of a therapeutic dose of 60 mg. In a majority of patients, CTx levels decline to below the detection limit of the assay. When denosumab is discontinued, BTMs rapidly increase to levels that transiently exceed the original baseline values. In addition, BMD falls quickly [6]. It has recently been shown that along with the marked reduction in BMD after denosumab is stopped, there is a rapid return of vertebral fracture risk to baseline levels and an apparent increase in the risk of multiple vertebral fractures [7].

Healthy postmenopausal women enrolled in a phase I dose-escalation study who received a single dose of denosumab demonstrated reduced NTx, another marker of bone resorption, as early as 12 hours post-dose (mean decrease: 77% in the 3.0 mg/kg group versus 46% in the placebo group) [8]. The timing of the nadir of NTx levels varied with the dose administered: as early as 2 weeks for the 0.01, 0.03, 0.3, and 1.0 mg/kg dose groups; 1 month for the 0.1 mg/kg group, and 3 months for the 3.0 mg/kg group. Patients in the higher dose groups showed the most prolonged suppression of NTx, with levels returning to baseline after 6 to 9 months. Although there was

Primer on the Metabolic Bone Diseases and Disorders of Mineral Metabolism, Ninth Edition. Edited by John P. Bilezikian.
© 2019 American Society for Bone and Mineral Research. Published 2019 by John Wiley & Sons, Inc.
Companion website: www.wiley.com/go/asbmrprimer

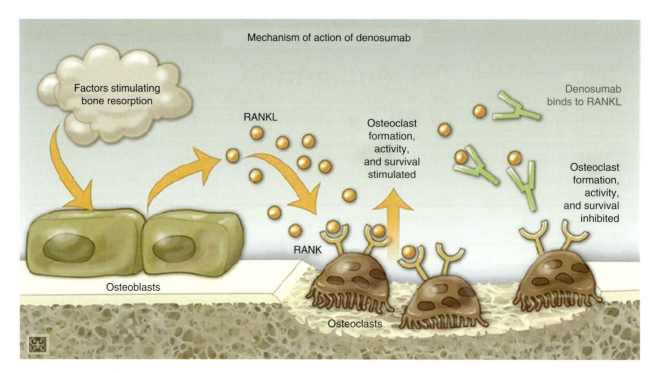

Fig. 72.1. Denosumab is an inhibitor of RANKL, the principal regulator of osteoclastic bone resorption. Source: [49]. Reproduced with permission of John Wiley & Sons.

a decrease in levels of the bone formation marker bone-specific alkaline phosphatase (BSAP), it fell less rapidly and not to the same extent as the changes in NTx.

The time course of bone resorption suppression shows a tendency for the reduced level to increase slightly during the last 2 months of the 6-month period [9]. Throughout the sequence of denosumab administration, repeated every 6 months, CTx levels tend to rise a bit earlier, the significance of which is not certain. In the pivotal registration trial, known as FREEDOM (Fracture REduction Evaluation of Denosumab in Osteoporosis every 6 Months), and its extension, BTMs remained below pretreatment baseline values throughout [10]. The rapid reversibility of BTMs and BMD values has been noted [11].

Similar to the general behavior of monoclonal antibodies, denosumab shows dose-dependent, nonlinear pharmacokinetics across a wide dose range. At the mean maximum denosumab concentration (C_{max}) of 6.75 µg/mL after a single subcutaneous 60 mg denosumab dose, the median time to maximum denosumab concentration (t_{max}) was 10 days. Serum denosumab concentrations declined over a period of 4 to 5 months [6]. A fixed dose of denosumab 60 mg administered every 6 months provided similar RANKL inhibition as using weight-based dosing, thus obviating the need for dosage adjustment based on weight [12]. Furthermore, there are no significant differences in pharmacokinetic or pharmacodynamic parameters in patients with impaired renal function, requiring no adjustment of the dose of denosumab in the treatment of such patients [13].

CLINICAL EFFICACY

The initial phase I clinical trial with healthy postmenopausal women exposed to a single subcutaneous dose of denosumab (0.01, 0.03, 0.1, 0.3, 1.0, or 3.0 mg/kg) showed, along with a suppression of NTs, a transient decrease in the serum calcium concentration (<10% on average) along with a transient increase in serum PTH (up to threefold after 4 days in the 3.0 mg/kg group).

In a phase II randomized, placebo-controlled, dose-ranging study [14], denosumab increased BMD (+9.4% to 11.8% compared with 2.4% decrease at the lumbar spine with placebo after 48 months). BTMs were suppressed in a rapid and sustained manner. Discontinuation of therapy led to a return towards baseline values for both BMD and BTMs. These major changes are associated with an increase in fractures in the year after stopping the drug [7].

In the FREEDOM pivotal phase III clinical trial, denosumab significantly reduced fracture incidence at vertebral (68%), hip (40%), and other nonvertebral sites (20%) as compared with placebo controls [15, 16]. Additionally, denosumab led to progressive increments in BMD at the lumbar spine (9.2%) and total hip (6%) at 3 years. The one-third radius showed a small increase also [17]. These results, along with a favorable safety profile, led to FDA approval of denosumab for the treatment of postmenopausal women with osteoporosis at high risk for fracture [18].

Further information has become available from those initial trials that have been extended for as long as

10 years [19]. In a rather unique manner that seems to characterize this drug, BMD at the lumbar spine and total hip was found to continuously increase for the entire duration of therapy. This observation is very different to that seen in all other classes of antiresorptive drugs, where the increment in BMD tends to reach a plateau after approximately 3 years of exposure. Moreover, at the lumbar spine, the slope of the increase seems to be maintained for the extended period of time. Compared with baseline, there were cumulative BMD gains over 10 years in the lumbar spine (21.7%) and in the total hip (9.2%) when compared with baseline. No declines were seen at the one-third radius site. BTMs continued to show values substantially below baseline.

Histomorphometric analyses of iliac crest bone biopsies showed that exposure to denosumab for 1 to 3 years resulted in a 50% to 80% reduction in median eroded surface [15]. Double tetracycline labels were detectable in only about 20% of patients, indicating markedly reduced bone formation. Osteoclasts were not visible. Indices of bone quality were normal. Additionally, micro-computed tomography analysis in a subset of the biopsies showed reduced porosity and increased volumetric BMD (vBMD) in cortical bone [20]. After 5 years of denosumab [21], these changes persisted [21]. More recently, Dempster and colleagues have evaluated transiliac bone biopsies in subjects who have been treated with up to 10 years of continuous exposure [22,23]. The following conclusions were drawn from this longitudinal analysis: cancellous and cortical structure was maintained, along with the antiresorptive effect. Bone remodeling remained low but stable. Trabecular bone showed an increased in tetracycline labels suggesting time-dependent increases in bone turnover of that compartment. Bone matrix mineralization increased through year 5 but not further over the next 5 years [23]. These histomorphometric effects are reversible when the drug is discontinued. Based upon FREEDOM and the Study of Transitioning from Alendronate to Denosumab (STAND) trials [24], there was nearly complete recovery of reduced bone formation with 14 or 15 biopsies showing clear double tetracycline labels over the course of 1 to 3 years after a 12-month treatment period.

In the head-to-head comparator study known as DECIDE (Determining Efficacy: Comparison of Initiating Denosumab versus alEndronate), the denosumab group at 12 months showed a significantly greater BMD increase at the total hip compared with alendronate (3.5% versus 2.6%; $p<0.0001$), with treatment differences of 0.6% at the femoral neck and 1.1% at lumbar spine [25]. Similar results were obtained in the STAND trial with the denosumab group showing a greater increase in BMD (+1.9%) compared with alendronate (+1.05%) ($p<0.0001$) [26]. Postmenopausal women previously treated with alendronate were randomized to receive 12 months of denosumab by subcutaneous injection every 6 months or oral risedronate 150 mg monthly [27]. Denosumab use was associated with greater BMD gains at the total hip (2.0% versus 0.5%), femoral neck (1.4% versus 0%), and lumbar spine (3.4% versus 1.1%) ($p<0.0001$) for comparisons at all sites. Similar comparisons favoring denosumab insofar as BMD was concerned, were made when ibandronate was the bisphosphonate [28]. Patients who did not respond well to at least 2 years of bisphosphonate therapy showed significant gains at the lumbar spine and total hip sites when switched to denosumab [29].

Results relevant to combination therapy with teriparatide for osteoporosis are promising. The DATA (Denosumab And Teriparatide Administration) study randomized 94 postmenopausal women with osteoporosis to receive either 20 μg teriparatide daily, 60 mg denosumab every 6 months, or combination therapy [30]. Unlike combination studies with BPs and teriparatide [31,32], combination therapy with denosumab and teriparatide was found to be more beneficial than monotherapy. At 12 months, lumbar spine BMD showed greater gains with combined therapy (+9.1%) than with teriparatide (+6.2%; $p=0.0139$) or denosumab alone (5.5%; $p=0.0005$). A similar pattern was observed for BMD changes at the total hip site (combination +4.9% versus teriparatide +0.7%; $p<0.0001$; denosumab +2.5%; $p=0.0011$) [30]. Beyond beneficial effects on BMD with combined therapy, skeletal microstructure as imaged by HRpQCT at the distal tibia and radius showed greater improvement with combination therapy than with monotherapy [33].

Denosumab has also been shown to be effective in men who are undergoing androgen deprivation therapy. The HALT (Hormone Ablation Bone Loss Trial) study addressed effects of denosumab in men who were being treated for nonmetastatic prostate cancer. Denosumab was associated with a 5.6% increase in lumbar spine BMD as compared with a loss of 1.0% in the placebo group ($p<0.001$). Furthermore, at 36 months, the incidence of new vertebral fractures was lower in the treated group (1.5% versus 3.9% with placebo; $p=0.006$) [16].

The study leading to the registration of denosumab in men with osteoporosis is known as ADAMO (Study to Compare the Efficacy and Safety of DenosumAb Versus Placebo in Males with Osteoporosis) [34,35]. Men treated with denosumab had BMD gains at 1 year of 5.7% at the lumbar spine, 2.4% at the total hip, and 2.1% at the femoral neck (adjusted $p \leq 0.0144$ for BMD percent differences at all sites compared with placebo) [34].

Women with hormone receptor-positive nonmetastatic breast cancer treated with adjuvant aromatase inhibitor therapy treated with denosumab for 2 years showed BMD gains at the lumbar spine of 5.5% at 12 months and 7.6% at 24 months, which was significantly greater than those in the placebo group. BMD increases were not influenced by duration of aromatase inhibitor therapy. This trial led to FDA approval of denosumab to increase bone mass in women at high risk for fracture receiving adjuvant aromatase inhibitor therapy for breast cancer [36]. Moreover, the risk of clinical fractures was reduced in postmenopausal patients with early hormone receptor-positive breast cancer receiving treatment with aromatase inhibitors treated with adjuvant denosumab 60 mg twice per year.

ADVERSE EVENTS

Since RANKL has an important role in immunocompetence, there has been concern that patients treated with denosumab might be a greater risk for infections [37]. Although the initial trials showed an imbalance regarding skin infections, particularly cellulitis as a serious adverse event and eczema in treated patients versus placebo, the overall incidence of infections was similar between the groups [16]. Additional data from the FREEDOM extension did not corroborate earlier observations about an imbalance in infections. The current view is that patients are not at greater risk for infections when treated with denosumab. [21].

There have been reports of osteonecrosis of the jaw (ONJ) in patients treated with the customary dose of denosumab for osteoporosis and when denosumab is used as an antimetastatic drug in cancer at much higher doses [38,39]. As is the case for BPs, ONJ is a rare event in patients receiving the twice-yearly dosage regimen for osteoporosis. Also, similar to the association between ONJ and BPs, risk factors include recent dental extraction, poor oral hygiene, diabetes mellitus, use of removable dental appliances, or chemotherapy [39].

There are also several reports of atypical femoral fractures in patients treated with denosumab [40–43]. During the extended FREEDOM trial two cases of atypical femoral fractures were confirmed. One patient suffered an atypical diaphyseal femoral fracture after 8 years of treatment and another patient developed a midshaft femur fracture in the third year after crossing over from placebo to denosumab [9].

Denosumab can lead to a reduction in the serum calcium concentration. It is therefore important that, prior to being treated, vitamin D sufficiency should be ensured (25-hydroxyvitamin D >30 ng/mL) [44]. This is particularly important in patients whose creatinine clearance is <30 mL/min. Anaphylaxis has been reported in five patients with no fatal outcomes [45]. The most common adverse reactions in postmenopausal women were back pain, pain in extremity, hypercholesterolemia, musculoskeletal pain, and cystitis. These changes were numerically different from the incidence of these events in the placebo control arm of the study.

DURATION OF TREATMENT

The rapid reversibility of denosumab with regard to BTMs, BMD, and, most recently, fracture protection [6,7] has raised concerns about stopping therapy without transitional pharmacological intervention. In patients whose therapy with denosumab is going to be discontinued, therefore, the current thinking is that patients should be treated, at least temporarily, with another agent such as intravenous zoledronic acid, administered 7–8 months after the last dose of denosumab [46]. In this way, gains appreciated with denosumab administration can be sustained [47]. The concept of a drug holiday can no longer be safely applied to denosumab [48].

DENOSUMAB IN PATIENTS WITH METASTATIC BONE DISEASE

Denosumab with a different brand name (Xgeva; Amgen, Inc.) and a different dose is approved for the prevention of skeletal-related events in: (i) patients with bone metastases from solid tumors; (ii) the treatment of adults and skeletally mature adolescents with giant cell tumor of bone that is unresectable or where surgical resection is likely to result in severe morbidity; and (iii) in the treatment of hypercalcemia of malignancy refractory to BP therapy [49]. This topic is covered in more detail elsewhere in this book.

CONCLUSION

The advent of denosumab as an effective therapy for postmenopausal osteoporosis, male osteoporosis, and the special situations of men and women with prostate and breast cancer, respectively, has widened options for maintaining and improving skeletal health among patients at risk for fracture.

REFERENCES

1. Costa AG, Bilezikian JP. How long to treat with denosumab. Curr Osteoporos Rep. 2015;13(6):415–20.
2. Lewiecki EM. Denosumab update. Curr Opin Rheumatol. 2009;21(4):369–73.
3. Bone HG, Bolognese MA, Yuen CK, et al. Effects of denosumab on bone mineral density and bone turnover in postmenopausal women. J Clin Endocrinol Metab. 2008;93(6):2149–57.
4. McClung MR, Lewiecki EM, Cohen SB, et al. Denosumab in postmenopausal women with low bone mineral density. N Engl J Med. 2006;354(8):821–31.
5. Lewiecki EM, Miller PD, McClung MR, et al. Two-year treatment with denosumab (AMG 162) in a randomized phase 2 study of postmenopausal women with low BMD. J Bone Miner Res. 2007;22(12):1832–41.
6. Bone HG, Bolognese MA, Yuen CK, et al. Effects of denosumab treatment and discontinuation on bone mineral density and bone turnover markers in postmenopausal women with low bone mass. J Clin Endocrinol Metab. 2011;96(4):972–80.
7. Cummings SR, Ferrari S, Eastell R, et al. Vertebral fractures after discontinuation of denosumab: a post hoc analysis of the randomized placebo-controlled FREEDOM Trial and Its extension. J Bone Miner Res. 2018;33:190–8.

8. Bekker P, Holloway D, Rasmussen A, et al. A single-dose placebo-controlled study of AMG 162, a fully monoclonal antibody to RANKL, in postmenopausal women. J Bone Miner Res. 2004;19(7):1059–66.
9. Papapoulos S, Lippuner K, Roux C, et al. The effect of 8 or 5 years of denosumab treatment in postmenopausal women with osteoporosis: results from the FREEDOM extension study. Osteoporos Int. 2015;26(12):2773–83.
10. Bone HG, Chapurlat R, Brandi ML, et al. The effect of three or six years of denosumab exposure in women with postmenopausal osteoporosis: results from the FREEDOM extension. J Clin Endocrinol Metab. 2013;98(11):4483–92.
11. Miller PD, Bolognese MA, Lewiecki EM, et al.; Amg Bone Loss Study Group. Effect of denosumab on bone density and turnover in postmenopausal women with low bone mass after long-term continued, discontinued, and restarting of therapy: a randomized blinded phase 2 clinical trial. Bone. 2008;43(2):222–9.
12. Sutjandra L, Rodriguez R, Doshi S, et al. Population pharmacokinetic meta-analysis of denosumab in healthy subjects and postmenopausal women with osteopenia or osteoporosis. Clin Pharmacokinet. 2011;50(12):793–807.
13. Block G, Bone H, Fang L, et al. A single-dose study of denosumab in patients with various degrees of renal impairment. J Bone Miner Res. 2012;27(7):1471–9.
14. Miller P, Bolognese M, Lewiecki E, et al. Effect of denosumab on bone density and turnover in postmenopausal women with low bone mass after long-term continued, discontinued, and restarting of therapy: a randomized blinded phase 2 clinical trial. Bone. 2008;43(2):222–9.
15. Cummings SR, San Martin J, McClung MR, et al. Denosumab for prevention of fractures in postmenopausal women with osteoporosis. N Engl J Med. 2009;361(8):756–65.
16. Smith MR, Egerdie B, Hernandez Toriz N, et al. Denosumab in men receiving androgen-deprivation therapy for prostate cancer. N Engl J Med. 2009;361(8):745–55.
17. Bolognese MA, Teglbjaerg CS, Zanchetta JR, et al. Denosumab significantly increases DXA BMD at both trabecular and cortical sites: results from the FREEDOM study. J Clin Densitom. 2013;16(2):147–53.
18. Rachner TD, Hadji P, Hofbauer LC. Novel therapies in benign and malignanct bone diseases. Pharmacol Ther. 2012;134(3):338–344.
19. Bone HG, Brandi ML, Chapurlat R, et al. Ten years of denosumab treatment in postmenopausal women with osteoporosis: results from the FREEDOM extension trial. J Bone Miner Res. 2015;30(suppl 1):S471.
20. Reid IR, Miller PD, Brown JP, et al.; Denosumab Phase 3 Bone Histology Study Group. Effects of denosumab on bone histomorphometry: the FREEDOM and STAND studies. J Bone Miner Res. 2010;25(10):2256–65.
21. Brown JP, Reid IR, Wagman RB, et al. Effects of up to 5 years of denosumab treatment on bone histology and histomorphometry: the FREEDOM study extension. J Bone Miner Res. 2014;29(9):2051–6.
22. Dempster DW, Daizadeh N, Fahrleitner-Pammer A, et al. Effect of 10 years of denosumab treatment on bone histology and histomorphometry in the FREEDOM extension study. J Bone Miner Res. 2016;31(suppl):S2.
23. Dempster DW, Brown JP, Yue S, et al. Effects of up to 10 years of denosumab treatment on bone mineral matrix mineralization: results from the FREEDOM extension. J Bone Miner Res. 2016;31(suppl):S53.
24. Brown JP, Dempster DW, Ding B, et al. Bone remodeling in postmenopausal women who discontinued denosumab treatment: off-treatment biopsy study. J Bone Miner Res. 2011;26(11):2737–44.
25. Brown J, Prince R, Deal C, et al. Comparison of the effect of denosumab and alendronate on BMD and biochemical markers of bone turnover in postmenopausal women with low bone mass: a randomized, blinded, phase 3 trial. J Bone Miner Res. 2009;24(1):153–61.
26. Kendler D, Roux C, Benhamou C, et al. Effects of denosumab on bone mineral density and bone turnover in postmenopausal women transitioning from alendronate therapy. J Bone Miner Res. 2009;25(1):72–81.
27. Roux C, Hofbauer L, Ho P, et al. Denosumab compared with risedronate in postmenopausal women suboptimally adherent to alendronate therapy: efficacy and safety results from a randomized open-label study. Bone. 2014;58:48–54.
28. Recknor C, Czerwinski E, Bone H, et al. Denosumab compared with ibandronate in postmenopausal women previously treated with bisphosphonate therapy: a randomized open-label trial. Obstet Gynecol. 2013;121(6):1291–9.
29. Kamimura M, Nakamura Y, Ikegami S, et al. Significant improvement of bone mineral density and bone turnover markers by denosumab therapy in bisphosphonate-unresponsive patients. Osteoporos Int. 2017;28(2):559–66.
30. Tsai J, Uihlein A, Lee H, et al. Teriparatide and denosumab, alone or combined, in women with postmenopausal osteoporosis: the DATA study randomised trial. Lancet. 2013;6;382(9886):50–6.
31. Cosman F, Eriksen E, Recknor C, et al. Effects of intravenous zoledronic acid plus subcutaneous teriparatide [rhPTH(1–34)] in postmenopausal osteoporosis. J Bone Miner Res. 2011;26(3):503–11.
32. Finkelstein J, Wyland J, Lee H, et al. Effects of teriparatide, alendronate, or both in women with postmenopausal osteoporosis. J Clin Endocrinol Metab. 2010;95(4):1838–45.
33. Tsai JN, Uihlein AV, Burnett-Bowie SM, et al. Effects of two years of teriparatide, denosumab, or both on bone microarchitecture and strength (DATA-HRpQCT study). J Clin Endocrinol Metab. 2016;101(5):2023–30.
34. Orwoll E, Teglbjaerg CS, Langdahl BL, et al. A randomized, placebo-controlled study of the effects of denosumab for the treatment of men with low bone mineral density. J Clin Endocrinol Metab. 2012;97(9):3161–9.
35. Langdahl BL, Teglbjaerg CS, Ho PR, et al. A 24-month study evaluating the efficacy and safety of denosumab for the treatment of men with low bone mineral density: results from the ADAMO trial. J Clin Endocrinol Metab. 2015;100(4):1335–42.
36. Ellis GK, Bone HG, Chlebowski R, et al. Randomized trial of denosumab in patients receiving adjuvant aro-

matase inhibitors for nonmetastatic breast cancer. J Clin Oncol. 2008;26:4875–82.
37. Zaheer S, LeBoff M, Lewiecki EM. Denosumab for the treatment of osteoporosis. Expert Opin Drug Metab Toxicol. 2015;11(3):461–70.
38. Khan AA, Morrison A, Hanley DA, et al.; International Task Force on Osteonecrosis of the Jaw. Diagnosis and management of osteonecrosis of the jaw: a systematic review and international consensus. J Bone Miner Res. 2015;30(1):3–23.
39. Boquete-Castro A, Gomez-Moreno G, Calvo-Guirado JL, et al. Denosumab and osteonecrosis of the jaw. A systematic analysis of events reported in clinical trials. Clin Oral Implants Res. 2016;27(3):367–75.
40. Schilcher J, Aspenberg P. Atypical fracture of the femur in a patient using denosumab—a case report. Acta Orthop. 2014;85(1):6–7.
41. Villiers J, Clark DW, Jeswani T, et al. An atraumatic femoral fracture in a patient with rheumatoid arthritis and osteoporosis treated with denosumab. Case Rep Rheumatol. 2013;2013:249872.
42. Khow KS, Yong TY. Atypical femoral fracture in a patient treated with denosumab. J Bone Miner Metab. 2015;33(3):355–8.
43. Thompson RN, Armstrong CL, Heyburn G. Bilateral atypical femoral fractures in a patient prescribed denosumab—a case report. Bone. 2014;61:44–7.
44. Holick MF, Binkley NC, Bischoff-Ferrari HA, et al. Evaluation, treatment, and prevention of vitamin D deficiency: an Endocrine Society clinical practice guideline. J Clin Endocrinol Metab. 2011;96(7):1911–30.
45. Geller M, Wagman R, Ho P, et al. Early findings from Prolia® post-marketing safety surveillance for atypical femur fracture, osteonecrosis of the jaw, severe symptomatic hypocalcemia, and anaphylaxis. Ann Rheum Dis. 2014;73(suppl 2):766–7.
46. Horne AM, Mihov B, Reid IR. Bone loss after romosozumab/denosumab: effects of bisphosphonates. Calcif Tissue Int. 2018;103:55–61.
47. Leder B, Jiang L, Tsai JN. Relative efficacy toprompt follow-up therapy in postmenopausal women completing the denosumab and teriparatide administration (DATA) study. J Bone Miner Res. 2016;31(suppl):S101.
48. McClung MR. Cancel the denosumab holiday. Osteoporos Int. 2016;27(5):1677–82.
49. Lewiecki EM, Bilezikian JP. Denosumab for the treatment of osteoporosis and cancer-related conditions. Clin Pharmacol Ther. 2012;91(1):123–33.

73

Parathyroid Hormone and Abaloparatide Treatment for Osteoporosis

Felicia Cosman[1] and Susan L. Greenspan[2]

[1]Columbia College of Physicians and Surgeons, Columbia University; and Clinical Research Center, Helen Hayes Hospita, West Haverstraw, New York, NY, USA
[2]Divisons for Geriatrics, Endocrinology, and Metabolism, University of Pittsburgh; and Osteoporosis Prevention and Treatment Center, University of Pittsburgh Medical Center, Pittsburgh, PA, USA

INTRODUCTION

As a result of its unique mechanism of action, parathyroid hormone, the only approved anabolic therapy for osteoporosis, produces larger increments in bone mass (particularly in the spine) than those seen with antiresorptive therapies. PTH treatment first stimulates bone formation and subsequently stimulates both resorption and formation; the balance remains positive for formation throughout [1–3]. In the iliac crest, the growth of new bone permits restoration of microarchitecture, including improved trabecular connectivity and enhanced cortical thickness [4, 5]. Bone formation is accelerated in cancellous and endocortical as well as periosteal envelopes [6–10]. A recent study of teriparatide (TPTD) versus placebo in patients undergoing total hip replacement indicated a rapid effect of TPTD to stimulate bone formation in the femoral neck of the human femur, suggesting a mechanism for improvement in hip strength with TPTD [11, 12].

This chapter highlights the most important clinical PTH studies. TPTD is the recombinant or synthesized aminoterminal PTH(1–34) fragment. PTH(1–84) is the intact human recombinant molecule. PTH without other designation denotes either compound. Abaloparatide, currently under US Food and Drug Administration (FDA) review, is a synthetic analogue of PTHrP, which also acts through the PTH1 receptor to produce potent anabolic activity.

CANDIDATES FOR ANABOLIC THERAPY

Patients at high risk of fractures, including those with prevalent vertebral fractures, other osteoporosis-related fractures with BMD in the low bone mass or osteoporosis range, or very low BMD, even in the absence of fractures (T-score$\leq$–3), are candidates for anabolic therapy. Anabolic therapy is optimally used first line but should also be recommended for individuals who incur incident fractures or active bone loss while on other therapies or who have persistent osteoporosis despite prior therapy. PTH is contraindicated in individuals at elevated risk for osteosarcoma (skeletal radiation and Paget disease) and those with primary or metastatic bone cancer, myeloma, hyperparathyroidism, or hypercalcemia. PTH treatment is given as a daily subcutaneous injection for 18 to 24 months [13].

POSTMENOPAUSAL OSTEOPOROSIS

Teriparatide monotherapy

Neer and colleagues randomized 1637 postmenopausal women with prevalent vertebral fractures to TPTD, 20 or 40 μg, or placebo [13]. After 19 months, TPTD 20 μg increased spine BMD by 9.7%, femoral neck (FN) BMD by 2.8%, and total hip (TH) BMD by 2.6% (all $p<0.001$)

Primer on the Metabolic Bone Diseases and Disorders of Mineral Metabolism, Ninth Edition. Edited by John P. Bilezikian.
© 2019 American Society for Bone and Mineral Research. Published 2019 by John Wiley & Sons, Inc.
Companion website: www.wiley.com/go/asbmrprimer

but decreased radius BMD (–2.1%; NS). Vertebral fracture risk reduction with TPTD 20 μg was 65% (absolute risk 4% versus 14% placebo), and height loss was reduced in patients with incident vertebral fractures. Compared with placebo, incident nonvertebral fractures and nonvertebral fragility fractures were reduced by 40% and by 50%, respectively (similar for TPTD 40 μg). Despite the decline in radius BMD, there were fewer wrist and hip fractures in TPTD-treated women (numbers too small to evaluate statistically).

Sustained serum calcium increases were rare (3%) with the approved 20 μg dose. Mean 24-hour urine calcium increased 40 mg and mean serum uric acid increased 25%, without clinical sequelae. Overall, new cancer diagnoses were significantly less frequent with TPTD versus placebo [13]. In rodents, long-term high-dose TPTD produced osteogenic sarcoma, dependent on dose and duration of administration [14, 15]. In almost 15 years of postmarketing experience there has been no evidence of increased osteosarcoma risk, and several long-term surveillance studies in the United States and in Europe have shown no association between TPTD exposure and osteosarcoma [16–19].

TPTD-induced BMD changes were not dependent on patient age, baseline BMD, or prior fracture history [20], but were related to baseline biochemical bone turnover levels [21]. Early PTH-induced changes in bone turnover markers (at 1 and 3 months) were predictive of ultimate change in spine BMD and bone structure [21, 22]. Longer duration of TPTD treatment (≥14 months) was associated with greater reduction in nonvertebral fracture incidence and reduced back pain [23, 24].

McClung and colleagues randomized 203 postmenopausal women with osteoporosis to TPTD (20 μg/day) or alendronate for 18 months [1]. In TPTD-treated women, bone turnover markers peaked within 6 months, suggesting developing resistance, as seen previously [13, 25, 26]. Spine BMD increased 10.3% (by DXA) and 19% (by QCT) with TPTD versus 5.5% and 3.8%, respectively, with alendronate. FN BMD measured by DXA increased similarly in both groups, though when measured by QCT, cortical volumetric BMD declined 1.2% with TPTD and increased 7.7% with alendronate. Clinical fracture incidence was similar between the groups. Moderate/severe back pain was less common with TPTD (15% versus 33% with alendronate; $p = 0.003$). Bone strength of the spine assessed by finite element modeling increased significantly more with TPTD than alendronate. Femoral strength increased with TPTD ($p = 0.06$) but without a group difference [27, 28].

Hadji and colleagues randomized 710 postmenopausal women with acute painful osteoporosis-related vertebral fractures to TPTD versus risedronate for 18 months [29]. Although there was no difference in the primary outcome (proportion of women who experienced reduction in worst back pain at 6 months), there was a significant reduction in the proportion of patients experiencing worsening back pain from 6 to 18 months with TPTD. BMD of the spine, TH, and FN increased significantly more with TPTD than with risedronate. Moreover, the incidence of new vertebral fractures was reduced by >50% in women on TPTD versus risedronate (4% TPTD versus 9% risedronate; $p < 0.01$) and fractures were milder ($p = 0.04$ group difference). There was no difference in nonvertebral fracture incidence.

PTH(1–84) monotherapy

In one study, 217 women were randomized to 50, 75, or 100 μg of PTH(1–84) or placebo. There was a dose-dependent increase in spine BMD, but no increase in TH or total body BMD [30]. In another study, 2532 postmenopausal women with osteoporosis (19% with prevalent vertebral fracture) were randomized to 100 mg of recombinant PTH(1–84) or placebo injection for 18 months [31]. Mean spine BMD increment was 7% with PTH(1–84) versus placebo. In the per protocol adherent population ($n = 1870$), new or worsened vertebral fracture incidence was 3.4% with placebo and 1.4% with PTH(1–84) (relative risk reduction 58%), but nonvertebral fracture incidence was not reduced. The incidence of hypercalcemia was 28.3% with PTH(1–84) and 4.5% with placebo [31]. PTH(1–84) therapy is currently available in Europe but not the United States.

One very small trial compared PTH(1–84) and TPTD for 18 months with HRpQCT of the tibia and radius. Both PTH peptides increased cortical porosity and decreased cortical density, while cortical thickness increased only with TPTD. Strength estimated by finite element modeling was maintained with TPTD, but declined with PTH(1–84) [32].

Discontinuation of parathyroid hormone and sequential treatment with antiresorptive therapy

Observational studies suggest that BMD is lost in individuals who do not take antiresorptive agents after cessation of PTH, whereas antiresorptive therapy can maintain PTH-induced gains or induce further BMD increments [25, 30, 33–36]. In a 30-month observational follow-up after TPTD discontinuation, nonvertebral fracture risk remained lower compared with the prior placebo group (difference insignificant) [13]; however 60% of women started another osteoporosis medication (most often a bisphosphonate) during the observational period. BMD changes were greatest in those who started antiresorptive agents within 6 months of stopping TPTD, and TH and FN BMD returned to baseline in those who did not start an antiresorptive [37]. In the subgroup of 549 women who had paired radiographs and did not use oteoporosis medications during 18 months of follow-up, 16% of the placebo group, and 10% of the 20 μg TPTD group, experienced one or more new vertebral fractures (37% relative risk reduction; $p = 0.08$). Risk reductions for new vertebral fractures were slightly greater in women who did start antiresorptives during follow-up (41%; $p = 0.004$). Spine

BMD gains were greatest in those who started antiresorptives within 6 months after TPTD cessation and declined substantially in those who did not [35].

Following 24 months of TPTD in men and postmenopausal women, after TPTD was discontinued, spine BMD decreased 4.1% in men and 7.1% in women over the next year [38]. TH bone mass was stable in men, but decreased in women.

Subjects originally randomized to 1 year of treatment with PTH(1–84) were subsequently randomized to receive alendronate or placebo for an additional year [39]. Over the 2-year trial, in women who received PTH(1–84) followed by alendronate, BMD increments were 12%, 4%, and 4% in the spine, TH, and FN compared with 4%, 0%, and 1%, respectively, in women who received PTH(1–84) followed by placebo.

Parathyroid hormone and antiresorptive combination therapy

Although PTH and antiresorptive agents could theoretically produce additive effects on bone strength, studies have shown different outcomes based on skeletal site (spine versus hip), type of measurement (DXA versus QCT), specific antiresorptive therapy, and whether patients are treatment-naïve or treatment-experienced. Furthermore, in treatment-experienced patients, there are differences based on whether the prior antiresorptive was continued or stopped when TPTD was initiated.

Combination treatment in treatment-naïve women: parathyroid hormone and bisphosphonates

Black and colleagues randomized 238 treatment-naïve women to PTH(1–84) with alendronate versus each agent alone [40]. Spine BMD (by DXA) increased similarly with PTH(1–84) and combination treatment (mean 6.3% and 6.1%, respectively). TH BMD (by DXA) increased with combination (1.9%) but not with PTH(1–84) alone (0.3%). Radial BMD declined more with PTH(1–84) alone (−3.4%) versus the combination (−1.1%). Integral QCT increments of the spine and TH were similar between PTH(1–84) and combination groups; however, trabecular spine BMD increased more with PTH(1–84) alone (25.5%) than with the combination (12.6%). In contrast, QCT-assessed cortical bone density declined in the hip (−1.7%) with PTH(1–84) alone, but was unchanged with the combination. The results demonstrated no clear evidence of additive effect in the spine with combination therapy; however, in the hip, combination therapy was superior to PTH(1–84) monotherapy.

In 93 women randomized to receive alendronate for 6 months prior to TPTD 40μg (versus either agent alone), BMD gains in both the spine and hip (by DXA) were lower in those given TPTD after alendronate, compared with those given TPTD alone [41]. However, the spine BMD difference was not significant if groups were restricted to those who did not discontinue medication prematurely. This is one of few PTH trials where treatment duration was a full 24 months and TH and FN BMD levels increased most markedly during the latter year; however the dose used was double the approved dose.

Cosman and colleagues randomized 412 treatment-naïve postmenopausal women with osteoporosis to receive TPTD, intravenous zoledronic acid (ZOL), or the combination [42]. With combination therapy, the spine BMD increase (7.5%) was similar to TPTD alone (7.0%), whereas the TH BMD and FN BMD increases were larger (2.3% and 2.2% combination versus 1.1% and 0.1% TPTD alone, respectively). Clinical fractures were reported as adverse events and adjudicated. Fractures occurred in 9.5% of patients with ZOL, 5.8% with TPTD, and 2.9% with combination treatment ($p<0.05$ versus ZOL alone).

Combination treatment in treatment-naïve women: teriparatide and raloxifene

Deal and colleagues randomized 137 postmenopausal women to receive TPTD or TPTD plus raloxifene for 6 months. Spine BMD increments were similar in the two groups, while TH BMD increased more with TPTD plus raloxifene [43].

Combination treatment in treatment-naïve women: teriparatide and denosumab

Leder and colleagues randomized 94 postmenopausal women with osteoporosis to TPTD 20μg daily, denosumab 60mg every 6 months, or combination treatment for 2 years [44]. Spine BMD increased more with the combination (12.9%) than with TPTD alone (9.5%; $p=0.01$). FN and TH BMD also increased significantly more with the combination (6.8% and 6.3%, respectively) than with TPTD monotherapy (2.8% and 2%, respectively). The biggest incremental gains with combination therapy above TPTD monotherapy were in the first year. BMD increments in the second year did not differ among groups [44].

Combination treatment in women on established antiresorptive therapy

Patients maintained on long-term antiresorptive treatment are a distinct, but clinically important, population since many of these patients have fractures or do not achieve BMD above the osteoporotic range, and thus might benefit from anabolic therapy. Possible explanations for differences between treatment-naïve and treatment-experienced women include: (i) reduced active bone surface in the treatment-experienced; (ii) the increase in endogenous PTH seen for up to 12 months when potent antiresorptive agents are administered to treatment-naïve individuals (which might produce a different response to exogenous PTH); and perhaps

(iii) unique effects on osteoclast and/or osteoblast activation in the treatment-experienced. Studies in these patients have followed two basic designs: antiresorptive agents are stopped when PTH is started (sequential monotherapy [45–48]) or antiresorptive agents are continued when PTH is started (combination therapy [49, 50]). In studies where bisphosphonates or denosumab are discontinued when PTH is started, hip BMD declines consistently over the first year, an effect not seen in protocols where TPTD is added to ongoing bisphosphonate.

Sequential studies where antiresorptive therapy is stopped when teriparatide is started

In an observational study where TPTD was initiated after long-term alendronate or raloxifene [45], a significant reduction in TH BMD was seen within 6 months in the group previously on alendronate (not in women switched from raloxifene). TH BMD returned to baseline at 18 months. In women previously treated with risedronate ($n = 146$) or alendronate ($n = 146$), TH BMD declined significantly in both groups of patients for 1 year [48]. In another multinational observational study, European Study of Forsteo (EUROFORS), in women who switched from bisphosphonates to TPTD, TH BMD was below baseline throughout the first year [47, 51].

Leder and colleagues followed 27 patients who received TPTD after original randomization to denosumab for 2 years [48]. After transition to TPTD, spine BMD increased, however TH and FN BMD declined precipitously during the first year of TPTD. At the end of the second TPTD year, TH BMD was still below and FN BMD just above the levels attained after the switch from denosumab. This study allows direct comparison of a 4-year sequence of TPTD for 2 years followed by denosumab for 2 years compared with denosumab for 2 years followed by TPTD for 2 years. Over 4 years, in the group that transitioned from TPTD to denosumab, TH and FN BMD increased 6.6% and 8.3%, respectively, whereas in those who switched from denosumab to TPTD, TH and FN increments were 2.8% and 4.9% (all differences significant). In the spine, 4-year BMD increments were not significantly different with the various sequences.

A transient decline in hip BMD upon sequential treatment with potent antiresorptive switch to PTH therapy is likely because of excessive cortical remodeling and increased cortical porosity, as seen in iliac crest bone biopsies.

Sequential/combination studies where antiresorptive therapy is continued when teriparatide is started

In 52 women on established hormone therapy (HT), spine BMD increased 14%, and total body and TH BMD both increased 4% over 3 years in women randomized to TPTD plus HT versus TPTD alone [25]. Vertebral fracture incidence was significantly reduced in patients receiving TPTD plus HT compared with HT alone [25]. Forty-two postmenopausal women on raloxifene for ≥1 year were randomized to continued raloxifene or raloxifene plus TPTD. With the combination, spine BMD increased 10% and TH BMD 3% [50]. Cosman and colleagues randomized 126 women on long-term alendronate to continue alendronate and receive TPTD, or to continue alendronate monotherapy [49]. Over 15 months, spine BMD rose 6.1% with TPTD monotherapy and TH BMD did not decline at any time point during this study. There were no TPTD monotherapy (switch) groups in these studies.

In a randomized trial, Cosman and colleagues compared continuing versus stopping the antiresorptive agent when TPTD was initiated in 102 women on prior alendronate and 96 women on prior raloxifene [52]. Women within each antiresorptive cohort were randomized to switch to or add TPTD. In the group switched from alendronate to TPTD, TH BMD declined in the first 6 months (as seen in other switch studies already described). Spine and TH BMD increases at both 6 and 18 months were greater in patients who added TPTD to ongoing alendronate compared with those who switched to TPTD, and at no time point did TH BMD decline with the combination. Differences between combination and switch protocols were minimal with raloxifene pretreatment.

Results of studies where patients were switched to TPTD, compared with those where TPTD was added in combination with ongoing antiresorptives, suggest that there may be a role for combination therapy. This might be particularly important in patients who have very low hip BMD and/or those in whom hip fractures have occurred whilst on the antiresorptive. In these patients, any decline in hip BMD might be detrimental and a greater increase in hip BMD with combination treatment favorable.

PARATHYROID HORMONE TREATMENT IN MEN WITH OSTEOPOROSIS

Orwoll and colleagues randomized 437 men with primary or hypogonadal osteoporosis to TPTD 20 or 40 µg versus placebo [53]. After 1 year, spine BMD rose 5.4% and 8.5% with 20 and 40 µg, respectively, versus no change with placebo. There were dose-dependent increases in TH and total body BMD. In a follow up of 355 men, lateral spine radiographs after 18 months (with antiresorptive therapy in many) showed a 50% reduction in vertebral fracture risk in those men initially assigned to TPTD versus placebo ($p = 0.07$) [33].

Finkelstein and colleagues randomized 83 men with osteoporosis to TPTD 40 µg (double the approved dose), alendronate alone, or TPTD after 6 months of alendronate pretreatment [54]. Many men on TPTD required dose adjustment (by 25% to 50%) due to hypercalcemia or side effects. After 24 months, the spine BMD increase was largest with TPTD monotherapy (18.1%) compared with combination (14.8%) or alendronate alone (7.9%). Similar trends were seen for the lateral spine and FN, but for TH and total body increases were similar in the three groups. In the radius, BMD declined with TPTD monotherapy versus slight increases in the other groups.

Walker and colleagues randomized 29 men to risedronate versus TPTD 20 µg or combination treatment for 18 months. Spine BMD increments did not differ between groups, however TH BMD increased to a greater extent with the combination compared with TPTD alone [55].

GLUCOCORTICOID-TREATED OSTEOPOROSIS

PTH could be a preferred treatment for glucocorticoid-induced osteoporosis because it improves the reduction in osteoblast function and life span. In the first 18 months of an active comparator trial in 428 glucocorticoid-treated women and men, TPTD increased spine BMD 7.2% and TH BMD 3.8%, both significantly greater than that of 3.4% and 2.4%, respectively, with alendronate [56]. Fewer new vertebral fractures occurred with TPTD compared with alendronate (0.6% versus 6.1%; $p=0.004$). At 36 months, TPTD compared with alendronate resulted in 11% versus 5.3% BMD increases at the lumbar spine, 5.2% versus 2.7% at the TH, 6.3% versus 3.4% at the FN (all $p=0.001$), and fewer vertebral fractures (1.7% TPTD versus 7.7% alendronate) [57]. There were no differences in nonvertebral fracture incidence at 18 or 36 months. Gluer and colleagues showed superior microstructure and density with teriparatide versus risedronate in male glucocorticoid-induced osteoporosis over 18 months [58].

OTHER PARATHYROID HORMONE ISSUES

Rechallenge with parathyroid hormone

Women who received TPTD in addition to ongoing alendronate were then followed for a year on continued alendronate alone, during which BMD remained stable [59]. Rechallenge with TPTD produced similar biochemical and BMD changes to those seen during the first TPTD course [59]. In women and men who completed a 2-year TPTD course and 1 year follow-up off therapy, rechallenge with TPTD increased BMD of the spine but less than that achieved during the first TPTD course [60]

Daily versus dyclic teriparatide treatment

To confirm prior findings in alendronate-treated women receiving TPTD, a recent study compared 3 month on/off cycles to daily TPTD over 24 months in a cohort of treatment-naïve women and a cohort on prior alendronate [61, 62]. In the latter group of women, results of cyclic treatment were similar to those of daily therapy, despite administration of only half of the cumulative dose.

However, in the women on no other therapy, cyclic treatment did not offer any anabolic advantage; in the cyclic group, BMD effect was half that of the daily group, consistent with the TPTD dose [61, 62].

Other delivery systems

Several transdermal and oral delivery systems have failed to reproduce the necessary pharmacokinetic profile to move forward in development. However, in one study, TPTD 40 µg delivered by transdermal microneedle patch increased spine BMD similarly and TH BMD more than subcutaneous TPTD [63].

Other possible applications

Both PTH(1–84) and TPTD have been utilized successfully for hypoparathyroidism [64, 65]. In postmenopausal women with radius fracture, TPTD 20 µg (but not 40 µg) shortened time to healing [66, 67] and PTH might also have accelerated pelvic fracture healing [68]. TPTD might also have a role in healing of atypical femur fractures and osteonecrosis of the jaw [69, 70]. Other potential orthopedic applications include spine fusion and improved prosthesis adherence.

ABALOPARATIDE FOR OSTEOPOROSIS TREATMENT

In the Abaloparatide Comparator Trial in Vertebral Endpoints (ACTIVE), 2463 postmenopausal women with osteoporosis were randomized to blinded daily subcutaneous abaloparatide versus placebo or open label TPTD [71]. At 18 months, spine BMD increase was similar with abaloparatide (11.2%) and TPTD (10.5%); the TH and FN BMD increments were significantly larger with abaloparatide (4.2% and 3.6%, respectively) compared with TPTD (3.3% and 2.7%, respectively). New vertebral fracture incidence was reduced by 86% with abaloparatide and 80% with TPTD compared with placebo (both $p<0.001$). Nonvertebral fractures were reduced by 43% with abaloparatide ($p=0.049$) and 28% with TPTD (NS, $p=0.22$). Major osteoporotic fractures were reduced by 70% with abaloparatide ($p<0.001$) and 33% with TPTD (NS, $p=0.14$); major osteoporotic fracture incidence was 55% lower with abaloparatide than TPTD ($p=0.03$). Time to first nonvertebral fracture revealed very early separation between abaloparatide and both placebo and TPTD groups. Prespecified evaluations of treatment effects on fracture and BMD by subgroups of age, baseline BMD, prevalent spine fracture, or prior nonvertebral fracture revealed no significant qualitative or quantitative interactions [72]. The ACTIVE study is in an ongoing extension where all participants from the abaloparatide and placebo arms transitioned to open label alendronate [73].

CONCLUSION

Because of the underlying effects it produces on microarchitecture and mass of bone, as well as cellular renewal, PTH may be able to ensure greater strength and more long-term protection against fracture than antiresorptive agents alone. Antiresorptive agents are needed after PTH to maintain benefits. For patients with severe osteoporosis, early anabolic therapy may help achieve therapeutic goals quicker than antiresorptive agents alone and may permit a shorter duration of overall therapy, hopefully minimizing risk of long-term adverse consequences.

REFERENCES

1. McClung MR, San Martin J, Miller PD, et al. Opposite bone remodeling effects of teriparatide and alendronate in increasing bone mass. Arch Intern Med. 2005;165(15):1762–8.
2. Arlot M, Meunier PJ, Boivin G, et al. Differential effects of teriparatide and alendronate on bone remodeling in postmenopausal women assessed by histomorphometric parameters. J Bone Miner Res. 2005;20(7):1244–53.
3. Lindsay R, Cosman F, Zhou H, et al. A novel tetracycline labeling schedule for longitudinal evaluation of the short-term effects of anabolic therapy with a single iliac crest bone biopsy: early actions of teriparatide. J Bone Miner Res. 2006;21(3):366–73.
4. Jiang Y, Zhao JJ, Mitlak BH, et al. Recombinant human parathyroid hormone (1-34) [teriparatide] improves both cortical and cancellous bone structure. J Bone Miner Res. 2003;18(11):1932–41.
5. Dempster DW, Cosman F, Kurland ES, et al. Effects of daily treatment with parathyroid hormone on bone microarchitecture and turnover in patients with osteoporosis: a paired biopsy study. J Bone Miner Res. 2001;16(10):1846–53.
6. Parfitt AM. Parathyroid hormone and periosteal bone expansion. J Bone Miner Res. 2002;17(10):1741–3.
7. Burr DB. Does early PTH treatment compromise bone strength? The balance between remodeling, porosity, bone mineral, and bone size. Curr Osteoporos Rep. 2005;3(1):19–24.
8. Lindsay R, Zhou H, Cosman F, et al. Effects of a one-month treatment with PTH(1-34) on bone formation on cancellous, endocortical, and periosteal surfaces of the human ilium. J Bone Miner Res. 2007;22(4):495–502.
9. Zanchetta JR, Bogado CE, Ferretti JL, et al. Effects of teriparatide [recombinant human parathyroid hormone (1-34)] on cortical bone in postmenopausal women with osteoporosis. J Bone Miner Res. 2003;18(3):539–43.
10. Uusi-Rasi K, Semanick LM, Zanchetta JR, et al. Effects of teriparatide [rhPTH (1-34)] treatment on structural geometry of the proximal femur in elderly osteoporotic women. Bone. 2005;36(6):948–58.
11. Cosman F, Dempster DW, Nieves JW, et al. Effect of teriparatide on bone formation in the human femoral neck. J Clin Endocrinol Metab. 2016;101(4):1498–505.
12. Eriksen EF, Keaveny TM, Gallagher ER, et al. Literature review: The effects of teriparatide therapy at the hip in patients with osteoporosis. Bone. 2014;67:246–56.
13. Neer RM, Arnaud CD, Zanchetta JR, et al. Effect of parathyroid hormone (1-34) on fractures and bone mineral density in postmenopausal women with osteoporosis. N Engl J Med. 2001;344(19):1434–41.
14. Vahle JL, Long GG, Sandusky G, et al. Bone neoplasms in F344 rats given teriparatide [rhPTH(1-34)] are dependent on duration of treatment and dose. Toxicol Pathol. 2004;32(4):426–38.
15. Vahle JL, Sato M, Long GG, et al. Skeletal changes in rats given daily subcutaneous injections of recombinant human parathyroid hormone (1-34) for 2 years and relevance to human safety. Toxicol Pathol. 2002;30(3):312–21.
16. Harper KD, Krege JH, Marcus R, et al. Osteosarcoma and teriparatide? J Bone Miner Res. 2007;22(2):334.
17. Subbiah V, Madsen VS, Raymond AK, et al. Of mice and men: divergent risks of teriparatide-induced osteosarcoma. Osteoporos Int. 2010;21(6):1041–5.
18. Andrews EB, Gilsenan AW, Midkiff K, et al. The US postmarketing surveillance study of adult osteosarcoma and teriparatide: study design and findings from the first 7 years. J Bone Miner Res. 2012;27(12):2429–37.
19. Von Scheele B, Martin RD, Gilsenan AW, et al. The European postmarketing adult Osteosarcoma Surveillance Study: characteristics of patients. A preliminary report. Acta Orthopaed. 2009;80(suppl 334):67–74.
20. Marcus R, Wang O, Satterwhite J, et al. The skeletal response to teriparatide is largely independent of age, initial bone mineral density, and prevalent vertebral fractures in postmenopausal women with osteoporosis. J Bone Miner Res. 2003;18(1):18–23.
21. Chen P, Satterwhite JH, Licata AA, et al. Early changes in biochemical markers of bone formation predict BMD response to teriparatide in postmenopausal women with osteoporosis. J Bone Miner Res. 2005;20(6):962–70.
22. Dobnig H, Sipos A, Jiang Y, et al. Early changes in biochemical markers of bone formation correlate with improvements in bone structure during teriparatide therapy. J Clin Endocrinol Metab. 2005;90(7):3970–7.
23. Lindsay R, Miller P, Pohl G, et al. Relationship between duration of teriparatide therapy and clinical outcomes in postmenopausal women with osteoporosis. Osteoporos Int. 2009;20(6):943–8.
24. Lindsay R, Krege JH, Marin F, et al. Teriparatide for osteoporosis: importance of the full course. Osteoporos Int. 2016;27(8):2395–410.
25. Cosman F, Nieves J, Woelfert L, et al. Parathyroid hormone added to established hormone therapy: effects on vertebral fracture and maintenance of bone mass after parathyroid hormone withdrawal. J Bone Miner Res. 2001;16(5):925–31.
26. Kurland ES, Cosman F, McMahon DJ, et al. Parathyroid hormone as a therapy for idiopathic osteoporosis in

men: effects on bone mineral density and bone markers. J Clin Endocrinol Metab. 2000;85(9):3069–76.
27. Keaveny TM, Donley DW, Hoffmann PF, et al. Effects of teriparatide and alendronate on vertebral strength as assessed by finite element modeling of QCT scans in women with osteoporosis. J Bone Miner Res. 2007;22(1): 149–57.
28. Keaveny TM, McClung MR, Wan X, et al. Femoral strength in osteoporotic women treated with teriparatide or alendronate. Bone. 2012;50(1):165–70.
29. Hadji P, Zanchetta JR, Russo L, et al. The effect of teriparatide compared with risedronate on reduction of back pain in postmenopausal women with osteoporotic vertebral fractures. Osteoporos Int. 2012;23(8):2141–50.
30. Rittmaster RS, Bolognese M, Ettinger MP, et al. Enhancement of bone mass in osteoporotic women with parathyroid hormone followed by alendronate. J Clin Endocrinol Metab. 2000;85(6):2129–34.
31. Greenspan SL, Bone HG, Ettinger MP, et al. Effect of recombinant human parathyroid hormone (1-84) on vertebral fracture and bone mineral density in postmenopausal women with osteoporosis: a randomized trial. Ann Intern Med. 2007;146(5):326–39.
32. Hansen S, Hauge EM, Beck Jensen JE, et al. Differing effects of PTH 1-34, PTH 1-84, and zoledronic acid on bone microarchitecture and estimated strength in postmenopausal women with osteoporosis: an 18-month open-labeled observational study using HR-pQCT. J Bone Miner Res. 2013;28(4):736–45.
33. Kaufman JM, Orwoll E, Goemaere S, et al. Teriparatide effects on vertebral fractures and bone mineral density in men with osteoporosis: treatment and discontinuation of therapy. Osteoporos Int. 2005;16(5):510–6.
34. Lane NE, Sanchez S, Modin GW, et al. Bone mass continues to increase at the hip after parathyroid hormone treatment is discontinued in glucocorticoid-induced osteoporosis: results of a randomized controlled clinical trial. J Bone Miner Res. 2000;15(5):944–51.
35. Lindsay R, Scheele WH, Neer R, et al. Sustained vertebral fracture risk reduction after withdrawal of teriparatide in postmenopausal women with osteoporosis. Arch Intern Med. 2004;164(18):2024–30.
36. Kurland ES, Heller SL, Diamond B, et al. The importance of bisphosphonate therapy in maintaining bone mass in men after therapy with teriparatide [human parathyroid hormone(1-34)]. Osteoporos Int. 2004;15(12):992–7.
37. Prince R, Sipos A, Hossain A, et al. Sustained nonvertebral fragility fracture risk reduction after discontinuation of teriparatide treatment. J Bone Miner Res. 2005;20(9):1507–13.
38. Leder BZ, Neer RM, Wyland JJ, et al. Effects of teriparatide treatment and discontinuation in postmenopausal women and eugonadal men with osteoporosis. J Clin Endocrinol Metab. 2009;94(8):2915–21.
39. Black DM, Bilezikian JP, Ensrud KE, et al. One year of alendronate after one year of parathyroid hormone (1-84) for osteoporosis. N Engl J Med. 2005;353(6):555–65.
40. Black DM, Greenspan SL, Ensrud KE, et al. The effects of parathyroid hormone and alendronate alone or in combination in postmenopausal osteoporosis. N Engl J Med. 2003;349(13):1207–15.
41. Finkelstein JS, Wyland JJ, Lee H, et al. Effects of teriparatide, alendronate, or both in women with postmenopausal osteoporosis. J Clin Endocrinol Metab. 2010;95(4):1838–45.
42. Cosman F, Eriksen EF, Recknor C, et al. Effects of intravenous zoledronic acid plus subcutaneous teriparatide [rhPTH(1-34)] in postmenopausal osteoporosis. J Bone Miner Res. 2011;26(3):503–11.
43. Deal C, Omizo M, Schwartz EN, et al. Combination teriparatide and raloxifene therapy for postmenopausal osteoporosis: results from a 6-month double-blind placebo-controlled trial. J Bone Miner Res. 2005;20(11):1905–11.
44. Leder BZ, Tsai JN, Uihlein AV, et al. Two years of denosumab and teriparatide administration in postmenopausal women with osteoporosis (The DATA Extension Study): a randomized controlled trial. J Clin Endocrinol Metab. 2014;99(5):1694–700.
45. Ettinger B, San Martin J, Crans G, et al. Differential effects of teriparatide on BMD after treatment with raloxifene or alendronate. J Bone Miner Res. 2004;19(5):745–51.
46. Miller PD, Delmas PD, Lindsay R, et al. Early responsiveness of women with osteoporosis to teriparatide after therapy with alendronate or risedronate. J Clin Endocrinol Metab. 2008;93(10):3785–93.
47. Boonen S, Marin F, Obermayer-Pietsch B, et al. Effects of previous antiresorptive therapy on the bone mineral density response to two years of teriparatide treatment in postmenopausal women with osteoporosis. J Clin Endocrinol Metab. 2008;93(3):852–60.
48. Leder BZ, Tsai JN, Uihlein AV, et al. Denosumab and teriparatide transitions in postmenopausal osteoporosis (the DATA-Switch study): extension of a randomised controlled trial. Lancet. 2015;386(9999):1147–55.
49. Cosman F, Nieves J, Zion M, et al. Daily and cyclic parathyroid hormone in women receiving alendronate. N Engl J Med. 2005;353(6):566–75.
50. Cosman F, Nieves JW, Zion M, et al. Effect of prior and ongoing raloxifene therapy on response to PTH and maintenance of BMD after PTH therapy. Osteoporos Int. 2008;19(4):529–35.
51. Obermayer-Pietsch BM, Marin F, McCloskey EV, et al. Effects of two years of daily teriparatide treatment on BMD in postmenopausal women with severe osteoporosis with and without prior antiresorptive treatment. J Bone Miner Res. 2008;23(10):1591–600.
52. Cosman F, Wermers RA, Recknor C, et al. Effects of teriparatide in postmenopausal women with osteoporosis on prior alendronate or raloxifene: differences between stopping and continuing the antiresorptive agent. J Clin Endocrinol Metab. 2009;94(10):3772–80.
53. Orwoll ES, Scheele WH, Paul S, et al. The effect of teriparatide [human parathyroid hormone (1-34)] therapy on bone density in men with osteoporosis. J Bone Miner Res. 2003;18(1):9–17.

54. Finkelstein JS, Hayes A, Hunzelman JL, et al. The effects of parathyroid hormone, alendronate, or both in men with osteoporosis. N Engl J Med. 2003;349(13):1216–26.
55. Walker MD, Cusano NE, Sliney J Jr, et al. Combination therapy with risedronate and teriparatide in male osteoporosis. Endocrine. 2013;44(1):237–46.
56. Saag KG, Shane E, Boonen S, et al. Teriparatide or alendronate in glucocorticoid-induced osteoporosis. N Engl J Med. 2007;357(20):2028–39.
57. Saag KG, Zanchetta JR, Devogelaer JP, et al. Effects of teriparatide versus alendronate for treating glucocorticoid-induced osteoporosis: thirty-six-month results of a randomized, double-blind, controlled trial. Arthritis Rheum. 2009;60(11):3346–55.
58. Gluer CC, Marin F, Ringe JD, et al. Comparative effects of teriparatide and risedronate in glucocorticoid-induced osteoporosis in men: 18-month results of the EuroGIOPs trial. J Bone Miner Res. 2013;28(6):1355–68.
59. Cosman F, Nieves JW, Zion M, et al. Retreatment with teriparatide one year after the first teriparatide course in patients on continued long-term alendronate. J Bone Miner Res. 2009;24(6):1110–5.
60. Finkelstein JS, Wyland JJ, Leder BZ, et al. Effects of teriparatide retreatment in osteoporotic men and women. J Clin Endocrinol Metab. 2009;94(7):2495–501.
61. Cosman F, Nieves JW, Zion M, et al. Daily or cyclical teriparatide treatment in women with osteoporosis on no prior therapy and women on alendronate. J Clin Endocrinol Metab. 2015;100(7):2769–76.
62. Dempster DW, Cosman F, Zhou H, et al. Effects of daily or cyclic teriparatide on bone formation in the iliac crest in women on no prior therapy and in women on alendronate. J Bone Miner Res. 2016;31(8):1518–26.
63. Cosman F, Lane NE, Bolognese MA, et al. Effect of transdermal teriparatide administration on bone mineral density in postmenopausal women. J Clin Endocrinol Metab. 2010;95(1):151–8.
64. Cusano NE, Rubin MR, McMahon DJ, et al. The effect of PTH(1-84) on quality of life in hypoparathyroidism. J Clin Endocrinol Metab. 2013;98(6):2356–61.
65. Cusano NE, Rubin MR, McMahon DJ, et al. Therapy of hypoparathyroidism with PTH(1-84): a prospective four-year investigation of efficacy and safety. J Clin Endocrinol Metab. 2013;98(1):137–44.
66. Aspenberg P, Johansson T. Teriparatide improves early callus formation in distal radial fractures. Acta Orthopaed. 2010;81(2):234–6.
67. Aspenberg P, Genant HK, Johansson T, et al. Teriparatide for acceleration of fracture repair in humans: a prospective, randomized, double-blind study of 102 postmenopausal women with distal radial fractures. J Bone Miner Res. 2010;25(2):404–14.
68. Peichl P, Holzer LA, Maier R, et al. Parathyroid hormone 1-84 accelerates fracture-healing in pubic bones of elderly osteoporotic women. J Bone Joint Surg Am. 2011;93(17):1583–7.
69. Shane E, Burr D, Abrahamsen B, et al. Atypical subtrochanteric and diaphyseal femoral fractures: second report of a task force of the American Society for Bone and Mineral Research. J Bone Miner Res. 2014;29(1):1–23.
70. Cheung A, Seeman E. Teriparatide therapy for alendronate-associated osteonecrosis of the jaw. N Engl J Med. 2010;363(25):2473–4.
71. Miller PD, Hattersley G, Riis BJ, et al.; ACTIVE Study Investigators. Effect of abaloparatide vs placebo on new vertebral fractures in postmenopausal women with osteoporosis: a randomized clinical trial. JAMA. 2016;316(7):722–33.
72. Cosman F, Hattersley G, Hu MY, et al. Effects of abaloparatide-SC on fractures and bone mineral density in subgroups of postmenopausal women with osteoporosis and varying baseline risk factors. J Bone Miner Res. 2017;32(1):17–23.
73. Cosman F, Miller PD, Williams GC, et al. Eighteen months of treatment with subcutaneous abaloparatide followed by 6 months of treatment with alendronate in postmenopausal women with osteoporosis: results of ACTIVExtend trial. Mayo Clin Proc. 2017;92:200–10.

74

Combination Anabolic and Antiresorptive Therapy for Osteoporosis

Joy N. Tsai and Benjamin Z. Leder

Massachusetts General Hospital; and Harvard Medical School, Boston, MA, USA

INTRODUCTION

Osteoporosis is a common, treatable disease that is projected to cost the US health care system US$25 billion annually by 2025 [1]. Unlike other chronic conditions such as hypertension or diabetes, the current standard of care for osteoporosis generally involves the use of a single medication at a single dose no matter how severe the disease. Because no currently available agent is able to fully restore bone strength in most patients with established osteoporosis, using a combination of drugs has been examined as a potential therapeutic approach, particularly in patients at very high risk of fracture. Early studies investigating the combination of two antiresorptive agents (most often a hormonal agent combined with a bisphosphonate) have generally not shown a significant benefit over monotherapy, an expected finding given the potency of the later generation nitrogen-containing bisphosphonates [2–12]. Conversely, the combination of antiresorptive and anabolic agents was hypothesized to provide unique benefits due to their distinct and potentially complementary mechanisms of action. Specifically, it was hoped that by administering these drugs together, bone formation could be stimulated without concomitant bone resorption, leading to greater gains in bone mass and strength. This chapter will review the clinical trials investigating the various combinations of antiresorptive and anabolic treatments, either in combination or in sequence, with a focus on their efficacy in postmenopausal osteoporotic women, the primary patient population studied.

CONCURRENT THERAPY

While daily subcutaneous injections of PTH analogs stimulate bone formation, increase BMD, and decrease fracture risk in high-risk populations, they also stimulate bone resorption. And while this stimulation of bone resorption may be an important intermediate mechanism by which these agents exert their anabolic effect, it also may limit their overall therapeutic potential, particularly in cortical bone [13]. The antiresorptive and anabolic combinations evaluated in clinical trials include PTH analogs (teriparatide and PTH(1–84)) combined with: (i) oral and parenteral bisphosphonates; (ii) estrogen or selective estrogen receptor modulators; or (iii), the receptor activator of NF-κB ligand (RANKL) inhibitor, denosumab [14–19].

Parathyroid hormone analog/bisphosphonate combinations

The most well studied combination therapy approach has been the coadministration of bisphosphonates and PTH analogs. Finkelstein and colleagues [14] and Black and colleagues [15] studied the effects of PTH analogs and oral alendronate in two separate studies. In the study conducted by Finkelstein and colleagues, 93 postmenopausal women were randomized to receive teriparatide 40 μg daily (double the US Food and Drug Administration [FDA] approved dose), alendronate 10 mg daily, or both for 30 months [14]. Although the combination group had been

---

Primer on the Metabolic Bone Diseases and Disorders of Mineral Metabolism, Ninth Edition. Edited by John P. Bilezikian.
© 2019 American Society for Bone and Mineral Research. Published 2019 by John Wiley & Sons, Inc.
Companion website: www.wiley.com/go/asbmrprimer

hypothesized to show the greatest benefit, teriparatide monotherapy increased spine, femoral neck, and total hip BMD more than combination treatment or alendronate monotherapy. Additionally, teriparatide monotherapy increased lumbar spine trabecular volumetric BMD (vBMD), assessed by QCT, more than combination treatment or alendronate. As expected, biochemical markers of bone formation (serum osteocalcin and P1NP) and resorption (CTx) increased and decreased in the teriparatide and alendronate groups, respectively. In the combination group, both bone formation and resorption markers increased to levels two- to threefold lower than the group receiving teriparatide monotherapy [14]. Of note, a similar study performed in men with idiopathic osteoporosis reported essentially the same findings [19, 20].

In the study by Black and colleagues (referred to as the PATH study), 238 postmenopausal women were randomized to receive PTH(1–84) 100μg daily, alendronate 10mg daily, or both for 1 year [15]. In this study, spine BMD increased similarly in all three groups. Total hip BMD increased similarly in the combination therapy and alendronate groups, which were both greater than the PTH(1–84) monotherapy group. Additionally, trabecular vBMD of the spine measured by QCT increased nearly twofold more in the PTH(1–84) group than the combination group. Trabecular total hip vBMD increased in all treatment groups, without statistically significant between-group differences, while cortical vBMD of the total hip and femoral neck decreased with PTH(1–84), did not change with combination therapy, and increased with alendronate. In women treated with both alendronate and PTH(1–84), P1NP increased transiently in the first month of therapy and then remained suppressed for the duration of the study. CTx did not show an early increase but was similarly suppressed throughout the treatment period (though this suppression was not as complete as the suppression in the alendronate monotherapy group).

The effects of combined teriparatide and intravenous zoledronic acid was studied in a 12-month randomized controlled trial of 412 postmenopausal women assigned to receive teriparatide 20μg daily, a single administration of zoledronic acid 5mg, or both [16]. While combined therapy did show some benefits at early time points, at the end of the 1-year treatment period spine BMD had increased similarly in the combination group and the teriparatide group while total hip and femoral neck BMD increased similarly in the combination group and the zoledronic acid group. Of note, serum CTx was initially suppressed in women treated with both drugs but then increased, a pattern that may explain why the early beneficial effects did not persist.

Finally, in the Parathyroid and Ibandronate Combination Study (PICS), 44 postmenopausal women were randomized to receive 3 months of PTH(1–84) 100μg followed by 9 months of oral ibandronate 150mg monthly for two cycles or 6 months of combined PTH(1–84) and ibandronate followed by 18 months of ibandronate alone [21]. Increases in both DXA-derived areal BMD and QCT-derived vBMD did not differ between groups at any measured site.

Taken together, while there are subtle difference in the distinct combinations of bisphosphonates and PTH analogs investigated, the cumulative results were disappointing in that no combination appears to offer a significant advantage over monotherapy. Moreover, the observed changes in biochemical markers of bone turnover in these studies suggest that bisphosphonates are only partially able to block the proresorptive effects of PTH analogs.

Parathyroid analog/estrogen or selective estrogen receptor modulator combinations

The combination of PTH analogs and estrogen has been studied in several small clinical trials [22, 23]. These studies generally demonstrate that adding PTH analogs to ongoing hormone replacement therapy results in continued increases in spine and hip BMD but the superiority of combination treatment cannot be assessed due to the lack of PTH monotherapy comparators. Deal and colleagues studied the combination of a SERM and teriparatide in 137 postmenopausal women randomized to receive raloxifene 60mg daily combined with either teriparatide 20μg daily or placebo for 6 months [24]. Total hip BMD increased more in the combination group, whereas spine and femoral neck BMD increased similarly in the two groups. Of note, although P1NP increased equally in the two groups, CTx levels increased more in the teriparatide monotherapy group than in the combination group.

Teriparatide/denosumab combination

In contrast to bisphosphonate-containing combinations, the combination of denosumab and teriparatide shows additive effects on BMD [17, 18]. Denosumab is a potent antiresorptive that blocks the binding of RANKL to RANK and therefore inhibits osteoclast differentiation, maturation, function, and survival [25]. In the 24-month DATA (Denosumab And Teriparatide Administration) study, 94 postmenopausal women were randomized to teriparatide 20μg daily, denosumab 60mg subcutaneously every 6 months, or both medications. Women receiving combination therapy experienced larger and more rapid increases in BMD at the total hip, femoral neck, and spine than either monotherapy group (Fig. 74.1). Additionally, denosumab prevented the loss in distal radius BMD that is consistently observed with teriparatide monotherapy. Interestingly, most of the benefit of combination therapy was manifest during the first 12 months of treatment, during which lumbar spine BMD increased by 9.1%, femoral neck BMD increased by 4.2%, and total hip BMD by 4.9%. Moreover, the BMD increases at the femoral neck and total hip exceeded the sum of the increases in the teriparatide and denosumab monotherapy groups.

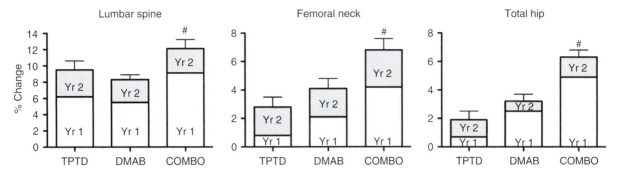

Fig. 74.1. Mean percent change (SE) in BMD from months 0–12 (white) and 12–24 (gray) in women treated with teriparatide (TPTD), denosumab (DMAB), or combination therapy (COMBO). # = $p < 0.05$ versus both other groups at 24 month. Source: [17]. Reproduced with permission of Oxford University Press.

The combination of denosumab and teriparatide also improved bone microarchitecture of the peripheral skeleton more than either drug alone (as assessed by HRpQCT of the distal tibia and distal radius) [26, 27]. Specifically, total vBMD, cortical vBMD, and cortical thickness increased more in the combination group than in either monotherapy group at the tibia, and estimated bone strength, assessed by finite element analysis, increased more with combination therapy than teriparatide at the radius and tibia. Finally, denosumab, when given along with teriparatide, fully prevented the increase in cortical porosity observed at the radius and tibia in women treated with teriparatide alone.

It is not yet clear why the denosumab/teriparatide combination additively increases BMD whereas bisphosphonate/teriparatide combinations do not. One potential explanation is suggested by the differential changes in bone turnover that are observed with these distinct approaches. Specifically, bone resorption markers are equally suppressed in patients treated with denosumab monotherapy and those treated with combined therapy whereas markers of bone formation are more suppressed in those treated with denosumab monotherapy than in those treated with both drugs, especially at the early time points. This pattern contrasts with the pattern observed when bisphosphonates and teriparatide are coadministered (during which teriparatide's proresorptive effects are only partially blocked). Thus, it is possible that the superior efficacy of combined denosumab and teriparatide may be due, at least in part, to denosumab's ability to fully block the proresorptive effects of teriparatide while still allowing for teriparatide-induced stimulation of bone formation even in the absence of ongoing bone resorption (modeling-based bone formation).

SWITCHING FROM ANTIRESORPTIVE THERAPY TO ANABOLIC THERAPY

The importance of the order in which anabolic and antiresorptive agents are prescribed is becoming increasingly clear. Several studies have investigated the effects of anabolic therapy when used after antiresorptive agents [28–32]. In a study by Ettinger and colleagues, postmenopausal women who had previously received alendronate or raloxifene for 18–36 months were then administered 18 months of teriparatide (20 µg daily) [28]. In the patients who had received raloxifene, BMD at both the total hip and spine increased more than in those who had received alendronate. In fact, women who received alendronate experienced only modest spine BMD increases (4.1%) and no gains in total hip BMD. Subsequent studies have confirmed this "blunting" of teriparatide's anabolic effects when given after potent bisphosphonates. Specifically, spine BMD generally increases less than would be expected if given de novo and total hip and femoral neck BMD either slightly decrease or remain stable in the first year of treatment and increase modestly thereafter [29–31].

A different and more concerning pattern is observed when denosumab therapy is transitioned to teriparatide therapy. In an extension to the DATA study discussed above (DATA-Switch), postmenopausal osteoporotic women who had received 2 years of denosumab monotherapy in the original DATA study were then assigned to receive 2 years of teriparatide therapy [32]. This transition from denosumab to teriparatide resulted in transient decreases in spine BMD and more sustained decreases in total hip and femoral neck BMD (Fig. 74.2). At the distal radius, a site made up of predominately cortical bone, switching from denosumab to teriparatide was associated with 24 months of sustained BMD decreases with concomitant increases in cortical porosity [33].

Notably, the decreases in BMD observed in the denosumab to teriparatide transition coincided with very large increases in biochemical markers of bone turnover. Specifically, both serum osteocalcin and CTx increased to levels two- to threefold higher than those typically observed when teriparatide is administered to patients without prior denosumab treatment [32, 34] (Fig. 74.3). Moreover, these accelerated rates of bone turnover were sustained throughout the entire 24 months of teriparatide therapy. Although the DATA-Switch study size precludes any conclusions concerning fracture risk, physicians should take into account

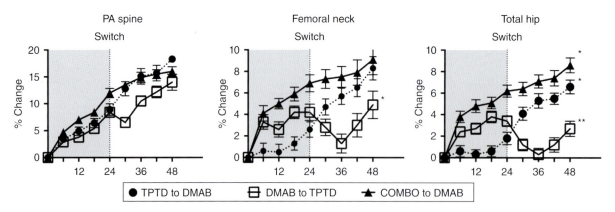

Fig. 74.2. Mean percent change (SE) in BMD from months 0–48 in subjects treated with teriparatide followed by denosumab (TPTD to DMAB), denosumab followed by teriparatide (DMAB to TPTD), and combination therapy followed by denosumab (COMBO to DMAB). * = $p<0.05$ versus both other groups at month 48. ** = $p<0.0005$ versus both other groups at month 48. Source: [32]. Reproduced with permission of Elsevier.

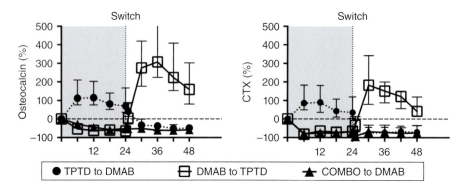

Fig. 74.3. Mean percent change (SE) in serum osteocalcin and c-telopeptide (CTx) from months 0–48 in subjects treated with teriparatide followed by denosumab (TPTD to DMAB), denosumab followed by teriparatide (DMAB to TPTD), and combination therapy followed by denosumab (COMBO to DMAB). Source: [32]. Reproduced with permission of Elsevier.

the potential effects of this rapid high turnover bone loss when considering anabolic therapy in denosumab-treated patients, especially given the recent case reports of multiple vertebral fractures occurring in the setting of the more modest acceleration in bone metabolism that occurs when denosumab is discontinued without any drug transition [35–38].

ADDING ANABOLIC THERAPY TO ONGOING ANTIRESORPTIVE THERAPY

In contrast to switching from an antiresorptive drug to an anabolic drug, adding an anabolic agent to ongoing antiresorptive therapy does not appear to induce transient or sustained decreases in BMD [31, 39]. In the study by Cosman and colleagues, postmenopausal women who had been treated with alendronate or raloxifene for at least 18 months were randomized to switch to teriparatide (20 μg daily) monotherapy or to have teriparatide added to the antiresorptive drug for an additional 18 months [31]. The BMD changes in the two treatment groups were similar in women in the raloxifene stratum. In the alendronate stratum, adding teriparatide increased spine and total hip BMD more than when switching to teriparatide.

SWITCHING FROM ANABOLIC THERAPY TO ANTIRESORPTIVE THERAPY

The importance of following a completed course of PTH analog therapy with antiresorptive therapy has been established in clinical trials. In an extension to the PATH study discussed above, postmenopausal women who had received PTH(1–84) for 12 months were then randomized to either placebo or alendronate for an additional 12 months [40]. While BMD at the femoral neck, total hip, and lumbar spine increased in women who received alendronate, those who received placebo experienced no gain in femoral neck or total hip BMD and a -1.7% decline in spine BMD. At the distal radius, there was no change in either group.

Similarly, in the European Study of Forsteo (EUROFORS), postmenopausal women with severe osteoporosis who had received 1 year of teriparatide (20 μg daily) were randomized to continue teriparatide, switch to raloxifene, or receive no further treatment during the second year [41]. Spine BMD decreased in the group who received no further treatment whereas it was stable in those receiving raloxifene and increased in those continuing teriparatide. Femoral neck BMD increased similarly in all groups and total hip BMD increased only in those who received active treatment.

Switching from teriparatide to denosumab appears to result in the largest gains in BMD of any sequential approach. In the DATA-Switch study previously discussed, postmenopausal women originally treated with teriparatide alone or combined teriparatide and denosumab were switched to 2 years of denosumab. In both of these treatment groups, spine BMD further increased resulting in 4-year gains of 15% to 18% (Fig. 74.2). In the teriparatide to denosumab group, total hip BMD increased by an additional 4.7% over 2 years, resulting in a total 4-year BMD increase of 6.6%. In the combination to denosumab group, total hip BMD increased by a more modest 2.2%, with a resulting 4-year increase of 8.6% [32].

ADDING ANTIRESORPTIVE THERAPY TO ONGOING ANABOLIC THERAPY

Muschitz and colleagues investigated the effects of adding an antiresorptive agent to ongoing anabolic therapy [42]. In this trial, 125 postmenopausal women, all of whom had previously received long-term bisphosphonate therapy, first received 9 months of teriparatide (20 μg daily) monotherapy and were then randomized to an additional 9 months of teriparatide monotherapy, teriparatide plus raloxifene, or teriparatide plus alendronate. After 18 months, spine BMD increased more in both combination groups than in the teriparatide monotherapy group whereas total hip BMD increased more in the combined teriparatide/alendronate group than in the other two groups.

SUMMARY

In summary, combined treatment with PTH analogs and oral or parenteral bisphosphonates does not appear to provide significant benefit when compared with monotherapy. Conversely, the combination of teriparatide and denosumab increases hip and spine BMD more than either drug alone and results in improved cortical microarchitecture and greater estimated bone strength at the distal radius and tibia. Furthermore, the initial use of an anabolic agent followed by an antiresorptive drug provides the largest net increases in bone mass of any sequential approach and this strategy should be strongly considered in patients with severe disease in whom the eventual use of anabolic agent is likely. Finally, the specific transition from denosumab to teriparatide appears to be uniquely associated with highly accelerated bone turnover and rapid bone loss and should be avoided in patients with significant skeletal fragility.

REFERENCES

1. Burge R, Dawson-Hughes B, Solomon DH, et al. Incidence and economic burden of osteoporosis-related fractures in the United States, 2005–2025. J Bone Miner Res. 2007;22(3):465–75.
2. Greenspan SL, Resnick NM, Parker RA. Combination therapy with hormone replacement and alendronate for prevention of bone loss in elderly women: a randomized controlled trial. JAMA. 2003;289(19):2525–33.
3. Sanad Z, Ellakwa H, Desouky B. Comparison of alendronate and raloxifene in postmenopausal women with osteoporosis. Climacteric. 2011;14(3):369–77.
4. Tseng LN, Sheu WH, Ho ES, et al. Effects of alendronate combined with hormone replacement therapy on osteoporotic postmenopausal Chinese women. Metabolism. 2006;55(6):741–7.
5. Harris ST, Eriksen EF, Davidson M, et al. Effect of combined risedronate and hormone replacement therapies on bone mineral density in postmenopausal women. J Clin Endocrinol Metab. 2001;86(5):1890–7.
6. Palomba S, Orio F Jr, Colao A, et al. Effect of estrogen replacement plus low-dose alendronate treatment on bone density in surgically postmenopausal women with osteoporosis. J Clin Endocrinol Metab. 2002;87(4):1502–8.
7. Tiras MB, Noyan V, Yildiz A, et al. Effects of alendronate and hormone replacement therapy, alone or in combination, on bone mass in postmenopausal women with osteoporosis: a prospective, randomized study. Hum Reprod. 2000;15(10):2087–92.
8. Bone HG, Greenspan SL, McKeever C, et al. Alendronate and estrogen effects in postmenopausal women with low bone mineral density. Alendronate/Estrogen Study Group. J Clin Endocrinol Metab. 2000;85(2):720–6.
9. Lindsay R, Cosman F, Lobo RA, et al. Addition of alendronate to ongoing hormone replacement therapy in the treatment of osteoporosis: a randomized, controlled clinical trial. J Clin Endocrinol Metab. 1999;84(9):3076–81.
10. Wimalawansa SJ. A four-year randomized controlled trial of hormone replacement and bisphosphonate, alone or in combination, in women with postmenopausal osteoporosis. Am J Med. 1998;104(3):219–26.
11. Wimalawansa SJ. Combined therapy with estrogen and etidronate has an additive effect on bone mineral density in the hip and vertebrae: four-year randomized study. Am J Med. 1995;99(1):36–42.
12. Evio S, Tiitinen A, Laitinen K, et al. Effects of alendronate and hormone replacement therapy, alone and in combination, on bone mass and markers of bone turnover in elderly women with osteoporosis. J Clin Endocrinol Metab. 2004;89(2):626–31.
13. Uihlein AV, Leder BZ. Anabolic therapies for osteoporosis. Endocrinol Metab Clin North Am. 2012;41(3):507–25.
14. Finkelstein JS, Wyland JJ, Lee H, et al. Effects of teriparatide, alendronate, or both in women with postmenopausal osteoporosis. J Clin Endocrinol Metab. 2010;95(4):1838–45.

15. Black DM, Greenspan SL, Ensrud KE, et al. The effects of parathyroid hormone and alendronate alone or in combination in postmenopausal osteoporosis. New Engl J Med. 2003;349(13):1207–15.
16. Cosman F, Eriksen EF, Recknor C, et al. Effects of intravenous zoledronic acid plus subcutaneous teriparatide [rhPTH(1-34)] in postmenopausal osteoporosis. J Bone Miner Res. 2011;26(3):503–11.
17. Leder BZ, Tsai JN, Uihlein AV, et al. Two years of denosumab and teriparatide administration in postmenopausal women with osteoporosis (the DATA Extension study): a randomized controlled trial. J Clin Endocrinol Metab. 2014;99(5):1694–700.
18. Tsai JN, Uihlein AV, Lee H, et al. Teriparatide and denosumab, alone or combined, in women with postmenopausal osteoporosis: the DATA study randomised trial. Lancet. 2013;382(9886):50–6.
19. Finkelstein JS, Leder BZ, Burnett SM, et al. Effects of teriparatide, alendronate, or both on bone turnover in osteoporotic men. J Clin Endocrinol Metab. 2006;91(8):2882–7.
20. Finkelstein JS, Hayes A, Hunzelman JL, et al. The effects of parathyroid hormone, alendronate, or both in men with osteoporosis. New Engl J Med. 2003;349(13): 1216–26.
21. Schafer AL, Sellmeyer DE, Palermo L, et al. Six months of parathyroid hormone (1-84) administered concurrently versus sequentially with monthly ibandronate over two years: the PTH and Ibandronate Combination Study (PICS) randomized trial. J Clin Endocrinol Metab. 2012;97(10):3522–9.
22. Ste-Marie LG, Schwartz SL, Hossain A, et al. Effect of teriparatide [rhPTH(1-34)] on BMD when given to postmenopausal women receiving hormone replacement therapy. J Bone Miner Res. 2006;21(2):283–91.
23. Cosman F, Nieves J, Woelfert L, et al. Parathyroid hormone added to established hormone therapy: effects on vertebral fracture and maintenance of bone mass after parathyroid hormone withdrawal. J Bone Miner Res. 2001;16(5):925–31.
24. Deal C, Omizo M, Schwartz EN, et al. Combination teriparatide and raloxifene therapy for postmenopausal osteoporosis: results from a 6-month double-blind placebo-controlled trial. J Bone Miner Res. 2005;20(11): 1905–11.
25. Lacey DL, Timms E, Tan HL, et al. Osteoprotegerin ligand is a cytokine that regulates osteoclast differentiation and activation. Cell. 1998;93(2):165–76.
26. Tsai JN, Uihlein AV, Burnett-Bowie SA, et al. Comparative effects of teriparatide, denosumab, and combination therapy on peripheral compartmental bone density, microarchitecture, and estimated strength: the DATA-HRpQCT Study. J Bone Miner Res. 2015;30(1): 39–45.
27. Tsai JN, Uihlein AV, Burnett-Bowie SM, et al. Effects of two years of teriparatide, denosumab, or both on bone microarchitecture and strength (DATA-HRpQCT study). J Clin Endocrinol Metab. 2016;101(5):2023–30.
28. Ettinger B, San Martin J, Crans G, et al. Differential effects of teriparatide on BMD after treatment with raloxifene or alendronate. J Bone Miner Res. 2004;19(5):745–51.
29. Miller PD, Delmas PD, Lindsay R, et al. Early responsiveness of women with osteoporosis to teriparatide after therapy with alendronate or risedronate. J Clin Endocrinol Metab. 2008;93(10):3785–93.
30. Boonen S, Marin F, Obermayer-Pietsch B, et al. Effects of previous antiresorptive therapy on the bone mineral density response to two years of teriparatide treatment in postmenopausal women with osteoporosis. J Clin Endocrinol Metab. 2008;93(3):852–60.
31. Cosman F, Wermers RA, Recknor C, et al. Effects of teriparatide in postmenopausal women with osteoporosis on prior alendronate or raloxifene: differences between stopping and continuing the antiresorptive agent. J Clin Endocrinol Metab. 2009;94(10):3772–80.
32. Leder BZ, Tsai JN, Uihlein AV, et al. Denosumab and teriparatide transitions in postmenopausal osteoporosis (the DATA-Switch study): extension of a randomised controlled trial. Lancet. 2015;386(9999):1147–55.
33. Tsai J, Uihlein AV, Burnett-Bowie SA, et al. (eds). Effect of denosumab (DMAB) and teriparatide (TPTD) transitions on peripheral bone mineral density (BMD) and microarchitecture: the DATA-Switch HR-pQCT Study. . J Bone Miner Res. 2015;30(suppl 1):S18.
34. Leder BZ, Uihlein AV, Neer RM, et al. The effects of combined denosumab and teriparatide administration on bone mineral density in postmenopausal women: the DATA (Denosumab and Teriparatide Administration) study. ASBMR 2012 Annual Meeting, Minneapolis, MN2012.
35. Anastasilakis AD, Makras P. Multiple clinical vertebral fractures following denosumab discontinuation. Osteoporos Int. 2016;27(5):1929–30.
36. Aubry-Rozier B, Gonzalez-Rodriguez E, Stoll D, et al. Severe spontaneous vertebral fractures after denosumab discontinuation: three case reports. Osteoporos Int. 2016;27(5):1923–5.
37. Lamy O, Gonzalez-Rodriguez E, Stoll D, et al. Severe rebound-associated vertebral fractures after denosumab discontinuation: nine clinical cases report. J Clin Endocrinol Metab. 2017;102(2):354–8.
38. Popp AW, Zysset PK, Lippuner K. Rebound-associated vertebral fractures after discontinuation of denosumab-from clinic and biomechanics. Osteoporos Int. 2016;27(5):1917–21.
39. Cosman F, Nieves JW, Zion M, et al. Effect of prior and ongoing raloxifene therapy on response to PTH and maintenance of BMD after PTH therapy. Osteoporos Int. 2008;19(4):529–35.
40. Black DM, Bilezikian JP, Ensrud KE, et al. One year of alendronate after one year of parathyroid hormone (1-84) for osteoporosis. New Engl J Med. 2005;353(6):555–65.
41. Eastell R, Nickelsen T, Marin F, et al. Sequential treatment of severe postmenopausal osteoporosis after teriparatide: final results of the randomized, controlled European Study of Forsteo (EUROFORS). J Bone Miner Res. 2009;24(4):726–36.
42. Muschitz C, Kocijan R, Fahrleitner-Pammer A, et al. Antiresorptives overlapping ongoing teriparatide treatment result in additional increases in bone mineral density. J Bone Miner Res. 2013;28(1):196–205.

75

Strontium Ranelate and Calcitonin

Leonardo Bandeira[1] and E. Michael Lewiecki[2]

[1] Department of Medicine, College of Physicians and Surgeons, Columbia University Medical Center, New York, NY, USA
[2] New Mexico Clinical Research & Osteoporosis Center, Albuquerque, NM, USA

STRONTIUM RANELATE

Strontium, an alkaline earth divalent cation, is a trace element in the human body (~0.00044% of body weight), mainly deposited in bone tissue. It resembles chemically the calcium molecule but has more than twice the atomic weight. Both compete for intestinal absorption and renal excretion, but strontium is less well absorbed and eliminated more than calcium. The dietary source of strontium comes mainly from vegetables and cereals [1, 2].

Strontium ranelate (SR) is composed of an organic molecule, ranelic acid, which is capable of binding to two stable strontium atoms, promoting an increase in element bioavailability [3]. The molecule has been used as treatment for osteoporosis in several countries but not in the United States. In Europe, it was approved for clinical use in 2004 [4]. The therapeutic dose is 2 g once a day, taken orally, available in sachets. It should be used between meals or at bedtime because the absorption is reduced by food, especially milk and dairy products, as strontium competes with calcium for intestinal absorption.

Pharmacology

Unlike most drugs used for osteoporosis, SR may stimulate bone formation while inhibiting resorption [3], which makes it an attractive medication from a physiological point of view. However, its exact mechanism of action remains unknown.

SR intestinal absorption is poor, but skeletal sites with active osteogenesis have greater uptake of the absorbed drug, thus increased deposition of the drug is observed on trabecular bone compared with cortical bone. Factors influencing bone incorporation of SR include treatment duration and dose [5]. When a metal with higher atomic weight than calcium, such as strontium, is incorporated in bone, there will be greater attenuation of photon beams emitted with DXA. Accordingly, patients treated with SR have an increase in BMD that is independent of its effect on fracture risk [6]. The drug is not metabolized in the body and excretion is mainly by the kidneys [5].

Clinical efficacy

The Spinal Osteoporosis Therapeutic Intervention (SOTI) was the first phase III trial that investigated the clinical efficacy of SR; its primary objective was to assess the medication effectiveness on vertebral fractures (VFs). In this study, 1649 postmenopausal women with osteoporosis and at least one VF were randomized to receive SR or placebo. The risk of new VF was reduced by 49% and 41% in the treatment group after 1 year and 3 years, respectively, compared with placebo ($p<0.001$ for both). In addition, there was a 52% and 38% reduction in risk of symptomatic VF after 1 year ($p=0.003$) and 3 years ($p<0.001$) of treatment, respectively. The number of patients needed to treat (NNT) to prevent one VF was nine after 3 years. The SR group had less loss of height ≥1 cm (30.1% versus 37.5%; $p=0.003$) and a trend to less back pain (17.7% versus 21.3%; $p=0.07$). There was a risk reduction in nonvertebral fractures that was not statistically significant [7].

In addition to showing reduction of VF risk, the study found increased BMD in the SR group compared with placebo at all skeletal sites (14.4% at lumbar spine, 7.2% at femoral neck, and 8.6% at total hip; $p<0.001$). Increased BMD was maintained even after adjustment for strontium content at the lumbar spine (8.1%; $p<0.001$). There were small changes in bone turnover markers but individuals in the SR group showed dissociation between formation and resorption. While the bone-specific alkaline phosphatase (BAP), a bone formation marker, was higher in the SR group compared with placebo ($p=0.003$), the bone resorption marker CTx) was lower ($p=0.006$) compared with placebo. These effects were more pronounced on the initiation of treatment but remained after 3 years [7]. An extension of SOTI showed that subjects who used the medication for 5 years continued to have increasing BMD while those who switched to placebo after 4 years had a decline. Subjects who were in the placebo group for 4 years and switched to SR in the fifth year had a similar BMD increase to that seen in the SR group during the first year of the trial [8].

Another study, the Treatment of Peripheral Osteoporosis (TROPOS), was designed to evaluate nonvertebral fractures in postmenopausal women with osteoporosis. After 3 years, patients who used SR had a 16% decrease in all nonvertebral fractures ($p=0.04$) and a 19% decrease in major nonvertebral fractures ($p=0.031$). In the subgroup with high fracture risk (age ≥74 years and femoral neck T-score ≤ −3), treatment was associated with a 36% reduction in risk of hip fractures ($p=0.046$). Patients who used SR also had BMD improvement compared with placebo (8.2% at femoral neck and 9.8% at total hip; $p<0.001$) [9].

Although it was not the primary objective of the TROPOS study, VF risk reduction was also observed. There were 45% and 39% less VFs after 1 year and 3 years, respectively, in the SR group. In patients who had never experienced a VF, there was a 45% reduction in the risk of having a first VF (incidence SR 7.7% versus 14% placebo; $p<0.001$). In those who had a previous VF, there was a 32% reduction in the risk of having another VF (SR 22.7% versus 31.5% placebo; $p<0.001$) [9].

In a subgroup of subjects over the age of 80 years from SOTI and TROPOS treated with SR for 5 years, there was a risk reduction of vertebral (31%; $p=0.01$), nonvertebral (27%; $p=0.018$), and major nonvertebral (33%; $p=0.0005$) fractures compared with placebo. There was also increased BMD at the lumbar spine (23.1%) and femoral neck (12.8%) [10]. In subjects previously treated with a bisphosphonate, use of SR led to less gain in BMD than in naïve patients [11].

SR also appears to have beneficial effect on quality of life by improving physical and emotional functions and back pain, as assessed by quality of life questionnaires [12].

Studies with micro-computed tomography and HRpQCT showed improvement in various parameters of bone microarchitecture in subjects using SR that was greater than that observed in subjects who used alendronate. Although SR is preferentially incorporated into trabecular bone, it also has an effect on cortical bone. Increase of cortical thickness, bone area and density, trabecular number and density, and decrease of trabecular separation were observed. The load to failure estimated by finite element analysis was also higher in patients who took SR. There was no difference in cortical porosity [5, 13–15].

Adverse events

In general, SR was well tolerated in clinical trials, with no major adverse events (AEs) reported in patients with good adherence to therapy. The main AEs were related to the gastrointestinal tract (nausea, vomiting, and diarrhea); headache, dermatitis, and eczema were also reported. The AEs occurred mainly at the initiation of treatment and were transient. Some patients showed increased serum creatine kinase (CK), but this was transient and without associated muscle symptoms. In patients aged over 80 years using SR, a few (less than 1%) had seizures, with an incidence that was statistically higher than in the placebo group [7, 9, 10].

Concerns arose about SR safety after reports of severe AEs. The drug has been associated with rare skin reactions (drug rash with eosinophilia and systemic symptoms, and toxic epidermal necrolysis) and cardiovascular diseases such as myocardial infarction and venous thromboembolism [16, 17]. As a result, the European Medicines Agency (EMA) limited its recommendation for SR to patients with severe osteoporosis without contraindications who cannot be treated with other medications. The agency added a contraindication for its use in patients with poorly controlled hypertension and/or history of ischemic heart disease, cerebrovascular disease, or peripheral arterial disease. Furthermore, it was recommended that patients undergo to a cardiovascular assessment prior to initiation of therapy and at regular periods while taking SR [17, 18].

In 2017 the company responsible for manufacturing SR decided to cease its production due to the restricted indications and limited use of the drug [19].

CALCITONIN

In humans, calcitonin (CT) is a 32 amino acid peptide produced by the C cells (parafollicular cells) of the thyroid gland [20]. It is initially synthesized as a pre-prohormone (15 kDa), which is converted to the mature form after enzymatic cleavage. The peptide was first described by Copp and Cameron in 1961 as a hormone secreted in a hypercalcemic state that caused rapid and transient decreases in serum calcium [21]. It is currently used in clinical practice as a medullary thyroid carcinoma (MTC) marker in the diagnosis, monitoring, and prognosis of this disease, since the hormone is produced by this type of tumor and correlates with its size [22].

Physiology and pharmacology

Osteoclasts express CT receptors and respond to stimulus of this peptide with a rapid decrease in their number and activity. By binding on these receptors, the hormone induces morphological flattening of the osteoclast ruffled border and withdrawal of the cell from sites of active bone resorption. It has also been demonstrated CT receptors are found on osteoblasts, suggesting that the peptide has a stimulatory effect on bone formation through an action on these cells [20]. In the kidneys CT is associated with mild phosphaturic and calciuric effect [20, 23]. The kidneys play the most important role on elimination of peptides, such as CT [24].

Unlike estrogens, whose deficiency causes more bone resorption and bone loss and increased fracture risk, there is no evidence of a significant physiological effect of CT in humans. Patients with deficiency or increase of serum levels of this hormone (eg, after total thyroidectomy and on MTC, respectively) have no skeletal or mineral homeostasis abnormalities [23, 25]. In other living beings, however, the hormone is important in the maintenance of calcium and sodium homeostasis, as in saltwater fish, for example [23, 26].

Formulations

CT was one of the first drugs shown to have an antiresorptive effect on the bone [20]. CT from fishes, reptiles, birds, and mammals have been tested as potential pharmacological agents. Synthetic salmon CT has been the most widely used in clinical practice because it has a higher potency (50 to 100 times more than human CT) [20].

The first salmon CT formulation available, in 1974, was by injectable route [20]. The following year it was approved for use subcutaneously in Paget disease. In 1980, it was approved for treatment of hypercalcemic emergencies subcutaneously, intramuscularly, or intravenously, and in 1984, for postmenopausal osteoporotic women by subcutaneous or intramuscular administration. The injectable dose for the treatment of osteoporosis is 50 to 100 IU (10 to 20 μg) daily or on alternate days [20, 27].

Given the inconvenience of the injectable route, an intranasal spray formulation of CT was released in 1987. A few years later, in 1995, it was approved for treatment of postmenopausal women with osteoporosis at a 200 IU/day dose [20]. The lower bioavailability of the intranasal CT explains the higher dose necessary when the drug in taken by this route [28].

Clinical efficacy

Despite the use of CT for decades in the treatment of postmenopausal osteoporotic women, its clinical effectiveness has never been strongly proved. Small studies using injectable salmon CT showed increased BMD at the lumbar spine, especially when the medication was combined with hormone-replacement therapy (HRT). Reduction of hip fractures has only been observed in a retrospective study when the drug was combined with calcium replacement [29–31]. Combination therapy with androgens seems to increase bone mass more than injectable CT taken alone [32]. Furthermore, sequential parenteral CT did not act synergistically with parathyroid hormone therapy and its use in this setting did not bring additional benefits in bone turnover markers and histomorphometric parameters [33, 34].

The registration trial with intranasal CT, the Prevent Recurrence of Osteoporotic Fractures (PROOF) study, showed approximately 30% reduction in VF risk with a 200 IU/day dose. The antifracture effect, however, was not observed with higher (400 IU/day) or lower (100 IU/day) doses. Changes in BMD and in bone turnover markers were mild. There was only a 1.5% increase in lumbar spine BMD and a 12% suppression on CTx [35]. This study was widely criticized, had a large withdrawal rate (744 of 1255 subjects), and did not show reduction in nonvertebral fractures, but it led to approval for intranasal CT [20].

The Qualitative Evaluation of Salmon Calcitonin Therapy (QUEST) study analyzed the bone microarchitecture in postmenopausal osteoporotic patients using MRI, 3D micro-computed tomography, and bone biopsy. There was an increase or maintenance of trabecular microarchitecture in the distal radius and lower trochanter, and a greater trabecular number at the os calcis, in patients who used intranasal CT compared with those who used placebo, regardless of changes in BMD [36]. Thus, its antifracture efficacy seems to be more related to reduction in bone resorption and preservation of trabecular microarchitecture than to an effect on bone density [20].

Although not included in the therapy indications, an analgesic effect of CT is suggested in some conditions such as Paget bone disease and diabetic neuropathy [37, 38]. This effect is possibly mediated by an increase of circulating endorphins. It appears to promote early mobilization and recovery of motor function after a VF [39]. Studies have shown that the use of eel CT with risedronate was more effective than risedronate alone on reduction of back pain in postmenopausal women with osteoporosis [40, 41].

Oral calcitonin

An oral formulation of CT is being studied as a possible treatment for osteoporosis. CTx suppression and a slight BMD increase were demonstrated, especially at the lumbar spine, in postmenopausal women with osteopenia or osteoporosis who used this formulation. When compared with patients who used the intranasal route, those in the oral CT group had greater suppression of bone turnover markers and BMD elevation; however, no antifracture effect has been demonstrated. Since the bioavailability is even lower by the oral route, the doses used in the studies were higher, ranging from 0.15 to 2.5 mg once a day [42, 46].

A randomized, double-blind, placebo-controlled, phase III study enrolled 4665 postmenopausal women with osteoporosis to assess the efficacy of oral CT in this population. Despite showing a slight increase in lumbar spine BMD compared with women on placebo (1.02 ± 0.12 versus $0.18 \pm 0.12\%$; $p < 0.0001$), there was no effect on preventing new fractures in the CT group after 3 years. Urinary CTx levels were 15% lower in the active group than in the placebo arm at year 1 and 2, but not at year 3 [46].

Adverse events

AEs vary according to the CT formulation. If subcutaneous, intramuscular, or intravenous, reactions at the injection site, nausea, and flushing occurs in up to 20% of the patients. The high incidence of AEs along with the inconvenience of these routes led to replacement of its use by intranasal CT once the latter was approved for the treatment of postmenopausal osteoporosis [20]. Intranasal CT is generally well tolerated and the few side effects associated with this formulation involve mild and transitional nasal reactions [20]. Oral CT was also well tolerated in studies. AEs were mild or moderate and more often related to the gastrointestinal tract (nausea, abdominal pain, constipation, diarrhea) [42, 43, 46].

In 2010, two phase III trials of oral CT in the treatment of male patients with osteoarthritis showed an imbalance of prostate cancer in those who used the medication. To evaluate this relationship, two meta-analyses were conducted incorporating studies with both intranasal and oral CT. An increased risk of developing any type of cancer, especially basal cell carcinoma, in patients using CT was found. Excluding this type of cancer, the risk difference of malignancy was not statistically significant between the CT and placebo groups. Furthermore, the risk was almost entirely attributable to a single study, the PROOF, and these meta-analyses were not prospectively designed to evaluate risk of malignancy [35, 47]. A prospective case-control study that enrolled more than 5000 participants showed increased lung and liver cancer risk in women who took high doses of intranasal CT. Interestingly this study also showed a protective effect of CT on breast cancer [48].

Considering the controversial and inconsistent data plus the known drug mechanism of action and the distribution of its receptors, it is still hard to advance a strong argument for a causal relationship between CT and cancer [47, 49]. Moreover, if CT was really oncogenic, patients with MTC would have a greater chance of acquiring secondary primary cancers. However, this has not been observed in these patients [50].

Despite the uncertainties regarding the malignancy risk, along with the weak evidence on fracture risk reduction, the EMA has recently withdrawn CT as an indication for the treatment of postmenopausal osteoporosis. The US Food and Drug Administration (FDA) has retained its approval for women who are more than 5 years postmenopausal [27, 51].

CONCLUSION

SR has been shown to increase BMD and reduce the risk of VFs and nonvertebral fractures in women with postmenopausal osteoporosis (Table 75.1). SR has the theoretical advantage of dissociating bone formation and resorption. Notwithstanding, its mechanism of action has never been fully elucidated and the medication was never approved in the United States. Concerns about an increased risk in cardiovascular disease and serious skin reactions have limited its use even in countries where the drug is approved.

CT was one of the first released medications for the treatment of postmenopausal osteoporosis. Nevertheless, its antifracture effect has only been proven in the spine and with only one dose of the intranasal route. This effect was demonstrated in a single study, which had many limitations. The effects of CT on BMD and bone turnover markers are minimal (Table 75.1). Moreover, a possible,

Table 75.1. Benefits and limitations of strontium ranelate and calcitonin.

	Benefits	Limitations
Strontium ranelate	Oral route Induces bone formation–resorption uncoupling Not metabolized in the body Decreases vertebral and nonvertebral fracture risk Increases LS and hip site BMD Beneficial effects on quality of life Improves trabecular and cortical bone Well tolerated	Not FDA approved Exact mechanism of action remains unknown Overestimates BMD Associated with rare skin reactions and cardiovascular diseases Production ceased in 2017
Calcitonin	Decreases VF risk Improves trabecular microarchitecture Has an analgesic effect Well tolerated by IN route	No oral formulation approved Clinical effectiveness was never strongly proved Mild changes in BMD and bone turnover markers No effects in nonvertebral fractures Possible association with cancer

BMD = bone mineral density; FDA = US Food and Drug Administration; IN = intranasal; LS = lumbar spine; VF = vertebral fracture.

although not very plausible, relationship between CT and cancer has been reported.

Therefore, considering safety issues and the existence of other more effective medications, both SR and CT are no longer included as first-line therapy for postmenopausal osteoporosis. Calcitonin use should be limited for patients with contraindications or intolerance to other drugs. SR is no longer produced by the manufacturer.

REFERENCES

1. Pors Nielsen S. The biological role of strontium. Bone. 2004;35(3):583–8.
2. Marcus CS, Lengemann FW. Absorption of Ca45 and Sr85 from solid and liquid food at various levels of the alimentary tract of the rat. J Nutr. 1962;77:155–60.
3. Marie PJ. Strontium ranelate: a physiological approach for optimizing bone formation and resorption. Bone. 2006;38(2 suppl 1):10–4.
4. European Medicines Agency. *Protelos Strontium Ranelate. EPAR Summary for the Public 2014*. http://www.ema.europa.eu/docs/en_GB/document_library/EPAR_-_Summary_for_the_public/human/000560/WC500045520.pdf (accessed May 2018).
5. Dahl SG, Allain P, Marie PJ, et al. Incorporation and distribution of strontium in bone. Bone. 2001;28(4):446–53.
6. Nielsen SP, Slosman D, Sorensen OH, et al. Influence of strontium on bone mineral density and bone mineral content measurements by dual X-ray absorptiometry. J Clin Densitom. 1999;2(4):371–9.
7. Meunier PJ, Roux C, Seeman E, et al. The effects of strontium ranelate on the risk of vertebral fracture in women with postmenopausal osteoporosis. N Engl J Med. 2004;350(5):459–68.
8. Meunier PJ, Roux C, Ortolani S, et al. Effects of long-term strontium ranelate treatment on vertebral fracture risk in postmenopausal women with osteoporosis. Osteoporos Int. 2009;20(10):1663–73.
9. Reginster JY, Seeman E, De Vernejoul MC, et al. Strontium ranelate reduces the risk of nonvertebral fractures in postmenopausal women with osteoporosis: Treatment of Peripheral Osteoporosis (TROPOS) study. J Clin Endocrinol Metab. 2005;90(5):2816–22.
10. Seeman E, Boonen S, Borgstrom F, et al. Five years treatment with strontium ranelate reduces vertebral and nonvertebral fractures and increases the number and quality of remaining life-years in women over 80 years of age. Bone. 2010;46(4):1038–42.
11. Middleton ET, Steel SA, Aye M, et al. The effect of prior bisphosphonate therapy on the subsequent BMD and bone turnover response to strontium ranelate. J Bone Miner Res. 2010;25(3):455–62.
12. Marquis P, Roux C, de la Loge C, et al. Strontium ranelate prevents quality of life impairment in post-menopausal women with established vertebral osteoporosis. Osteoporos Int. 2008;19(4):503–10.
13. Rizzoli R, Chapurlat RD, Laroche JM, et al. Effects of strontium ranelate and alendronate on bone microstructure in women with osteoporosis. Results of a 2-year study. Osteoporos Int. 2012;23(1):305–15.
14. Rizzoli R, Laroche M, Krieg MA, et al. Strontium ranelate and alendronate have differing effects on distal tibia bone microstructure in women with osteoporosis. Rheumatol Int. 2010;30(10):1341–8.
15. Arlot ME, Jiang Y, Genant HK, et al. Histomorphometric and microCT analysis of bone biopsies from postmenopausal osteoporotic women treated with strontium ranelate. J Bone Miner Res. 2008;23(2):215–22.
16. Musette P, Brandi ML, Cacoub P, et al. Treatment of osteoporosis: recognizing and managing cutaneous adverse reactions and drug-induced hypersensitivity. Osteoporos Int. 2010;21(5):723–32.
17. Reginster JY. Cardiac concerns associated with strontium ranelate. Expert Opin Drug Safety. 2014;13(9):1209–13.
18. European Medicines Agency. *Strontium Ranelate. PSUR Assessment Report 2013*. http://www.ema.europa.eu/docs/en_GB/document_library/EPAR_-_Assessment_Report_-_Variation/human/000560/WC500147168.pdf (accessed May 2018).
19. Bolland MJ, Gray A. Cessation of strontium ranelate supply. BMJ. 2017;357:j2580.
20. Chesnut CH 3rd, Azria M, Silverman S, et al. Salmon calcitonin: a review of current and future therapeutic indications. Osteoporos Int. 2008;19(4):479–91.
21. Copp DH, Cameron EC. Demonstration of a hypocalcemic factor (calcitonin) in commercial parathyroid extract. Science. 1961;134(3495):2038.
22. Brutsaert EF, Gersten AJ, Tassler AB, et al. Medullary thyroid cancer with undetectable serum calcitonin. J Clin Endocrinol Metab. 2015;100(2):337–41.
23. Lima JG, Nobrega LHC, Mendonca RP. Metabolismo osseo e mineral. In: Bandeira F, Mancini M, Graf H, et al (eds) *Endocrinologia e Diabetes*. Rio de Janeiro: Medbook, 2015.
24. Hysing J, Gordeladze JO, Christensen G, et al. Renal uptake and degradation of trapped-label calcitonin. Biochem Pharmacol. 1991;41(8):1119–26.
25. The ESHRE Capri Workshop Group. Bone fractures after menopause. Hum Reprod Update. 2010;16(6):761–73.
26. Wales NA, Barrett AL. Depression of sodium, chloride and calcium ions in the plasma of goldfish (Carassius auratus) and immature freshwater- and seawater-adapted eels (Anguilla anguilla L.) after acute administration of salmon calcitonin. J Endocrinol. 1983;98(2):257–61.
27. US Food and Drug Administration, Joint Meeting of the Advisory Committee for Reproductive Health Drugs and the Drug Safety and Risk Management Advisory Committee. *Calcitonin Salmon for the Treatment of Postmenopausal Osteoporosis*. 2013. https://www.medscape.com/viewarticle/780323 (accessed May 2018).
28. Ozsoy Y, Gungor S, Cevher E. Nasal delivery of high molecular weight drugs. Molecules. 2009;14(9):3754–79.

29. Kanis JA, Johnell O, Gullberg B, et al. Evidence for efficacy of drugs affecting bone metabolism in preventing hip fracture. BMJ. 1992;305(6862):1124–8.
30. Mazzuoli GF, Passeri M, Gennari C, et al Effects of salmon calcitonin in postmenopausal osteoporosis: a controlled double-blind clinical study. Calcif Tissue Int. 1986;38(1):3–8.
31. Meschia M, Brincat M, Barbacini P, et al. Effect of hormone replacement therapy and calcitonin on bone mass in postmenopausal women. Eur Obstet Gynecol Reprod Biol. 1992;47(1):53–7.
32. Szucs J, Horvath C, Kollin E, et al. Three-year calcitonin combination therapy for postmenopausal osteoporosis with crush fractures of the spine. Calcif Tissue Int. 1992;50(1):7–10.
33. Hodsman AB, Fraher LJ, Ostbye T, et al. An evaluation of several biochemical markers for bone formation and resorption in a protocol utilizing cyclical parathyroid hormone and calcitonin therapy for osteoporosis. J Clin Invest. 1993;91(3):1138–48.
34. Hodsman AB, Steer BM, Fraher LJ, et al. Bone densitometric and histomorphometric responses to sequential human parathyroid hormone (1-38) and salmon calcitonin in osteoporotic patients. Bone Miner. 1991;14(1):67–83.
35. Chesnut CH 3rd, Silverman S, Andriano K, et al. A randomized trial of nasal spray salmon calcitonin in postmenopausal women with established osteoporosis: the prevent recurrence of osteoporotic fractures study. Am J Med. 2000;109(4):267–76.
36. Chesnut CH 3rd, Majumdar S, Newitt DC, et al. Effects of salmon calcitonin on trabecular microarchitecture as determined by magnetic resonance imaging: results from the QUEST study. J Bone Miner Res. 2005;20(9):1548–61.
37. Altman RD, Collins-Yudiskas B. Synthetic human calcitonin in refractory Paget's disease of bone. Arch Intern Med. 1987;147(7):1305–8.
38. Quatraro A, Minei A, De Rosa N, et al. Calcitonin in painful diabetic neuropathy. Lancet. 1992;339(8795):746–7.
39. Lyritis GP, Paspati I, Karachalios T, et al. Pain relief from nasal salmon calcitonin in osteoporotic vertebral crush fractures. A double blind, placebo-controlled clinical study. Acta Orthop Scand Suppl. 1997;275:112–4.
40. Hongo M, Miyakoshi N, Kasukawa Y, et al. Additive effect of elcatonin to risedronate for chronic back pain and quality of life in postmenopausal women with osteoporosis: a randomized controlled trial. J Bone Miner Res. 2015;33(4):432–9.
41. Takakuwa M, Iwamoto J. Elcatonin in combination with risedronate is more effective than risedronate alone for relieving back pain in postmenopausal women with osteoporosis. Biol Pharm Bull. 2012;35(7):1159–65.
42. Binkley N, Bone H, Gilligan JP, et al. Efficacy and safety of oral recombinant calcitonin tablets in postmenopausal women with low bone mass and increased fracture risk: a randomized, placebo-controlled trial. Osteoporos Int. 2014;25(11):2649–56.
43. Binkley N, Bolognese M, Sidorowicz-Bialynicka A, et al. A phase 3 trial of the efficacy and safety of oral recombinant calcitonin: the Oral Calcitonin in Postmenopausal Osteoporosis (ORACAL) trial. J Bone Miner Res. 2012;27(8):1821–9.
44. Tanko LB, Bagger YZ, Alexandersen P, et al. Safety and efficacy of a novel salmon calcitonin (sCT) technology-based oral formulation in healthy postmenopausal women: acute and 3-month effects on biomarkers of bone turnover. J Bone Miner Res. 2004;19(9):1531–8.
45. Hamdy RC, Daley DN. Oral calcitonin. Int J Women Health. 2012;4:471–9.
46. Henriksen K, Byrjalsen I, Andersen JR, et al. A randomized, double-blind, multicenter, placebo-controlled study to evaluate the efficacy and safety of oral salmon calcitonin in the treatment of osteoporosis in postmenopausal women taking calcium and vitamin D. Bone. 2016;91:122–9.
47. Wells G, Chernoff J, Gilligan JP, et al. Does salmon calcitonin cause cancer? A review and meta-analysis. Osteoporos Int. 2016;27:13–9.
48. Sun LM, Lin MC, Muo CH, et al. Calcitonin nasal spray and increased cancer risk: a population-based nested case-control study. J Clin Endocrinol Metab. 2014;99(11):4259–64.
49. Hill AB. The environment and disease: association or causation? 1965. J Roy Soc Med. 2015;108(1):32–7.
50. Ronckers CM, McCarron P, Ron E. Thyroid cancer and multiple primary tumors in the SEER cancer registries. Int J Cancer. 2005;117(2):281–8.
51. Overman RA, Borse M, Gourlay ML. Salmon calcitonin use and associated cancer risk. Ann Pharmacother. 2013;47(12):1675–84.

76

Adverse Effects of Drugs for Osteoporosis

Bo Abrahamsen[1] and Daniel Prieto-Alhambra[2]

[1]Odense Patient Data Explorative Network, Department of Clinical Research, University of Southern Denmark, Odense; and Department of Medicine, Holbæk Hospital, Holbæk, Denmark
[2]Pharmaco- and Device Epidemiology, Oxford NIHR Biomedical Research Centre, Nuffield Department of Orthopaedics, Rheumatology, and Musculoskeltal Sciences (NDORMS), University of Oxford, Oxford, UK; and GREMPAL Research Group – Idiap Jordi Gol and CIBERFes, Universitat Autònoma de Barcelona and Instituto Carlos III (FEDER Research Funds), Barcelona, Spain

INTRODUCTION

In the following chapter we will address key safety aspects of current pharmacotherapy for osteoporosis and briefly also new drugs in phase III trials. The chapter is primarily aimed at clinicians and focuses more on the risk estimates than on pathophysiology although this will be addressed briefly where clinically important. We shall draw on data from both clinical trials and post-licensing observational studies. The review will not address reproductive safety as none of the therapies discussed here can be considered safe in pregnancy or lactation at this stage. Prescribers should refer to the official prescriber information from the producers and government agencies before prescribing because, due to space restraints, we do not cover all caveats and contraindications. Strontium ranelate, which has been the subject of significant concerns over cardiovascular safety in patients with hypertension or heart disease and which has never been licensed in the United States will not be covered here. For similar reasons, we shall not be reviewing calcitonin, which was withdrawn as an osteoporosis drug in Europe after the European Medicines Agency concluded in 2012 that risks including cancer risks could outweigh antifracture benefits (see Chapter 75). The literature review was completed in the second half of 2016.

SOURCES OF DRUG SAFETY DATA

Need for observational data in drug safety research

Where well-conducted, randomized controlled trials (RCTs) exist, they provide gold standard evidence on drug efficacy. There are, however, some limitations to RCTs that make them less useful for the study of some unwanted safety issues:

1. *External validity/generalizability*. It is known that RCT participants are not necessarily representative of all actual patients [1]. Indeed, a recent report [2] has demonstrated that only half of the real world users of alendronate would have been eligible for the pivotal Fracture Intervention Trial (FIT) RCT [3].
2. *Power*. RCTs are costly and usually powered for efficacy outcomes. Unwanted effects are most often secondary outcomes, and serious safety issues are—fortunately—most often rare and can have a long latency period (see the section on atypical femur fractures for an illustrative example).
3. *Ethics and logistics*. Even once potential drug side effects are known—or suspected based on observational data—it is difficult to imagine that one could

---

Primer on the Metabolic Bone Diseases and Disorders of Mineral Metabolism, Ninth Edition. Edited by John P. Bilezikian.
© 2019 American Society for Bone and Mineral Research. Published 2019 by John Wiley & Sons, Inc.
Companion website: www.wiley.com/go/asbmrprimer

set up an RCT for the study of safety outcomes. Both ethics (are we allowed to expose patients to such suspected risk/s?) and logistics (would we be able to recruit for such a study?) make such studies unfeasible.

All these issues make a case for observational postauthorization safety studies (PASS). Most drug regulators recognize this need, and do sometimes request large PASS as part of ongoing safety surveillance. More often than not, these will use broad inclusion criteria to assess the risk–benefit of drug use in the wider community and in actual practice settings (for some examples, see http://www.encepp.eu/encepp/studiesDatabase.jsp; accessed May 2018). In many cases routinely collected data sources will be used (such as electronic medical records or claims databases), whilst other studies assemble ad hoc drug or disease registry data.

Strengths and limitations of observational drug safety studies

Observational PASS are designed to assess the public health burden of any potential unwanted effect/s when drugs are used in the community after regulatory approval. Generalizability and large statistical power are the main strengths of safety studies, but there are also some limitations that need to be acknowledged:

1. *Information bias.* With few exceptions, PASS increasingly use data that were not collected in research settings but in actual practice conditions. Although this is to some extent one of the key strengths of such studies, it is recognized that such routinely collected data is only as good as the coders' recording quality. Effort is hence needed for the identification of good quality data sources, as well as, sometimes, for the identification of certain safety events, especially when no specific diagnostic code is available, as for atypical femur fracture (AFF) or osteonecrosis of the jaw (ONJ) [4].
2. *Confounding by indication.* Treatment allocation/prescription is, in actual practice, the result of clinical judgment, naturally resulting in differences between drug users and nonusers (and, similarly, in users of drug A versus drug B). This so-called "confounding by indication" is the main source of bias in observational drug safety studies—and the opposite to what we see in randomized experiments. Advanced methods (including multivariable regression, propensity score methods, case-only studies, and others) are often used to minimize such confounding and related bias, which is—despite all these—always to be considered when interpreting observational drug (pharmaco-epidemiologic) studies.
3. *Differential ascertainment.* In addition, it is also common that treated patients (compared with nontreated), as well as those treated with more costly (compared with first line/cheaper) medications, have more frequent contacts with their prescriber, even if only for repeat prescriptions. This can result in differential coding and hence ascertainment bias.

SKELETAL/BONE SIDE EFFECTS

Atypical femur fractures

Atypical femur fractures are low-energy or spontaneous fractures of the subtrochanteric femur or the femoral shaft characterized by a set of specific radiological criteria first defined by the ASBMR Task Force on AFF in 2010 [5]. Following new data [6, 7], the criteria were refined in order to make the diagnosis more specific [8]. In short, AFFs are substantially transverse fractures that originate at the lateral femoral cortex. They are always noncomminuted, and usually have localized or general cortical thickening. There is only limited information on BMD pretreatment in AFF cases but case series have indicated that many patients may have initiated treatment with a relatively high BMD (osteopenia or milder degrees of osteoporosis). Fractures may be bilateral, and if so they are usually symmetrical in position, and incomplete fractures very often require pinning due to delayed healing [9, 10]. AFFs are rare, but not unknown, in persons who have not used antiresorptive treatment [11]. Before 2018, AFFs did not have an International Classification of Diseases (ICD) diagnosis code that can be used in register-based pharmacovigilance but because they are confined to the proximal femur and femoral shaft it is possible to at least monitor the combined rate of such fractures for harm/benefit considerations.

Bisphosphonates

In 2012, Dell and colleagues [12] reviewed femur fracture X-rays in the Kaiser California radiology database to estimate the incidence rate of AFFs as a function of duration of bisphosphonate use. Where the incidence rate was 0.2 per 10,000 patient-years with up to 2 years of exposure, the authors calculated a rate of 10.7 per 10,000 with more than 10 years of use, suggesting an exponential relationship between duration of use and the risk of AFF. A more rapid increase in risk was suggested by Swedish researchers who reported a similar rate (11 per 10,000) already after 4+ years of use, a rate that seems large but not impossible when compared with the total number of subtrochanteric and shaft fractures observed in Danish health care data after a similar length of exposure [13]. The short look-back for drug history in the Swedish prescription register could have inflated the risk estimate, as discussed elsewhere [14]. The high rate of AFFs after 4+ years of use reported in Sweden is still low compared with the reported rate of classic hip fractures (Fig. 76.1), which was found to be 151 per 10,000 patient years, so 12 times higher than the rate of AFFs [15]. However, if an exponential increase

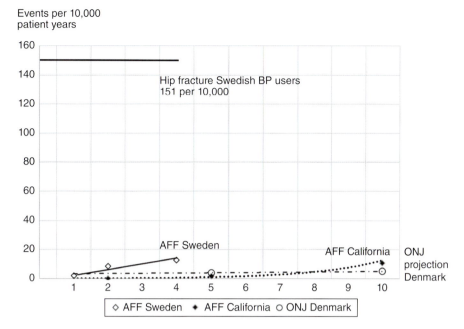

Fig. 76.1. Rates of atypical femur fractures (AFFs) [12, 13] and severe osteonecrosis of the jaw (ONJ) [30] as a function of treatment duration with oral bisphosphonates (BP) based on observational studies, compared with the rate of hip fractures in Swedish patients treated with oral bisphosphonates [52].

in risk with time were to be assumed then AFFs could outnumber classic hip fractures after 8 to 10 years of treatment, although this contrasts with the low number of AFFs seen even after 10 years in the Dell study.

Accordingly, a recent Danish study [16] did not find a net increase in the risk of subtrochanteric or shaft fractures with the use of 10 dose-years of alendronate use or over, suggesting that any increase in AFFs was offset by a reduction in the risk of conventional fractures (non-AFFs) in the same skeletal region. It is unclear if the risk of AFF is confined to a specific subgroup of patients or whether it is a risk posed to all patients treated with bisphosphonates. Specific risk factors that appear to further increase the likelihood of bisphosphonate-induced AFFs include Asian heritage [17], glucocorticoid and proton pump inhibitor exposure [18, 19], anterolateral femoral bowing, and low muscle mass [20].

Other antiresorptive drugs

AFFs have also been observed with denosumab; the cumulative exposure-adjusted incidence rate in the FREEDOM (Fracture Reduction Evaluation of Denosumab in Osteoporosis every 6 Months) extension study was 1.0 per 10,000 person-years [21], hence within the lower end of the range reported for bisphosphonates earlier. AFFs were also reported as a potential adverse event in the phase II clinical trial of the cathepsin K inhibitor odanacatib. The development program is no longer being taken forward because of a less favourable harm/benefit profile than expected. Other classes of anti-resorptive drugs have not been associated with AFF in the literature.

Osteonecrosis of the jaw

This condition is sometimes referred to as bisphosphonate-related osteonecrosis of the jaw. This term is best avoided because it conflates the outcome and the exposure, which makes association studies ambiguous, and because ONJ is also seen after exposure to non-bisphosphonates such as denosumab [22] and even drugs from outside the bone arena such as angiogenesis inhibitors used in oncology [23]. Radiation-induced osteonecrosis or osteo-radionecrosis, is regarded as a different entity so that ONJ can be defined by an area of exposed bone in the maxillofacial region that does not heal within 8 weeks after identification by a health provider. The updated American Association of Oral and Maxillofacial Surgeons (AAOMS) definition of medication-related ONJ stipulates that the patient should not have received radiotherapy to the head or face and also requires exposure to antiresorptive or anti-angiogenesis agent/s [24].

ONJ is thought to arise through a degraded mucosal barrier to infection combined with compromised local bone healing and reduced angiogenesis [25]. The pathophysiology and clinical management of ONJ is covered in a specific chapter in this book (see Chapter 120).

The absolute risk of ONJ is low with the antiresorptive regimens used in treating osteoporosis but is higher in an oncology setting. This may be influenced by the higher intensity intravenous administration of antiresorptives employed to prevent or treat skeletal metastases but also by the coadministration of antineoplastic drugs and high-dose glucocorticoids in this therapeutic area [26].

Oral bisphosphonates

The estimates for the incidence of ONJ in oral bisphosphonate users range from below 0.1 per 10,000 patient-years in the Ontario survey of oral and maxillofacial surgeons [27], to 1.5 per 10,000 patient-years in the US HealthCore claims database [28], and up to 6.9 per 10,000 patient years in a Swedish oral and maxillofacial clinical survey [29]. Preliminary observations from Denmark suggest that the rate of surgically treated osteonecrosis and osteomyelitis of the jaw and oral cavity increases from 4.0 (3.1 to 5.0) per 10,000 patient-years for a treatment duration below 5 years to 4.8 (3.2 to 7.0) with 5 to 10 years of treatment in adherent users [30]. When adjusted for differences in baseline characteristics, use for 5+ years was associated with a higher risk than use for a shorter period of time. This analysis does not include milder cases that were conservatively treated.

Intravenous bisphosphonates

An incidence rate of up to 9 per 10,000 patient-years has been reported in the osteoporosis setting compared with rates up to 1222 per 10,000 patient years in the oncology setting, as recently reviewed in detail by the International Task force on ONJ [25]. The incidence of ONJ in cancer populations is critically dependent on the type of cancer and on any coadministration of angiogenesis inhibitors [26].

Denosumab

In the FREEDOM extension trial, five adjudicated ONJ events were reported in the active treatment arm for up to 8 years of treatment [21], corresponding to an incidence of 4.2 per 10,000 patient-years and hence comparable with the event rate seen with bisphosphonates. In oncology trials in prostate and breast cancer, ONJ event rates were 340 and 290, respectively, per 10,000 for patients treated for a year or longer [31].

There are no data on ONJ for other classes of antiresorptive drugs than those referenced here; romosozumab will be addressed in the new bone anabolic agents section later in this chapter.

EXTRASKELETAL SIDE EFFECTS OF ANTIRESORPTIVE THERAPY

Gastrointestinal side effects

Oral bisphosphonates have been associated with gastrointestinal (GI) side effects such as esophageal ulcers, esophagitis, and upper GI hemorrhage. The summary of product characteristics (SmPC) for alendronate recognizes that "Alendronate can cause local irritation of the upper gastro-intestinal mucosa," and lists the following as "precautions of use": upper GI problems such as dysphagia, esophageal disease, gastritis, duodenitis, ulcers, or with a recent history (within the previous year) of major GI disease such as peptic ulcer, or active GI bleeding.

Some (but not all) pivotal RCTs found an increased risk of GI symptoms: studies of oral ibandronate found a 1.7% to 7.4% excess risk of GI issues (compared with placebo) [32]. Interestingly, the HORIZON (Health Outcomes and Reduced Incidence with Zoledronic Acid Once Yearly) RCT of parenteral zolendronate have also suggested an increased incidence of GI symptoms including nausea, vomiting, diarrhea, and dyspepsia among patients in the treatment arm [33]. Conversely, no increase in risk of GI side effects was seen in alendronate users enrolled in the Fracture Intervention Trial (FIT) [3] or even in the extension (up to 10 years follow-up) FLEX [34] RCTs. Similarly, no differences in risk of such side effects were seen between the risedronate and the placebo arms in the 5-year [35] or in the extension (totaling 7 years exposure) Vertebral Efficacy with Risedronate Therapy (VERT) Multinational (VERT-MN) trial [36].

Atrial fibrillation and other cardiovascular events

At least one recent systematic review and meta-analysis [37] of RCT data has cleared concerns on the overall cardiovascular safety of bisphosphonate therapy (Fig. 76.2).

Similarly, this and other similar studies [38] have not found an increased risk of atrial fibrillation amongst bisphosphonate users in a number of RCTs where such data were available or reported (Fig. 76.3). Other systematic reviews including observational studies have, however, found a worrying increase in risk of atrial fibrillation particularly amongst users of intravenous bisphosphonates [39]. Ascertainment bias may arise if physicians are more likely to monitor for arrhythmias in patients receiving intravenous bisphosphonates than other anti-osteoporosis drugs but there are no data to establish if this is the case. It is hence still unclear whether such excess risk (of atrial fibrillation but not of stroke or other cardiovascular events) exists in patients underrepresented in RCTs, or whether such findings are the result of unresolved confounding/bias in observational studies.

Other anti-osteoporosis drugs such as strontium ranelate, raloxifene, and more recently, in pre-marketing studies, odanacatib have been associated with an increased risk of thromboembolic and cardiovascular (myocardial infarction or stroke) events. These are not covered here but must be considered before prescribing any of these drugs, particularly in patients at high cardiovascular (or thromboembolic) risk. Raloxifene and strontium ranelate are contraindicated in patients with a history of previous thromboembolic events, and the latter is also to be avoided in patients with previous cardiovascular disease or certain risk factors. See the SmPC for details.

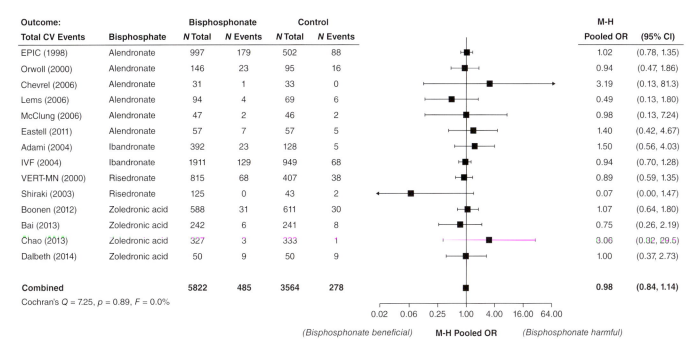

Fig. 76.2. Overall cardiovascular (CV) safety of oral and intravenous bisphosphonates. Source [53]. Reproduced under Creative Commons license 4.0.

Renal disease and hypocalcemia

Moderate or severe renal impairment is a contraindication or precaution of use for most anti-osteoporosis medications with the exception of denosumab. Extra caution is recommended with the use of intravenous bisphosphonates (eg, zolendronate) given reports of acute kidney injury.

Although no formal contraindication or need for dose adjustment is mentioned with denosumab, recent safety data have led to concerns regarding the risk of severe hypocalcemia amongst users of this drug, particularly in those with previous renal disease [40]. In addition, hypocalcemia is a known side effect and/or contraindication of most anti-osteoporosis therapies including bisphosphonates. Guidance on the monitoring and prevention of drug-related hypocalcemia is available in the SmPC documentation for each of these drugs.

Reducing bone turnover (with any antiresorptive drug such as bisphosphonates or denosumab) further in patients who may have developed adynamic bone disease as a consequence of severe CKD is unlikely to be of benefit to the patient.

Other side effects

Other potential side effects of anti-osteoporosis drugs are mentioned in their respective SmPC documentation (see the full SmPC documentation for updated safety profiles and indication/s). Some examples include:

- Bisphosphonates: musculoskeletal pain (and even related flu-like syndrome) and osteonecrosis of the external auditory canal.
- Denosumab: skin infections and allergic reaction to dry, natural rubber, needle cover.
- Raloxifene: hot flushes, leg cramps, headache (including migraine), and increased blood pressure.
- Estrogens: breast cancer (may be limited to combinations with progestogens [41]), endometrial cancer (preventable by appropriately administered progestagens), and deep vein thrombosis (see Chapter 70).

NOTE: The safety profile of bone formation (anabolic) agents is discussed in the next section and is thus not covered in detail here.

BONE FORMATION AGENTS

Teriparatide

Teriparatide has been approved for human use for over a decade. The most serious safety concern—which led to alerts from both the US and European regulators—was related to a dose- and time-dependent increase in risk of osteosarcoma in rat studies [42]. These data led to the contraindication of use of teriparatide in subjects at risk of such bone malignancy, including: (i) a previous history of bone cancer; (ii) Paget disease of bone; (iii) previous radiotherapy (to the skeleton or other organs if skeletal tissue is exposed); and (iv) in children and young adults

584 Adverse Effects of Drugs for Osteoporosis

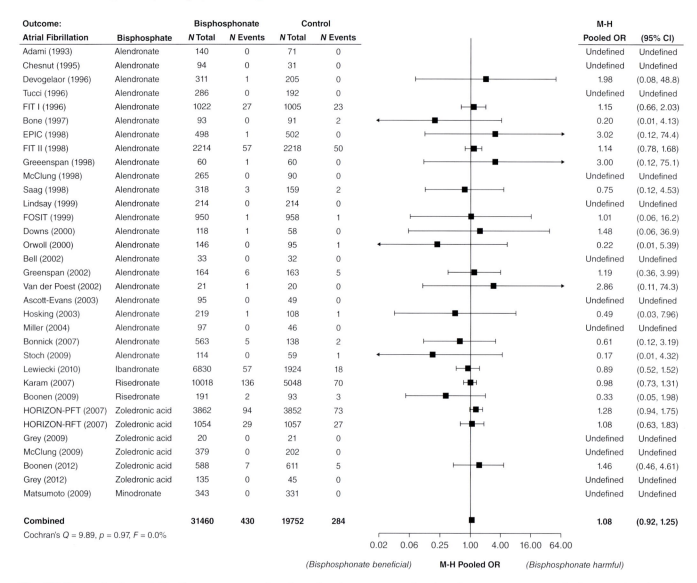

Fig. 76.3. Risk of atrial fibrillation associated with oral and intravenous bisphosphonate use. Source: [53]. Reproduced under Creative Commons license 4.0.

with open epiphyses [43]. Reassuringly though, primate experiments have not found any increase in risk of osteosarcoma with exposure to teriparatide [44, 45]. More importantly, human studies (both RCTs [46–48] and postmarketing observational studies [49]) have not reported an excess risk of bone malignancies with the use of teriparatide.

More common, and less worrying, side effects reported with teriparatide include the following (see the SmPC document for details):

- Transient orthostatic hypotension.
- Transient elevation of serum calcium concentration/s.
- Commonly reported adverse reactions such as nausea, limb pain, headache, and dizziness.

New bone anabolic agents

Two new bone formation agents (abaloparatide and romosozumab) have been recently tested in phase III RCTs and are hence are likely to obtain approval for clinical use in the coming months/years.

Abaloparatide

In the recently reported phase III RCT, the most common adverse events leading to therapy discontinuation in the abaloparatide arm were nausea (1.6%), dizziness (1.2%), headache (1.0%), and palpitations (0.9%) [50]. Hypercalcemia was significantly more common in abaloparatide users (3.4%) than in placebo users (0.4%), but less than in patients allocated to teriparatide in the trial (6.4%).

Romosozumab

In the 12-month double-blinded, placebo-controlled part of this phase III RCT serious adverse events potentially related to hypersensitivity appeared in seven subjects in the treatment arm, as were 187 (5.2%) injection site reactions (compared with 104 [2.9%] in the placebo arm) [51]. In addition, two cases of attributed ONJ (one after switching to denosumab) and one of AFF were seen during the trial, all of them in the romosozumab arm. Cardiovascular safety is under review.

More data are needed on the safety profiles of both these agents when used in clinical (rather than research/RCT controlled) settings.

CONCLUSION

In the majority of patients with osteoporosis, the benefit/harm ratio of anti-osteoporosis drugs is overwhelmingly favourable, certainly in the short and medium term. However, scenarios based on primary rather than secondary prevention and scenarios that include hip fracture benefits alone without considering other fractures prevented, will lead to high number needed to treat (NNT) estimates. This contributes to a widening treatment gap for osteoporosis by signaling to decision makers that we are not treating the right patients.

Intriguingly, it is more than anything the very low rate of the most feared adverse events ONJ and AFF that has slowed the process of obtaining exact absolute risk estimates that can be used for evidence-based clinical guidelines. One may say that these risks were too rare to have been detectable in the RCTs but it stands to reason that adjudication procedures would not have been in place for the newest safety outcomes. Another challenge is that while the benefit of antiresorptives may decline with longer use as fewer resorption cycles remain active and open to targeting, we lack data to refute that rare skeletal adverse events may increase exponentially in frequency with time. Hence the risk/benefit ratio is not likely to be constant over time. Further, the risk of adverse events may be higher in groups of patients who would not have been eligible for inclusion in the licensing trials but who we would not like to deny osteoporosis treatment. In practice, patients with RCT exclusion criteria make up a large proportion of real world users of osteoporosis therapy[2].

In contrast to the long-term safety outcomes, short-term outcomes such as atrial fibrillation, GI symptoms, acute phase reactions, and musculoskeletal pain are events that can usually be detected early in the course of treatment. They often pose less of a difficulty clinically as patients can often be managed on other anti-osteoporosis drugs with a different adverse event profile, something that is not necessarily possible for adverse events that follow directly from the skeletal action of anti-osteoporosis drugs.

REFERENCES

1. Rothwell PM. External validity of randomised controlled trials: 'to whom do the results of this trial apply?' Lancet. 2005;365(9453):82–93.
2. Reyes C, Pottegård A, Schwarz P, et al. Real-life and RCT participants: alendronate users versus FITs' trial eligibility criterion. Calcif Tissue Int. 2016;99(3):243–9.
3. Black DM, Cummings SR, Karpf DB, et al. Randomised trial of effect of alendronate on risk of fracture in women with existing vertebral fractures. Lancet. 1996;348(9041):1535–41.
4. Acquavella J, Ehrenstein V, Schiødt M, et al. Design and methods for a Scandinavian pharmacovigilance study of osteonecrosis of the jaw and serious infections among cancer patients treated with antiresorptive agents for the prevention of skeletal-related events. Clin Epidemiol. 2016;8:267–72.
5. Shane E, Burr D, Ebeling PR, et al. Atypical subtrochanteric and diaphyseal femoral fractures: report of a task force of the American Society for Bone and Mineral Research. J Bone Miner Res. 2010;25(11):2267–94.
6. Schilcher J, Koeppen V, Ranstam J, et al. Atypical femoral fractures are a separate entity, characterized by highly specific radiographic features. A comparison of 59 cases and 218 controls. Bone. 2013;52(1):389–92.
7. Ng AC, Png MA, Mohan PC, et al. Atypical femoral fractures: transverse morphology at lateral cortex is a critical feature. J Bone Miner Res. 2014;29(3):639–43.
8. Shane E, Burr D, Abrahamsen B, et al. Atypical subtrochanteric and diaphyseal femoral fractures: second report of a task force of the American Society for Bone and Mineral Research. J Bone Miner Res. 2014;(29):1–23.
9. Banffy MB, Vrahas MS, Ready JE, et al. Nonoperative versus prophylactic treatment of bisphosphonate-associated femoral stress fractures. Clin Orthop Relat Res. 2011;469(7):2028–34.
10. Ha YA, Cho MA, Park KH, et al. Is surgery necessary for femoral insufficiency fractures after long-term bisphosphonate therapy? Clin Orthop Relat Res. 2010;468(12):3393–8.
11. Lian K, Trollip J, Sandhu S, et al. Audit of atypical femoral fractures and a description of some of their features. Can Assoc Radiol J. 2016;67(1):69–75.
12. Dell RM, Adams AL, Greene DF, et al. Incidence of atypical nontraumatic diaphyseal fractures of the femur. J Bone Miner Res. 2012;27(12):2544–50.
13. Schilcher J, Koeppen V, Aspenberg P, et al. Risk of atypical femoral fracture during and after bisphosphonate use. Acta Orthop. 2015;86(1):100–7.
14. Reyes C, Hitz M, Prieto-Alhambra D, et al. Risks and benefits of bisphosphonate therapies. J Cell Biochem. 2016;117(1):20–8.

15. Schilcher J. Bisphosphonate use and atypical fractures of the femoral shaft. N Engl J. 2011;364(18):1728–37.
16. Abrahamsen B, Eiken P, Prieto-Alhambra D, et al. Risk of hip, subtrochanteric, and femoral shaft fractures among mid and long term users of alendronate: nationwide cohort and nested case-control study. BMJ. 2016;353(24):i3365.
17. Lo JC, Zheng P, Grimsrud CD, et al. Racial/ethnic differences in hip and diaphyseal femur fractures. Osteoporos Int. 2014;25(9):2313–8.
18. Giusti A, Hamdy NA, PapapoulosSE. Atypical fractures of the femur and bisphosphonate therapy: a systematic review of case/case series studies. Bone. 2010;47(2):169–80.
19. Kim D, Sung Y-K, Cho S-K, et al. Factors associated with atypical femoral fracture. Rheumatol Int. 2016;36(1):65–71.
20. Shin WC, Moon NH, Jang JH, et al. Anterolateral femoral bowing and loss of thigh muscle are associated with occurrence of atypical femoral fracture: effect of failed tension band mechanism in mid-thigh. J Orthop Sci. 2017;22(1):99–104.
21. Papapoulos S, Lippuner K, Roux C, et al. The effect of 8 or 5 years of denosumab treatment in postmenopausal women with osteoporosis: results from the FREEDOM Extension study. Osteoporos Int. 2015;26(12):2773–83.
22. Rachner TD, Platzbecker U, Felsenberg D, et al. Osteonecrosis of the jaw after osteoporosis therapy with denosumab following long-term bisphosphonate therapy. Mayo Clin Proc. 2013;88(4):418–9.
23. Ramírez L, López-Pintor RM, Casañas E, et al. New non-bisphosphonate drugs that produce osteonecrosis of the jaws. Oral Health Prev Dent. 2015;;13(5):385–93.
24. Ruggiero SL, Dodson TB, Fantasia J; and American Association of Oral and Maxillofacial Surgeons. American Association of Oral and Maxillofacial Surgeons position paper on medication-related osteonecrosis of the jaw—2014 update. J Oral Maxillofac Surg. 2014;72(10):1938–56.
25. Khan AA, Morrison A, Hanley DA, et al. Diagnosis and management of osteonecrosis of the jaw: a systematic review and international consensus. J Bone Miner Res. 2015;30(1):3–23.
26. Christodoulou C, Pervena A, Klouvas G, et al. Combination of bisphosphonates and antiangiogenic factors induces osteonecrosis of the jaw more frequently than bisphosphonates alone. Oncology. 2009;76(3):209–11.
27. Khan AA, Rios LP, Sándor GKB, et al. Bisphosphonate-associated osteonecrosis of the jaw in Ontario: a survey of oral and maxillofacial surgeons. J Rheumatol. 2011;38(7):1396–402.
28. Tennis P, Rothman KJ, Bohn RL, et al. Incidence of osteonecrosis of the jaw among users of bisphosphonates with selected cancers or osteoporosis. Pharmacoepidemiol Drug Saf. 2012;21(8):810–7.
29. Ulmner M, Jarnbring F, Törring O. Osteonecrosis of the jaw in Sweden associated with the oral use of bisphosphonate. J Oral Maxillofac Surg. 2014;72(1):76–82.
30. Abrahamsen B, Eiken P, Prieto-Alhambra D, et al. Surgically treated osteonecrosis and osteomyelitis of the jaw and oral cavity in patients highly adherent to alendronate treatment. ASMBR Annual Meeting, 2016, abstract FR0297.
31. Stopeck AT, Fizazi K, Body J-J, et al. Safety of long-term denosumab therapy: results from the open label extension phase of two phase 3 studies in patients with metastatic breast and prostate cancer. Support Care Cancer. 2016;24(1):447–55.
32. Eriksen EF, Díez-Pérez A, Boonen S. et al. Update on long-term treatment with bisphosphonates for postmenopausal osteoporosis: a systematic review. Bone. 2014;58:126–35.
33. Pazianas M, Cooper C, Ebetino FH, et al. Long-term treatment with bisphosphonates and their safety in postmenopausal osteoporosis. Ther Clin Risk Manag. 2010;6:325–43.
34. Bone HG, Hosking D, Devogelaer JP, et al. Ten years' experience with alendronate for osteoporosis in postmenopausal women. N Engl J Med. 2004;350(12):1189–99.
35. Sorensen OH, Crawford GM, Mulder H, et al. Long-term efficacy of risedronate: a 5-year placebo-controlled clinical experience. Bone. 2003;32(2):120–6.
36. Mellström DD, Sörensen OH, Goemaere S, et al. Seven years of treatment with risedronate in women with postmenopausal osteoporosis. Calcif Tissue Int. 2004;75(6):462–8.
37. Kim DH, Rogers JR, Fulchino LA, et al. Bisphosphonates and risk of cardiovascular events: a meta-analysis. PLoS One. 2015;10(4):e0122646.
38. Barrett-Connor E, Swern AS, Hustad CM, et al. Alendronate and atrial fibrillation: a meta-analysis of randomized placebo-controlled clinical trials. Osteoporos Int. 2012;23(1):233–45.
39. Sharma A, Einstein AJ, Vallakati A, et al. Risk of atrial fibrillation with use of oral and intravenous bisphosphonates. Am J Cardiol. 2014;113(11):1815–21.
40. Dave V, Chiang CY, Booth J, et al. Hypocalcemia post denosumab in patients with chronic kidney disease stage 4-5. Am J Nephrol. 2015;41(2):129–37.
41. Gurney EP, Nachtigall MJ, Nachtigall LE, et al. The Women's Health Initiative trial and related studies: 10 years later: a clinician's view. J Steroid Biochem Mol Biol. 2014;142:4–11.
42. Vahle JL, Sato M, Long GG, et al. Skeletal changes in rats given daily subcutaneous injections of recombinant human parathyroid hormone (1-34) for 2 years and relevance to human safety. Toxicol Pathol. 2002;30(3):312–21.
43. Cipriani C, Capriani C, Irani D, et al. Safety of osteoanabolic therapy: a decade of experience. J Bone Miner Res. 2012;27(12):2419–28.
44. Jerome CP, Johnson CS, Vafai HT, et al. Effect of treatment for 6 months with human parathyroid hormone (1-34) peptide in ovariectomized cynomolgus monkeys (Macaca fascicularis). Bone. 1999;25(3):301–9.
45. Vahle JL, Zuehlke U, Schmidt A, et al. Lack of bone neoplasms and persistence of bone efficacy in cynomolgus

macaques after long-term treatment with teriparatide [rhPTH(1-34)]. J Bone Miner Res. 2008;23(12):2033–9.
46. Neer RM, Arnaud CD, Zanchetta JR, et al. Effect of parathyroid hormone (1-34) on fractures and bone mineral density in postmenopausal women with osteoporosis. N Engl J Med. 2001;344(19):1434–41.
47. Orwoll ES, Scheele WH, Paul S, et al. The effect of teriparatide [human parathyroid hormone (1-34)] therapy on bone density in men with osteoporosis. J Bone Miner Res. 2003;18(1):9–17.
48. Saag KG, Zanchetta JR, Devogelaer J-P, et al. Effects of teriparatide versus alendronate for treating glucocorticoid-induced osteoporosis: thirty-six-month results of a randomized, double-blind, controlled trial. Arthritis Rheum. 2009;60(11):3346–55.
49. Andrews EB, Gilsenan AW, Midkiff K, et al. The US postmarketing surveillance study of adult osteosarcoma and teriparatide: study design and findings from the first 7 years. J Bone Miner Res. 2012;27(12):2429–37.
50. Miller PD, Hattersley G, Riis BJ, et al.; and ACTIVE Study Investigators. Effect of abaloparatide vs placebo on new vertebral fractures in postmenopausal women with osteoporosis: a randomized clinical trial. JAMA. 2016;316(7):722–33.
51. Cosman F, Crittenden DB, Adachi JD, et al. Romosozumab treatment in postmenopausal women with osteoporosis. N Engl J Med. 2016;375(16):1532–43.
52. Schilcher J, Michaelsson K, Aspenberg P. Bisphosphonate use and atypical fractures of the femoral shaft. N Engl J Med. 2011;364(18):1728–37.
53. Kim DH, Rogers JR, Fulchino LH, et al. Bisphosphonates and risk of cardiovascular events: a meta-analysis. PLoS One 2015;10(4):e0122646.

77

Orthopedic Principles of Fracture Management

Manoj Ramachandran[1] and David G. Little[2]

[1]Royal London Hospital, Barts Health NHS Trust, London, UK
[2]Paediatrics and Child Health, University of Sydney, Sydney; andOrthopaedic Research and Biotechnology, The Children's Hospital at Westmead, Westmead, NSW, Australia

INTRODUCTION

A broad range of fractures and associated injuries present to orthopedic and trauma surgeons for management. The principle that fracture immobilization supports both alignment and union of the fracture, while minimizing discomfort, has been known since antiquity [1]. More modern concepts involve internal fixation of fracture fragments to allow early joint motion wherever possible. The energy of the injury, associated soft tissue injury, and displacement of the fracture usually guide the actual intervention.

The initial management of fractures consists of realignment of the broken limb segment and immobilization of the fractured extremity once the initial assessment, evaluation, and management of any life-threatening injury are completed [2]. The aim of fracture treatment is to obtain union of the fracture in the most anatomical position compatible with maximal functional recovery of the extremity or the spine with minimal complications. This is accomplished by obtaining and subsequently maintaining reduction of the fracture with an immobilization technique that allows the fracture to heal and, at the same time, provides the patient with functional aftercare. Either nonoperative or operative means may be used. Any surgical technique, if chosen, should minimize additional soft tissue and bone injury, which can delay fracture healing. Modern fracture fixation methods are thus designed to be minimally invasive.

In open fractures, in addition to the treatment aims outlined, the prevention of infection is vital [3]. This is achieved by urgent wound irrigation and debridement (with serial irrigations and debridements every 24 to 48 hours, often with the use of negative pressure dressings until the wounds are clean and closed), antibiotic administration, and tetanus vaccination. If soft tissue coverage over the injury is inadequate, soft tissue transfers or free flaps are performed when the wound is clean and the fracture is definitively fixed.

BIOLOGY OF FRACTURE HEALING

Fracture healing involves a combination of inflammatory, vascular, anabolic, and catabolic events [4]. Briefly, the acute fracture sets off an inflammatory response with the recruitment of cells of multiple lineages. In the absence of infection, the profile of the cellular response rapidly involves the recruitment of macrophages and osteoclasts to deal with debris, while an immature blastema of undifferentiated mesenchymal cells is recruited from the surrounding periosteum, soft tissues, and possibly from the circulation. An active restoration of the damaged blood supply via angiogenesis is a major part of any successful wound repair, and its deficiency may lead to failure of repair in patients with severe local tissue damage or in smokers or those with systemic disease such as diabetes, and several other factors [5]. When there is motion at the fracture site, the central region of the fracture callus differentiates into a cartilage model. This model undergoes endochondral ossification to achieve boney union, which is also a vascular event.

Primer on the Metabolic Bone Diseases and Disorders of Mineral Metabolism, Ninth Edition. Edited by John P. Bilezikian.
© 2019 American Society for Bone and Mineral Research. Published 2019 by John Wiley & Sons, Inc.
Companion website: www.wiley.com/go/asbmrprimer

Absolutely rigid fracture healing conditions may prevent this cartilage model from forming, resulting in healing by intramembranous ossification [6]. In the longer term, the woven bone initially repairing the fracture is remodelled into lamellar bone. This process may take years.

Fracture healing may be delayed by local or systemic factors. Open and high-energy fractures, particularly of the tibia, are prone to delay in healing, as are fractures fixed with a large interfragmentary gap. Smoking, diabetes, chronic corticosteroid use, and poor nutrition are some of the systemic factors known to affect the speed of fracture healing. Osteoporosis has not been definitively shown to be associated with delayed union—the above comorbidities may be more important. The US Food and Drug Administration (FDA) define nonunion as more than 9 months without healing and no progress over the last 3 months. The term delayed union does not have a definitive time frame as it differs depending on the site involved. A delayed union is simply a delay in fracture healing beyond what would normally be appropriate for any given fracture. Delayed union is clearly a risk for nonunion. Another definition of nonunion is a fracture that will not heal without intervention.

TREATMENT PRINCIPLES

Fracture management can be divided into nonoperative and operative techniques. The nonoperative technique consists of a closed reduction (required if the fracture is significantly displaced or angulated), achieved by applying traction to the long axis of the injured limb and then reversing the mechanism of injury [7]. This is followed by a period of immobilization with casting. Casts are made from plaster of Paris or synthetic materials. Complications of casts can include the development of pressure ulcers, thermal burns during plaster hardening, and joint stiffness. Traction alone (skin or skeletal) is rarely used nowadays for definitive fracture management.

If the fracture cannot be reduced (eg, due to soft tissue interposition), surgical intervention is likely to be required. Indications for surgery include failed nonoperative management, unstable fractures that cannot be adequately maintained in a reduced position, displaced intra-articular fractures (>2 mm), impending pathological fractures, unstable or complicated open fractures, fractures in growth areas in skeletally immature individuals that have increased risk for growth arrest, and nonunions or malunions that have failed to respond to nonoperative treatment [8]. Contraindications to internal fixation include active infection (local or systemic), soft tissues that compromise the overlying fracture or the surgical approach (eg, burns and previous surgical scars), medical conditions that contraindicate surgery or anesthesia (eg, recent myocardial infarction), and cases where amputation would be more appropriate (eg, severe neurovascular injury).

SURGICAL OPTIONS

The treatment goals for surgical fracture management, as outlined by the Association for the Study of Internal Fixation (known internationally as AO), are anatomical reduction of the fracture fragments (for the correction of length, angulation, and rotation, and for intra-articular fractures for the restoration of the joint surface), stable internal fixation sufficient to cope with physiological biomechanical demands, preservation of blood supply, and active, pain-free mobilization of adjacent muscles and joints [9]. The objectives of open reduction and internal fixation include exposure and reduction of the fracture, followed by maintenance of the reduction by stabilization using one or a combination of the methods below.

Kirschner wires

Kirschner wires, or K-wires, placed percutaneously or through a mini-open approach, are commonly used for temporary and definitive treatment of fractures. However, they have poor resistance to torque and bending forces, and rely on friction with the bone for maintenance of reduction. Therefore, when they are used as the sole form of fixation, casting or splinting is used in conjunction. K-wires are commonly used as adjunctive fixation for plates and screws that involve fractures around joints.

Plates and screws

Plates and screws are commonly used in the management of articular fractures as they provide strength and stability to neutralize the forces on the injured limb for functional postoperative aftercare. Plate designs vary, depending on the anatomical region and size of the bone the plate is used for. All plates should be applied with minimal stripping of the soft tissue. Five main plate designs exist:

- *Buttress (antiglide) plates* counteract the compression and shear forces that commonly occur with fractures that involve the metaphysis and epiphysis. These plates are commonly used with interfragmentary screw fixation and require anatomical contouring to achieve stable fixation.
- *Compression plates* counteract bending, shear, and torsional forces by providing compression across the fracture site via the eccentrically loaded holes in the plate. Compression plates are commonly used in the long bones, especially the fibula, radius, and ulna, and in nonunion or malunion surgery.
- *Neutralization plates* are used in combination with interfragmentary screw fixation. The interfragmentary compression (lag) screws provide compression at the fracture site. This plate function neutralizes bending,

shear, and torsional forces on the lag screw fixation, as well as increasing the stability of the construct. These plates are commonly used for fractures involving the fibula, radius, ulna, and humerus.
- *Bridge plates* are useful in the management of multi-fragmented diaphyseal and metaphyseal fractures. Indirect reduction techniques are preferred in bridge plating without disrupting the soft tissue attachments to the bone fragments.
- *Tension band plates* convert tension forces into compressive forces, thereby providing stability (eg, in oblique olecranon fractures).

Locking plates are now commonplace in fixation constructs. A locking plate acts like an internal fixator. There is no need to exactly anatomically contour the plate onto the bone, thus reducing bone necrosis and allowing for a minimally invasive technique. Locking screws directly anchor and lock onto the plate, thereby providing angular and axial stability. These screws are incapable of toggling, sliding, or becoming dislodged, thus reducing the possibility of a secondary loss of reduction, as well as eliminating the possibility of intraoperative overtightening of the screws. The locking plate is indicated for osteoporotic fractures, for short and metaphyseal segment fractures, and for bridging comminuted areas. These plates are also appropriate for metaphyseal areas where subsidence may occur or prostheses are involved.

A range of anatomically contoured plates for each bone are now available. The technique of minimally invasive percutaneous plate osteosynthesis (MIPPO) with indirect reduction is becoming increasingly popular. This involves the use of anatomically preshaped plates and instrumentation to safely and effectively insert the plate percutaneously or through limited incisions [10]. Advantages of MIPPO may include faster bone healing, reduced infection rates, decreased need for bone grafting, less postoperative pain, faster rehabilitation, and more esthetic results. Some disadvantages include difficulty with indirect reduction, increased radiation exposure due to more intraoperative radiographs being utilized, malunion, pseudoarthrosis through diastases, and delayed union with flexible fixation in simple fractures.

Intramedullary nails

These nails operate like an internal splint that shares the load with the bone and can be flexible or rigid, locked or unlocked, and reamed or unreamed. Locked intramedullary nails provide relative stability to maintain bone alignment and length and to limit rotation. Ideally, the intramedullary nail allows for compressive forces at the fracture site, which stimulates bone healing.

Intramedullary nails are commonly used for femoral and tibial diaphyseal fractures and can be employed to stabilize humeral diaphyseal fractures. The advantages of intramedullary nails include minimally invasive procedures, early postoperative ambulation, and early range of motion being permitted in adjacent joints. Reaming may also increase union rates, possibly by providing the equivalent of a bone graft to the fractured region. The realization that reaming may increase union rates has led to a decrease in the use of unreamed nails.

External fixation

External fixation provides fracture stabilization at a distance from the fracture site, without interfering with the soft tissue structures that are near the fracture. This technique not only provides stability for the extremity and maintains bone length, alignment, and rotation without requiring casting, but it also allows for inspection of the soft tissue structures that are vital for fracture healing. Indications for external fixation (temporarily or as definitive care) are as follows [10]:

- Open fractures that have significant soft tissue disruption (eg, type II or III open fractures).
- Soft tissue injury (eg, burns).
- Pelvic fractures.
- Severely comminuted and unstable fractures.
- Fractures associated with bony defects.
- Limb-lengthening and bone transport procedures.
- Fractures associated with infection or nonunion.

Complications of external fixation include pin tract infection, pin loosening or breakage, interference with joint motion due to the pins transfixing soft tissues, neurovascular damage during pin placement, malalignment caused by poor placement of the fixator, delayed union, and malunion. Modern external fixators, such as the Taylor Spatial Frame, allow rapid placement on the day of presentation and subsequent postoperative adjustment to effect more anatomical reduction of the fracture in the weeks following intervention.

FRACTURES IN OSTEOPOROTIC PATIENTS

Vertebral compression fractures, Colles (distal radius) fractures, hip fractures, and other peripheral (nonvertebral) fractures all occur in patients with osteoporosis. These fractures may indicate the need for treatment of osteoporosis for secondary fracture prevention.

Osteoporotic fractures are usually low-energy injuries. Some fractures such as fatigue fractures of the pelvis have no displacement. Colles fractures and hip fractures are usually displaced and require reduction and fixation. Colles fractures are sometimes held in a cast, but may require wire fixation or low-profile plating to maintain reduction. Colles fractures nearly always heal, but significant malunion can interfere with function. Intertrochanteric hip fractures are usually fixed with sliding hip screw devices that allow compression of the fracture fragments on weight bearing; intramedullary devices can provide similar results [9]. Subcapital neck of femur fractures require joint arthroplasty of some form in the majority of cases because of the high incidence of nonunion and avascular necrosis. The functional status of the patient may determine whether a hemi-, bipolar or total hip arthroplasty are required. In recent times the development of locking plate technology has improved the surgeon's ability to internally fix fractures in osteoporotic bone, but further research is needed as in many cases optimal fixation cannot be achieved. Augmentation techniques such as adding poly(methyl methacrylate) (PMMA) to aid in screw fixation have also been introduced [11]. It is a principle of management for all osteoporotic fractures to institute load bearing and functional tasks as quickly as possible to minimize loss of function or mobility.

Intervention for vertebral fractures in osteoporotic patients

Acute vertebral compression fractures can be painful and lead to disability, whereas multiple "silent" compression fractures lead to kyphosis and loss of height. While in many clinically apparent fractures the pain settles in a few weeks, it has been estimated that one-third of fractures can remain chronically painful. While most vertebral compression fractures are managed nonoperatively, there has been an increase in intervention for painful acute fracture to minimize morbidity. These techniques are known as vertebroplasty and kyphoplasty.

In vertebroplasty, a percutaneous approach to the vertebral body is made under fluoroscopic control, either through or adjacent to the pedicles. Bone cement, usually PMMA, is injected into the fracture under pressure via a cannula while the cement is in a fluid state. This acutely stabilizes the fracture and results in an immediate reduction of pain which is significant enough in many cases to allow immediate return to activities of daily living. Pain reduction is thought to be from stabilization of the fracture, although heat necrosis of nerve endings from the exothermic setting of the cement has also been suggested.

Vertebroplasty makes little difference to the spinal alignment as the injection of cement usually does little to change the wedging deformity of the fracture. Kyphoplasty is designed to address these limitations. In this technique, cannulae are placed usually bilaterally to allow the introduction of balloon tamps, which are expanded with radioopaque saline. Some elevation of the end plate and thus correction of deformity can be achieved by this method. The balloon expansion creates a void into which cement is injected in a slightly more viscous state than in vertebroplasty. The literature available suggests that while correction of vertebral morphology is achievable, overall spinal balance is usually not affected as alterations in shape can be accommodated by disk spaces and deformity at other levels [12, 13].

When one considers vertebroplasty and kyphoplasty versus conservative care, both procedures appear to be effective in relieving pain in a few days in the majority of individuals. One nonrandomized study showed decreases in pain, rapid return to function, and decreased hospital stay in vertebroplasty versus conservatively treated patients [14]. A systematic review favored a therapeutic effect and some superiority for kyphoplasty over nonoperative treatment [15]. There is a suggestion from this meta-analysis that kyphoplasty may be associated with fewer cement leakage events than vertebroplasty. Serious complications include neurological sequelae and have been reported to run at about 1%.

Recent randomized trials of vertebroplasty have placed these apparent benefits into doubt. Two separate sham-controlled vertebroplasty trials showed no benefit of active over sham treatment [16, 17]. These trials have been criticized for including patients who were not "acute," however a study using the data of both trials found no difference even in the acute subset [18]. There has been a significant decline in the use of vertebroplasty and kyphoplasty in the United States over a recent 10-year period according to one study [19]. Further study of the best evidence-based treatment for vertebral fractures is ongoing.

REFERENCES

1. Smith Papyrus. https://www.nlm.nih.gov/news/turn_page_egyptian.html (accessed May 2018).
2. American College of Surgeons. *ATLS Student Course Manual: Advanced Trauma Life Support* (9th ed.). Chicago: American College of Surgeons, 2012
3. Gustilo RB, Merkow RL, Templeman D. The management of open fractures. J Bone Joint Surg Am. 1990;72(2):299–304.
4. Schindeler A, McDonald MM, Bokko P, et al. Bone remodeling during fracture repair: The cellular picture. Semin Cell Dev Biol. 2008;19(5):459–66.
5. Zura R, Xiong Z, Einhorn T, et al. Epidemiology of fracture nonunion in 18 human bones. JAMA Surg. 2016;151(11):e162775
6. Thompson Z, Miclau T, Hu D, et al. A model for intramembranous ossification during fracture healing. J Orthop Res. 2002;20(5):1091–8.
7. Charnley, J. *The Closed Treatment of Common Fractures* (4th ed.). Cambridge, UK: Greenwich Medical Media, 1999.
8. Canale ST. *Campbell's Operative Orthopaedics* (10th ed.). St. Louis: Mosby-Year Book; 2003.
9. Ruedi TP, Buckley R, Moran C (eds). *AO Principles of Fracture Management* (2nd ed.). New York: Thieme Medical Publishers, 2007.
10. Krettek C, Schandelmaier P, Miclau T, et al. Minimally invasive percutaneous plate osteosynthesis (MIPPO) using the DCS in proximal and distal femoral fractures. Injury. 1997;28(suppl 1):A20–30.

11. Bucholz RW, Heckman JD, Court-Brown C, et al. (eds). *Rockwood and Green's Fractures in Adults* (6th ed.). Philadelphia: Lippincott Williams and Wilkins, 2005.
12. Kammerlander C, Neuerburg C, Verlaan JJ, et al. The use of augmentation techniques in osteoporotic fracture fixation. Injury. 2016;47(suppl 2):S36–43.
13. Pradhan BB, Bae HW, Kropf MA, et al. Kyphoplasty reduction of osteoporotic vertebral compression fractures: correction of local kyphosis versus overall sagittal alignment. Spine. 2006;31:435–41.
14. Diamond TH, Bryant C, Browne L, et al. Clinical outcomes after acute osteoporotic vertebral fractures: a 2-year non-randomised trial comparing percutaneous vertebroplasty with conservative therapy. Med J Aust. 2006;184:113–7.
15. Taylor RS, Fritzell P, Taylor RJ. Balloon kyphoplasty in the management of vertebral compression fractures: an updated systematic review and meta-analysis. Eur Spine J. 2007;16:1085–100.
16. Buchbinder R, Osborne RH, Ebeling PR, et al. A randomized trial of vertebroplasty for painful osteoporotic vertebral fractures. N Engl J Med. 2009;361(6):557–68.
17. Kallmes DF, Comstock BA, Heagerty PJ, et al. A randomized trial of vertebroplasty for osteoporotic spinal fractures. N Engl J Med. 2009;361(6):569–79.
18. Staples MP, Kallmes DF, Comstock BA, et al. Effectiveness of vertebroplasty using individual patient data from two randomised placebo controlled trials: meta-analysis. BMJ. 2011;343:d3952.
19. Hirsch JA, Chandra RV, Pampati V, et al. Analysis of vertebral augmentation practice patterns: a 2016 update. J Neurointerv Surg. 2016;pii: neurintsurg-2016-012767 (epub ahead of print).

78

Adherence to Osteoporosis Therapies

Stuart L. Silverman[1] and Deborah T. Gold[2]

[1]Department of Medicine, Division of Rheumatology, Cedars-Sinai Medical Center, UCLA David Geffen School of Medicine and the OMC Clinical Research Center, Los Angeles, CA, USA
[2]Departments of Psychiatry and Behavioral Sciences, Sociology, and Psychology and Neuroscience, Center for the Study of Aging and Human Development, Duke University Medical Center, Durham, NC, USA

INTRODUCTION

Osteoporosis therapies reduce the risk of osteoporotic-related fractures [1]. To be effective, pharmaceutical treatment needs to be started in a timely fashion and consistently maintained [2, 3]. Real-world clinical practice studies have shown that patients who continue (ie, are persistent) with osteoporosis therapy have a reduced risk of fracture, hospitalization, and morbidity [4–6]. Compared with persistent patients, risk for fractures can increase up to 45% when patients are not persistent [4, 5, 7–9].

Poor persistence to all medications is an overwhelming public health problem and cause for serious concern. The poor persistence seen with osteoporosis medications is no different from that of medications for other asymptomatic or symptomatic conditions such as hypertension, diabetes, or asthma [10]. Patients with poor persistence to osteoporosis medications also are nonpersistent with nutritional interventions such as calcium [11] or with medications such as statins for other asymptomatic diseases [12].

DEFINING NONPERSISTENCE

Research on medication behaviors has been inconsistent in its terminology. In order to minimize confusion, we use the terminology suggested by the International Society for Pharmacoeconomics and Outcomes Research in 2008; we define persistence as the duration of time from initiation to discontinuation of therapy [13]. Given the complexity of this literature and the many different ways in which terms are used throughout that literature, it seems important to specify this definition. More specifically, there are two ways in which a patient can be identified as nonpersistent.

The patient who starts a medication and then stops it without physician recommendation engages in *secondary* nonpersistence. It is secondary nonpersistence that has received the bulk of the research attention in the last decade. But recently, several studies have suggested that primary nonpersistence may be as important as or more important than secondary nonpersistence for the overall treatment effectiveness of osteoporosis medications. *Primary* nonpersistence occurs when the patient refuses to even start the medication or does not even fill the prescription [14]. According to Reynolds and colleagues, primary nonadherence occurs when a provider orders a medication for a patient and the order is never picked up or dispensed. Primary nonpersistence is more important than has been recognized. In a close-ended system where prescriptions had no cost, up to 30% of patients did not fill the first prescription [14].

OPTIMIZING PERSISTENCE

Historically, optimizing compliance with all osteoporosis therapies has been challenging, especially with the oral bisphosphonates which require shorter dosing intervals and more complex regimens (eg, taking them first thing in the morning with 8oz of plain water and not eating or drinking for 30 to 60 minutes afterward). Patients found these requirements intrusive. More recently, research has found that overall persistence appears improved with the introduction of new medications and different delivery

---

Primer on the Metabolic Bone Diseases and Disorders of Mineral Metabolism, Ninth Edition. Edited by John P. Bilezikian.
© 2019 American Society for Bone and Mineral Research. Published 2019 by John Wiley & Sons, Inc.
Companion website: www.wiley.com/go/asbmrprimer

systems that have longer intervals between doses [15–19]. However, others report that—regardless of dosing options—approximately half of women discontinue osteoporosis treatment in their first year of therapy [20]; and only 44% of patients in one study taking weekly or monthly oral bisphosphonates were compliant with dosing instructions [21]. Numerous studies have examined persistence with oral bisphosphonates, reporting 12-month persistence of between 16% and 61% in postmenopausal women with osteoporosis [9, 22]. Long-term persistence data beyond 12 months are sparse. In one long-term study of Dutch patients, cumulative persistence rates for oral osteoporosis medications (including alendronate, risedronate, ibandronate, etidronate, raloxifene, and strontium ranelate) declined over time from 71% after 6 months, 60% after 1 year, to 27% after 5 years in women [23]. In a study using US claims databases, persistence with oral bisphosphonates was 20% during a 24-month study period [5].

When injectable osteoporosis medications administered by a health care provider became available (i.v. zoledronic acid, i.v. ibandronate, and s.c. denosumab), many thought that persistence would improve because medication administration would be managed by the health care professional. Injectables do have better persistence than oral medications but have not significantly improved medication behaviors. In two separate studies of zoledronic acid administered as intravenous infusions annually, persistence (defined as returning for a second infusion) was only 36% in elderly patients in Korea [24] and 68% in a US Medicare retrospective database analysis [25]. Intravenous ibandronate was shown to have better adherence than an oral bisphosphonate [26]. In two other separate observational studies of adherence in the United States and Europe, 12-month adherence to denosumab was better than historical data on oral bisphosphonates, but still suboptimal [27, 28]. In a Swedish study, persistence with denosumab was better than that with an oral bisphosphonate [29]. Surprisingly, osteoporosis medication self-injected daily by the patient did have good adherence. In two observational studies of self-infected teriparatide (Forteo; Eli Lilly, Indianapolis, IN, USA), 86% and 87% of patients in the United States [30] and United Kingdom [31], respectively, were persistent at 12 months. This may relate to the availability of patient support programs or may have included patients whose motivation to take the medications had a higher self-perceived risk of fracture.

However, poor persistence is a major problem that will not solve itself or be solved with a quick fix. Persistence with osteoporosis medications is a complex and dynamic process that can only change over time. Patients may stop and restart the medication for no apparent reasons [32]; or they may stop one medication and switch to another. Unfortunately switching medications may lower overall persistence with osteoporosis therapy [33, 34]. And, importantly, it has become clear that reasons for early discontinuation may differ from those of later discontinuation [35].

REASONS FOR NONPERSISTENCE

Researchers have sought for over a decade to explain nonpersistence with osteoporosis medications. These authors feel strongly that nonpersistence is not simply forgetfulness but is a choice made by patients for a multitude of reasons [36]. Physicians must not assume persistence in a patient for whom an osteoporosis medication is prescribed without some effort to determine whether that patient is taking the medication as directed. This choice to take medications as prescribed is based on a multifaceted social construct [37] where understanding, choice, risk/benefit ratio, and perceived need all interact to result in unpredictable patterns of usage and acceptability. We are now perhaps closer to understanding this social construct than ever before. We now understand that there are both external factors (media, social media, the internet, family and peer groups) and internal factors (individual beliefs regarding risks and benefits regarding taking osteoporosis medications).

Patient perception of risk/benefit is influenced by both perceived and actual side effects. Many side effects such as gastrointestinal adverse events lower the likelihood of persistence [38]. Perception of the risk/benefit ratio of osteoporosis medication also depends on the individual patient's severity of perceived need and evaluation of personal risk. Many patients believe that osteoporotic fractures are a natural consequence of aging rather than the manifestation of a chronic and treatable disease. Fractures are not seen as the consequence of a fragile skeleton but as the result of falls.

Further, it is critical that, as health care professionals, we make every effort to understand the health care beliefs and motivations of our patients [39]. Many may believe that they can prevent or treat osteoporosis with a more natural approach to treatment (eg, just calcium and vitamin D); they may also believe that they as individuals are more prone to medication side effects. Some patients may be resistant to any change. Furthermore, treatment satisfaction may play a role in a patient's willingness to continue treatment [40].

Patients are influenced by not only media (print, television, radio) [41], but also by social media, and may rely on the internet and others without health care training for information about risks and benefits [42]. Unfortunately, we lack the tools to assess the health benefits of these patients. No simple algorithm to identify the barriers to persistence in these patients or to correct them exists.

SUMMARY

How do we solve the problem of nonpersistence with osteoporosis medication? Interventions to improve persistence have typically failed. Single variable interventions have frequently failed [43, 44]. Positive results seen in the study by Clowes and colleagues [45] with monitoring have not been replicated since [46, 47]. One future point of intervention may be in communication content and style between health care provider and patient. If the patient lacks trust in the health care provider, information provided on risks and benefits from that source will not be accepted by the patient [48]. Yet we know that it is important for the health care providers to make strong positive recommendations to patients about the need for therapy in general and for the specific therapy that they prescribed.

Solving major public health problems such as nonpersistence with osteoporosis medication is truly an intellectual and pragmatic challenge. We will need time, sometimes a long time, to effect this level of change. Humans do not alter health behaviors overnight. For example, cigarette smoking was deemed a significant public health problem, and powerful efforts to change this behavior began in 1965. Yet it has taken over 50 years to significantly reduce the proportion of smokers in the US population. Despite the clear relationship between smoking and major health problems such as lung cancer and heart disease, the American public moved slowly in embracing this change. Similarly, despite the close relationship between low bone density and/or history of fracture and future risk of fracture, the majority of patients at risk have not yet embraced this relationship and begun to take osteoporosis medications as directed [49].

CONCLUSION

Factors associated with non-persistence with osteoporosis medications are better known and understood now than they were a decade ago but some remain a black box. Improving persistence remains a challenge in view of the often conflicting information that patients receive from healthcare professionals, media and social media. We must accept that we will not achieve 100% persistence in our patients. Some of our patients are simply resistant to our concepts or recommendations. Instead of trying to improve the persistence of all patients with osteoporosis, we need to focus our interventions on those patients willing and likely to accept change.

Follow-up with knowledgeable and trusted health care providers to provide reinforcement and to talk to the patient to identify and surmount barriers is helpful.

REFERENCES

1. Silverman S, Christiansen C. Individualizing osteoporosis therapy. Osteoporos Int. 2012;23:797–809.
2. Sampalis JS, Adachi JD, Rampakakis E, et al. Long-term impact of adherence to oral bisphosphonates on osteoporotic fracture incidence. J Bone Miner Res. 2012;27:202–10.
3. Meijer WM, Penning-van Beest FJ, Olson M, et al. Relationship between duration of compliant bisphosphonate use and the risk of osteoporotic fractures. Curr Med Res Opin. 2008;11:3217–22.
4. Halpern R, Becker L, Iqbal SU, et al. The association of adherence to osteoporosis therapies with fracture, all-cause medical costs, and all-cause hospitalizations: a retrospective claims analysis of female health plan enrollees with osteoporosis. J Manag Care Pharm. 2011;17:25–39.
5. Siris ES, Harris ST, Rosen CJ, et al. Adherence to bisphosphonate therapy and fracture rates in osteoporotic women: relationship to vertebral and nonvertebral fractures from 2 US claims databases. Mayo Clin Proc. 2006;81:1013–22.
6. Mikyas Y, Agodoa I, Yurgin N. A systematic review of osteoporosis medication adherence and osteoporosis-related fracture costs and men. Appl Health Econ Health Policy. 2014; 12:267–77.
7. Ross S, Samuels E, Gairy K, et al. A meta-analysis of osteoporotic fracture risk with medication nonadherence. Value Health. 2011;14:571–81.
8. Siris ES, Pasquale MK, Wang Y, et al. Estimating bisphosphonate use and fracture reduction among US women aged 45 years and older, 2001–2008. J Bone Miner Res. 2011;26:3–11.
9. Siris ES, Selby PL, Saag KG, et al. Impact of osteoporosis treatment adherence on fracture rates in North America and Europe. Am J Med. 2009;122:S3–13.
10. Rolnick SJ, Pawloski PA, Hedblom BD, et al. Patient characteristics associated with medication adherence. Clin Med Res. 2013;11:54–65.
11. Barrett-Connor E, Wade SW, Downs RW, et al. Self-reported calcium use in a cohort of postmenopausal women receiving osteoporosis therapy: Results from POSSIBLE US. Osteoporos Int. 2015;26:2175–84.
12. Curtis JR, Xi J, Westfall AO, et al. Improving the prediction of medication compliance: the example of bisphosphonates for osteoporosis. Med Care. 2009;47(3):334–41.
13. Cramer JA, Roy A, Burrell A, et al. Medication compliance and persistence: terminology and definitions. Value Health. 2008;11(1):44–7.
14. Reynolds K, Muntner P, Cheetham TC, et al. Primary non-adherence to bisphosphonates in an integrated healthcare setting. Osteoporos Int. 2013;24(9):2509–17.
15. Cramer JA, Lynch NO, Gaudin AF, et al. The effect of dosing frequency on compliance and persistence with bisphosphonate therapy in postmenopausal women: a comparison of studies in the United States, the United Kingdom, and France. Clin Ther. 2006;28:1686–94.
16. Lappe JM. Nonadherence to osteoporosis medications. Clin Rev Bone Miner Metab. 2006;4:25–32.
17. Kamatari M, Koto S, Ozawa N, et al. Factors affecting long-term compliance of osteoporotic patients with bisphosphonate treatment and QOL assessment in actual practice: alendronate and risedronate. J Bone Miner Metab. 2007;25:302–9.
18. Payer J, Killinger A, Sulkova I, et al. Preferences of patients receiving bisphosphonates how to influence the therapeutic adherence. Biomed Pharmacother. 2008;62:122–4.
19. Iglay K, Cao X, Mavros P, et al. Systematic literature review and meta-analysis and medication adherence with once weekly versus once daily therapy. Clin Ther. 2015;37:1813–21.

20. McHorney CA, Schousboe JT, Cline RR, et al. The impact of osteoporosis medication beliefs and side-effect experiences on non-adherence to oral bisphosphonates. Curr Med Res Opin. 2007;23:3137–52.
21. Wtrisalova M, Touskova T, Ladova K, et al. Adherence to oral bisphosphonates: 30 more minutes in dosing instructions matter. Climacteric. 2015;18:608–16.
22. Rabenda V, Mertens R, Fabri V, et al. Adherence to bisphosphonates therapy and hip fracture risk in osteoporotic women. Osteoporos Int. 2008;19:811–8.
23. vanBoven JF, de Boer PT, Postma MJ, et al. Persistence with osteoporosis medication among newly-treated osteoporotic patients. J Bone Miner Metab. 2013;31:562–70.
24. Lee YK, Nho JH, Ha YC, et al. Persistence with intravenous zoledronate in elderly patients with osteoporosis. Osteoporos Int. 2012;23:2329–33.
25. Curtis JR, Yun H, Matthews R, et al. Adherence with intravenous zoledronate and intravenous ibandronate in the United States Medicare population. Arthritis Care Res. 2012;64:1855–63.
26. Hadji P, Felsenberg D, Amling M, et al. The non-interventional BonViva Intravenous Versus Alendronate (VIVA) study: real-world adherence and persistence to medication, efficacy, and safety, in patients with postmenopausal osteoporosis. Osteoporos Int. 2014;25:339–47.
27. Silverman SL, Siris E, Kendler DL, et al. Persistence at 12 months with denosumab in postmenopausal women with osteoporosis: Interim results from a perspective observational study. Osteoporos Int. 2015;26:361–7.
28. Hadji P, Papaioannou N, Gielen E, et al. Persistence, adherence and medication taking behavior in women with postmenopausal osteoporosis receiving denosumab in routine practice in Germany, Austria, Greece and Belgium: 12 month results from a European non-interventional study. Osteoporos Int. 2015;26:2479–89.
29. Karlsson L, Lundkvist J, Psachoulia E, et al. Persistence with denosumab and persistence with oral bisphosphonates for the treatment of postmenopausal osteoporosis: a retrospective, observational study and a meta-analysis. Osteoporos Int. 2015;26:2401–11.
30. Gold DT, Weinstein DL, Pohl G, et al. Factors associated with persistence with teriparatide therapy: results from the DANCE observational study. J Osteoporos. : 2011; article ID 314970.
31. Arden NK, Earl S, Fisher DJ, et al. Persistence with teriparatide in patients with osteoporosis: the UK experience. Osteoporos Int. 2006;17:1626–9.
32. Klop C, Welsing PM, Elders PJ, et al. Long-term persistence with anti-osteoporosis drugs after fracture. Osteoporos Int. 2015;26(6):1831–40.
33. Li L, Roddam A, Ferguson S, et al. Switch patterns of osteoporosis medication and its impact on persistence among postmenopausal women in the UK general practice research database. Menopause. 2014;21:1106–13.
34. Wade SW, Satram-Hoang S, Stolshek B. Long-term persistence and switching patterns mold women using osteoporosis therapies: 24 and 36 months results from POSSIBLE US. Osteoporos Int. 2014;25:2279–90.
35. Carbonell-Abella C, Pages-Castella A, Javaid MK, et al. Early (1-year) discontinuation of different anti-osteoporosis medications compared: a population-based cohort study. Calcif Tissue Int. 2015;97(6):535–41.
36. Silverman SL, Schousboe JT, Gold DT. Oral bisphosphonate compliance and persistence: a matter of choice? Osteoporos Int. 2011;22:21–6.
37. Salter C, McDaid L, Bhattacharya D, et al. Abandoned acid? Understanding adherence to bisphosphonate medications for the prevention of osteoporosis among older women: a qualitative longitudinal study. PLoS One. 2014;9:e83552.
38. Modi A, Gold DT, Yang X, et al. Association between gastrointestinal events and healthcare resource utilization among patients with osteoporosis: analysis of a managed care population in the US. J Manag Care Spec Pharm. 2015;21(9):811–23.
39. Lindsay BR, Olufade T, Bauer J, et al. Patient reported barriers to osteoporosis therapy. Arch Osteoporos. 2016;11:19.
40. Palacios S, Agodoa I, Bonnick S, et al. Treatment satisfaction in postmenopausal women suboptimally inherent to bisphosphonates who transitioned to denosumab compared with risedronate or ibandronate. Clin Endocrinol Metab. 2015;100(3):E487–92.
41. Barozzi N, Peeters GG, Tett SE. Actions following adverse drug events—how do these influence uptake and utilisation of newer and/or similar medications? BMC Health Serv Res. 2015;15:498.
42. Silverman S, Calderon A, Kaw K, et al. Patient weighting of osteoporosis medication attributes across racial and ethnic groups: a study of osteoporosis medication preferences using conjoint analysis. Osteoporos Int. 2013;24:2067–77.
43. Hiligsmann M, Salas M, Hughes DA, et al. Interventions to improve osteoporosis medication adherence and persistence: a systematic review and literature by the ISPOR medication adherence and persistence special interest group. Osteoporos Int. 2013;24:2907–18.
44. Solomon DH, Iversen MD, Avorn J, et al. Osteoporosis telephonic intervention to improve medication adherence: a large pragmatic randomized controlled trial. Arch Int Med. 2012;26:477–83.
45. Clowes JA, Peel NF, Eastell R. The impact of monitoring on adherence and persistence with antiresorptive treatment for postmenopausal osteoporosis: a randomized controlled trial. J Clin Endocrinol Metab. 2004;89(3):1117–23.
46. Bianchi ML, Duca P, Vai S, et al. Improving adherence and persistence with oral therapy of osteoporosis. Osteoporos Int. 2015;26:1629–38.
47. Tuzun S, Akyuz G, Eskiyurt N, et al. Impact of the training on the compliance and persistence of weekly bisphosphonate treatment in postmenopausal osteoporosis: a randomized controlled study. Int J Med Sci. 2013;10:1880–7.
48. Prose NS. Paying attention. JAMA. 2000;283:2763.
49. Solomon DH, Johnston SS, Boytsov NN, et al. Osteoporosis medication use after hip fracture in U.S. patients between 2002 and 2011. J Bone Miner Res. 2014;29(9):1929–37.

79
Cost-Effectiveness of Osteoporosis Treatment

Anna N.A. Tosteson

Multidisciplinary Clinical Research Center in Musculoskeletal Diseases, The Dartmouth Institute for Health Policy and Clinical Practice; and Department of Medicine, Geisel School of Medicine, Dartmouth College, Lebanon, NH, USA

INTRODUCTION

Osteoporosis affects a large proportion of the elderly population and is associated with fractures that are costly in both human and economic terms [1]. In 2005, the US population sustained an estimated 2 million incident fractures at a cost of US$16.9 billion [2]. With annual fracture-related expenditures projected to increase to US$25.3 billion within the next 10 years, there is widespread recognition that rising elderly populations and constrained health care budgets will continue to challenge health care systems to find cost-effective approaches to osteoporosis care. Cost-effectiveness analysis is a form of economic evaluation that estimates the value of an intervention by weighing the expected net increase in cost of an intervention against its expected net gain in health [3]. The rationale for cost-effectiveness analysis is that when health care resources are limited, expenditures should be planned to maximize health outcomes within available resources. The cost-effectiveness of new treatments relative to current care standards is one attribute that policy-makers may consider when making formulary coverage decisions. In this chapter, the methodology of cost-effectiveness analysis is described, recent developments in the cost-effectiveness of osteoporosis care are discussed, and key findings are highlighted.

OVERVIEW OF METHODS FOR COST-EFFECTIVENESS ANALYSIS

Cost-effectiveness ratio

The incremental cost-effectiveness ratio, or ICER, which estimates expected cost per unit of health gained, is the primary outcome measure used to characterize value in cost-effectiveness studies. Consider two alternative treatments, A and B, where the average cost of A is higher than the average cost of B, then the ICER is defined as follows:

$$\text{ICER} = \frac{(\text{Cost}_A - \text{Cost}_B)}{(\text{Effectiveness}_A - \text{Effectiveness}_B)}$$

Using this definition, the value of each more costly intervention is judged relative to the improvement in health that it provides over and above health outcomes associated with the less costly alternative.

Choice of comparator

When assessing the cost-effectiveness of a new osteoporosis intervention, the standard of care that is used as the basis for comparison (ie, the comparator) may have a

---

Primer on the Metabolic Bone Diseases and Disorders of Mineral Metabolism, Ninth Edition. Edited by John P. Bilezikian.
© 2019 American Society for Bone and Mineral Research. Published 2019 by John Wiley & Sons, Inc.
Companion website: www.wiley.com/go/asbmrprimer

marked impact on the intervention's estimated value. Cost-effectiveness analyses of osteoporosis prevention conducted prior to 2002, when the Women's Health Initiative findings were published [4], typically included hormone therapy as a comparator. The choice of comparator today depends on whether treatment is being considered for a man or woman and whether or not the individual has established osteoporosis. Unless cost-effectiveness is measured relative to a reasonable alternative, the estimated ICER may not provide a meaningful estimate of an intervention's value. While ICERs computed relative to "no intervention" have meaning for the minority of patients who have no other viable treatment option, for the majority of patients in whom less costly treatments are possible, they are potentially misleading estimates of value. In general, when new interventions are compared with "no intervention" (technically an average rather than an incremental cost-effectiveness ratio), they will have more favorable value than when compared with active treatment comparators.

Table 79.1. Components of direct medical costs to consider when assessing the cost-effectiveness of osteoporosis treatment.

Cost Component
Medication
Acquisition
Health care services for routine monitoring
Health care services for treatment side effects/sequelae
Fracture
Acute care services
Rehabilitation services
Ongoing disability services
Extended life years
Health care services

Model-based analyses

Estimating an ICER typically requires mathematical modeling to project expected health and cost implications of alternative treatments over a longer time horizon than can be observed in any clinical trial [5] and/or to expand the treatments and/or population subgroups considered. Most analyses utilize Markov state-transition models [6], which are comprised of a discrete number of health states, each with an associated cost and health state value (ie, health utility), along with annual probabilities of transition among the health states. Other modeling methods that detail the biological processes related to bone health have also been proposed [7].

Estimating the cost of osteoporosis treatment

To estimate the net difference in cost of a new treatment relative to a comparator, several types of direct medical costs should be considered (Table 79.1). The cost of medical care in future years of life may also be included. Against these costs, potential savings that may accrue due to fracture prevention are considered and include the cost of acute fracture care, rehabilitation services (if required), and costs of ongoing fracture-related disability (if present). Differences in the cost of providing health care from country to country make generalizability of cost-effectiveness findings across countries challenging.

Indirect costs of an illness are those that are associated with a loss in productivity due to morbidity and mortality. Such costs may be incurred by the individual who sustains a fracture and/or by their care givers. However, there is scant evidence available to address the latter. The human capital approach, which values productivity changes based on lost earnings [8], has been applied to assess the cost of fractures in some US cost-of-illness studies [9–11], but to date such costs have not been included in cost-effectiveness analyses of osteoporosis treatment. Some argue that productivity costs are adequately reflected in the denominator of the cost-effectiveness ratio when quality-adjusted life-years (QALYs) are used to measure effectiveness [3].

Whether or not each of these potential costs/savings are included in the analysis depends on the perspective that is taken. For informing public policy decision-makers, the societal perspective is generally most desirable. The marked impact that perspective may have on the cost-effectiveness of osteoporosis treatment under a health care system such as in the United States, where different payors are responsible for health care at different ages, is shown by an example that underscores the disparity between who pays for the prevention and who realizes the potential long-term savings. Consider the 5-year treatment of high-risk 55-year-old women from two perspectives: (i) a private insurer who pays for health care services up until age 65; and (ii) a government insurer who pays after age 65 (eg, Medicare in the United States). For the private insurer who pays for the costs of the treatment and monitoring, but will realize limited savings due to fractures averted, treatment does not appear cost-effective. In contrast, the government payor only benefits from fractures averted and sees treatment as cost-saving. This simplistic example suggests that optimal decisions for public health require a broad perspective that considers the full time horizon of costs and benefits.

Estimating the effectiveness of osteoporosis treatment

Quality-adjusted life-years

The recommended measure for assessing the effectiveness of health interventions, is the QALY [3], which takes both length of life and quality of life into account. The use of QALYs facilitates comparisons of economic value across disease areas (eg, interventions to control diabetes can be compared with osteoporosis treatments). Cost-effectiveness studies that report ICERs as cost per QALY gained are often referred to as cost-utility analyses, because to estimate QALYs, health state values or "utilities" that reflect preferences for various health states are used.

Although QALYs have the potential to incorporate the intangible fracture-related costs of pain and suffering, data on health state values for fracture-related health outcomes are also required. Evidence on the impact of fractures on QALYs has been summarized in several reviews [12, 13]. The absolute QALY losses associated with fractures vary based on who is asked (eg, a patient who sustained a vertebral fracture versus a patient imagining a vertebral fracture) and how they are asked (eg, visual analog scale, time trade-off), yet published studies consistently report health state values for fracture-related outcomes that are significantly below ideal health. Although a growing literature addresses preference-based measures of health in osteoporosis, many cost-effectiveness studies continue to rely on expert opinion regarding the quality-of-life impact that fractures have both initially and in the long term [14–16].

When evaluating the value of osteoporosis treatment, it is important to consider the potential adverse impact that treatment side effects may have on estimates of quality-adjusted life expectancy. The potential for side effects to offset quality-of-life gains due to fracture prevention was first highlighted in studies of the role of hormone therapy in osteoporosis prevention [17, 18].

Number of fractures prevented

The value of osteoporosis treatment is sometimes reported in disease-specific terms as the number of fractures prevented, which is problematic for two reasons. First, some osteoporosis interventions, such as raloxifene, have extraskeletal health effects that go unaccounted for when value is reported in terms of cost per fracture prevented. Second, inherent differences in human and economic costs of different fracture types (eg, wrist versus hip) make interpreting cost per fracture prevented challenging. To address this, analysts sometimes report costs in terms of specific fracture types (eg, cost per hip fracture prevented or per vertebral fracture prevented) or in "hip fracture equivalent units" [19].

COST-EFFECTIVENESS OF OSTEOPOROSIS TREATMENT

Cost-effectiveness analysis and clinical practice guidelines

As constrained health care budgets are increasingly felt, guideline developers recognize that costs cannot be entirely ignored [20]. One approach to setting treatment thresholds that has seen growing application in the osteoporosis literature, is to identify the absolute fracture risk at which the cost per QALY gained falls below a "willingness to pay" per QALY gained threshold [12, 14, 15, 19, 21, 22]. This approach was utilized by the National Osteoporosis Foundation (NOF) to identify a 10-year absolute hip fracture risk at which a treatment cost of US$60,000 or lower per QALY gained for treatment was compared with cost for no intervention [22]. The Fracture Risk Assessment Tool (FRAX) facilitates such risk predictions for previously untreated populations [23] and a report from the NOF guide committee provided insight into specific clinical factors that meet the intervention thresholds (3% for 10-year hip fracture risk or 20% for hip, wrist, spine, and shoulder fracture risks combined) based on an adaptation of FRAX for the US population [24, 25]. While such analyses could be revisited for newer pharmacological agents and/or changing evidence on treatment side effects and treatment costs, the NOF analysis highlighted the importance of absolute fracture risk in identifying cost-effective treatment thresholds.

Cost-effectiveness of osteoporosis treatment

Fracture risk, treatment cost, the impact that fractures have on health-related quality of life, treatment persistence, and the durability of treatment [26] all influence the value of osteoporosis treatment. A US analysis of an unspecified treatment that reduces fracture incidence by 35% relative to no intervention [22] demonstrated the marked improvement in cost-effectiveness that average-risk women attain with advancing age due to their higher absolute fracture risk (Fig. 79.1). For example, a treatment costing US$900 per year costs in excess of US$580,000 per QALY gained for a 50-year-old woman

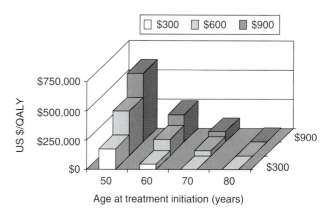

Fig. 79.1. The impact that annual treatment cost (US$300, 600, or 900) has on cost per QALY gained at different ages of treatment initiation when treatment reduces fracture incidence by 35% and 5-year loses in health-related quality of life following fracture are modeled. Source: [22]. Reproduced with permission of Springer.

whose 10-year hip fracture risk is 2.5%, compared with only US$4000 per QALY gained for an 80-year-old woman whose 10-year hip fracture risk is 4%.

Prior to 1993, most osteoporosis cost-effectiveness studies assessed the value of hormone therapy [27, 28]. Several reviews and technology assessments have addressed the value of other osteoporosis treatments with evaluation of bisphosphonates being a frequent focus [28–32]. Most recently, a systematic review of cost-effectiveness studies addressing pharmacological postmenopausal osteoporosis treatments published between 2008 and 2013 called for higher quality and more consistent cost-effectiveness reporting [33]. The review noted that treatments were generally cost-effective when directed to women older than 60 to 65 years of age with low bone mass, especially if they have had a prior vertebral fracture. Studies also address the value of calcium and vitamin D [34], raloxifene [14, 35–39], teriparatide [40, 41], calcitonin [42], strontium ranelate [43, 44], vitamin K [45], and denosumab [46] as well as the cost-effectiveness of selective treatment strategies [47–49] and special populations [50].

Bisphosphonate treatment is generally cost-effective when used in moderately high-risk populations such as women over age 65 with osteoporosis [30]. While treatment of elderly populations with calcium and vitamin D is potentially cost-saving [34], studies of raloxifene differ in their findings depending on the comparators included, the side effects considered and individual women's risks [35–38,51]. Discrepant findings have also been reported for teriparitide use among women at high risk of fracture [40, 41]. Cost-effectiveness studies that distinguish between agents on the basis of treatment persistence are of relevance due to the advent of agents with differing modes of administration (eg, weekly oral agent versus annual injection) [52]. Likewise, there is increasing interest in programs to support treatment adherence [53].

SUMMARY

Due to the size of the elderly population that is at risk for complications of osteoporosis-related fractures, it is imperative that cost-effective approaches to osteoporosis management be identified. While a growing literature addresses the value of specific treatments for various population subgroups, clinical practice guidelines identify cost-effective intervention thresholds on the basis of absolute 10-year fracture risk. For the US population, cost-effective treatment intervention thresholds of 3% or greater for 10-year hip fracture risk, or 20% or greater hip, wrist, spine, and shoulder fracture risk combined, have been recommended [25]. Risk assessment tools for predicting 10-year fracture risk [54] provide a tool for efficiently targeting therapy to those individuals who stand to benefit most from osteoporosis treatment.

ACKNOWLEDGMENTS

The author thanks Ms. Loretta Pearson MPhil and Ms. Rebecca Smith for research assistance and editorial support.

REFERENCES

1. Office of the Surgeon General. *Bone Health and Osteoporosis: A Report of the Surgeon General.* Rockville, MD: US Department of Health and Human Services, 2004.
2. Burge R, Dawson-Hughes B, Solomon DH, et al. Incidence and economic burden of osteoporosis-related fractures in the United States, 2005–2025. J Bone Miner Res. 2007;22(3):465–75.
3. Neumann PJ, Sanders GD, Russell LB et al. *Cost-effectiveness in Health and Medicine* (2nd edn.). New York: Oxford University Press, 2017.
4. Rossouw JE, Anderson GL, Prentice RL, et al. Risks and benefits of estrogen plus progestin in healthy postmenopausal women: principal results from the Women's Health Initiative randomized controlled trial. JAMA. 2002;288(3):321–33.
5. Tosteson AN, Jonsson B, Grima DT, et al. Challenges for model-based economic evaluations of postmenopausal osteoporosis interventions. Osteoporos Int. 2001;12: 849–57.
6. Sonnenberg FA, Beck JR. Markov models in medical decision making: a practical guide. Med Decis Making. 1993;13(4):322–38.
7. Vanness DJ, Tosteson AN, Gabriel SE, et al. The need for microsimulation to evaluate osteoporosis interventions. Osteoporos Int. 2005;16(4):353–8.
8. Hodgson TA, Meiners MR. Cost-of-illness methodology: a guide to current practices and procedures. Milbank Mem Fund Q Health Soc. 1982;60(3):429–62.
9. Holbrook T, Grazier K, Kelsey J, et al. *The Frequency of Occurrence, Impact, and Cost of Musculoskeletal Conditions in the United States.* Rosemont, IL: American Academy of Orthopaedic Surgeons, 1984.
10. Praemer A, Furner S, Rice D. *Musculoskeletal Conditions in the United States.* Rosemont, IL: American Academy of Orthopaedic Surgeons, 1992.
11. Praemer A, Furner S, Rice D. *Musculoskeletal Conditions in the United States.* Rosemont, IL: American Academy of Orthopaedic Surgeons, 1999.
12. Brazier JE, Green C, Kanis JA. A systematic review of health state utility values for osteoporosis-related conditions. Osteoporos Int. 2002;13(10):768–76.
13. Tosteson ANA, Hammond CS. Quality of life assessment in osteoporosis: health status and preference-based instruments. Pharmacoeconomics. 2002;20(5):289–303.
14. Kanis JA, Borgstrom F, Zethraeus N, et al. Intervention thresholds for osteoporosis in the UK. Bone. 2005;36(1): 22–32.

15. Kanis JA, Johnell O, Oden A, et al. Intervention thresholds for osteoporosis in men and women: a study based on data from Sweden. Osteoporos Int. 2005;16(1):6–14.
16. Kanis JA, Johnell O, Oden A, et al. Intervention thresholds for osteoporosis. Bone. 2002;31(1):26–31.
17. Weinstein MC. Estrogen use in postmenopausal women—costs, risks, and benefits. N Engl J Med. 1980;303(6):308–16.
18. Weinstein MC, Schiff I. Cost-effectiveness of hormone replacement therapy in the menopause. Obstet Gynecol Surv. 1983;38(8):445–55.
19. Kanis JA, Oden A, Johnell O, et al. The burden of osteoporotic fractures: a method for setting intervention thresholds. Osteoporos Int. 2001;12(5):417–27.
20. Guyatt G, Baumann M, Pauker S, et al. Addressing resource allocation issues in recommendations from clinical practice guideline panels: suggestions from an American College of Chest Physicians task force. Chest. 2006;129(1):182–7.
21. Borgstrom F, Johnell O, Kanis JA, et al. At what hip fracture risk is it cost-effective to treat? International intervention thresholds for the treatment of osteoporosis. Osteoporos Int. 2006;17(10):1459–71.
22. Tosteson AN, Melton LJ 3rd, Dawson-Hughes B, et al. Cost-effective osteoporosis treatment thresholds: the United States perspective. Osteoporos Int. 2008;19(4):437–47.
23. Kanis JA, McCloskey EV, Johansson H, et al. Case finding for the management of osteoporosis with FRAX—assessment and intervention thresholds for the UK. Osteoporos Int. 2008;19(10):1395–408.
24. Dawson-Hughes B. A revised clinician's guide to the prevention and treatment of osteoporosis. J Clin Endocrinol Metab. 2008;93(7):2463–5.
25. Foundation NO. *Clinician's Guide to Prevention and Treatment of Osteoporosis*. Washington, DC: National Osteoporosis Foundation, 2008.
26. Jonsson B, Kanis J, Dawson A, et al. Effect and offset of effect of treatments for hip fracture on health outcomes. Osteoporos Int. 1999;10(3):193–9.
27. Anon. Osteoporosis: review of the evidence for prevention, diagnosis and treatment and cost-effectiveness analysis. Introduction. Osteoporos Int. 1998;8(suppl 4): S7–80.
28. Fleurence RL, Iglesias CP, Torgerson DJ. Economic evaluations of interventions for the prevention and treatment of osteoporosis: a structured review of the literature. Osteoporos Int. 2006;17(1):29–40.
29. Zethraeus N, Borgstrom F, Strom O, et al. Cost-effectiveness of the treatment and prevention of osteoporosis—a review of the literature and a reference model. Osteoporos Int. 2007;18(1):9–23.
30. Fleurence RL, Iglesias CP, Johnson JM. The cost effectiveness of bisphosphonates for the prevention and treatment of osteoporosis: a structured review of the literature. Pharmacoeconomics. 2007;25(11):913–33.
31. Urdahl H, Manca A, Sculpher MJ. Assessing generalisability in model-based economic evaluation studies: a structured review in osteoporosis. Pharmacoeconomics. 2006;24(12):1181–97.
32. Stevenson M, Jones ML, De Nigris E, et al. A systematic review and economic evaluation of alendronate, etidronate, risedronate, raloxifene and teriparatide for the prevention and treatment of postmenopausal osteoporosis. Health Technol Assess. 2005;9(22):1–160.
33. Hiligsmann M, Evers SM, Sedrine WB, et al. A systematic review of cost-effectiveness for postmenopausal osteoporosis. Pharmacoeconomics. 2015;33:205–24.
34. Torgerson D, Kanis J. Cost-effectiveness of preventing hip fracture in the elderly using vitamin D and calcium. Q J Med. 1995;88:135–9.
35. Ivergard M, Strom O, Borgstrom F, et al. Identifying cost-effective treatment with raloxifene in postmenopausal women using risk algorithms for fractures and invasive breast cancer. Bone. 2010;47(5):966–74.
36. Goeree R, Blackhouse G, Adachi J. Cost-effectiveness of alternative treatments for women with osteoporosis in Canada. Curr Med Res Opin. 2006;22(7):1425–36.
37. Mobley LR, Hoerger TJ, Wittenborn JS, et al. Cost-effectiveness of osteoporosis screening and treatment with hormone replacement therapy, raloxifene, or alendronate. Med Decis Making. 2006;26(2):194–206.
38. Borgstrom F, Johnell O, Kanis JA, et al. Cost effectiveness of raloxifene in the treatment of osteoporosis in Sweden: an economic evaluation based on the MORE study. Pharmacoeconomics. 2004;22(17):1153–65.
39. Armstrong K, Chen TM, Albert D, et al. Cost-effectiveness of raloxifene and hormone replacement therapy in postmenopausal women: impact of breast cancer risk. Obstet Gynecol. 2001;98(6):996–1003.
40. Lundkvist J, Johnell O, Cooper C, et al. Economic evaluation of parathyroid hormone (PTH) in the treatment of osteoporosis in postmenopausal women. Osteoporos Int. 2006;17(2):201–11.
41. Liu H, Michaud K, Nayak S, et al. The cost-effectiveness of therapy with teriparatide and alendronate in women with severe osteoporosis. Arch Intern Med. 2006;166(11):1209–17.
42. Coyle D, Cranney A, Lee KM, et al. Cost effectiveness of nasal calcitonin in postmenopausal women: use of Cochrane Collaboration methods for meta-analysis within economic evaluation. Pharmacoeconomics. 2001;19(5 Pt 2):565–75.
43. Borgstrom F, Jonsson B, Strom O, et al. An economic evaluation of strontium ranelate in the treatment of osteoporosis in a Swedish setting: based on the results of the SOTI and TROPOS trials. Osteoporos Int. 2007;17(12):1781–93.
44. Stevenson M, Davis S, Lloyd-Jones M, et al. The clinical effectiveness and cost-effectiveness of strontium ranelate for the prevention of osteoporotic fragility fractures in postmenopausal women. Health Technol Assess. 2007;11(4):1–134.
45. Stevenson M, Lloyd-Jones M, Papaioannou D. Vitamin K to prevent fractures in older women: systematic review and economic evaluation. Health Technol Assess. 2009;13(45):iii–xi, 1–134.

46. Jonsson B, Strom O, Eisman JA, et al. Cost-effectiveness of denosumab for the treatment of postmenopausal osteoporosis. Osteoporos Int. 2011;22(3):967–82.
47. Schousboe JT, Nyman JA, Kane RL, et al. Cost-effectiveness of alendronate therapy for osteopenic postmenopausal women. Ann Intern Med. 2005;142(9):734–41.
48. Schousboe JT, Bauer DC, Nyman JA, et al. Potential for bone turnover markers to cost-effectively identify and select post-menopausal osteopenic women at high risk of fracture for bisphosphonate therapy. Osteoporos Int. 2007;18(2):201–10.
49. Schousboe JT, Taylor BC, Fink HA, et al. Cost-effectiveness of bone densitometry followed by treatment of osteoporosis in older men. JAMA. 2007;298(6):629–37.
50. Kanis JA, Stevenson M, McCloskey EV, et al. Glucocorticoid-induced osteoporosis: a systematic review and cost-utility analysis. Health Technol Assess. 2007;11(7):iii–iv, ix–xi, 1–231.
51. Kanis JA, Borgstrom F, De Laet C, et al. Assessment of fracture risk. Osteoporos Int. 2005;16(6):581–9.
52. Kanis JA, Cooper C, Hiligsmann M, et al. Partial adherence: a new perspective on health economic assessment in osteoporosis. Osteoporosis Int. 2011;22(10):2565–73.
53. Hiligsmann M, Rabenda V, Bruyere O, et al. The clinical and economic burden of non-adherence with oral bisphosphonates in osteoporotic patients. Health Policy. 2010;96(2):170–7.
54. Kanis JA, McCloskey EV, Johansson H, et al. Case finding for the management of osteoporosis with FRAX—assessment and intervention thresholds for the UK. Osteoporosis Int. 2008;19(10):1395–408.

80
Future Therapies

Michael R. McClung

Institute of Health and Ageing, Australian Catholic University, Melbourne, VIC, Australia;
and Oregon Osteoporosis Center, Portland, OR, USA

INTRODUCTION

The mechanisms of action of the several classes of drugs currently available for treating osteoporosis, as well as their efficacy and safety, have been reviewed in detail in the previous chapters. Several of these drugs are very effective in reducing the risk of vertebral (morphometric and clinical) fractures and of hip fractures in women with postmenopausal osteoporosis [1]. The ability of these agents to protect patients from nonvertebral fracture is more limited. In general, the drugs are well tolerated and, while potential and sometimes serious safety concerns exist with each drug, the benefit/risk ratio of current treatments is very favorable if treatment is targeted toward and confined to patients at high risk for fracture. [2]

However, the dosing regimen of some of the agents is considered by patients to be awkward or inconvenient; intolerance to oral bisphosphonates is common; adherence to oral therapies is poor; and concerns about theoretical or rare adverse events limit the acceptance of several therapies. These limitations to our current therapies explain, in some part, the observation that most patients who are identified as candidates for therapy are not being treated [3]. This provides both a need and justification for the development of new therapeutic agents to address these unmet needs. New agents would ideally be more effective than current treatments, especially in the reduction of nonvertebral fracture. They should improve or normalize bone strength by stimulating the formation of new bone, rebuilding the trabecular microarchitecture, and correcting the cortical thinning and porosity that characterize severe osteoporosis. New agents should also be well tolerated without serious safety concerns such as the skeletal effects of long-term bisphosphonate therapy.

They should be easily and conveniently administered and perhaps be able to be used in combination with existing therapies, either simultaneously or in sequence, to optimize skeletal benefits.

This chapter will focus on two classes of emerging therapies, inhibitors of cathepsin K and of sclerostin, and will focus on single drugs in each of these classes that have recently completed phase III registration trials. A third new drug, abaloparatide, has recently received regulatory approval in the United States and has been discussed in Chapter 73. While some additional therapies targeting different molecular pathways are being considered, none have entered clinical trials and will not be discussed in this review.

CATHEPSIN K INHIBITORS

Cathepsin K (CatK) is the major osteoclast-derived proteolytic enzyme in bone, hydrolyzing type I collagen and other bone matrix proteins [4]. The enzyme is highly expressed in osteoclasts with quite limited expression in other tissues. Genetic deficiency of CatK in humans results in the syndrome of pyknodysostosis, characterized by an osteoclast-rich form of osteopetrosis with high bone mass and reduced bone resorption but with preservation of bone formation. Targeted disruption of the *CatK* gene in mice results in a phenotype with an osteosclerotic skeleton with increased trabecular and cortical bone mass of good quality [5]. Histomorphometry demonstrates a high rate of bone formation, normal bone mineralization, and normal or increased osteoclast numbers, but a decrease in the ability of osteoclasts to resorb bone matrix. Compared with normal animals, CatK-deficient mice have increased bone strength at the vertebral body and femoral midshaft.

Primer on the Metabolic Bone Diseases and Disorders of Mineral Metabolism, Ninth Edition. Edited by John P. Bilezikian.
© 2019 American Society for Bone and Mineral Research. Published 2019 by John Wiley & Sons, Inc.
Companion website: www.wiley.com/go/asbmrprimer

Several small molecular weight inhibitors of human CatK have been developed. The preclinical studies with these inhibitors have been restricted to monkeys and rabbits. Relacatib is a relatively nonspecific CatK inhibitor. Balicatib is a basic, highly selective, nitrile-based CatK inhibitor in enzyme-based assays, but the selectivity is largely lost in whole cell assays due its accumulation in lysosomes (a property called lysosomotropism) where high concentrations exert effects on non-K cathepsins. ONO-5334 is a potent hydrazine-based non-lysosomotropic inhibitor of CatK with high selectivity. Odanacatib is a nonbasic, non-lysosomotropic, nitrile-based molecule that retains its high enzyme selectivity in cell-based assay systems [6].

In preclinical studies in ovariectomized rabbits and monkeys, CatK inhibitors induced dose-dependent decreases bone resorption at both trabecular and cortical sites and preservation of areal BMD [7–13]. Treatment increased osteoclast numbers when assessed by histomorphometry, and increased serum levels of tartrate-resistant acid phosphatase 5b, a marker of osteoclast number. The effects of CatK inhibition on bone formation are more complex. Like other resorption inhibitors, CatK inhibitors reduced trabecular bone formation in ovariectomized monkeys. However, the reduction in cortical bone formation rate was less with CatK inhibition than with alendronate. In contrast to the effects of bisphosphonates and denosumab, CatK inhibition is associated with maintenance of endocortical bone formation and increased periosteal bone formation in the femur, resulting in increased cortical bone thickness and volume in the hip and improved bending strength of long bones.

Preclinical studies demonstrated the unique actions that CatK inhibition has on bone remodeling. While inhibiting CatK decreases the capacity of viable osteoclasts to resorb bone, the number and other functions of osteoclasts is preserved. Communication between these functioning osteoclasts and osteoblasts remains intact in the presence of CatK inhibition [14.] Preservation of osteoblast function might also be due to reduced degradation by CatK of matrix-derived growth proteins such as TGF-β and IGF-1.

The favorable preclinical results led to clinical trials with the four CatK inhibitors. The development of relacatib was discontinued after a phase I study because of drug–drug interactions with paracetamol, ibuprofen, and atorvastatin. In a phase II study, treatment with balicatib led to reduced markers of bone resorption and increased BMD but, because of several cases of morphea-like skin lesions and of serious respiratory infections, development of this compound was halted [15]. These toxicities might have been because of inhibition of CatK in skin or lung or to the lysosomotropism described earlier, causing off-target inhibition of other cathepsins expressed in skin and pulmonary fibroblasts [16].

ONO-5334, administered in oral doses of 50 mg twice daily, 100 mg once daily, or 300 mg once daily, was compared with placebo and alendronate 70 mg weekly in postmenopausal women with osteoporosis in a phase II trial [17]. Compared with alendronate, markers of bone resorption were similarly reduced by the 300 mg daily dose of ONO-5334 over 12 months of therapy, but bone formation markers were inhibited less with ONO-5334 than with alendronate. As in animal studies, levels of tartrate-resistant acid phosphatase 5b (TRAP5b), a marker of osteoclast number, were reduced by alendronate but increased by the CatK inhibitor. The effects of ONO-5334 300 mg daily and alendronate on BMD of the lumbar spine were comparable. No clinically relevant safety concerns, including the rate of skin adverse events, were identified. However, the development of this compound was also halted.

Odanacatib is the only CatK inhibitor that has been evaluated in a phase III fracture endpoint trial. Pharmacokinetic single dose studies in young healthy subjects demonstrated a long half-life consistent with once-weekly dosing and good bioavailability in either the fasting or fed state [18, 19]. Markers of bone resorption and formation decreased and promptly returned to baseline. Pharmacodynamics and pharmacokinetics were similar in older men and women compared with young adults [20].

The phase II study evaluated doses of odanacatib ranging from 3 to 50 mg given orally once weekly (QW) [21]. Except for the 3 mg QW dose, the early decreases in markers of bone turnover and the increases in BMD were dose-dependent. Bone resorption markers were reduced and remained below baseline throughout the 2-year study. Bone formation markers, however, decreased soon after treatment was begun but returned to baseline values by the end of the second year. At 24 months, BMD at the lumbar spine and total hip increased by 5.5% and 3.2%, respectively, compared with baseline. Similar results were observed in Japanese patients with osteoporosis [22]. On the basis of these results, the 50 mg QW dose was selected for phase III studies. Complete, rapid reversibility of the effects on bone turnover and BMD was demonstrated [23] Open label therapy with odanacatib 50 mg QW demonstrated impressive progressive gains in BMD at the spine and hip regions for as long as 8 years [24, 25]. In patents who had previously been treated with bisphosphonates, odanacatib therapy increased BMD and decreased markers of bone resorption while increasing markers of bone formation, suggesting that odanacatib would be an attractive agent in patients who remained at high risk of fracture after several years of bisphosphonate therapy [26]. Imaging studies in patients receiving odanacatib revealed improvements in bone geometry, microarchitecture, and estimated bone strength in cortical and trabecular skeletal compartments [27–29].

The effect of odanacatib on fracture risk was tested in the phase III Long-term Odanacatib Fracture Trial (LOFT) [30]. Several aspects of this trial were unique. It is the largest and most complex osteoporosis treatment trial. More than 16,000 patients in 348 study sites in 40 countries were randomized to receive odanacatib 50 mg QW or placebo, assigned in a 1:1 ratio, for up to 5 years. It was a placebo-controlled, randomized, event-driven trial targeting hip fractures. Coprimary endpoints of new morphometric vertebral fracture, nonvertebral fracture, and hip fracture were specified. The large number of subjects was necessitated by having hip fracture as a coprimary endpoint and by the ethics of not enrolling very high-risk patients. An open label extension with treatment continued for 10 years was planned to follow the blinded, placebo-controlled portion of the study.

At a planned interim analysis when 70% of the hip fracture target was reached, all of the efficacy endpoints had been met, and the Data Monitoring Board recommended that the study be terminated but that the patients continue to be followed to obtain further safety data. All subjects returned for study close out between August and October 2012. Those who completed that part of LOFT (the base study) and who had not experienced a fragility fracture, were invited to participate in the blinded extension study. Some study sites were either not allowed by ethical committees to continue patients on placebo treatment or chose not to participate in the extension study. Patients who had experienced a decrease in BMD of more than 7% at any study visit were not allowed to enroll in the extension. As a result, more patients in the placebo group than in the active treatment group were not eligible for the extension. For these reasons, the base study results should be considered as the primary dataset.

Several separate committees of experts were established to adjudicate adverse events of special interest, including those previously associated with CatK inhibitor therapy (morphea-like skin lesions, serious respiratory infections) and adverse events possibly associated with anti-remodeling therapy (osteonecrosis of the jaw, fractures of the femur with atypical features, poor fracture healing). All cardiovascular adverse events were adjudicated because of preclinical data suggesting a possible treatment benefit.

In the base study, significant reductions in the risk of new morphometric vertebral fracture (54%), clinical nonvertebral (47%), and hip fracture (23%) were reported [31]. There was a positive interaction between nonvertebral fracture risk and time on therapy; the longer on therapy, the greater the reduction in fracture risk. These endpoints were similar after all patients had reached 5 years of placebo or odanacatib therapy [32]. Rare cases of morphea-like skin lesions, not associated with systemic features and fractures of the femoral shaft, with atypical features occurred more often in the treatment group than with placebo. After adjudication, a small but statistically significant increase in stroke risk was observed [33], leading Merck to halt the clinical development of odanacatib and to not file for registration [34]

ANTISCLEROSTIN THERAPY

As discussed in previous chapters, sclerostin is an osteoblast inhibiting glycoprotein expressed primarily in osteocytes and modulated by hormonal influences and skeletal loading. Sclerostin inhibits osteoblast activity by interrupting the Wnt signaling pathway. Genetic deficiency of sclerostin results in diseases characterized by very high bone mass, bone tissue of good quality (unlike the brittle bones of patients with osteopetrosis), and very low risk for fracture [35]. Homozygous patients experience cranial and facial distortion during growth and, as adults, bony overgrowth such as cranial basilar stenosis. Heterozygotes, who have intermediate levels of sclerostin, have a normal phenotype except for high bone mass, suggesting that inhibiting sclerostin has promise as a treatment strategy to activate bone formation and to improve bone mass as a treatment for osteoporosis and other bone disorders. Sclerostin-deficient mice also exhibit high bone mass with increased rates of bone formation in both trabecular and cortical bone [36]. These effects result in substantially stronger skeletons than in genetically normal mice.

Preclinical studies

Inhibiting sclerostin activity with an antisclerostin antibody in animals demonstrated the potential of this strategy to improve bone mass and structure. In aged, ovariectomized rats given a sclerostin-inhibiting antibody, increased bone formation was observed on trabecular, endocortical, intracortical, and periosteal bone surfaces [37]. Trabecular and cortical bone thickness was increased, and cortical porosity was reduced. Treatment for 5 weeks resulted in bone mass and bone strength that exceeded the sham-operated control animals and restored the architectural abnormalities induced by ovariectomy. In gonad-intact female cynomolgus monkeys, treatment with a humanized sclerostin-neutralizing antibody once monthly for 2 months transiently increased markers of bone formation, increased BMD in the lumbar spine, femoral neck, proximal tibia, and distal radius [37]. Increased modeling-based bone formation was observed on all skeletal surfaces. Such therapy also increased the thickness of trabeculae in all three spatial orientations (axial, oblique, and transverse) and converted rod-like structures into more mechanically sound plate-like structures [38, 39]. Substantial increase in bone strength occurred, with a strong correlation between bone mineral content (measured by QCT) and peak load observed in both the lumbar spine and femoral diaphysis.

Preclinical studies with particular relevance to the potential clinical use of antisclerostin therapy include the observation that the anabolic response to sclerostin antibody therapy was as robust in aged mice as in younger mice [40]. Additionally, the anabolic effect of therapy was unaffected by previous or simultaneous treatment with a bisphosphonate and could be restored after a short treatment-free interval [38, 41]. Relevant to the phase III study design, the increases in bone mass achieved with sclerostin antibody treatment were maintained or improved when that therapy was followed by a RANKL inhibitor [38]. In a rat lifetime carcinogenic toxicity study with romosozumab, an antisclerostin antibody evaluated in humans, no treatment-related effects on tumor incidence were observed [42].

Clinical trials

Blosozumab

Blosozumab, a humanized IgG4 monoclonal sclerostin-binding antibody, administered subcutaneously in doses as high as 270 mg every 2 weeks (Q2W), resulted in progressive dose-dependent increases in BMD over a 12-month treatment period [43]. Compared with placebo, the average increase in BMD was 17.7% in the lumbar spine, 6.2% in the total hip, and 7.3% in the total body with the 270 mg Q2W dose. Markers of bone formation increased rapidly; serum P1NP increased to a peak of 160% above baseline at 4 weeks with the 270 mg Q2W dose. Despite ongoing therapy, P1NP values then gradually decreased, remaining above baseline at 6 months but falling to near baseline at 12 months. Serum CTx decreased to as much as 40% below baseline after 2 weeks of treatment and remained below baseline and lower than values in the placebo group for the remainder of the treatment interval. Upon discontinuing blosozumab therapy after 12 months, BMD in the lumbar spine and total hip decreased to or toward baseline in all treatment groups [44]. Bone formation and resorption markers remained near baseline during the year off therapy. The planned phase III study with blosozumab was not initiated, and further development of the drug by Lilly has been halted, perhaps because of the occurrence of more frequent or severe injection site reactions with a more concentrated preparation of the drug.

Romosozumab

Romosozumab is a humanized monoclonal IgG2 antibody with high specificity for human sclerostin. Phase I single and multiple dose studies demonstrated rapid and marked increases in biochemical markers of bone formation, decreases in bone resorption markers, and increases in BMD [45, 46]. The effects on trabecular and cortical structural parameters were also assessed by HRpQCT scans of the lumbar spine in 48 subjects (32 women, 16 men) with low bone mass in a placebo-controlled phase Ib study [47]. Women received active treatment of 1 or 2 mg/kg Q2W or 2 or 3 mg/kg Q4W, while men were given 1 mg/kg Q2W or 3 mg/kg Q2W for 3 months. All active treatment groups were combined for analyses. At 3 months, significant increases were observed in trabecular BMD (9.5%) and apparent density-weighted cortical thickness and cortical stiffness (26.9%). These improvements in structural parameters were maintained during a 3-month off-treatment follow-up period.

The phase II romosozumab dose-ranging study (NCT00896532) evaluated treatment effects in 419 postmenopausal women ages 55 to 85 years with low bone mass who were randomly assigned to receive subcutaneous doses of romosozumab ranging from 70 mg every 3 months (Q3M) to 210 mg QM or placebo injections [48]. Other patients were randomly assigned to receive open label teriparatide 20 µg s.c. daily or alendronate 70 mg weekly. At 12 months, the average increases in BMD at the lumbar spine and total hip were 11.3% and 4.1%, respectively, with the 210 mg QM dose of romosozumab. These increases were significantly greater than those observed with teriparatide or alendronate. BMD at the one-third radial site decreased by 0.9% with placebo, by 1.3% with romosozumab, and by 1.7% with teriparatide.

Bone formation markers rose promptly, peaked at 1 to 3 months, returned to baseline by 6 months, and were then below baseline for the remainder of the 12-month treatment interval. During the second year of the study, markers of bone formation and resorption remained below baseline in patients who continued romosozumab [48]. Consistent with these results, lumbar spine BMD increased 3.8% during the second year of romosozumab therapy, a smaller amount than had occurred in the first year. After 2 years, romosozumab therapy was discontinued, and patients were randomized to receive denosumab 60 mg s.c. Q6M or placebo for 12 months. In those who received placebo, BMD values decreased to or toward baseline values. Formation markers also returned to baseline values, while serum CTx levels rose to values significantly greater than baseline before falling toward pretreatment values. In patients who were switched to denosumab, BMD increased in the lumbar spine (3.7%) and total hip (1.1%), increments that were similar to the increases during the second year of romosozumab therapy. Volumetric BMD was assessed by QCT in a subset of patients from the romosozumab phase II study [49]. Increases in integral volumetric BMD in both the lumbar spine and total hip were significantly greater with romosozumab compared with teriparatide.

The Fracture Study in Postmenopausal Women with Osteoporosis (FRAME) trial (NCT01575834) evaluated the effectiveness of romosozumab 210 mg QM compared with placebo for 12 months followed in both treatment groups by an additional 12 months of denosumab 60 mg Q6M in 7180 women with postmenopausal osteoporosis whose average age was 71 years [50]. About 18% had prevalent vertebral fracture (almost all were mild or grade 1 deformities), and 21.7% had a history of prior nonvertebral fracture. The geographic distribution of patients in this study was different to that in previous fracture endpoint trials. Forty-three percent of patients were from Latin America, and 29% were from Central or Eastern Europe. Eighty-nine percent of patients completed 12 months of the trial, and 83.9% completed the 2-year follow-up. During 12 months of romosozumab therapy, new vertebral fractures occurred in 1.8% of the placebo group and 0.8% of the group on romosozumab, a relative risk reduction of 73% (95% CI, 53% to 84%; $p<0.001$). The treatment effect was particularly evident during the second 6 months of therapy. At the end of the second year during which all patients had received open label denosumab therapy, new vertebral fracture risk was reduced by 75% in the patents who had initially received romosozumab compared with the group that had received placebo. During the second year, 25 of 3327 in the placebo group experienced a new fracture while this occurred in only five of 3325 patients in the romosozumab group.

Clinical fracture risk was reduced by 36% (2.5% with placebo and 1.6% with romosozumab; $p=0.008$) at 12 months. Nonvertebral fracture risk was reduced by 25%, from 2.1% with placebo to 1.6% with romosozumab (hazard ratio, 0.75; 95% CI, 0.53 to 1.05; $p=0.10$). In a preplanned subgroup analysis, significant interaction of nonvertebral fracture risk reduction with geography was observed. This was explored in more detail in a post hoc analysis that demonstrated very low fracture risk (assessed by the Fracture Risk Assessment Tool, FRAX) and fracture incidence in the Latin American subgroup in whom no treatment effect was observed. In the remainder of the study population, a significant 42% risk reduction was observed. Romosozumab was associated with BMD increases of 13.3% and 6.8% in the lumbar spine and total hip, respectively, compared with baseline and with placebo at 12 months. After 12 additional months of denosumab therapy, the total increases in lumbar spine and total hip BMD from the original baseline were 17.6% and 8.8%, respectively. These results of the FRAME study have been filed with American and European regulatory agencies, and decisions about registration are expected soon. A second phase III fracture endpoint study in postmenopausal women with osteoporosis (NCT01631214) will compare the effects of subcutaneous romosozumab 210 mg QM with oral alendronate 70 mg weekly for 12 months, followed by another year in which both treatment groups will receive alendronate. Results of a phase III registration study evaluating the safety and effectiveness of romosozumab in men with osteoporosis will be available soon (NCT02186171).

Safety

Except for mild injection site reactions, romosozumab has been well tolerated. One patient developed transient but symptomatic hepatitis 1 day after receiving a dose of 10 mg/kg in a phase I study [45]. The frequency of abnormal liver function tests was not greater with romosozumab compared with placebo. Mild, asymptomatic, transient decreases in serum calcium with reciprocal increases in PTH have been noted with the highest doses of romosozumab, likely related to the combination of the rapid formation of new bone matrix and inhibition of bone resorption induced by treatment.

In the phase III FRAME study, the frequencies of mortality and of adverse and serious adverse events, including cardiovascular disease, were balanced between the treatment and control groups [50]. Injection-site reactions, usually mild in severity, were observed in 5.2% of the romosozumab group and in 2.9% of the placebo group. All oral adverse events and femur fractures in the FRAME study were adjudicated by panels of experts to identify those events that met established criteria for osteonecrosis of the jaw or femoral fractures with atypical features. Two patients in the romosozumab group had adverse events consistent with the definition of osteonecrosis of the jaw, both of whom had recognized risk factors. An event consistent with an atypical femoral fracture occurred 3.5 months after the first dose of romosozumab in a patient with a history of prodromal pain at the site of fracture that was present at the time of enrollment.

Antiromosozumab antibodies were detected in 20% of patients during the first year of therapy in the phase II study, and 3% of patients had antibodies with neutralizing activity in vitro [48]. Binding and neutralizing antidrug antibodies were found in 18% and 7%, respectively, of patients receiving romosozumab in the FRAME study [20]. The presence of antibodies had no detectable effect on efficacy and did not correlate with injection site reactions or adverse events. The occurrence of antidrug antibodies and injections site reactions were also observed with blosozumab [43]. One patient in the phase II blosozumab study developed neutralizing antibodies at week 24 of treatment, and titers continued to rise during the 12-month treatment phase. The BMD response in this patient appeared to be blunted, perhaps related to the presence of the antibodies.

Summary

The novel anabolic effects of antisclerostin antibody therapy offer great promise for the treatment of osteoporosis. The results of the second year of the FRAME study are the first demonstration that sequential therapy with an anabolic agent followed by a potent antiremodeling agent is superior to monotherapy with an antiremodeling drug [50]. These data, coupled with the results of the extension of the abaloparatide phase III study [51] provide strong justification for the use of such sequential therapy to treat patients with severe osteoporosis. There will also be great interest in exploring the use of antisclerostin therapy in disorders of impaired bone formation.

CONCLUSION

The results of the studies of new compounds with novel targets demonstrate both the potential and the risks of drug development for osteoporosis. Both human and animal genetic models suggested that inhibiting CatK and sclerostin would have salutary skeletal effects. This promise was bolstered by very thorough and robust preclinical programs demonstrating impressive increases in bone mass and strength and was confirmed in clinical trials. Despite this background, five of the six drugs discussed here failed to become available for clinical use. As of this writing, the regulatory status of romosozumab and its place in the treatment paradigm, if approved, are still uncertain. The failures were most often related to off-target safety issues, a point of vulnerability for all drugs developed for the treatment of chronic diseases requiring long-term therapy where the mortality of the population considered for therapy is low.

There are no other new drugs in active clinical trials. New molecular pathways will be discovered that will provide interesting new targets for therapy. It is hoped that new approaches to the design and conduct of registration trials, using validated surrogates for fracture, will simplify the process of drug development. In the meantime, our clinical research focus must turn away from the wait for new drugs toward taking optimal advantage of the tools for evaluation and the broad menu of treatment options that are already available to us.

REFERENCES

1. Black DM, Rosen CJ. Postmenopausal osteoporosis. N Engl J Med. 2016;374:2096–7.
2. McClung M, Harris ST, Miller PD, et al. Bisphosphonate therapy for osteoporosis: benefits, risks, and drug holiday. Am J Med. 2013;126:13–20.
3. Khosla S, Shane E. A crisis in the treatment of osteoporosis. J Bone Miner Res. 2016;31:1485–7.
4. Rodan SB, Duong LT. Cathepsin K—a new molecular target for osteoporosis. IBMS BoneKEy. 2008;5:16–24.
5. Pennypacker B, Shea M, Liu Q, et al. Bone density, strength, and formation in adult cathepsin K (−/−) mice. Bone. 2009;44:199–207.
6. Gauthier JY, Chauret N, Cromlish W, et al. The discovery of odanacatib (MK-0822), a selective inhibitor of cathepsin K. Bioorg Med Chem Lett. 2008;18:923–98.
7. Stroup GB, Kumar S, Jerome CP. Treatment with a potent cathepsin K inhibitor preserves cortical and trabecular bone mass in ovariectomized monkeys. Calcif Tissue Int. 2009;85:344–55.
8. Jerome C, Missbach M, Gamse R. Balicatib, a cathepsin K inhibitor, stimulates periosteal bone formation in monkeys. Osteoporos Int. 2011;22:3001–11.
9. Cusick T, Chen CM, Pennypacker BL, et al. Odanacatib treatment increases hip bone mass and cortical thickness by preserving endocortical bone formation and stimulating periosteal bone formation in the ovariectomized adult rhesus monkey. J Bone Miner Res. 2010;27:524–37.
10. Pennypacker BL, Chen CM, Zheng H, et al. Inhibition of cathepsin K increases modeling-based bone formation, and improves cortical dimension and strength in adult ovariectomized monkeys. J Bone Miner Res. 2014;29: 1847–58.
11. Duong LT, Crawford R, Scott K, et al. Odanacatib, effects of 16-month treatment and discontinuation of therapy on bone mass, turnover and strength in the ovariectomized rabbit model of osteopenia. Bone. 2016;93:86–96.
12. Duong LT, Pickarski M, Cusick T, et al. Effects of long term treatment with high doses of odanacatib on bone mass, bone strength, and remodeling/modeling in newly ovariectomized monkeys. Bone. 2016;88:113–24.
13. Cabal A, Williams DS, Jayakar RY, et al. Long-term treatment with odanacatib maintains normal trabecular biomechanical properties in ovariectomized adult monkeys as demonstrated by micro-CT-based finite element analysis. Bone Rep. 2017;6:26–33.
14. Lotinun S, Kiviranta R, Matsubara T, et al. Osteoclast-specific cathepsin K deletion stimulates S1P-dependent bone formation. J Clin Invest. 2013;123(2):666–81.
15. Rünger TM, Adami S, Benhamou CL, et al. Morphea-like skin reactions in patients treated with the cathepsin K inhibitor balicatib. J Am Acad Dermatol. 2012;66: e89–96.
16. Falgueyret JP, Desmarais S, Oballa R, et al. Lysosomotropism of basic cathepsin K inhibitors contributes to increased cellular potencies against off-target cathepsins and reduced functional selectivity. J Med Chem. 2005;48:7535–43.
17. Eastell R, Nagase S, Ohyama M, et al. Safety and efficacy of the cathepsin K inhibitor ONO-5334 in postmenopausal osteoporosis: the OCEAN study. J Bone Miner Res. 2011;26:1303–12.
18. Stoch SA, Zajic S, Stone J, et al. Effect of the cathepsin K inhibitor odanacatib on bone resorption biomarkers in healthy postmenopausal women: two double-blind, randomized, placebo-controlled phase I studies. Clin Pharmacol Ther. 2009;86:175–82.
19. Stoch SA, Zajic S, Stone JA, et al. Odanacatib, a selective cathepsin K inhibitor to treat osteoporosis: safety, tolerability, pharmacokinetics and pharmacodynamics—results from single oral dose studies in healthy volunteers. Br J Clin Pharmacol. 2013;75:1240–54.
20. Anderson MS, Gendrano IN, Liu C, et al. Odanacatib, a selective cathepsin K inhibitor, demonstrates comparable pharmacodynamics and pharmacokinetics in older men and postmenopausal women. J Clin Endocrinol Metab. 2014;99:552–60.
21. Bone HG, McClung MR, Roux C, et al. Odanacatib, a cathepsin-K inhibitor for osteoporosis: a two-year study in postmenopausal women with low bone density. J Bone Miner Res. 2010;25:937–47.
22. Nakamura T, Shiraki M, Fukunaga M, et al. Effect of the cathepsin K inhibitor odanacatib administered once weekly on bone mineral density in Japanese patients with osteoporosis—a double-blind, randomized, dose-finding study. Osteoporos Int. 2014;25:367–76.
23. Eisman JA, Bone HG, Hosking DJ, et al. Odanacatib in the treatment of postmenopausal women with low bone mineral density: three-year continued therapy and resolution of effect. J Bone Miner Res. 2011;26:242–51.
24. Langdahl B, Binkley N, Bone H, et al. Odanacatib in the treatment of postmenopausal women with low bone mineral density: five years of continued therapy in a phase 2 study. J Bone Miner Res. 2012;27:2251–8.
25. Rizzoli R, Benhamou CL, Halse J, et al. Continuous treatment with odanacatib for up to 8 years in postmenopausal women with low bone mineral density: a phase 2 study. Osteoporos Int. 2016;27:2099–107.

26. Bonnick S, De Villiers T, Odio A, et al. Effects of odanacatib on BMD and safety in the treatment of osteoporosis in postmenopausal women previously treated with alendronate: a randomized placebo-controlled trial. J Clin Endocrinol Metab. 2013;98:4727–35.
27. Brixen K, Chapurlat R, Cheung AM, et al. Bone density, turnover, and estimated strength in postmenopausal women treated with odanacatib: a randomized trial. J Clin Endocrinol Metab. 2013;98:571–80.
28. Cheung AM, Majumdar S, Brixen K, et al. Effects of odanacatib on the radius and tibia of postmenopausal women: improvements in bone geometry, microarchitecture, and estimated bone strength. J Bone Miner Res. 2014;29:1786–94.
29. Engelke K, Fuerst T, Dardzinski B, et al. Odanacatib treatment affects trabecular and cortical bone in the femur of postmenopausal women: results of a two-year placebo-controlled trial. J Bone Miner Res. 2015;30: 30–8.
30. Bone HG, Dempster DW, Eisman JA, et al. Odanacatib for the treatment of postmenopausal osteoporosis: development history and design and participant characteristics of LOFT, the Long-Term Odanacatib Fracture Trial. Osteoporos Int. 2015;26:699–712.
31. McClung MR, Langdahl B, Papapoulos S, et al. Odanacatib antifracture-efficacy and safety in postmenopausal women with osteoporosis. Results from the phase III Long-term Odanacatib Fracture Trial (LOFT). J Bone Miner Res. 2014;29(S1):S51.
32. McClung MR, Langdahl B, Papapoulos S, et al. Odanacatib efficacy and safety in postmenopausal women with osteoporosis: 5-year data from the extension of the phase 3 Long-term Odanacatib Fracture Trial. J Bone Miner Res. 2016;31(S1):S50.
33. O'Donoghue M, Cavallari I, Bonaca M, et al. The Long-term Odanacatib Fracture Trial (LOFT): cardiovascular safety results. J Bone Miner Res. 2016;31(S1):S32.
34. Merck. Merck provides update on odanacatib development program. 2016. http://www.mrknewsroom.com/news-release/research-and-development-news/merck-provides-update-odanacatib-development-program (accessed May 2018).
35. van Lierop AH, Appelman-Dijkstra NM, Papapoulos SE. Sclerostin deficiency in humans. Bone. 2017;96:51–62.
36. Li X, Ominsky MS, Niu QT, et al. Targeted deletion of the sclerostin gene in mice results in increased bone formation and bone strength. J Bone Miner Res. 2008;23: 860–9.
37. Ominsky MS, Boyce RW, Li X, Ke HZ. Effects of sclerostin antibodies in animal models of osteoporosis. Bone. 2017;96:63–75.
38. Ominsky MS, Boyd SK, Varela A, et al. Romosozumab improves bone mass and strength while maintaining bone quality in ovariectomized cynomolgus monkeys. J Bone Miner Res. 2017;32:788–801.
39. Matheny JB, Torres AM, Ominsky MS, et al. Romosozumab treatment converts trabecular rods into trabecular plates in male cynomolgus monkeys. Calcif Tissue Int. 2017;101:82–91.
40. Thompson ML, Chartier SR, Mitchell SA, et al. Preventing painful age-related bone fractures: antisclerostin therapy builds cortical bone and increases the proliferation of osteogenic cells in the periosteum of the geriatric mouse femur. Mol Pain. 2016;12:pii: 1744806916677147.
41. Li X, Ominsky MS, Warmington KS, et al. Increased bone formation and bone mass induced by sclerostin antibody is not affected by pretreatment or cotreatment with alendronate in osteopenic, ovariectomized rats. Endocrinology. 2011;152:3312–22.
42. Chouinard L, Felx M, Mellal N, et al. Carcinogenicity risk assessment of romosozumab: a review of scientific weight-of-evidence and findings in a rat lifetime pharmacology study. Regul Toxicol Pharmacol. 2016;81: 212–22.
43. Recker RR, Benson CT, Matsumoto T, et al. A randomized, double-blind phase 2 clinical trial of blosozumab, a sclerostin antibody, in postmenopausal women with low bone mineral density. J Bone Miner Res. 2015;30:216–24.
44. Recknor CP, Recker RR, Benson CT, et al. The effect of discontinuing treatment with blosozumab: follow-up results of a phase 2 randomized clinical trial in postmenopausal women with low bone mineral density. J Bone Miner Res. 2015;30:1717–25.
45. Padhi D, Jang G, Stouch B, et al. Single-dose, placebo-controlled, randomized study of AMG 785, a sclerostin monoclonal antibody. J Bone Miner Res. 2011;26:19–26.
46. Padhi D, Allison M, Kivitz AJ, et al. Multiple doses of sclerostin antibody romosozumab in healthy men and postmenopausal women with low bone mass: a randomized, double-blind, placebo-controlled study. J Clin Pharmacol. 2014;54:168–78.
47. McClung MR, Grauer A, Boonen S, et al. Romosozumab in postmenopausal women with low bone mineral density. N Engl J Med. 2014;370:412–20.
48. McClung MR, Chines A Brown JP, et al. Effects of 2 years of treatment with romosozumab followed by 1 year of denosumab or placebo in postmenopausal women with low bone mineral density. ASBMR Annual Meeting 2014, abstract 1152.
49. Genant HK, Engelke K, Bolognese MA, et al. Effects of romosozumab compared with teriparatide on bone density and mass at the spine and hip in postmenopausal women with low bone mass. J Bone Miner Res. 2017;32:181–7.
50. Cosman F, Crittenden DB, Adachi JD, et al. Romosozumab treatment in postmenopausal women with osteoporosis. N Engl J Med. 2016;375:1532–43.
51. Cosman F, Miller PD, Williams GC, et al. Eighteen months of treatment with subcutaneous abaloparatide followed by 6 months of treatment with alendronate in postmenopausal women with osteoporosis: results of the ACTIVExtend Trial. Mayo Clin Proc. 2017;92:200–10.

Section VII
Metabolic Bone Diseases

Section Editors: Suzanne M. Jan de Beur and Peter R. Ebeling

Chapter 81. Approach to Parathyroid Disorders   613
John P. Bilezikian

Chapter 82. Primary Hyperparathyroidism   619
Shonni J. Silverberg, Francisco Bandeira, Jianmin Liu, Claudio Marcocci, and Marcella D. Walker

Chapter 83. Familial States of Primary Hyperparathyroidism   629
Andrew Arnold, Sunita K. Agarwal, and Rajesh V. Thakker

Chapter 84. Non-Parathyroid Hypercalcemia   639
Mara J. Horwitz

Chapter 85. Hypocalcemia: Definition, Etiology, Pathogenesis, Diagnosis, and Management   646
Anne L. Schafer and Dolores M. Shoback

Chapter 86. Hypoparathyroidism   654
Tamara Vokes, Mishaela R. Rubin, Karen K. Winer, Michael Mannstadt, Natalie E. Cusano, Harald Jüppner, and John P. Bilezikian

Chapter 87. Pseudohypoparathyroidism   661
Agnès Linglart, Michael A. Levine, and Harald Jüppner

Chapter 88. Disorders of Phosphate Homeostasis   674
Mary D. Ruppe and Suzanne M. Jan de Beur

Chapter 89. Rickets and Osteomalacia   684
Michaël R. Laurent, Nathalie Bravenboer, Natasja M. Van Schoor, Roger Bouillon, John M. Pettifor, and Paul Lips

Chapter 90. Pathophysiology and Treatment of Chronic Kidney Disease–Mineral and Bone Disorder   695
Mark R. Hanudel, Sharon M. Moe, and Isidro B. Salusky

Chapter 91. Disorders of Mineral Metabolism in Childhood   705
Thomas O. Carpenter and Nina S. Ma

Chapter 92. Paget Disease of Bone   713
Julia F. Charles, Ethel S. Siris, and G. David Roodman

Chapter 93. Epidemiology, Diagnosis, Evaluation, and Treatment of Nephrolithiasis   721
Murray J. Favus and David A. Bushinsky

Chapter 94. Immobilization and Burns: Other Conditions Associated with Osteoporosis   730
William A. Bauman, Christopher Cardozo, and Gordon L. Klein

81

Approach to Parathyroid Disorders

John P. Bilezikian

Metabolic Bone Diseases Unit, Division of Endocrinology, Department of Medicine,
College of Physicians and Surgeons, Columbia University, New York, NY, USA

INTRODUCTION

Disorders of the parathyroid glands are an important consideration in disorders of mineral metabolism. In this section, chapters focus upon many of these disorders either caused by excessive or inadequate secretion of PTH. The section also features disorders in which the parathyroid glands overproduce or underproduce PTH as a normal physiological adjustment to other inciting pathophysiological events that lead either to hypercalcemia or hypocalcemia. The primary parathyroid disorders, in which parathyroid glandular activity is intrinsically abnormal (eg, primary hyperparathyroidism [PHPT], hypoparathyroidism) and the secondary parathyroid disorders, in which increased or decreased parathyroid glandular activity is a normal adjustment to another pathophysiological process (eg, vitamin D deficiency, chronic renal disease), have given us new insights into the importance of PTH not only in the regulation of the serum calcium concentration but also in terms of skeletal health. Not covered in this section, but elsewhere, is a consequence of intense interest in and investigation of PTH, namely the use of PTH as a treatment for osteoporosis. Based upon our expanded knowledge base about the parathyroids, I offer, in this chapter, an approach to the primary and secondary parathyroid disorders. More detailed information will be found in the individual chapters that follow.

OVERSECRETION OF PTH CAUSED BY INTRINSIC FUNCTIONAL ABNORMALITIES OF THE PARATHYROID GLAND(S)

Primary oversecretion of PTH: PHPT

PHPT is the classic endocrine disorder associated with parathyroid gland hyperfunction (see Chapter 82 and [1]). It is most often a sporadic occurrence with only one of the four parathyroid glands involved in a benign, adenomatous process of excessive synthetic and secretory activity. Arnold and colleagues (Chapter 83) provide a comprehensive discussion of the genetics of the hyperparathyroid diseases when they present in their many other manifestations such as familial hypercalciuric hypercalcemia (FHH), hyperparathyroidism–jaw tumor syndrome, and in multiple glandular syndromes such as MEN1 and MEN2.

Although known since the late 1920s, PHPT has undergone a rather dramatic change in its clinical phenotype from a symptomatic disorder of "bones and stones" to one that in many parts of the world is asymptomatic. It is typically discovered incidentally during the course of a calcium measurement on a routine biochemical screening test [2]. Assisted by new technologies with which the skeleton can be evaluated, these and other aspects of PHPT have spawned greater interest than ever before in this disease.

Primer on the Metabolic Bone Diseases and Disorders of Mineral Metabolism, Ninth Edition. Edited by John P. Bilezikian.
© 2019 American Society for Bone and Mineral Research. Published 2019 by John Wiley & Sons, Inc.
Companion website: www.wiley.com/go/asbmrprimer

Hypercalcemia is the major clinical clue to the diagnosis of PHPT. Several highly useful PTH assays expedite the diagnosis and clearly distinguish this disease from nonparathyroid etiologies of hypercalcemia (see Chapter 84). Even with a PTH level that is not frankly elevated, PTH levels in the normal range establishes the diagnosis. The exquisite physiological regulation of PTH by calcium indicates that readily detectable levels of PTH in the context of hypercalcemia essentially rule out most other causes of hypercalcemia. Lithium, thiazide diuretics, and the exceedingly rare example of true ectopic PTH secretion are exceptions to this useful rule.

The skeleton, one of the major target organs in PHPT, has been a rich source of knowledge about PTH action. First shown by DXA and then followed by histomorphometric analyses of bone biopsies, the trabecular skeleton is relatively well preserved [3, 4]. DXA and bone biopsies also have established that the skeletal compartment that preferentially shows deterioration is cortical. These findings suggest that the nonvertebral skeleton, comprised substantially of cortical bone, should be a greatest risk for fracture in this disease. Epidemiological and more recent fracture data, however, argue that the fracture risk with PHPT is increased at all major sites [5, 6], suggesting that the trabecular compartment is also involved in PHPT. Confirmatory microarchitectural data from high resolution peripheral CT [7] and, indirectly, from the trabecular bone score index [8] support the epidemiological data. With successful surgery, both cortical and trabecular elements improve as shown by DXA [9] and also by high resolution peripheral CT [10].

Whereas these skeletal features are noteworthy, kidney stones are the most common overt complications of PHPT, with incidence figures in the neighborhood of 20% in most series [11, 12].

With appreciation that PHPT historically displays protean manifestations, it has been exceedingly difficult to identity nontraditional target organs that are specifically affected by the hyperparathyroid process in asymptomatic PHPT. It is still uncertain whether, and to what extent, putative neurocognitive and cardiovascular manifestations can be directly linked to the disease and/or whether they are reversible upon successful parathyroidectomy [13]. The most recent guidelines for surgery in PHPT do not include these putative nontraditional aspects of PHPT, but they are acknowledged for their potential importance. Rather, the guidelines are directed to more easily measurable and traditional end points, such as bone density, kidney stones, and renal function [14].

PHPT is being increasingly recognized in subjects whose serum total and ionized calcium are consistently normal. "Normocalcemic PHPT" is associated with levels of PTH that are consistently elevated [15]. Secondary causes of an elevated PTH level must be ruled out before normocalcemic PHPT can be seriously considered. An important consideration is the 25-hydroxyvitamin D level which is the index of body stores of vitamin D. Reduced 25-hydroxyvitamin D can account for an elevated PTH. The controversial matter of how vitamin D adequacy is defined is covered in this section of the Primer (see Chapter 89). For the purposes of establishing the diagnosis of normocalcemic PHPT, many experts require a 25-hydroxyvitamin D level that is greater than 30 ng/mL [16]. This level helps to ensure that the patient is not showing a subtle form of secondary hyperparathyroidism with levels that some people might regard as normal, such as between 20 and 30 ng/mL.

The normal serum calcium level in normocalcemic PHPT might lead one to expect weaker evidence for target organ involvement than in subjects whose PHPT is accompanied by overt hypercalcemia. However, in the experience so far with normocalcemic PHPT, many subjects show reduced BMD [17]. This may be caused by the fact that most of the published literature on normocalcemic PHPT has dealt with a referral population that is being evaluated for a specific reason, such as reduced bone mineral density. Screening an unselected population for normocalcemic PHPT might lead to the identification of a normocalcemic cohort whose parathyroid disease is minimal [18].

Approach to the patient

After the patient with PHPT is evaluated, parathyroidectomy may be recommended. The recommendation for surgical intervention is based usually on meeting one or more of the criteria for parathyroid surgery as set forth by the latest International Workshop [14]. However, these guidelines are not rules. Patients who meet one or more criteria may decide against surgery; patients who do not meet any criteria might opt for the operation [19]. This latter view is acceptable if there are no medical contraindications to surgery. Virtually all parathyroid surgeons now require successful preoperative localization of the abnormal parathyroid gland. Advances in imaging with high resolution modalities such as 4D CT have made it possible to identify abnormal parathyroid tissue in the vast majority of patients with PHPT [20, 21]. With preoperative localization and a highly experienced surgeon, parathyroidectomy can be performed under local anesthesia and conscious sedation with an outstanding outcome record [22, 23]. Successful parathyroidectomy is defined by a fall in intraoperative PTH levels by more than 50%, into the normal range, after removal of the offending adenoma. In patients who do not meet or refuse the recommendation of parathyroidectomy, observation is needed with annual serum calcium measurements and regular BMD monitoring. Pharmacological approaches to these individuals include the use of bisphosphonate [24] or the calcimimetic, cinacalcet [25].

Undersecretion of PTH: hypoparathyroidism

In contrast to primary hypersecretion of PTH, which is relatively common, the hypoparathyroid states in which PTH is undersecreted are uncommon. In fact, hypoparathyroidism is defined as an orphan disease because there are fewer than 200,000 affected individuals in the USA [26]. Hypoparathyroidism presents a contrast with its more common counterpart in another way, namely in terms of symptomatology. Asymptomatic PHPT is the most common way the hypersecretion syndrome presents. The widespread use of screening biochemical panels is responsible, in large part, for this observation. In hypoparathyroidism, on the other hand, subjects generally are not asymptomatic. They are invariably discovered only when classical signs and symptoms of hypocalcemia are present (see Chapter 85 and [26]). In hypoparathyroidism, the serum calcium is typically below normal limits and the PTH level is either undetectable or inappropriately low for the hypocalcemic state [27]. Similar to the physiology that governs suppression of PTH in hypercalcemic states not caused by PHPT, the physiology of hypocalcemia indicates that the PTH should be elevated, if not caused by an intrinsic functional abnormality or absence of the parathyroids. When hypocalcemia is not associated with elevated PTH levels, the diagnosis of a hypoparathyroidism is clear. Sequelae of neck surgery and autoimmune destruction of the parathyroids are the two most common causes of hypoparathyroidism. The genetics of the autoimmune form and other much less common hereditary manifestations of hypoparathyroidism, either in isolated form or involving multiple organ systems, are reviewed in this section of the Primer (Chapter 86).

In virtually all forms of hypoparathyroidism, the condition is permanent. The one exception is the setting of severe hypomagnesemia. In this situation, the marked magnesium deficiency impairs parathyroid secretory function, mimicking a hypoparathyroidism (see Chapter 23). However, with severe magnesium deficiency, the secretory abnormality is reversible, after magnesium is administered. Thus, in the evaluation of anyone who has hypocalcemia and low PTH levels, the serum magnesium should be measured.

The skeletal manifestations of hypoparathyroidism present an opportunity to address questions related to the role of PTH in skeletal metabolism and structure. The detailed knowledge of skeletal abnormalities in PHPT provides a counterpoint and permits the delineation of some of the findings to PTH itself. PHPT is a high turnover disease; hypoparathyroidism is a low turnover disease. Although the two ends of the bone turnover spectrum, high and low, are not always appreciated by the measurement of circulating or urinary bone turnover markers, such as P1NP, osteocalcin, CTX, or urinary NTX, it is clear by dynamic histomorphometry of bone biopsies that the two disorders are diametrically different from each other. By double tetracycline labeling, there is very little bone turnover in hypoparathyroidism [28], whereas bone turnover is generally high in PHPT (see Chapter 82 and [1]). In hypoparathyroidism, BMD is above average, using age-specific norms (Z-scores) or young normative databases (T-scores). It is not unusual for these individuals with hypoparathyroidism to show BMD values that are one- to threefold higher than normal (Z- and T-scores +1 to +3). By bone biopsy, there appears to be a cortical predominance in hypoparathyroidism that contrasts with the trabecular predominance of the hyperparathyroid skeleton [29]. These insights suggest a role for PTH in helping to regulate the distribution of cortical and trabecular bone at given sites. Clearly, overall site specificity to the distribution of cortical and trabecular bone is unlikely to be governed by PTH, but within a given site (lumbar spine, hip, or forearm), PTH might be a key modulator. Improvement of the abnormalities in these skeletal compartments after parathyroidectomy in PHPT and after PTH administration in hypoparathyroidism helps to assign this modulatory role to PTH [30].

Besides skeletal abnormalities, subjects with hypoparathyroidism are prone to deposition of calcium–phosphate complexes in the kidney, basal ganglia, and other soft tissues [31, 32]. Similar to PHPT, in hypoparathyroidism, there is a spectrum on nonspecific observations that are not clearly related to the disease. In many nonspecific ways, individuals with hypoparathyroidism do not feel well. They complain of lack of energy, easy fatigability, arthritic symptoms (without frank arthritis) and "brain fog." To what extent, these nonspecific symptoms can be reversed by PTH administration is not yet clear.

The administration of PTH represents a logical step in the definitive treatment of a disease characterized by lack of PTH. Although for many years, this goal was unrealized, the situation has changed because of studies that have led to the FDA approval of rhPTH(1-84) for the treatment of hypoparathyroidism. Early promising results of studies in hypoparathyroidism with teriparatide [PTH(1-34)] and the full length PTH molecule itself [PTH(1-84)] [30, 33, 34], led to a definitive clinical trial in which rhPTH(1-84) was shown to be efficacious in reducing the need for supplemental calcium and active vitamin D while maintaining the serum calcium level [35]. These advances led to an international symposium in which guidelines for the diagnosis and management of hypoparathyroidism were published recently [36, 37].

Approach to the patient

Hypoparathyroidism usually presents as a symptomatic condition with complaints ranging from mild paresthesias to tetany, and even seizures. In the patients who have not ever had neck surgery (note, the neck surgery may have occurred decades before), it is important to consider multiglandular autoimmune endocrine deficiency syndromes (see Chapter 85). It is also important to evaluate the patient for the possibility of ectopic calcifications (basal ganglia and other brain sites), renal calcifications, and if symptomatic, joint calcifications. The mainstay of treatment is calcium and active vitamin D (1,25-dihydroxyvitamin D). Although patients can be controlled with calcium and active vitamin D, and thiazide diuretics, in some cases, they can present a challenge because of unexplained wide swings in the level of control. In addition, many patients require very large doses of calcium and 1,25-dihydroxyvitamin D that raise additional concerns about the long-term sequelae of such chronic treatment. The use of calcium and active vitamin D (I prefer the judicious use of both parent vitamin D — either ergocalciferol or cholecalciferol — and 1,25-dihydroxyvitamin D) is clearly not equivalent to replacement therapy with PTH. Since PTH replacement therapy has become available for the treatment of hypoparathyroidism, this is now an option for those who cannot be well controlled without it. The international conference on the diagnosis and management of hypoparathyroidism has offered several situations in which the use of rhPTH(1-84) might be well advised [37].

HYPERCALCEMIC AND HYPOCALCEMIC DISORDERS NOT CAUSED BY AN INTRINSIC ABNORMALITY OF THE PARATHYROID GLANDS

Nonparathyroid-dependent hypercalcemia: reduced PTH levels

As already noted, the differential diagnosis of hypercalcemia caused by PTH or other cause is straightforward because of the PTH assay. An undetectable PTH level, the normal physiological response to hypercalcemia, argues in a compelling manner that the hypercalcemia is caused by a non-PTH dependent mechanism (see Chapter 84). When the hypercalcemia is associated with suppressed PTH, the next task is to establish the cause of hypercalcemia. The initial focus is usually malignancy, particularly if the patient presents with constitutional signs. If the malignancy is lung, breast, renal, or is myeloma, the diagnosis is usually readily established. If the PTHrP level is elevated, a squamous cell cancer becomes most suspect. Other malignancies such as pancreatic cancer or lymphoma may require extensive diagnostic testing which is generally not recommended unless there are clinical or biochemical clues. For example, a lymphomatous or granulomatous process might be suspected if the 1,25-dihydroxyvitamin D level is elevated. If the 25-hydroxyvitamin D is elevated, on the other hand, the diagnostic possibility focuses upon exogenous ingestion of vitamin D. There are times when after an extensive search, the etiology of the non-PTH-dependent hypercalcemia is not clear. In that scenario, time usually declares the condition.

Hypocalcemia: elevated PTH levels

The PTH assay helps to distinguish between hypocalcemia caused by a hypoparathyroid state and hypocalcemic conditions associated with a normal physiological response to hypocalcemia. If the PTH level is elevated, the search begins for the cause of the hypocalcemia. Often this is readily apparent such as a malabsorption syndrome, renal, or liver failure. In these so-called secondary hyperparathyroid states, the serum calcium can be low, but it is often in the lower range of normal. The therapeutic approach depends upon adequate control of the stimulus for PTH secretion. In the secondary hyperparathyroidism of renal disease, the pathophysiology and subsequent therapeutic approaches can be complex and is covered elsewhere in this Primer (see Chapter 90). A major goal in the secondary hyperparathyroidism of renal failure is to ascertain what the PTH level is and then to control PTH so that it is unlikely to be associated with unwanted effects of the secondary hyperparathyroid state. There are various official guidelines that suggest goals for maintaining PTH levels within a range [38, 39]. All guidelines acknowledge, however, that an acceptable PTH level in renal failure will be higher than the normal range. This point takes into account the fact that there are circulating inactive fragments of PTH that accumulate in renal failure and are detected by the commonly used intact assay for PTH. A new chapter in this Primer focuses upon chronic renal disease and fracture risk (see Chapter 65).

The dictum that in PHPT, the serum calcium is elevated, and that in the secondary hyperparathyroidism states, such as renal failure, the serum calcium is not elevated has to be qualified by several points. First, we now recognize a PHPT (normocalcemic PHPT) that is characterized by normal serum calcium levels, as described above. This is the reason why secondary causes for an elevated PTH must be ruled out before considering the diagnosis of normocalcemic PHPT. Second, with prolonged stimulation of the parathyroid glands caused by a renal or severe gastrointestinal disease, the serum calcium can rise to levels above normal. In this setting, the chronic stimulation of PTH secretion leads to the emergence of a semiautonomous state caused by the selection and proliferation of a clone of parathyroid cells into a single adenomatous gland. Thus, a prolonged secondary hyperparathyroidism can "morph" into a PHPT. Whenever the serum calcium becomes chronically elevated in someone who has had a prolonged stimulus for PTH secretion, this possibility should be considered.

In the absence of renal disease, liver disease, and overt vitamin D deficiency, identifying the stimulus for high PTH levels can be a challenge. The phosphorus level, if elevated, can be a clue to the diagnosis of pseudohypoparathyroidism, a classic genetic disorder associated with PTH resistance [40]. If the physical signs of pseudohypoparathyroidism are present (the Type 1a variant), the diagnosis is straightforward. However, there are other forms of pseudohypoparathyroidism, in which the classical physical phenotype is not present. Such variants of pseudohypoparathyroidism can present without any physical findings. This topic is covered elsewhere in this Primer (Chapter 87).

ACKNOWLEDGMENT

NIH grants: DK 32333 and DK 069350.

REFERENCES

1. Bilezikian JP. *Primary hyperparathyroidism*. In: DeGroot L (ed.), Singer F (section ed.) www.ENDOTEXT.org. South Dartmouth: MDTEXT.COM, Inc, 2017.
2. Silverberg SJ, Bilezikian JP. Primary hyperparathyroidism. In: Jameson JL, DeGroot LJ (eds) *Endocrinology* (7th edn). Philadelphia: Saunders, 2015, pp. 1105–24.
3. Silverberg SJ, Shane E, De La Cruz L, et al. Skeletal disease in primary hyperparathyroidism. J Bone Min Res. 1989;4:283–91.
4. Dempster DW, Silverberg SJ, Shane E, et al. Bone histomorphometry and bone quality in primary hyperparathyroidism. In: Bilezikian JP (editor-in-chief) *The Parathyroids: Basic and Clinical Concepts* (3rd edn). Cambridge, MA: Academic Press, 2015, pp. 429–45.
5. Khosla S, Melton LJ III, Wermers RA, et al. Primary hyperparathyroidism and the risk of fracture: a population based study. J Bone Min Res. 1999;14:1700–7.
6. Vignali E, Viccica G, Diacinti D, et al. Morphometic vertebral fractures in postmenopausal women with primary hyperparathyroidism. J Clin Endocrinol Metab. 2009;94:2306–12.
7. Silva BC, Cusano NE, Hans D, et al. Skeletal imaging in primary hyperparathyroidism. In: Bilezikian JP (editor-in-chief) *The Parathyroids: Basic and Clinical Concepts* (3rd edn). Cambridge, MA: Academic Press, 2015, pp. 447–54.
8. Silva BC, Boutroy S, Zhang C, et al. Trabecular bone score (TBS) – a novel method to evaluate bone microarchitectural texture in patients with primary hyperparathyroidism. J Clin Endocrinol Metab. 2013;98:1963–70.
9. Rubin MR, Bilezikian JP, McMahon DJ, et al. The natural history of primary hyperparathyroidism with or without parathyroid surgery after 15 years. J Clin Endocrinol Metab. 2008;93:3462–70.
10. Hansen S, Hauge EM, Rasmussen L, et al. Parathyroidectomy improves bone geometry and microarchitecture in female patients with primary hyperparathyroidism: a 1-year prospective controlled study using high resolution peripheral quantitative computed tomography. J Bone Miner Res. 2012;27:1150–8.
11. Rejnmark L, Vestergaard P, Mosekilde L. Nephrolithiasis and renal calcifications in primary hyperparathyroidism. J Clin Endocrinol Metab. 2011;96:2377–85.
12. Peacock M. Primary hyperparathyroidism and the kidney. In: Bilezikian JP (editor-in-chief) *The Parathyroids: Basic and Clinical Concepts* (3rd edn). Cambridge, MA: Academic Press, 2015, pp. 455–67.
13. Walker MD, Silverberg SJ. Non-traditional manifestations of primary hyperparathyroidism. In: Bilezikian JP (editor-in-chief) *The Parathyroids: Basic and Clinical Concepts* (3rd edn). Cambridge, MA: Academic Press, 2015, pp. 469–80.
14. Bilezikian JP, Brandi ML, Eastell R, et al. Consensus Statement: Guidelines for the Management of Asymptomatic Primary Hyperparathyroidism: Summary Statement from the Fourth International Workshop. J Clin Endocrinol Metab. 2014;99:3561–9.
15. Lowe H, McMahon DJ, Rubin MR, et al. Normocalcemic primary hyperparathyroidism: further characterization of a new clinical phenotype. J Clin Endocrinol Metab. 2007;92:3001–5.
16. Silverberg SJ, Clarke BL, Peacock M, et al. Consensus Statement: Current Issues in the Presentation of Asymptomatic Primary Hyperparathyroidism: Proceedings of the Fourth International Workshop. J Clin Endocrinol Metab. 2014;99:3580–94.
17. Cusano NE, Silverberg SJ, Bilezikian JP. Normocalcemic PHPT. In: Bilezikian JP (editor-in-chief) *The Parathyroids: Basic and Clinical Concepts* (3rd edn). Cambridge, MA: Academic Press, 2015, pp. 331–40.
18. Cusano NE, Maalouf N, Wang PY, et al. Normocalcemic hyperparathyroidism and hypoparathyroidism in two community-based non-referral populations. J Clin Endocrinol Metab. 2013;98:2734–41.
19. Marcocci C, Getani F. Primary hyperparathyroidism. N Eng J Med. 2011;365:2389–97
20. Hunter GJ, Schellinger D, Vu TH, et al. Accuracy of four-dimensional CT for the localization of abnormal parathyroid glands in patients with primary hyperparathyroidism. Radiology. 2012;264:789–95.
21. Khan AA, Hanley DA, Rizzoli R, et al. Primary hyperparathyroidism: review and recommendations on evaluation, diagnosis and management. A Canadian and international perspective. Osteoporosis Int. 2017;28:1–19.
22. Van Udelsman B, Udelsman R. Surgery in primary hyperparathyroidism: extensive personal experience. J Clin Densitom. 2013:16:54–9.
23. Udelsman R, Donovan P, Shaw C. Cure predictability during parathyroidectomy. World J Surg. 2014;38: 525–33.
24. Khan AA, Bilezikian JP, Kung AWC, et al. Alendronate in primary hyperparathyroidism: a double-blind, randomized, placebo-controlled trial. J Clin Endocrinol Metab. 2004;89:3319–25.
25. Peacock M, Bilezikian JP, Bolognese MA et al. Cinacalcet HCl reduces hypercalcemia in primary hyperparathyroidism across a wide spectrum of disease severity. J Clin Endocrinol Metab. 2011;96:E9–E18.
26. Clarke BL, Brown EM, Collins MT, et al. Epidemiology and diagnosis of hypoparathyroidism. J Clin Endocrinol Metab. 2016;101:2284–99.
27. Shoback D. Hypoparathyroidism. N Eng J Med. 2008;359: 391–403.
28. Rubin MR, Dempster DW, et al. Dynamic and structural properties of the skeleton in hypoparathyroidism. J Bone Miner Res. 2008;23:2018–24.
29. Rubin MR, Dempster DW, Kohler T, et al. Three-dimensional cancellous bone structure in hypoparathyroidism. Bone. 2009;46:190–5.

30. Rubin MR, Dempster DW, Sliney J, et al. PTH (1-84) administration reverses abnormal bone remodeling dynamics and structure in hypoparathyroidism. J Bone Miner Res. 2011;26:2727–36.
31. Mitchell DM, Regan S, Cooley MR, et al. Long-term follow-up of patients with hypoparathyroidism. J Clin Endocrinol Metab. 2012;97:4507–14.
32. Shoback DM, Bilezikian JP, Costa AG, et al. Presentation of hypoparathyroidism: etiologies and clinical features. J Clin Endocrinol Metab. 2016;101:2300–12.
33. Winer KK, Zhang B, Shrader JA, et al. Synthetic human parathyroid hormone 1-34 replacement therapy: a randomized crossover trial comparing pump versus injections in the treatment of chronic hypoparathyroidism. J Clin Endocrinol Metab. 2012;97:391–9.
34. Sikjaer T, Rejnmark L, Thomsen JS, et al. Changes in 3-dimensional bone structure indices in hypoparathyroid patients treated with PTH(1-84): a randomized controlled study. J Bone Miner Res. 2012;27:781–8.
35. Mannstadt M, Clarke BL, Vokes T, et al. Efficacy and safety of recombinant human parathyroid hormone (1-84) in hypoparathyroidism (REPLACE): a double-blind, placebo-controlled, randomized, phase 3 study. Lancet Diabetes Endocrinol. 2013;1:275–83.
36. Bilezikian JP, Brandi ML, Cusano NE, et al. Management of hypoparathyroidism: present and future. J Clin Endocrinol Metab. 2016;101:2313–24.
37. Brandi ML, Bilezikian JP, Shoback D, et al. Management of hypoparathyroidism: summary statement and guidelines. J Clin Endocrinol Metab. 2016;101:2273–83.
38. KDOQI US commentary on the 2009 KDIGO clinical practice guideline for the diagnosis, evaluation, and treatment of CKD-mineral and bone disorder (CKD-MBD). Am J Kidney Dis. 2010;55:773–99.
39. Kidney Disease: Improving Global Outcomes (KDIGO) CKDMBD Work Group. KDIGO clinical practice guideline for the diagnosis, evaluation, prevention, and treatment of chronic kidney disease-mineral and bone disorder (CKD-MBD) Kidney Int Suppl. 2009;113:S1–130.
40. Levine MA. Molecular and clinical aspects of pseudohypoparathyroidism. In: Bilezikian JP (editor-in-chief) *The Parathyroids: Basic and Clinical Concepts* (3rd edn). Cambridge, MA: Academic Press, 2015, pp. 781–807.

82

Primary Hyperparathyroidism

Shonni J. Silverberg[1], Francisco Bandeira[2], Jianmin Liu[3], Claudio Marcocci[4], and Marcella D. Walker[1]

[1]Department of Medicine, Columbia University College of Physicians and Surgeons, New York, NY, USA
[2]Division of Endocrinology, Diabetes and Metabolic Bone Diseases, Agamenon Magalhães Hospital, Brazilian Ministry of Health, University of Pernambuco Medical School, Recife, Brazil
[3]Department of Endocrine and Metabolic Diseases, Rui-jin Hospital, Shanghai Jiao-tong University School of Medicine, Shanghai, China
[4]Department of Clinical and Experimental Medicine, University of Pisa, Pisa, Italy

INTRODUCTION

Primary hyperparathyroidism (PHPT) is a disorder resulting from excessive secretion of PTH from one or more of the four parathyroid glands. In 90% of cases, hypercalcemia can be explained by PHPT or malignancy. The differential diagnosis of hypercalcemia is considered in Chapter 84.

PHPT is a relatively common endocrine disease. The disease was classically symptomatic, characterized by skeletal, renal, and neuromuscular manifestations, until the advent of the multichannel autoanalyzer in the 1970s revealed the presence of an asymptomatic phenotype of the disease. Once previously unrecognized cases were diagnosed in the 1970s and 1980s, incidence seemed to decline until a rise in the mid-1990s [1]. Availability and accuracy of data varies around the world (see A Global View later), but one recent report cites a prevalence of 0.86% in the USA, whereas another reports a threefold increase in incidence over the past 15 years [2, 3]. Women are affected more often than men. There are also ethnic variations in incidence in the USA (ie, higher incidence in black than white people) [2]. Although the majority of patients are postmenopausal women, the disease can present at all ages. When diagnosed in childhood, an unusual event, it is important to consider the possibility that the disease is a harbinger of a genetic endocrinopathy, such as multiple endocrine neoplasia (MEN) type 1 (MEN1) or 2 (MEN2).

ETIOLOGY AND PATHOGENESIS

PHPT occurs as a sporadic disease in over 90% of patients, and in 85% to 90% of cases is caused by a benign, solitary adenoma [4]. A parathyroid adenoma is a collection of chief cells surrounded by a rim of normal tissue at the outer perimeter of the gland. In those with a parathyroid adenoma, the remaining three parathyroid glands are usually normal. Multigland involvement occurs in approximately 10% of cases, with either multiple adenomas or four hyperplastic glands. Very rarely (<1%) there is pathological evidence of parathyroid carcinoma [5] in which the malignant tissue might show mitoses, vascular, or capsular invasion, and fibrous trabeculae. Unless gross local or distant metastases are present, the diagnosis of parathyroid cancer is difficult to make histologically. Specific genetic studies and immunohistochemical analyses *(CDC73* gene and parafibromin staining) may help to distinguish benign from malignant parathyroid tissue when standard approaches are not clear [5].

In about 10% of cases, PHPT may be part of hereditary syndromic (MEN1, MEN2, and MEN4), hereditary hyperparathyroidism–jaw tumor syndrome (HPT-JT), or non-syndromic (familial isolated hyperparathyroidism, familial hypocalciuric hypercalcemia [FHH], and neonatal severe hyperparathyroidism) forms of the disease, in which multigland involvement is more common.

The pathophysiology of PHPT relates to the loss of normal control of PTH synthesis and secretion. In

---

Primer on the Metabolic Bone Diseases and Disorders of Mineral Metabolism, Ninth Edition. Edited by John P. Bilezikian.
© 2019 American Society for Bone and Mineral Research. Published 2019 by John Wiley & Sons, Inc.
Companion website: www.wiley.com/go/asbmrprimer

adenomas, the parathyroid cell becomes less sensitive to the inhibitory effect of extracellular calcium (the "set-point" is shifted to the right), with increased secretion of PTH by individual cells [6]. In hyperplasia, the "set point" for calcium is not changed for a given parathyroid cell; instead, an increase in the number of cells, each secreting a normal amount of PTH, accounts for the increased PTH in the circulation. In virtually all other hypercalcemic conditions, there is feedback suppression of the parathyroid glands, and PTH levels are very low or undetectable.

The etiology of PHPT is apparent in only a small minority of patients. These include those who received external neck irradiation in childhood or through exposure during a radiation leak over 20 years previously. Chernobyl victims show a markedly increased risk of PHPT (Odds Ratio 63, CI 36–113) [7]. Long-term lithium therapy, which decreases parathyroid sensitivity to calcium, can be associated with hypercalcemia and hyperparathyroidism in a small percentage of patients [8].

The molecular basis for PHPT remains elusive in the vast majority of patients (see Chapter 83 for more extensive discussion). The clonal origin of most parathyroid adenomas suggests a defect at the level of the gene controlling growth of the parathyroid cell or the expression of PTH [9]. Several genes have been shown to be involved (Table 82.1). Germline mutations are found in hereditary forms. Somatic mutations usually occur in sporadic forms, even though in some cases germline mutations have been described. Cyclin-D1 overexpression has been found in 20% to 40% of sporadic parathyroid adenomas. In approximately 8% of cases a pericentromeric inversion causing a rearrangement of *CCND1* (the gene encoding for cyclin-D1) with the *PTH* gene accounts for transcriptional activation and overexpression of cyclin-D1 [9, 10]. Inactivating mutations of the *MEN1* gene have been reported in 12% to 35% of cases of sporadic parathyroid adenomas [9, 11]. The role of this tumor suppressor gene and menin, its gene product, in the pathogenesis of these sporadic adenomas is unknown. Other genetic abnormalities have been detected in a small proportion of parathyroid adenomas, including mutations in *CDC73*, *CDKN1B* (the *MEN4* encoding p27Kip1), and the aryl hydrocarbon receptor-interacting protein (AIP) genes [9, 12]. Contradictory results have been reported on the role of mutations in the *CTNNB1*, the gene encoding for β-catenin. Inactivating germline *CDC73* mutations have been described in the HPT–JT syndrome, in which they are associated with an increased risk of parathyroid cancer. Somatic mutations of this gene have also been reported in up to 70% of patients with sporadic parathyroid carcinoma [5, 9]. Germline and somatic mutations of the prune homolog 2 (Drosophila) (*PRUNE2*) gene have been detected in four out of 22 (18%) patients with parathyroid carcinoma [13]. Two studies have also provided evidence for a potential role of microRNA 296 as a novel suppressor gene in parathyroid carcinoma [14]. The genetic basis of familial hyperparathyroidism syndromes are incompletely understood as well. FHH and neonatal severe hyperparathyroidism are caused

Table 82.1. Genes involved in hereditary and sporadic primary hyperparathyroidism.

| Gene[a] | Chromosomal Locus | Hereditary | | Sporadic[c] |
		Disorder[b]	Pattern of Inheritance	
MEN1	11q13.1	MEN1, FIHP	Autosomal dominant	PA
RET	10q21.2	MEN2A	Autosomal dominant	–
CDKN1B	12p13.1	MEN4, FIHP	Autosomal dominant	PA
CDC73	1q31.2	HPT–JT, FIHP	Autosomal dominant	PA, PC
CASR	3q21.1	FHH1, FIHP	Autosomal dominant	–
		NSHPT	Autosomal recessive	–
GNA11	19p13.3	FHH2	Autosomal dominant	–
AP2S1	19q13.2-q13.3	FHH3	Autosomal dominant	–
CCND1	11q13.3	–	–	PA
AIP	11.q13.2	–	–	PA
CTNNB1	3p22.1	–	–	PA
PRUNE2	9q21.2	–	–	PC

[a]Abbreviations: *MEN1* = multiple endocrine neoplasia types 1; *RET* = rearranged during transfection; *CDKN1B* = cyclin-dependent kinase inhibitor 1B; *CDC73* = cell division cycle 73; *CASR* = calcium sensing receptor; *GNA11* = G protein subunit alpha 11; *AP2S1* = adaptor related protein complex 2 sigma 1 subunit; *CCND1* = cyclin D1; *AIP* = aryl hydrocarbon receptor interacting protein; *CTNNB1* = catenin beta 1; *PRUNE2* = prune homolog 2.
[b]Abbreviations: MEN1, 2A, 4 = multiple endocrine neoplasia types 1, 2A and 4; MEN2A = FIHP = familial isolated hyperparathyroidism; HPT–JT = hyperparathyroidism–jaw tumor; FHH1, 2 and 3 = familial hypocalciuric hypercalcemia; NSHPT = neonatal severe primary hyperparathyroidism.
[c]PA = parathyroid adenoma; PC = parathyroid carcinoma.

by inactivating mutations of the calcium-sensing receptor (*CASR*), whereas mutations of *MEN1*, *CDC73*, *CASR*, and *CDKN1B* have been described in some kindreds with familial isolated hyperparathyroidism.

SIGNS AND SYMPTOMS: FROM "CLASSICAL" TO "ASYMPTOMATIC" DISEASE

"Classical" PHPT was a symptomatic disease associated with a typical skeletal disorder (osteitis fibrosa cystica), nephrolithiasis, and neuromuscular complaints. Today the overwhelming majority of patients in the USA lack these symptoms [15]. Osteitis fibrosa cystica, characterized by subperiosteal resorption of the distal phalanges, tapering of the distal clavicles, a "salt and pepper" appearance of the skull, bone cysts, and brown tumors of the long bones, is now found in fewer than 5% of American patients.

The incidence of kidney stones has also declined, from 33% in the 1960s to 15% to 20% now. Nephrolithiasis remains the most common overt complication, and with increased abdominal imaging, clinically silent stones are also increasingly recognized [16]. Other renal features of PHPT include nephrocalcinosis and hypercalciuria (calcium excretion >250 mg/day for women or 300 mg/day for men), which is found in up to 30% of patients. The incidence of nephrocalcinosis is unknown. PHPT may also be associated with reduced creatinine clearance.

The classic neuromuscular syndrome of PHPT included a definable myopathy that has virtually disappeared [15]. Many patients do, however, complain of easy fatiguability, weakness, and mental weariness. These complaints lead some experts to conclude that the term "asymptomatic" PHPT is a misnomer. Some psychiatric complaints (depression and anxiety) are also more common in these patients, and there is evidence of specific reversible areas of mild cognitive impairment (ie, verbal memory and nonverbal abstraction) [17]. However, the cumulative results of three randomized studies of surgery in patients with mild disease do not suggest that an individual patient can expect specific reversible changes in quality of life (QOL) or psychiatric symptoms after parathyroidectomy [18–20]. Therefore, surgical intervention specifically for improving QOL, neuropsychological, or psychiatric symptoms is not recommended [21].

In classical PHPT, cardiovascular features included myocardial, valvular, and vascular calcification, with subsequent increased cardiovascular mortality. The limited available data suggests that cardiovascular mortality is not increased in mild disease [22]. However, there is evidence for increased vascular stiffness (both of aorta and carotids), and increased aortic valve calcification [23, 24]. A recent meta-analysis showing that parathyroidectomy reduced left ventricular mass in PHPT also supports subtle cardiovascular effects of the hyperparathyroid state [25]. In the absence of MEN, randomized clinical trial data suggest that hypertension is not corrected or improved after successful surgery [26].

Other organ systems previously affected by PHPT are rarely involved today. Gastrointestinal manifestations included peptic ulcer disease and pancreatitis. Peptic ulcer disease is not linked in a pathophysiological way to PHPT unless MEN1 or MEN4 is present. Pancreatitis is no longer a complication of PHPT because the hypercalcemia tends to be so mild. Gout and pseudogout, anemia, band keratopathy, and loose teeth are not part of the picture of mild disease.

CLINICAL FORMS OF PHPT

In the USA today, classical PHPT is rarely observed [15]. The most common clinical presentation of PHPT is characterized by mild asymptomatic hypercalcemia, often discovered on a routine multichannel screening test. Most patients do not have specific complaints and do not show evidence for any target organ complications.

Normocalcemic PHPT is now recognized as a specific phenotype of the disease [15, 27, 28]. These patients have consistently normal total and ionized calcium serum calcium concentrations with elevated PTH levels in the absence of an identifiable cause for secondary hyperparathyroidism. The diagnosis is often made in individuals who are undergoing skeletal screening or evaluation for low BMD. In order to make this diagnosis, alternative causes for PTH elevation must be ruled out (ie, vitamin D deficiency, renal failure, hypercalciuria, calcium malabsorption). Vitamin D deficiency can lower calcium levels in hypercalcemic PHPT into the normal range. In some patients this constellation of findings represents the earliest manifestation of hypercalcemic PHPT, when PTH alone is elevated, and serum calcium is still normal. Whereas some patients quickly progress to hypercalcemic disease, some remain normocalcemic for many years. It is unknown whether surgical criteria for hypercalcemic disease apply to patients with normocalcemic disease [21].

Rarely, a patient will present with life-threatening hypercalcemia, so-called acute PHPT, or parathyroid crisis [29]. These patients are invariably symptomatic. This diagnosis should be considered in any patient who presents with acute hypercalcemia of unclear etiology. Other unusual clinical presentations of PHPT include multiple endocrine neoplasia, familial PHPT not associated with other endocrine disorders, familial cystic parathyroid adenomatosis, and neonatal PHPT.

DIAGNOSIS AND EVALUATION OF PHPT

The physical examination rarely gives any clear indications of PHPT but a history and physical may suggest an alternative cause for hypercalcemia. The diagnosis of PHPT is instead established by laboratory tests [28].

Biochemistries

The biochemical hallmarks of PHPT are hypercalcemia (except in patients with normocalcemic PHPT) and elevated levels of PTH. In the presence of hypercalcemia, an elevated level of PTH virtually establishes the diagnosis. A PTH level in the mid or upper end of the normal range in the face of hypercalcemia is also consistent with the diagnosis. PTH levels may be measured using the second generation immunoradiometric (IRMA) or immunochemiluminometric (ICMA) or by the third generation assay specific for PTH(1-84). Second generation assays also measure large carboxyterminal fragments [PTH(7-84)] of PTH and will overestimate the amount of bioactive hormone, particularly in renal disease. Most data suggest that either second or third generation PTH assays can be used in the diagnosis of the disease [28]. All nonparathyroid-mediated causes of hypercalcemia (including malignancy) are associated with suppressed PTH levels. There is no cross-reactivity between PTH and PTHrP (the major causative factor in humoral hypercalcemia of malignancy) in the above assays for PTH. PTH concentration and serum calcium levels are also elevated in FHH [30]. Some cases of reversible PHPT are related to lithium or thiazide diuretic use [8, 31]. More commonly, thiazide diuretics unmask PHPT by inhibiting calcium excretion. If it is safe to withdraw the medication, the diagnosis is confirmed by persistence of hypercalcemia and elevated PTH levels several months later.

Serum phosphorus tends to be in the lower normal range and frankly low in approximately one-third of patients. Bone turnover markers are high normal or elevated in those with active bone disease. 1,25-dihydroxyvitamin D concentration is elevated in 25% of patients whereas 25-hydroxyvitamin D levels tend to be in the lower end of the normal range. Until recently most patients with PHPT had levels that were consistently below 30 ng/mL and often less than 20 ng/mL [32]. Newer data shows less vitamin D deficiency in American PHPT patients, likely due to self-supplementation with vitamin D [33].

Bone

Although osteitis fibrosa cystica is uncommon, imaging studies show that the skeleton remains an important target organ in PHPT. Routine bone mineral densitometry is indicated in all patients with PHPT. DXA generally shows preferential effects on the cortical as opposed to the cancellous skeleton (Fig. 82.1) with reduced BMD of the distal third of the forearm, a site enriched in cortical bone, whereas the lumbar spine, a site enriched in cancellous bone, is relative preserved [34]. The preferential reduction of bone density at the distal forearm underscores the importance of measuring BMD at that site in PHPT. A small subset of patients (15%) present with an atypical BMD profile, characterized by vertebral osteopenia or osteoporosis, whereas occasionally patients may show reduced BMD at all sites [35].

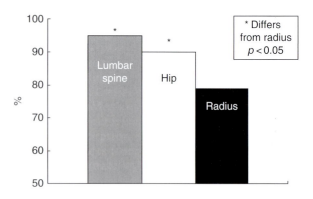

Fig. 82.1. Bone mineral density in primary hyperparathyroidism. Data are shown as percent of expected by site. Source: [34]. Reproduced with permission of John Wiley & Sons.

Table 82.2. Guidelines for use in patients with asymptomatic primary hyperparathyroidism.

Measure	Surgical Criteria	Guidelines for Follow-Up of Nonsurgical Patients
Serum calcium	>1 mg/dL above normal	Annually
Skeletal		
DXA	≤T-score −2.5	Every 1–2 years at spine, hip and forearm
Vertebral imaging	Vertebral fracture	If clinically indicated (eg, height loss, back pain)
Renal		
Estimated GFR	<60 mL/min	Annually
Serum creatinine		Annually
24-hour urine calcium	>400 mg/day	If nephrolithiasis suspected
Biochemical stone risk	Increased risk	If nephrolithiasis suspected
Renal imaging	Stone present	If nephrolithiasis suspected
Age	Age <50	

Guidelines describe patients in whom surgical intervention is desirable and criteria for follow-up of nonsurgical patients. It is recommended that all symptomatic patients be sent for parathyroidectomy.

Despite data showing sparing of cancellous (lumbar) BMD by bone densitometry in PHPT, several studies indicate an increased risk of vertebral fractures in PHPT and/or a reduction in risk after parathyroidectomy [36, 37]. For this reason, the most recent guidelines recommend imaging of the spine to assess for clinically silent vertebral compression fracture (Table 82.2). Data regarding fracture risk at nonvertebral skeletal sites are inconsistent [15].

New imaging techniques provide insight into this paradoxical increase in vertebral fracture risk in PHPT despite relatively preserved BMD. High resolution peripheral quantitative tomography shows that both trabecular as well as cortical microarchitecture is degraded in PHPT [38]. Trabecular bone score (TBS) assessment of the lumbar spine also shows values in the partially degraded range in PHPT [39]. Thus it is clear that even asymptomatic disease can have deleterious effects on both cancellous and cortical bone.

Renal

Evaluation for renal involvement includes obtaining a history for nephrolithiasis, an estimate of glomerular filtration rate, and a 24-hour urinary calcium measurement (to rule out FHH and assess kidney stone risk) [21]. Once the diagnosis of PHPT is confirmed, renal imaging for occult stones is indicated [21].

Other

Cardiac work-up and neuropsychological testing are not part of the routine evaluation of patients with PHPT.

TREATMENT OF PHPT

Parathyroid surgery

Surgery provides the only option for cure of PHPT. Although surgery is indicated in all patients with classical symptoms of PHPT, four national and international conferences (1990, 2002, 2008, 2013) have updated guidelines for surgical intervention in those with no clear signs or symptoms of their disease (Table 82.2) [21, 40–42].

Surgical guidelines (Table 82.2) [21]

Asymptomatic patients are advised to have surgery if they have: (i) serum calcium more than 1 mg/dL above the upper limit of normal; (ii) renal indications – creatinine clearance <60 mL/min, nephrocalcinosis or nephrolithiasis identified on imaging, 24 hour urine calcium >400 mg/day or increased stone risk by biochemical stone risk analysis; (iii) skeletal indications – bone density in the osteoporotic range at any site (T-score ≤ –2.5) or, recognizing that spine DXA may not accurately reflect fracture risk, evidence of vertebral fracture by imaging; (iv) patients younger than 50 years of age who are at greater risk for disease progression than older patients. Surgery is an acceptable approach even in patients who do not meet surgical guidelines. Physician and patient preference are clearly important in this decision. Advances in surgical technique, efficacy, and safety may also shift the balance in favor of intervention in the eyes of some patients and their physicians [43, 44].

Preoperative localization is used to identify candidates for minimally invasive parathyroidectomy (MIP) as well as in disease that is recurrent or persistent after surgery [43, 44]. These techniques should not be used to make the diagnosis; rather they should guide the surgeon once a diagnosis is made. The most widely used localization modalities are technetium-99m-sestamibi (with or without SPECT – single photon emission computed tomography) or ultrasound. The former is excellent in single gland disease, but is often inaccurate in multigland disease. Four-dimensional CT scanning (the 4th dimension is time because multiple images are acquired over time after contrast) provides excellent anatomical detail. Arteriography and selective venous studies are reserved for those individuals in whom the noninvasive studies have not been successful. In patients who have undergone prior unsuccessful surgery, localization by two different modalities is suggested. MRI can be considered in this situation but is not an early choice for localization.

Surgery

Even without localization, an experienced parathyroid surgeon will find the abnormal parathyroid gland(s) 95% of the time in the patient who has not had previous neck surgery. The glands are notoriously variable in location, requiring the surgeon's knowledge of typical ectopic sites (intrathyroidal, retroesophageal, lateral neck, mediastinum). Four-gland exploration, long considered the gold standard surgical approach, remains the procedure of choice in patients with no suggestive localization studies and those with hereditary disease or lithium-induced disease, in whom multigland involvement is common. Today, MIP is the procedure of choice in patients in whom preoperative localization has localized single gland disease [43, 44] in centers with the capability to measure intraoperative PTH levels. Taking advantage of the short half-life of PTH (3–5 minutes), an intraoperative PTH level is drawn shortly after resection [44]. If the PTH level falls by 50% and is within the normal range, the adenoma that has been removed is considered to be the only source of abnormal glandular activity. There is some concern about a 50% decline alone, because if the PTH level remains elevated, other glandular sources of PTH may remain. In the case of multiglandular disease, the approach is to remove all tissue except for a parathyroid tissue fragment that is left in situ or autotransplanted in the nondominant forearm. Potential complications of surgery include damage to the recurrent laryngeal nerve, which can lead to hoarseness and reduced voice volume, and permanent hypoparathyroidism in those who have had previous neck surgery or who undergo subtotal parathyroidectomy (for multiglandular disease).

Postoperative transient hypocalcemia, when the remaining suppressed parathyroid glands are regaining their sensitivity to calcium, can be prevented in most cases by providing patients with several grams of calcium daily during the first postoperative week. Prolonged postoperative symptomatic hypocalcemia as a result of rapid deposition of calcium into bone ("hungry bone syndrome") is rare today, but may require treatment with parenteral calcium.

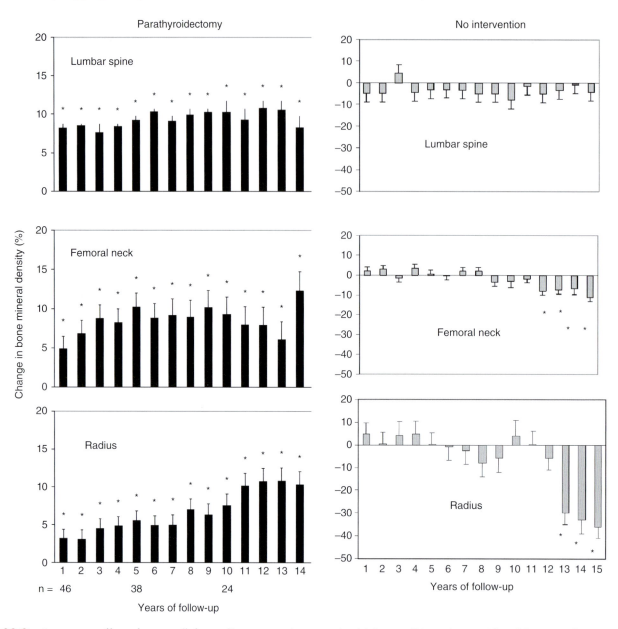

Fig. 82.2. Long-term effect of surgery (left panel) versus nonintervention (right panel) in patients with mild primary hyperparathyroidism. Source: [46]. Reproduced with permission of Oxford University Press.

After successful surgery, the patient is cured. Serum biochemistries and PTH levels normalize. Long-term observational data confirm short-term randomized trial findings showing that bone density improves in the first several years after surgery [18–20, 45, 46]. The cumulative increase in bone mass at the lumbar spine and femoral neck can be as high as 12%, an increase that is sustained for years after parathyroidectomy (Fig. 82.2). The substantial improvement found at the lumbar spine is even more impressive in patients with vertebral osteopenia or osteoporosis. These patients should be routinely referred for surgery, regardless of the severity of their hypercalcemia.

Nonsurgical management

Most patients who are not surgical candidates for parathyroidectomy do well when they are managed conservatively. In most such patients biochemical indices and BMD remain stable for close to a decade [45]. However, patients followed for longer than that period begin to show evidence of bone loss, particularly at the more cortical sites (hip and radius; Fig. 82.2) [46]. In a 15-year observational cohort, 37% of patients with asymptomatic PHPT had biochemical or bone densitometric evidence of disease progression. Those under the age of 50 years have a far higher incidence of progressive disease than do older patients (65% versus 23%), supporting the notion that younger patients should be referred for parathyroidectomy [47]. Finally, today, as in the day of classical PHPT, patients with symptomatic disease do poorly when observed without surgery. Thus, the data support the safety of observation without surgery only in selected patients with asymptomatic PHPT, and even in those, indefinite observation is not clearly desirable.

A set of general medical guidelines is recommended for patients who do not undergo surgery (Table 82.2) [21]. Serum calcium levels should be measured once yearly with annual assessment of serum creatinine and annual or biannual bone densitometry at the spine, hip, and distal one-third site of the forearm. Spine and renal imaging should be repeated if there is suspicion for a new vertebral fracture or nephrolithiasis. Adequate hydration and ambulation are always encouraged. Thiazide diuretics and lithium should be avoided if possible. Dietary calcium intake should therefore be moderate, because low calcium diets could theoretically lead to further stimulation of PTH secretion. High calcium intake (>1 g/day) should be avoided in patients whose 1,25-dihydroxyvitamin D levels are elevated. Current guidelines recommend treatment with vitamin D to maintain 25-hydroxyvitamin D levels higher than 20 ng/mL [21]. However, a recent RCT indicates higher vitamin D levels could be beneficial [48]. Treatment with 2800 IU cholecalciferol daily resulted in an increase in vitamin D from 25 to 37 ng/mL, a decline in PTH, and an increase in lumbar spine BMD without a significant change in serum or urine calcium.

Medical therapy

We still lack an effective and safe therapeutic agent approved for the medical management of PHPT in most patients [49]. *Oral phosphate* will lower the serum calcium by 0.5 to 1 mg/dL, by interfering with absorption of dietary calcium, inhibiting bone resorption and renal production of 1,25-dihydroxyvitamin D. Currently, oral phosphate is not recommended because of concerns for ectopic soft tissue calcification (increased calcium–phosphate product), the possibility that it could further elevate PTH levels, and its limited gastrointestinal tolerance. *Estrogen replacement therapy* remains an option in postmenopausal women desiring hormone replacement for treatment of symptoms of menopause [50, 51]. The rationale for estrogen use is based on the known antagonism by estrogen of PTH-mediated bone resorption. Although the serum calcium concentration does tend to decline after estrogen administration (by 0.5 mg/dL), PTH levels and the serum phosphorous concentration do not change. Estrogen replacement may improve BMD in these patients as well. Preliminary data suggest that the *selective estrogen receptor modulator*, raloxifene, may have a similar effect on serum calcium levels in postmenopausal women with PHPT [52]. *Bisphosphonates* are an option for low bone density in affected patients [49, 53]. Alendronate inhibits bone resorption and improves vertebral BMD, but it does not affect the underlying disorder. *Calcimimetic* agents increase the sensitivity of the extracellular calcium-sensing receptor to calcium, leading to a subsequent reduction in PTH synthesis and secretion, and ultimately a fall in the serum calcium. Trials of cinacalcet HCl have shown normalization of serum calcium for up to 5.5 years, and sustained decreases in serum calcium levels across a wide range of disease severity [54, 55]. However, this agent does not provide the equivalent of a "medical parathyroidectomy" because it does not improve bone density or reduce urinary calcium. There are no data on the effect of calcimimetics on constitutional or neuropsychological symptoms, fracture, or nephrolithiasis. Cinacalcet is approved for use in parathyroid cancer, and for PHPT in Europe. The FDA approved its limited use in PHPT in 2011 for the treatment of severe hypercalcemia in patients who are unable to undergo parathyroidectomy.

PRIMARY HYPERAPARATHYROIDISM: A GLOBAL VIEW

The presentation of PHPT differs in various geographic locales, and in many countries asymptomatic PHPT is not the prevailing form of the disease [15]. A brief summary of available data follows.

Europe

As in the USA, asymptomatic PHPT has become the predominant form of presentation in Europe [15, 56]. Population-based studies showed gender differences in the prevalence of PHPT (0.3% of adult Swedish men versus 1.6% of women) and, as in the USA, an increasing prevalence over the past decades (from 1.82 to 6.72 per 1000 population in Scotland) [56–58]. Incidence has also increased in the Czech Republic (0.08 to 0.24/1000 persons/year over four decades) [59].

Several studies provide clinical comparisons with PHPT in the USA. A retrospective study showed similar ionized calcium and PTH concentrations, but higher total calcium levels and lower BMI-adjusted hip BMD in Italian versus American patients [56]. Higher rates of kidney stones (55%) and vertebral fractures (35%), especially when discovered by imaging in asymptomatic patients, have also been reported from Europe [16]. Turkish patients had higher calcium and PTH levels than are generally found in the USA or Western Europe (mean PTH: 467 pg/mL; mean calcium: 11.9 mg/dL) [60]. Finally, European data on normocalcemic PHPT are very limited and have not used the rigorous diagnostic criteria recommended in international guidelines.

Latin America

Changes in the presentation of PHPT have also been observed in Latin America. Although population wide prevalence data are not available, in Recife, Brazil the prevalence of PHPT among 4207 endocrinology clinic patients was 0.78% of whom 82% were asymptomatic [61]. The female/male ratio was 7:1, and 90% of these women were postmenopausal.

Both symptomatic and asymptomatic disease are observed [62]. In severely affected patients, BMD is lower, PTH and bone markers are higher, vitamin D deficiency more common, and adenomas larger than in asymptomatic patients [63]. Patients with osteitis fibrosa cystica have large increases in BMD after parathyroidectomy and vitamin D supplementation [63]. Kidney stones are more common than in the USA [64, 65]. Over 80% of surgical patients had a single adenoma [62, 63, 65]. Normocalcemic PHPT is also increasingly recognized in Latin America. One study reported similar frequency of kidney stones and fractures, and no differences in BMD except for the forearm, which was higher in normocalcemic than in hypercalcemic patients [66].

Asia and Africa

In contrast to Western countries, PHPT in most Asian countries (China, India, Iran, Pakistan, Saudi Arabia, and Thailand) and South Africa still presents with classical hypercalcemic manifestations, including osteitis fibrosa cystica, osteoporosis, fractures, and renal stones [67–70]. The biochemical profile is similarly more severe than is found in Western patients. Recently, asymptomatic disease has been reported in China and India, probably due to the increased frequency of routine biochemistry testing and neck ultrasonography [70, 71]. Parathyroid carcinoma is more frequent in Asian than Western patients, with prevalence of 6% and 12% in Chinese and Japanese PHPT patients, respectively [70, 72]. Parathyroidectomy increases BMD and improves quality of life in Asian patients compared with those without surgery [73]. In parathyroid cancer, en-bloc tumor resection at initial or even second surgery has provided a chance of cure in Asian patients [74].

ACKNOWLEDGMENT

This work was supported in part by NIH Grants DK32333, DK074457, DK084986, and DK104105.

REFERENCES

1. Wermers RS, Khosla S, Atkinson EJ, et al. Incidence of primary hyperparathyroidism in Rochester, Minnesota, 1993–2001: an update on the changing epidemiology of the disease. J Bone Miner Res. 2006;21(1):171–7.
2. Press DM, Siperstein AE, Berber E, et al. The prevalence of undiagnosed and unrecognized primary hyperparathyroidism: a population-based analysis from the electronic medical record. Surgery. 2013;154(6):1232–7.
3. Yeh MW, Ituarte PH, Zhou HC, et al. Incidence and prevalence of primary hyperparathyroidism in a racially mixed population. J Clin Endocrinol Metab. 2013;98(3):1122–9.
4. Marcocci C, Cetani F. Clinical practice. Primary hyperparathyroidism. N Engl J Med. 2011;365:2389–97.
5. Cetani F, Marcocci C. Parathyroid carcinoma. In: Bilezikian JP (editor-in-chief) *The Parathyroids: Basic and Clinical Concepts* (3rd edn). Cambridge, MA: Academic Press, 2015, pp. 409–22.
6. Brown EM. Role of the calcium-sensing receptor in extracellular calcium homeostasis. Best Pract Res Clin Endocrinol Metab. 2013;27(3):333–43.
7. Boehm BO, Rosinger S, Belyi D, et al. The parathyroid as a target for radiation damage. N Engl J Med. 2011;365(7):676–8.
8. Szalat A, Mazeh H, Freund HR. Lithium-associated hyperparathyroidism: report of four cases and review of the literature. Eur J Endocrinol. 2009;160:317–23.
9. Arnold A, Levine MA. Molecular basis of primary hyperparathyroidism. In: Bilezikian JP (editor-in-chief) *The Parathyroids: Basic and Clinical Concepts* (3rd edn). Cambridge, MA: Academic Press, 2015, pp. 279–97.
10. Yi Y, Nowak NJ, Pacchia AL, et al. Chromosome 11 genomic changes in parathyroid adenoma and hyperplasia: array CGH, FISH, and tissue microarray. Genes Chromosomes Cancer. 2008;47(8):639–48.
11. Cromer MK, Starker LF, Choi M, et al. Identification of somatic mutations in parathyroid tumors using whole-exome sequencing. J Clin Endocrinol Metab. 2012;97(9): E1774–81.
12. Pardi E, Marcocci C, Borsari S, et al. Aryl hydrocarbon receptor interacting protein mutations occur rarely in sporadic parathyroid adenomas J Clin Endocrinol Metab. 2013;98(7):2800–10.
13. Yu W, McPherson JR, Stevenson M, et al. Whole-exome sequencing studies of parathyroid carcinomas reveal novel PRUNE2 mutations, distinctive mutational spectra related to APOBEC-catalyzed DNA mutagenesis and mutational enrichment in kinases associated with cell migration and invasion. J Clin Endocrinol Metab. 2015;100(2): E360–4.
14. Corbetta S, Vaira V, Guarnieri V, et al. Differential expression of microRNAs in human parathyroid carcinomas compared with normal parathyroid tissue. Endocr Relat Cancer 2010;17(1):135–46.
15. Silverberg SJ, Clarke BL, Peacock M, et al. Current issues in the presentation of asymptomatic primary hyperparathyroidism: proceedings of the Fourth International Workshop. J Clin Endocrinol Metab. 2014; 99:3580–94.
16. Cipriani C, Biamonte F, Costa AG, et al. Prevalence of kidney stones and vertebral fractures in primary hyperparathyroidism using imaging technology. J Clin Endocrinol Metab. 2015;100(4):1309–15.

17. Walker MD, McMahon DJ, Inabnet WB, et al. Neuropsychological features in primary hyperparathyroidism: a prospective study. J Clin Endocrinol Metab. 2009;94(6):1951–8.
18. Bollerslev J, Jannson S, Mollerup CL, et al. Medical observation, compared with parathyroidectomy, for asymptomatic primary hyperparathyroidism: a prospective, radomized trial. J Clin Endocrinol Metab. 2007;92(5):1687–92.
19. Ambrogini E, Centani F, Cianferotti L, et al. Surgery or surveillance for mild asymptomatic primary hyperparathyroidism: a prospective, randomized clinical trial. J Clin Endocrinol Metab. 2007;92(8):3114–21.
20. Rao DS, Phillips ER, Divine GW, et al. Randomized controlled clinical trial of surgery versus no surgery in patients with mild asymptomatic primary hyperparathyroidism. J Clin Endocrinol Metab. 2004;89(11):5415–22.
21. Bilezikian JP, Brandi ML, Eastell R, et al. Guidelines for the management of asymptomatic primary hyperparathyroidism: summary statement from the Fourth International Workshop. J Clin Endocrinol Metab. 2014;99:3561–9.
22. Wermers RA, Khosla S, Atkinson EJ, et al. Survival after the diagnosis of hyperparathyroidism: a population-based study. Am J Med 1998;104(2):115–22.
23. Walker MD, Fleischer J, Rundek T, et al. Carotid vascular abnormalities in primary hyperparathyroidism. J Clin Endocrinol Metab. 2009;94:3849–56.
24. Iwata S, Walker MD, Di Tullio MR, et al. Aortic valve calcification in mild primary hyperparathyroidism. J Clin Endocrinol Metab. 2012;97(1):132–7.
25. McMahon DJ, Carrelli A, Palmeri N, et al. Effect of parathyroidectomy upon left ventricular mass in primary hyperparathyroidism: a meta-analysis. J Clin Endocrinol Metab. 2015;100:4399–407.
26. Bollerslev J, Rosen T, Mollerup CL, et al.; SIPH Study Group. Effect of surgery on cardiovascular risk factors in mild primary hyperparathyroidism. J Clin Endocrinol Metab. 2009;94(7):2255–61.
27. Cusano NE, Silverberg SJ, Bilezikian JP. Normocalcemic primary hyperparathyroidism. J Clin Densitom 2013;16(1):33–9.
28. Eastell R, Brandi ML, Costa AG, et al. Diagnosis of asymptomatic primary hyperparathyroidism: proceedings of the Fourth International Workshop. J Clin Endocrinol Metab. 2014;99(10):3570–9.
29. Fitzpatrick LA. Acute primary hyperparathyroidism. In: Bilezikian JP (editor-in-chief) *The Parathyroids: Basic and Clinical Concepts* (3rd edn). Cambridge, MA: Academic Press, 2015, pp. 401–8.
30. Fuleihan GEH, Brown EM. Familial Hypocalciuric Hypercalcemia and Neoanatal Severe Hyperparathyroidism. In: Bilezikian JP (editor-in-chief) *The Parathyroids: Basic and Clinical Concepts* (3rd edn). Cambridge, MA: Academic Press, 2015, pp. 365–87.
31. Wermers RA, Kearns AE, Jenkins GD, et al. Incidence and clinical spectrum of thiazide-associated hypercalcemia. Am J Med 2007;120(10):911e–e915.
32. Boudou P, Ibrahim F, Cormier C, et al. A very high incidence of low 25 hydroxy-vitamin D serum concentration in a French population of patients with primary hyperparathyroidism. J Endocrinol Invest 2006;29(6):511–15.
33. Walker MD, Cong E, Lee JA, et al. Low vitamin D levels have become less common in primary hyperparathyroidism. Osteoporos Int. 2015;26:2837–43.
34. Silverberg SJ, Shane E, de la Cruz L, et al. Skeletal disease in primary hyperparathyroidism. J Bone Miner Res. 1989;4(3):283–91.
35. Silverberg SJ, Locker FG, Bilezikian JP. Vertebral osteopenia: a new indication for surgery in primary hyperparathyroidism. J Clin Endocrinol Metab. 1996;81(11):4007–12.
36. Vignali E, Viccica G, Diacinti D, et al. Morphometric vertebral fractures in postmenopausal women with primary hyperparathyroidism. J Clin Endocrinol Metab. 2009; 94:2306–12.
37. Lundstam K, Heck A, Mollerup C, et al. Effects of parathyroidectomy versus observation on the development of vertebral fractures in mild primary hyperparathyroidism. J Clin Endocrinol Metab. 2015;100:1359–67.
38. Stein EM, Silva BC, Boutroy S, et al. Primary hyperparathyroidism is associated with abnormal cortical and trabecular microstructure and reduced bone stiffness in postmenopausal women. J Bone Miner Res. 2013;28:1029–40.
39. Walker MD, Saeed I, Lee JA, et al. Effect of concomitant vitamin D deficiency or insufficiency on lumbar spine volumetric bone mineral density and trabecular bone score in primary hyperparathyroidism. Osteoporos Int. 2016;27:3063–71.
40. National Institutes of Health. Consensus development conference statement on primary hyperparathyroidism. J Bone Miner Res. 1991;6(Suppl 2):S9–S13.
41. Bilezikian JP, Potts JT Jr, Fuleihan Gel-H, et al. Summary statement from a workshop on asymptomatic primary hyperparathyroidism: a perspective for the 21st century. J Clin Endocrinol Metab. 2002;87(12):5353–61.
42. Bilezikian JP, Khan AA, Potts JT. Guidelines for the management of asymptomatic primary hyperparathyroidism: summary statement from the third international workship. J Clin Endocrinol Metab. 2009;94(2):335–9.
43. Udelsman R, Åkerström G, Biagini C, et al. The surgical management of asymptomatic primary hyperparathyroidism: proceedings of the Fourth International Workshop. J Clin Endocrinol Metab. 2014;99(10):3595–606
44. Wilhelm SM, Wang TS, Ruan DT, et al. The American Association of Endocrine Surgeons Guidelines for Definitive Management of Primary Hyperparathyroidism. JAMA Surg. 2016;151(10):959–68
45. Silverberg SJ, Shane E, Jacobs TP, et al. A 10-year prospective study of primary hyperparathyroidism with or without parathyroid surgery. N Engl J Med. 1999; 341(17):1249–55.
46. Rubin MR, Bilezikian JP, McMahon DJ, et al. The natural history of primary hyperparathyroidism with or without parathyroid surgery after 15 years. J Clin Endocrinol Metab. 2008;93(9):3462–70.

47. Silverberg SJ, Brown I, Bilezikian JP. Age as a criterion for surgery in primary hyperparathyroidism. Am J Med 2002;113(8):681–4.
48. Rolighed L, Rejnmark L, Sikjaer T, et al. Vitamin D treatment in primary hyperparathyroidism: a randomized placebo controlled trial. J Clin Endocrinol Metab. 2014; 99:1072–80.
49. Marcocci C, Bollerslev J, Khann AA, et al. Medical management of primary hyperparathyroidism: proceedings from the fourth International Workshop on the Management of Asymptomatic Primary Hyperparathyroidisim. J Clin Endocrinol Metab. 2014;99(10):3607–18.
50. Marcus R, Madvig P, Crim M, et al. Conjugated estrogens in the treatment of postmenopausal women with hyperparathyroidism. Ann Intern Med 1984;100(5):633–40.
51. Grey AB, Stapleton JP, Evans MC, et al. Effect of hormone replacement therapy on BMD in post-menopausal women with primary hyperparathyroidism. A randomized controlled trial. Ann Intern Med 1996;125(5):360–8.
52. Rubin MR, Lee K, Silverberg SJ. Raloxifene lowers serum calcium and markers of bone turnover in primary hyperparathyroidism. J Clin Endocrinol Metab. 2003; 88(3):1174–8.
53. Kahn AA, Bilezikian JP, Kung AW, et al. Alendronate in primary hyperparathyroidism: A double-blind, randomized, placebo-controlled trial. J Clin Endocrinol Metab. 2004;89(7):3319–25.
54. Peacock M, Bilezikian JP, Klassen P, et al. Cinacalcet hydrochloride maintains long-term normocalcemia in patients with primary hyperparathyroidism. J Clin Endocrinol Metab. 2005;90(1):135–41.
55. Peacock M, Bolognese MA, Borofsky M, et al. Cinacalcet treatment of primary hyperparathyroidism: biochemical and bone densitometric outcomes in a five-year study. J Clin Endocrinol Metab. 2009;94(12):4860–7.
56. De Lucia F, Minisola S, Romagnoli E, et al. Effect of gender and geographic location on the expression of primary hyperparathyroidism. J Endocrinol Invest 2013;36(2):123–6.
57. Lundgren E, Rastad J, Thurfjell E, et al. Population-based screening for primary hyperparathyroidism with serum calcium and parathyroid hormone values in menopausal women. Surgery 1997;121:287–94.
58. Yu N, Donnan PT, Murphy MJ, et al. Epidemiology of primary hyperparathyroidism in Tayside, Scotland, UK. Clin Endocrinol 2009;71:485–93.
59. Broulík P, Adámek S, Libánský P, et al. Changes in the pattern of primary hyperparathyroidism in Czech Republic. Prague Med Rep 2015;116(2):112–21.
60. Usta A, Alhan E, Cinel A, et al. A 20-year study on 190 patients with primary hyperparathyroidism in a developing country: Turkey experience. Int Surg 2015;100(4):648–55.
61. Eufrasino CS, Veras A, Bandeira F. Epidemiology of primary hyperparathyroidism and its nonclassical manisfestations in the city of Recife, Brazil. Clin Med Ins Endocrinol Diabetes 2013;6:69–74.
62. Bandeira F, Cassibba S. Hyperparathyroidism and bone health. Curr Rheumatol Rep 2015;17(7):48.
63. Bandeira F, Griz L, Caldas G, et al. From mild to severe primary hyperparathyroidism: the Brazilian experience. Arq Bras Endocrinol Metab 2006; 50(4):657–63.
64. Oliveira U, Ohe M, Santos R, et al. Analysis of the diagnostic presentation profile, patathyroidectomy indication and bone mineral density follow-up of Brazilian patients with primary hyperparathyroidism. Braz J Med Biol Res 2007;40:519–26.
65. Spivacow F, Martinez C, Polonsky A. Primary hyperparathyroidism: postoperative long-term evolution. Medicina (B Aires) 2010;70:408–14.
66. Amaral L, Queiroz D, Marques T, et al. Normocalcemic versus hypercalcemic primary hyperparathyroidism: more stone than bone? *J Osteoporos* 2012;128352.
67. Shah VN, Bhadada S, Bhansali A, et al. Changes in clinical and biochemical presentations of primary hyperparathyroidism in India over a period of 20 years. The Indian J Med Res. 2014;139(5):694–9.
68. Liu JM, Cusano NE, Silva BC, et al. Primary hyperparathyroidism: a tale of two cities revisited - new york and shanghai. Bone Res. 2013;1(2):162–9.
69. Paruk IM, Esterhuizen TM, Maharaj S, et al. Characteristics, management and outcome of primary hyperparathyroidism in South Africa: a single-centre experience. Postgrad Med J. 2013;89(1057):626–31.
70. Zhao L, Liu JM, He XY, et al. The changing clinical patterns of primary hyperparathyroidism in Chinese patients: data from 2000 to 2010 in a single clinical center. J Clin Endocrinol Metab. 2013;98(2):721–8.
71. Mithal A, Kaur P, Singh VP, et al. Asymptomatic primary hyperparathyroidism exists in North India: retrospective data from 2 tertiary care centers. Endocrine Practice 2015;21(6):581–5.
72. Kobayashi T, Sugimoto T, Chihara K. Clinical and biochemical presentation of primary hyperparathyroidism in Kansai district of Japan. Endocr J. 1997;44(4):595–601.
73. Ramakant P, Verma AK, Chand G, et al. Salutary effect of parathyroidectomy on neuropsychiatric symptoms in patients with primary hyperparathyroidism: evaluation using PAS and SF-36v2 scoring systems. J Postgrad Med. 2011;57(2):96–101.
74. Xue S, Chen H, Lv C, et al. Preoperative diagnosis and prognosis in 40 parathyroid carcinoma patients. Clin Endocrinol. 2016;85(1):29–36.

83

Familial States of Primary Hyperparathyroidism

Andrew Arnold[1], Sunita K. Agarwal[2], and Rajesh V. Thakker[3]

[1]*Division of Endocrinology and Metabolism; Departments of Molecular Medicine and Medicine and Genetics; and Center for Molecular Oncology, University of Connecticut School of Medicine, Farmington, CT, USA*
[2]*Metabolic Diseases Branch, The National Institute of Diabetes and Digestive and Kidney Diseases, National Institutes of Health, Bethesda, MD, USA*
[3]*University of Oxford, Nuffield Department of Clinical Medicine, OCDEM Churchill Hospital, Oxford, UK*

INTRODUCTION

Persons with familial primary hyperparathyroidism (PHPT), defined by the combination of hypercalcemia and elevated or nonsuppressed serum PTH, are a small and important subgroup of all cases with PHPT (about 5%) [1, 2]. Their familial syndromes include multiple endocrine neoplasia (MEN) (types 1, 2A, and 4), familial (benign) hypocalciuric hypercalcemia (FHH), neonatal severe primary hyperparathyroidism (NSHPT), hyperparathyroidism–jaw tumor syndrome (HPT–JT), and familial isolated primary hyperparathyroidism (FIHPT) [1, 3–5]. These syndromes exhibit Mendelian inheritance patterns, and for most the main genes have been identified (Table 83.1). As more knowledge accumulates on genetic contributions to complex phenotypes, contributions are being identified from additional genes, including some for less penetrant and more subtle predispositions to PHPT.

FAMILIAL HYPOCALCIURIC HYPERCALCEMIA

Familial benign hypocalciuric hypercalcemia (FHH) (OMIM #145980) is an autosomal dominant syndrome [6, 7]. The prevalence of FHH is similar to multiple endocrine neoplasia type 1 (MEN1); either accounts for about 2% of PHPT cases.

Clinical expressions

Persons with FHH usually have mild or no symptoms [6, 7]. Easy fatigue, weakness, thought disturbance, or polydipsia are less common and less severe than in typical PHPT. Nephrolithiasis or hypercalciuria are as uncommon as in those unaffected. Bone radiographs are usually normal. Either chondrocalcinosis or premature vascular calcification can be present but are generally silent clinically. Bone mass and susceptibility to fracture are normal. Hypercalcemia has virtually 100% penetrance at all ages. Onset of PHPT is otherwise rare in infancy and thus can be useful for diagnosis. The range of serum calcium levels is similar to that in typical PHPT with a normal ratio of free to bound calcium in serum [6, 8, 9]. Serum magnesium is typically in the high range of normal or modestly elevated, and serum phosphate is modestly depressed. Urinary excretion of calcium is normal, with hypercalcemic and unaffected family members showing a similar distribution of values. Normal urine calcium with high serum calcium explains the concept of relative hypocalciuria in FHH. Parathyroid function, including serum PTH and $1,25(OH)_2D$, is usually normal, with modest elevations of either index in 5% to 10% of cases [8, 9]. Such "normal" parathyroid function indices in the presence of lifelong hypercalcemia are inappropriate, diagnostically useful, and reflect the primary role for the parathyroids in causing this hypercalcemia.

The parathyroid glands in FHH may be of normal size or may be mildly enlarged. Standard subtotal

Table 83.1. Outline of syndromes of familial primary hyperparathyroidism with emphasis on major features that distinguish among the syndromes.

Syndrome	Main Genes and Mutation/Variant Types[a]	Parathyroid Gland Aspects	Aspects Outside of the Parathyroids
FHH	CASR[-,b]	Hypercalcemia begins at birth Increased secretion not growth Persist after subtotal PTX Avoid PTX	Relative hypocalciuria
NSHPT	CASR[=]	Hypercalcemia begins at birth Ca above 16 mg % Four very large PT glands Needs urgent total PTX	Relative hypocalciuria
MEN1	MEN[-]	Begins at average age 20 years Asymmetric adenomas Recur 12 years after successful subtotal PTX	Tumors among more than 20 tissues (pituitary, pancreaticoduodenal, foregut, carcinoid, adrenal cortex, dermis, etc.)
MEN2A	RET[+]	Like MEN1 but later, less intense, and less symmetrical	C-cell cancer that is preventable Find and treat pheochromocytoma(s) before thyroid or parathyroid surgery
MEN4	CDKN1B[-]	Like MEN1	Tumors among pituitary, adrenal cortex, neuroendocrine tissues, thyroid, and uterus
HPT–JT	HRPT2[-]	Hypercalcemia can occur by age 10, often later Parathyroid cancer in 15% Benign or malignant Microcystic histology sometimes	Benign jaw tumors, renal cysts, and/or uterine tumors
FIHPT	MEN1[-], CASR[-], or HRPT2[-] in 30%, GCM2[+,c] Other gene(s) not identified	No specific features	None by definition Another occult syndrome may emerge later

FHH = familial hypocalciuric hypercalcemia; FIHPT = familial isolated primary hyperparathyroidism; HPT–JT:NSHPT = neonatal severe primary hyperparathyroidism; hyperparathyroidism jaw tumor syndrome; MEN = multiple endocrine neoplasia (can be type 1, 2A, or 4); PTX = parathyroidectomy.
[a]Mutation types in germline: [-] = heterozygous inactivating; [=] = homozygous inactivating; [+] = heterozygous activating.
[b]Other genes identified for similar or identical syndromes are: FHH from AP2S1[-] and GNA11[-]; MEN1 from a cyclin-dependent kinase inhibitor (CDKI) mutation (p27[-], p15[-], p18[-], or p21[-]). MEN1 from p27[-] mutation was termed MEN4 by OMIM (Online Mendelian Inheritance in Man).
[c]Specific associated variants of undetermined penetrance, role in clinical management not yet established.

parathyroidectomy in FHH results in only a very transient lowering of serum calcium, followed by persistence of the hypercalcemia within a few days after surgery [6].

Pathogenesis/genetics

Most cases result from heterozygous inactivating mutation of the *CASR* gene, which encodes a calcium-sensing receptor (CaSR) [1, 3, 6, 9, 10]. From the surface of the parathyroid cell, this CaSR "reports" the level of ionized calcium in serum. About 30% of probands and kindreds with FHH lack a detectable mutation in *CASR* [6, 9]. Some of these express mutation in *AP2S1* or *GNA11*, but others do not have mutation detectable in these identified genes [8, 11]. One rare case had homozygous *CASR* mutation despite only a mild form of FHH; the parents and other relatives with heterozygous *CASR* mutation were normocalcemic, pointing to a milder part of the spectrum of heterozygous (and homozygous) mutation [12].

The parathyroid cell in FHH shows decreased sensitivity to elevations of extracellular calcium, caused by the inactivating mutation in its CaSR. This results in impaired calcium suppression of PTH secretion, usually with little or no increase of parathyroid cell proliferation [6].

There also is a disturbed calcium-sensing function intrinsic to the kidneys in FHH. Normally, the CaSR functions in the kidney to maintain tubular calcium reabsorption in the direction that would correct for changes in the serum calcium. The tubular reabsorption of calcium is also increased by rises of PTH; in FHH, it is high and remains high even after an intended or unintended total parathyroidectomy [6, 9]. CaSRs are normally expressed in additional tissues outside of the parathyroids and kidneys, but clinical dysfunction there has not been reported in FHH or even in NSHPT [6, 9].

The usual distinctions between typical PHPT and the PHPT of *CASR* mutation can be blurred in some cases [13] or, rarely, in all carriers in an entire kindred. In particular, affected members in one large kindred with a germline missense mutation in the *CASR* had a more typical PHPT syndrome, unlike the distinctive PHPT of FHH [14]. Several other small families with *CASR* loss-of-function mutations have contained some members with one or more features resembling typical PHPT. Hypocalciuric hypercalcemia can also be caused by antibodies against the CaSR and can then be associated with other autoimmune features, but this is without *CASR* mutation; this is rare and generally not familial [6].

Diagnosis of the family and the carrier

In the presence of hypercalcemia, a normal PTH, just like a relatively low urine calcium, warns about possible FHH. The family diagnosis usually is made from typical clinical features in one or more members of a family such as hypercalcemia, relative hypocalciuria, and failed parathyroidectomy [1, 6]. Recognition of hypercalcemia before age 10 years is almost specific for the diagnosis of FHH in that kindred.

Family screening for FHH traits can be important to establish the syndrome in a proband, in an entire family, and eventually to diagnose additional relatives [1, 9]. Because of high penetrance for hypercalcemia in all FHH carriers, an accurate assignment for each relative at risk can usually be made from one determination of serum calcium (preferably ionized or albumin-adjusted).

Because urinary calcium excretion in a fixed interval depends heavily on glomerular filtration rate (GFR) and collection interval, total calcium excretion is not a valid index to distinguish a case of FHH from typical PHPT. The ratio of renal calcium clearance to creatinine clearance is an empirical and useful index for this specific comparison [1, 6].

$$Ca_{Cl}/Cr_{Cl} = [Ca_u \times V/Ca_s]/[Cr_u \times V/Cr_s] = [Ca_u \times Cr_s]/[Cr_u \times Ca_s]$$

In hypercalcemic individuals with FHH, this clearance ratio averages one-third of that in typical PHPT, and values below 0.01 (valid units will all cancel out) are suggestive of FHH [1].

CASR mutation analysis has an occasional role in the diagnosis of the syndrome, particularly with an inconclusive clinical evaluation of the family [1, 9]. *CASR* mutation may be undetectable if located outside the tested coding exons; this may account for much of the 30% lack of identified *CASR* mutation in typical families with FHH [3, 6, 10].

Management

Despite lifelong hypercalcemia, FHH can be compatible with survival into the ninth decade. Chronic hypercalcemia in FHH should rarely be treated, and it has been resistant to several types of drugs (diuretics, bisphosphonates, phosphates, or estrogens) [6]. Calcimimetics, for example cinacalcet, are a type of drug that acts like Ca^{2+} to stimulate the normal or even the mutated CaSR on the parathyroid cell and thereby decrease release of PTH. They might be effective (if used off-label) in rare cases of FHH for which treatment is appropriately contemplated [9, 15]. This potential for efficacy would be expected to depend on the domain of the mutated residues within the CaSR.

Because of their generally benign course and lack of response to subtotal parathyroidectomy, very few cases should undergo parathyroidectomy. In rare situations, such as relapsing pancreatitis, very high PTH, or very high serum calcium (persistently higher than 14 mg/dL), debulking parathyroidectomy, and even total parathyroidectomy may be indicated.

Sporadic hypocalciuric hypercalcemia

Without a positive family history or mutations of *CASR*, *GNA11*, or *AP2S1*, the management of sporadic hypocalciuric hypercalcemia is challenging. This should generally be managed as if it is typical FHH, unless the features of another PHPT syndrome become more prominent [6, 9].

NEONATAL SEVERE PRIMARY HYPERPARATHYROIDISM

Clinical expressions

NSHPT (OMIM #239200) is an extremely rare neonatal state of life-threatening, severe hypercalcemia, very high PTH, rib fractures, hypotonia, respiratory distress, and massive enlargement of all parathyroid glands [6, 9, 10]. The rare case to survive without early surgery is likely to show general impairments to development. The main relevance of NSHPT here is toward understanding some severe ways that *CASR* mutation can disrupt the parathyroids selectively.

Pathogenesis/genetics

This disorder typically results from homozygous or compound heterozygous *CASR* inactivating mutation [6, 9]. Initial molecular genetic analyses have indicated that the hypercellular parathyroids in NSHPT are polyclonal, generalized hyperplastic expansions rather than monoclonal neoplasms [16].

Diagnosis

Diagnosis is usually based on the unique clinical features, often combined with parental consanguinity and/or FHH in first-degree relatives [6, 9].

Management

Urgent total parathyroidectomy for severe symptoms and signs can be lifesaving [6, 12]. This may allow the patient to have a normal life, with treated hypoparathyroidism. Persistence of the intrinsic renal defect makes treatment of this hypoparathyroidism simpler. Cinacalcet has been used to correct the hypercalcemia, whilst the child is awaiting surgery [9].

MULTIPLE ENDOCRINE NEOPLASIA TYPE 1

MEN1 (OMIM #131100) is a rare and often heritable disorder with an estimated prevalence of 2–3 per 100,000 in unselected persons. Approximately 2% of all cases of PHPT are caused by MEN1. It is defined by consensus as tumors in two of its three main tissues (parathyroids, pituitary, and pancreaticoduodenal endocrine); affected persons are also predisposed to tumors in many other hormonal and nonhormonal tissues. By extension, familial MEN1 is defined as MEN1 with a first-degree relative showing tumor in at least one of the three main tissues [4, 5].

Clinical expressions

PHPT is the most penetrant hormonal component of MEN1 and is the initial clinical manifestation of the disorder in most cases [4, 17]. In MEN1, PHPT has several features different from the common sporadic (nonfamilial) forms of PHPT. The female-to-male ratio is about 1.0 in MEN1, in contrast to about threefold female predominance in sporadic PHPT [4]. PHPT presents about 30 years earlier in MEN1, typically in the second to fourth decade of life and has been found as early as 8 years of age [4]. Much earlier onset of primary hyperparathyroidism is likely to explain the much earlier onset of osteoporosis in MEN1.

Multiple parathyroid tumors are typical in MEN1; these tumors may vary widely in size, with an average 10:1 ratio between the largest and smallest tumor. A powerful drive to parathyroid tumorigenesis exists in MEN1, reflected by an impressively high rate of recurrent PHPT; this averages about 50% at 12 years after parathyroidectomy [17]. Because of tumors in multiple parathyroid glands, an ectopic location is more likely in MEN1 than in common adenoma [18].

Some of the other tumors associated with MEN1 include duodenal gastrinomas, insulinoma, nonhormonal islet tumors, bronchial or thymic carcinoids, gastric enterochromaffin-like tumors, adrenocortical adenomas, lipomas, facial angiofibromas, and truncal collagenomas [4, 5]. In a family with few and/or mainly young affected members, tumors of MEN1 may be expressed in only the parathyroids, or in the parathyroids plus only in one additional tissue. Such families should be followed for development of other tumors of MEN1 [4].

Pathogenesis/genetics

Familial MEN1 shows an autosomal dominant inheritance pattern, and the main genetic basis is an inactivating germline heterozygous mutation of the *MEN1* tumor suppressor gene [1, 4, 5]. *MEN1* encodes menin, whose molecular pathways and detailed functions remain under study. Individuals with MEN1 have typically inherited one inactivated copy of the *MEN1* gene from an affected parent, but up to 10% may have a spontaneous or new germline mutation. The outgrowth of a tumor follows from the subsequent somatic (i.e., acquired) inactivation of the normal, remaining copy of the *MEN1* gene.

MEN1 mutation is not identified in 30% of probands and families with MEN1 [4, 5]. Several cases among this group have MEN1 from mutation in *CDKN1B*, encoding the p27KIP1 cyclin-dependent kinase inhibitor (CDKI) [19]. This combination has been termed MEN4. Few among the remainder have mutation in *p15*, *p18*, or *p21*, encoded by three other CDKI genes [20]; mutations in these genes can also be found, apparently with lower penetrance, in some patients with sporadic parathyroid adenomas [21]. The yield of detectable *MEN1* or CDKI mutation in cases with a sporadic but true MEN1 phenotype, limited to parathyroid plus pituitary tumor, is much lower (about 7%); this suggests the existence of other predisposing gene(s) in this subgroup [5, 22].

Diagnosis of carriers

Direct sequencing for germline *MEN1* mutations is commercially available, albeit costly, and the indications for such testing remain under discussion [1, 4]. Gene analyses, typically limited to the coding region and near to it, fail to detect pathogenic *MEN1* mutation in about 30% of typical MEN1 kindreds. Some of these kindreds may have unrecognized *MEN1* defects, such as large deletions or small noncoding mutations [4]. Only when a detectable mutation is found in a proband can DNA testing be applied in that same family for presymptomatic carrier diagnosis. Presymptomatic diagnosis, although not established to broadly improve mortality/morbidity in MEN1, can importantly impact patient management [4]. For example, *MEN1* gene testing can be helpful for diagnosis or rarely for intervention in an MEN1-like proband, when the clinical diagnosis is inconclusive but a suspicion of MEN1 exists: for example, in a young adult with sporadic or familial isolated multigland PHPT, or in Zollinger

Ellison Syndrome (ZES). In the latter, *MEN1* mutation occurs in one-quarter of cases, and identifying it can lead to avoidance of abdominal surgery that would otherwise be indicated. Also, the capacity for DNA testing to definitively exclude the possibility that a clinically unaffected, at-risk member of an MEN1 kindred is a carrier can be much appreciated [4].

Periodic screening with serum calcium, PTH, etc., can provide a non-DNA-based alternative for carrier ascertainment [4, 5]. When *MEN1* mutation is not identified or sought in a proband or family with, or suspicious for, clinical MEN1, classical ascertainment testing with physical or biochemical traits can be used, e.g. detecting facial angiofibromas or periodic screening for serum calcium and PTH.

Diagnosis of other tumors

For MEN1-like probands or for established MEN1 carriers or other family members at risk for developing manifestations of MEN1, screening for tumor indices at baseline or at follow-up for the emergence of tumors is recommended, because a benefit seems likely [4]. Tumor screening and other management of the pituitary, pancreaticoduodenal, and other MEN1 tumors outside of the parathyroids is mostly beyond the scope of this chapter.

Management

Once the biochemical diagnosis is established, the indications for referral to surgery are similar to those in sporadic PHPT [4]. Osteoporosis is a surgical indication that is frequent in women with MEN1 by 35 years of age.

Because of multiple parathyroid gland involvement, minimally invasive parathyroidectomy is not recommended for MEN1 patients [4]. Instead, the initial operation most frequently performed in MEN1 is 3.5 gland subtotal parathyroidectomy with transcervical near-total thymectomy. A parathyroid remnant is usually left on its native vascular pedicle in the neck and may be marked with a clip. Some centers may consider it beneficial to autotransplant parathyroids to the forearm during an intended complete parathyroidectomy (see below). Transcervical thymectomy in MEN1 has an unproven benefit but is widely used, because it may prevent or cure thymic carcinoids (mainly in males); in addition, the thymus is a common site for parathyroid tumors in MEN1 patients with recurrent PHPT. Involvement of an experienced parathyroid surgical team is crucial to optimal outcome [4]. In patients in whom parathyroid surgery has failed or who have co-morbidity that contradict surgery, cinacalcet, an allosteric modulator of the CaSR has been used effectively to control hypercalcemia [23].

It should be emphasized that MEN1-associated cancers in non-parathyroid tissues cause fully one-third of the deaths in MEN1 cases. For most of these cancers, no effective prevention or cure currently exists, partly because of their problematic locations. New drugs and other promising treatments, including some drugs already FDA approved for common islet tumors, still need more evaluation in similar tumors of MEN1.

MULTIPLE ENDOCRINE NEOPLASIA TYPE 2A

Multiple endocrine neoplasia type 2 (MEN2) is subclassified into three clinical syndromes with mutations in the same gene, *RET* [4, 10, 24]. These are MEN2A, MEN2B, and familial medullary thyroid cancer (FMTC). Of these three syndromes, MEN2A (OMIM #171400) is the most common and the only one that manifests PHPT.

Clinical expressions

MEN2A is a heritable predisposition to medullary thyroid or C-cell cancer (MTC), pheochromocytoma, and PHPT [24]. The frequencies of these tumors in an adult carrier of MEN2A are above 90% for MTC, 40% to 50% for pheochromocytoma, and 20% for PHPT.

This intermediate penetrance of PHPT in MEN2A contrasts with the higher penetrance found in every other familial PHPT syndrome.

Pathogenesis/genetics

MEN2A is inherited in an autosomal dominant pattern, with both genders affected in equal proportions; the gene defect is heterozygous germline gain of function mutation of the *RET* proto-oncogene [24]. Germline RET mutation is detectable in more than 95% of MEN2A families. *RET* mutation at codon 634 accounts for 85% of MEN2A and is more highly associated with the expression of PHPT [24].

There are both differences and much overlap in the specific *RET* gene mutations underlying MEN2A and FMTC; in contrast, MEN2B is caused by one of two entirely distinct *RET* mutations [24]. Why parathyroid disease fails to develop in the latter two MEN2 syndromes remains unclear. Unlike the numerous and seemingly random different inactivated codons of MEN1 that are typical of a tumor suppressor mechanism, the mutated *RET* codons in MEN2A are limited in number, reflecting the need for highly specific changes in selected domains of the RET protein to activate this oncoprotein [24].

The RET protein is a plasma membrane spanning receptor tyrosine kinase (RTK) that normally transduces growth and differentiation signals in developing tissues, including around the neural crest. Knowledge of its molecular pathway has resulted in evaluation of RTK inhibitors (e.g. vandetanib and cabozantinib) in phase III clinical trials of patients with advanced MTC, which have shown tumor responses and improvements in progression free survival [24].

Diagnosis of a kindred, of carriers, and of parathyroid tumors

Gene sequencing for germline *RET* mutations is central to clinical management of MEN2A, particularly to guide thyroidectomy in management/prevention of MTC [24]. PHPT in MEN2A is often asymptomatic and diagnosed and treated incidentally to thyroid surgery. Otherwise, its biochemical diagnosis, as well as indications for surgery, parallel those in sporadic PHPT.

Management

PHPT is a less urgent expression of MEN2A than MTC or pheochromocytoma. Conceptually consistent with its genetics, PHPT in MEN2A is multiglandular, but fewer than four overtly enlarged glands may be present at one time. Thus, bilateral neck exploration to identify all abnormal parathyroid glands is advisable in known or suspected MEN2A; resection of enlarged parathyroid glands (up to 3.5 glands) is the most common operation. Issues of preoperative tumor localization in unoperated MEN2A patients expressing PHPT are similar to MEN1.

RET testing during childhood can secure a preventive or curative thyroidectomy (ie, sufficiently early in childhood as to minimize the likelihood that extracapsular metastases of C-cell cancer will have already occurred) [24].

HYPERPARATHYROIDISM–JAW TUMOR SYNDROME

Clinical expressions

HPT–JT (OMIM #145001) is a rare, autosomal dominant combination of PHPT, ossifying or cementifying fibromas of the mandible and maxilla, renal manifestations including cysts, hamartomas, or Wilms tumors, and uterine tumors [3, 25, 26]. Among adults of "classical" HPT–JT kindreds, PHPT is the most highly penetrant manifestation at 80%, next to ossifying fibromas of the maxilla or mandible at 30%, and a renal lesion slightly less frequently. A wide range of penetrance values occurs for uterine tumors [25].

PHPT in HPT–JT may develop as early as during the first decade of life [3, 25]. Although all parathyroids are at risk, surgical exploration can show a solitary parathyroid tumor (even solitary atypical adenoma or solitary cancer) rather than multigland disease. Parathyroid neoplasms can be macro- or microcystic, and, whereas most tumors are classified as adenomas, parathyroid carcinoma (15% to 20% of PHPT) is markedly overrepresented in HPT–JT cases [3, 25]. In contrast, parathyroid cancer almost never occurs in MEN1, MEN2, or FHH. Dissemination of parathyroid cancer to the lungs can occur in the early 20s in HPT–JT. After a period of euparathyroidism, operated cases may manifest recurrent PHPT, and a solitary tumor asynchronously originating in a different parathyroid gland may prove responsible.

Pathogenesis/genetics

Germline mutation of the *HRPT2* gene (also called *CDC73*) causes HPT–JT [1, 3, 25, 26]. The yield of *HRPT2* mutation in HPT–JT kindreds is about 60% to 70%; some or perhaps many of the remaining kindreds may have other *HRPT2* defects, such as large deletions or small noncoding mutations that evade detection by standard genetic testing methods. Mutations of *HRPT2* cause tumors, including parathyroid cancer, by inactivating or eliminating its protein product, parafibromin, consistent with a classical "two-hit" tumor suppressor mechanism. Both parafibromin's normal cellular roles and its influence (via loss of function) in tumorigenesis are under study.

Importantly, many cases with seemingly sporadic presentations of parathyroid carcinoma (OMIM #608266) also harbor a germline mutation in *HRPT2*, thus representing newly ascertained HPT–JT, occult HPT–JT, or another variant syndrome [3, 27].

Diagnosis of carriers and cancers

The biochemical diagnosis of PHPT in HPT–JT parallels that in sporadic PHPT. Recognition of germline *HRPT2* mutation in classic or variant HPT–JT has opened the door to DNA-based diagnosis in probands or individuals with apparently sporadic parathyroid carcinoma, and for carrier identification in at-risk family members, aimed at preventing or curing parathyroid malignancy [1, 3, 25].

Before parathyroid surgery of a known or likely carrier of HPT–JT, the surgeon should be alerted to the possibility of parathyroid cancer.

Management

Management in HPT–JT focuses on monitoring and surgery to address the high risk for current or future parathyroid malignancy [1, 25, 28]. The finding of biochemical PHPT should lead promptly to surgery. All parathyroids should be identified at operation, signs of malignancy sought, and resection of abnormal glands performed. Because of the potential for malignancy, the consideration of prophylactic total parathyroidectomy (perhaps even for euparathyroid carriers) has been raised as an alternative approach. This is not currently favored in view of the difficulty in removing all parathyroid tissue, the burdens and adverse consequences of lifelong hypoparathyroidism, the incomplete penetrance of parathyroid cancer in the syndrome, and the likelihood that close biochemical monitoring for recurrent PHPT will promote successful management or prevention of cancer [3, 28].

FAMILIAL ISOLATED HYPERPARATHYROIDISM

Clinical expressions and diagnosis

FIHPT (OMIM #145000) is clinically defined as familial PHPT without the extraparathyroid manifestations of another syndromal category [1, 4, 10]. The diagnosis of FIHPT should therefore be changed, if features of another PHPT syndrome develop. Partly because this category is likely to encompass several occult or even totally unknown causes, the spectrum of its PHPT is broad.

Pathogenesis/genetics

FIHPT is genetically heterogeneous and can be caused by incomplete expression from germline mutation in *MEN1*, *HRPT2*, or *CASR* [3]. Recently, activating variants of the glial cells missing 2 (*GCM2*) gene, which encodes a parathyroid-specific transcription factor, have also been reported in some patients with FIHPT [29, 30]. However, the majority of families do not have a detectable mutation in any of these genes. One unidentified gene may be on the short arm of chromosome 2 (OMIM #610071); one or more other genes likely accounts for other families [3].

Mutation testing should be considered, e.g. when results might impact on the advisability of, or approach to, management of parathyroid and other tumors or to more gene testing in relatives [4].

Management

Management is very similar to that in common PHPT. Monitoring and management must recognize that additional features of a genetically defined PHPT syndrome or a previously unidentified PHPT syndrome could become detectable [4]. For example, the heightened risk of parathyroid carcinoma must be borne in mind in FIHPT, when occult HPT–JT is possible.

OVERLAPPING CONSIDERATIONS AMONG ALL FORMS OF FAMILIAL PHPT

Multifocal parathyroid gland hyperfunction

For three hereditary PHPT syndromes (MEN1, MEN2A, and HPT–JT), the germline mutation in the parathyroid cell causes susceptibility to postnatal and gradual overgrowth of mono- or oligoclonal parathyroid tumor [1]. For two others (FHH and NSHPT), the phenotype is fully expressed around birth (ie, with no postnatal delay). Stated differently, an underlying feature of these five multiorgan syndromes is that every parathyroid cell carries the same syndromal germline mutation; this is either sufficient to cause the overfunction phenotype in all parathyroid cells immediately or to put each parathyroid cell at risk to yield clonal proliferation many years later.

Detection of asymptomatic carriers

Once a PHPT syndrome has been diagnosed, testing for carriers among asymptomatic relatives should be considered (Fig. 83.1) [1, 4]. The concept of the carrier must include disease predisposition in a relative, even without an identifiable syndromal mutation in that family. Testing of germline DNA is often the gold standard for detection of carriers; however, carrier testing by use of traits that are expressed early and with high penetrance (such as hypercalcemia for the early carrier in FHH) or lower penetrance (such as hypercalcemia for delayed parathyroid cancer in HPT–JT) is sometimes a useful alternative.

The possible benefits of germline mutation testing include providing information to the subject, family, and physician [1, 4].

Among all PHPT syndromes, carrier testing only for MEN2 can lead to a major intervention of almost certain efficacy in reducing mortality (from medullary thyroid carcinoma) [24]. Testing for HPT–JT can lead to management that may lessen mortality associated with parathyroid malignancy [1, 3]. Testing in other PHPT syndromes is mainly for information to the physician and patient, and is less urgent. Such information about silent or even affected carriers is widely used to plan screening for tumors at baseline and during follow-up.

Monitoring for tumors

Tumor monitoring is best performed with a syndrome-specific protocol; it should be followed in each carrier of a syndrome of PHPT [1, 4]. Monitoring for parathyroid and other tumors should address tumors present at the time of initial ascertainment of the carrier and the tumors that emerge during periodic follow-up. The plan for monitoring must deal with issues of cost and effectiveness.

Special approaches in surgery for familial parathyroid tumors

Many aspects of surgery for parathyroid adenoma require modification, when multiple parathyroid glands have the potential to be overactive [1, 4, 28]. Efforts are made intraoperatively to know when sufficient pathological tissue has been removed. Traditionally, the identification of all four parathyroid glands has been pursued for this reason. Rapid measurement of PTH intraoperatively to see a major drop from high values also can be useful immediately to judge if any overactive parathyroid tissue remains in vivo, but must be interpreted with greater caution when multigland disease is expected.

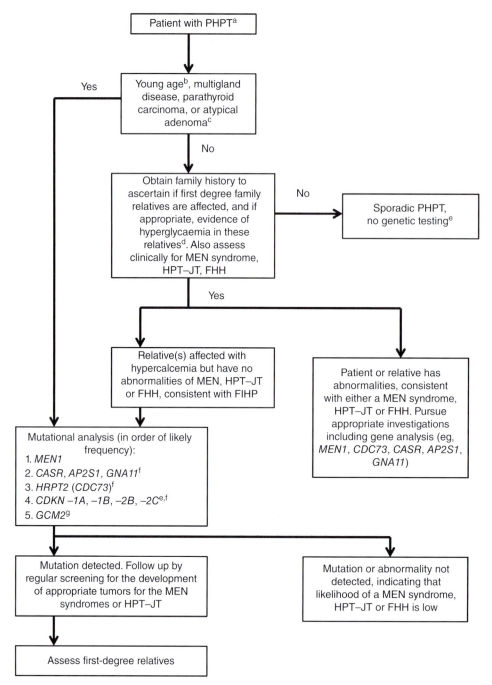

Fig. 83.1. Suggested clinical approach to genetic testing in a patient with PHPT. AP2S1 = adaptor protein 2 sigma subunit; CASR = calcium sensing receptor; CDC73 = cell division cycle 73; CDKN = cyclin dependent kinase inhibitor; FIHPT = familial isolated primary hyperparathyroidism; FHH = familial hypocalciuric hypercalcemia; GCM2 = glial cells missing 2, a parathyroid-specific transcription factor; GNA11 = G-protein alpha 11 subunit; HPT–JT = hyperparathyroid jaw-tumor syndrome; HRPT2 = hyperparathyroidism type 2; MEN = multiple endocrine neoplasia; PHPT = primary hyperparathyroidism; PTH = parathyroid hormone; RET = rearranged during transfection proto-oncogene. Source: [10]. Reproduced with permission of Elsevier.

[a] PHPT presenting without manifestations of MEN-associated tumors, or tumors associated with HPT–JT.

[b] Guidelines for MEN1 recommend *MEN1* mutational analysis in patients with PHPT occurring before the age of 30 years, and ~10% of PHPT patients below the age of 45 years have been reported to have a germline mutation involving the *MEN1*, *CASR* or *HRPT2* (*CDC73*) genes.

[c] Atypical parathyroid adenoma may have cysts or fibrous bands.

[d] PHPT may be the first manifestation of MEN1 and HPT–JT in ~90% and ~95% of patients, respectively, with these disorders.

[e] <5% of patients presenting with nonfamilial (sporadic) and nonsyndromic PHPT, due to solitary parathyroid adenomas in the sixth–ninth decade of life may have rare variants/mutations of *CDKN1A*, *CDKN2B*, or *CDKN2C*.

[f] *CASR*, *AP2S1*, *GNA11*, *HRPT2* (*CDC73*), *CDKN1B*, and *RET* mutations are associated with FHH1, FHH3, FHH2, HPT–JT, MEN4, and MEN2, respectively.

[g] GCM2 activating variants may occur in some FIHPT patients without *MEN1*, *CDC73*, or *CASR* mutations. Specific associated variants are of undetermined penetrance, with their role in clinical management not yet established.

Efforts to identify all four parathyroid glands and remove several glands result in increased frequency of postoperative hypoparathyroidism [4]. To minimize permanent hypoparathyroidism, in some centers, small fragments of the most normal-appearing parathyroid tumor are autografted immediately to the nondominant forearm. In other centers, because of difficulty in establishing such grafts, the surgeon leaves a small remnant of parathyroid in the neck, attached to its own vascular pedicle. In either case, one may cryopreserve fragments of the most normal-appearing tissue for possible delayed autograft for late postoperative hypoparathyroidism. However, such cryopreservation is not permitted in many centers because of concerns about legal liabilities. A normal-sized gland is not sufficiently large to give a satisfactory fresh or cryopreserved autograft; thus, tumor tissue is used. Fortunately, it can sustain euparathyroidism for many years. The potential for malignancy of any parathyroid tissue in HPT–JT argues against the autografting option there.

Beyond hypoparathyroidism, other complications are also more frequent after surgery for multigland PHPT, including familial PHPT, than after surgery for adenoma. These include complications to surrounding tissues (such as a recurrent laryngeal nerve) and postoperative persistent PHPT. The most obvious reason for the latter is an incomplete exploration for multigland disease.

True recurrent PHPT is a late complication that is much more frequent in MEN1 and other familial PHPT, than in common adenoma [4, 17]. True recurrence is defined for convenience as PHPT after a 3- to 6-month postoperative period of documented euparathyroidism. It could arise when a small tumorous remnant becomes overactive or when a previously normal parathyroid cell progresses into a tumor clone.

ACKNOWLEDGMENTS

This work was supported in part by: the intramural program of NIDDK (SKA); the Medical Research Council UK, the Wellcome Trust and National Institute of Health Research (NIHR) UK (RVT); and the Murray-Heilig Fund in Molecular Medicine (AA). We wish to acknowledge Dr Stephen J. Marx for his important authorship role in previous editions of this chapter.

REFERENCES

1. Eastell R, Brandi ML, Costa AG, et al. Diagnosis of asymptomatic primary hyperparathyroidism: Proceedings of the Fourth International Workshop. J Clin Endocrinol Metab. 2014;99:3570–9.
2. Khan AA, Hanley DA, Rizzoli R, et al. Primary hyperparathyroidism: review and recommendations on evaluation, diagnosis, and management. A Canadian and international consensus. Osteoporosis. 2017. Int. 28:1–19.
3. Arnold A, Lauter K. Genetics of hyperparathyroidism including parathyroid cancer. In: Weiss RE, Refetoff S (eds) *Genetic Diagnosis of Endocrine Disorders* (2nd ed.). Cambridge, MA: Academic Press, 2016, pp. 165–72.
4. Thakker RV, Newey PJ, Walls GV, et al. Clinical practice guidelines for multiple endocrine neoplasia type 1 (MEN1). J Clin Endocrinol Metab. 2012;97:2990–3011.
5. Thakker RV. Multiple endocrine neoplasia type 1 (MEN1) and type 4 (MEN4). Mol Cell Endocrinol. 2014;386:2–15.
6. Egbuna OI, Brown EM. Genetically determined disorders of the calcium-sensing receptor. In: Thakker RV, Whyte MP, Eisman JA, et al. (eds) *Genetics of Bone Biology and Skeletal Disease* (2nd ed.). Cambridge, MA: Academic Press, 2018, pp. 459–478.
7. Hannan FM, Babinsky VN, Thakker RV. Disorders of the calcium-sensing receptor and partner proteins: insights into the molecular basis of calcium homeostasis. J Mol Endocrinol. 2016;57:R127–42.
8. Nesbit MA, Hannan FM, Howles SA, et al. Mutations affecting G-protein subunit α11 in hypercalcemia and hypocalcemia. N Engl J Med. 2013;368:2476–86.
9. Hannan FM, Howles SA, Rogers A, et al. Adaptor protein-2 sigma subunit mutations causing familial hypocalciuric hypercalcaemia type 3 (FHH3) demonstrate genotype-phenotype correlations, codon bias and dominant-negative effects. Hum Mol Genet. 2015;24:5079–92.
10. Thakker RV. Familial and hereditary forms of primary hyperparathyroidism. In: Bilezikian JP (editor-in-chief) *The Parathyroids: Basic and Clinical Concepts* (3rd edn). Cambridge, MA: Academic Press, 2015, pp. 341–63.
11. Nesbit MA, Hannan FM, Howles SA, et al. Mutations in AP2S1 cause familial hypocalciuric hypercalcemia type 3. Nat Genet. 2013;45:93–7.
12. Pearce SHS, Trump D, Wooding C, et al. Calcium-sensing receptor mutations in familial benign hypercalcemia and neonatal hyperparathyroidism. J Clin Invest. 1995;96:2683–92.
13. Starker LF, Akerström T, Long WD, et al. Frequent germline mutations of the MEN1, CASR, and HRPT2/CDC73 genes in young patients with clinically non-familial primary hyperparathyroidism. Horm Cancer. 2012;3:44–51.
14. Carling T, Szabo E, Bai M, et al. Familial hypercalcemia and hypercalciuria caused by a novel mutation in the cytoplasmic tail of the calcium receptor. J Clin Endocrinol Metab. 2000;85:2042–7.
15. Howles SA, Hannan FM, Babinsky VN, et al. Cinacalcet for symptomatic hypercalcaemia caused by AP2S1 mutations. N Engl J Med. 2016;374:1396–8.
16. Corrado KR, Andrade SC, Bellizzi J, et al. Polyclonality of parathyroid tumors in neonatal severe hyperparathyroidism. J Bone Miner Res. 2015;30:1797–802.
17. Marx SJ. Multiple endocrine neoplasia type 1. In: Bilezikian JP (editor-in-chief) *The Parathyroids: Basic and Clinical Concepts* (3rd edn). Cambridge, MA: Academic Press, 2015, pp. 535–84.

18. Kivlen MH, Bartlett DL, Libutti SK, et al. Reoperation for hyperparathyroidism in multiple endocrine neoplasia type 1. Surgery. 2001;130:991–8.
19. Pellegata NS, Quintanilla-Martinez L, Siggelkow H, et al. Germ-line mutations in p27Kip1 cause a multiple endocrine neoplasia syndrome in rats and humans. Proc Natl Acad Sci U S A 2006;103:15558–63. Erratum in: Proc Natl Acad Sci U S A. 2006;103:19213.
20. Agarwal SK, Mateo C, Marx SJ. Rare germline mutations in cyclin-dependent kinase inhibitor genes in MEN1 and related states. J Clin Endocrinol Metab. 2009;94:1826–34.
21. Costa-Guda J, Soong CP, Parekh V, et al. Germline and somatic mutations in cyclin-dependent kinase inhibitor genes CDKN1A, CDKN2B, and CDKN2C in sporadic parathyroid adenomas. Horm Cancer 2013;4:301–7.
22. de Laat JM, van der Luijt RB, Pieterman CR, et al. MEN1 redefined, a clinical comparison of mutation-positive and mutation-negative patients. BMC Med. 2016;14:182.
23. Giusti F, Cianferotti L, Gronchi G, et al. Cinacalcet therapy in patients affected by primary hyperparathyroidism associated to Multiple Endocrine Neoplasia Syndrome type 1 (MEN1). Endocrine. 2016;52:495–506.
24. American Thyroid Association Guidelines Task Force, Kloos RT, Eng C, Evans DB, et al. Medullary thyroid cancer: management guidelines of the American Thyroid Association. Thyroid 2009;19:565–612.
25. Newey PJ, Bowl MR, Cranston T, Thakker RV. Cell division cycle protein 73 homolog (CDC73) mutations in the hyperparathyroidism-jaw tumor syndrome (HPT-JT) and parathyroid tumors. Hum Mutation. 2010;31:295–307.
26. Sharretts JM, Simonds WF. Clinical and molecular genetics of parathyroid neoplasms. Best Pract Res Clin Endocrinol Metab. 2010;24:491–502.
27. Shattuck TM, Valimaki S, Obara T, et al. Somatic and germ-line mutations of the HRPT2 gene in sporadic parathyroid carcinoma. N Engl J Med. 2003;349:1722–29.
28. El-Hajj Fuleihan G, Arnold A. Parathyroid carcinoma. In: Post TW (ed.) *UpToDate*. Waltham, MA: UpToDate, Inc., 2017.
29. Guan B, Welch JM, Sapp JC, et al. GCM2-activating mutations in familial isolated hyperparathyroidism. Am J Hum Genet 2016;99:1–11.
30. Guan B, Welch JM, Vemulapalli M, et al. Ethnicity of patients with germline GCM2-activating variants and primary hyperparathyroidism. J Endocr Soc 2017;1:488–99.

84

Non-Parathyroid Hypercalcemia

Mara J. Horwitz

Division of Endocrinology and Metabolism, The University of Pittsburgh School of Medicine, Pittsburgh, PA, USA

PATHOPHYSIOLOGY OF HYPERCALCEMIA

Hypercalcemia is defined as a serum calcium greater than two SD above the normal mean in a given laboratory, commonly 10.5 mg/dL for total serum calcium, and 1.25 mmol/L for ionized serum calcium. There is no formal grading system for defining the severity of hypercalcemia. In general, however, serum calcium concentrations less than 12 mg/dL can be considered mild, those between 12 and 14 mg/dL moderate, and those greater than 14 mg/dL severe.

The serum calcium concentration is tightly regulated by the flux of serum ionized calcium to and from four physiological compartments: the skeleton, the intestine, the kidney, and serum binding proteins. Hypercalcemia can result only from abnormal binding of calcium to serum proteins, or abnormal flux of calcium into extracellular fluid. Understanding hypercalcemia in a mechanistic construct is critical for accurate diagnosis, and for effective treatment. For example, hypercalcemia from vitamin D intoxication or milk-alkali syndrome commonly arises from increased gastrointestinal absorption of calcium, and would not be expected to respond to antiresorptive agents such as bisphosphonates. Conversely, humoral hypercalcemia of malignancy results principally from increased skeletal resorption and renal calcium reabsorption, and therefore is not influenced by restricting dietary calcium intake.

CLINICAL SIGNS AND SYMPTOMS OF HYPERCALCEMIA

Hypercalcemia raises the electrical potential difference across cell membranes, and increases the depolarization threshold. Clinically, this is manifest as a spectrum of neurological symptoms ranging from mild tiredness, to obtundation, to coma. The presence or absence and degree of neurological symptoms depends not only on the calcium level but also on the abruptness of onset of hypercalcemia, the age and underlying neurological status of the patient, and comorbidities and medications.

Hypercalcemia acts directly at the nephron to prevent normal reabsorption of water, leading to a functional form of nephrogenic diabetes and polyuria. This may lead to thirst, prerenal azotemia, and significant dehydration which are common clinical features of hypercalcemia. Hypercalcemia may also cause precipitation of calcium phosphate salts in the renal interstitium (nephrocalcinosis), vasculature, cardiac conduction system, the cornea ("band keratopathy"), and the gastric mucosa. Hypercalcemia may lead to renal failure.

Hypercalcemia can also lead to electrocardiographic abnormalities, the most specific of which is a prolonged Q-Tc interval. Hypercalcemia increases the depolarization threshold of skeletal and smooth muscle making them more refractory to neuronal activation which manifest clinically as skeletal muscle weakness and constipation. Nausea, anorexia, vomiting, and flushing are also common. Finally, hypercalcemia may lead to pancreatitis.

DISORDERS THAT LEAD TO HYPERCALCEMIA

The complete differential diagnosis of hypercalcemia is shown in Table 84.1. In this chapter, nonparathyroid causes of hypercalcemia are considered. The PTH-dependent disorders, including primary and tertiary hyperparathyroidism and familial benign hypocalciuric hypercalcemia (FHH), are discussed in Chapter 83.

Primer on the Metabolic Bone Diseases and Disorders of Mineral Metabolism, Ninth Edition. Edited by John P. Bilezikian.
© 2019 American Society for Bone and Mineral Research. Published 2019 by John Wiley & Sons, Inc.
Companion website: www.wiley.com/go/asbmrprimer

Table 84.1. Differential diagnosis of hypercalcemia.

PTH-dependent hypercalcemia (see Chapter 81)
Cancer
    Humoral hypercalcemia of malignancy (HHM) [1–9, 50]
    Local osteolytic hypercalcemia (LOH) [1, 2, 6–9]
    1,25(OH)2 vitamin D and lymphoma/dysgerminoma [10]
    Authentic ectopic PTH secretion [11, 12]
    Other
Granulomatous disorders [10, 13–15]
Endocrine disorders [16–21]
Immobilization [22–24]
Milk-alkali syndrome [25–27]
Total parenteral nutrition [28, 29]
Abnormal protein binding [30, 31]
Medications [32–43]
    Vitamin D or vitamin D analogs
    Thiazide diuretics
    Vitamin A
    Lithium
    Parathyroid hormone
    Estrogen/SERMs
    Aminophylline and theophylline
    Foscarnet
    Growth hormone
    8-chloro-cyclic AMP
Chronic and acute renal failure [44, 45]
End stage liver disease [46]
Manganese intoxication
Fibrin glue [47]
Hypophosphatemia
Pediatric syndromes (see Chapter 91) [48]

Cancer

Malignancy-associated hypercalcemia (MAHC) accounts for ~90% of hypercalcemia among hospitalized patients. It is usually found in patients with advanced and clinically obvious malignant disease and denotes a poor prognosis. MAHC can be subdivided into four mechanistic subtypes: humoral hypercalcemia of malignancy (HHM); local osteolytic hypercalcemia (LOH); 1,25(OH)$_2$-vitamin D-induced hypercalcemia; and, authentic ectopic hyperparathyroidism. Occasionally patients with MAHC also have previously undiagnosed primary hyperparathyroidism that preceded the malignancy.

Humoral hypercalcemia of malignancy

HHM is the most common form of MAHC, accounting for ~80% of subjects in large series of unselected patients with MAHC. HHM results from the secretion of PTHrP by HHM-associated tumors. The most common type of tumors causing HHM are squamous, breast, and renal carcinomas.

The continuous secretion of PTHrP by tumors leads to a dramatic uncoupling of bone resorption from formation, by activating osteoclastic bone resorption and suppressing osteoblastic bone formation. As a result, enormous net quantities of calcium (700–1000 mg/day) leave the skeleton, causing marked hypercalcemia. In addition, the anticalciuric effects of PTHrP prevent or restrict effective renal calcium clearance. Interestingly, HHM is associated with reductions in circulating 1,25(OH)$_2$D levels, which in turn limit intestinal calcium absorption. Thus, in pathophysiological terms, HHM results from enhanced skeletal resorption coupled with an inability to clear calcium through the kidney.

These observations stand in striking contrast to primary hyperparathyroidism (HPT), in which 1,25(OH)$_2$D is increased, and both osteoblast and osteoclast activity are increased but remain coupled. Bone scans, bone biopsies, and autopsy reveal few or no skeletal metastases in patients with HHM. This finding emphasizes the humoral nature of the syndrome and stands in contrast to patients with LOH, described later.

It has been reported that in very rare instances, benign neoplastic lesions also may lead to hypercalcemia by systemic overproduction of PTHrP, a condition that has been referred to as "humoral hypercalcemia of benignancy." Examples of this syndrome include benign uterine fibroids, benign ovarian tumors, insulinomas, and pheochromocytomas. In addition, hypercalcemia caused by physiological production of PTHrP without malignancy has been reported during pregnancy and lactation.

Local Osteolytic Hypercalcemia

The tumors that most commonly produce LOH are breast cancer and hematological neoplasms (myeloma, lymphoma, leukemia) in subjects with widespread skeletal involvement. They account for ~20% of patients with MAHC. Indeed, evidence suggests that the frequency of MAHC caused by LOH may be declining with the widespread use of bisphosphonates to prevent skeletal fractures, metastases, and pain in patients with myeloma and breast cancer.

Patients with LOH are characterized at bone biopsy or autopsy by extensive skeletal metastases or marrow infiltration. The "osteoclast-activating factors" (OAFs) responsible for LOH are reviewed in Chapter 6. Bone scintigraphic scans are generally intensely and widely positive in patients with metastatic disease from solid tumors, but may be completely negative despite extensive marrow involvement in patients with multiple myeloma, reflecting a reduction in bone formation.

In mechanistic terms, LOH can be thought of as primarily a resorptive (skeletally-derived) form of hypercalcemia in which massive removal of calcium from the skeleton exceeds the normal ability of the kidney to clear calcium. As the dehydration associated with such marked hypercalcemia occurs, the hypercalcemia is also exacerbated by a typical decline in renal function.

1,25(OH)$_2$D-induced hypercalcemia

In the 1970s, reports began to appear describing patients with lymphomas in whom hypercalcemia occurred as a result of increased production of 1,25(OH)$_2$D. The primary pathophysiological abnormality in this syndrome is that the malignant cells or adjacent normal cells overexpress the enzyme 1-α-hydroxylase leading to elevated circulating concentrations of active 1,25(OH)$_2$D. Because 1,25(OH)$_2$D activates intestinal calcium absorption, this syndrome is principally an absorptive form of hypercalcemia, although decreased renal clearance of calcium may also develop as a consequence of the dehydration caused by the hypercalcemia. In addition, 1,25(OH)$_2$D may increase osteoclast-mediated bone resorption by directly activating the RANKL pathway, and therefore worsening hypercalcemia. It is of note that circulating levels of 1,25(OH)$_2$D may not always reflect the high local concentrations observed in patients with marrow tumor involvement.

Authentic ectopic hyperparathyroidism

There are rare descriptions in the literature in which production of authentic PTH, and not PTHrP, by tumors was shown to cause MAHC. Approximately 20 cases have been reported in which convincing evidence exists for hypercalcemia resulting from ectopic secretion of PTH from malignant tumors. In addition, there is one report of a gastric carcinoma producing both PTH and PTHrP.

Other mechanisms for MAHC

The four categories described above account for more than 99% of patients with MAHC. Occasionally, however, patients who do not fit any of these categories have been described. For example, there are rare case reports in which elevated circulating concentrations of prostaglandin E2 may have been responsible, caused by its ability to induce RANKL expression, which in turn activates osteoclastic activity.

Granulomatous diseases

Almost every granulomatous disease has been reported to cause hypercalcemia. The most common is sarcoidosis, but tuberculosis, berylliosis, histoplasmosis, coccidoimycosis, *Pneumocystis*, and others have all been associated with the syndrome. In the case of sarcoidosis, ~10% of patients become hypercalcemic, and 20% hypercalciuric, during the course of their disease.

The mechanism in most cases is inappropriate production of 1,25(OH)$_2$D by the granulomas, as a result of increased activity of 1-α-hydroxylase in macrophages and lymphocytes. This 1-α-hydroxylase is distinct from the renal enzyme and appears to be stimulated by IFN-γ rather than PTH. This results in elevated circulating concentrations of 1,25(OH)$_2$D, which in turn lead to intestinal hyperabsorption of calcium, hypercalciuria, and ultimately hypercalcemia.

The syndrome reverses with the eradication of granulomas (eg, by glucocorticoids or antituberculosis medications), and by oral or intravenous hydration coupled with lowering dietary intake of vitamin D and calcium. Ketoconazole also may inhibit the 1-α-hydroxylase in macrophages in sarcoid patients. Because sunlight is a source of vitamin D, sunlight exposure should be reduced and exogenous vitamin D$_2$ or D$_3$ should not be used.

Other endocrine disorders

Hyperparathyroidism is the classical endocrine disorder associated with hypercalcemia, but there are four other endocrine disorders that may also cause hypercalcemia.

1. Hyperthyroidism has been reported to cause mild increases in ionized or total serum calcium in up to 50% of affected patients. It is believed to result from increases in osteoclastic bone resorption caused by thyroid hormone-induced increases in RANKL, a key regulator of osteoclast function.
2. Addisonian crisis has also been reported to cause hypercalcemia with increases in both ionized as well as total calcium. In general, hypercalcemia is mild, and responds to standard therapy for hypoadrenalism (fluid resuscitation and intravenous glucocorticoids). The cause is not known.
3. Pheochromocytoma has been associated with hypercalcemia. In some cases, the hypercalcemia is caused by primary hyperparathyroidism in the setting of multiple endocrine neoplasia type 2. However in others, the tumors have been shown to secrete PTHrP or catecholamine secretion by the pheochromocytoma is sufficient to activate bone resorption.
4. The VIP-oma syndrome is caused by vasoactive intestinal polypeptide (VIP) secretion by pancreatic islet or other neuroendocrine tumors, and is associated with severe watery diarrhea ("pancreatic cholera"), hypokalemia, and achlorhydria (the WDHA syndrome). Interestingly, 90% of patients with this rare syndrome have been reported to be hypercalcemic, although the mechanism is unknown. VIP has been shown to stimulate osteoclastic bone resorption in vitro, suggesting at least one potential mechanism.

Immobilization

Immobilization in association with another cause of high bone turnover (such as the high turnover associated with youth, hyperparathyroidism, myeloma or breast cancer with bone metastases, and Paget disease) may cause hypercalcemia.

Immobilization in patients with high bone turnover suppresses osteoblastic bone formation and markedly increases osteoclastic bone resorption, leading to complete uncoupling of these two normally tightly coupled processes. The result is massive loss of calcium from the skeleton, with resultant hypercalcemia and reductions in bone mineral density. PTH levels are generally low or undetectable in immobilization-induced hypercalcemia. It has been suggested that this process is mediated by sclerostin which is elevated in patients who are immobilized and appears to inhibit bone formation. The process is most effectively reversed by restoration of normal weight bearing.

Milk-alkali syndrome

Originally reported in 1949, this syndrome initially described patients who developed moderate or severe hypercalcemia when treated with large amounts of milk (several quarts or gallons per day) and absorbable antacids (eg, baking soda, or sodium bicarbonate) for peptide ulcer disease. Additional features of the syndrome were a metabolic alkalosis caused by antacid ingestion and renal failure caused by hypercalcemia.

Contemporary reports are primarily of patients taking large amounts of calcium carbonate for peptic ulcer or esophageal reflux symptoms in excess of 4000 mg/day, which can induce hypercalciuria and hypercalcemia in normal adults. The hypercalcemia reverses with hydration and correction of excessive calcium ingestion. Renal damage, however, may be permanent.

Abnormal protein binding

Hypercalcemia may be "artifactual" or "factitious" in some settings. In general, this refers to situations in which the total serum calcium is elevated, but the ionized serum calcium is normal. In one example, severe dehydration may lead to increases in serum albumin concentration and in the albumin-bound component of total serum calcium. This results in an increase in total but not ionized serum calcium.

An analogous situation has been reported in subjects with multiple myeloma or Waldenstrom's macroglobulinemia whose monoclonal immunoglobulin specifically recognizes calcium ion. In these cases, patients have displayed severe increases in total serum calcium, in the absence of symptoms or signs of hypercalcemia. In these cases, the ionized serum calcium and urinary calcium excretion was found to be normal.

Medications

A number of medications may cause hypercalcemia. Calcium-containing antacids are included above in the section on milk-alkali syndrome.

Vitamin D intoxication from standard vitamin D preparations has been reported in association with inappropriate addition by dairies or manufacturers of vitamin D to milk or to infant formula. Vitamin D intoxication may also occur with use of doses of vitamin D in excess of 50,000 units two or three times per week. Vitamin D analogues such as calcitriol [1,25(OH)$_2$D] used in the treatment of hypoparathyroidism, chronic renal failure, and metabolic bone disease may also cause hypercalcemia. The mechanism in all of the above is a combination of increased intestinal calcium absorption and bone resorption induced by vitamin D, together with reductions in renal ability to clear calcium as a result of dehydration.

Vitamin A intoxication can cause hypercalcemia. This may occur through the excessive use of vitamin supplements. The use of retinoic acid derivatives for the treatment of dermatological disorders or as chemotherapy agents has also been associated with induction of hypercalcemia.

Thiazide diuretics such as hydrochlorothiazide or chlorthalidone commonly cause mild hypercalcemia. This has been ascribed to their ability to increase distal renal tubular calcium reabsorption independently of PTH. There may an association between thiazide-induced hypercalcemia and PHPT.

Lithium has been reported to cause hypercalcemia in as many as 15% of patients. It has been suggested that lithium may actually induce parathyroid hyperplasia or induce parathyroid adenomas by binding to the calcium sensing receptor in the parathyroid gland. It is also possible that the coincidence of lithium use with hyperparathyroidism may also represent the simultaneous occurrence of two common clinical syndromes.

Parathyroid hormone, both the PTH(1-34) and PTH(1-84) forms used for treatment of osteoporosis, are associated with hypercalcemia in a substantial minority of patients so treated. In general, it is mild and requires little or no treatment, or a reduction in the dose of PTH or supplemental calcium, but it can be severe and require discontinuation of PTH therapy.

Other medications that are known to cause hypercalcemia with no known mechanism are listed in Table 84.1.

Acute and chronic renal failure

The recovery phase from acute renal failure caused by rhabdomyolysis has been associated with hypercalcemia. Typically, this follows an episode of severe hyperphosphatemia and hypocalcemia in the acute, oliguric phase, accompanied by severe secondary hyperparathyroidism. It has been ascribed to residual effects of PTH on bone turnover, as well as release of calcium phosphate precipitated into soft tissues such as skeletal muscle during the early hypocalcemic, hyperphosphatemic phase.

Chronic renal failure and dialysis are associated with hypercalcemia. Frequently, it is associated with the use of calcitriol or other vitamin D-analogues used to prevent secondary hyperparathyroidism, or with the use of oral calcium binding agents and supplements. Hypercalcemia in this population may also result from tertiary hyperparathyroidism, as discussed in Chapter 82.

Hypercalcemia has also been observed after kidney transplant particularly in cases of moderate to severe secondary hyperparathyroidism.

Miscellaneous

Several other diseases and syndromes associated with hypercalcemia are listed in Table 84.1.

APPROACH TO DIAGNOSIS

Although space precludes a detailed description of the specific approach to the differential diagnosis of each of the causes of hypercalcemia in Table 84.1, it is helpful to consider several broad guidelines. First, the causes of hypercalcemia can be divided into two broad categories: those associated with elevated PTH values (eg, primary and tertiary hyperparathyroidism, ectopic hyperparathyroidism, FHH, and occasionally lithium treatment) and those in which PTH is low normal or frankly suppressed.

Second, common diseases are common. Thus, the most common cause of hypercalcemia among outpatients is primary hyperparathyroidism and the most common cause among inpatients is cancer. Thus, it is reasonable to begin a diagnostic strategy with these two disorders in mind.

Third, most patients with MAHC have large, bulky tumors which are obvious on initial screening examinations and CT scans: thus, if no tumor is apparent after a careful physical examination and appropriate imaging procedures, attention should be focused on the less common items in Table 84.1. Exceptions to this guideline include small neuroendocrine tumors such as pheochromocytomas, bronchial carcinoids, and pancreatic islet tumors, which may be small and difficult to find.

Fourth, although it is tempting initially to select what seems an obvious diagnosis and manage the patient with this diagnosis in mind, because many of the less common items in Table 84.1 are overlooked unless specifically considered and because so many of these are easily treatable, it is critical to consider every diagnosis in Table 84.1 in every patient. For example, a patient with breast cancer can also have primary hyperparathyroidism, which is easily treated and may change the overall prognostic perception. Similarly, patients with lung cancer may also have tuberculosis, and hypercalcemia in this setting may reverse with appropriate antitubercular treatment, again, altering the overall prognostic perception. Another common example is milk-alkali syndrome occurring in a patient whose hypercalcemia has inappropriately been attributed to a coexisting cancer.

Fifth, carefully documenting the duration of hypercalcemia is helpful. Most hypercalcemia syndromes are unstable, and rapidly become more severe if left untreated (MAHC, immobilization, and vitamin D intoxication are examples), whereas long-term (>6 months), stable hypercalcemia has a relatively short differential diagnosis that includes largely primary and tertiary hyperparathyroidism, FHH, thiazide and lithium use, and occasional cases of sarcoid.

Sixth, it is useful to consider the principal underlying pathophysiological mechanism responsible for hypercalcemia before completely committing to a diagnosis or therapeutic plan. Is the hypercalcemia mainly postprandial (ie, gastrointestinal in origin) as might occur in sarcoid or vitamin D intoxication? Or is it equally apparent in fasting conditions, which may suggest inability of the kidney to clear calcium (as in FHH or thiazide use) or excessive bone resorption (as in MAHC)? And between these two, is the urinary calcium/creatinine ratio very high (suggesting gastrointestinal causes [such as sarcoid, milk alkali syndrome or vitamin D intoxication], or bone resorption [from cancer, immobilization, etc.]), or is it normal? Is the patient taking oral calcium supplements or antacids? These considerations help to narrow the diagnostic possibilities and suggest specific laboratory tests, such as PTH, PTHrP measurements, vitamin D metabolites, thyroid indices, serum ACE, serum/urine protein electrophoresis, lithium levels, bone marrow biopsy, liver biopsy, etc.

Considering these tenets while considering each of the items in Table 84.1 will facilitate and accelerate an accurate diagnosis.

MANAGEMENT

Management of hypercalcemia optimally targets the underlying pathophysiology. Of course, sometimes these are not possible, or therapy must be started before a definitive diagnosis is made. In these cases, targeting the underlying pathophysiology is most appropriate. Thus, in patients whose hypercalcemia is primarily based on accelerated bone resorption (eg, LOH, HHM, immobilization), therapy should include agents that block bone resorption, such as zoledronate, pamidronate, or denosumab [49]. In patients whose hypercalcemia is principally GI in origin (eg, sarcoidosis, milk-alkali syndrome, vitamin D intoxication, 1,25(OH)2D-secreting lymphomas), reducing or eliminating oral calcium and vitamin D intake and sunlight exposure may be most appropriate. For those with important renal contributions to hypercalcemia (eg, dehydration), increasing renal calcium clearance by increasing the GFR using saline infusions is critical. Blocking renal calcium absorption using loop diuretics such as furosemide may further enhance calcium excretion but is somewhat controversial as it may worsen intravascular depletion and should only be used after the patient is fully rehydrated and in a monitored setting. Of course, many patients have contributions from several sources and optimal therapy targets each of these components.

For hyperparathyroidism and its variants, specific therapy is discussed in Chapter 82. For patients with cancer, the most effective long-term therapy is tumor eradication. If this is not possible, or while waiting for a response to chemotherapy, aggressive hydration with saline, keeping a careful watch for signs of congestive heart failure, accompanied by a loop diuretic such as furosemide, are appropriate. Limiting oral calcium intake is not important in HHM and LOH, because intestinal calcium absorption is already low as a result of the low 1,25$(OH)_2$D concentrations in these patients, and because cachexia is a common feature of these patients. On the other hand, in patients with 1,25$(OH)_2$D-induced hypercalcemia from lymphoma, reducing oral calcium and vitamin D intake is important. Whereas some physicians wait to see the magnitude of the decline in serum calcium induced by hydration and diuresis, when the serum calcium exceeds 12.0 mg/dL, many recommend instituting antiresorptive therapy with an intravenous bisphosphonate such as zoledronate or denosumab concurrently with hydration.

For the granulomatous diseases, correcting the underlying cause is critical, where possible (eg, tuberculosis). In sarcoid, limiting calcium and vitamin D intake and sun exposure are important, together with oral or parenteral hydration. Glucocorticoid therapy may be necessary to treat the granulomas, to lower intestinal calcium absorption, and to lower 1,25$(OH)_2$D concentrations.

For immobilization-induced hypercalcemia, weight-bearing ambulation is the mainstay of therapy. Often, however, this is not possible because of spinal cord injury or pain. Here, aggressive hydration and intravenous bisphosphonates are effective and important.

For the remainder of the diagnoses in Table 84.1, correcting the underlying disorder or withdrawing or reducing the dose of the offending medication corrects the serum calcium.

ACKNOWLEDGMENTS

This work was supported by NIH grants DK51081 and DK073039.

REFERENCES

1. Hodak S, Stewart AF. *Disorders of Serum Minerals. CECIL Essentials of Medicine* (9th ed.). Philadelphia: WB Saunders, 2016, pp. 741–9.
2. Goltzman D. Approach to hypercalcemia (updated 8 August 2016). In: De Groot LJ, Chrousos G, Dungan K, et al. (eds). *Endotext* [Internet]. South Dartmouth, MA: MDText.com, Inc.
3. Burtis WJ, Brady TG, Orloff JJ, et al. Immunochemical characterization of circulating parathyroid hormone-related protein in patients with humoral hypercalcemia of malignancy. N Engl J Med. 1990;322:1106–12.
4. Stewart AF, Vignery A, Silvergate A, et al. Quantitative bone histomorphometry in humoral hypercalcemia of malignancy: uncoupling of bone cell activity. J Clin Endo Metab. 1982;55:219–27.
5. Horwitz MJ, Tedesco MB, Sereika SK, et al. Modeling hyperparathyroidism, humoral hypercalcemia of malignancy and lactation in humans: continuous infusion of PTH or PTHrP suppresses bone formation and uncouples bone turnover. J Bone Min Research (e-pub online 5–11) 2011.
6. Dean T, Vilardaga J-P, Potts JT, Gardella TJ. Altered selectivity of parathyroid hormone (PTH) and PTH-related protein (PTHrP) for distinct conformations of the PTH/PTHrP receptor. Mol Endo. 2008;22:156–66.
7. Knecht TP, Behling CA, Burton DW, et al. The humoral hypercalcemia of benignancy. A newly appreciated syndrome. Am J Clin Pathol. 1996;105:487–92.
8. Mirrakhimov AE. Hypercalcemia of malignancy: an update on pathogenesis and management. N Am J Med Sci. 2015;7(11):483–93.
9. Waning DL, Guise TA. Molecular mechanisms of bone metastasis and associated muscle weakness. Clin Cancer Res. 2014;20(12):3071–7.
10. Tebben PJ, Singh RJ, Kumar R. Vitamin D-mediated hypercalcemia: mechanisms, diagnosis and treatment. Endocr Rev. 2016;37(5):521–47.
11. Doyle M, Malcolm JC. An usual case of malignancy-related hypercalcemia. Int J Gen Med. 2014;7:21–7.
12. VanHouten JN, Yu N, Rimm D, et al. Hypercalcemia of malignancy due to ectopic transactivation of the parathyroid hormone gene. J Clin Endocrinol Metab. 2006;91:580–3.
13. Donovan PJ, Sundac L, Pretorius CJ, et al. Calcitriol-mediated hypercalcemia: causes and course in 101 patients. J Clin Endocrinol Metab. 2013;98(10):4023–9.
14. Adams JS, Gacad MA. Characterization of 1 hydroxylation of vitamin D_3 sterols by cultured alveolar macrophages from patients with sarcoidosis. J Exp Med. 1985;161:755–65.
15. Gkonos PJ, London R, Hendler ED. Hypercalcemia and elevated 1,25-dihydroxyvitamin D levels in a patient with end stage renal disease and active tuberculosis. N Engl J Med. 1984;311:1683–5.
16. Ross DS, Nussbaum SR. Reciprocal changes in parathyroid hormone and thyroid function after radioiodine treatment of hyperthyroidism. J Clin Endocrinol Metab. 1989;68:1216–19.
17. Rosen HN, Moses AC, Gundberg C, et al. Therapy with parenteral pamidronate prevents thyroid hormone-induced bone turnover in humans. J Clin Endocrinol Metab. 1993;77:664–9.
18. Muls E, Bouillon R, Boelaert J, et al. Etiology of hypercalcemia in a patient with Addison's disease. Calcif Tissue Int 1982;34:523–6.
19. Vasikaran SD, Tallis GA, Braund WJ. Secondary hypoadrenalism presenting with hypercalcaemia. Clin Endocrinol. 1994;41:261–5.
20. Ghaferi AA, Chojnacki KA, Long, WD, et al. Pancreatic VIPomas: subject review and one institutional experience. J Gastrointest Surg. 2008;12:382–93.
21. Mune T, Katakami H, Kato Y, et al. Production and secretion of parathyroid hormone-related protein in

pheochromocytoma: participation of an α-adrenergic mechanism. J Clin Endocrinol Metab. 1993;76:757–62.
22. Stewart AF, Adler M, Byers CM, et al. Calcium homeostasis in immobilization: an example of resorptive hypercalciuria. N Engl J Med. 1982;306:1136–40.
23. Cano-Torres EA, Gonzalex-Cantu A, Hinojosa-Garza G, et al. Immobilization induced hypercalcemia. Clin Cases Miner Bone Metab. 2016;13(1):46–7.
24. Gaudino A, Pennisi P, Bratengeier C, et al. Increased sclerostin serum levels associated with bone formation and resorption markers in patients with immobilization-induced bone loss. J Clin Endocrinol Metab. 2010;95(5):2248–53.
25. Orwoll ES. The milk-alkali syndrome: current concepts. Ann Intern Med 1982;97:242–48.
26. Beall DP, Scofield RH. Milk-alkali syndrome associated with calcium carbonate consumption. Medicine. 1995;74:89–96.
27. Holick MF, Shao Q, Liu WW, et al. The vitamin D content of fortified milk and infant formula. N Engl J Med. 1992;326:1178–81.
28. Ott SM, Maloney NA, Klein GL, et al. Aluminum is associated with low bone formation in patients receiving chronic parenteral nutrition. Ann Intern Med. 1983;96:910–14.
29. Klein GL, Horst RL, Norman AW, et al. Reduced serum levels of 1a, 25-dihydroxyvitamin D during long-term total parenteral nutrition. Ann Intern Med. 1981;94:638–43.
30. Merlini G, Fitzpatrick LA, Siris ES, et al. A human myeloma immunoglobulin G binding four moles of calcium associated with asymptomatic hypercalcemia. J Clin Immunol. 1984;4:185–96.
31. Elfatih A, Anderson NR, Fahie-Wilson MN, et al. Pseudo-pseudohypercalcaemia, apparent primary hyperparathyroidism and Waldenström's macroglobulinaemia. J Clin Pathol. 2007;60:436–7.
32. Twigt BA, Houweling BM, Vriens MR, et al. Hypercalcemia in patients with bipolar disorder treated with lithium: a cross-sectional study. Int J Bipolar Disord 2013;1:18.
33. Porter RH, Cox BG, Heaney D, et al. Treatment of hypoparathyroid patients with chlorthalidone. N Engl J Med. 1978;298:577.
34. Griebeler ML, Kearns AE, Ryu E, et al. Thiazide-associated hypercalcemia: incidence and association with primary hyperparathyroidism over two decades. J Clin Endocrinol Metab. 2016;101(3):1166–73.
35. McPherson ML, Prince SR, Atamer E, et al. Theophylline-induced hypercalcemia. Ann Intern Med. 1986;105:52–4.
36. Gayet S, Ville E, Durand JM, et al. J. Foscarnet-induced hypercalemia in AIDS. AIDS. 1997;11:1068–70.
37. Knox JB, Demling RH, Wilmore DW, et al. Hypercalcemia associated with the use of human growth hormone in an adult surgical intensive care unit. Arch Surg. 1995;130:442–5.
38. Sakoulas G, Tritos NA, Lally M, et al. Hypercalcemia in an AIDS patient treated with growth hormone. AIDS. 1997;11:1353–6.
39. Miller PD, Bilezikian JP, Diaz-Curiel M, et al. Occurrence of hypercalciuria in patients with osteoporosis treated with teriparatide. J Clin Endocrinol Metab. 2007;92:3535–41.
40. Valente JD, Elias AN, Weinstein GD. Hypercalcemia associated with oral isotretinoin in the treatment of severe acne. JAMA. 1983;250:1899.
41. Villablanca J, Khan AA, Avramis VI, et al. Phase I trial of 13-cis-retinoic acid in children with neuroblastoma following bone marrow transplantation. J Clin Oncol. 1995;13:894–901.
42. Ellis MJ, Gao F, Dehdashti F, et al. Lower-dose vs. high-dose oral estradiol therapy of hormone receptor-positive, aromatase inhibitor- resistant advance breast cancer. JAMA. 2009;302:774–80.
43. Jacobus CH, Holick MF, Shao Q, et al. Hypervitaminosis D associated with drinking milk. N Engl J Med. 1992;326:1173–7.
44. Llach F, Felsenfeld AJ, Haussler MR. The pathophysiology of altered calcium metabolism in rhabdomyolysis-induced acute renal failure. N Engl J Med. 1981;305:117–23.
45. Messa P, Cafforio C, Alfieri C. Calcium and phosphate changes after renal transplantation. J Nephrol. 2010:23 (S16);175–81.
46. Kuchay MS, Mishra SK, Farooqui KJ, et al. Hypercalcemia of advanced chronic liver disease: a forgotten clinical entity! Clin Cases Min Bone Metab. 2016;13(1):15–18.
47. Sarkar S, Hussain N, Herson V. Fibrin glue for persistent pneumothorax in neonates. J Perinatology. 2003;23:82–4.
48. Marks BE, Doyle DA. Idiopathic infantile hypercalcemia: case report and review of the literature. J Pediatr Endocrinol Metab. 2016;29(2):127–32.
49. Hu MI, Glezerman IG, Leboulleux S, et al. Denosumab for treatment of hypercalcemia of malignancy. J Clin Endocrinol Metab. 2014;99(9):3144–52.
50. Moseley JM, Kubota M, Diefenbach-Jagger H, et al. Parathyroid hormone-related protein purified from a human lung cancer cell line. Proc Natl Acad Sci U S A. 1987;84(14):5048–52.

85
Hypocalcemia: Definition, Etiology, Pathogenesis, Diagnosis, and Management

Anne L. Schafer and Dolores M. Shoback

University of California San Francisco and San Francisco Department of Veterans Affairs Medical Center, San Francisco, CA, USA

DEFINITION

Hypocalcemia, defined as an ionized calcium concentration that falls below the lower limit of the normal range, is a commonly encountered clinical problem with multiple causes. A normal level of ionized calcium (usually 1.00 to 1.25 mM) is critical for many vital cellular functions including hormonal secretion, skeletal and cardiac muscle contraction, cardiac conduction, blood clotting, and neurotransmission. Approximately 50% of the total serum calcium is ionized, with the remainder being protein-bound (45% to 50%, predominantly to albumin), or complexed to circulating anions such as phosphate (<5%). Total serum calcium is usually the only value a clinician has when making an initial determination of the state of serum calcium homeostasis in a patient, because ionized calcium determinations are not routine measurements in most clinical settings. Therefore, the clinician must make the first assessment based on total serum calcium levels.

The total serum calcium concentration is a reliable indicator of the serum ionized calcium concentration under most but not all circumstances. One important common situation where total serum calcium poorly reflects the ionized calcium concentration is when hypoalbuminemia is present. When serum albumin is depressed, total serum calcium often falls to subnormal levels. This can be mistaken for hypocalcemia. A bedside estimation of the corrected serum total calcium should be performed in the hypoalbuminemic patient to determine whether there is real concern for hypocalcemia. This estimation is usually performed using the following formula: adjusted total calcium (mg/dL) = measured total calcium (mg/dL) + [0.8 × (4.0 − measured serum albumin [g/dL])]. It is far better, however, when there is any question, to establish that the ionized calcium is truly low by making a direct measurement. Estimates of the ionized calcium are poor surrogates for actual measurements because, in addition to albumin, disturbances in pH and other circulating substances (eg, citrate, phosphate, paraproteins) can influence the serum total calcium, and these confounding factors are not considered in this estimation. Further, when serum albumin is very low, the correction equation may underperform. It is imperative that the clinician establishes that the ionized calcium concentration is indeed reduced, before an exhaustive search for an etiology of hypocalcemia is undertaken. Full evaluation may be costly and is unjustified if there is no or only weak evidence that the serum ionized calcium level is subnormal.

ETIOLOGY AND PATHOGENESIS

There are numerous etiologies of low serum ionized calcium. The disorders can be classified broadly as ones in which there is inadequate PTH or vitamin D, PTH, or vitamin D resistance, or a miscellaneous cause (Table 85.1). The last category encompasses a large and diverse spectrum of conditions that the clinician encounters in the course of practice. It is incumbent on the clinician to be aware of the etiologies, the pathogenic mechanisms, the intricacies of diagnostic testing

---

Primer on the Metabolic Bone Diseases and Disorders of Mineral Metabolism, Ninth Edition. Edited by John P. Bilezikian.
© 2019 American Society for Bone and Mineral Research. Published 2019 by John Wiley & Sons, Inc.
Companion website: www.wiley.com/go/asbmrprimer

Table 85.1. Etiologies of hypocalcemia.

Inadequate PTH production
  Postsurgical
  Constitutively active *CaSR* mutations (OMIM #145980)
  Autoimmune
    Isolated
    Polyendocrine failure syndrome type 1 (OMIM #240300 and OMIM #607358)
    Acquired antibodies that activate the CaSR
  Parathyroid gland agenesis
    GCMB mutations (OMIM #603716)
  PTH gene mutations
    Autosomal recessive (OMIM #168450.0002)
    Autosomal dominant (OMIM #168450.0001)
  X-linked hypoparathyroidism
  Postradiation therapy
  Secondary to infiltrative processes
    Iron overload: hemochromatosis, thalassemia after transfusions
    Wilson's disease
    Metastatic tumor
  Magnesium excess
  Magnesium deficiency

Syndromes with component of hypoparathyroidism
  DiGeorge syndrome (OMIM #188400)
  HDR (hypoparathyroidism, deafness, renal anomalies) syndrome (OMIM #146255 and OMIM #256340)
  Blomstrand lethal chondrodysplasia (OMIM #215045)
  Kenney-Caffey syndrome (OMIM #244460)
  Sanjad-Sakati syndrome (OMIM #241410)
  Kearns-Sayre syndrome (OMIM #530000)

Vitamin D or calcium deficiency
  Nutritional deficiency
  Lack of sunlight exposure
  Malabsorption
  Postgastric bypass surgery
  End-stage liver disease and cirrhosis
  Chronic kidney disease

PTH resistance
  Pseudohypoparathyroidism
  Magnesium depletion

Vitamin D resistance
  Pseudovitamin D deficiency rickets (vitamin D-dependent rickets type 1)
  Vitamin D-resistant rickets (vitamin D-dependent rickets type 2)

Miscellaneous
  Acute pancreatitis
  Hyperphosphatemia
    Phosphate retention caused by acute or chronic renal failure
    Excess phosphate absorption caused by enemas, oral supplements
    Massive phosphate release caused by tumor lysis or crush injury
  Drugs
    Bisphosphonate or denosumab therapy — especially in patients with vitamin D insufficiency or deficiency
    Foscarnet
    Imatinib mesylate — lowers both calcium and phosphate
  "Hungry bone syndrome" or recalcification tetany
    Postthyroidectomy for Grave's disease
    Postparathyroidectomy
  Interference of gadolinium-containing contrast agents with the assay for total calcium in patients with chronic renal failure (pseudohypocalcemia)
  Rapid transfusion of large volumes of citrate-containing blood
  Osteoblastic metastases
  Rhadomyolysis
  Acute critical illness — multiple contributing etiologies

including sequencing for mutations and other genetic analysis, and the best approaches to therapies in patients with hypocalcemia.

Hypoparathyroidism is a rare diagnosis. It is most commonly the sequela of thyroid, parathyroid, or laryngeal surgery during which most or all functioning parathyroid tissues are damaged, devitalized, and/or inadvertently removed. Perhaps next in frequency is the mild hypocalcemia and hypoparathyroidism caused by constitutively activating mutations of the calcium-sensing receptor (CaSR). These mutations lead to the inappropriate suppression of PTH secretion at subnormal serum calcium levels. The disorder presents as autosomal dominant hypocalcemia (ADH) in families and may go unrecognized because the hypocalcemia is often mild. The biochemical hallmark of ADH is often hypercalciuria, which worsens with attempts to treat the hypocalcemia with calcium salts and activated vitamin D metabolites. Exacerbation of hypercalciuria with nephrocalcinosis and renal failure can result from these efforts. Such renal complications occur because the constitutively active CaSRs in the kidney misperceive prevailing serum calcium concentrations as higher than they are, and this enhances renal excretion of calcium. It has been appreciated that patients can develop antibodies that activate parathyroid and renal CaSRs. This produces an acquired form of hypocalcemia with low PTH and elevated urinary calcium levels, thus mimicking the genetic disorder. These rare individuals often have other autoimmune disorders. Destruction of the parathyroid glands on an immune basis can occur in isolation or as part of the type 1 autoimmune polyendocrine syndrome (APS1). This is an autosomal recessive disorder caused by mutations in the autoimmune regulator (AIRE-I) gene. APS1 includes mucocutaneous candidiasis and adrenal insufficiency most commonly, in addition to hypoparathyroidism, as well as other autoimmune manifestations. APS1 typically presents in childhood or adolescence. The most important autoantibodies in APS1 are those directed against type 1 interferons (alpha and omega, for example) and can appear long before endocrine function has been lost. Patients with APS1 as well as those with isolated autoimmune hypoparathyroidism often have antibodies reactive with CaSR epitomes. However, it is undetermined what role, if any, the CaSR plays in the pathogenesis of the tissue destruction.

There are multiple modes of inheritance of hypoparathyroidism, depending on the molecule involved. Autosomal recessive mutations in *glial cell missing B (GCMB)*, a transcription factor essential for parathyroid gland development, are a rare cause of hypoparathyroidism. Mutations in a gene near *SOX3* on the X chromosome underlie the pathogenesis of X-linked hypoparathyroidism. Another syndrome, HDR (hypoparathyroidism, deafness, renal anomalies), is caused by mutations in the transcription factor *GATA3*. Variable penetrance of the renal anomalies and hearing deficits has been observed. The more common DiGeorge syndrome, a result of multiple developmental anomalies in tissues arising from the third and fourth branchial pouches, includes a spectrum of hypoparathyroidism, thymic aplasia and immunodeficiency, cardiac defects, cleft palate, and abnormal facies. A variety of other very rare genetic syndromes is worth considering when a patient presents with a constellation of features that includes hypoparathyroidism (eg, Kenney-Caffey, Kearns-Sayre, Sanjad-Sakati, and other syndromes; Table 85.1).

The different forms of PTH resistance, or pseudohypoparathyroidism, are very rare. They are described, as are disorders of vitamin D resistance.

In contrast to the rarity of hypoparathyroidism, vitamin D deficiency and disordered vitamin D metabolism are more common causes of hypocalcemia. Hypocalcemia may also be a presenting sign of nutritional rickets caused by severely impaired calcium intake or absorption. Whereas vitamin D deficiency and insufficiency occur in multiple clinical settings (elderly patients, postmenopausal women with fractures, nursing home residents, and so forth), it is uncommon to see frankly low ionized calcium values in such patients, particularly when 25-hydroxyvitamin D (25OHD) levels are just mildly depressed. Generally, low serum ionized calcium values result from longstanding severe vitamin D deficiency and chronically low serum 25OHD levels, and are accompanied by a significant degree of secondary hyperparathyroidism. Nevertheless, it is essential in the evaluation of patients with low serum ionized calcium values that one carefully considers vitamin D inadequacy, disorders of vitamin D activation by the kidney (including CKD, discussed below), and reduced vitamin D-mediated signaling as possible contributors to the etiology of the hypocalcemia.

Disorders of magnesium homeostasis are worth mentioning because both magnesium excess and deficiency can produce hypocalcemia that is generally mild and caused by functional (and reversible) hypoparathyroidism. Hypomagnesemia, often of a transient and correctable nature, accompanies a vast number of clinical situations particularly in ill and hospitalized patients (eg, malnutrition, pancreatitis, chronic alcohol abuse, diarrhea, diuretic and antibiotic therapy, and chemotherapeutic agents such as cisplatin derivatives). Low serum magnesium levels found in conjunction with these clinical entities require evaluation and often at least short-term therapy. Primary renal magnesium wasting states such as Gitelman syndrome, caused by mutations in the renal thiazide-sensitive NaCl cotransporter, are more persistent and require long-term magnesium and other electrolyte replacement therapy to correct the biochemical parameters and clinical symptoms. Other rare entities that involve primary renal magnesium wasting include autosomal recessive disorders caused by mutations in the *paracellin-1* or in the Na-K ATPase subunit (*FXYDZ* gene). Hypomagnesemia interferes with PTH action at the PTH target organs, bone and kidney, particularly by interfering with receptor-mediated activation of adenylate cyclase through the stimulatory G protein alpha subunit (Gs α). Magnesium is a cofactor for the adenylate cyclase enzyme complex. Hence, chronic hypomagnesemia produces a functional state of PTH resistance. Importantly, the normal physiological response to hypocalcemia is lacking in the patient with magnesium depletion. Once the magnesium depletion is corrected, parathyroid function returns to normal. Hypermagnesemia, in contrast, activates parathyroid CaSRs, thereby suppressing PTH secretion directly. Magnesium levels high enough to stimulate the CaSR tend to occur only in patients with CKD, or in the rare instance when magnesium is used for tocolytic therapy for preterm labor.

Hypocalcemia may occur in the conditions listed in the category "Miscellaneous" (Table 85.1). In terms of frequency, pancreatitis is the most common disorder and is often associated with a low serum calcium. This has been ascribed to the precipitation of calcium-containing salts in the inflamed pancreatic tissue and the presence of excess free fatty acids in the circulation. Hypocalcemia in patients with pancreatitis often correlates with illness severity, because pancreatitis may progress rapidly with hemorrhage, hypotension, and sepsis as complicating features.

Acute and chronic hyperphosphatemia can cause low serum total calcium. The most common cause for chronic hyperphosphatemia is CKD. Hypocalcemia in CKD has many contributing factors, including low 1,25-dihydroxyvitamin D production and poor nutrition. Acute changes in phosphate balance can also lower serum calcium. In any situation where large amounts of phosphate are rapidly absorbed into the intravascular compartment, there is the potential for the serum ionized calcium to fall, even to symptomatic levels. This can be found with phosphate-containing enemas and supplements, especially when the latter are given intravenously for the treatment of hypophosphatemia. Also, in the setting of acute tumor lysis caused by cytolytic therapy for high-grade lymphomas, sarcomas, leukemias, and solid tumors, cell breakdown with the rapid release of phosphate from intracellular nucleotides can quickly depress serum ionized calcium.

Treating normocalcemic patients with aminobisphosphonates (particularly intravenous bisphosphonates such as zoledronic acid and pamidronate) or denosumab, which block bone resorption dramatically, has the potential to cause a low serum ionized calcium. However, this is relatively infrequent unless concomitant vitamin D deficiency/insufficiency has gone unaddressed, or unless calcium absorption is markedly impaired (eg, after malabsorptive bariatric surgery). Denosumab may be more likely to cause hypocalcemia in the setting of advanced CKD, even in the case of vitamin D sufficiency and recommended calcium intake. The drug foscarnet, used to treat immunocompromised patients with refractory cytomegalovirus or herpes infections, can lower both serum calcium and magnesium to symptomatic levels. The tyrosine kinase inhibitor imatinib mesylate, used to treat chronic myeloid leukemia and gastrointestinal stromal tumors, can cause both hypocalcemia and hypophosphatemia, likely caused by its direct skeletal effects.

The "hungry bone syndrome," or recalcification tetany, can occur after parathyroidectomy for any form of hyperparathyroidism or after thyroidectomy for hyperthyroidism. Skeletal uptake of calcium and phosphate is intense because of the presence of a mineral-depleted bone matrix and the sudden removal by surgery of the stimulus for maintaining high rates of bone resorption (either PTH or thyroid hormones). Depending on the severity of the bone hunger, hypocalcemia and hypophosphatemia can persist for weeks and require large doses of calcium and vitamin D metabolites. If there has been no permanent damage to the parathyroid glands, intact PTH levels should rise appropriately to supranormal levels. In some cases, however, damage to the remaining parathyroid glands or suppression of function of the remaining glands by a previously dominant adenoma may confuse the picture. Careful management and repeated mineral and PTH analyses will usually allow the diagnosis to become clear over time.

The entity of pseudohypocalcemia caused by gadolinium-containing MRI agents received considerable attention because of the frequency of performing MR angiography in patients with CKD. The clearance of gadolinium in CKD patients is very prolonged. Total serum calcium levels, as measured by standard arsenazo III reagents, will appear to be low in patients with gadolinium-containing contrast agents in the circulation. Gadolinium complexes with this calcium-sensitive dye and blocks the colorimetric detection of calcium. Because ionized calcium is measured in a completely different manner, serum ionized calcium levels will be normal in these individuals, and there are no symptoms of hypocalcemia. To avert this problem, most radiology departments have shifted to the use of contrast agents that do not complex with arsenazo III.

Acute and critical illness, often in the intensive care unit setting, is frequently accompanied by hypocalcemia, including frankly low serum ionized calcium values. This entity is typically multifactorial with poor nutrition, vitamin D insufficiency, renal dysfunction, acid-base disturbances, cytokines, and other factors contributing. It is prudent to follow the serum ionized calcium values and treat as deemed appropriate based on clinical circumstances.

SIGNS AND SYMPTOMS

Patients with low serum ionized calcium values can present with no symptoms or with significant morbidity (Table 85.2). Their presentation depends on the severity and chronicity of the disturbance. Chronic hypocalcemia, even with very low levels of serum ionized calcium, can be asymptomatic. The only clue may be the presence of Chvostek's sign. The most frequent cause of symptoms is neuromuscular irritability, including tetany, carpopedal spasms, muscle twitching and cramping, circumoral tingling, abdominal cramps, and in severe

Table 85.2. Signs and symptoms of hypocalcemia.

Symptoms
    Paresthesias
        Circumoral and acral tingling
    Increased neuromuscular irritability
        Tetany
        Muscle cramping and twitching
        Muscle weakness
        Abdominal cramping
    Laryngospasm
    Bronchospasm
    Altered central nervous system function
        Seizures of all types: grand mal, petit mal, focal
        Altered mental status and sensorium
        Impaired concentration
        Papilledema, pseudotumor cerebri
        Choreoathetoid movements
        Depression
        Coma
    Generalized fatigue
    Cataracts
    Congestive heart failure
Signs
    Chvostek's sign
    Trousseau's sign
    Prolongation of the QT-c interval
    Basal ganglia and other intracerebral calcifications

cases, laryngospasm, bronchospasm, seizures, and even coma. Basal ganglia and other intracerebral calcifications can be observed on imaging studies. Ocular findings include cataracts, particularly when there are longstanding elevations in the calcium-phosphate product, and pseudotumor cerebri may be present. Longstanding hypocalcemia can also cause cardiomyopathy and congestive heart failure, which reverses with management of the usually very low serum ionized calcium levels. Hypocalcemia is well known for its effects on cardiac conduction, which are manifested on the ECG as prolongation of the QTc interval. In severe hypocalcemia, there is an increased risk of ventricular tachycardia and Torsades de Pointes. In addition, the patient often feels a sense of generalized weakness, fatigue, and depression that often lifts as the mineral disturbance, and vitamin D deficiency, if present, are successfully treated

DIAGNOSIS: TESTS AND INTERPRETATION

The mainstays of diagnostic testing are determination of serum total calcium and albumin or ionized calcium, total magnesium, intact PTH, 25OHD, and phosphate values (Fig. 85.1). CKD is typically already known from a patient's history. In patients without severe CKD, measuring 25OHD level is the best way to exclude vitamin D deficiency. Considerable attention has been directed to the reliability of contemporary 25OHD assays and clinically relevant cut-points for diagnosing vitamin D deficiency/insufficiency. These important clinical issues are discussed in Chapter 90. Intact PTH measured in a reliable two-site assay will readily disclose inappropriately low, low normal, or even undetectable values in hypoparathyroid states and generally normal levels (but inappropriately so) in patients with Mg^{2+} depletion. In marked contrast, patients with vitamin D deficiency or pseudohypoparathyroidism have elevated levels of PTH (secondary hyperparathyroidism). Serum phosphate levels are low in vitamin D deficiency and elevated or at the high end of the normal range in patients with hypoparathyroidism and pseudohypoparathyroidism, an important distinguishing measurement to make.

Accurate determination of 24-hour urinary calcium or magnesium excretion in hypocalcemia and hypomagnesemia, respectively, can be extremely helpful. Strikingly elevated urinary calcium levels in the mildly hypocalcemic patient suggest ADH. Milder elevations or even low urinary calcium levels are expected in hypoparathyroid states, depending on the serum calcium level. In contrast, vitamin D deficiency with secondary hyperparathyroidism classically produces hypocalciuria, in an effort by the kidney, under the influence of PTH, to conserve calcium for systemic needs. The presence of significant magnesium in the urine in a hypomagnesemic patient strongly suggests primary renal magnesium wasting and not gastrointestinal losses.

MANAGEMENT OF HYPOCALCEMIA: ACUTE AND CHRONIC

The goals of treatment are to alleviate symptoms, maintain an acceptable serum ionized calcium or total serum calcium, heal demineralized bones when osteomalacia is present, and (depending on the etiology of the hypocalcemia) avoid hypercalciuria. Avoiding hypercalciuria (urine calcium >300mg/24h) is critical for the prevention of renal dysfunction, stones, and nephrocalcinosis, although this has never been established experimentally. When clinical circumstances dictate urgent treatment, intravenous calcium salts are used. Seizures, severe tetany, laryngospasm, bronchospasm, or altered mental status are strong indicators for intravenous therapy. In contrast, if the patient is minimally symptomatic despite low numbers, the oral regimen outlined below can be used.

For urgent treatment, the preferred intravenous salt is calcium gluconate (10mL [10% solution] = 1g calcium gluconate = 90mg elemental calcium = 4.65mEq). The first 10mL is infused slowly over 10 min with ECG monitoring, and this can be followed by a second 10mL infused over 10 min with the same close monitoring. Generally, an infusion is begun. One method is to prepare 11g calcium gluconate in D5W to provide a final volume of 1000mL (~1mg elemental calcium/mL), and to infuse

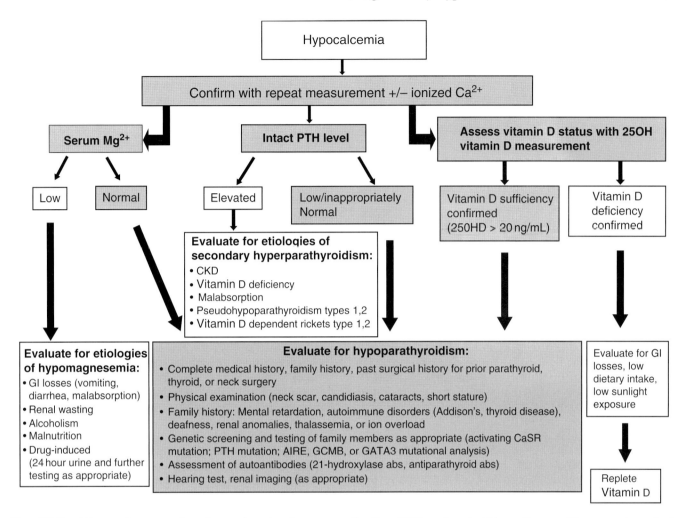

Fig. 85.1. Diagnostic approach to hypocalcemia. abs = autoantibodies; CKD = chronic kidney disease; GI = gastrointestinal; PTH = parathyroid hormone. Source: Bilezikian, et al. (2011). Reproduced with permission of John Wiley & Sons.

it at a rate of 0.5 to 2.0 mg elemental calcium/kg/h. The goals are to control symptoms, to restore the ionized calcium to the lower end of the normal range (~1.0 mM), and to normalize the QT_c interval. Higher rates (up to 2.0 mg/kg/h) may sometimes be needed to stabilize the patient. Serum ionized calcium level should be monitored closely (eg, after 1 hour and then every 4 hours). Once the serum ionized calcium is stabilized, a chronic oral regimen is begun. Infusion rates are tapered down as serum ionized calcium reaches the target, symptoms resolve, and oral medications are tolerated.

Chronic management of hypocalcemia uses oral calcium supplements, activated vitamin D metabolites (calcitriol and alfacalcidol), and sometimes thiazide diuretics. When magnesium depletion is responsible for hypocalcemia, magnesium deficits are generally large and poorly reflected by the serum magnesium level, because magnesium is predominantly an intracellular cation. Supplementation with magnesium salts over an extended period of time will usually be needed to replenish total body magnesium stores. Serum calcium and PTH secretory capacity will return to normal in nearly all cases unless there are ongoing and unaddressed losses.

Calcium supplements of all types work to treat hypocalcemia. A few general principles are worth emphasizing. It is best to divide the supplements throughout the day and to time their administration to coincide with meals, because this will enhance absorption (especially of calcium carbonate, which requires the acidic environment of the gastrointestinal tract for full absorption). The most efficient means of supplementation is in the form of calcium carbonate or citrate salts. The former is ~40% and the latter ~21% elemental calcium by weight. Generally, 500 to 1000 mg elemental calcium twice or three times a day is a reasonable starting dose and can be escalated upward. This is performed based on patient tolerance, compliance, and clinical goals.

When calcium supplements are insufficient to reach the target serum calcium, activated vitamin D metabolites are prescribed. Calcitriol (1,25-dihydroxyvitamin D; 0.25 to 1.0 μg once or twice daily) is necessary in patients with inadequate 1-α hydroxylase activity. Furthermore, most clinicians prefer calcitriol to vitamin D_2 or D_3 for hypocalcemia conditions because of its rapid onset and offset of action (1 to 3 days) and ease of titration. That said, when renal function is intact and some PTH is present for vitamin D activation, ergocalciferol (vitamin D_2) or cholecalciferol (vitamin D_3) may be used. Doses in the range of 25,000 to 50,000 IU daily (occasionally even more) may be needed to treat hypoparathyroidism, pseudohypoparathyroidism, or vitamin D deficiency in patients with malabsorption. Care must be exercised, because these forms of vitamin D have a long tissue half-life (weeks to months) caused by long-term storage in fat, and toxicity may be difficult to predict and to treat.

In patients who develop hypercalciuria on these regimens and/or have difficulty achieving the serum calcium goal safely, one can take advantage of the calcium-retaining actions of thiazide diuretics. Effective doses of hydrochlorthiazide typically range from 50 to 100 mg/day, although lower doses can be tried. Serum calcium, phosphorus, potassium, and creatinine levels should be monitored regularly, along with 25OHD levels if vitamin D_2 or D_3 therapy is used, to avoid toxicity.

For the treatment of hypoparathyroidism, the opportunity exists to treat with the missing hormone. Studies of PTH therapy have been performed in both adults and children. In 2015, the US Food and Drug Administration approved recombinant human PTH(1-84) in adults with hypoparathyroidism who cannot be well controlled on conventional management with calcium and active forms of vitamin D alone. Clinical trial evidence supporting the use of PTH(1-84) therapy in adults with hypoparathyroidism includes that from two randomized, placebo-controlled trials and one open-label study of 6 years duration. In the first trial, 62 patients on chronic calcium and activated vitamin D were treated with PTH(1-84) 100 μg/day (fixed dose) or placebo injections for 24 weeks. In the second trial, 134 patients on chronic calcium and activated vitamin D were treated with PTH(1-84) or placebo injections for 24 weeks, with PTH(1-84) dose starting at 50 μg/day and escalating to 75 then 100 μg/day. In the open-label trial, 33 patients were treated for 6 years initially with every other day injections (100 μg) and later up-titrated to 50 to 100 μg/day. The treatment of hypoparathyroidism is fully detailed in Chapter 86.

SUGGESTED READING

Abu-Alfa AK, Younes A. Tumor lysis syndrome and acute kidney injury: evaluation, prevention, and management. Am J Kidney Dis. 2010;55(5 Suppl 3):S1–13.

Agus ZS. Mechanisms and causes of hypomagnesemia. Curr Opin Nephrol Hypertens. 2016;25:301–7.

Akirav EM, Ruddle NH, Herold KC. The role of AIRE in human autoimmune disease. Nat Rev Endocrinol. 2010;7:25–33.

Ali A, Christie PT, Grigorieva IV, et al. Functional characterization of GATA3 mutations causing the hypoparathyroidism-deafness-renal (HDR) dysplasia syndrome: insight into mechanisms of DNA binding by the GATA3 transcription factor. Hum Mol Genet. 2007;16:265–75.

Antakia R, Edafe O, Uttley L, et al. Effectiveness of preventative and other surgical measures on hypocalcemia following bilateral thyroid surgery: a systematic review and meta-analysis. Thyroid. 2015;25:95–106.

Arabi A, El Rassi R, El-Hajj Fuleihan G. Hypovitaminosis D in developing countries-prevalence, risk factors and outcomes. Nat Rev Endocrinol. 2010;6:550–61.

Ayuk J, Gittoes NJ. How should hypomagnesaemia be investigated and treated? Clin Endocrinol. 2011;75:743–746.

Berman E, Nicolaides M, Maki RG, et al. Altered bone and mineral metabolism in patients receiving imatinib mesylate. N Engl J Med. 2006;354:2006–13.

Betterle C, Garelli S, Presotto F. Diagnosis and classification of autoimmune parathyroid disease. Autoimmun Rev. 2014;13:417–22.

Bilezikian JP, Khan A, Potts JT, et al. Hypoparathyroidism in the adult: epidemiology, diagnosis, pathophysiology, target-organ involvement, treatment, and challenges for future research. J Bone Miner Res. 2011;26:2317–37.

Bowl MR, Nesbit MA, Harding B, et al. An interstitial deletion-insertion involving chromosomes 2p25.3 and Xq27.1, near SOX3, causes X-linked recessive hypoparathyroidism. J Clin Invest. 2005;115:2822–31.

Clarke BL, Brown EM, Collins MT, et al. Epidemiology and diagnosis of hypoparathyroidism. J Clin Endocrinol Metab. 2016;101:2284–99.

Cooper MS, Gittoes NJ. Diagnosis and management of hypocalcaemia. BMJ. 2008;336:1298–302.

Forsythe RM, Wessel CB, Billiar TR, et al. Parenteral calcium for intensive care unit patients. Cochrane Database Syst Rev. 2008;8:CD006163.

Hannan FM, Babinsky VN, Thakker RV. Disorders of the calcium-sensing receptor and partner proteins: insights into the molecular basis of calcium homeostasis. J Mol Endocrinol. 2016;57:R127–42.

Hannan FM, Nesbit MA, Zhang C, et al. Identification of 70 calcium-sensing receptor mutations in hyper- and hypocalcaemic patients: evidence for clustering of extracellular domain mutations at calcium-binding sites. Hum Mol Genet. 2012;21:2768–78.

Kelly A, Levine MA. Hypocalcemia in the critically ill patient. J Intensive Care Med. 2013;28:166–77.

Kemp EH, Gavalas NG, Krohn KJ, et al. Activating autoantibodies against the calcium-sensing receptor detected in two patients with autoimmune polyendocrine syndrome type 1. J Clin Endocrinol Metab. 2009;94:4749–56.

Khan MI, Waguespack SG, Hu MI. Medical management of postsurgical hypoparathyroidism. Endocr Pract. (Suppl 1) 2011;17:18–25.

Konrad M, Schlingmann KP. Inherited disorders of renal hypomagnesaemia. Nephrol Dial Transplant. 2014;29:iv63–iv71.

Lemos MC, Thakker RV. GNAS mutations in pseudohypoparathyroidism type 1a and related disorders. Hum Mutat. 2015;36:11–19.

Liamis G, Milionis HJ, Elisaf M. A review of drug-induced hypocalcemia. J Bone Miner Metab. 2009;27:635–42.

Malloy PJ, Feldman D. Genetic disorders and defects in vitamin D action. Rheum Dis Clin N Am. 2012;38:93–106.

Mannstadt M, Clarke BL, Vokes T, et al. Efficacy and safety of recombinant human parathyroid hormone (1-84) in hypoparathyroidism (REPLACE): a double-blind, placebo-controlled, randomized phase 3 study. Lancet Diabetes Endocrinol. 2014;1:275–83.

McDonald-McGinn DM, Sullivan KE. Chromosome 22q11.2 deletion syndrome (DiGeorge syndrome/velocardiofacial syndrome). Medicine. 2011;90:1–18.

Munns CF, Shaw N, Kiely M, et al. Global consensus recommendations on prevention and management of nutritional rickets. J Clin Endocrinol Metab. 2016;101:394–415.

Prince MR, Erel HE, Lent RW, et al. Gadodiamide administration causes spurious hypocalcemia. Radiology. 2003;227:639–46.

Rosen CJ, Brown S. Severe hypocalcemia after intravenous bisphosphonate therapy in occult vitamin D deficiency. N Engl J Med. 2003;348:1503–4.

Rubin MR, Sliney J, McMahon DJ, et al. Therapy of hypoparathyroidism with intact parathyroid hormone. Osteoporos Int. 2010;21:1927–34.

Shah M, Bancos I, Thompson GB, et al. Teriparatide therapy and reduced postoperative hospitalization for postsurgical hypoparathyroidism. JAMA Otolaryngol Head Neck Surg. 2015;141:822–7.

Shoback D. Clinical practice. Hypoparathyroidism. N Engl J Med. 2008;359:391–403.

Shoback DM, Bilezikian JP, Costa AG, et al. Presentation of hypoparathyroidism: etiologies and clinical features. J Clin Endocrinol Metab. 2016;101:2300–12.

Sikjaer T, Amstrup AK, Rolighed L, et al. PTH(1-84) replacement therapy in hypoparathyroidism: a randomized controlled trial on pharmacokinetic and dynamic effects after 6 months of treatment. J Bone Miner Res. 2013;28:2232–43.

Thacher TD, Clarke BL. Vitamin D insufficiency. Mayo Clin Proc. 2011;86:50–60.

Thomee C, Schubert SW, Parma J, et al. GCMB mutation in familial isolated hypoparathyroidism with residual secretion of parathyroid hormone. J Clin Endocrinol Metab. 2005;90:2487–92.

Turan S, Bastepe M GNAS spectrum of disorders. Curr Osteoporos Rep. 2015;13:146–58.

Vandyke K, Fitter S, Dewar AL, et al. Dysregulation of bone remodeling by imatinib mesylate. Blood. 2010;115:766–74

Vivien B, Langeron O, Morel E, et al. Early hypocalcemia in severe trauma. Crit Care Med. 2006;33:1946–52.

Walker Harris V, Jan De Beur S. Postoperative hypoparathyroidism: medical and surgical therapeutic options. Thyroid. 2009;19:967–73.

Winer KK, Sinaii N, Reynolds J, et al. Long-term treatment of 12 children with chronic hypoparathyroidism: a randomized trial comparing synthetic human parathyroid hormone 1-34 *versus* calcitriol and calcium. J Clin Endocrinol Metab. 2010;85:2680–8.

Winer KK, Ko CW, Reynolds JC, et al. Long-term treatment of hypoparathyroidism: a randomized controlled study comparing parathyroid hormone-(1-34) *versus* calcitriol and calcium. J Clin Endocrinol Metab. 2003;88:4214–20.

Witteveen JE, van Thiel S, Romijn JA, et al. Hungry bone syndrome: still a challenge in the post-operative management of primary hyperparathyroidism: a systematic review of the literature. Eur J Endocrinol. 2013;168:R45–53.

86
Hypoparathyroidism

Tamara Vokes[1], Mishaela R. Rubin[2], Karen K. Winer[3], Michael Mannstadt[4], Natalie E. Cusano[5], Harald Jüppner[6], and John P. Bilezikian[2]

[1]Department of Medicine, University of Chicago, Chicago, IL, USA
[2]Metabolic Bone Diseases Unit, Division of Endocrinology, Department of Medicine, College of Physicians and Surgeons, Columbia University, New York, NY, USA
[3]NICHD, National Institutes of Health, Bethesda, MD, USA
[4]Endocrine Unit, Massachusetts General Hospital and Harvard Medical School, Boston, MA, USA
[5]Department of Medicine, Lenox Hill Hospital, New York, NY, USA
[6]Endocrine Unit and Pediatric Nephrology Unit, Departments of Medicine and Pediatrics, Harvard Medical School, Massachusetts General Hospital, Boston, MA, USA

INTRODUCTION

Hypoparathyroidism is a rare endocrine disorder characterized by low serum calcium caused by absent or inappropriately low levels of PTH. In 2015, the first international conference on hypoparathyroidism was held in Florence, Italy. The guidelines and supporting papers resulting from that conference were published in 2016 [1–4] and provide detailed information on the presentation and management of this rare condition. In the majority of patients, hypoparathyroidism is caused by thyroid and less commonly, parathyroid or other head and neck surgery [1]. The remaining cases are caused by autoimmune, genetic, and infiltrative disorders. Although genetic causes are relatively uncommon (particularly in adults), they provide important insight into the mechanisms of PTH regulation and may lead to future therapeutic targets. Consequently, they will be discussed in greater detail.

GENETIC CAUSES OF HYPOPARATHYROIDISM

Genetic causes of hypoparathyroidism are important in several ways: genetic testing can confirm the clinical diagnosis, reveal the presence of an unsuspected complex disease, and allow for carrier screening to be performed. Genetic etiologies of hypoparathyroidism have also provided invaluable information about the development of parathyroid glands and mechanism of PTH secretion. Although genetic causes of hypoparathyroidism comprise only a small subgroup of patients with nonsurgical hypoparathyroidism, precise prevalence data are not available. In these patients, hypoparathyroidism can either be part of a complex disease (syndromic form) or occur as isolated hypoparathyroidism without other syndromic features. Among the complex diseases (Table 86.1), 22q11 deletion syndrome, also called DiGeorge syndrome, is probably the most common [5]. Caused by a microdeletion of part of chromosome 22q11, patients present with the typical triad of congenital heart disease, immunodeficiency, and hypoparathyroidism, but may have many other manifestations of the disease. Up to about 50% of patients with the 22q11 deletion syndrome develop hypoparathyroidism [5]. The deleted gene responsible for hypoparathyroidism is Tbx1, which codes for a transcription factor that plays a role in the development of parathyroid glands. Mice with homozygous deletion of Tbx1 develop hypoparathyroidism [6].

The autoimmune polyglandular syndrome (APS1), also referred to as the autoimmune polyendocrinopathy-candidiasis-ectodermal dystrophy syndrome (APECED), is another complex disease. It is the only monogenetic autoimmune disease, and can be accompanied by hypoparathyroidism [7]. Patients affected by APECED develop

---

Primer on the Metabolic Bone Diseases and Disorders of Mineral Metabolism, Ninth Edition. Edited by John P. Bilezikian.
© 2019 American Society for Bone and Mineral Research. Published 2019 by John Wiley & Sons, Inc.
Companion website: www.wiley.com/go/asbmrprimer

Table 86.1. Complex diseases with hypoparathyroidism.

Syndrome	Gene(s)
22q11 deletion syndrome	Microdeletion of several genes including Tbx1
APS1 (APECED)	AIRE
HDR	GATA3
Kenney-Caffey, Sanjad-Sakati	TBCE, FAM111A
Kearns-Sayre, MELAS, MTPDS	Mitochondrial genes

Table 86.2. Diseases with isolated familial hypoparathyroidism.

Disease	Gene
ADH1	CASR
ADH2	Gα11
Hypoparathyroidism (autosomal dominant or recessive)	GCM2
Hypoparathyroidism (autosomal dominant or recessive)	PTH
X-linked hypoparathyroidism	Deletion/insertion involving SOX3

mucocutaneous candidiasis, Addison disease, hypoparathyroidism, and several other autoimmune diseases.

Isolated familial hypoparathyroidism (Table 86.2) is a genetically heterogeneous group of diseases. For simplicity, the disease mechanism can be grouped into three distinct categories. In the first one, parathyroid glands do not form, typically because of the loss-of-function of an essential transcription factor. For example, patients with homozygous loss-of-function or heterozygous dominant-negative mutations in GCM2 (glial cells missing 2) develop hypoparathyroidism and do not develop parathyroid glands [8–10]. Patients with X-linked recessive hypoparathyroidism harbor a deletion-insertion near the Sox3 gene, a transcription factor that also plays a role in parathyroid development [11].

In the second group, parathyroid glands do form normally but have a functional defect in PTH secretion caused by a genetic mutation. The most common example is autosomal-dominant hypocalcemia (ADH) type 1, caused by heterozygous activating mutations in the calcium-sensing receptor (CASR) [12]. The CASR on the surface of the parathyroid cells detects and is activated by rising concentrations of extracellular calcium, leading to reduced PTH secretion. Activating mutations in the CASR cause a leftward shift in the calcium-PTH curve and therefore lead to a reduction in PTH at normal and reduced calcium levels. In addition, patients with ADH1 have hypercalciuria, caused by low circulating levels of PTH, and the activation of the CASR in the renal tubules.

More recently, heterozygous gain-of-function mutations in Gα11, one of the signal transducers downstream of the CASR, have been identified in patients with ADH type 2, a similar type of familial hypocalcemia [13]. It came as a surprise that patients with a mutation in this ubiquitously expressed G protein only had hypoparathyroidism and possibly short stature, without derangements in other organ function. Further studies will be needed to increase our understanding of this disease.

Finally, familial isolated hypoparathyroidism can be caused by a rare autosomal recessive or dominant mutation in the gene coding for PTH. These mutations can cause incorrect processing and secretion of PTH or can be associated with a less active hormone [1].

Current therapies of patients with hypoparathyroidism, including those with genetic forms of the disease, typically consist of oral calcium, calcitriol, or PTH injections. Patients with ADH1 carry a higher risk of developing nephrocalcinosis because of high urinary calcium excretion and should therefore be treated only if symptomatic. The elucidation of the mechanism of genetic forms of the disease may facilitate the development of targeted drug therapies, such as calcilytics or Gα11 inhibitors, in the future.

CLINICAL FEATURES

Many signs and symptoms of hypoparathyroidism are attributable to hypocalcemia. Most commonly, the clinical presentation of hypoparathyroidism reflects the effects of chronic hypocalcemia which lead to neuromuscular symptoms (paresthesia and numbness) of varying severity. Occasionally, minimally symptomatic patients with hypoparathyroidism are identified by the discovery of hypocalcemia from a biochemical screening test. In contrast, the hypocalcemia of hypoparathyroidism can also present acutely after anterior neck surgery or in individuals with established hypoparathyroidism whose needs for supplemental calcium and active vitamin D change or who are noncompliant. Such acute hypocalcemic states can be a medical emergency because seizures and laryngospasm, which can occur under these circumstances, are each potentially lethal.

Hypocalcemia can affect the function of most organs, but in hypoparathyroidism, the most apparent organ systems that become dysfunctional are neurological, cognitive, muscular, and cardiac [14]. The calcium imbalance predisposes these systems to an irritability that can be subtle (paresthesia, prolonged cQT interval on electrocardiogram) or dangerous.

Renal manifestations and extraskeletal calcification

Although hypercalciuria and hypophosphaturia are expected in hypoparathyroidism because PTH effects on renal tubular functions are absent, this is not generally the case in the untreated disease because the filtered load of calcium is usually lower than normal. Similarly, hyperphosphatemia leads to a greater filtered load of phosphate. Urinary calcium and phosphate excretion may therefore be normal in untreated hypoparathyroid subjects. However, once treated, many of these patients become hypercalciuric because they require large amounts of calcium and/or active vitamin D for control of the serum calcium concentration. They are also at risk for renal calcium deposition, either as kidney stones or within the renal parenchyma, because the elevated calcium-phosphate product (>55 mg^2/dL^2) is often worsened by the need for large amounts of calcium and active vitamin D [15]. Ectopic calcium deposition, however, does not require an elevated calcium-phosphate product. Renal function is also at risk over time. The risk to renal function in hypoparathyroidism is particularly high in those with activating CaSR mutations [15, 16].

Over time, soft tissue calcifications can develop as a result of chronic hypocalcemia and hyperphosphatemia with an elevated serum calcium-phosphate product. The calcifications can occur either without treatment or as a result of large amounts of supplemental calcium and active vitamin D. These calcifications are typically found in the kidney (nephrolithiasis and nephrocalcinosis) and in the brain (basal ganglia in particular) [17] but can also be found in joints, eyes, skin, and vasculature [1, 14–17]. Ectopic calcifications can be found in patients whose calcium-phosphate product does not exceed 55 mg^2/dL^2. Therapeutic goals are to maintain this product as close to normal as possible.

TREATMENT OF HYPOPARATHYROIDISM

Treating hypoparathyroidism with calcium and active vitamin D

Conventional therapy of hypoparathyroidism consists of calcium and active vitamin D supplementation. The goals of therapy are: (i) to ameliorate symptoms of hypocalcemia; (ii) to maintain serum calcium in the low–normal range; (iii) to reduce or minimize hypercalciuria; (iv) to maintain serum phosphorus in the high–normal range; and (v) to maintain the calcium phosphate product well below the upper range of normal (55 mg^2/dL^2). During dose adjustments, albumin-corrected serum calcium, creatinine, and phosphorus should be measured weekly to monthly, and at least twice annually with a stable regimen. Urinary calcium and creatinine excretion should be considered during dose adjustments and should be measured at least annually with a stable regimen to evaluate for nephrotoxicity [3].

In the setting of acute hypocalcemia, calcium gluconate is commonly used. Calcium gluconate is first given as a 1 to 2 gr bolus (90 to 180 mg elemental calcium), followed by a slower infusion of 0.5 to 1.5 mg/kg/h (see Chapter 85). For long-term management, oral calcium carbonate and calcium citrate are the two most common forms of calcium supplementation, containing 40% and 21% elemental calcium, respectively. Calcium carbonate is less expensive but may be associated with gastrointestinal discomfort in some patients and must be taken with protein-based meals for proper absorption. Calcium citrate can be taken without respect to meals and should be used in patients on acid-suppressing medications that can inhibit absorption of calcium carbonate. Patients typically require calcium supplementation of 500 to 2000 mg 2 to 3 times daily, although higher doses or a more frequent dosing interval may be necessary [3]. "Natural" formulations of calcium, such as dolomite, are not recommended as they can contain lead [18].

PTH is the major inducer of 1-α-hydroxylase activity in the kidney. In hypoparathyroidism, it is, therefore, also necessary to supplement patients with active vitamin D, 1,25-dihydroxyvitamin D (calcitriol), to facilitate calcium absorption [19]. Patients typically require supplementation with 0.25 to 2.0 μg daily. Doses greater than 0.75 μg daily should be administered in two divided doses. Calcitriol has a relatively short half-life of 5 to 8 hours and toxicity is thus easily managed. Outside the USA, alphacalcidol and dihydrotachysterol are also used. Alphacalcidol has a similar onset of action as calcitriol of 1 to 3 days but a slightly longer offset [20]. Doses of 0.5 to 3.0 μg are typically used.

Use of parent vitamin D (ergo- or cholecalciferol) is also recommended to maintain 25-hydroxyvitamin D levels at 30 ng/mL or above. Use of very high doses of parent vitamin D in lieu of active vitamin D is no longer recommended given the availability and safety of the above analogs. Thiazide diuretics may be helpful in the case of hypercalciuria. Relatively high doses (50 to 100 mg daily) are usually needed to decrease urinary calcium excretion. Low phosphate diets or use of phosphate binders are usually not required but may be recommended for certain patients [4].

The use of large doses of calcium and active vitamin D often required to maintain serum calcium within the goal low–normal range raises concerns for potential complications. Conventional therapy can lead to nephrocalcinosis or nephrolithiasis within the renal system and deposition of calcium within soft tissues. Conventional therapy also does not directly address concerns caused by lack of PTH in hypoparathyroid patients, in particular regarding quality of life/neurocognitive complaints and decreased skeletal turnover. Consequently, when PTH formulations became available, several groups explored its utility in hypoparathyroidism.

Treating hypoparathyroidism with PTH(1-34)

PTH(1-34) is not approved for treatment of hypoparathyroidism. However, initial investigations of PTH use for treatment of this disease employed PTH(1-34) and provided a proof of concept that facilitated subsequent developments of therapeutic options such as PTH(1-84). Winer and colleagues studied PTH(1-34) replacement therapy given by subcutaneous injection in adults and children with hypoparathyroidism of various etiologies. Patients who received PTH(1-34) did not concurrently receive calcitriol, thiazide diuretics, or phosphate binders. In a series of randomized controlled studies, daily PTH(1-34) injections were first compared with conventional therapy [21] and then to twice daily PTH injections [22, 23]. In separate studies, adults and children receiving twice-daily injections or conventional therapy were compared in a randomized parallel trial over 3 years [24, 25]. These studies showed that twice daily subcutaneous PTH(1-34) maintained stable serum and urine calcium levels in the normal range with stable renal function over a 3 years and no impairment of linear growth in children [24, 25].

Simultaneous normalization of serum and urine calcium with normal levels of bone turnover markers was ultimately achieved with continuous administration of PTH(1-34) by pump (Omnipod by Insulet) [26, 27]. This method of hormonal delivery showed clear advantages over injection therapy in patients of all etiologies, including adults with postsurgical hypoparathyroidism and children with CaR or autoimmune polyendocrinopathy. Moreover, pump delivery resulted in a more than 60% reduction in the daily PTH(1-34) dose compared with twice-daily injection delivery. Additionally, by raising serum magnesium levels, pump delivery of PTH permitted a significant reduction in needs for supplemental magnesium. This proved the most physiological approach to replacement therapy as pump delivery of PTH(1-34) has the potential to mimic normal parathyroid secretory dynamics.

Treating hypoparathyroidism with PTH(1-84)

In January of 2015, PTH(1-84) was approved by the US Food and Drug Administration (FDA) for treatment of adults with chronic hypoparathyroidism. Several earlier studies provided a proof of concept for once daily use of PTH(1-84) in this disease. An investigator initiated study from Columbia University in New York reported initially 2 year results of treating 30 adults with PTH(1-84) at a fixed dose of 100 μg every other day [28]. In response to PTH therapy, patients were able to reduce serum phosphate and maintain serum calcium at a target level while reducing oral calcium and calcitriol significantly. The same group has reported the results of PTH(1-84) treatment for 4 years [29] and recently for 6 years [30]. Over time, the dosing regimens changed from a fixed every other day dose in the initial study to individualized dosing and (for 90% of patients) a daily injection [30]. Among the 33 patients who completed 6 years of investigation 48% were taking 50 μg per day, 20% taking 100 μg/day, 10% taking 75 μg/day with only a few patients taking the injection every other or every third day [30]. This illustrated the need for individualized dosing of the drug in this condition. PTH(1-84) therapy was associated with a reduction in calcium and calcitriol dose by an average of 53% and 67% respectively. Urinary calcium excretion also decreased and remained below baseline for the duration of the study [30]. Interestingly, the adverse events seem to diminish after the first year of the study with hypercalcemia (serum calcium >10.2 mg/dL) observed in only 2.5% of all the values over 6 years.

A second investigator-initiated study was conducted in Denmark where 62 patients were randomized in a double-blind fashion to receive a daily injection of 100 μg of PTH(1-84) or a matching placebo for 6 months [31]. Calcium and active vitamin D doses were reduced only if patients developed hypercalcemia or hypercalciuria. Plasma phosphate decreased and calcium increased in the PTH-treated group, with 19% of measurements falling above the upper limit of normal and 17 episodes of symptomatic hypercalcemia (one requiring hospitalization) in 11 PTH-treated patients and one such episode in the placebo group [31]. This experience underscores the need for individualized regimens so that the appropriate dose of PTH(1-84) can be achieved for a given patient.

The registration trial of PTH(1-84), known as REPLACE, was conducted in 134 adults with hypoparathyroidism (78% female and 74% having surgical etiology) randomized in 2:1 ratio to active drug or placebo given as daily injection in the thigh at escalating doses starting at 50 μg and increasing to 75 μg and then 100 μg per day over 6 months [32]. The doses of calcium and/or active vitamin D were reduced with each up-titration of the drug and increased if necessary to maintain calcium at the target range. The triple primary endpoint, defined as a reduction in *both* calcium and active vitamin D dose of at least 50% while maintaining serum calcium in the target range, was achieved in 53% of PTH-treated patients and in one (2%) placebo-treated patient [32]. The secondary endpoint defined as independence from active vitamin D and a reduction in oral calcium to less than 500 mg/day was achieved in 41% of PTH and 2% of placebo group. The PTH(1-84) dose at the end of the study was 100 μg/day in 52%, 75 μg/day in 27% and 50 μg/day in 21% of the PTH-treated patients. Serum calcium was maintained in the target range whereas serum phosphate decreased in the PTH but not in the placebo group [32]. The incidence of adverse events during the maintenance phase was similar between groups. However, during the 4-week follow-up phase when the study drug was stopped and patients returned to their prestudy supplement doses, the incidence of hypocalcemia was significantly greater in patients who had received PTH during the study. Consequently, if PTH therapy is stopped for any reason, close monitoring of serum calcium and prospective increase in supplement doses is mandatory.

AREAS OF SPECIAL INTEREST

Skeletal manifestations

BMD, as determined by dual energy X-ray absorptiometry, is typically above average as compared with age- and sex-matched controls at all measurement sites [30, 31, 33]. Imaging with peripheral quantitative computed tomography and high-resolution peripheral computed tomography, as well as direct histomorphometric analysis of bone by transiliac bone biopsy, show that both cortical and trabecular compartments of bone are altered, with increased cortical volumetric BMD and trabecular bone volume fraction and decreased cortical porosity [34–36]. Lower bone turnover is typical, as shown by dynamic histomorphometry studies of the iliac crest bone biopsy.

Reduced bone remodeling [1, 33, 37] is associated with positive bone balance [38], which accounts for the above-average features of skeletal density and microstructure. The abnormally low bone remodeling in hypoparathyroidism and dense bone suggest that hypoparathyroid bone is overly mature and, therefore, theoretically more vulnerable to fracture than euparathyroid bone. Fracture data, however, are limited, because this is a rare disease. A small cohort showed an increase in morphometric vertebral fractures in postmenopausal women with hypoparathyroidism; however, larger registry studies in Denmark did not detect a difference in overall fracture rate between hypoparathyroid patients and controls [39–41].

PTH(1-84) treatment affects skeletal parameters. In the longest study to date, PTH(1-84) treatment for 6 years induced site-specific densitometric changes [30] with increases in lumbar spine (3.8%) and total hip (2.4%), no change in femoral neck, and a decrease at the distal one third radius BMD (−4.4 %). Bone turnover markers increased significantly, reaching a threefold peak above baseline values at 1 year and subsequently declining but remaining higher than pretreatment values. Histomorphometric data in the same cohort after 2 years of treatment showed an increase in cancellous bone volume that was characterized by greater trabecular number and tunneling, along with an increase in cortical porosity [37]. Simulated bone strength by microCT was shown to transiently increase [42], whereas mineralization density distribution was shown to decrease [43]. Overall, although fracture data are lacking, these observations show that PTH stimulates the hypoparathyroid skeleton and perhaps reverses abnormal bone parameters.

Quality of life in hypoparathyroidism

Many patients with hypoparathyroidism treated with conventional therapy with calcium and active vitamin D have complaints suggestive of reduced quality of life (QoL) [1, 14, 16]. These including physical (fatigue), neuromuscular (weakness, cramps, paresthesia, seizures), cognitive ("brain fog"), and emotional difficulties (anxiety, depression, personality disorders). Only recently, systematic efforts to define the nature and incidence of QoL impairments in hypoparathyroidism have been reported [44–48]. When compared with healthy controls or to patients who had thyroid surgery but retained normal parathyroid function, patients with surgical hypoparathyroidism had significantly higher global complaint scores [44], lower physical summary scores on SF36 and muscle function [48], and lower QoL scores than anticipated by healthy subjects given the description of the disease or experienced (endocrine) surgeons [46].

Introduction of PTH therapy has raised hopes that replacing the missing hormone would restore QoL. The findings from the studies, however, have been inconsistent. In an open label study from Columbia University in New York described in references 28–30, QoL as assessed by SF36 was low in all domains at baseline despite control of serum calcium that was in an acceptable range for hypoparathyroidism [49, 50]. All domains improved significantly in response to PTH(1-84) at 1 year [49] with improvements maintained in subjects who completed 5 years of therapy [50]. Similar findings were reported in an Italian study with PTH(1-34) [51], although in that study many patients had hypocalcemia at baseline which improved during the study [51]. Different conclusions resulted from the above-mentioned Danish double-blinded placebo-controlled study [31] which enrolled relatively well-controlled hypoparathyroid patients and found that PTH(1-84) treated patients had less improvement in SF36 scores compared with placebo and actually had worse performance on at least some muscle function tests [52]. It should be remembered, however, that many PTH(1-84)-treated patients had hypercalcemia which may have negatively affected their well-being. Finally, QoL was assessed by SF36 during REPLACE with preliminary analysis revealing improved scores in the PTH(1-84)-treated but not in placebo-treated patients with no statistically significant between-group differences [53].

FUTURE DIRECTIONS

With the advent of PTH(1-84) as an FDA-approved therapy for hypoparathyroidism, it is likely that the therapeutic direction of this disease will change. The indications for treatment with PTH(1-84) are directed towards those who cannot be well controlled on conventional therapy. The International Conference on Hypoparathyroidism has interpreted this directive broadly to include the following categories: serum calcium cannot be reliably controlled with conventional therapy; daily oral calcium needs exceed 2.5 g/day; or need for calcitriol exceeds 1.5 μg/day (with the analogs, >3.0 μg/day); major gastrointestinal dysfunction (eg, gastric bypass surgery); high serum phosphate or high calcium-phosphate product; renal manifestations such as hypercalciuria, renal calcifications; reduced creatinine clearance (<60 cm^3/min); reduced quality of life. These guidelines are likely to be modified as more experience is gained with the use of PTH(1-84) in hypoparathyroidism. The availability of PTH(1-84) as a treatment of hypoparathyroidism adds this disease to the list of hormone deficiency states for which the replacement hormone is now available.

REFERENCES

1. Shoback DM, Bilezikian JP, Costa AG, et al. Presentation of hypoparathyroidism: etiologies and clinical features. J Clin Endocrinol Metab. 2016;101(6):2300–12.
2. Clarke BL, Brown EM, Collins MT, et al. Epidemiology and diagnosis of hypoparathyroidism. J Clin Endocrinol Metab. 2016;101(6):2284–99.
3. Brandi ML, Bilezikian JP, Shoback D, et al. Management of hypoparathyroidism: summary statement and guidelines. J Clin Endocrinol Metab. 2016;101(6):2273–83.
4. Bilezikian JP, Brandi ML, Cusano NE, et al. Management of hypoparathyroidism: present and future. J Clin Endocrinol Metab. 2016;101(6):2313–24.
5. McDonald-McGinn DM, Sullivan KE, Marino B, et al. 22q11.2 deletion syndrome. Nat Rev Dis Primers. 2015;1:15071.
6. Merscher S, Funke B, Epstein JA, et al. TBX1 is responsible for cardiovascular defects in velo-cardio-facial/DiGeorge syndrome. Cell. 2001;104(4):619–29.
7. Akirav EM, Ruddle NH, Herold KC. The role of AIRE in human autoimmune disease. Nat Rev Endocrinol. 2011;7(1):25–33.
8. Ding C, Buckingham B, Levine MA. Familial isolated hypoparathyroidism caused by a mutation in the gene for the transcription factor GCMB. J Clin Invest. 2001;108(8):1215–20.
9. Günther T, Chen ZF, Kim J, et al. Genetic ablation of parathyroid glands reveals another source of parathyroid hormone. Nature. 2000;406(6792):199–203.
10. Mannstadt M, Bertrand G, Muresan M, et al. Dominant-negative GCMB mutations cause an autosomal dominant form of hypoparathyroidism. J Clin Endocrinol Metab. 2008;93(9):3568–76.
11. Bowl MR, Nesbit MA, Harding B, et al. An interstitial deletion-insertion involving chromosomes 2p25.3 and Xq27.1, near SOX3, causes X-linked recessive hypoparathyroidism. J Clin Invest. 2005;115(10):2822–31.
12. Hannan FM, Babinsky VN, Thakker RV. Disorders of the calcium-sensing receptor and partner proteins: insights into the molecular basis of calcium homeostasis. J Mol Endocrinol 2016;57(3):R127–42.
13. Roszko KL, Bi RD, Mannstadt M. Autosomal dominant hypocalcemia (hypoparathyroidism) types 1 and 2. Front Physiol. 2016;7:458.
14. Bilezikian JP, Khan A, Potts JT Jr, et al. Hypoparathyroidism in the adult: epidemiology, diagnosis, pathophysiology, target-organ involvement, treatment, and challenges for future research. J Bone Miner Res. 2011;26(10):2317–37.
15. Mitchell DM, Regan S, Cooley MR, et al. Long-term follow-up of patients with hypoparathyroidism. J Clin Endocrinol Metab. 2012;97(12):4507–14.
16. Shoback D. Clinical practice. Hypoparathyroidism. N Engl J Med. 2008;359(4):391–403.
17. Goswami R, Millo T, Mishra S, et al. Expression of osteogenic molecules in the caudate nucleus and gray matter and their potential relevance for basal ganglia calcification in hypoparathyroidism. J Clin Endocrinol Metab. 2014;99(5):1741–8.
18. Bourgoin BP, Evans DR, Cornett JR, et al. Lead content in 70 brands of dietary calcium supplements. Am J Public Health. 1993;83(8):1155–60.
19. Neer RM, Holick MF, DeLuca HF, et al. Effects of 1alpha-hydroxy-vitamin D3 and 1,25-dihydroxy-vitamin D3 on calcium and phosphorus metabolism in hypoparathyroidism. Metabolism 1975;24(12):1403–13.
20. Haussler MR, Cordy PE. Metabolites and analogues of vitamin D. Which for what? JAMA. 1982;247(6):841–4.
21. Winer KK, Yanovski JA, Cutler GB Jr. Synthetic human parathyroid hormone 1-34 vs calcitriol and calcium in the treatment of hypoparathyroidism. JAMA. 1996;276(8):631–6.
22. Winer KK, Yanovski JA, Sarani B, et al. A randomized, cross-over trial of once-daily versus twice-daily parathyroid hormone 1-34 in treatment of hypoparathyroidism. J Clin Endocrinol Metab. 1998;83(10):3480–6.
23. Winer KK, Sinaii N, Peterson D, et al. Effects of once versus twice-daily parathyroid hormone 1-34 therapy in children with hypoparathyroidism. J Clin Endocrinol Metab. 2008;93(9):3389–95.
24. Winer KK, Ko CW, Reynolds JC, et al. Long-term treatment of hypoparathyroidism: a randomized controlled study comparing parathyroid hormone-(1-34) versus calcitriol and calcium. J Clin Endocrinol Metab. 2003;88(9):4214–20.
25. Winer KK, Sinaii N, Reynolds J, et al. Long-term treatment of 12 children with chronic hypoparathyroidism: a randomized trial comparing synthetic human parathyroid hormone 1-34 versus calcitriol and calcium. J Clin Endocrinol Metab. 2010;95(6):2680–8.
26. Winer KK, Zhang B, Shrader JA, et al. Synthetic human parathyroid hormone 1-34 replacement therapy: a randomized crossover trial comparing pump versus injections in the treatment of chronic hypoparathyroidism. J Clin Endocrinol Metab. 2012;97(2):391–9.
27. Winer KK, Fulton KA, Albert PS, et al. Effects of pump versus twice-daily injection delivery of synthetic parathyroid hormone 1-34 in children with severe congenital hypoparathyroidism. J Pediatr. 2014;165(3):556–63e1.
28. Rubin MR, Sliney J Jr, McMahon DJ, et al. Therapy of hypoparathyroidism with intact parathyroid hormone. Osteoporos Int. 2010;21(11):1927–34.
29. Cusano NE, Rubin MR, McMahon DJ, et al. Therapy of hypoparathyroidism with PTH(1-84): a prospective four-year investigation of efficacy and safety. J Clin Endocrinol Metab. 2013;98(1):137–44.
30. Rubin MR, Cusano NE, Fan WW, et al. Therapy of hypoparathyroidism with PTH(1-84): a prospective six year investigation of efficacy and safety. J Clin Endocrinol Metab. 2016;101(7):2742–50.
31. Sikjaer T, Rejnmark L, Rolighed L, et al. The effect of adding PTH(1-84) to conventional treatment of hypoparathyroidism: a randomized, placebo-controlled study. J Bone Miner Res. 2011;26(10):2358–70.
32. Mannstadt M, Clarke BL, Vokes T, et al. Efficacy and safety of recombinant human parathyroid hormone

(1-84) in hypoparathyroidism (REPLACE): a double-blind, placebo-controlled, randomised, phase 3 study. Lancet Diabetes Endocrinol. 2013;1(4):275–83.
33. Rubin MR, Dempster DW, Zhou H, et al. Dynamic and structural properties of the skeleton in hypoparathyroidism. J Bone Miner Res. 2008;23(12):2018–24.
34. Chen Q, Kaji H, Iu MF, et al. Effects of an excess and a deficiency of endogenous parathyroid hormone on volumetric bone mineral density and bone geometry determined by peripheral quantitative computed tomography in female subjects. J Clin Endocrinol Metab. 2003;88(10):4655–8.
35. Cusano NE, Nishiyama KK, Zhang C, et al. Noninvasive assessment of skeletal microstructure and estimated bone strength in hypoparathyroidism. J Bone Miner Res. 2016;31(2):308–16.
36. Rubin MR, Dempster DW, Kohler T, et al. Three dimensional cancellous bone structure in hypoparathyroidism. Bone. 2010;46(1):190–5.
37. Rubin MR, Dempster DW, Sliney J Jr, et al. PTH(1-84) administration reverses abnormal bone-remodeling dynamics and structure in hypoparathyroidism. J Bone Miner Res. 2011;26(11):2727–36.
38. Langdahl BL, Mortensen L, Vesterby A, et al. Bone histomorphometry in hypoparathyroid patients treated with vitamin D. Bone. 1996;18(2):103–8.
39. Mendonca ML, Pereira FA, Nogueira-Barbosa MH, et al. Increased vertebral morphometric fracture in patients with postsurgical hypoparathyroidism despite normal bone mineral density. BMC Endocr Disord. 2013;13:1.
40. Underbjerg L, Sikjaer T, Mosekilde L, et al. Cardiovascular and renal complications to postsurgical hypoparathyroidism: a Danish nationwide controlled historic follow-up study. J Bone Miner Res. 2013;28(11):2277–85.
41. Underbjerg L, Sikjaer T, Mosekilde L, et al. Postsurgical hypoparathyroidism – risk of fractures, psychiatric diseases, cancer, cataract, and infections. J Bone Miner Res. 2014;29(11):2504–10.
42. Rubin MR, Zwahlen A, Dempster DW, et al. Effects of parathyroid hormone administration on bone strength in hypoparathyroidism. J Bone Miner Res. 2016;31(5):1082–8.
43. Misof BM, Roschger P, Dempster DW, et al. PTH(1-84) Administration in hypoparathyroidism transiently reduces bone matrix mineralization. J Bone Miner Res. 2016;31(1):180–9.
44. Arlt W, Fremerey C, Callies F, et al. Well-being, mood and calcium homeostasis in patients with hypoparathyroidism receiving standard treatment with calcium and vitamin D. Eur J Endocrinol. 2002;146(2):215–22.
45. Astor MC, Løvås K, Debowska A, et al. Epidemiology and health-related quality of life in hypoparathyroidism in Norway. J Clin Endocrinol Metab. 2016;101(8):3045–53.
46. Cho NL, Moalem J, Chen L, et al. Surgeons and patients disagree on the potential consequences from hypoparathyroidism. Endocr Pract. 2014;20(5):427–46.
47. Hadker N, Egan J, Sanders J, et al. Understanding the burden of illness associated with hypoparathyroidism reported among patients in the paradox study. Endocr Pract. 2014;20(7):671–9.
48. Sikjaer T, Moser E, Rolighed L, et al. Concurrent hypoparathyroidism is associated with impaired physical function and quality of life in hypothyroidism. J Bone Miner Res. 2016;31(7):1440–8.
49. Cusano NE, Rubin MR, McMahon DJ, et al. The effect of PTH(1-84) on quality of life in hypoparathyroidism. J Clin Endocrinol Metab. 2013;98(6):2356–61.
50. Cusano NE, Rubin MR, McMahon DJ, et al. PTH(1-84) is associated with improved quality of life in hypoparathyroidism through 5 years of therapy. J Clin Endocrinol Metab. 2014;99(10):3694–9.
51. Santonati A, Palermo A, Maddaloni E, et al. PTH(1-34) for surgical hypoparathyroidism: a prospective, open-label investigation of efficacy and quality of life. J Clin Endocrinol Metab. 2015;100(9):3590–7.
52. Sikjaer T, Rolighed L, Hess A, et al. Effects of PTH(1-84) therapy on muscle function and quality of life in hypoparathyroidism: results from a randomized controlled trial. Osteoporos Int. 2014;25(6):1717–26.
53. Vokes T, Mannstadt M, Levine M, et al. Recombinant human parathyroid hormone (rhPTH [1–84]) therapy in hypoparathyroidism and improvement in quality of life. Poster presented at the ASBMR 37th Annual Meeting, SU0018, 2015.

87

Pseudohypoparathyroidism

Agnès Linglart[1], Michael A. Levine[2], and Harald Jüppner[3]

[1]APHP, Hôpital Bicêtre Paris Sud, Service d'endocrinologie et diabétologie pour enfants, Centre de référence des Maladies Rares du métabolisme du calcium et du phosphate, Plateforme d'Expertise Paris Sud des maladies rares, filière OSCAR, Le Kremlin Bicêtre, France
[2]Division of Endocrinology and Diabetes, The Children's Hospital of Philadelphia and Department of Pediatrics, University of Pennsylvania Perelman School of Medicine, Philadelphia, PA, USA
[3]Endocrine Unit and Pediatric Nephrology Unit, Departments of Medicine and Pediatrics, Harvard Medical School, Massachusetts General Hospital, Boston, MA, USA

INTRODUCTION

The term pseudohypoparathyroidism (PHP) refers to a historical nomenclature for a group of related metabolic and developmental disorders in which the most prominent biochemical features, ie, hypocalcemia and hyperphosphatemia, are the result of target organ resistance to PTH rather than to deficiency of PTH. Therefore, a key factor that distinguishes PHP from the different forms of hypoparathyroidism is that serum PTH levels are elevated in the former condition and usually low or inappropriately normal in the latter. PHP and related disorders are caused by epigenetic and/or genetic defects that impair accumulation or responsiveness to the second messenger cAMP. Consequently, a reduced or markedly impaired cAMP effect in the proximal renal tubules leads to increased reabsorption of phosphate and decreased conversion of 25-hydroxyvitamin D to 1,25-dihydroxyvitamin D.

The broad clinical and biochemical phenotypes of PHP and related disorders with impaired cAMP production reflect primarily an underlying pathophysiology in which molecular defects reduce expression or function of proteins that are downstream of the heptahelical PTH/PTHrP receptor (PTH1R), most notably the alpha-subunit of the heterotrimeric stimulatory G protein (Gsα) that couples the PTH1R and numerous other heptahelical receptors to activation of adenylyl cyclase (Table 87.1).

Overall, the prevalence of PHP has been estimated to be between 0.34 and 1.1/100,000 inhabitants by Japanese [1] and Danish [2] studies, respectively. However, the prevalence may be largely underestimated because of the extreme variability of the disease presentation, and often unrecognized symptoms in early life.

Several systems of classification have been developed to distinguish among forms of PHP and its variants. Historically, these distinctions have been based upon the urinary excretion of cAMP and phosphate in responses to PTH administration (PHP type 1 or 2); the presence or absence of the somatic features of Albright hereditary osteodystrophy (see later); and/or in vitro analyses of Gsα function or expression (PHP type 1A, B, or C) [3–5]. These various classifications are challenged today by advances in the genetic and epigenetic pathophysiology of these disorders and have encouraged attempts to develop a contemporary system of classification that is more reflective of the underlying disease mechanism and the molecular alterations [6, 7]. A European network recently proposed a new system of classification that reflects these concerns [8], yet has limitations because patients affected by PHP and related disorders usually have metabolic phenotypes that are unrelated to abnormalities in PTH or PTHrP signaling events.

PSEUDOHYPOPARATHYROIDISM AND RELATED DISORDERS

PHP type 1 refers to a form of PHP in which there is a markedly blunted or absent increase in urinary excretion of nephrogenous cAMP and phosphate in response to PTH administration. Not only the biochemical, but also

---

Primer on the Metabolic Bone Diseases and Disorders of Mineral Metabolism, Ninth Edition. Edited by John P. Bilezikian.
© 2019 American Society for Bone and Mineral Research. Published 2019 by John Wiley & Sons, Inc.
Companion website: www.wiley.com/go/asbmrprimer

Table 87.1. PHP, related disorders, and main differential diagnosis.

Disease	Frequent Clinical Features	Main Biochemical Features	Additional Features That May Be Present	Principal Epigenetic/Genetic Defects	Differential Diagnosis
PHP1A	Brachydactyly E Adult short stature, yet normal growth during infancy/childhood Subcutaneous ossifications, osteoma cutis Early-onset obesity Cognitive/behavior impairment	PTH resistance with hypocalcemia and/or hyperphosphatemia TSH resistance GH deficiency Calcitonin resistance	Sleep apnea Reactive airways disease Otitis media Carpal tunnel syndrome Craniosynostosis, Chiari 1 Metabolic acidosis	Loss of function mutations involving GNAS exons encoding Gsα (maternal allele) Rare: GNAS imprinting defect	Monogenic and syndromic obesities BDMR TRPS Turner syndrome
PPHP	Brachydactyly E Short stature Subcutaneous ossifications, osteoma cutis Small for gestational age		In rare cases, elevated PTH levels	Loss of function mutations involving GNAS exons encoding Gsα (paternal allele)	PTHLH mutations TRPS
POH	Extensive ectopic ossifications Small for gestational age at birth Leanness			Loss of function mutations involving GNAS exons encoding Gsα (paternal allele)	FOP
Osteoma cutis	Superficial ectopic ossification			Loss of function mutations involving GNAS exons encoding Gsα (paternal allele)	
PHP1B	Macrosomia Early-onset obesity Normal adult height	PTH resistance, hypocalcemia, hyperphosphatemia Mild TSH resistance Calcitonin resistance	Brachydactyly, in some cases Weight above the reference range	Loss of maternal GNAS methylation imprints; deletions in STX16 or GNAS, inversion involving GNAS regions	Vitamin D deficiency Magnesium deficiency
Acrodysostosis	Early-onset and extensive brachydactyly Short stature Cognitive/behavior impairment[a] Obesity	Mild PTH resistance[b] Mild TSH resistance[b] Calcitonin resistance[b]	Craniosynostosis, Chiari 1 Sleep apnea Otitis media Hearing loss	Mutations involving exons encoding PRKAR1A or PDE4D	PTHLH mutations TRPS

BDMR = brachydactyly mental retardation syndrome; FOP = fibrodysplasia ossificans progressiva; GH = growth hormone; PTHLH = parathyroid-hormone-like hormone; TRPS = tricho-rhino-phalangeal syndrome; TSH = thyroid stimulating hormone.
[a] Associated with PDE4D mutations.
[b] Associated with PRKAR1A mutations.

the clinical phenotypes of the PHP type 1 subtypes, can show considerable overlap, although distinct genetic and/or epigenetic defects decrease expression or function of Gsα and reduce heptahelical receptor-dependent activation of adenylyl cyclase.

GNAS LOCUS

Gsα is encoded by *GNAS*, a highly complex locus that encodes not only Gsα (exons 1–13), but also generates several additional transcripts through the use of alternative first exons that all splice onto *GNAS* exons 2–13. These alternative transcripts include A/B (also termed 1A), the extra-large stimulatory G protein (XLαs), and the transcript coding the neuroendocrine secretory protein 55 (NESP55) [9]. Furthermore, an antisense transcript (AS) is also derived from *GNAS*. NESP55 and XLαs are protein-coding transcripts whose exact functions in humans are still debated. XLαs can mimic or enhance Gsα action in vitro [10] and in vivo [11,12]. The A/B transcript may encode a truncated form of Gsα that can inhibit the actions of the full-length signaling protein [13].

Transcription of these different mRNAs is controlled by genomic imprinting, a regulatory process that suppresses transcription from one parental allele. Except for *GNAS* exon 1, which is required for Gsα transcription, imprinting of all other first *GNAS* exons is regulated through differentially methylated regions (DMR) that are located in the promoter sequences for each respective exon. Thus the promoters for exons XL, A/B, and AS are methylated on the maternal *GNAS* allele and transcription occurs predominantly from the paternal allele; transcripts encoding NESP55 arise from the maternal allele because of methylation of the DMR in the *NESP55* promoter on the paternal allele [14]. Although Gsα is biallelically expressed in most cells, expression of this ubiquitous signaling protein is restricted to the maternal *GNAS* allele in several tissues/cells, including renal proximal tubules, thyroid, pituitary somatotropes, and gonads [15–17]. The predominantly maternal expression of Gsα is likely controlled by cell-specific proteins that act on yet to be determined regulatory sequences in the paternal *GNAS* allele. The establishment and/or maintenance of methylation of *GNAS* DMRs has been localized in humans to two genomic regions that are within or close to *GNAS*. One region is located within the *STX16* gene and controls the A/B DMR only [18, 19], the other region, which encompasses AS exons 3 and 4, controls methylation of all DMRs within the entire *GNAS* locus [20–27].

PHP1A (OMIM #103580)

PHP1A is an autosomal dominant disease caused by inactivating mutations that involve the maternal *GNAS* exons encoding Gsα. Because Gsα expression is derived exclusively or preferentially from the maternal *GNAS* allele in some tissues, such as the proximal renal tubules, heterozygous *GNAS* mutations on the maternal allele are sufficient to cause PTH resistance. In addition to PTH resistance, patients affected by PHP1A typically manifest resistance to additional hormones (eg, thyroid stimulating hormone [TSH], gonadotropins, calcitonin, growth hormone releasing hormone [GHRH], and possibly others), whose target tissues also show predominant expression of Gsα from the maternal *GNAS* allele. By contrast, responsiveness to other hormones (eg, ACTH, vasopressin) is usually normal because the cells that express the receptors for these hormones express Gsα from both parental alleles.

The mean age at diagnosis of PHP1A is 8.5 + 3.8 years (mean ± SD) [2.8–41], unless a positive family history prompts earlier investigations [42]. PTH resistance is usually not present at birth, but develops slowly during the first years of life [42]. Phosphate and PTH levels increase first, followed later by a decrease in serum calcium levels. Delayed onset of PTH resistance may be explained by a progressive reduction of Gsα expression from the paternal allele in renal proximal tubules as shown in mice with ablation of *Gnas* exon 1 on either the maternal or the paternal allele [28]. Hypocalcemia is thought to result from decreased production of 1,25-dihydroxyvitamin D (1,25D), although most patients do not show low 1,25D levels. By contrast to the proximal tubule, it appears that Gsα is likely expressed from both parental alleles in the distal renal tubules. Therefore, urinary calcium reabsorption remains responsive to PTH.

Hypothyroidism is common in PHP1A, and is the result of TSH resistance. Hence, although levels of TSH are elevated, the thyroid gland is not enlarged. The elevation in serum levels of TSH is a common, if not a constant feature of PHP1A and it is often detected by newborn screening programs. Clinically significant hypothyroidism is unusual. In older children, TSH resistance is mild and asymptomatic, with TSH ranging between 5 and 50 mIU/L and free T4 in the low normal range. Resistance to calcitonin is also common, but rarely assessed, and does not appear to cause any clinical manifestations. Elevated serum levels of calcitonin in patients with PHP1A should not be construed as evidence of medullary thyroid hyperplasia/cancer in the absence of additional testing [15, 16, 43, 44].

Growth hormone deficiency resulting from GHRH resistance is present in 50 to 80% of PHP1A children [45, 46]. Although common, growth hormone deficiency is not the principle cause of short stature in PHP1A. Girls and women with PHP1A also manifest partial or complete resistance to gonadotrophins manifesting as amenorrhea or oligomenorrhea. Many very overweight women with PHP1A will have menstrual abnormalities that resemble typical polycystic ovary disease. Data on fertility are lacking. Cryptorchidism is common in males [47–49].

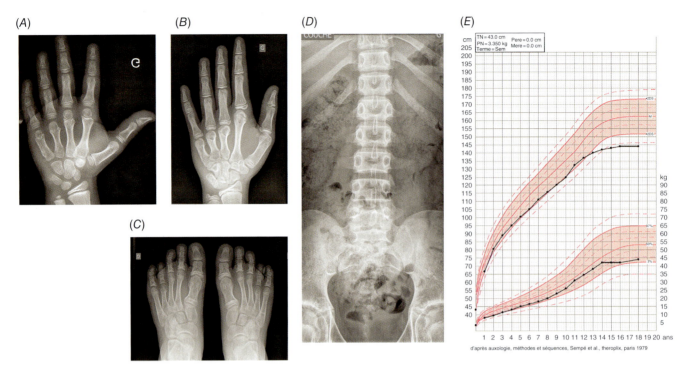

Fig. 87.1. Features of Albright osteodystrophy (AHO). *A, B, C,* Brachymetacarpy at the age of 5.5 and 11 years and brachymetatarsy in a PHP1A patient caused by a loss-of-function mutation in *GNAS* (A102V). *D,* Narrowed intervertebral space of the lumbar spine in the same patient. *E,* Typical pattern of growth and weight gain in a girl affected with PHP1A and loss-of-function mutation in *GNAS* (p.P144Sfs*5).

Albright hereditary osteodystrophy

Haploinsufficiency in chondrocytes where paternal Gsα expression is not reduced likely leads to diminished auto/paracrine PTHrP-dependent signaling events. Patients present with clinical features that resemble PTHrP haploinsufficiency (gene name: PTHLH, parathyroid hormone-like hormone) ie, brachydactyly type E of hands and feet (particularly shortening of III, IV, and V metacarpals and I distal phalanx), round facies, short femoral necks and narrowed lumbar spine (Fig. 87.1A). Brachydactyly is not present at birth, but evolves over time caused by premature closure of epiphyseal plates. PHP1A patients are born with mild growth retardation (46 cm [38–51], [n = 29] [31]), and are usually of normal stature until they undergo rapid and premature closure of the epiphyses between 10 and 15 years of age (Fig. 87.1B). Bone age is usually not advanced in infancy. Short stature is found in more than 75% of the adult PHP1A patients, which results presumably from a combination of accelerated hypertrophic chondrocyte differentiation leading to premature growth plate closure, an absent pubertal spurt, and growth hormone (GH) deficiency [46, 50, 51].

Subcutaneous ossifications including *osteoma cutis* are specific features of PHP1A that are found in 30% to 60% of patients, and can be present at birth and before any other signs of Albright hereditary osteodystrophy (AHO) become apparent. These are usually micronodular and superficial, although, in rare cases, ossified plaques may occur in PHP1A (Fig. 87.2) [52]. In patients with *GNAS* defects, ectopic ossification represents an anomalous process of cellular differentiation that leads to the formation of intramembraneous bone (see later). The development of ectopic bone in AHO is most likely unrelated to preceding trauma, unlike the process in patients with fibrodysplasia ossificans progressiva (FOP, OMIM #135100), in which even minor tissue damage can stimulate extensive ectopic endochondral bone formation. Nonetheless, precautions are advised to prevent pressure-dependent ossifications and to avoid surgical interventions.

Obesity

Several factors have been identified as contributing to obesity in PHP1A, such as resistance to epinephrine and other hormones in peripheral tissues, as well as reduced resting energy expenditure and increased food intake that could be related to decreased cAMP-dependent events downstream of MC4R in regions of the hypothalamus in which Gsα is transcribed predominantly from the maternal allele [32, 50, 53, 54]. Patients develop early-onset obesity before the age of 2 years, which can occur without obvious hyperphagia. About two-thirds of

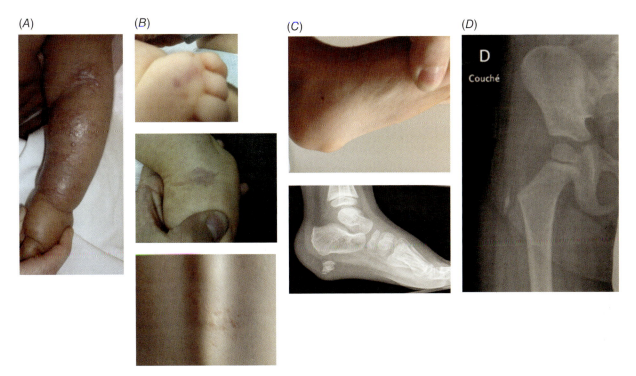

Fig. 87.2. Subcutaneous ossifications in patients with loss-of-function mutation in *GNAS*. *A,* Extensive ossifications on the arm of a neonate affected with PHP1A. *B,* Patterns and evolution of micronodular ossifications in patients with PHP1A. *C,* Nodular painful ossification of the heel that can be removed surgically. *D,* Radiographic appearance of subcutaneous ossifications.

PHP1A adults are obese; the average Z-score of their body mass index is 1.7 ± 0.2 [29, 32, 37].

Cognitive function

Developmental delay and learning disability are present in 40% to 70% of patients, and are often associated with an increased incidence of psychiatric manifestations.

Other features

Recently, several reports have highlighted the frequency of carpal tunnel syndrome, sleep apnea, atypical asthma, otitis media, craniosynostosis, Chiari 1 malformation, metabolic acidosis and diarrhea [55, 56]. Anomalies of tooth structure with delayed or absent eruption of teeth, might be caused by defective PTHrP signaling [57].

In summary

The diagnosis of PHP1A should be suspected in: (i) infants with early onset obesity, subcutaneous ossifications, and mildly elevated TSH; (ii) children with symptoms of hypocalcemia, brachydactyly, obesity, motor and/or cognitive delay, and mildly elevated TSH; and (iii) adolescents/adults with hypocalcemia, brachydactyly, short stature, motor and/or cognitive delay, and mildly elevated TSH.

Molecular diagnosis of PHP1A

PHP1A results from heterozygous inactivating mutations involving the maternal *GNAS* allele (OMIM #610540, 20q13.2-q13.3) that can involve any of the 13 exons encoding Gsα, although mutations in exon 13 are more likely to cause PHP1C (see later). All types of mutations can be found [41]. In some cases, it may be desirable or necessary to confirm that the mutation is on the maternally derived *GNAS* allele. In these cases, the parental origin of the mutant allele can often be assigned through genetic analysis of the parents. Parental DNA studies will be negative in cases where the patient carries a de novo mutation or a parent is mosaic, in which case research studies of patient's RNA are usually able to assign the parental origin of the defective allele. Some patients with PTH-resistance and limited features of AHO do not harbor typical *GNAS* mutations in the coding regions but show methylation changes at DMRs of *GNAS* (Fig. 87.3) [58–60] (see later, PHP1B).

PHP1C (OMIM #612462)

Patients with PHP1C are clinically and biochemically indistinguishable from patients with PHP1A save for subtle differences in Gsα function that are observed when patient-derived Gsα protein activity is analyzed in

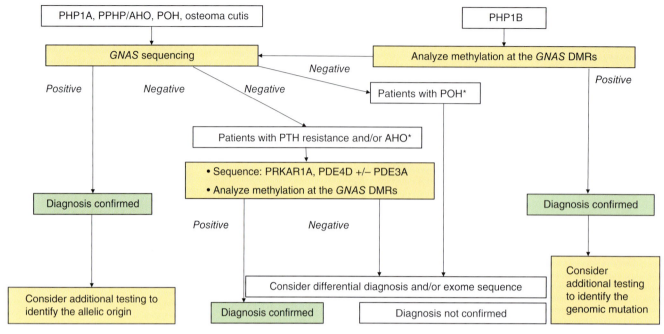

Fig. 87.3. Decision tree for the molecular analysis of pseudohypoparathyroidism of pseudohypoparathyroidism (PHP) and related disorders. AHO = Albright hereditary osteodystrophy; DMR = differentially methylated region; POH = progressive osseous heteroplasia; PPHP = pseudopseudohypoparathyroidism.

vitro. The basis for this unique pattern of Gsα activity is thought to be the location of the mutation within exon 13 of the maternally derived *GNAS* allele. We propose that the term PHP1C be eliminated from the nomenclature.

PSEUDOPSEUDOHYPOPARATHYROIDISM OR ISOLATED AHO

Some years after the initial description of PHP1, asymptomatic relatives of patients with PHP1A were identified who showed typical features of AHO but lacked PTH resistance. This form of the disorder was termed pseudopseudohypoparathyroidism (PPHP). By contrast to patients with PHP1A, patients with PPHP have completely normal hormone responsiveness despite a similar 50% reduction in Gsα protein activity/levels in tissues where Gsα is biallelically expressed [3, 5]. The basis for this paradox was solved when it was observed that these patients carry heterozygous mutations on the *paternal GNAS* exons encoding Gsα, and therefore have normal levels of Gsα in cells and tissues where Gsα is preferentially expressed from the maternal *GNAS* allele. Additional clinical features have been recently identified in patients with PPHP, including low birth weight and length (ie, small for gestational age) that may reflect decreased expression of XLαs [31, 61].

PROGRESSIVE OSSEOUS HETEROPLASIA (OMIM #166350)

Progressive osseous heteroplasia (POH) is characterized by extensive and clinically significant ossifications that are present in the skin as well as in the deep connective tissue. There is usually at least one bony plate. Heterotopic ossification is progressive and can be debilitating. POH patients display few if any other AHO features, and typically do not develop PTH resistance [52]. Characteristic features of POH include early onset of ectopic ossification in the first year of life, leanness, and dermomyotomal pattern of the lesions. Inactivating mutations involving the paternal allele of *GNAS* (mainly exons 2–13) are found in 60% to 70% of cases [62].

Remarkably, similar mutations on paternally derived *GNAS* alleles can also cause a very limited form of ectopic ossification termed osteoma cutis, and patients with this disorder do not manifest any other features of AHO.

The striking differences in the phenotypes caused by identical mutations on the paternal *GNAS* allele, osteoma cutis, PPHP, and POH, remains unknown. Heterotopic bone likely reflects the impaired differentiation of mesenchymal stem cells. Heterozygous *Gnas* exon 1-knockout mice develop, although slowly, heterotopic bone formation initiated in subcutaneous adipose tissue that is prone to injury and pressure [63]. The bone development may be accompanied by accelerated osteoblast transdifferentiation of adipose stromal cells, at least when investigated in vitro. Gsα appears to be a key factor in the switch between osteoblast and adipocyte fate. Gsα downregulation facilitates the expression of Runx2/Cbfa1, collagen type I, osteopontin, and alkaline phosphatase. Overall, GNAS defects provide a sensitized background for ectopic osteoblast differentiation [64].

PHP1B (OMIM #603233)

PTH resistance is the principal manifestation of this PHP variant. As in PHP1A, PTH resistance develops over time [21], and hypocalcemia is usually the initial manifestation of the disorder. Mean age at diagnosis is 16.7 ± 10.0 (mean ± SD) [18, 35, 41, 58, 65–75]. Other features include occasional, mild elevation of TSH with usually normal serum concentrations of thyroid hormones as evidence for partial TSH resistance. Furthermore, elevated calcitonin levels, early-onset obesity, macrosomia, and brachydactyly have been described in some patients. Micronodular ossifications are extremely rare [67]. It is noteworthy that these patients do not appear to have resistance to GHRH or gonadotropins [65, 73]. Similar to patients with PHP1A, patients with PHP1B retain PTH responsiveness in the distal renal tubule where calcium reabsorption occurs. Hypercalciuria is unusual in calcitriol-treated children unless serum levels of PTH are suppressed, yet urinary calcium is higher in adult PHP1B compared with adult PHP1A patients, respectively (personal data and [76]).

Molecular diagnosis

Loss of methylation at GNAS exon A/B:TSS-DMR of the maternal GNAS allele is a consistent finding in PHP1B patients, which accounts for decreased Gsα expression from the affected allele, and hence the PTH resistance [18, 65, 77].

Fifteen percent to 20% of PHP1B cases are familial in which maternal transmission of the mutations leads to dominant inheritance of the disorder (AD-PHP1B). In most AD-PHP1B cases, the loss of methylation is restricted to the GNAS exon A/B:TSS-DMR and to the GNAS AS2:DMR, which are caused by a recurrent 3-kilobase deletion removing a genomic region upstream of GNAS that comprises STX16 exons 4–6 [18, 19, 21, 23, 78, 79]. Another deletion has provided evidence for a cis-acting control element(s) located within GNAS exon NESP or the region centromeric thereof that regulates exon A/B methylation [22]. Other control elements that regulate methylation of all differentially methylated GNAS regions are located within the NESP and AS3-4 region [20, 25, 26, 72, 80, 81].

Most cases of PHP1B are sporadic, however, and these nonfamilial cases of PHP1B have more extensive methylation defects than patients with most familial PHP1B forms, such that loss of methylation of GNAS exon A/B:TSS-DMR and GNAS AS2:DMR are typically associated with loss of methylation at the XL and AS1 DMR as well as gain of methylation at the NESP DMR. In 8% to 10% of sporadic cases, paternal uniparental disomy involving the long arm of chromosome 20 (patUPD20q) that includes the GNAS locus is the cause of deficient maternal Gsα expression and hence PHP1B [70, 74, 82]. Methylation changes, assessed by analysis of DNA isolated from peripheral circulating mononuclear cells, are usually complete but may be partial in some cases, thereby complicating the definition of the molecular defect. The identification of partial methylation defects in circulating cells has suggested that the molecular defect may have occurred postzygotically, at least in some patients, thus leading to mosaicism [83].

More extensive imprinting defects occur at multiple genetic loci (ie, multilocus imprinting disorder, MLID) in 8% to 10% of PHP1B patients, but these additional methylation abnormalities do not seem to cause obvious clinical consequences [70, 83]. Only two patients with MLID have been reported who presented with features of PHP1B and the Beckwith–Wiedemann syndrome [84, 85].

The familial and sporadic forms of PHP1B appear to have similar clinical and biochemical phenotypes [65, 68]. However, the birth weights of neonates with a STX16 deletion were higher if they were born to mothers who are carriers of the same molecular defect, but are unaffected because their deletion resides on the paternal allele [86]. Given the role of the other GNAS-derived transcripts, specifically XLαs and NESP, and the differential expression induced by the methylation anomalies at this locus, future characterization of growth and behavior may reveal subtle differences between these subtypes.

In summary

The molecular characterization of PHP1B may include: (i) at a minimum, demonstration of loss of methylation at the GNAS exon A/B:TSS-DMR; (ii) optimally, comprehensive analysis of methylation at all GNAS DMRs; (iii) identification of a genetic defect in a patient with familial PHP1B, including STX16 or NESP/AS deletions or inversions; (iv) in case of methylation abnormalities at several DMRs within the GNAS locus (complete or partial), a search for deletions within GNAS should be considered; (v) for sporadic PHP1B cases, patUPD20q should be excluded. It should be noted that for most sporadic PHP1B patients, disease-causing mutations have not been identified.

ACRODYSOSTOSIS

Acrodysostosis has been included in this chapter because of the clinical overlap with PHP1A and because at least one variant is consistent with the biochemical definition of PHP type 2. PHP2 is defined by preserved excretion of nephrogenous cAMP response to PTH, but an absent phosphaturic response, which suggested that the molecular defect is in the proximal renal tubules distal to the generation of cAMP downstream of the PTH1R signaling pathway. This biochemical profile was initially described in a patient, who lacked AHO features [87], and subsequent studies identified severe hypocalcemia caused by profound vitamin D deficiency as the basis for a reversible form of PTH resistance [88].

Acrodysostosis is a chondrodysplasia that includes severe brachydactyly, facial dysostosis, and nasal hypoplasia, features that can also be found in AHO. By contrast to brachydactyly E in AHO, the brachydactyly in acrodysostosis usually involves all phalanges, as well as metacarpals and metatarsals, and it is present at birth or shortly thereafter. Bone maturation progresses rapidly leading to a premature fusion of epiphyses. Short femoral necks and narrowed lumbar spine have been described. As in PHP1A, affected patients may have a normal height during infancy and early childhood, then as growth velocity decreases and in absence of pubertal spurt, short stature is common, and more severe than that of PHP1A (−3.5 SD on average) [89–95].

Most newborns affected by acrodysostosis are short for their gestational age [93, 96, 97]. Obesity has been reported as a frequent feature, but the incidence of ear infections, sleep apnea, cryptorchidism, Chiari 1, and craniosynostosis is still unknown. Ectopic ossifications have not been observed in these patients.

Patients with acrodysostosis harbor heterozygous germline mutations in the genes encoding PRKAR1A, the cAMP-dependent regulatory subunit of protein kinase A [94, 96], or PDE4D [96], which encodes a class IV cAMP-specific phosphodiesterase that regulates intracellular cAMP concentrations. Patients carrying a heterozygous PRKAR1A mutation have elevated blood and urine cAMP concentrations that adequately respond to PTH administration and thus correspond to the definition of PHP type 2.

The molecular cause of acrodysostosis is still unknown in 10% to 20% of the patients.

Acrodysostosis caused by PRKAR1A mutations is frequently associated with resistance to PTH and/or other hormones that mediate their actions through the cAMP signaling pathway. These resistances are usually mild and no episodes of hypocalcemia have been described so far; thyroid hormones are usually in the low range of normal. In contrast, patients with acrodysostosis caused by PDE4D mutations harbor no — or extremely minor — endocrine anomalies but manifest behavioral difficulties [92, 93].

DIFFERENTIAL DIAGNOSES

The differential diagnoses depend on the age and predominant features at time of referral (Table 87.1). The *brachydactyly mental retardation syndrome* (BDMR) (OMIM #600430) is caused by a deletion at the 2q37 chromosomic region that leads to a loss of one copy of *HDAC4*. It is characterized by short stature, obesity, developmental delay, behavioral anomalies, autism spectrum disorder, brachydactyly, and skeletal and craniofacial abnormalities reminiscent of AHO, but no ectopic ossification [98]. The *tricho-rhino-phalangeal syndrome* (TRPS) — TRPS I (OMIM #190350), TRPS II (OMIM #150230) and TRPS III (OMIM #190351) — is caused by the haploinsufficiency of *TRPS1*. TRPS is characterized by ectodermal anomalies, intellectual disability in some patients, dysmorphic features, and skeletal anomalies resembling AHO (brachydactyly, cone-shaped epiphyses, and short stature).

MANAGEMENT

Endocrine defects

The objectives of therapy for PTH-resistance are: (i) maintain serum levels of calcium within the normal range (2.3 to 2.5 mmol/L); (ii) avoid hypercalciuria; (iii) prevent excessive bone resorption caused by chronically elevated PTH; and (iv) prevent chronically elevated phosphate levels and an elevated calcium×phosphate product. To achieve these goals, we recommend that serum levels of PTH are maintained in the high normal or slightly elevated range, ie, below 150 pg/ml, because oversuppression of PTH will lead to excessive urinary excretion of calcium and an increased risk of renal calcifications. Treatment consists of administration of active forms of vitamin D (calcitriol or alfacalcidol), oral calcium supplements, and vitamin D to maintain normal circulating levels of 25(OH)D [99]. Unlike patients with PTH deficient hypoparathyroidism, treatment with vitamin D analogs rarely leads to hypercalciuria. As with treatment of other forms of hypocalcemia, the dose of active forms of vitamin D may need to be increased during periods of high growth velocity such as infancy or puberty. In contrast, levels of active vitamin D may need to be reduced during late pregnancy. Calcium supplements are recommended if dietary calcium intake is insufficient and/or to reverse or prevent hyperphosphatemia. Any deficiency of magnesium should be corrected. Patients with acrodysostosis caused by a PRKAR1A mutation usually do not develop hypocalcemia and therefore do not require treatment with vitamin D analogs except if PTH is elevated or growth velocity decreased.

Patients should be screened for associated endocrine defects such as hypothyroidism, hypogonadism, and GH deficiency, and hormone replacement with levothyroxine, sex steroids, and growth hormone should be instituted as appropriate. The efficacy of GH therapy to improve adult height in PHP1A patients has not been fully established. In one study, Mantovani and colleagues have shown a significant increase in growth velocity during GH treatment; however, near-adult height was not improved in six patients [46]. A prospective trial

(NCT00209235) is currently ongoing to evaluate the efficacy of GH in children with PHP1A.

The treatment of early-onset obesity is the most challenging aspect of PHP1A management, and is limited to the usual recommendations of increasing physical activity, replacing hormone deficiencies, and reducing caloric intake. An analysis of resting energy expenditure may provide useful guidance for a recommended daily caloric intake. Patients who are obese are at increased risk for sleep apnea and atypical asthma, and may benefit from appropriate screening examinations. In patients with cognitive and/or motor delay, it is important to rule out craniosynostosis and Chiari 1, to perform a neuropsychological evaluation, and provide supportive care.

Another challenge is the management of heterotopic ossifications. Asymptomatic ossifications should not be surgically resected, because they will recur unless removed entirely. Similarly, infiltrative ossifications, plaques, and osteoma cutis should not be resected because of the risk of recurrence and damaging sequelae. On the other hand, it may be reasonable to consider surgical resection of ossifications that are painful or interfere with function, particularly if these are limited in size (Fig. 87.2C). Pharmacological attempts to limit the extension of ossification using bisphosphonates, or nonsteroidal anti-inflammatory drugs have been unsuccessful so far [100]. New hope for patients include the manipulation of the intracellular cAMP concentrations using phosphodiesterase inhibitors.

CONCLUSION

PHP and related disorders encompass a series of disorders that impair the cAMP signaling pathway downstream of the receptor for PTH/PTHrP, and several other Gsα-coupled receptors. Patients present with a variable combination of symptoms of AHO and endocrine resistances reflecting the tissue-specific and time-dependent expression of factors involved in the pathway (eg, Gsα, PRKAR1A, PDE4D). It is nowadays widely recommended to establish the underlying genetic defect, which allows improved care, monitoring, and follow up of patients, and appropriate genetic counseling [101].

REFERENCES

1. Nakamura Y, Matsumoto T, Tamakoshi A, et al. Prevalence of idiopathic hypoparathyroidism and pseudohypoparathyroidism in Japan. J Epidemiol. 2000;10(1):29–33.
2. Underbjerg L, Sikjaer T, Mosekilde L, et al. Pseudohypoparathyroidism – epidemiology, mortality and risk of complications. Clin Endocrinol. 2016;84(6):904–11.
3. Chase LR, Melson GL, Aurbach GD. Pseudohypoparathyroidism: defective excretion of 3′,5′-AMP in response to parathyroid hormone. J Clin Invest. 1969;48(10):1832–44.
4. Levine MA, Downs RW, Singer M, et al. Deficient activity of guanine nucleotide regulatory protein in erythrocytes from patients with pseudohypoparathyroidism. Biochem Biophys Res Commun. 1980;94(4):1319–24.
5. Albright F, Forbes AP, Henneman PH. Pseudopseudohypoparathyroidism. Trans Assoc Am Physicians. 1952;65:337–50.
6. Mantovani G, Spada A, Elli FM. Pseudohypoparathyroidism and Gsα-cAMP-linked disorders: current view and open issues. Nat Rev Endocrinol. 2016;12(6):347–56.
7. Mantovani G, Elli FM, Spada A. GNAS epigenetic defects and pseudohypoparathyroidism: time for a new classification? Horm Metab Res Horm Stoffwechselforschung Horm Métabolisme. 2012;44(10):716–23.
8. Thiele S, Mantovani G, Barlier A, et al. From pseudohypoparathyroidism to inactivating PTH/PTHrP signalling disorder (iPPSD), a novel classification proposed by the EuroPHP network. Eur J Endocrinol. 2016;175(6):P1–17.
9. Hayward BE, Moran V, Strain L, et al. Bidirectional imprinting of a single gene: GNAS1 encodes maternally, paternally, and biallelically derived proteins. Proc Natl Acad Sci U S A. 1998;95(26):15475–80.
10. Linglart A, Mahon MJ, Kerachian MA, et al. Coding GNAS mutations leading to hormone resistance impair in vitro agonist- and cholera toxin-induced adenosine cyclic 3′,5′-monophosphate formation mediated by human XLalphas. Endocrinology. 2006;147(5):2253–62.
11. Liu Z, Turan S, Wehbi VL, et al. Extra-long Gαs variant XLαs protein escapes activation-induced subcellular redistribution and is able to provide sustained signaling. J Biol Chem. 2011;286(44):38558–69.
12. Mariot V, Wu JY, Aydin C, et al. Potent constitutive cyclic AMP-generating activity of XLαs implicates this imprinted GNAS product in the pathogenesis of McCune-Albright syndrome and fibrous dysplasia of bone. Bone. 2011;48(2):312–20.
13. Freson K, Jaeken J, Van Helvoirt M, et al. Functional polymorphisms in the paternally expressed XLalphas and its cofactor ALEX decrease their mutual interaction and enhance receptor-mediated cAMP formation. Hum Mol Genet. 2003;12(10):1121–30.
14. Hayward BE, Kamiya M, Strain L, et al. The human GNAS1 gene is imprinted and encodes distinct paternally and biallelically expressed G proteins. Proc Natl Acad Sci U S A. 1998;95(17):10038–43.
15. Liu J, Erlichman B, Weinstein LS. The stimulatory G protein alpha-subunit Gs alpha is imprinted in human thyroid glands: implications for thyroid function in pseudohypoparathyroidism types 1A and 1B. J Clin Endocrinol Metab. 2003;88(9):4336–41.
16. Mantovani G, Ballare E, Giammona E, et al. The gsalpha gene: predominant maternal origin of transcription in human thyroid gland and gonads. J Clin Endocrinol Metab. 2002;87(10):4736–40.
17. Li T, Vu TH, Zeng ZL, et al. Tissue-specific expression of antisense and sense transcripts at the imprinted Gnas locus. Genomics. 2000;69(3):295–304.

18. Bastepe M, Fröhlich LF, Hendy GN, et al. Autosomal dominant pseudohypoparathyroidism type Ib is associated with a heterozygous microdeletion that likely disrupts a putative imprinting control element of GNAS. J Clin Invest. 2003;112(8):1255–63.
19. Grigelioniene G, Nevalainen PI, Reyes M, et al. A large inversion involving GNAS exon A/B and all exons encoding Gsα Is associated with autosomal dominant pseudohypoparathyroidism type Ib (PHP1B). J Bone Miner Res. 2017;32(4):776–83.
20. Bastepe M, Fröhlich LF, Linglart A, et al. Deletion of the NESP55 differentially methylated region causes loss of maternal GNAS imprints and pseudohypoparathyroidism type Ib. Nat Genet. 2005;37(1):25–7.
21. Linglart A, Gensure RC, Olney RC, et al. A novel STX16 deletion in autosomal dominant pseudohypoparathyroidism type Ib redefines the boundaries of a cis-acting imprinting control element of GNAS. Am J Hum Genet. 2005;76(5):804–14.
22. Richard N, Abeguilé G, Coudray N, et al. A new deletion ablating NESP55 causes loss of maternal imprint of A/B GNAS and autosomal dominant pseudohypoparathyroidism type Ib. J Clin Endocrinol Metab. 2012;97(5):E863–7.
23. Elli FM, de Sanctis L, Peverelli E, et al. Autosomal dominant pseudohypoparathyroidism type Ib: a novel inherited deletion ablating STX16 causes loss of imprinting at the A/B DMR. J Clin Endocrinol Metab. 2014;99(4):E724–8.
24. Nakamura A, Hamaguchi E, Horikawa R, et al. Complex genomic rearrangement within the GNAS region associated with familial pseudohypoparathyroidism type 1b. J Clin Endocrinol Metab. 2016;jc20161725.
25. Chillambhi S, Turan S, Hwang D-Y, et al. Deletion of the noncoding GNAS antisense transcript causes pseudohypoparathyroidism type Ib and biparental defects of GNAS methylation in cis. J Clin Endocrinol Metab. 2010;95(8):3993–4002.
26. Rezwan FI, Poole RL, Prescott T, et al. Very small deletions within the NESP55 gene in pseudohypoparathyroidism type 1b. Eur J Hum Genet. 2015;23(4):494–9.
27. Perez-Nanclares G, Velayos T, Vela A, et al. Pseudohypoparathyroidism type Ib associated with novel duplications in the GNAS locus. PloS One. 2015;10(2):e0117691.
28. Turan S, Fernandez-Rebollo E, Aydin C, et al. Postnatal establishment of allelic Gαs silencing as a plausible explanation for delayed onset of parathyroid hormone resistance owing to heterozygous Gαs disruption. J Bone Miner Res. 2014;29(3):749–60.
29. Kayemba-Kay's S, Tripon C, Heron A, et al. Pseudohypoparathyroidism type IA subclinical hypothyroidism and rapid weight gain weight as early clinical signs: a clinical study of 10 cases. J Clin Res Pediatr Endocrinol. 2016;8(4):432–8.
30. Mouallem M, Shaharabany M, Weintrob N, et al. Cognitive impairment is prevalent in pseudohypoparathyroidism type Ia, but not in pseudopseudohypoparathyroidism: possible cerebral imprinting of Gsalpha. Clin Endocrinol. 2008;68(2):233–9.
31. Richard N, Molin A, Coudray N, et al. Paternal GNAS mutations lead to severe intrauterine growth retardation (IUGR) and provide evidence for a role of XLαs in fetal development. J Clin Endocrinol Metab. 2013;98(9):E1549–56.
32. Roizen JD, Danzig J, Groleau V, et al. Resting energy expenditure is decreased in pseudohypoparathyroidism type 1A. J Clin Endocrinol Metab. 2015;101(3):880–8.
33. Linglart A, Carel JC, Garabédian M, et al. GNAS1 lesions in pseudohypoparathyroidism Ia and Ic: genotype phenotype relationship and evidence of the maternal transmission of the hormonal resistance. J Clin Endocrinol Metab. 2002;87(1):189–97.
34. Thiele S, de Sanctis L, Werner R, et al. Functional characterization of GNAS mutations found in patients with pseudohypoparathyroidism type Ic defines a new subgroup of pseudohypoparathyroidism affecting selectively Gsα-receptor interaction. Hum Mutat. 2011;32(6):653–60.
35. Kinoshita K, Minagawa M, Anzai M, et al. Characteristic height growth pattern in patients with pseudohypoparathyroidism: comparison between Type 1a and Type 1b. Clin Pediatr Endocrinol. 2007;16(1):31–6.
36. Cho SY, Yoon YA, Ki C-S, et al. Clinical characterization and molecular classification of 12 Korean patients with pseudohypoparathyroidism and pseudopseudohypoparathyroidism. Exp Clin Endocrinol Diabetes. 2013;121(9):539–45.
37. Thiele S, Werner R, Grötzinger J, et al. A positive genotype-phenotype correlation in a large cohort of patients with pseudohypoparathyroidism Type Ia and pseudopseudohypoparathyroidism and 33 newly identified mutations in the GNAS gene. Mol Genet Genomic Med. 2015;3(2):111–20.
38. Elli FM, deSanctis L, Ceoloni B, et al. Pseudohypoparathyroidism type Ia and pseudo-pseudohypoparathyroidism: the growing spectrum of GNAS inactivating mutations. Hum Mutat. 2013;34(3):411–6.
39. Garin I, Elli FM, Linglart A, et al. Novel microdeletions affecting the GNAS locus in pseudohypoparathyroidism: characterization of the underlying mechanisms. J Clin Endocrinol Metab. 2015;100(4):E681–7.
40. Wu Y-L, Hwang D-Y, Hsiao H-P, et al. Mutations in pseudohypoparathyroidism 1a and pseudopseudohypoparathyroidism in ethnic Chinese. PloS One. 2014;9(3):e90640.
41. Elli FM, Linglart A, Garin I, et al. The prevalence of GNAS deficiency-related diseases in a large cohort of patients characterized by the EuroPHP network. J Clin Endocrinol Metab. 2016;101(10):3657–68.
42. Usardi A, Mamoune A, Nattes E, et al. Progressive development of PTH resistance in patients with inactivating mutations on the maternal allele of GNAS. J Clin Endocrinol Metab. 2017;102(6):1844–50.
43. Balavoine A-S, Ladsous M, Velayoudom F-L, et al. Hypothyroidism in patients with pseudohypoparathyroidism type Ia: clinical evidence of resistance to TSH and TRH. Eur J Endocrinol. 2008;159(4):431–7.

44. Vlaeminck-Guillem V, D'herbomez M, Pigny P, et al. Pseudohypoparathyroidism Ia and hypercalcitoninemia. J Clin Endocrinol Metab. 2001;86(7):3091–6.
45. Germain-Lee EL, Groman J, Crane JL, et al. Growth hormone deficiency in pseudohypoparathyroidism type 1a: another manifestation of multihormone resistance. J Clin Endocrinol Metab. 2003;88(9):4059–69.
46. Mantovani G, Ferrante E, Giavoli C, et al. Recombinant human GH replacement therapy in children with pseudohypoparathyroidism type Ia: first study on the effect on growth. J Clin Endocrinol Metab. 2010;95(11):5011–7.
47. Namnoum AB, Merriam GR, Moses AM, et al. Reproductive dysfunction in women with Albright's hereditary osteodystrophy. J Clin Endocrinol Metab. 1998;83:824–9.
48. Linglart A, Carel JC, Garabédian M, et al. GNAS1 lesions in pseudohypoparathyroidism Ia and Ic: genotype phenotype relationship and evidence of the maternal transmission of the hormonal resistance. J Clin Endocrinol Metab. 2002;87(1):189–97.
49. Mantovani G, Spada A. Resistance to growth hormone releasing hormone and gonadotropins in Albright's hereditary osteodystrophy. J Pediatr Endocrinol Metab. 2006;19 Suppl 2:663–70.
50. Long DN, McGuire S, Levine MA, et al. Body mass index differences in pseudohypoparathyroidism type 1a versus pseudopseudohypoparathyroidism may implicate paternal imprinting of Galpha(s) in the development of human obesity. J Clin Endocrinol Metab. 2007;92(3):1073–9.
51. de Sanctis L, Giachero F, Mantovani G, et al., Study Group Endocrine diseases due to altered function of Gsα protein of the Italian Society of Pediatric Endocrinology and Diabetology (ISPED). Genetic and epigenetic alterations in the GNAS locus and clinical consequences in pseudohypoparathyroidism: Italian common healthcare pathways adoption. Ital J Pediatr. 2016;42(1):101.
52. Pignolo RJ, Ramaswamy G, Fong JT, et al. Progressive osseous heteroplasia: diagnosis, treatment, and prognosis. Appl Clin Genet. 2015;8:37–48.
53. Shoemaker AH, Lomenick JP, Saville BR, et al. Energy expenditure in obese children with pseudohypoparathyroidism type 1a. Int J Obes 2005;37(8):1147–53.
54. Wang L, Shoemaker AH. Eating behaviors in obese children with pseudohypoparathyroidism type 1a: a cross-sectional study. Int J Pediatr Endocrinol. 2014;2014(1):21.
55. Shoemaker AH, Jüppner H. Nonclassic features of pseudohypoparathyroidism type 1A. Curr Opin Endocrinol Diabetes Obes. 2017;24(1):33–8.
56. Kashani P, Roy M, Gillis L, et al. The association of pseudohypoparathyroidism type Ia with Chiari malformation type I: a coincidence or a common link? Case Rep Med. 2016;2016:7645938.
57. Reis MTA, Matias DT, Faria MEJ de, et al. Failure of tooth eruption and brachydactyly in pseudohypoparathyroidism are not related to plasma parathyroid hormone-related protein levels. Bone. 2016;85:138–41.
58. Mantovani G, de Sanctis L, Barbieri AM, et al. Pseudohypoparathyroidism and GNAS epigenetic defects: clinical evaluation of albright hereditary osteodystrophy and molecular analysis in 40 patients. J Clin Endocrinol Metab. 2010;95(2):651–8.
59. de Nanclares GP, Fernández-Rebollo E, Santin I, et al. Epigenetic defects of GNAS in patients with pseudohypoparathyroidism and mild features of Albright's hereditary osteodystrophy. J Clin Endocrinol Metab. 2007;92(6):2370–3.
60. Mariot V, Maupetit-Méhouas S, Sinding C, et al. A maternal epimutation of GNAS leads to Albright osteodystrophy and parathyroid hormone resistance. J Clin Endocrinol Metab. 2008;93(3):661–5.
61. Turan S, Thiele S, Tafaj O, et al. Evidence of hormone resistance in a pseudo-pseudohypoparathyroidism patient with a novel paternal mutation in GNAS. Bone. 2015;71:53–7.
62. Elli FM, Barbieri AM, Bordogna P, et al. Screening for GNAS genetic and epigenetic alterations in progressive osseous heteroplasia: first Italian series. Bone. 2013;56(2):276–80.
63. Pignolo RJ, Xu M, Russell E, et al. Heterozygous inactivation of Gnas in adipose-derived mesenchymal progenitor cells enhances osteoblast differentiation and promotes heterotopic ossification. J Bone Miner Res. 2011;26(11):2647–55.
64. Elli FM, Boldrin V, Pirelli A, et al. The complex GNAS imprinted locus and mesenchymal stem cells differentiation. Horm Metab Res. 2017;49(4):250–8.
65. Elli FM, de Sanctis L, Bollati V, et al. Quantitative analysis of methylation defects and correlation with clinical characteristics in patients with pseudohypoparathyroidism type I and GNAS epigenetic alterations. J Clin Endocrinol Metab. 2014;99(3):E508–17.
66. Brix B, Werner R, Staedt P, et al. Different pattern of epigenetic changes of the GNAS gene locus in patients with pseudohypoparathyroidism type Ic confirm the heterogeneity of underlying pathomechanisms in this subgroup of pseudohypoparathyroidism and the demand for a new classification of GNAS-related disorders. J Clin Endocrinol Metab. 2014;99(8):E1564–70.
67. Maupetit-Méhouas S, Mariot V, Reynès C, et al. Quantification of the methylation at the GNAS locus identifies subtypes of sporadic pseudohypoparathyroidism type Ib. J Med Genet. 2011;48(1):55–63.
68. Linglart A, Bastepe M, Jüppner H. Similar clinical and laboratory findings in patients with symptomatic autosomal dominant and sporadic pseudohypoparathyroidism type Ib despite different epigenetic changes at the GNAS locus. Clin Endocrinol. 2007;67(6):822–31.
69. Yuno A, Usui T, Yambe Y, et al. Genetic and epigenetic states of the GNAS complex in pseudohypoparathyroidism type Ib using methylation-specific multiplex ligation-dependent probe amplification assay. Eur J Endocrinol. 2013;168(2):169–75.
70. Perez-Nanclares G, Romanelli V, Mayo S, et al., Spanish PHP Group. Detection of hypomethylation syndrome

among patients with epigenetic alterations at the GNAS locus. J Clin Endocrinol Metab. 2012;97(6):E1060–7.
71. Kinoshita K, Minagawa M, Takatani T, et al. Establishment of diagnosis by bisulfite-treated methylation-specific PCR method and analysis of clinical characteristics of pseudohypoparathyroidism type 1b. Endocr J. 2011;58(10):879–87.
72. Takatani R, Molinaro A, Grigelioniene G, et al. Analysis of multiple families with single individuals affected by pseudohypoparathyroidism type Ib (PHP1B) reveals only one novel maternally inherited GNAS deletion. J Bone Miner Res. 2016;31(4):796–805.
73. Mantovani G, Bondioni S, Linglart A, et al. Genetic analysis and evaluation of resistance to thyrotropin and growth hormone-releasing hormone in pseudohypoparathyroidism type Ib. J Clin Endocrinol Metab. 2007;92(9):3738–42.
74. Takatani R, Minagawa M, Molinaro A, et al. Similar frequency of paternal uniparental disomy involving chromosome 20q (patUPD20q) in Japanese and Caucasian patients affected by sporadic pseudohypoparathyroidism type Ib (sporPHP1B). Bone. 2015;79:15–20.
75. Fernandez-Rebollo E, García-Cuartero B, Garin I, et al. Intragenic GNAS deletion involving exon A/B in pseudohypoparathyroidism type 1A resulting in an apparent loss of exon A/B methylation: potential for misdiagnosis of pseudohypoparathyroidism type 1B. J Clin Endocrinol Metab. 2010;95(2):765–71.
76. Mizunashi K, Furukawa Y, Sohn HE, et al. Heterogeneity of pseudohypoparathyroidism type I from the aspect of urinary excretion of calcium and serum levels of parathyroid hormone. Calcif Tissue Int. 1990;46(4):227–32.
77. Liu J, Litman D, Rosenberg MJ, et al. A GNAS1 imprinting defect in pseudohypoparathyroidism type IB. J Clin Invest. 2000;106(9):1167–74.
78. Turan S, Ignatius J, Moilanen JS, et al. De novo STX16 deletions: an infrequent cause of pseudohypoparathyroidism type Ib that should be excluded in sporadic cases. J Clin Endocrinol Metab. 2012;97(12):E2314–2319.
79. Jüppner H, Linglart A, Fröhlich LF, et al. Autosomal-dominant pseudohypoparathyroidism type Ib is caused by different microdeletions within or upstream of the GNAS locus. Ann N Y Acad Sci. 2006;1068:250–5.
80. Fröhlich LF, Mrakovcic M, Steinborn R, et al. Targeted deletion of the Nesp55 DMR defines another Gnas imprinting control region and provides a mouse model of autosomal dominant PHP-Ib. Proc Natl Acad Sci U S A. 2010;107(20):9275–80.
81. Jüppner H, Bastepe M. Different mutations within or upstream of the GNAS locus cause distinct forms of pseudohypoparathyroidism. J Pediatr Endocrinol Metab 2006;19 Suppl 2:641–6.
82. Dixit A, Chandler KE, Lever M, et al. Pseudohypoparathyroidism type 1b due to paternal uniparental disomy of chromosome 20q. J Clin Endocrinol Metab. 2013;98(1):E103–8.
83. Maupetit-Méhouas S, Azzi S, Steunou V, et al. Simultaneous hyper- and hypomethylation at imprinted loci in a subset of patients with GNAS epimutations underlies a complex and different mechanism of multilocus methylation defect in pseudohypoparathyroidism type 1b. Hum Mutat. 2013;34(8):1172–80.
84. Sano S, Matsubara K, Nagasaki K, et al. Beckwith-Wiedemann syndrome and pseudohypoparathyroidism type Ib in a patient with multilocus imprinting disturbance: a female-dominant phenomenon? J Hum Genet. 2016;61(8):765–9.
85. Bakker B, Sonneveld LJH, Woltering MC, et al. A girl with Beckwith-Wiedemann syndrome and pseudohypoparathyroidism type 1B due to multiple imprinting defects. J Clin Endocrinol Metab. 2015;100(11):3963–6.
86. Bréhin A-C, Colson C, Maupetit-Méhouas S, et al. Loss of methylation at GNAS exon A/B is associated with increased intrauterine growth. J Clin Endocrinol Metab. 2015;100(4):E623–31.
87. Drezner M, Neelon FA, Lebovitz HE. Pseudohypoparathyroidism type II: a possible defect in the reception of the cyclic AMP signal. N Engl J Med. 1973;289(20):1056–60.
88. Akın L, Kurtoğlu S, Yıldız A, et al. Vitamin D deficiency rickets mimicking pseudohypoparathyroidism. J Clin Res Pediatr Endocrinol. 2010;2(4):173–5.
89. Lynch DC, Dyment DA, Huang L, et al. Identification of novel mutations confirms pde4d as a major gene causing acrodysostosis. Hum Mutat. 2013;34(1):97–102.
90. Kaname T, Ki C-S, Niikawa N, et al. Heterozygous mutations in cyclic AMP phosphodiesterase-4D (PDE4D) and protein kinase A (PKA) provide new insights into the molecular pathology of acrodysostosis. Cell Signal. 2014;26(11):2446–59.
91. Muhn F, Klopocki E, Graul-Neumann L, et al. Novel mutations of the PRKAR1A gene in patients with acrodysostosis. Clin Genet. 2013;84(6):531–8.
92. Lindstrand A, Grigelioniene G, Nilsson D, et al. Different mutations in PDE4D associated with developmental disorders with mirror phenotypes. J Med Genet. 2014;51(1):45–54.
93. Linglart A, Fryssira H, Hiort O, et al. PRKAR1A and PDE4D mutations cause acrodysostosis but two distinct syndromes with or without GPCR-SIGNALING HORMONE RESISTANce. J Clin Endocrinol Metab. 2012;97(12):E2328–38.
94. Linglart A, Menguy C, Couvineau A, et al. Recurrent PRKAR1A mutation in acrodysostosis with hormone resistance. N Engl J Med. 2011;364(23):2218–26.
95. Mitsui T, Kim O-H, Hall CM, et al. Acroscyphodysplasia as a phenotypic variation of pseudohypoparathyroidism and acrodysostosis type 2. Am J Med Genet A. 2014;164A(10):2529–34.
96. Michot C, Le Goff C, Goldenberg A, et al. Exome sequencing identifies PDE4D mutations as another cause of acrodysostosis. Am J Hum Genet. 2012;90(4):740–5.
97. Elli FM, Bordogna P, de Sanctis L, et al. Screening of PRKAR1A and PDE4D in a large Italian series of patients clinically diagnosed with Albright hereditary osteodystrophy and/or pseudohypoparathyroidism. J Bone Miner Res. 2016;31(6):1215–24.

98. Leroy C, Landais E, Briault S, et al. The 2q37-deletion syndrome: an update of the clinical spectrum including overweight, brachydactyly and behavioural features in 14 new patients. Eur J Hum Genet. 2013;21(6):602–12.
99. Kooh SW, Fraser D, DeLuca HF, et al. Treatment of hypoparathyroidism and pseudohypoparathyroidism with metabolites of vitamin D: evidence for impaired conversion of 25-hydroxyvitamin D to 1 alpha,25-ihydroxyvitamin D. N Engl J Med. 1975;293(17):840–4.
100. Macfarlane RJ, Ng BH, Gamie Z, E et al. Pharmacological treatment of heterotopic ossification following hip and acetabular surgery. Expert Opin Pharmacother. 2008;9(5):767–86.
101. Mantovani G, Bastepe M, Monk D, et al. Diagnosis and management of pseudohypoparathyroidism and related disorders: first international Consensus Statement. Nat Rev Endocrinol. 2018;14(8):476–500.

88

Disorders of Phosphate Homeostasis

Mary D. Ruppe[1] and Suzanne M. Jan de Beur[2]

[1]Division of Endocrinology, Department of Medicine, Houston Methodist Hospital, Houston, TX, USA
[2]Division of Endocrinology and Metabolism, Department of Medicine, The Johns Hopkins School of Medicine, Baltimore, MD, USA

INTRODUCTION

Phosphorus is a critical element in skeletal development, bone mineralization, membrane composition, nucleotide structure, and cellular signaling. Serum phosphorus concentration is regulated by diet, hormones, pH, and changes in renal, skeletal, and intestinal function. The focus of this chapter is the molecular basis of human disorders of phosphate homeostasis.

HYPOPHOSPHATEMIA

Clinical consequences

Hypophosphatemia is observed in up to 5% of hospitalized patients [1]. Among alcoholic patients and those with severe sepsis, the prevalence is up to 30% to 50%. The clinical manifestations of hypophosphatemia are dependent on the severity and chronicity of the phosphorus depletion. Severe hypophosphatemia is observed in a variety of clinical settings including chronic alcoholism, refeeding syndrome, diabetic ketoacidosis, and critical illness.

The symptoms of hypophosphatemia are a consequence of intracellular phosphorus depletion: (i) tissue hypoxia from reduced 2,3-diphosphoglycerate (2,3-DPG) in the erythrocyte that increases affinity of hemoglobin for oxygen and (ii) compromised cellular function from diminished tissue content of ATP.

Causes of hypophosphatemia

The three major mechanisms by which hypophosphatemia occur are redistribution of phosphorus from extracellular fluid into cells, increased urinary excretion, and decreased intestinal absorption (Table 88.1). The diagnosis is often evident from the history. If the diagnosis remains obscure, the tubular reabsorption of phosphate from a 2 hour, second urine void after an overnight fast should be calculated. Tubular maximum of phosphate/glomerular filtration rate (TmP/GFR) can be calculated by the equation $TmP/GFR = P - PE/100.1 \log_e(P/PE)$ where $PE = UP \times SeCr/UCr$ (P = serum phosphate; Cr = creatinine; Se = serum; U = urine). TmP/GFR is age and gender specific but generally falls between 2.6 and 4.4 mg/dL.

Hypophosphatemia secondary to renal phosphate wasting can result from primary renal transporter defects, excess phosphaturic hormones such as PTH and fibroblast growth factor 23 (FGF23), or medications (Fig. 88.1). The remainder of the section will focus on renal phosphate wasting syndromes.

TUMOR-INDUCED OSTEOMALACIA

Tumor-induced osteomalacia (TIO), or oncogenic osteomalacia, is an acquired, paraneoplastic syndrome of renal phosphate wasting [2].

Clinical and biochemical manifestations

TIO may present at any age, although most patients are diagnosed in the sixth decade and present with progressive muscle weakness, bone pain, and fractures. Children display rachitic features including gait disturbances, growth retardation, and skeletal deformities. The average time from onset of symptoms to a correct diagnosis often exceeds 2.5 years [3]. Once the syndrome is recognized, an average of 5 years elapses until identification of the underlying tumor [4]. Until the underlying tumor is

---

Primer on the Metabolic Bone Diseases and Disorders of Mineral Metabolism, Ninth Edition. Edited by John P. Bilezikian.
© 2019 American Society for Bone and Mineral Research. Published 2019 by John Wiley & Sons, Inc.
Companion website: www.wiley.com/go/asbmrprimer

Table 88.1. Causes of hypophosphatemia.

Decreased Intestinal absorption
  Vitamin D deficiency or resistance
    Nutritional Deficiency
      Low sun exposure, low dietary intake
    Malabsorption
      Celiac disease, Crohn disease
      Gastrectomy, bowel resection, gastric bypass
      Pancreatitis
      Chronic diarrhea
    Chronic liver disease
    Chronic renal disease
    Increased catabolism
      Anticonvulsant therapy
    Vitamin D receptor defects
      Vitamin D-dependent rickets, type 2
    Vitamin D synthetic defects
      CYP27B1 (vitamin D-dependent rickets, type 1)
      CYP27A1
      CYP2R1
    Nutritional deficiencies
      Alcoholism, anorexia, starvation
  Antacids containing aluminum or magnesium

Increased urinary losses
  Renal phosphate wasting disorders (Table 88.2)
  Primary and secondary hyperparathyroidism
  Diabetic ketoacidosis (osmotic diuresis)
  Medications
    Calcitonin, diuretics, glucocorticoids, bicarbonate
  Acute volume expansion

Intracellular shifts
  Increased insulin
    Refeeding, treatment of diabetic ketoacidosis, insulin therapy
  Hungry bone syndrome
  Acute respiratory alkalosis
  Tumor consumption
    Leukemia blast crisis, lymphoma
  Sepsis
  Sugars
    Glucose, fructose, glycerol
  Recovery from metabolic acidosis

identified, other renal phosphate wasting syndromes must be considered. Documentation of a previously normal serum phosphorus level supports the diagnosis of TIO, although in rare instances patients with autosomal dominant hypophosphatemic rickets (ADHR) can present in adulthood. In situations when inherited hypophosphatemic rickets must be excluded, genetic testing is indicated.

The biochemical hallmarks of TIO are low serum phosphate, phosphaturia (secondary to reduced proximal renal tubular phosphate reabsorption), and frankly low or inappropriate normal levels of serum $1,25(OH)_2D_3$ that is expected to be elevated in the face of hypophosphatemia (Table 88.2). Calcium and PTH are typically normal. Bone histomorphometry shows severe osteomalacia with clear evidence of a mineralization defect with increased mineralization lag time and excessive osteoid (Fig. 88.2). The defect of renal phosphate wasting in concert with impaired $1,25(OH)_2D_3$ synthesis results in poor bone mineralization and fractures [5].

Over 90% of the associated tumors are phosphaturic mesenchymal tumor, mixed connective tissue type [6] (PMTMCT; Fig. 88.2). Characterized by an admixture of spindle cells, osteoclast-like giant cells, prominent blood vessels, cartilage-like matrix, and metaplastic bone, these tumors occur equally in soft tissue and bone. Although typically benign, malignant variants of PMTMCT have been described. These mesenchymal tumors ectopically express and secrete FGF23 and other phosphaturic proteins [7].

FGF23, a circulating fibroblast growth factor produced primarily by osteocytes [8], has two main physiological functions: (i) FGF23 promotes internalization of sodium phosphate cotransporters (NaPiIIa, NaPiIIc) from the renal brush border membrane reducing reabsorption of urinary phosphate resulting in hypophosphatemia [9]; (ii) it diminishes protein expression of the 25 hydroxy-l-α-hydroxylase enzyme (CYP27B1) that converts vitamin D to $1,25(OH)_2D_3$ [10] while increasing the activity of the vitamin D 24-hydroxylase enzyme which catabolizes 25-hydroxyvitamin D to 24,25-dihydroxyvitamin D, an inactive form. Together, this profoundly disrupts the compensatory increase in $1,25(OH)_2D_3$ triggered by hypophosphatemia [11]. Production of FGF23 is controlled by circulating factors (PTH, $1,25(OH)_2D_3$, iron, dietary phosphorous) and bone derived factors (DMP1, PHEX, ENPP1, hypoxia-inducible factor-1α) [12]. Circulating levels of FGF23 are elevated in most patients with TIO [13]. After surgical resection, FGF23 levels plummet. Other secreted proteins such as MEPE (matrix extracellular phosphoglycoprotein), FGF7, and sFRP4 (secreted frizzled related protein 4) are highly expressed in mesenchymal tumors associated with TIO, but the role of these proteins in the disease process is unknown.

Treatment

Detection and localization of the underlying tumor in TIO is imperative because surgical resection is curative. However, the tumors are often small, slow growing, and frequently found in a variety of anatomical locations, including the long bones, the distal extremities, the nasopharynx, the sinuses, and the groin. A thorough physical examination should be performed to detect any palpable masses because the tumors have been found in the subcutaneous tissues. The tumors can be difficult to localize with conventional imaging techniques frequently requiring a combination of functional and anatomical approaches. A recent study evaluating ^{68}Ga-DOTATATE PET/CT, Octreoscan SPECT/CT and ^{18}F deoxyglucose-PET scanning in TIO showed ^{68}Ga-DOTATATE PET/CT has the greatest sensitivity and specificity [14]. Because in vitro

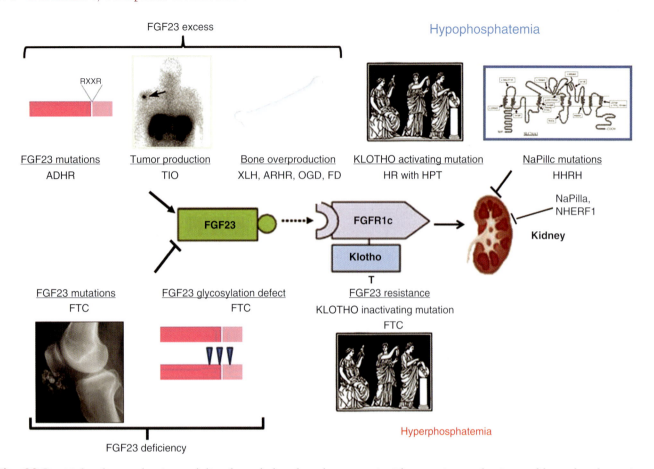

Fig. 88.1. Molecular mechanisms of disorders of phosphate homeostasis. Three major mechanisms of hypophosphatemia are FGF23 excess owing to ectopic production as in tumor-induced osteomalacia (TIO), excess bone production found in X-linked hypophosphatemic rickets (XLH), autosomal recessive hypophosphatemic rickets (ARHR), autosomal dominant hypophosphatemic rickets (ADHR), fibrous dysplasia (FD), and osteoglophonic dysplasia (OGD), and mutation in the FGF23 gene that renders the protein resistant to inactivation. Hypophosphatemia may also be secondary to excess KLOTHO, the cofactor necessary for FGF23 signaling as found in a patient with hypophosphatemic rickets with hyperparathyroidism. Finally, homozygous inactivating mutations in SLC34A3, that encodes NaPiIIc, or dominant negative mutations in SLC34A2 that encodes NaPiIIa, result in phosphate wasting caused by absence of sodium–phosphate cotransporters. Hyperphosphatemia is caused by FGF23 deficiency, either through inactivating mutations in FGF23, aberrant glycosylation of FGF23, or FGF23 resistance caused by inactivating KLOTHO mutations. FTC = familial tumoral calcinosis.

studies show that many mesenchymal tumors express somatostatin receptor, ^{111}In-pentetreotide scintigraphy (octreotide scan; Fig. 88.2), a scanning technique that uses a radiolabeled somatostatin analog, has been used to localize these tumors in some patients [15, 16]. Venous sampling for FGF23 has been used for tumor localization but seems to be more suited for confirmation that a mass observed on imaging is producing FGF23 than the de novo localization of the tumor [17].

The definitive treatment is complete tumor resection with wide margins, which results in rapid correction of the biochemical perturbations and remineralization of bone. Radiofrequency ablation has been reported to have been beneficial [18]. However, even after the diagnosis of TIO is made, the tumor often remains obscure or incompletely resected. Therefore, medical management is frequently necessary. The current practice is to treat TIO with phosphorus and calcitriol. The phosphorus supplementation serves to replace ongoing renal phosphate loss, and the calcitriol replaces insufficient renal production of $1,25(OH)_2D_3$ which enhances renal and gastrointestinal phosphorus reabsorption. Generally, patients are treated with phosphorus (1 to 2 g/day), divided into three to four doses daily, and calcitriol (1 to 3 μg/day). In some cases, administration of calcitriol alone may improve the biochemical abnormalities and heal the osteomalacia. Therapy and dosing should be tailored to improve symptoms and normalize alkaline phosphatase. With appropriate treatment, muscle and bone pain will improve, and healing of the osteomalacia will ensue.

Monitoring for therapeutic complications is important to prevent unintended hypercalcemia, hypercalciuria, nephrocalcinosis, and nephrolithiasis. The incidence

Table 88.2. Characteristics of renal phosphate wasting disorders.

Disease (OMIM)	Defect	Pathogenesis
Tumor-induced osteomalacia	Mesenchymal tumor	Ectopic, unregulated production of FGF23 and other phosphatonins sFRP4, MEPE, FGF7
X-linked hypophosphatemia (307800)	PHEX mutation	Inappropriate FGF23 synthesis from bone
Autosomal dominant hypophoshatemic rickets (193100)	FGF23 mutation	Increased circulating intact FGF23 caused by mutations that render it resistant to cleavage
Hereditary hypophosphatemic rickets with hypercalcuria (241530)	SLC34A3 mutation	Loss of function NaPiIIc mutations that result in renal phosphate wasting without a defect in 1,25(OH)$_2$D synthesis
Autosomal recessive hypophosphatemic rickets type 1 (241520)	DMP1 mutation	Loss of DMP1 causes impaired osteocyte differentiation and increased production of FGF23
Autosomal recessive hypophosphatemic rickets type 2 (613312)	ENPP1 mutation	Increased production of FGF23
Hypophosphatemic rickets and hyperparathyroidism (612089)	α-KLOTHO translocation	Increased KLOTHO, FGF23 and downstream FGF23 signaling
Fibrous dysplasia (139320)	GNAS mutation	Increased FGF23 production from the dysplastic bone
Cutaneous skeletal hypophosphatemic syndrome	Excess FGF23 production	Increased FGF23 production from the dysplastic bone and from the nevi
Osteoglophonic dysplasia (166250)	FGFR1 mutation	Increased FGF23 production from the dysplastic bone
NPHLOP1 (612286)	SLC34A1 mutation	Renal phosphate wasting without a defect in 1,25(OH)$_2$D synthesis
NPHLOP2 (612287)	SLC9A3R1 mutation	Renal phosphate wasting through potentiation of PTH-mediated cAMP production
Fanconi syndrome and hypoposphatemic rickets (613388)	SLC34A1 mutation	Renal phosphate wasting without a defect in 1,25(OH)$_2$D synthesis
Raine syndrome (259775)	FAM20C mutation	Increased FGF23 production

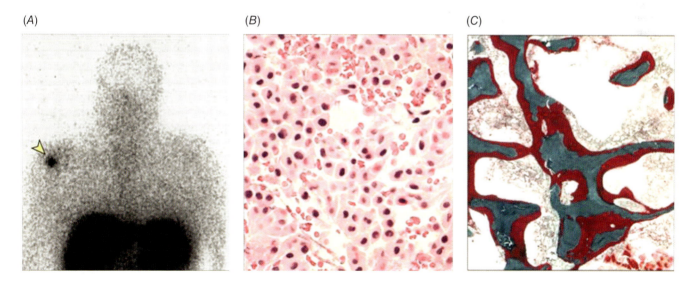

Fig. 88.2. Radiographic and histological features in TIO. (*A*) Octreotide scan showing small mesenchymal tumor in the head of the humerus. (*B*) Hemiangiopericytoma with numerous pericytes and vascular channels (H&E stain). (*C*) Bone biopsy with Goldner stain. Excessive osteoid or unmineralized bone matrix composed mainly of collagen stains pink. Mineralized bone stains blue. This bone biopsy shows severe osteomalacia.

of hyperparathyroidism with treatment is unknown. To assess safety and efficacy, monitoring of serum and urine calcium, renal function, and parathyroid status is recommended monthly at the initiation of treatment and every 3 months when on chronic therapy. There have been reports of the use cinacalcet in patients who did not tolerate medical therapy with phosphorus and calcitriol [19]. This allowed a decrease in the phosphorus dose to one that was tolerated. Octreotide treatment has shown a mixed response [15, 20]. Peptide receptor radionuclide therapy use has also been reported in a single patient [21]. There is a current clinical trial evaluating the use of a monoclonal antibody to FGF23 for the treatment of TIO in adults (http://www.clinicaltrials.gov, KRN23).

X-LINKED HYPOPHOSPHATEMIC RICKETS

First described by Albright in 1939 [22], X-linked hypophosphatemic rickets (XLH) is characterized by growth retardation, rachitic and osteomalacic bone disease, and dental abscesses. It is the most common disorder of renal phosphate wasting, occurring in 3.9 to 5 per 100,000 live births.

Genetics

Although the X-linked inheritance was first detailed in 1958 [23], it was not until the 1990s that the genetic basis of XLH was discovered to be mutations in *PHEX* (phosphate-regulating gene with homologies to endopeptidases on the X chromosome) [24]. To date, more than 330 mutations have been described (PHEX database: www.PHEXdb.mcgill.ca). The *PHEX* gene codes for a protein of unknown function that is a member of the M13 family of membrane-bound metalloproteases that is present in osteoblasts, osteocytes, and odontoblasts but not in kidney tubules [25].

Clinical and biochemical manifestations

Before children begin to walk, clinical findings are limited. In infants, the majority of testing that is performed is because of a known family history. After the child is ambulatory, progressive lower extremity bowing becomes apparent with a decrease in height velocity and with variable bone and/or joint pain. Dental manifestations include abscessed noncarious teeth, enamel defects, enlarged pulp chambers, and taurodontism. Cranial abnormalities with frontal bossing and an increased anteroposterior skull length are variably present. Adults exhibit short stature, bone and joint pain, pseudofractures, enthesopathy (calcification of ligaments and tendons), and dental manifestations. Typically, patients with XLH have increased spine BMD which may reflect calcific enthesopathy. It is unclear if there is a change in long-term fracture rates.

As in TIO, laboratory findings in XLH (Table 88.1) show low serum phosphate, phosphaturia, and low or inappropriately normal levels of serum 1,25(OH)2D3. Calcium and PTH are typically normal. In infants the diagnosis may be difficult to establish because phosphate levels may initially be normal. In addition, the infant normal range is substantially higher than that of older children and so the low phosphate may go unrecognized. Excess FGF23 derived from increase bone production is the etiology of hypophosphatemia in XLH [13]. How PHEX deficiency leads to increased FGF23 is the subject of active investigation.

The diagnosis of XLH is based on a consistent history and physical, radiological evidence of rachitic disease and appropriate biochemical findings. Family history may indicate multigenerational or sporadic occurrence of the disorder. Mutational analysis of the *PHEX* gene is available; however, studies have shown that mutations can only be found in 50% to 70% of affected individuals [26].

Treatment and complications

Treatment consists of oral phosphorus administered three to five times daily and calcitriol. The treatment is generally started at a low dose to avoid gastrointestinal side effects. The doses are then titrated to a weight-based dose of calcitriol at 20 to 30 ng/kg/day along with phosphorus at 20 to 40 mg/kg/day [27].

The treatment leads to resolution of radiographic rickets and improved but not normal growth. Age at initiation of therapy, height at initiation of therapy, and possibly sex of the patient influence peak height attainment. Despite pharmacological therapy some patients require surgical intervention to correct lower extremity deformities. Therapeutic requirements drop dramatically as children enter adulthood because of lower bone turnover and the closure of the epiphyseal plates. In adults, treatment is variable. Some are not treated, some are treated with a low dose of calcitriol alone, or with both calcitriol and phosphorus. Treatment is indicated in adults with spontaneous insufficiency fractures, pending orthopedic procedures, biochemical evidence of osteomalacia, or disabling skeletal pain [27]. Adults that are treated have less severe dental disease but enthesopathy is unaffected [28]. However, treatment with calcitriol and phosphorus can lead to an unwanted increase in serum FGF23 levels [29]. The clinical significance of this is unknown, although there is an association between elevation in serum FGF23 and an increased risk of vascular and nonvascular mortality in the general population [30]. In a small study of short-term treatment of adult XLH patients with subcutaneous calcitonin there was a transient increase in serum phosphorus with a decrease in serum FGF23 levels [31]. Further studies are needed to determine if this is an efficacious treatment for XLH. Early phase clinical trials evaluating a neutralizing monoclonal antibody against FGF23 (KRN23) in adults and children with XLH showed that KRN23 increases serum phosphate, TmP/GFR and 1,25(OH)$_2$D$_3$ in treated subjects [32, 33] as well as improved physical functioning and stiffness on quality of life testing in adults [34]. As in TIO, treatment with phosphorus and calcitriol need to be monitored frequently to avoid complications.

AUTOSOMAL DOMINANT HYPOPHOSPHATEMIC RICKETS

ADHR is a rare form of hypophosphatemic rickets with clinical characteristic similar to XLH.

Genetics

Early reports documented an inherited renal phosphate wasting syndrome with a pattern that included male-to-male transmission that distinguished it from XLH [35]. Missense mutations in one of two arginine residues in the *FGF23* gene at positions 176 or 179 have been identified in affected members of ADHR families [36]. The mutated arginine residues, located in the consensus proprotein convertase cleavage RXXR motif, prevent inactivation of FGF23 and thus result in prolonged or enhanced FGF23 action [37].

Clinical and biochemical manifestations

The clinical and biochemical findings in ADHR are similar to those in XLH (Table 88.1). In contrast to XLH, there are instances of delayed onset and rarely, resolution of the phosphate wasting. FGF23 levels have been shown to vary with disease status in ADHR [38]. Within the same family, there can be variable presentations with two subgroups of affected individuals described. Those with childhood onset have biochemical and clinical similarities to XLH, whereas those presenting later often lack lower extremity deformities, presumably because of fusion of the growth plate before the development of hypophosphatemia. Recent studies have implicated iron deficient states in the late manifestation of ADHR with a case report detailing normalization of the biochemical abnormalities after iron supplementation [39]. Furthermore, low levels of iron in ADHR subjects are associated with higher FGF23 levels [40].

Treatment

As in XLH, treatment consists of phosphate and calcitriol. Patients who do not respond to pharmacological therapy may require surgical intervention to straighten bowed limbs.

Hereditary hypophosphatemic rickets with hypercalciuria

Hereditary hypophosphatemic rickets with hypercalciuria (HHRH) is a rare genetic form of hypophosphatemic rickets characterized by hypophosphatemia, renal phosphate wasting, and preserved responsiveness of 1,25(OH)$_2$D$_3$ to hypophosphatemia (Table 88.1). This appropriate increase in 1,25(OH)$_2$D$_3$ leads to increased calcium absorption from the gastrointestinal tract and thus to hypercalciuria and nephrolithiasis. The genetic defect in HHRH is loss of function mutations in the gene that encodes NaPiIIc (SLC34A3) [41, 42]. The mutations can be either heterozygous or homozygous loss of function mutations. Biochemical testing in HHRH distinguishes it from XLH in that 1,25(OH)$_2$D$_3$ is elevated and hypercalciuria is present. Treatment consists of phosphate supplements alone.

Autosomal recessive hypophosphatemic rickets

Autosomal recessive hypophosphatemic rickets (ARHR) type 1 is caused by loss of function mutations in dentin matrix protein 1 (DMP-1) [43]. DMP-1 has two functions: it translocates into the nucleus to regulate gene transcription early in osteocyte proliferation and, likely in response to calcium fluxes, becomes phosphorylated and is exported to the extracellular matrix to facilitate mineralization by hydroxyapatite in a process that requires appropriate cleavage of the full-length protein. Loss of DMP-1 function in ARHR leads to modestly and variably increased serum FGF23, dramatically increased expression of FGF23 in bone, defects in osteocyte differentiation, and impaired skeletal mineralization.

Mutations in ectonucleotide pyrophosphatase/phosphodiesterase 1 (ENPP1) cause a second type of ARHR (ARHR2) [44, 45]. ENPP1 regulates extracellular pyrophosphate and is required for bone mineralization. In addition to the clinical rickets, those with ENPP1 mutations exhibit progressive early onset hearing loss [46]. As in XLH, treatment consists of phosphorus and calcitriol and close monitoring for complications is required.

Fibrous dysplasia

Polyostotic fibrous dysplasia (FD) and McCune Albright syndrome (MAS) are caused by activating missense mutations in *GNAS* which leads to hormone-independent activation of G-protein (Gsα)-coupled signaling. In FD, there is replacement of medullary bone and bone marrow with undermineralized bone and fibrosis. In MAS, FD is part of a triad of features such as precocious puberty, as well as other hyperfunctioning endocrine disorders and *café au lait* lesions. In some instances when FD burden is high, patients with FD and MAS exhibit FGF23-mediated phosphate wasting. The degree of FD correlates with the degree of phosphate wasting. There is debate over whether the hypophosphatemia should be treated if there is no evidence for pathological rickets. When it is treated, treatment and complication considerations are as for XLH [47].

Cutaneous skeletal hypophosphatemia syndrome

Cutaneous skeletal hypophosphatemia syndrome is disorder of phosphate wasting that has been associated with a group of disorders with skin manifestations, including linear nevus sebaceous syndrome (also known as epidermal nevus syndrome). Affected individuals have clinical evidence of multiple cutaneous nevi with radiological evidence of FD. It has been associated with somatic RAS mutations which have been referred to as RASopathies. Elevated levels of FGF23 are responsible for the renal phosphate wasting [48]. As in XLH, treatment consists of phosphorus and calcitriol and similar monitoring for complications of therapy is indicated.

OTHER DISORDERS OF RENAL PHOSPHATE WASTING

In a case report of hypophosphatemic rickets and hyperparathyroidism [49], the patient had renal phosphate wasting, inappropriately normal $1,25(OH)_2D_3$, and hyperparathyroidism secondary to a genetic translocation resulting in increased levels of α-KLOTHO, the cofactor necessary for FGF23 to bind and activate its receptor (Table 88.1). Unexpectedly, FGF23 serum levels are also markedly elevated in this disorder implicating α-KLOTHO in the regulation of serum phosphate, of FGF23 expression, and of parathyroid function.

Osteoglophonic dysplasia is a rare autosomal dominantly inherited form of dwarfism caused by activating mutations in FGF receptor 1 (FGFR1; Table 88.1) [50]. As in FD, these patients exhibit a high burden of nonossifying bony lesions that produce FGF23 with resultant renal phosphate wasting and lower than expected $1,25(OH)_2D_3$ levels. As in FD and MAS, the extent of bone lesions correlates with FGF23 levels and the degree of phosphate wasting.

Two patients with hypophosphatemia secondary to renal phosphate wasting and osteopenia or nephrolithiasis (NPHLOP1) were found to have heterozygous, dominant negative, mutations in the renal type IIa sodium–phosphate cotransporter gene (SLC34A1) [51]. Bone pain and muscle weakness were absent in the affected patients.

Hypophosphatemia associated with nephrolithiasis and low bone mineral density (NPHLOP2) has also been found in connection with mutations in the sodium/hydrogen exchanger regulatory factor 1 (SLC9A3R1) [52].

Two patients with autosomal recessive renal Fanconi syndrome and hypophosphatemic rickets (FRTS2) harbored in-frame duplication of SLC34A1. The patients presented with bone deformities, fractures, and severe short stature. As in HHRH, hypercalciuria and $1,25(OH)_2D_3$ levels are observed [53].

Raine syndrome (RAS) is an autosomal recessive osteosclerotic bone dysplasia that has been associated with FAM20C mutations. There is a severe form that is lethal shortly after birth. Recently, a milder form that is associated with hypophosphatemia has been described. The mutant FAM20C likely decreases DMP-1 activity leading to increased FGF23 production [54].

In summary, either through mutations of the sodium phosphate transporters themselves, damage to the proximal renal tubule or through aberrant regulation of FGF23, decreased expression or function of the renal sodium–phosphate cotransporters likely represent the common pathway in renal phosphate wasting observed in these syndromes (Fig. 88.1).

HYPERPHOSPHATEMIA

Serum inorganic phosphorus levels are maintained between 2.5 and 4.5 mg/dL (0.81–1.45 mmol/L) in adults and between 6 and 7 mg/dL in children under 2. Oral phosphate loads of up to 4000 mg/day can be efficiently excreted by the kidneys with minimal rise in serum phosphorus through increased FGF23 production leading to downregulation of the sodium phosphate cotransporters in the proximal renal tubules. PTH secretion also contributes to renal phosphate excretion. Excess phosphorus complexes with calcium, resulting in a decrease in ionized calcium, which stimulates PTH secretion. There are four general mechanisms whereby phosphate entry into the extracellular fluid can outstrip the rate of renal excretion: acute exogenous phosphate loads, redistribution of intracellular phosphate to the extracellular space, decreased renal excretion, and pseudohyperphosphatemia caused by interference with analytical detection methods.

Clinical manifestations of hyperphosphatemia

With rapid elevations in phosphate load, hypocalcemia and tetany can occur. Hypocalcemia results when hyperphosphatemia suppresses the renal 1α-hydroxylase enzyme, reducing circulating $1,25(OH)_2D_3$, impairing intestinal calcium absorption.

Chronic hyperphosphatemia can lead to soft tissue calcifications. In the case of renal failure, secondary hyperparathyroidism and renal osteodystrophy can occur. Hyperphosphatemia stimulates vascular cells to undergo osteogenic differentiation leading to calcification of coronary arteries and heart valves. Medial calcification of peripheral arteries may lead to calciphylaxis, a disorder with high mortality and morbidity.

Causes of hyperphosphatemia

Major causes of hyperphosphatemia are listed in Table 88.3.

Table 88.3.	Causes of hyperphosphatemia.
Mechanism	Etiology
Decreased renal excretion	Renal insufficiency/failure
	Hypoparathyroidism
	Pseudohypoparathyroidism
	Tumoral calcinosis
	Acromegaly
	Bisphosphonates
Acute phosphate load	Phosphate-containing laxatives
	Fleet's phosphosoda enemas
	Intravenous phosphate
	Parental nutrition
Redistribution to the extracellular space	Tumor lysis
	Rhabdomyolysis
	Acidosis
	Hemolytic anemia
	Severe hyperthermia
	Fulminant hepatitis
	Systemic infections
Pseudohyperphosphatemia	Hyperglobulinemia
	Hyperlipidemia
	Hemolysis
	Hyperbilirubinemia

FAMILIAL TUMORAL CALCINOSIS

Hyperphosphatemic familial tumoral calcinosis (HFTC; OMIM #211900) sometimes referred to as hyperostosis–hyperphosphatemia syndrome is an autosomal recessive disorder with progressive deposition of calcium phosphate crystals in periarticular spaces and soft tissues. There are both hyperphosphatemic and normophosphatemic forms (NFTC; OMIM #610455) of this disorder.

Genetics

In the normophosphatemic form, mutations in sterile α motif domain-containing-9 protein (SAMD9; OMIM #610456) have been described. For the hyperphosphatemic form, inactivating mutations have been found in UDP-N-acetyl-α-D-galactosamine: polypeptide N-acetylgalactosaminyltransferase 3 (GALNT3; OMIM #601756), FGF23, and KLOTHO [55] (Fig. 88.1). The mutations lead to inadequate FGF23 protein levels or FGF23 action.

Clinical and biochemical manifestations

Patients with FTC have heterotopic calcifications that are typically painless. Pain and other clinical complications occur if there is infiltration of skin, marrow, teeth, blood vessels, and nerves. Range of motion is not affected unless the masses become large. Dental disease characterized by short bulbous roots, pulp stones, and radicular dentin deposited in swirls may be present. Biochemically, along with the hyperphosphatemia, there is increased $1,25(OH)_2D_3$ with normal calcium and alkaline phosphatase levels. Urinary phosphate excretion is frequently low. Radiographs show large aggregates of irregularly dense calcified lobules [56].

TREATMENT OF HYPERPHOSPHATEMIA

Treatment of hyperphosphatemia should address the underlying etiology. In acute exogenous phosphorus overload, prompt discontinuation of supplemental phosphorus and hydration allow for rapid correction of hyperphosphatemia. With transcellular shifts as in tumor lysis and rhabdomyolysis, dietary phosphorus restriction and diuresis are often successful. In diabetic ketoacidosis, treatment with insulin and treatment of acidosis reverses the hyperphosphatemia. In renal failure, phosphate binders (calcium salts, sevelamer, and lanthanum carbonate), along with dietary restriction, are indicated. Hemodialysis may be indicated in acute hyperphosphatemia in renal dysfunction.

In FTC, medical therapy with aluminum hydroxide along with dietary phosphate and calcium deprivation has been utilized.

Use of sevelamer, acetazolamide, sodium thiosulfate and IL-1 antagonists have been reported to successfully decrease tumor burden. Surgical intervention is an option in patients when the masses are painful, interfere with function, or are cosmetically unacceptable.

REFERENCES

1. Halevy J, Bulvik S. Severe hypophosphatemia in hospitalized patients. Arch Intern Med. 1988;148:153–5.
2. McCance R. Osteomalacia with Looser's nodes (milkman's syndrome) due to a raised resistance to vitamin d acquired about the age of 15 years. Q J Med. 1947;16:33–46.
3. Drezner MK. Tumor-induced osteomalacia. In: Favus MJ (ed.) *Primer on Metabolic Bone Diseases and Disorders of Mineral Metabolism*. Philadelphia: Lippincott-Raven, 1999, pp. 331–7.
4. Jan de Beur SM. Tumor-induced osteomalacia. JAMA. 2005;294:1260–7.
5. Kumar R. Tumor-induced osteomalacia and the regulation of phosphate homeostasis. Bone. 2000;27:333–8.
6. Folpe, AL, Fanburg-Smith JC, Billings SD, et al. Most osteomalacia-associated mesenchymal tumors are a single histopathologic entity: an analysis of 32 cases and a comprehensive review of the literature. Am J Surg Pathol. 2004;28:1–30.
7. De Beur SM, Finnegan RB, Vassiliadis J, et al. Tumors associated with oncogenic osteomalacia express genes important in bone and mineral metabolism. J Bone Miner Res. 2002;17:1102–10.

8. Sitara D, Razzaque MS, Hesse M, et al. Homozygous ablation of fibroblast growth factor-23 results in hyperphosphatemia and impaired skeletogenesis, and reverses hypophosphatemia in Phex-deficient mice. Matrix Biol. 2004;23:421–32.
9. Gattineni J, Bates C, Twombley K, et al. FGF23 decreases renal NaPi-2a and NaPi-2c expression and induces hypophosphatemia in vivo predominantly via FGF receptor 1. Am J Physiol Renal Physiol. 2009;297: F282–91.
10. Shimada T, Hasegawa H, Yamazaki Y, et al. FGF-23 is a potent regulator of vitamin D metabolism and phosphate homeostasis. J Bone Miner Res. 2004;19:429–35.
11. Strom, TM, Juppner H. PHEX, FGF23, DMP1 and beyond. Curr Opin Nephrol Hyperten. 2008;17:357–62.
12. Zhang Q, Doucet M, Tomlinson RE, et al. The hypoxia-inducible factor-1α activates ectopic production of fibroblast growth factor 23 in tumor-induced osteomalacia. Bone Res. 2016;4:1–6.
13. Jonsson KB, Zahradnik R, Larsson T, et al. Fibroblast growth factor 23 in oncogenic osteomalacia and X-linked hypophosphatemia. N Engl J Med. 2003;348:1656–63.
14. El-Maouche D, Sadowski SM, Papadakis GZ, et al. 68Ga-DOTATATE for tumor localizationin tumor-induced osteomalacia. J Clin Endocrinol Metab. 2016;101:3575–81.
15. Jan de Beur SM, Streeten EA, Civelek AC, et al. Localisation of mesenchymal tumours by somatostatin receptor imaging. Lancet. 2002;359:761–3.
16. Seufert J, Ebert K, Muller J, et al. Octreotide therapy for tumor-induced osteomalacia. N Engl J Med. 2001;345:1883–8.
17. Andreopoulou P, Dumitrescu CE, Kelly MH, et al. Selective venous catheterization for the localization of phosphaturic mesenchymal tumors. J Bone Min Res. 2011;26:1295–302.
18. Hesse E, Rosenthal H, Bastian L. Radiofrequency ablation of a tumor causing oncogenic osteomalacia. N Engl J Med. 2007;357:422–4.
19. Geller JL, Khosravi A, Kelly MH, et al. Cinacalcet in the management of tumor-induced osteomalacia. J Bone Miner Res. 2007;22:931–7.
20. Paglia F, Dionisi S, Minisola S. Octreotide for tumor-induced osteomalacia. N Engl J Med. 2002;346:1748–9; author reply 1748–9.
21. Basu, S, Fargose P. 177Lu-DOTATATE PPRT in recurrrent skull-base phosphaturic mesenchymal tumor causing paraneoplastic oncogenic osteomalacia: a potential therapeutic application of PRRT beyond neuroendrine tumors. J Nucl Med Technol. 2016;44:248–50.
22. Albright F, Butler A, Bloomberg E. Rickets resistant to vitamin D therapy. Am J Dis Child. 1939;54:529–47.
23. Winters RW, Graham JB, Williams TF, et al. A genetic study of familial hypophosphatemia and vitamin D resistant rickets with a review of the literature. Medicine. 1958;37:97–142.
24. The HYP Consortium. A gene (PEX) with homologies to endopeptidases is mutated in patients with X-linked hypophosphatemic rickets. Nat Genet. 1995;11:130–6.
25. Beck L, Soumounou Y, Martel J, et al. Pex/PEX tissue distribution and evidence for a deletion in the 3' region of the Pex gene in X-linked hypophosphatemic mice. J Clin Invest. 1997;99:1200–9.
26. Ruppe MD. X-linked hypophosphatemia. In: Pagen RA, Adam MP, Ardinger HH, et al. (eds) GeneReviews. Seattle: University of Washington, 2012 (updated 2014, 2016).
27. Carpenter TO, Imel EA, Holm IA, et al. A clinician's guide to X-linked hypophosphatemia. J Bone Miner Res. 2011;26:1381–8.
28. Connor J, Olear E, Insogna KL, et al. Conventional therapy in adults with X-linked hypophosphatemia effects enthesopathy and dental disease. J Clin Endocrinol Metab. 2015;100:3625–32.
29. Imel, EA, DiMeglio LA, Hui SL, et al. Treatment of X-linked hypophosphatemia with calcitriol and phosphate increases circulating fibroblast growth factor 23 concentrations. J Clin Endocrinol Metab. 2010;95:1846–50.
30. Souma N, Isakova T, Lipiszko D, et al. Fibroblast growth factor 23 and cause-specific mortality in the general population: the northen Manhattan study. J Clin Endocrinol Metab. 2016;101:3779–86.
31. Liu, ES, Carpenter TO, Gundberg CM, et al. Calcitonin administration in X-linked hypophosphatemia. N Engl J Med. 2011;364:1678–80.
32. Carpenter TO, Imel EA, Ruppe MD, et al. Randomized trial of the anti-FGF23 antibody KRN23 in X-linked hypophosphatemia. J Clin Invest. 2014;124:1587–97.
33. Imel EA, Zhang X, Ruppe MD, et al. Prolonged correction of serum phosphorus in adults with X-linked hypophosphatemia using monthly doses of KRN23. J Clin Endocrinol Metab. 2015;100:2565–73.
34. Ruppe MD, Zhang X, Imel EA, et al. Effect of four montly dose of a human monoclonal anti-FGF23 antibody (KRN23) on quality of life in X-linked hypophosphatemia. Bone Reports. 2016;5:158–62.
35. Wilson DR, York SE, Jaworski ZF, et al. Studies in hypophosphatemic vitamin D-refractory osteomalacia in adults. Medicine. 1965;44:99–134.
36. The ADHR Consortium. Autosomal dominant hypophosphataemic rickets is associated with mutations in FGF23. Nat Genet. 2000;26:345–8.
37. Shimada T, Muto T, Urakawa I, et al. Mutant FGF-23 responsible for autosomal dominant hypophosphatemic rickets is resistant to proteolytic cleavage and causes hypophosphatemia in vivo. Endocrinology. 2002;143: 3179–82.
38. Imel, EA, Hui SL, Econs MJ. FGF23 concentrations vary with disease status in autosomal dominant hypophosphatemic rickets. J Bone Miner Res. 2007;22:520–6.
39. Kapelari K, Köhle J, Kotzot D et al. Iron supplementation associated with loss of phenotype in autosomal dominant hypophaetmic rickets. J Clin Endocrinol Metab. 2015;100:3388–92.
40. Imel, EA, Peacock M, Gray AK, et al. Iron modifies plasma FGF23 differently in autosomal dominant hypophosphatemic rickets and healthy humans. J Clin Endocrinol Metab. 2011;96:3541–9.

41. Bergwitz C, Roslin NM, Tieder M, et al. SLC34A3 mutations in patients with hereditary hypophosphatemic rickets with hypercalciuria predict a key role for the sodium-phosphate cotransporter NaPi-IIc in maintaining phosphate homeostasis. Am J Hum Genet. 2006;78:179–92.
42. Lorenz-Depiereux B, Benet-Pages A, Eckstein G, et al. Hereditary hypophosphatemic rickets with hypercalciuria is caused by mutations in the sodium-phosphate cotransporter gene SLC34A3. Am J Hum Genet. 2006;78:193–201.
43. Lorenz-Depiereux B, Bastepe M, Benet-Pages A, et al. DMP1 mutations in autosomal recessive hypophosphatemia implicate a bone matrix protein in the regulation of phosphate homeostasis. Nat Genet. 2006;38:1248–50.
44. Levy-Litan V, Hershkovitz E, Avizov L, et al. Autosomal-recessive hypophosphatemic rickets is associated with an inactivation mutation in the ENPP1 gene. Am J Hum Genet. 2010;86:273–8.
45. Lorenz-Depiereux, B, D Schnabel, D Tiosano, et al. Loss-of-function ENPP1 mutations cause both generalized arterial calcification of infancy and autosomal-recessive hypophosphatemic rickets. Am J Hum Genet. 2010;86:267–72.
46. Steichen-Gersdorf E, Lorenz-Depiereux B, Stron TM et al. Early onset hearing loss in autosomal recessive hypophosphatemic rickets caused by loss of function mutations in ENPP1. J Pediatr Endocrinol Metab. 2015;7–8:967–70.
47. Boyce AM, Collins MT. Fibrous dysplasia/McCune-Albright syndrome. In: Pagen RA, Adam MP, Ardinger HH, et al. (eds) *GeneReviews*. Seattle: University of Washington, 2015.
48. Ovejaro D, Lim YH, Boyce AM, et al. Cutaneous skeletal hypophosphatemia syndrome: clincal spectrum, natural history and treatment. Osteoporos Int. 2016;12:3615–26.
49. Brownstein, CA, Adler F, Nelson-Williams C, et al. A translocation causing increased alpha-klotho level results in hypophosphatemic rickets and hyperparathyroidism. Proc Natl Acad Sci U S A. 2008;105:3455–60.
50. White, KE, Cabral JM, Davis SI, et al. Mutations that cause osteoglophonic dysplasia define novel roles for FGFR1 in bone elongation. Am J Hum Genet. 2005;76:361–7.
51. Prie D, Huart V, Bakouh N, et al. Nephrolithiasis and osteoporosis associated with hypophosphatemia caused by mutations in the type 2a sodium-phosphate cotransporter. N Engl J Med. 2002;347:983–91.
52. Karim Z, Gerard B, Bakouh N, et al. NHERF1 mutations and responsiveness of renal parathyroid hormone. N Engl J Med. 2008;359:1128–35.
53. Magen D, Berger L, Coady MJ, et al. A loss-of-function mutation in NaPi-IIa and renal Fanconi's syndrome. N Engl J Med. 2010;362:1102–9.
54. Kinoshita Y, Hori M, Taguchi M, et al. Functional analysis of mutant FAM20C in Raine syndrome with FGF23-related hypophosphatemia. Bone. 2014;67:145–51.
55. Folsom LJ, Imel EA. Hyperphosphatemic familial tumoral calcinosis: genetic models of deficienct FGF23 action. Curr Osteoporos Rep. 2015;2:78–87.
56. Ramnitz MS, Gourh P, Goldbach-Mansky R, et al. Phenotypic and genotypic characterization and treatment of a cohort with familial tumoral calcinosis/hyperostosis-hyperphosphatemia syndrome. J Bone Miner Res. 2016;10:1845–54.

89
Rickets and Osteomalacia

Michaël R. Laurent[1,2], Nathalie Bravenboer[3], Natasja M. Van Schoor[4], Roger Bouillon[5], John M. Pettifor[6], and Paul Lips[7]

[1]Centre for Metabolic Bone Diseases, University Hospitals Leuven, Leuven, Belgium
[2]Gerontology and Geriatrics, Department of Clinical and Experimental Medicine, KU Leuven, Leuven, Belgium
[3]Department of Clinical Chemistry, VU University Medical Center, MOVE Research Institute, Amsterdam, The Netherlands
[4]Amsterdam Public Health Research Institute, Department of Epidemiology and Biostatistics, VU University Medical Center, Amsterdam, The Netherlands
[5]Laboratory of Clinical and Experimental Endocrinology, Department of Chronic Diseases, Metabolism and Ageing, KU Leuven, Leuven, Belgium
[6]MRC Developmental Pathways for Health Research Unit, Department of Pediatrics, Faculty of Health Sciences, University of the Witwatersrand, Johannesburg, South Africa
[7]Department of Internal Medicine, Endocrine Section, VU University Medical Center, Amsterdam, The Netherlands

INTRODUCTION

Rickets and osteomalacia (from Greek *osteon* and *malakia*, bone softness) are diseases characterized by hypomineralization of bone matrix. Rickets occurs only in children (before epiphyseal closure) and additionally leads to abnormal growth plate development, stunting, and bone deformities. The underlying causes of both disorders are related to severe deficiency of vitamin D, calcium, phosphate, and/or to direct inhibition of the mineralization process. Their histological hallmark is hyperosteoidosis and a prolonged mineralization lag time. Clinical features, which in adults may be misdiagnosed as osteoporosis, frequently include low BMD and sometimes fractures. However, bone pain, muscle weakness, and biochemical abnormalities (particularly raised serum alkaline phosphatase, ALP) should raise suspicion of the diagnosis of osteomalacia. Treatment of osteomalacia and rickets is aimed at correcting the underlying mineral deficit. In this chapter, we will review the epidemiology, etiology, pathophysiology, clinical features, diagnosis, and treatment of rickets and osteomalacia.

DEFINITION

Nutritional rickets can be defined as "a disorder of defective chondrocyte differentiation and mineralization of the growth plate and defective osteoid mineralization, … caused by vitamin D deficiency and/or low calcium intake in children" [1]. Rickets related to calcium and/or vitamin D deficiency is also termed "calciopenic" rickets, to distinguish it from hypophosphatemic ("phosphopenic") forms which are primarily caused by phosphate deficiency (see Chapter 88).

Osteomalacia is defined by the presence of both hyperosteoidosis and delayed mineralization. However, this histological definition is problematic because: (i) the exact histomorphometric criteria for osteomalacia remain somewhat debated (see Bone histology later);

Primer on the Metabolic Bone Diseases and Disorders of Mineral Metabolism, Ninth Edition. Edited by John P. Bilezikian.
© 2019 American Society for Bone and Mineral Research. Published 2019 by John Wiley & Sons, Inc.
Companion website: www.wiley.com/go/asbmrprimer

(ii) bone biopsy is an invasive procedure; and (iii) there is a lack of availability and expertise to perform bone histomorphometry in most clinical settings.

EPIDEMIOLOGY

Rickets

In the middle of the seventeenth century, rickets became known in Europe as the English disease. During the industrial revolution, high levels of atmospheric pollution, narrow sunless streets, child labor, and poor diets all contributed to the development of vitamin D deficiency and probably dietary calcium deficiency. In the late nineteenth and early twentieth centuries, rickets was almost ubiquitous among underprivileged infants in industrialized regions in the USA and Europe [2]. With the discovery of vitamin D and its role in the prevention and treatment of rickets, nutritional rickets was almost eradicated in the USA and a number of countries in Europe. Recently, however, the prevalence of rickets in these regions has increased, particularly among children of immigrants from the Indian subcontinent, the Middle East and Africa, and among African-Americans; amongst all of these groups maternal vitamin D deficiency during pregnancy and lactation is common [3–5].

Studies from Africa and the Indian subcontinent have highlighted the role of low dietary calcium intakes in the pathogenesis of rickets. Calcium intakes of affected children are typically <300 mg/day. Diets in such (sub)tropical countries are characteristically free of dairy products and high in phosphate [6].

Osteomalacia

In adults in Asia and the Middle East, calcium intake often is low and severe vitamin D deficiency is surprisingly common, particularly in subjects with little sunshine exposure [7]. Severe vitamin D deficiency is associated with low BMD, raised serum ALP, and secondary hyperparathyroidism and/or hypocalcemia, although only a minority of those with severe vitamin D deficiency develop osteomalacia [7, 8]. One study estimated a 2% to 3.6% prevalence of clinically diagnosed osteomalacia in young adult women in Pakistan [9].

In Western countries, dark-skinned and/or veiled immigrants remain at high risk of developing severe vitamin D deficiency, low BMD, and raised ALP [10,11]. This is of particular concern in pregnant women [12] because of the risk of severe vitamin D deficiency in their newborns, which may rarely lead to congenital rickets. The elderly are another risk group for osteomalacia, with a 2% to 5% prevalence reported in older studies [13, 14]. Osteoporotic fractures in the elderly were historically associated with osteomalacia but currently this is rarely the case [15, 16]. Nevertheless, a large German autopsy series found a 4.9% prevalence of hyperosteoidosis (>5% osteoid volume/bone volume) [17]. The exact level of 25-hydroxyvitamin D (25OHD) below which rickets and osteomalacia develop remains unclear, which may in part be because of inaccuracy of immunoassays at low 25OHD concentrations. Finally, osteomalacia is commonly found in patients suffering from gastrointestinal disorders such as celiac disease [18]. Also, after gastric bypass surgery, the majority of patients are at risk of developing vitamin D deficiency, calcium malabsorption, and secondary hyperparathyroidism [19, 20]. However, only ~25% of bariatric patients in whom osteomalacia is clinically suspected will be confirmed using rigorous histomorphometric criteria [21, 22].

ETIOLOGY

The causes of osteomalacia and rickets are similar and can be classified according to underlying mechanisms: (i) vitamin D deficiency or resistance; (ii) calcium deficiency independent of vitamin D; (iii) hypophosphatemic disorders; and (iv) mineralization inhibitors (Table 89.1). Additionally, metabolic acidosis in certain gastrointestinal or renal diseases may contribute to impaired mineralization. More recently, the osteocyte-derived hormone FGF23 has been implicated in specific forms of rickets or osteomalacia (see Chapter 88). Finally, unmineralized bone matrix may be observed in connective tissue disorders such as type VI osteogenesis imperfecta [23]. Similarly, the ultra-rare fibrogenesis imperfecta ossium and axial osteomalacia feature osteomalacia on bone biopsy, although their clinical presentation is that of sclerosing bone disorders [24, 25] (see Chapter 107).

PATHOPHYSIOLOGY

Rickets

The mechanisms underlying growth plate abnormalities in rickets are distinct from those involved in hypomineralization of osteoid. Hypophosphatemia plays a key role in the pathogenesis of the growth plate abnormalities, which are characterized histologically by an increase in the width of the zone of hypertrophied chondrocytes with disruption of the columnar pattern, an accumulation of growth plate cartilage, a failure of mineralization of the cartilage, and a lack of vascular invasion. Normal phosphate levels are required for normal growth plate maturation through regulation of chondrocyte apoptosis and matrix mineralization [27].

In pediatric practice there is no global consensus on the serum 25OHD cut-offs to define vitamin D status, although a recent consensus statement recommends classifying serum 25OHD lower than 30 nmol/L as deficient and 30 to 50 nmol/L as insufficient [1]. In the same consensus, dietary calcium intake lower than 300 mg/day during childhood and adolescence is defined as deficient.

Table 89.1. Causes of rickets and osteomalacia.

Vitamin D-related rickets/osteomalacia
Severe vitamin D deficiency:
- Low sunshine exposure, low dietary intake
- Malabsorption (calcium absorption may also be impaired): bariatric surgery (derivative types), bowel resection, celiac disease, cholestatic liver disease, exocrine pancreatic insufficiency, gastrectomy, inflammatory bowel disease
- Impaired hepatic 25-hydroxylation: severe cirrhosis (rare)
- Impaired renal 1α-hydroxylation: chronic kidney disease (renal osteomalacia), hypoparathyroidism
- Increased renal losses: nephrotic syndrome
- Increased catabolism: enzyme-inducing drugs (anticonvulsants, rifampicin, St John's wort)

Vitamin D-dependent or -resistant rickets:
- Type 1A: 1α-hydroxylase (CYP27B1) deficiency (OMIM #264700)
- Type 1B: 25-hydroxylase (CYP2R1) deficiency (OMIM #600081)
- Type 2A: hereditary vitamin D-resistant rickets (VDR mutations) (OMIM #277440)
- Type 2B: vitamin D-dependent rickets with normal VDR (hnRNP overexpression) (OMIM #600785)
- Type 3: activating CYP3A4 mutation [26]

Calcium deficiency (with normal vitamin D status)
Nutritional: very low dietary calcium intake
Calcium malabsorption (similar causes as vitamin D malabsorption, see above)
Hypercalciuria: in combination with renal phosphate wasting (see below)

Hypophosphatemic rickets/osteomalacia
Gastrointestinal causes: poor nutritional intake (eg, breastfed very low birth weight infants), chronic diarrhea, excessive phosphate binders
Renal phosphate wasting:
- Tumor-induced (oncogenic) osteomalacia
- Fanconi syndrome (medications like tenofovir, adefovir, ifosfamide, light chain gammopathy, Sjögren syndrome)
- X-linked dominant hypophosphatemic rickets (PHEX mutations) (OMIM #307800)
- X-linked recessive hypophosphatemic rickets (CLCN5 mutations) (OMIM #300554)
- Autosomal dominant hypophosphatemic rickets (FGF23 mutations) (OMIM #193100)
- Autosomal recessive hypophosphatemic rickets type 1 (DMP1 mutations) (OMIM #241520), type 2 (ENPP1 mutations) (OMIM #613312)
- Hereditary hypophosphatemic rickets with hypercalciuria (SLC34A3) (OMIM #241530)
- Dent disease-1 (CLCN5 mutations) (OMIM #300009,) Dent disease-2 (OCRL mutations) (OMIM #300555), Lowe oculocerebrorenal syndrome (OMIM #309000)

Mineralization inhibitors
Metabolic acidosis: renal insufficiency, renal tubular acidosis (± renal phosphate wasting), ileostomy and urinary diversion (also intestinal calcium losses)
Aluminum toxicity (eg, from antacids, dialysis fluid)
Fluorosis (including endemic fluorosis from borehole water)
Iron (in dialysis patients, or with FGF23 mediated hypophosphatemia)
Etidronate overdose (in Paget disease)
Environmental intoxication with cadmium (Itai Itai disease, ± renal phosphate wasting), strontium, etc.
Hypophosphatasia (inorganic pyrophosphate accumulation) (OMIM #146300)

Matrix abnormalities
Type VI osteogenesis imperfecta (SERPINF1 mutations) (OMIM #613982)
Fibrogenesis imperfecta ossium
Axial osteomalacia

Osteomalacia

Vitamin D, through its active metabolite 1,25-dihydroxyvitamin D [1,25(OH)$_2$D] maintains plasma calcium concentrations primarily by promoting intestinal calcium absorption [28, 29]. Vitamin D deficiency and decreased calcium absorption (caused by low intake and/or low fractional absorption) as may occur in the elderly, provoke secondary hyperparathyroidism (on average ~15% increase in PTH, Table 89.2) and increased bone turnover, whereas 1,25(OH)$_2$D, plasma calcium, and calciuria usually remain normal (Fig. 89.1A).

Table 89.2. Proposed classification of vitamin D-deficient status.

	25OHD (nmol/L)	1,25(OH)$_2$D	PTH Increase	Bone Histology
Severe deficiency	<12.5	(Relatively) low	↑↑↑	Incipient or overt osteomalacia
Deficiency	12.5–25	Normal	↑	High turnover, risk of osteomalacia
Insufficiency	25–50	Normal	= or (↑)	Normal or high turnover
Replete	>50	Normal	=	Normal

Source: [14]. Reproduced with permission of Oxford University Press.

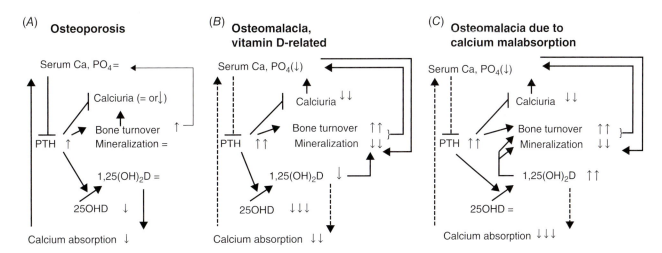

Fig. 89.1. Pathophysiology of calcium and phosphate homeostasis in different calcium- and/or vitamin D-deficient states. (A) In osteoporosis (eg, in the elderly), moderate vitamin D deficiency and/or impaired calcium absorption lead to high–normal PTH levels, which increases bone turnover whereas 1,25(OH)$_2$D and calciuria usually remain in the normal range. (B) In vitamin D deficiency-related rickets/osteomalacia, severe 25OHD deficiency compromises 1,25(OH)$_2$D and intestinal calcium absorption, leading to low–normal or frankly decreased serum calcium and phosphate and more pronounced secondary hyperparathyroidism, which partially restores 1,25(OH)$_2$D but also increases bone resorption, reduces mineralization and calciuria. (C) In rickets/osteomalacia with normal 25OHD and primarily impaired calcium absorption, low–normal or frankly decreased serum calcium and phosphate also trigger more pronounced secondary hyperparathyroidism, which together with increased 1,25(OH)$_2$D increases bone resorption, reduces mineralization, and calciuria.

In severe vitamin D and/or calcium deficiency, however, 1,25(OH)$_2$D and intestinal calcium absorption are compromised [29]. The calciotropic hormones then maintain plasma calcium concentrations at the expense of skeletal integrity, because hypocalcemia dysregulates vital physiology like neuromuscular and cardiac excitation and blood coagulation. In rickets and osteomalacia, secondary hyperparathyroidism is more pronounced, which contributes to hypocalciuria, hypophosphatemia, and increased bone resorption. Vitamin D also promotes mineralization by osteoblasts in vitro. Thus, impaired mineralization in severe vitamin D deficiency may be mediated by: (i) compromised 1,25(OH)$_2$D; (ii) hypocalcemia; and (iii) hypophosphatemia (Fig. 89.1B). However, patients with very low intestinal calcium absorption may also develop osteomalacia despite normal serum 25OHD concentrations. Indeed, correction of vitamin D deficiency alone may be insufficient to restore calcium absorption, not only in gastrointestinal diseases or after bariatric surgery [30] but also in nutritional osteomalacia and rickets [31, 32]. In situations of low calcium absorption with (relatively) normal serum 25OHD concentrations, 1,25(OH)$_2$D not only increases bone resorption and calcium release from bone but also upregulates mineralization inhibitors, reducing calcium incorporation into bone (Fig. 89.1C) [33]. Calcemia can be maintained in the low–normal range in osteomalacia by: (i) net efflux of calcium from skeletal reservoirs by increased resorption and decreased mineralization; (ii) very efficient renal calcium handling; and (iii) any small remaining amount of calcium absorbed from the intestine. However, inadvertent treatment with antiresorptive or osteoanabolic drugs may provoke hypocalcemia.

CLINICAL PICTURE

Rickets

The clinical features of rickets depend on the age of presentation. They typically include decreased longitudinal growth, widening of the metaphyseal zones (Fig. 89.2), and painful swelling around these zones [34].

Rickets in the first 3 months of life (congenital rickets) is very rare, because the fetus is to a large extent protected from the effects of maternal vitamin D deficiency. During the first 6 months of life, clinical vitamin D deficiency typically presents with symptoms of hypocalcemia (convulsions, apneic spells, twitching), rather than bone deformities (see Chapter 91). Classical vitamin D-deficiency rickets with bone deformities is commonest between 6 and 18 months of age, although the disease may occur in older infants and children with a second peak occurring during the adolescent growth spurt.

In infants older than 6 months, skull deformities with delayed closure of the sutures, enlargement of the costochondral junctions (rachitic rosary), chest deformities such as Harrison's sulcus and the violin case deformity, widening of the wrists, and bone deformities of the appendicular skeleton become prominent. The types of deformity depend on the motor development of the child and on which bones are weight bearing at that time. Associated with these bone signs are delayed eruption of teeth, hypotonia and delay in motor developmental milestones, increased sweating, and a propensity to lower respiratory tract infections. Children who develop rickets later, for example those with low dietary calcium intakes or X-linked hypophosphatemic rickets, tend to have more severe lower limb deformities than chest or upper limb deformities.

Osteomalacia

The clinical features of osteomalacia include bone pain, muscular weakness, and difficult walking. The muscular weakness is mainly localized in the proximal muscles around the shoulder and pelvic girdles. It manifests with difficulties in standing up from a chair or stair climbing. It may be so severe that the patient is completely bedridden. When plasma calcium concentrations are very low, symptoms of hypocalcemia such as occult or overt tetany and convulsions, may prevail. Fractures may occur, including stress fractures, but also typically "osteoporotic" fractures such as femoral or vertebral fractures.

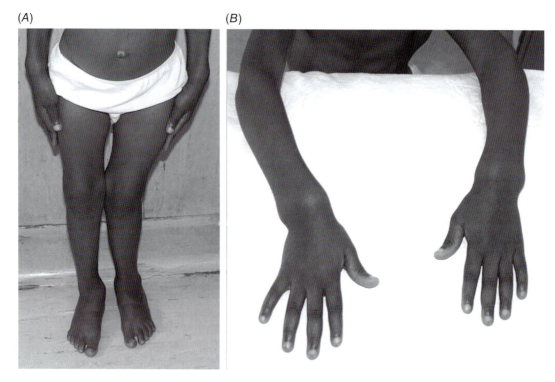

Fig. 89.2. Clinical findings in rickets. (A) Windswept deformities in the legs of a young child. Note the valgus deformity at the left knee and the varus deformity at the right knee. (B) Obvious clinical widening of the metaphyseal region of the distal forearms in a young child with severe rickets.

DIAGNOSIS

A high index of suspicion for clinical, biochemical, or radiographic features of osteomalacia or rickets should be maintained in patients with known predisposing conditions (Table 89.1). Clinical signs and symptoms alone are highly sensitive but unfortunately not specific for diagnosis [35]. Bone biopsy remains the gold standard to diagnose osteomalacia, but this is problematic as explained previously (Definition). There is great need for clinical definitions for probable or definite rickets and osteomalacia, such as recently suggested in Japanese guidelines [36]. A similar clinical scoring system for screening children (>18 months of age) with suspected dietary calcium deficiency rickets has also been proposed [37].

Laboratory findings

Patients with rickets and osteomalacia related to vitamin D deficiency typically have very low serum 25OHD concentrations, that is below 15 to 30 nmol/L [29]. However, only a subgroup of subjects with severe vitamin D deficiency develops rickets or osteomalacia whereas others may develop rickets/osteomalacia while having normal 25OHD concentrations, when calcium intake or intestinal absorption is severely compromised [6, 31]. Increased serum ALP activity is the best single predictor of biopsy-proven osteomalacia [35, 38]. However, other causes of increased ALP concentrations should be excluded, for example liver cholestasis, Paget disease, or bone metastasis. A combination of raised ALP and raised PTH with low calcium or phosphate has the best diagnostic properties [35, 38]. Hypocalcemia in osteomalacia is a late feature; most subjects will have low–normal serum calcium or intermittent hypocalcemia. Secondary hyperparathyroidism is common [35] but not specific. In infants, hypocalcemia may be the presenting feature early in the disease; however, this may disappear as secondary hyperparathyroidism increases calcium mobilization from bone and reduces renal losses. In the last stages of rickets, hypocalcemia reappears caused by decompensation of the calcium homeostatic mechanisms [39]. Hypophosphatemia may cause or contribute to rickets and osteomalacia, particularly in phosphate-related conditions (see Table 89.1). The fractional tubular reabsorption of phosphate (TRP) and the renal phosphate threshold (TmP/GFR) from a fasting 2-hour morning urine (second morning void) and a blood sample at the same time should be assessed in chronic hypophosphatemia to distinguish between renal phosphate wasting or gastrointestinal losses. Hypocalciuria (defined as <100 mg but often <50 mg Ca^{2+}/g creatinine on a 24-hour urine collection, or <2 mg/kg body weight/24-h in children) is a sensitive indicator of calcium deficit and may be used to monitor therapy. Although bone turnover markers are usually increased, this does not contribute to the diagnostic process, and neither do measurements of $1,25(OH)_2D$ (except in vitamin D-dependent or -resistant rickets).

Radiographic findings

Rickets

A recent consensus statement recommends the use of radiographic changes at the growth plates of the wrist and knee to confirm the diagnosis of rickets [1]. Early changes of rickets include radiolucent bones with thin cortices and loss of the normal trabecular pattern. Loss of the calcification zone and widening of the growth plate ensues. As the disease becomes more severe, the growth plate widens with fraying of the metaphyseal border (Fig. 89.3). The growth plate abnormalities are associated with delayed epiphyseal development and closure (delayed bone age) and deformities of the shafts of the long bones (especially those that are weight bearing). The Rickets Severity Score (RSS) is a validated scale to assess the radiographic abnormalities [40]. In the infant and young child varus deformities of the lower limbs are common, but in older children, valgus or windswept deformities of the lower limb become more frequent. In longstanding and severe calciopenic rickets, radiographic features of secondary hyperparathyroidism may be found. Fractures may be associated with severe rickets in children, and Looser zones may also be noted (see next section).

Osteomalacia

In osteomalacia, radiographs show less contrast and are less sharp as if the patient had moved during the X-ray. The classical sign of osteomalacia is the pseudofracture or Looser zone. It is a radiolucent line through one cortical plate, often with sclerosis at the margins (Fig. 89.4).

Bone scintigraphy may show multiple hot spots at the ribs, pelvis, in metaphyseal regions, and at pseudofracture sites, which may be mistaken for bone metastases. BMD T-scores may be (very) low, similar to or more severe than in osteoporosis [18]. The amount of nonmineralized bone (osteoid) may be high, leading to large and rapid increases of BMD (even more than 50%) when appropriate therapy has been instituted.

Bone histology

A transiliac bone biopsy after tetracycline double labeling (see Chapter 39) provides a definitive diagnosis of osteomalacia or can exclude it. Although there are no formal guidelines, most experts agree that osteomalacia requires the presence of three criteria: osteoid volume greater than 10%, uncorrected osteoid thickness greater

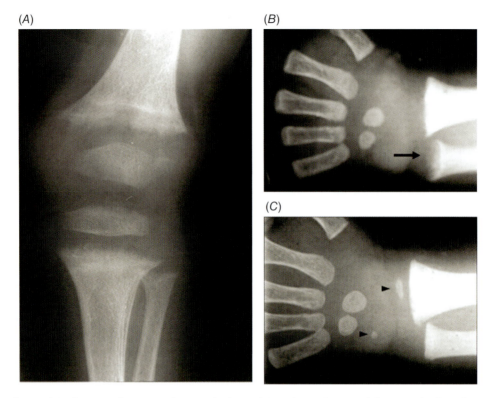

Fig. 89.3. (A) Radiographic changes of severe rickets at the knee. Note the widening of the growth plate, loss of provisional zone of calcification and fraying and splaying at the metaphysis. Subperiosteal new bone formation is apparent along the tibial cortex. (B) Florid nutritional rickets in a 1.5-year-old child. Wrist X-ray shows typical blurry concave margin of the ulna (arrow). (C) Healing of rickets and appearance of new ossification centers (arrow heads) after initiation of treatment.

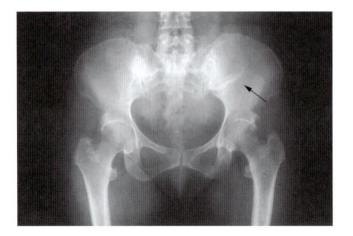

Fig. 89.4. Pseudofracture (arrow) in the left os ilium in a patient with osteomalacia.

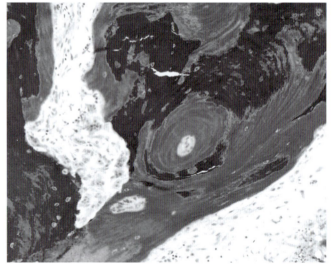

Fig. 89.5. Bone histology findings in osteomalacia in a patient with celiac disease. Goldner stain, mineralized bone is black, osteoid tissue is grey. Besides thick osteoid seams, increased bone resorption by multinucleated osteoclasts is visible.

than 15 μm (or >12.5 μm after correction for obliquity), and mineralization lag time of more than 100 days [41]. Tetracycline double labels are usually absent, the osteoid surface is extended to more than 70%, and osteoid seams contain more than four lamellae. Bone resorption is usually increased. Resorption lacunae are visible with many multinucleated osteoclasts (Fig. 89.5). The mineralization index, based on measurements of osteoid thickness, osteoid volume, mineralization rate, and bone formation rate, may be helpful [42].

TREATMENT AND PREVENTION

The treatment of rickets and osteomalacia primarily involves identification and treatment of the underlying causes.

Treatment of rickets

Calciopenic rickets should be treated with both vitamin D and calcium. Vitamin D deficiency rickets responds to small doses of vitamin D, but it is recommended that infants younger than 12 months of age should receive 2000 IU/day for 3 months, whereas older children should receive 3000 to 6000 IU/day [1]. Oral calcium supplements of at least 500 mg/day are recommended, although doses of 1000 to 2000 mg/day may produce more rapid radiographic healing of rickets [43]. In cases where the child is symptomatically hypocalcemic, a calcium loading dose should be given intravenously followed by a slow infusion of maintenance calcium until symptomatic hypocalcemia is controlled. Nutritional rickets also responds to vitamin D boluses, which may be useful in situations where compliance cannot be assured. Oral bolus therapy is preferred over intramuscular administration. The dosages recommended by expert opinion are as follows: age 3 to 12 months 50,000 IU; 1 to 12 years 150,000 IU; and older than 12 years 300,000 IU as single doses [1]. Larger boluses are not recommended because of the risk of developing hypercalcemia. Although there are some pharmacological reasons (longer half-life) to suggest that vitamin D_3 might be preferable to vitamin D_2 for the management of nutritional rickets, there is no clinical evidence to suggest that vitamin D_2 is inferior to vitamin D_3 in the treatment or prevention of rickets.

Prevention of rickets

Vitamin D deficiency rickets is eminently preventable, yet vitamin D deficiency and its clinical consequences remain a major public health problem throughout the world. The recent global consensus recommends that all infants younger than 12 months of age should receive vitamin D supplements at a dose of 400 IU/day (irrespective of the method of feeding). Thereafter all children and adults need to meet their nutritional requirements for vitamin D (600 IU/day) through diet or supplements [1, 44]. Specific attention should be paid to children in at-risk groups to ensure that their requirements are met. In regions in which vitamin D deficiency and nutritional rickets are common, governments should give serious consideration to fortifying commonly used foods with vitamin D.

Treatment of osteomalacia

Nutritional osteomalacia may be treated with remarkably low doses of calcium (eg, 1000 mg of elementary calcium) and vitamin D (eg, 800 to 1200 IU/day). A higher initial vitamin D dose (eg, 2000 IU/day) or a loading dose (eg 25,000 to 50,000 IU once) may be considered to bring the serum 25OHD concentration into the sufficient range more quickly, but these are not required for maintenance. A calcium supplement should always be added [31, 33]. It is usually prescribed as calcium citrate or carbonate. Although calcium citrate may be better absorbed in certain conditions, for example in patients with achlorhydria, calcium carbonate may have advantages, for example when metabolic acidosis is simultaneously present.

In contrast, when gastrointestinal absorption is severely impaired, some patients may require high to very high doses of oral calcium (eg, 1000 to 4000 mg/day) and vitamin D (eg, 4000 to 10,000 IU/day of vitamin D_3). In these circumstances it is important to monitor tolerance, compliance, and biochemical response to therapy more frequently to avoid both over- and undertreatment. High doses of calcium are often poorly tolerated, particularly after bariatric surgery. Increased sunlight exposure or UV lamp therapy may provide an additional source of vitamin D [45], although compliance may be more difficult [46, 47]. Megadoses of vitamin D (eg, a 500,000 IU intramuscular or oral dose) should be avoided if possible because of the risk of hypervitaminosis D in the first months, which may be associated with an increased risk of falls and fractures [48]. In countries where it is available, calcifediol (25OHD) may be useful in patients with fat malabsorption, because it is a more polar metabolite absorbed mainly via the portal venous system. This also applies to calcitriol and alphacalcidol, but these should generally be avoided in disorders in which 1α-hydroxylation is normal because of their narrower therapeutic range. The underlying cause should always be treated if possible, for example adherence to a gluten-free diet in celiac disease.

The aim of treatment is to improve symptoms, normalize plasma calcium and phosphate concentrations, correct secondary hyperparathyroidism, and to bring serum 25OHD within the target range of 20 to 30 ng/mL (50 to 75 nmol/L). Regular monitoring is recommended during follow-up. ALP may initially increase but gradually normalizes over several months, as do hypocalciuria and imaging abnormalities (in contrast to rickets where radiological improvement may be more rapid). BMD may improve quickly and substantially, although a certain degree of osteopenia may remain in women after menopause or in older men. Importantly, antiresorptive or osteoanabolic drugs are contraindicated because they may precipitate hypocalcemia.

HEREDITARY VITAMIN D-DEPENDENT AND RESISTANT RICKETS

The observation that some forms of rickets could not be cured by regular doses of vitamin D led to the discovery of rare inherited abnormalities of vitamin D metabolism or the vitamin D receptor (VDR). These inborn errors of

Hereditary vitamin D-dependent rickets (HVDDR) type 1A

After the discovery of 1,25(OH)$_2$D, it became apparent that some patients with congenital rickets had very low serum concentrations of 1,25(OH)$_2$D that did not increase after vitamin D supplementation [49]. It turned out that these patients have inactivating mutations of the 1α-hydroxylase gene. The disease is autosomal recessive. Patients with vitamin D-dependent rickets type 1A generally present during infancy with rickets and signs of hypocalcemia, tetany, or convulsions. Serum 25OHD is within normal range, but serum 1,25(OH)$_2$D is very low or undetectable. Radiologic examination and bone biopsy show features indistinguishable from nutritional rickets. These patients should be treated with 1,25(OH)$_2$D or 1α-hydroxyvitamin D to restore serum 1,25(OH)$_2$D to normal levels, which cures their rickets within a few months.

Hereditary vitamin D-dependent rickets type 1B

This type of rickets/osteomalacia is caused by a lack of 25-hydroxylase activity in the liver (CYP2R1 mutation) and can be cured by treatment with low doses of 25OHD [50].

Hereditary vitamin D-dependent rickets type 2A

True resistance to 1,25(OH)$_2$D was discovered as some children with hereditary rickets did not respond to treatment with calcitriol [51]. In fact, these children had a high serum 1,25(OH)$_2$D, leading to the suspicion of a (post)receptor defect. After cloning of the VDR, mutations have been identified at the DNA binding domain, the ligand binding domain, and other domains [52]. Hereditary vitamin D-dependent rickets type 2A is an autosomal recessive disease.

Affected children born to normal heterozygote parents present early in life with rickets and signs of hypocalcemia, including tetany and convulsions. Most kindreds have associated alopecia, but the disease may rarely occur in children without alopecia. The degree of hormone resistance (ie, hypocalcemia), may also vary. Serum PTH is increased and serum 1,25(OH)$_2$D is markedly elevated.

The success of treatment is variable and depends on the degree of hormone resistance. When some VDR function is present, a pharmacological dose of calcitriol or alphacalcidol can improve calcium absorption and heal the rickets. In case of complete resistance, active vitamin D metabolites are ineffective. Such severely affected individuals can be treated with calcium infusion that overcomes the defective calcium absorption [53]. Because calcium absorption from the intestine also has a passive component by diffusion — which is independent of vitamin D — very high doses of oral calcium can also be effective.

Hereditary vitamin D-dependent rickets type 2B

This rare variant of HVDDR is caused by overexpression of heterogeneous nuclear ribonucleoproteins (hnRNPs) which act as vitamin D response element-binding proteins, exerting a dominant-negative influence on VDR-mediated gene transactivation [54]. The disorder mimics the clinical and biochemical findings of type 2A, but without VDR mutations.

Hereditary vitamin D-dependent rickets type 3

Type 3 HVDDR has recently been recognized in children from two unrelated families with vitamin-D responsive rickets caused by an identical, heterozygous activating mutation in CYP3A4, which enhances 25OHD catabolism in a similar way as enzyme-inducing drugs do [26]. This appears to be an autosomal dominant disorder, treatable using high doses of 25OHD.

CONCLUSION

Rickets and osteomalacia are not "forgotten" disorders as they are commonly perceived, even in developed countries. More research into their modern epidemiology, diagnostic criteria, and treatment is required, particularly in populations such as immigrant children, the elderly, bariatric surgery patients, and rare genetic forms of these diseases. For the diagnosis of osteomalacia, the availability of bone biopsies should be increased, but at the same time noninvasive diagnostic alternatives are a high research priority [55].

REFERENCES

1. Munns CF, Shaw N, Kiely M, et al. Global consensus recommendations on prevention and management of nutritional rickets. J Clin Endocrinol Metab. 2016;101: 394–415.
2. Rajakumar K. Vitamin D, cod-liver oil, sunlight, and rickets: a historical perspective. Pediatrics. 2003;112: e132–5.
3. Munns CF, Simm PJ, Rodda CP, et al. Incidence of vitamin D deficiency rickets among Australian children: an Australian Paediatric Surveillance Unit study. Med J Aust. 2012;196:466–8.
4. Thacher TD, Fischer PR, Tebben PJ, et al. Increasing incidence of nutritional rickets: a population-based study in Olmsted County, Minnesota. Mayo Clin Proc. 2013;88:176–83.
5. Goldacre M, Hall N, Yeates DG. Hospitalisation for children with rickets in England: a historical perspective. Lancet. 2014;383:597–8.
6. Pettifor JM. Calcium and vitamin D metabolism in children in developing countries. Ann Nutr Metab. 2014;64 Suppl 2:15–22.

7. Garg MK, Tandon N, Marwaha RK, et al. The relationship between serum 25-hydroxy vitamin D, parathormone and bone mineral density in Indian population. Clin Endocrinol. 2014;80:41–6.
8. Islam MZ, Shamim AA, Kemi V, et al. Vitamin D deficiency and low bone status in adult female garment factory workers in Bangladesh. Br J Nutr. 2008;99:1322–9.
9. Herm FB, Killguss H, Stewart AG. Osteomalacia in Hazara District, Pakistan. Trop Doct. 2005;35:8–10.
10. van der Meer IM, Middelkoop BJ, Boeke AJ, et al. Prevalence of vitamin D deficiency among Turkish, Moroccan, Indian and sub-Sahara African populations in Europe and their countries of origin: an overview. Osteoporos Int. 2011;22:1009–21.
11. Islam MZ, Viljakainen HT, Kärkkäinen MU, et al. Prevalence of vitamin D deficiency and secondary hyperparathyroidism during winter in pre-menopausal Bangladeshi and Somali immigrant and ethnic Finnish women: associations with forearm bone mineral density. Br J Nutr. 2012;107:277–83.
12. van der Meer IM, Karamali NS, Boeke AJ, et al. High prevalence of vitamin D deficiency in pregnant non-Western women in The Hague, Netherlands. Am J Clin Nutr 2006;84:350–3;quiz 468–9.
13. Campbell GA, Kemm JR, Hosking DJ, et al. How common is osteomalacia in the elderly? Lancet. 1984;2:386–8.
14. Lips P. Vitamin D deficiency and secondary hyperparathyroidism in the elderly: consequences for bone loss and fractures and therapeutic implications. Endocr Rev. 2001;22:477–501.
15. Compston JE, Vedi S, Croucher PI. Low prevalence of osteomalacia in elderly patients with hip fracture. Age Ageing. 1991;20:132–4.
16. Robinson CM, McQueen MM, Wheelwright EF, et al. Changing prevalence of osteomalacia in hip fractures in southeast Scotland over a 20-year period. Injury. 1992;23:300–2.
17. Priemel M, von Domarus C, Klatte TO, et al. Bone mineralization defects and vitamin D deficiency: histomorphometric analysis of iliac crest bone biopsies and circulating 25-hydroxyvitamin D in 675 patients. J Bone Miner Res. 2010;25:305–12.
18. Rabelink NM, Westgeest HM, Bravenboer N, et al. Bone pain and extremely low bone mineral density due to severe vitamin D deficiency in celiac disease. Arch Osteoporos. 2011;6:209–13.
19. Karefylakis C, Näslund I, Edholm D, et al. Vitamin D status 10 years after primary gastric bypass: gravely high prevalence of hypovitaminosis D and raised PTH levels. Obes Surg. 2014;24:343–8.
20. Chakhtoura MT, Nakhoul NN, Shawwa K, et al. Hypovitaminosis D in bariatric surgery: A systematic review of observational studies. Metabolism. 2016;65:574–85.
21. Parfitt AM, Podenphant J, Villanueva AR, et al. Metabolic bone disease with and without osteomalacia after intestinal bypass surgery: a bone histomorphometric study. Bone. 1985;6:211–20.
22. Bisballe S, Eriksen EF, Melsen F, et al. Osteopenia and osteomalacia after gastrectomy: interrelations between biochemical markers of bone remodelling, vitamin D metabolites, and bone histomorphometry. Gut. 1991;32:1303–7.
23. Homan EP, Rauch F, Grafe I, et al. Mutations in SERPINF1 cause osteogenesis imperfecta type VI. J Bone Miner Res. 2011;26:2798–803.
24. Frame B, Frost HM, Pak CY, et al. Fibrogenesis imperfecta ossium. A collagen defect causing osteomalacia. N Engl J Med. 1971;285:769–72.
25. Whyte MP, Fallon MD, Murphy WA, et al. Axial osteomalacia. Clinical, laboratory and genetic investigation of an affected mother and son. Am J Med. 1981;71:1041–9.
26. Roizen J, Li D, O'Lear L, et al. Vitamin D deficiency due to a recurrent gain-of-function mutation in cyp3a4 causes a novel form of vitamin d dependent rickets. ASBMR 2016, September 27 2016, Atlanta, GA, USA.
27. Demay MB, Sabbagh Y, Carpenter TO. Calcium and vitamin D: what is known about the effects on growing bone. Pediatrics. 2007;119 Suppl 2:S141–4.
28. Christakos S, Dhawan P, Verstuyf A, et al. Vitamin D: metabolism, molecular mechanism of action, and pleiotropic effects. Physiol Rev. 2016;96:365–408.
29. Need AG, O'Loughlin PD, Morris HA, et al. Vitamin D metabolites and calcium absorption in severe vitamin D deficiency. J Bone Miner Res. 2008;23:1859–63.
30. Schafer AL, Weaver CM, Black DM, et al. intestinal calcium absorption decreases dramatically after gastric bypass surgery despite optimization of vitamin D status. J Bone Miner Res. 2015;30:1377–85.
31. Lafage-Proust MH, Lieben L, Carmeliet G, et al. High bone turnover persisting after vitamin D repletion: beware of calcium deficiency. Osteoporos Int 2013;24:2359–63.
32. Andersen R, Molgaard C, Skovgaard LT, et al. Effect of vitamin D supplementation on bone and vitamin D status among Pakistani immigrants in Denmark: a randomised double-blinded placebo-controlled intervention study. Br J Nutr. 2008;100:197–207.
33. Lieben L, Masuyama R, Torrekens S, et al. Normocalcemia is maintained in mice under conditions of calcium malabsorption by vitamin D-induced inhibition of bone mineralization. J Clin Invest 2012;122:1803–15.
34. Pettifor JM. Vitamin D deficiency and nutritional rickets in children. In: Feldman D, Pike JW, Glorieux FH (eds) Vitamin D (2nd ed.). London: Academic Press, 2005, p. 1065–83.
35. Nisbet JA, Eastwood JB, Colston KW, et al. Detection of osteomalacia in British Asians: a comparison of clinical score with biochemical measurements. Clin Sci. (Lond) 1990;78:383–9.
36. Fukumoto S, Ozono K, Michigami T, et al. Pathogenesis and diagnostic criteria for rickets and osteomalacia — proposal by an expert panel supported by the Ministry of Health, Labour and Welfare, Japan, the Japanese Society for Bone and Mineral Research, and the Japan Endocrine Society. J Bone Miner Metab. 2015;33:467–73.

37. Thacher TD, Fischer PR, Pettifor JM. The usefulness of clinical features to identify active rickets. Ann Trop Paediatr. 2002;22:229–237.
38. Peach H, Compston JE, Vedi S, et al. Value of plasma calcium, phosphate, and alkaline phosphatase measurements in the diagnosis of histological osteomalacia. J Clin Pathol. 1982;35:625–30.
39. Fraser D, Kooh SW, Scriver CR. Hyperparathyroidism as the cause of hyperaminoaciduria and phosphaturia in human vitamin D deficiency. Pediatr Res. 1967;1: 425–35.
40. Thacher TD, Fischer PR, Pettifor JM, et al. Radiographic scoring method for the assessment of the severity of nutritional rickets. J Trop Pediatr. 2000;46:132–9.
41. Dempster DW, Compston JE, Drezner MK, et al. Standardized nomenclature, symbols, and units for bone histomorphometry: a 2012 update of the report of the ASBMR Histomorphometry Nomenclature Committee. J Bone Miner Res. 2013;28(1):2–17.
42. Parfitt AM, Qiu S, Rao DS. The mineralization index — a new approach to the histomorphometric appraisal of osteomalacia. Bone 2004;35:320–5.
43. Thacher TD, Smith L, Fischer PR, et al. Optimal dose of calcium for treatment of nutritional rickets: a randomized controlled trial. J Bone Miner Res. 2016;31:2024–31.
44. Ross AC, Manson JE, Abrams SA, et al. The 2011 report on dietary reference intakes for calcium and vitamin D from the Institute of Medicine: what clinicians need to know. J Clin Endocrinol Metab. 2011;96(1):53–8.
45. Chel VG, Ooms ME, Popp-Snijders C, et al. Ultraviolet irradiation corrects vitamin D deficiency and suppresses secondary hyperparathyroidism in the elderly. J Bone Miner Res. 1998;13(8):1238–42.
46. Khazai NB, Judd SE, Jeng L, Wolfenden LL, et al. Treatment and prevention of vitamin D insufficiency in cystic fibrosis patients: comparative efficacy of ergocalciferol, cholecalciferol, and UV light. J Clin Endocrinol Metab. 2009;94:2037–43.
47. Sundbom M, Berne B, Hultin H. Short-term UVB treatment or intramuscular cholecalciferol to prevent hypovitaminosis D after gastric bypass-a randomized clinical trial. Obes Surg. 2016;26:2198–203.
48. Sanders KM, Stuart AL, Williamson EJ, et al. Annual high-dose oral vitamin D and falls and fractures in older women: a randomized controlled trial. JAMA. 2010;303:1815–22.
49. Fraser D, Kooh SW, Kind HP, et al. Pathogenesis of hereditary vitamin-D-dependent rickets. An inborn error of vitamin D metabolism involving defective conversion of 25-hydroxyvitamin D to 1 alpha,25-dihydroxyvitamin D. N Engl J Med. 1973;289:817–22.
50. Thacher TD, Levine MA. CYP2R1 mutations causing vitamin D-deficiency rickets. J Steroid Biochem Mol Biol. 2017;173:333–6.
51. Brooks MH, Bell NH, Love L, et al. Vitamin-D-dependent rickets type II. Resistance of target organs to 1,25-dihydroxyvitamin D. N Engl J Med. 1978;298:996–9.
52. Malloy PJ, Hochberg Z, Tiosano D, et al. The molecular basis of hereditary 1,25-dihydroxyvitamin D3 resistant rickets in seven related families. J Clin Invest. 1990;86:2071–9.
53. Balsan S, Garabedian M, Larchet M, et al. Long-term nocturnal calcium infusions can cure rickets and promote normal mineralization in hereditary resistance to 1,25-dihydroxyvitamin D. J Clin Invest. 1986;77:1661–7.
54. Chen H, Hewison M, Hu B, et al. Heterogeneous nuclear ribonucleoprotein (hnRNP) binding to hormone response elements: a cause of vitamin D resistance. Proc Natl Acad Sci U S A. 2003;100:6109–14.
55. Kronenberg HM. Bone and mineral metabolism: where are we, where are we going, and how will we get there? J Clin Endocrinol Metab. 2016;101(3):795–8.

90

Pathophysiology and Treatment of Chronic Kidney Disease–Mineral and Bone Disorder

Mark R. Hanudel[1], Sharon M. Moe[2], and Isidro B. Salusky[1]

[1]Department of Pediatrics, Division of Nephrology, David Geffen School of Medicine at UCLA, Los Angeles, CA, USA
[2]Department of Medicine, Division of Nephrology, Indiana University School of Medicine, Indianapolis, IN, USA

INTRODUCTION

The kidney plays an important role in bone and mineral homeostasis, regulating calcium, phosphate, PTH, fibroblast growth factor 23 (FGF23), and calcitriol (1,25-dihydroxyvitamin D, 1,25(OH)$_2$D) metabolism. Early in the course of CKD, dysregulation of mineral metabolism occurs, resulting in pathological alterations of bone modeling, remodeling, and growth. Such changes are termed "renal osteodystrophy," which specifically refers to alterations in bone morphology associated with CKD and is characterized by the histomorphometric parameters of bone turnover, mineralization, and volume. These parameters are measured via bone biopsy, the gold standard diagnostic assessment of renal osteodystrophy.

Renal osteodystrophy

The five traditional types of renal osteodystrophy (mild, osteitis fibrosa, osteomalacia, adynamic, mixed) are classified on the basis of bone turnover and mineralization [1]. Mild disease, osteitis fibrosa, and mixed disease are characterized by increased turnover, but mild disease and osteitis fibrosa have normal mineralization, whereas mixed disease has abnormal mineralization. Both osteomalacia and adynamic disease are characterized by decreased turnover, with abnormal mineralization in osteomalacia and decreased cellularity in adynamic disease. High turnover bone disease is primarily caused by secondary hyperparathyroidism, whereas low turnover bone disease may be precipitated by oversuppression of PTH secretion and/or resistance of the bone to PTH effects. Besides turnover and mineralization, the final bone biopsy parameter that can be measured is bone volume which, although useful for assessing bone fragility, has not traditionally been included in renal osteodystrophy characterization.

Over time, the epidemiology of renal osteodystrophy has changed. Historically, high turnover hyperparathyroid bone disease was predominant; however, over the past two decades, more low turnover disease, and especially adynamic bone disease, has been observed [2, 3]. Many factors may contribute to the pathogenesis of the low turnover state, including PTH resistance, decreased calcitriol levels, sex hormone deficiency, diabetes, and decreased osteocyte Wnt (Wingless-related integration site) signaling secondary to increased expression of Wnt antagonists such as sclerostin and Dickkopf-1 [4]. However, as renal osteodystrophy treatment paradigms shift from aggressive PTH suppression to caution against PTH oversuppression, less low turnover disease and more high turnover disease may again be observed in the future. Also, the type of renal osteodystrophy observed in individual CKD patients is not immutable, but rather may change over time with CKD progression [4]. Low bone turnover and a mineralization defect may be observed early in the course of CKD [3, 5]; however, as CKD progresses and increasing PTH levels overcome peripheral PTH resistance, high turnover disease may develop [4].

Primer on the Metabolic Bone Diseases and Disorders of Mineral Metabolism, Ninth Edition. Edited by John P. Bilezikian.
© 2019 American Society for Bone and Mineral Research. Published 2019 by John Wiley & Sons, Inc.
Companion website: www.wiley.com/go/asbmrprimer

Chronic kidney disease–mineral and bone disorder

Renal osteodystrophy is one measure of the skeletal component of the systemic disorder termed CKD–mineral and bone disorder (CKD–MBD). CKD–MBD describes a broader clinical syndrome that develops as a systemic disorder of mineral and bone metabolism caused by CKD, which is manifested by abnormalities in biochemical measures of bone and mineral metabolism, disordered bone (renal osteodystrophy, poor linear growth, and increased fractures), and/or extraskeletal calcification [6]. CKD–MBD focuses on the interrelatedness of disordered mineral metabolism, abnormal bone, and soft tissue/vascular calcification, and how these pathological processes contribute to cardiovascular morbidity and mortality in CKD patients (Fig. 90.1). For example, the increasingly predominant form of renal osteodystrophy, adynamic bone disease, is associated with arterial calcification [7], which is a strong predictor of cardiovascular and all-cause mortality in CKD patients [8]. Cardiovascular disease is the leading cause of death in both adult [9] and pediatric [10] CKD patients. Strikingly, relatively young dialysis patients (20 to 30 years old) have similar rates of cardiovascular mortality as octogenarians in the general population [11]. In addition to the traditional risk factors for cardiovascular disease found in the general population (ie, hypertension, diabetes, dyslipidemia), abnormal mineral metabolism is an established risk factor for cardiovascular disease in CKD patients. Therefore, optimizing CKD–MBD treatment paradigms may improve clinical outcomes and life expectancy in the CKD population.

PATHOGENESIS OF CKD–MBD

CKD–MBD pathogenesis involves a complex interplay among the kidney, bone, and parathyroid glands. As functional nephrons are lost and glomerular filtration rate (GFR) declines, a cascade of maladaptive events develops that results in bone disease, extraskeletal calcification, and adverse cardiovascular outcomes. Different factors have been implicated in the pathogenesis of this maladaptive response, but the primary trigger remains to be defined.

Traditional paradigm

The traditional paradigm of disordered bone and mineral metabolism in CKD posits that reduced functional renal mass causes phosphate retention and decreased renal 1α-hydroxylase (CYP27B1) activity. Phosphate retention decreases $1,25(OH)_2D$ levels in order to lessen enteral phosphate absorption, and decreased renal 1α-hydroxylase activity inhibits conversion of 25-hydroxyvitamin D (25OHD) to $1,25(OH)_2D$. Decreased $1,25(OH)_2D$ levels reduce intestinal calcium absorption, and low free calcium concentrations stimulate the calcium-sensing receptor (CaSR) in the parathyroid glands, increasing PTH expression. Furthermore, as PTH induces renal 1α-hydroxylase activity, decreased $1,25(OH)_2D$ levels promote PTH secretion as part of a classical feedback loop. Also, increased phosphate stimulates PTH production. Increased PTH activity on bone results in increased bone

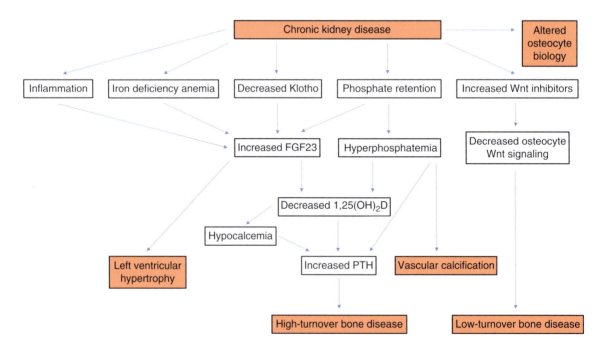

Fig. 90.1. CKD–mineral and bone disorder (CKD–MBD) pathophysiology. Changes precipitated by chronic kidney disease lead to increased circulating levels of FGF23, phosphate, and Wingless-related integration site (Wnt) inhibitors, which contribute to the bone, vascular, and cardiac disease of CKD–MBD.

turnover and resorption, which weakens bones and increases the calcium–phosphate cross-product, which may promote vascular calcification. Low 1,25(OH)$_2$D levels, low ionized calcium concentrations, and phosphate retention were identified as key elements that contribute to the development of CKD-associated secondary hyperparathyroidism, and therapies have been designed to address these abnormalities.

Role of FGF23

The identification of FGF23, its function, and its regulation have altered the traditional paradigm, and studies suggest that changes in FGF23 precede the development of CKD-associated secondary hyperparathyroidism. FGF23 levels in the bone [12] and blood [13] increase very early in the course of adult and pediatric CKD, before alterations in other mineral metabolism parameters occur (Fig. 90.2A). FGF23 physiologically functions to lower phosphate levels by two mechanisms. First, FGF23 directly affects phosphate concentrations by decreasing expression of the kidney proximal tubule type II sodium–phosphate cotransporters (NaPi-2a and NaPi-2c), reducing urinary phosphate reabsorption. Secondly, FGF23 indirectly affects phosphate levels by altering renal vitamin D metabolism. FGF23 decreases expression of renal 1α-hydroxylase, which converts 25OHD to active 1,25(OH)$_2$D, and increases expression of renal 24-hydroxylase, which converts 25OHD and 1,25(OH)$_2$D to inactive metabolites. Reduced renal 1,25(OH)$_2$D production results in decreased enteral phosphate absorption. The observation that FGF23 levels increase very early in the course of CKD suggests a role for FGF23 in the "trade-off" hypothesis of disordered mineral metabolism in CKD. The "trade-off" hypothesis states that, in the setting of decreased functional nephrons, maintenance of phosphate balance requires increased secretion of phosphaturic humoral factors to increase per nephron phosphate excretion, but with the trade-off of higher

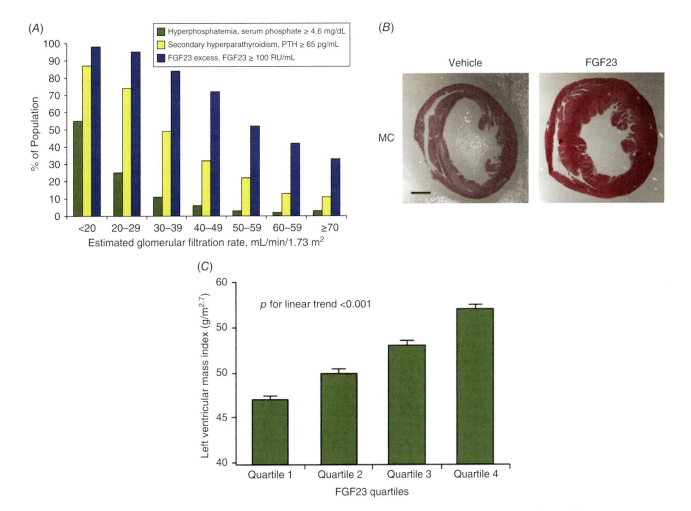

Fig. 90.2. (A) FGF23 levels increase early in the course of CKD, before alterations in other mineral metabolism parameters are observed [13]. (B) Mice injected with FGF23 develop left ventricular hypertrophy [22]. (C) In a large cohort of adult CKD patients, higher FGF23 levels were associated with increased left ventricular mass index [22].

circulating hormone levels which may ultimately cause adverse effects [14]. Indeed, progressively increasing FGF23 levels help to maintain normophosphatemia until late in the CKD course [13], but may also have adverse, "off-target" cardiovascular effects, as described below.

As GFR decreases in CKD, further phosphate retention progressively increases FGF23 production. FGF23 decreases renal $1,25(OH)_2D$ production, resulting in secondary hyperparathyroidism. Thus, in the updated FGF23-centric model of early CKD disordered mineral metabolism, FGF23-mediated inhibition of renal 1α-hydroxylase activity contributes as much, if not more so, to decreased $1,25(OH)_2D$ levels as reduced functional nephron mass. Although in the absence of kidney dysfunction, FGF23 itself directly acts on the parathyroid glands to decrease PTH expression, in the setting of CKD, downregulation of fibroblast growth factor receptor 1 (FGFR1) and the FGF23 coreceptor Klotho in the parathyroid glands induces parathyroid gland resistance to FGF23 such that elevated FGF23 levels do not decrease PTH production [15].

Klotho, which exists in both a membrane-bound form and a soluble form [16], also affects FGF23 levels. Membrane Klotho in the kidney functions as an obligate coreceptor for FGF23, and soluble Klotho functions as a pleiotropic endocrine factor [16]. Klotho deficiency increases FGF23 levels [16], and CKD has been characterized as a state of Klotho deficiency, with decreased renal Klotho expression and decreased circulating soluble Klotho levels [16]. As $1,25(OH)_2D$ induces Klotho expression, FGF23-mediated decreases in $1,25(OH)_2D$ inhibit Klotho production, which further increases FGF23 levels, engendering a pathological downward spiral [16].

Nonmineral metabolism factors also contribute to increased FGF23 production in CKD. Inflammation increases bone and circulating FGF23 levels [17]. Iron deficiency, which is common in CKD, also increases FGF23 expression. Iron chelation increases FGF23 expression in vitro [18], and iron-deficient mice with normal [17–19] and impaired kidney function [17, 19] have increased osteocytic FGF23 expression. In murine models, both absolute iron deficiency, induced by low-iron diets [17–19], and "functional" iron deficiency, induced by inflammation or administration of exogenous hepcidin [17], increase bone FGF23 expression. In a small study of iron-deficient dialysis patients, iron supplementation decreased circulating FGF23 levels [20]. In nondialysis CKD patients, the use of ferric citrate both lowered serum phosphate levels and improved iron parameters, contributing to a reduction in FGF23 concentrations [21].

FGF23 contributes to CKD-associated morbidity and mortality not only by serving as the catalyst for the development of CKD–MBD, but also via its "off-target" effects on the heart. In vitro and in vivo studies have shown that FGF23 directly causes cardiac myocyte hypertrophy [22] (Fig. 90.2B). Physiologically, FGF23, with its obligate coreceptor Klotho, binds to and activates renal proximal tubule FGFR1 to induce phosphaturia; however, in the development of pathological cardiac hypertrophy, FGF23 acts through FGFR4 [23] and does so independently of Klotho [22]. In human CKD cohorts, higher circulating FGF23 levels are associated with increased left ventricular mass [22] (Figure 90.2C) and increased mortality [24]. Although these data are associative, it has also been shown that in dialysis patients treated with cinacalcet, which lowers FGF23 (but also PTH, calcium, and phosphate), a reduction of FGF23 by 30% reduced cardiovascular mortality, heart failure events, and sudden cardiac death [25]. Besides adverse cardiovascular effects, higher FGF23 levels are also independently associated with CKD progression [24].

Inhibition of Wnt signaling

In addition to FGF23 contributing to CKD–MBD and cardiac hypertrophy, other circulating factors produced by primary kidney injury/repair mechanisms directly affect the vasculature, the myocardium, and the skeleton [26]. Specifically, local kidney repair mechanisms involving Wnt signaling pathway reactivation produce autoregulatory Wnt inhibitors, including sclerostin and Dickkopf-1 (Dkk1) [26]. In animal models of early CKD, increased circulating levels of such Wnt inhibitors were observed in the circulation, with concomitant decreased bone formation rates and increased vascular calcification; Dkk1 neutralization increased bone formation rates and decreased vascular calcification, demonstrating the effect of circulating Wnt inhibitors on CKD–MBD pathogenesis [27].

Vascular calcification

Abnormal mineral metabolism, including hypercalcemia and hyperphosphatemia, is an established risk factor for vascular calcification in CKD patients. However, in a study of nondialysis adult CKD patients without hypercalcemia or hyperphosphatemia, 40% were found to have coronary artery calcification [28]. Changes in carotid artery wall thickness are also apparent in children with early CKD [29]. These children with early CKD lack not only the mineral metabolism-associated risk factors common in the dialysis population, but also the traditional adult risk factors for vascular calcification, which include hypertension, diabetes, and dyslipidemia. This observation suggests that factors unique to CKD, and independent of circulating mineral metabolism factors, contribute to vascular calcification.

The pathophysiology of calcification in CKD patients clearly differs from that observed in the general population, although the mechanisms by which vascular calcification develop remain to be fully elucidated. In contrast to the calcified atherosclerotic plaques that develop in the vascular intima of aging individuals with normal kidney function, uremia facilitates calcification of the tunica media. Indeed, in patients with advanced CKD, calcification can occur both in atherosclerotic plaques, caused by traditional risk factors (such as hypertension and diabetes), and in the medial layer caused by CKD. In CKD, the entire smooth muscle layer surrounding arteries may be replaced not only by calcium deposits, but also by tissue that resembles bone. Indeed, osteoblasts and vascular smooth muscle cells have a common mesenchymal origin and, in CKD patients, core binding factor-1 (Cbfa1/Runx2) is thought to trigger mesenchymal cell to osteoblast transformation. Mice deficient in Cbfa1 fail to mineralize bone [30], and arteries obtained from patients undergoing renal transplantation show increased Cbfa1 mRNA expression [31]. Also, in the uremic milieu, there is upregulation of promineralization factors such as bone sialoprotein, osteonectin, alkaline phosphatase, type I collagen, and bone morphogenic protein-2, as well as downregulation of calcification inhibitors, including fetuin A and matrix Gla protein [32].

THERAPEUTIC APPROACHES IN CKD–MBD

Current therapeutic approaches in CKD–MBD focus primarily on treating hyperphosphatemia and lowering elevated PTH levels. The goal of therapy is that optimization of these parameters limits the contribution of CKD–MBD to adverse CKD-associated outcomes, including fractures, CKD progression, and cardiovascular morbidity and mortality. In pediatric CKD–MBD, a further therapeutic aim is to minimize complications to the growing skeleton and maximize linear growth. Although the current therapeutic approaches have been utilized for many years, the incidence rates of fractures and cardiovascular disease in CKD patients remain largely unchanged. Thus, there is a critical need to develop new therapeutic strategies, as related to our evolving understanding of CKD–MBD pathogenesis.

Hyperphosphatemia

The 2017 Kidney Disease Improving Global Outcomes (KDIGO) Clinical Practice Guideline Update for the Diagnosis, Evaluation, Prevention, and Treatment of CKD–MBD suggests that, for patients with CKD stages 3 (estimated GFR <60 mL/min/1.73 m^2) through 5D (estimated GFR <15 ml/min/1.73 m^2 and on dialysis; end-stage kidney disease), elevated serum phosphate levels be lowered toward the normal range [33]. It is noted that there is an absence of data supporting maintenance of phosphate within the normal range [33], and that trial data demonstrating improvement in patient-centered outcomes with phosphate-lowering treatments are still lacking [33]. In children with CKD, it must be noted that the normal ranges of phosphate vary substantially by age, and that normal serum phosphate concentrations in infants and toddlers are higher, because positive phosphate balance is required for optimal early bone growth.

Dietary phosphate restriction

Maintaining serum phosphate concentrations at near-normal levels often requires dietary phosphate restriction and enteral phosphate binders. The 2003 Kidney Disease Outcomes Quality Initiative (KDOQI) guidelines recommend that, when serum phosphate levels are elevated, dietary phosphate intake should be restricted to 800-1000 mg daily [34]. In pediatric CKD patients with hyperphosphatemia, it is suggested that dietary phosphate intake be reduced to 80% of the age-specific daily adequate intake (AI) for infants, or 80% of the recommended dietary allowances (RDA) for older children. However, dietary phosphate restriction must be balanced with adequate protein intake. Especially in larger adults, providing adequate protein while concurrently limiting dietary phosphate intake to less than one gram daily is very difficult. Per the KDOQI guidelines, multiplying the recommended daily protein intake (in grams) times 10-12 mg phosphate per gram of protein provides an estimate of a reasonable range for daily phosphate intake [34]. Protein sources with the least amount of phosphate per gram of protein should be favored (examples of which are listed in KDOQI tables) [34]. Besides the amount of dietary phosphate consumed, the protein source of the phosphate should also be considered [33]. In a crossover study of CKD patients fed vegetarian and meat diets with equivalent protein and phosphate concentrations, the vegetarian diet was associated with lower serum phosphate and FGF23 levels, suggesting some benefit of plant versus animal protein [35].

Enteral phosphate binders

Patients with progressively or persistently elevated serum phosphate, despite dietary restriction, may require the use of phosphate-binding agents [33]. Enteral binders taken with meals complex with phosphate in the intestinal tract, limiting phosphate absorption by blocking passive diffusion across a paracellular gradient. Several phosphate binders are available for clinical use, and KDIGO recommends that the choice of phosphate binder takes into account CKD stage, the presence of other components of CKD–MBD, concomitant therapies, and side effect profile [1].

Calcium-based phosphate binders, including calcium carbonate and calcium acetate preparations (containing 40% and 25% elemental calcium, respectively), are widely prescribed and are effective in lowering serum phosphate levels. However, the benefit of calcium-based phosphate binders, especially when used concurrently with active vitamin D sterols, must be weighed against the possible adverse effects of hypercalcemia and/or extraskeletal calcification. Indeed, the KDOQI guidelines recommend that the total dose of elemental calcium provided by calcium-based phosphate binders should not exceed 1500 mg daily, and the total intake of elemental calcium (including dietary calcium) should not exceed

2000 mg daily [34]. A recent crossover calcium balance study illustrates the effects of exceeding this recommended elemental calcium intake. In this study, adult patients with CKD stages 3 and 4 received a controlled diet (957 mg elemental calcium daily) with or without calcium carbonate supplementation (1500 mg elemental calcium daily, providing a total of 2457 mg elemental calcium daily) for three weeks [36]. A daily elemental calcium intake of 2457 mg resulted in a positive calcium balance of 508 mg daily, which was 447 mg higher than the placebo group [36]. Moreover, calcium kinetic modeling suggested that some of the retained calcium may have been deposited in extraskeletal tissue and not incorporated into bone [36]. In children with CKD, it is suggested that the total dose of elemental calcium provided by calcium-based phosphate binders and by dietary calcium not exceed twice the age-specific AI or RDA for calcium; however, calcium balance studies are still needed in children in order to define the optimal calcium intake. As with phosphate, normal ranges of serum calcium in children vary by age, and normal serum calcium concentrations in infants and toddlers are higher, because positive calcium balance is required for optimal early bone growth.

Given concerns for hypercalcemia and possible extraskeletal calcification with the use of calcium-based phosphate binders, alternative, non-calcium-based phosphate binders may be used. One of the most commonly used non-calcium-based binders is sevelamer, a phosphate-binding resin that is formulated as sevelamer hydrochloride and sevelamer carbonate. Sevelamer effectively controls hyperphosphatemia without increasing the incidence of hypercalcemia. In some, but not all, clinical trials (several of which are summarized in [1]), compared with calcium-based binders, sevelamer attenuated the progression of arterial calcification in CKD patients. A recent meta-analysis of randomized controlled trials of calcium-based binders versus non-calcium-based binders, which included 25 trials and approximately 8000 patients, found that the use of calcium-based binders resulted in higher all-cause mortality than sevelamer in particular and non-calcium-based binders in general [37].

Other non-calcium-based phosphate binders include lanthanum, aluminum, and iron salts. Lanthanum carbonate also effectively controls hyperphosphatemia without increasing the incidence of hypercalcemia. However, lanthanum can be absorbed and may accumulate in tissues, including bone and liver [38], limiting its use as a first-line binder. In children, the use of lanthanum is not recommended, because animal studies have shown that lanthanum is deposited in developing bone, including the growth plates [38]. Although evidence of a direct toxic effect of lanthanum on bone is lacking [38], in the absence of longer term safety data [38], as well as efficacy in clinical outcomes, awareness of lanthanum accumulation in bone is prudent.

Aluminum hydroxide is a very effective phosphate binder; however, its use is limited by the risk of aluminum toxicity, which may manifest as neurotoxicity and impairment of bone mineralization. Although some of the most severe cases of aluminum toxicity occurred in patients exposed to dialysate contaminated with aluminum [1], it has been shown that aluminum retention and bone toxicity can occur when using recommended doses of aluminum hydroxide [39]. Because there is no ability to predict a safe aluminum dose, KDIGO recommended that the long-term use of aluminum-based binders be avoided [1].

Iron-based binders, such as ferric citrate and sucroferric oxyhydroxide, are effective in controlling phosphate levels [21, 40]. Ferric citrate has also been shown to improve iron parameters [21]. Given recent evidence that iron deficiency increases FGF23 production [17–19], the use of ferric citrate, especially in iron-deficient CKD patients, may affect FGF23 levels by both limiting phosphate absorption and delivering iron.

Inhibition of the enteral type II sodium-phosphate cotransporter (NPT2b)

Enteral phosphate binders limit passive, paracellular phosphate absorption. However, such binder-induced effects lead to upregulation of the intestinal sodium-phosphate NPT2b transporter, increasing active enteral phosphate absorption, offsetting some of the beneficial binder effect [41]. In animal studies, compared with wild type mice with CKD, NPT2b knockout mice with CKD had significantly lower serum phosphate concentrations (and a blunted increase in FGF23) [41]. Moreover, treating the NPT2b-deficient mice with sevelamer carbonate further reduced serum phosphate levels [41]. In a study of adult CKD patients, extended release niacin, a NPT2b inhibitor, reduced both serum phosphate and FGF23 [42]. Therefore, maximal suppression of enteral phosphate absorption in CKD patients may possibly be best achieved using NPT2b inhibitors in combination with binders. The ongoing CKD Optimal Management with BInders and NicotinamidE (COMBINE) study is designed to assess whether the addition of NPT2b blockade to phosphate binders improves phosphate and FGF23 levels, as well as surrogate measures of cardiovascular disease and CKD progression, in pre-dialysis CKD patients [43].

Inhibition of the enteral sodium/hydrogen exchanger isoform 3 (NHE3)

Tenapanor, a small molecule inhibitor of the intestinal sodium/hydrogen exchanger isoform 3 (NHE3), has been shown to effect dose-dependent reductions in serum phosphate concentrations in hyperphosphatemic hemodialysis patients [44]. The mechanism by which tenapanor reduces enteral phosphate uptake is currently being studied.

Frequent hemodialysis

Lastly, the KDIGO guidelines suggest that, in dialysis patients with persistent hyperphosphatemia, increased dialytic phosphate removal may be employed [1]. Compared with the usual thrice-weekly hemodialysis regimen, daily hemodialysis significantly reduces serum phosphate levels and the enteral phosphate binder dose [45]. Because the daily pill burden associated with enteral phosphate binders is often very high, many patients have difficulty adhering to binder treatment regimens. In these patients, more intensive dialysis regimens may be useful. However, in frequent hemodialysis patients, serum phosphate levels should be closely monitored, because many patients may actually require the addition of phosphate into the dialysate to prevent hypophosphatemia [45].

Secondary hyperparathyroidism

Secondary hyperparathyroidism is common in CKD–MBD and can lead to increased bone turnover, which may be associated with marrow fibrosis and abnormal mineralization, fractures, worsening anemia, hypercalcemia, and cardiovascular morbidity and mortality [1]. As such, lowering elevated circulating PTH levels has been a focus of CKD–MBD treatment for decades. The KDIGO guidelines state that, in patients with CKD stages 3–5 not on dialysis, the optimal PTH level is not known [1]. However, in such patients with intact PTH levels that are progressively rising or persistently above the upper limit of normal for the assay, an evaluation for potentially modifiable factors (including hyperphosphatemia, hypocalcemia, high phosphate intake, and vitamin D deficiency) should be undertaken [33]. In dialysis patients, it is suggested that intact PTH levels be maintained in the range of approximately two to nine times the upper limit of normal for the assay [1]. In the Dialysis Outcomes and Practice Patterns Study (DOPPS), a prospective cohort study including 25,588 dialysis patients, as compared with the reference group of patients with PTH levels of 101 to 300 pg/mL, the group of patients with PTH levels of >600 pg/mL had a 21% increased risk for all-cause mortality [46].

Patients with CKD stages 3–5 not on dialysis who have persistently elevated intact PTH levels are evaluated for hyperphosphatemia, hypocalcemia, and deficiency of 25OHD [1]. Normalization of serum phosphate (with dietary phosphate restriction and/or enteral binders) and serum calcium (with calcium supplementation) may help to decrease high PTH levels. Furthermore, because these patients still have some functioning renal parenchyma with intact renal 1α-hydroxylase activity, normalization of 25OHD levels will provide adequate substrate for conversion to 1,25(OH)$_2$D, which inhibits PTH secretion. Indeed, in vitamin D deficient CKD patients, ergocalciferol therapy delays the development of secondary hyperparathyroidism [47]. The 2017 KDIGO update suggests that calcitriol and vitamin D analogs not be routinely used to lower PTH levels in patients with CKD stages 3–5 not on dialysis, but rather reserved for patients with CKD stages 4–5 and severe or progressive hyperparathyroidism [33]. In dialysis patients requiring PTH-lowering therapy, active vitamin D sterols (calcitriol and its analogs) and/or calcimimetics may be used [33].

Active vitamin D sterols

Active vitamin D sterols (calcitriol and its analogs) lower PTH levels, but also promote intestinal calcium and phosphate absorption. Calcitriol is a commonly used active vitamin D sterol; however, it is associated with an increased risk of hypercalcemia and hyperphosphatemia. As such, vitamin D analogs with less calcemic and/or phosphatemic effects have been developed, including paricalcitol and doxercalciferol. However, all active vitamin D sterols have the ability to increase serum calcium and phosphate levels. Therefore, judicious use of calcitriol and its analogs, with careful monitoring, is warranted. Also, switching from calcium-based to noncalcium-based phosphate binders, and/or increasing the dose of noncalcium-based phosphate binders, may help to prevent hypercalcemia and hyperphosphatemia, respectively, allowing for better optimization of active vitamin D sterol treatment.

Although active vitamin D sterols effectively treat secondary hyperparathyroidism, they also increase bone and circulating FGF23 levels [48]. Therefore, the beneficial effects of these medications may come at the cost of higher FGF23 levels, which are independently associated with increased mortality in CKD patients [24], as well as hypercalcemia. Indeed, the 2017 KDIGO update cites two recent randomized controlled trials of paricalcitol versus placebo in nondialysis CKD patients that showed an increased risk of hypercalcemia, without beneficial effects on cardiac endpoints, in the paricalcitol-treated groups [33].

Calcimimetics

Another medication used for treatment of secondary hyperparathyroidism is cinacalcet, an oral tablet which acts as a calcimimetic by allosteric activation of the calcium-sensing receptor (CaSR) on the parathyroid glands. Cinacalcet effectively lowers serum PTH [49], as well as FGF23 [25]. However, the EVOLVE study, a large randomized trial of cinacalcet versus placebo in dialysis patients, found no significant difference in the risk of death or major cardiovascular events in the primary, unadjusted intention-to-treat analysis [49]. Yet, after adjustment for baseline characteristics (mean age was higher in the cinacalcet group), there was a 12% reduction in the primary composite end point in the cinacalcet group [49]. Cinacalcet is approved for use in adults on dialysis but not children. The long-term effects of cinacalcet in pediatric patients have not been studied and, because the calcium-sensing receptor is expressed on growth plate cartilage, the effects of cinacalcet on growth must be elucidated before cinacalcet use can be recommended in children. Hypocalcemia may occur with cinacalcet, so serum calcium levels must be frequently monitored and dosage changes made accordingly.

Recently, an intravenous calcimimetic, etelcalcetide, has been approved for the treatment of secondary hyperparathyroidism in adult dialysis patients. As compared with cinacalcet, PTH lowering was noninferior with etelcalcetide [50]. Etelcalcetide also reached the secondary superiority end point (52% of patients randomized to etelcalcetide versus 40% of patients randomized to cinacalcet experienced a 50% reduction in PTH concentrations from baseline), and was associated with greater FGF23 reduction [50]. Further studies of etelcalcetide will be needed to assess clinical outcomes as well as longer-term efficacy and safety.

Regarding treatment approaches to secondary hyperparathyroidism, the use of active vitamin D sterols alone versus cinacalcet plus low-dose active vitamin D sterols has been compared. In the ACHIEVE study, hemodialysis patients were randomized to receive either cinacalcet with low doses of paricalcitol or doxercalciferol, or flexible, escalating doses of paricalcitol or doxercalciferol, with the primary endpoint being the proportion of subjects who simultaneously achieved a mean PTH level of 150–300 pg/mL and a mean calcium–phosphate cross-product value of $<55\,mg^2/dL^2$ [51]. No difference in the primary endpoint was observed between groups, which was attributed to oversuppression of PTH in the cinacalcet group [51]. In the IMPACT SHPT study, hemodialysis patients were randomized to receive either paricalcitol alone or cinacalcet plus low-dose active vitamin D sterols, with the primary endpoint being the proportion of subjects who achieved a mean PTH level of 150–300 pg/mL [52]. Paricalcitol versus cinacalcet plus low-dose active vitamin D resulted in a higher percentage of patients achieving goal PTH levels [52], but also induced higher FGF23 levels [53]. Notably, the incidence of hypocalcemia in the cinacalcet group (~50%) was higher than the incidence of hypercalcemia in the paricalcitol group (<10%) [52]. Based on the current data, one treatment approach to secondary hyperparathyroidism cannot be favored over another. Therefore, the benefits and risks of each medication, or combination of medications, must be weighed. The KDIGO guidelines suggest that initial drug selection be based on serum calcium, serum phosphate, and other aspects of CKD–MBD; and that calcium or noncalcium-based phosphate binders be adjusted so that secondary hyperparathyroidism treatment does not compromise calcium and phosphate levels [1].

Parathyroidectomy

Lastly, when secondary hyperparathyroidism is severe and refractory to medical management, parathyroidectomy is suggested [1]. Severe secondary hyperparathyroidism may be defined as intact PTH levels persistently greater than 800 to 1000 pg/mL [1, 34]. Effective surgical therapy can be accomplished by subtotal parathyroidectomy, or total parathyroidectomy with parathyroid tissue autotransplantation [34]. Postparathyroidectomy, serum calcium and phosphate must be monitored very closely, as "hungry bone syndrome," a condition characterized by acutely increased skeletal calcium and phosphate uptake, may cause marked hypocalcemia and/or hypophosphatemia. Treatment consists of large doses of active vitamin D and calcium. Phosphate supplementation may worsen hypocalcemia, and is generally not recommended unless the hypophosphatemia is severe.

ACKNOWLEDGMENTS

The authors' work has been supported by an NIH/NIDDK K08 Mentored Clinical Scientist Research Career Development Award (DK111980, to MRH), an NIH/NICHD K12 Child Health Research Career Development Award (HD034610, to MRH), a UCLA Clinical and Translational Science Institute, and UCLA Children's Discovery and Innovation Institute Children's Health Team Science Award (to MRH and IBS), an NIH/NIDDK R01 grant (DK35423, to IBS), an NIH/NIDDK U34 grant (DK104619, to IBS), and NIH R01 grants (DK11087 and DK100306, to SMM).

REFERENCES

1. KDIGO clinical practice guideline for the diagnosis, evaluation, prevention, and treatment of Chronic Kidney Disease-Mineral and Bone Disorder (CKD–MBD). Kidney Int Suppl. 2009(113):S1–130.
2. Martin KJ, Olgaard K, Coburn JW, et al. Diagnosis, assessment, and treatment of bone turnover abnormalities in renal osteodystrophy. Am J Kidney Dis. 2004;43(3):558–65.
3. Graciolli FG, Neves KR, Barreto F, et al. The complexity of chronic kidney disease-mineral and bone disorder across stages of chronic kidney disease. Kidney Int. 2017;91(6):1436–46.
4. Drueke TB, Massy ZA. Changing bone patterns with progression of chronic kidney disease. Kidney Int. 2016;89(2):289–302.
5. Wesseling-Perry K, Pereira RC, Tseng CH, et al. Early skeletal and biochemical alterations in pediatric chronic kidney disease. Clin J Am Soc Nephrol. 2012;7(1):146–52.
6. Moe S, Drüeke T, Cunningham J, et al. Definition, evaluation, and classification of renal osteodystrophy: a position statement from Kidney Disease: Improving Global Outcomes (KDIGO). Kidney Int. 2006;69(11):1945–53.
7. Bover J, Ureña P, Brandenburg V, et al. Adynamic bone disease: from bone to vessels in chronic kidney disease. Semin Nephrol. 2014;34(6):626–40.
8. London GM, Guérin AP, Marchais SJ, et al. Arterial media calcification in end-stage renal disease: impact on all-cause and cardiovascular mortality. Nephrol Dial Transplant. 2003;18(9):1731–40.

9. United States Renal Data System. 2016 USRDS annual data report: Epidemiology of kidney disease in the United States. 2016, National Institutes of Health, National Institute of Diabetes and Digestive and Kidney Diseases, Bethesda, MD, USA.
10. Mitsnefes MM. Cardiovascular disease in children with chronic kidney disease. J Am Soc Nephrol. 2012;23(4):578–85.
11. Foley RN, Parfrey PS, Sarnak MJ. Clinical epidemiology of cardiovascular disease in chronic renal disease. Am J Kidney Dis. 1998;32(5 Suppl 3):S112–9.
12. Pereira RC, Juppner H, Azucena-Serrano CE, et al. Patterns of FGF-23, DMP1, and MEPE expression in patients with chronic kidney disease. Bone. 2009;45(6):1161–8.
13. Isakova T, Wahl P, Vargas GS, et al. Fibroblast growth factor 23 is elevated before parathyroid hormone and phosphate in chronic kidney disease. Kidney Int. 2011;79(12):1370–8.
14. Gutierrez OM. Fibroblast growth factor 23 and disordered vitamin D metabolism in chronic kidney disease: updating the "trade-off" hypothesis. Clin J Am Soc Nephrol. 2010;5(9):1710–6.
15. Galitzer H, Ben-Dov IZ, Silver J, et al. Parathyroid cell resistance to fibroblast growth factor 23 in secondary hyperparathyroidism of chronic kidney disease. Kidney Int. 2010;77(3):211–8.
16. Hu MC, Kuro-o M, Moe OW. Klotho and chronic kidney disease. Contrib Nephrol. 2013;180:47–63.
17. David V, et al. Inflammation and functional iron deficiency regulate fibroblast growth factor 23 production. Kidney Int. 2016;89(1):135–46.
18. Farrow EG, Yu X, Summers LJ, et al. Iron deficiency drives an autosomal dominant hypophosphatemic rickets (ADHR) phenotype in fibroblast growth factor-23 (Fgf23) knock-in mice. Proc Natl Acad Sci U S A. 2011;108(46):E1146–55.
19. Hanudel MR, Chua K, Rappaport M, et al. Effects of dietary iron intake and chronic kidney disease on fibroblast growth factor 23 metabolism in wild-type and hepcidin knockout mice. Am J Physiol Renal Physiol. 2016;311(6):F1369–F1377.
20. Iguchi A, Kazama JJ, Yamamoto S, et al. Administration of ferric citrate hydrate decreases circulating FGF23 levels independently of serum phosphate levels in hemodialysis patients with iron deficiency. Nephron, 2015;131(3):161–6.
21. Block GA, Fishbane S, Rodriguez M, et al. A 12-week, double-blind, placebo-controlled trial of ferric citrate for the treatment of iron deficiency anemia and reduction of serum phosphate in patients with CKD Stages 3-5. Am J Kidney Dis. 2015;65(5):728–36.
22. Faul C, Amaral AP, Oskouei B, et al. FGF23 induces left ventricular hypertrophy. J Clin Invest. 2011;121(11):4393–408.
23. Grabner A. Amaral AP, Schramm K, et al. Activation of cardiac fibroblast growth factor receptor 4 causes left ventricular hypertrophy. Cell Metab. 2015;22(6):1020–32.
24. Isakova T, Xie H, Yang W, et al. Fibroblast growth factor 23 and risks of mortality and end-stage renal disease in patients with chronic kidney disease. JAMA. 2011;305(23):2432–9.
25. Moe SM, Chertow GM, Parfrey PS, et al. Cinacalcet, fibroblast growth factor-23, and cardiovascular disease in hemodialysis: the Evaluation of Cinacalcet HCl Therapy to Lower Cardiovascular Events (EVOLVE) Trial. Circulation. 2015;132(1):27–39.
26. Hruska KA, Seifert M, Sugatani T. Pathophysiology of the chronic kidney disease-mineral bone disorder. Curr Opin Nephrol Hypertens. 2015;24(4):303–9.
27. Fang Y, Ginsberg C, Seifert M, et al. CKD-induced wingless/integration1 inhibitors and phosphorus cause the CKD-mineral and bone disorder. J Am Soc Nephrol. 2014;25(8):1760–73.
28. Russo D, Palmiero G, De Blasio AP, et al. Coronary artery calcification in patients with CRF not undergoing dialysis. Am J Kidney Dis. 2004;44(6):1024–30.
29. Litwin M, Wühl E, Jourdan C, et al. Altered morphologic properties of large arteries in children with chronic renal failure and after renal transplantation. J Am Soc Nephrol. 2005;16(5):1494–500.
30. Ducy P, Zhang R, Geoffroy V, et al. Osf2/Cbfa1: a transcriptional activator of osteoblast differentiation. Cell. 1997;89(5):747–54.
31. Moe SM, Duan D, Doehle BP, et al. Uremia induces the osteoblast differentiation factor Cbfa1 in human blood vessels. Kidney Int. 2003;63(3):1003–11.
32. Wesseling-Perry K, Juppner H. The osteocyte in CKD: new concepts regarding the role of FGF23 in mineral metabolism and systemic complications. Bone. 2013;54(2):222–9.
33. KDIGO clinical practice guideline update for the diagnosis, evaluation, prevention, and treatment of Chronic Kidney Disease–Mineral and Bone Disorder (CKD–MBD). Kidney Int Suppl. 2017;7(1):1–59.
34. K/DOQI clinical practice guidelines for bone metabolism and disease in chronic kidney disease. Am J Kidney Dis. 2003(42):S1–201.
35. Moe SM, Zidehsarai MP, Chambers MA, et al. Vegetarian compared with meat dietary protein source and phosphorus homeostasis in chronic kidney disease. Clin J Am Soc Nephrol. 2011;6(2):257–64.
36. Hill KM, Martin BR, Wastney ME, et al. Oral calcium carbonate affects calcium but not phosphorus balance in stage 3-4 chronic kidney disease. Kidney Int. 2013;83(5):959–66.
37. Sekercioglu N, Thabane L, Díaz Martínez JP, et al. Comparative effectiveness of phosphate binders in patients with chronic kidney disease: a systematic review and network meta-analysis. PLoS One, 2016;11(6):e0156891.
38. Drueke TB. Lanthanum carbonate as a first-line phosphate binder: the "cons". Semin Dial. 2007;20(4):329–32.
39. Salusky IB, Foley J, Nelson P, et al. Aluminum accumulation during treatment with aluminum hydroxide and dialysis in children and young adults with chronic renal disease. N Engl J Med. 1991;324(8):527–31.

40. Greig SL, Plosker GL. Sucroferric oxyhydroxide: a review in hyperphosphataemia in chronic kidney disease patients undergoing dialysis. Drugs. 2015;75(5):533–42.
41. Schiavi SC, Tang W, Bracken C, et al. Npt2b deletion attenuates hyperphosphatemia associated with CKD. J Am Soc Nephrol. 2012;23(10):1691–700.
42. Rao M, Steffes M, Bostom A, et al. Effect of niacin on FGF23 concentration in chronic kidney disease. Am J Nephrol. 2014;39(6):484–90.
43. Isakova T, x JH, Sprague SM, et al. Rationale and approaches to phosphate and fibroblast growth factor 23 reduction in CKD. J Am Soc Nephrol. 2015;26(10):2328–39.
44. Block GA, Rosenbaum DP, Leonsson-Zachrisson M, et al. Effect of tenapanor on serum phosphate in patients receiving hemodialysis. J Am Soc Nephrol. 2017;28(6):1933–42.
45. Copland M, Komenda P, Weinhandl ED, et al. Intensive hemodialysis, mineral and bone disorder, and phosphate binder use. Am J Kidney Dis. 2016;68(5s1):S24–s32.
46. Tentori F, Blayney MJ, Albert JM, et al. Mortality risk for dialysis patients with different levels of serum calcium, phosphorus, and PTH: the Dialysis Outcomes and Practice Patterns Study (DOPPS). Am J Kidney Dis. 2008;52(3):519–30.
47. Shroff R, Wan M, Gullett A, et al. Ergocalciferol supplementation in children with CKD delays the onset of secondary hyperparathyroidism: a randomized trial. Clin J Am Soc Nephrol. 2012;7(2):216–23.
48. Wesseling-Perry K, Pereira RC, Sahney S, et al. Calcitriol and doxercalciferol are equivalent in controlling bone turnover, suppressing parathyroid hormone, and increasing fibroblast growth factor-23 in secondary hyperparathyroidism. Kidney Int. 2011;79(1):112–9.
49. EVOLVE Trial Investigators, Chertow GM, Block GA, Correa-Rotter R, et al. Effect of cinacalcet on cardiovascular disease in patients undergoing dialysis. N Engl J Med. 2012;367(26):2482–94.
50. Block GA, Bushinsky DA, Cheng S, et al. Effect of etelcalcetide vs cinacalcet on serum parathyroid hormone in patients receiving hemodialysis with secondary hyperparathyroidism: a randomized clinical trial. JAMA. 2017;317(2):156–64.
51. Fishbane S, Shapiro WB, Corry DB, et al. Cinacalcet HCl and concurrent low-dose vitamin D improves treatment of secondary hyperparathyroidism in dialysis patients compared with vitamin D alone: the ACHIEVE study results. Clin J Am Soc Nephrol. 2008;3(6):1718–25.
52. Ketteler M, Martin KJ, Wolf M, et al. Paricalcitol versus cinacalcet plus low-dose vitamin D therapy for the treatment of secondary hyperparathyroidism in patients receiving haemodialysis: results of the IMPACT SHPT study. Nephrol Dial Transplant. 2012;27(8):3270–8.
53. Cozzolino M, Ketteler M, Martin KJ, et al. Paricalcitol- or cinacalcet-centred therapy affects markers of bone mineral disease in patients with secondary hyperparathyroidism receiving haemodialysis: results of the IMPACT-SHPT study. Nephrol Dial Transplant. 2014;29(4):899–905.

91

Disorders of Mineral Metabolism in Childhood

Thomas O. Carpenter[1] and Nina S. Ma[2]
[1]Yale University School of Medicine, New Haven, CT, USA
[2]Boston Children's Hospital, Harvard Medical School, Boston, MA, USA

INTRODUCTION

Disorders of mineral homeostasis may present differently in children than in adults. This chapter outlines disorders of mineral metabolism emphasizing specific features of the childhood age group.

DISORDERS OF CALCIUM HOMEOSTASIS

Hypocalcemia

Clinical presentation

In the newborn with acute hypocalcemia, jitteriness, hyperacusis, irritability, and limb-jerking may occur, with progression to generalized or focal clonic seizures. Laryngospasm may lead to a misdiagnosis of croup. Atrioventricular heart block occurs in premature infants with hypocalcemia, and electrocardiograms should be performed in newborns with significant bradycardia [1]. Apnea, tachycardia, tachypnea, cyanosis, edema, and vomiting have been reported in newborns with hypocalcemia. Paresthesias, increased neuromuscular irritability, and positive Chvostek's and Trousseau's signs may be evident in older children.

Transient hypocalcemia of the newborn

Early neonatal hypocalcemia occurs during the first 3 days of life, and is found in premature infants, infants of diabetic mothers, and asphyxiated infants. The premature infant has an exaggerated postnatal depression in circulating calcium (Ca), such that total Ca levels may drop below 7.0 mg/dL, but the proportional drop in ionized Ca is less. PTH insufficiency may contribute to early neonatal hypocalcemia in premature infants.

Late neonatal hypocalcemia presents as tetany between 5 and 10 days of life, occurs more frequently in term than in premature infants, is usually not correlated with birth trauma or asphyxia, and may be associated with maternal vitamin D insufficiency.

Hypocalcemia associated with magnesium (Mg) deficiency may present as late neonatal hypocalcemia. Severe hypomagnesemia (circulating levels of Mg below 0.8 mg/dL) may occur in congenital defects of intestinal Mg absorption or renal tubular reabsorption [2]. Hypocalcemia in this setting may be refractory to therapy unless Mg levels are corrected.

Maternal hyperparathyroidism may result in neonatal hypocalcemia. The child's serum phosphorus (Pi) is often more than 8 mg/dL and symptoms may be exacerbated by high Pi intake. Maternal hypercalcemia results in increased Ca delivery to the fetus, thereby suppressing parathyroid gland responsivity. As a result, eucalcemia in the infant is not maintained postnatally because persistent parathyroid gland suppression

Hypocalcemia presenting in childhood

Hypoparathyroidism

Persistent hypocalcemia in childhood may be caused by congenital hypoparathyroidism. Mutations in genes involved in parathyroid gland development, PTH processing, PTH secretion, PTH structure, and PTH resistance have been identified (see Chapters 27, 86 and 87). The most frequently identified disorder of parathyroid gland

---

Primer on the Metabolic Bone Diseases and Disorders of Mineral Metabolism, Ninth Edition. Edited by John P. Bilezikian.
© 2019 American Society for Bone and Mineral Research. Published 2019 by John Wiley & Sons, Inc.
Companion website: www.wiley.com/go/asbmrprimer

development is the DiGeorge anomaly (OMIM #188400), microdeletions of chromosome 22q11.2 [3], presenting with hypoparathyroidism, T-cell incompetence caused by a partial or absent thymus, and conotruncal heart defects (tetralogy of Fallot, truncus arteriosus) or aortic arch abnormalities. Cleft palate and facial dysmorphism may occur. Mild parathyroid defects not apparent in infancy may present with hypocalcemia later in life during periods of stress or increased metabolic demands [4, 5]. Deletion of *TBX1*, which encodes a T-box transcription factor, is sufficient to cause the cardiac, parathyroid, thymic, facial, and vellopharyngeal features of the syndrome. However, variability in phenotype with similar genetic defects occurs [6]. A similar phenotype identified as DiGeorge 2 Syndrome has been attributed to a deletion of distal end of chromosome 10p [7] (601362); *GATA3* may be the responsible gene for the hypoparathyroidism accompanying this deletion, which likely represents a variant of HDR (hypoparathyroidism, sensorineural deafness, and renal dysplasia syndrome) (146255). Other genetic defects result in disrupted parathyroid gland development (eg, loss of the tubulin chaperone E, *TBCE* [Sanjad-Sakati and Kenny-Caffey syndromes, 241410 and 244460], *GCM2* [603716], *SOX3* [307700]), abnormal PTH processing and molecular structure (*PTH*, 168450), abnormal PTH secretory dynamics (*CaSR*, 601198 and *GNA11*, 615361), and resistance to PTH action (*GNAS*, 103580 and 603233). Individuals with classic PTH resistance (pseudohypoparathyroidism), caused by loss of function of the Gs alpha protein, often do not develop clinically evident hypocalcemia until a few years of age. Mitochondrial defects (eg, Kearns-Sayre syndrome [530000], mutations in mitochondrial trifunctional protein [*HADHB*, 143450]) and Kenny Caffey Syndrome Type 2 (mutations in *FAM111A*) (615292) are more recently described causes of hypoparathyroidism.

Acquired hypoparathyroidism in children is most commonly caused by autoimmune destruction of the glands (autoimmune polyendocrinopathy syndrome type 1 [APS1, 240300]). Manifestations of APS1 include adrenal insufficiency, mucocutaneous candidiasis, and hypoparathyroidism; loss of function mutations in the *AIRE* gene, which encodes an autoimmune regulator with features of a transcription factor, is often identified in such cases. Autoantibodies against parathyroid-specific antigens (NALP5, CaSR) have been identified [8, 9]. Thyroidectomy may result in inadvertent removal of parathyroid tissue, resulting in acquired hypoparathyroidism, as can heavy metal deposition associated with thalassemia and Wilson's disease.

Vitamin D-related hypocalcemia

Vitamin D deficiency has been most commonly observed in African-American infants that are breastfed or have limited dietary intake of dairy products. Older age groups may be affected [10]. Chronically low intake of vitamin D and/or sunlight exposure may result in rickets. Dietary Ca deficiency can result in a similar clinical picture; some children have a combined Ca/vitamin D deficiency [11].

Inherited defects in vitamin D metabolism may mimic nutritional rickets. Mutations in the major vitamin D 25-hydroxylase (*CYP2R1*, 60081) are extremely rare [12]. Mutations in 1-alpha-hydroxylase (*CYP27B1*) (1-alpha-hydroxylase deficiency, or vitamin D-dependent rickets, type 1, 264700) are well described; mutations may result in reduced enzymatic activity caused by disruption of binding to substrate or to the adrenodoxin moiety of the enzyme complex [13]. Mutations in the vitamin D receptor (*VDR*) (hereditary vitamin D resistance, or vitamin D-dependent rickets, type 2, 277440) are also well described.

Other causes of hypocalcemia

Severe hypocalcemia may occur in children after rectal or oral administration of Pi enema preparations [14]. The resultant hyperphosphatemia can be extreme (up to 20 mg/dL), with accompanying severe hypocalcemia and hypomagnesemia. Such preparations should never be administered to infants less than 2 years of age. Hyperphosphatemia with rhabdomyolysis has resulted in hypocalcemia. Rotavirus infections and other enteropathies may induce malabsorption-related hypocalcemia [15]. Hungry bone syndrome after treatment of severe rickets or after parathyroidectomy can result in transient, severe hypocalcemia [16].

Infantile osteopetrosis, may present with hypocalcemia caused by impaired bone resorption [17]. Decreases in ionized Ca occur in infants undergoing exchange transfusions with citrated blood products or receiving lipid infusions. Citrate and fatty acids may complex with ionized Ca, reducing free Ca levels. Likewise, EDTA chelation therapy has resulted in acute hypocalcemia associated with fatality [18]. Alkalosis secondary to adjustments in ventilatory assistance may provoke a shift from ionized to protein-bound Ca. Prolonged pharmacological inhibition of bone resorption, as with potent bisphosphonate therapy, may also precipitate hypocalcemia [19].

Treatment of hypocalcemia

Early neonatal hypocalcemia is usually treated when total serum Ca is below 6 mg/dL (1.25 to 1.50 mmol/L) (or ionized Ca below 3 mg/dL, 0.62 to 0.72 mmol/L) in premature infants or below 7 mg/dL (1.75 mmol/L) in term infants. Therapy of acute tetany consists of intravenous (never intramuscular) Ca gluconate (10% solution) given slowly (<1 mL/min); 1 to 3 mL will usually arrest seizure activity. Doses should not exceed 20 mg of elemental Ca/kg body weight, and may be repeated up to four times per 24 hours. After successful management of acute emergencies, maintenance therapy is achieved by intravenous administration of 20 to 50 mg of elemental Ca/kg body weight/24 hours. Ca glubionate is a commonly used oral supplement (most preparations provide

115 mg of elemental Ca/5 mL). Management of late neonatal tetany should include low-phosphate formula such as Similac PM 60/40, in addition to Ca supplements. Therapy can usually be discontinued after several weeks.

The place of vitamin D in the management of transient hypocalcemia is less clear. A significant portion of intestinal Ca absorption in newborns occurs by facilitated diffusion and is not vitamin D dependent. Thus, vitamin D metabolites may not be as useful for the short-term management of transient hypocalcemia as added Ca.

In persistent hypoparathyroidism, calcitriol [1,25(OH)$_2$D] is used in the long term. In the older child, Ca and active vitamin D metabolites are usually titrated to maintain serum Ca in an asymptomatic range without incurring hypercalciuria. In hypoparathyroidism, a target serum Ca of 7.5 to 9.0 mg/dL is recommended, because higher serum Ca levels often result in hypercalciuria. In autosomal dominant hypocalcemia, caused by activating mutations in the calcium-sensing receptor or its associated coupling protein (CaSR,GNA11) (601199, 139313), thiazide diuretics can be helpful if symptomatic hypocalcemia coexists with hypercalciuria. Hypercalciuria is not typically observed in patients with pseudohypoparathyroidism type 1a when serum Ca is maintained in the usual normal range. The use of PTH for treatment of hypoparathyroidism is an appealing option that is discussed in Chapter 86.

Treatment of vitamin D deficiency rickets is accomplished by provision of adequate oral vitamin D (1000 to 10,000 units daily depending on the age of the child), and providing adequate dietary calcium (30 to 70 mg elemental calcium/kg body weight/day in three divided doses) [20]. The clinical response to pharmacological vitamin D therapy in infants and small children requires monitoring to avoid hypervitaminosis D [21]. A daily dose of 400 to 800 units of vitamin D will prevent vitamin D deficiency in premature and exclusively breastfed infants.

Therapy with calcitriol is effective for 1-alpha-hydroxylase deficiency, but in hereditary vitamin D resistance exceptionally high dosages may be required, and even complete lack of response may be evident. Intravenous calcium has been useful in such settings, particularly until secondary hyperparathyroidism is corrected, with gradual progression to enteral calcium treatment [22, 23].

Disorders of hypercalcemia

Infants with mild to moderate hypercalcemia (11.0 to 12.5 mg/dL) are usually asymptomatic. More severe hypercalcemia may lead to failure to thrive, poor feeding, hypotonia, vomiting, seizures, lethargy, polyuria, dehydration, and hypertension. Hypercalcemia is discussed in detail in Chapters 81 and 84. Specific childhood syndromes are described below.

Severe neonatal hyperparathyroidism (SNHP, 239200) presents in the first few days of life. Serum Ca levels may reach 30 mg/dL. Serum Pi is low, and serum PTH is elevated. Nephrocalcinosis may be present on ultrasonagraphic examination. SNHP is a rare autosomal recessive disorder caused by homozygous CaSR loss-of-function mutations [24], occurring in families with familial hypocalciuric hypercalcemia (FHH, 145980); SNHP is life-threatening, usually requiring emergency extirpation of the parathyroid glands. Significant hypercalcemia requiring therapeutic intervention has been reported in newborns with heterozygous mutations of the CaSR as well. In addition, FHH2 and FHH3 caused by inactivating mutations in GNA11 (145981) or AP2S1 (600740) have been recently described [25, 26]; infants with FHH3 may have significant hypercalcemia requiring therapy.

Hypercalcemia occurs in infants with Williams syndrome (194050), often associated with a deletion at the elastin gene locus. Growth failure, characteristic facies, cardiovascular abnormalities (usually supravalvular aortic stenosis or peripheral pulmonic stenosis), delayed psychomotor development, and selective mental deficiency may be present. Hypercalcemia usually subsides spontaneously by a year of age, but may rarely persist for longer. Treatment has consisted of a vitamin D-free, low Ca diet with the addition of high doses of corticosteroids if necessary. We find pamidronate safer and more effective than steroid therapy if medication is indicated.

Subcutaneous fat necrosis is a self-limited disorder of infancy presenting with hypercalcemia and erythematous or violacious skin. Pamidronate is useful when significant hypercalcemia is unresponsive to dietary Ca and vitamin D restriction.

Vitamin D intoxication, is evidenced by increased circulating 25-OHD levels. Total 1,25(OH)$_2$D levels are usually normal whereas free 1,25(OH)$_2$D is elevated [27]. Vitamin A intoxication may manifest with bone pain, hypercalcemia, headache, pseudotumor cerebri, and an exfoliative erythematous rash. Alopecia and ear discharge may be present. Hypercalcemia is mediated by increased bone resorption. Serum retinyl ester levels should be determined to establish the diagnosis.

Other conditions in children in which hypercalcemia may occur include Down syndrome, skeletal dysplasias (such as Jansen's, 156400), hypophosphatasia (241500, 241510), SHORT syndrome (269880) [28] and osteogenesis imperfecta (120150 and others). Endogenous overproduction of 1,25(OH)$_2$D occurs in granulomatous diseases, such as cat-scratch disease. Inflammatory disorders may generate hypercalcemia via increased bone resorption as in Crohn disease. Other causes include those commonly encountered in adults: immobilization, malignancy, and acquired hyperparathyroidism. Iatrogenic causes include parenteral nutrition, dialysis, and drugs (eg, thiazides, antifungals) [29].

Increasing reports of "idiopathic infantile hypercalcemia" attributed to loss of function mutations in CYP24A1 (126065), the vitamin D-24 hydroxylase, has invoked impaired catabolism of 1,25(OH)$_2$D as one potential mechanism for this syndrome [30, 31]. Some reports suggest that a significant proportion of patients with unexplained hypercalcemia and suppressed circulating PTH levels are affected [31]. Generally, a severe phenotype is evident in biallelic mutations, and mild forms may be evident with certain heterozygous mutations. Recently, mutations in *SLC34A1* (616963), encoding the renal sodium phosphate cotransporter (NaPi2a) have been identified as another cause of this syndrome [32].

Finally, hypercalcemia may accompany primary hypophosphatemia in certain situations.

Treatment of hypercalcemia

Management of acute hypercalcemia consists of administration of intravenous saline with attention to restoration of euvolemia. Furosemide (1 mg/kg body weight) has been used intravenously at 6 to 8 hour intervals. However, its use has fallen into disfavor in the setting of normal heart function and risk of further compromise in volume status. The renal handling of other solutes (eg, potassium and magnesium) may be altered. Bisphosphonate therapy for unremitting hypercalcemia in children has become widely accepted [33]. Calcitonin is often effective in the short term, but patients may become refractory to this drug with repetitive use. High dose glucocorticoid therapy has been used in the past, but complications with this approach may be greater than that found with bisphosphonates.

DISORDERS OF PHOSPHATE HOMEOSTASIS

Disorders of hypophosphatemia

Serum Pi is greater in young children than in adults. Unfortunately lapses in diagnosis of childhood hypophosphatemia occur because this clinical difference is not always recognized (Table 91.1).

Table 91.1. Normative values in mg/dL (mmol/L) for serum phosphate by age.

Age (y)	Mean	2.5th Percentile	97.5th Percentile
0–0.5	6.7 (2.15)	5.8 (1.88)	7.5 (2.42)
2	5.6 (1.81)	4.4 (1.43)	6.8 (2.20)
4	5.5 (1.77)	4.3 (1.38)	6.7 (2.15)
6	5.3 (1.72)	4.1 (1.33)	6.5 (2.11)
8	5.2 (1.67)	4.0 (1.29)	6.4 (2.06)
10	5.1 (1.63)	3.8 (1.24)	6.2 (2.01)
12	4.9 (1.58)	3.7 (1.19)	6.1 (1.97)
14	4.7 (1.53)	3.6 (1.15)	6.0 (1.92)
16	4.6 (1.49)	3.4 (1.10)	5.8 (1.88)
20	4.3 (1.39)	3.1 (1.01)	5.5 (1.78)
Adult	3.6 (1.15)	2.7 (0.87)	4.4 (1.41)

Source: Brodehl J, Gellissen K, Weber HP. Postnatal development of tubular phosphate reabsorption. Clin Nephrol. 1982 Apr;17(4):163–71.

Hypophosphatemia may result from decreased Pi supply, excessive renal losses, or intracellular/extracellular compartmental movement of Pi. "Supply" problems result from dietary deficiency or limited intestinal absorption of Pi. Reduced dietary intake occurs in breast-fed premature infants, because human milk is relatively low in Pi content. Fortifiers have been developed to restore mineral content to human milk; these may result in hypercalcemia, so monitoring may be indicated when used. Hypophosphatemia caused by inadequate dietary Pi can be treated with 20 to 25 mg of elemental phosphorus/kg body weight/day, given orally in three to four divided doses. Children with complex gastrointestinal disorders fed amino acid-based formulas (particularly Neocate), may develop hypophosphatemia which can be associated with fractures or rickets. Bioavailability of Pi appears to be compromised, despite normal Pi content of the formula. Thus, serum Pi levels should be monitored in children fed with these formulas. Treatment consists of providing phosphate supplementation or implementing a change in formula under careful observation.

Hypophosphatemia secondary to renal losses occurs in several primary Pi wasting disorders, of which X-linked hypophosphatemia (XLH, 307800) is the most common [34]. XLH typically presents in the second or third year of life with progressive leg bowing. Children may be incorrectly diagnosed with other disorders (typically metaphyseal dysplasias). A delayed diagnosis of XLH may result in the child missing early medical therapy, which is beneficial for growth and leg alignment. XLH is an FGF23-mediated form of hypophosphatemia, and the elevated circulating FGF23 further limits $1,25(OH)_2D$ production and increases its turnover.

Medical treatment consists of the combined administration of Pi with calcitriol. Dose adjustments based on clinical monitoring result in a wide range of doses. Calcitriol is usually given in two daily doses, at 20 to 50 ng/kg body weight/day, and phosphorus at 20 to 50 mg/kg body weight/day, in three to five divided doses. Detailed treatment guidelines for XLH are available [35]. Hyperparathyroidism or hypercalcemia may complicate therapy, and monitoring of serum Ca, Pi, alkaline phosphatase, PTH, and urinary Ca and creatinine is suggested at 3 to 4 month intervals in children. Height and assessment of bow defects should be routinely monitored. Radiographs of the epiphyses of the distal femur and proximal tibia are obtained every 2 years, or more frequently when no improvement or progressive skeletal disease is evident. Nephrocalcinosis (per ultrasonogram) may result from this therapy, but significant clinical sequelae in patients with mild nephrocalcinosis are not generally observed.

Treatment with an inhibitory anti-FGF23 antibody, burosumab, has been recently investigated in adults and children with XLH, demonstrating correction of serum Pi and $1,25(OH)_2D$, as well as radiographic improvement in rickets in children previously treated with conventional therapy [36–38]. Burosumab was recently approved by the US FDA for the treatment of XLH in children and adults in the US and by the European Medical Agency (EMA) for the treatment of children with XLH in Europe. Other

FGF23-mediated hypophosphatemic disorders include tumor-induced osteomalacia (caused by neoplastic overproduction of FGF23) (see Chapter 83), autosomal dominant hypophosphatemic rickets (193100) caused by mutations that disrupt proteolytic cleavage of FGF23 [39], and in fibrous dysplasia/McCune Albright syndrome (174800) [40]. Autosomal recessive hypophosphatemic rickets (ARHR, 241520) is caused by mutations in dentin matrix protein 1 (DMP1), in Raine syndrome due to mutations in FAM20C, and in ARHR2 (613312), due to mutations in ectonucleotide pyrophosphatase/phophodiesterase 1 (ENPP1) [41–43]. This group of disorders is discussed in detail in Chapter 83. Finally, this constellation of biochemical findings in concert with parathyroid hyperplasia are described with overexpression of Klotho (612089) [44].

Renal Pi loss independent of FGF23 occurs in hereditary hypophosphatemic rickets with hypercalciuria (HHRH, 241530) secondary to mutations in the renal NaPi2c cotransporter (SLC34A3) [45]. In contrast to XLH and other FGF23-mediated disorders, circulating 1,25(OH)$_2$D is elevated in HHRH; hypercalciuria is common and renal stones may occur. Osteoporosis may develop. HHRH is treated with oral phosphate, without vitamin D metabolites. Generalized tubular dysfunction (Fanconi syndrome, 134600) may occur in cystinosis (219800), Lowe syndrome (309000), and Wilson disease (277900). Hypophosphatemia may occur in Dent disease (300009), an X-linked recessive disorder caused by mutations in CLCN5, which encodes a renal tubular chloride channel.

Acute movement of extracellular Pi to intracellular compartments results in hypophosphatemia, as occurs during correction of diabetic ketoacidosis with insulin-induced Pi uptake, or with refeeding after long-term nutritional deprivation, as found during rehabilitation of patients with anorexia nervosa. Pi levels reach a nadir within a week of refeeding; slow oral feeds minimize the severity of this phenomenon. Targeting a minimum 4-day weight gain between 0.36 kg and 0.55 kg has been recommended to reduce complications of refeeding in anorexic adolescents. [46].

Hyperphosphatemia

Hyperphosphatemia is uncommon in children with normal renal function. It may occur with use of phosphate-containing enemas often accompanied by a reciprocal decrease in serum Ca, precipitating tetany and seizures. Serum Pi may increase with rapid lysis of bulky tumors (tumor lysis syndrome).

Hyperphosphatemia is a biochemical hallmark of hypoparathyroidism and pseudohypoparathyroidism, and is observed in chronic kidney disease, where progressive loss of nephrons results in limited excretion of Pi. Hyperphosphatemic tumoral calcinosis (HTC, 211900) results from loss of function mutations in FGF23 [47], GALNT3, which encodes an enzyme that initiates O-glycosylation of FGF23, important for preventing proteolytic processing and secretion [48], or Klotho, a membrane protein necessary for FGF23 signaling through FGF receptors [49]. As expected from the impaired FGF23 activity, HTC has the converse biochemical phenotype of XLH (hyperphosphatemia caused by an increased tubular maximum threshold for Pi reabsorption and increased circulating 1,25(OH)$_2$D). When serum Pi and/or Ca concentrations result in chronic elevations in the Ca×Pi product, risk of soft tissue calcification may ensue. National Kidney Foundation guidelines recommend maintaining the Ca×Pi product less than 65 mg^2/dL^2 in children under 12 years of age, and less than 55 mg^2/dL^2 in older children [50].

DISORDERS OF MAGNESIUM

Disorders of hypomagnesemia

Familial hypomagnesemia with secondary hypocalcemia (602014) is an autosomal recessive disease caused by mutations in the TRPM6 ion channel, resulting in electrolyte abnormalities in the newborn period [51], presenting with tetany or seizures. Hypocalcemia may be refractory to therapy unless Mg levels are corrected. Mutations in a renal tubular paracellular transport protein, paracellin, of the claudin family, and encoded by CLDN16, may also cause hypomagnesemia, hypocalcemia, and hypercalciuria (248250) [52]. CLDN19, another member of the claudin family, has been implicated as causal to heritable hypomagnesemia (248190) [53]. *Gitelman syndrome* (263800) is an autosomal recessive disorder of Mg and potassium wasting with metabolic alkalosis and hypocalciuria, caused by mutations in the gene encoding a thiazide-sensitive Na-Cl cotransporter (SLC12A3) [2]. Mutations in HNF1B, known to be important in renal development, have been shown to manifest hypomagnesemia caused by renal tubular wasting [54]. Hypomagnesemia may accompany renal losses caused by the tubulopathy observed with certain medications: chemotherapeutic agents, aminoglycosides, and cyclosporine are typical culprits in children. Refeeding syndrome, which may result in hypophosphatemia and hypokalemia, may also involve hypomagnesemia [55]. Low serum magnesium levels may also accompany hypoparathyroidism.

Treatment of hypomagnesemia

For acute symptomatic hypomagnesemia, Mg sulfate is given intravenously using cardiac monitoring or intramuscular as a 50% solution at a dose of 0.1 to 0.2 mL/kg body weight. One or two doses may treat transient hypomagnesemia: a dose may be repeated after 12 to 24 hours. Patients with primary defects in Mg metabolism require long-term oral Mg, best given in several divided doses through the day to avoid diarrhea. We begin oral supplementation at 5 mg of elemental Mg/kg body weight/day. A variety of salts are available for oral use; we have had limited complications with Mg oxide.

Hypermagnesemia

Hypermagnesemia, unusual in pediatrics, may occur transiently after fetal exposure to maternal Mg infusions used in the management of eclampsia/preeclampsia, or with excessive use of cathartic preparations [56]. Severe hypermagnesemia can result in apnea, respiratory depression, and cardiac arrhythmias. Hypocalcemia may also result from hypermagnesemia.

SKELETAL MANIFESTATIONS OF DISORDERS OF CALCIUM AND PHOSPHATE

The typical skeletal abnormality in the growing child with a paucity of available Ca or Pi is rickets. The clinical use of the term "rickets" refers to growth plate cartilage abnormalities observed in the long bones, and manifest radiographically by widened metaphyses, irregular or "frayed" metaphyseal edges, and "cupped" metaphyseal deformations (Fig. 91.1).

The histological correlate of the radiographic findings at the growth plate is the expansion of the hypertrophic zone of chondrocytes [57]. Rickets is usually accompanied by osteomalacia in bone tissue. Weight bearing on the undermineralized skeleton results in characteristic bowing. A child with overt rickets may have minimal leg deformity before walking. However, enlarged wrists or costochondral junctions ("rachitic rosary") are typical. An osteomalacic skull (craniotabes) may be present.

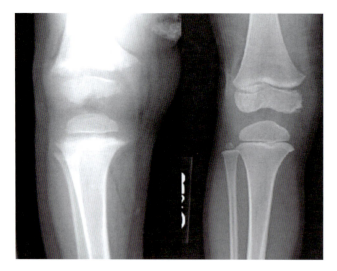

Fig. 91.1. *Left*, Radiograph of the right knee in an infant with vitamin D-deficiency rickets. Classic epiphyseal deformities are observed, with metaphyseal flaring and frayed edges of the metaphyseal–growth plate junction. *Right*, a normal knee is shown for comparison.

REFERENCES

1. Stefanaki E, Koropuli M, Stefanaki S, et al. Atrioventricular block in preterm infants caused by hypocalcemia: a case report and review of the literature. Eur J Obst Gyn Reproduct Biol. 2005;120(1):115–6.
2. Schlingmann KP, Konrad M, Seyberth HW. Genetics of hereditary disorders of magnesium homeostasis. Pediatr Nephrol. 2004;19(1):13–25.
3. Webber SA, Hatchwell E, Barber JC, et al. Importance of microdeletions of chromosomal region 22q11 as a cause of selected malformations of the ventricular outflow tracts and aortic arch: a three-year prospective study. J Pediatr. 1996;129(1):26–32.
4. Sykes KS, Bachrach LK, Siegel-Bartelt J, et al. Velocardiofacial syndrome presenting as hypocalcemia in early adolescence. Archiv Pediatr Adolescent Med. 1997;151(7):745–7.
5. Hiéronimus S, Bec-Roche M, Pedeutour F, et al. The spectrum of parathyroid gland dysfunction associated with the microdeletion 22q11. Eur J Endocrinol. 2006;155(1):47–52.
6. Thakker RV. Genetics of endocrine and metabolic disorders: parathyroid. Revs Endocr Metabol Disorders. 2004;5(1):37–51.
7. Daw SC, Taylor C, Kraman M, et al. A common region of 10p deleted in DiGeorge and velocardiofacial syndromes. Nat Genet. 1996;13(4):458–60.
8. Alimohammadi M, Björklund P, Hallgren A, et al. Autoimmune polyendocrine syndrome type 1 and NALP5, a parathyroid autoantigen. N Engl J Med. 2008;358(10):1018–28.
9. Brown EM. Anti-parathyroid and anti-calcium sensing receptor antibodies in autoimmune hypoparathyroidism. Endocrinol Metab Clin North Am. 2009;38(2):437–45.
10. Gordon CM, DePeter KC, Feldman HA, et al. Prevalence of vitamin D deficiency among healthy adolescents. Arch Pediatr Adolescent Med. 2004;158(6):531–7.
11. DeLucia MC, Mitnick ME, Carpenter TO. Nutritional rickets with normal circulating 25-hydroxyvitamin D: a call for re-examining the role of dietary calcium intake in North American children. J Clin Endocrinol Metab. 2003;88(8):3539–45.
12. Cheng JB, Levine MA, Bell NH, et al. Genetic evidence that the human CYP2R1 enzyme is a key vitamin D 25-hydroxylase. Proc Nat Acad Sci USA. 2004;101(20):7711–5.
13. Zalewski A, Ma NS, Legeza B, et al. Vitamin D-dependent rickets type 1 caused by mutations in CYP27B1 affecting protein interactions with adrenodoxin. J Clin Endocrinol Metab. 2016;101(9):3409–18.
14. Walton DM, Thomas DC, Aly HZ, et al. Morbid hypocalcemia associated with phosphate enema in a six-week-old infant. Pediatrics. 2000;106(3):E37.
15. Foldenauer A, Vossbeck S, Pohlandt F. Neonatal hypocalcaemia associated with rotavirus diarrhoea. Eur J Peds. 1998;157(10):838–42.

16. Yesilkaya E, Cinaz P, Bideci A, et al. Hungry bone syndrome after parathyroidectomy caused by an ectopic parathyroid adenoma. J Bone Miner Metab. 2009;27(1):101–4.
17. Martinez C, Polgreen LE, DeFor TE, et al. Characterization and management of hypercalcemia following transplantation for osteopetrosis. Bone Marrow Transplantat. 2010;45(5):939–44.
18. Baxter AJ, Krenzelok EP. Pediatric fatality secondary to EDTA chelation. Clin Toxicol (Phila). 2008;46(10):1083–4.
19. Perman MJ, Lucky AW, Heubi JE, et al. Severe symptomatic hypocalcemia in a patient with RDEB treated with intravenous zoledronic acid. Arch Dermatol. 2009;145(1):95–6.
20. Misra M, Pacaud D, Petryk A, et al.; Drug and Therapeutics Committee of the Lawson Wilkins Pediatric Endocrine Society. Vitamin D deficiency in children and its management: review of current knowledge and recommendations. Pediatrics. 2008;122(2):398–417.
21. Vanstone MS, Oberfield SE, Shader L, et al. Hypercalcemia in children receiving pharmacologic doses of vitamin D. Pediatrics. 2012;129(4):1060–3.
22. Balsan S, Garabédian M, Larchet M, et al. Long-term nocturnal calcium infusions can cure rickets and promote normal mineralization in hereditary resistance to 1,25-dihydroxyvitamin D. J Clin Invest. 1986;77(5):1661–7.
23. Ma NS, Malloy PJ, Pitukcheewanont P, et al. Hereditary vitamin D resistant rickets: identification of a novel splice site mutation in the vitamin D receptor gene and successful treatment with oral calcium therapy. Bone. 2009;45(4):743–6.
24. Pidasheva S, D'Souza-Li L, Canaff L, et al. CASRdb: calcium-sensing receptor locus-specific database for mutations causing familial (benign) hypocalciuric hypercalcemia, neonatal severe hyperparathyroidism, and autosomal dominant hypocalcemia. Hum Mutat. 2004;24(2):107–11.
25. Nesbit MA, Hannan FM, Howles SA, et al. Mutations affecting G-protein subunit α11 in hypercalcemia and hypocalcemia. N Engl J Med. 2013;368(26):2476–86.
26. Nesbit MA, Hannan FM, Howles SA, et al. Mutations in AP2S1 cause familial hypocalciuric hypercalcemia type 3. Nat Genet. 2013;45(1):93–7.
27. Pettifor JM, Bikle DD, Caveleros M, et al. Serum levels of free 1,25-dihydroxyvitamin D in vitamin D toxicity. Ann Intern Med. 1995;122(7):511–3.
28. Reardon W, Temple IK. Nephrocalcinosis and disordered calcium metabolism in two children with SHORT syndrome. Am J Med Genet. 2008;146A(10):1296–8.
29. Smith PB, Steinbach WJ, Cotten CM, et al. Caspofungin for the treatment of azole resistant candidemia in a premature infant. J Perinatol. 2007;27(2):127–9.
30. Schlingmann KP, Kaufmann M, Weber S, et al. Mutations in CYP24A1 and idiopathic infantile hypercalcemia. N Eng J Med. Epub 2011 June 15.
31. Dauber A, Nguyen TT, Sochett E, et al. Genetic defect in CYP24A1, the vitamin D 24-hydroxylase gene, in a patient with severe infantile hypercalcemia. J Clin Endocrinol Metab. 2012;97(2):E268–74.
32. Schlingmann KP, Ruminska J, Kaufmann M, et al. Autosomal-recessive mutations in SLC34A1 encoding sodium-phosphate cotransporter 2A cause idiopathic infantile hypercalcemia. J Am Soc Nephrol. 2016;27(2):604–14.
33. Lteif AN, Zimmerman D. Bisphosphonates for treatment of childhood hypercalcemia. Pediatrics. 1998;102(4):990–3.
34. Holm IA, Econs MJ, Carpenter TO. Familial hypophosphatemia and related disorders. In: Glorieux FH, Juppner H, Pettifor JM (eds) Pediatric Bone: Biology and Diseases (2nd ed.). San Diego: Elsevier, 2012, pp. 699–726.
35. Carpenter TO, Imel EA, Holm IA, et al. A clinician's guide to X-linked hypophosphatemia. J Bone Min Res. 2011;26(7):1381–8.
36. Carpenter TO, Imel EA, Ruppe MD, et al. Randomized trial of the anti-FGF23 antibody KRN23 in X-linked hypophosphatemia. J Clin Invest. 2014;124(4):1587–97.
37. Imel EA, Zhang X, Ruppe MD, et al. Prolonged correction of serum phosphorus in adults with X-linked hypophosphatemia using monthly doses of KRN23. J Clin Endocrinol Metab. 2015;100(7):2565–73.
38. Carpenter TO, Whyte MP, Imel EA, et al. Burosumab therapy in children with X-linked hypophosphatemia. N Engl J Med. 2018;378(21):1987–98.
39. White KE, Jonsson KB, Carn G, et al. The autosomal dominant hypophosphatemic rickets (ADHR) gene is a secreted polypeptide overexpressed by tumors that cause phosphate wasting. J Clin Endocrinol Metab. 2001;86(2):497–500.
40. Riminucci M, Collins MT, Fedarko NS, et al. FGF-23 in fibrous dysplasia of bone and its relationship to renal phosphate wasting. J Clin Invest. 2003;112(5):683–92.
41. Feng JQ, Ward LM, Liu S, et al. Loss of DMP1 causes rickets and osteomalacia and identifies a role for osteocytes in mineral metabolism. Nat Genet. 2006;38(11):1310–5.
42. Lorenz-Depiereux B, Schnabel D, Tiosano D, et al. Loss-of-function ENPP1 mutations cause both generalized arterial calcification of infancy and autosomal-recessive hypophosphatemic rickets. Am J Hum Genet. 2010;86(2):267–72.
43. Levy-Litan V, Hershkovitz E, Avizov L, et al. Autosomal-recessive hypophosphatemic rickets is associated with an inactivation mutation in the ENPP1 gene. Am J Hum Genet. 2010;86(2):273–78.
44. Brownstein CA, Adler F, Nelson-Williams C, et al. A translocation causing increased α–Klotho level results in hypophosphatemic rickets and hyperparathyroidism. Proc Nat Acad Sci U S A. 2008;105(9):3455–60.
45. Bergwitz C, Roslin NM, Tieder M, et al. SLC34A3 mutations in patients with hereditary hypophosphatemic rickets with hypercalciuria (HHRH) predict a key role for the sodium-phosphate cotransporter NaPi-IIc in maintaining phosphate homeostasis and skeletal function. Am J Hum Gen. 2006;78(2):179–92.
46. Fisher M. Treatment of eating disorders in children, adolescents, and young adults. Pediatr Rev. 2006;27(1):5–16.

47. Benet-Pagès A, Orlik P, Strom TM, et al. An FGF23 missense mutation causes familial tumoral calcinosis with hyperphosphatemia. Hum Mol Genet. 2005;14(3):385–90.
48. Topaz O, Shurman DL, Bergman R, et al. Mutations in GALNT3, encoding a protein involved in O-linked glycosylation, cause familial tumoral calcinosis. Nat Genet. 2004;36(6):579–81.
49. Ichikawa S, Imel EA, Kreiter ML, et al. A homozygous missense mutation in human KLOTHO causes severe tumoral calcinosis. J Clin Invest. 2007;117(9):2684–91.
50. National Kidney Foundation. K/DOQI clinical practice guidelines for bone metabolism and disease in children with chronic kidney disease. Am J Kidney Dis. 2005;46(Suppl 1):S1.
51. Schlingmann KP, Weber S, Peters M, et al. Hypomagnesemia with secondary hypocalcemia is caused by mutations in TRPM6, a new member of the TRPM gene family. Nat Genet. 2002;31(2):166–70.
52. Simon DB, Lu Y, Choate KA, et al. Paracellin-1, a renal tight junction protein required for paracellular Mg2+ resorption. Science. 1999;285(5424):103–6.
53. Konrad M, Schaller A, Seelow D, et al. Mutations in the tight-junction gene claudin 19 (CLDN19) are associated with renal magnesium wasting, renal failure, and severe ocular involvement. Am J Hum Genet. 2006;79(5):949–57.
54. Adalat S, Woolf AS, Johnstone KA, et al. HNF1B mutations associate with hypomagnesemia and renal magnesium wasting. J Am Soc Nephrol. 2009;20(5):1123–31.
55. Fuentebella J, Kerner JA. Refeeding syndrome. Ped Clin North Am. 2009;56(5):1201–10.
56. Kutsal E, Aydemir C, Eldes N, et al. Severe hypermagnesemia as a result of excessive cathartic ingestion in a child without renal failure. Pediatr Emerg Care. 2007;23(8):570–2.
57. Sabbagh Y, Carpenter TO, Demay MB. Hypophosphatemia leads to rickets by impairing caspase-mediated apoptosis of hypertrophic chondrocytes. Proc Nat Acad Sci USA. 2005;102(27):9637–42.

92

Paget Disease of Bone

Julia F. Charles[1], Ethel S. Siris[2], and G. David Roodman[3]

[1]*Brigham and Women's Hospital, Boston, MA, USA*
[2]*Department of Medicine, Columbia University, New York, NY, USA*
[3]*Division of Hematology and Oncology, Indiana University School of Medicine, Indianapolis, IN, USA*

INTRODUCTION

Paget disease of bone is a localized disorder of bone remodeling and the second most common bone disease after osteoporosis. The process is initiated by increases in osteoclast-mediated bone resorption, with compensatory increases in new bone formation. Affected skeletal sites develop a disorganized mosaic of woven and lamellar bone resulting in bone that is expanded in size, less compact, more vascular, and more susceptible to deformity or fracture than normal bone. Clinical signs and symptoms vary from one patient to the next depending on the number and location of affected skeletal sites, as well as on the degree and extent of the abnormal bone turnover. Most patients are asymptomatic, but a substantial minority may experience symptoms, including bone pain or deformity, secondary arthritis, fracture, excessive warmth over bone from hypervascularity, and compression of neural tissues adjacent to pagetic bone.

ETIOLOGY

Both genetic and environmental factors have been implicated in the pathogenesis of Paget disease. Paget disease occurs commonly in families and can be transmitted vertically in an autosomal dominant pattern. Fifteen percent to 30% of Paget disease patients have positive family histories of the disorder [1], and familial aggregation studies in a US population suggest that first-degree relatives of pagetic subjects have a seven times greater risk for developing Paget disease than someone without an affected relative [2]. Recent genome-wide association studies have identified multiple susceptibility loci for Paget disease, including variants in *CSF-1*, *RANK*, *PML* and other genes [3–5].

The most frequent mutations linked to Paget disease are in the gene encoding the ubiquitin binding protein, sequestasome-1, *SQSTM1/p62*. Mutations in *SQSTM1* occur in 30% of patients with familial Paget disease with the P392L mutation being the most frequent [6]. *SQSTM1* gene mutations have been associated with severity of Paget disease, with carriers having an earlier age of onset and more commonly requiring surgery and bisphosphonate therapy [7]. Sequestasome-1 plays an important role in the NFκB signaling pathway. The clinical phenotype of patients with *SQSTM1* mutations can be variable clinical phenotype (including no evidence of Paget disease in at least one or two carriers) and no gene dose effect can be observed between heterozygotes and homozygotes individuals. Recent studies have reported that the *SQSTM1*[P392L] mutation is either a predisposing mutation for Paget disease or can result in Paget disease in experimental animal models [8, 9]. However, mice in which the normal *SQSTM1* gene has been replaced with *SQSTM1*[P392L] either develop pagetic-like lesions predominantly in their femurs [8] or do not develop Paget disease [9].

There is a restricted geographical distribution for the occurrence of Paget disease. Paget disease is most common in Europe, North America, Australia, and New Zealand in persons of Anglo-Saxon descent and is extremely uncommon in Asia, Africa, and Scandinavia. Some recent studies have reported an apparent decline in the frequency and severity of Paget disease in both Great Britain and New Zealand [10, 11]. The basis for this decline is unknown, but the changes are too rapid to be explained by genetic factors and cannot be explained by migration patterns of persons with a predisposition to Paget disease.

For more than 30 years, studies have suggested that Paget disease may result from a chronic paramyxoviral infection. This is based on ultrastructural studies by Rebel and colleagues [12] who showed that nuclear and, less commonly, cytoplasmic inclusions similar to nucleocapsids from paramyxoviruses were present in osteoclasts from Paget disease patients. Mills and colleagues [13] also reported that the measles virus nucleocapsid antigen was present in osteoclasts from patients with Paget disease, but not from patients with other bone diseases. In some specimens, both measles virus and respiratory syncytial virus nucleocapsid proteins were shown by immunocytochemistry on serial sections. Others reported that osteoclasts from patients with Paget disease expressed canine distemper virus (CDV) nucleocapsid transcripts [14, 15].

Kurihara and colleagues [16] provided evidence for a pathophysiological role for measles virus in the abnormal osteoclast activity in Paget disease both in vitro and in vivo. Transfection of the gene encoding the measles virus nucleocapsid protein (MVNP) into normal human osteoclast precursors resulted in formation of osteoclasts that expressed many characteristics of pagetic osteoclasts. However, other workers have been unable to confirm the presence of measles virus or CDV in pagetic osteoclasts [17]. Kurihara and colleagues also targeted the MVNP gene to cells in the osteoclast lineage in transgenic mice, and found that 29% of these mice develop localized bone lesions that are similar to lesions observed in patients with Paget disease [18], whereas mice expressing both the MVNP gene and the $SQSTM1^{P392L}$ mutation develop exuberant pagetic lesions [19]. They found that many of the effects of MVNP observed in these mice were moderated by IL-6 [19]. Recently, Teramachi and colleagues [20] determined if osteoclasts expressing MVNP, $SQSTM1^{P392L}$, or normal osteoclasts stimulated osteoblast differentiation, because treatments that target osteoclasts block both pagetic bone resorption and formation (see later). They found that the high IL-6 levels induced by MVNP increased IGF1 production by osteoclasts and the expression of coupling factors, specifically ephrinB2 on osteoclasts and EphB4 on osteoblasts. IGF1 further enhanced ephrinB2 expression on osteoclasts as well as osteoblast differentiation. Importantly, ephrinB2 and IGF1 levels were increased in MVNP-expressing osteoclasts from patients with Paget disease and MVNP-transduced human osteoclasts compared with levels detected in controls.

Among the many questions that still remain to be explained to understand the contributions of environmental and genetic factors to Paget disease are: (i) since paramyxoviral infections such as measles virus occur worldwide, why does Paget disease have a very restricted geographic distribution?; (ii) how does the virus persist in osteoclasts in patients who are immunocompetent for such long periods of time, because measles virus infections generally occur in children rather than adults, and Paget disease is usually diagnosed in patients over the age of 55?; (iii) why does Paget disease remain so highly localized in patients after diagnosis? (iv) what is the explanation for the variable phenotypic presentation of patients with familial Paget disease, especially those patients who carry the mutated gene yet do not have Paget disease even though they are over 70 years of age?

PATHOLOGY

The initiating lesion in Paget disease is an increase in bone resorption caused by an abnormality in osteoclasts found at affected sites. Pagetic bone contains more numerous osteoclasts than normal, and these osteoclasts contain substantially more nuclei than normal osteoclasts. In response to the increase in bone resorption, numerous osteoblasts are recruited to pagetic sites where rapid new bone formation occurs. It is generally believed that the osteoblasts are intrinsically normal [21, 22].

In the earliest phases of Paget disease, increased bone resorption dominates, and lytic changes can be found on radiographs. After this, there is a combination of increased resorption and relatively tightly coupled new bone formation, produced by large numbers of osteoblasts present at these sites. During this phase, presumably because of the accelerated nature of bone formation, the new bone that is made is abnormal. Newly deposited collagen fibers are laid down in a haphazard rather than a linear fashion, creating more primitive woven bone. The end product is the so-called mosaic pattern of woven bone plus irregular sections of lamellar bone linked in a disorganized way by numerous cement lines representing the extent of previous areas of bone resorption. The bone marrow becomes infiltrated by excessive fibrous connective tissue and by an increased number of blood vessels, explaining the hypervascular state of the bone. Bone matrix is typically normally mineralized, and tetracycline labeling shows increased mineralization rates. It is not unusual, however, to find areas of pagetic biopsies in which widened osteoid seams are apparent, perhaps reflecting locally inadequate calcium/phosphorus products.

In time, the hypercellularity at a locus of affected bone may diminish, leaving the end product of a sclerotic, pagetic mosaic without evidence of active bone turnover, so-called burned out Paget disease. Typically, all phases of the pagetic process can be observed at the same time at different sites in a particular subject. The chaotic architectural changes that occur in pagetic bone contribute to the loss of structural integrity. Figure 92.1 compares the appearances of normal and pagetic bone by scanning electron microscopy.

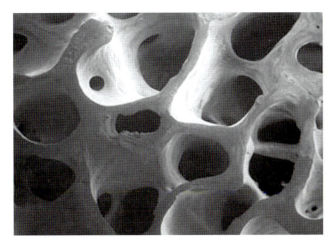

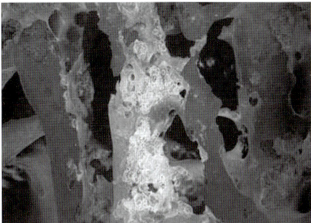

Fig. 92.1. Scanning electron micrographs with sections of normal bone (*top*) and pagetic bone (*bottom*). Both samples were taken from the iliac crest. The normal bone shows the trabecular plates and marrow spaces to be well preserved, whereas the pagetic bone has completely lost this architectural appearance. Extensive pitting of the pagetic bone is apparent, due to dramatically increased osteoclastic bone resorption. Source: Photographs courtesy of Dr David Dempster.

BIOCHEMICAL INDICES IN PAGET DISEASE

Measurements of biochemical markers of bone turnover are useful clinically in the assessment of the extent and severity of disease in the untreated state and for monitoring the response to treatment [23, 24]. Increased osteoclast-mediated bone resorption is reflected by increased levels of biomarkers of bone resorption, such as C- and N-terminal telopeptides of collagen, CTX, and NTX. Secondary increases in osteoblastic activity are associated with elevated levels of bone formation markers including serum total alkaline phosphatase (SAP), bone specific alkaline phosphatase, and procollagen type-1 N-terminal propeptide (P1NP). In untreated patients, the values of serum CTX or urine NTX and SAP rise in proportion to each other, reflecting the preserved coupling of resorption and formation. The magnitude of the increase in markers offers an estimate of the extent or severity of the abnormal bone turnover, with higher levels reflecting a more active, ongoing localized metabolic process. Active monostotic disease may have lower SAP values than polyostotic disease. Lower values (eg, <3 times the upper limit of normal) may indicate fewer pagetic sites or a lesser degree of increased bone turnover at affected sites. However, mild elevations in a patient with limited and highly localized disease (eg, the proximal tibia) may still be associated with symptoms and clear progression of disease at that site. Even a so-called "normal" SAP (eg, at the upper limit of the normal range) may not truly be normal for the pagetic patient. To be confident that the SAP reflects quiescent disease, a result in the middle of the normal range is probably required.

Monitoring biochemical markers is helpful in assessing treatment effects. All markers of bone turnover have moderate to strong correlation with disease activity assessed by scintigraphy [25]. Potent bisphosphonates are capable of normalizing biochemical markers in a majority of patients and bringing the markers to near normal in most others. CTX or NTX may become normal in days to a few weeks after bisphosphonate therapy is initiated. It is often adequate, however, to monitor SAP alone, with a baseline measure pretreatment, a posttreatment test 1 to 3 months after treatment is completed and at 6 to 12 month intervals thereafter to determine duration of the effect of that treatment course. Assessment and monitoring of disease activity by biochemical markers, specifically SAP, is recommended by the Endocrine Society Clinical Practice Guidelines [24].

Serum calcium is typically normal in untreated Paget disease, but secondary hyperparathyroidism and transient decreases in serum calcium can occur in some patients being treated with potent bisphosphonates. This results from the early suppression of bone resorption in the setting of not yet reduced new bone formation [26]. As restoration of coupling occurs with time, PTH levels fall. The problem can be largely avoided by being certain that such patients are and remain replete in both calcium and vitamin D.

CLINICAL FEATURES

Paget disease affects both men and women, with most series describing a slight male predominance. It is rarely observed in individuals younger than age 25 years, and is thought to develop after the age of 40 in most instances. In a survey of over 800 selected patients in the USA, 600 of whom had symptoms, the average age at diagnosis was 58 years [27]. Paget disease is often an incidental finding and it seems likely that many patients have the disorder for a period of time before diagnosis.

Paget disease may be monostotic, affecting only a single bone or portion of a bone (Fig. 92.2), or may be polyostotic, involving two or more bones. Sites of disease are often asymmetric. The most common sites of

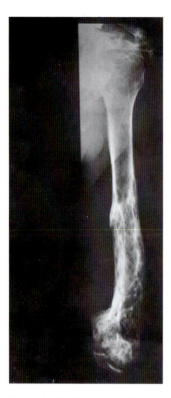

Fig. 92.2. Radiograph of a humerus showing typical pagetic change in the distal half, with cortical thickening, expansion, and mixed areas of lucency and sclerosis, contrasted with normal bone in the proximal half.

involvement include the pelvis, femur, spine, skull, and tibia. The humerus, clavicle, scapula, ribs, and facial bones are less commonly involved, and the hands and feet are only rarely affected. Clinical observation suggests that in most instances, sites affected with Paget disease when the diagnosis is made are the only ones that will show pagetic change over time. Although progression of disease within a given bone may occur, the sudden appearance of new sites of involvement years after the initial diagnosis is uncommon.

Most patients with Paget disease are asymptomatic and diagnosis is most often made when an elevated SAP is noted on routine screening or when a radiograph taken for an unrelated problem reveals typical skeletal changes. The development of symptoms or complications of Paget disease is influenced by the particular areas of involvement, the interrelationship between affected bone and adjacent structures, the extent of metabolic activity, and presence or absence of disease progression within an affected site.

Signs and symptoms

Bone pain from a site of pagetic involvement, experienced either at rest or with motion, is probably the most common symptom. Pagetic bone associated with a high turnover state has an increased vascularity, leading to a sensation of warmth of the skin overlying bone (eg, skull or tibia) that some patients perceive as an unpleasant sensation. Small transverse lucencies along the expanded cortices of involved weight-bearing bones or advancing, lytic, blade-of-grass lesions sometimes cause pain.

A bowing deformity of the femur or tibia can cause clinical problems. A bowed limb is typically shortened, resulting in specific gait abnormalities that can lead to abnormal mechanical stresses. Clinically severe secondary arthritis can occur at joints adjacent to pagetic bone (eg, the hip, knee, or ankle).

Back pain may result from enlarged pagetic vertebrae. Vertebral compression fractures can occur because the bone is of suboptimal quality. Lumbar spinal stenosis with neural impingement may arise, producing radicular pain and possibly motor impairment. Kyphosis may occur, or there may be a forward tilt of the upper back, particularly when a compression fracture or spinal stenosis is present. Paget disease in the thoracic spine may rarely cause direct spinal cord compression with motor and sensory changes. Several cases of apparent direct cord compression have been documented to have resulted from a vascular steal syndrome, whereby hypervascular pagetic bone "steals" blood from the neural tissue [28].

Paget disease of the skull may be asymptomatic, but common complaints in up to one-third of patients with diffuse skull involvement may include an increase in head size with or without frontal bossing or deformity, or headache, sometimes described as a band-like tightening around the head. Hearing loss may occur as a result of isolated or combined conductive or neurosensory abnormalities; cochlear damage from pagetic involvement of the temporal bone with loss of bone density in the cochlear capsule may be an important component [29]. Cranial nerve palsies (such as in nerves II, VI, and VII) occur rarely. With extensive skull involvement, a softening of the base of the skull may produce flattening and basilar invagination, so that the odontoid process begins to extend upward as the skull sinks downward upon it. Rarely basilar invagination can produce direct brainstem compression or an obstructive hydrocephalus and increased intra-cranial pressure caused by blockage of cerebrospinal fluid flow. Pagetic involvement of the facial bones may cause facial deformity, dental problems, and, rarely, narrowing of the airway.

Fracture through pagetic bone can occur, particularly in long bones with active areas of advancing lytic disease; the most common sites are the femoral shaft or subtrochanteric area [30]. Increased vascularity of high turnover pagetic bone may lead to substantial blood loss in the presence of fractures caused by trauma. Fractures may also occur in areas of malignant degeneration, a rare complication of Paget disease. Far more common are small fissure fractures along the convex surfaces of bowed lower extremities. These fissure fractures may be asymptomatic, stable, and persistent for years, but sometimes a more extensive transverse lucent area extends medially from the cortex, typically with symptoms of discomfort, and may lead to a clinical fracture with time. These painful lesions warrant treatment and careful radiographic follow-up over time. Fracture through pagetic bone usually heals normally, although some groups have reported as high as a 10% rate of nonunion.

Neoplastic degeneration is a relatively rare event, occurring with an incidence of less than 1%. This lesion typically presents as severe new pain at a pagetic site and has a grave prognosis. The majority of the tumors are classified as osteosarcomas, although fibrosarcomas and chondrosarcomas have been reported. The most common site of sarcomatous change appears to be the pelvis, with the femur and humerus next in frequency [31]. Although these lesions involve cells of osteoblastic lineage, osteosarcomas are typically osteolytic [32].

Less commonly, benign giant-cell tumors may develop in pagetic bone. These may present as localized masses with lytic changes on radiographs. Biopsy reveals clusters of large osteoclast-like cells, which some authors believe represent reparative granulomas [33]. These tumors usually show a remarkable sensitivity to high dose glucocorticoids, and the mass will shrink or even disappear after treatment with prednisone or dexamethasone [34], although some will grow back after treatment ends.

DIAGNOSIS

When Paget disease is suspected, the diagnostic evaluation should include a careful medical history, including family history of the condition and symptom history, and a focused physical examination. The physical examination should note the presence or absence of warmth, tenderness, or bone deformity in the skull, spine, pelvis, and extremities, as well as evidence of loss of range of motion at major joints or leg length discrepancy.

Laboratory tests include measurement of SAP, and serum calcium and 25-hydroxyvitamin D if bisphosphonate treatment is considered. Markers of bone resorption, as described earlier, may be useful in patients with limited disease burden and normal SAP [24]. Radiographic studies (bone scans and conventional radiographs) complete the initial evaluation. Bone biopsy is not usually indicated, because the characteristic radiographic and laboratory findings are diagnostic in most instances.

Bone scans are the most sensitive means of identifying possible pagetic sites but are nonspecific, and can be positive in degenerative joints or, more ominously, metastatic disease. Plain radiographs of areas of increased activity on bone scan provide the most specific information, because radiographic findings are usually characteristic to the point of being pathognomonic. Enlargement or expansion of bone, cortical thickening, coarsening of trabecular markings, and typical lytic and sclerotic changes may be observed. Radiographs also show the condition of joints adjacent to involved sites, indicate the degree to which lytic or sclerotic lesions predominate, and show the presence or absence of deformity or fracture, including fissure fractures.

Unless new symptoms develop or current symptoms become significantly worse, raising concern for fracture or sarcomatous change, repeated imaging is usually unnecessary. CT or MRI imaging is rarely required in routine cases. CT imaging may be helpful in assessment of a fracture where radiographs are not sufficient, and MRI scans are useful in assessing the possibility of sarcoma, giant cell tumor, or metastatic disease at a site of Paget disease. Anecdotal data suggest that PET scans of sclerotic lesions in patients with Paget disease may help distinguish pagetic lesions from bone metastases, because the former are likely to be minimally to nonmetabolic as compared with marked hypermetabolic changes observed with bone metastases [35].

The characteristic X-ray and clinical features of Paget disease usually eliminate problems with differential diagnosis. However, an older patient may occasionally present with severe bone pain, elevated SAP, a positive bone scan, and less-than-characteristic radiographic areas of lytic or blastic change. Here the possibility of metastatic disease to bone or some other form of metabolic bone disease (eg, osteomalacia with secondary hyperparathyroidism) must be considered. Old radiographs and laboratory tests are very helpful in this setting, because normal studies a year earlier would make a diagnosis of Paget disease less likely. A similar dilemma occurs when someone with known and established Paget disease develops multiple painful new sites; here, too, the likelihood of metastatic disease must be carefully considered, and bone biopsy for a tissue diagnosis may be indicated.

TREATMENT

Specific antipagetic therapy consists of those agents capable of suppressing the activity of pagetic osteoclasts. The goal of medical treatment is to relieve symptoms and to prevent future complications. Current guidelines suggest treatment of most patients with active disease at risk for complications with high potency bisphosphonates, either intravenous zolendronic acid, or oral alendronate or risedronate [24]. Other bisphosphonates may be used, however, and a detailed review of bisphosphonate treatment, including information on dosing regimens, clinical trial results, and side effects, has been published [36]. Case reports suggest that denosumab, an antiresorptive agent that blocks RANKL, the cytokine responsible for osteoclast formation, could have utility in managing Paget disease in patients with contraindications to bisphosphonates [37].

It has been shown that suppression of the pagetic process by any of the available agents can effectively ameliorate certain symptoms in the majority of patients. Bone aches or pain, excessive warmth over bone, headache caused by skull involvement, low-back pain secondary to pagetic vertebral changes, and some syndromes of neural compression (eg, radiculopathy and some examples of slowly progressive brainstem or spinal cord compression) are the most likely to be relieved. Pain caused by a secondary arthritis from pagetic bone involving the spine, hip, knee, ankle, or shoulder may or may not respond to antipagetic treatment. Filling in of osteolytic blade-of-grass lesions in weight-bearing bones has been reported in some cases after either calcitonin or bisphosphonate treatment. On the other hand, a bowed extremity or other bone deformity will not change after treatment, and clinical experience indicates that deafness is unlikely to improve, although limited studies suggest that progression of hearing loss may be slowed [38] or even, in one case with pamidronate, reversed [39].

A second indication for treatment is to prevent the development of late complications in those patients deemed to be at risk, based on their sites of involvement and evidence of active disease. While it has not been proven that suppression of pagetic bone turnover prevents future complications, normal patterns of new bone deposition are restored in biopsy specimens after suppression of pagetic activity [40, 41]. Untreated active disease with persistent abnormal bone turnover for many years could potentially lead to severe bone deformity over time. Indeed, substantial (eg, 50%) but incomplete suppression of elevated indices of bone turnover with older and less effective therapies has been associated with disease progression [42]. Thus, some treatment guides recommend treatment of asymptomatic but active disease (ie, SAP above normal) at sites where the potential for later problems or complications exists (eg, weight-bearing bones, areas near major joints, vertebral bodies, extensively involved skull) [24, 36]. Others argue that the evidence does not support such use; in the PRISM clinical trial disease suppression with bisphosphonates failed to reduce short-term complications or improve quality of life. However, the median duration of observation was only 3 years [43].

Although controlled studies are not available to prove efficaciousness in this situation, the use of a potent bisphosphonate before elective surgery on metabolically active pagetic bone also is recommended [24, 44]. The goal is to reduce the hypervascularity associated with moderately active disease (eg, a threefold or more elevation in SAP) to minimize blood loss at operation.

Recommendations for the management of Paget disease have been published as guideline or management documents by consensus panels in the USA [24, 36], UK [45] and Canada [46].

Bisphosphonates

Studies with alendronate [47], risedronate [48], pamidronate [49], and zoledronic acid (also referred to as zoledronate) [50] have all showed the efficacy of these agents in suppressing the localized bone turnover abnormality and in improving many symptoms in patients with Paget disease. In most instances the drug of choice, based on the efficaciousness of the agent and patient preferences regarding an intravenous or an oral regimen, is intravenous zoledronate or oral risedronate. Generic alendronate, 40 mg per day for 6 months (with the potential to repeat after a drug-free interval) and generic pamidronate, with several possible dosing approaches based on the patient's status [36], are also available at lower cost but with less convenient dosing regimens.

Risedronate is prescribed as a daily oral dose of 30 mg for 2 months — note that this is a different dosing regimen than that for osteoporosis. The pill is taken after an overnight fast upon arising each morning with 225 mL of plain water. The patient must remain upright and take nothing else by mouth for 30 min, after which he or she should eat. A follow-up measurement of SAP 1 to 2 months after completing the course is useful; if the value is not yet normal or near normal, a third or fourth month of risedronate could be offered with a good likelihood of normalcy or near normalcy of indices thereafter. In the pivotal clinical trial 80% of the patients had achieved a normal SAP 6 months after initiation of 2 months of treatment, with a period of subsequent disease suppression of up to 18 months [49]. Periodic measurement of SAP (every 6 to 12 months) should be performed and retreatment is suggested, if indicated, if and when SAP rises above normal or increases by >25% of the nadir value if full remission was not achieved.

Zoledronic acid at a dose of 5 mg is administered as a single 15 min intravenous infusion. In the pivotal clinical trial comparing one 5 mg infusion of zoledronic acid with 2 months of 30 mg per day oral risedronate, a normal SAP was achieved by 96% of zoledronic acid subjects compared with 74% of risedronate subjects [50]. In practice, if a patient has a very high SAP pretreatment that fails to reach normal or near normal by a few months after the infusion, a second infusion can be provided. Biochemical remission after zolendronic acid treatment may be prolonged with 87% of patients maintaining response at 5 to 6 years [41]. For patients who enter a biochemical remission or near remission after one (or two) doses, 6 to 12 month follow-up SAP measurements are suggested, and once the SAP begins to rise above normal or more than 25% above nadir levels if remission was not achieved, and if treatment is again indicated based on symptoms or concerns about complications, another dose can be provided. Again, note that treating at variable intervals based on biochemical remission and relapse differs from the regimen that is used when zoledronic acid is given for osteoporosis.

It is important to emphasize the need for full repletion of both calcium and vitamin D before and during treatment with potent bisphosphonates to avoid hypocalcemia and secondary hyperparathyroidism. Calcium and vitamin D repletion should be maintained thereafter in these patients as a general principle.

Side effects with alendronate and risedronate include upper gastrointestinal symptoms consistent with esophageal irritation in a minority of individuals. The initial dose of either pamidronate or zoledronic acid in a patient who has not previously received a nitrogen containing bisphosphonate can be associated with a flu-like reaction for 1 to 2 days after treatment with fever, headache, myalgia, and arthralgia, ameliorated by using acetaminophen or an NSAID; this reaction is unlikely to occur with subsequent doses. Finally, relatively rare cases of uveitis or iritis have been described with nitrogen-containing bisphosphonates. In such patients, a 6-month course of etidronate can be given, because these compounds do not contain the nitrogen atom.

Osteonecrosis of the jaw has been described as a complication of bisphosphonate therapy. At least seven patients with Paget disease have been reported to have had this complication, most of whom were given very high doses for prolonged periods of time outside the usual prescribing guidelines [51]. This topic is discussed in detail elsewhere in the Chapter 120.

Calcitonin

Synthetic salmon calcitonin is available as a subcutaneous injection. It is less effective than the nitrogen-containing bisphosphonates and is most useful in the rare patient who is intolerant of all bisphosphonates or if bisphosphonate therapy is contraindicated. The usual starting dose is 100 U (0.5 mL; the drug is available in a 2 mL vial), generally self-injected subcutaneously, initially on a daily basis. Symptomatic benefit may be apparent in a few weeks, and the biochemical benefit (typically about a 50% reduction from baseline in SAP) is usually observed after 3 to 6 months of treatment. After this period, many clinicians reduce the dose to 50 to100U every other day or three times weekly. Escape from the efficacy of salmon calcitonin may sometimes occur after a variable period of benefit. The main side effects of parenteral salmon calcitonin include, in a minority of patients, the development of nausea or queasiness, with or without flushing of the skin of the face and ears. Intranasal calcitonin is not indicated for use in Paget disease, but anecdotal experience suggests it may relieve some symptoms and lower elevated bone turnover markers in patients with mild disease.

Other therapies

Analgesics such as acetaminophen, aspirin, and nonsteroidal anti-inflammatory agents (NSAIDs) may be tried empirically with or without antipagetic therapy to relieve pain. In particular, pain from pagetic arthritis (ie, osteoarthritis caused by deformed pagetic bone at a joint space) is often helped by some of these agents.

Surgery on pagetic bone [52] may be necessary in the setting of established or impending fracture. Elective joint replacement, more complex with Paget disease than with typical osteoarthritis, is often very successful in relieving refractory pain. Rarely, osteotomy is performed to alter a bowing deformity in the tibia. Neurosurgical intervention is sometimes required in cases of spinal cord compression, spinal stenosis, or basilar invagination with neural compromise. Although medical management may be beneficial and adequate in some instances, all cases of serious neurological compromise require immediate neurological and neurosurgical consultation to allow the appropriate plan of management to be developed.

REFERENCES

1. Morales-Piga AA, Rey-Rey JS, Corres-Gonzalez J, et al. Frequency and characteristics of familial aggregation of Paget's disease of bone. J Bone Miner Res. 1995;10(4):663–70.
2. Siris ES, Ottman R, Flaster E, et al. Familial aggregation of Paget's disease of bone. J Bone Miner Res. 1991;6(5):495–500.
3. Albagha OM, Visconti MR, Alonso N, et al. Genome-wide association study identifies variants at CSF1, OPTN and TNFRSF11A as genetic risk factors for Paget's disease of bone. Nat Genet. 2010;42(6):520–4.
4. Albagha OM, Wani SE, Visconti MR, et al. Genome-wide association identifies three new susceptibility loci for Paget's disease of bone. Nat Genet. 2011;43(7):685–9.
5. Singer FR. Paget's disease of bone-genetic and environmental factors. Nat Rev Endocrinol. 2015;11(11):662–71.
6. Hocking LJ, Herbert CA, Nicholls RK, et al. Genomewide search in familial Paget disease of bone shows evidence of genetic heterogeneity with candidate loci on chromosomes 2q36, 10p13, and 5q35. Am J Hum Genet. 2001;69(5):1055–61.
7. Visconti MR, Langston AL, Alonso N, et al. Mutations of SQSTM1 are associated with severity and clinical outcome in paget disease of bone. J Bone Miner Res. 2010;25(11):2368–73.
8. Daroszewska A, van 't Hof RJ, Rojas JA, et al. A point mutation in the ubiquitin-associated domain of SQSMT1 is sufficient to cause a Paget's disease-like disorder in mice. Hum Mol Genet. 2011;20(14):2734–44.
9. Kurihara N, Hiruma Y, Zhou H, et al. Mutation of the sequestosome 1 (p62) gene increases osteoclastogenesis but does not induce Paget disease. J Clin Invest. 2007;117(1):133–42.
10. Cooper C, Schafheutle K, Dennison E, et al. The epidemiology of Paget's disease in Britain: is the prevalence decreasing? J Bone Miner Res. 1999;14(2):192–7.
11. Cundy T, McAnulty K, Wattie D, et al. Evidence for secular change in Paget's disease. Bone. 1997;20(1):69–71.
12. Rebel A, Malkani K, Basle M, et al. Is Paget's disease of bone a viral infection? Calcif Tissue Res. 1977;22(Suppl):283–6.
13. Mills BG, Singer FR, Weiner LP, et al. Evidence for both respiratory syncytial virus and measles virus antigens in the osteoclasts of patients with Paget's disease of bone. Clin Orthop Relat Res. 1984;183:303–11.
14. Gordon MT, Mee AP, Sharpe PT. Paramyxoviruses in Paget's disease. Semin Arthritis Rheum. 1994;23(4):232–4.
15. Mee AP, Dixon JA, Hoyland JA, et al. Detection of canine distemper virus in 100% of Paget's disease samples by in situ-reverse transcriptase-polymerase chain reaction. Bone. 1998;23(2):171–5.

16. Kurihara N, Reddy SV, Menaa C, et al. Osteoclasts expressing the measles virus nucleocapsid gene display a pagetic phenotype. J Clin Invest. 2000;105(5):607–14.
17. Ooi CG, Walsh CA, Gallagher JA, et al. Absence of measles virus and canine distemper virus transcripts in long-term bone marrow cultures from patients with Paget's disease of bone. Bone. 2000;27(3):417–21.
18. Kurihara N, Zhou H, Reddy SV, et al. Expression of measles virus nucleocapsid protein in osteoclasts induces Paget's disease-like bone lesions in mice. J Bone Miner Res. 2006;21(3):446–55.
19. Kurihara N, Hiruma Y, Yamana K, et al. Contributions of the measles virus nucleocapsid gene and the SQSTM1/p62(P392L) mutation to Paget's disease. Cell Metab. 2011;13(1):23–34.
20. Teramachi J, Nagata Y, Mohammad K, et al. Measles virus nucleocapsid protein increases osteoblast differentiation in Paget's disease. J Clin Invest. 2016;126(3):1012–22.
21. Rebel A, Basle M, Pouplard A, et al. Bone tissue in Paget's disease of bone. Ultrastructure and immunocytology. Arthritis Rheum. 1980;23(10):1104–14.
22. Singer FR, Mills BG, Gruber HE, et al. Ultrastructure of bone cells in Paget's disease of bone. J Bone Miner Res. 2006;21(Suppl 2):51–4.
23. Shankar S, Hosking DJ. Biochemical assessment of Paget's disease of bone. J Bone Miner Res. 2006;21(Suppl 2):22–7.
24. Singer FR, Bone HG, 3rd, Hosking DJ, et al. Paget's disease of bone: an endocrine society clinical practice guideline. J Clin Endocrinol Metab. 2014;99(12):4408–22.
25. Al Nofal AA, Altayar O, BenKhadra K, et al. Bone turnover markers in Paget's disease of the bone: a systematic review and meta-analysis. Osteoporos Int. 2015;26(7):1875–91.
26. Siris ES, Canfield RE. The parathyroids and Paget's disease of bone. In: Bilezikian J, Levine M, Marcus R (eds) The Parathyroids. New York: Raven Press; 1994, pp. 823–8.
27. Siris ES. Indications for medical treatment of Paget's disease of bone. In: Singer FR, Wallach S (eds) Paget's Disease of Bone: Clinical Assessment, Present and Future Therapy. New York: Elsevier, 1991, pp. 44–56.
28. Herzberg L, Bayliss E. Spinal-cord syndrome due to non-compressive Paget's disease of bone: a spinal-artery steal phenomenon reversible with calcitonin. Lancet. 1980;2(8184):13–5.
29. Monsell EM. The mechanism of hearing loss in Paget's disease of bone. Laryngoscope. 2004;114(4):598–606.
30. Barry HC. Orthopedic aspects of Paget's disease of bone. Arthritis Rheum. 1980;23(10):1128–30.
31. Wick MR, Siegal GP, Unni KK, et al. Sarcomas of bone complicating osteitis deformans (Paget's disease): fifty years' experience. Am J Surg Pathol. 1981;5(1):47–59.
32. Hansen MF, Seton M, Merchant A. Osteosarcoma in Paget's disease of bone. J Bone Miner Res. 2006;21(Suppl 2):58–63.
33. Upchurch KS, Simon LS, Schiller AL, et al. Giant cell reparative granuloma of Paget's disease of bone: a unique clinical entity. Ann Intern Med. 1983;98(1):35–40.
34. Jacobs TP, Michelsen J, Polay JS, et al. Giant cell tumor in Paget's disease of bone: familial and geographic clustering. Cancer. 1979;44(2):742–7.
35. Sundaram M. Imaging of Paget's disease and fibrous dysplasia of bone. J Bone Miner Res. 2006;21(Suppl 2):28–30.
36. Siris ES, Lyles KW, Singer FR, et al. Medical management of Paget's disease of bone: indications for treatment and review of current therapies. J Bone Miner Res. 2006;21(Suppl 2):94–8.
37. Reid IR, Sharma S, Kalluru R, et al. Treatment of Paget's disease of bone with denosumab: case report and literature review. Calcif Tissue Int. 2016;99(3):322–5.
38. el Sammaa M, Linthicum FH, Jr., House HP, et al. Calcitonin as treatment for hearing loss in Paget's disease. Am J Otol. 1986;7(4):241–3.
39. Murdin L, Yeoh LH. Hearing loss treated with pamidronate. J R Soc Med. 2005;98(6):272–4.
40. Meunier PJ. Bone histomorphometry and skeletal distribution of Paget's disease of bone. Semin Arthritis Rheum. 1994;23(4):219–21.
41. Reid IR, Lyles K, Su G, et al. A single infusion of zoledronic acid produces sustained remissions in Paget disease: data to 6.5 years. J Bone Miner Res. 2011;26(9):2261–70.
42. Meunier PJ, Vignot E. Therapeutic strategy in Paget's disease of bone. Bone. 1995;17(5 Suppl):489S–91S.
43. Langston AL, Campbell MK, Fraser WD, et al. Randomized trial of intensive bisphosphonate treatment versus symptomatic management in Paget's disease of bone. J Bone Miner Res. 2010;25(1):20–31.
44. Kaplan FS. Surgical management of Paget's disease. J Bone Miner Res. 1999;14(Suppl 2):34–8.
45. Selby PL, Davie MW, Ralston SH, et al. Guidelines on the management of Paget's disease of bone. Bone. 2002;31(3):366–73.
46. Drake WM, Kendler DL, Brown JP. Consensus statement on the modern therapy of Paget's disease of bone from a Western Osteoporosis Alliance symposium. Biannual Foothills Meeting on Osteoporosis, Calgary, Alberta, Canada, September 9–10, 2000. Clin Ther. 2001;23(4):620–6.
47. Siris E, Weinstein RS, Altman R, et al. Comparative study of alendronate versus etidronate for the treatment of Paget's disease of bone. J Clin Endocrinol Metab. 1996;81(3):961–7.
48. Miller PD, Brown JP, Siris ES, et al. A randomized, double-blind comparison of risedronate and etidronate in the treatment of Paget's disease of bone. Paget's Risedronate/Etidronate Study Group. Am J Med. 1999;106(5):513–20.
49. Harinck HI, Papapoulos SE, Blanksma HJ, et al. Paget's disease of bone: early and late responses to three different modes of treatment with aminohydroxypropylidene bisphosphonate (APD). Br Med J (Clin Res Ed). 1987;295(6609):1301–5.
50. Reid IR, Miller P, Lyles K, et al. Comparison of a single infusion of zoledronic acid with risedronate for Paget's disease. N Engl J Med. 2005;353(9):898–908.
51. Khosla S, Burr D, Cauley J, et al. Bisphosphonate-associated osteonecrosis of the jaw: report of a task force of the American Society for Bone and Mineral Research. J Bone Miner Res. 2007;22(10):1479–91.
52. Parvizi J, Klein GR, Sim FH. Surgical management of Paget's disease of bone. J Bone Miner Res. 2006;21(Suppl 2):75–82.

93

Epidemiology, Diagnosis, Evaluation, and Treatment of Nephrolithiasis

Murray J. Favus[1] and David A. Bushinsky[2]

[1]Department of Medicine, The University of Chicago, Chicago, IL, USA
[2]Department of Medicine, University of Rochester Medical Center, Rochester, NY, USA

INTRODUCTION

The signs and symptoms of a patient actively passing a kidney stone may be indistinguishable from that of another patient symptomatic with a kidney stone but with a stone of different composition and treatment requirements (Table 93.1). Therefore, the physician must be mindful of the several types of stones found in the urinary tract, their prevalence, and the methods available to identify the stone's chemical composition from urinary crystals and blood and urine chemistries (Table 93.2).

For example, in Western cultures 70% to 75% of all stones contain Ca as either pure Ca oxalate, or Ca phosphate, or some combination of both. Ca oxalate and Ca phosphate crystal formation and aggregation in the final urine are dictated by the concentrations of a number of ion species including hypercalciuria and hyperoxaluria, hypocitraturia, and hyperuricosuria, and urine pH and urine volume [1, 2]. Another 10% to 15% of stones are composed of struvite (magnesium ammonium phosphate) that form in a urine infected with gram-negative urease-producing bacteria, usually Proteus [1, 3]. Uric acid stones account for 5% to 10% of stones and are usually found with hyperuricosuria in urine of low volume and low pH that reduces the solubility of uric acid [1, 4]. Cystinuria accounts for about 1% of all kidney stones and is caused by the high levels of urine cystine that result from an inherited defect in renal tubule cystine transport. Cystine kidney stones are usually discovered in childhood and often progress to renal failure at a young age [5].

Rarely, stones result from crystal formation of excreted pharmacological agents or metabolites such as alkali, carbonic anhydrase inhibitor, and topiramate.

GEOGRAPHIC DISTRIBUTION OF PREVALENCE AND INCIDENCE

In many countries in the Americas, Europe, and Asia the prevalence and incidence of kidney stones have continued to increase over the past 20 to 30 years [6]. In the USA the prevalence of kidney stones varies with gender, race, and geographic location. For men, rates vary from 4% to 9% with new stone appearance rates increasing from 78.5 to 123.6 cases per 100,000 per year during this time. For women, stone rates range from almost 2% to 4%. In 1974 the rate was 36 cases per 100,000 per year [7]. For whites, the risk for stone formation is several times that of African-Americans. The incidence of nephrolithiasis tracked by the Mayo Clinic indicates an increase in stone frequency since 1950 [8, 9]. The prevalence of a history of nephrolithiasis in the US population reported in the NHANES III (1988 to 1994) survey increased from 3.8% (1976 to 1980) to 5.2%. In the USA the lifetime risk for stone formation is 12% in men and 5% in women. Recurrence rates of new stone formation are high. If untreated, stones will recur at the rate of 50% in 5 to 10 years [9].

In the recent NHANES US population survey analyzed for kidney stones, occurrence between 2007 and 2010 was 8.8%, with 10.6% for men compared with 7.1% for women [10]. Kidney stones were more common among obese subjects (11.2%) compared with 6.1% for nonobese adults. Black, non-Hispanic, and Hispanic adults had lower stone rates than white, non-Hispanics (black, non-Hispanic: 0.37%; Hispanic: 0.60%). A strong association of obesity and diabetes with a history of kidney stones was also reported. Recurrence rates of stones show rates

---

Primer on the Metabolic Bone Diseases and Disorders of Mineral Metabolism, Ninth Edition. Edited by John P. Bilezikian.
© 2019 American Society for Bone and Mineral Research. Published 2019 by John Wiley & Sons, Inc.
Companion website: www.wiley.com/go/asbmrprimer

Table 93.1. Properties and manifestations of stones by composition.

Stone event	Ca Oxalate	Uric Acid	Struvite	Cystine
Stone passage	+ +	+ +	− −	+ +
Crystalluria	+ +	+ +	− −	+ +
Small, separate stone	+ +	+ +	− −	+
Radiodense	+ +	− −	+	+ +
Staghorn	− −	+ +	+ + +	+ +
Nephrocalcinosis	+ +	− −	− −	− −
Sludge and obstruction	− −	+ +	− −	+ +

Table 93.2. Stone type and frequency by composition.

Composition	Frequency (%)	Crystal Shape
Calcium oxalate	15–35	Dumbbell shape for calcium oxalate monohydrate; and bipyramidal for calcium oxalate dihydate
Calcium phosphate[a]	5–20	Elongate, narrow
Mixed Ca oxalate/phosphate	40–45	Mixed
Uric acid	2–13	Flat rhomboidal
Struvite[b]	20–30	Rectangular prisms
Cystine	1–3	Hexagonal plates
Ammonium urate	0.5–1.0	Flat rhomboidal
Mixed calcium oxalate/uric acid	2–5	Mixed

[a]CaP also known as brushite or apatite; [b]Struvite is magnesium ammonium phosphate. Frequency is the incidence of the crystals found in all stones reported from five series with a total of 2668 patients.

as high as 75%; with 40% to 50% of the recurrences occurring within 5 years of the initial stone event [11]. Patients who have formed two or more stones tend to have successive shorter intervals between each new stone event [12]. The factors that determine the accelerating pace of stone formation in recurrent stone formers are not known. Therefore, in any single stone-former, one cannot predict who will relapse. However, the natural history of stone disease and the high rate of recurrence require careful diagnostic evaluation, early treatment, and long-term follow-up.

NUTRITION AND LIFESTYLE

Dietary selections may increase the risk of kidney stone formation through excessive consumption of animal proteins, salt, rapidly absorbed monosaccharides, and low intakes of fruits and vegetables rich in potassium. Urine supersaturation (SS) for Ca oxalate, uric acid, and Ca phosphate may increase with changing urine concentration of solutes. For example, excess protein intake may contribute to low urine pH, resulting in high urine Ca and uric acid and low urine citrate. Increased urine solute concentration can result from high caloric intake and low fluid intake.

Climate, occupation, and fluid intake

High fluid intake can reduce Ca oxalate kidney stones by diluting the urine and lowering the critical urine concentrations of Ca, oxalate, and other important urinary ion species [13]. The hot and humid climate of the southeast USA is an example of climate contributing to a high local/regional prevalence of kidney stones.

Calcium intake

The role of Ca intake on the appearance of new kidney stones is complex. High intake of dietary Ca appears to decrease the risk of symptomatic kidney stones, whereas intake of supplemental Ca may increase stone risk [14]. As dietary Ca reduces the absorption of oxalate, the apparently different effects caused by the form of Ca may be associated with the timing of Ca ingestion relative to the amount of oxalate consumed. High Ca intake is an infrequent cause of kidney stones. However, in a large study population of postmenopausal women participating in the National Institutes of Health sponsored Women's Health Initiative, women given 700 mg of Ca supplement daily in addition to their usual Ca intake experienced a 17% increase in kidney stone formation compared with those not given Ca supplementation [15]. Thus, there is evidence for an inverse relationship between dietary Ca and stone appearance in contrast to a positive relationship of Ca intake versus appearance of kidney stone when Ca supplements are used.

Salt

Increased dietary salt ingestion in Western countries has had multiple effects on health including increased urinary Ca excretion and predisposition for first-time stone formation in women and men.

Oxalate

Variations in dietary oxalate have only a modest effect on the risk of Ca oxalate kidney stone formation. However, urinary oxalate excretion is an important contributor to Ca oxalate stone formation [16]. Further, urine oxalate excretion may not be solely determined by dietary oxalate. According to a small pilot study, a recently discovered potential regulator of oxalate regulation is the gut microbiome [17]. Individuals with oxalate-containing kidney stones have a unique gut microbiome (GMB) compared with those without kidney stones. The abundant GMB *Eubacterium* content was inversely correlated with urine oxalate levels but not with samples from patients with uric acid stones or Ca-based stones [17]. If reproduced in other laboratories, then GMB manipulation might represent a novel treatment or preventative for kidney stone disease.

Protein

Population surveys on the role of protein intake on kidney stone formation have yielded variable results. Protein intake lowers urine pH, which alters the concentration of soluble and insoluble salts that promote crystalluria and growth of kidney stones [16].

GENETIC CONTRIBUTIONS

Several lines of evidence support a genetic basis for some stone disorders. A prospective study of men with a family history of kidney stones had a risk of incident stones double those without a family history of stones [18]. A twin study showed a concordance of kidney stones in monozygotic twins approached twice the rate observed in dizygotic twins. Based on this data, heritability was estimated to account for 56% of kidney stone preference. Another large twin study showed that a concordance of kidney stones in monozygotic twins was almost twice the rate observed in dizygotic twins [19]. From this data, heritability was estimated to account for 56% of kidney stone prevalence.

EVALUATION OF THE STONE PATIENT

Clinical

In the USA signs and symptoms of a kidney stone are the ninth most common reason for a visit to an emergency department. The acute onset of flank pain may be the initial sign of an intraabdominal medical or surgical emergency caused by a kidney stone. Clinical signs suggesting a stone causing an acute intrarenal or ureteral obstruction include flank pain with an intermittent crescendo that increases with time, followed by clearing, and then a return (renal colic). The pain may move lower in the flank toward the groin, suggesting the movement of an obstructing ureteral calculus more distally toward the bladder. Gross or microscopic hematuria is common. Prompt imaging of the kidney beds and ureters using CT or ultrasound is now the most common approach to making the diagnosis of kidney stone.

Imaging: CT and ultrasound

All patients suspected of harboring a stone in the urinary tract should undergo an imaging procedure to determine whether the new stone is located within the kidney parenchyma, renal pelvis, upper or lower ureter, or bladder, and whether there is ureteral obstruction. Localization of stones is also important in choosing medications, surgery, or lithotripsy. Compared with previous techniques which used plain radiography of the abdomen, intravenous pyelography (IVP), the more recent noninfused CT of the abdomen with 5 mm cuts is the most sensitive imaging technique for determining the number and location of stones within the renal parenchyma or along the upper or lower urinary tract (Fig. 93.1 and Fig. 93.2) [20]. Using this technique, sensitivity and specificity were both at and above 95% [21], and stones can be distinguished from nonstone structures such as kidney tissue or blood clots.

Nephrocalcinosis can be identified as a myriad of tiny, almost microscopic specks of radiodense calcium arrayed along the calyces (Table 93.1). Small, separate, radiodense stones of less than 1 cm in diameter suggest Ca or less commonly, cystine stones.

Radiodense stones suggest either Ca or struvite (magnesium ammonium phosphate, infection stone) composition, but struvite stones are usually large and fill the calyceal system (Table 93.1). Cystine stones appear to

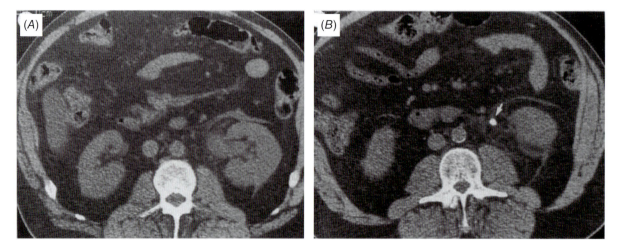

Fig. 93.1. Unenhanced CT of a 48-year-old man presenting with left flank pain. (A) Enlargement of left kidney, dilatations of the collecting stem, and perinephric stranding indicates urinary obstruction. (B) Stone (7 mm) in proximal left ureter (arrow). Inflammatory change is present in fat surrounding the lower pole of left kidney and proximal left ureter. After unsuccessful ESWL a nephrostomy catheter was placed and nephrolithotomy was performed. Contrast enhanced studies were not performed. Source: [20].

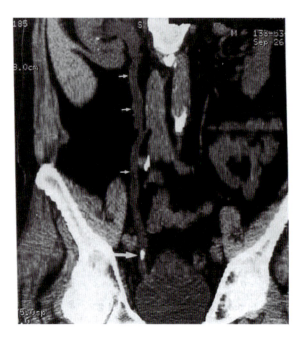

Fig. 93.2. Helical CT data in a 71-year-old man with right flank pain allowed reconstruction of curved reformatted image in the coronal plane. Unenhanced helical CT axial images (not shown) revealed a stone in the distal right ureter. Dilated right ureter (small arrows) was revealed to level of obstructing calculus (large arrow). Source: [20].

be radiodense, but less dense than Ca-containing stones. Small, radiolucent stones suggest uric acid composition (Table 93.1). Uric acid stones appear as filling defects on IVP. Filling defects that occupy the renal pelvis are staghorn stones and may be of struvite, uric acid, or cystine composition. Sludge may be of either uric acid or cystine and can fill the renal pelvis and cause obstruction (Table 93.1). Plain radiographs of the abdomen can identify large stones of greater than 3 mm. Ultrasound may not accurately visualize all stones and therefore cannot be used for follow-up to determine the appearance of new stones. When nephrolithiasis is absent, other causes of acute flank or abdominal pain should be sought.

The rising incidence of kidney stones in the USA has increased the numbers of visits to emergency rooms (from 178 to 340 visits per 100,000 individuals from 1992 to 2009) and with it the use of imaging [22]. At the same time, use of imaging, especially CT, in stone patients, increased over threefold from 21% to 71%. Much of the increase in CT use was to evaluate recurrent stone formers who represent over 20% of all stone formers. In the emergency room setting, evaluating patients with acute abdominal pain caused by nephrolithiasis were recruited into a multicenter study that randomized patients to either ultrasound vs CT. The results concluded that although ultrasound was less sensitive than CT for the diagnosis of nephrolithiasis, using ultrasound as the initial test in patients with suspected nephrolithiasis (and using other imaging as needed) resulted in no need for CT in most patients, lower cumulative radiation exposure, and no significant differences in the risk of subsequent serious adverse events, pain scores, return emergency department visits, or hospitalizations [23].

Urine crystal analysis

Direct inspection of the crystal structures in urine at low magnifications can readily be obtained using standard clinical microscopy (Fig. 93.3 and Fig. 93.4).

Analytical methods used on freshly collected early morning urine sediment includes polarization microscopy. This identifies crystals using standard microscopy with polarized light, which is inexpensive and sufficient for clinical diagnosis [24]. Polarization microscopy is

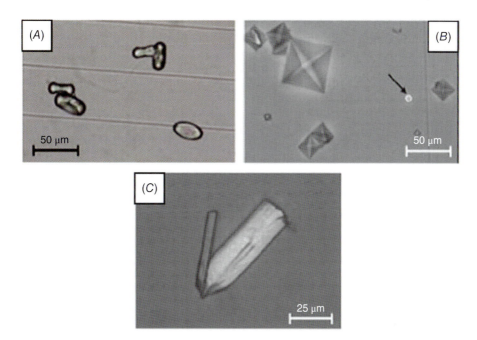

Fig. 93.3. Calcium oxalate and Ca phosphate crystals. (A) Ca oxalate monohydrate (whewellite) crystals with the typical ovoid shape. The main crystal in Ca-containing kidney stones is composed of $Ca(C_2O_4)H_2O$. (B) Typical octahedral (bipyramidal) crystals of Ca oxalate dihydrate (weddellite) composed of $CaC_2O_4 \cdot 2H_2O$. (C) Rod-shaped pale yellow crystals of Ca phosphate dihydrate (brushite) with chemical formula of $CaHPO_4 \cdot 2H_2O$. Source: [24].

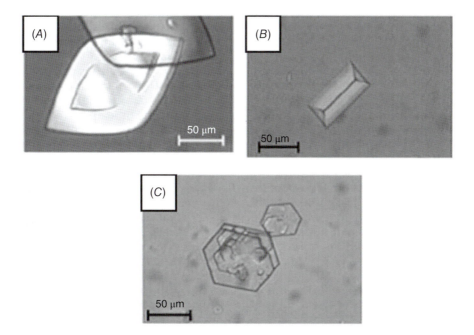

Fig. 93.4. Uric acid, struvite, and cystine crystals. (A) Typical lozenges of uric acid dihydrate crystals. (B) A coffin-shaped crystal of magnesium ammonium phosphate (struvite). (C) Cystine crystals. Source: [24].

effective in identification of Ca oxalate and uric acid crystals but less reliable in identifying Ca phosphate crystals and stones of mixed composition. Infrared spectroscopy is a highly specialized technique for identification of unusual crystals and is based on the interaction of infrared light and the molecular structure of the stone. The procedure is more time consuming, but is able to identify small amounts of minor crystals (5% of a solitary stone). X-ray diffraction crystallography depends upon the interaction of X-rays and the crystalline structure of the stone, but is not able to identify noncrystalline material in the stone. Chemical analysis does not provide the unique crystal structure and therefore crystals with similar molecular weight but different structures may not be distinguished one from the other.

LABORATORY BIOCHEMICAL TESTING

Initial laboratory testing within hours of the onset of acute symptoms may reveal primary hyperparathyroidism, gout, dehydration, and status of renal function. The tests should include serum calcium, electrolytes, uric acid, creatinine, and estimated glomerular filtration rate (eGFR).

Tests after recovery of acute stone passage

Blood and urine collections for laboratory tests may be conducted after recovery of the acute stone event and any surgical procedure and after 4 to 6 weeks of resumption of the usual diet when nutritional steady-state has been re-established.

24-hour urine collection

Ca intake from food sources should be about 900 to 1000 mg daily. Vitamin D intake should be based on intake to maintain serum 25OHD levels within the target range. Blood draw should be fasting upon completion of the 24-hour urine collection. Measurements on blood serum samples are should include: Ca, phosphate, magnesium, creatinine, PTH, 25OHD, uric acid, electrolytes, and albumin.

For the 24-hour urine collection one needs a proper container with appropriate preservatives. Measurements on aliquots of the urine include: volume, pH, Ca, phosphorus, magnesium, sodium, potassium, chloride, bicarbonate, uric acid, oxalate, citrate, ammonium, sulfate, and the calculated supersaturation for Ca oxalate, Ca phosphate, and uric acid.

Supersaturation

Urine crystals vary with respect to concentrations of ionic species, pH, proteins, and naturally occurring crystal inhibitors and promoters. The many ionic species in the urine with varying solubilities of complexes formed predict formation of many potential unique crystals. Each ionic species in the urine may at any one time exist in the soluble or solid phase. Each salt has its unique ion activity product (AP). The solubility product (SP) of an ion complex is the AP below which crystallization does not occur, and is also known as the formation product (FP) or the AP at which the solid phase or crystallization occurs. Supersaturation is the ratio of AP/SP. At supersaturation (SS) values less than 1.0, crystals of a substance will dissolve; at SS values greater than one, crystals can spontaneously form and grow. Phase changes are driven by SS which is usually approximated for such salts by the ratio of their concentration in the urine to their solubilities. Computer algorithms designed to solve 23 simultaneous equations are used to calculate SS for Ca oxalate and Ca phosphates including apatite, and uric acid [25]. FP may be influenced by the presence of inhibitors or promoters in urine, usually small, highly charged molecules. The promoters and inhibitors create a metastable state of supersaturation between the SP and the FP. For Ca ion species, promoters of crystal formation include Ca, oxalate, sodium, pH, and low urine volume whereas inhibitors of crystal formation include high urine volume, citrate, magnesium, potassium, and pyrophosphate.

Stones containing Ca ultimately arise because of a phase change of Ca reacting with ionic species that may form soluble (citrate) or insoluble (oxalate, phosphate) salts. Insoluble complexes may change from the liquid (soluble) phase of urine to a solid (potentially crystalline) phase that may form crystal aggregations of Ca oxalate, brushite, or apatite resulting in a significant mass capable of anchoring along the upper urinary tract, growing and becoming a clinical stone [26]. Thus, patients form crystals and stones that correlate with SS values measured in their urine.

For Ca oxalate SS, the major determinants are urine volume, and Ca and oxalate concentrations. For Ca phosphate (brushite) SS, urine Ca concentration and pH are the main determinants. Measurements of the state of SS for these Ca ion species are clinically informative as to the cause of the stone formed and for assessing therapy to reduce SS. The quantitative assessment of a patient's SS is a powerful diagnostic tool for determining stone composition and for providing a rational approach to management of stone prevention.

Hypercalciuria

Hypercalciuria is one of the most common causes of Ca oxalate and Ca phosphate kidney stones. The definition of hypercalciuria is based upon the measurements of urine Ca excretion by men and women who are ingesting self-selected diets of average Ca intake and are free of osteoporosis and other bone, parathyroid, or vitamin D disorders. Because of the wide range of individual efficiencies of Ca absorption caused by age, dietary Ca intake, and body requirements for Ca, the ranges for 24-hour urine Ca excretion rates vary widely in a population. Children have lower urine Ca excretion rates and have their own table of normative values.

Hodgkinson and Pyrah proposed that urine Ca excretion rates in men of more than 300 mg/24 h and more than 250 mg/24 h for women define hypercalciuria [27, 28]. Coe suggested a more stringent definition of hypercalciuria for men and women eating ad libitum of more than 4.0 mg/kg body weight per day independent of sex or age [29]. A third definition of hypercalciuria is urine Ca excretion per lean body mass expressed as milligrams of urine Ca per gram urine creatinine (normal is less than 140). This estimate is useful in slightly built men and women especially. A petite woman weighing 49 kg with a 24 hour urine Ca of 160 mg/24 h (normal using urine Ca in mg/24 h) may be elevated at more than140 mg Ca/g urine creatinine when calculated as lean body mass.

In healthy adults, the relative increase in urine Ca as diet Ca increases (slope) averages about 8%. In contrast, in patients with idiopathic hypercalciuria, urine Ca excretion increases twofold across a similar range of Ca intakes. Using the same clinical studies, there was a continuum of daily urine Ca excretion values without a bimodal distribution when kidney stone patients and healthy adults were compared. Included is data from 175 kidney stone formers, and 373 healthy adults. About 40% of Ca stone formers have daily Ca excretion of 300 mg/24 h or above, whereas only 10% of healthy adults have such elevated rates of excretion Also, the distribution of urine Ca excretion rates is higher and shifted to the right in stone formers compared with healthy adults [27, 28].

Treatment of idiopathic hypercalciuria

Lowering urine Ca by enhancing renal tubule reabsorption with a thiazide diuretic agent has been widely used to stimulate Ca retention and lower urine Ca excretion. Ca balance studies have shown that thiazide (chlorthalidone) improves Ca balance from negative to or toward positive [30]. During thiazide therapy, Ca intake should be from food sources and Ca supplements avoided. Ca intake should be adequate at 800 to 1000 mg daily. Ca intake should not be restricted to very low intakes, and high Ca intakes should be avoided. Adequate hydration is important because dehydration may raise serum Ca. Because of the potassium-wasting effects of chlorthalidone and other thiazide diuretics, serum potassium should be monitored and low potassium levels treated with potassium supplements. Salt intake should be limited to 100 mEq daily for optimal effect of chlorthalidone. High salt intake increases urine Ca excretion and can nullify the thiazide-lowering effects on urine Ca [31]. While lowering urine Ca, chlorthalidone and other thiazide diuretics can improve bone mineral density in idiopathic hypercalciuria patients with low bone mass.

Hyperoxaluria

Normal urinary oxalate excretion is taken as below 45 mg/24 h. Most patients with hyperoxaluria have urine oxalate excretion well above 45 mg/24 h primarily caused by excess ingestion of dietary oxalate or excess oxalate production from intermediary metabolism [1, 16]. Oxalate-rich foods include teas, chocolate, dark green leafy vegetables, soy, ripe rhubarb, legumes, nuts, and berries. Accidental ingestion of antifreeze (ethylene glycol) results in the generation of large amounts of oxalate that can cause urine oxalate excretion to approach 100 mg/24 h which is sufficient to promote Ca oxalate kidney stones and Ca oxalate deposition in the kidney causing acute renal failure [16, 26].

Treatment to prevent subsequent stones is focused on lowering SS Ca oxalate by reducing dietary oxalate. If hypercalciuria is also present, then a thiazide diuretic agent may be added.

Hypocitraturia

Hypocitraturia is defined as less than 500 mg/24 h in women and less than 350 mg/24 h in men. Low urine citrate is common among women with Ca stone disease. In the urine Ca citrate is soluble and reduces the amount of Ca available to bind oxalate and form insoluble Ca oxalate complexes [32]. Systemic acidosis reduces citrate excretion, whereas alkalosis enhances urine citrate. Other causes of low urine citrate include high dietary protein, acid loading, hypokalemia, distal renal tubular acidosis, diarrhea, infection, and the use of acetazolamide. High levels of urine citrate lower urine Ca oxalate SS and reduce the risk of Ca oxalate stone formation.

Urine pH

A number of ion complexes in the urine are pH-sensitive, including Ca phosphate and uric acid crystals. Chronic alkaline urine with hypocitraturia suggests distal renal tubular acidosis [33].

Attempts to raise urine pH should be performed with caution because high urine pH increases Ca phosphate SS, raises CAP supersaturation, and increases the risk for Ca phosphate $[Ca_3(PO_4)_2]$ stone formation [34]. High dietary protein intake indicated by high urinary sulfate, urine urea nitrogen, and protein catabolic rate are indicators of dietary acid loading. Bone is a major buffer of acid loading, and bone buffering may contribute to hypercalciuria and bone loss.

Hyperuricosuria

Hyperuricosuria is defined as urine uric acid greater than 750 mg/24 h in women and greater than 800 mg/24 h in men. High urine uric acid excretion raises uric acid SS and contributes to uric acid kidney stone formation. Increased Ca oxalate stone formation may also occur in the presence of elevated urine uric acid SS, forming mixed Ca oxalate–uric acid stones [35]. Uric acid SS is sensitive to urine pH, because acid urine lowers urine uric acid concentration and its solubility.

Infection stone

Urine ammonium levels are increased during urinary tract infection with urease-producing gram-negative bacteria. Persistent infection and high urine pH promote formation of magnesium ammonium phosphate crystals that grow, promote stone formation, and fill the renal calyceal systems in both kidneys with large staghorn calculi [36].

As infection within staghorn calculi is difficult to eradicate with antibiotics, and there is a high risk for renal failure, complete surgical removal of all infected stone fragments is required to sterilize the renal pelvis.

INVASIVE AND NONINVASIVE TECHNIQUES FOR STONE REMOVAL

Stones that cause obstruction, bleeding, severe pain, or serious infection may require surgical intervention. Removal of such stones depends upon their size and location using cystoscopy, lithotripsy, or percutaneous nephrolithotomy. Stones less than 5 mm in diameter may pass through the upper and lower urinary tract spontaneously and require no procedure, whereas stones 7 mm and greater in diameter are unlikely to pass. Extracorporeal shock wave lithotripsy (ESWL) fragments stones in the renal pelvis and upper and lower urinary tracts into small pieces that can pass without difficulty. ESWL is effective for removing stones less than 2 cm in diameter, with success rates highest in kidneys that contain one stone. Also, stones less than 2 cm in the upper two-thirds of the ureter can be fragmented by ESWL. Cystoscopy is effective in removing stones located in the lower one-third of the ureter [37].

Stones lodged at the ureteropelvic junction or within calyceal diverticulae in the upper tract are best removed through endourological techniques. Because too many shock waves may result in renal damage, very large stones and staghorn calculi are not treated with ESWL. Ureterolithotomy rather than ESWL is the procedure of choice for removal of stones in the lower segment of the ureter. Stones larger than 2 cm require both ESWL and percutaneous nephrolithotomy. Overall, 35% to 55% of ESWL procedures fail to remove stone fragments. Indications for open surgical procedures include: removal of large infected staghorn calculi; when anatomy is complex; when obstructions resist ESWL; or with limited access via a lower ureteral approach.

REFERENCES

1. Worcester EM, Coe FL. Calcium kidney stones. N Engl J Med. 2010;363:954–63.
2. Moe OW. Kidney stones: pathophysiology and medical management. Lancet. 2006;367:333–44.
3. Healy KA, Ogan K. Pathophysiology and management of infectious staghorn calculi. Urol Clin North Am. 2007;34:363–74.
4. Shekarriz B, Stoller ML. Uric acid nephrolithiasis: current concepts and controversies. J Urol. 2002;168:1307–14.
5. Knoll T, Schubert AB, Fahlenkamp D, et al. Urolithiasis through the ages: data on more than 200,000 urinary stone analyses. J Urol. 2011;185:1304–11.
6. Favus MJ. Epidemiology of nephrolithiasis. In: Rosen CJ (ed.) *Primer on the Metabolic Bone Diseases and Disorders of Mineral Metabolism* (8th ed.) American Society for Bone and Mineral Research. Oxford: Wiley-Blackwell, 2013, pp. 857.
7. Romero V, Akpinar H, Assimos DG. Kidney stones: a global picture of prevalence, incidence, and associated risk factors. Rev Urol. 2010;12:e86–e96.
8. Lieske JC, Pena de la Vegas LS, Slezak JM, et al. Renal stone epidemiology in Rochester, Minnesota. An Update. Kidney Int. 2006;69:760–4.
9. Stamatelou KK, Francis ME, Jones CA, et al. Time trends in reported prevalence of kidney stones in the USA: 1976–1994. Kidney Int. 2003;64:1817–23.
10. Scales CD Jr, Smith AC, Hanley JM, et al. Urologic diseases in America project. Prevalence of kidney stones in the United States. Eur Urol. 2012;62:160–5.
11. Sutherland JW, Parks JH, Coe FL. Recurrence after a single renal stone in a community practice. Miner Electrolyte Metab. 1985;11:267–9.
12. Coe FL, Keck J, Norton ER. The natural history of calcium nephrolithiasis. JAMA. 1977;238:1519–23.
13. Borghi L, Meschi T, Amato F, et al. Urinary volume, water and recurrence in idiopathic calcium nephrolithiasis: a 5-year randomized prospective study. J Urol. 1996;155:839–43.
14. Curhan GC, Willett WW, Speizer FE, et al. Comparison of dietary calcium with supplemental calcium and other nutrients as factors affecting the risk for kidney stones in women. Ann Int Med 1997;126:497–504.
15. Jackson RD, LaCroix AZ, Gass M, et al. Calcium plus vitamin D supplementation and the risk of fractures. N Engl J Med. 2006;354:669–83.
16. Taylor EN, Curhan GC. Oxalate and the risk for nephrolithiasis. Am J Soc Nephrol. 2007;18:2198–204.
17. Stern JM, Moazami S, Qiu Y, et al. Evidence for a distinct gut microbiome in kidney stone formers compared to non-stone formers. Urolithiasis. 2016;44:399–407.
18. Ljunghall S, Danielson BG, Fellstrom B, et al. Family history of renal stones in recurrent stone patients Br J Urol. 1985;57:370–4.
19. Goldfarb DS, Fischer ME, Keich Y, et al. A twin study of genetic and dietary influences on nephrolithiasis: a report from the Vietnam Era Twin (VET) registry. Kidney Int. 2005;67:1053–61.
20. Dalrymple NC, Verga M, Anderson KR, et al. The value of unenhanced helical computerized tomography in the management of acute flank pain. J Urol. 1988;159:735–40.
21. Portis AJ, Sundaram CP. Diagnosis and initial management of kidney stones. Am Fam Physicians. 2001;63:1329–38.
22. Fwu CW, Eggers PW, Kimmel PL, et al. Emergency department visits, use of imaging, and drugs for urolithiasis have increased in the United States. Kidney Int. 2013;83(3):479–86.

23. Smith-Bindman R, Aubin C, Bailitz J, et al. Ultrasonography versus computed tomography for suspected nephrolithiasis. N Engl J Med. 2014;371:1100–10.
24. Daudon M, Frochot V. Crystalluria. Clin Chem Lab Med 2015;53(Suppl):S1479–S1487.
25. Parks JH, Coward M, Coe FL. Correspondence between stone composition and urine supersaturation in nephrolithiasis. Kidney Int. 1997;51(3):894–900.
26. Coe FL, Evan A, Worcester E. Kidney stone disease. J Clin Invest. 2005;115(10):2598–608.
27. Hodgkinson A, Pyrah LN. The urinary excretion of calcium and inorganic phosphate in 344 patients with calcium stone of renal origin. Br J Surg. 1958;46(195):10–18.
28. Robertson WG, Morgan DB. The distribution of urinary calcium excretions in normal persons and stone-formers. Clin Chim Acta. 1972;37:503–8.
29. Coe FL. Treated and untreated recurrent calcium nephrolithiasis in patients with idiopathic hypercalciuria, hyperuricosuria, or no metabolic disorder. Ann Int Med. 1977;87(4):404–10.
30. Coe FL, Parks JH, Bushinsky DA, et al. Chlorthalidone promotes mineral retention in patients with idiopathic hypercalciuria. Kidney Int. 1988;33(6):1140–6.
31. Shah O, Assimos DG, Holmes RP. Genetic and dietary factors in urinary citrate excretion. J Endourol. 2005;19(2):177–82.
32. Pinheiro VB, Baxmann AC, Tiselius HG, et al. The effect of sodium bicarbonate upon urinary citrate excretion in calcium stone formers. Urology. 2013;82(1):33–7.
33. Arampatzis S, Röpke-Rieben B, Lippuner K, et al. Prevalence and densitometric characteristics of incomplete distal renal tubular acidosis in men with recurrent calcium nephrolithiasis. Urol Res. 2012;40:53–9.
34. Halperin ML, Dhadli SC, Kamel KS. Physiology of acid-base balance: links with kidney stone prevention. Semin Nephrol. 2006;26:441–6.
35. Friedlander JI, Moreira DM, Hartman C, et al. Comparison of the metabolic profile of mixed calcium oxalate/uric acid stone formers to that of pure calcium oxalate and pure uric acid stone formers. Urology. 2014;84:289–94.
36. Abdulwahab OM, Ahmed A, Chaparala H, et al. Does stone removal help patients with recurrent urinary tract infections? J Urol. 2015;194(4):997–1001.
37. Donaldson JF, Lardas M, Scrimgeour D, et al. Systematic review and meta-analysis of the clinical effectiveness of shock wave lithotripsy, retrograde intrarenal surgery, and percutaneous nephrolithotomy for lower-pole renal stones. Eur Urol. 2015;67:612–6.

94

Immobilization and Burns: Other Conditions Associated with Osteoporosis

William A. Bauman[1,2], Christopher Cardozo[1–3], and Gordon L. Klein[4]

[1]Department of Veterans Affairs Rehabilitation Research and Development Service National Center for the Medical Consequences of Spinal Cord Injury, James J. Peters Veterans Affairs Medical Center, Bronx, NY, USA
[2]Departments of Medicine and Rehabilitation Medicine, Icahn School of Medicine at Mount Sinai, New York, NY, USA
[3]Department of Pharmacological Sciences, Icahn School of Medicine at Mount Sinai, New York, NY, USA
[4]Department of Orthopaedic Surgery, University of Texas Medical Branch, Galveston, TX, USA

NEUROLOGICAL DISEASE AND OSTEOPOROSIS OF IMMOBILIZATION

Immobilization osteoporosis occurs in several conditions that represent a spectrum of inactivity ranging from mild gait disturbances to absolute bed rest. The etiology and severity of the immobilizing condition will determine the location, magnitude, and characteristics of the skeletal deterioration. However, the determinants common to all forms of bone loss secondary to immobilization are the reduction in mechanical load on the skeleton and the duration of immobilization. Other factors that may serve to determine the degree of bone loss include age, gender, genetic factors, nutritional intake, vitamin D sufficiency, anabolic hormonal status, nervous system integrity, alcohol intake, and muscle strength and function. The risk of fracture in more severe forms of immobilization is increased in the short term in proportion to the degree of bone loss; if the immobilization is less severe or of limited duration, then a heightened risk of fracture is less likely with advancing age. Interventions that have been shown to maintain bone mass and integrity after stroke or Parkinson disease (PD) may not be prescribed because of lack of awareness or enthusiasm for prophylactic approaches to fracture. In persons who have extreme neurological compromise caused by spinal cord injury (SCI), several pharmacological and/or rehabilitation treatment approaches have been tried but have not proven to be efficacious at the knee, which is the anatomical site most prone to fracture. However, newer classes of agents hold promise to be safe and effective in preventing or, perhaps, reversing bone loss caused by motor-complete paralysis.

SPINAL CORD INJURY

Bone loss after SCI

The most severe form of immobilization-related bone deterioration occurs after motor-complete SCI. The distal femur and proximal tibia, regions which are composed predominantly of trabecular bone, have been reported to lose as much as 1% per week during the initial months after paralysis, resulting in losses of 50% to 60% of bone at these sites over the first 2 years of SCI [1]; this represents an astounding degree of bone loss when compared with other conditions associated with bone loss, including those of spaceflight, bed rest, or postmenopausal osteoporosis. Skeletal regions below the level of lesion also have marked loss of BMD. As would be expected, spinal lesions that are less neurologically complete, especially those that permit weight-bearing activities, are associated with lesser degrees of bone deterioration [2]. Loss of

---

Primer on the Metabolic Bone Diseases and Disorders of Mineral Metabolism, Ninth Edition. Edited by John P. Bilezikian.
© 2019 American Society for Bone and Mineral Research. Published 2019 by John Wiley & Sons, Inc.
Companion website: www.wiley.com/go/asbmrprimer

BMD in the femoral and tibial diaphysis of 35% and 25%, respectively, has been reported, with this loss occurring by thinning of the cortical envelope by approximately 0.25 mm/year over the initial 5 to 7 years after paralysis [3]; it should be appreciated that 80% of total long-bone mass resides in the cortex. The severity of immobilization is the most obvious and, probably, most important reason for the extreme magnitude of bone loss. However, other mechanisms may also contribute. In men after traumatic SCI, there is usually a rapid and persistent fall in serum testosterone levels, as well as a concomitant fall in serum estrogen levels, and in women there is often a temporary disruption in menstrual cycles, which one may assume would also lead to a relative hypogonadal state [4, 5]. Serum growth hormone has been also reported to be depressed after SCI, thus exacerbating the catabolic environment [6].

Because of an increase in plasma ionized calcium concentration caused by heightened bone resorption, the PTH–vitamin D axis is profoundly suppressed [7], thereby reducing PTH-mediated renal tubular resorption of calcium and gut absorption of calcium, as well as possibly contributing to increased sclerostin production. Another probable factor is local and systemic production of inflammatory mediators after a major catastrophic event. Furthermore, myokines and other muscle-derived entities, such as miRNAs, which may have osteocatabolic properties, may be postulated to be released from muscle tissue undergoing rapid atrophy. Finally, high dose glucocorticoid administration is occasionally employed immediately after acute SCI in an effort to preserve neurological function.

Fracture after SCI

In 2014, there were estimated to be approximately 280,000 individuals with SCI in the USA [8], with this population increasing by ~12,000 cases per year. The risk of fracture during the first year after SCI has been reported to be 1% per year but rises to 4.6% per year after 20 years of SCI [9]. The most common sites of fracture, regardless of gender, are the distal femur and proximal tibia, with other sites in the lower extremities also at higher risk. In a large study that used the Veterans Affairs' patient database, men and women with SCI sustained incident fractures with a frequency of 10.5% and 11.5%, respectively [10]. The risk of fracture is correlated to the BMD of the epiphysis [11], and may be associated with completeness of motor lesion (eg, motor-complete is greater than motor-incomplete), lower level of lesion (eg, paraplegia is greater than tetraplegia; the level of activity is generally greater with use of the arms/hands and better trunk balance), longer duration of injury (>10 years), low BMI (<19 kg/m^2), alcohol intake, prescription of anticonvulsant medications, and a prior history of fracture [9, 12].

Thus, sublesional BMD in conjunction with SCI-specific risk factors (although the actual extent to which these ancillary risk factors contribute to fracture risk independent of BMD has not yet been determined) should be used by physicians and therapists to guide treatment decisions. Because an individual with SCI often lacks sensation below the level of lesion, the presenting complaint after the occurrence of fracture may be unexplained swelling at the site of fracture, increased spasticity, fever, or autonomic dysreflexia (eg, a condition of uncontrolled sympathetic response to noxious external or bodily stimuli, that generally occurs in patients with SCI at or above thoracic level six). Delayed union and nonunion of fractures have been observed in SCI patients, but there is no epidemiological information concerning incidence of occurrence. Long-term complications of lower extremity fractures include increased pain and spasticity, reduced range of motion, and decreased quality of life. As is the case for elderly able-bodied men, hip fractures in older men with SCI have been reported to be associated with a greatly increased mortality

For each additional point in the Charlson comorbidity index in persons with SCI who sustain a fracture, there was a 10% increase in the hazard of death [10], an observation that supports the concept that both the fracture and comorbidities are contributors to mortality.

Rehabilitation and pharmacological approaches for bone health

Preservation of bone mass and architecture at time of acute SCI has proven to be a difficult medical problem to address. The initial enthusiasm for the use of bisphosphonates has ebbed after reports suggested that this class of agents lacks efficacy at the knee region in persons with motor-complete SCI [13, 14]. A few case series suggest that early administration of bisphosphonates may reduce bone loss in the lower extremities in those who have motor-incomplete lesions and can ambulate. Some of the difficulty interpreting the literature stems from the fact that the completeness of motor lesions and functional ability of subjects has not been adequately controlled in experimental designs [15, 16].

Pamidronate or zoledronic acid administered shortly after paralysis lacked efficacy in preventing bone loss at the knee in individuals who had motor-complete SCI or were nonambulatory [13, 14]. Denosumab, a human monoclonal anti-RANKL antibody, is a potent antiresorptive agent that was administered for 12 months to persons with SCI who had an average duration of injury of 15 months and were wheelchair-bound (13 of 14 subjects had motor-complete lesions); BMD was significantly increased at the total hip (2.4±3.6%), femoral neck (3.0±3.6%), and lumbar spine (7.8±3.7%)

BMD significantly increased when compared with baseline values [17], but BMD at the knee was not evaluated. A clinical trial in persons within 3 months of acute motor-complete SCI is currently underway to test the safety and efficacy of denosumab to prevent loss of BMD at the distal femur and proximal tibia, as well as preserve trabecular microarchitecture (ClinicalTrial.gov Identifier: NCT01983475). Little data are available after SCI regarding the efficacy of teriparatide (recombinant PTH 1-34; Forteo, Eli Lilly, Indianapolis, IN, USA) or abaloparatide (a recombinant parathyroid hormone-related peptide analogue that received approval by the FDA in 2017; Radius Health, Waltham, MA, USA). Teriparatide administration in conjunction with robotically-assisted gait training for 6 and 12 months in 12 nonambulatory subjects with chronic SCI did not improve BMD values at the hip or spine [18], but this work had limitations in study design. Preclinical studies in rodent models of SCI have shown that sclerostin antagonism is protective of BMD loss at the knee and mid-femoral shaft, despite concomitant profound muscle atrophy of the limbs [19]. Antisclerostin human monoclonal antibodies (romosozumab, Amgen, Inc., Thousand Oaks, CA, USA) are in phase 3 clinical trials in postmenopausal osteoporosis, and preliminary results have been quite encouraging.

Reloading of the skeleton by a variety of methods has been evaluated as a means to prevent or reverse osteoporosis after SCI. Reloading of the sublesional skeleton after SCI by standing or partial body weight supported treadmill training have not been efficacious in reducing loss of sublesional long-bone BMD. However, when used to elicit cyclical muscle contraction, electrical stimulation (ES) has been shown to be beneficial to bone mass. When initiated within several months of acute SCI, isometric contraction of the soleus muscle over 4 to 6 years partially preserved trabecular bone along the posterior aspect of the tibia where the forces applied by ES were greatest [20]. Another study performed in those with early SCI showed that when ES was combined with standing, this dual rehabilitation approach had greater beneficial effect on BMD at the hip and knee than ES alone (unpublished observation). ES has been less effective to increase lower extremity long-bone BMD in individuals with chronic SCI. Despite the promise of ES, its translation to clinical care has been difficult because it is labor intensive, and also because bone accrued as a result of ES is rapidly lost when ES training is reduced or terminated.

Low intensity, high frequency vibration (LIV) is being evaluated as a mechanical intervention for osteoporosis. In a trial with nine subjects with motor-complete SCI with durations of injury of 2 or more years, LIV that was applied for 20 min per day for 5 days a week over 6 months had no effect on BMD or trabecular bone architecture [21]. Additional work is required to determine the possible beneficial effects of LIV in the SCI population.

STROKE

Incidence of stroke and related bone loss

The 2016 Executive Summary from the American Heart Association reported that stroke was the fifth leading cause of death in the USA, with approximately 800,000 people having a new or recurrent stroke [22]. For many patients, stroke predisposes to falls because of persistent gait impairments which, in the elderly, may be exacerbated by poor balance, visual impairments, and the possible occurrence of seizures after stroke. Most of the bone loss occurs on the side of paresis, with the more severe functional impairments resulting in greater bone loss and increased risk of fracture because of falls [23]. Those with higher Functional Ambulation Category scores have relatively less reduction in BMD at the femoral neck on the paretic side, and those with greater functional loss also have bone loss on the side contralateral to paresis [24]. BMD of the proximal femur on the paretic side has been reported to be decreased by as much as 14% in association with bone loss of the nonparetic limb secondary to reduced ambulation [23]. In more severe cases as categorized by the Scandinavian Stroke Scale, the BMD of the humerus falls by as much as 27% on the paretic side [25]. In 33 ambulatory stroke patients who took a greater number of steps per day and had higher vertical ground reaction forces also had higher BMD values of the proximal femur than those who were less active [26]. In cross-sectional studies, the longer the duration of paresis, the greater the bone loss, with a greater magnitude of effect observed for the upper than the lower extremity [27]. To date, though, there have been no longitudinal studies of bone loss in stroke victims with duration of paresis longer than 12 months. The risk of fracture in stroke survivors is usually reported to be two to four times higher than that of an age- and gender-matched healthy population [28]. Over a period of 2.5 years, there were 24,263 fractures in a cohort of 273,288 stroke patients followed in the Swedish National Register; 14,263 of these fractures (59% of the total number observed) occurred at the hip on the paretic side [29]. Hip fractures in the general population are appreciated to be associated with disability, increased morbidity, and mortality, and this association should also hold true in those with stroke. As such, preservation of hip BMD after stroke should be considered as standard medical practice, which is seldom the case despite the literature suggesting its importance to continued health and wellbeing.

Rehabilitation and pharmacological approaches to preserve bone

Efforts to mobilize stroke victims early after paresis appear to improve bone mass of the lower extremities. A community fitness and mobility exercise program maintained BMD of the hip after stroke compared with a more sedentary control group [30]. Pharmacological therapy with bisphosphonates appears to be efficacious in two studies [31, 32]. Stroke patients, regardless of gender (male = 280; female = 374), who received risedronate rather than placebo had a significantly reduced incidence of hip fractures [31]. In another limited study (n = 14), zoledronic acid administered to patients shortly after hemiparesis reduced loss of BMD at the hip of the paretic side compared with placebo [32]. Thus, it would seem that rehabilitation and/or pharmacological intervention, especially if initiated early after the catastrophic neurological event, preserves bone health and should be considered in all patients after stroke associated with motor impairment of the limbs.

PARKINSON DISEASE

Prevalence of PD and bone loss

The prevalence of PD is estimated to be from 1.5 to 22 per 100,000 population per year, with differences in the range of disease because of populations at risk, diagnostic criteria, and methods employed to identify cases [33]. A population sample that includes 20% of the medical center admissions in the USA from 1988 to 2007 found that 3.6% of patients who were hospitalized with a hip fracture had PD, and the prevalence was fourfold higher than that in an age- and gender-matched general population sample [34].

Bone loss over time in persons with PD is greater than in healthy controls. In one study of 5937 men who were older than 65 years and followed on average for 4.6 years, the annual age-adjusted loss of BMD at the hip was −1.08% in those with PD compared with 0.36% in controls [35]. Women older than 65 years with PD had comparable findings to that of men with respect to the rate of bone loss, with those with PD losing 1.3% versus 0.6% per year for controls [36]. Other smaller studies have reported a more accelerated bone loss at the hip of 2.5% and 3.9% in those who had PD for an average of 3.0 or 4.6 years, respectively [37, 38].

The etiology of bone loss in individuals with PD is likely reduced and/or impaired ambulation caused by neuromuscular dysfunction, malnutrition, vitamin D deficiency, and anti-PD medications, specifically L-dopa. Body weight has emerged as a reliable and potent factor associated with hip BMD in PD patients and directly related to the severity of the disease; body weight was reported to account for 58% of the age-adjusted and 72% of the multivariate-adjusted difference in BMD at the hip between groups with PD [36]. Patients with PD, perhaps caused by diminished sunlight exposure, have been reported to have lower serum vitamin D levels than patients with Alzheimer disease or healthy controls. In the general population, hyperhomocysteinemia has been reported to be associated with lower BMD values, possibly caused by an effect on bone remodeling by reducing bone formation and increasing bone resorption. Higher doses of L-dopa administered in the treatment of PD result in increased homocysteine production, which is assumed to be caused by its dopa decarboxylase activity; higher plasma homocysteine levels have been observed in PD patients being treated with L-dopa compared with patients with PD or healthy controls not on L-dopa [39].

Pharmacological and rehabilitation approaches to preserve bone

There is some evidence supporting the use of antiresorptive therapy in patients with PD, although the literature is quite limited. In a study of women who were treated with risedronate plus supplemental vitamin D [40], a significant reduction in fracture risk was observed. A study performed in 272 men with PD treated with risedronate (again both groups received supplemental vitamin D) showed a significant increase in BMD and a decrease in markers of bone resorption, but a nonsignificant decrease in the number of hip fractures [41]. Because of esophageal dysfunction and associated difficulties in swallowing in those with PD, it would seem to be appropriate to consider treatment with an intravenous antiresorptive agent, such as zoledronic acid or denosumab.

Thus, PD patients have a markedly increased risk of fracture, particularly at the hip, caused by osteoporosis and an increased risk of falls. The morbidity and mortality associated with hip fractures would be anticipated to be increased in the PD population compared with a matched cohort. Care providers should be aware of and identify secondary risk factors for osteoporosis that lend themselves to treatment. Strategies for fall prevention should be instituted, including aggressive physical rehabilitation and adequate nutrition. Pharmacological intervention with antiresorptive agents, possibly by intravenous delivery, should be considered early in the course of the disease to prevent insidious bone deterioration with increasing severity of disease.

ABNORMALITIES IN BONE AND CALCIUM METABOLISM AFTER BURNS

Introduction

Because humans did not evolve a specific means of protection against burn injury, the responses are nonspecific and have unintended consequences. The two major adaptive responses in question are the inflammatory response and the stress response.

The inflammatory response

A systemic inflammatory response occurs within 24 hours of severe burn injury and includes high circulating levels of the proinflammatory cytokines interleukin (IL)-1β and IL-6 [42]. Both cytokines stimulate osteoblast RANKL production with subsequent marrow stem cell differentiation into osteoclasts, thus increasing bone resorption. That this occurs early on after burns can be inferred from the success of acute administration of the bisphosphonate pamidronate within the first 10 days after burn injury in preventing both total body and lumbar spine bone loss, both acutely [43] and for up to 2 years after this acute intervention [44]. In contrast, those not receiving pamidronate lost approximately 3% of total body BMC in the ensuing 6 months and 7% of lumbar spine BMD in the first 6 weeks postburn [43].

Moreover, children with severe burns develop acute sustained hypocalcemia and hypoparathyroidism with urinary calcium wasting [45] suggesting cytokine-mediated upregulation of the parathyroid calcium-sensing receptor (CaR). A 50% upregulation of the CaR has been shown to occur within 48h in a sheep model of burn injury [46] and suggests that the CaR upregulation may serve to clear the excess calcium that enters the blood after cytokine-mediated bone resorption.

The stress response

The three- to eightfold increase in urine free cortisol [42, 47] within the first 24 hours of the burn injury could well act synergistically with the inflammatory cytokines during the first 2 weeks in stimulating osteoblast RANKL production and subsequent bone resorption. However, by the second week postburn there are no visible surface osteoblasts on bone biopsy [45], markedly reduced surface uptake of tetracycline label, and disappearance of double labeling [42, 47], as well as a failure of the marrow stromal cells cultured from the bone biopsies of burned children to exhibit normal quantities of markers of osteoblast differentiation [47]. Along with the marked reduction in bone formation, urinary deoxypyridinoline, as a marker of resorption, also fell [42] creating an adynamic bone despite persistent high circulating levels of resorptive cytokines.

Thus, there would appear to be two stages of postburn bone loss: a primary resorption mediated by both inflammatory cytokines and endogenous glucocorticoids, and a secondary adynamic stage mediated primarily by endogenous glucocorticoids.

Bone remodeling resumes by 12 months post-burn [44] but the lumbar spine BMD Z-scores in burned children who did not receive acute bisphosphonate therapy remained significantly lower than in those who did [44].

Other possible contributing factors

Vitamin D deficiency, beginning sometime in the first year postburn, is caused by at least two factors: lack of routine supplementation [48] and failure of the skin to convert normal quantities of 7-dehydrocholesterol to vitamin D_3 [49]. The deficiency is progressive such that at 2 years, whereas all serum levels of 25OHD are low, all levels of $1,25(OH)_2D$ are normal. By 7 years postburn, not only are all serum levels of 25OHD low but half of all subjects, levels of $1,25(OH)_2D$ are also low [48].

Immobilization after burn injury has not been adequately studied with regard to bone loss. Between weekly skin grafts the patient's mobility is restricted, but the effects of immobilization are mediated by the sympathetic nervous system via beta adrenergic receptors on the osteoblast [50]. By 2 weeks postburn, in the presence of osteoblast apoptosis, it is unclear what kind of effect sympathetic drive can have on bone.

Treatment

Current treatment consists of a 1-year course of oral oxandrolone, which increases both lean body mass followed by an increase in bone mineral content and bone area, though not BMD [51, 52]. It acts via IGF-1 stimulation, although the remainder of the signaling pathway is unclear at present [51]. Whereas pamidronate has been used to prevent postburn bone loss, it is not universally used to treat patients after burn injury.

REFERENCES

1. Biering-Sorensen F, Bohr HH, Schaadt OP. Longitudinal study of bone mineral content in the lumbar spine, the forearm and the lower extremities after spinal cord injury. Eur Clin Invest. 1990;20(3):330–5.
2. Garland DE, Adkins RH, Kushwaha V, et al. Risk factors for osteoporosis at the knee in the spinal cord injury population. J Spinal Cord Med. 2004;27(3):202–6.
3. Eser P, Frotzler A, Zehnder Y, et al. Relationship between the duration of paralysis and bone structure: a pQCT study of spinal cord injured individuals. Bone. 2004;34(5):869–80.
4. Axel SJ. Spinal cord injured women's concerns: menstruation and pregnancy. Rehabil Nurs. 1982;7(5):10–15.
5. Clark MJ, Schopp LH, Mazurek MO, et al. Testosterone levels among men with spinal cord injury: relationship between time since injury and laboratory values. Am J Phys Med Rehabil. 2008;87(9):758–67.
6. Bauman WA, Spungen AM, Flanagan S, et al. Blunted growth hormone response to intravenous arginine in subjects with a spinal cord injury. Horm Metab Res. 1994;26(3):152–6.
7. Stewart AF, Adler M, Byers CM, et al. Calcium homeostasis in immobilization: an example of resorptive hypercalciuria. N Engl J Med. 1982;306(19):1136–40.
8. Devivo MJ. Epidemiology of traumatic spinal cord injury: trends and future implications. Spinal Cord. 2012;50(5):365–72.
9. Zehnder Y, Luthi M, Michel D, et al. Long-term changes in bone metabolism, bone mineral density, quantitative ultrasound parameters, and fracture incidence after spinal cord injury: a cross-sectional observational study in 100 paraplegic men. Osteoporos Int. 2004;15(3):180–9.
10. Carbone LD, Chin AS, Burns SP, et al. Mortality after lower extremity fractures in men with spinal cord injury. J Bone Miner Res. 2014;29(2):432–9.
11. Eser P, Frotzler A, Zehnder Y, et al. Fracture threshold in the femur and tibia of people with spinal cord injury as determined by peripheral quantitative computed tomography. Arch Phys Med Rehabil. 2005;86(3):498–504.

12. Morse LR, Battaglino RA, Stolzmann KL, et al. Osteoporotic fractures and hospitalization risk in chronic spinal cord injury. Osteoporos Int. 2009;20(3):385–92.
13. Bauman WA, Cirnigliaro C, LaFountaine MF, et al. Zoledronic acid administration failed to prevent bone loss at the knee in persons with acute spinal cord injury. J Bone Min Metab. 2015;33(4):410–21.
14. Bauman WA, Wecht JM, Kirshblum S, et al. Effect of pamidronate administration on bone in patients with acute spinal cord injury. J Rehabil Res Dev. 2005;42(3):305–13.
15. Shapiro J, Smith B, Beck T, et al. Treatment with zoledronic acid ameliorates negative geometric changes in the proximal femur following acute spinal cord injury. Calcif Tissue Int. 2007;80(5):316–22.
16. Bubbear JS, Gall A, Middleton FR, et al. Early treatment with zoledronic acid prevents bone loss at the hip following acute spinal cord injury. Osteoporos Int. 2011;22(1):271–9.
17. Gifre, L, Vidal J, Carrasco JL, et al. Denosumab increases sublesional bone mass in osteoporotic individuals with recent spinal cord injury. Osteoporos Int. 2016;27(1):405–10.
18. Gordon KE, Wald MJ, Schnitzer TJ. Effect of parathyroid hormone combined with gait training on bone density and bone architecture in people with chronic spinal cord injury. PMR. 2013;5(8):633–71.
19. Qin W, Li X, Peng Y, et al. Sclerostin antibody preserves the morphology and structure of osteocytes and blocks the severe skeletal deterioration after motor-complete spinal cord injury in rats. J Bone Miner Res. 2015;30(11):1994–2004.
20. Shields RK, Dudley-Javoroski S. Musculoskeletal adaptations in chronic spinal cord injury: effects of long-term soleus electrical stimulation training. Neurorehabil Neural Repair. 2007;21(2):169–79.
21. Wuermser LA, Beck LA, Lamb JL, et al. The effect of low-magnitude whole body vibration on bone density and microstructure in men and women with chronic motor complete paraplegia. J Spinal Cord Med. 2015;38(2):178–86.
22. Mozaffarian D, Benjamin EJ, Go AS, et al. Executive Summary: Heart Disease and Stroke Statistics — 2016 Update: A Report from the American Heart Association. Circulation. 2016;133(4):447–54.
23. Lazoura O, Groumas N, Antoniadou E, et al. Bone mineral density alterations in upper and lower extremities 12 months after stroke measured by peripheral quantitative computed tomography and DXA. J Clin Densitom. 2008;11(4):511–7.
24. Jorgensen L, Jacobsen BK, Wilsgaard T, et al. Walking after stroke: does it matter? Changes in bone mineral density within the first 12 months after stroke. A longitudinal study. Osteoporos Int. 2000;11(5):381–7.
25. Jorgensen L, Jacobsen BK. Functional status of the paretic arm affects the loss of bone mineral in the proximal humerus after stroke: a 1-year prospective study. Calcif Tissue Int. 2001;68(1):11–5.
26. Worthen LC, Kim CM, Kautz SA, et al. Key characteristics of walking correlate with bone density in individuals with chronic stroke. J Rehabil Res Dev. 2005;42(6):761–8.
27. del Puente A, Pappone N, Mandes MG, et al. Determinants of bone mineral density in immobilization: a study on hemiplegic patients. Osteoporos Int. 1996;6(1):50–4.
28. Ramnemark A, Nyberg L, Borssen B, et al. Fractures after stroke. Osteoporos Int. 1998;8(1):92–5.
29. Kanis J, Oden A, Johnell O. Acute and long-term increase in fracture risk after hospitalization for stroke. Stroke. 2001;32(3):702–6.
30. Pang MY, Eng JJ, Dawson AS, et al. A community-based fitness and mobility exercise program for older adults with chronic stroke: a randomized, controlled trial. J Am Geriatr Soc. 2005;53(10):1667–74.
31. Sato Y, Iwamoto J, Kanoko T, et al. Risedronate sodium therapy for prevention of hip fracture in men 65 years or older after stroke. Arch Intern Med. 2005;165(15):1743–8.
32. Poole KE, Loveridge N, Rose CM, et al. A single infusion of zoledronate prevents bone loss after stroke. Stroke. 2007;38(5):1519–25.
33. Wirdefeldt K, Adami HO, Cole P, et al. Epidemiology and etiology of Parkinson's disease: a review of the evidence. Eur J Epidemiol. 2011;26(Suppl 1):S1–58.
34. Bhattacharya RK, Dubinsky RM, Lai SM, et al. Is there an increased risk of hip fracture in Parkinson's disease? A nationwide inpatient sample. Mov Disord. 2012;27(11):1440–3.
35. Fink HA, Kuskowski MA, Cauley JA, et al. Association of stressful life events with accelerated bone loss in older men: the osteoporotic fractures in men (MrOS) study. Osteoporos Int. 2014;25(12):2833–9.
36. Schneider JL, Fink HA, Ewing SK, et al. The association of Parkinson's disease with bone mineral density and fracture in older women. Osteoporos Int. 2008;19(7):1093–7.
37. Lee SH, Kim MJ, Kim BJ, et al. Homocysteine-lowering therapy or antioxidant therapy for bone loss in Parkinson's disease. Mov Disord. 2010;25(3):332–40.
38. Lorefalt B, Toss G, Granerus AK. Bone mass in elderly patients with Parkinson's disease. Acta Neurol Scand. 2007;116(4):248–54.
39. Hu XW, Qin SM, Li D, et al. Elevated homocysteine levels in levodopa-treated idiopathic Parkinson's disease: a meta-analysis. Acta Neurol Scand. 2013;128(2):73–82.
40. Sato Y, Iwamoto J, Honda Y. Once-weekly risedronate for prevention of hip fracture in women with Parkinson's disease: a randomised controlled trial. J Neurol Neurosurg Psych. 2011;82(12):1390–3.
41. Sato Y, Honda Y, Iwamoto J. Risedronate and ergocalciferol prevent hip fracture in elderly men with Parkinson disease. Neurology. 2007;68(12):911–5.
42. Klein GL, Herndon DN, Goodman WG, et al. Histomorphometric and biochemical characterization of bone following acute severe burns in children. Bone. 1995;17(5):455–60.
43. Klein GL, Wimalawansa SJ, Kulkarni G, et al. The efficacy of acute administration of pamidronate on the

conservation of bone mass following severe burn injury in children: a double-blind, randomized, controlled study. Osteoporos Int. 2005;16(6):631–5.
44. Przkora R, Herndon DN, Sherrard DJ, et al. Pamidronate preserves bone mass for at least 2 years following acute administration for pediatric burn injury. Bone. 2007;41(2):297–302.
45. Klein GL, Nicolai M, Langman CB, et al. Dysregulation of calcium homeostasis after severe burn injury in children: possible role of magnesium depletion. J Pediatr. 1997;131(2):246–51.
46. Murphey ED, Chattopadhyay N, Bai M, et al. Up-regulation of the parathyroid calcium-sensing receptor after burn injury in sheep: a potential contributory factor to postburn hypocalcemia. Crit Care Med. 2000;28(12):3885–90.
47. Klein GL, Bi LX, Sherrard DJ, et al. Evidence supporting a role of glucocorticoids in short-term bone loss in burned children. Osteoporos Int. 2004;15(6):468–74.
48. Klein GL, Langman CB, Herndon DN. Vitamin D depletion following burn injury in children: a possible factor in post-burn osteopenia. J Trauma. 2002;52(2):346–50.
49. Klein GL, Chen TC, Holick MF, et al. Synthesis of vitamin D in skin after burns. Lancet. 2004;363(9405):291–2.
50. Takeda S, Elefteriou F, Levasseur R, et al. Leptin regulates bone formation via the sympathetic nervous system. Cell. 2002;111(3):305–17.
51. Porro LJ, Herndon DN, Rodriguez NA, et al. Five-year outcomes after oxandrolone administration in severely burned children: a randomized clinical trial of safety and efficacy. J Am Coll Surg. 2012;214(4):489–502; discussion 502–4.
52. Murphy KD, Thomas S, Mlcak RP, et al. Effects of long-term oxandrolone administration in severely burned children. Surgery. 2004;136(2):219–24.

Section VIII
Cancer and Bone

Section Editors: Theresa Guise and G. David Roodman

Chapter 95. Mechanisms of Osteolytic and Osteoblastic Skeletal Lesions   739
G. David Roodman and Theresa Guise

Chapter 96. Clinical and Preclinical Imaging in Osseous Metastatic Disease   743
Siyang Leng and Suzanne Lentzsch

Chapter 97. Metastatic Tumors and Bone   752
Julie A. Sterling and Rachelle W. Johnson

Chapter 98. Myeloma Bone Disease and Other Hematological Malignancies   760
Claire M. Edwards and Rebecca Silbermann

Chapter 99. Osteogenic Osteosarcoma   768
Yangjin Bae, Huan-Chang Zeng, Linchao Lu, Lisa L. Wang, and Brendan Lee

Chapter 100. Skeletal Complications of Breast and Prostate Cancer Therapies   775
Catherine Van Poznak and Pamela Taxel

Chapter 101. Bone Cancer and Pain   781
Denis Clohisy and Lauren M. MacCormick

Chapter 102. Radiotherapy-Induced Osteoporosis   788
Laura E. Wright

Chapter 103. Skeletal Complications of Childhood Cancer   793
Manasa Mantravadi and Linda A. DiMeglio

Chapter 104. Medical Prevention and Treatment of Bone Metastases   799
Catherine Handforth, Stella D'Oronzo, and Janet Brown

Chapter 105. Radiotherapy of Skeletal Metastases   809
Srinivas Raman, K. Liang Zeng, Oliver Sartor, Edward Chow, and Øyvind S. Bruland

Chapter 106. Concepts and Surgical Treatment of Metastatic Bone Disease   816
Kristy Weber and Scott L. Kominsky

Primer on the Metabolic Bone Diseases and Disorders of Mineral Metabolism, Ninth Edition. Edited by John P. Bilezikian.
© 2019 American Society for Bone and Mineral Research. Published 2019 by John Wiley & Sons, Inc.
Companion website: www.wiley.com/go/asbmrprimer

95
Mechanisms of Osteolytic and Osteoblastic Skeletal Lesions

G. David Roodman[1] and Theresa Guise[2]

[1]Division of Hematology and Oncology, Indiana University School of Medicine, Indianapolis, IN, USA
[2]Department of Oncology, Medicine, and Pharmacology, Division of Endocrinology, Indiana University School of Medicine, Indianapolis, IN, USA

INTRODUCTION

Bone is a very frequent site for solid tumor metastasis and for involvement with multiple myeloma (MM). Approximately 60% to 70% of breast and prostate cancer patients and 85% of MM patients having bone involvement with advanced disease [1, 2]. More than 450,000 patients in the United States suffer from cancer in bone (CIB) [3, 4]. Tumor metastasis to bone can be purely lytic, as in MM and breast cancer (BCa), or blastic, as in prostate cancer (PCa), although patients with PCa bone metastasis can have very high bone resorption marker levels [5]. Lytic and blastic bone metastases represent extremes of a continuum, with most patients having a mixture of osteolytic and osteoblastic components in their lesions. This marked imbalance in osteoclast and osteoblast numbers and activity in CIB causes debilitating skeletal-related events (SREs) that have catastrophic sequelae for patients. These sequelae include excruciating bone pain, pathological fractures, spinal cord and nerve compression syndromes, and derangements of calcium and phosphate homeostasis [6]. Bone involvement is the major cause of severe cancer-related pain in patients with advanced malignancies [7]. SREs not only increase morbidity and mortality, but also diminish the quality of life for patients [7]. Importantly, once cancers involve bone, the majority of patients are incurable. Thus, to improve outcomes for patients with CIB, new mechanistic-based therapies are needed that can control the growth of cancer cells in bone and prevent or inhibit the terrible sequelae associated with CIB. This chapter will provide an overview of the mechanisms responsible for osteolytic and osteoblastic metastasis, and provide examples of how their identification has resulted in the development of new treatments for CIB patients. Details of these mechanisms are provided in the chapters that follow in this section.

BONE AS THE PREFERRED SITE FOR METASTASIS

Stephen Paget originally proposed the "seed and soil" hypothesis, in which tumor cells preferentially metastasize to specific sites such as bone because of their unique properties rather than just to sites contiguous to the primary tumor [8]. Multiple mechanisms contribute to tumor cells homing to bone and the subsequent development of bone lesions. Kang and colleagues found that expression of only three to four genes in BCa cells was required for them to preferentially metastasize to bone [9]. These genes included *interleukin 11*, *connective tissue growth factor*, *CXCR4* or *osteopontin*, and *MMP1*, which encode for an osteolytic factor, an angiogenic factor, an adhesive factor, and a metalloprotease, respectively. TGF is released and activated from bone matrix by the increased osteoclast activity that is induced by tumor cells in bone, which further upregulates expression of these genes. In addition, multiple interactions between tumor cells and cells in the metastatic sites induce the secretion of a variety of factors, such as lysyl oxidase, that direct the migration, proliferation, and differentiation of tumor cells to metastatic sites [10]. Further, exosomes released by the tumor cells can prepare future metastatic sites for tumor metastasis [11].

Primer on the Metabolic Bone Diseases and Disorders of Mineral Metabolism, Ninth Edition. Edited by John P. Bilezikian.
© 2019 American Society for Bone and Mineral Research. Published 2019 by John Wiley & Sons, Inc.
Companion website: www.wiley.com/go/asbmrprimer

Adhesion molecules expressed on tumor cells and cellular products from tumor cells also increase bone metastasis. Shiozawa and colleagues showed that adhesion molecules on marrow stromal cells and osteoblasts in the endosteal hematopoietic stem cell niche allow hematopoietic stem cells and PCa cells that express receptors for these adhesion molecules to home to the stem cell niche [12]. Importantly, when cancer cells home to the stem cell niche, they displace and induce hematopoietic stem cells to terminally differentiate. Recently, Lawson and colleagues showed that when myeloma cells bind to the stem cell niche, they become dormant and can be activated and released later to become active tumor cells by osteoclastic bone resorption [13].

BONE REMODELING IN OSTEOLYTIC METASTASIS

Under normal conditions, bone resorption and formation are tightly linked, with bone formation occurring at discrete sites of previous bone resorption [14]. This "coupled" bone remodeling is largely dependent on communication between osteoclasts and osteoblasts, via bidirectional signaling between Eph receptors expressed on osteoblast precursors and ephrins on osteoclasts [15]. In osteolytic metastasis and MM, the bone remodeling process is imbalanced, or uncoupled, with increased osteoclastic bone resorption driven by osteoclast-activating factors produced by the tumor cells, or by cells in the bone microenvironment in response to the tumor cells [16]. For example, RANKL and IL-6 levels are increased in the bone microenvironment by tumor cells, while osteoprotegerin (OPG) is suppressed. In addition, multiple osteoblast inhibitors are produced in purely lytic tumors like MM, including the Wnt inhibitors DKK1 and secreted frizzled-related protein 2 (sFRP2), IL-7, and sclerostin that block bone repair [17].

BONE REMODELING IN OSTEOBLASTIC METASTASIS

Bone lesions such as those with PCa are classified as osteoblastic, primarily due to their characteristic osteosclerotic appearance on X-ray or CT scan. Bone formation markers, including bone-specific alkaline phosphatase and procollagen type 1 C propeptide, are elevated in PCa patients with bone disease. However, the new bone that is formed is of poor quality. Osteoblastic tumor cells secrete factors such as BMPs, PDGF, IGF-1, endothelin-1 (ET-1), FGF, and VEGF, that directly and indirectly increase osteoblast activity [18]. ET-1 has also been shown to activate the Wnt signaling pathway and suppress expression of the Wnt antagonist DKK1 [19], to further increase bone formation.

THE VICIOUS CYCLE HYPOTHESIS

The "vicious or feed-forward cycle" hypothesis proposes that tumor cells in bone stimulate osteoclastic bone resorption by secreting factors that directly or indirectly induce osteoclast formation. The increased bone resorption in turn releases and activates immobilized growth factors from the bone matrix, such as TGF-β and other growth factors, to stimulate the growth of tumor cells [20]. As noted earlier, some tumors, in particular PCa, also produce osteoblast-stimulating factors that increase bone formation. Release of cytokines and chemokines from the increased numbers of osteoblasts further stimulates tumor growth. In addition, adhesive interaction between marrow stromal cells and tumor cells enhance chemoresistance of tumor cells to further prevent tumor cell death in bone.

OSTEOCYTES AND BONE METASTASIS

In addition to the contribution of osteoclasts and osteoblasts to tumor growth in bone, recent studies have shown that osteocytes also contribute to tumor growth in bone [21, 22]. Osteocytes comprise more than 95% of the cells in bone and although embedded into the bone matrix, can physically interact with tumor cells in the bone via the osteocytic lacunar–cannalicular network. Osteocytes regulate normal osteoblast and osteoclast formation and activity through the production of sclerostin, an osteoblast inhibitor, and RANKL, M-CSF, and OPG, cytokines that regulate osteoclastogenesis and bone resorption. Further, Sottnik and colleagues [21] reported that tumor-induced pressure in the bone microenvironment in PCa bone metastasis upregulated CCL5 and matrix metalloproteinases in osteocytes, which also enhanced tumor growth in bone. In addition, recent studies have shown that apoptotic osteocytes are increased in patients with myeloma and produce increased levels of IL-11 that stimulate osteoclast formation [23]. Delgado-Calle and colleagues showed that direct cell to cell interaction between MM cells and osteocytes resulted in bidirectional Notch signaling between the osteocytes and MM cells. This triggered osteocyte apoptosis, upregulating *Sost* (the gene encoding sclerostin), decreased Wnt signaling, downregulating OPG and increasing RANKL production by osteocytes, and increased tumor cell growth [22, 24].

ROLE OF IMMUNE CELLS IN BONE METASTASIS

Immune cells present in the marrow, including mesenchymal stromal cells, hematopoietic cells, T cells, B cells, and macrophage-derived cells, stimulate bone destruction or enhance tumor cell homing to bone [25]. Dendritic cells and myeloid suppressor cells can differentiate into osteoclasts in the presence of MM cells in bone [26]. Further, Th17 T cells are an important subset of CD4+ T helper cells that contribute to the growth of cancer in bone. In normal bone marrow, the population of Th1 cells (interferon-producing T cells) that can protect patients with cancer from metastasis [27], exceeds the population of Th17 cells (IL-17-producing T cells). However, in myeloma, the ratio of Th1 to Th17 cells in the bone marrow is reversed, resulting in a 10-fold excess of Th17 cells compared with Th1 cells [28]. This change in T-cell subset distribution creates a microenvironment that enhances osteoclast activation. The increased IL-17 levels in the marrow also support myeloma cell growth and increased osteoclast formation, since IL-17 can induce RANKL production by T cells. IL-17 is also a growth factor for prostate cancer cells [29] and plays a role in bone metastasis from BCa [30]. Lin and colleagues demonstrated that TNF-α, IL-6, and IL-17 generated by host immune cells or tumor cells also led to loss of antitumor immunity and enhanced tumor growth [31]. Mesenchymal stromal cells also contribute to tumor growth in bone. They produce large amounts of IL-6 that enhance the growth and prevent apoptosis of MM cells, and stimulate osteoclast formation, as well as producing RANKL. Finally, immune cell–tumor cell interactions can suppress antitumor immune responses, and are an emerging therapeutic target for modulating the growth of tumors [32]. Further, antibodies that target cytokines and surface receptors on tumor cells, and small molecule inhibitors that bind cytokine receptors that enhance both tumor growth and bone destruction, are in development.

SUMMARY

Multiple mechanisms contribute to bone metastasis and make bone a preferential site for metastasis. Agents that target the mechanisms that mediate the effects of cancer cells in bone are now being developed or are in clinical trial to treat patients with CIB. These treatments include novel antibodies, immune modulators, and microRNAs that suppress the expression of genes that contribute to bone metastasis. They should allow patients with CIB to have an improved quality of life and increased survival.

REFERENCES

1. Greenberg AJ, Rajkumar SV, Therneau TM, et al. Relationship between initial clinical presentation and the molecular cytogenetic classification of myeloma. Leukemia. 2014;28(2):298–403.
2. Mundy GR. Metastasis to bone: causes, consequences and therapeutic opportunities. Nat Rev Cancer. 2002;2(8): 584–93.
3. Mundy GR. Mechanisms of bone metastasis. Cancer. 1997;80(suppl 8):1546–56.
4. Coleman RE. Skeletal complications of malignancy. Cancer. 1997;80(suppl 8):1588–94.
5. Coleman RE, Major P, Lipton A, et al. Predictive value of bone resorption and formation markers in cancer patients with bone metastases receiving the bisphosphonate zoledronic acid. J Clin Oncol. 2005;23(22):4925–35.
6. Roodman GD. Machanisms of bone metastasis. Discov Med. 2004;4(22):144–8.
7. Mantyh P. Bone cancer pain: causes, consequences, and therapeutic opportunities. Pain. 2013;154(suppl 1): S54–62.
8. Paget S. The distribution of secondary growths in cancer of the breast. 1889. Cancer Metastasis Rev. 1989;8(2):98–101.
9. Kang Y, Siegel PM, Shu W, et al. A multigenic program mediating breast cancer metastasis to bone. Cancer Cell. 2003;3(6):537–49.
10. Erler JT, Bennewith KL, Cox TR, et al. Hypoxia-induced lysyl oxidase is a critical mediator of bone marrow cell recruitment to form the premetastatic niche. Cancer Cell. 2009;15(1):35–44.
11. Roccaro AM, Sacco A, Maiso P, et al. BM mesenchymal stromal cell-derived exosomes facilitate multiple myeloma progression. J Clin Invest. 2013;123(4):1542–55.
12. Shiozawa Y, Pedersen EA, Havens AM, et al. Human prostate cancer metastases target the hematopoietic stem cell niche to establish footholds in mouse bone marrow. J Clin Invest. 2011;121(4):1209–312.
13. Lawson MA, McDonald MM, Kovacic N, et al. Osteoclasts control reactivation of dormant myeloma cells by remodeling the endosteal niche. Nature Commun. 2015;6:8983.
14. Sims NA, Gooi JH. Bone remodeling: multiple cellular interactions required for coupling of bone formation and resorption. Semin Cell Dev Biol. 2008;19(5):444–51.
15. Zhao C, Irie N, Takada Y, et al. Bidirectional ephrineB2-EphB4 signaling controls bone homeostasis. Cell Metab. 2006;4(2):111–21.
16. Esteve FR, Roodman GD. Pathophysiology of myeloma bone disease. Best Pract Res Clin Haematol. 2007;20(4): 613–24.
17. Giuliani N, Rizzoli V, Roodman GD. Multiple myeloma bone disease: pathophysiology of osteoblast inhibition. Blood. 2006;108(13):3992–6.
18. Roodman DG, Silbermann R. Mechanisms of osteolytic and osteoblastic skeletal lesions. BoneKEy Rep. 2015;4:753.
19. Clines GA, Mohammad KS, Bao Y, et al. Dickkopf homog 1 mediates endothelin-1-stimulated new bone formation. Mol Endocrinol. 2007;21(2):486–98.
20. Roodman GD. Mechanisms of bone metastasis. N Engl J Med. 2004;350(16):1655–64.
21. Sottnik JL, Dai J, Zhang H, et al. Tumor-induced pressure in the bone microenvironment causes osteocytes to promote the growth of prostate cancer bone metastases. Cancer Res. 2015;75(11):2151–8.
22. Delgado-Calle J, Bellido T, Roodman GD. Role of osteocytes in multiple myeloma bone disease. Curr Opin Support Palliat Care. 2014;8(4):407–13.

23. Giuliani N, Ferretti M, Bolzoni M, et al. Increased osteocyte death in multiple myeloma patients: role in myeloma-induced osteoclast formation. Leukemia. 2012;26(6):1391–401.
24. Delgado-Calle J, Anderson J, Cregor MD, et al. Bidirectional Notch signaling and osteocyte-derived factors in the bone marrow microenvironment promote tumor cell proliferation and bone destruction in multiple myeloma. Cancer Res. 2016;76(5):1089–100.
25. Mori G, D'Amelio P, Faccio R, et al. Bone-immune cell crosstalk: bone diseases. J Immunol Res. 2015;2015:108451.
26. Danilin S, Merkel AR, Johnson JR, et al. Myeloid-derived suppressor cells expand during breast cancer progression and promote tumor-induced bone destruction. Oncoimmunology. 2012;1(9):1484–94.
27. Bidwell BN, Slaney CY, Withana NP, et al. Silencing of Irf7 pathways in breast cancer cells promotes bone metastasis through immune espace. Nat Med. 2012;18(8):1224–31.
28. Noonan K, Marchionni L, Anderson J, et al. A novel role of IL-17-producing lymphocytes in mediating lytic bone disease in multiple myeloma. Blood. 2010;116(18):35654–63.
29. Zhang Q, Liu S, Ge D, et al. Interleukin-17 promotes formation and growth of prostate adenocarcinoma in mouse models. Cancer Res. 2012;72(10):2589–99.
30. Bian G, Zhao WY. Il-17, an important prognostic factor and potential therapeutic target for breast cancer? Eur J Immunol. 2014;44(2):604–5.
31. Lin WW, Karin M. A cytokine-mediated link between innate immunity, inflammation, and cancer. J Clin Invest. 2007;117(5):1175–83.
32. Malas S, Harrasser M, Lacy KE, et al. Antibody therapies for melanoma: new and emerging opportunities to activate immunity (Review). Oncology Rep. 2014;32(3):875–86.

96

Clinical and Preclinical Imaging in Osseous Metastatic Disease

Siyang Leng and Suzanne Lentzsch
Division of Hematology and Oncology, Columbia University Medical Center, New York, NY, USA

INTRODUCTION

Bone metastases are frequent complications in both solid and hematological malignancies, with a greater than 95% incidence in advanced myeloma, greater than 65% incidence in advanced breast or prostate cancer, and frequent occurrence in thyroid, lung, melanoma, and kidney cancers [1]. The occurrence of bone metastases is associated with a negative prognosis, and they cause substantial morbidity in the form of pain, fracture, hypercalcemia, spinal cord compression, and impaired mobility [2, 3].

Bone metastases can be classified into osteolytic, osteoblastic/sclerotic, or mixed phenotypes. Their formation depends on disruption of the dynamic between osteoblasts, osteoclasts, and the bone marrow stroma. A review of the pathophysiology of bone metastases is beyond the scope of this chapter, and the reader is directed to excellent reviews regarding this topic elsewhere [3, 4].

Imaging plays a vital role in the detection, evaluation, and monitoring of osseous metastatic disease. Common uses for imaging include staging, assessment of the risk for pathological fracture, and monitoring of the response to systemic therapy. Imaging modalities can be broadly divided into those that detect morphological or anatomical alterations, and those that assess for aberrant metabolic activity. Metabolic-based approaches have the advantage that, in addition to direct clinical utility, they can also be used to study processes such as tumor growth and metastasis, angiogenesis, and pharmacokinetics.

MORPHOLOGICAL/ANATOMICAL IMAGING TECHNIQUES

Plain film radiography

Plain film radiographymeasures the decline of X-rays passing though tissues. Plain radiographs can be used to assess the alignment of bone/vertebrae, the presence of fractures, and—by measuring cortical thickness—the risk for fracture [5]. They are frequently used in the initial evaluation of bone-related symptoms, due to their quick acquisition times and lower costs than other modalities. Additionally, they remain the standard screening test for bone disease in multiple myeloma. An advantage over bone scans (discussed further later in this chapter) is that they can identify lytic lesions as well as blastic lesions.

In other malignancies, plain radiographs are not typically used to screen for bone disease. This is because of poor sensitivity—up to 30% to 75% of bone must be destroyed in order for detection to occur—which is worse than that of CT, radionuclide bone scintigraphy (RBS), MRI, or PET [6]. Plain films, by virtue of being 2-dimensional (2D), do not offer spatial or structural detail about the location of tumors or bone damage, and additional imaging is often required.

Dual-energy X-ray absorptiometry

DXA is a method of measuring bone mineral density that is in widespread use for osteoporosis detection but also has emerging cancer-related applications. In this technique, a radiograph tube generates two photon beams of different energy levels, and a detector on the opposite side of the patient measures the attenuation of the beams as they pass through body tissue. The difference in the attenuation of the two beams allows for the differentiation of bone from other tissues, and measurement of mineral density.

DXA has an established role in bone density monitoring in certain cancer settings: women with breast cancer receiving aromatase inhibitors or who develop therapy-induced ovarian failure, and men with prostate cancer on androgen deprivation therapy [7]. While not established currently, DXA is under continued investigation for the risk stratification of patients with multiple myeloma. Individuals with monoclonal gammopathy of undetermined significance and smoldering multiple myeloma, both precursors to multiple myeloma, have a higher risk for fractures and have more bone imaging abnormalities on MRI or PET than the normal population [8]. There is interest in using DXA and other imaging techniques to identify patients with osteopenia and others at high risk for developing myeloma bone disease. In select cases, DXA is used to determine multiple myeloma-associated bone disease.

Computed tomography

CT entails the use of a rotating focal radiation source, and detectors opposite the source that measure the attenuation of the X-rays as they pass through tissues, with different rates of decline depending on the density of the tissue. The 2D images collected can be combined to form a 3-dimensional (3D) structure. Due to the natural contrast between bone and soft tissue, CT is well suited for both bone and soft tissue imaging. In myeloma, for example, it demonstrates bone lesions as well as extramedullary metastases. It can detect lytic as well as blastic lesions without the use of contrast.

CT is frequently used when results are desired urgently, when detailed information about the extent or location of bone destruction is desired, and for preoperative or preprocedural planning. It is widely available, and generally less expensive than other modalities discussed here. A key limitation of CT is its insufficient visualization of spinal cord pathologies; MRI offers better visualization of the spinal cord, the adjacent spaces and meninges, and the bone marrow [6]. Another important limitation is radiation exposure and increased risk of secondary cancer, a significant problem for patients who survive for years and undergo repeated scans, and for those who have been effectively cured and are now under surveillance [9].

Several modifications of CT are also in current clinical use. QCT utilizes CT to measure bone mineral density, where Hounsfield units are converted into bone density values. QCT delivers more radiation than DXA and is less widely available, but also has several advantages over DXA including 3D representation of bone density, and ability to avoid inaccuracies from hyperostosis or extraosseous calcium deposits. A modification of QCT called HRpQCT is restricted to peripheral bones such as the radius and tibia. It allows for accurate measurement of the cortical and trabecular compartments separately. Currently, the main application of QCT is osteoporosis, but there is significant clinical and research interest in the use of QCT for cancers that have a significant bone component [10, 11].

Whole body low-dose CT is a protocol developed for use in multiple myeloma which utilizes a radiation dose comparable to skeletal survey, is easy to implement, has a quick acquisition time (75 seconds), does not require contrast, and offers a 3D view of the skeletal system. It has significantly higher sensitivity than a skeletal survey as it identifies more lytic lesions. However, compared with MRI, it has lower specificity due to its inability to detect abnormalities in the bone marrow signal [12].

Micro-CT (µCT) is a modality mainly used to image small animals and tissue specimens, with an identical set-up compared with CT. µCT has high spatial resolution (50 to 100 µm^3 of voxels) and rapid data acquisition time (minutes). A variety of bone parameters including mineral density, volume, surface ratio, and trabecular thickness can be measured (Fig. 96.1). Recent innovations have centered on the development of noniodinated contrast agents, such as gold or bismuth nanoparticles, that produce greater enhancement of tumors and vasculature. Nanoparticles can have therapeutic agents added to them—for example, gold nanoparticles absorb radiation readily, and can be used to both diagnose tumors and deliver radiation to them. Similar to CT, a key limitation of µCT is radiation exposure, which is generally limited but which for high-resolution scans can become biologically significant, altering tumor growth and the surrounding biological and immunological pathways [13–15].

Magnetic resonance imaging

MRI allows visualization of various tissues based upon differences in their absorption and emission of electromagnetic energy. This is typically performed through excitation of hydrogen atoms in an oscillating magnetic field. As the hydrogen atoms return to equilibrium, they emit radiofrequency signals which are captured by a receiving coil. Different tissues can be distinguished by their differential return to equilibrium. Positional information is determined through the use of gradients in the magnetic field [16].

MRI is particularly useful in evaluating for preclinical bone disease or small bone lesions—when abnormalities are present within the marrow and/or cortex but before symptoms develop [17]. As discussed earlier, MRI is superior to other modalities for imaging the spine, where visualization of both bone and soft tissue is important [18].

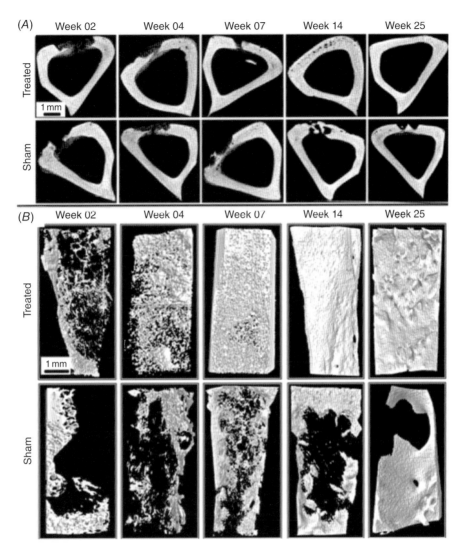

Fig. 96.1. Micro-CT imaging of a tibial bone defect before and after treatment with an osteoinductive gel scaffold. New Zealand White male rabbits underwent a surgical procedure which created a defect in their tibia. μCT was performed to evaluate healing. Axial (A) and 3D reconstruction (B) views are provided, showing significantly more healing of bone in the arm that was treated with an osteoinductive gel scaffold. Source: [50]. Reproduced with permission of Sagar, http://journals.plos.org/plosone/article?id=10.1371/journal.pone.0077578.

Additionally, MRI can help distinguish between stress and pathological fractures, particularly important in patients who are not known to have cancer. A well-defined, low-signal T1-weighted abnormality around a fracture has been suggested to be the discriminating feature of a pathological fracture [19].

The most significant advantage for use of MRI to detect osseous metastatic disease is its good performance. Studies have generally shown MRI to be more sensitive and specific than CT for detecting a variety of bone metastases. For example, a pooled sensitivity of 90.4% for MRI versus 77.1% for CT and specificity of 96.0% versus 83.2% were found in one study [20]. A meta-analysis of whole body MRI found a pooled sensitivity of 89.9% and specificity of 91.8% for detecting bone metastases [21]. Another advantage with MRI is the lack of radiation exposure—again most important for cancer patients who require repeat imaging and have extended survival.

If not for a number of limitations, MRI would likely be even more widely used. MRI is contraindicated for patients with certain pacemakers, cochlear implants, and other metal objects. It is sensitive to movement and, not infrequently, results are compromised by patient movement. Patients with severe claustrophobia may not be able to tolerate the machine. MRI has longer acquisition times, and is generally more costly than CT. The use of marrow-stimulating factors may alter the marrow signal, making scan results noninterpretable [18].

Micro-MRI (μMRI), similar to μCT, is mainly used to image small animals and tissue specimens, although it can be used to image the extremities of humans as well. It has

Table 96.1. Comparison of different imaging modalities including source, possible applications, and the benefits and limitations of each technique. *Italics* denotes preclinical applications.

Modality	Source	Possible Applications	Advantages	Limitations
X-ray	Radiation	Lytic or blastic lesions Structural, anatomical information	Low cost Wide availability	Low sensitivity Patient discomfort from needing to reposition in between images
MRI	Electromagnetic waves	Lytic or blastic lesions Visualization of spinal cord and canal Assesses bone marrow compartment Structural, anatomical information Procedure planning	High sensitivity and specificity High spatial resolution Good soft tissue contrast No radiation exposure	Low throughput (eg, 1–2 hours for MRI of entire spine versus 10–20 minutes for CT of entire spine) Subject to motion artifact Patients with claustrophobia or implanted devices may not be able to undergo procedure Costly
CT, WBLDCT	Radiation (X-rays)	Lytic or blastic lesions Bone density and structure analysis Structural, anatomical information Measurement and monitoring of tumor burden Bone structure and health Procedure planning	Fast acquisition time Widely available High spatial resolution High sensitivity compared with X-ray Depicts extramedullary disease Less costly than PET or MRI Comfortable	Radiation exposure* (CT chest: 5–10 mSv, CT abdomen and pelvis: 10 mSv; WBLDCT: 4–7 mSv) More costly than X-ray
PET	Radiation [high-energy γ-rays]	Lytic or blastic lesions Highlights areas of active tumor Measurement and monitoring of tumor burden Prognostication value in some cancers *Visualization of molecular and metabolic processes (eg, whole body pharmacokinetics or receptor expression)* *Real-time signaling pathway activity (reporter constructs)*	High sensitivity High penetration depth Depicts extramedullary disease	Limited spatial resolution Radiation exposure* (PET/CT: 25 mSv) Potential for false positives High cost Limited availability; usually not available as inpatient
SPECT	Radiation [low-energy γ-rays]	Blastic lesions only Measurement and monitoring of tumor burden *Visualization of molecular and metabolic processes (eg, apoptosis or receptor expression levels)*	High sensitivity (but lower than PET) High penetration depth	Limited spatial resolution Radiation exposure* (4 mSv) Not quantitative
BLI	Bioluminescent light	*Early cell localization and bone homing, small cell numbers* *Long-term monitoring of cell viability and metastatic bone tumor burden (functional and drug response studies)* *Real-time signaling pathway activity (reporter constructs)* *Visualization of molecular and metabolic processes (eg, VEGFR activity)*	High sensitivity High throughput Low costs Small cell number imaging possible	Limited spatial resolution Limited penetration depth Not clinically applicable currently
FI	Fluorescent light	*Early cell localization and bone homing* *Long-term monitoring of cell viability and metastatic bone tumor burden (functional and drug response studies NIRFs)* *Real-time signaling pathway activity (reporter constructs)* Real-time monitoring of processes involved in bone metastasis (smart probes, eg, MMP sense) Image-guided surgery	High sensitivity High throughput Low costs	Limited spatial resolution Limited penetration depth Autofluorescence Not clinically applicable currently

*Radiation exposure doses provided are estimates given in millisieverts (mSv). Risk of developing cancer is thought to significantly increase when exposure exceeds 50–100 mSv.

BLI = bioluminescence imaging; CT = computed tomography; FI = fluorescence imaging; MMP = matrix metalloproteinase; MRI = magnetic resonance imaging; NIRF = near infrared fluorescent; PET = positron emission tomography; SPECT = single-photon emission computed tomography; WBLDCT = whole body low-dose computed tomography.

good spatial resolution, and is adept at identifying bone and soft tissue lesions [22]. With μMRI there is no use of radiation and thus no concern for the tissue-modifying properties of radiation. Despite this advantage, μMRI is used less frequently than μCT because it is more costly and has longer image acquisition times. Animals have to be anesthetized, which may entail alterations in body temperature and other physiology, that in turn limit the use of μMRI to anatomical studies only [13].

MOLECULAR/FUNCTIONAL IMAGING TECHNIQUES

Radionuclide bone scintigraphy

RBS comprises the intravenous administration of a radioactive bone-seeking agent, ^{99m}technetium (^{99m}Tc), complexed with a phosphate compound, most commonly methylene diphosphonate (MDP). Typically 3 hours after administration, patients are imaged using a planar gamma camera, which captures the 140 keV γ-rays produced by the radioactive decay of the ^{99m}Tc tracer. Anterior and posterior views are collected [2].

RBS demonstrates areas of active bone remodeling through the phosphate groups in the administered radiotracers binding to hydroxyapatite in the bone matrix. Osteoblastic metastases, which display locally increased bone turnover, accumulate more Tc^{99m}-MDP than normal bone. Current evidence gives strongest support to the use of RBS for the detection of bone metastases in common solid tumors, and its use is also recommended in the staging of breast and prostate carcinomas in certain clinical scenarios [20].

Studies examining the use of RBS in predominantly solid tumor malignancies have reported both sensitivities and specificities that are typically in the range of 80% to 90% [20]. RBS is sensitive for small changes, even a 5% change in bone turnover can be detected, whereas plain film radiography or CT requires up to 40% to 50% loss of mineral before an abnormal signal is displayed. Disadvantages include lack of utility for tumors that are predominantly lytic (eg, multiple myeloma), very aggressive tumors wherein bone destruction rapidly outpaces remodeling, and false positives from degenerative change, inflammation, trauma, or Paget disease [5].

Single-photon emission computed tomography

Single-photon emission computed tomography (SPECT) is a 3D variant of RBS in which detectors rotate around the patient to record the volumetric distribution of the radiotracer. It is often used as a complementary technique to RBS, to increase sensitivity and localize vertebral lesions. It can be combined with CT (SPECT-CT) for additional sensitivity [23]. A recent meta-analysis for the detection of vertebral metastases found good overall sensitivity and specificity, with an overall performance second only to MRI (and higher than RBS, PET, and CT) [24]. Despite this, SPECT is not as frequently used as other modalities for detecting bone metastases in the United States.

Micro-SPECT (μSPECT) is a preclinical version of SPECT that is used in animals. Due to high sensitivity, only tiny amounts of radiotracer are needed, and multiple tracers can be used to image several molecular events simultaneously. Limitations include cost, exposure to radiation, and limited resolution, currently as low as 1000 μm [25, 26].

Positron emission tomography

PET is an imaging technique that detects pairs of high energy γ-rays travelling in opposite directions emitted indirectly by a positron-emitting radioactive atom. Patients are injected with a radiotracer, and are passed through a ring of detectors. A signal is recorded when detector elements on opposite ends of a ring are triggered at the same time, and a 3D map of the decay events is generated [2]. The predominant radiotracer in use for PET is ^{18}fluoride-fluorodeoxyglucose (^{18}F-FDG). Cancer cells have altered metabolic pathways, relying upon aerobic glycolysis (the Warburg effect) to sustain their proliferative rate [27]. They preferentially accumulate ^{18}F-FDG, which upon phosphorylation becomes trapped within cells [28].

FDG-PET has broad clinical and preclinical utility. When combined with CT, it becomes very useful for cancer staging in many solid and hematological malignancies, due to its ability to demonstrate distant and extramedullary metastases. It is useful for detecting lytic bone disease, a key limitation of RBS, and is increasingly the preferred imaging in smoldering myeloma (whole body MRI is also acceptable), and in multiple myeloma (to assess for active disease and extramedullary metastases, neither of which is reflected by skeletal survey). It is also used to help distinguish between benign versus malignant bone lesions [29]. The sensitivity and specificity of FDG-PET varies by histology, with generally better performance in cancers that have predominantly lytic bone disease, and when FDG-PET is combined with CT. The previously cited meta-analysis reported a pooled sensitivity for PET (includes studies that examined either PET or PET/CT) of 86.9% (compared with 90.4% for MRI, 77.1% for CT, and 75.1% for RBS), and specificity of 97.0% (compared with 96.0% for MRI, 83.2% for CT, and 93.6% for RBS) [20].

Despite its overall good performance, PET has a number of limitations that reduce its utility. It may be less sensitive than RBS at detecting bone lesions that are blastic or mixed [29]. It may miss lesions that are less than 8 to 10 mm, unless they are very hypermetabolic. Findings may be difficult to interpret—for example, low levels of uptake can be from treated disease, early relapse, chronic inflammation, infection, or physiological uptake—and there are false positives. It can be significantly more costly than other modalities. Careful scheduling is required, because radionuclides have time-dependent decay, and because patient glucose levels can compete with FDG for uptake by tumor (patients should be euglycemic with a 6- to 8-hour fast prior). Marrow rebound after chemotherapy or use of marrow-stimulating factors can cause false positives. There is radiation exposure, particularly if PET is combined with CT [30].

Radiotracers based upon alternative metabolic substrates have been developed, such as those involved in lipid (cell membrane) metabolism: acetate and choline, complexed to either ^{11}carbon or ^{18}F [31]. The acetate and choline-based tracers have been predominantly studied in prostate cancer, and remain under investigational use due to challenges in synthesis, limitations in half-life, and limited clinical applicability compared with ^{18}F-FDG [2]. Another radiotracer under investigation is ^{18}F-sodium fluoride (^{18}F-NaF), which, like ^{99m}Tc-MDP, is incorporated into the bone during the mineralization process. Emerging evidence suggests that it is superior to ^{99m}Tc-MDP RBS for detecting bone metastases in prostate cancer [32]. Its performance compared with FDG is unclear, and may differ by histology [33]. Multiple other radiotracers that target tumor metabolism and angiogenesis are in development.

Micro-PET (μPET) is a preclinical version of PET for use in animals. Advantages are quick acquisition time and good performance characteristics similar to that of PET. Like μSPECT, only tiny amounts of radiotracer are generally needed. Limitations include cost, short half-lives of commonly used tracers, and exposure to radiation. Resolution suffers compared with μCT or μMRI, at 1000 μm [26, 34].

Optical imaging

Optical imaging encompasses a set of emerging techniques that use visible, ultraviolet, and infrared light to examine anatomy and cellular processes. All techniques are currently preclinical, although clinical trials are underway. Optical imaging is divided broadly based upon contrast agent into bioluminescence imaging (BLI), which uses a natural light-emitting protein such as luciferase to trace the location of certain cells or processes in vivo, and fluorescence imaging (FI), which uses fluorochromes that emit light after being stimulated by an external light source. The choice of BLI or FI is dictated by the research question, with BLI favored when in vivo metabolic activity is being assessed. In addition to the choice of contrast agent, there is also choice in the technology used to create the image. Many commercial systems exist, and vary by characteristics such as wavelength illuminated and detected, detection technology, and support for multimodal imaging [25, 35, 36].

The popularity of optical methods is increasing due to several advantages: the methods tend to have high

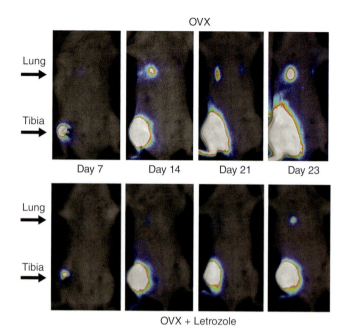

Fig. 96.2. Bioluminescence imaging demonstrating metastatic progression of 4T1 breast cancer cells in BALB/c mice. BALB/c mice were injected with 4T1 murine mammary cancer cells in their right tibia. They were then exposed to different treatments and followed with serial bioluminescence imaging. Representative images of one animal in each group are shown here, showing progressive tumor growth. OVX = ovarectomized. Source: [37]. Reproduced with permission of Springer.

overall sensitivity and specificity, are noninvasive, do not involve radiation, are economical, and have fairly straightforward interpretation. Key limitations are a low depth of penetration—signal detection is only possible for tissues within a few centimeters of the surface—and limited spatial resolution, reaching 1000 μm [35].

BLI is based on the detection of photons emitted by an enzymatic reaction in which a substrate is oxidized by a luciferase. Many applications of this system have been described. As an example, in one study, breast cancer cells were tagged with luciferase and introduced into the tibia of mice. The mice were then randomized to different treatment arms, and after therapy were injected with luciferin and imaged with BLI, which detected tumor growth in both the bone and lung (Fig. 96.2) [37]. Systems examining bone metastasis in hepatocellular carcinoma, prostate cancer, and other malignancies have been described [38].

BLI has several advantages, including high sensitivity, a low background signal, a high signal/noise ratio, and a short acquisition time. However, it mainly provides planar imaging with limited ability to localize signal and determine depth. In addition, the technique is restricted to small animals because it would work only very superficially in larger objects. The color of the animals is also important because dark skin pigmentation or fur color can attenuate the bioluminescent light (the same holds true for fluorescent light) [39].

FI utilizes synthesized fluorescent compounds that are targeted to specific cell compartments or molecules. After administration of the fluorophore, light of a specific wavelength is applied, and a detector records light emitted by the excited fluorophore. Multiple classes of fluorophores exist but, for bone, tetracyclines (eg, BoneTag) and fluorescently labeled bisphosphonates (eg, OsteoSense) are most frequently used [36, 40, 41]. ProSense is a cathepsin-activated fluorophore that can be used as a marker of osteoclast activity. Increased osteoclast activity increases cathepsin, which cleaves the fluorophores and frees them, allowing them to be detected [42]. MMPSense is a fluorophore that is activated by matrix metalloproteinases (MMP) 2 and 9 and thus can be used to detect MMP activity [43].

FI does not require a metabolically active organism in order to produce photons, meaning that it can be used in vitro or ex vivo. Drawbacks of FI include autofluorescence, tissue absorbance, limited depth information (restricting the technique to small animals), and poor spatial resolution [35]. Using fluorescent proteins that are in the far-red or near-infrared area of the spectrum can help allow for deeper imaging, because autofluorescence is less prominent at these wavelengths [44].

Due to the advantageous characteristics of optical imaging discussed here, there is significant interest in finding clinical applications, one of which is surgery. Fluorescence-guided surgery, in which FI is used to detect fluorophore-expressing cancer cells, has been shown to help identify metastases and improve survival in mouse models [45, 46]. Many clinical trials in humans are under way, and validation is eagerly awaited.

MULTIMODALITY IMAGING

Given that each imaging modality has advantages and limitations, combining two or more modalities may yield better performance, albeit at greater economic cost and logistical inconvenience. Currently in the clinic, PET/CT is the predominant multimodal imaging technique. In 2010, PET/MRI emerged as an alternate multimodal technique, with potential advantages of improved soft tissue imaging, and functional MRI capability. Adoption has been limited, and national guidelines have not been established. Current studies suggest general equivalency between PET/MRI and PET/CT, except for lung lesions, in which some studies suggest CT is better, and for prostate cancer and other cancers with bone lesions, in which MRI may be better [47].

In the preclinical setting, μCT has been combined with optical imaging to provide complimentary information on changes in the bone matrix and in osteolytic activity [48]. This combination can also be used to avoid the use of radiotracers. A limitation is that this combination does not overcome the limited tissue penetration of optical imaging.

CONCLUSION

Several modalities exist for imaging osseous metastatic lesions in both clinical and preclinical settings. Each comes with limitations, and sometimes multiple modalities are required to help answer difficult clinical questions. Because imaging is not a perfect reflection of function and presence of cancer, biopsies and correlation with clinical features and lab studies continue to be necessary.

Bone disease is a frequent occurrence in cancer which often requires urgent attention, and continued refinement of the costs, acquisition times, and performance characteristics of the imaging techniques will help to expedite cancer care. However, we are also realizing that the use of imaging does not by itself improve patient outcomes [49]. Finding the appropriate balance between clinical precision and overuse will continue to be a key challenge for cancer care.

REFERENCES

1. Coleman RE. Skeletal complications of malignancy. Cancer 1997;80:1588–94.
2. Ulmert D, Solnes L, Thorek DLJ. Contemporary approaches for imaging skeletal metastasis. Bone Res. 2015;3:15024.
3. Roodman GD. Mechanisms of bone metastasis. New Engl J Med. 2004;350:1655–64.
4. Weidle UH, Birzele F, Kollmorgen G, et al. Molecular mechanisms of bone metastasis. Cancer Genomics Proteomics. 2016;13:1–12.
5. Rosenthal DI. Radiologic diagnosis of bone metastases. Cancer. 1997;80:1595–607.
6. Hamaoka T, Madewell JE, Podoloff DA, et al. Bone imaging in metastatic breast cancer. J Clin Oncol. 2004;22:2942–53.
7. Gralow JR, Biermann JS, Farooki A, et al. NCCN Task Force report: bone health in cancer care. J Natl Compr Cancer Network. 2013;11(suppl 3):S1–50; quiz S1.
8. Kristinsson SY, Minter AR, Korde N, et al. Bone disease in multiple myeloma and precursor disease: novel diagnostic approaches and implications on clinical management. Expert Rev Mol Diagn. 2011;11:593–603.
9. Brenner DJ, Hall EJ. Computed tomography—an increasing source of radiation exposure. New Engl J Med. 2007;357:2277–84.
10. Nishiyama KK, Shane E. Clinical Imaging of bone microarchitecture with HR-pQCT. Curr Osteoporos Rep. 2013;11:147–55.
11. Hui SK, Khalil A, Zhang Y, et al. Longitudinal assessment of bone loss from diagnostic CT scans in gynecologic cancer patients treated with chemotherapy and radiation. Am J Obstet Gynecol. 2010;203:353.e1–.e7.
12. Pianko MJ, Terpos E, Roodman GD, et al. Whole-body low-dose computed tomography and advanced imaging

techniques for multiple myeloma bone disease. Clin Cancer Res. 2014;20:5888–97.
13. de Kemp RA, Epstein FH, Catana C, et al. Small-animal molecular imaging methods. J Nucl Med. 2010;51(suppl 1):18s–32s.
14. Ashton JR, West JL, Badea CT. In vivo small animal micro-CT using nanoparticle contrast agents. Front Pharmacol. 2015;6:256.
15. Schambach SJ, Bag S, Schilling L, et al. Application of micro-CT in small animal imaging. Methods. 2010;50:2–13.
16. Callaghan PT. *Principles of Nuclear Magnetic Resonance Microscopy*. Oxford: Oxford University Press, 1994.
17. Messiou C, Cook G, deSouza NM. Imaging metastatic bone disease from carcinoma of the prostate. Br J Cancer. 2009;101:1225–32.
18. Lecouvet FE, Larbi A, Pasoglou V, et al. MRI for response assessment in metastatic bone disease. Eur Radiol. 2013;23:1986–97.
19. Fayad LM, Kamel IR, Kawamoto S, et al. Distinguishing stress fractures from pathologic fractures: a multimodality approach. Skeletal Radiol. 2005;34:245–59.
20. Yang HL, Liu T, Wang XM, et al. Diagnosis of bone metastases: a meta-analysis comparing (1)(8)FDG PET, CT, MRI and bone scintigraphy. Eur Radiol. 2011;21:2604–17.
21. Wu LM, Gu HY, Zheng J, et al. Diagnostic value of whole-body magnetic resonance imaging for bone metastases: a systematic review and meta-analysis. J Magn Reson Imaging. 2011;34:128–35.
22. Graham TJ, Box G, Tunariu N, et al. Preclinical evaluation of imaging biomarkers for prostate cancer bone metastasis and response to cabozantinib. J Natl Cancer Inst. 2014;106:dju033.
23. Sedonja I, Budihna NV. The benefit of SPECT when added to planar scintigraphy in patients with bone metastases in the spine. Clin Nucl Med. 1999;24:407–13.
24. Liu T, Wang S, Liu H, et al. Detection of vertebral metastases: a meta-analysis comparing MRI, CT, PET, BS and BS with SPECT. J Cancer Res Clin Oncol. 2017;143(3):457–65.
25. Massoud TF, Gambhir SS. Molecular imaging in living subjects: seeing fundamental biological processes in a new light. Genes Dev. 2003;17:545–80.
26. Chatziioannou AF. Instrumentation for molecular imaging in preclinical research: micro-PET and Micro-SPECT. Proc Am Thorac Soc. 2005;2:533–6, 10–11.
27. Vander Heiden MG, Cantley LC, Thompson CB. Understanding the Warburg effect: the metabolic requirements of cell proliferation. Science. 2009;324:1029–33.
28. Wahl RL, Herman JM, Ford E. The promise and pitfalls of positron emission tomography and single-photon emission computed tomography molecular imaging-guided radiation therapy. Semin Radiat Oncol. 2011;21:88–100.
29. Fogelman I, Cook G, Israel O, et al. Positron emission tomography and bone metastases. Semin Nucl Med. 2005;35:135–42.
30. Griffeth LK. Use of PET/CT scanning in cancer patients: technical and practical considerations. Proceedings (Bayl Univ Med Cent). 2005;18:321–30.
31. Jadvar H. Prostate cancer: PET with (18)F-FDG, (18)F- or (11)C-acetate, and (18)F- or (11)C-choline. J Nucl Med. 2011;52:81–9.
32. Poulsen MH, Petersen H, Hoilund-Carlsen PF, et al. Spine metastases in prostate cancer: comparison of technetium-99m-MDP whole-body bone scintigraphy, [(18)F]choline positron emission tomography(PET)/computed tomography (CT) and [(18) F]NaF PET/CT. BJU Int. 2014;114:818–23.
33. Ota N, Kato K, Iwano S, et al. Comparison of (1)(8)F-fluoride PET/CT, (1)(8)F-FDG PET/CT and bone scintigraphy (planar and SPECT) in detection of bone metastases of differentiated thyroid cancer: a pilot study. Br J Radiol. 2014;87:20130444.
34. Koba W, Jelicks LA, Fine EJ. MicroPET/SPECT/CT imaging of small animal models of disease. Am J Pathol. 2013;182:319–24.
35. Ventura M, Boerman OC, de Korte C, et al. Preclinical imaging in bone tissue engineering. Tissue Eng B Rev. 2014;20:578–95.
36. Snoeks TJ, Khmelinskii A, Lelieveldt BP, et al. Optical advances in skeletal imaging applied to bone metastases. Bone. 2011;48:106–14.
37. Wang W, Belosay A, Yang X, et al. Effects of letrozole on breast cancer micro-metastatic tumor growth in bone and lung in mice inoculated with murine 4T1 cells. Clin Exp Metastasis. 2016;33:475–85.
38. Elshafae SM, Hassan BB, Supsavhad W, et al. Gastrin-releasing peptide receptor (GRPr) promotes EMT, growth, and invasion in canine prostate cancer. Prostate. 2016;76:796–809.
39. O'Neill K, Lyons SK, Gallagher WM, et al. Bioluminescent imaging: a critical tool in pre-clinical oncology research. J Pathol. 2010;220:317–27.
40. van Gaalen SM, Kruyt MC, Geuze RE, et al. Use of fluorochrome labels in in vivo bone tissue engineering research. Tissue Eng B Rev. 2010;16:209–17.
41. Wen D, Qing L, Harrison G, et al. Anatomic site variability in rat skeletal uptake and desorption of fluorescently labeled bisphosphonate. Oral Dis. 2011;17:427–32.
42. Kozloff KM, Quinti L, Patntirapong S, et al. Non-invasive optical detection of cathepsin K-mediated fluorescence reveals osteoclast activity in vitro and in vivo. Bone. 2009;44:190–8.
43. Bremer C, Tung CH, Weissleder R. In vivo molecular target assessment of matrix metalloproteinase inhibition. Nat Med. 2001;7:743–8.
44. Hilderbrand SA, Weissleder R. Near-infrared fluorescence: application to in vivo molecular imaging. Curr Opin Chem Biol. 2010;14:71–9.
45. Miwa S, De Magalhaes N, Toneri M, et al. Fluorescence-guided surgery of human prostate cancer experimental bone metastasis in nude mice using anti-CEA DyLight 650 for tumor illumination. J Orthopaed Res. 2016;34:559–65.

46. Keereweer S, Kerrebijn JD, Mol IM, et al. Optical imaging of oral squamous cell carcinoma and cervical lymph node metastasis. Head Neck. 2012;34:1002–8.
47. Spick C, Herrmann K, Czernin J. 18F-FDG PET/CT and PET/MRI perform equally well in cancer: evidence from studies on more than 2,300 patients. J Nucl Med. 2016;57:420–30.
48. Lim E, Modi K, Christensen A, et al. Monitoring tumor metastases and osteolytic lesions with bioluminescence and micro CT Imaging. J Visu Exp. 2011:50;e2775.
49. Healy MA, Yin H, Reddy RM, et al. Use of positron emission tomography to detect recurrence and associations with survival in patients with lung and esophageal cancers. J Natl Cancer Instit. 2016;108(7).
50. Sagar N, Pandey AK, Gurbani D, et al. In-vivo efficacy of compliant 3D nano-composite in critical-size bone defect repair: a six month preclinical study in rabbit. PloS One. 2013;8:e77578.

97
Metastatic Tumors and Bone

Julie A. Sterling[1,2,3,4] and Rachelle W. Johnson[2,3]

[1]*Department of Veterans Affairs, Tennessee Valley Healthcare System, Nashville, TN, USA*
[2]*Center for Bone Biology, Department of Medicine, Division of Clinical Pharmacology, Vanderbilt University Medical Center, Nashville, TN, USA*
[3]*Department of Cancer Biology, Vanderbilt University, Nashville, TN, USA*
[4]*Department of Biomedical Engineering, Vanderbilt University, Nashville, TN, USA*

IMPORTANCE OF THE PROBLEM

Despite decades of research into bone metastatic disease, bone metastases remain a significant complication faced by patients with some of the most common cancers including breast, prostate, lung, and renal. For example, approximately 70% of breast and prostate cancer patients and 20% to 40% of renal cell carcinoma, melanoma, and lung cancer patients who die from their disease will develop bone metastases [1, 2]. Once established in bone, many factors within the microenvironment (cellular interactions, physical interactions, stress conditions, etc.) stimulate tumor cells to secrete factors such as PTHrP, that increase osteoblast production of RANKL, thus stimulating osteoclast-mediated bone destruction [3, 4] (Fig. 97.1). In patients, the bone disease can include evidence of both osteolytic (bone-destructive) and osteoblastic (bone-forming) activity, but in either case leads to an increase in pain and skeletal-related events (SREs) including spinal cord compression and fracture. While palliative treatments such as bisphosphonates have been very effective for reducing pain and time to SREs, they do not eliminate disease. Therefore, many groups are continuing to study tumor-induced bone disease (TIBD) in order to develop a better understanding of the processes that lead to tumor establishment in bone and subsequent bone disease. Developing a better understanding of TIBD may help develop therapeutic strategies to inhibit both tumor growth in bone and the resulting destruction.

CONCEPT OF TUMOR-INDUCED BONE DISEASE

TIBD was originally described as the "vicious cycle," which centered on tumor cell secretion of PTHrP, interleukins, and other factors that stimulated osteoblasts to produce RANKL and activate the osteoclasts [5, 6]. As the osteoclasts degraded bone, growth factors (eg, TGF-β, Wnt, BMP) were released from the bone, which further stimulated the growth of tumor cells and the secretion of tumor-produced factors. These in turn stimulated bone destruction through enhanced osteoclastogenesis and the release of more growth factors. While this theory has largely held up, we now know that it is a far more complex and dynamic process and as such is often referred to as tumor-induced or cancer-induced bone disease. Current views of TIBD include interactions with the bone and bone marrow microenvironment, both cellular and physical, as key players in tumor growth and bone destruction. While the interactions between the tumors and the bone/bone marrow microenvironment are acknowledged as important, there is still much to be learned about these interactions to produce more effective treatments for TIBD.

Primer on the Metabolic Bone Diseases and Disorders of Mineral Metabolism, Ninth Edition. Edited by John P. Bilezikian.
© 2019 American Society for Bone and Mineral Research. Published 2019 by John Wiley & Sons, Inc.
Companion website: www.wiley.com/go/asbmrprimer

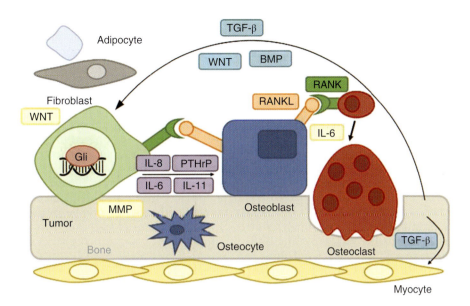

Fig. 97.1. Updated concept of tumor-induced bone disease. Tumor cells disseminate to the bone marrow where they express/secrete signaling molecules (eg, PTHrP, IL-8, IL-6, and IL-11) that modify the bone microenvironment, including osteoblasts and osteocytes. Signals to and from osteoblast lineage cells, myocytes, fibroblasts, and adipocytes in the bone microenvironment enable tumor cells to colonize the bone and induce tumor-induced bone destruction, releasing growth factors such as TGF-β, Wnt, and BMPs from the bone matrix, which in turn act directly and indirectly (such as through adipocytes, fibroblasts, and myocytes) on tumor cells. For all secreted molecules: blue = bone-derived; purple = tumor-derived; yellow = mesenchymal lineage-derived.

TUMOR INTERACTIONS WITH THE BONE/BONE MARROW MICROENVIRONMENT

From the early descriptions of TIBD, the fact that bone metastatic tumor cells required interactions with the bone microenvironment to colonize and invade the bone were clear, but due to the complexity of the bone/bone marrow microenvironment, fully characterizing these interactions has been exceedingly challenging. In fact, early studies solely focused on the interactions of tumor cells with osteoblasts and osteoclasts, and were able to demonstrate the importance of the interactions between these cell types. Through these studies the primary signaling pathways were established to include TGF-β, Wnt, BMP, and other signaling pathways that appear to be important in the interactions between tumors and other cells/features in the bone microenvironment. Many of these early signaling pathways were noted to regulate PTHrP, which was identified early on as a key driver of TIBD [7]. These studies continue today, and have revealed more molecules that mediate tumor–osteoclast interactions, such as tumor secretion of soluble intercellular adhesion molecule 1 (ICAM1), which stimulates osteoclastogenesis through repression of microRNAs [8].

Tumor interactions with myeloid lineage cells

The bone marrow is a rich reservoir of immune cells and other progenitor cells. While these cell types were largely ignored for many years, many groups have more recently begun to investigate the role of the immune progenitor cells in TIBD. One of the first immune cell types to be investigated in TIBD was the myeloid-derived suppressor cells (MDSCs), because they had been seen in many other tumors to help establish the premetastatic niche [9, 10]. They have since been demonstrated to expand and contribute to tumor growth and bone destruction in bone metastatic breast and prostate cancer [11, 12]. In prostate cancer, tumor-derived PTHrP was shown to indirectly increase MDSC angiogenic potential, thus promoting tumor metastasis and growth in bone [11]. In breast cancer, Danilin and colleagues demonstrated that MDSCs contributed to bone destruction by activating tumor expression of Gli2 and PTHrP, thus stimulating bone destruction [12]. Two studies have also demonstrated that MDSCs from a tumor-bearing mouse have the potential to differentiate into osteoclasts, therefore the expansion of this population leads to a larger pool of osteoclast progenitors, which may explain the ability of MDSCs to contribute to bone destruction [12, 13]. While the signaling pathways involved in the cross-talk between MDSCs and tumor cells is still being investigated, cross-talk between PLCγ2 and β-catenin has been shown to promote MDSC expansion in bone [14].

Another cell type that has been demonstrated to play an important role in TIBD is that of the macrophages. For example, bone marrow macrophages have been shown to support prostate cancer growth in bone [15] and to modulate osteoblasts [16]. Another group demonstrated that tumor-associated macrophages (TAMs) may take up bisphosphonates bound to calcified nodules in the primary site [17]. Whether this uptake alters the behavior of the TAMS or metastasis to the bone remains to be determined.

Tumor interactions with mesenchymal lineages

Tumor cells interact with many mesenchymal lineage cells in the bone marrow, including osteoblasts, osteocytes, myocytes, fibroblasts, and adipocytes. These interactions can influence both the morbidity and pathophysiology of TIBD, and a large amount of data characterizing these interactions has been produced in recent years. Of particular interest is the finding that tumor cells in the bone marrow can disrupt muscle function and promote cachexia, or muscle wasting, through oxidization of the ryanodine receptor 1 (RyR1) [18]. The finding that TGF-β is a driver of cancer-associated muscle weakness, in addition to TIBD, is particularly important because TGF-β inhibitors are currently in clinical trials for metastatic breast cancer (http://clinicaltrials.gov, accessed May 2018) and may therefore target multiple aspects of the disease.

Great progress has been made in understanding the molecular interactions between tumor cells and the mesenchymal lineage, particularly with regard to tumor–osteoblast interactions. For example, we now know that increased flow of the sympathetic nervous system stimulates RANKL expression on osteoblast lineage cells, and that the direct interactions of RANKL^{osteoblast} and RANK$^{tumor\ cell}$ promotes breast cancer bone colonization [19]. Interestingly, senescent osteoblasts also appear to play a role in disease progression and can promote bone colonization by breast cancer cells [20]. This was proposed to be mediated through secretion of the senescence-associated secretory phenotype factor IL-6, which is secreted by senescent and active osteoblasts, and can stimulate osteoclastogenesis. These IL-6-mediated effects were previously confirmed by an independent group [21], who showed a similar positive effect on prostate cancer-induced bone destruction with IL-6 neutralization. IL-6 produced by human breast cancer cells also stimulates tumor production of RANK and stimulates further IL-6 secretion and TIBD [22]. Bone marrow fibroblasts have also been shown to promote breast cancer bone colonization, which was mediated in part by Wnt3a secreted by fibroblasts [23].

In addition to osteoblasts, there is an increasing appreciation of tumor–osteocyte interactions in the field. As breast cancer cells expand within the bone marrow microenvironment, they are proposed to increase intraosseous pressure, and therefore promote osteocyte MMP and RANTES (regulated on activation, normal T cell expressed and secreted; also known as CCL5) signaling to further promote the outgrowth of prostate cancer cells [24]. Multiple myeloma cells also interact with and can induce osteocyte apoptosis and osteolysis, fueling osteolytic bone disease [25].

There has been a surge in interest in adipocyte–tumor cell interactions [26, 27], and while many groups are currently investigating the significance of the marrow adipocyte in bone colonization and disease progression, as of yet little has been published. Bone marrow adipocytes have been shown to enhance the Warburg effect in prostate cancer cells via hypoxia-inducible factor 1α (HIF-1α) activation, and tumor cells in turn promote adipocyte lipolysis [28]. This was tested extensively in vitro, but has not been functionally tested in vivo, and thus its role in promoting bone colonization remains unknown. In support of a role for adipocytes in disease progression, marrow adipose tissue (MAT) and serum adiponectin increase during cancer therapy [29], suggesting marrow adipocytes may play an endocrine role in cancer progression. There is great interest in this particular area of the field, and there is still much to be learned about these interactions.

Tumor interactions with the physical microenvironment

In addition to interactions between tumor cells and bone/bone marrow cells, several groups have demonstrated that interactions with the physical bone microenvironment also influence TIBD [30, 31]. For example, the rigidity of bone has been shown to mediate cross-talk between integrin β3 and TGF-β signaling pathways [32, 33], leading to an increase in Gli2 and PTHrP expression and subsequent bone destruction. Interestingly, the rigidity of the bone not only influences gene expression within the tumor cells, but also affects gene expression of fibroblasts [23] and mesenchymal stem cells [34]. In addition to the influence of the bone on tumor cell behavior, tumor cells and drug treatments are both known to alter bone quality and the structure of bone [35–37], suggesting that there may be a bidirectional interaction between tumor cells and the physical properties of bone. Thus several groups are focusing on how tumor cells alter the biomechanical properties of bone. The current publications are just the beginning of this area of research, since the bone is a complex 3D environment with many structural parameters beyond rigidity that likely influence tumor cell behavior and gene expression [38, 39]. More research is needed to fully understand these interactions, but many of the signaling pathways appear to be similar to those required for mechanical loading responses known to be critical for bone healing and turnover.

TUMOR DORMANCY

Tumor dormancy is the concept that tumor cells disseminate from a primary tumor and enter a quiescent state in other organs, where they may later re-emerge as a clinically detectable metastasis. While there is no standardized definition of a "dormant" tumor cell, it is generally accepted that these cells are either in a slow-growing state, a truly quiescent/nonreplicating state, or a balanced turnover state, in which tumor micrometastases replicate and die off at the same rate [40]. Tumor cells may enter a dormant state in many tissues, including the bone marrow, which provides a unique and supportive microenvironment for tumor cells. There are a number of signaling pathways and environmental cues that have been implicated in promoting tumor dormancy [41], but as reviewed in detail by Croucher and colleagues [42], there is relatively little known about the cellular and molecular mechanisms that regulate tumor dormancy in bone. These mechanisms may in fact be distinct from those mechanisms that facilitate establishment of tumor cells in bone.

Based on data from multiple groups, it is clear that the osteoblast may be a key mediator of tumor cell dormancy in prostate and breast cancer, as well as multiple myeloma, and these interactions have revealed that the Gas6/Axl signaling pathway [43, 44], CXCR4/CXCL12 axis [45], and leukemia inhibitory factor (LIF) produced by osteoblasts in the bone marrow may provide paracrine and autocrine signals to maintain tumor cells in a dormant state in the bone marrow [46]. The endosteal niche in particular, where osteoblasts reside, has also been shown to harbor myeloma cells that are low cycling or quiescent, and modulation of the endosteal niche pushes these dormant tumor cells to a proliferative state [47]. Similarly, the perivascular niche has been implicated in maintaining breast cancer cells in a dormant state, where thrombospondin 1 produced by the endothelial cells is responsible for maintenance of the dormant state, and TGF-β signaling and periostin promote the outgrowth of these tumor cells [48]. In contrast to myeloma–osteoblast interactions, which maintain cells in a dormant state, osteoblasts have been shown to promote breast cancer cell proliferation via adheren junctions in the context of breast cancer dormancy [49]. Thus, there may be competing mechanisms at play in dormancy. There does, however, appear to be some overlap in the pathways and gene expression signatures across different tissue types, as evidenced by an overlapping dormancy signature between prostate cancer, head and neck squamous cell carcinoma, and breast cancer [50].

One of the most interesting findings to date is that the therapies that target the primary tumor may act differently on disseminated tumor cells (DTCs) in the context of the tumor–bone microenvironment. For example, in prostate cancer we now know that, paradoxically, androgen ablation therapy promotes the outgrowth of bone marrow DTCs, and that this may be prevented by adjuvant bisphosphonate treatment [51], and mitotic quiescence is associated with a more tumorigenic phenotype in prostate cancer cells disseminated to the bone marrow [49]. These findings highlight the complicated nature of the dormancy problem and indicate that the therapies used to target the primary tumor may impact DTCs in alarmingly different ways. These data are further supported by the finding that LIF receptor signaling via STAT3 maintains a dormancy phenotype in breast cancer cells, and that downregulation of STAT3 signaling, while beneficial in primary tumors, promotes the outgrowth of breast cancer cells disseminated to the bone marrow [46]. Interestingly, there appears to be some overlap between dormant prostate DTCs in the bone marrow and cancer stem cells (CSCs) [49, 52]. Greater clarification of the link between CSCs ad DTCs in the bone marrow across different tumor types is needed. The issue of targeting dormant cancer cells remains complicated, and it is still widely debatable whether it is of greater clinical benefit to treat dormant tumor cells as a chronic disease by maintaining their dormant state, or to drive tumor cells out of dormancy with the hopes of eradicating them [48]. This will likely remain a debate for some time to come.

NEW MODELS OF TUMOR-INDUCED BONE DISEASE

Studies of TIBD have and continue to rely heavily on both animal and in vitro models. Probably the most widely accepted in vivo model is the intracardiac injection model where tumor cells are injected into the left cardiac ventricle and circulate throughout the body [53, 54]. Using this model, tumor cells frequently home to, but do not truly metastasize to the hindlimbs, where they can easily be quantified using faxitron and fluorescent/luminescent imaging. The most common cell lines used in this model are the bone metastatic MDA-MB-231 human breast cancer cells, which have been developed by several groups [54–59]. A bone metastatic variant of MDA-MB-231 cells has also been developed in which cells are injected into the tail vein and induce bone metastases more rapidly than by intracardiac injection [60], although the intracardiac model is more common in the field. Other cells that can be used in this model include bone metastatic variants of the prostate (PC-3 [61, 62]), lung (RWGT2 [59]), and mouse mammary fat pad (4T1 [53, 63]). Most models do not easily home to the bone, especially osteoblastic prostate cancer, and therefore intratibial injections are also a commonly used model (including the LN series, ACE-1, and others) [62]. This model is a good surrogate for replicating tumor establishment in bone, but like the intracardiac model is not a true metastasis model. A handful of cell lines do metastasize to bone from their primary site, including the 4T1-5T2 (and other similar 4T1 models) [12, 19, 63, 64]. These orthotopic metastasis models are ideal both because they replicate all the steps of metastasis and because they can be used in immune competent mice, but there are no established human models that will metastasize to bone from the primary site. Another issue with the current breast cancer models is that those cell lines that aggressively colonize the bone have been exclusively estrogen receptor negative (ER−) models, while patients with ER+ and luminal breast cancer subtypes develop bone metastases at a higher rate than those with any other breast cancer subtype [65, 66]. Therefore, several groups are working to develop ER+ models that will metastasize to bone. For example, Capietto and colleagues developed an ER+ variant of the SSM2 cells that can metastasize to bone [14], and several groups are now using the MCF7 human breast cancer cell line, which is ER+ and serves as a good model of tumor dormancy in bone following intracardiac [46, 67] and intrailiac [68] inoculation.

While patient-derived xenograft (PDX) models have been extremely successful in other fields for studying patient tumor cells, bone metastatic tumors do not typically grow well as a PDX. One group has successfully propagated bone metastatic prostate cancer cells in PDX models by growing cells in a 3D hydrogel with patient tumor cells and osteoblasts [69]. Other groups have demonstrated that the physical rigidity and compression forces of the bone are an important contributor to TIBD. One group showed that hindlimb loading (compression) could stimulate the expression of PTHrP and induce more bone destruction and tumor growth [70, 71]. Another group has shown that the rigidity of the bone microenvironment can stimulate PTHrP expression leading to more bone destruction [32, 72]. This suggests that the structural microenvironment is important for allowing tumor cells to grow. Thus, several other groups have been developing more bone-like models that replicate the structure, rigidity, and fluid flow parameters of tumors in the bone microenvironment. For example, one group is using silk scaffolds to grow myeloma cells in vitro [73], because these scaffolds resemble the trabecular structure and rigidity of bone. Another group is using polyurethane scaffolds of varying pore sizes and rigidities to replicate bone [74]. These studies [32, 69–73] and those from Lynch and colleagues have shown that the rigidity, pore sizes, and fluid sheer rates all contribute to bone formation and TIBD [75].

An interesting conceptual model for tumor metastasis and dormancy in the metastatic niche is the Pienta laboratory's proposal of the body as an ecosystem and of the tumor cell as creating its own niche within this ecosystem, ultimately leading to collapse of the ecosystem, or death [76–78]. This concept, although not a new physical model, is an interesting take on tumor–microenvironment interactions because it takes into account the impact of the metastatic niche and the tumor–bone microenvironment on the rest of the systems in the body, and vice versa, and thus promotes thinking about tumor cells in the context of their larger ecosystem.

PROMISING CLINICAL TARGETS

Osteoclast inhibitors such as bisphosphonates (eg, zoledronic acid) or RANKL inhibitors (denosumab) remain the standard of care for treating patients with bone metastases [79, 80]. While these treatments have been very successful for the reduction of SREs and development of bone metastases, they do not consistently extend disease-free survival [80]. Therefore, many groups are investigating other potential therapeutics that may be more effective than osteoclast inhibitors alone. Many of these are still in the preclinical phases and will likely come to the forefront over the next several years. Therapies that are being tested clinically in multiple tumor types, but not yet in bone metastatic breast cancer patients, include the TGF-β inhibitors, which have been shown by multiple groups to block TIBD and improve bone quality in tumor-bearing animals [35, 37, 81–84]. Integrin inhibition also looks promising and seems to both reduce metastasis to bone and osteoclast-mediated bone destruction [32, 57]. Additionally, integrin inhibition appears to increase immune suppression in tumor-bearing mice [85], suggesting that it may be a strong multipotential therapeutic approach to reduce tumor and bone destruction. Some of these inhibitors have reached clinical trials for cancer patients, but have not yet been tested in patients with bone metastases [86]. Also at the forefront are inhibitors targeting sclerostin, src kinase signaling, activin A, and cathepsin K, which have been tested extensively in preclinical models of low bone mass [87]. The investigation of cathepsin K inhibitors in blocking bone metastases may be stopped because the cathepsin K inhibitor odanacatib was recently evaluated in clinical trials for osteoporosis and was found to carry an increased risk of stroke, as reported at the ASBMR 2016 Annual Meeting. The makers of this drug have since decided not to seek US Food and Drug Administration (FDA) approval.

Other groups are focusing on improving drug delivery using nanoparticle or other targeting approaches in order to improve the efficacy of current or novel therapeutic options [88–90]. While these delivery approaches have made rapid progress, it is still an emerging area with many new developments likely to be made over the next decade.

ACKNOWLEDGMENTS

The authors acknowledge the following funding: W81XWH-15-1-0622 (JAS), 1I01BX001957 (JAS), 1R01CA163499 (JAS), and 4R00CA194198 (RWJ).

REFERENCES

1. Johnson RW, Schipani E, Giaccia AJ. HIF targets in bone remodeling and metastatic disease. Pharmacol Ther. 2015;150:169–77.
2. Suva LJ, Washam C, Nicholas RW, et al. Bone metastasis: mechanisms and therapeutic opportunities. Nat Rev Endocrinol. 2011;7(4):208–18.
3. Buenrostro D, Mulcrone PL, Owens P, et al. The bone microenvironment: a fertile soil for tumor growth. Curr Osteoporos Rep. 2016;14(4):151–8.
4. Weilbaecher KN, Guise TA, McCauley LK. Cancer to bone: a fatal attraction. Nat Rev Cancer. 2011;11(6):411–25.

5. Mundy GR. Mechanisms of bone metastasis. Cancer. 1997;80(suppl 8):1546–56.
6. Sterling JA, Edwards JR, Martin TJ, et al. Advances in the biology of bone metastasis: how the skeleton affects tumor behavior. Bone. 2011;48(1):6–15.
7. Guise TA, Mundy GR. Physiological and pathological roles of parathyroid hormone-related peptide. Curr Opin Nephrol Hypertens. 1996;5(4):307–15.
8. Ell B, Mercatali L, Ibrahim T, et al. Tumor-induced osteoclast miRNA changes as regulators and biomarkers of osteolytic bone metastasis. Cancer Cell. 2013;24(4):542–56.
9. Yang L, DeBusk LM, Fukuda K, et al. Expansion of myeloid immune suppressor Gr+CD11b+ cells in tumor-bearing host directly promotes tumor angiogenesis. Cancer Cell. 2004;6(4):409–21.
10. Yang L, Huang J, Ren X, et al. Abrogation of TGF beta signaling in mammary carcinomas recruits Gr-1+CD11b+ myeloid cells that promote metastasis. Cancer Cell. 2008;13(1):23–35.
11. Park SI, Lee C, Sadler WD, et al. Parathyroid hormone-related protein drives a CD11b+Gr1+ cell-mediated positive feedback loop to support prostate cancer growth. Cancer Res. 2013;73(22):6574–83.
12. Danilin S, Merkel AR, Johnson JR, et al. Myeloid-derived suppressor cells expand during breast cancer progression and promote tumor-induced bone destruction. Oncoimmunology. 2012;1(9):1484–94.
13. Sawant A, Deshane J, Jules J, et al. Myeloid-derived suppressor cells function as novel osteoclast progenitors enhancing bone loss in breast cancer. Cancer Res. 2013;73(2):672–82.
14. Capietto AH, Kim S, Sanford DE, et al. Down-regulation of PLCgamma2-beta-catenin pathway promotes activation and expansion of myeloid-derived suppressor cells in cancer. J Exp Med. 2013;210(11):2257–71.
15. Soki FN, Cho SW, Kim YW, et al. Bone marrow macrophages support prostate cancer growth in bone. Oncotarget. 2015;6(34):35782–96.
16. Michalski MN, Koh AJ, Weidner S, et al. Modulation of osteoblastic cell efferocytosis by bone marrow macrophages. J Cell Biochem. 2016;117(12):2697–706.
17. Junankar S, Shay G, Jurczyluk J, et al. Real-time intravital imaging establishes tumour-associated macrophages as the extraskeletal target of bisphosphonate action in cancer. Cancer Discov. 2015;5(1):35–42.
18. Waning DL, Mohammad KS, Reiken S, et al. Excess TGF-beta mediates muscle weakness associated with bone metastases in mice. Nat Med. 2015;21(11):1262–71.
19. Campbell JP, Karolak MR, Ma Y, et al. Stimulation of host bone marrow stromal cells by sympathetic nerves promotes breast cancer bone metastasis in mice. PLoS Biol. 2012;10(7):e1001363.
20. Luo X, Fu Y, Loza AJ, et al. Stromal-initiated changes in the bone promote metastatic niche development. Cell Rep. 2016;14(1):82–92.
21. Zheng Y, Basel D, Chow SO, et al. Targeting IL-6 and RANKL signaling inhibits prostate cancer growth in bone. Clin Exp Metastasis. 2014;31(8):921–33.
22. Zheng Y, Chow SO, Boernert K, et al. Direct crosstalk between cancer and osteoblast lineage cells fuels metastatic growth in bone via auto-amplification of IL-6 and RANKL signaling pathways. J Bone Miner Res. 2014;29(9):1938–49.
23. Johnson RW, Merkel A, Page JM, et al. Wnt signaling induces gene expression of factors associated with bone destruction in lung and breast cancer. Clin Exp Metastasis. 2014;31(8):945–59.
24. Sottnik JL, Dai J, Zhang H, et al. Tumor-induced pressure in the bone microenvironment causes osteocytes to promote the growth of prostate cancer bone metastases. Cancer Res. 2015;75(11):2151–8.
25. Delgado-Calle J, Anderson J, Cregor MD, et al. Bidirectional Notch signaling and osteocyte-derived factors in the bone marrow microenvironment promote tumor cell proliferation and bone destruction in multiple myeloma. Cancer Res. 2016;76(5):1089–100.
26. McDonald MM, Fairfield H, Falank C, et al. Adipose, bone, and myeloma: contributions from the microenvironment. Calcif Tissue Int. 2016;100(5):433–48.
27. Morris EV, Edwards CM. Bone marrow adipose tissue: a new player in cancer metastasis to bone. Front Endocrinol. 2016;7:90.
28. Diedrich JD, Rajagurubandara E, Herroon MK, et al. Bone marrow adipocytes promote the Warburg phenotype in metastatic prostate tumors via HIF-1alpha activation. Oncotarget. 2016;7(40):64854–77.
29. Cawthorn WP, Scheller EL, Learman BS, et al. Bone marrow adipose tissue is an endocrine organ that contributes to increased circulating adiponectin during caloric restriction. Cell Metab. 2014;20(2):368–75.
30. Kostic A, Lynch CD, Sheetz MP. Differential matrix rigidity response in breast cancer cell lines correlates with the tissue tropism. PLoS One. 2009;4(7):e6361.
31. Provenzano PP, Inman DR, Eliceiri KW, et al. Matrix density-induced mechanoregulation of breast cell phenotype, signaling, and gene expression through a FAK-ERK linkage. Oncogene. 2009;28(49):4326–43.
32. Page JM, Merkel AR, Ruppender NS, et al. Matrix rigidity regulates the transition of tumor cells to a bone-destructive phenotype through integrin beta3 and TGF-beta receptor type II. Biomaterials. 2015;64:33–44.
33. Ruppender NS, Merkel AR, Martin TJ, et al. Matrix rigidity induces osteolytic gene expression of metastatic breast cancer cells. PLoS One. 2010;5(11):e15451.
34. Guo R, Ward CL, Davidson JM, et al. A transient cell-shielding method for viable MSC delivery within hydrophobic scaffolds polymerized in situ. Biomaterials. 2015;54:21–33.
35. Biswas S, Nyman JS, Alvarez J, et al. Anti-transforming growth factor ss antibody treatment rescues bone loss and prevents breast cancer metastasis to bone. PLoS One. 2011;6(11):e27090.
36. Edwards JR, Nyman JS, Lwin ST, et al. Inhibition of TGF-beta signaling by 1D11 antibody treatment increases bone mass and quality in vivo. J Bone Miner Res. 2010;25(11):2419–26.

37. Nyman JS, Merkel AR, Uppuganti S, et al. Combined treatment with a transforming growth factor beta inhibitor (1D11) and bortezomib improves bone architecture in a mouse model of myeloma-induced bone disease. Bone. 2016;91:81–91.
38. Guelcher SA, Sterling JA. Contribution of bone tissue modulus to breast cancer metastasis to bone. Cancer Microenviron. 2011;4(3):247–59.
39. Sterling JA, Guelcher SA. Bone structural components regulating sites of tumor metastasis. Curr Osteoporos Rep. 2011;9(2):89–95.
40. Zhang XH, Giuliano M, Trivedi MV, et al. Metastasis dormancy in estrogen receptor-positive breast cancer. Clin Cancer Res. 2013;19(23):6389–97.
41. Quayle L, Ottewell PD, Holen I. Bone metastasis: molecular mechanisms implicated in tumour cell dormancy in breast and prostate cancer. Curr Cancer Drug Targets. 2015;15(6):469–80.
42. Croucher PI, McDonald MM, Martin TJ. Bone metastasis: the importance of the neighbourhood. Nat Rev Cancer. 2016;16(6):373–86.
43. Shiozawa Y, Pedersen EA, Patel LR, et al. GAS6/AXL axis regulates prostate cancer invasion, proliferation, and survival in the bone marrow niche. Neoplasia. 2010;12(2):116–27.
44. Taichman RS, Patel LR, Bedenis R, et al. GAS6 receptor status is associated with dormancy and bone metastatic tumor formation. PLoS One. 2013;8(4):e61873.
45. Wang N, Docherty FE, Brown HK, et al. Prostate cancer cells preferentially home to osteoblast-rich areas in the early stages of bone metastasis: evidence from in vivo models. J Bone Miner Res. 2014;29(12):2688–96.
46. Johnson RW, Finger EC, Olcina MM, et al. Induction of LIFR confers a dormancy phenotype in breast cancer cells disseminated to the bone marrow. Nat Cell Biol. 2016;18(10):1078–89.
47. Lawson MA, McDonald MM, Kovacic N, et al. Osteoclasts control reactivation of dormant myeloma cells by remodelling the endosteal niche. Nat Commun. 2015;6:8983.
48. Ghajar CM, Peinado H, Mori H, et al. The perivascular niche regulates breast tumour dormancy. Nat Cell Biol. 2013;15(7):807–17.
49. Wang N, Docherty F, Brown HK, et al. Mitotic quiescence, but not unique "stemness," marks the phenotype of bone metastasis-initiating cells in prostate cancer. FASEB J. 2015;29(8):3141–50.
50. Chery L, Lam HM, Coleman I, et al. Characterization of single disseminated prostate cancer cells reveals tumor cell heterogeneity and identifies dormancy associated pathways. Oncotarget. 2014;5(20):9939–51.
51. Ottewell PD, Wang N, Meek J, et al. Castration-induced bone loss triggers growth of disseminated prostate cancer cells in bone. Endocr Relat Cancer. 2014;21(5):769–81.
52. Shiozawa Y, Berry JE, Eber MR, et al. The marrow niche controls the cancer stem cell phenotype of disseminated prostate cancer. Oncotarget. 2016;5(27):41217–32.
53. Yoneda T, Sasaki A, Mundy GR. Osteolytic bone metastasis in breast cancer. Breast Cancer Res Treat. 1994;32(1):73–84.
54. Campbell JP, Merkel AR, Masood-Campbell SK, et al. Models of bone metastasis. J Visual Exp. 2012;67:e4260.
55. Johnson RW, Nguyen MP, Padalecki SS, et al. TGF-beta promotion of Gli2-induced expression of parathyroid hormone-related protein, an important osteolytic factor in bone metastasis, is independent of canonical Hedgehog signaling. Cancer Res. 2011;71(3):822–31.
56. Wright LE, Ottewell PD, Rucci N, et al. Murine models of breast cancer bone metastasis. BoneKEy Rep. 2016;5:804.
57. Zhao Y, Bachelier R, Treilleux I, et al. Tumor alphav-beta3 integrin is a therapeutic target for breast cancer bone metastases. Cancer Res. 2007;67(12):5821–30.
58. Guise TA, Yin JJ, Taylor SD, et al. Evidence for a causal role of parathyroid hormone-related protein in the pathogenesis of human breast cancer-mediated osteolysis. J Clin Invest. 1996;98(7):1544–9.
59. Guise TA, Yoneda T, Yates AJ, et al. The combined effect of tumor-produced parathyroid hormone-related protein and transforming growth factor-alpha enhance hypercalcemia in vivo and bone resorption in vitro. J Clin Endocrinol Metab. 1993;77(1):40–5.
60. Garcia T, Jackson A, Bachelier R, et al. A convenient clinically relevant model of human breast cancer bone metastasis. Clin Exp Metastasis. 2008;25(1):33–42.
61. Hansen AG, Arnold SA, Jiang M, et al. ALCAM/CD166 is a TGF-beta-responsive marker and functional regulator of prostate cancer metastasis to bone. Cancer Res. 2014;74(5):1404–15.
62. Simmons JK, Elshafae SM, Kellar ET, et al. Review of animal models of prostate cancer bone metastasis. Vet Sci Rev. 2014;1:16–39.
63. Hiraga T, Williams PJ, Ueda A, et al. Zoledronic acid inhibits visceral metastases in the 4T1/luc mouse breast cancer model. Clin Cancer Res. 2004;10(13):4559–67.
64. Rose AA, Pepin F, Russo C, et al. Osteoactivin promotes breast cancer metastasis to bone. Mol Cancer Res. 2007;5(10):1001–14.
65. Kennecke H, Yerushalmi R, Woods R, et al. Metastatic behavior of breast cancer subtypes. J Clin Oncol. 2010;28(20):3271–7.
66. Smid M, Wang Y, Zhang Y, et al. Subtypes of breast cancer show preferential site of relapse. Cancer Res. 2008;68(9):3108–14.
67. Thomas RJ, Guise TA, Yin JJ, et al. Breast cancer cells interact with osteoblasts to support osteoclast formation. Endocrinology. 1999;140(10):4451–8.
68. Wang H, Yu C, Gao X, et al. The osteogenic niche promotes early-stage bone colonization of disseminated breast cancer cells. Cancer Cell. 2015;27(2):193–210.
69. Fong EL, Wan X, Yang J, et al. A 3D in vitro model of patient-derived prostate cancer xenograft for controlled interrogation of in vivo tumor-stromal interactions. Biomaterials. 2016;77:164–72.
70. Lynch ME, Brooks D, Mohanan S, et al. In vivo tibial compression decreases osteolysis and tumor formation

in a human metastatic breast cancer model. J Bone Miner Res. 2013;28(11):2357–67.
71. Lynch ME, Fischbach C. Biomechanical forces in the skeleton and their relevance to bone metastasis: biology and engineering considerations. Adv Drug Deliv Rev. 2014;79–80:119–34.
72. Page JM, Merkel AR, Ruppender NS, et al. Altering adsorbed proteins or cellular gene expression in bone-metastatic cancer cells affects PTHrP and Gli2 without altering cell growth. Data Brief. 2015;4:440–6.
73. Abbott RD, Wang RY, Reagan MR, et al. The use of silk as a scaffold for mature, sustainable unilocular adipose 3D tissue engineered systems. Adv Healthcare Mat. 2016;5(13):1667–77.
74. Guo R, Lu S, Page JM, et al. Fabrication of 3D scaffolds with precisely controlled substrate modulus and pore size by templated-fused deposition modeling to direct osteogenic differentiation. Adv Healthcare Mat. 2015;4(12):1826–32.
75. Lynch ME, Chiou AE, Lee MJ, et al. Three-dimensional mechanical loading modulates the osteogenic response of mesenchymal stem cells to tumor-derived soluble signals. Tissue Eng A. 2016;22(15–16):1006–15.
76. Amend SR, Pienta KJ. Ecology meets cancer biology: the cancer swamp promotes the lethal cancer phenotype. Oncotarget. 2015;6(12):9669–78.
77. Yang KR, Mooney SM, Zarif JC, et al. Niche inheritance: a cooperative pathway to enhance cancer cell fitness through ecosystem engineering. J Cell Biochem. 2014;115(9):1478–85.
78. Chen KW, Pienta KJ. Modeling invasion of metastasizing cancer cells to bone marrow utilizing ecological principles. Theor Biol Med Model. 2011;8:36.
79. Coleman R, Body JJ, Aapro M, et al. Bone health in cancer patients: ESMO Clinical Practice Guidelines. Ann Oncol. 2014;25(suppl 3):iii124–37.
80. Coleman R, Cameron D, Dodwell D, et al. Adjuvant zoledronic acid in patients with early breast cancer: final efficacy analysis of the AZURE (BIG 01/04) randomised open-label phase 3 trial. Lancet Oncol. 2014;15(9):997–1006.
81. Balooch G, Balooch M, Nalla RK, et al. TGF-beta regulates the mechanical properties and composition of bone matrix. Proc Natl Acad U S A. 2005;102(52):18813–8.
82. Nyman JS, Uppuganti S, Makowski AJ, et al. Predicting mouse vertebra strength with micro-computed tomography-derived finite element analysis. BoneKEy Rep. 2015;4:664.
83. Mohammad KS, Javelaud D, Fournier PG, et al. TGF-beta-RI kinase inhibitor SD-208 reduces the development and progression of melanoma bone metastases. Cancer Res. 2011;71(1):175–84.
84. Yin JJ, Selander K, Chirgwin JM, et al. TGF-beta signaling blockade inhibits PTHrP secretion by breast cancer cells and bone metastases development. J Clin Invest. 1999;103(2):197–206.
85. Su X, Esser AK, Amend SR, et al. Antagonizing Integrin beta3 Increases Immunosuppression in Cancer. Cancer Res. 2016;76(12):3484–95.
86. Goodman SL, Picard M. Integrins as therapeutic targets. Trends Pharmacol Sci. 2012;33(7):405–12.
87. Terpos E, Confavreux CB, Clezardin P. Bone antiresorptive agents in the treatment of bone metastases associated with solid tumours or multiple myeloma. BoneKEy Rep. 2015;4:744.
88. Adjei IM, Sharma B, Peetla C, et al. Inhibition of bone loss with surface-modulated, drug-loaded nanoparticles in an intraosseous model of prostate cancer. J Control Release. 2016;232:83–92.
89. Sun W, Han Y, Li Z, et al. Bone-targeted mesoporous silica nanocarrier anchored by zoledronate for cancer bone metastasis. Langmuir. 2016;32(36):9237–44.
90. Yin Q, Tang L, Cai K, et al. Pamidronate functionalized nanoconjugates for targeted therapy of focal skeletal malignant osteolysis. Proc Natl Acad Sci U S A. 2016;113(32):E4601–9.

98

Myeloma Bone Disease and Other Hematological Malignancies

Claire M. Edwards[1] and Rebecca Silbermann[2]

[1]*Nuffield Department of Surgical Sciences; and Nuffield Department of Orthopaedics, Rheumatology and Musculoskeletal Sciences, Botnar Research Centre, University of Oxford, Oxford, UK*
[2]*Department of Medicine, Division of Hematology-Oncology, Indiana University School of Medicine, Indianapolis, IN, USA*

INTRODUCTION

Hematological malignancies can have multiple direct and indirect effects on bone including pathological fractures, bone pain, and hypercalcemia. The frequency of bone involvement in association with these malignancies varies widely depending on the underlying diagnosis. Bone involvement in multiple myeloma occurs in nearly 80% of patients over their disease course, and significantly impacts quality of life, morbidity, performance status, and survival. Skeletal complications in the much rarer human T-cell leukemia virus type 1 (HTLV-1) associated adult T-cell leukemia and lymphoma—the second most common hematological malignancy to affect bone—most frequently manifest as hypercalcemia, which affects approximately 70% of patients over the course of their disease. In contrast, bone involvement from Hodgkin disease and non-Hodgkin lymphoma is rare.

Under normal physiological conditions, interactions between bone-resorbing cells, osteoclasts (OCs), and bone-forming cells, osteoblasts (OBs), are balanced, allowing for coupled bone remodeling and normal hematopoiesis. Dysregulation of the normal bone remodeling process can occur in the setting of malignancy, resulting in osteolytic, osteoblastic, or mixed osteolytic/osteoblastic lesions. In addition, increased osteoclastic bone resorption results in increased renal tubular calcium resorption. This impairs glomerular filtration and can result in hypercalcemia.

This chapter discusses the pathophysiology of skeletal lesions in hematological malignancies, with a focus on the current understanding of their associated bone disease, and includes brief discussions of radiological imaging of skeletal lesions, pharmacological treatments, and other hematological malignancies that affect bone.

MULTIPLE MYELOMA

Multiple myeloma (MM) is a plasma cell malignancy characterized by monoclonal paraprotein production from terminally differentiated plasma cells and lytic bone disease. Laboratory findings include an elevated monoclonal paraprotein in the serum and/or urine and decreased normal immunoglobulin levels. Expansion of the plasma cell population in the bone marrow leads to leukopenia, anemia, and thrombocytopenia.

MM has the highest incidence of bone involvement amongst malignant diseases, and is the second most common hematological malignancy, accounting for approximately 15% of all hematological malignancies. The American Cancer Society estimated that approximately 30,000 new cases of MM would be diagnosed in the United States in 2016, with an estimated 12,650 deaths [1]. While recent advances in the treatment of MM have resulted in an improvement in median overall survival from 4.6 years for patients diagnosed between 2001 and 2005 to 6.1 years for patients diagnosed between 2006 and 2010, the disease remains incurable [2]. Approximately 70% of patients present with bone pain at diagnosis, while 85% develop bone lesions during their disease course, and up to 60% develop pathological

---

Primer on the Metabolic Bone Diseases and Disorders of Mineral Metabolism, Ninth Edition. Edited by John P. Bilezikian.
© 2019 American Society for Bone and Mineral Research. Published 2019 by John Wiley & Sons, Inc.
Companion website: www.wiley.com/go/asbmrprimer

fractures [3]. While the clinical presentation of myeloma is quite variable, with approximately 20% of patients asymptomatic at presentation (disease in these patients is generally identified through routine lab studies), bone pain is the most common symptom at presentation and is frequently centered on the chest or back and exacerbated by movement. Skeletal manifestations of MM, particularly osteolytic bone lesions, represent the most prominent source of pain and disability, however MM bone disease can also lead to diffuse osteopenia. Lytic lesions most often involve the axial skeleton, skull, and femur. Importantly, these lesions rarely heal, even when patients achieve a complete remission. Therefore, management of myeloma bone disease remains a crucial component of the long-term care of MM patients. In contrast to other tumors that involve bone, myeloma is rarely associated with osteosclerotic lesions except in the POEMS syndrome (polyneuropathy, organomegaly, endocrinopathy, monoclonal gammopathy and skin changes), a multisystem disease occurring rarely in the setting of plasma cell dyscrasias [4].

Fifteen percent to 20% of newly diagnosed myeloma patients have hypercalcemia, (defined as a corrected serum calcium level greater than 11.5 mg/dL), due to increased bone resorption, decreased bone formation, and impaired renal function, all of which are often exacerbated by immobility. Unlike other malignancies resulting in metastatic disease to the bone, PTHrP is rarely overproduced by myeloma cells. The severity of hypercalcemia in patients with myeloma is not correlated with serum PTHrP levels and instead reflects tumor burden [3]. Symptomatic hypercalcemia can result in anorexia, nausea, vomiting, confusion, fatigue, constipation, renal stones, depression, and polyuria and is suggestive of a high tumor burden.

The bone marrow microenvironment in MM also contributes to tumor growth and the destructive process. It is comprised of cellular and extracellular elements including OBs, OCs, osteocytes, endothelial cells, immune cells, adipocytes, and MM cells. Interactions between MM cells and their bone marrow microenvironment are tightly regulated. Under normal physiological conditions, balanced interactions within the marrow microenvironment result in coupled bone remodeling (Fig. 98.1). In MM, bone remodeling is uncoupled and is characterized by generalized OC activation and suppressed OB function with decreased bone formation. In addition, the bone destructive process releases growth factors from the bone matrix that increase the proliferation of MM cells. This results in a "vicious cycle" of bone destruction, leading to increased tumor mass and further bone destruction (Fig. 98.2).

Bone marrow biopsies from MM patients demonstrate a correlation between tumor burden, OC number, and resorptive surface [5]. MM cells do not directly resorb bone, but instead produce or induce multiple osteoclastogenic factors in the bone marrow microenvironment that directly increase OC formation and activity. This is in part achieved by increasing expression of receptor activator of NF-κB ligand (RANKL), a critical OC differentiation factor produced by marrow stromal cells and OB, and decreasing stromal production of osteoprotegerin (OPG), a soluble decoy receptor for RANKL [6]. Myeloma cells adhere to bone marrow stromal cells (which are pre-osteoblasts [preOBs]) via binding of surface very late antigen 4 (VLA-4; α4β1 integrin) to vascular cell adhesion molecule 1 (VCAM-1) expressed on stromal cells. This results in production of marrow stromal cell-derived osteoclastogenic cytokines such as RANKL, M-CSF, IL-11, and IL-6, and MM cell-derived osteoclastogenic cytokines such as macrophage inflammatory protein 1α (MIP-1α) and IL-3 [7–10]. In addition, apoptosis of osteocytes, the most abundant bone cells, is linked to both normal physiological and pathological bone remodeling.

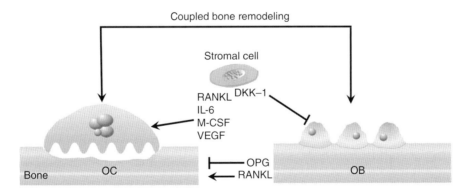

Fig. 98.1. Physiological bone remodeling is marked by balanced interactions between osteoclasts (OC) and osteoblasts (OB) within the bone marrow microenvironment. Locally produced cytokines and systemic hormones regulate the formation and activation of OCs. Systemic hormones (not pictured) stimulate OC formation by inducing the expression of RANKL on marrow stromal cells and OBs. Stromal cells also produce OC-stimulating factors including IL-6, M-CSF, and VEGF that induce OC formation. In addition, stromal cells produce Dickkopf-1 (DKK1), an OB inhibitory factor. Coupling factors produced by OCs such as ephrins (not shown), also drive OB differentiation while suppressing further OC formation and activity. OBs produce OPG, a soluble RANKL inhibitor. Under physiological conditions, OB and OC activity is balanced, in part due to the OPG/RANKL ratio. Source: [56]. Reproduced with permission of American Association for Cancer Research.

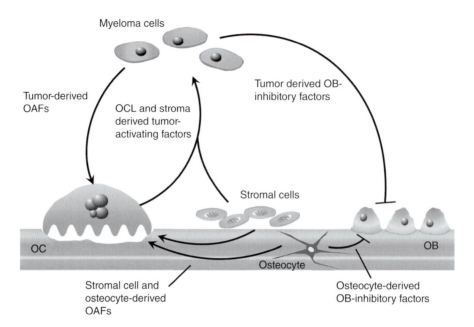

Fig. 98.2. The "vicious cycle" of myeloma bone disease. In myeloma bone disease osteoclastogenesis is favored and osteoblastogenesis is inhibited. Myeloma cells produce osteoclast (OC) activating factors (OAFs) that directly or indirectly activate OCs and stimulate marrow stromal cells and T cells to further increase production of OAFs and decrease production of OC inhibitory factors. Myeloma cell-derived OAFs include MIP-1α, IL-3, RANKL, and TNF-α. Myeloma cells also induce marrow stromal cell production of OAFs such as RANKL, M-CSF, IL-6, and TNF-α, and decrease expression of OPG. This results in an increased RANKL/OPG ratio, which promotes OC development. Tumor growth in the myeloma bone marrow microenvironment is stimulated by OC and stromal cell-derived soluble factors such as IL-6, osteopontin, BAFF, and APRIL. In addition, myeloma cells induce changes in marrow stromal cells that increase production of factors that support myeloma cells such as IL-6, VEGF, and IGF-1, in part via adhesive interactions through VCAM-1 on marrow stromal cells and $\alpha_4\beta_1$ on myeloma cells. The bone destructive process itself releases growth factors from the bone matrix that further increase the growth of myeloma cells. These include IGFs, FGFs, PDGFs, BMPs, and TGF-β (not shown on diagram). Suppression of osteoblast (OB) differentiation is mediated by soluble tumor-derived factors such as sclerostin, DKK1, IL-7, hepatocyte growth factor (HGF), TNF-α, the secreted frizzled-related proteins (sFRPs), and IL-3. The bone matrix-derived OC stimulatory factor TGF-β also inhibits OB differentiation. Myeloma cells induce other cells in the bone marrow microenvironment to increase production of OB suppressors. Examples include osteocyte production of sclerostin and DKK1, and stromal cell production of TNF-α. Finally, bidirectional signaling between ephrin B2 in OCs, and its receptor, EphB4, in bone marrow stromal cells and OBs (not illustrated) negatively regulates OC formation and promotes OB differentiation. Both ephrin B2 and EphB4 are decreased in myeloma. Source: [57]. Reproduced with permission of Massachusetts Medical Society.

Osteocyte apoptosis can induce OC precursor recruitment and differentiation, and bone biopsies from MM patients with skeletal disease have been reported to have fewer viable osteocytes and more osteocyte apoptosis than healthy controls or MM patients without bone lesions. This suggests that there is a correlation between the number of apoptotic osteocytes and OC number [11].

The ratio of RANKL to OPG is critical for the regulation of lytic activity under both normal physiological conditions and in MM [12]. RANKL increases OC formation and survival by binding to its receptor RANK on OC precursor cells and OCs. Soluble RANKL levels are increased in MM patients and correlate with disease activity, however the primary source of RANKL in the MM bone marrow microenvironment is not clear. While MM cells themselves express RANKL and induce RANKL expression in stromal cells (through adhesive interactions, soluble factors produced by MM cells, including TNF-α and Dickkopf1 [DKK1]), emerging evidence also suggests that MM cells upregulate RANKL expression in osteocytes [13]. The effects of increased RANKL in MM are further enhanced by the decreased production of OPG by marrow stromal cells which occurs as a result of OB suppression [14, 15]. MIP-1α acts as a chemotactic factor for OC precursors and can induce differentiation of OC progenitors contributing to OC formation, while potentiating the effects of RANKL [16]. MIP-1α also directly promotes the growth, survival, and migration of MM cells by inducing the activation of multiple signaling pathways crucial for MM cell growth and survival [17].

The increased OC activity in MM bone disease is accompanied by highly suppressed or absent OB activity, with decreased bone formation and calcification despite increased bone resorption [5, 18]. As a result, serum alkaline phosphatase and osteocalcin are normal or decreased in patients with myeloma bone involvement. Osteocyte-derived molecules, such as sclerostin, a canonical Wnt signaling antagonist, and DKK1, also regulate OB activity, possibly contributing to OB inhibition in MM bone disease [19]. OBs also affect myeloma cell growth indirectly via their regulation of OCs. Co-culture experiments have demonstrated reduced myeloma cell proliferation in the presence of OBs as compared with OC or marrow stromal cells [20], a finding that has been confirmed in murine models of myeloma bone disease [21].

Mechanisms of OB suppression in MM are of significant clinical interest as there are currently no generally effective anabolic bone agents available for MM patients. Mechanisms currently under investigation include: (i) downregulation of the osteogenic transcription factor runt-related transcription factor 2 (Runx2) via direct cell to cell contact between MM and OB progenitor cells; (ii) Wnt signaling inhibitors such as DKK1, soluble frizzled receptor-like protein 3 (sFRP3) and sclerostin, that inhibit osteogenesis [22]; and (iii) modulation of OB differentiation by TGF-β superfamily members including BMP-2 [23], activin A [24], and TGF-β itself [25]. DKK1 is of particular interest as it is highly expressed in the bone marrow of MM patients with bone lesions, appears to be involved in early bone disease, and has roles in the regulation of both OC and OB function. Importantly, lytic bone disease in MM patients persists in the absence of active disease, suggesting that persistent OB suppression is regulated by soluble inhibitors of OB differentiation as well as by MM cell-induced epigenetic changes in pre-OB. Consistent with this, Gfi1, a transcriptional repressor of Runx2, directly binds the *Runx2* promoter, recruiting chromatin corepressors in pre-OB to induce epigenetic repression of Runx2. Gfi1 is elevated in marrow stromal cells from MM patients and induced in stromal cells by MM cells [26, 27].

Interactions between myeloma cells and bone cells are also now thought to be instrumental in the control of dormancy [28]. Following myeloma cell colonization of the endosteal niche, myeloma cells enter a reversible dormant state in which they are relatively chemoresistant. Dormancy is induced by contact with bone-lining OBs and switched off (reactivating the myeloma cells) by OC remodeling of the endosteal niche. As such, interactions between myeloma cells and bone cells drive not only development of the bone disease, but also key steps in dormancy and disease relapse.

Imaging in multiple myeloma bone disease

Multiple imaging modalities may be used to identify MM bone disease. While conventional whole body skeletal radiography remains the traditional gold standard, this method can underestimate lytic bone disease (identification of lytic bone lesions requires loss of at least 30% of trabecular bone volume). The utility of ^{18}F-fluorodeoxyglucose (FDG) PET/CT has been evaluated in MM and has a sensitivity of approximately 85% and a specificity of approximately 90% [29]. The International Myeloma Working Group acknowledges the utility of newer and highly sensitive imaging modalities, including low-dose whole body CT, MRI, and PET/CT for the evaluation of myeloma patients [30]. Importantly, however, osteoporosis and vertebral compression fractures cannot be used in isolation to meet diagnostic criteria for active myeloma. Bone density studies, such as DXA scans, are not routinely used as a part of the standard diagnostic workup for MM patients. Traditional technetium bone scintigraphy scans underestimate the extent of bone disease and are thus not recommended in patients with MM [31].

Models of multiple myeloma bone disease

In order to study MM bone disease and identify new mechanisms of disease pathogenesis and therapeutic targets, preclinical models that mimic human disease, including the osteolytic bone diseases, are invaluable [32]. The 5T Radl myeloma model is a syngenic model that effectively mimics tumor growth within bone, osteolytic bone disease, and anemia, and has been highly informative for many years. Genetically engineered mice such as the Vk*MYC and Eu-xbp-1s models are associated with tumor development within bone and evidence of bone disease. Xenograft models that allow the inoculation of human cell lines such as JJN-3 and U266 are also associated with an osteolytic bone disease, but are limited by a compromised immune system. Implantations of fetal human bone, rabbit bone, or synthetic bone scaffolds in immune-deficient mice allow the growth of primary myeloma cells within the implanted bone, but also have several limitations. Recently, advances have been made where genetically humanized mice support the growth of primary plasma cells from both patients with myeloma and monoclonal gammopathy of undetermined significance (MGUS) in vivo [33].

Management of multiple myeloma bone disease

Current pharmacological management of MM bone disease relies on the combination of systemic antimyeloma therapy and antiresorptive therapy to block ongoing osteoclastic bone destruction. Bisphosphonates, potent inhibitors of osteoclast activity, remain the standard of care for the prevention and control of bone destruction in MM. Bisphosphonates inhibit farnesyl diphosphate synthase, which inhibits protein prenylation, a process required for OC formation and survival. This reduces bone turnover and decreases both the incidence of skeletal-related events (SREs), including the development of osteolytic lesions, pathological fractures, hypercalcemia, and bone pain, and the time to the first SRE [34]. In addition, bisphosphonate treatment can be used to manage pain related to bone disease [35]. Importantly, the Medical Research Council IX trial reported that administration of intravenous bisphosphonates (zoledronic acid as compared with oral clodronate) in combination with systemic myeloma therapy prolonged median progression-free and overall survival in MM patients without radiological evidence of bone disease [36]. This suggests that bisphosphonates have a direct antimyeloma effect, a hypothesis supported by in vitro data. Multiple international organizations including the American Society of Clinical Oncology, the European Myeloma Network, the International Myeloma Working Group, and the National Comprehensive Cancer Center Network have published guidelines advising the initiation of bisphosphonate therapy for MM patients regardless of whether lytic bone lesions are identified [37]. There are limited data to date regarding the ideal duration of bisphosphonate therapy in MM or the utility of following bone resorption markers to determine the ideal dosing schedule of bisphosphonate therapy.

Denosumab, a human monoclonal antibody that binds to RANKL with high affinity and specificity, inhibits bone resorption and prevents SREs in patients refractory to bisphosphonate therapy [38, 39]. Denosumab was approved by the US Food and Drug Administration (FDA) for the prevention of SREs in patients with bone metastases from solid tumors in 2010, and is under investigation for use in MM bone disease.

Bortezomib, a proteasome inhibitor active against MM, directly alters OB and OC activity by decreasing RANKL and DKK1 levels in the sera of myeloma patients [40]. In clinical studies of both newly diagnosed and relapsed myeloma patients, bortezomib therapy, either alone or in combination with other agents, demonstrated improvement in markers of OB activity and OC inhibition [41]. Bortezomib's effects on OB differentiation have been extensively studied. Several clinical trials showed increased bone-specific alkaline phosphatase (ALP), a marker for osteoblast activation, in myeloma patients whose tumor responded to the drug. Some authors have interpreted these findings as evidence that bortezomib directly stimulates osteoblasts and inhibits osteoclasts. Others have suggested that biochemical markers of bone formation peak after 6 weeks of bortezomib treatment due to a direct inhibitory effect on bone resorption by OCs that counteract bortezomib's initial direct OB stimulatory effect. Alternatively, bortezomib's direct inhibition of myeloma cells in the bone marrow microenvironment allows for normalization of OB and OC function because these effects are only seen in patients whose myeloma responds to bortezomib treatment. Limited data on the bone effects of carfilzomib, a second generation proteasome inhibitor, or ixazomib, an oral proteasome inhibitor, are available at this time.

MONOCLONAL GAMMOPATHY OF UNDETERMINED SIGNIFICANCE

MGUS is a premalignant monoclonal plasma cell disorder that produces no overt symptoms, but is associated with an annual risk of progression to myeloma of approximately 1%. There is increasing evidence that MGUS is associated with a disruption of skeletal health and bone metabolism, with an increased fracture risk in patients with MGUS, and an increased prevalence of MGUS in patients with osteoporosis. Recent studies using high resolution QCT have identified significant differences in the microstructural architecture of bone in patients with MGUS, associated with a decrease in vertebral BMD and an increase in cortical porosity and deficits in biomechanical bone strength [42]. Furthermore, patients with MGUS have elevated circulating levels of factors associated with osteolysis, including DKK1 and MIP-1α [43].

ADULT T-CELL LEUKEMIA/LYMPHOMA

Adult T-cell leukemia/lymphoma (ATLL) is a malignancy of CD4+ T cells caused by infection with human T-cell leukemia virus type 1 (HTLV-1). ATLL was initially reported in southern Japan, and has been reported sporadically in areas where HTLV-1 infection is rare, including the United States. The additive lifetime risk of developing ATLL among HTLV-1 carriers ranges between 1% and 5% in Japan and Jamaica. Aggressive presentations of ATLL, and leukemic or lymphomatous presentations, occur in approximately 80% of patients, and can include lytic bone lesions and symptoms associated with severe hypercalcemia in approximately 70% of patients [44]. In contrast to the hypercalcemia that typically develops in myeloma, the hypercalcemia associated with ATLL is mediated by PTHrP and IL-1 [45]. It is hypothesized that the HTLV-1 and HTLV-11 tax proteins transactivate PTHrP via the cellular transcription factors activator protein 2 (AP-2) and AP-1 [46, 47]. However, increased transcription of PTHrP also occurs in a tax-independent manner [48].

ATLL cells produce chemokines that affect bone remodeling, including IL-1, IL-6, TNF-α, and MIP-1α/MIP-1β. Circulating ATLL cells infiltrate a variety of tissues, mediated by MIP-1α induction of integrin-mediated adhesion to the endothelium and subsequent transmigration [46]. As in myeloma, MIP-1α in ATLL is important for the chemotaxis of monocytes, including OC progenitor cells, and the production of the osteoclastogenic factors IL-6, PTHrP, and RANKL by osteoblasts or stromal cells [49, 50]. MIP-1α has been proposed as a mediator for the hypercalcemia in ATLL, by enhancing OC formation and inducing RANKL expression on ATLL cells in an autocrine fashion [51]. In addition, IL-1 and PTHrP have also been reported to mediate bone destruction in ATLL, with elevated PTHrP levels in patients and increased concentrations of IL-1 and PTHrP in media conditioned by ATLL cells in vitro [45].

Bone involvement has also been infrequently reported in more classic forms of acute lymphoid leukemia and is thought to be similarly mediated by PTHrP production by malignant cells [52].

NON-HODGKIN LYMPHOMA

Bone involvement with non-Hodgkin lymphoma (NHL), the most frequently diagnosed hematological malignancy, is rare. Less than 10% of NHL patients present with bone involvement, while 7% to 25% of all patients with NHL eventually develop bone findings during the course of their disease. The most common histological subtypes of NHL that present with bone manifestations include histiocytic, undifferentiated, and poorly differentiated NHL. In addition, lytic bone lesions are more often seen in patients with diffuse rather than nodular patterns of lymph node involvement, and frequently involve the axial skeleton [53]. As in ATLL, serum levels of PTHrP are elcvated in NHL patients with hypercalcemia.

HODGKIN DISEASE

Bone involvement in Hodgkin disease (HD) is also uncommon and seldom encountered at diagnosis. Sites of involvement include the spine, pelvis, femur, humerus, ribs, sternum, scapula, and base of the skull; however, as with NHL, vertebral and femoral involvement is most common [54]. Bone disease in patients with HD can be lytic, blastic, or mixed. Increased new bone formation by tumor cell stimulation of OB activity occurs at sites of previous osteoclastic activity. Hypercalcemia does occur in HD, and is associated with excess production of $1,25(OH)_2D_3$ or PTHrP by the lymphoma cells [55].

The most frequent presentation is that of a localized, solitary, osteoblastic mass in a patient with mixed cellularity, nodular sclerosing disease. Radiological findings can include a vertebral sclerotic pattern along with a periosteal reaction and hypertrophic pulmonary osteoarthropathy [54]. As in NHL, radiographic patterns cannot predict histological type or the prognosis of HD, and must be used with clinical staging to predict prognosis. Bone biopsies often show fibrosis and a mixed inflammatory infiltrate with rare atypical cells.

SUMMARY

Dysregulation of physiological bone remodeling in the setting of malignancy results in osteolytic, osteoblastic, or mixed lesions. PTHrP frequently mediates nonmyelomatous bone disease. With the exception of multiple myeloma, bone involvement in hematological malignancies is rare, however bone lesions can significantly contribute to patient morbidity and pain. Thus, consideration of the potential consequences of skeletal manifestations of hematological malignancies is an important component of the care of these patients.

REFERENCES

1. Siegel RL, Miller KD, Jemal A. Cancer statistics, 2016. CA Cancer J Clin. 2016;66(1):7–30.
2. Kumar SK, Dispenzieri A, Lacy MQ, et al. Continued improvement in survival in multiple myeloma: changes in early mortality and outcomes in older patients. Leukemia. 2014;28(5):1122–8.
3. Roodman GD. Pathogenesis of myeloma bone disease. J Cell Biochem. 2010;109(2):283–91.
4. Dispenzieri A. POEMS syndrome: 2011 update on diagnosis, risk-stratification, and management. Am J Hematol. 2011;86(7):591–601.
5. Taube T, Beneton MN, McCloskey EV, et al. Abnormal bone remodelling in patients with myelomatosis and normal biochemical indices of bone resorption. Eur J Haematol. 1992;49(4):192–8.
6. Pearse RN, Sordillo EM, Yaccoby S, et al. Multiple myeloma disrupts the TRANCE/osteoprotegerin cytokine axis to trigger bone destruction and promote tumor progression. Proc Natl Acad Sci U S A. 2001;98(20):11581–6.
7. Choi SJ, Cruz JC, Craig F, et al. Macrophage inflammatory protein 1-alpha is a potential osteoclast stimulatory factor in multiple myeloma. Blood. 2000;96(2):671–5.
8. Lee JW, Chung HY, Ehrlich LA, et al. IL-3 expression by myeloma cells increases both osteoclast formation and growth of myeloma cells. Blood. 2004;103(6):2308–15.
9. Giuliani N, Colla S, Rizzoli V. New insight in the mechanism of osteoclast activation and formation in multiple myeloma: focus on the receptor activator of NF-kappaB ligand (RANKL). Exp Hematol. 2004;32(8):685–91.
10. Gunn WG, Conley A, Deininger L, et al. A crosstalk between myeloma cells and marrow stromal cells stimulates production of DKK1 and interleukin-6: a potential role in the development of lytic bone disease and tumor progression in multiple myeloma. Stem Cells. 2006;24(4):986–91.
11. Giuliani N, Ferretti M, Bolzoni M, et al. Increased osteocyte death in multiple myeloma patients: role in myeloma-induced osteoclast formation. Leukemia. 2012;26(6):1391–401.
12. Kobayashi Y, Udagawa N, Takahashi N. Action of RANKL and OPG for osteoclastogenesis. Crit Rev Eukaryot Gene Expr. 2009;19(1):61–72.
13. Delgado-Calle J, Anderson J, Cregor MD, et al. Bidirectional Notch signaling and osteocyte-derived factors in the bone marrow microenvironment promote tumor cell proliferation and bone destruction in multiple myeloma. Cancer Res. 2016;76(5):1089–100.
14. Giuliani N, Bataille R, Mancini C, et al. Myeloma cells induce imbalance in the osteoprotegerin/osteoprotegerin ligand system in the human bone marrow environment. Blood. 2001;98(13):3527–33.
15. Qiang YW, Chen Y, Stephens O, et al. Myeloma-derived Dickkopf-1 disrupts Wnt-regulated osteoprotegerin and RANKL production by osteoblasts: a potential mechanism underlying osteolytic bone lesions in multiple myeloma. Blood. 2008;112(1):196–207.

16. Oyajobi BO, Franchin G, Williams PJ, et al. Dual effects of macrophage inflammatory protein-1alpha on osteolysis and tumor burden in the murine 5TGM1 model of myeloma bone disease. Blood. 2003;102(1):311–9.
17. Lentzsch S, Gries M, Janz M, et al. Macrophage inflammatory protein 1-alpha (MIP-1 alpha) triggers migration and signaling cascades mediating survival and proliferation in multiple myeloma (MM) cells. Blood. 2003;101(9):3568–73.
18. Bataille R, Chappard D, Marcelli C, et al. Mechanisms of bone destruction in multiple myeloma: the importance of an unbalanced process in determining the severity of lytic bone disease. J Clin Oncol. 1989;7(12):1909–14.
19. Delgado-Calle J, Bellido T, Roodman GD. Role of osteocytes in multiple myeloma bone disease. Curr Opin Support Palliat Care. 2014;8(4):407–13.
20. Yaccoby S, Wezeman MJ, Zangari M, et al. Inhibitory effects of osteoblasts and increased bone formation on myeloma in novel culture systems and a myelomatous mouse model. Haematologica. 2006;91(2):192–9.
21. Edwards CM, Edwards JR, Lwin ST, et al. Increasing Wnt signaling in the bone marrow microenvironment inhibits the development of myeloma bone disease and reduces tumor burden in bone in vivo. Blood. 2008;111(5):2833–42.
22. Gaur T, Lengner CJ, Hovhannisyan H, et al. Canonical WNT signaling promotes osteogenesis by directly stimulating Runx2 gene expression. J Biol Chem. 2005;280(39):33132–40.
23. Ryoo HM, Lee MH, Kim YJ. Critical molecular switches involved in BMP-2-induced osteogenic differentiation of mesenchymal cells. Gene. 2006;366(1):51–7.
24. Vallet S, Mukherjee S, Vaghela N, et al. Activin A promotes multiple myeloma-induced osteolysis and is a promising target for myeloma bone disease. Proc Natl Acad Sci U S A. 2010;107(11):5124–9.
25. Lee MH, Kwon TG, Park HS, et al. BMP-2-induced Osterix expression is mediated by Dlx5 but is independent of Runx2. Bioche Biophys Res Commun. 2003;309(3):689–94.
26. D'Souza S, del Prete D, Jin S, et al. Gfi1 expressed in bone marrow stromal cells is a novel osteoblast suppressor in patients with multiple myeloma bone disease. Blood. 2011;118(26):6871–80.
27. Silbermann R, Adamik J, Zhou D, et al. p62-ZZ domain signaling inhibition rescues MM-induced epigenetic repression at the *Runx2* promoter and allows osteoblast differentiation of MM patient pre-osteoblasts *in vitro*. Blood. 2016;128:4410.
28. Lawson MA, McDonald MM, Kovacic N, et al. Osteoclasts control reactivation of dormant myeloma cells by remodelling the endosteal niche. Nat Commun. 2015;6:8983.
29. Bredella MA, Steinbach L, Caputo G, et al. Value of FDG PET in the assessment of patients with multiple myeloma. AJR Am J Roentgenol. 2005;184(4):1199–204.
30. Rajkumar SV, Dimopoulos MA, Palumbo A, et al. International Myeloma Working Group updated criteria for the diagnosis of multiple myeloma. Lancet Oncol. 2014;15(12):e538–48.
31. Dimopoulos M, Terpos E, Comenzo RL, et al. International Myeloma Working Group consensus statement and guidelines regarding the current role of imaging techniques in the diagnosis and monitoring of multiple myeloma. Leukemia. 2009;23(9):1545–56.
32. Lwin ST, Edwards CM, Silbermann R. Preclinical animal models of multiple myeloma. BoneKEy Rep. 2016;5:772.
33. Das R, Strowig T, Verma R, et al. Microenvironment-dependent growth of preneoplastic and malignant plasma cells in humanized mice. Nat Med. 2016;22(11):1351–7.
34. Berenson JR, Rosen LS, Howell A, et al. Zoledronic acid reduces skeletal-related events in patients with osteolytic metastases. Cancer. 2001;91(7):1191–200.
35. Terpos E, Roodman GD, Dimopoulos MA. Optimal use of bisphosphonates in patients with multiple myeloma. Blood. 2013;121(17):3325–8.
36. Morgan G, Davies F, Gregory W, et al. Evaluating the effects of zoledroic acid (ZOL) on overall survival (OS) in patients (Pts) with multiple myeloma (MM): results of the Medical Research Council (MRC) Myeloma IX study. J Clin Oncol. 2010;28(7s):Abstract 8021.
37. Anderson KC, Alsina M, Atanackovic D, et al. Multiple Myeloma, Version 2.2016: Clinical Practice Guidelines in Oncology. J Natl Compr Cancer Netw. 2015;13(11):1398–435.
38. Fizazi K, Lipton A, Mariette X, et al. Randomized phase II trial of denosumab in patients with bone metastases from prostate cancer, breast cancer, or other neoplasms after intravenous bisphosphonates. J Clin Oncol. 2009;27(10):1564–71.
39. Body JJ, Facon T, Coleman RE, et al. A study of the biological receptor activator of nuclear factor-kappaB ligand inhibitor, denosumab, in patients with multiple myeloma or bone metastases from breast cancer. Clin Cancer Res. 2006;12(4):1221–8.
40. Terpos E, Heath DJ, Rahemtulla A, et al. Bortezomib reduces serum dickkopf-1 and receptor activator of nuclear factor-kappaB ligand concentrations and normalises indices of bone remodelling in patients with relapsed multiple myeloma. Br J Haematol. 2006;135(5):688–92.
41. Zangari M, Esseltine D, Lee CK, et al. Response to bortezomib is associated to osteoblastic activation in patients with multiple myeloma. Br J Haematol. 2005;131(1):71–3.
42. Ng AC, Khosla S, Charatcharoenwitthaya N, et al. Bone microstructural changes revealed by high-resolution peripheral quantitative computed tomography imaging and elevated DKK1 and MIP-1alpha levels in patients with MGUS. Blood. 2011;118(25):6529–34.
43. Drake MT. Unveiling skeletal fragility in patients diagnosed with MGUS: no longer a condition of undetermined significance? J Bone Miner Res. 2014;29(12):2529–33.
44. Kiyokawa T, Yamaguchi K, Takeya M, et al. Hypercalcemia and osteoclast proliferation in adult T-cell leukemia. Cancer. 1987;59(6):1187–91.

45. Roodman GD. Mechanisms of bone lesions in multiple myeloma and lymphoma. Cancer. 1997;80(suppl 8):1557–63.
46. Raza S, Naik S, Kancharla VP, et al. Dual-positive (CD4+/CD8+) acute adult T-cell leukemia/lymphoma associated with complex karyotype and refractory hypercalcemia: case report and literature review. Case Rep Oncol. 2010;3(3):489–94.
47. Shu ST, Martin CK, Thudi NK, et al. Osteolytic bone resorption in adult T-cell leukemia/lymphoma. Leuk Lymphoma. 2010;51(4):702–14.
48. Richard V, Lairmore MD, Green PL, et al. Humoral hypercalcemia of malignancy: severe combined immunodeficient/beige mouse model of adult T-cell lymphoma independent of human T-cell lymphotropic virus type-1 tax expression. Am J Pathol. 2001;158(6):2219–28.
49. Tanaka Y, Maruo A, Fujii K, et al. Intercellular adhesion molecule 1 discriminates functionally different populations of human osteoblasts: characteristic involvement of cell cycle regulators. J Bone Miner Res. 2000;15(10):1912–23.
50. Han JH, Choi SJ, Kurihara N, et al. Macrophage inflammatory protein-1alpha is an osteoclastogenic factor in myeloma that is independent of receptor activator of nuclear factor kappaB ligand. Blood. 2001;97(11):3349–53.
51. Okada Y, Tsukada J, Nakano K, et al. Macrophage inflammatory protein-1alpha induces hypercalcemia in adult T-cell leukemia. J Bone Miner Res. 2004;19(7):1105–11.
52. Inukai T, Hirose K, Inaba T, et al. Hypercalcemia in childhood acute lymphoblastic leukemia: frequent implication of parathyroid hormone-related peptide and E2A-HLF from translocation 17;19. Leukemia. 2007;21(2):288–96.
53. Pear BL. Skeletal manifestations of the lymphomas and leukemias. Semin Roentgenol. 1974;9(3):229–40.
54. Franczyk J, Samuels T, Rubenstein J, et al. Skeletal lymphoma. Can Assoc Radiol J. 1989;40(2):75–9.
55. Seymour JF, Gagel RF. Calcitriol: the major humoral mediator of hypercalcemia in Hodgkin's disease and non-Hodgkin's lymphomas. Blood. 1993;82(5):1383–94.
56. Raje N, Roodman GD. Advances in the biology and treatment of bone disease in multiple myeloma. Clin Cancer Res. 2011;17(6):1278–86.
57. Roodman GD. Pathogenesis of myeloma bone disease. N Engl J Med. 2004;32:290–2.

99

Osteogenic Osteosarcoma

Yangjin Bae[1], Huan-Chang Zeng[1], Linchao Lu[2], Lisa L. Wang[2], and Brendan Lee[1]

[1]*Department of Molecular and Human Genetics, Baylor College of Medicine, Texas Children's Hospital, Houston, TX, USA*
[2]*Department of Pediatrics, Section of Hematology/Oncology, Baylor College of Medicine, Texas Children's Hospital, Houston, TX, USA*

INTRODUCTION

Osteogenic sarcoma (also known as osteosarcoma or OS) is the most common primary malignant tumor of bone [1, 2]. Nevertheless, it is a rare disease, with only about 900 new cases diagnosed annually in the United States, accounting for less than 1% of all cancers [3]. Osteosarcoma affects individuals of all ages, but its incidence is bimodal, with the first peak in adolescents (eight per million at age group of 15- to 19-year-olds) and the second peak in the elderly (six per million at age group of 75- to 79-year-olds) [1, 4]. Accordingly, osteosarcoma accounts for 5% of pediatric cancers overall (about 400 new cases each year) [5]. There is a high percentage of osteosarcoma associated with Paget disease and occurring as a second or later cancer among the elderly [1, 4]. Although osteosarcoma can arise in any bone, it preferentially affects anatomical sites where rapid bone remodeling occurs such as the metaphyses of long bones (distal femur > proximal tibia > proximal humerus) [6]. In children and adolescents, these anatomical regions account for the majority of primary tumors. However, in the elderly, the distribution of anatomical sites is more variable and can include the axial skeleton and skull [1].

Biopsy is required for the diagnosis of osteosarcoma. There are several different histological subtypes including conventional, telangiectatic, small cell, high-grade surface, low-grade central, periosteal, and parosteal osteosarcoma [7]. Conventional osteosarcoma is the most common subtype in childhood and adolescence, comprising about 85% of all cases, and is subdivided into osteoblastic, chondroblastic, and fibroblastic types based on the predominant histological characteristics of the malignant cells [8, 9]. Osteosarcoma is characterized by the production of osteoid by tumor cells. Approximately 20% of patients will have detectable metastatic disease at the time of initial presentation, with the lungs being the most common sites of metastasis followed by other bones. Treatment consists of surgery to remove the primary tumor and high-dose chemotherapy to treat micrometastatic disease. In general, the 5-year survival rate for nonmetastatic disease is about 70% in younger patients aged <25 years and approximately 45% in patients aged 60 and above. Patients with distant metastases have much poorer 5-year survival rates of the order of 30% or less [1, 4]. Strikingly, there have been no substantial improvements in survival rates for either group of patients over the past several decades [10]. Clearly, new treatment strategies and drugs are needed.

The etiology of osteosarcoma remains largely unknown. However, there has been renewed effort in understanding the molecular biology and pathogenesis of osteosarcoma [2]. Recent reviews on studies of familial syndromes, specimens, and cell lines derived from human osteosarcoma patients describe the genetic factors and signaling pathways that may be involved in several key pathogenic processes, including initiation, progression, invasion, and metastasis [2, 5, 11]. Several genetic factors have been evaluated as potential diagnostic and prognostic biomarkers of disease as well as potential therapeutic targets. Based on these findings, novel agents have been tested in several phase I and II clinical trials [12]. More recently, genetically engineered osteosarcoma mouse models have been generated in an attempt to recapitulate the human disease [13, 14]. Understanding these models will broaden our knowledge of the molecular basis of osteosarcoma and will also advance preclinical studies for new therapeutic strategies. This chapter updates the

---

Primer on the Metabolic Bone Diseases and Disorders of Mineral Metabolism, Ninth Edition. Edited by John P. Bilezikian.
© 2019 American Society for Bone and Mineral Research. Published 2019 by John Wiley & Sons, Inc.
Companion website: www.wiley.com/go/asbmrprimer

current understanding of osteosarcoma biology and reviews recent discovery of osteosarcoma-driver genes through next generation sequencing and animal models.

CHALLENGES IN THE TREATMENT OF OSTEOSARCOMA

Current standard treatments

Surgery and chemotherapy are two essential components of therapy for osteosarcoma. Unlike other high-grade sarcomas, osteosarcoma is relatively resistant to radiotherapy. Prior to the 1970s, before the introduction of chemotherapy for the treatment of osteosarcoma, when surgery (most often amputation) was the sole treatment, the 5-year survival rates were only 10% to 20% [8]. Most patients developed distant metastatic disease postoperatively despite complete removal of the tumor [9]. Thus although only 20% of patients initially present with clinically detectable metastatic disease—as seen by modern imaging technologies including computerized tomography, bone scintigraphy, and magnetic resonance imaging—virtually all patients already have micrometastases at the time of diagnosis [8]. After the introduction of adjuvant (postoperative) chemotherapy in the 1970s in addition to surgery, the survival rates increased dramatically, thus establishing the critical role of chemotherapy in the treatment of osteosarcoma [15, 16].

The most effective chemotherapeutic agents include doxorubicin (or adriamycin), methotrexate, cisplatin, and ifosfamide [10]. In the late 1970s, the concept of neoadjuvant (preoperative) chemotherapy was introduced, which offered several advantages such as early eradication of micrometastases, shrinkage of tumor bulk (making surgery more feasible), and importantly the ability to determine the degree of tumor necrosis at the time of definitive resection [17]. The percent tumor necrosis (or histological response) has been found to be a prognostic factor, with >90% tumor necrosis being considered a good response and favorable prognostic factor [9]. In addition to assessment of histological response, administration of neoadjuvant chemotherapy allows time for the orthopedic surgeon to plan for limb salvage surgery [8], which has largely replaced amputation for extremity tumors.

While chemotherapy is important for treating osteosarcoma, surgery is still a mainstay of therapy and is essential for survival. Complete resection may be difficult depending on the tumor location, for example in the spine and pelvic bones, where the risk of local recurrence is high. A dismal outcome was reported for those patients with pelvic osteosarcoma where the 5-year survival rate was only about 19% [18, 19]. A major challenge is curing patients who are not eligible for metastasectomy; in these cases, less effective radiotherapy and palliative chemotherapy can be applied toward management of disease burden [8]. Thus, despite improvements in survival from use of intensive chemotherapy and improved surgical techniques, there is a continued need for new therapeutic approaches.

Limitations of current standard treatments

Despite the relative success of current treatments, about 40% of all patients will relapse mostly within 2 years, and half of them will die in less than 5 years [2, 8]. This is attributable in large part to the chemoresistance of tumor cells. Another limitation to current treatment regimens is the toxicity associated with chemotherapy. Chemotherapeutic drugs kill not only tumor cells but normal tissues as well, causing major renal, hematological, and cardiac toxicities [2]. Some of these toxicities occur during administration of the drugs, while others, such as cardiotoxicity from doxorubicin, can occur many years later. The recent Childhood Cancer Survivor Study (CCSS) performed on 733 long-term survivors (>5 years) of childhood cancer with a mean follow-up 21.6 years, showed that 86.9% of osteosarcoma survivors experienced at least one chronic medical condition. Prospective evaluation of survivors will be important to assess both acute and long-term effects of current treatments and their impact on survivorship [20]. Another potential late effect of chemotherapy is secondary malignancy, particularly with alkylating agents such as ifosfamide. The majority of secondary malignant neoplasms occur around 10 years from diagnosis, and the incidence is 3% to 5% [20, 21].

Given the problems of chemoresistance, organ toxicity, and secondary malignancies associated with medical treatment for osteosarcoma, researchers have investigated other therapeutic avenues. The identification of new active agents for this disease has been a challenge, partially due to the fact that the primary endpoint of clinical trials has traditionally been radiographic response using WHO or Response Evaluation Criteria in Solid Tumors (RECIST) criteria [22]. Given that osteosarcoma is a bone-forming tumor with mineralization of stromal tissue, radiographic appearance may underestimate therapeutic activity. Thus, several phase II studies are currently being conducted by the Children's Oncology Group with the primary endpoint being prolongation of progression-free survival compared with historical controls [23]. Agents under investigation include a RANKL inhibitor (denosumab), anti-GD2 antibody (dinutuximab), and antibody–drug conjugate directed toward type 1 transmembrane glycoprotein NMB (glembatumumab vedotin).

One agent that has shown promise in the treatment of osteosarcoma is the immune modulating agent muramyl tripeptide phosphatidyl ethanolamine (MTP-PE). It has been safely administered along with standard adjuvant chemotherapy agents and has shown some benefit in overall survival; however, it has not been shown to improve event-free survival in osteosarcoma patients [24, 25]. MTP-PE is an activator of macrophages and monocytes and remains a promising agent for incorporation into standard of care; however, it requires further study and is not currently approved for use in the United States [26]. A more recent immune approach involves the use of host T cells to recognize and eradicate osteosarcoma cells expressing specific tumor-associated antigens. A phase I study of chimeric antigen receptor T cells specifically recognizing the human epidermal growth factor receptor 2 (Her2) antigen in sarcoma patients showed that these cells could be safely administered, and several osteosarcoma patients had encouraging responses [27]. T-cell immunotherapeutic and other immune-based approaches such as PD1 and PDL1 immune checkpoint inhibitors continue to be investigated and offer promise for the treatment of osteosarcoma.

GENETIC FACTORS OF OSTEOSARCOMA

Germline mutations in familial cancer predisposition syndromes

Germline mutations in *TP53*, *RB1*, *RECQL4*, *BLM*, and *WRN* genes cause Li–Fraumeni syndrome, hereditary retinoblastoma, Rothmund–Thomson syndrome, Bloom syndrome, and Werner syndrome, respectively, and all of these syndromes are predisposed to osteosarcoma [5]. Also, mutations in *EXT1* and *EXT2* and *SQSTM1* are associated with hereditary multiple exostoses and Paget disease, respectively. These genes are also expected to contribute to osteosarcoma development [28, 29]. Mutations in the *TP53* gene have only been observed in about 15% to 20% of sporadic osteosarcomas [30]. An additional 10% to 20% of sporadic osteosarcomas are due to inactivation of p53 function by amplification or overexpression of *MDM2* and *COPS3* [31, 32]. More recent next generation sequencing studies of osteosarcoma tumors have revealed frequent structural alterations (deletions and genomic rearrangement) of *TP53* [33, 34]. About 70% of sporadic osteosarcomas contain genetic alterations of *RB1*, but few point mutations have been found, and deletions or structural alterations are seen in 40% of cases [35]. Inactivation of Rb1 alone in mice does not cause osteosarcoma, so it is likely that Rb is an enhancer during osteosarcomagenesis [15]. The RECQ family of proteins plays a role in maintaining genomic integrity. While less is known about the direct role of these proteins in osteosarcomagenesis, patients with Rothmund–Thomson syndrome with *RECQL4* mutations have an extremely high rate of occurrence of osteosarcoma as well as skeletal dysplasias [36, 37]. Two recent studies showed that complete loss of function of Recql4 in mouse skeletal cells leads to developmental bone abnormalities and osteoporosis, but does not induce the development of osteosarcoma [38, 39]. Lu and colleagues suggested that RECQL4 is critical for skeletal development by modulating p53 activity in vivo [38]. Ng and colleagues further suggested that mutant, not null, alleles of *RECQL4* may account for tumor suppression and susceptibility to osteosarcoma [39]. Further investigation of how RECQL4 functionally interacts with p53 in the context of bone tumor formation may provide more detailed molecular mechanistic information regarding osteosarcomagenesis.

Discovery of somatic mutations in osteosarcoma by next generation sequencing

Next generation whole genome sequencing studies allowed the discovery of multiple genetic alterations in sporadic osteosarcomas. A large proportion of *TP53* genetic alterations were due to somatic structural variations such as deletions and translocations with breakpoints within *TP53* [33]. Interestingly, these genetic events seen in *TP53* are not typical for most other cancers, and are a result of genomic instability in osteosarcoma. In addition to the *TP53* gene, *RB1* and RB1-interacting protein genes are also inactivated in osteosarcoma [35]. Other recurrently altered genes include *ATRX* and *DLG2* [33]. Perry and colleagues found that *PTEN*, *TSC2*, and *AKT1* genes involved in the PI3K/mammalian target of rapamycin (mTOR) pathway were altered in osteosarcoma [40]. In addition, other genes that were affected included growth factor receptor tyrosine kinases (*PDGFRA*, *PDGFRB*, *JAK1*, *ALK*, *KDR*, *FGFR4*), Wnt pathway members, cell cycle regulatory molecules, and genes involved in DNA repair [40]. Furthermore, a study using a *Sleeping Beauty* transposon forward genetic screen in mice identified hundreds of potential osteosarcoma driver genes. A significant portion of these genes was enriched in the PI3/AKT (serine/threonine kinase 1)/mTOR, MAPK, and ERbB (receptor tyrosine kinase) pathways [41]. Among these, *PTEN* was the most frequently identified gene in the screen, and the contribution of PTEN (phosphatase and tensin homolog) to osteosarcoma pathogenesis was further demonstrated through cooperation with *TP53* in a genetic mouse model [41]. In addition, axon guidance genes *Sema4d* and *Sema6d* were validated as oncogenic.

Overall, the recent discovery of multiple somatic mutations and structural variations in genes driving osteosarcoma by next generation sequencing indicates the genetic complexity of osteosarcoma. Delineating the molecular mechanisms underlying the functions of these genes and their roles specifically in osteosarcoma pathogenesis will require rigorous in vivo validation.

Altered signaling pathways contributing to osteosarcomagenesis

Several evolutionarily conserved signaling pathways have been linked to the pathogenesis of osteosarcoma and metastasis. They include Wnt (wingless-type mouse mammary tumor virus integration site), Notch, TGF/BMPs, SHH (Sonic hedgehog), and GFs (growth factors) pathways. Thus far, the Wnt and Notch pathways have been the most extensively studied. Elevated levels of cytoplasmic and/or nuclear localized β-catenin, a critical mediator of the canonical Wnt pathway, have been detected in the majority of osteosarcomas, as well as sporadic mutations of β-catenin [11]. Ectopic expression of the Wnt agonist Dkk3 suppresses invasion and motility of osteosarcoma cell lines [42]. Inactivation of Wif1, a secreted Wnt antagonist, increases β-catenin levels and accelerates the development of osteosarcomas in mice [43]. Altered Notch signaling has been associated with several human cancers. A recent study showed that deletion of RBPJk, a transcription factor in the canonical Notch pathway, completely blocked tumor formation in osteoblast-specific cNICD mice, but rarely affected tumor formation in p53 mutant mice [14]. Also, Wif1 expression was drastically decreased in this mouse model, indicating the crosstalk between Wnt and Notch pathways in osteosarcoma progression. Along with other genetic mouse models of loss of TP53, RB, and TP53/RB, the gain of Notch signaling in osteoblasts results in spontaneous osteosarcoma. Based on complementary genomic approaches using whole exome, whole genome, and RNA sequencing, the PI3K/mTOR pathway was identified as a therapeutic target pathway for osteosarcoma [38]. The Hedgehog pathway was found to be associated with osteosarcoma metastasis. Silencing GLI2, a critical downstream Hedgehog transcription factor, inhibited metastasis [44]. Thus, multiple pathways have been implicated in osteosarcoma and this highlights the diverse mechanisms involved in osteosarcoma pathogenesis.

Noncoding RNAs and osteosarcoma

Recent studies showed that the long noncoding RNAs (lncRNAs) play roles in tumor development in various types of cancers including osteosarcoma. LncRNAs are a group of noncoding transcripts of about 200 nucleotides in length. Li and colleagues found a group of differentially regulated lncRNAs by microarray using osteosarcoma tissues and paired adjacent nontumor tissues [45]. The expression of MALAT1, initially found as a predictive biomarker for metastasis in the early stage of nonsmall cell lung cancer [46], was repressed in the MG-63 osteosarcoma cell line by the tumor suppressive function of MYC-6 and suppressed cell proliferation [46]. Dong and colleagues demonstrated that the expression of MALAT1 was elevated in osteosarcoma tumors and positively correlated with pulmonary metastasis [47]. Furthermore, MALAT1 knockdown suppressed tumor growth in vivo and inhibited the RhoA and coiled-coil-containing Rho-associated protein kinase (ROCK) expression [48]. Increasing evidence suggests that H19 plays an important role in the development and progression of cancers. H19 was aberrantly expressed and induced through upregulated Hedgehog signaling and Yap1 expression in osteosarcoma [49].

MicroRNAs (miRNAs) are also implicated in osteosarcoma biology by functioning as oncogenes or tumor suppressors by regulating downstream targets. The miR-34 family identified as a direct target of p53 shares a similar function as p53 by inducing apoptosis, cell cycle arrest, and senescence. Further analysis of miR-34 genes showed the epigenetic silencing of their promoter in primary osteosarcoma. Also, miR-34 genes underwent minimal deletions in primary osteosarcoma. These genetic dysregulations of miR-34 genes are associated with decreased expression of miR-34s in osteosarcoma samples [50]. MiR-143 was found to be downregulated in osteosarcoma cell lines and primary tumor samples, and restoration of miR-143 promoted cell apoptosis and suppressed tumorigenesis by targeting B-cell lymphoma 2 (Bcl-2), an antiapoptotic factor [51]. Interestingly, it was demonstrated that downregulation of miR-143 correlated with pulmonary metastasis of human osteosarcoma cells by cellular invasion, probably through elevated expression of MMP13 [52]. It has been suggested in various studies that miRNAs can be potential therapeutic targets for osteosarcoma. However, further studies are needed to dissect the molecular function of miRNAs in osteosarcoma, their efficacy as treatment, and delivery of miRNAs to tumor cells.

CELL OF ORIGIN AND CANCER STEM CELLS IN OSTEOSARCOMA

Cell of origin

Dysregulation of the proper osteogenic differentiation process, such as through an increase in proliferation or block of differentiation, at early stages can lead to bone cancer [11]. Conditional knockout mouse models of tumor suppressor p53 or Rb1 genes with Cre recombinase controlled by different lineage-specific promoters have suggested that the cells of origin in osteosarcoma are mesenchymal stem cell (MSC) derived bone-forming cells. Homozygous deletion of p53 and Rb1 in the early mesenchymal tissues of embryonic limb buds developed sarcomas at significantly high rates [53]. Deletion of p53 and Rb1 in preosteoblasts resulted in early-onset osteosarcomas with high metastatic rates [54]. In addition, deletion of p53 in committed osteoblasts also resulted in high incidence rates of osteosarcomas [13]. Transgenic mice with a constitutively active Notch1 intracellular domain in committed osteoblasts have an elevated incidence of osteosarcoma formation when p53 is inactivated [14]. Upregulation of Hedgehog signaling by removal of Patched1, a negative regulator, in committed osteoblasts also induces osteosarcoma formation in a p53 heterozygous background [49]. These studies suggest that cells from either undifferentiated MSCs or osteogenic lineage cells have the potential to develop osteosarcoma, and the key factors determining tumor formation may lie in which cell types undergo dysregulation of oncogenes or tumor suppressor genes.

Cancer stem cells

In the past decades, chemotherapy treatments for various cancers led to the discovery of the cancer stem cell (CSC), which is a subtype of cancer cells resistant to drugs and is associated with tumor recurrence and metastasis. It was first identified in acute myeloid leukemia and later found in different types of cancers including osteosarcoma [55]. These cells possess several properties including high expression of drug efflux transporters, abnormal cellular metabolism, extensive DNA repair mechanisms, and deregulation of self-renewal pathways.

The putative CSCs of osteosarcoma are able to reinitiate the full repertoire of the tumorigenesis. They undergo not only osteogenic, but also chondrogenic and adipogenic, lineage differentiation in response to appropriate environmental cues. Several CSC-specific surface markers including CD117, CD133, CD248, CD271, and STOR1 are used to isolate CSCs from osteosarcomas [56]. These CSCs show increased expression of the pluripotent stem cell markers, such as SOX2, OCT3/4, and NANOG, and confer the property of self-renewal and drug resistance [56].

Microenvironmental niches in osteosarcoma provide an environment for CSCs to maintain stemness. Various growth factor-stimulated signaling pathways such as IGF-1, BMP, FGF, and TGF-β are deregulated in the microenvironmental niches [57]. Similar to other types of solid tumors, hypoxic condition in the bone also allow stemness of CSCs in osteosarcomas by the hypoxia inducible factor 1 (HIF1) signaling pathway.

Overall, the important properties of CSCs in osteosarcoma are associated with chemoresistance and metastasis, which directly affect patient survival. Therefore, further research to characterize CSCs of osteosarcoma is critical to develop targeted therapies.

DISCUSSION

Osteosarcoma is an extremely complex and aggressive malignancy that develops through multiple functional interactions of many genetic factors and altered signaling pathways during tumor initiation, progression, and metastasis. In recent years, great progress has been made in the discovery of osteosarcoma driver genes and their pathogenic roles in osteosarcomagenesis by utilizing genomic and transcriptomic analyses and genetic mouse models. One of these discoveries has been the identification of a high degree of genomic instability, a hallmark of osteosarcoma, in patients and mouse models. Structural variations (deletions and translocations) involving TP53 and RB1 were confirmed as unique in osteosarcoma compared with other types of cancers. This phenomenon suggests that p53-independent preexisting genomic instability may account for the complex genetic mechanism in osteosarcoma development. Therefore, uncovering upstream factors (pathways) of p53 or p53-independent mechanisms governing the genome instability will be critical to delineating the complexity of osteosarcomagenesis. Overall, our new understanding of the role of driver genes in osteosarcoma development and metastasis will facilitate tailoring genetic-based targeted therapy for osteosarcoma. Also, a growing body of evidence suggests the importance of noncoding RNAs (lncRNAs and miRNAs) in osteosarcomas. Because of the versatile function of noncoding RNAs, they are attractive as therapeutic tools. Despite the vast amount of genomic data now available, translation into direct patient care is still in the early stages, and continued development of genetic mouse models and large animal models will help to elucidate the pathological roles of driver genes and also provide the platform for preclinical testing for new therapeutic approaches.

ACKNOWLEDGMENTS

This work was supported by the BCM Intellectual and Developmental Disabilities Research Center (HD024064) from the Eunice Kennedy Shriver National Institute of Child Health and Human Development, the BCM Advanced Technology Cores with funding from the NIH (AI036211, CA125123, and RR024574), the Cancer Prevention and Research Initiative of Texas (CPRIT grant RP170488), the Rolanette and Berdon Lawrence Bone Disease Program of Texas, and the BCM Center for Skeletal Medicine and Biology.

REFERENCES

1. Mirabello L, Troisi RJ, Savage SA. Osteosarcoma incidence and survival rates from 1973 to 2004. Cancer. 2009;115(7):1531–43.
2. Rickel K, Fang F, Tao J. Molecular genetics of osteosarcoma. Bone. 2017;102:69–79.
3. Gurney J, Swensen A, Bulterys M. Malignant bone tumors. In: Ries L, Smith M, Gurney J, et al. (eds) *Cancer Incidence and Survival Among Children and Adolescents: United States SEER Program 1975–1995, National Cancer Institute, SEER Program*. NIH Pub. No. 99-4649. Bethesda, MD: National Cancer Institute 1999, pp. 99–110.
4. Jawad M, Cheung M, Clarke J, et al. Osteosarcoma: improvement in survival limited to high-grade patients only. J Cancer Res Clin Oncol. 2011;137(4):597–607.
5. Wang LL. Biology of osteogenic sarcoma. Cancer J. 2005;11(4):294–305.
6. Unni KK, Inwards CY. *Dahlin's Bone Tumors: General Aspects and Data on 10,165 Cases* (6th ed.). Philadelphia: Lippincott Williams & Wilkins, 2009, p. 416.
7. Yarmish G, Klein MJ, Landa J, et al. Imaging characteristics of primary osteosarcoma: nonconventional subtypes 1. Radiographics. 2010;30(6):1653–72.

8. Ta HT, Dass CR, Choong PF, et al. Osteosarcoma treatment: state of the art. Cancer Metastasis Rev. 2009;28(1–2):247–63.
9. Kim HJ, Chalmers PN, Morris CD. Pediatric osteogenic sarcoma. Curr Opin Pediatr. 2010;22(1):61–6.
10. Marina NM, Smeland S, Bielack SS, et al. Comparison of MAPIE versus MAP in patients with a poor response to preoperative chemotherapy for newly diagnosed high-grade osteosarcoma (EURAMOS-1): an open-label, international, randomised controlled trial. Lancet Oncol. 2016;17(10):1396–408.
11. Wagner ER, Luther G, Zhu G et al. Defective osteogenic differentiation in the development of osteosarcoma. Sarcoma. 2011;2011:325238.
12. Subbiah V, Kurzrock R. Phase 1 clinical trials for sarcomas: the cutting edge. Curr Opin Oncol. 2011;23(4):352–60.
13. Walkley CR, Qudsi R, Sankaran VG, et al. Conditional mouse osteosarcoma, dependent on p53 loss and potentiated by loss of Rb, mimics the human disease. Genes Dev. 2008;22(12):1662–76.
14. Tao J, Jiang MM, Jiang L, et al. Notch activation as a driver of osteogenic sarcoma. Cancer Cell. 2014;26(3):390–401.
15. Cores EP, Holland JF, Wang JJ, et al. Doxorubicin in disseminated osteosarcoma. JAMA. 1972;221(10):1132–8.
16. Jaffe N, Paed D, Farber S, et al. Favorable response of metastatic osteogenic sarcoma to pulse high-dose methotrexate with citrovorum rescue and radiation therapy. Cancer. 1973;31(6):1367–73.
17. Rosen G, Caparros B, Huvos AG, et al. Preoperative chemotherapy for osteogenic sarcoma: selection of postoperative adjuvant chemotherapy based on the response of the primary tumor to preoperative chemotherapy. Cancer. 1982;49(6):1221–30.
18. Jawad MU, Haleem AA, Scully SP. Malignant sarcoma of the pelvic bones. Cancer. 2011;117(7):1529–41.
19. Saab R, Rao BN, Rodriguez-Galindo C, et al. Osteosarcoma of the pelvis in children and young adults: the St. Jude Children's Research Hospital experience. Cancer. 2005;103(7):1468–74.
20. Nagarajan R, Kamruzzaman A, Ness KK, et al. Twenty years of follow-up of survivors of childhood osteosarcoma. Cancer. 2011;117(3):625–34.
21. Goldsby R, Burke C, Nagarajan R, et al. Second solid malignancies among children, adolescents, and young adults diagnosed with malignant bone tumors after 1976. Cancer. 2008;113(9):2597–604.
22. Gehan EA, Tefft MC. Will there be resistance to the RECIST (Response Evaluation Criteria in Solid Tumors)? J Natl Cancer Inst. 2000;92(3):179–81.
23. Lagmay JP, Krailo MD, Dang H, et al. Outcome of patients with recurrent osteosarcoma enrolled in seven phase II trials through Children's Cancer Group, Pediatric Oncology Group, and Children's Oncology Group: learning from the past to move forward. J Clin Oncol. 2016;34(25):3031–8.
24. Meyers PA, Schwartz CL, Krailo MD, et al.; Children's Oncology Group. Osteosarcoma: the addition of muramyl tripeptide to chemotherapy improves overall survival—a report from the Children's Oncology Group. J Clin Oncol. 2008;26(4):633–8.
25. Chou AJ, Kleinerman ES, Krailo MD, et al.; Children's Oncology Group. Addition of muramyl tripeptide to chemotherapy for patients with newly diagnosed metastatic osteosarcoma: a report from the Children's Oncology Group. Cancer. 2009;115(22):5339–48.
26. Kager L, Pötschger U, Bielack S. Review of mifamurtide in the treatment of patients with osteosarcoma. Ther Clin Risk Manag. 2010;6:279–86.
27. Ahmed N, Brawley VS, Hegde M, et al. Human epidermal growth factor receptor 2 (HER2)-specific chimeric antigen receptor-modified T cells for the immunotherapy of HER2-positive sarcoma. J Clin Oncol. 2015;33(15):1688–96.
28. Wuyts W, Van Hul W. Molecular basis of multiple exostoses: mutations in the EXT1 and EXT2 genes. Hum Mutat. 2000;15(3):220–7.
29. Rea SL, Majcher V, Searle MS, et al. SQSTM1 mutations—bridging Paget disease of bone and ALS/FTLD. Exp Cell Res. 2014;325(1):27–37.
30. Wunder JS, Gokgoz N, Parkes R, et al. TP53 mutations and outcome in osteosarcoma: a prospective, multicenter study. J Clin Oncol. 2005;23(7):1483–90.
31. Florenes VA, Maelandsmo GM, Forus A, et al. MDM2 gene amplification and transcript levels in human sarcomas: relationship to TP53 gene status. J Natl Cancer Inst. 1994;86(17):1297–302.
32. Henriksen J, Aagesen TH, Maelandsmo GM, et al. Amplification and overexpression of COPS3 in osteosarcomas potentially target TP53 for proteasome-mediated degradation. Oncogene. 2003;22(34):5358–61.
33. Chen X, Bahrami A, Pappo A, et al. Recurrent somatic structural variations contribute to tumorigenesis in pediatric osteosarcoma. Cell Rep. 2014;7(1):104–12.
34. Ribi S, Baumhoer D, Lee K, et al. TP53 intron 1 hotspot rearrangements are specific to sporadic osteosarcoma and can cause Li-Fraumeni syndrome. Oncotarget. 2015;6(10):7727–40.
35. Miller CW, Aslo A, Won A, et al. Alterations of the p53, Rb and MDM2 genes in osteosarcoma. J Cancer Res Clin Oncol. 1996;122(9):559–65.
36. Hicks MJ, Roth J, Kozinetz CA, et al. Clinicopathologic features of osteosarcoma in patients with Rothmund-Thomson syndrome. J Clin Oncol. 2007;25:370–5.
37. Meholin-Ray AR, Kozinetz CA, Schlesinger AE, et al. Radiographic abnormalities and genotype-phenotype correlation with RECQL4 mutation status in Rothmund-Thomson syndrome. AJR Am J Roentgenol. 2008;191(2):W62–6.
38. Lu L, Harutyunyan K, Jin W, et al. RECQL4 regulates p53 function in vivo during skeletogenesis. J Bone Miner Res. 2015;30(6):1077–89.
39. Ng AJ, Walia MK, Smeets MF, et al. The DNA helicase recql4 is required for normal osteoblast expansion and osteosarcoma formation. PLoS Genet. 2015;11(4):e1005160.
40. Perry JA, Kiezun A, Tonzi P, et al. Complementary genomic approaches highlight the PI3K/mTOR pathway

as a common vulnerability in osteosarcoma. Proc Natl Acad Sci U S A. 2014;111(51):E5564–73.
41. Moriarity BS, Otto GM, Rahrmann EP. A Sleeping Beauty forward genetic screen identifies new genes and pathways driving osteosarcoma development and metastasis. Nat Genet. 2015;47(6):615–24.
42. Hoang BH, Kubo T, Healey JH, et al. Dickkopf 3 inhibits invasion and motility of Saos-2 osteosarcoma cells by modulating the Wnt-beta-catenin pathway. Cancer Res. 2004;64(8):2734–9.
43. Kansara M, Tsang M, Kodjabachian L, et al. Wnt inhibitory factor 1 is epigenetically silenced in human osteosarcoma, and targeted disruption accelerates osteosarcomagenesis in mice. J Clin Invest. 2009;119(4):837–51.
44. Nagao-Kitamoto H, Nagata M, Nagano S. GLI2 is a novel therapeutic target for metastasis of osteosarcoma. Int J Cancer. 2015;136(6):1276–84.
45. Li JP, Liu LH, Li J, et al. Microarray expression profile of long noncoding rnas in human osteosarcoma. Biochem Biophys Res Commun. 2013;433(2):200–6.
46. Schmidt LH, Spieker T, Koschmieder S, et al. The long noncoding malat-1 rna indicates a poor prognosis in non-small cell lung cancer and induces migration and tumor growth. J Thorac Oncol. 2011;6(12):1984–92.
47. Dong Y, Liang G, Yuan B, et al. Malat1 promotes the proliferation and metastasis of osteosarcoma cells by activating the pi3k/akt pathway. Tumour Biol. 2015;36(3):1477–86.
48. Cai X, Liu Y, Yang W, et al. Long noncoding rna malat1 as a potential therapeutic target in osteosarcoma. J Orthop Res. 2015;34(6):932–41.
49. Chan LH, Wang W, Yeung W, et al. Hedgehog signaling induces osteosarcoma development through yap1 and h19 overexpression. Oncogene. 2014;33(40):4857–66.
50. He C, Xiong J, Xu X, et al. Functional elucidation of MiR-34 in osteosarcoma cells and primary tumor samples. Biochem Biophys Res Commun. 2009;388(1):35–40.
51. Zhang H, Cai X, Wang Y, et al. MicroRNA-143, down-regulated in osteosarcoma, promotes apoptosis and suppresses tumorigenicity by targeting Bcl-2. Oncol Rep. 2010;24(5):1363–9.
52. Osaki M, Takeshita F, Sugimoto Y, et al. MicroRNA-143 regulates human osteosarcoma metastasis by regulating matrix metalloprotease-13 expression. Mol Ther. 2011;19(6):1123–30.
53. Lin PP, Pandey MK, Jin F, et al. Targeted mutation of p53 and Rb in mesenchymal cells of the limb bud produces sarcomas in mice. Carcinogenesis. 2009;30(10):1789–95.
54. Berman SD, Calo E, Landman AS, et al. Metastatic osteosarcoma induced by inactivation of Rb and p53 in the osteoblast lineage. Proc Natl Acad Sci U S A. 2008;105(33):11851–6.
55. Borah A, Raveendran S, Rochani A, et al. Targeting self-renewal pathways in cancer stem cells: clinical implications for cancer therapy. Oncogenesis. 2015;4:e177.
56. Abarrategi A, Tornin J, Martinez-Cruzado L, et al. Osteosarcoma: cells-of-origin, cancer stem cells, and targeted therapies. Stem Cells Int. 2016;2016:3631764.
57. Alfranca A, Martinez-Cruzado L, Tornin J, et al. Bone microenvironment signals in osteosarcoma development. Cell Mol Life Sci. 2015;72(16):3097–113.

100

Skeletal Complications of Breast and Prostate Cancer Therapies

Catherine Van Poznak[1] and Pamela Taxel[2]

[1]*Department of Internal Medicine—Hematology/Oncology, University of Michigan, Ann Arbor, MI, USA*
[2]*Department of Medicine, University of Connecticut Health Center, Farmington, CT, USA*

INTRODUCTION

In early-stage breast cancer (BCA) and prostate cancer (PCA), adjuvant anticancer therapies may be administered in an attempt to eradicate occult tumor cells and increase the likelihood of cure. In advanced disease, anticancer therapies are used to control the known tumor burden, which may improve quality and quantity of life. Anticancer therapies for BCA and PCA often include chemotherapies and hormonal manipulations. The impact of estrogen and androgen on bone metabolism has been well defined [1], as are the effects of steroids on bone [2]. Anticancer therapies for BCA and PCA often incorporate interventions that negatively impact bone metabolism and bone health. This patient population may have unique risks for loss of bone mass.

BCA and PCA tumors frequently express receptors for estrogen and progesterone, and androgen, respectively. When the tumor is considered hormone-sensitive, these hormone receptors are frequently targeted as part of the anticancer therapy in both early- and late-stage BCA and PCA. The suppression of these sex hormones during cancer therapy often has the unintended consequence of accelerating loss of bone mass as reflected by BMD and increasing the risk of fragility fracture. Chemotherapy and the supportive therapies used with them may also accelerate loss of BMD. Radiotherapy whether used in the curative or palliative setting is associated with a small increase risk for fractures occurring within the treated field. The BCA and PCA population is of significant size; cancer survivors in the United States include approximately 3.5 million people with BCA and 3.3 million people with PCA [3]. This chapter reviews skeletal complications of therapies used to manage BCA and PCA.

NONMETASTASTIC BREAST CANCER

Approximately one in eight women will be diagnosed with BCA, and the median age of diagnosis of BCA in the United States is 61 years [4]. In 2018, over 268,000 new cases of invasive BCA are expected to be diagnosed [3]. In early-stage (nonmetastatic) BCA, the goals of cancer care are curative. Therapies used in this setting may include surgery, chemotherapy, radiotherapy, and/or endocrine therapy. In general, chemotherapy may reduce the risk of BCA recurrence by approximately one-third, and antiestrogen therapy reduces the risk of recurrence by approximately one-half. Approximately 75% of all invasive BCAs express the estrogen receptor (ER) or the progesterone receptor (PR) and are considered hormone receptor positive (HR+).

The Women's Health Initiative Observational Study demonstrated that women with a history of BCA have a greater rate of fractures compared with women without a history of BCA [5], and studies have associated chemotherapy with loss of BMD [6]. Premenopausal women treated with chemotherapy for BCA have a 21% to 100% risk of amenorrhea [7]. Rates of amenorrhea are affected by age and chemotherapy regimen used. In the setting of chemotherapy-induced ovarian dysfunction, the rate of lumbar spine BMD loss has been shown to be approximately 6.7% at 1 year [8]. Similarly rapid rates of lumbar spine BMD loss are seen 2 years after surgical ovarian ablation with a 17.7% loss, and a 5% decline after chemical ovarian suppression with goserelin, a leutinizing hormone-releasing hormone (LHRH) agonist [9]. The impact of these BCA-associated hormonal changes on fracture risk are less well defined.

Primer on the Metabolic Bone Diseases and Disorders of Mineral Metabolism, Ninth Edition. Edited by John P. Bilezikian.
© 2019 American Society for Bone and Mineral Research. Published 2019 by John Wiley & Sons, Inc.
Companion website: www.wiley.com/go/asbmrprimer

The vast majority of women diagnosed with early-stage BCA have HR+ disease, and endocrine therapy with either tamoxifen or an aromatase inhibitor (AI) will be used as part of the cancer care regimen. The American Society of Clinical Oncology (ASCO) guideline on the use of adjuvant endocrine therapy for HR+ BCA recommends the use of endocrine therapy for 5 to 10 years based on the clinical situation. Endocrine therapy may include tamoxifen, an AI, or these drugs used in sequence [10]. The recent publication of the phase III clinical trial known as MA-17-R [11] demonstrated that the use of an AI for 10 years rather than 5 years improved BCA outcomes but without a survival benefit at median follow-up of 6.3 years. However, the risk of fracture was 14% in those treated with an AI for 10 years, and 9% in those treated for 5 years ($p = 0.001$).

Table 100.1. Estimated annual bone loss at the lumbar spine with different cancer therapies compared with normal ranges (in italic).

Therapy/Population	Bone loss
Normal aging adult male bone loss	*0.5%*
Men on androgen-deprivation therapy for PCA	4.6%
Normal aging early menopausal bone loss	*2%*
Normal aging late menopausal bone loss	*1%*
Postmenopausal woman on aromatase inhibitor	2.6%
LHRH agonist therapy plus aromatase inhibitor	7.0%
Ovarian failure related to chemotherapy	7.6%

LHRH = luteinizing hormone-releasing hormone; PCA = prostate cancer.

NONMETASTASTIC PROSTATE CANCER

Approximately 12.9% of men will be diagnosed with PCA during their lifetime, which represents 10.7% of all new cancer cases in the United States and the second most common cancer in men worldwide. It is the second leading cause of cancer deaths in American men, with over 26,000 deaths annually. It is estimated that over 164,000 new cases will be diagnosed in 2018 [3]. Prostate cancer affects men primarily between the ages of 55 and 75 years, with a median age at diagnosis of 66 years. It is more likely to occur in African American men, and in those with a family history of prostate cancer. Over 90% of cases are diagnosed at the local stage, with a 5-year survival of 100.0% [12].

PCA cell growth and proliferation is stimulated by androgens via the androgen receptor, and thus can be mitigated by the inhibition of androgen production. This can be accomplished by antigen deprivation therapy (ADT), which includes both castration as well as antiandrogen therapy. Surgical castration is no longer commonly used since the advent of chemical castration with agents known as LHRH agonists or gonadotrophin-releasing hormone (GnRH) agonists. These agents act at the level of the pituitary to downregulate gonadotropin receptors, and therefore lead to the suppression of both testosterone and estrogen within several weeks [13, 14].

IMPACT OF CANCER ENDOCRINE THERAPIES ON THE LUMBAR SPINE

In general, it can be said that the effects of antiestrogen and antiandrogen therapies have a negative impact on BMD, except for tamoxifen in postmenopausal women. Table 100.1 illustrates the general trends in lumbar spine BMD changes across standard interventions used in postmenopausal women with early-stage BCA and in men receiving antiandrogen treatment for PCA. The studies selected for representation here are of short duration and are not all inclusive. Other than the study known as MA-27-B [15], which compared anastrozole with exemestane, cross comparisons of drug effects on BMD cannot be made. Note that tamoxifen, a selective estrogen receptor modulator, is expected to have a modest improvement on BMD in postmenopausal women. However, tamoxifen has a negative impact on BMD of premenopausal women, as does ovarian suppression by any means [16]. With the use of tamoxifen, one study demonstrated that the total body BMD at 2 years decreased by 1.5% in contrast to the control group loss of 0.3% [9].

Beyond demonstrating changes in BMD, there are data illustrating the increased risk of any fracture with endocrine therapies used to treat BCA or PCA. The meta-analysis by Amir and colleagues demonstrated the use of an AI to be associated with a 7.5% absolute risk of fracture while tamoxifen was associated with a 5.2% risk [17]. Analysis of 1778 men with PCA treated with ADT versus 1157 men with PCA but not treated with ADT demonstrated an incidence of fragility fractures of 9.0% versus 5.9%, respectively (adjusted hazard ratio [HR], 1.65; 95% CI, 1.53 to 1.78, $p = 0.0001$) [18]. Of note, this analysis did not specify the stage or indications for the use of ADT and these data may not be directly applicable to early-stage PCA.

SCREENING FOR OSTEOPOROSIS IN BREAST CANCER AND PROSTATE CANCER

The National Comprehensive Cancer Network (NCCN) Task Force for Bone Health in Cancer Care [16] recommends considering bone health in those with cancer and who are increased risk for bone loss or fracture due to cancer therapy or age. This assessment includes obtaining a history that covers bone-related features, a physical exam, laboratory data (if indicated), use of the Fracture Risk Assessment Tool (FRAX), and a measurement of the BMD. Similarly, the ASCO guidelines on BCA survivorship [19] and PCA survivorship [20] point out that the rate and magnitude of bone loss associated with cancer therapy is higher than that for normal age-related bone loss. Hence, assessment of bone health is a component of survivorship care. These recommendations are consistent with those of other national specialty societies, health care agencies, and individual investigators—with special attention to the cancer therapy as a secondary risk factor for osteoporosis and fragility fracture.

These guidelines, and the work of other investigators, will hopefully increase the frequency of screening for osteoporosis in this patient population. Several large database studies have demonstrated a low use of bone density testing. In women with BCA in one study the rate tested was found to be below 20% [21], and in men receiving ADT for PCA at baseline or within 1 year of therapy with ADT the rate ranged from 10% to 15% [22].

Standard therapies that are US Food and Drug Administration (FDA) approved for managing osteopenia or osteoporosis in populations without a cancer diagnosis have been shown to have efficacy in the setting of cancer therapy for BCA or PCA. In addition, there have been studies using standard or more frequent doses of antiresorptive therapy that have shown similar effects (Fig. 100.1) [8, 24–28]. There are limited fracture data on the effects of antiresorptive treatments in the setting of antihormonal therapy in patients with BCA or PCA. One randomized controlled study, known as AZURE, used frequent dose zoledronic acid monthly for six doses, then every 3 months for eight doses, followed by every 6 months for five doses, for a total treatment period of 5 years as an experimental adjuvant therapy to decrease the risk of BCA recurrence. Patients treated with a wide range of anticancer therapies were randomized to the above regimen with zoledronic acid or to the control group. At 84 months, the zoledronic acid-treated group had a fracture rate of 6.2% while the control group had a fracture rate of 8.3% [23].

Two randomized controlled clinical trials examined the use of denosumab 60 mg every 6 months in BCA and PCA patients and reported that the rate of vertebral fractures decreased [24]. In the BCA study, patients were treated with an AI with or without denosumab. The denosumab group experienced a new vertebral fracture rate of 3.2% while the control group had a rate of 6.1% with an odds ratio of 0.53 (95% CI, 0.33 to 0.85; $p = 0.009$) [24]. A similar magnitude of effect was seen in the second study where patients with PCA were treated with ADT with or without denosumab. The denosumab-treated group had a rate of vertebral fractures of 1.5% at 36 months, while 3.9% of the control group experienced new vertebral fractures with a relative risk of 0.72 (95% CI, 0.19 to 0.78; $p = 0.006$) [25].

In addition to the therapies that have received FDA approval for managing BMD or bone metastases in men with PCA, there are other agents that have been studied in this patient population. The SERM toremifene demonstrated antifracture efficacy in the setting of ADT therapy. In a phase III trial, over 800 men on ADT were randomized to receive daily toremifene versus placebo for 24 months. Vertebral fracture, a primary endpoint, was assessed morphometrically. One percent of the toremifene group versus 4.8% of the placebo group sustained new vertebral fractures. This represented a relative risk reduction of 79.5% ($p < 0.005$) and an absolute risk reduction of 3.8% of new vertebral fractures with toremifene. Bone density increased at all sites in the toremifene group [29]. Thus, denosumab and toremifene are the only agents that have shown fracture efficacy in men with PCA. Toremifene is not FDA approved for the management of bone mass.

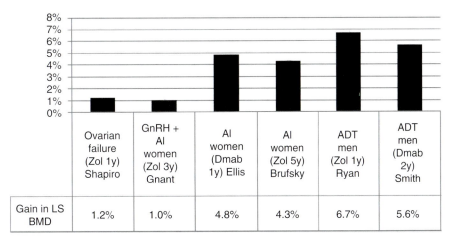

Fig. 100.1. Antiresorptive therapy maintains or improves lumbar spine BMD when used with endocrine therapy [8, 24–28].

METASTATIC BREAST CANCER

When BCA metastasizes, bone is often the first site of spread, and it is the most common distant organ involved, with approximately 70% of patients with metastatic BCA having osseous lesions [30]. Bone metastases are associated with skeletal-related events (fracture, spinal cord compression, hypercalcemia of malignancy, as well as the need for surgery or radiation to the bone). In addition, bone metastases are associated with significant morbidity and mortality. To reduce the risk of skeletal complications from osseous metastases, patients with bone metastases from BCA are routinely treated with bone-modifying agents such as denosumab, pamidronate, or zoledronic acid [31]. Discussion of the management of bone metastases is outside of the scope of this chapter and is covered elsewhere.

Anticancer therapies for metastatic BCA include the interventions used in nonmetastatic BCA (chemotherapy, endocrine therapy, radiotherapy, and surgery) as well as additional medications specifically FDA-labeled for metastatic disease. Therapy for metastatic disease is palliative. Endocrine therapy in the metastatic setting may include tamoxifen, or the AIs, as well as fulvestrant, an estrogen receptor downregulator. Novel endocrine combination therapy regimens include everolimus (a mammalian target of rapamycin or mTOR inhibitor) in combination with exemestane (an AI) and palbociclib with either letrozole or fulvestrant [32]

The effects of systemic therapies on BMD and fracture in the metastatic setting are not well defined. These patients are likely to have bone metastases impacting bone integrity, making it difficult to separate bone events secondary to treatment from events related to the tumor. The use of high-dose osteoclast inhibition used to prevent skeletal-related events far outpaces the dosing used for the prevention and treatment of osteoporosis. In patients with advanced BCA but no frank evidence of osseous metastases, an increased risk of fracture has been identified with a hazard ratio of 22.7 (95% CI, 9.1 to 57.1; $p < 0.0001$) [33]. However, at the time of this study, the methods to identify bone metastases were much less sensitive then those currently used. Radiotherapy whether used in the curative or palliative setting is associated with a small increase in risk for fractures occurring within the treated field.

ROLE OF ADT AND LHRH AGONISTS AND ANTAGONISTS IN PROSTATE CANCER

Approximately 33% to 70% of men with prostate cancer receive GnRH agonist treatment with or without antiandrogens for localized disease before (neoadjuvant) or after (adjuvant) other treatment modalities, or in those with a rising prostate-specific antigen (PSA) after primary therapy with surgery or radiation has failed [34]. Neoadjuvant therapy as well as adjuvant ADT have both shown benefit in terms of disease-specific survival, time to progression, and all-cause mortality [35].

Degarelix is a third-generation GnRH antagonist, and can be used as first-line treatment of androgen-dependent advanced PCA. It acts directly on pituitary receptors and blocks the action of LHRH. With this agent there is no unwanted surge in gonadotropin and testosterone levels as is seen with GnRH agonists [36]. Clinical trials have demonstrated that degarelix treatment results in superior disease control when compared with a GnRH agonist as measured by PSA progression-free survival. Degarelix may likely delay progression to castration-resistant disease, and is generally well tolerated, with limited toxicity [37]. It was approved by the FDA in 2008.

ADDITIONAL ANTICANCER THERAPIES FOR ADVANCED PROSTATE CANCER

Oral estrogen therapy as the mainstay of medical castration was historically used to treat men with PCA prior to the advent of LHRH agonist therapy. However, a large Veterans Administration study determined that oral estrogen, although able to achieve medical castration and efficacious in the treatment of PCA, resulted in an increased risk of thromboembolic phenomena, and therefore this therapy was abandoned [38]. Estrogen delivered via transdermal patch has been studied; however, the outcomes on bone density and fractures are awaiting further follow-up [39].

Abiraterone is an inhibitor of adrenal hormone synthesis and is currently approved for use in the setting of castrate-resistant PCA—defined as disease progression despite a low testosterone level (<50 ng/dL). It inhibits cytochrome P450c17, an enzyme that is critical in extragonadal and testicular androgen synthesis [40]. It has been shown to increase survival in patients with castrate-resistant PCA, and is currently approved for use after treatment with docetaxel, a chemotherapeutic agent [41], or in patients who are chemotherapy-naïve [40]. Due to inhibition of adrenal hormone biosynthesis, prednisone is given in combination with it at 5 mg twice per day.

With regard to the metabolic effects on bone turnover, a small trial of 49 men receiving abiraterone demonstrated a significant decrease of serum C teleopeptide values at 3, 6, and 9 months compared with baseline values, and an increase of alkaline phosphatase at 3 months in abiraterone-treated patients [42]. However, 20 of 49 subjects had received zoledronic acid during the observation period, and the outcomes (effect) might be difficult to determine. To date, studies on BMD and fractures have not been reported.

Enzalutamide is a targeted androgen receptor inhibitor approved by the FDA for the treatment of castration-resistant PCA. This agent can be used before or after chemotherapy. A large placebo-controlled double blind trial in men with metastatic castrate-resistant PCA and no prior chemotherapy demonstrated that radiographic progression was delayed by enzalutamide, and overall survival was improved with a 30% reduction in the risk of death compared with placebo, and a delay in initiation of chemotherapy by a median of 17 months [43]. In a murine study both enzalutamide and orchidectomy decreased bone mass in the axial skeleton as shown by a reduced areal BMD at the lumbar spine [44]. No human studies are available currently.

Radium-223 is an α-particle-emitting radiopharmaceutical medication that has been recently approved in the setting of castrate-resistant PCA with symptomatic bone metastases without visceral involvement. It is a calcimimetic that directly targets bone, and delivers cytotoxic radiation to the sites of bone metastases. In a recent phase III trial, radium 223 reduced the risk of mortality by 30% and prolonged time to first symptomatic skeletal event by 5.8 months [45]. Further, it demonstrated an overall good safety profile and was not associated with significant myelosuppression. This is the first treatment for castrate-resistant PCA that has shown an overall survival benefit [46]. The impact of radium on bone integrity outside of skeletal-related events is not yet defined.

CONCLUSION

By the year 2026 it is anticipated that the number of BCA and PCA survivors will number approximately 4.6 million and 4.5 million. respectively [47]. An approximate 90,000 women have bone metastases from BCA and approximately 63,000 men have bone metastases from PCA [48]. Hence, the bone health of this population is a public health concern. Standard interventions can be made to improve bone integrity in early-stage as well as advanced BCA and PCA. There is a need for additional therapies and practices to reduce the risk of skeletal complications. There is an increasing awareness of the importance of bone health in cancer survivorship and additional data are needed to define the optimal care of this vulnerable population.

REFERENCES

1. Riggs BL, Khosla S, Melton LJ 3rd. Sex steroids and the construction and conservation of the adult skeleton. Endocr Rev. 2002;23(3):279–302.
2. Weinstein, RS. Glucocorticoid-induced bone disease. N Engl J Med. 2011;365:62–70.
3. Siegel RL, Miller KD, Jemal A. Cancer statistics, 2018. CA Cancer J Clin. 2018;68(1):7–30.
4. National Cancer Institute. Breast cancer risk in American Women. https://www.cancer.gov/types/breast/risk-fact-sheet (accessed May 2018).
5. Chen Z, Maricic M, Bassford TL, et al Fracture risk among breast cancer survivors: results from the Women's Health Initiative Observational Study. Arch Intern Med. 2005;165(5):552–8.
6. Greep NC, Giuliano AE, Hansen NM, et al. The effects of adjuvant chemotherapy on bone density in postmenopausal women with early breast cancer. Am J Med. 2003;114(8):653–9.
7. Minton SE, Munster PN. Chemotherapy-induced amenorrhea and fertility in women undergoing adjuvant treatment for breast cancer. Cancer Control. 2002;9(6):466–72.
8. Shapiro CL, Halabi S, Hars V, et al. Zoledronic acid preserves bone mineral density in premenopausal women who develop ovarian failure due to adjuvant chemotherapy: final results from CALGB trial 79809. Eur J Cancer. 2011;47(5):683–9.
9. Sverrisdóttir A, Fornander T, Jacobsson H, et al. Bone mineral density among premenopausal women with early breast cancer in a randomized trial of adjuvant endocrine therapy. J Clin Oncol. 2004;22(18):3694–9.
10. Burstein HJ, Temin S, Anderson H, et al. Adjuvant endocrine therapy for women with hormone receptor-positive breast cancer: American Society of Clinical Oncology Clinical Practice Guideline focused update. J Clin Oncol. 2014;32(21):2255–69.
11. Goss PE, Ingle JN, Pritchard KI, et al. Extending aromatase-inhibitor adjuvant therapy to 10 years. N Engl J Med. 2016;375(3):209–19.
12. National Cancer Institute. Stat Facts: Prostate Cancer. http://seer.cancer.gov/statfacts/html/prost.html (accessed May 2018).
13. Limonta P, Montagnani Marelli M, Moretti RM. LHRH analogues as anticancer agents: pituitary and extrapituitary sites of action. Expert Opin Investig Drugs. 2001;10(4):709–20.
14. Wilson HC, Shah SI, Abel PD, et al. Contemporary hormone therapy with LHRH agonists for prostate cancer: avoiding osteoporosis and fracture. Cent European J Urol. 2015;68(2):165–8.
15. Goss PE, Hershman DL, Cheung AM, et al. Effects of adjuvant exemestane versus anastrozole on bone mineral density for women with early breast cancer (MA.27B): a companion analysis of a randomised controlled trial. Lancet Oncol. 2014;15(4):474–82.
16. Gralow JR, Biermann JS, Farooki A, et al. NCCN Task Force Report: bone health in cancer care. J Natl Compr Canc Netw. 2013;11(suppl 3):S1–50.
17. Amir E, Seruga B, Niraula S, et al. Toxicity of adjuvant endocrine therapy in postmenopausal breast cancer patients: a systematic review and meta-analysis. J Natl Cancer Inst. 2011;103(17):1299–309.
18. Alibhai SM, Duong-Hua M, Cheung AM, et al. Fracture types and risk factors in men with prostate cancer on androgen deprivation therapy: a matched cohort study of 19,079 men. J Urol. 2010;184(3):918–23.
19. Runowicz CD, Leach CR, Henry NL, et al. American Cancer Society/American Society of Clinical Oncology breast cancer survivorship care guideline. J Clin Oncol. 2016;34(6):611–35.

20. Resnick MJ, Lacchetti C, Bergman J, et al. Prostate cancer survivorship care guideline: American Society of Clinical Oncology Clinical Practice Guideline endorsement. J Clin Oncol. 2015;33(9):1078–85.
21. Snyder CF, Frick KD, Kantsiper ME, et al. Prevention, screening, and surveillance care for breast cancer survivors compared with controls: changes from 1998 to 2002. J Clin Oncol. 2009;27(7):1054–61
22. Tsang DS, Alibhai SM. Bone health care for patients with prostate cancer receiving androgen deprivation therapy. Hosp Pract. 2014;42(2):89–102.
23. Coleman R, Cameron D, Dodwell D, et al.; AZURE Investigators. Adjuvant zoledronic acid in patients with early breast cancer: final efficacy analysis of the AZURE (BIG 01/04) randomised open-label phase 3 trial. Lancet Oncol. 2014;15(9):997–1006.
24. Gnant MF, Mlineritsch B, Luschin-Ebengreuth G, et al.; Austrian Breast and Colorectal Cancer Study Group. Zoledronic acid prevents cancer treatment-induced bone loss in premenopausal women receiving adjuvant endocrine therapy for hormone-responsive breast cancer: a report from the Austrian Breast and Colorectal Cancer Study Group. J Clin Oncol. 2007;25(7):820–8.
25. Smith MR, Egerdie B, Hernández Toriz N, et al.; Denosumab HALT Prostate Cancer Study Group. Denosumab in men receiving androgen-deprivation therapy for prostate cancer. N Engl J Med. 2009;361(8):745–55.
26. Ellis GK, Bone HG, Chlebowski R, et al. Randomized trial of denosumab in patients receiving adjuvant aromatase inhibitors for nonmetastatic breast cancer. J Clin Oncol. 2008;26(30):4875–82.
27. Brufsky AM, Harker WG, Beck JT, et al. Final 5-year results of Z-FAST trial: adjuvant zoledronic acid maintains bone mass in postmenopausal breast cancer patients receiving letrozole. Cancer. 2012;118(5):1192–201.
28. Ryan CW, Huo D, Demers LM, et al. Zoledronic acid initiated during the first year of androgen deprivation therapy increases bone mineral density in patients with prostate cancer. J Urol. 2006;176(3):972–8; discussion 978.
29. Smith MR, Malkowicz SB, Brawer MK, et al. Toremifene decreases vertebral fractures in men younger than 80 years receiving androgen deprivation therapy for prostate cancer. J Urol. 2011; 186(6):2239–44.
30. Coleman RE, Smith P, Rubens RD. Clinical course and prognostic factors following recurrence from breast cancer. Br J Cancer. 1998;77:336–40.
31. Van Poznak CH, Temin S, Yee GC, et al.; American Society of Clinical Oncology. American Society of Clinical Oncology executive summary of the clinical practice guideline update on the role of bone-modifying agents in metastatic breast cancer. J Clin Oncol. 2011;29(9):1221–7.
32. Rugo HS, Rumble RB, Macrae E, et al. Endocrine therapy for hormone receptor-positive metastatic breast cancer: American Society of Clinical Oncology guideline. J Clin Oncol. 2016;34(25):3069–103.
33. Kanis JA, McCloskey EV, Powles T, et al. A high incidence of vertebral fracture in women with breast cancer. Br J Cancer. 1999;79(7–8):1179–81.
34. Msaouel P, Diamanti E, Tzanela M, et al. Luteinising hormone-releasing hormone antagonists in prostate cancer therapy. Expert Opin Emerg Drugs. 2007;12(2):285–99.
35. Payne H, Mason M. Androgen deprivation therapy as adjuvant/neoadjuvant to radiotherapy for high-risk localised and locally advanced prostate cancer: recent developments. Br J Cancer. 2011;105(11):1628–34.
36. Rick FG, Block NL, Schally AV. An update on the use of degarelix in the treatment of advanced hormone-dependent prostate cancer. Onco Targets Ther. 2013;6: 391–402.
37. Klotz L, Boccon-Gibod L, Shore ND, et al. The efficacy and safety of degarelix: a 12-month, comparative, randomized, open-label, parallel-group phase III study in patients with prostate cancer. BJU Int. 2008;102(11):1531–8.
38. Byar DP, Corle DK. Hormone therapy for prostate cancer: results of the Veterans Administration Cooperative Urological Research Group studies. NCI Monogr. 1988;7:165–70.
39. Langley RE, Cafferty FH, Alhasso AA, et al. Cardiovascular outcomes in patients with locally advanced and metastatic prostate cancer treated with luteinising-hormone-releasing-hormone agonists or transdermal oestrogen: the randomised, phase 2 MRC PATCH trial (PR09). Lancet Oncol. 2013;14(4):306–16.
40. Ryan CJ, Smith MR, de Bono JS, et al.; COU-AA-302 Investigators. Abiraterone in metastatic prostate cancer without previous chemotherapy. N Engl J Med. 2013;368(2):138–48.
41. de Bono JS, Logothetis CJ, Molina A, et al.; COU-AA-301 Investigators. Abiraterone and increased survival in metastatic prostate cancer. N Engl J Med. 2011;364(21): 1995–2005.
42. Iuliani M, Pantano F, Buttigliero C, et al. Biological and clinical effects of abiraterone on anti-resorptive and anabolic activity in bone microenvironment. Oncotarget. 2015;6(14):12520–8.
43. Beer TM, Armstrong AJ, Rathkopf DE, et al.; PREVAIL Investigators. Enzalutamide in metastatic prostate cancer before chemotherapy. New Engl J Med. 2014;371(5): 424–33.
44. Wu J, Movérare-Skrtic S, Börjesson AE, et al. Enzalutamide reduces the bone mass in the axial but not the appendicular skeleton in male mice. Endocrinology. 2016;157(2):969–77.
45. Shore ND. Radium-223 dichloride for metastatic castration-resistant prostate cancer: the urologist's perspective. Urology. 2015;85(4):717–24.
46. Shirley M, McCormack PL. Radium-223 dichloride: a review of its use in patients with castration-resistant prostate cancer with symptomatic bone metastases. Drugs. 2014;74(5):579–86.
47. Miller KD, Siegel RL, Lin CC, et al. Cancer treatment and survivorship statistics, 2016. CA Cancer J Clin. 2016;66(4):271–89.
48. Li S, Peng Y, Weinhandl ED, et al. Estimated number of prevalent cases of metastatic bone disease in the US adult population. Clin Epidemiol. 2012;4:87–93.

101

Bone Cancer and Pain

Denis Clohisy and Lauren M. MacCormick

Orthopaedic Surgery, University of Minnesota, Minneapolis, MN, USA

EPIDEMIOLOGY OF BONE CANCER PAIN

Bone cancer pain is one of the most difficult pain entities to treat and significantly affects the quality of life for many patients with bone cancer. Pain is the most common presenting symptom in patients with skeletal metastases and is directly proportional to its impact on the cancer patient's quality of life [1, 2]. Two main types of cancer pain exist: ongoing pain and breakthrough pain. Ongoing pain is typically described as a dull and aching pain that is constant in nature and progresses in accordance with the disease process. Breakthrough pain is most commonly associated with bone metastases and is characterized by sharp pain, intermittent in nature, and exacerbated by movement. Breakthrough pain is difficult to treat and is often recalcitrant in nature, but can be found in as many as 80% of patients with advanced disease [3, 4]. Significant insight into understanding bone cancer pain and the development of new therapeutic strategies for bone cancer pain are due to the development of novel animal models and recent clinical trials.

MECHANISMS OF BONE CANCER PAIN

More is now known about the causes of bone cancer pain because of the development of novel animal models; rodent and canine models of bone cancer pain have been described in the literature. Each model differs in the route of inoculation of tumor cells, type of tumor, immunocompetency of the host, and species [5]. Despite these differences, a wealth of information has been generated regarding the pathophysiological mechanisms that drive bone cancer pain. Ultimately, such pain is part of a multifactorial process initiated by a complex interaction between the host cells of the affected bone and the tumor cells that affect the peripheral and central nervous system.

Bone cancer pain has nociceptive and neuropathic components. Nociceptive pain generally occurs during tissue damage as a result of the release of neurotransmitters, cytokines, and factors from damaged cells, adjacent blood vessels, and nerve terminals. Pain is transduced at the level of the primary afferent nerve fiber that innervates peripheral tissues. Bone is densely innervated by both sensory and sympathetic nerve fibers within the bone marrow, mineralized bone, and periosteum (Figs 101.1 and 101.2) [6, 7]. Neuropathic pain is secondary to tumor growth that directly destroys the distal ends of nerve fibers that normally innervate bone as well as the induction of pathological sprouting of sympathetic and sensory nerve fibers [8–11].

The majority of metastatic skeletal malignancies are destructive in nature and produce regions of osteolysis (bone destruction). This occurs through activation, recruitment, and proliferation of osteoclasts and is characterized by an increased number and size of osteoclasts found in tumor-bearing sites [12–15]. The activation and proliferation of osteoclasts is mediated by the interaction between RANK expressed on osteoclasts and RANKL expressed on osteoblasts. An increased expression of both RANK and RANKL has been found in tumor-bearing sites. Selective inhibition of osteoclasts using either bisphosphonates or the soluble decoy receptor for RANKL, osteoprotegerin (OPG), results in inhibition of cancer-induced osteolysis, cancer pain behaviors, and neurochemical markers of peripheral and central sensitization [16–20].

Tumor-derived cytokines, growth factors, and peptides have been shown to activate primary afferent nerve fibers that innervate bone. Prostaglandins, interleukins, protons, bradykinin, chemokines, tumor necrosis factor α, nerve growth factor (NGF), and endothelins are all examples of chemical mediators released from tumor cells or the host immune system that sensitize nerve terminals,

---

Primer on the Metabolic Bone Diseases and Disorders of Mineral Metabolism, Ninth Edition. Edited by John P. Bilezikian.
© 2019 American Society for Bone and Mineral Research. Published 2019 by John Wiley & Sons, Inc.
Companion website: www.wiley.com/go/asbmrprimer

782 Bone Cancer and Pain

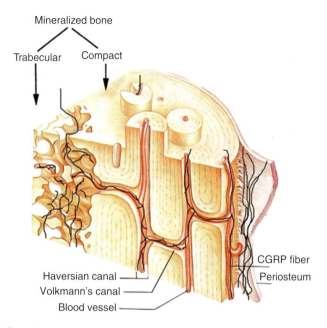

Fig. 101.1. Schematic diagram demonstrating the innervation within the periosteum, mineralized bone, and bone marrow. All three tissues may be sensitized during the various stages of bone cancer pain. CGRP = calcitonin gene-related peptide. Source: [6]. Reproduced with permission of Elsevier.

resulting in cancer pain [5, 21–23]. Each mediator has a specialized receptor that converts the chemical signal to an electrical signal (Fig. 101.3). In bone cancer pain, chemical mediators are released that bind to respective receptors causing pain transduction. Peripheral sensitization occurs when constant nerve stimulation leads to decreased excitation thresholds, upregulation of receptors in nerve terminals, or recruitment of previously silent pain receptors [9, 10, 24, 25].

Central sensitization

Central sensitization refers to the heightened reactivity of central nervous system neurons in the face of sustained peripheral neural input. Whereas this may occur within the thalamus and cortex, research has been primarily focused on the dorsal horn of the spinal cord. Electrophysiological and anatomical studies have shown a change in the activity and responsiveness of dorsal horn neurons in response to persistent painful stimulation. Allodynia is a condition where normally non-noxious stimulation is painful and is a condition that can result from central sensitization [26].

Persistent stimulation of unmyelinated C fibers results in the increased activity and responsiveness of spinal

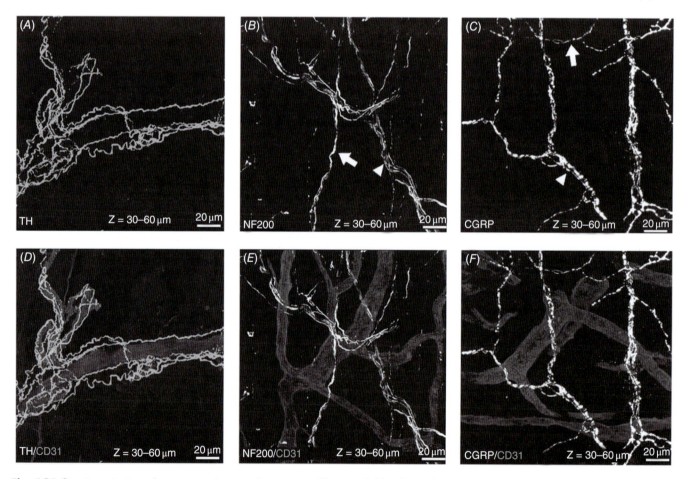

Fig. 101.2. Association of sensory and sympathetic nerve fibers with blood vessels in the bone periosteum shown by high-power computed tomography scans of bone in cross-section overlaid by confocal images. (A) Sympathetic nerve fibers wrapping around CD31+ blood vessels of the periosteum (D). (B) Neuronal fibers (NF200+) neurofilament-positive and calcitonin gene-related peptide-positive (CGRP+) sensory nerve fibers (C) do not associate with CD31+ blood vessels, as seen in (E) and (F), respectively. Source: [7]. Reproduced with permission of Elsevier.

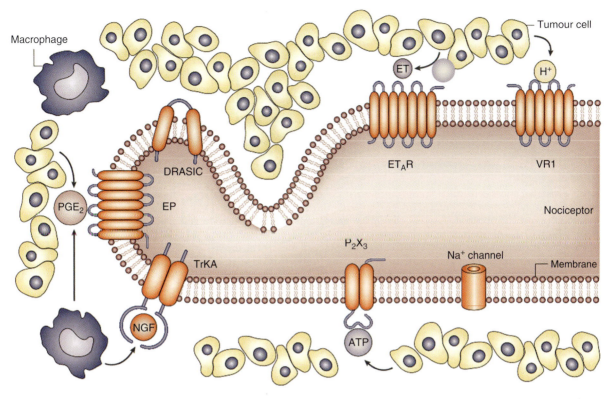

Fig. 101.3. Schematic diagram of a peripheral pain fiber expressing receptors and ion channels. Interaction between neurotransmitters and chemical mediators and their cognate receptor results in pain transduction and signaling. ATP = adenosine triphosphate; DRASIC= dorsal root acid sensing ion channel; EP = prostaglandin E receptor; ET = endothelin; ET_AR = endothelin A receptor; H+ = protons; NA+ = sodium; NGF = nerve growth factor; PGE_2 = prostaglandin E_2; P_2X_3 = purinergic ion-gated receptor; TrKA = high affinity nerve growth factor tyrosine kinase receptor A; VR1 = vanilloid receptor-1. Source: [51]. Reproduced with permission.

neurons that receive input from the stimulated unmyelinated C fiber. This increased responsiveness is of short duration and is known as wind up [27]. Sensitization can also occur when persistent stimulation results in phenotypical changes in neurons that do not receive, but are adjacent to, neurons that receive the persistent painful stimulation. Typically, these neurons receive input from Aβ fibers that normally do not transmit painful stimuli. However, once sensitized, these neurons are capable of transmitting both nonpainful and painful information. Central sensitization is mediated in part by glutamate, substance P, prostaglandins, and growth factors. Associated receptors or channels are known as N-methyl-D-aspartate (NMDA), neurokinin-1 (NK-1), prostaglandin E receptor (EP), and tyrosine kinase B (trkB), respectively. Upregulation of transient receptor potential (TRPV1) and sodium channels have also been reported in central sensitization [28, 29].

Reorganization of the peripheral and central nervous systems in response to cancer pain

Several studies have demonstrated that nociceptor peripheral sensitization occurs in animals with cancer pain in experimental models [5, 9, 24]. In normal mice, the neurotransmitter substance P is synthesized by nociceptors and released in the spinal cord when noxious mechanical stress is applied to the femur. Substance P, in turn, binds to and activates the NK-1 receptor that is expressed by a subset of spinal cord neurons, eliciting a response. In mice with bone cancer, the reorganization of nociceptive nerve fibers causes mechanical allodynia where nonpainful levels of mechanical stress induce the release of substance P, making the stimuli noxious [24].

Phenotypic alterations with extensive neurochemical reorganization in the innervation of tumor-bearing bones, occur during the sensitization of peripheral nerves. Specific neural changes that may mediate pain include astrocyte hypertrophy and decreased expression of glutamate reuptake transporters. Increased extracellular glutamate levels result in central nervous system excitotonicity, which is the pathological process by which neurons are damaged by overactivation of their receptor, and contributes to the development of neuropathic pain [30, 31].

Several studies examining mouse models of breast and prostate cancer have demonstrated significant sprouting of new sensory and sympathetic nerve fibers that display a unique morphology and high density of fibers [8, 9] (Fig. 101.4). To further evaluate the driving force for the new nociceptive fibers, reverse transcriptase polymerase chain reaction analysis for NGF showed the surrounding tumor-associated inflammatory, immune,

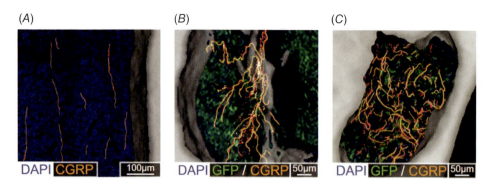

Fig. 101.4. Prostate cancer cells cause sprouting of sensory nerve fibers in bone shown by high-power computer tomography scans of bone in cross-section overlaid by confocal images. 4,6-Diamidino-2-phenylindole (DAPI) stained nuclei appear blue, green fluorescent protein (GFP) expressing prostate cancer cells appear green, and calcitonin gene-related peptide-positive (CGRP+) sensory nerve fibers appear yellow/red. (A) Sham femur showing control level of nerve sprouting seen in characteristic liner morphology. (B) Prostate tumor-bearing femur from mouse killed at an early stage of metastatic disease showing tumor colonies and markedly highly branched sensory nerve sprouting. (C) Prostate tumor-bearing femur from mouse killed at an advanced stage of metastatic disease with a high density of sensory nerve fibers. Source: [9].

and stromal cells are the major source of NGFs in painful tumors [9]. Local anti-NGFs can block ectopic sprouting and pathological reorganization of these nociceptive fibers, suggesting that prophylactic treatment may be capable of preventing a significant aspect of bone cancer pain [10].

THERAPEUTIC STRATEGIES

Pain research has significantly improved our understanding of acute and chronic pain mechanisms. By highlighting key molecular mechanisms involved in pain transmission, new drugs are currently being studied as potential novel and selective therapies. Currently available medications such as the opioids are fraught with side-effect profiles that may limit their clinical efficacy and patient quality of life. Research is now focused on specific receptor or channel targets within the nervous system that limit systemic complications.

Growth factors and cytokines

NGF modulates inflammatory and neuropathic pain states. In chronic pain, NGF levels are elevated in peripheral tissues. Neutralizing antibodies against NGF are effective in reducing and in some cases preventing chronic pain [32]. In vitro studies have shown that growth and differentiation of NGF-dependent sensory nerve cell lines can be inhibited by naturally neutralizing antibodies. These same antibodies have been shown to inhibit the in vitro migration and metastasis of prostate cancer cells [25]. In animal models, anti-NGF antibodies reduce continuous and breakthrough pain by blocking the nociceptive stimuli associated with the sensitization in the peripheral or central nervous system (Fig. 101.5) [32]. In breast, prostate, and sarcoma models of bone cancer pain, anti-NGF therapy has been found to be more effective in reducing pain-related behaviors when compared with administration of intravenous morphine sulfate [32, 33]. Recently, a study evaluating the efficacy of an anti-NGF antibody, tanezumab, in cancer patients with bone metastases found a slight decrease in pain over 8 weeks when compared with placebo. This effect did not reach statistical significance, but improvement in pain was increased in patients with higher baseline pain and lower baseline opioid use [34]. Additionally, a phase II trial to test the efficacy and safety of tanezumab as an adjunct to opioids in patients with pain related to bone metastases over a 16-week period has been conducted, but no results have been reported from this trial (Identifier NCT00545129).

In addition to NGF, glial-derived growth factor (GDGF) and brain-derived growth factor (BDGF) play roles in cancer-related bone pain. BDGF has been implicated in the modulation of central sensitization because its expression is increased in nociceptive neurons in models of peripheral neuropathy. BDGF sensitizes C-fiber activity resulting in hyperalgesia and allodynia. Inhibition of BDGF and its cognate receptor, trkB, results in decreased C-fiber firing and a reduction in pain behaviors [35, 36]. GDGF is important in the survival of sensory neurons and supporting cells. Neuropathic pain behaviors commonly observed in animal models of chronic pain are prevented or reversed after GDGF administration. The analgesic effects of GDGF show strong temporal regulation, and the timing of administration determined whether treatment was protective or therapeutic in nature [36, 37].

Endothelins are a family of vasoactive peptides that are expressed by several tumors, and levels appear to correlate with pain severity. Direct application of endothelin to peripheral nerves induces activation of primary afferent fibers and pain-specific behaviors [38]. Selective blockade of endothelin receptors blocks bone cancer pain-related behaviors and spinal changes indicative of peripheral and central sensitization [39, 40]. Recently, a phase III clinical

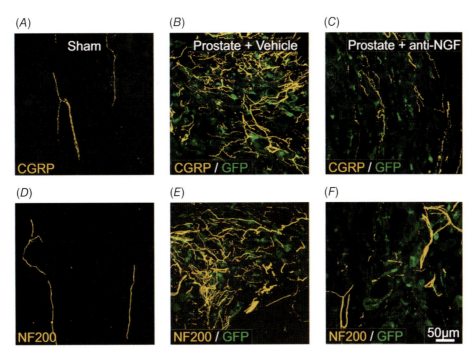

Fig. 101.5. The mesh-like network of nociceptive nerve sprouting in prostate cancer is inhibited by anti-nerve growth factor (anti-NGF) therapy shown by high-power computed tomography scans of bone in cross-section overlaid by confocal images. Calcitonin gene-related peptide-positive (CGRP+) and neuronal fibers (NF200+) nerve fibers appear orange and yellow, and green fluorescent protein (GFP) expressing prostate cancer cells appear green. (A,D) Sham-operated mice show regular innervation of bone by two types of nerve fibers: (A) CGRP+ and (D) NF200+. (B,E) GFP-transfected prostate cancer cells growing in bone after 26 days, with the CGRP+ and NF200+ nerve fibers. (C,F) Prevention of CGRP+ and NF200+ nerve fiber sprouting due to anti-NGF antibody therapy. Source: [9].

trial involving patients with hormone-resistant prostate cancer evaluated the effects of an endothelin A antagonist in addition to standard chemotherapy compared with standard chemotherapy alone, and found no improvement in health-related quality of life or decrease in progression in pain (Identifier NCT00617669).

Ion channels

The TRPV1 family of channels is located on unmyelinated C fibers and spinal nociceptive neurons that mediate pain transmission. TRPV1 channels can be activated by noxious heat, capsaicin, and acid. Mice that lack the channel are unable to develop chronic pain states, and when antagonists to TRPV1 are administered orally or into the intrathecal space, chronic pain is markedly reduced [22, 41, 42]. Interestingly, intrathecal administration of resinifera toxin, a potent capsaicin analog, in a canine bone cancer model resulted in reductions in pain behaviors and selective destruction of small sensory neurons [43]. At sites of osteolytic bone cancer, the acidic microenvironment at the osteoclast–bone interface mediates pain. Acid signals received by TRPV1 channels residing within bone are thought to stimulate intracellular nociceptive signaling. TRPV1 antagonists serve as a method of inhibiting pain transmission by forming a molecular blockade of the activated intracellular transcription factors in these signaling pathways [23, 41]. A human trial evaluating TRPV1 antagonists demonstrated a potent antihyperalgesic effect of this agent [44].

Osteoclasts

Osteoclasts play an essential role in cancer-induced bone loss and contribute to the etiology of bone cancer pain. Treatment with bisphosphonates and denosumab, which both reduce bone resorption, are the standard of care for patients with bone metastases. Both agents have demonstrated patient-reported improvement in quality of life in clinical trials in patients with lung, breast, and prostate cancer [19, 20, 45–47]. Denosumab was evaluated against zoledronic acid (bisphosphonate) in a randomized clinical trial evaluating the prevention of skeletal-related events in patients with bone metastases from breast cancer. While both therapies were well tolerated and delayed or prevented skeletal-related events, denosumab trended toward superior reductions in patient-reported pain and improved quality of life [48]. Both agents are recommended in patients with solid tumors and bone metastases with and without associated pain [49]. Ongoing evaluation is needed to determine dose and dosing intervals with consideration for patient-specific factors [50].

REFERENCES

1. Coleman RE. Clinical features of metastatic bone disease and risk of skeletal morbidity. Clin Cancer Res. 2006;12(20 Pt 2):s6243–9.
2. Mantyh PW. Cancer pain and its impact on diagnosis, survival and quality of life. Nat Rev Neurosci. 2006;7(10):797–809.
3. Swarm R, Abernethy AP, Anghelescu DL, et al. Adult cancer pain. J Natl Compr Cancer Netw. 2010;8(9):1046–86.
4. Mercadante S, Fulfaro F. Management of painful bone metastases. Curr Opin Oncol. 2007;19(4):308–14.
5. Jaggi AS, Jain V, Singh N. Animal models of neuropathic pain. Fundam Clin Pharmacol. 2011;25(1):1–28.
6. Mach DB, Rogers SD, Sabino MC, et al. Origins of skeletal pain: sensory and sympathetic innervation of the mouse femur. Neuroscience. 2002;113(1):155–66.
7. Martin CD, Jimenez-Andrade JM, Ghilardi JR, et al. Organization of a unique net-like meshwork of CGRP+ sensory fibers in the mouse periosteum: implications for the generation and maintenance of bone fracture pain. Neurosci Lett. 2007;427(3):148–52.
8. Bloom AP, Jimenez-Andrade JM, Taylor RN, et al. Breast cancer-induced bone remodeling, skeletal pain, and sprouting of sensory nerve fibers. J Pain. 2011;12(6):698–711.
9. Jimenez-Andrade JM, Bloom AP, Stake JI, et al. Pathological sprouting of adult nociceptors in chronic prostate cancer-induced bone pain. J Neurosci. 2010;30(44):14649–56.
10. Mantyh WG, Jimenez-Andrade JM, Stake JI, et al. Blockade of nerve sprouting and neuroma formation markedly attenuates the development of late stage cancer pain. Neuroscience. 2010;171(2):588–98.
11. Mantyh WG. Bone cancer pain: from mechanism to therapy. Curr Opin Support Palliat Care. 2014;8(2):83–90.
12. Taube T, Elomaa I, Blomqvist C, et al. Histomorphometric evidence for osteoclast-mediated bone resorption in metastatic breast cancer. Bone. 1994;15(2):161–6.
13. Clohisy DR, Ramnaraine ML. Osteoclasts are required for bone tumors to grow and destroy bone. J Orthop Res. 1998;16(6):660–6.
14. Sterling JA, Edwards JR, Martin TJ, et al. Advances in the biology of bone metastasis: how the skeleton affects tumor behavior. Bone. 2011;48(1):6–15.
15. Zhang Y, Ma B, Fan Q. Mechanisms of breast cancer bone metastasis. Cancer Lett. 2010;292(1):1–7.
16. Clohisy DR, Ramnaraine ML, Scully S, et al. Osteoprotegerin inhibits tumor-induced osteoclastogenesis and bone tumor growth in osteopetrotic mice. J Orthop Res. 2000;18(6):967–76.
17. Roudier MP, Bain SD, Dougall WC. Effects of the RANKL inhibitor, osteoprotegerin, on the pain and histopathology of bone cancer in rats. Clin Exp Metastasis. 2006.23(3–4):167–75.
18. Lamoureux F, Moriceau G, Picarda G, et al. Regulation of osteoprotegerin pro- or anti-tumoral activity by bone tumor microenvironment. Biochim Biophys Acta. 2010;1805(1):17–24.
19. Diel IJ. Effectiveness of bisphosphonates on bone pain and quality of life in breast cancer patients with metastatic bone disease. A review. Support Care Cancer. 2007;15(11):1243–9.
20. Rodrigues P, Hering F, Campagnari JC. Use of bisphosphonates can dramatically improve pain in advanced hormone-refractory prostate cancer patients. Prostate Cancer Prostatic Dis. 2004;7(4):350–4.
21. White FA, Jung H, Miller RJ. Chemokines and the pathophysiology of neuropathic pain. Proc Natl Acad Sci U S A. 2007;104(51):20151–8.
22. White JP, Urban L, Nagy I. TRPV1 function in health and disease. Curr Pharm Biotechnol. 2011;12(1):130–44.
23. Yoneda T, Hata K, Nakanishi M, et al. Involvement of acidic microenvironment in the pathophysiology of cancer-associated bone pain. Bone. 2011;48(1):100–5.
24. Schmidt BL, Hamamoto DT, Simone DA, et al. Mechanism of cancer pain. Mol Interv. 2010;10(3):164–78.
25. Warrington RJ, Lewis KE. Natural antibodies against nerve growth factor inhibit in vitro prostate cancer cell metastasis. Cancer Immunol Immunother. 2011;60(2):187–95.
26. Latremoliere A, Woolf CJ. Central sensitization: a generator of pain hypersensitivity by central neural plasticity. J Pain. 2009;10(9):895–926.
27. Woolf CJ. 2011. Central sensitization: implications for the diagnosis and treatment of pain. Pain. 2011;152(suppl 3):S2–15.
28. Xiaoping G, Xiaofang Z, Yaguo Z, et al. Involvement of the spinal NMDA receptor/PKCgamma signaling pathway in the development of bone cancer pain. Brain Res. 2010;1335:83–90.
29. Yanagisawa Y, Furue H, Kawamata T, et al. Bone cancer induces a unique central sensitization through synaptic changes in a wide are of the spinal cord. Mol Pain. 2010;6:38.
30. Schwei MJ, Honore P, Rogers SD, et al. Neurochemical and cellular reorganization of the spinal cord in a murine model of bone cancer pain. J Neurosci. 1999;19(24):10886–97.
31. Gao YJ, Ji RR. Targeting astrocyte signaling for chronic pain. Neurotherapeutics. 2010;7(4):482–93.
32. Sevcik MA, Ghilardi JR, Peters CM, et al. Anti-NGF therapy profoundly reduces bone cancer pain and the accompanying increase in markers of peripheral and central sensitization. Pain. 2005;115(1–2):128–41.
33. Halvorson KG, Kubota K, Sevcik MA, et al. A blocking antibody to nerve growth factor attenuates skeletal pain induced by prostate tumor cells growing in bone. Cancer Res. 2005;65(20):9426–35.
34. Sopata M, Katz N, Carey W, et al. Efficacy and safety of tanezumab in the treatment of pain from bone metastases. Pain. 2015;156(9):1703–13.
35. Wright MA, Ribera AB. Brain-derived neurotrophic factor mediates non-cell-autonomous regulation of sensory neuron position and identity. J Neurosci. 2010;30(43):14513–21.

36. Hunt SP, Mantyh PW. The molecular dynamics of pain control. Nat Rev Neurosci. 2001;2(2):83–91.
37. Patil SB, Brock JH, Colman DR, et al. Neuropathic pain- and glial derived neurotrophic factor-associated regulation of cadherins in spinal circuits of the dorsal horn. Pain. 2011;152(4):924–35.
38. Hans G, Deseure K, Adriaensen H. Endothelin-1-induced pain and hyperalgesia: a review of pathophysiology, clinical manifestations and future therapeutic options. Neuropeptides. 2008;42(2):119–32.
39. Peters CM, Lindsay TH, Pomonis JD, et al. Endothelin and the tumorigenic component of bone cancer pain. Neuroscience. 2004;126(4):1043–52.
40. Davar G. Endothelin-1 and metastatic cancer pain. Pain Med. 2001;2(1):24 7.
41. Ghilardi JR, Rohrich H, Lindsay TH, et al. Selective blockade of the capsaicin receptor TRPV1 attenuates bone cancer pain. Neuroscience. 2005;25(12):3126–31.
42. Cui M, Honore P, Zhong C, et al. TRPV1 receptors in the CNS play a key role in broad-spectrum analgesia of TRPV1 antagonists. Neuroscience. 2006;26(37):9385–93.
43. Brown DC, Iadarola MJ, Perkowski SZ, et al. Physiologic and antinociceptive effects of intrathecal resiniferatoxin in a canine bone cancer model. Anesthesiology. 2005;103(5):1052–9.
44. Arendt-Nielsen L, Harris S, Whiteside GT, et al. A randomized, double blind, positive-controlled, 3-way crossover human experimental pain study of a TRPV1 antagonist (8V116517) in healthy volunteers and comparison with preclinical profile. Pain. 2010;157(9):2057–67.
45. Saad F, Eastham J. Zoledronic acid improves clinical outcomes when administered before onset of bone pain in patients with prostate cancer. Urology. 2010.76(5):1175–81.
46. Broom R, Du H, Clemons M, et al. Switching breast cancer patients with progressive bone metastases to third-generation bisphosphonates: measuring impact using the Functional Assessment of Cancer Therapy-Bone Pain. J Pain Symptom Manage. 2009;38(2):244–57.
47. Namazi H. Zoledronic acid and survival in patient with metastatic bone disease from lung cancer and elevated markers of osteoclast activity: a novel molecular mechanism. J Thorac Oncol. 2008;3(8):943–4.
48. Stopeck AT, Lipton A, Body JJ, et al. Denosumab compared with zoledronic acid for the treatment of bone metastases in patients with advanced breast cancer: a randomized, double-blind study. J Clin Oncol. 2010;28(35):5132–9.
49. Nabal M, Librada S, Redondo MJ, et al. The role of paracetamol and non-steroidal anti-inflammatory drugs in addition to WHO step III opioids in the control of pain in advanced cancer. A systematic review of the literature. Palliative Med. 2012;26(4):305–12.
50. Kuchuk I, Clemons M, Addison C. Time to put an end to the "one size fits all" approach to bisphosphonate use in patients with metastatic breast cancer? Curr Oncol. 2012;19(5):e303–4.
51. Mantyh PW, Clohisy DR, Koltzenburg M, et al. 2002. Molecular mechanisms of cancer pain. Nat Cancer Rev. 2002;2(3):201–9.

102
Radiotherapy-Induced Osteoporosis

Laura E. Wright

Department of Medicine, Division of Endocrinology, Indiana University, Indianapolis, IN, USA

INTRODUCTION

Radiotherapy (RT), either as monotherapy or in combination with surgery or chemotherapy, is utilized in over 50% of cancer patients during the course of disease treatment [1]. As patient survival continues to improve with effective cancer treatment approaches, the long-term side effects of radiotherapy (RT) on the skeletal system have emerged. Clinical studies report that cancer patients receiving RT are at increased risk of osteoporosis and pathological fracture. Many factors can influence the response of skeletal tissue to radiation exposure, including the total absorbed dose, radiation energy, dose per fraction, and the developmental stage of the patient. RT-induced osteoporosis and the resulting increase in fracture risk appear to result from an acute period of bone resorption, followed by long-term suppression of bone formation that ultimately prevents the recovery of bone mass and impairs bone quality. This review highlights our current understanding of the skeletal consequences of radiation exposure in cancer patients, and the potential cellular and molecular mechanisms responsible for radiation-induced bone loss.

FRACTURE RISK FOLLOWING RADIOTHERAPY

Healthy by-standing bone adjacent to an irradiated tumor is estimated to absorb up to half of the RT dose, which can exceed 50 Gy for gynecological malignancies [2]. Despite efforts to minimize dose-limiting side effects by protecting healthy tissues, the incidence of pathological fracture at sites in the direct path of therapeutic irradiation is increased relative to nonirradiated skeletal sites in cancer patients and survivors [3–10]. Patients receiving RT for pelvic tumors including cervical, rectal, and anal cancers have increased risk of hip fracture relative to cancer patients who undergo surgery or chemotherapy alone [3–7]. Among men treated for prostate cancer, fracture incidence was shown to be approximately 7%, with a median time to diagnosis of 20 months [8]. Likewise, a dose-dependent relationship between radiation and rib fracture incidence has been identified in breast cancer patients [9], with reported rib fracture rates of up to 19% in irradiated populations [10]. In addition to evidence for the direct effects of radiation therapy on bone, radiation-treated breast cancer patients are reported to have hip fracture rates up to 20 times higher than average reported fracture rates for breast cancer patients 4 years after diagnosis [11–13]. Taken together, these striking clinical findings implicate both direct and systemic mechanisms at play in the pathology of radiation-induced bone loss.

CHANGES IN BONE QUANTITY AND QUALITY AFTER IRRADIATION

Systemic reduction in BMD has been detected in cancer patients within the first year of radiotherapy [4, 14, 15]. The radiation-induced deficits in bone that have been documented include demineralization, thinning, sclerosis, and loss of trabecular bone volume [4, 15–17]. A relatively comprehensive and prospective study by Nishiyama and colleagues assessed changes in BMC in the third lumbar vertebra (L3) by QCT in a cohort of patients with uterine or cervical malignancies [18]. In this study, patients that did not receive RT had no significant change in BMC over the course of the year assessed. Patients receiving RT experienced a mean loss of BMC of 32% at 5 weeks, 40% at 3 months, 47% at 6 months, and 49% at 12 months relative to pre-RT qCT scan measurements. These findings highlight RT-induced skeletal

---

Primer on the Metabolic Bone Diseases and Disorders of Mineral Metabolism, Ninth Edition. Edited by John P. Bilezikian.
© 2019 American Society for Bone and Mineral Research. Published 2019 by John Wiley & Sons, Inc.
Companion website: www.wiley.com/go/asbmrprimer

damage as an acute event that translates into prolonged deficits.

In animal models, acute deterioration of trabecular bone has been detected following ionizing radiation (2 Gy) as early as three days postexposure [19]. Trabecular bone microarchitecture is compromised and volume reduced in the proximal tibia, distal femur, and the fifth lumbar vertebrae of mice 1 week after exposure to as little as 2 Gy [19, 20]. The early loss of trabecular bone appears to translate into prolonged deficits in bone quantity and quality: loss of trabecular bone persists for months after both photon and charged particle exposures [21, 22]. Additionally, single-site irradiation of the hindlimb (2 Gy) of mice causes a reduction in trabecular bone volume at both irradiated and contralateral-shielded sites, analogous to the system-wide osteopenia observed in many cancer patients following RT [20].

LOSS OF BONE STRENGTH IN ANIMAL IRRADIATION MODELS

Photon radiation appears to produce greater damage to the trabecular network than to cortical bone [21]. However, at relatively low doses (50 cGy), high linear energy transfer (LET) heavy ion radiation does appear to increase cortical bone porosity, cortical area, and polar moment of inertia [23]. To date, the most direct assessment of radiation-induced reduction in bone strength comes from rodent and rabbit models, although studies are limited. Reduced ultimate strength of cortical bone from rabbit tibiae has been observed at 4 and 12 months after exposure to a 50 Gy total dose [24]. Compressive testing of mouse distal femora showed a reduction of strength at 12 weeks after 5 and 12 Gy acute doses of X-ray [25]. The subsequent loss of strength as determined by compressive testing and estimated by finite element analysis (FEA) occurred despite a transient increase in bone volume and a sustained elevation of cortical BMC indicating that the bone appeared to be more brittle. Changes in bone strength after irradiation may be influenced by both architectural and material properties. Additionally, a 2 Gy dose of heavy ion radiation induces a loss of vertebral stiffness when tested by compression loading and determined by FEA [26].

RADIATION EFFECTS ON OSTEOCLASTS

A heightened state of osteoclastic bone resorption is thought to play a substantial role in the acute phase of RT-induced bone loss. *In vitro* studies have shown that a radiation dose of 8 Gy was capable of stimulating murine macrophage RAW264.7 cell differentiation into multinucleated osteoclasts [20], suggesting that ionizing radiation may have a direct effect on osteoclast precursor cells. In preclinical studies, proresorptive inflammatory cytokines including interleukin (IL)-1, IL-6, IL-17, and TNF-α are known to be elevated within 24 to 48 hours following radiation exposure [27] and these factors may drive a systemic increase in bone resorption [28, 29]. Preceding measurable bone loss, serum tartrate-resistant acid phosphatase (TRAP5b) is increased 24 hours after total-body irradiation of mice, and histological assessment of bone reveals an increase in osteoclast number and activity as early as 3 days post-RT exposure [19, 28, 29]. Other preclinical studies have similarly shown a rapid increase in TRAP-positive osteoclast numbers at irradiated sites (2 Gy) 7 days post-RT [20]. Use of the antiresorptive bisphosphonates risedronate and zoledronic acid effectively block radiation-induced osteoclast activation and prevent bone loss at multiple skeletal sites [29, 30], indicating that osteoclasts are a player in the acute phase of bone loss following RT.

RADIATION EFFECTS ON OSTEOBLASTS

The suppression of bone formation through apoptosis of osteoblasts is thought to be a major contributor to radiation-induced damage to bone long term. The overall decrease in the number and activity of osteoblasts following radiation exposure has been widely reported along with attenuated bone matrix formation [20, 29, 31]. In vitro and in vivo data confirm impaired bone formation, which can be attributed to osteoblast cell-cycle arrest, DNA damage, reduced collagen synthesis, and increased apoptosis [20, 32–35]. Furthermore, recent in vitro studies have shown that activation of the Wnt/β-catenin pathway can overcome radiation-induced DNA damage and osteoblast apoptosis [36], and that the anabolic drug teriparatide (rhPTH 1-34) protects trabecular bone following focal radiation in rats by accelerating the repair of DNA double-strand breaks in osteoblasts [36, 37]. These studies provide a mechanistic basis for RT-induced impaired osteoblastic bone formation, and support the potential use of anabolic agents in the prevention of RT-induced bone loss.

RADIATION EFFECTS ON OSTEOCYTES

Doses of radiation as low as 2 to 4 Gy have been reported to induce osteocyte apoptosis [4, 16, 20, 38], providing evidence that the osteocyte is highly sensitive to radiation exposure. Impaired osteocyte function and/or cell death after radiation exposure could blunt the mechanosensitivity of bone and its response to dynamic loads, slowing bone growth, and retarding bone repair, thus compromising its material strength. In vitro, osteocytes appear more radiosensitive than osteoblasts when compared head-to-head in proliferation and apoptosis assays [20]. Importantly, the focal radiation of bone in vivo has recently been found to increase osteocyte production of the Wnt antagonist sclerostin, and monoclonal sclerostin antibody (Scl-Ab) treatment protects osteoblasts from radiation-induced apoptosis by accelerating DNA repair [37], similar to what was found with teriparatide treatment [36]. Moreover, Scl-Ab improves osteocyte survival and restores osteocyte canaliculi structure after irradiation [37]. Radiation-induced osteocyte expression of sclerostin may be a previously unappreciated contributor to the etiology of long-term bone loss and increased risk of fracture in cancer patients. Collectively, these data support the continued investigation of Scl-Ab as a potential treatment in cancer patients at risk for RT-induced osteoporosis.

RADIATION EFFECTS ON BONE MARROW ADIPOSITY

A reciprocal relationship has been identified clinically between bone marrow adiposity and bone volume in cancer patients treated with radiotherapy [3, 4, 39]. Consistent with these clinical findings, preclinical studies confirm the rapid increase in marrow adiposity in irradiated bone [20]. Radiation-induced DNA damage to mesenchymal stem cells (MSCs) has been linked to cellular senescence resulting in reduced osteogenic potential and increased adipogenic potential [40]. Expression levels of the Wnt/β-catenin pathway transcription factors Runx2 and Osx, critical mediators of osteoblastic lineage commitment and osteoblast differentiation, respectively, are reduced in irradiated MSCs [40–44]. This radiation-induced MSC lineage commitment switch away from the osteoblastic progeny via inhibition of the Wnt/β-catenin pathway favors the adipogenic lineage [40, 43] and likely contributes to reduced osteoblastic bone formation after RT. In line with these findings, Scl-Ab treatment partially attenuates radiation-induced MSC lineage commitment toward adipocytes and protects bone [37]. While the underlying mechanisms responsible for fatty infiltration of the marrow after radiation exposure are still being clarified, it is clear that this marrow defect is strongly associated with bone loss and increased fracture risk [40], and that damage to MSCs after radiotherapy could drive this defect in human cancer patients.

RADIATION EFFECTS ON BONE VASCULATURE

Since the turn of the twentieth century, radiation-induced bone damage (previously termed "osteitis") was thought to be initiated primarily by the reduction of vascularity in bone [45]. This vascular damage is characterized by swelling, vacuolization of endothelial cells within the osteon, the deposition of sclerotic connective tissue within the marrow cavity, and fibrosis [15, 16, 46]. These changes can result in the constriction of the vessel lumen and subsequent localized hypoxia. Bones of the skull and jaw are considered especially at risk for vascular injury because of their superficial location and relative lack of blood vessels [46, 47]. Vascular damage in both the marrow cavity and Haversian system, inflammation, and the generation of reactive oxygen species (ROS) have been reported in animal models after irradiation and are thought to contribute to deleterious effects of radiation at skeletal sites [27, 31, 46].

CONCLUSION

In summary, the etiology of radiation-induced osteoporosis and increased fracture risk appears multifactorial. The combination of an acute increase in osteoclast activity leading to rapid bone loss, followed by a lengthy reduction in bone formation resulting from osteocyte expression of sclerostin, impaired osteoblast function, and an MSC lineage commitment switch toward adipogenesis are likely contributing to skeletal complications after clinical RT. Protection of bone in cancer patients by targeting these pathways with antiresorptives (eg, bisphosphonates) or bone-forming agents (eg, teriparatide, Scl-Ab) may reduce fracture risk in osteoporotic cancer patients and survivors.

ACKNOWLEDGMENT

The Department of Defense Breast Cancer Research Program supported this work (BC134025 LEW).

REFERENCES

1. Delaney G, Jacob S, Featherstone C, Barton M. The role of radiotherapy in cancer treatment: estimating optimal utilization from a review of evidence-based clinical guidelines. Cancer. 2005;104(6):1129–37.
2. Willey JS, Lloyd SAJ, Nelson GA, et al. Ionizing radiation and bone loss: space exploration and clinical therapy applications. Clin Rev Bone Miner Metab. 2011;9(1):54–62.
3. Baxter NN, Habermann EB, Tepper JE, et al. Risk of pelvic fractures in older women following pelvic irradiation. JAMA. 2005;294(20):2587–93.
4. Mitchell JM, Logan PH. Radiation-induced changes in bone. Radiographics. 1998;18(5):1125–36.
5. Williams HJ, Davies AM. The effect of X-rays on bone: a pictorial review. Eur Radiol. 2006;16(3):619–33.
6. Kwon JW, Huh SJ, Yoon YC, et al. Pelvic bone complications after radiation therapy of uterine cervical cancer: evaluation with MRI. Am J Roentgenol. 2008;191:987–94.
7. Ikushima H, Osaki K, Furutani S, et al. Pelvic bone complications following radiation therapy of gynecological malignancies: clinical evaluation of radiation-induced pelvic insufficiency fractures. Gynecol Oncol. 2006;103:1100–4.

8. Igdem S, Alco G, Ercan T, et al. Insufficiency fractures after pelvic radiotherapy in patients with prostate cancer. Int J Radiat Oncol Bio Phys. 2010;77:818–23.
9. Pierce SM, Recht A, Lingos TI, et al. Long-term radiation complications following conservative surgery (CS) and radiation therapy (RT) in patients with early stage breast cancer. Int J Radiat Oncol Biol Phys. 1992;23: 912–23.
10. Overgaard M. Spontaneous radiation-induced rib fractures in breast cancer patients treated with postmastectomy irradiation. A clinical radiobiological analysis of the influence of fraction size and dose-response relationships on late bone damage. Acta Oncol. 1988;27(2): 117–22.
11. Kirstensen B, Ejlertsen B, Mouridsen HT, et al. Femoral fractures in postmenopausal breast cancer patients treated with adjuvant tamoxifen. Breast Cancer Res Treat. 1996;39:321–6.
12. Chen A, Maraicic M, Aragaki AK, et al. Fracture risk increases after diagnosis of breast or other cancers in postmenopausal women: results from the Women's Health Initiative. Osteoporos Int. 2009;20:527–36.
13. Jia D, Gaddy D, Suva LJ, et al. Rapid loss of bone mass and strength in mice after abdominal irradiation. Radiat Res. 2011;176(5):624–35.
14. Nishiyama K, Inaba F, Higashihara T, et al. Radiation osteoporosis — an assessment using single energy quantitative computed tomography. Eur Radiol. 1992;2(4): 322–5.
15. Hopewell JW. Radiation-therapy effects on bone density. Med Pediatr Oncol. 2003;41(3):208–11.
16. Ergun H, Howland WJ. Postradiation atrophy of mature bone. CRC Crit Rev Diagn Imaging. 1980;12(3):225–43.
17. Howland W, Loeffler RK, Starchman DE. Post-irradiation atrophic changes of bone and related complications. Radiology. 1975;117:677–85.
18. Nishiyama K, Inaba F, Higashirara T, et al. Radiation osteoporosis — an assessment using single energy quantitative computed tomography. Eur Radiol 1992;2:322–5.
19. Kondo H, Searby ND, Mojarrab R, et al. Total-body irradiation of postpubertal mice with (137)Cs acutely compromises the microarchitecture of cancellous bone and increases osteoclasts. Radiat Res. 2009;171(3):283–9.
20. Wright LE, Buijs JT, Kim HS, et al. Single-limb irradiation induces local and systemic bone loss in a murine model. J Bone Miner Res. 2015;30(7):1268–79.
21. Bandstra ER, Pecaut MJ, Anderson ER, et al. Long-term dose response of trabecular bone in mice to proton radiation. Radiat Res. 2008;169(6):607–14.
22. Hamilton SA, Pecaut MJ, Gridley DS, et al. A murine model for bone loss from therapeutic and space-relevant sources of radiation. J Appl Physiol. 2006;101(3): 789–93.
23. Bandstra ER, Thompson RW, Nelson GA, et al. Musculoskeletal changes in mice from 20–50 cGy of simulated galactic cosmic rays. Radiat Res. 2009;172(1):21–9.
24. Sugimoto M, Takahashi S, Toguchida J, et al. Changes in bone after high-dose irradiation. Biomechanics and histomorphology. J Bone Joint Surg. 1991;73(3):492–7.
25. Wernle JD, Damron TA, Allen MJ, et al. Local irradiation alters bone morphology and increases bone fragility in a mouse model. J Biomech. 2010;43(14):2738–46.
26. Alwood JS, Yumoto K, Mojarrab R, et al. Heavy ion irradiation and unloading effects on mouse lumbar vertebral microarchitecture, mechanical properties and tissue stresses. Bone. 2010;47(2):248–55.
27. Lorimore SA, Coates PJ, Scobie GE, et al. Inflammation-type responses after exposure to ionizing radiation in vivo: a mechanism for radiation-induced bystander effects? Oncogene. 2001;20(48):7085–95.
28. Willey JS, Lloyd SA, Robbins ME, et al. Early increase in osteoclast number in mice after whole-body irradiation with 2 Gy X rays. Radiat Res. 2008;170(3): 388–92.
29. Willey JS, Livingston EW, Robbins ME, et al. Risedronate prevents early radiation-induced osteoporosis in mice at multiple skeletal locations. Bone. 2010;46(1):101–11.
30. Keenawinna L, Oest ME, Mann KA, et al. Zoledronic acid prevents loss of trabecular bone after focal irradiation in mice. Radiat Res. 2013;180(1):89–99.
31. Cao X, Wu X, Frassica D, Yu B, et al. Irradiation induces bone injury by damaging bone marrow microenvironment for stem cells. Proc Natl Acad Sci U S A. 2011;108(4):1609–14.
32. Gal TJ, Munoz-Antonia T, Muro-Cacho CA, et al. Radiation effects on osteoblasts in vitro: a potential role in osteoradionecrosis. Arch Otolaryngol Head Neck Surg. 2000;126(9):1124–8.
33. Dudziak ME, Saadeh PB, Mehrara BJ, et al. The effects of ionizing radiation on osteoblast-like cells in vitro. Plast Reconstr Surg. 2000;106(5):1049–61.
34. Szymczyk KH, Shapiro IM, Adams CS. Ionizing radiation sensitizes bone cells to apoptosis. Bone. 2004;34(1): 148–56.
35. Sakurai T, Sawada Y, Yoshimoto M, et al. Radiation-induced reduction of osteoblast differentiation in C2C12 cells. J Radiat Res. 2007;48(6):515–21.
36. Chandra A, Lin T, Tribble MB, et al. PTH1-34 alleviates radiotherapy-induced local bone loss by improving osteoblast and osteocyte survival. Bone. 2014; 67:33–40.
37. Chandra A, Lin T, Young T, et al. Suppression of sclerostin alleviates radiation-induced bone loss by protecting bone-forming cells and their progenitors through distinct mechanisms. J Bone Miner Res. 2017;32(2): 360–72.
38. Rabelo GD, Beletti ME, Dechichi P. Histological analysis of the alterations on cortical bone channels network after radiotherapy: a rabbit study. Microsc Res Tech. 2010;73(11):1015–8.
39. Hui SK, Khalil A, Zhang Y, et al. Longitudinal assessment of bone loss from diagnostic computer tomography scans in gynecologic cancer patients treated with chemotherapy and radiation. Am J Obstet Gynecol. 2010;203:353–7.
40. Georgiou KR, Hui SK, Xian CJ. Regulatory pathways associated with bone loss and bone marrow adiposity by aging, chemotherapy, glucocorticoid therapy and radiotherapy. Am J Stem Cell. 2012;1(3):205–24.

41. Despars G, Caronneau CL, Bardeau P, et al. Loss of the osteogenic differentiation potential during senescence is limited to bone progenitor cell and is dependent on p53. PLOS ONE. 2013;8(8):1–11.
42. Carbonneau CL, Despars G, Rojas-Sutterlin S, et al. Ionizing radiation-induced expression of INK4a/ARF in murine bone marrow-derived stromal cell populations interferes with bone marrow homeostasis. Blood. 2012;119:717–26.
43. Su W, Chen Y, Zeng W, et al. Involvement of Wnt signaling in the injury of murine mesenchymal stem cells exposed to X-radiation. Int J Rad Biol. 2012;88(9): 635–41.
44. Komori T. Regulation of bone development and maintenance by Runx2. Front Biosci. 2008;13:898–903.
45. Ewing J. Radiation osteitis. Acta Radiol. 1926;6: 399–412.
46. Rohrer MD, Kim Y, Fayos JV. The effect of cobalt-60 irradiation on monkey mandibles. Oral Surg Oral Med Oral Pathol. 1979;48(5):424–40.
47. Williams HJ, Davies AM. The effect of X-rays on bone: a pictorial review. Eur Radiol. 2006;16(3):619–33.

103

Skeletal Complications of Childhood Cancer

Manasa Mantravadi and Linda A. DiMeglio
Department of Pediatrics, Indiana University School of Medicine, Indianapolis, IN, USA

INTRODUCTION

With improvements in therapy since the 1970s, the survival rate for children with cancer has steadily increased. Currently, the 5-year rate is 83.5%, with almost 45% of children surviving 20 years or more [1]. The survival rate for acute lymphocytic leukemia (ALL), the most common childhood malignancy, is near 90% [1]. Data from national registries show an estimated 388,501 survivors of childhood cancer living in the USA as of January 2011 [2]. Many of these individuals have long-term sequelae of cancer and its therapy. Among survivors of ALL, the cumulative incidence of a chronic health condition reached 73.4% (95% CI, 69.0 to 77.9) 30 years after the cancer diagnosis, with a cumulative incidence of 42.4% (95% CI, 33.7 to 51.2) for severe, disabling, or life-threatening conditions or death caused by a chronic condition [3]. This chapter focuses on the skeletal complications of childhood cancers.

Childhood cancer is often associated with a variety of factors that have detrimental effects on bone health [4]. Cancer can cause prolonged nutritional deficiencies, suboptimal physical activity, and pubertal disruption during critical years of growth and bone accrual. Cancer treatments can also more directly be associated with skeletal complications such as decreased BMD, altered epiphyseal growth, and avascular necrosis [5–8]. Hormonal deficiencies secondary to cranial irradiation causing growth hormone deficiency and/or central hypogonadism, or gonadal irradiation leading to secondary hypogonadism, may further compromise bone accrual [4]. Moreover, direct bone irradiation can be detrimental to bone health as it is cytotoxic to epiphyseal chondrocytes [9], increases hypovascularity, and results in reduced bone strength [10, 11]. Finally, chemotherapeutic agents, including glucocorticoids and methotrexate, interfere with calcium uptake, bone mineral accrual, and skeletal development [4, 12]. The remainder of this chapter will expand upon several of these factors/complications individually.

DECREASED BONE MINERAL DENSITY

Deficits in BMD among survivors of childhood cancer have been well documented [13–15]. The most data on the influences of cancer on childhood BMD come from populations of children with ALL. Decreased BMD and reduced markers of bone formation have been shown in patients with pediatric ALL at diagnosis [4, 16]. Between 13% and 21% of newly diagnosed patients already have a decreased BMD [16]. In the Canadian STeroid-Associated Osteoporosis in the Pediatric Population (STOPP) research program, vertebral compression fractures were noted in 16% of newly diagnosed ALL patients, and children with fractures had lower lumbar spine BMD (LSBMD) Z-scores [13]. This low bone density is likely in part caused by direct leukemic cell infiltration; these cells expand into the medullary cavity and can damage the spongiosa and cause bone pain [16, 17]. Additionally, factors secreted by leukemic cells (such as the osteoclast-stimulating interleukins 6 and 8) and decreases in the conversion of 25OHD to 1,25 vitamin D are thought to have a deleterious effect on bone mineralization [18, 19].

During treatment for ALL, BMD also decreases significantly [20, 21]. Numerous factors that can be associated with BMD deficits including high cumulative steroid doses, methotrexate use, hematopoietic cell transplant (HCT), cranial irradiation, and testicular irradiation, are often inherent to ALL treatment [12]. Persons with ALL who received corticosteroid doses greater than 9080 mg/m² (prednisone equivalent) over 3 years, or greater than 3000 mg/m²/year for protocols of other lengths, seem to be at risk for decreased BMD [18, 22] Antimetabolite chemotherapy such as methotrexate, has a cytotoxic effect on osteoblasts and results in reduced bone volume and formation of new bone; a total methotrexate dose of greater than 40,000 mg/m² is shown to increase risk [18]. Alkylating agents, such as ifosfamide, can reduce BMD through toxicity to the gonads decreasing sex steroid production [12].

Children with brain tumors may also be particularly susceptible to reduced BMD caused by growth hormone deficiency (GHD) and hypogonadism caused by primary tumor effects of pituitary and/or cranial irradiation. Patients exposed to cranial radiation doses of 18 Gy or more are at increased risk for GHD. Fractionated total body radiation doses of more than 12 Gy for HCT also increase risk of GHD [4]. Growth hormone (GH) therapy can improve BMD in persons with GHD as the cause of decreased BMD [4].

Direct tumor irradiation is part of the treatment for solid tumors such as rhabdomyosarcoma, Ewing's sarcoma, Wilms' tumor, and neuroblastoma. Radiation not only causes hypovascularity but has a direct cytotoxic effect on the epiphyseal chondrocytes [9–11]. In pediatric patients with bone sarcomas, bone mineral deficits may not be present initially but may develop with time. Patients with newly diagnosed Ewing sarcoma and osteosarcoma showed no deficits in LSBMD after completion of neoadjuvant chemotherapy, although in patients with a lower extremity tumor, local BMD deficits were found, with lower femoral neck BMD in the affected compared with the nonaffected limb [23]. However, at an average of over 5 years after remission, bone mineral deficits were reported in bone sarcoma survivors, including survivors of Ewing sarcoma and osteosarcomas [24].

In addition to treatment-related risk and concurrent endocrinopathies such as GHD and hypogonadism, factors associated with a lower BMD include white race, male gender, younger age at diagnosis, and vitamin D deficiency [25–27]. Poor nutrition, alcohol consumption, low physical activity, and smoking are also risk factors [27–30].

Recent data from the St. Jude Lifetime Cohort study of 862 childhood cancer survivors treated at St. Jude Children's Research Hospital (mean age 31.3 years) showed that both low BMD (measured by vertebral qCT) and frailty/prefrailty were associated with GHD, smoking, and alcohol [31]. Long-term ALL survivors are less likely to meet the Centers for Disease Control and Prevention physical activity recommendations and are more likely to report less leisure-time physical activity than the general population [32]. Lack of physical activity may play a greater role in the low BMD found in pediatric solid tumor survivors than in other cancers. In the Childhood Cancer Survivor Study (CCSS), a retrospectively ascertained cohort of 35,923 childhood cancer survivors diagnosed between 1970 and 1999, the highest prevalence of physical performance limitations were found in survivors of brain tumors (36.9%), bone tumors (26.6%), and Hodgkin disease (23.3%) [33, 34]. Marinovic and colleagues observed a significant BMD increase in a 1-year longitudinal study in children with ALL during the first 3 years after completion of therapy without cranial irradiation [35]. These findings suggest a positive effect of long-term completion therapy and increase in physical activity on BMD, body composition, and bone metabolism in patients who have been treated for ALL [35]. Nutrition is known to play an important role in bone mineral accrual. Yet, less than 30% of ALL survivors at least 5 years after therapy met the recommended dietary intakes for vitamin D and calcium [36].

These findings suggest that ALL survivors should receive lifestyle counseling and undergo screening for hormonal deficits to minimize the risk of low BMD and frailty [31]. The Long-Term Follow-up Guidelines from the Children's Oncology Group (a National Cancer Institute supported clinical trials group) recommend that all patients treated with agents that predispose to reduced BMD (glucocorticoids, cranial radiation, methotrexate, or HCT) have a quantitative BMD measure 2 years after cancer chemotherapy completion [4]. Measurement should be with either DXA or QCT [4].

Cancer survivors should be assessed for hypogonadism at their annual follow-up visits. Clinicians should also include diet, exercise, and lifestyle histories as essential components of these annual visits. It is likely important to ensure the best nutrition possible during and immediately after therapy, as cholecalciferol and calcium supplementation provided no added benefit to nutritional counseling for improving LSBMD among adolescent and young adult survivors of ALL who had been in remission at least 5 years [37].

FRACTURES

Failure to accrue sufficient bone mineral during childhood and adolescence increases the risk for early onset osteoporosis among childhood cancer survivors and places them at risk for fractures, particularly later in life. Sixteen percent of children with ALL enrolled in STOPP developed incident vertebral fractures 12 months after therapy initiation [13]. The presence of low LSBMD Z-score or vertebral fractures of any grade were associated with significantly increased odds of incident vertebral fractures [13]. Children with incident vertebral fractures had greater increases in LSBMD Z-scores between 6 and 12 months suggesting fractures may have occurred early in the observation period, and were then followed by a degree of recovery with enhanced BMD accrual [13]. A similar study found the proportion of children with incident vertebral fractures in the 4 years after ALL diagnosis was 26.4%, that most of the incident fractures happen during the first year (when GC exposure is highest), and that discrete clinical predictors could be ascertained around the time of diagnosis (including vertebral fractures at diagnosis, low LSBMD Z-scores, and younger age) and during chemotherapy (GC exposure and low LSBMD Z-scores) [38].

There are conflicting data regarding future fracture risk in childhood cancer survivors. Older studies reported fracture rates were increased sixfold in survivors compared with healthy controls, tending to occur during or shortly after discontinuation of chemotherapy [8, 16]. These were small studies that followed patients for only 1 year after therapy. Recent data from the CCSS cohort showed an increased risk for osteoporosis in ALL survivors when compared with siblings, yet the fracture prevalence was not increased [27]. The median age at follow-up among survivors was 36.2 years [27]. In one study, low LSBMD at diagnosis and during treatment, rather than the treatment-related decline of LSBMD, determined the increased fracture risk of 17.8% in children with ALL [39]. More long-term data are needed to evaluate whether these children have increased rates of fracture in older adulthood.

ALTERED GROWTH

Growth deficits are another important endocrine sequelae of childhood cancer. They can be short term (with subsequent catch up growth) or lifelong depending on the cancer and its treatment modalities. Chemotherapy is generally associated with some growth deceleration during treatment; although many children then experience a period of rapid catch-up growth after treatment cessation [6, 40], sometimes chemotherapies and cancers can be associated with an attenuation of growth during this period as well.

Chemotherapy directly interferes with bone growth in several ways including inhibition of chondrocyte destruction (chondrocyte cell death and the associated release of growth factors are an essential part of longitudinal bone growth) [12, 41–43]. In most cases, trunk length is affected more than standing height (leading to disproportionate stature) because of the additive effects of attenuated growth at each of the many vertebral column epiphyses [44]. Long-term glucocorticoid administration as part of treatment of ALL also limits growth by inhibiting pituitary GH secretion [11, 41, 43].

Of all the chemotherapeutic agents, the antimetabolites 6-mercaptopurine and methotrexate are most associated with attenuated catch-up growth after treatment for ALL [12, 41–43]. Intensive chemotherapy with agents such as doxorubicin, actinomycin D, cisplatin, or high dose glucocorticoids can significantly decrease height in the long term [6, 12, 40].

Cranial radiation can have pronounced adverse effects on linear growth, primarily through effects on the hypothalamic–pituitary axis. As noted above, hypothalamic–pituitary radiation in doses of 18 Gy or more, and total body irradiation (TBI) in doses with single fractions 10 Gy or more or fractionated doses of 12 Gy or more are associated with risk for GHD [20, 45, 46]. Surgical resection for suprasellar tumors also increases GHD risk. GH replacement improves final height in childhood cancer survivors with GHD [47–49].

Direct irradiation of long bones and the vertebral column can also retard bone growth. Children previously treated with radiation doses of more than 20 Gy to the spine respond less well to GH therapy because deficits in linear growth are related to radiation-induced skeletal dysplasia [49].

Cranial radiation of more than 18 Gy can cause central precocious puberty (CPP) which induces accelerated epiphyseal closure and can lead to reduced height accrual [47, 50, 51]. Gonadotropin-releasing hormone (GnRH) agonists can be used to delay pubertal progression in children with CPP to improve final adult height, especially when combined with GH therapy for concurrent GHD. Hypothyroidism, as thyroid-stimulating hormone (TSH) deficiency from cranial radiotherapy or primary hypothyroidism from local radiotherapy, TBI, or ^{131}I-MIBG and ^{131}I-labeled monoclonal antibody, can also contribute to growth failure [52–54].

Because of all the above risks for growth failure for children with cancer, the Children's Oncology Group recommends physical examinations during treatment and yearly for survivors, with particular attention to growth velocity and pubertal progression. In addition, yearly TSH and T4 screening is recommended [25]. In those whose clinical findings are worrisome, additional hormonal assessments and/or imaging (e.g. bone age) should be pursued.

AVASCULAR NECROSIS

Avascular necrosis (AVN), also known as aseptic osteonecrosis, is a known complication of chemotherapy and glucocorticoid use. AVN is characterized by ischemia-related destruction of bone tissue leading to significant morbidity including compromise of joint function, pain, and disruption in skeletal growth and development [55]. It is most typically found in the femoral head, although it can affect other joints. Often it is initially asymptomatic, and not until the disease advances do significant clinical manifestations present. At that point, typically, AVN presents as nonspecific bone or joint pain. While the exact pathogenesis remains unclear, glucocorticoid exposure, particularly dexamethasone, has been consistently associated with AVN [5, 56, 57]. Proposed mechanisms include an increase in intramedullary adipocytosis causing interference in vascular bone glucocorticoid-induced hypoperfusion from downregulation of vascular endothelial growth factor [58, 59].

While longitudinal studies are limited, the reported incidence varies from 1% to 72% [5, 55–57, 60, 61]. Overall rates are much greater in studies that employ screening

MRI and thereby identify asymptomatic disease or early symptomatic disease. In patients with ALL, the reported incidence of symptomatic AVN ranges from a 1.8% 5-year cumulative incidence [60] to a 3-year life-table incidence of 9.3% [56]. After HCT, reported incidences of AVN in any joint range from 3.9% to 44.2% [62–64].

Recently, a growing recognition of the possibility of treatment-induced AVN has led to a high index of suspicion and prompted prospective monitoring of at-risk patient populations [55]. MRI is the most sensitive and specific method for detection and monitoring of osteonecrosis [65]. Sansgiri and colleagues describe subtle signal changes that precede the earliest reported MRI signs of osteonecrosis and also appeared to predict subsequent development of extensive osteonecrosis on follow-up MRI examinations in leukemia patients [66]. In another prospective study, approximately 15% of children with ALL or advanced-stage non-Hodgkin lymphoma whose therapy included large cumulative doses of prednisone had MRI changes consistent with AVN, primarily in the knees, after the first year of therapy [67]. No published recommendations exist regarding monitoring of asymptomatic patients for AVN. However, based on recent literature, Kaste and colleagues suggests that older patients (>10 years) exposed to glucocorticoids or those who undergo bone marrow transplantation may need to be monitored for osteonecrosis, even in the absence of symptoms [55]. Adolescents are at higher risk than younger children because of rapid skeletal maturation and bone turnover during the pubertal years [5, 68].

Current treatment options range from conservative management with physical activity restriction to surgical intervention for more severe cases. Bisphosphonate therapy may have a limited role in preserving joint shape; studies of this therapy are ongoing.

CONCLUSION

As outlined above, detrimental effects on bone health are an important long-term consequence of childhood cancers and their treatments. Deficits in bone mass accrual that persist into adulthood will likely increase children's lifetime risk of osteoporosis and fractures. Patients should be followed in long-term survivorship clinics and an endocrine referral is suggested for children with decreased BMD by DXA (Z-score <-2), recurrent fractures, GHD, or suspected hypogonadism [25].

The glucocorticoids, chemotherapies and radiation therapies used to treat these cancers are in large part responsible for the increased risk of a low BMD. However, there are a number of modifiable contributing factors such as nutrition, exercise, smoking, and alcohol consumption among others. Understanding the time course and long-term effects on BMD through each stage of therapy can help guide further screening and prevention strategies along with treatment options for skeletal morbidity associated with childhood cancer survivorship.

REFERENCES

1. Phillips SM, Padgett LS, Leisenring WM, et al. Survivors of childhood cancer in the United States: prevalence and burden of morbidity. Cancer Epidemiol Biomarkers Prev. 2015;24(4):653–63.
2. Howlader N, Krapcho M, Miller D, et al. SEER Cancer Statistics Review, 1975–2013, National Cancer Institute. Bethesda, MD, based on November 2015 SEER data submission. posted to SEER website April 2016, http://seer.cancer.gov/csr/1975_2013/
3. Essig S, Li Q, Chen Y, Hitzler J, et al. Risk of late effects of treatment in children newly diagnosed with standard-risk acute lymphoblastic leukaemia: a report from the Childhood Cancer Survivor Study cohort. Lancet Oncol. 2014;15(8):841–51.
4. Wasilewski-Masker K, Kaste SC, Hudson MM, et al. Bone mineral density deficits in survivors of childhood cancer: long-term follow-up guidelines and review of the literature. Pediatrics. 2008;121(3):e705–13.
5. Lackner H, Benesch M, Moser A, et al. Aseptic osteonecrosis in children and adolescents treated for hemato-oncologic diseases: a 13-year longitudinal observational study. J Pediatr Hematol Oncol. 2005;27(5):259–63.
6. Viana MB, Vilela MI. Height deficit during and many years after treatment for acute lymphoblastic leukemia in children: a review. Pediatr Blood Cancer. 2008;50 (2 Suppl):509–16; discussion 17.
7. van der Sluis IM, van den Heuvel-Eibrink MM, et al. Bone mineral density, body composition, and height in long-term survivors of acute lymphoblastic leukemia in childhood. Med Pediatr Oncol. 2000;35(4):415–20.
8. van der Sluis IM, van den Heuvel-Eibrink MM, Hahlen K, et al. Altered bone mineral density and body composition, and increased fracture risk in childhood acute lymphoblastic leukemia. J Pediatr. 2002;141(2):204–10.
9. Brennan BM, Rahim A, Adams JA, et al. Reduced bone mineral density in young adults following cure of acute lymphoblastic leukaemia in childhood. Br J Cancer. 1999;79(11–12):1859–63.
10. Nyaruba MM, Yamamoto I, Kimura H, et al. Bone fragility induced by X-ray irradiation in relation to cortical bone-mineral content. Acta Radiol. 1998;39(1):43–6.
11. Wagner LM, Neel MD, Pappo AS, et al. Fractures in pediatric Ewing sarcoma. J Pediatr Hematol Oncol. 2001;23(9):568–71.
12. van Leeuwen BL, Kamps WA, Jansen HW, et al. The effect of chemotherapy on the growing skeleton. Cancer Treat Rev. 2000;26(5):363–76.
13. Halton J, Gaboury I, Grant R, et al. Advanced vertebral fracture among newly diagnosed children with acute lymphoblastic leukemia: results of the Canadian Steroid-Associated Osteoporosis in the Pediatric Population (STOPP) research program. J Bone Miner Res. 2009;24(7):1326–34.
14. Sala A, Barr RD. Osteopenia and cancer in children and adolescents: the fragility of success. Cancer. 2007;109(7):1420–31.

15. Arikoski P, Komulainen J, Riikonen P, et al. Alterations in bone turnover and impaired development of bone mineral density in newly diagnosed children with cancer: a 1-year prospective study. J Clin Endocrinol Metab. 1999;84(9):3174–81.
16. Halton JM, Atkinson SA, Fraher L, et al. Altered mineral metabolism and bone mass in children during treatment for acute lymphoblastic leukemia. J Bone Miner Res. 1996;11(11):1774–83.
17. Samuda GM, Cheng MY, Yeung CY. Back pain and vertebral compression: an uncommon presentation of childhood acute lymphoblastic leukemia. J Pediatr Orthoped. 1987;7(2):175–8.
18. Mandel K, Atkinson S, Barr RD, et al. Skeletal morbidity in childhood acute lymphoblastic leukemia. J Clin Oncol. 2004;22(7):1215–21.
19. Jayanthan A, Miettunen PM, Incoronato A, et al. Childhood acute lymphoblastic leukemia (ALL) presenting with severe osteolysis: a model to study leukemia-bone interactions and potential targeted therapeutics. Pediatr Hematol Oncol. 2010;27(3):212–27.
20. Chow EJ, Friedman DL, Yasui Y, et al. Timing of menarche among survivors of childhood acute lymphoblastic leukemia: a report from the Childhood Cancer Survivor Study. Pediatr Blood Cancer. 2008;50(4):854–8.
21. Huma Z, Boulad F, Black P, et al. Growth in children after bone marrow transplantation for acute leukemia. Blood. 1995;86(2):819–24.
22. Strauss AJ, Su JT, Dalton VM, et al. Bony morbidity in children treated for acute lymphoblastic leukemia. J Clin Oncol. 2001;19(12):3066–72.
23. Muller C, Winter CC, Rosenbaum D, et al. Early decrements in bone density after completion of neoadjuvant chemotherapy in pediatric bone sarcoma patients. BMC Musculoskelet Disord 2010;11:287.
24. Ruza E, Sierrasesumaga L, Azcona C, et al. Bone mineral density and bone metabolism in children treated for bone sarcomas. Pediatr Res. 2006;59(6):866–71.
25. Hudson MM, Landier W, Bhatia S. Long-term follow-up guidelines for survivors of childhood, adolescent, and young adult cancers. In: Children's Oncology Group (ed.) Version 2. www.survivorshipguidelines.org
26. Makitie O, Heikkinen R, Toiviainen-Salo S, et al. Long-term skeletal consequences of childhood acute lymphoblastic leukemia in adult males: a cohort study. Eur J Endocrinol. 2013;168(2):281–8.
27. Wilson CL, Dilley K, Ness KK, et al. Fractures among long-term survivors of childhood cancer: a report from the Childhood Cancer Survivor Study. Cancer. 2012;118(23):5920–8.
28. Hoorweg-Nijman JJ, Kardos G, Roos JC, et al. Bone mineral density and markers of bone turnover in young adult survivors of childhood lymphoblastic leukaemia. Clin Endocrinol. 1999;50(2):237–44.
29. Kaste SC, Ahn H, Liu T, et al. Bone mineral density deficits in pediatric patients treated for sarcoma. Pediatr Blood Cancer. 2008;50(5):1032–8.
30. Kaste SC, Rai SN, Fleming K, et al. Changes in bone mineral density in survivors of childhood acute lymphoblastic leukemia. Pediatr Blood Cancer. 2006;46(1):77–87.
31. Wilson CL, Chemaitilly W, Jones KE, et al. Modifiable factors associated with aging phenotypes among adult survivors of childhood acute lymphoblastic leukemia. J Clin Oncol. 2016;34(21):2509–15.
32. Florin TA, Fryer GE, Miyoshi T, et al. Physical inactivity in adult survivors of childhood acute lymphoblastic leukemia: a report from the Childhood Cancer Survivor Study. Cancer Epidemiol Biomarkers Prev. 2007;16(7):1356–63.
33. Ness KK, Hudson MM, Ginsberg JP, et al. Physical performance limitations in the Childhood Cancer Survivor Study cohort. J Clin Oncol. 2009;27(14):2382–9.
34. Ness KK, Morris EB, Nolan VG, et al. Physical performance limitations among adult survivors of childhood brain tumors. Cancer. 2010;116(12):3034–44.
35. Marinovic D, Dorgeret S, Lescoeur B, et al. Improvement in bone mineral density and body composition in survivors of childhood acute lymphoblastic leukemia: a 1-year prospective study. Pediatrics. 2005;116(1):e102–8.
36. Tylavsky FA, Smith K, Surprise H, et al. Nutritional intake of long-term survivors of childhood acute lymphoblastic leukemia: evidence for bone health interventional opportunities. Pediatr Blood Cancer. 2010;55(7):1362–9.
37. Kaste SC, Qi A, Smith K, et al. Calcium and cholecalciferol supplementation provides no added benefit to nutritional counseling to improve bone mineral density in survivors of childhood acute lymphoblastic leukemia (ALL). Pediatr Blood Cancer. 2014;61(5):885–93.
38. Cummings EA, Ma J, Fernandez CV, et al. Incident vertebral fractures in children with leukemia during the four years following diagnosis. J Clin Endocrinol Metab. 2015;100(9):3408–17.
39. te Winkel ML, Pieters R, Hop WC, et al. Bone mineral density at diagnosis determines fracture rate in children with acute lymphoblastic leukemia treated according to the DCOG–ALL9 protocol. Bone. 2014;59:223–8.
40. Vilela MI, Viana MB. Longitudinal growth and risk factors for growth deficiency in children treated for acute lymphoblastic leukemia. Pediatr Blood Cancer. 2007;48(1):86–92.
41. Groot-Loonen JJ, Otten BJ, van t' Hof MA, et al. Chemotherapy plays a major role in the inhibition of catch-up growth during maintenance therapy for childhood acute lymphoblastic leukemia. Pediatrics. 1995;96(4):693–5.
42. Rainsford KD. Doxorubicin is a potent inhibitor of interleukin 1 induced cartilage proteoglycan resorption invitro. J Pharm Pharmacol 1989;41(1):60–3.
43. Samuelsson BO, Marky I, Rosberg S, et al. Growth and growth hormone secretion after treatment for childhood non-Hodgkin's lymphoma. Med Pediatr Oncol. 1997;28(1):27–34.
44. Davies HA, Didcock E, Didi M, et al. Disproportionate short stature after cranial irradiation and combination chemotherapy for leukaemia. Arch Dis Child. 1994;70(6):472–5.

45. Chemaitilly W, Boulad F, Heller G, et al. Final height in pediatric patients after hyperfractionated total body irradiation and stem cell transplantation. Bone Marrow Transplant. 2007;40(1):29–35.
46. Darzy KH. Radiation-induced hypopituitarism after cancer therapy: who, how and when to test. Nat Clin Pract Endcrinol Metab. 2009;5(2):88–99.
47. Gleeson HK, Stoeter R, Ogilvy-Stuart AL, et al. Improvements in final height over 25 years in growth hormone (GH)-deficient childhood survivors of brain tumors receiving GH replacement. J Clin Endocrinol Metab. 2003;88(8):3682–9.
48. Adan L, Sainte-Rose C, Souberbielle JC, et al. Adult height after growth hormone (GH) treatment for GH deficiency due to cranial irradiation. Med Pediatr Oncol. 2000;34(1):14–9.
49. Chemaitilly W, Sklar CA. Endocrine complications in long-term survivors of childhood cancers. Endocr Relat Cancer. 2010;17(3):R141–59.
50. Oberfield SE, Soranno D, Nirenberg A, et al. Age at onset of puberty following high-dose central nervous system radiation therapy. Arch Pediatr Adolesc Med. 1996;150(6):589–92.
51. Brownstein CM, Mertens AC, Mitby PA, et al. Factors that affect final height and change in height standard deviation scores in survivors of childhood cancer treated with growth hormone: a report from the childhood cancer survivor study. J Clin Endocrinol Metab. 2004;89(9):4422–7.
52. Rose SR, Leong GM, Yanovski JA, et al. Thyroid function in non-growth hormone-deficient short children during a placebo-controlled double blind trial of recombinant growth hormone therapy. J Clin Endocrinol Metab. 1995;80(1):320–4.
53. Sklar C, Whitton J, Mertens A, et al. Abnormalities of the thyroid in survivors of Hodgkin's disease: data from the Childhood Cancer Survivor Study. J Clin Endocrinol Metab. 2000;85(9):3227–32.
54. Chow EJ, Friedman DL, Stovall M, et al. Risk of thyroid dysfunction and subsequent thyroid cancer among survivors of acute lymphoblastic leukemia: a report from the Childhood Cancer Survivor Study. Pediatr Blood Cancer. 2009;53(3):432–7.
55. Kaste SC, Karimova EJ, Neel MD. Osteonecrosis in children after therapy for malignancy. Am J Roentgenol. 2011;196(5):1011–8.
56. Mattano LA, Jr., Sather HN, Trigg ME, et al. Osteonecrosis as a complication of treating acute lymphoblastic leukemia in children: a report from the Children's Cancer Group. J Clin Oncol. 2000;18(18):3262–72.
57. Kadan-Lottick NS, Dinu I, Wasilewski-Masker K, et al. Osteonecrosis in adult survivors of childhood cancer: a report from the childhood cancer survivor study. J Clin Oncol. 2008;26(18):3038–45.
58. Drescher W, Schneider T, Becker C, et al. Selective reduction of bone blood flow by short-term treatment with high-dose methylprednisolone. An experimental study in pigs. J Bone Joint Surg. 2001;83(2):274–7.
59. Wang G, Zhang CQ, Sun Y, et al. Changes in femoral head blood supply and vascular endothelial growth factor in rabbits with steroid-induced osteonecrosis. J Int Med Res. 2010;38(3):1060–9.
60. Burger B, Beier R, Zimmermann M, et al. Osteonecrosis: a treatment related toxicity in childhood acute lymphoblastic leukemia (ALL) – experiences from trial ALL-BFM 95. Pediatr Blood Cancer. 2005;44(3):220–5.
61. Kawedia JD, Kaste SC, Pei D, et al. Pharmacokinetic, pharmacodynamic, and pharmacogenetic determinants of osteonecrosis in children with acute lymphoblastic leukemia. Blood. 2011;117(8):2340–7; quiz 556.
62. Kaste SC, Shidler TJ, Tong X, et al. Bone mineral density and osteonecrosis in survivors of childhood allogeneic bone marrow transplantation. Bone Marrow Transplant. 2004;33(4):435–41.
63. Faraci M, Calevo MG, Lanino E, et al. Osteonecrosis after allogeneic stem cell transplantation in childhood. A case-control study in Italy. Haematologica. 2006;91(8):1096–9.
64. Sharma S, Yang S, Rochester R, et al. Prevalence of osteonecrosis and associated risk factors in children before allogeneic BMT. Bone Marrow Transplant. 2011;46(6):813–9.
65. Bassounas AE, Karantanas AH, Fotiadis DI, et al. Femoral head osteonecrosis: volumetric MRI assessment and outcome. Eur J Radiol. 2007;63(1):10–5.
66. Sansgiri RK, Neel MD, Soto-Fourier M, et al. Unique MRI findings as an early predictor of osteonecrosis in pediatric acute lymphoblastic leukemia. Am J Roentgenol. 2012;198(5):W432–9.
67. Ribeiro RC, Fletcher BD, Kennedy W, et al. Magnetic resonance imaging detection of avascular necrosis of the bone in children receiving intensive prednisone therapy for acute lymphoblastic leukemia or non-Hodgkin lymphoma. Leukemia. 2001;15(6):891–7.
68. Mont MA, Jones LC, Hungerford DS. Nontraumatic osteonecrosis of the femoral head: ten years later. J Bone Joint Surg. 2006;88(5):1117–32.

104
Medical Prevention and Treatment of Bone Metastases

Catherine Handforth, Stella D'Oronzo, and Janet Brown
Academic Unit of Clinical Oncology, University of Sheffield, Weston Park Hospital, Sheffield, UK

INTRODUCTION

The presence of metastatic bone disease is a devastating complication that has significant impact on morbidity and mortality in cancer patients. Although the type, incidence, and consequences of bone metastases (BM) may vary between primary cancer sites, all patients with BM require consideration as to how both their primary tumour and their bone disease should be managed in terms of avoiding complications and maximizing both quality of life and survival. This chapter will describe normal bone turnover, the pathophysiology of BM, and the most common clinical sequelae. We will then explain the current evidence for the prevention and treatment of BM for the most osteotropic tumours (breast, prostate, lung, and renal).

NORMAL BONE TURNOVER

Normal bone is in a dynamic metabolic state and undergoes constant remodelling. Resorption of old bone by osteoclasts is balanced with osteoblast-mediated synthesis of new bone matrix. Key to this process is the interaction between the cell-surface receptor activator of nuclear factor kappa-B (RANK) expressed by osteoclast precursor cells, which interacts with its ligand (RANKL) (predominantly made by osteoblasts) and results in osteoclast activation [1]. Negative regulation exists in the form of osteoprotegrin (OPG) produced by osteoblasts which acts as a soluble decoy receptor to RANKL [2].

PATHOPHYSIOLOGY OF BM

Current understanding of cancer metastasis to bone is based upon Paget's 'seed and soil' hypothesis, where cancer cells (seeds) preferentially interact with the bone microenvironment (soil) and cells within it to facilitate their growth and survival. The primary tumour is able to prepare a premetastatic niche in bone, recruiting stromal and bone-marrow derived cells through the secretion of exosomes and growth factors (such as vascular endothelial growth factor and placental growth factor) [3]. This cross-talk is essential to create a favourable microenvironment, where disseminated tumour cells (DTCs) may survive in a quiescent state for some time, before BM are detectable.

A self-propagating 'vicious cycle' takes effect when DTCs become active in bone. DTCs secrete a variety of factors such as PTHrP, which increases osteoblast expression of RANKL, reduces OPG levels, and results in osteoclastogenesis [4]. Breakdown of type 1 collagen releases growth factors such as TGFβ, PDGF, and IGF, which promote further tumour cell proliferation (and PTHrP release) [4]. In contrast, the bone-forming metastases commonly associated with prostate cancer lead to a predominantly sclerotic appearance. Although there is increased bone resorption, there is also upregulation in signalling molecules which activate osteoblasts, such as BMPs; TGFβ; FGF; Wnt-family members and endothelin-1 [5].

BIOMARKERS OF BONE TURNOVER

During normal bone turnover, proteins, protein fragments, and bone mineral components are released directly into the blood and urine. Metastatic bone disease accelerates this process altering the levels of bone turnover biomarkers (BTM). BTM are usually classified according to whether they are produced from bone formation or resorption [6, 7]. Examples of resorption BTM include the N- and C-terminal cross-linked telopeptide breakdown products of type 1 collagen (NTX and CTX), and markers of bone formation include bone-derived alkaline phosphatase (BALP) and the N- and C-terminal propeptides formed from cleavage of the type 1 procollagen molecule (P1NP and P1CP).

CLINICAL SEQUELAE OF BM

BM are associated with significant morbidity, which is an increasingly important consideration given improvements in cancer survival. Clinical sequelae are collectively termed skeletal related events (SREs) and include; pathological fracture; radiotherapy for bone pain; surgical intervention to treat or prevent an impending fracture; spinal cord or nerve root compression and hypercalcaemia. Not only do SREs negatively impact on quality of life, but they are also associated with further SREs and with worse survival outcomes [8, 9]. Additional psychosocial consequences may include increased risk of depression/anxiety, mobility or independence loss, and increased demand upon health and social care services.

BONE-TARGETED AGENTS

Several bone-targeted agents (BTAs) may be used in the management of BM. Bisphosphonates have a high affinity for mineralized bone matrix, where they bind selectively to hydroxyapatite and are released during resorption. Ingestion of biphosphonates by osteoclasts results in osteoclast inhibition, either through induction of apoptosis (non-nitrogen-containing biphosphonates such as clodronate) or inhibition of the HMG CoA reductase pathway, required for osteoclastogenesis (nitrogen-containing biphosphonates such as zoledronate, ibandronate, and pamidronate). Most biphosphonates are administered orally or as an intravenous infusion. The most commonly used intravenous biphosphonate is zoledronate, whose usual dose is 4 mg but requires adjustment in patients with a creatinine clearance of less than 60 mL/min.

Denosumab is a fully humanized monoclonal IgG$_2$ antibody that targets RANKL and prevents its interaction with RANK on osteoclast precursors. Consequent inhibition of osteoclast differentiation and activation causes a rapid reduction in bone resorption. In patients with BM, denosumab is given as a subcutaneous injection every 4 weeks, and the usual dose is 120 mg.

BTA are generally well tolerated. The commonest adverse effects are a mild and self-limiting flu-like syndrome, hypocalcaemia (may be avoided by supplementation with calcium and vitamin D), and renal impairment. The most serious adverse event is osteonecrosis of the jaw (ONJ), which occurs both with bisphosphonates and with denosumab. For example, in a prostate cancer study, denosumab caused ONJ with a similar frequency to zoledronate (1.1% versus 0.7% per 100 patient years) in the blinded phase of the study, but subsequent analysis of the open-label extension phase reported that the incidence rate rises with continued denosumab treatment to around 4.1% per 100 patient years [10]. The risk of ONJ increases with higher frequency and longer duration of BTA treatment, poor dental hygiene, invasive dental procedures (such as extractions), and concomitant medications (such as steroids) [11]. Guidelines to reduce the risk of ONJ are available and emphasize the importance of patient education, regular dental assessment, and avoidance of dental procedures during treatment with biphosphonates or denosumab.

In addition to the BTA described above, a range of alpha- and beta-emitting bone-seeking radiopharmaceuticals are available for the treatment of BM. Examples include radium-223 (alpha), strontium-89 and samarium-153 (beta). They cause apoptosis mainly via double strand DNA breaks and have selective toxicity in view of their affinity for areas of high bone turnover. The most important and often limiting toxicity associated with their use is profound myelosuppression, especially in cancer patients who may have received or be receiving other cytotoxic therapies. Myelosuppression is much less of an issue with radium-223 than with strontium-89 or samarium-153, because the alpha particles have only a short range of penetration, which minimizes the toxicity to healthy tissue including bone marrow.

BM IN PROSTATE CANCER

Prostate cancer is the most common noncutaneous malignancy in men, with more than 420,000 new cases and 90,000 deaths occurring annually in Europe. More than half of cases occur in men aged over 70 [10] and the incidence is projected to increase considerably over the next two decades as a result of the aging population. Bone is the most common site of metastasis; between 70% and 80% of men with advanced prostate cancer and 90% of men who die from prostate cancer will have bone involvement [11, 12]. Without specific bone-targeted treatment, around 50% of men with BM from prostate cancer will experience a skeletal-related event (SRE) within 2 years of diagnosis [8]. Although the recent introduction of newer therapies has led to improved survival, these agents do not specifically address the local consequences of BM, and metastatic prostate cancer remains incurable.

Biomarkers of bone turnover

BTM are frequently raised in men with prostate cancer and BM. Several studies have suggested that they may correlate with the extent of bone involvement, determine the need for bone-targeted treatment, and predict bone-related adverse events and survival [13]. However, there is not currently sufficient evidence to warrant their use in routine clinical practice, and they require further evaluation in large prospective studies.

Bone-targeted agents for the treatment of BM in prostate cancer

Castration-sensitive prostate cancer

Two large studies sought to determine the role of biphosphonates in patients with hormone-sensitive prostate cancer with BM. The first of these compared zoledronate with placebo, and the primary outcome was the time to first SRE [14]. The study was terminated early because of withdrawal of sponsor support, but of the 299 recorded SREs, the median time to first SRE did not significantly differ between the zoledronate and placebo groups (31.9 versus 29.8 months, $p=0.39$). Results from the ongoing multiarm multistage STAMPEDE trial (all men have hormone sensitive locally advanced, high risk or metastatic prostate cancer and receive androgen deprivation therapy [ADT] +/− experimental treatment) have shown that the addition of zoledronate to first line ADT does not improve overall survival (OS) [15]. Therefore, no evidence supports the use of biphosphonates before the development of castration resistance in men with prostate cancer.

Castration-resistant prostate cancer

Zoledronate was compared with placebo in a randomized controlled trial (RCT) of 643 patients with metastatic castration resistant prostate cancer (mCRPC), with a primary endpoint of SRE incidence [16]. There were significantly more SREs in the placebo group compared with the zolendronate groups (49% versus 38%, $p=0.028$). Secondary endpoints of time to first SRE and rate of SREs per year were also significantly less frequent in the zoledronate arms. Where zoledronate was continuously administered after a SRE, it reduced the risk of subsequent SREs, and was also associated with a marked decrease in BTM. A subsequent RCT compared chemotherapy alone with zoledronate and with strontium in men with mCRPC, and found that zoledronate did not improve either clinical progression-free survival (PFS) or OS, but did prolong the median SRE-free interval, suggesting that it may have a role as maintenance after chemotherapy [17].

Amongst men with localised CRPC, a prostate-specific antigen (PSA) greater than 8 ng/mL and/or a PSA doubling time of less than 10 months, denosumab was found to be superior to placebo in terms of BM-free survival (29.5 versus 25.2 months, $p=0.028$) and reduction in BTM [18]. However, there was no improvement in PFS or OS, and a high incidence of adverse effects reported in this study.

A large RCT compared zoledronate with denosumab in 1904 patients with mCRPC [19]. The time to first SRE was the primary endpoint, and those in the denosumab group were found to have a significantly longer time to first SRE than those in the zoledronate arm (20.7 versus 17.1 months, HR 0.82, $p=0.0008$ for superiority). Those in the denosumab arm experienced fewer SREs, and greater decrease in BTM than those randomized to zoledronate. OS and PFS did not differ between the groups. Hypocalcaemia (usually asymptomatic) and ONJ were more frequently observed with denosumab.

Currently, based upon the results of the trials mentioned above, patients with mCRPC who are at risk of SREs are often treated with BTA every 3 to 4 weeks. Treatment choice relies upon availability, ease of administration, potential adverse events, and patient and clinician preference.

Radiopharmaceuticals in the treatment of BM

Strontium-89 and Samarium-153 are radiopharmaceuticals licensed for the treatment of bone pain but do not improve survival in prostate cancer [20]. They are predominantly used in the palliative setting, where a range of contraindications (eg, recent radiotherapy or renal impairment) limit their use.

The ALSYMPCA trial compared radium-223 with placebo in 921 men with symptomatic BM from CRPC. Radium was associated with significantly improved OS compared with placebo (14.9 versus 11.3 months, $p<0.001$), with an equal incidence of myelosuppression [21]. Secondary endpoints were also favourable in terms of radium treatment, with longer time to the first symptomatic skeletal event (SSE) (15.6 versus 9.8 months with placebo, HR 0.66, $p<0.001$) and longer time to increase in levels of alkaline phosphatase and PSA. There were no safety issues when radium-223 was given alongside biphosphonates (continued as standard care in 41% patients who received biphosphonates before trial entry). The trial was terminated early for efficacy, and radium-223 received fast track approval.

A number of clinical trials involving radium-223 are currently open to recruitment (Table 104.1) [22]. These include phase II and III studies in combination with abiraterone, enzalutamide, external beam radiotherapy, sipuleucel-T, and ADT. A number of observational studies

Table 104.1. Ongoing phase II/III studies of radium-223 in prostate cancer.

Trial Identifier	Location	Trial Design	Trial Size	Main Endpoints	Start and Estimated Completion Dates
NCT02194842 mCRPC PEACE III	International multicentre	Randomized phase III open-label trial comparing enzalutamide versus a combination of radium-223 and enzalutamide in asymptomatic or mildly symptomatic CRPC with BM	560	Primary: rPFS Secondary: OS, prostate cancer-specific survival, first SSE, time to initiation of next systemic therapy, treatments elected after first disease progression, second PFS interval in sequential regimen, pain, time to pain progression, occurrence of AEs, time to first use of opioid analgesics, QoL	Oct 2015- April 2021
NCT02043678 (ERA223)	International multicentre	Phase III randomized, double-blind, placebo-controlled trial of radium-223 in combination with abiraterone and prednisolone in asymptomatic or mildly symptomatic chemotherapy-naïve mCRPC	800	Primary: SSE-free survival Secondary: OS, time to opiate use, time to pain progression, time to cytotoxic chemotherapy, rPFS, AEs	March 2014-Dec 2017
NCT02346526	USA	Phase II single arm open-label biomarker study of radium-223 in mCRPC	22	Primary: change from baseline in bone scan index at 2 months Secondary: mean percentage change in bone lesion area by 18-month survival status, changes in CTC and BTM number	May 2015- July 2021
NCT02023697	International multicentre	Three-arm, randomized, open-label phase II trial of radium-223 50 versus 80kBq/kg, and versus 50kBq/kg in an extended dosing schedule in CRPC patients with BM	389	Primary: SSE-free survival Secondary: OS, time to first SSE, rPFS, time to radiological progression, pain improvement and pain progression, AEs, change in analgesic use/24 hours	March 2014-July 2018
NCT02463799	USA	Phase II, randomized trial of sipuleucel-T with or without radium-223 in men with asymptomatic or minimally symptomatic CRPC and BM	34	Primary: immune response to treatment with sipuleucel-T measured by peripheral PA2024 T-cell proliferation Secondary: various immunological endpoints, safety of combined use of radium-223 and sipuleucel-T, time to PSA/ALP/pain progression, time to first opioid use, time to radiographic or clinical progression, time to first SRE, time to first chemotherapy use	Dec 2015-Dec 2020
NCT02225704	Ireland	Phase II interventional trial of radium-223 and enzalutamide in mCRPC	44	Primary: AEs Secondary: time to clinical and PSA progression, PSA response, time to first SRE, pain assessment, OS	June 2015- Dec 2017

NCT	Country	Study	N	Endpoints	Dates
NCT02278055	USA	Phase II observational trial of radium-223 in symptomatic mCRPC	63	Primary: Change in pain response Secondary: Change in BTM/ALP	Oct 2014-Oct 2018
NCT02582749	USA	Phase II randomized trial of ADT versus ADT and radium-223 in newly diagnosed prostate cancer with BM	204	Primary: rPFS Secondary: AE incidence, time to first SRE, secondary malignancy, PSA CR and PR, time to castration resistance, OS, pain score and analgesic use	April 2016-Jan 2020
NCT02484339	Germany	Phase II open-label trial of radium-223 + EBRT versus EBRT alone in CRPC with limited BM	274	Primary: Time to rPFS Secondary: Time to local and distant BM progression, OS, time to SRE, pain control, PSA response; time to ALP response	Dec 2014- Dec 2017
NCT02803437	Japan	Phase II prospective observational cohort study of the safety of radium-223 in mCRPC	300	Primary: AEs Secondary: Change in BTM, change in analgesic use	July 2016- Dec 2018
NCT02199197	USA	Phase II randomized trial of radium-223 + enzalutamide versus enzalutamide alone in mCRPC	50	Change in BTM	June 2014- June 2019
NCT02141438 REASSURE	International multicentre	Observational prospective cohort study of the safety of radium-223 in mCRPC patients and to evaluate the risk of developing second primary cancers	1334	Primary: incidence of developing second primary malignancies, incidence of SAEs, bone marrow suppression Secondary: OS, pain score and pain interference score using the Brief Pain Inventory questionnaire	Aug 2014- Dec 2023
NCT02450812 (URANIS)	Germany	Observational, prospective, single-arm cohort study of radium-223 in chemotherapy-naive mCRPC	500	Primary: OS Secondary: SSE-free survival, time to next tumour treatments, AE incidence, QoL, ADLs, and function	May 2015- Jan 2020
NCT02398526 PARABO	Germany	An observational, prospective, single-arm cohort trial for pain evaluation in radium-223-treated patients with bone metastases	300	Primary: pain response Secondary: OS, time to first SSE, AEs, various other pain endpoints, QoL, opioid use, relation between bone lesions and pain palliation, time to next tumour treatment	March 2015-Dec 2019

ADT = androgen deprivation therapy; AE: adverse event; ALP = bone-specific alkaline phosphatase; BM = bone metastases; BTM = biomarkers of bone turnover; CR = complete response; CRPC = castration resistant prostate cancer; CTCs = circulating tumour cells; EBRT = external beam radiotherapy; mCRPC = metastatic castration resistant prostate cancer; OS: overall survival, PFS = progression-free survival PR = partial response; PSA = prostate specific antigen; QoL = quality of life; rPFS = radiological progression-free survival; SAE = serious adverse event; SRE = skeletal related events; SSE = symptomatic skeletal event.

are also seeking to determine the safety and efficacy of radium-223 predominantly in the CRPC BM setting. The most common study endpoints are: OS, PFS, incidence of adverse effects, time to SRE, PSA response, change in BTM, quality of life and pain responses.

BTA in the prevention of BM

Studies have investigated the ability of both clodronate and zoledronate to prevent the development of BM. Clodronate was no more efficacious than placebo in reducing the time to development of BM, and was associated with significantly more adverse events [23].

Zoledronate was compared with placebo in men who had recently developed CRPC. A low event rate caused the study to fail to recruit to target, but comparison of the partial cohorts revealed that there was no significant difference in the time to development of first BM [24]. Due to the frequency of imaging, important insights were gained into the natural history of CRPC. After 2 years of follow-up, 33% of the 201 patients that received placebo developed BM, and the median time to BM was 30 months. Low baseline PSA and PSA velocity were found to be predictive of metastasis-free survival. A subsequent randomized phase III trial compared zoledronate with an observation arm in 1433 men with high risk localized disease, and failed to show any benefit from zoledronate in the prevention of BM [25].

Assessment of bone health

Overall bone health is an important consideration in all men with prostate cancer. Aside from BM, various factors are associated with loss of BMD and increased risk of SREs. The first of these is ADT, which is a cornerstone of therapy in prostate cancer treatment. It is most frequently achieved by the use of luteinizing hormone releasing hormone (LHRH) agonists (such as goserelin and leuprorelin), LHRH antagonists (such as degarelix) and antiandrogens, and leads to rapid decrease in circulating androgens and oestrogens to castration levels within 2 to 4 weeks. Loss of BMD is most rapid in the first year of treatment, with studies reporting between 5% and 10% BMD loss [26].

In addition to ADT, other prostate cancer treatments which directly affect bone include: glucocorticoids which are administered alongside abiraterone and with chemotherapy; comorbid conditions and medications used to treat them; and normal age-related BMD loss (usually between 0.5% and 1.0% per year). A 10% to 15% loss of BMD doubles the fracture risk [27], which predisposes men to subsequent fractures and increases the risk of mortality.

In men with prostate cancer, bone health assessment should be carried out by DXA scans, along with fracture risk assessment calculation using a validated tool such as FRAX [28]. Bone health should be optimized through lifestyle modifications (such as stopping smoking and minimizing alcohol intake), calcium and vitamin D supplementation, and regular exercise. Those found to be at high risk of fracture should be referred for consideration of BTA.

> **Summary box 1** Treatment and prevention of BM in prostate cancer
>
> - Evidence does not yet support the routine use of BTM in men with metastatic prostate cancer
> - SREs are common in men with prostate cancer and have an adverse effect on quality of life, function, and independence
> - There is no evidence to support the use of BTA before the development of castration resistance in men with prostate cancer and BM
> - Either zoledronate or denosumab may be used to reduce the risk of SREs in men with BM from CRPC
> - Radium-223 may be used to treat bone pain, and may improve survival and reduce SREs in men with BM from CRPC
> - No BTAs are currently recommended for the prevention of BM
> - Assessment and optimization of overall skeletal health is important to reduce the risk of SREs

BM IN BREAST CANCER

Epidemiology

Breast cancer is the most common cancer and the leading cause of cancer-related mortality in women, with more than 460,000 new cases and 130,000 deaths annually in Europe [29]. Owing to the recent advances in diagnostic and therapeutic strategies, both incidence and prevalence continue to increase at the expense of mortality. Longer life expectancy enables the development of metastatic disease including skeletal involvement; the latter affects up to 70% of patients with breast cancer and constitutes a major clinical, social, and economical concern [30].

Prevention of BM in breast cancer

Biphosphonates have been investigated for their ability to prevent BM in breast cancer, because of their potential to disrupt different stages of BM development. Biphosphonates can reduce tumour cell homing towards bone [31] and are internalized into breast cancer cells where they exert a pro-apoptotic effect [32–34]. They are also able to disrupt osteoclast activity and deprive tumour cells of bone-derived growth factors.

Several trials have investigated the role of biphosphonates in the adjuvant breast cancer setting [35], but the major contribution came from the ABCSG-12 and AZURE studies. The ABCSG-12 trial reported a disease-free survival (DFS) advantage after adjuvant zoledronate in premenopausal women undergoing ovarian suppression plus endocrine therapy [36]. Results from AZURE also showed improved DFS in those patients with established menopause, but did not find this to be the case in premenopausal women [37]. This could be partially explained by the loss of oestrogen in postmenopausal women which is necessary to promote DTC survival and proliferation in the BM niche. A large meta-analysis of 22,982 patients from 36 trials also found that zoledronate was associated with a significant reduction in distant recurrence in postmenopausal women compared with controls (18.4% versus 21.9%) [38].

Current evidence-based guidelines suggest that adjuvant biphosphonates (zoledronate, clodronate, or ibandronate) should be offered, along with vitamin D and calcium supplementation, in postmenopausal women with intermediate–high risk of breast cancer recurrence and premenopausal patients undergoing adjuvant ovarian suppression [35]. The treatment should be continued for between 3 and 5 years, depending on both recurrence and SRE risk.

Denosumab is also currently under investigation for the prevention of BM in women with breast cancer. Two ongoing trials (ABCSG 18 and D-CARE) are seeking to evaluate different schedules of adjuvant denosumab, in terms of bone-specific endpoints.

Treatment of BM in breast cancer

The aims of BM treatment in breast cancer are to control bone pain and prevent SREs. Biphosphonates are currently administered in this clinical setting and, in particular, zoledronate has proven its effectiveness in the presence of lytic lesions [39]. Treatment is initially given on a 4-weekly basis for the first year, and can subsequently be reduced to 12 weekly without loss of efficacy [40]. It is generally recommended that treatment is limited to 2 years to reduce the risk of adverse effects. However, treatment may be prolonged if it is felt to be of clinical benefit and the patient is closely monitored. Concurrent administration of calcium and vitamin D is recommended to reduce the risk of hypocalcemia.

A comparison of zoledronate and denosumab found that denosumab was superior in terms of time to first SRE in women with breast cancer and BM (HR 0.82, $p=0.01$. OS and adverse event rates were similar [41]. However, there is no current consensus as to the optimal duration of treatment.

Radium-223 has been investigated in preclinical models of breast cancer, where it was found to prevent cancer-related cachexia and reduce both the whole-body tumour burden and the number of osteolytic lesions [42]. A subsequent phase II nonrandomized trial investigated the efficacy and safety of radium-223 in 23 women with metastatic breast cancer with bone-predominant disease. There was a consistent reduction of BTM and in BM metabolic activity on FDG PET-CT [43]. There was also a significant reduction in pain, and the treatment was well tolerated.

Several clinical trials are currently on-going to investigate the efficacy and safety of radium-223 in combination with a range of other breast cancer treatments (such as hormonal therapy and chemotherapy with everolimus and capecitabine).

> **Summary box 2** BM prevention and treatment in breast cancer
>
> - BM affect up to 70% of patients with advanced breast cancer, are mostly lytic, and localized at the axial skeleton
> - Adjuvant zoledronate or clodronate are recommended for BM prevention in postmenopausal women and premenopausal patients undergoing ovarian suppression (along with vitamin D and calcium supplementation)
> - Either zoledronate or denosumab may be used with vitamin D and calcium in patients with BM from breast cancer
> - Adjuvant denosumab treatment is still under investigation
> - Phase II trials are ongoing to investigate the role of radium-223 in the BM setting

BM IN LUNG CANCER

Therapeutic advances in lung cancer have improved survival and led to more frequent bone involvement. Approximately 30% to 40% of patients with lung cancer will develop BM during the course of their disease.

A large retrospective study of 661 patients with nonsmall cell lung cancer (NSCLC) found that BM affected 57.7% of patients at diagnosis and were associated with a median survival of 9.5 months [44]. The osteolytic pattern prevailed (74.3%) over mixed (14.3%) and osteoblastic (11.4%) lesions. The study authors proposed a score to predict the OS after BM diagnosis, identifying four prognostic factors: age over 65 years, concomitant visceral metastases, ECOG performance status (>2), and nonadenocarcinoma histology. The presence of more than two of these factors was associated with worse prognosis (median survival 5 versus 8 months, $p=0.001$) [44].

BTA in the treatment of BM in lung cancer

A phase III randomized trial of 773 patients with bone-metastatic solid malignancies including lung cancer compared zoledronate to placebo [45]. After 21 months, zoledronate reduced the risk of SRE by 31% (RR 0.693), delayed the median time to first SRE (236 versus 155 days, $p=0.009$), and decreased annual SRE incidence (1.74/year versus 2.71/year, $p=0.012$), as compared with placebo. A small retrospective study of pamidronate reported improvement in OS (15.4 versus 2.1 months,

$p=0.001$) and favorable toxicity profile when compared with no treatment [46].

Comparison of zoledronate and denosumab in a subgroup analysis of 702 NSCLC patients in a larger study revealed an OS improvement (9.5 versus 8 months, $p=0.01$) in the denosumab group [47]. The efficacy of denosumab in addition to first-line chemotherapy in NSCLC is currently being investigated.

> **Summary box 3** BM in lung cancer
> - Lung cancer is the third most common malignancy spreading to bone
> - BM are involved in up to 30% to 40% of lung cancer patients during their lifetime, are mainly lytic, and affect the axial skeleton in 75% of NSCLC patients.
> - Zoledronate is effective in preventing SREs and delaying their first occurrence
> - Denosumab is noninferior to zoledronate in SRE prevention
> - Systemic treatments for lung cancer, including chemotherapies and targeted therapies, may contribute to bone health maintenance

BM IN RENAL CELL CARCINOMA

Bone is the second most common site of distant metastatic spread in patients with advanced renal cell carcinoma (RCC). Approximately one-third of patients will have BM at diagnosis or will develop BM [9]. BM in RCC are highly lytic, and tend to be more aggressive than BM from other tumours. Advances in targeted therapies and immunotherapy have improved survival in RCC, but survival outcomes remain considerably poorer for those with BM [48].

Biomarkers of bone turnover

In various other tumour types, BTM have been found to be diagnostic of BM and been able to predict outcomes, including SREs and survival [49–51]. Studies in RCC have included small numbers, and reported conflicting results. BTM are not currently used in the management of patients with RCC, and require further validation in larger prospective studies.

BTA for the management of BM in RCC

There is no evidence to support the use of BTA in the prevention of BM in RCC. The only biphosphonate licensed for use in RCC is zoledronate. Its clinical benefit was shown by a large RCT with placebo, which included 74 patients with RCC. Amongst these 74 patients, there was a reduction in the incidence of SRE incidence with zoledronate (37% versus 74% for placebo, $p=0.014$). Treatment with zoledronate also prolonged the time to first SRE and significantly extended the median time to BM progression [52]. Despite this, zoledronate remains underused in patients with RCC BM compared with patients with other tumours [53].

A large study compared zoledronate to denosumab, and found that denosumab was noninferior to zoledronate with respect to SRE, disease progression, adverse events, and OS [54]. Denosumab is now licensed for the treatment of RCC BM in the USA and Europe.

Consideration of treatment toxicity is especially important when BTA are used in patients with RCC. Renal impairment is common and therefore biphosphonates should be used with caution, with appropriate dose reductions and close monitoring. Denosumab does not exacerbate pre-existing renal impairment and may be a safer treatment option. It has also been reported that the concurrent administration of antiangiogenic therapies with biphosphonates increases the risk of ONJ [55].

Novel treatments in RCC such as cabozantanib have recently been shown to have survival benefits, and may also offer an advantage in terms of specific bone endpoints [56, 57], and it is hoped that these findings will be confirmed by larger studies. There are a number of early phase trials of immune checkpoint inhibitors (such as PD-L1 inhibitors) that are currently being undertaken in RCC. It remains to be shown as to whether these will have a future role in the management of metastatic bone disease.

> **Summary box 4** BM in renal cell carcinoma
> - BM in RCC are aggressive and frequently result in SREs
> - Evidence does not support the use of BTM in RCC, or the use of BTA to prevent BM
> - Zoledronate is licensed for the prevention of SRE, but is relatively underused in RCC
> - Denosumab is noninferior to zoledronate, and has advantages in terms of ease of administration and less renal impairment
> - The effects of various novel treatments for metastatic RCC on bone endpoints are yet to be determined

REFERENCES

1. Boyle WJ, Simonet WS Lacey DL. Osteoclast differentiation and activation. Nature. 2003;423:337–42.
2. Hofbauer LC, Khosla S, Dunstan CR, et al. Estrogen stimulates gene expression and protein production of osteoprotegerin in human osteoblastic cells. Endocrinology. 1999;140:4367–70.
3. Scenay J, Smyth MJ, Möller A. The pre-metastatic niche: finding common ground. Cancer Metastasis Rev. 2013;32:449–64.

4. Sims NA, Gooi JH. Bone remodeling: multiple cellular interactions required for coupling of bone formation and resorption. Semin Cell Dev Biol. 2008;19:444–51.
5. Wood SL, Westbrook JA, Brown JE. Omic-profiling in breast cancer metastasis to bone: implications for mechanisms, biomarkers and treatment. Cancer Treat Rev. 2014;40(1):139–52.
6. Fohr B, Dunstan CR Seibel MJ. Markers of bone remodelling in metastatic bone disease. J Clin Endocrinol Metab. 2003;88:5059–75.
7. Coleman R, Brown J, Terpos E et al. Bone markers and their prognostic value in metastatic bone disease: clinical evidence and future directions. Cancer Treat Rev. 2008;34:629–39.
8. Saad F, Gleason DM, Murray R et al. Long-term efficacy of zoledronic acid for the prevention of skeletal complications in patients with metastatic hormone-refractory prostate cancer. J Natl Cancer Inst. 2004;96:879–82.
9. Woodward E, Jagdev S, McParland L, et al. Skeletal complications and survival in renal cancer patients with bone metastases. Bone. 2011;48(1):160–6.
10. CRUK. Cancer statistics. http://www.cancerresearchuk.org/health-professional/cancer-statistics/statistics-by-cancer-type/prostate-cancer/incidence#heading-One. Published 2013. Accessed April 15, 2016.
11. So A, Chin J, Ont O, et al. Management of skeletal-related events in patients with advanced prostate cancer and bone metastases: incorporating new agents into clinical practice. J Can Urol Assoc. 2012;6(6):465–70.
12. Bubendorf L, Schöpfer A, Wagner U, et al. Metastatic patterns of prostate cancer: an autopsy study of 1,589 patients. Hum Pathol. 2000;31(5):578–83.
13. Brown JE, Sim S. Evolving role of bone biomarkers in castration-resistant prostate cancer. Neoplasia. 2010;9:685–96.
14. Smith MR, Halabi S, Ryan CJ, et al. Randomized controlled trial of early zoledronic acid in men with castration-sensitive prostate cancer and bone metastases: results of CALGB 90202 (alliance). J Clin Oncol. 2014;32(11):1143–50.
15. James ND, Sydes MR, Clarke NW, et al. Addition of docetaxel, zoledronic acid, or both to first-line long-term hormone therapy in prostate cancer (STAMPEDE): survival results from an adaptive, multiarm, multistage, platform randomised controlled trial. Lancet. 2016;387:1163–77.
16. Saad F, Gleason DM, Murray R, et al. A randomized, placebo-controlled trial of zoledronic acid in patients with hormone-refractory metastatic prostate carcinoma. J Natl Cancer Inst. 2002;94:1458–68.
17. James ND, Pirrie SJ, Pope AM, et al. Clinical outcomes and survival following treatment of metastatic castrate-refractory prostate cancer with docetaxel alone or with strontium-89, zoledronic acid, or both: the TRAPEZE Randomized Clinical Trial. JAMA Oncol. 2016;2(4):493–9.
18. Smith MR, Saad F, Coleman R, et al. Denosumab and bone-metastasis-free survival in men with castration-resistant prostate cancer: results of a phase 3, randomised, placebo-controlled trial. Lancet. 2012;379:39–46.
19. Fizazi K, Carducci M, Smith M, et al. Denosumab versus zoledronic acid for treatment of bone metastases in men with castration-resistant prostate cancer: a randomised, double-blind study. Lancet. 2011;377:813–22.
20. El-Amm J, Aragon-Ching JB. Targeting bone metastases in metastatic castration-resistant prostate cancer. Clin Med insights Oncol. 2016;10(Suppl 1):7–9.
21. Parker C, Nilsson S, Heinrich D, et al. Alpha emitter radium-223 and survival in metastatic prostate cancer. N Engl J Med. 2013;369(3):213–23.
22. National Institutes of Health. Clinicaltrials.gov.
23. Mason MD, Sydes MR, Glaholm J, et al. Oral sodium clodronate for nonmetastatic prostate cancer — results of a randomized double-blind placebo-controlled trial: Medical Research Council PR04. J Natl Cancer Inst. 2007;99(10):765–76.
24. Smith MR, Kabbinavar F, Saad F, et al. Natural history of rising serum prostate-specific antigen in men with castrate non-metastatic prostate cancer. J Clin Oncol. 2005;23(13):2918–25.
25. Wirth M, Tammela T, Cicalese V, et al. Prevention of bone metastases in patients with high-risk nonmetastatic prostate cancer treated with zoledronic acid: efficacy and safety results of the Zometa European Study (ZEUS). Eur Urol. 2015;67(3):482–91.
26. Daniell HW, Dunn SR, Ferguson DW et al. Progressive osteoporosis during androgen deprivation therapy for prostate cancer. J Urol. 2000;163:181–6.
27. Faulkner KG. Bone matters: are density increases necessary to reduce fracture risk? J Bone Miner Res. 2000;15:183.
28. University of Sheffield. FRAX online calculator. https://www.sheffield.ac.uk/FRAX/. Accessed June 7, 2018.
29. Cancer Research UK (CRUK) and National cancer intelligence network (NCIN). CRUK breast cancer statistics. http://www.cancerresearchuk.org/health-professional/cancer-statistics/statistics-by-cancer-type/breast-cancer#KSe8ZK5GxJiqGX75.99. Accessed June 7, 2018.
30. Senkus E, Kyriakides S, Penault-Llorca F, et al. Primary breast cancer: ESMO clinical practice guidelines for diagnosis, treatment and follow-up. Ann Oncol. 2013;24(Suppl.6).
31. Hoffmann O, Aktas B, Goldnau C, et al. Effect of ibandronate on disseminated tumor cells in the bone marrow of patients with primary breast cancer: a pilot study. Anticancer Res. 2011;31:3623–8.
32. Van der Pluijm G, Vloedgraven H, van Beek E van der W-P, et al. Bisphosphonates inhibit the adhesion of breast cancer cells to bone matrices in vitro. J Clin Invest. 1996;13:698–705.
33. Boissier S, Ferreras M, Peyruchaud O, et al. Bisphosphonates inhibit breast and prostate carcinoma cell invasion, an early event in the formation of bone metastases. Cancer Res. 2000;13:2949–54.
34. Senaratne SG, Pirianov G, Mansi JL, et al. Bisphosphonates induce apoptosis in human breast cancer cell lines. Br J Cancer. 2000;82:1459–68.
35. Hadji P, Coleman RE, Wilson C, et al. Adjuvant bisphosphonates in early breast cancer: consensus guidance for

clinical practice from a European Panel. Ann Oncol. 2016;27:379–90.
36. Gnant M, Mlineritsch B, Schippinger W, et al. Endocrine therapy plus zoledronic acid in premenopausal breast cancer. N Eng J Med 2009;360:679–91.
37. Coleman R, Cameron D, Dodwell D, et al. Adjuvant zoledronic acid in patients with early breast cancer: final efficacy analysis of the AZURE (BIG 01/04) randomised open- label phase 3 trial. Lancet Oncol. 2014;15:997–1006.
38. Coleman R, Gnant M, Paterson A, et al. Effects of bisphosphonate treatment on recurrence and cause-specific mortality in women with early breast cancer: a meta-analysis of individual patient data from randomized trials. In: San Antonio Breast Cancer Symposium, 2013:Abstract S4–07.
39. Rosen LS, Gordon DH, Dugan W, et al. Zoledronic acid is superior to pamidronate for the treatment of bone metastases in breast carcinoma patients with at least one osteolytic lesion. Cancer. 2004;100:36–43.
40. Hortobagyi GN, Lipton A, Chew HK, et al. Efficacy and safety of continued zoledronic acid every 4 weeks versus every 12 weeks in women with bone metastases from breast cancer: results of the OPTIMIZE-2 trial. J Clin Oncol. 2014;32:5s.
41. Stopeck AT, Lipton A, Body JJ, et al. Denosumab compared with zoledronic acid for the treatment of bone metastases in patients with advanced breast cancer: a randomized, double- blind study. J Clin Oncol. 2010;28:5132–9.
42. Suominen MI, Rissanen JP, Käkönen R, et al. Survival benefit with radium-223 Dichloride in a mouse model of breast cancer bone metastasis. J Natl Cancer Inst. 2013;105:908–16.
43. Coleman R, Aksnes AK, Naume B, et al. A phase IIa, nonrandomized study of radium-223 dichloride in advanced breast cancer patients with bone-dominant disease. Breast Cancer Res Treat. 2014;(145):411–18.
44. Santini D, Barni S, Intagliata S, et al. Natural history of non-small- cell lung cancer with bone metastases. Sci Rep. 2015;5:Article 18670. http://www.nature.com/articles/srep18670. Accessed June 7, 2018.
45. Rosen L, Gordon D, Tchekmedyian N, et al. Long term efficacy and safety of zoledronic acid in the treatment of skeletal metastases in patients with non small cell lung carcinoma and other solid tumors: a randomized, phase III, double blind, placebo- controlled trial. Cancer. 2004;100:2613–21.
46. Spizzo G, Seeber A, Mitterer M. Routine use of pamidronate in NSCLC patients with bone metastasis: results from a retrospective analysis. Anticancer Res. 2009;29:5245–50.
47. Henry DH, Costa L, Goldwasser F, et al. Randomized, double-blind study of denosumab versus zoledronic acid in the treatment of bone metastases in patients with advanced cancer (excluding breast and prostate cancer) or multiple myeloma. J Clin Oncol. 2011;29(9):1125–32.
48. Riechelmann RP, Chin S, Wang L, et al. Sorafenib for metastatic renal cancer: the Princess Margaret Experience. Am J Clin Oncol. 2008;31:182–7.
49. Armstrong AJ, Eisenberger MA, Halabi S, et al. Biomarkers in the management and treatment of men with metastatic castration-resistant prostate cancer. Eur Urol. 2012;61(3):549–9.
50. Zhao H, Han KL, Wang ZY, et al. Value of C-telopeptide-cross-linked Type I collagen, osteocalcin, bone-specific alkaline phosphatase and procollagen Type I N-terminal propeptide in the diagnosis and prognosis of bone metastasis in patients with malignant tumors. Med Sci Monit. 2011;17(11):CR626–633.
51. Coleman RE, Major P, Lipton A, et al. Predictive value of bone resorption and formation markers in cancer patients with bone metastases receiving the bisphosphonate zoledronic acid. J Clin Oncol. 2005;23(22):4925–35.
52. Lipton A, Zheng M, Seaman J. Zoledronic acid delays the onset of skeletal related events and progression of skeletal disease in patients with advanced renal cell carcinoma. Cancer. 2003;98;962–9.
53. Wood SL, Brown JE. Skeletal metastasis in renal cell carcinoma: current and future management options. Cancer Treat Rev. 2012;38(4):284–91.
54. Henry DH, Costa L, Goldwasser FJ et al. Randomized, double-blind study of denosumab versus zoledronic acid in the treatment of bone metastases in patients with advanced cancer (excluding breast and prostate cancer) or multiple myeloma. J Clin Oncol. 2011;29(9):1125–32.
55. Christodoulou C, Pervena A, Klouvas G, et al. Combination of bisphosphonates and antiangiogenic factors induces osteonecrosis of the jaw more frequently than bisphosphonates alone. Oncology. 2009;76:209–11.
56. Quoix E, Zalcman G, Oster J-P, et al. Carboplatin and weekly paclitaxel doublet chemotherapy compared with monotherapy in elderly patients with advanced non-small-cell lung cancer: IFCT-0501 randomised, phase 3 trial. Lancet. 2011;378(9796):1079–88.
57. Choueiri C, Escudier B, Powles T, et al. Cabozantinib versus everolimus in advanced renal-cell carcinoma. N Engl J Med. 2015;373:1814–23.

105
Radiotherapy of Skeletal Metastases

Srinivas Raman[1], K. Liang Zeng[1], Oliver Sartor[2], Edward Chow[1], and Øyvind S. Bruland[3]

[1]Department of Radiation Oncology, Odette Cancer Centre, Sunnybrook Health Sciences Centre, University of Toronto, Toronto, ON, Canada
[2]Departments of Medicine and Urology, Tulane Medical School, New Orleans, LA, USA
[3]Department of Oncology, The Norwegian Radium Hospital, University of Oslo, Oslo, Norway

INTRODUCTION

Bone is the most common site of symptomatic metastases. Two-thirds to three-quarters of patients with advanced disease from breast and prostate carcinomas have skeletal metastases [1]. Pain is the most common presenting symptom [1, 2]. Additional complications including pathological fracture, nerve entrapment/spinal cord compression (SCC), bone marrow insufficiency, and hypercalcaemia can have a devastating impact on quality of life [1, 3]. SCC is of particular concern to cancer patients [4], as these patients can have debilitating consequences from neurological injury.

Optimal management of bone metastases involves a combination of medical treatment, radiation therapy, surgery, bone-targeted radiopharmaceuticals, bisphosphonates, and denosumab depending on the biology of the disease, extent of the involvement, and the life expectancy of the patient. This chapter discusses the importance of external beam radiotherapy (EBRT) and the emerging role of radiopharmaceuticals in the management of bone metastases.

EXTERNAL BEAM RADIOTHERAPY

Skeletal metastases are the most frequent indication for palliative radiotherapy. EBRT relieves pain from localized skeletal metastases [5, 6]. However, the lack of tumor-only selectivity limits its clinical use. Further, because skeletal metastases usually are multiple and throughout the axial skeleton [1, 2, 4], larger or multiple fields are often necessary. Table 105.1 outlines considerations when prescribing palliative radiotherapy for bone metastases.

Pain palliation

From more than 25 randomized clinical trials (RCTs) and three recent meta-analyses, single-fraction (SF) EBRT provides equivalent pain relief compared with multifraction (MF) EBRT for uncomplicated bone metastases [6–9].

In one of the first RCTs, 90% experienced some degree of pain relief and 54% achieved complete pain palliation [10]. The trial initially concluded that the low-dose, short-course schedules were as effective as the high-dose protracted schedules. The UK Bone Pain Trial Working Party randomized 765 patients with bone metastases to either a SF or a MF regimen [11]. There were no significant differences in the time to first improvement in pain, time to complete pain relief, time to first increase in pain at any time up to 12 months after randomization, and no differences in the incidence of nausea, vomiting, spinal cord compression, or pathological fracture between the two groups. Retreatment was, however, twice as common after SF than after MF radiotherapy. The study concluded that a SF is as safe and effective for the palliation of metastatic bone pain for at least 12 months with greater convenience and lower cost than MF treatment.

The large Dutch Bone Metastases Study included 1171 patients and found similar results [12]. In this trial, the retreatment rates were 25% in the single

Primer on the Metabolic Bone Diseases and Disorders of Mineral Metabolism, Ninth Edition. Edited by John P. Bilezikian.
© 2019 American Society for Bone and Mineral Research. Published 2019 by John Wiley & Sons, Inc.
Companion website: www.wiley.com/go/asbmrprimer

Table 105.1. Factors to consider when prescribing palliative radiotherapy (EBRT) for bone metastases.

EBRT — Single fraction (SF)	EBRT — Fractionated (MF)
Indication: "pain relief" Short life expectancy Concomitant visceral metastases Poor performance status Inflammatory pain Aspects of cost and inconvenience	Indication: "local tumor control" Expected long-term survival Predominantly bone or bone only metastasis Good performance status Neuropathic pain Spinal cord compression Postoperative EBRT after an orthopedic procedure in selected cases Impending fractures where surgery is not indicated

8 Gy arm, 7% in the MF arm, and more pathological fractures were observed in the SF group, but the absolute percentage was low. In a cost-utility analysis of this RCT, there was no difference in life expectancy or quality-adjusted life expectancy. The estimated cost of radiotherapy, including retreatments and nonmedical costs, was significantly lower for the SF than for the MF schedule [13].

A Scandinavian RCT planned to recruit 1000 patients with painful bone metastases randomized to single 8 Gy or 30 Gy in 10 fractions [14]. The data monitoring committee recommended closure after 376 patients because interim analyses indicated that the treatment groups had similar outcomes. Equivalent pain relief within the first 4 months was experienced, and no differences were found for fatigue, global quality of life, and survival between the groups [14].

Two meta-analyses published in 2003 showed no significant difference in complete and overall pain relief between SF and MF EBRT [6, 7]. Results were similar, with Wu and colleagues reporting a complete response rate of 33% and 32% after SF and MF EBRT, respectively, compared with 34% and 32% for Sze and colleagues. Overall response rates from the two meta-analyses were 62% and 59% [7], compared with 60% and 59% [6], for SF and MF, respectively. Most patients experienced pain relief in the first 2 to 4 weeks after EBRT [7]. Side effects were similar and generally consisted of nausea and vomiting.

An updated meta-analysis reviewed 25 RCTs involving a total of 2818 randomizations to SF arms and 2799 to MF arms [8]. The overall response rate to SF was 60%, and the complete response rate was 23%, not significantly different from the 61% and 24% experienced by patients randomized to MF. No differences in acute toxicity, pathological fracture, or spinal cord compression were found, similar to the conclusions of the 2003 systematic reviews. However, SF was associated with a higher rate of retreatment when compared with MF (20% versus 8%). Recent therapeutic guidelines for the treatment of bone metastases from the American College of Radiology (ACR) Appropriateness Criteria Expert Panel on Radiation Oncology and American Society for Radiation Oncology (ASTRO) endorsed these findings [15–17].

Neuropathic pain and spinal cord compression

Certain groups of patients with a neuropathic pain component could benefit from a protracted schedule. In a comparison of single 8 Gy versus 20 Gy in five fractions for 272 patients with a neuropathic pain component [18], SF was not as effective as MF; however, it was also not significantly worse. The authors recommended MF as standard for patients with neuropathic pain but those with short survival or poor performance status, as well as when cost/inconvenience of MF is relevant, SF could be used [18]. The addition of pregabalin, an antineuropathic agent, to radiotherapy has not been found to be effective in cancer bone pain [19]. However, this has not been tested in a selected population with neuropathic pain.

In the treatment of neoplastic SCC, at least three RCTs have shown that in patients with a poor prognosis, short course RT can provide equivalent palliation to longer treatment schedules. Maranzano and colleagues showed that short course RT (16 Gy in two fractions) was equivalent to split-course RT (30 Gy in eight fractions; 5 Gy × 3 + 3 Gy × 5) in a patient population with a survival prognosis of less than 6 months [20]. In a similar trial, Maranzano and colleagues showed that a single 8 Gy fraction was equivalent to 16 Gy in two fractions in a population with a survival prognosis of less than 3 months [21]. Recently, Rades and colleagues showed that 20 Gy in five fractions was equivalent to 30 Gy in 10 fractions [22]. Single fraction or short-course radiotherapy should be considered in patients with SCC and poor prognosis. Longer-course radiotherapy may lead to better local control of SCC [23] and may be considered in those with better prognosis.

Patchell and colleagues showed that functional outcomes after surgery and EBRT was superior to EBRT alone in a landmark trial [24]. This paradigm has been challenged because of limitations in the original study design and small sample size. More recent case-matched comparisons suggest an equivalency between both modalities of treatment [25]. The ASTRO guidelines recommend that an individual approach must be taken considering many factors including tumor histology, extent of involvement, patient prognosis, mobility, spine stability, and previous treatments [26].

Impending fracture and risk prediction

An impending fracture has a significant likelihood of fracture. Although some believe that all patients with proximally located femoral metastases should undergo preventive surgery, this would result in a large number of unnecessary surgical procedures [27]. Furthermore, a proportion of patients will not be candidates for an operative procedure, or will refuse surgical intervention. Often minimum life expectancy, a reasonable performance status, manageable comorbidities, and adequate remaining bone to support the implanted hardware are required in order to justify the morbidity and mortality risk [28]. In candidates for surgery, postoperative radiotherapy has a role in reducing subsequent local tumor growth and avoidance of further orthopedic surgeries for loosening of the fixation(s) [29].

If orthopedic intervention is not appropriate, patients may receive EBRT alone. Although EBRT can provide pain relief and tumour control, it does not restore bone stability, and remineralization will take weeks to months [30]. Patients should be warned of the increased risk of fracture in the periradiation period caused by an induced hyperemic response at the periphery of the tumour that temporarily weakens the adjacent bone. Pain relief may allow the patient to be more mobile and, hence, at greater risk for fracture. As such, measures to reduce anatomical forces across the lesion (crutches, a sling, or a walker) are routinely introduced during this time.

Although there is no consensus on appropriate dose fractionation, most authors recommend a MF course of EBRT in a patient with an impending or established fracture [31]. One retrospective series analyzed 27 pathological fractures in various sites treated with doses of 40 to 50 Gy over 4 to 5 weeks. Healing with remineralization was observed in 33%, with pain relief in 67% [31]. Koswig and Budach reported that the recalcification showed a significant difference between patients treated with 30 Gy in 10 fractions (173%) and those treated with a single 8 Gy (120%, $p<0.0001$) [32].

Reirradiation

Subsets of patients with metastatic disease have longer life expectancies than in the past because of advances in systemic therapy and may therefore outlive the duration of benefit provided by their initial palliative EBRT. This may require reirradiation of previously treated sites [33].

Retreatment rates after SF varied from 18% to 25% compared with 7% to 9% after MF [11, 12, 14, 34]. Sande and colleagues showed that patients in the SF arm received significantly more reirradiations as compared with the MF arm (27% versus 9%, $p=0.002$) [35]. The Dutch Bone Metastases Study Group found similar findings [36]. Of patients not responding to initial radiation, 66% who initially received a single 8 Gy responded to retreatment, compared with 33% of patients who initially received a MF course. Retreatment in patients after pain progression was successful in 70% of those who received SF initially, compared with 57% of those who received more than one fraction. Overall, reirradiation was effective in 63% of all such treated patients. Jeremic and colleagues also noticed the efficacy of the second single 4 Gy reirradiation for patients with painful bone metastases who had already twice received single fraction radiation [37].

The radiation dose, efficacy, and safety of reirradiation was assessed in NCIC CTG SC.20 [38]. Patients who had previously received in-field radiation ($n=850$) were randomized between 8 Gy in one fraction and 20 Gy in multiple fractions. There was a 28% response rate in the 8 Gy arm and a 32% response rate in the 20 Gy arm; this difference was not statistically significant and was within the predetermined noninferiority limits. The difference in response rates between 8 Gy (45%) and 20 Gy (51%) was also not statistically significant but greater than the predetermined noninferiority limits. The overall toxicity rates were acceptable with both treatment schedules. It is important to consider reirradiation of sites of metastatic bone pain, especially when this follows an initial period of response after EBRT. However, data from the same trial also shows that a proportion of initial nonresponders will respond to reirradiation and therefore, reirradiation must be considered in all patients whenever safe and feasible [38].

Pain flare

Palliative radiotherapy is not without side effects. Pain flare is a very common with reported incidence ranging from 2% to 44% after conventional EBRT and 10% to 68% for stereotactic body radiation therapy (SBRT) [39]. One published definition of pain flare is a two-point increase in the worst pain score (0–10) compared with baseline worst pain with no decrease in analgesic intake, or a 25% increase in analgesic intake employing daily oral morphine equivalent dose (OMED) with no decrease in worst pain score [40].

Dexamethasone as prophylaxis for radiation-induced pain flare has shown to be effective from the results of a recent phase III trial [41]. Dexamethasone (8 mg/day for 5 days) reduced the incidence of pain flare from 35% to 26%. A phase II observational study similarly supports the use of dexamethasone in SBRT [42].

Survival prediction

The development of bone metastases is generally associated with incurable disease. Accurate prediction of life expectancy of patients with advanced cancer can guide appropriate clinical decisions, direct planning of supportive services, help patients and families plan end-of-life issues, steer the allocation of resources, and determine eligibility for hospice referrals or enrollment into clinical trials.

Models have been developed to specifically predict the life expectancy of patients with bone metastases. Chow and colleagues developed a life expectancy model for predicting survival in patients attending a palliative radiotherapy clinic [43]. Six factors were found to statistically impact survival: primary cancer site, site of metastases,

Karnofsky Performance Status (KPS), and fatigue, appetite, and shortness of breath scores. This model was later reduced to three factors without loss of predictive ability: primary cancer site, site of metastases, and KPS [44]. Both a survival prediction score (SPS) and number of risk factors (NRF) method were used to predict survival based on three prognostic groups. This model for predicting survival was externally validated at other institutions [45, 46]. Krishnan and colleagues also developed the TEACHH model for predicting life expectancy at different ends of the prognostic spectrum: more than 1 year and less than 3 months [47]. The abbreviation represents factors predictive for a shorter life expectancy: primary tumor (T), ECOG status (E), age (A), previous palliative chemotherapy courses (C), recent hospitalizations (H), and hepatic metastasis (H). Three prognostic groups with different median survivals of 19.9 months, 5 months, and 1.7 months were created based on the presence or absence of these risk factors.

Similar models for predicting survival have been developed for various other clinical scenarios including spinal metastases [48, 49], SCC [50, 51], and reirradiation of bone metastases [52], and can help guide clinical decision making in patients with bone metastases.

BONE-SEEKING RADIOPHARMACEUTICALS

Treatment with bone-seeking radiopharmaceuticals (BSRs) is an intriguing alternative for selected patients. The preferential delivery of ionizing radiation after intravenous injection in areas of amplified osteoblastic activity will target multiple (symptomatic and asymptomatic) metastases simultaneously. The target is Ca-OH-apatite (hydroxyapatite) particularly abundant in sclerotic metastases from prostate cancer but also present, although more heterogeneously distributed, in mixed sclerotic/osteolytic metastases from breast cancer. This is evident from the biodistribution image common to all BSRs – exemplified as "hot-spots" visualized on a routine diagnostic bone scan (by ^{99m}Tc-MDP; a radiolabelled bisphosphonate). BSRs effectively relieve pain, and have been thoroughly reviewed [53–57]. In their commercially available formulations, the radioisotopes involved used to be beta-emitters: strontium-89 dichloride (Metastron, GE Healthcare, Cardiff, UK) or ^{153}Sm-EDTMP (Quadramet, Schering AG, Berlin, Germany, and Cytogen Co., Princeton, NJ, USA), but a new era has emerged with the approval of Xofigo, a bone-targeted alpha radiopharmaceutical comprising radium-223 (Bayer Healthcare, Whippany, NJ, USA).

Because of the millimeter range of the emitted electrons, the crossirradiation of the bone marrow represents a concern. Following intravenous injection, bone marrow is the dose-limiting organ, and disease-associated bone marrow suppression already present in these patients often results in delayed and unpredictable recovery. This limits the usefulness of beta-emitting BSRs, especially when dosages are increased to deliver potential antitumor radiation levels and/or repeated treatments are attempted. Some clinical studies to date have reported on the feasibility of combining BRSs and chemotherapy [58–61].

The bone-seeking calcium mimetic, radium-223 is bound into the hydroxyapatite within the stroma of osteoblastic or sclerotic metastases. Emitted alpha-particle radiation induces double-strand DNA breaks resulting in a potent and highly localized cytotoxic effect in the target areas containing metastatic cancer cells and their microenvironment, and the short path length of the high-energy alpha-particles implies that toxicity to adjacent healthy tissue and particularly the normal bone marrow is minimized.

In a phase I study of single-dosage administration of escalating amounts of the natural bone-seeker 223Ra in 25 patients with bone metastases from breast and prostate cancer [62], dose-limiting hematological toxicity was not observed. A phase II RCT of EBRT plus either saline or 223Ra injections (given four times at 4-week intervals) in patients with bone metastases from castrate-resistant prostate cancer resulted in a statistically significant decrease from baseline compared with placebo both in bone alkaline phosphatase and prostate-specific antigen [63]. A favorable adverse event profile was observed with minimal bone marrow toxicity for patients who received 223Ra. Importantly, and surprisingly, survival analyses from this phase II trial indicated a survival benefit for 223Ra [63]. The phase III ALSYMPCA trial then established 223Ra efficacy and safety in CRPC patients with bone metastases and no visceral disease [64]. Importantly, all patients received standard of care including various second-line hormonal therapies with placebo or 223Ra. In this setting, in the intent to treat analysis, 223Ra improved OS significantly (median 14.9 versus 11.3 months; HR = 0.70; 95% CI: 0.58, 0.83) and time to first symptomatic skeletal event was significantly prolonged (median 15.6 versus 9.8 months; HR = 0.66; 95% CI: 0.52, 0.83). OS was improved regardless if given to docetaxel naïve patients, or those patients randomized to treatment post-docetaxel. The safety profile was favorable, and secondary end points (including time to tALP increase, tALP response rate, and tALP normalization rate) favored radium-223 treatment [64, 65]. This trial was pivotal for multiple regulatory agencies throughout the world to approve commercial use of this 223Ra.

The combination of radiotherapy or radionuclide therapy with bisphosphonates has been reviewed [66] and examined in ALSYMPCA [65]. Given mechanistic similarity, denosumab may act similarly to bisphosphonates. Interestingly, nonrandomized trials suggest that OS may be improved with combinations of denosumab and 223Ra compared with 223Ra alone [67]. Overall, one can hypothesize that agents such as bisphosphonates and denosumab provide synergy with 223Ra, but this needs to be vigorously tested in prospective phase III randomized settings. There is a strong theoretical basis for synergistic activity between the two modalities, with enhanced healing in metastatic disease in animal studies resulting in better biochemical strength, stability, and bone microarchitecture. Further, the enhanced deposition of hydroxyapatite may improve the 223Ra deposition at sites of bone metastatic disease.

REFERENCES

1. Coleman RE. Clinical features of metastatic bone disease and risk of skeletal morbidity. Clin Cancer Res. 2006;12(20):6243s–9s.
2. Hage WD, Aboulafia AJ, Aboulafia DM. Incidence, location, and diagnostic evaluation of metastatic bone disease. Orthop Clin North Am. 2000;31(4):515–28.
3. British Association of Surgical Oncology Guidelines. The management of metastatic bone disease in the United Kingdom. The Breast Specialty Group of the British Association of Surgical Oncology. Eur J Surg Oncol. 1999;25(1):3–23.
4. Coleman RE, Smith P, Rubens RD. Clinical course and prognostic factors following bone recurrence from breast cancer. Br J Cancer. 1998;77(2):336–40.
5. Chow E, Wong R, Hruby G, et al. Prospective patient-based assessment of effectiveness of palliative radiotherapy for bone metastases. Radiother Oncol. 2001;61(1):77–82.
6. Sze WM, Shelley MD, Held I, et al. Palliation of metastatic bone pain: single fraction versus multifraction radiotherapy — a systematic review of randomised trials. Clin Oncol. 2003;15(6):345–52.
7. Wu JS-Y, Wong R, Johnston M, et al., Cancer Care Ontario Practice Guidelines Initiative Supportive Care Group. Meta-analysis of dose-fractionation radiotherapy trials for the palliation of painful bone metastases. Int J Radiat Oncol Biol Phys. 2003;55(3):594–605.
8. Chow E, Zeng L, Salvo N, et al. Update on the systematic review of palliative radiotherapy trials for bone metastases. Clin Oncol. 2012;24(2):112–24.
9. Chow E, Harris K, Fan G, et al. Palliative radiotherapy trials for bone metastases: a systematic review. J Clin Oncol. 2007;25(11):1423–36.
10. Tong D, Gillick L, Hendrickson FR. The palliation of symptomatic osseous metastases: final results of the Study by the Radiation Therapy Oncology Group. Cancer. 1982;50(5):893–9.
11. Bone Pain Trial Working Party. 8 Gy single fraction radiotherapy for the treatment of metastatic skeletal pain: randomised comparison with a multifraction schedule over 12 months of patient follow-up. Radiother Oncol. 1999;52(2):111–21.
12. Steenland E, Leer JW, van Houwelingen H, et al. The effect of a single fraction compared with multiple fractions on painful bone metastases: a global analysis of the Dutch Bone Metastasis Study. Radiother Oncol. 1999;52(2):101–9.
13. Hout WB van den, Linden YM van der, Steenland E, et al. Single-versus multiple-fraction radiotherapy in patients with painful bone metastases: cost–utility analysis based on a randomised trial. J Natl Cancer Inst. 2003;95(3):222–9.
14. Kaasa S, Brenne E, Lund J-A, et al. Prospective randomised multicenter trial on single fraction radiotherapy (8 Gy x 1) versus multiple fractions (3 Gy x 10) in the treatment of painful bone metastases. Radiother Oncol. 2006;79(3):278–84.
15. Janjan N, Lutz ST, Bedwinek JM, et al. Therapeutic guidelines for the treatment of bone metastasis: a report from the American College of Radiology Appropriateness Criteria Expert Panel on Radiation Oncology. J Palliat Med. 2009;12(5):417–26.
16. Janjan N, Lutz ST, Bedwinek JM, et al. Clinical trials and socioeconomic implication in the treatment of bone metastasis: a report from the American College of Radiology Appropriateness Criteria Expert Panel on Radiation Oncology. J Palliat Med. 2009;12(5):427–31.
17. Lutz S, Balboni T, Jones J, et al. Palliative radiation therapy for bone metastases: update of an ASTRO Evidence-Based Guideline. Pract Radiat Oncol. 2017;7(1):4–12.
18. Roos DE, Turner SL, O'Brien PC, et al. Randomized trial of 8 Gy in 1 versus 20 Gy in 5 fractions of radiotherapy for neuropathic pain due to bone metastases (Trans-Tasman Radiation Oncology Group, TROG 96.05). Radiother Oncol. 2005;75(1):54–63.
19. Fallon M, Hoskin PJ, Colvin LA, et al. Randomized double-blind trial of pregabalin versus placebo in conjunction with palliative radiotherapy for cancer-induced bone pain. J Clin Oncol. 2016;34(6):550–6.
20. Maranzano E, Bellavita R, Rossi R. Radiotherapy alone or surgery in spinal cord compression? The choice depends on accurate patient selection. J Clin Oncol. 2005;23(32):8270–4.
21. Maranzano E, Trippa F, Casale M, et al. 8Gy single-dose radiotherapy is effective in metastatic spinal cord compression: results of a phase III randomized multicentre Italian trial. Radiother Oncol. 2009;93(2):174–9.
22. Rades D, Šegedin B, Conde-Moreno AJ, et al. Radiotherapy with 4 Gy × 5 Versus 3 Gy × 10 for metastatic epidural spinal cord compression: final results of the SCORE-2 Trial (ARO 2009/01). J Clin Oncol. 2016;34(6):597–602.
23. Rades D, Lange M, Veninga T, et al. Final results of a prospective study comparing the local control of short-course and long-course radiotherapy for metastatic spinal cord compression. Int J Radiat Oncol Biol Phys. 2011;79(2):524–30.
24. Patchell RA, Tibbs PA, Regine WF, et al. Direct decompressive surgical resection in the treatment of spinal cord compression caused by metastatic cancer: a randomised trial. Lancet. 2005;366(9486):643–8.
25. Rades D, Huttenlocher S, Dunst J, et al. Matched pair analysis comparing surgery followed by radiotherapy and radiotherapy alone for metastatic spinal cord compression. J Clin Oncol. 2010;28(22):3597–604.
26. Lutz S, Berk L, Chang E, et al. Palliative radiotherapy for bone metastases: an ASTRO evidence-based guideline. Int J Radiat Oncol Biol Phys. 2011;79(4):965–76.
27. van der Linden YM, Kroon HM, Dijkstra SP, et al. Simple radiographic parameter predicts fracturing in metastatic femoral bone lesions: results from a randomised trial. Radiother Oncol. 2003;69(1):21–31.

28. Healey JH, Brown HK. Complications of bone metastases. Cancer 2000;88(S12):2940–51.
29. Bickels J, Dadia S, Lidar Z. Surgical management of metastatic bone disease. J Bone Jt Surg Am. 2009;91(6):1503–16.
30. Agarawal JP, Swangsilpa T, van der Linden Y, et al. The role of external beam radiotherapy in the management of bone metastases. Clin Oncol. 2006;18(10):747–60.
31. Rieden K, Kober B, Mende U, et al. [Radiotherapy of pathological fractures and skeletal lesions in danger of fractures]. Strahlenther Onkol Organ Dtsch Rontgengesellschaft Al. 1986;162(12):742–9.
32. Koswig S, Budach V. [Remineralization and pain relief in bone metastases after after different radiotherapy fractions (10 times 3 Gy vs. 1 time 8 Gy). A prospective study]. Strahlenther Onkol Organ Dtsch Rontgengesellschaft Al. 1999;175(10):500–8.
33. Morris DE. Clinical experience with retreatment for palliation. Semin Radiat Oncol. 2000;10(3):210–21.
34. Hartsell WF, Scott CB, Bruner DW, et al. Randomized trial of short- versus long-course radiotherapy for palliation of painful bone metastases. J Natl Cancer Inst. 2005;97(11):798–804.
35. Sande TA, Ruenes R, Lund JA, et al. Long-term follow-up of cancer patients receiving radiotherapy for bone metastases: results from a randomised multicentre trial. Radiother Oncol. 2009;91(2):261–6.
36. van der Linden YM, Lok JJ, Steenland E, et al. Single fraction radiotherapy is efficacious: a further analysis of the Dutch Bone Metastasis Study controlling for the influence of retreatment. Int J Radiat Oncol Biol Phys. 2004;59(2):528–37.
37. Jeremic B, Shibamoto Y, Igrutinovic I. Second single 4 Gy reirradiation for painful bone metastasis. J Pain Symptom Manage. 2002;23(1):26–30.
38. Chow E, Linden YM van der, Roos D, et al. Single versus multiple fractions of repeat radiation for painful bone metastases: a randomised, controlled, non-inferiority trial. Lancet Oncol. 2014;15(2):164–71.
39. McDonald R, Chow E, Rowbottom L, et al. Incidence of pain flare in radiation treatment of bone metastases: a literature review. J Bone Oncol. 2014;3(3–4):84–9.
40. Hird A, Chow E, Zhang L, et al. Determining the incidence of pain flare following palliative radiotherapy for symptomatic bone metastases: results from three Canadian cancer centers. Int J Radiat Oncol Biol Phys. 2009;75(1):193–7.
41. Chow E, Meyer RM, Ding K, et al. Dexamethasone in the prophylaxis of radiation-induced pain flare after palliative radiotherapy for bone metastases: a double-blind, randomised placebo-controlled, phase 3 trial. Lancet Oncol. 2015;16(15):1463–72.
42. Khan L, Chiang A, Zhang L, et al. Prophylactic dexamethasone effectively reduces the incidence of pain flare following spine stereotactic body radiotherapy (SBRT): a prospective observational study. Support Care Cancer. 2015;23(10):2937–43.
43. Chow E, Fung K, Panzarella T, et al. A predictive model for survival in metastatic cancer patients attending an outpatient palliative radiotherapy clinic. Int J Radiat Oncol Biol Phys. 2002;53(5):1291–302.
44. Chow E, Abdolell M, Panzarella T, et al. Predictive model for survival in patients with advanced cancer. J Clin Oncol. 2008;26(36):5863–9.
45. Glare P, Shariff I, Thaler HT. External validation of the number of risk factors score in a palliative care outpatient clinic at a comprehensive cancer center. J Palliat Med. 2014;17(7):797–802.
46. Angelo K, Dalhaug A, Pawinski A, et al. Survival prediction score: a simple but age-dependent method predicting prognosis in patients undergoing palliative radiotherapy. Int Sch Res Not. 2014:e912865.
47. Krishnan MS, Epstein-Peterson Z, Chen Y-H, et al. Predicting life expectancy in patients with metastatic cancer receiving palliative radiotherapy: the TEACHH model. Cancer. 2014;120(1):134–41.
48. van der Linden YM, Dijkstra SPDS, Vonk EJA, et al., Dutch Bone Metastasis Study Group. Prediction of survival in patients with metastases in the spinal column: results based on a randomized trial of radiotherapy. Cancer. 2005;103(2):320–8.
49. Westhoff PG, de Graeff A, Monninkhof EM, et al. An easy tool to predict survival in patients receiving radiation therapy for painful bone metastases. Int J Radiat Oncol Biol Phys. 2014;90(4):739–47.
50. Rades D, Rudat V, Veninga T, et al. Prognostic factors for functional outcome and survival after reirradiation for in-field recurrences of metastatic spinal cord compression. Cancer. 2008;113(5):1090–6.
51. Rades D, Douglas S, Veninga T, et al. Validation and simplification of a score predicting survival in patients irradiated for metastatic spinal cord compression. Cancer. 2010;116(15):3670–3.
52. Chow E, Ding K, Parulekar WR, et al. Predictive model for survival in patients having repeat radiation treatment for painful bone metastases. Radiother Oncol. 2016;118(3):547–51.
53. Lewington VJ. Bone-seeking radionuclides for therapy. J Nucl Med. 2005;46 Suppl 1:38S–47S.
54. Silberstein EB. Systemic radiopharmaceutical therapy of painful osteoblastic metastases. Semin Radiat Oncol. 2000;10(3):240–9.
55. Finlay IG, Mason MD, Shelley M. Radioisotopes for the palliation of metastatic bone cancer: a systematic review. Lancet Oncol. 2005;6(6):392–400.
56. Bauman G, Charette M, Reid R, et al. Radiopharmaceuticals for the palliation of painful bone metastasis — a systemic review. Radiother Oncol. 2005;75(3):258–70.
57. Reisfield GM, Silberstein EB, Wilson GR. Radiopharmaceuticals for the palliation of painful bone metastases. Am J Hosp Palliat Med. 2005;22(1):41–6.
58. Tu S-M, Kim J, Pagliaro LC, et al. Therapy tolerance in selected patients with androgen-independent prostate cancer following strontium-89 combined with chemotherapy. J Clin Oncol. 2005;23(31):7904–10.
59. Pagliaro LC, Delpassand ES, Williams D, et al. A phase I/II study of strontium-89 combined with gemcitabine in the treatment of patients with androgen independent

prostate carcinoma and bone metastases. Cancer. 2003;97(12):2988–94.
60. Sciuto R, Festa A, Rea S, et al. Effects of low-dose cisplatin on 89Sr therapy for painful bone metastases from prostate cancer: a randomized clinical trial. J Nucl Med. 2002;43(1):79–86.
61. Akerley W, Butera J, Wehbe T, et al. A multiinstitutional, concurrent chemoradiation trial of strontium-89, estramustine, and vinblastine for hormone refractory prostate carcinoma involving bone. Cancer. 2002;94(6): 1654–60.
62. Nilsson S, Larsen RH, Fosså SD, et al. First clinical experience with alpha-emitting radium-223 in the treatment of skeletal metastases. Clin Cancer Res. 2005;11(12): 4451–9.
63. Nilsson S, Franzén L, Parker C, et al. Bone targeted radium-223 in symptomatic, hormone-refractory prostate cancer: a randomised, multicentre, placebo-controlled phase II study. Lancet Oncol. 2007;8(7):587–94.
64. Parker C, Nilsson S, Heinrich D, et al. Alpha emitter radium-223 and survival in metastatic prostate cancer. N Engl J Med. 2013;369(3):213–23.
65. Sartor O, Coleman R, Nilsson S, et al. Effect of radium-223 dichloride on symptomatic skeletal events in patients with castration-resistant prostate cancer and bone metastases: results from a phase 3, double-blind, randomised trial. Lancet Oncol. 2014;15(7): 738–46.
66. Vassiliou V, Bruland O, Janjan N, et al. Combining systemic bisphosphonates with palliative external beam radiotherapy or bone-targeted radionuclide therapy: interactions and effectiveness. Clin Oncol. 2009;21(9): 665–7.
67. Saad F, Carles J, Gillessen S, et al. Radium-223 and concomitant therapies in patients with metastatic castration-resistant prostate cancer: an international, early access, open-label, single-arm phase 3b trial. Lancet Oncol. 2016;17(9):1306–16.

106
Concepts and Surgical Treatment of Metastatic Bone Disease

Kristy Weber[1] and Scott L. Kominsky[2]

[1]Department of Orthopaedic Surgery, University of Pennsylvania, Philadelphia, PA, USA
[2]Departments of Orthopaedic Surgery and Oncology, Johns Hopkins University, Baltimore, MD, USA

INTRODUCTION

Over 1.6 million people are diagnosed with cancer each year [1] and approximately 50% of those will develop bone metastasis. As treatments improve for primary and metastatic disease, patients are living longer with their disease. This often causes them to experience the morbidities of related bone disease. Although the most worrisome clinical problem is progressive disease in the skeleton, patients can also experience treatment-related osteoporosis. Additional physiological disruptions in patients with bone metastasis include anemia and hypercalcemia. The bone lesions themselves can cause extreme pain and put the patient at risk for pathological fractures. Patients become less mobile and may function at a lower level. Prolonged immobilization due to pain or risk of fracture creates potential problems with thromboembolic disease or decubitus ulcers. Lesions in the vertebral region can cause progressive neurological deficits. Overall quality of life is often markedly diminished.

Comprehensive treatment of bone metastasis is beyond the scope of this chapter. Importantly, advances in systemic chemotherapy, targeted biological therapy, and immunotherapy have been increasingly effective for some underlying primary cancers. Different forms and varying doses of radiation are used to target metastatic cancer cells within the bone to provide palliative pain relief and potentially abrogate the need for surgical intervention. Minimally invasive treatments such as cryoablation and kyphoplasty are effective at relieving pain when surgery and/or further radiation is not feasible. This chapter focuses on treatment that affects the neoplastic process as well as the bone microenvironment. A brief review of the molecular events related to metastatic bone disease will be discussed. The use of bisphosphonate or anti-RANKL therapy as well as surgical stabilization in these settings will be summarized.

BIOLOGY OF METASTATIC BONE LESIONS

Tumor–bone interactions

Colonization and growth of tumor cells in the bone can be thought of as occurring in three phases: colonization, dormancy, and expansion (Fig. 106.1).

First, upon arrival in the bone vasculature, disseminated tumor cells are drawn into the bone microenvironment by signals, such as SDF-1, produced by mesenchymal cells of the osteoblastic niche [2, 3]. Second, consistent with the long latency that frequently occurs between the treatment of primary tumors and relapse in bone, evidence suggests that initial tumor–bone cell interaction leads to tumor cell quiescence/dormancy. In prostate cancer, stimulation of the annexin II receptor on the surface of prostate cancer cells by annexin II on the surface of bone lining cells induces tumor cell expression of the receptor tyrosine kinase AXL [4, 5]. AXL can in turn be activated by its ligand growth arrest-specific protein 6 (GAS6) on the surface of bone lining cells, though GAS6 may also be found on various mesenchymal cells within the bone environment. Additional mechanisms of tumor

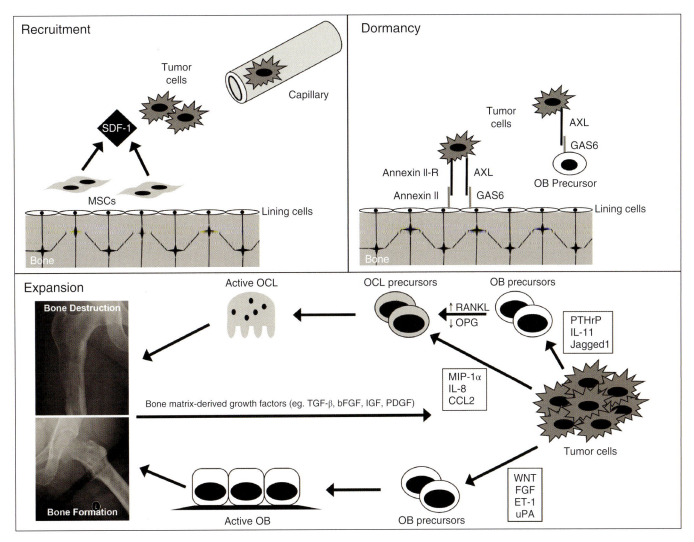

Fig. 106.1. Tumor cell recruitment, dormancy, and the general cycle of bone destruction (osteolytic metastasis – breast, lung, thyroid, and kidney cancers) and bone formation (osteoblastic metastasis – prostate cancer). MSC = mesenchymal stromal cells; OB = osteoblast; OCL = osteoclast; OPG = osteoprotegerin.

cell dormancy likely exist and are currently the subject of investigation. In the third phase of bone metastasis, tumor cells awake from dormancy and form an expanding mass. Though the mechanism whereby tumor cells are released from quiescence remains unknown, it is believed to be an extrinsic event that alters the growth-suppressive environment, such as the initiation of bone remodeling.

Once actively dividing, tumor cells alter the bone microenvironment, skewing the normally balanced process of bone remodeling toward either net bone destruction or formation, causing significant morbidity [6–11]. The majority of research aimed at elucidating tumor–bone interactions has been performed in the field of breast cancer bone metastasis, a predominantly osteolytic disease. Studies using mouse models have provided evidence of a "vicious cycle" of tumor growth and bone destruction driven by TGF-β. In the course of normal bone remodeling, TGF-β is released, stimulating breast cancer cells to secrete PTHrP [12]. PTHrP then stimulates osteoblast precursors to increase receptor activator for nuclear factor κB ligand (RANKL), which increases osteoclast differentiation. An increased number of active osteoclasts then destroy more bone. This releases numerous growth factors in addition to TGF-β including basic FGF (bFGF), IGF, and PDGF, fueling tumor growth and restarting the cycle. Strikingly, along with PTHrP, numerous other TGF-β-regulated genes have been shown to promote osteolytic bone metastasis including IL-11, CTGF, COX-2, Jagged1, and ADAMTS1 [13–16]. Given its widespread effect on bone metastasis, the TGF-β signaling pathway has become the target of various experimental therapeutic drugs [17]. In addition to TGF-β-regulated genes, studies indicate that the chemokines IL-8, CCL2, and MIP-1α promote osteolytic metastasis using mouse models of breast cancer and myeloma [18–20].

Rather than increasing osteoclast numbers via effects on osteoblasts, these factors exert direct effects on osteoclast precursors, stimulating their recruitment and differentiation into mature osteoclasts.

As opposed to other common cancer types affecting the skeleton, prostate cancer causes predominantly osteoblastic lesions in bone. Utilizing mouse models of prostate cancer, several tumor-produced paracrine factors that stimulate osteoblasts to form new bone have been identified. Endothelin-1 (ET-1) has been shown to directly stimulate the proliferation and differentiation of osteoblast precursors [21]. Beyond its direct effects, ET-1 enhances the activation of WNT signaling in osteoblast precursors by decreasing expression of the WNT inhibitor, Dickkopf homolog 1 (Dkk1) [21]. Additionally, production of the osteoblast precursor mitogen, FGF, was reported to promote prostate cancer growth and sclerotic lesions in the skeleton [22]. Urokinase plasminogen activator (uPA) has also been reported to influence prostate cancer bone metastasis, wherein it induces proteases capable of performing multiple functions including the degradation of PTHrP and liberation of IGF-1 from its inhibitory binding proteins, thus favoring osteoblast differentiation and activity [23, 24].

Pharmacological treatment of bone metastasis

Currently, there are several pharmacological agents used in the treatment of patients with bone metastasis which are covered in more detail elsewhere in the Primer. Bisphosphonates and RANKL inhibitors are currently used to decrease the incidence of skeletally related events in patients with metastatic bone disease. The most commonly used bisphosphonate in the USA is zoledronic acid. Alternatively, denosumab is a RANKL inhibitor that is approved for treatment of patients with bone metastasis and shown to be both safe and effective. In systematic reviews of patients with metastatic bone disease, denosumab is often more effective than zoledronic acid at preventing skeletally-related events [25–28].

Bisphosphonates and denosumab have been used with published success to treat bone pain and hypercalcemia in breast and prostate cancer and myeloma and are most efficacious when used as an adjunct to systemic cancer therapies [25–28]. They are given when bone metastasis is first diagnosed, because they significantly increase the time to the first skeletal-related event. Osteonecrosis of the jaw is a rare but serious complication in patients with bone metastasis on bisphosphonates and is estimated to occur with a frequency of 0.6% to 6.2% [29]. It is recommended that patients have a routine dental examination before starting on bisphosphonates, and risk factors include high cumulative doses, poor oral health, and a history of dental extractions. There is also a risk of atypical femoral fractures from using antiresorptive agents, and there are now guidelines of how long to continue treatment to mitigate this potential complication [30].

Radium-223 dichloride is a radiopharmaceutical approved for use in the treatment of prostate cancer bone metastasis. Radium-223 acts as a calcium mimetic, accumulating in areas of new bone formation, with a half-life of 11.4 days. In clinical trials, radium-223 has been shown to increase patient survival and delay the time to the first skeletally-related event [31].

SURGICAL TREATMENT OF METASTATIC BONE DISEASE

Impending fractures

As patients with bone metastasis are unlikely to be surgically cured, the primary focus of orthopedic oncologists is to improve quality of life. If a bone lesion is discovered at an early stage, radiation or systemic medical treatment may prevent further destruction and allow surgery to be avoided. However, if a lesion progresses despite nonsurgical treatment or is initially discovered after there has been extensive cortical destruction and pain, surgical stabilization should be considered [32]. Patients who have prophylactic fixation of their extremity before it actually fractures have a shorter hospitalization, quicker return to premorbid function, and less hardware complications [33]. Elective stabilization also allows the medical oncologist and surgeon to coordinate operative treatment and systemic chemotherapy. The difficulty lies in reliably determining which bone lesions will eventually result in a fracture. Several classifications have been proposed which involve determination of pain, cortical destruction, and size/location of the bone lesion [34, 35]. CT-based structural rigidity analysis to predict fracture risk in patients with bone metastasis was 100% sensitive and 90% specific in a recent study [36].

Surgical treatment

The goals of surgical treatment of patients with bone metastasis are to improve function and decrease pain. Treatment of impending or actual fractures secondary to metastatic bone disease utilizes different principles than those used for routine traumatic fractures [32]. The underlying bone quality is often poor, and the patient may have progressive osseous destruction despite treatment.

Upper extremity bone metastases are less common than those in the lower extremity and can often be treated nonoperatively. However, if patients require their upper extremities for weight bearing (ie, have lower extremity lesions that cause pain or are unable to bear weight without assistive devices), then surgical treatment should be considered in order to improve mobility. Lesions in the scapula and clavicle are generally treated nonoperatively with radiation or minimally invasive options, because most surgical options do not improve function in these areas. Extensive bone destruction in the proximal humerus is either treated by a proximal humeral prosthetic replacement or an intramedullary device with or without supplemental methylmethacrylate if secure fixation can be achieved. Diaphyseal humeral lesions are treated with intramedullary devices or occasionally intercalary metal spacers [37, 38]. Distal humeral lesions are less common and are stabilized with crossed intramedullary pins, dual plating, or segmental distal humeral prosthetic reconstruction. Bone metastases distal to the elbow are extremely rare and are treated on an individual basis.

Lower extremity metastasis is more common than in the upper extremities and has a larger impact on quality of life due to the need for weight bearing. Pelvic lesions are usually treated nonoperatively or with minimally invasive techniques if the acetabulum is not affected. Acetabular lesions are treated according to specific classification schemes depending on the extent and location of bone loss [39]. Patients with severe bone loss in the acetabulum should have a reasonably long predicted life span and good performance status in order to make the procedure and recovery worthwhile. Metastases to the femoral neck are common and often lead to hip fractures [40]. The treatment is either a bipolar hemiarthroplasty or a total hip replacement depending on the status of the acetabular disease [41]. Internal fixation with plates and screws is not indicated, because there is a high risk of hardware failure with disease progression in this area. In the intertrochanteric and subtrochanteric regions, options for prosthetic reconstruction or intramedullary fixation are available depending on the extent and location of bone loss as well as the tumor histology (Figs 106.2 and 106.3).

Tumors that are less responsive to systemic treatment or radiation (ie, renal cell carcinoma) are often treated more aggressively with surgical resection (Fig. 106.3). Femoral diaphyseal lesions are treated with intramedullary fixation [41]. It is important that the intramedullary device includes stabilization of the femoral neck to avoid a future hip fracture. Distal femoral lesions are treated by intramedullary fixation, plate fixation or prosthetic reconstruction. Lesions distal to the knee are uncommon and treated on an individual basis.

The most common site of bone metastasis is the thoracic spine. If patients are neurologically intact and there are no fracture fragments impinging on the spinal cord, radiation is often the treatment of choice. If patients have intractable pain, progressive neurological deficits, or deformity progression, surgical stabilization should be considered [42, 43].

Minimally invasive options

In selected patients, minimally invasive procedures provide an alternative to surgery and can produce long-lasting pain relief. Kyphoplasty and vertebroplasty are commonly used techniques for patients who have osteolytic spine metastasis without neurological compromise. Both techniques can be performed safely, stabilize the collapsed vertebral body, and yield quick pain relief

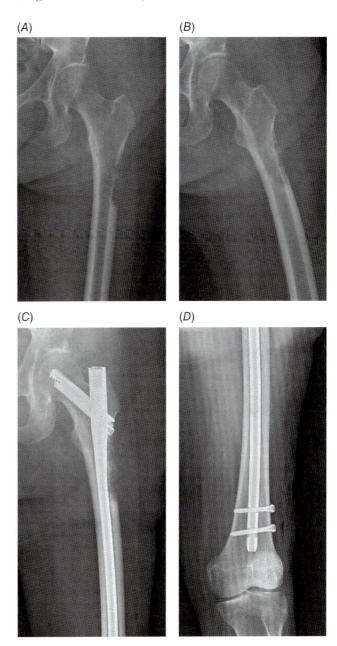

Fig. 106.2. (A) and (B) Anteroposterior and lateral radiographs of the left proximal femur in a 62-year-old woman with metastatic lung carcinoma reveals an osteolytic lesion involving the lateral cortex in the subtrochanteric region at high risk for pathological fracture. There were no additional lesions in the remaining femur. (C) and (D) Anteroposterior radiographs after stabilization of the femur with an intramedullary reconstruction nail. The patient received postoperative radiation to the femur after the incisions healed.

[44, 45]. Radiofrequency ablation, cryoablation, or similar techniques are used to treat metastasis in multiple bony sites with most patients achieving some measure of pain relief [46]. Cyberknife treatment is a type of minimally invasive outpatient radiosurgery used for spine metastasis. A comparison with external beam radiation (EBRT)

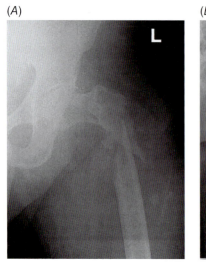
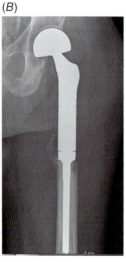

Fig. 106.3. (A) A radiograph of the left hip in a 68-year-old man with metastatic renal cell carcinoma shows a pathological fracture through an osteolytic lesion in the subtrochanteric femoral region. (B) Postoperative left hip radiograph after a cemented proximal femoral megaprosthesis to reconstruct and stabilize the hip. The patient received postoperative radiation that encompassed the entire reconstruction.

showed that EBRT was more cost effective but with more acute toxicities and need for further subsequent interventions at the same vertebral level [47].

CONCLUSION

In summary, the more that is discovered about the mechanisms of cancer-induced bone destruction, the more likely there will be effective pharmacological agents to maintain skeletal integrity and decrease tumor growth. For patients who are at risk for pathological fracture or have sustained an actual fracture, there are a multitude of surgical options to stabilize the skeleton with a focus on pain control and return to function.

REFERENCES

1. Siegel RL, Miller KD, Jemal A. Cancer statistics, 2016. CA Cancer J Clin. 2016; 66:7–30.
2. Kaplan RN, Psaila B, Lyden D. Niche-to-niche migration of bone-marrow-derived cells. Trends Mol Med. 2007;13:72–81.
3. Wang N, Docherty FE, Brown HK, et al. Prostate cancer cells preferentially home to osteoblast-rich areas in the early stages of bone metastasis: evidence from in vivo models. J Bone Miner Res. 2014;29:2688–96.
4. Shiozawa Y, Havens AM, Jung Y, et al. Annexin II/annexin II receptor axis regulates adhesion, migration, homing, and growth of prostate cancer. J Cell Biochem. 2008;105:370–80.
5. Shiozawa Y, Pedersen EA, Patel LR, et al. GAS6/AXL axis regulates prostate cancer invasion, proliferation, and survival in the bone marrow niche. Neoplasia 2010;12:116–27.
6. Chirgwin JM, Mohammad KS, Guise TA. Tumor-bone cellular interactions in skeletal metastases. J Musculoskelet Neuronal Interact. 2004;4:308–18.
7. Kominsky S, Doucet M, Brady K, et al. TGF-β influences the development of renal cell carcinoma bone metastasis. J Bone Miner Res. 2007;22:37–44.
8. Mundy GR. Metastasis to bone: causes, consequences and therapeutic opportunities. Nat Rev Cancer. 2002;2:584–93.
9. Park JI, Lee MG, Cho K, et al. Transforming growth factor-beta1 activates interleukin-6 expression in prostate cancer cells through the synergistic collaboration of the Smad2, p38-NF-kappaB, JNK, and Ras signaling pathways. Oncogene. 2003;22:4314–32.
10. Roodman GD. Role of cytokines in the regulation of bone resorption. Calcif Tissue Int. 1993;53(Suppl. 1):S94–8.
11. Kwan Tat S, Padrines M, Théoleyre S. IL-6, RANKL, TNF-alpha/IL-1: interrelations in bone resorption pathophysiology. Cytokine Growth Factor Rev. 2004;15:49–60.
12. Kakonen SM, Selander KS, Chirgwin JM, et al. Transforming growth factor-beta stimulates parathyroid hormone-related protein and osteolytic metastases via Smad and mitogen-activated protein kinase signaling pathways. J Biol Chem. 2002;277:24571–8.
13. Kang Y, Siegel PM, Shu W, et al. A multigenic program mediating breast cancer metastasis to bone. Cancer Cell 2003;3:537–49.
14. Singh B, Berry JA, Shoher A, et al. COX-2 involvement in breast cancer metastasis to bone. Oncogene 2007;26:3789–96.
15. Sethi N, Dai X, Winter CG, et al. Tumor-derived Jagged1 promotes osteolytic bone metastasis of breast cancer by engaging notch signaling in bone cells. Cancer Cell 2011;19:192–205.
16. Lu X, Wang Q, Hu G, et al. ADAMTS1 and MMP1 proteolytically engage EGF-like ligands in an osteolytic signaling cascade for bone metastasis. Genes Dev. 2009;23:1882–94.
17. Neuzillet C, Tijeras-Raballand A, Cohen R, et al. Targeting the TGF-β pathway for cancer therapy. Pharmacol Therapeut. 2015;147:22–31.
18. Bendre MS, Margulies AG, Walser B, et al. Tumor-derived interleukin-8 stimulates osteolysis independent of the receptor activator of nuclear factor-kappa B ligand pathway. Cancer Res. 2005;65:11001–19.
19. Lu X, Kang Y. Chemokine (C-C motif) ligand 2 engages CCR2+ stromal cells of monocytic origin to promote breast cancer metastasis to lung and bone. J Biol Chem. 2009;284:29087–96.
20. Han JH, Choi SJ, Kurihara N, et al. Macrophage inflammatory protein-1 alpha is an osteoclastogenic factor in myeloma that is independent of receptor activator of nuclear factor kappa B ligand. Blood 2001;97:3349–53.

21. Clines GA, Mohammad KS, Grunda JM, et al. Regulation of postnatal trabecular bone formation by the osteoblast endothelin A receptor. J Bone Miner Res. 2011;26(10): 2523–36.
22. Valta MP, Tuomela J, Bjartell A, et al. FGF-8 is involved in bone metastasis of prostate cancer. Int J Cancer. 2008;123:22–31.
23. Cramer SD, Chen Z, Peehl DM. Prostate specific antigen cleaves parathyroid hormone-related protein in the PTH-like domain: inactivation of PTHrP-stimulated cAMP accumulation in mouse osteoblasts. J Urol. 1996;156: 526–31.
24. Cohen P, Peehl DM, Graves HC, et al. Biological effects of prostate specific antigen as an insulin-like growth factor binding protein-3 protease. J Endocrinol. 1994;142: 407–15.
25. Vignani F, Bertaglia V, Buttigliero C, et al. Skeletal metastases and impact of anticancer and bone-targeted agents in patients with castration-resistant prostate cancer. Cancer Treat Rev. 2016;44:61–73.
26. Smith MR, Saad F, Shore ND, et al. Effect of denosumab on prolonging bone metastasis-free survival (BMFS) in men with nonmetastatic castrate-resistant prostate cancer (CRPC) presenting with aggressive PSA kinetics. J Clin Oncol. 2012;30:6
27. Wang Z, Qiao D, Lu Y, et al. Systematic literature review and network meta-analysis comparing bone-targeted agents for the prevention of skeletal-related events in cancer patients with bone metastasis. Oncologist. 2015;20:440–9.
28. Raje N, Vadhan-Raj S, Willenbacher W, et al. Evaluating results from the multiple myeloma patient subset treated with denosumab or zoledronic acid in a randomized phase 3 trial. Blood Cancer J. 2016;6:e378.
29. Hoff AO, Toth B, Hu M, et al. Epidemiology and risk factors for osteonecrosis of the jaw in cancer patients. Ann N Y Acad Sci. 2011;1218:47–54.
30. Gedmintas L, Solomon DH, Kim SC. Bisphosphonates and risk of subtrochanteric, femoral shaft, and atypical femur fracture: a systematic review and meta-analysis. J Bone Miner Res. 2013;28:1729–37.
31. Humm JL, Sartor O, Parker C, et al. Radium-223 in the treatment of osteoblastic metastases: a critical clinical review. Int J Radiation Oncol Biol Phys. 2015;91: 898–906.
32. Biermann JS, Holt GE, Lewis VO, et al. Metastatic bone disease: diagnosis, evaluation, and treatment. Instr Course Lect. 2010;59:593–606.
33. Katzer A, Meenen NM, Grabbe F, et al. Surgery of skeletal metastases. Arch Orthop Trauma Surg. 2002;122(5): 251–8.
34. Beals RK, Lawton GD, Snell WE. Prophylatic internal fixation of the femur in metastatic breast cancer. Cancer. 1971;28:1350–4.
35. Mirels H. Metastatic disease in long bones: a proposed scoring system for diagnosing impending pathological fractures. Clin Orthop. 1989;249:256–65.
36. Nazarian A, Entezari V, Zurakowski D, et al. Treatment planning and fracture prediction in patients with skeletal metastasis with CT-based rigidity analysis. Clin Cancer Res. 2015:21:2514–19.
37. Redmond BJ, Biermann JS, Blasier RB. Interlocking intramedullary nailing of pathological fractures of the shaft of the humerus. J Bone Joint Surg Am. 1996;78:891–6.
38. Damron TA, Sim FH, Shives TC, et al. Intercalary spacers in the treatment of segmentally destructive diaphyseal humeral lesions in disseminated malignancies. Clin Orthop. 1996;324:233–43.
39. Marco RA, Sheth DS, Boland PJ, et al. Functional and oncological outcome of acetabular reconstruction for the treatment of metastatic disease. J Bone Joint Surg Am. 2000;82:642–51.
40. Schneiderbauer MM, Von Knoch M, Schleck CD, et al. Patient survival after hip arthroplasty for metastatic disease of the hip. J Bone Joint Surg. 2004;86:1684–9.
41. O'Connor M, Weber K. Indications and operative treatment for long bone metastasis with a focus on the femur. Clinical Orthop Rel Res. 2003;415S:276–8.
42. Bohm P, Huber J. The surgical treatment of bony metastasis of the spine and limbs. J Bone Joint Surg. 2002;84B:521–9.
43. Holman PJ, Suki D, McCutcheon I, et al. Surgical management of metastatic disease of the lumbar spine: experience with 139 patients. J Neurosurg Spine. 2005;2:550–63.
44. Qian Z, Sun Z, Yang H, et al. Kyphoplasty for the treatment of malignant vertebral compression fractures caused by metastases. J Clin Neurosci. 2011;18: 763–7.
45. Kassamali RH, Ganeshan A, Hoey ET, et al. Pain management in spinal metastases: the role of percutaneous vertebral augmentation. Ann Oncol. 2011;22:782–6.
46. Kurup AN, Callstrom MR. Ablation of skeletal metastases: current status. J Vasc Interv Radiol. 2010;8(Suppl) :S242–50.
47. Haley ML, Gerszten PC, Heron DE, et al. Efficacy and cost-effectiveness analysis of external beam and stereotactic body radiation therapy in the treatment of spine metastases: a matched-pair analysis. J Neurosurg Spine. 2011;14:537–42.

Section IX
Sclerosing and Dysplastic Bone Diseases

Section Editor: Michael P. Whyte

Chapter 107. Sclerosing Bone Disorders 825
Michael P. Whyte

Chapter 108. Fibrous Dysplasia 839
Michael T. Collins, Alison M. Boyce, and Mara Riminucci

Chapter 109. The Osteochondrodysplasias 848
Fabiana Csukasi and Deborah Krakow

Chapter 110. Ischemic and Infiltrative Disorders of Bone 853
Michael P. Whyte

Chapter 111. Tumoral Calcinosis – Dermatomyositis 861
Nicholas J. Shaw

Chapter 112. Genetic Disorders of Heterotopic Ossification: Fibrodysplasia Ossificans Progressiva and Progressive Osseous Heteroplasia 865
Frederick S. Kaplan, Robert J. Pignolo, Mona Al Mukaddam, and Eileen M. Shore

Chapter 113. Osteogenesis Imperfecta 871
Joan C. Marini

Chapter 114. Fibrillinopathies: Skeletal Manifestations of Marfan Syndrome and Marfan-Related Conditions 878
Gary S. Gottesman and Michael P. Whyte

Chapter 115. Hypophosphatasia and Other Enzyme Deficiencies Affecting the Skeleton 886
Michael P. Whyte

Primer on the Metabolic Bone Diseases and Disorders of Mineral Metabolism, Ninth Edition. Edited by John P. Bilezikian.
© 2019 American Society for Bone and Mineral Research. Published 2019 by John Wiley & Sons, Inc.
Companion website: www.wiley.com/go/asbmrprimer

107

Sclerosing Bone Disorders

Michael P. Whyte

Center for Metabolic Bone Disease and Molecular Research, Shriners Hospital for Children;
and Division of Bone and Mineral Diseases, Department of Internal Medicine,
Washington University School of Medicine at Barnes-Jewish Hospital, St Louis, MO, USA

INTRODUCTION

Increased skeletal mass is caused by many rare osteochondrodysplasias [1] and by various dietary, metabolic, endocrine, hematological, infectious, and neoplastic disorders (Table 107.1). This chapter reviews principally the key Mendelian diseases.

OSTEOPETROSIS

Osteopetrosis (OPT) (OMIM #: 166600, 259700, 259710, 259720, 259730, 607634, 611490, 611497) [2], sometimes called "marble bone disease," was identified in 1904 by Albers-Schönberg [3]. All true OPTs are the consequence of failed osteoclast (OC)-mediated resorption of the skeleton during growth. Traditionally, two major clinical forms are discussed: the autosomal recessive (AR) infantile (malignant) type typically fatal in early childhood if untreated [4], and the autosomal dominant (AD) adult (benign) type associated with relatively few symptoms [5]. For 30 years, two types of AD "OPT" have confused the literature [6] because so-called type 1 (ADO 1) is actually the high bone mass disorder caused by LRP5 gene activation and its anabolic effect on bone formation, leaving ADO 2 as the genuine AD OPT perhaps better called Albers-Schönberg disease (A-SD). Traditionally, some OPTs have also been designated "intermediate," "lethal," "transient infantile," "post infectious," or with neuronal storage disease [7]. OPT with renal tubular acidosis and cerebral calcification, the first OPT understood molecularly, is the inborn-error-of-metabolism carbonic anhydrase II deficiency (see later). OPT, lymphedema, anhydrotic ectodermal dysplasia, and immunodeficiency (OL-EDA-ID) is X-linked and affects boys [8]. Now, the genetic defects are known for most OPTs (see later) and underpin a nosology that reflects OC pathobiology [9]. Nongenetic (iatrogenic) OPT from bisphosphonate exposure (see later) was first described in 2003 [10]. Persistence of primary spongiosa (calcified cartilage deposited during endochondral bone formation) provides this as the diagnostic histopathological hallmark [10]. Understandably, "osteopetrosis" has been used generically for many years to describe radiodense skeletons. However, precise diagnosis is now crucial because therapies for genuine OPTs (eg, bone marrow transplantation [BMT]) are inappropriate for other sclerosing bone disorders.

Clinical presentation

Infantile OPT manifests during the first year of life [4]. Nasal stuffiness from underdeveloped sinuses is an early sign. Cranial foramina do not widen and optic, oculomotor, and facial nerves may become paralyzed. Hearing loss is common. Blindness can occur also from raised intracranial pressure or sometimes retinal degeneration. Some patients develop hydrocephalus or sleep apnea. Eruption of teeth is delayed, and there is failure to thrive. Bones are dense yet fragile. Physical findings include macrocephaly, frontal bossing, "adenoid" appearance, nystagmus, hepatosplenomegaly, *genu valgum*, and short stature. Recurrent infection and spontaneous bruising and bleeding reflect myelophthisis caused by excessive bone, abundant OCs, and fibrous tissue crowding marrow

---

Primer on the Metabolic Bone Diseases and Disorders of Mineral Metabolism, Ninth Edition. Edited by John P. Bilezikian.
© 2019 American Society for Bone and Mineral Research. Published 2019 by John Wiley & Sons, Inc.
Companion website: www.wiley.com/go/asbmrprimer

Table 107.1. Conditions that cause focal or generalized increases in skeletal mass.

Dysplasias and Dysostoses	Metabolic	Other
Autosomal dominant osteosclerosis	Carbonic anhydrase II deficiency	Axial osteomalacia
Central osteosclerosis with ectodermal dysplasia	Fluorosis	Diffuse idiopathic skeletal hyperostosis (DISH)
Craniodiaphyseal dysplasia	Heavy metal poisoning	Erdheim–Chester disease
Congenital sclerosing osteomalacia with cerebral calcification (Raine syndrome)	Hepatitis C-associated osteosclerosis	Fibrogenesis imperfecta ossium
Craniometaphyseal dysplasia	Hypervitaminosis A, D	Hypertrophic osteoarthropathy
Dysosteosclerosis	Hyper-, hypo-, and pseudohypoparathyroidism	Ionizing radiation
Endosteal hyperostosis (van Buchem disease and sclerosteosis)	Hypophosphatemic osteomalacia	Leukemia
Frontometaphyseal dysplasia	LRP5 and 6 activation (high-bone-mass phenotype)	Lymphoma
Infantile cortical hyperostosis (Caffey disease)	Milk-alkali syndrome	Mastocytosis
Juvenile Paget's disease (osteoectasia with hyperphosphatasia)	Renal osteodystrophy	Multiple myeloma
Lenz–Majewski syndrome	X-linked hypophosphatemia	Myelofibrosis
Melorheostosis		Osteomyelitis
Metaphyseal dysplasia (Pyle disease)		Osteonecrosis
Mixed sclerosing bone dystrophy		Paget's bone disease
Oculodento-osseous dysplasia		Sarcoidosis
Osteodysplasia of Melnick and Needles		Sickle cell disease
Osteopathia striata		Skeletal metastases
Osteopetrosis		Systemic lupus erythematosus
Osteopoikilosis		Tuberous sclerosis
Progressive diaphyseal dysplasia (Camurati–Engelmann disease)		
Pycnodysostosis		
Trichodentoosseous dysplasia		
Tubular stenosis (Kenny–Caffey syndrome)		

spaces. Hypersplenism causing hemolysis may exacerbate the anemia. Without treatment patients usually die in the first decade of life from pneumonia, hemorrhage, severe anemia, or sepsis [4].

Adult OPT (A-SD) emerges radiographically during childhood, yet some "carriers" have only biochemical findings in adult life [5, 11]. Potential complications include facial palsy, compromised vision or hearing, psychomotor delay, osteomyelitis of the mandible [11], carpal tunnel syndrome, slipped capital femoral epiphysis, and osteoarthritis. The long bones become brittle and may fracture.

Intermediate OPT during childhood causes short stature, cranial nerve deficits, ankylosed teeth that predispose to osteomyelitis of the jaw, recurrent fractures, and mild or occasionally moderately severe anemia. The prognosis is poorly understood. Neuronal storage disease with OPT is especially severe and features epilepsy and neurodegeneration [7]. Lethal OPT manifests in utero and causes stillbirth. Transient infantile OPT inexplicably resolves during the first months of life.

Radiological features

All three components of skeletal development are disrupted in OPT: growth, modeling, and remodeling [10]. Diffusely increased mineralized bone mass is the major finding. The thickened trabecular and cortical bone are called "osteosclerosis" and "hyperostosis," respectively. In malignant OPT, hypocalcemia with secondary hyperparathyroidism can cause rachitic-like changes in growth plates ("osteopetrorickets"). Paranasal and mastoid sinuses are underpneumatized. Vertebrae can show a "bone-in-bone" (endobone) configuration. In A-SD, alternating sclerotic and lucent bands may parallel the iliac crest and long bone physes, and the spine has a "rugger jersey" appearance [6]. Metaphyses widen and can develop a club shape or "Erlenmeyer flask" deformity (Fig. 107.1). Rarely, distal phalanges are eroded. Pathological ("chalk stick") fractures of long bones reflect the skeletal brittleness. The skull is thickened and dense, especially its base in A-SD. Skeletal scintigraphy helps show fractures and osteomyelitis. MRI can

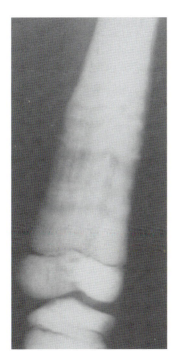

Fig. 107.1. Osteopetrosis. Anteroposterior radiograph of the distal femur of a 10-year-old boy shows a widened metadiaphysis with characteristic alternating dense and lucent bands.

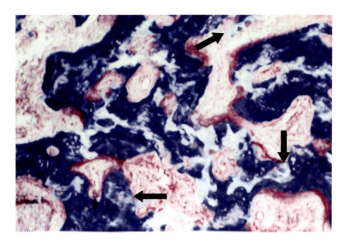

Fig. 107.2. Osteopetrosis. Characteristic areas of lightly stained calcified primary spongiosa (arrows) are found within darkly stained mineralized bone.

assess BMT if engraftment enlarges medullary spaces. Cranial CT and MRI findings have been detailed for pediatric patients [12].

Laboratory findings

In infantile OPT, failed bone resorption can cause hypocalcemia and circulating calcium (Ca) levels depend on dietary intake [13]. Secondary hyperparathyroidism with elevated serum calcitriol is common. In A-SD, this is mild [11]. Increased serum acid phosphatase and the brain isoenzyme of creatine kinase (BB-CK) from defective yet excessive OCs [14] are biomarkers for most OPTs [14]. Several LDH isoenzymes are elevated in A-SD [14].

Histopathological findings

The radiographic features of the OPTs are distinctive, but the OC failure during endochondral bone formation provides a pathognomonic histological finding. Unresorbed remnants of primary spongiosa persist as "islands" or "bars" of calcified cartilage embedded within trabecular bone (Fig. 107.2). OC numbers may be increased, normal, or (rarely) decreased or absent [15]. In infantile OPT, OCs are typically on bone surfaces, their nuclei are especially numerous, and ruffled borders or clear zones are absent. Fibrous tissue often crowds marrow spaces. A-SD may show increased osteoid and few OCs also lacking ruffled borders, or OCs can be especially numerous and large [6]. Immature "woven" bone is common. Rounded, hypermultinucleated OCs off bone surfaces characterize bisphosphonate-induced OPT [10].

Etiology and pathogenesis

The etiology and pathogenesis of the OPTs is complex. Defects have been proposed for the stem cell or its microenvironment that enable osteoclastogenesis, the mononuclear precursor cell, or mature heterokaryon. Osteoblast (OB) effects have also been proposed, and theoretically the skeletal matrix could resist resorption. Defective production of interleukin (IL)-2 or superoxide has been considered because leukocyte function may be abnormal in infantile OPT. In OPT with neuronal storage disease involving ceroid lipofuscin, lysosomes could be defective [7]. Virus-like inclusions of uncertain significance have been found in the OCs of mild OPT. Ultimately, impaired skeletal resorption causes fragility because unresorbed cartilage accumulates, collagen fibers do not interconnect osteons, woven bone remodels poorly to compact bone, and microcracks fail to heal.

Now, the genetic basis is known for most OPTs and patients [9]. Mutation analysis usually reveals *TCIRG1* or *CLCN7* defects [15]. A-SD is caused by disruption of chloride channel 7 caused by heterozygous mutation in *CLCN7* [16]. Rarely, bi-allelic *CLCN7* defects cause malignant or intermediate OPT. However, malignant OPT typically involves bi-allelic mutations in *TCIRG1* (*ATP6I*) encoding the α3 subunit of the vacuolar proton pump. *CA II* mutation explains carbonic anhydrase (CA) II deficiency OPT. Loss of function of the gene encoding OPT-associated transmembrane protein 1 (*OSTM1*) causes especially severe OPT. Notably, mutations of these four genes disrupt hydrochloric acid generation by OCs to resorb bone. OL-EDA-ID reflects inactivation of *IKBKG*, a key modulator (NEMO) of NF-κB [8]. Other OPTs are from defective *SNX10* or *PLEKHM1*, and especially rare infantile OPT resistant to BMT reflects deactivation of *TNFSF11* that encodes RANKL [15]. Other OPTs with few OCs include deactivation of RANK (*TNFRSF11A*) and dysosteosclerosis involving *SLC29A2* [17, 18].

Diagnosis

Traditionally, diagnosis of OPT has considered the complications, progression, and investigation of the family. Now, it includes mutation analysis offered by commercial laboratories [9].

Treatment

Because the etiology, OC disruption, and outcome differ among the OPTs, a precise diagnosis is crucial for therapy. Radiographic studies can occasionally detect malignant OPT late in pregnancy. Early prenatal diagnosis by sonography has generally been unsuccessful.

Bone marrow transplantation

BMT from HLA-identical donors has essentially cured some patients with infantile OPT [19]. Success using BMT from HLA-nonidentical donors has improved [19, 20], including administration of progenitor cells from parental blood. Importantly, BMT will not benefit all OPTs because the fundamental defect is sometimes extrinsic to the OC lineage (eg, RANKL deficiency) [15]. Histomorphometry of bone helps to predict the outcome of BMT. Patients with severely crowded medullary space seem less likely to engraft, and BMT early on seems best [19]. Hypercalcemia can occur as OC function begins. Severe, acute, pulmonary hypertension has been a not infrequent complication [21].

Hormonal, dietary, and other therapy

Some success has been reported with a Ca-deficient diet, but Ca supplementation may be necessary if there is "osteopetrorickets." High doses of calcitriol to stimulate quiescent OCs, while dietary Ca is limited to prevent hypercalciuria and hypercalcemia, has been largely abandoned. The observation that OPT leukocytes produce insufficient superoxide led in the USA to recombinant human interferon gamma-1b (Actimmune, Horizon Pharma, Inc., USA) treatment as a means to slow progression of malignant OPT. High-dose glucocorticoid treatment stabilizes pancytopenia and hepatomegaly, and one case report describes reversal of malignant OPT after prednisone treatment [22]. Prednisone and a low Ca/high phosphate diet was discussed as an alternative when BMT was first used as a treatment [23]. Small interfering RNA may one day help A-SD. Gene therapy is being studied for malignant OPT [24]. Delivery of soluble RANKL is being tested in vitro.

Supportive

Surgical decompression of the optic and facial nerves and auditory canal [25] may benefit some patients. Hyperbaric oxygenation can help osteomyelitis of the jaw [26]. Joint replacement is difficult but possible [27]. Concerning fractures, internal fixation may be necessary, but challenging [28].

CARBONIC ANHYDRASE II DEFICIENCY

In 1983, AR OPT, renal tubular acidosis (RTA), and cerebral calcification was identified as carbonic anhydrase II (CA II) deficiency [29] (OMIM #611492) [2].

Clinical presentation

Severity varies considerably. In infancy or early childhood, patients can suffer fractures, failure to thrive, developmental delay, short stature, optic nerve compression with blindness, and dental malocclusion, and RTA may explain hypotonia, apathy, and muscle weakness. Periodic hypokalemic paralysis can occur. Mental subnormality is common, but not inevitable. Recurrent long bone fractures, although uncommon, can cause significant morbidity. Life expectancy does not seem shortened, but the oldest reported cases are young adults.

Radiological features

Skeletal findings can be subtle at birth. The osteosclerosis and modeling defects during childhood can gradually diminish. Also, cerebral calcification appears between ages 2 and 5 years, affects cortical and basal ganglia gray matter, resembles this finding in hypoparathyroidism, and increases during childhood.

Laboratory findings

Metabolic acidosis manifests as early as birth. Both proximal and distal RTA are reported; distal (type I) RTA seems better documented. Any anemia is generally mild.

Etiology and pathogenesis

CAs occur in many tissues including brain, kidney, erythrocytes, cartilage, lung, and gastric mucosa and accelerate the first step in $CO_2 + H_2O <-> H_2CO_3 <-> H^+ + HCO_3^-$. CA II deficiency OPT reveals its significance in bone, kidney, and perhaps brain [30]. Heterozygous carrier erythrocytes have ~50% of normal CA II levels [29, 31]. There is a knockout mouse model.

Treatment

Transfusion of healthy erythrocytes did not improve the systemic acidosis [31]. The skeletal impact of long-term treatment with HCO_3^- for the RTA is unknown. Reportedly, BMT corrected the OPT, slowed the cerebral calcification, yet the RTA persisted [32].

PYCNODYSOSTOSIS

Pycnodysostosis is the AR disorder that perhaps affected painter Henri de Toulouse-Lautrec (1864–1901) [33]. More than 100 patients have been described since 1962, mostly from Europe or the USA, with others from Israel, Indonesia, India, Africa, and especially Japan. Parental consanguinity is recorded in fewer than 30% of cases.

Clinical presentation

Pycnodysostosis is typically diagnosed during infancy or early childhood because of disproportionate short stature and a relatively large cranium, fronto-occipital prominence, small facies and chin, obtuse mandibular angle, high-arched palate, dental malocclusion with retained deciduous teeth, proptosis, and a beaked and pointed nose. The anterior fontanel and cranial sutures are usually open. Sclerae can be blue. Fingers are short and clubbed from acro-osteolysis or aplasia of terminal phalanges, and the hands are small and square. The thorax is narrow sometimes with *pectus excavatum*, kyphoscoliosis, and lumbar hyperlordosis. Recurrent fractures typically involve the lower limbs causing *genu valgum*. Rickets has been described. Adult height ranges from 4' 3" to 4' 11". Mental retardation affects fewer than 10% of cases. Recurrent respiratory infections and right heart failure may complicate chronic upper airway obstruction caused by micrognathia.

Radiographic features

Pycnodysostosis can be considered an OPT (see later). Uniform osteosclerosis appears in childhood and increases with age. The calvarium and skull base are sclerotic, and the orbital ridges are radiodense. However, the marked modeling defects of OPT do not occur, although long bones have narrow medullary canals. Additional findings include delayed closure of cranial sutures and fontanels (prominently the anterior), obtuse mandibular angle, wormian bones, gracile clavicles with hypoplastic ends, partial absence of the hyoid bone, and hypoplasia of the distal phalanges and ribs. Endobones and radiodense striations are absent. Recurrent fractures can occur.

Laboratory findings

Serum levels of Ca, phosphorus (P), and alkaline phosphatase (ALP) are usually unremarkable. There is no anemia.

Electron microscopy has suggested defective degradation of bone collagen. Inclusions in chondrocytes and virus-like inclusions in OCs have been reported. Diminished growth hormone secretion with low circulating insulin-like growth factor 1 levels is reported [34].

Etiology and pathogenesis

In 1996, loss of function mutation was discovered within the CTSK gene that encodes cathepsin K [35]. Cathepsin K, a lysosomal cysteine protease, is highly expressed in OCs [36]. Impaired collagen degradation seems a pathogenetic flaw [35].

Treatment

There is no established medical treatment. BMT has not yet seemed appropriate [37]. Replacement therapy for growth hormone deficiency is beneficial [37]. Fractures of long bones are typically transverse and heal satisfactorily, although delayed union with massive callus formation is possible [38]. Skeletal hardness challenges internal fixation of long bones or extraction of teeth. Jaw fracture has occurred. Osteomyelitis of the mandible may require antibiotics and surgery.

PROGRESSIVE DIAPHYSEAL DYSPLASIA (CAMURATI–ENGELMANN DISEASE)

Progressive diaphyseal dysplasia (PDD) (OMIM #131300) [2] was characterized by Cockayne in 1920 [39]. Camurati recognized its AD transmission. Engelmann described severe involvement in 1929. Hyperostosis occurs gradually on both the periosteal and endosteal surfaces of long bones. When severe, the axial skeleton and skull become involved. The clinical and laboratory features and responsiveness to glucocorticoid treatment have suggested severe PDD be considered an inflammatory connective tissue disease [40]. In 2001, mutations were identified within a specific region of the gene that encodes TGF-β1 [41].

Clinical presentation

All races are affected, and severity is quite variable. PDD typically presents during childhood with limping or a broad-based and waddling gait, leg pain, and muscle wasting with decreased subcutaneous fat in the extremities mimicking a muscular dystrophy. Severely affected individuals have a characteristic body habitus that includes tall stature, a large head with prominent forehead, proptosis, and thin limbs with thickened painful bones and little muscle mass. Cranial nerve palsies may develop when the skull is involved. Puberty is sometimes delayed. Raised intracranial pressure can occur. Physical findings include palpably widened and tender bones. Some patients have hepatosplenomegaly, Raynaud's phenomenon, and other findings suggestive of vasculitis [40]. The course is variable, sometimes with improvement during adult life [42]. Some mutation carriers lack radiographic changes, but bone scintigraphy is abnormal.

Radiological features

The tibias and femurs are most commonly affected; less frequently the radii, ulnas, humeri, scapulae, clavicles, and pelvis, and occasionally short tubular bones. Typically, skeletal disease progresses. Hyperostosis of major long bone is fairly symmetrical, involves both the periosteal and endosteal surfaces, and gradually spreads to metaphyses but spares epiphyses (Fig. 107.3). Diaphyses slowly widen with irregular surfaces. Age-of-onset, rate of progression, and degree of bony change are highly variable. With mild disease, especially in adolescents or young adults, radiographic and scintigraphic abnormalities may involve only the lower limbs. In severely affected children, regional osteopenia is possible.

Clinical, radiographic, and scintigraphic findings are generally concordant. Bone scanning typically shows focally increased radionuclide accumulation, but in advanced quiescent disease can be unremarkable [43]. Conversely, markedly increased radioisotope accumulation with minimal radiographic findings can represent early disease [43]. The cranial MRI and CT findings have been delineated.

Laboratory findings

Serum ALP and urinary hydroxyproline levels are elevated in some PDD patients. Mild hypocalcemia and hypocalciuria sometimes occur in severe disease, probably reflecting positive Ca balance [42]. Other biochemical parameters of skeletal and mineral homeostasis are typically normal. Mild anemia and leukopenia and elevated erythrocyte sedimentation rate are reported [40]. Diaphyseal histopathology shows nascent woven bone undergoing centripetal maturation and incorporation into the cortex. Electron microscopy of muscle has disclosed myopathic and vascular changes.

Etiology and pathogenesis

PDD reflects mutation within one region of *TGFβ1*. Consequently, its latency-associated peptide remains bound keeping TGF-β1 active in the skeletal matrix. PDD has been described as increasingly severe in ensuing generations ("anticipation") [44], but the *TGFβ1* mutations argue otherwise. However, there may be locus heterogeneity for PDD [45].

Treatment

The course is somewhat unpredictable. Symptoms may remit during adolescence or adult life. Prednisone in small doses on alternate days is effective for bone pain and weakness and can correct bone histopathology [46]. Resection creating a cortical "window" has relieved localized bone pain. Bisphosphonate therapy may be beneficial, but has transiently increased pain. Losartan can be helpful, but its side effects may limit use [47]. Intranasal calcitonin diminished pain in one patient.

ENDOSTEAL HYPEROSTOSIS

In 1955, van Buchem and colleagues described AR *hyperostosis corticalis generalisata* [48]. Subsequently, it and two additional (also remarkably instructive) disorders [49] were classified as endosteal hyperostoses. The second AR and especially severe entity is sclerosteosis. The AD

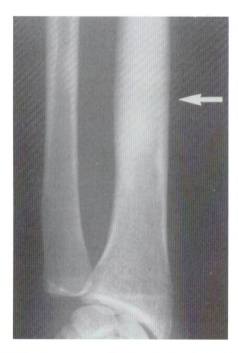

Fig. 107.3. Progressive diaphyseal dysplasia (Camurati–Engelmann disease). The distal radius of this 20-year-old woman has characteristic patchy thickening (arrow) of the periosteal and endosteal surfaces of the diaphysis.

relatively mild form was originally called Worth disease [50], later ADO 1 [6], and now "high bone mass."

Van Buchem disease

Van Buchem disease (OMIM #239100) [2] predominantly has Dutch ancestry [48].

Clinical presentation

Progressive asymmetrical enlargement and marked thickening of the jaw occurs during puberty but without prognathism. Dental malocclusion is uncommon. Recurrent facial nerve palsy, deafness, and optic atrophy from narrowing of cranial foramina are common and can present as early as infancy. Long bones may be painful to pressure but are not fragile, and joint range-of-motion is not compromised. Van Buchem suggested the excessive bone was essentially of normal quality [48]. Sclerosteosis (see later) is even more severe with in utero onset of syndactyly and then excessive height.

Radiological features

Endosteal hyperostosis produces dense cortices with a narrow medullary canal. Long bones are, however, properly modeled. Osteosclerosis also affects the skull base, facial bones, mandible, vertebrae, pelvis, and ribs.

Laboratory findings

Serum ALP (bone isoform) may be increased whereas Ca and P are normal.

Etiology and pathogenesis

Initially, van Buchem disease and sclerosteosis were regarded as allelic disorders with different modifying genes. Subsequently, loss of function mutation of the gene *SOST* encoding sclerostin explained sclerosteosis [51], whereas van Buchem disease was from a 52 kilobase deletion of a downstream enhancer of *SOST* [52]. In health, sclerostin binds to LRP5/6, antagonizes canonical Wnt signaling, and promotes OB apoptosis to inhibit bone formation.

Treatment

There is no specific medical therapy. Decompression of narrowed foramina may help cranial nerve palsies. Surgery can recontour the mandible [53].

Sclerosteosis

Sclerosteosis (OMIM #269500) [2], like van Buchem disease, affects primarily Afrikaners or others of Dutch ancestry but differs by excessive height and syndactyly [51, 52].

Clinical presentation

At birth, only syndactyly of variable severity (from either cutaneous or bony fusion of the middle and index fingers) may be noted. In early childhood, patients become tall and heavy with skeletal overgrowth, especially of the skull causing facial disfigurement. Deafness and facial palsy are prominent. The mandible has a square configuration. Bones are resistant to fracture. The small cranial cavity may raise intracranial pressure and cause headache with brainstem compression. If untreated, this often significantly shortens life expectancy [54].

Radiological features

Except for syndactyly, the skeleton appears normal in early childhood. Then, progressive bone acquisition widens the skull and mandible (Fig. 107.4). Long bones develop thickened cortices. Vertebral pedicles, ribs, tubular bones, and the pelvis can also appear dense. Auditory ossicles may fuse and the internal canals and cochlear aqueducts become narrow.

Histopathological findings

Dynamic histomorphometry of nondecalcified skull showed accelerated bone formation with thickened trabeculae and osteoidosis, whereas resorption appeared quiescent [55].

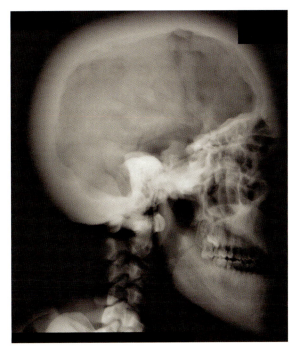

Fig. 107.4. Sclerosteosis. Lateral radiograph of the skull of an affected young man shows diffuse marked thickening.

Etiology and pathogenesis

Loss of function mutations in *SOST* cause sclerosteosis type 1 [51]. Missense mutation of *LRP4* underlies extremely rare sclerosteosis type 2 [56]. Enhanced OB activity with failure of OCs to compensate explains the dense bone [55]. No abnormality of Ca homeostasis or of pituitary gland function has been documented. The pathogenesis of the neurological defects has been described [55].

Treatment

There is no established medical treatment. Surgery for syndactyly is difficult if involving the bony fusion of sclerosteosis type 2. Management of the neurological complications has been reviewed [55].

High bone mass disorder

Certain mutations of the low-density lipoprotein receptor-related protein 5 gene (*LRP5*) and *LRP6* enhance Wnt signaling, stimulate OBs, and increase skeletal mass with good quality bone inherited as an AD trait (OMIM #607636) [2, 57]. Some patients have torus palatinus [57], and others oropharyngeal exostoses, and cranial nerve palsies [58].

OSTEOPOIKILOSIS

Osteopoikilosis (OPK) (OMIM #166700) [2] is typically an AD radiographic curiosity of "spotted bones." When connective tissue nevi are present, the disorder is the Buschke–Ollendorff syndrome (BOS) [59]. Joint contractions and limb length inequality can occur, especially if there are also changes of melorheostosis (see later). In 2004, deactivating mutations in *LEMD3* explained OPK and BOS [60].

Clinical presentation

OPK is usually an incidental finding. The bony lesions are asymptomatic, but if misunderstood can precipitate investigation for metastatic disease to the skeleton [61]. Family members should be screened with a radiograph of a wrist and knee in early adult life. The nevi of BOS usually involve the lower trunk or proximal extremities and are small asymptomatic papules, but sometimes are yellow or white discs or plaques, deep nodules, or streaks [61].

Radiological features

Numerous, small, usually round or oval foci of osteosclerosis are common in the ends of the short tubular bones, metaepiphyses of long bones, and tarsal, carpal, and pel-

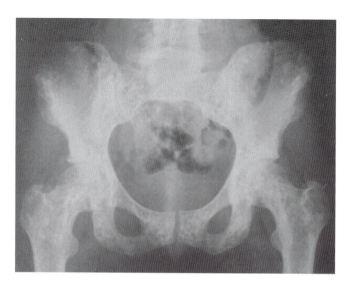

Fig. 107.5. Osteopoikilosis. The characteristic feature is the spotted appearance shown here in the pelvis and metaepiphyseal regions of the femora.

vic bones, and remain unchanged for decades (Fig. 107.5). Bone scanning is not abnormal [61].

Histopathological studies

Foci of osteosclerosis are thickened trabeculae that merge with surrounding normal bone or are islands of cortical bone that include Haversian systems. Mature lesions appear to be remodeling slowly. The skin lesion, *dermatofibrosis lenticularis disseminata*, consists of unusually broad, markedly branched, interlacing elastin fibers in the dermis; the epidermis is normal [59].

OSTEOPATHIA STRIATA

Osteopathia striata (OMIM #166500) features linear striations at the ends of long bones and in the ileum. Like OPK or BOS, it is usually a radiographic curiosity, but can occur in important disorders including osteopathia striata with cranial sclerosis (OMIM #300373) [2] and osteopathia striata with focal dermal hypoplasia (Goltz syndrome) (OMIM #305600) [2]. Histopathological studies of bone are lacking.

Clinical presentation

Osteopathia striata alone is an AD trait. Symptoms that may have led to its discovery are probably unrelated. However, when it accompanies sclerosis of the skull caused by WTX gene mutation [62], cranial nerve palsies are common. Osteopathia striata with focal dermal hypoplasia is a serious disorder of males, transmitted as an X-linked recessive trait, that features widespread linear areas of dermal hypoplasia through which adipose tissue can herniate and a variety of bony defects in the limbs.

Radiological features

Gracile linear striations are found in cancellous bone, particularly within metaepiphyses of major long bones and the periphery of the iliac bones. Carpal, tarsal, and tubular bones of the hands and feet are less often and more subtly affected. The striations appear unchanged for years. Radionuclide accumulation is not increased during bone scintigraphy [61].

Treatment

Radiographic screening of a knee of family members during young adult life may be prudent. Osteopathia striata with cranial sclerosis has been detected prenatally by sonography, and *WTX* mutation is present.

MELORHEOSTOSIS

Melorheostosis (MEL) (OMIM #155950) [2], from the Greek, refers to "flowing hyperostosis." The radiographic appearance of mature lesions resembles wax that has dripped down a candle. Since discovered in 1922 [63], about 200 cases have been described. A retrospective review of 24 patients at Mayo Clinic was published in 2016 [64].

Clinical presentation

MEL occurs sporadically, including when it accompanies AD osteopoikilosis (see previously). Presentation is typically during childhood. Monomelic involvement is usual; bilateral disease is characteristically asymmetrical. Cutaneous changes may overlie the skeletal lesions and include linear scleroderma-like patches as well as hypertrichosis. Soft tissue abnormalities including fibromas, fibrolipomas, capillary hemangiomas, and lymphangiectasia are often noted before the hyperostosis. There can be arterial aneurysms. Pain and stiffness are the major symptoms. Leg-length inequality can follow premature fusion of epiphyses. Affected joints may contract. Bone lesions seem to advance most rapidly during childhood. In adult life, MEL may or may not progress. Nevertheless, pain becomes worse especially if subperiosteal bone formation continues.

Radiological features

Dense, irregular, and eccentric hyperostosis of both the periosteal and endosteal surfaces of a single bone, or several adjacent bones in a sclerotomal distribution, is the hallmark of MEL (Fig. 107.6). Any bone may be

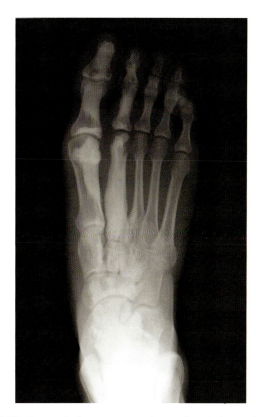

Fig. 107.6. Melorheostosis. Radiograph of the right foot shows characteristic areas of sclerosis and hyperosteosis.

affected, but most commonly in the lower extremities. Ossification can also develop in soft tissues near skeletal lesions, particularly near joints. MEL bone is hyperemic and "hot" during bone scanning.

Laboratory features

Serum CA, P, and ALP levels are normal.

Histopathological features

MEL features endosteal thickening during growth and periosteal new bone formation during adult life. Affected bones are sclerotic with thickened, irregular lamellae. Marrow fibrosis may be present. In the skin, unlike in true scleroderma, the collagen of the scleroderma-like lesions appears normal and has therefore been called linear melorheostotic scleroderma.

Etiology and pathogenesis

The etiology and pathogenesis of MEL is becoming understood. The distribution of the bone and soft tissue lesions suggested a segmentary embryonic defect likely from postzygotic mosaicism. Mutation of *KRAS* and *MAP2K1* has been found in some patients [65, 66]. Linear scleroderma may be a fundamental abnormality that extends into and alters the skeleton. In affected skin, there may be altered expression of several adhesion proteins. However, the germline loss of function mutations in *LEMD3* that cause osteopoikilosis and the Buschke–Ollendorff syndrome (see previously) do not explain typical sporadic MEL [60].

Treatment

Surgical correction of contractures can be difficult and counterproductive; recurrent or worse deformity is common. Limb shortening typically involves the longer extremity. Distraction techniques seem promising [67].

AXIAL OSTEOMALACIA

Axial osteomalacia (OMIM #109130) [2] features coarsened trabecular bone in the axial, but not the appendicular, skeleton [68]. Fewer than 20 patients are described. Except for an affected mother and son [69], all have been sporadic cases.

Clinical presentation

Most patients have been middle-aged or elderly men. Dull, vague, chronic, axial bone pain (often in the cervical spine) usually prompts radiographic discovery.

Radiological features

Abnormalities are confined essentially to the spine and pelvis where the coarsened trabecular pattern resembles an osteomalacia. The cervical spine and ribs seem most severely affected. However, Looser's zones characteristic of osteomalacia are not reported. The appendicular skeleton is unremarkable. Several patients have had features of ankylosing spondylitis.

Laboratory studies

Serum ALP (bone isoform) may be increased. In a few patients, circulating P levels tended to be low. For others, osteomalacia occurred despite normal levels of Ca, P, 25-hydroxyvitamin D, and 1,25-dihydroxyvitamin D.

Histopathological findings

Iliac crest specimens have distinct corticomedullary junctions, but the cortices can be especially wide and porous. Total bone volume may be increased. There is excess osteoid, but collagen has a normal lamellar pattern. Tetracycline labeling confirms defective mineralization. OBs are flat "lining" cells, but stain intensely for ALP. Changes of secondary hyperparathyroidism are absent.

Etiology and pathogenesis

Axial osteomalacia may reflect an OB defect.

Treatment

There is no established medical therapy, but the course seems relatively benign. One long-term follow-up showed unchanging symptoms and radiographic findings. Methyltestosterone and stilbestrol or vitamin D_2 (as much as 20,000 U/day for 3 years) was not helpful. Slightly improved skeletal histology, but not symptoms, was reported after Ca and vitamin D_2 supplementation.

FIBROGENESIS IMPERFECTA OSSIUM

Fibrogenesis imperfecta ossium (FIO) was identified in 1950. Approximately 10 cases have been reported. Radiographic studies suggest generalized osteopenia, but coarse and dense appearing trabecular bone explains its inclusion among the osteosclerotic disorders. The clinical, biochemical, radiological, and histopathological features have been carefully contrasted with axial osteomalacia [68].

Clinical presentation

Presentation occurs during middle age or later. Both sexes are affected. Characteristically, gradual onset of intractable skeletal pain precedes rapid deterioration. The course becomes debilitating with progressive immobility. Spontaneous fracturing is prominent. Physical examination shows marked bony tenderness.

Radiological features

Skeletal changes spare only the skull. Initially, there may be osteopenia and slightly abnormal appearing trabecular bone. Subsequent findings are more consistent with osteomalacia showing further alterations of the trabecular bone pattern, heterogeneous bone density, and cortical thinning. Corticomedullary junctions become indistinct. Areas of the skeleton may have a mixed lytic and sclerotic appearance. Remaining trabeculae appear coarse and dense in a "fish-net" pattern. Pseudofractures may develop. Some patients have a "rugger jersey" spine. Diaphyses may show periosteal reaction. The findings in Waldenstrom's macroglobulinemia can resemble FIO [70]. Generalized versus axial changes distinguish FIO from axial osteomalacia, respectively. The histopathological findings also differ [68].

Laboratory findings

Serum Ca and P levels and renal tubular function are typically normal, but serum ALP is increased. Urine hydroxyproline is normal or elevated. Acute agranulocytosis and macroglobulinemia have been reported.

Histopathological findings

This is an osteomalacia. At sites of defective mineralization, collagen lacks birefringence. Electron microscopy reveals thin and randomly organized collagen fibrils in a "tangled" pattern. Cortical bone in the femurs and tibias may show the least abnormality. OBs and OCs can be abundant. In some regions, 300–500nm diameter matrix structures have been observed [70]. Unless bone is viewed with polarized-light or electron microscopy, FIO can be mistaken for osteoporosis or other forms of osteomalacia [70].

Etiology and pathogenesis

The etiology is unknown. Genetic factors have not been implicated.

Treatment

There is no recognized therapy. Temporary improvement can occur spontaneously. Treatment with vitamin D (or active metabolite), Ca, salmon calcitonin, or sodium fluoride has not been helped. In fact, ectopic calcification complicated high-dose vitamin D therapy. Treatment with melphalan and prednisolone seemed beneficial for one patient, and more recently plasmapheresis [71].

PACHYDERMOPERIOSTOSIS

Pachydermoperiostosis (hypertrophic osteoarthropathy: primary) causes clubbing of the digits, thickening and hyperhidrosis of the skin especially on the face and forehead (*cutis verticis gyrata*), and periosteal new bone formation particularly in the distal extremities. AD (OMIM #167100) [2] and AR inheritance with variable expression is established [72].

Clinical presentation

Black men seem most severely affected. Age at presentation is variable, but is usually during adolescence [72]. All principal features (clubbing, periostitis, and pachydermia) trouble some patients; others manifest just one or two. Clinical problems emerge over a decade and then sometimes abate [72]. Progressive enlargement of the hands and feet may cause a paw-like appearance with excessive perspiration. Acro-osteolysis can occur. Fatigue and arthralgias of the elbows, wrists, knees, and ankles are common. Stiffness of the appendicular and the axial skeleton may develop. Compression of cranial or spinal nerves has occurred. Cutaneous changes include coarsening, thickening, furrowing, pitting, and oiliness of the skin, especially the scalp and face. Myelophthisic anemia with extramedullary hematopoiesis may occur. Life expectancy is not compromised.

Radiological features

Severe periostitis thickens tubular bones distally, typically the radius, ulna, tibia, and fibula, but sometimes also the metacarpals, tarsals/metatarsals, clavicles, pelvis, skull base, and phalanges. The spine is rarely involved. Clubbing is obvious, and acro-osteolysis can occur. Ankylosis of joints, especially in the hands and feet, may trouble older patients. The major diagnostic challenge involves secondary hypertrophic osteoarthropathy (pulmonary or otherwise). Here, however, the radiographic findings typically feature smooth and undulating periosteal reactions whereas in pachydermoperiostosis periosteal proliferation is exuberant, irregular, and often involves epiphyses. Bone scanning for either condition reveals symmetrical, diffuse, regular uptake along the cortical margins of long bones, especially in the legs, causing a "double stripe" sign.

Histological findings

Nascent periosteal bone roughens cortical bone surfaces while undergoing cancellous compaction with blending to the original cortex. There may also be osteopenia of trabecular bone from quiescent formation. Mild cellular hyperplasia and thickening of blood vessels is found near synovial membranes, but synovial fluid is unremarkable. Electron microscopy shows layered basement membranes.

Etiology and pathogenesis

A controversial hypothesis suggested some circulating factor acts on the vasculature causing hyperemia and soft tissues changes, but later reduced blood flow. In 2008, AR pachydermoperiostosis was explained by loss of function mutation within *HPGD* that encodes 15-hydroxyprostaglandin dehydrogenase [72]. Subsequently, *SLCO2A1* mutation was identified [73]. These defects would increase prostaglandin E2 levels.

Treatment

The treatment has been recently reviewed [74]. Painful synovial effusions may respond to nonsteroidal anti-inflammatory drugs. Colchicine reportedly helped one patient. Contractures or neurovascular compression by osteosclerotic lesions can require surgical intervention. A variety of treatments may help the hyperhidrosis.

HEPATITIS C-ASSOCIATED OSTEOSCLEROSIS

In 1992, severe generalized osteosclerosis and hyperostosis was reported in former drug abusers [75]. Approximately 20 cases have been reported, and hepatitis C virus infection proved common to all. Periosteal, endosteal, and trabecular bone thickening occurs throughout the skeleton except the cranium. During active disease, the forearms and legs are painful. Gradual, spontaneous remission [76] with decreasing bone density can occur. The IGF system features distinctive increases in circulating levels of IGF binding protein 2 and "big" IGF II [77]. Remodeling of excessive good quality bone seems accelerated during active disease, and pain may respond or become transiently worse from antiresorptive therapy. Antiviral therapy improved one patient [78].

OTHER SCLEROSING BONE DISORDERS

Table 107.1 lists many additional conditions associated with focal or generalized increases in skeletal mass. Sarcoidosis characteristically causes cysts within coarsely reticulated bone. However, sclerotic areas occasionally appear in the axial skeleton or long bones. Although multiple myeloma typically features generalized osteopenia and osteolytic lesions, indolent forms can manifest widespread osteosclerosis. Lymphoma, myelofibrosis, and mastocytosis may also increase bone mass. Metastatic carcinoma, especially from the prostate, can cause dense bones. Diffuse osteosclerosis reflects secondary but not primary hyperparathyroidism (eg, renal disease).

REFERENCES

1. Boulet C, Madani H, Lenchik, et al. Sclerosing bone dysplasias: genetic, clinical and radiology update of hereditary and non-hereditary disorders. Br J Radiol. 2016;89:549.
2. Online Mendelian Inheritance in Man, OMIM. *McKusick-Nathans Institute of Genetic Medicine*, Johns Hopkins University, Baltimore, MD, January 1, 2017. https://omim.org/.
3. Albers-Schönberg H. Roentgen bilder einer seltenen Kochennerkrankung. Munch Med Wochenschr. 1904; 51:365.
4. Loría-Cortés R, Quesada-Calvo E, Cordero-Chaverri C. Osteopetrosis in children: a report of 26 cases. J Pediatr. 1977;91:43–7.
5. Johnston CC Jr, Lavy N, Lord T, et al. Osteopetrosis: a clinical, genetic, metabolic, and morphologic study of the dominantly inherited, benign form. Medicine. 1968;47:149–67.
6. Bollerslev J, Steiniche T, Melsen F, et al. Structural and histomorphometric studies of iliac crest trabecular and cortical bone in autosomal dominant osteopetrosis: a study of two radiological types. Bone. 1986;10:19–24.
7. Jagadha V, Halliday WC, Becker LE, et al. The association of infantile osteopetrosis and neuronal storage disease in two brothers. Acta Neuropathol. 1988;75:233–40.
8. Dupuis-Girod S, Corradini N, Hadj-Rabia S, et al. Osteopetrosis, lymphedema, anhidrotic ectodermal dysplasia, and immunodeficiency in a boy and incontinentia pigmenti in his mother. Pediatrics. 2002;109:e97.
9. Shamriz O, Shaag A, Yaacov B, et al. The use of whole exome sequencing for the diagnosis of autosomal recessive malignant infantile osteopetrosis. Clin Genet. 2016;92(1):80–5.
10. Whyte MP, McAlister WH, Novack DV, et al. Bisphosphonate-induced osteopoetrosis: novel bone modeling defects, metaphyseal osteopenia, and osteosclerosis fractures after drug exposure ceases. J Bone Miner Res. 2008;23:1698–707.
11. Waguespack SG, Hui SL, DiMeglio LA, et al. Autosomal dominant osteopetrosis: clinical severity and natural history of 94 subjects with a chloride channel 7 gene mutation. J Clin Endocrinol Metab. 2007;92:771–8.
12. Elster AD, Theros EG, Key LL, et al. Cranial imaging in autosomal recessive osteopetrosis. Skull base and brain. Radiology. 1992;183:137–44.
13. Key L, Carnes D, Cole S, et al. Treatment of congenital osteopetrosis with high dose calcitriol. N Engl J Med. 1984;310:409–15.
14. Whyte MP, Kempa LG, McAlister WH, et al. Elevated serum lactate dehydrogenase isoenzymes and aspartate transaminase distinguish Albers-Schönberg disease (chloride channel 7 deficiency osteopetrosis) among the sclerosing bone disorders. J Bone Miner Res. 2010;25:2515–26.
15. Sobacchi C, Frattini A, Guerrini MM, et al. Osteoclast-poor human osteopetrosis due to mutations in the gene encoding RANKL. Nat Genet. 2007;39:960–62.
16. Cleiren E, Bénichou O, Van Hul E, et al. Albers-Schönberg disease (autosomal dominant osteopetrosis, type II) results from mutations in the CLCN7 chloride channel gene. Hum Mol Genet. 2001;10:2861–7.
17. Whyte MP, Wenkert D, McAlister WH, et al. Dysosteosclerosis presents as an "osteoclast-poor" form of osteopetrosis: comprehensive investigation of a 3-year-old girl and literature review. J Bone Miner Res. 2010;25:2527–39.
18. Campeau PM, Lu JT, Sule G, et al. Whole exome sequencing identifies mutations in the nucleoside transporter gene SLC29A2 in dysosteosclerosis, a form of osteopetrosis. Hum Mol Genet. 2012;21:4904–9.
19. Orchard PJ, Fasth AL, Le Rademacher J, et al. Hematopoietic stem cell transplantation for infantile osteopetrosis. Blood. 2015;126:270–6.
20. Chiesa R, Ruggeri A, Paviglianiti A, et al., on behalf of Eurocord, Inborn Errors Working Part, Cell Therapy, Immunobiology Working Party of the European Group for Blood and Marrow Transplantation. Outcomes after unrelated umbilical cord blood transplantation for children with osteopetrosis. Biol Blood Marrow Transplant. 2016;22:1997–2002.
21. Steward CG, Pellier I, Mahajan A, et al.; Working Party on Inborn Errors of the European Blood and Marrow Transplantation Group. Severe pulmonary hypertension: a frequent complication of stem cell transplantation for malignant infantile osteopetrosis. Br J Haematol. 2004;124:63–71.
22. Iacobini M, Migliaccio S, Roggini M, et al. Case report: apparent cure of a newborn with malignant osteopetrosis

using prednisone therapy. J Bone Miner Res. 2001;16: 2356–60.
23. Dorantes LM, Mejia AM, Dorantes S. Juvenile osteopetrosis: effects of blood and bone of prednisone and low calcium, high phosphate diet. Arch Dis Child. 1986;61:666–70.
24. Thudium CS, Moscatelli H, Lofvall H, et al. Regulation and function of lentviral vector-mediated TCIRG1 expression in osteoclasts from patients with infantile malignant osteopetrosis: implications for gene therapy. Calcif Tissue Int. 2016;99:638–48.
25. Dozier TS, Duncan IM, Klein AJ, et al. Otologic manifestations of malignant osteopetrosis. Otol Neurotol. 2005;26:762–6.
26. Sun H J, Xue L, Wu C-B, et al. Clinical characteristics and treatment of osteopetrosis complicated by osteomyelitis of the mandible. J Craniofacial Surg. 2016;27:e728–e730.
27. Strickland JP, Berry DJ. Total joint arthroplasty in patients with osteopetrosis: a report of 5 cases and review of the literature. J Arthoplasty. 2005;20:815–20.
28. Farfan MA, Olarte CM, Pesantez RF, et al. Recommendations for fracture management in patients with osteopetrosis: case report. Arch Orthop Trauma Surg. 2015;135:351–6.
29. Sly WS, Hewett-Emmett D, Whyte MP, et al. Carbonic anhydrase II deficiency identified as the primary defect in the autosomal recessive syndrome of osteopetrosis with renal tubular acidosis and cerebral calcification. Proc Natl Acad Sci U S A. 1983;80:2752–6.
30. Shah GN, Bonapace G, Hu PY, et al. Carbonic anhydrase II deficiency syndrome (osteopetrosis with renal tubular acidosis and brain calcification): novel mutations in CA2 identified by direct sequencing expand the opportunity for genotype- phenotype correlation. Hum Mutat. 2004;24:272.
31. Whyte MP, Hamm LL 3rd, Sly WS. Transfusion of carbonic anhydrase-replete erythrocytes fails to correct the acidification defect in the syndrome of osteopetrosis, renal tubular acidosis, and cerebral calcification (carbonic anhydrase-II deficiency). J Bone Miner Res. 1988;3:385–8.
32. McMahon C, Will A, Hu P, et al. Bone marrow transplantation corrects osteopetrosis in the carbonic anhydrase II deficiency syndrome. Blood. 2001;97:1947–50.
33. Maroteaux P, Lamy M. The malady of Toulouse-Lautrec. JAMA. 1965;191:715–7.
34. Soliman AT, Rajab A, AlSalmi I, et al. Defective growth hormone secretion in children with pycnodysostosis and improved linear growth after growth hormone treatment. Arch Dis Child. 1996;75:242–4.
35. Gelb BD, Brijmme D, Desnick RJ. Pycnodysostosis: cathepsin K deficiency. In: Scriver CR, Beaudet AL, Sly WS, et al. (eds) *The Metabolic and Molecular Bases of Inherited Disease* (8th ed.) New York: McGraw-Hill Book Company, 2001, pp. 3453–68.
36. Fratzl-Zelman N, Valenta A, Roschger P, et al. Decreased bone turnover and deterioration of bone structure in two cases of pycnodysostosis. J Clin Endocrinol Metab. 2004;89:1538–47.
37. Turan S. Current research on pyconodysostosis. Intract Rare Dis Res 2014;3:91–93.
38. Marti PR, Font RU. Orthopaedic disorders of pycnodysostosis: a report of five clinical cases. Int Orthopaed. 2016;40:2221–31.
39. Engelmann G. Ein fall von osteopathia hyperostotica (sclerotisans) multiplex infantilis. Fortschr Geb Roentgen. 1929;39:1101–6.
40. Crisp AJ, Brenton DP. Engelmann's disease of bone: a systemic disorder? Ann Rheum Dis. 1982;41:183–8.
41. Saito T, Kinoshita A, Yoshiura K, et al. Domain-specific mutations of a transforming growth factor (TGF)-beta 1 latency-associated peptide cause Camurati-Engelmann disease because of the formation of a constitutively active form of TGF-beta 1. J Biol Chem. 2001;276: 11469–72.
42. Smith R, Walton RJ, Corner BD, et al. Clinical and biochemical studies in Engelmann's disease (progressive diaphyseal dysplasia). Q J Med. 1977;46:273–94.
43. Kumar B, Murphy WA, Whyte MP. Progressive diaphyseal dysplasia (Engelmann's disease): scintigraphic-radiologic-clinical correlations. Radiology. 1981;140: 87–92.
44. Saraiva JM. Anticipation in progressive diaphyseal dysplasia. J Med Genet 2000;37:394–5.
45. Hecht JT, Blanton SH, Broussard S, et al. Evidence for locus heterogeneity in the Camurati-Engelmann (DPD1) syndrome. Clin Genet. 2001;59:198–200.
46. Naveh Y, Alon U, Kaftori JK, et al. Progressive diaphyseal dysplasia: evaluation of corticosteroid therapy. Pediatrics. 1985;75:321–3.
47. Ayyavoo A, Derraik JGB, Cutfield WS, et al. Elimination of pain and improvement of exercise capacity in Camurati-Engelmann disease with Losartan. J Clin Endocrinol Metab. 2014;99:3978–82.
48. Van Buchem FSP, Prick JJG, Jaspar HHJ. *Hyperostosis Corticalis Generalisata Familiaris (Van Buchem's Disease)*. Amsterdam: Excerpta Media, 1976.
49. Appelman-Dijkstra NM, Papapoulos SE. From disease to treatment: from rare skeletal disorders to treatments for osteoporosis. Endocrine. 2016;52:414–26.
50. Perez-Vicente JA, Rodriguez de Castro E, Lafuente J, et al. Autosomal dominant endosteal hyperostosis. Report of a Spanish family with neurological involvement. Clin Genet. 1987;31:161–9.
51. Brunkow ME, Gardner JC, Van Ness J, et al. Bone dysplasia sclerosteosis results from loss of the SOST gene product, a novel cystine knot-containing protein. Am J Hum Genet. 2001;68:577–89.
52. Loots GG, Kneissel M, Keller H, et al. Genomic deletion of a long-range bone enhancer misregulates sclerostin in Van Buchem disease. Genome Res. 2005;15:928–35.
53. Schendel SA. Van Buchem disease: surgical treatment of the mandible. Ann Plast Surg. 1988;20:462–7.
54. Hamersma H, Gardner J, Beighton P. The natural history of sclerosteosis. Clin Genet. 2003;63:192–7.
55. Stein SA, Witkop C, Hill S, et al. Sclerosteosis, neurogenetic and pathophysiologic analysis of an American kinship. Neurology. 1983;33:267–77.

56. Leupin O, Piters E, Halleux C, et al. Bone overgrowth-associated mutations in the LRP4 gene impair sclerostin facilitator function. J Biol Chem. 2011;286:19489–500.
57. Boyden LM, Mao J, Belsky J, et al. High bone density due to a mutation in LDL-receptor-related protein 5. N Engl J Med. 2002;345:1513–21.
58. Rickels MR, Zhang X, Mumm S, et al. Oropharyngeal skeletal disease accompanying high bone mass and novel LRP5 mutation. J Bone Miner Res. 2005;20(5):878–85.
59. Uitto J, Santa Cruz DJ, Starcher BC, et al. Biochemical and ultrastructural demonstration of elastin accumulation in the skin of the Buschke-Ollendorff syndrome. J Invest Dermatol. 1981;76:284–7.
60. Mumm S, Wenkert D, Zhang X, et al. Deactivating germline mutations in LEMD3 cause osteopoikilosis and Buschke-Ollendorff syndrome, but not sporadic melorheostosis. J Bone Miner Res. 2007;22:243–50.
61. Whyte MP, Murphy WA, Seigel BA. 99m Tc-pyrophosphate bone imaging in osteopoikilosis, osteopathia striata, and melorheostosis. Radiology. 1978;127:439–43.
62. Jenkins ZA, van Kogelenberg M, Morgan T, et al. Germline mutations in WTX cause a sclerosing skeletal dysplasia but do not predispose to tumorigenesis. Nat Genet. 2009;41:55–100.
63. Leri A, Joanny J. Une affection non decrite des os. Hyperostose "en coulee" sur toute la longueur d'un membre ou "melorheostose." Bull Mem Soc Med Hop Paris 1922;46:1141–5.
64. Smith GC, Pingree MJ, Freeman LA, et al. Melorheostosis: a retrospective clinical analysis of 24 patients at the Mayo Clinic. PM R. 2017;9(3):283–8.
65. Whyte MP, Griffith M, Trani L, et al. Melorheostosis: exome sequencing of an associated dermatosis implicates postzygotic mosaicism of mutated KRAS. Bone. 2017;101:145–55.
66. Kang H, Jha S, Deng Z, et al. Somatic activating mutations in MAP2K1 cause melorheostosis. Nat Commun. doi: 10.1038/s41467-018-03720-z.
67. Atar D, Lehman WB, Grant AD, et al. The Ilizarov apparatus for treatment of melorheostosis: case report and review of the literature. Clin Orthop. 1992;281:163–7.
68. Christmann D, Wenger JJ, Dosch JC, et al. (Axial osteomalacia. Comparative analysis with fibrogenesis imperfecta ossium (author's transl)). (Article in French). J Radiol. 1981;62:37–41.
69. Whyte MP, Fallon MD, Murphy WA, et al. Axial osteomalacia: clinical, laboratory and genetic investigation of an affected mother and son. Am J Med. 1981;71:1041–9.
70. Epperla N, McKiernan FE, Kenney CV. Radiographic findings in Waldenstrom's macroglobulinemia resembling fibrogenesis imperfecta ossium (FIO): a case report. Skeletal Radiol. 2014;43:381–5.
71. Bakos B, Lukats A, Lakatos P, et al. Report on a case of fibrogenesis imperfecta ossium and a possible new treatment option. Osteoporos Int. 2014;25:1643–6.
72. Uppal S, Diggle CP, Carr IM, et al. Mutations in 15-hydroxyprostaglandin dehydrogenase cause primary hypertrophic osteoarthropathy. Nat Genet. 2008;40:789–93.
73. Yuan L, Chen L, Liao R-X, et al. A common mutation and a novel mutation in the HPGD gene in nine patients with primary hypertrophic osteoarthropathy. Calcif Tissue Int. 2015;97:336–42.
74. Giancane G, Diggle CP, Legger EG, et al. Primary hypertrophic osteoarthropathy: an update on patient features and treatment. J Rheumatol. 2015;42:2211–2214.
75. Whyte MP, Teitelbaum SL, Reinus WR. Doubling skeletal mass during adult life: the syndrome of diffuse osteosclerosis after intravenous drug abuse. J Bone Miner Res. 1996;11:554–8.
76. Serraino C, Melchio R, Silvestri A, et al. Hepatitis C-associated osteosclerosis: a new case with long-term follow-up and a review of the literature. Intern Med 2015;54:777–83.
77. Khosla S, Ballard FJ, Conover CA. Use of site-specific antibodies to characterize the circulating form of big insulin-like growth factor II in patients with hepatitis C-associated osteosclerosis. J Clin Endocrinol Metab. 2002;87(8):3867–70.
78. Shadado S, Akehi Y, Yotsumoto K, et al. A case of hepatitis C-associated osteosclerosis that was improved with the combination therapy of peginterferon alfa-2b and ribavirin. Clin J Gastroenterol. 2011;4:255–61.

108

Fibrous Dysplasia

Michael T. Collins[1], Alison M. Boyce[1], and Mara Riminucci[2]

[1]Section of Skeletal Diseases and Mineral Homeostasis, National Institute of Dental and Craniofacial Research, National Institutes of Health, Department of Health and Human Services, Bethesda, MD, USA
[2]Dipartimento di Medicina Molecolare, Sapienza Università, Rome, Italy

INTRODUCTION

Fibrous dysplasia of bone (FD) (OMIM #174800) is an uncommon skeletal disorder with a broad spectrum of clinical presentation. On one end of the spectrum, patients may present in adulthood with an incidentally discovered, asymptomatic radiographic finding of no clinical significance. On the other end of the spectrum, patients present early in life with disabling disease. The disease may involve one bone (monostotic FD), multiple bones (polyostotic FD), or the entire skeleton (panostotic FD) [1–3]. FD may be associated with a wide range of extraskeletal manifestations, the most common of which is patches of cutaneous hyperpigmentation commonly referred to as café-au-lait macules. These lesions vary widely in size but typically have characteristic features that include jagged, "coast of Maine" borders, some relationship with the midline, and sometimes follow the developmental lines of Blashcko (Fig. 108.1A–E). Other extraskeletal manifestations include hyperfunctioning endocrinopathies, such as precocious puberty, hyperthyroidism, growth hormone excess, and Cushing's syndrome. FD in combination with one or more extraskeletal manifestations is known as McCune–Albright syndrome (MAS) [4–7]. Renal phosphate wasting caused by overproduction of FGF23 by dysplastic bone cells, is one of the more common and clinically significant extraskeletal manifestations [8]. An association of FD with possibly premalignant, cystic lesions of the pancreas (intraductal papillary mucinous neoplasms) has recently been reported, and is emerging as one of the more common extraskeletal manifestations [9]. More rarely, FD may be associated with intramuscular myxomas (Mazabraud's syndrome) [10] or dysfunction of the heart, liver, or other organs within the context of the MAS [11].

ETIOLOGY AND PATHOGENESIS

Genetically, FD is a somatic mosaic disease caused by missense mutations of the alpha subunit of the stimulatory G protein (Gαs), encoded at the *GNAS* locus on chromosome 20q13.3 [12–14]. The mutation most often occurs at exon 8, likely as a consequence of abnormal methylation of a CpG dinucleotide during early embryonic development [15], in which the arginine 201 is converted to a histidine or a cysteine (p.R201H, p.R201C). Rarely, other substitutions (eg, p.R201S) or mutations in other codons, (p.Q227L) have been reported [16–19]. Gαs is a central component of the cAMP-dependent signal transduction pathway. All mutations associated with FD are gain of function mutations that impair the intrinsic GTPase activity of the protein leading to the generation of excess intracellular cAMP [20]. Based on the lack of verified vertical transmission, discordance in monozygotic twins [21], and presence of wild-type Gαs in many tissues in a severely affected infant who died from severe MAS (unpublished data), it is assumed somatic mosaicism is necessary for the survival of the mutated embryo. However, germ-line transmission of Gαs[R201C] has been reported recently in transgenic mouse models of FD that express the mutated sequence under the control of constitutive promoters [22]. The survival of these murine strains suggest that germ line gain of function mutations of Gαs are not necessarily embryonically lethal, at least in mice. Interestingly, although the transgene was ubiquitously expressed in developing embryos, skeletal lesions in this mouse model of FD were not evident until the postnatal period; first detected at approximately 2 months of age (equivalent to approximately 20 years of age in humans). This could be consistent with downstream epigenetic/nongenetic determinants operating in

---

Primer on the Metabolic Bone Diseases and Disorders of Mineral Metabolism, Ninth Edition. Edited by John P. Bilezikian.
© 2019 American Society for Bone and Mineral Research. Published 2019 by John Wiley & Sons, Inc.
Companion website: www.wiley.com/go/asbmrprimer

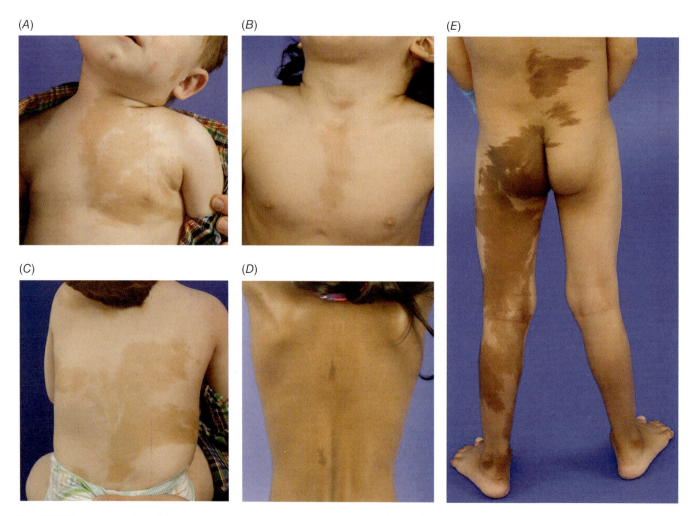

Fig. 108.1. Cafe-au-lait skin pigmentation. (A) Large pigmented macules with a classic appearance on the face, neck, and chest shows jagged "coast of Maine" borders, and the tendency to respect the midline of the body. (B) A smaller lesion located in the midline of the chest also shows irregular borders. Note the goiter associated with McCune–Albright thyroid disease. (C) Large lesions on the back following developmental lines of Blashcko. Note that the lesion crosses the midline. (D) Small but typical lesions distributed along the midline of the spine. (E) Extensive pigmentation involving the back, buttock, and left lower extremity show the tendency of lesions to originate in a central location and extend distally.

this model that modulate the development of the phenotype. It is possible that the same determinants play a role in the human disease, and in addition to genetic mosaicism, could explain the variability of the clinical phenotype observed in FD/MAS patients.

The constitutive activation of Gαs in the skeleton results in the growth of fibrous-like tissue that expands in the marrow space at the expense of marrow adipocytes, hematopoiesis, and normal trabecular bone. Within the fibrous tissue, pathological, mechanically unsound bone with abnormal architecture and structure is deposited. At first glance, the microscopic appearance of typical FD lesions may recall many different neoplastic and nonneoplastic fibrous–osseous diseases that affect the skeleton. However, specific patterns of distribution of bone trabeculae at different skeletal sites (Fig. 108.2A,B), as well as some recurrent, site-independent microscopic features of FD bone (Fig. 108.2C) [23] may assist in the differential diagnosis, especially in patients with isolated, monostotic lesions. Increased bone resorption and bone matrix hypomineralization (Fig. 108.2D) are commonly observed and likely contribute to the mechanical instability of the bone that leads to recurrent fractures and bone deformities. Although the majority of FD lesions are characterized by these histopathological features, rare histological variants of the disease have also been reported, such as fibrocartilaginous dysplasia (Fig. 108.2E,F) [24].

The recognition that FD is a disorder of the whole bone/bone marrow organ, and the development of experimental models based on the use of skeletal stem cells has led to major advances in our understanding of the pathogenesis of FD [25, 26]. FD lesions reflect the abnormal differentiation of Gαs mutation-bearing marrow osteoprogenitor cells from which defective osteoblasts

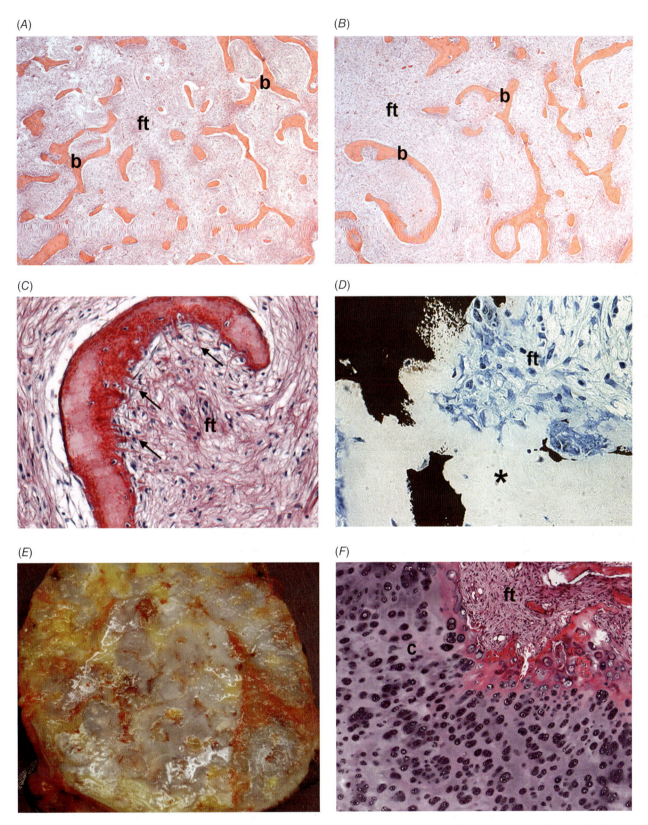

Fig. 108.2. Histopathology of fibrous dysplasia. (A,B) Typical "chinese writing" histological pattern of FD. (C) Abnormally oriented collagen fibers on the surface of FD bone trabeculae (sharpey fibers, arrows). (D) Excess unmineralized osteoid in FD bone (asterisk). (E,F) Macroscopic and microscopic features of the cartilaginous variant of FD. A,B,F = HE stain; C = Sirius red stain; D = von Kossa stain; b = bone; ft = fibrous tissue; c = cartilage.

are generated [27, 28]. Support for the concept of deranged osteoblast differentiation as pathogenic in FD is supported experimentally by both studies of osteoprogenitor cells isolated from FD [29], as well as normal human skeletal stem cells that have been transduced with $G\alpha s^{R201C}$ and exhibit pathological features consistent with FD cells/tissue patients [26]. Most of the essential histopathological features of FD may be ascribed to the osteogenic nature of the lesional tissue, including the osteomalacic changes of FD bone, which result from excess production of the phosphaturic hormone, FGF23, by osteogenic tissue [14, 30]. The inappropriate osteoclastogenesis detected within FD lesions depends on osteogenic cells through the secretion of IL-6 [31], and likely through the upregulation of RANKL [26]. In addition, the clinical expression of FD may be modulated by biological mechanisms specifically related to osteoprogenitor cells. For example, the asymmetric and random expression of the two Gαs alleles [32] that physiologically occurs in skeletal clonogenic cells may affect the level of expression of the mutated allele and, therefore, the likelihood of developing the disease. In addition, the progressive, age-related, apoptosis-dependent loss of mutation-bearing clonogenic progenitors that takes place in FD lesions [33] may contribute to the decrease in lesion activity ("burn-out") that is frequently observed clinically in FD lesions as patients age.

Recently, transgenic murine lines expressing the $G\alpha s^{R201C}$ sequence under the control of different promoters have been developed. Mice strains with constitutive expression of the transgene produce a FD-like tissue that bears histological similarities to the human disease, including evolution through distinct histopathological phases [22]. In contrast, murine lines with transgene expression targeted to mature osteoblasts do not reproduce FD and develop a high bone mass phenotype [34, 35]. These results further support the concept of FD as a disease of the entire skeletal lineage, demonstrating that the essential, morbidity-causing features of FD are not dependent on the direct impact of Gαs mutations in osteoblasts, per se. However, the exact identity of the cell population(s) in the bone/bone marrow organ from which FD lesions emanate still remains unknown. Additional tissue-specific transgenic models are necessary to address this point and clarify the cellular (and molecular) pathogenesis of the disease.

CLINICAL FEATURES

The clinical presentation of FD is variable and depends upon the location and extent of the skeletal disease. The proximal metaphysis of the femora and the skull base are the two sites most commonly involved. However, any part or combination of bones may be affected [36]. FD in the appendicular skeleton typically presents with limp, fracture, and/or pain. Deformities are common, and in particular affect weight-bearing bones such as the proximal femur, which classically develops a coxa vara "shepherd's crook" deformity (Fig. 108.3B,C) [37]. In patients with extensive involvement of the lower extremities, recurrent fractures and severe deformity may result in impaired ambulation that necessitates assistive devices. FD in the craniofacial skeleton leads to bony expansion, resulting in variable degrees of facial deformity and infrequently functional impairment. Involvement of the optic canals is common, but only rarely leads to visual deficits [38]. Similarly, temporal bone FD occurs frequently, but uncommonly results in hearing and/or otologic dysfunction [39]. Sinonasal involvement commonly leads to chronic nasal congestion [40], and involvement of the alveolar bones of the maxilla and mandible may affect dental occlusion [41]. Scoliosis as a result of axial involvement is common and in rare cases may be progressive and even potentially fatal, emphasizing the importance of detection, monitoring, and treatment of FD-associated scoliosis [42, 43].

The sites of skeletal involvement (the "map" of affected tissues) are established early in patients with FD. While bone disease is typically not clinically apparent at birth, 90% of craniofacial FD lesions are established before the age of five, and 75% of all sites are evident by 15 years of age [44] The clinical implication is that essentially all significant disease is apparent early in life, probably by the age of 5 [3]. Functional impairment is also established relatively early; most patients who require assistive ambulation will develop this need during childhood or adolescence [44].

Pain is a common feature of FD. Sites commonly affected include the ribs, long bones, and craniofacial bones, whereas lesions in the spine and pelvis are typically less painful [36]. FD pain is typically more common in adults; however, pain during childhood is often underdiagnosed and undertreated [45].

The radiographic appearance of FD lesions is also characterized by site- and age-specific features (Fig. 108.3A–E). In toddlers and young children, FD lesions in the craniofacial and appendicular skeleton often appear heterogeneous on plain films and computed tomography (Fig. 108.3A) [46]. Lesions in older children and adolescents develop a characteristic homogeneous appearance, classically defined as a "ground glass" (Fig. 108.3B,D). However, in older individuals the appearance tends to become "sclerotic" and heterogeneous, which likely reflects an age-related decrease in disease activity (Fig. 103.3C,E). Lesions in the spine, ribs, and pelvis are common, but difficult to appreciate on plain radiographs. These lesions are more readily detected by CT, or bone scintigraphy, which is a sensitive imaging technique for the detection of FD lesions (Fig. 108.3F,G).

Secondary lesions may arise in association with preexisting FD. Examples include intramuscular myxomas (Mazabraud's syndrome, Fig. 108.3H), fluid-filled cysts (aneurysmal bone cysts, Fig. 108.3I), and malignancies, consistent with the small increased risk of oncogenesis associated with *GNAS* mutations [47]. Malignancies are rare (<1% incidence) and are more likely in patients exposed to high-dose external beam radiation, including pituitary radiation for treatment of McCune–Albright-associated growth hormone excess [48, 49]. Rapid lesion expansion and disruption of the cortex on radiographs should alert the

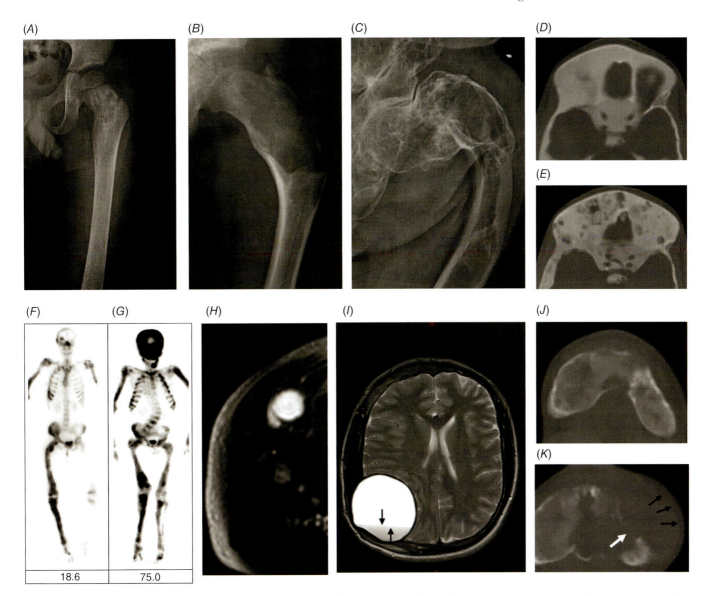

Fig. 108.3. Representative radiographic features of fibrous dysplasia (FD). (A–E) The age-related changes in the appearance of FD. In early childhood (A – age 3.5), there is an inhomogeneous, streaky appearance. In childhood (B – age 12) there is the classic ground glass appearance. As patients age (C – age 60), significant sclerosis occurs, which has the effect of increasing measures of bone density. In addition, B and C show the development of the classic "shepherd's crook" that commonly occurs. (D,E) The age-related changes that occur in craniofacial FD. In children (D – age 4) the dominant characteristic appearance is classic ground glass. However, with age (E – same patient, age 15) craniofacial lesions become more lytic in appearance. (F,G) ^{99}Tc-MDP bone scan images demonstrating radiotracer uptake at areas of FD. The patchy uptake shows the mosaic nature of the disease. Numbers below the panels indicate quantification of the skeletal disease burden as described in reference [50]. (H) Intramuscular myxomas can also be observed in patients with FD. They can be single (white arrow), or multiple, and are usually asymptomatic and found incidentally. (I) Fluid-filled cystic lesions occur infrequently in existing FD. The fluid-filled nature of the lesion is best shown by MRI; the black arrows outline the fluid/fluid level that is diagnostic of the aneurysmal bone cyst-like lesion. (J,K) Sarcomatous transformation from preexisting FD. (J) is a rare but serious feature of FD. The hallmark radiographic features are shown in (K): a soft tissue mass (black arrows) that has breached the cortex (white arrow).

clinician to the possibility of sarcomatous change (Fig. 108.3J,K). Osteogenic sarcoma is the most common, but not the only type of bone tumor that may complicate FD. The clinical course is usually aggressive, surgery is the primary treatment, and chemotherapeutic regimens do not appear to improve prognosis significantly.

MANAGEMENT AND TREATMENT

The diagnosis of FD is established based on expert assessment of clinical, radiographic, and histopathological features. All patients should undergo a complete skeletal evaluation at the time of diagnosis to determine disease burden. This is best achieved with total body bone scintigraphy to identify areas of FD (Fig. 108.3 F,G), followed by radiographs to evaluate lesions anatomically [50]. Screening and treatment of extraskeletal disease is a critical component of management in FD patients, because untreated endocrinopathies are associated with poorer outcomes [38, 42, 51, 52]. In particular, growth hormone excess has been associated with craniofacial morbidity [38, 51], and hypophosphatemia increases fractures and bone pain [42].

Genetic testing may be helpful in distinguishing isolated, monostotic FD from unrelated fibro-osseous lesions of the skeleton, which may mimic FD both clinically and radiographically (osteofibrous dysplasia, ossifying fibromas of jawbones; reviewed in [53]). Isolated lesions of the proximal femur may be improperly diagnosed and classified as distinct fibro-osseous lesions. For example, all cases of so-called "liposclerosing myxofibrous tumor," which is radiographically similar to FD, in which DNA was analyzed for *GNAS* mutations, were mutation positive and therefore represent monostotic FD [54]. Multiple nonossifying fibromas, skeletal angiomatosis, and Ollier's disease may sometimes enter the differential diagnosis. Distinction from these entities relies on histology and mutation analysis. Given the mosaic nature of the disease, tissue from lesions must be used for genomic analysis.

Management is focused on treating endocrinopathies, preventing fracture and deformities, optimizing function, and treating pain. Proximal femur FD is often best managed by internal fixation with intramedullary nails, which in children may require specially designed nails or modification of existing devices [55–57]. Clinicians should be aware that techniques used in other disorders may be unsuccessful in FD. In particular bone grafting and external fixation are commonly used techniques that are now known to be frequently ineffective in FD [58, 59]. Indications for surgery of the craniofacial bones may include cosmesis, or documented vision, or hearing loss. In all cases, operations should be undertaken with caution and only by experienced surgeons. Contouring or "shaving" procedures for cosmesis often invoke a rapid regrowth, especially in children and patients with growth hormone excess [60]. Operations for vision loss are frequently unsuccessful at restoring vision [61, 62], and prophylactic surgery to prevent vision loss is not advocated [38, 62].

Treatment with bisphosphonates (pamidronate, zoledronic acid, etc.) has been advocated based on observational studies, with claims of reduced pain, decreased serum and urine markers of bone metabolism, and improvement in the radiographic appearance of the disease [63–65]. However, an open label, prospective study of pamidronate with appropriate histological, radiographic, and clinical end points showed pain relief but no benefit radiographically or histologically [66]. A placebo-controlled trial of oral alendronate did not show a benefit on pain or radiographic appearance of FD lesions. Of concern, recent evidence shows that patients with FD are at risk for bisphosphonate-related osteonecrosis of the jaw (ONJ) [67]. Therefore, based on the current literature, intravenous bisphosphonates are likely beneficial for treatment of FD-related bone pain; however, they should be used judiciously and at the lowest therapeutic dose and longest interval between treatments to minimize the risk of ONJ.

FUTURE TREATMENT

Currently available medical and surgical therapies are not satisfactory. Advances in surgical techniques, especially large, limb salvage techniques, and devices for older patients with advanced, disabling disease are needed, as are size-appropriate devices for children. Medical therapy that includes adaptation of existing drugs, such as tocilizumab [68] (results of a controlled, blinded trial should be available soon), denosumab, and novel drugs that either target pain pathways or target the mutated Gαs [69, 70] should be considered. Efforts are being made to elucidate the effects of silencing the disease-causing gene in skeletal stem cells in in vitro and in vivo models [71], on devising strategies for using stem cells as either a tool or a target of therapies [71], the development of interfering allele-specific oligonucleotides, and on identifying novel drugs that specifically target the constitutively active Gsα or its functional effects. As part of these approaches, it will be important to elucidate additional molecular mediators of the disease phenotype downstream of the mutated Gsα. The establishment of an international consortium of clinician investigators, the coalescence of patient support groups, and public/private and academic/industry partnerships will lead to improved understanding and better treatment for patients with FD/MAS in the future.

ACKNOWLEDGMENT

This work was supported in part by the Division of Intramural Research, National Institute of Dental and Craniofacial Research, National Institutes of Health, Bethesda, MD, USA. The work was also supported by research grants from the University of Pennsylvania Orphan Disease Center in partnership with the Fibrous Dysplasia Foundation (MDBR17-114 and MDBR18-114), and Telethon (GGP15198).

REFERENCES

1. Lichtenstein L, Jaffe HL. Fibrous dysplasia of bone: a condition affecting one, several or many bones, the graver cases of which may present abnormal pigmentation of skin, premature sexual development, hyperthyroidism or still other extraskeletal abnormalities. Arch Pathol. 1942;33:777–816.

2. Collins MT. Spectrum and natural history of fibrous dysplasia of bone. J Bone Miner Res. 2006;21(Suppl. 2):P99–P104.
3. Hart ES, Kelly MH, Brillante B, et al. Onset, progression, and plateau of skeletal lesions in fibrous dysplasia, and the relationship to functional outcome. J Bone Miner Res. 2007;22:1468–74.
4. McCune DJ Osteitis fibrosa cystica; the case of a nine year old girl who also exhibits precocious puberty, multiple pigmentation of the skin and hyperthyroidism. Am J Dis Child. 1936;52:743–4.
5. Albright F, Butler AM, Hampton AO, et al. Syndrome characterized by osteitis fibrosa disseminata, areas of pigmentation and endocrine dysfunction, with precocious puberty in females, report of five cases. N Engl J Med. 1937;216:727–46.
6. Danon M, Crawford JD. The McCune-Albright syndrome. Ergeb Inne Med Kinderheilkd. 1987;55:81–115.
7. Dumitrescu CE, Collins MT. McCune-Albright syndrome. Orphanet J Rare Dis. 2008;3:12.
8. Collins MT, Chebli C, Jones J, et al. Renal phosphate wasting in fibrous dysplasia of bone is part of a generalized renal tubular dysfunction similar to that seen in tumor- induced osteomalacia. J Bone Miner Res. 2001;16(5):806–13.
9. Gaujoux S, Salenave S, Ronot M, et al. Hepatobiliary and pancreatic neoplasms in patients with McCune-Albright syndrome. J Clin Endocrinol Metab 2014;99(1):E97–101.
10. Cabral CE, Guedes P, Fonseca T, et al. Polyostotic fibrous dysplasia associated with intramuscular myxomas: Mazabraud's syndrome. Skeletal Radiol. 1998;27(5):278–82.
11. Shenker A, Weinstein LS, Moran A, et al. Severe endocrine and nonendocrine manifestations of the McCune-Albright syndrome associated with activating mutations of stimulatory G protein GS. J Pediatr. 1993;23(4):509–18.
12. Weinstein LS, Shenker A, Gejman PV, et al. Activating mutations of the stimulatory G protein in the McCune-Albright syndrome. N Engl J Med. 1991;325(24):1688–95.
13. Shenker A, Weinstein LS, Sweet DE, et al. An activating Gs alpha mutation is present in fibrous dysplasia of bone in the McCune-Albright syndrome. J Clin Endocrinol Metab. 1994;79(3):750–5.
14. Bianco P, Riminucci M, Majolagbe A, et al. Mutations of the GNAS1 gene, stromal cell dysfunction, and osteomalacic changes in non-McCune-Albright fibrous dysplasia of bone. J Bone Miner Res. 2000;15(1):120–8.
15. Riminucci M, Saggio I, Robey PG, et al. Fibrous dysplasia as a stem cell disease. J Bone Miner Res. 2006;21(Suppl. 2):P125–31.
16. Riminucci M, Fisher LW, Majolagbe A, et al. A novel GNAS1 mutation, R201G, in McCune-albright syndrome. J Bone Miner Res. 1999;14(11):1987–9.
17. Idowu BD, Al-Adnani M, O'Donnell P, et al. A sensitive mutation-specific screening technique for GNAS1 mutations in cases of fibrous dysplasia: the first report of a codon 227 mutation in bone. Histopathology 2007;50(6):691–704.
18. Jour G, Oultache A, Sadowska J, et al. GNAS mutations in fibrous dysplasia: a comparative study of standard sequencing and locked nucleic acid PCR sequencing on decalcified and nondecalcified formalin-fixed paraffin-embedded tissues. Appl Immunohistochem Mol Morphol. 2016;24(9):660–7.
19. Lee SE, Lee EH, Park H, et al. The diagnostic utility of the GNAS mutation in patients with fibrous dysplasia: meta-analysis of 168 sporadic cases. Human Pathol. 2012;43(8):1234–42.
20. Landis CA, Masters SB, Spada A, et al. GTPase inhibiting mutations activate the alpha chain of Gs and stimulate adenylyl cyclase in human pituitary tumours. Nature 1989;340(6236):692–6.
21. Lemli L. Fibrous dysplasia of bone. Report of female monozygotic twins with and without the McCune-Albright syndrome. J Pediatr. 1977;91:947–9.
22. Saggio I, Remoli C, Spica E, et al. Constitutive expression of Gsalpha(R201C) in mice produces a heritable, direct replica of human fibrous dysplasia bone pathology and demonstrates its natural history. J Bone Miner Res. 2014;29(11):2357–68.
23. Riminucci M, Liu B, Corsi A, et al. The histopathology of fibrous dysplasia of bone in patients with activating mutations of the Gs alpha gene: site-specific patterns and recurrent histological hallmarks. J Pathol. 1999;187(2):249–58.
24. Ishida T, Dorfman HD. Massive chondroid differentiation in fibrous dysplasia of bone (fibrocartilaginous dysplasia). Am J Surg Pathol. 1993;17(9):924–30.
25. Bianco P, Kuznetsov SA, Riminucci M, et al. Reproduction of human fibrous dysplasia of bone in immunocompromised mice by transplanted mosaics of normal and Gsalpha-mutated skeletal progenitor cells. J Clin Invest. 1998;101(8):1737–44.
26. Piersanti S, Remoli C, Saggio I, et al. Transfer, analysis, and reversion of the fibrous dysplasia cellular phenotype in human skeletal progenitors. J Bone Miner Res. 2010;25(5):1103–16.
27. Riminucci M, Fisher LW, Shenker A, et al. Fibrous dysplasia of bone in the McCune-Albright syndrome: abnormalities in bone formation. Am J Pathol. 1997;151(6):1587–600.
28. Robey PG, Kuznetsov S, Riminucci M, et al. The role of stem cells in fibrous dysplasia of bone and the McCune-Albright syndrome. Pediatr Endocrinol Rev. 2007;4(Suppl. 4):386–94.
29. Regard JB, Cherman N, Palmer D, et al. Wnt/beta-catenin signaling is differentially regulated by Galpha proteins and contributes to fibrous dysplasia. Proc Natl Acad Sci U S A 2011;108(50):20101–6.
30. Riminucci M, Collins MT, Fedarko NS, et al. FGF-23 in fibrous dysplasia of bone and its relationship to renal phosphate wasting. J Clin Invest. 2003;112(5):683–92.
31. Riminucci M, Kuznetsov SA, Cherman N, et al. Osteoclastogenesis in fibrous dysplasia of bone: in situ and in vitro analysis of IL-6 expression. Bone 2003;33(3):434–42.
32. Michienzi S, Cherman N, Holmbeck K, et al. GNAS transcripts in skeletal progenitors: evidence for random asymmetric allelic expression of Gs alpha. Hum Mol Genet. 2007;16(16):1921–30.

33. Kuznetsov SA, Cherman N, Riminucci M, et al. Age-dependent demise of GNAS-mutated skeletal stem cells and "normalization" of fibrous dysplasia of bone. J Bone Miner Res. 2008;23(11):1731–40.
34. Remoli C, Michienzi S, Sacchetti B, et al. Osteoblast-specific expression of the fibrous dysplasia (FD)-causing mutation Gsalpha(R201C) produces a high bone mass phenotype but does not reproduce FD in the mouse. J Bone Miner Res. 2015;30(6):1030–43.
35. Hsiao EC, Boudignon BM, Halloran BP, et al. Gs G protein-coupled receptor signaling in osteoblasts elicits age-dependent effects on bone formation. J Bone Miner Res. 2010;25(3):584–93.
36. Kelly MH, Brillante B, Collins MT. Pain in fibrous dysplasia of bone: age-related changes and the anatomical distribution of skeletal lesions. Osteoporos Int. 2008;19(1):57–63.
37. Ippolito E, Farsetti P, Boyce AM, et al. Radiographic classification of coronal plane femoral deformities in polyostotic fibrous dysplasia. Clin Orthop Relat Res. 2014;472(5):1558–67.
38. Lee JS, FitzGibbon E, Butman JA, et al. Normal vision despite narrowing of the optic canal in fibrous dysplasia. N Engl J Med. 2002;347(21):1670–6.
39. Frisch CD, Carlson ML, Kahue CN, et al. Fibrous dysplasia of the temporal bone: a review of 66 cases. Laryngoscope. 2015;125(6):1438–43.
40. DeKlotz TR, Kim HJ, Kelly M, et al. Sinonasal disease in polyostotic fibrous dysplasia and McCune-Albright Syndrome. Laryngoscope. 2013;123(4):823–8.
41. Burke AB, Collins MT, Boyce AM. Fibrous dysplasia of bone: craniofacial and dental implications. Oral Dis. 2017;23(6):697–708.
42. Leet AI, Chebli C, Kushner H, et al. Fracture incidence in polyostotic fibrous dysplasia and the McCune-Albright syndrome. J Bone Miner Res. 2004;19(4):571–7.
43. Mancini F, Corsi A, De Maio F, et al. Scoliosis and spine involvement in fibrous dysplasia of bone. Eur Spine J. 2009;18(2):196–202.
44. Hart ES, Kelly MH, Brillante B, et al. Onset, progression, and plateau of skeletal lesions in fibrous dysplasia and the relationship to functional outcome. J Bone Miner Res. 2007;22(9):1468–74.
45. Kelly MH, Brillante B, Collins MT. Pain in fibrous dysplasia of bone: age-related changes and the anatomical distribution of skeletal lesions. Osteoporos Int. 2008;19(1):57–63.
46. Leet AI, Collins MT. Current approach to fibrous dysplasia of bone and McCune-Albright syndrome. J Child Orthop. 2007;1(1):3–17.
47. Ruggieri P, Sim FH, Bond JR, et al. Malignancies in fibrous dysplasia. Cancer 1994;73(5):1411–24.
48. Saglik Y, Atalar H, Yildiz Y, et al. Management of fibrous dysplasia. A report on 36 cases. Acta Orthop Belg. 2007;73(1):96–101.
49. Liu F, Li W, Yao Y, et al. A case of McCune-Albright syndrome associated with pituitary GH adenoma: therapeutic process and autopsy. J Pediatr Endocrinol Metab. 2011;24(5–6):283–7.
50. Collins MT, Kushner H, Reynolds JC, et al. An instrument to measure skeletal burden and predict functional outcome in fibrous dysplasia of bone. J Bone Miner Res. 2005;20(2):219–26.
51. Boyce AM, Glover M, Kelly MH, et al. Optic neuropathy in McCune-Albright syndrome: effects of early diagnosis and treatment of growth hormone excess. J Clin Endocrinol Metab. 2013;98(1):E126–34.
52. Boyce AM, Collins MT. Fibrous dysplasia/McCune–Albright syndrome. In: Pagon RA, Adam MP, Ardinger HH, et al. (eds.) GeneReviews. Seattle: University of Washington, 1993.
53. Bianco P, Gehron Robey P, Wientroub S. Fibrous dysplasia. In: Glorieux F, Pettifor J, Juppner H (eds) Pediatric Bone: Biology and Disease. New York: Academic Press, 2003, pp. 509–39.
54. Corsi A, De Maio F, Ippolito E, et al. Monostotic fibrous dysplasia of the proximal femur and liposclerosing myxofibrous tumor: which one is which? J Bone Miner Res. 2006;21(12):1955–8.
55. Stanton RP. Surgery for fibrous dysplasia. J Bone Miner Res. 2006;21(Suppl. 2):P105–9.
56. Ippolito E, Bray EW, Corsi A, et al. Natural history and treatment of fibrous dysplasia of bone: a multicenter clinico-pathologic study promoted by the European Pediatric Orthopaedic Society. J Pediatr Orthop B. 2003;12(3):155–77.
57. Keijser LC, Van Tienen TG, Schreuder HW, et al. Fibrous dysplasia of bone: management and outcome of 20 cases. J Surg Oncol. 2001;76(3):157–166; discussion 167–58.
58. Stanton RP, Ippolito E, Springfield D, et al. The surgical management of fibrous dysplasia of bone. Orphanet J Rare Dis. 2012;24(7):Suppl 1:S1.
59. Leet AI, Boyce AM, Ibrahim KA, et al. Bone-grafting in polyostotic fibrous dysplasia. J Bone Joint Surg Am. 2016;98(3):211–9.
60. Boyce AM, Burke A, Cutler Peck C, et al. Surgical management of polyostotic craniofacial fibrous dysplasia: long-term outcomes and predictors for postoperative regrowth. Plast Reconstr Surg. 2016;37(6):1833–9.
61. Chen YR, Chang CN, Tan YC. Craniofacial fibrous dysplasia: an update. Chang Gung Med J. 2006;29(6):543–9.
62. Cutler CM, Lee JS, Butman JA, et al. Long-term outcome of optic nerve encasement and optic nerve decompression in patients with fibrous dysplasia: risk factors for blindness and safety of observation. Neurosurgery 2006;59(5):1011–1017; discussion 1017–18.
63. Liens D, Delmas PD, Meunier PJ. Long-term effects of intravenous pamidronate in fibrous dysplasia of bone. Lancet 1994;343(8903):953–4.
64. Chapurlat RD, Delmas PD, Liens D, et al. Long-term effects of intravenous pamidronate in fibrous dysplasia of bone. J Bone Miner Res. 1997;12(10):1746–52.
65. Majoor BC, Appelman-Dijkstra NM, Fiocco M, et al. Outcome of long-term bisphosphonate therapy in McCune-Albright syndrome and polyostotic fibrous dysplasia. J Bone Miner Res. 2017;32(2):264–76.
66. Plotkin H, Rauch F, Zeitlin L, et al. Effect of pamidronate treatment in children with polyostotic fibrous dysplasia of bone. J Clin Endocrinol Metab. 2003;88(10):4569–75.

67. Metwally T, Burke A, Tsai JY, et al. Fibrous dysplasia and medication-related osteonecrosis of the jaw. J Oral Maxillofac Surg. 2016;74(10):1983–99.
68. de Boysson H, Johnson A, Hablani N, et al. Tocilizumab in the treatment of a polyostotic variant of fibrous dysplasia of bone. Rheumatology (Oxford) 2015;54(9):1747–9.
69. Chapurlat RD, Gensburger D, Jimenez-Andrade JM, et al. Pathophysiology and medical treatment of pain in fibrous dysplasia of bone. Orphanet J Rare Dis. 2012;7(Suppl. 1):S3.
70. Bhattacharyya N, Hu X, Chen CZ, et al. A high throughput screening assay system for the identification of small molecule inhibitors of gsp. PloS One 2014;9(3):e90766.
71. Piersanti S, Remoli C, Saggio I, et al. Transfer, analysis and reversion of the fibrous dysplasia cellular phenotype in human skeletal progenitors. J Bone Miner Res. 2010;25(5):1103–16.

109
The Osteochondrodysplasias

Fabiana Csukasi[1] and Deborah Krakow[2]

[1]Orthopaedic Surgery, University of California at Los Angeles, CA, USA
[2]Orthopaedic Surgery and Human Genetics, University of California at Los Angeles, CA, USA

INTRODUCTION

The skeletal dysplasias or osteochondrodysplasias are a group of at least 450 well-delineated genetic disorders that affect primarily bone and cartilage, but can also have significant effects on muscle, tendons, and ligaments [1]. By definition, skeletal dysplasias have generalized abnormalities in cartilage and bone, whereas the dysostoses are characterized by abnormalities in a single or group of bones [2]. However, deeper clinical phenotyping and molecular discoveries have shown that the distinction between these classifications is quite blurred. Advances in genetic technologies have contributed to identifying the molecular basis in at least 75% of these disorders, providing us the opportunities to translate molecular findings into clinical service. Understanding the genes that produce these disorders: (i) allows us to delineate the spectrum of a disease; (ii) provides diagnostic service for families at risk for recurrence; (iii) gives clues to anticipating the natural history and choosing care; as well as (iv) advances our comprehension of pathways underlying the development and maintenance of the skeleton.

Skeletal development involves bone growth, modeling, and remodeling and occurs by two distinct processes, endochondral and membranous ossification. Endochondral ossification forms the appendicular skeleton and some parts of the axial skeleton and involves a sequence of carefully orchestrated developmental cascades. These include limb bud initiation, specification of mesenchymal cells that then condense triggering cartilage differentiation, ensuing ossification of developing bones, and finally growth and maturation of the cartilage growth plate [3, 4]. Membranous ossification results from condensing mesenchymal cells that progress almost directly to bone. The bones of the skull, lateral clavicle, and pubis form via mesenchymal ossification. Endochondral bone growth is mediated by the cartilage growth plate in which certain resting chondrocytes proliferate, undergo hypertrophy, and then apoptosis (as well as potential transdifferentiation) to become the growing scaffold of bone [5]. Multiple molecular mechanisms (genes and pathways) act in concert to regulate skeletogenesis, and often genetic perturbations to these highly orchestrated processes leading to skeletal dysplasias [6].

Osteochondrodysplasias are inherited in an autosomal recessive, autosomal dominant, X-linked recessive, X-linked dominant, and Y-linked manner [7]. Many present as sporadic cases. Recognition of the mode of inheritance is important because it imparts information to families regarding future recurrences. An uncommon pattern of inheritances of some skeletal dysplasias is somatic mosaicism in which one of the parents is mildly affected and their offspring is more severely affected when the gene defect is transmitted from the germline [8]. Gonadal mosaicism can also cause familial recurrence of a known dominant disorder because one parent is clinically unaffected and heterozygosity for a mutation in one of the cell lineages that comprise the pool of his/her progenitor germ cells is passed on [9]. These are rare occurrences, but explains recurrences of "sporadic" disorders in families. Osteogensis imperfecta type II is a good example (see Chapter 113)

The explosion in knowledge from molecular genetics has allowed for gene identification in more than two-thirds of the well-delineated skeletal dysplasias [1]. This technical advancement has allowed for more precise diagnoses and improved patient care based on the established natural history of each disorder. It has also imparted molecular information on the repertoire of genes that can affect the skeleton. The historical approach to the osteochondrodysplasias that involved clinical diagnosis through physical evaluation and radiographic review, then directed molecular testing, has now in many centers been augmented by use of the skeletal dysplasia

---

Primer on the Metabolic Bone Diseases and Disorders of Mineral Metabolism, Ninth Edition. Edited by John P. Bilezikian.
© 2019 American Society for Bone and Mineral Research. Published 2019 by John Wiley & Sons, Inc.
Companion website: www.wiley.com/go/asbmrprimer

gene panel or exome analyses that are somewhat unbiased to clinical findings. The limitations to this approach can include delay in diagnosis, cost, and potential nondiagnosis. However, the benefits include molecular conformation for this group of disorders that are rare, hard to diagnose, and sometimes have multiple responsible genes (locus heterogeneity). As technology advances, its cost, precision, and availability should improve.

Molecular diagnosis can be important particularly for entitics associated with both allelic and locus heterogeneity. For some disorders, the type and location of the mutation within the disease-producing gene (protein) can impart long-term natural history information, for example nonsense or loss of protein mutations (haploinsufficiency) can lead to a different severity of disease relative to missense mutations. This is illustrated by the 16 distinct chondrodysplasias attributed to heterozygosity for mutations in the gene encoding type II collagen range from uniformly lethal to early-onset osteoarthritis *COL2A1* ("type II collagenopathies") [10]. Haploinsufficiency or loss of function for type II collagen causes the relatively mild Stickler syndrome, whereas a glycine substitution within the triple helix can produce the perinatal lethal achondrogenesis II (Fig. 109.1) [10, 11]. While the molecular and phenotypic spectrum of type II collagen disorders is quite large, most of the disorders share the common radiographic feature of abnormal epiphyseal centers.

CLINICAL APPROACH TO THE INITIAL EVALUATION OF AN INDIVIDUAL WITH A SKELETAL DYSPLASIA

Many individuals who present for evaluation of a skeletal disorder are newborns or children. Newborns with lethal skeletal disorders (there are at least 50) [1] should be provided with palliative care. Individuals with skeletal dysplasias usually present with some degree of disproportion. Dependent on the disorder they can have relative macrocephaly, frequently either small or narrow chest appearance relative to the abdomen, rhizomelia (short upper portion extremity), mesomelia (short mid portion extremity), and frequently brachydactyly (short hands including phalanges). Frequently the facies are normal, but numerous disorders have a flat nasal bridge and midface hypoplasia including many of the lethal disorders, as well as achondroplasia, campomelic dysplasia, chondrodysplasia punctata (all forms), type II collagen disorders, Larsen syndrome, and the mucopolysaccharidoses (most forms). Micrognathia occurs in some of the following: type II collagen disorders, acrofacial dysostoses, Robinow syndrome and, again, many of the lethal skeletal dysplasias.

Diagnosis is based on clinical, radiographic, and any available molecular data. Evaluation includes a thorough physical examination describing any distinct features, particularly the facies, and key measurements that

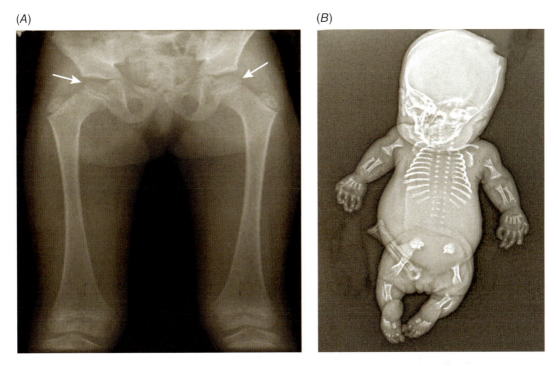

Fig. 109.1. Radiographs of COL2A1 mutational spectrum. (A) Anteroposterior (AP) view of the pelvis of a 15-year-old girl with Stickler syndrome caused by heterozygosity for a *COL2A1* mutation. Arrows point to small irregular hip epiphyses. (B) AP view of an ex utero 33-week fetus with achondrogenesis II caused by a heterozygous *COL2A1* mutation.

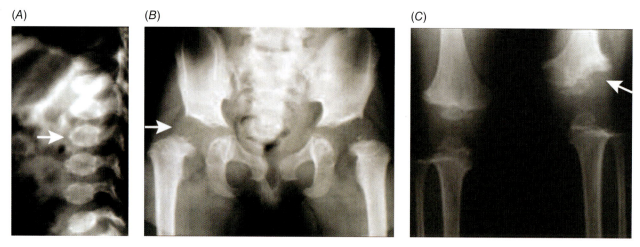

Fig. 109.2. Radiographs depicting a spondyloepimetaphyeal dysplasia. (A) Lateral spine shows rounded vertebral bodies with end plate erosion (arrow). (B) Anteroposterior (AP) vies of the pelvis showing delayed and abnormal epiphyseal development at the hips (arrow), as well as irregular metaphyseal borders at the acetabular roof and proximal femora. (C) AP view of the knee showing irregular metaphyses (arrow) and small, irregular epiphyses.

include head circumference, weight, height/length, chest circumference if it appears small, and palm and middle phalange lengths [12]. Careful delineation of any dysmorphic facial features should be performed and include evaluation of the fontanels, nasal bridge, midface, philtrum, mandible, palate, and ears. Frequently, the neck will appear short and if the chest is small the nipples may appear widely spaced. In many disorders the hands will appear small, and the phalanges will appear to have wide spaces between them because they are short. Attention should be paid to the proportion of the upper and mid sections of the arm. In skeletal disorders, there is frequently disproportion. Rhizomelia is found in achondroplasia and spondyloepiphyseal dysplasia, two of the common nonlethal disorders. Very significant mesomelia relative to rhizomelic segment suggests a group of specific disorders, the mesomelic dysplasias. Frequently when there is rhizomelia and mesomelia, increased skin creases and skin folds reflect abnormal underlying bone length. A significantly curved long bone may appear substantially shorter than observed on radiographs.

After clinical evaluation, it is critical to obtain a genetic skeletal survey [13]. This includes images of the skull, extremities (including hands and feet) and the spine. Radiographic evaluation should start with overall assessment of epiphyseal ossification to determine if they are delayed or irregular for age, then there should be consideration for an epiphyseal dysplasia. If the metaphyses are widened, flared, or irregular then the diagnosis of a metaphyseal chondrodysplasia should be entertained. If diaphyseal abnormalities are present, such as widening and/or cortical thickening, or marrow space expansion, then a diaphyseal dysplasia is implied. Any combinations of the aforementioned abnormalities help categorize and diagnose the disorder (eg, epimetaphyseal dysplasia). If the vertebral bodies are affected, then there is a "spondylo-" component, further categorizing the disorder. Once the extent of radiographic abnormalities is determined and generates a category (eg, spondylometaphyseal dysplasia), classification schemes can refine the diagnosis [1] (Fig. 109.2).

Organ system abnormalities beyond the skeleton are occasionally encountered and can direct the diagnosis. Abnormal formed genitalia and congenital heart defects are common in the skeletal ciliopathies (chondroectodermal dysplasia, asphyxiating thoracic dysplasia, and short rib polydactyly syndromes), campomelic dysplasia, omodysplasia, Robinow dysplasia, Antley–Bixler, and severe disorders of cholesterol metabolism [14–19]. Immune deficiency and Hirschprung disease occur in metaphyseal chondrodysplasia, McKusick type (cartilage hair hypoplasia) [20, 21].

LONG-TERM FOLLOW-UP FOR AFFECTED INDIVIDUALS

Patients with skeletal dysplasias are best managed by a multidisciplinary approach that includes pediatricians, neonatologists, medical geneticists, internists, endocrinologists, neurosurgeons, otolaryngologists, and orthopaedic surgeons. Affected individuals and their families should be encouraged to seek out support from organizations dedicated to the well-being of individuals with dwarfism (eg, Little People of America). Pediatricians and internists should be informed that children and adults with normal immune systems should receive standard doses of vaccines on routine schedules. Exception to live vaccines include skeletal disorders in which the affected individual has an established immunodeficiency [20]. The specific diagnosis should direct subspeciality care based on the known natural history. Many of these disorders have respiratory complications, early onset osteoarthritis, spinal deformities as well as progressive stenosis, and management should be predicated on the medical and surgical complications.

MOLECULAR PATHWAYS INVOLVED IN THE OSTEOCHONDRODYSPLASIAS

With more than 450 distinct skeletal dysplasias, multiple classification schemes have been proposed [1]. Varying methods have classified these disorders based on clinical, radiographic, pathological, biochemical, molecular, and developmental criteria. No scheme can fully address their large molecular diversity. Clinical and radiological nosologies do not address the cartilage and bone pathobiology, whereas pathological, biochemical, and molecular classifications do not provide insight to explore the differential diagnoses if there are shared phenotypic findings. Currently, the classification of the genetic skeletal disorders has 42 distinct skeletal groups according to genetic loci, but also based on clinical and radiographic findings; thus, a blended classification scheme [1].

The mutated genes underlying osteochondrodysplasias exemplify the complexity of cartilage and bone biology [1]. They encode defects in: (i) proteins in the extracellular matrix; (ii) metabolic pathways (enzymes, ion channels, transporters); (iii) processes involved in folding, transporting, and degradation of macromolecules; (iv) regulating ciliary function; (v) having nuclear roles (transcription factors); (vi) RNA processing; (vii) controlling cytoskeleton activity; (viii) encoding hormones, growth factors, receptors, and signaling pathway components and lastly, those of unclear function. Many of the involved genes have multiple roles, for example SOX9 is crucial to patterning of the skeleton, and after development regulates transition from proliferating to hypertrophic chondrocytes. Grouping of these disorders based on gene or protein function provides insight into disease. For example, autosomal dominant multiple epiphyseal dysplasia (MED) (Fig. 109.3) causes mild short stature, epiphyseal abnormalities, and early onset arthritis [22]. Dominantly inherited mutations in the genes that encode the extracellular matrix proteins that include cartilage oligomatrix protein, matrilin 3, type IX collagen (COL9A1, COL9A2, and COL9A3), and COL2A1 all produce indistinguishable forms of multiple epiphyseal dysplasia [22, 23]. Recessively inherited MED involves mutations in *SLC26A2*, but has some findings distinguishing it from the dominant forms. These dominantly inherited MED mutations encode proteins that interact in the cartilage extracellular matrix to form higher-order macromolecules, implicating locus heterogeneity producing phenotype similarities based on shared function. Similar biology is found in the genes producing the skeletal ciliopathies, Ellis van Creveld syndrome, asphyxiating thoracic syndrome, short-rib polydactyly syndromes, and cranioectodermal dysplasia. These disorders share clinical findings such as a high degree of lethality and polydactyly, and in many cases overlapping radiographic features such as long narrow rib cages, shortening, and bending of the appendicular skeleton (Fig. 109.4). All of these disorders have marked locus heterogeneity yet all result from mutations in genes involved in ciliary function.

Disease gene discovery in the skeletal dysplasias has provided deep insights into the functions of hundreds of molecules important in the skeleton. Modeling in mice of both loss and gain of function in candidate genes has also advanced our understanding of genetic control of skeletogenesis and maintenance of cartilage and bone, thus providing candidate genes for the osteochondrodysplasias while revealing key biological processes.

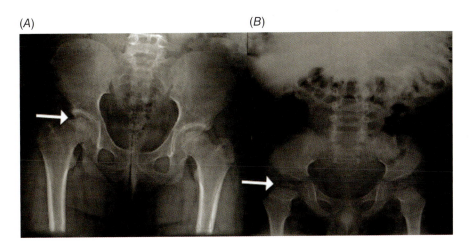

Fig. 109.3 Radiographs in multiple epiphyseal dysplasia (MED). (A, B) Anteroposterior view of hips in two individuals with MED showing small or delayed hip epiphyses for age (arrows). (A) is an adult, (B) is a 12-year-old child.

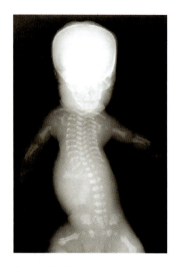

Fig. 109.4 Radiograph of a 34-week gestation newborn with short rib polydactyly type II showing long narrow chest with short ribs, shortening of all the appendicular bones, with irregular metaphyseal borders.

ACKNOWLEDGMENTS

DK is supported by NIH grants RO1 AR066124, DE019567, AR062651 and the Orthopaedic Institute for Children/Orthopaedic Hospital Research Center.

REFERENCES

1. Bonafe L, Cormier-Daire V, Hall C, et al. Nosology and classification of genetic skeletal disorders: 2015 revision. Am J Med Genet A. 2015;167A(12):2869–92.
2. Krakow D, Rimoin DL. The skeletal dysplasias. Genet Med. 2010;12(6):327–41.
3. Provot S, Schipani E. Molecular mechanisms of endochondral bone development. Biochem Biophys Res Comm. 2005;328(3):658–65.
4. Hall BK, Miyake T. All for one and one for all: condensations and the initiation of skeletal development. Bioessays. 2000;22(2):138–47.
5. Ballock RT, O'Keefe RJ. The biology of the growth plate. J Bone Joint Surg. 2003;85(4):715–26.
6. Kronenberg HM. Developmental regulation of the growth plate. Nature. 2003;423(6937):332–6.
7. Warman ML, Cormier-Daire V, Hall C, et al. Nosology and classification of genetic skeletal disorders: 2010 revision. Am J Med Genet A. 2011;155A(5):943–68.
8. Edwards MJ, Wenstrup RJ, Byers PH, et al. Recurrence of lethal osteogenesis imperfecta due to parental mosaicism for a mutation in the COL1A2 gene of type I collagen. The mosiac parent exhibits phenotypic features of a mild form of the disease. Human Mutat. 1992;1(1):47–54.
9. Cohn D, Starman B, Blumberg B, et al. Recurrence of lethal osteogenesis imperfecta due to parental mosaicism for a dominant mutation in a human type I collagen gene (COL1A1). Am J Hum Genet. 1990;46(3):591.
10. Deng H, Huang X, Yuan L. Molecular genetics of the COL2A1-related disorders. Mutat Res Rev Mutat Res. 2016;768:1–13.
11. Barat-Houari M, Sarrabay G, Gatinois V, et al. Mutation update for COL2A1 gene variants associated with Type II collagenopathies. Hum Mutat. 2016;37(1):7–15.
12. Garn SM, Hertzog KP, Poznanski AK, et al. Metacarpophalangeal length in the evaluation of skeletal malformation. Radiology. 1972;105(2):375–81.
13. Krakow D. Skeletal dysplasias. Clin Perinatol. 2015;42(2):301–19.
14. Mansour S, Hall C, Pembrey M, et al. A clinical and genetic study of campomelic dysplasia. J Med Genet. 1995;32(6):415–20.
15. Kelley RI, Kratz LE, Glaser RL, et al. Abnormal sterol metabolism in a patient with Antley-Bixler syndrome and ambiguous genitalia. Am J Med Genet. 2002;110(2):95–102.
16. Patton M, Afzal A. Robinow syndrome. J Med Genet. 2002;39(5):305–10.
17. Ho NC, Francomano CA, van Allen M. Jeune asphyxiating thoracic dystrophy and short-rib polydactyly type III (Verma-Naumoff) are variants of the same disorder. Am J Med Genet. 2000;90(4):310–4.
18. Maroteaux P, Sauvegrain J, Chrispin A, et al. Omodysplasia. Am J Med Genet. 1989;32(3):371–5.
19. Baujat G, Le Merrer M. Ellis-van Creveld syndrome. Orphanet J Rare Dis. 2007;2(6):27.
20. Lux SE, Johnston Jr RB, August CS, et al. Chronic neutropenia and abnormal cellular immunity in cartilage-hair hypoplasia. N Engl J Med. 1970;282(5):231–6.
21. Mäkitie O, Kaitila I. Cartilage-hair hypoplasia — clinical manifestations in 108 Finnish patients. Eur J Pediatr. 1993;152(3):211–7.
22. Jackson GC, Mittaz-Crettol L, Taylor JA, et al. Pseudoachondroplasia and multiple epiphyseal dysplasia: a 7-year comprehensive analysis of the known disease genes identify novel and recurrent mutations and provides an accurate assessment of their relative contribution. Hum Mutat. 2012;33(1):144–57.
23. Terhal PA, Nievelstein RJ, Verver EJ, et al. A study of the clinical and radiological features in a cohort of 93 patients with a COL2A1 mutation causing spondyloepiphyseal dysplasia congenita or a related phenotype. Am J Med Genet A. 2015;167A(3):461–75.

110
Ischemic and Infiltrative Disorders of Bone

Michael P. Whyte

Division of Bone and Mineral Diseases, Department of Internal Medicine, Washington University School of Medicine at Barnes-Jewish Hospital; and Center for Metabolic Bone Disease and Molecular Research, Shriners Hospital for Children, St Louis, MO, USA

INTRODUCTION

Interruption of blood flow to the skeleton can cause ischemic (avascular or aseptic) necrosis (IN), a focal disorder that disrupts bone and cartilage [1–5]. Ischemia, if sufficiently severe and prolonged, will kill osteocytes and osteoblasts. Significant clinical problems arise if skeletal repair is inadequate as necrotic bone undergoes resorption [3]. Skeletal strength may be compromised enough to engender fracture [1, 3]. Several additional important skeletal disorders are caused by proliferation or infiltration of specific cell types within marrow spaces [6, 7]. Reviewed briefly herein are systemic mastocytosis (SM) [6] and histiocytosis-X — now called Langerhans cell histiocytosis (LCH) [7].

ISCHEMIC NECROSIS

Various conditions are associated with IN (Table 110.1) [1, 5], and a considerable number of clinical presentations are possible based primarily on the affected skeletal site (Table 110.2) [2, 4]. Focal change in skeletal density is the principal radiographic feature [4], but may take several months to appear. Characteristic radiographic signs include patchy areas of sclerosis and osteopenia, crescent-shaped subchondral radiolucencies, bone collapse, and diaphyseal periostitis [4]. Initially, joint space is preserved despite epiphyseal damage, but may become compromised if there is underlying fracture [3]. Legg–Calvé–Perthes disease (LCPD), an archetypal form of IN, is discussed in the next section. Osteonecrosis of the jaw is reviewed in Chapter 120.

Legg–Calvé–Perthes disease

LCPD is a relatively common, complex, and controversial problem of IN of the capital femoral epiphysis in children [8, 9]. Boys are affected more often than girls (~5:1). Familial occurrence varies from 1% to 20% [10, 11], but twin studies indicate no genetic basis despite any familial clustering [12]. LCPD typically presents between ages 2 and 12 years (mean age 7 years). When it manifests later in childhood, the term "adolescent IN" indicates the poorer prognosis compared with onset during adult life. Usually one hip is involved, but ~20% of patients have bilateral disease and epiphyseal dysplasia should be considered in the differential diagnosis.

Although the etiology of LCPD is not always apparent [13], the pathogenesis is fairly well understood [3]. Interruption of blood flow to the capital femoral epiphysis can have many causes including raised intracapsular pressure caused by congenital or developmental abnormalities, synovitis, venous thrombosis, or perhaps increased blood viscosity [8, 10, 11]. Consequently, most if not all of the capital femoral epiphysis is rendered ischemic and osteoblasts, osteocytes, and marrow cells may die. Furthermore, endochondral ossification ceases temporarily because blood flow to the growth plate is impaired. However, articular cartilage remains intact initially because the synovial fluid provides nourishment. Subsequently, revascularization of necrosed areas proceeds from the periphery to the center of the epiphysis. New bone is deposited on central trabecular osseous debris and subchondral cortical bone. However, as removal of necrotic osseous tissue begins, bone resorption can exceed reparative bone formation. If so, subchondral bone is weakened. When no fracture occurs at the site of reparative bone resorption, healing can follow

---

Primer on the Metabolic Bone Diseases and Disorders of Mineral Metabolism, Ninth Edition. Edited by John P. Bilezikian.
© 2019 American Society for Bone and Mineral Research. Published 2019 by John Wiley & Sons, Inc.
Companion website: www.wiley.com/go/asbmrprimer

Table 110.1. Causes of ischemic necrosis of cartilage and bone.

Endocrine/metabolic
    Alcohol abuse
    Bisphosphonate therapy
    Glucocorticoid therapy
    Cushing's syndrome
    Gout
    Osteomalacia
Storage diseases (eg, Gaucher's disease)
Hemoglobinopathies (eg, sickle cell disease)
Trauma (eg, dislocation, fracture)
Dysbaric conditions
Collagen vascular disorders
HIV infection
Irradiation
Pancreatitis
Renal transplantation
Idiopathic, familial

Table 110.2. Common sites of osteochondrosis and ischemic necrosis of bone.

Adult skeleton
    Osteochondritis dissecans (König)
    Osteochondrosis of lunate (Kienböck)
    Fractured head of femur (Axhausen, Phemister)
    Proximal fragment of fractured carpal scaphoid
    Fractured head of humerus
    Fractured talus
    Osteonecrosis of the knee (spontaneous or idiopathic ischemic necrosis)
    Idiopathic ischemic necrosis of the femoral head
Developing skeleton
    Osteochondrosis of femoral head (Legg–Calvé–Perthes)
    Slipped femoral epiphysis
    Vertebral epiphysitis affecting secondary ossification centers (Scheuermann)
    Vertebral osteochondrosis of primary ossification centers (Calvé)
    Osteochondrosis of tibial tuberosity (Osgood–Schlatter)
    Osteochondrosis of tarsal scaphoid (Köhler)
    Osteochondrosis of medial tibial condyle (Blount)
    Osteochondrosis of primary ossification center of patella (Köhler) and of secondary ossification center (Sinding Larsen)
    Osteochondrosis of os calcis (Sever)
    Osteochondrosis of head of second metatarsal (Freiberg) and of other metatarsals and metacarpals
    Osteochondrosis of the humeral capitellum (Panner)

Reproduced with permission from Edeiken J, Dalinka M, Karasic D (eds). 1990 Edeiken's Roentgen Diagnosis of Diseases of Bone, 4th ed. Williams & Wilkins, Baltimore, MD, USA, p. 937.

without further symptoms. If fracture occurs, there will be symptoms and trabecular bone collapse may cause a second episode of ischemia [8, 10, 11]. Proximal femur growth can be stunted because of physeal dysfunction, including premature closure of the growth plate.

Children suffering LCPD typically limp, complain of knee or anterior thigh pain, and have limited hip mobility (especially with abduction or internal rotation). Trendelenburg sign may be positive. If treatment (see later) is unsuccessful, adduction and flexion contractures of the hip can lead to muscle atrophy in the thigh.

Radiographs for diagnosis and follow-up should include anteroposterior and "frog" lateral views. Often, bone age is 1 to 3 years delayed [4]. Sequential studies typically show cessation of growth of the capital femoral epiphysis, resorption of necrotic bone, reossification, and finally healing but sometimes with subchondral fracture (Fig. 110.1). MRI is important because findings change with circulatory compromise, soft tissues and bone are visualized, and femoral head containment (see below) can be assessed [14]. Generally, the more extensive the involvement of the capital femoral epiphysis, the worse the prognosis. Girls seem to fare worse than boys because they tend to have greater disruption of the capital femoral epiphysis and they mature earlier providing less time for femoral head modeling before closure of the growth plates. Similarly, onset of LCPD at 2 to 6 years of age leads to the least femoral head deformity, but not always a good outcome [15], whereas onset after 10 years of age has a poor outcome [8, 10, 11, 16]. The short-term prognosis for LCPD reflects the severity of femoral head deformity at completion of the healing phase. The long-term outcome depends upon how much degenerative osteoarthritis develops.

Importantly, as epiphyseal reossification proceeds, the femoral head will shape according to impacting mechanical forces [4, 8, 10, 11], and orthopedic assessment and management is crucial [17, 18]. Prevention of femoral head deformity is a major goal. Significant distortion predisposes to osteoarthritis. This seems worse for children who lose "containment" of the femoral head by the acetabulum. Therefore, coverage of the femoral head by the acetabulum is sought as a mold during reparative reossification [8, 10, 11]. Management may be observation, intermittent treatment of symptoms with periodic bed rest, stretching exercises to maintain hip range-of-motion, and early or late surgical prevention or correction of deformity [8, 10, 11, 17, 19]. Radiologic follow-up is essential, and arthrography, bone scintigraphy, and especially MRI are useful [10, 11, 14]. However, the long-term outcome from various treatment approaches remains controversial [17, 19]. Osteoarthritis of the hip is likely [20], and arthroplasty may become necessary [21].

Other presentations of ischemic necrosis

Among the considerable number and variety of perturbations that cause IN (Table 110.1) — now including HIV infection in adults and children [22, 23], bone antiresorptives

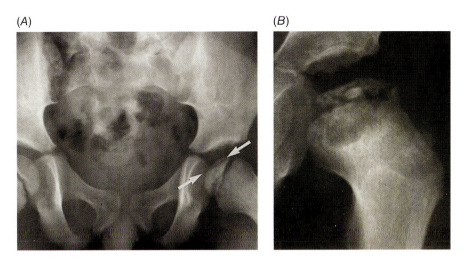

Fig. 110.1. Legg-Calvé-Perthes disease. (A) The affected left capital femoral epiphysis of this 4-year-old boy is denser and smaller than the contralateral normal side. A radiolucent area "crescent sign" (arrows) indicates subchondral bone collapse. (B) Seven months later, there is flattening of the capital femoral epiphysis with widening and irregularity of the femoral neck.

in adults, and regional migratory osteoporosis [24] — numerous distinctive presentations are possible (Table 110.2) [4, 5]. The specific diagnosis depends on the patient's age, involved site, and size of the area of affected bone. Mechanisms for vascular insufficiency include traumatic rupture, internal obstruction, or external pressure compromising blood flow. Arteries, veins, or sinusoids may be involved causing "ischemic," "avascular," "aseptic," or "idiopathic" necrosis [1, 4, 5]. The pathogenesis of disrupted blood flow may, however, be incompletely understood [1, 4, 5]. When nontraumatic, the predisposed sites seem to reflect the physiological marrow blood flow decrease that proceeds distally to proximally in the appendicular skeleton as red marrow is converted to fatty marrow with aging [4]. Accordingly, disorders that increase the size and/or number of adipocytes within critical areas of medullary space (eg, alcohol abuse, Cushing's syndrome) may ultimately compress sinusoids and infarct bone. Additionally, fat embolization, hemorrhage, and abnormalities in the quality of the bone tissue itself may also cause traumatic or nontraumatic osteonecrosis [1, 3]. Over months or years, dead bone may, or may not, slowly resorb. Osteosclerosis will occur if new bone encases dead bone and/or if there is bony collapse.

After infarction, necrotic bone does not change radiographic density for at least 10 days. Currently, MRI is the most sensitive way to detect IN of the skeleton, and is particularly useful early on although false negatives can occur [25–27]. Bone scintigraphy, although nonspecific, can also be positive before radiographic changes are apparent [28–30]. Prior to revascularization, infarcted areas show decreased radioisotope uptake. Later, there is increased tracer accumulation. CT is especially helpful for detecting osteonecrosis of the femoral head, because the central bony structure has an "asterisk" shape that will be distorted by new bone formation [31]. Histopathological studies confirm this pathogenesis. Various processes of skeletal death and repair are focal

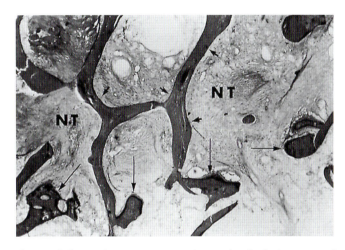

Fig. 110.2. Ischemic necrosis. This undecalcified section of an affected femoral head shows a typical area of dead bone (arrows) having a smooth acellular surface. A band of necrotic tissue (NT) is present. Reparative bone formation is occurring in adjacent areas where darkly stained, newly synthesized, osteoid is covered by osteoblasts (arrowheads; Goldner stain × 160).

and may be occurring simultaneously (Fig. 110.2).

As for LCPD, symptoms result primarily from skeletal disintegration. The various clinical presentations of IN (Table 110.2) [2, 4] are sometimes divided into two major anatomical categories: diaphysometaphyseal and epiphysometaphyseal [2]. Diaphysometaphyseal ischemia can reflect dysbaric disorders, hemoglobinopathies, collagen vascular diseases, thromboembolic problems, gout, storage disorders (eg, Gaucher's disease), acute or chronic pancreatitis, pheochromocytoma, and other conditions. Typically, large bones (especially the distal femur or proximal tibia) are involved where radiographic changes extend into the metaphysis. Lesions are often symmetrical; however, their size can vary considerably. Small bones may be affected (eg, hands and feet of infants with sickle cell anemia). New bone deposition delineates infarcted bone especially well on radiographic study. Epiphysometaphyseal infarcts can reflect dysbaric conditions, sickle cell disease, Cushing's syndrome, gout, trauma, storage problems, and other disorders. When the lesions are small, they typically involve children or young adults and occur without a history of injury, although occult trauma may be important. Thrombosis, disease of arterial walls, or abnormalities within adjacent bone (eg, Gaucher's disease, LCH) may cause this type of IN.

Osteochondrosis refers to atraumatic IN typically affecting an ossification ("growth") center. Osteochondritis dissecans describes a small epiphysometaphyseal infarct that can cause fracture adjacent to the joint space. This lesion appears as a small, dense, button-like area of osseous tissue separated by a radiolucent band from the intact bone. This fragment can heal in place or become loose and enter the joint. Larger infarcts too are often idiopathic, occur frequently in adults, and typically involve the hip and the femoral condyles. Large areas of ischemic bone can collapse flattening joint surfaces and destroying articular cartilage. Ultimately, this will lead to osteoarthritis (Fig. 110.3). Very extensive epiphysometaphyseal infarction results from trauma or systemic disease and frequently involves the femoral head [2].

Eponyms for specific presentations of IN or osteochondrosis of the skeleton remain numerous and popular (eg, Blount disease, Scheuermann disease). However, classification according to the involved anatomical site is more informative. Table 110.2 matches the eponym with the affected skeletal region and helps to show that the patient's age is an important factor for at risk sites [2].

Treatment of IN varies according to the site and size of the lesion and the patient's age, but various aspects remain controversial. Conservative or surgical approaches may be appropriate [5, 32, 33]. Bone morphogenetic protein utility is uncertain [34].

INFILTRATIVE DISORDERS

Systemic mastocytosis

Systemic mastocytosis (SM) comprises several disorders featuring increased numbers of mast cells [6, 35, 36]. It is one of eight myeloproliferative neoplasms [37]. The viscera (principally the liver, spleen, gastrointestinal tract, and lymph nodes) become involved [6, 37–41]. The skin can contain numerous hyperpigmented macules that reflect dermal mast cell accumulation, a condition called urticaria pigmentosa (Fig. 110.4). Bone marrow is also typically involved and can cause skeletal pathology.

Symptoms of SM result primarily from release of mediator substances by the mast cells and include generalized pruritus, urticaria, flushing, episodic hypotension, syncope, diarrhea, weight loss, and peptic ulcer [37–41]. With cutaneous involvement, histamine release occurs from stroking the skin causing swelling, itchiness, and redness (Darier sign). Skeletal complications develop relatively infrequently but include bone pain or tenderness from deformity resulting from fracture [37–45]. Serum tryptase elevation is a good, but not disease-specific, marker for SM [35]. Urinary N-methylhistamine increase signifies bone marrow involvement [38]. Diagnostic investigation can be complex [6].

Radiographic abnormalities of the skeleton are common in SM (~70% of patients). The findings are well delineated [46, 47]. Classically, there are diffuse, poorly

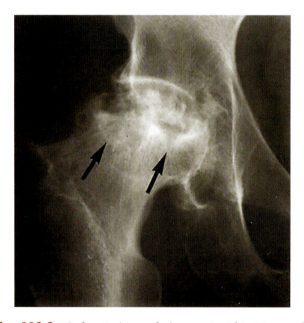

Fig. 110.3. Ischemic (avascular) necrosis. This 50-year-old man has advanced avascular necrosis of the femoral head. Much of the femoral head has been resorbed, causing collapse of the articular surface. The necrotic area is fragmented. A sclerotic zone of reparative tissue (arrows) demarcates viable versus necrotic tissues. The acetabular cartilage is focally thin, indicating developing secondary osteoarthritis.

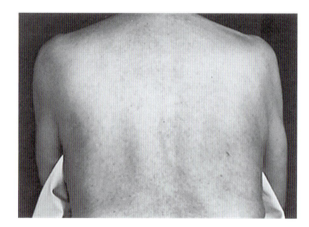

Fig. 110.4. Systemic mastocytosis. Numerous characteristic hyperpigmented macules (urticaria pigmentosa) cover the back of this 61-year-old woman.

demarcated, sclerotic, and lucent areas within red marrow (ie, axial skeleton) (Fig. 110.5). Circumscribed lesions can occur especially in the skull and extremities. Such focal findings may be mistaken for metastatic disease. Lytic areas are often small with a surrounding rim of osteosclerosis; rarely, they are large and lead to fracture. Progression of the radiographic changes can signify generalized involvement [46, 47], but focal bony changes may be absent despite such involvement. Elevated serum IL-6 levels may be a marker for disease progression [48]. Generalized osteopenia (without discrete bony abnormalities) is also a common presentation [42, 43, 45], yet has a relatively benign prognosis [49]. Bone scintigraphy helps detect involved skeletal areas [50] and can provide information regarding disease activity and prognosis [51]. Reportedly, hip bone density correlates with urinary excretion of the histamine metabolite, methylimidazoleacetic acid [44].

Histopathological correlates of SM within the skeleton are also well characterized [38, 42, 52, 53]. Examination of metachromatically stained undecalcified sections of bone can be an especially effective way to establish the diagnosis. Transiliac crest biopsy may be better than bone marrow aspiration or biopsy [42, 52, 53]. Undecalcified sections of iliac crest show nodules 150 to 450 μm in diameter that resemble granulomas (ie, mast cell granulomas). They contain eosinophils, lymphocytes, plasma cells, and characteristic oval or spindle-shaped cells that resemble histiocytes or fibroblasts but contain granules that stain metachromatically and are actually a type of mast cell (Fig. 110.6). Additionally, the marrow contains increased numbers of these mast cells individually or in small aggregates [42, 52, 53]. Tetracycline-based histomorphometry shows rapid bone remodeling [52, 53]. High serum levels of C-telopeptide of type I collagen correlate with reduction in BMD [54].

The etiology of SM is unknown [6, 55]. Its persistence after bone marrow transplantation (for an additional condition) suggested that a defective myeloid precursor cell was not at fault [56, 57]. SM seems to be a multitopic monoclonal proliferation of cytologically and/or functionally abnormal tissue mast cells [39]. Patients often succumb to a granulocytic neoplasm [35, 38, 39, 58, 59]. Many harbor a mutation in the C-KIT proto-oncogene in their aberrant mast cells [60]. Bone marrow investigation is mandatory for all adults highly suspected of SM regardless of *KIT* mutation status [6, 61]. Somatic mutation of other genes is found in advanced SM [62].

Treatment of SM is discussed in recent reviews [35, 38–41, 62–66], and must be "tailored" in individual patients [65, 66]. Reportedly, severe bone pain from advanced skeletal disease responds to radiotherapy [67]. Bisphosphonates have controlled pain and improved bone density in early trials [68–70]. Denosumab may also be useful [71]. *KIT* mutational status may help select candidates for imatinib therapy [72]. Midostaurin can have efficacy for advanced SM [73].

Langerhans cell histiocytosis

Histiocytosis-X is the term coined in 1953 to unify what had been regarded as three distinct entities: Letterer–Siwe disease, Hand–Schüller–Christian disease, and eosinophilic granuloma [7, 74]. An immature clonal Langerhans cell is considered the pathognomonic and

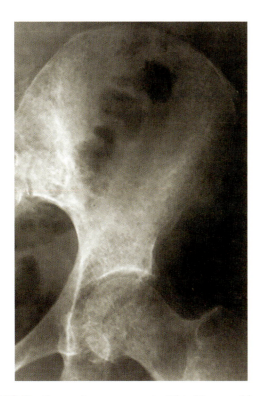

Fig. 110.5. Systemic mastocytosis. This 81-year-old woman has characteristic diffuse, punctuate radiolucencies within her left hemipelvis and hip that indicate a permeative process of the bone marrow.

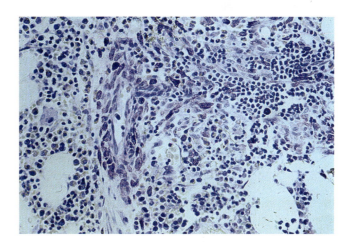

Fig. 110.6. Mast cell granuloma. This nondecalcified, metachromatically stained, section of iliac crest shows a characteristic mast cell granuloma that contains numerous spindle-shaped mast cells (Geimsa stain; magnification ×250).

linking feature now called LCH [7, 74]. About 1200 cases are diagnosed yearly in the USA. Sex incidence is equal. Northern Europeans are affected more commonly than Hispanics, and the condition is rare in Blacks. The tripartite distinction persists because of the generally different clinical courses and prognoses [7, 74].

Letterer–Siwe disease presents between several weeks and 2 years of age with lymphadenopathy, hepatosplenomegaly, anemia, hemorrhagic tendency, fever, failure to grow, and skeletal lesions. It has ended fatally after just several weeks [7, 74].

Hand–Schüller–Christian disease is a chronic condition that begins in early childhood, although symptoms may not manifest until the third decade [7, 74]. The classic triad of findings consists of exophthalmos, diabetes insipidus, and bony lesions. However, this presentation occurs in only 10% of cases. The most common skeletal manifestation is osteolytic lesions in the skull, with overlying soft tissue nodules (Fig. 110.7) [7, 75]. Proptosis is associated with destruction of orbital bones. Spontaneous remissions and exacerbations may occur. Soft tissue nodules may remit without treatment.

Eosinophilic granuloma occurs most frequently in children between 3 and 10 years of age, and is rare after age 15 years [74]. A solitary and painful lesion in a flat bone is the most common finding [7, 75]. There may be a soft tissue mass. The calvarium is usually affected, although any bone can be involved. The prognosis is excellent, with monostotic lesions healing spontaneously or responding well to X-ray therapy.

The radiographic findings in the skeleton are similar in the three disorders [7, 46, 75]. Single bony foci are most prevalent. Nevertheless, multiple areas can be affected and show progressive enlargement. Individual lesions are well defined (ie, "punched-out," osteolytic, and destructive with scalloped edges). They vary from a few millimeters to several centimeters in diameter. Fewer than 50% of these radiolucencies show marginal reactive osteosclerosis. Membranous bones as well as long bones can be affected. In the long bones, defects occur in the medullary canal where there is erosion of the endosteal cortex (commonly in the metaphyseal or epiphyseal regions). Periosteal reaction is frequent and produces a solid layer of new bone. In the skull, the bony tables can be eroded. Destruction of orbital bones may or may not be associated with exophthalmos. Vertebra plana (ie, flattened vertebra) can result from spinal involvement in young children. Radionuclide accumulation is poor during bone scanning [46]. Biochemical parameters of mineral homeostasis are usually normal.

LCH seems to reflect some poorly understood dysfunction of the immune system [74]. TNF-alpha and other cytokines are increased in the lesions [7]. LCH is an extremely heterogeneous disorder that can include major congenital malformations. Many tissues and organs can be involved, including brain, lung, oropharynx, gastrointestinal tract, skin, and bone marrow. Diabetes insipidus is common because of pituitary infiltration. Prognosis is age-related; infants and the elderly have poorer outcomes. The signs and symptoms of the three principal clinical forms of LCH also differ.

LCH tends to be benign and self-limiting when there is no systemic involvement. Treatment for severe disease includes chemotherapy, radiation therapy, and immunotherapy [76–78]. Central nervous system involvement is often treated by radiation therapy. Allogeneic bone marrow transplantation was reportedly successful in a severe case with poor prognosis [79]. Methylprednisolone injected into lesions is effective [75]. Denosumab has seemed useful for adult multisystem LCH [80].

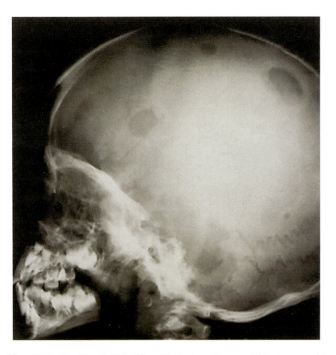

Fig. 110.7. Hand–Schüller–Christian disease. This 2-year-old boy has multiple, well-defined, beveled-edge, lucent lesions within his skull. Note the extensive destruction of the paranasal sinuses and at the base of his skull.

REFERENCES

1. Jones LC, Hungerford DS. Osteonecrosis: etiology, diagnosis, and treatment. Curr Opin Rheumatol. 2004;16:443–9.
2. Steinberg ME, Steinberg DR. Classification systems for osteonecrosis: an overview. Orthop Clin North Am. 2004;35:273–83.
3. Assouline-Dayan Y, Chang C, Greenspan A, et al. Pathogenesis and natural history of osteonecrosis. Semin Arthritis Rheum. 2002;32:94–124.
4. Resnick D, Niwayama G. *Diagnosis of Bone and Joint Disorders* (4th ed.). Philadelphia: WB Saunders, 2002.
5. Pavelka K. Osteonecrosis. Clin Rheumatol. 2000;14:399–414.
6. González-de-Olano D, Matito A, Orfao A, et al. Advances in the understanding and clinical management of mastocytosis and clonal mast cell activation syndromes. F1000Res. 2016;5:2666.

7. Coppes-Zantinga A, Egeler RM. The Langerhans cell histiocytosis X files revealed. Br J Haematol. 2002;116:3–9.
8. Divi SN, Bielski RJ. Legg-Calvé-Perthes disease. Pediatr Ann. 2016;45:e144–9.
9. Chaudhry S, Phillips D, Feldman D. Legg-Calvé-Perthes disease: an overview with recent literature. Bull Hosp Joint Dis. 2014;72:18–27.
10. Conway JJ. A scintigraphic classification of Legg-Calvé-Perthes disease. Semin Nucl Med. 1993;23:274–95.
11. Thompson GH, Price CT, Roy D, et al. Legg-Calve-Perthes disease: current concepts. Instr Course Lect. 2002;51:367–84.
12. Metcalfe D, Van Dijck S, Parsons N, et al. A twin study of Perthes disease. Pediatrics. 2016;137(3):e20153542.
13. Perry DC, Skellorn PJ, Bruce CE. The lognormal age of onset distribution in Perthes' disease. Bone Joint J. 2016;98-B:710–14.
14. Kotoura Y, Kim W-C, Hosokawa M, et al. Assessment of lateral subluxation in Legg-Calvé-Perthes disease: a time-sequential study of magnetic resonance imaging and plain radiography. J Pediatr Orthop B. 2015;24:493–506.
15. Nakamura J, Kamegaya M, Saisu T, et al. Outcome of patients with Legg-Calvé-Perthes onset before 6 years of age. J Pediatr Orthop. 2015;35:144–50.
16. Mukherjee A, Fabry G. Evaluation of the prognostic indices in Legg-Calvé-Perthes disease: statistical analysis of 116 hips. J Pediatr Orthop. 1990;10:153–8.
17. Kim SS, Lee CW, Kim HJ, et al. Treatment of late-onset Legg-Calve-Perthes disease by arthrodiastasis. Clin Orthop Surg. 2016;8:452–7.
18. Mazloumi SM, Ebrahimzadeh MH, Kachooei AR. Evolution in diagnosis and treatment of Legg-Calve-Perthes disease. Arch Bone Joint Surg. 2014;2:86–92.
19. Mosow N, Vettorazzi E, Breyer S, et al. Outcome after combined pelvic and femoral osteotomies in patients with Legg-Calve-Perthes disease. J Bone Joint Surg Am. 2017;99:207–13.
20. Heesakkers N, van Kempen R, Feith R, et al. The long-term prognosis of Legg-Calvé-Perthes disease: a historical prospective study with a median follow-up of forty one years. Int Orthop. 2015;39:859–63.
21. Seufert CR, McGrory BJ. Treatment of arthritis associated with Legg-Calve-Perthes disease with modular total hip arthroplasty. J Arthrop. 2015;30:1743–6.
22. Gaughan DM, Mofenson LM, Hughes MD, et al., Pediatric AIDS Clinical Trials Group Protocol 219 Team. Osteonecrosis of the hip (Legg-Calvé-Perthes disease) in human immunodeficiency virus–infected children. Pediatrics. 2002;109:1–8.
23. Qaqish RB, Sims KA. Bone disorders associated with the human immunodeficiency virus: pathogenesis and management. Pharmacotherapy. 2004;24:1331–46.
24. Trevisan C, Ortolani S, Monteleone M, et al. Case report: regional migratory osteoporosis. a pathogenetic hypothesis based on three cases and a review of the literature. Clin Rheumatol. 2002;21:418–25.
25. Watson RM, Roach NA, Dalinka MK. Avascular necrosis and bone marrow edema syndrome. Radiol Clin North Am. 2004;42:207–19.
26. Saini A, Saifuddin A. MRI of osteonecrosis. Clin Radiol. 2004;59:1079–93.
27. Mitchell DG, Rao VM, Dalinka MK, et al. Femoral head avascular necrosis: correlation of MR imaging, and clinical findings. Radiology. 1987;162:709–15.
28. Mitchell MD, Kundel HL, Steinberg ME, et al. Avascular necrosis of the hip: comparison of MR, CT, and scintigraphy. Am J Roentgenol. 1986;147:67–71.
29. Bonnarens F, Hernandez A, D'Ambrosia RD. Bone scintigraphic changes in osteonecrosis of the femoral head. Orthop Clin North Am. 1985;16:697–703.
30. Spencer JD, Maisey M. A prospective scintigraphic study of avascular necrosis of bone in renal transplant patients. Clin Orthop. 1985;194:125–35.
31. Dihlmann W. CT analysis of the upper end of the femur: the asterisk sign and ischemic bone necrosis of the femoral head. Skeletal Radiol. 1982;8:251–8.
32. Canale ST. *Campbell's Operative Orthopaedics* (9th ed.). St Louis: CV Mosby, 1998.
33. Smith SW, Fehring TK, Griffin WL, et al. Core decompression of the osteonecrotic femoral head. J Bone Joint Surg Am. 1995;77:674–80.
34. Krishnakumar GS, Roffi A, Reale D, et al. Clinical application of bone morphogenetic proteins for bone healing: a systemic review. Int Orthop. 2017;41:1073–83.
35. Akin C, Metcalfe DD. Systemic mastocytosis. Annu Rev Med. 2004;55:419–32.
36. Abid A, Malone MA, Curci K. Mastocytosis. Prim Care Clin Office Pract. 2016;43:505–18.
37. Pardanani A. Systemic mastocytosis in adults: 2017 update on diagnosis, risk stratification and management. Am J Hematol. 2016;91:1147–59.
38. Pardanani A. Systemic mastocytosis: bone marrow pathology, classification, and current therapies. Acta Haematol. 2005;114:41–51.
39. Akin C. Clonality and molecular pathogenesis of mastocytosis. Acta Haematol. 2005;114:61–9.
40. Valent, P, Akin C, Sperr WR, et al. Diagnosis and treatment of systemic mastocytosis: state of the art. Br J Haematol. 2003;122:695–717.
41. Valent P, Akin C, Sperr WR, et al. Mastocytosis: pathology, genetics, and current options for therapy. Leuk Lymphoma. 2005;46:35–48.
42. Fallon MD, Whyte MP, Teitelbaum SL. Systemic mastocytosis associated with generalized osteopenia: histopathological characterization of the skeletal lesion using undecalcified bone from two patients. Hum Pathol. 1981;12:813–20.
43. Harvey JA, Anderson HC, Borek D, et al. Osteoporosis associated with mastocytosis confined to bone: report of two cases. Bone. 1989;10:237–41.
44. Cook JV, Chandy J. Systemic mastocytosis affecting the skeletal system. J Bone Joint Surg Br. 1989;71:536.
45. Lidor C, Frisch B, Gazit D, et al. Osteoporosis as the sole presentation of bone marrow mastocytosis. J Bone Miner Res. 1990;5:871–6.

46. Edeiken J, Dalinka M, Karasick D. *Edeiken's Roentgen Diagnosis of Diseases of Bone* (4th ed.). Baltimore: Williams and Wilkins, Baltimore, 1990.
47. Resnick D, Niwayama G. *Diagnosis of Bone and Joint Disorders* (4th ed.). Philadelphia: WB Saunders, 2002.
48. Mayado A, Teodosio C, Garcia-Montero AC, et al. Increased IL6 plasma levels in indolent systemic mastocytosis patients are associated with high risk of disease progression. Leukemia. 2016;30:124–30.
49. Andew SM, Freemont AJ. Skeletal mastocytosis. J Clin Pathol. 1993;46:1033–5.
50. Arrington ER, Eisenberg B, Hartshorne MF, et al. Nuclear medicine imaging of systemic mastocytosis. J Nucl Med. 1989;30:2046–8.
51. Chen CC, Andrich MP, Mican JM, et al. A retrospective analysis of bone scan abnormalities in mastocytosis: correlation with disease category and prognosis. J Nucl Med. 1994;35:1471–5.
52. De Gennes C, Kuntz D, de Vernejoul MC. Bone mastocytosis. A report of nine cases with a bone histomorphometric study. Clin Orthop. 1991;279:281–91.
53. Chines A, Pacifici R, Avioli LV, et al. Systemic mastocytosis presenting as osteoporosis: a clinical and histomorphometric study. J Clin Endocrinol Metab. 1991;72:140–4.
54. Artuso A, Caimmi C, Tripi G, et al. Longitudinal evaluation of bone mineral density and bone metabolism markers in patients with indolent systemic mastocytosis without osteoporosis. Calcif Tissue Int. 2017;100:40–6.
55. Oranje AP, Riezebos P, van Toorenenbergen AW, et al. Urinary N-methylhistamine as an indicator of bone marrow involvement in mastocytosis. Clin Exp Dermatol. 2002;27:502–6.
56. Van Hoof A, Criel A, Louwagie A, et al. Cutaneous mastocytosis after autologous bone marrow transplantation. Bone Marrow Transplant. 1991;8:151–3.
57. Ronnov-Jessen D, Nielsen PL, Horn T. Persistence of systemic mastocytosis after allogeneic bone marrow transplantation in spite of complete remission of the associated myelodysplastic syndrome. Bone Marrow Transplant. 1991;8:413–15.
58. Lawrence JB, Friedman BS, Travis WD, et al. Hematologic manifestations of systemic mast cell disease: a prospective study of laboratory and morphologic features and their relation to prognosis. Am J Med. 1991;91:612–24.
59. Valent P, Akin C, Hartmann K, et al. Advances in the classification and treatment of mastocytosis: current status and outlook toward the future. Cancer Res. 2017;77:1261–70.
60. Fritsche-Polanz R, Jordan JH, Feix Al, et al. Mutation analysis of *C-KIT* in patients with myelodysplastic syndromes without mastocytosis and cases of systemic mastocytosis. Br J Haematol. 2001;113:357–64.
61. Kristensen T, Vestergaard H, Bindslev-Jensen C, et al. Prospective evaluation of the diagnostic value of sensitive KIT D816V mutation analysis of blood in adults with suspected systemic mastocytosis. Allergy. 2017;72(11):1737–43.
62. Ustun C, Arock M, Kluin-Nelemans HC, et al. Advanced systemic mastocytosis: from molecular and genetic progress to cinical practice. Haematologica. 2016;101(10):1133–43.
63. Gasior-Chrzan B, Falk ES. Systemic mastocytosis treated with histamine H1 and H2 receptor antagonists. Dermatology 1992;184:149–52.
64. Metcalfe DD. The treatment of mastocytosis: an overview. J Invest Dermatol. 1991;96:55S–59S.
65. Valent, P, Ghannadan M, Akin C, et al. On the way to targeted therapy of mast cell neoplasms: identification of molecular targets in neoplastic mast cells and evaluation of arising treatment concepts. Eur J Clin Invest. 2004;34(Suppl 2):41–52.
66. Krokowski M, Sotlar K, Krauth MT, et al. Delineation of patterns of bone marrow mast cell infiltration in systemic mastocytosis: value of CD25, correlation with subvariants of the disease, and separation from mast cell hyperplasia. Am J Clin Pathol. 2005;124:560–68.
67. Johnstone PA, Mican JM, Metcalfe DD, et al. Radiotherapy of refractory bone pain due to systemic mast cell disease. Am J Clin Oncol. 1994;17:328–30.
68. Brumsen C, Hamady NAT, Papapoulos SE. Osteoporosis and bone marrow mastocytosis: Dissociation of skeletal responses and mast cell activity during long-term bisphosphonate therapy. J Bone Miner Res. 2002;17:567–9.
69. Greene LW, Asadipooya K, Corradi PF, et al. Endocrine manifestations of systemic mastocytosis in bone. Rev Endocr Metab Disord. 2016;17:419–31.
70. Rossini M, Zanotti R, Orsolini G, et al. Prevalence, pathogenesis, and treatment options for mastocytosis-related osteoporosis. Osteoporos Int 2016;27:2411–21.
71. Orsolini G, Gavioli I, Tripi G, et al. Denosumab for the treatment of mastocytosis-related osteoporosis: a case series. Calcif Tissue Int 2017;100:595–8.
72. Alvarez-Twose I, Matito A, Morgado JM, et al. Imatinib in systemic mastocytosis: a phase IV clinical trial in patients lacking exon 17 *KIT* mutations and review of the literature. Oncotarget. 2016;8(40):68950–63.
73. Gotlib J, Kluin-Nelemans HC, George TI, et al. Efficacy and safety of midostaurin in advanced systemic mastocytosis. N Engl J Med. 2016;374;2530–41.
74. Lam KY. Langerhans cell histiocytosis (histiocytosis X). Postgrad Med J. 1997;73:391–4.
75. Alexander JE, Seibert JJ, Berry DH, et al. Prognostic factors for healing of bone lesions in histiocytosis X. Pediatr Radiol. 1988;18:326–32.
76. Bollini G, Jouve JL, Gentet JC, et al. Bone lesions in histiocytosis X. J Pediatr Orthop. 1991;11:469–77.
77. Greenberger JS, Crocker AC, Vawter G, et al. Results of treatment of 127 patients with systemic histiocytosis (Letterer-Siwe syndrome, Schüller-Christian syndrome and multifocal eosinophilic granuloma). Medicine (Baltimore). 1981;60:311–88.
78. Haupt R, Minkov M, Astigarraga I, et al. Euro Histio Network. Langerhans cell histiocytosis (LCH): guidelines for diagnosis, clinical work-up, and treatment for patients till the age of 18 years. Pediatr Blood Cancer. 2013;60:175–84.
79. Ringden O, Aohstrom L, Lonnqvist B, et al. Allogeneic bone marrow transplantation in a patient with chemotherapy-resistant progressive histiocytosis X. N Engl J Med. 1987;316:733–5.
80. Makras P, Tsoli M, Anastasilakis AD, et al. Denosusmab for the treatment of adult multisystem Langerhans cell histiocytosis. Metabolism. 2017;69:107–11.

111

Tumoral Calcinosis – Dermatomyositis

Nicholas J. Shaw

Department of Endocrinology and Diabetes, Birmingham Children's Hospital, Birmingham, UK

TUMORAL CALCINOSIS

Tumoral calcinosis (TC) is a rare metabolic disorder characterized by progressive deposition of calcium phosphate crystals in periarticular spaces and soft tissues. Its biochemical hallmark is hyperphosphatemia caused by increased renal tubular reabsorption of phosphate. The form referred to as hyperphosphatemic familial TC (HFTC) (OMIM #211900) is an autosomal recessive disorder. TC is also described in the absence of elevated plasma phosphate, referred to as normophosphatemic familial TC (NFTC) (OMIM #610455). Although the first description of this condition was in 1898, the term tumoral calcinosis was not used until 1943 [1].

Clinical features

Mineral deposition presents as soft tissue masses around major joints. In one report, the order of frequency for the first lesions is hips, elbows, shoulders, and scapulae [2]. Onset can vary from 22 months to adulthood with the majority manifesting by 20 years of age. Black ancestry is common with many cases from Africa. The soft tissue masses are usually painless and can enlarge to be orange or grapefruit size. Although periarticular they do not usually impair range of movement because they are extracapsular. They can compress adjacent neural structures such as the sciatic nerve and may also cause ulceration of overlying skin including a sinus tract that leaks a chalky fluid and may become infected. Some patients have features of pseudoxanthoma elasticum (ie, skin changes, vascular calcification, and angioid streaks of the retina). A specific dental abnormality may occur as hypoplastic teeth containing short bulbous roots and almost complete obliteration of pulp cavities with pulp stones. A related condition, hyperostosis–hyperphosphataemia syndrome, features recurrent bone pain and swelling particularly affecting the long bones. However, recent evidence indicates the two conditions represent a spectrum of the same disease despite the same mutations in GALNT3 [3]. In two subjects with bony involvement, TC destroyed a shoulder joint and ulnar growth plate. Three subjects had evidence of systemic inflammation with elevated serum C-reactive protein levels.

Radiographic findings

Radiographs show early and small lesions near bursae and are often distributed along the extensor surfaces of large joints [4]. They comprise multiple globular amorphous calcific components separated by radiolucent fibrous septae. Occasionally, fluid levels indicate a cystic component (Fig. 111.1). An inflammatory process, "diaphysitis," may also be observed on radiographs, CT, or MRI usually occurring in the middle of long bones. Vascular calcification has also been reported on radiographs or CT. A bone scan is the most reliable and simplest method for detection, localization, and assessment of extension of the calcific masses. Periarticular masses that are radiologically indistinguishable from those described in TC may be found in chronic renal failure. Hyperostosis–hyperphosphatemia causes periosteal reaction and cortical hyperostosis.

Biochemical findings

Many subjects with TC have elevated levels of serum phosphate and 1,25 dihydroxyvitamin D levels are either inappropriately elevated or normal [5]. The tubular maximum of phosphate reabsorption ($TmPO_4$/glomerular filtration rate)

Primer on the Metabolic Bone Diseases and Disorders of Mineral Metabolism, Ninth Edition. Edited by John P. Bilezikian.
© 2019 American Society for Bone and Mineral Research. Published 2019 by John Wiley & Sons, Inc.
Companion website: www.wiley.com/go/asbmrprimer

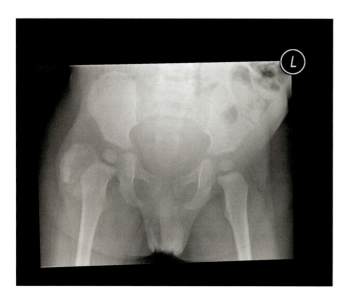

Fig. 111.1. An area of tumoral calcinosis overlying the right hip in a young child.

is elevated but renal function is otherwise normal. However, TC is also described with normal levels of serum phosphate. Serum calcium, alkaline phosphatase, and parathyroid hormone levels are usually normal. Positive calcium and phosphorus balances are caused by increased gastrointestinal absorption and reduced renal excretion.

Histopathology

It is suggested that their early lesions are triggered by bleeding followed by aggregation of foamy histiocytes which then become transformed into cystic cavities lined by osteoclast-like giant cells and histiocytes. Movement and friction because of their periarticular location seems key to the transformation. In a review of 111 cases from Zaire, histology identified exuberant cellular proliferation adjacent to the classical cystic form [6]. This consisted either of ill-defined reactive-like perivascular solid cell nests admixed with mononuclear and iron-loaded macrophages or well-organized fibrohistiocytic nodules of variable size embedded in a dense collagenous stroma. Mature lesions are filled with calcareous material in a viscous milky fluid.

Etiology and pathogenesis

Elucidation of the genetic basis for TC began in 2004 when investigation of large Druze and African American kindreds mapped to 2q24-q31 with biallelic mutations were discovered in the *GALNT3* gene [7]. Pseudoautosomal dominant inheritance was shown in one family in which individuals expressing the full phenotype harboured biallelic mutations in *GALNT3* whilst those heterozygous for the mutations showed incomplete expression with increased serum phosphate or 1,25 dihydroxyvitamin D levels but no calcified deposits [8]. Thus, autosomal recessive inheritance was confirmed. Subsequently, loss of function mutations in the gene for *FGF23* were identified in affected individuals who were negative for mutations in *GALNT3* [9, 10]. Elevated plasma levels of C-terminal FGF23 occur in these individuals with low plasma levels of intact FGF23 indicating increased processing of biologically active intact FGF23.

GALNT3 is now known to produce an enzyme that selectively o-glycosylates a furin-like convertase recognition sequence in FGF23 thus preventing proteolytic processing of FGF23 and allowing secretion of intact FGF23. Thus, loss of function mutations in *GALNT3* result in defective secretion of intact FGF23 causing hyperphosphatemia and increased synthesis of 1,25 dihydroxyvitamin D. In addition, a mutation in the *KLOTHO* gene was reported in a girl with TC who in addition to elevated serum phosphate and 1,25 dihydroxyvitamin D levels also had hypercalcemia and a high serum parathyroid hormone level [11]. There was evidence of serum C-terminal and intact FGF23 yet reduced FGF23 bioactivity. KLOTHO is a cofactor required by FGF23 to bind and signal through its FGF receptors [12]. Thus, mutations in three genes, *GALNT3*, *FGF23*, and *KLOTHO* cause the clinical and biochemical features of TC caused by functional deficiency or resistance to intact FGF23. Individuals with hyperostosis–hyperphosphatemia syndrome have homozygous or compound heterozygous mutations in the *GALNT3* gene [13, 14], thus confirming that the two conditions represent different aspects of the same disease. There is now evidence that normophosphatemic familial TC can be caused by mutations in the gene encoding the sterile alpha motif domain-containing-9 protein (*SAMD9*) [15].

Treatment

Surgical removal of the calcified masses may be required if they are painful, affect function, or for cosmetic reasons. Several different medical approaches to treatment have been reported, although usually as case reports. Aluminium hydroxide combined with dietary phosphate and calcium restriction has been reported to be successful [16]. Calcitonin has been used to induce phosphaturia [17]. The combination of acetazolamide with aluminium hydroxide, used for 14 years for one patient, reportedly improved the lesions [18]. Bisphosphonate therapy with alendronate reportedly alleviated symptoms within 12 weeks in one patient [19]. Phosphate-binding sevelamer, used for end stage renal failure, may be a possible treatment option. It was used in conjunction with acetazolamide and a low phosphate diet for a young girl with a large elbow mass caused by a homozygous mutation in FGF23 [20]. This lowered serum phosphate and shrunk the mass. Several members of the recently reported cohort were treated with a low phosphate diet, sevelamer, and acetazolamide with variable responses, but complete resolution of TC occurred in one subject [3].

It may be that the use of FGF23 or a means of enhancing its activity will be useful for treatment of TC in the future.

DERMATOMYOSITIS IN CHILDREN

Juvenile dermatomyositis (JD) is a rare idiopathic inflammatory disorder of the skin and muscle featuring progressive weakness predominantly of the proximal muscles and a rash which particularly affects the face and the extremities. JD differs from adult onset dermatomyositis in that it is frequently associated with small vessel vasculitis in the skin, muscle, and gastrointestinal tract and there is no association with malignancy. Dystrophic soft tissue calcification or "calcinosis" damages or devitalizes tissues despite normal calcium/phosphorus metabolism (Fig. 111.2).

Clinical presentation

JD is estimated to affect 1.9 to 2.5 per million children younger than 16 years and is more common in girls than boys (2:1 ratio). In a UK survey, the median age of onset was 6.8 years with two peaks in the girls at 6 and 11 years [21]. Of the reported cases were, 88% were White. Calcinosis usually manifests 1 to 3 years after the disease onset and reportedly occurs in 20% to 40% of affected individuals [22]. The duration of untreated JD is associated with pathological calcifications, thus demonstrating a clear link with chronic inflammation [23, 24]. The dystrophic calcification can cause pain, skin ulceration, limited joint mobility, contractures, and predispose to abscess formation. The calcification once present typically remains stable, but rarely some spontaneous resolution is reported. The clinical course of JD in children is variable with some having long-term relapsing or persistent disease whereas others recover.

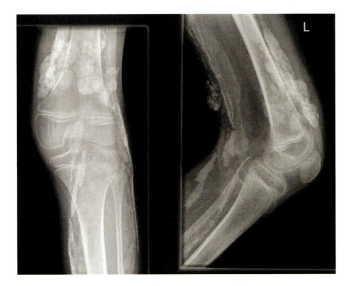

Fig. 111.2. Subcutaneous calcification in a child with juvenile dermatomyositis.

Biochemical and histological features

Serum levels of calcium, phosphate, and alkaline phosphatase are usually normal. Urinary levels of γ-carboxyglutamic acid have been reported to be elevated particularly if there is calcinosis. The mineral in the calcified deposits is poorly crystallized hydroxyapatite [22] or carbonate apatite containing relatively more mineral than matrix with a composition more similar to enamel than bone. Bone matrix proteins such as osteopontin, sialoprotein, and osteonectin are present within the calcifications with more osteonectin than is found in bone.

Radiographic features

Four types of dystrophic calcification can occur:

1. Superficial masses within the skin
2. Deep, discrete, subcutaneous nodular masses near joints that can impair movement (calcinosis circumscripta)
3. Deep, linear, sheet-like deposits within intramuscular fascial planes (calcinosis universalis)
4. Lacy, reticular subcutaneous deposits that encase the trunk to form a generalized "exoskeleton".

"Milk of calcium" fluid collections are a rare complication of the calcinosis [25]. Although established calcification can be readily seen radiographically, MRI is a sensitive method for detection and localization of myositis and oedema. It is also an excellent modality for monitoring progression or remission of the disease [26].

Treatment

High-dose corticosteroids soon after the onset of symptoms remains the mainstay of treatment and reduces the risk of calcinosis by suppressing the inflammatory process. Additional agents include methotrexate and infliximab. Several therapies suggesting benefit including bisphosphonates, diltiazem, and surgical extirpation are evaluated in case reports [27, 28]. However, a review of the published literature over a 32-year period concluded that no treatment has convincingly prevented or reduced calcinosis, with a lack of systematic study and clinical therapeutic trials [29].

REFERENCES

1. Inclan A, Leon P, Camejo MG. Tumoral calcinosis. JAMA. 1943;121:490–5.
2. Slavin RE, Wen J, Kumar D, et al. Familial tumoral calcinosis: a clinical, histopathologic, and ultrastructural study with an analysis of its calcifying process and pathogenesis. Am J Surg Pathol. 1993;17:788–802.

3. Ramnitz MS, Gourh P, Goldbach-Mansky R, et al. Phenotypic and genotypic characterization and treatment of a cohort with familial tumoral calcinosis/hyperostosis-hyperphosphatemia syndrome. J Bone Miner Res. 2016;31:1845–54.
4. Martinez S, Vogler JB, Harrelson JM, et al. Imaging of tumoral calcinosis: new observations. Radiology. 1990;174:215–22.
5. Lyles KW, Halsey DL, Friedman NE, et al. Correlations of serum concentrations of 1,25-dihydroxyvitamin D, phosphorous and parathyroid hormone in tumoral calcinosis. J Clin Endocrinol Metab. 1988;67:88–92.
6. Pakasa NM, Kalengayi RM. Tumoral calcinosis: a clinicopathological study of 111 cases with emphasis on the earliest changes. Histopathology. 1997;31:18–24.
7. Topaz O, Shurman DL, Bergman R, et al. Mutations in GALNT3, encoding a protein involved in o-linked glycosylation, cause familial tumoral calcinosis. Nature Genet. 2004;36:579–81.
8. Ichikawa S, Lyles KW, Econs MJ. A novel GALNT3 mutation in a pseudoautosomal dominant form of tumoral calcinosis: evidence that the disorder is autosomal recessive. J Clin Endocrinol Metab. 2005;90:2420–3.
9. Benet-Pages A, Orlik P, Strom TM, et al. An FGF23 missense mutation causes familial tumoral calcinosis with hyperphosphataemia. Hum Mol Genet. 2005;14:385–90.
10. Larsson T, Yu X, Davis SI, et al. A novel recessive mutation in fibroblast growth factor-23 causes familial tumoral calcinosis. J Clin Endocrinol Metab. 2005;90:2424–7.
11. Ichikawa S, Imel EA, Kreiter ML, et al. A homozygous missence mutation in human KLOTHO causes severe tumoral calcinosis. J Clin Invest. 2007;117:2684–91.
12. Urakawa I, Yamazaki Y, Shimada T, et al. Klotho converts canonical FGF receptor into a specific receptor for FGF23. Nature. 2006;444:770–7.
13. Ichikawa S, Baujat G, Seyahi A, et al. Clinical variability of familial tumoral calcinosis caused by novel GALNT3 mutations. Am J Med Genet A. 2010;152A:896–903.
14. Joseph L, Hing SN, Presneau N, et al. Familial tumoral calcinosis and hyperostosis-hyperphosphataemia syndrome are different manifestations of the same disease: novel missense mutations in GALNT3. Skeletal Radiol. 2010;39:63–8.
15. Topaz O, Indelman M, Chefetz I, et al. A deleterious mutation in SAMD9 causes normophosphatemic familial tumoral calcinosis. Am J Hum Genet. 2006;79:759–64.
16. Gregosiewicz A, Warda E. Tumoral calcinosis: successful medical treatment. J Bone Joint Surg Am. 1989;71:1244–9.
17. Salvi A, Cerudelli B, Cimino A, et al. Phosphaturic action of calcitonin in pseudotumoral calcinosis. Horm Metab Res. 1983;15:260.
18. Yamaguchi T, Sugimoto T, Imai Y, et al. Successful treatment of hyperphosphatemic tumoral calcinosis with long term acetazolimide. Bone. 1995;16:247S–50S.
19. Jacob JJ, Mathew K, Thomas N. Idiopathic sporadic tumoral calcinosis of the hip: successful oral bisphosphonate therapy. Endocr Pract. 2007;13:182–6.
20. Lammoglia JJ, Mericq V. Familial tumoral calcinosis caused by a novel FGF23 mutation: response to induction of tubular renal acidosis with acetazolamide and the non-calcium phosphate binder sevelamer. Horm Res. 2009;71:178–84.
21. Symmons DPM, Sills JA, Davis SM. The incidence of juvenile dermatomyositis: results from a nation-wide study. Br J of Rheumatol. 1995;34:732–6.
22. Pachman LM, Veis A, Stock S, et al. Composition of calcifications in children with juvenile dermatomyositis. Arthritis Rheum. 2006;54:3345–50.
23. Pachman LM, Abbott K, Sinacore JM, et al. Duration of illness is an important variable for untreated children with untreated dermatomyositis. J Pediatr. 2006;148:247–53.
24. Saini I, Kalaivani M, Kabra SK. Calcinosis in juvenile dermatomyositis: frequency, risk factors and outcome. Rheumatol Int. 2016;36:961–5.
25. Samson C, Soulen RL, Gursel E. Milk of calcium fluid collections in juvenile dermatomyositis: MR characteristics. Pediatr Radiol. 2000;30:28–9.
26. Park JH, Vital TL, Ryder NM, et al. Magnetic resonance imaging and P-31 magnetic spectroscopy provide unique quantitative data useful in the longitudinal management of patients with dermatomyositis. Arthritis Rheum. 1994;37:736–46.
27. Mukamel M, Horev G, Mimouni M. New insights into calcinosis of juvenile dermatomyositis: a study of composition and treatment. J Pediatr. 2001;138:763–6.
28. Oliveri MB, Palermo R, Mautalen C, et al. Regression of calcinosis during diltiazem treatment in juvenile dermatomyositis. J Rheumatol. 1996;23:2152–5.
29. Boulman N, Slobodin G, Rozenbaum M, et al. Calcinosis in rheumatic diseases. Semin Arthritis Rheum. 2005;34:805–12.

112

Genetic Disorders of Heterotopic Ossification: Fibrodysplasia Ossificans Progressiva and Progressive Osseous Heteroplasia

Frederick S. Kaplan[1], Robert J. Pignolo[2], Mona Al Mukaddam[1], and Eileen M. Shore[1]

[1]Center for Research in FOP and Related Disorders, Department of Orthopaedic Surgery, The University of Pennsylvania School of Medicine, Philadelphia, PA, USA
[2]Division of Geriatric Medicine and Gerontology; and Division of Endocrinology, Diabetes, Metabolism, Nutrition, Department of Internal Medicine, Mayo Clinic College of Medicine, Rochester, MN, USA

INTRODUCTION

Fibrodysplasia ossificans progressiva (FOP: OMIM #135100) is a rare heritable disorder of connective tissue characterized by congenital malformations of the great toes, and progressive heterotopic endochondral ossification (HEO) in characteristic anatomical patterns [1]. HEO may also occur sporadically following joint replacement, central nervous system trauma, athletic injury, war wounds, atherosclerosis, and valvular heart disease [2].

FOP is among the rarest of human afflictions, with an estimated incidence of one per two million individuals. All races are affected. Autosomal dominant transmission with complete penetrance but variable expression is established. However, reproductive fitness is low and most occurrences are sporadic. Gonadal mosaicism has also been reported [3].

CLINICAL PRESENTATION

Malformations of the great toes are present at birth in all classically affected individuals (Fig. 112.1). Typically, episodes of soft-tissue swelling ("flare-ups") leading to HEO begin during the first decade of life (Fig. 112.1) [4]. FOP is usually diagnosed when radiographic evidence of heterotopic ossification is noted; however, misdiagnosis is common and can lead to unnecessary biopsies and other invasive procedures that cause permanent harm [5].

The severity of FOP differs greatly among patients [1, 3]. Most become immobilized and confined to a wheelchair by the third decade of life [6]. Wide variability in the rate of disease progression, even among identical twins, attests to the importance of environmental factors [7].

Flare-ups appear spontaneously or after muscle fatigue, minor trauma, intramuscular injections, or influenza-like viral illnesses [1, 8, 9]. Swellings develop rapidly during the course of several hours. Aponeuroses, fascia, tendons, ligaments, and connective tissue of voluntary muscles may be affected. Although some lesions regress spontaneously, most mature by an endochondral pathway to form heterotopic bone with marrow elements [10]. Flare-ups are unpredictable and episodic. Once ossification develops, it is permanent. Disability is cumulative [4, 6].

Bony masses from FOP immobilize joints and cause contractures and deformity. Ossification around the hips, typically present by the third decade of life, often prevents ambulation [6]. Involvement of the muscles of mastication (frequently following injection of local anesthetic or overstretching of the jaw during dental procedures) can ankylose the jaw, impair nutrition, and diminish quality

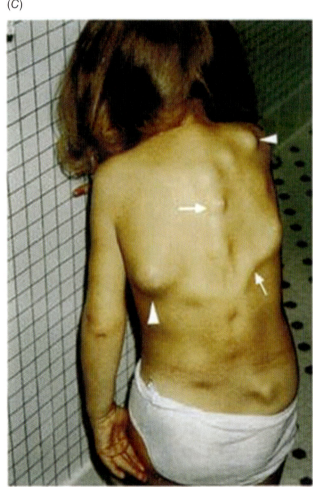

Fig. 112.1. Fibrodysplasia ossificans progressiva. Characteristics of FOP are seen in early childhood. (A) Short malformed great toes at birth (arrows) herald (B) later spontaneous appearance of the preosseous soft tissue lesions on the neck and back (arrow heads) and should invoke FOP even before their transformation to heterotopic bone (arrows). (C) Inspection of the toes for hallux valgus and/ or genetic DNA sequence analysis of *ACVR1* would then confirm the diagnosis obviating a lesional biopsy (trauma) that could exacerbate the condition Source: Kaplan FS, Smith RM. Clinical vignette – fibrodysplasia ossificans progressiva (FOP). J Bone Miner Res. 1997;12:855. Reproduced with permission of John Wiley & Sons.

of life [11, 12]. Scoliosis is common and associated with malformation of the ribs and costovertebral joints or with heterotopic bone that asymmetrically connects the ribs to the pelvis [13]. Ankylosis of the spine and ribs as well as chin-on-chest deformity further restrict mobility and may imperil cardiopulmonary function (Fig. 112.1) [14]. Restrictive disease of the chest wall may lead to early mortality [15]. Vocal, smooth, diaphragm, and extraocular muscles as well as the tongue and heart are spared [1]. Hearing impairment is common [16].

RADIOGRAPHIC FEATURES

Skeletal anomalies and soft-tissue ossification are radiological hallmarks of FOP [17]. The malformations almost always involve the great toe, although other skeletal anomalies commonly occur. In some cases, the thumbs are strikingly short [1]. Osteochondromas are frequent [18]. Progressive fusion of cervical vertebrae may be confused with Klippel–Feil syndrome [19]. Malformation of the temporomandibular joint and fusion of costovertebral joints are common [11, 15]. The femoral necks may be broad and short. Early degenerative arthritis is common [1, 20].

Radiographs and bone scans suggest normal modeling and remodeling of HEO [21]. Bone fractures are not increased, but when they occur are repaired involving normal processes in either heterotopic or normotopic bone [22].

Bone scintigraphy detects abnormalities in soft tissue before HEO can be demonstrated radiographically [22].

LABORATORY FINDINGS

Routine biochemical studies are usually normal, although serum prostanoids and alkaline phosphatase and urinary basic fibroblast growth factor levels may be increased during the inflammatory, fibroproliferative, and osteogenic phases of flare-ups, respectively [23, 24]. Elevated numbers of circulating osteoprogenitor cells have been noted during early flare-ups [25].

HISTOPATHOLOGY

Early preosseous FOP lesions consist of an intense aggregation of mononuclear inflammatory cells including lymphocytes, macrophages, and mast cells in the perivascular spaces of edematous muscle [26, 27]. Following the catabolic phase of muscle cell death, a highly anabolic fibroproliferative phase (often mistaken for aggressive juvenile fibromatosis) consists, in part, of mesenchymal-like stem cells derived from Tie2+ cells that differentiate through an endochondral pathway into mature heterotopic bone [28, 29].

ETIOLOGY AND PATHOGENESIS

The BMP signaling pathway is highly dysregulated in FOP [30–33]. FOP cells overexpress BMP4, cannot upregulate expression of multiple BMP antagonists in response to a BMP challenge [30, 31], and exhibit a defect in BMP receptor internalization and increased activation of downstream targets, suggesting that altered BMP receptor signaling participates in HEO formation in FOP [32].

In 2006, genome-wide linkage analysis localized the causative gene for FOP to chromosome 2q23-24, a locus containing the activin A type I receptor/activin-like kinase 2 (ACVR1/ALK2) gene encoding a BMP type I receptor [34]. A recurrent heterozygous missense mutation (c.617G>A; R206H) in the glycine-serine (GS) activation domain of ACVR1/ALK2 was identified in all affected individuals with classic features of either sporadic or inherited FOP, making molecular confirmation possible [34, 35]. Subsequently, a knock-in mouse model of the identified mutation recapitulated the FOP phenotype [36]. Protein modeling predicted destabilization of the GS domain, consistent with enhanced activation of the ACVR1/ALK2 receptor as the underlying pathogenesis of the ectopic chondrogenesis, osteogenesis, and joint fusion of FOP [34]. The GS domain is a specific binding site for FKBP1A (also known as FKBP12), a highly conserved inhibitory protein that prevents leaky activation of type I receptors in the absence of ligand. The ACVR1/ALK2 (R206H) protein interacts less with FKBP1A in the absence of BMP, suggesting this impaired FKBP1A-ACVR1/ALK2 interaction contributes in part to BMP-independent pathway signaling [37]. Basal and ligand-stimulated dysregulation of BMP pathway signaling characterize connective tissue progenitor cells from FOP patients, and in vitro and in vivo FOP models [37–43]. The ACVR1 (R206H) receptor causes FOP, in part, by being hyperresponsive to BMP ligands as well as by responding to the normally antagonistic ligand activin A [44, 45]. Additionally, early FOP lesions dramatically amplify BMP pathway signaling through an intracellular ligand-independent HIF-1α mechanism [46]. Individuals with rare variants and atypical forms of FOP have been described, and all have activating mutations of ACVR1/ALK2 that, like the common R206H mutation, cause gain of receptor activity [47, 48].

TREATMENT

There is no established medical treatment for FOP; its rarity, variability, and fluctuating clinical course pose substantial uncertainties when evaluating experimental therapies. Management is currently supportive [49]. High-dose glucocorticoids have limited use but are most effective in the management of the early inflammatory flare-ups. Bone marrow transplantation is ineffective, as even the normal immune system has triggered FOP flare-ups in one genetically susceptible chimeric patient [24]. Research to develop treatments for FOP [50, 51] has mainly focused on targeted inhibition of ACVR1/ALK2 receptor and BMP pathway signaling [52], and/or inhibition of the preosseous chondrogenic anlagen of HEO [53, 54], and offers hope for the future. Activation of the retinoid signaling pathway and retinoic acid receptor gamma (RARγ) inhibits chondrogenesis and HEO [53], and the RARγ agonist palovarotene is being used in FDA-approved clinical trials for FOP. Information about clinical trials on FOP can be found at http://clinicaltrials.gov/

Removal of FOP lesions is often followed by significant recurrence of HEO. Surgical release of joint contractures is unsuccessful, and risks new trauma-induced HEO [49]. Spinal bracing is ineffective and surgical intervention is associated with numerous complications [13]. Dental therapy should preclude mandibular blocks and stretching of the jaw [11]. Guidelines for general anesthesia have been reported [55]. While physical therapy to maintain joint mobility may be harmful by provoking or exacerbating lesions, occupational therapy evaluations are often helpful [56]. Intramuscular injections should be avoided [8, 49]. Prevention of falls, influenza, recurrent pulmonary infections, and complications of restrictive chest wall disease is important [49].

PROGNOSIS

Despite widespread HEO and severe disability, some patients live productive lives into the seventh decade. Most, however, die earlier from cardiopulmonary complications of severe restrictive chest wall involvement [15].

PROGRESSIVE OSSEOUS HETEROPLASIA

Research investigating FOP led to the discovery of progressive osseous heteroplasia (POH: OMIM #166350), a distinct developmental disorder of heterotopic ossification [57, 58]. As with FOP, POH is an autosomal dominant genetic disorder of heterotopic ossification. However, unlike in FOP, the heterotopic ossification of POH commonly begins in the dermis and progresses to deeper tissues by an intramembranous, rather than an endochondral, pathway [59]. In 2000, identification of two patients with POH and features of Albright hereditary osteodystrophy suggested a shared genetic etiology for the two conditions [60], later confirmed in a third patient [61]. These discoveries rapidly led to the identification of paternally-inherited inactivating mutations of the GNAS gene as the cause of POH [62]. A recent report that somatic disruption of GNAS in chick embryos mimics the induction of POH-like ossification suggested that somatic cell inactivation of GNAS could be responsible for the mosaic distribution of lesions in POH [63].

There are no specific phenotype–genotype correlations that distinguish POH from the more benign forms of limited dermal ossification [64]. Reduced expression of Gs-alpha, one of several proteins encoded by GNAS, can induce an osteoblast-like phenotype in mice and in human and mouse mesenchymal stem cells [65–68]. Heterozygous inactivation of Gnas by disruption of the Gs-alpha-specific exon 1 alters osteoblast differentiation in $Gnas^{+/-}$ mice, and manifests as subcutaneous heterotopic ossification by an intramembranous process [67]. GNAS-encoded G-proteins and downstream cAMP signaling seem to regulate cell fate lineage decisions at an early cell commitment stage, and appear to regulate osteogenesis, at least in part, through interactions with the BMP signaling pathway [66], and raises consideration whether treatments for FOP could also be applicable to POH.

Gs-alpha restricts bone formation by inhibiting Hedgehog signaling in mesenchymal progenitor cells, and Hedgehog signaling is upregulated in ectopic osteoblasts and progenitor cells in animal models of POH and in human POH osteoblasts and progenitor cells, suggesting that Hedgehog inhibitors could be repurposed for treating POH [68].

Presently, treatment of POH is supportive [69]. Removal of deep POH lesions is often followed by significant recurrence of heterotopic ossification. Surgical release of joint contractures is unsuccessful, and risks new trauma-induced heterotopic ossification.

REFERENCES

1. Kaplan FS, Glaser DL, Shore EM, et al. The phenotype of fibrodysplasia ossificans progressiva. Clin Rev Bone Miner Metab. 2005;3:183–8.
2. Pignolo RJ, Foley KL. Nonhereditary heterotopic ossification. Clin Rev Bone Miner Metab. 2005;3:261–6.
3. Shore EM, Feldman GJ, Xu M, et al. The genetics of fibrodysplasia ossificans progressiva. Clin Rev Bone Miner Metab. 2005;3:201–4.
4. Pignolo RJ, Bedford-Gay C, Liljesthrom M, et al. The natural history of flare-ups in fibrodysplasia ossificans progressiva: a comprehensive global assessment. J Bone Miner Res. 2016;31:650–6.
5. Kitterman JA, Kantanie S, Rocke DM, et al. Iatrogenic harm caused by diagnostic errors in fibrodysplasia ossificans progressiva. Pediatrics. 2005;116:654–61.
6. Rocke DM, Zasloff M, Peeper J, et al. Age and joint-specific risk of initial heterotopic ossification in patients who have fibrodysplasia ossificans progressiva. Clin Orthop. 1994;301:243–8.
7. Hebela N, Shore EM, Kaplan FS. Three pairs of monozygotic twins with fibrodysplasia ossificans progressiva: the role of environment in the progression of heterotopic ossification. Clin Rev Bone Miner Metab. 2005;3:205–8.
8. Lanchoney TF, Cohen RB, Rocke DM, et al. Permanent heterotopic ossification at the injection site after diphtheria-tetanus-pertussis immunizations in children who have fibrodysplasia ossificans progressiva. J Pediatr. 1995;126:762–4.

9. Scarlett RF, Rocke DM, Kantanie S, et al. Influenza-like viral illnesses and flare-ups of fibrodysplasia ossificans progressiva (FOP). Clin Orthop. Rel Res. 2004;423: 275–9.
10. Kaplan FS, Tabas JA, Gannon FH, et al. The histopathology of fibrodysplasia ossificans progressiva: an endochondral process. J Bone Joint Surg Am. 1993;75: 220–30.
11. Luchetti W, Cohen RB, Hahn GV, et al. Severe restriction in jaw movement after routine injection of local anesthetic in patients who have progressiva. Oral Surg Oral Med Oral Pathol Oral Radiol Endod. 1996;81:21–5.
12. Janoff HB, Zasloff M, Kaplan FS. Submandibular swelling in patients with fibrodysplasia ossificans progressiva. Otolaryngol Head Neck Surg. 1996;114:599–604.
13. Shah PB, Zasloff MA, Drummond D, et al. Spinal deformity in patients who have fibrodysplasia ossificans progressiva. J Bone Joint Surg Am. 1994;76:1442–50.
14. Kaplan FS, Glaser DL. Thoracic insufficiency syndrome in patients with fibrodysplasia ossificans progressiva. Clin Rev Bone Miner Metab. 2005;3:213–16.
15. Kaplan FS, Zasloff MA, Kitterman JA, et al. Early mortality and cardiorespiratory failure in patients with fibrodysplasia ossificans progressiva. J Bone Joint Surg Am. 2010;92:686–91.
16. Levy CE, Lash AT, Janoff HB, et al. Conductive hearing loss in individuals with fibrodysplasia ossificans progressiva. Am J Audiol. 1999;8:29–33.
17. Mahboubi S, Glaser DL, Shore EM, et al. Fibrodysplasia ossificans progressiva (FOP). Pediatr Radiol. 2001;31: 307–14.
18. Deirmengian GK, Hebela NM, O'Connell M, et al. Proximal tibial osteochondromas in patients with fibrodysplasia ossificans progressiva. J Bone Joint Surg Am. 2008;90:366–74.
19. Schaffer AA, Kaplan FS, Tracy MR, et al. Developmental anomalies of the cervical spin in patients with fibrodysplasia ossificans progressiva are distinctly different from those in patients with Klippel-Feil syndrome. Spine. 2005;30:1379–85.
20. Kaplan FS, Groppe JC, Seemann P, et al. Fibrodysplasia ossificans progressiva: developmental implications of a novel metamorphogene. In: Bronner F, Farach-Carson MC, Roach HI (eds).Bone and Development. London: Springer Verlag, 2010.
21. Kaplan FS, Strear CM, Zasloff MA. Radiographic and scintigraphic features of modeling and remodeling in the heterotopic skeleton of patients who have fibrodysplasia ossificans progressiva. Clin Orthop. 1994;304:238–47.
22. Einhorn TA, Kaplan FS. Traumatic fractures of heterotopic bone in patients who have fibrodysplasia ossificans progressiva. Clin Orthop. 1994;308:173–7.
23. Kaplan F, Sawyer J, Connors S, et al. Urinary basic fibroblast growth factor: a biochemical marker for preosseous fibroproliferative lesions in patients with FOP. Clin Orthop. 1998;346:59–65.
24. Kaplan FS, Glaser DL, Shore EM, et al. Hematopoietic stem-cell contribution to ectopic skeletogenesis. J Bone Joint Surg Am. 2007;89:347–57.
25. Suda RK, Billings PC, Egan KP, et al. Circulating osteogenic precursor cells in heterotopic bone formation. Stem Cells. 2009;27:2209–19.
26. Gannon FH, Valentine BA, Shore EM, et al. Acute lymphocytic infiltration in an extremely early lesion of fibrodysplasia ossificans progressiva. Clin Orthop. 1998;346:19–25.
27. Gannon FH, Glaser D, Caron R, et al. Mast cell involvement in fibrodysplasia ossificans progressiva. Hum Pathol. 2001;32:842–8.
28. Lounev V, Ramachandran R, Wosczyna MN, et al. Identification of progenitor cells that contribute to heterotopic skeletogenesis. J Bone Joint Surg Am. 2009;91:652–63.
29. Wosczyna MN, Biswas AA, Cogswell CA, et al. Multipotent progenitors resident in the skeletal muscle interstitium exhibit robust BMP-dependent osteogenic activity and mediate heterotopic ossification. J Bone Miner Res. 2012;27(5):1004–17.
30. Shafritz AB, Shore EM, Gannon FH, et al. Dysregulation of bone morphogenetic protein 4 (BMP4) gene expression in fibrodysplasia ossificans progressiva. N Engl J Med. 1996;335:555–61.
31. Ahn J, Serrano de La Peña L, Shore EM, et al. Paresis of a bone morphogenetic protein antagonist response in a genetic disorder of heterotopic skeletogenesis. J Bone Joint Surg Am. 2003;85:667–74.
32. Serrano de la Peña L, Billings PC, Fiori JL, et al. Fibrodysplasia ossificans progressiva (FOP), a disorder of ectopic osteogenesis, misregulates cell surface expression and trafficking of BMPRIA. J Bone Miner Res. 2005;20:1168–76.
33. Fiori JL, Billings PC, Serrano de la Peña L, et al. Dysregulation of the BMP-p38 MAPK signaling pathway in cells from patients with fibrodysplasia ossificans progressiva (FOP). J Bone Miner Res. 2006;21:902–9.
34. Shore EM, Xu M, Feldman GJ, et al. A recurrent mutation in the BMP type I receptor ACVR1 causes inherited and sporadic fibrodysplasia ossificans progressiva. Nature Genet. 2006;38:525–7.
35. Kaplan FS, Xu M, Glaser DL et al. Early diagnosis of fibrodysplasia ossificans progressiva. Pediatrics. 2008;121: e1295–300.
36. Chakkalakal SA, Zhang D, Culbert AL, et al. An Acvr1 Knock-in mouse has fibrodysplasia ossificans progressiva. J Bone Miner Res. 2012;27:1746–56.
37. Shen Q, Little SC, Xu M, et al. The fibrodysplasia ossificans progressiva R206H ACVR1 mutation activates BMP-independent chondrogenesis and zebrafish embryo ventralization. J Clin Invest. 2009;119:3462–72.
38. Billings PC, Fiori JL, Bentwood JL, et al. Dysregulated BMP signaling and enhanced osteogenic differentiation of connective tissue progenitor cells from patients with fibrodysplasia ossificans progressiva (FOP). J Bone Miner Res. 2008;23:305–13.
39. Kaplan FS, Pignolo RJ, Shore EM. The FOP metamorphogene encodes a novel type I receptor that dysregulates BMP signaling. Cytokine Growth Factor Rev. 2009;20: 399–407.

40. Culbert AL, Chakkalakal SA, Theosmy EG, et al. Alk2 regulates early chondrogenic fate in fibrodysplasia ossificans progressiva heterotopic endochondral ossification. Stem Cells. 2014;32:1289–300.
41. Fukuda T, Kohda M, Kanomata K, et al. Constitutively activated ALK2 and increased SMAD1/5 cooperatively induce bone morphogenetic protein signaling in fibrodysplasia ossificans progressiva. J Biol Chem. 2009;284:7149–56.
42. van Dinther M, Visser N, de Gorter DJ, et al. ALK2 R206H mutation linked to fibrodysplasia ossificans progressiva confers constitutive activity to the BMP type I receptor and sensitizes mesenchymal cells to BMP-induced osteoblasts differentiation and bone formation. J Bone Miner Res. 2010;25:1208–15.
43. Song GA, Kim HJ, Woo KM, et al. Molecular consequences of the ACVR1 (R206H) mutation of fibrodysplasia ossificans progressiva. J Biol Chem. 2010;285:22542–53.
44. Hatsell SJ, Idone V, Wolken DM, et al. ACVR1(R206H) receptor mutation causes fibrodysplasia ossificans progressiva by imparting responsiveness to activin A. Sci Transl Med. 2015;7(303)ra137.
45. Hino K, Ikeya M, Horigome K, et al. Neofunction of ACVR1 in fibrodysplasia ossificans progressiva. Proc Natl Acad Sci U S A. 2015;112:15438–43.
46. Wang H, Lindborg C, Lounev V, et al. Cellular hypoxia promotes heterotopic ossification by amplifying BMP signaling. J Bone Miner Res. 2016;31:1652–65.
47. Kaplan FS, Xu M, Seemann P, et al. Classic and atypical fibrodysplasia ossificans progressiva (FOP) phenotypes are caused by mutations in the bone morphogenetic protein (BMP) type I receptor ACVR1. Hum Mutat. 2009;30:379–90.
48. Chaikuad A, Alfano I, Kerr G, et al. Structure of the bone morphogenetic protein receptor ALK2 and implications for fibrodysplasia ossificans progressiva. J Biol Chem. 2012;287:36990–8.
49. Pignolo RJ, Shore EM, Kaplan FS. Fibrodysplasia ossificans progressiva: diagnosis, management, and therapeutic horizons. In Emerging Concepts in Pediatric Bone Disease. Pediatr Endocrinol Rev. 2013;10(S-2):437–48.
50. Kaplan FS, Pignolo RJ, Shore EM. From mysteries to medicines: drug development for fibrodysplasia ossificans progressive. Expert Opin Orphan Drugs. 2013;1:637–49.
51. Pacifici M, Shore EM 2016. Common mutations in ALK2/ACVR1, a multi-faceted receptor, have roles in distinct pediatric musculoskeletal and neural disorders. Cytokine Growth Factor Rev. 2016;27:93–104.
52. Hong CC, Yu PB. Applications of small molecule BMP inhibitors in physiology and disease. Cytokine Growth Factor Rev. 2009;20:409–18.
53. Shimono K, Tung WE, Macolino C, et al. Potent inhibition of heterotopic ossification by nuclear retinoic acid receptor-gamma agonists. Nat Med. 2011;17:454–60.
54. Chakkalakal SA, Uchibe K, Convente MR, et al. Palovarotene inhibits heterotopic ossification and maintains limb mobility and growth in mice with the human ACVR1 (R206H) fibrodysplasia ossificans progressiva (FOP) mutation. J Bone Miner Res. 2016;31:1666–75.
55. Kilmartin E, Grunwald Z, Kaplan FS, et al. General anesthesia for dental procedures in patients with fibrodysplasia ossificans progressiva: a review of 42 cases in 30 patients. Anesth Analg. 2014;118:298–301.
56. Levy CE, Berner TF, Bendixen R. Rehabilitation for individuals with fibrodysplasia ossificans progressiva. Clin Rev Bone Miner Metab. 2005;3:251–6.
57. Kaplan FS, Craver R, MacEwen GD, et al. Progressive osseous heteroplasia: a distinct developmental disorder of heterotopic ossification. J Bone Joint Surg Am. 1994;76:425–36.
58. Kaplan FS, Shore EM. Progressive osseous heteroplasia. J Bone Miner Res. 2000;15:2084–94.
59. Shore EM, Kaplan FS. Inherited human diseases of heterotopic bone formation. Nat Rev Rheumatol. 2010;6:518–27.
60. Eddy MC, Jan De Beur SM, Yandow SM, et al. Deficiency of the alpha-subunit of the stimulatory G protein and severe extraskeletal ossification. J Bone Miner Res. 2000;15:2074–83.
61. Yeh GL, Mathur S, Wivel A, et al. GNAS1 mutation and Cbfa1 misexpression in a child with severe congenital platelike osteoma cutis. J Bone Miner Res. 2000;15:2063–73.
62. Shore EM, Ahn J, Jan de Beur S, et al. Paternally-inherited inactivating mutations of the GNAS1 gene in progressive osseous heteroplasia. N Engl J Med. 2002;346:99–106.
63. Cairns DM, Pignolo RJ, Uchimura T, et al. Somitic disruption of GNAS in chick embryos mimics progressive osseous heteroplasia. J Clin Invest. 2013;123:3624–33.
64. Adegbite NS, Xu M, Kaplan FS, et al. Diagnostic and mutational spectrum of progressive osseous heteroplasia (POH) and other forms of GNAS-based heterotopic ossification. Am J Med Genet A. 2008;146A:1788–96.
65. Leitman SA, Ding C, Cooke DW, et al. Reduction in Gs-alpha induces osteogenic differentiation in human mesenchymal stem cells. Clin Orthop Rel Res. 2005;434:231–38.
66. Zhang S, Xu M, Kaplan FS, et al. Different roles of Gnas and cAMP signaling during early and late stages of osteogenic differentiation. Hormone Metab. Res. 2012;44:724–31.
67. Pignolo RJ, Xu M, Russell E, et al. Heterozygous inactivation of Gnas confers enhances osteoblast differentiation and formation of heterotopic ossification J Bone Miner Res. 2011;26:2647–55.
68. Regard JB, Malhotra D, Gvozdanovic-Jeremic J, et al. Activation of hedgehog signaling by loss of GNAS causes heterotopic ossification. Nat Med. 2013;19:1505–12.
69. Pignolo RJ, Ramaswamy G, Fong JT, et al. Progressive osseous heteroplasia: diagnosis, treatment, and prognosis. Application Clin Genet. 2015;8:37–48.

113
Osteogenesis Imperfecta

Joan C. Marini

The Eunice Kennedy Shriver National Institute of Child Health and Human Development,
Section on Heritable Disorders of Bone and Extracellular Matrix,
National Institutes of Health, Bethesda, MD, USA

INTRODUCTION

Osteogenesis imperfecta (OI), or "brittle bone disease," is a genetic disorder of connective tissue characterized by fragile bones and susceptibility to fracture from mild trauma [1]. OI is both clinically and genetically heterogeneous. The phenotype of OI ranges from perinatal lethality to subtle presentation as early osteoporosis. Individuals with OI have varying combinations of growth deficiency, defective tooth formation (dentinogenesis imperfecta), hearing loss, blue sclerae, macrocephaly, scoliosis, barrel chest, and ligamentous laxity. Classical OI is an autosomal dominant condition caused by defects in either gene (*COL1A1/COL1A2*) encoding the chains comprising type I collagen, the major structural protein of bone and skin matrix [2]. Classical OI nosology uses the Sillence classification from 1979, proposed before OI gene defects were known [3]. Subsequent studies showed type I OI, which is relatively mild, is caused by deficiency of type I collagen [4], whereas moderate and severe OI types are caused by structural defects in type I collagen [2]. Recurrence of classical OI types among offspring of clinically healthy parents is caused by parental mosaicism [5]. Beginning in 2006, the genes causing rare forms of (mostly) recessive OI were identified, shifting the OI paradigm of a collagen structural disorder to a collagen-related disorder [6]. Currently, 18 OI types span lethal to mild and account for nearly all cases (Table 113.1). They can be organized into type I collagen deficiency and structural defects plus four functional groups comprising genetic defects in bone mineralization (*IFITM5, SERPINF1*), collagen modification (*CRTAP, P3H1, PPIB*), collagen processing and crosslinking (*SERPINH1, FKBP10, BMP1*), and osteoblast differentiation and function (*SP7, TMEM38B, WNT1, CREB3L1, SPARC,* and *MBTPS2*).

CLINICAL PRESENTATION

Because these 18 types of OI vary widely in symptoms and age at presentation, the differential diagnosis can vary [7]. A positive family history is usually absent, because most mutations are recessive or occur de novo. Prenatally, severe OI types II, III, VII, VIII IX, X, XIII, XVI, or XVIII may be difficult to distinguish from thanatophoric dysplasia, campomelic dysplasia, and achondrogenesis type I. Neonatally, type III or VIII OI and perinatal hypophosphatasia may have overlapping features, but perinatal hypophosphatasia has the biochemical distinction of low serum alkaline phosphatase activity. In childhood, for milder forms of OI, the major distinctions are with juvenile hypophosphatasia and idiopathic osteoporosis and child abuse.

The key diagnostic feature for OI is the generalized nature of the connective tissue defect, with characteristic facial features, relative macrocephaly, thoracic configuration (barrel chest or pectus excavatum), joint laxity, vertebral compressions, and growth deficiency present in variable combinations. Among the recessive OIs, those caused by defects in the collagen 3-hydroxylation complex overlap clinically with types II and III, whereas those caused by defects in *SERPINH1, SERPINF1, FKBP10, TMEM38B, WNT1, CREB3L1,* and *MBTPS2* overlap clinically with types II, III, and IV. Recessive OI types almost always have white sclerae and unremarkable dentition, whereas dominant forms may have blue or white sclerae and dentinogenesis imperfecta [6]. Nevertheless, the definitive diagnosis of OI is now molecular, using DNA sequencing panels of dominant and recessive OI genes, sometimes supplemented by collagen biochemical studies.

Table 113.1. Osteogenesis imperfecta (OI) nosology.

OI type	Inheritance	Defective gene	Defective protein
Defects in collagen synthesis and structure			
Type I, II, III, IV	AD	COL1A1 or COL1A2	_1(I) or _2(I) collagen
Defects in bone mineralization			
Type V	AD	IFITM5	BRIL
Type VI	AR	SERPINF1	PEDF
Defects in collagen modification			
Type VII	AR	CRTAP	CRTAP
Type VIII	AR	LEPRE1	P3H1
Type IX	AR	PPIB	PPIB (CyPB)
Defects in collagen processing and crosslink			
Type X	AR	SERPINH1	HSP47
Type XI	AR	FKBP10	FKBP65
Unclassified	AR	PLOD2	LH2
Type XII	AR	BMP1	BMP1
Defects in osteoblast differentiation and function			
Type XIII	AR	SP7	SP7 (OSTERIX)
Type XIV	AR	TMEM38B	TRIC-B
Type XV	AR/AD	WNT1	WNT1
Type XVI	AR	CREB3L1	OASIS
Type XVII	AR	SPARC	SPARC (osteonectin)
Type XVIII	XR	MBTPS2	S2P

AD - = autosomal dominant; AR = autosomal recessive; XR = X-linked recessive.

CLINICAL PHENOTYPES

The Sillence classification for OI (Table 113.1) utilized clinical and radiographic criteria [3]. Now, the current genetic classification assigns the original types I–IV to collagen structural mutations, and extends the numeration based on discoveries of new OI genes.

Type I OI is the most common and mildest form of the disorder. Fracture onset is postnatal, usually after attaining ambulation, but sometimes presenting in middle age as early onset osteoporosis. Fracture incidence decreases markedly after puberty. Individuals with type I OI usually have blue sclerae and often bruise easily. They may suffer hearing loss (onset as early as late childhood, but usually in their 20s), dentinogenesis imperfecta, and joint hyperextensibility. Growth deficiency and long bone deformity are generally mild.

Type II OI is usually lethal, from respiratory causes, in the perinatal period, although survival for months is not uncommon. These babies are often premature and small for gestational age. Legs are usually held in the frog-leg position with hips abducted and knees flexed. Radiographically, long bones are extremely osteopenic, with in utero fractures and undertubulated modeling. The skull is severely undermineralized with wide anterior and posterior fontanels. Scleral hue is blue-gray. Their bones comprise predominantly "woven bone."

Type III OI, the "severe progressive deforming" type, presents at birth and overlaps with the mild end of the type II OI spectrum. Affected individuals have extremely fragile bones and, over a lifetime, will have dozens to hundreds of fractures. Their soft long bones deform from normal muscle tension and fractures. Extreme growth deficiency results in final stature in the range of a prepubertal child. Almost all type III patients develop scoliosis. Radiographically, there is metaphyseal flaring and "popcorn" calcification at growth plates [8]. These patients require intensive physical rehabilitation and orthopedic care to attain assisted ambulation; many use wheelchairs for mobility. Type III OI is compatible with a full lifespan, although many individuals have respiratory insufficiency and *cor pulmonale* in middle age, and some die in childhood.

Type IV is the moderately severe Sillence form. The disorder may be apparent at birth or during school age. Scleral hue is variable. These children have several fractures yearly and bowing of their long bones. Fracture incidence decreases after puberty. Final stature is in the range of pubertal children; many respond to growth hormone with significant additional height and improved bone histology [9]. Radiographically, they have osteopenia and mild modeling abnormalities. Platybasia, vertebral compressions, and scoliosis may be present. With rehabilitation intervention and orthopedic management, independent mobility is generally attained. Type IV OI is compatible with a full lifespan.

OI/EDS is a discrete subgroup of patients who have both the skeletal manifestations of OI (types IV or III) plus the joint laxity of Ehlers–Danlos syndrome [10]. Hip dysplasia and early progressive scoliosis often occur. Bone tissue is soft and friable, requiring extra care for spinal fixation. Mutations are found in the amino terminal region of the type I collagen chains and interfere with collagen N-propeptide processing [10].

Mutations in the type I collagen C-propeptide also cause OI, although the propeptide is processed before collagen is incorporated into bone matrix [11]. These cases range from mild to lethal. Mutations in the C-propeptide cleavage site itself or in the C-propeptidase lead to a paradoxically high bone mass form of OI with elevated DXA BMD Z-scores and bone hypermineralization [12].

Type V and VI OI are related functionally because their causative genes influence each other and impact bone mineralization, but they are quite distinct clinically [13]. Type V OI is defined by its bone histology and a clinical triad of hypertrophic callus, dense metaphyseal bands, and ossification of interosseus membranes [14]. Autosomal recessive type VI OI features white sclerae, distinctive fish-scale bony lamellae, and broad osteoid bands, and is defined both genetically as defects in SERPINF1 and biochemically as low to absent PEDF in serum. Type VI OI is not apparent at birth, but progresses in childhood to severe OI with scoliosis.

Types VII, VIII, and IX OI have autosomal recessive inheritance and are caused by deficiency of components of the procollagen prolyl 3-hydroxylation complex, CRTAP [15], P3H1 [16], and CyPB [17], respectively. Types VII and VIII present indistinguishably, with severe to lethal bone dysplasia, normal cranial size, rhizomelia, and white sclerae. Survivors of type VIII OI in their second or third decade have extreme growth deficiency, white sclerae and DXA BMD Z-scores of less than –6. Type VIII OI caused by a West African founder mutation is almost invariably lethal perinatally [18]. Fewer than 10 cases of Type IX OI have been reported [6]. Most are lethal perinatally but two were moderate in severity; none have rhizomelia.

Types X and XI OI are caused by deficiency of collagen chaperones, HSP47 [19] and FKBP10 [20], respectively. Type X features severe skeletal dysplasia, blue sclerae, and dentinogenesis imperfecta in childhood. Defects in FKBP10 encompass a spectrum with severe OI alone, OI plus congenital contractures (Bruck syndrome), and congenital contractures alone (Kuskokwim syndrome) [21–23]. Contractures are a variable feature, even among siblings. Affected individuals have deforming OI with fractures, vertebral compressions and short stature. Teeth and sclerae are normal.

Type XII OI is a severe autosomal recessive form, caused by deficiency of the C-propeptide processing protease, BMP1 [24]. Patients have multiple fractures, kyphoscoliosis, and short stature with rhizomelia. Some have elevated DXA BMD, overlapping with high bone mass OI [12].

The most recent group of genes found to cause OI leads to defects in osteoblast differentiation and function. There is wide variability of phenotype among them, although they share normal teeth, hearing, and scleral hue. An interesting defect in an endoplasmic reticulum cation channel that affects calcium flux kinetics causes moderately severe OI that is difficult to distinguish clinically from type IV OI, although cardiac defects may be an additional concern [25]. Defects in WNT1 may be dominant, causing osteoporosis in adults, or recessive, causing progressive deforming OI with kyphoscoliosis [6, 26]. Two individuals with SPARC defects had progressive deforming OI with muscle hypotonia [27]. And two genes in the RIP (regulated intramembrane proteolyisis) pathway, CREB3L1 [28] and X-linked MBTPS2 [29], each cause a skeletal phenotype that resembles type III OI.

RADIOGRAPHIC AND DXA FEATURES

The skeletal survey in classical OI shows generalized osteopenia. Long bones have thin cortices and a gracile appearance. In moderately to severely affected patients, long bones have bowing and modeling deformities, especially undertubulation, metaphyseal flaring, and "popcorn" appearance at the metaphyses [8]. Long bones of the upper extremity often seem milder than those of the lower extremity. Vertebrae often have central compressions even in mild type I OI which often appear first at the T_{12}-L_1 level. In moderate to severe OI, vertebrae will have central and anterior compressions and may appear compressed throughout. The compressions are generally consistent with the patient's L_1-L_4 DXA BMD Z-score but often not with scoliosis [30]. In the lateral spine, it is not easy to assess asymmetric vertebral collapse, which, along with paraspinal ligamentous laxity, underlies OI scoliosis. The skull of OI patients, despite a wide phenotypic range of severity, has wormian bones. Patients with type III and IV OI may also have platybasia, which should be followed with CT studies for basilar impression and invagination.

Distinctive features are noted in some OI types. In type V OI, dense metaphyseal bands, ossification of the interosseus membrane of the forearm, and hypertrophic callus are found [14]. Types VII and VIII OI long bones have a cystic and disorganized appearance with "popcorn" calcifications at epiphyses; limbs show rhizomelia [15, 16].

Bone densitometry using DXA (L_1-L_4) generally correlates with the severity of OI [7], and is useful to follow its course. Importantly, the Z-score compares the mineral quantity of OI bone to bone with normal matrix structure and crystal alignment [31]. In OI, many mutations result in irregular crystal alignment within abnormal matrix. DXA does not measure bone quality, which reflects bone geometry, histomorphometry, and mechanical properties.

LABORATORY FINDINGS

Serum chemistries related to bone and mineral metabolism are generally normal in OI. Alkaline phosphatase may be elevated after a fracture and is significantly elevated during childhood in type VI [32]. PEDF has been reported to be low to absent in type VI, which provides a useful screening test [33]. Acid phosphatase is elevated in type VIII OI and would be expected to be elevated in type VII [34]. Hormones of the growth axis have normal levels [35]. Bone histomorphometry shows defects in bone modeling and in number of trabeculae [36]. Cortical width and cancellous bone volume are decreased in all types. The rate of bone remodeling is increased, as are osteoblast and osteoclast surfaces. When viewed under polarized light, the lamellae of OI bone are thinner and less smooth than in controls; those in type VI have a fish-scale appearance [32], whereas type V bone has a mesh-like appearance [14]. Mineral apposition rate is normal: crystal disorganization may contribute to bone weakness.

ETIOLOGY AND PATHOGENESIS

About 80% to 85% of patients who have OI have dominantly inherited abnormalities of type I collagen, the major structural protein of the bone extracellular matrix [37]. Of the rare OI types, only type V, which has a recurrent mutation at the 5′-end of *IFITM5*, has dominant inheritance [38]. The recessive types of OI can be divided into four functional groups, those with mineralization defects, those with deficiency of a component of the procollagen prolyl 3-hydroxylation complex, those with defects in collagen processing and crosslinking, and those with defects in osteoblast differentiation and function. OI is now understood more widely to be a type I collagen-related disorder because all of the deficient/defective proteins either interact with collagen directly or affect its synthesis, crosslinking, or level of synthesis [6].

The pathophysiology of classical type I collagen-based OI encompasses multiple levels of dysfunction, ranging from abnormal collagen folding and overmodification and ER stress, to abnormal crosslinking in matrix, disturbed interactions with noncollagenous proteins, altered bone cellular differentiation and function, and hypermineralization of bone, all leading to brittle bone tissue [7]. Murine studies have implicated bone cell dysfunction (osteoblast ER stress, with increased expression of CHOP and GRP78, cytoskeletal dysfunction, and impaired matrix production, as well as increased osteoclast numbers and activity) [7, 39].

Patients with type I OI, who synthesize reduced amounts of structurally normal type I collagen because of a null *COL1A1* allele, display a relative increase in the COL3/COL1 ratio, typically detected using cultured fibroblast studies [4]. Probands with the clinically significant types II, III, and IV OI synthesize a mixture of normal collagen and collagen with a structural defect [2]. With rare exceptions, the structural defects are either amino acid substitutions for one of the glycine residues that occur at every third position along the chain and are essential for proper helix folding (80%) or alternative splicing of an exon (20%), resulting more frequently in out-of-frame than in-frame alternative transcripts. Structural abnormalities delay the folding of the triple helix, expose the constituent chains to modifying enzymes for a longer time, and result in overmodification detectable as slower electrophoretic migration. However, the biochemical test does not accurately detect abnormalities in the amino one-third of the α1(I) or amino half of the α2(I) chain [40]. Cultured fibroblasts in types VII [15] and VIII OI [16], and some in type IX OI [17, 41], also produce collagen with overmodification of the helical regions of the chains, suggesting that deficiency of the 3-hydroxylation complex delays helix folding [7].

Genotype-phenotype modeling of the collagen helical mutations indicates that the α1(I) and α2(I) chains play distinct roles in maintaining matrix integrity [2]. About one-third of the amino acid substitutions in α1(I) are lethal, especially if a branched or charged side chain is introduced. Two exclusively lethal regions coincide with the proposed major ligand-binding regions for the collagen monomer with integrins, matrix metalloproteinases (MMPs), fibronectin, and cartilage oligomeric matrix protein (COMP). For the α2(I) chain, only one-fifth of substitutions are lethal; these cluster in eight regularly spaced regions along the chain, coinciding with the proteoglycan binding regions on the collagen fibril.

The phenotypes resulting from failure to process either N- or C-terminal propeptides have distinct etiologies and pathogenesis. At the amino end of the helix, the combined OI/EDS syndrome is caused by the incorporation of uncleaved pN-collagen into fibrils, which results in decreased fibril diameter and compromised matrix integrity [10]. At the C-terminal end, incorporation of pC-collagen into fibrils causes bone fragility from a seemingly paradoxical increase in bone mineralization, causing high bone mass OI [12]. Mutations in the C-propeptide itself have a range of phenotypes depending on where they are located in the "flower-like" structure defined by X-ray crystallography [11]. These mutations delay chain incorporation and may interfere with processing.

The genes underlying types V and VI OI, *IFITM5* and *SERPINF1*, are connected by a pathway related to bone mineralization [13]. The unique type V OI mutation at the 5′ end of *IFITM5* causes a gain of function addition to BRIL, resulting in increased osteoblast *SERPINF1* expression and PEDF secretion, plus increased expression of multiple osteoblast markers [42]. Conversely, a heterozygous BRIL p.S40L substitution causes atypical type VI OI, with decreased osteoblasts, *SERPINF1* transcripts, and PEDF synthesis, and decreased expression of other osteoblast markers [13]. However, both BRIL mutations decrease the expression and secretion of type I collagen. The pathway apparently connects back from PEDF to BRIL, since PEDF levels affect *IFITM5* expression (Kang and Marini, unpublished observation).

The causal genes for types VII, VIII, and IX OI encode the components of the ER-localized procollagen prolyl 3-hydroxylation complex, CRTAP, Prolyl 3-hydroxlase 1 (P3H1), and CyPB [7]. CRTAP and P3H1 are mutually protective in the complex, so null mutations in one component lead to absence of both proteins [43]. The complex modifies specific proline residues; α1(I)P986 is normally fully hydroxylated whereas α2(I)P707 is partially modified [7]. Defects in the complex lead to severely reduced or absent 3-hydroxylation, but also to overmodification of the full collagen helix, indicating that folding is also slowed in the absence of the chaperone function of the complex. The 3-hydroxylation modification itself has been proposed to function to fine-tune the alignment of collagen monomers in the fibril [6, 44].

The genes underlying types X and XI OI, SERPINF1 and FKBP10, also have protein products, HSP47 and FKBP65, functionally connected to each other [45]. HSP47 is a collagen specific chaperone that binds preferentially to folded collagen helices, preventing procollagen aggregation during secretory transport. HSP47 supports the stabilization of FKBP65, and FKBP65 may regulate HSP47 localization [45]. Furthermore, FKBP65 serves as the PPIase for lysyl hydroxylase 2 (LH2) and supports dimerization to its active form [46]. In its absence, LH2 hydroxylation of the telopeptide lysine required for crosslinking of collagen in the extracellular matrix is severely reduced, impairing matrix incorporation of collagen.

The most recently delineated group of six OI genes (SP7, TMEM38B, WNT1, CREB3L1, SPARC, and MBTPS2) cause defects in osteoblast differentiation and function. Although their protein products do not interact directly with collagen, they indirectly impact collagen secretion and, in one case, modification. Type XIV OI is caused by null mutation in TMEM38B, which encodes an ER membrane channel for potassium and impacts calcium flux kinetics [25]. Multiple collagen-related enzymes are calcium dependent, and these mutations globally disregulate collagen synthesis. In addition, helical lysyl hydroxylation is reduced, and collagen is underhydroxylated. Two other genes in this functional category, MBTPS2[29] and CREB3L1 [28], are components of the regulated intramembrane proteolysis (RIP) pathway, which also impacts cholesterol metabolism. These OI types show RIP is critical for skeletal development and suggest that there are cell-specific RIP substrates as well as general pathways involving an ER stress response.

TREATMENT

Early and consistent rehabilitation intervention is the basis for maximizing the physical potential of individuals with OI [47, 48]. Physical therapy should begin in infancy for the severest types, promoting muscle strengthening, and if possible, protected ambulation. Programs to assure that children have muscle strength to lift a limb against gravity should continue between any orthopedic interventions using isotonic and aerobic conditioning. Swimming should be encouraged for aerobic conditioning.

Orthopedic care should come from surgeons experienced in OI. To prevent loss of function, fractures should not be allowed to heal without reduction. The goals of orthopedic surgery are to correct deformity for ambulation and to interrupt a cycle of fracture and refracture. Osteotomy requires intramedullary rod fixation. The hardware available includes telescoping rods (Bailey–Dubow [49] or Fassier–Duval [50] rods) and nonelongating rods (Rush rods). Selection of the rod with the smallest diameter suited to the situation helps to avoid cortical atrophy. Use of extensible rods may lead to fewer revisions for patients with significant growth potential.

The complications of OI, including abnormal pulmonary functions, hearing loss, and basilar invagination, are best managed in a specialized and coordinated care program. The severe growth deficiency of OI is responsive to exogenous growth hormone (rGH) administration in about one half of cases of type IV OI [9] and most type I OI [51]; some treated children can attain normal heights. Responders to recombinant rGH also had increased L_1-L_4 DXA BMD, and bone biopsy increased bone volume per total volume (BV/TV) and bone formation rate (BFR). Growth hormone (rGH) remains under study for its effects on the OI skeleton.

For the past two decades, antiresorptive bisphosphonates have featured prominently in the treatment of OI in children, with the rationale that increased bone volume and BMD will increase resistance to fractures. Four controlled trials and several recent meta-analyses [52] were in agreement that bisphosphonate treatment increased vertebral DXA BMD Z-scores and improved vertebral geometry [6, 53, 54]. However, any decrease in long-bone fracture rate was equivocal, with reduced relative risk representing an interplay of positive effects of increased bone stiffness and load bearing versus a decline in material quality (increased brittleness). Furthermore, the improved vertebral geometry did not diminish the prevalence of scoliosis at maturity, even in children who received bisphosphonate before age 5 years [30]. Functional changes in ambulation, muscle strength, and bone pain reported in the uncontrolled trials have not been supported in controlled trials. Studies have shown that the maximum gains in BMD and histology are achieved in 2 to 4 years of treatment [6, 53, 54]. Current use of bisphosphonates at the NICHD for classical OI is to treat for 2 to 3 years and then to discontinue the drug while continuing to follow the patient.

The prolonged half-life and recirculation of pamidronate in children up to 8 years after treatment cessation [55] may pose pediatric specific skeletal and reproductive risks [55]. High cumulative doses of bisphosphonate may induce adynamic bone with defective modeling and may lead to accumulation of bone microdamage [6]. Delayed osteotomy healing and tooth eruption were noted at Montreal protocol doses. This has engendered interest in short-acting antiresorptives, such as the RANKL-antibody denosumab, and anabolic drugs, such as investigational antisclerostin antibody.

REFERENCES

1. Marini JC. Osteogenesis imperfecta. In: Kliegman RM, Stanton, BF, St. Geme JW, et al. (eds) Nelson's Textbook of Pediatrics (20th ed.). Philadelphia: Elsevier, 2015, pp. 3380–4.
2. Marini JC, Forlino A, Cabral WA, et al. Consortium for osteogenesis imperfecta mutations in the helical domain of type I collagen: regions rich in lethal mutations align with collagen binding sites for integrins and proteoglycans. Hum Mutat. 2007;28:209–21.
3. Sillence DO, Senn A, Danks DM. Genetic heterogeneity in osteogenesis imperfecta. J Med Genet. 1979;16:101–16.
4. Willing MC, Pruchno CJ, Byers PH. Molecular heterogeneity in osteogenesis imperfecta type I. Am J Med Genet. 1993;45:223–7.
5. Cohn DH, Starman BJ, Blumberg B, et al. Recurrence of lethal osteogenesis imperfecta due to parental mosaicism for a dominant mutation in a human type I collagen gene (COL1A1). Am J Hum Genet. 1990;46:591–601.
6. Forlino A, Marini JC. Osteogenesis imperfecta. Lancet. 2016;387:1657–71.
7. Forlino A, Cabral WA, Barnes AM, Marini JC. New perspectives on osteogenesis imperfecta. Nat Rev Endocrinol. 2011;7:540–57.
8. Obafemi AA, Bulas DI, Troendle J, et al. Popcorn calcification in osteogenesis imperfecta: incidence, progression, and molecular correlation. Am J Med Genet Part A. 2008;146A:2725–32.
9. Marini JC, Hopkins E, Glorieux FH, et al. Positive linear growth and bone responses to growth hormone treatment in children with types III and IV osteogenesis imperfecta: high predictive value of the carboxyterminal propeptide of type I procollagen. J Bone Min Res. 2003;18:237–43.
10. Cabral WA, Makareeva E, Colige A, et al. Mutations near amino end of alpha1(I) collagen cause combined osteogenesis imperfecta/Ehlers-Danlos syndrome by interference with N-propeptide processing. J Biol Chem. 2005;280:19259–69.
11. Symoens S, Hulmes DJ, Bourhis JM, et al. Type I procollagen C-propeptide defects: study of genotype-phenotype correlation and predictive role of crystal structure. Hum Mutat. 2014;35:1330–41.
12. Lindahl K, Barnes AM, Fratzl-Zelman N, et al. COL1 C-propeptide cleavage site mutations cause high bone mass osteogenesis imperfecta. Hum Mutat. 2011;32:598–609.
13. Farber CR, Reich A, Barnes AM, et al. A novel IFITM5 mutation in severe atypical osteogenesis imperfecta type VI impairs osteoblast production of pigment epithelium-derived factor. J Bone Miner Res. 2014;29:1402–11.
14. Glorieux FH, Rauch F, Plotkin H, et al. Type V osteogenesis imperfecta: a new form of brittle bone disease. J Bone Miner Res. 2000;15:1650–8.
15. Barnes AM, Chang W, Morello R, et al. Deficiency of cartilage-associated protein in recessive lethal osteogenesis imperfecta. N Engl J Med. 2006;355:2757–64.
16. Cabral WA, Chang W, Barnes AM, et al. Prolyl 3-hydroxylase 1 deficiency causes a recessive metabolic bone disorder resembling lethal/severe osteogenesis imperfecta. Nat Genet. 2007;39:359–65.
17. Barnes AM, Carter EM, Cabral WA, et al. Lack of cyclophilin B in osteogenesis imperfecta with normal collagen folding. N Engl J Med. 2010;362:521–8.
18. Cabral WA, Barnes AM, Adeyemo A, et al. A founder mutation in LEPRE1 carried by 1.5% of West Africans and 0.4% of African Americans causes lethal recessive osteogenesis imperfecta. Genet Med. 2012;14:543–51.
19. Christiansen HE, Schwarze U, Pyott SM, et al. Homozygosity for a missense mutation in SERPINH1, which encodes the collagen chaperone protein HSP47, results in severe recessive osteogenesis imperfecta. Am J Hum Genet. 2010;86(3):389–98.
20. Alanay Y, Avaygan H, Camacho N, et al. Mutations in the gene encoding the RER protein FKBP65 cause autosomal-recessive osteogenesis imperfecta. Am J Hum Genet. 2010;86:551–9.
21. Barnes AM, Cabral WA, Weis M, et al. Absence of FKBP10 in recessive type XI osteogenesis imperfecta leads to diminished collagen cross-linking and reduced collagen deposition in extracellular matrix. Hum Mutat. 2012;33:1589–98.
22. Schwarze U, Cundy T, Pyott SM, et al. Mutations in FKBP10, which result in Bruck syndrome and recessive forms of osteogenesis imperfecta, inhibit the hydroxylation of telopeptide lysines in bone collagen. Hum Mol Genet. 2013;22:1–17.
23. Barnes AM, Duncan G, Weis M, et al. Kuskokwim syndrome, a recessive congenital contracture disorder, extends the phenotype of FKBP10 mutations. Hum Mutat. 2013;34:1279–88.
24. Martinez-Glez V, Valencia M, Caparros-Martin JA, et al. Identification of a mutation causing deficient BMP1/mTLD proteolytic activity in autosomal recessive osteogenesis imperfecta. Hum Mutat. 2012;33:343–50.
25. Cabral WA, Ishikawa M, Garten M, et al. Absence of the ER cation channel TMEM38B/TRIC-B disrupts intracellular calcium homeostasis and dysregulates collagen synthesis in recessive osteogenesis imperfecta. PLoS Genet 2016;12:e1006156.
26. Keupp K, Beleggia F, Kayserili H, et al. Mutations in WNT1 cause different forms of bone fragility. Am J Hum Genet. 2013;92:565–74.
27. Mendoza-Londono R, Fahiminiya S, Majewski J, et al. Recessive osteogenesis imperfecta caused by missense mutations in SPARC. Am J Hum Genet. 2015;96:979–85.
28. Symoens S, Malfait F, D'Hondt S, et al. Deficiency for the ER-stress transducer OASIS causes severe recessive osteogenesis imperfecta in humans. Orphanet J Rare Dis. 2013;8:154.
29. Lindert U, Cabral WA, Ausavarat S, et al. MBTPS2 mutations cause defective regulated intramembrane

30. Sato A, Ouellet J, Muneta T, et al. Scoliosis in osteogenesis imperfecta caused by COL1A1/COL1A2 mutations — genotype-phenotype correlations and effect of bisphosphonate treatment. Bone 2016;86:53–7.
31. Fratzl P, Paris O, Klaushofer K, et al. Bone mineralization in an osteogenesis imperfecta mouse model studied by small-angle x-ray scattering. J Clin Invest. 1996;97: 396–402.
32. Glorieux FH, Ward LM, Rauch F, et al. Osteogenesis imperfecta type VI: a form of brittle bone disease with a mineralization defect. J Bone Miner Res. 2002;17:30–8.
33. Rauch F, Husseini A, Roughley P, et al. Lack of circulating pigment epithelium-derived factor is a marker of osteogenesis imperfecta type VI. J Clin Endcrinol Metabol. 2012;97:E1550–6.
34. Fratzl-Zelman N, Barnes AM, Weis M, et al. Non-lethal type VIII osteogenesis imperfecta has elevated bone matrix mineralization. J Clin Endcrinol Metabol. 2016;101:3516–25.
35. Marini JC, Bordenick S, Heavner G, et al. The growth hormone and somatomedin axis in short children with osteogenesis imperfecta. J Clin Endcrinol Metabol. 1993;76:251–6.
36. Rauch F, Travers R, Parfitt AM, et al. Static and dynamic bone histomorphometry in children with osteogenesis imperfecta. Bone 2000;26:581–9.
37. Bardai G, Moffatt P, Glorieux FH, et al. DNA sequence analysis in 598 individuals with a clinical diagnosis of osteogenesis imperfecta: diagnostic yield and mutation spectrum. Osteoporos Int. 2016;27(12):3607–13.
38. Marini JC, Reich A, Smith SM. Osteogenesis imperfecta due to mutations in non-collagenous genes: lessons in the biology of bone formation. Curr Opin Pediatr 2014;26:500–7.
39. Bianchi L, Gagliardi A, Maruelli S, et al. Altered cytoskeletal organization characterized lethal but not surviving Brtl+/- mice: insight on phenotypic variability in osteogenesis imperfecta. Hum Mol Genet. 2015;24: 6118–33.
40. Cabral WA, Milgrom S, Letocha AD, et al. Biochemical screening of type I collagen in osteogenesis imperfecta: detection of glycine substitutions in the amino end of the alpha chains requires supplementation by molecular analysis. J Med Genet. 2006;43:685–90.
41. Pyott SM, Schwarze U, Christiansen HE, et al. Mutations in PPIB (cyclophilin B) delay type I procollagen chain association and result in perinatal lethal to moderate osteogenesis imperfecta phenotypes. Hum Mol Genet. 2011;20:1595–609.
42. Reich A, Bae AS, Barnes AM, et al. Type V OI primary osteoblasts display increased mineralization despite decreased COL1A1 expression. J Clin Endcrinol Metabol. 2015;100:E325–32.
43. Chang W, Barnes AM, Cabral WA, et al. Prolyl 3-hydroxylase 1 and CRTAP are mutually stabilizing in the endoplasmic reticulum collagen prolyl 3-hydroxylation complex. Hum Mol Genet. 2010;19:223–34.
44. Hudson DM, Kim LS, Weis M, et al. Peptidyl 3-hydroxyproline binding properties of type I collagen suggest a function in fibril supramolecular assembly. Biochemistry. 2012;51:2417–24.
45. Duran I, Nevarez L, Sarukhanov A, et al. HSP47 and FKBP65 cooperate in the synthesis of type I procollagen. Hum Mol Genet. 2015;24:1918–28.
46. Gjaltema RA, van der Stoel MM, Boersema M, et al. Disentangling mechanisms involved in collagen pyridinoline cross-linking: The immunophilin FKBP65 is critical for dimerization of lysyl hydroxylase 2. Proc Natl Acad U S A. 2016;113:7142–7.
47. Binder H, Conway A, Hason S, et al. Comprehensive rehabilitation of the child with osteogenesis imperfecta. Am J Med Genet. 1993;45:265–9.
48. Gerber LH, Binder H, Weintrob J, et al. Rehabilitation of children and infants with osteogenesis imperfecta. A program for ambulation. *Clin Orthop Rel Res.* 1990:254–62.
49. Zionts LE, Ebramzadeh E, Stott NS. Complications in the use of the Bailey-Dubow extensible nail. Clin Orthop Rel Res. 1998:186–95.
50. Fassier F. Experience with the Fassier-Duval rod: effectiveness and complications. *Abstract at: 9th International Conference on Osteogenesis imperfecta*, 2005, Annapolis, MD, USA.
51. Antoniazzi F, Bertoldo F, Mottes M, et al. Growth hormone treatment in osteogenesis imperfecta with quantitative defect of type I collagen synthesis. J Paediatr 1996;129:432–9.
52. Hald JD, Evangelou E, Langdahl BL, et al. Bisphosphonates for the prevention of fractures in osteogenesis imperfecta: meta-analysis of placebo-controlled trials. J Bone Min Res. 2015;30:929–33.
53. Land C, Rauch F, Munns CF, et al. Vertebral morphometry in children and adolescents with osteogenesis imperfecta: effect of intravenous pamidronate treatment. Bone. 2006;39:901–6.
54. Letocha AD, Cintas HL, Troendle JF, et al. Controlled trial of pamidronate in children with types III and IV osteogenesis imperfecta confirms vertebral gains but not short-term functional improvement. J Bone Min Res. 2005;20:977–86.
55. Papapoulos SE, Cremers SC. Prolonged bisphosphonate release after treatment in children. N Engl J Med. 2007;356:1075–6.

114

Fibrillinopathies: Skeletal Manifestations of Marfan Syndrome and Marfan-Related Conditions

Gary S. Gottesman[1] and Michael P. Whyte[1,2]

[1]*Center for Metabolic Bone Disease and Molecular Research, Shriners Hospital for Children, St Louis, MO, USA*
[2]*Division of Bone and Mineral Diseases, Department of Internal Medicine, Washington University School of Medicine at Barnes-Jewish Hospital, St Louis, MO, USA*

INTRODUCTION

The fibrillinopathies are connective tissue disorders that share a constellation of features yet differ by specific clinical and molecular findings. This chapter discusses the phenotype, molecular basis, pathogenesis, and treatment of the most prevalent types while emphasizing their skeletal manifestations.

Marfan syndrome

Marfan syndrome (MFS; OMIM #154700) affects at least 1 in 5000 people and is an autosomal dominant (AD) disorder caused by defects in fibrillin-1 (FBN1) leading to pleiotropic complications primarily involving the cardiovascular, musculoskeletal, integumentary, and visual systems. MFS nosology has been honed during the past three decades, last updated in 2010 (Table 114.1) [1–3] allowing MFS to be distinguished from more than 20 other phenotypically-related disorders (OMIM numbers specified in Table 114.2) including: (i) congenital contractural arachnodactyly; (ii) familial aortic aneurysm; (iii) mitral valve prolapse with aortic, skeletal, and skin manifestations; (iv) mitral valve prolapse syndrome; (v) isolated (familial) ectopia lentis; and (vi) recently described Marfan lipodystrophy syndrome (Table 114.2). Several additional disorders that resemble MFS phenotypically reflect altered protein interactions with FBN1 (eg, the Loeys-Dietz syndromes 1–5 with altered TGF-β signaling) [4, 5]. Still other FBN1 fibrillinopathies have distinctive clinical features (eg, short stature, round facies, craniosynostosis, acromelia, broad phalanges, decreased joint mobility, thick skin, hepatomegaly, neurological developmental delay, hoarse voice, muscle weakness, and neuropathy) (Table 114.3). Congenital contractural arachnodactyly (also known as distal arthrogryposis type 9) is a FBN2 fibrillinopathy that includes several features of MFS (eg, Marfanoid body habitus, dolichostenomelia, mitral valve prolapse, ectopia lentis), but also osteopenia and congenital kyphosis with decreased diaphyseal cortical thickness, sparse trabeculae, and with decreased mineralization of the axial compared with the appendicular skeleton, along with hypoplastic calf muscles, motor development delay, and a characteristic crumpled ear helix [6, 7] (Tables 114.2 and 114.3).

The newest diagnostic criteria for MFS require strict adherence and depend on specific anatomical, family history, and systemic score (SS) findings (Table 114.1) [3]. The SS reflects physical examination and clinical findings requiring at least 7 points (maximum of 20) to document systemic involvement. Errant diagnosis risks inappropriate medical intervention, restriction of physical activity compromising its health benefits, isolation, increased financial burden, and altered career or reproductive choices [3]. Characteristic skeletal features of MFS figure importantly in the SS and include wrist (1 point) and thumb (1 point) signs, with added import (3 points total) if both are present [3] (Fig. 114.1). The "wrist sign" (resulting from disproportionately long digits and diminished skeletal

---

Primer on the Metabolic Bone Diseases and Disorders of Mineral Metabolism, Ninth Edition. Edited by John P. Bilezikian.
© 2019 American Society for Bone and Mineral Research. Published 2019 by John Wiley & Sons, Inc.
Companion website: www.wiley.com/go/asbmrprimer

Table 114.1. Revised diagnostic criteria of Marfan syndrome (MFS).

I. In the *absence* of family history (FH)
1. AoD (Z ≥2) *and* EL: MFS
2. AoD (Z ≥2) *and* FBN1: MFS
3. AoD (Z ≥2) *and* systemic score (≥7): MFS[a]
4. EL *and* FBN1 with known AoD: MFS
   - EL with **or** without systemic score (≥7) *and* no *FBN1* **or** with AoD *and* *FBN1* unknown: EL syndrome
   - AoD (Z <2) *and* systemic score (≥5) with at least one skeletal feature without EL: MASS
   - MVP *and* AoD (Z <2) *and* systemic score (>5) without EL: MVPS

II. In the *presence* of family history (FH)[b]
5. EL *and* FH of MFS: MFS
6. Systemic score (≥7 points) *and* FH of MFS: MFS[a]
7. AoD (Z ≥2 above 20 years of age *or* ≥3 below 20 years of age) *and* FH of MFS = MFS[a]

III. Systemic score (SS) (score ≥7 points indicates systemic involvement; maximum = 20 points)
Axial skeleton:
- Facial features (any 3/5): **1 point** (dolichocephaly, enophtalmos, downslanting palpebral fissures, malar hypoplasia, retrognathia)
- Pectus carinatum deformity: **2 points**
- Pectus excavatum or chest asymmetry: **1 point**
- Scoliosis or thoracolumbar kyphosis: **1 point**
- Dural ectasia: **2 points**
- Protrusio acetabuli: **2 points**

Appendicular skeleton:
- Wrist *and* thumb sign: **3 points**; wrist *or* thumb sign: **1 point**
- Reduced elbow extension: **1 point**
- Hindfoot deformity: **2 points**
- Plain pes planus: **1 point**
- Reduced US/LS *and* increased arm/height *and* no severe scoliosis: **1 point**

Other Systems:
Visual:
- Myopia >3 diopters: **1 point**

Pulmonary:
- Pneumothorax: **2 points**

Cardiac:
- Mitral valve prolapse (all types): **1 point**

Integumentary:
- Skin striae: **1 point**

AoD = aortic diameter at the sinuses of Valsalva; EL = ectopia lentis; *FBN1* = fibrillin-1 mutation; MASS = myopia, mitral valve prolapse, aortic root dilation, skeletal findings, striae syndrome; MVPS = mitral valve prolapse syndrome; US/LS = upper segment/lower segment ratio; Z = Z-score.
[a]If discriminating features of Shprintzen–Golberg syndrome, Loeys–Dietz syndromes or vascular Ehlers-Danlos syndrome are present, then *TGFBR1/2*, *TGFB2/3*, *SMAD3*, *SKI* testing, collagen biochemistry, and *COL3A1* testing are indicated.
[b]When a family member was independently diagnosed using the above criteria.
Source: adapted from [3] and [9].

muscle and subcutaneous fat) [8] is positive if a thumb and fifth digit can encircle the opposite wrist with the thumb reaching over the fifth fingernail at least to the cuticle. A positive "thumb sign" (resulting from a long thumb and hypermobile hand) [8] occurs if the entire distal phalanx of the thumb can be folded beyond the ulnar border of the palm. Additional points are given if there is lowering of the midfoot (ankle pronation) together with significant hindfoot valgus and forefoot abduction (2 points), which distinguishes this criterion from a typical flat foot (1 point), and congenital elbow contractures causing reduced extension (≤170°) (1 point). Axial skeleton criteria include pectus deformities (from longitudinal rib overgrowth) [9], where pectus carinatum (2 points) outweighs pectus excavatum (1 point), and scoliosis or thoracolumbar kyphosis (1 point). Protrusio acetabuli (radiographically evident medial protrusion at least 3 mm beyond the ilio-ischial line) is 2 points. Dural ectasia (2 points) is expansion of the meningeal sac perhaps from weakening of the most dependent section of the dura from pressures generated by CSF. It may widen the neural canal, thin the vertebral bodies and posterior elements, dilate neural foramina, and

Table 114.2. Marfan syndrome and phenotypically-related disorders.

Disorder	Inheritance	OMIM	Locus	Mutant Gene
Marfan syndrome (MFS)	AD	154700	15q21.1	FBN1
Acromicric dysplasia (ACMICD)	AD	102370	15q21.1	FBN1
Shprintzen-Goldberg craniosynostosis syndrome (SGS)	AD	182212	15q21.1	FBN1
Stiff skin syndrome (SSKS)	AD	184900	15q21.1	FBN1
Weill-Marchesani syndrome 2 (WMS2)	AD	608328	15q21.1	FBN1
Geleophysic dysplasia 2 (GPHYSD2)	AD	614185	15q21.1	FBN1
Marfan lipodystrophy syndrome (MFLS)	AD	616914	15q21.1	FBN1
Arthrogryposis, distal, type 9 (DA9)	AD	121050	5q23-q31	FBN2
Geleophysic dysplasia 1 (GPHYSD1)	AR	231050	9q34.2	ADAMTSL2
Weill–Marchesani syndrome 1 (WMS1)	AR	277600	19p13.2	ADAMTS10
Weill–Marchesani-Like syndrome (WMLS)	AR	613195	15q26.3	ADAMTS17
Weill–Marchesani syndrome 3 (WMS3)	AR	614819	14q24.3	LTBP2
Dental anomalies and short stature (DASS)	AR	601216	11q13.1	LTBP3
Cutis Laxa, AR, Type IC (ARCL1C)	AR	613177	19q13.1-q13.2	LTBP4
Shprintzen-Goldberg craniosynostosis syndrome (SGS)	AD	182212	1p36.33-1p36.32	SKI
Camurati–Engelmann disease (CED)	AD	131300	19q13.2	TGFB1
Loeys–Dietz syndrome 1 (LDS1)	AD	609192	9q22.33	TGFBR1
Loeys–Dietz syndrome 2 (LDS2)	AD	610168	3p24.1	TGFBR2
Loeys–Dietz syndrome 3 (LDS3)	AD	613795	15q22.33	SMAD3
Loeys–Dietz syndrome 4 (LDS4)	AD	614816	1q41	TGFB2
Loeys–Dietz syndrome 5 (LDS5)	AD	610380	3p22	TGFB3

AD = autosomal dominant; AR = autosomal recessive.
Arthrogryposis, Distal, Type 9 = Congenital contractural arachnodactyly = Beals syndrome.
Cutis Laxa, AR, Type IC, ARCL1C = Urban-Davis-Rifkin syndrome (URDS).
Dental anomalies and short stature (DASS) = tooth agenesis syndrome (STHAG).
OMIM number reassignments:
Loeys-Dietz 1B 610168 moved to LDS2 610168.
Loeys-Dietz 2A 608967 moved to LDS1 609192.
Loeys-Dietz 2B 610380 moved to LDS2 610168.
Tooth Agenesis syndrome 613097 moved to DASS 601216.
Source: [9]. Reproduced with permission of John Wiley & Sons.

protrude the dura beyond the neural canal [10]. Dural ectasia, best detected by CT or MRI but sometimes evident radiographically, distinguishes MFS from all other fibrillinopathies and Marfan-related disorders except TGF-β-related disorders, the Loeys-Dietz syndromes, and vascular Ehlers-Danlos syndrome (COL3A1) [3]. Symptoms include headache, low back and proximal lower extremity pain, and lower extremity radicular findings of numbness and weakness [10, 11] with an incidence as high as 90% [10–15]. Lastly, axial/appendicular skeletal disproportion (ie, reduced upper-segment-to-lower-segment ratio of <0.85 for white adults, <0.78 for black adults, Asian adult data not available) coupled with an increased ratio of arm span to height without severe scoliosis (>1.05 for adults) contributes 1 point to the SS [3].

Also contributing to the SS are facial features: dolichocephaly, enophthalmos, downslanting palpebral fissures, malar hypoplasia, and retrognathia likely resulting from changes in the cranial, orbital, maxillary and mandibular components of the skull [16, 17]. Three of these five common facial features are required to add 1 point to the SS [3]. In addition, the cardiorespiratory findings of pneumothorax and mitral valve prolapse add 2 and 1 point(s), respectively. Significant myopia (ie, >3 diopters) adds 1 point. Skin striae add 1 point [3].

Definitive diagnosis of MFS requires aortic dilatation at the sinuses of Valsalva (diameter Z-score ≥2) or past aortic dissection (Table 114.1). Ectopia lentis is another crucial finding (with or without aortic involvement). FBN1 mutation analysis is now an integral part of MFS diagnosis (Table 114.1) [3].

Any association of MFS and osteopenia or low BMD has been difficult to assess, especially in children. Conventional radiographs are typically said to show "osteopenia" [18]. However, limitations of the areal measurements of DXA, rather than volumetric BMD (g/cm^2, not g/cm^3), compromise this assessment. Pediatric bones change size and shape during growth, posing challenges for DXA [19, 20]. Tall and short subjects artifactually have more or less dense bones, respectively by DXA. Newer DXA studies in MFS adjust for height, weight, age, gender and ethnicity, and have shown significantly lower whole body and lumbar spine BMC and BMD [21, 22]. One MFS DXA study correcting BMD for

Table 114.3. Skeletal features of Marfan syndrome and related disorders.

	MFS	MFLS	SGS	AD	GD	WMS	DA9	LDS
Axial skeletal abnormalities								
Kyphosis	++	+	++	−	−	−	+++	++
Pectus deformity	+++	++	+++	−	−	−	++	++
Scoliosis	+++	++	++	++	−	−	+++	+++
Dural ectasia	+++	++	−	−	−	−	−	+++
Appendicular skeletal abnormalities								
Arachnodactyly	+++	+++	+++	−	−	−	+++	+++
Brachydactyly	−	−	−	+++	+++	+++	−	−
Camptodactyly	+	+	++	−	−	++	+++	++
Club foot deformity	−	−	+	−	−	−	−	+++
Dolichostenomelia	+++	+++	+++	−	−	−	++	+
Joint hypermobility	++	++	++	−	−	−	+/−	+++
Joint contracture	+/−	++	−	+	+++	−	++	−
Pes planus	++	++	−	−	−	−	++	++
Protusio acetabuli	+++	+/−	−	−	−	−	+	+
Short stature	−	−	−	+++	+++	+++	−	−
Low bone mineral density	++	−	+	−	−	+/−	++	+/−
Craniofacial abnormalities								
Macrocephaly	−	++	−	−	−	−	−	−
High-arched palate	+++	+++	+++	−	−	−	++	+++
Bifid uvula	−	−	+	−	−	−	−	←→
Cleft palate	−	−	+	−	−	−	+/−	←→
Craniosynostosis	−	+	++	−	−	−	+/−	←→
Dolichocephaly	+++	+++	−	−	−	−	−	+
Down-slant of palpebral fissures	++	+++	+++	−	+	−	−	++
Enophthalmos	++	−	−	−	+/−	−	−	+
Exophthalmos	−	+++	+++	−	−	−	−	−
Hypertelorism	−	−	+++	−	++	−	+/−	←→
Long upper lip	−	−	−	−	+++	−	−	−
Malar hypoplasia	+++	++	++	−	++	+	−	+++
Micrognathia/retrognathia	++	+++	+++	−	+	−	+	+++

− = not associated; +/− = rare or subtle; + = occasionally observed; ++ = commonly observed; +++ = generally observed; ←→ = phenotypic spectrum varies across genotypes.
AD = acromicric dysplasia; DA9 = distal arthrogryposis type 9 (also known as congenital contractural arachnodactyly); LDS = Loeys–Dietz syndromes 1–5.
Source: [9].

height–age (BMD$_{HAZ}$), showed significantly diminished BMD$_{HAZ}$ at the lumbar spine and femur. Follow-up showed BMD$_{HAZ}$ was essentially stable in the spine, increased modestly in the femoral neck, and decreased significantly in the total femur [22]. Those aged under 10 years had lower femoral neck BMD$_{HAZ}$ (−1.41), whereas those older had markedly lower total femur BMD$_{HAZ}$ (−1.94). Why was not clear [22, 23]. In 2015, children, adolescents, and young adults with MFS were compared with healthy controls. In children with MFS, axial and appendicular bone and muscle mass were diminished and decreased into adulthood culminating in compromised peak bone mass [23]. However, the fracture incidence in adult MFS is not well understood. Diminished muscle mass may have been more evident had they used height-matched controls [23]. It is a possibility that less physical activity from concerns about aortic dilatation negatively impacted muscle accrual [23]. Alternatively, the decreased muscle mass may maintain mechanical strain on the bone at a constant set point. Patients with premature termination codons in *FBN1* likely causing haplo-insufficiency had reduced bone mass compared with those with in-frame mutations predicted to have a dominant/negative effect [23].

Congenital contractural arachnodactyly

Congenital contractural arachnodactyly (CCA) is a rare AD disorder delineated from MFS by crumpled ear helices, multiple congenital flexion contractures, and typically absence of dolichocephaly, myopia, ectopia lentis, and aortic enlargement [24, 25]. Common to both disorders are arachnodactyly, dolichostenomelia, scoliosis, progressive kyphoscoliosis (sometimes present at birth in CCA), and decreased muscle mass. Diminished bone mass is considered more common in CCA [26] (Table 114.3). Contractures may resolve, but camptodactyly persists [26]. Lifespan is reportedly normal unless significant cardiac or spinal deformity is present. In severe neonatal cases, the clinical phenotypes of CCA and MFS are similar, showing why molecular diagnosis is essential [27].

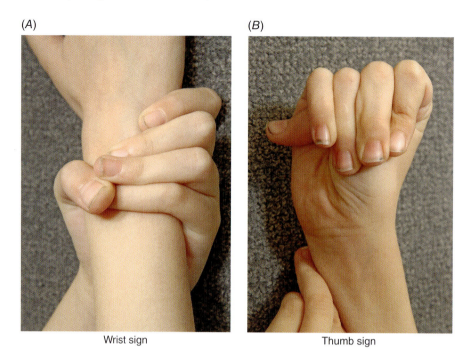

Fig. 114.1. Characteristic skeletal features of Marfan syndrome included in the systemic score are the (A) wrist sign (1 point) and the (B) thumb sign (1 point).

Marfan syndrome-related disorders

The various clinical phenotypes associated with FBN1 fibrillinopathies seem to reflect mutations in specific domains of *FBN1* [28]. Shprintzen–Golberg syndrome (SGS) is a rare AD condition with skeletal features that overlap with MFS and Loeys–Dietz syndrome (Table 114.3). Craniofacial abnormalities (tall or prominent forehead, proptosis, posteriorly rotated ears, and craniosynostosis) and neurodevelopmental delay with mild to moderate intellectual disability are hallmarks of SGS. Recent reports link most SGS patients to SKI (a proto-oncoprotein that negatively regulates SMAD-dependent TGF-β signaling) [29, 30]. Two SGS patients had *FBN1* mutations in a specific *FBN1* EGF-like domain. Both SKI and FBN1 regulate TGF-β function and their overlapping phenotypes may result from specific protein alterations that impact TGF-β signaling similarly [29, 31, 32].

Marfan lipodystrophy syndrome (MLS), the newest FBN1 fibrillinopathy, features premature birth, congenital lipodystrophy, accelerated linear growth that outstrips weight gain, and a progeroid facies together with MFS features that meet the newest criteria for MFS diagnosis [33, 34]. The skeletal phenotype can be severe featuring craniofacial abnormalities, kyphosis, scoliosis, protrusio acetabuli, and dural ectasia, but aortic arch dilatation and/or ectopia lentis are reported in only ~50% of cases [34].

Still other FBN1 fibrillinopathies have distinctive clinical features (Table 114.3). Weill–Marchesani syndrome shows AD short stature, brachydactyly, camptodactyly, early and severe myopia, and ectopia lentis. Acromicric dysplasia and geleophysic dyplasia 2 (GD2) feature brachydactly and short stature. GD2 is further characterized by facial abnormalities and cardiac involvement. Stiff skin syndrome (SSKS) presents with generalized thick, indurated skin that may limit joint mobility and cause contractures, without any skeletal features of MFS. The mutations in *FBN1* causing SSKS involve an RGD motif, a putative integrin binding site. Each of these allelic disorders results from mutations in *FBN1* domains predicted to manifest near each other within the encoded microfibril, suggesting they alter integrin binding and microfibril function [28, 35].

Added to the FBN1 and FBN2 fibrillinopathies, several disorders significantly overlap the MFS phenotype and mimic the FBN1 Marfan-related disorders. The Loeys–Dietz syndromes (LDS 1–5) are genetically heterogeneous yet share a triad of clinical features: arterial tortuosity and aneurysms, hypertelorism, and a bifid uvula or cleft palate. Craniosynostosis is also common. LDS can cause immunological-related disorders with an increased prevalence of food allergies, asthma, rhinitis, and eczema [4].

Mutations in any of five genes, *TGFBR1* or *2*, *TGFB2* or *3*, and *SMAD3* explain LDS, resulting from altered TGF-β signaling [4, 5, 36]. Specific clinical criteria seem wanting, and the diagnosis relies primarily upon molecular testing [4]. The most concerning feature is rapidly progressive aortic aneurysmal disease potentially at younger ages and smaller aortic dimensions than in MFS.

Homocystinuria is an inborn error of metabolism that deserves mention because of phenotypic similarities with MFS. This autosomal recessive (AR) disorder is caused by mutations in the cystathionine β-synthase gene (*CBS*) resulting in deficient enzyme activity.

Untreated homocystinuria causes myopia, ectopia lentis, mental retardation, skeletal changes (excessive height and extremity length), and risks thrombosis and thromboembolic events. Approximately 50% of patients respond to pharmacological doses of vitamin B_6, stressing the importance of distinguishing homocystinuria from MFS [37].

PATHOGENESIS

The pathogenesis of MFS is complex. MFS results from alterations in the structure or expression of FBN1, a major structural element of extracellular matrix 10-nM microfibrils [8, 9, 28] encoded by *FBN1* localized at chromosome 15q21.1 [8, 37, 38]. FBN1 has multiple functional domains with intact, repetitive cysteine-rich motifs. Microfibril assembly depends on the length of the fibrillin monomer [8]. Fibrillin-containing microfibrils, as a component of elastic fibers or elastin-free structural components, help organize various tissues. FBN1 also modifies cellular behavior by interacting with integrin receptors as well as by complexing with TGF-β and BMP [9]. MFS and related phenotypes represent more than 1000 FBN1 pathogenic variants [37]. FBN1 is incorporated into large microfibrillar structures found in both elastic and nonelastic tissues and acts in: (i) the formation and homeostasis of the elastic matrix; (ii) matrix-cell attachments; and (iii) may regulate select growth factors. MFS animal models have elucidated microfibrillar modulation, matrix sequestration, and activation of TGF-β [37]. TGF-β signaling is elevated in FBN1-deficient mice and medical therapy with the angiotensin 1 receptor blocker, losartan, or a TGF-β neutralizing antibody, mitigated or prevented changes in the lung, mitral valve, skeletal muscle, and vasculature. Noncanonical TGF-β signaling (Smad-independent), through mitogen-activated protein kinase cascades, may be another mechanism driving development of aortic aneurysm and dissection [39], and may be amenable to pharmacological intervention [39]. The relevance of this signaling to additional manifestations of MFS and matrix-degrading enzymes in MFS aortic disease is being studied [37].

Mutated FBN1 is believed to dominant-negatively interfere with the normal protein product. Haploinsufficiency of *FBN1* has also been implicated pathogenetically because some patients harbor premature termination codons as their only *FBN1* mutation [37].

MEDICAL MANAGEMENT AND TREATMENT

The pleiotropic nature of MFS and the MFS-related disorders necessitates multidisciplinary management including regular surveillance. Identifying a primary coordinating physician for the multiple specialists is crucial for success. Often a medical geneticist, pediatrician, internist, or cardiologist assumes this lead role. Others should include an ophthalmologist, orthopedist, and cardiothoracic surgeon. Vision problems can usually be managed with eyeglasses. Surgical substitution of a prosthetic lens is rarely needed, but best performed after growth has ceased. Scoliosis bracing in MFS is of limited utility if curvatures exceed 25°, but surgical intervention risks complications from the underlying disorder itself [40, 41]. Spondylolisthesis also occurs in MFS. Simple interventions like supportive orthotics or wedge inserts in shoes can reduce fatigue and muscle discomfort from pes planus. Pectus deformity is rarely severe enough to warrant surgical repair, but evidence of cardiopulmonary compromise deserves consultation with a cardiothoracic surgeon.

Long-term survival in MFS patients has improved dramatically during the past four decades, primarily from new surgical techniques to prevent aortic dissection. Surgical intervention for aortic enlargement is recommended when the aortic root approaches a diameter of 5 cm in older children and adults, but may be warranted earlier if enlargement is progressing rapidly (0.5 to 1.0 cm yearly) [42]. Additionally, aortic valve dysfunction with progressive and severe aortic regurgitation may require repair before the 5 cm threshold. A family history of aortic dissection intensifies surveillance for aortic root enlargement. Severe mitral valve involvement is an indication for early intervention in children [42].

Pharmacological therapy may improve cardiovascular function for congestive heart failure. Prevention of vascular complications includes therapy to decrease aortic wall injury from hemodynamic stress. Beta-blockers are often recommended at diagnosis and for progressive aortic dilatation [37]. Other antihypertensives can be used if beta blockers are not tolerated; however, their efficacy and safety in MFS are under investigation [37]. Antibiotic prophylaxis is required for dental work to prevent subacute bacterial endocarditis [37].

Routine surveillance of MFS patients should include annual ophthalmologic examination and echocardiography to monitor the ascending aorta when aortic root enlargement is slight or slowly increasing. More frequent noninvasive evaluations are indicated when: (i) aortic root diameter approaches 4.5 cm in adults; (ii) aortic dilation rate exceeds ~0.5 cm yearly; and (iii) significant aortic regurgitation is present. Surveillance of the entire aorta with CT or MR angiography is recommended beginning in young adulthood [37].

Contact or competitive sports or isometric exercise should be avoided to reduce stress on the cardiovascular system and modified to avoid trauma, joint injury, and pain. Medicinals that stimulate the cardiovascular system (decongestants and caffeine) or cause vasoconstriction (including triptans) are contraindicated. Additionally, LASIK correction of refractive errors is not recommended because refractive corneal surgery may cause transient elevation of intraocular pressure leading to iatrogenic myopia [37, 43]. For those at risk of pneumothorax, breathing against resistance or positive pressure ventilation should be avoided. To prevent devastating cardiovascular events, relatives of MFS patients suspected of having the disorder should undergo echocardiography. If clinical findings are subtle in the index case, all close relatives should be evaluated with echocardiography. Pregnant women with MFS warrant intensive surveillance by a high-risk obstetrician including through the immediate postpartum period [37].

Ongoing investigations of medical therapy of MFS with angiotensin receptor blockers like losartan continue in children and adults (see clinicaltrials.gov) after animal models showed dramatic protection against aortic enlargement [39]. Multiple prospective trials have shown that beta blockers combined with losartan provides better protection against aortic root enlargement in both children and adults with MFS than either pharmaceutical alone. However, interventions that neutralize TGF-β may have a detrimental effect on aortic growth (found in a murine model) [7]. New medical therapies will hopefully correct the pathogenic effects of excessive TGF-β signaling on the cardiovascular and musculoskeletal systems while preserving TGF-β activity during growth and development of the aorta.

Medical surveillance and therapy for the MFS-related conditions are not as well codified, but are similar in scope because of the pleotropic nature of these diseases.

REFERENCES

1. Beighton P, de Paepe A, Danks D, et al. International nosology of heritable disorders of connective tissue, Berlin, 1986. Am J Med Genet. 1988;29:581–94.
2. De Paepe A, Devereux RB, Dietz HC, et al. Revised diagnostic criteria for the Marfan syndrome. Am J Med Genet. 1996;62:417–26.
3. Loeys BL, Dietz HC, Braverman AC, et al. The revised Ghent nosology for the Marfan syndrome. J Med Genet. 2010;47:476–85.
4. MacCarrick G, Black JH, Bowdin S, et al. Loeys-Dietz syndrome: a primer for diagnosis and management. Genet Med. 2014;16:576–87.
5. Bertoli-Avella AM, Gillis E, Morisaki H, et al. Mutations in a TGF-β ligand, TGFB3, cause syndromic aortic aneurysms and dissections. J Am Coll Cardiol. 2015;65:1324–36.
6. Epstein CJ, Graham CB, Hodgkin WE, et al. Hereditary dysplasia of bone with kyphoscoliosis, contractures, and abnormally shaped ears. J Pediatr. 1968;73(3):379–86.
7. Smaldone S, Ramirez F. Fibrillin microfibrils in bone physiology. Matrix Biol. 2016;52–54:191–7.
8. Dietz HC, Pyeritz RE. Marfan syndrome and related disorders. In: Valle D, Beaudet AL, Vogelstein B, et al. (eds) *The Online Metabolic and Molecular Bases of Inherited Disease*. New York: McGraw-Hill, 2014.
9. Arteaga-Solis E, Ramirez F. Skeletal manifestations of Marfan syndrome and related disorders of the connective tissue. In: Rosen JR (ed.) *Primer on the Metabolic Bone Disease and Disorders of Mineral Metabolism* (8th ed.). Hoboken: John Wiley & Sons, Inc., 2013.
10. Pyeritz RE, Fishman EK, Bernhardt BA, et al. Dural ectasia is a common feature of the Marfan syndrome Am J Hum Genet. 1988;43:726–32.
11. Foran JR, Pyeritz RE, Dietz HC, et al. Characterization of the symptoms associated with dural ectasia in the Marfan patient. Am J Med Genet A. 2005;134:58–65.
12. Fattori R, Nienaber CA, Descovich B, et al. Importance of dural ectasia in phenotypic assessment of Marfan's syndrome. Lancet. 1999;354:910–13.
13. Ahn NU, Sponseller PD, Ahn UM, et al. Dural ectasia in the Marfan syndrome: MR and CT findings and criteria. Genet Med. 2000;2:173–9.
14. Ahn NU, Nallamshetty L, Ahn UM, et al. Dural ectasia and conventional radiography in Marfan lumbosacral spine. Skeletal Radiol. 2001;30:338–45.
15. Habermann CR, Weiss F, Schoder V, et al. MR evaluation of dural ectasia in Marfan syndrome: reassessment of the established criteria in children, adolescents, and young adults. Radiology. 2005;234:535–41.
16. Baran S, Ignys A, Ignys I. Respiratory dysfunction in patients with Marfan syndrome. J Physiol Pharmacol. 2007;58(Suppl 5):37–41.
17. De Coster P, De Pauw G, Martens L, et al. Craniofacial structure in Marfan syndrome: a cephalometric study. Am J Med Genet A. 2004;131(3):240–8.
18. Magid D, Pyeritz RE, Fishman EK. Musculoskeletal manifestations of the Marfan syndrome: radiologic features. Am J Roentgenol. 1990;155(1):99–104.
19. Binkovitz LA, Henwood MJ. Pediatric DXA: technique and interpretation. Pediatr Radiol. 2007;37:21–31.
20. Zhang F, Whyte MP, Wenkert D. Dual-energy X-ray absorptiometry interpretation: a simple equation for height correction in preteenage children. J Clin Densitom. 2012;15:267–74.
21. Grover M, Brunetti-Pierri N, Belmont J, et al. Assessment of bone mineral status in children with Marfan syndrome. Am J Med Genet A. 2012;158A:2221–4.
22. Trifiro G, Marelli S, Viecca M, et al. Areal bone mineral density in children and adolescents with Marfan syndrome: evidence of an evolving problem. Bone. 2015;73:176–80.
23. Haine E, Salles J-P, Khau Van Kien K, et al. Muscle and bone impairment in children with Marfan syndrome: correlation with age and FBN genotype. J Bone Miner Res. 2015; 30:1369–76.
24. Hecht F, Beals RK. 'New' syndrome of congenital contractural arachnodactyly originally described by Marfan in 1896. Pediatrics. 1972;49:574–9.
25. Putnam EA, Zhang H, Ramirez F, et al. Fibrillin-2 (FBN2) mutations result in the Marfan-like disorder, congenital contractural arachnodactyly. Nat Genet. 1995;11:456–8.
26. Tunçbilek E, Alanay Y. Congenital contractural arachnodactyly (Beals syndrome). Orphanet J Rare Dis. 2006;1:20–2.
27. Woolnough R, Dhawan A, Dow K, et al. Are patients with Loeys-Dietz syndrome misdiagnosed with Beals syndrome? Pediatrics. 2017;139(3):e20161281.

28. Sakai LY, Keene DR, Renard M, et al. FBN1: the disease-causing gene for Marfan syndrome and other genetic disorders. Gene 2016;591:279–91.
29. Doyle AJ, Doyle JJ, Bessling SL, et al. Mutations in the TGF-β repressor SKI cause Shprintzen-Goldberg syndrome with aortic aneurysm. Nat Genet. 2012;44:1249–54.
30. Greally MT. Shprintzen-Goldberg syndrome, January 13, 2006 (updated June 13, 2013). In: Pagon RA, Adam MP, Ardinger HH, et al. (eds) GeneReviews.® Seattle: University of Washington, 1993–2017. Available from: https://www.ncbi.nlm.nih.gov/books/NBK1277/. Accessed June 12, 2018.
31. Sood S, Eldadah ZA, Krause WL, et al. Mutation in fibrillin-1 and the Marfanoid-craniosynostosis (Shprintzen-Goldberg) syndrome. Nat Genet. 1996;12:209–11.
32. Kosaki K, Takahashi D, Udaka T, et al. Molecular pathology of Shprintzen-Goldberg syndrome. Am J Hum Genet A. 2006;140A:104–8.
33. Graul-Neumann LM, Kienitz T, Robinson PN, et al. Marfan syndrome with neonatal progeroid syndrome-like lipodystrophy associated with a novel frameshift mutation at the 3′ terminus of the FBN1-gene. Am J Med Genet A. 2010;152A:2749–55.
34. Garg A, Xing C. De novo heterozygous FBN1 mutations in the extreme C-terminal region cause progeroid fibrillinopathy. Am J Med Genet A. 2014;164A:1341–5.
35. Loeys BL, Gerber EE, Riegert-Johnson D, et al. Mutations in fibrillin-1 cause congenital scleroderma: stiff skin syndrome. Sci Transl Med. 2010;2:23ra20.
36. Verstraeten A, Alaerts M, Van Laer L, et al. Marfan syndrome and related disorders: 25 years of gene discovery. Hum Mutat. 2016;37:524–31.
37. Dietz HC. Marfan syndrome. 18 April, 2001 (updated February 2, 2017). In: Pagon RA, Adam MP, Ardinger HH, et al. (eds) GeneReviews.® Seattle: University of Washington, 1993–2017. Available from: https://www.ncbi.nlm.nih.gov/books/NBK1277/. Accessed June 12, 2018.
38. Sakai LY, Keene DR, Engvall E. Fibrillin, a new 350-kD glycoprotein, is a component of extracellular microfibrils. J Cell Biol. 1986;103:2499.
39. Holm TM, Habashi JP, Doyle JJ, et al. Noncanonical TGFβ signaling contributes to aortic aneurysm progression in Marfan syndrome mice. Science. 2011;332:358–61.
40. Sponseller PD, Bhimani M, Solacoff D, et al. Results of brace treatment of scoliosis in Marfan syndrome. Spine. 2000;25:2350–4.
41. Jones KB, Erkula G, Sponseller PD, et al. Spine deformity correction in Marfan syndrome. Spine 2002;27:2003–12.
42. Miyahara S, Okita Y. Overview of current surgical strategies for aortic disease in patients with Marfan syndrome. Surg Today. 2016;46:1006–18.
43. Alcorn D, Milewicz D, Maumenee, IH. Ocular management in Marfan syndrome. https://www.marfan.org/resource/fact-sheet/overview-ocular-management-marfan-syndrome#.WOOwQE_fPIU. Accessed June 12, 2018.
44. Fuhrhop SK, McElroy MJ, Dietz HC 3rd, et al. High prevalence of cervical deformity and instability requires surveillance in Loeys-Dietz syndrome. J Bone Joint Surg Am. 2015;97:411–19.
45. Watanabe K, Okada E, Kosaki K, et al. Surgical treatment for scoliosis in patients with Shprintzen-Goldberg syndrome. J Pediatr Orthop. 2011;31:186–93.
46. Goldblatt J, Hyatt J, Edwards C, et al. Further evidence for a marfanoid syndrome with neonatal progeroid feature and severe generalized lipodystrophy due to frameshift mutations near the 30 end of the FBN1 gene. Am J Med Genet A. 2011;155:717–20.
47. Horn D, Robinson PN. Progeroid facial features and lipodystrophy associated with a novel splice site mutation in the final intron of the FBN1 gene. Am J Med Genet A. 2011;155:721–4.
48. Takenouchi T, Hida M, Sakamoto Y, et al. Severe congenital lipodystrophy and a progeroid appearance: mutation in the penultimate exon of FBN1 causing a recognizable phenotype. Am J Med Genet A. 2013;161:3057–62.

115

Hypophosphatasia and Other Enzyme Deficiencies Affecting the Skeleton

Michael P. Whyte

Center for Metabolic Bone Disease and Molecular Research, Shriners Hospital for Children; and Division of Bone and Mineral Diseases, Department of Internal Medicine, Washington University School of Medicine at Barnes-Jewish Hospital, St Louis, MO, USA

INTRODUCTION

Inborn errors of metabolism featuring enzyme deficiencies can importantly compromise the skeleton [1]. Five such entities are reviewed here.

HYPOPHOSPHATASIA

Hypophosphatasia (HPP) is the rare heritable rickets or osteomalacia (OMIM #146300, #241500, #241510) [1] characterized biochemically by subnormal activity of the tissue-nonspecific (bone/liver) isoenzyme of alkaline phosphatase (TNSALP) [2, 3]. Although in health some TNSALP seems present in all tissues [4], HPP disturbs mainly the skeleton and teeth but with the greatest range of severity of all skeletal diseases [5]. Many case and several family reports have established its clinical, radiological, biochemical, and histopathological features, and a nosology of six principal forms reflects patient age when skeletal disease or other significant complications present: perinatal, infantile, childhood (mild or severe), adult, and odonto HPP [6]. Generally, the earlier the presentation the more severe the clinical course [2, 3].

Perinatal HPP manifests in utero [7–9]. At birth, extreme skeletal hypomineralization features caput membraneceum and short deformed limbs. Respiratory compromise proves rapidly fatal. Survival is very rare [7–9]. The radiographic features are pathognomonic [2]. Sometimes the skeleton is so poorly calcified that merely a few bones can be found [10], or the calvarium is ossified only centrally, segments of the spinal column appear missing, and severe rachitic changes are present in the limbs [2, 3]. Importantly, patients with a "benign prenatal" form of HPP manifest bowing in utero that corrects postnatally and then ranges broadly from infantile to odonto HPP (see later) [9].

Infantile HPP presents before 6 months of age [3, 10, 11]. Development appears normal until rachitic deformities, poor feeding, inadequate weight gain, hypotonia, and wide fontanels are noted. Vitamin B_6-dependent seizures predict a lethal course [12]. Hypercalcemia can cause recurrent vomiting, and hypercalciuria sometimes nephrocalcinosis and renal compromise [10, 13]. A hypomineralized calvarium may give the illusion of widely open fontanels, yet there is functional craniosynostosis. A flail chest predisposes to pneumonia [8]. The radiographic changes are pathognomonic, but less striking than in perinatal HPP [2]. Abrupt transition from relatively normal diaphyses to hypomineralized metaphyses can suggest sudden metabolic deterioration. Spontaneous improvement or progressive skeletal deterioration are possible [10], with ~50% of these babies dying in infancy [7, 8, 10]. Progressive skeletal demineralization with fractures and advancing thoracic deformity herald a fatal outcome [7, 10, 11].

Childhood HPP (mild and severe forms) [6] causes premature loss of deciduous teeth (before 5 years of age), without root resorption, explained by hypomineralization of dental cementum [14]. Lower incisors are typically lost first, but all teeth can be affected. Permanent teeth fare better. Short stature, dolichocephaly, muscle weakness, and delayed walking with a waddling gait

Primer on the Metabolic Bone Diseases and Disorders of Mineral Metabolism, Ninth Edition. Edited by John P. Bilezikian.
© 2019 American Society for Bone and Mineral Research. Published 2019 by John Wiley & Sons, Inc.
Companion website: www.wiley.com/go/asbmrprimer

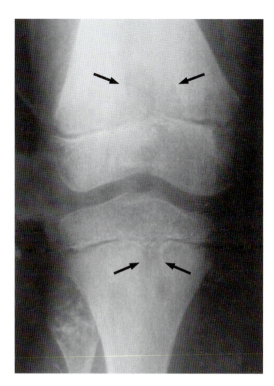

Fig. 115.1. The metaphysis of the proximal tibia of this 10-year-old boy with mild childhood hypophosphatasia shows a subtle but characteristic "tongue" of radiolucency (arrows). Note, however, that his rickets does not manifest with widening of the growth plate.

represent typical complications [15]. The clinical course during childhood is unusually unchanging [16]. However, after puberty patients can seem improved, but skeletal complications sometimes recur in middle age [2, 3, 17]. Radiographs show characteristic "tongues" of lucency projecting from growth plates into metaphyses (Fig. 115.1) [2]. True premature fusion of cranial sutures with craniosynostosis can lead to a "beaten-copper" appearance of the skull [2, 3].

Adult HPP usually presents during middle age with recurrent, poorly healing, metatarsal stress fractures [17]. Sometimes patients recount rickets and/or premature loss of deciduous teeth during childhood [2, 3, 17]. Subsequently, discomfort in the hips or thighs may indicate femoral pseudofractures [17–20]. Chondrocalcinosis from calcium pyrophosphate dihydrate crystal deposition or pyrophosphate arthropathy including pseudogout may occur [21]. Radiographs can also show osteopenia [2, 3, 17–20].

Odonto HPP represents dental manifestations alone [2, 3].

HPP is a remarkable rickets/osteomalacia because circulating calcium (Ca) or phosphorus (P) levels are not low, and serum ALP activity is low not high [2, 3]. In fact, hypercalcemia occurs frequently in infantile HPP [10, 11, 13] caused by dyssynergy between gut absorption of Ca and the defective skeletal growth and mineralization [2, 3]. If there is hypercalcemia, serum parathyroid hormone and $1,25(OH)_2D$ concentrations can be suppressed. In the childhood and adult forms of HPP, many patients are hyperphosphatemic caused by enhanced renal reclamation of P (increased TmP/GFR) [2, 3]. Nondecalcified HPP bone shows rickets or osteomalacia without secondary hyperparathyroidism [22].

Three TNSALP phosphocompound substrates accumulate extracellularly in HPP [2, 3]: phosphoethanolamine (PEA), inorganic pyrophosphate (PPi), and pyridoxal 5′-phosphate (PLP). If vitamin B_6 is not being supplemented, elevated plasma PLP is a sensitive and specific biochemical marker for HPP [2, 3]. In fact, the hypophosphatasemia (low serum ALP activity) and elevated plasma PLP level reflect the clinical severity of HPP [2, 3].

Perinatal and infantile HPP are autosomal recessive traits [2, 9, 10]; milder forms typically represent autosomal dominant inheritance [6]. Carrier parents and siblings often have low or low-normal serum ALP activity and sometimes mildly elevated plasma PLP levels. Pyridoxine given orally causes exaggerated increments in plasma PLP for all patients and some carriers [2, 3].

HPP is diagnosed from a consistent clinical history and physical findings, radiographic or histopathological evidence of rickets or osteomalacia, and hypophosphatasemia together with TNSALP substrate (PLP) accumulation [2, 3, 22]. Mutation analysis of *TNSALP* (*ALPL*) is available commercially. At least 340 mutations (~80% missense) are known [23].

Fetal sonography and radiography have detected HPP in the second trimester [9]. Typically, however, prenatal diagnosis requires *TNSALP* mutation analysis [2, 3, 5].

Aberrant vitamin B_6 metabolism in HPP revealed that TNSALP is a cell surface enzyme [2, 3]. Extracellular accumulation of PPi, an inhibitor of hydroxyapatite crystal growth, impairs skeletal and tooth cementum mineralization [2–4]. PPi is currently assayed only in research laboratories.

Unless deficiencies are documented, avoiding traditional treatments for rickets or osteomalacia seem best in HPP because circulating levels of Ca, P, and 25OHD are usually not low [2, 3]. In fact, supplementation could provoke or exacerbate hypercalcemia or hypercalciuria [13]. Hypercalcemia in perinatal or infantile HPP responds to restriction of dietary Ca [10], and perhaps to salmon calcitonin and/or glucocorticoid therapy [2, 3, 13]. Fractures can mend spontaneously, but healing may be delayed, including after osteotomy. In adults with HPP, load-sharing intramedullary rods, rather than load-sparing plates, seem best for fractures and pseudofractures [20]. Expert dental care is important. Soft foods and dentures may be necessary even for some pediatric patients.

Marrow cell transplantation seemed to rescue two severely affected infants [11]. Teriparatide stimulated TNSALP biosynthesis by osteoblasts and healed fractures in adult HPP [18, 19]. *TNSALP* knockout mice manifest infantile HPP and have helped elucidate the role of *TNSALP* and have provided a means to test treatment of the TNSALP deficiency [4]. Enzyme-replacement therapy for HPP using a recombinant bone-targeted TNSALP (asfotase alfa) showed excellent efficacy and safety, first in severely affected infants and young children [10], and then in severely affected older children [15]. In 2015, asfotase alfa was approved internationally typically for pediatric-onset HPP [5]. This offers new opportunities, but also poses new challenges [5].

MUCOPOLYSACCHARIDOSES

Mucopolysaccharidoses (eg, Hunter, Hurler, Morquio disease) are caused by diminished activity of the lysosomal enzymes that degrade glycosaminoglycans [24, 25]. Accumulation of these complex carbohydrates within marrow cells somehow distorts the skeleton in a characteristic radiographic pattern called "dysostosis multiplex" featuring macrocephaly, J-shaped sella turcica, widened clavicles, oval or hook-shaped vertebrae, oar-shaped ribs, dysplasia of the capital femoral epiphyses, coxa valga, epiphyseal and metaphyseal dysplasia, proximal tapering of the second and fifth metacarpals, and osteoporosis with coarsened trabeculae [25–27]. Joint contractures are also common [24, 25]. BMP [28] or other growth factor signaling [29] may be disrupted [30]. The clinical spectrum and pathophysiology of the skeletal complications has been reviewed [29], with the severity and manifestations differing among these enzymopathies and causal gene mutations [31]. In fact, each disorder manifests broad-ranging severity [24, 25].

Enzyme assays and genetic tests are available [1]. Increasingly, treatment is by marrow cell transplantation or by specific enzyme replacement [32–34] sometimes used initially [35]. Management of the skeletal disease has been reviewed [36, 37].

HOMOCYSTINURIA

Homocystinuria is a rare autosomal recessive disorder (OMIM #236200) caused by cystathionine β-synthase deficiency [38]. Consequently, homocystine, an intermediate in methionine metabolism, accumulates endogenously. This can cause thrombosis and embolism and modification of connective tissue proteins including fibrillin within periosteum and perichondrium. Major complications involve the eyes, central nervous system, vasculature, and skeleton [38]. Dislocation of the ocular lens can occur first, followed by mental subnormality and thrombotic events. Patients seem "Marfanoid," but joint mobility is limited. There may be pectus excavatum or carinatum, arachnodactyly, and genu valgum [38]. Generalized osteoporosis occurs with "codfish" vertebrae and kyphoscoliosis [26, 27]. Bones are elongated and overtubulated (narrow) [26, 27]. Osteopenia occurs in affected children and adults [39]. Mild manifestations are said to predict responsiveness (including the skeletal disease) to pyridoxine (vitamin B_6) therapy, but this is controversial [40, 41]. Treatment may instead include a low methionine–cysteine diet and betaine [40, 42, 43]. Homocystine is increased in plasma. Other causes of homocystinemia are being studied in relationship to common osteoporoses [41–45].

ALKAPTONURIA

Alkaptonuria is an autosomal recessive disorder caused by a deficiency of homogentisic acid oxidase due to loss of function mutations within the AKU gene [46]. Its prevalence is fewer than one in 250,000 (OMIM #203500) [1] Phenylalanine and tyrosine degradation is blocked, leading to homogentisic acid accumulation in tissues. Homogentisic acid may inhibit collagen synthesis by compromising lysyl hydroxylase [46, 47]. Oxidation and polymerization of excess homogentisic acid explains the characteristic black urine when exposed to air and discoloration of connective tissues [46]. "Ochronosis" refers to the pigmentation in the sclera, bucchal mucosa, teeth, skin, nails, endocardium, intima of large vessels, hyaline cartilage of major joints, and intervertebral disks [46]. In elderly patients, such pigmentation is also striking in costal, laryngeal, and tracheal cartilage, and in fibrocartilage, tendons, and ligaments. Radiographic changes in the spine are nearly pathognomonic, showing dense calcification of remaining disk material. Calcification of ear cartilage can also occur. The shoulders and hips are most likely to develop osteoarthritis. Although its pathogenesis is not well understood, severe degenerative disease from tissue fragility occurs in the spinal column where disks calcify and vertebrae fuse, and in major appendicular joints, especially the hips and knees [47]. Patients may present with low back pain that progresses, followed by appendicular joint pain accompanying kyphoscoliosis [48].

There is no established medical treatment [49], but a low-protein or other special diets seem worthwhile. Ascorbic acid may block homogentisic acid polymerization [46].

DISORDERS OF COPPER TRANSPORT

Wilson disease (OMIM #277900) and Menkes disease (OMIM #309400) [1] are genetic disorders of copper ($Cu2^+$) metabolism caused by deficiencies of $Cu2^+$-transporting ATPases in the trans-Golgi of different tissues [50].

Wilson disease affects approximately one in 55,000 people in the USA, causes impaired biliary excretion of $Cu2^+$, and leads to hepatic injury and $Cu2^+$ storage in additional tissues. Severity is broad-ranging. Kayser–Fleischer rings in the eyes, hepatitis and cirrhosis, renal tubular dysfunction and calculi, neurological disease, and hypoparathyroidism are potential complications [50]. The skeletal disease includes osteoporosis, osteomalacia, and chondrocalcinosis with osteoarthritis and joint hypermobility. Hypercalciuria, hyperphosphaturia, and sometimes renal tubular acidosis can occur [51]. DXA generally shows normal BMD [52] or predominantly axial osteopenia [53] likely involving a complex pathogenesis [54, 55]. Loss of function mutation disrupts the ATPase, Cu^{2+} transporting, beta-polypeptide gene, *ATP7B* [1]. Cu^{2+} chelation using penicillamine is usually effective.

In Menkes disease, an X-linked recessive disorder [1], boys develop Cu^{2+} deficiency causing kinky sparse hair and central nervous system disease including mental retardation, seizures, and intracranial hemorrhage [50]. The skeletal sequelae include short stature, microcephaly, brachycephaly, wormian bones, metaphyseal dysplasia featuring widening and spurs, joint laxity, and osteoporosis [26, 27]. Death usually occurs by age 3 years. A mild form is called "occipital horn syndrome" [50]. Serum Cu^{2+} and ceruloplasmin levels are low because of mutation within the ATPase, Cu^{2+} transporting, alpha-polypeptide gene, *ATP7A* [1].

DISCLOSURE

The author discloses prior research grant support, honoraria, and travel from Alexion Pharmaceuticals, New Haven, CT, USA.

REFERENCES

1. McKusick-Nathans Institute of Genetic Medicine, Johns Hopkins University. 2008. Online Mendelian Inheritance in Man. http://www.ncbi.nlm.nih.gov/omim/. Accessed June 12, 2018.
2. Whyte MP. Hypophosphatasia: aetiology, nosology, pathogenesis, diagnosis and treatment. Nat Rev Endocrinol 2016;12:233–46.
3. Whyte, MP. Hypophosphatasia and how alkaline phosphatase promotes mineralization. In: Thakker RV, Whyte MP, Eisman J, et al. (eds.) *Genetics of Bone Biology and Skeletal Disease*. 2nd ed. San Diego: Academic Press, 2018, pp. 481–504.
4. Millan JL, Whyte MP. Alkaline phosphatase and hypophosphatasia. Calcified Tissue Int. 2016;98:398–416.
5. Whyte MP. Hypophosphatasia: enzyme replacement therapy brings new opportunities and new challenges (perspective). J Bone Miner Res.. 2017;32:667–75.
6. Whyte MP, Zhang F, Wenkert D, et al. Hypophosphatasia: validation and expansion of the clinical nosology for children from 25 years experience with 173 pediatric patients. Bone. 2015;75:229–39.
7. Leung ECW, Mhanni AA, Reed M, et al. Outcome of perinatal hypophosphatasia in Manitoba Mennonites: a retrospective cohort analysis. J Inherit Metabol Dis Rep 2013;11:73–8.
8. Whyte MP, Greenberg CR, Ozono K, et al. Asfotase alfa treatment improves survival for perinatal and infantile hypophosphatasia. J Clin Endocrinol Metab.ol. 2016; 101:334–42.
9. Wenkert D, McAlister WH, Coburn SP, et al. Hypophosphatasia: non-lethal disease despite skeletal presentation in utero (17 new cases and literature review). J Bone Miner Res. 2011;26:2389–98.
10. Whyte MP, Greenberg CR, Salman NJ, et al. Enzyme replacement therapy for life-threatening hypophosphatasia. N Engl J Med. 2012;366(10):904–13.
11. Cahill RA, Wenkert D, Perman SA, et al. Infantile hypophosphatasia: transplantation therapy trial using bone fragments and cultured osteoblasts. J Clin Endocrinol Metab. 2007;95:2923–30.
12. Baumgartner-Sigl SB, Haberlandt E, Mumm S, et al. Pyridoxine-responsive seizures as the first symptom of infantile hypophosphatasia caused by two novel missense mutations (c.677T>C, p.M226T; c.1112C>T, p.T371I) of the tissue-nonspecific alkaline phosphatase gene. Bone. 2007;40:1655–61.
13. Barcia JP, Strife CF, Langman CB. Infantile hypophosphatasia: treatment options to control hypercalcemia, hypercalciuria, and chronic bone demineralization. J Pediatr. 1997;130:825–8.
14. Van den Bos T, Handoko G, Niehof A, et al. Cementum and dentin in hypophosphatasia. J Dent Res. 2005;84: 1021–5.
15. Whyte MP, Madson KL, Phillips D, et al. Asfotase alfa therapy for children with hypophosphatasia. JCI Insight 2016;1:e85971;1–10.
16. Whyte MP, Mumm S, McAlister WH, et al. Hypophosphatasia: natural history study of 101 affected children studied at a single research center. Bone. 2016;93:125–38.
17. Khandwala HM, Mumm S, Whyte MP. Low serum alkaline phosphatase activity with pathologic fracture: case report and brief review of adult hypophosphatasia. Endocr Pract 2006;12:676–81.
18. Whyte MP, Mumm S, Deal C. Adult hypophosphatasia treated with teriparatide. J Clin Endocrinol Metab. 2007;92:1203–8.
19. Camacho PM, Mazhari AM, Wilczynski C, et al. Adult hypophosphatasia treated with teriparatide: report of two patients and review of the literature. Endocr Pract 2016;22:941–50.
20. Coe JD, Murphy WA, Whyte MP. Management of femoral fractures and pseudofractures in adult hypophosphatasia. J Bone Joint Surg Am. 1986;68:981–90.
21. Guañabens N, Mumm S, Möller I, et al. Calcific periarthritis as the only clinical manifestation of hypophosphatasia in middle-aged sisters. J Bone Miner Res. 2014;29:929–34.
22. Fallon MD, Weinstein RS, Goldfischer S, et al. Hypophosphatasia: clinicopathologic comparison of the infantile, childhood, and adult forms. Medicine (Baltimore) 1984;63:12–24.
23. Mornet E. Tissue nonspecific alkaline phosphatase gene mutations database. 2005. http://www.sesep.uvsq.fr/Database.html Accessed June 12, 2018.
24. Neufeld EF, Muenzer J. The mucopolysaccharidoses. In: Scriver CR, Beaudet AL, Sly WS, et al. (eds) *The Metabolic and Molecular Bases of Inherited Disease* (8th ed.). New York: McGraw-Hill, 2001, pp. 3421–52.

25. Leroy JG, Wiesmann U. Disorders of lysosomal enzymes. In: Royce PM, Steinmann B (eds) *Connective Tissue and Its Heritable Disorders*. New York: Wiley-Liss, 2001, pp. 8494–9.
26. Taybi H, Lachman RS. *Radiology of Syndromes, Metabolic Disorders, and Skeletal Dysplasias* (5th ed.). St. Louis: Mosby, 2006.
27. Resnick D, Niwayama G. *Diagnosis of Bone and Joint Disorders* (4th ed.). Philadelphia: WB Saunders, 2002.
28. Khan SA, Nelson MS, Pan C, et al. Endogenous heparan sulfate and heparin modulate bone morphogenetic protein-4 signaling and activity. Am J Physiol Cell Physiol. 2008;294:C1387–C1397.
29. Kingma SDK, Wagemans T, Ijlst L, et al. Altered interaction and distribution of glycosaminoglycans and growth factors in mucopolysaacharidosis type I bone disease. Bone. 2016;88:92–100.
30. Clarke LA, Hollak CEM. The clinical spectrum and pathophysiology of skeletal complications in lysosomal storage disorders. Best Pract Res Clin Endocrinol Metabol. 2015;29:219–35.
31. Muenzer J. The mucopolysaccharidoses: a heterogeneous group of disorders with variable pediatric presentations. J Pediatr. 2004;144:S27–S34.
32. Schiffmann R, Brady RO. New prospects for the treatment of lysosomal storage diseases. Drugs. 2002;62: 733–42.
33. Braunlin EA, Stauffer NR, Peters CH, et al. Usefulness of bone marrow transplantation in the Hurler syndrome. Am J Cardiol. 2003;92(7):882–6.
34. Aldenhoven M, Kurtzberg J. Cord blood is the optimal graft source for the treatment of pediatric patients with lysosomal storage diseases: clinical outcomes and future directions. Cytotherapy. 2015;17:765–74.
35. Ghosh A, Miller W, Orchard PJ, et al. Enzyme replacement therapy prior to haematopoietic stem cell transplantation in mucopolysaccharidosis type I: 10 year combined experience of 2 centres. Mol Genet Metab. 2016;117:373–7.
36. Tomatsu S, Almeciga-Diaz CJ, Montano AM, et al. Therapies for the bone in mucopolysaccharidoses. Mol Genet Metab. 2015;114:94–109.
37. Kurtzberg J. Early HSCT corrects the skeleton in MPS. Blood. 2015;125:1518–19.
38. Mudd SH, Levy HL, Kraus JP. Disorders of transsulfuration. In: Scriver CR, Beaudet AL, Sly WS, et al. (eds) *The Metabolic and Molecular Bases of Inherited Disease* (8th ed.). New York: McGraw-Hill, 2001, pp. 2007–56.
39. Weber DR, Coughlin C, Brodsky JL, et al. Low bone mineral density is a common finding in patients with homocystinuria. Mol Genet Metab. 2016;117:351–4.
40. Lim JS, Lee DH. Changes in bone mineral density and body composition of children with well-controlled homocystinuria caused by CBS deficiency. Osteoporos Int. 2013;24:2535–8.
41. Green TJ, McMahon JA, Skeaff CM, et al. Lowering homocysteine with B vitamins has no effect on biomarkers of bone turnover in older persons: a 2-y randomized controlled trial. Am J Clin Nutr. 2007; 85: 460–4.
42. Cagnacci A, Bagni B, Zini A, et al. Relation of folates, vitamin B12 and homocysteine to vertebral bone mineral density change in postmenopausal women. A five-year longitudinal evaluation. Bone. 2008;42: 314–20.
43. Kumar T, Sharma GS, Singh LR. Homocystinuria: therapeutic approach. Clinica Chimica Acta 2016;458: 55–62.
44. Herrmann W, Herrmann M. Is hyperhomocysteinemia a risk factor for osteoporosis? Expert Rev Endocrinol Metab. 2008;3:309–13.
45. Salari P, Larijani B, Abdollahi M. Association of hyperhomocysteinemia with osteoporosis: a systematic review. Therapy. 2008;5:215–22.
46. La Du BN. Alkaptonuria. In: Scriver CR, Beaudet AL, Sly WS, et al. (eds) *The Metabolic and Molecular Bases of Inherited Disease* (8th ed.). New York: McGraw-Hill, 2001, pp. 2109–23.
47. Mannoni A, Selvi E, Lorenzini S, et al. Alkaptonuria, ochronosis, and ochronotic arthropathy. Sem Arthrit Rheum. 2004;33: 239–48.
48. Furuncuoglu Y, Demir MK, Toktas ZO, et al. Alkaptonuria and ochronotic spondyloarthropathy: a delayed imaging diagnosis. Spine J. 2016;16: e159–e161.
49. Gallagher JA, Dillon JP, Sireau N, et al. Alkaptonuria: an example of a "fundamental disease" – a rare disease with important lessons for more common disorders. Sem Cell Dev Biol. 2016;52:53–7.
50. Culotta VC, Gitlin JD. Disorders of copper transport. In: Scriver CR, Beaudet AL, Sly WS, et al. (eds) *The Metabolic and Molecular Bases of Inherited Disease* (8th ed.). New York: McGraw-Hill, 2001, pp. 3105–26.
51. Subrahmanyam DKS, Vadiveian M, Giridharan S, et al. Wilson's disease – a rare cause of renal tubular acidosis with metabolic bone disease. Indian J Nephrol. 2014;24: 171–4.
52. Quemeneur AS, Trocello J mea HK, Ostertag A, et al. Bone status and fractures in 85 adults with Wilson's disease. Osteoporos Int. 2014;25:2573–80.
53. Weiss KH, Van de Moortele M, Gotthardt DN, et al. Bone demineralisation in a large cohort of Wilson disease patients. J Inherit Metab Dis. 2015;38:949–56.
54. Shin JJ, Lee J-P, Rah J-H. Fracture in a young male patient leading to the diagnosis of Wilson's disease: a case report. J Bone Metab. 2015;22:33–7.
55. Wang H, Zhou A, Hu J, et al. Renal impairment in different phenotypes of Wilson disease. Neurol Sci. 2015;36:2111–15.

Section X
Oral and Maxillofacial Biology and Pathology

Section Editor: Laurie McCauley

Chapter 116. Craniofacial Morphogenesis 893
Erin Ealba Bumann and Vesa Kaartinen

Chapter 117. Development and Structure of Teeth and Periodontal Teeth 901
Petros Papagerakis and Thimios Mitsiadis

Chapter 118. Genetic Craniofacial Disorders Affecting the Dentition 911
Yong-Hee Patricia Chun, Paul H. Krebsbach, and James P. Simmer

Chapter 119. Pathology of the Hard Tissues of the Jaws 918
Paul C. Edwards

Chapter 120. Osteonecrosis of the Jaw 927
Sotirios Tetradis, Laurie McCauley, and Tara Aghaloo

Chapter 121. Alveolar Bone Homeostasis in Health and Disease 933
Chad M. Novince and Keith L. Kirkwood

Chapter 122. Oral Manifestations of Metabolic Bone Diseases 941
Erica L. Scheller, Charles Hildebolt, and Roberto Civitelli

Chapter 123. Dental Implants and Osseous Healing in the Oral Cavity 949
Takashi Matsuura and Junro Yamashita

Primer on the Metabolic Bone Diseases and Disorders of Mineral Metabolism, Ninth Edition. Edited by John P. Bilezikian.
© 2019 American Society for Bone and Mineral Research. Published 2019 by John Wiley & Sons, Inc.
Companion website: www.wiley.com/go/asbmrprimer

116
Craniofacial Morphogenesis

Erin Ealba Bumann[1] and Vesa Kaartinen[2]

[1]*Department of Oral and Craniofacial Sciences, University of Missouri-Kansas City School of Dentistry, Kansas City, MO, USA*
[2]*Department of Biologic and Materials Sciences, University of Michigan School of Dentistry, Ann Arbor, MI, USA*

INTRODUCTION

Formation and growth of craniofacial skeletal elements is a long and complicated process, which starts during early embryo development and is completed when the facial and cranial growth ends in late adolescence. It involves coordinated interactions of cells with different tissue origins including ectoderm, mesoderm, endoderm, and neural crest. Each individual has unique facial features, which result from subtle variations in growth and patterning of cranial and facial structures. Disturbances in these developmental processes often result in congenital craniofacial malformations, which are among the most common birth defects in humans. Here we review key developmental steps of craniofacial morphogenesis and summarize common birth defects resulting in a failure of these processes.

ANTERIOR-POSTERIOR AXIS FORMATION AND HEAD INDUCTION

A seed for the craniofacial development is planted during early postimplantation embryogenesis, when an embryo consists of three cell types: trophoectoderm, primitive endoderm, and epiblast [1]. Epiblast gives rise to embryonal tissues, which form the embryo proper, while primitive endoderm and trophoectoderm give rise to extra-embryonal tissues. In mouse embryogenesis, early postimplantation embryos elongate to form an egg cylinder, in which the epiblast is surrounded by the primitive endoderm-derived visceral endoderm (VE) distally and trophoectoderm-derived extra-embryonic ectoderm proximally [2]. This arrangement of different tissue types in close proximity to each other is critical for early patterning events. While the egg cylinder appears initially symmetrical, the VE displays regional gene expression soon after implantation. An important morphogen Wnt3 is expressed in prospective posterior VE, while the gene encoding Hhex, a homeobox transcription factor, is expressed in the distal end of the embryo [3, 4]. The Hhex-expressing VE cells quickly migrate to the prospective anterior region and give rise to the anterior visceral endoderm (AVE) [5], which forms an important head organizer (Fig. 116.1A). By secreting Nodal (Lefty1, Cer1) and Wnt (Dkk1) antagonists in the prospective anterior embryo, the AVE establishes a gradient of Nodal and Wnt signals along the anterior–posterior (A-P) axis, which confines the site of primitive streak formation diagonally to the AVE in the prospective posterior end of the embryo [6]. The importance of the AVE in head induction has been highlighted by ablation of AVE cells from the mouse embryo, which resulted in loss of the forebrain and other critical head structures [7].

Primitive streak (PS), which is induced by Wnt and Nodal signals from the VE and/or posterior epiblast, is the first morphologically distinct structure to form in the epiblast [8]. During a process called gastrulation, epiblast cells in the PS undergo epithelial to mesenchymal transdifferentiation to form mesodermal cells that migrate into the space between the epiblast and VE [8, 9]. PS elongates along the A-P axis, gradually forming different mesodermal cell populations. The last population to emerge forms the axial mesendoderm that gives rise to the prechordal plate, notochord, and definitive endoderm. The prechordal plate and anterior definitive endoderm function as important signaling centers controlling brain, head, and jaw patterning during neurulation and early craniofacial development [10, 11] (Fig. 116.1B,C).

Primer on the Metabolic Bone Diseases and Disorders of Mineral Metabolism, Ninth Edition. Edited by John P. Bilezikian.
© 2019 American Society for Bone and Mineral Research. Published 2019 by John Wiley & Sons, Inc.
Companion website: www.wiley.com/go/asbmrprimer

894 Craniofacial Morphogenesis

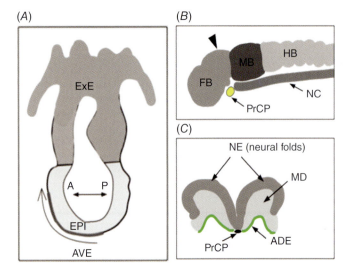

Fig. 116.1. Head organizers. (A) Anterior visceral endoderm (AVE) in the mouse embryo at the pre-streak stage. ExE = extraembryonic ectoderm; A = anterior; P = posterior; EPI = epiblast. Curved arrow depicts AVE migration to the prospective anterior region, where it functions as a head organizer. (B) Prechordal plate (PrCP) in the mouse embryo during neurulation. FB = forebrain; MB = midbrain; HB = hindbrain; NC = notochord. Black arrowhead depicts the level of the cross-section shown in C. (C) Anterior definitive endoderm (ADE) in the mouse embryo during neurulation (green). NE = neuroectoderm; MD = mesoderm. (B) Source: [10]. Reproduced with permission of Elsevier. (C) Source: [11]. Reproduced with permission of John Wiley & Sons.

FORMATION OF CRANIOFACIAL ECTOMESENCHYME (NEURAL CREST)

A critical step in early craniofacial development is the formation, migration, and differentiation of the neural crest [12]. This multipotent cell population forms in the interphase between neural and non-neural ectoderm at all axial levels. Upon induction, neural crest cells undergo epithelial to mesenchymal transdifferentiation, delaminate from dorsal ridges of the neural tube just before it closes, and migrate ventrolaterally along stereotypical migration paths to their target tissues, where they differentiate to multiple different cell types (eg, melanocytes, ganglia of peripheral nerves, and cells of the adrenal medulla) [13, 14]. Neural crest cells formed in the anterior embryo from the midbrain to the hindbrain level migrate to the cranial, facial, and neck region, where they give rise to the cranial neural crest (CNC) (Fig. 116.2) [15]. These cells, which are collectively called ectomesenchyme, populate the mesenchyme of the frontonasal process (FNP), maxillary process (MXP), and mandibular process of the first pharyngeal arch (Figs 116.2 and 116.3), and form most of the cranial and facial connective tissues (eg, chondrocytes, osteocytes, pericytes, and dermal cells). Most of the cranial and facial bones are derived from the CNC (Fig. 116.2). Congenital defects in tissues derived from the neural crest form a diverse class of disorders called neurocristopathies [16].

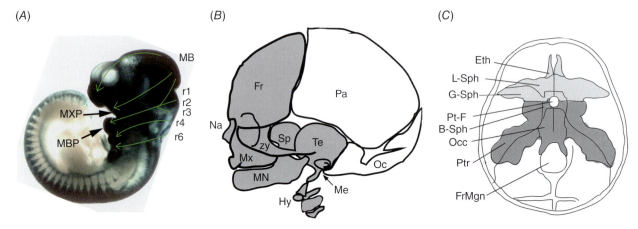

Fig. 116.2. (A) Cranial neural crest cell migration to the prospective cranial, facial, and neck region in a mouse embryo at embryonal day 10 (lineage tracing using R26 reporter assay). Blue staining illustrates neural crest cells, visualized using lacZ staining. MB = midbrain; r = rhombomere; MXP = maxillary process of the first pharyngeal arch; MBP = mandibular process of the first pharyngeal arch. Green arrows depict the patterns of cranial neural crest (CNC) migration. (B) Schematic illustration of newborn (human) craniofacial bones. Gray color depicts the bones derived from the CNC. Fr = frontal bone; Na = nasal articulation; Mx = maxillary bone; Mn = mandibular bone; Zy = zygomatic arch; Sp = Sphenoid bone; Te = temporal bone; Pa = parietal bone; Oc = occipital bone, Me = middle ear; Hy = hyoid. (C) Schematic presentation of the cranial base. Light gray color depicts the bones derived from the CNC, dark gray color depicts the bones derived from the paraxial mesoderm. Eth = ethmoid; L-Sph = lesser wing of sphenoid; G-Sph = greater wing of sphenoid; Pt-F = pituitary foramen; B-Sph = body of sphenoid; Occ = base of occipital bone; Ptr = petrous bone; FrMgn = foramen magnum.

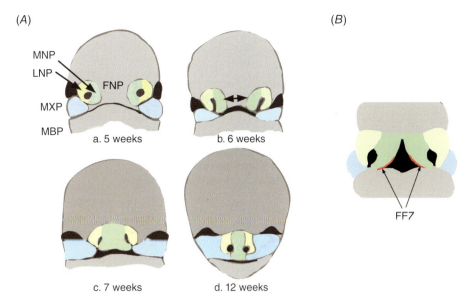

Fig. 116.3. Facial fusion. (A) Schematic presentation of human facial fusion at 5, 6, 7, and 12 weeks (of gestation). FNP = frontonasal process; MNP = medial nasal process (green); LNP = lateral nasal process (yellow); MXP = maxillary process of the first pharyngeal arch (blue); MBP = mandibular process of the first pharyngeal arch (grey). (B) Frontonasal ectodermal zone (FEZ, red).

MIDFACIAL DEVELOPMENT AND PALATOGENESIS

Midfacial development, maxillary patterning, and lip fusion take place during embryonal weeks 5 to 8 of human development. CNC cells migrating from the midbrain/anterior hindbrain region surround the forebrain and form the FNP (Fig. 116.3). At inferior–lateral frontal regions, CNC cells give rise to the mesenchyme of MXP and the horseshoe-shaped medial nasal process (MNP) and lateral nasal process (LNP). These processes grow in a stereotypical manner and form appropriate contacts. Fusion between the MNP and MXP form the upper lip and jaw, while fusion between the MNPs (double arrow in Fig. 116.3A, 6 weeks) results in formation of the philtrum, middle portion of the nose, and primary palate (Fig. 116.3A, 12 weeks).

Midfacial growth is controlled by reciprocal interactions between the ventral forebrain, FNP ectoderm, and neural crest cells [17]. *Shh* expression in the forebrain neuroectoderm makes the adjacent FNP ectoderm competent to express *Shh* upon ingress of neural crest cells. These events induce a formation of a signaling center called the frontonasal ectodermal zone (FEZ) in the facial ectoderm adjacent to *Shh*-expressing cells (Fig. 116.3B). FEZ regulates expression of the key morphogens (eg, Bmps and Fgfs) in the underlying mesenchyme to control appropriate growth and patterning of the FNP in the midfacial region, resulting in appropriate development of the upper jaw [18]. Moreover, antagonistic interactions between Shh and Wnt control growth and patterning of the MXP and MNP during lip fusion [19, 20]. During later stages of upper jaw development, bone formation of the maxilla, like that of most of the craniofacial bones, takes place via intramembraneous ossification, which is regulated by Notch (Jagged1) and BMP signaling [21].

The secondary palate separates the oral and nasal cavities. Its formation (palatogenesis) starts at week 7 of human gestation, when bilateral ridges called palatal shelves emerge from the maxillary process [22]. They first grow down along the sides of the tongue, then rapidly elevate, become adherent, form a contact in the midline, and eventually fuse by week 11 of pregnancy in humans (Fig. 116.4). Palatal shelf growth and patterning are controlled by complex signaling interactions between the palatal epithelium and mesenchyme [23]. The anterior palate, the so-called hard palate (bony palate) differs both anatomically and functionally from the posterior palate (soft palate), which is muscular and functions during speech and swallowing. In concordance with the different structure and function, molecular mechanisms controlling anterior palate formation differ from those of posterior palate formation. In the anterior prefusion palate, Shh in the epithelium stimulates *Bmp2* and *Osr2* expression in the underlying mesenchyme. Osr2 positively regulates *Fgf10* expression in the mesenchyme, which in turn controls *Shh* expression. In the posterior palate, a different set of transcription factors (eg, Tbx22 and Meox) regulate palatal shelf growth and patterning.

The palatal shelf fusion, the last step of palatogenesis, has been intensely studied and several different alternative processes have been proposed to explain a mechanism of fusion, such as apoptosis, epithelial–mesenchymal transdifferentiation, and migration [24–27]. However, recent studies using in vivo animal models and live imaging have

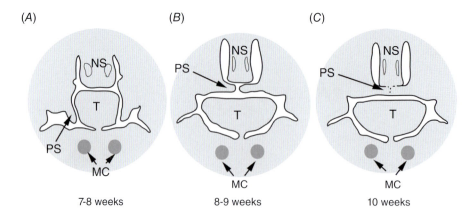

Fig. 116.4. Schematic presentation depicting growth (A, 7 to 8 weeks), elevation (B, 8 to 9 weeks), and fusion (C, 10 weeks of gestation) of palatal shelves (frontal orientation). NS = nasal septum; T = tongue; PS = palatal shelves; MC = Meckel cartilage.

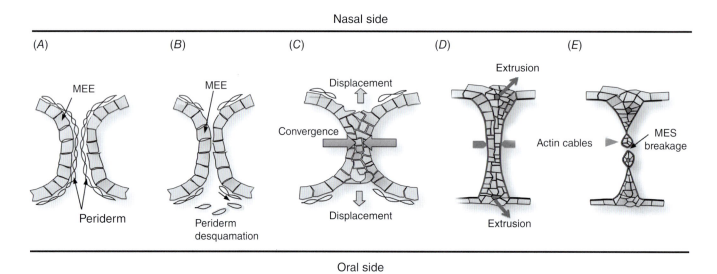

Fig. 116.5. Schematic model for palatal shelf fusion. (A) Apposing palatal shelves are covered by periderm cells. (B) Palatal shelves become adherent after periderm desquamation. (C) Medial edge epithelium (MEE) cells converge and displace MEE cells to the oral and nasal epithelial triangles. (D) Concomitant actin cable contraction and cell extrusion further narrow the midline epithelial seam (MES) resulting in MES breakage (E).

shown that establishment of palatal shelf adherence and subsequent fusion involve coordinated loss of periderm cells and epithelial cell intercalation, convergence, and extrusion (as explained later in this chapter). During early oral development, the oral epithelium, including the epithelium of palatal shelves, is covered by a single cell layer of flattened periderm cells. These cells have been shown to play a critical role in oral development by protecting the underlying epithelium and by preventing aberrant (unwanted) fusion events [28]. However, recent studies have suggested that in the tips of the palatal shelves, TGF-β3-triggered desquamation of periderm cells covering the medial edge epithelium (MEE) is required for appropriate adherence of the apposing palatal shelves [29, 30] (Fig. 116.5). Once adherence has been established, MEE cells intercalate and form the midline epithelial seam (MES) [31]. Multicellular actin cables contract and push MES cells into oral and nasal epithelial triangles, where some of them undergo programmed cell death as a result of cell extrusion. These contractile cellular events, which eventually lead to complete degradation of the midline seam, require rho-kinase and myosin light chain kinase resulting in myosin IIA activation [31] (Fig. 116.5).

Malformations affecting midfacial development are among the most common birth defects in humans. Failure of the forebrain to divide into cerebral hemispheres results in holoprosencephaly (HPE) (one in 250 pregnancies), which is characterized by severe brain, skull, and midfacial defects [32]. It can be caused by a failure in several morphogenic processes involved in head, brain, or face development. These include Nodal and Cripto signaling, which are important in prechordal plate growth and differentiation [33, 34]. In addition, mutations in genes encoding Shh or its signaling proteins have been shown to cause HPE both in animal models and in humans; however, they explain only a small subset of human HPE cases [32]. Other important pathways include retinoids and fibroblast growth factors (FGFs) [35]. In contrast, excessive hedgehog signaling in the anterior forebrain has been shown to result in aberrant proliferation of CNC cells in the FNP and the expanded facial midline, a condition called frontonasal dysplasia. Recent studies have shown that mutations in genes encoding intraflagellar transport proteins, which result in functional defects of primary cilia and gain of Shh function, play a major role in frontonasal dysplasia pathogenesis [36].

Orofacial clefts (eg, cleft lip with or without cleft palate [CL/P] and palate only [CPO]) affect about one in 1000 live births [37]. Their etiology is complex and involves both environmental and genetic factors. Moreover, the etiology of CL/P differs from that of CPO. About 70% of the cases are isolated (ie, they are not associated with known syndromes), and the affected individuals show no signs of other defects such as cognitive impairment or cardiac malformations. Close to 300 syndromes are known in which orofacial clefting is one of the primary features [38]. About 75% of them are triggered by known genetic causes, most inherited by Mendelian fashion. Together with animal models, the CL/P and CPO syndromes have played a crucial role in unraveling the molecular mechanisms of lip and palate fusion. Many genes that are causally related to orofacial clefting in syndromic cases have been shown to play a role in nonsyndromic cases as well [38]. The other risk factors include environmental and lifestyle causes such as alcohol consumption, cigarette smoking, and recreational and some prescription drugs.

CRANIAL BASE DEVELOPMENT

The cranial base, or ventral part of the cranium, is composed of the basioccipital, sphenoid, ethmoid, and frontal bones. Developmentally, the cranial base is split into an anterior and posterior cranial base due to differences in their embryological origins. The anterior cranial base (greater and lesser wings of the sphenoid, ethmoid, and frontal bones) arises from the CNC and the posterior cranial base (body of the sphenoid and basioccipital bones) develops from the paraxial mesoderm [39] (Fig. 116.2C). These different origins suggest unique genetic control over their morphogenesis, development, and growth.

The cranial base angulation is a unique cranial feature of modern humans, thought to support larger brain volumes, adapt the head for bipedalism, and influence facial shapes [40]. The flex of the cranial base is seen at the sella turcica, a saddle-shaped depression in the body of the sphenoid bone that holds the pituitary gland, which corresponds to the junction of the anterior and posterior cranial bases. The anterior cranial base is known to play a role in the development of the upper/midface and growth in the facial skeleton, particularly influencing the development of the maxilla. Growth of the anterior cranial base is found through adulthood [40]. The posterior cranial base grows at approximately half of the rate of the anterior cranial base and articulates with the mandible. So, the posterior cranial base plays a larger role in the position of the mandible [40]. Anterior and posterior cranial base abnormalities have been noted in syndromes and developmental defects.

Three synchondroses, or cartilaginous unions, are present at the midline of the cranial base. They are named the spheno-occipital, intersphenoidal, and spheno-ethmoidal synchondrosis [41]. The intersphenoidal synchondrosis fuses prenatally and therefore does not play a role in postnatal growth. The spheno-ethmoidal synchondrosis fuses during adolescence and is derived from CNC. The spheno-occipital synchondrosis is the last to fuse and is derived from both CNC and paraxial mesoderm [42]. The spheno-occipital synchondrosis is thought to play the most important role in directing cranial growth [41].

Cranial base abnormalities have been identified that affect only the anterior or posterior cranial bases. Specifically, syndromes with anterior cranial base abnormalities include the Pfeiffer, Down, and Williams syndromes. Posterior cranial base abnormalities are seen in the Saethre–Chotzen and Turner syndromes. There are also syndromes that include both anterior and posterior cranial base abnormalities, such as Crouzon syndrome [40]. Interestingly, certain genes play a large role in proper formation in either the anterior or the posterior cranial base. PTHrP is known to play a critical role in chondrocytic differentiation [43]. PTHrP-deficient mice show a significant difference from the wildtype mice in the posterior cranial base while hardly any changes were noted in the anterior cranial base [43]. This suggests that the cranial base might play a primary role in craniofacial anomalies and that different therapies could be developed to target specific anterior and posterior craniofacial defects.

MANDIBULAR DEVELOPMENT

The first pharyngeal arch divides into the maxillary and mandibular processes. The mandible, or lower jaw, is formed from the mandibular process and is innervated by the mandibular division of the trigeminal nerve [44]. It differs in multiple ways from the maxilla, including that the mandible is not fixed, it can move in a number of directions through the temporomandibular joint, and its proper formation is critical to allow for proper mastication and vocalization [44]. The mandible is also the largest and strongest bone in the craniofacial skeleton.

The osteoblasts and chondrocytes of the mandible originate from CNC cells, while osteoclasts and myocytes originate from the paraxial mesoderm. Meckel cartilage begins to form in the sixth week of embryonic development [44]. It is formed in close proximity to the mandible but does not make a direct contribution to the formation of the mandible; rather it forms the sphenomandibular ligament and the incus and malleus of the inner ear. The primary ossification center of the mandible begins lateral to Meckel cartilage at the sixth week of development and intramembranous ossification of the mandible begins at the seventh week [44].

Mandibles vary in size ranging from micrognathia, a small mandible, to macrognathia, a large mandible. Classically, Darwin's finches have been studied to analyze differences in jaw length. Differences in mandibular length are seen in natural variation, but can become amplified in syndromic conditions. One such condition is Pierre Robin sequence, which is characterized by micrognathia that leads to the development of a cleft palate and glossoptosis [45]. These patients have life-threating conditions related to breathing and feeding. The disruptions in the growth and elongation of the Meckel cartilage are thought to be the primary cause of the micrognathia. This is supported by the fact that multiple syndromes with Pierre Robin sequence have a mutation in genes known to be important in cartilage formation. Most notably, mutations in *SOX9* and its enhancer region have been identified in patients with Pierre Robin sequence and confirmed in mouse models [46, 47].

The three secondary cartilages of the mandible are the condylar, coronoid, and symphyseal cartilages. The condylar cartilage begins to form in the ninth week of development [44]. The condylar cartilage is important in the postnatal growth of the mandible and formation of the temporomandibular joint. The coronoid cartilage begins to form in the 10th week of development and disappears before birth [44]. Additionally, the symphyseal cartilage disappears in the first year of life and allows the two halves of the mandible to completely fuse.

CALVARIAL DEVELOPMENT

Almost the entire calvaria, except the supraocciptial bone, forms by intramembranous ossification. The frontal, parietal, temporal, and interparietal occipital bones begin to ossify by the ninth week of development. The fontal and temporal bones are derived from the CNC, and the parietal and occipital bones originate from the paraxial mesoderm (Fig. 116.2). The ossification of each bone advances until they come in close proximity to each other at the suture, a fibrous band that connects the bones of the skull. Sutures are needed to support brain growth and are the primary site of osteoblast differentiation for further bone growth. There are four sutures visible from the dorsal view of the calvaria, including the lambdoid (separating the parietal bones from the occipital bone), coronal (separating the frontal and parietal bones), sagittal (separating the parietal bones), and metopic (separating the frontal bones). At birth the human brain is approximately 40% of its final adult volume and grows to about 80% of this volume by 3 years old. This continued postnatal growth of the calvaria is important and continues until the child is around 7 years old [48].

The dura mater is first identified in the seventh week of development. It surrounds the developing brain and is in direct contact with the bone of the calvaria. The dura mater has been shown to play a critical role in establishing cranial development and in the development of a suture, yet once the suture is formed it is much less dependent on the dura mater to maintain normal growth [48]. This suggests that there are different signaling pathways between the dura mater and sutures throughout development of the calvaria.

Craniosynostosis is the premature fusion of one or more of the cranial sutures, resulting in an abnormal head shape. Craniosynostosis occurs in 1:2000 live births. Most commonly only one suture is fused, the specific etiology is unknown, and these babies tend to be otherwise healthy. When more than one suture is closed, the craniosynostosis is usually part of a syndrome, which occurs in approximately 20% cases of craniosynostosis [48]. Many of the syndromes have a mutation that leads to constitutive activation in the FGF receptor, including the Apert, Crouzon, and Pfeiffer syndromes [49]. The current treatment for craniosynostosis to correct the abnormal face and cranium shapes, as well as intracranial pressure, is invasive surgery.

SUMMARY

Craniofacial morphogenesis is a complex developmental process involving all morphogen signaling pathways and all three germ cell layers. While most of the key fundamental steps in craniofacial development (eg, head induction and midfacial fusion) take place during the first trimester of gestation, some critical processes (eg, cranial suture fusion and cranial and jaw growth) occur during the postnatal period, being finally completed in early adulthood. Failure of normal developmental events, whether in early embryogenesis or in postnatal maturation, often results in congenital craniofacial malformations, which are among the most common birth defects in humans.

REFERENCES

1. Nowotschin S, Hadjantonakis AK. Cellular dynamics in the early mouse embryo: from axis formation to gastrulation. Curr Opin Genet Dev. 2010;20(4):420–7.
2. Ang SL, Constam DB. A gene network establishing polarity in the early mouse embryo. Semin Cell Dev Biol. 2004;15(5):555–61.
3. Martinez-Barbera JP, Beddington RS. Getting your head around Hex and Hesx1: forebrain formation in mouse. Int J Dev Biol. 2001;45(1):327–36.
4. Liu P, Wakamiya M, Shea MJ, et al. Requirement for Wnt3 in vertebrate axis formation. Nat Genet. 1999;22(4):361–5.

5. Beddington RS, Robertson EJ. Anterior patterning in mouse. Trends Genet. 1998;14(7):277–84.
6. Mesnard D, Filipe M, Belo JA, et al. The anterior-posterior axis emerges respecting the morphology of the mouse embryo that changes and aligns with the uterus before gastrulation. Curr Biol. 2004;14(3):184–96.
7. Thomas P, Beddington R. Anterior primitive endoderm may be responsible for patterning the anterior neural plate in the mouse embryo. Curr Biol. 1996;6(11):1487–96.
8. Tam PP, Loebel DA, Tanaka SS. Building the mouse gastrula: signals, asymmetry and lineages. Curr Opin Genet Dev. 2006;16(4):419–25.
9. Tam PP, Beddington RS. Establishment and organization of germ layers in the gastrulating mouse embryo. Ciba Found Symp. 1992;165:27–41; discussion 2–9.
10. Robb L, Tam PP. Gastrula organiser and embryonic patterning in the mouse. Semin Cell Dev Biol. 2004;15(5):543–54.
11. Billmyre KK, Klingensmith J. Sonic hedgehog from pharyngeal arch 1 epithelium is necessary for early mandibular arch cell survival and later cartilage condensation differentiation. Dev Dyn. 2015;244(4):564–76.
12. Knecht AK, Bronner-Fraser M. Induction of the neural crest: a multigene process. Nat Rev Genet. 2002;3(6):453–61.
13. Le Douarin NM. The neural crest in the neck and other parts of the body. Birth Defects Orig Artic Ser. 1975;11(7):19–50.
14. LeDouarin. *The Neural Crest*. Cambridge: Cambridge University Press, 1982.
15. Trainor PA, Krumlauf R. Patterning the cranial neural crest: hindbrain segmentation and Hox gene plasticity. Nat Rev Neurosci. 2000;1(2):116–24.
16. Bolande RP. Neurocristopathy: its growth and development in 20 years. Pediatr Pathol Lab Med. 1997;17(1):1–25.
17. Hu D, Marcucio RS, Helms JA. A zone of frontonasal ectoderm regulates patterning and growth in the face. Development. 2003;130(9):1749–58.
18. Foppiano S, Hu D, Marcucio RS. Signaling by bone morphogenetic proteins directs formation of an ectodermal signaling center that regulates craniofacial development. Dev Biol. 2007;312(1):103–14.
19. Kurosaka H, Iulianella A, Williams T, et al. Disrupting hedgehog and WNT signaling interactions promotes cleft lip pathogenesis. J Clin Invest. 2014;124(4):1660–71.
20. Song L, Li Y, Wang K, et al. Lrp6-mediated canonical Wnt signaling is required for lip formation and fusion. Development. 2009;136(18):3161–71.
21. Hill CR, Yuasa M, Schoenecker J, et al. Jagged1 is essential for osteoblast development during maxillary ossification. Bone. 2014;62:10–21.
22. Bush JO, Jiang R. Palatogenesis: morphogenetic and molecular mechanisms of secondary palate development. Development. 2012;139(2):231–43.
23. Lane J, Kaartinen V. Signaling networks in palate development. WIREs Syst Biol Med. 2014;6(3):271–8.
24. Shuler CF, Guo Y, Majumder A, et al. Molecular and morphologic changes during the epithelial-mesenchymal transformation of palatal shelf medial edge epithelium in vitro. Int J Dev Biol. 1991;35(4):463–72.
25. Kaartinen V, Cui XM, Heisterkamp N, et al. Transforming growth factor-beta3 regulates transdifferentiation of medial edge epithelium during palatal fusion and associated degradation of the basement membrane. Dev Dyn. 1997;209(3):255–60.
26. Mori C, Nakamura N, Okamoto Y, et al. Cytochemical identification of programmed cell death in the fusing fetal mouse palate by specific labelling of DNA fragmentation. Anat Embryol (Berl). 1994;190(1):21–8.
27. Carette MJ, Ferguson MW. The fate of medial edge epithelial cells during palatal fusion in vitro: an analysis by DiI labelling and confocal microscopy. Development. 1992;114(2):379–88.
28. Richardson RJ, Hammond NL, Coulombe PA, et al. Periderm prevents pathological epithelial adhesions during embryogenesis. J Clin Invest. 2014;124(9):3891–900.
29. Hu L, Liu J, Li Z, et al. TGFbeta3 regulates periderm removal through DeltaNp63 in the developing palate. J Cell Physiol. 2015;230(6):1212–25.
30. Lane J, Yumoto K, Pisano J, et al. Control elements targeting Tgfb3 expression to the palatal epithelium are located intergenically and in introns of the upstream Ift43 gene. Front Physiol. 2014;5:258.
31. Kim S, Lewis AE, Singh V, et al. Convergence and extrusion are required for normal fusion of the mammalian secondary palate. PLoS Biol. 2015;13(4):e1002122.
32. Petryk A, Graf D, Marcucio R. Holoprosencephaly: signaling interactions between the brain and the face, the environment and the genes, and the phenotypic variability in animal models and humans. WIREsDev Biol. 2015;4(1):17–32.
33. Anderson RM, Lawrence AR, Stottmann RW, et al. Chordin and noggin promote organizing centers of forebrain development in the mouse. Development. 2002;129(21):4975–87.
34. McKean DM, Niswander L. Defects in GPI biosynthesis perturb Cripto signaling during forebrain development in two new mouse models of holoprosencephaly. Biol Open. 2012;1(9):874–83.
35. Roessler E, Muenke M. The molecular genetics of holoprosencephaly. Am J Med Genet C Semin Med Genet. 2010;154C(1):52–61.
36. Brugmann SA, Allen NC, James AW, et al. A primary cilia-dependent etiology for midline facial disorders. Hum Mol Genet. 2010;19(8):1577–92.
37. Vanderas AP. Incidence of cleft lip, cleft palate, and cleft lip and palate among races: a review. Cleft Palate J. 1987;24(3):216–25.
38. Leslie EJ, Marazita ML. Genetics of cleft lip and cleft palate. Am J Med Genet C Semin Med Genet. 2013;163C(4):246–58.
39. Couly GF, Coltey PM, Le Douarin NM. The triple origin of skull in higher vertebrates: a study in quail-chick chimeras. Development. 1993;117(2):409–29.

40. Nie X. Cranial base in craniofacial development: developmental features, influence on facial growth, anomaly, and molecular basis. Acta Odontol Scand. 2005;63(3):127–35.
41. Pan A, Chang L, Nguyen A, et al. A review of hedgehog signaling in cranial bone development. Front Physiol. 2013;4:61.
42. McBratney-Owen B, Iseki S, Bamforth SD, et al. Development and tissue origins of the mammalian cranial base. Dev Biol. 2008;322(1):121–32.
43. Ishii-Suzuki M, Suda N, Yamazaki K, et al. Differential responses to parathyroid hormone-related protein (PTHrP) deficiency in the various craniofacial cartilages. Anat Rec. 1999;255(4):452–7.
44. Lee SK, Kim YS, Oh HS, et al. Prenatal development of the human mandible. Anat Rec. 2001;263(3):314–25.
45. Tan TY, Farlie PG. Rare syndromes of the head and face—Pierre Robin sequence. WIREs Dev Biol. 2013;2(3):369–77.
46. Benko S, Fantes JA, Amiel J, et al. Highly conserved non-coding elements on either side of SOX9 associated with Pierre Robin sequence. Nat Genet. 2009;41(3):359–64.
47. Mori-Akiyama Y, Akiyama H, Rowitch DH, et al. Sox9 is required for determination of the chondrogenic cell lineage in the cranial neural crest. Proc Natl Acad Sci U S A. 2003;100(16):9360–5.
48. Tubbs RS, Bosmia AN, Cohen-Gadol AA. The human calvaria: a review of embryology, anatomy, pathology, and molecular development. Childs Nerv Syst. 2012;28(1):23–31.
49. Rice DP, Aberg T, Chan Y, et al. Integration of FGF and TWIST in calvarial bone and suture development. Development. 2000;127(9):1845–55.

117

Development and Structure of Teeth and Periodontal Teeth

Petros Papagerakis[1] and Thimios Mitsiadis[2]

[1]College of Dentistry; College of Medicine (Anatomy and Cell Biology); College of Pharmacy and Nutrition, Toxicology and Biomedical Engineering Graduate Programs, University of Saskatchewan, Saskatoon, SK, Canada; and School of Dentistry, Department of Orthodontics and Pediatric Dentistry, Center for Computational Medicine, University of Michigan, Ann Arbor, MI, USA
[2]Institute of Oral Biology, Medical Faculty of the University of Zurich, Zurich, Switzerland

INTRODUCTION

Tooth development or odontogenesis is the complex process by which dental mineralized tissues are formed from embryonic cells which differentiate into ameloblasts that secrete enamel, odontoblasts that produce dentin, and cementoblasts that make cementum. Enamel is of epithelial origin and covers the crown of each tooth. In contrast, dentin and cementum are of mesenchymal origin. Dentin forms the bulk of the tooth and extends within both the crown and root. It has a yellow color, in contrast to the much whiter and harder enamel. Cementum is deposited only in the root area on the recently mineralized dentin matrix. The tooth is anchored onto its socket (alveolar bone) by the periodontal ligament (PDL), a connective tissue structure that surrounds the tooth root and connects each tooth to the alveolar bone through a specialized set of collagen fibers.

STAGES OF TOOTH MORPHOGENESIS AND THEIR MOLECULAR CONTROL

Mammalian teeth have distinctive crown and root morphologies, which are highly adapted to their particular masticatory function. Generation of individual teeth relies upon interactions between the oral epithelium and mesenchymal cells that are derived from the cranial neural crest cells (CNCCs). Their formation involves a precisely orchestrated series of molecular and morphogenetic events. Although many diverse types of teeth exist in different species, nonhuman tooth development is largely the same as in humans. Therefore, we have used here examples of mouse tooth development as a prototype model for understanding human tooth formation. When possible we have tried to make connections to human pathologies.

Morphologically, tooth development begins with a thickening of the oral epithelium that forms a structure known as the dental lamina. Within the dental lamina cells start to proliferate and to invaginate the underlying mesenchyme in precise positions to form the dental placodes (which define where the teeth will be positioned into the jaws). Tooth development proceeds through a series of morphological stages that necessitate sequential and reciprocal interactions between the oral epithelium and the underlying cranial neural crest-derived mesenchyme. In mice, the oral epithelium starts thickenning at embryonic day 10.5 (E10.5) and progressively acquires bud (E13.5), cap (E14.5), and bell (E16.5) configurations (Figure 117.1). At the bell stage, two mesenchymal cell populations can be distinguished: the dental follicle and dental pulp. Dental pulp cells adjacent to the dental epithelium differentiate into odontoblasts, while epithelial cells juxtaposing the dental pulp differentiate into ameloblasts [1, 2]. Dental follicle gives rise to specialized periodontal cells.

Signaling molecules control all the steps of tooth formation by coordinating cell proliferation, differentiation, apoptosis, extracellular matrix synthesis, and mineral

---

Primer on the Metabolic Bone Diseases and Disorders of Mineral Metabolism, Ninth Edition. Edited by John P. Bilezikian.
© 2019 American Society for Bone and Mineral Research. Published 2019 by John Wiley & Sons, Inc.
Companion website: www.wiley.com/go/asbmrprimer

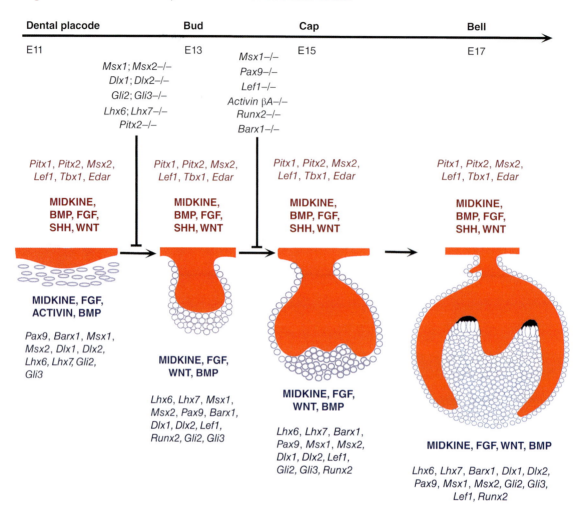

Fig. 117.1. Schematic representation of the various stages of embryonic mouse molar development. The most important signaling molecules (in bold capitals) and transcription factors (in italics) that are expressed in tooth epithelium (red) and mesenchyme (blue) are shown.

deposition. The same molecules are repetitively used during the different stages of tooth development and are regulated according to a precise timing mechanism [3, 4]. Signals produced at a wrong time lead to abnormal cell proliferation, differentiation, and apoptosis, thus affecting the overall tooth development and shape.

Numerous studies have shown that BMPs regulate epithelial–mesenchymal interactions during tooth initiation [3]; Wnts [4] and sonic hedgehog (shh) [5] regulate cell proliferation, migration, and differentiation; and FGFs regulate tooth-specific gene expression and cell proliferation [6] (Fig. 117.1).

The territory of the mammalian dentition is defined early, before any obvious sign of tooth development. The earliest marker that defines the oral epithelial area where teeth will grow is the transcription factor Pitx2 [7]. Mutations of *PITX2* in humans result in Rieger syndrome, which is characterized by eye and tooth defects including anodontia (lack of teeth).

The mesenchyme of teeth is derived from CNCCs that form a pool of multipotent progenitors. CNCCs migrate from the dorsal part of the neural tube and subsequently generate craniofacial structures of unique morphology and function, such as teeth [8]. Malformations and syndromes that arise due to defects in neural crest cell development are collectively called neurocristopathies in humans.

Dental fields within the oral epithelium are established by epithelium-derived signals that form morphogenic gradients, providing positional information [1]. These signals determine the display and fate of the CNCCs, leading to the generation of distinct tooth shapes. As an example, ectodysplasin A (EDA) signaling molecules have been shown to be involved in the determination of the size of the dental fields in the oral epithelium, and, thus, the proportion of the size and number of teeth [9]. Nonsyndromic hypodontia or congenital absence of one or more permanent teeth is a common anomaly of dental development in humans. Mutations in the *EDA* gene are related with X-linked recessive hypohidrotic ectodermal dysplasia (XLHED). XLHED is a genetic diseases characterized by the defective morphogenesis of teeth, hair, and sweat glands [10].

SIGNALS CONTROLLING DENTAL CELL DIFFERENTIATION

Dental cell differentiation results in the formation of the three dental mineralized tissues (enamel, dentin, and cementum), which are connected through the PDL to the alveolar bone.

The specification of the various dental cell types during dental cell differentiation (ie, stratum intermedium, stellate reticulum, outer and inner dental epithelial cells, ameloblasts, dental pulp fibroblasts, odontoblasts, and PDL fibroblasts) involves differential expression of specialized genes with restricted developmental and circadian patterns during odontogenesis. It is possible that the determination of cell fates in teeth occurs via inhibitory interactions between adjacent dental cells. These interactions seem to be mediated mainly through the Notch signaling pathway.

BMPs and FGFs have opposite effects on the expression of Notch receptors and ligands (ie, Delta and Jagged) in dental tissues [11], indicating that cell fate choices during odontogenesis are under the concomitant control of the Notch and BMP/FGF signaling pathways. Notch-mediated lateral inhibition has a pivotal role in the establishment of the tooth morphology, as shown in *Jagged2* mutant mice where the overall development and structure of their teeth is severely affected [12]. Mutations in Notch signaling pathway members cause developmental phenotypes that affect the liver, skeleton, heart, eye, face, kidney, and vasculature in humans [13]. The tooth phenotype is still unclear in these patients. In addition, genetic findings have shown that Tbx1, a Notch signalling target, plays a significant role for the early determination of epithelial cells to adopt the ameloblast fate. Indeed, hypoplastic incisors that lack enamel are observed in mice where the *Tbx1* gene was deleted [14]. Furthermore, *TBX1* mutations in humans are associated with the DiGeorge syndrome which is characterized by abnormal cell differentiation resulting in defects of many organs including the heart and teeth [15].

DENTAL MINERALIZED TISSUE FORMATION

Dentin

During tooth development, dental pulp cells start to differentiate into odontoblasts at the dental–enamel junction (DEJ) under the future cusp tip. Young odontoblasts are columnar cells that secrete a mantle matrix, called predentin, which is rich in type I collagen and matrix vesicles. After predentin deposition, the basal lamina associated with the inner enamel epithelium disintegrates [16], followed by a major upregulation in the production and secretion of enamel matrix proteins and MMP-20 [17].

Dentin is tough and elastic, and its prime feature is its penetration by odontoblast tubules, that radiate out from the dental pulp to the periphery (Fig. 117.2). These, with their many side branches that remain in the tubules within the dentin, are analogous to the canaliculi that house osteocyte processes in bone. Functional odontoblasts are highly polarized cells with a specialized cellular process, the odontoblast process, which transverses the heterogeneous layers of primary dentin within the dentin tubules [18].

Primary dentin forms most of the tooth prior to root eruption. Primary dentin is composed of peritubular or intratubular dentin (which creates the wall of the dentinal tubule); intertubular dentin found between the tubules, mantle dentin (which is the first predentin that forms within the tooth and is devoid of tubules); and circumpulpal dentin (which is the inner layer of dentin around the outer pulpal wall). The continuous secretion of dentin matrix (called orthodentin) is associated with the progressive lengthening of the odontoblast cell process and retraction of the odontoblasts toward the dental pulp. In contrast, during physiopathological situations in humans several other types of dentin are formed, such as sclerotic/reactionary orthodentin, where tubules may be obliterated, fibrodentin, and also osteodentin.

Odontoblasts secrete tooth-related proteins, such as dentin sialoprotein (DSP) dentin phosphoprotein (DPP), and dentin glycoprotein (DGP), which are encoded by a single gene called *dentin sialophosphoprotein* (*DSPP*) [19]. However, the majority of dentin is composed of proteins common to both dentin and bone. These proteins include type I, III, and V collagens, bone sialoprotein (BSP), osteopontin (OPN), dentin matrix protein-1 (DMP-1), osteocalcin (OC), and osteonectin (ON) [20]. Several proteins involved in calcium and phosphate handling are also synthesized jointly by osteoblasts and odontoblasts, including calbindin-D28k [21], calcium pump [22], and alkaline phosphatase [23].

Dentin continues to be slowly deposited even after tooth eruption and complete root formation. This dentin could be either the regularly deposited secondary dentin or irregular tertiary dentin, as a response of the pulp–dentin complex to attrition or disease [24]. Nerves pass from the dental pulp between the odontoblasts and extend into the dentin tubules for variable distances.

The dental pulp shows signs of aging [25], which may include diffuse or local calcifications and the formation of dental stones as well as limited reparative dentin formation.

Enamel

Dental enamel is the hardest mineralized tissue found in mammals. Mature enamel contains less than 1% organic material. It is acellular and contains no collagen. Enamel forms in an extracellular space lined by ameloblasts, which control both the ionic and the organic contents of the enamel extracellular space [26]. Enamel mineral is mainly formed by calcium hydroxyapatite, exhibiting peculiar dimensions and organization of its crystallites.

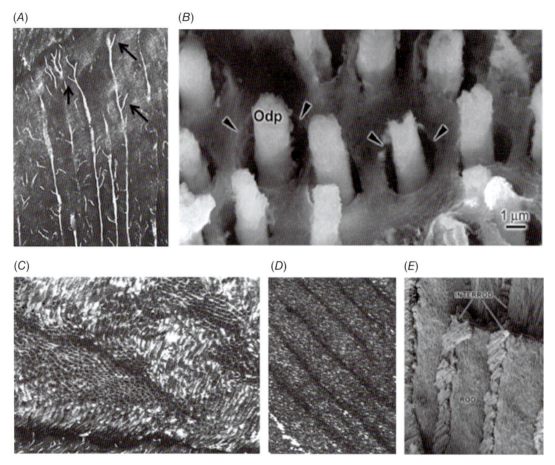

Fig. 117.2. (A) Dentin processes transverse the whole primary dentin matrix. Many ramifications are observed (arrows). (B) Close capture of odontoblast processes (Odp) showing peritubular (arrowheads) and intertubular dentin. (C) Decussation (crossing in an X fashion) of the enamel prisms, with zones of prisms with contrasting 3D courses forming the Hunter-Schreger bands can be seen. (D) Enamel prisms. (E) Enamel rods and inter-rod enamel. Crystals in rod and inter-rod enamel are structurally similar, but have different orientations. Source: [32]. Reproduced with permission of John Wiley & Sons.

Enamel crystals are only ~25 nm thick and 65 nm wide, but are believed to extend uninterrupted from the DEJ to the surface of the tooth [27]. The secretory end of the ameloblast ends in a six-sided pyramid-like projection known as the Tomes' process. Tomes' processes organize the enamel crystallites into rods (prisms). The angulation of the Tomes' process is significant in the orientation of enamel rods. Enamel formation is divided into secretory, transition, and maturation stages [28, 29]. The stages are shown in Fig. 117.3 [30].

During the secretory stage, enamel matrix deposition is orchestrated by both pre-ameloblasts and secretory ameloblasts. Ameloblast differentiation initiates at the cusps of the tooth germ, where the first epithelial cells differentiate into pre-ameloblasts. The signals driving ameloblast differentiation derive from the newly differentiated odontoblasts [18]. Pre-ameloblasts are secretory cells that have not yet formed a Tomes' process and they deposit a thin layer of aprismatic enamel on the dentin surface. Thereafter, secretory ameloblasts initiate prismatic enamel deposition. During the secretory stage, mineral is deposited rapidly on the tips of the crystals and very slowly on the crystals sides (for review see [31]). As the crystals extend, the enamel layer expands. During the secretory stage, ameloblasts secrete mostly amelogenin, enamelin, and ameloblastin [32]. These enamel-specific proteins catalyze the extension of ribbon-like enamel crystallites comprised of the mineral calcium hydroxyapatite. Amelogenins are the predominant enamel matrix proteins and they assemble into spheres that occupy the spaces between the crystal ribbons, serving to separate and support them [33]. Ameloblastin is thought to be important for ameloblast attachment to the mineralizing matrix [34]. Enamelin is probably responsible for the initiation of mineralization at the mineralization front [35]. As ameloblasts secrete enamel proteins and extend the mineral ribbons, they retreat from the existing enamel surface, thus increasing the thickness of the enamel extracellular space. Unlike collagen-based mineralization processes, which are two-step processes (secretion of an organic matrix followed by matrix mineralization), enamel secretion and mineralization occurs in a single step.

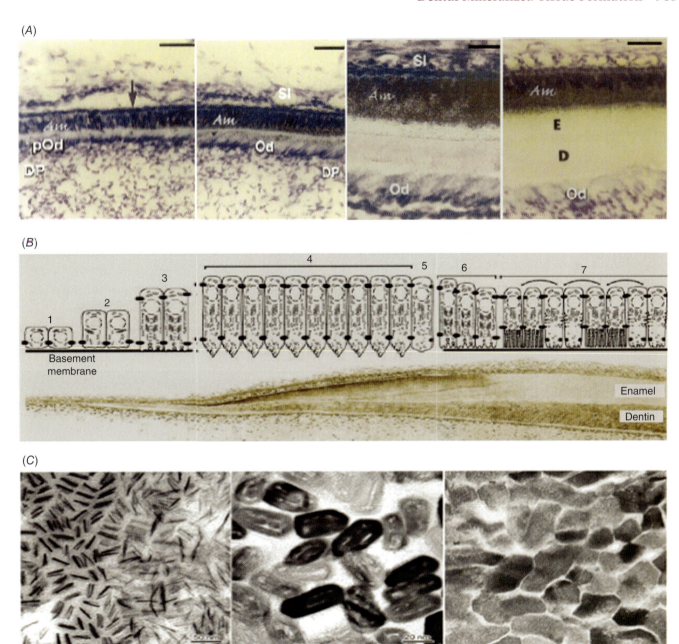

Fig. 117.3. (A) Stage-specific gene expression is a characteristic of amelogenesis. Here, as an example, amelogenin mRNA shows low expression in pre-ameloblasts and maturation stage ameloblasts. In contrast, secretory stage ameloblasts show strong expression of amelogenin mRNA using in situ hybridization (black spots). Am = ameloblasts; D = dentin; DP = dental pulp; E = enamel; Od = odontoblasts; pOd = pre-odontoblasts; SI = stratum intermedium. (B) Ameloblast changes during enamel formation. The epithelial cells of the inner enamel epithelium (1) rest on a basement membrane. These cells increase in length as the ameloblasts differentiate above the predentin matrix (2). Presecretory ameloblasts send processes through the degenerating basement membrane as they initiate the secretion of enamel proteins on the dentin surface (3). After establishing the dental–enamel junction and mineralizing a thin layer of aprismatic enamel, secretory ameloblasts develop a secretory specialization, or Tomes' process. Along the secretory face of the Tomes' process, in place of the absent basement membrane, ameloblasts secrete proteins at a mineralization front where the enamel crystals grow in length (4). At the end of the secretory stage, ameloblasts lose their Tomes' processes and produce a thin layer of aprismatic enamel (5). At this point the enamel has achieved its final thickness. During the transition stage of amelogenesis, ameloblasts undergo a major restructuring that diminishes their secretory activity and changes the types of proteins secreted (6). Proteinases are secreted in the matrix to degrade the accumulated enamel proteins and new basement membrane is formed. During the maturation stage, ameloblasts modulate between ruffled and smooth-ended phases (7). Ameloblasts continue to harden the enamel layer by promoting deposition of mineral on the sides of enamel crystals laid down during the secretory stage. (C) Enamel crystals are very thin at the secretory stage (left panel). Enamel crystals thicken during the maturation stage (central and right panels). Source: [54]. Reproduced with permission of Elsevier.

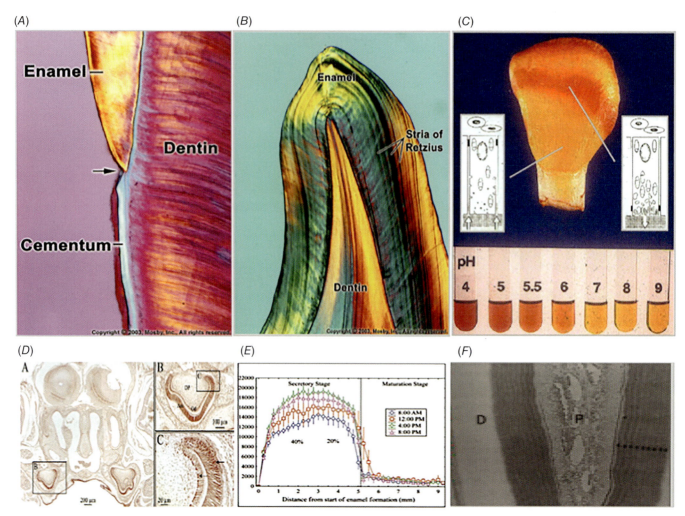

Fig. 117.4. (A) Enamel, dentin, and cementum constitute the three dental mineralized tissues. All three tissues are controlled by circadian rhythms. (B) Striae of Retzius (SR) lines (also called long-period incremental lines) are seen in enamel. About seven cross-striations are observed as vertical lines between the SR. (C) Maturation ameloblasts cycle between ruffle-ended and smooth-ended phases. Beneath the ruffle-ended ameloblasts the crystals rapidly mineralize and the pH drops below 6 (red). Beneath the smooth-ended ameloblasts the enamel is neutralized and rises above 7.0 (orange). (D) Immunohistochemistry of clock protein in a postnatal 4-day-old mouse. (A) Clock protein expression was detected in the developing first molars. (B,C) Higher magnifications showing that the nuclei (arrows) of ameloblasts (Am) and odontoblasts (Od) have strong clock expression relative to the dental pulp (DP) cells. (E) Daily variation in ameloblast secretion of proteins containing methionine. Substantially greater amounts of secretory activity for enamel proteins occur in the late afternoon (4:00 p.m., green) compared with early morning (8:00 a.m., blue) throughout the secretory stage. These differences are noticeably larger (up to 40%) for inner enamel formation (distance, 0.5 to 3.0 mm) than for outer enamel formation (20%). (F) Similar to enamel, dentin (D) also contains marks of short- and long-period growth lines. Lines of Owen, the equivalent of SR, can be traced over considerable distances and are deposited 6 to 10 days apart in different individuals. This radiograph of a transverse dentin section shows nine densely labeled circumpulpal bands after infusion with labeled proline for 10 days. P = pulp. Reproduced with permission of Elsevier.

Amelogenesis is characterized by a maturation stage, during which mineral is deposited exclusively on the sides of the crystals. Crystals grow in width and thickness until further growth is prevented by contact with adjacent crystals. Maturation stage ameloblasts oscillate between smooth-ended and ruffle-ended morphologies (Fig. 117.4C). Rapid mineralization occurs beneath ruffle-ended ameloblasts, which is associated with a substantial drop in pH (acidity is generated by mineral deposition) [36]. Smooth-ended ameloblasts neutralize this acid by secreting bicarbonate.

During the maturation stage, the ameloblasts create the space necessary for a continued increase in the volume of mineral by gradually removing enamel proteins progressively degraded by stage-specific proteases (ie, MMP-20 at the secretory stage and kallikrein 4 [klk4] at the maturation stage) [37]. Maturation-stage ameloblasts secrete and assemble a specialized basal lamina that securely attaches them to the enamel surface to assure its mineralization [38]. Two ameloblast-specific basal lamina proteins have recently been characterized: amelotin (AMTN) and Apin (ODAM) [39]. Transition- and maturation-stage ameloblasts also express genes necessary for enamel mineralization such as carbonic anhydrase II (CA2), which allows the cells to secrete bicarbonate to neutralize the acid generated by hydroxyapatite formation, calcium binding protein (Calb1), calcium sensing receptor (CaSR), anion exchanger Ae2 (SLC4a4), and alkaline phosphatase (TNAP). The expression of several growth factors, hormones, and transcription factors is also characterized during amelogenesis [40]. Of interest, genes initially found to be expressed only in ameloblasts were also detected in odontoblasts [41], and vice versa, odontoblast-specific genes were found to be expressed in ameloblasts [42].

As ameloblasts move away from the dentin, they travel in groups across the surface that they make. This results in decussation (crossing in an X fashion) of the enamel prisms, with zones of prisms with contrasting 3D courses forming the Hunter-Schreger bands. The unerupted enamel is protected from resorption by a layer of cells termed the reduced enamel epithelium, generated from mature ameloblasts remnants. These cells disappear once the tooth erupts.

Cementum and periodontal ligament

The dental follicle, a sac of loose connective tissue that separates the developing tooth from its bony crypt, is essential for eruption and will become the PDL on tooth eruption.

The cementum is an avascular and unnerved mineralized tissue with ultrastructural similarity to bone and covers the entire root surface. It is the interface between the dentin and the PDL and contributes to periodontal tissue repair and regeneration after damage. The organic extracellular matrix of cementum contains proteins that selectively enhance the attachment and proliferation of cell populations residing within the PDL space [43].

Cementum is deposited initially on the newly mineralized dentin matrix of the root by cells derived from the dental follicle and/or by epithelial–mesenchymal transition of root sheath epithelial cells [44]. Secretory proteins from the cells of the epithelial root sheath may be included in the first-formed cementum matrix. Embryologically, there are two types of cementum. These are the primary cementum (which is acellular and develops slowly as the tooth erupts) and the secondary cementum (which is formed after the tooth is in occlusion). Where cementum is deposited very rapidly, it is cellular, containing cementocytes that resemble the osteocytes of bone.

The main collagen of both the extrinsic and intrinsic fibers is type I. The noncollagenous proteins of cementum identified are similar to bone matrix proteins, making difficult to distinguish cementum from other calcified connective tissues. So far only the cementum attachement protein (CAP) may be specific to cementum but its specificity is controversial.

During root development, the dental follicle becomes rapidly organized into the PDL, which supports the tooth, provides nutrition and mechano-sensation, and allows physiological tooth movement. The PDL is unique among the various ligament and tendon systems of the body, in the sense that it is the only soft tissue to span between two distinct hard tissues, namely the cementum of the dental root and the alveolar bone [45].

Through the groups of fibers of the PDL, comprising type I and type III collagen, functioning teeth are linked to each other, the gingiva, and the alveolar bone. On either side of the ligament, its principal fibers are incorporated within cementum and alveolar bone. Within the ligament, there is constant adaptive remodeling of the soft tissue. The PDL collagen fibers are categorized according to their orientation and location along the tooth. The completeness and vitality of the PDL fibers are essential for the functioning of the tooth.

The alveolar bone is a part of the periodontal tissues, functioning as an anchorage of the tooth root to the alveoli and resorbing the forces generated by the function of mastication. Progenitor cells, which are responsible for alveolar bone formation, lie in the periosteal region, the PDL, or around the blood vessels. Alveolar bone marrow is considered to be a useful and easily accessible source of progenitor cells, as they have similar osteogenic potential to those derived from the iliac crest [46]. The periosteum is also considered as a suitable cell source for bone regeneration.

As the permanent tooth erupts, the alveolar bone is resorbed to allow its passage, its root develops, and the crown transverses the oral mucosa that contributes to a tight ring seal of epithelial cells on the enamel close to the junction of the crown and root. The complex molecular signaling cascades controlling eruption and root growth are unclear. At eruption, the root of the tooth is not yet fully formed, and root completion in humans takes 18 months more in the deciduous teeth and up to 3 years in the permanent teeth. Cementum in permanent teeth sees little remodeling, but the surface of alveolar bone is continually resorbing and forming to allow the tooth to move in response to eruption, growth drift, or changing functional forces. Resorption of deciduous tooth roots begins shortly after their completion, appearing first and most extensively on the aspect adjacent to the successional tooth.

Circadian formation of dental mineralized tissues

Dental mineralized tissues form by additive modes of growth that preserve within the hard tissues short- and long-period lines of incremental growth. In dental enamel there are two regularly occurring incremental markers: daily cross-striations and long-period striae of Retzius. These lines correspond to what was, at precise points in time during the secretory stage of amelogenesis, the enamel surface (Fig. 117.4B). Daily incremental lines (called von Ebner's lines) are also observed in dentin. Cross-striations in enamel and von Ebner's lines in dentin delineate the amount of mineral deposited in a single day. Circadian rhythms have been demonstrated using ^{3}H-proline tracers that label collagen in dentin formation [47]. Twice as much collagen is secreted during the daylight 12 hours as during the night time 12 hours (Fig. 117.4E). Similarly to collagen, amelogenin shows circadian rhythms [48, 49]. Consistently, ameloblasts and odontoblasts strongly express clock genes [50].

Although cementum is laid down centrifugally from the cementum–dentin junction and is marked by incremental lines, little is known regarding the circadian control of cementum formation.

Elucidating the role of circadian control on dental mineralized tissue formation may help in understanding the phenotypic differences in tooth structure and form among individuals. Altered expression or polymorphisms of clock genes may also be related with dental disease predisposition, as has shown for other pathological situations such as diabetes and cancer.

STEM CELLS DURING TOOTH REPAIR

Stem cells play a critical role in tissue homeostasis and repair. Their fate is regulated by cell intrinsic determinants and signals from a specialized microenvironment. The reparative mechanisms following dental injury involve a series of highly conserved processes that share genetic programs that occur throughout embryogenesis (reviewed in [51]). In a severe injury, the dying odontoblasts are replaced by stem/progenitor cells, which differentiate into a new generation of odontoblasts that produce the reparative dentin [49]. Signaling molecules released at the injury site may attract dental pulp stem cells and thus initiate the healing process. Notch molecules and nestin are involved in the dynamic processes triggered by pulp injury [51]. Notch expression is activated in cells close to the injury site, as well as in cells located at the root apex, suggesting that these sites represent stem cell niches within the dental pulp. Activation of the Notch molecules in endothelial cells after injury may reflect another pool of stem cells. Epithelial stem cells also exist in human teeth in the root area (Athanassiou et al., unpublished data) and may be implicated in the regeneration of cementum and PDL tissues [52].

CONCLUSION

Elucidating the controls of tooth initiation, pattern, and mineralization necessitates a thorough understanding of the cellular and molecular levels. Understanding when and how signaling molecules control these events will open new horizons and create new challenges. Although much has been discovered about the signaling pathways that dictate tooth morphogenesis, very little is known about the control of dental cell differentiation and subsequent matrix formation. Mutliscale mathematical modeling of complex pathway interactions during tooth morphogenesis and differentiaion may result in a better understanding of development and diseases [53]. Novel scientific knowledge together with tissue engineering approaches are likely to instruct the development of novel therapies in dentistry.

ACKNOWLEDGMENTS

We would like to thank all the members of the Laboratory of Dental Research at the University of Michigan and in particular Drs. Hu, Simmer, and Yamakoshi for scientific interactions, discussions, and material shared. The research presented here is partially supported by the NIH grant DE018878-01A1 to Petros Papagerakis. The work of Thimios Mitsiadis has been supported by funds from the University of Zurich.

REFERENCES

1. Mitsiadis TA, Graf D. Cell fate determination during tooth development and regeneration. Birth Defects Res C Embryo Today. 2009;87:199–211.
2. Bluteau G, Luder HU, De Bari C, et al. Stem cells for tooth engineering. Eur Cell Mater. 2008;16:1–9.
3. Vainio S, Karavanova I, Jowett A, et al. Identification of BMP-4 as a signal mediating secondary induction between epithelial and mesenchymal tissues during early tooth development. Cell. 1993;75:45–58.
4. Dassule HR, McMahon AP. Analysis of epithelial-mesenchymal interactions in the initial morphogenesis of the mammalian tooth. Dev Biol. 1998;202:215–27.
5. Khan M, Seppala M, Zoupa M, et al. Hedgehog pathway gene expression during early development of the molar tooth root in the mouse. Gene Expr Patterns. 2007;7:239–43.
6. Bei M. Moleculer genetics of ameloblast cell lineage. J Exp Zool B Mol Dev Evol. 2009;312B:437–44.
7. Mucchielli ML, Mitsiadis TA, Raffo S, et al. Mouse Otlx2/RIEG expression in the odontogenic epithelium precedes tooth initiation and requires mesenchyme-derived signals for its maintenance. Dev Biol. 1997;189(2):275–84.
8. Trainor PA, Krumlauf R. Patterning the cranial neural crest: hindbrain segmentation and Hox gene plasticity. Nat Rev Neurosci. 2000;1:116–24.
9. Mikkola ML. TNF superfamily in skin appendage development. Cytokine Growth Factor Rev. 2008;19:219–30.
10. Zhang J, Han D, Song S, et al. Correlation between the phenotypes and genotypes of X-linked hypohidrotic ectodermal dysplasia and non-syndromic hypodontia caused by ectodysplasin-A mutations. Eur J Med Genet. 2011;54:e377–82.
11. Mitsiadis TA, Hirsinger E, Lendahl U, et al. Delta-Notch signaling in odontogenesis: correlation with cytodifferentiation and evidence for feedback regulation. Dev Biol. 1998;204:420–31.

12. Mitsiadis TA, Graf D, Luder H, et al. BMPs and FGFs target Notch signalling via jagged 2 to regulate tooth morphogenesis and cytodifferentiation. Development. 2010;137:3025–35.
13. Penton A, Leonard L, Spinner N. Notch signaling in human development and disease. Semin Cell Dev Biol. 2012;23(4):450–7.
14. Caton J, Luder HU, Zoupa M, et al. Enamel-free teeth: Tbx1 deletion affects amelogenesis in rodent incisors. Dev Biol. 2009;328:493–505.
15. Toka O, Kari M, Dittrich S, et al. Dental aspects in patients with DiGeorge syndrome. Quintessence Int. 2010;41:551–6.
16. Reith EJ. The early stage of amelogenesis as observed in molar teeth of young rats. J Ultrastruct Res. 1967;17:503–26.
17. Inai T, Kukita T, Ohsaki Y, et al. Immunohistochemical demonstration of amelogenin penetration toward the dental pulp in the early stages of ameloblast development in rat molar tooth germs. Anat Rec. 1991;229:259–70.
18. Sasaki T, Garant PR. Structure and organization of odontoblasts. Anat Rec. 1996;245:235–49.
19. Yamakoshi Y, Hu JC, Fukae M, et al. Dentin glycoprotein: the protein in the middle of the dentin sialophosphoprotein chimera. J Biol Chem. 2005;280:17472–9.
20. Butler WT, Ritchie H. The nature and functional significance of dentin extracellular matrix proteins. Int J Dev Biol. 1995;39:169–79.
21. Bailleul-Forestier I, Davideau JL, Papagerakis P, et al. Immunolocalization of vitamin D receptor and calbindin-D28k in human tooth germ. Pediatr Res. 1996;39:636–42.
22. Borke JL, Zaki AE, Eisenmann DR, et al. Expression of plasma membrane Ca pump epitopes parallels the progression of mineralization in rat incisor. J Histochem Cytochem. 1993;41:175–81.
23. Goseki M, Oida S, Nifuji A, et al. Properties of alkaline phosphatase of the human dental pulp. J Dent Res. 1990;69:909–12.
24. Smith AJ, Cassidy N, Perry H, et al. Reactionary dentinogenesis. Int J Dev Biol. 1995;39:273–80.
25. Mitsiadis TA, De Bari C, About I. Apoptosis in developmental and repair-related human tooth remodeling: a view from the inside. Exp Cell Res. 2008;314:869–77.
26. Simmer JP, Fincham AG. Molecular mechanisms of dental enamel formation. Crit Rev Oral Biol Med. 1995;6:84–108.
27. Daculsi G, Menanteau J, Kerebel LM, et al. Length and shape of enamel crystals. Calcif Tissue Int. 1984;36:550–5.
28. Nanci A. Enamel: composition, formation, and structure. In: Nanci A (ed.) *Ten Cate's Oral Histology Development, Structure, and Function*. St. Louis, MO: Mosby, 2003, pp. 145–91.
29. Smith CE, Nanci A. Overview of morphological changes in enamel organ cells associated with major events in amelogenesis. Int J Dev Biol. 1995;39:153–61.
30. Hu JC, Chun YH, Al Hazzazzi T, et al. Enamel formation and amelogenesis imperfecta. Cells Tissues Organs. 2007;186:78–85.
31. Simmer JP, Papagerakis P, Smith CE, et al. Regulation of dental enamel shape and hardness. J Dent Res. 2010;89:1024–38.
32. Robinson C, Brookes SJ, Shore RC, et al. The developing enamel matrix: nature and function. Eur J Oral Sci. 1998;106(suppl 1):282–91.
33. Fincham AG, Moradian-Oldak J, Diekwisch TG, et al. Evidence for amelogenin "nanospheres" as functional components of secretory-stage enamel matrix. J Struc Biol. 1995;115:50–9.
34. Fukumoto S, Kiba T, Hall B, et al. Ameloblastin is a cell adhesion molecule required for maintaining the differentiation state of ameloblasts. J Cell Biol. 2004;167:973–83.
35. Hu JC, Hu Y, Smith CE, et al. Enamel defects and ameloblast-specific expression in enamelin knockout/LACZ knockin mice. J Biol Chem. 2008;283:10858–71.
36. Smith CE. Cellular and chemical events during enamel maturation. Crit Rev Oral Biol Med. 1998;9:128–61.
37. Lu Y, Papagerakis P, Yamakoshi Y, et al. Functions of KLK4 and MMP-20 in dental enamel formation. Biol Chem. 2008;389:695–700.
38. Al Kawas S, Warshawsky H. Ultrastructure and composition of basement membrane separating mature ameloblasts from enamel. Arch Oral Biol. 2008;53:310–7.
39. Moffatt P, Smith CE, St-Arnaud R, et al. Characterization of Apin, a secreted protein highly expressed in tooth-associated epithelia. J Cell Biochem. 2008;103:941–56.
40. Davideau JL, Papagerakis P, Hotton D, et al. In situ investigation of vitamin D receptor, alkaline phosphatase, and osteocalcine gene expression in oro-facial mineralized tissues. Endocrinology. 1996;137:3577–85.
41. Papagerakis P, MacDougall M, Bailleul-Forestier I, et al. Expression of amelogenin in odontoblasts. Bone. 2003;32:228–40.
42. Papagerakis P, Berdal A, Mesbah M, et al. Investigation of osteocalcin, osteonectin, and dentin sialophosphoprotein in developing human teeth. Bone. 2002;30:377–85.
43. MacNeil RL, Somerman MJ. Molecular factors regulating development and regeneration of cementum. J Periodontal Res. 1993;28:550–9.
44. Huang X, Bringas P Jr, Slavkin HC, et al. Fate of HERS during tooth root development. Dev Biol. 2009;334:22–30.
45. McCulloch CA, Lekic P, McKee MD. Role of physical forces in regulating the form and function of the periodontal ligament. Periodontol 2000. 2000;24:56–72.
46. Matsubara T, Suardita K, Ishii M, et al. Alveolar bone marrow as a cell source for regenerative medicine: differences between alveolar and iliac bone marrow stromal cells. J Bone Miner Res. 2005;20:399–409.
47. Ohtsuka M, Saeki S, Igarashi K, et al. Circadian rhythms in the incorporation and secretion of 3H-proline by odontoblasts in relation to incremental lines in rat dentin. J Dent Res. 1998;77:1889–95.

48. Athanassiou-Papaefthymiou M, Kim D, Harbron L, et al. Molecular and circadian controls of ameloblasts. Eur J Oral Sci. 2011;119(suppl 1):35–40.
49. Zheng L, Seon YJ, Mourão MA, et al. Circadian rhythms regulate amelogenesis. Bone. 2013;55(1):158–65.
50. Zheng L, Papagerakis S, Schnell SD, et al. Expression of clock proteins in developing tooth. Gene Expr Patterns. 2011;11:202–6.
51. Mitsiadis TA, Rahiotis C. Parallels between tooth development and repair: conserved molecular mechanisms following carious and dental injury. J Dent Res. 2004;83:896–902.
52. Mitsiadis TA, Papagerakis P. Regenerated teeth: the future of tooth replacement? Regen Med. 2011;6:135–9.
53. Papagerakis S, Zheng L, Schnell S, et al. The circadian clock in oral health and diseases. J Dent Res. 2014;93(1):27–35.
54. Hu JC, Chun YH, Al Hazzazzi T, et al. Enamel formation and amelogenesis imperfecta. Cells Tissues Organs. 2007;186:78–85.

118

Genetic Craniofacial Disorders Affecting the Dentition

Yong-Hee Patricia Chun[1], Paul H. Krebsbach[2], and James P. Simmer[3]

[1]Department of Periodontics; and Department of Cell Systems and Anatomy, University of Texas Health Science Center at San Antonio, San Antonio, TX, USA
[2]UCLA School of Dentistry, Los Angeles, CA, USA
[3]Department of Biological and Materials Sciences, University of Michigan Dental Research Laboratory, Ann Arbor, MI, USA

GENETIC DISORDERS AFFECTING THE DENTITION

Regulation of early development of ectodermal organs (eg, hair, teeth, and many exocrine glands) originates from similar sequential and reciprocal interactions between the epithelium and mesenchyme [1]. Four conserved signaling pathways mediate epithelial–mesenchymal interactions during tooth development: BMPs, FGFs, sonic hedgehog (Shh), and wingless-related (Wnt). Additionally, the ectodermal-specific ectodysplasin A (EDA) signaling pathway controls tooth number and tooth shape through regulation of *Fgf20* expression in the dental epithelium [2]. Genetic disorders that alter early events in tooth development may be limited in their phenotypes to the dentition or compromise the development of other ectodermal organs, as in syndromes.

The skeleton contains two of the five mineralized tissues in the body: bone and calcified cartilage. The other three mineralized tissues, dentin, enamel, and cementum, are found in teeth. The mineral in each of these hard tissues is a biological apatite structurally resembling calcium (Ca) hydroxyapatite $Ca_{10}(PO_4)_6(OH)_2$, with the most common substitutions being carbonate (CO_3^{2-}) for phosphate (PO_4^{3-}), and fluoride (F^-) for the hydroxyl (OH^-) group. Therefore, disorders of Ca and inorganic phosphate (Pi) metabolism potentially affect multiple hard tissues. In some conditions, the oral manifestations may be the earliest or most obvious sign of a broader problem involving bone and mineral metabolism and may lead to the diagnosis.

This chapter provides a concise overview of dental genetic malformations that offer early evidence of undiagnosed syndromic or systemic conditions of bone and mineral.

Genetic diseases affecting number of teeth: familial tooth agenesis and supernumerary teeth

Familial tooth agenesis is common and can be an isolated finding or occur in syndromes [3]. The most prominent causes of familial tooth agenesis are genetic defects in the Wnt signaling system—ie, wingless-type MMTV integration site family, member 10A (*WNT10A*, 2q35), low-density lipoprotein receptor-related protein 6 (*LRP6*, 12p12.3) [4], and axis inhibition protein 2 (*AXIN2*, 17q23-q24) [5]. Individuals with mutations in a single *WNT10A* allele exhibit molar root taurodontism and mild tooth agenesis with incomplete penetrance in their permanent dentitions, whereas mutations in both alleles cause severe tooth agenesis and fewer cusps on their molars [6]. Although rare, specific *AXIN2* mutations cause an autosomal dominant (AD) form of familial tooth agenesis associated with gastrointestinal polyps that turn malignant by the fourth decade [5].

Msh homeobox 1 (*MSX1*, 4p16.1) and paired box 9 (*PAX9*, 14q12-q13) express interacting transcription factors that are critical for the progression of tooth development beyond the bud stage [7, 8]. Mutations in *MSX1* and *PAX9* cause similar AD patterns of familial tooth agenesis, with *PAX9* mutations being more likely to include second molars, whereas *MSX1* mutations are more likely to include maxillary first bicuspids among the teeth missing [9].

Hypohidrotic ectodermal dysplasia (HED) is a heritable condition that features missing teeth, thin and sparse hair, absent sweat glands, and defective nails and salivary glands. X-linked recessive hypohidrotic ectodermal dysplasia (XLHED) affects only males and is caused by mutations in ectodysplasin (*EDA*, Xq12-q13.1). In mild cases, hypodontia is the only manifestation [10]. Mutations in the genes that encode the ectodysplasin A receptor (*EDAR*, 2q13) and the EDAR-associated death domain (*EDARADD*, 1q42.3) cause ectodermal dysplasia. Additionally, combinations of heterozygous mutations in *EDAR*, *EDARADD*, and/or *WNT10A* (digenic inheritance) greatly increase the severity of tooth agenesis (increases the number of missing teeth) compared with siblings with a heterozygous defect in only one of these genes [11].

Syndromic tooth agenesis can result from mutations in *FGF10* (5p13) and its receptors (ie, *FGFR3*, 4p16.3), and involve aplasia of the lacrimal and salivary glands (ALSG), [12] or cause lacrimo-auriculo-dento-digital (LADD) syndrome [13]. LADD syndrome features aplasia or hypoplasia of the lacrimal and salivary systems associated with a variety of dental phenotypes: agenesis of the lateral maxillary incisors (hypodontia), or small (microdontia) and peg-shaped laterals, mild enamel dysplasia, and delayed tooth eruption.

Variations in tooth number can also include extra teeth. Nonsyndromic supernumerary teeth are found in ~1% of primary dentitions and ~2% of permanent dentitions, and tend to be familial, but of unknown etiology. Supernumerary teeth also occur in syndromes [14]. Cleidocranial dysplasia is an AD disorder caused by mutations in runt-related transcription factor 2 (*RUNX2*, 6p21). The most notable syndromic feature is hypoplasia or aplasia of the clavicles permitting abnormal apposing of the shoulders on physical examination. Multiple supernumerary teeth form, and fail to erupt, but respond well to orthodontic movement [15].

Familial adenomatous polyposis is a disorder caused by truncation mutations in adenomatous polyposis of the colon (*APC*, 5q21-q22) and is characterized by radioopaque lesions in the jaw comprised of clumped toothlets (odontomas) and gastrointestinal polyps that usually undergo malignant change by the fourth decade [16]. The discovery of either odontomas or familial tooth agenesis should raise concern about gastrointestinal polyps, especially when the oral findings appear to have arisen spontaneously (ie, not observed in either parent), as the family will not manifest intestinal cancer.

Genetic diseases affecting dentin: osteogenesis imperfecta, dentinogenesis imperfecta, and dentin dysplasia

Inherited dentin defects can be isolated or syndromic [17]. The most common disorder with dentin malformations is AD osteogenesis imperfecta (OI), which is caused by defects in the genes that encode type I collagen [18]. In some instances of OI, the dentin defects are the only prominent manifestation [19,20]. Nonsyndromic dentin defects are predominantly caused by dominant-negative AD mutations in dentin sialophosphoprotein (*DSPP*, 4q21.3) [21].

The Shields classification of inherited dentin defects divides genetic dentin malformations into two disease groups with five subtypes: dentinogenesis imperfecta (DGI, types I to III) and dentin dysplasia (DD, types I and II), with all five forms showing AD inheritance [22]. Type I DGI is a collective designation for OI with DGI, now largely abandoned in deference to the current OI nosology. Type II DGI, however, is more prevalent. Clinically, the teeth of type II DGI feature an amber-like appearance (Fig. 118.1). Also, they are narrower at the cervical margins and thus exhibit a bulbous or bell-shaped crown. Type III DGI is a rare form, also called the "Brandywine isolate," after the prototype kindred identified in Brandywine, Maryland. This form features in the deciduous teeth with multiple pulp exposures with considerable variation radiographically, ranging from shell teeth to normal pulp chambers to pulpal obliteration. The permanent teeth are the same as in type II DGI. In type I DD both the permanent and deciduous teeth appear to have normal shapes and color. Dental radiographs, however, show short roots with periapical radiolucencies in noncarious teeth. The primary teeth show total obliteration of the pulp. Type II DD appears to be a mild form of type II DGI, featuring amber tooth coloration with total pulpal obliteration in

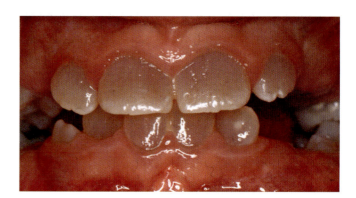

Fig. 118.1. Dental features of dentinogenesis imperfecta. These permanent teeth of this patient exhibit the characteristic blue-gray or opalescent appearance associated with dentinogenesis imperfecta. Courtesy of Dr. Jan C-C. Hu.

the primary teeth, and a thistle-tube pulp configuration with ubiquitous pulp stones and normal to near normal coloration in the permanent teeth.

It is important for patients with DGI displaying attrition of the permanent first molars that the dentist intervene by placing stainless steel crowns over the deteriorating first molars to maintain vertical dimension, thereby preventing rapid attrition of the anterior teeth. The stainless steel crowns are later replaced with cast gold crowns when the pulp chambers have receded sufficiently to avoid exposure during crown preparation.

Genetic diseases affecting enamel: amelogenesis imperfecta

Amelogenesis imperfecta (AI) is a heterogeneous group of isolated inherited defects in the enamel layer of teeth. The enamel may be thin, soft, rough, and/or pigmented. The pattern of inheritance can be AD, autosomal recessive (AR), X-linked dominant (XLD), or X-linked recessive (XLR) [23]. When the various enamel phenotypes and the pattern of inheritance are considered, 14 subtypes are recognized [23]. Isolated enamel defects have been associated with defects in *AMELX*, *ENAM*, *AMBN*, *AMTN*, *MMP20*, *FAM83H*, *KLK4*, *SLC24A4*, *WDR72*, *ITGB6*, *C4orf26*, *GPR68*, *ACP4*, *COL17A1*, *LAMA3*, and *LAMB3*. The enamel in patients with AI can be thin (hypoplastic) or absent, of normal thickness but soft and stained (hypomaturation), or both thin and soft (hypocalcification). Amelogenin (*AMELX*, Xp22.3) mutations cause XLD AI with heterozygous females often showing vertical bands of normal and defective enamel, while hemizygous males usually have little or no enamel on their teeth [24].

AI also refers to the enamel defects in syndromes. Persons with AD enamel malformations are carriers of AR junctional epidermolysis bullosa when the causative mutations are in *COL17A1*, *LAMA3*, or *LAMB3*. Mutations in both alleles of the cyclin and CBS domain divalent metal cation transport mediator 4 gene (*CNNM4*, 2q11), which encodes a protein that transports magnesium ions out of ameloblasts, cause cone–rod dystrophy and AI [25]. The cone–rod dystrophy leads to blindness. Mutations in family with sequence similarity 20 member A (*FAM20A*, 17q24.2) cause enamel renal syndrome [26]. The distinctive combination of oral findings in this disorder include severe enamel hypoplasia, unerupted molars associated with pericoronal radiolucencies, intrapulpal calcifications, and enlarged gingiva [27]. Mutations in porcupine (*PORCN*, Xp11.23) cause focal dermal hypoplasia, an XLD condition with in utero lethality in males. The enamel defects show vertical banding [28, 29] due to random X-chromosome inactivation causing mosaicism for expression of the defective copy of the gene (lyonization).

Genetic diseases causing cleft lip and palate

Cleft lip and/or palate (CL/P) are relatively common craniofacial malformations (one in 700 births) that can profoundly impact on nutritional, speech, dental, and psychological development. Most facial clefting birth defects are multifactorial and nonsyndromic, although there are nearly 300 recognized syndromes that may include a facial cleft as a manifestation. Between 15% and 50% of instances of CL/P occur in defined syndromes. The more common syndromes with cleft palates include Apert (*FGFR2*, 10q26.13), Stickler (*COL2A1*, 12q13.11), and Treacher Collins (*TCOF1*, 5q32). Van der Woude (*IRF6*, 1q32.2) and Waardenberg (*NECTIN1*, 11q23.3) syndromes are associated with CL/P. In Van der Woude syndrome, mutation analysis can identify variants in the interferon regulatory factor 6 gene that correlate with an increased risk of facial clefting [30].

Some of the genes that cause syndromic CLP are also implicated in nonsyndromic CL/P, such as *TBX22* (Xq21.1), *NECTIN1*, and *IRF6* [31]. While most nonsyndromic CL/P cases have a complex etiology involving both genetic and environmental factors, in some cases the CL/P is caused by highly penetrant deleterious variations in specific genes and show a familial pattern of inheritance [32].

ORAL MANIFESTATIONS OF METABOLIC BONE DISEASES OF GENETIC ORIGIN

Metabolic diseases of bone are disorders of bone remodeling that characteristically involve the entire skeleton, and are often manifest in the oral cavity, which can lead to the diagnosis of the underlying systemic disease. Numerous studies suggest that subclinical derangements in calcium homeostasis and bone metabolism may also contribute to a variety of dental abnormalities including alveolar ridge resorption and periodontal bone loss in predisposed individuals. The significance of this spectrum of diseases and their overall impact on oral health and dental management are likely to increase as the elderly proportion of the population increases in the coming decades [33].

Hypophosphatemia

Phosphorus homeostasis, is incompletely understood (see Chapter 25). Though regulated by many hormones including PTH) calcitonin, and vitamin D, now Pi-regulating factors have been discovered. FGF23 is secreted by osteocytes into the circulation and inhibits Pi reabsorption and $1,25(OH)_2D$ production by the kidney (see Chapter 27). Specific defects in the FGF23 gene (*ADHR*, 12p13) cause AD hypophosphatemic rickets. Phosphate-regulating neutral endopeptidase (PHEX), an endopeptidase, and dentin matrix protein 1 (DMP1), a proteoglycan involved in bone mineralization, both inhibit FGF23 expression. Loss of function mutations in *PHEX* (Xp22.2)

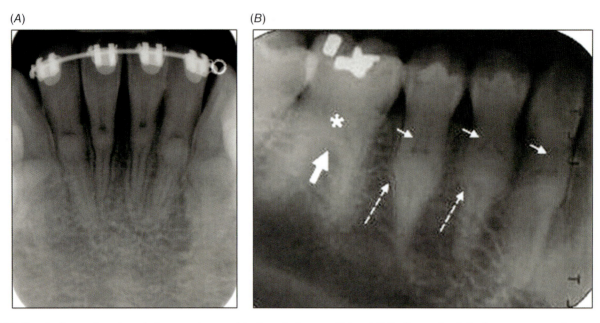

Fig. 118.2. Radiographs of teeth from patients with dental abnormalities. (A) Radiograph of the lower incisors from a patient with a homozygous missense mutation (p.Ser71Gly) in *FGF23* showing round calcareous deposits (arrows) within the pulp chambers. (B) Radiograph of the lower incisors from a patient with hyperphosphatemic familial tumoral calcinosis with compound heterozygous mutations (p.Arg438Cys and p.Gln592*) in *GALNT3* with bulbous tooth roots (dashed arrows), thistle-shaped pulps (small arrows), long root trunk (thick arrow), and pulp stones (asterisk). (A) Source [48]. (B) Source: [46] Reproduced with permission of Springer.

and *DMP1* (4q21) cause XLD and AR hypophosphatemia, respectively [34–37].

In familial hypophosphatemia, dental findings are often the presenting sign and resemble those seen generally in rickets and osteomalacia. Patients may present with enamel hypoplasia, enamel discoloration, poorly mineralized dentin, enlarged pulp chambers and root canals, and abscessed primary or permanent teeth that have no signs of dental caries [38–40]. Microbial infection of the pulp is thought to occur through invasion of dentinal tubules exposed by attrition of enamel or through enamel microfractures [41]. Improving understanding of the mechanisms of phosphate homeostasis is providing new insights into the genetic etiologies of hypophosphatemias [42].

Hyperphosphatemia

Hyperphosphatemic familial tumoral calcinosis (HFTC) is an AR metabolic disorder characterized by hyperphosphatemia, tooth root defects, and the progressive deposition of Ca/P$_i$ crystals in periarticular spaces, soft tissues, and sometimes bone (see Chapter 115). HFTC can be caused by inactivating mutations in polypeptide N-acetylgalactosaminyltransferase 3 (*GALNT3*, 2q24-q31), *FGF23*, or Klotho (*KL*, 13q12) [43]. An early sign of the resulting hyperphosphatemia is a distinctive, localized thickening of dental tooth roots, pulp stones, and partial obliteration of the pulp chamber [44–46] (Fig. 118.2). These root deformities complicate dental treatment, if root canal therapy becomes necessary.

Hypophosphatasia

Hypophosphatasia (HPP) is caused by loss of function mutation of the skeletal alkaline phosphatase (*ALPL*, 1p36.1-p34) gene (see Chapter 115). Osteoblasts show the highest level of ALPL expression, and profound skeletal hypomineralization occurs in the severest forms of HPP. However, the hard tissue that seems to be the most sensitive to the deficiency is cementum [47]. The classic oral presentation of childhood hypophosphatasia is premature loss of fully-rooted deciduous teeth (Fig. 118.3). Histological examination shows lack of cementum on their root surface, so that the attachment apparatus fails because periodontal ligament fibers do not connect the alveolar bone to the root. In the permanent teeth, large pulp spaces, late eruption, and delayed apical closure are often observed. Bone loss uniformly affects areas adjacent to the teeth (horizontal pattern), and in the adult form of HPP there may be widespread dental caries.

ACKNOWLEDGMENTS

The authors thank Dr. Jan C.-C. Hu, University of Michigan for Figs 118.1 and 118.3. This work was supported by awards from the National Institutes of Health,

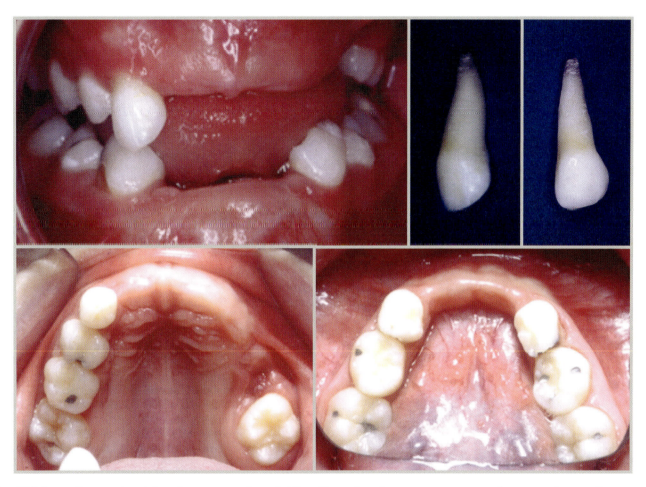

Fig. 118.3. Fully-rooted, exfoliated primary teeth in childhood hypophosphatasia in a patient aged 6 years. The maxillary cuspid and incisor (upper right) brought in by the parents had fallen out without trauma 2 years previously. This patient also showed periodontal attachment problems in her primary posterior teeth, which were mobile. Some childhood hypophosphatasia patients present with enamel hypoplasia. These dental findings are diagnostic of childhood hypophosphatasia and are often the first sign leading to the diagnosis. Courtesy of Dr. Jan C-C. Hu.

National Institutes for Dental and Craniofacial Research DE015846 (JPS), DE022800 (YPC), DE025758 (YPC), and DE026769 (YPC).

The URLs for data presented herein are as follows: Online Mendelian Inheritance in Man (OMIM) http://www.ncbi.nlm.nih.gov/Omim/(accessed May 2018).

REFERENCES

1. Pispa J, Thesleff I. Mechanisms of ectodermal organogenesis. Dev Biol. 2003;262(2):195–205.
2. Lan Y, Jia S, Jiang R. Molecular patterning of the mammalian dentition. Semin Cell Dev Biol. 2014;25–26:61–70.
3. Nieminen P. Genetic basis of tooth agenesis. J Exp Zool B Mol Dev Evol. 2009;312B(4):320–42.
4. Massink MP, Creton MA, Spanevello F, et al. loss-of-function mutations in the WNT co-receptor LRP6 cause autosomal-dominant oligodontia. Am J Hum Genet. 2015;97(4):621–6.
5. Lammi L, Arte S, Somer M, et al. Mutations in AXIN2 cause familial tooth agenesis and predispose to colorectal cancer. Am J Hum Genet. 2004;74(5):1043–50.
6. Yang J, Wang SK, Choi M, et al. Taurodontism, variations in tooth number, and misshapened crowns in Wnt10a null mice and human kindreds. Mol Genet Genomic Med. 2015;3(1):40–58.
7. Stockton DW, Das P, Goldenberg M, et al. Mutation of PAX9 is associated with oligodontia. Nature Genet. 2000;24(1):18–9.
8. Vastardis H, Karimbux N, Guthua SW, et al. A human MSX1 homeodomain missense mutation causes selective tooth agenesis. Nat Genet. 1996;13(4):417–21.
9. Kim JW, Simmer JP, Lin BP, et al. Novel MSX1 frameshift causes autosomal-dominant oligodontia. J Dent Res. 2006;85(3):267–71.
10. Han D, Gong Y, Wu H, et al. Novel EDA mutation resulting in X-linked non-syndromic hypodontia and the pattern of EDA-associated isolated tooth agenesis. Eur J Med Genet. 2008;51(6):536–46.

11. Arte S, Parmanen S, Pirinen S, et al. Candidate gene analysis of tooth agenesis identifies novel mutations in six genes and suggests significant role for WNT and EDA signaling and allele combinations. PLoS One. 2013;8(8):e73705.
12. Entesarian M, Matsson H, Klar J, et al. Mutations in the gene encoding fibroblast growth factor 10 are associated with aplasia of lacrimal and salivary glands. Nat Genet. 2005;37(2):125–7.
13. Milunsky JM, Zhao G, Maher TA, et al. LADD syndrome is caused by FGF10 mutations. Clin Genet. 2006;69(4):349–54.
14. Lubinsky M, Kantaputra PN. Syndromes with supernumerary teeth. Am J Med Genet A. 2016;170(10):2611–6.
15. Proffit WR, Vig KW. Primary failure of eruption: a possible cause of posterior open-bite. Am J Orthod. 1981;80(2):173–90.
16. Oner AY, Pocan S. Gardner's syndrome: a case report. Br Dent J. 2006;200(12):666–7.
17. Kim JW, Simmer JP. Hereditary dentin defects. J Dent Res. 2007;86(5):392–9.
18. O'Connell AC, Marini JC. Evaluation of oral problems in an osteogenesis imperfecta population. Oral Surg Oral Med Oral Path Oral Radiol Endod. 1999;87(2):189–96.
19. Pallos D, Hart PS, Cortelli JR, et al. Novel COL1A1 mutation (G559C) [correction of G599C] associated with mild osteogenesis imperfecta and dentinogenesis imperfecta. Arch Oral Biol. 2001;46(5):459–70.
20. Wang SK, Chan HC, Makovey I, et al. Novel PAX9 and COL1A2 missense mutations causing tooth agenesis and OI/DGI without skeletal abnormalities. PLoS One. 2012;7(12):e51533.
21. Yang J, Kawasaki K, Lee M, et al. The dentin phosphoprotein repeat region and inherited defects of dentin. Mol Genet Genomic Med. 2016;4(1):28–38.
22. Shields ED, Bixler D, el-Kafrawy AM. A proposed classification for heritable human dentine defects with a description of a new entity. Arch Oral Biol. 1973;18(4):543–53.
23. Witkop CJ Jr. Amelogenesis imperfecta, dentinogenesis imperfecta and dentin dysplasia revisited: problems in classification. J Oral Pathol. 1988;17(9–10):547–53.
24. Wright JT, Hart PS, Aldred MJ, et al. Relationship of phenotype and genotype in X-linked amelogenesis imperfecta. Connect Tissue Res. 2003;44 (suppl 1):72–8.
25. Polok B, Escher P, Ambresin A, et al. Mutations in CNNM4 cause recessive cone-rod dystrophy with amelogenesis imperfecta. Am J Hum Genet. 2009;84(2):259–65.
26. Wang SK, Aref P, Hu Y, et al. FAM20A mutations can cause enamel-renal syndrome (ERS). PLoS Genet. 2013;9(2):e1003302.
27. O'Sullivan J, Bitu CC, Daly SB, et al. Whole-exome sequencing identifies FAM20A mutations as a cause of amelogenesis imperfecta and gingival hyperplasia syndrome. Am J Hum Genet. 2011;88(5):616–20.
28. Gysin S, Itin P. Blaschko linear enamel defects—a marker for focal dermal hypoplasia: case report of focal dermal hypoplasia. Case Rep Dermatol. 2015;7(2):90–4.
29. Tejani Z, Batra P, Mason C, et al. Focal dermal hypoplasia: oral and dental findings. J Clin Pediatr Dent. 2005;30(1):67–72.
30. Zuccati G. Implant therapy in cases of agenesis. J Clin Orthod. 1993;27(7):369–73.
31. Kohli SS, Kohli VS. A comprehensive review of the genetic basis of cleft lip and palate. J Oral Maxillofac Pathol. 2012;16(1):64–72.
32. Pengelly RJ, Arias L, Martinez J, et al. Deleterious coding variants in multi-case families with non-syndromic cleft lip and/or palate phenotypes. Sci Rep. 2016;6:30457.
33. Solt DB. The pathogenesis, oral manifestations, and implications for dentistry of metabolic bone disease. Curr Opin Dent. 1991;1(6):783–91.
34. Feng JQ, Ward LM, Liu S, et al. Loss of DMP1 causes rickets and osteomalacia and identifies a role for osteocytes in mineral metabolism. Nat Genet. 2006;38(11):1310–5.
35. Lorenz-Depiereux B, Bastepe M, Benet-Pages A, et al. DMP1 mutations in autosomal recessive hypophosphatemia implicate a bone matrix protein in the regulation of phosphate homeostasis. Nat Genet. 2006;38(11):1248–50.
36. Roetzer KM, Varga F, Zwettler E, et al. Novel PHEX mutation associated with hypophosphatemic rickets. Nephron Physiol. 2007;106(1):8–12.
37. Alizadeh Naderi AS, Reilly RF. Hereditary disorders of renal phosphate wasting. Nat Rev Nephrol. 2010;6(11):657–65.
38. Goodman JR, Gelbier MJ, Bennett JH, et al. Dental problems associated with hypophosphataemic vitamin D resistant rickets. Int J Paediatr Dent. 1998;8(1):19–28.
39. Pereira CM, de Andrade CR, Vargas PA, et al. Dental alterations associated with X-linked hypophosphatemic rickets. J Endod. 2004;30(4):241–5.
40. Baroncelli GI, Angiolini M, Ninni E, et al. Prevalence and pathogenesis of dental and periodontal lesions in children with X-linked hypophosphatemic rickets. Eur J Paediatr Dent. 2006;7(2):61–6.
41. Hillmann G, Geurtsen W. Pathohistology of undecalcified primary teeth in vitamin D-resistant rickets: review and report of two cases. Oral Surg Oral Med Oral Pathol Oral Radiol Endod. 1996;82(2):218–24.
42. Bastepe M, Juppner H. Inherited hypophosphatemic disorders in children and the evolving mechanisms of phosphate regulation. Rev Endocr Metab Disord. 2008;9(2):171–80.
43. Folsom LJ, Imel EA. Hyperphosphatemic familial tumoral calcinosis: genetic models of deficient FGF23 action. Curr Osteoporos Rep. 2015;13(2):78–87.
44. Burkes EJ Jr, Lyles KW, Dolan EA, et al. Dental lesions in tumoral calcinosis. J Oral Pathol Med. 1991;20(5):222–7.
45. Chefetz I, Heller R, Galli-Tsinopoulou A, et al. A novel homozygous missense mutation in FGF23 causes familial tumoral calcinosis associated with disseminated visceral calcification. Hum Genet. 2005;118(2):261–6.

46. Dumitrescu CE, Kelly MH, Khosravi A, et al. A case of familial tumoral calcinosis/hyperostosis-hyperphosphatemia syndrome due to a compound heterozygous mutation in GALNT3 demonstrating new phenotypic features. Osteoporos Int. 2009;20(7):1273–8.
47. Whyte MP. Physiological role of alkaline phosphatase explored in hypophosphatasia. Ann N Y Acad Sci. 2010;1192:190–200.
48. Benet-Pages A, Orlik P, Strom TM, et al. An FGF23 missense mutation causes familial tumoral calcinosis with hyperphosphatemia. Hum Mol Genet. 2005;14(3):385–90.

119
Pathology of the Hard Tissues of the Jaws

Paul C. Edwards

Department of Oral Pathology, Medicine and Radiology, Indiana University School of Dentistry, Indianapolis, IN, USA

INTRODUCTION

Unique to the bones of the jaw is the presence of teeth, specialized hard tissue organs that are supported by the alveolar processes. The teeth are composed of unique hard tissues (enamel, dentin, and cementum) that are found nowhere else in the body and develop through a process involving sequential and reciprocal interactions between oral epithelium and ectomesenchyme. In humans, this comprises the development of two dentitions, the primary (or deciduous) teeth and the permanent teeth, occurring over a time frame from the fetal period to the late teens, in order to accommodate the growth of the jaws.

As a result of the unique developmental processes involved in the development of the teeth and the fact that these specialized mineralized tissues are continuously exposed to the harsh oral environment, the mandible and maxilla are home to a distinctive set of pathological entities. This chapter provides a brief review of some of the more common of these entities: tooth demineralization, dental caries, root resorption, odontogenic cysts, odontogenic neoplasms, and nonodontogenic tumors of the jaw.

TOOTH DEMINERALIZATION, DENTAL CARIES, AND ROOT RESORPTION

Dental caries ("tooth decay") is one of the most prevalent chronic diseases affecting modern society, with estimates of untreated caries of the permanent teeth affecting 2.4 billion people worldwide [1]. Once viewed primarily as a disease of childhood, since the introduction of water fluoridation, fluoride toothpaste, and access to dental care, there has been a dramatic decrease in tooth loss among many children in Western Europe and the USA [2–4]. As a result of this shift towards keeping more teeth into adulthood, caries has shifted from an acute disease of childhood to a slowly progressing chronic disease of adulthood.

The caries process represents the end result of a multifaceted interaction between transmissible cariogenic oral microflora, primarily *Streptococcus mutans* and *Lactobacillus* species, and fermentable dietary carbohydrates [5]. Oral microflora, vertically transmitted to the child through the mother, colonize the teeth through a process involving bacterial surface proteins and salivary constituents on the enamel surface [6]. Through the action of cariogenic oral microflora, an adherent extracellular polysaccharide matrix, the dental biofilm, is produced. The metabolism of refined carbohydrates within this biofilm produces lactic acid, resulting in a drop in pH at the biofilm–tooth interface, leading to dissolution of the mineral component of the tooth. This demineralization is countered by remineralization, involving diffusion of phosphate and calcium ions back into the hydroxyapatite mineral component of the tooth as the local pH level rises through the buffering capacity of saliva. Caries results when the rate of demineralization exceeds that of remineralization (Fig. 119.1).

In addition to the frequency and duration of exposure to fermentable dietary carbohydrate, other factors involved in an individual's risk of developing caries include the virulence of the specific pathogenic genotypes of colonizing *S. mutans*, host immune response, and the protein and mineral composition and buffering capacity of saliva [7]. By disrupting the tooth surface biofilm, oral hygiene practices can reduce the caries risk.

Primer on the Metabolic Bone Diseases and Disorders of Mineral Metabolism, Ninth Edition. Edited by John P. Bilezikian.
© 2019 American Society for Bone and Mineral Research. Published 2019 by John Wiley & Sons, Inc.
Companion website: www.wiley.com/go/asbmrprimer

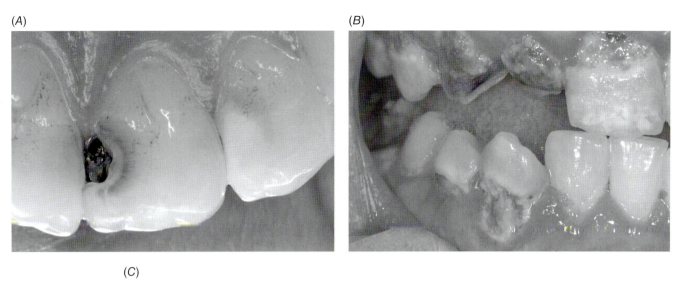

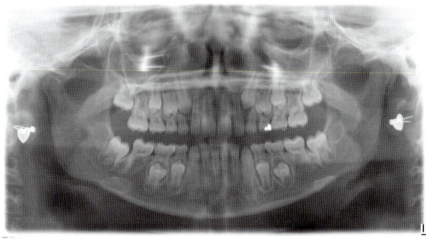

Fig. 119.1. (A) Localized "cavitated" carious lesion of the maxillary central incisor extending into the dentine. (B) Patient with rampant caries. This patient was a chronic methamphetamine abuser. (C) A 10-year-old girl with diminished root formation of the mesial roots of all four first permanent molars. Also note the narrowing of the pulp chambers in the affected teeth. Radiographically, this closely resembles external root resorption. However, there was no history of orthodontic treatment or trauma. (B) Source: [56]. (C) Courtesy of Dr. Charles Scanlon.

Fluoridated hydroxyapatite and fluorapatite are formed on exposure of enamel to trace quantities (~1 ppm) of fluoride ion, increasing enamel resistance to acid dissolution. The overall magnitude of caries reduction afforded by fluoride averages 25% in both children and adults, whether delivered professionally, self-administered in toothpaste, or by means of community water fluoridation.

Although the caries process (Fig. 119.1A,B) in the heavily mineralized enamel outer layer of the tooth is primarily a physicochemical process, the underlying dentin contains 20% organic matrix, primarily type I collagen and a smaller component of noncollagenous proteins. Continued acid dissolution of the mineralized component of dentin results in exposure of the organic matrix to enzymatic degradation by bacterial collagenases and host-derived matrix metalloproteinases [8]. Because of the presence of odontoblast cell processes extending from the dental pulp through tubules within the dentin, inflammation within this tightly integrated "dentin–pulp complex," either as a result of direct extension of bacteria into the pulp tissue or by secondary strangulation of venous blood flow, predisposes to pulpal necrosis. Subsequent to pulpal necrosis, production of inflammatory mediators derived from degradation of bacteria and necrotic pulp tissue leads to the formation of a mass of chronically inflamed granulation tissue and progressive bone destruction at the apex of the nonvital tooth root (periapical granuloma).

External root resorption is a very common, almost universal, process characterized by loss of root cementum and dentin. Most individuals exhibit minimal root resorption of no clinical significance. However, significant root resorption can be seen in a small percentage of patients following orthodontic therapy. Other etiological factors include trauma, periradicular inflammation, excess occlusal forces, direct effect from odontogenic cysts and tumors, and hyperparathyroidism. External root resorption is mediated by macrophage-derived osteoclasts, odontoclasts, periodontal ligament-derived cells, cementum, inflammatory cytokines, and dental pulp [9].

A newly defined dental phenotype, molar root–incisor malformation (MRIM), that can be clinically mistaken for root resorption, has recently been described [10]. MRIM is a developmental dental defect, characterized by diminished root formation of the first permanent molars with narrowing of affected pulp chambers (Fig. 119.1C). MRIM appears to be more common in individuals with a history of renal disease, meningomyelocele, or meningitis, although healthy individuals are also affected.

CYSTS AND NEOPLASMS OF THE JAWS

Origin

Cysts and neoplasms of the jaws are categorized as being of odontogenic origin, recapitulating structures involved in the development of the teeth, or of nonodontogenic origin. By definition, odontogenic cysts and neoplasms are unique to the oral and maxillofacial region.

During tooth development, a linear epithelial structure, termed the dental lamina, from which the individual teeth ultimately form, arises from an ingrowth of surface epithelium into the underlying connective tissue [11]. Following tooth development, these epithelial structures undergo apoptosis. However, residual "epithelial rests" are believed to be the source of epithelium from which several developmental odontogenic cysts arise. The molecular events involved in the development of these odontogenic lesions remain poorly understood [12, 13].

Clinical and radiographic presentation

Odontogenic cysts and neoplasms originate in the tooth-bearing areas of the jaws (eg, in the alveolar processes that support the teeth and above the inferior alveolar nerve canal in the mandible), and are characterized by replacement of bone by soft tissue or, less commonly, a mixture of soft and hard tissue. In the absence of secondary infection or significant expansion, odontogenic cysts and tumors typically cause few symptoms and are usually identified during routine dental radiographic examination (Fig. 119.2).

The radiographic and clinical presentations of these lesions, though often characteristic, are not pathognomonic. As with extragnathic bone lesions, correlation of radiographic features with histopathology is often required to arrive at a definitive diagnosis.

Odontogenic cysts

Odontogenic cysts can be further subclassified into inflammatory cysts and those of developmental origin.

Inflammatory odontogenic cysts: periapical granuloma and periapical cyst

The periapical granuloma and periapical (radicular) cyst are common, closely related, slow-growing lesions that develop at the apex or midroot area of teeth exhibiting pulpal necrosis, usually representing the end result of caries extending into the pulp or previous trauma to the dental pulp. Continued bone destruction can lead to cortical bone perforation (Fig. 119.3A) and the formation of a mass of acute and chronically inflamed granulation tissue in the oral cavity (parulis; Fig. 119.3B).

The periapical granuloma, comprising chronically inflamed granulation tissue, represents the precursor to the periapical cyst. It is believed that inflammatory mediators, cytokines, and growth factors released from mononuclear inflammatory cells and neighboring stromal cells from the degradation of necrotic pulp tissue stimulate residual epithelial cell rests to proliferate [14]. Although the mechanism is not thoroughly understood, the end result is the formation of a cyst lining (periapical cyst).

When sufficient tooth structure remains to allow for restoration of the tooth, nonvital teeth with periapical radiolucent lesions suggestive of a periapical granuloma or cyst (Fig. 119.2A) should receive endodontic treatment, a procedure in which the degradation products are mechanically removed from the pulp chamber and canals. Alternatively, these lesions can be treated definitively by extraction of the causative tooth with conservative curettage of any cyst lining. Failure to excise the cyst lining, if present, can lead to continued expansion of the lesion, resulting in a residual cyst. In extremely rare cases, the epithelial lining of a periapical cyst or residual cyst can undergo malignant transformation [15].

Developmental odontogenic cysts

Dentigerous cyst

The dentigerous (follicular) cyst is a slow-growing developmental cyst seen in association with the crown of an unerupted (impacted) tooth. If untreated, the dentigerous cyst is capable of causing significant bone destruction. The third molar and maxillary canine teeth are the teeth most commonly associated with the development of dentigerous cysts, reflecting the fact that these teeth are most prone to impaction as a result of being the last teeth to come into the mouth in the normal tooth eruption sequence [16].

Treatment involves enucleation of the cyst lining with extraction of the associated impacted tooth. Rarely, the cyst lining can undergo transformation into an odontogenic neoplasm, most commonly a cystic ameloblastoma [17], or squamous cell carcinoma [18].

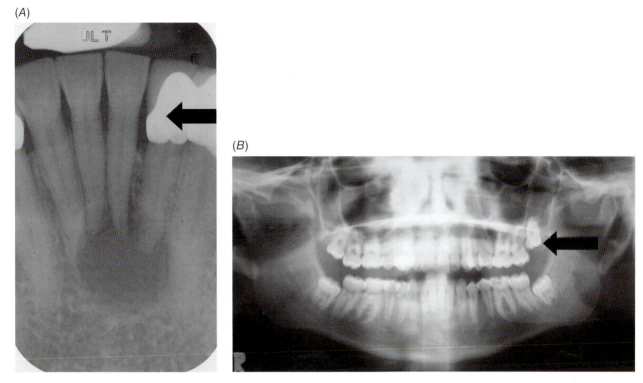

Fig. 119.2. The dentist's typical armamentarium of radiographic techniques includes intraoral dental radiographs and extraoral panoramic imaging. (A) Intraoral radiograph (periapical film) demonstrating the outline of the mandibular anterior teeth and surrounding alveolar bone. A unilocular radiolucent lesion with a well-defined corticated border is evident in the middle, at the apices of the incisors. The left lateral incisor (arrow) was nonvital. The differential diagnosis was periapical cyst versus periapical granuloma. (B) Extraoral panoramic radiography provides a complete overview of the maxillary and mandibular bones and neighboring structures. This radiographic technique is widely used as a screening tool. This panoramic radiograph, taken on an 18-year-old patient, reveals the presence of unerupted third molars (wisdom teeth). A small well-defined radiolucent area with corticated borders is evident around the crown of the maxillary left third molar (arrow). As the width of the radiolucent area is less than 4 mm and the associated third molar has not completed its root development, this most likely represents a normal dental follicle surrounding the developing tooth.

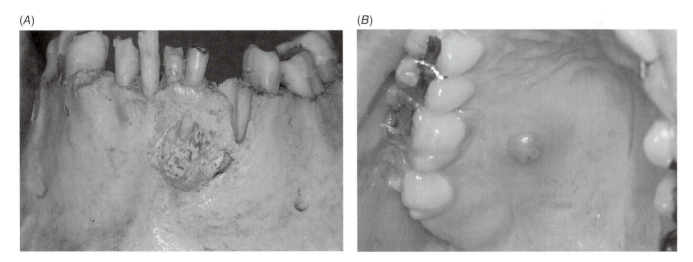

Fig. 119.3. (A) Cadaveric mandible with missing buccal cortical plate. The crowns of the mandibular incisors are fractured, presumably resulting in pulpal necrosis and the subsequent development of an inflammatory periapical granuloma or cyst. (B) A 25-year-old male with a necrotic right maxillary first molar presented with a soft tissue nodule composed of inflamed epithelial-lined granulation tissue and abscess, consistent with a parulis, overlying the palatal root. Note that the palate is an uncommon location for the development of a parulis, as these are typically noted on the buccal aspect of the alveolar processes.

Odontogenic keratocyst (keratocystic odontogenic tumor)

The odontogenic keratocyst (OKC) demonstrates a preference for the posterior mandible. It is characterized by a propensity to cause significant bone destruction (Fig. 119.4A–C) and exhibits a high recurrence rate after conservative treatment [19]. OKCs range in size from small, unilocular radiolucent lesions, sometimes associated with an impacted tooth, to large, destructive, multilocular radiolucent lesions.

Approximately 5% of OKCs are associated with the nevoid basal cell carcinoma syndrome (NBCCS or Gorlin syndrome), an autosomal dominantly inherited condition. Additional stigmata include the development of multiple basal cell carcinomas involving sun-exposed skin at an early age, the presence of small pitlike developmental defects in the palms of the hands and plantar surfaces of the feet, and an increased incidence of neoplasms including medulloblastoma and meningioma [20]. NBCCS is associated with germline loss of function mutations in the patched-1 (*PTCH1*) gene, a tumor suppressor gene that is a component of the sonic hedgehog pathway [21]. Even in the absence of these stigmata, development of a single OKC in a patient under 25 years of age or multiple OKCs in a patient of any age warrants further investigation to rule out NBCCS.

PTCH1 mutations have also been documented in conventional (sporadic) nonsyndrome associated OKCs [22, 23]. A World Health Organization working group has renamed this entity the "keratocystic odontogenic tumor" to emphasize its aggressive, neoplastic-like

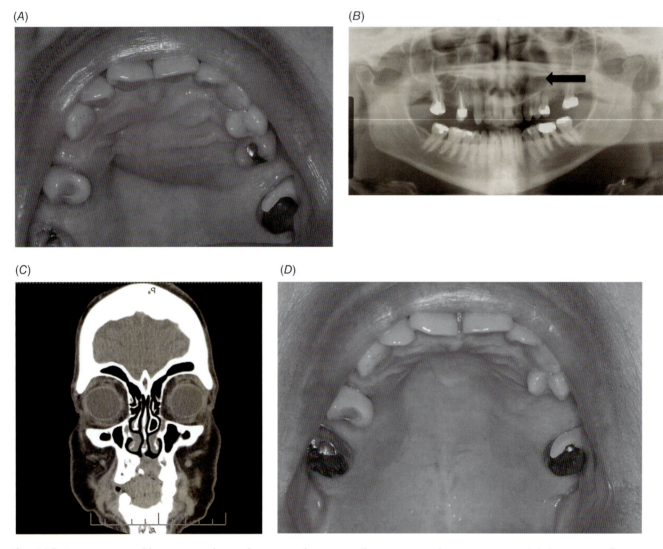

Fig. 119.4. A 56-year-old woman with an odontogenic keratocyst (keratocystic odontogenic tumor). (*A*) Note significant expansion of the maxillary anterior palatal bone. (*B*) Panoramic radiography reveals a large, well-defined, corticated radiolucent lesion extending from the maxillary right canine to the left second premolar area (arrow). (*C*) Cone beam computed tomography reveals extensive destruction of the palatal bone, extending to the nasal floor. (*D*) Significant remodeling of the palatal bone is evident 6 months after marsupialization. Despite the bone remodeling, definitive surgical treatment must still be performed following marsupialization.

behavior [24]. Additional findings supportive of a neoplastic nature include the demonstration of increased proliferative activity in the epithelial lining [25] and loss of heterozygosity at loci associated with the *p16* and *p53* tumor suppressor genes [26]. However, this new terminology is not universally accepted. Arguments offered against reclassifying the OKC as a "benign cystic neoplasm" is the documentation of *PTCH1* mutations in other developmental odontogenic cysts [27], the fact that molecular genetic criteria alone are not necessarily sufficient to define neoplasia, and, most notably, the fact that histologically the OKC remains an epithelial-lined pathological cavity of developmental origin (ie, a cyst).

A recent study identified two distinct molecular subtypes of non NBCCS associated OKC [28]: a more prevalent subtype exhibiting a gene expression profile resembling the secretory ameloblast, and a smaller subset sharing an expression profile more closely associated with the odontoblast. PTCH1 expression was found to be downregulated in both molecular subtypes.

Treatment options vary from surgical curettage with adjuvant chemical fixation, to aggressive surgical resection [29]. In select cases in which surgical removal would be disfiguring, marsupialization and decompression can be attempted to try to reduce the size of the lesion prior to definitive treatment (Fig. 119.4D). However, it must be clearly explained to the patient that definitive surgical intervention will still be needed following marsupialization.

Regardless of the treatment approach selected, patients with a history of OKC should be followed radiographically for an indefinite period, as recurrences have been documented even decades after treatment [30].

Lateral periodontal cyst

The lateral periodontal cyst (LPC) is an uncommon developmental cyst exhibiting limited growth potential that is often overlooked in the clinical differential diagnosis of a unilocular radiolucency occurring along the lateral root surface of an anterior tooth [31]. In contrast to the periapical cyst, there is no causal relationship to pulpal necrosis. Treatment involves conservative debridement of the lesion with preservation of the associated tooth. The botryoid odontogenic cyst (BOC) represents a histologically identical variant to the LPC, presenting as a multilocular radiolucent lesion. The BOC exhibits a higher recurrence rate than its unilocular counterpart, the LPC.

In order to avoid unnecessary endodontic therapy or tooth extraction, assessment of pulpal vitality of all teeth with radiolucent lesions involving the roots of the teeth is necessary.

Glandular odontogenic cyst

The glandular odontogenic cyst (GOC) is a rare lesion [32], with one study documenting only 11 cases among 55,000+ oral cavity biopsies [33]. The GOC exhibits histological features and clinical behavior varying along a spectrum from a benign process akin to the lateral periodontal cyst to a destructive lesion with features more suggestive of a malignant process, akin to the low-grade central mucoepidermoid carcinoma. The majority of GOCs occur in the anterior mandible, often crossing the midline [34].

Odontogenic neoplasms

Benign odontogenic neoplasms

The ameloblastoma is a locally destructive neoplasm with a propensity to cause significant cortical expansion and is characterized by a high rate of recurrence. Ameloblastomas expressing BRAF-V600E, as assessed by immunohistochemistry, may exhibit more aggressive clinical behavior [35].

Although the ameloblastom exhibits a marked predilection for the posterior mandible [36], they can occur anywhere within the tooth-bearing areas of the jaws and, especially at an early stage, can be mistaken clinically and radiographically for less aggressive lesions (Fig. 119.5).

The incidence of ameloblastomas is estimated at 0.3 to 2.3 new cases per million persons per year [37].

The unicystic ameloblastoma, seen predominantly in teenagers at an average 20 years earlier than the conventional ameloblastoma, represents a cystic variant of the conventional ameloblastoma, possibly exhibiting a lower risk of recurrence [38]. Often associated with an impacted mandibular third molar, the unicystic ameloblastoma is frequently mistaken radiographically for a dentigerous cyst.

Malignant odontogenic neoplasms

Malignant odontogenic tumors are extremely rare [39]. Radiographically, they usually present as destructive lesions with irregular, poorly defined radiographic margins. Pain, paresthesia, and a tendency for early lymph node metastasis are characteristic. More commonly, jaw malignancies originate by direct extension from neighboring soft tissue (eg, oral squamous cell carcinoma), originate from extraoral sites (especially metastatic breast, colon, and prostate carcinoma), or are of primary nonodontogenic origin. Rarely, maxillary sinus malignancies can mimic the pain of dental origin [40].

NONODONTOGENIC TUMORS OF THE JAWS: CENTRAL GIANT CELL LESION

A substantial number of additional nonodontogenic cysts, pseudocysts, and tumors can also occur in the jaws. Among these is a process unique to the jaws: the central giant cell lesion (CGCL). Interestingly, histologically identical lesions are seen in association with hyperparathyroidism (the "brown tumor" of hyperparathyroidism).

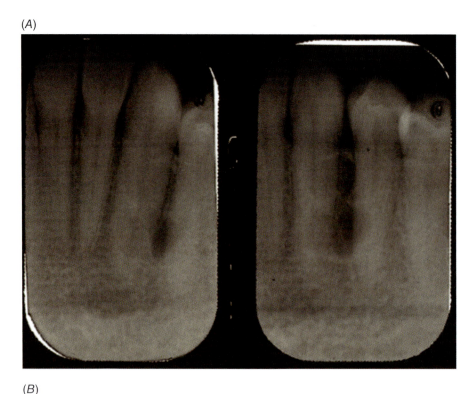

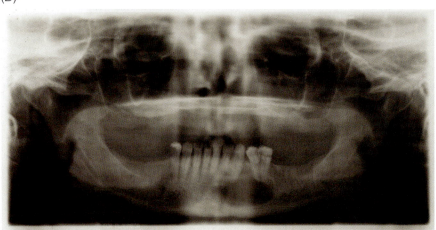

Fig. 119.5. A 48-year-old woman with an ameloblastoma of the anterior mandible. (A) On initial presentation, a multilocular radiolucent lesion was noted involving the midroot area between the mandibular left canine and first premolar. All associated teeth where vital. The radiographic differential diagnosis included botryoid odontogenic cyst, odontogenic keratocyst, and central giant cell lesion. The patient failed to follow through with the recommended surgical excision and biopsy. (B) Panoramic radiograph taken 3 years later showing significant increase in the size of the lesion. Note also the displacement of the neighboring teeth. On biopsy, this was an ameloblastoma.

The classic CGCL is a variably aggressive non-neoplastic reactive lesion with an estimated incidence of 1.1 per million persons per year [41]. The CGCL is characterized histologically by the presence of multi-nucleated giant cells (MGCs), typically concentrated in areas of hemorrhage, and believed to develop from the fusion of mononuclear phagocytes [42], in a background of spindle-shaped mesenchymal cells. Although these MCGs share similarities with the osteoclast [43], phenotypic differences exist [44]. CGCLs are occasionally noted in association with other centrally occurring jaw lesions, principally central odontogenic fibroma [45], raising the specter of intraosseous hemorrhage and abnormal repair of bone as potential etiological factors [46].

Isolated lesions are commonly treated by surgical curettage. Treatment options for large or multiple lesions include the injection of intralesional corticosteroids [47], subcutaneous or intranasal administration of calcitonin [48], and therapy with IFNα-2a [49].

CGCL-like lesions of the jaws are also the hallmark of cherubism, an autosomal dominantly inherited condition caused by mutations in the *Sh3bp2* gene. However, *Sh3bp2* mutations do not appear to play a role in the pathogenesis of sporadic CGCLs [50].

Several syndromes with overlapping facial and skeletal features have been identified as conferring an increased propensity to the development of CGCL-like lesions. These syndromes include Noonan syndrome (*PTPN11, KRAS, NRAS, SOS1, RAF1, BRAF, MAP2K1, SHOC2,* and *CBL* mutations) [51, 52], neurofibromatosis type I (*NF1* mutations) [53, 54], cardiofacio-cutaneous syndrome (*BRAF, MAP2K1, MAP2K2,* and *KRAS* mutations), and Noonan syndrome with multiple lentigines (*PTPN11, RAF1,* and *BRAF* mutations). These entities have recently been recognized to represent a related group of syndromes, the RAS/MAPK syndromes, or RASopathies, resulting from mutations at different points along the Ras/MAPK pathway. Although the RAS/MAPK pathway plays an important role in bone homeostasis, the exact mechanism linking alterations in Ras/MAPK signal transduction to the development of CGCL-like lesions is not entirely clear [55]. It is likely that the CGCL represents a diverse group of lesions, all characterized by osteoclastic overactivity, occurring at a greater incidence in a group of predisposed individuals, in response to other, as of yet unidentified, precipitating factors (eg, trauma and vascular compromise).

REFERENCES

1. Lagerweij MD, van Loveren C. Declining caries trends: are we satisfied? Curr Oral Health Rep. 2015;2: 212–7.
2. Kassebaum NJ, Bernabé E, Dahiya M, et al. Global burden of untreated caries: a systematic review and metaregression. J Dent Res. 2015;94:650–8.
3. Griffin SO, Regnier E, Griffin PM, et al. Effectiveness of fluoride in preventing caries in Adults. J Dent Res. 2007;86:410–5.
4. Edwards PC, Kanjirath P. Recognition and management of common acute conditions of the oral cavity resulting from tooth decay, periodontal disease and trauma: an update for the family physician. J Am Board Fam Med. 2010;23:285–94.
5. Loesche WJ. Role of *Streptococcus mutans* in human dental decay. Microbiol Rev. 1986;50:353–80.
6. Napimoga MH, Hofling JF, Klein MI, et al. Transmission, diversity and virulence factors of *Streptococcus mutans* genotypes. J Oral Sci. 2005;47:59–64.
7. Selwitz RH, Ismail AI, Pitts NB. Dental caries. Lancet. 2007;369:51–9.
8. Chaussain-Miller C, Fioretti F, Goldberg M, et al. The role of matrix metalloproteinases (MMPs) in human caries. J Dent Res. 2006;85:22–32.
9. Wang Z, McCauley LK. Osteoclasts and odontoclasts: signaling pathways to development and disease. Oral Dis. 2011;17:129–42.
10. Wright JT, Curran A, Kim KJ, et al. Molar root-incisor malformation: considerations of diverse developmental and etiologic factors. Oral Surg Oral Med Oral Pathol Oral Radiol. 2016;121:164–72.
11. Cobourne MT, Sharpe PT. Tooth and jaw: molecular mechanisms of patterning in the first branchial arch. Arch Oral Biol. 2003;48:1–14.
12. Kumamoto H. Molecular pathology of odontogenic tumors. J Oral Pathol Med. 2006;35:65–74.
13. Gomes CC, Duarte AP, Diniz MP, et al. Current concepts of ameloblastoma pathogenesis. J Oral Path Med. 2010;39:585–91.
14. Lin LM, Huang GT, Rosenberg PA. Proliferation of epithelial cell rests, formation of apical cysts, and regression of apical cysts after periapical wound healing. J Endod. 2007;33:908–16.
15. Whitlock RI, Jones JH. Squamous cell carcinoma of the jaw arising in a simple cyst. Oral Surg Oral Med Oral Pathol. 1967;24:530–6.
16. Daley TD, Wysocki GP. The small dentigerous cyst: a diagnostic dilemma. Oral Surg Oral Med Oral Pathol Oral Radiol Endod. 1995;79:77–81.
17. Holmlund HA, Anneroth G, Lundquist G, et al. Ameloblastoma originating from odontogenic cysts. J Oral Pathol Med. 1991;20:318–21.
18. Bodner L, Manor E, Shear M, et al. Primary intraosseous squamous cell carcinoma arising in an odontogenic cyst: a clinicopathologic anaylsis. J Oral Med Pathol. 2011;40(10):733–8.
19. Myoung H, Hong SP, Hong SD, et al. Odontogenic keratocyst: review of 256 cases for recurrence and clinicopathologic parameters. Oral Surg Oral Med Oral Pathol Radiol Endod. 2001;91:328–333.
20. Kimonis VE, Goldstein AM, Pastakia B, et al. Clinical manifestations in 105 persons with nevoid basal cell carcinoma syndrome. Am J Med Genet. 1997;69:299–308.
21. Hahn H, Wicking C, Zaphiropoulous PG, et al. Mutations of the human homolog of Drosophila *patched* in the nevoid basal cell carcinoma syndrome. Cell. 1996;85: 841–51.
22. Gu XM, Zhao HS, Sun LS, et al. *PTCH* mutations in sporadic and Gorlin-syndrome-related odontogenic keratocysts. J Dent Res. 2006;85:859–63.
23. Li TJ. The odontogenic keratocyst: a cyst, or a cystic neoplasm. J Dent Res. 2011;90:133–42.
24. Philipsen HP. Keratocystic odontogenic tumor. In: Barnes L, Eveson JW, Reichart P, et al. (eds) *Pathology and Genetics of Head and Neck Tumors*. Lyons, France: IARC Press, 2005, pp. 306–7.
25. Slootweg PJ. p53 protein and Ki-67 reactivity in epithelial odontogenic lesions. An immunohistochemical study. J Oral Pathol Med. 1995;24:393–7.
26. Henley J, Summerlin DJ, Tomich C, et al. Molecular evidence supporting the neoplastic nature of odontogenic keratocyst: a laser capture microdissection study of 15 cases. Histopathology. 2005;47:582–6.

27. Pavelic B, Levanat S, Crnic I, et al. *PTCH* gene altered in dentigerous cysts. J Oral Pathol Med. 2001;30:569–76.
28. Hu S, Divaris K, Parker J, et al. Transcriptome variability in keratocystic odontogenic tumor suggests distinct molecular subtypes. Sci Rep. 2016;12;6:24236.
29. Blanas N, Freund B, Schwartz M, et al. Systematic review of the treatment and prognosis of the odontogenic keratocyst. Oral Surg Oral Med Oral Pathol Radiol Endod. 2000;90:553–8.
30. Kolokythas A, Fernandes RP, Pazoki A, et al. Odontogenic keratocyst: to decompress or not to decompress? A comparative study of decompression and enucleation versus resection/peripheral ostectomy J Oral Maxillofac Surg. 2007;65:640–4.
31. Fantasia JE. Lateral periodontal cyst. An analysis of forty-six cases. Oral Surg Oral Med Oral Pathol Radiol Endod. 1979;48:237–43.
32. Gardner DG, Kessler HP, Morency R, et al. The glandular odontogenic cyst: an apparent entity. J Oral Pathol. 1988;17:359–66.
33. Jones AV, Craig GT, Franklin CD. Range and demographics of odontogenic cysts diagnosed in a UK population over a 30-year period. J Oral Pathol Med. 2006;35:500–7.
34. Hussain K, Edmondson HD, Browne RM. Glandular odontogenic cysts: diagnosis and treatment. Oral Surg Oral Med Oral Pathol Oral Radiol Endod. 1995;79:593–602.
35. Fregnani ER, da Cruz Perez DE, Paes de Almeida O, et al. Braf- V600E expression correlates with ameloblastoma aggressiveness. Histopathology. 2016;70(3):473–84.
36. Reichart PA, Philipsen HP, Sonner S. Ameloblastoma: biological profile of 3677 cases. Eur J Cancer B Oral Oncol. 1995;31B:86–99.
37. Shear M, Singh S. Age-standardized incidence rates of ameloblastoma and dentigerous cyst on the Witwatersrand, South Africa. Community Dent Oral Epidemiol. 1978;6:195–9.
38. Philipsen HP, Reichart PA. Unicystic ameloblastoma. A review of 193 cases from the literature. Oral Oncol. 1998;34:317–25.
39. Slootweg PJ. Malignant odontogenic tumors: an overview. Mund Kiefer Gesichts Chir.2002; 6:295–302.
40. Edwards PC, Hess S, Saini T. Sinonasal undifferentiated carcinoma of the maxillary sinus. J Can Dent Assoc. 2006;72:161–5.
41. de Lange J, van den Akker HP. Clinical and radiological features of central giant cell lesions of the jaw. Oral Surg Oral Med Oral Pathol Oral Radiol Endod. 2005;99:464–70.
42. Abe E, Mocharla H, Yamate T, et al. Meltrin-alpha, a fusion protein involved in multinucleated giant cell and osteoclast formation. Calcif Tissue Int. 1999;64:508–15.
43. Liu B, Yu SF, Li TJ. Multinucleated giant cells in various forms of giant cell containing lesions of the jaws express features of osteoclasts. J Oral Pathol Med. 2003;32:367–75.
44. Tobon-Arroyave SI, Franco-Gonzalez LM, Isaza-Guzman DM, et al. Immunohistochemical expression of RANK, GR-alpha and CTR in central giant cell granulomas of the jaws. Oral Oncol. 2005;41:480–8.
45. Odell EW, Lombardi T, Barrett AW, et al. Hybrid central giant cell granuloma and central odontogenic fibroma-like lesions of the jaws. Histopathology. 1997;30:165–71.
46. Dorfman HD, Czerniak B. Giant cell lesions. In: Dorfman HD, Czerniak B (eds) *Bone Tumors.* St. Louis, MO: Mosby, 1998, pp. 559–606.
47. Terry BC, Jacoway JR. Management of central giant cell lesion: an alternative to surgical therapy. Oral Maxillofac Surg Clin North Am. 1994;6:579–600.
48. de Lange J, Rosenberg AJ, Van den Akker HP, et al. Treatment of central giant cell granuloma of the jaw with calcitonin. Int J Oral Maxillofac Surg. 1999;28:372–6.
49. Kaban LB, Mulliken JB, Ezekowitz RA, et al. Antiangiogenic therapy of a recurrent giant cell tumor of the mandible with interferon alfa-2a. Pediatrics. 1999;103:1145–9.
50. Teixeira RC, Horz HP, Damante JH, et al. SH3BP2-encoding exons involved in cherubism are not associated with central giant cell granuloma. Int J Oral Maxillofac Surg. 2011;40:851–5.
51. Cohen MM Jr, Gorlin RJ. Noonan-like/multiple giant cell lesion syndrome. Am J Med Genet. 1991;40:159–66.
52. Edwards PC, Fox J, Fantasia JE, et al. Bilateral central giant cell granulomas of the mandible in an eight year-old girl with Noonan syndrome (Noonan-like/multiple giant cell lesions syndrome). Oral Surg Oral Med Oral Pathol Oral Radiol Endod. 2005;99:334–40.
53. Ruggieri M, Pavone V, Polizzi A, et al. Unusual form of recurrent giant cell granuloma of the mandible and lower extremities in a patient with neurofibromatosis type 1. Oral Surg Oral Med Oral Pathol Oral Radiol Endod. 1999;87:67–72.
54. Edwards PC, Fantasia JE, Saini T, et al. Clinically aggressive central giant cell granulomas in two patients with neurofibromatosis 1. Oral Surg Oral Med Oral Pathol Oral Radiol Endod. 2006;102:765–72.
55. Edwards PC. Insight into the pathogenesis and nature of central giant cell lesions of the jaws. Med Oral Patol Oral Cir Bucal. 2015;20:e196–8.
56. Shaner JW, Kimmes N, Saini T, et al. "Meth mouth": rampant caries in methamphetamine abusers. AIDS Patient Care STDs. 2006;20:4–8.

120

Osteonecrosis of the Jaw

Sotirios Tetradis[1], Laurie McCauley[2], and Tara Aghaloo[1]

[1] Division of Diagnostic and Surgical Sciences, UCLA School of Dentistry, Los Angeles, CA, USA
[2] Periodontics and Oral Medicine, University of Michigan School of Dentistry, Ann Arbor, MI, USA

DEFINITION

Although osteonecrosis of the jaw (ONJ) was first described nearly 15 years ago [1, 2], the disease process is poorly understood, patient diagnosis is not based on objective measures or laboratory tests but on clinical observations, and patient treatment is mostly empirical [3]. Thus it is not surprising that there is considerable disagreement in the definition of ONJ. Most proposals agree that a patient is considered to suffer from ONJ if all the following attributes are present: (i) presence of exposed bone in the maxillofacial region that does not heal within 8 weeks after identification by a health care provider; (ii) current or previous treatment with antiresorptive medication; and (iii) no history of radiation therapy to the craniofacial region [4, 5].

There are two notable modifications to this definition:(i) inclusion of patients without overt bone exposure but with bone that can be probed through an intraoral or extraoral fistula, and (ii) inclusion of patients who have not been exposed to antiresorptive medications but have a history of treatment with antiangiogenic agents [4]. Addition of the first group of patients seems reasonable because bone, although not clinically visible, is clearly accessible to the oral environment. However, whether the same mechanisms contribute to bone exposure in patients on antiresorptives versus antiangiogenics, and whether ONJ management requires similar intervention in patients on these different medications is unknown [5].

Furthermore, inclusion of patients on antiresorptives in the absence of bone exposure but with nonspecific symptoms, clinical and radiographic findings of no odontogenic origin as stage 0 ONJ was proposed [4], based on findings that up to 50% of such patients progress to bone exposure and clinical ONJ [6], yet this remains controversial [5]. Interestingly, ONJ can occur in patients not exposed to antiresorptive or antiangiogenic agents [4, 7]. The lack of a universally accepted ONJ definition not only creates difficulties in the identification and management of patients with the disease, but also leads to uncertainty in identification and adjudication of patients with ONJ in clinical epidemiologic studies [3]. For details regarding the epidemiologic aspects of ONJ, see Chapter 76.

ETIOLOGY

The precise etiology of ONJ is unknown and likely complex. Several theories have been proposed with the most widely supported described here. Antiresorptive agents, notably bisphosphonates and denosumab, are associated with ONJ. Other pharmacological modifiers including immunomodulators and antiangiogenic agents increase the risk of ONJ. A multi-step trajectory to clinical ONJ likely includes (i) underlying osseous compromise; (ii) trauma exposing the osseous tissue, and (iii) infection. There are several studies that suggest antiresorptive treatment may lead to areas of bone necrosis and/or altered biomechanical integrity long before clinical ONJ presents [8–10]. Notably, empty osteocyte lacunae are found in cortical bone after long-term treatment with zoledronate in the absence of clinical ONJ. This is one of the underlying premises supporting the categorization of stage 0 ONJ [4], yet similar findings have not been convincingly verified in humans.

Suppression of bone turnover, through inhibition of osteoclastic activity, is at the core of the underlying necrosis and altered biomechanical integrity; however, suppression of bone turnover alone is not sufficient to

---

Primer on the Metabolic Bone Diseases and Disorders of Mineral Metabolism, Ninth Edition. Edited by John P. Bilezikian.
© 2019 American Society for Bone and Mineral Research. Published 2019 by John Wiley & Sons, Inc.
Companion website: www.wiley.com/go/asbmrprimer

cause ONJ. It is likely that inhibition of osteoclastic function compromises the ability of the alveolar bone to respond to extrinsic local factors and maintain a normal homeostasis [11].

Trauma features prominently in the pathogenesis of ONJ with 46% to 79% of ONJ patients having dentoalveolar trauma preceding ONJ clinical presentation, and specifically dental extraction [12]. Spontaneous incidence of ONJ may also be attributed to the trauma associated with normal masticatory function or dental prosthetic devices coupled with a very thin oral mucosa overlying the bone. Once the osseous tissue is exposed, infection is likely a key factor [13]. More than 750 species of bacteria are found in the oral cavity and comprise complex communities existing primarily in biofilms on the surfaces of teeth, prostheses, gingiva, and tongue [14]. The immune system reaction to oral microflora that are deleterious to wound healing coupled with compromised osseous integrity and the lack of osteoclastic remodeling may be particularly devastating for the development of clinical ONJ. Several cell types other than osteoclasts have been implicated as contributors to ONJ including $\gamma\delta T$ cells, macrophages, epithelial cells, and others [15]. Recent evidence suggests that bisphosphonate therapy results in significant differences in the expression of genes regulating immune function, barrier functions, tissue remodeling, and lymphangiogenesis [16, 17]. Hence, the convergence of alterations in bone structure, the incitement for wound healing that deviates from normal, and the activation of inflammatory and immune reactions likely contributes to the compromised osseous and mucosal clinical presentation of ONJ.

RISK FACTORS

In general, local factors that disrupt oral soft and hard tissue integrity or systemic factors that compromise healing have been shown to contribute to ONJ development, progression, and severity. Proposed risk factors are [4, 5]:

- Local
  - invasive dental procedure
  - denture use
  - periodontal or periapical disease
  - poor oral hygiene
  - anatomical factors
- Systemic
  - antiresorptive therapy
  - antiangiogenic agents
  - diabetes
  - glucocorticoid therapy
  - smoking.

Potent antiresorptive agents such as zoledronate, alendronate, and denosumab are risk factors for ONJ because of the potential negative impact on the underlying osseous tissue. Patients who have had invasive dental procedures have marked trauma. Denture wearers experience trauma when abnormal forces, often resulting from poorly fitting dentures, are transmitted through the thin mucosa overlying the osseous tissues. Patients with periodontal disease have increased local bone turnover and the opportunity for pathogenic microflora to colonize the inflamed mucosal and osseous tissues. Individuals with diabetes, smokers, and patients on other therapeutics such as antiangiogenic agents and cancer chemotherapies may be more prone to compromised wound healing and/or altered immune responses after infection and hence predisposed to atypical exposed bone that necroses and restricts the essential epithelialization for repair of the exposed bone.

CLINICAL APPEARANCE AND STAGING

ONJ patients present with a spectrum of findings from mild to severe (Fig. 120.1). Patient symptomatology ranges from nearly absent to mild tenderness to severe pain and discomfort [18]. Clinically, the common denominator is the presence of bone not covered by oral mucosa but exposed to the oral cavity. Although both jaws can be affected, the mandible is a more frequent site: about 70% to 75% of the cases affect the mandible, and only 20% to 25% the maxilla, while in 4% to 5% of patients both jaws are affected [19]. The exposed bone area can range from very small to quite extensive encompassing a large area of the alveolar ridge. Presence of infection and inflammation, intraoral or extraoral fistula, mobile bone sequestra, oro-antral communication, and even pathological fracture might be observed [4, 5, 18].

Specific staging of ONJ has been a source of controversy and discussion since the original description of the disease. Because the clinical presentation was how ONJ was originally discovered, it is not surprising that staging is based primarily on clinical criteria. However, radiographic findings contribute greatly to identifying the extent of disease and proximity to vital structures. Multiple groups, most notably the ASBMR and the American Association of Oral and Maxillofacial Surgeons (AAOMS), have published comprehensive position papers that have been revised to characterize ONJ [4, 5]. Stage 1 is defined as exposed, necrotic bone or bone that can be probed through a fistula, but without symptoms or signs of infection (Fig. 120.1A). Stage 2 is characterized by exposed, necrotic bone or bone that can be probed through a fistula with pain or clinical evidence of infection such as swelling, erythema, or purulence. Stage 3 ONJ includes patients with extensive areas of exposed, necrotic bone with pain and evidence of infection. Here, the exposed, necrotic bone extends beyond the alveolar bone to structures including: (i) the maxillary sinus creating an oroantral communication; (ii) nasal cavity creating an oronasal communication; (iii) inferior alveolar nerve canal manifesting as paresthesia; (iv) inferior border of the mandible causing a pathological fracture or extraoral fistula; or

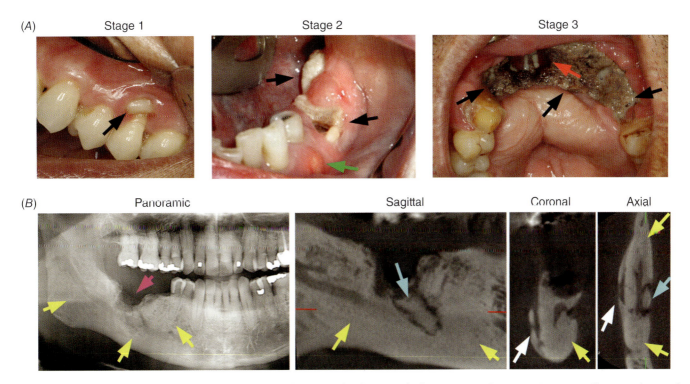

Fig. 120.1. (A) Typical examples of various stages of ONJ. In the first panel of a patient with stage 1 ONJ a small exposed area of bone at the buccal surface of the second premolar (black arrow) without evidence of infection is noted. In the second panel, a larger area of exposed bone (black arrow) with edema of the surrounding soft tissues and fistula track formation (green arrow) are seen. In the last panel, extensive exposed bone that encompasses the right and left maxillary alveolar ridge (black arrows) and extends into the nasal cavity with a naso-oral communication (red arrow) and exposure of the nasal septum is noted. (B) Typical radiographic findings in an ONJ patient. The first panel is part of a panoramic radiograph and shows osteosclerosis of the right posterior alveolar ridge and body of the mandible (yellow arrows) and a nonhealed extraction socket (magenta arrow). Sagittal, coronal, and axial sections of a cone beam CT scan of the same patient capture the osteosclerotic changes in more detail (yellow arrows) and also demonstrate sequestrum formation (cyan arrows) and periosteal bone formation (white arrows).

(v) zygoma. Although the diagnostic criteria are largely clinical, it is widely recognized that utilizing only clinical criteria to stage ONJ may greatly underestimate its severity [4, 5]. Therefore, radiographic evaluation plays a major role in evaluating the extent of disease.

Radiographically, incomplete healing of extraction sockets, thickening of the lamina dura and cortical outlines, irregular areas of osteolysis, erosions or destruction of cortices, diffused and extensive sclerosis of the trabecular bone, exuberant periosteal bone formation, and sequestration might be observed [20, 21] (Fig. 120.1B). These findings are present in all radiographic modalities but are easier to appreciate in advanced 3D imaging modalities such as cone beam CT (CBCT) and multidetector CT (MDCT) and less obvious in 2D projections such as panoramic and periapical radiographs [20–22]. Hence, advanced 3D imaging with CBCT or MDCT of ONJ patients is recommended [5]. Other advanced imaging modalities, such as MRI, scintigraphy, single-photon emission computed tomography (SPECT), or PET can further characterize bone and soft tissue changes around ONJ lesions and help in early diagnosis and treatment planning of patients with ONJ [20, 23, 24].

TREATMENT

Therapy for ONJ remains controversial; treatment protocols range from conservative oral microbial mouthrinses to aggressive segmental resection with microvascular reconstruction [4, 5]. First and foremost, disease prevention with regular dental visits, adequate hygiene procedures, and restorative and periodontal maintenance is recommended [25]. Multiple studies demonstrate that dental screening and preventive therapy reduce the risk of ONJ [19, 26–28]. Both periodontal and periapical disease increase the risk of ONJ and are the most common reasons for tooth extraction or surgical dentoalveolar intervention, which reinforce the general dentist's role in ONJ prevention and management [4, 29, 30].

If dental disease cannot be contained and adequate oral health cannot be maintained, ONJ may develop and require treatment. Therapy is generally performed according to disease severity or staging. Well-accepted guidelines advocate conservative management including antibacterial mouthrinse, close clinical follow-up, and patient education with periodic evaluation for continued antiresorptive or antiangiogenic therapy. Unfortunately, as disease severity increases, conservative management is less predictable in improving signs or symptoms of pain or infection [4, 31–34].

For patients with stage 1 and 2 disease, local wound care with an antimicrobial swab on the area of exposed, necrotic bone to treat or prevent pain or infection before considering surgical intervention is recommended [4].

Stage 2 disease that includes signs or symptoms of infection requires conservative management as above that also includes systemic antibiotics and pain control. Published guidelines advocate for superficial debridement to relieve soft tissue irritation and facilitate infection control in more advanced disease [4] so it is important to consider if patients are medically able to undergo a surgical and anesthetic intervention. Similar to conservative, nonsurgical therapy, success rates to provide pain relief and/or mucosal closure vary from 15% to 100% with sequestrectomy and debridement [35–40]. Therefore, conservative management is often attempted as first-line therapy before any surgical intervention [41].

Stage 3 ONJ is the most severe form and may require advanced surgical debridement or resection to provide long-term palliative care [4]. Surgical resection improves pain and leads to soft tissue healing in 72% to 95% of cases [39, 42–45]. However, resection may pose technical challenges because it is often difficult to identify healthy bone to serve as surgical margins. Furthermore, resective surgery requires extended hospitalization, and risks potential morbidity in medically compromised patients. For these reasons, alternative therapies continue to be investigated, such as PTH [46, 47] and stem cell therapy [48, 49]. However, these interventions are still in early stages of development with only isolated case reports demonstrating improved treatment outcomes, and they need to be validated in larger clinical studies.

RESEARCH DIRECTIONS

Several aspects of ONJ pathophysiology, progression, and management remain largely unknown. Research efforts on multiple fronts are needed to advance characterization of the disease and improve clinical outcomes.

Clinical studies have provided valuable information towards understanding ONJ incidence in various patient cohorts and identifying local and systemic risk factors [19, 50–52]. In the future, utilization of big data "omics" methodologies could lead to the development of markers that will identify patients at higher risk and diagnose the disease at early stages. Recent advances in understanding of the oral microbiome and saliva diagnostics offer great opportunities for a personalized approach towards this goal [53, 54]. Prospective epidemiologic studies would be required to validate these findings, as well as to establish effective treatment observations and best practices in the management of patients with ONJ. However, such clinical approaches are limited because of the relatively low incidence of ONJ in patients on antiresorptive medications.

Translational investigations are needed to offer controlled experimental designs that will shed light on the mechanisms of ONJ establishment and progression, will assess treatment interventions, and will ultimately inform and complement clinical efforts. ONJ animal models that capture several of the clinical, radiographic, and histological findings of the disease have been established. These models point to a central role for bone turnover suppression through osteoclastic inhibition by bisphosphonates or RANKL inhibitors and the contribution of dentoalveolar trauma and particularly tooth extractions, and periodontal or periapical disease with associated infection/inflammation in ONJ pathogenesis [55–57]. Utilizing these models, the dysregulation of immune responses and defective oral wound healing in the presence of antiresorptive medications have been implicated [58]. Importantly, potential pharmacological or stem cell therapeutic approaches have been proposed [59]. Further refinement and expansion of such studies will reveal the sequence of events and uncover the molecular and cellular players in the multifaceted process of ONJ development and establishment that implicates an attenuated osteoclastic function, an altered immune response, and a compromised mucosal and submucosal wound healing.

ONJ involves the abnormal function and defective interplay and communication of multiple cell types and tissues. Although in vitro experiments might seem unlikely to capture the complexity and intricacy of ONJ pathological processes, such approaches have provided important understanding of the molecular effects of bisphosphonates on various cell types involved in oral tissue function and integrity [55, 60]. Further in vitro studies could provide significant insight into the molecular signals and transduction cascades that lead to altered cellular function and thus complement translational and clinical studies towards a comprehensive understanding of ONJ pathophysiology and establishment of effective interventional methodologies.

REFERENCES

1. Marx RE. Pamidronate (Aredia) and zoledronate (Zometa) induced avascular necrosis of the jaws: a growing epidemic. J Oral Maxillofac Surg. 2003;61(9):1115–7.
2. Ruggiero SL, Mehrotra B, Rosenberg TJ, et al. Osteonecrosis of the jaws associated with the use of bisphosphonates: a review of 63 cases. J Oral Maxillofac Surg. 2004;62(5):527–34.
3. Kim HY, Kim JW, Kim SJ, et al. Uncertainty of current algorithm for bisphosphonate-related osteonecrosis of the jaw in population-based studies: a systematic review. J Bone Miner Res. 2017;32(3):584–91.
4. Ruggiero SL, Dodson TB, Fantasia J, et al.; American Association of Oral and Maxillofacial Surgeons.

American Association of Oral and Maxillofacial Surgeons position paper on medication-related osteonecrosis of the jaw—2014 update. J Oral Maxillofac Surg. 2014; 72(10):1938–56.
5. Khan AA, Morrison A, Hanley DA, et al.; International Task Force on Osteonecrosis of the Jaw. Diagnosis and management of osteonecrosis of the jaw: a systematic review and international consensus. J Bone Miner Res. 2015;30(1):3–23.
6. Fedele S, Porter SR, D'Aiuto F, et al. Nonexposed variant of bisphosphonate-associated osteonecrosis of the jaw: a case series. Am J Med. 2010;123(11):1060–4.
7. Aghaloo TL, Tetradis S. Osteonecrosis of the Jaw in the Absence of Antiresorptive or Antiangiogenic Exposure: A Series of 6 Cases. J Oral Maxillofac Surg. 2017;75(1): 129–42.
8. Allen MR, Burr DB. Mandible matrix necrosis in beagle dogs after 3 years of daily oral bisphosphonate treatment. J Oral Maxillofac Surg. 2008;66(5):987–94.
9. Kim JW, Landayan ME, Lee JY, et al. Role of microcracks in the pathogenesis of bisphosphonate-related osteonecrosis of the jaw. Clin Oral Investig. 2016;20(8): 2251–8.
10. De Ponte FS, Catalfamo L, Micali G, et al. Effect of bisphosphonates on the mandibular bone and gingival epithelium of rats without tooth extraction. Exp Ther Med. 2016;11(5):1678–84.
11. Aghaloo T, Hazboun R, Tetradis S. Pathophysiology of Osteonecrosis of the Jaws. Oral Maxillofac Surg Clin North Am. 2015;27(4):489–96.
12. Yazdi PM, Schiodt M. Dentoalveolar trauma and minor trauma as precipitating factors for medication-related osteonecrosis of the jaw (ONJ): a retrospective study of 149 consecutive patients from the Copenhagen ONJ Cohort. Oral Surg Oral Med Oral Pathol Oral Radiol. 2015;119(4):416–22.
13. Wei X, Pushalkar S, Estilo C, et al. Molecular profiling of oral microbiota in jawbone samples of bisphosphonate-related osteonecrosis of the jaw. Oral Dis. 2012;18(6): 602–12.
14. Jenkinson HF, Lamont RJ. Oral microbial communities in sickness and in health. Trends Microbiol. 2005;13(12): 589–95.
15. Landesberg R, Woo V, Cremers S, et al. Potential pathophysiological mechanisms in osteonecrosis of the jaw. Ann N Y Acad Sci. 2011;1218:62–79.
16. Kalyan S, Wang J, Quabius, ES, et al. Systemic immunity shapes the oral microbiome and susceptibility to bisphosphonate-associated osteonecrosis of the jaw. J Transl Med. 2015;13:212.
17. Yamashita J, Koi K, Yang DY, et al. Effect of zoledronate on oral wound healing in rats. Clin Cancer Res. 2011;17(6):1405–14.
18. Ruggiero SL. Diagnosis and Staging of Medication-Related Osteonecrosis of the Jaw. Oral Maxillofac Surg Clin North Am. 2015;27(4):479–87.
19. Saad F, Brown JE, Van Poznak C, et al. Incidence, risk factors, and outcomes of osteonecrosis of the jaw: integrated analysis from three blinded active-controlled phase III trials in cancer patients with bone metastases. Ann Oncol. 2012;23(5):1341–7.
20. Bedogni A, Blandamura S, Lokmic Z, et al. Bisphosphonate-associated jawbone osteonecrosis: a correlation between imaging techniques and histopathology. Oral Surg Oral Med Oral Pathol Oral Radiol Endod. 2008;105(3):358–64.
21. Arce K, Assael LA, Weissman JL, et al. Imaging findings in bisphosphonate-related osteonecrosis of jaws. J Oral Maxillofac Surg. 2009;67(5 suppl):75–84.
22. Hutchinson M, O'Ryan F, Chavez V, et al. Radiographic findings in bisphosphonate-treated patients with stage 0 disease in the absence of bone exposure. J Oral Maxillofac Surg. 2010;68(9): 2232–40.
23. Krishnan A, Arslanoglu A, Yildirm N, et al. Imaging findings of bisphosphonate-related osteonecrosis of the jaw with emphasis on early magnetic resonance imaging findings. J Comput Assist Tomogr. 2009;33(2):298–304.
24. Catalano L, Del Vecchio S, Petruzziello F, et al. Sestamibi and FDG-PET scans to support diagnosis of jaw osteonecrosis. Ann Hematol. 2007;86(6):415–23.
25. Ruggiero SL, Fantasia J, Carlson E. Bisphosphonate-related osteonecrosis of the jaw: background and guidelines for diagnosis, staging and management. Oral Surg Oral Med Oral Pathol Oral Radiol Endod. 2006;102(4): 433–41.
26. Dimopoulos MA, Kastritis E, Bamia C, et al. Reduction of osteonecrosis of the jaw (ONJ) after implementation of preventive measures in patients with multiple myeloma treated with zoledronic acid. Ann Oncol. 2009; 20(1):117–20.
27. Ripamonti CI, Maniezzo M, Campa T, et al. Decreased occurrence of osteonecrosis of the jaw after implementation of dental preventive measures in solid tumour patients with bone metastases treated with bisphosphonates. The experience of the National Cancer Institute of Milan. Ann Oncol. 2009;20(1):137–45.
28. Bonacina R, Mariani U, Villa F, et al. Preventive strategies and clinical implications for bisphosphonate-related osteonecrosis of the jaw: a review of 282 patients. J Can Dent Assoc. 2011;77:b147.
29. Pichardo SE, van Merkesteyn JP. Bisphosphonate related osteonecrosis of the jaws: spontaneous or dental origin? Oral Surg Oral Med Oral Pathol Oral Radiol. 2013; 116(3):287–92.
30. Marx RE, Sawatari Y, Fortin M, et al. Bisphosphonate-induced exposed bone (osteonecrosis/osteopetrosis) of the jaws: risk factors, recognition, prevention, and treatment. J Oral Maxillofac Surg. 2005;63(11):1567–75.
31. Melea PI, Melakopoulos I, Kastritis E, et al. Conservative treatment of bisphosphonate-related osteonecrosis of the jaw in multiple myeloma patients. Int J Dent. 2014; 2014:427273.
32. Van den Wyngaert T, Claeys T, Huizing MT, et al. Initial experience with conservative treatment in cancer patients with osteonecrosis of the jaw (ONJ) and predictors of outcome. Ann Oncol. 2009;20(2):331–6.
33. Scoletta M, Arduino PG, Dalmasso P, et al. Treatment outcomes in patients with bisphosphonate-related

osteonecrosis of the jaws: a prospective study. Oral Surg Oral Med Oral Pathol Oral Radiol Endod. 2010;110(1): 46–53.
34. Nicolatou-Galitis O, Papadopoulou E, Sarri T, et al. Osteonecrosis of the jaw in oncology patients treated with bisphosphonates: prospective experience of a dental oncology referral center. Oral Surg Oral Med Oral Pathol Oral Radiol Endod. 2011;112(2):195–202.
35. Vescovi P, Merigo E, Meleti M, et al. Bisphosphonates-related osteonecrosis of the jaws: a concise review of the literature and a report of a single-centre experience with 151 patients. J Oral Pathol Med. 2012;41(3):214–21.
36. Thumbigere-Math V, Sabino MC, Gopalakrishnan R, et al. Bisphosphonate-related osteonecrosis of the jaw: clinical features, risk factors, management, and treatment outcomes of 26 patients. J Oral Maxillofac Surg. 2009;67(9):1904–13.
37. Williamson RA. Surgical management of bisphosphonate induced osteonecrosis of the jaws. Int J Oral Maxillofac Surg. 2010;39(3):251–5.
38. Vescovi P, Merigo E, Meleti M, et al. Conservative surgical management of stage I bisphosphonate-related osteonecrosis of the jaw. Int J Dent. 2014;2014:107690.
39. Graziani F, Vescovi P, Campisi G, et al. Resective surgical approach shows a high performance in the management of advanced cases of bisphosphonate-related osteonecrosis of the jaws: a retrospective survey of 347 cases. J Oral Maxillofac Surg. 2012;70(11):2501–7.
40. Wutzl A, Biedermann E, Wanschitz F, et al. Treatment results of bisphosphonate-related osteonecrosis of the jaws. Head Neck. 2008;30(9):1224–30.
41. Rodriguez-Lozano FJ, Onate-Sanchez RE. Treatment of osteonecrosis of the jaw related to bisphosphonates and other antiresorptive agents. Med Oral Patol Oral Cir Bucal. 2016;21(5):e595–600.
42. Carlson ER, Basile JD. The role of surgical resection in the management of bisphosphonate-related osteonecrosis of the jaws. J Oral Maxillofac Surg. 2009;67(5 suppl): 85–95.
43. Bedogni A, Saia G, Bettini G, et al. Long-term outcomes of surgical resection of the jaws in cancer patients with bisphosphonate-related osteonecrosis. Oral Oncol. 2011; 47(5):420–4.
44. Voss PJ, Joshi J, Oshero A, et al. Surgical treatment of bisphosphonate-associated osteonecrosis of the jaw: technical report and follow up of 21 patients. J Craniomaxillofac Surg. 2012;40(8):719–25.
45. Schubert M, Klatte I, Linek W, et al. The saxon bisphosphonate register - therapy and prevention of bisphosphonate-related osteonecrosis of the jaws. Oral Oncol. 2012;48(4):349–54.
46. Cheung A, Seeman E. Teriparatide therapy for alendronate-associated osteonecrosis of the jaw. N Engl J Med. 2010;363(25):2473–4.
47. Neuprez A, Rompen E, Crielaard JM, et al. Teriparatide therapy for denosumab-induced osteonecrosis of the jaw in a male osteoporotic patient. Calcif Tissue Int. 2014; 95(1):94–6.
48. Cella L, Oppici A, Arbasi M, et al. Autologous bone marrow stem cell intralesional transplantation repairing bisphosphonate related osteonecrosis of the jaw. Head Face Med. 2011;7:16.
49. Gonzalvez-Garcia M, Rodriguez-Lozano FJ, Villanueva V, et al. Cell therapy in bisphosphonate-related osteonecrosis of the jaw. J Craniofac Surg. 2013;24(3):e226–8.
50. Barasch A, Cunha-Cruz J, Curro F, et al. Dental risk factors for osteonecrosis of the jaws: a CONDOR case-control study. Clin Oral Investig. 2013;17(8):1839–45.
51. Barasch A, Cunha-Cruz J, Curro FA, et al. Risk factors for osteonecrosis of the jaws: a case-control study from the CONDOR dental PBRN. J Dent Res. 2011;90(4):439–44.
52. Yamazaki T, Yamori M, Ishizaki T, et al. Increased incidence of osteonecrosis of the jaw after tooth extraction in patients treated with bisphosphonates: a cohort study. Int J Oral Maxillofac Surg. 2012;41(11):1397–403.
53. Takahashi N. Oral Microbiome Metabolism: From "Who Are They?" to "What Are They Doing?". J Dent Res. 2015;94(12):1628–37.
54. Cuevas-Cordoba B, Santiago-Garcia J. Saliva: a fluid of study for OMICS. OMICS, 2014;18(2):87–97.
55. Allen MR. Medication-Related Osteonecrosis of the Jaw: Basic and Translational Science Updates. Oral Maxillofac Surg Clin North Am. 2015;27(4):497–508.
56. Aghaloo TL, Cheong S, Bezouglaia O, et al. RANKL inhibitors induce osteonecrosis of the jaw in mice with periapical disease. J Bone Miner Res. 2014;29(4):843–54.
57. Williams DW, Lee C, Kim T, et al. Impaired Bone Resorption and Woven Bone Formation Are Associated with Development of Osteonecrosis of the Jaw-Like Lesions by Bisphosphonate and Anti-Receptor Activator of NF-kappaB Ligand Antibody in Mice. Am J Pathol. 2014;184(11):3084–93.
58. Zhang Q, Atsuta I, Liu S, et al. IL-17-mediated M1/M2 macrophage alteration contributes to pathogenesis of bisphosphonate-related osteonecrosis of the jaws. Clin Cancer Res. 2013;19(12):3176–88.
59. Li Y, Xu J, Mao L, et al. Allogeneic mesenchymal stem cell therapy for bisphosphonate-related jaw osteonecrosis in Swine. Stem Cells Dev. 2013;22(14):2047–56.
60. Landesberg R, Cozin M, Cremers S, et al. Inhibition of oral mucosal cell wound healing by bisphosphonates. J Oral Maxillofac Surg. 2008;66(5):839–47.

121

Alveolar Bone Homeostasis in Health and Disease

Chad M. Novince[1] and Keith L. Kirkwood[2]

[1]*Department of Oral Health Sciences and the Center for Oral Health Research, Medical University of South Carolina, Charleston, SC, USA*
[2]*Department of Oral Biology, University at Buffalo, The State University of New York, Buffalo, NY, USA*

INTRODUCTION

Periodontal diseases constitute a variety of inflammatory conditions affecting the health of the periodontium, with the primary etiological factor being microbial dental plaque biofilms. Plaque-induced periodontal diseases are generally divided into gingivitis and periodontitis [1]. Gingivitis is the presence of gingival inflammation without loss of connective tissue attachment. Periodontitis is the presence of gingival inflammation with pathological detachment of periodontal ligament collagen fibers from cementum, which can progress to alveolar bone loss [2, 3]. Although it is critical to recognize that plaque-induced inflammatory periodontal diseases are influenced by genetic, systemic, and environmental factors, the scope of the chapter is focused on pathological changes in the periodontal tissues resulting from dental plaque biofilm local etiological factors. Alveolar bone metabolism and homeostasis will be addressed considering osteoimmunomodulatory implications of the normal commensal oral flora in periodontal health and effects of pathogenic shifts in the oral microbiota during periodontitis disease states.

ANATOMY OF THE PERIODONTIUM

Alveolar bone is a unique osseous tissue because of its integration with teeth and close proximity to the resident oral microbiota. The periodontium, which is the supporting structure of the teeth, is comprised of alveolar bone, cementum (covering the root surfaces), and the intervening periodontal ligament (Fig. 121.1).

The periodontal ligament fibrous attachment, extending from the root-covered cementum to the alveolar bone, suspends the tooth in the alveolar bony socket, acting as a cushion to occlusal forces [4]. Alveolar bone is under a constant state of robust physiologic remodeling, which has been attributed to masticatory occlusal forces transmitted through the fibrous osseous junction of the periodontal ligament to the alveolar bone [4].

The spatial relationship of the oral microbiota to alveolar bone is unique in that no other microbiome community colonizes an external body surface directly integrated with osseous tissue. The alveolar bone is separated from direct contact with oral biofilms by superficial gingival connective tissue and epithelium [4, 5]. Whereas epithelial tissues typically act as an impermeable barrier to microbiota colonizing external body surfaces, the junctional epithelial attachment at the tooth is unique in that it is highly porous. The periodontal innate immune system defends against microbial invasion of the permeable junctional epithelium, in both health and disease, via the coordinated expression of chemokine gradients which lead to the continuous transit of polymorphonuclear leukocytes into the gingival crevice [4, 5]. Expression of chemokines including E-selectin, intercellular adhesion molecules (ICAMs), and interleukin-8 (IL-8) drives polymorphonuclear leukocyte homing and transit from the highly vascularized gingival connective tissue to the junctional epithelium [5, 6]. Calculated to transit through the gingival tissue at a rate of approximately 30,000 neutrophils per minute [6, 7],

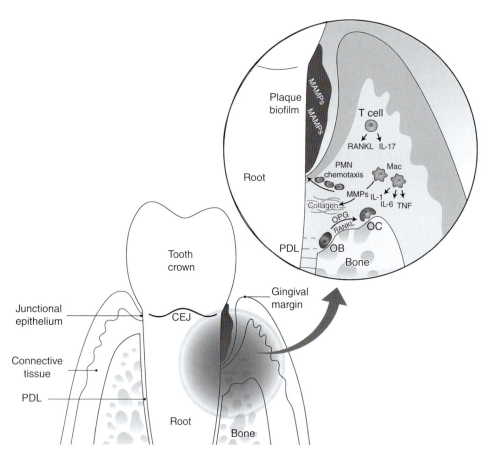

Fig. 121.1. Anatomy of periodontal tissues supporting the tooth, in periodontal health (left) and periodontal disease (right). The subgingival plaque biofilm upregulates host immune response mechanisms, resulting in periodontitis-mediated destruction of supporting connective tissue and alveolar bone (right). The inset depicts stimulation of the host immune response by microbe-associated molecular patterns (MAMPs) derived from subgingival plaque biofilm. Polymorphonuclear leukocytes (PMNs) home from the connective tissue towards the epithelium lining the periodontal pocket. Macrophage (Mac) cells secrete matrix metalloproteinases (MMPs) and proinflammatory cytokines such as tumor necrosis factor (TNF), which have procatabolic effects disrupting normal connective tissue remodeling/homeostasis. Osteoblast (OB) cells and lymphocytic cells secrete increased RANKL and other inflammatory signaling molecules, which drive osteoclast (OC) differentiation and exacerbate the proresorptive nature of the periodontitis microenvironment. IL-1 = interleukin-1; IL-6 = interleukin-6; IL-17 = interleukin-17; OPG = osteoprotegerin; PDL = periodontal ligament; CEJ = cementoenamel junction.

polymorphonuclear leukocytes passing through the junctional epithelium into the gingival crevice form a barrier wall between the host tissue and dental plaque biofilm. In periodontal health, it is important to note that neutrophils transiting through the periodontal tissues do not reside in the extracellular matrix [6, 8].

PATHOGENESIS OF PERIODONTITIS

Contradictory to prior empirical thought that periodontal tissue destruction is secondary to pathogenic bacteria direct catabolic actions on periodontal tissue matrices, extensive molecular research has demonstrated that the host immune response is the primary mediator of periodontal tissue destruction [9–11]. Although experimental periodontitis studies have clearly shown that pathogenic oral bacteria induce host immune response mechanisms driving pathophysiologic alveolar bone loss [9–11], recent investigations in the germ-free versus specific-pathogen-free mouse model have intriguingly revealed that the normal commensal oral flora has immunomodulatory effects significantly impacting alveolar bone homeostasis in health [12, 13].

Historical and current perspectives

Theories of periodontal disease pathogenesis have evolved over time because of advances in research technologies that have furthered the understanding of microbial biofilm ecosystems and the host immune response. Early observations that increased bacterial plaque accumulations were associated with periodontal disease sites shaped the *nonspecific plaque hypothesis*, which stated

Table 121.1. Host cell pattern-recognition receptor (PRR) binding of periodontal bacteria-derived microbe-associated molecular pattern (MAMP) ligands.

Host PRR	Receptor Localization	MAMP Ligand	Periodontal Bacteria
TLR2/1	Plasma membrane	Lipoproteins	*P. gingivalis*
or		Lipoproteins	*T. forsythia*
TLR2/6		Lipoproteins	*A. viscosus*
		Peptidoglycan	*A. naeslundii*
		Lipoproteins Lipoteichoic acid Peptidoglycan	*S. gordonii*
TLR4	Plasma membrane	Lipopolysaccharide (LPS)	*P. gingivalis*
			A. actinomycetemcomitans
			F. nucleatum
TLR9	Endolysosome	CpG-DNA	*P. gingivalis*
			T. forsythia
NOD1	Cytoplasm	γ-D-glutamyl-mesodiaminopimelic acid	*P. gingivali*
NOD2	Cytoplasm	Muramyl dipeptide	*A. actinomycetemcomitans*
			F. nucleatum

that increased bacterial quantity, irrespective of the presence of specific bacteria, mediates periodontal tissue destruction [14, 15]. Bacterial culture techniques and the application of whole genome DNA probes, enabling investigators to identify prominent bacteria present in plaque deposits isolated from clinical periodontal pocket disease sites, led to the *specific plaque hypothesis.* This classical periodontal pathogenesis theory dictates that infection by a specific putative periopathogenic bacterium (eg, *Porphyromonas gingivalis, Aggregatibacter actinomycetemcomitans*) or a small cluster of interacting putative periopathogenic bacteria (eg, "red complex"— *Porphyromonas gingivalis, Tannerella forsythia, Treponema denticola*) drives periodontitis-associated alveolar bone loss [15, 16]. Advances in bacterial genome sequencing, enabling the characterization of the commensal bacterial communities (microbiota) and corresponding genome (microbiome), have more recently called into question whether periodontitis is secondary to infection by the empirically defined periopathogenic bacteria. Recognition that the oral microbiota is a diverse community consisting of over 700 known bacterial species [17], and knowledge that empirically defined periopathogenic bacteria are present in both health and disease [15, 18], implies that the pathogenesis of periodontitis is secondary to ecological shifts in the normal oral flora. Periodontitis is currently understood to be a polymicrobial disease, characterized by a shift from predominantly Gram-positive bacteria in health to predominantly Gram-negative bacteria in disease [19], which is speculated to favor the putative periopathogenic bacteria [5, 15].

Molecular basis of host–microbe interactions

The host distinguishes between self and among commensal/pathogenic microbiota via the direct recognition of microbe-associated molecular patterns (MAMPs) at pattern-recognition receptors (PRRs) [20–22]. Not present in higher eukaryotes, MAMPs include microbial cell wall macromolecules, nucleic acids, and other evolutionary conserved molecular motifs uniquely found in microorganisms [20–22]. MAMPs function as distinct molecular ligands having high affinity for corresponding PRRs, which are expressed by host periodontal innate immune cells (neutrophils, monocytes, macrophages, dendritic cells, natural killer cells), adaptive immune cells (T lymphocytes, B lymphocytes), and extracellular matrix cells (epithelial cells, fibroblasts, cementoblasts, osteoblasts) [23–27]. MAMP-PRR recognition induces host cell signaling, resulting in the expression of proinflammatory cytokines and type I interferons, which facilitates the host mounting an immune defense response to colonizing/invading microorganisms [20–22]. Whereas MAMP-PRR signaling immunomodulation plays a critical role in the balanced regulation of commensal microbes in health and homeostasis, periodontal research has focused on periopathogenic bacteria associated MAMP recognition at host cell PRRs [23–27] (Table 121.1).

The two PRR families studied most extensively in the context of periodontitis are the Toll-like receptors (TLRs) and the nucleotide-binding oligomerization domain (NOD)-like receptors (NLRs). TLRs are transmembrane receptors that recognize extracellular MAMPs, whereas NLRs are cytosolic receptors that recognize intracellular MAMPs [20–22]. TLR2 and TLR4 recognize extracellular bacterial cell wall components at the cell surface, whereas TLR9 senses bacterial nucleic acids (CpG-DNA) within endosomes. Specialized NLRs that recognize bacterial peptidoglycan structures of invading pathogens in the cytoplasm include NOD1 (senses γ-D-glutamyl-mesodiaminopimelic acid) and NOD2 (senses muramyl dipeptide) [20–22] (Table 121.1).

Unlike TLR4, which is specific for lipopolysaccharide (LPS) at the outer membrane of Gram-negative bacteria, TLR2 recognizes diverse macromolecules because it forms heterodimer protein complexes with TLR1 and TLR6 [21, 28, 29]. TLR2/6 complexes recognize peptidoglycan, lipoteichoic acid, and diacylated lipoproteins derived from Gram-positive bacteria whereas TLR2/1 complexes recognize triacylated lipoproteins derived from Gram-negative bacteria [21, 28, 29]. Considering the plaque biofilm shifts from a predominantly Gram-positive microbiota in periodontal health to a predominantly Gram-negative microbiota in periodontitis disease states [15, 19], Gram-negative bacteria-induced catabolic actions are likely mediated through increased signaling at the TLR2/1 versus TLR2/6 receptor complex, as well as the concomitant activation of the TLR4 receptor by LPS. Given that periopathogenic bacteria express a multitude of heterogeneous MAMPs recognized at distinct TLRs and NLRs (Table 121.1), which commonly activate downstream mitogen-activated protein kinase (MAPK) and NF-κB signal transduction pathways [20, 21], ongoing research is indicated to elucidate how synergistic PRR signaling cross-talk contributes to the pathophysiologic expression of proinflammatory cytokines driving periodontal tissue destruction.

Host immune response in periodontal tissues

PRRs are expressed in the periodontium in health, and notably have been reported to be upregulated in periodontal tissues afflicted by periodontitis [30–32]. Whereas the commensal oral flora–host defense response has been postulated to be in a dynamic steady state that supports homeostasis of the periodontal tissues in health [33], Gram-negative periopathogenic shifts in the oral microbiota can pathophysiologically upregulate PRR signaling. Chronic supraphysiologic PRR stimulation leads to the excessive production of proinflammatory mediators, which can ultimately have catabolic effects on balanced tissue remodeling processes [5, 23]. Periodontitis-driven supraphysiologic immune and inflammatory response processes result in pathophysiologic levels of various prostaglandins, cytokines, proteases, matrix metalloproteinases (MMPs), and other host enzymes being released from epithelial cells, fibroblasts, osteoblasts, polymorphonuclear leukocytes, monocytes, macrophages, or other host cells [11, 34]. In periodontal health, balanced expression of extracellular matrix degrading MMPs and their endogenous tissue inhibitors of metalloproteinases (TIMPs) facilitates fibroblast-mediated homeostatic remodeling of connective tissues. To the contrary, in periodontitis disease states, increased MMPs/TIMPs ratios drive connective tissue degradation [11, 34]. Periodontal ligament collagen fiber detachment from the root cementum results in apical migration of the junctional epithelium, which facilitates the apical extension of subgingival biofilms (Fig. 121.1). As the periodontal pocket deepens, the anaerobic microenvironment favors the propagation of Gram-negative periopathogenic bacteria, increasing the site's risk for disease progression [2, 3].

Periodontitis-driven alveolar bone loss is secondary to the host cell inflammatory infiltrate radiating from the apical extent of the dental plaque biofilm [35, 36]. The subgingival plaque range of effectiveness in generating alveolar bone loss has been estimated to be about 2.5 mm [3], and is attributed to the localized catabolic disruption of balanced osteoclast–osteoblast mediated bone remodeling processes [37, 38].

The inflammatory infiltrate within periodontitis-afflicted connective tissue can initiate alveolar bone destruction through the modulation of local factors and cytokines stimulating osteoclastogenesis and/or inhibiting osteoblastogenesis. Periodontitis has been reported to alter the expression of RANKL and/or its decoy receptor known as osteoprotegerin (OPG), resulting in a higher RANKL/OPG ratio favoring osteoclastic bone resorption [39, 40]. Although there have been conflicting reports, the source of upregulated RANKL expression in the periodontitis-afflicted microenvironment has been primarily attributed to lymphocytic cells [41–43]. A multitude of cells in the periodontal tissues express potent proinflammatory cytokines known to enhance RANKL-mediated osteoclastogenesis—interleukin (IL)-1β, tumor necrosis factor-α (TNF-α), IL-6, and IL-17 [11, 34]—exacerbating osteoclast bone resorptive processes (Fig. 121.1). Although periodontitis research has largely centered on enhanced osteoclastogenesis, studies have clearly demonstrated that periopathogenic bacteria–host immune response mechanisms impair osteoblastogenesis [44, 45]. In light of Baron and Saffar's [37] seminal experimental periodontitis–alveolar bone remodeling investigation, revealing that periodontitis-induced alveolar bone loss is secondary to enhanced osteoclast-mediated bone resorption and suppressed osteoblast-mediated bone formation, periodontal bone destruction is caused by the catabolic disruption of normal osteoclast–osteoblast-mediated bone remodeling processes.

DIAGNOSIS

Periodontitis-afflicted persons lacking access to preventive dental care commonly do not seek professional treatment until the most severely progressed stages of disease. The fact that periodontitis is a painless chronic inflammatory condition, and that symptoms of alveolar bone loss (ie, tooth mobility, tooth migration) do not typically manifest clinically until the terminal stages of disease progression, underscores the importance of early diagnosis and intervention. Highlighting the prevalence of periodontitis and need for periodontal therapy, the 2009–2012 National Health and Nutrition Examination Survey (NHANES) revealed that 46% of adults in the United States are afflicted by periodontitis [46].

Information routinely collected during periodontal examination consists of a medical history and dental history (including prior/current periodontal disease), along with a comprehensive clinical examination. Despite advances in our understanding of the molecular/cellular pathophysiology of periodontal diseases, diagnosis still relies on clinical assessments. Clinical probing depth measurements are performed with a periodontal probe, which is used to measure the linear distance from the gingival margin to the periodontal attachment (Fig. 121.1). Clinical attachment level (CAL) measurements, which measure the linear distance from the cementoenamel junction (CEJ) to the periodontal attachment, more accurately reflect lost periodontal support because they evaluate gingival recession/excess in addition to probing depth. Radiographs are assessed to determine the extent, severity, and morphology of periodontal bone loss, which should correlate with clinical probing depth and CAL findings. Clinical parameters of plaque control (plaque/calculus accretions) are evaluated as local risk factors, and clinical parameters of inflammation (bleeding upon probing) are assessed as risk markers, for periodontal disease progression.

PERIODONTAL TREATMENT

Whereas genetic and environmental factors can modify susceptibility to the initiation and progression of plaque-induced periodontal diseases, periodontal therapy is primarily directed towards the disruption of local dental plaque biofilms in order to resolve inflammation. It is important to understand that plaque-induced gingivitis and periodontitis are largely preventable conditions, dependent on personal and professional measures that disrupt dental plaque biofilm maturation. Failure to perform thorough personal plaque control and/or present for regular professional debridements (necessary for removal of calcified plaque deposits known as tartar/calculus) results in microbial plaque communities shifting towards a more prominent Gram-negative composition, which ultimately elicits a pathophysiologic proinflammatory host immune response.

Gingivitis treatment

Gingivitis is the presence of gingival inflammation without loss of connective tissue attachment or alveolar bone. Gingivitis is a reversible disease, and therapy is aimed at reducing etiological plaque biofilm factors, which facilitates healing of the gingival tissues. Therapy for gingivitis is usually limited to professional supragingival plaque/calculus removal (ie, dental prophylaxis) and personal plaque control (ie, oral hygiene) instructions [47].

Periodontitis treatment

Periodontitis is the presence of gingival inflammation with irreversible detachment of periodontal ligament collagen fibers from cementum, which can progress to alveolar bone loss [2, 3]. Because plaque-induced periodontitis-mediated attachment/alveolar bone loss is primarily driven by local dental plaque biofilms, the primary objective of both nonsurgical and surgical therapeutic interventions is directed towards reducing local plaque etiological factors. The basis for surgical interventions (typically performed after unsatisfactory nonsurgical treatment outcomes) is to increase access and removal of residual subgingival plaque biofilms, in order to further resolve inflammation and abate disease progression [47].

Nonsurgical periodontitis treatment

The most effective nonsurgical therapy for periodontitis is scaling and root planing, which is supra/subgingival mechanical debridement (using hand instruments and/or ultrasonic powered instruments) of plaque/calculus accretions. Clear benefits have been validated from scaling and root planing combined with personal plaque control, including reduction of inflammation, shifts to a less pathogenic subgingival flora, decreased probing depths, gain of clinical attachment, and decreased disease progression [47].

The US Food and Drug Administration (FDA) has approved several controlled release local delivery antimicrobial agents as adjuncts to nonsurgical scaling and root planing therapy in the treatment of periodontitis. Antimicrobial agents for local delivery to periodontal pockets, currently available in the US, include 1.0 mg minocycline microspheres, 2.5 mg chlorhexidine in gelatin matrix, and 10% doxycycline hyclate in a bioabsorbable polymer. The use of local antimicrobial delivery as an adjunct to scaling and root planing has been shown to provide defined but marginal improvements in clinical periodontal parameters [48].

Systemically administered antibiotics have been used for periodontitis as adjuncts to scaling and root planing, and for patients who do not respond to conventional therapy despite adherence to plaque control protocols [47]. While microbial culturing and DNA probes can be applied to identify the presence of specific periopathogenic bacteria, these technique are not routinely applied due to the current understanding that periodontitis is driven by polymicrobial shifts in dental plaque biofilms. Antibiotic drug selection, dosage, and interval of therapy are largely empirically based. Combination antibiotic therapies have been reported to be effective, which is likely due to broadening the antimicrobial spectrum targeting the Gram-negative flora shift associated with periodontitis. Judicious use of systemic antibiotics is recommended to avoid the emergence of resistant microorganisms, and adverse side effects such as *Clostridium difficile* infection and drug-related toxicities.

Systemic administration of subantimicrobial dose doxycycline hyclate (20 mg, BID) is FDA approved as an adjunctive host immunomodulatory therapy to scaling and root planing for the treatment of periodontitis. The subantimicrobial doxycycline therapy acts as a host collagenase inhibitor. Another systemic therapy geared towards modulating the host immune response includes the administration of nonsteroidal anti-inflammatory drugs. Although the above host immunomodulation therapies have demonstrated beneficial effects on clinical parameters of periodontitis, long-term administration of any systemic agent has potential undesirable negative side effects [47].

Surgical periodontitis treatment

Periodontal surgical therapy is utilized to access subgingival plaque/calculus deposits on root surfaces that cannot be effectively debrided via nonsurgical instrumentation. Although the primary objective of periodontal surgery is to provide access facilitating thorough debridement of subgingival bacterial etiological factors, surgical therapy approaches vary considerably depending on the anatomical location, severity and morphology of attachment/bone loss, and the therapeutic objectives of correcting the diseased periodontal anatomy. Flap access surgeries provide access to debride subgingival root surfaces, but do not surgically alter the periodontal anatomy. Gingivectomy surgeries reduce the depth of periodontal pockets by resecting excess suprabony gingival tissue. Osseous resective surgeries correct pathological alterations in the alveolar bone architecture, in combination with resecting excess gingival tissues, in order to more effectively reduce periodontal pocket depth. Regenerative periodontal surgeries include bone grafting (autografts, allografts, xenografts, alloplasts) and guided tissue regeneration (GTR). GTR employs barrier membranes (resorbable, nonresorbable), with or without bone grafts and/or biological agents (platelet-derived growth factor, bone matrix proteins, enamel matrix derived proteins), with the objective of restoring lost supporting periodontal tissues. Although the different surgical therapeutic approaches differ in their immediate reduction of periodontal pocket depth and gain in clinical attachment levels, the long-term success of all surgical therapies is dependent on thorough patient oral hygiene measures and regular professional maintenance therapy [47].

Periodontal maintenance therapy

Both personal plaque control and regular professional periodontal maintenance therapy are critical in preventing the recurrence of inflammation and progressive destruction of periodontal tissues in prior periodontitis patients. Periodontal maintenance therapy is initiated after initial periodontal therapy. Periodontal maintenance entails updating medical and dental records, thorough clinical evaluation, radiographic review, debridement of supra/subgingival plaque/calculus accretions with root planing as indicated, and review of patient oral hygiene technique. Depending on the periodontitis patient's history and risk for periodontal disease progression, periodontal maintenance therapy intervals vary. Typically periodontal maintenance therapy is carried out every 3 months, which is in line with research demonstrating that subgingival plaque biofilms can shift towards pretreatment pathogenic flora levels 2 to 3 months after treatment [49].

FUTURE PERSPECTIVES

Periodontal research is currently centered on advancing knowledge about how polymicrobial shifts in plaque biofilms alter microbial community interactions, and the host immune response. Germ-free versus specific-pathogen-free animal investigations are actively delineating how the commensal flora impacts normal host immune response mechanisms, knowledge which is critical in understanding how pathogenic flora shifts disrupt periodontal tissue health and homeostasis. Advanced microbial genomic sequencing technologies, identifying specific microorganisms associated with periodontal health versus disease, will provide opportunities for noninvasive therapeutic interventions in the oral microbiome. Clinical salivary diagnostics geared towards measuring host cellular/molecular biomarkers, predictive of periodontal disease susceptibility and/or progression, are also actively being pursued.

Another emerging area of periodontal research is centered in the rapidly developing field of osteoimmunology, which is the study of immune cell interactions with bone cells. Delineating osteoimmunological mechanisms having catabolic effects on alveolar bone remodeling could provide avenues for novel therapeutic interventions aimed at the prevention and treatment of periodontitis-associated alveolar bone destruction. Notably, administration of recombinant PTH 1-34, an anabolic drug having osteoimmunomodulatory actions, was recently shown to improve clinical outcomes in the treatment of periodontitis. Patients treated with intermittent PTH 1-34 showed greater resolution of alveolar bone defects and accelerated osseous wound healing in the oral cavity [50]. Tissue engineering and regeneration remain active areas of investigation, as new biologicals undergo clinical trials to enhance predictability of regenerative therapies.

ACKNOWLEDGMENTS

The authors thank Johannes Aartun for graphic design support.

REFERENCES

1. Armitage GC; Research, Science and Therapy Committee of the American Academy of Periodontology. Position paper: Diagnosis of periodontal diseases. J Periodontol. 2003;74(8):1237–47.
2. Listgarten MA. Pathogenesis of periodontitis. J Clin Periodontol. 1986;13(5):418–30.
3. Page RC, Schroeder HE. *Periodontitis in man and other animals: a comparative review*. New York: Karger, 1982.
4. Nanci A, Bosshardt DD. Structure of periodontal tissues in health and disease. Periodontol 2000. 2006;40:11–28.

5. Darveau RP. Periodontitis: a polymicrobial disruption of host homeostasis. Nat Rev Microbiol. 2010;8(7):481–90.
6. Tonetti MS, Imboden MA, Lang NP. Neutrophil migration into the gingival sulcus is associated with transepithelial gradients of interleukin-8 and ICAM-1. J Periodontol. 1998;69(10):1139–47.
7. Schiott CR, Loe H. The origin and variation in number of leukocytes in the human saliva. J Periodontal Res. 1970;5(1):36–41.
8. Garant PR. Plaque-neutrophil interaction in monoinfected rats as visualized by transmission electron microscopy. J Periodontol. 1976;47(3):132–8.
9. Baker PJ. The role of immune responses in bone loss during periodontal disease. Microbes Infect. 2000;2(10):1181–92.
10. Taubman MA, Valverde P, Han X, et al. Immune response: the key to bone resorption in periodontal disease. J Periodontol. 2005;76(11 suppl):2033–41.
11. Garlet GP. Destructive and protective roles of cytokines in periodontitis: a re-appraisal from host defense and tissue destruction viewpoints. J Dent Res. 2010;89(12):1349–63.
12. Hajishengallis G, Liang S, Payne MA, et al. Low-abundance biofilm species orchestrates inflammatory periodontal disease through the commensal microbiota and complement. Cell Host Microbe. 2011;10(5):497–506.
13. Irie K, Novince CM, Darveau RP. Impact of the Oral Commensal Flora on Alveolar Bone Homeostasis. J Dent Res. 2014;93(8):801–6.
14. Theilade E. The non-specific theory in microbial etiology of inflammatory periodontal diseases. J Clin Periodontol. 1986;13(10):905–11.
15. Berezow AB, Darveau RP. Microbial shift and periodontitis. Periodontol 2000. 2011;55(1):36–47.
16. Socransky SS, Haffajee AD, Cugini MA, et al. Microbial complexes in subgingival plaque. J Clin Periodontol. 1998;25(2):134–44.
17. Aas JA, Paster BJ, Stokes LN, et al. Defining the normal bacterial flora of the oral cavity. J Clin Microbiol. 2005;43(11):5721–32.
18. Teles R, Teles F, Frias-Lopez J, et al. Lessons learned and unlearned in periodontal microbiology. Periodontol 2000. 2013;62(1):95–162.
19. Marsh PD. Microbial ecology of dental plaque and its significance in health and disease. Adv Dent Res. 1994;8(2):263–71.
20. Takeuchi O, Akira S. Pattern recognition receptors and inflammation. Cell. 2010;140(6):805–20.
21. Kawai T, Akira S. Toll-like receptors and their crosstalk with other innate receptors in infection and immunity. Immunity. 2011;34(5):637–50.
22. Cao X. Self-regulation and cross-regulation of pattern-recognition receptor signalling in health and disease. Nat Rev Immunol. 2016;16(1):35–50.
23. Hans M, Hans VM. Toll-like receptors and their dual role in periodontitis: a review. J Oral Sci. 2011;53(3):263–71.
24. Huang N, Gibson FC, 3rd. Immuno-pathogenesis of Periodontal Disease: Current and Emerging Paradigms. Curr Oral Health Rep. 2014;1(2):124–32.
25. Song B, Zhang Y, Chen L, et al. The role of Toll-like receptors in periodontitis. Oral Dis. 2017;23(2):168–80.
26. Crump KE, Sahingur SE. Microbial Nucleic Acid Sensing in Oral and Systemic Diseases. J Dent Res. 2016;95(1):17–25.
27. Teng YT. Protective and destructive immunity in the periodontium: Part 1—innate and humoral immunity and the periodontium. J Dent Res. 2006;85(3):198–208.
28. Kawai T, Akira S. The role of pattern-recognition receptors in innate immunity: update on Toll-like receptors. Nat Immunol. 2010;11(5):373–84.
29. O'Neill LA, Golenbock D, Bowie AG. The history of Toll-like receptors - redefining innate immunity. Nat Rev Immunol. 2013;13(6):453–60.
30. Beklen A, Hukkanen M, Richardson R, et al. Immunohistochemical localization of Toll-like receptors 1-10 in periodontitis. Oral Microbiol Immunol. 2008;23(5):425–31.
31. Sugawara Y, Uehara A, Fujimoto Y, et al. Toll-like receptors, NOD1, and NOD2 in oral epithelial cells. J Dent Res. 2006;85(6):524–9.
32. Uehara A, Takada H. Functional TLRs and NODs in human gingival fibroblasts. J Dent Res. 2007;86(3):249–54.
33. Dixon DR, Reife RA, Cebra JJ, et al. Commensal bacteria influence innate status within gingival tissues: a pilot study. J Periodontol. 2004;75(11):1486–92.
34. Liu YC, Lerner UH, Teng YT. Cytokine responses against periodontal infection: protective and destructive roles. Periodontol 2000. 2010;52(1):163–206.
35. Waerhaug J. The angular bone defect and its relationship to trauma from occlusion and downgrowth of subgingival plaque. J Clin Periodontol. 1979;6(2):61–82.
36. Garant PR, Cho MI. Histopathogenesis of spontaneous periodontal disease in conventional rats. I. Histometric and histologic study. J Periodontal Res. 1979;14(4):297–309.
37. Baron R, Saffar JL. A quantitative study of bone remodeling during experimental periodontal disease in the golden hamster. J Periodontal Res. 1978;13(4):309–15.
38. McCauley LK, Nohutcu RM. Mediators of periodontal osseous destruction and remodeling: principles and implications for diagnosis and therapy. J Periodontol. 2002;73(11):1377–91.
39. Belibasakis GN, Bostanci N. The RANKL-OPG system in clinical periodontology. J Clin Periodontol. 2012;39(3):239–48.
40. Kajiya M, Giro G, Taubman MA, et al. Role of periodontal pathogenic bacteria in RANKL-mediated bone destruction in periodontal disease. J Oral Microbiol. 2010;2:5532.
41. Teng YT, Nguyen H, Gao X, et al. Functional human T-cell immunity and osteoprotegerin ligand control alveolar bone destruction in periodontal infection. J Clin Invest. 2000;106(6):R59–67.
42. Kawai T, Matsuyama T, Hosokawa Y, et al. B and T lymphocytes are the primary sources of RANKL in the bone resorptive lesion of periodontal disease. Am J Pathol. 2006;169(3):987–98.

43. Pacios S, Xiao W, Mattos M, et al. Osteoblast Lineage Cells Play an Essential Role in Periodontal Bone Loss Through Activation of Nuclear Factor-Kappa B. Sci Rep. 2015;5:16694.
44. Algate K, Haynes DR, Bartold PM, et al. The effects of tumour necrosis factor-alpha on bone cells involved in periodontal alveolar bone loss; osteoclasts, osteoblasts and osteocytes. J Periodontal Res. 2016;51(5):549–66.
45. Herbert BA, Novince CM, Kirkwood KL. Aggregatibacter actinomycetemcomitans, a potent immunoregulator of the periodontal host defense system and alveolar bone homeostasis. Mol Oral Microbiol. 2016;31(3):207–27.
46. Eke PI, Dye BA, Wei L, et al. Update on Prevalence of Periodontitis in Adults in the United States: NHANES 2009 to 2012. J Periodontol. 2015;86(5):611–22.
47. Rosen PS; Research, Science and Therapy Committee of the American Academy of Periodontology. Treatment of plaque-induced gingivitis, chronic periodontitis, and other clinical conditions. J Periodontol. 2001;72(12):1790–800.
48. Greenstein G, Tonetti M; Research, Science and Therapy Committee of the American Academy of Periodontology. Position Paper: The role of controlled drug delivery for periodontitis. J Periodontol. 2000;71(1):125–40.
49. Cohen RE; Research, Science and Therapy Committee of the American Academy of Periodontology. Position paper: Periodontal maintenance. J Periodontol. 2003;74(9):1395–401.
50. Bashutski JD, Eber RM, Kinney JS, et al. Teriparatide and osseous regeneration in the oral cavity. N Engl J Med. 2010;363(25):2396–405.

122

Oral Manifestations of Metabolic Bone Diseases

Erica L. Scheller[1], Charles Hildebolt[2], and Roberto Civitelli[1]

[1]*Division of Bone and Mineral Diseases, Department of Internal Medicine, Washington University in St. Louis, MO, USA*
[2]*Mallinckrodt Institute of Radiology, Washington University in St. Louis, MO, USA*

MORPHOLOGY OF THE JAWS

The oral skeleton consists of two main bones, the mandible and the maxilla. The mandible, or lower jaw, is formed by intramembranous ossification and is the only mobile bone of the facial skeleton. The maxilla, or upper jaw, forms the roof of the oral cavity and houses the maxillary sinuses. The architecture of the mandibular and maxillary jaw bones is similar to that of the rest of the skeleton: the spongy cancellous bone is surrounded by a cortical shell. However, the jaws are also unique, functioning to support the teeth and their associated periodontal ligaments. Thus, in addition to biomechanics and bone density, tooth retention is considered an essential component of skeletal health in the oral cavity.

CLINICAL ASSESSMENT OF THE JAWS IN HUMANS

Dentists typically use clinical measures to assess the loss of the tooth attachment apparatus: gingiva, periodontal ligament, alveolar bone, and cementum. To measure attachment loss (AL), a periodontal probe is inserted alongside the tooth until it encounters firm resistance from the base of the gingival sulcus (Fig. 122.1). On the probe, the distance from the base of the sulcus to the tooth's cementoenamel junction (CEJ) is measured as AL. This measurement is often used as a surrogate measure for bone loss in the jaw because a reduction in alveolar bone height will result in a higher AL. However, AL may also increase as a consequence of inflammation, or overeruption of the tooth. After tooth loss, alveolar bone resorption accelerates, a phenomenon called residual ridge resorption.

Although it is useful as a clinical tool, AL is an indirect measurement of alveolar bone loss. A more direct assessment of oral bone health can be obtained by radiographic methods, although many of these are not applicable to routine clinical settings. The most commonly used index is alveolar crestal height (ACH), which is determined by measuring the distance from the CEJ to the top of the alveolar bone crest on dental bitewing radiographs (Fig. 122.2). The index is typically reported as the average ACH of all teeth measured. Advanced positioning techniques have been developed to track radiographic bone loss over time. Using a custom repositioning device, two images taken at different times can be registered by minimizing trabecular noise and subtracted (Figs 122.3 and 122.4). With this approach, it is possible to achieve a least significant change (threshold that represents real biological change with 95% confidence) of 0.06 mm for crest height change, compared with 0.49 mm for CEJ-AC on conventional radiographs [1].

Panoramic radiographs that show the entire jaw have also been used to clinically assess bone loss. Most studies quantify some aspects of cortical thickness along the inferior border of the mandible relative to other radiographic landmarks, for example at the gonion, the antegonial notch, or the mental foramen [2, 3]. An automated, computer-aided system has also been developed [4]. However, panoramic radiographs are not as widely used as are

---

Primer on the Metabolic Bone Diseases and Disorders of Mineral Metabolism, Ninth Edition. Edited by John P. Bilezikian.
© 2019 American Society for Bone and Mineral Research. Published 2019 by John Wiley & Sons, Inc.
Companion website: www.wiley.com/go/asbmrprimer

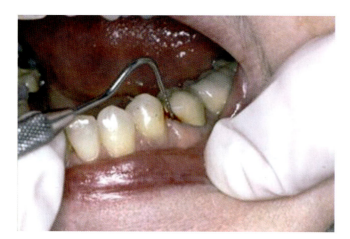

Fig. 122.1. Clinical measurement of attachment loss. A periodontal probe (with mm markings) is inserted into the gingival sulcus until it encounters firm resistance from the base of the sulcus, and the distance from the base of the sulcus to the cementoenamel junction (CEJ) is measured as attachment loss. Probing depth is the distance from the depth of the sulcus to the crest of the gingiva.

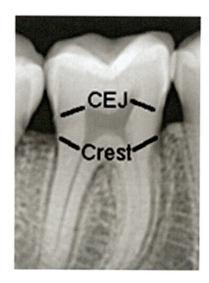

Fig. 122.2. Cementoenamel junction (CEJ) and alveolar crest as used to measure alveolar crest height (ACH) on a dental radiographic image. Logarithmic-transformed image, enhanced for display.

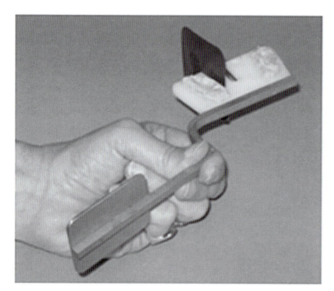

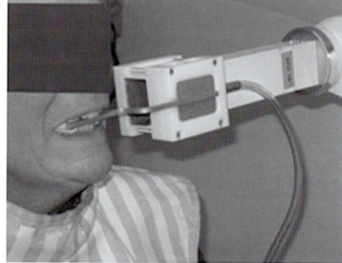

Fig. 122.3. A cross-arch, precision patient positioning bar for the molar–premolar region. The device is positioned with bite-registration material; the positioning ring is rigidly aligned with the X-ray tube by vacuum coupling. A phosphor screen, backed by two lead backscatter foils in a septic barrier envelope, is held in a vertical slot.

periapical and bitewing radiographs, and measurements vary with subject positioning.

BMD has been assessed using a variety of techniques including DXA [5], single-photon absorptiometry, dual-photon absorptiometry, and CT. BMD by DXA varies between jaw sites and is highest in the mandible [5]. However, positioning, reproducibility, and obstruction by teeth in dentate subjects make BMD measurement by DXA difficult. QCT allows assessment of BMD even in regions that contain teeth. However, it is expensive and involves exposure to relatively high radiation doses. Cone beam CT has become widely available in dentistry. Indeed, linear measurements using this technology are accurate and reliable, and images of oral bone can be studied in three dimensions. However, there is a large amount of scattered radiation, which negatively affects determination of BMD. All imaging techniques are limited by cost, accuracy, reliability, precision, and/or practicality.

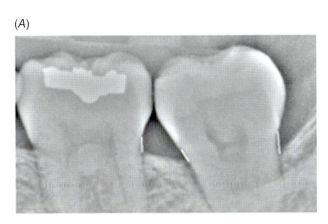
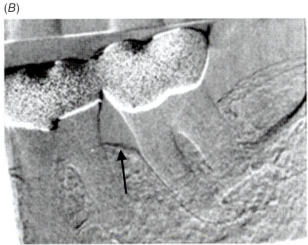

Fig. 122.4. Determination of alveolar bone loss using digital subtraction imaging. (A) High-pass-filtered image (features >40 pixels removed). Subsequent images are registered by minimizing trabecular noise. (B) Subtraction image. The arrow points to a dark area representing a change of −0.084 mm in crest ridge height between the two original images. Adapted from [1]. Reproduced with permission from Elsevier.

MORPHOLOGY AND ASSESSMENT OF THE JAWS IN LABORATORY ANIMALS

Unlike humans, rodents have continuously erupting central incisors that extend into the jaw underneath the roots of the molars. This presents a unique challenge when defining appropriate analysis regions for cortical and cancellous bone. Larger animals such as dogs and nonhuman primates more closely recapitulate the human dentition, but are not as commonly studied.

Micro–computed tomography is the most widely applied method for analysis of the jaws in laboratory animals and has been used to study alveolar remodeling, periodontal ligament thickness, morphology of cortical and trabecular bone, and skeletal regeneration in the mandible and maxilla (reviewed in [6]). Standardized analysis protocols, however, remain scarce. The most comprehensive study published to date provides recommendations for cancellous analysis in the region of the mandibular molars in mice, rats, rabbits, dogs, and nonhuman primates [7].

RELATIONSHIP BETWEEN ORAL AND POSTCRANIAL BONE MASS

With aging, the jaws undergo decreased trabeculation, cortical thinning, and increased porosity [8, 9]. The severity of this process depends upon the area of oral bone studied. For example, porosity of the buccal cortex and trabecular density of the mandible are dependent upon the presence or absence of teeth.

Earlier studies established that in postmenopausal women oral BMD is significantly associated with total body calcium and BMD at other skeletal sites (ie, forearm and vertebrae) [10]. Furthermore, good correlations between appendicular (forearm) or central (vertebral, proximal femur) BMD measured by DXA and oral bone measurements derived from dental radiographs have been consistently reported [8, 11]. Similarly, CT-based methods have confirmed a significant correlation between mandible and lumbar vertebrae bone mass [12] and longitudinal studies have shown that ACH and alveolar radio-densitometric parameters were correlated with changes in femoral and lumbar BMD [13–15]. Thus, oral bone density and microarchitecture reflect, in large part, the rest of the skeleton.

Several studies have also reported significant correlations between a mandibular cortical index and biochemical markers of bone turnover, specifically serum total alkaline phosphatase and urinary N-telopeptide crosslinks of type I collagen, in both women and men [16]. Such correlation is important in that bone turnover fluctuates widely with age, menopausal status, and even within a day; yet the correlation indicates that oral bone loss is associated with active bone remodeling.

MECHANISMS OF ORAL BONE LOSS

In principle, the same systemic processes that cause age-dependent bone loss in the postcranial skeleton occur in the jaws. This process adds to the effect of periodontal disease, resulting in progressive reduction in ACH and loosening of the tooth attachment apparatus [17]. Genetic factors that predispose a person to systemic bone loss also predispose to alveolar bone loss. Likewise, certain lifestyle factors, such as cigarette smoking and suboptimal calcium and/or vitamin D intake, and pathological conditions that facilitate local infections, such as diabetes, may increase the risk of systemic bone loss and deterioration of the alveolar bone. Periodontal disease is a key contributor to alveolar bone loss. In this condition secretion of inflammatory cytokines, including IL-1, IL-6, and TNF-α, activates osteoclastogenesis and bone resorption in response to bacterial infection [18]. Thus, the fundamental pathogenetic mechanism of alveolar bone resorption in periodontal disease is activation of a normal biological process (osteoclast formation and activity) by an abnormal local condition (infection-driven inflammation).

ASSESSMENT OF ORAL BONE AS A SCREENING TOOL FOR OSTEOPOROSIS OR FRACTURE RISK

In the United States, 62% of adults and 83% of children visit the dentist at least once per year [19]. This visit typically includes a review of the patient's medical history, an oral health screening, and dental radiographs. Thus, the use of parameters of dental health to estimate the risk of systemic bone loss and/or fragility fractures has been looked upon as a potential screening tool for bone health in the general population.

Measurements of cortical bone from panoramic radiographs have been proposed. However, initial studies revealed poor reproducibility, probably because of the need for manual assessment [20]. Specific training for dental practitioners marginally improved the predictive value of the oral measures [21]. Indeed, when good reproducibility is achieved, the sensitivity for predicting osteoporosis from mandibular radiometric measurements is greater than 80% [22]. A semi-automated, computer-aided system has also been developed to analyze panoramic radiographs. Although such a system does not significantly improve the predictive value relative to manual measurements, it does offer improved applicability to routine clinical settings [4, 23].

Data from the OSTEODENT consortium indicate that the diagnostic value of panoramic mandibular cortical thickness measures at the mental foramen was only modestly lower in predicting osteoporosis than algorithms based upon clinical risk factors [24]. Combining the two methods increases specificity but lowers sensitivity. Different radiometric parameters have been studied in the OSTEODENT project and, overall, measurement of cortical thickness appears to perform best for screening patients who would be recommended to undergo DXA testing for osteoporosis [25]. Inclusion of clinical data might improve the diagnostic performance of these measurements [26].

Trabecular patterns have also been used to assess oral bone. Trabecular coarseness on periapical radiographs was highly correlated with DXA measures of the forearm [27]. With periapical radiographs, morphologic features of trabecular oral bone had an accuracy of 92% in distinguishing osteoporotic and control patients [28]. In an expanded study of 598 women, the accuracy of trabecular features for predicting hip fractures was similar to that obtained with standard risk assessment tools [29]. These methods were not as successful when used with panoramic radiographs. An OSTEODENT-based collaboration also found that trabecular patterns determined on panoramic radiographs were accurate in identifying osteoporotic women with slightly better success than with periapical radiographs [30].

More recently, automated determinations of cortical width were combined with other clinical risk factors in an OSTEODENT index that was tested in 339 women against the WHO fracture risk assessment tool (FRAX) [31]. The OSTEODENT index had the same predictive value for identifying subjects who should be treated for osteoporosis as had FRAX without inclusion of DXA [32]. It is possible that with further refinements the OSTEODENT index's predictive value may increase, thus offering a new platform for osteoporosis screening in dental clinics [33].

OSTEOPOROSIS AND TOOTH LOSS

Loss of jaw BMD, tooth attachment, and alveolar bone height occur in osteoporosis [8, 34], but whether systemic bone loss directly contributes to tooth loss remains controversial [8, 17]. The extent to which osteoporosis contributes to tooth loss is inherently difficult to determine because teeth are lost primarily to decay and trauma. Also, dentists use various thresholds for determining when teeth should be extracted. Nevertheless, the available data would suggest that at least in elderly people low BMD is often associated with a lower number of teeth [35], even though this may not be the case in younger people and in early postmenopausal women.

Dental implants are becoming more frequent after tooth loss. During this procedure a bioengineered post is placed into the bone and then capped with a replacement tooth or crown. A common concern is whether implant osteointegration and osseous regeneration are compromised in patients with osteoporosis. A case-control study of 98 patients, half with osteoporosis, found no correlation between BMD by DXA and implant failure [36]. Implant success rates in osteoporotic patients are also high, about 97% in some studies [37]. Thus, the diagnosis of osteoporosis is not currently a contraindication for dental implant therapy.

PAGET DISEASE OF BONE

Up to 17% of patients with Paget disease of bone have manifestations in the oral cavity, primarily in the maxilla, less frequently in the mandible [38]. Enlargement of the alveolar ridge and in some cases the middle third of the face is frequently observed, leading to tooth spreading and an abnormal occlusal pattern. Such alveolar ridge enlargement may require edentulous patients with Paget disease to have their dentures fitted more frequently. Another complication of Paget disease is hypercementosis (excessive deposition of the mineralized cementum structure of the tooth root), which may result in tooth ankylosis. Conversely, Paget disease may lead to loosening of the teeth during its osteolytic phase [38]. The reduced bone radiodensity may resemble cemento-osseous dysplasia. At a more advanced osteosclerotic stage, numerous irregular radiopaque areas become evident, in the typical "cotton wool" appearance of Paget disease of bone. Histologically, the early lytic phase is associated with increased osteoclast number, and in the later stage, the exuberant osteoblast activity results in osteosclerotic areas next to the osteolytic areas in the typical "mosaic" pattern of pagetic bone. There is little information on the effect of systemic therapy for Paget disease of bone with oral localization. One case report suggested that a 6-month course with alendronate in a patient with Paget disease had no obvious negative effects on placement of dental implants [39].

PRIMARY HYPERPARATHYROIDISM

Several manifestations of hyperparathyroidism occur in the oral cavity. These include partial or complete loss of lamina dura, increased periodontal ligament width, decreased alveolar bone density, and, at more advanced stages, brown tumor formation [40]. Loss of the lamina dura is not a pathognomonic sign of the disease because it is also seen in Cushing syndrome and osteomalacia. However, acro-osteolysis, subperiosteal bone resorption with scalloping, typical of parathyroid bone disorders, occurs in the alveolar bone as well as in the phalanges. Brown tumors are now rarely observed because primary hyperparathyroidism is typically diagnosed well before the long-term consequences of PTH excess become manifest. In a recent study, parameters of periodontal health, such as AL, probing depth, and bleeding on probing were normal in patients with primary hyperparathyroidism; however, there was a positive correlation between increased widening of the periodontal ligament and serum PTH levels. Decreased cortical bone density and an increased presence of tori have also been reported [41].

RENAL OSTEODYSTROPHY

The oral manifestations of renal osteodystrophy are primarily the consequence of secondary hyperparathyroidism and share many features of primary hyperparathyroidism including loss of lamina dura, "ground glass" appearance of the bone, loss of trabeculation, and brown tumor formation [42]. Probably as a consequence of the stimulatory effect of PTH on periosteal bone formation, an enlargement of the jaws may occur in renal osteodystrophy, associated with cementum resorption [43]. However, renal osteodystrophy does not lead to widening of the periodontal ligament and indices of periodontal disease are unchanged in patients with secondary hyperparathyroidism from chronic renal failure [44].

IMPACT OF OSTEOPOROSIS THERAPY ON ORAL BONE HEALTH

One modifiable factor that may contribute to age-dependent bone loss is inadequate intake of vitamin D. A number of studies point to beneficial effects of vitamin D on periodontal health (reviewed in [45]) In a 12-year study of 562 men (mean age 63), daily intake of 800 IU or more of vitamin D was associated with lower odds of periodontal disease and bone loss relative to those whose intake was below 400 IU/day [46]. As an adjunct to scaling and root planing, supplementation with 500 mg calcium and 250 IU vitamin D improved gingival and oral health indices but failed to improve probing depth, AL, or bone density beyond scaling and root planing alone [47]. In another study, subjects taking ≥1000 mg oral calcium and ≥400 IU vitamin D daily had better clinical parameters of periodontal health and less bone loss relative to individuals not taking supplements [48]. However, consistent periodontal care (scaling and root planing) minimizes the apparent benefit of vitamin D supplementation [49].

Clinical studies have shown that estrogen deficiency decreases alveolar bone density [50]. A study in ovariectomized monkeys revealed eroded endosteal surfaces and decreased cortical bone density in histological sections of mandibular bone, associated with enlarged Haversian canals [51]. Accordingly, estrogen replacement therapy is beneficial to oral bone health. In the Leisure World Cohort and the Nurses' Health Study, estrogen use reduced tooth loss [52, 53], and data from the Third United States National Health and Nutrition Examination Survey (NHANES III) demonstrated that women who had taken estrogen had less AL relative to nonusers [54]. Furthermore, in a double-blind, randomized, 3-year, controlled trial hormone/estrogen replacement therapy improved alveolar bone mass as well as femoral and vertebral bone density [13, 55].

Bisphosphonates, potent inhibitors of bone resorption, should in theory protect from both alveolar and systemic bone loss, because the two processes are fundamentally similar. In a study on periodontal disease, patients who received periodontal maintenance therapy and were randomized to take either oral alendronate (10 mg daily) or risedronate (5 mg daily) for one year experienced significantly greater improvements in AL, probing depth, and gingival bleeding relative to a placebo group [56]. However, in another randomized trial alendronate (70 mg once weekly) did not significantly improve alveolar bone loss in subjects of both genders (71% had periodontal disease) who also received periodontal care, although there was a detectable positive effect in a subgroup of subjects with low alveolar bone mass at baseline [57]. Intriguingly, alendronate delivered locally as a 1% gel to patients with aggressive, chronic periodontal disease significantly improved clinical parameters of periodontal health relative to placebo, as an adjunct therapy to scaling and root planing [58]. However, the use of bisphosphonates for alveolar bone loss is at present overshadowed by the concerns engendered by the reported association between bisphosphonate use and osteonecrosis of the jaw. Such a severe side effect seems to be linked primarily to the use of very high doses of bisphosphonates in cancer patients, whereas the incidence of this event in patients treated with bisphosphonates for osteoporosis is quite low [59]. There is also concern that bisphosphonates increase the risk of implant failure; however, studies indicate that the risk is low [60].

SUMMARY

Several important questions remain regarding the correlation between systemic and oral bone mass. A better understanding of the rate of bone loss in the jaws compared with other skeletal regions with age, and the effect of menopause and other systemic conditions is still needed. Longitudinal progression of alveolar bone loss and the effects of different therapies on postcranial bone density compared with other skeletal sites also remain to be determined. Methodologies to assess oral density and alveolar bone loss need to be further refined and improved, especially those based on radiodensitometric approaches. Application of relatively simple and reliable methods to assess oral bone status in routine clinical settings could also be helpful in identifying subjects with or at risk of osteoporosis. Additional studies on this potential application of oral bone mass assessment would be of high impact. Metabolic bone diseases and periodontal disease are major health concerns in the United States, especially in older populations. Thus, studies that improve our understanding of the mechanisms by which metabolic bone diseases impact oral bone health would be highly relevant to improving quality of life in subjects affected by these highly prevalent disorders.

REFERENCES

1. Hildebolt CF, Couture R, Garcia NM, et al. Alveolar bone measurement precision for phosphor-plate images. Oral Surg Oral Med Oral Pathol Oral Radiol Endod. 2009;108(3):e96–107.
2. Dagistan S, Bilge OM. Comparison of antegonial index, mental index, panoramic mandibular index and mandibular cortical index values in the panoramic radiographs of normal males and male patients with osteoporosis. Dentomaxillofac Radiol. 2010;39(5):290–4.
3. Valerio CS, Trindade AM, Mazzieiro ET, et al. Use of digital panoramic radiography as an auxiliary means of low bone mineral density detection in post-menopausal women. Dentomaxillofac Radiol. 2013;42(10):20120059.
4. Nakamoto T, Taguchi A, Ohtsuka M, et al. A computer-aided diagnosis system to screen for osteoporosis using dental panoramic radiographs. Dentomaxillofac Radiol. 2008;37(5):274–81.
5. Gulsahi A, Paksoy CS, Ozden S, et al. Assessment of bone mineral density in the jaws and its relationship to radiomorphometric indices. Dentomaxillofac Radiol. 2010;39(5):284–9.
6. Faot F, Chatterjee M, de Camargos GV, et al. Micro-CT analysis of the rodent jaw bone micro-architecture: A systematic review. Bone Reports. 2015;2:14–24.
7. Bagi CM, Berryman E, Moalli MR. Comparative bone anatomy of commonly used laboratory animals: implications for drug discovery. Comp Med. 2011;61(1):76–85.
8. Hildebolt CF. Osteoporosis and oral bone loss. Dentomaxillofac Radiol. 1997;26(1):3–15.
9. Von Wowern N. Microradiographic and histomorphometric indices of mandibles for diagnosis of osteopenia. Scand J Dent Res. 1982;90(1):47–63.
10. Kribbs PJ, Chesnut CH, Ott SM, et al. Relationships between mandibular and skeletal bone in an osteoporotic population. J Prosthet Dent. 1989;62(6):703–7.
11. White SC. Oral radiographic predictors of osteoporosis. Dentomaxillofac Radiol. 2002;31(2):84–92.
12. Taguchi A, Tanimoto K, Suei Y, et al. Relationship between the mandibular and lumbar vertebral bone mineral density at different postmenopausal stages. Dentomaxillofac Radiol. 1996;25(3):130–5.
13. Civitelli R, Pilgram TK, Dotson M, et al. Alveolar and postcranial bone density in postmenopausal women receiving hormone/estrogen replacement therapy: a randomized, double-blind, placebo-controlled trial. Arch Intern Med. 2002;162(12):1409–15.
14. Hildebolt CF, Pilgram TK, Yokoyama-Crothers N, et al. The pattern of alveolar crest height change in healthy postmenopausal women after 3 years of hormone/estrogen replacement therapy. J Periodontol. 2002;73(11):1279–84.
15. Jacobs R, Ghyselen J, Koninckx P, et al. Long-term bone mass evaluation of mandible and lumbar spine in a group of women receiving hormone replacement therapy. Eur J Oral Sci. 1996;104(1):10–16.
16. Deguchi T, Yoshihara A, Hanada N, et al. Relationship between mandibular inferior cortex and general bone metabolism in older adults. Osteoporos Int. 2008; 19(7):935–40.
17. Oh T, Bashutski J, Giannobile WV. The Interrelationship Between Osteoporosis and Oral Bone Loss. Grand Rounds Oral-Sys Med. 2007;2:10–21.
18. Offenbacher S. Periodontal diseases: pathogenesis. Ann Periodontol. 1996;1(1):821–78.
19. U.S. Department of Health and Human Services. *Health, United States, 2015: With Special Feature on Racial and Ethnic Disparities.* Hyattsville, MD: National Center for Health Statistics, 2016.

20. Devlin H, Horner K. Mandibular radiomorphometric indices in the diagnosis of reduced skeletal bone mineral density. Osteoporos Int. 2002;13(5):373–8.
21. Sutthiprapaporn P, Taguchi A, Nakamoto T, et al. Diagnostic performance of general dental practitioners after lecture in identifying post-menopausal women with low bone mineral density by panoramic radiographs. Dentomaxillofac Radiol. 2006;35(4):249–52.
22. Taguchi A, Asano A, Ohtsuka M, et al.; OSPD International Collaborative Group. Observer performance in diagnosing osteoporosis by dental panoramic radiographs: results from the osteoporosis screening project in dentistry (OSPD). Bone. 2008;43(1):209–13.
23. Arifin AZ, Asano A, Taguchi A, et al. Computer-aided system for measuring the mandibular cortical width on dental panoramic radiographs in identifying postmenopausal women with low bone mineral density. Osteoporos Int. 2006;17(5):753–9.
24. Karayianni K, Horner K, Mitsea A, et al. Accuracy in osteoporosis diagnosis of a combination of mandibular cortical width measurement on dental panoramic radiographs and a clinical risk index (OSIRIS): the OSTEODENT project. Bone. 2007;40(1):223–9.
25. Horner K, Karayianni K, Mitsea A, et al. The mandibular cortex on radiographs as a tool for osteoporosis risk assessment: the OSTEODENT Project. J Clin Densitom. 2007;10(2):138–46.
26. Nackaerts O, Jacobs R, Devlin H, et al. Osteoporosis detection using intraoral densitometry. Dentomaxillofac Radiol. 2008;37(5):282–7.
27. Jonasson G, Bankvall G, Kiliaridis S. Estimation of skeletal bone mineral density by means of the trabecular pattern of the alveolar bone, its interdental thickness, and the bone mass of the mandible. Oral Surg Oral Med Oral Pathol Oral Radiol Endod. 2001;92(3):346–52.
28. White SC, Rudolph DJ. Alterations of the trabecular pattern of the jaws in patients with osteoporosis. Oral Surg Oral Med Oral Pathol Oral Radiol Endod. 1999;88(5):628–35.
29. White SC, Atchison KA, Gornbein JA, et al. Change in mandibular trabecular pattern and hip fracture rate in elderly women. Dentomaxillofac Radiol. 2005;34(3):168–74.
30. Geraets WG, Verheij JG, van der Stelt PF, et al. Prediction of bone mineral density with dental radiographs. Bone. 2007;40(5):1217–21.
31. Kanis JA, Oden A, Johansson H, et al. FRAX and its applications to clinical practice. Bone. 2009;44(5):734–43.
32. Horner K, Allen P, Graham J, et al. The relationship between the OSTEODENT index and hip fracture risk assessment using FRAX. Oral Surg Oral Med Oral Pathol Oral Radiol Endod. 2010;110(2):243–9.
33. Taguchi A. Triage screening for osteoporosis in dental clinics using panoramic radiographs. Oral Dis. 2010;16(4):316–27.
34. Jeffcoat MK, Lewis CE, Reddy MS, et al. Post-menopausal bone loss and its relationship to oral bone loss. Periodontol 2000. 2000;23:94–102.
35. Klemetti E, Vainio P, Lassila V, et al. Cortical bone mineral density in the mandible and osteoporosis status in postmenopausal women. Scand J Dent Res. 1993;101(4):219–23.
36. Becker W, Hujoel PP, Becker BE, et al. Osteoporosis and implant failure: an exploratory case-control study. J Periodontol. 2000;71(4):625–31.
37. Von Wowern N, Gotfredsen K. Implant-supported overdentures, a prevention of bone loss in edentulous mandibles? A 5-year follow-up study. Clin Oral Implants Res. 2001;12(1):19–25.
38. Smith BJ, Eveson JW. Paget's disease of bone with particular reference to dentistry. J Oral Pathol. 1981;10(4):233–47.
39. Pirih FQ, Zablotsky M, Cordell K, et al. Case report of implant placement in a patient with Paget's disease on bisphosphonate therapy. J Mich Dent Assoc. 2009;91(5):38–43.
40. Silverman S, Gordan G, Grant T, et al. The dental structures in primary hyperparathyroidism. Studies in forty-two consecutive patients. Oral Surg Oral Med Oral Pathol. 1962;15:426–36.
41. Padbury AD, Tözüm TF, Taba M, et al. The impact of primary hyperparathyroidism on the oral cavity. J Clin Endocrinol Metab. 2006;91(9):3439–45.
42. Silverman S, Ware WH, Gillooly C. Dental aspects of hyperparathyroidism. Oral Surg Oral Med Oral Pathol. 1968;26(2):184–9.
43. Goultschin J, Eliezer K. Resorption of cementum in renal osteodystrophy. J Oral Med. 1982;37(3):84–6.
44. Frankenthal S, Nakhoul F, Machtei EE, et al. The effect of secondary hyperparathyroidism and hemodialysis therapy on alveolar bone and periodontium. J Clin Periodontol. 2002;29(6):479–83.
45. Hildebolt CF. Effect of vitamin D and calcium on periodontitis. J Periodontol. 2005;76(9):1576–87.
46. Alshouibi EN, Kaye EK, Cabral HJ, et al. Vitamin D and periodontal health in older men. J Dent Res. 2013;92(8):689–93.
47. Perayil J, Menon KS, Kurup S, et al. Influence of Vitamin D & Calcium Supplementation in the Management of Periodontitis. J Clin Diagn Res. 2015;9(6):ZC35–8.
48. Miley DD, Garcia MN, Hildebolt CF, et al. Cross-sectional study of vitamin D and calcium supplementation effects on chronic periodontitis. J Periodontol. 2009;80(9):1433–9.
49. Garcia MN, Hildebolt CF, Miley DD, et al. One-year effects of vitamin D and calcium supplementation on chronic periodontitis. J Periodontol. 2011;82(1):25–32.
50. Payne JB, Reinhardt RA, Nummikoski PV, et al. Longitudinal alveolar bone loss in postmenopausal osteoporotic/osteopenic women. Osteoporos Int. 1999;10(1):34–40.
51. Tanaka M, Yamashita E, Anwar RB, et al. Radiological and histologic studies of the mandibular cortex of ovariectomized monkeys. Oral Surg Oral Med Oral Pathol Oral Radiol Endod. 2011;111(3):372–80.

52. Grodstein F, Colditz GA, Stampfer MJ. Post-menopausal hormone use and tooth loss: a prospective study. J Am Dent Assoc. 1996;127(3):370–7, quiz 392.
53. Paganini-Hill A. The benefits of estrogen replacement therapy on oral health. The Leisure World cohort. Arch Intern Med. 1995;155(21):2325–9.
54. Ronderos M, Jacobs DR, Himes JH, et al. Associations of periodontal disease with femoral bone mineral density and estrogen replacement therapy: cross-sectional evaluation of US adults from NHANES III. J Clin Periodontol. 2000;27(10):778–86.
55. Hildebolt CF, Pilgram TK, Dotson M, et al. Estrogen and/or calcium plus vitamin D increase mandibular bone mass. J Periodontol. 2004;75(6):811–6.
56. Lane N, Armitage GC, Loomer P, et al. Bisphosphonate therapy improves the outcome of conventional periodontal treatment: results of a 12-month, randomized, placebo-controlled study. J Periodontol. 2005;76(7):1113–22.
57. Jeffcoat MK, Cizza G, Shih WJ, et al. Efficacy of bisphosphonates for the control of alveolar bone loss in periodontitis. J Int Acad Periodontol. 2007;9(3):70–6.
58. Pradeep AR, Kanoriya D, Singhal S, et al. Comparative evaluation of subgingivally delivered 1% alendronate versus 1.2% atorvastatin gel in treatment of chronic periodontitis: a randomized placebo-controlled clinical trial. J Investig Clin Dent. 2017;8(3).
59. Ruggiero SL, Dodson TB, Fantasia J, et al.; American Association of Oral and Maxillofacial Surgeons. American Association of Oral and Maxillofacial Surgeons position paper on medication-related osteonecrosis of the jaw—2014 update. J Oral Maxillofac Surg. 2014;72(10):1938–56.
60. Madrid C, Sanz M. What impact do systemically administered bisphosphonates have on oral implant therapy? A systematic review. Clin Oral Implants Res. 2009;20(suppl 4):87–95.

123
Dental Implants and Osseous Healing in the Oral Cavity

Takashi Matsuura and Junro Yamashita

Department of Oral Rehabilitation, Fukuoka Dental College, Fukuoka, Japan

INTRODUCTION

Osseointegration occurs when a dental implant, which is typically made of titanium or titanium alloy, is placed in the jawbone and successfully integrated with the living bone via osseous wound healing (Fig. 123.1). Osseous trauma caused during implant placement triggers a sequence of biological events: protein adsorption to the implant surfaces, blood clot formation, inflammatory responses, woven bone formation, maturation of woven bone to lamellar bone, and subsequent bone remodeling. The wound healing pattern around dental implants is for the most part comparable with typical osseous wound healing that occurs during bone fractures and after tooth extraction. However, because osseous wound healing around implants involves host bone on one side and a biocompatible metal or ceramic on the other, differences do exist and there are factors that contribute to the differences. These include, but are not limited to, the space between the implant surface and the host bone, the implant surface topography and chemical composition, the surface treatment, the implant micromovement, and the implant macro design. Bone remodeling is crucial for the maintenance of long-term stable osseointegration. Mechanical stimuli from implants during mastication play a major role in the maintenance of osseointegration and adjacent bone through bone remodeling. In this chapter the focus will be on osseous wound healing around dental implants.

WOUND HEALING AROUND IMPLANTS

Osseous wound healing around implants can be divided into four major phases: hemostatic, inflammatory, proliferative, and remodeling. Although these phases are sequential, there is some overlap.

Hemostatic phase

Osteotomy during a dental implant placement surgery prompts bleeding. As soon as implants are exposed to blood, plasma proteins such as albumin, vitronectin, and fibronectin are adsorbed onto the implant surface. This protein adsorption provides a foundation for blood cells, such as leukocytes, red blood cells, and platelets, to adhere. Upon bleeding, thrombin converts fibrinogen into insoluble fibrin. Factor XIII then stimulates cross-linking of fibrin strands to form a stable 3D fibrin mesh matrix. Fibronectin adheres to this fibrin matrix to support cell attachment. In this way the fibrin matrix serves as a scaffold for cells to build osseous tissue. It is crucial to have stable blood clot formation on the implant surface for successful osseous wound healing to take place [1]. Upon the arrival of blood cells at the site, they are activated and multiple growth and differentiation factors are released. Serotonin is released from platelets to promote hemostasis [2]. PDGF, TGF-β, VEGF, and IGF are released to chemo-attract immune cells and

---

Primer on the Metabolic Bone Diseases and Disorders of Mineral Metabolism, Ninth Edition. Edited by John P. Bilezikian.
© 2019 American Society for Bone and Mineral Research. Published 2019 by John Wiley & Sons, Inc.
Companion website: www.wiley.com/go/asbmrprimer

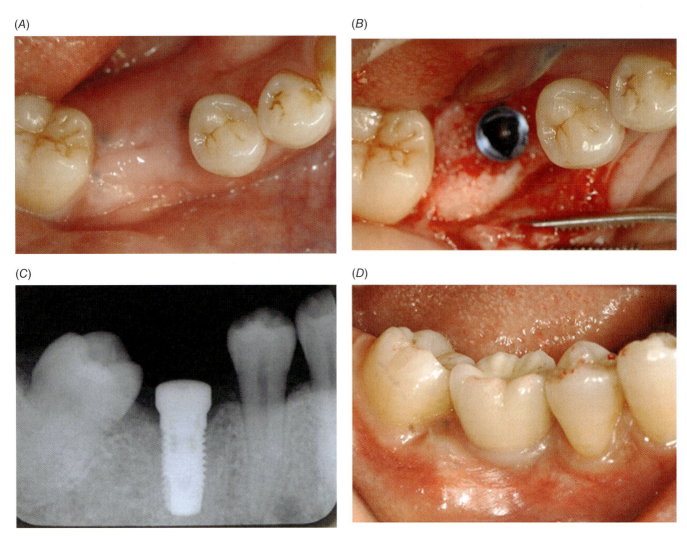

Fig. 123.1. Dental implant treatment to replace the mandibular first molar. (A) The mandibular first molar was missing. (B) Osteotomy was performed and a dental implant was placed in the alveolar bone. (C) A radiograph 3 months after implant placement. Osseointegration was achieved. (D) A porcelain crown was placed on the implant. Osseointegration is well maintained and the implant-supported crown is functioning well.

mesenchymal progenitors and stimulate their mitosis for subsequent tissue repair.

Inflammatory phase

Following the hemostatic phase, vasodilation occurs through the action of bradykinin, histamine, and substance P released from activated platelets [3]. Increased vascular permeability stimulates influx of serum fluids, proteins, and leukocytes. Polymorphonuclear leukocytes (PMNs) are chemotactically attracted to a wound site via the concentration gradient of bacterial proteins, fibrinopeptides, and proinflammatory interleukins [4]. PMNs combat bacteria with phagocytosis, the release of ROS and granule enzymes, the secretion of tissue digestive enzymes such as MMPs, and the generation of neutrophil extracellular traps (NETs). PMNs also stimulate macrophages. Activated macrophages take over and further eliminate tissue debris and bacteria. Macrophages release proinflammatory cytokines to further orchestrate the inflammatory process and later secrete cytokines and growth factors, such as tissue inhibitors for MMPs (TIMPs), VEGF, and FGF to stop matrix degradation and to promote angiogenesis and extracellular matrix formation.

Proliferative phase

The proliferative phase is characterized by the formation of granulation tissue and blood vessels. Fibroblasts are chemo-attracted to the implant surface in a blood clot by a concentration gradient of cytokines such as PDGF, TGF-β, bFGF, and connective tissue growth factor (CTGF). As fibroblasts migrate, they degrade the fibrin clot by secreting MMPs, and then deposit fibronectin, collagen, and decorin [5]. The low-oxygen tissue environment attracts macrophages, which are also stimulated to express VEGF [6]. Perivascular cells migrate from the outer wall of blood vessels to the hypoxia site and initiate blood vessel formation responding to the VEGF gradient [7]. Newly formed blood vessels restore the oxygen supply in the vicinity, which is a prerequisite for osteogenesis to happen because osteoblasts need blood vessels close by to survive. Hence, new bone formation occurs in close approximation to blood vessels. Burkhardt et al. reported that significantly increased VEGF secretion was noted in a blood clot at the implant surface [1], thereby contributing to osseointegration at the implant surface. Implant osteotomy traumatizes the surrounding bone up to 500 μm from the cutting surface, resulting in necrotic bone formation [8]. Such necrotic bone is repaired by the team work of osteoclasts and osteoblasts. Osteoclasts are derived from cells in the monocyte/macrophage lineage. Osteoclast precursors leave blood vessels and migrate towards the implant surface in response to cytokines such as stromal cell-derived factor 1 (SDF-1), IL-8, and monocyte chemotactic protein 1 (MCP-1/CCL2) [9]. Osteoclasts are formed by the fusion of precursors and resorb the damaged bone at the osteotomy site to initiate bone repair. This early bone resorption reduces the implant's primary stability [10]. As osteoclastic bone resorption proceeds, growth factors such as BMPs, TGF-β, and PDGF are released from the bone matrix. This in turn attracts osteoblasts and resultant bone formation occurs. Osteoblasts deposit an organic matrix mostly composed of collagen type I at the implant surface and at the osteoclast-resorbed areas. The matrix is subsequently mineralized to form woven bone, which is considered less organized immature bone.

Remodeling phase

Again a team of osteoblasts and osteoclasts works together to replace woven bone with lamellar bone where the collagen matrix and mineral structures are well organized and matured [11]. The lamellar bone around dental implants is subjected to repeated mechanical stress during mastication and is constantly remodeled throughout life. Because the alveolar bone, where teeth and implants are housed, receives mechanical stimuli during mastication, the alveolar bone remodeling is considered more robust compared to other bones in dogs [12]. However, this has yet to be validated in humans. Osteocytes play a major role in mechanotransduction, by which bone remodeling is largely governed. Functional stress causes interstitial pericellular fluid shifts in the cytoplasmic processes of osteocytes [13]. Such fluid shifts trigger intracellular signaling [14]. Thus mechanical stress during function perceived by local osteocytes is converted to molecular signals, transferred to neighboring osteocytes through gap junctions, and eventually shared with osteoblasts and osteoclasts that reside on distant bone surfaces [15]. This communication by osteocytes involves the trafficking of small messenger molecules, such as nitric oxide and prostaglandin signaling [16]. In this way the bone condition around implants, such as stress levels and microcrack formation, is conveyed to distant osteoclasts and osteoblasts which, after receiving the call, will then initiate bone remodeling.

ULTRASTRUCTURE AT THE BONE–IMPLANT INTERFACE AFTER OSSEOUS HEALING

Osseointegration is defined as a process in which a clinically asymptomatic fixation of alloplastic material is achieved and maintained in bone during functional healing. Direct contact between bone and the implant surface is seen at the light microscopic level. However, at the electron microscopic level, there are two thin layers between the bone and the implant surface. The first layer on the implant surface is approximately 20 to 40 nm in thickness, collagen-free, and low in mineral. It is surrounded by a 100- to 500-nm layer with randomly arranged collagen filaments and mineral gradient [17]. The first layer can be stained by ruthenium red. Treatment with chondroitinases, which degrades the glycosaminoglycan (GAG) chains of proteoglycans, weakens the interfacial staining intensity of ruthenium red between bone and the implant surface, suggesting the presence of proteoglycans in the first layer. However, the presence of proteoglycans in this layer has not been verified by immunochemical analysis [18]. The second layer consists of genuine collagen fibrils rather than fine collagen filaments [19]. The collagen fibrils are fewer, thinner, and less densely packed than those in surrounding bone. However, calcification is at least observed in the second layer. From these findings, it is clear that collagen fibrils are much thinner at the implant surface than in the surrounding bone.

Extensive research has been performed and many hypotheses made to clarify the composition of the first layer. It has been proposed that the first layer at the implant surface may be composed of collagen-binding small leucine-rich proteoglycans (SLRPs) such as decorin, fibromodulin, and lumican. SLRPs consist of core protein and GAG chains. SLRPs bind specifically to collagen fibrils by their core proteins. The binding site is a hydroxyapatite nucleation site where mineral is preferentially deposited in the early stage of mineralization [20]. On the other hand, the GAG chains of SLRPs are outside of fibrils, suppress collagen fibril growth [21], and interact with other proteoglycans and/or adsorbed proteins at the implant surface [22]. Among the SLRPs, decorin is the most abundant proteoglycan in bone. In fact, differentiated osteoblasts express decorin more highly than other SLRPs [23]. It has been demonstrated that osteoblasts induced to overexpress decorin deposit thin collagen fibrils and inhibit mineralization [24]. For these reasons decorin is likely a major component of the thin layers, especially the first layer, between the implant surface and bone. Another hypothesis is that the overhydroxylation of collagen lysine residues makes collagen fibrils thin and suppresses mineralization. Specific lysine residue is subjected to enzymatic hydroxylation followed by glycosylation, resulting in the generation of galactose-hydroxylysine (G-Hyl) and glucose-G-Hyl (GG-Hyl) [25]. Therefore, the overhydroxylation of lysine causes the overproduction of G-Hyl and GG-Hyl. Because both G-Hyl and GG-Hyl interfere with collagen fibrillogenesis [26], the overhydroxylation of lysine may be a mechanism of thin collagen fibrils at the second layer of the implant interface.

ROUGH IMPLANT SURFACE AND OSSEOINTEGRATION

Until the 1990s dental implants had predominantly smooth machined surfaces. In an effort to boost the success rate of osseointegration, microtextured (rough) implant surfaces were developed. The surface roughness is typically in a range of 0.5 to 8.5 μm of RA (profile roughness average), with which bone-to-implant contact (BIC) values increase [27]. Acid-etching, anodization, sandblasting, grit-blasting, and different coating procedures are employed to manufacture the rough implant surfaces [28]. Osseous wound healing is generally promoted on the rough surfaces because they exhibit better serum protein adsorption, more stable fibrin clot formation, and enhanced osteoblast function (adhesion, proliferation, differentiation, extracellular matrix formation, and mineralization) when compared to smooth surfaces [1]. As a result of these biological advantages, dental implants with rough surfaces achieve osseointegration more quickly and result in an improved BIC than those with smooth surfaces [27].

Another important factor that has a significant impact on the success of osseointegration is the hydrophilicity of the implant surfaces. Titanium implant surfaces are predominantly covered by a titanium dioxide (TiO_2) layer. Although the surface energy of this layer is initially good, it does diminish because the TiO_2 layer absorbs hydrocarbons and carbonates from ambient air. This hydrocarbon deposition hinders serum protein adsorption and suppresses osteoblastic function [29]. Furthermore, the rough surface treatment of implants reduces the surface energy as well [30]. In vitro studies revealed that hydrophilic implant surfaces are responsible for thicker blood clot formation with a more organized and layered architecture compared to hydrophobic surfaces [31]. Blood clots on hydrophilic surfaces exhibit a high number of activated platelets and strong complement activation. In these blood clots, blood cells synergistically interact with fibroblasts to produce a provisional collagen matrix and both osseous healing and angiogenesis are promoted [1], thereby enhancing osseointegration. In vivo studies confirmed that hydrophilic surfaces exhibit higher bone-to-implant contact 2 to 4 weeks after implant placement than hydrophobic surfaces [32]. In fact, higher torque is required to remove implants with hydrophilic surfaces than implants with hydrophobic surfaces, indicating improved osseointegration [33]. Once hydrophobic surfaces are treated with alkali or ultraviolet-mediated photofunctionalization to convert them to hydrophilic surfaces, high serum protein adsorption and subsequent bone formation are regained [34]. Thus, hydrophilic surfaces are beneficial for osseointegration.

ORAL MUCOSAL HEALING

Soft tissue laceration is inevitable during dental implant placement, and the repair of such traumatized tissue begins immediately after surgery. The reparative process of soft tissue wound healing around implants is essentially similar to that around teeth; coagulation, inflammatory responses, re-epithelialization, granulation tissue formation, matrix synthesis, and tissue remodeling. Because dental implants are biomaterials with no cementum, the attachment of soft tissue to the implant surface is different from that to teeth. In the case of teeth, the junctional epithelium is a crucial seal between the inside and outside of the body. The junctional epithelium adheres tightly to the tooth surface by means of hemidesmosomes. This biological seal develops around implants as well. Peri-implant junctional epithelium is formed around the transmucosal neck of an implant. It is structurally similar to the junctional epithelium around teeth [35], however the seal is much weaker around an implant. Moreover, the connective tissue attachment differs between teeth and implants. In teeth, the connective tissue adheres to the cementum through special fibers, such as dentogingival, dentoperiosteal, trans-septal, and circumferential fibers. Blood supply comes from the periodontal ligament and periosteum. In implants, the fibers run primarily parallel to the implant surface and the microvasculature is poorly developed in the peri-implant connective tissue. The healed peri-implant connective tissue is more or less a scar tissue cuff. Thus, although the soft tissue seal around implants is macroscopically similar to that around teeth, the tissue microstructure is quite different.

MECHANOBIOLOGY

Primary stability is a prerequisite for the success of osseointegration when implants are placed [36]. Primary stability is equivalent to proper mechanical stability at implant placement which comes from pure mechanical engagement of implant threads with the surrounding bone. Mechanical stress as a response to the implant surface during function influences early osseous healing. When primary stability is not obtained and implants are mechanically not secure, fibrous tissue formation occurs

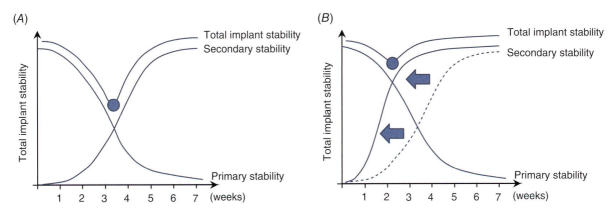

Fig. 123.2. Total implant stability results from the balance between primary stability and secondary stability. (A) Primary stability begins to drop soon after implant placement, whereas secondary stability gradually increases approximately 2 weeks after implant placement. The solid circle denotes a stability dip where the two curves meet. (B) A stability dip moves upward when the secondary stability curve shifts left. Source: [37]. Reproduced with permission from Wolters Kluwer.

around the implants, resulting in osseointegration failure. Mechanical stability diminishes as healing proceeds, because osteoclastic resorption of the bone which engages the implant threads takes place. During implant osteotomy, necrosis occurs in bone within approximately 500 μm of the cutting surface. This necrotic bone is the very bone that engages the implant threads. Because the necrotic bone is subjected to resorption for repair, the implant loses mechanical primary stability as bone resorption proceeds. The secondary stability secured by peri-implant bone formation begins approximately 2 weeks after implant placement. Hence the loss of primary mechanical stability is faster than the development of the secondary stability established by new bone formation. This causes a stability dip between these two processes (Fig. 123.2) [37]. When the establishment of secondary stability takes even more time because of bone degenerative conditions, such as osteoporosis and diabetes, the second curve shifts right and successful osseointegration may not be obtained. On the other hand, when the establishment of secondary stability is promoted, the second curve shifts left and successful osseointegration is achieved. Therefore, a good strategy for successful osseointegration would be to prevent a stability dip or to lessen the dip. The control of micromovement of implants and a reduction of stress in wounds is clinically important as well [38]. However, it is also known that adapted mechanical deformation, which promotes osseointegration during osseous wound healing, does exist although its magnitude, direction, and frequency are as yet undefined [39].

During the maintenance of osseointegration, it is accepted that appropriate mechanical stimuli have a positive influence on bone remodeling [40]. Mechanical stress generated around implants during function is sensed by osteocytes, converted to molecular signals, and transmitted to distant osteoclasts and osteoblasts via the osteocyte network [41]. It is known that an appropriate range of mechanical stress (deformation) is required to maintain bone health. Frost proposed that this deformation would be within a window of 400 to 3000 με [42]. Indeed, a previous study reported that bone apposition was observed in animals in which mechanical stimuli through implants exerted 3400 to 6600 με of strain [43]. Strain values above 6700 με resulted in accelerated bone resorption. Thus, for continuous maintenance of osseointegration, adapted and appropriate mechanical stimuli are essential.

FAILURE OF OSSEOINTEGRATION

The long-term success rate of dental implants is reported to be over 96% but the success rate drops when ailing implants are taken into consideration [44]. This is because there is a high prevalence of peri-implantitis [45]. Strictly, peri-implantitis is not a failure entirely of osseointegration. It is a chronic inflammatory disease that causes gradual marginal bone loss (Fig. 123.3). Although the short-term success rate of titanium dental implants is even higher [46], there are instances where dental implant loss does occur as a result of osseointegration failure. An early failure in osseointegration is noted during osseous wound healing where implants encapsulated by granulated fibrous tissue are seen. The causes of failure are related to surgical procedures, excessive biomechanical stress, and host biological responses [47]. A late failure in osseointegration occurs typically within a year after implant placement. Biomechanical overstress from parafunctional habits, such as clenching and grinding, is likely responsible for late osseointegration failure [48]. To prevent a failure in osseointegration, surgical procedures and the control of biomechanical stress can be modified and improved. Although at present our comprehension of host biological responses is not rich, progress has been made regarding implant surface modifications by using the application of bioactive molecules and altering topography. Currently

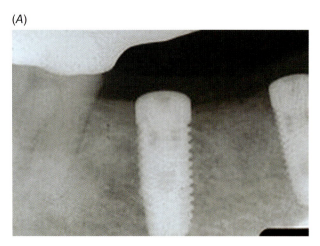 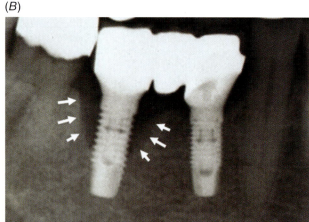

Fig. 123.3. Peri-implantitis. (A) Dental implants were placed in the mandibular posterior edentulous site. Osseointegration was achieved and the implant-supported prosthesis was in function. (B) After 3 years of service one implant developed peri-implantitis. There is extensive bone resorption around the implant (arrows).

these modifications promote a positive outcome in osseointegration but clearly more understanding is crucial to cope with host biological responses [49].

SUMMARY

In the past, titanium and titanium alloys have been considered bioinert materials. However, it is now clear that titanium implants are recognized as foreign bodies to the host. After placement, the surrounding living tissue interacts with implants. The surface topography and chemistry of these implants exert a strong influence on host biological responses during osseous wound healing. Mechanical stimuli, which are converted to molecular signals, modulate wound healing and the maintenance of osseointegration. Because our knowledge of molecular events during osseointegration and its maintenance is still scarce, further extensive research on biological responses to dental implants at the cellular and molecular levels is warranted.

REFERENCES

1. Burkhardt MA, Waser J, Milleret V, et al. Synergistic interactions of blood-borne immune cells, fibroblasts and extracellular matrix drive repair in an in vitro peri-implant wound healing model. Sci Rep. 2016;6:21071.
2. Lopez-Vilchez I, Diaz-Ricart M, White JG, et al. Serotonin enhances platelet procoagulant properties and their activation induced during platelet tissue factor uptake. Cardiovasc Res. 2009;84:309–16.
3. Brain SD, Williams TJ. Inflammatory oedema induced by synergism between calcitonin gene-related peptide (CGRP) and mediators of increased vascular permeability. Br J Pharmacol. 1985;86:855–60.
4. Cybulsky MI, Colditz IG, Movat HZ. The role of interleukin-1 in neutrophil leukocyte emigration induced by endotoxin. Am J Pathol. 1986;124:367–72.
5. Zhang Y, Lin Z, Foolen J, Schoen I, et al. Disentangling the multifactorial contributions of fibronectin, collagen and cyclic strain on MMP expression and extracellular matrix remodeling by fibroblasts. Matrix Biol. 2014;40: 62–72.
6. Xiong M, Elson G, Legarda D, et al. Production of vascular endothelial growth factor by murine macrophages: regulation by hypoxia, lactate, and the inducible nitric oxide synthase pathway. Am J Pathol. 1998;153:587–98.
7. Yamagishi S, Yonekura H, Yamamoto Y, et al. Vascular endothelial growth factor acts as a pericyte mitogen under hypoxic conditions. Lab Invest. 1999;79:501–9.
8. Ohtsu A, Kusakari H, Maeda T, et al. A histological investigation on tissue responses to titanium implants in cortical bone of the rat femur. J Periodontol. 1997;68: 270–83.
9. Niwa T, Mizukoshi K, Azuma Y, et al. Fundamental study of osteoclast chemotaxis toward chemoattractants expressed in periodontitis. J Periodontal Res. 2013; 48:773–80.
10. Gomes JB, Campos FE, Marin C, et al. Implant biomechanical stability variation at early implantation times in vivo: an experimental study in dogs. Int J Oral Maxillofac Implants. 2013;28:e128–134.
11. Andersen TL, Sondergaard TE, Skorzynska KE, et al. A physical mechanism for coupling bone resorption and formation in adult human bone. Am J Pathol. 2009;174:239–47.
12. Huja SS, Fernandez SA, Hill KJ, et al. Remodeling dynamics in the alveolar process in skeletally mature dogs. Anat Rec A Discov Mol Cell Evol Biol. 2006;288:1243–9.
13. Price C, Zhou X, Li W, et al. Real-time measurement of solute transport within the lacunar-canalicular system of mechanically loaded bone: direct evidence for load-induced fluid flow. J Bone Miner Res. 2011;26:277–85.

14. Ajubi NE, Klein-Nulend J, Alblas MJ, et al. Signal transduction pathways involved in fluid flow-induced PGE2 production by cultures osteocytes. Am J Physiol. 1999;276:E171–8.
15. Moriishi T, Fukuyama R, Ito M, et al. Osteocyte network; a negative regulatory system for bone mass augmented by the induction of Rankl in osteoblasts and Sost in osteocytes at unloading. PLoS One. 2012;7:e40143.
16. Chow JW, Fox SW, Lean JM, et al. Role of nitric oxide and prostaglandins in mechanically induced bone formation. J Bone Miner Res. 1998;13:1039–44.
17. Linder L, Albrektsson T, Brånemark PI, et al. Electron microscopic analysis of the bone-titanium interface. Acta Orthop Scand. 1983;54:45–52.
18. Nakamura H, Shim J, Butz F, et al. Glycosaminoglycan degradation reduces mineralized tissue-titanium interfacial strength. J Biomed Mater Res A. 2006;77:478–86.
19. Listgarten MA, Buser D, Steinemann SG, et al. Light and transmission electron microscopy of the intact interfaces between non-submerged titanium-coated epoxy resin implants and bone or gingiva. J Dent Res. 1992;71:364–71.
20. Landis WJ, Song MJ, Leith A, et al. Mineral and organic matrix interaction in normally calcifying tendon visualized in three dimensions by high-voltage electron microscopic tomography and graphic image reconstruction. J Struct Biol. 1993;110:39–54.
21. Rühland C, Schönherr E, Robenek H, et al. The glycosaminoglycan chain of decorin plays an important role in collagen fibril formation at the early stages of fibrillogenesis. FEBS J. 2007;274:4246–55.
22. Gubbiotti MA, Vallet SD, Ricard-Blum S, et al. Decorin interacting network: a comprehensive analysis of decorin-binding partners and their versatile functions. Matrix Biol. 2016;55:7–21.
23. Matsuura T, Tsubaki S, Tsuzuki T, et al. Differential gene expression of collagen-binding small leucine-rich proteoglycans and lysyl hydroxylases, during mineralization by MC3T3-E1 cells cultured on titanium implant material. Eur J Oral Sci. 2005;113:225–31.
24. Mochida Y, Parisuthiman D, Pornprasertsuk-Damrongsri S, et al. Decorin modulates collagen matrix assembly and mineralization. Matrix Biol. 2009;28:44–52.
25. Sricholpech M, Perdivara I, Nagaoka H, et al. Lysyl hydroxylase 3 glucosylates galactosylhydroxylysine residues in type I collagen in osteoblast culture. J Biol Chem. 2011;286:8846–56.
26. Notbohm H, Nokelainen M, Myllyharju J, et al. Recombinant human type II collagens with low and high levels of hydroxylysine and its glycosylated forms show marked differences in fibrillogenesis in vitro. J Biol Chem. 1999;274:8988–92.
27. Shalabi MM, Gortemaker A, Van't Hof MA, et al. Implant surface roughness and bone healing: a systematic review. J Dent Res. 2006;85:496–500.
28. Jarmar T, Palmquist A, Brånemark R, et al. Characterization of the surface properties of commercially available dental implants using scanning electron microscopy, focused ion beam, and high-resolution transmission electron microscopy. Clin Implant Dent Relat Res. 2008;10:11–22.
29. Hayashi R, Ueno T, Migita S, et al. Hydrocarbon deposition attenuates osteoblast activity on titanium. J Dent Res. 2014;93:698–703.
30. Zhao G, Schwarz Z, Wieland M, et al. High surface energy enhances cell response to titanium substrate microstructure. J Biomed Mater Res A. 2005;74:49–58.
31. Milleret V, Tugulu S, Schlottig F, et al. Alkali treatment of microrough titanium surfaces affects macrophage/monocyte adhesion, platelet activation and architecture of blood clot formation. Eur Cell Mater. 2011;21:430–44.
32. Buser D, Broggini N, Wieland M, et al. Enhanced bone apposition to a chemically modified SLA titanium surface. J Dent Res. 2004;83:529–33.
33. Ferguson SJ, Broggini N, Wieland M, et al. Biomechanical evaluation of the interfacial strength of a chemically modified sandblasted and acid-etched titanium surface. J Biomed Mater Res A. 2006;78:291–7.
34. Aita H, Hori N, Takeuchi M, et al. The effect of ultraviolet functionalization of titanium on integration with bone. Biomaterials. 2009;30:1015–25.
35. Atsuta I, Yamaza T, Yoshinari M, et al. Changes in the distribution of laminin-5 during peri-implant epithelium formation after immediate titanium implantation in rats. Biomaterials. 2005;26:1751–60.
36. Lioubavina-Hack N, Lang NP, Karring T. Significance of primary stability for osseointegration of dental implants. Clin Oral Implants Res. 2006;17:244–50.
37. Suzuki S, Kobayashi H, Ogawa T. Implant stability change and osseointegration speed of immediately loaded photofunctionalized implants. Implant Dent. 2013;22:481–90.
38. Schnitman PA, Hwang JW. To immediately load, expose, or submerge in partial edentulism: a study of primary stability and treatment outcome. Int J Oral Maxillofac Implants. 2011;26:850–9.
39. Leucht P, Kim JB, Wazen R, et al. Effect of mechanical stimuli on skeletal regeneration around implants. Bone. 2007;40:919–30.
40. Zhang X, Duyck J, Vandamme K, et al. Ultrastructural characterization of the implant interface response to loading. J Dent Res 2014;93:313–8.
41. Barros RR, Degidi M, Novaes AB, et al. Osteocyte density in the peri-implant bone of immediately loaded and submerged dental implants. J Periodontol. 2009;80:499–504.
42. Frost HM. Bone's mechanostat: a 2003 update. Anat Rec A Discov Mol Cell Evol Biol. 2003;275:1081–101.
43. Melsen B, Lang NP. Biological reactions of alveolar bone to orthodontic loading of oral implants. Clin Oral Implants Res. 2001;12:144–52.
44. Zupnik J, Kim SW, Ravens D, et al. Factors associated with dental implant survival: a 4-year retrospective analysis. J Periodontol. 2011;82:1390–5.
45. Derks J, Tomasi C. Peri-implant health and disease. A systematic review of current epidemiology. J Clin Periodontol. 2015;42(suppl 16):S158–71.

46. Alsaadi G, Quirynen M, Komárek A, et al. Impact of local and systemic factors on the incidence of oral implant failure, up to abutment connection. J Clin Periodontol. 2007;34:610–17.
47. Olmedo-Gaya MV, Manzano-Moreno FJ, Cañaveral-Cavero E, et al. Risk factors associated with early implant failure: A 5-year retrospective clinical study. J Prosthet Dent. 2016;115:150–5.
48. Chrcanovic BR, Kisch J, Albrektsson T, et al. Bruxism and dental implant failures: a multilevel mixed effects parametric survival analysis approach. J Oral Rehabil. 2016;43:813–23.
49. Boyan BD, Cheng A, Olivares-Navarrete R, et al. Implant surface design regulates mesenchymal stem cell differentiation and maturation. Adv Dent Res. 2016;28:10–17.

Section XI
Integrative Physiology of the Skeleton

Section Editors: Mone Zaidi and Clifford J. Rosen

Chapter 124. Integrative Physiology of the Skeleton   959
Clifford J. Rosen and Mone Zaidi

Chapter 125. The Hematopoietic Niche and Bone   966
Stavroula Kousteni, Benjamin J. Frisch, Marta Galan-Diez, and Laura M. Calvi

Chapter 126. Adipocytes and Bone   974
Clarissa S. Craft, Natalie K. Wee, and Erica L. Scheller

Chapter 127. The Vasculature and Bone   983
Marie Hélène Lafage-Proust and Bernard Roche

Chapter 128. Immunobiology and Bone   992
Roberto Pacifici and M. Neale Weitzmann

Chapter 129. Cellular Bioenergetics of Bone   1004
Wen-Chih Lee and Fanxin Long

Chapter 130. Endocrine Bioenergetics of Bone   1012
Patricia F. Ducy and Gerard Karsenty

Chapter 131. Central Neuronal Control of Bone Remodelling   1020
Hiroki Ochi, Paul Baldock, and Shu Takeda

Chapter 132. Peripheral Neuronal Control of Bone Remodeling   1028
Katherine J. Motyl and Mary F. Barbe

Chapter 133. The Pituitary–Bone Axis in Health and Disease   1037
Mone Zaidi, Tony Yuen, Wahid Abu-Amer, Peng Liu, Terry F. Davies, Maria I. New, Harry C. Blair, Alberta Zallone, Clifford J. Rosen, and Li Sun

Chapter 134. Neuropsychiatric Disorders and the Skeleton   1047
Madhusmita Misra and Anne Klibanski

Chapter 135. Interactions Between Muscle and Bone   1055
Marco Brotto

Primer on the Metabolic Bone Diseases and Disorders of Mineral Metabolism, Ninth Edition. Edited by John P. Bilezikian.
© 2019 American Society for Bone and Mineral Research. Published 2019 by John Wiley & Sons, Inc.
Companion website: www.wiley.com/go/asbmrprimer

124

Integrative Physiology of the Skeleton

Clifford J. Rosen[1] and Mone Zaidi[2]

[1] *Maine Medical Center, Scarborough, ME, USA*
[2] *The Mount Sinai Bone Program, Department of Medicine, Icahn School of Medicine at Mount Sinai, New York, NY, USA*

INTRODUCTION

In the 21st century, two sharply contrasting disorders of body composition, obesity and osteoporosis, have reached near epidemic proportions and challenged our scientific resources. The growing prevalence of these diseases has also enabled investigators to explore new perspectives on the molecular, cellular, and genetic determinants that regulate body composition and energy homeostasis shared by both disorders [1]. This, in turn, has resulted in a sea-change in our understanding of the multiple mechanisms controlling body mass and energy utilization, particularly as they relate to the skeleton. Shared regulation of bone and fat occurs at several levels, including the skeleton. Acquisition and maintenance of bone and fat are also mediated through central and peripheral mechanisms including multiple endocrine and paracrine determinants, some that arise from the skeleton [2]. Importantly, the hypothalamus is the principal controller of these target tissues through neuronal (eg, sympathetic nervous system—SNS) and hormonal mediators [3]. In the case of the former, the hypothalamus modulates fat and bone tissue via two major pathways: (i) SNS activation that modulates appetite, insulin sensitivity, energy utilization, and skeletal remodeling; and (ii) paracrine and endocrine hormonal factors secreted by the hypothalamus. In respect to the latter, paracrine pituitary activating hormones such as corticotropin-releasing hormone and thyrotropin-releasing hormone can modify metabolic homeostasis through their secondary target glands, the adrenals and thyroid. Similarly the hypothalamus can secrete factors such as FGF-21, brain-derived neurotrophic factor (BDNF), oxytocin, neuromedin U, neuropeptide Y (NPY), melatonin, and IL-6 that impact both bone and fat in a systemic fashion [4, 5]. Circulating serum factors, such as TNF-α and PTHrP, may also play a role in determining how the body utilizes fuel and distributes it to other organs, particularly in states of cachexia [6]. Muscle-derived growth factors can have an impact on bone and adipose tissues, particularly inguinal depots (or subcutaneous depots in humans). Currently one of the most exciting myokines is irisin, which is a cleaved protein product from muscle after exercise that induces beiging of white adipose tissue and also positively impacts cortical bone [7].

In addition to shared regulatory determinants between bone and fat, as well as a common central signal from the hypothalamus, bone and fat cells arise from the same mesenchymal derived progenitor cell, primarily in the bone marrow [8] (Fig. 124.1). There is emerging evidence of significant plasticity between these two lineages, and some have suggested that there may be transdifferentiation of these terminally differentiated adult cells. Cell fate decisions in the bone marrow are critical for defining total fat and bone mass, particularly because recent studies have hinted that around 10% of adipocytes in peripheral depots may originate from the marrow after bone marrow transplantation [9]. Cell fate within the marrow is also particularly relevant in cases where fuel supplies are low, for example in anorexia nervosa [10]. New insights from genetic engineering in mice, and clinical disorders such as anorexia nervosa, where paradoxically fat mass is absent in the periphery but high in the marrow and is accompanied by an impairment in osteogenesis, have allowed investigators to examine the physiological mechanisms that link acquisition and maintenance of BMD to energy utilization and adipose remodeling. On the other hand, genetic determinants of bone and fat in humans using genomewide association

Integrative Physiology of the Skeleton

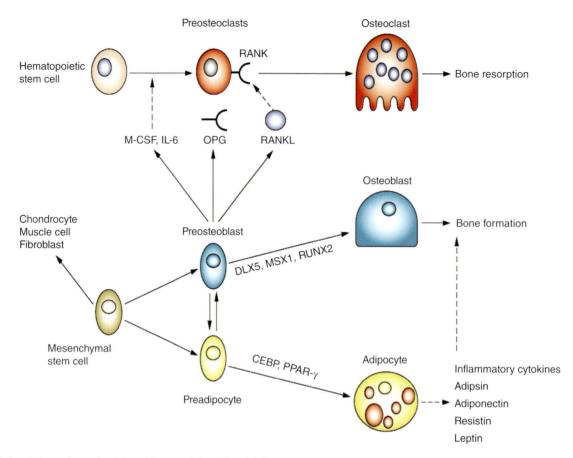

Fig. 124.1. Mesenchymal origin of bone and fat cells. OPG = osteoprotegerin.

study have reinforced the intimate relationship between body composition and the skeleton. Finally, and probably most importantly, the seminal discoveries by the Karsenty group that bone-specific proteins regulate glucose homeostasis, testosterone production, and muscle function (see Chapter 130) provide proof that the skeleton is a critical element of integrative physiology. In this overview we will focus on several advances in adipose and bone biology that have revised our thinking about the role of the skeleton in whole body homeostasis.

A COMMON ORIGIN FOR FAT AND BONE CELLS

Bone marrow surrounds trabecular elements in the skeleton and is composed of multiple cells that are pluripotent [11]. In addition to the hematopoietic elements that include red and white blood cells, platelets, and their progenitors, the bone marrow stroma or connective tissue contains mesenchymal stromal cells, an adult stem cell population capable of extensive self-renewal and plasticity [12]. Stromal cells are regulated by endocrine, paracrine, and autocrine signals, and in response can enter bone, cartilage, or fat lineages, depending upon the activation of intrinsic transcription factors [12]. The allocation of stem cells into bone-forming osteoblasts, some of which may enter the circulation, is accelerated after skeletal injury as well as during the rapid growth phase of puberty and with administration of PTH 1–34 [13,14]. Runx2/Cbfa1 and osterix are two of several bone-specific transcription factors that are required for this process [15]. In contrast, entry of stromal cells into the fat lineage occurs through activation of the nuclear receptor PPAR-γ2 by endogenous fatty acids or exogenous ligands (see Fig. 124.1) [16]. Stem cell specification into the fat or bone pathway is often considered as an either-or-paradigm, that is, commitment of these cells is exclusive to one lineage or the other, based principally on PPAR-γ2 complex activation by a class of drugs called the thiazolidinediones (TZDs: rosiglitazone or pioglitazone) [17,18]. These agents improve insulin sensitivity and enhance bone resorption while shifting mesenchymal cell fate from preosteoblasts to preadipocytes [19]. The clinical outcome of treatment with the TZDs is often manifested as improved glucose tolerance, associated with modest weight gain and increased marrow adiposity at the expense of skeletal mass [18-,19]. But in other circumstances enhanced marrow adipogenesis can coexist with active bone formation, such as in the inbred mouse strain, C3H/HeJ, or after treatment with certain PPAR-γ agonists [18–20]. In that vein, emerging evidence suggests there may be a mixed population of cells that have transcriptional characteristics found in both fat and bone cells. Westendorf and colleagues have noted that some Runx2-positive cells have large perilipin-positive lipid droplets in HDAC3 null mice consistent with the hypothesis that these cells are "osteo-adipocytes" [21]. Lanske and colleagues have shown that some marrow adipocytes express RANKL, a stromal cell cytokine characteristic of preosteoblasts [22]. Hence, the progression or "switch" of stem cells into either the fat or bone lineage may not be mutually exclusive, and there is certain to be plasticity during fate decisions. Moreover, genetic determinants are likely to be important at this critical junction for both physiological and pathological states (see later).

Although much of what we have learned about the "switching mechanism" in mesenchymal stromal cells has come from basic investigations, clinical experiences have also been very illustrative. For example, glucocorticoids are remarkable for their ability to enhance marrow adipogenesis at the expense of osteoblast differentiation [23]. In the syndrome of glucocorticoid-induced osteoporosis, bone loss is quite rapid at a time when fat mass is enhanced, particularly in central depots. Similarly, the TZDs reduce bone mass and increase the risk of peripheral fractures primarily because of the switch in stem cell allocation into the adipocyte lineage [24]. Another less well-known example of "switching" occurs with aging. Fatty infiltration of the vertebral marrow is a characteristic feature of older individuals and can be recognized on MRI, usually as an incidental finding. Its presence, however, is inversely related to BMD and skeletal integrity in many circumstances [25]. Importantly, cell fate decisions are also related to their metabolic program (see later in this chapter), which is tied to energy availability and illustrated by the syndrome of anorexia nervosa. Thus, bone and fat cells share a common origin, and their fates are interwoven in a context-specific manner. Pharmacological manipulation of mesenchymal stem cells to reduce adipogenesis or enhance osteogenesis is a promising new but yet untested avenue for therapeutic investigations.

BIOENERGETICS OF CELLS IN THE BONE MARROW NICHE IN RELATION TO ENERGY NEEDS AND WHOLE BODY METABOLISM

It is likely that cell fate in the niche relates to metabolic flexibility and programming of progenitor cells. Each of the cells in the bone marrow niche has its own specific nutrient requirements in order to survive, particularly in a hypoxic environment, and these, in turn, are linked to adenosine diphosphate (ATP) demands. In general, the more differentiated the cell, the greater the energy needs. However, within that context there are major differences between mature osteoblasts and adipocytes. First, when considering survival and maintenance of MSC stemness in the relative hypoxia of the marrow, there is a need for metabolic adaptation. Stemness allows a stable pool of progenitors that can be called on in numerous circumstances, particularly injury and inflammatory states [26]. Hypoxia induces the stabilization of HIF1α, a major transcription factor for stem cells and progenitors as well as multiple downstream target genes, particularly VEGFα [27].

Metabolic reprogramming of quiescent cells is necessary to prevent differentiation, and this occurs through a shift from oxidative phosphorylation to glycolysis. Importantly, glycolysis, although less efficient in generating ATP than mitochondrial oxidation, reduces oxidative stress and ROS generation, key elements that drive stem cell differentiation. Although glycolysis is the major driver of ATP generation in MSCs, entrance into a specific differentiation program, either adipogenic or osteogenic, requires distinct metabolic requirements that are very context-specific [28]. For adipocytic differentiation, several studies have suggested that mitochondrial oxidation of fatty acids and the generation of ROS are essential to achieve full maturation. The process of glucose entry and fatty acid oxidation through the Krebs cycle generates more molecules of ATP per mol of glucose (36:1) than glycolysis (2:1), but it comes at a cost because mitochondrial respiration leads to the generation of ROS from the electron transport chain (ETC). ROS (eg, H_2O_2, superoxides) can further suppress mitochondrial respiration and promote an adipogenic program that is associated with more insulin resistance and less lipolysis [29,30]. Excess ROS in adipocytes may also cause mitochondrial DNA damage or further changes to complex I in the ETC, leading to metabolic dysfunction [31]. Although some ROS are generated during normal adipocyte differentiation, much less is produced during early stages of osteoblast maturation.

During osteoblastic differentiation, glycolysis is the predominant ATP generator, even though it is less efficient per mol of glucose [32]. Glycolysis can occur rapidly and can happen in both hypoxic and normoxic states (ie, the Warburg effect). Two key proteins, Glut1, the principal glucose transporter, and lactate dehydrogenase (LDH), the enzyme that converts pyruvate to lactate, are essential during glycolysis. Karsenty and colleagues showed that in osteoblasts, glucose uptake through the Glut 1 transporter inhibits 5′ adenosine monophosphate-activated protein kinase (AMPK), which in turn prevents ubiquitination of runt-related transcription factor 2 (Runx2) [16]. In a feed-forward system, Runx2 begins the differentiation program in osteoblasts and increases Glut1 expression. Furthermore, preosteoblasts differentiate under the influence of various ligands, particularly the Wnts and IGFs. Long and colleagues demonstrated that glycolysis is a major feature of Wnt3a-induced osteoblast differentiation [33]. More recently the Long group reported that HIF1α, which is important for stemness, is also a critical transcriptional regulator of glycolysis, triggered in part by relative hypoxia in the bone marrow niche [34]. Remarkably, much older ex vivo studies from Neuman and colleagues demonstrated that PTH treatment produced lactic acid in calvarial osteoblasts, supporting the tenet that osteoblasts utilize glycolysis to generate lactate and fuel collagen synthesis and mineralization [35]. Guntur et al. also showed that glycolysis was essential for terminal differentiation of osteoblasts and that oxidative phosphorylation was more important early in the differentiation scheme [36]. Thus it is likely there is a distinct metabolic program that features a transient phase of oxidative phosphorylation following glycolysis and that is then switched off as glycolysis re-emerges as a predominant driver of ATP. Indeed, recent studies suggest that both oxidative phosphorylation and glycolysis are occurring in differentiating cells, and the relative proportion that contributes to ATP determines the final energy production. Too much oxidative phosphorylation or glycolysis can inhibit the other, so there clearly must be a fine balance, which is very context-specific.

The transient phase of mitochondrial respiration is very time-sensitive and may occur in vitro between days 3 and 9 of osteoblastic differentiation. During this time period, AMPK is activated and this may induce lipophagy as well as oxidative phosphorylation [37]. Other studies have shown that metformin, which upregulates AMPK, can enhance differentiation but only during specific time periods. In a similar vein, others have demonstrated that glutaminolysis is also essential for osteoblast differentiation through the Wnt signaling system [34]. Hence, there are at least three substrates for ATP generation and differentiation of MSCs: glutamine, which enters the Krebs cycle via alpha ketoglutarate; glucose, which through glycolysis can generate lactate as well as ribose nucleotides via the pentose phosphate shunt; and fatty acids, which are metabolized via acetyl coenzyme A (acetyl CoA) in the mitochondria. Generation of acetyl CoA can also lead to enhanced nuclear acetylation that in turn can impact transcriptional processes unifying the processes of protein production and energy utilization. Finally autophagy cannot be overlooked as a fueling mechanism for the cell, particularly during times of stress or nutrient deficiency. AMPK stimulates autophagy as well as glycolysis and inhibits mammalian target of rapamycin (mTOR) and overall protein synthesis [37,38]. This would also lead to increased fatty acid entry into the mitochondria for ATP generation because lipophagy is also stimulated.

In summary, the bone marrow niche has significant energy requirements that are tissue- and time-specific. Allocation into the osteoblast or adipocyte lineage is dictated by multiple transcription factors, which in turn must be governed by specific metabolic programs and their inherent flexibility. Changes in energy availability are certain to alter cell fate decisions and these in turn can impact metabolic homeostasis. Taken together, the determination of the identity of marrow adipocytes as osteoblasts in disguise, or a novel fat cell, is a critical question because of the intimate relationship between adipogenesis and skeletal remodeling, particularly in states of energy insufficiency. Genetic determinants of metabolic programs are certain to link bone and fat metabolism.

CONTROL OF SKELETAL AND ADIPOSE TISSUE REMODELING

Bone and adipose tissue remodeling are functionally related through a complex neuroendocrine circuit that involves the brain, adipose depots, and the skeleton. In mammals, the SNS can drive lipolysis through release of norepinephrine, which activates beta-adrenergic receptors, of which there are three (β1–3 AR) in adipocytes. The effects of the SNS on bone are more complex and are driven primarily through activation of the β2AR. Interrelated to the neural modulation of bone and fat, several hormonal factors are also regulatory, particularly during puberty. For example, during linear growth and the acquisition of peak bone mass, surges in growth hormone and gonadal steroid secretion provide a stimulus for skeletal expansion and stem cell recruitment into cartilage and bone. These processes are fueled through lipolysis of white adipose tissue. As such, malnutrition or undernutrition can cause short stature and a severe reduction in peak bone mass. Thus, when considering the regulation of fat and bone by systemic and local factors, it is helpful to classify these as peripheral or central mediators. The peripheral mediators are adipokines that act via the central nervous system (CNS) to regulate sympathetic outflow to both fat and bone tissue. Central mediators arise from the CNS as either neural peptides or paracrine factors.

Leptin is an adipokine produced by fat cells [39]. The circulating levels of leptin are directly related to total fat mass. Leptin regulates appetite, reproduction, and energy utilization by crossing the blood–brain barrier and binding to a receptor in the hypothalamus. In the ventromedial nucleus of the hypothalamus, leptin triggers activation of the SNS. Deficiencies of, or resistance to leptin can cause obesity, impaired fertility, and changes in appetite, in both rodents and humans [40,41]. Surprisingly, the absence of leptin results in high bone mass, even though gonadal steroids are markedly suppressed in the animal model of total leptin deficiency, the ob/ob mouse. Studies of these mice demonstrated that the high bone mass of leptin deficiency is a result of reduced sympathetic tone innervating β2-adrenergic receptors in osteoblasts (see Fig. 124.2). Those observations provided evidence that bone remodeling is regulated by a hypothalamic relay via the SNS, and is primed by leptin, the sensor of peripheral fuel status in fat depots.

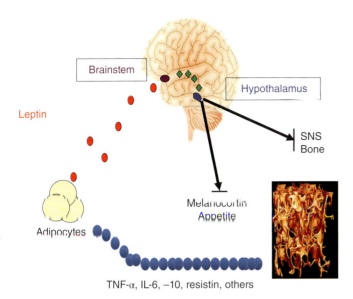

Fig. 124.2. Afferent signal phase I: leptin controls bone mass and energy metabolism by inhibiting serotonin synthesis and increasing sympathetic nervous system (SNS) activity.

Not surprisingly, the system is more complex than originally perceived, and includes other hypothalamic networks such as the neuropeptide cocaine- and amphetamine-regulated transcript (CART), melanocortin 4 receptors, the neuropeptide Y receptor system, and neuromedin U [42,43]. Cannabinoid receptors that regulate appetite and energy balance also modulate bone turnover centrally and peripherally, principally by blocking sympathetic innervation. Karsenty et al. have reported that the bone-specific protein, osteocalcin, when undercarboxylated, regulates insulin secretion, gonadal status, and muscle activity [44]. If validated in humans, this connection provides another step in a complex integrative circuit that regulates body composition, although in this case, the master regulator is the skeleton.

Adiponectin is a secretory peptide made by differentiated adipocytes, primarily from white adipose tissue. Its expression occurs late in adipocyte differentiation and is regulated by the master transcription factor PPARG. Adiponectin can induce insulin sensitivity and is thought to be a modulator of glucose transport. It is upregulated by the TZD class of antidiabetic agents. Cawthorn and colleagues demonstrated that adiponectin secretion also occurs in marrow adipose tissue and this may regulate glucose levels in anorexia nervosa, a condition associated with low energy intake but high insulin sensitivity [45]. In experimental animals, adiponectin has been shown to have dual effects on bone mass—direct and indirect. In respect to the former, adiponectin has been shown to inhibit osteoblast differentiation in the bone marrow; on the other hand, adiponectin can downregulate sympathetic tone, thereby indirectly enhancing bone mass. Karsenty and colleagues have suggested that leptin and adiponectin act in a yin and yang mode in the CNS [46]. Hence, it is apparent from experimental evidence that one locus of control over body composition is modulated through the SNS.

SYMPATHETIC NERVOUS SYSTEM CONTROL OVER FAT AND BONE REMODELING

Several lines of evidence point towards an efferent sympathetic pathway from the brain controlling skeletal metabolism. Genetically engineered mice with both a global and conditional deletion of the β2-adrenergic receptor (B2AR) have high bone mass at 8 and 16 weeks of age. Mice treated with isoproterenol, a beta-adrenergic receptor agonist, lose bone mass whereas treatment with the beta-adrenergic receptor antagonist propranolol may protect against ovariectomy-induced bone loss [5,46,47]. Patients with reflex sympathetic dystrophy, a disease characterized by high sympathetic tone, are prone to low bone mass that at least in some cases can be mitigated by beta-blockers [48]. Sympathetic overactivity has also been proposed as a contributing mechanism for microgravity-induced bone loss during space flight although long-term studies have not been done [49].

The link between leptin and bone via a sympathetic efferent was established by Elefteriou et al. through failure to reverse the high bone mass phenotype in B2AR-deficient mice following intracerebroventricular infusion of leptin [50]. The expression of B2AR (Adrβ2) in osteoblasts provided another link in the pathway between hypothalamus and osteoblast, and mice with selective knockout of Adrβ2 on osteoblasts have a high bone mass from increased bone formation and reduced bone resorption (see Fig. 124.2). Studies also indicated that sympathetic signaling in osteoblasts is responsible for regulatory control of osteoblast function through inhibition of osteoblast proliferation via circadian clock genes and that the SNS also favors bone resorption by increasing the expression of RANKL [51–53]. The balance of bone formation and resorption from chronic stress-induced sympathetic activity may shift to favor bone resorption as seen with chronic stimulation of β-AR with low-dose agonist treatment in mice which induces bone loss mainly via enhanced bone resorption [54], suggesting the control of each cell type by the SNS is temporal. Second-generation antipsychotics, such as risperidone, also upregulate sympathetic tone and uncouple remodeling; these effects can also be blocked with propranolol [55].

CONCLUSION

There is emerging evidence that the skeleton is closely integrated with other tissues, particularly adipose depots and the CNS. New evidence is emerging to support a bidirectional messaging system that allows communications between cell types. Further studies are certain to support the premise that the skeleton is a major component of integrative physiology.

REFERENCES

1. Kawai M, de Paula FJA, Rosen CJ. New insights into osteoporosis: The bone-fat connection. J Intern Med. 2012;272:317–29.
2. Guntur AR, Rosen CJ. The skeleton: A multi-functional complex organ. New insights into osteoblasts and their role in bone formation: The central role of PI3Kinase. J Endocrinol. 2011;211:123–30.
3. Shaikh MG, Crabtree N, Kirk JMW, et al. The relationship between bone mass and body composition in children with hypothalamic and simple obesity. Clin Endocrinol (Oxf). 2014;80(1):85–91.
4. Baldock PA, Sainsbury A, Allison S, et al. Hypothalamic control of bone formation: distinct actions of leptin and y2 receptor pathways. J Bone Miner Res. 2005;20(10):1851–7.
5. Sato S, Hanada R, Kimura A, et al. Central control of bone remodeling by neuromedin U. Nat Med. 2007;13(10):1234–40.
6. Kir S, White JP, Kleiner S, et al. Tumour-derived PTH-related protein triggers adipose tissue browning and cancer cachexia. Nature. 2014;513(7516):100–4.
7. Colaianni G, Cuscito C, Mongelli T, et al. The myokine irisin increases cortical bone mass. Proc Natl Acad Sci U S A. 2015;112(39):12157–62.
8. Bianco P. "Mesenchymal" stem cells. Annu Rev Cell Dev Biol. 2014;30:677–704.
9. Rydén M, Uzunel M, Hård JL, et al. Transplanted Bone Marrow-Derived Cells Contribute to Human Adipogenesis. Cell Metab. 2015;22(3):408–17.
10. Miller KK, Lee EE, Lawson EA, et al. Determinants of skeletal loss and recovery in anorexia nervosa. J Clin Endocrinol Metab. 2006;91(8):2931–7.
11. Bianco P, Kuznetsov SA, Riminucci M, et al. Postnatal Skeletal Stem Cells. Methods Enzymol. 2006;419:117–48.
12. Kuznetsov SA, Riminucci M, Ziran N, et al. The interplay of osteogenesis and hematopoiesis: expression of a constitutively active PTH/PTHrP receptor in osteogenic cells perturbs the establishment of hematopoiesis in bone and of skeletal stem cells in the bone marrow. J Cell Biol. 2004;167(6):1113–22.
13. Bianco P. Minireview: The stem cell next door: skeletal and hematopoietic stem cell "niches" in bone. Endocrinology. 2011;152(8):2957–62.
14. Undale A, Srinivasan B, Drake M, et al. Circulating osteogenic cells: characterization and relationship to rates of bone loss in postmenopausal women. Bone. 2010;47(1):83–92.
15. Eghbali-Fatourechi GZ, Mödder UIL, Charatcharoenwitthaya N, et al. Characterization of circulating osteoblast lineage cells in humans. Bone. 2007;40(5):1370–7.
16. Wei J, Shimazu J, Makinistoglu MP, et al. Glucose Uptake and Runx2 Synergize to Orchestrate Osteoblast Differentiation and Bone Formation. Cell. 2015;161(7):1576–91.
17. Ackert-Bicknell C, Rosen C. The genetics of PPARG and the skeleton. PPAR Res. 2006;2006:93258.
18. Ackert-Bicknell CL, Shockley KR, Horton LG, et al. Strain-specific effects of rosiglitazone on bone mass, body composition, and serum insulin-like growth factor-I. Endocrinology. 2009;150:1330–40.
19. Lazarenko OP, Rzonca SO, Hogue WR, et al. Rosiglitazone induces decreases in bone mass and strength that are reminiscent of aged bone. Endocrinology. 2007;148:2669–80.
20. Stechschulte LA, Czernik PJ, Rotter ZC, et al. PPARG Post-translational Modifications Regulate Bone Formation and Bone Resorption. EBioMedicine. 2016;10:174–84.
21. McGee-Lawrence ME, Carpio LR, Schulze RJ, et al. Hdac3 Deficiency Increases Marrow Adiposity and Induces Lipid Storage and Glucocorticoid Metabolism in Osteochondroprogenitor Cells. J Bone Miner Res. 2016;31(1):116–28.
22. Fan Y, Bi R, Densmore MJ, et al. Parathyroid hormone 1 receptor is essential to induce FGF23 production and maintain systemic mineral ion homeostasis. FASEB J. 2016;30(1):428–40.
23. Mazziotti G, Delgado A, Maffezzoni F, et al. Skeletal Fragility in Endogenous Hypercortisolism. Front Horm Res. 2016;46:66–73.
24. Rzonca SO, Suva LJ, Gaddy D, et al. Bone Is a Target for the Antidiabetic Compound Rosiglitazone. Endocrinology. 2004;145:401–6.
25. Moerman EJ, Teng K, Lipschitz DA, et al. Aging activates adipogenic and suppresses osteogenic programs in mesenchymal marrow stroma/stem cells: the role of PPAR-gamma2 transcription factor and TGF-beta/BMP signaling pathways. Aging Cell. 2004;3(6):379–89.
26. Kuhn NZ, Tuan RS. Regulation of stemness and stem cell niche of mesenchymal stem cells: Implications in tumorigenesis and metastasis. J Cell Physiol. 2010;222(2):268–77.
27. Imanirad P, Dzierzak E. Hypoxia and HIFs in regulating the development of the hematopoietic system. Blood Cells Mol Dis. 2013;51(4):256–63.
28. Esen E, Long F. Aerobic glycolysis in osteoblasts. Curr Osteoporos Rep. 2014;12(4):433–8.
29. Ushio-Fukai M, Rehman J. Redox and metabolic regulation of stem/progenitor cells and their niche. Antioxid Redox Signal. 2014;21(11):1587–90.
30. Wang T, Si Y, Shirihai OS, et al. Respiration in adipocytes is inhibited by reactive oxygen species. Obesity (Silver Spring). 2010;18(8):1493–502.
31. Dong X, Bi L, He S, et al. FFAs-ROS-ERK/P38 pathway plays a key role in adipocyte lipotoxicity on osteoblasts in co-culture. Biochimie. 2014;101:123–31.
32. Chen J, Tu X, Esen E, et al. WNT7B promotes bone formation in part through mTORC1. PLoS Genet. 2014;10(1):e1004145.
33. Esen E, Chen J, Karner CM, et al. WNT-LRP5 signaling induces Warburg effect through mTORC2 activation during osteoblast differentiation. Cell Metab. 2013;17(5):745–55.

34. Regan JN, Lim J, Shi Y, et al. Up-regulation of glycolytic metabolism is required for HIF1α-driven bone formation. Proc Natl Acad Sci U S A. 2014;111(23):8673–8.
35. Nichols FC, Neuman WF. Lactic acid production in mouse calvaria in vitro with and without parathyroid hormone stimulation: lack of acetazolamide effects. Bone. 1987;8(2):105–9.
36. Guntur AR, Le PT, Farber CR, et al. Bioenergetics during calvarial osteoblast differentiation reflect strain differences in bone mass. Endocrinology. 2014;155(5):1589–95.
37. Xi G, Rosen CJ, Clemmons DR. IGF-I and IGFBP-2 Stimulate AMPK Activation and Autophagy, Which Are Required for Osteoblast Differentiation. Endocrinology. 2016;157(1):268–81.
38. Pantovic A, Krstic A, Janjetovic K, et al. Coordinated time-dependent modulation of AMPK/Akt/mTOR signaling and autophagy controls osteogenic differentiation of human mesenchymal stem cells. Bone. 2013;52(1):524–31.
39. Friedman JM, Mantzoros CS. 20 years of leptin: from the discovery of the leptin gene to leptin in our therapeutic armamentarium. Metabolism. 2015;64(1):1–4.
40. Ducy P, Amling M, Takeda S, et al. Leptin inhibits bone formation through a hypothalamic relay: a central control of bone mass. Cell. 2000;100:197–207.
41. Holloway WR, Collier FM, Aitken CJ, et al. Leptin inhibits osteoclast generation. J Bone Miner Res. 2002;17(2):200–9.
42. Zhang Y, Proenca R, Maffei M, et al. Positional cloning of the mouse obese gene and its human homologue. Nature. 1994;372(6505):425–32.
43. Tartaglia LA, Dembski M, Weng X, et al. Identification and expression cloning of a leptin receptor, OB-R. Cell. 1995;83(7):1263–71.
44. Ferron M, Hinoi E, Karsenty G, et al. Osteocalcin differentially regulates beta cell and adipocyte gene expression and affects the development of metabolic diseases in wild-type mice. Proc Natl Acad Sci U S A. 2008;105(13):5266–70.
45. Cawthorn WP, Scheller EL, Learman BS, et al. Bone marrow adipose tissue is an endocrine organ that contributes to increased circulating adiponectin during caloric restriction. Cell Metab. 2014;20(2):368–75.
46. Kajimura D, Lee HW, Riley KJ, et al. Adiponectin regulates bone mass via opposite central and peripheral mechanisms through foxo1. Cell Metab. 2013;17(6):901–15.
47. Takeda S, Karsenty G. Molecular bases of the sympathetic regulation of bone mass. Bone. 2008;42(5):837–40.
48. Sandhu HS, Herskovits MS, Singh IJ. Effect of surgical sympathectomy on bone remodeling at rat incisor and molar root sockets. Anat Rec. 1987;219(1):32–8.
49. Schwartzman RJ. New treatments for reflex sympathetic dystrophy. N Engl J Med. 2000;343(9):654–6.
50. Navasiolava NM, Custaud M-A, Tomilovskaya ES, et al. Long-term dry immersion: review and prospects. Eur J Appl Physiol. 2011;111(7):1235–60.
51. Elefteriou F, Ahn JD, Takeda S, et al. Leptin regulation of bone resorption by the sympathetic nervous system and CART. Nature. 2005;434(7032):514–20.
52. Kondo H, Togari A. Continuous treatment with a low-dose β-agonist reduces bone mass by increasing bone resorption without suppressing bone formation. Calcif Tissue Int. 2011;88(1):23–32.
53. Ding Y, Arai M, Kondo H, et al. Effects of capsaicin-induced sensory denervation on bone metabolism in adult rats. Bone. 2010;46(6):1591–6.
54. Togari A, Kondo H, Hirai T, et al. [Regulation of bone metabolism by sympathetic nervous system]. Nihon Yakurigaku Zasshi. 2015;145(3):140–5.
55. Motyl KJ, DeMambro VE, Barlow D, et al. Propranolol Attenuates Risperidone-Induced Trabecular Bone Loss in Female Mice. Endocrinology. 2015;156(7):2374–83.

125

The Hematopoietic Niche and Bone

Stavroula Kousteni[1], Benjamin J. Frisch[2],
Marta Galan-Diez[1], and Laura M. Calvi[3]

[1] *Columbia University, New York, NY, USA*
[2] *University of Rochester, Rochester, NY, USA*
[3] *University of Rochester Multidisciplinary Neuroendocrinology Clinic, Rochester, NY, USA*

INTRODUCTION—HETEROGENEITY OF HSCs AND THEIR NICHES

One of the critical and unique functions of the skeleton in terrestrial vertebrates is to provide the anatomical spaces for storing and facilitating differentiation of hematopoietic progenitors and precursors. The most immature cells capable of generating the entire hematopoietic system are hematopoietic stem cells (HSCs) that are also found in bone marrow (BM). A true HSC is defined as a cell that can generate the entire hematopoietic system throughout the lifetime of an adult. HSCs represent a functionally heterogeneous cell population in terms of self-renewal, life span, and differentiation. Self-renewal heterogeneity is manifested by the fact that long-term and short-term repopulating HSCs show differential abilities to engraft into irradiated hosts and to maintain multilineage hematopoiesis for extended periods of time or by serial transplantation [1]. Single purified HSCs are characterized by large fluctuations in their contributions to myeloid and lymphoid lineages [2]. Studies in mice have demonstrated a differential propensity of HSCs to give rise to lymphoid or myeloid cells or platelets [3, 4]. The latter subpopulation may also be primed toward the megakaryocytic lineage.

At this time, little is known about the extrinsic regulation of HSC subpopulations. However, HSCs grown in culture without a supportive stromal cell layer demonstrate loss of long-term engraftment capacity. In addition, several studies have shown that disruptions in the microenvironment can lead to aberrant hematopoiesis and even hematopoietic malignancies in mice [5–8]. These observations have led to the hypothesis that a specific microenvironment or niche is necessary for the maintenance and regulation of normal adult hematopoiesis and HSC functions. As a result of the singular homing of HSCs to BM spaces it has long been suspected that non-hematopoietic cells contained in the skeleton, as well as their secreted products, including the matrix, may provide cellular and molecular components that are critical for the regulation of hematopoiesis and HSCs. Moreover, the functional heterogeneity of HSCs may indicate a matching heterogeneity in the niches for lymphoid-, myeloid-, or megakaryocyte-biased HSCs that exert differential influences to support the function and behavior of these HSC subsets.

This chapter will review the principal cell types and skeletal signals that have been implicated as regulatory cellular components of the HSC niche. These components not only illustrate the extraordinary complexity of skeletal tissue, but also provide critical clues to novel therapeutic targets for HSC expansion in conditions of myeloablative injury or in cases of malignant transformation of HSCs in hematological cancers.

THE HSC NICHE

Osteoblastic cells

Cells of the osteoblast lineage were implicated early in the regulation of HSC function (Fig. 125.1). More recent studies may suggest that their effects on the support of

Primer on the Metabolic Bone Diseases and Disorders of Mineral Metabolism, Ninth Edition. Edited by John P. Bilezikian.
© 2019 American Society for Bone and Mineral Research. Published 2019 by John Wiley & Sons, Inc.
Companion website: www.wiley.com/go/asbmrprimer

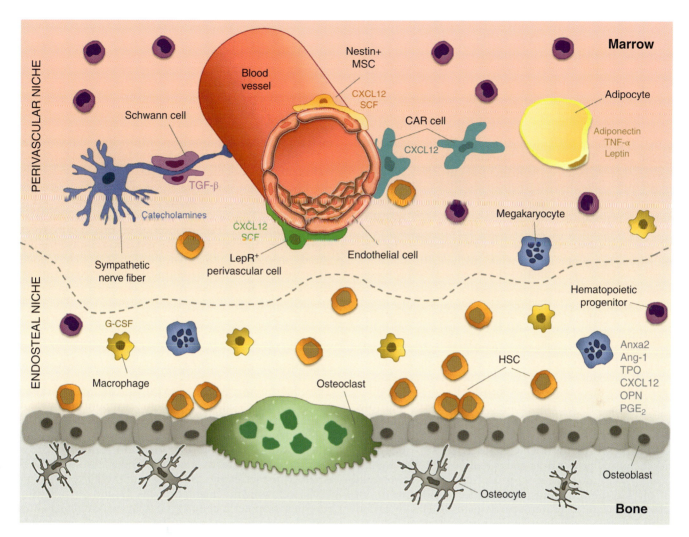

Fig. 125.1. The bone marrow niche. The principal cell types and skeletal signals involved in the control of the bone marrow microenvironment are represented here along the two main niches, the endosteal and the perivascular niche. The growth factors, cytokines, and other short-lived secreted signals are depicted close to their sources of origin. TGF-β = transforming growth factor beta; G-CSF = granulocyte colony-stimulating factor; CXCL12 = CXC-chemokine ligand 12; SCF = stem cell factor; LepR+ = leptin-receptor expressing perivascular cells; MSC = nestin+ mesenchymal stem cells; CAR cells = CXCL12 abundant reticular cells; TNF-α = tumor necrosis factor alpha; Anxa2 = annexin 2; Ang-1 = angiopoietin-1; TPO = thrombopoietin; OPN = osteopontin; PGE_2 = prostaglandin E_2.

hematopoiesis vary depending on the stage of osteoblast differentiation. Studies in mice have demonstrated that transplanted HSCs preferentially engraft at the endosteal surface, and in close contact with osteoblastic cells [9–11] and that HSCs with high proliferative and long-term engrafting potential are found tightly adhered to the endosteal surface [12]. In addition, when cotransplanted with HSCs, osteoblasts can increase their engraftment rate [13].

Alterations in the number of cells of the osteoblast lineage can either stimulate [10, 14, 15] or limit HSC expansion [16], promote quiescence and HSC mobilization [17, 18], support expansion of the erythroid lineage [10, 15], and regulate B lymphopoiesis [14]. Osteocytes expand the myeloid lineage through disruption of $G\alpha_s$ signaling [19]. Similarly, osteoblast dysfunction results in pancytopenia via distinct mechanisms. Osteoblast-depleted mice show reduced normal long-term HSC self-renewal [20] and develop a hematological phenotype that favors myeloid but suppresses lymphoid and erythroid expansion [8]. In the presence of leukemia this phenotype may contribute to marrow failure by allowing the replacement of healthy hematopoietic cells with leukemic blasts. In contrast, osteoclasts, the bone-resorbing cells, appear not to be involved in the maintenance and mobilization of HSCs [21].

Osteoblasts produce many growth factors and cytokines that are known to be important for hematopoiesis [15, 22] (Fig. 125.1). Treatment with PTH is capable of stimulating self-renewal of HSCs and increasing the HSC population despite the fact that HSCs do not express the PTH receptor, PTH1R, indicating a microenvironmentally mediated effect [15]. In addition to increasing HSCs, studies utilizing either PTH treatment or the constitutive activation of PTH1R on osteoblastic cells revealed that PTH signaling increases bone volume and the expression of the Notch ligand Jagged1 on osteoblastic cells [15, 23]. Notch signaling is known to support stem cell self-renewal in other stem cell systems and has been implicated in the support of HSCs [24]. Further the inhibition of Notch signaling ablates the PTH-dependent increase in HSC number. Additional studies have further suggested that Notch signaling between osteoblasts and HSCs plays a role in maintaining repopulating potential [25] and, in the context of activated β-catenin, in inducing myeloid malignancies [7, 26]. A separate study questioned the importance of canonical Notch signaling in the HSC niche [27] and work remains to be done elucidating the precise role Notch signaling plays in the maintenance of HSCs.

Annexin 2 (Anxa2) is expressed by both osteoblasts and endothelial cells in the BM, and treatment of mice with Anxa2 inhibitors impairs HSC homing and engraftment [28]. Angiopoietin-1 (Ang-1) is expressed by osteoblasts and promotes stronger adhesion and quiescence in HSCs through interactions with the receptor tyrosine kinase Tie2 [17]. Genetic deletion of both Tie1 and Tie2 in mice results in a loss of HSC maintenance in the marrow, again suggesting a role of Ang-1 in the regulation of hematopoiesis [29]. Osteoblastic cells also express thrombopoietin, a regulator of HSC quiescence [18]. Moreover, mesenchymal progenitors and osteolineage cells produce local C-X-C motif ligand 12 (CXCL12), a critical regulator of HSC homing and maintenance. Osteopontin, an osteoblast-secreted protein, participates in HSC location and is a negative regulator of their proliferation [16].

Short-lived secreted signals are also potential osteoblastic-dependent HSC regulators. Prostaglandin E_2 (PGE_2) is an arachidonic acid derivative that is produced by multiple cell types within the BM including osteoblastic cells. Treatment of mice with PGE_2 increases short-term repopulating HSCs [30] and ex vivo treatment of HSCs with dimethyl-PGE_2 increases HSC repopulating potential both in mouse BM and in human cord blood samples [31, 32], one of the first illustrations of how identification of HSC regulatory signals can be used pharmacologically for therapeutic purposes.

Although numerous studies using both in vivo and in vitro experimental models firmly implicate cells of the osteoblastic lineage in HSC maintenance and regulation, the differentiation stage of the osteolineage cell which supports HSCs has begun to emerge, and this supports the hypothesis that immature cells in the osteoblastic lineage are critical for HSC regulation. Self-renewing osteoprogenitor cells in the marrow can form supportive HSC niches [33]. Nestin-positive putative mesenchymal stem cells (MSCs) express high levels of genes implicated in HSC maintenance. Additionally, these nestin-positive cells are spatially colocated with HSCs and depletion of the nestin-positive population results in HSC mobilization [34]. Conversely, activation of the PTH1R in terminally mature osteocytes, in spite of expansion in the osteoblastic pool, did not increase microenvironmental support for HSCs [35]. Additional studies have suggested that mature osteoblasts do not directly affect HSC maintenance in the BM [36]. Accordingly, deletion of *Scf* or *Cxcl12* using the *osterix-Cre* or *Col2.3-cre* mice did not alter HSC numbers but, instead, reduced the numbers of B lymphoid progenitors [37, 38]. Similarly, osteoblast ablation in *Col2.3-TK* transgenic mice depleted pre-pro-B and pro-B cells from the BM without affecting HSCs. Differentiating osteoblasts are therefore emerging as having a dual function in regulating HSC activity; whereas the most immature subset influences HSC proliferation, the mature osteoblast subset tailors HSC differentiation along the lymphoid, myeloid, and erythroid lineages. The precise characteristics of the osteolineage cells responsible for HSC support are of great interest because they could predict whether therapeutic strategies that stimulate osteoblastic cells could also achieve beneficial HSC effects.

Endothelial cells

The endothelium gives rise to the first definitive HSCs in the developing embryo and there is mounting evidence that endothelial cells also play an important role in the maintenance of HSCs in the adult marrow (Fig. 125.1). HSCs and hematopoietic progenitor cells localize to endothelial structures near the endosteal surface [39]. Angiocrine factors produced by endothelial cells support HSCs following myeloablation [40, 41] and expand immature hematopoietic cells ex vivo. In vivo imaging shows HSC localization to areas of close spatial relationship between vascular endothelial cells and endosteal osteoblasts suggesting that both structures may be necessary components of a singular niche [9]. More recently, data have suggested that endothelial cells may not directly support HSC populations, because deletion of a critical HSC-regulatory signal, stem cell factor (SCF), does not result in loss of HSC function [37]. However, other endothelial signals, such as the Notch ligand Jagged1, have been proven necessary for maintenance of HSC quiescence [42]. In fact, either arteriolar or sinusoidal cell populations may orchestrate mesenchymal populations found in their proximity, that are then responsible for HSC support (reviewed in [43]).

CXCL12 abundant reticular cells

CXCL12, also known as stromal cell-derived factor 1 (SDF1), is a chemokine produced by osteoblasts and endothelial cells. CXCL12 signals through its receptor, C-X-C motif receptor 4 (CXCR4), on HSCs to induce migration to and retention in the BM (reviewed in [43]). HSCs were shown to be closely associated with CXCL12 abundant reticular (CAR) cells that are in turn closely associated with endothelial cells (Fig. 125.1). Recently, in vivo experiments crossing transgenic mice expressing GFP under the control of the CXCL12 promoter (which originally led to the identification of CAR cells) with lines expected to target mature osteoblasts (osteocalcin-Cre) showed CXCL12-GFP expression in a majority of CAR cells, suggesting that a significant subset of these cells may be in the osteoblastic lineage [44].

Sympathetic nervous system neuronal cells and glia

HSCs and cells in the BM microenvironment can receive and respond to signals from the sympathetic nervous system (SNS) (Fig. 125.1). Catecholamines, produced by sympathetic nerves, are delivered to the BM niche through the blood circulation or by secretion from the nerve endings acting in a paracrine mode. At the same time, HSCs express catecholaminergic receptors, suggesting that they are able to directly respond to signals from the SNS. Indeed, treatment of HSCs with dopamine agonists or norepinephrine enhances colony formation in vitro, provided that G-CSF is present, and promotes the ability of HSCs to engraft in vivo [45, 46]. The SNS also affects the migration of HSCs from the BM to the bloodstream through direct actions on HSCs and also indirectly through the microenvironment. SNS neurons coordinate the circadian oscillation of HSC numbers in the marrow, and ablation of SNS neurons results in a loss of circadian controlled release of HSCs into the periphery. In turn, osteoblastic cells in the BM can influence HSC egress through a process that is regulated by the SNS and involves suppression of CXCL12 production. Sympathectomy of one tibia in a mouse downregulates CXCL12 expression while the sham-operated contralateral tibia is unaffected [47], and although it does not affect HSC numbers it impairs mobilization in response to G-CSF [48]. Inhibition of CXCL12 may promote HSC migration by inducing critical pathways for cytokine-induced mobilization [48].

In the BM, glial cells are closely associated with HSCs, and produce numerous factors previously identified as playing a role in the HSC niche [49] (Fig. 125.1). These include activated TGF-β which regulates HSC dormancy ex vivo. As a result, ablation of this population in the BM results in loss of HSC dormancy and ultimately of HSC numbers [49]. These studies suggest that the SNS is an important contributor in regulating the function of the HSC niche.

Adipocytes

A regulatory role for adipocytes in the BM has emerged with recent studies (Fig. 125.1). BM adipocytes secrete cytokines, fatty acids, and hormones that have a potential to influence the function of other neighboring cells in the BM microenvironment via paracrine mechanisms that are still being unraveled. Marrow adipocytes preserve the HSC pool via secretion of adiponectin and TNF-α, which increase proliferation of HSCs while retaining their repopulating potential [50]. Another adipokine, leptin, has been proposed to sustain myelopoiesis and lymphopoiesis, whereas its increased levels in the BM in mice with diet induced obesity correlate with enhanced hematopoiesis [51, 52]. At the same time, adipocytes prevent HSC expansion and may inhibit HSC function and hematopoietic reconstitution [53]. Genetic or pharmacological ablation of marrow adipocytes enhances engraftment of HSCs and improves hematopoietic recovery following myeloablative injury. Lipid-filled BM adipocytes have been connected to the repression of growth and differentiation of HSCs. Lastly, pharmacological inhibition of adipogenesis in mice improves hematopoietic recovery following chemotherapy [54]. In addition to better elucidating their mechanisms of action in the niche it is interesting to understand whether the role of BM adipocytes differs under various conditions, such as obesity or aging.

Macrophages and monocytes

Macrophages and monocytes are recent additions to the HSC niche's cellular milieu (Fig. 125.1). They are involved in the G-CSF-dependent mobilization of HSCs out of the marrow, and depletion of the macrophage population results in loss of osteoblastic cells and increased HSC mobilization into the periphery [55, 56]. In addition, reduction of monocytic phagocytes from the marrow reduces the expression of genes associated with HSC retention in nestin-positive cells, and specific ablation of macrophage populations resulted in the egress of HSCs from the marrow [57]. In addition, macrophages negatively regulate HSCs in response to interferon [58], a mechanism by which infection may modulate the HSC pool. It is unclear whether macrophages and monocytes specifically interact with HSCs or function only in support of other niche components.

NICHE FOR B LYMPHOPOIESIS

In addition to HSCs there is emerging evidence that B lymphopoiesis is regulated by microenvironmental factors in the BM (Fig. 125.1). B-cell precursors are in direct contact with stromal cells in the marrow space that express CXCL12. Osteoblasts support the development of B lymphocytes from HSCs in vitro and are necessary and sufficient for B-cell commitment and maturation [59] whereas ablation of osteoblasts results in a loss of B lymphocytes before the loss of HSCs [60, 59]. Osteoblastic support for B lymphopoiesis appears to involve Gα$_s$ signaling because the loss of Gα$_s$ in osteolineage cells results in a decrease in B-cell precursors in the BM [14]. Similarly, human BM stromal cells in coculture with human CD34+

cord blood cells can support B lymphopoiesis [61]. More recently, G-CSF was proposed to reprogram BM stromal cell properties resulting in suppression of B lymphopoiesis in mice [62]. Interestingly, in the BM, stroma osteoblasts are the main cells with a myelopoietic supportive capacity which is mediated by sustained G-CSF release [63, 64]. Therefore, BM stromal cells, and in particular osteoblasts, support B lymphopoiesis and inhibiting this capacity could favor myelopoiesis.

MALIGNANCY AND THE HSC NICHE

Although cancer metastasis is described elsewhere in this primer, it is important to note that recent data have suggested that metastatic cancers home to benign HSC niches [65]. Therefore, as therapeutic targets are identified for HSC expansion through manipulation of niche components, the effect of their stimulation on malignant cells should be monitored, particularly in the setting of a prior cancer diagnosis.

Both xenograft and syngeneic models have suggested that leukemic cells compete with benign HSCs for access to the HSC niche and disrupt normal interactions between HSCs and their microenvironment [66, 67]. Further, these interactions with the HSC niche can protect leukemic cells from chemotherapy leading to relapse of the disease [68, 69]. In models of acute myelogenous leukemia (AML) this protection can be abrogated by inhibition of the CXCL12/CXCR4 axis [70, 71], whereas in models of acute lymphoblastic leukemia (ALL) it was demonstrated that, following chemotherapy, residual leukemic cells resided in a specialized niche that was dependent on leukemic production of CCL3 and TGF-β1 [72].

The mesenchymal cell lineage is of particular interest for its role in regulating both benign HSCs and their malignant counterparts. Ablation of osteoblastic cells leads to accelerated leukemia progression in several murine models of leukemia that include AML, chronic myeloid leukemia (CML), myelomonocytic leukemia, and lymphoblastic leukemia [8]. Further, there is evidence in a syngeneic model of blast crisis CML that leukemic cells actively inhibit osteoblastic cells, likely through inflammatory mediators such as the chemokine CCL3 [73, 74]. In fact, CML remodels the endosteal niche so as to reinforce and self-perpetuate the blast crisis [74]. Similarly, osteoblast numbers are decreased in patients with AML and myelodysplastic syndrome (MDS) and in mouse models of AML [8]. Pharmacological maintenance of osteoblast numbers with inhibitors of gut-derived serotonin synthesis reduces AML burden and prolongs life span [8]. Under certain conditions osteolineage cells have also been demonstrated to function as leukemia-initiating cells. Genetic alterations in osteolineage cells such as deletion of *Dicer1*, a main regulator of the microRNA biosynthesis machinery, using the osterix promoter [5], overexpression of a constitutively active (ca) β-catenin using the 2.3 kb collagen 1 promoter [7, 26], or activating mutations of the protein tyrosine phosphatase SHP2 in mesenchymal stem/progenitor cells and osteoprogenitors using the nestin promoter [75] have all been shown to induce hematological malignancies. In the case of overexpression of constitutively active β-catenin, the development of AML-like hematopoietic malignancy was shown to be completely dependent on overexpression of the notch ligand jagged1 on osteoblastic cells and it leads to clonal expansion characterized by recurrent chromosomal aberrations and somatic mutations [7, 26]. As a result, AML can be transferred to healthy irradiated recipients. Activation of the β-catenin/Jagged1 leukemogenic pathway by osteoblasts is detected in 38% of AML and MDS/AML patients. In the case of the myeloproliferative neoplasm that develops in mice with activating mutations of the protein tyrosine phosphatase SHP2 in mesenchymal stem/progenitor cells and osteoprogenitors, the disease can be completely reversed by antagonizing CCL3 signaling through administration of small molecule inhibitors of its receptors CCR1 and CCR5 [75]. These studies highlight both the importance of osteolineage cells for the maintenance of a benign HSC population, and the potential that the primary defect in many hematological malignancies may lie in the supportive BM microenvironment.

Further discoveries of the cellular and molecular components of normal and malignant niches are therefore needed to safely manipulate the microenvironment in the context of malignancy.

CONCLUSION

The importance of the skeletal microenvironment in hematopoietic regulation is beginning to be elucidated and holds great promise for translation in hematopoietic recovery and in the treatment of hematopoietic malignancies as well as metastatic disease.

Previous and emerging studies show that multiple cell populations within the BM microenvironment contribute to the complex regulation of HSC function. Many cell types likely coordinate to provide one regulatory niche, and equally as likely there are separate niches providing support for distinct and separate populations of HSCs, and/or different potential HSC fates. Our increased understanding of healthy HSC niches should foster studies on the altered HSC niches in BM disorders. The balance of extrinsic influences from the supportive niche may also vary under different physiological conditions. Newborn, adult, and aged HSCs have different physiological demands. The exploration of how the BM microenvironment ages will reveal essential information for the treatment of age-related BM illnesses. Likewise, understanding how the niche controls HSC function during stress situations, such as infections, radiotherapy, and chemotherapy, is needed. Further study in this area has benefited greatly from interactions of the field of hematopoiesis with bone biology, and will likely continue to progress rapidly to increase our understanding and therapeutic use of the complex cellular relationships within the BM. Targeting the niche itself is an attractive potential possibility for the treatment of hematological disorders. A big challenge in this endeavor will be to translate animal research into humans. Improving the availability of human tissue samples will be essential to reach this goal and will advance our understanding of the importance and the complexity of the BM microenvironment to HSC function.

REFERENCES

1. Yamamoto R, Morita Y, Ooehara J, et al. Clonal analysis unveils self-renewing lineage-restricted progenitors generated directly from hematopoietic stem cells. Cell. 2013;154(5):1112–26.
2. Yang L, Bryder D, Adolfsson J, et al. Identification of Lin(-)Sca1(+)kit(+)CD34(+)Flt3- short-term hematopoietic stem cells capable of rapidly reconstituting and rescuing myeloablated transplant recipients. Blood. 2005;105(7):2717–23.
3. Dykstra B, Kent D, Bowie M, et al. Long-term propagation of distinct hematopoietic differentiation programs in vivo. Cell Stem Cell. 2007;1(2):218–29.
4. Benz C, Copley MR, Kent DG, et al. Hematopoietic stem cell subtypes expand differentially during development and display distinct lymphopoietic programs. Cell Stem Cell. 2012;10(3):273–83.
5. Raaijmakers MH, Mukherjee S, Guo S, et al. Bone progenitor dysfunction induces myelodysplasia and secondary leukaemia. Nature. 2010;464(7290):852–57.
6. Walkley CR, Olsen GH, Dworkin S, et al. A microenvironment-induced myeloproliferative syndrome caused by retinoic acid receptor gamma deficiency. Cell. 2007;129(6):1097–110.
7. Kode A, Manavalan JS, Mosialou I, et al. Leukaemogenesis induced by an activating beta-catenin mutation in osteoblasts. Nature. 2014;506(7487):240–44.
8. Krevvata M, Silva BC, Manavalan JS, et al. Inhibition of Leukemia Cell Engraftment and Disease Progression in Mice by Osteoblasts. Blood. 2014;124(18):2834–46.
9. Lo Celso C, Fleming HE, Wu JW, et al. Live-animal tracking of individual haematopoietic stem/progenitor cells in their niche. Nature. 2009;457(7225):92–6.
10. Zhang J, Niu C, Ye L, et al. Identification of the haematopoietic stem cell niche and control of the niche size. Nature. 2003;425(6960):836–41.
11. Xie Y, Yin T, Wiegraebe W, et al. Detection of functional haematopoietic stem cell niche using real-time imaging. Nature. 2009;457(7225):97–101.
12. Haylock DN, Williams B, Johnston HM, et al. Hemopoietic stem cells with higher hemopoietic potential reside at the bone marrow endosteum. Stem Cells. 2007;25(4):1062–9.
13. El-Badri NS, Wang BY, Cherry, et al. Osteoblasts promote engraftment of allogeneic hematopoietic stem cells. Exp Hematol. 1998;26(2):110–16.
14. Wu JY, Purton LE, Rodda SJ, et al. Osteoblastic regulation of B lymphopoiesis is mediated by Gs(alpha)-dependent signaling pathways. Proc Natl Acad Sci U S A. 2008;105(44):16976–81.
15. Calvi LM, Adams GB, Weibrecht KW, et al. Osteoblastic cells regulate the haematopoietic stem cell niche. Nature. 2003;425(6960):841–6.
16. Stier S, Ko Y, Forkert R, et al. Osteopontin is a hematopoietic stem cell niche component that negatively regulates stem cell pool size. J Exp Med. 2005;201(11):1781–91.
17. Arai F, Hirao A, Ohmura M, et al. Tie2/angiopoietin-1 signaling regulates hematopoietic stem cell quiescence in the bone marrow niche. Cell. 2004;118(2):149–61.
18. Yoshihara H, Arai F, Hosokawa K, et al. Thrombopoietin/MPL signaling regulates hematopoietic stem cell quiescence and interaction with the osteoblastic niche. Cell Stem Cell. 2007;1(6):685–97.
19. Fulzele K, Krause DS, Panaroni C, et al. Myelopoiesis is regulated by osteocytes through Gsalpha-dependent signaling. Blood. 2013;121(6):930–9.
20. Bowers M, Zhang B, Ho Y, et al. Osteoblast ablation reduces normal long-term hematopoietic stem cell self-renewal but accelerates leukemia development. Blood. 2015;125(17):2678–88.
21. Miyamoto K, Yoshida S, Kawasumi M, et al. Osteoclasts are dispensable for hematopoietic stem cell maintenance and mobilization. J Exp Med. 2011;208(11):2175–81.
22. Marusic A, Kalinowski JF, Jastrzebski S, et al. Production of leukemia inhibitory factor mRNA and protein by malignant and immortalized bone cells. J Bone Miner Res. 1993;8(5):617–24.
23. Weber JM, Forsythe SR, Christianson CA, et al. Parathyroid hormone stimulates expression of the Notch ligand Jagged1 in osteoblastic cells. Bone. 2006;39(3):485–93.
24. Liu J, Sato C, Cerletti M, et al. Notch signaling in the regulation of stem cell self-renewal and differentiation. Curr Top Dev Biol. 2010;92:367–409.
25. Chitteti BR, Cheng YH, Poteat B, et al. Impact of interactions of cellular components of the bone marrow microenvironment on hematopoietic stem and progenitor cell function. Blood. 2010;115(16):3239–48.
26. Kode A, Mosialou I, Manavalan JS, et al. FoxO1-dependent induction of acute myeloid leukemia by osteoblasts in mice. Leukemia. 2016;30(1):1–13.
27. Maillard I, Koch U, Dumortier A, et al. Canonical notch signaling is dispensable for the maintenance of adult hematopoietic stem cells. Cell Stem Cell. 2008;2(4):356–66.
28. Jung Y, Wang J, Song J, et al. Annexin II expressed by osteoblasts and endothelial cells regulates stem cell adhesion, homing, and engraftment following transplantation. Blood. 2007;110(1):82–90.
29. Puri MC, Bernstein A. Requirement for the TIE family of receptor tyrosine kinases in adult but not fetal hematopoiesis. Proc Natl Acad Sci U S A. 2003;100(22):12753–8.

30. Frisch BJ, Porter RL, Gigliotti BJ, et al. In vivo prostaglandin E2 treatment alters the bone marrow microenvironment and preferentially expands short-term hematopoietic stem cells. Blood. 2009;114(19):4054–63.
31. North TE, Goessling W, Walkley CR, et al. Prostaglandin E2 regulates vertebrate haematopoietic stem cell homeostasis. Nature. 2007;447(7147):1007–11.
32. Goessling W, Allen RS, Guan X, et al. Prostaglandin E2 enhances human cord blood stem cell xenotransplants and shows long-term safety in preclinical nonhuman primate transplant models. Cell Stem Cell. 2011;8(4):445–58.
33. Sacchetti B, Funari A, Michienzi S, et al. Self-renewing osteoprogenitors in bone marrow sinusoids can organize a hematopoietic microenvironment. Cell. 2007;131(2):324–36.
34. Mendez-Ferrer S, Michurina TV, Ferraro F, et al. Mesenchymal and haematopoietic stem cells form a unique bone marrow niche. Nature. 2010;466(7308):829–34.
35. Calvi LM, Bromberg O, Rhee Y, et al. Osteoblastic expansion induced by parathyroid hormone receptor signaling in murine osteocytes is not sufficient to increase hematopoietic stem cells. Blood. 2012;119(11):2489–99.
36. Kunisaki Y, Bruns I, Scheiermann C, et al. Arteriolar niches maintain haematopoietic stem cell quiescence. Nature. 2013;502(7473):637–43.
37. Ding L, Morrison SJ. Haematopoietic stem cells and early lymphoid progenitors occupy distinct bone marrow niches. Nature. 2013;495(7440):231–5.
38. Greenbaum A, Hsu YM, Day RB, et al. CXCL12 in early mesenchymal progenitors is required for haematopoietic stem-cell maintenance. Nature. 2013;495(7440):227–30.
39. Kiel MJ, Yilmaz OH, Iwashita T, et al. SLAM family receptors distinguish hematopoietic stem and progenitor cells and reveal endothelial niches for stem cells. Cell. 2005;121(7):1109–21.
40. Butler JM, Nolan DJ, Vertes EL, et al. Endothelial cells are essential for the self-renewal and repopulation of Notch-dependent hematopoietic stem cells. Cell Stem Cell. 2010;6(3):251–64.
41. Kobayashi H, Butler JM, O'Donnell R, et al. Angiocrine factors from Akt-activated endothelial cells balance self-renewal and differentiation of haematopoietic stem cells. Nat Cell Biol. 2010;12(11):1046–56.
42. Poulos MG, Guo P, Kofler NM, et al. Endothelial Jagged-1 is necessary for homeostatic and regenerative hematopoiesis. Cell Rep. 2013;4(5):1022–34.
43. Boulais PE, Frenette PS. Making sense of hematopoietic stem cell niches. Blood. 2015;125(17):2621–9.
44. Zhang J, Link DC. Targeting of Mesenchymal Stromal Cells by Cre-Recombinase Transgenes Commonly Used to Target Osteoblast Lineage Cells. J Bone Miner Res. 2016;31(11):2001–7.
45. Kalinkovich A, Spiegel A, Shivtiel S, et al. Blood-forming stem cells are nervous: direct and indirect regulation of immature human CD34+ cells by the nervous system. Brain Behav Immun. 2009;23(8):1059–65.
46. Spiegel A, Shivtiel S, Kalinkovich A, et al. Catecholaminergic neurotransmitters regulate migration and repopulation of immature human CD34+ cells through Wnt signaling. Nat Immunol. 2007;8(10):1123–31.
47. Mendez-Ferrer S, Lucas D, Battista M, et al. Haematopoietic stem cell release is regulated by circadian oscillations. Nature. 2008;452(7186):442–7.
48. Katayama Y, Battista M, Kao WM, et al. Signals from the sympathetic nervous system regulate hematopoietic stem cell egress from bone marrow. Cell. 2006;124(2):407–21.
49. Yamazaki S, Ema H, Karlsson G, et al. Nonmyelinating Schwann cells maintain hematopoietic stem cell hibernation in the bone marrow niche. Cell. 2011;147(5):1146–58.
50. DiMascio L, Voermans C, Uqoezwa M, et al. Identification of adiponectin as a novel hemopoietic stem cell growth factor. J Immunol. 2007;178(6):3511–20.
51. Claycombe K, King LE, Fraker PJ. A role for leptin in sustaining lymphopoiesis and myelopoiesis. Proc Natl Acad Sci U S A. 2008;105(6):2017–21.
52. Trottier MD, Naaz A, Li Y, et al. Enhancement of hematopoiesis and lymphopoiesis in diet-induced obese mice. Proc Natl Acad Sci U S A. 2012;109(20):7622–9.
53. Naveiras O, Nardi V, Wenzel PL, et al. Bone-marrow adipocytes as negative regulators of the haematopoietic microenvironment. Nature. 2009;460(7252):259–63.
54. Zhu RJ, Wu MQ, Li ZJ, et al. Hematopoietic recovery following chemotherapy is improved by BADGE-induced inhibition of adipogenesis. Int J Hematol. 2013;97(1):58–72.
55. Christopher MJ, Rao M, Liu F, et al. Expression of the G-CSF receptor in monocytic cells is sufficient to mediate hematopoietic progenitor mobilization by G-CSF in mice. J Exp Med. 2011;208(2):251–60.
56. Winkler IG, Sims NA, Pettit AR, et al. Bone marrow macrophages maintain hematopoietic stem cell (HSC) niches and their depletion mobilizes HSCs. Blood. 2010;116(23):4815–28.
57. Chow A, Lucas D, Hidalgo A, et al. Bone marrow CD169+ macrophages promote the retention of hematopoietic stem and progenitor cells in the mesenchymal stem cell niche. J Exp Med. 2011;208(2):261–71.
58. McCabe A, Zhang Y, Thai V, et al. Macrophage-Lineage Cells Negatively Regulate the Hematopoietic Stem Cell Pool in Response to Interferon Gamma at Steady State and During Infection. Stem Cells. 2015;33(7):2294–305.
59. Zhu J, Garrett R, Jung Y, et al. Osteoblasts support B-lymphocyte commitment and differentiation from hematopoietic stem cells. Blood. 2007;109(9):3706–12.
60. Visnjic D, Kalajzic Z, Rowe DW, et al. Hematopoiesis is severely altered in mice with an induced osteoblast deficiency. Blood. 2004;103(9):3258–64.
61. Ichii M, Oritani K, Yokota T, et al. Regulation of human B lymphopoiesis by the transforming growth factor-beta

superfamily in a newly established coculture system using human mesenchymal stem cells as a supportive microenvironment. Exp Hematol. 2008;36(5):587–97.
62. Day RB, Bhattacharya D, Nagasawa T, et al. Granulocyte colony-stimulating factor reprograms bone marrow stromal cells to actively suppress B lymphopoiesis in mice. Blood. 2015;125(20):3114–17.
63. Taichman RS, Emerson SG. Human osteoblasts support hematopoiesis through the production of granulocyte colony-stimulating factor. J Exp Med. 1994;179(5):1677–82.
64. Morad V, Pevsner-Fischer M, Barnees S, et al. The myelopoietic supportive capacity of mesenchymal stromal cells is uncoupled from multipotency and is influenced by lineage determination and interference with glycoxylation. Stem Cells. 2008;26(9):2275–86.
65. Shiozawa Y, Pedersen EA, Havens AM, et al. Human prostate cancer metastases target the hematopoietic stem cell niche to establish footholds in mouse bone marrow. J Clin Invest. 2011;121(4):1298–312.
66. Colmone A, Amorim M, Pontier AL, et al. Leukemic cells create bone marrow niches that disrupt the behavior of normal hematopoietic progenitor cells. Science. 2008;322(5909):1861–5.
67. Glait-Santar C, Desmond R, Feng X, et al. Functional Niche Competition Between Normal Hematopoietic Stem and Progenitor Cells and Myeloid Leukemia Cells. Stem Cells. 2015;33(12):3635–42.
68. Lane SW, Wang YJ, Lo Celso C, et al. Differential niche and Wnt requirements during acute myeloid leukemia progression. Blood. 2011;118(10):2849–56.
69. Ishikawa F, Yoshida S, Saito Y, et al. Chemotherapy-resistant human AML stem cells home to and engraft within the bone-marrow endosteal region. Nat Biotechnol. 2007;25(11):1315–21.
70. Nervi B, Ramirez P, Rettig MP, et al. Chemosensitization of acute myeloid leukemia (AML) following mobilization by the CXCR4 antagonist AMD3100. Blood. 2009;113(24):6206–14.
71. Zeng Z, Shi YX, Samudio IJ, et al. Targeting the leukemia microenvironment by CXCR4 inhibition overcomes resistance to kinase inhibitors and chemotherapy in AML. Blood. 2009;113(24):6215–24.
72. Duan CW, Shi J, Chen J, et al. Leukemia propagating cells rebuild an evolving niche in response to therapy. Cancer Cell. 2014;25(6):778–93.
73. Frisch BJ, Ashton JM, Xing L, et al. Functional inhibition of osteoblastic cells in an in vivo mouse model of myeloid leukemia. Blood. 2012;119(2):540–50.
74. Schepers K, Pietras EM, Reynaud D, et al. Myeloproliferative neoplasia remodels the endosteal bone marrow niche into a self-reinforcing leukemic niche. Cell Stem Cell. 2013;13(3):285–99.
75. Dong L, Yu WM, Zheng H, et al. Leukaemogenic effects of Ptpn11 activating mutations in the stem cell microenvironment. Nature. 2016;539(7628):304–8.

126

Adipocytes and Bone

Clarissa S. Craft, Natalie K. Wee, and Erica L. Scheller

Division of Bone and Mineral Diseases, Department of Internal Medicine, Washington University, St Louis, MO, USA

INTRODUCTION

Within the skeleton there is an extensive population of adipocytes that are collectively referred to as the bone marrow adipose tissue (abbreviated MAT or BMAT). Both acronyms have gained traction in recent years, reflecting the realization that the adipocytes within the skeleton are more than just nodules of lipid and can behave, in many ways, like peripheral adipose tissues [1]. This chapter will review the evolution of MAT nomenclature, from *yellow marrow* to *marrow fat* to *marrow adipose tissue*, and discuss how this shift has been driven by changes in our developing understanding of MAT's function both within and beyond the skeleton.

MARROW ADIPOSE TISSUE DISTRIBUTION AND DEVELOPMENT

The histology and distribution of MAT within adult bones was largely defined in the 1930s [2–5] (Fig. 126.1A, B). Early work aptly recognized that the color of the bone marrow is different, red or yellow, depending on its location in the body—thus the terms *red marrow* and *yellow marrow* came into use [6]. Red marrow is primarily made up of blood-forming cells with scattered adipocytes (Fig. 126.1C). It is concentrated in the axial skeleton: the skull, vertebrae, ribs, sternum, and pelvis, and the proximal portions of the appendicular skeleton [7,8]. By contrast, the yellow marrow is filled with MAT adipocytes. Yellow marrow is contained primarily in the appendicular skeleton (hands, feet, tibia, radius/ulna, and distal humerus and femur). This distribution paints a centripetal pattern where the yellow, adipocyte-rich marrow is concentrated toward the periphery and red, hematopoietic marrow predominates in the center of the body (Fig. 126.1A). The extent of red marrow, balanced by the quantity of yellow, varies from person to person and with disease state, as will be discussed. On average, however, approximately two-thirds of the bone marrow is made up of adipose tissue in the adult human skeleton [7].

In humans, lipid-filled MAT adipocytes begin to form at or slightly before birth in the distal extremities such as the toes [9]. Fatty conversion of distal bone marrow accelerates between 4 and 8 weeks of age with continued red to yellow marrow conversion in the limbs, appearing to progress from peripheral to central sites, until age 20 to 25 [6,8,9]. At this point MAT expansion slows, subsequently undergoing more gradual age-related accumulation in areas of red marrow throughout life (Fig. 126.1B). This pattern of development is recapitulated in many vertebrate species including mice [10], rats [11], and rabbits [12]. However, the timing of MAT development varies widely. In rabbits, mature MAT conversion of the marrow occurs by age 4 to 6 months [12]. In mice, this process occurs as early as 12 to 16 weeks of age [10]. The absolute volume of MAT within the red marrow also varies markedly between species, as low as 0.7% by volume in mice and as high as 70% in humans [1]. The density of MAT adipocytes within the red marrow appears to be directly proportional to the size of the animal, likely reflecting underlying differences in metabolic rate and/or hematopoietic demand (ie, humans > rabbits > rats > mice) [1,2].

Primer on the Metabolic Bone Diseases and Disorders of Mineral Metabolism, Ninth Edition. Edited by John P. Bilezikian.
© 2019 American Society for Bone and Mineral Research. Published 2019 by John Wiley & Sons, Inc.
Companion website: www.wiley.com/go/asbmrprimer

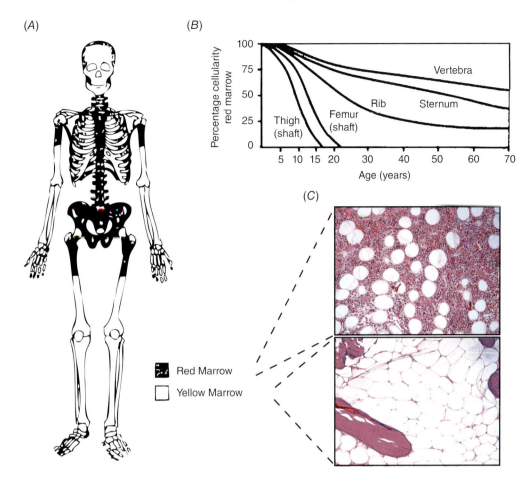

Fig. 126.1. Bone marrow adipose tissue in humans. (A) Distribution of red and yellow marrow in an adult human skeleton (reprinted from [8] with permission from Springer). (B) Conversion of red, hematopoietic, cellular marrow to yellow, fatty marrow over time at five skeletal sites. Higher percentage cellularity of the red marrow indicates more hematopoietic, blood-forming cells (reprinted from [8] with permission from Springer). Lower percentage cellularity indicates that the hematopoietic cells have been replaced with adipocytes. (C) Representative histology of human red marrow and yellow marrow, H&E stain.

MARROW ADIPOSE TISSUE EXPANSION AND METABOLIC DISEASE

It was not until the 1950s and 1960s that MAT became recognized as a fat depot with the potential for adipose tissue-like characteristics [2, 13]. This caused the term *yellow marrow* to fall into disuse in favor of the more popular terms *marrow fat* or *fatty marrow*. Like peripheral adipose tissues, MAT has the potential to undergo increases in cell number and/or size [14], referred to as hyperplasia or hypertrophy respectively. MAT expansion occurs in a wide variety of conditions, including aging, diabetes, obesity, anorexia, estrogen deficiency, osteoporosis, and with glucocorticoid use (reviewed in [1, 15]). It is particularly notable that MAT accumulates both in states of high-fat diet-induced obesity [16] and during anorexia in humans or caloric restriction in mice [17, 18], two conditions that are seemingly opposite to one another. This highlights the potential for MAT to behave differently than peripheral white adipose tissue in some contexts. Mechanistically, emerging clinical evidence suggests that MAT formation may be driven by increases in circulating lipids such as triglycerides [19, 20]. Insulin resistance [21], estrogen depletion, increases in circulating glucocorticoids [22, 23], and gonadal dysfunction may also drive MAT expansion in metabolic disease.

Regarding function, it is unclear whether increases in MAT are uniformly physiologically adaptive or in some contexts pathologically destructive. For example, current work suggests that MAT formation promotes metabolic adaptation during caloric restriction [18], however, by contrast, increases in MAT have been repeatedly associated with bone loss in conditions such as osteoporosis [24]. This may be explained, at least in part, by the ability of the MAT adipocyte to express distinct phenotypic characteristics depending on its location within the skeleton [10, 25].

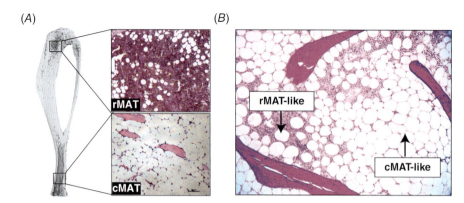

Fig. 126.2. Regulated and constitutive marrow adipocytes. (A) Osmium-stained mouse tibia showing marrow fat in dark gray and bone in light gray with corresponding representative histology. Regulated MAT (rMAT) adipocytes in the proximal tibia are defined as single adipocytes interspersed within red marrow at sites of active hematopoiesis. Constitutive MAT (cMAT) adipocytes, by contrast, exist as confluent sheets of cells and are located in the more distal portions of the skeleton. (B) In larger species, these phenotypes can exist side by side. Human biopsy specimen, H&E stain.

SITE-SPECIFIC VARIATION OF THE MARROW ADIPOSE TISSUE ADIPOCYTE

In the 1970s, research on MAT took a significant step forward, largely as a result of the work of Dr. Mehdi Tavassoli. Based on experiments in rabbits, he proposed that MAT exists in not one, but two unique forms that undergo differential regulation by stimulated hematopoietic demand [25]. This work was recently confirmed and extended in mice, rats, rabbits, and humans [1, 10]. The terms regulated MAT (rMAT) and constitutive MAT (cMAT) have been proposed to capture this complexity (Fig. 126.2A). Regulated MAT is currently defined as single adipocytes interspersed within red marrow at sites of active hematopoiesis. By contrast, constitutive MAT adipocytes make up confluent "depots" of cells within the yellow marrow. In larger species such as rabbits and humans it is worth noting that rMAT- and cMAT-like adipocytes may exist side by side (Fig. 126.2B).

Regulated MAT adipocytes are, in essence, identical to the adipocytes within the red marrow that were discussed previously. They accumulate gradually throughout life in the proximal and central regions of the skeleton. The lipid composition and transcription factor expression in rMAT adipocytes mimics peripheral white adipose tissue (WAT) [10]. Regulated MAT adipocytes, as suggested by their name, readily undergo changes in cell number and/or size when acted on by external forces such as prolonged cold exposure in rodents [10], induction of sympathetic tone with leptin in the brain [26, 27], and stimulated hemolysis [25]. By contrast, cMAT adipocytes in the yellow marrow are refractory to change, have increased lipid unsaturation, are larger in size, and have higher expression of transcription factors *Cebpa/Cebpb* [10]. It is unclear whether differences between rMAT and cMAT are cell autonomous or dictated by their surrounding microenvironment. Future work is needed to pinpoint the relationship of MAT to bone turnover and hematopoietic function within the context of these differences.

MARROW ADIPOSE TISSUE EVOLUTION PROVIDES CLUES ABOUT ITS FUNCTION

Examination of MAT within the context of vertebrate evolution can inform hypotheses as to its function relative to bone, adipose, and hematopoietic tissues (reviewed in [28]) (Fig. 126.3). The first evidence of MAT occurs in the bony fishes. Bony fish such as the zebrafish do not have a hematopoietic bone marrow, instead delegating blood cell formation to the spleen, kidney, intestinal submucosa, and thymus [29]. However, even in the absence of a hematopoietic marrow, osteoclasts contribute to endochondral bone resorption and allometric growth of the skeleton. The spaces created by this process in zebrafish lack hematopoietic activity but are uniformly filled with cMAT-like adipocytes [30]. By contrast, the ray-finned fish *Garra congoensis* has hematopoietic bone marrow with associated adipocytes that histologically resemble the "rMAT" type [28]. These findings suggest that rMAT and cMAT adipocytes evolved at roughly the same time and supports the hypothesis that any differences in function are, at least in part, dictated by their surrounding microenvironment.

Unlike most fish, the majority of amphibians and reptiles have a hematopoietic bone marrow. Unilocular MAT adipocytes have been reported in species including the salamander, newt, leopard frog, and gecko [28]. The leopard frog, in particular, provides an intriguing example of the ability of MAT to undergo seasonal variation. Specifically, fatty yellow marrow converts to red hematopoietic marrow, generally in the early summer [31]. A similar pattern of marrow conversion occurs in the

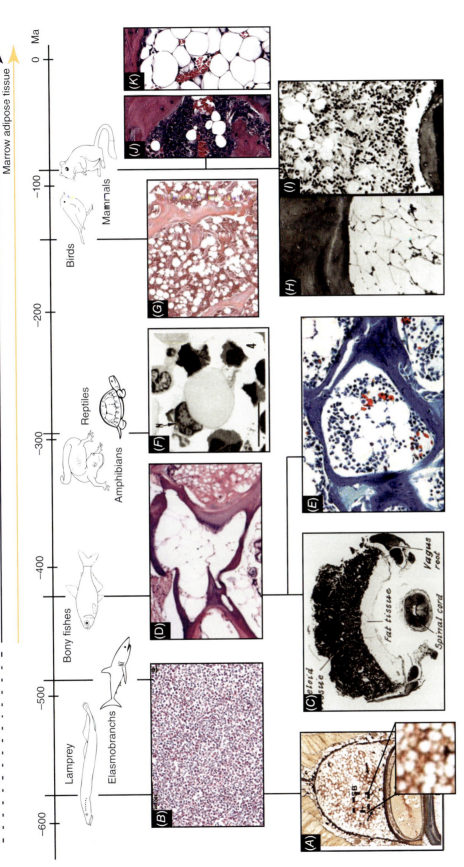

Fig 126.3. Evolution of marrow adipose tissue in vertebrates relative to bone marrow and peripheral adipose tissue. The first evidence of white adipose tissue (WAT) in vertebrates occurs in the sea lamprey, before the evolution of an ossified skeleton. The Elasmobranchs, a group of cartilaginous fishes, have a partially ossified skeleton but no bone marrow, marrow adipose tissue (MAT), or WAT. MAT first becomes apparent in the bony fishes, after the evolution of bone resorption, both in the presence and absence of hematopoietic cells in the adipose tissue of a postmetamorphic sea lamprey. (*B*) Normal adipose tissue is only present in mammals. (*A*) Groups of hematopoietic cells in the adipose tissue of the supraneural body of a postmetamorphic sea lamprey. (*B*) Normal epigonal organ composed of a relatively uniform sheet of mature and developing granulocytes without adipocytes. (*C*) Cross-section through the myeloid organ of *Amia* bony fish with adjacent fat tissue. (*D*) Cross-section of zebrafish bone showing fatty filling of the skeletal space (middle) and adjacent cartilage (right). (*E*) Cancellous bone of the lower jaw in the bony fish *Garra congoensis* with numerous marrow adipocytes (Image credit: Franck Genten). (*F*) Slimy salamander *Plethodon glutinosus*. Semi-thin plastic section of a fat cell in marrow. The densely stained nucleus (N) is located eccentrically in the cell, and the cytoplasm is filled with lightly stained lipid material. (*G*) Bone marrow of a female Leghorn chicken showing trabeculae, hematopoietic marrow, and abundant adipocytes. (*H*) Armadillo dermal plate marrow in December. Few hematopoietic cells are present but fat cells are abundant. (*I*) Armadillo dermal plate marrow in October. Active hematopoietic cells interspersed with adipocytes in a typical dermal plate marrow. (*J*) Red, hematopoietic bone marrow of a C3H/HeJ mouse femur containing rMAT adipocytes. (*K*) Yellow, fatty bone marrow of a C3H/HeJ mouse caudal vertebra containing cMAT adipocytes. Source: [28]. Reprinted with permission from Springer.

dermal plates of the nine-banded armadillo (Fig. 126.3). The volume of dermal plate marrow is highest in the spring, summer, and fall and contains highly active regions of hematopoiesis [32]. In the winter, however, the marrow becomes dull in color and filled with fat cells. Thus, in some contexts, MAT appears to balance seasonal variations in hematopoiesis. This is in line with more recent work in mice demonstrating that sites of high MAT (ie, tail vertebrae) have increased quiescence of hematopoietic progenitors and that inhibition of MAT expansion can accelerate hematopoietic reconstitution after irradiation [33]. This evolutionary and experimental evidence supports the hypothesis that MAT actively interacts with blood-forming cells within the bone marrow.

MARROW ADIPOSE TISSUE AND BONE LOSS

But what about MAT and bone? As discussed previously, current evidence suggests that MAT evolved coincident with the appearance of endochondral bone resorption [28]. Indeed, MAT is present even before hematopoietic colonization of the skeleton in species such as zebrafish. Thus, filling bone with MAT, rather than fluid, may offer some evolutionary advantage for skeletal biomechanics or regulation of bone turnover.

In the 1980s, research on the relationship between MAT and skeletal health dramatically expanded, casting MAT in a negative light—a putative "villain" in diseases such as osteoporosis [34–36]. Indeed, many studies have since shown a negative correlation between MAT volume and bone volume or density in humans [24, 37–41]. However, in recent years, we have realized that the relationship between MAT expansion and bone loss is significantly more nuanced, subject to variations in both skeletal site and disease state. For example, reports that correlate MAT accumulation with decreases in BMD or formation and increased bone loss are generally based on rMAT-enriched sites such as the proximal femur, hip, and lumbar spine (reviewed in [2]). Conversely, studies demonstrating resistance to bone loss at sites of high MAT have all selected cMAT-enriched sites as their area of interest (eg, distal tibia and tail vertebrae) [42–44].

Indeed, more MAT does not necessarily mean less bone. Distal, yellow marrow regions are filled with cMAT-like adipocytes, yet they actually have comparable or even higher basal cancellous bone volume and increased trabecular thickness relative to proximal red marrow sites [42–44]. Consideration of mouse strains is also informative. For example, C3H/HeJ mice have significantly more MAT than do C57BL/6J [10]. C3H/HeJ mice also have more bone, counterintuitive to the hypothesis that MAT causes bone loss [10] (Fig. 126.4).

Furthermore, MAT accumulation is not necessary for loss of skeletal integrity during disease. Analysis of transiliac biopsy specimens from premenopausal women with idiopathic osteoporosis found that even though the patients had increased MAT adipocyte number and size, this failed to correlate with measures of bone formation and volume [39]. Similarly, in rats with ovariectomy-induced osteoporosis, bone loss precedes MAT expansion. In one study, at 4 weeks after ovariectomy a 47% decrease in trabecular bone volume was observed at the tibial metaphysis without a corresponding increase in MAT adipocyte volume, which was not present until the 12-week time point [45]. In mice, ovariectomy [46], type 1 diabetes [47], and high-fat diet feeding [16] cause MAT expansion and bone loss. However, in all three situations it has been shown that the bone loss can occur independent of MAT accumulation. Thus, MAT expansion is not a necessary precursor for bone loss in these conditions. More work is clearly needed to determine at which point and in which contexts MAT has the capacity to directly regulate bone turnover and homeostasis.

Lastly, there are insights about MAT and bone that can be learned from patients with congenital generalized lipodystrophy (CGL) (reviewed in [2]). Patients with CGL1 and CGL2 do not have MAT. There is a high incidence of rapid growth and osteosclerosis during infancy with advanced skeletal age and high bone density persisting through adolescence [2]. During adolescence, which is generally a time of rapid MAT expansion in humans, up to 70% of patients with CGL1 or CGL2 develop bone cysts [2]. Conversely, patients with CGL3 and CGL4 appear to retain MAT and do not develop osteosclerosis or cysts, though it is worth noting that current case reports are limited in number and scope [48–50]. Taken together, this shows that MAT is not a necessary component of initial skeletal development and patterning. However, MAT expansion may be important for maintenance of skeletal homeostasis during growth.

There are many diverse mechanisms underlying the relationships between MAT and bone. Overexpression of RANKL can drive MAT expansion in mice, implying that osteoblasts may have the capacity to stimulate both osteoclastogenesis and adipogenesis through a conserved mechanism [51]. Conversely, release of soluble signaling mediators and lipids by the adipocytes may directly impact the function of skeletal cells [52]. The actions and content of lipids in the skeleton are reviewed here [53].

PERIPHERAL ADIPOCYTES AND BONE

In addition to MAT interacting with bone, peripheral adipose depots, such as WAT, can influence skeletal homeostasis. Mechanically, increased adiposity can generate a loading response in bone and is thus commonly perceived as a positive regulator of bone density (reviewed in [54]). Adipocyte-secreted factors may also independently influence the relationship between fat and bone. For example, adipokines such as leptin and adiponectin have been shown to act on distant tissues including the skeleton.

Leptin acts as a satiety signal, regulating food intake and energy homeostasis. The influence of leptin on the

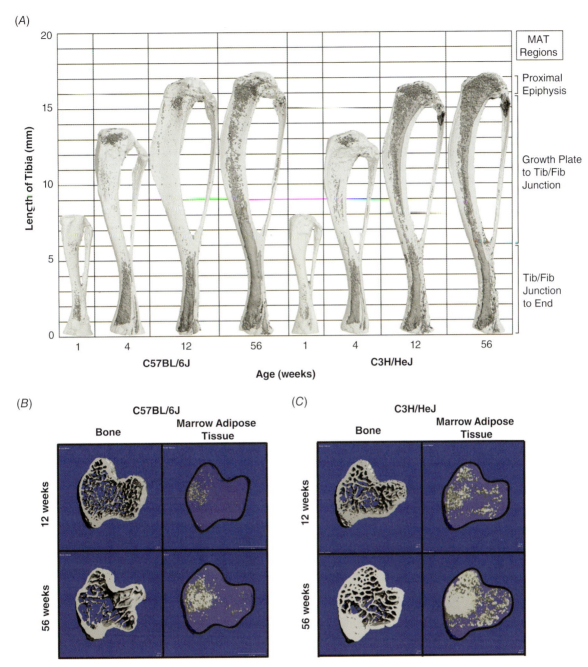

Fig. 126.4. Marrow adipose tissue and trabecular bone in C3H/HeJ and C57BLK/6J mice. (A) Osmium-stained tibias were scanned by micro–computed tomography and reconstructed with the decalcified bone overlaid. Marrow fat is dark gray and bone is light gray. Development of MAT in the distal tibia occurs as early as 1 week of age in both C57BL/6J and C3H/HeJ mice. However, rates of MAT expansion in the proximal tibia vary markedly between strains. (B, C) Representative images of the proximal tibial metaphysis both before decalcification and after osmium staining. Marrow adipose tissue is in white. C3H/HeJ mice have more marrow adipocytes and more bone than C57BL/6J. Quantified data for this figure are available in [10]. Source: Scheller, https://www.nature.com/articles/ncomms8808?WT.ec_id=NCOMMS-20150812&spMailingID=49302896&spUserID=MzcwNDE0MDAxODMS1&spJobID=741904452&spReportId=NzQxOTA0NDUyS0, licensed under: CC BY 4.0.

skeleton was first identified in 2000, when Ducy and colleagues reported that leptin acts in the brain to increase sympathetic tone, leading to reductions in cancellous bone mass [55, 56]. By contrast, leptin has an anabolic effect in the periphery [57–59]. Indeed, the actions of leptin on the skeleton are pleiotropic with evidence for both central and peripheral pathways regulating bone mass. In clinical studies, the relationship between circulating leptin and bone density in humans has been mixed. Some report positive associations between leptin and the skeleton, others show no relationship, and a few suggest leptin is detrimental (summarized in [60]). Altogether, the relationship between leptin and bone mass is complex and context-dependent (reviewed in [61]); appreciatively, other factors are often modulated with changes in adiposity and this may account for the reported differences in leptin's actions on the skeleton.

Adiponectin is another adipokine secreted from WAT. Both osteoblasts and osteoclasts express adiponectin receptors (AdipoR1, AdipoR2). The adiponectin knockout mouse has a high bone mass phenotype at 12 to 14 weeks of age, indicative of a detrimental role of adiponectin on the skeleton through increasing RANKL expression from osteoblasts [62, 63]. However, the actions of adiponectin on bone are highly dynamic and alter with age: older animals (6 to 9 months of age) are observed to have a low bone mass phenotype [63]. Both central signaling via FOXO1 and the sympathetic nervous system and direct peripheral pathways are implicated in contributing to the bone phenotype.

MARROW FAT AS AN ADIPOSE TISSUE

Histologically, MAT adipocytes resemble unilocular cells, comparable to those in WAT. However, unlike WAT which can increase seemingly without limit, MAT is constrained spatially by the skeleton. Recent evidence supports the notion that the origin and function of MAT, although related, is distinct from that of peripheral adipose tissues [1]. Evolutionarily, with the exception of the cartilaginous fish, WAT-like structures are present in the majority of vertebrate species [64–66]. By contrast, brown adipose tissue (BAT) is a purely mammalian adaptation and has not been found in birds, reptiles, or fish despite significant effort [66]. On the evolutionary timeline, MAT represents an adipocyte population present in bony fish, reptiles, birds, and mammals that likely appeared around the same time or slightly after WAT and well before thermogenesis was relegated from muscle to BAT in mammals [28]. There is also recent evidence that MAT has a unique progenitor when compared to WAT and BAT (reviewed in [1]). This does not rule out functional overlap between MAT and other adipose tissues, which is likely present, but rather prompts consideration of MAT as an adipose tissue depot with the potential for unique functional characteristics, particularly those related to bone turnover and hematopoiesis.

CURRENT STANDARDS FOR IMAGING AND ANALYSIS OF MARROW ADIPOSE TISSUE

MAT has been imaged with tools including PET, MRI, and CT. In humans, assessment of skeletal marrow fat content with water–fat MRI correlates highly with histology [67]. Whole body assessment with MRI reveals that the amount of total MAT in humans is approximately 1.35 kg, ranging from 0.5 to 3.0 kg [38, 68]. Similar estimations can be derived from PET/CT-based calculations of marrow volume (average MAT content of 1.03 ± 0.37 kg) [1]. This means that in a person of average size, ~8% of their total fat is contained within the skeleton. Depending on peripheral body composition, however, this can range from 1% to 30%. In addition to total MAT volume, MRI can be used to estimate the saturation of the marrow lipids based on the presence of carbon–carbon double bonds. This may be particularly relevant for bone because a recent study revealed that MAT saturation was an independent predictor of fracture risk in postmenopausal women with diabetes [69]. In rodents, MAT volume is generally quantified using osmium tetroxide staining and micro–computed tomography (Figs 126.2 and 126.4) [10, 70]. The osmium technique highlights the asymmetric distribution of the MAT within rodent bones, a point which must be carefully considered if one attempts quantification by serial histology in lieu of CT-based techniques.

SUMMARY

The nomenclature for MAT has undergone steady evolution over the course of the past century [2]. However, because the word "fat" can be used to refer to lipid accumulation in non-adipocytes, the terms *marrow adipocyte*, *marrow adipose tissue*, and *bone marrow adipose tissue* (in lieu of *marrow fat* or *yellow marrow*) are most accurate for current use. MAT development is conserved across mammalian species, occurring shortly after birth in a well-defined centripetal pattern. In some skeletal regions, MAT undergoes regulation and expansion with disease or pharmacotherapy. In others, MAT remains stubbornly persistent, perhaps pointing to its necessary role in skeletal or metabolic homeostasis. For now, we recognize that MAT has the potential to be physiologically adaptive—perhaps helping the body to partition energy, regulate bone turnover, or fuel hematopoiesis. By contrast, MAT expansion may also be pathological, contributing to bone loss in some contexts. More work is clearly needed to determine when and how this occurs and whether it is related to the phenotype of the MAT adipocyte itself.

ACKNOWLEDGMENTS

This work was supported by NIH grant R00-DE024178. Thank you to Deb Novack and Hero Robles for their help with the human bone marrow images.

REFERENCES

1. Scheller EL, Cawthorn WP, Burr AA, et al. Marrow Adipose Tissue: Trimming the Fat. Trends Endocrinol Metab. 2016;27(6):392–403.

2. Scheller EL, Rosen CJ. What's the matter with MAT? Marrow adipose tissue, metabolism, and skeletal health. Ann N Y Acad Sci. 2014;1311:14–30.
3. Custer RP. Studies on the Structure and Function of Bone Marrow Part I. J Lab and Clin Med. 1932;17:951–60.
4. Custer RP, Ahlfeldt FE. Studies on the Structure and Function of Bone Marrow II. J Lab and Clin Med. 1932;17:960–2.
5. Huggins C, Blocksom BH, Jr. Changes in Outlying Bone Marrow Accompanying a Local Increase of Temperature Within Physiological Limits. J Exp Med. 1936;64:253–74.
6. Piney A. The Anatomy of the Bone Marrow. The British Medical Journal. 1922;2:792–5.
7. Vogler JB, Murphy WA. Bone marrow imaging. Radiology. 1988;168(3):679–93.
8. Kricun ME. Red-yellow marrow conversion: its effect on the location of some solitary bone lesions. Skeletal Radiol. 1985;14(1):10–19.
9. Emery JL, Follett GF. Regression of bone-marrow haemopoiesis from the terminal digits in the foetus and infant. Br J Haematol. 1964;10:485–9.
10. Scheller EL, Doucette CR, Learman BS, et al. Region-specific variation in the properties of skeletal adipocytes reveals regulated and constitutive marrow adipose tissues. Nat Commun. 2015;6:7808.
11. Tavassoli M, Watson LR, Khademi R. Retention of hemopoiesis in tail vertebrae of newborn rats. Cell Tissue Res. 1979;200(2):215–22.
12. Bigelow CL, Tavassoli M. Fatty involution of bone marrow in rabbits. Acta Anat (Basel). 1984;118(1):60–4.
13. Zakaria E, Shafrir E. Yellow bone marrow as adipose tissue. Proc Soc Exp Biol Med. 1967;124(4):1265–8.
14. Rozman C, Feliu E, Berga L, et al. Age-related variations of fat tissue fraction in normal human bone marrow depend both on size and number of adipocytes: a stereological study. Exp Hematol. 1989;17(1):34–7.
15. Fazeli PK, Horowitz MC, MacDougald OA, et al. Marrow fat and bone—new perspectives. J Clin Endocrinol Metab. 2013;98(3):935–45.
16. Scheller EL, Khoury B, Moller KL, et al. Changes in Skeletal Integrity and Marrow Adiposity during High-Fat Diet and after Weight Loss. Front Endocrinol (Lausanne). 2016;7:102.
17. Devlin MJ, Cloutier AM, Thomas NA, et al. Caloric restriction leads to high marrow adiposity and low bone mass in growing mice. J Bone Miner Res. 2010;25(9):2078–88.
18. Cawthorn WP, Scheller EL, Learman BS, et al. Bone marrow adipose tissue is an endocrine organ that contributes to increased circulating adiponectin during caloric restriction. Cell Metab. 2014;20(2):368–75.
19. Bredella MA, Gill CM, Gerweck AV, et al. Ectopic and serum lipid levels are positively associated with bone marrow fat in obesity. Radiology. 2013;269(2):534–41.
20. Slade JM, Coe LM, Meyer RA, et al. Human bone marrow adiposity is linked with serum lipid levels not T1-diabetes. J Diabetes Complicat. 2012;26(1):1–9.
21. Walji TA, Turecamo SE, DeMarsilis AJ, et al. Characterization of metabolic health in mouse models of fibrillin-1 perturbation. Matrix Biol. 2016;55:63–76.
22. Vande Berg BC, Malghem J, Lecouvet FE, et al. Fat conversion of femoral marrow in glucocorticoid-treated patients: a cross-sectional and longitudinal study with magnetic resonance imaging. Arthritis Rheum. 1999;42(7):1405–11.
23. Li GW, Xu Z, Chen QW, et al. The temporal characterization of marrow lipids and adipocytes in a rabbit model of glucocorticoid-induced osteoporosis. Skeletal Radiol. 2013;42(9):1235–44.
24. Schwartz AV. Marrow fat and bone: review of clinical findings. Front Endocrinol (Lausanne). 2015;6:40.
25. Tavassoli M. Marrow adipose cells. Histochemical identification of labile and stable components. Arch Pathol Lab Med. 1976;100(1):16–18.
26. Ambati S, Li Q, Rayalam S, et al. Central leptin versus ghrelin: effects on bone marrow adiposity and gene expression. Endocrine. 2010;37(1):115–123.
27. Lindenmaier LB, Philbrick KA, Branscum AJ, et al. Hypothalamic Leptin Gene Therapy Reduces Bone Marrow Adiposity in ob/ob Mice Fed Regular and High-Fat Diets. Front Endocrinol (Lausanne). 2016;7:110.
28. Craft CS, Scheller EL. Evolution of the Marrow Adipose Tissue Microenvironment. Calcif Tissue Int. 2017;100(5):461–75.
29. Bennett CM, Kanki JP, Rhodes J, et al. Myelopoiesis in the zebrafish, Danio rerio. Blood. 2001;98(3):643–51.
30. Witten PE, Hansen A, Hall BK. Features of mono- and multinucleated bone resorbing cells of the zebrafish Danio rerio and their contribution to skeletal development, remodeling, and growth. J Morphol. 2001;250(3):197–207.
31. Jordan HE. The histology of the blood and the red bone-marrow of the leopard frog, Rana pipiens. American Journal of Anatomy. 1919;25(4):436–80.
32. Weiss LP, Wislocki GB. Seasonal variations in hematopoiesis in the dermal bones of the nine-banded armadillo. Anat Rec. 1956;126(2):143–63.
33. Naveiras O, Nardi V, Wenzel PL, et al. Bone-marrow adipocytes as negative regulators of the haematopoietic microenvironment. Nature. 2009;460(7252):259–63.
34. Wronski TJ, Smith JM, Jee WS. Variations in mineral apposition rate of trabecular bone within the beagle skeleton. Calcif Tissue Int. 1981;33(6):583–6.
35. Burkhardt R, Kettner G, Böhm W, et al. Changes in trabecular bone, hematopoiesis and bone marrow vessels in aplastic anemia, primary osteoporosis, and old age: a comparative histomorphometric study. Bone. 1987;8(3):157–64.
36. Nelson-Dooley C, Della-Fera MA, Hamrick M, et al. Novel treatments for obesity and osteoporosis: targeting apoptotic pathways in adipocytes. Curr Med Chem. 2005;12(19):2215–25.
37. Wren TA, Chung SA, Dorey FJ, et al. Bone marrow fat is inversely related to cortical bone in young and old subjects. J Clin Endocrinol Metab. 2011;96(3):782–6.
38. Shen W, Chen J, Punyanitya M, et al. MRI-measured bone marrow adipose tissue is inversely related to

DXA-measured bone mineral in Caucasian women. Osteoporos Int. 2007;18(5):641–7.
39. Cohen A, Dempster DW, Stein EM, et al. Increased marrow adiposity in premenopausal women with idiopathic osteoporosis. J Clin Endocrinol Metab. 2012;97(8):2782–91.
40. Yeung DK, Griffith JF, Antonio GE, et al. Osteoporosis is associated with increased marrow fat content and decreased marrow fat unsaturation: a proton MR spectroscopy study. J Magn Reson Imaging. 2005;22(2):279–85.
41. Griffith JF, Yeung DK, Antonio GE, et al. Vertebral bone mineral density, marrow perfusion, and fat content in healthy men and men with osteoporosis: dynamic contrast-enhanced MR imaging and MR spectroscopy. Radiology. 2005;236(3):945–51.
42. Miyakoshi N, Sato K, Abe T, et al. Histomorphometric evaluation of the effects of ovariectomy on bone turnover in rat caudal vertebrae. Calcif Tissue Int. 1999;64(4):318–24.
43. Li M, Shen Y, Qi H, et al. Comparative study of skeletal response to estrogen depletion at red and yellow marrow sites in rats. Anat Rec. 1996;245(3):472–80.
44. Ma YF, Ke HZ, Jee WS. Prostaglandin E2 adds bone to a cancellous bone site with a closed growth plate and low bone turnover in ovariectomized rats. Bone. 1994;15(2):137–46.
45. Martin RB, Zissimos SL. Relationships between marrow fat and bone turnover in ovariectomized and intact rats. Bone. 1991;12(2):123–31.
46. Iwaniec UT, Turner RT. Failure to generate bone marrow adipocytes does not protect mice from ovariectomy-induced osteopenia. Bone. 2013;53(1):145–53.
47. Motyl KJ, McCabe LR. Leptin treatment prevents type I diabetic marrow adiposity but not bone loss in mice. J Cell Physiol. 2009;218(2):376–84.
48. Premkumar A, Chow C, Bhandarkar P, et al. Lipoatrophic-lipodystrophic syndromes: the spectrum of findings on MR imaging. AJR Am J Roentgenol. 2002;178(2):311–18.
49. Rajab A, Heathcote K, Joshi S, et al. Heterogeneity for congenital generalized lipodystrophy in seventeen patients from Oman. Am J Med Genet. 2002;110(3):219–25.
50. Kim CA, Delépine M, Boutet E, et al. Association of a homozygous nonsense caveolin-1 mutation with Berardinelli-Seip congenital lipodystrophy. J Clin Endocrinol Metab. 2008;93(4):1129–34.
51. Rinotas V, Niti A, Dacquin R, et al. Novel genetic models of osteoporosis by overexpression of human RANKL in transgenic mice. J Bone Miner Res. 2014;29(5):1158–69.
52. Gunaratnam K, Vidal C, Gimble JM, et al. Mechanisms of palmitate-induced lipotoxicity in human osteoblasts. Endocrinology. 2014;155(1):108–16.
53. During A, Penel G, Hardouin P. Understanding the local actions of lipids in bone physiology. Prog Lipid Res. 2015;59:126–46.
54. Chan MY, Frost SA, Center JR, et al. Relationship between body mass index and fracture risk is mediated by bone mineral density. J Bone Miner Res. 2014;29(11):2327–35.
55. Takeda S, Elefteriou F, Levasseur R, et al. Leptin regulates bone formation via the sympathetic nervous system. Cell. 2002;111(3):305–17.
56. Ducy P, Amling M, Takeda S, et al. Leptin inhibits bone formation through a hypothalamic relay: a central control of bone mass. Cell. 2000;100(2):197–207.
57. Scheller EL, Song J, Dishowitz MI, et al. Leptin functions peripherally to regulate differentiation of mesenchymal progenitor cells. Stem Cells. 2010;28(6):1071–80.
58. Turner RT, Kalra SP, Wong CP, et al. Peripheral leptin regulates bone formation. J Bone Miner Res. 2013;28(1):22–34.
59. Yue R, Zhou BO, Shimada IS, et al. Leptin Receptor Promotes Adipogenesis and Reduces Osteogenesis by Regulating Mesenchymal Stromal Cells in Adult Bone Marrow. Cell Stem Cell. 2016;18(6):782–96.
60. Scheller EL, Song J, Dishowitz MI, et al. A potential role for the myeloid lineage in leptin-regulated bone metabolism. Horm Metab Res. 2012;44(1):1–5.
61. Wee NK, Baldock PA. The Skeletal Effects of Leptin. In: Blum EL (ed.) *Leptin: Biosynthesis, Functions and Clinical Significance*. Nova Science Publishers, 2014, pp 129–40.
62. Williams GA, Wang Y, Callon KE, et al. In vitro and in vivo effects of adiponectin on bone. Endocrinology. 2009;150(8):3603–10.
63. Kajimura D, Lee HW, Riley KJ, et al. Adiponectin regulates bone mass via opposite central and peripheral mechanisms through FoxO1. Cell Metab. 2013;17(6):901–15.
64. Gesta S, Tseng YH, Kahn CR. Developmental origin of fat: tracking obesity to its source. Cell. 2007;131(2):242–56.
65. Vague J, Fenasse R. Comparative anatomy of adipose tissue. In: Terjung R (ed.) *Comprehensive Physiology*. Hoboken, NJ: John Wiley & Sons, 2010.
66. Pond CM. The Evolution of Mammalian Adipose Tissue. In: Symonds ME (ed.) *Adipose Tissue Biology*. New York: Springer, 2012, pp 227–69.
67. Arentsen L, Yagi M, Takahashi Y, et al. Validation of marrow fat assessment using noninvasive imaging with histologic examination of human bone samples. Bone. 2015;72:118–22.
68. Shen W, Chen J, Gantz M, et al. Ethnic and sex differences in bone marrow adipose tissue and bone mineral density relationship. Osteoporos Int. 2012;23(9):2293–301.
69. Patsch JM, Li X, Baum T, et al. Bone marrow fat composition as a novel imaging biomarker in postmenopausal women with prevalent fragility fractures. J Bone Miner Res. 2013;28(8):1721–8.
70. Scheller EL, Troiano N, Vanhoutan JN, et al. Use of osmium tetroxide staining with microcomputerized tomography to visualize and quantify bone marrow adipose tissue in vivo. Meth Enzymol. 2014;537:123–39.

127

The Vasculature and Bone

Marie Hélène Lafage-Proust and Bernard Roche

SAINBIOSE Inserm, Université de Lyon, Saint-Étienne, France

INTRODUCTION

All the functions of bone, including locomotion, hematopoiesis, calcium/phosphate metabolism, and endocrine secretions, depend on a common blood supply [1]. Not only do bone blood vessels bring oxygen, nutrients, and regulatory factors and remove metabolic waste products, they also transport or bear bone precursor cells. Throughout life, these functional relationships are supported by a complex network of reciprocal signals in which vascular cells stimulate bone cells, and bone cells, in turn, elicit cues to modulate blood vessels.

ANATOMY OF BONE VESSELS

In long bones, the principal nutrient artery enters the marrow cavity through a diaphysis foramen and divides early into two branches, heading towards the metaphyses. Straight arterioles abruptly emerge from these arteries which feed the bone marrow (BM) and contribute to cortical and trabecular bone vascularization. An additional blood supply is provided by epiphyseal, metaphyseal, and periosteal arteries which irrigate the subchondral bone and the growth plate, the trabecular bone, and the cortex outer part, respectively (Fig. 127.1) [2]. Arterial capillaries enter cortical bone through Volkmann's channels and connect perpendicularly to vessels of the Haversian canals, thereby connecting endosteum and periosteum. Thus, cortical bone is irrigated by a double blood supply, the inner two-thirds depending on medullary arteries and the outer third on periosteal vessels. The respective contribution of these two networks varies with age, with a predominant centrifugal perfusion in young bones and a periosteal centripetal one in older bones [3]. Similarly, a double venous network drains blood out of the BM cavity. Cortical bone capillaries drain in the periosteal or medullary venous networks while BM sinusoids join the large central venous sinus (Fig. 127.1) which, in turn, drains into the principal nutrient veins.

HETEROGENEOUS HISTOLOGY OF BONE MICROVESSELS

Feeding arteries are small (30 to 70 μm diameter), with a layer of smooth muscle cells (SMC) found up to the metarterioles. SMCs constitute a precapillary sphincter able to bypass the downstream capillary network by establishing a shunt with postcapillary venules, ensuring territorial blood redistribution. Vasoregulation takes place in straight arterioles and metarterioles which account for most of the bone vascular resistance. Marrow microvessels are heterogeneous. Arterial capillaries are smaller (10 μm) than venous sinusoids (20 to 30 μm). In the diaphysis, capillaries are parallel while sinusoids are perpendicular to the bone long axis (Figs 127.1 and 127.2). Arterial capillaries are composed of a monolayer of endothelial cells disposed on a continuous basal lamina and highly express adhesion and junction molecules [4]. The sinusoid basal lamina is discontinuous and delineates pores, through which most of the cell traffic to and from the BM [3, 5] takes place. Notably, lymphatic vessels are found along the periosteum, occasionally in cortical bone and almost never in the BM [6].

PERIVASCULAR CELLS AND PERICYTES

Endothelial cells and surrounding pericytes share the same basal lamina. Pericyte coverage is denser on the arterial network [7] and contributes to the stabilization of vessel walls. Perivascular cells can express neural/glial antigen (NG)-2, alpha smooth muscle actin (α-SMA), PDGFRβ, desmin, high levels of CXCL12-abundant reticular cells

---

Primer on the Metabolic Bone Diseases and Disorders of Mineral Metabolism, Ninth Edition. Edited by John P. Bilezikian.
© 2019 American Society for Bone and Mineral Research. Published 2019 by John Wiley & Sons, Inc.
Companion website: www.wiley.com/go/asbmrprimer

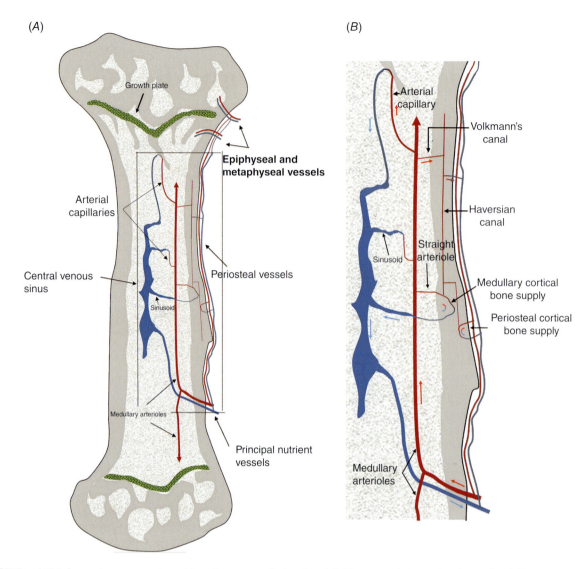

Fig. 127.1. (A) Schematic presentation of long bone vascularization. (B) Close-up of the region framed in (A).

(CXCL12: C-X-C motif chemokine 12), leptin receptor [8], nestin [9], transgelin [10], or CD146 (in humans) [11]. These markers are not mutually exclusive and illustrate the heterogeneity of this cell population. Subsets of pericytes strongly regulate the hematopoietic stem cell niche [12]. For example, in vivo transplantation assays showed that CD146+ human BM pericytes are committed to bone and hematopoietic supportive stroma formation including adipocytes, but not to cartilage [11].

OXYGEN TENSION IN THE BONE MARROW AND BONE PERFUSION

Oxygen tension, which depends on bone perfusion, is a key regulatory factor in bone. The overexpression of hypoxia inducible factor 1 alpha (HIF-1α) in osteoblasts leads to increased angiogenesis and osteogenesis [13]. Studies modeling oxygen distribution [14] and in vivo experiments concluded that BM is rather hypoxic. However, the BM regional pattern of pO_2 remains controversial. Using a marker of tissue hypoxia, Levesque et al. concluded that the endocortical regions are more hypoxic than the BM center [15]. In contrast, Spencer et al. [16], who developed direct in vivo pO_2 measurements in mouse BM, reported that the lowest pO_2 was found in the central perisinusoidal area whereas the endocortical regions were less hypoxic.

Bone perfusion evaluation can be achieved through arteriolar blockade by injecting fluorescent microspheres [17] which provides a single instantaneous blood flow measurement. Laser Doppler allows longer and multiple measurements of bone perfusion in mouse tibia [18]. It is of note that structural (vessel density) and functional (perfusion) parameters are usually poorly correlated [19].

fluorescent acetylated low density lipoprotein (Dil-Ac-LDL; DIL) or isolectin binding, immunohistochemistry (IHC), or endothelial-specific promoter protocols using Tie2 [22] or vascular endothelial cadherin promoters. Interestingly, double IHC staining of BM vessels distinguishes arterial capillaries, which express only CD31, from venous sinusoids, which express both CD31 and endomucin [23] (Fig. 127.2). Similarly, Sca-1 and Tie2 are expressed in arterial capillaries only, whereas only sinusoids are able to internalize DIL [22]. Furthermore, mouse bone vessels and circulating cells can be visualized through the calvaria by intravital multi-photon confocal microscopy (IVM) [24, 25]. Using IVM, Ishii et al. observed GFP-labeled preosteoclasts trafficking in and out of vessels through a glass window inserted in a mouse femur [26]. As for cortical bone, micro-computed tomography (μCT) assessment of pore number constitutes a surrogate method for evaluating the cortical vascular network in mice [27] and humans [28].

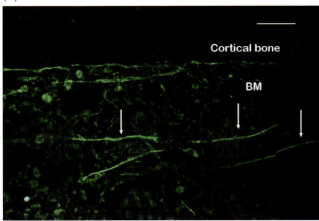

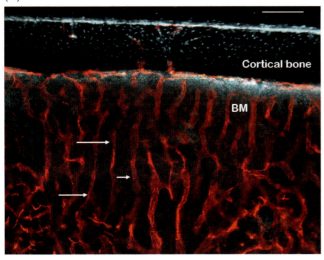

Fig. 127.2. Immunohistochemistry staining of bone vessels in mouse tibia on 70 μm-thick cryosections (personal data). (A) CD31 immunostaining of the arterial capillary network. Bar: 100 μm. Some cells in the BM express CD31. (B) Endomucin immunostaining of the veinous sinusoids. Nuclei are stained with 4,6-diamidino-2-phenylindole (DAPI). Bar: 100 μm.

Thus, a higher number of vessels does not necessarily result in better bone perfusion [19].

ASSESSMENT OF BONE VESSELS (see [20] for review)

Bone vessels are visualized by either filling the vascular bed by a contrast product (India ink, lead or barium-containing compounds) or labeling the vessel wall. Then, vessel density and size can be quantified on histology sections or X-ray-based 2D or 3D images [21]. Vessel wall labeling can be performed via intravenous injection of molecules internalized by endothelial cells such as

VESSEL ROLES DURING OSSIFICATION, MODELING, AND GROWTH

Whether it be during endochondral ossification [29], bone growth [30], or fracture repair [31, 32], vessels grow along a gradient of cues synthesized by hypertrophic chondrocytes, penetrate the cartilage, bring osteoblast precursors, and promote their migration towards the bone template, the growth plate, or the injury site, respectively. Vessel growth responds to neural [33] and angiogenic factors including VEGF [34], placental growth factor (PlGF), FGFs, PDGF, and BMPs. Overall, de novo osteogenesis and angiogenesis are tightly coupled in primary ossification and bone modeling situations. Subsequently, stimulating adequate angiogenesis in bone constructs has become a technical challenge in order to improve bone tissue engineering [35] and fracture repair [36].

THE ROLE OF VESSELS IN BONE REMODELING

Bone remodeling basic multicellular units (BMU) are functionally linked to a microvessel [37]. Bone vessels bring osteoclast precursors to the front of the BMU whereas the osteoprogenitors are recruited at the back from circulating cells [38] or differentiate from pericytes [39] (Fig. 127.3). In trabecular bone, the BMU is separated from the marrow by a thin canopy of cells [40] which delineates the bone remodeling compartment (BRC) (Figs 127.3 and 127.4). Kristensen et al. observed that the presence and spatial orientation of the vessels neighboring the BRC depended on the remodeling activity of the bone surface [41]. In cortical bone, remodeling secondary osteons constituting the Haversian canals are also centered by vessels. Of note, vascularization in long bone

986   The Vasculature and Bone

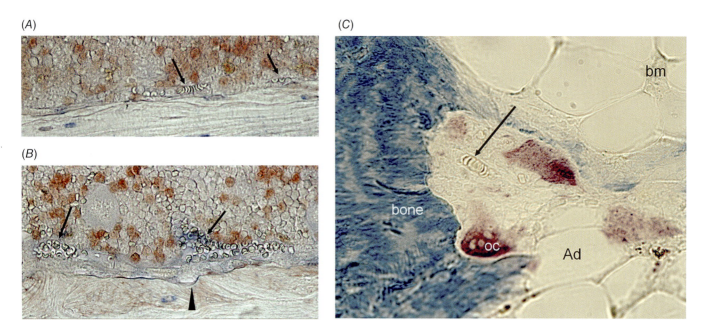

Fig. 127.3. (A, B) Endocortical region of mouse tibia. (A) Capillaries orthogonal to non-remodeling bone surfaces (arrows). (B) Capillary leaning on a bone remodeling compartment (arrow) facing a resorption lacuna (arrow head). Bar: 50 μm. (C) Human trabecular bone embedded in metacrylate, stained with aniline blue with tartrate-resistant acid phosphatase (TRAP)-positive osteoclasts (oc). The arrow indicates a vessel filled with red blood cells. Ad = adipocyte; bm = bone marrow.

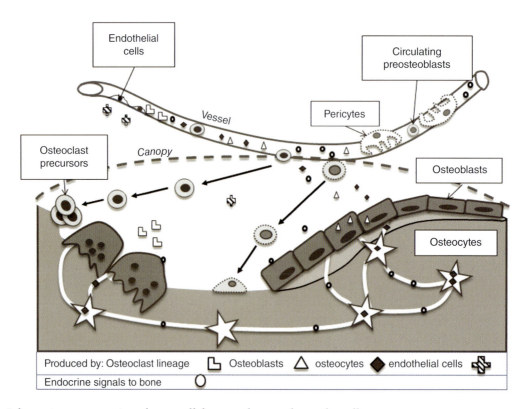

Fig. 127.4. Schematic representation of cross-talk between bone and vascular cells.

cortices differs among species with major differences between mice [42] and humans [43]. The mechanisms that control the bone remodeling/vessels coupling remain obscure. Bone cell traffic through the vessel wall to and from the remodeling bone surface involves, inter alia, sphingosine-phosphate 1 [44] or CXCR4/CXCL12 signaling [45, 46]. Kusumbe et al. [23] showed that osteoprogenitors are borne on the wall of a subtype of bone vessels, namely "type H vessels," which express high levels of both CD31 and endomucin. They demonstrated that activating the HIF-1α signaling pathway in endothelial cells only, increased both the number of type H vessels and bone formation. Furthermore, PDGF-BB released by preosteoclasts induces growth of type H vessels and stimulates bone formation in ovariectomized mice [47]. Thus, vessels are indispensable for bone remodeling. However, the tight spatial and temporal regulations that allow adequate growth and regression of the capillary that accompanies the BMU throughout its life span are not fully understood. For instance, overexpression of VEGF in the chondro-osteoblast lineage in adult mice is angiogenic and increases trabecular bone mass but it also induces fibrotic marrow and augments cortical porosity [48]. Many other cues are involved in the cross-talk between bone and vascular cells (reviewed in Chim et al. [49]), which are briefly illustrated in Table 127.1 and Fig. 127.4.

BONE VESSELS IN HUMAN BONE DISEASES

Bone primary tumors, metastases, and osteonecrosis of the jaw are not mentioned here.

Local and regional pathologies

Vascular obstruction (Fig. 127.5)

As for other organs, vessel obstruction in bone induces ischemia, often followed by necrosis of the downstream vascular territory. However, the clinical consequences differ according to the location of the vascular thrombosis.

Epiphyseal osteonecrosis (ON) may lead to subsequent collapse of subchondral bone with flattening of the articular surface, followed by joint destruction requiring prosthesis replacement when possible. ON occurs at epiphyses in which collateral vascular supply is scarce: humeral and femoral head, medial femoral condyle and tibial plateau, talus, lunate, etc. Vascular lesions induced by Garden type IV hip fracture can cause femoral head ON, demonstrating that impaired blood perfusion is actually at its origin. However, ON may arise from many other risk factors such as glucocorticoid treatment, alcohol consumption, hyperlipidemia, sickle cell disease, caisson disease, and hyperviscosity

Table 127.1. Examples of molecular cross-talk between bone and vessels cells (+ means "expressed by").

	Osteoclasts	Osteocytes	Osteoblasts	Endothelial Cells	Pericytes and Smooth Muscle Cells
Nitric oxide		+	+	+	Vasodilation
	←Effects on bone depend on dose→				
VEGF	+	+	+	Angiogenesis	Vasodilation
	↑Resorption		Migration		
BMP7	+	+	+ Osteogenesis	Angiogenesis	
Endothelin-1			+	+	Vasoconstriction
				Angiogenesis	
RANKL	↑Resorption	+	+	+	
				Angiogenesis	
FGF-2	↑Resorption		+		Vasodilation
			Osteogenesis	Angiogenesis	
PTHrP	↑Resorption		+	Angiogenesis	Vasodilation
PDGF-BB	+			Angiogenesis	Vasodilation
	←Effects on bone depend on dose→				
PEDF	↓Resorption		+	Anti-angiogenesis	
PGE₂	↓Resorption		+	Angiogenesis	Vasodilation
			↑Formation		
Periostin			+	Angiogenesis	Migration
			↑Formation		

PDGF-BB = platelet-derived growth factor-BB; PEDF = pigmented epithelium-derived factor; PGE₂ = prostaglandin E₂.

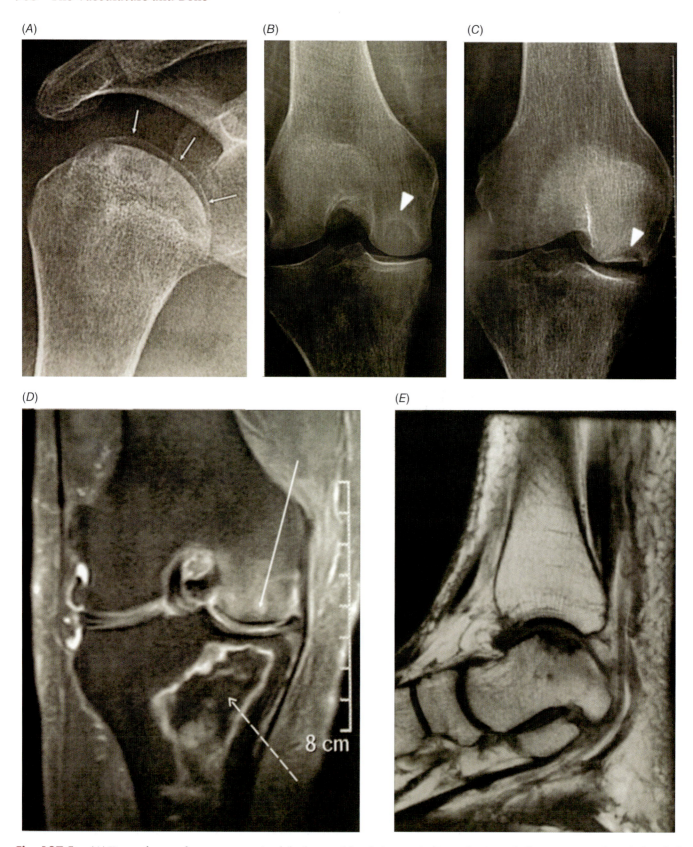

Fig. 127.5. *(A)* X-ray of avascular osteonecrosis of the humeral head. Arrows indicate the "egg shell sign." Note the subchondral fracture. *(B, C)* X-ray follow-up (1 month apart) of osteonecrosis of medial condyle (arrow). *(D)* T2 MRI image of a diabetic patient with severe ischemic lower limb arteriopathy combining osteonecrosis of the medial femoral condyle (arrow) and infarction of the medial region of tibia metaphysis (dotted arrow). *(E)* T1 MRI image of osteonecrosis of the talus.

syndromes. In these cases, the pathophysiology is more complex and involves an increase in adipocyte content, lipotoxicity, and BM edema leading to a rise in marrow pressure, all of which contribute to vascular obstruction.

Bone infarction occurs at the center of the metaphysis or the diaphysis of long bones. It does not necessarily induce pain when small and isolated. At the final stage the lesion is calcified and appears on plain X-rays as irregular and sclerotic at the periphery with normal density at the center. Beside steroid use, bone infarcts are common in Gaucher disease and sickle cell anemia.

Diseases associated with changes in bone perfusion

Some nonmalignant bone diseases involve changes in bone vascularization via increase in perfusion. Paget disease and "regional pain syndrome" at its early stage [50] both feature a significant increase in bone blood flow and accelerated resorption. However, the former leads to bigger bones whereas the latter induces bone loss. In Paget disease, the increase in perfusion is clearly linked to the high turnover rate because anti-osteoclastic treatments dramatically reduce the perfusion of the affected bones.

Osteoporosis, vascular diseases, and aging

Aging is associated with both osteoporosis and vascular calcifications, a complication of atherosclerosis. Whether this is the result of a common pathophysiological mechanism (the so-called bone–vascular axis [51]), or the result of independent processes linked to aging remains a matter of debate. For instance, Hyder et al. showed, in a large population of aging subjects, that lower BMD was independently associated with structural and functional measures of atherosclerosis in men and with more severe and calcified carotid plaques in both sexes [52]. A number of molecular studies support the idea of mechanistic links between bone loss and vascular disease. Vascular SMCs can undergo differentiation into osteoblast-like cells as a result of inflammatory cues involved in the "inflammaging" process or the loss of sex steroid associated with aging. For instance, estrogen deficiency induces RANKL expression in endothelial cells, which, in turn, increases production of BMP2, a potent stimulator of osteogenesis, and decreases expression of matrix Gla protein, an inhibitor of vessel media mineralization [53]. In addition, MRI-based studies showed reduced bone perfusion in postmenopausal osteoporotic women compared to age-matched healthy controls [54], suggesting that bone loss could be partly explained by a dysfunction of bone vasculature.

CONCLUSION

The precise mechanisms that regulate bone–vessel crosstalk remain unclear. However, accumulating data make bone vascularization a potential therapeutic target.

REFERENCES

1. Marenzana M, Arnett TR. The Key Role of the Blood Supply to Bone. Bone Res. 2013;1(3):203–15.
2. Kelly PJ. Anatomy, physiology, and pathology of the blood supply of bones. J Bone Joint Surg Am. 1968;50(4):766–83.
3. Bridgeman G, Brookes M. Blood supply to the human femoral diaphysis in youth and senescence. J Anat. 1996;188:611–21.
4. Itkin T, Gur-Cohen S, Spencer JA, et al. Distinct bone marrow blood vessels differentially regulate haematopoiesis. Nature. 2016;532(7599):323–8.
5. Inoue S, Osmond DG. Basement membrane of mouse bone marrow sinusoids shows distinctive structure and proteoglycan composition: a high resolution ultrastructural study. Anat Rec. 2001;264(3):294–304.
6. Edwards JR, Williams K, Kindblom LG, et al. Lymphatics and bone. Hum Pathol. 2008;39(1):49–55.
7. Armulik A, Genové G, Betsholtz C. Pericytes: developmental, physiological, and pathological perspectives, problems, and promises. Dev Cell. 2011;21(2):193–215.
8. Zhou BO, Yue R, Murphy MM, et al. Leptin Receptor-expressing mesenchymal stromal cells represent the main source of bone formed by adult bone marrow. Cell Stem Cell. 2014;15(2):154–68.
9. Kunisaki Y, Bruns I, Scheiermann C, et al. Arteriolar niches maintain haematopoietic stem cell quiescence. Nature. 2013;502(7473):637–43.
10. Zhang J, Link DC. Targeting of Mesenchymal Stromal Cells by Cre-Recombinase Transgenes Commonly Used to Target Osteoblast Lineage Cells. J Bone Miner Res. 2016;31(11):2001–7.
11. Sacchetti B, Funari A, Remoli C, et al. No Identical "Mesenchymal Stem Cells" at Different Times and Sites: Human Committed Progenitors of Distinct Origin and Differentiation Potential Are Incorporated as Adventitial Cells in Microvessels. Stem Cell Rep. 2016;6(6):897–913.
12. Acar M, Kocherlakota KS, Murphy MM, et al. Deep imaging of bone marrow shows non-dividing stem cells are mainly perisinusoidal. Nature. 2015;1;526(7571).
13. Wan C, Shao J, Gilbert SR, et al. Role of HIF-1alpha in skeletal development. Ann N Y Acad Sci. 2010;1192:322–6.
14. Chow DC, Wenning LA, Miller WM, et al. Modeling pO(2) distributions in the bone marrow hematopoietic compartment. I. Krogh's model. Biophys J. 2001;81(2):675–84.
15. Levesque JP, Winkler IG, Hendy J, et al. Hematopoietic progenitor cell mobilization results in hypoxia with increased hypoxia-inducible transcription factor-1 alpha and vascular endothelial growth factor A in bone marrow. Stem Cells. 2007;25(8):1954–65.
16. Spencer JA, Ferraro F, Roussakis E, et al. Direct measurement of local oxygen concentration in the bone marrow of live animals. Nature. 2014;508(7495):269–73.
17. Serrat MA. Measuring bone blood supply in mice using fluorescent microspheres. Nat Protoc. 2009;4(12):1749–58.

18. Roche B, Vanden-Bossche A, Normand M, et al. Validated Laser Doppler protocol for measurement of mouse bone blood perfusion - response to age or ovariectomy differs with genetic background. Bone. 2013; 55(2):418–26.
19. Roche B, Vanden-Bossche A, Malaval L, et al. Parathyroid hormone 1-84 targets bone vascular structure and perfusion in mice: impacts of its administration regimen and of ovariectomy. J Bone Miner Res. 2014;29:1608–18.
20. Lafage-Proust MH, Roche B, Langer M, et al. Assessment of bone vascularization and its role in bone remodeling. Bonekey Rep. 2015;4:662.
21. Roche B, David V, Vanden-Bossche A, et al. Structure and quantification of microvascularisation within mouse long bones: what and how should we measure? Bone. 2012;50(1):390–9.
22. Li XM, Hu Z, Jorgenson ML, et al. High levels of acetylated low-density lipoprotein uptake and low tyrosine kinase with immunoglobulin and epidermal growth factor homology domains-2 (Tie2) promoter activity distinguish sinusoids from other vessel types in murine bone marrow. Circulation. 2009;120(19):1910–8.
23. Kusumbe A, Ramasamy S, Adams RH. Coupling of angiogenesis and osteogenesis by a specific vessel subtype in bone. Nature. 2014; 507:325–8.
24. Lo Celso C, Fleming HE, Wu JW, et al. Live-animal tracking of individual haematopoietic stem/progenitor cells in their niche. Nature. 2009;457:92–6.
25. Huang C, Ness VP, Yang X, et al. Spatiotemporal Analyses of Osteogenesis and Angiogenesis via Intravital Imaging in Cranial Bone Defect Repair. J Bone Miner Res. 2015;30(7):1217–30.
26. Ishii M, Egen JG, Klauschen F, et al. Sphingosine-1-phosphate mobilizes osteoclast precursors and regulates bone homeostasis. Nature. 2009;458:524–8.
27. Schneider P, Krucker T, Meyer E, et al. Simultaneous 3D visualization and quantification of murine bone and bone vasculature using micro-computed tomography and vascular replica. Microsc Res Tech. 2009;72:690–701.
28. Cooper DM, Thomas CD, Clement JG, et al. Age-dependent change in the 3D structure of cortical porosity at the human femoral midshaft. Bone. 2007;40(4):957–65.
29. Maes C, Kobayashi T, Selig MK, et al. Osteoblast precursors, but not mature osteoblasts, move into developing and fractured bones along with invading blood vessels. Dev Cell. 2010;19(2):329–44.
30. Gerber HP, Vu TH, Ryan AM, et al. VEGF couples hypertrophic cartilage remodeling, ossification and angiogenesis during endochondral bone formation. Nat Med. 1999;5(6):623–8.
31. Stegen S, van Gastel N, Carmeliet G. Bringing new life to damaged bone: the importance of angiogenesis in bone repair and regeneration. Bone. 2015;70:19–27.
32. Boerckel JD, Uhrig BA, Willett NJ, et al. Mechanical regulation of vascular growth and tissue regeneration in vivo. Proc Natl Acad Sci. 2011;108:674–80.
33. Tomlinson RE, Li Z, Zhang Q, et al. NGF-TrkA Signaling by Sensory Nerves Coordinates the Vascularization and Ossification of Developing Endochondral Bone. Cell Rep. 2016; 16(10):2723–35.
34. Hu K, Olsen BR. Vascular endothelial growth factor control mechanisms in skeletal growth and repair. Dev Dyn. 2017;246(4):227–34.
35. Nguyen LH, Annabi N, Nikkhah M, et al. Vascularized bone tissue engineering: approaches for potential improvement. Tissue Eng Part B Rev. 2012;18(5):363–82.
36. Street J, Bao M, de Guzman L, et al. Vascular endothelial growth factor stimulates bone repair by promoting angiogenesis and bone turnover. Proc Natl Acad Sci U S A. 2002;99:9656–61.
37. Parfitt AM. The mechanism of coupling: a role for the vasculature. Bone. 2000;26(4):319–23.
38. Eghbali-Fatourechi GZ, Lamsam J, Fraser D, et al. Circulating osteoblast-lineage cells in humans. N Engl J Med. 2005;352:1959–66.
39. Bianco P, Sacchetti B, Riminucci M. Osteoprogenitors and the hematopoietic microenvironment. Best Pract Res Clin Haematol. 2011;24:37–47.
40. Kristensen HB, Andersen TL, Marcussen N, et al. Osteoblast recruitment routes in human cancellous bone remodeling. Am J Pathol. 2014;184:778–89.
41. Kristensen HB Andersen TL, Marcussen N, et al. Increased presence of capillaries next to remodeling sites in adult human cancellous bone. J Bone Miner Res. 2013;28(3):574–85.
42. Schneider P, Voide R, Stampanoni M, et al. The importance of the intracortical canal network for murine bone mechanics. Bone. 2013;53:120–8.
43. Palacio-Mancheno PE, Larriera AI, Doty SB, et al. 3D Assessment of Cortical Bone Porosity and Tissue Mineral Density Using High-Resolution μCT: Effects of Resolution and Threshold Method. J Bone Miner Res. 2014;29:142–50.
44. Ishii M, Egen JG, Klauschen F, et al. Sphingosine-1-phosphate mobilizes osteoclast precursors and regulates bone homeostasis. Nature. 2009;458:524–8.
45. Zhang Q, Guo R, Schwarz EM, et al. TNF inhibits production of stromal cell-derived factor 1 by bone stromal cells and increases osteoclast precursor mobilization from bone marrow to peripheral blood. Arthritis Res Ther. 2008;10:R37.
46. Otsuru S, Tamai K, Yamazaki T, et al. Circulating bone marrow-derived osteoblast progenitor cells are recruited to the bone-forming site by the CXCR4/stromal cell-derived factor-1 pathway. Stem Cells. 2008;26(1):223–34.
47. Xie H, Cui Z, Wang L, et al. PDGF-BB secreted by preosteoclasts induces angiogenesis during coupling with osteogenesis. Nat Med. 2014;20:1270–8.
48. Maes C, Goossens S, Bartunkova S, et al. Increased skeletal VEGF enhances beta-catenin activity and results in excessively ossified bones. EMBO J. 2010;29(2):424–41.
49. Chim SM, Tickner J, Chow ST, et al. Angiogenic factors in bone local environment. Cytokine Growth Factor Rev. 2013;24(3):297–310.

50. Driessens M. Circulatory aspects of reflex sympathetic dystrophy. In: Schoutens A, Arlet J, Gardeniers JW, et al. (eds) *Bone Circulation and Vascularization in Normal and Pathological Conditions*. New York: Plenum Press, 1993, pp 217–31.
51. Thompson B, Towler DA. Arterial calcification and bone physiology: role of the bone-vascular axis. Nat Rev Endocrinol. 2012;8:529–43.
52. Hyder JA, Allison MA, Barrett-Connor E, et al. Atherosclerosis. 2010;209(1):283–9.
53. Osako MK, Nakagami H, Koibuchi N, et al. Estrogen inhibits vascular calcification via vascular RANKL system: common mechanism of osteoporosis and vascular calcification. Circ Res. 2010;107(4):466–75.
54. Wang YX, Griffith JF, Kwok AW, et al. Reduced bone perfusion in proximal femur of subjects with decreased bone mineral density preferentially affects the femoral neck. Bone. 2009;45:711–5.

128

Immunobiology and Bone

Roberto Pacifici[1,2] and M. Neale Weitzmann[1,3]

[1]*Division of Endocrinology, Metabolism and Lipids, Department of Medicine, Emory University, Atlanta, GA, USA*
[2]*Immunology and Molecular Pathogenesis Program, Emory University, Atlanta, GA, USA*
[3]*Atlanta Department of Veterans Affairs Medical Center, Decatur, GA, USA*

IMMUNE CELLS RELEVANT TO BONE

T cells

The lymphocytes critical for bone are those that reside in the bone marrow (BM). T cells are highly mobile cells that account for ~5% of the BM cells [1]. BM is also a niche for central memory CD8+ cells which have a higher activation state than peripheral blood CD8+ cells and thus secrete higher levels of effector cytokines. Disease states may further alter the number and phenotype of BM T cells. For example, postmenopausal women with osteoporotic fractures have a higher proportion of TNF-producing CD8+ cells as compared to age-matched control women [2]. Some T-cell lineages (eg, Th17 cells) stimulate bone resorption whereas others (eg, regulatory T cells) inhibit osteoclastogenesis. Moreover, there is increasing recognition that T cells may stimulate bone formation by secreting Wnt ligands that activate Wnt signaling in osteoblastic cells [3, 4]. The effects of T cells on bone remodeling are also determined by their activation state. Activated CD4+ and CD8+ T cells tend to promote bone loss whereas resting CD4+ T cells may dampen bone resorption in vivo. Accordingly, T-cell null mice have a significant increase in bone resorption and reduced bone density as compared to controls. Conversely, T cells rendered anergic though CD28 costimulation blockade secrete Wnt10b and promote bone formation [4].

Among the most osteoclastogenic subsets of T cells are Th17 cells, which are defined as CD4+ cells and have the capacity to produce IL-17. Th17 cells abound in the intestine and BM. BM is a large reservoir of TGFβ and IL-6, factors essential for Th17 cell differentiation. Th17 cells potently induce osteoclastogenesis by secreting IL-17, RANKL, TNF, IL-1, and IL-6, along with low levels of IFNγ. IL-17 also potentiates the osteoclastogenic activity of RANKL by upregulating RANK. Th17 cells have been linked to postmenopausal osteoporosis because the differentiation of Th17 cells is induced by ovariectomy (OVX) [5], while the bone loss induced by OVX is prevented by silencing of IL-17R [6] and treatment with anti-IL-17 Ab [7]. Importantly, elevated levels of IL-17 have been found in postmenopausal women with osteoporosis [8].

The most bone-sparing population of T cells are regulatory T cells (Tregs), a suppressive population of mostly CD4+ T cells defined by the expression of the transcription factor FoxP3 and the ability to block conventional T-cell proliferation and production of effector cytokines. Tregs reside in close proximity to endosteal bone surfaces and osteoclasts and prevent OVX-induced bone loss [9]; estrogen increases the relative number of Tregs [10].

Tregs downregulate osteoclast formation and blunt bone resorption through the secretion of IL-4, IL-10, and TGFβ. IL-4 is a potent inhibitor of osteoclastogenesis [11]. In addition to secreting pro- and anti-inflammatory factors, T cells are also armed with surface costimulatory molecules such as RANKL and CD40L, which activate the cognate receptors RANK and CD40 in osteoclast precursors and osteoblastic cells, respectively [12]. CD40L has been linked to postnatal skeletal maturation because children affected by X-linked hyper-IgM syndrome, a condition in which CD40L production is impaired because of a mutation of the CD40L gene, have low bone density [13]. Two mechanisms have been identified to link the CD40L/CD40 to bone. First, activation of CD40 signaling in B cells by T cell–expressed CD40L promotes production of the anti-osteoclastogenic factor osteoprotegerin

---

Primer on the Metabolic Bone Diseases and Disorders of Mineral Metabolism, Ninth Edition. Edited by John P. Bilezikian.
© 2019 American Society for Bone and Mineral Research. Published 2019 by John Wiley & Sons, Inc.
Companion website: www.wiley.com/go/asbmrprimer

(OPG) by B cells [14], thereby decreasing bone resorption. In addition, activation of CD40 signaling in stromal cells (SCs) by T cell-expressed CD40L provides proliferative and survival cues to SCs in vitro and in vivo [15]. CD40L also increases the commitment of SCs to the osteoblastic lineage while activation of CD40 signaling in osteoblastic cells increases their osteoclastogenic activity [15]. Attesting to the relevance of the CD40L signaling pathway, CD40 is required for expansion of SCs and to increase their osteoclastogenic activity in ovariectomy- and continuous PTH-induced bone loss [15, 16].

B cells

B cells regulate bone resorption by secreting OPG, a soluble decoy receptor for RANKL. The major sources of OPG were initially thought to be osteoblasts and their stromal cell precursors. However, an analysis of the bone phenotype of B-cell knockout mice has shown that B cells, their precursors, and plasma cells are the dominant producers of OPG in the murine bone microenvironment in vivo [14].

Human tonsil B cells have further been demonstrated to secrete OPG, which is significantly upregulated by the activation of the CD40 receptor [17]. CD40 is a costimulatory molecule constitutively expressed by professional antigen-presenting cells (APC) such as macrophages, dendritic cells, and B cells, and partners with a receptor that is transiently upregulated on the surface of activated T cells. Mouse splenic B cells likewise produced upregulated concentrations of OPG in response to a recombinant soluble ligand to CD40 (sCD40L). In line with these data both CD40 and CD40L knockout mice displayed an osteoporotic phenotype and a significant deficiency in BM OPG concentrations. This deficiency in total OPG further correlated with a B-cell-specific deficiency in OPG production [18]. Thus, the emerging data suggest that the B lineage is likely a major source of OPG in the bone microenvironment and that T cell to B cell signaling, through the costimulatory molecules CD40L and CD40, plays an important role in regulating basal osteoclast formation and in regulating bone homeostasis.

Mice lacking RANKL in B lymphocytes are partially protected from the bone loss caused by OVX because of a failure of OVX to increase osteoclast numbers and bone resorption [19]. By contrast, deletion of RANKL from B cells had no impact on bone mass in estrogen-replete mice.

These findings may provide in part a novel explanation for the propensity for osteopenia and osteoporosis development in numerous pathological conditions in which altered immune function or immunodeficiency in B cells and/or T cells results. Such conditions include solid organ and BM transplantation, patients treated with immunosuppressive agents, and in the acquired immune deficiency syndrome (AIDS) associated with human immunodeficiency virus (HIV) infection.

Monocytes/macrophages

Monocytes and macrophages play key roles in both innate and adaptive immunity. Parabiosis studies performed in the 1970s identified the monocyte as the likely precursor of the osteoclasts; however, it was only in 1997 and 1998 that the key downstream regulators of osteoclastogenesis were identified. The first to be identified was OPG, which was found to suppress osteoclast differentiation [20]. The following year a potent inducer of osteoclastogenesis was identified and named osteoprotegerin ligand (OPGL) [21], now referred to as RANKL. It is now recognized that the osteoclast precursor is a cell of monocytic origin expressing the receptor for RANKL, RANK. When permissive concentrations of M-CSF are present, binding of RANKL to RANK on osteoclast precursors catalyzes their differentiation into pre-osteoclasts and the fusion of preosteoclasts into mature bone-resorbing osteoclasts. As a potent decoy receptor of RANKL, OPG functions as a negative regulator of osteoclast differentiation [21]. Macrophages exert complex effects in bone. Depletion of macrophage precursors results in osteopenia and blunted iPTH anabolic activity. By contrast, depletion of mature phagocytic macrophages potentiates iPTH-dependent anabolism by activating efferocytosis, a process that stimulates other BM cells to secrete Wnt10b, Wnt3a, and TGFβ.

IMMUNE CELLS IN BONE DISEASE

Bone loss secondary to sex steroid deficiency

The central mechanism by which estrogen deficiency induces bone loss is an increase in osteoclast formation and osteoclast life span driven by enhanced production of RANKL and TNF [22, 23]. Therefore, estrogen-dependent bone loss is regarded as a form of inflammatory bone loss [22, 23]. In humans, estrogen deficiency is associated with an expansion of RANKL- and TNF-expressing T cells and B cells [2, 24]. Menopause increases the levels of both IL-1 and TNF [22] whereas treatment with TNF and IL-1 inhibitors prevents the increase in bone resorption that results from estrogen deficiency [25]. Indeed, the causal role of TNF in OVX-induced bone loss in mice has been demonstrated in multiple models. Mechanistically, TNF stimulates bone resorption through potentiation of RANKL activity and induction of Th17 cells. Indeed, postmenopausal women with osteoporosis exhibit elevated levels of serum IL-17 [26]. In mice, OVX expands Th17 cells via TGFβ, IL-6, and IL-1β and TNF overproduction induced by estrogen deficiency [27] whereas Th17 differentiation is inhibited by estrogen via a direct effect on CD4+ T cells mediated by estrogen receptor alpha (ERα) [28]. The importance of IL-17 in bone loss is highlighted by the fact that silencing of IL-17R [6], or treatment with anti-IL-17 antibody [7], prevents OVX-induced bone loss.

Evidence now suggests that T cell-produced TNF plays a pivotal role in the mechanism of OVX-induced bone loss because no bone loss occurred, and no increased TNF production was detected, in T-cell null mice, or mice depleted of T cells [29]. Furthermore, bone loss did not occur in mice lacking T-cell TNF production [30], or lacking the costimulatory molecule CD40L [16], as well as mice treated with CTLA4-Ig, an agent that transmits an inhibitory signal to T cells [31]. These findings have been confirmed by other laboratories [32, 33]. However, the role of T cells in OVX-induced bone loss remains controversial because not all studies have corroborated a key role of T cells [34, 35]. In recent years additional evidence has emerged from human studies in favor of a role for T cell–produced TNF in postmenopausal bone loss [2] and T cell and B cell-produced RANKL [24]. Interestingly, OVX induces proliferation of T cells via an antigen-dependent process thereby increasing the number of CD4+ and CD8+ T cells in the BM and enhancing their production of TNF [30]. This process is driven by enhanced antigen presentation by macrophages and dendritic cells [31, 36], although the nature of the involved antigen remains unknown. Because the T cells of OVX mice have similar features to T cells exposed to bacteria we hypothesized that increased exposure to microorganisms provides the antigens required for T-cell activation and ensuing systemic immune responses required for sex steroid deficiency–induced bone loss.

Studies in germ free (GF) mice revealed that the absence of microbiota confers complete protection against the loss of trabecular bone induced by sex steroid depletion [37]. By contrast, sex steroid deficiency results in a similar loss of cortical bone in GF and conventionally-raised mice, indicating that sex steroid deprivation induces cortical bone loss via a microbiota-independent mechanism [37]. This study revealed that the microbiota is required for sex steroid deprivation to induce bone loss because it drives the expansion of conventional T cells and Th17 cells, thereby increasing their production of TNF, RANKL, and IL-17. Still unclear is whether bone loss is induced by gut cytokines that reach the BM, by immune cells activated in the gut that home to the BM, or by BM cells that are activated by foreign antigens of intestinal origin.

The intestinal epithelium forms a tight physiological barrier to separate systemic tissue compartments from the resident flora of the gut lumen. A breakdown in barrier integrity leads to increased bacterial translocation, often resulting in chronic immunogenic disease states. In response to sex steroid deprivation, we detected an increased gut permeability. Our proposed model is that increased permeability allows an expanded range of molecules and potential antigenic load to enter epithelial submucosa, initiating aberrant intestinal and systemic immune responses. Sex steroids maintain a tight physiological barrier between the systemic tissue compartments and the resident flora of the gut lumen. Depletion of sex steroid levels leads to increased gut permeability, and offers a positive link between a weakened intestinal barrier and the signature osteoclastogenic cytokine profile associated with osteoporosis.

Importantly several laboratories [38, 39], including ours [37], have shown that probiotics, defined as viable microorganisms that confer a health benefit when administered in adequate quantities, prevent the bone loss induced by OVX by decreasing inflammatory cytokine production in the gut and BM. The roles of the microbiota and gut permeability in the bone loss induced by sex steroid deficiency as well as the mechanism of action of probiotics are summarized in Fig. 128.1.

Role of T cells in primary hyperparathyroidism

Primary hyperparathyroidism (PHPT) is mimicked by continuous PTH (cPTH) infusion. PHPT and cPTH treatment cause cortical bone loss by enhancing endosteal resorption, and severe chronic elevations of PTH levels may also lead to trabecular bone loss [40], although PHPT and cPTH treatment often induce a modest increase in cancellous bone [41]. The bone effects of cPTH result from its binding to the PTH/PTH-related protein (PTHrP) receptor (PPR or PTHR1), expressed on BM stromal cells (SCs), osteoblasts, and osteocytes, but also on T cells and macrophages. SCs and osteoblasts were the first targets of PTH to be identified and earlier consensus held that the catabolic effect of cPTH is mostly mediated with enhanced production of RANKL and decreased production of OPG by SCs and osteoblasts [42]. More recent studies in mice with deletion and/or overexpression of PPR and RANKL in osteocytes [43–45] led to the recognition that osteocytes represent essential targets of PTH in bone, and that increased production of RANKL by osteocytes plays an important role in cPTH-induced bone loss [44, 45].

The discovery that T lymphocytes express functional PPR [3] and respond to PTH prompted investigations on the role of T cells as mediators of the effects of cPTH in bone. Early studies revealed that levels of PTH typically found in PHPT require the presence of T cells to induce bone loss, whereas conditions that cause extreme elevations in PTH levels induce bone loss via T-cell-independent mechanisms. PPR signaling in T cells stimulates the release of TNF [46] that drives bone resorption. Thus, deletion of T cells or T-cell production of TNF [15, 46] blocks the effects of cPTH as effectively as deletion of PPR signaling in osteocytes. Because of these reports, T cells are now recognized as a second critical target of PTH in bone.

Conditional silencing of the PTH receptor PPR in T cells blunts the stimulation of bone resorption by blocking TNF release, thus preventing cortical bone loss and converting the effects of cPTH in trabecular bone from catabolic to anabolic [46]. cPTH stimulates bone cells and immune cells to release growth factors and cytokines. TGFβ and IL-6 direct the differentiation of naïve CD4+ cells into Th17 cells. IL-17 provides an important connection between T cells and osteocytes because IL-17 regulates osteocytic RANKL production [47], which is one key effect of PTH on osteocytes. cPTH treatment

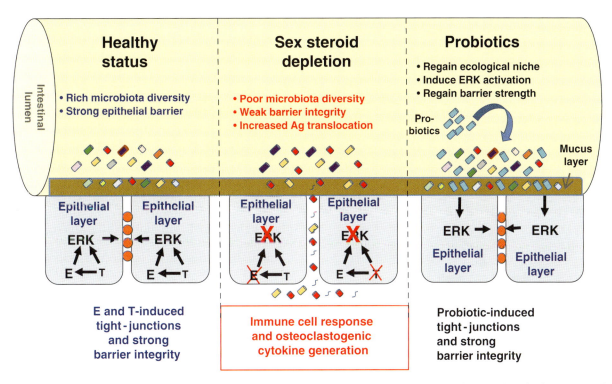

Fig. 128.1. Summary of the role of gut microbiota, gut permeability, and probiotics in the mechanism by which sex steroids cause bone loss. Estrogen (E) or testosterone (T) deficiency weakens barrier integrity by blunting expression of tight junction proteins via an ERK-dependent mechanism. The resulting increased bacterial translocation along with decreased microbiota diversity induces a local and systemic immune response that causes an increased production of osteoclastogenic cytokines. Probiotics increase microbiota diversity and decrease gut permeability, thus reducing peri-intestinal inflammation. Reproduced with permission.

increases the relative and absolute frequency of Th17+ cells and the levels of IL-17 in peripheral blood, spleen, and BM [47].

PTH receptor signaling activates $G\alpha_s$, which further promotes Th17 cell differentiation via Ca^{2+} influx [47]. Attesting to the relevance of $G\alpha_s$ signaling in CD4+ cells for Th17 cell generation, cPTH was found not to expand BM and splenic Th17 cells and not to exerts its bone catabolic activity in $G\alpha_s^{\Delta CD4,8}$ mice, demonstrating that silencing of $G\alpha_s$ in T cells prevents the expansion of Th17 cells and the bone loss induced by cPTH [47]. Consistent with this model, the L-type calcium channel blocker diltiazem blocks the expansion of Th17 cells, the increase in bone resorption, and the loss of cortical and trabecular bone induced by cPTH [47]. The finding may suggest a potential therapeutic role for L-type calcium channel blockers in the treatment of hyperparathyroidism.

To demonstrate the relevance of IL-17, mice were treated with cPTH and an IL-17 antibody (IL-17 Ab). These studies revealed that IL-17 Ab completely prevents the loss of cortical and trabecular bone induced by cPTH by decreasing bone resorption [47].

The fact that silencing of PPR signaling in T cells and osteocytes induces similar bone-sparing effects is in keeping with a "serial circuit" regulatory model, where signals from one population affect the response to cPTH of the other. IL-17A is a probable candidate because it is a potent inducer of RANKL. In support of this hypothesis, neutralization of IL-17, via treatment with IL-17 Ab and deletion of IL-17RA, blocks the capacity of cPTH to increase the production of RANKL by osteocytes and osteoblasts [47] and the bone catabolic activity of cPTH. Therefore, IL-17 mediates the bone catabolic activity of cPTH by upregulating the production of RANKL by osteocytes and osteoblasts (Fig. 128.2).

Although numerous studies have investigated the role of immune cells and cytokines in the mechanism of action of PTH in animal models, little information is available in humans.

To investigate the effects of PHPT on the production of cytokines, unfractionated peripheral blood nucleated cells were obtained from healthy controls and similar subjects affected by PHPT. In PHPT patients, blood samples were obtained before surgery and 1 month after successful resolution of PHPT by parathyroidectomy. These studies revealed that the mRNA levels of IL-17A in peripheral blood nucleated cells were approximately threefold higher in PHPT patients than in healthy controls [47]. Moreover, surgical restoration of normal parathyroid function was associated with the normalization of IL-17A levels. Furthermore, the mRNA levels of the IL-17-inducing transcription factor RORC were approximately threefold higher in PHPT patients before surgery than in healthy controls and parathyroidectomy was

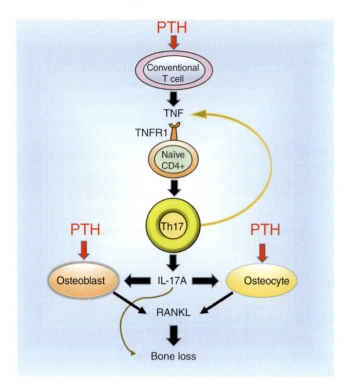

Fig. 128.2. Summary of the role of IL-17 in the bone loss induced by continuous PTH (cPTH) treatment. PTH binds to the PTH receptor PPR expressed in conventional CD4+ and CD8+ T cells and induces the secretion of TNF. This cytokine promotes the differentiation of naïve CD4+ cells into Th17 cells via TNFR1 signaling. Th17 cells release additional TNF, which further stimulates Th17 differentiation. Th17 cells secrete IL-17, which targets osteocytes and osteoblasts, thus increasing their sensitivity to TNF. In the presence of IL-17, PPR activation in osteocytes and osteoblasts stimulates these cells to release RANKL, which stimulates bone resorption and induces bone loss. Silencing of IL-17 or IL-17RA signaling blocks the capacity of cPTH to stimulate the production of RANKL by osteocytes and osteoblasts. Reproduced with permission.

followed by a decrease in RORC mRNA levels. PTH levels were directly correlated with IL-17A level. These findings suggest that increased IL-17A gene expression in PHPT patients is caused by increased levels of circulating PTH.

Role of T cells in the anabolic activity of intermittent PTH (iPTH) treatment

Mice lacking T cells exhibit a blunted increase in bone formation and trabecular bone volume in response to iPTH [3]. With regard to the mechanism by which T cells potentiate the bone anabolic activity of iPTH, it is now clear that in the absence of T cells, iPTH is unable to increase the commitment of stem cells to the osteoblastic lineage, induce osteoblast proliferation and differentiation, and mitigate osteoblast apoptosis. All of these actions of iPTH were found to hinge on the capacity of T cells to activate Wnt signaling in osteoblastic cells [3]. BM CD8+ T cells produce large amounts of the osteogenic Wnt ligand Wnt10b [3]. The pivotal role of T cell-produced Wnt10b was revealed by the dampened effect of iPTH on bone volume in global and T-cell-specific wnt10b null mice [3, 48]. These data indicate that CD8+ T cells potentiate the anabolic activity of PTH by providing Wnt10b, which is a critical Wnt ligand required for activating Wnt signaling in osteoblastic cells. Importantly, a recent report in humans has shown that treatment with teriparatide, a form of iPTH treatment, increases the BM levels of Wnt10b [49]. That study has also shown that T cells represent the main source of Wnt10b in humans treated with teriparatide [49]. By contrast, patients affected by PHPT did not exhibit increased Wnt10b expression [49].

Additional interactions between T cells and SCs may contribute to the anabolic activity of iPTH. An important mediator of T cell–SC interaction is the T-cell costimulatory ligand CD40L [50]. Attesting to relevance, silencing of CD40L blocks the effects of iPTH on SC proliferation, differentiation, and life span, resulting in a blunted anabolic activity of iPTH on trabecular bone. Together, the data suggest a requirement for a dual regulatory interaction between T cells and SCs. According to this model, silencing of either Wnt10b or CD40L is sufficient to blunt the responsiveness of SCs and their osteoblastic progeny to iPTH.

The residual bone anabolic activity of PTH observed in T cell-deficient mice is presumably the result of suppressed production of sclerostin [51]. Indeed, mice treated with anti-sclerostin antibody maintain a partial anabolic response to iPTH. The sclerostin-independent activity of iPTH has shown to be caused completely by increased production of Wnt10b by T cells [52].

In summary, the available data are consistent with a complex modality of action of iPTH that includes suppression of sclerostin production and increased T-cell production of Wnt10b. In conditions of normal baseline bone turnover and bone mass and partial sclerostin blockade that mimics the repressive activity of the hormone on sclerostin production, iPTH stimulates osteoblastogenesis, bone density, and trabecular bone volume independently of sclerostin through a Wnt10b-mediated mechanism.

Osteoblasts, osteocytes, and T cells are all key targets of PTH

Robust evidence shows that silencing of PPR signaling in either osteoblasts, osteocytes, or T cells results in the partial or total blockade of the bone anabolic and catabolic effects of PTH. These findings may sound surprising and perhaps contradictory, if one envisions PTH acting using a "parallel circuit" regulatory modality. In this model, PTH targets osteoblasts, osteocytes, and T cells independently, and interactions between immune cells and bone cells do not play a significant role. This

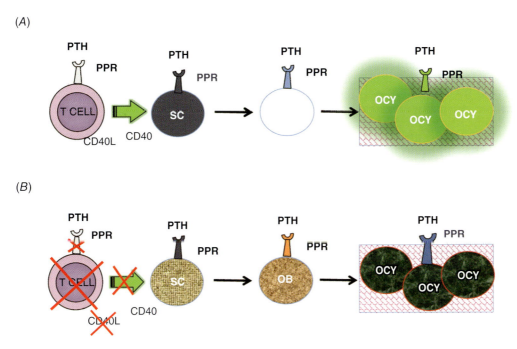

Fig. 128.3. Silencing of PTH/PTH-related protein (PTHrP) receptor (PPR) signaling in either T cells, osteoblasts (OB), or osteocytes (OCY) leads to decreased or absent responses to cPTH and iPTH. This type of response is consistent with a serial circuit regulatory mechanism. SC = stem cell. Reproduced with permission.

model forces investigators to focus on the effects of PTH on one cell population, for example the capacity of the hormone to stimulate the osteocytic production of RANKL, and underestimate the contributions of the effects of PTH on another lineage, for example the capacity to induce T-cell TNF production. However, the fact that silencing of PPR signaling in one lineage (eg, T cells) induces the same effects as silencing of PPR signaling in another (eg, osteocytes) is in keeping with a "serial circuit" regulatory mode (Fig. 128.3). In this model, BM T cells provide cell surface signals and secrete cytokines that direct the differentiation of SCs toward osteoblasts characterized by a high sensitivity to PTH. Conversely, SCs that differentiate in a microenvironment devoid of T cells acquire a permanent phenotype characterized by reduced sensitivity to PTH. Because osteoblasts differentiate into osteocytes, these osteoblasts will give rise to osteocyte populations that are also characterized by diminished responsiveness to PTH. Thus, by acting on SCs and early osteoblast precursors, T cells have the capacity to set the responsiveness of all osteoblast-lineage cells to PTH, including osteocytes. In addition, cytokines secreted by T cells, such as TNF and IL-17, target mature osteoblasts and osteocytes, regulating their production of RANKL [47]. We thus propose the existence of a serial, rather than a parallel, connection between T cells, osteoblasts, and osteocytes with reciprocal, bidirectional regulatory interactions. T cells are "upstream" targets of PTH whereas osteoblasts and osteocytes are "downstream targets." Therefore, T cells, osteoblasts, and osteocytes are all critical targets of PTH.

The role of immune cells in the bone loss associated with HIV infection and antiretroviral therapy

Given the important role of immune cells in regulating basal bone homeostasis through OPG-mediated control of bone resorption [14], damage to the immune system would be predicted to cause a RANKL/OPG imbalance and hence bone loss in the context of immunodeficiency. In fact, clinical studies have reported that bone loss, and osteopenia and osteoporosis, are common in HIV-infected subjects [53], leading to an overall 2% to 4% increase in fracture rate [54]. Hip fracture risk may be increased as much as ninefold [55] and is associated with significant morbidity and a 24% to 32% mortality rate within the first year of fracture.

The factors that contribute to HIV-induced bone loss, however, have been difficult to delineate given that HIV infection leads to AIDS, a complex disease state associated with high incidence of low BMI, muscle wasting, kidney disease, hypogonadism, vitamin D deficiency, and other conditions that are themselves often directly associated with bone disease. Furthermore, high rates of smoking and recreational drug and alcohol use are common and may further contribute to bone loss [56]. However, recent work by our group now demonstrates that, despite this complexity, a fundamental defect in the immuno-skeletal interface (ISI) may contribute significantly to dysregulated bone turnover and bone loss in HIV infection [53, 57]. To control for confounders associated with human populations, we and others have utilized an animal model, the HIV transgenic (Tg) rat. Consistent with human studies, the skeletons of the HIV-Tg rats have significantly denuded BMD, and trabecular and cortical bone structure [58]. Mechanistically, we identified B-cell defects causing a decline in B-cell OPG expression and increase in B-cell RANKL expression. This RANKL/OPG imbalance was consistent with a significant increase in osteoclasts and bone resorption [59].

Importantly, we recently confirmed the existence of this inversion in the B-cell RANKL/OPG ratio in human HIV infection in a translational clinical study involving 58 HIV-negative control and 62 HIV-infected subjects [57]. Importantly, the B-cell RANKL/OPG ratio remained significantly associated with HIV status following multivariable analysis adjusting for multiple covariates with known effect on the skeleton including age, sex, race, BMI, smoking, alcohol consumption, and fracture history. The B-cell RANKL/OPG ratio further correlated significantly with BMD (and with T- and Z-scores) at the TH and FN, but not LS [57]. Although levels of serum cytokines are not always predictive of bone turnover, serum OPG (but not RANKL) has previously been reported to be significantly associated with LS, TH, and FN Z-scores in HIV-infected subjects [60]. Taken together, these studies have revealed an underlying defect in the ISI that may significantly contribute to bone loss and increased fracture risk in HIV-infected subjects.

Antiretroviral therapy-induced inflammation exacerbates HIV-induced bone loss

Another confounder in HIV bone loss is that the antiretroviral drugs used to combat HIV appear to further contribute to skeletal deterioration, with most studies showing a BMD loss of between 2% and 6% within the first 2 years of antiretroviral therapy (ART) initiation [61]. Surprisingly, bone loss seems to occur with all drug classes, although the magnitude of bone loss is variable, and it is now recognized that a significant component of the bone loss induced by ART is independent of ART regimen [62, 63], suggesting that bone cells are regulated by an indirect mechanism. Chronic inflammation has long been associated with bone loss in conditions as diverse as rheumatoid arthritis, periodontal infection, and estrogen deficiency [53]. Because inflammation is often documented in patients on ART [64], we recently hypothesized a common mechanism for ART-induced bone loss involving inflammation associated with the homeostatic reconstitution of T cells and the reactivation of adaptive immunity, following viral suppression by ART. This hypothesis is further consistent with a recent clinical study in which we observed an increase in plasma levels of the osteoclastogenic/inflammatory cytokines RANKL and TNF following ART initiation in some subjects [65]. Bone resorption markers and the magnitude of CD4+ T-cell recovery further correlated significantly with ART, and enhanced bone loss was documented in patients with the lowest CD4 T-cell count at time of ART initiation.

The latter observations are consistent with, and provide a mechanistic explanation for, a previous report of increased bone loss in subjects with low baseline CD4 nadir before ART [66]. Taken together, these observations support the notion of an inflammatory bone loss associated with T-cell repopulation after ART initiation [65].

Recently we developed a murine model to further explore this hypothesis using T cell–deficient mice that were reconstituted with T cells in a syngeneic adoptive transfer [65]. As predicted, homeostatic reconstitution of T cells caused an increase in indices of bone resorption and a dramatic loss of BMD and of trabecular and cortical bone mass, along with production of RANKL and TNF by activated immune cells including T cells, B cells, and macrophages. Taken together, these data support the concept that in addition to any direct effects of ART on bone turnover, a potent indirect mode of bone loss, common to all ART, may drive bone resorption through an inflammatory state arising from T-cell reconstitution and adaptive immune reactivation, in response to viral suppression by ART [65].

Therapies and interventions to prevent bone loss and fracture associated with HIV/ART remain conservative, although published studies reveal efficacy with standard anti-osteoporotic medications such as the bisphosphonate alendronate, with or without vitamin D and calcium supplementation in osteopenic and/or osteoporotic subjects [67] and with zoledronic acid [68]. However, despite enhanced guidelines [69], most subjects are not screened for bone status and/or do not receive interventions at the time of ART initiation to protect from bone loss during this time of intense bone resorption.

We recently proposed a strategy by which all subjects initiating ART would automatically receive a single dose of long-acting bisphosphonate such as zoledronic acid, irrespective of bone status, as a protective measure, and we tested this approach in a double-blind placebo-controlled phase IIb clinical trial. Initial data from baseline through the first 48 weeks reveal complete protection from bone loss at the LS, hip, and FN [70]. Although advanced time points remain to be assessed, this outcome suggests that ART-associated bone loss can indeed be controlled through early pharmacological intervention.

Bone loss in chronic inflammatory states

Clinicians have recognized for many decades that chronic inflammation is associated with bone loss and fractures (Fig. 128.4). Rheumatoid arthritis (RA) is an inflammatory autoimmune disease responsible for joint destruction and crippling disability in approximately 2% of adults. In RA, inflammation in the synovial membrane of affected joints leads to chronic pain and fatigue and ultimately to permanent disability and increased mortality. Furthermore, systemic disruption of bone remodeling leads to loss of bone mass and development of a generalized systemic osteoporosis [71]. RA is one of the best studied disease states

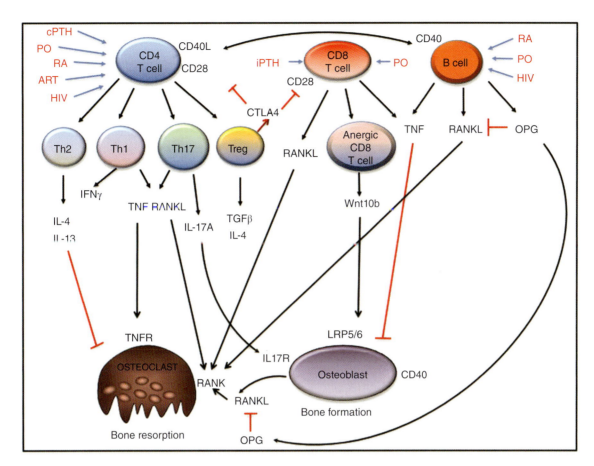

Fig. 128.4. Schematic of immune regulation of bone under different inflammatory states. Multiple pathological events drive bone loss by converging on the adaptive immune response. Rheumatoid arthritis (RA), estrogen deficiency associated with postmenopausal osteoporosis (PO), hyperparathyroidism-induced bone loss (modeled by cPTH treatment), and antiretroviral therapy (ART) all initiate inflammatory states leading to production of inflammatory mediators by T and/or B cells that drive up concentrations of the key osteoclastogenic cytokine RANKL, directly or indirectly. In addition, lymphocytes are a key source of the potent inflammatory cytokine TNF, that promotes RANKL production by osteoblast-lineage cells, suppresses the RANKL decoy receptor osteoprotegerin (OPG), and synergizes with RANKL to upregulate bone resorption and suppress bone formation. Disruption of CD4 T-cell function and costimulation by human immunodeficiency virus (HIV) infection further disrupt T cell to B cell communication leading to B-cell decline in OPG and production by B cells of RANKL. The net effect of these actions is increased osteoclastic bone resorption and bone loss. In addition to these proresorptive effects of the adaptive immune system, iPTH and the CTLA4-induced anergy of CD8 T cells promote Wnt10b production, driving up osteoblast differentiation and bone formation.

involving disruption of the ISI and how chronic T-cell and B-cell activation drives local and systemic bone resorption. Although RANKL is the key osteoclastogenic cytokine and RANKL production by activated T cells has been demonstrated to play significant roles in animal models of RA [72], other inflammatory cytokines likely contribute significantly to both inflammation and bone resorption, including IL-1, IL-6, IL-7, IL-17, and TNF. As with estrogen deficiency, a key protagonist of inflammation and bone loss in RA is TNF. Indeed, TNF overexpression in mice is considered one of the most accurate animal models of human RA [73] and TNF antagonists are routinely used for clinical management of RA in humans. Although current therapies focus on alleviation of inflammatory cascades, the same cytokines are key players in driving bone loss and hence likely reap a double benefit.

Inflammatory bowel diseases, and especially Crohn's disease, are another typical example of inflammatory diseases associated with bone loss. Interestingly, patients with Crohn's disease were recognized to be at risk of osteoporosis because of intermittent steroid use and altered micronutrient absorption. It is now clear that the mechanism is much more complex and involves increased production of inflammatory cytokines (RANKL, TNF, IL-17, IL-6) by activated immune cells in the intestine and systemically. Alterations of the gut microbiota and increased intestinal permeability are features of Crohn's disease, and are responsible for activating the immune system in the intestinal wall. Migration of activated cells and cytokines from the intestinal wall to the BM plays a

Table 128.1. Effects of T and B cells in bone.

Cell Lineage	Activity/Factor Secreted	Effect of OVX	Effect of cPTH	Effect of iPTH	Effect of HIV Infection	Effect of ART
B cells	OPG, RANKL, TNF	Expand B lineage Increase OPG production	?	?	Diminished number Decreased OPG production Increased RANKL production	Increase RANKL and TNF production
CD4+ (Th1)	TNF, RANKL, IFNγ, IL-1, IL-6 Increase bone resorption	Increase number, activation state, and cytokine secretion	Increase TNF production	None	Severely depleted number	Partial Th1 recovery Increase RANKL and TNF production
CD4+ (Th2)	IL-4, IL-13 Decrease bone resorption	None	None	None	Severely depleted number Decreased IL-4	Partial Th2 and IL-4 recovery
CD4+IL-17+ (Th17 cells)	TNF, RANKL, IFNγ, IL-1, IL-6, IL-17 Increase bone resorption	Increase number, activation state, and cytokine secretion	Expand Th17 pool	None	Severely depleted number	Partial Th17 recovery
CD4+ FoxP3+ (Tregs)	IL-10, IL-35, TGFβ Block osteoclast formation Decrease bone resorption	Decrease number and activation state	None	Increase number Stimulate bone formation	Severely depleted number	Partial recovery of Tregs
CD8+	TNF, RANKL, Wnt10b Increase bone resorption and/or bone formation	Increase number, activation state, and TNF production	Increase TNF production	Increase Wnt10b production	Depleted number	Partial CD8 recovery RANKL and TNF production

OVX = ovariectomy; c/iPTH = continuous/intermittent PTH; HIV = human immunodeficiency virus; ART = antiretroviral therapy; OPG = osteoprotegerin.

Table 128.2. Effects of immune cells and immune cell factor modulation on bone.

Mouse Strain	Baseline Phenotype	Effect of OVX	Effect of cPTH	Effect of iPTH
WT mice	Normal	Bone loss	Bone loss	Bone growth
T-cell$^{-/-}$ mice (nude mice)	Low bone mass Increased bone resorption	No bone loss (protection limited to cortical bone in some studies)	No bone loss	Blunted/no bone anabolism
αβ null mice (TCRβ$^{-/-}$ mice)	Low bone mass Increased bone resorption	No bone loss	No bone loss	Blunted bone anabolism
μMT/μMT mice (mature B-cell$^{-/-}$)	Low bone mass Increased bone resorption	Bone loss	?	?
CD40L$^{-/-}$ mice	Low bone mass Increased bone resorption Decreased osteoblast number and life span	No bone loss	No bone loss	Blunted bone anabolism
Wnt10b$^{-/-}$ mice	Very low bone mass. Decreased bone formation	No compensatory increase in bone formation	?	No bone anabolism
IL-17R$^{-/-}$ mice	Low bone mass Increased bone resorption	No bone loss	No bone loss	?
Treg transgenic mice	High bone mass Decreased bone resorption	No bone loss	?	No bone anabolism

OVX = ovariectomy; c/iPTH = continuous/intermittent PTH.

pivotal role in the systemic bone loss associated with inflammatory bowel disease.

A similar paradigm is present in periodontal disease, a condition that is also initiated and aggravated by changes in the oral microbiome that lead to local and systemic inflammation with aberrant production of LPS, RANKL and TNF and other factor responsible for local and systemic bone loss [74].

The effects of T and B cells in bone and of immune cells and immune cell factors are summarized in Tables 128.1 and 128.2 respectively.

REFERENCES

1. Di Rosa F, Pabst R. The bone marrow: a nest for migratory memory T cells. Trends Immunol. 2005;26(7):360–6.
2. D'Amelio P, Grimaldi A, Di Bella S, et al. Estrogen deficiency increases osteoclastogenesis up-regulating T cells activity: a key mechanism in osteoporosis. Bone. 2008;43(1):92–100.
3. Terauchi M, Li JY, Bedi B, et al. T lymphocytes amplify the anabolic activity of parathyroid hormone through Wnt10b signaling. Cell Metab. 2009;10(3):229–40.
4. Roser-Page S, Vikulina T, Zayzafoon M, et al. CTLA-4Ig-induced T cell anergy promotes Wnt-10b production and bone formation in a mouse model. Arthritis Rheumatol. 2014;66(4):990–9.
5. Tyagi AM, Srivastava K, Mansoori MN, et al. Estrogen deficiency induces the differentiation of IL-17 secreting Th17 cells: a new candidate in the pathogenesis of osteoporosis. PLoS One. 2012;7(9):e44552.
6. DeSelm CJ, Takahata Y, Warren J, et al. IL-17 mediates estrogen-deficient osteoporosis in an Act1-dependent manner. J Cell Biochem. 2012;113(9):2895–902.
7. Tyagi AM, Mansoori MN, Srivastava K, et al. Enhanced immunoprotective effects by anti-IL-17 antibody translates to improved skeletal parameters under estrogen deficiency compared with anti-RANKL and anti-TNF-alpha antibodies. J Bone Miner Res. 2014;29(9):1981–92.
8. Molnar I, Bohaty I, Somogyine-Vari E. IL-17A-mediated sRANK ligand elevation involved in postmenopausal osteoporosis. Osteoporos Int. 2014;25(2):783–6.
9. Zaiss MM, Sarter K, Hess A, et al. Increased bone density and resistance to ovariectomy-induced bone loss in FoxP3-transgenic mice based on impaired osteoclast differentiation. Arthritis Rheum. 2010;62(8):2328–38.
10. Tai P, Wang J, Jin H, et al. Induction of regulatory T cells by physiological level estrogen. J Cell Physiol. 2008;214(2):456–64.
11. Wei S, Wang MW, Teitelbaum SL, et al. Interleukin-4 reversibly inhibits osteoclastogenesis via inhibition of NF-kappa B and mitogen-activated protein kinase signaling. J Biol Chem. 2002;277(8):6622–30.
12. Pacifici R. Osteoimmunology and its implications for transplantation. Am J Transplant. 2013;13(9):2245–54.
13. Lopez-Granados E, Temmerman ST, Wu L, et al. Osteopenia in X-linked hyper-IgM syndrome reveals a regulatory role for CD40 ligand in osteoclastogenesis. Proc Natl Acad Sci U S A. 2007;104(12):5056–61.
14. Li Y, Toraldo G, Li A, et al. B cells and T cells are critical for the preservation of bone homeostasis and attainment of peak bone mass in vivo. Blood. 2007;109(9):3839–48.
15. Gao Y, Wu X, Terauchi M, et al. T cells potentiate PTH-induced cortical bone loss through CD40L signaling. Cell Metab. 2008;8(2):132–45.
16. Li JY, Tawfeek H, Bedi B, et al. Ovariectomy disregulates osteoblast and osteoclast formation through the T-cell receptor CD40 ligand. Proc Natl Acad Sci U S A. 2011;108(2):768–73.
17. Yun TJ, Chaudhary PM, Shu GL, et al. OPG/FDCR-1, a TNF receptor family member, is expressed in lymphoid cells and is up-regulated by ligating CD40. J Immunol. 1998;161(11):6113–21.
18. Li Y, Li A, Yang X, et al. Ovariectomy-induced bone loss occurs independently of B cells. J Cell Biochem. 2007;100(6):1370–5.
19. Onal M, Xiong J, Chen X, et al. Receptor activator of nuclear factor kappaB ligand (RANKL) protein expression by B lymphocytes contributes to ovariectomy-induced bone loss. J Biol Chem. 2012;287(35):29851–60.
20. Simonet WS, Lacey DL, Dunstan CR, et al. Osteoprotegerin: a novel secreted protein involved in the regulation of bone density. Cell. 1997;89(2):309–19.
21. Lacey DL, Timms E, Tan HL, et al. Osteoprotegerin ligand is a cytokine that regulates osteoclast differentiation and activation. Cell. 1998;93:165–76.
22. Weitzmann MN, Pacifici R. Estrogen deficiency and bone loss: an inflammatory tale. J Clin Invest. 2006;116(5):1186–94.
23. Khosla S, Pacifici R. Estrogen deficiency, postmenopausal osteoporosis, and age-related bone loss. In: Marcus R, Feldman D, Dempster DW, et al. (eds) Osteoporosis. 2. Amsterdam: Elsevier, 2013, pp 1113–38.
24. Eghbali-Fatourechi G, Khosla S, Sanyal A, et al. Role of RANK ligand in mediating increased bone resorption in early postmenopausal women. J Clin Invest. 2003;111(8):1221–30.
25. Charatcharoenwitthaya N, Khosla S, Atkinson EJ, et al. Effect of blockade of TNF-alpha and interleukin-1 action on bone resorption in early postmenopausal women. J Bone Miner Res. 2007;22(5):724–9.
26. Zhang J, Fu Q, Ren Z, et al. Changes of serum cytokines-related Th1/Th2/Th17 concentration in patients with postmenopausal osteoporosis. Gynecol Endocrinol. 2015;31(3):183–90.
27. Pacifici R. Role of T cells in ovariectomy induced bone loss-revisited. J Bone Miner Res. 2012;27(2):231–9.
28. Lelu K, Laffont S, Delpy L, et al. Estrogen receptor alpha signaling in T lymphocytes is required for estradiol-mediated inhibition of Th1 and Th17 cell differentiation and protection against experimental autoimmune encephalomyelitis. J Immunol. 2011;187(5):2386–93.
29. Cenci S, Weitzmann MN, Roggia C, et al. Estrogen deficiency induces bone loss by enhancing T-cell production of TNF-alpha. J Clin Invest. 2000;106(10):1229–37.

30. Roggia C, Gao Y, Cenci S, et al. Up-regulation of TNF-producing T cells in the bone marrow: A key mechanism by which estrogen deficiency induces bone loss in vivo. Proc Natl Acad Sci U S A. 2001;98(24):13960–5.
31. Grassi F, Tell G, Robbie-Ryan M, et al. Oxidative stress causes bone loss in estrogen-deficient mice through enhanced bone marrow dendritic cell activation. Proc Natl Acad Sci U S A. 2007;104(38):15087–92.
32. Yamaza T, Miura Y, Bi Y, et al. Pharmacologic stem cell based intervention as a new approach to osteoporosis treatment in rodents. PLoS One. 2008;3(7):e2615.
33. Tyagi AM, Srivastava K, Kureel J, et al. Premature T cell senescence in Ovx mice is inhibited by repletion of estrogen and medicarpin: a possible mechanism for alleviating bone loss. Osteoporos Int. 2012;23(3):1151–61.
34. Sass DA, Liss T, Bowman AR, et al. The role of the T-lymphocyte in estrogen deficiency osteopenia. J Bone Miner Res. 1997;12(3):479–86.
35. Lee SK, Kadono Y, Okada F, et al. T lymphocyte-deficient mice lose trabecular bone mass with ovariectomy. J Bone Miner Res. 2006;21(11):1704–12.
36. Cenci S, Toraldo G, Weitzmann MN, et al. Estrogen deficiency induces bone loss by increasing T cell proliferation and lifespan through IFN-gamma-induced class II transactivator. Proc Natl Acad Sci U S A. 2003;100(18):10405–10.
37. Li JY, Chassaing B, Tyagi AM, et al. Sex steroid deficiency-associated bone loss is microbiota dependent and prevented by probiotics. J Clin Invest. 2016;126(6):2049–63.
38. Ohlsson C, Engdahl C, Fak F, et al. Probiotics protect mice from ovariectomy-induced cortical bone loss. PLoS One. 2014;9(3):e92368.
39. Britton RA, Irwin R, Quach D, et al. Probiotic L. reuteri treatment prevents bone loss in a menopausal ovariectomized mouse model. J Cell Physiol. 2014;229(11):1822–30.
40. Potts J. Primary hyperparathyroidism. In: Avioli LV, Krane S (eds). Metabolic Bone Diseases. 1. (3rd ed.). San Diego: Academic Press, 1998, pp 411–42.
41. Parisien M, Dempster DW, Shane E, et al. Histomorphometric analysis of bone in primary hyperparathyroidism. In: Bilezikian JP, Levine RMM (eds) The Parathyroids. Basic and Clinical Concepts (2nd ed.). San Diego: Academic Press, 2001, pp 423–36.
42. Weir EC, Lowik CW, Paliwal I, et al. Colony stimulating factor-1 plays a role in osteoclast formation and function in bone resorption induced by parathyroid hormone and parathyroid hormone-related protein. J Bone Miner Res. 1996;11(10):1474–81.
43. Powell WF, Jr, Barry KJ, Tulum I, et al. Targeted ablation of the PTH/PTHrP receptor in osteocytes impairs bone structure and homeostatic calcemic responses. J Endocrinol. 2011;209(1):21–32.
44. Saini V, Marengi DA, Barry KJ, et al. Parathyroid hormone (PTH)/PTH-related peptide type 1 receptor (PPR) signaling in osteocytes regulates anabolic and catabolic skeletal responses to PTH. J Biol Chem. 2013;288(28):20122–34.
45. Xiong J, Piemontese M, Thostenson JD, et al. Osteocyte-derived RANKL is a critical mediator of the increased bone resorption caused by dietary calcium deficiency. Bone. 2014;66C:146–54.
46. Tawfeek H, Bedi B, Li JY, et al. Disruption of PTH Receptor 1 in T Cells Protects against PTH-Induced Bone Loss. PLoS One. 2010;5(8):e12290.
47. Li JY, D'Amelio P, Robinson J, et al. IL-17A Is Increased in Humans with Primary Hyperparathyroidism and Mediates PTH-Induced Bone Loss in Mice. Cell Metab. 2015;22(5):799–810.
48. Bedi B, Li JY, Tawfeek H, et al. Silencing of parathyroid hormone (PTH) receptor 1 in T cells blunts the bone anabolic activity of PTH. Proc Natl Acad Sci U S A. 2012;109(12):E725–33.
49. D'Amelio P, Sassi F, Buondonno I, et al. Treatment with intermittent PTH increases Wnt10b production by T cells in osteoporotic patients. Osteoporos Int. 2015;26(12):2785–91.
50. Grewal IS, Flavell RA. CD40 and CD154 in cell-mediated immunity. Annu Rev Immunol. 1998;16:111–35.
51. Keller H, Kneissel M. SOST is a target gene for PTH in bone. Bone. 2005;37(2):148–58.
52. Li JY, Walker LD, Tyagi AM, et al. The sclerostin-independent bone anabolic activity of intermittent PTH treatment is mediated by T-cell-produced Wnt10b. J Bone Miner Res. 2014;29(1):43–54.
53. Weitzmann MN, Ofotokun I. Physiological and pathophysiological bone turnover - role of the immune system. Nat Rev Endocrinol. 2016;12(9):518–32.
54. Triant VA, Brown TT, Lee H, et al. Fracture prevalence among human immunodeficiency virus (HIV)-infected versus non-HIV-infected patients in a large U.S. healthcare system. J Clin Endocrinol Metab. 2008;93(9):3499–504.
55. Prieto-Alhambra D, Guerri-Fernandez R, De Vries F, et al. HIV Infection and Its Association With an Excess Risk of Clinical Fractures: A Nationwide Case-Control Study. J Acquir Immune Defic Syndr. 2014;66(1):90–5.
56. Ofotokun I, Weitzmann MN. HIV and bone metabolism. Discov Med. 2011;11(60):385–93.
57. Titanji K, Vunnava A, Sheth AN, et al. Dysregulated B cell expression of RANKL and OPG correlates with loss of bone mineral density in HIV infection. PLoS Pathog. 2014;10(10):e1004497.
58. Lafferty MK, Fantry L, Bryant J, et al. Elevated suppressor of cytokine signaling-1 (SOCS-1): a mechanism for dysregulated osteoclastogenesis in HIV transgenic rats. Pathog Dis. 2014;71(1):81–9.
59. Vikulina T, Fan X, Yamaguchi M, et al. Alterations in the immuno-skeletal interface drive bone destruction in HIV-1 transgenic rats. Proc Natl Acad Sci U S A. 2010;107(31):13848–53.
60. Brown TT, Chen Y, Currier JS, et al. Body composition, soluble markers of inflammation, and bone mineral density in antiretroviral therapy-naive HIV-1-infected individuals. J Acquir Immune Defic Syndr. 2013;63(3):323–30.

61. McComsey GA, Tebas P, Shane E, et al. Bone disease in HIV infection: a practical review and recommendations for HIV care providers. Clin Infect Dis. 2010;51(8):937–46.
62. Brown TT, McComsey GA, King MS, et al. Loss of bone mineral density after antiretroviral therapy initiation, independent of antiretroviral regimen. J Acquir Immune Defic Syndr. 2009;51(5):554–61.
63. Piso RJ, Rothen M, Rothen JP, et al. Markers of bone turnover are elevated in patients with antiretroviral treatment independent of the substance used. J Acquir Immune Defic Syndr. 2011;56(4):320–4.
64. Deeks SG, Tracy R, Douek DC. Systemic effects of inflammation on health during chronic HIV infection. Immunity. 2013;39(4):633–45.
65. Ofotokun I, Titanji K, Vikulina T, et al. Role of T-cell reconstitution in HIV-1 antiretroviral therapy-induced bone loss. Nat Commun. 2015;6:8282.
66. Grant PM, Kitch D, McComsey GA, et al. Low Baseline CD4+ Count Is Associated With Greater Bone Mineral Density Loss After Antiretroviral Therapy Initiation. Clin Infect Dis. 2013;57(10):1483–8.
67. McComsey GA, Kendall MA, Tebas P, et al. Alendronate with calcium and vitamin D supplementation is safe and effective for the treatment of decreased bone mineral density in HIV. Aids. 2007;21(18):2473–82.
68. Bolland MJ, Grey A, Horne AM, et al. Effects of intravenous zoledronate on bone turnover and bone density persist for at least five years in HIV-infected men. J Clin Endocrinol Metab. 2012;97(6):1922–8.
69. Brown TT, Hoy J, Borderi M, et al. Recommendations for evaluation and management of bone disease in HIV. Clin Infect Dis. 2015;60(8):1242–51.
70. Ofotokun I, Titanji K, Lahiri CD, et al. A Single-dose Zoledronic Acid Infusion Prevents Antiretroviral Therapy-induced Bone Loss in Treatment-naive HIV-infected Patients: A Phase IIb Trial. Clin Infect Dis. 2016;63(5):663–71.
71. Weitzmann MN, Ofotokun I. Physiological and pathophysiological bone turnover - role of the immune system. Nat Rev Endocrinol. 2016;12(9):518–32.
72. Kong YY, Feige U, Sarosi I, et al. Activated T cells regulate bone loss and joint destruction in adjuvant arthritis through osteoprotegerin ligand. Nature. 1999;402(6759):304–9.
73. Li P, Schwarz EM. The TNF-alpha transgenic mouse model of inflammatory arthritis. Springer Semin Immunopathol. 2003;25(1):19–33.
74. Hienz SA, Paliwal S, Ivanovski S. Mechanisms of Bone Resorption in Periodontitis. J Immunol Res. 2015;2015:615486.

129
Cellular Bioenergetics of Bone

Wen-Chih Lee[1] and Fanxin Long[1,2]

[1]*Department of Orthopedic Surgery, Washington University School of Medicine, St Louis, MO, USA*
[2]*Departments of Medicine and Developmental Biology, Washington University School of Medicine, St Louis, MO, USA*

INTRODUCTION

Bone remodeling occurs asynchronously throughout the skeleton at anatomically distinct sites and replaces ~10% of the adult human skeleton each year [1]. The remodeling cycle begins with osteoclasts pumping molar quantities of hydrochloric acid to dissolve the bone minerals, and this process requires adenosine triphosphate (ATP) hydrolysis by vacuolar H$^+$ (V)-ATPase [2]. On the other hand, osteoblasts build bone mass by synthesizing and secreting bone matrix proteins [3]. Thus, both osteoclasts and osteoblasts consume large amounts of energy to sustain their normal function. Elucidating the bioenergetic mechanisms in the bone cells is not only essential for basic bone biology but may also open new avenues for developing bone therapies.

GLUCOSE METABOLISM IN OSTEOBLASTS

Glucose is a major nutrient for both energy production and biomass synthesis in mammalian cells. In most cell types, glucose transportation across the plasma membrane is mediated by the Glut family of facilitative transporters including 14 isoforms (also known as solute carrier family 2A) [4]. The Glut proteins transport glucose along a concentration gradient without a requirement for energy. Once inside the cell, glucose is phosphorylated by members of the hexokinase family to glucose-6-phosphate (G6P), which is then either converted to glycogen for storage or metabolized further to release energy and to produce building blocks for biosynthesis. A majority of G6P in most cell types enters the core glycolysis pathway to generate pyruvate that is then further metabolized through either the tricarboxylic acid (TCA) cycle or lactate production. Complete oxidation of pyruvate through the TCA cycle coupled with oxidative phosphorylation (OXPHOS) extracts considerably more energy from glucose than does the conversion to lactate (>30 versus 2 ATP per glucose molecule). On the other hand, the lactate pathway consumes glucose at a faster pace and produces energy without the need for oxygen. Besides the core glycolysis pathway, a number of glycolytic intermediates can be diverted for further metabolism through other mechanisms without direct production of ATP. These include shunting of G6P through the pentose phosphate pathway (PPP) critical for nucleotide and lipid synthesis, and conversion of fructose-6-P for protein glycosylation via the hexosamine biosynthetic pathway (HBP) [5]. In addition, 3-P-glycerate can be used for de novo synthesis of serine and glycine, whereas glyceraldehyde-3-P is a precursor for glycerol, which forms the backbone of triglycerides and phospholipids (Fig. 129.1). The distribution of glucose among the various metabolic fates is dictated by the specific energy and biosynthesis requirements of each cell.

Glucose is a major nutrient for osteoblasts. A series of studies in the early 1960s demonstrated that bone slices in culture consumed glucose at a brisk rate [6, 7]. Similar observations were made with primary cultures of calvarial osteoblasts [8]. Recently, both direct measurements and PET/CT imaging of radiolabeled glucose analogs confirmed a significant uptake of glucose by bone in the mouse [9, 10]. Glucose uptake in the osteoblast lineage is mainly mediated by the Glut family of transporters. Expression of both Glut1 and Glut3 was detected in rodent osteoblastic cells (PyMS) or the osteosarcoma cell line UMR 106-01 [11–13]. Glut1 mRNA was recently shown to express at a much higher level than Glut3 in primary osteoblast cultures, and selective deletion of

---

Primer on the Metabolic Bone Diseases and Disorders of Mineral Metabolism, Ninth Edition. Edited by John P. Bilezikian.
© 2019 American Society for Bone and Mineral Research. Published 2019 by John Wiley & Sons, Inc.
Companion website: www.wiley.com/go/asbmrprimer

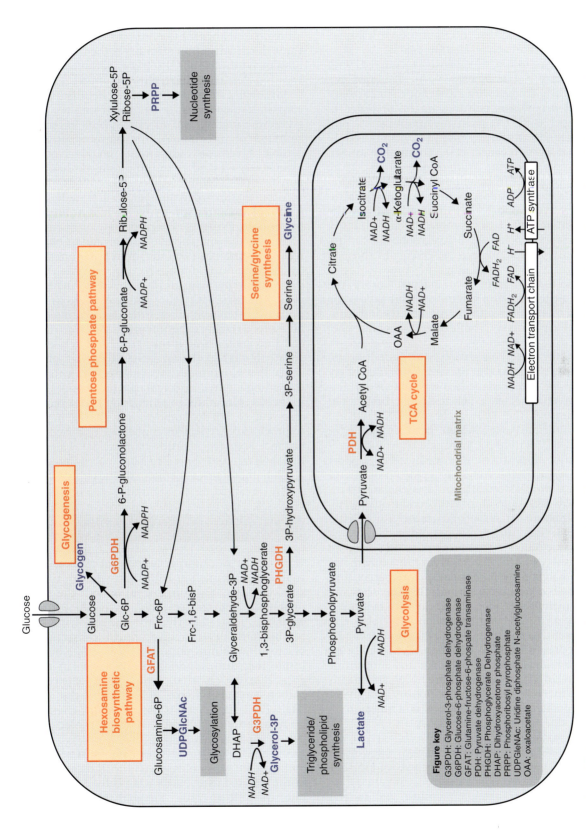

Fig. 129.1. Metabolic fates of glucose in mammalian cells. Principal metabolites from glucose are shown in blue, and certain key enzymes in red. Note that many reactions of the TCA cycle are reversible but are depicted here as unidirectional for simplicity.

Glut1 in osteoblast precursors suppressed osteoblast differentiation [9]. In addition, although normally detected at very low levels in neonatal calvarial osteoblasts, Glut4 mRNA was shown to increase after culture with β-glycerophosphate and ascorbate, or insulin [14].

The brisk consumption of glucose is coupled with rapid production of lactate by osteoblasts. This was evident from the early studies with either bone slices or primary calvarial osteoblasts cultured in the presence of abundant oxygen [6–8]. The phenomenon of lactate production from glucose in the presence of oxygen is akin to that observed in certain cancer cells, commonly known as aerobic glycolysis or the Warburg effect [15]. More recent studies have confirmed aerobic glycolysis as a predominant mode of glucose metabolism in primary calvarial osteoblasts even though OXPHOS increases as the cells further differentiate to form mineralized nodules in response to ascorbic acid and β-glycerophosphate [16, 17]. Functionally, mimicking aerobic glycolysis through stabilization of Hif-1α signaling in preosteoblasts increases bone formation in vivo independent of the increase in angiogenesis [18]. Overall, aerobic glycolysis is a prominent metabolic feature of osteoblasts and stimulation of aerobic glycolysis promotes the osteoblast phenotype [19].

The mechanism for aerobic glycolysis to benefit osteoblasts is not fully understood at present. From the energy viewpoint, aerobic glycolysis is less efficient in extracting ATP from glucose on a per molecule basis than metabolism through the TCA cycle and OXPHOS. In cancer cells, multiple benefits have been proposed for aerobic glycolysis, including a rapid rate of ATP production, provision of biosynthetic intermediates in support of cell proliferation, establishment of a pro-tumor microenvironment, as well as modulation of signaling transduction through ROS and chromatin modification [20, 21]. Although mature osteoblasts generally exhibit little proliferation in vivo, they produce and secrete large amounts of extracellular matrix proteins [3, 22]. Aerobic glycolysis therefore may be necessary for providing metabolic intermediates as building blocks for matrix proteins. This view is consistent with an early study showing significant contributions of glucose carbons to amino acids in collagen [23]. In addition, enhanced aerobic glycolysis was recently shown to reduce nuclear acetyl-coA levels and suppress histone acetylation, hence favoring osteoblast differentiation from a bipotential progenitor cell population [24]. Further studies are necessary to elucidate the full mechanism through which aerobic glycolysis promotes osteoblast differentiation and function.

GLUTAMINE METABOLISM IN OSTEOBLASTS

Glutamine is the most abundant amino acid in the plasma, normally present at a concentration of 500 to 750 μM in humans [25]. Besides being a direct building block of proteins, glutamine is an important energy source and an essential carbon and nitrogen donor for the synthesis of amino acids, nucleotides, glutathione, and hexosamine [26–28] (Fig. 129.2). In addition, glutamine efflux in exchange for the import of essential amino acids through the antiporter stimulates the master regulator of protein synthesis mTORC1 [29]. Although it can be synthesized in mammalian cells, glutamine is classified as a conditional essential amino acid because demand may surpass its synthesis. It has been long recognized that glutamine metabolism in cancer cells exceeds the consumption of other nonessential amino acids [30, 31]. Like aerobic glycolysis, glutamine metabolism has been extensively studied as an important mechanism for supporting cancer cell growth [32, 33].

Although limited in number, several studies have begun to reveal important roles for glutamine metabolism in osteoblasts. Explants of calvaria and long bones exhibited active uptake and metabolism of glutamine [34]. More recently, glutamine was shown to be required in calvarial osteoblast cultures for matrix mineralization [35]. In addition, glutamine consumption by bone marrow stromal cells decreases with aging and appears to associate with impaired osteoblast differentiation [36]. Importantly, stable isotope tracing experiments in a bone marrow stromal cell line demonstrated that glutamine was converted to citrate through oxidation in the TCA cycle, providing evidence that glutamine contributes to energy production in the mitochondria in osteoblast precursors [37]. Finally, increased glutathione production from glutamine in response to Hif-1α stabilization has been proposed to improve the survival of implanted osteogenic cells and enhance bone regeneration in mice [38]. Thus, glutamine metabolism contributes to both energy production and redox homeostasis in osteoblast-lineage cells.

FATTY ACID METABOLISM IN OSTEOBLASTS

Fatty acids are another important carbon and energy source in mammalian cells. Fatty acids can be synthesized de novo or acquired from outside the cell through membrane transporters [39]. To produce energy, fatty acids are first transported into the mitochondrial matrix in the form of acyl-coA through consecutive actions of carnitine palmitoyltransferase 1 (CPT1) and 2 (CPT2). Inside the mitochondrial matrix, acyl-coA then undergoes β-oxidation that sequentially cleaves off two carbons as acetyl-coA that enters the TCA cycle (Fig. 129.3). Complete oxidation of long-chain fatty acids by β-oxidation yields more ATP per molecule than glucose or amino acids.

A number of studies have demonstrated fatty acid metabolism by osteoblasts in vitro. Rat calvarial explants or the isolated osteoblasts were shown to oxidize palmitate in culture [40]. Supplementation of carnitine increased palmitate oxidation and osteoblast markers in primary cultures of porcine osteoblasts [41]. Similarly, carnitine stimulated both proliferation and activity of

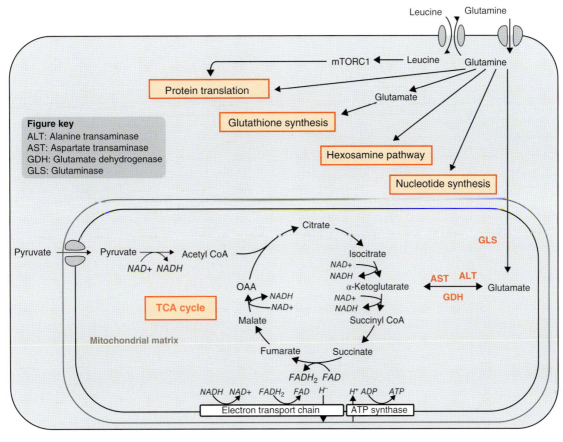

Fig. 129.2. Multiple roles of glutamine in mammalian cells. Red boxes denote major metabolic pathways. Certain key enzymes are shown in red.

human osteoblasts in culture [42]. In addition, lipid supplementation of serum-free medium was found to support proliferation of calvarial osteoblasts, even though the mechanism for the proliferative effect was not explored [43]. More recently, palmitate oxidation was observed to increase during further differentiation of murine calvarial osteoblasts in the presence of β-glycerophosphate and ascorbic acid [44]. However, genetic studies are necessary to determine the physiological importance of fatty acid oxidation in osteoblast differentiation and function in vivo.

REGULATION OF OSTEOBLAST METABOLISM BY BONE ANABOLIC SIGNALS

Wnt signaling, a major mechanism for stimulating bone accrual in both mice and humans, has been shown to promote aerobic glycolysis during osteoblast differentiation [45]. In particular, Wnt3a, Wnt7b, and Wnt10b, which are known to promote osteoblast differentiation, all stimulate glucose consumption and lactate production in the bone marrow stromal cell line ST2 cells. However, Wnt5a does not induce either osteoblastogenesis or glycolysis in those cells. Mechanistically, Wnt3a signals through mTORC2 and Akt to increase the protein levels of a number of metabolic enzymes including Hk2, Pfk1, Pfkfb3, and Ldha without changes in their mRNA abundance. In addition, Wnt3a increases the protein level of Pdk1, a negative regulator of pyruvate dehydrogenase activity, and therefore reduces the flux of glucose-derived pyruvate entering the TCA cycle [24]. Consistent with in vitro observations, bone protein extracts from mice either lacking the Wnt coreceptor Lrp5 or expressing a hyperactive mutant *Lrp5* allele showed a lower or higher level of glycolytic enzymes, respectively, than the normal control [45]. Thus, bone anabolic Wnt signaling directly stimulates aerobic glycolysis, but future experiments are necessary to elucidate the contribution of metabolic reprogramming to Wnt-induced bone accrual in vivo.

Wnt signaling also promotes glutamine metabolism during osteoblast differentiation. In ST2 cells, Wnt3a increases glutamine oxidation in the TCA cycle by rapidly increasing glutaminase (Gls) level and activity downstream of mTORC1 [37]. Strikingly, the increase in glutamine catabolism leads to activation of Gcn2 by lowering intracellular glutamine levels. This in turn induces Atf4 that activates transcription of genes responsible for amino acid uptake or synthesis to promote protein synthesis. Importantly, pharmacological inhibition of Gls with BPTES ameliorates the excessive bone formation

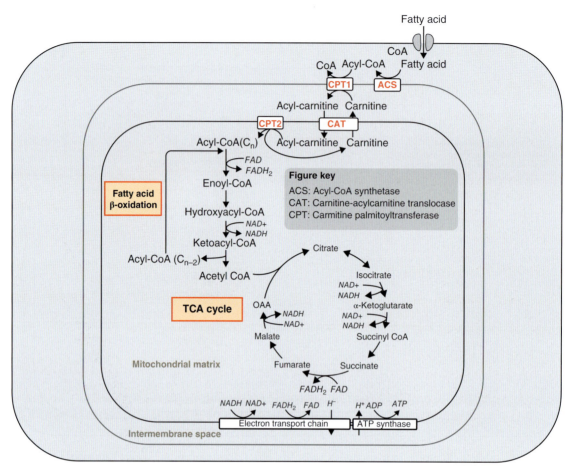

Fig. 129.3. A schematic for fatty acid oxidation in mammalian cells. Red boxes denote major metabolic pathways. Certain key enzymes are shown in red.

caused by hyperactive Wnt signaling in mice expressing a mutant *Lrp5* allele [46]. Although BPTES did not affect bone mass under basal conditions in those experiments, future genetic studies are warranted to determine the potential contribution of glutamine catabolism to normal bone formation in vivo.

In addition, Wnt has been shown to stimulate fatty acid metabolism in osteoblasts. Murine calvarial osteoblasts lacking Lrp5 express lower levels of fatty acid metabolism genes and exhibit less oxidation of oleate in vitro [44]. Conversely, Wnt activation in calvarial osteoblasts through either expression of a hyperactive mutant *Lrp5* allele or stimulation with Wnt10b increases the expression of fatty acid metabolism genes as well as oleate oxidation. Moreover, Gsk3β inhibition by LiCl or β-catenin overexpression exhibits a similar effect, thus implicating β-catenin as a potential mediator for Wnt signaling in promoting fatty acid oxidation. Intriguingly, however, deletion of Lrp6 in osteoblasts did not have the same effect on fatty acid metabolism in that study, even though Lrp6 has been shown to activate β-catenin signaling more effectively than Lrp5 [47]. The physiological function of fatty acid oxidation in bone formation also remains to be demonstrated in vivo. Nonetheless, the evidence to date supports the view that bone anabolic Wnt signaling directly reprograms cellular metabolism by stimulating energy production from fatty acids and glutamine in the mitochondria, while promoting aerobic glycolysis that may be necessary for providing biosynthetic intermediates.

Intermittent use of PTH is well known to stimulate bone formation and has been an effective anabolic therapy for osteoporosis [48]. Interestingly, long before its bone anabolic use, PTH was shown to stimulate aerobic glycolysis in long bone or calvarium explants, as well as in isolated calvarial osteoblasts [49–52]. Similarly, PTH induced glucose uptake in PyMS [13]. More recently, studies in MC3T3-E1 have uncovered a main mechanism through which PTH stimulates aerobic glycolysis [53]. Specifically, PTH induces Igf signaling which in turn activates the PI3K-mTORC2 cascade, resulting in upregulation of metabolic enzymes such as Hk2, Ldha, and Pdk1. Interestingly, although PTH reduces the flux of glucose-derived pyruvate into the TCA cycle, it increases the oxygen consumption rate in Seahorse assays, indicating an increased use of alternative substrates fuelling mitochondrial OXPHOS in response to PTH. The alternative fuel source however remains to be identified. Regardless, pharmacological suppression of glycolysis with the Pdk1 inhibitor dichloroacetate (DCA) diminishes the bone anabolic effect of intermittent PTH in the mouse. Thus, PTH stimulates bone formation in part through activation of aerobic glycolysis downstream of Igf signaling.

BIOENERGETICS IN OSTEOCLASTS

During bone resorption, osteoclasts secrete large amounts of protons and proteolytic enzymes to remove both minerals and the matrix proteins. The high energy demand is likely met by robust OXPHOS because osteoclasts are known to contain abundant mitochondria. Deletion of the mitochondrial transcription factor A (Tfam) in the mouse with cathepsin K-Cre resulted in lower intracellular ATP levels in osteoclasts and accelerated their apoptosis [54]. Moreover, disruption of mitochondrial complex I through deletion of Ndufs4 impaired osteoclast differentiation and function resulting in osteopetrosis in the mouse [55]. During osteoclast differentiation, Rankl has been shown to stimulate mitochondrial biogenesis involving upregulation of Ppargc1b [56]. However, the mechanism for mitochondrial biogenesis during osteoclastogenesis is likely complex, as alternative NF-κB signaling via RelB and NIK is required downstream of Rankl but this requirement cannot be obviated by Ppargc1b overexpression [57]. Confirming the energy requirement of osteoclast differentiation, mTOR inhibition or AMPK activation, both mimicking energy deficiency, inhibited osteoclast formation [58]. Conversely, AMPKa1 or AMPKa2 deletion, presumably mirroring energy-replete states, increased osteoclast differentiation [59].

The nutrient source for energy production in osteoclasts remains to be fully elucidated. Studies with chicken osteoclasts have demonstrated that glucose, instead of fatty acids or ketone bodies, is the principal energy source for bone resorption [60]. In those studies, resorption increased with glucose concentration but plateaued at 7 through 25 mM. Also in chicken osteoclasts, glucose was shown to stimulate both protein and mRNA levels of V-ATPase, likely through transcriptional regulation involving p38 MAPK [61]. In keeping with the chicken studies, Rankl-induced osteoclast differentiation from either RAW264.7 cells or murine bone marrow macrophages was coupled with increased glycolysis and osteoclast differentiation was maximal at 5 mM but decreased significantly at 20 mM [62]. Moreover, Glut1 and glycolytic enzymes increased during osteoclastogenesis in response to Rankl, and deprivation of glucose inhibited osteoclast differentiation [58]. Glycolysis here may be driven by lactate production because Ldhb increased during Rankl-induced osteoclast differentiation, and deletion of Ldha or Ldhb suppressed osteoclastogenesis [63]. In addition, the glutamine transporter Slc1a5 and glutaminase 1 increased during osteoclast differentiation and pharmacological inhibition of glutamine metabolism diminished osteoclastogenesis [58]. Thus, both glucose and glutamine appear to contribute to energy production in osteoclasts.

CONCLUSION

Energy is a common requirement for all cells, but the bioenergetic pathways employed by various cell types likely differ depending on their biological functions. Although both osteoblasts and osteoclasts are expected to consume large amounts of energy, a major function of osteoblasts is to synthesize collagen whereas osteoclasts specialize in pumping acids to the extracellular milieu. Our knowledge about how the bone cells integrate energy production and their specific function is rudimentary at present. However, the data so far indicate that osteoblasts metabolize glucose mainly through aerobic glycolysis, a less efficient but more rapid mechanism than OXPHOS for energy production from glucose. The predominance of aerobic glycolysis in the face of high energy demands creates an apparent "energy paradox" for osteoblasts. It is possible that the brisk rate of glucose consumption provides sufficient energy for collagen production. Alternatively, fatty acids and amino acids such as glutamine may provide additional energy through OXPHOS. Beyond energy production, aerobic glycolysis may be necessary for generating metabolic intermediates from glucose to support the synthesis of bone matrix proteins. Similar to osteoblasts, osteoclasts also use glucose as a major energy source. Unlike osteoblasts, however, osteoclasts appear to derive energy mainly through OXPHOS although some lactate production may still be necessary for sustaining a proper glycolytic flux feeding into the TCA cycle. Further investigation is necessary to clarify the precise contributions of glycolysis versus OXPHOS to bioenergetics in both osteoblasts and osteoclasts.

Other aspects of bioenergetics in bone cells also warrant further attention. Compared to glucose, fatty acids and amino acids as energy sources have been understudied, in part because of their greater complexity. How the bone cells achieve their unique metabolic features during differentiation is also just beginning to be explored. In addition, advances in imaging techniques and in situ metabolomics are necessary to achieve optimal resolution and sensitivity for studies of bone cell metabolism in vivo [64]. Finally, illuminating the relationship between dysregulation of bioenergetics and aging or diabetes may reveal new opportunities for combating bone fragility associated with those conditions.

REFERENCES

1. Sims NA, Martin, TJ. Coupling the activities of bone formation and resorption: a multitude of signals within the basic multicellular unit. BoneKEy Reports. 2014;3:481.

2. Blair HC, Teitelbaum SL, Ghiselli R, et al. Osteoclastic bone resorption by a polarized vacuolar proton pump. Science. 1989;245(4920):855–7.
3. Long F. Building strong bones: molecular regulation of the osteoblast lineage. Nature reviews. Mol Cell Biol. 2012; 13(1):27–38.
4. Augustin R. The protein family of glucose transport facilitators: It's not only about glucose after all. IUBMB Life. 2010;62(5):315–33.
5. Bouche C, Serdy S, Kahn CR, et al. The cellular fate of glucose and its relevance in type 2 diabetes. Endocr Rev. 2004;25(5):807–30.
6. Borle AB, Nichols N, Nichols, Jr G. Metabolic studies of bone in vitro. I. Normal bone. J Biol Chem. 1960;235:1206–10.
7. Cohn DV, Forscher BK. Aerobic metabolism of glucose by bone. J Biol Chem. 1962;237:615–8.
8. Peck WA, Birge, Jr SJ, Fedak SA. Bone Cells: Biochemical and Biological Studies after Enzymatic Isolation. Science. 1964;146(3650):1476–7.
9. Wei J, Shimazu J, Makinistoglu MP, et al. Glucose Uptake and Runx2 Synergize to Orchestrate Osteoblast Differentiation and Bone Formation. Cell. 2015;161(7): 1576–91.
10. Zoch ML, Abou DS, Clemens TL, et al. In vivo radiometric analysis of glucose uptake and distribution in mouse bone. Bone Res.2016;4:16004.
11. Thomas DM, Rogers SD, Ng KW, et al. Dexamethasone modulates insulin receptor expression and subcellular distribution of the glucose transporter GLUT 1 in UMR 106-01, a clonal osteogenic sarcoma cell line. J Mol Endocrinol. 1996;17(1):7–17.
12. Thomas DM, Maher F, Rogers SD, et al. Expression and regulation by insulin of GLUT 3 in UMR 106-01, a clonal rat osteosarcoma cell line. Biochem Biophys Res Commun. 1996;218(3):789–93.
13. Zoidis E, Ghirlanda-Keller C, Schmid C. Stimulation of glucose transport in osteoblastic cells by parathyroid hormone and insulin-like growth factor I. Mol Cell Biochem. 2011;348(1-2):33–42.
14. Li Z, Frey JL, Wong GW, et al. Glucose Transporter-4 Facilitates Insulin-stimulated Glucose Uptake in Osteoblasts. Endocrinology. 2016;157(11):4094–103.
15. Warburg O. On the origin of cancer cells. Science. 1956;123(3191):309–14.
16. Guntur AR, Le PT, Farber CR, et al. Bioenergetics during calvarial osteoblast differentiation reflect strain differences in bone mass. Endocrinology. 2014;155(5):1589–95.
17. Komarova SV, Ataullakhanov FI, Globus RK. Bioenergetics and mitochondrial transmembrane potential during differentiation of cultured osteoblasts. Am J Physiol Cell Physiol. 2000;279(4):C1220–9.
18. Regan JN, Lim J, Shi Y, et al. Up-regulation of glycolytic metabolism is required for HIF1alpha-driven bone formation. Proc Natl Acad Sci U S A. 2014;111(23):8673–8.
19. Esen E, Long F. Aerobic glycolysis in osteoblasts. Curr Osteoporos Rep. 2014;12(4):433–8.
20. Vander Heiden MG, Cantley LC, Thompson CB. Understanding the Warburg effect: the metabolic requirements of cell proliferation. Science. 2009;324 (5930):1029–33.
21. Liberti MV, Locasale JW. The Warburg Effect: How Does it Benefit Cancer Cells? Trends Biochem Sci. 2016;41(3):211–8.
22. Owen M, Macpherson S. Cell Population Kinetics of an Osteogenic Tissue. II. J Cell Biol. 1963;19:33–44.
23. Flanagan B, Nichols, Jr G. Metabolic studies of bone in vitro. V. Glucose metabolism and collagen biosynthesis. J Biol Chem. 1964;239:1261–5.
24. Karner CM, Esen E, Chen J, et al. Wnt Protein Signaling Reduces Nuclear Acetyl-CoA Levels to Suppress Gene Expression during Osteoblast Differentiation. J Biol Chem. 2016;291(25):13028–39.
25. Walsh NP, Blannin AK, Robson PJ, et al. Glutamine, exercise and immune function. Links and possible mechanisms. Sports Med. 1998;26(3):177–91.
26. Slawson C, Copeland RJ, Hart GW. O-GlcNAc signaling: a metabolic link between diabetes and cancer? Trends Biochem Sci. 2010;35(10):547–55.
27. Stegen S, van Gastel N, Eelen G, et al. HIF-1α Promotes Glutamine-Mediated Redox Homeostasis and Glycogen-Dependent Bioenergetics to Support Postimplantation Bone Cell Survival. Cell Metab. 2016;23(2):265–79.
28. Wise DR, Thompson CB. Glutamine addiction: a new therapeutic target in cancer. Trends Biochem Sci. 2010;35(8):427–33.
29. Nicklin P, Bergman P, Zhang B, et al. Bidirectional transport of amino acids regulates mTOR and autophagy. Cell. 2009;136(3):521–34.
30. Coles NW, Johnstone RM. Glutamine metabolism in Ehrlich ascites-carcinoma cells. Biochem J. 1962;83: 284–91.
31. Reitzer LJ, Wice BM, Kennell D. Evidence that glutamine, not sugar, is the major energy source for cultured HeLa cells. J Biol Chem. 1979;254(8):2669–76.
32. DeBerardinis RJ, Mancuso A, Daikhin E, et al. Beyond aerobic glycolysis: transformed cells can engage in glutamine metabolism that exceeds the requirement for protein and nucleotide synthesis. Proc Natl Acad Sci U S A 2007;104(49):19345–50.
33. Yuneva MO, Fan TW, Allen TD, et al. The metabolic profile of tumors depends on both the responsible genetic lesion and tissue type. Cell Metab. 2012;15(2):157–70.
34. Biltz RM, Letteri JM, Pellegrino ED, et al. Glutamine metabolism in bone. Miner Electrolyte Metab. 1983;9(3):125–31.
35. Brown PM, Hutchison JD, Crockett JC. Absence of glutamine supplementation prevents differentiation of murine calvarial osteoblasts to a mineralizing phenotype. Calcif Tissue Int. 2011;89(6):472–82.
36. Huang T, Liu R, Fu X, et al. Aging Reduces an ERRalpha-Directed Mitochondrial Glutaminase Expression Suppressing Glutamine Anaplerosis and Osteogenic Differentiation of Mesenchymal Stem Cells. Stem Cells. 2017;35(2):411–24.
37. Karner CM, Esen E, Okunade AL, et al. Increased glutamine catabolism mediates bone anabolism in response to WNT signaling. J Clin Invest. 2015;125(2):551–62.

38. Stegen S, van Gastel N, Eelen G, et al. HIF-1α Promotes Glutamine-Mediated Redox Homeostasis and Glycogen-Dependent Bioenergetics to Support Postimplantation Bone Cell Survival. Cell Metab. 2016;23(2):265–79.
39. Glatz JF, Luiken JJ, Bonen A. Membrane fatty acid transporters as regulators of lipid metabolism: implications for metabolic disease. Physiol Rev. 2010;90(1):367–417.
40. Adamek G, Felix R, Guenther HL, et al. Fatty acid oxidation in bone tissue and bone cells in culture. Characterization and hormonal influences. Biochem J. 1987;248(1):129–37.
41. Chiu KM, Keller ET, Crenshaw TD, et al. Carnitine and dehydroepiandrosterone sulfate induce protein synthesis in porcine primary osteoblast-like cells. Calcif Tissue Int. 1999;64(6):527–33.
42. Colucci S, Mori G, Vaira S, et al. L-carnitine and isovaleryl L-carnitine fumarate positively affect human osteoblast proliferation and differentiation in vitro. Calcif Tissue Int. 2005;76(6):458–65.
43. Catherwood BD, Addison J, Chapman G, et al. Growth of rat osteoblast-like cells in a lipid-enriched culture medium and regulation of function by parathyroid hormone and 1,25-dihydroxyvitamin D. J Bone Miner Res. 1988;3(4):431–8.
44. Frey JL, Li Z, Ellis JM, et al. Wnt-Lrp5 signaling regulates fatty acid metabolism in the osteoblast. Mol Cell Biol. 2015;35(11):1979–91.
45. Esen E, Chen J, Karner CM, et al. WNT-LRP5 signaling induces Warburg effect through mTORC2 activation during osteoblast differentiation. Cell Metab. 2013;17(5):745–55.
46. Karner CM, Esen E, Okunade AL, et al. Increased glutamine catabolism mediates bone anabolism in response to WNT signaling. J Clin Invest. 2015;125(2):551–62.
47. MacDonald BT, Semenov MV, Huang H, et al. Dissecting molecular differences between Wnt coreceptors LRP5 and LRP6. PLoS One. 2011;6(8):e23537.
48. Hodsman AB, Bauer DC, Dempster DW, et al. Parathyroid hormone and teriparatide for the treatment of osteoporosis: a review of the evidence and suggested guidelines for its use. Endocr Rev. 2005;26(5):688–703.
49. Borle AB, Nichols N, Nichols, Jr G. Metabolic studies of bone in vitro. II. The metabolic patterns of accretion and resorption. J Biol Chem. 1960;235:1211–4.
50. Neuman WF, Neuman MW, Brommage R. Aerobic glycolysis in bone: lactate production and gradients in calvaria. Am J Physiol 1978;234(1):C41–50.
51. Rodan GA, Rodan SB, Marks, Jr SC. Parathyroid hormone stimulation of adenylate cyclase activity and lactic acid accumulation in calvaria of osteopetrotic (ia) rats. Endocrinology. 1978;102(5):1501–5.
52. Felix R, Neuman WF, Fleisch H. Aerobic glycolysis in bone: lactic acid production by rat calvaria cells in culture. Am J Physiol. 1978;234(1):C51–5.
53. Esen E, Lee SY, Wice BM, et al. PTH Promotes Bone Anabolism by Stimulating Aerobic Glycolysis Via IGF Signaling. J Bone Miner Res 2015;30(11):1959–68.
54. Miyazaki T, Iwasawa M, Nakashima T, et al. Intracellular and extracellular ATP coordinately regulate the inverse correlation between osteoclast survival and bone resorption. J Biol Chem. 2012;287(45):37808–23.
55. Jin Z, Wei W, Yang M, et al. Mitochondrial complex I activity suppresses inflammation and enhances bone resorption by shifting macrophage-osteoclast polarization. Cell Metab. 2014;20(3):483–98.
56. Ishii KA, Fumoto T, Iwai K, et al. Coordination of PGC-1beta and iron uptake in mitochondrial biogenesis and osteoclast activation. Nat Med. 2009;15(3):259–66.
57. Zeng R, Faccio R, Novack DV. Alternative NF-kappaB Regulates RANKL-Induced Osteoclast Differentiation and Mitochondrial Biogenesis via Independent Mechanisms. J Bone Miner Res. 2015;30(12):2287–99.
58. Indo Y, Takeshita S, Ishii KA, et al. Metabolic regulation of osteoclast differentiation and function. J Bone Miner Res. 2013;28(11):2392–9.
59. Kang H, Viollet B, Wu D. Genetic deletion of catalytic subunits of AMP-activated protein kinase increases osteoclasts and reduces bone mass in young adult mice. J Biol Chem. 2013;288(17):12187–96.
60. Williams JP, Blair HC, McDonald JM, et al. Regulation of osteoclastic bone resorption by glucose. Biochem Biophys Res Commun. 1997;235(3):646–51.
61. Larsen KI, Falany ML, Ponomareva LV, et al. Glucose-dependent regulation of osteoclast H(+)-ATPase expression: potential role of p38 MAP-kinase. J Cell Biochem. 2002;87(1):75–84.
62. Kim JM, Jeong D, Kang HK, et al. Osteoclast precursors display dynamic metabolic shifts toward accelerated glucose metabolism at an early stage of RANKL-stimulated osteoclast differentiation. Cell Physiol Biochem. 2007;20(6):935–46.
63. Ahn H, Lee K, Kim JM, et al. Accelerated Lactate Dehydrogenase Activity Potentiates Osteoclastogenesis via NFATc1 Signaling. PLoS One. 2016;11(4):e0153886.
64. Miura D, Fujimura Y, Wariishi H. In situ metabolomic mass spectrometry imaging: recent advances and difficulties. J Proteomics. 2012;75(16):5052–60.

130
Endocrine Bioenergetics of Bone

Patricia F. Ducy[1] and Gerard Karsenty[2]

[1]*Department of Pathology & Cell Biology, College of Physicians and Surgeons, Columbia University, New York, NY, USA*
[2]*Department of Genetics & Development, College of Physicians and Surgeons, Columbia University, New York, NY, USA*

INTRODUCTION

Bone physiology is an expensive process in energetic terms. Not only does bone synthesize daily a massive amount of proteins during bone growth and modeling but the constant succession of destruction and formation that defines remodeling requires a remarkable supply of energy to bone cells. The extent of this demand is only magnified by the fact that the skeleton is one of the largest organs in the body. Although the survival advantage conferred to vertebrates by the skeleton, ie, the ambulatory function, justifies this metabolic expense it also suggests that energy and bone metabolisms may regulate each other. For instance, when food supply is scarce and energy stores are depleted there should be mechanisms slowing down bone growth. Conversely, bone cells should have means to regulate the flow of energy intake and/or storage to fulfill their needs. Given that the skeleton is an intrinsic player of movement, even more so during exercise, the existence of a coregulation between bone and muscle activity seems with hindsight logical.

We proposed 15 years ago a hypothesis contending that there should be a common regulation, endocrine in nature, of bone growth/remodeling and energy metabolism. This hypothesis implied that bone should receive signals from organs not classically associated with its physiology, such as fat or pancreas, but perhaps more provocatively it also inferred that bone could be a fully fledged endocrine organ regulating energy metabolism. Demonstrating the existence of the first tenet of this hypothesis opened our field to novel regulatory mechanisms influencing bone biology including the central control of bone mass. Validating its second principle identified osteocalcin, a circulating peptide secreted by osteoblasts, as a multifaceted hormone.

OSTEOCALCIN, A HORMONE CONTRIBUTING TO ENERGY HOMEOSTASIS

The global and osteoblast-specific inactivation of the *Osteocalcin* genes results, in mice fed a normal chow diet, in hyperglycemia, low β-cell mass, decreased insulin secretion and sensitivity, increased fat mass, and decreased energy expenditure [1]. Conversely, inactivation of *Esp*, the gene encoding a phosphatase activity present in osteoblasts that negatively regulates the function of osteocalcin (see later in this chapter), generates mice that are lean, hypoglycemic, hyperinsulinemic, and have increased energy expenditure, increased glucose tolerance, and enhanced insulin sensitivity [1]. None of these mutant mouse models display changes in food intake.

Consistent with the phenotype of the *Esp*-deficient mice, osteocalcin infusions or injections in wild-type mice improve their glucose tolerance, increase their β-cell mass and insulin secretion, decrease their fat mass, serum triglyceride levels, and the expression of lipolysis-inducing genes (*Tgl* and *Perilipin*), and increase the expression of genes involved in thermogenesis (*Pgc1α* and *Ucp1*) in brown adipose tissue [2, 3]. Such metabolic improvements are observed in mice on a normal diet or in mice in which glucose intolerance and insulin resistance have been induced by high fat diet feeding or by a treatment with gold thioglucose, a chemical destroying neurons that control satiety [2–4].

The majority of clinical studies available indicate that osteocalcin's actions on energy and glucose metabolism are conserved between human and mice; some went as far as to propose that osteocalcin serum levels could have a predictive value for glycemia, insulin resistance, or the development of diabetes. These studies have been

---

Primer on the Metabolic Bone Diseases and Disorders of Mineral Metabolism, Ninth Edition. Edited by John P. Bilezikian.
© 2019 American Society for Bone and Mineral Research. Published 2019 by John Wiley & Sons, Inc.
Companion website: www.wiley.com/go/asbmrprimer

extensively reviewed previously [5–7] therefore only a few examples that illustrate particular features will be cited here. Serum levels of osteocalcin are negatively correlated with fasting plasma glucose and hemoglobin 1c (HbA$_1$c) [8–10]. Improved fasting insulin levels and insulin resistance, as estimated by the model assessment of insulin resistance HOMA-IR, in adult men and women, whether or not diabetic, are associated with elevated osteocalcin levels [11–14]. Levels of circulating osteocalcin are inversely correlated with BMI, fat mass, and a number of lipid abnormalities [11, 12, 15–18]. Lastly, from a genomic point of view, the nonsynonymous SNP rs34702397 (R94Q) in the human *Osteocalcin* gene appears to correlate with type 2 diabetes in African-American patients, the non-rare rs1800247 SNP is associated with HOMA-IR, and the G–C–G haplotype composed of three SNPs (rs2758605–rs1543294–rs2241106) is significantly associated with an age- and sex-adjusted lower BMI [19–21].

Regulation of pancreas development and physiology by osteocalcin

The mode of action of osteocalcin on the pancreas has been defined through the analysis of cell-specific gene deletion mouse models and cell-based assays using rodent and human pancreatic islets (Fig. 130.1). Insulin secretion, β-cell proliferation, and β-cell mass are decreased in osteocalcin-deficient mice, whereas the opposite is true in Esp$^{-/-}$ mice, the osteocalcin gain-of-function model, or in mice treated with recombinant osteocalcin [1, 2]. These effects on β-cell biology are at least in part direct because osteocalcin enhances the expression of the *Ins1* and *Ins2* insulin genes as well as of *Cyclin-dependent kinase 4*, *CyclinD1*, and *CyclinD2* [1, 2, 22] in cultured islets or β-cell lines. Perfusion experiments on isolated mouse islets showed that osteocalcin also triggers insulin secretion because of its ability to increase cytosolic Ca^{2+} levels [22].

Likewise, osteocalcin enhances glucose-stimulated insulin secretion and insulin content in rat islets exposed to high glucose compared to untreated islets [23]. A direct role for osteocalcin signaling in the regulation of human β-cell proliferation and function has also been documented. Treatment of human pancreatic islets with recombinant osteocalcin in vitro induces a significant increase in glucose-stimulated insulin content and upregulates the expression of SUR1, an ATP-sensitive potassium channel (KATP) subunit playing a crucial role in glucose homeostasis and insulin secretion [24]. Likewise, osteoblast-conditioned medium increases insulin content and expression of genes related to survival in human islets but these changes are abolished when an anti-osteocalcin antibody is added to the medium [25]. In vivo, human islets transplanted under the kidney capsule of nondiabetic immunodeficient (NOD-scid) mice dosed for a month with recombinant osteocalcin show a threefold increase in the number of proliferating β cells and when subjected to a glucose stimulation test, these mice experience increased plasma levels of human insulin and C-peptide compared to vehicle-treated mice [24]. These results echo the positive correlation observed between osteocalcin serum levels and the homeostasis model assessment of β-cell function (HOMA-%B) in normal or diabetic patients before and after glycemic control [14, 18, 26, 27].

Evidence of a direct effect of osteocalcin on β cells in vivo came from the inactivation of the osteocalcin receptor in peripheral tissues, a G-protein coupled receptor termed Gprc6a [28, 29]. Indeed, specific inactivation of *Gprc6a* in the β-cell lineage causes glucose intolerance, low insulin secretion, and low β-cell mass in the presence of normal insulin sensitivity [29–31]. The presence of Gprc6a is necessary for osteocalcin action on pancreatic islets because recombinant osteocalcin can induce insulin secretion and *Ins2* and *CyclinD1* expression in wild-type but not in Gprc6a-deficient islets [30]. Likewise, recombinant osteocalcin can increase BrdU incorporation

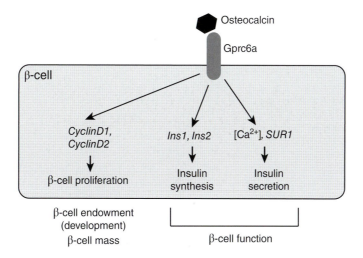

Fig. 130.1. Role of osteocalcin/Gprc6a signaling in pancreatic β cells.

in control but not in Gprc6a-deficient β cells [30]. The function of Gprc6a as an osteocalcin receptor mediating its metabolic function is most likely conserved between mice and humans because it was recently shown that two patients harboring a F464Y dominant negative mutation in *GPRC6A* display altered insulin secretion and glucose intolerance, and that the non-rare polymorphism rs2274911 in this gene is associated with features of insulin resistance [32, 33].

Inactivation of *Gprc6a* in the β-cell lineage revealed an unexpected role for osteocalcin signaling during pancreas morphogenesis. During late development (ie, at a stage when osteocalcin levels rapidly rise in the embryo), β cells undergo a transient peak of proliferation that significantly contributes to the rapid increase in β-cell mass observed during the neonatal period [30, 34]. This process is impaired in the absence of osteocalcin or of Gprc6a in β cells, already causing more than a 20% decrease in proportional β-cell area by embryonic day 17.5 in mutant mice [30]. This developmental function of osteocalcin signaling in the β-cell lineage, which involves CyclinD1 but does not affect β-cell differentiation, identifies this endocrine pathway as a determinant of β-cell endowment.

Regulation of muscle bioenergetics by osteocalcin

Remarkably, osteocalcin serum levels significantly increase during exercise in mice and humans [35–37]. This increase is needed to favor uptake and utilization of glucose and fatty acids, the main nutrients of the myofibers [35] (Fig. 130.2). Osteocalcin directly promotes glucose uptake in myoblasts by triggering the translocation of the glucose transporter GLUT4 to the plasma membrane and favors the production of ATP through the tricarboxylic acid (TCA) cycle. In parallel, in part by enhancing phosphorylation of the hormone-sensitive lipase (HSL), osteocalcin favors the conversion of intramyocellular triglycerides into free fatty acids while increasing their utilization by promoting the phosphorylation of the cellular energy sensor adenosine monophosphate-activated protein kinase (AMPK) and thereby the accumulation of fatty acids transporters. Independently of these cell metabolism changes, osteocalcin also signals in myofibers to increase the expression and secretion of IL-6, a myokine that enhances the production of nutrients upon exercise by promoting liver gluconeogenesis and lipolysis in white adipose tissue [35, 38]. Thus, the body's adaptation to

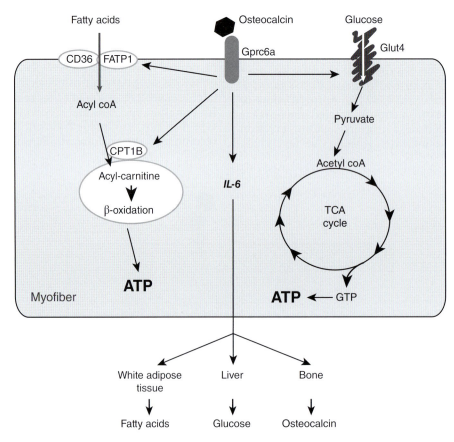

Fig. 130.2. Mechanism of osteocalcin/Gprc6a action in myofibers during exercise.

exercise involves an integrated regulation of energy metabolism in which osteocalcin, via its induction of IL-6 secretion by muscle, mobilizes nutrients stored in fat and liver and promotes their intake and utilization in myofibers (Fig. 130.2). Osteocalcin signaling in myofibers is also necessary to maintain muscle mass in adult mice [39]. That levels of total osteocalcin in serum are positively associated with fat-free mass in premenopausal women and both osteocalcin and IL-6 serum levels rise in young women during exercise suggest that osteocalcin positively regulates muscle physiology in human [35, 40].

This cross-talk between bone and muscle has been validated in vivo through the analysis of mutant mice deficient for osteocalcin either globally or only in osteoblasts postnatally, or lacking osteocalcin's receptor, Gprc6a, specifically in myofibers. By 3 months of age, these mutant mice run 20% to 30% less time and distance than control littermates [35]. The use of conditional models of *Osteocalcin* and *Gprc6a* inactivation rules out that this phenotype results from non-muscle-related metabolic abnormalities or a developmental defect. Furthermore, when administered immediately before exercise, osteocalcin increases the time and distance run by 3-month-old wild-type mice on a treadmill by over 20%. This injection, however, has no beneficial effect when Gprc6a is inactivated in myofibers, confirming that this receptor mediates all effects of osteocalcin in muscle cells [35].

In mice, rhesus monkeys, and humans serum levels of total and bioactive osteocalcin drop around mid-life [35, 41] (ie, when exercise capacity begins to decline), and neither osteocalcin nor IL-6 levels increase during exercise to the same extent in old mice as they do in 3-month-old mice. In addition, injections of recombinant osteocalcin in 12-month-old mice confer them with an exercise capacity seen only in 3-month-old mice [35]. This improvement results from an increase in nutrient uptake in oxidative muscles while circulating IL-6 levels also rise. This observation indicates that osteocalcin is sufficient to increase muscle function in mice as they age.

Cell-based evidence suggests that IL-6 may upregulate osteocalcin production in bone via its stimulation of bone resorption [42], explaining in part why serum levels of bioactive osteocalcin increase during exercise. In support of this notion, bone resorption and serum levels of bioactive osteocalcin cannot increase in IL-6-deficient mice during exercise as they do in wild-type mice [35]. Additional studies are needed to understand the cellular basis of this regulation.

REGULATION OF OSTEOCALCIN SECRETION AND BIOACTIVITY

Abiding by the basic principle of endocrinology that a hormone needs to be both quickly produced in response to the body's needs and strictly kept in check, osteocalcin secretion and activity are under tight positive and negative controls. Several hormones, secreted molecules, nutrients, neuromediators known to regulated energy metabolism, and/or bone homeostasis have been shown to control osteocalcin production and/or activity at different levels (Fig. 130.3).

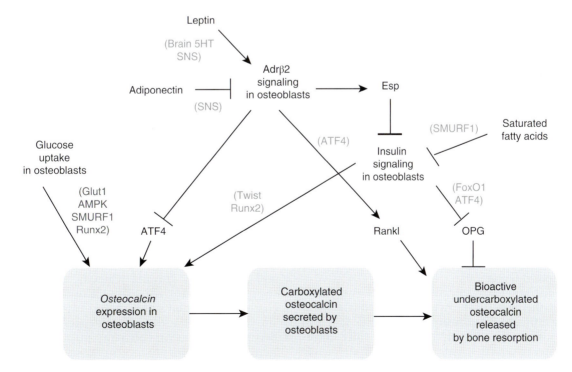

Fig. 130.3. Regulation of osteocalcin production and activation.

Regulation of osteocalcin activation by insulin and bone resorption

Like most peptide hormones, osteocalcin is produced as a pre-pro-molecule that is sequentially cleaved in osteoblasts so that only the mature protein is secreted. In addition, osteocalcin is posttranslationally modified on three glutamic acid (GLU) residues that are carboxylated by a vitamin K-dependent γ-carboxylase to form three γ-carboxyglutamic acid (GLA) residues [43]. This modification confers on osteocalcin its affinity for mineral ions and therefore explains why large amounts of carboxylated osteocalcin are stored in the bone matrix. Along with carboxylated osteocalcin, a form of osteocalcin decarboxylated on residue 13 in mice and residue 17 in humans (ie, an undercarboxylated osteocalcin) circulates in blood. Genetic evidence, cell-based assays, and association studies have established that it is this blood-borne undercarboxylated osteocalcin that is responsible for its metabolic functions [1–3,22,44], and one needs to dissociate this undercarboxylated form of osteocalcin from the fully uncarboxylated form marginally produced by osteoblasts. This latter form, which results from an absence of γ-carboxylation, is produced at very low levels in normal conditions (less than 5% of total osteocalcin in mouse serum [45]; less than 7% in human bone extracts [46]). Of note, although this uncarboxylated osteocalcin is active (it is actually equivalent to recombinant osteocalcin, which is bacterially produced uncarboxylated) its concentration is so low compared to that of undercarboxylated osteocalcin (less than one-tenth in mouse serum and less than one-fifth in human bone [45, 46]) that reducing its concentration by supplementation with vitamin K cannot have a metabolic effect [47].

Once carboxylated osteocalcin is secreted by osteoblasts and bound to the mineralized bone matrix it can be released during osteoclastic resorption, which also activates this hormone by causing the decarboxylation of its first GLA residue because of the low pH existing in the resorption lacuna [48, 49]. This process is enhanced by insulin signaling in osteoblasts that decreases the expression of osteoprotegerin (OPG) in a FoxO1-dependent manner and thereby enhances osteoclast function [48, 50, 51]. Conversely, the product of the *Esp* gene, a protein tyrosine phosphatase that dephosphorylates and inactivates the insulin receptor, eventually inhibits osteocalcin decarboxylation. This explains why *Esp*-deficient mice [1, 48] present a phenotype of a gain-of-osteocalcin function. Insulin signaling in osteoblasts may also stimulate the Runx2-mediated regulation of *Osteocalcin* expression by inhibiting the negative action of Twist on this transcription factor [50].

The demonstration that undercarboxylated osteocalcin in humans is the bioactive form of this hormone is hampered by the difficulty in quantifying its specific levels in blood. Although it can quantify the level of fully uncarboxylated osteocalcin, the hydroxyapatite affinity assay developed in the 1980s [52] cannot accurately measure the undercarboxylated form because it lacks the specificity to distinguish between forms having two or three GLA residues. More importantly, there is no immunodetection assay commercially available that can quantify undercarboxylated osteocalcin concentration in an unbiased manner. Most assays recognize total osteocalcin, without differentiating its carboxylation status, except for a non-carboxylated ELISA detection kit (Takara), which mostly reacts with fully decarboxylated osteocalcin (M. Ferron, personal communication), and for a specific ELISA that was developed specifically to analyze mouse serum [45]. This latter ELISA was most recently used to demonstrate that undercarboxylated osteocalcin levels rise during exercise and are associated with an increase in bone resorption as measured by circulating levels of the bone resorption marker CTx [35]. In contrast, hindlimb immobilization in rats causes a 30% reduction of blood levels of undercarboxylated osteocalcin that correlates with muscle loss and muscle weakness [53].

Regulation of osteocalcin decarboxylation by adipokines and the sympathetic nervous system

The positive regulation of osteocalcin's bioactivation by insulin signaling in osteoblasts combined with osteocalcin's ability to increase insulin production requires that other mechanisms exist that can put a break to this feed-forward loop. Although other molecules may do the same, leptin, the adipocyte-specific hormone that negatively controls bone mass accrual by upregulating sympathetic output (SNS) through an inhibition of central serotonin synthesis, plays such a role [22, 54–57]. By increasing SNS signaling through the β_2-adrenergic receptor (Adrβ2) expressed by osteoblasts, leptin indirectly increases the expression of *Rankl* and thus stimulates bone resorption [56, 58]. Sympathetic signaling also strongly favors *Esp* expression and therefore OPG secretion [22, 48] and the ATF4-mediated upregulation of *Osteocalcin* expression [51, 59]. The sum of all these effects of leptin/SNS signaling in osteoblasts is a decrease in bioactive osteocalcin, as inferred by increased serum levels of undercarboxylated osteocalcin and insulin existing in Adrβ2-deficient mice [22].

Adiponectin is another adipocyte-derived hormone that affects bone remodeling and osteocalcin production. Adiponectin acts both through a central relay and directly on osteoblasts [60]. In young animals, when osteocalcin production is high, adiponectin signaling in osteoblasts seems predominant and blunts the production of total and undercarboxylated osteocalcin by limiting proliferation and favoring apoptosis of osteoblasts. Later on, a second mechanism prevails whereby adiponectin crosses the blood–brain barrier and binds to neurons of the locus coeruleus to inhibit sympathetic output, ie, opposing leptin central action [60]. At that age (6 months old and beyond) adiponectin-deficient mice become glucose intolerant and poor producers of insulin, but normalizing their sympathetic tone rescues this phenotype [60]. Thus, the adipocyte exerts both positive and negative feedbacks on osteocalcin biology, establishing a functional link between fat, a classical player in energy storage and homeostasis, and bone.

Regulation of osteocalcin production by nutrients

Glucose uptake in osteoblasts also regulates *Osteocalcin* expression. GLUT1-mediated glucose uptake stabilizes Runx2, a potent regulator of osteocalcin expression [61], in part by inhibiting AMPK's activation of the ubiquitinase SMURF1 [62]. Because Runx2 also increases *Glut1* expression, glucose uptake by osteoblasts is further enhanced [62]. This feed-forward loop allows the secretion of large amounts of carboxylated, ready to be activated, osteocalcin that can be stored in the bone matrix.

Free saturated fatty acids, whose levels rise upon high fat diet feeding, are also potent regulators of osteocalcin production. Increased levels of fatty acids in osteoblasts trigger insulin receptor ubiquitination by SMURF1 and its degradation; this hampers bone resorption and the production of undercarboxylated osteocalcin [63]. This decrease in bioactive osteocalcin production contributes to whole organism insulin resistance.

CONCLUSIONS AND PERSPECTIVES

The complete story of the role of bone in general and of osteocalcin in particular in the regulation of energy metabolism is still being written. For instance, osteocalcin favors *Adiponectin* expression by adipocytes and influences fat mass and enhances energy expenditure but the cellular and molecular bases for these activities are still unknown. In addition, conditional deletion of its receptor in particular tissues such as liver or the gut could identify new target tissues and metabolic-related function(s) for this hormone. Perhaps more importantly, there is evidence that the role of the skeleton in regulating energy metabolism is not completely understood and most likely involves yet unknown hormones secreted by bone cells. The best illustration for this assumption is the fact that ablation of osteoblasts in mice causes an effect on appetite independent of osteocalcin [64]. Hence, there must be at least one other hormone made by the osteoblast that controls this particular aspect.

REFERENCES

1. Lee NK, Sowa H, Hinoi E, et al. Endocrine regulation of energy metabolism by the skeleton. Cell. 2007;130(3):456–69.
2. Ferron M, Hinoi E, Karsenty G, et al. Osteocalcin differentially regulates beta cell and adipocyte gene expression and affects the development of metabolic diseases in wild-type mice. Proc Natl Acad Sci U S A. 2008;105(13):5266–70.
3. Ferron M, McKee MD, Levine RL, et al. Intermittent injections of osteocalcin improve glucose metabolism and prevent type 2 diabetes in mice. Bone. 2012;50(2):568–75.
4. Zhou B, Li H, Liu J, et al. Autophagic dysfunction is improved by intermittent administration of osteocalcin in obese mice. Int J Obes (Lond). 2016;40(5):833–43.
5. Liu J-M, Rosen CJ, Ducy P, et al. Regulation of glucose handling by the skeleton: Insights from mouse and human studies. Diabetes. 2016;65(11):3225–32.
6. Ducy P. The role of osteocalcin in the endocrine cross-talk between bone remodelling and energy metabolism. Diabetologia. 2011;54(6):1291–7.
7. Ferron M, Lacombe J. Regulation of energy metabolism by the skeleton: Osteocalcin and beyond. Arch Biochem Biophys. 2014;561:137–46.
8. Liu D-M, Guo X-Z, Tong H-J, et al. Association between osteocalcin and glucose metabolism: A meta-analysis. Osteoporos Int. 2015;26(12):2823–33.
9. Jung KY, Kim KM, Ku EJ, et al. Age- and sex-specific association of circulating osteocalcin with dynamic measures of glucose homeostasis. Osteoporos Int. 2016;27(3):1021–29.
10. González-García ZM, Kullo IJ, Coletta DK, et al. Osteocalcin and type 2 diabetes risk in latinos: A life course approach. Am J Hum Biol. 2015;27(6):859–61.
11. Pittas AG, Harris SS, Eliades M, et al. Association between serum osteocalcin and markers of metabolic phenotype. J Clin Endocrinol Metab. 2009;94(3):827–32.
12. Kanazawa I, Yamaguchi T, Yamamoto M, et al. Serum osteocalcin level is associated with glucose metabolism and atherosclerosis parameters in type 2 diabetes mellitus. J Clin Endocrinol Metab. 2009;94(1):45–9.
13. Im JA, Yu BP, Jeon JY, et al. Relationship between osteocalcin and glucose metabolism in postmenopausal women. Clin Chim Acta. 2008;396(1-2):66–9.
14. Bao YQ, Zhou M, Zhou J, et al. Relationship between serum osteocalcin and glycemic variability in Type 2 diabetes. Clin Exp Pharmacol Physiol. 2011;38(1):50–4.
15. Kindblom JM, Ohlsson C, Ljunggren O, et al. Plasma osteocalcin is inversely related to fat mass and plasma glucose in elderly Swedish men. J Bone Miner Res. 2009;24(5):785–91.
16. Lee YJ, Lee H, Jee SH, et al. Serum osteocalcin is inversely associated with adipocyte-specific fatty acid-binding protein in the Korean metabolic syndrome research initiatives. Diabetes Care. 2010;33(7):e90.
17. Saleem U, Mosley TH, Jr, Kullo IJ. Serum osteocalcin is associated with measures of insulin resistance, adipokine levels, and the presence of metabolic syndrome. Arterioscler Thromb Vasc Biol. 2010;30(7):1474–8.
18. Zhou M, Ma X, Li H, et al. Serum osteocalcin concentrations in relation to glucose and lipid metabolism in Chinese individuals. Eur J Endocrinol. 2009;161(5):723–9.
19. Das SK, Sharma NK, Elbein SC. Analysis of osteocalcin as a candidate gene for Type 2 Diabetes (T2D) and

intermediate traits in Caucasians and African Americans. Dis Markers. 2010;28(5):281–6.
20. Korostishevsky M, Malkin I, Trofimov S, et al. Significant association between body composition phenotypes and the osteocalcin genomic region in normative human population. Bone. 2012;51(4):688–94.
21. Ling Y, Gao X, Lin H, et al. A common polymorphism rs1800247 in osteocalcin gene is associated with hypertension and diastolic blood pressure levels: The Shanghai Changfeng study. J Hum Hypertens. 2016;30(11):679–84.
22. Hinoi E, Gao N, Jung DY, et al. The sympathetic tone mediates leptin's inhibition of insulin secretion by modulating osteocalcin bioactivity. J Cell Biol. 2008;183(7):1235–42.
23. Kover K, Yan Y, Tong PY, et al. Osteocalcin protects pancreatic beta cell function and survival under high glucose conditions. Biochem Biophys Res Commun. 2015;462(1):21–26.
24. Sabek OM, Nishimoto SK, Fraga D, et al. Osteocalcin effect on human beta-cells mass and function. Endocrinology. 2015;156(9):3137–46.
25. Kover K, Tong PY, Pacicca D, et al. Bone marrow cavity: A supportive environment for islet engraftment. Islets. 2011;3(3):93–101.
26. Ma X, Chen F, Hong H, et al. The relationship between serum osteocalcin concentration and glucose and lipid metabolism in patients with type 2 diabetes mellitus - the role of osteocalcin in energy metabolism. Ann Nutr Metab. 2015;66(2-3):110–16.
27. Bulló M, Moreno-Navarrete JM, Fernández-Real JM, et al. Total and undercarboxylated osteocalcin predict changes in insulin sensitivity and β cell function in elderly men at high cardiovascular risk. Am J Clin Nutr. 2012;95(1):249–55.
28. Oury F, Sumara G, Sumara O, et al. Endocrine regulation of male fertility by the skeleton. Cell. 2011;144(5):796–809.
29. Pi M, Chen L, Huang MZ, et al. GPRC6A null mice exhibit osteopenia, feminization and metabolic syndrome. PLoS One. 2008;3(12):e3858.
30. Wei J, Hanna T, Suda N, et al. Osteocalcin promotes beta-cell proliferation during development and adulthood through Gprc6a. Diabetes. 2014;63(3):1021–31.
31. Pi M, Kapoor K, Ye R, et al. Evidence for osteocalcin binding and activation of GPRC6A in beta-cells. Endocrinology. 2016:en20152010.
32. Oury F, Ferron M, Huizhen W, et al. Osteocalcin regulates murine and human fertility through a pancreas-bone-testis axis. J Clin Invest. 2013;123(6):2421–33.
33. Di Nisio A, Rocca MS, Fadini GP, et al. The rs2274911 polymorphism in GPRC6A gene is associated with insulin resistance in normal weight and obese subjects. Clin Endocrinol (Oxf). 2017;86(2):185–91.
34. Ackermann AM, Gannon M. Molecular regulation of pancreatic beta-cell mass development, maintenance, and expansion. J Mol Endocrinol. 2007;38(1-2):193–206.
35. Mera P, Laue K, Ferron M, et al. Osteocalcin signaling in myofibers is necessary and sufficient for optimum adaptation to exercise. Cell Metab. 2016;23(6):1078–92.
36. Kim YS, Nam JS, Yeo DW, et al. The effects of aerobic exercise training on serum osteocalcin, adipocytokines and insulin resistance on obese young males. Clin Endocrinol (Oxf). 2015;82(5):686–94.
37. Levinger I, Zebaze R, Jerums G, et al. The effect of acute exercise on undercarboxylated osteocalcin in obese men. Osteoporos Int. 2011;22(5):1621–6.
38. Pedersen BK, Febbraio MA. Muscles, exercise and obesity: Skeletal muscle as a secretory organ. Nat Rev Endocrinol. 2012;8(8):457–65.
39. Mera P, Laue K, Wei J, et al. Osteocalcin is necessary and sufficient to maintain muscle mass in older mice. Mol Metab. 2016;5(10):1042–47.
40. Liu JM, Zhao HY, Zhao L. An independent positive relationship between the serum total osteocalcin level and fat-free mass in healthy premenopausal women. J Clin Endocrinol Metab 2013;98(5):2146–52.
41. Gundberg CM, Lian JB, Gallop PM. Measurements of gamma-carboxyglutamate and circulating osteocalcin in normal children and adults. Clin Chim Acta. 1983;128(1):1–8.
42. Jilka RL, Hangoc G, Girasole G, et al. Increased osteoclast development after estrogen loss: Mediation by interleukin-6. Science. 1992;257(5066):88–91.
43. Hauschka PV, Lian JB, Cole DE, et al. Osteocalcin and matrix Gla protein: vitamin K-dependent proteins in bone. Physiol Rev. 1989;69(3):990–1047.
44. Ferron M, Lacombe J, Germain A, et al. GGCX and VKORC1 inhibit osteocalcin endocrine functions. J Cell Biol. 2015;208(6):761–76.
45. Ferron M, Wei J, Yoshizawa T, et al. An ELISA-based method to quantify osteocalcin carboxylation in mice. Biochem Biophys Res Commun. 2010;397(4):691–6.
46. Cairns JR, Price PA. Direct demonstration that the vitamin K-dependent bone Gla protein is incompletely gamma-carboxylated in humans. J Bone Miner Res. 1994;9(12):1989–97.
47. Shea MK, Dawson-Hughes B, Gundberg CM, et al. Reducing undercarboxylated osteocalcin with vitamin K supplementation does not promote lean tissue loss or fat gain over 3 years in older women and men: A randomized controlled trial. J Bone Miner Res. 2017;32(2):243–9.
48. Ferron M, Wei J, Yoshizawa T, et al. Insulin signaling in osteoblasts integrates bone remodeling and energy metabolism. Cell. 2010;142(2):296–308.
49. Ivaska KK, Hentunen TA, Vaaraniemi J, et al. Release of intact and fragmented osteocalcin molecules from bone matrix during bone resorption in vitro. J Biol Chem. 2004;279(18):18361–9.
50. Fulzele K, Riddle RC, DiGirolamo DJ, et al. Insulin receptor signaling in osteoblasts regulates postnatal bone acquisition and body composition. Cell. 2010;142(2):309–19.
51. Kode A, Mosialou I, Silva BC, et al. FoxO1 protein cooperates with ATF4 protein in osteoblasts to control glucose homeostasis. J Biol Chem. 2012;287(12):8757–68.
52. Gundberg CM, Hauschka PV, Lian JB, et al. Osteocalcin: Isolation, characterization, and detection. Methods Enzymol. 1984;107:516–44.

53. Lin X, Hanson E, Betik AC, et al. Hindlimb immobilization, but not castration, induces reduction of undercarboxylated osteocalcin associated with muscle atrophy in rats. J Bone Miner Res. 2016;31(11):1967–78.
54. Ducy P, Amling M, Takeda S, et al. Leptin inhibits bone formation through a hypothalamic relay: A central control of bone mass. Cell. 2000;100(2):197–207.
55. Takeda S, Elefteriou F, Levasseur R, et al. Leptin regulates bone formation via the sympathetic nervous system. Cell. 2002;111(3):305–17.
56. Elefteriou F, Ahn JD, Takeda S, et al. Leptin regulation of bone resorption by the sympathetic nervous system and cart. Nature 2005;434(7032):514–20.
57. Yadav VK, Oury F, Suda N, et al. A serotonin-dependent mechanism explains the leptin regulation of bone mass, appetite, and energy expenditure. Cell. 2009;138(5):976–89.
58. Yoshizawa T, Hinoi E, Jung DY, et al. The transcription factor ATF4 regulates glucose metabolism in mice through its expression in osteoblasts. J Clin Invest. 2009;119(9):2807–17.
59. Yang X, Matsuda K, Bialek P, et al. ATF4 is a substrate of RSK2 and an essential regulator of osteoblast biology; implication for Coffin-Lowry Syndrome. Cell. 2004;117(3):387–98.
60. Kajimura D, Lee HW, Riley KJ, et al. Adiponectin regulates bone mass via opposite central and peripheral mechanisms through FoxO1. Cell Metab. 2013;17(6):901–15.
61. Ducy P, Zhang R, Geoffroy V, et al. Osf2/Cbfa1: A transcriptional activator of osteoblast differentiation. Cell. 1997;89(5):747–54.
62. Wei J, Shimazu J, Makinistoglu MP, et al. Glucose uptake and Runx2 synergize to orchestrate osteoblast differentiation and bone formation. Cell. 2015;161(7):1576–91.
63. Wei J, Ferron M, Clarke CJ, et al. Bone specific insulin resistance disrupts whole-body glucose homeostasis via decreased osteocalcin activation. J Clin Invest. 2014;124(4):1–13.
64. Yoshikawa Y, Kode A, Xu L, et al. Genetic evidence points to an osteocalcin-independent influence of osteoblasts on energy metabolism. J Bone Miner Res. 2011;26(9):2012–25.

131
Central Neuronal Control of Bone Remodeling

Hiroki Ochi[1], Paul Baldock[2], and Shu Takeda[3]

[1]*Department of Physiology and Cell Biology, Tokyo Medical and Dental University, Tokyo, Japan*
[2]*Bone and Mineral Research Program, Garvan Institute of Medical Research, St Vincent's Hospital, Sydney, NSW, Australia*
[3]*Division of Endocrinology, Toranomon Hospital, Tokyo, Japan*

INTRODUCTION

All homeostatic functions, including those in bone, are controlled by the brain. Indeed, clinical evidence that traumatic brain injury (TBI) accelerates the healing of fractures suggests that there is a link between the central nervous system and bone remodeling [1]. The discovery that leptin regulates bone formation through the central nervous system initiated a new research field: neuronal control of bone remodeling [2]. Since then, other neuropeptides and neurotransmitters, such as neuropeptide Y (NPY) [3], cocaine- and amphetamine-regulated transcript (CART) [4], neuromedin U [5], and, more recently, serotonin [6, 7], as well as sensory nerve innervation of bone have been demonstrated to possess bone-regulating activity (Fig. 131.1).

LEPTIN AND THE SYMPATHETIC NERVOUS SYSTEM (SNS)

Leptin is a 16-kDa peptide hormone that is synthesized by adipocytes [8]. Leptin affects appetite and energy metabolism through the pro-opiomelanocortin (POMC) pathway, which increases food intake and energy expenditure, and the Agouti-related peptide (AgRP)/NPY pathway, which decreases food intake and energy expenditure [8]; both pathways are active in the arcuate nucleus. *ob/ob* mice that lack functional leptin are obese and sterile [8]. In spite of their hypogonadism, the most common cause of osteoporosis, *ob/ob* mice and *db/db* mice that lack a functional leptin receptor exhibit high bone mass [9]. Intracerebroventricular (ICV) infusion of leptin in *ob/ob* mice or wild-type mice at a minimal dose, without any detectable leakage into the general circulation, results in reduced bone mass [9]. This observation was subsequently verified in both rats and sheep [10,11]. Moreover, mice lacking leptin receptors specifically in the central nervous system demonstrate an identical bone phenotype to mice lacking leptin receptors in the entire body (*ob/ob* mice), whereas mice lacking leptin receptors only in osteoblasts have normal bone metabolism [12]. Thus, leptin acts on the central nervous system to regulate bone mass.

Leptin stimulates activity of the sympathetic nervous system, causing a decrease in bone mass mainly through the adrenergic 2 receptor (adrb2), the most strongly expressed adrenergic receptor in osteoblasts [13]. Treatment of wild-type mice with the nonselective beta-agonist, isoproterenol, or the adrb2-selective agonist, clenbuterol or salbutamol, decreases bone mass [13,14]. In contrast, mice with decreased sympathetic activity (*dopamine-beta-hydroxylase*$^{-/-}$ mice [13], *adrb2*$^{-/-}$ mice[4], and mice treated with nonselective beta-blockers [13]) show high bone mass because of an increase in bone formation and a decrease in bone resorption. In addition, these mice are resistant to the effect of leptin on decreasing bone mass, demonstrating that the main downstream pathway of leptin in bone metabolism is the sympathetic nervous system. Moreover, osteoblast-specific *adrb2*$^{-/-}$ mice recapitulate the bone abnormality of *adrb2*$^{-/-}$ mice [15], indicating that adrb2 in osteoblasts,

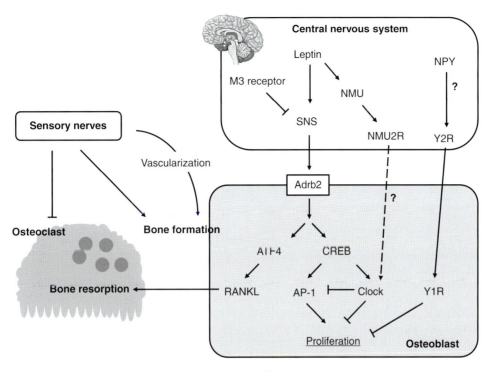

Fig. 131.1. A schematic diagram of the central neuronal control of bone remodeling.

not in other tissue, is responsible for the effect of the sympathetic nervous system on bone.

Sympathetic signaling in osteoblasts involves two different pathways to control bone formation and bone resorption. The former is via cAMP response element-binding protein (CREB) and c-myc transcription factors [15,16]. Upon stimulation of the SNS–adrb2 pathway in osteoblasts, CREB phosphorylation is inhibited by an unidentified mechanism [15] that results in the inhibition of further downstream effectors, ie, molecular clocks such as Per, Cry, and AP-1 transcription factors [16]. Molecular experiments have revealed that Per1 and Per2 negatively regulate the expression of c-myc and G_1 cyclins and, thus, osteoblast proliferation and that the absence of Per1 and Per2 or Cry1 and Cry2 favors bone mass accrual [16]. In contrast, AP-1 stimulates the expression of c-myc and G_1 cyclins [16]. As a combined effect of these antagonistic proteins, the SNS decreases bone formation [16]. The latter pathway, which regulates bone resorption, is also mediated through osteoblasts [4]. Upon stimulation of the SNS–adrb2 pathway, osteoblastic ATF4 is phosphorylated, inducing *Rankl* expression [4]. Because the bone resorption abnormality in *adrb2-/-* mice cannot be corrected by leptin ICV infusion, leptin signaling has been suggested to be dependent on the SNS to regulate bone resorption; however, given that osteoclasts do exist in *adrb2-/-* mice, SNS signaling is not essential for osteoclast differentiation.

Various mouse models of osteoporosis, such as ovariectomy-induced [13], unloading-induced [17], or depression-induced bone loss [18], are all ameliorated by concomitant treatment with beta-blockers, though there are some conflicting reports. These discrepancies may be related to the amount of beta-blocker used in the study: a low dose of propranolol that does not affect any cardiovascular functions was sufficient to increase bone formation parameters, and increasing the doses of propranolol progressively decreases its beneficial effect on bone formation [19].

Recently, the norepinephrine content in bone was shown to be decreased by TBI via cannabinoid receptor 1 signaling, and TBI-induced stimulation of osteogenesis was restrained by a beta agonist.[20] Many epidemiologic studies have also confirmed the effect of a beta-blocker on bone mass or fracture [21,22]. Though there are some conflicting results showing either beneficial or indifferent effects of beta-blockers in the prevention of osteoporotic fractures, a meta-analysis of eight studies demonstrated that the use of beta-blockers is associated with reduction of hip fracture risk and the risk of any fracture [22]. Considering the widespread usage of beta-blockers in clinical medicine, beta-blockers can also be easily applied to the treatment of osteoporosis. However, because most of the studies addressing the relationship between beta-blockers and osteoporotic fracture are observational studies, randomized clinical trials are strongly needed.

Other adrenergic receptors and muscarinic receptor are also involved in bone remodeling. *M3 muscarinic receptor-/-* mice and neuron-specific *M3 muscarinic receptor-/-* mice, both of which show increases in sympathetic nervous activity, demonstrate a low bone mass phenotype because of a decrease in bone formation and an increase in bone resorption [23], whereas osteoblast-specific

$M3$ *muscarinic receptor*$^{-/-}$ mice have no bone abnormality [23]. These results demonstrate that the para-sympathetic nervous system affects bone mass by targeting neurons and that the balance between the autonomic nervous systems defines the bone mass. Moreover, *α2A/α2C adrenergic receptor*$^{-/-}$ mice exhibit high bone mass despite an increase in sympathetic nervous system activity, and selective α2R agonists increase osteoclast formation [24], indicating that α-adrenergic receptors are also involved in bone remodeling. Whether these observations may be applied to human beings remains unknown.

SEROTONIN

Serotonin is a monoamine compound produced exclusively by the action of Tph2 in the central nervous system [25]. *Tph2*$^{-/-}$ mice develop an osteoporotic phenotype because of an increase in SNS activity, and *ob/ob* mice exhibit an increase in the concentration of serotonin in the brain [6]. The anti-osteogenic and anorexigenic actions of leptin are hampered in mice lacking leptin receptors exclusively in serotonergic neurons located in the brainstem [6], whereas mice lacking leptin receptors specifically in the arcuate nucleus or ventromedial nucleus of the hypothalamus (VMH), which are known to be indispensable for the anorexigenic action of leptin [8], have normal bone mass [6], demonstrating a pivotal role of serotonin in the action of leptin. Thus, brainstem-derived serotonin favors bone mass accrual through its binding to the 5-HT2c receptor [6]. However, contradictory observations using identical mouse models are also reported [26]: serotonergic neurons do not express leptin receptors, and hence serotonergic neuron-specific leptin receptor$^{-/-}$ mice have normal bone mass and normal body weight. Currently, the cause of this discrepancy is not known, although it has been proposed that the serotonergic neuron-specific leptin receptor$^{-/-}$ mice reported by Yadav et al. might have been globally leptin receptor$^{-/-}$ mice. Moreover, selective serotonin reuptake inhibitors (SSRIs), which inhibit serotonin uptake and are believed to stimulate the serotonin signaling pathway, increase the risk of fracture [27]. A recent study has reported that fluoxetine, one of the most-prescribed SSRIs, affects bone remodeling through two different mechanisms. Peripherally, fluoxetine inhibits osteoclast differentiation via decreases in intracellular Ca^{2+} levels and Nfatc1 transcription. This mechanism does not depend on a serotonin-reuptake mechanism. However, fluoxetine also induces a brain-serotonin-dependent rise in sympathetic tone that increases bone resorption depending on the administration time of fluoxetine. This mechanism is sufficient to counteract the local effect of fluoxetine, suggesting that chronic SSRI use promotes bone loss through a central serotonin-sympathetic tone activation [28].

In the periphery, serotonin is mostly produced by the action of Tph1 by enterochromaffin cells in the gastrointestinal tract [25]. *Tph1*$^{-/-}$ mice, whose serum concentration of serotonin is markedly decreased, present high bone mass [7]. Peripheral serotonin acts on HTr1b and inhibits CREB phosphorylation and osteoblast proliferation [7]. Unexpectedly, gut-derived serotonin is shown to be a downstream mediator of the skeletal effect of LRP5: the gut-specific inactivation of LRP5 or the activation of LRP5 signaling only in the gut fully recapitulates low bone mass in *LRP5*$^{-/-}$ mice or high bone mass in mice carrying an LRP5-activating mutation in the entire body [7]. Moreover, a low-tryptophan diet normalizes the high serum serotonin concentration and low bone mass in *LRP5*$^{-/-}$ mice [7].

A negative association between BMD and serum serotonin levels is also reported in humans [29], and importantly an inhibitor of Tph1, which does not affect Tph2, favors bone mass accrual in rodents to a similar extent as PTH injection [30]. Thus, serotonin and Tph1 form an attractive target for a novel bone anabolic therapy for osteoporosis. However, contradictory observations using identical mouse models have been reported [31]: osteocyte-specific *LRP5*$^{-/-}$ mice show a decrease in bone mass, which is identical to observations in global *LRP5*$^{-/-}$ mice, whereas gut-specific *LRP5*$^{-/-}$ mice and *Tph1*$^{-/-}$ mice have a normal bone mass. Moreover, LRP5 does not affect *Tph1* expression or serum serotonin concentration in mice, whereas treatment with a different Tph1 inhibitor does not affect bone mass in mice, and hence LRP5 regulates bone metabolism exclusively acting in bone. Currently, the cause of this discrepancy is unknown and requires further corroboration.

NEUROMEDIN U

Neuromedin U is a neuropeptide produced in the gastrointestinal tract and brain, and NMU inhibits food intake by a leptin-independent mechanism [32]. *Nmu*$^{-/-}$ mice present a high bone mass phenotype with an isolated increase in bone formation [5], similar to that observed in *ob/ob* mice. This phenotype is not cell-autonomous, because *Nmu*$^{-/-}$ osteoblasts are indistinguishable from wild-type osteoblasts in vitro [5]. In contrast, ICV infusion of NMU in *Nmu*$^{-/-}$ mice and wild-type mice decreases bone formation and bone mass [5]. Importantly, leptin ICV infusion or isoproterenol treatment does not decrease bone mass in *Nmu*$^{-/-}$ mice, demonstrating that NMU mediates the action of leptin and SNS in the regulation of bone formation [5]. Further analysis revealed that NMU in the hypothalamus affects only the negative regulator of osteoblast proliferation, namely, the molecular clock [5,16].

SENSORY NERVES AND BONE METABOLISM

The effects of sensory denervation on bone metabolism have been examined in animals treated with capsaicin. Previous studies showed that sensory denervation induces a reduction in BMD and bone mass by increasing the number and activity of osteoclasts [33,34]. These results suggest that the sensory innervation of bone maintains bone metabolism.

Calcitonin gene-related peptide (CGRP), an efferent neurotransmitter in sensory nerves, inhibits osteoclast differentiation in vitro [35–37], and αCGRP knockout mice showed low bone mass via decreased bone formation [38]. The continuous administration of magnesium ions by a magnesium-rich intramedullary implant increased the expression of CGRP protein in the lumbar dorsal root ganglions (DRGs), and the CGRP-positive sensory nerve innervation of bone and bone periosteum subsequently accelerated cortical bone formation [39]. These results suggest that CGRP derived from sensory nerve endings promotes bone formation in vivo.

A recent study reported that neuron-derived semaphorin 3A (Sema3A) regulates the sensory innervation of bone. Neuron-specific Sema3A knockout mice had low bone mass via the suppression of osteoblast activity. Interestingly, whole body sensory function and bone innervation were significantly decreased in this conditional knockout mouse [40].

The types of sensory nerve fibers in the bone were reported in previous studies [41,42], and more than 80% of all sensory nerve fibers innervating bone are TrkA positive. TrkA is a receptor of nerve growth factor (NGF). Tomlinson et al. reported that NGF was expressed in perichondrial cells close to the long bone primary ossification center, and TrkA-positive sensory nerves were observed in this region [43]. TrkA-NGF signaling-disrupted mice exhibited impaired sensory innervation and abnormality in bone development during embryogenesis. Moreover, the inhibition of TrkA-NGF signaling also delayed the vascular invasion of the primary ossification center, suggesting that TrkA-NGF signaling coordinately regulates sensory innervation, vascularization, and ossification [43].

THE NEUROPEPTIDE Y SYSTEM

The NPY system comprises three ligands, NPY, peptide YY (PYY), and pancreatic polypeptide (PP), mediating their actions through five Y receptor subtypes, NPY1R, NPY2R, NPY4R, NPY5R, and NPY6R [44,45]. NPY is predominantly neural, produced by central and peripheral neurons, and is often cosecreted with noradrenaline [46]. NPYergic neurons are abundant in the brain, with high levels in several hypothalamic nuclei (the arcuate nucleus and VMH) [47–49]. Early studies identified NPY-immunoreactive fibers in bone associated with blood vessels [50–53] but also cells in the periosteum and bone-lining cells [50,51]. Central NPY treatment was associated with a reduction in bone mass [9]. Additionally, NPY treatment in osteoblastic cell lines inhibited the cAMP response to PTH and norepinephrine [54,55], suggesting the presence of functional Y receptors and a possible regulatory role for NPY in bone-forming cells.

A recent publication has confirmed the role of NPY in skeletal metabolism. NPY-null mice demonstrated a generalized bone anabolic phenotype [56] without significant changes in body weight. Despite early reports of no effect [57], the negative relationship between hypothalamic NPY and bone formation is consistent with previous reports of reduced bone formation after overexpression of NPY in hypothalamic neurons [58] or CSF [9] of wild-type mice, as well increased bone mass after loss of NPY receptors (discussed later in this chapter). Interestingly, central NPY overexpression represents a model of forced central starvation, similar to that evident in leptin-deficient ob/ob mice [59]. Importantly, elevation of central NPY (mimicking the conditions encountered in the hypothalamus during starvation [60]) decreased bone mass despite marked increases in body weight, as evident in ob/ob mice [59]. In this manner, weight may be matched to bone mass; calorie restriction reduces body weight and increases central NPY [61], which inhibits bone formation as a component of the whole body energy conservation response. Conversely, excessive calorie intake increases body weight but reduces NPY expression, which stimulates bone formation, thereby correlating bone mass to increases in body mass. Thus, the central perception of body weight, ie, as evident by alterations in central NPY, may act to correlate bone mass to changes in body weight [56], a process that occurs in addition to the well-described mechanical responses to altered body weight. Interestingly, NPY is also expressed in osteoblasts and osteocytes, and expression is reduced in vitro by mechanical loading [56,62]. Thus, NPY signaling may regulate multiple processes as part of a system to coordinate bone and energy homeostasis.

Hypothalamic NPY2 receptor effects on bone

Two NPY receptors have been connected with skeletal homeostasis, NPY1R and NPY2R. Both receptors are expressed in the hypothalamus as well as in peripheral nerves [63–65]. Analysis of the distal femur of germline $NPY2r^{-/-}$ mice revealed a greater cancellous bone volume associated with a greater rate of bone formation resulting from increased osteoblast activity, without an increase in the mineralizing surface [3,66]. Parameters of bone resorption were unchanged. Critically, the bone phenotype of germline $NPY2r^{-/-}$ mice was recapitulated in adult mice following selective deletion of NPY2R solely from the hypothalamus, showing a role for central NPY2R in this pathway. Moreover, the skeletal changes observed in germline and conditional $NPY2r^{-/-}$ mice occurred in the absence of measurable changes in bone-active endocrine factors. Thus, these findings indicated that the anabolism resulting from NPY2R deletion was mediated through a neural mechanism originating within the hypothalamus.

Importantly, one study showed that ablation of hypothalamic NPY2R specifically from NPYergic neurons produced only moderate increases in cancellous bone volume and no effect on cortical bone mass [67]. This result indicates that, within the hypothalamus, the NPY2R-mediated regulation of bone mass is mediated through neuronal populations other than NPY neurons. Preliminary indications suggest that sympathetic neurons emanating from the paraventricular nucleus, the target region for arcuate NPY, may be responsible for the efferent pathway (unpublished observation).

Osteoblastic NPY1 receptor effects on bone

The NPY1 receptor has recently been confirmed to be a second Y receptor active in the regulation of bone. Similar to that observed in NPY2R-deficient mice, loss of NPY1R expression resulted in a generalized anabolic phenotype with greater bone mass and formation [68], although with an additional increase in bone resorption. The bone phenotype, however, differed from that of the $NPY2r^{-/-}$ mice in several critical aspects. Most importantly, conditional deletion of hypothalamic NPY1R receptors had no effect on bone homeostasis, indicating a noncentral mechanism for NPY1R action in bone. The existence of a direct NPY1R-mediated effect on anabolism was suggested after identification of NPY1R expression in osteoblastic cells in vivo [68]. Deletion of NPY1R from osteoblastic cells in vitro recapitulated the bone anabolic changes evident in germline $NPY1r^{-/-}$ mice, although bone resorption was not different from that in wild-type mice [69]. Moreover, treatment of wild-type osteoblast-like cultures with NPY resulted in a decrease in cell number, a response that was completely absent in $NPY1r^{-/-}$ cultures, indicating functional osteoblastic NPY1R.

Moreover, this osteoblastic NPY1R expression may be directly involved in the $NPY2r^{-/-}$ phenotype. $NPY1r^{-/-}$ $NPY2r^{-/-}$ mice do not display an additive phenotype in bone, and NPY1R expression is substantially reduced in osteoblast-like cultures from $NPY2r^{-/-}$ mice. These studies indicate that, although its role in the control of bone homeostasis is yet to be fully elucidated, NPY1R signaling may be a critical downstream component of the neural regulation of bone mass.

THE CANNABINOID RECEPTORS

The endocannabinoid system mediates its actions via two cannabinoid receptors, CB1-R and CB2-R, which couple to inhibitory G proteins [70]. CB1-R is primarily found within the CNS [71], whereas CB2-R is predominantly expressed in peripheral tissue [20]. Cannabinoid receptors are also expressed in osteoblasts and osteoclasts and play a role in the control of bone homeostasis by a centrally mediated and direct mechanism.

The CB1 receptor plays a significant role in regulating BMD [72]. Mice with inactivation of CB1-R have been demonstrated to have increased BMD and additionally are protected against ovariectomy-induced bone loss [72]. Furthermore, synthetic cannabinoid receptor antagonists inhibit osteoclast formation and bone resorption in vitro and protect against ovariectomy-induced bone loss in vivo [72].

There is limited evidence concerning CB2-R action on bone mass in humans. Karsak et al. have provided evidence that the *CNR2* gene, encoding the CB2-R, exhibits a significant association of a single polymorphism and haplotypes encompassing the *CNR2* gene on human chromosome 1p36 with low BMD [73].

CB2-R-deficient mice display accelerated age-related cancellous bone loss and cortical expansion, albeit with unaltered cortical thickness [74]. Despite the loss of bone, $CB2r^{-/-}$ mice exhibit an increased mineral appositional rate and bone formation rate. This low bone mass, associated with high bone turnover, is another phenotypic parallel with postmenopausal osteoporosis [74]. The presence of functional CB2-Rs has been demonstrated in both the osteoblast and osteoclast lineages [74]. Combined, these studies indicate that CB2-R signaling contributes to the maintenance of bone mass by two mechanisms: (i) stimulating stromal cells/osteoblasts directly; and (ii) inhibiting monocytes/osteoclasts, both directly and by inhibiting osteoblast/stromal cell RANKL expression. Jointly, these data suggest that the cannabinoid system plays an important role in the regulation and maintenance of bone mass through the signaling of both the CB1-R and CB2-R.

THE MELANOCORTIN SYSTEM

Melanocortins are a complex family comprising a number of endogenous agonists that are all derived from a single precursor, pro-opiomelanocortin (POMC), and α-, β- and γ-melanocyte-stimulating hormone (MSH) and adrenocorticotropic hormone (ACTH) elicit the action of melanocortins by interacting with five melanocortin receptors (MCRs), identified as G-protein coupled receptors MCR1–5 [75,76]. In addition to the melanocortin agonists, AgRP has been identified as a high-affinity antagonist [77].

The regulation of bone homeostasis by this system centers around the action of the melanocortin 4 receptor (MC4R) expressed in hypothalamic neurons. Patients deficient in MC4R are known to exhibit high BMD, resulting from a decrease in bone resorption [78]. Importantly, the greater BMD is still evident following correction of the obesity that is characteristic of MC4R deficiency [78]. Mechanistic studies in mice have enabled dissection of this pathway to bone and, interestingly, have implicated another hypothalamic neuropeptide, CART. Hypothalamic CART expression is increased in $MC4R^{-/-}$ mice, which display a high bone mass phenotype resulting from decreased osteoclast number and function [4,79], as evident in human studies. Additionally, MC4R-mutant mice lacking one or two copies of CART exhibited a significantly lower bone mass [4,79], demonstrating that increased CART signaling is critical for the low-bone-resorption/high-bone-mass phenotype observed in MC4R-deficient mice.

REFERENCES

1. Perkins R, Skirving AP. Callus formation and the rate of healing of femoral fractures in patients with head injuries. J Bone Joint Surg Br. 1987;69(4):521–4.
2. Takeda S, Karsenty G. Molecular bases of the sympathetic regulation of bone mass. Bone. 2008;42(5):837–40.
3. Baldock PA, Sainsbury A, Couzens M, et al. Hypothalamic Y2 receptors regulate bone formation. J Clin Invest. 2002;109(7):915–21.
4. Elefteriou F, Ahn JD, Takeda S, et al. Leptin regulation of bone resorption by the sympathetic nervous system and CART. Nature. 2005;434(7032):514–20.
5. Sato S, Hanada R, Kimura A, et al. Central control of bone remodeling by neuromedin U. Nat Med. 2007;13(10):1234–40.
6. Yadav VK, Oury F, Suda N, et al. A serotonin-dependent mechanism explains the leptin regulation of bone mass, appetite, and energy expenditure. Cell. 2009;138(5):976–89.
7. Yadav VK, Ryu JH, Suda N, et al. Lrp5 controls bone formation by inhibiting serotonin synthesis in the duodenum. Cell. 2008;135(5):825–37.
8. Gautron L, Elmquist JK. Sixteen years and counting: an update on leptin in energy balance. J Clin Invest 2011;121(6):2087–93.
9. Ducy P, Amling M, Takeda S, et al. Leptin inhibits bone formation through a hypothalamic relay: a central control of bone mass. Cell. 2000;100(2):197–207.
10. Pogoda P, Egermann M, Schnell JC, et al. Leptin inhibits bone formation not only in rodents, but also in sheep. J Bone Miner Res. 2006;21(10):1591–9.
11. Guidobono F, Pagani F, Sibilia V, et al. Different skeletal regional response to continuous brain infusion of leptin in the rat. Peptides. 2006;27(6):1426–33.
12. Shi Y, Yadav VK, Suda N, et al. Dissociation of the neuronal regulation of bone mass and energy metabolism by leptin in vivo. Proc Natl Acad Sci U S A. 2008;105(51):20529–33.
13. Takeda S, Elefteriou F, Levasseur R, et al. Leptin regulates bone formation via the sympathetic nervous system. Cell. 2002;111(3):305–17.
14. Bonnet N, Brunet-Imbault B, Arlettaz A, et al. Alteration of trabecular bone under chronic beta2 agonists treatment. Med Sci Sports Exerc. 2005;37(9):1493–501.
15. Kajimura D, Hinoi E, Ferron M, et al. Genetic determination of the cellular basis of the sympathetic regulation of bone mass accrual. J Exp Med. 2011;208(4):841–51.
16. Fu L, Patel MS, Bradley A, et al. The molecular clock mediates leptin-regulated bone formation. Cell. 2005;122(5):803–15.
17. Kondo H, Nifuji A, Takeda S, et al. Unloading induces osteoblastic cell suppression and osteoclastic cell activation to lead to bone loss via sympathetic nervous system. J Biol Chem. 2005;280(34):30192–200.
18. Yirmiya R, Goshen I, Bajayo A, et al. Depression induces bone loss through stimulation of the sympathetic nervous system. Proc Natl Acad Sci U S A. 2006;103(45):16876–81.
19. Bonnet N, Laroche N, Vico L, et al. Dose effects of propranolol on cancellous and cortical bone in ovariectomized adult rats. J Pharmacol Exp Ther. 2006;318(3):1118–27.
20. Tam J, Trembovler V, Di Marzo V, et al. The cannabinoid CB1 receptor regulates bone formation by modulating adrenergic signaling. FASEB J. 2008;22(1):285–94.
21. Pasco JA, Henry MJ, Sanders KM, et al. Beta-adrenergic blockers reduce the risk of fracture partly by increasing bone mineral density: Geelong Osteoporosis Study. J Bone Miner Res. 2004;19(1):19–24.
22. Wiens M, Etminan M, Gill SS, et al. Effects of antihypertensive drug treatments on fracture outcomes: a meta-analysis of observational studies. J Intern Med. 2006;260(4):350–62.
23. Shi Y, Oury F, Yadav VK, et al. Signaling through the M(3) muscarinic receptor favors bone mass accrual by decreasing sympathetic activity. Cell Metab. 2010;11(3):231–8.
24. Fonseca TL, Jorgetti V, Costa CC, et al. Double disruption of alpha2A- and alpha2C-adrenoceptors results in sympathetic hyperactivity and high-bone-mass phenotype. J Bone Miner Res. 2011;26(3):591–603.
25. Ducy P, Karsenty G. The two faces of serotonin in bone biology. J Cell Biol. 2010;191(1):7–13.
26. Lam DD, Leinninger GM, Louis GW, et al. Leptin does not directly affect CNS serotonin neurons to influence appetite. Cell Metab. 2011;13(5):584–91.
27. Wu Q, Bencaz AF, Hentz JG, et al. Selective serotonin reuptake inhibitor treatment and risk of fractures: a meta-analysis of cohort and case-control studies. Osteoporos Int. 2012;23(1):365–75.
28. Ortuno MJ, Robinson ST, Subramanyam P, et al. Serotonin-reuptake inhibitors act centrally to cause bone loss in mice by counteracting a local anti-resorptive effect. Nat Med. 2016;22(10):1170–9.
29. Modder UI, Achenbach SJ, Amin S, et al. Relation of serum serotonin levels to bone density and structural parameters in women. J Bone Miner Res. 2010;25(2):415–22.
30. Yadav VK, Balaji S, Suresh PS, et al. Pharmacological inhibition of gut-derived serotonin synthesis is a potential bone anabolic treatment for osteoporosis. Nat Med. 2010;16(3):308–12.
31. Cui Y, Niziolek PJ, MacDonald BT, et al. Lrp5 functions in bone to regulate bone mass. Nat Med. 2011;17(6):684–91.
32. Brighton PJ, Szekeres PG, Willars GB. Neuromedin U and its receptors: structure, function, and physiological roles. Pharmacol Rev. 2004;56(2):231–48.
33. Ding Y, Arai M, Kondo H, et al. Effects of capsaicin-induced sensory denervation on bone metabolism in adult rats. Bone. 2010;46(6):1591–6.
34. Offley SC, Guo TZ, Wei T, et al. Capsaicin-sensitive sensory neurons contribute to the maintenance of trabecular bone integrity. J Bone Miner Res. 2005;20(2):257–67.
35. Cornish J, Callon KE, Bava U, et al. Effects of calcitonin, amylin, and calcitonin gene-related peptide on osteoclast development. Bone. 2001;29(2):162–8.

36. Ishizuka K, Hirukawa K, Nakamura H, et al. Inhibitory effect of CGRP on osteoclast formation by mouse bone marrow cells treated with isoproterenol. Neurosci Lett. 2005;379(1):47–51.
37. Wang L, Shi X, Zhao R, et al. Calcitonin-gene-related peptide stimulates stromal cell osteogenic differentiation and inhibits RANKL induced NF-kappaB activation, osteoclastogenesis and bone resorption. Bone. 2010;46(5):1369–79.
38. Schinke T, Liese S, Priemel M, et al. Decreased bone formation and osteopenia in mice lacking alpha-calcitonin gene-related peptide. J Bone Miner Res. 2004;19(12):2049–56.
39. Zhang Y, Xu J, Ruan YC, et al. Implant-derived magnesium induces local neuronal production of CGRP to improve bone-fracture healing in rats. Nat Med. 2016;22(10):1160–9.
40. Fukuda T, Takeda S, Xu R, et al. Sema3A regulates bone-mass accrual through sensory innervations. Nature. 2013;497(7450):490–3.
41. Castaneda-Corral G, Jimenez-Andrade JM, Bloom AP, et al. The majority of myelinated and unmyelinated sensory nerve fibers that innervate bone express the tropomyosin receptor kinase A. Neuroscience. 2011;178:196–207.
42. Jimenez-Andrade JM, Mantyh WG, Bloom AP, et al. A phenotypically restricted set of primary afferent nerve fibers innervate the bone versus skin: therapeutic opportunity for treating skeletal pain. Bone. 2010;46(2):306–13.
43. Tomlinson RE, Li Z, Zhang Q, et al. NGF-TrkA Signaling by Sensory Nerves Coordinates the Vascularization and Ossification of Developing Endochondral Bone. Cell Rep. 2016;16(10):2723–35.
44. Blomqvist AG, Herzog H. Y-receptor subtypes—how many more? Trends Neurosci. 1997;20(7):294–8.
45. Lin S, Boey D, Couzens M, et al. Compensatory changes in [125I]-PYY binding in Y receptor knockout mice suggest the potential existence of further Y receptor(s). Neuropeptides. 2005;39(1):21–8.
46. Grundemar L, Hakanson R. Multiple neuropeptide Y receptors are involved in cardiovascular regulation. Peripheral and central mechanisms. Gen Pharmacol. 1993;24(4):785–96.
47. Chronwall BM, DiMaggio DA, Massari VJ, et al. The anatomy of neuropeptide-Y-containing neurons in rat brain. Neuroscience. 1985;15(4):1159–81.
48. Hokfelt T, Broberger C, Zhang X, et al. Neuropeptide Y: some viewpoints on a multifaceted peptide in the normal and diseased nervous system. Brain Res Brain Res Rev. 1998;26(2-3):154–66.
49. Lindefors N, Brene S, Herrera-Marschitz M, et al. Regulation of neuropeptide Y gene expression in rat brain. Ann N Y Acad Sci. 1990;611:175–85.
50. Ahmed M, Bjurholm A, Kreicbergs A, et al. Neuropeptide Y, tyrosine hydroxylase and vasoactive intestinal polypeptide-immunoreactive nerve fibers in the vertebral bodies, discs, dura mater, and spinal ligaments of the rat lumbar spine. Spine. 1993;18(2):268–73.
51. Hill EL, Turner R, Elde R. Effects of neonatal sympathectomy and capsaicin treatment on bone remodeling in rats. Neuroscience. 1991;44(3):747–55.
52. Lindblad BE, Nielsen LB, Jespersen SM, et al. Vasoconstrictive action of neuropeptide Y in bone. The porcine tibia perfused in vivo. Acta Orthop Scand. 1994;65(6):629–34.
53. Sisask G, Bjurholm A, Ahmed M, et al. The development of autonomic innervation in bone and joints of the rat. J Auton Nerv Syst. 1996;59(1-2):27–33.
54. Bjurholm A. Neuroendocrine peptides in bone. Int Orthop. 1991;15(4):325–9.
55. Bjurholm A, Kreicbergs A, Schultzberg M, et al. Neuroendocrine regulation of cyclic AMP formation in osteoblastic cell lines (UMR-106-01, ROS 17/2.8, MC3T3-E1, and Saos-2) and primary bone cells. J Bone Miner Res. 1992;7(9):1011–9.
56. Baldock PA, Lee NJ, Driessler F, et al. Neuropeptide Y knockout mice reveal a central role of NPY in the coordination of bone mass to body weight. PLoS One. 2009;4(12):e8415.
57. Elefteriou F, Takeda S, Liu X, et al. Monosodium glutamate-sensitive hypothalamic neurons contribute to the control of bone mass. Endocrinology. 2003;144(9):3842–7.
58. Baldock PA, Sainsbury A, Allison S, et al. Hypothalamic control of bone formation: distinct actions of leptin and y2 receptor pathways. J Bone Miner Res. 2005;20(10):1851–7.
59. Sainsbury A, Schwarzer C, Couzens M, et al. Y2 receptor deletion attenuates the type 2 diabetic syndrome of ob/ob mice. Diabetes. 2002;51(12):3420–7.
60. de Rijke CE, Hillebrand JJ, Verhagen LA, et al. Hypothalamic neuropeptide expression following chronic food restriction in sedentary and wheel-running rats. J Mol Endocrinol. 2005;35(2):381–90.
61. Lauzurica N, Garcia-Garcia L, Pinto S, et al. Changes in NPY and POMC, but not serotonin transporter, following a restricted feeding/repletion protocol in rats. Brain Res. 2010;1313:103–12.
62. Igwe JC, Jiang X, Paic F, et al. Neuropeptide Y is expressed by osteocytes and can inhibit osteoblastic activity. J Cell Biochem. 2009;108(3):621–30.
63. Kishi T, Elmquist JK. Body weight is regulated by the brain: a link between feeding and emotion. Mol Psychiatry. 2005;10(2):132–46.
64. Kopp J, Xu ZQ, Zhang X, et al. Expression of the neuropeptide Y Y1 receptor in the CNS of rat and of wild-type and Y1 receptor knock-out mice. Focus on immunohistochemical localization. Neuroscience. 2002;111(3):443–532.
65. Naveilhan P, Neveu I, Arenas E, et al. Complementary and overlapping expression of Y1, Y2 and Y5 receptors in the developing and adult mouse nervous system. Neuroscience. 1998;87(1):289–302.
66. Baldock PA, Allison SJ, McDonald MM, et al. Hypothalamic regulation of cortical bone mass: opposing activity of Y2 receptor and leptin pathways. J Bone Miner Res. 2006; 21:1600–7.

67. Shi YC, Lin S, Wong IP, et al. NPY neuron-specific Y2 receptors regulate adipose tissue and trabecular bone but not cortical bone homeostasis in mice. PLoS One. 2010;5(6):e11361.
68. Baldock PA, Allison SJ, Lundberg P, et al. Novel role of Y1 receptors in the coordinated regulation of bone and energy homeostasis. J Biol Chem. 2007;282(26):19092–102.
69. Lee NJ, Nguyen AD, Enriquez RF, et al. Osteoblast specific Y1 receptor deletion enhances bone mass. Bone. 2010;48(3):461–7.
70. Howlett AC, Barth F, Bonner TI, et al. International Union of Pharmacology. XXVII. Classification of cannabinoid receptors. Pharmacol Rev. 2002;54(2):161–202.
71. Mackie K. Signaling via CNS cannabinoid receptors. Mol Cell Endocrinol. 2008;286(1-2 Suppl 1):S60–5.
72. Idris AI, van't Hof RJ, Greig IR, et al. Regulation of bone mass, bone loss and osteoclast activity by cannabinoid receptors. Nat Med. 2005;11(7):774–9.
73. Karsak M, Cohen-Solal M, Freudenberg J, et al. Cannabinoid receptor type 2 gene is associated with human osteoporosis. Hum Mol Genet. 2005;14(22):3389–96.
74. Ofek O, Karsak M, Leclerc N, et al. Peripheral cannabinoid receptor, CB2, regulates bone mass. Proc Natl Acad Sci U S A. 2006;103(3):696–701.
75. Beltramo M, Campanella M, Tarozzo G, et al. Gene expression profiling of melanocortin system in neuropathic rats supports a role in nociception. Brain Res Mol Brain Res. 2003;118(1-2):111–8.
76. Nijenhuis WA, Oosterom J, Adan RA. AgRP(83-132) acts as an inverse agonist on the human-melanocortin-4 receptor. Mol Endocrinol. 2001;15(1):164–71.
77. Emmerson PJ, Fisher MJ, Yan LZ, et al. Melanocortin-4 receptor agonists for the treatment of obesity. Curr Top Med Chem. 2007;7(11):1121–30.
78. Farooqi IS, Yeo GS, Keogh JM, et al. Dominant and recessive inheritance of morbid obesity associated with melanocortin 4 receptor deficiency. J Clin Invest. 2000;106(2):271–9.
79. Ahn JD, Dubern B, Lubrano-Berthelier C, et al. Cart overexpression is the only identifiable cause of high bone mass in melanocortin 4 receptor deficiency. Endocrinology. 2006;147(7):3196–202.

132
Peripheral Neuronal Control of Bone Remodeling

Katherine J. Motyl[1] and Mary F. Barbe[2]

[1]*Maine Medical Center Research Institute, Maine Medical Center, Scarborough, ME, USA*
[2]*Department of Anatomy and Cell Biology, School of Medicine, Lewis Katz School of Medicine, Temple University, Philadelphia, PA, USA*

INTRODUCTION

Nerves have long been established to terminate within bone, with significant innervation observed in the periosteum, medullary cavity, and cortical bone. Both autonomic motor and sensory neurons innervate bone. This pattern of innervation changes during fracture repair and in pathological conditions, such as bone cancers [1], suggestive of a functional link between the skeleton and nervous system [2]. Furthermore, discrete defects in bone have been found in patients with neurological disorders, such as complex regional pain syndrome, and there is increasing evidence that the central nervous system (CNS) plays an important role in regulating bone turnover. Despite the field making great strides within the last 20 years, little is known about the basic interactions between bone cells and peripheral nerves, and how acutely and specifically this innervation modulates bone turnover. We will examine the existing anatomical evidence for innervation of bone, explore how innervation is modulated under physiological and pathological conditions, and then focus on recent literature that has expanded our understanding of how bone cells are influenced by local neuronal signals. It is important to point out that the nervous system also interacts with bone indirectly via endocrine pathways; however, we will focus primarily on evidence for peripheral nerves interacting directly with bone to modulate remodeling.

PERIPHERAL NERVES IN BONE

The autonomic nervous system contains both sympathetic and parasympathetic nerves that, in general, send a response from the CNS or spinal cord to the target tissue after an afferent sensory input. The cell bodies of postsynaptic sympathetic and sensory neurons are housed in sympathetic and dorsal root ganglia (DRG), respectively, located segmentally along the vertebral column. For example, sensory neurons that innervate the hindlimbs of mice originate in the L4 and L5 DRG. In contrast, postsynaptic parasympathetic neurons typically originate in small parasympathetic ganglia near the target tissue. Sensory neurons are not limited to afferent functions; some secrete neuropeptides like calcitonin gene-related peptide (CGRP) or substance P that induce an effect in the target tissue. An interaction between the nervous system and bone lies in evidence that nerves terminate in bone marrow (BM), cortical bone, and periosteum in both the axial and appendicular skeleton [3, 4]. Myelinated and nonmyelinated nerve terminals are also associated with blood vessels in bone or in close proximity to osteoblasts and osteoclasts [5]. Furthermore, developmental studies in rats and mice have revealed autonomic and sensory innervation of long bones as early as embryonic day 14.5, with timing similar to tissue vascularization [6]. A variety of work in both animal models and humans has demonstrated a general positive role for innervation in bone development and maintenance of

Primer on the Metabolic Bone Diseases and Disorders of Mineral Metabolism, Ninth Edition. Edited by John P. Bilezikian.
© 2019 American Society for Bone and Mineral Research. Published 2019 by John Wiley & Sons, Inc.
Companion website: www.wiley.com/go/asbmrprimer

homeostasis. For example, denervation in rats and mice, either surgically or chemically (with capsaicin, which destroys capsaicin-sensitive sensory neurons), reduces bone growth, volume, and fracture healing [7, 8]. Mice with global and neuron-specific deletions of Semaphorin 3A, an axonal chemorepellent, show impaired bone development and reduced sensory innervation of bone [9]. More recently, nerve growth factor (NGF) signaling through neurotrophic tyrosine kinase receptor type 1 (TrkA) was shown to be required for innervation of the femur and vascularization, and induction of primary and secondary ossification [6].

INNERVATION OF BONE IS DYNAMIC

Like bone tissue, the innervation of bone tissue and marrow is highly dynamic; however, changes in innervation do not necessarily correspond to changes in bone remodeling. Bone fracture is associated with a dramatic increase in innervation of sympathetic and sensory neurons during the healing process [10]. Interestingly, inhibition of NGF/TrkA signaling via anti-NGF or anti-TrkA treatment reduces fracture pain, but does not appear to delay healing as complete denervation does [11, 12]. Cancer metastases to bone are associated with bone loss, but also an increase in innervation and bone pain (which can also be relieved in animal models by anti-NGF therapy) [13].

There are also reports of reduced nerve fiber density in certain diseases and conditions in which bone loss occurs. For example, 14 days after ovariectomy (OVX) in rats, there is a significant reduction of nerve fiber density in the tibia in association with bone loss [14]. Furthermore, neuropathy is a common side effect of both type 1 and type 2 diabetes mellitus (T1DM and T2DM). Mouse models of these diseases (streptozotocin and *ob/ob* mice) are associated with reduced sympathetic innervation of the BM compartment, which impairs hematopoietic stem cell mobilization in the BM niche [15, 16]. It is not known whether sensory or sympathetic neuropathy is necessary for reduced bone formation in T1DM, or if other complications (ie, microvascular dysfunction and hyperglycemia) are more direct contributors. Interestingly, in a study of aging rats, there were no differences in CGRP or neurofilament 200-positive sensory nerve fiber density in the rat periosteum, or pain response to fracture, between 4-, 13-, and 36-month-old rats, despite a drastic decrease in trabecular and cortical bone volume [17]. It remains unclear how aging modifies sympathetic and sensory innervation of the BM compartment.

Accumulating evidence supports a strong modulatory role for the peripheral nervous system in bone tissues in response to mechanical loading, whether during healing post fracture or cyclical loading. A dense network of nerve endings that are CGRP, substance P, neuropeptide Y (NPY), vasoactive intestinal peptide (VIP), and/or glutamate immunopositive is present in periosteum, BM, and on metaphyseal trabecular bone, although more abundant in the first two structures [3, 4, 18]. This network has mechanosensing responses, showing site-specific sprouting after bone fracture. For example, both CGRP and substance P immunoreactive nerve endings increase during the healing phase in callus sites with maximum bone loading and formation, ie, the bony callus on the concave side of angulated fractures, compared to the less loaded convex side or concave side of a straight fracture [19]. NPY shows temporal and side-specific changes during angular fracture healing [20]. It increases early (during inflammation) in the concave side, as well as later (during remodeling) in the convex side in parallel with bone resorption and callus reduction [20]. Other functions of neurotransmitters in bone will be discussed further in the next section.

The way in which peripheral nerves modulate bone homeostasis is circumstance-dependent, with bone innervation being responsible for bone pain, maintenance of the hematopoietic niche, and response to mechanical loading. It is also important to note that innervation also plays an important role in regulating vascular tone, which could in turn impact the microenvironment to which bone cells are exposed. Preclinical studies have described important roles for specific neurotransmitters in everyday bone homeostasis, and in response to some of the paradigms described previously (ie, loading, aging, OVX). Context-dependent functions of neurotransmitters in bone are outlined in the following sections and in Table 132.1.

FUNCTION OF NEUROTRANSMITTERS IN BONE

Acetylcholine

Although the evidence for a sympathetic nervous system (SNS)-mediated effect on bone is clear (see later in this chapter), the extent to which the parasympathetic nervous system (PSNS) plays a regulatory role in bone is less clear. Acetylcholine (ACh) is the neurotransmitter associated with the PSNS that binds to nicotinic ACh receptors (nAChR) and muscarinic receptors. Several of the five subtypes of muscarinic and twelve subunits of nicotinic receptors are expressed in osteoblasts, osteoclasts, and osteocytes [21–24]. Deletion of the α2nAChR results in low trabecular bone mass in the femur and spine with increased bone resorption [21]. ACh either does not change or suppresses osteoclast differentiation [25, 26] and suppresses osteoblast proliferation and alkaline phosphatase activity [23]. However, PSNS innervation of bone is less clear and changes associated with fracture, OVX, cancer, aging, etc., are unknown.

Calcitonin gene-related peptide

Calcitonin gene-related peptide (CGRP) is abundant in sensory neurons, with roles in pain transmission and vasodilation [3]. CGRP has been studied extensively in

Table 132.1. Neurochemical signaling systems present in bone.

Neurochemical	Positive Nerves in Bone	Expression of Receptors and Transporters on Bone Cells	Function
Acetylcholine (ACh)	Potentially vesicular ACh transporter expression in bone	Both muscarinic and nicotinic receptors found in osteoblasts, osteoclasts, and osteocytes	Reduced/unchanged osteoclast differentiation, reduced osteoblast proliferation, increased viability of MLO-Y4 osteocytes
Calcitonin gene-related peptide (CGRP)	CGRP-positive nerve endings in the periosteum and bone marrow	Calcitonin receptors (CTR) and calcitonin receptor-like receptors (CRLR) on osteoblasts, preosteoblasts, osteoclasts, osteocytes, and bone-lining cells	Activates osteoblast-induced bone formation and inhibits osteoclast resorptive activity through the RANKL/OPG pathway
Dopamine (DA)	TH-positive nerve endings in bone marrow and periosteum. DA itself identified in bone marrow	D1, D2, D3, D4, and D5 mRNA expression in MC3T3-E1 cells by nested PCR and protein detected on the surface of human osteoclast precursors by flow cytometry	DA inhibits MSC migration but modestly improves osteoblast mineralization. DA suppresses osteoclastogenesis through D2-like signaling
Glutamate	No	Ionotropic glutamate receptors (eg, NMDA) are present on osteoblasts and bone lining cells; metabotropic receptors (eg, GluR8) are present on osteoclasts	Mechanical loading reduces GluR2/3, GluR4, GluR5,6,7, and NMDAR2a in osteoclasts and bone lining cells in response to 4 days of cyclical compressive loading
Neuropeptide Y (NPY)	Peripheral nerves containing NPY identified in early postnatal and adult long bones and calvaria	Y1 receptors, but not Y2 receptors, detected in osteoblasts in vivo	Y1$^{-/-}$ mice have more bone formation through nonhypothalamic pathways, whereas osteoblast/osteocyte overexpression of NPY reduces bone formation
Norepinephrine (NE)	Sympathetic, tyrosine hydroxylase positive nerve endings found in bone marrow and periosteum	Expression of α1AR, all α2AR, β2AR, and to a lesser extent β1AR and β3AR, shown on osteoblasts. Osteoclasts express βARs and αARs	Activation of βARs directly inhibits osteoblast activity and increases osteoclastogenesis in RAW 264.7 cells and human osteoclast-like cells
Purines and pyrimidines	Released by osteoblasts from shear stress, released from neurons, but not yet shown in bone itself	Osteoblasts express the P2X2R, P2X5R, and P2Y2R subtypes; osteoclasts express the P2X2R, P2X4R, and P2X7R subtypes	ATP and ADP can stimulate osteoclastogenesis and bone resorption through the P2Y$_1$ receptor. ATP can stimulate resorption through the P2X$_2$ receptor. ATP and UTP suppress bone formation through the P2Y$_2$ receptor
Serotonin (5-HT)	Identified in DRG only by Tph1 mRNA, not yet in bone itself	5-HT$_{2A,2B,2C}$ mRNA expression in MC3T3-E1 cells and fetal chicken primary osteoblasts, osteocytes, and periosteal fibroblasts MLO-Y4 osteocytes and MC3T3-E1 osteoblasts express 5-HTT, 5-HT$_{1A}$, and 5-HT$_{2A}$ by western blot, IHC, and mRNA. 5-HT$_{2A, 2B, 2C}$ and 5-HTT mRNA in human PBMCs differentiated with RANKL and M-CSF	Serotonin receptor agonist α-methyl-5-HT stimulated proliferation of primary chicken periosteal cells but not osteoblasts Serotonin increased PBMC osteoclast differentiation and RAW264.7 cell proliferation, as well as increased proliferation of normal human osteoblasts and MC3T3-E1 cells [70]. Serotonin stimulated OPG and suppressed RANKL expression in MC3T3-E1 cells
Substance P	Substance P nerves found in periosteum, especially post fracture, and in bone marrow and on trabeculae	Neurokinin-1 receptor (NK-1) on osteoclasts > osteoblasts and osteocytes	Stimulates late-stage osteoblastic bone formation, and post-fracture bone formation, but may be involved in inhibiting bone resorption during callus remodeling
Vasoactive intestinal peptide (VIP)/ pituitary adenylate cyclase activating peptides (PACAP)	Positive nerves can be found in periosteum (VIP/PACAP), synovial joint membranes (VIP), and bone marrow (PACAP)	Osteoblasts and osteogenic cells express VPAC1 and/or PAC1, dependent on the cell line; osteoclasts express VPAC1 and PAC1 receptors	Dependent on the osteoblastic cell type/line, VIP/PACAP can increase or inhibit osteogenesis markers and differentially alter expression of components of the RANKL/OPG pathway (thus enhancing or suppressing osteoclastogenesis) VIP/PACAP inhibits osteoclast formation and resorptive activity

ADP = adenosine diphosphate; ATP = adenosine triphosphate; DRG = dorsal root ganglia; IHC = immunohistochemistry; M-CSF = macrophage colony-stimulating factor; MSC = mesenchymal stem cell; NMDA = N-methyl-D-aspartate; OPG = osteoprotegerin; UTP = uridine triphosphate.

bone because of its presence in nerve endings in periosteum and bone marrow [3, 18], its differentiative effects on osteoblasts [27], and inductive effects on osteogenic molecules such as bone morphogenetic protein 2 (BMP2) [27]. CGRP activates osteoblast-induced bone formation and inhibits osteoclast resorptive activity through the receptor activator of NF-κB ligand/osteoprotegerin (RANKL/OPG) pathway, in a manner similar to mechanotransduction through oscillatory fluid flow (OFF)-induced shear stress [28]. CGRP levels increase in bone in an ulna end-loading model, concomitant with increased periosteal and total bone area in the loaded bones [29]. Adaptive modeling responses were also observed in distant bones in this model, such as in the contralateral unloaded ulna, suggestive of neuronal signaling crosstalk between the limbs [30]. These responses were ameliorated when peripheral nerve signaling was temporarily blocked with brachial plexus anesthesia (BPA). Load-induced bone formation is also decreased in CGRP-α knockout mice, but not in CGRP-β knockout mice, and even further with BPA [29]. Thus, CGRP-α release from nerve fibers in bone is implicated in osteoblast-induced bone formation and inhibition of osteoclast resorptive activity.

Dopamine

Dopamine (DA) is a precursor to epinephrine and norepinephrine yet also plays a significant role as a neurotransmitter involved in a variety of physiological functions, including hormone secretion, reward, mood, and movement [31, 32]. Tyrosine hydroxylase (TH), the enzyme responsible for converting tyrosine to L-dopa, upstream of DA and norepinephrine (NE), is used as a marker for dopaminergic neurons in the CNS and as a marker of peripheral sympathetic nerves. The effects of DA are elicited by a family of G-protein coupled receptors (D1–5) which signal through distinct pathways, with D1-like receptors (D1 and D5) stimulating adenylate cyclase (AC) to produce cyclic adenosine monophosphate (cAMP), and D2-like receptors (D2, D3, and D4) inhibiting AC [33]. DA is present in BM and receptor expression has been demonstrated in osteoblasts and osteoclasts [34, 35]. DA concentrations similar to those in a synapse negatively regulate murine BM mesenchymal stem cell (BM-MSC) actin polymerization and migration in response to VEGF [36, 37]. However, another study found high concentrations (50 μM) of DA moderately improved mineralization of MC3T3-E1 cells [38]. DA and a D2R agonist suppress osteoclast differentiation in human CD14+ cells isolated from peripheral blood mononuclear cells [35]. Similarly, the antipsychotic drug risperidone (which is largely a D2R antagonist) increased osteoclast differentiation from primary murine BM-MSCs [39]. Interestingly, mice with a deletion of the DA transporter (DAT) gene, which is responsible for dopamine reuptake into presynaptic terminals, have low trabecular bone mass [40]. Whether these findings resulted from changes in formation or resorption is unclear and no studies have determined whether the effects of DAT deletion on bone could have been mediated within the bone itself or through the CNS.

Glutamate

Glutamate is best known as the major excitatory transmitter in the brain. In bone, glutamatergic signaling mechanisms respond to very fast stimulatory signals and involve multiple glutamatergic receptor subtypes and transporters [41]. The initiating stimulus for glutamate release from osteoblasts remains unclear [41]. Mason proposed in 2004 that mechanical loading may open stretch-sensitive calcium channels in osteocytes, triggering the release of glutamate from osteocytes that then activates osteoblast receptors such as N-methyl-D-aspartate (NMDA, an ionotropic glutamate receptor) [42]. Osteoblastic NMDA receptors show classical voltage-sensitive Mg^{2+} blockade and may function as detectors of coincident receptor activation and membrane depolarization [43]. Transcripts for metabotropic glutamate receptors, such as *mGluR1b*, *mGluR4*, and *mGluR8*, have been detected in osteoblasts and osteoclasts, and the mGluR8 protein has been detected in osteoclasts. Mechanical loading of bone regulates glutamate receptor expression, with immunoexpression of GluR2/3, GluR4, GluR5,6,7, and NMDAR2a decreasing in osteoblasts and bone-lining cells in response to 4 days of cyclical compressive loading [44]. On the flipside, disuse-induced bone loss is also associated with a downregulation of NMDA receptors [45]. Thus, changes in bone mass have been linked to changes in glutamate signaling components in bone cells, although studies have also identified glutamatergic-containing nerves in the periosteum [4].

Norepinephrine

Sympathetic fibers are generally associated with increasing heart rate, vasoconstriction, and increased respiration rate. Norepinephrine (NE) is the neurotransmitter responsible for transmitting sympathetic signals to target tissues after release from sympathetic nerve terminals, which are immunopositive for TH. NE binds to α- and β-adrenergic receptors (AR), and the response is dependent upon the specific receptors present. Osteoblasts express β2AR and to a lesser extent β1AR and β3AR, each of which interacts with $Gα_s$ to stimulate AC and cAMP production [46]. Downstream effects of selective and nonselective β-agonists are generally downregulation of bone formation and increased *Rankl* production, whereas inhibition of βARs with antagonists or genetic deletion is protective of bone loss from aging or other conditions in which sympathetic activity is high, eg, leptin treatment, stress, and exposure to the atypical antipsychotic drug risperidone [47–50]. Osteoblasts express α1AR, which modulates clock gene expression and regulates cyclic expression of *Bmp4* and *Osteoprotegerin* [34, 51]. Furthermore,

$\alpha_{2A,2B\&2C}$ARs are also expressed on osteoblasts and α_{2C}AR deletion in particular has site-specific effects on bone remodeling [52]. Osteoclasts also express adrenergic receptors, and stimulation directly increases osteoclastogenesis [53, 54]. Early studies did not support a major influence of the SNS on functional adaptation of bone to mechanical loading [55, 56], although the SNS appeared to modulate unloading-induced bone loss [55]. More recently, deletion of the β1AR, but not β2AR, has been shown to impair bone formation after axial compression loading of the tibia [57]. The norepinephrine transporter (NET) is responsible for reuptake of NE in neurons and is expressed in several osteoblast models [58]. Deletion of NET and treatment with an antagonist both cause bone loss via reduced bone formation and increased resorption.

Several clinical studies support a role for the SNS in bone turnover. One study found that although unloading of bone was similar among all paraplegic patients studied, patients with a high level (T4 to T7) spinal cord injury (SCI) had a significant inverse correlation between duration of paralysis and stress–strain index. These patients, unlike those with lower thoracic SCI (T8 to T12), were susceptible to autonomic dysreflexia, which could lead to heightened SNS output to bone [59]. Similarly, patients with pheochromocytomas, which are rare catecholamine-producing tumors, have higher levels of CTx, but not P1NP, levels that normalize after resection of the tumor [60]. Complex regional pain syndrome type I is also associated with increased bone resorption [61] and higher resting heart rate is also an independent predictor of fracture risk [62]. Additionally, some but not all studies find a reduced risk of osteoporotic fractures in patients taking β-adrenergic receptor antagonists [63, 64].

Neuropeptide Y

NPY has diverse functions, including controlling feeding behavior, memory formation, induction of vasoconstriction, and energy metabolism. NPY is produced at several sites, both centrally and peripherally, including the hypothalamus, adrenal medulla, pancreatic cells, sympathetic nerves (eg, those innervating skeleton [20] in which it is costored and released with NE), and cells of osteoblastic lineage. It signals through G-protein linked Y receptors. Y2 receptors are responsible for the neuroinhibitory effects of NPY. Their activation in the hypothalamus suppresses whole body bone formation, whereas global or hypothalamic deletion of Y2 receptors leads to a generalized increase in bone volume, mineral apposition rate, and osteoprogenitor cells [65]. A global deletion of the Y1 receptor, a receptor identified on osteocytes and osteoblasts, also increases trabecular and cortical bone, and osteoblast activity [66]. Its deletion in osteoblasts or BM-MSCs prevents any effects of NPY and leads to increased osteoblast activity and bone mass [5, 66]. Release of NPY from skeletal nerves has direct, local inhibitory effects on bone formation by acting on Y1 receptors to inhibit cAMP and ERK pathways, leading to downregulation of osteoblast gene expression and activity [5, 67]. Most notably, during stress, peripheral sources of NPY signal Y1 receptors on osteoblasts, reducing their response to mechanical loading-induced bone formation, a finding suggesting that NPY's central actions are integrated with local bone activity. In summary, osteogenic cells respond to NPY in an inhibitory manner, acting to oppose CGRP, substance P, and VIP, which primarily increase bone formation and reduce bone resorption.

Serotonin

Serotonin is a monoamine neurotransmitter produced in neurons and in enterochromaffin cells of the gut by tryptophan hydroxylase 2 and 1, respectively. Clinical studies suggest that selective serotonin reuptake inhibitors (SSRIs) are associated with reduced BMD and increased fracture risk [68]. The majority of evidence regarding the role of serotonin in regulating bone remodeling points to: (i) serotonin acting as a neurotransmitter in the brain to inhibit SNS output to bone; and (ii) serotonin production in the gut by enterochromaffin cells expressing tryptophan hydroxylase 1 (Tph1) negatively regulates bone accrual through endocrine pathways [69]. However, some suggest that bone cells themselves produce serotonin, because both osteoclasts and osteoblasts have been shown to express Tph1 [2, 70, 71]. Serotonin transporter (5-HTT), 5-HT$_{1A,1B}$, and 5-HT$_{2A,2B,2C}$ mRNA expression (and in some cases protein) have been shown in a variety of osteoblast lineage models, including MC3T3-E1 cells, fetal chicken primary osteoblasts, osteocytes, and periosteal fibroblasts, and MLO-Y4 osteocytes [71, 72]. In some models, direct serotonin treatment increases proliferation and OPG expression while suppressing RANKL [70]. In vivo, 5-HT$_{1B}$ knockout mice (global and osteoblast-specific) have high bone mass resulting from high bone formation, whereas global deletion of 5-HT$_{2A}$ and osteoblast-specific deletion of 5-HT$_{2B}$ have no effect [69, 73]. 5-HT$_{2A,2B,2C}$ and 5-HTT mRNA are also expressed in human PBMCs differentiated with RANKL and macrophage colony-stimulating factor (M-CSF), and serotonin directly increased PBMC osteoclast differentiation as well as RAW264.7 cell proliferation [70]. However, whether peripherally derived serotonin has direct neurotransmitter capabilities in bone is less clear. There is evidence of Tph2, the neuron-specific tryptophan hydroxylase, within the L4 to L5 DRG, which house neuronal cell bodies responsible for sensory innervation of leg bones [74]. Thus, an additional unexplored pathway through which serotonin could impact bone is through direct innervation of bone by serotonergic neurons.

Substance P

Substance P is an undecapeptide and tachykinin. In addition to its role as a neurotransmitter involved in pain transmission in sensory nerves, it has essential roles in gut motility, exacerbation of inflammation in central and peripheral tissues (eg, skin, gastrointestinal tract, and lungs), and activation of endothelial cell retraction and vascular smooth muscle dilation. In bone, its preferred receptor, the neurokinin-1 receptor (NK-1R), is expressed by osteoblasts, osteocytes, and osteoclasts, and substance P-NK-1 binding can stimulate late-stage osteoblastic bone formation [75]. After capsaicin treatment of the sciatic nerve in adult rats, substance P (and CGRP) levels decreased in the nerve and tibia [8], as did tibial BMD and strength as a result of reduced bone formation and increased resorption [8]. Bones that are denervated as a consequence of peripheral nerve injury also show significant bone loss and increased fracture risk [76]. Interestingly, in a fracture study in rats, the density of substance P immunopositive nerves altered in temporally different manners in areas of bone formation versus bone resorption [19]. There was increased sprouting of substance P+ nerves that coexpressed GAP-43 (a nerve growth marker) into the fracture hematoma on day 3 post fracture (presumably as a consequence of inflammatory processes). The density of substance P+/GAP-43+ nerves remained high in the periosteum until cortical bridging had occurred on the concave side of angulated fractures (day 21). In contrast, substance P+ nerves that did not coexpress GAP-43 increased on the convex side on day 35 and remained increased until resorptive remodeling of the callus had completed. Combined, this suggests that substance P has both a stimulatory role on bone formation [8, 75] and a resorptive role during bone remodeling [19].

Purines and pyrimidines

Extracellular nucleotides, such as adenosine 5'-triphosphate (ATP) and uridine triphosphate (UTP), are soluble factors that are key components of bone's mechanotransduction system. ATP and UTP are released from neurons and osteocytes in response to mechanical stimulation such as fluid shear stress [77, 78], and bind to purinoreceptors (P2Rs) that are known modulators of osteoblast function. P2 receptors can be divided into subfamilies, in which seven metabotropic P2X and eight ionotropic P2Y receptors have been identified. Osteoblasts express P2X2R, P2X5R, and P2Y2R, whereas osteoclasts express P2X2R, P2X4R, and P2X7R [79]. Past studies suggest that ATP stimulates osteoclastogenesis and bone resorption through P2Y1R, and resorption through P2X2R [77]. ATP and UTP mediate inhibition of osteoblast bone mineralization through the osteoclastic P2X receptors that then activate ERK1/2 signaling [77, 80]. A recent study shows that P2X7R and pannexin 1 (Panx1), a mechanosensitive channel, form a functional complex that may provide a pathway for OFF-induced ATP release from osteocytes [78]. Interestingly, exposure to high glucose typically encountered with T1DM was associated with blunted load-induced ATP signaling (reduced P2R and Panx1 expression, and lowered flow-induced calcium signaling responses and ATP release from osteocytes) [78].

Vasoactive intestinal peptide and pituitary cyclase activating polypeptide

VIP and pituitary cyclase activating polypeptide (PACAP) are both in the VIP–secretin– growth hormone–releasing hormone (GHRH)–glucagon superfamily with 68% homology in function. Both are present on skeletal nerves located in the periosteum [20]. Their biological actions are mediated by three G-protein coupled transmembrane receptors: PACAP type 1 receptors (PAC1), and VIP receptors 1 and 2 (VPAC1 and VPAC2). Both VIP and PACAP bind equally to VIP receptors, yet PACAP binds with higher affinity to PAC1. Osteoblasts, osteoblast cell lines, and osteoclasts express different functional subtypes of these receptors. Mouse calvarial osteoblasts express the VPAC2 receptor, which when stimulated by VIP elevates intracellular cAMP and causes increased expression of osteogenesis markers (increased alkaline phosphatase and enhanced mineralization) [81]. Similarly, stimulation of UMR-106 osteoblastic tumor-like cells (which express only PAC1 [82]) increases alkaline phosphatase and other osteogenic proteins, albeit through noncanonical PACAP signaling pathways (BMPs/Smad1 and Hedgehog) [82]. In contrast, MC3T3-E1 cells express only the VPAC2 receptor, which when activated by VIP or PACAP causes cAMP accumulation, inhibition of alkaline phosphatase RNA expression (thus, inhibition of osteoblastic differentiation rather than enhancement), and increased release of IL-6, a stimulator of osteoclast resorptive activity [83].

There is also variation across cell types and studies in the induction pattern of osteoclastogenic factors in osteoblasts by VIP and PACAP. In one study [84], VIP increased *Rankl* and decreased *Opg* mRNA expression in mouse calvarial osteoblasts via cAMP and ERK pathways, changes that would enhance osteoclastic activity [84]. MC3T3-E1 cells showed similar changes in *Rankl* and *Opg* expression, whereas BM stromal cells and UMR-106 cells showed no changes in these factors after VIP treatment [84]. However, in another study, levels of RANKL decreased and those of OPG increased in MC3T3-E1 cells after VIP or CGRP treatment (changes that would suppress bone resorption), matching their response to OFF-induced shear stress [28]. VIP/PACAP also bind to VPAC1 and PAC1 receptors on osteoclasts (which lack VPAC2). Such binding inhibits osteoclast formation and resorptive activity, cholinergic agonist carbamylcholine-induced apoptosis, and cathepsin K activity, via increased AC and cAMP signaling. In summary, although more studies are needed to fully understand the role of VIP and PACAP in bone, they are clearly involved in regulating the activity of both osteoblasts and osteoclasts, and thus play a vital role in bone remodeling.

CONCLUSION

Innervation of bone is required for normal development and fracture healing. There is a clear role for the SNS (through NE and NPY) in causing bone loss in clinical situations as well as in animal models, and some evidence for an opposing role of the PSNS in bone. Sensory neurotransmitters (such as CGRP, substance P, and VIP) as well as extracellular nucleotides are important for response to loading and maintenance of normal bone mass. However, data supporting direct, peripheral neuronal roles for neurotransmitters like dopamine, glutamate, and serotonin are only beginning to emerge. Despite the breadth of research in this field, we are limited in our ability to understand how peripheral nerves directly interact with bone cells for a variety of reasons. Neuron-specific deletions are not bone-nerve-specific, thus any neuron-specific deletion could be indirectly affecting bone through the CNS or another peripheral tissue. Another limitation is that despite histological evidence for innervation of bone, synapses with osteoblasts, osteocytes, or osteoclasts have not been identified. However, some investigators have shown direct interactions between neurons in culture and osteoblasts and osteoclasts, which may prove to be a useful tool in discerning cellular and molecular mechanisms [6, 85, 86]. Additionally, basic roles for other neurotransmitters, such as endogenous opioids, have not yet been established in bone. Furthermore, the effects of some neurotransmitters (like VIP/PACAP) are varied from study to study, with cell model, animal model, and experimental design likely contributing to differences. Despite these limitations, there is an ever-growing number of publications to support a homeostatic and pathophysiological role of peripheral bone innervation in modulating bone remodeling, and further research investigating these interactions could lead to novel therapeutics and treatment strategies.

REFERENCES

1. Mach DB, Rogers SD, Sabino MC, et al. Origins of skeletal pain: sensory and sympathetic innervation of the mouse femur. Neuroscience. 2002;113(1):155–66.
2. Warden SJ, Bliziotes MM, Wiren KM, et al. Neural regulation of bone and the skeletal effects of serotonin (5-hydroxytryptamine). Mol Cell Endocrinol. 2005;242(1-2):1et al9.
3. Martin CD, Jimenez-Andrade JM, Ghilardi JR, et al. Organization of a unique net-like meshwork of CGRP+ sensory fibers in the mouse periosteum: implications for the generation and maintenance of bone fracture pain. Neurosci Lett. 2007;427(3):148–52.
4. Serre CM, Farlay D, Delmas PD, et al. Evidence for a dense and intimate innervation of the bone tissue, including glutamate-containing fibers. Bone. 1999;25(6):623–9.
5. Franquinho F, Liz MA, Nunes AF, et al. Neuropeptide Y and osteoblast differentiation—the balance between the neuro-osteogenic network and local control. FEBS J. 2010;277(18):3664–74.
6. Tomlinson RE, Li Z, Zhang Q, et al. NGF-TrkA Signaling by Sensory Nerves Coordinates the Vascularization and Ossification of Developing Endochondral Bone. Cell Rep. 2016;16(10):2723–35.
7. Heffner MA, Anderson MJ, Yeh GC, et al. Altered bone development in a mouse model of peripheral sensory nerve inactivation. J Musculoskelet Neuronal Interact. 2014;14(1):1–9.
8. Offley SC, Guo TZ, Wei T, et al. Capsaicin-sensitive sensory neurons contribute to the maintenance of trabecular bone integrity. J Bone Miner Res. 2005;20(2):257–67.
9. Fukuda T, Takeda S, Xu R, et al. Sema3A regulates bone-mass accrual through sensory innervations. Nature. 2013;497(7450):490–3.
10. Jimenez-Andrade JM, Bloom AP, Mantyh WG, et al. Capsaicin-sensitive sensory nerve fibers contribute to the generation and maintenance of skeletal fracture pain. Neuroscience. 2009;162(4):1244–54.
11. Majuta LA, Longo G, Fealk MN, et al. Orthopedic surgery and bone fracture pain are both significantly attenuated by sustained blockade of nerve growth factor. Pain. 2015;156(1):157–65.
12. Rapp AE, Kroner J, Baur S, et al. Analgesia via blockade of NGF/TrkA signaling does not influence fracture healing in mice. J Orthop Res. 2015;33(8):1235–41.
13. McCaffrey G, Thompson ML, Majuta L, et al. NGF blockade at early times during bone cancer development attenuates bone destruction and increases limb use. Cancer Res. 2014;74(23):7014–23.
14. Burt-Pichat B, Lafage-Proust MH, Duboeuf F, et al. Dramatic decrease of innervation density in bone after ovariectomy. Endocrinology. 2005;146(1):503–10.
15. Albiero M, Poncina N, Tjwa M, et al. Diabetes causes bone marrow autonomic neuropathy and impairs stem cell mobilization via dysregulated p66Shc and Sirt1. Diabetes. 2014;63(4):1353–65.
16. Busik JV, Tikhonenko M, Bhatwadekar A, et al. Diabetic retinopathy is associated with bone marrow neuropathy and a depressed peripheral clock. J Exp Med. 2009;206(13):2897–906.
17. Jimenez-Andrade JM, Mantyh WG, Bloom AP, et al. The effect of aging on the density of the sensory nerve fiber innervation of bone and acute skeletal pain. Neurobiol Aging. 2012;33(5):921–32.
18. Hill EL, Elde R. Distribution of CGRP-, VIP-, D beta H-, SP-, and NPY-immunoreactive nerves in the periosteum of the rat. Cell Tissue Res. 1991;264(3):469–80.
19. Li J, Ahmed M, Bergstrom J, et al. Occurrence of substance P in bone repair under different load comparison of straight and angulated fracture in rat tibia. J Orthop Res. 2010;28(12):1643–50.
20. Long H, Ahmed M, Ackermann P, et al. Neuropeptide Y innervation during fracture healing and remodeling. A study of angulated tibial fractures in the rat. Acta Orthop. 2010;81(5):639–46.
21. Bajayo A, Bar A, Denes A, et al. Skeletal parasympathetic innervation communicates central IL-1 signals regulating bone mass accrual. Proc Natl Acad Sci U S A. 2012;109(38):15455–60.
22. Liu PS, Chen YY, Feng CK, et al. Muscarinic acetylcholine receptors present in human osteoblast and bone tissue. Eur J Pharmacol. 2011;650(1):34–40.

23. Sato T, Abe T, Chida D, et al. Functional role of acetylcholine and the expression of cholinergic receptors and components in osteoblasts. FEBS Lett. 2010;584(4):817–24.
24. Ma Y, Li X, Fu J, et al. Acetylcholine affects osteocytic MLO-Y4 cells via acetylcholine receptors. Mol Cell Endocrinol. 2014;384(1-2):155–64.
25. Mandl P, Hayer S, Karonitsch T, et al. Nicotinic acetylcholine receptors modulate osteoclastogenesis. Arthritis Res Ther. 2016;18:63.
26. Ternes S, Trinkaus K, Bergen I, et al. Impact of acetylcholine and nicotine on human osteoclastogenesis in vitro. Int Immunopharmacol. 2015;29(1):215–21.
27. Tian G, Zhang G, Tan YH. Calcitonin gene-related peptide stimulates BMP-2 expression and the differentiation of human osteoblast-like cells in vitro. Acta Pharmacol Sin. 2013;34(11):1467–74.
28. Yoo YM, Kwag JH, Kim KH, et al. Effects of neuropeptides and mechanical loading on bone cell resorption in vitro. Int J Mol Sci. 2014;15(4):5874–83.
29. Sample SJ, Hao Z, Wilson AP, et al. Role of calcitonin gene-related peptide in bone repair after cyclic fatigue loading. PLoS One 2011;6(6):e20386.
30. Sample SJ, Heaton CM, Behan M, et al. Role of calcitonin gene-related peptide in functional adaptation of the skeleton. PLoS One. 2014;9(12):e113959.
31. Pijl H. Reduced dopaminergic tone in hypothalamic neural circuits: expression of a "thrifty" genotype underlying the metabolic syndrome? Eur J Pharmacol. 2003;480(1-3):125–31.
32. Parekh PK, Ozburn AR, McClung CA. Circadian clock genes: effects on dopamine, reward and addiction. Alcohol. 2015;49(4):341–9.
33. Missale C, Nash SR, Robinson SW, et al. Dopamine receptors: from structure to function. Physiol Rev. 1998;78(1):189–225.
34. Hirai T, Tanaka K, Togari A. α1B-Adrenergic receptor signaling controls circadian expression of Tnfrsf11b by regulating clock genes in osteoblasts. Biol Open. 2015;4(11):1400–9.
35. Hanami K, Nakano K, Saito K, et al. Dopamine D2-like receptor signaling suppresses human osteoclastogenesis. Bone. 2013;56(1):1–8.
36. Chakroborty D, Chowdhury UR, Sarkar C, et al. Dopamine regulates endothelial progenitor cell mobilization from mouse bone marrow in tumor vascularization. J Clin Invest. 2008;118(4):1380–9.
37. Shome S, Dasgupta PS, Basu S. Dopamine regulates mobilization of mesenchymal stem cells during wound angiogenesis. PLoS One. 2012;7(2):e31682.
38. Lee DJ, Tseng HC, Wong SW, et al. Dopaminergic effects on in vitro osteogenesis. Bone Res. 2015;3:15020.
39. Motyl KJ, Dick-de-Paula I, Maloney AE, et al. Trabecular bone loss after administration of the second-generation antipsychotic risperidone is independent of weight gain. Bone. 2012;50(2):490–8.
40. Bliziotes M, McLoughlin S, Gunness M, et al. Bone histomorphometric and biomechanical abnormalities in mice homozygous for deletion of the dopamine transporter gene. Bone. 2000;26(1):15–19.
41. Brakspear KS, Mason DJ. Glutamate signaling in bone. Front Endocrinol (Lausanne). 2012;3:97.
42. Mason DJ. Glutamate signalling and its potential application to tissue engineering of bone. Eur Cell Mater. 2004;7:12–25; discussion 25–6.
43. Gu Y, Genever PG, Skerry TM, et al. The NMDA type glutamate receptors expressed by primary rat osteoblasts have the same electrophysiological characteristics as neuronal receptors. Calcif Tissue Int. 2002;70(3):194–203.
44. Szczesniak AM, Gilbert RW, Mukhida M, et al. Mechanical loading modulates glutamate receptor subunit expression in bone. Bone. 2005;37(1):63–73.
45. Ho ML, Tsai TN, Chang JK, et al. Down regulation of N-methyl D-aspartate receptor in rat-modeled disuse osteopenia. Osteoporos Int. 2005;16(12):1780–8.
46. Nuntapornsak A, Wongdee K, Thongbunchoo J, et al. Changes in the mRNA expression of osteoblast-related genes in response to beta(3)-adrenergic agonist in UMR106 cells. Cell Biochem Funct. 2010;28(1):45–51.
47. Takeda S, Elefteriou F, Levasseur R, et al. Leptin regulates bone formation via the sympathetic nervous system. Cell. 2002;111(3):305–17.
48. Bouxsein ML, Devlin MJ, Glatt V, et al. Mice lacking beta-adrenergic receptors have increased bone mass but are not protected from deleterious skeletal effects of ovariectomy. Endocrinology. 2009;150(1):144–52.
49. Motyl KJ, DeMambro VE, Barlow D, et al. Propranolol attenuates risperidone-induced trabecular bone loss in female mice. Endocrinology. 2015;156(7):2374–83.
50. Yirmiya R, Goshen I, Bajayo A, et al. Depression induces bone loss through stimulation of the sympathetic nervous system. Proc Natl Acad Sci U S A. 2006;103(45):16876–81.
51. Hirai T, Tanaka K, Togari A. alpha1-adrenergic receptor signaling in osteoblasts regulates clock genes and bone morphogenetic protein 4 expression through up-regulation of the transcriptional factor nuclear factor IL-3 (Nfil3)/E4 promoter-binding protein 4 (E4BP4). J Biol Chem. 2014;289(24):17174–83.
52. Cruz Grecco Teixeira MB, Martins GM, Miranda-Rodrigues M, et al. Lack of alpha2C-Adrenoceptor Results in Contrasting Phenotypes of Long Bones and Vertebra and Prevents the Thyrotoxicosis-Induced Osteopenia. PLoS One. 2016;11(1):e0146795.
53. Arai M, Nagasawa T, Koshihara Y, et al. Effects of beta-adrenergic agonists on bone-resorbing activity in human osteoclast-like cells. Biochim Biophysica Acta. 2003;1640(2-3):137–42.
54. Kondo H, Takeuchi S, Togari A. β-Adrenergic signaling stimulates osteoclastogenesis via reactive oxygen species. Am J Physiol Endocrinol Metab. 2013;304(5):E507–15.
55. Marenzana M, Chenu C. Sympathetic nervous system and bone adaptive response to its mechanical environment. J Musculoskelet Neuronal Interact. 2008;8(2):111–20.
56. de Souza RL, Pitsillides AA, Lanyon LE, et al. Sympathetic nervous system does not mediate the load-induced cortical

new bone formation. J Bone Miner Res. 2005;20(12): 2159–68.
57. Pierroz DD, Bonnet N, Bianchi EN, et al. Deletion of beta-adrenergic receptor 1, 2, or both leads to different bone phenotypes and response to mechanical stimulation. J Bone Miner Res. 2012;27(6):1252–62.
58. Ma Y, Krueger JJ, Redmon SN, et al. Extracellular norepinephrine clearance by the norepinephrine transporter is required for skeletal homeostasis. J Biol Chem. 2013;288(42):30105–13.
59. Dionyssiotis Y, Trovas G, Galanos A, et al. Bone loss and mechanical properties of tibia in spinal cord injured men. J Musculoskelet Neuronal Interact. 2007;7(1): 62–8.
60. Veldhuis-Vlug AG, El Mahdiui M, Endert E, et al. Bone resorption is increased in pheochromocytoma patients and normalizes following adrenalectomy. J Clin Endocrinol Metab. 2012;97(11):E2093–7.
61. Manicourt DH, Brasseur JP, Boutsen Y, et al. Role of alendronate in therapy for posttraumatic complex regional pain syndrome type I of the lower extremity. Arthritis Rheum. 2004;50(11):3690–7.
62. Kado DM, Lui LY, Cummings SR, Study Of Osteoporotic Fractures Research Group. Rapid resting heart rate: a simple and powerful predictor of osteoporotic fractures and mortality in older women. J Am Geriatr Soc. 2002;50(3):455–60.
63. Ruths S, Bakken MS, Ranhoff AH, et al. Risk of hip fracture among older people using antihypertensive drugs: a nationwide cohort study. BMC Geriatr. 2015;15:153.
64. Wiens M, Etminan M, Gill SS, et al. Effects of antihypertensive drug treatments on fracture outcomes: a meta-analysis of observational studies. J Intern Med. 2006;260(4):350–62.
65. Baldock PA, Allison S, McDonald MM, et al. Hypothalamic regulation of cortical bone mass: opposing activity of Y2 receptor and leptin pathways. J Bone Miner Res. 2006;21(10):1600–7.
66. Baldock PA, Allison SJ, Lundberg P, et al. Novel role of Y1 receptors in the coordinated regulation of bone and energy homeostasis. J Biol Chem. 2007;282(26): 19092–102.
67. Horsnell H, Baldock PA. Osteoblastic Actions of the Neuropeptide Y System to Regulate Bone and Energy Homeostasis. Curr Osteoporos Rep. 2016;14(1): 26–31.
68. Tsapakis EM, Gamie Z, Tran GT, et al. The adverse skeletal effects of selective serotonin reuptake inhibitors. Eur Psychiatry. 2012;27(3):156–69.
69. Ducy P. 5-HT and bone biology. Curr Opin Pharmacol. 2011;11(1):34–8.
70. Gustafsson BI, Thommesen L, Stunes AK, et al. Serotonin and fluoxetine modulate bone cell function in vitro. J Cell Biochem 2006;98(1):139–51.
71. Bliziotes M, Eshleman A, Burt-Pichat B, et al. Serotonin transporter and receptor expression in osteocytic MLO-Y4 cells. Bone. 2006;39(6):1313–21.
72. Westbroek I, van der Plas A, de Rooij KE, et al. Expression of serotonin receptors in bone. The Journal of biological chemistry 2001;276(31):28961–8.
73. Yadav VK, Ryu JH, Suda N, et al. Lrp5 controls bone formation by inhibiting serotonin synthesis in the duodenum. Cell. 2008;135(5):825–37.
74. Tegeder I, Costigan M, Griffin RS, et al. GTP cyclohydrolase and tetrahydrobiopterin regulate pain sensitivity and persistence. Nat Med. 2006;12(11):1269–77.
75. Goto T, Nakao K, Gunjigake KK, et al. Substance P stimulates late-stage rat osteoblastic bone formation through neurokinin-1 receptors. Neuropeptides. 2007;41(1): 25–31.
76. Apel PJ, Crane D, Northam CN, et al. Effect of selective sensory denervation on fracture-healing: an experimental study of rats. J Bone Joint Surg Am. 2009;91(12): 2886–95.
77. Hoebertz A, Arnett TR, Burnstock G. Regulation of bone resorption and formation by purines and pyrimidines. Trends Pharmacol Sci. 2003;24(6):290–7.
78. Seref-Ferlengez Z, Maung S, Schaffler MB, et al. P2X7R-Panx1 Complex Impairs Bone Mechanosignaling under High Glucose Levels Associated with Type-1 Diabetes. PLoS One 2016;11(5):e0155107.
79. Hoebertz A, Townsend-Nicholson A, Glass R, et al. Expression of P2 receptors in bone and cultured bone cells. Bone. 2000;27(4):503–10.
80. Li W, Wei S, Liu C, Song M, et al. Regulation of the osteogenic and adipogenic differentiation of bone marrow-derived stromal cells by extracellular uridine triphosphate: The role of P2Y2 receptor and ERK1/2 signaling. Int J Mol Med. 2016;37(1):63–73.
81. Lundberg P, Bostrom I, Mukohyama H, et al. Neurohormonal control of bone metabolism: vasoactive intestinal peptide stimulates alkaline phosphatase activity and mRNA expression in mouse calvarial osteoblasts as well as calcium accumulation mineralized bone nodules. Regul Pept. 1999;85(1):47–58.
82. Juhasz T, Matta C, Katona E, et al. Pituitary adenylate cyclase-activating polypeptide (PACAP) signalling enhances osteogenesis in UMR-106 cell line. J Mol Neurosci. 2014;54(3):555–73.
83. Nagata A, Tanaka T, Minezawa A, et al. cAMP activation by PACAP/VIP stimulates IL-6 release and inhibits osteoblastic differentiation through VPAC2 receptor in osteoblastic MC3T3 cells. J Cell Physiol. 2009;221(1):75–83.
84. Persson E, Lerner UH. The neuropeptide VIP regulates the expression of osteoclastogenic factors in osteoblasts. J Cell Biochem. 2011;112(12):3732–41.
85. Suga S, Goto S, Togari A. Demonstration of direct neurite-osteoclastic cell communication in vitro via the adrenergic receptor. J Pharmacol Sci. 2010;112(2):184–91.
86. Obata K, Furuno T, Nakanishi M, et al. Direct neurite-osteoblastic cell communication, as demonstrated by use of an in vitro co-culture system. FEBS Lett. 2007; 581(30):5917–22.

133

The Pituitary–Bone Axis in Health and Disease

Mone Zaidi[1], Tony Yuen[1], Wahid Abu-Amer[1], Peng Liu[1], Terry F. Davies[1], Maria I. New[1], Harry C. Blair[2], Alberta Zallone[3], Clifford J. Rosen[4], and Li Sun[1]

[1]*The Mount Sinai Bone Program, Department of Medicine, Icahn School of Medicine at Mount Sinai, New York, NY, USA*
[2]*The Pittsburgh VA Medical Center and Departments of Pathology and of Cell Biology, University of Pittsburgh School of Medicine, Pittsburgh, PA, USA*
[3]*Department of Histology, University of Bari, Bari, Italy*
[4]*Maine Medical Center, Scarborough, ME, USA*

INTRODUCTION

Traditionally, a specific, limited function has been ascribed to each anterior and posterior pituitary hormone. However, the recent use of mouse genetics has led to the realization that these hormones and their receptors have more ubiquitous functions in integrative physiology. They are particularly abundant in organs such as the skeleton that are regulated by, and respond to, both local factors and systemic signals related to central metabolism and reproduction. Notably, although the skeleton expresses steroid-family receptors that play major roles in regulation, major pituitary hormones also have critical direct actions in skeletal homeostasis.

The skeletal expression of pituitary glycoprotein receptors further reflects that their function in endocrine control is evolutionarily more recent [1]. Thus, growth hormone (GH), follicle-stimulating hormone (FSH), thyroid-stimulating hormone (TSH), adrenocorticotrophic hormone (ACTH), prolactin (PRL), oxytocin (OXT), and vasopressin (AVP) all affect bone and, in mice, the haploinsufficiency of either the ligand and/or receptor often yields a skeletal phenotype with the primary target organ remaining unperturbed. Recognition and in-depth analysis of the mechanism of action of each pituitary hormone has improved our understanding of bone pathophysiology and opens new avenues for therapy. In this chapter, we discuss the interaction of each pituitary hormone with bone and the potential it holds in understanding and treating osteoporosis.

PITUITARY HORMONE RECEPTORS AND LIGANDS IN BONE

ACTH is the clearest example of a pituitary hormone being part of a widely distributed G-protein coupled receptor (GPCR) system that is known to participate in local cell differentiation in several contexts. Yet, this distributed function is overshadowed by its pituitary–adrenal signaling function. There are five melanocortin receptors, including the ACTH receptor (MC2R), which regulate cellular functions, including pigment production, appetite, and sexual function. All are controlled by ligands processed from a single large pro-hormone, pro-opiomelanocortin (POMC). Hormone production occurs by tissue-specific regulated proteolysis, with ACTH being the predominant product in the anterior pituitary. At other sites, POMC, three melanotropins, and β-endorphin are synthesized from the same precursor. There are reports of ACTH production by human macrophage/monocyte cells [2], making it possible that MC2Rs in

---

Primer on the Metabolic Bone Diseases and Disorders of Mineral Metabolism, Ninth Edition. Edited by John P. Bilezikian.
© 2019 American Society for Bone and Mineral Research. Published 2019 by John Wiley & Sons, Inc.
Companion website: www.wiley.com/go/asbmrprimer

bone may be activated by local, instead of pituitary-derived ACTH. Such decentralized control is also exemplified by corticotropin-releasing factor (CRF), which, in the adult, stimulates pituitary ACTH production, whereas in the fetus it stimulates cortisol synthesis directly [3]. This fetal system shows that evolution for centralization of ACTH as the CRF second messenger has not yet completely supplanted an ancestral regulatory system.

TSH and FSH are two of a group of hormones, along with chorionic gonadotropin (hCG) and luteinizing hormone (LH), that are heterodimeric proteins that share a common α-chain. Their specificity, however, depends on their differing β-chains. These hormones are particularly interesting in that simpler phyla have distributed functions. In coelenterates, which have a primitive nervous system but no endocrine glands, a TSHR family gene is readily identifiable, widely expressed, and shows the intron–exon structure found in mammals [4]. In lower vertebrates, such as in bony fish, the TSHR is abundant in the thyroid, but is also detectable in ovaries, heart, muscle, and brain [5]. In fish, the gonadal expression of the receptors, LH receptor (LHR) and FSH receptor (FSHR), is established, and all higher orders retain this. In fact, multiple differently processed forms of the FSHR occur in fish [6]; this may reflect isoforms with differing functions (see later in this chapter). Further, in fish the FSHR binds both FSH and LH, whereas the LHR recognizes only LH [7]. Although high-level FSHR expression is restricted to gonads, low-level expression is seen in the spleen [8], quite similarly to findings in human cells (see later in this chapter).

Low-level TSH production by bone marrow cells has likewise been reported [9]. A newly identified splice variant activates the TSHR in both murine and human bone, and to an extent exerts local osteoprotection [10]. Locally produced TSHβv is positively regulated by thyroid hormones [10], unlike the negative feedback exerted on TSH secretion from the pituitary. Lymphocytes also express TSH, but such production is unlikely to affect circulating levels. There is no evidence, however, for bone or marrow cell production of FSH, although coproduction of TSHβ and FSH is noted in CD11β cells from mouse thyroid [11]. Overall, therefore, the presence of GPCRs in tissues other than traditional endocrine targets, such as the skeleton, and in some cases, coexistence of their ligands, comes as no surprise. What does, however, come as a surprise is that the skeleton appears to be more sensitive to GPCR stimulation than the primary target organs, at least in mouse genetic and limited human studies.

GROWTH HORMONE

GH, a single-chain polypeptide, plays a vital role in skeletal homeostasis. It directly affects bone through a GPCR, but its primary action occurs via its release of insulin-like growth factors (IGFs). The predominant IGF, IGF-1, is synthesized mainly in the liver and approximately 80% that circulates is bound to IGF binding protein-3 (IGFBP3) and the acid-labile subunit (ALS). The importance of IGF-1 in skeletal homeostasis is testified by the demonstration that growth retardation and osteoporosis in GHR-deficient mice are arrested by the overexpression of IGF-1 [12]. Furthermore, and importantly, despite elevated GH levels, mice lacking both liver IGF-1 (LID) and ALS, with depleted serum IGF-1, show reduced bone growth and bone strength [13]. These results suggest that the skeletal effects of GH require IGF-1. In fact, the induction of osteoclastic activity by GH also appears to require IGF-1 made from bone marrow stromal cells, which then activates bone resorption by acting on osteoclastic receptors, as well as by altering RANKL expression [14].

There is also evidence to suggest that GH can act independently of IGF. For example, GH replacement reverses the increased adiposity in hypophysectomized rats, whereas IGF-1 replacement does not. Furthermore, in ovariectomized LID mice, GH reverses osteopenia [15]. Although these findings point to a direct action of GH on bone and other tissues, selective deletion of this GPCR in osteoblasts and other cells should provide further clarity.

FOLLICLE-STIMULATING HORMONE

We discovered that FSH acts on an FSHR in the osteoclasts and directly stimulates bone resorption [16]. Mice haploinsufficient in FSHβ showed evidence of increased bone mass in the face of normal ovarian function, suggesting a fundamental role of the FSH/FSHR interaction in bone physiology [16]. Several studies have now confirmed direct effects of FSH on the skeleton in rodents and humans. Notably, amenorrheic women with a higher mean serum FSH (approximately 35 IU/L) have greater bone loss than those with lower levels (approximately 8 IU/L) in the face of near-equal estrogen levels [17]. Likewise, patients with functional hypothalamic amenorrhea, in whom both FSH and estrogen were low, showed slight to moderate skeletal defects. Importantly, women harboring an activating *FSHR* polymorphism, *rs6166*, have lower bone mass and high resorption markers [18]. In fact, digenic combinations between wild-type genotype of the 3′UTR marker for the *CYP19A1* gene, the *IVS4* marker of the same gene, and the *BMP15* and *FSHR* genes have been described as being osteoprotective [19]. These studies attest to a role for FSHRs in human physiology and the pathophysiology of human postmenopausal osteoporosis. Consistent with these human studies, exogenous administration of FSH to rats augments ovariectomy-induced bone loss, and an FSH antagonist reduces bone loss after ovariectomy or FSH injection [20].

Clinical correlations between bone loss and serum FSH levels are also consistent with genetic studies. Most impressive is the Study of Women's Health Across the

Nation (SWAN), a longitudinal cohort of 2375 perimenopausal women. Not only was there a strong correlation between serum FSH levels and markers of bone resorption, a change in FSH levels over 4 years predicted decrements in bone mass [21]. Analyses of data from Chinese women showed similar trends: a significant association between bone loss and high serum FSH. In a group of southern Chinese women aged between 45 and 55 years, those in the highest quartile of serum FSH lost bone at a 1.3- to 2.3-fold higher rate than those in the lowest quartile. A more recent analysis of perimenopausal Chinese women aged between 45 and 50 years of age revealed a strong correlation between serum C-telopeptide and FSH levels [22]. Importantly, C-telopeptide levels were greater when serum FSH levels were greater than 40 mIU/mL [22].

Likewise, examination of a National Health and Nutrition Evaluation Survey III (NHANES III) cohort of women between the ages of 42 and 60 years showed a strong correlation between serum FSH and femoral neck bone mineral density (BMD) [23]. A cross-sectional analysis of 92 postmenopausal women found that serum osteocalcin and C-telopeptide cross-linked (CTx) were both positively correlated with FSH, but not with estradiol [24]. The Bone Turnover Range of Normality (BONTURNO) Study group likewise showed that women with serum FSH levels greater than 30 IU/mL had significantly higher bone turnover markers than age-matched women, despite having normal menses [25]. Consistent with this, lower serum FSH levels and higher serum estrogen levels have been associated with lower rates of lumbar spine bone loss in some but not all phases of the menopausal transition.

In contrast, Gourlay et al. failed to show a strong relationship between bone mass and FSH or, indeed, estrogen [26]. Interestingly, however, the same authors documented an independent correlation between FSH and lean mass. This latter association makes biological sense inasmuch as FSHRs are present on mesenchymal stem cells [16], known to have the propensity for adipocytic and/or myocytic differentiation. Nonetheless, the weight of the evidence prompts the use of FSH at least as a serum marker for identifying "fast bone losers" during the early phases of the menopausal transition [27].

Mechanistically, FSH increases osteoclast formation, function, and survival through a distinct FSHR isoform [16, 28]. Wu et al. further showed that the osteoclastogenic response to FSH was abolished in mice lacking immunoreceptor tyrosine-based activation motif (Itam) adapter signaling molecules [28]. This suggests an interaction between FSH and immune receptor complexes, although the significance of this remains unclear. In a separate study, FSHR activation was shown to enhance RANK receptor expression [29]. In addition, FSH indirectly stimulates osteoclast formation by releasing osteoclastogenic cytokines, namely IL-1β, TNF-α, and IL-6 in proportion to the surface expression of FSHRs [30]. In a study of 36 women between the ages of 20 and 50, serum FSH levels correlated with circulating cytokine concentrations [30].

A group has, however, failed to identify FSHRs on osteoclasts, having likely used primers targeted to the ovarian isoform [31]. We very consistently find FSHR in human CD14+ cells and osteoclasts using nested primers and sequencing to verify the specificity of the reaction, and amplifying regions that contain an intron to avoid the pitfall of genomic DNA contamination [32]. Furthermore, cellular responsiveness to FSH appears also to be determined by the level of FSH glycosylation [33], with the prediction that the fully glycosylated isoform is more active on the bone receptor.

It has further been difficult to tease out the action of FSH from that of estrogen in vivo, as FSH releases estrogen, and the actions of FSH and estrogen on the osteoclast are opposed. The injection of FSH into mice with intact ovaries, or its transgenic overexpression [31], even in *hpg* mice, is unlikely to reveal proresorptive actions of FSH. This is because direct effects of FSH on the osteoclast will invariably be masked by the antiresorptive and anabolic actions of the ovarian estrogen so released in response to FSH.

As noted previously, there is evidence that women with low FSH levels undergo less bone loss [17], and that the effectiveness of estrogen therapy is related to the degree of FSH suppression [34]. With that said, patients with pituitary hypogonadism lose bone. Leuprolide treatment, and hence the lowering of FSH, has not been shown to prevent hypogonadal hyper-resorption [35]. Although this proves that low estrogen is a *cause* of acute hypogonadal bone loss, it does not exclude a role for FSH in human skeletal homeostasis [35].

Rapid and profuse bone loss begins 3 years before the last menstrual period, when serum estrogen is relatively normal [36]. This is when the rate of bone loss is maximal, and therefore cannot be attributed to changes in serum estrogen [21, 36]. Addressing the importance of FSH elevations in this period, Lukefahr and coworkers used a unique rat model of osteoporosis during the perimenopausal transition [37]. This perimenopausal rodent-equivalent, in which the ovotoxin 4-vinylcyclohexene diepoxide (VCD) is administered to rats, was characterized by a prolonged estrogen-replete period when serum FSH levels were elevated. Longitudinal measurements revealed that significant decreases in BMD (5% to 13%) occurred during periods of increased FSH and decreased inhibins [37].

To leverage this selective increase in FSH early in the menopausal transition, an antibody to a 13-amino-acid-long peptide sequence within the receptor-binding domain of the FSH β-subunit was generated [38]. The FSH antibody bound FSH specifically and blocked its action on osteoclast formation in vitro. When injected into ovariectomized mice, the FSH antibody attenuated bone loss not only by inhibiting bone resorption, but also by stimulating bone formation, a yet uncharacterized action of FSH [38]. Notably, stromal cells isolated from mice treated with the FSH antibody showed greater osteoblast precursor colony counts, similarly to stromal cells isolated from *Fshr*-/- mice. This suggested that FSH negatively regulates osteoblast differentiation via signaling-efficient FSHRs present on mesenchymal stem cells. There is recent direct evidence for FSH action on osteoblast precursors, albeit in the opposite direction [39]. Overall, the data prompt the future development of a novel FSH-blocking agent as a means of uncoupling bone formation and bone resorption to a therapeutic advantage in humans. An interesting alternative strategy, for which a proof-of-principle study is available, is the use of an FSHβ vaccine. It has been shown that immunizing ovariectomized rats with the GST-FSHβ antigen significantly prevents trabecular bone loss and increases bone strength [40].

THYROID-STIMULATING HORMONE

TSH is a direct inhibitor of osteoclasts [41]. The haploinsufficiency of TSHRs in heterozygotic $Tshr^{-/-}$ mice results in osteoporosis in the face of unaltered thyroid hormone levels [41]. Furthermore, $Tshr^{-/-}$ mice are osteoporotic, a phenotype that cannot be explained by the known pro-osteoclastic action of thyroid hormones, particularly as $Tshr^{-/-}$ mice are hypothyroid. Furthermore, skeletal runting, but not the osteoporotic phenotype, is reversed upon rendering $Tshr^{-/-}$ mice euthyroid by thyroid hormone replacement [41]. Thus, TSH acts on bone independently of thyroid hormones, and the osteoporosis of hyperthyroidism may, in part, be a result of low TSH [42].

The osteoporosis of TSHR deficiency is of the high-turnover variety. $Tshr^{-/-}$ mice showed evidence of increased osteoclastic activity, similarly to hyt/hyt mice that have defective TSHR signaling [43]. Studies show that recombinant TSH attenuates the genesis, function, and survival of osteoclasts in vitro in bone marrow [44] and murine embryonic stem cell cultures [45]. In contrast, the overexpression of constitutively activated TSHR in osteoclast precursor cells [46] or transgenically, in mouse precursors [43], inhibits osteoclastogenesis. In postmenopausal women, a single subcutaneous injection of TSH drastically lowers serum C-telopeptide to premenopausal levels within 2 days, with recovery at day 7 [47]. In none of the studies with TSH replacement did thyroid hormones increase, exemplifying again that the pituitary–bone axis is more primitive than the pituitary–thyroid axis. Furthermore, an activating TSHR antibody was shown to inhibit osteoclastogenesis in vitro [48].

The anti-osteoclastogenic action of TSH is mediated by reduced NF-κB and Janus N-terminus kinase (JNK) signaling, and TNF-α production [41, 46]. The effect of TSH on TNF-α synthesis is mediated transcriptionally by binding of two high-mobility group box proteins, HMGB1 and HMGB2, to the TNF-α gene promoter. TNF-α production is expectedly upregulated in osteoporotic $Tshr^{-/-}$ mice [41], and the genetic deletion of TNF-α in these mice reverses the osteoporosis and the bone formation and resorption defects, proving that the $Tshr^{-/-}$ phenotype is mediated by TNF-α, at least in part [49].

The role of TSH in osteoblast regulation is less explicit, although increasing evidence suggests an anabolic action. It inhibits osteoblastogenesis in bone marrow-derived cell cultures [41], but stimulates differentiation and mineralization in murine cell cultures through a Wnt5a-dependent mechanism [50]. Likewise, in vivo, intermittently administered TSH is anabolic in both rats and mice [43]. In rats, TSH, injected up to once every 2 weeks, inhibits ovariectomy-induced bone loss, 28 weeks following ovariectomy [43]. Calcein-labeling studies are consistent with a direct anabolic action of intermittent TSH. Furthermore, in humans, Martini et al. [51] showed an increase in N-terminal propeptide of human procollagen type I (PINP), a marker of bone formation, validating the conclusion that a bolus dose of TSH is indeed anabolic. Likewise, it has been shown recently that antibody-activated TSH signaling contributes to high bone formation, independent of the actions of thyroid hormone.

Epidemiologically, a 4.5-fold increase in the risk of vertebral fractures and a 3.2-fold increase in the risk of nonvertebral fractures is seen at TSH levels lower than 0.1 IU/L [52]. There is also a strong negative correlation between low serum TSH and high C-telopeptide levels, without an association with thyroid hormone [53]. In patients on L-thyroxine, significantly greater bone loss has been noted in those with suppressed TSH than those without suppression [54]. The Tromso study supports this: participants with serum TSH below 2 SD had a significantly lower BMD, those with TSH above 2 SD had a significantly increased BMD, whereas there was no association between TSH and BMD at normal TSH levels [55]. In patients taking suppressive doses of thyroxine for thyroid cancer, the serum level of cathepsin K, a surrogate but yet unvalidated resorption marker, was elevated [56], and the HUNT 2 study found a positive correlation between TSH and BMD at the distal forearm [57]. In fact, it is now evident that in patients presenting with an elevated TSH, the long-term risk of hip and other osteoporotic fractures is related to the cumulative duration of periods with low TSH, which likely result from excessive thyroid hormone replacement [58]. In the OPENTHYRO study, hip and major osteoporosis-related fracture risk was shown to be a function of the duration of hyperthyroidism [59]. Likewise, TSH suppression increased the risk of postoperative osteoporosis in low- and intermediate-risk thyroid cancer patients [60]. There is also evidence for reduced bone density and/or increased bone resorption that correlates with low TSH levels in hypothyroid patients undergoing thyroid hormone therapy [61].

Analysis of data from NHANES has shown that the odds ratio for correlations between TSH and bone mass ranged between 2 and 3.4 [44]. There is increasing new evidence that elderly women with low-normal TSH levels in the euthyroid range display lower BMD and, in some instances, weaker femoral structure [62]. Such women also have a higher incidence of vertebral fractures, independent of age, BMD, and thyroid hormones [63].

In a retrospective study in women with postmenopausal osteoporosis, there was a positive correlation between serum TSH and BMD; furthermore, serum TSH level was associated with a protective effect in a regression model [64]. Another study found significant positive correlations between serum TSH level within normal range and the BMD of the lumbar spine [65]. This was not obvious in a group of perimenopausal women [66]. In elderly men, lower TSH levels within the normal range were associated with an approximately 30% increase in the risk of hip fracture [67].

In genetic studies, patients harboring the *TSHR-D727E* polymorphism have high bone mass [68], and allelic associations have been reported in the Rotterdam study [69]. Another polymorphism, *T+140974TC*, seen in the Korean population, is also associated with increased BMD, more so in patients with an elevated TSH, again substantiating the role of TSH in protecting against bone loss [69].

Physiologically, therefore, TSH uncouples bone remodeling by inhibiting osteoclastic bone resorption and stimulating osteoblastic bone formation, particularly when given intermittently. Furthermore, absent TSH signaling stimulates bone remodeling directly, and through TNF-α production, causing net bone loss. Low TSH levels may thus contribute to the pathophysiology of osteoporosis of hyperthyroidism, which has traditionally been attributed to high thyroid hormone levels alone. This latter hypothesis was studied in wild-type and *Tshr$^{-/-}$* mice that were rendered either euthyroid or hyperthyroid by implanting T4 pellets [70]. Not surprisingly, it was noted that hyperthyroid *Tshr$^{-/-}$* mice suffered greater bone loss than wild-type mice that were rendered equally hyperthyroid [70]. This suggested that absent TSH signaling in *Tshr$^{-/-}$* mice itself contributes to hyperthyroid bone loss, which has been thought solely to arise from elevated thyroid hormones.

ADRENOCORTICOTROPHIC HORMONE

Glucocorticoids, under natural regulation mainly by ACTH, are important coregulators of many processes including vascular tone, central metabolism, and immune response. At higher, pharmacological levels, they become anti-inflammatory and immunosuppressant drugs, with incident complications including diabetes, osteoporosis, and osteonecrosis. Osteonecrosis is a painful, debilitating condition that affects metabolically active bone, typically the femoral head [71], and invariably requires surgical treatment. The underlying mechanisms of glucocorticoid-induced osteonecrosis are poorly understood, although a key finding is that osteonecrosis occurs before macroscopic vascular changes [72].

Isales et al. [73] discovered that bone-forming units strongly express melanocortin receptor 2 (MC2R). We showed that, as with the adrenal cortex, ACTH induces vascular endothelial growth factor (VEGF) production in osteoblasts through its action on MC2Rs [74]. This likely translates into the protection by ACTH of glucocorticoid-induced osteonecrosis in a rabbit model [74]. We speculate that VEGF suppression secondary to ACTH suppression may contribute to bone damage with long-term glucocorticoid therapy. Much needs to be done to validate this idea towards a therapeutic advantage, considering that ACTH analogs are already approved for human use.

PROLACTIN

PRL, a peptide hormone secreted by the anterior pituitary, primarily acts to induce and maintain lactation and prevent another pregnancy by suppressing folliculogenesis and libido. During pregnancy, it increases the calcium bioavailability for milk production and fetal skeletogenesis by promoting intestinal calcium absorption and skeletal mobilization. Accelerated bone turnover and bone loss is noted in hyperprolactinemic adults [75]. Antagonism of PRL by bromocriptine, a dopamine agonist, reverses the bone loss.

This osteoclastic action of PRL is traditionally thought to arise from the accompanying hypoestrogenemia. However, it has been shown that osteoblasts express PRL receptors (PRLRs) [76], suggesting a direct interaction between PRL and the osteoblast. In fact, the pattern of bone loss is distinct in PRL-exposed and ovariectomized rats [77].

Ex vivo, PRL decreases osteoblast differentiation markers [77], in part through the PI3K signaling pathway [78]. In vivo, PRL accelerates bone resorption in adult mice through an indirect action on osteoclasts, notably by increasing the RANK/OPG (osteoprotegerin) ratio [77]. Osteoclasts themselves do not possess PRLRs [76]. In contrast, in infant rats, PRL causes net bone gain and increased osteocalcin expression. Likewise, in human fetal osteoblast cells, PRL decreases the RANKL/OPG ratio [78]. It appears therefore that the net effect of PRL on bone depends on the biological maturity of the organism. In the fetal stage, it promotes bone growth and mineralization, while accelerating bone resorption in the mother to make nutrients available. Further insight is needed to clarify the role of PRL in bone metabolism and determine the cellular pathways.

OXYTOCIN

OXT is a nonapeptide synthesized in the hypothalamus and released into circulation via the posterior pituitary. Its primary function is to mediate the milk ejection reflex in nursing mammals. It also stimulates uterine contraction during parturition; however, OXT is not a requirement for this function. Thus, *Oxt*-null mice can deliver normally, but are unable to nurse. Subcutaneous OXT injection completely rescues the milk ejection phenotype, attesting to this being a peripheral, as opposed to a central, action [79]. Central actions of OXT include the regulation of social behavior, including sexual and maternal behavior, affiliation, and social memory, as well as penile erection and ejaculation. OXT also controls food, predominantly carbohydrate, intake centrally. Thus, the social amnesia, aggressive behavior, and overfeeding observed in $Oxt^{-/-}$ and $Oxtr^{-/-}$ mice are reversed on intracerebroventricular OXT injection.

OXT acts on a GPCR, present in abundance on osteoblasts [80], osteoclasts, and their precursors [81]. In line with the ubiquitous distribution of OXT receptors (OXTRs), cells of bone marrow also synthesize OXT, suggesting the existence of autocrine and paracrine interactions [81]. In vitro, OXT stimulates osteoblast differentiation and bone formation [81]. Thus, $Oxt^{-/-}$ and $Oxtr^{-/-}$ mice, including the haploinsufficient heterozygotes with normal lactation, display severe osteoporosis caused by a bone-forming defect [81]. This not only indicates that the osteoblast is the target for OXT, but also that bone is more sensitive to OXT than the breast, hitherto considered its primary target. Once again, the finding emphasizes a relatively primitive pituitary–bone axis. Effects of OXT on bone resorption in vivo appear minimal: OXT stimulates osteoclastogenesis but inhibits the activity of mature osteoclasts, with a net zero effect on resorption [81].

In vivo gain-of-function studies document a direct effect of OXT on bone. Intraperitoneal OXT injections result in increased BMD and ex vivo osteoblast formation [81]. In contrast, short-term intracerebroventricular OXT does not affect bone turnover markers. OXT injections in wild-type rats alter the RANKL/OPG ratio in favor of bone formation, again attesting to an anabolic action [82].

Although unproven, OXT may have a critical role in bone anabolism during pregnancy and lactation. Both are characterized by excessive bone resorption in favor of fetal and postpartum bone growth, respectively [83]. This bone loss is, however, completely reversed upon weaning by an as yet unidentified mechanism. OXT peaks in blood during late pregnancy and lactation, and although its pro-osteoclastogenic action may contribute to intergenerational calcium transfer, its anabolic action could enable the restoration of the maternal skeleton. That $Oxt^{-/-}$ pups show hypomineralized skeletons, and $Oxt^{-/-}$ moms display reduced bone formation markers are suggestive of such an action. Estrogen positively regulates osteoblastic OXT production as well as OXTR expression, and therefore can synergize this action through a local feed-forward loop [84]. There is evidence that these osteoblastic actions are in part mediated via the nuclear localization of the OXTR following its activation by ligand [85].

In postmenopausal women, there is evidence that serum OXT levels correlate strongly with BMD, particularly at the hip, in the 6-year-long prospective OPUS study [86]. Moreover, low serum OXT levels are associated with severe osteoporosis, independently of estrogen [87]. In separate studies, the low nocturnal oxytocin secretion in amenorrheic athletes is associated with site-dependent microarchitectural impairments.

It is also possible that the skeletal anabolic actions of OXT could be used to a therapeutic advantage. For example, systemic administration of OXT has been shown to enhance osteointegration of titanium implants in ovariectomized rats [88]. Likewise, in mice and rabbits, OXT reversed the bone loss after ovariectomy [89], but not after orchiectomy [89]. The latter finding emphasizes possible sex differences that might arise from the regulation of OXT action by estrogen, as noted previously [85]. The MINOS study confirms that serum OXT in men is not associated with BMD, bone turnover rate, or prevalence of fractures [90].

VASOPRESSIN

AVP is a posterior pituitary hormone that regulates mineral excretion. However, two AVP receptors, AVPR1α and AVPR2, coupled to ERK activation, have been shown to be present on osteoblasts and osteoclasts [91]. In contrast to the anabolic action of OXT noted above [81], AVP injected into wild-type mice has been found to reduce osteoblast formation while enhancing osteoclastogenesis [91]. Conversely, the exposure of osteoblast precursors to AVPR1α or AVPR2 antagonists, namely SR49059 or ADAM, results in increased osteoblastogenesis [91]. The effect of these inhibitors is phenocopied in mice lacking the Avpr1α. In contrast, osteoclast formation and bone resorption are both reduced in $Avpr1α^{-/-}$ cultures [91]. This increased bone formation and reduced resorption results in a profound enhancement of bone mass in $Avpr1α^{-/-}$ mice and in wild-type mice injected with SR49059 [91]. It also reverses the bone loss in compound $Avpr1α^{-/-}:Oxtr^{-/-}$ mice [92]. These data not only establish a primary role for AVP signaling in bone mass regulation, but also call for further studies on the skeletal actions of AVPR inhibitors used commonly in hyponatremic patients.

A long-standing dogma has centered round the mechanism(s) underlying the osteoporosis that accompanies hyponatremia. It is quite possible that, in a case report, the documented bone loss in a patient with syndrome of inappropriate antidiuretic hormone (SIADH) had arisen from a 30-fold elevation in serum AVP, although the accompanying high aldosterone could be an additional culprit. Hyperaldosteronism has been linked to bone loss in rodents [93], and the use of spironolactone in patients with secondary hyperaldosteronism resulting from heart failure reduces fracture risk [94]. Nonetheless, the complementary cell-based, pharmacological, and mouse genetic data noted previously argue for a role for AVP in causing hyponatremia-induced bone loss. There is thus a strong rationale for the routine evaluation of skeletal health in patients with chronic hyponatremia, as well as for the potential of therapeutic osteoprotection.

However, because AVPR2 is expressed by osteoblasts and osteoclasts, and is active functionally in regulating bone mass [91], it is possible that highly selective AVPR2 inhibitors, such as tolvaptan [95], when used as aquaretics in the therapy of chronic hyponatremia, could, in fact, themselves offer osteoprotection. This was not found to be the case [92], suggesting that AVPR1α is the major functional skeletal receptor for AVP. Interestingly, although AVPR1α and OXTR have opposing effects on bone mass, they share receptors to a limited extent [92]. Whereas the OXTR is not indispensable for AVP action in inhibiting osteoblastogenesis, AVP-stimulated gene expression is inhibited when the OXTR is deleted in $Avpr1\alpha^{-/-}$ cells. In contrast, OXT does not interact with AVPRs in vivo in a model of lactation-induced bone loss where OXT levels are high [92].

PITUITARY HORMONES AND BODY COMPOSITION

In women, FSH secretion begins to increase during the perimenopausal transition, preceding declines in estrogen by about 2–3 years [21]. During this period, and concurrent with the sharp decline in BMD, there is the onset of increasing visceral and bone marrow adiposity and a decrease in lean mass. This clinical phenotype is associated with disrupted energy metabolism and decreased physical activity. Considering that FSHRs are present on bone and fat tissue [16, 96–98], a question has arisen whether a single agent can be used to treat two aging-related medical conditions — osteoporosis and obesity — simultaneously.

A polyclonal antibody to FSHβ was used to examine the effect of FSH blockade in mice pair-fed on a high-fat diet, ovariectomized and sham-operated mice, and mice on a normal chow but allowed to eat *ad libitum* [98]. In all instances, the FSH antibody prevented the development of visceral and subcutaneous obesity and induced energy-producing "beige" adipocytes. This beiging phenotype was evident from immunolabeling for uncoupling protein 1 (UCP1), measuring the expression of Ucp1 and other brown adipose tissue (BAT) genes in white adipose tissue, and examining UCP1 expression in vivo in live UCP1 reporter, Thermo mice [98]. Furthermore, the FSH antibody enhanced mitochondrial density in the photo-activatable mitochondria (PhAM) mouse and enhanced basal energy expenditure and oxygen consumption, independently of increased activity. Studies with a corresponding monoclonal FSH antibody to human FSH recapitulated the data, as did mice haploinsufficient in the *Fshr*. Importantly, the FSH antibody failed to prevent obesity in $Fshr^{+/-}$ mice, testifying to its action via the FSH axis in vivo. Finally, there was an important yet unexplained increase in lean mass in all groups treated with the FSH antibody. This phenotype is concordant with a strong association between reduced lean mass and high FSH levels in post-menopausal women.

CONCLUSION

The discovery of the direct regulation of bone by glycoprotein hormones helps explain some of the inconsistencies of older models that assumed that pituitary signaling was mediated entirely via endocrine organs through steroid-family signals. Important direct responses include actions of TSH, FSH, ACTH, PRL, OXT, and AVP in bone. It is important, in evaluating these new signaling mechanisms, to consider that the skeletal responses may or may not have similar mechanisms to the responses of the traditional endocrine targets, and that the signals may vary in importance because of secondary endocrine and paracrine control. The discovery of direct skeletal, and more recently adipose-specific, responses of pituitary hormones nevertheless offers a new set of therapeutic opportunities. As has been noted with OXT [87], it is also possible that actions of pituitary hormones may extend beyond bone to be exerted via GPCRs expressed in other vital tissues, such as muscle.

ACKNOWLEDGMENTS

Mone Zaidi, Li Sun, Terry F. Davies, and Harry C. Blair are supported by grants from the National Institutes of Health. Alberta Zallone is supported by the Ministry of Education, Italy.

DISCLOSURES

Mone Zaidi consults for Merck, Roche, and Shire, and is a named inventor of an issued US patent related to osteoclastic bone resorption filed by the Icahn School of Medicine at Mount Sinai (ISMMS). In the event the issued patent is licensed, he would be entitled to a share of any proceeds ISMMS receives from the licensee. All other authors have nothing to disclose.

REFERENCES

1. Blair HC, Robinson LJ, Sun L, et al. Skeletal receptors for steroid-family regulating glycoprotein hormones: A multilevel, integrated physiological control system. Ann N Y Acad Sci. 2011;1240:26–31.
2. Pallinger E, Csaba G. A hormone map of human immune cells showing the presence of adrenocorticotropic hormone, triiodothyronine and endorphin in immunophenotyped white blood cells. Immunology. 2008;123: 584–9.
3. Sirianni R, Rehman KS, Carr BR, et al. Corticotropin-releasing hormone directly stimulates cortisol and the cortisol biosynthetic pathway in human fetal adrenal cells. J Clin Endocrinol Metab. 2005;90:279–85.
4. Vibede N, Hauser F, Williamson M, et al. Genomic organization of a receptor from sea anemones, structurally and evolutionarily related to glycoprotein hormone receptors from mammals. Biochem Biophys Res Commun. 1998;252:497–501.
5. Kumar RS, Ijiri S, Kight K, et al. Cloning and functional expression of a thyrotropin receptor from the gonads of a

vertebrate (bony fish): potential thyroid-independent role for thyrotropin in reproduction. Mol Cell Endocrinol. 2000;167:1–9.
6. Kobayashi T, Andersen O. The gonadotropin receptors FSH-R and LH-R of Atlantic halibut (Hippoglossus hippoglossus), 1: isolation of multiple transcripts encoding full-length and truncated variants of FSH-R. Gen Comp Endocrinol. 2008;156:584–94.
7. Bogerd J, Granneman JC, Schulz RW, et al. Fish FSH receptors bind LH: how to make the human FSH receptor to be more fishy? Gen Comp Endocrinol. 2005;142:34–43.
8. Kumar RS, Ijiri S, Trant JM. Molecular biology of the channel catfish gonadotropin receptors: 2. Complementary DNA cloning, functional expression, and seasonal gene expression of the follicle-stimulating hormone receptor. Biol Reprod. 2001;65:710–7.
9. Vincent BH, Montufar-Solis D, Teng BB, et al. Bone marrow cells produce a novel TSHbeta splice variant that is upregulated in the thyroid following systemic virus infection. Genes Immun. 2009;10:18–26.
10. Baliram R, Chow A, Huber AK, et al. Thyroid and bone: macrophage-derived TSH-beta splice variant increases murine osteoblastogenesis. Endocrinology. 2013;154:4919–26.
11. Klein JR, Wang HC. Characterization of a novel set of resident intrathyroidal bone marrow-derived hematopoietic cells: potential for immune-endocrine interactions in thyroid homeostasis. J Exp Biol. 2004;207:55–65.
12. De Jesus K, Wang X, Liu JL. A general IGF-I overexpression effectively rescued somatic growth and bone deficiency in mice caused by growth hormone receptor knockout. Growth Factors. 2009;27:438–47.
13. Yakar S, Rosen CJ, Beamer WG, et al. Circulating levels of IGF-1 directly regulate bone growth and density. J Clin Invest. 2002;110:771–81.
14. Rubin J, Ackert-Bicknell CL, Zhu L, et al. IGF-I regulates osteoprotegerin (OPG) and receptor activator of nuclear factor-kappaB ligand in vitro and OPG in vivo. J Clin Endocrinol Metab. 2002;87:4273–9.
15. Fritton JC, Emerton KB, Sun H, et al. Growth hormone protects against ovariectomy-induced bone loss in states of low circulating insulin-like growth factor (IGF-1). J Bone Miner Res. 2010;25:235–46.
16. Sun L, Peng Y, Sharrow AC, et al. FSH directly regulates bone mass. Cell. 2006;125:247–60.
17. Devleta B, Adem B, Senada S. Hypergonadotropic amenorrhea and bone density: new approach to an old problem. J Bone Miner Metab. 2004;22:360–4.
18. Rendina D, Gianfrancesco F, De Filippo G, et al. FSHR gene polymorphisms influence bone mineral density and bone turnover in postmenopausal women. Eur J Endocrinol. 2010;163:165–72.
19. Mendoza N, Quereda F, Presa J, et al. Estrogen-related genes and postmenopausal osteoporosis risk. Climacteric. 2012;15:587–93.
20. Liu S, Cheng Y, Xu W, et al. Protective effects of follicle-stimulating hormone inhibitor on alveolar bone loss resulting from experimental periapical lesions in ovariectomized rats. J Endod. 2010;36:658–63.
21. Sowers MR, Greendale GA, Bondarenko I, et al. Endogenous hormones and bone turnover markers in pre- and perimenopausal women: SWAN. Osteoporos Int. 2003;14:191–7.
22. Wang B, Song Y, Chen Y, et al. Correlation analysis for follicle-stimulating hormone and C-terminal cross-linked telopetides of type i collagen in menopausal transition women with osteoporosis. Int J Clin Exp Med. 2015;8:2417–22.
23. Gallagher CM, Moonga BS, Kovach JS. Cadmium, follicle-stimulating hormone, and effects on bone in women age 42-60 years, NHANES III. Environ Res. 2010;110:105–11.
24. Garcia-Martin A, Reyes-Garcia R, Garcia-Castro JM, et al. Role of serum FSH measurement on bone resorption in postmenopausal women. Endocrine. 2012;41:302–8.
25. Adami S, Bianchi G, Brandi ML, et al.; BONTURNO study group. Determinants of bone turnover markers in healthy premenopausal women. Calcif Tissue Int. 2008;82:341–7.
26. Gourlay ML, Preisser JS, Hammett-Stabler CA, et al. Follicle-stimulating hormone and bioavailable estradiol are less important than weight and race in determining bone density in younger postmenopausal women. Osteoporos Int. 2011;22:2699–708.
27. Zaidi M, Turner CH, Canalis E, et al. Bone loss or lost bone: rationale and recommendations for the diagnosis and treatment of early postmenopausal bone loss. Curr Osteoporos Rep. 2009;7:118–26.
28. Wu Y, Torchia J, Yao W, et al. Bone microenvironment specific roles of ITAM adapter signaling during bone remodeling induced by acute estrogen-deficiency. PLoS One. 2007;2:e586.
29. Cannon JG, Kraj B, Sloan G. Follicle-stimulating hormone promotes RANK expression on human monocytes. Cytokine. 2011;53:141–4.
30. Cannon JG, Cortez-Cooper M, Meaders E, et al. Follicle-stimulating hormone, interleukin-1, and bone density in adult women. Am J Physiol Regul Integr Comp Physiol. 2010;298:R790–8.
31. Allan CM, Kalak R, Dunstan CR, et al. Follicle-stimulating hormone increases bone mass in female mice. Proc Natl Acad Sci U S A. 2010;107:22629–34.
32. Tourkova IL, Witt MR, Li L, et al. Follicle stimulating hormone receptor in mesenchymal stem cells integrates effects of glycoprotein reproductive hormones. Ann N Y Acad Sci. 2015;1335:100–9.
33. Meher BR, Dixit A, Bousfield GR, et al. Glycosylation Effects on FSH-FSHR Interaction Dynamics: A Case Study of Different FSH Glycoforms by Molecular Dynamics Simulations. PLoS One. 2015;10:e0137897.
34. Kawai H, Furuhashi M, Suganuma N. Serum follicle-stimulating hormone level is a predictor of bone mineral density in patients with hormone replacement therapy. Arch Gynecol Obstet. 2004;269:192–5.

35. Drake MT, McCready LK, Hoey KA, et al. Effects of suppression of follicle-stimulating hormone secretion on bone resorption markers in postmenopausal women. J Clin Endocrinol Metab. 2010;95:5063–8.
36. Randolph JF, Jr, Sowers M, Gold EB, et al. Reproductive hormones in the early menopausal transition: relationship to ethnicity, body size, and menopausal status. J Clin Endocrinol Metab. 2003.88:1516–22.
37. Lukefahr AL, Frye JB, Wright LE, et al. Decreased bone mineral density in rats rendered follicle-deplete by an ovotoxic chemical correlates with changes in follicle-stimulating hormone and inhibin A. Calcif Tissue Int. 2012;90:239–49.
38. Zhu LL, Tourkova I, Yuen T, et al. Blocking FSH action attenuates osteoclastogenesis. Biochem Biophys Res Commun. 2012;422:54–8.
39. Su XY, Zou X, Chen QZ, et al. Follicle-Stimulating Hormone β-subunit Potentiates Bone Morphogenetic Protein 9-induced Osteogenic Differentiation in Mouse Embryonic Fibroblasts. J Cell Biochem. 2017;118(7):1792–1802.
40. Geng W, Yan X, Du H, et al. Immunization with FSHbeta fusion protein antigen prevents bone loss in a rat ovariectomy-induced osteoporosis model. Biochem Biophys Res Commun. 2013;434:280–6.
41. Abe E, Marians RC, Yu W, et al. TSH is a negative regulator of skeletal remodeling. Cell. 2003;115:151–62.
42. Zaidi M, Sun L, Davies TF, et al. Low TSH triggers bone loss: fact or fiction? Thyroid. 2006;16:1075–6.
43. Sun L, Vukicevic S, Baliram R, et al. Intermittent recombinant TSH injections prevent ovariectomy-induced bone loss. Proc Natl Acad Sci U S A. 2008;105:4289–94.
44. Morris MS. The association between serum thyroid-stimulating hormone in its reference range and bone status in postmenopausal American women. Bone. 2007;40:1128–34.
45. Ma R, Latif R, Davies TF. Thyrotropin-independent induction of thyroid endoderm from embryonic stem cells by activin A. Endocrinology. 2009;150:1970–5.
46. Hase H, Ando T, Eldeiry L, et al. TNFalpha mediates the skeletal effects of thyroid-stimulating hormone. Proc Natl Acad Sci U S A. 2006;103:12849–54.
47. Mazziotti G, Sorvillo F, Piscopo M, et al. Recombinant human TSH modulates in vivo C-telopeptides of type-1 collagen and bone alkaline phosphatase, but not osteoprotegerin production in postmenopausal women monitored for differentiated thyroid carcinoma. J Bone Miner Res. 2005;20:480–6.
48. Ma R, Morshed S, Latif R, et al. The influence of thyroid-stimulating hormone and thyroid-stimulating hormone receptor antibodies on osteoclastogenesis. Thyroid. 2011;21:897–906.
49. Sun L, Zhu LL, Lu P, et al. Genetic confirmation for a central role for TNFalpha in the direct action of thyroid stimulating hormone on the skeleton. Proc Natl Acad Sci U S A. 2013;110:9891–6.
50. Baliram R, Latif R, Berkowitz J, et al. Thyroid-stimulating hormone induces a Wnt-dependent, feed-forward loop for osteoblastogenesis in embryonic stem cell cultures. Proc Natl Acad Sci U S A 2011;108:16277–82.
51. Martini G, Gennari L, De Paola V, et al. The effects of recombinant TSH on bone turnover markers and serum osteoprotegerin and RANKL levels. Thyroid. 2008;18:455–60.
52. Bauer DC, Ettinger B, Nevitt MC, et al.; Study of Osteoporotic Fractures Research Group. Risk for fracture in women with low serum levels of thyroid-stimulating hormone. Ann Intern Med. 2001;134:561–8.
53. Zofkova I, Hill M. Biochemical markers of bone remodeling correlate negatively with circulating TSH in postmenopausal women. Endocr Regul. 2008;42:121–7.
54. Kim MK, Yun KJ, Kim MH, et al. The effects of thyrotropin-suppressing therapy on bone metabolism in patients with well-differentiated thyroid carcinoma. Bone. 2015;71:101–5.
55. Grimnes G, Emaus N, Joakimsen RM, et al. The relationship between serum TSH and bone mineral density in men and postmenopausal women: the Tromso study. Thyroid. 2008;18:1147–55.
56. Mikosch P, Kerschan-Schindl K, Woloszczuk W, et al. High cathepsin K levels in men with differentiated thyroid cancer on suppressive L-thyroxine therapy. Thyroid. 2008;18:27–33.
57. Svare A, Nilsen TI, Bjoro T, et al. Hyperthyroid levels of TSH correlate with low bone mineral density: the HUNT 2 study. Eur J Endocrinol. 2009;161:779–86.
58. Abrahamsen B, Jorgensen HL, Laulund AS, et al. The excess risk of major osteoporotic fractures in hypothyroidism is driven by cumulative hyperthyroid as opposed to hypothyroid time: an observational register-based time-resolved cohort analysis. J Bone Miner Res. 2015;30:898–905.
59. Abrahamsen B, Jorgensen HL, Laulund AS, et al. Low serum thyrotropin level and duration of suppression as a predictor of major osteoporotic fractures-the OPENTHYRO register cohort. J Bone Miner Res. 2014;29:2040–50.
60. Wang LY, Smith AW, Palmer FL, et al. Thyrotropin suppression increases the risk of osteoporosis without decreasing recurrence in ATA low- and intermediate-risk patients with differentiated thyroid carcinoma. Thyroid. 2015;25:300–7.
61. Karimifar M, Esmaili F, Salari A, et al. Effects of Levothyroxine and thyroid stimulating hormone on bone loss in patients with primary hypothyroidism. J Res Pharm Pract. 2014;3:83–7.
62. Ding B, Zhang Y, Li Q, et al. Low Thyroid Stimulating Hormone Levels Are Associated with Low Bone Mineral Density in Femoral Neck in Elderly Women. Arch Med Res. 2016;47:310–4.
63. Mazziotti G, Porcelli T, Patelli I, et al. Serum TSH values and risk of vertebral fractures in euthyroid postmenopausal women with low bone mineral density. Bone. 2010;46:747–51.
64. Acar B, Ozay AC, Ozay OE, et al. Evaluation of thyroid function status among postmenopausal women with and without osteoporosis. Int J Gynaecol Obstet. 2016;134: 53–7.
65. Noh HM, Park YS, Lee J, et al. A cross-sectional study to examine the correlation between serum TSH levels

65. and the osteoporosis of the lumbar spine in healthy women with normal thyroid function. Osteoporos Int. 2015;26:997–1003.
66. van Rijn LE, Pop VJ, Williams GR. Low bone mineral density is related to high physiological levels of free thyroxine in peri-menopausal women. Eur J Endocrinol. 2014;170:461–8.
67. Waring AC, Harrison S, Fink HA, et al.; Osteoporotic Fractures in Men (MrOS) Study. A prospective study of thyroid function, bone loss, and fractures in older men: The MrOS study. J Bone Miner Res. 2013;28:472–9.
68. Liu RD, Chen RX, Li WR, et al. The Glu727 Allele of Thyroid Stimulating Hormone Receptor Gene is Associated with Osteoporosis. N Am J Med Sci. 2012;4:300–4.
69. van der Deure WM, Uitterlinden AG, Hofman A, et al. Effects of serum TSH and FT4 levels and the TSHR-Asp727Glu polymorphism on bone: the Rotterdam Study. Clin Endocrinol (Oxf). 2008;68:175–81.
70. Baliram R, Sun L, Cao J, et al. Hyperthyroid-associated osteoporosis is exacerbated by the loss of TSH signaling. J Clin Invest. 2012;122:3737–41.
71. Mankin HJ. Nontraumatic necrosis of bone (osteonecrosis). N Engl J Med. 1992;326:1473–9.
72. Eberhardt AW, Yeager-Jones A, Blair HC. Regional trabecular bone matrix degeneration and osteocyte death in femora of glucocorticoid- treated rabbits. Endocrinology. 2001;142:1333–40.
73. Isales CM, Zaidi M, Blair HC. ACTH is a novel regulator of bone mass. Ann N Y Acad Sci. 2010;1192: 110–6.
74. Zaidi M, Sun L, Robinson LJ, et al. ACTH protects against glucocorticoid-induced osteonecrosis of bone. Proc Natl Acad Sci U S A. 2010;107:8782–7.
75. Naylor KE, Iqbal P, Fledelius C, et al. The effect of pregnancy on bone density and bone turnover. J Bone Miner Res. 2000;15:129–37.
76. Coss D, Yang L, Kuo CB, et al. Effects of prolactin on osteoblast alkaline phosphatase and bone formation in the developing rat. Am J Physiol Endocrinol Metab. 2000;279:E1216–25.
77. Seriwatanachai D, Thongchote K, Charoenphandhu N, et al. Prolactin directly enhances bone turnover by raising osteoblast-expressed receptor activator of nuclear factor kappaB ligand/osteoprotegerin ratio. Bone. 2008;42:535–46.
78. Seriwatanachai D, Charoenphandhu N, Suthiphongchai T, et al. Prolactin decreases the expression ratio of receptor activator of nuclear factor kappaB ligand/osteoprotegerin in human fetal osteoblast cells. Cell Biol Int. 2008;32:1126–35.
79. Nishimori K, Young LJ, Guo Q, et al. Oxytocin is required for nursing but is not essential for parturition or reproductive behavior. Proc Natl Acad Sci U S A. 1996;93:11699–704.
80. Copland JA, Ives KL, Simmons DJ, et al. Functional oxytocin receptors discovered in human osteoblasts. Endocrinology. 1999;140:4371–4.
81. Tamma R, Colaianni G, Zhu LL, et al. Oxytocin is an anabolic bone hormone. Proc Natl Acad Sci U S A. 2009;106:7149–54.
82. Elabd SK, Sabry I, Hassan WB, et al. Possible neuroendocrine role for oxytocin in bone remodeling. Endocr Regul. 2007;41:131–41.
83. Wysolmerski JJ. The evolutionary origins of maternal calcium and bone metabolism during lactation. J Mammary Gland Biol Neoplasia. 2002;7:267–76.
84. Colaianni G, Sun L, Di Benedetto A, et al. Bone marrow oxytocin mediates the anabolic action of estrogen on the skeleton. J Biol Chem. 2012;287:29159–67.
85. Di Benedetto A, Sun L, Zambonin CG, et al. Osteoblast regulation via ligand-activated nuclear trafficking of the oxytocin receptor. Proc Natl Acad Sci U S A. 2014; 111:16502–7.
86. Breuil V, Panaia-Ferrari P, Fontas E, et al. Oxytocin, a new determinant of bone mineral density in postmenopausal women: analysis of the OPUS cohort. J Clin Endocrinol Metab. 2014;99:E634–41.
87. Breuil V, Amri EZ, Panaia-Ferrari P, et al. Oxytocin and bone remodelling: relationships with neuropituitary hormones, bone status and body composition. Joint Bone Spine. 2011;78:611–5.
88. Wang M, Lan L, Li T, et al. The effect of oxytocin on osseointegration of titanium implant in ovariectomized rats. Connect Tissue Res. 2016;57:220–5.
89. Beranger GE, Djedaini M, Battaglia S, et al. Oxytocin reverses osteoporosis in a sex-dependent manner. Front Endocrinol (Lausanne). 2015;6:81.
90. Breuil V, Fontas E, Chapurlat R, et al. Oxytocin and bone status in men: analysis of the MINOS cohort. Osteoporos Int. 2015;26:2877–82.
91. Tamma R, Sun L, Cuscito C, et al. Regulation of bone remodeling by vasopressin explains the bone loss in hyponatremia. Proc Natl Acad Sci U S A. 2013;110: 18644–9.
92. Sun L, Tamma R, Yuen T, et al. Functions of vasopressin and oxytocin in bone mass regulation. Proc Natl Acad Sci U S A. 2016;113:164–9.
93. Chhokar VS, Sun Y, Bhattacharya SK, et al. Loss of bone minerals and strength in rats with aldosteronism. Am J Physiol Heart Circ Physiol. 2004;287: H2023–6.
94. Carbone LD, Cross JD, Raza SH, et al. Fracture risk in men with congestive heart failure risk reduction with spironolactone. J Am Coll Cardiol. 2008;52:135–8.
95. Hori M. Tolvaptan for the treatment of hyponatremia and hypervolemia in patients with congestive heart failure. Future Cardiol. 2013;9:163–76.
96. Cui H, Zhao G, Liu R, et al. FSH stimulates lipid biosynthesis in chicken adipose tissue by upregulating the expression of its receptor FSHR. J Lipid Res. 2012; 53:909–17.
97. Liu XM, Chan HC, Ding GL, et al. FSH regulates fat accumulation and redistribution in aging through the Gai/Ca(2+)/CREB pathway. Aging Cell. 2015;14: 409–20.
98. Lui P, Ji Y, Yuen T, et al. Blocking FSH induces thermogenic adipose tissue and reduces body fat. Nature. 2017;546:107–12.

134

Neuropsychiatric Disorders and the Skeleton

Madhusmita Misra[1,2] and Anne Klibanski[3]

[1]Department of Pediatrics, Harvard Medical School, Boston, MA, USA
[2]Pediatric Endocrinology, Massachusetts General Hospital and Harvard Medical School, Boston, MA, USA
[3]Neuroendocrine Unit, Massachusetts General Hospital and Harvard Medical School, Boston, MA, USA

INTRODUCTION

Studies have provided evidence for the control of bone metabolism by the central nervous system (CNS) and through autocrine/paracrine mechanisms via neurotransmitters released by bone cells and their microenvironment. Bone is densely innervated by autonomic and sensory nerve fibers [1]. Bone cells, mainly osteoblasts, express receptors for neurotransmitters and neuropeptides including acetylcholine [2], norepinephrine (NE) [3], endocannabinoids (ECs) [4], neuropeptide Y [5], calcitonin gene-related peptide, and substance P [1]. Central interleukin (IL)-1 signaling also has been implicated in the regulation of bone metabolism [6]. To date, the best experimentally characterized brain-to-bone pathway is the sympathetic nervous system (SNS), which mediates the skeletal effects of leptin and serotonin via a hypothalamic serotonergic relay [7]. Sympathetic nerve terminals form a synaptic-like junction with osteoblasts, which express β2-adrenergic receptors (β2AR). These receptors are activated by norepinephrine (NE) released from sympathetic terminals, tonically restraining bone formation [8]. Activation of β2AR in osteoblasts stimulates RANKL expression, which increases osteoclast number and activity [9]. Skeletal sympathetic tone is downregulated by brainstem-derived serotonin and M3 acetylcholine muscarinic receptors expressed in the brain [7, 10]. NE discharge by skeletal sympathetic terminals is attenuated by endocannabinoid 2-arachidonoylglycerole (2-AG) released into the sympatho-osteoblast junction by osteoblasts and activates CB1 cannabinoid receptors in the prejunctional membrane [3].

Neuropsychiatric diseases are attributable to organic nervous system pathology, mainly brain pathology. It is now well established that diseases such as anorexia nervosa (AN) and major depressive disorder (MDD) are associated with low BMD. Schizophrenia and Alzheimer disease (AD) have also been associated with low BMD.

EATING DISORDERS

Anorexia nervosa

AN is characterized by severe low weight and a distorted body image, associated with cognitive biases that modify body self-perception [11]. Although amenorrhea is no longer required for the diagnosis according to the *Diagnostic and Statistical Manual of Mental Disorders* (DSM-5), it remains a common feature. AN patients exhibit a high incidence of comorbidities such as anxiety and depression, and AN has the highest mortality rate of any psychiatric disorder. AN typically begins in adolescence and young adulthood, affecting 0.2% to 4% of individuals in this age range [12]. Although the condition is much more common in females, 5% to 15% of all patients have been reported to be male [13].

AN is highly heritable with multiple polymorphisms in more than 40 genes involved in eating behavior, motivation and reward, personality traits, and emotion. Some polymorphisms have been identified in genes related to skeletal remodeling such as brain-derived neurotrophic factor and the NE transporter. In addition, an abnormal

---

Primer on the Metabolic Bone Diseases and Disorders of Mineral Metabolism, Ninth Edition. Edited by John P. Bilezikian.
© 2019 American Society for Bone and Mineral Research. Published 2019 by John Wiley & Sons, Inc.
Companion website: www.wiley.com/go/asbmrprimer

brain response to the anorexic effects of estrogen, and serotonin dysregulation have been reported [14,15]. Autoantibodies to melanocortin, a neuropeptide implicated in personality traits associated with eating disorders and the control of bone mass, have been demonstrated in AN [16]. However, a causal relationship between these findings and AN has not been established.

Impact on bone

Low BMD is a common finding in AN [17]. One study in adult women with AN reported that 92% and 38% had T-scores of < –1 or < –2 respectively at one or more sites [18]. One study reported Z-scores of < –1 in 52% and < –2 in 11% at one or more sites in adolescents with AN [19]. Bone accrual rates are markedly decreased compared to healthy adolescents, raising concerns regarding suboptimal peak bone mass acquisition and increased fracture risk in later life. Both trabecular and cortical sites are impacted, and use of QCT techniques has revealed lower trabecular and cortical volumetric BMD, increased cortical porosity, reduced cortical and trabecular thickness, decreased trabecular number, and greater trabecular separation at the non-weight-bearing radius in adults and adolescents with AN compared with controls (reviewed in [20]). Assessment of biochemical markers of bone turnover suggests reduced bone turnover in adolescents with AN, versus an uncoupled increase in bone resorption in adults with AN [21,22]. Importantly, an increased risk for fracture has been reported, with a 31% fracture prevalence in adolescents and young adults with AN compared to 19% in controls [23], and a twofold increase in fracture risk in adults with AN compared to controls, and a cumulative risk of 57% at 40 years [24,25].

Pathophysiology of low bone density

Low BMD is a consequence of reduced muscle mass, hypogonadism, low levels of insulin-like growth factor-1 (IGF-1), relative hypercortisolemia, and high peptide YY (PYY) levels resulting from profound alterations in energy metabolism (reviewed in [26]). This is only partially reversible with weight regain [26]. Further, low BMD is associated with higher brown fat activity and marrow fat in women with AN [20]. Specifically, women with AN have higher spine and femur marrow adiposity than controls associated with lower BMD at the spine and hip. Further, marrow fat and levels of preadipocyte factor-1, a regulator of osteoblast and adipocyte differentiation, are higher in women with active AN compared with those who have recovered from AN and controls [27]. Certain medications commonly used in patients with AN have also been implicated in impaired bone metabolism. In particular, a longer duration of use of selective serotonin reuptake inhibitors (SSRIs) is associated with lower BMD, even after controlling for age and duration since diagnosis [28].

Evaluation of bone status

A diagnosis of AN should prompt assessment of BMD using DXA. In one study, more than 50% of those with Z-scores < –2 had been diagnosed with the condition for less than 6 months [29], suggesting that the effect on bone can be rapid. Of note, onset of subclinical or frank AN can precede its diagnosis by months to years. Adolescents with AN should have BMD assessments at the spine and whole body, whereas adults should have assessments of the spine and hip [30]. Serial BMD assessment is important in adolescents to capture suboptimal bone accrual rates resulting in decreasing BMD Z-scores over time, and in adults to capture ongoing bone loss. Also, weight gain can lead to stabilization of bone loss or increases in bone mass, and DXA assessment may assist in determining the need for therapeutic intervention. When early onset of AN has stunted statural growth, it is essential to adjust for height, because DXA measures of areal BMD can be artifactually low in those with shorter stature.

Strategies to manage low bone density

The best strategy to improve low BMD in AN is weight regain and resumption of menses. Weight regain is associated with increases in lean mass, and an improvement in many hormonal changes that contribute to low BMD. Increases in weight alone result in increases in hip BMD, whereas menses recovery alone results in increases in spine BMD; women who regain weight and also resume menses have an improvement in BMD at both sites [31]. However, the magnitude of increase in BMD is small and is usually inconsistent across patients. In adolescents, an increase in BMI of at least 10% with resumption of menses over a 1-year period was associated with some improvement in BMD measures at the spine and whole body, but not to the extent seen in controls [32]. Thus residual deficits may persist, at least in the short term. This is particularly concerning in adolescents given the narrow window of time available for optimizing accrual.

One strategy to improve BMD in AN is to address the deleterious hormonal changes that contribute to low BMD. Multiple studies have shown that estrogen replacement with an oral estrogen–progesterone combination pill is not effective in increasing BMD in adolescents or adults with AN, possibly related to the IGF-1 suppressive effects of oral estrogen, and use of non-physiological forms of estrogen in most combination pills [33,34]. In adolescents 12 to 18 years old, one randomized controlled trial (RCT) showed that estrogen replacement using the 17-β estradiol patch (100 μg), with 2.5 mg of medroxyprogesterone acetate given for 10 days of every month, led to increased bone accrual rates over 18 months (3.5% at the spine and 2.9% at the hip compared to placebo after adjusting for baseline age and weight changes) to approximate rates observed in normal-weight controls, although BMD did not reach values in healthy controls [35]. In contrast, physiological testosterone replacement was not effective in increasing BMD in adult women with AN [36]. In another study, compared to placebo, daily administration of 50 mg of dehydroepiandrosterone (DHEA) orally with a low-dose estrogen–progesterone combination pill maintained BMD Z-scores in females 13 to 27 years old [37]. Further, low IGF-1 levels contribute to decreased bone formation in AN, and physiological replacement doses of recombinant human (rh) IGF-1 increase bone formation markers in adolescents and adults [21,38]. When rhIGF-1 was administered in physiological doses with an oral estrogen–progesterone combination pill, it increased spine BMD by 2.8% in adults with AN compared to no therapy, but had no effect alone [39].

Finally, a few studies have examined the effect of teriparatide and bisphosphonates on bone in AN. One small 6-month study in older AN women reported an increase in antero-posterior and lateral spine BMD (by 6% to 10%) after the use of teriparatide (20 mg daily) compared to placebo [40]. Another RCT in adult women with AN reported a significant increase in spine and hip BMD by 4% and 2% respectively following risedronate use (35 mg weekly) [36]. However, a study of alendronate (10 mg daily) versus placebo in adolescents with AN showed no improvement in spine BMD, but did demonstrate a small increase at the femoral neck [41].

Current recommendations to manage low BMD in AN focus on weight regain and menses recovery. It can take 6 to 12 months of being at "normal" weight (>90% median BMI for age) before menses resume. In one study, participants were about 2 kg heavier at the time that they resumed menses compared to the weight at which they lost menses [42]. In those with osteoporosis, strategies include physiological estrogen replacement (and cyclic progesterone). We use the 100 μg 17-β estradiol patch applied once or twice weekly, with 100 to 200 mg of micronized progesterone given for 12 days of every month. This strategy should also be considered in adolescent girls with AN demonstrating a decrease in BMD Z-scores on serial DXA assessments, given the narrow window available to optimize bone accrual. Other strategies include oral DHEA (50 mg daily) given with oral estrogen–progesterone combination pills, or a bisphosphonate (in adults with AN and osteoporosis). The latter should be used very cautiously in women of reproductive age given the long half-life of bisphosphonates and concerns of teratogenicity (although data thus far are mostly reassuring). Use of bisphosphonates should be reserved for those with osteoporosis and a history of fracture for whom other strategies, such as efforts at weight regain, are not effective. Treatment studies for low BMD in males with AN are lacking, and the recommended strategy is weight regain.

Other eating disorders

Fewer studies have examined the impact of other eating disorders on bone; however, one meta-analysis reported that bulimia nervosa is also associated with low BMD [17]. Another study reported a higher fracture risk in bulimia nervosa [24]. It is unclear whether this risk applies to women who remain eumenorrheic and have never been underweight.

MAJOR DEPRESSIVE DISORDER

MDD is characterized by all-encompassing low mood accompanied by loss of interest or pleasure in normally enjoyable activities, as well as alterations in appetite, body weight, sleep patterns, psychomotor activity, fatigue, worthlessness and low self-esteem, cognitive impairments, and suicidal ideation. More than half of the individuals who commit suicide have MDD or another mood disorder. Several central systems and processes have been implicated in depression, including alterations in serotonergic and adrenergic transmission, hormonal dysregulation, reduced neurotrophins and neurogenesis, as well as neuroinflammation and IL-1 secretion [43,44]. MDD is the second leading cause of years of life lived with disability [45].

Impact on bone

Several meta-analyses (reviewed in [46]) have demonstrated an association between MDD and low BMD. One of the more comprehensive of these included 23 studies comparing a total of 2327 depressed with 21,141 non-depressed subjects [47]. It showed lower BMD at the spine, femoral neck, and distal radius in patients with MDD, and higher bone resorption markers than in non-depressed subjects. However, the overall effect size for the entire population (adult males and females) was small [47]. Women were significantly more vulnerable to depression-associated low BMD [47,48], which may be related to the greater responsiveness of depressed women to various stressors (reviewed in [19]). Another meta-analysis reported a sex-specific impact, with lower spine BMD noted in women with MDD and lower femoral BMD in men with MDD [49]. Of note, premenopausal women display a greater MDD-associated decrease in BMD compared to postmenopausal subjects [47], consistent with a decreased rate of bone accrual in female teenagers with MDD [50], resulting in reduced peak bone mass. However, one study in adolescents 12 to 18 years old with MDD found no impact of the disorder on bone in females with the disorder, but did report a deleterious effect in males [51]. A study in older adolescents and young adults (mean age 19 years) reported lower BMD at the spine and whole body [52], however, sex-specific changes were not examined. In postmenopausal women, the association of MDD with low BMD may be masked by multiple factors contributing to bone loss, including estrogen depletion, reduced physical activity, and nutritional disturbances (reviewed in [10]). Williams et al. have also reported a 68% increased risk of incident fractures in women with depression, although the association was attenuated after adjusting for use of psychotropic medication [53]. In a meta-analysis, depression was associated with a 17% increase in fracture risk reported as hazard ratios, and a 52% increase in risk reported as risk ratios [54].

Pathophysiology of low bone density

Body weight, number of previous depressive episodes, total duration of disease, history of estrogen treatment, and race do not appear to modulate the association between MDD and BMD (reviewed in [47]). In addition, serum 25-hydroxyvitamin D, PTH, free T3, IGF-1, and TSH levels do not differ in depressed versus non-depressed subjects (reviewed in [47]). Studies using antidepressant therapy as a covariate found no evidence of an effect of these medications on BMD in MDD, although the Geelong Osteoporosis Study reported an association with lower BMD in lower weight men (<75 to 110 kg depending on bone site) [55]. (See "Neuropsychiatric Medications" for the effects of these medications on bone.)

Of note, in MDD, there is a well-established increase in endogenous cortisol production and NE, known to reduce bone formation and which may contribute to low BMD. Loss of bone mass and impaired microarchitecture have been shown in mice with chronic mild stress (CMS), an established rodent model for depression [56]. These mice show reduced bone formation and generalized trabecular bone loss, and these skeletal deficits and depressive symptoms are prevented by imipramine [56]. The depressive-like state is associated with increased NE levels in bone and elevated serum corticosterone. Furthermore, CMS-induced bone loss, but not the depressive-like state, is prevented using the β-adrenergic antagonist propranolol, suggesting that bone sympathetic innervation allows for communication of depressive signals to the skeleton.

Other systems implicated in the association between MDD and low BMD include the endocannabinoid system and inflammatory cytokines such as IL-1, IL-6, and TNF-α (reviewed in [46]). Another possible contributing factor is cigarette smoking, which has repeatedly been shown to negatively influence bone mass in cross-sectional studies in men and women. Likewise, depression and excessive alcohol consumption are common comorbidities and alcohol abuse is a recognized risk factor for osteoporosis. Finally, changes in exercise activity and food consumption occur with MDD, and certain nutrients reported to be deficient in patients with MDD are necessary for optimal bone health (reviewed in [46, 57]). Bone density assessment may be considered in adults with MDD. Optimizing nutrition, physical activity, and vitamin D status, and treating the underlying MDD are important.

SCHIZOPHRENIA

Schizophrenia is characterized by abnormal thinking and emotional responsiveness, and commonly manifests as auditory hallucinations, paranoia or delusions, and/or disorganized speech and thinking. Onset of symptoms typically occurs in adolescence or young adulthood, with a global lifetime prevalence of about 0.5%. A positive relationship between schizophrenia, low BMD [58,59], and osteoporotic fractures [60,61]) has been attributed to schizophrenia-associated factors such as undernutrition, inadequate exercise, smoking, alcohol use, and use of antipsychotic medications [62, 63]. In particular, neuroleptic-induced hyperprolactinemia and the resultant hypogonadism have received much attention [58,59]. Of note, a comparison of 965 adult schizophrenia patients and 405 non-schizophrenic community members in the same district reported that young male and female schizophrenia patients have lower BMD than non-schizophrenics by the age of 20 years, indicative of impaired bone accrual before that age. Interestingly, they did not exhibit age-related bone loss, with females even showing some gain, such that BMD in schizophrenics older than 60 years was higher than in controls [64]. This "protection" may result from mutations at the $Dgcr8$ gene involved in the 22q11.2 microdeletions shown in schizophrenia, because silencing of $Dgcr8$ decreases osteoclastogenesis and causes a mild increase in BMD [65, 66]. Bone changes may be linked to comorbidities and behavioral patterns, and disease-induced impairment of the brain-to-bone communication.

ALZHEIMER DISEASE

Dementia syndromes affect 10% of the population aged 65 and older. AD is the commonest cause of dementia, and accounts for 65% of dementia cases either alone or in combination with vascular dementia. Although osteoporosis is observed already at the age of 50, its consequences become apparent from the seventh decade and onwards. Expectedly, this circumstantial relationship results in a significant correlation between the two conditions. This is enhanced by common risk factors shared by the two diseases such as low body mass, reduced physical activity and exposure to sunlight, and nutritional deficiencies. That this relationship may not be merely circumstantial is suggested by (i) the higher incidence of osteoporosis in AD patients compared to those in the same age group without memory impairment (reviewed in [67]), (ii) a similar incidence of osteoporotic fractures in women and men with AD compared to several-fold higher incidence of fractures in females in the general population [68], and (iii) studies in mice that indicate a causal role [69,70]. Importantly, treatment and rehabilitation in post-fracture AD patients is more difficult than in non-AD patients with less than half regaining pre-fracture functional status. Also, dementia patients fall more frequently, resulting in an increased incidence of femoral neck fractures.

NEUROPSYCHIATRIC MEDICATIONS

SSRIs

Osteoblasts, osteocytes, and osteoclasts express functional serotonin (5-hydroxytryptamine, 5-HT) receptors and serotonin transporter (5-HTT) (reviewed in [71]). In osteoblasts, 5-HT receptor agonists influence cell proliferation, potentiate the PTH-induced increase in AP-1 activity, and modulate cellular response to mechanical stimulation. In osteocytes, 5-HT increases whole cell cAMP and PGE_2, involved in the transduction of mechanical stimuli [72]. In osteoclasts, 5-HT and 5-HTT affect differentiation, but not activity [71].

The CNS does not appear to be a likely source of 5-HT available to bone, because the blood–brain barrier is impermeable to 5-HT and serotonergic innervation has not been demonstrated in the skeleton. 5-HT could be synthesized and released by bone cells and act in an autocrine/paracrine manner. Indeed, mRNA transcripts for tryptophan hydroxylase-1 (Tph1), a rate-limiting enzyme in 5-HT synthesis, have been detected in osteoblast and osteocyte cell lines [73]. Most 5-HT is produced in the gastrointestinal (GI) tract and stored in dense granules in platelets. Because 5-HT from this source is released only upon platelet activation, it is an unlikely activator of bone cell 5-HT receptors. However, a small fraction of GI-derived 5-HT remains in the circulation, and serum 5-HT has been suggested as a negative regulator of osteoblast proliferation and bone formation [74]. Disruption of the 5-HTT gene or pharmacological inhibition of 5-HTT by SSRIs leads to a low bone mass phenotype in growing mice [71]. SSRIs also have deleterious effects on trabecular and cortical bone in adult mice [75]. Many studies report that antidepressants in general, but mainly SSRIs, are associated with low BMD and a dose-dependent increase in the risk of fractures and low bone mass in children (reviewed in [76]). This may be linked to dysregulation of the balance between the skeletal, GI, and central serotonergic systems, especially after prolonged administration [77].

Neuroleptics

Many typical and atypical antipsychotics induce dose-dependent hyperprolactinemia and may result in hypogonadism and thus bone loss. Prolactin production is controlled by dopaminergic and serotonergic systems. Dopamine, the predominant regulator of prolactin, tonically inhibits its production by acting on D2 receptors on lactotrophs. In contrast, 5-HT stimulates prolactin secretion by activating $5HT_{1A}R$ and $5HT_2R$ [78]. Neuroleptics antagonize receptors in the brain dopamine pathways, including differential inhibition of D2 and 5-HT receptors, causing a dose-dependent increase in serum prolactin (reviewed in [41]). In addition, prolactin may affect the skeleton directly through prolactin receptors expressed in osteoblasts and their precursors. In vitro studies suggest that osteoblastic processes affected depend on the level of hyperprolactinemia. Prolactin concentrations typically recorded during lactation (100 ng/mL) inhibit preosteoblast proliferation, osteoblast number, and production of a mineralized extracellular matrix. This concentration stimulates RUNX2 and alkaline phosphatase expression in early-stage osteoblast differentiation and inhibits the same genes in late stages [79]. Higher concentrations (up to 500 ng/mL) increase osteoblastic expression of osteoclastogenic factors such as RANKL, MCP-1, Cox-2, TNF-α, IL-1, and ephrin-B1 [80]. However, clinically, hyperprolactinemia has not been shown to cause bone mass reduction unless it results in amenorrhea.

Overall, the highest and most consistent rates of hyperprolactinemia are reported with antipsychotics such as amisulpride, risperidone, and paliperidone, whereas quetiapine and aripiprazole are the most favorable in this regard, with the former causing no change, and the latter, because of a partial dopamine agonist effect, causing a reduction in prolactin levels. Newly approved antipsychotics perform similarly to clozapine (asenapine and iloperidone), ziprasidone, and olanzapine (lurasidone) in this regard [81].

SUMMARY

The association between neuropsychiatric and skeletal disorders is emerging as an important theme in bone and psychiatric pathophysiology. Comorbidities shared by the two disciplines impact millions of male and female patients of all ages and present a wealth of unanswered basic mechanistic questions and unresolved clinical issues. Perhaps the most urgent of these issues is the conflict between the beneficial effects of psychiatric medications, particularly in mood disorders and schizophrenia, and their deleterious effects on the skeleton.

REFERENCES

1. Imai S, Matsusue Y. Neuronal regulation of bone metabolism and anabolism: calcitonin gene-related peptide-, substance P-, and tyrosine hydroxylase-containing nerves and the bone. Microsc Res Tech. 2002;58:61–9.
2. Sato T, Abe T, Chida D, et al. Functional role of acetylcholine and the expression of cholinergic receptors and components in osteoblasts. FEBS Lett. 2010;584:817–24.
3. Tam J, Trembovler V, Di Marzo V, et al. The cannabinoid CB1 receptor regulates bone formation by modulating adrenergic signaling. FASEB J. 2008;22:285–94.
4. Bab I, Ofek O, Tam J, et al. Endocannabinoids and the regulation of bone metabolism. J Neuroendocrinol. 2008;20(suppl 1):69–74.
5. Franquinho F, Liz MA, Nunes AF, et al. Neuropeptide Y and osteoblast differentiation—the balance between the neuro-osteogenic network and local control. FEBS J. 2010;277:3664–74.
6. Bajayo A, Goshen I, Feldman S, et al. Central IL-1 receptor signaling regulates bone growth and mass. Proc Natl Acad Sci U S A. 2005;102:12956–61.
7. Yadav VK, Oury F, Suda N, et al. A serotonin-dependent mechanism explains the leptin regulation of bone mass, appetite, and energy expenditure. Cell. 2009;138:976–89.

8. Takeda S, Elefteriou F, Levasseur R, et al. Leptin regulates bone formation via the sympathetic nervous system. Cell. 2002;111:305–17.
9. Elefteriou F, Ahn JD, Takeda S, et al. Leptin regulation of bone resorption by the sympathetic nervous system and CART. Nature. 2005;434:514–20.
10. Shi Y, Oury F, Yadav VK, et al. Signaling through the M(3) muscarinic receptor favors bone mass accrual by decreasing sympathetic activity. Cell Metab. 2010;11:231–8.
11. Brooks S, Prince A, Stahl D, et al. A systematic review and meta-analysis of cognitive bias to food stimuli in people with disordered eating behaviour. Clin Psychol Rev. 2011;31:37–51.
12. Lucas AR, Beard CM, O'Fallon WM, et al. 50-year trends in the incidence of anorexia nervosa in Rochester, Minn.: a population-based study. Am J Psychiatry. 1991;148:917–22.
13. Attia E. Anorexia nervosa: current status and future directions. Annu Rev Med. 2010;61:425–35.
14. Rask-Andersen M, Olszewski PK, Levine AS, et al. Molecular mechanisms underlying anorexia nervosa: focus on human gene association studies and systems controlling food intake. Brain Res Rev. 2010;62:147–64.
15. Yamashiro T, Fukunaga T, Yamashita K, et al. Gene and protein expression of brain-derived neurotrophic factor and TrkB in bone and cartilage. Bone. 2001;28:404–9.
16. Ahn JD, Dubern B, Lubrano-Berthelier C, et al. Cart overexpression is the only identifiable cause of high bone mass in melanocortin 4 receptor deficiency. Endocrinology. 2006;147:3196–202.
17. Robinson L, Aldridge V, Clark EM, et al. A systematic review and meta-analysis of the association between eating disorders and bone density. Osteoporos Int. 2016;27:1953–66.
18. Grinspoon S, Thomas E, Pitts S, et al. Prevalence and predictive factors for regional osteopenia in women with anorexia nervosa. Ann Intern Med. 2000;133:790–4.
19. Misra M, Aggarwal A, Miller KK, et al. Effects of anorexia nervosa on clinical, hematologic, biochemical, and bone density parameters in community-dwelling adolescent girls. Pediatrics. 2004;114:1574–83.
20. Misra M, Klibanski A. Anorexia nervosa and bone. J Endocrinol. 2014;221:R163–76.
21. Grinspoon S, Baum H, Lee K, et al. Effects of short-term recombinant human insulin-like growth factor I administration on bone turnover in osteopenic women with anorexia nervosa. J Clin Endocrinol Metab. 1996;81:3864–70.
22. Misra M, Soyka LA, Miller KK, et al. Serum osteoprotegerin in adolescent girls with anorexia nervosa. J Clin Endocrinol Metab. 2003;88:3816–22.
23. Faje AT, Fazeli PK, Miller KK, et al. Fracture risk and areal bone mineral density in adolescent females with anorexia nervosa. Int J Eat Disord. 2014;47:458–66.
24. Vestergaard P, Emborg C, Stoving RK, et al. Fractures in patients with anorexia nervosa, bulimia nervosa, and other eating disorders—a nationwide register study. Int J Eat Disord. 2002;32:301–8.
25. Lucas AR, Melton LJ, 3rd, Crowson CS, et al. Long-term fracture risk among women with anorexia nervosa: a population-based cohort study. Mayo Clin Proc. 1999;74:972–7.
26. Misra M, Klibanski A. Endocrine consequences of anorexia nervosa. Lancet Diabetes Endocrinol. 2014;2:581–92.
27. Fazeli PK, Bredella MA, Freedman L, et al. Marrow fat and preadipocyte factor-1 levels decrease with recovery in women with anorexia nervosa. J Bone Miner Res 2012;27:1864–71.
28. Misra M, Le Clair M, Mendes N, et al. Use of SSRIs may Impact Bone Density in Adolescent and Young Women with Anorexia Nervosa. CNS Spectr. 2010;15:579–86.
29. Bachrach LK, Guido D, Katzman D, et al. Decreased bone density in adolescent girls with anorexia nervosa. Pediatrics. 1990;86:440–7.
30. Crabtree NJ, Arabi A, Bachrach LK, et al. Dual-energy X-ray absorptiometry interpretation and reporting in children and adolescents: the revised 2013 ISCD Pediatric Official Positions. J Clin Densitom. 2014;17:225–42.
31. Miller KK, Lee EE, Lawson EA, et al. Determinants of skeletal loss and recovery in anorexia nervosa. J Clin Endocrinol Metab. 2006;91:2931–7.
32. Misra M, Prabhakaran R, Miller KK, et al. Weight gain and restoration of menses as predictors of bone mineral density change in adolescent girls with anorexia nervosa-1. J Clin Endocrinol Metab. 2008;93:1231–7.
33. Klibanski A, Biller B, Schoenfeld D, et al. The effects of estrogen administration on trabecular bone loss in young women with anorexia nervosa. J Clin Endocrinol Metab. 1995;80:898–904.
34. Strokosch GR, Friedman AJ, Wu SC, et al. Effects of an oral contraceptive (norgestimate/ethinyl estradiol) on bone mineral density in adolescent females with anorexia nervosa: a double-blind, placebo-controlled study. J Adolesc Health. 2006;39:819–27.
35. Misra M, Katzman D, Miller KK, et al. Physiologic estrogen replacement increases bone density in adolescent girls with anorexia nervosa. J Bone Miner Res. 2011;26:2430–8.
36. Miller KK, Meenaghan E, Lawson EA, et al. Effects of risedronate and low-dose transdermal testosterone on bone mineral density in women with anorexia nervosa: a randomized, placebo-controlled study. J Clin Endocrinol Metab. 2011;96:2081–8.
37. Divasta AD, Feldman HA, Giancaterino C, et al. The effect of gonadal and adrenal steroid therapy on skeletal health in adolescents and young women with anorexia nervosa. Metabolism. 2012;61:1010–20.
38. Misra M, McGrane J, Miller KK, et al. Effects of rhIGF-1 administration on surrogate markers of bone turnover in adolescents with anorexia nervosa. Bone. 2009;45:493–8.
39. Grinspoon S, Thomas L, Miller K, et al. Effects of recombinant human IGF-I and oral contraceptive administration on bone density in anorexia nervosa. J Clin Endocrinol Metab. 2002;87:2883–91.

40. Fazeli PK, Wang IS, Miller KK, et al. Teriparatide increases bone formation and bone mineral density in adult women with anorexia nervosa. J Clin Endocrinol Metab. 2014;99:1322–9.
41. Golden NH, Iglesias EA, Jacobson MS, et al. Alendronate for the treatment of osteopenia in anorexia nervosa: a randomized, double-blind, placebo-controlled trial. J Clin Endocrinol Metab. 2005;90:3179–85.
42. Golden NH, Jacobson MS, Schebendach J, et al. Resumption of menses in anorexia nervosa. Arch Pediatr Adolesc Med. 1997;151:16–21.
43. Levinson DF. The genetics of depression: a review. Biol Psychiatry. 2006;60:84–92.
44. Goshen I, Kreisel T, Ben Menachem Zidon O, et al. Brain interleukin-1 mediates chronic stress-induced depression in mice via adrenocortical activation and hippocampal neurogenesis suppression. Mol Psychiatry. 2008;13:717–28.
45. Murray CJ, Lopez AD. Alternative projections of mortality and disability by cause 1990-2020: Global Burden of Disease Study. Lancet. 1997;349:1498–504.
46. Bab I, Yirmiya R. Depression, selective serotonin reuptake inhibitors, and osteoporosis. Curr Osteoporos Rep. 2010;8:185–91.
47. Yirmiya R, Bab I. Major depression is a risk factor for low bone mineral density: a meta-analysis. Biol Psychiatry. 2009;66:423–32.
48. Wu Q, Magnus JH, Liu J, et al. Depression and low bone mineral density: a meta-analysis of epidemiologic studies. Osteoporos Int. 2009;20:1309–20.
49. Schweiger JU, Schweiger U, Huppe M, et al. Bone density and depressive disorder: a meta-analysis. Brain Behav. 2016;6:e00489.
50. Dorn LD, Susman EJ, Pabst S, et al Association of depressive symptoms and anxiety with bone mass and density in ever-smoking and never-smoking adolescent girls. Arch Pediatr Adolesc Med. 2008;162:1181–8.
51. Fazeli PK, Mendes N, Russell M, et al. Bone density characteristics and major depressive disorder in adolescents. Psychosom Med. 2013;75:117–23.
52. Calarge CA, Butcher BD, Burns TL, et al. Major depressive disorder and bone mass in adolescents and young adults. J Bone Miner Res. 2014;29:2230–7.
53. Williams LJ, Pasco JA, Jackson H, et al. Depression as a risk factor for fracture in women: A 10 year longitudinal study. J Affect Disord. 2016;192:34–40.
54. Wu Q, Liu J, Gallegos-Orozco JF, et al. Depression, fracture risk, and bone loss: a meta-analysis of cohort studies. Osteoporos Int. 2010;21:1627–35.
55. Rauma PH, Pasco JA, Berk M, et al. The association between major depressive disorder, use of antidepressants and bone mineral density (BMD) in men. J Musculoskelet Neuronal Interact. 2015;15:177–85.
56. Yirmiya R, Goshen I, Bajayo A, et al. Depression induces bone loss through stimulation of the sympathetic nervous system. Proc Natl Acad Sci U S A. 2006;103:16876–81.
57. Rosenblat JD, Gregory JM, Carvalho AF, et al. Depression and Disturbed Bone Metabolism: A Narrative Review of the Epidemiological Findings and Postulated Mechanisms. Curr Mol Med. 2016;16:165–78.
58. Tseng PT, Chen YW, Yeh PY, et al. Bone Mineral Density in Schizophrenia: An Update of Current Meta-Analysis and Literature Review Under Guideline of PRISMA. Medicine (Baltimore). 2015;94:e1967.
59. Gomez L, Stubbs B, Shirazi A, et al. Lower Bone Mineral Density at the Hip and Lumbar Spine in People with Psychosis Versus Controls: a Comprehensive Review and Skeletal Site-Specific Meta-analysis. Curr Osteoporos Rep. 2016;14:249–59.
60. Wu CS, Chang CM, Tsai YT, et al. Antipsychotic treatment and the risk of hip fracture in subjects with schizophrenia: a 10-year population-based case-control study. J Clin Psychiatry. 2015;76:1216–23.
61. Tsai KY, Lee CC, Chou YM, et al. The risks of major osteoporotic fractures in patients with schizophrenia: a population-based 10-year follow-up study. Schizophr Res. 2014;159:322–8.
62. Misra M, Papakostas GI, Klibanski A. Effects of psychiatric disorders and psychotropic medications on prolactin and bone metabolism. J Clin Psychiatry. 2004;65:1607–18; quiz 590, 760–1.
63. Kishimoto T, De Hert M, Carlson HE, et al. Osteoporosis and fracture risk in people with schizophrenia. Curr Opin Psychiatry. 2012;25:415–29.
64. Renn JH, Yang NP, Chueh CM, et al. Bone mass in schizophrenia and normal populations across different decades of life. BMC Musculoskelet Disord. 2009;10:1.
65. Stark KL, Xu B, Bagchi A, et al. Altered brain microRNA biogenesis contributes to phenotypic deficits in a 22q11-deletion mouse model. Nat Genet. 2008;40:751–60.
66. Sugatani T, Hruska KA. Impaired micro-RNA pathways diminish osteoclast differentiation and function. J Biol Chem. 2009;284:4667–78.
67. Francis PT, Palmer AM, Snape M, et al. The cholinergic hypothesis of Alzheimer's disease: a review of progress. J Neurol Neurosurg Psychiatry. 1999;66:137–47.
68. Cumming RG, Nevitt MC, Cummings SR. Epidemiology of hip fractures. Epidemiol Rev. 1997;19:244–57.
69. Cui S, Xiong F, Hong Y, et al. APPswe/Abeta regulation of osteoclast activation and RAGE expression in an age-dependent manner. J Bone Miner Res. 2011;26:1084–98.
70. Yang MW, Wang TH, Yan PP, et al. Curcumin improves bone microarchitecture and enhances mineral density in APP/PS1 transgenic mice. Phytomedicine. 2011;18:205–13.
71. Warden SJ, Bliziotes MM, Wiren KM, et al. Neural regulation of bone and the skeletal effects of serotonin (5-hydroxytryptamine). Mol Cell Endocrinol. 2005;242:1–9.
72. Cherian PP, Cheng B, Gu S, et al. Effects of mechanical strain on the function of Gap junctions in osteocytes are mediated through the prostaglandin EP2 receptor. J Biol Chem. 2003;278:43146–56.
73. Bliziotes M, Eshleman A, Burt-Pichat B, et al. Serotonin transporter and receptor expression in osteocytic MLO-Y4 cells. Bone. 2006;39:1313–21.

74. Yadav VK, Ryu JH, Suda N, et al. Lrp5 controls bone formation by inhibiting serotonin synthesis in the duodenum. Cell. 2008;135:825–37.
75. Bonnet N, Bernard P, Beaupied H, et al. Various effects of antidepressant drugs on bone microarchitectecture, mechanical properties and bone remodeling. Toxicol Appl Pharmacol. 2007;221:111–8.
76. Williams LJ, Pasco JA, Jacka FN, et al. Depression and bone metabolism. A review. Psychother Psychosom. 2009;78:16–25.
77. Ziere G, Dieleman JP, van der Cammen TJ, et al. Selective serotonin reuptake inhibiting antidepressants are associated with an increased risk of nonvertebral fractures. J Clin Psychopharmacol. 2008;28:411–7.
78. Durham RA, Johnson JD, Eaton MJ, et al. Opposing roles for dopamine D1 and D2 receptors in the regulation of hypothalamic tuberoinfundibular dopamine neurons. Eur J Pharmacol. 1998;355:141–7.
79. Seriwatanachai D, Charoenphandhu N, Suthiphongchai T, et al. Prolactin decreases the expression ratio of receptor activator of nuclear factor kappaB ligand/osteoprotegerin in human fetal osteoblast cells. Cell Biol Int. 2008;32:1126–35.
80. Wongdee K, Tulalamba W, Thongbunchoo J, et al. Prolactin alters the mRNA expression of osteoblast-derived osteoclastogenic factors in osteoblast-like UMR106 cells. Mol Cell Biochem. 2011;349:195–204.
81. Peuskens J, Pani L, Detraux J, et al. The effects of novel and newly approved antipsychotics on serum prolactin levels: a comprehensive review. CNS Drugs. 2014;28:421–53.

135

Interactions Between Muscle and Bone

Marco Brotto

Bone-Muscle Collaborative Sciences, College of Nursing and Health Innovation,
University of Texas at Arlington, Arlington, TX, USA

INTRODUCTION: THE VISIBLE MECHANICAL COUPLING

The mechanical coupling between bones and skeletal muscles occurs through the sometimes forgotten or much less studied component of the musculoskeletal system—tendons—and is visible. Muscle contractions apply load to the bone, and bone adjusts its mass and architecture to changes in mechanical load as contraction of skeletal muscle is required for locomotion. This mechanical perspective has naturally led to the observation and immediate interpretation that, as muscle function declines, the result would be decreased loading of the skeleton, leading to a decrease in bone mass [1]. However, mechanical factors alone cannot explain the totality of the changes in the musculoskeletal system during development, aging, and pathological conditions [2]. As reviewed in Novotny, there is a clear mismatch between muscle width and bone mineral mass with the latter dropping significantly more with aging in both men and women (Fig. 135.1). Furthermore, in women this mismatch is greater.

In fact, even the reduction in force in skeletal muscle during aging, commonly referred to as sarcopenia, cannot be explained by the decrease in muscle mass alone, suggesting that even at the organ level there are more than physical interactions (Fig. 135.2) [3–5]. These observations imply that beyond the mechanical coupling there is also a biochemical coupling.

One of our goals in this chapter is to show that muscle and bone conform to the classical definition of endocrine organs. Furthermore, we remind readers of the exceptional complexity of the musculoskeletal (MSK) system. In contrast to the simplistic view that only bones and muscles are part of the MSK, the system is composed of bones, blood vessels, cartilage, joints, ligaments, nerves, skeletal muscles, tendons, and other connective tissues. Of particular interest to this chapter, we shall call attention to a rarely discussed member of the connective tissues, Sharpey's fibers or perforated fibers, which may provide a conduit for transport of factors between bones and muscles. We shall attempt to answer the commonly asked question of how factors from one tissue reach another tissue if there are impermeable barriers to transport of molecules in the organism. We shall draw from both in vitro cell and in vivo animal models of bone–muscle cross-talk as well as clinical human situations that exemplify the applicability of bone–muscle interactions.

THE INVISIBLE BIOCHEMICAL COUPLING: BONES AND MUSCLES AS ENDOCRINE ORGANS

The physiological roles of the MSK in movement and mechanical support along with the close proximity of its chief components have largely defined the connection between bones and muscles. The mechanostat model, which maintained that bone strength and density was largely a function of imposed mechanical forces [6], derived from an impressive body of evidence in support of the biomechanical relationship between bones and muscles. Intriguingly, even the strongest supporters of the model acknowledged the possibility of other agents locally and systemically influencing the skeletal architecture [6].

It took many years of research to establish the essential role of skeletal muscles in the maintenance of body glucose homeostasis [7, 8], which is also demonstrated by the fact that by having a larger volume of muscle

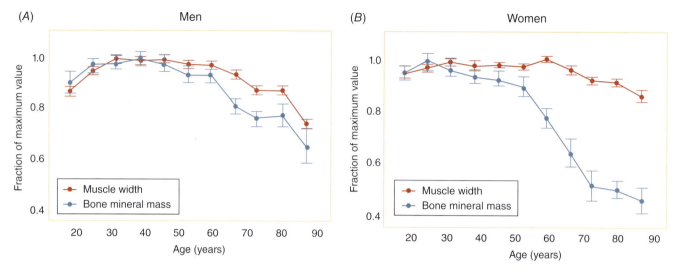

Fig. 135.1. Muscle width and bone mineral mass mismatch during aging. Relationship between bone mineral mass of the radius and muscle width in the forearm in adult men (A) and women (B) during aging. Reproduced from [2].

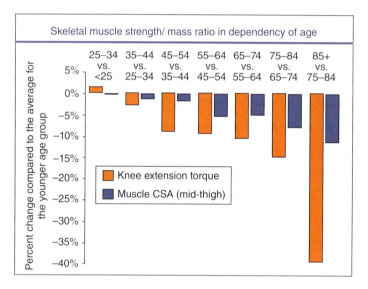

Fig. 135.2. Skeletal muscle strength does not match muscle mass. Relationship between loss of muscle mass and loss of strength per major decade of life shows a clear uncoupling of muscle mass and force, suggesting additional mechanisms beyond mechanical. Reproduced from [5] with permission from John Wiley & Sons.

and/or being more active physically a person can help prevent diabetes mellitus type 2 [9–11]. However, a series of fascinating studies have also shown a role of bone in regulating glycemia and body metabolism, which further brings these tissues together and illustrates the close biochemical coupling between them. The implications of these findings for diabetes, obesity, and cardiovascular diseases have not even begun to be explored. Interventions (ie, new exercise regimes, specific diets, new targeting drugs, etc.) that simultaneously stimulate glucose uptake by bone and muscle cells could for example be far more effective for diabetes and preventing obesity.

BONE AS AN ENDOCRINE ORGAN: BONE SIGNALING TO MUSCLE

In the 1990s, support for the endocrine function of osteocytes began to gain momentum with the works of Marotti et al. [12] and Urist [13] that led to the identification and naming of bone morphogenetic protein (BMP). Using fluid flow shear stress experiments, Klein-Nulend et al. reported that osteocytes secrete large amounts of prostaglandins [14]. Furthermore, osteocytes seem to exert a self-regulatory role by secreting sclerostin, a protein that inhibits bone formation [15]. This discovery has led to a new class of anti-sclerostin antibodies for the treatment of osteoporosis [16]. Another protein produced mainly by osteocytes is FGF23, which regulates vitamin D metabolism and systemic phosphate levels [17]. It appears that osteocyte expression of FGF23 is modulated by DMP1, PHEX, and MEPE [18].

We recently reported that in the *DMP1* knockout mouse, an animal model of rickets and osteomalacia that has elevated serum levels of FGF23, there is severely compromised skeletal muscle function [19]. Furthermore, we found that muscle quality was reduced; muscle mass alone could not explain the decrease in muscle performance in these mice [19]. These results suggest that a disease primarily of bone can lead to muscle phenotype. These and other observations have led us to propose that bone–muscle signaling is essential for optimal muscle function [20].

Osteocalcin is mainly synthesized by osteoblasts and has been shown to contribute to calcium ion homeostasis, energy metabolism, and male fertility [21]. A vitamin K-dependent posttranslational carboxylation is required for the activation of osteocalcin [22]. Osteocalcin has gained significant clinical attention since serum undercarboxylated osteocalcin has been identified as a biomarker for high risk of hip fracture in elderly women [23, 24]. Interestingly, osteoblasts express the *Esp* (osteo-testicular phosphatase gene) gene that inhibits osteocalcin [25] and the *Esp* knockout mouse has increased muscle mass, suggesting that an osteoblast-derived factor can directly modulate bone mass. In addition, supplementation with osteocalcin is able to restore reduced exercise capacity in Esp knockout mice.

As aforementioned, prostaglandins (PGs) are another major class of osteocyte-secreted factors with fundamental roles in bone metabolism because they are able to stimulate both bone resorption and formation. PGs may also contribute to fracture healing and heterotopic ossification [26]. Obviously, these data only point to autocrine/paracrine effects of PGE_2 in bone itself. However, experiments our group performed using osteocyte and muscle cell lines have revealed that PGE_2 secretion from osteocytes is more than 100 times greater than PGE_2 secretion from muscle cells. This excess amount of PGE_2 from osteocytes could interplay with injured muscles, and would aid in muscle regeneration and repair. Interestingly, our recent in vitro studies have provided support for a role of osteocyte-secreted PGE_2 in modulating the process of myogenesis [27, 28]. We believe that the study of the roles of prostaglandins and lipid mediators in bone–muscle interactions is only in its infancy. Recently, several specific lipids derived from membrane phospholipids have been identified as important mediators in bone and muscle [29, 30] physiology and the number of such molecules will increase in the near future. The deciphering of the molecular mechanisms underlying the actions of these lipid mediators, as well as how they might interact to promote optimal function of bone and muscle, could lead to both new biomarkers for bone–muscle health/disease and new treatments for the MSK unit.

An interesting and feasible assumption is that structural components important for the release of factors could affect bone–muscle biochemical cross-talk. Bonewald and colleagues have demonstrated that connexin 43 is an integral membrane protein in gap junctions in bone, with essential roles for the release of PGE_2 from osteocytes in response to loading [31]. Shen et al. specifically targeted the knockdown of connexin 43 in osteoblasts/osteocytes, and observed an expected bone phenotype (ie, reduced cortical bone thickness), but in clear confirmation of bone to muscle signaling via secreted factors from bone acting in muscles, a defective muscle phenotype in fast-twitch extensor digitorum longus (EDL) muscle was also observed [32].

Our group has also recently reported a study in which the conditional deletion of membrane bound transcription factor peptidase-site 1 (*MBTPS1*) in osteocytes did not lead to major phenotypic changes in bone structure or density. Nevertheless, *MBTPS1* deletion in bone resulted in a 30% increase in contractile force and a 12% increase in muscle mass of soleus, a slow-twitch muscle, but not in the fast-twitch EDL muscle, indicating a highly sophisticated signaling role of bone cells to muscle cells, capable of influencing or regulating one muscle fiber type but not the other [33]. These new data could have profound implications for the treatment of muscle atrophy/weakness, including sarcopenia.

Other factors worth mentioning are WNTs and the specific signaling pathway they activate, the Wnt/β-catenin signaling pathway [34]. Studies by Johnson's laboratory were essential to demonstrate the role of the Wnt/β-catenin signaling pathway in humans and mice. His laboratory discovered causal mutations in the low-density lipoprotein receptor-related protein 5 (*LRP5*) gene underlying conditions of altered bone mass, which steered in a new era in bone research [35]. Our research groups recently found that WNT3a is secreted at 10 times greater levels by MLO-Y4 osteocytes when compared to C2C12 myotubes. Furthermore, WNT3a when added to myoblasts dramatically accelerated myogenic differentiation (Fig. 135.3) (Huang & Brotto, unpublished observations, 2017). WNT3a is found in the extracellular matrix of different cell types [36], is expressed in C2C12 cells and 2T3 osteoblasts [37, 38], and is detected at relatively high levels in serum of both humans (16.9 ± 2.4 ng/mL) and mice (0.225 to 3.74 ng/mL), suggesting a potential role of WNT3a in bone–muscle cross-talk. Our groups are currently conducting muscle- and bone-targeted and timed specific deletions of β-catenin to investigate the role of this pathway in bone–muscle cross-talk in vivo.

One last factor that deserves mention is TFG-β, which has attracted attention because of its catabolic, rather than anabolic effects on muscle. Work by Guise and colleagues has demonstrated a new role for bone-derived TGF-β as a cause of skeletal muscle weakness in the setting of osteolytic cancer in the bone [39, 40]. An interesting question that arises from these studies is whether there might be a balance of anabolic and catabolic stimulants from bone signaling to muscles and whether this

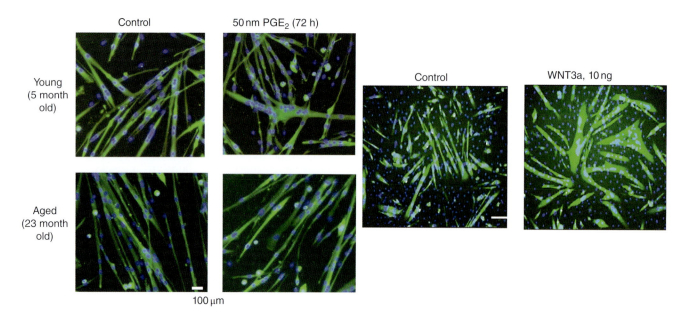

Fig. 135.3. Effects of bone-secreted factors on myogenesis. PGE_2, which is secreted 100 to 1000 more times by osteocytes than muscle cells, accelerates myogenesis in myoblasts from young mice but not from aged mice. WNT3a, which is secreted at 10 times higher levels by osteocytes as compared to muscle cells, dramatically increases myogenesis of C2C12 myoblasts. (Unpublished results from the Brotto Research Laboratory.)

balance alters with disease state and aging for example. As these questions find answers, potentially new treatments for MSK diseases will develop.

It is impossible in this summary to provide details of the ever-growing number of bone-secreted factors (adenosine triphosphate [ATP], calcium, DKK1, DMP1, FGF23, matrix extracellular phosphoglycoprotein [MEPE], nitric oxide, osteoprotegerin, osteocalcin, prostaglandins [particularly PGE_2], RANKL, sclerostin, sclerostin [SOST], TFG-β, WNTs, etc.). These factors exemplify numerous biochemical structures that range from simple organic molecules, to lipid mediators, to complex proteins, all of which help illustrate the diversity and far-reaching impact of bone as an endocrine organ.

MUSCLE AS AN ENDOCRINE ORGAN: MUSCLE SIGNALING TO BONE

Skeletal muscle represents the largest organ system, comprising 40% to 50% of the body. Skeletal muscle develops from myogenic precursor cells often referred to as satellites or myoblasts. Muscle cell proliferation and differentiation (ie, myogenesis) occur in the embryonic and early stages of development, and continue throughout the life span. Skeletal muscle cells undergo myogenesis repeatedly as muscles regenerate in response to injury [41, 42]. Diseases such as muscular dystrophy show that there seems to be a limit to the number of cycles in which muscles can undergo myogenesis before they start showing signs of deterioration due to loss of plasticity, characterized by gradual replacement of muscle by fibrotic and necrotic tissue.

The functional role of skeletal muscle to move and support the body has long been acknowledged; it was about one decade ago with the pioneering and truly innovative work of Pedersen et al. that the endocrine-like function of skeletal muscle was appreciated.

In recent years, results from different research groups have supported the concept that muscle functions as an endocrine organ. Nonetheless, the pioneering work of Pedersen and colleagues first highlighted this new role of muscles [43]. They coined the word myokines to refer to how muscles regulate systemic metabolism by releasing these secretory factors to interact with other tissues or organs, including brain, liver, fat, bone, and muscles themselves. The myokines discovered include myostatin, leukemia inhibitory factor (LIF), IL-6, IL-7, brain-derived neurotrophic factor (BDNF), IGF-1, FGF2, and follistatin-like protein 1 (FSTL-1), and some of them have been shown to be important for bone development, such as IGF-1 and FGF2 [43]. It is reasonable to postulate that many myokines could influence bone directly through binding to specific receptors. It is also possible that myokines prime bone cells for the action of the local factors released by bone cells or other tissues. Moreover, indirect effects could also be involved in bone–muscle interaction, because myokines can affect pancreas and adipose and other tissues, which have been shown in previous studies to cross-talk with bone biochemically. For example, irisin, a newly identified myokine, can induce the switch of white fat to brown-fat-like tissue, which could subsequently exert effects on bone [44]. In addition, recent data also suggest that irisin can directly regulate osteogenesis in osteoblast and bone marrow stromal cell cultures [45, 46].

Furthermore, some myokines, including prostaglandins, IL-6, and LIF, have been shown to enhance myoblast/myotube differentiation after injury [47–50], which can affect bone function. Other factors secreted by injured skeletal muscle are TGF-α and TGF-β1 [47, 51]. It is interesting that these myokines inhibit muscle cell differentiation and proliferation. Again, there seems to be intrinsic homeostatic balance. As mentioned in "Bone as an Endocrine Organ," TGF-β released from bone seems to exert catabolic effects in skeletal muscles in the context of osteolytic lesions/cancers, pointing to the complexity of bone–muscle interactions.

Insulin resistance in peripheral tissues characterizes obesity and type 2 diabetes. A high caloric intake combined with a sedentary lifestyle is the leading cause of these conditions, which have reached worldwide epidemic proportions. Because myokines seem to act metabolically in most tissues in the body, including bone, and are stimulated by physical activity/exercise [52], sedentary lifestyles may favor for example conditions where both muscle and bone cells are not working to control glycemia.

As with bone-secreted factors, which in another review article we were tempted to refer to as "osteokynes," myokines are also expanding as new methods and models interrogate their existence. IL-5 is an anabolic myokine with potential roles in muscle–fat cross-talk; IL-7 seems to modulate satellite cell function during myogenesis [53]; and IL-8 acts to increase angiogenesis [54]. Angiogenesis is crucial for bone survival and maintenance and it is noteworthy that a secreted myokine could influence bone angiogenesis.

Recently identified myokines and their mechanisms of action have been reviewed in [55]. They are: (i) irisin, which regulates the conversion of white fat into brown fat (or browning of adipose tissue); (ii) meteorin-like 1, which apparently has a similar function to irisin but appears to be released by injured muscle; (iii) myonectin, which stimulates improved uptake of fatty acids by the liver; (iv) musclin, which enhances mitochondrial biogenesis; and (v) SPARC, which has beneficial actions on colon cancer.

As in bone to muscle signaling, animal models can add important insights. Zimmers et al. and colleagues used the myostatin-deficient mouse model to investigate the effects of increased muscle mass on bone mineral content and density. Despite the overwhelming muscle growth in this model [56], a consistent correlation was not found in all regions of the skeletal system; the authors reported increased cortical BMD in the distal femur and increased periosteal circumference along the humerus [57]. This strongly suggests that higher muscle mass and forces alone were not sufficient to increase bone mass in the entire body, but rather in specific areas of the skeletal system. Arguably, the two regions where they reported increased bone mass/function are some of the higher metabolically active regions regarding muscle activity. Was it the more constant contractions in these regions or the higher rate of release of myokines because of the higher activity that led to the bone phenotype? Likely, it is a combination of both effects.

As the body of knowledge related to myokines continues to expand, researchers are investigating the role of certain myokines, or the lack thereof, because of a sedentary lifestyle, the Standard American Diet (SAD), and the connection to these chronic diseases [52].

With a deeper understanding of skeletal muscle myokines and their specific roles in bone–muscle cross-talk and, more specifically, in muscle to bone signaling, comes the promise of innovative approaches to the prevention and treatment of MSK diseases and a host of other disorders throughout the body.

ADDITIONAL EVIDENCE FOR BONE-MUSCLE CROSS-TALK FROM CLINICAL EXAMPLES

Additional evidence comes to light as we examine pathological conditions, for example medial tibial stress syndrome, one of the most prevalent syndromes in military personnel and runners [58, 59]. In this syndrome, there is inflammation in the posterior tibial muscle as well as inflammation of periosteum and the bone underneath the affected muscle region. Despite many studies, the question remains unanswered: does muscle induce the inflammation to the periosteum or does the periosteum induce it to the muscle [58, 59]? It is certain that the mechanisms are common in both tissues. It is very interesting that, to date, treatments for this condition have been inadequate, likely because effective approaches to treat both tissues simultaneously are not yet in place.

Equally intriguing and yet lacking an understanding of the molecular mechanisms, is the current therapeutic approach utilizing muscle flaps to accelerate the healing of compound bone fractures as well as for the treatment of chronic osteomyelitis [60]. Our recent studies may provide insights into the protective roles of muscles for bone function. We reported that muscle-secreted factors protect osteocytes from dexamethasone-induced cell death [61]. Accordingly, muscle flaps might work as a reservoir of myokines and mechanical stimuli that lead to improvement in bone healing and function.

HOW CAN THE "KINES" CROSS BARRIERS?

A commonly asked question is how the factors secreted by one tissue might reach another tissue. Although this is interesting, in general, we do not have problems in accepting that hormones released by glands reach into all cells of the body and act even in nuclear receptors, but there seems to be an unexplained resistance in the field to the concept that bone factors can reach muscle and muscle factors can reach bone.

The anatomical proximity and, furthermore, the intricate complexity of this proximity provides confidence for the concept that bones and muscles influence each other in a paracrine manner. Elegant studies by Schnitzler's laboratory revealed for the first time Sharpey's or perforating fibers originating from muscle fibers penetrating deep into the periosteum and beyond [62]. These fibers are rich in collagen and extracellular matrix material, and therefore one could postulate their potential function in the transmission of bone–muscle signaling, including mechanical and biochemical signals, but this possibility remains unexplored.

Experiments performed by Beno and colleagues [63] showed that mouse tail vein injection of small dyes and molecules up to 70 kDa can permeate the osteocyte-lacunar–canalicular network in just a few minutes; this is bigger than the size of WNT3a (~40 kDa), and in fact the majority of WNTs. This demonstrates that the canalicular fluid has ready access to the circulation and suggests that factors secreted by osteocytes could enter the blood and have effects on distant target cells, including muscle and vice versa.

Lai et al. [64] in a very innovative study investigated the permeability of the periosteum in intact mouse tibia by tracking and modeling fluorescent tracers that mimic myokines under confocal microscopy. They found that the periosteum was semipermeable with a cut-off MW of approximately 40 kDa. Therefore, a number of known myokines, based on simple diffusion equations through the periosteum, should be able to cross this barrier and reach bone cells.

Both bone and muscle are highly vascularized and factors released into the circulation could simply reach the other tissue systematically. Furthermore, both tissues release exosomes and macrovesicles, which can carry a host of molecules to other tissues for the exchange of information. Intriguingly, circulatory extracellular vesicle content increases in an intensity-dependent manner in response to endurance exercise, and the term "exerkines" has been created to refer to this phenomenon [65].

CLOSING REMARKS: BEYOND BONE–MUSCLE INTERACTIONS

The endocrine communication revealed by bone–muscle interactions seems to be more far-reaching because molecules from these tissues also interact with the adipose tissue, brain, gut, and the immune system. The field of tissue cross-talk is only in its infancy. Detailed profiling of "kines" in health, disease, and exercise does not yet exist. We do not know how aging affects the production, secretion, and interaction of these molecules. According to the Centers for Disease Control and Prevention, the prevalence of obesity in the United States among adults 65 years of age and older is nearly 35% [66]. The American Diabetes Association reports that nearly 25% to 30% of US adults aged 60 and over have diabetes, and the prevalence of metabolic syndrome increases with age [67]. There can be no doubt that healthier MSK systems lead to healthier humans.

Therefore we should welcome a new era of basic, translational, and clinical research aimed at addressing some of the questions in this chapter and many others raised by the biochemical interactions between bone and muscle. These answers will lead to innovative therapeutic approaches and unprecedented advances in the fight against chronic diseases such as osteoporosis, sarcopenia, diabetes, obesity, and many MSK diseases that afflict humanity.

REFERENCES

1. Frost HM. Bone "mass" and the "mechanostat": a proposal. Anat Rec. 1987;219:1–9.
2. Novotny SA, Warren GL, Hamrick MW. Aging and the muscle-bone relationship. Physiology (Bethesda). 2015;30:8–16.
3. Manring H, Abreu E, Brotto L, et al. Novel excitation-contraction coupling related genes reveal aspects of muscle weakness beyond atrophy-new hopes for treatment of musculoskeletal diseases. Front Physiol. 2014;5:37.
4. Manini TM, Clark BC. Dynapenia and aging: an update. J Gerontol A Biol Sci Med Sci. 2012;67:28–40.
5. Moore AZ, Caturegli G, Metter EJ, et al. Difference in muscle quality over the adult life span and biological correlates in the Baltimore Longitudinal Study of Aging. J Am Geriatr Soc. 2014;62:230–6.
6. Frost HM. Perspectives: a proposed general model of the "mechanostat" (suggestions from a new skeletal-biologic paradigm). Anat Rec. 1996;244:139–47.
7. Brozinick JT, Jr, Etgen GJ, Jr, Yaspelkis BB, III, et al. The effects of muscle contraction and insulin on glucose-transporter translocation in rat skeletal muscle. Biochem J. 1994;297:539–45.
8. Katz A, Sahlin K, Broberg S. Regulation of glucose utilization in human skeletal muscle during moderate dynamic exercise. Am J Physiol. 1991;260:E411–5.
9. Yang J. Enhanced skeletal muscle for effective glucose homeostasis. Prog Mol Biol Transl Sci. 2014;121:133–63.
10. Toledo FG, Menshikova EV, Ritov VB, et al. Effects of physical activity and weight loss on skeletal muscle mitochondria and relationship with glucose control in type 2 diabetes. Diabetes. 2007;56:2142–7.
11. Toledo FG, Goodpaster BH. The role of weight loss and exercise in correcting skeletal muscle mitochondrial abnormalities in obesity, diabetes and aging. Mol Cell Endocrinol. 2013;379:30–34.
12. Marotti G, Ferretti M, Muglia MA, et al. A quantitative evaluation of osteoblast-osteocyte relationships on growing endosteal surface of rabbit tibiae. Bone. 1992;13:363–8.
13. Urist MR. Bone: formation by autoinduction. Science. 1965;150:893–9.
14. Klein-Nulend J, Burger EH, Semeins CM, et al. Pulsating fluid flow stimulates prostaglandin release and inducible

prostaglandin G/H synthase mRNA expression in primary mouse bone cells. J Bone Miner Res. 1997;12:45–51.
15. Winkler DG, Sutherland MK, Geoghegan JC, et al. Osteocyte control of bone formation via sclerostin, a novel BMP antagonist. EMBO J. 2003;22:6267–76.
16. McClung MR. Clinical utility of anti-sclerostin antibodies. Bone. 2017;96:3–7.
17. Shimada T, Kakitani M, Yamazaki Y, et al. Targeted ablation of Fgf23 demonstrates an essential physiological role of FGF23 in phosphate and vitamin D metabolism. J Clin Invest. 2004;113:561–8.
18. Martin A, Liu S, David V, et al. Bone proteins PHEX and DMP1 regulate fibroblastic growth factor Fgf23 expression in osteocytes through a common pathway involving FGF receptor (FGFR) signaling. FASEB J. 2011;25:2551–62.
19. Wacker MJ, Touchberry CD, Silswal N, et al. Skeletal Muscle, but not Cardiovascular Function, Is Altered in a Mouse Model of Autosomal Recessive Hypophosphatemic Rickets. Front Physiol. 2016;7:173.
20. Brotto M, Bonewald L. Bone and muscle: Interactions beyond mechanical. Bone. 2015;80:109–14.
21. Karsenty G, Wagner EF. Reaching a genetic and molecular understanding of skeletal development. Dev Cell. 2002;2:389–406.
22. Hauschka PV, Lian JB, Cole DE, et al. Osteocalcin and matrix Gla protein: vitamin K-dependent proteins in bone. Physiol Rev. 1989;69:990–1047.
23. Szulc P, Chapuy MC, Meunier PJ, et al. Serum undercarboxylated osteocalcin is a marker of the risk of hip fracture: a three year follow-up study. Bone. 1996;18:487–8.
24. Szulc P, Chapuy MC, Meunier PJ, et al. Serum undercarboxylated osteocalcin is a marker of the risk of hip fracture in elderly women. J Clin Invest. 1993;91:1769–74.
25. Coiro V, Volpi R, Cataldo S, et al. Effect of physiological exercise on osteocalcin levels in subjects with adrenal incidentaloma. J Endocrinol Invest. 2012;35:357–8.
26. Blackwell KA, Raisz LG, Pilbeam CC. Prostaglandins in bone: bad cop, good cop? Trends Endocrinol Metab. 2010;21:294–301.
27. Mo C, Romero-Suarez S, Bonewald L, et al. Prostaglandin E2: from clinical applications to its potential role in bone- muscle crosstalk and myogenic differentiation. Recent Pat Biotechnol. 2012;6:223–9.
28. Mo C, Zhao R, Vallejo J, et al. Prostaglandin E2 promotes proliferation of skeletal muscle myoblasts via EP4 receptor activation. Cell Cycle. 2015;14:1507–16.
29. During A, Penel G, Hardouin P. (2015) Understanding the local actions of lipids in bone physiology. Prog Lipid Res 59, 126–46.
30. Goto-Inoue N, Yamada K, Inagaki A, et al. Lipidomics analysis revealed the phospholipid compositional changes in muscle by chronic exercise and high-fat diet. Sci Rep. 2013;3:3267.
31. Cherian PP, Siller-Jackson AJ, Gu S, et al. Mechanical strain opens connexin 43 hemichannels in osteocytes: a novel mechanism for the release of prostaglandin. Mol Biol Cell. 2005;16:3100–6.
32. Shen H, Grimston S, Civitelli R, et al. Deletion of connexin43 in osteoblasts/osteocytes leads to impaired muscle formation in mice. J Bone Miner Res. 2015;30:596–605.
33. Gorski JP, Huffman NT, Vallejo J, et al. Deletion of Mbtps1 (Pcsk8, S1p, Ski-1) Gene in Osteocytes Stimulates Soleus Muscle Regeneration and Increased Size and Contractile Force with Age. J Biol Chem. 2016;291:4308–22.
34. Regard JB, Zhong Z, Williams BO, et al. Wnt signaling in bone development and disease: making stronger bone with Wnts. Cold Spring Harb Perspect Biol. 2012;4.
35. Johnson ML. LRP5 and bone mass regulation: Where are we now? Bonekey Rep. 2012;1:1.
36. Willert K, Brown JD, Danenberg E, et al. Wnt proteins are lipid-modified and can act as stem cell growth factors. Nature. 2003;423:448–52.
37. Vaes BL, Dechering KJ, van Someren EP, et al. Microarray analysis reveals expression regulation of Wnt antagonists in differentiating osteoblasts. Bone. 2005;36:803–11.
38. Kalajzic I, Staal A, Yang WP, et al. Expression profile of osteoblast lineage at defined stages of differentiation. J Biol Chem. 2005;280:24618–26.
39. Waning DL, Mohammad KS, Reiken S, et al. Excess TGF-beta mediates muscle weakness associated with bone metastases in mice. Nat Med. 2015;21:1262–71.
40. Regan JN, Trivedi T, Guise TA, et al. The Role of TGFbeta in Bone-Muscle Crosstalk. Curr Osteoporos Rep. 2017;15:18–23.
41. Nag AC, Foster JD. Myogenesis in adult mammalian skeletal muscle in vitro. J Anat. 1981;132:1–18.
42. Shefer G, Van de Mark DP, Richardson JB, et al. Satellite-cell pool size does matter: defining the myogenic potency of aging skeletal muscle. Dev Biol. 2006;294:50–66.
43. Pedersen BK, Febbraio MA. Muscles, exercise and obesity: skeletal muscle as a secretory organ. Nat Rev Endocrinol. 2012;8:457–65.
44. Zhang Y, Li R, Meng Y, et al. Irisin stimulates browning of white adipocytes through mitogen-activated protein kinase p38 MAP kinase and ERK MAP kinase signaling. Diabetes. 2014;63:514–25.
45. Qiao X, Nie Y, Ma Y, et al. Irisin promotes osteoblast proliferation and differentiation via activating the MAP kinase signaling pathways. Sci Rep. 2016;6:18732.
46. Colaianni G, Cuscito C, Mongelli T, et al. The myokine irisin increases cortical bone mass. Proc Natl Acad Sci U S A. 2015;112:12157–62.
47. Kurek JB, Bower JJ, Romanella M, et al. The role of leukemia inhibitory factor in skeletal muscle regeneration. Muscle Nerve 1997;20:815–22.
48. Pedersen BK, Steensberg A, Fischer C, et al. Searching for the exercise factor: is IL-6 a candidate? J Muscle Res Cell Motil. 2003;24:113–19.
49. Pedersen BK, Febbraio MA. Muscle as an endocrine organ: focus on muscle-derived interleukin-6. Physiol Rev. 2008;88:1379–406.
50. Goldspink DF, Goldspink G. (1986) The role of passive stretch in retarding muscle atrophy. In: Nix WA, Vrbová G (eds) *Electrical Stimulation and Neuromuscular Disorders*. Berlin: Springer, pp 91–100.

51. Li Y, Huard J. Differentiation of muscle-derived cells into myofibroblasts in injured skeletal muscle. Am J Pathol. 2002;161:895–907.
52. Pedersen L, Olsen CH, Pedersen BK, et al. Muscle-derived expression of the chemokine CXCL1 attenuates diet-induced obesity and improves fatty acid oxidation in the muscle. Am J Physiol Endocrinol Metab. 2012;302:E831–40.
53. Pedersen BK, Akerstrom TC, Nielsen AR, et al. Role of myokines in exercise and metabolism. J Appl Physiol 2007;103:1093–8.
54. Nielsen AR, Pedersen BK. The biological roles of exercise-induced cytokines: IL-6, IL-8, and IL-15. Appl Physiol Nutr Metab. 2007;32:833–9.
55. Whitham M, Febbraio MA. The ever-expanding myokinome: discovery challenges and therapeutic implications. Nat Rev Drug Discov. 2016;15:719–29.
56. Zimmers TA, Davies MV, Koniaris LG, et al. Induction of cachexia in mice by systemically administered myostatin. Science. 2002;296:1486–8.
57. Elkasrawy MN, Hamrick MW. Myostatin (GDF-8) as a key factor linking muscle mass and bone structure. J Musculoskelet Neuronal Interact. 2010;10:56–63.
58. Tweed JL, Avil SJ, Campbell JA, et al. Etiologic factors in the development of medial tibial stress syndrome: a review of the literature. J Am Podiatr Med Assoc. 2008;98:107–11.
59. Reshef N, Guelich DR. Medial tibial stress syndrome. Clin Sports Med. 2012;31:273–90.
60. Chan JK, Harry L, Williams G, et al. Soft-tissue reconstruction of open fractures of the lower limb: muscle versus fasciocutaneous flaps. Plast Reconstr Surg. 2012;130:284e–295e.
61. Jahn K, Lara-Castillo N, Brotto L, et al. Skeletal muscle secreted factors prevent glucocorticoid-induced osteocyte apoptosis through activation of beta-catenin. Eur Cell Mater. 2012;24:197–209; discussion 209–10.
62. Schnitzler CM. Childhood cortical porosity is related to microstructural properties of the bone-muscle junction. J Bone Miner Res. 2015;30:144–55.
63. Beno T, Yoon YJ, Cowin SC, et al. Estimation of bone permeability using accurate microstructural measurements. J Biomech. 2006;39:2378–87.
64. Lai X, Price C, Lu XL, et al. Imaging and quantifying solute transport across periosteum: implications for muscle-bone crosstalk. Bone. 2014;66:82–9.
65. Safdar A, Saleem A, Tarnopolsky MA. The potential of endurance exercise-derived exosomes to treat metabolic diseases. Nat Rev Endocrinol. 2016;12:504–17.
66. Fakhouri TH, Ogden CL, Carroll MD, et al. Prevalence of obesity among older adults in the United States, 2007-2010. NCHS data brief. 2012:1–8.
67. Beltran-Sanchez H, Harhay MO, Harhay MM, et al. Prevalence and trends of metabolic syndrome in the adult U.S. population, 1999-2010. J Am Coll Cardiol. 2013;62:697–703.

Index

Entries in **bold** indicate tables; page locators in **bold** indicate figures.

abaloparatide
   adverse effects 584–585
   immobilization and burns 732
   overview and historical development 396
   parathyroid hormone treatment for osteoporosis 563
abiraterone 778
aBMD *see* areal bone mass/mineral density
abnormal protein binding 642
abnormal vertebral segmentation (AVS) 4
ABQ *see* algorithm-based qualitative
AC *see* articular cartilage
ACAN *see* aggrecan
acetazolamide 862
acetylcholine (Ach) 1029, **1030**
acetyl coenzyme A (acetylCoA) 1006
aCGH *see* microarray comparative genomic hybridization
Ach *see* acetylcholine
ACH *see* achondroplasia; alveolar crestal height
achondrogenesis II 849, *849*
achondroplasia (ACH) 8, 376
acid–base balance 536
acid-suppressive medications 483
ACPA *see* anticitrullinated protein antibodies
ACR *see* American College of Rheumatology
acrodysostosis 667–668
ACTH *see* adrenocorticotropic hormone
actin ring 47
activation frequency 312
activator protein-1 (AP-1) 415, 1021
ACTIVE study 563
activin A type I receptor/activin-like kinase 2 (ACVR1/
   ALK2) 63–64, 867–868
activins 60
acute lymphoblastic leukemia (ALL) 970
acute lymphocytic leukemia (ALL) 793–795
acute myelogenous leukemia (AML) 970
acute primary hyperparathyroidism 621
acute renal failure 642–643
ACVR1/ALK2 and *ACVR1/ALK2 see* activin A type I receptor/
   activin-like kinase 2
AD *see* Alzheimer disease
ADAMO study 555
Addisonian crisis 641

adenomatous polyposis coli (Apc) protein 68
5′-adenosine monophosphate-activated protein kinase
   (AMPK) 961–962, 1009, 1014, 1017
adenosine triphosphate (ATP)
   cellular bioenergetics of bone 1004, 1009
   endocrine bioenergetics of bone 1013
   integrative physiology of the skeleton 961–962
   mechanotransduction 79–80
   peripheral neuronal control of bone remodeling 1033
ADH *see* autosomal dominant hypocalcemia
adherence to osteoporosis therapies 593–596
   concepts and definitions 593
   nonpersistence 593
   optimizing persistence 593–594
   reasons for nonpersistence 594
   summary and conclusion 594–595
ADHR *see* autosomal dominant hypophosphatemic rickets
adinopectin 963
adipocytes 974–982
   bone histomorphometry 312
   bone loss and MAT 978, *979*
   concepts and definitions 974
   current standards for imaging and analysis of MAT 980
   distribution and development of MAT 974, *975*
   evolution and function of MAT 976–978, *977*
   expansion and metabolic disease of MAT 975
   hematopoietic stem cell niche 969
   integrative physiology of the skeleton 959–963
   marrow fat as an adipose tissue 980
   peripheral adipocytes and bone 978–980
   regulated and constitutive marrow adipocytes 976, *976*
   site-specific variation of MAT adipocytes 976
adipokines 1016–1017
adiponectin 980, 1016–1017
adrb2/*adrb2 see* adrenergic 2 receptor
Adrβ2 *see* β2-adrenergic receptor
adrenergic 2 receptor (adrb2) 1020–1021
adrenocorticotropic hormone (ACTH) 513,
   1037–1038, 1041
ADT *see* androgen deprivation therapy; antigen deprivation
   therapy
adult hypophosphatasia (HPP) 887
adult T-cell leukemia/lymphoma (ATLL) 764

Primer on the Metabolic Bone Diseases and Disorders of Mineral Metabolism, Ninth Edition. Edited by John P. Bilezikian.
© 2019 American Society for Bone and Mineral Research. Published 2019 by John Wiley & Sons, Inc.
Companion website: www.wiley.com/go/asbmrprimer

advanced glycation end-products (AGE) 289–290
  diabetes and bone loss 488
adverse effects of drugs for osteoporosis 579–587
  antisclerostin therapy 607
  atrial fibrillation and other cardiovascular events 582, *584*
  atypical femur fractures 580–581, *581*, 585
  bisphosphonates 549–550, 580–583, *583*, *584*
  bone formation agents 583–585
  concepts and definitions 579
  denosumab 556, 582, 583
  extraskeletal side effects of antiresorptive therapy 582–583
  gastrointestinal side effects 582
  need for observational data 579–580
  osteonecrosis of the jaw 581–582, 585
  renal disease and hypocalcemia 583
  skeletal/bone side effects 580–582
  sources of drug safety data 579–580
  strengths and limitations of observational data 580
  strontium ranelate and calcitonin 574, 576
  teriparatide 583–584
adynamic bone disease, transplantation osteoporosis 424
AED *see* antiepileptic drugs
AER *see* apical ectoderm ridge
aerobic conditioning 875
aerobic glycolysis 1006, 1007
AF *see* atrial fibrillation
AFF *see* atypical femoral fractures
AGE *see* advanced glycation end-products
aggrecan (ACAN), endochondral ossification 12, 13
aging
  bone vasculature 989
  epidemiology of osteoporotic fractures 399
  ethnicity and age-related loss of bone strength 131–134
  exercise for osteoporotic fracture prevention and management 521
  Fracture Risk Assessment Tool 331–332, *332*
  gonadal steroids 194, 200, *201*
  gonadal steroids and osteoporosis pathogenesis 412–413, *413*
  mechanical loading 141–142, 143
  menopause and age-related bone loss 155–161, 194, 200
  muscle–bone interactions 1055, *1056*
  nonsex steroid hormone changes with aging 416–417
  osteoporosis in men 443, 447
  pituitary–bone axis 1040–1041
  prevention of falls 526–533
  sarcopenia and osteoporosis 499–500
  skeletal healing/repair 104
  transcriptional profiling 369
AHO *see* Albright hereditary osteodystrophy
AI *see* amelogenesis imperfecta; aromatase inhibitors
Albers-Schönberg disease (A-SD) 825–827
Albright hereditary osteodystrophy (AHO), pseudohypoparathyroidism 664, *664*, *665*, *666*, 668
alcohol, neuropsychiatric disorders 1050
alendronate 395, 429–430
  combination therapy 567–571
  oral manifestations of metabolic bone diseases 945–946
  Paget disease of bone 717–719
  postmenopausal osteoporosis 548, 550
algorithm-based qualitative (ABQ) assessment, vertebral fracture 325
alkaline phosphatase, bone matrix composition 86

alkaline phosphatase (ALP)
  bone turnover markers 293–297
  genetic craniofacial disorders affecting dentition 914
  Paget disease of bone 715, 717–718
  rickets and osteomalacia 684, 685, 689, 691
  sclerosing bone disorders 829, 830
alkaptonuria 888
ALL *see* acute lymphoblastic leukemia; acute lymphocytic leukemia
allometric approach, bone mass measurement in children with osteoporosis risk factors 247
allosteric modulators, calcium-sensing receptor 224–225
ALM *see* appendicular lean mass
ALP *see* alkaline phosphatase
αKlotho (αKL) 188–189, 191
α-SMA *see* α-smooth muscle actin
α-smooth muscle actin (α-SMA) 26
ALSG *see* aplasia of the lacrimal and salivary glands
aluminium hydroxide 700
  tumoral calcinosis 862
alveolar bone homeostasis 933–940
  anatomy of the periodontium 933–934, *934*
  concepts and definitions 933
  diagnosis of periodontal diseases 936–937
  future perspectives 938
  gingivitis treatment 937
  historical and current perspectives 934–935
  host immune response in periodontal tissues 936
  maintenance therapy 938
  molecular basis of host–microbe interactions 935–936, **935**
  oral manifestations of metabolic bone diseases 945–946
  pathogenesis of periodontitis 934–936
  periodontitis treatment 937–938
  treatment of periodontal diseases 937–938
alveolar crestal height (ACH) 941, 943
Alx4 6
Alzheimer disease (AD) 1050
ameloblastoma 923, *924*
ameloblasts 904–907
amelogenesis 904–906, *905*, *906*
amelogenesis imperfecta (AI) 913
American College of Rheumatology (ACR) 469, 471
American Society of Clinical Oncology (ASCO) 776–777
AML *see* acute myelogenous leukemia
amniotic fluid, fetal calcium metabolism 180
AMPK *see* 5′-adenosine monophosphate-activated protein kinase
AN *see* anorexia nervosa
anabolic therapy
  adverse effects 583–585
  chronic kidney disease 507–508
  combination therapy 567–572
  parathyroid hormone treatment for osteoporosis 559–563
androgen deprivation therapy (ADT) 801
  osteoporosis in men 445
  skeletal effects of medications for nonskeletal disorders 482
androgens *see* gonadal steroids
angiogenesis, skeletal healing/repair 111–112
angiopoietin-1 (Ang-1) 968
angiotensin receptor blockers 884
animal models: allelic determinants for BMD 359–366
  clustering 362
  concepts and definitions 359
  concordance with human data 363
  covariation 362

dimensions 361
dynamic phenotypes 361
experimental design and genetic architecture 363
future directions 363–365
gene expression 361
gene networks 363, 364
heritability 361–362
intersite discordance 363
mechanical performance 361
phenotypes 359–361, 360
pleiotropy 362
principal components and other composite phenotypes 361
sex limitation 362–363
themes of existing data 361–363
trabecular structure 361
animal models: genetic manipulation 351–358
advantages and disadvantages of gene targeting 352–353
advantages and disadvantages of overexpression approaches 352
cartilage 354
chondrocytes 351
concepts and definitions 351
considerations when using inducible knockouts 356
Cre drivers and lineage tracers for Cre activity **355–356**
CRISPR/Cas9 genomic engineering 357
drivers for gene recombination 354
gene targeting 352–356
lineage tracing and activity reporters **355–356**, 356–357
osteoblasts/osteocytes 351–352, 354
osteoclasts 352, 354–355
overexpression of target genes 351–352
resources and repositories 351
tendon and ligament 352
tissue-specific and inducible knockout and overexpression 353–354, 353
uncondensed mesenchyme and mesenchymal condensations 354
ankylosing spondylitis (AS), inflammation-induced bone loss 459–460, 462–463
annexin 2 (Anxa2) 968
anorexia nervosa
integrative physiology of the skeleton 959–960
premenopausal osteoporosis 439
anorexia nervosa (AN), neuropsychiatric disorders 1047–1049
anterior visceral endoderm (AVE) 893
antibiotics
alveolar bone homeostasis 937–938
osteonecrosis of the jaw 929–930
anticitrullinated protein antibodies (ACPA), inflammation-induced bone loss 460
antidiabetic agents, skeletal effects of medications for nonskeletal disorders 483
antiepileptic drugs (AED), skeletal effects of medications for nonskeletal disorders 483–484
anti-FGF23 antibody, childhood disorders of mineral metabolism 708–709
antigen deprivation therapy (ADT) 776–778
antiglide plates 589
antihormonal drugs 482–483
antimyostatin antibody 502
antipsychotics 1051
antiresorptive therapy
adverse effects of drugs for osteoporosis 580–583
combination therapy 561–562, 567–572
parathyroid hormone treatment for osteoporosis 560–562
antiretroviral therapy (ART) 474–477, 997–998

antisclerostin therapy 605–607
Anxa2 see annexin 2
AP-1 see activator protein-1
Apc see adenomatous polyposis coli
APECED see autoimmune polyglandular syndrome
apical ectoderm ridge (AER) 5–6, 6
aplasia of the lacrimal and salivary glands (ALSG) 912
apoptosis
bisphosphonates for postmenopausal osteoporosis 546
bone histomorphometry 312
glucocorticoid-induced osteoporosis 468
gonadal steroids 199
osteocytes 40–41, 43
osteosarcoma 771
sarcopenia and osteoporosis 498
signal transduction cascades controlling osteoblast differentiation 56
appendicular lean mass (ALM) 528
APS1 see autoimmune polyendocrinopathy syndrome type 1
ARE see hormone response elements
areal bone mass/mineral density (aBMD) 93–94
diagnostic potential of DXA 247–248
gonadal steroids and osteoporosis pathogenesis 412–413
human immunodeficiency virus 475
measurement in adults 266–267
measurement in children with osteoporosis risk factors 243–248
mechanical loading 141, 144
pregnancy and lactation 148–149, 151, 152
reporting aBMD by DXA 247
vertebral fracture assessment 248, 248
ARHR see autosomal recessive hypophosphatemic rickets
aromatase inhibitors (AI)
bone turnover markers 295
skeletal complications of anticancer therapies 776
skeletal effects of medications for nonskeletal disorders 482–483
ART see antiretroviral therapy
arthritis of inflammatory bowel disease 459
articular cartilage (AC) 13
AS see ankylosing spondylitis
ASCO see American Society of Clinical Oncology
A-SD see Albers-Schönberg disease
ATF4/Atf4 17, 34
ATLL see adult T-cell leukemia/lymphoma
ATP see adenosine diphosphate; adenosine triphosphate
atrial fibrillation (AF) 582, 584
atypical femoral fractures (AFF) 451, 580–581, 581, 585
autoimmune polyendocrinopathy syndrome type 1 (APS1) 648, 654–655, 706
autosomal dominant hypocalcemia (ADH) 175
calcium-sensing receptor 225
diagnostic tests and interpretation 650
etiology and pathogenesis 648
hypoparathyroidism 655
parathyroid hormone 208
autosomal dominant hypophosphatemic rickets (ADHR)
clinical and biochemical manifestations 679
disorders of phosphate homeostasis 675, 679–680
FGF23 and the regulation of phosphorus metabolism 189
genetics 679
treatment 679
autosomal recessive hypophosphatemic rickets (ARHR) 43, 189–190, 679
autosomal recessive renal Fanconi syndrome and hypophosphatemic rickets (FRTS2) 680

avascular necrosis (AVN) 795–796
AVE *see* anterior visceral endoderm
AVN *see* avascular necrosis
AVP *see* vasopressin
AVS *see* abnormal vertebral segmentation
axial osteomalacia 834
axis inhibition protein 2 (AXIN2) 68, 911
AXL 816–817

balanced structure variants 374
base-pair substitutions 373–374
basic multicellular units (BMU) 985–986
bazedoxifene 543
BD *see* brachydactyly
BDGF *see* brain-derived growth factor
BDMR *see* brachydactyly mental retardation syndrome
Beckwith–Wiedemann syndrome 667
bending tests of long bones 97–99, *97*, 142–143
benign odontogenic neoplasms 923, *924*
β2-adrenergic receptor (Adrβ2) 962–963
beta-blockers 484, 883–884, 1020–1021
BGLAP/*Bglap see* osteocalcin
bigylcan 86
bioactive dietary constituents 537
bioenergetics *see* endocrine bioenergetics of bone
bioluminescence imaging (BLI) 748, *748*
biopsy
   adipocytes 978
   chronic kidney disease 507
   hematological malignancies 761–762
   ischemic and infiltrative disorders of bone 857
   juvenile osteoporosis 421
   osteosarcoma 768
   Paget disease of bone 717
   premenopausal osteoporosis 438–439
   rickets and osteomalacia 685, 689–690
bisphosphonates (BP)
   adherence to osteoporosis therapies 594
   adverse effects 549–550, 580–583, *583*, *584*
   antifracture efficacy 546–548, *547*
   atypical fractures of the femur 549, *549*
   bone turnover markers 296–298
   chronic kidney disease 507
   combination therapy 561, 567–568
   concepts and definitions 545
   cost-effectiveness of osteoporosis treatment 600
   diabetes and bone loss 489
   excessive suppression of bone remodeling 548–549
   fibrous dysplasia of bone 844
   fracture risk 546, *547*
   glucocorticoid-induced osteoporosis 470
   hematological malignancies 763–764
   immobilization and burns 731–733
   immunobiology 998
   intermittent administration 547–548
   juvenile dermatomyositis 863
   juvenile osteoporosis 421–422
   long-term effects on bone fragility 548
   medical prevention and treatment of metastatic bone disease 800–801, 804–806
   neuropsychiatric disorders 1049
   non-parathyroid hypercalcemia 643–644
   obesity and skeletal health 495
   oral manifestations of metabolic bone diseases 945–946
   osteogenesis imperfecta 875
   osteonecrosis of the jaw 549, 927, 930
   osteoporosis in men 446–447
   overview and historical development 395–396
   Paget disease of bone 717–719
   pharmacological treatment of bone metastasis 818
   pharmacology 545–546
   postmenopausal osteoporosis 545–552
   premenopausal osteoporosis 439
   primary hyperparathyroidism 625
   pseudohypoparathyroidism 669
   radiotherapy of skeletal metastases 812–813
   sclerosing bone disorders 825, 830
   secondary causes of osteoporosis 512
   special issues related to osteoporosis treatment 548–549
   structures *546*
   trabecular bone score 282
   translational genetics of osteoporosis 389
   transplantation osteoporosis 426, **427–428**, 429–432
   tumoral calcinosis 862
   tumor-induced bone disease 753–754
BLC *see* bone lining cells
BLI *see* bioluminescence imaging
Blomstrand's chondrodysplasia 217
blosozumab 606
B lymphopoiesis 969–970
BMAD *see* bone mineral apparent density
BMAT *see* adipocytes
BMC *see* bone mineral content
BMD *see* bone mass/mineral density
BMI *see* body mass index
BMP/*Bmp see* bone morphogenic proteins
BMPR1/*Bmpr1* and BMPR2/*Bmpr2* 64
BMS *see* bone mineral strength
BMSC *see* bone marrow stromal cells
BMSi *see* bone material strength index
BMT *see* bone marrow transplantation
BMU *see* basic multicellular units
body composition 256–257, **257**
body mass index (BMI)
   glucocorticoid-induced osteoporosis 468
   obesity and skeletal health 492–495
   sarcopenia and osteoporosis 500
   trabecular bone score 278, 282–283
   transplantation osteoporosis 424–425
bone cancer pain 781–787
   central sensitization 782–783
   epidemiology 781
   growth factors and cytokines 781–782, *783*, 784–785, *785*
   innervation and sensitization *782*
   ion channels 783, *785*
   mechanisms 781–784, *782*
   nociceptive pain and neuropathic pain 781
   ongoing pain and breakthrough pain 781
   osteoclasts 785
   radiotherapy of skeletal metastases 809–812, **810**
   reorganization of PNS and CNS in response to 783–784, *784*
   therapeutic strategies 784–785
bone histomorphometry 310–318
   basic histomorphometric variables 311–312
   biopsy procedure 315
   bone (re)modeling 310–313, *311*
   chronic kidney disease 506
   concepts and definitions 310

denosumab 555
findings in metabolic bone disease 313–315, **314**
fluorochrome labeling 315
indications for bone biopsy and histomorphometry 316, **316**
intermediary organization of the skeleton 310
interpretation of findings 312–313
kinetic features 312
obtaining the specimen 315
organization and function of bone cells 310–311
reference values 312
rickets and osteomalacia 684–685, 689–690
sclerosing bone disorders 827, 831–835
specimen processing and analysis 316
structural features from the transilial biopsy 311
bone lining cells (BLC) 20, *21*, 23–25
bone loss
   adipocytes 978, *979*
   bone turnover markers 296
   bone vasculature 989
   denosumab 555
   diabetes mellitus 487–491
   ethnicity and age-related loss of bone strength 133
   gonadal steroids 198, 200
   gonadal steroids and osteoporosis pathogenesis 412–416
   immobilization and burns 730–734
   immunobiology 993–994, *995*, 997–1001
   inflammation-induced bone loss 459–466
   juvenile osteoporosis 420–421
   menopause and age-related bone loss 155–161
   oral manifestations of metabolic bone diseases 943–944
   pituitary–bone axis 1042–1043
   radiotherapy-induced osteoporosis 788–790
   secondary causes of osteoporosis 512
   skeletal effects of medications for nonskeletal disorders 482–484
   transplantation osteoporosis **425**, 430–431
   see also osteoporosis
bone marrow (BM)
   adipocytes 974–982
   bone histomorphometry 312
   bone vasculature 983–985
   hematopoietic stem cell niche 966–971, *967*
   immunobiology 992–994
   integrative physiology of the skeleton 959–961
   peripheral neuronal control of bone remodeling 1028–1029, 1031
   radiotherapy-induced osteoporosis 790
bone marrow stromal cells (BMSC) 21–27, *21*, 370
bone marrow transplantation (BMT)
   fibrodysplasia ossificans progressiva 867–868
   sclerosing bone disorders 825, 828
   transplantation osteoporosis 425, 426, 431
bone mass/mineral density (BMD)
   animal models: allelic determinants for BMD 359–366
   animal models: genetic manipulation 357
   antisclerostin therapy 606–607
   bisphosphonates for postmenopausal osteoporosis 547–548
   bone mass, structure, and quality assessment 93–95
   bone turnover markers 296–297
   cathepsin K inhibitors 604–605
   central neuronal control of bone remodeling 1022–1024
   childhood cancer 793–795
   chronic kidney disease 508
   combination therapy 568–571
   denosumab 553–555
   diagnostic potential of DXA 247–248
   diet and nutrition 135–138
   dual-energy X-ray absorptiometry 243–249, *244*, *245*
   epidemiology of osteoporotic fractures 398, 402
   ethnicity and age-related loss of bone strength 132–133
   exercise for osteoporotic fracture prevention and management 518–523
   fetal and neonatal bone development 118, 119
   fibrillinopathies 880–881
   Fracture Liaison Service 407–408
   Fracture Risk Assessment Tool 331–337
   genome-wide association studies 378, 380–383
   glucocorticoid-induced osteoporosis 467, 470
   gonadal steroids and osteoporosis pathogenesis 199–200, 412–413, *413*
   human immunodeficiency virus 474–475, 477–478
   hypoparathyroidism 657–658
   immobilization and burns 730–734
   immunobiology 998
   inflammation-induced bone loss 461–463
   integrative physiology of the skeleton 959–960
   International Society for Clinical Densitometry 253–254, **254**, 256–257
   juvenile osteoporosis 420–422
   low bone mass in children 402, *403*
   magnetic resonance imaging 274–275
   measurement in adults 252–257, 260–268
   measurement in children with osteoporosis risk factors 243–249
   mechanical loading 141, 144
   neuropsychiatric disorders 1047–1051
   nutritional support for osteoporosis 534–536, 537
   obesity and skeletal health 493–495
   oral manifestations of metabolic bone diseases 942–943
   osteogenesis imperfecta 873, 875
   osteoporosis in men 445–446
   overview and historical development 395
   parathyroid disorders 614
   parathyroid hormone treatment for osteoporosis 559–563
   peripheral neuronal control of bone remodeling 1032–1033
   peripheral quantitative computed tomography 247, 248–249
   pituitary–bone axis 1039–1042
   pregnancy and lactation 148–149, 151, 152
   premenopausal osteoporosis 436–440, *437*
   primary hyperparathyroidism 622–623, *622*, 626
   radiotherapy-induced osteoporosis 788–789
   reference point indentation 289–290
   reporting aBMD by DXA 247
   rickets and osteomalacia 685
   sarcopenia and osteoporosis 499–501, 502
   sclerosing bone disorders 825
   secondary causes of osteoporosis 511–513
   skeletal complications of anticancer therapies 775–778
   skeletal effects of medications for nonskeletal disorders 482–484
   skeletal growth 123–128
   strontium ranelate 573–577
   trabecular bone score 277–284
   translational genetics of osteoporosis 385–390, **386**, **387**
   transplantation osteoporosis 424–426, 429–430
   vertebral fracture 248, *248*, 319, **327**
   World Health Organization classification 252, **253**
bone mass, structure, and quality assessment 93–100
   advanced techniques 260–271
   allometric approach 247
   animal models: allelic determinants for BMD 361
   areal bone mineral density 93–94

bone mass, structure, and quality assessment (cont'd)
  assessment of fracture risk 255
  bending tests of long bones 97–99, 97, 142–143
  body composition 256–257, **257**
  bone lesions and metastases 268
  bone mass measurement in adults 252–259, 260–271
  bone mass measurement in children with osteoporosis risk factors 243–251
  bone mineral apparent density 246, 247
  bone mineral content for height or height for age 246
  clinical applications 265–268
  compression tests of vertebral bodies 95–97
  concepts and definitions 93, 243, 252, 260
  developing techniques 249
  diagnosis of osteoporosis 255
  diagnostic potential of DXA 247–248
  dual-energy X-ray absorptiometry 94, 243–249, *244*, *245*, 252–257, 260–261, 265–268
  ethnicity and age-related loss of bone strength 131–134
  finite element analysis 264–265, *264*, *265*, 267–268
  fracture toughness testing 98–99, *98*
  high-resolution peripheral quantitative computed tomography 261–264, **262**
  hip geometry 256, *256*
  indications for DXA **246**
  magnetic resonance imaging 249, 274–275
  measurement of bone mineral density 253–255
  mechanical loading 141–144
  mechanostat or functional model 247
  medical prevention and treatment of metastatic bone disease 804
  micro-computed tomography 93, 94–95, *96*
  opportunistic screening 266
  peripheral quantitative computed tomography 247, 248–249
  precision errors for QCT, HRpQCT, and FEA **263**
  quality standards 253–254, **254**
  quantitative computed tomography of spine and hip 260–261, *261*, *262*
  quantitative ultrasound 257
  radiotherapy-induced osteoporosis 788–789
  reference point indentation 287–290
  reporting aBMD by DXA 247
  segmentation and analysis 95
  standard techniques 252–259
  statistical parameter mapping 268
  vertebral fracture assessment 248, *248*, 255–256, *255*, **256**
  World Health Organization classification 252, **253**
bone material strength index (BMSi) 288–289
bone matrix composition 84–92
  bone as a composite 84–90
  collagen-related genes and proteins **85**
  concepts and definitions 84
  extracellular matrix 84
  fetal and neonatal bone development 119–120
  gla-containing proteins 90, **90**
  glycosylated proteins 86–88, **88**
  mineral 84–85
  noncollagenous proteins 85–90
  other components 91
  Paget disease of bone 714
  proteoglycans 86, **87**
  serum-derived proteins 85–86, **86**
  SIBLING family, N-glycosylated proteins and other glycoproteins 88–90, **89**
  water and lipids 91

bone matrix volume 156
bone metastasis *see* metastatic bone disease
bone mineral apparent density (BMAD) 246, 247
bone mineral content (BMC)
  BMC for height or height for age 246
  bone mass measurement in children with osteoporosis risk factors 246
  diet and nutrition 135–138
  immobilization and burns 734
  mechanical loading 141–144, *142*
  radiotherapy-induced osteoporosis 788–789
  skeletal growth *124*
bone mineral strength (BMS) 487–488
bone morphogenic proteins (BMP)
  craniofacial morphogenesis 895
  endochondral ossification 14–17
  fibrodysplasia ossificans progressiva 867
  gonadal steroids 197
  inflammation-induced bone loss 462
  muscle–bone interactions 1056–1057
  odontogenesis 902–903
  osteoblasts 33, 54–56
  peripheral neuronal control of bone remodeling 1031
  skeletal healing/repair 103–105
  skeletal morphogenesis in embryonic development 5–8
  TGF-β superfamily 60, 62–65
  transcriptional profiling 371
bone morphology
  bone size, shape and microarchitecture 125–127, *126*
  ethnicity and age-related loss of bone strength 132–133
  maturational stage-specificity of illness effects 128–129
  puberty and sex differences 127–129, *127*, *128*
bone–muscle interactions *see* muscle–bone interactions
bone perfusion 984–985, 989
bone (re)modeling
  bisphosphonates for postmenopausal osteoporosis 548–549
  bone histomorphometry 310–313, *311*
  bone stress injuries 450–451, *451*
  bone turnover markers 293
  bone vasculature 985–987, *986*
  cellular bioenergetics of bone 1004
  central neuronal control of bone remodeling 1020–1027
  dental implants and osseous healing 951
  ethnicity and age-related loss of bone strength 133
  fetal and neonatal bone development 117
  gonadal steroids 198–199, *199*
  hematological malignancies 761, *761*
  hypoparathyroidism 658
  inflammation-induced bone loss 462
  integrative physiology of the skeleton 963
  menopause and age-related bone loss 155, 157–158
  metastatic bone disease 817–818
  osteocytes 40–41, *41*, 43
  osteolytic and osteoblastic skeletal lesions 740
  peripheral neuronal control of bone remodeling 1028–1036
  regulation of calcium homeostasis 169–170
  secondary causes of osteoporosis 512
  TGF-β superfamily *61*
  see also skeletal healing/repair
bone resorption
  central neuronal control of bone remodeling 1020–1024
  endocrine bioenergetics of bone 1016
  gonadal steroids 415–416, *416*

mechanism 47, *48*
osteoclasts 46–53, *47*
bone-seeking radiopharmaceuticals (BSR)
    medical prevention and treatment of metastatic bone disease 779, 800–805, **802–803**
    pharmacological treatment of bone metastasis 818
    radiotherapy of skeletal metastases 812–813
bone-specific alkaline phosphatase (BSAP) 506, 554
bone stress injuries (BSI) 450–458
    bone (re)modeling 450–451, *451*
    classification/grading 454–456, **455**
    concepts and definitions 450
    diagnosis 454
    epidemiology 451–452
    management 455–456
    mechanical loading 452–456, *452*
    pathophysiology 450–451, *451*
    risk factors 452–454, *452*
    site-specific susceptibilities 452–453, **453**
bone-targeted agents (BTA) 800–801, 804–806
bone turnover markers (BTM) 293–301
    analytical and preanalytical variability 293–294, **294**
    biochemical markers of bone formation and resorption 293, **294**
    bone lesions and metastases 295
    bone loss 296
    chronic kidney disease 506–507
    concepts and definitions 293
    conditions that impact BTM measurements 294–295
    denosumab 553–555
    diabetes mellitus 295
    fracture risk 296
    levels after discontinuation of antifracture treatments 297
    levels and therapeutic efficacy of anti-osteoporotic treatment 297
    medical prevention and treatment of metastatic bone disease 800–801, 806
    medications 295
    men 298
    metabolic effect 296–297
    monitoring 296–297
    multiple myeloma 295
    osteolytic and osteoblastic skeletal lesions 740
    osteoporosis 293–301
    Paget disease of bone 294–295, 715
    parathyroid disorders 615
    primary hyperparathyroidism 294, 622
    reference values 295–296
    renal osteodystrophy 295
    start of treatment for osteoporosis 297
    treatment monitoring at the individual level 297
bone vasculature 983–991
    anatomy of bone vessels 983, *984*
    assessment of bone vessels 985, *985*
    bone remodeling 985–987, *986*
    childhood cancer 793–796
    concepts and definitions 983
    cross-talk between bone and vascular cells *986*, **987**
    diseases associated with bone perfusion changes 989
    heterogeneous histology of bone microvessels 983, *985*
    human bone diseases 987–989
    ossification, modeling, and growth 985
    osteoporosis, vascular diseases, and aging 989
    oxygen tension in bone marrow and bone perfusion 984–985
    perivascular cells and pericytes 983–984
    radiotherapy-induced osteoporosis 790
    vascular obstruction/epiphyseal osteonecrosis 987–989, *988*
bortezomib 764
BOS *see* Buschke–Ollendorff syndrome
bowing deformities 321, *322*, 716
BP *see* bisphosphonates
brachydactyly (BD) 65
brachydactyly mental retardation syndrome (BDMR) 668
brain-derived growth factor (BDGF) 784
breast cancer
    bone cancer pain 783–785
    concepts and definitions 775
    endocrine therapy 776
    epidemiology 804
    medical prevention and treatment of metastatic bone disease 804–805
    metastatic breast cancer 778
    nonmetastatic breast cancer 775–776
    osteolytic and osteoblastic skeletal lesions 739
    parathyroid hormone-related protein 217
    screening for osteoporosis 776–777, *777*
    skeletal complications of anticancer therapies 775–780
breastfeeding 137–138
bridge plates 590
brittle bone 119
Bruck syndrome 873
BSAP *see* bone-specific alkaline phosphatase
BSI *see* bone stress injuries
BSR *see* bone-seeking radiopharmaceuticals
BTA *see* bone-targeted agents
BTM *see* bone turnover markers
bulimia nervosa 1049
burns *see* immobilization and burns
Buschke–Ollendorff syndrome (BOS) 832
buttress plates 589
BZA 543

CAL *see* clinical attachment level
calbindin 235–236
calcimimetics 625, 701–702
calcineurin inhibitors 425
calcitonin 574–577
    adverse effects 576
    benefits and limitations **576**
    calcium-sensing receptor 221
    clinical efficacy 575
    fetal and neonatal bone development 118
    fetal calcium metabolism 180
    formulations 575
    glucocorticoid-induced osteoporosis 471
    oral formulation 575–576
    Paget disease of bone 719
    physiology and pharmacology 574–575
    transplantation osteoporosis 429
calcitonin gene-related peptide (CGRP) 1023, 1028–1031, **1030**
calcitriol
    childhood disorders of mineral metabolism 707
    chronic kidney disease–mineral and bone disorder 701–702
    disorders of phosphate homeostasis 676, 678
    hypoparathyroidism 655–657
    sclerosing bone disorders 828
    transplantation osteoporosis 429, 432

calcium
  actions of key hormones in target tissues 168–169
  blood levels 165–166
  bone remodeling and mineralization 169–170
  bone stress injuries 454, 456
  bone turnover markers 294
  calcium balance 166, *166*
  cell levels 165
  childhood cancer 793–794
  childhood disorders of mineral metabolism 705–708, 710
  chronic kidney disease–mineral and bone disorder 698–702
  cost-effectiveness of osteoporosis treatment 600
  diet and nutrition 135–138
  familial states of primary hyperparathyroidism 629–631
  fetal calcium metabolism 179–186
  genetic craniofacial disorders affecting dentition 911
  glucocorticoid-induced osteoporosis 469–470
  hormonal regulation of calcium homeostasis 166–170, *167*
  human immunodeficiency virus 476–477
  hypoparathyroidism 655–657
  immobilization and burns 733–734
  intestinal calcium transport 168, **169**
  management of hypocalcemia 650–652
  medical prevention and treatment of metastatic bone disease 805
  nephrolithiasis 721–723, 726–727
  non-parathyroid hypercalcemia 642–644
  nutritional support for osteoporosis 535–538, *535*
  oral manifestations of metabolic bone diseases 943–944
  osteoporosis in men 446
  Paget disease of bone 718
  parathyroid disorders 614–616
  parathyroid hormone treatment for osteoporosis 560
  pregnancy and lactation 147–152
  premenopausal osteoporosis 438
  primary hyperparathyroidism 621–622, 625
  pseudohypoparathyroidism 668
  regulation of calcium homeostasis 165–172
  regulation of hormone production 167–168
  renal calcium handling 168–169, **169**
  rickets and osteomalacia 685–689, **686**, *687*, 691–692
  sclerosing bone disorders 829, 834
  temporal sequence of homeostasis regulation 170
  total body distribution 165
  transplantation osteoporosis 426, 429–430
  vitamin D 233–236
  see also hypercalcemia; hypocalcemia
calcium-based phosphorus binders 699–700
calcium gluconate 650–651
calcium-sensing receptor (CaSR) 167, 169–170, 221–229
  allosteric modulators 224–225
  binding of Ca^{2+} and activation of CaSR-mediated signaling 223–224
  bone and cartilage 226
  breast and placenta 226
  CaSR activation 222–223, *223*
  CaSR expression 223–224
  CASR gene 119, 221, *222*
  CaSR ligands 223
  CaSR protein 222
  C-cells 225
  chronic kidney disease–mineral and bone disorder 696
  concepts and definitions 221
  familial states of primary hyperparathyroidism 630–631, 635
  fetal and neonatal bone development 119
  fetal calcium metabolism 180
  hypocalcemia 648
  hypoparathyroidism 655
  immobilization and burns 734
  intestine 226
  intracellular signalling 223–225, *224*
  kidney 225–226
  magnesium homeostasis 174–175
  parathyroid 225
  parathyroid hormone 207–208
  parathyroid hormone-related protein 212
  primary hyperparathyroidism 621, 625
  regulation of calcium homeostasis 225–226
  structure and function 221–225
callus formation 108–111
calmodulin-dependent kinase (CaMK) 48
calvarial development 898
CaMK *see* calmodulin-dependent kinase
cAMP response element-binding protein (CREB) 1021–1022
Camurati–Engelmann disease 62, 829–830
canagliflozin 483, 489
cancellous bone volume 311, 312
cancer-associated cachexia 217
cancer in bone (CIB) 739
cancer stem cells (CSC) 755, 772
candidate gene studies 386, **386**
canine distemper virus (CDV) 714
cannabinoid receptors 1021, 1024, 1047
capsaicin 1022–1023
CAR *see* CXCL12-abundant reticular cells
carbohydrates, dietary 537
carbonic anhydrase II deficiency 828–829
cardiac transplantation (CT) 429
cardiovascular system 216, 582, *584*
carnitine palmitoyltransferase (CPT1/2) 1006
cART *see* combination antiretroviral therapy
cartilage 354
cartilage oligomeric matrix protein (COMP) 874
CART signaling 1024
CaSR/*CASR see* calcium-sensing receptor
catecholamines 969
cathepsin K
  bone turnover markers in osteoporosis 296
  osteoclasts 46–48, 50
  sclerosing bone disorders 829
cathepsin K inhibitors 603–605, 756
CBCT *see* cone beam computed tomography
Cbfa1 *see* core binding factor-1
CCA *see* congenital contractural arachnodactyly
CCD *see* cleidocranial dysplasia
CCSS *see* Childhood Cancer Survivor Study
CD4+ T cells 992–994
CD8+ T cells 992–994, 996
CD40/CD40L 992–994, 996
CD106 *see* vascular cell adhesion molecule-1
CD146 *see* melanoma-associated cell adhesion molecule
CDC73 620–621
CDKN1B 620–621
CDMP1/*Cdmp1* 63
CDV *see* canine distemper virus
CEJ *see* cementoenamel junction
celiac disease 438–439
cell adhesion molecules 12, 22

cell–cell communication 3–9, 42–43, 51–52
cellular bioenergetics of bone 1004–1011
  bioenergetics in osteoclasts 1009
  concepts and definitions 1004
  fatty acid metabolism in osteoblasts 1006–1007, *1008*
  glucose metabolism in osteoblasts 1004–1006, *1005*
  glutamine metabolism in osteoblasts 1006, *1007*
  integrative physiology of the skeleton 961–962
  regulation of osteoblast metabolism by bone anabolic signals 1007–1009
cementoenamel junction (CEJ) 941, *942*
cementum 907
central giant cell lesion (CGCL) 923–925
central nervous system (CNS)
  bone cancer pain 782–785
  cannabinoid receptors 1021, 1024
  central neuronal control of bone remodeling 1020–1027, *1021*
  concepts and definitions 1020
  hypothalamic NPY2 receptor effects on bone 1023–1024
  integrative physiology of the skeleton 962–963
  leptin and the sympathetic nervous system 1020–1022
  melanocortin system 1024
  neuromedin U 1022
  neuropeptide Y system 1023–1024
  neuropsychiatric disorders 1047, 1051
  osteoblastic NPY1 receptor effects on bone 1024
  schematic diagram *1021*
  sensory nerves and bone metabolism 1022–1023
  serotonin 1022
central precocious puberty (CPP) 795
CFU-F *see* colony-forming unit-fibroblasts
CGCL *see* central giant cell lesion
CGL *see* congenital generalized lipodystrophy
CGRP *see* calcitonin gene-related peptide
chemokine C-X-C motif ligand 12-abundant reticular cells *see* CXCL12
chemotherapy
  childhood cancer 793–796
  osteosarcoma 769–770
  skeletal complications of anticancer therapies 775–780
  transplantation osteoporosis 425, 430–431
CHF *see* congestive heart failure
childhood cancer 793–798
  altered growth 795
  avascular necrosis 795–796
  bone mass/mineral density 793–795
  concepts and definitions 796
  fracture risk 794–795
Childhood Cancer Survivor Study (CCSS) 794–795
childhood disorders of mineral metabolism 705–712
  calcium homeostasis 705–708
  concepts and definitions 705
  magnesium homeostasis 709–710
  phosphate homeostasis 708–709
  skeletal manifestations of calcium and phosphate disorders 710, *710*
childhood hypophosphatasia (HPP) 886–887
childhood obesity 402
children with human immunodeficiency virus 477, *477*
CHL *see* crown-heel length
chlorthalidone 642
chondrocytes
  animal models: genetic manipulation 351
  endochondral ossification 12–13, *13*

fetal calcium metabolism 181
  parathyroid hormone-related protein 212
  transcriptional profiling 368
chondrodysplasias 217
chondrogenesis 12–13, *13*, 61
Chrohn's disease 999
chromosomal abnormalities 346–348, **347**
chromosomal microarray (CMA) analysis 375
chronic kidney disease (CKD)
  anabolic agents 507–508
  antiresorptive patients 507
  bone biopsy for quantitative purposes 507
  bone turnover markers 506–507
  concepts and definitions 505
  diagnosis of osteoporosis in renal impairment patients 506
  FGF23 and the regulation of phosphorus metabolism 191–192
  fracture risk 505–506, 508
  glomerular filtration rate 505, 507–508
  hypocalcemia 649–650
  management of osteoporosis in CKD patients 505–509
  transplantation osteoporosis 424, 426
  treatment of osteoporosis in renal impairment patients 507–508
chronic kidney disease–mineral and bone disorder (CKD–MBD) 695–704
  active vitamin D sterols 701
  bone turnover markers in osteoporosis 295
  calcimimetics 701–702
  concepts and definitions 695–696
  FGF23 activity 697–698, *697*, 700–702
  hemodialysis 701–702
  hyperphosphatemia 699–701
  inhibition of enteral sodium/hydrogen exchanger isoform 3 (NHE3) 700
  inhibition of enteral type II sodium-phosphate cotransporter 700
  inhibition of Wnt signaling 695, 698
  management of osteoporosis in CKD patients 505–508
  parathyroidectomy 702
  pathogenesis 696–699
  pathophysiology 696, *696*
  renal osteodystrophy 695
  secondary hyperparathyroidism 701–702
  therapeutic approaches 699–702
  traditional paradigm 696–697
  transplantation osteoporosis 424, 426
  vascular calcification 698–699
chronic myeloid leukemia (CML) 970
chronic obstructive pulmonary disease (COPD) 468, 511–512
chronic renal failure (CRF) 237, 642–643
chronic respiratory failure 424–425
Chvostek's sign 649
CIB *see* cancer in bone
circulating osteogenic precursor (COP) cells 27
CKD *see* chronic kidney disease
CKD–MBD *see* chronic kidney disease–mineral and bone disorder
CLCN7 120, 827–828
cleft lip and/or palate (CL/P) 897, 913
cleft palate only (CPO) 897
cleidocranial dysplasia (CCD) 54, 56, 120
CLIA *see* Clinical Laboratory Improvement Amendments
clinical attachment level (CAL) 937
Clinical Laboratory Improvement Amendments (CLIA) 376
Clinical Practice Research Datalink (CPRD) 402
clinical risk factors (CRF) Tool 331
clodronate 804

CL/P *see* cleft lip and/or palate
cluster analysis 362, 400
CMA *see* chromosomal microarray
cMAT *see* constitutive marrow adipocytes
CML *see* chronic myeloid leukemia
CNC *see* cranial neural crest
CNCC *see* cranial neural crest cells
CNNM2/*CNNM2* 175
CNR *see* contrast/noise ratio
CNS *see* central nervous system
coenzyme A (acetyl CoA) 962
cognitive function 665
collagens
   animal models: genetic manipulation 351–352, 354
   bone matrix composition **85**
   dental implants and osseous healing 951–952
   diabetes and bone loss 487
   endochondral ossification 12–16
   fetal and neonatal bone development 119
   genetic testing 375–376
   inflammation-induced bone loss 462
   juvenile osteoporosis 421
   medical prevention and treatment of metastatic bone disease 799–800
   osteoblasts 31–32, 35
   osteochondrodysplasias 849, 851
   osteogenesis imperfecta 871, 874
   osteoprogenitor cells 20, 23
   pathology of the hard tissues of the jaws 919
   TGF-β superfamily 63–64
   transcriptional profiling 368–369
   translational genetics of osteoporosis 385
Colles fractures 590–591
colony-forming unit-fibroblasts (CFU-F) 22, 24, 25
combination antiretroviral therapy (cART) 474–475, 476
combination therapy 567–572
   adding anabolic therapy to ongoing antiresorptive therapy 561–562, 570
   adding antiresorptive therapy to ongoing anabolic therapy 571
   calcitonin 575
   concepts and definitions 567
   concurrent therapy 567–569
   parathyroid analog/estrogen or SERM combination 568
   parathyroid hormone analog/bisphosphonate combinations 561, 567–568
   parathyroid hormone for postmenopausal osteoporosis 561–562
   switching from anabolic to antiresorptive therapy 570–571
   switching from antiresorptive to anabolic therapy 562, 569–570, *570*
   teriparatide/denosumab combination 555, 561–562, 568–571, *569, 570*
COMP *see* cartilage oligomeric matrix protein
compression plates 589
compression tests of vertebral bodies 95–97
computed tomography-based structural rigidity analysis (CTRA) 268
computed tomography (CT)
   adipocytes 980
   bone cancer pain *782, 784, 785*
   bone stress injuries 454
   hematological malignancies 763
   ischemic and infiltrative disorders of bone 855
   metastatic bone disease 744, *745*, 747
   nephrolithiasis 723–724, *724*
   nuclear medicine 303–308
   oral manifestations of metabolic bone diseases 942–943
   osteonecrosis of the jaw 929
   Paget disease of bone 717
   parathyroid disorders 614
   primary hyperparathyroidism 623
   reference point indentation 287
   skeletal healing/repair 110, *110*
   tumoral calcinosis 861
   vertebral fracture 327–329, *328*
cone beam computed tomography (CBCT) 929, 942
confounding by indication 580
congenital contractural arachnodactyly (CCA) 881
congenital generalized lipodystrophy (CGL) 978
congestive heart failure (CHF) 424
conjugated estrogen 543
connexins (Cx) 43, 79–80
constitutive marrow adipocytes (cMAT) 976, *976*
continuous parathyroid hormone (cPTH) infusion 994–996, *996*
contrast/noise ratio (CNR) 94–95
COP *see* circulating osteogenic precursor
COPD *see* chronic obstructive pulmonary disease
copper transport disorders 888–889
copy number variants (CNV) 345–348, **347**, 374–375
core binding factor-1 (Cbfa1) 699, 960
core width 311
cortical bone
   animal models: allelic determinants for BMD 363
   bone histomorphometry 311, 312
   bone mass measurement in adults 263–264
   magnetic resonance imaging 273–274, 275
cortical porosity
   bone histomorphometry 311
   ethnicity and age-related loss of bone strength 132
   genome-wide association studies 381–382
   hypoparathyroidism 658
   menopause and age-related bone loss 156–158, *159*
   radiotherapy-induced osteoporosis 789
   skeletal growth 128
   strontium ranelate 574
cortical thick ascending limb of the loop of Henle (CTAL) 167, 169, 225
corticosteroids 511, 863
corticotropin-releasing factor (CRF) 1038
cost-effectiveness of osteoporosis treatment 597–602
   choice of comparator 597–598
   concepts and definitions 597
   cost-effectiveness analysis and clinical practice guidelines 599
   estimating the cost of osteoporosis treatment 598, **598**
   estimating the effectiveness of osteoporosis treatment 598–599
   findings 599–600, *599*
   incremental cost-effectiveness ratio 597–598
   methods for cost-effectiveness analysis 597–599
   model-based analyses 598
   number of fractures prevented 599
   quality-adjusted life years 598–600
covariation 362
COX2 *see* cyclooxygenase 2
CPO *see* cleft palate only
CPP *see* central precocious puberty

CPRD *see* Clinical Practice Research Datalink
CPT1/2 *see* carnitine palmitoyltransferase
cPTH *see* continuous parathyroid hormone
cranial base development 897
cranial neural crest cells (CNCC) 901–902
cranial neural crest (CNC) 894, *894*, 897–898
cranial suture closure 119
craniofacial morphogenesis 893–900
    anterior-posterior axis formation and head induction 893, *894*
    calvarial development 898
    concepts and definitions 893
    cranial base development 897
    facial fusion *895*
    formation of craniofacial ectomesenchyme (neural crest) 894, *894*, 897–898
    mandibular development 897–898
    midfacial development and palatogenesis 895–897, *896*
craniosynostosis 8, 898
CRD *see* cysteine-rich domain
CREB *see* cAMP response element-binding protein
Cre-loxP 50
Cre recombinase 353–356, **355–356**
CRF *see* chronic renal failure; clinical risk factors; corticotropin-releasing factor
Cripto signaling pathway 896–897
CRISPR/Cas9 genomic engineering 357
CRM1 214
Crohn's disease 511, 707
cross-talk
    bone vasculature *986*, **987**
    muscle–bone interactions 1057, 1059
crown-heel length (CHL) 123, *124*
cRPI *see* cyclic reference point indentation
cryoablation 819
cryptorchidism 663
CsA *see* cyclosporine
CSC *see* cancer stem cells
CSMI *143*, 144
c-Src 48
CT *see* cardiac transplantation; computed tomography
CTAL *see* cortical thick ascending limb of the loop of Henle
CTRA *see* computed tomography-based structural rigidity analysis
CTX
    bone turnover markers 293, 295–298
    calcitonin 575–576
    combination therapy 568–570
    denosumab 553–554
    medical prevention and treatment of metastatic bone disease 800
    Paget disease of bone 715
Cupid's bow deformity 321, *322*
Cushing syndrome 513
cutaneous skeletal hypophosphatemia syndrome 680
Cx *see* connexins
CXCL12 25, 27, 983–984, 987
CXCL12-abundant reticular cells (CAR) 968–969, 983–984
cyberknife treatment 819
cyclic adenosine monophosphate (cAMP)
    fibrous dysplasia of bone 839
    peripheral neuronal control of bone remodeling 1031–1032
    pseudohypoparathyroidism 661–664, 668–669
cyclic reference point indentation (cRPI) 287–290
CyclinD1/2 620, 1013–1014

cyclooxygenase 2 (COX2) 81, 104
cyclosporine (CsA) 425, 426, 431
CYP2R1 230–231
CYP24A1 231–232
CYP27B1 231–232, 234–236
cysteine-rich domain (CRD) 70–71

DA *see* dopamine
DAA *see* direct-acting antiviral agents
DATA study 555, 569
DBP *see* serum vitamin D binding protein
DCA *see* dichloroaceteate
DCT *see* distal convoluted tubule
DD *see* dentin dysplasia
DECIDE trial 555
degarelix 778
degenerative-type scoliosis 321
denosumab 553–558
    adherence to osteoporosis therapies 594
    adverse effects 556, 582, 583
    bone cancer pain 785
    bone histomorphometry 555
    bone turnover markers in osteoporosis 296–298
    chronic kidney disease 506, 507
    clinical efficacy 554–555
    combination therapy 555, 561, 568–571, *569*, *570*
    concepts and definitions 553
    duration of treatment 556
    glucocorticoid-induced osteoporosis 470–471
    hematological malignancies 764
    immobilization and burns 731–732, *733*
    mechanism of action 553, *554*
    medical prevention and treatment of metastatic bone disease 800–801, 805–806
    metastatic bone disease 556
    non-parathyroid hypercalcemia 643–644
    osteogenesis imperfecta 875
    osteonecrosis of the jaw 927
    osteoporosis in men 447
    osteosarcoma 769
    overview and historical development 395, *396*
    pharmacodynamics, pharmacokinetics, and metabolism 553–554
    pharmacological treatment of bone metastasis 818
    premenopausal osteoporosis 440
    skeletal complications of anticancer therapies 777
    transplantation osteoporosis 432
dental caries 918–920, *919*
dental implants and osseous healing 949–956
    concepts and definitions 949, *950*
    failure of osseointegration 953–954, *954*
    hemostatic phase 949–950
    inflammatory phase 950
    mechanobiology and implant stability 952–953, *953*
    oral mucosal healing 952
    proliferative phase 950–951
    remodeling phase 951
    rough implant surface and osseointegration 952
    ultrastructure at bone–implant interface 951–952
    wound healing around implants 949–951
dentigerous cyst 920
dentin 903, *904*, 912–913, *912*, 919
dentin dysplasia (DD) 912–913
dentin matrix protein 1 (DMP1) 38, 43, 189–190, 352, 679

dentinogenesis imperfecta (DGI) 912–913, *912*
dentures 928
dermatomyositis in children *see* juvenile dermatomyositis
desert hedgehog (DHH) 56–57
dexamethasone 811
DGI *see* dentinogenesis imperfecta
diabetes mellitus (DM) 487–491
   bone turnover markers 295
   concepts and definitions 487
   efficacy and safety of osteoporosis therapies in diabetic patients 489
   endocrine bioenergetics of bone 1012–1013
   epidemiology of osteoporotic fractures 488
   evaluation of fracture risk 488
   pathophysiology of bone fragility 487–488
   peripheral neuronal control of bone remodeling 1029
   reference point indentation 289–290
   skeletal effects of glucose-lowering medications 488–489
   skeletal healing/repair 104–105
   trabecular bone score 281, 283
diabetic ketoacidosis 681
diaphysometaphyseal ischemia 855–856
Dicer1 225
dichloroaceteate (DCA) 1009
Dickkopfs (Dkk) family
   chronic kidney disease–mineral and bone disorder 695, 698
   hematological malignancies 762–764
   inflammation-induced bone loss 461–463
   osteosarcoma 771
   skeletal healing/repair 105
   Wnt/β-catenin signaling pathway 71, 72
diet and nutrition
   acid–base balance 536
   bioactive dietary constituents 537
   bone stress injuries 454, 456
   calcium 135–138
   chronic kidney disease–mineral and bone disorder 699
   concepts and definitions 534
   dietary patterns 536–537
   epidemiology of osteoporotic fractures 402–403
   fetal and neonatal bone development 118
   food 536
   Fracture Liaison Service 408
   fracture risk 537–538
   fruit and vegetables 136
   lactation/breastfeeding 137–138, 150–152, *151*
   milk supplementation 535, *535*, 536
   muscle strength, balance, and falling 537
   nephrolithiasis 722–723
   nutritional support for osteoporosis 534–540
   obesity and skeletal health 494–495
   pregnancy 136–137
   pregnancy and lactation 147–150
   recommendations for supplementation 538
   rickets and osteomalacia 684, 688, 691
   role of diet in building peak bone mass 534–535, **535**
   role of nutrition in maintaining bone mass 535–536
   role of supplementation in pharmacotherapy 538
   safety of supplementation 538
   salt 138, 536
   sarcopenia and osteoporosis 502
   sclerosing bone disorders 828
   skeletal effects of calcium and vitamin D supplementation 537–538
   skeletal growth 135–140
   soft drinks and milk avoidance 138
   vitamin D 136–137, 236–237
differential ascertainment 580
DiGeorge syndrome 648, 654, 903
1,25-dihydroxyvitamin D *see* calcitriol; vitamin D
dipeptidyl peptidase 4 (DPP-4) inhibitors 489
direct-acting antiviral agents (DAA) 478
disease-targeted gene panels 348
disseminated tumor cells (DTC) 755, 799–800, 805
distal convoluted tubule (DCT) 174–175
distal forearm fracture *399*, *400*, 413
distraction osteogenesis 111
Dkk *see* Dickkopfs
DM *see* diabetes mellitus
DMP1/*Dmp1* *see* dentin matrix protein 1
dopamine (DA) **1030**, 1031
dorsal root ganglia (DRG) 1028
Down syndrome 707
doxorubicin 769
doxycycline hyclate 356, 937–938
DPP-4 *see* dipeptidyl peptidase 4
DRG *see* dorsal root ganglia
DTC *see* disseminated tumor cells
dual-energy X-ray absorptiometry (DXA)
   best practices **254**
   body composition 256–257, **257**
   bone mass measurement in adults 252–257, 260–261, 265–268
   bone mass measurement in children with osteoporosis risk factors 243–249, *244*, *245*
   bone mass, structure, and quality assessment 94, 143
   chronic kidney disease 506
   diagnostic potential 247–248
   fibrillinopathies 880–881
   Fracture Liaison Service 407–408
   Fracture Risk Assessment Tool 332, 334–335
   genome-wide association studies 380
   glucocorticoid-induced osteoporosis 469
   gonadal steroids 199
   hematological malignancies 763
   indications **246**
   magnetic resonance imaging 274–275
   metastatic bone disease 743–744
   oral manifestations of metabolic bone diseases 942–944
   osteogenesis imperfecta 873, 875
   overview and historical development 395
   parathyroid disorders 614
   parathyroid hormone treatment for osteoporosis 560
   pregnancy and lactation 148–149, 151
   premenopausal osteoporosis 436–437, *437*
   quality standards 253–254, **254**
   reference point indentation 287, 290
   reporting aBMD by DXA 247
   sarcopenia and osteoporosis 500–501
   trabecular bone score 277, 282–283
   T-scores 252–254, **253**
   vertebral fracture 323, 325, 326–327, *326*
   vertebral fracture assessment 248, *248*, 255–256, *255*, **256**
   Z-scores *244*, 245–249, *245*, 253–254
dual task paradigm 518, **520**
Duchenne muscular dystrophy *244*, *245*

dwarfism 8
DXA *see* dual-energy X-ray absorptiometry
dysmobility syndrome 500–502, *501*

E11/gp38 38
EBRT *see* external beam radiotherapy
ECF *see* extracellular fluid
ECG *see* electrocardiography
ECM *see* extracellular matrix
ectodysplasin A (EDA) 902, 911–912
ectonucleotide pyrophosphatase/phosphodiesterase-1 (ENPP1) 190
ectopic hyperparathyroidism 640, 641, 643
ectopic ossifications 664, 666, 668
EDA *see* ectodysplasin A
EDA/*EDA see* ectodysplasin A
EDS *see* Ehlers–Danlos syndrome
eGFR *see* estimated glomerular filtration rate
Ehlers–Danlos syndrome (EDS) 873, 880
electrical stimulation (ES) 732
electrocardiography (ECG) 650
electron transport chain (ETC) 961
ELISA *see* enzyme-linked immunosorbent assay
enamel 903–907, *905*, 913, 919
Encyclopedia of DNA Elements (ENCODE) Project 383
endochondral ossification 12–19
    cell differentiation and signaling regulators 14–15, *14*
    chondrogenesis and chondrocyte hypertrophy 12–13, *13*
    concepts and definitions 12
    inflammation-induced bone loss 462
    molecular modulators of cartilage development 13–16
    orthopedic principles of fracture management 588–589
    osteoblast differentiation and bone formation 16–17
    osteochondrodysplasias 848
    primary and secondary ossification centers 13
    skeletal healing/repair 112
    skeletal morphogenesis in embryonic development 3–9
    transcriptional profiling 368–369
    translational genetics of osteoporosis 388
endocrine bioenergetics of bone 1012–1019
    concepts and definitions 1012
    muscle–bone interactions 1055–1059
    osteocalcin and energy homeostasis 1012–1015
    regulation of muscle bioenergetics by osteocalcin 1014–1015, *1014*
    regulation of osteocalcin activation by insulin and bone resorption 1016
    regulation of osteocalcin decarboxylation by adipokines and the SNS 1016–1017
    regulation of osteocalcin production by nutrients 1017
    regulation of osteocalcin secretion and bioactivity 1015–1017, *1015*
    regulation of pancreas development and physiology by osteocalcin 1013–1014, *1013*
endocrine disorders
    disorders of phosphate homeostasis 679–680
    growth hormone secretion disturbances 512–513
    hypercorticolism/Cushing syndrome 513
    hyperthyroidism 512
    non-parathyroid hypercalcemia 641
    pseudohypoparathyroidism 668–669
    secondary causes of osteoporosis 512–513
    see also individual disorders
endocrine therapy 776, 777
endosteal hyperostosis 830–832, *831*

endothelial cells 968, 985
endothelins 781–782, 784–785, 818
end-stage liver disease 424
end-stage renal disease (ESRD) 314
ENPP1/*ENPP1 see* ectonucleotide pyrophosphatase/phosphodiesterase-1
environmental toxins 118–119
enzalutamide 778–779
enzyme deficiencies 886–890
    alkaptonuria 888
    concepts and definitions 886
    disorders of copper transport 888–889
    homocystinuria 888
    hypophosphatasia 375, 454, 886–888
    mucopolysaccharidoses 888
enzyme-linked immunosorbent assay (ELISA) 189–191, 1016
eosinophilic granuloma 857–858
ephrins (EPH) 32, 714
epidemiology of osteoporotic fractures 398–404
    age, ethnicity and gender 399, 402
    bone stress injuries 451–452
    clustering of fractures in individuals 400
    concepts and definitions 398
    diabetes and bone loss 488
    diet and nutrition 402–403
    distal forearm fracture 399, *400*
    early life influences on adult fragility fracture 402–403
    Fracture Liaison Service 405–406
    geographical variation 401–402, *402*
    glucocorticoid-induced osteoporosis 467–468
    hip fracture 398–400, *399*
    human immunodeficiency virus 474
    low bone mass in children 402, *403*
    obesity and skeletal health 492–493
    osteoporosis in men 444, *444*
    prevention of falls 526
    socioeconomic status 399, 402
    time trends and future projections 401, *401*
    vertebral fracture 399, *400*
epidermis 236
epigenetic variants 374
epiphyseal osteonecrosis (ON) 987–989, *988*
epiphyseal reossification 854
epiphysometaphyseal ischemia 855–856
ERE *see* hormone response elements
ES *see* electrical stimulation
ESRD *see* end-stage renal disease
estimated glomerular filtration rate (eGFR) 429
estrogens 541–544
    animal models: genetic manipulation 354
    combination therapy 568
    concepts and definitions 541–542
    Global Consensus Statement 541–542
    gonadal steroids and osteoporosis pathogenesis 414–416
    menopausal hormone therapy 541–543
    molecular mechanism of action 194–195, *196*
    selective estrogen receptor modulators 201, 389, 542–543, 568
    tissue-selective estrogen complex 543
    see also gonadal steroids
estrogen therapy 625, 778
    see also hormone replacement therapy
ESWL *see* extracorporeal shock wave lithotripsy
ETC *see* electron transport chain
etelcalcetide 702

ethics 579–580
ethnicity
  acquisition and age-related loss of bone strength 131–134
  bone loss and remodeling imbalance 133
  bone mass and geometry in adulthood 132–133
  bone mass and geometry in childhood 132
  concepts and definitions 131
  epidemiology of osteoporotic fractures 402
  fracture rates in adulthood 131
  fracture rates in childhood and adolescence 131
  genome-wide association studies 380
  skeletal growth 127–128, *128*
etidronate 395, 547–548
EUROFORS study 570–571
exercise 517–525
  bone stress injuries 456
  concepts and definitions 517
  dual task paradigm 518, **520**
  endocrine bioenergetics of bone 1014–1015, *1014*
  fibrillinopathies 883–884
  hyperkyphosis 521
  imaging technologies 518, *519*
  neuropsychiatric disorders 1050
  obesity and skeletal health 494–495
  osteoporosis and fracture management 519–523
  osteoporosis, falls, and fracture prevention 518–519, 521
  prevention of falls 528
  progressive resistance training 518–521, **520**, 523
  sarcopenia and osteoporosis 502
  spine sparing strategies 521, *522*
  theoretical basis and principles of loading important to bone 517–518
  transplantation osteoporosis 430
  walking 518
  weight-bearing impact exercise 518, **520**
  whole body vibration 521–523
external beam radiotherapy (EBRT) 809–812, 819–820
external fixation 590
external validity 579
extracellular domain (ECD) 214, *214*, 222–223
extracellular fluid (ECF) 165–170, *167*
extracellular matrix (ECM) 31–32, 84–85
extracorporeal shock wave lithotripsy (ESWL) 728

facial fusion *895*
falls
  concepts and definitions 526
  epidemiology and cost of falls 526
  evidence for fracture reduction 529–530, **529**
  exercise for osteoporotic fracture prevention and management 518–519, 521
  fall mechanics and risk of fracture 527, *527*
  fracture risk 526–527
  nutritional support for osteoporosis 537
  obesity and skeletal health 493
  prevention of falls 526–533
  risk factors for falls 528
  sarcopenia and osteoporosis 501–502, 527–528
  strategies for fall prevention 528–530
  vitamin D 529–530, **529**
FAM20C/*Fam20c* 190
FAM210A 381
familial adenomatous polyposis 912
familial cancer predisposition syndromes 770

familial hypocalciuric hypercalcemia (FHH) 629–631
  calcium-sensing receptor 225
  childhood disorders of mineral metabolism 707
  clinical expressions 629–630
  diagnosis of the family and the carrier 631
  fetal and neonatal bone development 119
  fetal calcium metabolism 181
  magnesium homeostasis 175
  management 631
  non-parathyroid hypercalcemia 643
  parathyroid hormone 207–208
  pathogenesis/genetics 630–631, **630**
  pregnancy and lactation 149
  primary hyperparathyroidism 619
  sporadic hypocalciuric hypercalcemia 631
familial hypomagnesemia with hypercalciuria and nephrocalcinosis (FHHNC) 174
familial hypomagnesemia with secondary hypocalcemia (fHSH) 173
familial hypophosphatemia 914
familial hypophosphatemic rickets (FHR) 375
familial isolated hyperparathyroidism (FIHPT) 619, **630**, 635
familial states of primary hyperparathyroidism 629–638
  detection of asymptomatic cancers 635, *636*
  familial hypocalciuric hypercalcemia 119, 175, 629–631
  familial isolated hyperparathyroidism 619, 635
  hyperparathyroidism–jaw tumor syndrome 619, 620, 634
  monitoring of tumors 635
  multifocal parathyroid gland hyperfunction 635
  multiple endocrine neoplasia type 1 619–621, 632–633
  multiple endocrine neoplasia type 2A 619–621, 633–634
  neonatal severe primary hyperparathyroidism 119, 175, 619, 631–632
  outlines of syndromes 629, **630**
  overlapping considerations among all forms of familial PHPT 635–637
  surgical procedures 635–637
familial tooth agenesis 911–912
familial tumoral calcinosis (TC) 190–191, 681
family medical history
  genetics 344–345
  osteoporosis in men 446
  premenopausal osteoporosis 438–439
farnesyl pyrophosphate synthase (FPPS) 546
fatty acid metabolism 1006–1007, *1008*
FBN1/*FBN1* and FBN2/*FBN2* 878, 881–883
FD *see* fibrous dysplasia of bone
FDT *see* fuzzy distance transform
FEA *see* finite element analysis
FEMg *see* fractional excretion of magnesium
femoral neck (FN)
  Fracture Risk Assessment Tool 331–333
  mechanical loading 142–143
  skeletal growth 125–126, *126*
  transplantation osteoporosis 426, 430–431
ferric citrate 700
fetal and neonatal bone development 117–122
  concepts and definitions 117
  defects in bone matrix production 119
  defects in cranial suture closure and osteogenesis 119
  defects in mineral deposition 119–120
  defects in mineral homeostasis 119
  defects in osteoclastic function 120
  environmental influences 118–119

epigenetic contributions 119
extrinsic factors 118–119
inherited disorders 119–120
mechanical influences 118
nutritional influences 118
physiology 117–118
skeletal growth 123–125, *124*
fetal calcium metabolism 179–186
blood calcium regulation 183
calcium sensing receptor 180
calcium sources 183, *183*
concepts and definitions 179
fetal kidneys and amniotic fluid 180
fetal parathyroid glands 180
fetal skeleton 181
integrated fetal calcium homeostasis 183, *183*, *184*
maternal hyperparathyroidism 181
maternal hypoparathyroidism 181–182
maternal vitamin D deficiency 182–183
mineral ions and calciotropic hormones 179–183
placental mineral ion transport 180–181, 183
skeletal mineralization 183, *184*
FEZ *see* frontonasal ectodermal zone
FGF23/*FGF23* *see* fibroblast growth factor 23
FGF *see* fibroblast growth factors
FGFR *see* fibroblast growth factor receptors
FHH *see* familial hypocalciuric hypercalcemia
FHHNC *see* familial hypomagnesemia with hypercalciuria and nephrocalcinosis
FHR *see* familial hypophosphatemic rickets
fHSH *see* familial hypomagnesemia with secondary hypocalcemia
FI *see* fluorescence imaging
fiber, dietary 537
fibrillinopathies 878–885
congenital contractural arachnodactyly 881
Marfan syndrome 878–884, **879–881**
Marfan syndrome-related disorders 882–883
medical management and treatment 883–884
pathogenesis 883
fibroblast growth factor 23 (FGF23)
childhood disorders of mineral metabolism 708–709
chronic kidney disease–mineral and bone disorder 697–698, *697*, 700–702
concepts and definitions 187
disorders associated with decreased FGF23 activity 190–192
disorders associated with increased FGF23 activity 189–190
disorders of phosphate homeostasis 674–681
FGF23 activity 188–192
FGF23-associated syndromes 189–192, **191**
FGF23 gene and protein 188
FGF23 receptors 188–189
fibrous dysplasia of bone 842
genetic craniofacial disorders affecting dentition 913–914
muscle–bone interactions 1057–1058
osteocytes 38, 43
parathyroid hormone 205
phosphate metabolism 187–188
regulation of calcium homeostasis 166–169
regulation of FGF23 *in vivo* 188
regulation of phosphorus metabolism 187–193
serum assays 189
therapy for FGF23-mediated hypophosphatemic disorders 190
tumoral calcinosis 862
vitamin D 235

fibroblast growth factor receptors (FGFR)
fetal and neonatal bone development 120
FGF23 and the regulation of phosphorus metabolism 188–189
genetic testing 376
osteoblasts 33
fibroblast growth factors (FGF)
craniofacial morphogenesis 897
genetic craniofacial disorders affecting dentition 912
odontogenesis 903
skeletal healing/repair 103
skeletal morphogenesis in embryonic development 4–6, 8
fibrodysplasia ossificans progressiva (FOP) 664, 865–868
clinical presentation 865–867, *866*
concepts and definitions 865
etiology and pathogenesis 867
histopathology 867
laboratory findings 867
prognosis 868
radiographic features 867
TGF-β superfamily 64
treatment 867–868
fibrogenesis imperfecta ossium (FIO) 834–835
fibrosarcoma 717
fibrous dysplasia of bone (FD) 839–847
clinical presentation 839, *840*, 842–843
concepts and definitions 839
etiology and pathogenesis 839–842
future treatment 844
histopathological features 840, *841*
management and treatment 843–844
phosphate homeostasis 679–680
radiographic features 842, *843*
scintigraphy and PET 304–305
FIHPT *see* familial isolated hyperparathyroidism
finite element analysis (FEA)
bone mass measurement in adults 264–265, *264*, *265*, 267–268
overview and historical development 396
precision errors **263**
radiotherapy-induced osteoporosis 789
vertebral fracture 329
FIO *see* fibrogenesis imperfecta ossium
FISH *see* fluorescence in situ hybridization
fissure fractures 716–717
FIT *see* Fracture Intervention Trial
FKBP10/*FKBP10* 873
flavonoids 537
FLEX *see* Fracture Intervention Trial
FLS *see* Fracture Liaison Service
fluid intake 722
fluorescence imaging (FI) 748–749
fluorescence in situ hybridization (FISH) 346
fluoride 918–919
fluorochrome labeling 315–316
fluoxetine 1022
FN *see* femoral neck
focal adhesions 78, *79*
folic acid/folate 137
follicle-stimulating hormone (FSH) 415, 1037–1040
FOP *see* fibrodysplasia ossificans progressiva
formation period 312
FOXO-mediated transcription 23
FPPS *see* farnesyl pyrophosphate synthase
fractional excretion of magnesium (FEMg) 173–174
fracture healing *see* bone (re)modeling; skeletal healing/repair

Fracture Intervention Trial (FIT/FLEX) 548–549
Fracture Liaison Service (FLS) 405–411
   administrative and clinical organization 405, *406*
   case finding 407
   concepts and definitions 405
   differential diagnosis 408
   epidemiology of osteoporotic fractures 405–406
   five-step decision plan 405, *406*
   implementation 408–410, **409**
   phenotype of patients with recent fracture 407
   randomized controlled trials 405–406
   risk evaluation 407–408
   therapy and follow-up 408
fracture risk
   antisclerostin therapy 607
   bisphosphonates for postmenopausal osteoporosis 546, *547*
   bone mass, structure, and quality assessment 255
   bone stress injuries 452–456, *452*, **455**
   bone turnover markers 296
   childhood cancer 794–795
   chronic kidney disease 505–506, 508
   cost-effectiveness of osteoporosis treatment 599–600
   denosumab 554–555
   diabetes and bone loss 488
   exercise for osteoporotic fracture prevention and management 521
   Fracture Liaison Service 407–408
   glucocorticoid-induced osteoporosis 468
   human immunodeficiency virus 474, 476
   immobilization and burns 731
   nutritional support for osteoporosis 537–538
   obesity and skeletal health 493–495
   oral manifestations of metabolic bone diseases 944
   osteoporosis in men 445–447
   overview and historical development 395–396
   Paget disease of bone 716–717
   parathyroid hormone treatment for osteoporosis 559
   pituitary–bone axis 1040–1041
   prevention of falls 526–527
   radiotherapy-induced osteoporosis 788
   radiotherapy of skeletal metastases 810–811
   sarcopenia and osteoporosis 499–501
   secondary causes of osteoporosis 511–512
   skeletal effects of medications for nonskeletal disorders 482–484
   trabecular bone score 277–282
   translational genetics of osteoporosis 388–389
   transplantation osteoporosis **425**
   vertebral fracture 319
Fracture Risk Assessment Tool (FRAX) 331–339
   bone mass measurement in adults 255
   chronic kidney disease 505
   concepts and definitions 331
   cost-effectiveness of osteoporosis treatment 599
   Fracture Liaison Service 408
   glucocorticoid-induced osteoporosis 469
   gradient of risk 332–333, **334**
   guidelines in the United Kingdom 335–336, *336*
   guidelines in the United States 336
   guidelines without BMD testing 336
   human immunodeficiency virus 476
   input and output 331–332, *332*
   intervention and assessment thresholds 335, *335*
   life expectancy 331–332, *332*
   limitations 333–335

   obesity and skeletal health 494
   oral manifestations of metabolic bone diseases 944
   osteoporosis in men 445–446
   other applications 336–337
   performance characteristics 332–333, **334**
   prevention of falls 527
   QFracture 337, **337**
   sarcopenia and osteoporosis 500–501, *500*
   secondary causes of osteoporosis 331–332, **333**, 511
   skeletal complications of anticancer therapies 777
   trabecular bone score 278–281, *283*
   translational genetics of osteoporosis 389
fracture toughness testing 98–99, *98*
FRAME trial 606–607
FRAX *see* Fracture Risk Assessment Tool
FREEDOM trial 554–556, 581–582
Frizzled (Fzd) family 70, 72
frontonasal ectodermal zone (FEZ) 895
FRTS2 *see* autosomal recessive renal Fanconi syndrome and hypophosphatemic rickets
fruit 136, 536
FSH *see* follicle-stimulating hormone
functional model *see* mechanostat model
fuzzy distance transform (FDT) technique 273
Fzd *see* Frizzled

G6P *see* glucose-6-phosphate
GAG *see* glycosaminoglycan
GALNT3/*GALNT3* 190–191, 862
Gα
   disorders of phosphate homeostasis 679
   fibrous dysplasia of bone 839–842
   hematopoietic stem cell niche 969–970
   hypoparathyroidism 655
   immunobiology 995
   parathyroid hormone 208–209
   progressive osseous heteroplasia 868
   pseudohypoparathyroidism 661–667
γ-carboxyglutamic acid (GLA) 1016
gap junctions 42–43, 79–80, *81*
Garvan nomogram 527
Gas6 *see* growth arrest-specific protein 6
gastrointestinal bone disease 314
gastrointestinal side effects 582
Gaucher's disease 855–856
GC *see* glucocorticoids
GCMB *see* glial cell missing B
GCS *see* Global Consensus Statement
G-CSF *see* granulocyte colony-stimulating factor
GD2 *see* geleophysic dyplasia 2
GDF/*Gdf* 63
GDGF *see* glial-derived growth factor
GEF *see* guanine nucleotide exchange factors
GEFOS *see* Genetic Factors of Osteoporosis
geleophysic dyplasia 2 (GD2) 882
gender *see* sex differences
gene expression profiling *see* transcriptional profiling
gene networks 363, *364*
gene ontology (GO) analysis 369
generalizability 579
gene targeting 352–356
genetic craniofacial disorders affecting dentition 911–917
   amelogenesis imperfecta 913
   cleft lip and/or palate 913

concepts and definitions 911
familial tooth agenesis and supernumerary teeth 911–912
hyperphosphatemia 914, *914*
hypophosphatasia 914, *915*
hypophosphatemia 913–914
oral manifestations of metabolic bone diseases of genetic origin 913–914
osteogenesis imperfecta, dentinogenesis imperfecta, and dentin dysplasia 912–913, *912*
genetic disorders of heterotopic ossification *see* fibrodysplasia ossificans progressiva; progressive osseous heteroplasia
Genetic Factors of Osteoporosis (GEFOS) Consortium 380–381
Genetic Information Nondiscrimination Act (GINA) 376
Genetic Markers for Osteoporosis (GENOMOS) Consortium 380
genetics 343–350
    animal models: allelic determinants for BMD 359–366
    animal models: genetic manipulation 351–358
    approaches to genetic testing 373–377
    chromosomal abnormalities, CNVs, and mutations 345–348, **347**
    clinical approach 343–345
    concepts and definitions 343
    data interpretation and incidental findings 348–349
    disorders of phosphate homeostasis 678, 679, 681
    familial states of primary hyperparathyroidism 629–635, **630**
    family medical history and mode of disease inheritance 344–345
    genetic tests, their clinical utility and interpretation 345–349
    hypoparathyroidism 654–655
    inheritance 343
    medical history and physical examination 344
    mosaicism 348
    nephrolithiasis 723
    osteochondrodysplasias 848–849
    osteogenesis imperfecta 874–875
    osteosarcoma 770–771
    prenatal diagnosis 348
    pretest considerations and test selection 345–346
    primary hyperparathyroidism 620–621, **620**
    transcriptional profiling 367–372
    translational genetics of osteoporosis 385–392
    value of genetic testing 345, **345**
genetic testing 373–377
    base-pair substitutions 373–374
    clinical utility and interpretation 345–349
    concepts and definitions 373
    diagnosis and when to order 376–377
    evolving approaches of DNA-based testing 374–375
    large-scale variants 374
    metabolic bone disease 375
    methods to detect small scale changes in the DNA 374
    overview of available tests 373–375
    reproductive counseling 376–377
    skeletal disorders and available tests 375–376
    skeletal dysplasias 375–376
    small-scale variants 373–374
genome-wide association studies (GWAS) 378–384
    combining data from different platforms 380
    concepts and definitions 378
    ethnicity 380
    heritability of bone mineral density 378
    imputation 379
    linkage disequilibrium 383
    Manhattan plots 380, *382*
    meta-analysis 379–383, *382*

    musculoskeletal field successes 380–383
    novel applications of GWAS meta-analyses summary data 383
    phenotyping harmonization 379
    power and sample size considerations 379–381, *381*
    principles 378–379
    skeletal phenotypes 380
    translational genetics of osteoporosis 386–387, **388**
GENOMOS *see* Genetic Markers for Osteoporosis
geographical variation 401–402, *402*
GFR *see* glomerular filtration rate
GH *see* growth hormone
GHD *see* growth hormone deficiency
GINA *see* Genetic Information Nondiscrimination Act
gingivitis 937
GIOP *see* glucocorticoid-induced osteoporosis
GIP *see* glucose-dependent insulinotropic polypeptide
Gitelman syndrome 709
GLA *see* γ-carboxyglutamic acid
gla-containing proteins 90, **90**
glandular odontogenic cyst (GOC) 923
glial cell missing B (GCMB) 648
glial-derived growth factor (GDGF) 784
Gli proteins 6–7, 56–57, 753–754
Global Consensus Statement (GCS) 541–542
glomerular filtration rate (GFR) 505, 507–508, 696, 699
GLOW study 492–495
glucagon-like peptide 1 (GLP-1) 488, 489
glucocorticoid-induced osteoporosis (GIOP) 467–473
    changes in bone density and fracture rates 467
    dose effects and routes of glucocorticoid administration 467
    epidemiology 467–468
    Fracture Risk Assessment Tool 331–332, 334, **334**
    guidelines for treatment 471
    history and physical examination 469
    integrative physiology of the skeleton 961
    osteoporosis in men 445, 447
    parathyroid hormone treatment for osteoporosis 563
    pathogenesis 468–469
    populations at high risk 468
    premenopausal osteoporosis 440
    treatment and prevention 469–471
glucocorticoids (GC)
    adipocytes 975
    bone histomorphometry 313
    bone turnover markers 295
    childhood cancer 793–796
    fibrodysplasia ossificans progressiva 867
    non-parathyroid hypercalcemia 644
    Paget disease of bone 717
    reference point indentation 289
    secondary causes of osteoporosis 510
    trabecular bone score 282
    transplantation osteoporosis 425, 426, 431
glucose-dependent insulinotropic polypeptide (GIP) 488, 489
glucose metabolism 1004–1006, *1005*
glucose-6-phosphate (G6P) 1004–1006
glutamate 1029, **1030**, 1031
glutamine metabolism 1006, *1007*
Glut family 961–962, 1004–1006, 1009, 1017
glycocalyx 76–77
glycogen synthase kinase 3 (GSK3) 68
glycosaminoglycan (GAG) 951–952
glycosylated proteins 86–88, **88**

GNAS/*GNAS*
　disorders of phosphate homeostasis 679
　fibrous dysplasia of bone 839, 842, 844
　progressive osseous heteroplasia 868
　pseudohypoparathyroidism 663–667
GnRH *see* gonadotropin-releasing hormone
GO *see* gene ontology
GOC *see* glandular odontogenic cyst
gonadal mosaicism 848
gonadal steroids 194–204
　bone loss in men 415–416
　bone loss in women 414–415
　bone resorption and formation 415–416, *416*
　cellular targets 195
　changes in bone mass and structure with aging 412–413, *413*, *414*
　concepts and definitions 194, 412
　glucocorticoid-induced osteoporosis 468–469
　hormone biosynthesis 194
　hormone replacement therapy 200–201
　immunobiology 993–994, *995*
　linear growth 197–198
　menopause and age-related bone loss 156–157, 194, 200, *201*
　molecular mechanism of action 194–195, *196*
　neuropsychiatric disorders 1048–1049
　nonsex steroid hormone changes with aging 416–417
　osteoclasts 51
　osteoporosis 199–201, 412–418
　periosteal expansion 198
　pituitary–bone axis 1042
　putative gene targets 195–197
　sex steroid deficiency 199–201
　skeletal development and growth 197–198
　skeletal growth 128
　skeletal maintenance 198–199, *199*
　transplantation osteoporosis 425
　see also estrogens; testosterone
gonadotropin-releasing hormone (GnRH)
　bone turnover markers 295
　childhood cancer 795
　pregnancy and lactation 151
　premenopausal osteoporosis 440
　skeletal effects of medications for nonskeletal disorders 482
gonadotropin-releasing hormone (GnRH) agonists 776, 778
Gorlin syndrome 922–923
goserelin 775
GP *see* growth plate
G-protein coupled receptors (GPCR)
　endocrine bioenergetics of bone 1013–1015, *1013*
　mechanotransduction 78
　parathyroid hormone 206–209
　parathyroid hormone-related protein 213–214, *214*
　pituitary–bone axis 1037–1038, 1041–1042
　see also calcium-sensing receptor
gradient of risk (GR) 332–333, **334**
graft-versus-host disease (GVHD) 431
granulocyte colony-stimulating factor (G-CSF) 969–970
granulomatous diseases 641, 644
growth arrest-specific protein 6 (Gas6) 755, 816–817
growth hormone deficiency (GHD) 663–665, 794, 795
growth hormone (GH)
　osteogenesis imperfecta 875
　pituitary–bone axis 1037–1038
　secondary causes of osteoporosis 512–513

growth plate (GP) cartilage 13
GSK3 *see* glycogen synthase kinase 3
GTR *see* guided tissue regeneration
guanine nucleotide exchange factors (GEF) 48
guided tissue regeneration (GTR) 938
GVHD *see* graft-versus-host disease
GWAS *see* genome-wide association studies

HAART *see* highly active antiretroviral therapy
hair follicles 236
HAL *see* hip axis length
HALT study 555
Hand–Schüller–Christian disease 857–858, *858*
HAP *see* hydroxyapatite
hazard ratio (HR) 278–281
HBP *see* hexosamine biosynthetic pathway
hCG *see* human chorionic gonadotropin
HCT *see* hematopoietic cell transplant
HCV *see* hepatitis C virus
HD *see* Hodgkin disease
HDAC/*Hdac see* histone deacetylase
HDR *see* hypoparathyroidism, deafness, renal abnormalities
head organizers *894*
HED *see* hypohidrotic ectodermal dysplasia
helper T cells (Th) 740–741, 992–995
hematological malignancies 760–767
　adult T-cell leukemia/lymphoma 764
　bone (re)modeling 761, *761*
　concepts and definitions 760
　Hodgkin disease 765
　hypercalcemia 760, 761, 765
　imaging in multiple myeloma bone disease 763
　management of multiple myeloma bone disease 763–764
　models of multiple myeloma bone disease 763
　monoclonal gammopathy of undetermined significance 764
　non-Hodgkin lymphoma 765
　pathophysiology of multiple myeloma 760–764
　vicious cycle hypothesis *762*
　see also multiple myeloma
hematopoietic cell transplant (HCT) 793–794
hematopoietic stem cell (HSC) niche 966–973
　adipocytes 969, 976–978, 980
　bone marrow 966–971, *967*
　cell types of the hematopoitic niche 966–969
　CXCL12 abundant reticular cells 968–969
　endothelial cells 968
　heterogeneity of HSCs and their niches 966
　macrophages and monocytes 969
　malignancy and the HSC niche 970
　niche for B lymphopoiesis 969–970
　osteoclasts 52
　research directions 970–971
　sympathetic nervous system neuronal cells and glia 969
hemichannels 42–43
hemodialysis 701–702
HEO *see* heterotopic endochondral ossification
heparin 484
hepatitis C-associated osteosclerosis 835–836
hepatitis C virus (HCV) 476–478
Her2 *see* human epidermal growth factor receptor 2
hereditary hypophosphatemic rickets with hypercalciuria (HHRH) 679, 709

hereditary vitamin D-dependent and resistant rickets
    (HVDDR) 691–692
heritability
    animal models: allelic determinants for
        BMD 361–362
    genome-wide association studies 378
    nephrolithiasis 723
    osteochondrodysplasias 848
    osteosarcoma 770
    translational genetics of osteoporosis 385
HES/Hes 15
heterotopic endochondral ossification (HEO) 865, 867–868
heterotopic ossification 666, 669
    see also fibrodysplasia ossificans progressiva; progressive osseous
        heteroplasia
hexosamine biosynthetic pathway (HBP) 1004–1006
hFEA see homogenized finite element analysis
HFTC see hyperphosphatemic familial tumoral calcinosis
HHex 893
HHM see humoral hypercalcemia of malignancy
HHRH see hereditary hypophosphatemic rickets with
    hypercalciuria
HHS see hyperostosis–hyperphosphatemia syndrome
high bone mass disorder 832
highly active antiretroviral therapy (HAART) 475
high-resolution peripheral quantitative computed tomography
    (HR-pQCT)
    bone mass measurement in adults 261–267, **262**, *263*
    bone mass measurement in children with osteoporosis risk
        factors 248–249
    chronic kidney disease 506
    ethnicity and age-related loss of bone strength 132–133
    finite element analysis 264–265, *264*, *265*, 267–268
    gonadal steroids 199
    magnetic resonance imaging 275
    metastatic bone disease 744
    precision errors **263**
    trabecular bone score 277
hip axis length (HAL) 256, *256*
hip fracture
    epidemiology of osteoporotic fractures 398–400, *399*
    Fracture Liaison Service 407–408
    orthopedic principles of fracture management 590–591
    radiotherapy-induced osteoporosis 788
histiocytosis-X 857–858, *858*
histomorphometry see bone histomorphometry
histone deacetylase (HDAC) 15–16, 35, 215
histone proteins 35
HIV see human immunodeficiency virus
Hodgkin disease (HD) 765
holoprosencephaly (HPE) 896
homocystinuria 882–883, 888
homogenized finite element analysis (hFEA) 264, 267–268
HORIZON trial 548
hormone receptor-positive nonmetastatic breast cancer 555
hormone replacement therapy (HRT)
    bone turnover markers 298
    calcitonin 575
    estrogens, SERMs, and TSEC 541–543
    gonadal steroids 200–201
    neuropsychiatric disorders 1048–1049
    primary hyperparathyroidism 625
    sclerosing bone disorders 828

translational genetics of osteoporosis 389
    see also testosterone replacement therapy
hormone response elements (ERE/ARE) 194–195, *196*
Hox 7, 181
HPE see holoprosencephaly
HPP see hypophosphatasia
HPT-JT see hyperparathyroidism-jaw tumor syndrome
HR see hazard ratio
HR-pQCT see high resolution peripheral quantitative computed
    tomography
HRPT2 634
HRT see hormone replacement therapy
HSC see hematopoietic stem cells
HTC see hyperphosphatemic tumoral calcinosis
HTLV-1 see human T-cell leukemia virus type 1
human chorionic gonadotropin (hCG) 1038
human epidermal growth factor receptor 2 (Her2) antigen 770
human fetal and neonatal bone development see fetal and neonatal
    bone development
human genome-wide association studies see genome-wide
    association studies
human immunodeficiency virus (HIV) 474–481
    antiretroviral therapy 474–477
    children and adolescents 477, *477*
    concepts and definitions 474
    epidemiology of osteoporotic fractures 474
    hepatitis C virus coinfection 476–478
    immunobiology 993, 997–998
    management 476–477
    putative mechanisms 474–475, *475*
    reference point indentation 289–290
    screening 475–476
    special populations 477–478
human T-cell leukemia virus type 1 (HTLV-1) 764
humoral hypercalcemia of malignancy (HHM) 640
hungry bone syndrome 702, 706
HVDDR see hereditary vitamin D-dependent and resistant rickets
HVO see hypovitaminosis D osteopathy
hydrochlorothiazide 642
hydrogen peroxide 200
hydroxyapatite (HAP)
    bone mass measurement in adults 260
    bone matrix composition 84–85, 90
    genetic craniofacial disorders affecting dentition 911
    odontogenesis 904
    osteoblasts 31–32
    pathology of the hard tissues of the jaws 918–919
11β-hydroxysteroid dehydrogenase (11β-HSD) 468
hyperaldosteronism 1042
hypercalcemia
    calcium-sensing receptor 225
    childhood disorders of mineral metabolism 707–708
    chronic kidney disease–mineral and bone disorder 699–700
    enzyme deficiencies 886–887
    familial states of primary hyperparathyroidism 629–631
    fetal calcium metabolism 181
    hematological malignancies 760, 761, 765
    non-parathyroid hypercalcemia 639–645
    parathyroid disorders 614–616
    parathyroid hormone 207–208
    primary hyperparathyroidism 622
    treatment 708
    vitamin D 232

1082 Index

hypercalciuria
   enzyme deficiencies 886–887
   familial hypomagnesemia with hypercalciuria and nephrocalcinosis 174
   familial states of primary hyperparathyroidism 629–631
   hereditary hypophosphatemic rickets with hypercalciuria 679, 709
   hypocalcemia 648, 650, 652
   hypoparathyroidism 656
   nephrolithiasis 721, 726–727
   premenopausal osteoporosis 438
   treatment of idiopathic hypercalciuria 727
hypercorticolism 513
hyperkyphosis 521
hypermagnesemia 174, 710
hyperostosis–hyperphosphatemia syndrome (HHS) 190–191
hyperoxaluria 721, 727
hyperparathyroidism
   calcium-sensing receptor 221
   childhood disorders of mineral metabolism 705, 707
   chronic kidney disease 506
   chronic kidney disease–mineral and bone disorder 701–702
   disorders of phosphate homeostasis 678, 680
   fetal calcium metabolism 181
   nuclear medicine 305–308, *306*
   osteocytes 41–42
   parathyroid hormone 208
   pathology of the hard tissues of the jaws 923–925
   rickets and osteomalacia 686
   transplantation osteoporosis 425, 432
   see also primary hyperparathyroidism
hyperparathyroidism–jaw tumor syndrome (HPT-JT) 634
   clinical expressions 634
   diagnosis of carriers and cancers 634
   management 634
   pathogenesis/genetics 619, 620, **630**, 634
hyperphosphatemia
   causes 680, **681**
   childhood disorders of mineral metabolism 709
   chronic kidney disease 505, 506
   chronic kidney disease–mineral and bone disorder 699–701
   clinical and biochemical manifestations 680, 681
   familial tumoral calcinosis 681
   FGF23 and regulation of phosphorus metabolism 190–191
   genetic craniofacial disorders affecting dentition 914, *914*
   genetics 681
   treatment 681
hyperphosphatemic familial tumoral calcinosis (HFTC) 914, *914*
hyperphosphatemic tumoral calcinosis (HTC) 709
hyperprolactinemia 1051
hypersplenism 825
hyperthyroidism 512, 641
hypertrophic osteoarthropathy 307, *307*
hyperuricosuria 721, 727
hypocalcemia 646–653
   adverse effects of drugs for osteoporosis 583
   calcium-sensing receptor 225
   childhood disorders of mineral metabolism 705–707
   clinical presentation 705
   concepts and definitions 646
   diagnostic tests and interpretation 650, *651*
   etiology and pathogenesis 646–649, **647**
   fetal and neonatal bone development 119–120
   fetal calcium metabolism 179
   hypoparathyroidism 655–656, 658
   magnesium homeostasis 173, 175
   management of acute and chronic hypocalcemia 650–652
   medical prevention and treatment of metastatic bone disease 800
   parathyroid disorders 616
   parathyroid hormone 208
   pseudohypoparathyroidism 663
   regulation of calcium homeostasis 166
   signs and symptoms 649–650, **650**
   transient hypocalcemia of the newborn 705
   treatment 706–707
hypocitraturia 721, 727
hypogonadism
   bone histomorphometry 313
   osteoporosis in men 445, 447
   pituitary–bone axis 1039
   pseudohypoparathyroidism 668
   secondary causes of osteoporosis 513
   transplantation osteoporosis 425, 426
hypohidrotic ectodermal dysplasia (HED) 912
hypomagnesemia 173–175
   childhood disorders of mineral metabolism 705–706, 709
   parathyroid disorders 615
   treatment 709
hyponatremia 1042–1043
hypoparathyroidism 654–660
   areas of special interest 657–658
   calcium and vitamin D supplementation 655–656
   childhood disorders of mineral metabolism 705–707, 709
   clinical features 655–656
   complex diseases with hypoparathyroidism 654–655, **655**
   concepts and definitions 654
   diseases associated with isolated familial hypoparathyroidism 655, **655**
   fetal calcium metabolism 181–182
   future directions 658
   genetic causes 654–655
   hypocalcemia **647**, 648
   parathyroid disorders 614–616
   parathyroid hormone treatment for osteoporosis 563
   pregnancy and lactation 149, 152
   PTH(1–34) 656–657
   PTH(1–84) 657, 658
   quality of life 658
   renal manifestations and extraskeletal calcification 655–656
   skeletal manifestations 657–658
   treatment 656–657
hypoparathyroidism, deafness, renal abnormalities (HDR) syndrome 648
hypophosphatasia (HPP)
   bone stress injuries 454
   enzyme deficiencies 886–888
   genetic craniofacial disorders affecting dentition 914, *915*
   genetic testing 375
hypophosphatemia 189–190
   causes 674, **675**
   childhood disorders of mineral metabolism 708–709, **708**
   clinical consequences 674
   genetic craniofacial disorders affecting dentition 913–914
   molecular mechanisms 676
   rickets and osteomalacia **686**, 687
   see also individual disorders

hypophosphatemic osteopathy 313–314
hypophosphatemic vitamin D resistant rickets 454
hypothalamus 959, 962–963, 1023–1024
hypothyroidism 663, 668, 795
hypovitaminosis D osteopathy (HVO) 313–314
hypoxia-inducible factor (HIF-1α)
   bone vasculature 984, 987
   cellular bioenergetics of bone 1006
   integrative physiology of the skeleton 961–962

ibandronate 506, 549, 594
IBD *see* inflammatory bowel disease
IBSP/*Ibsp see* integrin-linked bone sialoprotein
ICER *see* incremental cost-effectiveness ratio
IDI *see* indentation distance increase
idiopathic infantile hypercalcemia 707–708
idiopathic juvenile osteoporosis (IJO) *see* juvenile osteoporosis
idiopathic osteoporosis (IOP)
   osteoporosis in men 444–445, 447
   premenopausal osteoporosis 438
   secondary causes of osteoporosis 512
IFITM5 874
IGF1R *see* insulin-like growth receptor 1 receptor
IGF *see* insulin-like growth factor
IHC *see* immunohistochemistry
Ihh/*Ihh see* Indian hedgehog
IJO *see* juvenile osteoporosis
IL *see* interleukins
IMAT *see* intramuscular adipose tissue
IMCD *see* inner medullary collecting duct
immobilization and burns 730–736
   abnormalities in bone and calcium metabolism after burns 733–734
   bone loss after SCI 730–731
   fracture risk 731
   incidence of stroke and related bone loss 732
   inflammatory response 733–734
   neurological disease 730
   non-parathyroid hypercalcemia 641–644
   Parkinson disease 733
   rehabilitation and pharmacological approaches 731–733
   spinal cord injury 730–732
   stress response 734
   stroke 732–733
   treatment of burns 734
immunobiology 992–1003
   ART-induced inflammation exacerbates HIV-induced bone loss 998
   B cells 993, 997–998, **1000**
   bone loss in chronic inflammatory states 998–1001, *999*
   bone loss secondary to sex steroid deficiency 993–994, *995*
   immune cells in bone disease 993–1001, **1000**
   immune cells relevant to bone 992–993
   monocytes and macrophages 993
   osteoblasts, osteocytes, and T cells as targets of PTH 996–997, *997*
   osteolytic and osteoblastic skeletal lesions 740–741
   role of immune cells in HIV/ART-associated bone loss 997–998
   role of T cells in anabolic activity of intermittent PTH treatment 996
   role of T cells in primary hyperparathyroidism 994–996, *996*
   T cells 992–997, **1000**
   tumor-induced bone disease 753–754
   vitamin D 235–236

immunohistochemistry (IHC) 985
immunoreceptor tyrosine-based activation motifs (ITAM) 50
immunosuppressive drugs 425–426
   see also glucocorticoids
IMO *see* infantile malignant osteopetrosis
impact reference point indentation (iRPI) 287 290
incremental cost-effectiveness ratio (ICER) 597–598
incretin-based therapies 489
indels 374
indentation distance increase (IDI) 288
Indian hedgehog (Ihh)
   endochondral ossification 16–17
   osteoblasts 33, 56–57
   parathyroid hormone-related protein 214–215, *215*
   skeletal morphogenesis in embryonic development 7–8
infantile hypophosphatasia (HPP) 886
infantile malignant osteopetrosis (IMO) 120
infantile osteopetrosis 706
infection stone 727–728
infiltrative disorders *see* ischemic and infiltrative disorders of bone
inflammation-induced bone loss 459–466
   ankylosing spondylitis 459–460, 462–463
   chronic obstructive pulmonary disease 511–512
   concepts and definitions 459–460
   immobilization and burns 733–734
   immunobiology 998–1001, *999*
   inflammatory bowel diseases 511
   rheumatoid arthritis 459–462, 511
   secondary causes of osteoporosis 510–512
   systemic lupus erythematosus 459, 463
inflammatory bowel disease (IBD) 511, 999
information bias 580
inner medullary collecting duct (IMCD) 226
inorganic pyrophosphate (PPi) 887
insulin and insulin resistance
   diabetes and bone loss 489
   endocrine bioenergetics of bone 1012–1014, 1016–1017
   integrative physiology of the skeleton 963
insulin-like growth factor (IGF)
   glucocorticoid-induced osteoporosis 468
   gonadal steroids 197–198
   gonadal steroids and osteoporosis pathogenesis 416–417
   muscle–bone interactions 1058
   nutritional support for osteoporosis 535
   Paget disease of bone 714
   pituitary–bone axis 1038
   sclerosing bone disorders 836
   secondary causes of osteoporosis 512–513
   skeletal healing/repair 103
insulin-like growth receptor 1 receptor (IGF1R) 78
integrative physiology of the skeleton 959–965
   bioenergetics of cells in bone marrow niche 961–962
   common origin for fat and bone cells 960–961
   concepts and definitions 959–960
   control of skeletal and adipose tissue remodeling 962–963, *963*
   mesenchymal progenitor cells in bone marrow 959–960, *960*
   role of hypothalamus 959
   sympathetic nervous system control over fat and bone remodeling 963
integrin-linked bone sialoprotein (IBSP) 16–17
integrins 48, 76, 78, 79
interferon-γ (IFN-γ) 50
interfragmentary strain theory 111–112

interleukins (IL)
    endocrine bioenergetics of bone 1014–1015
    gonadal steroids and osteoporosis pathogenesis 414–415
    hematological malignancies 761, 764
    immunobiology 992–996, *996*, *999*
    inflammation-induced bone loss 460
    muscle–bone interactions 1058–1059
    osteoclasts 48, 50
    osteolytic and osteoblastic skeletal lesions 741
    skeletal healing/repair 104
    tumor-induced bone disease 754
intermediary organization (IO) 310
intermittent parathyroid hormone (iPTH) infusion 996
International Osteoporosis Foundation (IOF) 437
International Society for Clinical Densitometry (ISCD)
    bone mass measurement in adults 253–254, **254**, 256–257
    bone mass measurement in children with osteoporosis risk factors 247–248
    chronic kidney disease 506
    premenopausal osteoporosis 436–437
    vertebral fracture **327**
intersphenoidal synchondrosis 897
intestinal transplantation 425
intracardiac injection model 755–756
intramedullary nails 590
intramedullary rod fixation 875
intramembranous ossification 112, 369
intramuscular adipose tissue (IMAT) 500
intravenous pyelography (IVP) 723–724
intravital multiphoton confocal microscopy (IVM) 985
IO *see* intermediary organization
IOF *see* International Osteoporosis Foundation
ion channels 79–80
IOP *see* idiopathic osteoporosis
iPTH *see* intermittent parathyroid hormone
iRPI *see* impact reference point indentation
ISCD *see* International Society for Clinical Densitometry
ischemic and infiltrative disorders of bone 853–860
    causes of ischemic necrosis **854**
    common sites of osteochondrosis and ischemic necrosis **854**
    concepts and definitions 853
    infiltrative disorders 856–858
    ischemic necrosis 853–856, **854**, *855*, *856*
    Langerhans cell histiocytosis 857–858, *858*
    Legg–Calvé–Perthes disease 853–854, *855*
    other presentations of ischemic necrosis 854–856
    systemic mastocytosis 856–857, *856*, *857*
isoflavones 537
isolated familial hypoparathyroidism 655, **655**
ITAM *see* immunoreceptor tyrosine-based activation motifs
IVM *see* intravital multiphoton confocal microscopy
IVP *see* intravenous pyelography

Jaccoud arthropathy 463
Jansen metaphyseal chondrodysplasia 217
juvenile dermatomyositis (JD) 863, *863*
juvenile-onset spondyloarthropathy 459
juvenile osteoporosis (IJO) 419–423
    biochemical findings 420
    bone biopsy 421
    clinical features 420, *420*
    concepts and definitions 419
    differential diagnosis 421, **421**, **422**

    pathophysiology 419–420
    prognosis 422
    radiological features 420, *420*
    treatment 421–422

karyotyping 346
KDIGO *see* Kidney Disease Improving Global Outcomes
keratocystic odontogenic tumor 922–923, *922*
Kidney Disease Improving Global Outcomes (KDIGO) 699–702
kidney–pancreas transplantation 429
kidney stones *see* nephrolithiasis
kidney transplantation 426, 429
killer osteoclasts 199
Kirschner wires (K-wires) 589
Klotho
    childhood disorders of mineral metabolism 709
    chronic kidney disease–mineral and bone disorder 698
    tumoral calcinosis 862
Kuskokwim syndrome 873
K-wires *see* Kirschner wires
kyphoplasty 591, 819

lacrimo-auriculo-dento-digital (LADD) syndrome 912
lactate dehydrogenase (LDH) 961
lactate production 1004–1006, 1009
lactation *see* breastfeeding; pregnancy and lactation
Lactobacillus spp. 918
lactose intolerance 536
lacunocanalicular network 38, *39*, 41–42
LADD *see* lacrimo-auriculo-dento-digital
Langerhans cell histiocytosis (LCH) 857–858, *858*
lanthanum carbonate 700
lasofoxifene 542–543
lateral periodontal cyst (LPC) 923
LCH *see* Langerhans cell histiocytosis
LCPD *see* Legg–Calvé–Perthes disease
LD *see* linkage disequilibrium
LDH *see* lactate dehydrogenase
L-dopa 733
LEF/TCF *see* lymphoid enhancer-binding factor/T-cell factor
Legg–Calvé–Perthes disease (LCPD) 853–854, *855*
leptin 962–963, 978–980, 1020–1022
leptin receptor (LepR) cells 25
LET *see* linear energy transfer
Letterer–Siwe disease 857–858
leucine-rich repeat (LRR) 86
leukemia inhibitory factor (LIF) 755, 1058–1059
leutenizing hormone-releasing hormone (LHRH) agonists 775–776, 778
LH *see* luteinizing hormone
LHRH *see* leutenizing hormone-releasing hormone
LIF *see* leukemia inhibitory factor
ligament 352
lineage tracers **355–356**, *356–357*
linear energy transfer (LET) 789
linear variable displacement transducer (LVDT) 95
linkage disequilibrium (LD) 383
linkage studies 386, **387**
lipids 91
lipopolysaccharide (LPS) 935–936
lipoprotein receptor-related protein (LRP)
    cellular bioenergetics of bone 1007–1008
    central neuronal control of bone remodeling 1022

genetic craniofacial disorders affecting dentition 911
    mechanotransduction 80–81
    sclerosing bone disorders 832
    signal transduction cascades controlling osteoblast
        differentiation 56
    translational genetics of osteoporosis 387
    Wnt/β-catenin signaling pathway 68–70, 71–72
LIV *see* low intensity, high frequency vibration
liver transplantation 430
Lmx1b gene 6
lncRNA *see* long noncoding RNA
local osteolytic hypercalcemia (LOH) 640
locking plates 590
Loeys–Dietz syndrome 62, 880, 882
LOFT trial 604–605
logistics 579–580
LOH *see* local osteolytic hypercalcemia; loss of heterozygosity
long noncoding RNA (lncRNA) 35, 771, 772
loss of heterozygosity (LOH) 374–375
low intensity, high frequency vibration (LIV) 732
LPC *see* lateral periodontal cyst
LPS *see* lipopolysaccharide
LRP/*LRP see* lipoprotein receptor-related protein
LRR *see* leucine-rich repeat
lumbar spine (LS)
    bone mass measurement in children with osteoporosis risk
        factors 243–246
    childhood cancer 793–795
    mechanical loading 142–143
    trabecular bone score 277–278, 282
lung cancer
    bone cancer pain 785
    medical prevention and treatment of metastatic bone
        disease 805–806
    parathyroid hormone-related protein 217
lung transplantation 429
luteinizing hormone (LH) 1037–1038
LVDT *see* linear variable displacement transducer
lymphoid enhancer-binding factor/T-cell factor (LEF/TCF)
    68–70, 80

M3 muscarinic receptor 1021–1022
McCune–Albright syndrome (MAS) 190, 679–680, 839–844
macrophage colony-stimulating factor (M-CSF) 46, 48–52,
    102–104
macrophage inflammatory protein 1α (MIP-1α) 761–764
macrophages 950–951, 969, 993
MAF *see* minor allele frequency
magnesium
    childhood disorders of mineral metabolism 705–706, 709–710
    concepts and definitions 173
    familial states of primary hyperparathyroidism 629
    homeostasis 173–178
    hypermagnesemia 174, 710
    hypocalcemia 648–650
    hypomagnesemia 173–175, 615, 705–706, 709
    inherited disorders of magnesium homeostasis 173–175, **176**
    intestinal absorption 174
    renal conservation 174–175
magnetic resonance imaging (MRI) 272–276
    adipocytes 980
    bone mass measurement in children with osteoporosis risk
        factors 249

bone strength, fracture, osteoporosis, and therapeutic
    response 274–275
bone stress injuries 450, 454
childhood cancer 796
concepts and definitions 272
cortical bone 273–274, 275
glucocorticoid-induced osteoporosis 469
hematological malignancies 763
ischemic and infiltrative disorders of bone 854–855
metabolic bone disease 303–305
metastatic bone disease 744–747
Paget disease of bone 717
pregnancy and lactation 149
signal-to-noise ratio 272–273, 274–275
spin-echo and gradient-echo sequences 272–273
trabecular bone 272–273, 273, 274, 275
vertebral fracture 327, 329, *328*
MAHC *see* malignancy-associated hypercalcemia
major depressive disorder (MDD) 1049–1050
major osteoporotic fractures (MOF) 278–281, *283*
MALAT1 771
malignancy-associated hypercalcemia (MAHC) 640, 641, 643
malignant odontogenic neoplasms 923
mammary gland 215–216, *216*
MAMP *see* microbe-associated molecular patterns
mandibular development 897–898
Manhattan plots 380, *382*
MAPK/ERK signaling pathway
    calcium-sensing receptor *224*
    FGF23 and the regulation of phosphorus metabolism 188–189
    parathyroid hormone 209
MAR *see* mineral apposition rate
Marfan lipodystrophy syndrome (MLS) 882
Marfan syndrome (MFS) 878–884
    congenital contractural arachnodactyly 881
    diagnostic criteria 878–880, **879**
    Marfan syndrome-related disorders 882–883
    medical management and treatment 883–884
    pathogenesis 883
    phenotypically-related disorders **880**
    skeletal features **881**, *882*
    TGF-β superfamily 62
marrow adipose tissue (MAT) *see* adipocytes
MAS *see* McCune Albright syndrome
mast cell granuloma 857, *857*
mastocytosis 512
MAT *see* adipocytes
maternal hyperparathyroidism 705
matrix abnormalities **686**
matrix-embedded osteocytes 20
matrix extracellular phosphoglycoprotein (MEPE) 38, 43, 88–90
matrix gla protein (MGP) 90
matrix metalloproteinases (MMP)
    alveolar bone homeostasis 936
    bone turnover markers 296, 298
    dental implants and osseous healing 950–951
    endochondral ossification 13, 15–16
    gonadal steroids 196–197
    metastatic bone disease 749
    odontogenesis 903, 906–907
    osteoclasts 47, *48*
    osteogenesis imperfecta 874
    pathology of the hard tissues of the jaws 919

Mazabraud's syndrome 842, *843*
MBTPS1 *see* membrane-bound transcription factor peptidase-site 1
MCAM *see* melanoma-associated cell adhesion molecule
M-CSF *see* macrophage colony-stimulating factor
MDCT *see* multidetector computed tomography
MDD *see* major depressive disorder
MDS *see* myelodysplastic syndrome
MDSC *see* myeloid-derived suppressor cells
measles virus nucleocapsid protein (MVNP) 714
mechanical loading 141–146
    age-related loss of bone strength 141–142, *143*
    bone stress injuries 452–456, *452*
    characteristics of effective loading prescription 142–143, *143*
    concepts and definitions 141
    exercise for osteoporotic fracture prevention and management 517–518
    peak bone mass/peak bone strength 141–142, *143*
    peak height velocity 141, *142*
    persistence of childhood bone adaptation 143–144
    transcriptional profiling 369–370
mechanosensation 42
mechanosome hypothesis 78, *79*
mechanostat theory 247, 517–518
mechanotransduction 75–83
    biochemical responses to mechanical stimuli 80–82
    concepts and definitions 75
    fluid flow and tissue strain in bone 76–77, *77*
    gap junctions/ion channels 79–80, *81*
    G-protein and receptor tyrosine kinase signaling 78
    integrins/focal adhesions 76, 78, *79*
    mechanosensors in bone 75–76
    molecular basis 77–80
    osteogenic mechanical stimuli 75–77
    prostaglandin and NO signaling 81–82
    Wnt/β-catenin signaling pathway 80–81, *81*
Meckel cartilage 897–898
MED *see* multiple epiphyseal dysplasia
medial edge epithelium (MEE) 896, *896*
medical history 344, 446
medication-induced hypercalcemia 642
medullary thyroid carcinoma (MTC) 574–575
MEE *see* medial edge epithelium
MEF2/*Mef2 see* myocyte enhancer factor
melanocortin system 1024
melanoma-associated cell adhesion molecule (MCAM) 22, 24–25
melorheostosis 833–834, *833*
membrane bound Klotho (mKL) 188–189
membrane-bound transcription factor peptidase-site 1 (MBTPS1) 1057
membranous ossification 848
MEN1 *see* multiple endocrine neoplasia type 1
MEN2A *see* multiple endocrine neoplasia type 2A
Menkes disease 888–889
menopausal hormone therapy (MHT) 541–543
menopause 155–161
    age-related bone loss 156
    bone histomorphometry 313
    bone (re)modeling 155, 157–158
    cortical porosity 156–158, *159*
    estrogens, SERMs, and TSEC 541–542
    ethnicity and age-related loss of bone strength 132–133
    gonadal steroids 194, 200
    hypoparathyroidism 658
    magnetic resonance imaging 274–275
    menopausal bone loss 156–159, *157*
    osteoclasts 51
    overview and historical development 395–396
    periosteal apposition 158, *159*
    surface area/bone matrix volume configuration 156
    trabecular bone 156–158, *158*
    trabecular bone score 282
    see also postmenopausal osteoporosis; premenopausal osteoporosis
MEPE *see* matrix extracellular phosphoglycoprotein
MES *see* midline epithelial seam; minimum effective strain
mesenchymal progenitor cells 3, 12–17, *14*
mesenchymal stromal/stem cells (MSC)
    hematopoietic stem cell niche 968
    integrative physiology of the skeleton 959–962
    mechanotransduction 76–77
    menopause and age-related bone loss 156–157
    osteoblasts 32–35, 54
    osteochondrodysplasias 848
    osteoprogenitor cells 21–22, *21*
    osteosarcoma 771
    peripheral neuronal control of bone remodeling 1031
    radiotherapy-induced osteoporosis 790
    sarcopenia and osteoporosis 498
    skeletal healing/repair 101–104
    TGF-β superfamily *61*
    transcriptional profiling 367–368
mesenchyme 354
meta-analysis
    calcitonin 576
    genome-wide association studies 379–383, *382*
    glucocorticoid-induced osteoporosis 470
    prevention of falls 529–530, **529**
metabolic bone diseases *see individual disorders*; oral manifestations of metabolic bone diseases
metabolic syndrome 500–501, *501*
metaphysis 128
metastatic bone disease
    assessment of bone health 804
    biology of metastatic bone lesions 816–818
    bone cancer pain 781–787
    bone mass measurement in adults 268
    bone metastases in breast cancer 804–805
    bone metastases in lung cancer 805–806
    bone metastases in prostate cancer 800–804
    bone metastases in renal cell carcinoma 806
    bone-targeted agents 800–801, 804
    bone turnover markers 295, 800–801, 806
    clinical and preclinical imaging 743–751
    clinical sequelae 800
    comparison of imaging modalities **746**
    computed tomography 744, *745*, 747
    concepts and definitions 743, 799, 816
    denosumab 556
    dual-energy X-ray absorptiometry 743–744
    fibrous dysplasia of bone 842–843
    hematopoietic stem cell niche 970
    impeding fractures 818
    magnetic resonance imaging 744–747
    medical prevention and treatment 799–808
    minimally invasive surgical options 819–820
    molecular/functional imaging techniques 747–749
    morphological/anatomical imaging techniques 743–747
    multimodality imaging 749

myeloma bone disease and other hematological
    malignancies 760–767
normal bone turnover 799
optical imaging 748–749, 748
osteolytic and osteoblastic skeletal lesions 739–741
osteosarcoma 768–774
pathophysiology of bone metastases 799–800
pharmacological treatment of bone metastasis 818
plain film radiography 743
positron emission tomography 747–748
radionuclide bone scintigraphy 747
radiotherapy-induced osteoporosis 788–792
radiotherapy of skeletal metastases 809–815
single-photon emission computed tomography 747
skeletal complications of breast and prostate cancer
    therapies 775–780
skeletal complications of childhood cancer 793–798
surgical treatment 818–820
tumor–bone interactions 816–818, 817
tumor-induced bone disease 752–759
vicious cycle hypothesis 817–818
see also individual diseases/disease types
metformin 489
methotrexate 793–795
MFS *see* Marfan syndrome
MGC *see* multinucleated giant cells
MGP *see* matrix gla protein
MGUS *see* monoclonal gammopathy of undetermined
    significance
MHT *see* menopausal hormone therapy
microarray comparative genomic hybridization (aCGH) 346
microarrays 368–369, 374–375
microbe-associated molecular patterns (MAMP) 935–936, **935**
microcomputed tomography (microCT)
    bone mass, structure, and quality assessment 93, 94–95, 96
    bone vasculature 985
    metastatic bone disease 744, 745
    oral manifestations of metabolic bone diseases 943
    skeletal healing/repair 110, 110
    trabecular bone score 277
microcrack density 312
microCT *see* micro-computed tomography
microindentation instruments 287–288, 288
micro-MRI 745–747
micronodular ossifications 665, 665, 667
micro-PET 748
microRNA (miRNA)
    osteoblasts 35
    osteosarcoma 771, 772
    transcriptional profiling 370–371
micro-SPECT 747
midfacial development 895–897, 896
midline epithelial seam (MES) 896, 896
milk 138
milk-alkali syndrome 642
mineral apposition rate (MAR) 312
mineralization inhibitors **686**
mineralization lag time 312
mineralization surface 312
mineral/matrix ratio (MMR) 65
minimally invasive parathyroidectomy (MIP) 623
minimally invasive percutaneous plate osteosynthesis
    (MIPPO) 590
minimum effective strain (MES) 517

minor allele frequency (MAF) 379, 380
MIP-1α *see* macrophage inflammatory protein 1α
MIP *see* minimally invasive parathyroidectomy
MIPPO *see* minimally invasive percutaneous plate osteosynthesis
MiR-34/*miR-34* 771
miRNA *see* microRNA
mKL *see* membrane bound Klotho
MLID *see* multilocus imprinting disorder
MLO-Y4 cell line 41, 43
MLPA *see* multiplex ligation-dependent probe amplification
MLS *see* Marfan lipodystrophy syndrome
MM *see* multiple myeloma
MMP/*Mmp see* matrix metalloproteases
MMR *see* mineral/matrix ratio
MOF *see* major osteoporotic fractures
molar root–incisor malformation (MRIM) 920
Molgaard model *see* allometric approach
monoclonal gammopathy of undetermined significance
    (MGUS) 764
monocytes 969, 993
mosaicism, genetics 348
MRI *see* magnetic resonance imaging
MRIM *see* molar root–incisor malformation
MrOS *see* Osteoporotic Fractures in Men Study
MSC *see* mesenchymal stromal/stem cells
Msh homeobox 1 (MSX1) 912
MSPC *see* multipotent mesenchymal stem or progenitor cells
MSX1 see Msh homeobox 1
MTC *see* medullary thyroid carcinoma
MTP-PE *see* muramyl tripeptide phosphatidyl ethanolamine
mucopolysaccharidoses 888
multidetector computed tomography (MDCT) 327, 328, 929
multifocal parathyroid gland hyperfunction 635
multilocus imprinting disorder (MLID) 667
multinucleated giant cells (MGC) 924
multiple endocrine neoplasia type 1 (MEN1) 632–633
    clinical expressions 632
    diagnosis of carriers 632–633
    diagnosis of other tumors 633
    management 633
    pathogenesis/genetics 619–621, **630**, 632
multiple endocrine neoplasia type 2A (MEN2A) 633–634
    clinical expressions 633
    diagnosis of kindred, carriers, and parathyroid tumors 634
    management 634
    pathogenesis/genetics 619–621, **630**, 633
multiple epiphyseal dysplasia (MED) 376, 851, *851*
multiple myeloma (MM)
    bone turnover markers 295
    hematological malignancies 760–764
    imaging in multiple myeloma bone disease 763
    management of multiple myeloma bone disease 763–764
    models of multiple myeloma bone disease 763
    non-parathyroid hypercalcemia 642
    osteoclasts 52
    osteolytic and osteoblastic skeletal lesions 739–741
    tumor-induced bone disease 754
multiplex ligation-dependent probe amplification (MLPA) 346,
    374–375
multipotent mesenchymal stem or progenitor cells (MSPC)
    20–27, *21*
muramyl tripeptide phosphatidyl ethanolamine (MTP-PE)
    769–770
muscle bioenergetics 1014–1015, *1014*

1088 *Index*

muscle–bone interactions 1055–1062
    aging 1055, *1056*
    bone signaling to muscle 1056–1058, *1058*
    cross-talk 1057, 1059
    endocrine bioenergetics of bone 1055–1059
    evidence of cross-talk from clinical examples 1059
    invisible biochemical coupling 1055–1056
    mechanism for kines crossing barriers 1059–1060
    muscle mass and muscle strength 1055, *1056*
    muscle signaling to bone 1058–1059
    myokines 1058–1060
    research directions 1060
    visible mechanical coupling 1055, *1056*
MVNP *see* measles virus nucleocapsid protein
myelodysplastic syndrome (MDS) 970
myeloid-derived suppressor cells (MDSC) 753
myelosuppression 800
myocyte enhancer factor (MEF2C/D) 15
myokines 1058–1060

NAM *see* negative allosteric modulators
National Bone Health Alliance (NBHA) 255
National Comprehensive Cancer Network (NCCN) 776–777
National Osteoporosis Foundation (NOF)
    bone mass measurement in adults 255
    cost-effectiveness of osteoporosis treatment 599
    Fracture Risk Assessment Tool 336
    nutritional support for osteoporosis 534
National Osteoporosis Guideline Group (NOGG) 335–336, *336*
NBCCS *see* nevoid basal cell carcinoma syndrome
NBHA *see* National Bone Health Alliance
N-cadherin 12
N-cam 12
NCCN *see* National Comprehensive Cancer Network
NCP *see* noncollagenous proteins
NE *see* norepinephrine
negative allosteric modulators (NAM) 224–225
neonatal hypocalcemia 705
neonatal severe primary hyperparathyroidism (NSHPT) 631–632
    clinical expressions 631
    diagnosis 632
    fetal and neonatal bone development 119
    magnesium homeostasis 175
    management 632
    parathyroid hormone 208
    pathogenesis/genetics 619, **630**, 631
neoplastic fractures 328–329, *329*, **329**
nephrocalcinosis
    clinical evaluation 723
    hypocalcemia 648, 650
    hypoparathyroidism 656
    magnesium homeostasis 174
    primary hyperparathyroidism 621
nephrolithiasis 721–729
    [a a]24-hour urine collection 726
    clinical evaluation 723
    computed tomography and ultrasound imaging 723–724, *724*
    concepts and definitions 721
    disorders of phosphate homeostasis 680
    evaluation of the stone patient 723–726
    familial states of primary hyperparathyroidism 629
    genetic contributions 723
    geographic distribution of prevalence and incidence 721–722
    hypercalciuria 721, 726–727
    hyperoxaluria 721, 727
    hyperuricosuria 721, 727
    hypocitraturia 721, 727
    hypoparathyroidism 656
    infection stone 727–728
    invasive and noninvasive stone removal techniques 728
    laboratory biochemical testing 726–728
    nutrition and lifestyle 722–723
    primary hyperparathyroidism 621, 623, 625
    stone properties and manifestations by composition **722**
    stone type and frequency by composition **722**
    supersaturation 726
    tests after recovery of acute stone passage 726
    treatment of idiopathic hypercalciuria 727
    urine crystal analysis 724–726, *725*
    urine pH 727
nerve growth factor (NGF)
    bone cancer pain 781–782, 783–784
    central neuronal control of bone remodeling 1023
    peripheral neuronal control of bone remodeling 1029
Nestin (Nes) cells 25–26
neural crest cells 3–4, *4*
neuroleptics 1051
neuromedin U (NMU) 1022
neuropeptide Y (NPY) 1023–1024, 1029, **1030**, 1032
neuropsychiatric disorders 1047–1054
    Alzheimer disease 1050
    anorexia nervosa 1047–1049
    concepts and definitions 1047
    eating disorders 1047–1049
    immobilization and burns 730
    major depressive disorder 1049–1050
    medications 1048, 1050–1051
    schizophrenia 1050
neurotransmitters *see individual substances*
neutralization plates 589–590
nevoid basal cell carcinoma syndrome (NBCCS) 922–923
next generation sequencing (NGS)
    chromosomal abnormalities, CNVs, and mutations 346–348
    genetic testing 374–375
    osteosarcoma 770
NFAT2 *see* nuclear factor for activated T cells-2
NFATc1 *see* nuclear factor of activated T cells, cytoplasmic 1
NGF *see* nerve growth factor
N-glycosylated proteins 88–90, **89**
NGS *see* next generation sequencing
NHE3 *see* sodium/hydrogen exchanger isoform 3
NHL *see* non-Hodgkin lymphoma
NIPD/NIPT *see* noninvasive prenatal genetic diagnosis/testing
nitric oxide (NO) signaling 81–82
NLR *see* nucleotide-binding oligomerization domain (NOD)-like receptors
NMU *see* neuromedin U
NO *see* nitric oxide
NOD *see* nucleotide-binding oligomerization domain
Nodal signaling pathway 893, 896–897
NOF *see* National Osteoporosis Foundation
NOGG *see* National Osteoporosis Guideline Group
noncollagenous proteins (NCP)
    bone matrix composition 85–90
    gla-containing proteins 90, **90**
    glycosylated proteins 86–88, **88**
    proteoglycans 86, **87**

serum-derived proteins 85–86, **86**
   SIBLING family, N-glycosylated proteins and other glycoproteins 88–90, **89**
non-Hodgkin lymphoma (NHL) 765
noninvasive prenatal genetic diagnosis/testing (NIPD/NIPT) 348
nonparathyroid-dependent hypercalcemia 616
non-parathyroid hypercalcemia 639–645
   abnormal protein binding 642
   acute and chronic renal failure 642–643
   cancer 640–641, 643–644
   clinical signs and symptoms 639
   diagnostic approach 643
   differential diagnosis **640**
   disorders that lead to hypercalcemia 639–643
   endocrine disorders 641
   granulomatous diseases 641, 644
   immobilization 641–644
   management 643–644
   medication-induced hypercalcemia 642
   milk-alkali syndrome 642
   pathophysiology 639
nonsmall cell lung cancer (NSCLC) 805–806
nonspecific plaque hypothesis 934–935
nonsteroidal anti-inflammatory drugs (NSAID) 104, 719
nonsynonymous variants 373
Noonan syndrome 925
norepinephrine (NE) **1030**, 1031–1032, 1047
Notch signaling pathway
   endochondral ossification 15–17
   hematopoietic stem cell niche 968
   odontogenesis 903, 908
   osteoblasts 33, 57
   osteosarcoma 770–771
   skeletal morphogenesis in embryonic development 4
Notum 71
NPT2a/b/c *see* sodium-phosphate cotransporter
NPY *see* neuropeptide Y
NRTI *see* nucleotide reverse transcriptase inhibitors
NSAID *see* nonsteroidal anti-inflammatory drugs
NSCLC *see* nonsmall cell lung cancer
NSHPT *see* neonatal severe primary hyperparathyroidism
NTX
   bone turnover markers 294–298
   denosumab 553–554
   medical prevention and treatment of metastatic bone disease 800
   Paget disease of bone 715
nuclear factor for activated T cells-2 (NFAT2) 54
nuclear factor-κB 460
nuclear factor of activated T cells, cytoplasmic 1 (NFATc1) 50
nuclear medicine
   applications 303–308
   hyperparathyroidism 305–307, *306*
   hypertrophic osteoarthropathy 307, *307*
   imaging systems 303
   metabolic bone disease 302–309
   osteomalacia and rickets 307–308, *308*
   osteoporosis 303–304, *304*
   Paget disease of bone 304–305, *305*
   positron emission tomography 303
   radiotracers 302, 303–308
   renal osteodystrophy 307
   scintigraphy 305, 307–308
   single-photon emission computed tomography 303

nucleotide-binding oligomerization domain (NOD)-like receptors (NLR) 935–936
nucleotide reverse transcriptase inhibitors (NRTI) 475
nutrition *see* diet and nutrition

OA *see* osteoarthritis
obesity 492–497
   body mass index/fracture relationship 492–493
   bone mineral density 493–494
   bone turnover and structure 494
   childhood obesity 402
   clinical management 494–495
   clinical risk factors 493
   concepts and definitions 492
   epidemiology of osteoporotic fractures 492–493
   falls and sarcopenia 493
   Fracture Risk Assessment Tool 494
   hematopoietic stem cell niche 969
   lifestyle measures 494
   morbidity and mortality of fractures in the obese 493
   muscle–bone interactions 1059–1060
   pathogenesis of fractures in the obese 493–494
   pharmacological intervention 495
   pseudohypoparathyroidism 664–665, 668–669
   sarcopenia and osteoporosis 500–501
   site-specificity 492–493
   weight loss 494–495
OCR *see* osteochondroreticular
odanacatib 604–605
odontogenesis 901–910
   amelogenesis 904–906, *905*, *906*
   cementum 907
   circadian formation of dental mineralized tissues *906*, 907–908
   concepts and definitions 901
   dental mineralized tissue formation 903–908
   dentin 903, *904*
   enamel 903–907, *905*
   periodontal ligament 901, 903, 907–908
   signals controlling dental cell differentiation 903
   stages of tooth morphogenesis and their molecular control 901–902, *902*
   stem cells during tooth repair 908
odontogenic cysts 920–923, *921*, *922*
odontogenic keratocyst (OKC) 922–923, *922*
odontogenic neoplasms 920, 923, *924*
odonto hypophosphatasia (HPP) 887
OI *see* osteogenesis imperfecta
OKC *see* odontogenic keratocyst
ON *see* osteonecrosis
oncogenic osteomalacia 308
ONJ *see* osteonecrosis of the jaw
ONO-5334 604
OPG *see* osteoprotegerin
OPK *see* osteopoikilosis
OPPG *see* osteoporosis pseudoglioma syndrome
opportunistic screening 266
OPT *see* osteopetrosis
optical imaging 748–749, *748*
oral contraceptives 295
oral manifestations of metabolic bone diseases 941–948
   clinical assessment of the jaws in humans 941–942
   genetic craniofacial disorders affecting dentition 913–914
   impact of osteoporosis therapy on oral bone health 945–946
   mechanisms of oral bone loss 943–944

oral manifestations of metabolic bone diseases (cont'd)
  morphology and assessment of the jaws in laboratory animals 943
  morphology of the jaws 941
  oral bone assessment to screen for osteoporosis/fracture risk 944
  osteoporosis and tooth loss 944
  Paget disease of bone 944–945
  primary hyperparathyroidism 945
  relationship between oral and postcranial bone mass 943
  renal osteodystrophy 945
orofacial clefts 897, 913
ORP150 38
orthopedic principles of fracture management 588–592
  biology of fracture healing 588–589
  Colles fractures and hip fractures 590–591
  concepts and definitions 588
  external fixation 590
  fractures in osteoporotic patients 590–591
  intramedullary nails 590
  Kirschner wires 589
  osteogenesis imperfecta 875
  plates and screws 589–590
  surgical options 589–590
  treatment principles 589
  vertebral fractures 591
OS *see* osteosarcoma
Osr2/*Osr2* 895
osseointegration *see* dental implants and osseous healing
osseous metastatic disease *see* metastatic bone disease
osteitis fibrosa cystica 181, 622, 626
osteoarthritis (OA)
  TGF-β superfamily 64–65
  trabecular bone score 282–283
  transcriptional profiling 370
osteoblasts 31–37
  animal models: genetic manipulation 351–352, 354
  BMP signaling 54–56
  bone histomorphometry 311
  cell biology 31, *32*
  cellular bioenergetics of bone 1004–1009
  central neuronal control of bone remodeling 1020–1022, 1024
  collagen production and extracellular matrix mineralization 31–32
  conclusion 57–58
  dental implants and osseous healing 951
  development 32–34
  differentiation 16–17, 32–35, 54–59
  endochondral ossification 16–17
  epigenetic posttranscriptional mechanisms of differentiation 35
  epigenetic transcription mechanisms of differentiation 34–35
  fetal calcium metabolism 181
  fibrous dysplasia of bone 842
  function 31–32
  glucocorticoid-induced osteoporosis 468
  gonadal steroids 195–196, 198–199
  gonadal steroids and osteoporosis pathogenesis 414–415, 416
  hedgehog signaling 56–57
  hematological malignancies 760–765
  hematopoietic stem cell niche 966–970
  immunobiology 996–997
  interactions with other cell types 32
  juvenile osteoporosis 421
  mechanical signals promoting differentiation 34
  mechanotransduction 76–77
  neuropsychiatric disorders 1051
  non-parathyroid hypercalcemia 641–642
  Notch signaling 57
  osteocytes 38–41
  osteogenic phenotype commitment of MSCs 32–33
  osteolytic and osteoblastic skeletal lesions 739–742
  osteoprogenitor cells 20, 22
  Paget disease of bone 714–715, 717
  peripheral neuronal control of bone remodeling 1028, 1031, 1033
  pituitary–bone axis 1042–1043
  radiotherapy-induced osteoporosis 789
  regulation 34–35
  regulation of calcium homeostasis 169–170
  Runx2 and Osterix transcription factors 54, *55*
  sarcopenia and osteoporosis 498
  sclerosing bone disorders 827
  signal transduction cascades 33, 54–59
  skeletal healing/repair 103
  TGF-β signaling 56
  TGF-β superfamily 61–62
  transcriptional control of differentiation 34
  transcriptional profiling 367–368
  tumor-induced bone disease 752, 756
  Wnt signaling 56
osteocalcin (OCN)
  activation by insulin and bone resorption 1016
  bone matrix composition 90
  decarboxylation by adipokines and the SNS 1016–1017
  endochondral ossification 16–17
  energy homeostasis 1012–1015
  muscle–bone interactions 1057
  osteocytes 38–40
  osteoprogenitor cells 20, *21*
  production by nutrients 1017
  regulation of muscle bioenergetics 1014–1015, *1014*
  regulation of pancreas development and physiology 1013–1014, *1013*
  secretion and bioactivity 1015–1017, *1015*
  skeletal healing/repair 103
  TGF-β superfamily 61–62
  vitamin D 234
osteochondral progenitor cells 3
osteochondrodysplasias 848–852
  clinical approach to initial evaluation 849–850
  concepts and definitions 848–849
  genetic testing 375–376
  heritability and genetics 848–849
  long-term follow-up 850–851
  molecular pathways involved 851
  multiple epiphyseal dysplasia 851, *851*
  short rib polydactyly type II 851, *852*
  spondyloepimetaphyseal dysplasia 850, *850*
  Stickler syndrome and achondrogenesis II 849, *849*
  see also individual disorders
osteochondroreticular (OCR) cells 22
osteochondrosis **854**, 856
osteoclasts
  actin ring 47
  alveolar bone homeostasis 936
  animal models: genetic manipulation 352, 354–355
  bisphosphonates for postmenopausal osteoporosis 546
  bone cancer pain 785
  bone disease 47
  bone histomorphometry 311

bone resorption 46–53, 47
  cell biology 46–47
  cell–cell interactions in bone marrow 51–52
  cellular bioenergetics of bone 1009
  dental implants and osseous healing 951
  factors regulating formation and/or function 50–52
  fetal and neonatal bone development 120
  fibrous dysplasia of bone 842
  gonadal steroids 196–197, 199
  gonadal steroids and osteoporosis pathogenesis 416
  hematological malignancies 760–764
  immunobiology 992–993
  juvenile osteoporosis 421
  mechanism of bone resorption 47, 48
  non-parathyroid hypercalcemia 641–642
  osteolytic and osteoblastic skeletal lesions 740–741
  osteonecrosis of the jaw 927–928, 930
  Paget disease of bone 714–715, 717
  peripheral neuronal control of bone remodeling 1028, 1033
  pituitary–bone axis 1040–1041
  proteins as regulators 50–51
  radiotherapy-induced osteoporosis 789
  regulation of calcium homeostasis 169–170
  sclerosing bone disorders 825, 827–828
  secondary causes of osteoporosis 510
  signals that regulate differentiation 48–50, 49
  signals that regulate function 48, 49
  small molecules as regulator 51
  transcriptional profiling 367–368
  tumor-induced bone disease 752
osteocyte halos 42
osteocytes 38–45
  animal models: genetic manipulation 351–352, 354
  bone disease 43
  bone (re)modeling 40–41, 41, 43
  cell biology 38
  cell death and apoptosis 40–41, 43
  dental implants and osseous healing 951
  gap junctions and hemichannels in communication 42–43
  glucocorticoid-induced osteoporosis 468
  gonadal steroids 195, 198–199
  gonadal steroids and osteoporosis pathogenesis 414–415
  immunobiology 996–997
  lacunocanalicular network 38, 39, 41–42
  markers 38–40, **40**
  mechanosensation and transduction 42
  mechanotransduction 75–80
  modification of microenvironment 41–42
  ontogeny 38–40
  osteolytic and osteoblastic skeletal lesions 740
  radiotherapy-induced osteoporosis 789–790
  sarcopenia and osteoporosis 498
  transcriptional profiling 367–368
OSTEODENT 944
osteogenesis 111–112, 119
osteogenesis imperfecta (OI) 871–877
  clinical phenotypes 872–873, **872**
  clinical presentation 871
  concepts and definitions 871
  etiology and pathogenesis 874–875
  genetic craniofacial disorders affecting dentition 912–913
  genetic testing 375–376
  juvenile osteoporosis 419, 421, **422**
  laboratory findings 874
  radiographic and DXA features 873
  treatment 875
osteogenic sarcoma *see* osteosarcoma
osteoglophonic dysplasis 680
osteoids 311, 313
osteolytic and osteoblastic skeletal lesions 739–742
  bone as preferred site for metastasis 739–740
  bone remodeling 740
  concepts and definitions 739
  osteocytes and bone metastasis 740
  role of immune cells in bone metastasis 740–741
  vicious cycle hypothesis 740
osteolytic osteolysis 41–42
osteoma cutis 664
osteomalacia 684–694
  axial osteomalacia 834
  bone histology 684–685, 689–690, 690
  childhood disorders of mineral metabolism 710
  clinical picture 688
  concepts and definitions 684–685
  diagnosis 689–690
  enzyme deficiencies 886–888
  epidemiology 685
  etiology 685, **686**
  genetic craniofacial disorders affecting dentition 914
  hereditary vitamin D-dependent and resistant rickets 691–692
  laboratory findings 689
  muscle–bone interactions 1057
  nuclear medicine 307–308, *308*
  pathophysiology 686–687, *687*
  radiographic findings 689, *690*
  treatment and prevention 691
  see also tumor-induced osteomalacia
osteonecrosis of the jaw (ONJ) 927–932
  adverse effects of drugs for osteoporosis 581–582, 585
  bisphosphonates for postmenopausal osteoporosis 549
  clinical appearance and staging 928–929, *929*
  concepts and definitions 927
  denosumab 556
  etiology 927–928
  fibrous dysplasia of bone 844
  medical prevention and treatment of metastatic bone disease 800–801, 806
  parathyroid hormone treatment for osteoporosis 563
  research directions 930
  risk factors 928
  translational genetics of osteoporosis 389–390
  treatment 929–930
osteonecrosis (ON)
  bone vasculature 987–989, *988*
  childhood cancer 796
  glucocorticoid-induced osteoporosis 470
  ischemic and infiltrative disorders of bone 855
  Paget disease of bone 719
  pharmacological treatment of bone metastasis 818
osteonectin 86–88
osteopathia striata 832–833
osteopenia
  disorders of phosphate homeostasis 680
  enzyme deficiencies 888
  inflammation-induced bone loss 462
  obesity and skeletal health 495
  secondary causes of osteoporosis 510, 511
  vertebral fracture 319

osteopetrorickets 826, 828
osteopetrosis (OPT) 825–828
  bone matrix composition 85
  cathepsin K inhibitors 603
  childhood disorders of mineral metabolism 706
  clinical presentation 825–826
  diagnosis 828
  etiology and pathogenesis 827–828
  histopathological findings 827, *827*
  laboratory findings 827
  osteoclasts and bone resorption 47
  radiological features 826–827, *827*
  treatment 828
osteopoikilosis (OPK) 832, *832*
osteopontin (OPN)
  bone matrix composition 88–90
  skeletal healing/repair 103
  vitamin D 234
osteoporosis 553–558
  adherence to osteoporosis therapies 593–596
  adipocytes 978
  adverse effects of drugs for osteoporosis 579–587
  antisclerostin therapy 605–607
  bisphosphonates for postmenopausal osteoporosis 545–552
  bone histomorphometry 313
  bone mass measurement in adults 255
  bone mass measurement in children with osteoporosis risk factors 243–251
  bone matrix composition 85
  bone stress injuries 450–458
  bone turnover markers 293–301
  bone vasculature 989
  cathepsin K inhibitors 603–605
  central neuronal control of bone remodeling 1020–1021, 1024
  chronic kidney disease 505–509
  combination therapy 561–562, 567–572
  cost-effectiveness of osteoporosis treatment 597–602
  diabetes and bone loss 487–491
  drugs associated with bone loss and fracture risk 482–484
  drugs that may protect against osteoporosis 484
  endocrine disorders 512–513
  enzyme deficiencies 888–889
  epidemiology of osteoporotic fractures 398–404
  estrogens, SERMs, and TSEC 541–544
  exercise for fracture prevention and management 517–525
  Fracture Liaison Service 405–411
  Fracture Risk Assessment Tool 331–337
  future therapies 603–609
  genome-wide association studies 380–383
  gonadal steroids 199–201, 412–418
  gradients of risk **334**
  human immunodeficiency virus 474–481
  immobilization and burns 730–736
  immunobiology 998–999
  inflammation-induced bone loss 459–466
  juvenile osteoporosis 419–423
  magnetic resonance imaging 274–275
  mastocytosis 512
  neuropsychiatric disorders 1050
  nuclear medicine 303–304, *304*
  nutritional support for osteoporosis 534–540
  obesity and skeletal health 492–497
  oral manifestations of metabolic bone diseases 944, 945–946
  orthopedic principles of fracture management 588–592
  osteoclasts 51
  osteocytes 43
  other secondary causes 510–516
  overview and historical development 395–397
  parathyroid hormone-related protein 217
  pituitary–bone axis 1039–1042
  pregnancy and lactation 149, 151–152
  premenopausal osteoporosis 436–442
  prevention of falls 518–519, 521, 526–533
  radiotherapy-induced osteoporosis 788–792
  reference point indentation 287
  sarcopenia 498–504
  screening for in breast cancer and prostate cancer 776–777, *777*
  screening for secondary causes for osteoporosis 513
  secondary causes 331–332, **333**
  skeletal effects of medications for nonskeletal disorders 482–486
  strontium ranelate and calcitonin 573–578
  systemic inflammatory disorders 510–512
  TGF-β superfamily 62
  trabecular bone score 278–283
  transcriptional profiling 370–371
  translational genetics of osteoporosis 385–392
  transplantation osteoporosis 424–435
  vertebral fracture 319–320, 328–329, *329*, **329**
  Wnt/β-catenin signaling pathway 71
  see also glucocorticoid-induced osteoporosis
osteoporosis in men 443–449
  age-related osteoporosis 447
  causes 444–445, **445**
  concepts and definitions 443
  effects of aging on the skeleton in men 443
  epidemiology of osteoporotic fractures 444, *444*
  evaluation 445–446, **446**
  glucocorticoid-induced osteoporosis 445, 447
  hypogonadism 445, 447
  idiopathic osteoporosis 444–445, 447
  parathyroid hormone treatment for osteoporosis 562–563
  prevention 446
  sarcopenia and osteoporosis 502
  secondary causes of osteoporosis 512–513
  skeletal development 443
  treatment 446–447
osteoporosis pseudoglioma syndrome (OPPG) 385
Osteoporotic Fractures in Men Study (MrOS) 502
osteoprogenitor cells 20–30
  bone development, growth, and homeostasis 22–24
  bone vasculature 987
  chemokine C-X-C motif ligand 12-abundant reticular cells 25, 27
  circulating osteogenic precursor cells 27
  concepts and definitions 20
  differentiation of osteoprogenitor cells and lineages 20, *21*
  leptin receptor cells 25
  Nestin cells 25–26
  osteocytes 38
  Osx+ cells 26
  PαS cells 25
  periosteum 26
  perivascular progenitors with osteogenic potential in BM environment 24–26, *24*
  recruitment in fracture healing 26
  skeletal stem cells 20, 21–22
osteoprotegerin (OPG)
  alveolar bone homeostasis 936
  bone cancer pain 781

   endocrine bioenergetics of bone 1016
   genome-wide association studies 381–382
   gonadal steroids 196
   gonadal steroids and osteoporosis pathogenesis 414–415
   hematological malignancies 761–762
   immunobiology 992–993, 997–998
   inflammation-induced bone loss 460, 462
   osteoblasts 32
   osteoclasts 46, 50–52
   osteolytic and osteoblastic skeletal lesions 740
   peripheral neuronal control of bone remodeling 1031–1032
   pituitary–bone axis 1041–1042
   secondary causes of osteoporosis 510
   skeletal healing/repair 102–104
   translational genetics of osteoporosis 387
osteosarcoma (OS) 768–774
   altered signaling pathways contributing to osteosarcomagenesis 770–771
   cell of origin and cancer stem cells 771–772
   concepts and definitions 768–769
   current standard treatments 769
   discovery of somatic mutations by NGS 770
   etiology 768–769
   fibrous dysplasia of bone 843
   genetics 770–771
   germline mutations in familial cancer predisposition syndromes 770
   incidence and prevalence 768
   limitations of current standard treatments 769–770
   noncoding RNAs 771, 772
   Paget disease of bone 717
   surgical resection 769
osteosarcopenia 499–502
osteosclerosis 855
osteotomy 875
Osterix (OSX)
   endochondral ossification 15, 17
   integrative physiology of the skeleton 960
   osteoblasts 34, 54, 55
   osteoprogenitor cells 22–23, 26
   skeletal morphogenesis in embryonic development 7–8
   TGF-β superfamily 65
OSTM1 120, 827–828
OSX/*Osx* *see* Osterix
ovariectomy (OVX)
   immunobiology 992–994
   peripheral neuronal control of bone remodeling 1029
   pituitary–bone axis 1040
overexpression of target genes 351–352
oxalate 723
oxandrolone 734
oxidative phosphorylation (OXPHOS) 1004–1006, 1008–1009
oxytocin (OXT) 1037, 1041–1043

P1NP
   bone turnover markers 293–298
   chronic kidney disease 506–507
   combination therapy 568
P2X7 receptors 79–81
p38 1009
p53/*p53* 771, 772
PACAP *see* pituitary cyclase activating polypeptide
pachydermoperiostosis 835
Paget disease of bone 713–720

   biochemical indices in Paget disease 715
   bone turnover markers 294–295
   bone vasculature 989
   clinical features 715–717, *716*
   concepts and definitions 713
   diagnosis 717
   etiology 713–714
   nuclear medicine 304–305, *305*
   oral manifestations of metabolic bone diseases 944–945
   pathology 714, *715*
   signs and symptoms 716–717
   treatment 717–719
paired box 9 (PAX9) 912
palatogenesis 895–897, *896*
palliative radiotherapy 809–811, **810**
PαS cells 25
PAM *see* positive allosteric modulators
pamidronate
   childhood disorders of mineral metabolism 707
   juvenile osteoporosis 421–422
   osteogenesis imperfecta 875
   transplantation osteoporosis 429–430
pancreas 236
pancreatic β-cells 1013–1014, *1013*
pancreatic islets 216–217
parasympathetic nervous system (PSNS) 1029
parathyroid disorders 613–618
   concepts and definitions 613
   extrinsic hypercalcemic and hypocalcemic disorders 616
   intrinsic functional abnormalities of parathyroid glands 613–616
   primary oversecretion of PTH 613–614
   undersecretion of PTH 614–616
   vitamin D 235
   see also individual disorders
parathyroidectomy 702
parathyroid hormone-like hormone (PTHLH) 664
parathyroid hormone (PTH) 205–211
   abaloparatide monotherapy 563
   amino acid sequences 206, *206*
   bone turnover markers 296–297
   calcium-sensing receptor 221, 226
   candidates for anabolic therapy 559
   cellular bioenergetics of bone 1008–1009
   childhood disorders of mineral metabolism 705–709
   chronic kidney disease 506, 508
   chronic kidney disease–mineral and bone disorder 695–698, 701–702
   combination therapy 561–562, 567–569
   concepts and definitions 205, 559
   daily versus cyclic teriparatide treatment 563
   discontinuation of antiresorptive therapy and sequential PTH 562
   discontinuation of PTH and sequential antiresorptive therapy 560–561
   familial states of primary hyperparathyroidism 629–631
   fetal and neonatal bone development 118
   fetal calcium metabolism 179–181, 183, *184*
   FGF23 and the regulation of phosphorus metabolism 190
   genes encoding human PTH 206, *207*, 212, *213*
   glucocorticoid-induced osteoporosis 468, 563
   gonadal steroids and osteoporosis pathogenesis 414
   hematopoietic stem cell niche 967–968
   hypocalcemia 646–652, **647**

parathyroid hormone (PTH) (cont'd)
  hypoparathyroidism 654–657
  immobilization and burns 731
  immunobiology 994–997, 996, 997
  magnesium homeostasis 174–175
  mechanism of action 208–209, 208
  non-parathyroid hypercalcemia 641–643
  nutritional support for osteoporosis 536
  oral manifestations of metabolic bone diseases 945
  osteoblasts 33, 57
  osteoclasts 50–51
  osteoporosis in men 562–563
  parathyroid disorders 613–618
  parathyroid hormone-related peptide 205–206
  parathyroid hormone treatment for osteoporosis 559–566
  postmenopausal osteoporosis 559–562
  pregnancy and lactation 147, 150–152
  premenopausal osteoporosis 440
  primary hyperparathyroidism 619–624
  pseudohypoparathyroidism 661–663, 666–669
  rechallenge with parathyroid hormone 563
  regulation of calcium homeostasis 166–170, 167
  regulation of PTH secretion 207–208
  rickets and osteomalacia 686, 689, 692
  skeletal healing/repair 105, 109–110
  synthesis and secretion 206–207
  teriparatide monotherapy 559–561
  transcriptional profiling 370–371
  transplantation osteoporosis 424, 432
  vitamin D 231–232
parathyroid hormone-related protein (PTHrP) 212–220
  amino acid sequences 206, 206
  calcium-sensing receptor 226
  cancer 217
  concepts and definitions 212
  craniofacial morphogenesis 897
  disease roles 217
  endochondral ossification 16
  fetal and neonatal bone development 118
  fetal calcium metabolism 180–181, 183, 184
  genes encoding human PTHrP 206, 207, 212, 213
  hematological malignancies 761, 764–765
  immunobiology 994, 996–997, 997
  mammary gland 215–216, 216
  medical prevention and treatment of metastatic bone disease 799–800
  metastatic bone disease 817–818
  non-parathyroid hypercalcemia 641, 643
  nuclear PTHrP 214
  osteoblasts 33, 57
  osteoclasts 50, 52
  osteoporosis 217
  pancreatic islets 216–217
  parathyroid hormone 205–206
  physiological functions 214–217
  placenta 216
  polyhormone 213
  pregnancy and lactation 147, 150–152
  pseudohypoparathyroidism 661–663
  PTHrP receptors 213–214, 214
  regulation of calcium homeostasis 169–170
  skeletal dysplasias 217
  skeleton 214–215, 215
  smooth muscle and the cardiovascular system 216
  teeth 216
  tumor-induced bone disease 752–754, 756
Parkinson disease (PD) 733
PASS see postauthorization safety studies
Patched-1 (Ptch1) 56–57, 922–923
pathology of the hard tissues of the jaws 918–926
  central giant cell lesion 923–925, 924
  concepts and definitions 918
  developmental odontogenic cysts 920–923, 922
  inflammatory odontogenic cysts 920, 921
  odontogenic neoplasms 920, 923, 924
  tooth demineralization, dental caries, and root resorption 918–920, 919
PATH study 570
patient-derived xenograft (PDX) models 756
pattern-recognition receptor (PRR) 935–936, **935**
Pax1 4
PAX9 see paired box 9
PBM *see* peak bone mass
PCA *see* principal component analysis
PCP *see* planar cell polarity
PCR *see* polymerase chain reaction
PD *see* Parkinson disease
PDD *see* progressive diaphyseal dysplasia
PDL *see* periodontal ligament
PDX *see* patient-derived xenograft
PEA *see* phosphoethanolamine
peak bone mass (PBM)
  epidemiology of osteoporotic fractures 402
  mechanical loading 141–142, 143
  nutritional support for osteoporosis 534–535, **535**
peak height velocity (PHV) 141, 142
pentose phosphate pathway (PPP) 1004–1006
PEPI trial 541
periapical cyst 920, 921
periapical granuloma 920, 921
pericytes 983–984
peri-implantitis 953–954, 954
perinatal and infantile hypophosphatasia 119–120
perinatal hypophosphatasia (HPP) 886
periodontal diseases
  alveolar bone homeostasis 933–940
  anatomy of the periodontium 933–934, 934
  concepts and definitions 933
  diagnosis 936–937
  future perspectives 938
  gingivitis treatment 937
  historical and current perspectives 934–935
  host immune response in periodontal tissues 936
  immunobiology 1001
  maintenance therapy 938
  molecular basis of host–microbe interactions 935–936, **935**
  pathogenesis of periodontitis 934–936
  periodontitis treatment 937–938
  treatment 937–938
periodontal ligament (PDL) 901, 903, 907–908
periosteal apposition 127–129, 127, 158, 159
periosteal expansion 198
periosteum 26
peripheral nervous system (PNS)
  bone cancer pain 783–785
  concepts and definitions 1028
  dynamic innervation of bone 1029
  function of neurotransmitters in bone 1029–1034, **1030**

peripheral nerves in bone 1028–1029
peripheral neuronal control of bone remodeling 1028–1036
  research directions 1033–1034
peripheral quantitative computed tomography (pQCT)
  animal models: allelic determinants for BMD 361
  bone mass measurement in children with osteoporosis risk factors 247
  diet and nutrition 138
  magnetic resonance imaging 274–275
  mechanical loading 143
perivascular cells 983–984
PET see positron emission tomography
PG see prostaglandins
pharmacogenetics 389–390
phenotyping
  animal models: allelic determinants for BMD 359–361, 360
  Fracture Liaison Service 407
  genome-wide association studies 379, 380
pheochromocytoma 641
Phex/Phex see phosphate-regulating neutral endopeptidase on the chromosome X
phosphatase and tensin homolog (PTEN) 770
phosphate-regulating neutral endopeptidase on the chromosome X (Phex)
  disorders of phosphate homeostasis 678
  FGF23 and the regulation of phosphorus metabolism 189
  genetic craniofacial disorders affecting dentition 913–914
  osteocytes 38, 43
phosphaturic mesenchymal tumor, mixed connective tissue type (PMTMCT) 675
phosphoethanolamine (PEA) 887
phosphorus
  autosomal dominant hypophosphatemic rickets 675, 679–680
  autosomal recessive hypophosphatemic rickets 679
  bone histomorphometry 313–314
  bone stress injuries 454
  characteristics of renal phosphate wasting disorders 677
  childhood disorders of mineral metabolism 708–709, 710
  chronic kidney disease 506
  chronic kidney disease–mineral and bone disorder 699–702
  concepts and definitions 187, 674
  cutaneous skeletal hypophosphatemia syndrome 680
  disorders associated with decreased FGF23 activity 190–192
  disorders associated with increased FGF23 activity 189–190
  disorders of phosphate homeostasis 674–683
  enzyme deficiencies 886–888
  familial states of primary hyperparathyroidism 629
  FGF23 activity 188–192, 674–681
  FGF23 and the regulation of phosphorus metabolism 187–193
  FGF23-associated syndromes 189–192, **191**
  FGF23 gene and protein 188
  FGF23 receptors 188–189
  fibrous dysplasia 679–680
  genetic craniofacial disorders affecting dentition 911
  hereditary hypophosphatemic rickets with hypercalciuria 679, 709
  molecular mechanisms of hypophosphatemia 676
  other disorders of renal phosphate wasting 680
  phosphate metabolism 187–188
  primary hyperparathyroidism 625
  pseudohypoparathyroidism 668
  regulation of FGF23 *in vivo* 188
  rickets and osteomalacia 686–687, 687
  sclerosing bone disorders 829

  serum assays 189
  therapy for FGF23-mediated hypophosphatemic disorders 190
  tumor-induced osteomalacia 674–678
  vitamin D 233, 235
  X-linked hypophosphatemic rickets 678–679
  see also hyperphosphatemia; hypophosphatemia
PHP see pseudohypoparathyroidism
PHPT see primary hyperparathyroidism
PHV see peak height velocity
physical examination 344
physical exercise see exercise
physiotherapy 422, 456
PI3k/Akt-dependent signaling pathway 33, 51, 57
PI3k/mTOR signaling pathway 770–771, 1006–1009
PICS study 568
Pierre Robin sequence 898
pioglitazone 960–961
pituitary–bone axis 1037–1046
  adrenocorticotrophic hormone 1037–1038, 1041
  concepts and definitions 1037
  follicle-stimulating hormone 1038–1040
  growth hormone 1038
  oxytocin 1041–1043
  pituitary hormone receptors and ligands in bone 1037–1038
  prolactin 1041
  thyroid-stimulating hormone 1038, 1040–1041
  vasopressin 1042–1043
pituitary cyclase activating polypeptide (PACAP) **1030**, 1033
Pitx2/*PITX2* 902
PKG see protein kinase G
placenta
  calcium-sensing receptor 226
  fetal calcium metabolism 180–181, 183
  parathyroid hormone-related protein 216
planar cell polarity (PCP) 8, 9
plasma membrane calcium pump (PMCA) 235
plates and screws 589–590
pleiotropy 362
PLP see pyridoxal 5′-phosphate
PMCA see plasma membrane calcium pump
PMN see polymorphonuclear leukocytes
PMTMCT see phosphaturic mesenchymal tumor, mixed connective tissue type
PNS see peripheral nervous system
POC see primary ossification center
POEMS syndrome 761
POH see progressive osseous heteroplasia
polymerase chain reaction (PCR) 374–375
polymorphonuclear leukocytes (PMN) 950
polymyalgia rheumatica 468
POMC see proopiomelanocortin
Porcupine 70
positive allosteric modulators (PAM) 224–225
positron emission tomography (PET)
  adipocytes 980
  hematological malignancies 763
  metabolic bone disease 303–308
  metastatic bone disease 747–748
postauthorization safety studies (PASS) 580
postmenopausal osteoporosis
  antisclerostin therapy 606–607
  bisphosphonates for postmenopausal osteoporosis 545–552
  central neuronal control of bone remodeling 1024
  combination therapy 569–571

postmenopausal osteoporosis (cont'd)
  denosumab 553–555
    parathyroid hormone treatment for osteoporosis 559–562
    pituitary–bone axis 1039–1042
    strontium ranelate and calcitonin 574, 576–577
PPARγ agonists 961
PPARγ pathway 488
PPHP see pseudopseudohypoparathyroidism
PPi see inorganic pyrophosphate
PPI see proton pump inhibitors
PPP see pentose phosphate pathway
pQCT see peripheral quantitative computed tomography
prebiotics 537
prednisone 468
pregnancy and lactation 147–154
  adolescent pregnancy and lactation 152
  calcium-sensing receptor 226
  cell biology in pregnancy 147–150, 148
  diet and nutrition 136–138, 147–152
  familial hypocalciuric hypercalcemia 149
  fetal calcium metabolism 179–186
  hypoparathyroidism and pseudohypoparathyroidism 149, 152
  implications 152
  intestinal absorption of calcium 147–148, 150
  lactation/breastfeeding 137–138, 150–152
  low calcium intake 150, 152
  mineral ions and calciotropic hormones 147, 150
  osteoporosis in pregnancy 149
  osteoporosis of lactation 151–152
  parathyroid hormone-related protein 215–216, 216
  premenopausal osteoporosis 437
  primary hyperparathyroidism 149
  renal handling of calcium 148, 150
  skeletal calcium metabolism 148–149, 150–151
  vitamin D 236
  vitamin D deficiency and insufficiency 149–150, 152
preimplantation genetic diagnosis 348
premenopausal osteoporosis 436–442
  bisphosphonates 439
  concepts and definitions 436
  denosumab 440
  evaluation 438–439, **439**
  glucocorticoid-induced osteoporosis 440
  human PTH(1–34) 440
  idiopathic osteoporosis 438
  management issues 439–440
  pregnancy and lactation 437
  secondary causes of osteoporosis in premenopausal women 437–438, **438**
  special issues related to BMD interpretation 437
  women with a history of low-trauma fracture 436
  women with low BMD 436–437, *437*
prenatal genetic testing 348
presomitic mesoderm (PSM) 4, *5*
primary hyperparathyroidism (PHPT) 619–628
  Asia and Africa 626
  biochemical hallmarks 622
  bone histomorphometry 313
  bone mass/mineral density 622–623, *622*, 626
  bone turnover markers 294
  classical and asymptomatic PHPT 621
  clinical forms 621
  concepts and definitions 619
  diagnosis and evaluation 621–623

  etiology and pathogenesis 619–620
  Europe 625
  genetics 620–621, **620**
  imaging guidelines **622**
  immunobiology 994–996, *996*
  Latin America 626
  multiple endocrine neoplasia 619–621
  nonsurgical management 624–625
  oral manifestations of metabolic bone diseases 945
  parathyroid disorders 613–615
  pregnancy and lactation 149
  premenopausal osteoporosis 439
  renal involvement 623
  signs and symptoms 621
  surgical procedures 623–624, *624*
  treatment 623–625
  see also familial states of primary hyperparathyroidism
primary hypertrophic osteoarthropathy 835
primary ossification center (POC) 13
primitive streak (PS) 893
principal component analysis (PCA) 361
PRKAR1A/*PRKAR1A* 668
PRL see prolactin
probiotics 994, *995*
progressive diaphyseal dysplasia (PDD) 829–830, *830*
progressive osseous heteroplasia (POH) 868
  pseudohypoparathyroidism 666–667
  signal transduction cascades controlling osteoblast differentiation 56–57
  skeletal morphogenesis in embryonic development 7
progressive resistance training (PRT) 518–521, **520**, 523
prolactin (PRL) 1037, 1041, 1051
proopiomelanocortin (POMC) 1037
prostaglandins (PG)
  hematopoietic niche 968
  mechanotransduction 79, 81–82
  muscle–bone interactions 1057, *1058*
prostate cancer
  additional therapies for advanced prostate cancer 778–779
  bone cancer pain 783–785
  castration-sensitive/resistant prostate cancer 801
  concepts and definitions 775
  endocrine therapy 776
  medical prevention and treatment of metastatic bone disease 800–804
  nonmetastatic prostate cancer 776
  osteolytic and osteoblastic skeletal lesions 739, 740
  radiotherapy-induced osteoporosis 788
  role of ADT and LHRH agonists and antagonists 778
  screening for osteoporosis 776–777, *777*
  skeletal complications of anticancer therapies 775–780
prostate-specific antigen (PSA) 778, 801
protein, dietary 536, 723
protein kinase G (PKG) 78
proteoglycans 86, **87**
proton pump inhibitors (PPI) 483
PRR see pattern-recognition receptor
PRT see progressive resistance training
PRUNE2 620
Prx1 26
PS see primitive streak
PSA see prostate-specific antigen
pseudohypocalcemia 649
pseudohypoparathyroidism (PHP) 661–673

acrodysostosis 667–668
Albright hereditary osteodystrophy 664, *664*, *665*, *666*, 668
classification systems 661
cognitive function 665
concepts and definitions 661
endocrine disorders 668–669
fetal calcium metabolism 182
GNAS locus 663
management 668–669
molecular diagnosis of PHP1A 665, *666*
molecular diagnosis of PHP1B 667
obesity 664–665, 668–669
other features 665
parathyroid disorders 616
parathyroid hormone 208
PHP1A and related disorders 663–665
PHP1B and related disorders 667
PHP1C and related disorders 665–666
pregnancy and lactation 149, 152
progressive osseous heteroplasia 666–667
pseudopseudohypoparathyroidism or isolated AHO 666
related disorders and differential diagnosis 661–668, **662**, *666*
pseudopseudohypoparathyroidism (PPHP) 666
PSNS *see* parasympathetic nervous system
psoriatic arthritis 459
Ptch1/*PTCH1 see* patched-1
PTEN/*PTEN see* phosphatase and tensin homolog
PTH(1-34)
    alveolar bone homeostasis 938
    hypoparathyroidism 656–657
    integrative physiology of the skeleton 960
    non-parathyroid hypercalcemia 642
PTH(1-84)
    combination therapy 568
    hypocalcemia 652
    hypoparathyroidism 657, 658
    non-parathyroid hypercalcemia 642
    osteoporosis monotherapy 560–561
    parathyroid disorders 615
    parathyroid hormone treatment for osteoporosis 560–561
PTH *see* parathyroid hormone
PTHLT *see* parathyroid hormone-like hormone
PTHrP/*Pthrp see* parathyroid hormone-related protein
puberty
    bone morphology 127–129, *127*, *128*
    gonadal steroids 197–198
purines **1030**, 1033
pyknodysostosis 47, 603, 829
Pyle disease 70
pyridoxal 5′-phosphate (PLP) 887
pyrimidines **1030**, 1033

QALY *see* quality-adjusted life years
QCT *see* quantitative computed tomography
QFracture **337**
QM *see* quantitative morphometry
QoL *see* quality of life
QTL *see* quantitative trait loci
quality-adjusted life years (QALY) 598–600
quality of life (QoL) 621, 658
quantitative computed tomography (QCT)
    bone mass measurement in adults 260–261, *261*, *262*, 265–266
    finite element analysis 264–265, *264*, *265*, 267–268
    Fracture Risk Assessment Tool 334–335

glucocorticoid-induced osteoporosis 469
gonadal steroids and osteoporosis pathogenesis 412–413
hematological malignancies 764
metastatic bone disease 744
oral manifestations of metabolic bone diseases 942
precision errors **263**
see also high-resolution peripheral quantitative computed tomography; peripheral quantitative computed tomography
quantitative morphometry (QM) 323–325, *325*
quantitative trait loci (QTL) 359, 361–363, 386–387, **388**
quantitative ultrasound (QUS) 257
QUEST study 575
QUS *see* quantitative ultrasound

RA *see* retinoic acid; rheumatoid arthritis
racial differences *see* ethnicity
radiofrequency ablation 676, 819
Radiographic Union Score for Tibial (RUST) 111, *111*
radiography
    bisphosphonates for postmenopausal osteoporosis *549*
    bone stress injuries 454
    bone vasculature *988*
    childhood disorders of mineral metabolism *710*
    enzyme deficiencies 886–887, *887*
    fibrillinopathies 879–881
    fibrodysplasia ossificans progressiva 867
    fibrous dysplasia of bone *843*
    genetic craniofacial disorders affecting dentition *914*
    ischemic and infiltrative disorders of bone 854, *855–858*, 856–858
    juvenile dermatomyositis 863, *863*
    juvenile osteoporosis 420, *420*
    metastatic bone disease 743, *818*, *819*
    oral manifestations of metabolic bone diseases 941–942, *942*
    osteochondrodysplasias 849–850, *849–852*
    osteogenesis imperfecta 873
    osteonecrosis of the jaw 929, *929*
    Paget disease of bone 716, 717
    pathology of the hard tissues of the jaws 920, *921*
    rickets and osteomalacia 689
    sclerosing bone disorders 826–835, *827*, *830–833*
    tumoral calcinosis 861, *862*
    vertebral fracture 319–321, *320–322*
radionuclide bone scintigraphy (RBS) 747
radiotherapy
    bone-seeking radiopharmaceuticals 812–813
    childhood cancer 793–796
    concepts and definitions 809
    external beam radiotherapy 809–812
    impending fracture and risk prediction 810–811
    neuropathic pain and spinal cord compression 810
    pain flare 811
    pain palliation 809–811, **810**
    reirradiation 811
    skeletal complications of anticancer therapies 775, 779
    skeletal metastases 809–815
    surgical treatment of metastatic bone disease 819–820
    survival prediction 811–812
radiotherapy-induced osteoporosis 788–792
    bone marrow adiposity 790
    bone vasculature 790
    changes in bone quantity and quality after radiation 788–789
    concepts and definitions 788
    fracture risk following radiotherapy 788

radiotherapy-induced osteoporosis (cont'd)
  loss of bone strength in animal models 789
    osteoblasts 789
    osteoclasts 789
    osteocytes 789–790
radiotracers 302, 303–308
radium-223 779, 800–805, **802–803**, 812, 818
Raine syndrome (RAS) 680
raloxifene 471, 542, 561–562, 569–571, 625
randomized controlled trials (RCT)
  adverse effects of drugs for osteoporosis 579–585
  antisclerostin therapy 605–608
  estrogens, SERMs, and TSEC 541
  exercise for osteoporotic fracture prevention and management 518
  Fracture Liaison Service 405–406
  nutritional support for osteoporosis 535
  prevention of falls 527–530
  transplantation osteoporosis **427–428**
Rank1 1016
RANKL *see* receptor activator of nuclear factor κ B ligand
RARγ *see* retinoic acid receptor γ
RAS *see* Raine syndrome
Ras-ERK MAPK signaling pathway 57, 925
RBPjk/*Rbpjk* 15–16
RBS *see* radionuclide bone scintigraphy
RCC *see* renal cell carcinoma
RCT *see* randomized controlled trials
reactive arthritis 459
reactive oxygen species (ROS)
  integrative physiology of the skeleton 961
  menopause and age-related bone loss 156–157
  radiotherapy-induced osteoporosis 790
receptor activator of nuclear factor κ B ligand (RANKL)
  adipocytes 978, 980
  alveolar bone homeostasis 936
  bone cancer pain 781
  bone vasculature 989
  cellular bioenergetics of bone 1009
  denosumab 553–554, *554*, 556
  fibrous dysplasia of bone 842
  genome-wide association studies 381
  glucocorticoid-induced osteoporosis 468, 470–471
  gonadal steroids 195–196
  gonadal steroids and osteoporosis pathogenesis 414–415
  hematological malignancies 761–762, 764
  immobilization and burns 733–734
  immunobiology 992–995, 997–1001
  inflammation-induced bone loss 460–462
  integrative physiology of the skeleton 961, 963
  medical prevention and treatment of metastatic bone disease 799–800
  metastatic bone disease 817–818
  non-parathyroid hypercalcemia 641
  osteoblasts 32–34
  osteoclasts 46–52
  osteocytes 40, 43
  osteolytic and osteoblastic skeletal lesions 740–741
  parathyroid hormone-related protein 215
  peripheral neuronal control of bone remodeling 1031–1032
  pituitary–bone axis 1041–1042
  regulation of calcium homeostasis 169–170
  sclerosing bone disorders 828
  secondary causes of osteoporosis 510, 512–513

  skeletal healing/repair 102–104
  translational genetics of osteoporosis 387
  tumor-induced bone disease 752, 754, 756
  vitamin D 234
receptor tyrosine kinase (RTK) 78
RECQL4/*RECQL4* 770
reference point indentation (RPI) 287–292
  applications 289–290
  clinical studies in humans 288–289
  concepts and definitions 287
  cyclic and impact modalities 287–290
  future needs and directions 290
  microindentation instruments 287–288, *288*
  potential advantages or RPI technology 290
  preclinical studies in animals 288
regional pain syndrome 989
regulated intramembrane proteolysis (RIP) pathway 875
regulated marrow adipocytes (rMAT) 976, *976*
regulatory T cells (Treg) 992–993
rehabilitation 731–733
reirradiation 811
renal cell carcinoma (RCC) 806
renal disease 583
renal osteodystrophy
  bone histomorphometry 314
  bone turnover markers 295
  chronic kidney disease–mineral and bone disorder 695
  nuclear medicine 307
  oral manifestations of metabolic bone diseases 945
renal tubular acidosis (RTA) 828–829
repeat expansions 374
REPLACE trial 657, 658
reproductive counseling 376–377
resistance exercise *see* exercise
retinoic acid (RA) 4, 5, 642
retinoic acid receptor γ (RARγ) 868
retinoid X receptor (RXR) 224, 233
RET/*RET* 633–634
RGD-containing proteins 88–90, **89**
rheumatoid arthritis (RA)
  glucocorticoid-induced osteoporosis 468
  immunobiology 998–999
  inflammation-induced bone loss 459–462
  secondary causes of osteoporosis 511
rhupus syndrome 459
rib fracture 788
rickets 684–694
  bone histology 684–685, 689–690
  bone stress injuries 454
  childhood disorders of mineral metabolism 709, 710, *710*
  clinical picture 688, *688*
  concepts and definitions 684–685
  diagnosis 689–690
  disorders of phosphate homeostasis 675, 678–680
  enzyme deficiencies 886–888
  epidemiology 685
  etiology 685, **686**
  FGF23 and the regulation of phosphorus metabolism 189–190
  genetic craniofacial disorders affecting dentition 914
  genetic testing 375
  hereditary vitamin D-dependent and resistant rickets 691–692
  laboratory findings 689
  muscle–bone interactions 1057
  nuclear medicine 307–308

osteocytes 42–43
pathophysiology 685, 687
radiographic findings 689, 690
sclerosing bone disorders 826, 828, 829
treatment and prevention 691
see also individual disorders
Rieger syndrome 902
ring finger 43 (Rnf43) 72
RIP *see* regulated intramembrane proteolysis
risedronate 547, 717–719, 945–946
rMAT *see* regulated marrow adipocytes
RNA sequencing 367–370
Rnf43 *see* ring finger 43
romosozumab 345, 396, 585, 606–607
root resorption 918–920, 919
ROS *see* reactive oxygen species
rosiglitazone 960–961
Rothmund–Thomson syndrome 770
RPI *see* reference point indentation
RTA *see* renal tubular acidosis
RTK *see* receptor tyrosine kinase
runt-related transcription factor (RUNX2/3)
chronic kidney disease–mineral and bone disorder 699
diabetes and bone loss 487
endochondral ossification 15–17
endocrine bioenergetics of bone 1016–1017
fetal and neonatal bone development 120
hematological malignancies 763
inflammation-induced bone loss 462
integrative physiology of the skeleton 960–962
osteoblasts 33–35, 54, 55
osteoprogenitor cells 22–23
skeletal healing/repair 103
skeletal morphogenesis in embryonic development 7
RUST *see* Radiographic Union Score for Tibial
RXR *see* retinoid X receptor

S100A8 196
Saethre–Chotzen syndrome (SCS) 6
salt, dietary 138, 536, 723
samarium-153 800–801
Sanger sequencing 346
sarcoidosis 641
sarcopenia
adverse health outcomes 501–502
clinical implications of sarcopenia and osteoporosis 498–501
combined therapeutic interventions 502
common mechanisms in sarcopenia and osteoporosis 498, 499
dysmobility syndrome and metabolic syndrome 500–502, 501
Fracture Risk Assessment Tool 500–501, 500
muscle–bone interactions 1055, 1056
obesity and skeletal health 493
osteoporosis 498–504
osteosarcopenia 499–502
prevention of falls 527–528
SARM *see* selective androgen receptor modulators
SBRT *see* stereotactic body radiation therapy
SBT *see* small bowel transplantation
scanning electron microscopy (SEM) 715
SCC *see* spinal cord compression
Scheuermann disease 320, 322
schizophrenia 1050
Schmorl's nodes 321, 322

SCI *see* spinal cord injury
scintigraphy
metabolic bone disease 305, 307–308
radionuclide bone scintigraphy 747
Scleraxis (Scx) 352, 356
sclerosing bone disorders 825–838
additional conditions 836
axial osteomalacia 834
carbonic anhydrase II deficiency 828–829
concepts and definitions 825
endosteal hyperostosis 830–832, 831
fibrogenesis imperfecta ossium 834–835
hepatitis C-associated osteosclerosis 835–836
melorheostosis 833–834, 833
osteopathia striata 832–833
osteopetrosis 825–828, 827
osteopoikilosis 832, 832
pachydermoperiostosis 835
progressive diaphyseal dysplasia 829–830, 830
pycnodysostosis 829
see also individual disorders
sclerosteosis 71–72, 831–832, 831
sclerostin (SOST)
bone turnover markers 297, 298
chronic kidney disease–mineral and bone disorder 695, 698
immobilization and burns 732
muscle–bone interactions 1056–1057
non-parathyroid hypercalcemia 642
osteocytes 38–41
radiotherapy-induced osteoporosis 790
skeletal healing/repair 105
Wnt/β-catenin signaling pathway 71, 72
SCS *see* Saethre–Chotzen syndrome
SCT *see* stem cell transplantation
Scx/*Scx* see Scleraxis
secondary hyperparathyroidism (SHPT) 424, 426
secondary ossification centers (SOC)
endochondral ossification 13
fetal and neonatal bone development 117
TGF-β superfamily 61
secreted frizzled-related proteins (Sfrp) 70–71
secular trends 401, 401
selective androgen receptor modulators (SARM) 201
selective estrogen receptor modulators (SERM) 542–543
combination therapy 568
gonadal steroids 201
primary hyperparathyroidism 625
translational genetics of osteoporosis 389
selective serotonin reuptake inhibitors (SSRI)
central neuronal control of bone remodeling 1022
neuropsychiatric disorders 1048, 1050–1051
peripheral neuronal control of bone remodeling 1032
skeletal effects of medications for nonskeletal disorders 484
semiquantitative (SQ) assessment 323, 324, 325, 329
sequestasome-1 (SQSTM1) 713–714
SERM *see* selective estrogen receptor modulators
seronegative spondyloarthritis 459–460, 462–463
serotonin
central neuronal control of bone remodeling 1022
neuropsychiatric disorders 1050–1051
peripheral neuronal control of bone remodeling **1030**, 1032
SERPINF1 874
serum-derived proteins 85–86, **86**
serum total alkaline phosphatase (SAP) 715, 717–718

serum vitamin D binding protein (DBP) 230–233
severe neonatal hyperparathyroidism (SNHP) 707
sex differences
	bone morphology 127–129, *127*, *128*
	bone stress injuries 454, 456
	epidemiology of osteoporotic fractures 399, 402
	ethnicity and age-related loss of bone strength 132
	gonadal steroids and osteoporosis pathogenesis 412–416
	nutritional support for osteoporosis 535
sex hormone-binding globulin (SHBG) 194, 415
sex limitation 362–363
sex steroids *see* gonadal steroids
Sfrp *see* secreted frizzled-related proteins
SGLT-2 *see* sodium glucose cotransporter 2
SGS *see* Shprintzen–Golberg syndrome
Sh3bp2 925
SHBG *see* sex hormone-binding globulin
Shh *see* Sonic hedgehog
short rib polydactyly type II 851, *852*
short vertebral body height (SVH) 320, *321*
Shprintzen–Golberg syndrome (SGS) 882
SHPT *see* secondary hyperparathyroidism
SIBLING *see* small integrin-binding ligand
signal-to-noise ratio (SNR) 272–273, 274–275
Sillence classification 872–873, **872**
single gene testing 346
single nucleotide polymorphism (SNP)
	chromosomal abnormalities, CNVs, and mutations 346
	genome-wide association studies 378–379
	translational genetics of osteoporosis 386, 389
single-photon emission computed tomography (SPECT) 303–305, 747
skeletal dysplasias *see* osteochondrodysplasias
skeletal growth 123–130
	bone size, shape and microarchitecture 125–127, *126*
	diet and nutrition 135–140
	gonadal steroids 197–198
	growth of metaphyses and fractures in childhood 128
	intrauterine growth 123
	maturational stage-specificity of illness effects on bone morphology 128–129
	mechanical loading 141–146
	postnatal growth 123–125
	prepubertal growth 125–127
	puberty and sex differences in bone morphology 127–129, *127*, *128*
	variances in bone trait 123–125, *124*
skeletal healing/repair 101–107
	aging 104
	biomechanical assessment 108–109, *109*
	biomechanical stages of fracture healing 109–110, *109*
	biomechanics of fracture healing 108–114
	cell biology 101, 108
	cellular contribution to healing tissues 101–102
	cigarette smoking 105
	conditions that impair fracture healing and therapeutic modalities 104–105
	diabetes mellitus 104–105
	gene expression profile during fracture healing 103
	mechanobiology of fracture healing 111–112, *112*
	molecular control of fracture healing 103–104
	molecular therapies to enhance bone healing 105
	noninvasive assessment of fracture healing 110–111, *110*
	orthopedic principles of fracture management 588–589
	osteoprogenitor cells 26
	process of skeletal healing 101–104
	Radiographic Union Score for Tibial scores 111, *111*
	tissue morphogenesis during bone repair *102*
	transcriptional profiling 371
skeletal morphogenesis in embryonic development
	cell lineage contribution 3, *4*
	concepts and definitions 3
	early skeletal patterning 3–7
	embryonic cartilage and bone formation 7–9, *8*
	embryonic development 3–11
	limb patterning and growth 5, *6*
	somitogenesis 3–4, *5*
skeletal-related events (SRE)
	hematological malignancies 763–764
	medical prevention and treatment of metastatic bone disease 800–801, 805–806
	osteolytic and osteoblastic skeletal lesions 739
	tumor-induced bone disease 752, 756
skeletal stem cells (SSC) 20, 21–22
SLC34A2/*SLC34A2* 187
SLE *see* systemic lupus erythematosus
SLRP *see* small leucine-rich proteoglycans
SM *see* systemic mastocytosis
SMAD/*Smad* 14–16, 60
small bowel transplantation (SBT) 430
small integrin-binding ligand (SIBLING) family proteins 88–90, **89**, 189
small leucine-rich proteoglycans (SLRP) 86, 951–952
SMC *see* smooth muscle cells
smoking 105, 468
Smoothened (Smo) 56–57
smooth muscle cells (SMC) 216, 983, 989
SMURF1 1017
SNHP *see* severe neonatal hyperparathyroidism
SNP *see* single nucleotide polymorphism
SNR *see* signal-to-noise ratio
SNS *see* sympathetic nervous system
SOC *see* secondary ossification center
socioeconomic status 399, 402
sodium chloride *see* salt, dietary
sodium fluoride 395
sodium glucose cotransporter 2 (SGLT-2) inhibitors 483, 489
sodium/hydrogen exchanger isoform 3 (NHE3) 700
sodium-phosphate cotransporter (NPT2a/b/c) 187–189, 700
SOF *see* Study of Osteoporotic Fractures
soft drinks 138
somatic mosaicism 848
somatostatin 676
somitogenesis 3–4, *5*
Sonic hedgehog (Shh)
	craniofacial morphogenesis 895–897
	odontogenesis 902
	osteoblasts 33, 56–57
	skeletal morphogenesis in embryonic development 4
SOST/*Sost see* sclerostin
SOTI trial 573–574
SOX/*Sox see* Sry-box
Sp7 *see* Osterix
specific plaque hypothesis 935
SPECT *see* single-photon emission computed tomography
spheno-ethmoidial synchondrosis 897
spheno-occipital synchondrosis 897
spinal cord compression (SCC) 810

spinal cord injury (SCI) 730–732
spine sparing strategies 521, 522
splicing variants 373–374
SPM *see* statistical parameter mapping
spondyloarthritis 282–283
spondyloepimetaphyseal dysplasia 850, *850*
sporadic hypocalciuric hypercalcemia 631
SQ *see* semiquantitative
SQSTM1 713–714
SRE *see* skeletal-related events
Sry-box (SOX) factors
   craniofacial morphogenesis 898
   endochondral ossification 13–17
   hypocalcemia 648
   osteoprogenitor cells 23
   skeletal morphogenesis in embryonic development 6–7, 9
SSC *see* skeletal stem cells
SSKS *see* stiff skin syndrome
SSRI *see* selective serotonin reuptake inhibitors
STAND trial 555
STAT3 signaling pathway 755
statins 484
statistical parameter mapping (SPM) 268
statistical power 579
stem cell transplantation (SCT) 425, 430–431
stereotactic body radiation therapy (SBRT) 811
Stickler syndrome 849, *849*
stiff skin syndrome (SSKS) 882
Streptococcus mutans 918
STRO-1 22
stroke 732–733
strontium-89 800–801
strontium ranelate 573–574, 576–577
   adverse effects 574
   benefits and limitations **576**
   clinical efficacy 573–574
   pharmacology 573
Study of Osteoporotic Fractures (SOF) 436
subcutaneous ossifications 664–669, *665*
substance P 1029, **1030**, 1032–1033
sucroferric oxyhydroxide 700
sulfonylureasmetformin 489
supernumerary teeth 911–912
surgical resection 769, 929–930
survival prediction 811–812
SVH *see* short vertebral body height
SWAN study 132
sympathetic nervous system (SNS)
   central neuronal control of bone remodeling 1020–1022
   endocrine bioenergetics of bone 1016–1017
   hematopoietic stem cell niche 969
   integrative physiology of the skeleton 959, 962–963, *963*
   neuropsychiatric disorders 1047
   peripheral neuronal control of bone remodeling 1032–1033
symphalangism (SYM) 65
synonymous variants 373
synovial inflammation 459–463
systemic inflammatory disorders *see individual disorders*; inflammation-induced bone loss
systemic lupus erythematosus (SLE) 459, 463
systemic mastocytosis (SM) 856–857, *856*, *857*

T2DM *see* type 2 diabetes mellitus
tacrolimus 425

TAF *see* tenofovir alafenamide
TAL *see* thick ascending limb of Henle's loop
TAM *see* tumor-associated macrophages
tamoxifen 356, 776
tanezumab 784
tartrate-resistant acid phosphatase (TRAP)
   bone turnover markers 293, 295
   cathepsin K inhibitors 604
   chronic kidney disease 506
   radiotherapy-induced osteoporosis 789
TBI *see* traumatic brain injury
TBLH *see* total body less head
TBS *see* trabecular bone score
Tbx1/*Tbx1* 654, 903
TC *see* tumoral calcinosis
TCA *see* tricarboxylic acid
TCIRG1 defects 827–828
TDF *see* tenofovir disoproxil fumarate
temporal arteritis 468
tenascin C (Tnc) 12
tendon 352
tenofovir alafenamide (TAF) 475, 476
tenofovir disoproxil fumarate (TDF) 475–477
tension band plates 590
teriparatide (TPTD) 298, 370–371, 395, 447, 470
   adverse effects 583–584
   chronic kidney disease 507–508
   combination therapy 555, 561–562, 568–571, *569*, *570*
   daily versus cyclic treatment 563
   enzyme deficiencies 887–888
   glucocorticoid-induced osteoporosis 563
   immobilization and burns 732
   osteoporosis in men 562–563
   parathyroid hormone treatment for osteoporosis 559–563
   postmenopausal osteoporosis 559–562
   radiotherapy-induced osteoporosis 789
testosterone 415–416, 425, 426, 429
testosterone replacement therapy (TRT)
   bone turnover markers 298
   glucocorticoid-induced osteoporosis 471
   osteoporosis in men 445, 447
tetracycline antibiotics 315
tetracycline double labeling 439, 689–690
TGF-β *see* transforming growth factor-β
Th *see* helper T cells
thiazide diuretics 484, 642, 656–657
thiazolidinediones (TZD) 483, 488, 960–961, 963
thick ascending limb of Henle's loop (TAL) 174, *175*
three-point bending test 97–98, *97*
thyroid-stimulating hormone (TSH) 663, 795, 1037–1038, 1040–1041
TIBD *see* tumor-induced bone disease
TID *see* total indentation distance
Tiki 71
TIMP *see* tissue inhibitors of metalloproteinases
TIO *see* tumor-induced osteomalacia
TIP39 *see* tuberoinfundibular peptide of 39 amino acids
tissue hypoxia 674
tissue inhibitors of metalloproteinases (TIMP) 936, 950
tissue-nonspecific isoenzyme of alkaline phosphatase (TNSALP) 120, 886–888
tissue-selective estrogen complex (TSEC) 543
titanium dioxide 952
TLR *see* toll-like receptors; Toll-like receptors

Tnc *see* tenascin C
TNF-α *see* tumor necrosis factor
TNF receptor-associated factor (TRAF) 48
TNFRSF11A/B 387
TNSALP/*TNSALP* *see* tissue-nonspecific isoenzyme of alkaline phosphatase
Toll-like receptors (TLR) 232, 935–936
Too Fit to Fracture initiative 521
tooth demineralization 918–920, *919*
tooth development *see* odontogenesis
toremifene 777
torque–twist curves 108–109, *109*
total body less head (TBLH) 243–245, *244*
total indentation distance (TID) 288
TP53 770
Tph1/*Tph1* and Tph2/*Tph2* 1022
trabecular bone
　adipocytes 978, *979*
　animal models: allelic determinants for BMD 361, 363
　bone histomorphometry 311
　bone mass measurement in adults 263–264
　magnetic resonance imaging 272–273, *273*, 274–275
　menopause and age-related bone loss 156–158, *158*
　radiotherapy-induced osteoporosis 789
trabecular bone score (TBS) 277–286
　clinical assessment 277–282
　clinical practice 282–284, *283*
　concepts and definitions 277
　cross-sectional studies 277–278
　diabetes mellitus 281, 283
　effect of different therapies 282
　fracture risk 277–282
　Fracture Risk Assessment Tool 278–281, 283
　longitudinal studies 278–281, **279–280**
　long-term glucocorticoid exposure 282
　primary hyperparathyroidism 623
　sarcopenia and osteoporosis 500, *500*
　technical aspects 277
TRAF *see* TNF receptor-associated factor
transcriptional profiling 367–372
　age- and location-dependent changes 369
　changes during bone development 368–369
　concepts and definitions 367
　fracture healing 371
　mechanical loading-induced changes 369–370
　osteoporosis 370–371
　skeletal cells and bone metabolism 367–368
transduction 42
transforming growth factor-β (TGF-β) 60–67
　activins 60
　ACVR1/ALK-2 63–64
　BMP-2 62–63
　BMP-4 63
　BMP-5/-6/-7 63, 65
　BMP-13/GDF-6 and BMP-14/GDF-5/CDMP-1 63, 65
　BMPR1A/ALK-3 64–65
　BMPR2 64
　bone development 60–65
　bone maintenance 62, 65
　bone morphogenic protein signaling 60, 62–65
　cell biology 60, *61*
　disease connections 62, 64–65
　endochondral ossification 14–15
　fibrillinopathies 878–880, 883–884
　hematological malignancies 763
　metastatic bone disease 817–818
　muscle–bone interactions 1057–1059
　osteoblasts 33, 35, 56
　sclerosing bone disorders 829–830
　skeletal healing/repair 103
　skeletal morphogenesis in embryonic development 8
　TGF-β1 61
　TGF-β2 61–62
　TGF-β3 62
　TGFβR1/ALK-5 62
　TGFβR2 62
　TG/KO phenotypes of ligands 60–63
　TG/KO phenotypes of receptors and signaling molecules 62, 63–64
　tumor-induced bone disease 754, 756
transient hypocalcemia of the newborn 705
transient receptor potential vanilloid (TRPV) family 168–169, 225–226, 235, 783, 785
translational genetics of osteoporosis 385–392
　candidate gene studies 386, **386**
　clinical application 388–390
　concepts and definitions 385
　genetics of osteoporosis 385–388
　genome-wide association studies 386–387, **388**
　individualized risk assessment 388–389
　linkage studies 386, **387**
　pathways 387–388
　pharmacogenetics 389–390
　whole genome sequencing 387
transplantation osteoporosis 424–435
　cardiac transplantation 429
　concepts and definitions 424
　diagnostic strategies 426, **427–428**
　intestinal transplantation 425
　kidney–pancreas transplantation 429
　kidney transplantation 426, 429
　liver transplantation 430
　lung transplantation 429
　management 426–431
　preexisting bone disease 424–425
　risk factors for bone loss and fractures **425**
　skeletal effects of immunosuppressive drugs 425–426
　small bowel transplantation 430
　stem cell transplantation 425, 430–431
traumatic brain injury (TBI) 1020–1021
Treg *see* regulatory T cells
tricarboxylic acid (TCA) cycle 1004–1006, 1008–1009, 1014–1015
tricho-rhino-phalangeal syndrome (TRPS) 668
TrkA *see* tyrosine kinase receptor type 1
TROPOS study 574
TRPM6/*TRPM6* 173–175
TRPS *see* tricho-rhino-phalangeal syndrome
TRPV *see* transient receptor potential vanilloid
TRT *see* testosterone replacement therapy
T-scores
　bisphosphonates for postmenopausal osteoporosis 548
　bone mass measurement in adults 252–254, **253**
　Fracture Risk Assessment Tool 335
　osteoporosis in men 445–446
　sarcopenia and osteoporosis 499–501
TSEC *see* tissue-selective estrogen complex
TSH *see* thyroid-stimulating hormone
tuberoinfundibular peptide of 39 amino acids (TIP39) 206, *206*, *207*

tumoral calcinosis (TC) 861–862
  biochemical and histopathological findings 861–862
  clinical features 861
  etiology and pathogenesis 862
  FGF23 and regulation of phosphorus metabolism 190–191
  radiographic findings 861, 862
  treatment 862
tumor-associated macrophages (TAM) 753–754
tumor dormancy 754–755
tumor-induced bone disease (TIBD) 752–759
  concepts and definitions 752–753
  importance of the problem 752, 753
  mesenchymal lineages 754
  myeloid lineage cells 753–754
  new models of TIBD 755–756
  physical microenvironment 754
  tumor dormancy 754–755
  tumor interactions with bone/bone marrow microenvironment 753–754
  vicious cycle hypothesis 752
tumor-induced osteomalacia (TIO)
  bone stress injuries 454
  clinical and biomedical manifestations 674–675
  disorders of phosphate homeostasis 674–678
  FGF23 and the regulation of phosphorus metabolism 189
  radiographic and histological features 677
  treatment 675–678
tumor necrosis factor-α (TNF-α)
  gonadal steroids and osteoporosis pathogenesis 415
  immunobiology 993–994, 999–1001
  inflammation-induced bone loss 460–462
  osteoclasts 50–51
  osteolytic and osteoblastic skeletal lesions 741
  secondary causes of osteoporosis 511
  skeletal healing/repair 101, 103–104
twin/family studies 359, 378, 385
type 1 diabetes mellitus (T1DM)
  bone turnover markers 295
  diabetes and bone loss 487–488
  peripheral neuronal control of bone remodeling 1029
type 2 diabetes mellitus (T2DM)
  bone turnover markers 295
  diabetes and bone loss 487–488
  endocrine bioenergetics of bone 1012–1013
  peripheral neuronal control of bone remodeling 1029
  reference point indentation 289
tyrosine kinase receptor type 1 (TrkA) 1023, 1029
TZD see thiazolidinediones

UCMA see unique cartilage matrix protein
ultrasound 257, 723–724
unique cartilage matrix protein (UCMA) 90
uridine triphosphate (UTP) 1033
urine crystal analysis 724–726, 725
urine pH 727
UTP see uridine triphosphate

van Buchem disease 71, 831
Vangl2 protein 8, 9
vascular calcification 698–699
vascular cell adhesion molecule-1 (VCAM-1) 22
vascular diseases 989
vascular endothelial growth factor (VEGF) 103–104, 951, 1041
vascular smooth muscle cells (VSMC) 216

vasculature see bone vasculature
vasoactive intestinal peptide (VIP) 1029, **1030**, 1033
vasoactive intestinal polypeptide (VIP)-oma syndrome 641
vasopressin (AVP) 1037, 1042–1043
vBMD see volumetric bone mass/mineral density
VCAM-1 see vascular cell adhesion molecule-1
VDDR see vitamin D-dependent rickets
VDRE see vitamin D response elements
VDR/Vdr 182
VE see visceral endoderm
vegetables 136, 536
VEGF see vascular endothelial growth factor
Venus flytrap (VFT) domain 222–223, 223
VERT see Vertebral Efficacy with Risedronate Therapy
vertebral fracture assessment (VFA)
  bone mass measurement in adults 255–256, 255, **256**
  bone mass measurement in children with osteoporosis risk factors 248, 248
  Fracture Liaison Service 407
  trabecular bone score 278
vertebral fractures 319–330
  algorithm-based qualitative assessment 325
  childhood cancer 794–795
  clinical detection 319
  clinical significance 319
  computed tomography 327–329, 328
  Cupid's bow deformity 321, 322
  defining a vertebral fracture 323
  detection 319–327
  diagnostic problems 320–321
  differentiating osteoporotic from neoplastic fractures 328–329, 329, **329**
  dual-energy X-ray absorptiometry 323, 325, 326–327, 326
  epidemiology of osteoporotic fractures 399, 400
  fracture risk 319
  ISCD recommendations **327**
  juvenile osteoporosis 420–422, 420
  magnetic resonance imaging 327–329, 328
  mild physiological wedging 320, 321, 420
  orthopedic principles of fracture management 591
  Paget disease of bone 716
  parathyroid hormone treatment for osteoporosis 560
  primary hyperparathyroidism 622
  quantitative morphometry 323–325, 325
  radiotherapy-induced osteoporosis 789
  Scheuermann disease 320, 322
  Schmorl's nodes 321, 322
  semiquantitative assessment 323, 324, 325, 329
  short vertebral body height 320, 321
  skeletal complications of anticancer therapies 777
  spinal radiography 319–321, 320–322
  strontium ranelate 573–574
vertebroplasty 591, 819
VERT study 547
VFA see vertebral fracture assessment
vFEA see voxel finite element analysis
VFT see Venus flytrap
VGCC see voltage-gated calcium channels
vicious cycle hypothesis 740, 752, 762, 817–818
VIP see vasoactive intestinal peptide; vasoactive intestinal polypeptide
visceral endoderm (VE) 893
vitamin A 642
vitamin B_6 886–887

vitamin D 230–240
  bone 233–234
  bone histomorphometry 313
  bone stress injuries 454, 456
  bone turnover markers 294
  calcium-sensing receptor 221, 223–224, 226
  childhood cancer 793–794
  childhood disorders of mineral metabolism 705–709
  chronic kidney disease 505
  chronic kidney disease–mineral and bone disorder 695–702
  classical target tissues 233–235
  cost-effectiveness of osteoporosis treatment 600
  denosumab 556
  diet and nutrition 136–137, 236–237
  disorders of phosphate homeostasis 675–679
  epidemiology of osteoporotic fractures 402–403
  epidermis and hair follicles 236
  fetal and neonatal bone development 118, 119
  fetal calcium metabolism 179–180, 182–183
  FGF23 and the regulation of phosphorus metabolism 188–189
  genetic craniofacial disorders affecting dentition 913–914
  genetics 344
  glucocorticoid-induced osteoporosis 468–470
  hematological malignancies 765
  human immunodeficiency virus 475–477
  hypocalcemia 646–652, **647**
  hypoparathyroidism 655–656, 658
  immobilization and burns 731, 733–734
  immune cells 235–236
  internalization of vitamin D metabolites 233
  intestine 234–235
  kidney 235
  medical prevention and treatment of metastatic bone disease 805
  metabolism 230–232, *232*
  molecular mechanism of action 233, *234*
  neuropsychiatric disorders 1050
  nonclassical target tissues 235–236
  non-parathyroid hypercalcemia 639–644
  nutritional support for osteoporosis 536, 537–538
  oral manifestations of metabolic bone diseases 943–944
  osteoclasts 51
  osteoporosis in men 446
  Paget disease of bone 718
  pancreas 236
  parathyroid disorders 615–616
  parathyroid glands 235
  parathyroid hormone 205
  parathyroid hormone-related protein 214
  pregnancy and lactation 147–148, 149–150, 152
  premenopausal osteoporosis 438
  prevention of falls 529–530, **529**
  primary hyperparathyroidism 621, 622, 625–626
  pseudohypoparathyroidism 661, 663, 668–669
  regulation of calcium and phosphate metabolism 233
  regulation of calcium homeostasis 166–170
  rickets and osteomalacia 684–692, **686**, **687**
  sarcopenia and osteoporosis 502
  sclerosing bone disorders 834–835
  secondary causes of osteoporosis 511, 513
  supplementation and treatment strategies 237
  transplantation osteoporosis 426, **427–428**, 429–432
  transport in the blood 232–233
  tumoral calcinosis 862
  vitamin D_3 production 230, *231*
vitamin D-dependent rickets (VDDR) 375, *710*
vitamin D-related hypocalcemia 706
vitamin D response elements (VDRE) 221, 233
voltage-gated calcium channels (VGCC) 80, 118, 235
volumetric bone mass/mineral density (vBMD) 94–95
  combination therapy 568–569
  ethnicity and age-related loss of bone strength 132–133
  gonadal steroids and osteoporosis pathogenesis 412–413, *413*
  measurement in adults 260–263, 266–267
  measurement in children with osteoporosis risk factors 248–249
  skeletal growth 123–128
voxel finite element analysis (vFEA) 264–265, 268
VSMC *see* vascular smooth muscle cells

Waldenstrom's macroglobulinemia 642
Warburg effect 1006, 1007
WAT *see* white adipose tissue
WBV *see* whole body vibration
weight-bearing impact exercise 518, **520**
Weill–Marchesani syndrome 882
WES *see* whole exome sequencing
WGS *see* whole genome sequencing
WHI *see* Women's Health Initiative
white adipose tissue (WAT) 976, 978, 980
WHO *see* World Health Organization
whole body vibration (WBV) 521–523
whole exome sequencing (WES) 348, 387
whole genome sequencing (WGS) 348, 387
Wilson disease 888–889
WLS/*Wls see* Wntless
Wnt/β-catenin signaling pathway 68–74
  cell biology 68, *69*
  cellular bioenergetics of bone 1007–1008
  chronic kidney disease–mineral and bone disorder 695, 698
  core signaling pathway 68
  craniofacial morphogenesis 893
  Dickkopfs and sclerostin 71, 72
  endochondral ossification 15–17
  fetal and neonatal bone development 119
  Frizzled 70, 72
  gene mutations and skeletal disease 70–72
  genetic craniofacial disorders affecting dentition 911–912
  genome-wide association studies 381–382
  immunobiology 996
  integrative physiology of the skeleton 961–962
  Lrp4 71–72
  Lrp5 and Lrp6 68–70, *71*
  mechanotransduction 80–81, *81*
  metastatic bone disease 818
  muscle–bone interactions 1057–1058, *1058*, 1060
  Notum and Tiki 71
  odontogenesis 902
  osteoblasts 33, 56
  osteocytes 40–41, 42
  osteoprogenitor cells 23
  osteosarcoma 770–771
  Porcupine/Wntless in skeletal tissues 70
  regulation of Frizzled receptor stability at cell surface 72
  regulation of Wnt production and secretion 68–70
  secreted frizzled-related proteins 70–71
  skeletal healing/repair 103, 105